Electronics in Action

MICROELECTRONIC CIRCUIT DESIGN

ELECTRONICS AND VLSI CIRCUITS RELATED TITLES

Senior Consulting Editor
Stephen W. Director, *University of Michigan, Ann Arbor*

Consulting Editor
Richard C. Jaeger, *Auburn University*

DeMicheli: *Synthesis and Optimization of Digital Circuits*
Franco: *Design with Operational Amplifiers and Analog Integrated Circuits*
Hodges, Jackson, and Saleh: *Analysis and Design of Digital Integrated Circuits*
Kang and Leblebici: *CMOS Digital Integrated Circuits Analysis and Design*
Kasap: *Principles of Electrical Engineering Materials and Devices*
Kovacs: *Micromachined Transducers Sourcebook*
Jaeger and Blalock: *Microelectronic Circuit Design*
Neamen: *Electronic Circuit Analysis and Design*
Neamen: *Semiconductor Physics and Devices*
Razavi: *Design of Integrated Circuits for Optical Communications*
Razavi: *Design of Analog CMOS Integrated Circuits*

THIRD EDITION

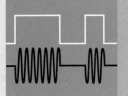

MICROELECTRONIC CIRCUIT DESIGN

RICHARD C. JAEGER
Auburn University

TRAVIS N. BLALOCK
University of Virginia

Boston   Burr Ridge, IL   Dubuque, IA   New York   San Francisco   St. Louis
Bangkok   Bogotá   Caracas   Kuala Lumpur   Lisbon   London   Madrid   Mexico City
Milan   Montreal   New Delhi   Santiago   Seoul   Singapore   Sydney   Taipei   Toronto

 **Higher Education**

MICROELECTRONIC CIRCUIT DESIGN, THIRD EDITION

Published by McGraw-Hill, a business unit of The McGraw-Hill Companies, Inc., 1221 Avenue of the Americas, New York, NY 10020.

Some ancillaries, including electronic and print components, may not be available to customers outside the United States.

This book is printed on acid-free paper.

3 4 5 6 7 8 9 0 DOW/DOW 0 9 8

ISBN 978–0–07–319163–8
MHID 0–07–319163–9

Senior Sponsoring Editor: *Michael S. Hackett*
Executive Marketing Manager: *Michael Weitz*
Senior Project Manager: *Kay J. Brimeyer*
Senior Production Supervisor: *Kara Kudronowicz*
Associate Media Producer: *Christina Nelson*
Designer: *Rick D. Noel*
Cover Designer: *Rick D. Noel*
(USE) Cover Image: *Cover image © Molecular Expressions, Photomicrograph-"The Little Engine That Could"*
Senior Photo Research Coordinator: *John C. Leland*
Photo Research: *LouAnn K. Wilson*
Compositor: *Interactive Composition Corporation*
Typeface: *10/12 Times Roman*
Printer: *R. R. Donnelley Willard, OH*

Library of Congress Cataloging-in-Publication Data

Jaeger, Richard C.
    Microelectronic circuit design / Richard C. Jaeger, Travis N. Blalock. – 3rd ed.
       p. cm.
    Includes bibliographical references and index.
    ISBN 978-0-07-319163-8—ISBN 0-07-319163-9 (hard copy : alk. paper)
    1. Integrated circuits–Design and construction. 2. Semiconductors–Design and construction. 3. Electronic circuit design. I. Blalock, Travis N. II. Title.

  TK7874.J333   2006
   621.3815—dc22
                                          2006032859

TO

BRIEF CONTENTS

CONTENTS

CHAPTER 7

COMPLEMENTARY MOS (CMOS) LOGIC DESIGN 352

CHAPTER 8

MOS MEMORY AND STORAGE CIRCUITS 398

PREFACE

Through study of this text, the reader will develop a comprehensive understanding of the basic techniques of modern electronic circuit design, analog and digital, discrete and integrated. Even though most readers may not ultimately be engaged in the design of integrated circuits (ICs) themselves, a thorough understanding of the internal circuit structure of ICs is prerequisite to avoiding many pitfalls that prevent the effective and reliable application of integrated circuits in system design.

Digital electronics has evolved to be an extremely important area of circuit design, but it is included almost as an afterthought in many introductory electronics texts. We present a more balanced coverage of analog and digital circuits. The writing integrates the authors' extensive industrial backgrounds in precision analog and digital design with their many years of experience in the classroom. A broad spectrum of topics is included, and material can easily be selected to satisfy either a two-semester or three-quarter sequence in electronics.

IN THIS EDITION

This revision has focused upon shortening the text and continuing to make the material more readable and accessible to the student. In shortening the text, deleted material has been moved to the Assessment, Review and Instruction System (ARIS) website including the treatment of dc-to-dc converters, JFET devices and circuits, the Ebers-Moll Model, the bandgap reference, the Gilbert multiplier, Gilbert mixer, D/A and A/D converters, and switched-capacitor circuits. Improved section transitions have been written and mathematical detail continues to be reduced.

The logic circuit material in Part II has been updated to use lower power supply voltages, and the treatment of delay in MOS logic circuits has been greatly simplified. The addition of Psuedo NMOS logic, that utilizes a PMOS load transistor, provides a transition to CMOS logic. The discussion of TTL logic circuit variations has been reduced.

The op-amp material in Part III has been reorganized into ideal operational amplifiers and op-amp applications (Chapter 11) and non-ideal limitations of real operational amplifiers (Chapter 12). Redundant material has been removed from the treatment of single transistor amplifiers in Chapters 13 and 14, as well as overlapping material on current mirrors in Chapters 4, 5 and 16. Chapters 15 and 16 have been merged into one new chapter. The discussion of multistage amplifiers concentrates on dc coupled circuits. Treatment of the high frequency response of multistage circuits has been completely revised.

The Structured Problem Solving Approach introduced in the Second Edition is used throughout the examples. The popular Electronics-in-Action features have been retained, and new ones added including include Direct Digital Synthesis, Dual Slope ADCs, Weighted resistor D/A Converter, Standard Cell Logic, and FPGAs.

Chapter Openers enhance the readers understanding of historical developments in electronics. Design notes highlight important ideas that the circuit designer should remember. The World Wide Web is viewed as an integral extension of the text, and a wide range of supporting materials and resource links are maintained and updated on the McGraw-Hill website (aris.mhhe.com).

Features of the book are outlined below.

The Structured Problem-Solving Approach is used throughout the examples

Electronics in Action features in each chapter

Chapter openers highlighting developments in the field of electronics

Design Notes and emphasis on practical circuit design

Broad use of SPICE throughout the text and examples

Integrated treatment of device modeling in SPICE

Numerous Exercises, Examples, and Design Examples

Large number of new problems

Integrated web materials

Continuously updated web resources and links

Placing the digital portion of the book first is also beneficial to students outside of electrical engineering, particularly computer engineering or computer science majors, who may only take the first course in a sequence of electronics courses.

The material in Part II deals primarily with the internal design of logic gates and storage elements. A comprehensive discussion of NMOS and CMOS logic design is presented in Chapters 6 and 7, and a discussion of memory cells and peripheral circuits appears in Chapter 8. Chapter 9 on bipolar logic design includes a detailed discussion of ECL and TTL. However, the material on bipolar logic has been reduced in deference to the import of MOS technology. This text does not include any substantial design at the logic block level, a topic that is fully covered in digital design courses.

Parts I and II of the text deal only with the large-signal characteristics of the transistors. This allows the reader to become comfortable with device behavior and i-v characteristics before they have to grasp the concept of splitting circuits into different pieces (and possibly different topologies) to perform dc and ac small-signal analyses. (The concept of a small-signal is formally introduced in Part III, Chapter 13.)

Although the treatment of digital circuits is more extensive than most texts, more than 50 percent of the material in the book, **Part III,** still deals with traditional analog circuits. The analog section begins in Chapter 10 with a discussion of concepts related to amplifiers and amplification. Chapters 11 and 12 present a comprehensive discussion of the operational amplifier and its many limitations. Chapter 13 presents a comprehensive development of the small-signal models for diode, BJT, and MOSFET. The hybrid-pi model and pi-models for the BJT and FET are used throughout.

Design concepts and device and circuit comparisons are emphasized wherever possible. A significantly stronger emphasis is given to MOS analog circuits than in many texts, and the treatment of bipolar and FET analog circuits is merged from Chapter 14 onward, permitting a continual comparison of design options and reasons for choosing one device over another in a particular circuit.

Chapters 13–15 provide an in-depth discussion of single-stage and multi-stage amplifier design using transistors. Chapter 15 discusses techniques that are important in IC design and studies the classic 741 operational amplifier.

Chapter 16 presents a detailed discussion of frequency response. In the final chapter, the classical two-port approach is taken in the presentation of feedback. However, a section is included that stresses the errors that can occur when the approach is incorrectly applied. A section discussing Blackman's Theorem shows how the problems associated with the two-port formulations can be avoided. Feedback amplifier stability and oscillators are discussed, as is the important method of determining loop-gain using successive voltage and current injection.

DESIGN

Design remains a difficult issue in educating engineers. The use of the well-defined problem-solving methodology presented in this text can significantly enhance an engineer's ability to understand the issues related to design. The design examples assist in building an understanding of the design process.

Part II launches directly into the issues associated with the design of NMOS and CMOS logic gates. The effects of device and passive-element tolerances are discussed throughout the text. In today's world, low-power, low-voltage design, often supplied from batteries, is playing an increasingly import role. In this edition the logic design examples have been moved away from 5 V to lower power supply levels. The use of the computer, including MATLAB, spreadsheets, or standard high-level languages to explore design options is a thread that continues throughout the text.

Methods for making design estimates and decisions are stressed throughout the analog portion of the text. Expressions for amplifier behavior are simplified beyond the standard hybrid-pi model expressions whenever appropriate. For example, the expression for the voltage gain of an amplifier in most texts is simply written as $|A_v| = g_m R_L$, which tends to hide the power supply voltage as the fundamental design variable. Rewriting this expression in approximate form as $g_m R_L \cong 10 V_{CC}$ for the BJT, or $g_m R_L \cong V_{DD}$ for the FET, explicitly displays the dependence of amplifier design on the choice of power supply voltage and provides a simple first-order design estimate for the voltage gain of the common-emitter and common-source amplifiers. These approximation techniques and methods for performance estimation are included as often as possible. Comparisons and design tradeoffs between the properties of BJTs and FETs are included throughout Part III.

Worst-case and Monte-Carlo analysis techniques are introduced at the end of the first chapter. These are not topics traditionally included in undergraduate courses. However, the ability to design circuits in the face of wide component tolerances and variations is a key component of electronic circuit design, and the design of circuits using standard components and tolerance assignment are discussed in examples and included in many problems.

PROBLEMS AND INSTRUCTOR SUPPORT

Specific design problems, computer problems, and SPICE problems are included at the end of each chapter. Design problems are indicated by ⚲, computer problems are indicated by 🖥, and SPICE problems are indicated by Ⓢ. The problems are keyed to the topics in the text and are also graded into three levels of difficulty with the more difficult or time-consuming problems indicated by * and **. The original problems and solutions can be found on the McGraw-Hill ARIS website. An Instructor's Manual containing solutions to all the problems is available from the authors. Solutions are also available as word processor files. In addition, copies of the original versions of all of the graphs and figures are available as PowerPoint files and can be retrieved from the world wide web. Instructor notes are available as PowerPoint slides.

COMPUTER USAGE AND SPICE

The computer is used as a tool throughout the text. The authors firmly believe that this means more than just the use of the SPICE circuit analysis program. In today's computing environment, it is often appropriate to use the computer to explore a complex design space rather than to try to reduce a complicated set of equations to some manageable analytic form. Examples of the process of setting up equations for iterative evaluation by computer through the use of spreadsheets, MATLAB and/or standard high-level language programs are illustrated in several places in the text. MATLAB is also used for Nyquist and Bode plot generation and is very useful for Monte Carlo analysis.

On the other hand, SPICE is used throughout the text. Results from SPICE simulation are included throughout and numerous SPICE problems are to be found in the problem sets. Wherever useful, a SPICE analysis is used with most examples. This edition also emphasizes the differences and utility of the dc, ac, transient, and transfer function analysis modes in SPICE. A discussion of SPICE device modeling is included following the introduction to each semiconductor device, and typical SPICE model parameters are presented with the models.

ACKNOWLEDGMENTS

We want to thank the large number of people have had an impact on the material in this text and on its preparation. Our students have helped immensely in polishing the manuscript and have managed to survive the many revisions of the manuscript. Our department heads, J. D. Irwin of Auburn University and L. R. Harriott of the University of Virginia, have always been highly supportive of faculty efforts to develop improved texts.

We want to thank all the reviewers including

Mohamed Bakr
McMaster University

Steve Bibyk
Ohio State University

Otmar Boser
Norwalk Community College

Daniel Bukofzer
California State University, Fresno

Marc Cahay
University of Cincinnati

Charles L. Carnal
Tennessee Technological University

Yung Jui Chen
University of Maryland, Baltimore County

Dale L. Critchlow
University of Vermont

Susan Earles
Florida Institute of Technology

Saleh Faruque
University of North Dakota

Patrick Fay
University of Notre Dame

Maysam Ghovanloo
North Carolina State University

George C. Giakos
The University of Akron

K. Gopalan
Purdue University Calumet

Sotoudeh Hamedi-Hagh
San Jose State University

Perry L. Heedley
California State University, Sacramento

Syed K. Islam
The University of Tennessee

Rajiv J. Kapadia
Minnesota State University

Aydin I. Karsilayan
Texas A&M University

Richard Kwor
University of Colorado at Colorado Springs

Daniel L. Lau
University of Kentucky

Colby Leider
University of Miami

Swaminathan Madhu
*Rochester Institute
of Technology*

Wagdy H. Mahmoud
*Tennessee Technological
University*

Oscar Moreira-Tamayo
*University of Texas
at Dallas
Texas Instruments, Inc.*

Kazutoshi Najita
*University of Hawaii
at Manoa*

Andrea Pacelli
Stony Brook University

Dennis L. Polla
University of Minnesota

Paulo F. Ribeiro
*Calvin College
Florida State University*

Bassam Shaer
*University of Minnesota
Duluth*

Hemchandra Shertukde
University of Hartford

Jose Silva-Martinez
Texas A&M University

Weizhong Wang
*University of Wisconsin
Milwaukee*

Finally, we want to be sure to thank the team at McGraw-Hill, including Kay Brimeyer, for a myriad of important suggestions and changes that helped polish the final version of the manuscript.

In developing this text, we have attempted to integrate our industrial backgrounds in precision analog and digital design with many years of experience in the classroom. We hope we have at least succeeded to some extent. Constructive suggestions and comments will be appreciated.

Richard C. Jaeger
Auburn University

Travis N. Blalock
University of Virginia

CHAPTER-BY-CHAPTER SUMMARY

PART I — SOLID-STATE ELECTRONICS AND DEVICES

Chapter 1 provides a historical perspective on the field of electronics beginning with vacuum tubes and advancing to giga-scale integration and its impact on the global economy. Chapter 1 also provides a classification of electronic signals and a review of some important tools from network analysis, including a review of the ideal operational amplifier. Because developing a good problem-solving methodology is of such import to an engineer's career, the comprehensive Structured Problem Solving Approach is used to help the students develop their problem solving skills. The structured approach is discussed in detail in the first chapter and used in all the subsequent examples in the text. Component tolerances and variations play an extremely important role in practical circuit design, and Chapter 1 closes with introductions to tolerances, temperature coefficients, worst-case design, and Monte Carlo analysis.

Chapter 2 deviates from the recent norm and discusses semiconductor materials including the covalent-bond and energy-band models of semiconductors. The chapter includes material on intrinsic carrier density, electron and hole populations, n- and p-type material, and impurity doping. Mobility, resistivity, and carrier transport by both drift and diffusion are included as topics. Velocity saturation is discussed, and an introductory discussion of microelectronic fabrication has been merged with Chapter 2.

Chapter 3 introduces the structure and i-v characteristics of solid-state diodes. Discussions of Schottky diodes, variable capacitance diodes, photo-diodes, solar cells, and LEDs are also included. This chapter introduces the concepts of device modeling and the use of different levels of modeling to achieve various approximations to reality. The SPICE model for the diode is discussed. The concepts of bias, operating point, and load-line are all introduced, and iterative mathematical solutions are also used to find the operating point with MATLAB and spreadsheets. Diode applications in rectifiers are discussed in detail and a

discussion of the dynamic switching characteristics of diodes is also presented.

Chapter 4 discusses MOS field-effect transistors, starting with a qualitative description of the MOS capacitor. Models are developed for the FET i-v characteristics, and a complete discussion of the regions of operation of the device is presented. Body effect is included. MOS transistor scaling and performance limits including scaling, cutoff frequency, and subthreshold conduction are discussed as well as basic Λ-based layout methods. Biasing circuits and load-line analysis are presented. The FET SPICE models and model parameters are discussed in Chapter 4.

Chapter 5 introduces the bipolar junction transistor and presents a heuristic development of the Transport (simplified Gummel-Poon) model of the BJT based upon superposition. The various regions of operation are discussed in detail. Common-emitter and common-base current gains are defined, and base transit-time, diffusion capacitance and cutoff frequency are all discussed. Bipolar technology and layout are introduced. Various bias circuits are discussed. The SPICE model for the BJT and the SPICE model parameters are discussed in Chapter 5.

PART II — DIGITAL CIRCUIT DESIGN

Chapter 6 begins with a compact introduction to digital electronics. Terminology discussed includes logic levels, noise margins, rise-and-fall times, propagation delay, fan out, fan in, and power-delay product. A short review of Boolean algebra is included. The introduction to MOS logic design is now merged with Chapter 6 and follows the historical evolution of NMOS logic gates focusing on the design of saturated-load, linear-load, and depletion-load circuit families. The impact of body effect on MOS logic circuit design is discussed in detail. The concept of reference inverter scaling is developed and employed to affect the design of other inverters, NAND gates, NOR gates, and complex logic functions throughout Chapters 6 and 7. Capacitances in MOS circuits are discussed, and methods for estimating the propagation delay and power-delay product of NMOS logic are presented. Details of several of the propagation delay analyses are moved to the MCD website, and the delay equation results for the various families have been collapsed into a much more compact form.

CMOS represents today's most important integrated circuit technology, and **Chapter 7** provides an in-depth look at the design of CMOS logic gates including inverters, NAND and NOR gates, and complex logic gates. In this case, the designs are based on simple scaling of a reference inverter design. Noise margin and latchup are discussed as well as a comparison of the power-delay products of various MOS logic families. Dynamic logic circuits and cascade buffer design are discussed in Chapter 7. A discussion of BiCMOS logic circuitry has been added to Chapter 9 after bipolar logic is introduced.

Chapter 8 ventures into the design of memory and storage circuits, including the six-transistor, four-transistor, and one-transistor memory cells. Basic sense-amplifier circuits are introduced as well as the peripheral address and decoding circuits needed in memory designs. ROMs and flip-flop circuitry are included in Chapter 8.

Chapter 9 discusses bipolar logic circuits including emitter-coupled logic and transistor-transistor logic. The use of the differential pair as a current switch and the large-signal properties of the emitter follower are introduced. Operation of the BJT as a saturated switch is included and followed by a discussion of low voltage and standard TTL. An introduction to BiCMOS logic now concludes the chapter on bipolar logic.

PART III—ANALOG CIRCUIT DESIGN

Chapter 10 provides a succinct introduction to analog electronics. The concepts of voltage gain, current gain power gain, and distortion are developed. Much care has been taken to be consistent in the use of the notation that defines these quantities as well as in the use of dc, ac, and total signal notation throughout the book. Bode plots are reviewed and amplifiers are classified by frequency response. MATLAB is utilized as a tool for producing Bode plots.

Chapter 11 provides a comprehensive introduction to the design of circuits involving ideal operational amplifiers. Classical op amp circuits including the inverting and non-inverting amplifiers, difference amplifier, instrumentation amplifier, summing amplifier, integrator and differentiator are all discussed. Basic active filter circuits are also introduced. Op amp based rectifiers, Schmitt triggers, and

multivibrators complete the discussions in Chapter 11. SPICE simulation of ideal op-amp circuits is discussed.

Chapter 12 now focuses on a comprehensive discussion of the characteristics and limitations of real operational amplifiers including the effects of finite gain and input resistance, non-zero output resistance, input offset voltage, input bias and offset currents, output voltage and current limits, finite bandwidth, and common-mode rejection. Cascade amplifiers are investigated including a discussion of the bandwidth of multistage amplifiers. The macro model concept is introduced and the discussion of SPICE simulation of op-amp circuits using various levels of models continues in Chapter 12.

Chapter 13 begins the general discussion of linear amplification using the BJT and FET as C-E and C-S amplifiers. Biasing for linear operation and the concept of small-signal modeling are both introduced, and small-signal models of the diode, BJT and FET are all developed. The limits for small-signal operation are all carefully defined. Appropriate points for signal injection and extraction are identified, and the use of coupling and bypass capacitors and inductors to separate the ac and dc designs is explored. The important $10V_{CC}$ and V_{DD} design estimates for the voltage gain of the C-E and C-S amplifiers are introduced, and the role of transistor amplification factor in bounding circuit performance is discussed. The role of Q-point design on power dissipation and signal range is also introduced.

Chapter 14 proceeds with an in-depth comparison of the characteristics of single-transistor amplifiers, including small-signal amplitude limitations. Amplifiers are classified as inverting amplifiers (C-E, C-S), non-inverting amplifiers (C-B, C-G), and followers (C-C, C-D). The treatment of MOS and bipolar devices is merged from Chapter 14 on, and design tradeoffs between the use of the BJT and the FET in amplifier circuits is an important thread that is followed through all of Part III. A new addition to this chapter is a detailed discussion of the design of coupling and bypass capacitors and the role of these capacitors in controlling the low frequency response of amplifiers.

Chapter 15 explores the design of multistage amplifiers, including ac- and dc-coupled circuits. An evolutionary approach to multistage op amp design is used. MOS and bipolar differential amplifiers are first introduced. Subsequent addition of a second gain stage and then an output stage convert the differential amplifiers into simple op amps. Class A, B, and AB operation are defined. Electronic current sources are designed and used for biasing of the basic operational amplifiers. Darlington, cascode, and cascade C-E circuits are presented in this chapter. Discussion of important FET-BJT design tradeoffs are included wherever appropriate.

Chapter 15 also introduces techniques that are of particular import in integrated circuit design. A variety of current mirror circuits are introduced and applied in bias circuits and as active loads in operational amplifiers. A wealth of circuits and analog design techniques are explored through the detailed analysis of the classic 741 operational amplifier.

Chapter 16 discusses the frequency response of analog circuits. The behavior of each of the three categories of single-stage amplifiers (C-E/C-S, C-B/C-G, and C-C/C-D) is discussed in detail, and BJT behavior is contrasted with that of the FET. The frequency response of the transistor is discussed, and the high frequency, small-signal models are developed for both the BJT and FET. Miller multiplication is used to obtain estimates of the lower and upper cutoff-frequencies of complex multistage amplifiers. Gain-bandwidth products and the gain-bandwidth tradeoffs in design are discussed. Cascode amplifier frequency response, and tuned amplifiers are included in this chapter.

Chapter 17 discusses feedback amplifier design using the classical two-port approach to account for the loading of the amplifier and feedback network on each other. Loop-gain calculations are discussed. A unique section on the use of the successive voltage and current injection technique for determining loop-gain is included. This method does not require the feedback loop to be broken and represents a useful technique in the laboratory as well as for SPICE simulation of high gain feedback amplifiers. Another unique section discusses errors that must be avoided when applying the two-port analysis methods to the shunt-series and series-series feedback topologies. This discussion is followed by a section that introduces Blackman's Theorem as a method to avoid the problems associated with the two-port calculations of input and output resistance.

Amplifier stability is also discussed in Chapter 3, and Nyquist diagrams and Bode plots (with MATLAB) are used to explore the phase and gain margin of amplifiers. Basic single-pole op amp compensation is discussed, and the unity gain-bandwidth product is related to amplifier slew rate. Design op amp compensation to achieve a desired phase margin is discussed. Relationships between the Nyquist and Bode techniques are explicitly discussed. The Barkhausen

criteria for oscillation are introduced, and the presentation of oscillator circuits includes RC, LC, and crystal implementations. The discussion of amplitude stabilization in oscillators includes techniques for calculating the amplitude of the oscillation.

Two **Appendices** include tables of standard component values (Appendix A), and summary of the device models and sample SPICE parameters (Appendix B). Data sheets for representative solid-state devices and operational amplifiers are now available via the WWW.

FLEXIBILITY

The chapters are designed to be used in a variety of different sequences, and there is more than enough material for a two-semester or three-quarter sequence in electronics. One can obviously proceed directly through the book. On the other hand, the material has been written so that the BJT chapter can be used immediately after the diode chapter if so desired (i.e., a 1-2-3-5-4 chapter sequence). At the present time, the order actually used at Auburn University is:

 1. Introduction
 2. Solid-State Electronics
 3. Diodes
 4. FETs
 5. Introduction to Digital Logic
 6. CMOS Logic
 7. Memory
 5. The BJT
 9. Bipolar Logic
10–17. Analog in sequence

The chapters have also been written so that Part II, Digital Circuit Design, can be skipped, and Part III, Analog Circuit Design, can be used directly after completion of the coverage of the solid-state devices in Part I. If so desired, many of the quantitative details of the material in Chapter 2 may be skipped. In this case, the sequence would be

 1. Introduction
 2. Solid-State Electronics
 3. Diodes
 4. FETs
 5. The BJT
10–17. Analog in sequence

MICROELECTRONIC
CIRCUIT DESIGN

PART ONE
SOLID STATE ELECTRONIC AND DEVICES

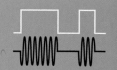

CHAPTER 1

INTRODUCTION TO ELECTRONICS

CHAPTER OUTLINE

CHAPTER GOALS

- Present a brief history of electronics.
- Quantify the explosive development of integrated circuit technology.
- Discuss initial classification of electronic signals.
- Review important notational conventions and concepts from circuit theory.
- Introduce methods for including tolerances in circuit analysis.
- Present the problem-solving approach used in this text.

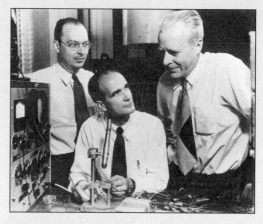

Figure 1.1 John Bardeen, William Shockley, and Walter Brattain in Brattain's laboratory in 1948.
Lucent Technologies Inc./Bell Labs

Figure 1.2 The first germanium bipolar transistor.
Lucent Technologies Inc./Bell Labs

November 1997 was the 50th anniversary of the 1947 discovery of the bipolar transistor by John Bardeen and Walter Brattain at Bell Laboratories, a seminal event that marked the beginning of the semiconductor age (see Figs. 1.1 and 1.2). The invention of the transistor and the subsequent development of microelectronics have done more to shape the modern era than any other event. The transistor and microelectronics have reshaped how business is transacted, machines are designed, information moves, wars are fought, people interact, and countless other areas of our lives.

This textbook develops the basic operating principles and design techniques governing the behavior of the devices and circuits that form the backbone of much of the infrastructure of our modern world. This knowledge will enable students who aspire to design and create the next generation of this technological revolution to build a solid foundation for more advanced design courses. In addition, students who expect to work in some other technology area will learn material that will help them understand microelectronics, a technology that will continue to have impact on how their chosen field develops. This understanding will enable them to fully exploit microelectronics in their own technology area. Now let us return to our short history of the transistor.

After the discovery of the transistor, it was only a few months until William Shockley developed a theory that described the operation of the bipolar junction transistor. Only 10 years later, in 1956, Bardeen, Brattain, and Shockley received the Nobel prize in physics for the discovery of the transistor.

In June 1948 Bell Laboratories held a major press conference to announce the discovery. In 1952 Bell Laboratories, operating under legal consent decrees, made licenses for the transistor available for the modest fee of $25,000 plus future royalty payments. About this time, Gordon Teal, another member of the solid-state group, left Bell Laboratories to work on the transistor at Geophysical Services, Inc., which subsequently became Texas Instruments (TI). There he made the first silicon transistors, and TI marketed the first all-transistor radio. Another early licensee of the transistor was Tokyo Tsushin Kogyo, which became the Sony Company in 1955. Sony subsequently sold a transistor radio with a marketing strategy based on the idea that everyone could now have a personal radio; thus was launched the consumer market for transistors. A very interesting account of these and other developments can be found in [1, 2] and their references.

Activity in electronics began more than a century ago with the first radio transmissions in 1895 by Marconi, and these experiments were followed after only a few years by the invention of the first electronic amplifying device, the triode vacuum tube. In this period, electronics—loosely defined as the design and application of electron devices—has had such a significant impact on our lives that we often overlook just how pervasive electronics has really become. One measure of the degree of this impact can be found in the gross domestic product (GDP) of the world. In 2004 the world GDP was approximately U.S. $55 trillion, and of this total, fully 10 percent, more than $5 trillion, was directly traceable to electronics. See Table 1.1 [3–5].

We commonly encounter electronics in the form of telephones, radios, televisions, and audio equipment, but electronics can be found even in seemingly mundane appliances such as vacuum cleaners, washing machines, and refrigerators. Wherever one looks in industry, electronics will be found. The corporate world obviously depends heavily on data processing systems to manage its operations. In fact, it is hard to see how the computer industry could have evolved without the use of its own products. In addition, the design process depends ever more heavily on computer-aided design (CAD) systems, and manufacturing relies on electronic systems for process control—in petroleum refining, automobile tire production, food processing, power generation, and so on.

TABLE 1.1
The Worldwide Electronics Market

CATEGORY	SHARE (%)
Data processing hardware	23
Data processing software and services	18
Professional electronics	10
Telecommunications	9
Consumer electronics	9
Active components	9
Passive components	7
Computer integrated manufacturing	5
Instrumentation	5
Office electronics	3
Medical electronics	2

1.1 A BRIEF HISTORY OF ELECTRONICS: FROM VACUUM TUBES TO ULTRA-LARGE-SCALE INTEGRATION

Because most of us have grown up with electronic products all around us, we often lose perspective of how far the industry has come in a relatively short time. At the beginning of the twentieth century, there were no commercial electron devices, and transistors were not invented until the late 1940s! Explosive growth was triggered by first the commercial availability of the bipolar transistor in the late 1950s, and then the realization of the integrated circuit (IC) in 1961. Since that time, signal processing using electron devices and electronic technology has become a pervasive force in our lives.

Table 1.2 lists a number of important milestones in the evolution of the field of electronics. The Age of Electronics began in the early 1900s with the invention of the first electronic two-terminal devices, called **diodes.** The **vacuum diode,** or diode **vacuum tube,** was invented by Fleming in 1904; in 1906 Pickard created a diode by forming a point contact to a silicon crystal. (Our study of electron devices begins with the introduction of the solid-state diode in Chapter 3.)

The invention of the three-element vacuum tube known as the **triode** was an extremely important milestone. The addition of a third element to a diode enabled electronic amplification to take place with good isolation between the input and output ports of the device. Silicon-based three-element devices now form the basis of virtually all electronic systems. Fabrication of tubes that could be used reliably in circuits followed the invention of the triode by a few years and enabled rapid circuit innovation. Amplifiers and oscillators were developed that significantly improved radio transmission and reception. Armstrong invented the super heterodyne receiver in 1920 and FM modulation in 1933. Electronics developed rapidly during World War II, with great advances in the field of radio communications and the development of radar. Although first demonstrated in 1930, television did not begin to come into widespread use until the 1950s.

An important event in electronics occurred in 1947, when John Bardeen, Walter Brattain, and William Shockley at Bell Telephone Laboratories invented the **bipolar transistor.**[1] Although field-effect devices had actually been conceived by Lilienfeld in 1925, Heil in 1935, and Shockley in 1952 [2], the technology to produce such devices on a commercial basis did not yet exist. Bipolar devices, however, were rapidly commercialized.

Then in 1958, the nearly simultaneous invention of the **integrated circuit (IC)** by Kilby at Texas Instruments and Noyce and Moore at Fairchild Semiconductor produced a new technology that would profoundly change our lives. The miniaturization achievable through IC technology made available complex electronic functions with high performance at low cost. The attendant characteristics of high reliability, low power, and small physical size and weight were additional important advantages.

In 2000, Jack St. Clair Kilby received a share of the Nobel prize for the invention of the integrated circuit. In the mind of the authors, this was an exceptional event as it represented one of the first awards to an electronic technologist.

Most of us have had some experience with personal computers, and nowhere is the impact of the integrated circuit more evident than in the area of digital electronics. For example, 1-gigabit (Gb) dynamic memory chips, similar to those in Fig. 1.3(c), contain more than 1 billion transistors. Creating this much memory using individual vacuum tubes [depicted in Fig. 1.3(a)] or even discrete transistors [shown in Fig. 1.3(b)] would be an almost inconceivable feat.

Levels of Integration

The dramatic progress of integrated circuit miniaturization is shown graphically in Figs. 1.4 and 1.5. The complexities of memory chips and microprocessors have grown exponentially with time. In the four decades since 1970, the number of transistors on a microprocessor chip has increased by a factor

---

[1] The term **transistor** is said to have originated as a contraction of "transfer resistor," based on the voltage-controlled resistance of the characteristics of the MOS transistor.

TABLE 1.2
Milestones in Electronics

YEAR	EVENT
1874	Ferdinand Braun invents the solid-state rectifier.
1884	American Institute of Electrical Engineers (AIEE) formed.
1895	Marconi makes first radio transmissions.
1904	Fleming invents diode vacuum tube—Age of Electronics begins.
1906	Pickard creates solid-state point-contact diode (silicon).
1906	Deforest invents triode vacuum tube (audion).
1910–1911	"Reliable" tubes fabricated.
1912	Institute of Radio Engineers (IRE) founded.
1907–1927	First radio circuits developed from diodes and triodes.
1920	Armstrong invents super heterodyne receiver.
1925	TV demonstrated.
1925	Lilienfeld files patent application on the field-effect device.
1927–1936	Multigrid tubes developed.
1933	Armstrong invents FM modulation.
1935	Heil receives British patent on a field-effect device.
1940	Radar developed during World War II—TV in limited use.
1947	Bardeen, Brattain, and Shockley at Bell Laboratories invent bipolar transistors.
1950	First demonstration of color TV.
1952	Shockley describes the unipolar field-effect transistor.
1952	Commercial production of silicon bipolar transistors begins at Texas Instruments.
1952	Ian Ross and George Dacey demonstrate the junction field-effect transistor.
1956	Bardeen, Brattain, and Shockley receive Nobel prize for invention of bipolar transistors.
1958	Integrated circuit developed simultaneously by Kilby at Texas Instruments and Noyce and Moore at Fairchild Semiconductor.
1961	First commercial digital IC available from Fairchild Semiconductor.
1963	AIEE and IRE merge to become the Institute of Electrical and Electronic Engineers (IEEE)
1967	First semiconductor RAM (64 bits) discussed at the IEEE International Solid-State Circuits Conference (ISSCC).
1968	First commercial IC operational amplifier—the μA709—introduced by Fairchild Semiconductor.
1970	One-transistor dynamic memory cell invented by Dennard at IBM.
1970	Low-loss optical fiber invented.
1971	4004 microprocessor introduced by Intel.
1972	First 8-bit microprocessor—the 8008—introduced by Intel.
1974	First commercial 1-kilobit memory chip developed.
1974	8080 microprocessor introduced.
1978	First 16-bit microprocessor developed.
1984	Megabit memory chip introduced.
1987	Erbium doped, laser-pumped optical fiber amplifiers demonstrated.
1995	Experimental gigabit memory chip presented at the IEEE ISSCC.
2000	Alferov, Kilby, and Kromer share the Nobel prize in physics for optoelectronics, invention of the integrated circuit, and heterostructure devices, respectively.

(a)   (b)

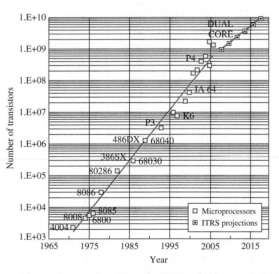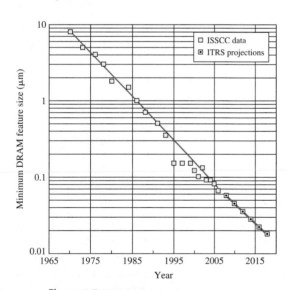

(c)   (d)

Figure 1.3 Comparison of (a) vacuum tubes, (b) individual transistors, (c) integrated circuits in dual-in-line packages (DIPs), and (d) ICs in surface mount packages.
Source: (a) Mark J. Wilson, American Radio Relay League, The ARRL Handbook, *1992.*

Figure 1.4 Microprocessor complexity versus time.

Figure 1.5 DRAM feature size versus year.

of one million as depicted in Fig. 1.4. Similarly, memory density has grown by a factor of more than 10 million from a 64-bit chip in 1968 to the demonstration of several experimental 1-Gbit chips in the late 1990s.

Since the commercial introduction of the integrated circuit, these increases in density have been achieved through a continued reduction in the minimum line width, or **minimum feature size,** that can be defined on the surface of the integrated circuit (see Fig. 1.5). Today most corporate semiconductor laboratories around the world are actively working on deep submicron processes with feature sizes below 0.1 μm—less than one one-hundredth the diameter of a human hair.

As the miniaturization process has continued, a series of commonly used abbreviations has evolved to characterize the various levels of integration. Prior to the invention of the integrated circuit, electronic systems were implemented in discrete form. Early ICs, with fewer than 100 components, were characterized as **small-scale integration,** or **SSI.** As density increased, circuits became identified as **medium-scale integration (MSI,** 100–1000 components/chip)**, large-scale integration (LSI,** $10^3 - 10^4$ components/chip)**,** and **very-large-scale integration (VLSI,** $10^4 - 10^9$ components/chip)**.** Today discussions focus on **ultra-large-scale integration (ULSI)** and **giga-scale integration (GSI,** above 10^9 components/chip)**.**

ELECTRONICS IN ACTION

Cellular Phone Evolution

The impact of technology scaling is ever present in our daily lives. One example appears visually in the pictures of cellular phone evolution below. Early mobile phones were often large and had to be carried in a relatively large pouch (hence the term "bag phone"). The next generation of analog phones could easily fit in your hand, but they had poor battery life caused by their analog communications technology. Implementations of second- and third-generation digital cellular technology are considerably smaller and have much longer battery life. As density continues to increase, additional functions such as a personal digital assistant (PDA) or cameras are integrated with the digital phone.

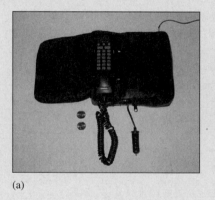

(a)   (b)   (c)

A decade of cellular phone evolution: (a) early Uniden "bag phone," (b) Nokia analog phone, and (c) Motorola miniature phone.

Source: © 2006 Motorola, Inc. Reprinted with permission from Motorola, Inc.

Cell phones also represent excellent examples of the application of **mixed-signal** integrated circuits that contain both analog and digital circuitry on the same chip. ICs in the cell phone contain analog radio frequency receiver and transmitter circuitry, analog-to-digital and digital-to-analog converters, CMOS logic and memory, and power conversion circuits.

1.2 CLASSIFICATION OF ELECTRONIC SIGNALS

The signals that electronic devices are designed to process can be classified into two broad categories: analog and digital. **Analog signals** can take on a continuous range of values, and thus represent continuously varying quantities; purely **digital signals** can appear at only one of several discrete levels. Examples of these types of signals are described in more detail in the next two subsections, along with the concepts of digital-to-analog and analog-to-digital conversion, which make possible the interface between the two systems.

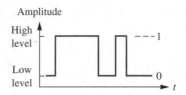

Figure 1.6 A time-varying binary digital signal.

1.2.1 DIGITAL SIGNALS

When we speak of digital electronics, we are most often referring to electronic processing of **binary digital signals,** or signals that can take on only one of two discrete amplitude levels as illustrated in Fig. 1.6. The status of binary systems can be represented by two symbols: a logical 1 is assigned to represent one level, and a logical 0 is assigned to the second level.[2] The two logic states generally correspond to two separate voltages—V_H and V_L—representing the high and low amplitude levels, and a number of voltage ranges are in common use. Although $V_H = 5$ V and $V_L = 0$ V represented the primary standard for many years, these have given way to lower voltage levels because of power consumption and semiconductor device limitations. Systems employing $V_H = 3.3, 2.5,$ and 1.5 V, with $V_L = 0$ V, are now used in many types of electronics.

However, binary voltage levels can also be negative or even bipolar. One high-performance logic family called ECL uses $V_H = -0.8$ V and $V_L = -2.0$ V, and the early standard RS-422 and RS-232 communication links between a small computer and its peripherals used $V_H = +12$ V and $V_L = -12$ V. In addition, the time-varying binary signal in Fig. 1.6 could equally well represent the amplitude of a current or that of an optical signal being transmitted down a fiber in an optical digital communication system. The more recent USB and Firewire standards have returned to the use of a single positive supply voltage.

Part II of this text discusses the design of a number of families of digital circuits using various semiconductor technologies. These include CMOS,[3] NMOS, and PMOS logic, which use field-effect transistors, and the TTL and ECL families, which are based on bipolar transistors.

1.2.2 ANALOG SIGNALS

Although quantities such as electronic charge and electron spin are truly discrete, much of the physical world is really analog in nature. Our senses of vision, hearing, smell, taste, and touch are all analog processes. Analog signals directly represent variables such as temperature, humidity, pressure, light intensity, or sound—all of which may take on any value, typically within some finite range. In reality, classification of digital and analog signals is largely one of perception. If we look at a digital signal similar to the one in Fig. 1.6 with an oscilloscope, we find that it actually makes a continuous transition between the high and low levels. The signal cannot make truly abrupt transitions between two levels. Designers of high-speed digital systems soon realize that they are really dealing with analog signals. The time-varying voltage or current plotted in Fig. 1.7 could be the electrical representation of temperature, flow rate, or pressure versus time, or the continuous audio output from a microphone. Some analog transducers produce output *voltages* in the range of 0 to 5 or 0 to 10 V, whereas others are designed to produce an output *current* that ranges between 4 and 20 mA. At the other extreme, signals brought in by a radio antenna can be as small as a fraction of a microvolt.

To process the information contained in these analog signals, electronic circuits are used to selectively modify the amplitude, phase, and frequency content of the signals. In addition, significant

---

[2] This assignment facilitates the use of Boolean algebra, reviewed in Chapter 6.

[3] For now, let us accept these initials as proper names without further definition. The details of each of these circuits are developed in Part II.

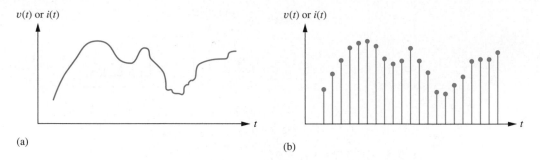

Figure 1.7 (a) A continuous analog signal; (b) sampled data version of signal in (a).

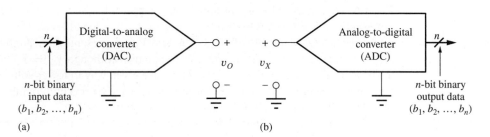

Figure 1.8 Block diagram representation for a (a) D/A converter and a (b) A/D converter.

increases in the voltage, current, and power level of the signal are usually needed. All these modifications to the signal characteristics are achieved using various forms of amplifiers, and Part III of this text discusses the analysis and design of a wide range of amplifiers using bipolar and field-effect transistors and operational amplifiers.

1.2.3 A/D AND D/A CONVERTERS—BRIDGING THE ANALOG AND DIGITAL DOMAINS

For analog and digital systems to be able to operate together, we must be able to convert signals from analog to digital form and vice versa. We sample the input signal at various points in time as in Fig. 1.7(b) and convert or quantize its amplitude into a digital representation. The quantized value can be represented in binary form or can be a decimal representation as given by the display on a digital multimeter. The electronic circuits that perform these translations are called digital-to-analog (D/A) and analog-to-digital (A/D) converters.

Digital-to-Analog Conversion
The **digital-to-analog converter,** often referred to as a **D/A converter** or **DAC,** provides an interface between the digital signals of computer systems and the continuous signals of the analog world. The D/A converter takes digital information, most often in binary form, as input and generates an output voltage or current that may be used for electronic control or analog information display. In the DAC in Fig. 1.8(a), an n-bit binary input word $(b_1, b_2, \ldots, b_n)$ is treated as a binary fraction and multiplied by a full-scale reference voltage V_{FS} to set the output of the D/A converter. The behavior of the DAC can be expressed mathematically as

$$v_O = (b_1 2^{-1} + b_2 2^{-2} + \cdots + b_n 2^{-n}) V_{FS} \qquad \text{for } b_i \in \{1, 0\} \qquad (1.1)$$

Examples of typical values of the full-scale voltage V_{FS} are 2.5, 5, 5.12, 10, and 10.24 V. The smallest voltage change that can occur at the output takes place when the **least significant bit** b_n, or **LSB,** in the digital word changes from a 0 to a 1. This minimum voltage change is also referred to as the

resolution of the converter and is given by

$$V_{\text{LSB}} = 2^{-n} V_{FS} \qquad (1.2)$$

At the other extreme, b_1 is referred to as the **most significant bit,** or **MSB,** and has a weight of one-half V_{FS}.

EXERCISE: A 10-bit D/A converter has $V_{FS} = 5.12$ V. What is the output voltage for a binary input code of (1100010001)? What is V_{LSB}? What is the size of the MSB?

ANSWERS: 3.925 V; 5 mV; 2.56 V

Analog-to-Digital Conversion

The **analog-to-digital converter (A/D converter** or **ADC)** is used to transform analog information in electrical form into digital data. The ADC in Fig. 1.8(b) takes an unknown continuous analog input signal, usually a voltage v_X, and converts it into an n-bit binary number that can be easily manipulated by a computer. The n-bit number is a binary fraction representing the ratio between the unknown input voltage v_X and the converter's full-scale voltage V_{FS}.

For example, the input–output relationship for an ideal 3-bit A/D converter is shown in Fig. 1.9(a). As the input increases from zero to full scale, the output digital code word stair-steps from 000 to 111. The output code is constant for an input voltage range equal to 1 LSB of the ADC. Thus, as the input voltage increases, the output code first underestimates and then overestimates the input voltage. This error, called **quantization error,** is plotted against input voltage in Fig. 1.9(b).

For a given output code, we know only that the value of the input voltage lies somewhere within a 1-LSB quantization interval. For example, if the output code of the 3-bit ADC is 100, corresponding to a voltage $V_{FS}/2$, then the input voltage can be anywhere between $\frac{7}{16}V_{FS}$ and $\frac{9}{16}V_{FS}$, a range of $V_{FS}/8$ V or 1 LSB. From a mathematical point of view, the ADC circuitry in Fig. 1.8(b) picks the values of the bits in the binary word to minimize the magnitude of the quantization error v_ε between the unknown input voltage v_X and the nearest quantized voltage level:

$$v_\varepsilon = |v_X - (b_1 2^{-1} + b_2 2^{-2} + \cdots + b_n 2^{-n}) V_{FS}| \qquad (1.3)$$

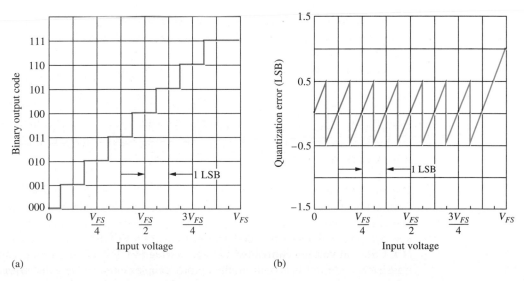

Figure 1.9 (a) Input-output relationship and (b) quantization error for 3-bit ADC.

1.3 NOTATIONAL CONVENTIONS

In many circuits we will be dealing with both dc and time-varying values of voltages and currents. The following standard notation will be used to keep track of the various components of an electrical signal. Total quantities will be represented by lowercase letters with capital subscripts, such as v_T and i_T in Eq. (1.4). The dc components are represented by capital letters with capital subscripts as, for example, V_{DC} and I_{DC} in Eq. (1.4); changes or variations from the dc value are represented by signal components v_{sig} and i_{sig}.

$$v_T = V_{DC} + v_{sig} \qquad \text{or} \qquad i_T = I_{DC} + i_{sig} \tag{1.4}$$

As examples, the total base-emitter voltage v_{BE} of a transistor or the total drain current i_D of a field-effect transistor are written as

$$v_{BE} = V_{BE} + v_{be} \qquad \text{and} \qquad i_D = I_D + i_d \tag{1.5}$$

Unless otherwise indicated, the equations describing a given network will be written assuming a consistent set of units: volts, amperes, and ohms. For example, the equation 5 V $= (10,000 \ \Omega)I_1 + 0.6$ V will be written as $5 = 10,000I_1 + 0.6$.

Resistance and Conductance Representations

In the circuits throughout this text, resistors will be indicated symbolically as R_x or r_x, and the values will be expressed in Ω, kΩ, MΩ, and so on. During analysis, however, it may be more convenient to work in terms of conductance with the following convention:

$$G_x = \frac{1}{R_x} \qquad \text{and} \qquad g_\pi = \frac{1}{r_\pi} \tag{1.6}$$

For example, conductance G_x always represents the reciprocal of the value of R_x, and g_π represents the reciprocal of r_π. The values next to a resistor symbol will always be expressed in terms of resistance (Ω, kΩ, MΩ).

Dependent Sources

In electronics, **dependent** (or **controlled**) **sources** are used extensively. Four types of dependent sources are summarized in Fig. 1.10, in which the standard diamond shape is used for controlled sources. The **voltage-controlled current source (VCCS), current-controlled current source (CCCS),** and **voltage-controlled voltage source (VCVS)** are used routinely in this text to model transistors and amplifiers or to simplify more complex circuits. Only the **current-controlled voltage source (CCVS)** sees limited use.

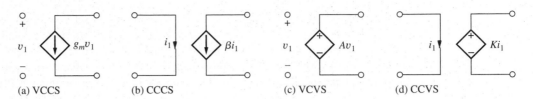

Figure 1.10 Controlled sources. (a) Voltage-controlled current source (VCCS). (b) Current-controlled current source (CCCS). (c) Voltage-controlled voltage source (VCVS). (d) Current-controlled voltage source (CCVS).

1.4 PROBLEM-SOLVING APPROACH

Solving problems is a centerpiece of an engineer's activity. As engineers, we use our creativity to find new solutions to problems that are presented to us. A well-defined approach can aid significantly in solving problems. The examples in this text highlight an approach that can be used in all facets of your career, as a student and as an engineer in industry. The method is outlined in the following nine steps:

1. State the **problem** as clearly as possible.
2. List the **known information and given data.**
3. Define the **unknowns** that must be found to solve the problem.
4. List your **assumptions.** You may discover additional assumptions as the analysis progresses.
5. Develop an **approach** from a group of possible alternatives.
6. Perform an **analysis** to find a solution to the problem. As part of the analysis, be sure to draw the circuit and label the variables.
7. **Check the results.** Has the problem been solved? Is the math correct? Have all the unknowns been found? Have the assumptions been satisfied? Do the results satisfy simple consistency checks?
8. **Evaluate the solution.** Is the solution realistic? Can it be built? If not, repeat steps 4–7 until a satisfactory solution is obtained.
9. **Computer-aided analysis.** SPICE and other computer tools are highly useful to check the results and to see if the solution satisfies the problem requirements. Compare the computer results to your hand results.

To begin solving a problem, we must try to understand its details. The first four steps, which attempt to clearly define the problem, can be the most important part of the solution process. Time spent understanding, clarifying, and defining the problem can save much time and frustration.

The first step is to write down a statement of the problem. The original problem description may be quite vague; we must try to understand the problem as well as, or even better than, the individual who posed the problem. As part of this focus on understanding the problem, we list the information that is known and unknown. Problem-solving errors can often be traced to imprecise definition of the unknown quantities. For example, it is very important for analysis to draw the circuit properly and to clearly label voltages and currents on our circuit diagrams. (Have you ever lost points on homework and exams as a result of poor labeling of variables in circuit diagrams?)

Often there are more unknowns than constraints, and we need engineering judgment to reach a solution. Part of our task in studying electronics is to build up the background for selecting between various alternatives. Along the way, we often need to make approximations and assumptions that simplify the problem or form the basis of the chosen approach. It is important to state these assumptions, so that we can be sure to check their validity at the end. Throughout this text you will encounter opportunities to make assumptions. Most often, you should make assumptions that simplify your computational effort yet still achieve useful results.

The exposition of the known information, unknowns, and assumptions helps us not only to better understand the problem but also to think about various alternative solutions. We must choose the approach that appears to have the best chance of solving the problem. There may be more than one satisfactory approach. Each person will view the problem somewhat differently, and the approach that is clearest to one individual may not be the best for another. Pick the one that seems best to you. As part of defining the approach, be sure to think about what computational tools are available to assist in the solution, including MATLAB®, Mathcad®, spreadsheets, SPICE, and your calculator.

Once the problem and approach are defined as clearly as possible, then we can perform any analysis required and solve the problem. After the analysis is completed we need to check the results. A number of questions should be resolved. First, have all the unknowns been found? (Have you ever lost credit on an exam because you forgot to work part of the problem?) Do the results make sense? Are they consistent with each other? Are the results consistent with assumptions used in developing the approach to the problem?

Then we need to evaluate the solution. Are the results viable? For example, are the voltage, current, and power levels reasonable? Can the circuit be realized with reasonable yield with real components? Will the circuit continue to function within specifications in the face of significant component variations? Is the cost of the circuit within specifications? If the solution is not satisfactory, we need to modify our approach and assumptions and attempt a new solution. An iterative solution is often required to meet the specifications in realistic design situations. SPICE and other computer tools are highly useful for checking results and ensuring that the solution satisfies the problem requirements.

The solutions to the examples in this text have been structured following the problem-solving approach introduced here. Although some examples may appear trivial, the power of the structured approach increases as the problem becomes more complex.

WHAT ARE REASONABLE NUMBERS?

Part of our results check should be to decide if the answer is "reasonable" and makes sense. Over time we must build up an understanding of what numbers are reasonable. Most solid-state devices that we will encounter are designed to operate from voltages ranging from a battery voltage of 1 V on the low end to no more than 40–50 V[4] at the high end. Typical power supply voltages will be in the 10- to 20-V range, and typical resistance values encountered will range from hundreds of Ω up to many MΩ.

Based on our knowledge of dc circuits, we should expect that the voltages in our circuits not exceed the power supply voltages. For example, if a circuit is operating from +8- and −5-V supplies, all of our calculated dc voltages must be between −5 and +8 V. In addition, the peak-to-peak amplitude of an ac signal should not exceed 13 V, the difference of the two supply voltages. With a 10-V supply, the maximum current that can go through a 100-Ω resistor is 100 mA; the current through a 10-MΩ resistor can be no more than 1 μA. Thus we should remember the following "rules" to check our results:

1. With few exceptions, the dc voltages in our circuits cannot exceed the power supply voltages. The peak-to-peak amplitude of an ac signal should not exceed the difference of the power supply voltages.
2. The currents in our circuits will range from microamperes to no more than a hundred milliamperes or so.

---

[4] The primary exception is in the area of power electronics, where one encounters much larger voltages and currents than the ones discussed here.

1.5 IMPORTANT CONCEPTS FROM CIRCUIT THEORY

Analysis and design of electronic circuits make continuous use of a number of important techniques from basic network theory. Circuits are most often analyzed using a combination of **Kirchhoff's voltage law,** abbreviated **KVL,** and **Kirchhoff's current law,** abbreviated **KCL.** Occasionally, the solution relies on systematic application of **nodal** or **mesh analysis. Thévenin** and **Norton circuit transformations** are often used to help simplify circuits, and the notions of voltage and current division also represent basic tools of analysis. Models of active devices invariably involve dependent sources, as mentioned in the last section, and we need to be familiar with dependent sources in all forms. Amplifier analysis also uses two-port network theory. A review of two-port networks is deferred until the introductory discussion of amplifiers in Chapter 10. If the reader feels uncomfortable with any of the concepts just mentioned, this is a good time for review. To help, a brief review of these important circuit techniques is given in the next several sections.

1.5.1 VOLTAGE AND CURRENT DIVISION

Voltage and current division are highly useful circuit analysis techniques that can be derived directly from basic circuit theory. They are both used routinely throughout this text, and it is very important to be sure to understand the conditions for which each technique is valid! Examples of both methods are provided next.

Voltage Division
Voltage division is demonstrated by the circuit in Fig. 1.11(a) in which the voltages v_1 and v_2 can be expressed as

$$v_1 = i_S R_1 \quad \text{and} \quad v_2 = i_S R_2 \tag{1.7}$$

Applying KVL to the single loop,

$$v_S = v_1 + v_2 = i_S(R_1 + R_2) \quad \text{and} \quad i_S = \frac{v_S}{R_1 + R_2} \tag{1.8}$$

Combining Eqs. (1.7) and (1.8) yields the basic voltage division formula:

$$v_1 = v_S \frac{R_1}{R_1 + R_2} \quad \text{and} \quad v_2 = v_S \frac{R_2}{R_1 + R_2} \tag{1.9}$$

For the resistor values in Fig. 1.11(a),

$$v_1 = 10 \text{ V} \frac{8 \text{ k}\Omega}{8 \text{ k}\Omega + 2 \text{ k}\Omega} = 8.00 \text{ V} \quad \text{and} \quad v_2 = 10 \text{ V} \frac{2 \text{ k}\Omega}{8 \text{ k}\Omega + 2 \text{ k}\Omega} = 2.00 \text{ V} \tag{1.10}$$

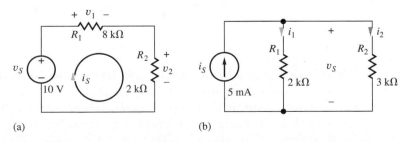

(a)   (b)

Figure 1.11   (a) A resistive voltage divider, (b) Current division in a simple network.

DESIGN NOTE   VOLTAGE DIVIDER RESTRICTIONS

Note that the voltage divider relationships in Eq. (1.9) can be applied only when the current through the two resistor branches is the same. Also, note that the formulas are correct if the resistances are replaced by complex impedances and the voltages are represented as **phasors.**

$$\mathbf{V_1} = \mathbf{V_S}\frac{Z_1}{Z_1 + Z_2} \qquad \text{and} \qquad \mathbf{V_2} = \mathbf{V_S}\frac{Z_2}{Z_1 + Z_2}$$

Current Division
Current division is also very useful. Let us find the currents i_1 and i_2 in the circuit in Fig. 1.11(b). Using KCL at the single node,

$$i_S = i_1 + i_2 \qquad \text{where } i_1 = \frac{v_S}{R_1} \text{ and } i_2 = \frac{v_S}{R_2} \tag{1.11}$$

and solving for v_S yields

$$v_S = i_S \frac{1}{\dfrac{1}{R_1} + \dfrac{1}{R_2}} = i_S\frac{R_1 R_2}{R_1 + R_2} = i_S(R_1 \| R_2) \tag{1.12}$$

in which the notation $R_1 \| R_2$ represents the parallel combination of resistors R_1 and R_2. Combining Eqs. (1.11) and (1.12) yields the current division formulas:

$$i_1 = i_S\frac{R_2}{R_1 + R_2} \qquad \text{and} \qquad i_2 = i_S\frac{R_1}{R_1 + R_2} \tag{1.13}$$

For the values in Fig. 1.11(b),

$$i_1 = 5 \text{ mA}\frac{3 \text{ k}\Omega}{2 \text{ k}\Omega + 3 \text{ k}\Omega} = 3.00 \text{ mA} \qquad i_2 = 5 \text{ mA}\frac{2 \text{ k}\Omega}{2 \text{ k}\Omega + 3 \text{ k}\Omega} = 2.00 \text{ mA}$$

DESIGN NOTE   CURRENT DIVIDER RESTRICTIONS

It is important to note that the same voltage must appear across both resistors in order for the current division expressions in Eq. (1.13) to be valid. Here again, the formulas are correct if the resistances are replaced by complex impedances and the currents are represented as **phasors.**

$$\mathbf{I_1} = \mathbf{I_S}\frac{Z_2}{Z_1 + Z_2} \qquad \text{and} \qquad \mathbf{I_2} = \mathbf{I_S}\frac{Z_1}{Z_1 + Z_2}$$

1.5.2 THÉVENIN AND NORTON CIRCUIT REPRESENTATIONS

Let us now review the method for finding **Thévenin** and **Norton equivalent circuits,** including a dependent source; the circuit in Fig. 1.12 serves as our illustration. Because the linear network in the dashed box has only two terminals, it can be represented by either the Thévenin or Norton equivalent circuits in Figs. 1.12(b) and 1.12(c). The work of Thévenin and Norton permits us to reduce complex circuits to a single source and equivalent resistance. We illustrate these two important techniques with the next four examples.

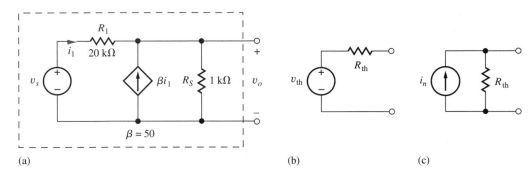

Figure 1.12 (a) Two-terminal circuit and its (b) Thévenin and (c) Norton equivalents.

EXAMPLE 1.1   THÉVENIN EQUIVALENT VOLTAGE

Let's practice finding the Thévenin equivalent voltage source for a circuit.

PROBLEM Find the Thévenin equivalent voltage for the circuit in Fig. 1.12(a).

SOLUTION **Known Information and Given Data:** Circuit topology and circuit values appear in Fig. 1.12(a).

Unknowns: Thévenin equivalent voltage v_{th}.

Approach: Voltage source v_{th} is defined as the open-circuit voltage at the output terminals of the circuit.

Assumptions: None.

Analysis: The open-circuit output voltage can be found by applying KCL at the output node:

$$\beta i_1 = \frac{v_o - v_s}{R_1} + \frac{v_o}{R_S} = G_1(v_o - v_s) + G_S v_o \tag{1.14}$$

by applying the notational convention for conductance from Sec. 1.3 ($G_S = 1/R_S$).
Current i_1 is given by

$$i_1 = G_1(v_s - v_o) \tag{1.15}$$

Substituting Eq. (1.15) into Eq. (1.14) and combining terms yields

$$G_1(\beta + 1)v_s = [G_1(\beta + 1) + G_S]v_o \tag{1.16}$$

The Thévenin equivalent output voltage is then found to be

$$v_o = \frac{G_1(\beta + 1)}{[G_1(\beta + 1) + G_S]}v_s = \frac{(\beta + 1)R_S}{[(\beta + 1)R_S + R_1]}v_s \tag{1.17}$$

where the second relationship was found by multiplying numerator and denominator by ($R_1 R_S$). For the values in this problem,

$$v_o = \frac{(50 + 1)1 \text{ k}\Omega}{[(50 + 1) \ 1 \text{ k}\Omega + 20 \text{ k}\Omega]}v_s = 0.718v_s \quad \text{and} \quad v_{th} = 0.718v_s \tag{1.18}$$

Check of Results: We have found the single requested unknown. A double check of the analysis shows it is correct.

EXAMPLE **1.2**   THÉVENIN EQUIVALENT RESISTANCE

Now let us find the Thévenin equivalent resistance for the circuit in Ex. 1.1.

PROBLEM   Find the Thévenin equivalent resistance for the circuit in Fig. 1.12(a).

SOLUTION   **Known Information and Given Data:** Circuit topology and circuit values appear in Fig. 1.12(a).

Unknowns: Thévenin equivalent resistance R_{th}.

Approach: R_{th} represents the equivalent resistance present at the output terminals with all independent sources set to zero. To find the **Thévenin equivalent resistance** R_{th}, we first set the independent sources in the network to zero. Remember, however, that any dependent sources must remain active. A test voltage or current source is then applied to the network terminals and the corresponding current or voltage calculated. In Fig. 1.13 v_s is set to zero, voltage source v_x is applied to the network, and the current i_x must be determined so that

$$R_{\text{th}} = \frac{v_x}{i_x} \tag{1.19}$$

can be calculated.

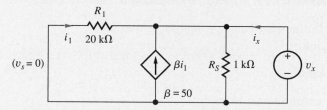

Figure 1.13 A test source v_x is applied to the network to find R_{th}.

Assumptions: None.

Analysis: Applying KCL,

$$i_x = -i_1 - \beta i_1 + G_S v_x \qquad \text{in which } i_1 = -G_1 v_x \tag{1.20}$$

Combining and simplifying these two expressions yields

$$i_x = [(\beta + 1)G_1 + G_S]v_x \qquad \text{and} \qquad R_{\text{th}} = \frac{v_x}{i_x} = \frac{1}{(\beta + 1)G_1 + G_S} \tag{1.21}$$

The denominator of Eq. (1.21) represents the sum of two conductances, which corresponds to the parallel combination of two resistances. Therefore, Eq. (1.21) can be rewritten as

$$R_{\text{th}} = \frac{1}{(\beta + 1)G_1 + G_S} = \frac{R_S \dfrac{R_1}{(\beta + 1)}}{R_S + \dfrac{R_1}{(\beta + 1)}} = R_S \left\| \frac{R_1}{(\beta + 1)} \right. \tag{1.22}$$

For the values in this example,

$$R_{\text{th}} = R_S \left\| \frac{R_1}{(\beta + 1)} \right. = 1\ \text{k}\Omega \left\| \frac{20\ \text{k}\Omega}{(50 + 1)} \right. = 1\ \text{k}\Omega \| 392\ \Omega = 282\ \Omega \tag{1.23}$$

Check of Results: We have found the single unknown requested. The value is less than 1 kΩ, which seems reasonable since we would not expect the resistance to exceed the value of R_S that appears in parallel with the output terminals.

We can also check the result by remembering that R_{th} is equal to the open-circuit voltage divided by the short-circuit current. If we short the output terminals in Fig. 1.12, we find the short-circuit current to be $i_{SC} = (\beta + 1)v_S/R_1$. Using the open-circuit voltage expression from Eq. (1.17), we find that Eq. (1.22) is correct.

EXERCISE: Show that the value of the short-circuit current given in Ex. 1.2 is correct, and verify the result in Eq. (1.22).

<table><tr><td>EXAMPLE **1.3**</td><td>**NORTON EQUIVALENT CIRCUIT**</td></tr></table>

Practice finding the Norton equivalent circuit for a network containing a dependent source.

PROBLEM   Find the Norton equivalent (Fig. 1.12(c)) for the circuit in Fig. 1.12(a).

SOLUTION   **Known Information and Given Data:** Circuit topology and circuit values appear in Fig. 1.12(a). The value of R_{th} was calculated in the previous example.

Unknowns: Norton equivalent current i_n.

Approach: The Norton equivalent current is found by determining the current coming out of the network when a short circuit is applied to the terminals.

Assumptions: None.

Analysis: For the circuit in Fig. 1.14, the output current will be

$$i_n = i_1 + \beta i_1 \qquad \text{and} \qquad i_1 = G_1 v_s \tag{1.24}$$

The short circuit across the output forces the current through R_S to be 0. Combining the two expressions in Eq. (1.24) yields

$$i_n = (\beta + 1)G_1 v_s = \frac{(\beta + 1)}{R_1} v_s \tag{1.25}$$

or

$$i_n = \frac{(50 + 1)}{20 \text{ k}\Omega} v_s = \frac{v_s}{392 \ \Omega} = (2.55 \text{ mS})v_s \tag{1.26}$$

The resistance in the Norton equivalent circuit also equals R_{th} found in Eq. (1.23).

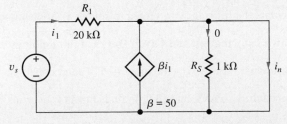

Figure 1.14 Circuit for determining short-circuit output current.

Check of Results: We have found the Norton equivalent current. Note that $v_{TH} = i_N R_{th}$ and this result can be used to check the calculations: $i_n R_{th} = (2.55 \text{ mS})v_s(282 \ \Omega) = 0.719 \, v_s$, which agrees within round-off error.

The Thévenin and Norton equivalent circuits for Fig. 1.12 calculated in the previous three examples appear for comparison in Fig. 1.15. Before closing this section, let us work one final example combining the use of Kirchhoff's voltage and current laws with a voltage-controlled current source in the circuit. For this problem, we will calculate the resistance presented to source v_s in the circuit of Fig. 1.16.

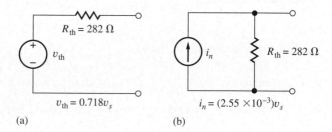

(a)   (b)

Figure 1.15 Completed (a) Thévenin and (b) Norton equivalent circuits for the two-terminal network in Fig. 1.12(a).

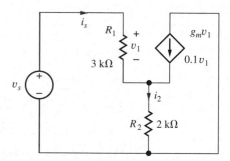

Figure 1.16 Circuit containing a voltage-controlled current source.

EXAMPLE 1.4   **THÉVENIN EQUIVALENT RESISTANCE**

Here we calculate the Thévenin equivalent resistance for a network containing a dependent voltage source.

PROBLEM Calculate the resistance presented to source v_s in the circuit of Fig. 1.16.

SOLUTION **Known Information and Given Data:** Circuit topology and element values appear in Fig. 1.16.

Unknowns: Input resistance presented to source v_s.

Approach: Use KVL and KCL to find current i_s in terms of v_s. The resistance presented to source v_s is $R = v_s/i_s$.

Assumptions: None.

Analysis: Applying KVL around the loop containing v_s yields

$$v_s = i_s R_1 + i_2 R_2 = i_s R_1 + (i_s + g_m v_1) R_2 \tag{1.27}$$

However, the voltage v_1 is related to the current i_s by

$$v_1 = i_s R_1 \tag{1.28}$$

Combining Eqs. (1.27) and (1.28) gives

$$v_s = i_s (R_1 + R_2 + g_m R_1 R_2) \tag{1.29}$$

and

$$R = \frac{v_s}{i_s} = R_1 + R_2(1 + g_m R_1) \tag{1.30}$$

For the values in this particular circuit, we find

$$R = 3000 \ \Omega + 2000 \ \Omega \ [1 + (0.1 \ \text{S})(3000 \ \Omega)] = 605 \ \text{k}\Omega \tag{1.31}$$

Check of Results: We have found the unknown resistance. However, at this point we should have reason to question the calculated value. It seems quite large compared to the various resistance values that appear in the circuit. Upon checking our calculations, we find no errors. In Chapters 13 and 14 we will find that this circuit is a model for the bipolar transistor operating as a common-collector amplifier, and one of its properties is to provide a significant impedance transformation.

1.6 FREQUENCY SPECTRUM OF ELECTRONIC SIGNALS

Fourier analysis and the **Fourier series** represent extremely powerful tools in electrical engineering. Results from Fourier theory show that complicated signals are actually composed of a continuum of sinusoidal components, each having a distinct amplitude, frequency, and phase. The **frequency spectrum** of a signal presents the amplitude and phase of the components of the signal versus frequency.

Nonrepetitive signals have continuous spectra with signals that may occupy a broad range of frequencies. For example, the amplitude spectrum of a television signal measured during a small time interval is depicted in Fig. 1.17. The TV video signal is designed to occupy the frequency range from 0 to 4.5 MHz.[5] Other types of signals occupy different regions of the frequency spectrum. Table 1.3 identifies the frequency ranges associated with various categories of common signals.

In contrast to the continuous spectrum in Fig. 1.17, Fourier series analysis shows that *any* *periodic* signal, such as the square wave of Fig. 1.18, contains spectral components only at discrete frequencies[6] that are related directly to the period of the signal. For example, the square wave of Fig. 1.18 having an amplitude V_O and period T can be represented by the Fourier series

$$v(t) = V_{DC} + \frac{2V_O}{\pi} \left(\sin \omega_o t + \frac{1}{3} \sin 3\omega_o t + \frac{1}{5} \sin 5\omega_o t + \cdots \right) \tag{1.32}$$

in which $\omega_o = 2\pi/T$ (rad/s) is the **fundamental radian frequency** of the square wave. We refer to $f_o = 1/T$ (Hz) as the **fundamental frequency** of the signal, and the frequency components at $2f_o$, $3f_o, 4f_o, \ldots$ are called the second, third, fourth, and so on **harmonic frequencies.**

---

[5] This signal is combined with a much higher carrier frequency prior to transmission.

[6] There are an infinite number of components, however.

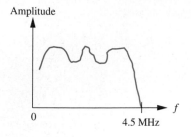

Figure 1.17 Spectrum of a TV signal.

TABLE 1.3
Frequencies Associated with Common Signals

CATEGORY	FREQUENCY RANGE
Audible sounds	20 Hz – 20 kHz
Baseband video (TV) signal	0 – 4.5 MHz
AM radio broadcasting	540 –1600 kHz
High-frequency radio communications	1.6 – 54 MHz
VHF television (Channels 2–6)	54 – 88 MHz
FM radio broadcasting	88 – 108 MHz
VHF radio communication	108 – 174 MHz
VHF television (Channels 7–13)	174 – 216 MHz
Maritime and government communications	216 – 450 MHz
Business communications	450 – 470 MHz
UHF television (Channels 14–69)	470 – 806 MHz
Fixed and mobile communications including	806 – 902 MHz
Allocations for analog and digital cellular	928 – 960 MHz
Telephones, personal communications, and other	1710 –1990 MHz
Wireless devices	2310 –2690 MHz
Satellite television	3.7 – 4.2 GHz
Wireless devices	5.0 – 5.5 GHz

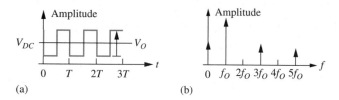

Figure 1.18 A periodic signal (a) and its amplitude spectrum (b).

1.7 AMPLIFIERS

The characteristics of analog signals are most often manipulated using linear amplifiers that affect the amplitude and/or phase of the signal without changing its frequency. Although a complex signal may have many individual components, as just described in Sec. 1.6, linearity permits us to use the **superposition principle** to treat each component individually.

For example, suppose the amplifier with voltage gain A in Fig. 1.19(a) is fed a sinusoidal input signal component v_s with amplitude V_s, frequency ω_s, and phase ϕ:

$$v_s = V_s \sin(\omega_s t + \phi) \tag{1.33}$$

Then, if the amplifier is linear, the output corresponding to this signal component will also be a sinusoidal signal at the same frequency but with a different amplitude and phase:

$$v_o = V_o \sin(\omega_s t + \phi + \theta) \tag{1.34}$$

Using phasor notation, the input and output signals would be represented as

$$\mathbf{V_s} = V_s \angle \phi \qquad \text{and} \qquad \mathbf{V_o} = V_o \angle (\phi + \theta) \tag{1.35}$$

The **voltage gain** of the amplifier is defined in terms of these phasors:

$$A = \frac{\mathbf{V_o}}{\mathbf{V_s}} = \frac{V_o \angle (\phi + \theta)}{V_s \angle \phi} = \frac{V_o}{V_s} \angle \theta \tag{1.36}$$

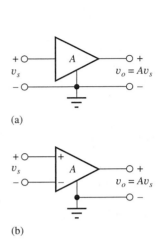

(a)

(b)

Figure 1.19 (a) Symbol for amplifier with single input and voltage gain A; (b) differential amplifier having two inputs and gain A.

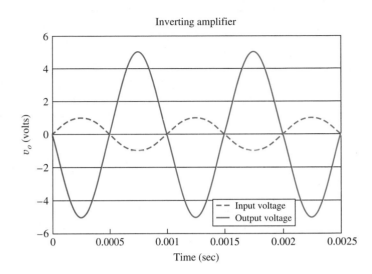

Figure 1.20 Input and output voltage waveforms for an amplifier with gain $A_v = -5$ and $v_s = 1 \sin 2000\pi t$ V.

This amplifier has a voltage gain with magnitude equal to V_o / V_s and a phase shift of θ. In general, both the magnitude and phase of the voltage gain will be a function of frequency. Note that amplifiers also often provide current gain and power gain as well as voltage gain, but these concepts will not be explored further until Chapter 10.

The curves in Fig. 1.20 represent the input and output voltage waveforms for an inverting amplifier with $A_v = -5$ and $v_s = 1 \sin 2000\pi t$ V. Both the factor of five increase in signal amplitude and the 180° phase shift (multiplication by -1) are apparent in the graph.

At this point, a note regarding the phase angle is needed. In Eqs. (1.33) and (1.34), ωt, ϕ, and θ must have the same units. With ωt normally expressed in radians, ϕ should also be in radians. However, in electrical engineering texts, ϕ is often expressed in degrees. We must be aware of this mixed system of units and remember to convert degrees to radians before making any numeric calculations.

EXERCISE: The input and output voltages of an amplifier are expressed as

$$v_s = 0.001 \sin(2000\pi t) \text{ V} \qquad \text{and} \qquad v_o = -5\cos(2000\pi t + 25°) \text{ V}$$

in which v_s and v_o are specified in volts when t is seconds. What are $\mathbf{V_S}$, $\mathbf{V_O}$, and the voltage gain of the amplifier?

ANSWERS: $0.001\angle 0°$; $5\angle{-65°}$; $5000\angle{-65°}$

1.7.1 IDEAL OPERATIONAL AMPLIFIERS

The **operational amplifier, "op amp"** for short, is a fundamental building block in electronic design and is discussed in most introductory circuit courses. A brief review of the ideal op amp is provided here; an in-depth study of the properties of ideal and nonideal op amps and the circuits used to build the op amp itself are the subjects of Chapters 11, 12, 15, and 16. Although it is impossible to realize the **ideal operational amplifier,** its use allows us to quickly understand the basic behavior to be expected from a given circuit and serves as a model to help in circuit design.

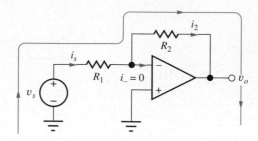

Figure 1.21 Inverting amplifier using op amp.

From our basic circuit courses, we may recall that op amps are differential (or difference) amplifiers that respond to the signal voltage that appears between the + and − input terminals of the amplifier depicted in Fig. 1.19(b). Ideal op amps are assumed to have infinite **voltage gain** and infinite **input resistance,** and these properties lead to two special assumptions that are used to analyze circuits containing ideal op amps:

1. The voltage difference across the input terminals is zero; that is, $v_- = v_+$.
2. Both input currents are zero.

Applying the Assumptions—The Inverting Amplifier The classic **inverting amplifier** circuit will be used to refresh our memory of the analysis of circuits employing op amps. The inverting amplifier is built by grounding the positive input of the operational amplifier and connecting resistors R_1 and R_2, called the **feedback network,** between the inverting input and the signal source and amplifier output node, respectively, as in Fig. 1.21. Note that the ideal op amp is represented by a triangular amplifier symbol without a gain A indicated.

Our goal is to determine the voltage gain A_v of the overall amplifier, and to find A_v, we must find a relationship between $\mathbf{v_s}$ and $\mathbf{v_o}$. One approach is to write an equation for the single loop shown in Fig. 1.21:

$$\mathbf{v_s} - \mathbf{i_s} R_1 - \mathbf{i_2} R_2 - \mathbf{v_o} = 0 \qquad (1.37)$$

Now we need to express $\mathbf{i_s}$ and $\mathbf{i_2}$ in terms of $\mathbf{v_s}$ and $\mathbf{v_o}$. By applying KCL at the inverting input to the amplifier, we see that $\mathbf{i_2}$ must equal $\mathbf{i_s}$ because Assumption 2 states that $\mathbf{i_-}$ must be zero:

$$\mathbf{i_s} = \mathbf{i_2} \qquad (1.38)$$

Current $\mathbf{i_s}$ can be written in terms of $\mathbf{v_s}$ as

$$\mathbf{i_s} = \frac{\mathbf{v_s} - \mathbf{v_-}}{R_1} \qquad (1.39)$$

where $\mathbf{v_-}$ is the voltage at the inverting input (negative input) of the op amp. But Assumption 1 states that the input voltage between the op amp terminals must be zero, so $\mathbf{v_-}$ must be zero because the positive input is grounded. Therefore

$$\mathbf{i_s} = \frac{\mathbf{v_s}}{R_1} \qquad (1.40)$$

Combining Eqs. (1.37)–(1.40), the voltage gain is given by

$$A_v = \frac{\mathbf{v_o}}{\mathbf{v_s}} = -\frac{R_2}{R_1} \qquad (1.41)$$

Referring to Eq. (1.41), we should note several things. The voltage gain is negative, indicative of an inverting amplifier with a 180° phase shift between its input and output signals. In addition, the magnitude of the gain can be greater than or equal to 1 if $R_2 \geq R_1$ (the most common case), but it can also be less than 1 for $R_2 < R_1$.

In the amplifier circuit in Fig. 1.21, the inverting-input terminal of the operational amplifier is at ground potential, 0 V, and is referred to as a **virtual ground.** The ideal operational amplifier adjusts its output to whatever voltage is necessary to force v_- to be zero.

DESIGN NOTE   VIRTUAL GROUND IN OP AMP CIRCUITS

Although the inverting input represents a virtual ground, it is *not* connected directly to ground (there is no direct dc path for current to reach ground). Shorting this terminal to ground for analysis purposes is a common error that must be avoided.

EXERCISE: The amplifier in Fig. 1.21 has a gain of -5 with $R_1 = 10$ kΩ. What is the value of R_2?

ANSWER: 50 kΩ

 ELECTRONICS IN ACTION

Amplifiers in a Familiar Electronic System—The FM Stereo Receiver

The block diagram of an FM radio receiver is an example of an electronic system that uses a number of amplifiers. The signal from the antenna can be very small, often in the microvolt range. The signal's amplitude and power level are increased sequentially by three groups of amplifiers: the radio frequency (RF), intermediate frequency (IF), and audio amplifiers. At the output, the amplifier driving the loudspeaker could be delivering a 100 W audio signal to the speaker, whereas the power originally available from the antenna may amount to only picowatts.

The local oscillator, which tunes the radio receiver to select the desired station, represents another special class of amplifiers; these are investigated at the end of Chapter 17. The mixer circuit actually changes the frequency of the incoming signal and is thus a nonlinear circuit. However, its design draws heavily on linear amplifier circuit concepts. Finally, the FM detector may be formed from either a linear or nonlinear circuit. Chapters 10 to 18 provide in-depth exploration of the design techniques used in linear amplifiers and oscillators and the foundation needed to understand more complex circuits such as mixers, modulators, and detectors.

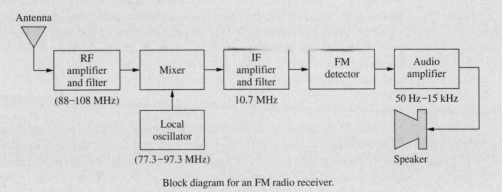

Block diagram for an FM radio receiver.

1.7.2 AMPLIFIER FREQUENCY RESPONSE

In addition to modifying the voltage, current, and/or power level of a given signal, amplifiers are often designed to selectively process signals of different frequency ranges. Amplifiers are classified into a number of categories based on their frequency response; five possible categories are shown in Fig. 1.22. The **low-pass amplifier,** Fig. 1.22(a), passes all signals below some upper cutoff frequency f_H, whereas the **high-pass amplifier,** Fig. 1.22(b), amplifies all signals above the lower cutoff frequency f_L. The **band-pass amplifier** passes all signals between the two cutoff frequencies f_L and f_H, as in Fig. 1.22(c). The **band-reject amplifier** in Fig. 1.22(d) rejects all signals having

Amplitude

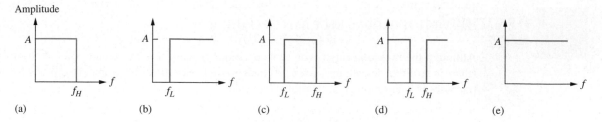

Figure 1.22 Ideal amplifier frequency responses: (a) low-pass, (b) high-pass, (c) band-pass, (d) band-reject, and (e) all-pass characteristics.

frequencies lying between f_L and f_H. Finally, the **all-pass amplifier** in Fig. 1.22(e) amplifies signals at any frequency. The all-pass amplifier is actually used to tailor the phase of the signal rather than its amplitude. Circuits that are designed to amplify specific ranges of signal frequencies are usually referred to as **filters.**

EXERCISE: (a) The band-pass amplifier in Fig. 1.22(c) has $f_L = 1.5$ kHz, $f_H = 2.5$ kHz, and $A = 10$. If the input voltage is given by

$$v_s = [0.5\sin(2000\pi t) + \sin(4000\pi t) + 1.5\sin(6000\pi t)] \text{ V}$$

what is the output voltage of the amplifier?
    (b) Suppose the same input signal is applied to the low-pass amplifier in Fig. 1.22(a), which has $A = 6$ and $f_H = 1.5$ kHz. What is the output voltage?

ANSWERS: $10.0\sin 4000\pi t$ V; $3.00\sin 2000\pi t$ V

1.8 ELEMENT VARIATIONS IN CIRCUIT DESIGN

Whether a circuit is built in discrete form or fabricated as an integrated circuit, the passive components and semiconductor device parameters will all have **tolerances** associated with their values. Discrete resistors can be purchased with a number of different tolerances including ±10 percent, ±5 percent, ±1 percent, or better, whereas resistors in ICs can exhibit even wider variations (±30 percent). Capacitors often exhibit asymmetrical tolerance specifications such as +20 percent/−50 percent, and power supply voltage tolerances are often specified in the range of 1–10 percent. For the semiconductor devices that we shall study in Chapters 3–5, device parameters may vary by 30 percent or more.

In addition to this initial value uncertainty due to tolerances, the values of the circuit components and parameters will vary with temperature and circuit age. It is important to understand the effect of these element changes on our circuits and to be able to design circuits that will continue to operate correctly in the face of such element variations. We will explore two analysis approaches, worst-case analysis and Monte Carlo analysis, that can help quantify the effects of tolerances on circuit performance.

1.8.1 MATHEMATICAL MODELING OF TOLERANCES

A mathematical model for symmetrical parameter variations is

$$P_{nom}(1 - \varepsilon) \leq P \leq P_{nom}(1 + \varepsilon) \tag{1.42}$$

in which P_{nom} is the nominal specification for the parameter such as the resistor value or independent source value, and ε is the fractional tolerance for the component. For example, a resistor R with

nominal value of 10 kΩ and a 5 percent tolerance could exhibit a resistance anywhere in the following range:

$$10{,}000 \ \Omega (1 - 0.05) \le R \le 10{,}000 \ \Omega (1 + 0.05)$$

or

$$9500 \ \Omega \le R \le 10{,}500 \ \Omega$$

EXERCISE: A 39-kΩ resistor has a 10 percent tolerance. What is the range of resistor values corresponding to this resistor? Repeat for a 3.6-kΩ resistor with a 1 percent tolerance.

ANSWERS: $35.1 \le R \le 42.9$ kΩ; $3.56 \le R \le 3.64$ kΩ.

1.8.2 WORST-CASE ANALYSIS

Worst-case analysis is often used to ensure that a design will function under a given set of component variations. Worst-case analysis is performed by choosing values of the various components that make a desired variable (such as voltage, current, power, gain, or bandwidth) as large or as small as possible. These limits are usually found by analyzing a circuit with the values of the various circuit elements pushed to their extremes. Although a design based on the worst case is often too conservative and represents "overdesign," it is important to understand the technique and its limitations. An easy way to explore worst-case analysis is with an example.

EXAMPLE 1.5   **WORST-CASE ANALYSIS**

Here we apply worst-case analysis to a simple voltage divider circuit.

PROBLEM   Find the nominal and worst-case values (highest and lowest) of output voltage V_O and source current I_S for the voltage divider circuit of Fig. 1.23.

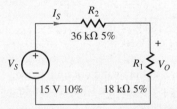

Figure 1.23 Resistor voltage divider circuit with tolerances.

SOLUTION   **Known Information and Given Data:** We have been given the voltage divider circuit in Fig. 1.23; the 15-V source V_S has a 10 percent tolerance; resistor R_1 has a nominal value of 18 kΩ with a 5 percent tolerance; resistor R_2 has a nominal value of 36 kΩ with a 5 percent tolerance. Expressions for V_O and I_S are

$$V_O = V_S \frac{R_1}{R_1 + R_2} \qquad \text{and} \qquad I_S = \frac{V_S}{R_1 + R_2} \qquad (1.43)$$

Unknowns: $V_O^{\text{nom}}, V_O^{\text{max}}, V_O^{\text{min}}, I_S^{\text{nom}}, I_S^{\text{max}}, I_S^{\text{min}}$

Approach: Find the nominal values of V_O and I_S with all circuit elements set to their nominal (ideal) values. Find the worst-case values by selecting the individual voltage and resistance values that force V_O and I_S to their extremes. Note that the values selected for the various circuit elements to produce V_O^{max} will most likely differ from those that produce I_S^{max}, and so on.

Assumptions: None.

Analysis: *(a) Nominal Values*
The nominal value of voltage V_O is found using the nominal values for all the parameters:

$$V_O^{\text{nom}} = V_S^{\text{nom}} \frac{R_1^{\text{nom}}}{R_1^{\text{nom}} + R_2^{\text{nom}}} = 15\,\text{V} \frac{18\,\text{k}\Omega}{18\,\text{k}\Omega + 36\,\text{k}\Omega} = 5\,\text{V} \tag{1.44}$$

Similarly, the nominal value of source current I_S is

$$I_S^{\text{nom}} = \frac{V_S^{\text{nom}}}{R_1^{\text{nom}} + R_2^{\text{nom}}} = \frac{15\,\text{V}}{18\,\text{k}\Omega + 36\,\text{k}\Omega} = 278\,\mu\text{A} \tag{1.45}$$

(b) Worst-Case Limits
Now let us find the worst-case values (the largest and smallest possible values) of voltage V_O and current I_S that can occur for the given set of element tolerances. First, the values of the components will be selected to make V_O as large as possible. However, it may not always be obvious at first to which extreme to push the individual component values. Rewriting Eq. (1.43) for voltage V_O will help:

$$V_O = V_S \frac{R_1}{R_1 + R_2} = \frac{V_S}{1 + R_2/R_1} \tag{1.46}$$

In order to make V_O as large as possible, the numerator of Eq. (1.46) should be large and the denominator small. Therefore, V_S and R_1 should be chosen to be as large as possible and R_2 as small as possible. Conversely, in order to make V_O as small as possible, V_S and R_1 must be small and R_2 must be large. Using this approach, the maximum and minimum values of V_O are

$$V_O^{\text{max}} = \frac{15\,\text{V}(1.1)}{1 + \dfrac{36\,\text{k}\Omega(0.95)}{18\,\text{k}\Omega(1.05)}} = 5.87\,\text{V} \quad \text{and} \quad V_O^{\text{min}} = \frac{15\,\text{V}(.90)}{1 + \dfrac{36\,\text{k}\Omega(1.05)}{18\,\text{k}\Omega(0.95)}} = 4.20\,\text{V} \tag{1.47}$$

The maximum value of V_O is 17 percent greater than the nominal value of 5 V, and the minimum value is 16 percent below the nominal value.

The worst-case values of I_S are found in a similar manner but require different choices for the values of the resistors:

$$I_S^{\text{max}} = \frac{V_S^{\text{max}}}{R_1^{\text{min}} + R_2^{\text{min}}} = \frac{15\,\text{V}(1.1)}{18\,\text{k}\Omega(0.95) + 36\,\text{k}\Omega(0.95)} = 322\,\mu\text{A}$$

$$\tag{1.48}$$

$$I_S^{\text{min}} = \frac{V_S^{\text{min}}}{R_1^{\text{max}} + R_2^{\text{max}}} = \frac{15\,\text{V}(0.9)}{18\,\text{k}\Omega(1.05) + 36\,\text{k}\Omega(1.05)} = 238\,\mu\text{A}$$

The maximum of I_S is 16 percent greater than the nominal value, and the minimum value is 14 percent less than nominal.

Check of Results: The nominal and worst-case values have been determined and range 14–17 percent above and below the nominal values. We have three circuit elements that are varying, and the sum of the three tolerances is 20 percent. Our worst-case values differ from the nominal case by somewhat less than this amount, so the results appear reasonable.

> **EXERCISE:** Find the nominal and worst-case values of the power delivered by source V_S in Fig. 1.23.
>
> **ANSWERS:** 4.17 mW, 3.21 mW, 5.31 mW.

DESIGN NOTE BE WARY OF WORST-CASE DESIGN

In a real circuit, the parameters will be randomly distributed between the limits, and it is unlikely that the various components will all reach their extremes at the same time. Thus the worst-case analysis technique will overestimate (often badly) the extremes of circuit behavior, and a design based on worst-case analysis usually represents an unnecessary overdesign that is more costly than necessary to achieve the specifications with satisfactory yield. A better, although more complex, approach is to attack the problem statistically using Monte Carlo analysis. However, if every circuit must work no matter what, worst-case analysis may be appropriate.

1.8.3 MONTE CARLO ANALYSIS

Monte Carlo analysis uses randomly selected versions of a given circuit to predict its behavior from a statistical basis. For Monte Carlo analysis, a value for each of the elements in the circuit is selected at random from the possible distributions of parameters, and the circuit is then analyzed using the randomly selected element values. Many such randomly selected realizations ("cases" or "instances") of the circuit are generated, and the statistical behavior of the circuit is built up from analysis of the many test cases. Obviously, this is a good use of the computer. Before proceeding, we need to refresh our memory concerning a few results from probability and random variables.

Uniformly Distributed Parameters

In this section, the variable parameters will be assumed to be uniformly distributed between the two extremes. In other words, the probability that any given value of the parameter will occur is the same. In fact, when the parameter tolerance expression in Eq. (1.42) was first encountered, most of us probably visualized it in terms of a uniform distribution.

In mathematical terms, the probability density function $p(r)$ for a uniformly distributed resistor r is represented graphically in Fig. 1.24(a). The probability that a resistor value lies between r and $(r + dr)$ is equal to $p(r)\, dr$. The total probability P must equal unity, so

$$P = \int_{-\infty}^{+\infty} p(r)\, dr = 1 \tag{1.49}$$

Using this equation with the uniform probability density of Fig. 1.24(a) yields $p(r) = \frac{1}{2\varepsilon R_{\text{nom}}}$ as indicated in the figure.

Monte Carlo analysis can be readily implemented with a spreadsheet, MATLAB®, Mathcad®, or another computer program using the **uniform random number generators** that are built into the software. Successive calls to these random number generators produce a sequence of pseudo-**random numbers** that are uniformly distributed between 0 and 1 with a mean of 0.5 as in Fig. 1.24(b).

For example, the Excel® spreadsheet contains the function called RAND( ) (used with a null argument), whereas MATLAB® uses rand,[7] and Mathcad® uses rnd(1). These functions generate random numbers with the distribution in Fig. 1.24(b). Other software products contain random

---

[7] In MATLAB®, rand generates a single random number, rand(n) is an $n \times n$ matrix of random numbers, and rand (n, m) is an $n \times m$ matrix of random numbers. In Mathcad®, rnd(x) returns a number uniformly distributed between o and x.

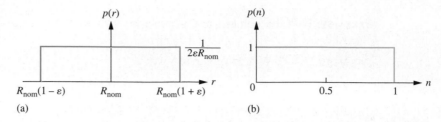

Figure 1.24   (a) Probability density function for a uniformly distributed resistor; (b) probability density function for a random variable uniformly distributed between 0 and 1.

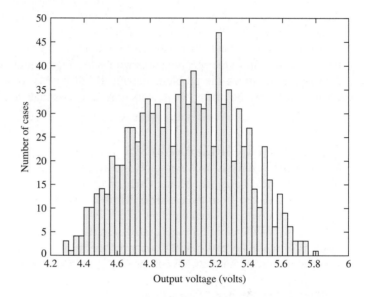

Figure 1.25   Histogram of a 1000-case simulation.

number generators with similar names. In order to use RAND( ) to generate the distribution in Fig. 1.24(a), the mean must be centered at R_{nom} and the width of the distribution set to $(2\varepsilon) \times R_{\text{nom}}$:

$$R = R_{\text{nom}}(1 + 2\varepsilon(\text{RAND}( ) - 0.5)) \tag{1.50}$$

Now let us see how we use Eq. (1.50) in implementing a Monte Carlo analysis.

EXAMPLE 1.6   MONTE CARLO ANALYSIS

Now we will apply Monte Carlo analysis to the voltage divider circuit.

PROBLEM   Perform a Monte Carlo analysis of the circuit in Fig. 1.23. Find the mean, standard deviation, and largest and smallest values for V_O, I_S, and the power delivered from the source.

SOLUTION   **Known Information and Given Data:** The voltage divider circuit appears in Fig. 1.23. The 15 V source V_S has a 10 percent tolerance, resistor R_1 has a nominal value of 18 kΩ with a 5 percent tolerance, and resistor R_2 has a nominal value of 36 kΩ with a 5 percent tolerance. Expressions for V_O, I_S, and P_S are

$$V_O = V_S \frac{R_1}{R_1 + R_2} \qquad I_S = \frac{V_S}{R_1 + R_2} \qquad P_S = V_S I_S$$

Unknowns: The mean, standard deviation, and largest and smallest values for V_O, I_S, and P_S.

Approach: To perform a Monte Carlo analysis of the circuit in Fig. 1.23, we assign randomly selected values to V_S, R_1, and R_2 and then use the values to determine V_O and I_S. Using Eq. (1.50) with the tolerances specified in Fig. 1.23, the power supply and resistor values are represented as

$$1. \quad V_S = 15(1 + 0.2(\text{RAND}() - 0.5))$$

$$2. \quad R_1 = 18{,}000(1 + 0.1(\text{RAND}() - 0.5)) \qquad (1.51)$$

$$3. \quad R_2 = 36{,}000(1 + 0.1(\text{RAND}() - 0.5))$$

Note that each variable must invoke a separate call of the function RAND( ) so that the random values will be independently selected. The random elements in Eqs. (1.51) are then used to evaluate the equations that characterize the circuit, including the power delivered from the source:

$$4. \quad V_O = V_S \frac{R_1}{R_1 + R_2}$$

$$5. \quad I_S = \frac{V_S}{R_1 + R_2} \qquad (1.52)$$

$$6. \quad P_S = V_S I_S$$

This example will utilize a spreadsheet. However, any number of computer tools could be used: MATLAB®, Mathcad®, C++, SPICE, or the like.

Assumptions: The parameters are uniformly distributed between their means. A 100-case analysis will be performed.

Analysis: The spreadsheet used in this analysis appears in Table 1.4. Equation sets (1.51) and (1.52) are entered into the first row of the spreadsheet, and then that row may be copied into as

TABLE 1.4

	V_S (V)	R_1 (Ω)	R_2 (Ω)	V_O (V)	I_S (A)	P (W)
TOLERANCE	**10.00%**	**5.00%**	**5.00%**			
Case 1	15.94	17,248	35,542	5.21	3.02E − 04	4.81E − 03
2	14.90	18,791	35,981	5.11	2.72E − 04	4.05E − 03
3	14.69	18,300	36,725	4.89	2.67E − 04	3.92E − 03
4	16.34	18,149	36,394	5.44	3.00E − 04	4.90E − 03
5	14.31	17,436	37,409	4.55	2.61E − 04	3.74E − 03
...						
95	16.34	17,323	36,722	5.24	3.02E − 04	4.94E − 03
96	16.38	17,800	35,455	5.47	3.08E − 04	5.04E − 03
97	15.99	17,102	35,208	5.23	3.06E − 04	4.89E − 03
98	14.06	18,277	35,655	4.76	2.61E − 04	3.66E − 03
99	13.87	17,392	37,778	4.37	2.51E − 04	3.49E − 03
Case 100	15.52	18,401	34,780	5.37	2.92E − 04	4.53E − 03
Avg	14.88	17,998	36,004	4.96	2.76E − 04	4.12E − 03
Nom.	15.00	18,000	36,000	5.00	2.78E − 04	4.17E − 03
Stdev	0.86	476	976	0.30	1.73E − 05	4.90E − 04
Max	16.46	18,881	37,778	5.70	3.10E − 04	5.04E − 03
WC-Max	16.50	18,900	37,800	5.87	3.22E − 04	—
Min	13.52	17,102	34,201	4.37	2.42E − 04	3.29E − 03
WC-Min	13.50	17,100	34,200	4.20	2.38E − 04	—

many additional rows as the number of statistical cases that are desired. The analysis is automatically repeated for the random selections to build up the statistical distributions, with each row representing one analysis of the circuit. At the end of the columns, the mean, standard deviation, and minimum and maximum values can all be calculated using built-in spreadsheet functions, and the overall spreadsheet data can be used to build histograms for the circuit performance. A portion of the spreadsheet output for 100 cases of the circuit of Fig. 1.23 is shown in Table 1.4.

Check of Results: The average values for V_O and I_S are 4.96 V and 276 μA, respectively, which are close to the values originally estimated from the nominal circuit elements. The averages will more closely approach the nominal values as the number of cases used in the analysis is increased. The standard deviations are 0.30 V and 17.3 μA, respectively.

A histogram (generated with MATLAB® hist(x, n)) of the results of a 1000-case simulation of the output voltage in the same problem appears in Fig. 1.25. Note that the overall distribution is becoming Gaussian in shape with the peak in the center near the mean value. The worst-case values calculated earlier are several standard deviations from the mean and lie outside the minimum and maximum values that occurred even in this 1000-case Monte Carlo analysis.

Some implementations of the SPICE circuit analysis program, PSPICE for example, actually contain a Monte Carlo option in which a full circuit simulation is automatically performed for any number of randomly selected test cases. These programs, which provide a powerful tool for much more complex statistical analysis than is possible by hand, can perform statistical estimates of delay, frequency response, and the like for circuits with many elements.

1.8.4 TEMPERATURE COEFFICIENTS

In the real world, all physical circuit elements change value as the temperature changes. Our circuit designs must continue to operate properly as the temperature changes. For example, the temperature range for commercial products is typically 0 to 70°C, whereas the standard military temperature range is −55 to +85°C. Other environments, such as the engine compartment of an automobile, can be even more extreme.

Mathematical Model

The basic mathematical model for incorporating element variation with temperature is

$$P = P_{\text{nom}}(1 + \alpha_1 \Delta T + \alpha_2 \Delta T^2) \qquad \text{and} \qquad \Delta T = T - T_{\text{nom}} \qquad (1.53)$$

Coefficients α_1 and α_2 represent the first- and second-order[8] **temperature coefficients,** and ΔT represents the difference between the actual temperature T and the temperature at which the nominal value is specified:

$$P = P_{\text{nom}} \qquad \text{for} \qquad T = T_{\text{nom}} \qquad (1.54)$$

Common values for the magnitude of α_1 range from 0 to plus or minus several thousand parts per million per degree C (1000 ppm/°C = 0.1%/°C). For example, nichrome resistors are highly stable and can exhibit a **temperature coefficient of resistance (TCR = α_1)** of only 50 ppm/°C. In contrast, diffused resistors in integrated circuits may have α_1 as large as several thousand ppm/°C. Most elements will also exhibit some curvature in their characteristics as a function of temperature, and α_2 will be nonzero, although small. We will neglect α_2 unless otherwise stated.

---

[8] Higher-order temperature dependencies can also be included.

SPICE Model

Most SPICE programs contain models for the temperature dependencies of many circuit elements. The temperature-dependent SPICE model for the resistor is equivalent to that given in Eq. (1.53):

$$R(\text{T}) = R(\text{TNOM}) * [1 + \text{TC1} * (\text{T} - \text{TNOM}) + \text{TC2} * (\text{T} - \text{TNOM})^2] \qquad (1.55)$$

in which the SPICE parameters are defined as follows:

TNOM = temperature at which the nominal resistor value is measured
T = temperature at which the simulation is performed
TC1 = first-order temperature coefficient
TC2 = second-order temperature coefficient

EXAMPLE 1.7   **TCR ANALYSIS**

Find the value of a resistor at various temperatures.

PROBLEM   A diffused resistor has a nominal value of 10 kΩ at a temperature of 25°C and has a TCR of + 1000 ppm/°C. Find its resistance at 40 and 75°C.

SOLUTION   **Known Information and Given Data:** The resistor's nominal value is 10 kΩ at $T = 25°C$. The TCR is 1000 ppm/°C.

Unknowns: The resistor values at 40 and 75°C.

Approach: Use the known values to evaluate Eq. (1.53).

Assumptions: Based on the TCR statement, $\alpha_1 = 1000$ ppm/°C and $\alpha_2 = 0$.

Analysis: The TCR of +1000 ppm/°C corresponds to

$$\alpha_1 = \frac{10^3}{10^6} \frac{1}{°C} = 10^{-3}/°C$$

The resistor value at 40°C would be

$$R = 10 \text{ k}\Omega \left[1 + \frac{10^{-3}}{°C}(40 - 25)°C\right] = 10.15 \text{ k}\Omega$$

and at 75°C the value would be

$$R = 10 \text{ k}\Omega \left[1 + \frac{10^{-3}}{°C}(75 - 25)°C\right] = 10.5 \text{ k}\Omega$$

Check of Results: 1000 ppm/°C corresponds to 0.1%/°C or 10 Ω/°C for the 10-kΩ resistor. A 15°C temperature change should shift the resistor value by 150 Ω, whereas a 50°C change should change the value by 500 Ω. Thus the answers appear correct.

EXERCISE: What will the resistor value in Ex. 1.7 be for $T = -55°C$ and $T = +85°C$?

ANSWERS: 9.20 kΩ, 10.6 kΩ.

1.9 NUMERIC PRECISION

Many numeric calculations will be performed throughout this book. Keep in mind that the circuits being designed can all be built in discrete form in the laboratory or can be implemented in integrated circuit form. In designing circuits, we will be dealing with components that have tolerances ranging from less than ± 1 percent to greater than ± 50 percent, and calculating results to a precision of more than three significant digits represents a meaningless exercise except in very limited circumstances. Thus, the results in this text are consistently represented with three significant digits: 2.03 mA, 5.72 V, 0.0436 μA, and so on. For example, see the answers in Eqs. (1.18), (1.23), and (1.31), and so on.

SUMMARY

- The age of electronics began in the early 1900s with Pickard's creation of the crystal diode detector, Fleming's invention of the diode vacuum tube, and then Deforest's development of the triode vacuum tube. Since that time, the electronics industry has grown to account for as much as 10 percent of the world gross domestic product.

- The real catalysts for the explosive growth of electronics occurred following World War II. The first was the invention of the bipolar transistor by Bardeen, Brattain, and Shockley in 1947; the second was the simultaneous invention of the integrated circuit by Kilby and by Noyce and Moore in 1958.

- Integrated circuits quickly became a commercial reality, and the complexity, whether measured in memory density (bits/chip), microprocessor transistor count, or minimum feature size, has changed exponentially since the mid-1960s. We are now in an era of ultra-large-scale integration (ULSI), having already put lower levels of integration—SSI, MSI, LSI, and VLSI—behind us.

- Electronic circuit design deals with two major categories of signals. Analog electrical signals may take on any value within some finite range of voltage or current. Digital signals, however, can take on only a finite set of discrete levels. The most common digital signals are binary signals, which are represented by two discrete levels.

- Bridging between the analog and digital worlds are the digital-to-analog and analog-to-digital conversion circuits (DAC and ADC, respectively). The DAC converts digital information into an analog voltage or current, whereas the ADC creates a digital number at its output that is proportional to an analog input voltage or current.

- Fourier demonstrated that complex signals can be represented as a linear combination of sinusoidal signals. Analog signal processing is applied to these signals using linear amplifiers; these modify the amplitude and phase of analog signals. Linear amplifiers do not alter the frequency content of the signal, other than changing the relative amplitudes and phases of the frequency components.

- Amplifiers are often classified by their frequency response into low-pass, high-pass, band-pass, band-reject, and all-pass categories. Electronic circuits that are designed to amplify specific ranges of signal frequencies are usually referred to as filters.

- Solving problems is one focal point of an engineer's career. A well-defined approach can help significantly in solving problems, and to this end, a structured problem-solving approach has been introduced in this chapter as outlined in these nine steps. Throughout the rest of this text, the examples will follow this problem-solving approach:

  1. State the **problem** as clearly as possible.
  2. List the **known information and given data.**

3. Define the **unknowns** that must be found to solve the problem.

4. List your **assumptions.** You may discover additional assumptions as the analysis progresses.

5. Develop an **approach** from a list of possible alternatives.

6. Perform an **analysis** to find a solution to the problem.

7. **Check the results.** Is the math correct? Have all the unknowns been found? Do the results satisfy simple consistency checks?

8. **Evaluate the solution.** Is the solution realistic? Can it be built? If not, repeat steps 4–7 until a satisfactory solution is obtained.

9. Use **computer-aided analysis** to check the results and to see if the solution satisfies the problem requirements.

- Our circuit designs will be implemented using real components whose initial values differ from those of the design and that change with time and temperature. Techniques for analyzing the influence of element tolerances on circuit performance include the worst-case analysis and statistical Monte Carlo analysis methods. Most circuit analysis programs include the ability to specify temperature dependencies for most circuit elements.

- In worst-case analysis, element values are simultaneously pushed to their extremes, and the resulting predictions of circuit behavior are often overly pessimistic.

- The Monte Carlo method analyzes a large number of randomly selected versions of a circuit to build up a realistic estimate of the statistical distribution of circuit performance. Random number generators in high-level computer languages, spreadsheets, Mathcad®, or MATLAB® can be used to randomly select element values for use in Monte Carlo analysis. Some circuit analysis packages such as PSPICE provide a Monte Carlo analysis option as part of the program.

KEY TERMS

All-pass amplifier
Analog signal
Analog-to-digital converter (A/D converter or ADC)
Band-pass amplifier
Band-reject amplifier
Binary digital signal
Bipolar transistor
Current-controlled current source (CCCS)
Current-controlled voltage source (CCVS)
Current division
Dependent (or controlled) source
Digital signal
Digital-to-analog converter (D/A converter or DAC)
Diode
Feedback network
Filters
Fourier analysis
Fourier series
Frequency spectrum

Fundamental frequency
Fundamental radian frequency
Giga-scale integration (GSI)
Harmonic frequency
High-pass amplifier
Ideal operational amplifier
Integrated circuit (IC)
Input resistance
Inverting amplifier
Kirchhoff's current law (KCL)
Kirchhoff's voltage law (KVL)
Large-scale integration (LSI)
Least significant bit (LSB)
Low-pass amplifier
Medium-scale integration (MSI)
Mesh analysis
Minimum feature size
Monte Carlo analysis
Most significant bit (MSB)
Nodal analysis
Nominal value

Norton circuit transformation
Norton equivalent circuit
Operational amplifier (op amp)
Phasor
Problem-solving approach
Quantization error
Random numbers
Resolution of the converter
Small-scale integration (SSI)
Superposition principle
Temperature coefficient
Temperature coefficient of resistance (TCR)
Thévenin circuit transformation
Thévenin equivalent circuit
Thévenin equivalent resistance

Tolerance
Transistor
Triode
Ultra-large-scale integration (ULSI)
Uniform random number generator
Vacuum diode
Vacuum tube
Very-large-scale integration (VLSI)
Virtual ground
Voltage-controlled current source (VCCS)
Voltage-controlled voltage source (VCVS)
Voltage division
Voltage gain
Worst-case analysis

REFERENCES

1. W. F. Brinkman, D. E. Haggan, and W. W. Troutman, "A History of the Invention of the Transistor and Where It Will Lead Us," *IEEE Journal of Solid-State Circuits,* vol. 32, no. 12, pp. 1858–65, December 1997.

2. www.pbs.org/transistor/sitemap.html.

3. 2006 *CIA Factbook,* www.cia.gov.

4. 2006 *Fortune* Global 500, www.fortune.com.

5. 2006 *Fortune* 500, www.fortune.com.

6. J. T. Wallmark, "The Field-Effect Transistor—An Old Device with New Promise," *IEEE Spectrum,* March 1964.

7. IEEE: www.ieee.org.

8. ISSCC: www.sscs.org.

9. IEDM: www.ieee.org.

10. International Technology Roadmap for Semiconductors: public.itrs.net.

11. Frequency allocations: www.fcc.org.

ADDITIONAL READING

Commemorative Supplement to the Digest of Technical Papers, 1993 IEEE International Solid-State Circuits Conference Digest, vol. 36, February 1993.

Digest of Technical Papers of the IEEE Custom Integrated International Circuits Conference, September of each year.

Digest of Technical Papers of the IEEE International Electronic Devices Meeting, December of each year.

Digest of Technical Papers of the IEEE International Solid-State Circuits Conference, February of each year.

Digest of Technical Papers of the IEEE International Symposia on VLSI Technology and Circuits, June of each year.

Electronics, Special Commemorative Issue, April 17, 1980.

Garratt, G. R. M. *The Early History of Radio from Faraday to Marconi.* London: Institution of Electrical Engineers (IEE), 1994.

"200 Years of Progress." *Electronic Design* 24, no. 4, February 16, 1976.

PROBLEMS

1.1 A Brief History of Electronics: From Vacuum Tubes to Ultra-Large-Scale Integration

1.1. Make a list of 20 items in your environment that contain electronics. A PC and its peripherals are considered one item. (Do not confuse electromechanical timers, common in clothes dryers or the switch in a simple thermostat, with electronic circuits.)

1.2. The change in memory density with time can be described by $B = 19.97 \times 10^{0.1977(\text{Year}-1960)}$. If a straight-line projection is made using this equation, what will be the number of memory bits/chip in the year 2020?

1.3. (a) How many years does it take for memory chip density to increase by a factor of 2, based on the equation in Prob. 1.2? (b) By a factor of 10?

1.4. The straight line in Fig. 1.4 is described by $N = 1610 \times 10^{0.1548(\text{Year}-1970)}$. Based on a straight-line projection of this figure, what will be the number of transistors in a microprocessor in the year 2020?

1.5. (a) How many years does it take for microprocessor circuit density to increase by a factor of 2, based on the equation in Prob. 1.4? (b) By a factor of 10?

1.6. If you make a straight-line projection from Fig. 1.5, what will be the minimum feature size in integrated circuits in the year 2025? The curve can be described by $F = 8.00 \times 10^{-0.05806(\text{Year}-1970)}$ μm. Do you think this is possible? Why or why not?

1.7. Based on Fig. 1.4, how many Pentium IV processors will we be able to place on one chip in the year 2020?

1.8. The filament of a small vacuum tube uses a power of approximately 1.5 W. Suppose that 75 million of these tubes are used to build the equivalent of a 64 Mb memory. How much power is required for this memory? If this power is supplied from a 220 V ac source, what is the current required by this memory?

1.2 Classification of Electronic Signals

1.9. Classify each of the following as an analog or digital quantity: (a) status of a light switch, (b) status of a thermostat, (c) water pressure, (d) gas tank level, (e) bank overdraft status, (f) light bulb intensity, (g) stereo volume, (h) full or empty cup, (i) room temperature, (j) TV channel selection, and (k) tire pressure.

1.10. A 12-bit D/A converter has a full scale voltage of 10.00 V. What is the voltage corresponding to the LSB? To the MSB? What is the output voltage if the binary input code is equal to (100100100100)?

1.11. A 10-bit D/A converter has a full scale voltage of 2.5 V. What is the voltage corresponding to the LSB? What is the output voltage if the binary input code is equal to (0101100101)?

1.12. An 8-bit A/D converter has $V_{FS} = 5$ V. What is the value of the voltage corresponding to the LSB? If the input voltage is 2.77 V, what is the binary output code of the converter?

1.13. A 15-bit A/D converter has $V_{FS} = 10$ V. What is the value of the LSB? If the input voltage is 6.83 V, what is the binary output code of the converter?

1.14. (a) A digital multimeter is being designed to have a readout with four decimal digits. How many bits will be required in its A/D converter? (b) Repeat for six decimal digits.

1.15. A 12-bit ADC has $V_{FS} = 5.12$ V and the output code is (101110111011). What is the size of the LSB for the converter? What range of input voltages corresponds to the ADC output code?

1.3 Notational Conventions

1.16. If $i_B = 0.002(1 + \cos 1000t)$ A, what are I_B and i_b?

1.17. If $v_{GS} = (4 + 0.5u(t - 1) + 0.2\cos 2000\pi t)$ V, what are V_{GS} and v_{gs}? [$u(t)$ is the unit step function.]

1.18. If $V_{CE} = 5$ V and $v_{ce} = (2\cos 5000t)$ V, write the expression for v_{CE}.

1.19. If $V_{DS} = 5$ V and $v_{ds} = (2\sin 2500t + 4\sin 1000t)$ V, write the expression for v_{DS}.

1.5 Important Concepts from Circuit Theory

1.20. Use voltage and current division to find V_1, V_2, I_2, and I_3 in the circuit in Fig. P1.20 if $V = 10$ V, $R_1 = 22$ kΩ, $R_2 = 47$ kΩ, and $R_3 = 180$ kΩ.

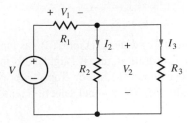

Figure P1.20

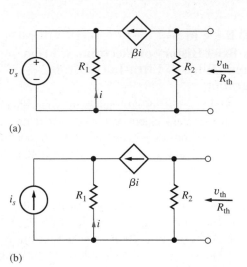

(a)

(b)

Figure P1.26

1.21. Use voltage and current division to find V_1, V_2, I_2, and I_3 in the circuit in Fig. P1.20 if $V = 18$ V, $R_1 = 56$ kΩ, $R_2 = 33$ kΩ, and $R_3 = 11$ kΩ.

1.22. Use current and voltage division to find I_1, I_2, and V_3 in the circuit in Fig. P1.22 if $I = 5$ mA, $R_1 = 2.4$ kΩ, $R_2 = 5.6$ kΩ, and $R_3 = 3.6$ kΩ.

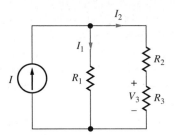

Figure P1.22

1.23. Use current and voltage division to find I_1, I_2, and V_3 in the circuit in Fig. P1.22 if $I = 250$ μA, $R_1 = 150$ kΩ, $R_2 = 68$ kΩ, and $R_3 = 82$ kΩ.

1.24. Find the Thévenin equivalent representation of the circuit in Fig. P1.24 if $g_m = 0.002$ S and $R_1 = 50$ kΩ.

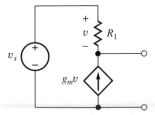

Figure P1.24

1.25. Find the Norton equivalent representation of the circuit in Fig. P1.24 if $g_m = 0.025$ S and $R_1 = 4$ kΩ.

1.26. Find the Thévenin equivalent representation of the circuit in Fig. P1.26(a) if $\beta = 120$, $R_1 = 100$ kΩ, and $R_2 = 39$ kΩ. (b) Repeat for the circuit in Fig. P1.26(b).

1.27. Find the Norton equivalent representation of the circuit in Fig. P1.26(a) if $\beta = 100$, $R_1 = 75$ kΩ, and $R_2 = 56$ kΩ.

1.28. What is the resistance presented to source v_s by the circuit in Fig. P1.26(a) if $\beta = 80$, $R_1 = 100$ kΩ, and $R_2 = 39$ kΩ?

1.29. Find the Thévenin equivalent representation of the circuit in Fig. P1.29 if $g_m = .0025$ S, $R_1 = 100$ kΩ, and $R_2 = 1$ MΩ.

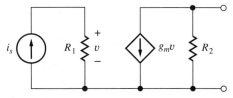

Figure P1.29

1.6 Frequency Spectrum of Electronic Signals

1.30. A signal voltage is expressed as $v(t) = (5 \sin 1000\pi t + 3 \cos 2000\pi t)$ V. Draw a graph of the amplitude spectrum for $v(t)$ similar to the one in Fig. 1.18(b).

*1.31. Voltage $v_1 = 2 \sin 20{,}000\pi t$ is multiplied by voltage $v_2 = 2 \sin 2000\pi t$. Draw a graph of the amplitude spectrum for $v = v_1 \times v_2$ similar to the one in Fig. 1.18(b). (Note that multiplication is a nonlinear mathematical operation. In electronics it is often called *mixing* because it produces a signal that contains output frequencies that are not in the input signal but depend directly on the input frequencies.)

1.7 Amplifiers

1.32. The input and output voltages of an amplifier are expressed as $v_s = 10^{-5} \sin(2 \times 10^7 \pi t)$ V and $v_o = 2 \sin(2 \times 10^7 \pi t + 36°)$ V. What are the magnitude and phase of the voltage gain of the amplifier?

*1.33. The input and output voltages of an amplifier are expressed as

$$v_s = [10^{-3} \sin(3000\pi t)$$
$$+ 2 \times 10^{-3} \sin(5000\pi t)] \text{ V}$$

and $\quad v_o = [10^{-1} \sin(3000\pi t - 12°)$
$$+ 10^{-2} \sin(5000\pi t - 45°)] \text{ V}$$

(a) What are the magnitude and phase of the voltage gain of the amplifier at a frequency of 2500 Hz? (b) At 1500 Hz?

1.34. What is the voltage gain of the amplifier in Fig. 1.21 if (a) $R_1 = 14 \text{ k}\Omega$ and $R_2 = 620 \text{ k}\Omega$? (b) For $R_1 = 18 \text{ k}\Omega$ and $R_2 = 180 \text{ k}\Omega$? (c) For $R_1 = 1.6 \text{ k}\Omega$ and $R_2 = 62 \text{ k}\Omega$?

1.35. Write an expression for the output voltage $v_o(t)$ of the circuit in Fig. 1.21 if $R_1 = 910 \,\Omega$, $R_2 = 8.2 \text{ k}\Omega$, and $v_s(t) = (0.01 \sin 750\pi t)$ V. Write an expression for the current $i_s(t)$.

1.36. Find an expression for the voltage gain $A_v = v_o/v_s$ for the amplifier in Fig. P1.36.

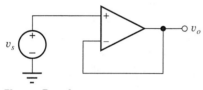

Figure P1.36

1.37. Find an expression for the voltage gain $A_v = v_o/v_s$ for the amplifier in Fig. P1.37.

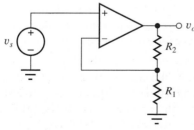

Figure P1.37

1.38. Write an expression for the output voltage $v_o(t)$ of the circuit in Fig. P1.38 if $R_1 = 2 \text{ k}\Omega$, $R_2 = 10 \text{ k}\Omega$, $R_3 = 51 \text{ k}\Omega$, $v_1(t) = (0.01 \sin 3770t)$ V, and

$v_2(t) = (0.05 \sin 10,000t)$ V. Write an expression for the voltage appearing at the inverting input (v_-).

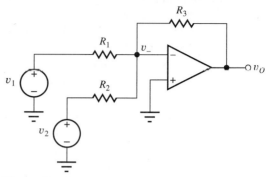

Figure P1.38

1.39. The circuit in Fig. P1.39 can be used as a simple 3-bit digital-to-analog converter (DAC). The individual bits of the binary input word $(b_1 \; b_2 \; b_3)$ are used to control the position of the switches, with the resistor connected to 0 V if $b_i = 0$ and connected to V_{REF} if $b_i = 1$. (a) What is the output voltage for the DAC as shown with input data of (011) if $V_{REF} = 5.0$ V? (b) Suppose the input data change to (100). What will be the new output voltage? (c) Make a table giving the output voltages for all eight possible input data combinations.

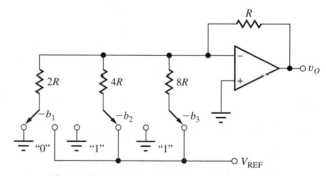

Figure P1.39

Amplifier Frequency Response

1.40. An amplifier has a voltage gain of 10 for frequencies below 6000 Hz, and zero gain for frequencies above 6000 Hz. Classify this amplifier.

1.41. An amplifier has a voltage gain of zero for frequencies below 1000 Hz, and zero gain for frequencies above 5000 Hz. In between these two frequencies the amplifier has a gain of 20. Classify this amplifier.

1.42. An amplifier has a voltage gain of 16 for frequencies above 10 kHz, and zero gain for frequencies below 10 kHz. Classify this amplifier.

1.43. The amplifier in Prob. 1.40 has an input signal given by $v_s(t) = (5 \sin 2000\pi t + 3 \cos 8000\pi t + 2 \cos 15000\pi t)$ V. Write an expression for the output voltage of the amplifier.

1.44. The amplifier in Prob. 1.41 has an input signal given by $v_s(t) = (0.5 \sin 2500\pi t + 0.75 \cos 8000\pi t + 0.6 \cos 12,000\pi t)$ V. Write an expression for the output voltage of the amplifier.

1.45. The amplifier in Prob. 1.42 has an input signal given by $v_s(t) = (0.5 \sin 2500\pi t + 0.75 \cos 8000\pi t + 0.8 \cos 12,000\pi t)$ V. Write an expression for the output voltage of the amplifier.

1.46. An amplifier has an input signal that can be represented as

$$v(t) = \frac{4}{\pi}\left(\sin \omega_o t + \frac{1}{3}\sin 3\omega_o t + \frac{1}{5}\sin 5\omega_o t\right) \text{V}$$

where $f_o = 1000$ Hz
(a) Use MATLAB to plot the signal for $0 \le t \le 5$ ms. (b) The signal $v(t)$ is amplified by an amplifier that provides a voltage gain of 5 at all frequencies. Plot the output voltage for this amplifier for $0 \le t \le 5$ ms. (c) A second amplifier has a voltage gain of 5 for frequencies below 2000 Hz but zero gain for frequencies above 2000 Hz. Plot the output voltage for this amplifier for $0 \le t \le 5$ ms. (d) A third amplifier has a gain of 5 at 1000 Hz, a gain of 3 at 3000 Hz, and a gain of 1 at 5000 Hz. Plot the output voltage for this amplifier for $0 \le t \le 5$ ms.

1.8 Element Variations in Circuit Design

1.47. (a) A 3.00-kΩ resistor is purchased with a tolerance of 1 percent. What is the possible range of values for this resistor? (b) Repeat for a 5 percent tolerance. (c) Repeat for a 10 percent tolerance.

1.48. The power supply voltage for a circuit must vary by no more than 50 mV from its nominal value of 2.5 V. What is its tolerance specification?

1.49. A 20,000 μF capacitor has an asymmetric tolerance specification of +20%/−50%. What is the possible range of values for this capacitor?

1.50. An 8200-Ω resistor is purchased with a tolerance of 10 percent. It is measured with an ohmmeter and found to have a value of 7905 Ω. Is this resistor within its specification limits? Explain your answer.

1.51. (a) The output voltage of a 5-V power supply is measured to be 5.30 V. The power supply has a 5 percent tolerance specification. Is the supply operating within its specification limits? Explain your answer. (b) The voltmeter that was used to make the measurement has a 1.5 percent tolerance. Does that change your answer? Explain.

1.52. A resistor is measured and found to have a value of 6066 Ω at 0°C and 6562 Ω at 100°C. What are the temperature coefficient and nominal value for the resistor? Assume $T_{\text{NOM}} = 27$°C.

1.53. Find the worst-case values of V_1, I_2, and I_3 for the circuit in Prob. 1.20 if the resistor tolerances are 10 percent and the voltage source tolerance is 5 percent.

1.54. Find the worst-case values of I_1, I_2, and V_3 for the circuit in Prob. 1.22 if the resistor tolerances are 5 percent and the current source tolerance is 2 percent.

1.55. Find the worst-case values for the Thévenin equivalent resistance for the circuit in Prob. 1.24 if the resistor tolerance is 20 percent and the tolerance on g_m is also 20 percent.

1.56. Perform a 200-case Monte Carlo analysis for the circuit in Prob. 1.53 and compare the results to the worst-case calculations.

1.57. Perform a 200-case Monte Carlo analysis for the circuit in Prob. 1.54 and compare the results to the worst-case calculations.

1.9 Numeric Precision

1.58. (a) Express the following numbers to three significant digits of precision: 3.2947, 0.995171, −6.1551. (b) To four significant digits. (c) Check these answers using your calculator.

1.59. (a) What is the voltage developed by a current of 1.763 mA in a resistor of 20.70 kΩ? Express the answer with three significant digits. (b) Express the answer with two significant digits. (c) Repeat for $I = 102.1$ μA and $R = 97.80$ kΩ.

SOLID-STATE ELECTRONICS

CHAPTER OUTLINE

Jack St. Clair Kilby.
Courtesy of Texas Instruments

CHAPTER GOALS

- Explore the characteristics of semiconductors and discover how engineers control semiconductor properties to fabricate electronic devices.

- Characterize resistivity and insulators, semiconductors, and conductors.

- Develop the covalent bond and energy band models for semiconductors.

- Understand the concepts of bandgap energy and intrinsic carrier concentration.

- Explore the behavior of the two charge carriers in semiconductors—electrons and holes.

- Discuss acceptor and donor impurities in semiconductors.

- Learn to control the electron and hole populations using impurity doping.

- Understand drift and diffusion currents in semiconductors.

- Explore the concepts of low-field mobility and velocity saturation.

- Discuss the dependence of mobility on doping level.

- Explore basic IC fabrication processes.

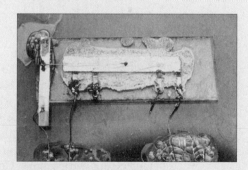

The Kilby integrated circuit.
Courtesy of Texas Instruments

Jack Kilby from Texas Instruments Inc. and Gordon Moore and Robert Noyce from Fairchild Semiconductor pioneered the nearly simultaneous development of the integrated circuit in the late 1950s. After years of litigation, the basic integrated circuit patents of Jack Kilby and Texas Instruments were upheld, and also finally recognized in Japan in 1994. Gorden E. Moore, Robert Noyce, and Andrew S. Grove founded the Intel Corporation in 1968. Kilby shared the 2000 Nobel prize in physics for invention of the integrated circuit.

Andy Grove, Robert Noyce, and Gordon Moore with Intel
8080 processor rubylith in 1978.
Courtesy of Intel Corporation

As discussed in Chapter 1, the evolution of solid-state materials and the subsequent development of
the technology for integrated circuit fabrication have revolutionized electronics and made possible
the modern information and technological revolution. Using silicon as well as other crystalline
semiconductor materials, we can now fabricate integrated circuits (ICs) that have more than a billion
electronic components on a single 2 cm × 2 cm die. Most of us have some familiarity with the
very high-speed microprocessor and memory components that form the building blocks for personal
computers and workstations. Consider for a moment the content of a 1-gigabit memory chip. The
memory array alone on this chip will contain more than 10^9 transistors and 10^9 capacitors—more
than 2 billion electronic components on a single die!

Our ability to build such phenomenal electronic system components is based on a detailed
understanding of solid-state physics as well as on development of fabrication processes necessary
to turn the theory into a manufacturable reality. Integrated circuit manufacturing is an excellent
example of a process requiring a broad understanding of many disciplines. IC fabrication requires
knowledge of physics, chemistry, electrical engineering, mechanical engineering, materials engineering, and metallurgy, to mention just a few disciplines. The breadth of understanding required is
a challenge, but it makes the field of solid-state electronics an extremely exciting and vibrant area
of specialization.

It is possible to explore the behavior of electronic circuits from a "black box" perspective, simply
trusting a set of equations that model the terminal voltage and current characteristics of each of the
electronic devices. However, understanding the underlying behavior of the devices leads a designer
to develop an intuition that extends beyond the simplified models of a black box approach. Building

our models from fundamental characteristics enables us to understand the limitations and appropriate uses of particular models. This is especially true when we experimentally observe deviations from our model predictions. One goal of this chapter is to develop a basic understanding of the underlying operational principles of semiconductor devices that enables us to place our simplified models in the appropriate context.

The material in this chapter provides the background necessary for understanding the behavior of the solid-state devices presented in subsequent chapters. We begin our study of solid-state electronics by exploring the characteristics of crystalline materials, with an emphasis on silicon, the most commercially important semiconductor. We look at electrical conductivity and resistivity and discuss the mechanisms of electronic conduction. The technique of impurity doping is discussed, along with its use in controlling conductivity and resistivity type.

2.1 SOLID-STATE ELECTRONIC MATERIALS

Electronic materials generally can be divided into three categories: **insulators, conductors,** and **semiconductors.** The primary parameter used to distinguish among these materials is the **resistivity** ρ, with units of $\Omega \cdot cm$. As indicated in Table 2.1, insulators have resistivities greater than $10^5 \ \Omega \cdot cm$, whereas conductors have resistivities below $10^{-3} \ \Omega \cdot cm$. For example, diamond, one of the highest quality insulators, has a very large resistivity, $10^{16} \ \Omega \cdot cm$. On the other hand, pure copper, a good conductor, has a resistivity of only $3 \times 10^{-6} \ \Omega \cdot cm$. Semiconductors occupy the full range of resistivities between the insulator and conductor boundaries; moreover, the resistivity can be controlled by adding various impurity atoms to the semiconductor crystal.

Elemental semiconductors are formed from a single type of atom (column IV of the periodic table of elements; see Table 2.2), whereas **compound semiconductors** can be formed from combinations of elements from columns III and V or columns II and VI. These later materials are often referred to as III–V (3–5) or II–VI (2–6) compound semiconductors. Table 2.3 presents some of the most useful possibilities. There are also ternary materials such as mercury cadmium telluride, gallium aluminum arsenide, gallium indium arsenide, and gallium indium phosphide.

Historically, germanium was one of the first semiconductors to be used. However, it was rapidly supplanted by silicon, which today is the most important semiconductor material. Silicon has a wider bandgap energy,[1] enabling it to be used in higher-temperature applications than germanium, and oxidation forms a stable insulating oxide on silicon, giving silicon significant processing advantages over germanium during fabrication of ICs. In addition to silicon, gallium arsenide, indium phosphide, and silicon carbide are commonly encountered today, although germanium is still used in some limited applications. The compound semiconductor materials gallium arsenide (GaAs) and indium phosphide (InP) are the most important material for optoelectronic applications, including light-emitting diodes (LEDs), lasers, and photodetectors.

TABLE 2.1
Electrical Classification of Solid Materials

MATERIALS	RESISTIVITY ($\Omega \cdot cm$)
Insulators	$10^5 < \rho$
Semiconductors	$10^{-3} < \rho < 10^5$
Conductors	$\rho < 10^{-3}$

---

[1] The meaning of bandgap energy is discussed in detail in Secs. 2.2 and 2.10.

TABLE 2.2
Portion of the Periodic Table, Including the Most Important Semiconductor Elements (shaded)

	IIIA	IVA	VA	VIA
	5 10.811 **B** Boron	6 12.01115 **C** Carbon	7 14.0067 **N** Nitrogen	8 15.9994 **O** Oxygen
IIB	13 26.9815 **Al** Aluminum	14 28.086 **Si** Silicon	15 30.9738 **P** Phosphorus	16 32.064 **S** Sulfur
30 65.37 **Zn** Zinc	31 69.72 **Ga** Gallium	32 72.59 **Ge** Germanium	33 74.922 **As** Arsenic	34 78.96 **Se** Selenium
48 112.40 **Cd** Cadmium	49 114.82 **In** Indium	50 118.69 **Sn** Tin	51 121.75 **Sb** Antimony	52 127.60 **Te** Tellurium
80 200.59 **Hg** Mercury	81 204.37 **Tl** Thallium	82 207.19 **Pb** Lead	83 208.980 **Bi** Bismuth	84 (210) **Po** Polonium

TABLE 2.3
Semiconductor Materials

SEMICONDUCTOR	BANDGAP ENERGY E_G (eV)
Carbon (diamond)	5.47
Silicon	1.12
Germanium	0.66
Tin	0.082
Gallium arsenide	1.42
Gallium nitride	3.49
Indium phosphide	1.35
Boron nitride	7.50
Silicon carbide	3.26
Cadmium selenide	1.70

Many research laboratories are exploring the formation of diamond, boron nitride, silicon carbide, and silicon germanium materials. Diamond and boron nitride are excellent insulators at room temperature, but they, as well as silicon carbide, can be used as semiconductors at much higher temperatures (600°C). Adding a small percentage (<10 percent) of germanium to silicon has been shown recently to offer improved device performance in a process compatible with normal silicon processing. Because of its performance advantages, this SiGe technology [1] has rapidly been introduced into fabrication of devices for RF (radio-frequency) applications, particularly for use in the telecommunications marketplace.

EXERCISE: What are the chemical symbols for antimony, arsenic, aluminum, boron, gallium, germanium, indium, phosphorus, and silicon?

ANSWERS: Sb, As, Al, B, Ga, Ge, In, P, Si

2.2 COVALENT BOND MODEL

Atoms can bond together in **amorphous, polycrystalline,** or **single-crystal** forms. Amorphous materials have a disordered structure, whereas polycrystalline material consists of a large number of small crystallites. Most of the useful properties of semiconductors, however, occur in high-purity, single-crystal material. Silicon—column IV in the periodic table—has four **electrons** in the outer shell. Single-crystal material is formed by the covalent bonding of each silicon atom with its four nearest neighbors in a highly regular three-dimensional array of atoms, as shown in Fig. 2.1. Much of the behavior we discuss can be visualized using the simplified two-dimensional **covalent bond model** of Fig. 2.2.

At temperatures approaching absolute zero, all the electrons reside in the covalent bonds shared between the atoms in the array, with no electrons free for conduction. The outer shells of the silicon atoms are full, and the material behaves as an insulator. As the temperature increases, thermal energy is added to the crystal and some bonds break, freeing a small number of electrons for conduction, as

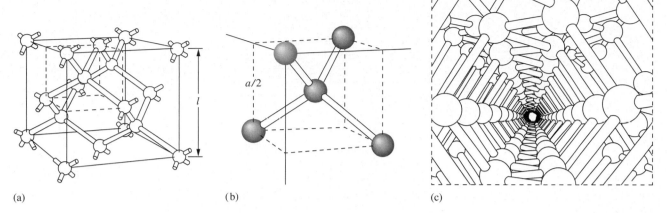

(a)   (b)   (c)

Figure 2.1 Silicon crystal lattice structure. (a) Diamond lattice unit cell. The cube side length $l = 0.543$ nm. (b) Enlarged top corner of the diamond lattice, showing the four nearest neighbors bonding within the structure. (c) View along a crystallographic axis.
Source: (a) and (b) Adapted from Electrons and Holes in Semiconductors *by William Shockley, © 1950 by Litton Educational Publishing. (c) Adapted from* The Architecture of Molecules *by Linus Pauling © 1964 by W. H. Freeman and Company, used with permission.*

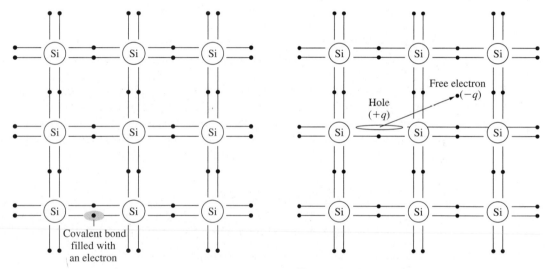

Figure 2.2 Two-dimensional silicon lattice with shared co-valent bonds. At temperatures approaching absolute zero, 0 K, all bonds are filled, and the outer shells of the silicon atoms are completely full.

Figure 2.3 An electron–hole pair is generated whenever a covalent bond is broken.

in Fig. 2.3. The density of these free electrons is equal to the **intrinsic carrier density** n_i (cm^{-3}), which is determined by material properties and temperature:

$$n_i^2 = BT^3 \exp\left(-\frac{E_G}{kT}\right) \qquad \text{cm}^{-6} \qquad (2.1)$$

where E_G = semiconductor bandgap energy in eV (electron volts)

k = Boltzmann's constant, 8.62×10^{-5} eV/K

T = absolute temperature, K

B = material-dependent parameter, 1.08×10^{31} K$^{-3} \cdot$ cm^{-6} for Si

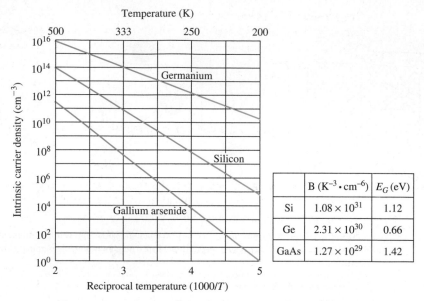

Figure 2.4 Intrinsic carrier density versus temperature from Eq. (2.1).

Bandgap energy E_G is the minimum energy needed to break a covalent bond in the semiconductor crystal, thus freeing electrons for conduction. Table 2.3 lists values of the bandgap energy for various semiconductors.

The *density of conduction (or free) electrons* is represented by the symbol n (electrons/cm^3), and for **intrinsic material** $n = n_i$. The term *intrinsic* refers to the generic properties of pure material. Although n_i is an intrinsic property of each semiconductor, it is extremely temperature-dependent for all materials. Figure 2.4 has examples of the strong variation of intrinsic carrier density with temperature for germanium, silicon, and gallium arsenide.

EXAMPLE 2.1   **INTRINSIC CARRIER CONCENTRATION**

Calculate the theoretical value of n_i in silicon at room temperature.

PROBLEM Calculate the value of n_i in silicon at room temperature (300 K).

SOLUTION **Known Information and Given Data:** Equation (2.1) defines n_i, B, and k. $E_G = 1.12$ eV from Table 2.3.

Unknowns: Intrinsic carrier concentration n_i.

Approach: Calculate n_i by evaluating Eq. (2.1).

Assumptions: $T = 300$ K at room temperature.

Analysis:

$$n_i^2 = 1.08 \times 10^{31} \ (\text{K}^{-3} \cdot \text{cm}^{-6})(300 \text{ K})^3 \exp\left[\frac{-1.12 \text{ eV}}{(8.62 \times 10^{-5} \text{ eV/K})(300 \text{ K})}\right]$$

$$n_i^2 = 4.52 \times 10^{19}/\text{cm}^6 \qquad \text{or} \qquad n_i = 6.73 \times 10^{9}/\text{cm}^3$$

Check of Results: The desired unknown has been found, and the value agrees with the results graphed in Fig. 2.4.

Discussion: For simplicity, in subsequent calculations we use $n_i = 10^{10}/cm^3$ as the room temperature value of n_i for silicon. The density of silicon atoms in the crystal lattice is approximately $5 \times 10^{22}/cm^3$. We see from this example that only one bond in approximately 10^{13} is broken at room temperature.

EXERCISE: Calculate the value of n_i in germanium at a temperature of 300 K.

ANSWER: $2.27 \times 10^{13}/cm^3$

A second charge carrier is actually formed when the covalent bond in Fig. 2.3 is broken. As an electron, which has charge $-q$ equal to -1.602×10^{-19} C, moves away from the covalent bond, it leaves behind a **vacancy** in the bond structure in the vicinity of its parent silicon atom. The vacancy is left with an effective charge of $+q$. An electron from an adjacent bond can fill this vacancy, creating a new vacancy in another position. This process allows the vacancy to move through the crystal. The moving vacancy behaves just as a particle with charge $+q$ and is called a **hole. Hole density** is represented by the symbol p (holes/cm^3).

As already described, two charged particles are created for each bond that is broken: one electron and one hole. For intrinsic silicon, $n = n_i = p$, and the product of the electron and hole concentrations is

$$pn = n_i^2 \qquad (2.2)$$

The **pn product** is given by Eq. (2.2) whenever a semiconductor is in **thermal equilibrium.** (This very important result is used later.) In thermal equilibrium, material properties are dependent only on the temperature T, with no other form of stimulus applied. Equation (2.2) does not apply to semiconductors operating in the presence of an external stimulus such as an applied voltage or current or an optical excitation.

EXERCISE: Calculate the intrinsic carrier density in silicon at 50 K and 325 K. On the average, what is the length of one side of the cube of silicon that is needed to find one electron and one hole at $T = 50$ K?

ANSWERS: $4.34 \times 10^{-39}/cm^3$; $4.01 \times 10^{10}/cm^3$; 6.13×10^{10} m

2.3 DRIFT CURRENTS AND MOBILITY IN SEMICONDUCTORS

2.3.1 DRIFT CURRENTS
Electrical resistivity ρ and its reciprocal, **conductivity** σ, characterize current flow in a material when an electric field is applied. Charged particles move or *drift* in response to the electric field, and the resulting current is called *drift current*. The **drift current density j** is defined as

$$j = Q\mathbf{v} \qquad (C/cm^3)(cm/s) = A/cm^2 \qquad (2.3)$$

where j = current density[2], the charge in coulombs moving through an area of unit cross section

Q = charge density[2], the charge in a unit volume

$\mathbf{v}$ = velocity of charge in an electric field

[2] Note that "density" has different meanings based on the context. Current density involves a cross-sectional area, whereas charge density is a volumetric quantity.

In order to find the charge density, we explore the structure of silicon using both the covalent bond model and (later) the energy band model for semiconductors. Next, we relate the velocity of the charge carriers to the applied electric field.

2.3.2 MOBILITY

We know from electromagnetics that charged particles move in response to an applied electric field. This movement is termed **drift,** and the resulting current flow is known as **drift current.** Positive charges drift in the same direction as the electric field, whereas negative charges drift in a direction opposed to the electric field. At low fields carrier drift velocity **v** (cm/s) is proportional to the electric field **E** (V/cm); the constant of proportionality is called the **mobility** μ:

$$\mathbf{v}_n = -\mu_n \mathbf{E} \qquad \text{and} \qquad \mathbf{v}_p = \mu_p \mathbf{E} \tag{2.4}$$

where $\mathbf{v}_n =$ velocity of electrons (cm/s)

$\mathbf{v}_p =$ velocity of holes (cm/s)

$\mu_n =$ **electron mobility,** 1350 cm^2/V · s in intrinsic Si

$\mu_p =$ **hole mobility,** 500 cm^2/V · s in intrinsic Si

Conceptually, holes are localized to move through the covalent bond structure, but electrons are free to move about the crystal. Thus, one might expect hole mobility to be less than electron mobility.

EXERCISE: Calculate the velocity of a hole in an electric field of 10 V/cm. What is the electron velocity in an electric field of 1000 V/cm? The voltage across a resistor is 1 V, and the length of the resistor is 2 μm. What is the electric field in the resistor?

ANSWERS: 5.00×10^3 cm/s; 1.35×10^6 cm/s; 5.00×10^3 V/cm

2.3.3 VELOCITY SATURATION

From physics, we know that the velocity of carriers cannot increase indefinitely, certainly not beyond the speed of light. In silicon, for example, the linear velocity-field relationship assumed in Eq. (2.4) is valid only for fields below approximately 5000 V/cm or 0.5 V/μm. As the electric field increases above this value, the velocity of both holes and electrons begins to saturate, as indicated in Fig. 2.5. At low fields, the slope of the characteristic represents the mobility, as defined by Eq. (2.4). For fields above approximately 3×10^4 V/cm in silicon, carrier velocity approaches the **saturated drift velocity** $\mathbf{v}_{\text{sat}}$. For electrons and holes in silicon, $\mathbf{v}_{\text{sat}}$ is approximately 10^7 cm/s. The velocity saturation phenomenon ultimately places an upper limit on the frequency response of solid-state devices.

EXERCISE: (a) What are the maximum drift velocities for electrons and holes in germanium. What are the low field mobilities? (b) What is the maximum drift velocity for electrons in gallium arsenide? What is the electron mobility in gallium arsenide?

ANSWERS: 6×10^6 cm/s, 4300 cm^2/V · s, 2100 cm^2/V · s; 2×10^7 cm/s, 7500 cm^2/V · s

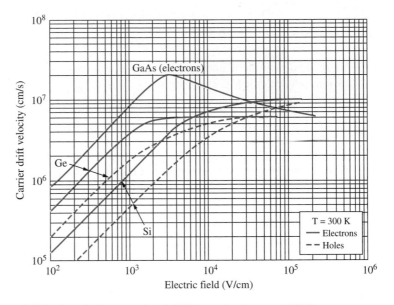

Figure 2.5 Carrier velocity versus electric field in semiconductors at 300 K.
Source: Semiconductor Devices: Physics and Technology by S. M. Sze, © 1985 by Bell Telephone Laboratories, by permission of John Wiley & Sons.

2.4 RESISTIVITY OF INTRINSIC SILICON

We are now in a position to calculate the electron and hole drift current densities j_n^{drift} and j_p^{drift}. For simplicity, we assume a one-dimensional current and avoid the vector notation of Eqs. (2.3) and (2.4):

$$j_n^{\text{drift}} = Q_n v_n = (-qn)(-\mu_n E) = qn\mu_n E \qquad \text{A/cm}^2$$
$$j_p^{\text{drift}} = Q_p v_p = (+qp)(+\mu_p E) = qp\mu_p E \qquad \text{A/cm}^2$$

(2.5)

in which $Q_n = (-qn)$ and $Q_p = (+qp)$ represent the charge densities (C/cm^3) of electrons and holes, respectively. The total drift current density is then given by

$$j_T^{\text{drift}} = j_n + j_p = q(n\mu_n + p\mu_p)E = \sigma E$$

(2.6)

This equation defines σ, the **electrical conductivity:**

$$\sigma = q(n\mu_n + p\mu_p) \qquad (\Omega \cdot \text{cm})^{-1}$$

(2.7)

Resistivity ρ is the reciprocal of conductivity:

$$\rho = \frac{1}{\sigma} \qquad (\Omega \cdot \text{cm})$$

(2.8)

The unit of resistivity, the Ohm-cm, may seem strange to many of us, but from Eq. (2.6), ρ represents the ratio of electric field to drift current density. The resistivity unit is therefore

$$\rho = \frac{E}{j_T^{\text{drift}}} \qquad \text{and} \qquad \frac{\text{V/cm}}{\text{A/cm}^2} = \Omega \cdot \text{cm}$$

(2.9)

EXAMPLE **2.2**   **RESISTIVITY OF INTRINSIC SILICON**

Here we determine if intrinsic silicon is an insulator, semiconductor, or conductor at room temperature by calculating its resistivity.

PROBLEM   Find the resistivity of intrinsic silicon at room temperature and classify it as an insulator, semiconductor or conductor.

SOLUTION   **Known Information and Given Data:** The room temperature mobilities for intrinsic silicon were given right after Eq. (2.4). For intrinsic silicon, the electron and hole densities are both equal to n_i.

Unknowns: Resistivity ρ and classification.

Approach: The conductivity and resistivity are defined by Eqs. (2.7) and (2.8), respectively.

Assumptions: Temperature is unspecified; assume "room temperature" with $n_i = 10^{10}/\text{cm}^3$.

Analysis: For intrinsic silicon, the charge density of electrons is given by $Q_n = -qn_i$, whereas the charge density for holes is $Q_p = +qn_i$. Substituting the given values into Eq. (2.7) yields

$$\sigma = (1.60 \times 10^{-19})[(10^{10})(1350) + (10^{10})(500)] \qquad (\text{C})(\text{cm}^{-3})(\text{cm}^2/\text{V} \cdot \text{s})$$

$$= 2.96 \times 10^{-6} \ (\Omega \cdot \text{cm})^{-1}$$

The resistivity ρ is equal to the reciprocal of the conductivity, so for intrinsic silicon

$$\rho = \frac{1}{\sigma} = 3.38 \times 10^5 \ \Omega \cdot \text{cm}$$

From Table 2.1, we see that intrinsic silicon can be characterized as an insulator, albeit near the low end of the insulator resistivity range.

Check of Results: The resistivity has been found, and intrinsic silicon is a poor insulator.

EXERCISE: Find the resistivity of intrinsic silicon at 400 K and classify it as an insulator, semiconductor, or conductor. Use the mobility values from Ex. 2.2.

ANSWERS: 1450 $\Omega \cdot$ cm, semiconductor

EXERCISE: Calculate the resistivity of intrinsic silicon at 50 K if the electron mobility is 6500 cm^2/V $\cdot$ s and the hole mobility is 2000 cm^2/V $\cdot$ s.

ANSWER: 1.69 $\times$ 10^{53} $\Omega \cdot$ cm

2.5 IMPURITIES IN SEMICONDUCTORS

The real advantages of semiconductors emerge when **impurities** are added to the material in minute but well-controlled amounts. This process is called **impurity doping,** or just **doping,** and the material that results is termed a **doped semiconductor.** Impurity doping enables us to change the resistivity over a very wide range and to determine whether the electron or hole population controls the resistivity

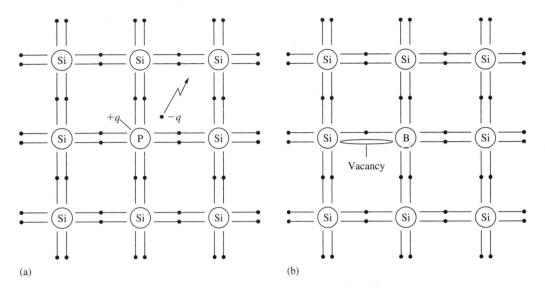

(a)   (b)

Figure 2.6 (a) An extra electron is available from a phosphorus donor atom. (b) Covalent bond vacancy from boron acceptor atom.

of the material. The following discussion focuses on silicon, although the concepts of impurity doping apply equally well to other materials. The impurities that we use with silicon are from columns III and V of the periodic table.

2.5.1 DONOR IMPURITIES IN SILICON

Donor impurities in silicon are from column V, having five valence electrons in the outer shell. The most commonly used elements are phosphorus, arsenic, and antimony. When a donor atom replaces a silicon atom in the crystal lattice, as shown in Fig. 2.6(a), four of the five outer shell electrons fill the covalent bond structure; it then takes very little thermal energy to free the extra electron for conduction. At room temperature, essentially every donor atom contributes (donates) an electron for conduction. Each donor atom that becomes ionized by giving up an electron will have a net charge of $+q$ and represents an immobile fixed charge in the crystal lattice.

2.5.2 ACCEPTOR IMPURITIES IN SILICON

Acceptor impurities in silicon are from column III and have one less electron than silicon in the outer shell. The primary acceptor impurity is boron, which is shown in place of a silicon atom in the lattice in Fig. 2.6(b). Because boron has only three electrons in its outer shell, a vacancy exists in the bond structure. It is easy for a nearby electron to move into this vacancy, creating another vacancy in the bond structure. This mobile vacancy represents a hole that can move through the lattice, as illustrated in Fig. 2.7(a) and (b), and the hole may simply be visualized as a particle with a charge of $+q$. Each impurity atom that becomes ionized by accepting an electron has a net charge of $-q$ and is immobile in the lattice, as in Fig. 2.7(a).

2.6 ELECTRON AND HOLE CONCENTRATIONS IN DOPED SEMICONDUCTORS

We now discover how to calculate the **electron** and **hole concentrations** in a semiconductor containing donor and acceptor impurities. In doped material, the electron and hole concentrations are no longer equal. If $n > p$, the material is called **n-type,** and if $p > n$, the material is referred to as

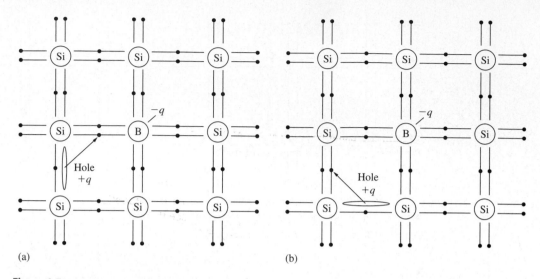

Figure 2.7 (a) Hole created after the boron atom accepts an electron. (b) Mobile hole moving through the silicon lattice.

p-type. The carrier with the larger population is called the **majority carrier,** and the carrier with the smaller population is termed the **minority carrier.**

To make detailed calculations of electron and hole densities, we need to keep track of the donor and acceptor impurity concentrations:

$$N_D = \textbf{donor impurity concentration} \quad \text{atoms/cm}^3$$
$$N_A = \textbf{acceptor impurity concentration} \quad \text{atoms/cm}^3$$

Two additional pieces of information are needed. First, the semiconductor material must remain charge neutral, which requires that the sum of the total positive charge and negative charge be zero. Ionized donors and holes represent positive charge, whereas ionized acceptors and electrons carry negative charge. Thus **charge neutrality** requires

$$q(N_D + p - N_A - n) = 0 \tag{2.10}$$

Second, the product of the electron and hole concentrations in intrinsic material was given in Eq. (2.2) as $pn = n_i^2$. It can be shown theoretically that $pn = n_i^2$ even for doped semiconductors in thermal equilibrium, and Eq. (2.2) is valid for a very wide range of doping concentrations.

2.6.1 n-TYPE MATERIAL ($N_D > N_A$)

Solving Eq. (2.2) for p and substituting into Eq. (2.10) yields a quadratic equation for n:

$$n^2 - (N_D - N_A)n - n_i^2 = 0$$

Now solving for n,

$$n = \frac{(N_D - N_A) \pm \sqrt{(N_D - N_A)^2 + 4n_i^2}}{2} \quad \text{and} \quad p = \frac{n_i^2}{n} \tag{2.11}$$

In practical situations $(N_D - N_A) \gg 2n_i$, and n is given approximately by $n \cong (N_D - N_A)$. The formulas in Eq. (2.11) should be used for $N_D > N_A$ (see Prob. 2.26).

2.6.2 *p*-TYPE MATERIAL ($N_A > N_D$)

For the case of $N_A > N_D$, we substitute for n in Eq. (2.10) and use the quadratic formula to solve for p:

$$p = \frac{(N_A - N_D) \pm \sqrt{(N_A - N_D)^2 + 4n_i^2}}{2} \quad \text{and} \quad n = \frac{n_i^2}{p} \qquad (2.12)$$

Again, the usual case is $(N_A - N_D) \gg 2n_i$, and p is given approximately by $p \cong (N_A - N_D)$. Equation (2.12) should be used for $N_A > N_D$.

Because of practical process-control limitations, impurity densities that can be introduced into the silicon lattice range from approximately 10^{14} to 10^{21} atoms/cm^3. Thus, N_A and N_D normally will be much greater than the intrinsic carrier concentration in silicon at room temperature. From the preceding approximate expressions, we see that the majority carrier density is set directly by the net impurity concentration: $p \cong (N_A - N_D)$ for $N_A > N_D$ or $n \cong (N_D - N_A)$ for $N_D > N_A$.

DESIGN NOTE — PRACTICAL DOPING LEVELS

In both *n*- and *p*-type semiconductors, the majority carrier concentrations are established "at the factory" by the engineer's choice of N_A and N_D and are independent of temperature over a wide range. In contrast, the minority carrier concentrations, although small, are proportional to n_i^2 and highly temperature dependent. For practical doping levels,

$$\text{For } n\text{-type } (N_D > N_A): \quad n \cong N_D - N_A \quad p = \frac{n_i^2}{N_D - N_A}$$

$$\text{For } p\text{-type } (N_A > N_D): \quad p \cong N_A - N_D \quad n = \frac{n_i^2}{N_A - N_D}$$

Typical values of doping fall in this range:

$$10^{14}/\text{cm}^3 \leq |N_A - N_D| \leq 10^{21}/\text{cm}^3$$

EXAMPLE 2.3 — ELECTRON AND HOLE CONCENTRATIONS

Calculate the electron and hole concentrations in a silicon sample containing both acceptor and donor impurities.

PROBLEM Find the type and electron and hole concentrations in a silicon sample at room temperature if it is doped with a boron concentration of $10^{16}/\text{cm}^3$ and a phosphorus concentration of $2 \times 10^{15}/\text{cm}^3$.

SOLUTION **Known Information and Given Data:** Boron and phosphorus doping concentrations and room temperature operation are specified.

Unknowns: Electron and hole concentrations (n and p).

Approach: Identify the donor and acceptor impurity concentrations and use their values to find n and p with Eq. (2.11) or Eq. (2.12), as appropriate.

Assumptions: At room temperature, $n_i = 10^{10}/\text{cm}^3$.

Analysis: Using Table 2.2 we find that boron is an acceptor impurity and phosphorus is a donor impurity. Therefore

$$N_A = 10^{16}/\text{cm}^3 \qquad \text{and} \qquad N_D = 2 \times 10^{15}/\text{cm}^3$$

Since $N_A > N_D$, the material is p-type, and we have $(N_A - N_D) = 8 \times 10^{15}/\text{cm}^3$. For $n_i = 10^{10}/\text{cm}^3$, $(N_A - N_D) \gg 2n_i$, and we can use the simplified form of Eq. (2.12):

$$p \cong (N_A - N_D) = 8.00 \times 10^{15} \text{ holes/cm}^3$$

$$n = \frac{n_i^2}{p} = \frac{10^{20}/\text{cm}^6}{8.00 \times 10^{15}/\text{cm}^3} = 1.25 \times 10^4 \text{ electrons/cm}^3$$

Check of Results: We have found the electron and hole concentrations. We can double check the pn product: $pn = 10^{20}/\text{cm}^6$, which is correct.

EXERCISE: Find the type and electron and hole concentrations in a silicon sample at a temperature of 400 K if it is doped with a boron concentration of $10^{16}/\text{cm}^3$ and a phosphorus concentration of $2 \times 10^{15}/\text{cm}^3$. Assume $n_i = 2 \times 10^{12}/\text{cm}^3$.

ANSWERS: $8.00 \times 10^{15}/\text{cm}^3$, $5.00 \times 10^8/\text{cm}^3$

EXERCISE: Silicon is doped with an antimony concentration of $2 \times 10^{16}/\text{cm}^3$. Is antimony a donor or acceptor impurity? Find the electron and hole concentrations at 300 K. Is this material n- or p-type?

ANSWERS: Donor; $2 \times 10^{16}/\text{cm}^3$; $5 \times 10^3/\text{cm}^3$; n-type

---

One might ask why we care about the minority carriers if they are so small in number. Indeed, we find shortly that semiconductor resistivity is controlled by the majority carrier concentration, and in Chapter 4 we find that field-effect transistors (FETs) are also majority carrier devices. However, the characteristics of diodes and bipolar junction transistors, discussed in Chapters 3 and 5, respectively, depend strongly on the minority carrier populations. Thus, to be able to design a variety of solid-state devices, we must understand how to manipulate both the majority and minority carrier concentrations.

2.7 MOBILITY AND RESISTIVITY IN DOPED SEMICONDUCTORS

The introduction of impurities into a semiconductor such as silicon actually degrades the mobility of the carriers in the material. Impurity atoms have slightly different sizes than the silicon atoms that they replace and hence disrupt the periodicity of the lattice. In addition, the impurity atoms are ionized and represent regions of localized charge that were not present in the original crystal. Both these effects cause the electrons and holes to scatter as they move through the semiconductor and reduce the mobility of the carriers in the crystal.

Figure 2.8 shows the dependence of mobility on the *total* impurity doping density $N_T = (N_A + N_D)$ in silicon. We see that mobility drops rapidly as the doping level in the crystal increases. Mobility in heavily doped material can be more than an order of magnitude less than that in lightly doped material. On the other hand, doping vastly increases the density of majority carriers in the semiconductor material and thus has a dramatic effect on resistivity that overcomes the influence of decreased mobility.

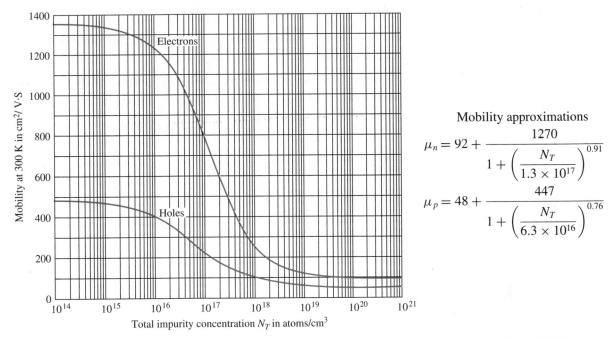

Mobility approximations

$$\mu_n = 92 + \frac{1270}{1 + \left(\dfrac{N_T}{1.3 \times 10^{17}}\right)^{0.91}}$$

$$\mu_p = 48 + \frac{447}{1 + \left(\dfrac{N_T}{6.3 \times 10^{16}}\right)^{0.76}}$$

Figure 2.8 Dependence of electron and hole mobility on total impurity concentration in silicon at 300 K.

EXERCISE: What are the electron and hole mobilities in a silicon sample with an acceptor impurity density of $10^{16}/cm^3$?

ANSWERS: 1250 cm^2/V·s; 400 cm^2/V·s

EXERCISE: What are the electron and hole mobilities in a silicon sample with an acceptor impurity density of $4 \times 10^{16}/cm^3$ and a donor impurity density of $6 \times 10^{16}/cm^3$?

ANSWERS: 800 cm^2/V·s, 230 cm^2/V·s

Remember that impurity doping also determines whether the material is n- or p-type, and simplified expressions can be used to calculate the conductivity of most extrinsic material. Note that $\mu_n n \gg \mu_p p$ in the expression for σ in Ex. 2.4. For doping levels normally encountered, this inequality will be true for n-type material, and $\mu_p p \gg \mu_n n$ will be valid for p-type material. The majority carrier concentration controls the conductivity of the material so that

$$\sigma \cong q\mu_n n \cong q\mu_n(N_D - N_A) \qquad \text{for } n\text{-type material}$$
$$\sigma \cong q\mu_p p \cong q\mu_p(N_A - N_D) \qquad \text{for } p\text{-type material}$$

(2.13)

We now explore the relationship between doping and resistivity with an example.

EXAMPLE 2.4   RESISTIVITY CALCULATION OF DOPED SILICON

This example contrasts the resistivity of doped silicon to that of pure silicon.

PROBLEM   Calculate the resistivity of silicon doped with a donor density $N_D = 2 \times 10^{15}/cm^3$. What is the material type? Classify the sample as an insulator, semiconductor, or conductor.

SOLUTION   **Known Information and Given Data:** $N_D = 2 \times 10^{15}/\text{cm}^3$.

Unknowns: Resistivity ρ, which also requires us to find the hole and electron concentrations (p and n) and mobilities (μ_p and μ_n); material type.

Approach: Use the doping concentration to find n and p and μ_n and μ_p; substitute these values into the expression for σ.

Assumptions: Since N_A is not mentioned, assume $N_A = 0$. Assume room temperature with $n_i = 10^{10}/\text{cm}^3$.

Analysis: In this case, $N_D > N_A$ and much much greater than n_i, so

$$n = N_D = 2 \times 10^{15} \text{ electron/cm}^3$$

$$p = \frac{n_i^2}{n} = 10^{20}/2 \times 10^{15} = 5 \times 10^4 \text{ holes/cm}^3$$

Because $n > p$, the silicon is n-type material. From Fig. 2.8, the electron and hole mobilities for an impurity concentration of $2 \times 10^{15}/\text{cm}^3$ are

$$\mu_n = 1320 \text{ cm}^2/\text{V} \cdot \text{s} \qquad \mu_p = 460 \text{ cm}^2/\text{V} \cdot \text{s}$$

The conductivity and resistivity are now found to be

$$\sigma = 1.6 \times 10^{19}[(1320)(2 \times 10^{15}) + (460)(5 \times 10^4)] = 0.422 \ (\Omega \cdot \text{cm})^{-1}$$

and

$$\rho = 1/\sigma = 2.37 \ \Omega \cdot \text{cm}$$

This silicon sample is a semiconductor.

Check of Results: We have found the required unknowns.

Discussion: Comparing these results to those for intrinsic silicon, we note that the introduction of a minute fraction of impurities into the silicon lattice has changed the resistivity by 5 orders of magnitude, changing the material in fact from an insulator to a midrange semiconductor. Based upon this observation, it is not unreasonable to assume that additional doping can change silicon into a conductor (see the exercise following Ex. 2.5). Note that the doping level in this example represents a replacement of less than 10^{-5} percent of the atoms in the silicon crystal.

EXAMPLE **2.5**   **WAFER DOPING—AN ITERATIVE CALCULATION**

Solutions to many engineering problems require iterative calculations as well as the integration of mathematical and graphical information.

PROBLEM   An n-type silicon wafer has a resistivity of 0.054 $\Omega \cdot \text{cm}$. What is the donor concentration N_D?

SOLUTION   **Known Information and Given Data:** The wafer is n-type silicon; resistivity is 0.054 $\Omega \cdot \text{cm}$.

Unknowns: Doping concentration N_D required to achieve the desired resistivity.

Approach: For this problem, an iterative trial-and-error solution is necessary. Because the resistivity is low, it should be safe to assume that

$$\sigma = q\mu_n n = q\mu_n N_D \qquad \text{and} \qquad \mu_n N_D = \frac{\sigma}{q}$$

We know that μ_n is a function of the doping concentration N_D, but the functional dependence is available only in graphical form. This is an example of a type of problem often encountered in engineering. The solution requires an iterative trial-and-error approach involving both mathematical and graphical evaluations. To solve the problem, we need to establish a logical progression of steps in which the choice of one parameter enables us to evaluate other parameters that lead to the solution. One method for this problem is

1. Choose a value of N_D.
2. Find μ_n from the mobility graph.
3. Calculate $\mu_n N_D$.
4. If $\mu_n N_D$ is not correct, go back to step 1.

Obviously, we hope we can make educated choices that will lead to convergence of the process after a few trials.

Assumptions: Assume the wafer contains only donor impurities.

Analysis: For this problem,

$$\frac{\sigma}{q} = (0.054 \times 1.6 \times 10^{-19})^{-1} = 1.2 \times 10^{20} \ (\text{V} \cdot \text{s} \cdot \text{cm})^{-1}$$

Choosing a first guess of $N_D = 1 \times 10^{16}/\text{cm}^3$:

TRIAL	N_D (cm^{-3})	μ_n (cm^2/V·s)	$\mu_n N_D$ (V·s·cm)$^{-1}$
1	1×10^{16}	1250	1.3×10^{19}
2	1×10^{18}	260	2.5×10^{20}
3	1×10^{17}	80	8.0×10^{19}
4	5×10^{17}	380	3.8×10^{20}
5	4×10^{17}	430	1.7×10^{20}
6	2×10^{17}	600	1.2×10^{20}

After six iterations, we find $N_D = 2 \times 10^{17}$ donor atoms/cm^3.

Check of Results: We have found the only unknown. $N_D = 2 \times 10^{17}/\text{cm}^3$ is in the range of practically achievable doping. See the Design Note in Sec. 2.6.

EXERCISE: What is the minimum value of donor doping required to convert silicon to a conductor at room temperature?

ANSWER: $6.25 \times 10^{19}/\text{cm}^3$ with $\mu_n \approx 100 \ \text{cm}^2/\text{V} \cdot \text{s}$

EXERCISE: Silicon is doped with a phosphorus concentration of $2 \times 10^{16}/\text{cm}^3$. What are N_A and N_D? What are the electron and hole mobilities? What are the mobilities if boron in a concentration of $3 \times 10^{16}/\text{cm}^3$ is added to the silicon?

ANSWERS: $N_A = 0/\text{cm}^3$; $N_D = 2 \times 10^{16}/\text{cm}^3$; $\mu_n = 1160 \ \text{cm}^2/\text{V} \cdot \text{s}$, $\mu_p = 370 \ \text{cm}^2/\text{V} \cdot \text{s}$; $\mu_n = 980 \ \text{cm}^2/\text{V} \cdot \text{s}$; $\mu_p = 290 \ \text{cm}^2/\text{V} \cdot \text{s}$

EXERCISE: Silicon is doped with a boron concentration of $4 \times 10^{18}/cm^3$. Is boron a donor or acceptor impurity? Find the electron and hole concentrations at 300 K. Is this material n-type or p-type? Find the electron and hole mobilities.

ANSWERS: Acceptor; $n = 25/cm^3$, $p = 4 \times 10^{18}/cm^3$; p-type; $\mu_n = 150$ cm^2/V·s and $\mu_p = 70$ cm^2/V·s

EXERCISE: Silicon is doped with an indium concentration of $7 \times 10^{19}/cm^3$. Is indium a donor or acceptor impurity? Find the electron and hole concentrations, the electron and hole mobilities, and the resistivity of this silicon material at 300 K. Is this material n- or p-type?

ANSWERS: Acceptor; $n = 1.4/cm^3$, $p = 7 \times 10^{19}/cm^3$; $\mu_n = 100$ cm^2/V·s and $\mu_p = 50$ cm^2/V·s; $\rho = 0.00179$ Ω·cm; p-type

2.8 DIFFUSION CURRENTS

As already described, the electron and hole populations in a semiconductor are controlled by the impurity doping concentrations N_A and N_D. Up to this point we have tacitly assumed that the doping is uniform in the semiconductor, but this need not be the case. Changes in doping are encountered often in semiconductors, and there will be gradients in the electron and hole concentrations. Gradients in these free carrier densities give rise to a second current flow mechanism, called **diffusion.** The free carriers tend to move (diffuse) from regions of high concentration to regions of low concentration in much the same way as a puff of smoke in one corner of a room rapidly spreads throughout the entire room.

A simple one-dimensional gradient in the electron or hole density is shown in Fig. 2.9. The gradient in this figure is positive in the $+x$ direction, but the carriers diffuse in the $-x$ direction, from high to low concentration. Thus the **diffusion current densities** are proportional to the negative of the carrier gradient:

$$
\begin{aligned}
j_p^{\text{diff}} &= (+q)D_p\left(-\frac{\partial p}{\partial x}\right) = -qD_p\frac{\partial p}{\partial x} \\
j_n^{\text{diff}} &= (-q)D_n\left(-\frac{\partial n}{\partial x}\right) = +qD_n\frac{\partial n}{\partial x}
\end{aligned}
\qquad \text{A/cm}^2 \qquad (2.14)
$$

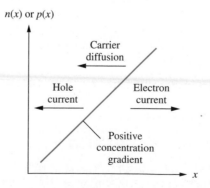

Figure 2.9 Carrier diffusion in the presence of a concentration gradient.

The proportionality constants D_p and D_n are the **hole** and **electron diffusivities,** with units (cm^2/s). Diffusivity and mobility are related by **Einstein's relationship:**

$$\frac{D_n}{\mu_n} = \frac{kT}{q} = \frac{D_p}{\mu_p} \qquad (2.15)$$

The quantity $(kT/q = V_T)$ is called the **thermal voltage V_T,** and its value is approximately 0.025 V at room temperature. We encounter the parameter V_T in several different contexts throughout this book. Typical values of the diffusivities (also referred to as the **diffusion coefficients**) in silicon are in the range 2 to 35 cm^2/s for electrons and 1 to 15 cm^2/s for holes at room temperature.

EXERCISE: Calculate the value of the thermal voltage V_T for $T = 50$ K, 300 K, and 400 K.

ANSWERS: 4.3 mV; 25.8 mV; 34.5 mV

DESIGN NOTE   THERMAL VOLTAGE V_T

$$V_T = kT/q = 0.0258 \text{ V at 300 K}$$

EXERCISE: What are the maximum values of the room temperature values (300 K) of the diffusion coefficients for electrons and holes in silicon based on the mobilities in Fig. 2.8?

ANSWERS: Using $V_T = 25.8$ mV; 35.1 cm^2/s, 12.8 cm^2/s

EXERCISE: An electron gradient of $+10^{16}/(\text{cm}^3 \cdot \mu\text{m})$ exists in a semiconductor. What is the diffusion current density at room temperature if the electron diffusivity $= 20$ cm^2/s? Repeat for a hole gradient of $+10^{20}/\text{cm}^4$ with $D_p = 4$ cm^2/s.

ANSWER: $+320$ A/cm^2; -64 A/cm^2

2.9 TOTAL CURRENT

Generally, currents in a semiconductor have both drift and diffusion components. The total electron and hole current densities j_n^T and j_p^T can be found by adding the corresponding drift and diffusion components from Eqs. (2.5) and (2.14):

$$j_n^T = q\mu_n n E + q D_n \frac{\partial n}{\partial x} \qquad \text{and} \qquad j_p^T = q\mu_p p E - q D_p \frac{\partial p}{\partial x} \qquad (2.16)$$

Using Einstein's relationship from Eq. (2.15), Eq. (2.16) can be rewritten as

$$j_n^T = q\mu_n n \left(E + V_T \frac{1}{n} \frac{\partial n}{\partial x} \right) \qquad \text{and} \qquad j_p^T = q\mu_p p \left(E - V_T \frac{1}{p} \frac{\partial p}{\partial x} \right) \qquad (2.17)$$

Equation (2.16) or (2.17) combined with Gauss' law

$$\nabla \cdot (\varepsilon E) = Q \tag{2.18}$$

where ε = permittivity (F/cm), E = electric field (V/cm), and Q = charge density (C/cm^3)

gives us a powerful mathematics approach for analyzing the behavior of semiconductors and forms the basis for many of the results presented in later chapters.

2.10 ENERGY BAND MODEL

This section discusses the **energy band model** for a semiconductor, which provides a useful alternative view of the electron–hole creation process and the control of carrier concentrations by impurities. Quantum mechanics predicts that the highly regular crystalline structure of a semiconductor produces periodic quantized ranges of allowed and disallowed energy states for the electrons surrounding the atoms in the crystal. Figure 2.10 is a conceptual picture of this band structure in the semiconductor, in which the regions labeled **conduction band** and **valence band** represent allowed energy states for electrons. Energy E_V corresponds to the top edge of the valence band and represents the highest permissible energy for a valence electron. Energy E_C corresponds to the bottom edge of the conduction band and represents the lowest available energy level in the conduction band. Although these bands are shown as continuums in Fig. 2.10, they actually consist of a very large number of closely spaced, discrete energy levels.

Electrons are not permitted to assume values of energy lying between E_C and E_V. The difference between E_C and E_V is called the *bandgap energy E_G*:

$$E_G = E_C - E_V \tag{2.19}$$

Table 2.3 listed examples of the bandgap energy for a number of semiconductors.

2.10.1 ELECTRON–HOLE PAIR GENERATION IN AN INTRINSIC SEMICONDUCTOR

In silicon at very low temperatures (≈ 0 K), the valence band states are completely filled with electrons, and the conduction band states are completely empty, as shown in Fig. 2.11. The semiconductor in

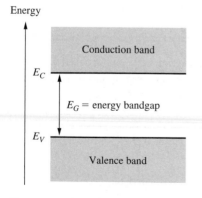

Figure 2.10 Energy band model for a semiconductor with bandgap E_G.

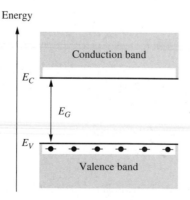

Figure 2.11 Semiconductor at 0 K with filled valence band and empty conduction band. This figure corresponds to the bond model in Fig. 2.2.

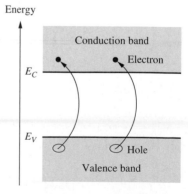

Figure 2.12 Creation of electron–hole pair by thermal excitation across the energy bandgap. This figure corresponds to the bond model of Fig. 2.3.

this situation does not conduct current when an electric field is applied. There are no free electrons in the conduction band, and no holes exist in the completely filled valence band to support current flow. The band model of Fig. 2.11 corresponds directly to the completely filled bond model of Fig. 2.2.

As temperature rises above 0 K, thermal energy is added to the crystal. A few electrons gain the energy required to surmount the energy bandgap and jump from the valence band into the conduction band, as shown in Fig. 2.12. Each electron that jumps the bandgap creates an electron–hole pair. This **electron–hole pair generation** situation corresponds directly to that presented in Fig. 2.3.

2.10.2 ENERGY BAND MODEL FOR A DOPED SEMICONDUCTOR

Figures 2.13 to 2.15 present the band model for **extrinsic material** containing donor and/or acceptor atoms. In Fig. 2.13, a concentration N_D of donor atoms has been added to the semiconductor. The donor atoms introduce new localized energy levels within the bandgap at a **donor energy level E_D** near the conduction band edge. The value of $(E_C - E_D)$ for phosphorus is approximately 0.045 eV, so it takes very little thermal energy to promote the extra electrons from the donor sites into the conduction band. The density of conduction-band states is so high that the probability of finding an electron in a donor state is practically zero, except for heavily doped material (large N_D) or at very low temperature. Thus at room temperature, essentially all the available donor electrons are free for conduction. Figure 2.13 corresponds to the bond model of Fig. 2.6.

In Fig. 2.14, a concentration N_A of acceptor atoms has been added to the semiconductor. The acceptor atoms introduce energy levels within the bandgap at the **acceptor energy level E_A** near the valence band edge. The value of $(E_A - E_V)$ for boron is approximately 0.044 eV, and it takes very little thermal energy to promote electrons from the valence band into the acceptor energy levels. At room temperature, essentially all the available acceptor sites are filled, and each promoted electron creates a hole that is free for conduction. Figure 2.14 corresponds to the bond model of Fig. 2.7.

2.10.3 COMPENSATED SEMICONDUCTORS

The situation for a **compensated semiconductor,** one containing both acceptor and donor impurities, is depicted in Fig. 2.15 for the case in which there are more donor atoms than acceptor atoms. Electrons seek the lowest energy states available, and they fall from donor sites, filling all the available acceptor sites. The remaining free electron population is given by $n = (N_D - N_A)$.

The energy band model just discussed represents a conceptual model that is complementary to the covalent bond model of Sec. 2.2. Together they help us visualize the processes involved in creating holes and electrons in doped semiconductors.

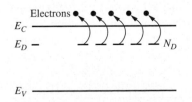

Figure 2.13 Donor level with activation energy $(E_C - E_D)$. This figure corresponds to the bond model of Fig. 2.6.

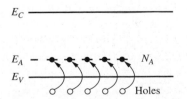

Figure 2.14 Acceptor level with activation energy $(E_A - E_V)$. This figure corresponds to the bond model of Fig. 2.7(b).

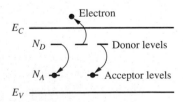

Figure 2.15 Compensated semiconductor containing both donor and acceptor atoms with $N_D > N_A$.

CCD Cameras

Modern astronomy is highly dependent on microelectronics for both the collection and analysis of astronomical data. Tremendous advancements in astronomy have been made possible by the combination of electronic image capture and computer analysis of the acquired images.

In the case of optical telescopes, the Charge-Coupled Device (CCD) camera converts photons to electrical signals that are then formed into a computer image. Like other photo-detector circuits, the CCD captures electrons that are generated when incident photons interact with the semiconductor material and create hole-electron pairs as in Fig. 2.12. A two-dimensional array of as many as several million CCD cells is formed on a single chip, similar to the one shown below. CCD imagers are especially important to astronomers because of their very high sensitivity and low electronic noise.

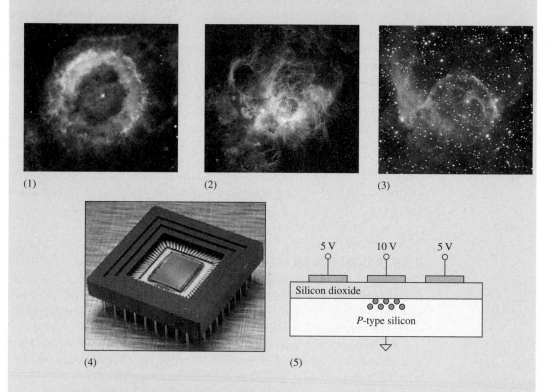

(1)   (2)   (3)

(4)   (5)

A simplified view of a CCD cell is shown here. A group of electrons have accumulated under the middle electrode due to the higher voltage present. The electrons are held within the semiconductor by the combination of the insulating silicon-dioxide layer and the fields created by the electrodes. The more incident light, the more electrons are captured. To read the charge out of the cell, the electrode voltages are manipulated to move the charge from electrode to electrode until it is converted to a voltage at the edge of the imaging array. The astronomical images were acquired with CCD cameras located on the Hubble Space Telescope.

Source: (1) NGC6369: The Little Ghost Nebula. Credit: Hubble Heritage Team, NASA; (2) NGC604: Giant Stellar Nursery. Credit: H. Yang (UIUC), HST, NASA; (3) NGC2359: Thors Helmet. Credits: Christine and David Smith, Steve Mandel, Adam Block (KPNO Visitor Program), NOAO, AURA, NSF. (4) The chip pictured above is a Kodak KAF-1401E CCD image sensor and is reprinted here with permission from Eastman Kodak Company.

2.11 OVERVIEW OF INTEGRATED CIRCUIT FABRICATION

Before we leave this chapter, we explore how an engineer uses selective control of semiconductor doping to form a simple electronic device. We do this by exploring the basic fabrication steps utilized to fabricate a solid-state diode. These ideas help us understand the characteristics of many electronic devices that depend strongly on the physical structure of the device.

Complex solid-state devices and circuits are fabricated through the repeated application of a number of basic IC processing steps including oxidation, photolithography, etching, ion implantation, diffusion, evaporation, sputtering, chemical vapor deposition, and epitaxial growth. **Silicon dioxide** (SiO_2) layers are formed by heating silicon wafers to a high temperature (1000 to 1200°C) in the presence of pure oxygen or water vapor. This process is called **oxidation.** Thin layers of metal films are deposited through **evaporation** by heating the metal to its melting point in a vacuum. Both conducting metal films and insulators can be deposited through a process called **sputtering,** which uses physical ion bombardment to effect transfer of atoms from a source target to the wafer surface.

Thin films of polysilicon, silicon dioxide, and **silicon nitride** can all be formed through **chemical vapor deposition** (CVD), in which the material is precipitated from a gaseous mixture directly onto the surface of the silicon wafer. Shallow *n*- and *p*-type layers are formed by **ion implantation,** where the wafer is bombarded by high-energy (50-keV to 1-MeV) acceptor or donor impurity atoms generated by a high-voltage particle accelerator. A greater depth of the impurity layers can be achieved by **diffusion** of the impurities at high temperatures, typically 1000 to 1200°C, in either an inert or oxidizing environment. Bipolar processes, as well as some CMOS processes, employ the **epitaxial growth** technique to form thin high-quality layers of crystalline silicon on top of the wafer. The epitaxial layer replicates the crystal structure of the original silicon substrate.

To build integrated circuits, localized *n*- and *p*-type regions must be formed selectively in the silicon surface. Silicon dioxide, silicon nitride, **polysilicon, photoresist,** and other materials can all be used to block out areas of the wafer surface to prevent penetration of impurity atoms during implantation and/or diffusion. **Masks** containing window patterns to be opened in the protective layers are produced using a combination of computer-aided design systems and photographic reduction techniques. The patterns are transferred from the mask to the wafer surface through the use of high-resolution optical photographic techniques, a process called **photolithography.** The windows defined by the masks are cut through the protective layers by wet-chemical **etching** using acids or by dry-plasma etching.

The fabrication steps just outlined can be combined in many different ways to form integrated circuits. A simple example is contained in Figs. 2.16 and 2.17. Here we wish to form, and make contact to, a localized *p*-type region in the surface of an *n*-type silicon wafer. In Fig. 2.17(a), a thin layer of silicon dioxide (1 μm) has been grown on the silicon wafer with a typical thickness of approximately 500 μm, and a layer of photoresist has been applied to the top of the SiO_2. The photoresist is exposed by shining light through a mask that contains patterns to be transferred to the wafer. After exposure and development, this photoresist (called positive resist) has an opening where it was exposed, as in Fig. 2.17(b). Next, the oxide is etched away using the photoresist as a barrier layer, leaving a window through both the photoresist and oxide layers, as in Fig. 2.17(c). Acceptor impurities are now implanted into the silicon through the window, but are blocked everywhere else by the barrier formed by the photoresist and oxide layers. After photoresist removal, a localized *p*-type region exists in the silicon below the window in the SiO_2, as in Fig. 2.17(d). The *p*-type region will extend from a few tenths of a micron to at most a few microns below the silicon surface.

Oxide is regrown on the wafer surface and coated with a new layer of photoresist, as indicated in Fig. 2.17(e). Contact windows are exposed through a second mask. The structure in Fig. 2.17(f) results following completion of the photolithography step and subsequent etching of the contact windows in the oxide. Contacts will be made to both the *n*-type substrate and the *p*-type region through these openings. Next, an aluminum layer is evaporated onto the silicon wafer and once again coated with photoresist as in Fig. 2.17(g). A third mask and photolithography step are used to transfer

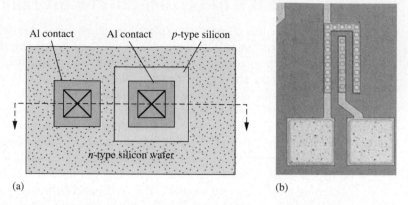

(a)   (b)

Figure 2.16 (a) Top view of the *pn* diode structure formed by fabrication steps in Fig. 2.17. (b) Photomicrograph of an actual diode.

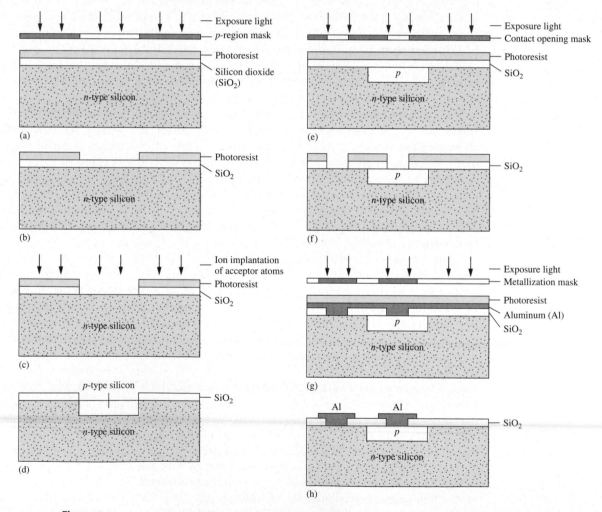

Figure 2.17 Silicon wafer (a) at first mask exposure step, (b) after exposure and development of photoresist, (c) following etching of silicon dioxide, and (d) after implantation/diffusion of acceptor impurity and resist removal. (e) Exposure of contact opening mask (f) after resist development and etching of contact openings. (g) Exposure of metal mask. (h) Final structure after etching of aluminum and resist removal.

the desired metallization pattern to the wafer surface, and then the aluminum is etched away wherever it is not coated with resist. The completed structure appears in Fig. 2.17(h) and corresponds to the top view in Fig. 2.16. Aluminum contacts have been made to both the *n*-type substrate and the *p*-type region. We have just stepped through the fabrication of our first solid-state device—a *pn* junction diode! Study of the characteristics, operation, and application of diodes is the topic of Chapter 3. Figure 2.16(b) is a photomicrograph of an actual diode.

ELECTRONICS IN ACTION

Lab-on-a-chip

The photo below[1] illustrates recent work in the integration of silicon microelectronic circuits, microfluidics, and a printed circuit board to realize a nanoliter DNA analysis device. DNA fluid samples are introduced at one end of the device, metered into nanoliter sized droplets, and propelled along a fluidic channel where the sample is mixed with other materials, heated, and optically stimulated. Integrated optical detectors are used to measure the resulting fluorescence for detection of target genetic bio-materials.

Devices such as the one below are revolutionizing health-care by improving our understanding of disease and disease mechanisms, enabling rapid diagnostics and providing for the screening of large numbers of potential treatments in a low-cost fashion. Bioengineering and in particular the application of microelectronics to health-care and life sciences is a rapidly growing and exciting field.

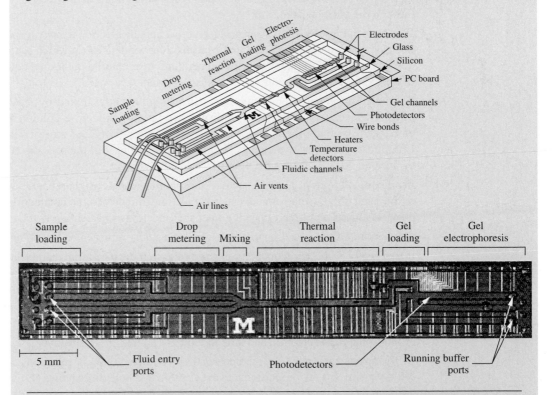

[1] Mark A. Burns, Brian N. Johnson, Sundaresh N. Brahmasandra, Kalyan Handique, James R. Webster, Madhavi Krishnan, Timothy S. Sammarco, Piu M. Man, Darren Jones, Dylan Heldsinger, Carlos H. Mastrangelo, David T. Burke, "An Integrated Nanoliter DNA Analysis Device," *Science*, vol. 282, no. 5388, 16 Oct 1998.

SUMMARY

- Materials are found in three primary forms: amorphous, polycrystalline, and crystalline. An amorphous material is a totally disordered or random material that shows no short range order. In polycrystalline material, large numbers of small crystallites can be identified. A crystalline material exhibits a highly regular bonding structure among the atoms over the entire macroscopic crystal.

- Electronic materials can be separated into three classifications based on their electrical resistivity. Insulators have resistivities above 10^5 $\Omega \cdot$ cm, whereas conductors have resistivities below 10^{-3} $\Omega \cdot$ cm. Between these two extremes lie semiconductor materials.

- Today's most important semiconductor is silicon (Si), which is used for fabrication of very-large-scale-integrated (VLSI) circuits. Two compound semiconductor materials, gallium arsenide (GaAs) and indium phosphide (InP), are the most important materials for optoelectronic applications including light-emitting diodes (LEDs), lasers, and photodetectors.

- The highly useful properties of semiconductors arise from the periodic nature of crystalline material, and two conceptual models for these semiconductors were introduced: the covalent bond model and the energy band model.

- At very low temperatures approaching 0 K, all the covalent bonds in a semiconductor crystal will be intact and the material will actually be an insulator. As temperature is raised, the added thermal energy causes a small number of covalent bonds to break. The amount of energy required to break a covalent bond is equal to the bandgap energy E_G.

- When a covalent bond is broken, two charge carriers are produced: an electron, with charge $-q$, that is free to move about the conduction band; and a hole, with charge $+q$, that is free to move through the valence band.

- Pure material is referred to as intrinsic material, and the electron density n and hole density p in an intrinsic material are both equal to the intrinsic carrier density n_i, which is approximately equal to 10^{10} carriers/cm^3 in silicon at room temperature. In a material in thermal equilibrium, the product of the electron and hole concentrations is a constant: $pn = n_i^2$.

- The hole and electron concentrations can be significantly altered by replacing small numbers of atoms in the original crystal with impurity atoms. Silicon, a column IV element, has four electrons in its outer shell and forms covalent bonds with its four nearest neighbors in the crystal. In contrast, the impurity elements (from columns III and V of the periodic table) have either three or five electrons in their outer shells.

- In silicon, column V elements such as phosphorus, arsenic, and antimony, with an extra electron in the outer shell, act as donors and add electrons directly to the conduction band. A column III element such as boron has only three outer shell electrons and creates a free hole in the valence band.

- The donor and acceptor impurity densities are usually represented by N_D and N_A, respectively.

- If n exceeds p, the semiconductor is referred to as n-type material, and electrons are the majority carriers and holes are the minority carriers. If p exceeds n, the semiconductor is referred to as p-type material, and holes become the majority carriers and electrons, the minority carriers.

- Electron and hole currents each have two components: a drift current and a diffusion current.

- Drift current is the result of carrier motion caused by an applied electric field. Drift currents are proportional to the electron and hole mobilities (μ_n and μ_p, respectively).

- Diffusion currents arise from gradients in the electron or hole concentrations. The magnitudes of the diffusion currents are proportional to the electron and hole diffusivities (D_n and D_p, respectively).

- Diffusivity and mobility are related by the Einstein relationship: $D/\mu = kT/q$. The expression kT/q has units of voltage and is often referred to as the thermal voltage V_T. Doping the semiconductor disrupts the periodicity of the crystal lattice, and the mobility—and hence diffusivity—both decrease monotonically as the impurity doping concentration is increased.

- The ability to add impurities to change the conductivity type and to control hole and electron concentrations is at the heart of our ability to fabricate high-performance, solid-state devices and high-density integrated circuits. In the next several chapters, we see how this capability is used to form diodes, field-effect transistors (FETs), and bipolar junction transistors (BJTs).

- Complex solid-state devices and circuits are fabricated through the repeated application of a number of basic IC processing steps, including oxidation, photolithography, etching, ion implantation, diffusion, evaporation, sputtering, chemical vapor deposition (CVD), and epitaxial growth.

- To build integrated circuits, localized *n*- and *p*-type regions must be formed selectively in the silicon surface. Silicon dioxide, silicon nitride, polysilicon, photoresist, and other materials can all be used to block out areas of the wafer surface to prevent penetration of impurity atoms during implantation and/or diffusion. Masks containing window patterns to be opened in the protective layers are produced using a combination of computer-aided design systems and photographic reduction techniques. The patterns are transferred from the mask to the wafer surface through the use of high-resolution photolithography.

KEY TERMS

Acceptor energy level	Hole
Acceptor impurities	Hole concentration
Acceptor impurity concentration	Hole density
Amorphous material	Hole diffusivity
Bandgap energy	Hole mobility
Charge neutrality	Impurities
Chemical vapor deposition	Impurity doping
Compensated semiconductor	Insulator
Compound semiconductor	Intrinsic carrier density
Conduction band	Intrinsic material
Conductivity	Ion implantation
Conductor	Majority carrier
Covalent bond model	Mask
Diffusion	Minority carrier
Diffusion coefficients	Mobility
Diffusion current density	*n*-type material
Donor energy level	Oxidation
Donor impurities	*p*-type material
Donor impurity concentration	Photolithography
Doped semiconductor	Photoresist
Doping	*pn* product
Drift current density	Polycrystalline material
Einstein's relationship	Polysilicon
Electrical conductivity	Resistivity
Electron	Saturated drift velocity
Electron concentration	Semiconductor
Electron diffusivity	Silicon dioxide
Electron–hole pair generation	Silicon nitride
Electron mobility	Single-crystal material
Elemental semiconductor	Sputtering
Energy band model	Thermal equilibrium
Epitaxial growth	Thermal voltage
Etching	Vacancy
Evaporation	Valence band
Extrinsic material	

REFERENCE

1. J. D. Cressler, "Re-Engineering Silicon: SiGe Heterojunction Bipolar Technology," *IEEE Spectrum,* pp. 49–55, March 1995.

ADDITIONAL READING

Jaeger, R. C. *Introduction to Microelectronic Fabrication,* 2d ed. Prentice-Hall, Reading, MA: 2001.
Campbell, S. A. *The Science and Engineering of Microelectronic Fabrication,* 2nd ed. Oxford University Press, New York: 2001.
Yang, E. S. *Microelectronic Devices.* McGraw-Hill, New York: 1988.
Pierret, R. F. *Semiconductor Fundamentals.* Addison-Wesley, Reading, MA: 1983.
Sze, S. M. *Physics of Semiconductor Devices.* Wiley, New York: 1981.

IMPORTANT EQUATIONS

$$n_1^2 = BT^3 \, \exp\left(-\frac{E_G}{kT}\right) \qquad \text{cm}^{-6} \tag{2.1}$$

where E_G = semiconductor bandgap energy in eV (electron volts)

k = Boltzmann's constant, 8.62×10^{-5} eV/K (1.38×10^{-23} J/K)

T = absolute temperature, K

B = material-dependent parameter, $1.08 \times 10^{31}/\text{K}^{-3} \cdot \text{cm}^{-6}$ for Si

$$\sigma = q(n\mu_n + p\mu_p) \qquad (\Omega \cdot \text{cm})^{-1} \tag{2.7}$$

Doped Semiconductors

$$q(N_D + p - N_A - n) = 0 \tag{2.10}$$

***n*-Type Material ($N_D > N_A$)**

$$n = \frac{(N_D - N_A) \pm \sqrt{(N_D - N_A)^2 + 4n_i^2}}{2} \qquad \text{and} \qquad p = \frac{n_i^2}{n} \tag{2.11}$$

***p*-Type Material ($N_A > N_D$)**

$$p = \frac{(N_A - N_D) \pm \sqrt{(N_A - N_D)^2 + 4n_i^2}}{2} \qquad \text{and} \qquad n = \frac{n_i^2}{p} \tag{2.12}$$

Currents

$$\begin{aligned}
j_n^{\text{drift}} &= Q_n v_n = (-qn)(-\mu_n E) = qn\mu_n E \qquad \text{A/cm}^2 \\
j_p^{\text{drift}} &= Q_p v_p = (+qp)(+\mu_p E) = qp\mu_p E \qquad \text{A/cm}^2
\end{aligned} \tag{2.5}$$

$$\begin{aligned}
j_p^{\text{diff}} &= (+q)D_p\left(-\frac{\partial p}{\partial x}\right) = -qD_p\frac{\partial p}{\partial x} \\
j_n^{\text{diff}} &= (-q)D_n\left(-\frac{\partial n}{\partial x}\right) = +qD_n\frac{\partial n}{\partial x}
\end{aligned} \qquad \text{A/cm}^2 \tag{2.14}$$

$$j_n^T = q\mu_n n E + qD_n\frac{\partial n}{\partial x} \qquad \text{and} \qquad j_p^T = q\mu_p p E - qD_p\frac{\partial p}{\partial x} \tag{2.16}$$

PROBLEMS

2.1 Solid-State Electronic Materials

2.1. Pure aluminum has a resistivity of $2.6\,\mu\Omega \cdot$ cm. Based on its resistivity, should aluminum be classified as an insulator, semiconductor, or conductor?

2.2. The resistivity of silicon dioxide is $10^{15}\ \Omega \cdot$ cm. Is this material a conductor, semiconductor, or insulator?

2.3. An aluminum interconnection line in an integrated circuit can be operated with a current density up to 10 mA/cm^2. If the line is 5 μm wide and 1 μm high, what is the maximum current permitted in the line?

2.2 Covalent Bond Model

2.4. Calculate the intrinsic carrier densities in silicon and germanium at (a) 100 K, (b) 300 K, and (c) 500 K. Use the information from the table in Fig. 2.4.

2.5. (a) At what temperature will $n_i = 10^{14}$/cm^3 in silicon? (b) Repeat the calculation for $n_i = 10^{16}$/cm^3.

2.6. Calculate the intrinsic carrier density in gallium arsenide at (a) 300 K, (b) 100 K, (c) 500 K. Use the information from the table in Fig. 2.4.

2.3 Drift Currents and Mobility in Semiconductors

2.7. Electrons and holes are moving in a uniform, one-dimensional electric field $E = +2500$ V/cm. The electrons and holes have mobilities of 700 and 250 cm^2/V $\cdot$ s, respectively. What are the electron and hole velocities? If $n = 10^{17}$/cm^3 and $p = 10^3$/cm^3, what are the electron and hole current densities?

2.8. Use Eq. (2.1) to calculate the actual temperature that corresponds to the value $n_i = 10^{10}$/cm^3 in silicon.

2.9. A current density of -1000 A/cm^2 exists in a semiconductor having a charge density of 0.01 C/cm^3. What are the carrier velocities?

2.10. The maximum drift velocity of electrons in silicon is 10^7 cm/s. If the silicon has a charge density of 0.4 C/cm^3, what is the maximum current density in the material?

2.11. A silicon sample is supporting an electric field of -2000 V/cm, and the mobilities of electrons and holes are 1000 and 400 cm^2/V $\cdot$ s, respectively. What are the electron and hole velocities?

If $p = 10^{17}$/cm^3 and $n = 10^3$/cm^3, what are the electron and hole current densities?

2.12. (a) A voltage of 5 V is applied across a 10-μm-long region of silicon. What is the electric field? (b) Suppose the maximum field allowed in silicon is 10^5 V/cm. How large a voltage can be applied to the 10-μm region?

2.13. The maximum drift velocity for holes in silicon is 10^7 cm/s. If the hole density in a sample is 10^{19}/cm^3, what is the maximum hole current density? If the sample has a cross section of 1 μm $\times$ 25 μm, what is the maximum current?

2.4 Resistivity of Intrinsic Silicon

2.14. At what temperature will intrinsic silicon become a conductor based on the definitions in Table 2.1? Assume that $\mu_n = 100$ cm^2/V $\cdot$ s and $\mu_p = 50$ cm^2/V $\cdot$ s. (Note that silicon melts at 1430 K.)

2.15. At what temperature will intrinsic silicon become an insulator, based on the definitions in Table 2.1? Assume that $\mu_n = 2000$ cm^2/V $\cdot$ s and $\mu_p = 750$ cm^2/V $\cdot$ s.

2.5 Impurities in Semiconductors

2.16. Draw a two-dimensional conceptual picture [similar to Fig. 2.6(b)] of the silicon lattice containing one donor atom and one acceptor atom in adjacent lattice positions. Are there any free electrons or holes?

2.17. GaAs is composed of equal numbers of atoms of gallium and arsenic in a lattice similar to that of silicon. (a) Suppose a silicon atom replaces a gallium atom in the lattice. Do you expect the silicon atom to behave as a donor or acceptor impurity? Why? (b) Suppose a silicon atom replaces an arsenic atom in the lattice. Do you expect the silicon atom to behave as a donor or acceptor impurity? Why?

2.18. Crystalline germanium has a lattice similar to that of silicon. (a) What are the possible donor atoms in Ge based on Table 2.2? (b) What are the possible acceptor atoms in Ge based on Table 2.2?

2.19. InP is composed of equal atoms of indium and phosphorus in a lattice similar to that of silicon. (a) Suppose a germanium atom replaces an indium atom in the lattice. Do you expect the germanium atom to behave as a donor or acceptor impurity? Why? (b) Suppose a germanium atom replaces a phosphorus atom in the lattice. Do you expect the germanium atom to behave as a donor or acceptor impurity? Explain.

2.20. A current density of 10,000 A/cm^2 exists in a 0.02-$\Omega \cdot$ cm n-type silicon sample. What is the electric field needed to support this drift current density?

2.21. The maximum drift velocity of carriers in silicon is approximately 10^7 cm/s. What is the maximum drift current density that can be supported in n-type silicon with a doping of 10^{16}/cm^3?

2.22. Silicon is doped with 10^{15} boron atoms/cm^3. How many boron atoms will be in a silicon region that is 1 μm long, 10 μm wide, and 0.5 μm deep?

2.6 Electron and Hole Concentrations in Doped Semiconductors

2.23. Suppose a semiconductor has $N_A = 10^{15}$/cm^3, $N_D = 10^{14}$/cm^3, and $n_i = 5 \times 10^{13}$/cm^3. What are the electron and hole concentrations?

2.24. Suppose a semiconductor has $N_D = 10^{16}$/cm^3, $N_A = 5 \times 10^{16}$/cm^3, and $n_i = 10^{11}$/cm^3. What are the electron and hole concentrations?

2.25. Suppose a semiconductor has $N_A = 2 \times 10^{17}$/cm^3, $N_D = 3 \times 10^{17}$/cm^3, and $n_i = 10^{17}$/cm^3. What are the electron and hole concentrations?

2.26. Equations (2.11) and (2.12) are actually valid for any combination of N_A and N_D. Try to calculate the hole and electron concentrations in silicon with your calculator using Eq. (2.11) for the case of $N_A = 2.5 \times 10^{18}$/cm^3 and $N_D = 0$. Did you get the correct result? If not, why not?

2.27. Silicon is doped with 6×10^{18} boron atoms/cm^3. (a) Is this n- or p-type silicon? (b) What are the hole and electron concentrations at room temperature? (c) What are the hole and electron concentrations at 200 K?

2.28. Silicon is doped with 3×10^{17} arsenic atoms/cm^3. (a) Is this n- or p-type silicon? (b) What are the hole and electron concentrations at room temperature? (c) What are the hole and electron concentrations at 250 K?

2.29. Silicon is doped with 2×10^{18} arsenic atoms/cm^3 and 8×10^{18} boron atoms/cm^3. (a) Is this n- or p-type silicon? (b) What are the hole and electron concentrations at room temperature?

2.30. Silicon is doped with 5×10^{17} boron atoms/cm^3 and 2×10^{17} phosphorus atoms/cm^3 (a) Is this n- or p-type silicon? (b) What are the hole and electron concentrations at room temperature?

2.7 Mobility and Resistivity in Doped Semiconductors

2.31. Silicon is doped with a donor concentration of 4×10^{16}/cm^3. Find the electron and hole concentrations, the electron and hole mobilities, and the resistivity of this silicon material at 300 K. Is this material n- or p-type?

2.32. Silicon is doped with an acceptor concentration of 10^{18}/cm^3. Find the electron and hole concentrations, the electron and hole mobilities, and the resistivity of this silicon material at 300 K. Is this material n- or p-type?

2.33. Silicon is doped with an indium concentration of 7×10^{19}/cm^3. Is indium a donor or acceptor impurity? Find the electron and hole concentrations, the electron and hole mobilities, and the resistivity of this silicon material at 300 K. Is this material n- or p-type?

2.34. A silicon wafer is uniformly doped with 5.5×10^{16} phosphorus atoms/cm^3 and 4.5×10^{16} boron atoms/cm^3. Find the electron and hole concentrations, the electron and hole mobilities, and the resistivity of this silicon material at 300 K. Is this material n- or p-type?

2.35. Repeat Example 2.5 for p-type silicon. Assume that the silicon contains only acceptor impurities. What is the acceptor concentration N_A?

*2.36. A p-type silicon wafer has a resistivity of 0.75 $\Omega \cdot$ cm. It is known that silicon contains only acceptor impurities. What is the acceptor concentration N_A?

2.37. A silicon sample is doped with 5.0×10^{19} donor atoms/cm^3 and 5.0×10^{19} acceptor atoms/cm^3. (a) What is its resistivity? (b) Is this an insulator, conductor, or semiconductor? (c) Is this intrinsic material? Explain your answers.

*2.38. n-type silicon wafers with a resistivity of 2.0 $\Omega \cdot$ cm are needed for integrated circuit fabrication. What donor concentration N_D is required in the wafers? Assume $N_A = 0$.

2.39. (a) What is the minimum donor doping required to convert silicon into a conductor based on the definitions in Table 2.1? (b) What is the minimum acceptor doping required to convert silicon into a conductor?

*2.40. It is conceptually possible to produce extrinsic silicon with a higher resistivity than that of intrinsic silicon. How would this occur?

*2.41. Measurements of a silicon wafer indicate that it is p-type with a resistivity of $1 \, \Omega \cdot \text{cm}$. It is also known that it contains only boron impurities. (a) What additional acceptor concentration must be added to the sample to change its resistivity to $0.25 \, \Omega \cdot \text{cm}$? (b) What concentration of donors would have to be added to the original sample to change the resistivity to $0.25 \, \Omega \cdot \text{cm}$? Would the resulting material be classified as n- or p-type silicon?

*2.42. A silicon wafer has a doping concentration of 1×10^{16} phosphorus atoms/cm^3. (a) Determine the conductivity of the wafer. (b) What concentration of boron atoms must be added to the wafer to make the conductivity equal to $5.0 \, (\Omega \cdot \text{cm})^{-1}$?

*2.43. A silicon wafer has a background concentration of 1×10^{16} boron atoms/cm^3. (a) Determine the conductivity of the wafer. (b) What concentration of phosphorus atoms must be added to the wafer to make the conductivity equal to $5.5 \, (\Omega \cdot \text{cm})^{-1}$?

2.8 Diffusion Currents

2.44. Make a table of the values of thermal voltage V_T for $T = 50$ K, 75 K, 100 K, 150 K, 200 K, 250 K, 300 K, 350 K, and 400 K.

2.45. The electron concentration in a region of silicon is shown in Fig. P2.45. If the electron mobility is $350 \, \text{cm}^2/\text{V} \cdot \text{s}$ and the width $W_B = 1 \, \mu\text{m}$, determine the electron diffusion current density. Assume room temperature.

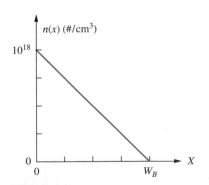

Figure P2.45

2.46. Suppose the hole concentration in silicon sample is described mathematically by

$$p(x) = 10^5 + 10^{19} \, \exp\left(-\frac{x}{L_p}\right) \text{holes/cm}^3, \, x \geq 0$$

in which L_p is known as the diffusion length for holes and is equal to $2.0 \, \mu\text{m}$. Find the diffusion current density for holes as a function of distance for $x \geq 0$ if $D_p = 15 \, \text{cm}^2/\text{s}$. What is the diffusion current at $x = 0$ if the cross-sectional area is $10 \, \mu\text{m}^2$?

2.9 Total Current

*2.47. A 5-μm-long block of p-type silicon has an acceptor doping profile given by $N_A(x) = 10^{14} + 10^{18} \, \exp(-10^4 x)$, where x is measured in cm. Use Eq. (2.17) to demonstrate that the material must have a nonzero internal electric field E. What is the value of E at $x = 0$ and $x = 5 \, \mu\text{m}$? (*Hint:* In thermal equilibrium, the total electron and total hole currents must each be zero.)

2.48. Figure P2.48 gives the electron and hole concentrations in a 2-μm-wide region of silicon. In addition, there is a constant electric field of 20 V/cm present in the sample. What is the total current density at $x = 0$? What are the individual drift and diffusion components of the hole and electron current densities at $x = 1.0 \, \mu\text{m}$? Assume that the electron and hole mobilities are 350 and $150 \, \text{cm}^2/\text{V} \cdot \text{s}$, respectively.

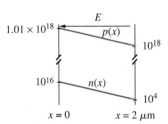

Figure P2.48

2.10 Energy Band Model

2.49. Draw a figure similar to Fig. 2.15 for the case $N_A > N_D$ in which there are two acceptor atoms for each donor atom.

*2.50. Electron–hole pairs can be created by means other than the thermal activation process as described in Figs. 2.3 and 2.12. For example, energy may be added to electrons through optical means by shining light on the sample. If enough optical energy is absorbed, electrons can jump the energy bandgap, creating electron–hole pairs. What is the maximum wavelength of light that we should expect silicon to be able to absorb? (*Hint:* Remember from physics that energy E is related to wavelength λ by $E = hc/\lambda$ in which Planck's constant $h = 6.626 \times 10^{-34}$ J $\cdot$ s and the velocity of light $c = 3 \times 10^{10}$ cm/s.)

2.11 Overview of Integrated Circuit Fabrication

2.51. Draw the cross section for a *pn* diode similar to that in Fig. 2.17(h) if the fabrication process utilizes a *p*-type substrate in place of the *n*-type substrate depicted in Fig. 2.17.

2.52. To ensure that a good ohmic contact is formed between aluminum and *n*-type silicon, an additional doping step is added to the diode in Fig. 2.17(h) to place an *n*+ region beneath the left-hand contact as in Fig. P2.52. Where might this step go in the process flow in Fig. 2.17? Draw a top and side view of a mask that could be used in the process.

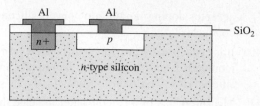

Figure P2.52

Miscellaneous

*2.53. Single crystal silicon consists of three-dimensional arrays of the basic unit cell in Fig. 2.1(a). (a) How many atoms are in each unit cell? (b) What is volume of the unit cell in cm^3? (c) Show that the atomic density of silicon is 5×10^{22} atoms/cm^3. (d) The density of silicon is 2.33 g/cm^3. What is the mass of one unit cell? (e) Based on your calculations here, what is the mass of a proton? Assume that protons and neutrons have the same mass and that electrons are much much lighter. Is your answer reasonable? Explain.

CHAPTER 3

SOLID-STATE DIODES AND DIODE CIRCUITS

CHAPTER OUTLINE

CHAPTER GOALS

- Understand diode structure and basic layout
- Develop electrostatics of the *pn* junction
- Explore various diode models including the mathematical model, the ideal diode model, and the constant voltage drop model
- Understand the SPICE representation and model parameters for the diode
- Define regions of operation of the diode, including forward and reverse bias and reverse breakdown
- Apply the various types of models in circuit analysis
- Explore different types of diodes including Zener, variable capacitance, and Schottky barrier diodes as well as solar cells and light emitting diodes (LEDs)

- Discuss the dynamic switching behavior of the *pn* junction diode
- Explore diode rectifiers
- Practice simulating diode circuits using SPICE

Photograph of an assortment of diodes

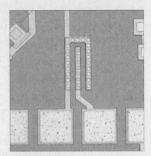

Fabricated Diode

The first electronic circuit element that we explore is the solid-state *pn* junction diode. The diode is an extremely important device in its own right with many important applications including ac-dc power conversion (rectification). In addition, the *pn* junction diode is a fundamental building block for other solid-state devices. In later chapters, we will find that two closely coupled diodes are used to form the bipolar junction transistor (BJT), two diodes are used in the metal-oxide-semiconductor field-effect transistor (MOSFET), and a single diode is used in formation of junction field-effect transistor (JFET). Gaining an understanding of diode characteristics is prerequisite to understanding the behavior of the field-effect and bipolar transistors that are used to realize both digital logic circuits and analog amplifiers.

The *pn* junction diode is formed by fabricating adjoining regions of *p*-type and *n*-type semiconductor material. Another type of diode, called the Schottky barrier diode, is formed by a non-ohmic contact between a metal such as aluminum, palladium, or platinum and an *n*-type or *p*-type semiconductor. Both types of solid-state diodes are discussed in this chapter. The vacuum diode, which was used before the advent of semiconductor diodes, still finds application in very high voltage situations.

The *pn* junction diode is a nonlinear element, and for many of us, this will be our first encounter with a nonlinear device. The diode is a two-terminal circuit element similar to a resistor, but its *i-v* characteristic, the relationship between the current through the element and the voltage across the element, is not a straight line. This nonlinear behavior allows electronic circuits to be designed to provide many useful operations, including rectification, mixing (a form of multiplication), and wave shaping. Diodes can also be used to perform elementary logic operations such as the AND and OR functions.

This chapter begins with a basic discussion of the structure and behavior of the *pn* junction diode and its terminal characteristics. Next is an introduction to the concept of modeling, and several different models for the diode are introduced and used to analyze the behavior of diode circuits. We begin to develop the intuition needed to make choices between models of various complexities in order to simplify electronic circuit analysis and design. Diode circuits are then explored, including the detailed application of the diode in rectifier circuits. The characteristics of Zener diodes, photo diodes, solar cells, and light-emitting diodes are also discussed.

3.1 THE *pn* JUNCTION DIODE

The ***pn* junction diode** is formed by fabrication of a *p*-type semiconductor region in intimate contact with an *n*-type semiconductor region, as illustrated in Fig. 3.1. The diode is constructed using the impurity doping process studied in Chapter 2.

An actual diode can be formed by starting with an *n*-type wafer with doping N_D and selectively converting a portion of the wafer to *p*-type by adding acceptor impurities with $N_A > N_D$. The point at which the material changes from *p*-type to *n*-type is called the metallurgical junction. The *p*-type region is also referred to as the **anode** of the diode, and the *n*-type region is called the **cathode** of the diode.

Figure 3.2 gives the circuit symbol for the diode, with the left-hand end corresponding to the *p*-type region of the diode and the right-hand side corresponding to the *n*-type region. We will see shortly that the "arrow" points in the direction of positive current in the diode.

3.1.1 *pn* JUNCTION ELECTROSTATICS

Consider a *pn* junction diode similar to Fig. 3.1 having $N_A = 10^{17}/\text{cm}^3$ on the *p*-type side and $N_D = 10^{16}/\text{cm}^3$ on the *n*-type side. The hole and electron concentrations on the two sides of the junction will be

$$\text{*p*-type side:} \quad p_p = 10^{17} \text{ holes/cm}^3 \qquad n_p = 10^3 \text{ electrons/cm}^3$$
$$\text{*n*-type side:} \quad p_n = 10^4 \text{ holes/cm}^3 \qquad n_n = 10^{16} \text{ electrons/cm}^3 \qquad (3.1)$$

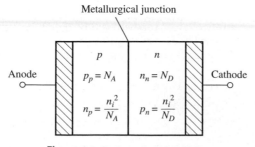

Figure 3.1 Basic *pn* junction diode.   **Figure 3.2** Diode circuit symbol.

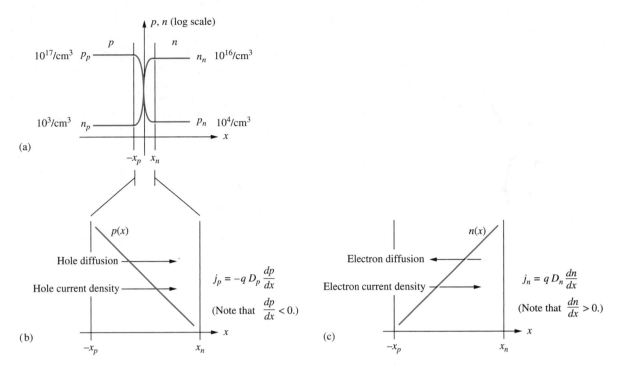

Figure 3.3 (a) Carrier concentrations; (b), (c) diffusion currents in the space charge region.

As shown in Fig. 3.3(a), a very large concentration of holes exists on the *p*-type side of the metallurgical junction, whereas a much smaller hole concentration exists on the *n*-type side. Likewise, there is a very large concentration of electrons on the *n*-type side of the junction and a very low concentration on the *p*-type side.

From our knowledge of diffusion from Chapter 2, we know that mobile holes will diffuse from the region of high concentration on the *p*-type side toward the region of low concentration on the *n*-type side and that mobile electrons will diffuse from the *n*-type side to the *p*-type side, as in Figs. 3.3(b) and (c). If the diffusion processes were to continue unabated, there would eventually be a uniform concentration of holes and electrons throughout the entire semiconductor region, and the *pn* junction would cease to exist. Note that the two diffusion current densities are both directed in the positive *x* direction, but this is inconsistent with zero current in the open-circuited terminals of the diode.

A second, competing process must be established to balance the diffusion current. The competing mechanism is a drift current, as discussed in Chapter 2, and its origin can be understood by focusing on the region in the vicinity of the **metallurgical junction** shown in Fig. 3.4. As mobile holes move out of the *p*-type material, they leave behind immobile negatively charged acceptor atoms. Correspondingly, mobile electrons leave behind immobile ionized donor atoms with a localized positive charge. A **space charge region (SCR),** depleted of mobile carriers, develops in the region immediately around the metallurgical junction. This region is also often called the **depletion region,** or **depletion layer.**

From electromagnetics, we know that a region of space charge ρ_c (C/cm^3) will be accompanied by an electric field E measured in V/cm through Gauss' law,

$$\nabla \cdot E = \frac{\rho_c}{\varepsilon_s} \tag{3.2}$$

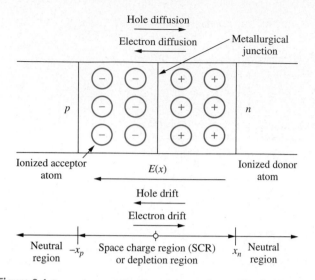

Figure 3.4 Space charge region formation near the metallurgical junction.

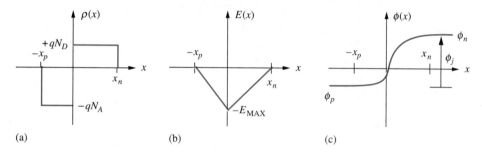

Figure 3.5 (a) Charge density (C/cm³), (b) electric field (V/cm), and (c) electrostatic potential (V) in the space charge region of a *pn* junction.

written assuming a constant semiconductor permittivity ε_s (F/cm). In one dimension, Eq. (3.2) can be rearranged to give

$$E(x) = \frac{1}{\varepsilon_s} \int \rho_c(x)\, dx \qquad (3.3)$$

Figure 3.5 illustrates the space charge and electric field in the diode for the case of uniform (constant) doping on both sides of the junction. As illustrated in Fig. 3.5(a), the value of the space charge density on the *p*-type side will be $-qN_A$ and will extend from the metallurgical junction at $x = 0$ to $-x_p$, whereas that on the *n*-type side will be $+qN_D$ and will extend from 0 to $+x_n$. The overall diode must be charge neutral, so

$$qN_Ax_p = qN_Dx_n \qquad (3.4)$$

The electric field is proportional to the first integral of the space charge density and will be zero in the (charge) neutral regions outside of the depletion region. Using this zero-field boundary condition yields the triangular electric field distribution in Fig. 3.5(b).

Figure 3.5(c) represents the integral of the electric field and shows that a **built-in potential** or **junction potential** ϕ_j, exists across the *pn* junction space charge region according to

$$\phi_j = -\int E(x)\, dx \qquad \text{V} \qquad (3.5)$$

ϕ_j represents the difference in the internal chemical potentials between the n and p sides of the diode, and it can be shown [1] to be given by

$$\phi_j = V_T \ln\left(\frac{N_A N_D}{n_i^2}\right) \tag{3.6}$$

where the **thermal voltage** $V_T = kT/q$ was originally defined in Chapter 2.

Equations (3.3) to (3.5) can be used to determine the total width of the depletion region w_{do} in terms of the built-in potential:

$$w_{do} = (x_n + x_p) = \sqrt{\frac{2\varepsilon_s}{q}\left(\frac{1}{N_A} + \frac{1}{N_D}\right)\phi_j} \quad \text{m} \tag{3.7}$$

From Eq. (3.7), we see that the doping on the more lightly doped side of the junction will be the most important in determining the **depletion-layer width.**

EXAMPLE 3.1   DIODE SPACE CHARGE REGION WIDTH

When diodes are actually fabricated, the doping levels on opposite sides of the *pn* junction tend to be quite asymmetric, and the resulting depletion layer tends to extend primarily on one side of the junction and is referred to as a "one-sided" step junction or one-sided abrupt junction. The *pn* junction that we analyze provides an example of the magnitudes of the distances involved in such a *pn* junction.

PROBLEM   Calculate the built-in potential and depletion-region width for a silicon diode with $N_A = 10^{17}/\text{cm}^3$ on the *p*-type side and $N_D = 10^{20}/\text{cm}^3$ on the *n*-type side.

SOLUTION   **Known Information and Given Data:** On the *p*-type side, $N_A = 10^{17}/\text{cm}^3$; on the *n*-type side, $N_D = 10^{20}/\text{cm}^3$. Theory describing the *pn* junction is given by Eqs. (3.4) through (3.7).

Unknowns: Built-in potential ϕ_j and depletion-region width w_{do}.

Approach: Find the built-in potential using Eq. (3.6); use ϕ_j to calculate w_{do} in Eq. (3.7)

Assumptions: Room temperature operation with $V_T = 0.025$ V. There are only donor impurities on the *n*-type side and acceptor impurities on the *p*-type side of the junction. The doping levels are constant on each side of the junction.

Analysis: The built-in potential is given by

$$\phi_j = V_T \ln\left(\frac{N_A N_D}{n_i^2}\right) = (0.025 \text{ V}) \ln\left[\frac{(10^{17}/\text{cm}^3)(10^{20}/\text{cm}^3)}{(10^{20}/\text{cm}^6)}\right] = 0.979 \text{ V}$$

For silicon, $\varepsilon_s = 11.7\varepsilon_o$, where $\varepsilon_o = 8.85 \times 10^{-14}$ F/cm represents the permittivity of free space.

$$w_{do} = \sqrt{\frac{2\varepsilon_s}{q}\left(\frac{1}{N_A} + \frac{1}{N_D}\right)\phi_j}$$

$$w_{do} = \sqrt{\frac{2 \cdot 11.7 \cdot (8.85 \times 10^{-14} \text{ F/cm})}{1.60 \times 10^{-19} \text{ C}}\left(\frac{1}{10^{17}/\text{cm}^3} + \frac{1}{10^{20}/\text{cm}^3}\right)0.979 \text{ V}} = 0.113 \text{ } \mu\text{m}$$

Check of Results: The built-in potential should be less than the bandgap of the material. For silicon the bandgap is approximately 1.2 V (see Table 2.3), so ϕ_j appears reasonable. The depletion-layer width seems quite small, but a double check of the numbers indicates that the calculation is correct.

Discussion: The numbers in this example are fairly typical of a *pn* junction diode. For the normal doping levels encountered in solid-state diodes, the built-in potential ranges between 0.5 V and 1.0 V, and the total depletion-layer width w_{do} can range from a fraction of 1 μm in heavily doped diodes to tens of microns in lightly doped diodes.

EXERCISE: Calculate the built-in potential and depletion-region width for a silicon diode if N_A is increased to 2×10^{18}/cm³ on the *p*-type side and $N_D = 10^{20}$/cm³ on the *n*-type side.

ANSWERS: 1.05 V; 0.0263 μm

3.1.2 INTERNAL DIODE CURRENTS

Remember that the electric field E points in the direction that a positive carrier will move, so electrons drift toward the positive x direction and holes drift in the negative x direction. The carriers drift in directions opposite the diffusion of the same carrier species. Because the terminal currents must be zero, a dynamic equilibrium is established in the junction region. Hole diffusion is precisely balanced by hole drift, and electron diffusion is exactly balanced by electron drift. This balance is stated mathematically in Eq. (3.8), in which the total hole and electron current densities must each be identically zero:

$$j_n^T = qn\mu_n E + qD_n\frac{\partial n}{\partial x} = 0 \quad \text{and} \quad j_p^T = qp\mu_p E - qD_p\frac{\partial p}{\partial x} = 0 \qquad \text{A/cm}^2 \qquad (3.8)$$

The difference in potential in Fig. 3.5(c) represents a barrier to both hole and electron flow across the junction. When a voltage is applied to the diode, the potential barrier is modified, and the delicate balances in Eq. (3.8) are disturbed, resulting in a current in the diode terminals.

EXAMPLE **3.2**   **DIODE ELECTRIC FIELD AND SPACE-CHARGE REGION EXTENTS**

Now we find the value of the electric field in the diode and the size of the individual depletion layers on either side of the *pn* junction.

PROBLEM   Find x_n, x_p, and E_{MAX} for the diode in Example 3.1.

SOLUTION   **Known Information and Given Data:** On the *p*-type side, $N_A = 10^{17}$/cm³; on the *n*-type side, $N_D = 10^{20}$/cm³. Theory describing the *pn* junction is given by Eqs. (3.4) through (3.7). From Ex. 3.1, $\phi_j = 0.979$ V and $w_{do} = 0.113$ μm.

Unknowns: x_n, x_p, and E_{MAX}

Approach: Use Eqs. (3.4) and (3.7) to find x_n and x_p; use Eq. (3.5) to find E_{MAX}

Assumptions: Room temperature operation

Analysis: Using Eq. (3.4), we can write

$$w_{do} = x_n + x_p = x_n\left(1 + \frac{N_D}{N_A}\right) \quad \text{and} \quad w_{do} = x_n + x_p = x_p\left(1 + \frac{N_A}{N_D}\right)$$

Solving for x_n and x_p gives

$$x_n = \frac{w_{do}}{\left(1 + \dfrac{N_D}{N_A}\right)} = \frac{0.113\ \mu\text{m}}{\left(1 + \dfrac{10^{20}/\text{cm}^3}{10^{17}/\text{cm}^3}\right)} = 1.13 \times 10^{-4}\ \mu\text{m}$$

and

$$x_p = \frac{w_{do}}{\left(1 + \dfrac{N_A}{N_D}\right)} = \frac{0.113\ \mu\text{m}}{\left(1 + \dfrac{10^{17}/\text{cm}^3}{10^{20}/\text{cm}^3}\right)} = 0.113\ \mu\text{m}$$

Equation (3.5) indicates that the built-in potential is equal to the area under the triangle in Fig. 3.5(b). The height of the triangle is $(-E_{\text{MAX}})$ and the base of the triangle is $x_n + x_p = w_{do}$:

$$\phi_j = \frac{1}{2} E_{\text{MAX}} w_{do} \quad \text{and} \quad E_{\text{MAX}} = \frac{2\phi_j}{w_{do}} = \frac{2(0.979\ \text{V})}{0.113\ \mu\text{m}} = 173\ \text{kV/cm}$$

Check of Results: From Eqs. (3.3) and (3.4), E_{MAX} can also be found from the doping levels and depletion-layer widths on each side of the junction. The equation in the next exercise can be used as a check of the answer.

EXERCISE: Using Eq. (3.3) and Fig. 3.5(a) and (b), show that the maximum field is given by

$$E_{\text{MAX}} = \frac{qN_A x_p}{\varepsilon_s} = \frac{qN_D x_n}{\varepsilon_s}$$

Use this formula to find E_{MAX}.

ANSWER: 175 kV/cm

EXERCISE: Calculate the maximum E_{MAX}, x_p, and x_n for a silicon diode if $N_A = 2 \times 10^{18}/\text{cm}^3$ on the p-type side and $N_D = 10^{20}/\text{cm}^3$ on the n-type side. Use $\phi_j = 1.05$ V and $w_{do} = 0.0263\ \mu\text{m}$.

ANSWERS: 799 kV/cm; $5.16 \times 10^{-4}\ \mu\text{m}$; 0.0258 μm

3.2 THE i-v CHARACTERISTICS OF THE DIODE

The diode is the electronic equivalent of a mechanical check valve—it permits current to flow in one direction in a circuit, but prevents movement of current in the opposite direction. We will find that this nonlinear behavior has many useful applications in electronic circuit design. To understand this phenomenon, we explore the relationship between the current in the diode and the voltage applied to the diode. This information, called the i-v characteristic of the diode, is first presented graphically and then mathematically in this section and Sec. 3.3.

The current in the diode is determined by the voltage applied across the diode terminals, and the diode is shown with a voltage applied in Fig. 3.6. The voltage v_D in Fig. 3.6 represents the voltage applied to the diode terminals; i_D is the current through the diode. The neutral regions of the diode represent a low resistance to current, and essentially all the external applied voltage is dropped across the space charge region.

The applied voltage disturbs the balance between the drift and diffusion currents at the junction specified in the two expressions in Eq. (3.8). A positive applied voltage reduces the potential barrier for electrons and holes, as in Fig. 3.7, and current easily crosses the junction. A negative voltage

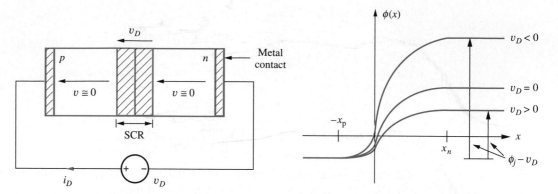

Figure 3.6 Diode with external applied voltage v_D.

Figure 3.7 Electrostatic junction potential for different applied voltages.

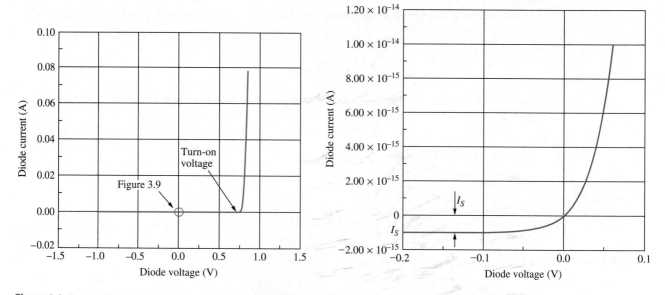

Figure 3.8 Graph of the i-v characteristics of a pn junction diode.

Figure 3.9 Diode behavior near the origin with $I_S = 10^{-15}$ A and $n = 1$.

increases the potential barrier, and although the balance in Eq. (3.8) is disturbed, the increased barrier results in a very small current.

The most important details of the diode i-v characteristic appear in Fig. 3.8. The diode characteristic is definitely not linear. For voltages less than zero, the diode is essentially nonconducting, with $i_D \cong 0$. As the voltage increases above zero, the current remains nearly zero until the voltage v_D exceeds approximately 0.5 to 0.7 V. At this point, the diode current increases rapidly, and the voltage across the diode becomes almost independent of current. The voltage required to bring the diode into significant conduction is often called either the **turn-on** or **cut-in voltage** of the diode.

Figure 3.9 is an enlargement of the region around the origin in Fig. 3.8. We see that the i-v characteristic passes through the origin; the current is zero when the voltage is zero. For negative voltages the current is not actually zero but reaches a limiting value labeled as $-I_S$ for voltages less than -0.1 V. I_S is called the **reverse saturation current,** or just **saturation current,** of the diode.

3.3 THE DIODE EQUATION: A MATHEMATICAL MODEL FOR THE DIODE

When performing both hand and computer analysis of circuits containing diodes, it is very helpful to have a mathematical representation, or model, for the i-v characteristics depicted in Fig. 3.8. In fact, solid-state device theory has been used to formulate a mathematical expression that agrees amazingly well with the measured the i-v characteristics of the pn junction diode. We study this extremely important formula called the **diode equation** in this section.

A voltage is applied to the diode in Fig. 3.10; in the figure the diode is represented by its circuit symbol from Fig. 3.2. Although we will not attempt to do so here, Eq. (3.8) can be solved for the hole and electron concentrations and the terminal current in the diode as a function of the voltage v_D across the diode. The resulting diode equation, given in Eq. (3.9), provides a **mathematical model** for the i-v characteristics of the diode:

$$i_D = I_S \left[\exp\left(\frac{q v_D}{nkT}\right) - 1 \right] = I_S \left[\exp\left(\frac{v_D}{n V_T}\right) - 1 \right] \qquad (3.9)$$

where I_S = reverse saturation current of diode (A) T = absolute temperature (K)

v_D = voltage applied to diode (V) n = nonideality factor (dimensionless)

q = electronic charge (1.60×10^{-19} C) $V_T = kT/q$ = thermal voltage (V)

k = Boltzmann's constant (1.38×10^{-23} J/K)

The total current through the diode is i_D, and the voltage drop across the diode terminals is v_D. Positive directions for the terminal voltage and current are indicated in Fig. 3.10. V_T is the thermal voltage encountered previously in Chapter 2 and will be assumed equal to 0.025 V at room temperature. I_S is the (reverse) saturation current of the diode encountered in Fig. 3.9, and n is a dimensionless parameter discussed in more detail shortly. The saturation current is typically in the range

$$10^{-18} \text{ A} \le I_S \le 10^{-9} \text{ A} \qquad (3.10)$$

From device physics, it can be shown that the diode saturation current is proportional to n_i^2, where n_i is the density of electrons and holes in intrinsic semiconductor material. After reviewing Eq. (2.1) in Chapter 2, we realize that I_S will be strongly dependent on temperature. Additional discussion of this temperature dependence is in Sec. 3.5.

The parameter n is termed the **nonideality factor.** For most silicon diodes, n is in the range 1.0 to 1.1, although it approaches a value of 2 in diodes operating at high current densities. From this point on, we assume that $n = 1$ unless otherwise indicated, and the diode equation will be written as

$$i_D = I_S \left[\exp\left(\frac{v_D}{V_T}\right) - 1 \right] \qquad (3.11)$$

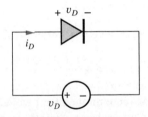

Figure 3.10 Diode with applied voltage v_D.

It is difficult to distinguish small variations in the value of n from an uncertainty in our knowledge in the absolute temperature. This is one reason that we will assume that $n = 1$ in this text. The problem can be investigated further by working the next exercise.

> **EXERCISE:** For $n = 1$ and $T = 300$ K, $n(KT/q) = 25.8$ mV. Verify this calculation. Now, suppose $n = 1.03$. What temperature gives the same value for nV_T?
>
> **ANSWER:** 291 K

The mathematical model in Eq. (3.11) provides a highly accurate prediction of the i-v characteristics of the pn junction diode. The model is useful for understanding the detailed behavior of diodes. It also provides a basis for understanding the i-v characteristics of the bipolar transistor in Chapter 5.

DESIGN NOTE

The static i-v characteristics of the diode are well-characterized by three parameters: Saturation current I_S, temperature via the thermal voltage V_T, and nonideality factor n.

$$i_D = I_S \left[\exp \left(\frac{v_D}{nV_T} \right) - 1 \right]$$

EXAMPLE 3.3   DIODE VOLTAGE AND CURRENT CALCULATIONS

In this example, we calculate some typical values of diode voltages for several different current levels and types of diodes.

PROBLEM   (a) Find the diode voltage for a silicon diode with $I_S = 0.1$ fA operating at room temperature at a current of 300 μA. What is the diode voltage if $I_S = 10$ fA? What is the diode voltage if the current increases to 1 mA?
(b) Find the diode voltage for a silicon power diode with $I_S = 10$ nA and $n = 2$ operating at room temperature at a current of 10 A.
(c) A silicon diode is operating with a temperature of 50°C and the diode voltage is measured to be 0.736 V at a current of 2.50 mA. What is the saturation current of the diode?

SOLUTION (a)   **Known Information and Given Data:** The diode currents are given and the saturation current parameter I_S is specified.

Unknowns: Diode voltage at each of the operating currents.

Approach: Solve Eq. (3.9) for the diode voltage and evaluate the expression at each operating current.

Assumptions: At room temperature, we will use $V_T = 0.025$ V $= 1/40$ V; assume $n = 1$, since it is not specified otherwise; assume dc operation: $i_D = I_D$ and $v_D = V_D$.

Analysis: Solving Eq. (3.9) for V_D with $I_D = 0.1$ fA yields

$$V_D = nV_T \ln\left(1 + \frac{I_D}{I_S}\right) = 1(0.025 \text{ V}) \ln\left(1 + \frac{3 \times 10^{-4} \text{ A}}{10^{-16} \text{ A}}\right) = 0.718 \text{ V}$$

For $I_S = 10$ fA:

$$V_D = nV_T \ln\left(1 + \frac{I_D}{I_S}\right) = 1(0.025 \text{ V}) \ln\left(1 + \frac{3 \times 10^{-4} \text{ A}}{10^{-14} \text{ A}}\right) = 0.603 \text{ V}$$

For $I_D = 1$ mA with $I_S = 0.1$ fA:

$$V_D = nV_T \ln\left(1 + \frac{I_D}{I_S}\right) = 1(0.025 \text{ V}) \ln\left(1 + \frac{10^{-3} \text{ A}}{10^{-16} \text{ A}}\right) = 0.748 \text{ V}$$

Check of Results: The diode voltages are all between 0.5 V and 1.0 V and are reasonable (the diode voltage should not exceed the bandgap for $n = 1$).

SOLUTION (b) **Known Information and Given Data:** The diode current is given and the values of the saturation current parameter I_S and n are both specified.

Unknowns: Diode voltage at the operating current

Approach: Solve Eq. (3.9) for the diode voltage and evaluate the resulting expression.

Assumptions: At room temperature, we will use $V_T = 0.025$ V $= 1/40$ V

Analysis: The diode voltage will be

$$V_D = nV_T \ln\left(1 + \frac{I_D}{I_S}\right) = 2(0.025 \text{ V}) \ln\left(1 + \frac{10 \text{ A}}{10^{-10} \text{ A}}\right) = 1.27 \text{ V}$$

Check of Results: Based on the comment at the end of part (a) and realizing that $n = 2$, voltages between 1 V and 2 V are reasonable for power diodes operating at high currents.

SOLUTION (c) **Known Information and Given Data:** The diode current is 2.50 mA and voltage is 0.736 V. The diode is operating at a temperature of 50°C.

Unknowns: Diode saturation current I_S

Approach: Solve Eq. (3.9) for the saturation current and evaluate the resulting expression. The value of the thermal voltage V_T will need to be calculated for $T = 50$°C.

Assumptions: The value of n is unspecified, so assume $n = 1$.

Analysis: Converting $T = 50$°C to Kelvins, $T = (273 + 50)$ K $= 323$ K, and

$$V_T = \frac{kT}{q} = \frac{(1.38 \times 10^{-23} \text{ J/K})(323 \text{ K})}{1.60 \times 10^{-19}°\text{C}} = 27.9 \text{ mV}$$

Solving Eq. (3.9) for I_S yields

$$I_S = \frac{I_D}{\exp\left(\dfrac{V_D}{nV_T}\right) - 1} = \frac{2.5 \text{ mA}}{\exp\left(\dfrac{0.736 \text{ V}}{0.0279 \text{ V}}\right) - 1} = 8.74 \times 10^{-15} \text{ A} = 8.74 \text{ fA}$$

Check of Results: The saturation current is within the range of typical values specified in Eq. (3.10).

EXERCISE: A diode has a reverse saturation current of 40 fA. Calculate i_D for diode voltages of 0.55 and 0.7 V. What is the diode voltage if $i_D = 6$ mA?

ANSWERS: 143 μA; 57.9 mA; 0.643 V

3.4 DIODE CHARACTERISTICS UNDER REVERSE, ZERO, AND FORWARD BIAS

When a dc voltage is applied to an electronic device, we say that we are providing a dc bias voltage or simply a **bias** to the device. As we develop our electronics expertise, choosing the bias will be important to all of the circuits that we analyze and design. We will find that bias determines device characteristics, power dissipation, voltage and current limitations, and other important circuit parameters. For a diode, there are two regions of operation, **reverse bias** and **forward bias,** corresponding to $v_D < 0$ V and $v_D > 0$ V, respectively. The **zero bias** condition, with $v_D = 0$ V, represents the boundary between the forward and reverse bias regions. When the diode is operating with reverse bias, we consider the diode "off" or nonconducting because the current is very small. For forward bias, the diode is usually in a highly conducting state and is considered "on."

3.4.1 REVERSE BIAS

For $v_D < 0$, the diode is said to be operating under reverse bias. Only a very small reverse leakage current, approximately equal to I_S, flows through the diode. This current is small enough that we usually think of the diode as being in the nonconducting or off state when it is reverse-biased. For example, suppose that a dc voltage $V = -4V_T = -0.1$ V is applied to the diode terminals so that $v_D = -0.1$ V. Substituting this value into Eq. (3.11) gives

$$i_D = I_S \left[\exp\left(\frac{v_D}{V_T} \right) - 1 \right] = I_S[\overset{\text{negligible}}{\exp(-4)} - 1] \approx -I_S \qquad (3.12)$$

because $\exp(-4) = 0.018$. For a reverse bias greater than $4V_T$—that is, $v_D \leq -4V_T = -0.1$ V—the exponential term $\exp(v_D/V_T)$ is much less than 1, and the diode current will be approximately equal to $-I_S$, a very small current. The current I_S was identified in Fig. 3.9.

EXERCISE: A diode has a reverse saturation current of 5 fA. Calculate i_D for diode voltages of −0.04 V and −2 V (see Sec. 3.6).

ANSWERS: −3.99 fA; −5 fA

The situation depicted in Fig. 3.9 and Eq. (3.12) actually represents an idealized picture of the diode. In a real diode, the reverse leakage current is several orders of magnitude larger than I_S due to the generation of electron–hole pairs within the depletion region. In addition, i_D does not saturate but increases gradually with reverse bias as the width of the depletion layer increases with reverse bias. (See Sec. 3.6.1).

3.4.2 ZERO BIAS

Although it may seem to be a trivial result, it is important to remember that the i-v characteristic of the diode passes through the origin. For zero bias with $v_D = 0$, we find $i_D = 0$. Just as for a resistor, there must be a voltage across the diode terminals in order for a nonzero current to exist.

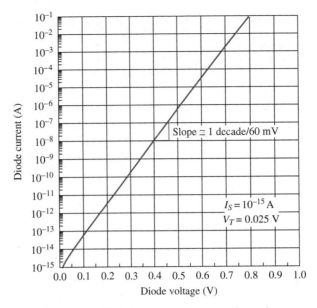

Figure 3.11 Diode i-v characteristic on semilog scale.

3.4.3 FORWARD BIAS

For the case $v_D > 0$, the diode is said to be operating under forward bias, and a large current can be present in the diode. Suppose that a voltage $v_D \geq +4V_T = +0.1$ V is applied to the diode terminals. The exponential term $\exp(v_D/V_T)$ is now much greater than 1, and Eq. (3.9) reduces to

$$i_D = I_S \left[\exp\left(\frac{v_D}{V_T}\right) \overset{\text{negligible}}{-\cancel{1}} \right] \cong I_S \exp\left(\frac{v_D}{V_T}\right) \tag{3.13}$$

The diode current grows exponentially with applied voltage for a forward bias greater than approximately $4V_T$.

The diode i-v characteristic for forward voltages is redrawn in semilogarithmic form in Fig. 3.11. The straight line behavior predicted by Eq. (3.13) for voltages $v_D \geq 4V_T$ is apparent. A slight curvature can be observed near the origin, where the -1 term in Eq. (3.13) is no longer negligible. The slope of the graph in the exponential region is very important. Only a 60-mV increase in the forward voltage is required to increase the diode current by a factor of 10. This is the reason for the almost vertical increase in current noted in Fig. 3.8 for voltages above the turn-on voltage.

EXAMPLE 3.4   **DIODE VOLTAGE CHANGE VERSUS CURRENT**

The slope of the diode i-v characteristic is an important number for circuit designers to remember.

PROBLEM   Use Eq. (3.13) to accurately calculate the voltage change required to increase the diode current by a factor of 10.

SOLUTION   **Known Information and Given Data:** The current changes by a factor of 10.

Unknowns: The diode voltage change corresponding to a one decade change in current; the saturation current has not been given.

Approach: Form an expression for the ratio of two diode currents using the diode equation. The saturation current will cancel out and is not needed.

Assumptions: Room temperature operation with $V_T = 25.0$ mV. Assume $I_D \gg I_S$

Analysis: Let

$$i_{D1} = I_S \exp\left(\frac{v_{D1}}{V_T}\right) \qquad \text{and} \qquad i_{D2} = I_S \exp\left(\frac{v_{D2}}{V_T}\right)$$

Taking the ratio of the two currents and setting it equal to 10 yields

$$\frac{i_{D2}}{i_{D1}} = \exp\left(\frac{v_{D2} - v_{D1}}{V_T}\right) = \exp\left(\frac{\Delta v_D}{V_T}\right) = 10 \qquad \text{and} \qquad \Delta v_D = V_T \ln 10 = 2.3 V_T$$

Therefore $\Delta V_D = 2.3 V_T = 57.5$ mV (or approximately 60 mV) at room temperature.

Check of Results: The result is consistent with the logarithmic plot in Fig. 3.11. The diode voltage changes approximately 60 mV for each decade change in forward current.

EXERCISE: A diode has a saturation current of 2 fA. (a) What is the diode voltage at a diode current of 40 μA (assume $V_T = 25.0$ mV)? Repeat for a diode current of 400 μA. What is the difference in the two diode voltages? (b) Repeat for $V_T = 25.8$ mV.

ANSWER: 0.593 V, 0.651 V, 57.6 mV; 0.612 V, 0.671 V, 59.4 mV

DESIGN NOTE

The diode voltage changes by *60 mV per decade* change in diode current. Sixty mV/decade often plays an important role in our thinking about the design of circuits containing both diode and bipolar transistors and is a good number to remember.

Figure 3.12 compares the characteristics of three diodes with different values of saturation current. The saturation current of diode A is 10 times larger than that of diode B, and the saturation current of diode B is 10 times that of diode C. The spacing between each pair of curves is

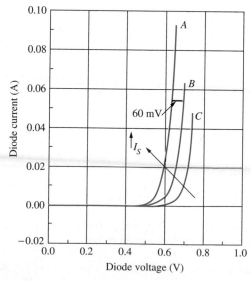

Figure 3.12 Diode characteristics for three different reverse saturation currents (a) 10^{-12} A, (b) 10^{-13} A, and (c) 10^{-14} A.

ELECTRONICS IN ACTION

The PTAT Voltage and Electronic Thermometry

The well-defined temperature dependence of the diode voltage discussed in Secs. 3.3–3.5 is actually used as the basis for most digital thermometers. We can build a simple electronic thermometer based on the circuit shown here in which two identical diodes are biased by current sources I_1 and I_2.

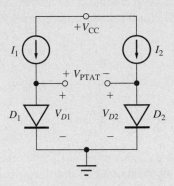

Wireless digital thermometer.

If we calculate the difference between the diode voltages using Eq. (3.14), we discover a voltage that is directly *proportional to absolute temperature* (PTAT), referred to as the PTAT voltage or V_{PTAT}:

$$V_{PTAT} = V_{D1} - V_{D2} = V_T \ln\left(\frac{I_{D1}}{I_S}\right) - V_T \ln\left(\frac{I_{D2}}{I_S}\right) = V_T \ln\left(\frac{I_{D1}}{I_{D2}}\right) = \frac{kT}{q} \ln\left(\frac{I_{D1}}{I_{D2}}\right)$$

The PTAT voltage has a temperature coefficient given by

$$\frac{dV_{PTAT}}{dT} = \frac{k}{q} \ln\left(\frac{I_{D1}}{I_{D2}}\right) = \frac{V_{PTAT}}{T}$$

By using two diodes, the temperature dependence of I_S has been eliminated from the equation. For example, suppose $T = 295$ K, $I_{D1} = 250$ μA, and $I_{D2} = 50$ μA. Then $V_{PTAT} = 40.9$ mV with a temperature coefficient of $+0.139$ mV/K.

This simple but elegant PTAT voltage circuit forms the heart of most of today's highly accurate electronic thermometers as depicted in the block diagram here. The analog PTAT voltage is amplified and then converted to a digital representation by an A/D converter. The digital output is scaled and offset to properly represent either the Fahrenheit or Celsius temperature scales and appears on an alphanumeric display. The scaling and offset shift can also be done in analog form prior to the A/D conversion operation.

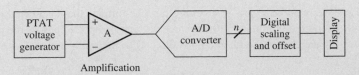

Block diagram of a digital thermometer.

approximately 60 mV. If the saturation current of the diode is reduced by a factor of 10, then the diode voltage must increase by approximately 60 mV to reach the same operating current level. Figure 3.12 also shows the relatively low sensitivity of the forward diode voltage to changes in the parameter I_S. For a fixed diode current, a change of two orders of magnitude in I_S results in a diode voltage change of only 120 mV.

3.5 DIODE TEMPERATURE COEFFICIENT

Another important number to keep in mind is the temperature coefficient associated with the diode voltage v_D. Solving Eq. (3.11) for the diode voltage under forward bias

$$v_D = V_T \ln\left(\frac{i_D}{I_S} + 1\right) = \frac{kT}{q} \ln\left(\frac{i_D}{I_S} + 1\right) \cong \frac{kT}{q} \ln\left(\frac{i_D}{I_S}\right) \qquad \text{V} \qquad (3.14)$$

and taking the derivative with respect to temperature yields

$$\frac{dv_D}{dT} = \frac{k}{q} \ln\left(\frac{i_D}{I_S}\right) - \frac{kT}{q} \frac{1}{I_S} \frac{dI_S}{dT} = \frac{v_D}{T} - V_T \frac{1}{I_S} \frac{dI_S}{dT} = \frac{v_D - V_{GO} - 3V_T}{T} \qquad \text{V/K} \quad (3.15)$$

where it is assumed that $i_D \gg I_S$ and $I_S \propto n_i^2$. In the numerator of Eq. (3.15), v_D represents the diode voltage, V_{GO} is the voltage corresponding to the silicon bandgap energy at 0 K, ($V_{GO} = E_G/q$), and V_T is the thermal potential. The last two terms result from the temperature dependence of n_i^2 as defined by Eq. (2.2). Evaluating the terms in Eq. (3.15) for a silicon diode with $v_D = 0.65$ V, $E_G = 1.12$ eV, and $V_T = 0.025$ V yields

$$\frac{dv_D}{dT} = \frac{(0.65 - 1.12 - 0.075) \text{ V}}{300 \text{ K}} = -1.82 \text{ mV/K} \qquad (3.16)$$

DESIGN NOTE

The forward voltage of the diode decreases as temperature increases, and the diode exhibits a temperature coefficient of approximately *−1.8 mV/°C* at room temperature.

EXERCISE: (a) Verify Eq. (3.15) using the expression for n_i^2 from Eq. (2.2). (b) A silicon diode is operating at $T = 300$ K, with $i_D = 1$ mA, and $v_D = 0.680$ V. Use the result from Eq. (3.16) to estimate the diode voltage at 275 K and at 350 K.

ANSWERS: 0.726 V; 0.589 V

3.6 DIODES UNDER REVERSE BIAS

We must be aware of several other phenomena that occur in diodes operated under reverse bias. As depicted in Fig. 3.13, the reverse voltage v_R applied across the diode terminals is dropped across the space charge region and adds directly to the built-in potential of the junction:

$$v_j = \phi_j + v_R \qquad \text{for } v_R > 0 \qquad (3.17)$$

The increased voltage results in a larger internal electric field that must be supported by additional charge in the depletion layer, as defined by Eqs. (3.2) to (3.5). Using Eq. (3.7) with the voltage

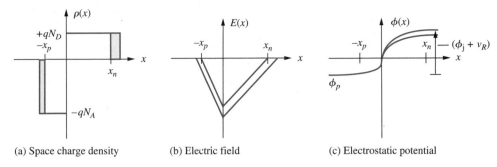

(a) Space charge density   (b) Electric field   (c) Electrostatic potential

Figure 3.13 The *pn* junction diode under reverse bias.

from Eq. (3.17), the general expression for the depletion-layer width w_d for an applied reverse-bias voltage v_R becomes

$$w_d = (x_n + x_p) = \sqrt{\frac{2\varepsilon_s}{q}\left(\frac{1}{N_A} + \frac{1}{N_D}\right)(\phi_j + v_R)}$$

or

$$(3.18)$$

$$w_d = w_{do}\sqrt{1 + \frac{v_R}{\phi_j}} \qquad \text{where } w_{do} = \sqrt{\frac{2\varepsilon_s}{q}\left(\frac{1}{N_A} + \frac{1}{N_D}\right)\phi_j}$$

The width of the space charge region increases approximately in proportion to the square root of the applied voltage.

EXERCISE: The diode in Example 3.1 had a zero-bias depletion-layer width of 0.113 μm and a built-in voltage of 0.979 V. What will be the depletion-layer width for a 10-V reverse bias? What is the new value of E_{MAX}?

ANSWERS: 0.378 μm; 581 kV/cm

3.6.1 SATURATION CURRENT IN REAL DIODES

The reverse saturation current actually results from the thermal generation of hole–electron pairs in the depletion region that surrounds the *pn* junction and is therefore proportional to the volume of the depletion region. Since the depletion-layer width increases with reverse bias, as described by Eq. (3.18), the reverse current does not truly saturate, as depicted in Fig. 3.9 and Eq. (3.9). Instead, there is gradual increase in reverse current as the magnitude of the reverse bias voltage is increased.

$$I_S = I_{SO}\sqrt{1 + \frac{v_R}{\phi_j}} \qquad (3.19)$$

Under forward bias, the depletion-layer width changes very little, and $I_S = I_{SO}$ for forward bias.

EXERCISE: A diode has $I_{SO} = 10$ fA and a built-in voltage of 0.8 V. What is I_S for a reverse bias of 10 V?

ANSWER: 36.7 fA

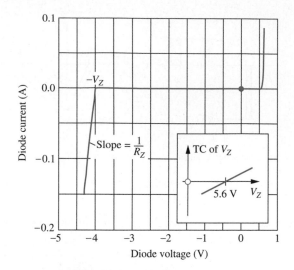

Figure 3.14 i-v characteristic of a diode including the reverse-breakdown region. The inset shows the temperature coefficient (TC) of V_Z.

3.6.2 REVERSE BREAKDOWN

As the reverse voltage increases, the electric field within the device grows, and the diode eventually enters the **breakdown region.** The onset of the breakdown process is fairly abrupt, and the current increases rapidly for any further increase in the applied voltage, as shown in the i-v characteristic of Fig. 3.14.

The magnitude of the voltage at which breakdown occurs is called the **breakdown voltage V_Z** of the diode and is typically in the range $2 \text{ V} \leq V_Z \leq 2000 \text{ V}$. The value of V_Z is determined primarily by the doping level on the more lightly doped side of the pn junction, but the heavier the doping, the smaller the breakdown voltage of the diode.

Two separate breakdown mechanisms have been identified: *avalanche breakdown* and *Zener breakdown.* These are discussed in the following two sections.

Avalanche Breakdown

Silicon diodes with breakdown voltages greater than approximately 5.6 V enter breakdown through a mechanism called **avalanche breakdown.** As the width of the depletion layer increases under reverse bias, the electric field increases, as indicated in Fig. 3.13. Free carriers in the depletion region are accelerated by this electric field, and as the carriers move through the depletion region, they collide with the fixed atoms. At some point, the electric field and the width of the space charge region become large enough that some carriers gain energy sufficient to break covalent bonds upon impact, thereby creating electron–hole pairs. The new carriers created can also accelerate and create additional electron–hole pairs through this **impact-ionization process,** as illustrated in Fig. 3.15.

Zener Breakdown

True **Zener breakdown** occurs only in heavily doped diodes. The high doping results in a very narrow depletion-region width, and application of a reverse bias causes carriers to tunnel directly between the conduction and valence bands, again resulting in a rapidly increasing reverse current in the diode.

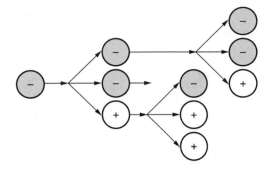

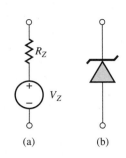

Figure 3.15 The avalanche breakdown process. (Note that the positive and negative charge carriers will actually be moving in opposite directions in the electric field in the depletion region.)

Figure 3.16 (a) Model for reverse-breakdown region of diode. (b) Zener diode symbol.

Breakdown Voltage Temperature Coefficient
We can differentiate between the two types of breakdown because the breakdown voltages associated with the two mechanisms exhibit opposite temperature coefficients (TC). In avalanche breakdown, V_Z increases with temperature; in Zener breakdown, V_Z decreases with temperature. For silicon diodes, a zero temperature coefficient is achieved at approximately 5.6 V. The avalanche breakdown mechanism dominates in diodes that exhibit breakdown voltages of more than 5.6 V, whereas diodes with breakdown voltages below 5.6 V enter breakdown via the Zener mechanism.

3.6.3 DIODE MODEL FOR THE BREAKDOWN REGION
In breakdown, the diode can be modeled by a voltage source of value V_Z in series with resistor R_Z, which sets the slope of the i-v characteristic in the breakdown region, as indicated in Fig. 3.16. The value of R_Z is normally small ($R_Z \leq 100 \ \Omega$), and the reverse current flowing in the diode must be limited by the external circuit or the diode will be destroyed.

From the i-v characteristic in Fig. 3.14 and the model in Fig. 3.16, we see that the voltage across the diode is almost constant, independent of current, in the reverse-breakdown region. Some diodes are actually designed to be operated in **reverse breakdown.** These diodes are called **Zener diodes** and have the special circuit symbol given in Fig. 3.16(b). Sample data sheets for a series of zener diodes can be found on the MCD website.

3.7 *pn* JUNCTION CAPACITANCE

Forward- and reverse-biased diodes also have a capacitance associated with the *pn* junction. This capacitance is important under dynamic signal conditions because it prevents the voltage across the diode from changing instantaneously.

3.7.1 REVERSE BIAS
Under reverse bias, w_d increases beyond its zero-bias value, as expressed by Eq. (3.18), and hence the amount of charge in the depletion region also increases. Because the charge in the diode is changing with voltage, a capacitance results. Using Eqs. (3.4) and (3.7), the total space charge on the n-side of the diode is given by

$$Q_n = qN_D x_n A = q \left(\frac{N_A N_D}{N_A + N_D} \right) w_d A \quad \text{C} \quad (3.20)$$

where A is the cross-sectional area of the diode and w_d is described by Eq. (3.18). The capacitance of the reverse-biased pn junction is given by

$$C_j = \frac{dQ_n}{dv_R} = \frac{C_{jo}A}{\sqrt{1 + \dfrac{v_R}{\phi_j}}} \qquad \text{where } C_{jo} = \frac{\varepsilon_s}{w_{do}} \qquad \text{F/cm}^2 \qquad (3.21)$$

in which C_{jo} represents the **zero-bias junction capacitance** per unit area of the diode.

Figure 3.17 Circuit symbol for the variable capacitance diode (varactor).

Equation (3.21) shows that the capacitance of the diode changes with applied voltage. The capacitance decreases as the reverse bias increases, exhibiting an inverse square root relationship. This voltage-controlled capacitance can be very useful in certain electronic circuits. Diodes can be designed with impurity profiles (called *hyper-abrupt profiles*) specifically optimized for operation as voltage-controlled capacitors. As for the case of Zener diodes, a special symbol exists for the variable capacitance diode, as shown in Fig. 3.17. Remember that this diode is to be operated under reverse bias. Sample data sheets for a series of variable capacitance diodes can be found on the MCD website.

> **EXERCISE:** What is the value of C_{jo} for the diode in Example 3.1? What is the zero bias value of C_j if the diode junction area is 100 μm $\times$ 125 μm? What is the capacitance at a reverse bias of 5 V?
>
> **ANSWERS:** 91.7 nf/cm^2; 11.5 pF; 4.64 pF

3.7.2 FORWARD BIAS

When the diode is operating under forward bias, additional charge is stored in the neutral regions near the edges of the space charge region. The amount of charge Q_D stored in the diode is proportional to the diode current:

$$Q_D = i_D \tau_T \qquad \text{C} \qquad (3.22)$$

The proportionality constant τ_T is called the diode **transit time** and ranges from 10^{-15} s to more than 10^{-6} s (1 fs to 1 μs) depending on the size and type of diode. Because we know that i_D is dependent on the diode voltage through the diode equation, there is an additional capacitance, the **diffusion capacitance C_D,** associated with the forward region of operation:

$$C_D = \frac{dQ_D}{dv_D} = \frac{(i_D + I_S)\tau_T}{V_T} \cong \frac{i_D \tau_T}{V_T} \qquad \text{F} \qquad (3.23)$$

in which V_T is the thermal voltage. The diffusion capacitance is proportional to current and can become quite large at high currents.

> **EXERCISE:** A diode has a transit time of 10 ns. What is the diffusion capacitance of the diode for currents of 10 μA, 0.8 mA, and 50 mA at room temperature?
>
> **ANSWERS:** 4 pF; 320 pF; 20 nF

3.8 SCHOTTKY BARRIER DIODE

In a p^+n junction diode, the p-side is a highly doped region (a conductor), and one might wonder if it could be replaced with a metallic layer. That is in fact the case, and in the **Schottky barrier diode,** one of the semiconductor regions of the pn junction diode is replaced by a non-ohmic rectifying

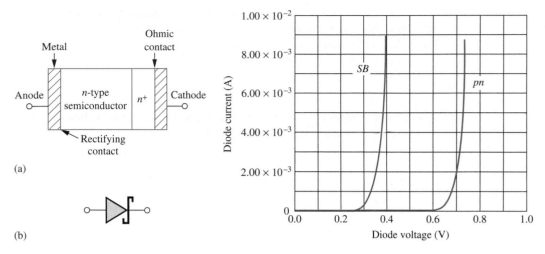

(a)

(b)

Figure 3.18 (a) Schottky barrier diode structure. (b) Schottky diode symbol.

Figure 3.19 Comparison of pn junction (pn) and Schottky diode (SB) i-v characteristics.

metal contact, as indicated in Fig. 3.18. It is easiest to form a Schottky contact to n-type silicon, and for this case the metal region becomes the diode anode. An n^+ region is added to ensure that the cathode contact is ohmic. The symbol for the Schottky diode appears in Fig. 3.18(b).

The Schottky diode turns on at a much lower voltage than its pn-junction counterpart, as indicated in Fig. 3.19. It also has significantly reduced internal charge storage under forward bias. We encounter an important use of the Schottky diode in bipolar logic circuits in Chapter 9. Schottky diodes also find important applications in high-power rectifier circuits and fast switching applications.

3.9 DIODE SPICE MODEL AND LAYOUT

The circuit in Fig. 3.20 represents the diode model that is included in SPICE programs. Resistance R_S represents the inevitable series resistance that always accompanies fabrication of, and making contacts to, a real device structure. The current controlled current source represents the ideal exponential behavior of the diode as described by Eq. 3.12 and **SPICE parameters IS, N,** and V_T. The model equation for i_D also includes a second term, not shown here, that models the effects of carrier generation in the space charge region in a manner similar to Eq. (3.19).

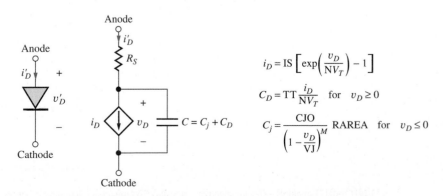

$$i_D = \text{IS}\left[\exp\left(\frac{v_D}{NV_T}\right) - 1\right]$$

$$C_D = \text{TT}\frac{i_D}{NV_T} \quad \text{for} \quad v_D \ge 0$$

$$C_j = \frac{\text{CJO}}{\left(1 - \dfrac{v_D}{\text{VJ}}\right)^M} \text{ RAREA} \quad \text{for} \quad v_D \le 0$$

Figure 3.20 Diode equivalent circuit and simplified versions of the model equations used in SPICE programs.

The capacitor specification includes the depletion-layer capacitance for the reverse-bias region modeled by **SPICE parameters CJO, VJ, and M,** as well as the diffusion capacitance associated with the junction under forward bias and defined by N and the transit-time parameter TT. In SPICE, the "junction grading coefficient" is an adjustable parameter. Using the typical value of $M = 0.5$ results in Eq. (3.21).

EXERCISE: Find the default values of the seven parameters in Table 3.1 for the SPICE program that you use in class. Compare to the values in Table 3.1.

TABLE 3.1
SPICE Diode Parameter Equivalences

PARAMETER	OUR TEXT	SPICE	TYPICAL DEFAULT VALUES
Saturation current	I_S	IS	10 fA
Ohmic series resistance	R_S	RS	0 Ω
Ideality factor or emission coefficient	n	N	1
Transit time	τ_T	TT	0 sec
Zero-bias junction capacitance for a unit area diode RAREA = 1	$C_{jo} \cdot A$	CJO	0 F
Built-in potential	ϕ_j	VJ	1 V
Junction grading coefficient	—	M	0.5
Relative junction area	—	RAREA	1

Diode Layout

Figure 3.21(a) shows the layout of a simple diode fabricated by forming a p-type diffusion in an n-type silicon wafer, as outlined in Chapter 2. This diode has a long rectangular p-type diffusion to increase the value of I_S, which is proportional to the junction area. Multiple contacts are formed to the p-type anode, and the p-region is surrounded by a collar of contacts to the n-type region. Both

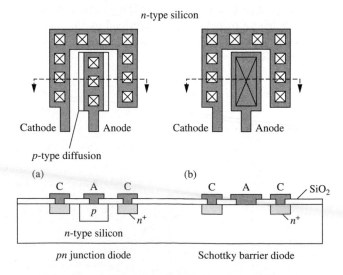

Figure 3.21 Layout of (a) a pn junction diode and (b) a metal-semiconductor Schottky diode. (See top view of diode in Chapter 3 opener.)

these sets of contacts are used to minimize the value of the extrinsic resistance R_S in series with the diode, as included in the model in Fig. 3.20. Identical contacts are used so that they all tend to etch open at the same time during the fabrication process. The use of multiple identical contacts also facilitates calculation of the overall contact resistance. Heavily doped n-type regions are placed under the n-region contacts to ensure formation of an ohmic contact by preventing formation of a Schottky barrier diode.

A conceptual drawing of a metal-semiconductor or Schottky diode also appears in Fig. 3.21(b) in which the aluminum metallization acts as the anode of the diode and the n-type semiconductor is the diode cathode. Careful attention to processing details is needed to form a diode rather than just an ohmic contact.

3.10 DIODE CIRCUIT ANALYSIS

We now begin our analysis of circuits containing diodes and introduce simplified circuit models for the diode. Figure 3.22 presents a series circuit containing a voltage source, resistor, and diode. Note that V and R may represent the Thévenin equivalent of a more complicated two-terminal network. Also note the notational change in Fig. 3.22. In the circuits that we analyze in the next few sections, the applied voltage and resulting diode voltage and current will all be dc quantities. (Recall that the dc components of the total quantities i_D and v_D are indicated by I_D and V_D, respectively.)

One common objective of diode circuit analysis is to find the **quiescent operating point,** or **Q-point,** for the diode. The Q-point consists of the dc current and voltage (I_D, V_D) that define the point of operation on the diode's i-v characteristic. We start the analysis by writing the loop equation for the circuit of Fig. 3.22:

$$V = I_D R + V_D \tag{3.24}$$

Equation (3.24) represents a constraint placed on the diode operating point by the circuit elements. The diode i-v characteristic in Fig. 3.8 represents the allowed values of I_D and V_D as determined by the solid-state diode itself. Simultaneous solution of these two sets of constraints defines the Q-point.

We explore several methods for determining the solution to Eq. (3.24), including graphical analysis and the use of models of varying complexity for the diode. These techniques will include

- Graphical analysis using the load-line technique
- Analysis with the mathematical model for the diode
- Simplified analysis with an ideal diode model
- Simplified analysis using the constant voltage drop model

3.10.1 LOAD-LINE ANALYSIS

In some cases, the i-v characteristic of the solid-state device may be available only in graphical form, as in Fig. 3.23. We must then use a graphical approach (**load-line analysis**) to find the simultaneous solution of Eq. (3.24) with the graphical characteristic. Equation (3.24) defines the **load line** for the

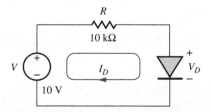

Figure 3.22 Diode circuit containing a voltage source and resistor.

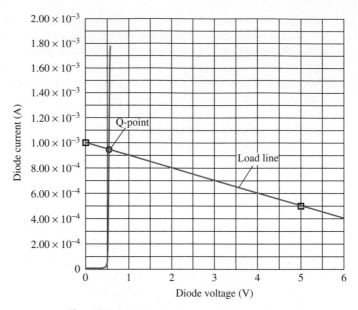

Figure 3.23 Diode i-v characteristic and load line.

diode. The Q-point can be found by plotting the graph of the load line on the i-v characteristic for the diode. The intersection of the two curves represents the quiescent operating point, or Q-point, for the diode.

EXAMPLE 3.5   **LOAD-LINE ANALYSIS**

The graphical load-line approach is an important concept for visualizing the behavior of diode circuits as well as for estimating the actual Q-point.

PROBLEM Use load-line analysis to find the Q-point for the diode circuit in Fig. 3.22 using the i-v characteristic in Fig. 3.23.

SOLUTION **Known Information and Given Data:** The Diode i-v characteristic is presented graphically in Fig. 3.23. Diode circuit is given in Fig. 3.22 with $V = 10$ V and $R = 10$ kΩ.

Unknowns: Diode Q-point (I_D, V_D)

Approach: Write the load-line equation and find two points on the load line that can be plotted on the graph in Fig. 3.23. The Q-point is at the intersection of the load line with the diode i-v characteristic.

Assumptions: Diode temperature corresponds to the temperature at which the graph in Fig. 3.23 was measured.

Analysis: Using the values from Fig. 3.22, Eq. (3.24) can be rewritten as

$$10 = I_D 10^4 + V_D \qquad (3.25)$$

Two points are needed to define the line. The simplest choices are

$$I_D = (10\ \text{V}/10\ \text{k}\Omega) = 1\ \text{mA} \quad \text{for} \quad V_D = 0 \quad \text{and} \quad V_D = 10\ \text{V} \quad \text{for} \quad I_D = 0$$

Unfortunately, the second point is not on the diode characteristic, as presented in Fig. 3.23, but we are free to choose any point that satisfies Eq. (3.25). Let's pick $V_D = 5$ V:

$$I_D = (10 - 5)\text{V}/10^4\ \Omega = 0.5\ \text{mA} \qquad \text{for } V_D = 5$$

These points and the resulting load line are plotted in Fig. 3.23. The Q-point is given by the intersection of the load line and the diode characteristic:

$$\text{Q-point} = (0.95\ \text{mA},\ 0.6\ \text{V})$$

Check of Results: We can double check our result by substituting the diode voltage found from the graph into Eq. (3.25) and calculating I_D. Using $V_D = 0.6$ V in Eq. (3.25) yields an improved estimate for the Q-point: (0.94 mA, 0.6 V). [Note that we could also substitute 0.95 mA into Eq. (3.25) and calculate V_D.]

Discussion: Note that the values determined graphically are not quite on the load line since they do not precisely satisfy the load-line equation. This is a result of the limited precision that we can obtain by reading the graph.

EXERCISE: Repeat the load-line analysis if $V = 5$ V and $R = 5$ kΩ.

ANSWERS: (0.88 mA, 0.6 V)

EXERCISE: Use SPICE to find the Q-point for the circuit in Fig. 3.22. Use the default values of parameters in your SPICE program.

ANSWERS: (935 μA, 0.653 V) for $I_S = 10$ fA

3.10.2 ANALYSIS USING THE MATHEMATICAL MODEL FOR THE DIODE

We can use our mathematical model for the diode to approach the solution of Eq. (3.25) more directly. The particular diode characteristic in Fig. 3.23 is represented quite accurately by diode Eq. (3.11), with $I_S = 10^{-13}$ A, $n = 1$, and $V_T = 0.025$ V:

$$I_D = I_s \left[\exp\left(\frac{V_D}{V_T} \right) - 1 \right] = 10^{-13}[\exp(40V_D) - 1] \tag{3.26}$$

Eliminating I_D by substituting Eq. (3.26) into Eq. (3.25) yields

$$10 = 10^4 \cdot 10^{-13}[\exp(40V_D) - 1] + V_D \tag{3.27}$$

The expression in Eq. (3.27) is called a *transcendental equation* and does not have a closed-form analytical solution, so we settle for a numerical answer to the problem.

One approach to finding a numerical solution to Eq. (3.27) is through simple trial and error. We can guess a value of V_D and see if it satisfies Eq. (3.27). Based on the result, a new guess can be formulated and Eq. (3.27) evaluated again. The human brain is quite good at finding a sequence of values that will converge to the desired solution.

On the other hand, it is often preferable to use a computer to find the solution to Eq. (3.27), particularly if we need to find the answer to several different problems or parameter sets. The computer, however, requires a much more well-defined iteration strategy than brute force trial and error.

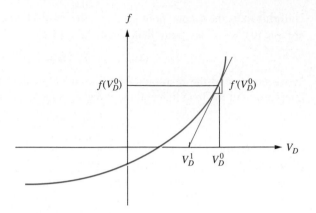

Figure 3.24 Newton update for finding a zero of the function f.

One possible numerical iteration strategy uses Newton's method for finding the zeros of an equation. Equation (3.27) can be rewritten as

$$f = 10 - 10^4 \cdot 10^{-13}[\exp(40V_D) - 1] - V_D \tag{3.28}$$

where we desire to find the value of V_D for which $f = 0$. Newton's method provides a numerical iteration strategy.

An initial guess V_D^0 is chosen for the value of V_D, and the value of f corresponding to V_D^0 is calculated. Unless we are extremely lucky, the value of f will not be zero. Referring to Fig. 3.24, the slope f' of the function f at $V_D = V_D^0$ is used to estimate an improved value V_D^1 that will force f closer to zero. It is desired that $f(V_D^1) = 0$, and

$$f'(V_D^0) = \frac{f(V_D^0) - 0}{V_D^0 - V_D^1} \qquad \text{or} \qquad V_D^1 = V_D^0 - \frac{f(V_D^0)}{f'(V_D^0)} \tag{3.29}$$

The iterative numerical procedure becomes:

1. Make an initial guess V_D^0.
2. Evaluate f and f' for this value of V_D.
3. Calculate a new guess for V_D using Eq. (3.29).
4. Repeat steps 2 and 3 until convergence is obtained.

This numerical iteration process can be done manually with a calculator or be easily implemented using a spreadsheet. Spreadsheets are quite useful in circuit design for finding solutions to iterative problems as well as for exploring the design space associated with many circuit design problems. As an example of the use of a spreadsheet, let us implement the preceding numeric iteration method. The function f, its derivative f', the iterative update for V_D, and the current I_D are all given by

$$f = 10 - 10^{-9}[\exp(40V_D) - 1] - V_D \qquad f' = -4 \times 10^{-8} \exp(40V_D) - 1$$

$$\tag{3.30}$$

$$V_D^{i+1} = V_D^i - \frac{f(V_D^i)}{f'(V_D^i)} \qquad \text{and} \qquad I_D = 10^{-13}[\exp(40V_D) - 1]$$

Table 3.2 gives the results of spreadsheet calculations. The initial guess of $V_D = 0.8$ V is shown in **bold** type in row 0. In any given row, f, f', and I_D are calculated from their respective expressions in Eq. (3.30) using the value of V_D in the same row. Except for the initial guess in row 0, the value of V_D in any row is calculated from V_D, f, and f' in the previous row. Once the formulas are entered into one row of the spreadsheet, the row may simply be copied into the cells in as many rows as desired, and the iteration process proceeds automatically. For the particular case in Table 3.2, the diode voltage and current calculations converge to within four significant digits in 13 iterations.

TABLE 3.2
Diode Iteration Problem Using a Spreadsheet

ITERATION #	V_D	f	f'	I_D
0	**0.8000**	-7.895×10^4	-3.159×10^6	7.896×10^0
1	0.7750	-2.904×10^4	-1.162×10^6	2.905×10^0
2	0.7500	-1.068×10^4	-4.276×10^5	1.069×10^0
3	0.7250	-3.927×10^3	-1.575×10^5	3.936×10^{-1}
$\vdots$				
11	0.5743	-5.989×10^{-2}	-3.804×10^2	9.486×10^{-4}
12	0.5742	-1.876×10^{-4}	-3.780×10^2	9.426×10^{-4}
13	0.5742	-1.858×10^{-9}	-3.780×10^2	9.426×10^{-4}
		$10 - 10^{-9} \times$ $[\exp(40V_D) - 1]$ $- V_D$	$-4 \times 10^{-8} \times$ $\exp(40V_D) - 1$	$10^{-13} \times$ $[\exp(40V_D) - 1]$

For these diode problems, it is best to choose an initial value of V_D that is larger than the expected solution. (See Prob. 3.61.)

Note that one can achieve answers to an almost arbitrary precision using the numerical approach. However, in most real circuit situations, we will not have an accurate value for the saturation current of the diode, and there will be significant tolerances associated with the sources and passive components in the circuit. For example, the saturation current specification for a given diode type may vary by factors ranging from 10:1 to as much as 100:1. In addition, resistors commonly have ± 5 percent to ± 10 percent tolerances, and we do not know the exact operating temperature of the diode (remember the -1.8 mV/K temperature coefficient) or the precise value of the parameter n. Hence, it does not make sense to try to obtain answers with a precision of more than two or three significant digits.

An alternative to the use of a spreadsheet is to write a simple program using a high-level language. The solution to Eq. (3.28) also can be found using the "solver" routines in many calculators, which use iteration procedures more sophisticated than that just described. MATLAB also provides the function fzero, which will calculate the zeros of a function as outlined in Example 3.6.

EXERCISE: An alternative expression (another transcendental equation) for the basic diode circuit can be found by eliminating V_D in Eq. (3.25) using Eq. (3.14). Show that the result is

$$10 = 10^4 I_D + 0.025 \ln\left(1 + \frac{I_D}{I_S}\right)$$

EXAMPLE **3.6**   SOLUTION OF THE DIODE EQUATION USING MATLAB

MATLAB is one example of a computer tool that can be used to find the solution to transcendental equations.

PROBLEM   Use MATLAB® to find the solution to Eq. (3.28).

SOLUTION   **Known Information and Given Data:** Diode circuit in Fig. 3.22 with $V = 10$ V, $R = 10$ kΩ, $I_S = 10^{-13}$ A, $n = 1$, and $V_T = 0.025$ V

Unknowns: Diode voltage V_D

Approach: Create a MATLAB "M-File" describing Eq. (3.28). Execute the program to find the diode voltage.

Assumptions: Room temperature operation with $V_T = 1/40$ V

Analysis: First, create an M-file for the function 'diode':

$$\text{function xd} = \text{diode(vd)}$$
$$\text{xd} = 10 - (10^{\wedge}(-9)) * (\exp(40 * \text{vd}) - 1) - \text{vd};$$

Then find the solution near 1 V:

$$\text{fzero('diode', 1)}$$

Answer: 0.5742 V

Check of Results: The diode voltage is positive and in the range of 0.5 to 0.8 V, which is expected for a diode. Substituting this value of voltage into the diode equation yields a current of 0.944 mA. This answer appears reasonable since we know that the diode current cannot exceed 10 V$/10$ k$\Omega =$ 1.0 mA, which is the maximum current available from the circuit.

EXERCISE: Use the MATLAB® to find the solution to

$$10 = 10^4 I_D + 0.025 \ln\left(1 + \frac{I_D}{I_S}\right) \qquad \text{for } I_S = 10^{-13} \text{ A}$$

ANSWER: 942.6 μA

EXAMPLE **3.7**   **EFFECT OF DEVICE TOLERANCES ON DIODE Q-POINTS**

Let us now see how sensitive our Q-point results are to the exact value of the diode saturation current.

PROBLEM   Suppose that there is a tolerance on the value of the saturation current such that the value is given by

$$I_S^{\text{nom}} = 10^{-15} \text{ A} \qquad \text{and} \qquad 2 \times 10^{-16} \text{ A} \leq I_S \leq 5 \times 10^{-15} \text{ A}$$

Find the nominal, smallest, and largest values of the diode voltage and current in the circuit in Fig. 3.22.

SOLUTION   **Known Information and Given Data:** The nominal and worst-case values of saturation current are given as well as the circuit values in Fig. 3.22.

Unknowns: Nominal and worst-case values for the diode Q-point: (I_D, V_D).

Approach: Use MATLAB or the solver on our calculator to find the diode voltages and then the currents for the nominal and worst-case values of I_S. Note from Eq. (3.24) that the maximum value of diode voltage corresponds to minimum current and vice versa.

Assumptions: Room temperature operation with $V_T = 0.025$ V. The voltage and resistance in the circuit do not have tolerances associated with them.

Analysis: For the nominal case, Eq. (3.28) becomes

$$f = 10 - 10^4(10^{-15})[\exp(40V_D) - 1] - V_D$$

for which the solver yields

$$V_D^{\text{nom}} = 0.689 \text{ V} \qquad \text{and} \qquad I_D^{\text{nom}} = \frac{(10 - 0.689) \text{ V}}{10^4 \, \Omega} = 0.931 \text{ mA}$$

For the minimum I_S case, Eq. (3.28) is

$$f = 10 - 10^4(2 \times 10^{-16})[\exp(40V_D) - 1] - V_D$$

and the solver yields

$$V_D^{\text{max}} = 0.729 \text{ V} \qquad \text{and} \qquad I_D^{\text{min}} = \frac{(10 - 0.729) \text{ V}}{10^4 \, \Omega} = 0.927 \text{ mA}$$

Finally, for the maximum value of I_S, Eq. (3.28) becomes

$$f = 10 - 10^4(5 \times 10^{-15})[\exp(40V_D) - 1] - V_D$$

and the solver gives

$$V_D^{\text{min}} = 0.649 \text{ V} \qquad \text{and} \qquad I_D^{\text{max}} = \frac{(10 - 0.649) \text{ V}}{10^4 \, \Omega} = 0.935 \text{ mA}$$

Check of Results: The diode voltages are positive and in the range of 0.5 to 0.8 V which is expected for a diode. The diode currents are all less than the short circuit current available from the voltage source (10 V/10 kΩ = 1.0 mA).

Discussion: Note that even though the diode saturation current in this circuit changes by a factor of 5:1 in either direction, the current changes by less than ±0.5%. As long as the driving voltage in the circuit is much larger than the diode voltage, the current should be relatively insensitive to changes in the diode voltage or the diode saturation current.

EXERCISE: Find V_D and I_D if the upper limit on I_S is increased to 10^{-14} A.

ANSWERS: 0.6316 V; 0.9368 A

EXERCISE: Use the Solver function in your calculator to find the solution to

$$10 = 10^4 I_D + 0.025 \ln\left(1 + \frac{I_D}{I_S}\right) \quad \text{for } I_S = 10^{-13} \text{ A} \qquad \text{and} \qquad I_S = 10^{-15} \text{ A}$$

ANSWER: 0.9426 mA; 0.9311 mA

3.10.3 THE IDEAL DIODE MODEL

Graphical load-line analysis provides insight into the operation of the diode circuit of Fig. 3.22, and the mathematical model can be used to provide more accurate solutions to the load-line problem. The next method discussed provides simplified solutions to the diode circuit of Fig. 3.22 by introducing simplified diode circuit models of varying complexity.

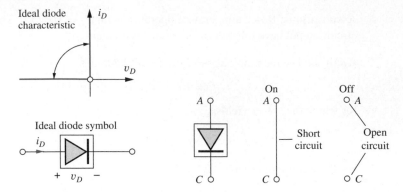

Figure 3.25 Ideal diode i-v characteristics and circuit symbol.

Figure 3.26 Circuit models for on and off states of the ideal diode.

The diode, as described by its i-v characteristic in Fig. 3.8 or by Eq. (3.11), is obviously a nonlinear device. However, most, if not all, of the circuit analysis that we have learned thus far assumed that the circuits were composed of linear elements. To use this wealth of analysis techniques, we will use **piecewise linear** approximations to the diode characteristic.

The **ideal diode model** is the simplest model for the diode. The i-v characteristic for the **ideal diode** in Fig. 3.25 consists of two straight-line segments. If the diode is conducting a forward or positive current (forward-biased), then the voltage across the diode is zero. If the diode is reverse-biased, with $v_D < 0$, then the current through the diode is zero. These conditions can be stated mathematically as

$$v_D = 0 \quad \text{for } i_D > 0 \qquad \text{and} \qquad i_D = 0 \quad \text{for } v_D \leq 0$$

The special symbol in Fig. 3.25 is used to represent the ideal diode in circuit diagrams.

We can now think of the diode as having two states. The diode is either conducting in the *on* state, or nonconducting and *off*. For circuit analysis, we use the models in Fig. 3.26 for the two states. If the diode is on, then it is modeled by a "short" circuit, a wire. For the off state, the diode is modeled by an "open" circuit, no connection.

Analysis Using the Ideal Diode Model

Let us now analyze the circuit of Fig. 3.22 assuming that the diode can be modeled by the ideal diode of Fig. 3.26. The diode has two possible states, and our analysis of diode circuits proceeds as follows:

1. Select a model for the diode.
2. Identify the anode and cathode of the diode and label the diode voltage v_D and current i_D.
3. Make an (educated) guess concerning the region of operation of the diode based on the circuit configuration.
4. Analyze the circuit using the diode model appropriate for the assumption in step 3.
5. Check the results to see if they are consistent with the assumptions.

For this analysis, we select the ideal diode model. The diode in the original circuit is replaced by the ideal diode, as in Fig. 3.27(b). Next we must guess the state of the diode. Because the voltage source appears to be trying to force a positive current through the diode, our first guess will be to assume that the diode is on. The ideal diode of Fig. 3.27(b) is replaced by its piecewise linear model for the on region in Fig. 3.28, and the diode current is given by

$$I_D = \frac{(10 - 0) \text{ V}}{10 \text{ k}\Omega} = 1.00 \text{ mA}$$

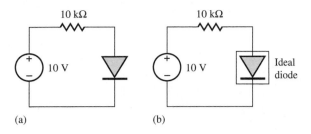

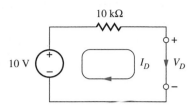

(a)   (b)

Figure 3.27 (a) Original diode circuit. (b) Circuit modeled by an ideal diode.

Figure 3.28 Ideal diode replaced with its model for the on state.

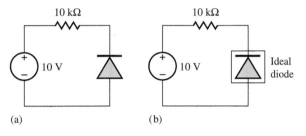

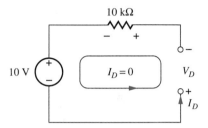

(a)   (b)

Figure 3.29 (a) Circuit with reverse-biased diode. (b) Circuit modeled by ideal diode.

Figure 3.30 Ideal diode replaced with its model for the off region.

The current $I_D \geq 0$, which is consistent with the assumption that the diode is on. The Q-point is therefore equal to (1 mA, 0 V). Based on the ideal diode model, we find that the diode is forward-biased and operating with a current of 1 mA.

Analysis of a Circuit Containing a Reverse-Biased Diode

A second circuit example in which the diode terminals have been reversed appears in Fig. 3.29; the ideal diode model is again used to model the diode [Fig. 3.29(b)]. The voltage source now appears to be trying to force a current backward through the diode. Because the diode cannot conduct in this direction, we assume the diode is off. The ideal diode of Fig. 3.29(b) is replaced by the open circuit model for the off region, as in Fig. 3.30.

Writing the loop equation for this case,

$$10 + V_D + 10^4 I_D = 0$$

Because $I_D = 0$, $V_D = -10$ V. The calculated diode voltage is negative, which is consistent with the starting assumption that the diode is off. The Q-point is (0, −10 V). The analysis shows that the diode in the circuit of Fig. 3.29 is indeed reverse-biased.

Although these two problems may seem rather simple, the complexity of diode circuit analysis increases rapidly as the number of diodes increases. If the circuit has N diodes, then the number of possible states is 2^N. A circuit with 10 diodes has 1024 different possible circuits that could be analyzed! Only through practice can we develop the intuition needed to avoid analysis of many incorrect cases. We analyze more complex circuits shortly, but first let's look at a slightly better piecewise linear model for the diode.

3.10.4 CONSTANT VOLTAGE DROP MODEL

We know from our earlier discussion that there is a small, nearly constant voltage across the forward-biased diode. The ideal diode model ignores the presence of this voltage. However, the piecewise linear model for the diode can be improved by adding a constant voltage V_{on} in series with the ideal diode, as shown in Fig. 3.31(b). This is the **constant voltage drop (CVD) model.** V_{on} offsets the

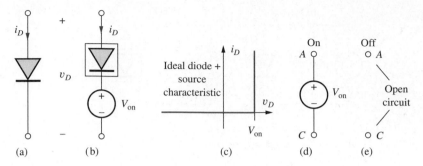

Figure 3.31 Constant voltage drop model for diode. (a) Actual diode. (b) Ideal diode plus voltage source V_{on}. (c) Composite i-v characteristic. (d) CVD model for the on state. (e) Model for the off state.

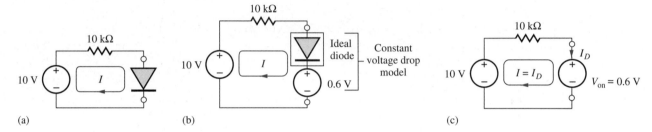

Figure 3.32 Diode circuit analysis using constant voltage drop model. (a) Original diode circuit. (b) Circuit with diode replaced by the constant voltage drop model. (c) Circuit with ideal diode replaced by the piecewise linear model.

i-v characteristic of the ideal diode, as indicated in Fig. 3.31(c). The piecewise linear models for the two states become a voltage source V_{on} for the on state and an open circuit for the off state. We now have

$$v_D = V_{on} \quad \text{for } i_D > 0 \qquad \text{and} \qquad i_D = 0 \quad \text{for } v_D \le V_{on}$$

We may consider the ideal diode model to be the special case of the constant voltage drop model for which $V_{on} = 0$. From the i-v characteristics presented in Fig. 3.8, we see that a reasonable choice for V_{on} is 0.6 to 0.7 V. We use a voltage of 0.6 V as the turn-on voltage for our diode circuit analysis.

Diode Analysis with the Constant Voltage Drop Model

Let us analyze the diode circuit from Fig. 3.22 using the CVD model for the diode. The diode in Fig. 3.32(a) is replaced by its CVD model in Fig. 3.32(b). The 10-V source once again appears to be forward biasing the diode, so assume that the diode is on, resulting in the simplified circuit in Fig. 3.32(c). The diode current is given by

$$I_D = \frac{(10 - V_{on}) \text{ V}}{10 \text{ k}\Omega} = \frac{(10 - 0.6) \text{ V}}{10 \text{ k}\Omega} = 0.940 \text{ mA} \tag{3.31}$$

which is slightly smaller than that predicted by the ideal diode model but quite close to the exact result found earlier.

3.10.5 MODEL COMPARISON AND DISCUSSION

We have analyzed the circuit of Fig. 3.22 using four different approaches; the various results appear in Table 3.3. All four sets of predicted voltages and currents are quite similar. Even the simple ideal diode model only overestimates the current by less than 10 percent compared to the mathematical model. We see that the current is quite insensitive to the actual choice of diode voltage. This is a result

TABLE 3.3
Comparison of Diode Circuit Analysis Results

ANALYSIS TECHNIQUE	DIODE CURRENT	DIODE VOLTAGE
Load-line analysis	0.94 mA	0.6 V
Mathematical model	0.942 mA	0.547 V
Ideal diode model	1.00 mA	0 V
Constant voltage drop model	0.940 mA	0.600 V

of the exponential dependence of the diode current on voltage as well as the large source voltage (10 V) in this particular circuit.

Rewriting Eq. (3.31),

$$I_D = \frac{10 - V_{on}}{10\ k\Omega} = \frac{10\ V}{10\ k\Omega}\left(1 - \frac{V_{on}}{10}\right) = (1.00\ \text{mA})\left(1 - \frac{V_{on}}{10}\right) \tag{3.32}$$

we see that the value of I_D is approximately 1 mA for $V_{on} \ll 10$ V. Variations in V_{on} have only a small effect on the result. However, the situation would be significantly different if the source voltage were only 1 V for example (see Prob. 3.65).

3.11 MULTIPLE-DIODE CIRCUITS

The load-line technique is applicable only to single-diode circuits, and the mathematical model, or numerical iteration technique, becomes much more complex for circuits with more than one nonlinear element. In fact, the SPICE electronic circuit simulation program referred to throughout this book is designed to provide numerical solutions to just such complex problems. However, we also need to be able to perform hand analysis to predict the operation of multidiode circuits as well as to build our understanding and intuition of diode circuit operation. In this section we discuss the use of the simplified diode models for hand analysis of more complicated diode circuits.

As the complexity of diode circuits increases, we must rely on our intuition to eliminate unreasonable solution choices. Even so, analysis of diode circuits often requires several iterations, Intuition can only be developed over time by working problems, and here we analyze a circuit containing three diodes.

Figure 3.33 is an example of a circuit with several diodes. In the analysis of this circuit, we will use the CVD model to improve the accuracy of our hand calculations.

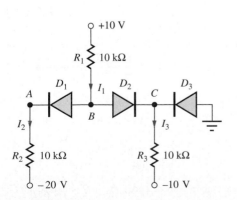

Figure 3.33 Example of a circuit containing three diodes.

TABLE 3.4
Possible Diode States for Circuit in Fig. 3.33

D_1	D_2	D_3
Off	Off	Off
Off	Off	On
Off	On	Off
Off	On	On
On	Off	Off
On	Off	On
On	On	Off
On	On	On

EXAMPLE **3.8**   **ANALYSIS OF A CIRCUIT CONTAINING THREE DIODES**

Now we will attempt to find the solution for a three-diode circuit. Our analysis will employ the CVD model.

PROBLEM   Find the Q-points for the three diodes in Fig. 3.33. Use the constant voltage drop model for the diodes.

SOLUTION   **Known Information and Given Data:** Circuit topology and element values in Fig. 3.33

Unknowns: (I_{D1}, V_{D1}), (I_{D2}, V_{D2}), (I_{D3}, V_{D3})

Approach: With three diodes, there are the eight On/Off combinations indicated in Table 3.4. A common method that we often use to find a starting point is to consider the circuit with all the diodes in the off state as in Fig 3.34(a). Here we see that the circuit produces large forward biases across D_1, D_2 and D_3. So our second step will be to assume that all the diodes are on.

Assumptions: Use the constant voltage drop model with $V_{\text{on}} = 0.6$ V.

Analysis: The circuit is redrawn using the CVD diode models in Fig. 3.34(b). Here we skipped the step of physically drawing the circuit with the ideal diode symbols but instead incorporated the piecewise linear models directly into the figure. Working from right to left, we see that the voltages at nodes C, B, and A are given by

$$V_C = -0.6 \text{ V} \qquad V_B = -0.6 + 0.6 = 0 \text{ V} \qquad V_A = 0 - 0.6 = -0.6 \text{ V}$$

With the node voltages specified, it is easy to find the current through each resistor:

$$I_1 = \frac{10 - 0}{10} \frac{\text{V}}{\text{k}\Omega} = 1 \text{ mA} \qquad I_2 = \frac{-0.6 - (-20)}{10} \frac{\text{V}}{\text{k}\Omega} = 1.94 \text{ mA}$$

$$I_3 = \frac{-0.6 - (-10)}{10} \frac{\text{V}}{\text{k}\Omega} = 0.94 \text{ mA} \tag{3.33}$$

Using Kirchhoff's current law, we also have

$$I_2 = I_{D1} \qquad I_1 = I_{D1} + I_{D2} \qquad I_3 = I_{D2} + I_{D3} \tag{3.34}$$

Combining Eqs. (3.33) and (3.34) yields the three diode currents:

$$I_{D1} = 1.94 \text{ mA} > 0 \qquad I_{D2} = -0.94 \text{ mA} < 0 \times \qquad I_{D3} = 1.86 \text{ mA} > 0 \tag{3.35}$$

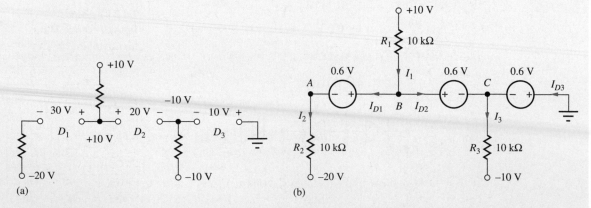

Figure 3.34 (a) Three diode circuit model with all diodes off. (b) Circuit model for circuit of Fig. 3.33 with all diodes on.

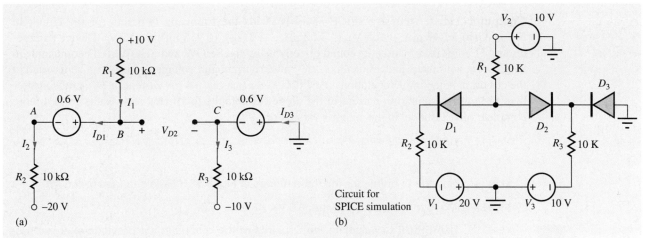

Figure 3.35 (a) Circuit with diodes D_1 and D_3 on and D_2 off.

Check of Results: I_{D1} and I_{D3} are greater than zero and therefore consistent with the original assumptions. However, I_{D2}, which is less than zero, represents a contradiction.

SECOND ITERATION   For our second attempt, let us assume D_1 and D_3 are on and D_2 is off, as in Fig. 3.35(a). We now have

$$+10 - 10{,}000 I_1 - 0.6 - 10{,}000 I_2 + 20 = 0 \qquad \text{with } I_1 = I_{D1} = I_2 \qquad (3.36)$$

which yields

$$I_{D1} = \frac{29.4}{20} \frac{\text{V}}{\text{k}\Omega} = 1.47\,\text{mA} > 0$$

Also

$$I_{D3} = I_3 = \frac{-0.6 - (-10)}{10} \frac{\text{V}}{\text{k}\Omega} = 0.940\,\text{mA} > 0$$

The voltage across diode D_2 is given by

$$V_{D2} = 10 - 10{,}000 I_1 - (-0.6) = 10 - 14.7 + 0.6 = -4.10\,\text{V} < 0$$

Check of Results: I_{D1}, I_{D3}, and V_{D2} are now all consistent with the circuit assumptions, so the Q-points for the circuit are

$$D_1\text{: } (1.47\,\text{mA}, 0.6\,\text{V}) \qquad D_2\text{: } (0\,\text{mA}, -4.10\,\text{V}) \qquad D_3\text{: } (0.940\,\text{mA}, 0.6\,\text{V})$$

Discussion: The Q-point values that we would have obtained using the ideal diode model are (see Prob. 3.76):

$$D_1\text{: } (1.50\,\text{mA}, 0\,\text{V}) \qquad D_2\text{: } (0\,\text{mA}, -5.00\,\text{V}) \qquad D_3\text{: } (1.00\,\text{mA}, 0\,\text{V})$$

The values of I_{D1} and I_{D3} differ by less than 6 percent. However, the reverse-bias voltage on D_2 differs by 20 percent. This shows the difference that the choice of models can make. The results from the circuit using the CVD model should be a more accurate estimate of how the circuit will actually perform than would result from the ideal diode case. Remember, however, that these calculations are both just approximations based on our models for the actual behavior of the real diode circuit.

Computer-Aided Analysis: SPICE analysis yields the following Q-points for the circuit in Fig. 3.35(b): (1.47 mA, 0.665 V), (−4.02 pA, −4.01 V), (0.935 mA, 0.653 V). Device parameter and Q-point information are found directly using the SHOW and SHOWMOD commands in SPICE. Or, voltmeters and ammeters (zero-valued current and voltage sources) can be inserted in the circuit in some implementations of SPICE. Note that the −4 pA current in D_2 is much larger than the reverse saturation current of the diode (IS defaults to 10 fA), and results from a more complete SPICE model in the author's version of SPICE.

EXERCISE: Find the Q-points for the three diodes in Fig. 3.33 if R_1 is changed to 2.5 kΩ.

ANSWERS: (2.13 mA, 0.6 V); (1.13 mA, 0.6 V); (0 mA, −1.27 V)

EXERCISE: Use SPICE to calculate the Q-points of the diodes in the previous exercise. Use $I_S = 1$ fA.

ANSWERS: (2.12 mA, 0.734 V); (1.12 mA, 0.718 V); (0 mA, −1.19 V)

3.12 ANALYSIS OF DIODES OPERATING IN THE BREAKDOWN REGION

Reverse breakdown is actually a highly useful region of operation for the diode. The reverse breakdown voltage is nearly independent of current and can be used as either a voltage regulator or voltage reference. Thus, it is important to understand the analysis of diodes operating in reverse breakdown.

Figure 3.36 is a single-loop circuit containing a 20-V source supplying current to a Zener diode with a reverse breakdown voltage of 5 V. The voltage source has a polarity that will tend to reverse-bias the diode. Because the source voltage exceeds the Zener voltage rating of the diode, $V_Z = 5$ V, we should expect the diode to be operating in its breakdown region.

3.12.1 LOAD-LINE ANALYSIS

The i-v characteristic for this Zener diode is given in Fig. 3.37, and load-line analysis can be used to find the Q-point for the diode, independent of the region of operation. The normal polarities for I_D and V_D are indicated in Fig. 3.36, and the loop equation is

$$-20 = V_D + 5000 I_D \tag{3.37}$$

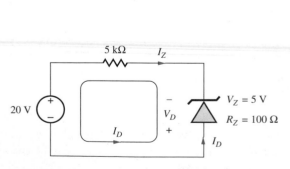

Figure 3.36 Circuit containing a Zener diode with $V_Z = 5$ V and $R_Z = 100$ Ω.

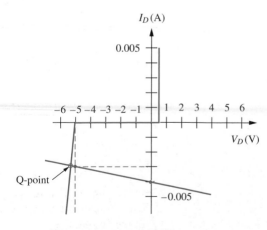

Figure 3.37 Load line for Zener diode.

In order to draw the load line, we choose two points on the graph:

$$V_D = 0, \; I_D = -4 \text{ mA} \qquad \text{and} \qquad V_D = -5 \text{ V}, \; I_D = -3 \text{ mA}$$

In this case the load line intersects the diode characteristic at a Q-point in the breakdown region: $(-2.9 \text{ mA}, -5.2 \text{ V})$.

3.12.2　ANALYSIS WITH THE PIECEWISE LINEAR MODEL

The assumption of reverse breakdown requires that the diode current I_D be less than zero or that the Zener current $I_Z = -I_D > 0$. We will analyze the circuit with the piecewise linear model and test this condition to see if it is consistent with the reverse-breakdown assumption.

In Fig. 3.38, the Zener diode has been replaced with its piecewise linear model from Fig. 3.16 in Sec. 3.6, with $V_Z = 5$ V and $R_Z = 100 \; \Omega$. Writing the loop equation this time in terms of I_Z:

$$20 - 5100 I_Z - 5 = 0 \qquad \text{or} \qquad I_Z = \frac{(20 - 5) \text{ V}}{5100 \; \Omega} = 2.94 \text{ mA} \qquad (3.38)$$

Because I_Z is greater than zero ($I_D < 0$), the solution is consistent with our assumption of Zener breakdown operation.

It is worth noting that diodes have three possible states when the breakdown region is included, further increasing analysis complexity.

3.12.3　VOLTAGE REGULATION

A useful application of the Zener diode is as a **voltage regulator,** as shown in the circuit of Fig. 3.39. The function of the Zener diode is to maintain a constant voltage across load resistor R_L. As long as the diode is operating in reverse breakdown, a voltage of approximately V_Z will appear across R_L. To ensure that the diode is operating in the Zener breakdown region, we must have $I_Z > 0$.

The circuit of Fig. 3.39 has been redrawn in Fig. 3.40 with the model for the Zener diode, with $R_Z = 0$. Using nodal analysis, the Zener current is expressed by $I_Z = I_S - I_L$. The currents I_S and

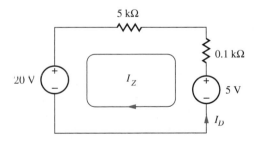

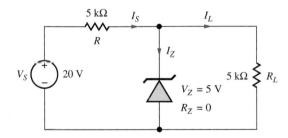

Figure 3.38 Circuit with piecewise linear model for Zener diode. Note that the diode model is valid only in the breakdown region of the characteristic.

Figure 3.39 Zener diode voltage regulator circuit.

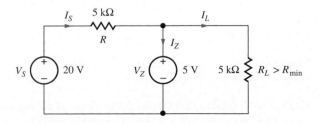

Figure 3.40 Circuit with a constant voltage model for the Zener diode.

I_L are equal to

$$I_S = \frac{V_S - V_Z}{R} = \frac{(20 - 5)\ \text{V}}{5\ \text{k}\Omega} = 3\ \text{mA} \qquad \text{and} \qquad I_L = \frac{V_Z}{R_L} = \frac{5\ \text{V}}{5\ \text{k}\Omega} = 1\ \text{mA} \qquad (3.39)$$

resulting in a Zener current $I_Z = 2$ mA. $I_Z > 0$, which is again consistent with our assumptions. If the calculated value of I_Z were less than zero, then the Zener diode no longer controls the voltage across R_L, and the voltage regulator is said to have "dropped out of regulation."

For proper regulation to take place, the Zener current must be positive,

$$I_Z = I_S - I_L = \frac{V_S}{R} - V_Z \left(\frac{1}{R} + \frac{1}{R_L} \right) > 0 \qquad (3.40)$$

Solving for R_L yields a lower bound on the value of load resistance for which the Zener diode will continue to act as a voltage regulator.

$$R_L > \frac{R}{\left(\dfrac{V_S}{V_Z} - 1 \right)} = R_{\min} \qquad (3.41)$$

EXERCISE: What is the value of $R_{\min}$ for the Zener voltage regulator circuit in Figs. 3.39 and 3.40? What is the output voltage for $R_L = 1\ \text{k}\Omega$? For $R_L = 2\ \text{k}\Omega$?

ANSWERS: 1.67 kΩ; 3.33 V; 5.00 V

3.12.4 ANALYSIS INCLUDING ZENER RESISTANCE

The voltage regulator circuit in Fig. 3.39 has been redrawn in Fig. 3.41 and now includes a nonzero Zener resistance R_Z. The output voltage is now a function of the current I_Z through the Zener diode. For small values of R_Z, however, the change in output voltage will be small.

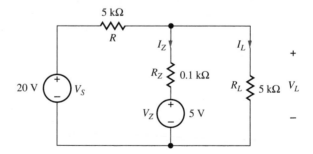

Figure 3.41 Zener diode regulator circuit, including Zener resistance.

EXAMPLE 3.9   **DC ANALYSIS OF A ZENER DIODE REGULATOR CIRCUIT**

Find the operating point for a Zener-diode-based voltage regulator circuit.

PROBLEM Find the output voltage and Zener diode current for the Zener diode regulator in Figs. 3.39 to 3.41 if $R_Z = 100\ \Omega$ and $V_Z = 5$ V.

SOLUTION **Known Information and Given Data:** Zener diode regulator circuit as modeled in Fig. 3.41 with $V_S = 20$ V, $R = 5$ kΩ, $R_Z = 0.1$ kΩ, and $V_Z = 5$ V

Unknowns: V_L, I_Z

Approach: The circuit contains a single unknown node voltage V_L, and a nodal equation can be written to find the voltage. Once V_L is found, I_Z can be determined using Ohm's law.

Assumptions: Use the piecewise linear model for the diode as drawn in Fig. 3.41.

Analysis: Writing the nodal equation for V_L yields

$$\frac{V_L - 20 \text{ V}}{5000 \ \Omega} + \frac{V_L - 5 \text{ V}}{100 \ \Omega} + \frac{V_L}{5000 \ \Omega} = 0$$

Multiplying the equation by 5000 Ω and collecting terms gives

$$52V_L = 270 \text{ V} \qquad \text{and} \qquad V_L = 5.19 \text{ V}$$

The Zener diode current is equal to

$$I_Z = \frac{V_L - 5 \text{ V}}{100 \ \Omega} = \frac{5.19 \text{ V} - 5 \text{ V}}{100 \ \Omega} = 1.90 \text{ mA} > 0$$

Check of Results: $I_Z > 0$ confirms operation in reverse breakdown. We see that the output voltage of the regulator is slightly higher than for the case with $R_Z = 0$, and the Zener diode current is reduced slightly. Both changes are consistent with the addition of R_Z to the circuit.

Computer-Aided Analysis: We can use SPICE to simulate the Zener circuit if we specify the breakdown voltage using SPICE parameters BV, IBV, and RS. BV sets the breakdown voltage, and IBV represents the current at breakdown. Setting BV = 5 V, and RS = 100 Ω and letting IBV default to 1 mA yields $V_L = 5.21$ V and $I_Z = 1.92$ mA, which agree well with our hand calculations. A transfer function analysis from V_S to V_L gives a yields a sensitivity of 21 mV/V and an output resistance of 108 Ω. The meaning of these numbers is in the next section.

EXERCISE: Find V_L, I_Z, and the Zener power dissipation in Fig. 3.41 if $R = 1$ kΩ.

ANSWERS: 6.25 V; 12.5 mA; 78.1 mW

3.12.5 LINE AND LOAD REGULATION

Two important parameters characterizing a voltage regulator circuit are **line regulation** and **load regulation.** Line regulation characterizes how sensitive the output voltage is to input voltage changes and is expressed as mV/V or as a percentage. Load regulation characterizes how sensitive the output voltage is to changes in the load current withdrawn from the regulator and has the units of Ohms.

$$\text{Line regulation} = \frac{dV_L}{dV_S} \qquad \text{and} \qquad \text{Load regulation} = \frac{dV_L}{dI_L} \qquad (3.42)$$

We can find expressions for these quantities from a straight forward analysis of the circuit in Fig. 3.41 similar to that in Ex. 3.9:

$$\frac{V_L - V_S}{R} + \frac{V_L - V_Z}{R_Z} + I_L = 0 \qquad (3.43)$$

For a fixed load current, we find the line regulation is

$$\text{Line regulation} = \frac{R_Z}{R + R_Z} \tag{3.44}$$

and for changes in I_L,

$$\text{Load regulation} = -(R_Z \| R) \tag{3.45}$$

The load regulation should be recognized as the Thévinen equivalent resistance looking back into the regulator from the load terminals.

> **EXERCISE:** What are the values of the load and line regulation for the circuit in Fig. 3.41?
>
> **ANSWERS:** 19.6 mV/V; 98.0 Ω. Note that these agree with the SPICE results in Ex. 3.9.

3.13 HALF-WAVE RECTIFIER CIRCUITS

Rectifiers represent an application of diodes that we encounter frequently every day, but they may not be recognized as such. The basic **rectifier circuit** converts an ac voltage to a pulsating dc voltage. A filter is then added to eliminate the ac components of the waveform and produce a nearly constant dc voltage output. Virtually every electronic device that is plugged into the wall utilizes a rectifier circuit to convert the 120-V, 60-Hz ac power line source to the various dc voltages required to operate electronic devices such as personal computers, audio systems, radio receivers, televisions, and the like. All of our battery chargers and "wall-warts" contain rectifiers. As a matter of fact, the vast majority of electronic circuits are powered by a dc source, usually based on some form of rectifier.

This section explores half-wave rectifier circuits with capacitor filters that form the basis for many dc power supplies. Up to this point, we have looked at only steady-state dc circuits in which the diode remained in one of its three possible states (on, off, or reverse breakdown). Now, however, the diode state will be changing with time, and a given piecewise linear model for the circuit will be valid for only a certain time interval.

3.13.1 HALF-WAVE RECTIFIER WITH RESISTOR LOAD

A single diode is used to form the **half-wave rectifier circuit** in Fig. 3.42. A sinusoidal voltage source $v_S = V_P \sin \omega t$ is connected to the series combination of diode D_1 and load resistor R. During the first half of the cycle, for which $v_S > 0$, the source forces a current through diode D_1 in the forward direction, and D_1 will be on. During the second half of the cycle, $v_S < 0$. Because a negative current cannot exist in the diode (unless it is in breakdown), it turns off. These two states are modeled in Fig. 3.43 using the ideal diode model.

When the diode is on, voltage source v_S is connected directly to the output and $v_O = v_S$. When the diode is off, the current in the resistor is zero, and the output voltage is zero. The input and

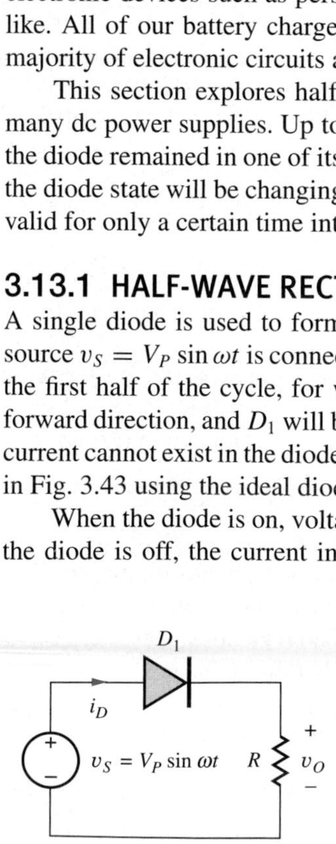

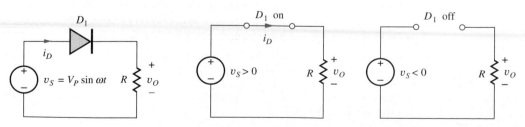

Figure 3.42 Half-wave rectifier circuit.

Figure 3.43 Ideal diode models for the two half-wave rectifier states.

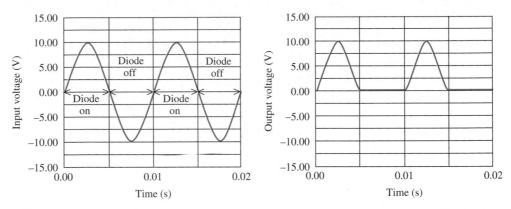

Figure 3.44 Sinusoidal input voltage v_S and pulsating dc output voltage v_O for the half-wave rectifier circuit.

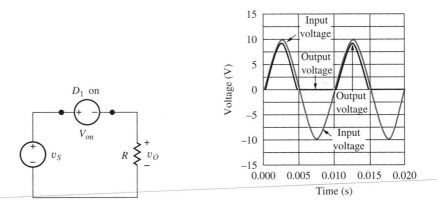

Figure 3.45 CVD model for the rectifier on state.

Figure 3.46 Half-wave rectifier output voltage with $V_P = 10$ V and $V_{on} = 0.7$ V.

output voltage waveforms are shown in Fig. 3.44(b), and the resulting current is called pulsating direct current. In this circuit, the diode is conducting 50 percent of the time and is off 50 percent of the time.

In some cases, the forward voltage drop across the diode can be important. Figure 3.45 shows the circuit model for the on-state using the CVD model. For this case, the output voltage is one diode-drop smaller than the input voltage during the conduction interval:

$$v_O = (V_P \sin \omega t) - V_{on} \qquad (3.46)$$

The output voltage remains zero during the off-state interval. The input and output waveforms for the half-wave rectifier, including the effect of V_{on}, are shown in Fig. 3.46 for $V_P = 10$ V and $V_{on} = 0.7$ V.

In many applications, a transformer is used to convert from the 120-V ac, 60-Hz voltage available from the power line to the desired ac voltage level, as in Fig. 3.47. The transformer can step the voltage up or down depending on the application; it also enhances safety by providing isolation from the power line. From circuit theory we know that the output of an ideal transformer can be represented by an ideal voltage source, and we use this knowledge to simplify the representation of subsequent rectifier circuit diagrams.

The unfiltered output of the half-wave rectifier in Fig. 3.42 or 3.47 is not suitable for operation of most electronic circuits because constant power supply voltages are required to establish proper bias for the electronic devices. A **filter capacitor** (or more complex circuit) can be added to filter the output of the circuit in Figs. 3.47 to remove the time-varying components from the waveform.

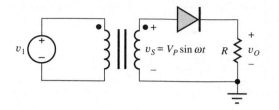

Figure 3.47 Transformer-driven half-wave rectifier.

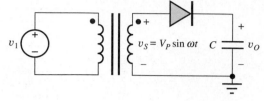

Figure 3.48 Rectifier with capacitor load (peak detector).

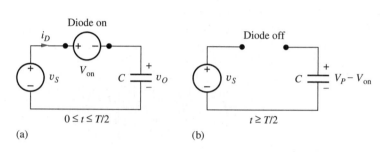

(a)   (b)

Figure 3.49 Peak-detector circuit models (constant voltage drop model). (a) The diode is on for $0 \leq t \leq T/2$. (b) The diode is off for $t > T/2$.

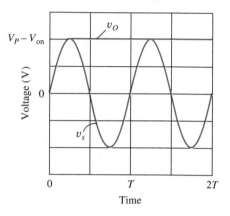

Figure 3.50 Input and output waveforms for the peak-detector circuit.

3.13.2 RECTIFIER FILTER CAPACITOR

To understand operation of the rectifier filter, we first consider the operation of the **peak-detector** circuit in Fig. 3.48. This circuit is similar to that in Fig. 3.47 except that the resistor is replaced with a capacitor C that is initially discharged [$v_O(0) = 0$].

Models for the circuit with the diode in the on and off states are in Fig. 3.49, and the input and output voltage waveforms associated with this circuit are in Fig. 3.50. As the input voltage starts to rise, the diode turns on and connects the capacitor to the source. The capacitor voltage equals the input voltage minus the voltage drop across the diode.

At the peak of the input voltage waveform, the current through the diode tries to reverse direction because $i_D = C[d(v_s - V_{on})/dt] < 0$, the diode cuts off, and the capacitor is disconnected from the rest of the circuit. There is no circuit path to discharge the capacitor, so the voltage on the capacitor remains constant. Because the amplitude of the input voltage source v_S can never exceed V_P, the capacitor remains disconnected from v_S for $t > T/2$. Thus, the capacitor in the circuit in Fig. 3.48 charges up to a voltage one diode-drop below the peak of the input waveform and then remains constant, thereby producing a dc output voltage

$$V_{dc} = V_P - V_{on} \qquad (3.47)$$

3.13.3 HALF-WAVE RECTIFIER WITH *RC* LOAD

To make use of this output voltage, a load must be connected to the circuit as represented by the resistor R in Fig. 3.51. Now there is a path available to discharge the capacitor during the time the diode is not conducting. Models for the conducting and nonconducting time intervals are shown in Fig. 3.52; the waveforms for the circuit are shown in Fig. 3.53. The capacitor is again assumed to be initially discharged. During the first quarter cycle, the diode conducts, and the capacitor is rapidly charged toward the peak value of the input voltage source. The diode cuts off at the peak of v_S, and

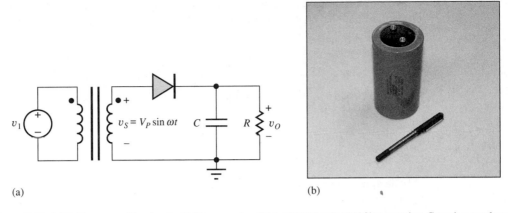

(a)   (b)

Figure 3.51 (a) Half-wave rectifier circuit with filter capacitor. (b) A-175,000-μF, 15-V filter capacitor. Capacitance tolerance is −10 percent, +75 percent.

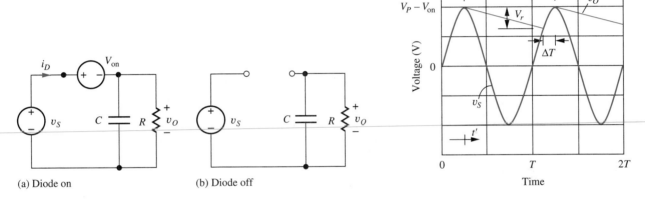

(a) Diode on   (b) Diode off

Figure 3.52 Half-wave rectifier circuit models.

Figure 3.53 Input and output voltage waveforms for the half-wave rectifier circuit.

the capacitor voltage then discharges exponentially through the resistor R, as governed by the circuit in Fig. 3.52(b). The discharge continues until the voltage $v_s - v_{on}$ exceeds the output voltage v_O, which occurs near the peak of the next cycle. The process is then repeated once every cycle.

3.13.4 RIPPLE VOLTAGE AND CONDUCTION INTERVAL

The output voltage is no longer constant as in the ideal peak-detector circuit but has a **ripple voltage** V_r. In addition, the diode only conducts for a short time ΔT during each cycle. This time ΔT is called the **conduction interval,** and its angular equivalent is the **conduction angle** θ_c. The variables ΔT and V_r are important values related to dc power supply design, and we will now develop expressions for these parameters.

During the discharge period, the voltage across the capacitor is described by

$$v_o(t') = (V_P - V_{on}) \exp\left(-\frac{t'}{RC}\right) \qquad \text{for } t' = \left(t - \frac{T}{4}\right) \geq 0 \tag{3.48}$$

We have referenced the t' time axis to $t = T/4$ to simplify the equation. The ripple voltage V_r is given by

$$V_r = (V_P - V_{on}) - v_o(t') = (V_P - V_{on})\left[1 - \exp\left(-\frac{T - \Delta T}{RC}\right)\right] \tag{3.49}$$

A small value of V_r is desired in most power supply designs; a small value requires RC to be much greater than $T - \Delta T$. Using $\exp(-x) \cong 1 - x$ for small x results in an approximate expression for the ripple voltage:

$$V_r \cong (V_P - V_{\text{on}}) \frac{T}{RC} \left(1 - \frac{\Delta T}{T} \right) \tag{3.50}$$

A small ripple voltage also requires $\Delta T \ll T$, and the final simplified expression for the ripple voltage becomes

$$V_r \cong \frac{(V_P - V_{\text{on}})}{R} \frac{T}{C} \tag{3.51}$$

The approximation of the exponential used in Eqs. (3.50) and (3.51) is equivalent to assuming that the capacitor is being discharged by a constant current so that the discharge waveform is a straight line. The ripple voltage V_R can be considered to be determined by an equivalent dc current equal to

$$I_{dc} = \frac{V_P - V_{\text{on}}}{R} \tag{3.52}$$

discharging the capacitor C for a time period T (that is, $\Delta V = (I_{dc}/C)\, T$).

An approximate expression can also be obtained for the conduction interval ΔT. At $t' = T - \Delta T$,

$$(V_P - V_{\text{on}}) \exp \left(-\frac{T - \Delta T}{RC} \right) = V_P \cos \omega (T - \Delta T) - V_{\text{on}} \tag{3.53}$$

Expanding the cosine term, recognizing that $\cos \omega T = 1$ and $\sin \omega T = 0$, and assuming that $\Delta T \ll T$ yields

$$(V_P - V_{\text{on}}) \left(1 - \frac{T}{RC} \right) = V_P \cos \omega \Delta T - V_{\text{on}} \tag{3.54}$$

For small $\omega \Delta T$, Eq. (3.54) can be reduced to

$$(V_P - V_{\text{on}}) \left(1 - \frac{T}{RC} \right) = V_P \left[1 - \frac{(\omega \Delta T)^2}{2} \right] - V_{\text{on}} \tag{3.55}$$

which can be solved for ΔT and simplified using Eq. (3.51):

$$\Delta T \cong \frac{1}{\omega} \sqrt{\frac{2T}{RC} \frac{(V_P - V_{\text{on}})}{V_P}} = \frac{1}{\omega} \sqrt{\frac{2V_r}{V_P}} \tag{3.56}$$

The conduction angle θ_c is given by

$$\theta_c = \omega \Delta T = \sqrt{\frac{2V_r}{V_P}} \tag{3.57}$$

EXAMPLE 3.10   HALF-WAVE RECTIFIER ANALYSIS

Here we see an illustration of numerical results for a half-wave rectifier with a capacitive filter.

PROBLEM Find the value of the dc output voltage, dc output current, ripple voltage, conduction interval, and conduction angle for a half-wave rectifier driven from a transformer having a secondary voltage of 12.6 V_{rms} (60 Hz) with $R = 15\ \Omega$ and $C = 25{,}000\ \mu\text{F}$. Assume the diode on-voltage $V_{\text{on}} = 1$ V.

SOLUTION **Known Information and Given Data:** Half-wave rectifier circuit with RC load as depicted in Fig. 3.51. Transformer secondary voltage is 12.6 V_{rms}, operating frequency is 60 Hz, $R = 15\ \Omega$, and $C = 25{,}000\ \mu F$.

Unknowns: dc output voltage V_{dc}, output current I_{dc}, ripple voltage V_r, conduction interval ΔT, conduction angle θ_C.

Approach: Given data can be used directly to evaluate Eqs. (3.47), (3.51), (3.56), and (3.57).

Assumptions: Diode on-voltage is 1 V. Remember that the derived results assume the ripple voltage is much less than the dc output voltage ($V_r \ll V_{dc}$) and the conduction interval is much less than the period of the ac signal ($\Delta T \ll T$).

Analysis: The ideal dc output voltage in the absence of ripple is given by Eq. (3.47):

$$V_{dc} = V_P - V_{on} = \left(12.6\sqrt{2} - 1\right)\ V = 16.8\ V$$

The nominal dc current delivered by the supply is

$$I_{dc} = \frac{V_P - V_{on}}{R} = \frac{16.8\ V}{15\ \Omega} = 1.12\ A$$

The ripple voltage is calculated using Eq. (3.51) with the discharge interval $T = 1/60$ s:

$$V_r \cong \frac{(V_P - V_{on})}{R}\frac{T}{C} = \frac{16.8\ V}{15\ \Omega}\frac{\frac{1}{60}\ s}{2.5 \times 10^{-2}\ F} = 0.747\ V$$

The conduction angle is calculated using Eq. (3.57)

$$\theta_c = \omega \Delta T = \sqrt{\frac{2V_r}{V_P}} = \sqrt{\frac{2 \cdot 0.75}{17.8}} = 0.290\ \text{rad or } 16.6°$$

and the conduction interval is

$$\Delta T = \frac{\theta_c}{\omega} = \frac{\theta_c}{2\pi f} = \frac{0.29}{120\pi} = 0.769\ mS$$

Check of Results: The ripple voltage represents 4.4 percent of the dc output voltage. Thus the assumption that the voltage is approximately constant is justified. The conduction time is 0.769 mS out of a total period $T = 16.7$ ms, and the assumption that $\Delta T \ll T$ is also satisfied.

Discussion: From this example, we see that even a 1-A power supply requires a significant filter capacitance C to maintain a low ripple percentage. In this case, $C = 0.025\ F = 25{,}000\ \mu F$.

EXERCISE: Find the value of the dc output voltage, dc output current, ripple voltage, conduction interval, and conduction angle for a half-wave rectifier that is being supplied from a transformer having a secondary voltage of 6.3 V_{rms} (60 Hz) with $R = 0.5\ \Omega$ and $C = 500{,}000\ \mu F$. Assume the diode on voltage $V_{on} = 1$ V.

ANSWERS: 7.91 V; 15.8 A; 0.527; 0.912 ms; 19.7°

EXERCISE: What are the values of the dc output voltage and dc output current for a half-wave rectifier that is being supplied from a transformer having a secondary voltage of 10 V$_{rms}$ (60 Hz) and a 2-Ω load resistor? Assume the diode on voltage $V_{on} = 1$ V. What value of filter capacitance is required to have a ripple voltage of no more than 0.1 V? What is the conduction angle?

ANSWERS: 13.1 V; 6.57 A; 1.10 F; 6.82°

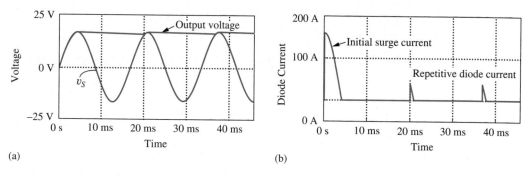

Figure 3.54 SPICE simulation of the half-wave rectifier circuit.

3.13.5 DIODE CURRENT

In rectifier circuits, a nonzero current is present in the diode for only a very small fraction of the period T, yet an almost constant dc current is flowing out of the filter capacitor to the load. The total charge lost from the capacitor during each cycle must be replenished by the current through the diode during the short conduction interval ΔT, which leads to high peak diode currents. Figure 3.54 shows the results of SPICE simulation of the diode current. The repetitive current pulse can be modeled approximately by a triangle of height I_P and width ΔT, as in Fig. 3.55.

Equating the charge supplied through the diode during the conduction interval to the charge lost from the filter capacitor during the complete period yields

$$Q = I_P \frac{\Delta T}{2} = I_{dc} T \qquad \text{or} \qquad I_P = I_{dc} \frac{2T}{\Delta T} \tag{3.58}$$

Here we remember that the integral of current over time represents charge Q. Therefore the charge supplied by the triangular current pulse in Fig. 3.56 is given by the area of the triangle, $I_P \Delta T / 2$.

For Ex. 3.10, the peak diode current would be

$$I_P = 1.12 \frac{2 \cdot 16.7}{0.769} = 48.6 \text{ A} \tag{3.59}$$

which agrees well with the simulation results in Fig. 3.55. The diode must be built to handle these high peak currents, which occur over and over. This high peak current is also the reason for the relatively large choice of V_{on} used in Ex. 3.10 (See Prob. 3.86.)

EXERCISE: (a) What is the forward voltage of a diode operating at a current of 48.6 A at 300 K if $I_S = 10^{-15}$ A? (b) At 50 C?

ANSWERS: 0.994 V; 1.07 V

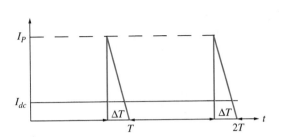

Figure 3.55 Triangular approximation to diode current pulse.

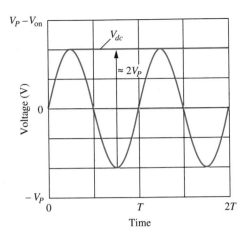

Figure 3.56 Peak reverse voltage across the diode in a half-wave rectifier.

3.13.6 SURGE CURRENT

When the power supply is first turned on, there can be an even larger current through the diode, as is visible in Fig. 3.54. During the first quarter cycle, the current through the diode is given approximately by

$$i_d(t) = i_c(t) \cong C \left[\frac{d}{dt} V_P \sin \omega t \right] = \omega C V_P \cos \omega t \qquad (3.60)$$

The peak value of this initial **surge current** occurs at $t = 0^+$ and is given by

$$I_{SC} = \omega C V_P = 2\pi (60 \text{ Hz})(0.025 \text{ F})(17.8 \text{ V}) = 168 \text{ A}$$

Using the numbers from Ex. 3.7 yields an initial surge current of almost 170 A! This value, again, agrees well with the simulation results in Fig. 3.54. If the input signal v_S does not happen to be crossing through zero when the power supply is turned on, the situation can be even worse, and rectifier diodes selected for power supply applications must be capable of withstanding very large surge currents as well as the large repetitive current pulses required each cycle.

In most practical circuits, the surge current will be large but cannot actually reach the values predicted by Eq. (3.60) because of series resistances in the circuit that we have neglected. The rectifier diode itself will have an internal series resistance (review the SPICE model in Sec. 3.9 for example), and the transformer will have resistances associated with both the primary and secondary windings. A total series resistance in the secondary of only a few tenths of an ohm will significantly reduce both the surge current and peak repetitive current in the circuit. In addition, the large time constant associated with the series resistance and filter capacitance causes the rectifier output to take many cycles to reach its steady state voltage. (See SPICE simulation problems at the end of this chapter.)

3.13.7 PEAK-INVERSE-VOLTAGE (PIV) RATING

We must also be concerned about the breakdown voltage rating of the diodes used in rectifier circuits. This is called the **peak-inverse-voltage (PIV)** rating of the rectifier diode. The worst-case situation for the half-wave rectifier is depicted in Fig. 3.56 in which it is assumed that the ripple voltage V_r is very small. When the diode is off, as in Fig. 3.52(b), the reverse bias across the diode is equal to $V_{dc} - v_S$. The worst case occurs when v_S reaches its negative peak of $-V_P$. The diode must therefore be able to withstand a reverse bias of at least

$$\text{PIV} \geq V_{dc} - v_S^{\min} = V_P - V_{on} - (-V_P) = 2V_P - V_{on} \cong 2V_P \qquad (3.61)$$

From Eq. (3.61), we see that diodes used in the half-wave rectifier circuit must have a PIV rating equal to twice the peak voltage supplied by the source v_S. The PIV value corresponds to the minimum value of Zener breakdown voltage for the rectifier diode. A safety margin of at least 25 to 50 percent is usually specified for the diode PIV rating in power supply designs.

3.13.8 DIODE POWER DISSIPATION

In high-current power supply applications, the power dissipation in the rectifier diodes can become significant. The average power dissipation in the diode is defined by

$$P_D = \frac{1}{T} \int_0^T v_D(t) i_D(t) \, dt \tag{3.62}$$

This expression can be simplified by assuming that the voltage across the diode is approximately constant at $v_D(t) = V_{on}$ and by using the triangular approximation to the diode current $i_D(t)$ shown in Fig. 3.55. Eq. (3.62) becomes

$$P_D = \frac{1}{T} \int_0^T V_{on} i_D(t) \, dt = \frac{V_{on}}{T} \int_{T-\Delta T}^T i_D(t) \, dt = V_{on} \frac{I_P}{2} \frac{\Delta T}{T} = V_{on} I_{dc} \tag{3.63}$$

Using Eq. (3.58) we see that the power dissipation is equivalent to the constant dc output current multiplied by the on-voltage of the diode. For the half-wave rectifier example, $P_D = (1 \text{ V})(1.1 \text{ A}) = 1.1$ W. This rectifier diode would probably need a heat sink to maintain its temperature at a reasonable level.

Another source of power dissipation is caused by resistive loss within the diode. Diodes have a small internal series resistance R_S, and the average power dissipation in this resistance can be calculated using

$$P_D = \frac{1}{T} \int_0^T i_D^2(t) R_S \, dt \tag{3.64}$$

Evaluation of this integral (left for Prob. 3.89) for the triangular current wave form in Fig. 3.55 yields

$$P_D = \frac{1}{3} I_P^2 R_S \frac{\Delta T}{T} = \frac{4}{3} \frac{T}{\Delta T} I_{dc}^2 R_S \tag{3.65}$$

Using the number from the rectifier example with $R_S = 0.20 \text{ }\Omega$ yields $P_D = 7.3$ W! This is significantly greater than the component of power dissipation caused by the diode on-voltage calculated using Eq. (3.63). The component of power dissipation described by Eq. (3.65) can be reduced by minimizing the peak current I_P through the use of the minimum required size of filter capacitor or by using the full-wave rectifier circuits, which are discussed in Sec. 3.14.

3.13.9 HALF-WAVE RECTIFIER WITH NEGATIVE OUTPUT VOLTAGE

The circuit of Fig. 3.51 can also be used to produce a negative output voltage if the top rather than the bottom of the capacitor is grounded, as depicted in Fig. 3.57(a). However, we usually draw the circuit as in Fig. 3.57(b). These two circuits are equivalent. In the circuit in Fig. 3.57(b), the diode conducts on the negative half cycle of the transformer voltage v_S, and the dc output voltage is $V_{dc} = -(V_P - V_{on})$.

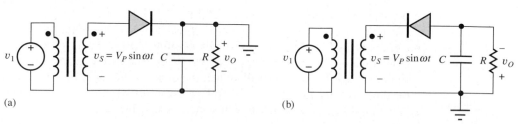

(a)   (b)

Figure 3.57 Half-wave rectifier circuits that develop negative output voltages.

ELECTRONICS IN ACTION

Power Cubes and Cell Phone Chargers

We actually encounter the unfiltered transformer driven half-wave rectifier circuit depicted in Fig. 3.47 frequently in our everyday lives in the form of "power cubes" and battery chargers for many portable electronic devices. An example is shown in the accompanying figure. The power cube contains only a small transformer and rectifier diode. The transformer is wound with small wire and has a significant resistance in both the primary and secondary windings. In the transformer in the photograph, the primary resistance is 600 Ω and the secondary resistance is 15 Ω, and these resistances actually help provide protection from failure of the transformer windings. Load resistance R in Fig. 3.47 represents the actual electronic device that is receiving power from the power cube and may often be a rechargable battery. In some cases, a filter capacitor may be included as part of the circuit that forms the load for the power cube.

The figure below shows a much more complex device used for recharging the batteries in a cell phone. The simplified schematic in part (c) of the figure utilizes a full-wave bridge rectifier with filter capacitor connected directly to the ac line. The rectifier's high voltage output is filtered by capacitor C_1 and feeds a switching regulator consisting of a switch, the transformer driving a half-wave rectifier with pi-filter (D_5, C_2, L, and C_3), and a feedback circuit that controls the output voltage by modulating the duty cycle of the switch. The transformer steps down the voltage and provides isolation from the high voltage ac line input. Diode D_6 and R clamp the inductor voltage when the switch opens. The feedback signal path is isolated from the input using an optical isolator. (See Electronics in Action in Chapter 5 for discussion of an optical isolator.) Note the wide range of input voltages accomodated by the circuit.

(a)

(b)

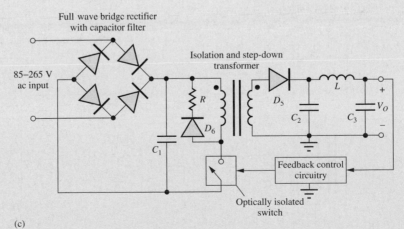

(c)

(a) Inside a simple power cube (b) Cell phone charger (c) Simplified schematic for the cell-phone charger.

3.14 FULL-WAVE RECTIFIER CIRCUITS

Full-wave rectifier circuits cut the capacitor discharge time in half and offer the advantage of requiring only one-half the filter capacitance to achieve a given ripple voltage. The full-wave rectifier circuit in Fig. 3.58 uses a **center-tapped transformer** to generate two voltages that have equal amplitudes but are 180 degrees out of phase. With the voltage v_S applied to the anode of D_1, and $-v_S$ applied to the anode of D_2, the two diodes form a pair of half-wave rectifiers operating on alternate half cycles of the input waveform. Proper phasing is indicated by the dots on the two halves of the transformer.

For $v_S > 0$, D_1 will be functioning as a half-wave rectifier, and D_2 will be off, as indicated in Fig. 3.59. The current exits the upper terminal of the transformer, goes through diode D_1, through the RC load, and returns back into the center tap of the transformer.

For $v_S < 0$, D_1 will be off, and D_2 will be functioning as a half-wave rectifier as indicated in Fig. 3.60. During this portion of the cycle, the current path leaves the bottom terminal of the transformer, goes through D_2, down through the RC load, and again returns into the transformer center tap. The current direction in the load is the same during both halves of the cycle; one-half of the transformer is utilized during each half cycle.

The load, consisting of the filter capacitor C and load resistor R, now receives two current pulses per cycle, and the capacitor discharge time is reduced to less than $T/2$, as indicated in the graph in Fig. 3.61. An analysis similar to that for the half-wave rectifier yields the same formulas for dc output voltage, ripple voltage, and ΔT, except that the discharge interval is $T/2$ rather than T. For a given capacitor value, the ripple voltage is one-half as large, and the conduction interval and peak current are reduced. The peak-inverse-voltage waveform for each diode is similar to the one shown

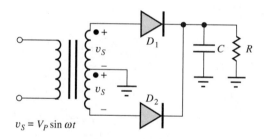

$v_S = V_P \sin \omega t$

Figure 3.58 Full-wave rectifier circuit using two diodes and a center-tapped transformer. This circuit produces a positive output voltage.

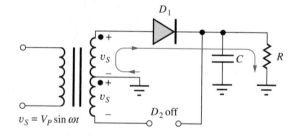

$v_S = V_P \sin \omega t$

Figure 3.59 Equivalent circuit for $v_S > 0$.

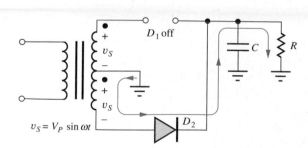

$v_S = V_P \sin \omega t$

Figure 3.60 Equivalent circuit for $v_S < 0$.

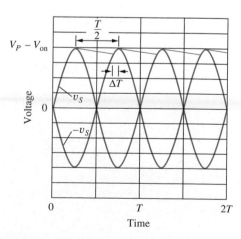

Figure 3.61 Voltage waveforms for the full-wave rectifier.

in Fig. 3.56 for the half-wave rectifier, with the result that the PIV rating of each diode is the same as in the half-wave rectifier. These results are summarized in Eqs. (3.66) to (3.70) for $v_S = V_P \sin \omega t$:

Full-wave Rectifier Equations:

$$V_{dc} = V_P - V_{on} \tag{3.66}$$

$$V_r = \frac{(V_P - V_{on})}{R} \frac{T}{2C} \tag{3.67}$$

$$\Delta T = \frac{1}{\omega} \sqrt{\frac{T}{RC} \frac{(V_P - V_{on})}{V_P}} = \frac{1}{\omega} \sqrt{\frac{2V_r}{V_P}} \tag{3.68}$$

$$\theta_c = \omega \Delta T = \sqrt{\frac{2V_r}{V_P}} \qquad I_P = I_{DC} \frac{T}{\Delta T} \tag{3.69}$$

$$\text{PIV} = 2V_P \tag{3.70}$$

3.14.1 FULL-WAVE RECTIFIER WITH NEGATIVE OUTPUT VOLTAGE

By reversing the polarity of the diodes, as in Fig. 3.62, a full-wave rectifier circuit with a negative output voltage is realized. Other aspects of the circuit remain the same as the previous full-wave rectifiers with positive output voltages.

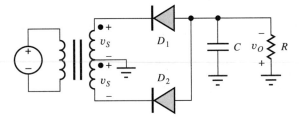

Figure 3.62 Full-wave rectifier with negative output voltage.

3.15 FULL-WAVE BRIDGE RECTIFICATION

The requirement for a center-tapped transformer in the full-wave rectifier can be eliminated through the use of two additional diodes in the **full-wave bridge rectifier circuit** configuration shown in Fig. 3.63. For $v_S > 0$, D_2 and D_4 will be on and D_1 and D_3 will be off, as indicated in Fig. 3.64. Current exits the top of the transformer, goes through D_2 into the RC load, and returns to the transformer through D_4. The full transformer voltage, now minus two diode voltage drops, appears across the load capacitor yielding a dc output voltage

$$V_{dc} = V_P - 2V_{on} \tag{3.71}$$

The peak voltage at node 1, which represents the maximum reverse voltage appearing across D_1, is equal to $(V_P - V_{on})$. Similarly, the peak reverse voltage across diode D_4 is $(V_P - 2V_{on}) - (-V_{on}) = (V_P - V_{on})$.

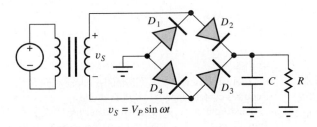

Figure 3.63 Full-wave bridge rectifier circuit with positive output voltage.

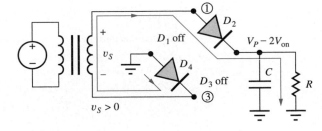

Figure 3.64 Full-wave bridge rectifier circuit for $v_S > 0$.

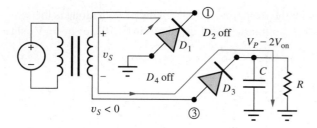

Figure 3.65 Full-wave bridge rectifier circuit for $v_S < 0$.

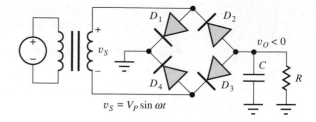

Figure 3.66 Full-wave bridge rectifier circuit with $v_O < 0$.

For $v_S < 0$, D_1 and D_3 will be on and D_2 and D_4 will be off, as depicted in Fig. 3.65. Current leaves the bottom of the transformer, goes through D_3 into the RC load, and back through D_1 to the transformer. The full transformer voltage is again being utilized. The peak voltage at node 3 is now equal to $(V_P - V_{on})$ and is the maximum reverse voltage appearing across D_4. Similarly, the peak reverse voltage across diode D_2 is $(V_P - 2V_{on}) - (-V_{on}) = (V_P - V_{on})$.

From the analysis of the two half cycles, we see that each diode must have a PIV rating given by

$$\text{PIV} = V_P - V_{on} \cong V_P \tag{3.72}$$

As with the previous rectifier circuits, a negative output voltage can be generated by reversing the direction of the diodes, as in the circuit in Fig. 3.66.

3.16 RECTIFIER COMPARISON AND DESIGN TRADEOFFS

Table 3.5 summarizes the characteristics of the half-wave, full-wave, and full-wave bridge rectifiers introduced in Secs. 3.13 to 3.15. The filter capacitor often represents a significant economic factor in terms of cost, size, and weight in the design of **rectifier circuits.** For a given ripple voltage, the value of the filter capacitor required in the full-wave rectifier is one-half that for the half-wave rectifier.

The reduction in peak current in the full-wave rectifier can significantly reduce heat dissipation in the diodes. The addition of the second diode and the use of a center-tapped transformer represent additional expenses that offset some of the advantage. However, the benefits of full-wave rectification usually outweigh the minor increase in circuit complexity.

The bridge rectifier eliminates the need for the center-tapped transformer and the PIV rating of the diodes is reduced, which can be particularly important in high-voltage circuits. The cost of the extra diodes is usually negligible, particularly since four-diode bridge rectifiers can be purchased in single-component form.

TABLE 3.5
Comparison of Rectifiers with Capacitive Filters

RECTIFIER PARAMETER	HALF-WAVE RECTIFIER	FULL-WAVE RECTIFIER	FULL-WAVE BRIDGE RECTIFIER
Filter capacitor	$C = \dfrac{V_P - V_{on}}{V_r}\dfrac{T}{R}$	$C = \dfrac{V_P - V_{on}}{V_r}\dfrac{T}{2R}$	$C = \dfrac{V_P - 2V_{on}}{V_r}\dfrac{T}{2R}$
PIV rating	$2V_P$	$2V_P$	V_P
Peak diode current (constant V_r)	Highest I_P	Reduced $\dfrac{I_P}{2}$	Reduced $\dfrac{I_P}{2}$
Comments	Least complexity	Smaller capacitor Requires center-tapped transformer Two diodes	Smaller capacitor Four diodes No center tap on transformer

DESIGN EXAMPLE 3.11

RECTIFIER DESIGN

Now we will use our rectifier theory to design a rectifier circuit that will provide a specified output voltage and ripple voltage.

PROBLEM Design a rectifier to provide a dc output voltage of 15 V with no more than 1 percent ripple at a load current of 2 A.

SOLUTION **Known Information and Given Data:** $V_{dc} = 15$ V, $V_r < 0.15$ V, $I_{dc} = 2$ A

Unknowns: Circuit topology, transformer voltage, filter capacitor, diode PIV rating, diode repetitive current rating, diode surge current rating.

Approach: Use given data to evaluate rectifier circuit equations. Let us choose a full-wave bridge topology that requires a smaller value of filter capacitance, a smaller diode PIV voltage, and no center tap in the transformer.

Assumptions: Assume diode on-voltage is 1 V. The ripple voltage is much less than the dc output voltage ($V_r \ll V_{dc}$), and the conduction interval should be much less than the period of the ac signal ($\Delta T \ll T$).

Analysis: The required transformer voltage is

$$V = \frac{V_P}{\sqrt{2}} = \frac{V_{dc} + 2V_{on}}{\sqrt{2}} = \frac{15 + 2}{\sqrt{2}} \text{ V} = 12.0 \text{ V}_{rms}$$

The filter capacitor is found using the ripple voltage, output current, and discharge interval:

$$C = I_{dc} \left(\frac{T/2}{V_r} \right) = 2 \text{ A} \left(\frac{1}{120} \text{ s} \right) \left(\frac{1}{0.15 \text{ V}} \right) = 0.111 \text{ F}$$

To find I_P, the conduction time is calculated using Eq. (3.56)

$$\Delta T = \frac{1}{\omega} \sqrt{\frac{2V_r}{V_P}} = \frac{1}{120\pi} \sqrt{\frac{2(0.15) \text{ V}}{17 \text{ V}}} = 0.352 \text{ ms}$$

and the peak repetitive current is found to be

$$I_P = I_{dc} \left(\frac{2}{\Delta T} \right) \left(\frac{T}{2} \right) = 2 \text{ A} \frac{(1/60) \text{ s}}{0.352 \text{ ms}} = 94.7 \text{ A}$$

The surge current estimate is

$$I_{surge} = \omega C V_P = 120\pi (0.111)(17) = 711 \text{ A}$$

The minimum diode PIV is $V_P = 17$ V. A choice with a safety margin would be PIV > 20 V. The repetitive current rating should be 95 A with a surge current rating of 710 A. Note that both of these calculations overestimate the magnitude of the currents because we have neglected series resistance of the transformer and diode. The minimum filter capacitor needs to be 111,000 μF. Assuming a tolerance of −30 percent, a nominal filter capacitance of 160,000 μF would be required.

Check of Results: The ripple voltage is designed to be 1 percent of the dc output voltage. Thus the assumption that the voltage is approximately constant is justified. The conduction time is 0.352 mS out of a total period $T = 16.7$ mS. Thus the assumption that $\Delta T \ll T$ is satisfied.

Computer-Aided Analysis: This design example represents an excellent place where simulation can be used to explore the magnitude of the diode currents and improve the design so that we don't over specify the rectifier diodes. A SPICE simulation with $R_S = 0.1$ Ω, $n = 2$, $I_S = 1$ μA, and a transformer series resistance of 0.1 Ω yields a number of unexpected results: $I_P = 11$ A, $I_{surge} = 70$ A, and $V_{dc} = 13$ V! The surge current and peak repetitive current are both reduced by almost an order of magnitude compared to our hand calculations! In addition the output voltage is lower than expected. If we think further, a peak current of 11 A will cause a peak voltage drop of 2.2 V across the total series resistance of 0.2 Ω, so it should not be surprising that the output voltage is 2 V lower than originally expected. The series resistances actually help to reduce the stress on the diodes. The time constant of the series resistance and the filter capacitor is 0.44 s, so the circuit takes many cycles to reach the steady state output voltage.

EXERCISE: Repeat the rectifier design assuming the use of a half-wave rectifier.

ANSWERS: $V = 11.3$ V$_{rms}$; $C = 222,000$ μF; $I_P = 184$ A; $I_{SC} = 1340$ A

ELECTRONICS IN ACTION

Three-Terminal IC Voltage Regulators

It is not easy to produce precise output voltages with rectifier circuits, particularly with changing load currents. Specially wound transformers may be required to produce the desired output voltages, and extremely large filter capacitances are required to reduce the output ripple voltage to very small values. A much better approach is to use integrated circuit voltage regulators to set the output voltage and remove the ripple. The IC regulators are available with a wide range of fixed output voltages as well as adjustable output versions.

An example of a rectifier circuit with a three-terminal 5-V regulator is shown in the accompanying figure. The regulator uses feedback with high-gain amplifier circuitry to greatly reduce the ripple voltage at the output. (We will study op amps and feedback amplifiers in detail in Part III of this text.) **IC voltage regulators** also provide outstanding line and load regulation, maintaining a constant output voltage even though the output current may change by many orders of magnitude. Capacitor C is the normal rectifier filter capacitor, and C_{B1} and C_{B2} (typically 0.001–0.01 μf) are bypass capacitors that provide a low-impedance path for high-frequency signals and are needed to ensure proper operation of the voltage regulator.

The regulator can reduce the ripple voltage by a factor of 100 to 1000 or more. To minimize power dissipation in the regulator, the rectifier can be designed with a relatively large ripple voltage at the input to the regulator, thus reducing the average input voltage to the regulator. The main design constraint is set by the input-output voltage differential V_{REG} across the regulator, which must not fall below a minimum "dropout voltage" value specified for the regulator, typically a few volts. The current I_{REG} needed to operate the IC regulator is only a few mA and typically represents a small percentage of the total current supplied by the rectifier: $I_S = I_L + I_{REG}$. An example of a voltage regulator family can be found on the MCD website.

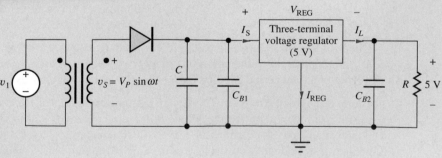

Half-wave rectifier and three-terminal IC voltage regulator.

3.17 DYNAMIC SWITCHING BEHAVIOR OF THE DIODE

Up to this point, we have tacitly assumed that diodes can turn on and off instantaneously. However, an unusual phenomenon characterizes the dynamic switching behavior of the *pn* junction diode. SPICE simulation is used to illustrate the switching of the diode in the circuit in Fig. 3.67, in which diode D_1 is being driven from voltage source v_1 through resistor R_1.

The source is zero for $t < 0$. At $t = 0$, the source voltage rapidly switches to $+1.5$ V, forcing a current into the diode to turn it on. The voltage remains constant until $t = 7.5$ ns. At this point the source switches to -1.5 V in order to turn the diode back off.

The simulation results are presented in Fig. 3.68. Following the voltage source change at $t = 0+$, the current increases rapidly. The internal capacitance of the diode prevents the diode voltage from changing instantaneously. The current actually overshoots its final value and then decreases as the diode turns on and the diode voltage increases to approximately 0.7 V. At any given time, the current flowing into the diode is given by

$$i_D(t) = \frac{v_1(t) - v_D(t)}{0.75 \text{ k}\Omega} \tag{3.73}$$

The initial peak of the current value occurs when v_1 reaches 1.5 V and v_D is still nearly zero:

$$i_{D\text{max}} = \frac{1.5 \text{ V}}{0.75 \text{ k}\Omega} = 2.0 \text{ mA} \tag{3.74}$$

After the diode voltage reaches its final value with $V_{\text{on}} \approx 0.7$ V, the current stabilizes at a forward current I_F of

$$I_F = \frac{1.5 - 0.7}{0.75 \text{ k}\Omega} = 1.1 \text{ mA} \tag{3.75}$$

At $t = 7.5$ ns, the input source rapidly changes polarity to -1.5 V. The diode current also rapidly reverses direction and is much greater than the reverse saturation current of the diode. The diode does

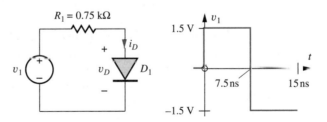

Figure 3.67 Circuit used to explore diode-switching behavior.

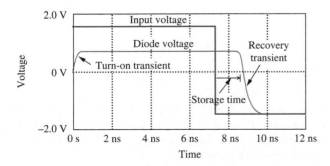
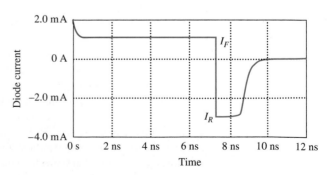

Figure 3.68 SPICE simulation results for the diode circuit in Fig. 3.67. (The diode transit time is equal to 5 ns.)

not turn off immediately. In fact, the diode actually remains forward-biased by the charge stored in the diode, with $v_D = V_{on}$, even though the current has changed direction! The reverse current I_R is equal to

$$I_R = \frac{-1.5 - 0.7}{0.75 \text{ k}\Omega} = -2.9 \text{ mA} \tag{3.76}$$

The current remains at -2.9 mA for a period of time called the diode **storage time τ_S**, during which the internal charge stored in the diode is removed. Once the stored charge has been removed, the voltage across the diode begins to drop and charges toward the final value of -1.5 V. The current in the diode drops rapidly to zero as the diode voltage begins to fall.

The turn-on time and recovery time are determined primarily by the charging and discharging of the nonlinear depletion-layer capacitance C_j through the resistance R_S. The storage time is determined by the diode transit time defined in Eq. (3.22) and by the values of the forward and reverse currents I_F and I_R:

$$\tau_S = \tau_T \ln\left[1 - \frac{I_F}{I_R}\right] \quad \text{and} \quad \tau_S = 5 \ln\left[1 - \frac{1.1 \text{ mA}}{-2.9 \text{ mA}}\right] \text{ ns} = 1.6 \text{ ns} \tag{3.77}$$

The SPICE simulation in results Fig. 3.68 agree well with this value.

Always remember that solid-state devices do not turn off instantaneously. The unusual storage time behavior of the diode is an excellent example of the switching delays that occur in *pn* junction devices in which carrier flow is dominated by the minority-carrier diffusion process. This behavior is not present in field-effect transistors, in which current flow is dominated by majority-carrier drift.

3.18 PHOTO DIODES, SOLAR CELLS, AND LIGHT-EMITTING DIODES

Several other important applications of diodes include photo detectors in communication systems, solar cells for generating electric power, and light-emitting diodes (LEDs). These applications all rely on the solid-state diode's ability to interact with optical photons.

3.18.1 PHOTO DIODES AND PHOTODETECTORS

If the depletion region of a *pn* junction diode is illuminated with light of sufficiently high frequency, the photons can provide enough energy to cause electrons to jump the semiconductor bandgap, creating electron–hole pairs. For photon absorption to occur, the incident photons must have an energy E_p that exceeds the bandgap of the semiconductor:

$$E_p = h\nu = \frac{hc}{\lambda} \geq E_G \tag{3.78}$$

where h = Planck's constant (6.626×10^{-34} J $\cdot$ s) λ = wavelength of optical illumination
       ν = frequency of optical illumination c = velocity of light (3×10^8 m/s)

The *i*-*v* characteristic of a diode with and without illumination is shown in Fig. 3.69. The original diode characteristic is shifted vertically downward by the photon-generated current. Photon absorption creates an additional current crossing the *pn* junction that can be modeled by a current source i_{PH} in parallel with the *pn* junction diode, as shown in Fig. 3.70.

Based on this model, we see that the incident optical signal can be converted to an electrical signal voltage using the simple **photodetector circuit** in Fig. 3.71. The diode is reverse-biased to enhance the width and electric field in the depletion region. The photon-generated current i_{PH} will flow through resistor R and produce an output signal voltage given by

$$v_o = i_{PH}R \tag{3.79}$$

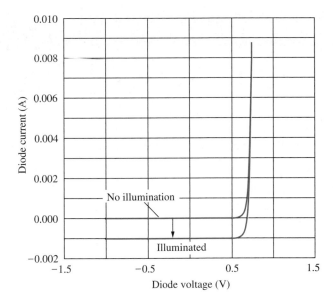

Figure 3.69 Diode i-v characteristic with and without optical illumination.

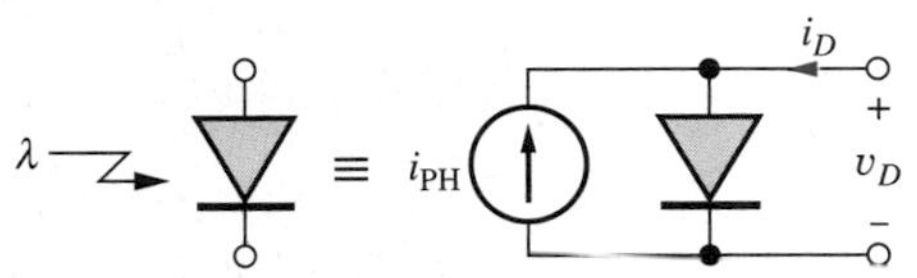

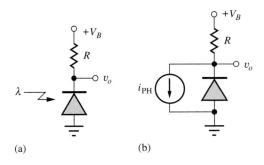

Figure 3.70 Model for optically illuminated diode. i_{PH} represents the current generated by absorption of photons in the vicinity of the pn junction.

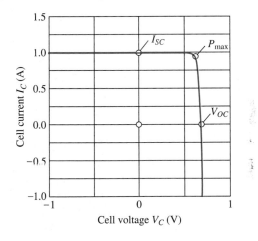

(a)   (b)

Figure 3.71 Basic photodetector circuit (a) and model (b).

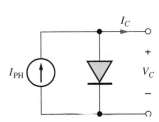

Figure 3.72 pn Diode under steady-state illumination as a solar cell.

Figure 3.73 Terminal characteristics for a pn junction solar cell.

In optical fiber communication systems, the amplitude of the incident light is modulated by rapidly changing digital data, and i_{PH} is a time-varying signal. The time-varying signal voltage at v_o is fed to additional electronic circuits to demodulate the signal and recover the original data that were transmitted down the optical fiber.

3.18.2 POWER GENERATION FROM SOLAR CELLS

In **solar cell** applications, the optical illumination is constant, and a dc current I_{PH} is generated. The goal is to extract power from the cell, and the i-v characteristics of solar cells are usually plotted in terms of the cell current I_C and cell voltage V_C, as defined in Fig. 3.72.

The i-v characteristic of the pn junction used for solar cell applications is plotted in terms of these terminal variables in Fig. 3.73. Also indicated on the graph are the short-circuit current I_{SC},

Solar Power for the Home

The following photo shows Auburn University's entry in the 2002 Solar Decathlon competition sponsored by the US Dept. of Energy and its private-sector partners BP Solar, The Home Depot, Electronic Data Systems (EDS), and the American Institute of Architects. Solar energy represents a clean and renewable source of power that significantly reduces pollutant emissions. For the competition, the solar energy available within the footprint of the house had to supply the total energy requirements for an entire home. The solar array on top of the house consists of 36 panels, connected as eighteen parallel strings of two panels each. Each solar panel (BP3160) is a series connection of 72 polycrystalline-silicon solar cells that can be represented by the simple model in Figs. 3.72 and 3.73, and each panel is specified to have an open-circuit voltage of 44.2 V and a short-circuit current of 4.8 A. The complete array produces a maximum power of 5.74 kW at an output voltage of 70 V and a current of 82 A. The solar cells charge batteries that drive ac inverters to supply 110/220-V 60-Hz power to the house. Note that two separate solar water heating panels are also visible on the roof of the building in the photograph.

Auburn University's award-winning entry in the 2002 Solar Decathlon.

the open-circuit voltage V_{OC}, and the maximum power point P_{max}. I_{SC} represents the maximum current available from the cell, and V_{OC} is the voltage across the open-circuited cell when all the photo current is flowing into the internal pn junction. For the solar cell to supply power to an external circuit, the product $I_C \times V_C$ must be positive, corresponding to the first quadrant of the characteristic. An attempt is made to operate the cell near the point of maximum output power P_{max}.

3.18.3 LIGHT-EMITTING DIODES (LEDS)

Light-emitting diodes, or **LEDs,** rely on the annihilation of electrons and holes through recombination rather than on the generation of carriers, as in the case of the photo diode. When a hole and electron recombine, an energy equal to the bandgap of the semiconductor can be released in the form of a photon. This recombination process is present in the forward-biased pn junction diode. In silicon, the recombination process actually involves the interaction of photons and lattice vibrations called phonons. The optical emission process in silicon is not nearly as efficient as that in the III–V compound semiconductor GaAs or the ternary materials such as $GaIn_{1-x}As_x$ and $GaIn_{1-x}P_x$. LEDs in these compound semiconductor materials provide visible illumination, and the color of the output can be controlled by varying the fraction x of arsenic or phosphorus in the material.

SUMMARY

In this chapter we investigated the detailed behavior of the solid-state diode.

- A *pn* junction diode is formed when *p*-type and *n*-type semiconductor regions are formed in intimate contact with each other. In the *pn* diode, large concentration gradients exist in the vicinity of the metallurgical junction, giving rise to large electron and hole diffusion currents.

- Under zero bias, no current can exist at the diode terminals, and a space charge region forms in the vicinity of the *pn* junction. The region of space charge results in both a built-in potential and an internal electric field, and the internal electric field produces electron and hole drift currents that exactly cancel the corresponding components of diffusion current.

- When a voltage is applied to the diode, the balance in the junction region is disturbed, and the diode conducts a current. The resulting i-v characteristics of the diode are accurately modeled by the diode equation:

$$i_D = I_S \left[\exp\left(\frac{v_D}{nV_T}\right) - 1 \right]$$

  where I_S = reverse saturation current of the diode

  n = nonideality factor (approximately 1)

  $V_T = kT/q$ = thermal voltage (0.025 V at room temperature)

- Under reverse bias, the diode current equals $-I_S$, a very small current.

- For forward bias, however, large currents are possible, and the diode presents an almost constant voltage drop of 0.6 to 0.7 V.

- At room temperature, an order of magnitude change in diode current requires a change of less than 60 mV in the diode voltage. At room temperature, the silicon diode voltage exhibits a temperature coefficient of approximately −1.8 mV/°C.

- One must also be aware of the reverse-breakdown phenomenon that is not included in the diode equation. If too large a reverse voltage is applied to the diode, the internal electric field becomes so large that the diode enters the breakdown region, either through Zener breakdown or avalanche breakdown. In the breakdown region, the diode again represents an almost fixed voltage drop, and the current must be limited by the external circuit or the diode can easily be destroyed.

- Diodes called Zener diodes are designed to operate in breakdown and can be used in simple voltage regulator circuits. Line regulation and load regulation characterize the change in output voltage of a power supply due to changes in input voltage and output current, respectively.

- As the voltage across the diode changes, the charge stored in the vicinity of the space charge region of the diode changes, and a complete diode model must include a capacitance. Under reverse bias, the capacitance varies inversely with the square root of the applied voltage. Under forward bias, the capacitance is proportional to the operating current and the diode transit time. These capacitances prevent the diode from turning on and off instantaneously and cause a storage time delay during turn-off.

- Direct use of the nonlinear diode equation in circuit calculations usually requires iterative numeric techniques. Several methods for simplifying the analysis of diode circuits were discussed, including the graphical load-line method and use of the ideal diode and constant voltage drop models.

- SPICE circuit analysis programs include a comprehensive built-in model for the diode that accurately reproduces both the ideal and nonideal characteristics of the diode and is useful for exploring the detailed behavior of circuits containing diodes.

- Important applications of diodes include half-wave, full-wave, and full-wave bridge rectifier circuits used to convert from ac to dc voltages in power supplies. Simple power supply circuits use capacitive filters, and the design of the filter capacitor determines power supply ripple voltage and diode conduction angle. Diodes used as rectifiers in power supplies must be able to withstand large peak repetitive currents as well as surge currents when the power supplies are first turned on. The reverse-breakdown voltage of rectifier diodes is referred to as the peak-inverse-voltage, or PIV, rating of the diode.

- Real diodes cannot turn on or off instantaneously because the internal capacitances of the diodes must be charged and discharged. The turn-on time is usually quite short, but diodes that have been conducting turn off much less abruptly. It takes time to remove stored charge within the diode, and this time delay is characterized by storage time τ_s. During the storage time, it is possible for large reverse currents to occur in the diode.

- Finally, the ability of the *pn* junction device to generate and detect light was discussed, and the basic characteristics of photo diodes, solar cells, and light-emitting diodes were presented.

KEY TERMS

Anode	Junction potential
Avalanche breakdown	Light-emitting diode (LED)
Bias current and voltage	Line regulation
Breakdown region	Load line
Breakdown voltage	Load-line analysis
Built-in potential (or voltage)	Load regulation
Cathode	Mathematical model
Center-tapped transformer	Metallurgical junction
Conduction angle	Nonideality factor (*n*)
Conduction interval	Peak detector
Constant voltage drop (CVD) model	Peak inverse voltage (PIV)
Cut-in voltage	Photodetector circuit
Depletion layer	Piecewise linear model
Depletion-layer width	*pn* junction diode
Depletion region	Q-point
Diffusion capacitance	Quiescent operating point
Diode equation	Rectifier circuits
Diode SPICE parameters (IS, RS, N, TT, CJO, VJ, M)	Reverse bias
	Reverse breakdown
Filter capacitor	Reverse saturation current (I_S)
Forward bias	Ripple current
Full-wave bridge rectifier circuit	Ripple voltage
Full-wave rectifier circuit	Saturation current
Half-wave rectifier circuit	Schottky barrier diode
Ideal diode	Solar cell
Ideal diode model	Space charge region (SCR)
Impact-ionization process	Storage time

Surge current

Thermal voltage (V_T)

Transit time

Turn-on voltage

Voltage regulator

Voltage transfer characteristic (VTC)

Zener breakdown

Zener diode

Zero bias

Zero-bias junction capacitance

REFERENCE

1. G. W. Neudeck, *The PN Junction Diode,* 2d ed. Pearson Education, Upper Saddle River, NJ: 1989.

ADDITIONAL READING

PSPICE, ORCAD, now owned by Cadence Design Systems, San Jose, CA.

T. Quarles, A. R. Newton, D. O. Pederson, and A. Sangiovanni-Vincentelli, *SPICE3 Version 3f3 User's Manual.* UC Berkeley: May 1993.

A. S. Sedra, and K. C. Smith. *Microelectronic Circuits.* 5th ed. Oxford University Press, New York: 2004.

PROBLEMS

3.1 The *pn* Junction Diode

3.1. A diode is doped with $N_A = 10^{19}$/cm^3 on the *p*-type side and $N_D = 10^{18}$/cm^3 on the *n*-type side. (a) What is the depletion-layer width w_{do}? (b) What are the values of x_p and x_n? (c) What is the value of the built-in potential of the junction? (d) What is the value of E_{MAX}? Use Eq. (3.3) and Fig. 3.5.

3.2. A diode is doped with $N_A = 10^{18}$/cm^3 on the *p*-type side and $N_D = 10^{15}$/cm^3 on the *n*-type side. (a) What are the values of p_p, p_n, n_p, and n_n? (b) What are the depletion-region width w_{do} and built-in voltage?

3.3. Repeat Prob. 3.2 for a diode with $N_A = 10^{18}$/cm^3 on the *p*-type side and $N_D = 10^{18}$/cm^3 on the *n*-type side.

3.4. Repeat Prob. 3.2 for a diode with $N_D = 10^{20}$/cm^3 on the *n*-type side and $N_A = 10^{18}$/cm^3 on the *p*-type side.

3.5. Repeat Prob. 3.2 for a diode with $N_A = 10^{16}$/cm^3 on the *p*-type side and $N_D = 10^{19}$/cm^3 on the *n*-type side.

3.6. A diode has $w_{do} = 0.4$ μm and $\phi_j = 0.85$ V. (a) What reverse bias is required to double the depletion-layer width? (b) What is the depletion-region width if a reverse bias of 5 V is applied to the diode?

3.7. A diode has $w_{do} = 1$ μm and $\phi_j = 0.6$ V. (a) What reverse bias is required to triple the depletion-layer width? (b) What is the depletion region width if a reverse bias of 10 V is applied to the diode?

3.8. Suppose a drift current density of 1000 A/cm^2 exists in the neutral region on the *n*-type side of a diode that has a resistivity of 0.5 $\Omega \cdot$ cm. What is the electric field needed to support this drift current density?

3.9. Suppose a drift current density of 5000 A/cm^2 exists in the neutral region on the *p*-type side of a diode that has a resistivity of 2 $\Omega \cdot$ cm. What is the electric field needed to support this drift current density?

3.10. The maximum velocity of carriers in silicon is approximately 10^7 cm/s. What is the maximum drift current density that can be supported in a region of *n*-type silicon with a doping of 4×10^{15}/cm^3?

3.11. The maximum velocity of carriers in silicon is approximately 10^7 cm/s. What is the maximum drift-current density that can be supported in a region of *p*-type silicon with a doping of 5×10^{17}/cm^3?

**3.12. Suppose that $N_A(x) = N_o \exp(-x/L)$ in a region of silicon extending from $x = 0$ to $x = 15$ μm, where N_o is a constant. Assume that $p(x) = N_A(x)$. Assuming that j_p must be zero in thermal equilibrium, show that a built-in electric field must exist and find its value for $L = 1$ μm and $N_o = 10^{18}$/cm^3.

3.13. What carrier gradient is needed to generate a diffusion current density of $j_n = 2000$ A/cm^2 if $\mu_n = 500$ cm^2/V · s?

3.14. Use the solver routine in your calculator to find the solution to Eq. (3.25) for $I_S = 10^{-16}$ A.

3.15. Use a spreadsheet to iteratively find the solution to Eq. 3.25 for $I_S = 10^{-13}$ A.

3.16. (a) Use MATLAB or MATHCAD to find the solution to Eq. 3.25 for $I_S = 10^{-13}$ A. (b) Repeat for $I_S = 10^{-15}$ A.

3.2 –3.4 The i-v Characteristics of the Diode; The Diode Equation: A Mathematical Model for the Diode; and Diode Characteristics Under Reverse, Zero, and Forward Bias

3.17. To what temperature does $V_T = 0.025$ V actually correspond?

3.18. What is the value of V_T for temperatures of $-55°$C, $0°$C, and $+85°$C?

3.19. (a) Plot a graph of the diode equation similar to Fig. 3.8 for a diode with $I_S = 10^{-12}$ A and $n = 1$. (b) Repeat for $n = 2$. (c) Repeat (a) for $I_S = 10^{-14}$ A.

3.20. A diode has $n = 1.04$ at $T = 300$ K. What is the value of $n \cdot V_T$? What temperature would give the same value of $n \cdot V_T$ if $n = 1.00$?

*3.21. What are the values of I_S and n for the diode in the graph in Fig. P3.21? Assume $V_T = 0.025$ V.

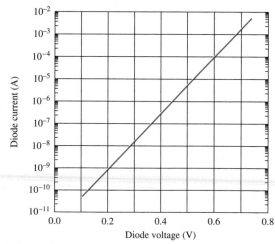

Figure P3.21

3.22. A diode has $I_S = 10^{-18}$ A and $n = 1.05$. (a) What is the diode voltage if the diode current is 70 μA? (b) What is the diode voltage if the diode current is

5 μA? (c) What is the diode current for $v_D = 0$ V? (d) What is the diode current for $v_D = -0.075$ V? (e) What is the diode current for $v_D = -5$ V?

3.23. A diode has $I_S = 10^{-17}$ A and $n = 1$. (a) What is the diode voltage if the diode current is 100 μA? (b) What is the diode voltage if the diode current is 10 μA? (c) What is the diode current for $v_D = 0$ V? (d) What is the diode current for $v_D = -0.06$ V? (e) What is the diode current for $v_D = -4$ V?

3.24. A diode has $I_S = 10^{-17}$ A and $n = 1$. (a) What is the diode current if the diode voltage is 0.675 V? (b) What will be the diode voltage if the current increases by a factor of 3?

3.25. A diode has $I_S = 10^{-10}$ A and $n = 2$. (a) What is the diode voltage if the diode current is 40 A? (b) What is the diode voltage if the diode current is 100 A?

3.26. A diode is operating with $i_D = 2$ mA and $v_D = 0.82$ V. (a) What is I_S if $n = 1$? (b) What is the diode current for $v_D = -5$ V?

3.27. A diode is operating with $i_D = 300$ μA and $v_D = 0.75$ V. (a) What is I_S if $n = 1$? (b) What is the diode current for $v_D = -3$ V?

3.28. The saturation current for diodes with the same part number may vary widely. Suppose it is known that 10^{-14} A $\leq I_S \leq 10^{-12}$ A. What is the range of forward voltages that may be exhibited by the diode if it is biased with $i_D = 1$ mA?

**3.29. The i-v characteristic for a diode has been measured under carefully controlled temperature conditions ($T = 307$ K), and the data are in Table P3.29. Use a spreadsheet or MATLAB to find the values of I_S and n that provide the best fit of the diode equation to the measurements in the least-squares sense. [That is,

TABLE P3.29

Diode i-v Measurements

DIODE VOLTAGE	DIODE CURRENT
0.500	6.591×10^{-7}
0.550	3.647×10^{-6}
0.600	2.158×10^{-5}
0.650	1.780×10^{-4}
0.675	3.601×10^{-4}
0.700	8.963×10^{-4}
0.725	2.335×10^{-3}
0.750	6.035×10^{-3}
0.775	1.316×10^{-2}

find the values of I_S and n that minimize the function $M = \sum_{m=1}^{n} (i_D^m - I_{Dm})^2$, where i_D is the diode equation from Eq. (3.1) and I_{Dm} are the measured data.] For your values of I_S and n, what is the minimum value of $M = \sum_{m=1}^{n} (i_D^m - I_{Dm})^2$?

3.5 Diode Temperature Coefficient

3.30. What is the value of V_T for temperatures of $-40°C$, $0°C$, and $+50°C$?

3.31. A diode with $I_S = 2.5 \times 10^{-16}$ A at $30°C$ is biased at a current of 1 mA. (a) What is the diode voltage? (b) If the diode voltage temperature coefficient is -1.8 mV/K, what will be the diode voltage at $50°C$?

3.32. A diode has $I_S = 10^{-15}$ A and $n = 1$. (a) What is the diode voltage if the diode current is 100 μA at $T = 25°C$? (b) What is the diode voltage at $T = 50°C$? Assume the diode voltage temperature coefficient is -2 mV/K at $0°C$.

3.33. A diode has $I_S = 10^{-15}$ A and $n = 1$. (a) What is the diode voltage if the diode current is 250 μA at $T = 25°C$? (b) What is the diode voltage at $T = 85°C$? Assume the diode voltage temperature coefficient is -1.8 mV/K at $55°C$.

*3.34. The temperature dependence of I_S is described approximately by

$$I_S = CT^3 \exp\left(-\frac{E_G}{kT}\right)$$

What is the diode voltage temperature coefficient based on this expression and Eq. (3.15) if $E_G = 1.21$ eV, $V_D = 0.7$ V, and $T = 300$ K?

3.35. The saturation current of a silicon diode is described by the expression in Prob. 3.34. (a) What temperature change will cause I_S to double? (b) To increase by 10 times? (c) To decrease by 100 times?

3.6 Diodes under Reverse Bias

3.36. A diode has $w_{do} = 1$ μm and $\phi_j = 0.8$ V. (a) What is the depletion layer width for $V_R = 5$ V? (b) For $V_D = -10$ V?

3.37. A diode has a doping of $N_D = 10^{15}/cm^3$ on the n-type side and $N_A = 10^{16}/cm^3$ on the p-type side. What are the values of w_{do} and ϕ_j? What is the value of w_d at a reverse bias of 10 V? At 100 V?

3.38. A diode has a doping of $N_D = 10^{20}/cm^3$ on the n-type side and $N_A = 10^{18}/cm^3$ on the p-type side.

What are the values of w_{do} and ϕ_j? What is the value of w_d at a reverse bias of 5 V? At 25 V?

*3.39. A diode has $w_{do} = 1$ μm and $\phi_j = 0.6$ V. If the diode breaks down when the internal electric field reaches 300 kV/cm, what is the breakdown voltage of the diode?

*3.40. Silicon breaks down when the internal electric field exceeds 300 kV/cm. At what reverse bias do you expect the diode of Prob. 3.2 to break down?

3.41. What are the breakdown voltage V_Z and Zener resistance R_Z of the diode depicted in Fig. P3.41?

**3.42. A diode is fabricated with $N_A \gg N_D$. What value of doping is required on the lightly doped side to achieve a reverse-breakdown voltage of 1000 V if the semiconductor material breaks down at a field of 300 kV/cm?

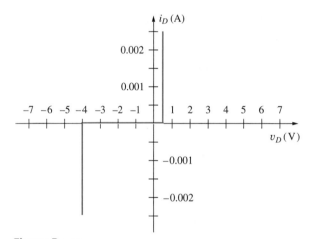

Figure P3.41

3.7 *pn* Junction Capacitance

3.43. What is the zero-bias junction capacitance/cm^2 for a diode with $N_A = 10^{15}/cm^3$ on the p-type side and $N_D = 10^{20}/cm^3$ on the n-type side? What is the diode capacitance with a 5-V reverse bias if the diode area is 0.05 cm^2?

3.44. What is the zero-bias junction capacitance per cm^2 for a diode with $N_A = 10^{18}/cm^3$ on the p-type side and $N_D = 10^{15}/cm^3$ on the n-type side. What is the diode capacitance with a 10 V reverse bias if the diode area is 0.02 cm^2?

3.45. A diode is operating at a current of 100 μA. (a) What is the diffusion capacitance if the diode transit time is 100 ps? (b) How much charge is stored in the diode? (c) Repeat for $i_D = 5$ mA.

3.46. A diode is operating at a current of 1 A. (a) What is the diffusion capacitance if the diode transit time is 10 ns? (b) How much charge is stored in the diode? (c) Repeat for $i_D = 100$ mA.

3.47. A *pn* junction diode has a cross-sectional area of $10^4 \ \mu m^2$. The *p*-type side has a doping concentration of $10^{19}/cm^3$ and the *n*-type side has a doping concentration of $10^{17}/cm^3$. What is the zero-bias capacitance of the diode? What is the capacitance at a reverse bias of 5 V?

3.48. A square *pn* junction diode is 5 mm on a side. The *p*-type side has a doping concentration of $10^{19}/cm^3$ and the *n*-type side has a doping concentration of $10^{16}/cm^3$. What is the zero-bias capacitance of the diode? What is the capacitance at a reverse bias of 3 V?

3.49. A variable capacitance diode with $C_{jo} = 39$ pF and $\phi_j = 0.75$ V is used to tune a resonant LC circuit as shown in Fig. P3.49. The impedance of the RFC (radio frequency choke) can be considered infinite. What are the resonant frequencies ($f_o = \frac{1}{2\pi\sqrt{LC}}$) for $V_{dc} = 1$ V and $V_{dc} = 10$ V?

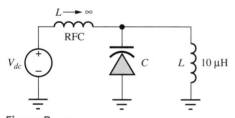

Figure P3.49

3.8 Schottky Barrier Diode

3.50. Suppose a Schottky barrier diode can be modeled by the diode equation in Eq. (3.11) with $I_S = 10^{-7}$ A. (a) What is the diode voltage at a current of 50 A? (b) What would be the voltage of a *pn* junction diode with $I_S = 10^{-15}$ A?

3.51. A Schottky barrier diode is modeled by the diode equation in Eq. (3.11) with $I_S = 10^{-11}$ A. (a) What is the diode voltage at a current of 4 mA? (b) What would be the voltage of a *pn* junction diode with $I_S = 10^{-14}$ A operating at the same current?

3.9 Diode SPICE Model and Layout

3.52. A *pn* diode has a resistivity of $2 \ \Omega \cdot cm$ on the *p*-type side and $0.01 \ \Omega \cdot cm$ on the *n*-type side. What is the value of R_S for this diode if the cross-sectional area of the diode is $0.01 \ cm^2$ and the lengths of the *p*- and *n*-sides of the diode are each 250 μm?

3.53. (a) A diode has $I_S = 5 \times 10^{-16}$ A and $R_S = 10 \ \Omega$ and is operating at a current of 1 mA at room temperature. What are the values of V_D and V_D'? (b) Repeat for $R_S = 100 \ \Omega$.

*3.54. A diode fabrication process has a specific contact resistance of $10 \ \Omega \cdot \mu m^2$. If the contacts are each $1 \ \mu m \times 1 \ \mu m$ in size, what are the total contact resistances associated with the anode and cathode contacts to the diode in Fig. 3.21(a).

3.55. (a) Estimate the area of the diode in Fig. 3.21(a) if the contact dimensions are $1 \ \mu m \times 1 \ \mu m$. (b) Repeat for $0.13 \ \mu m \times 0.13 \ \mu m$ contacts.

3.10 Diode Circuit Analysis Load-Line Analysis

3.56. (a) Plot the load line and find the Q-point for the diode circuit in Fig. P3.56 if $V = 5$ V and $R = 10 \ k\Omega$. Use the *i-v* characteristic in Fig. P3.41. (b) Repeat for $V = -6$ V and $R = 3 \ k\Omega$. (c) Repeat for $V = -3$ V and $R = 3 \ k\Omega$.

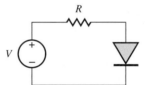

Figure P3.56

3.57. (a) Plot the load line and find the Q-point for the diode circuit in Fig. P3.56 if $V = 10$ V and $R = 5 \ k\Omega$. Use the *i-v* characteristic in Fig. P3.41. (b) Repeat for $V = -10$ V and $R = 5 \ k\Omega$. (c) Repeat for $V = -2$ V and $R = 2 \ k\Omega$.

3.58. Simulate the circuit in Prob. 3.56 with SPICE and compare the results to those in Prob. 3.56. Use $I_S = 10^{-15}$ A.

3.59. (a) Plot the load line and find the Q-point for the diode circuit in Fig. P3.56 if $V = -10$ V and $R = 10 \ k\Omega$. Use the *i-v* characteristic in Fig. P.3.41. (b) Repeat for $V = 10$ V and $R = 10 \ k\Omega$. (c) Repeat for $V = -4$ V and $R = 2 \ k\Omega$.

3.60. Use the *i-v* characteristic in Fig. P3.41. (a) Plot the load line and find the Q-point for the diode circuit in Fig. P3.56 if $V = 6$ V and $R = 4 \ k\Omega$. (b) For $V = -6$ V and $R = 3 \ k\Omega$. (c) For $V = -3$ V and $R = 3 \ k\Omega$. (d) For $V = 12$ V and $R = 8 \ k\Omega$. (e) For $V = -25$ V and $R = 10 \ k\Omega$.

Iterative Analysis and the Mathematical Model

3.61. Repeat the iterative procedure used in the spread-sheet in Table 3.2 for initial guesses of 0.7 V, 0.5 V, and 0.2 V. How many iterations are required for each case? Did any problem arise? If so, what is the source of the problem?

3.62. (a) Use direct trial and error to find the solution to the diode circuit in Fig. 3.22 using Eq. (3.27).

3.63. Use a spreadsheet or write a program in a high-level language to numerically find the Q-point for the circuit in Fig. 3.22 using the equation in the exercise on page 99.

3.64. Use MATLAB or MATHCAD to numerically find the Q-point for the circuit in Fig. 3.22 using the equation in the exercise on page 99.

Ideal Diode and Constant Voltage Drop Models

*3.65. Find the Q-point for the circuit in Fig. 3.22 using the same four methods as in Sec. 3.10 if the voltage source is 1 V. Compare the answers in a manner similar to Table 3.3.

3.66. Find the Q-point for the diode in Fig. P3.66 using (a) the ideal diode model and (b) the constant voltage drop model with $V_{on} = 0.6$ V. (c) Discuss the results. Which answer do you feel is most correct?

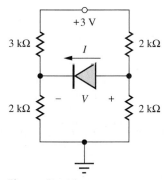

Figure P3.66

3.67. Find the worst-case values of the Q-point current for the diode in Fig. P3.66 using the ideal diode model if the resistors all have 10 percent tolerances.

3.68. Simulate the circuit of Fig. P3.66 and find the diode Q-point. Compare the results to those in Prob. 3.66.

3.69. (a) Find I and V in the four circuits in Fig. P3.69 using the ideal diode model. (b) Repeat using the constant voltage drop model with $V_{on} = 0.7$ V.

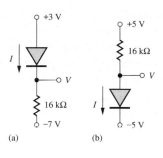

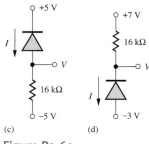

Figure P3.69

3.70. (a) Find I and V in the four circuits in Fig. P3.69 using the ideal diode model if the resistor values are changed to 100 kΩ. (b) Repeat using the constant voltage drop model with $V_{on} = 0.6$ V.

3.11 Multiple Diode Circuits

3.71. Find the Q-points for the diodes in the four circuits in Fig. P3.71 using (a) the ideal diode model and (b) the constant voltage drop model with $V_{on} = 0.75$ V.

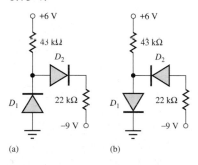

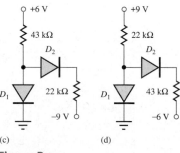

Figure P3.71

3.72. Find the Q-points for the diodes in the four circuits in Fig. P3.71 if the values of all the resistors are changed to 15 kΩ using (a) the ideal diode model and (b) the constant voltage drop model with $V_{on} = 0.75$ V.

3.73. Find the Q-point for the diodes in the circuits in Fig. P3.72 using the ideal diode model.

3.74. Find the Q-point for the diodes in the circuits in Fig. P3.72 using the constant voltage drop model with $V_{on} = 0.6$ V.

3.75. Simulate the diode circuits in Fig. P3.72 and compare your results to those in Prob. 3.73.

3.76. Verify that the values presented in Ex. 3.8 using the ideal diode model are correct.

3.77. Simulate the circuit in Fig. 3.33 and compare to the results in Ex. 3.8.

3.12 Analysis of Diodes Operating in the Breakdown Region

3.78. Find the Q-point for the Zener diode in Fig. P3.78.

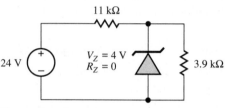

Figure P3.78

3.79. Draw the load line for the circuit in Fig. P3.78 on the characteristics in Fig. P3.41 and find the Q-point.

3.80. What is maximum load current I_L that can be drawn from the Zener regulator in Fig. P3.80 if it is to maintain a regulated output? What is the minimum value of R_L that can be used and still have a regulated output voltage?

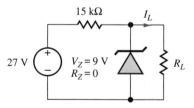

Figure P3.80

3.81. What is power dissipation in the Zener diode in Fig. P3.80 for $R_L = \infty$?

3.82. Load resistor R_L in Fig. P3.80 is 10 kΩ. What are the nominal and worst-case values of Zener diode current and power dissipation if the power supply voltage, Zener breakdown voltage and resistors all have 5 percent tolerances?

3.83. What is power dissipation in the Zener diode in Fig. P3.83 for (a) $R_L = 100$ Ω? (b) $R_L = \infty$?

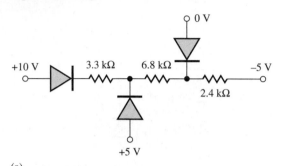

(a)

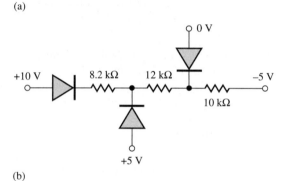

(b)

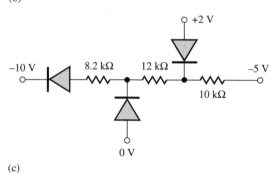

(c)

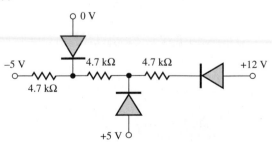

(d)

Figure P3.72

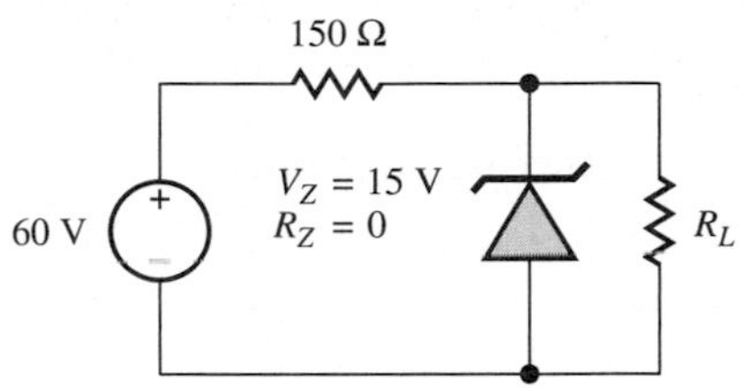

Figure P3.83

150 Ω

60 V

$V_Z = 15$ V
$R_Z = 0$

R_L

Figure P3.83

3.84. Load resistor R_L in Fig. P3.83 is 100 Ω. What are the nominal and worst-case values of Zener diode current and power dissipation if the power supply voltage, Zener breakdown voltage, and resistors all have 10 percent tolerances?

3.13 Half-Wave Rectifier Circuits

*3.85. (a) Use a spreadsheet or MATLAB or write a computer program to find the numeric solution to the conduction angle equation (Eq. 3.53) for a 60 Hz half-wave rectifier circuit that uses a filter capacitance of 100,000 μF. The circuit is designed to provide 5 V at 5 A. {That is, solve $[(V_P - V_{on}) \exp(-t/RC) = V_P \cos \omega t - V_{on}]$. Be careful! There are an infinite number of solutions to this equation. Be sure your algorithm finds the desired answer to the problem.} Assume $V_{on} = 1$ V. (b) Compare to calculations using Eq. (3.57).

3.86. A power diode has a reverse saturation current of 10^{-9} A and $n = 2$. What is the forward voltage drop at the peak current of 48.6 A that was calculated in the example in Sec. 3.13.5?

3.87. A power diode has a reverse saturation current of 10^{-8} A and $n = 1.6$. What is the forward voltage drop at the peak current of 100 A? What is the power dissipation in the diode in a half-wave rectifier application operating at 60 Hz if the series resistance is 0.01 Ω and the conduction time is 1 ms?

3.88. What is the actual average value (the dc value) of the rectifier output voltage for the waveform in Fig. P3.88 if V_r is 5 percent of $V_P - V_{on} = 18$ V?

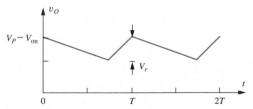

Figure P3.88

*3.89. Show that evaluation of Eq. (3.64) will yield the result in Eq. (3.65).

3.90. Draw the voltage waveforms, similar to those in Fig. 3.53, for the negative output rectifier in Fig. 3.57(b).

3.91. The half-wave rectifier in Fig. P3.91 is operating at a frequency of 60 Hz, and the rms value of the transformer output voltage is 6.3 V. (a) What is the value of the dc output voltage V_O if the diode voltage drop is 1 V? (b) What is the minimum value of C required to maintain the ripple voltage to less than 0.25 V if $R = 0.5$ Ω? (c) What is the PIV rating of the diode in this circuit? (d) What is the surge current when power is first applied? (e) What is the amplitude of the repetitive current in the diode?

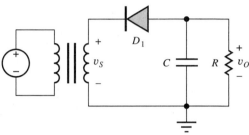

v_S D_1 C R v_O

Figure P3.91

3.92. The half-wave rectifier in Fig. P3.91 is operating at a frequency of 60 Hz, and the rms value of the transformer output voltage v_S is 12.6 V ± 10%. What are the nominal and worst case values of the dc output voltage V_O if the diode voltage drop is 1 V?

3.93. Simulate the behavior of the half-wave rectifier in Fig. P3.91 for $v_S = 10 \sin 120\pi t$, $R = 0.025$ Ω and $C = 0.5$ F. (Use IS $= 10^{-10}$ A, RS $= 0$, and RELTOL $= 10^{-6}$.) Compare the simulated values of dc output voltage, ripple voltage, and peak diode current to hand calculations. Repeat simulation with $R_S = 0.02$ Ω.

3.94. (a) Repeat Prob. 3.91 for a frequency of 400 Hz. (b) Repeat Prob. 3.91 for a frequency of 100 kHz.

3.95. For the Zener regulated power supply in Fig. P3.95, the rms value of v_s is 15 V, the operating ferquency is 60 Hz, $R = 100$ Ω, $C = 1000$ μF, the on-voltage of diodes D_1 and D_2 is 0.75 V, and the Zener voltage of diode D_3 is 15 V. (a) What type of rectifier is used in this power supply circuit? (b) What is the dc voltage at V_1? (c) What is the dc output voltage V_O? (d) What is the magnitude of the ripple voltage at V_1? (e) What is the minimum PIV rating for the rectifier diodes? (f) Draw a new version of the circuit that will produce an output voltage of -15 V.

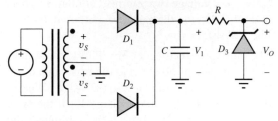

Figure P3.95

3.96.   A 3000-V, 1-A, dc power supply is to be designed with a ripple voltage ≤ 1 percent. Assume that a half-wave rectifier circuit (60 Hz) with a capacitor filter is used. (a) What is the size of the filter capacitor C? (b) What is the minimum PIV rating for the diode? (c) What is the rms value of the transformer voltage needed for the rectifier? (d) What is the peak value of the repetitive current in the diode? (e) What is the surge current at $t = 0^+$?

3.97.   A 3.3-V, 30-A dc power supply is to be designed with a ripple of less than 2.5 percent. Assume that a half-wave rectifier circuit (60 Hz) with a capacitor filter is used. (a) What is the size of the filter capacitor C? (b) What is the PIV rating for the diode? (c) What is the rms value of the transformer voltage needed for the rectifier? (d) What is the value of the peak repetitive diode current in the diode? (e) What is the surge current at $t = 0^+$?

*3.98.   Draw the voltage waveforms at nodes v_O and v_1 for the "voltage-doubler" circuit in Fig. P3.98 for the first two cycles of the input sine wave. What is the steady-state output voltage if $V_P = 17$ V?

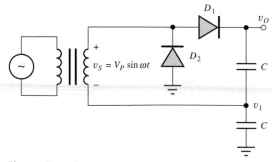

Figure P3.98

3.99.   Simulate the voltage-doubler rectifier circuit in Fig. P3.98 for $C = 500$ μF and $v_S = 1500 \sin 2\pi(60)t$ with a load resistance of $R_L =$

3000 Ω added between v_O and ground. Calculate the ripple voltage and compare to the simulation.

3.14 Full-Wave Rectifier Circuits

3.100.   The full-wave rectifier in Fig. P3.100 is operating at a frequency of 60 Hz, and the rms value of the transformer output voltage is 15 V. (a) What is the value of the dc output voltage if the diode voltage drop is 1 V? (b) What is the minimum value of C required to maintain the ripple voltage to less than 0.25 V if $R = 0.5$ Ω? (c) What is the PIV rating of the diode in this circuit? (d) What is the surge current when power is first applied? (e) What is the amplitude of the repetitive current in the diode?

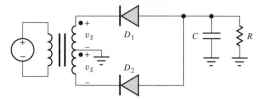

Figure P3.100

3.101.   Repeat Prob. 3.100 if the rms value of the transformer output voltage v_S is 9 V.

3.102.   Simulate the behavior of the full-wave rectifier in Fig. P3.100 for $R = 3$ Ω and $C = 22,000$ μF. Assume that the rms value of v_S is 10.0 V and the frequency is 400 Hz. (Use IS $= 10^{-10}$ A, RS $= 0$, and RELTOL $= 10^{-6}$.) Compare the simulated values of dc output voltage, ripple voltage, and peak diode current to hand calculations. Repeat simulation with $R_S = 0.25$.

3.103.   Repeat Prob. 3.97 for a full-wave rectifier circuit.

3.104.   Repeat Prob. 3.96 for a full-wave rectifier circuit.

*3.105.   The full-wave rectifier circuit in Fig. P3.105(a) was designed to have a maximum ripple of approximately 1 V, but it is not operating properly. The measured waveforms at the three nodes in the circuit are shown in Fig. P3.105(b). What is wrong with the circuit?

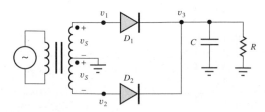

Figure P3.105(a)

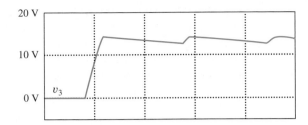

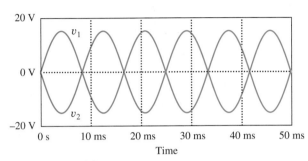

Figure P3.105(b) Waveforms for the circuit in Fig. P3.105(a).

3.15 Full-Wave Bridge Rectification

3.106. Repeat Prob. 3.100 for a full-wave bridge rectifier circuit.

3.107. Repeat Prob. 3.96 for a full-wave bridge rectifier circuit.

3.108. Repeat Prob. 3.97 for a full-wave bridge rectifier circuit.

*3.109. What are the dc output voltages V_1 and V_2 for the rectifier circuit in Fig. P3.109 if $v_S = 35 \sin 377t$ and $C = 10,000$ μF?

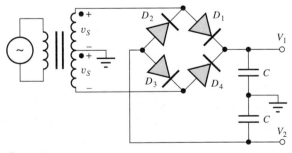

Figure P3.109

3.110. Simulate the rectifier circuit in Fig. P3.109 for $C = 100$ mF and $v_S = 35 \sin 2\pi(60)t$ with a 500-Ω load connected between each output and ground.

3.111. Repeat Prob. 3.100 if the full-wave bridge circuit in Fig. 3.66 is used instead of the rectifier in Fig. P3.100.

3.16 Rectifier Comparison and Design Tradeoffs

3.112. A 3.3-V, 15-A dc power supply is to be designed to have a ripple voltage of no more the 10 mV. Compare the pros and cons of implementating this power supply with half-wave, full-wave, and full-wave bridge rectifiers.

3.113. A 200-V, 3-A dc power supply is to be designed with less than a 2 percent ripple voltage. Compare the pros and cons of implementing this power supply with half-wave, full-wave, and full-wave bridge rectifiers.

3.114. A 3000-V, 1-A dc power supply is to be designed with less than a 4 percent ripple voltage. Compare the pros and cons of implementing this power supply with half-wave, full-wave, and full-wave bridge rectifiers.

3.17 Dynamic Switching Behavior of the Diode

*3.115. (a) Calculate the current at $t = 0^+$ in the circuit in Fig. P3.115. (b) Calculate I_F, I_R, and the storage time expected when the diode is switched off if $\tau_T = 7$ ns.

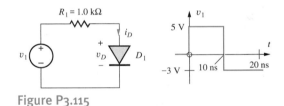

Figure P3.115

3.116. (a) Simulate the switching behavior of the circuit in Fig. P3.115. (b) Compare the simulation results to the hand calculations in Prob. 3.115.

*3.117. (a) Calculate the current at $t = 0^+$ in the circuit in Fig. P3.115 if R_1 is changed to 5 Ω. (b) Calculate I_F, I_R, and the storage time expected when the diode is switched off at $t = 10$ μs if $\tau_T = 250$ nS.

**3.118. The simulation results presented in Fig. 3.68 were performed with the diode transit time $\tau_T = 5$ ns. (a) Repeat the simulation of the diode circuit in Fig. 3.118(a) with the diode transit time changed to $\tau_T = 50$ ns. Does the storage time that you observe change in proportion to the value of τ_T in your simulation? Discuss. (b) Repeat the simulation with the input voltage changed to the one in Fig. P3.118(b), in which it is assumed that v_1 has been at 1.5 V for a long time, and compare the results to those obtained in (a). What is the reason for the difference between the results in (a) and (b)?

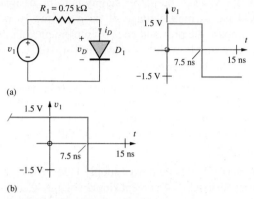

(a)

(b)

Figure 3.118

3.18 Photo Diodes, Solar Cells, and LEDs

*3.119. The output of a diode used as a solar cell is given by

$$I_C = 1 - 10^{-15}[\exp(40V_C) - 1] \text{ amperes}$$

What operating point corresponds to P_{max}? What is P_{max}? What are the values of I_{SC} and V_{OC}?

*3.120. Three diodes are connected in series to increase the output voltage of a solar cell. The individual outputs of the three diodes are given by

$$I_{C1} = 1.05 - 10^{-15}[\exp(40V_{C1}) - 1] \text{ A}$$
$$I_{C2} = 1.00 - 10^{-15}[\exp(40V_{C2}) - 1] \text{ A}$$
$$I_{C3} = 0.95 - 10^{-15}[\exp(40V_{C3}) - 1] \text{ A}$$

(a) What are the values of I_{SC} and V_{OC} for the series connected cell? (b) What is the value of P_{max}?

**3.121. The bandgaps of silicon and gallium arsenide are 1.12 eV and 1.42 eV, respectively. What are the wavelengths of light that you would expect to be emitted from these devices based on direct recombination of holes and electrons? To what "colors" of light do these wavelengths correspond?

CHAPTER 4

FIELD-EFFECT TRANSISTORS

CHAPTER GOALS

- Develop a qualitative understanding of the operation of the MOS field-effect transistor
- Define and explore FET characteristics in the cutoff, triode, and saturation regions of operation
- Develop mathematical models for the current-voltage (i-v) characteristics of MOSFETs
- Introduce the graphical representations for the output and transfer characteristic descriptions of electron devices
- Catalog and contrast the characteristics of both NMOS and PMOS enhancement-mode and depletion-mode FETs
- Learn the symbols used to represent FETs in circuit schematics
- Investigate circuits used to bias the transistors into various regions of operation
- Learn the basic structure and mask layout for MOS transistors and circuits
- Explore the concept of MOS device scaling

- Contrast three- and four-terminal device behavior
- Understand sources of capacitance in MOSFETs
- Explore FET Modeling in SPICE

In this chapter we begin to explore the field-effect transistor or FET. The FET has emerged as the dominant device in modern integrated circuits and is present in the vast majority of semiconductor circuits produced today. The ability to dramatically shrink the size of the FET device has made possible handheld computational power unimagined just 20 years ago.

As noted in Chapter 1, various versions of the field-effect device were conceived by Lilienfeld in 1928, Heil in 1935, and Shockley in 1952, well before the technology to produce such devices existed. The first successful metal-oxide-semiconductor field-effect transistors, or MOSFETs, were fabricated in the late 1950s, but it took nearly a decade to develop reliable commercial fabrication processes for MOS devices. Because of fabrication-related difficulties, MOSFETs with a p-type conducting region, PMOS devices, were the first to be commercially available in IC form, and the first microprocessors were built using PMOS processes. By the late 1960s, understanding and control of fabrication processes had improved to the point that devices with an n-type conducting region, NMOS transistors, could be reliably fabricated in large numbers, and NMOS rapidly supplanted PMOS technology because the improved mobility of the NMOS device translated directly into higher circuit performance. By the mid 1980s, power had become a severe problem, and the low-power characteristics of complementary MOS or CMOS devices caused a rapid shift to that technology even though it was a more complex and costly process. Today CMOS technology, which utilizes both NMOS and PMOS transistors, is the dominant technology in the electronics industry.

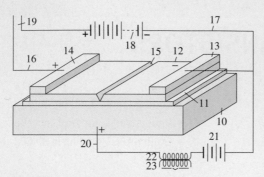

Drawing from Lillienfeld Patent [1]

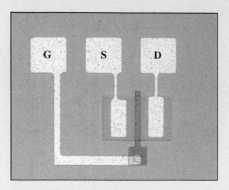

Top View of a Simple MOSFET

Chapter 4 explores the characteristics of the **metal-oxide-semiconductor field-effect transistor (MOSFET)** that is without doubt the most commercially successful solid-state device. It is the primary component in high-density VLSI chips, including microprocessors and memories. A second type of FET, the **junction field-effect transistor (JFET),** is based on a *pn* junction structure and finds application particularly in analog and RF circuit design. Its discussion can be found on the CDROM.

P-channel MOS (PMOS) transistors were the first MOS devices to be successfully fabricated in large-scale integrated (LSI) circuits. Early microprocessor chips used PMOS technology. Greater performance was later obtained with the commercial introduction of *n*-channel MOS (NMOS) technology, using both enhancement-mode and ion-implanted depletion-mode devices.

This chapter discusses the qualitative and quantitative *i*-*v* behavior of MOSFETs and investigates the differences between the various types of transistors. Techniques for biasing the transistors in various regions of operation are also presented.

Early integrated circuit chips contained only a few transistors, whereas today, the National Technology Roadmap for Semiconductors (NTRS [2]) projects the existence of chips with billions of transistors by the year 2010! This phenomenal increase in transistor density has been the force behind the explosive growth of the electronics industry outlined in Chapter 1 that has been driven by our ability to reduce (scale) the dimensions of the transistor without compromising its operating characteristics.

Although the bipolar junction transistor or BJT was successfully reduced to practice before the FET, the FET is conceptually easier to understand and is by far the most commercially important device. Thus, we consider it first. The BJT is discussed in detail in Chapter 5.

4.1 CHARACTERISTICS OF THE MOS CAPACITOR

At the heart of the MOSFET is the **MOS capacitor** structure depicted in Fig. 4.1. Understanding the qualitative behavior of this capacitor provides a basis for understanding operation of the MOSFET. The MOS capacitor is used to induce charge at the interface between the semiconductor and oxide. The top electrode of the MOS capacitor is formed of a low-resistivity material, typically aluminum or heavily doped polysilicon (polycrystalline silicon). We refer to this electrode as the **gate (*G*)** for reasons that become apparent shortly. A thin insulating layer, typically silicon dioxide, isolates the gate from the substrate or body—the semiconductor region that acts as the second electrode of the capacitor. Silicon dioxide is a stable, high-quality electrical insulator readily formed by thermal oxidation of the silicon substrate. The ability to form this stable high-quality insulator is one of the basic reasons that silicon is the dominant semiconductor material today. The semiconductor region may be *n*- or *p*-type. A *p*-type substrate is depicted in Fig. 4.1.

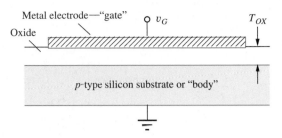

Figure 4.1 MOS capacitor structure on p-type silicon.

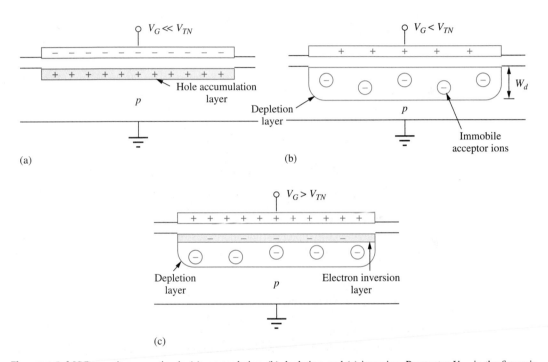

Figure 4.2 MOS capacitor operating in (a) accumulation, (b) depletion, and (c) inversion. Parameter V_{TN} in the figure is called the threshold voltage and represents the voltage required to just begin formation of the inversion layer.

The semiconductor forming the bottom electrode of the capacitor typically has a substantial resistivity and a limited supply of holes and electrons. Because the semiconductor can therefore be depleted of carriers, as discussed in Chapter 2, the capacitance of this structure is a nonlinear function of voltage. Figure 4.2 shows the conditions in the region of the substrate immediately below the gate electrode for three different bias conditions: accumulation, depletion, and inversion.

4.1.1 ACCUMULATION REGION

The situation for a large negative bias on the gate with respect to the substrate is indicated in Fig. 4.2(a). The large negative charge on the metallic gate is balanced by positively charged holes attracted to the silicon-silicon dioxide interface directly below the gate. For the bias condition shown, the hole density at the surface exceeds that which is present in the original p-type substrate, and the surface is said to be operating in the **accumulation region** or just in **accumulation.** This majority carrier accumulation layer is extremely shallow, effectively existing as a charge sheet directly below the gate.

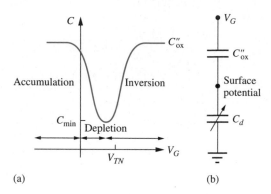

Figure 4.3 (a) Low frequency capacitance-voltage (C-V) characteristics for a MOS capacitor on a p-type substrate. (b) Series capacitance model for the C-V characteristic.

4.1.2 DEPLETION REGION

Now consider the situation as the gate voltage is slowly increased. First, holes are repelled from the surface. Eventually, the hole density near the surface is reduced below the majority-carrier level set by the substrate doping level, as depicted in Fig. 4.2(b). This condition is called **depletion** and the region, the **depletion region.** The region beneath the metal electrode is depleted of free carriers in much the same way as the depletion region that exists near the metallurgical junction of the pn junction diode. In Fig. 4.2(b), positive charge on the gate electrode is balanced by the negative charge of the ionized acceptor atoms in the depletion layer. The depletion-region width w_d can range from a fraction of a micron to tens of microns, depending on the applied voltage and substrate doping levels.

4.1.3 INVERSION REGION

As the voltage on the top electrode increases further, electrons are attracted to the surface. At some particular voltage level, the electron density at the surface exceeds the hole density. At this voltage, the surface has inverted from the p-type polarity of the original substrate to an n-type **inversion layer,** or **inversion region,** directly underneath the top plate as indicated in Fig. 4.2(c). This inversion region is an extremely shallow layer, existing as a charge sheet directly below the gate. In the MOS capacitor, the high density of electrons in the inversion layer is supplied by the electron–hole generation process within the depletion layer.

   The positive charge on the gate is balanced by the combination of negative charge in the inversion layer plus negative ionic acceptor charge in the depletion layer. The voltage at which the surface inversion layer just forms plays an extremely important role in field-effect transistors and is called the **threshold voltage** V_{TN}.

   Figure 4.3 depicts the variation of the capacitance of the NMOS structure with gate voltage. At voltages well below threshold, the surface is in accumulation, corresponding to Fig. 4.2(a), and the capacitance is high and determined by the **oxide thickness.** As the gate voltage increases, the surface depletion layer forms as in Fig. 4.2(b), the effective separation of the capacitor plates increases, and the capacitance decreases. The total capacitance can be modeled as the series combination of the fixed oxide capacitance C''_{ox} and the voltage dependent depletion-layer capacitance C_d, as in Fig. 4.3(b). The inversion layer forms at the surface as V_G exceeds threshold voltage V_{TN}, as in Fig. 4.2(c), and the capacitance rapidly increases back to the value determined by the oxide layer thickness.

4.2 THE NMOS TRANSISTOR

The MOSFET is formed by adding two heavily doped n-type (n^+) diffusions to the cross section of Fig. 4.1, resulting in the structure in Fig. 4.4. The diffusions provide a supply of electrons that can readily move under the gate as well as terminals that can be used to apply a voltage and cause a current in the channel region of the transistor.

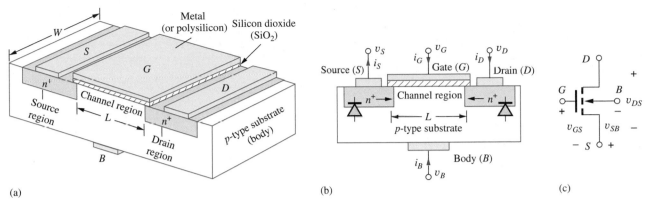

Figure 4.4 (a) NMOS transistor structure; (b) cross section; and (c) circuit symbol.

Figure 4.4 shows a planar view, cross section, and circuit symbol of an *n*-channel **MOSFET,** usually called an **NMOS transistor,** or **NMOSFET.** The central region of the NMOSFET is the MOS capacitor discussed in Sec. 4.1, and the top electrode of the capacitor is called the gate. The two heavily doped *n*-type regions (n^+ regions), called the **source (S)** and **drain (D),** are formed in the *p*-type substrate and aligned with the edge of the gate. The source and drain provide a supply of carriers so that the inversion layer can rapidly form in response to the gate voltage. The substrate of the NMOS transistor represents a fourth device terminal and is referred to synonymously as the **substrate terminal,** or the **body terminal (B).**

The terminal voltages and currents for the NMOS device are defined in Figs. 4.4(b) and (c). Drain current i_D, source current i_S, gate current i_G, and body current i_B are all defined, with the positive direction of each current indicated for an NMOS transistor. The important terminal voltages are the gate-source voltage $v_{GS} = v_G - v_S$ the drain-source voltage $v_{DS} = v_D - v_S$, and the source-bulk voltage $v_{SB} = v_S - v_B$. These voltages are all positive during normal operation of the NMOSFET.

Note that the source and drain regions form *pn* junctions with the substrate. These two junctions are kept reverse-biased at all times to provide isolation between the junctions and the substrate as well as between adjacent MOS transistors. Thus, the bulk voltage must be less than or equal to the voltages applied to the source and drain terminals to ensure that these *pn* junctions are properly reverse-biased.

The semiconductor region between the source and drain regions directly below the gate is called the **channel region** of the FET, and two dimensions of critical import are defined in Fig. 4.4. **L** represents the **channel length,** which is measured in the direction of current in the channel. **W** is the **channel width,** which is measured perpendicular to the direction of current. In this and later chapters we will find that choosing the values for W and L is an important aspect of the digital and analog IC designer's task.

4.2.1 QUALITATIVE *i-v* BEHAVIOR OF THE NMOS TRANSISTOR

Before attempting to derive an expression for the current-voltage characteristic of the NMOS transistor, let us try to develop a qualitative understanding of what we might expect by referring to Fig. 4.5. In the figure, the source, drain, and body of the NMOSFET are all grounded.

For a dc gate-source voltage, $v_{GS} = V_{GS}$, well below threshold voltage V_{TN}, as in Fig. 4.5(a), back-to-back *pn* junctions exist between the source and drain, and only a small leakage current can flow between these two terminals. For V_{GS} near but still below threshold, a depletion region forms beneath the gate and merges with the depletion regions of the source and drain, as indicated in Fig. 4.5(b). The depletion region is devoid of free carriers, so a current still does not appear between the source and drain. Finally, when the gate-channel voltage exceeds the threshold voltage V_{TN}, as

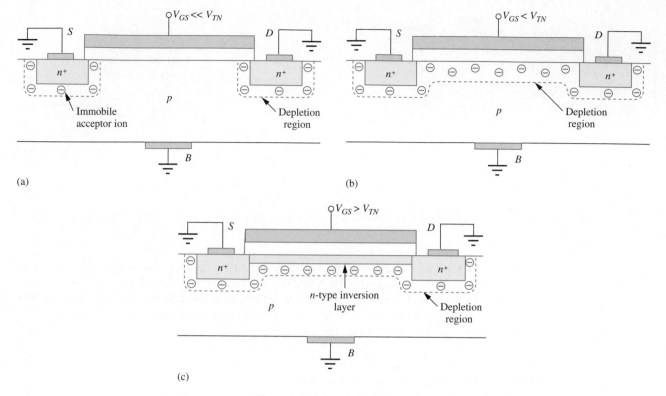

Figure 4.5 (a) $V_{GS} \ll V_{TN}$. (b) $V_{GS} < V_{TN}$. (c) $V_{GS} > V_{TN}$.

in Fig. 4.5(c), electrons flow in from the source and drain to form an inversion layer that connects the n^+ source region to the n^+ drain. A resistive connection, the channel, exists between the source and drain terminals.

If a positive voltage is now applied between the drain and source terminals, electrons in the channel inversion layer will drift in the electric field, creating a current in the terminals. Positive current in the NMOS transistor enters the drain terminal, travels down the channel, and exits the source terminal, as indicated by the polarities in Fig. 4.4(b). The gate terminal is insulated from the channel; thus, there is no dc gate current, and $i_G = 0$. The drain-bulk and source-bulk (and induced channel-to-bulk) pn junctions must be reverse-biased at all times to ensure that only a small reverse-bias leakage current exists in these diodes. This current is usually negligible with respect to the channel current i_D and is neglected. Thus we assume that $i_B = 0$.

In the device in Fig. 4.5, a channel must be induced by the applied gate voltage for conduction to occur. The gate voltage "enhances" the conductivity of the channel; this type of MOSFET is termed an **enhancement-mode device.** Later in this chapter we identify an additional type of MOSFET called a **depletion-mode device.** In Sec. 4.2.2, we develop a mathematical model for the current in the terminals of the NMOS device in terms of the applied voltages.

4.2.2 TRIODE[1] REGION CHARACTERISTICS OF THE NMOS TRANSISTOR

We saw in Sec. 4.2.1 that both i_G and i_B are zero. Therefore, the current entering the drain must be equal to the current leaving the source:

$$i_S = i_D \tag{4.1}$$

---

[1] This region of operation is also referred to as the "linear region." We will use triode region to avoid confusion with the concept of linear amplification introduced later in the text.

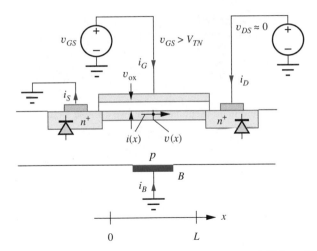

Figure 4.6 Model for determining i-v characteristics of the NMOS transistor.

An expression for the drain current i_D can be developed by considering the transport of charge in the channel in Fig. 4.6, which is depicted for a small value of v_{DS}. The electron charge per unit length (a line charge — C/cm) at any point in the channel is given by

$$Q' = -W C''_{\text{ox}}(v_{\text{ox}} - V_{TN}) \qquad C\text{/cm for } v_{\text{ox}} \geq V_{TN} \tag{4.2}$$

where $C''_{\text{ox}} = \varepsilon_{\text{ox}}/T_{\text{ox}}$, the oxide capacitance per unit area (F/cm^2)

$\varepsilon_{\text{ox}} = $ oxide permittivity (F/cm) $\qquad T_{\text{ox}} = $ oxide thickness (cm)

For silicon dioxide, $\varepsilon_{\text{ox}} = 3.9\varepsilon_o$, where $\varepsilon_o = 8.854 \times 10^{-14}$ F/cm.

The voltage v_{ox} represents the voltage across the oxide and will be a function of position in the channel:

$$v_{\text{ox}} = v_{GS} - v(x) \tag{4.3}$$

where $v(x)$ is the voltage at any point x in the channel referred to the source.

Note that v_{ox} must exceed V_{TN} for an inversion layer to exist, so Q' will be zero until $v_{\text{ox}} > V_{TN}$. At the source end of the channel, $v_{\text{ox}} = v_{GS}$, and it decreases to $v_{\text{ox}} = v_{GS} - v_{DS}$ at the drain end of the channel.

The electron drift current at any point in the channel is given by the product of the charge per unit length times the velocity v_x:

$$i(x) = Q'(x)v_x(x) \tag{4.4}$$

The charge Q' is represented by Eq. (4.2), and the velocity v_x of electrons in the channel is determined by the electron mobility and the transverse electric field in the channel:

$$i(x) = Q'v_x = [-W C''_{\text{ox}}(v_{\text{ox}} - V_{TN})](-\mu_n E_x) \tag{4.5}$$

The transverse field is equal to the negative of the spatial derivative of the voltage in the channel

$$E_x = -\frac{dv(x)}{dx} \tag{4.6}$$

Combining Eqs. (4.3) to (4.6) yields an expression for the current at any point in the channel:

$$i(x) = -\mu_n C''_{\text{ox}} W[v_{GS} - v(x) - V_{TN}]\frac{dv(x)}{dx} \tag{4.7}$$

We know the voltages applied to the device terminals are $v(0) = 0$ and $v(L) = v_{DS}$, and we can integrate Eq. (4.7) between 0 and L:

$$\int_0^L i(x)\,dx = -\int_0^{v_{DS}} \mu_n C''_{ox} W[v_{GS} - v(x) - V_{TN}]\,dv(x) \tag{4.8}$$

Because there is no mechanism to lose current as it goes down the channel, the current must be equal to the same value i_D at every point x in the channel, $i(x) = i_D$, and Eq. (4.8) finally yields

$$i_D = \mu_n C''_{ox} \frac{W}{L}\left(v_{GS} - V_{TN} - \frac{v_{DS}}{2}\right)v_{DS} = K'_n\left(\frac{W}{L}\right)\left(v_{GS} - V_{TN} - \frac{v_{DS}}{2}\right)v_{DS} \tag{4.9}$$

The value of $\mu_n C''_{ox}$ is fixed for a given technology and cannot be changed by the circuit designer. For circuit analysis and design purposes, Eq. (4.9) is therefore most often written as

$$i_D = K_n\left(v_{GS} - V_{TN} - \frac{v_{DS}}{2}\right)v_{DS} \tag{4.10}$$

where $K_n = K'_n W/L$ and $K'_n = \mu_n C''_{ox}$. Parameters K_n and K'_n are called **transconductance parameters** and both have units of A/V^2.

Equation (4.10) represents the classic expression for the drain-source current for the NMOS transistor in its **triode region** of operation, in which a resistive channel directly connects the source and drain. This resistive connection will exist as long as the voltage across the oxide exceeds the threshold voltage at every point in the channel:

$$v_{GS} - v(x) \geq V_{TN} \qquad \text{for } 0 \leq x \leq L \tag{4.11}$$

The voltage in the channel is maximum at the drain end where $v(L) = v_{DS}$. Thus, Eqs. (4.9) and (4.10) are valid as long as

$$v_{GS} - v_{DS} \geq V_{TN} \qquad \text{or} \qquad v_{GS} - V_{TN} \geq v_{DS} \tag{4.12}$$

Recapitulating for the triode region,

$$i_D = K'_n \frac{W}{L}\left(v_{GS} - V_{TN} - \frac{v_{DS}}{2}\right)v_{DS} \quad \text{for} \quad v_{GS} - V_{TN} \geq v_{DS} \geq 0 \quad \text{and} \quad K'_n = \mu_n C''_{ox} \tag{4.13}$$

Equation (4.13) is used frequently in the rest of this text. Commit it to memory!

Some additional insight into the mathematical model can be gained by regrouping the terms in Eq. (4.13):

$$i_D = \left[C''_{ox} W\left(v_{GS} - V_{TN} - \frac{v_{DS}}{2}\right)\right]\left(\mu_n \frac{v_{DS}}{L}\right) \tag{4.14}$$

For small drain-source voltages, the first term represents the average charge per unit length in the channel because the average channel voltage $\overline{v(x)} = v_{DS}/2$. The second term represents the drift velocity in the channel, where the average electric field is equal to the total voltage v_{DS} across the channel divided by the channel length L.

We should note that the term *triode region* is used because the drain current of the FET depends on the drain voltage of the transistor, and this behavior is similar to that of the electronic vacuum triode that appeared many decades earlier (see Table 1.2—Milestones in Electronics).

Note also that the **quiescent operating point** or **Q-point** of the FET is given by (I_D, v_{DS}).

EXERCISE: Calculate K_n' for a transistor with $\mu_n = 500$ cm²/v · s and $T_{ox} = 25$ nm.

ANSWER: 69.1 μA/V²

EXERCISE: An NMOS transistor has $K_n' = 50$ μA/V². What is the value of K_n if $W = 20$ μm, $L = 1$ μm? If $W = 60$ μm, $L = 3$ μm? If $W = 10$ μm, $L = 0.25$ μm?

ANSWERS: 1000 μA/V²; 1000 μA/V²; 2000 μA/V²

EXERCISE: Calculate the drain current in an NMOS transistor for $V_{GS} = 0$, 1 V, 2 V, and 3 V, with $V_{DS} = 0.1$ V, if $W = 10$ μm, $L = 1$ μm, $V_{TN} = 1.5$ V, and $K_n' = 25$ μA/V². What is the value of K_n?

ANSWERS: 0; 0; 11.3 μA; 36.3 μA; 250 μA/V²

4.2.3 ON RESISTANCE

The i-v characteristics in the triode region generated from Eq. (4.13) are drawn in Fig. 4.7 for the case of $V_{TN} = 1$ V and $K_n = 250$ μA/V². The curves in Fig. 4.7 represent a portion of the common-source **output characteristics** for the NMOS device. The output characteristics for the MOSFET are graphs of drain current i_D as a function of drain-source voltage v_{DS}. A family of curves is generated, with each curve corresponding to a different value of gate-source voltage v_{GS}. The output characteristics in Fig. 4.7 appear to be a family of nearly straight lines, hence the alternate name linear region (of operation). However, some curvature can be noted in the characteristics, particularly for $V_{GS} = 2$ V.

Let us explore the triode region behavior in more detail using Eq. (4.9). For small drain-source voltages such that $v_{DS}/2 \ll v_{GS} - V_{TN}$, Eq. (4.9) can be reduced to

$$i_D \cong \mu_n C_{ox}'' \frac{W}{L}(v_{GS} - V_{TN})v_{DS} \tag{4.15}$$

in which the current i_D through the MOSFET is directly proportional to the voltage v_{DS} across the MOSFET. The FET behaves much like a resistor connected between the drain and source terminals, but the resistor value can be controlled by the gate-source voltage. It has been said that this voltage-controlled resistance behavior originally gave rise to the name transistor, a contraction of "transfer-resistor."

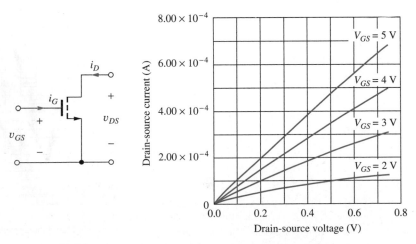

Figure 4.7 NMOS i-v characteristics in the triode region ($V_{SB} = 0$).

The resistance of the FET in the triode region near the origin, called the **on-resistance R_{on}**, is defined in Eq. (4.16) and can be found by taking the derivative of Eq. (4.13):

$$
R_{\text{on}} = \left[\left. \frac{\partial i_D}{\partial v_{DS}} \right|_{v_{DS} \to 0} \right]_{Q\text{-}pt}^{-1} = \left. \frac{1}{K'_n \dfrac{W}{L}(V_{GS} - V_{TN} - V_{DS})} \right|_{V_{DS} \to 0} = \frac{1}{K'_n \dfrac{W}{L}(V_{GS} - V_{TN})} \tag{4.16}
$$

We will find that the value of R_{on} plays a very important role in the operation of MOS logic circuits in Chapters 6–8. Note that R_{on} is also equal to the ratio v_{DS}/i_D from Eq. (4.15).

Near the origin, the i-v curves are indeed straight lines. However, curvature develops as the assumption $v_{DS} \ll v_{GS} - V_{TN}$ starts to be violated. For the lowest curve in Fig. 4.7, $V_{GS} - V_{TN} = 2 - 1 = 1$ V, and we should expect linear behavior only for values of v_{DS} below 0.1 to 0.2 V. On the other hand, the curve for $V_{GS} = 5$ V exhibits quasi-linear behavior throughout most of the range of Fig. 4.7.

> **EXERCISE:** Calculate the on-resistance of an NMOS transistor for $V_{GS} = 2$ V and $V_{GS} = 5$ V if $V_{TN} = 1$ V and $K_n = 250$ μA/V^2. What value of V_{GS} is required for an on-resistance of 2 kΩ?
>
> **ANSWERS:** 4 kΩ; 1 kΩ; 3v

4.2.4 SATURATION OF THE i-v CHARACTERISTICS

As discussed, Eq. (4.13) is valid as long as the resistive channel region directly connects the source to the drain. However, an unexpected phenomenon occurs in the MOSFET as the drain voltage increases above the triode region limit in Eq. (4.13). The current does not continue to increase, but instead saturates at a constant value. This unusual behavior is depicted in the i-v characteristics in Fig. 4.8 for several fixed gate-source voltages.

We can try to understand the origin of the current saturation by studying the device cross sections in Fig. 4.9. In Fig. 4.9(a), the MOSFET is operating in the triode region with $v_{DS} < v_{GS} - V_{TN}$, as discussed previously. In Fig. 4.9(b), the value of v_{DS} has increased to $v_{DS} = v_{GS} - V_{TN}$, for which the channel just disappears at the drain. Figure 4.9(c) shows the channel for an even larger value of v_{DS}. The channel region has disappeared, or *pinched off,* before reaching the drain end of the channel, and the resistive channel region is no longer in contact with the drain. At first glance, one may be

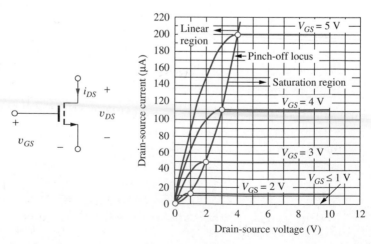

Figure 4.8 Output characteristics for an NMOS transistor with $V_{TN} = 1$ V and $K_n = 25 \times 10^{-6}$ A/V^2.

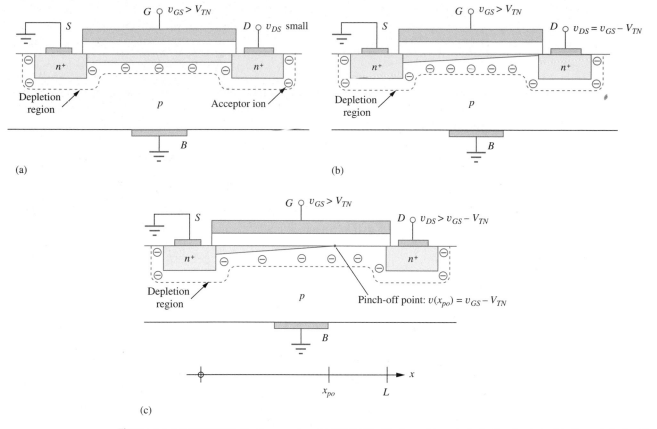

Figure 4.9 (a) MOSFET in the linear region. (b) MOSFET with channel just pinched off at the drain. (c) Channel pinch-off for $v_{DS} > v_{GS} - V_{TN}$.

inclined to expect that the current should become zero in the MOSFET. However, this is not the case. As depicted in Fig. 4.9(c), the voltage at the **pinch-off point** in the channel is always equal to

$$v_{GS} - v(x_{po}) = V_{TN} \qquad \text{or} \qquad v(x_{po}) = v_{GS} - V_{TN}$$

There is still a voltage equal to $v_{GS} - V_{TN}$ across the inverted portion of the channel, and electrons will be drifting down the channel from left to right. When the electrons reach the pinch-off point, they are injected into the depleted region between the end of the channel and the drain, and the electric field in the depletion region then sweeps these electrons on to the drain. Once the channel has reached pinch-off, the voltage drop across the inverted channel region is constant; hence, the drain current becomes constant and independent of drain-source voltage.

This region of operation of the MOSFET is often referred to as either the **saturation region** or the **pinch-off region** of operation. However, we will learn a different meaning for saturation when we discuss bipolar transistors in the next chapter. On the other hand, operation beyond pinchoff is the regime that we most often use for analog amplification, and in Part III we will use the term **active region** to refer to this region for both MOS and bipolar devices.

4.2.5 MATHEMATICAL MODEL IN THE SATURATION (PINCH-OFF) REGION

Now let us find an expression for the MOSFET drain current in the pinched-off channel. The drain-source voltage just needed to pinch off the channel at the drain is $v_{DS} = v_{GS} - V_{TN}$, and substituting this value into Eq. (4.13) yields an expression for the NMOS current in the saturation region of operation:

$$i_D = \frac{K_n' W}{2 L}(v_{GS} - V_{TN})^2 \qquad \text{for } v_{DS} \geq (v_{GS} - V_{TN}) \geq 0 \tag{4.17}$$

This is the classic square-law expression for the drain-source current for the n-channel MOSFET operating in pinch-off. The current depends on the square of $v_{GS} - V_{TN}$ but is now independent of the drain-source voltage v_{DS}. Equation (4.17) is *also used* frequently in the rest of this text. Be sure to commit it to memory!

The value of v_{DS} for which the transistor saturates is given the special name v_{DSAT} defined by

$$v_{\text{DSAT}} = v_{GS} - V_{TN} \tag{4.18}$$

and v_{DSAT} is referred to as the **saturation voltage,** or **pinch-off voltage,** of the MOSFET. Equation (4.17) can be interpreted in a manner similar to that of Eq. (4.14):

$$i_D = \left(C_{\text{ox}}'' W \frac{v_{GS} - V_{TN}}{2} \right) \left(\mu_n \frac{v_{GS} - V_{TN}}{L} \right) \tag{4.19}$$

The inverted channel region has a voltage of $v_{GS} - V_{TN}$ across it, as depicted in Fig. 4.9(c). Thus, the first term represents the magnitude of the average electron charge in the inversion layer, and the second term is the magnitude of the velocity of electrons in an electric field equal to $(v_{GS} - V_{TN})/L$.

An example of the overall output characteristics for an NMOS transistor with $V_{TN} = 1$ V and $K_n = 25$ μA/V^2 appeared in Fig. 4.8, in which the locus of pinch-off points is determined by $v_{DS} = v_{\text{DSAT}}$. To the left of the **pinch-off locus,** the transistor is operating in the triode region, and it is operating in the saturation region for operating points to the right of the locus. For $v_{GS} \leq V_{TN} = 1$ V, the transistor is cut off, and the drain current is zero. As the gate voltage is increased in the saturation region, the curves spread out due to the square-law nature of Eq. (4.17).

Figure 4.10 gives an individual output characteristic for $V_{GS} = 3$ V, showing the behavior of the individual triode and saturation region equations. The triode region expression given in Eq. (4.13) is represented by the inverted parabola in Fig. 4.10. Note that it does not represent a valid model for the i-v behavior for $V_{DS} > V_{GS} - V_{TN} = 2$ V for this particular device. Note also that the maximum drain voltage must never exceed the Zener breakdown voltage of the drain-substrate pn junction diode.

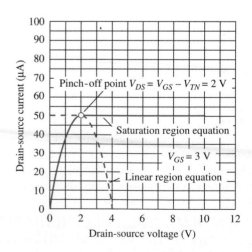

Figure 4.10 Output characteristic showing intersection of the linear region and saturation region equations at the pinch-off point.

EXERCISE: Calculate the drain current for an NMOS transistor operating with $V_{GS} = 5$ V and $V_{DS} = 10$ V if $V_{TN} = 1$ V and $K_n = 1$ mA/V^2. What is the W/L ratio of this device if $K'_n = 40$ μA/V^2? What is W if $L = 0.35$ μm?

ANSWERS: 8.00 mA; 25/1; 8.75 μm

4.2.6 TRANSCONDUCTANCE

An important characteristic of transistors is the **transconductance** given the symbol g_m. The transconductance of the MOS devices relates the change in drain current to a change in gate-source voltage:

$$g_m = \frac{di_D}{dv_{GS}}\bigg|_{Q-pt} = K'_n\frac{W}{L}(V_{GS} - V_{TN}) = \frac{2I_D}{V_{GS} - V_{TN}} \qquad (4.20)$$

where we have taken the derivative of Eq. (4.17) and evaluated the result at the Q-point. We encounter g_m frequently in electronics, particularly during our study of analog circuit design. It is interesting to note that g_m is the reciprocal of the on-resistance defined in Eq. (4.16).

EXERCISE: Find the drain current and transconductance for an NMOS transistor operating with $V_{GS} = 2.5$ V, $V_{TN} = 1$ V, and $K_n = 1$ mA/V^2.

ANSWERS: 1.13 mA; 1.5 mS

4.2.7 CHANNEL-LENGTH MODULATION

The output characteristics of the device in Fig. 4.8 indicate that the drain current is constant once the device enters the saturation region of operation. However, this is not quite true. Rather, the i-v curves have a small positive slope, as indicated in Fig. 4.11(a). The drain current increases slightly as the drain-source voltage increases. The increase in drain current visible in Fig. 4.11 is the result of a phenomenon called **channel-length modulation,** which can be understood by referring

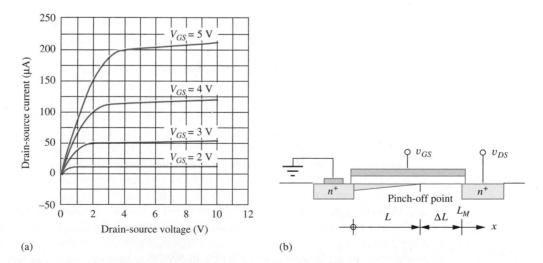

(a)   (b)

Figure 4.11 (a) Output characteristics including the effects of channel-length modulation. (b) Channel-length modulation.

to Fig. 4.11(b), in which the channel region of the NMOS transistor is depicted for the case of $v_{DS} > v_{DSAT}$. The channel pinches off before it makes contact with the drain. Thus, the actual length of the resistive channel is given by $L = L_M - \Delta L$. As v_{DS} increases above v_{DSAT}, the length of the depleted channel region ΔL also increases, and the effective value of L decreases. Therefore, the value of L in the denominator of Eq. (4.17) actually has a slight inverse dependence on v_{DS}, leading to an increase in drain current increases as v_{DS} increases. The expression in Eq. (4.17) can be heuristically modified to include this drain-voltage dependence as

$$i_D = \frac{K'_n}{2} \frac{W}{L} (v_{GS} - V_{TN})^2 (1 + \lambda v_{DS}) \qquad (4.21)$$

in which λ is called the **channel-length modulation parameter.** The value of λ is dependent on the channel length, and typical values are $0.001 \text{ V}^{-1} \le \lambda \le 0.10 \text{ V}^{-1}$. In Fig. 4.11, λ is approximately 0.01 V^{-1}, which yields a 10 percent increase in drain current for a drain-source voltage change of 10 V.

EXERCISE: Calculate the drain current for an NMOS transistor operating with $V_{GS} = 5$ V and $V_{DS} = 10$ V if $V_{TN} = 1$ V, $K_n = 1 \text{ mA/V}^2$, and $\lambda = 0.02 \text{ V}^{-1}$. What is I_D for $\lambda = 0$?

ANSWERS: 9.60 mA; 8.00 mA

EXERCISE: Calculate the drain current for the NMOS transistor in Fig. 4.11 operating with $V_{GS} = 4$ V and $V_{DS} = 5$ V if $V_{TN} = 1$ V, $K_n = 25 \text{ μA/V}^2$, and $\lambda = 0.01 \text{ V}^{-1}$. Repeat for $V_{GS} = 5$ V and $V_{DS} = 10$ V.

ANSWERS: 118 μA; 220 μA

4.2.8 TRANSFER CHARACTERISTICS AND DEPLETION-MODE MOSFETS

The output characteristics in Figs. 4.7 and 4.11 represented our first look at graphical representations of the i-v characteristics of the transistor. The output characteristics plot drain current versus drain-source voltage for fixed values of the gate-source voltage. The second commonly used graphical format, called the **transfer characteristic,** plots drain current versus gate-source voltage for a fixed drain-source voltage. An example of this form of characteristic is given in Fig. 4.12 for two NMOS transistors in the pinch-off region. Up to now, we have been assuming that the threshold voltage of the NMOS transistor is positive, as in the right-hand curve in Fig. 4.12. This curve corresponds to an enhancement-mode device with $V_{TN} = +2$ V. Here we can clearly see the turn-on of the transistor as v_{GS} increases. The device is off (nonconducting) for $v_{GS} \le V_{TN}$, and it starts to conduct as v_{GS} exceeds V_{TN}. The curvature reflects the square-law behavior of the transistor in the saturation region as described by Eq. (4.17).

However, it is also possible to fabricate NMOS transistors with values of $V_{TN} \le 0$. These transistors are called **depletion-mode MOSFETs,** and the transfer characteristic for such a device with $V_{TN} = -2$ V is depicted in the left-hand curve in Fig. 4.12(a). Note that a nonzero drain current exists in the depletion-mode MOSFET for $v_{GS} = 0$; a negative value of v_{GS} is required to turn the device off.

The cross section of the structure of a depletion-mode NMOSFET is shown in Fig. 4.12(b). A process called *ion implantation* is used to form a built-in n-type channel in the device so that the source and drain are connected through the resistive channel region. A negative voltage must be applied to the gate to deplete the n-type channel region and eliminate the current path between the source and drain (hence the name depletion-mode device). In Chapter 6 we will see that the ion-implanted depletion-mode device played an important role in the evolution of MOS logic circuits. The addition of the depletion-mode MOSFET to NMOS technology provided substantial performance improvement, and it was a rapidly accepted change in technology in the mid 1970s.

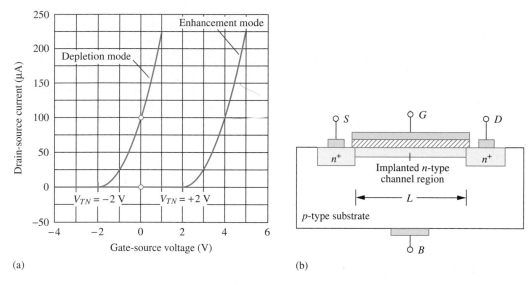

Figure 4.12 (a) Transfer characteristics for enhancement-mode and depletion-mode NMOS transistors. (b) Cross section of a depletion-mode NMOS transistor.

EXERCISE: Calculate the drain current for the NMOS depletion-mode transistor in Fig. 4.12 for $V_{GS} = 0$ V if $K_n = 50$ μA/V². Assume the transistor is in the pinch-off region. What value of V_{GS} is required to achieve the same current in the enhancement-mode transistor in the same figure?

ANSWERS: 100 μA; 4 V

EXERCISE: Calculate the drain current for the NMOS depletion-mode transistor in Fig. 4.12 for $V_{GS} = +1$ V if $K_n = 50$ μA/V². Assume the transistor is in the pinch-off region.

ANSWER: 225 μA

4.2.9 BODY EFFECT OR SUBSTRATE SENSITIVITY

Thus far, it has been assumed that the source-bulk voltage v_{SB} is zero. With $v_{SB} = 0$, the MOSFET behaves as if it were a three-terminal device. However, we find many circuits, particularly in ICs, in which the bulk and source of the MOSFET must be connected to different voltages so that $v_{SB} \neq 0$. A nonzero value of v_{SB} affects the i-v characteristics of the MOSFET by changing the value of the threshold voltage. This effect is called **substrate sensitivity,** or **body effect,** and can be modeled by

$$V_{TN} = V_{TO} + \gamma \left(\sqrt{v_{SB} + 2\phi_F} - \sqrt{2\phi_F} \right) \tag{4.22}$$

where V_{TO} = **zero-substrate-bias value for** V_{TN} (V)

γ = **body-effect parameter** ($\sqrt{V}$)

$2\phi_F$ = **surface potential parameter** (V)

Parameter γ determines the intensity of the body effect, and its value is set by the relative sizes of the oxide and depletion-layer capacitances C''_{ox} and C_d in Fig. 4.3. The surface potential represents the approximate voltage across the depletion layer at the onset of inversion. For typical NMOS transistors, -5 V $\leq V_{TO} \leq +5$ V, $0 \leq \gamma \leq 3\sqrt{V}$, and 0.3 V $\leq 2\phi_F \leq 1$ V.

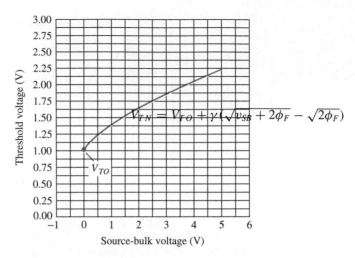

Figure 4.13 Threshold variation with source-bulk voltage for an NMOS transistor, with $V_{TO} = 1$ V, $2\phi_F = 0.6$ V and $\gamma = 0.75\sqrt{V}$.

We use $2\phi_F = 0.6$ V throughout the rest of this text, and Eq. (4.22) will be represented as

$$V_{TN} = V_{TO} + \gamma \left(\sqrt{v_{SB} + 0.6} - \sqrt{0.6} \right) \tag{4.23}$$

Figure 4.13 plots an example of the threshold-voltage variation with source-bulk voltage for an NMOS transistor, with $V_{TO} = 1$ V and $\gamma = 0.75\sqrt{V}$. We see that $V_{TN} = V_{TO} = 1$ V for $v_{SB} = 0$ V, but the value of V_{TN} more than doubles for $v_{SB} = 5$ V. In Chapter 6, we will see that this behavior can have a significant impact on the design of MOS logic circuits.

NMOS TRANSISTOR MATHEMATICAL MODEL SUMMARY

Equations (4.24) through (4.28) represent the complete model for the i-v behavior of the NMOS transistor.

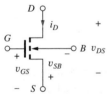

NMOS transistor

For all regions,

$$K_n = K_n' \frac{W}{L} \qquad K_n' = \mu_n C_{ox}'' \qquad i_G = 0 \qquad i_B = 0 \tag{4.24}$$

Cutoff region:

$$i_D = 0 \qquad \text{for } v_{GS} \leq V_{TN} \tag{4.25}$$

Triode region:

$$i_D = K_n \left(v_{GS} - V_{TN} - \frac{v_{DS}}{2} \right) v_{DS} \qquad \text{for } v_{GS} - V_{TN} \geq v_{DS} \geq 0 \tag{4.26}$$

Saturation region:

$$i_D = \frac{K_n}{2} (v_{GS} - V_{TN})^2 (1 + \lambda v_{DS}) \qquad \text{for } v_{DS} \geq (v_{GS} - V_{TN}) \geq 0 \tag{4.27}$$

Threshold voltage:

$$V_{TN} = V_{TO} + \gamma \left(\sqrt{v_{SB} + 2\phi_F} - \sqrt{2\phi_F} \right) \tag{4.28}$$

4.3 PMOS TRANSISTORS

MOS transistors with p-type channels (PMOS transistors) can also easily be fabricated. In fact, as mentioned earlier, the first commercial MOS transistors and integrated circuits used PMOS devices because it was easier to control the fabrication process of the PMOS technology. The PMOS device is built by forming p-type source and drain regions in an n-type substrate, as depicted in the device cross section in Fig. 4.14(a).

The qualitative behavior of the transistor is essentially the same as that of an NMOS device except that the normal voltage and current polarities are reversed. The normal directions of current in the **PMOS transistor** are indicated in Fig. 4.14. A negative voltage on the gate relative to the source ($v_{GS} < 0$) is required to attract holes and create a p-type inversion layer in the channel region. To initiate conduction in the enhancement-mode PMOS transistor, the gate-source voltage must be more negative than the threshold voltage of the p-channel device, denoted by V_{TP}. To keep the source-substrate and drain-substrate junctions reverse-biased, v_{SB} and v_{DB} must also be less than zero. This requirement is satisfied by $v_{DS} \leq 0$.

An example of the output characteristics for an enhancement-mode PMOS transistor is given in Fig. 4.14(b). For $v_{GS} \geq V_{TP} = -1$ V, the transistor is off. For more negative values of v_{GS}, the drain current increases in magnitude. The PMOS device is in the triode region for small values of V_{DS}, and the saturation of the characteristics is apparent at larger V_{DS}. The curves look just like those for

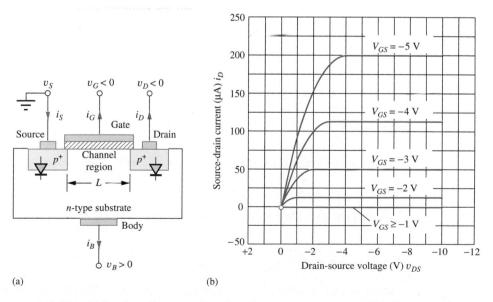

Figure 4.14 (a) Cross section of an enhancement-mode PMOS transistor. (b) Output characteristics for a PMOS transistor with $V_{TP} = -1$ V.

the NMOS device except for sign changes on the values of v_{GS} and v_{DS}. This is a result of assigning the positive current direction to current exiting from the drain terminal of the PMOS transistor.

PMOS TRANSISTOR MATHEMATICAL MODEL SUMMARY

The mathematical model for the PMOS transistor is summarized below in Eqs. (4.29) through (4.33).

For all regions,

$$K_p = K_p' \frac{W}{L} \qquad K_p' = \mu_p C_{ox}'' \qquad i_G = 0 \qquad i_B = 0 \tag{4.29}$$

Cutoff region:

$$i_D = 0 \qquad \text{for } V_{GS} \geq V_{TP} \tag{4.30}$$

Triode region:

$$i_D = K_p \left(v_{GS} - V_{TP} - \frac{v_{DS}}{2} \right) v_{DS} \qquad \text{for } 0 \leq |v_{DS}| \leq |v_{GS} - V_{TP}| \tag{4.31}$$

Saturation region:

$$i_D = \frac{K_p}{2} (v_{GS} - V_{TP})^2 (1 + \lambda |v_{DS}|) \qquad \text{for } |v_{DS}| \geq |v_{GS} - V_{TP}| \geq 0 \tag{4.32}$$

Threshold voltage:

$$V_{TP} = V_{TO} - \gamma \left(\sqrt{v_{BS} + 2\phi_F} - \sqrt{2\phi_F} \right) \tag{4.33}$$

For the enhancement-mode PMOS transistor, $V_{TP} < 0$. Depletion-mode PMOS devices can also be fabricated; $V_{TP} \geq 0$ for these devices.

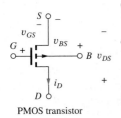

PMOS transistor

Various authors have different ways of writing the equations that describe the PMOS transistor. Our choice attempts to avoid as many confusing minus signs as possible. The drain-current expressions for the PMOS transistor are written in similar form to those for the NMOS transistor except that the drain-current direction is reversed and the values of v_{GS} and v_{DS} are now negative quantities. A sign must still be changed in the expressions, however. The parameter γ is normally specified as a positive value for both n- and p-channel devices, and a positive bulk-source potential will cause the PMOS threshold voltage to become more negative.

An important parametric difference appears in the expressions for K_p and K_n. In the PMOS device, the charge carriers in the channel are holes, so current is proportional to hole mobility μ_p. Hole mobility is typically only 40 percent of the electron mobility, so for a given set of voltage bias conditions, the PMOS device will conduct only 40 percent of the current of the NMOS device! Higher current capability leads to higher frequency operation in both digital and analog circuits. Thus, NMOS devices are preferred over PMOS devices in many applications.

EXERCISE: What is the region of operation and drain current of a PMOS transistor having $V_{TP} = -1$ V, $K_p = 0.4$ mA/V^2, and $\lambda = 0.02$ V^{-1} for (a) $V_{GS} = 0$ V, $V_{DS} = -1$ V; (b) $V_{GS} = -2$ V, $V_{DS} = -0.5$ V; (c) $V_{GS} = -2$ V, $V_{DS} = -2$ V?

ANSWERS: (a) cutoff, 0 A; (b) triode, 150 μA; (c) saturation, 208 μA

4.4 MOSFET CIRCUIT SYMBOLS

Standard circuit symbols for four different types of MOSFETs are given in Fig. 4.15: (a) NMOS enhancement-mode, (b) PMOS enhancement-mode, (c) NMOS depletion-mode, and (d) PMOS depletion-mode transistors. The four terminals of the MOSFET are identified as source (S), drain (D),

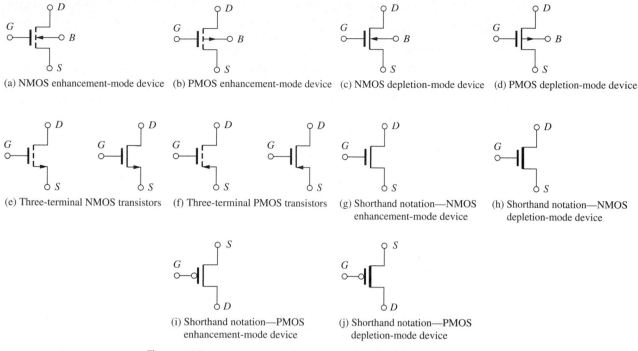

(a) NMOS enhancement-mode device (b) PMOS enhancement-mode device (c) NMOS depletion-mode device (d) PMOS depletion-mode device

(e) Three-terminal NMOS transistors (f) Three-terminal PMOS transistors (g) Shorthand notation—NMOS enhancement-mode device (h) Shorthand notation—NMOS depletion-mode device

(i) Shorthand notation—PMOS enhancement-mode device (j) Shorthand notation—PMOS depletion-mode device

Figure 4.15 (a)–(f) IEEE Standard MOS transistor circuit symbols. (g)–(j) Other commonly used symbols.

gate (G), and bulk (B). The arrow on the **bulk terminal** indicates the polarity of the bulk-drain, bulk-source, and bulk-channel *pn* junction diodes; the arrow points inward for an NMOS device and outward for the PMOS transistor. Enhancement-mode devices are indicated by the dashed line in the channel region, whereas depletion-mode devices have a solid line, indicating the existence of the built-in channel. The gap between the gate and channel represents the insulating oxide region. Table 4.1 summarizes the threshold-voltage values for the four types of NMOS and PMOS transistors.

In many circuit applications, the MOSFET substrate terminal is connected to its source. The shorthand notation in Fig. 4.15(e) and 4.15(f) is often used to represent these three-terminal MOS-FETs. The arrow identifies the source terminal and points in the direction of normal positive current.

To further add to the confusing array of symbols that the circuit designer must deal with, a number of additional symbols are used in other texts and reference books and in papers in technical journals. The wide diversity of symbols is unfortunate, but it is a fact of life that circuit designers must accept. For example, if one tires of drawing the dashed line for the enhancement-mode device as well as the substrate arrow, one arrives at the NMOS transistor symbol in Fig. 4.15(g); the channel line is then thickened to represent the NMOS depletion-mode device, as in Fig. 4.15(h). In a similar vein, the symbol in Fig. 4.15(i) represents the enhancement-mode PMOS transistor, and the corresponding depletion-mode PMOS device appears in Fig. 4.15(j). In the last two symbols, the circles represent a carry-over from logic design and are meant to indicate the logical inversion operation. We explore

TABLE 4.1
Categories of MOS transistors

	NMOS DEVICE	PMOS DEVICE
Enhancement-mode	$V_{TN} > 0$	$V_{TP} < 0$
Depletion-mode	$V_{TN} \leq 0$	$V_{TP} \geq 0$

ELECTRONICS IN ACTION

CMOS Camera on a Chip

Earlier in this text we examined the CCD image sensor widely used in astronomy. Although the CCD imager produces very high quality images, it requires an expensive specialized manufacturing process, complex control circuitry, and consumes a substantial amount of power. In the early 1990s, designers began developing techniques to integrate photo-detection circuitry onto inexpensive mainstream digital CMOS processes. In 1993, Dr. Eric Fossum's group at the Jet Propulsion Laboratory announced a CMOS digital camera on a chip. Since that time, many companies have designed camera chips that are based on mainstream CMOS processes, allowing the merging of many camera functions onto a single chip.

Pictured here is a photo of such a chip from Micron Technology.[1] The device produces full color images and has 1.3 million pixels in a 1280×1024 imaging array.

1.3 MegaPixel CMOS active-pixel image sensor.[1]

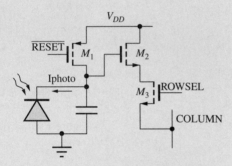

Typical photo diode pixel architecture.

A typical photodiode-based imaging pixel is also shown above. After asserting the $\overline{\text{RESET}}$ signal, the storage capacitor is fully charged to V_{DD} through transistor M_1. The reset signal is then removed, and light incident on the photodiode generates a photo current that discharges the capacitor. Different light intensities produce different voltages on the capacitor at the end of the light integration time. To read the stored value, the row select (ROWSEL) signal is asserted, and the capacitor voltage is driven onto the COLUMN bus via transistors M_2 and M_3.

In many designs random variations in the device characteristics will cause variations in the signal produced by each pixel for the same intensity of incident light. To correct for many of these variations, a technique known as *correlated double sampling* is used. After the signal level is read from a pixel, the pixel is reset and then read again to acquire a baseline signal. The baseline signal is subtracted from the desired signal, thereby removing the non-uniformities and noise sources which are common to both of the acquired signals.

Chips like this one are now common in digital cameras and digital camcorders. These now-common and inexpensive portable devices are enabled by the integration of analog photosensitive pixel structures with mainstream CMOS processes.

[1] The chip pictured above is a Micron Technology MI-MV13 image sensor and is reprinted here with permission from Micron Technology, Inc.

this more fully in Part II of this book. The symbols in Figs. 4.15(g) and (i) commonly appear in books discussing VLSI logic design.

The symmetry of MOS devices should be noted in the cross sections of Figs. 4.4 and 4.14. The terminal that is acting as the drain is actually determined by the applied potentials. Current can traverse the channel in either direction, depending on the applied voltage. For NMOS transistors, the n^+ region that is at the highest voltage will be the drain, and the one at the lowest voltage will be the source. For the PMOS transistor, the p^+ region at the lowest voltage will be the drain, and the one at the highest voltage will be the source. In later chapters, we shall see that this symmetry is highly useful in certain applications, particularly in MOS logic and dynamic random-access memory (DRAM) circuits.

DESIGN NOTE MOS DEVICE SYMMETRY

The MOS transistor terminal that is acting as the drain is actually determined by the applied potentials. Current can traverse the channel in either direction, depending on the applied voltage.

4.5 MOS TRANSISTOR FABRICATION AND LAYOUT DESIGN RULES[2]

In addition to choosing the circuit topology, the MOS integrated circuit designer must pick the values of the W/L ratios of the transistors and develop a layout for the circuit that ensures that it will achieve the performance specifications. Design of the layout of transistors and circuits in integrated form is constrained by a set of rules termed the **design rules** or **ground rules.** These rules are technology specific and specify minimum sizes, spacings and overlaps for the various shapes that define transistors. The sets of rules are different for MOS and bipolar processes, for MOS processes designed specifically for logic and memory, and even for similar processes from different companies.

4.5.1 MINIMUM FEATURE SIZE AND ALIGNMENT TOLERANCE

Processes are defined around a **minimum feature size F**, which represents the width of the smallest line or space that can be reliably transferred to the surface of a wafer using a given generation of lithographic manufacturing tools. To produce a basic set of ground rules, we must also know the maximum misalignment which can occur between two mask levels during fabrication. For example, Fig. 4.16(a) shows the nominal position of a metal line aligned over a contact window (indicated by the box with and × in it). The metal overlaps the contact window by at least one **alignment tolerance T** in all directions. During the fabrication process, alignment will not be perfect, and the actual structure may exhibit misalignment in the x or y directions or both. Figures 4.16(b) through 4.16(d) show the

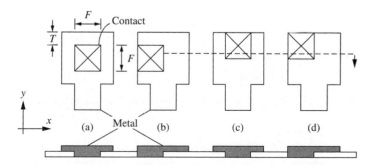

Figure 4.16 Misalignment of a metal pattern over a contact opening: (a) desired alignment, (b) one possible worst-case misalignment in the x direction, (c) one possible worst-case misalignment in the y direction, and (d) misalignment in both directions.

[2] Reproduced with permission from *Introduction to Microelectronic Fabrication*, Second Edition, by Richard C. Jaeger, Prentice Hall, 2000.

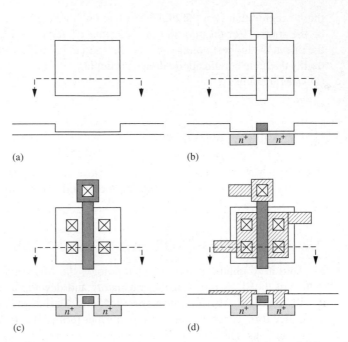

Figure 4.17 (a) Active area mask, (b) gate mask, (c) contact opening mask, (d) metal mask.

result of one possible set of worst-case alignments of the patterns in the x, y, and both directions simultaneously. Our set of design rules assume that T is the same in both directions. Transistors designed with our ground rules will fail to operate properly if the misalignment exceeds tolerance T.

4.5.2 MOS TRANSISTOR LAYOUT

Figure 4.17 outlines the process and mask sequence used to fabricate a basic polysilicon-gate transistor. The first mask defines the active area, or thin oxide region of the transistor, and the second mask defines the polysilicon gate of the transistor. The channel region of the transistor is actually produced by the intersection of these first two mask layers; the source and/or drain regions are formed wherever the active layer (mask 1) is *not* covered by the gate layer (mask 2). The third and fourth masks delineate the contact openings and the metal pattern. The overall mask sequence is

Active area mask	Mask 1
Polysilicon-gate mask	Mask 2 — align to mask 1
Contact window mask	Mask 3 — align to mask 2
Metal mask	Mask 4 — align to mask 3

The alignment sequence must be specified to properly account for alignment tolerances in the ground rules. In this particular example, each mask is aligned to the one used in the preceding step, but this is not always the case.

We will now explore a set of design rules similar in concept to those developed by Mead and Conway [3]. These ground rules were designed to permit easy translation of a design from one generation of technology to another by simply changing the size of one parameter Λ. To achieve this goal, the rules are quite forgiving in terms of the mask-to-mask alignment tolerance.

A composite set of rules for a transistor is shown graphically in Fig. 4.18 in which the minimum feature size $F = 2\,\Lambda$ and the alignment tolerance $T = F/2 = \Lambda$. (Parameter Λ could be 0.5, 0.25, or 0.1 μm, for example.) Note that an alignment tolerance equal to one-half the minimum feature size is a very forgiving alignment tolerance.

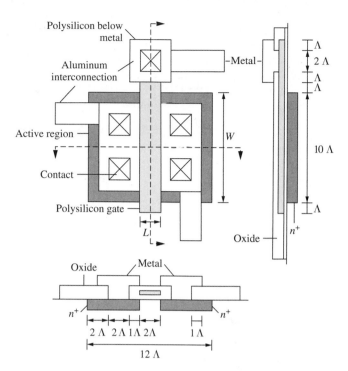

Figure 4.18 Composite top view and cross sections of a transistor with $W/L = 5/1$ demonstrating a basic set of ground rules.

For the transistor in Fig. 4.18, all linewidths and spaces must be a minimum feature size of $2\,\Lambda$. Square contacts are a minimum feature size of $2\,\Lambda$ in each dimension. To ensure that the metal completely covers the contact for worst-case misalignment, a $1\,\Lambda$ border of metal is required around the contact region. The polysilicon gate must overlap the edge of the active area and the contact openings by $1\,\Lambda$. However, because of the potential for tolerance accumulation during successive misalignments of masks 2 and 3, the contacts must be inside the edges of the active area by $2\,\Lambda$.

The transistor in Fig. 4.18 has a W/L ratio of $10\,\Lambda/2\,\Lambda$ or $5/1$, and the total active area is $120\,\Lambda^2$. Thus the active channel region represents approximately 17 percent of the total area of the transistor. Note that the polysilicon gate defines the edges of the source and/or drain regions and results in "self-alignment" of the edges of the gate to the edges of the channel region. Self-alignment of the gate to the channel reduces the size of the transistor and minimizes the "overlap capacitances" associated with the transistor. We will explore these capacitances in more detail in Sec. 4.6.

EXERCISE: What is the active area of the transistor in Fig. 4.18 if $\Lambda = 0.5\ \mu$m? What are the values of W and L for the transistor. What is the area of the transistor gate region? How many of these transistors could be packed together on a 1 cm $\times$ 1 cm integrated circuit die if the active areas of the individual transistors must be spaced apart by a minimum of $4\,\Lambda$?

ANSWERS: 30 μm^2; 1 μm; 5 μm; 5 μm^2; 1.79 million

4.6 CAPACITANCES IN MOS TRANSISTORS

Every electronic device has internal capacitances that limit the high-frequency performance of the particular device. In logic applications, these capacitances limit the switching speed of the circuits, and in amplifiers, the capacitances limit the frequency at which useful amplification can be

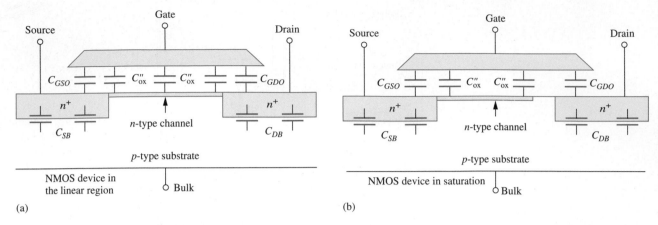

Figure 4.19 (a) NMOS capacitances in the linear region. (b) NMOS capacitances in the active region.

obtained. Thus knowledge of the origin and modeling of these capacitances is quite important, and an introductory discussion of the capacitances of the MOS transistor appears in this section.

4.6.1 NMOS TRANSISTOR CAPACITANCES IN THE TRIODE REGION

Figure 4.19(a) shows the various capacitances associated with the MOS field-effect transistor operating in the triode region, in which the resistive channel region connects the source and drain. A simple model for these capacitances was presented by Meyer [4]. The total gate-channel capacitance C_{GC} is equal to the product of the **gate-channel capacitance** per unit area C''_{ox} (F/m^2) and the area of the gate:

$$C_{GC} = C''_{ox}WL \tag{4.34}$$

In the Meyer model for the triode region, C_{GC} is partitioned into two equal parts. The **gate-source capacitance C_{GS}** and the **gate-drain capacitance C_{GD}** each consist of one-half of the gate-channel capacitance plus the overlap capacitances C_{GSO} and C_{GDO} associated with the gate-source or gate-drain regions:

$$C_{GS} = \frac{C_{GC}}{2} + C_{GSO}W = C''_{ox}\frac{WL}{2} + C_{GSO}W$$

$$C_{GD} = \frac{C_{GC}}{2} + C_{GDO}W = C''_{ox}\frac{WL}{2} + C_{GDO}W \tag{4.35}$$

The overlap capacitances arise from two sources. First, the gate is actually not perfectly aligned to the edges of the source and drain diffusion but overlaps the diffusions somewhat. In addition, fringing fields between the gate and the source and drain regions contribute to the values of C_{GSO} and C_{GDO}.

The **gate-source** and **gate-drain overlap capacitances** C_{GSO} and C_{GDO} are normally specified as oxide capacitances per unit width (F/m). Note that C_{GS} and C_{GD} each have a component that is proportional to the area of the gate and one proportional to the width of the gate.

The capacitances of the reverse-biased pn junctions, indicated by the **source-bulk** and **drain-bulk capacitances** C_{SB} and C_{DB}, respectively, exist between the source and drain diffusions and the substrate of the MOSFET. Each capacitance consists of a component proportional to the junction bottom area of the source (A_S) or drain (A_D) region and a component proportional to the perimeter of the source (P_S) or drain (P_D) junction region:

$$C_{SB} = C_J A_S + C_{JSW} P_S \qquad C_{DB} = C_J A_D + C_{JSW} P_D \tag{4.36}$$

Here C_J is called the junction bottom capacitance per unit area (F/m^2), and C_{JSW} is the junction sidewall capacitance per unit length. C_{SB} and C_{DB} will be present regardless of the region of operation. Note that the junction capacitances are voltage dependent [see Eq. (3.21)].

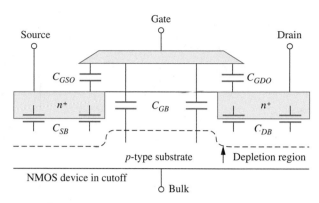

Figure 4.20 NMOS capacitances in the cutoff region.

4.6.2 CAPACITANCES IN THE SATURATION REGION

In the saturation region of operation, depicted in Fig. 4.19(b), the portion of the channel beyond the pinch-off point disappears. The Meyer models for the values of C_{GS} and C_{GD} become

$$C_{GS} = \frac{2}{3}C_{GC} + C_{GSO}W \qquad \text{and} \qquad C_{GD} = C_{GDO}W \qquad (4.37)$$

in which C_{GS} now contains two-thirds of C_{GC}, but only the overlap capacitance contributes to C_{GD}. Now C_{GD} is directly proportional to W, whereas C_{GS} retains a component dependent on $W \times L$.

4.6.3 CAPACITANCES IN CUTOFF

In the cutoff region of operation, depicted in Fig. 4.20, the conducting channel region is gone. The values of C_{GS} and C_{GD} now contain only the overlap capacitances:

$$C_{GS} = C_{GSO}W \qquad \text{and} \qquad C_{GD} = C_{GDO}W \qquad (4.38)$$

In the cutoff region, a small capacitance C_{GB} appears between the gate and bulk terminal, as indicated in Fig. 4.20.

$$C_{GB} = C_{GBO}W \qquad (4.39)$$

in which C_{GBO} is the gate-bulk **capacitance per unit width**.

It should be clear from Eqs. (4.34) to (4.39) that MOSFET capacitances depend on the region of operation of the transistor and are nonlinear functions of the voltages applied to the terminals of the device. In subsequent chapters we analyze the impacts of these capacitances on the behavior of digital and analog circuits. Complete models for these nonlinear capacitances are included in circuit simulation programs such as SPICE, and circuit simulation is an excellent tool for exploring the detailed impact of these capacitances on circuit performance.

> **EXERCISE:** Calculate C_{GS} and C_{GD} for a transistor operating in the triode and saturation regions if $C''_{ox} = 200$ μF/m², $C_{GSO} = C_{GDO} = 300$ pF/m, $L = 0.5$ μm, and $W = 5$ μm.
>
> **ANSWERS:** 1.75 fF, 1.75 fF; 1.83 fF, 1.5 fF

4.7 MOSFET MODELING IN SPICE

The SPICE circuit analysis program is used to simulate more complicated circuits and to make much more detailed calculations than we can perform by hand analysis. The circuit representation for the MOSFET model that is implemented in SPICE is given in Fig. 4.21, and as we can observe, the model

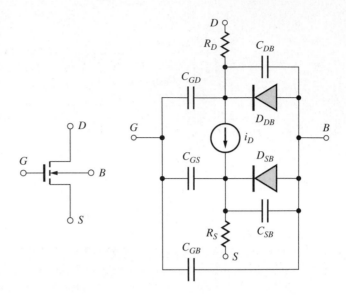

Figure 4.21 SPICE model for the NMOS transistor.

uses quite a number of circuits elements in an attempt to accurately represent the characteristics of a real MOSFET. For example, small resistances R_S and R_D appear in series with the external MOSFET source and drain terminals, and diodes are included between the source and drain regions and the substrate. The need for the power of the computer is clear here. It would be virtually impossible for us to use this sophisticated a model in our hand calculations.

As many as 20 different MOSFET models [5] of varying complexity are built into various versions of the SPICE simulation program, and they are denoted by "Level=Model_Number". The levels each have a unique mathematical formulation for current source i_D and for the various device capacitances. The model we have studied in this chapter is the most basic model and is referred to as the Level-1 model (LEVEL=1). Largely because of a lack of standard parameter usage at the time SPICE was first written, as well as the limitations of the programming languages originally used, the parameter names that appear in the models differ from those used in this text and throughout the literature. The LEVEL=1 model is coded into SPICE using the following formulas, which are the similar to those we have already studied.

Table 4.2 contains the equivalences of the **SPICE model** parameters and our equations summarized in Sec. 4.2. Typical and default values of the SPICE model parameters can be found in Table 4.2. A similar model is used for the PMOS transistor, but the polarities of the voltages and currents, and the directions of the diodes, are reversed.

$$\text{Triode region:} \qquad i_D = \text{KP}\frac{W}{L}\left(v_{GS} - \text{VT} - \frac{v_{DS}}{2}\right)v_{DS}(1 + \text{LAMBDA} \cdot v_{DS})$$

$$\text{Saturation region:} \quad i_D = \frac{\text{KP}}{2}\frac{W}{L}(v_{GS} - \text{VT})^2(1 + \text{LAMBDA} \cdot v_{DS}) \qquad (4.40)$$

$$\text{Threshold voltage: VT} = \text{VTO} + \gamma\left(\sqrt{v_{SB} + \text{PHI}} - \sqrt{\text{PHI}}\right)$$

Notice that the SPICE level-1 description includes the addition of channel-length modulation to the triode region expression. Also, be sure not to confuse SPICE threshold voltage **VT** with thermal voltage V_T.

TABLE 4.2
SPICE Parameter Equivalences

PARAMETER	OUR TEXT	SPICE	DEFAULT
Transconductance	K_n or K_p	KP	$20\ \mu A/V^2$
Threshold voltage	V_{TN} or V_{TP}	VT	—
Zero-bias threshold voltage	V_{TO}	VTO	1 V
Surface potential	$2\phi_F$	PHI	0.6 V
Body effect	γ	GAMMA	0
Channel length modulation	λ	LAMBDA	0
Mobility	μ_n or μ_p	UO	$600\ cm^2/V\cdot s$
Gate-drain capacitance per unit width	C_{GDO}	CGDO	0
Gate-source capacitance per unit width	C_{GSO}	CGSO	0
Gate-bulk capacitance per unit width	C_{GBO}	CGBO	0
Junction bottom capacitance per unit area	C_J	CJ	0
Grading coefficient	MJ	MJ	$0.5\ V^{0.5}$
Sidewall capacitance	C_{JSW}	CJSW	0
Sidewall grading coefficient	MJSW	MJSW	$0.5\ V^{0.5}$
Oxide thickness	T_{ox}	TOX	100 nm
Junction saturation current	I_S	IS	10 fA
Built-in potential	ϕ_j	PB	0.8 V
Ohmic drain resistance	—	RD	0
Ohmic source resistance	—	RS	0

The junction capacitances are modeled in SPICE by a generalized form of the capacitance expression in Eq. (3.21)

$$C_J = \frac{CJO}{\left(1 + \dfrac{v_R}{PB}\right)^{MJ}} \quad \text{and} \quad C_{JSW} = \frac{CJSWO}{\left(1 + \dfrac{v_R}{PB}\right)^{MJSW}} \tag{4.41}$$

in which v_R is the reverse bias across the *pn* junction.

EXERCISE: What are the values of SPICE model parameters **KP, LAMBDA, VTO, PHI,** *W*, and *L* for a transistor with the following characteristics: $V_{TN} = 1$ V, $K_n = 150\ \mu A/V^2$, $W = 1.5\ \mu m$, $L = 0.25\ \mu m$, $\lambda = 0.0133\ V^{-1}$, and $2\phi_F = 0.6$ V?

ANSWERS: $150\ \mu A/V^2$; $0.0133\ V^{-1}$; 1 V; 0.6 V; $1.5\ \mu m$; $0.25\ \mu m$ (specified in SPICE as 150U; 0.0133; 1; 0.6; 1.5U; 0.25U)

4.8 BIASING THE NMOS FIELD-EFFECT TRANSISTOR

As stated before, the MOS circuit designer has the flexibility to choose the circuit topology and W/L ratios of the devices in the circuit, and to a lesser extent, the voltages applied to the devices. As designers, we need to develop a mental catalog of useful circuit configurations, and we begin by looking at several basic circuits for biasing the MOSFET.

4.8.1 WHY DO WE NEED BIAS?

We have found that the MOSFET has three regions of operation: cutoff, triode, and saturation. For circuit applications, we want to establish a well-defined **quiescent operating point,** or **Q-point,** for the MOSFET in a particular region of operation. The Q-point for the MOSFET is represented by the

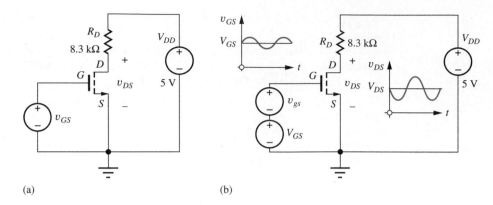

Figure 4.22 (a) Circuit for a logic inverter. (b) The same transistor used as a linear amplifier.

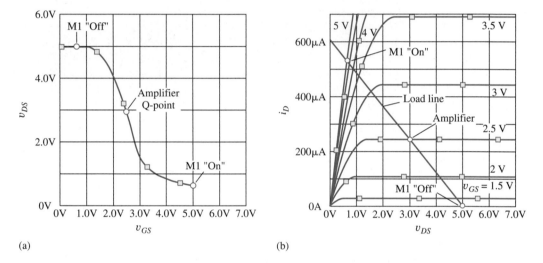

Figure 4.23 (a) Voltage transfer characteristic (VTC) with quiescent operating points (Q-points) corresponding to an "on-switch," an amplifier, and an "off switch." (b) The same three operating points located on the transistor output characteristics.

dc values (I_D, V_{DS}) that locate the operating point on the MOSFET output characteristics. [In reality, we need the three values (I_D, V_{DS}, V_{GS}), but two are enough if we know the region of operation of the device.]

For binary logic circuits investigated in detail in Part II of this text, the transistor acts as an "on-off" switch, and the Q-point is set to be in either the cutoff region ("off") or the triode region ("on"). For example, let us explore the circuit in Fig. 4.22(a) that can be used as either a logic inverter or a linear amplifier depending upon our choice of operating points. The voltage transfer characteristic (VTC) for the circuit appears in Fig. 4.23(a). For low values of v_{GS}, the MOSFET is off, and the output voltage is 5 V, corresponding to a binary "1" in a logic application. As v_{GS} increases, the output begins to drop and finally reaches its "on-state" voltage of 0.65 V for $v_{GS} = 5$ V. This voltage would correspond to a "0" in binary logic. These two logic states are also shown on the transistor output characteristics in Fig. 4.23(b). When the transistor is "on," it conducts a substantial current, and v_{DS} falls to 0.65 V. When the transistor is off, v_{DS} equals 5 V. We study the design of logic gates in detail in Chapters 6–9.

For amplifier applications, the Q-point is located in the region of high slope (high gain) near the center of the voltage transfer characteristic, also indicated in Fig. 4.23(a). At this operating point, the transistor is operating in saturation, the region in which high voltage, current and/or power gain can be achieved. To establish this Q-point, a **dc bias** V_{GS} is applied to the gate as in Fig. 4.22(b), and a small ac signal v_{gs} is added to vary the gate voltage around the bias value.[3] The variation in total gate-source voltage v_{GS} causes the drain current to change, and an amplified replica of the ac input voltage appears at the drain. Our study of the design of transistor amplifiers begins in Chapter 12.

For hand analysis and design of Q-points, channel-length modulation is usually ignored by assuming $\lambda = 0$. A review of Fig. 4.11 indicates that including λ changes the drain current by less than 10 percent. Generally, we do not know the values of transistor parameters to this accuracy, and the tolerances on both discrete or integrated circuit elements may be as large as 30 to 50 percent. If you explore some transistor specification sheets (see CD ROM or MCD website), you will discover parameters that have a 4 or 5 to 1 spread in values. You will also find parameters with only a minimum or maximum value specified. Thus, neglecting λ will not significantly affect the validity of our analysis. Also, many bias circuits involve feedback which further reduces the influence of λ. On the other hand, in Part III we will see that λ can play an extremely important role in limiting the voltage gain of analog amplifier circuits, and the effect of λ must often be included in the analysis of these circuits.

To analyze circuits containing MOSFETs, we must first assume a region of operation, just as we did to analyze diode circuits in Chapter 3. The bias circuits that follow will most often be used to place the transistor Q-point in the saturation region, and by examining Eq. (4.27) with $\lambda = 0$, we see that we must know the gate-source voltage V_{GS} to calculate the drain current I_D. Then, once we know I_D, we can find V_{DS} from the constraints of Kirchhoff's voltage law. Thus our most frequently used analysis approach will be to first find V_{GS} and then to use its value to find the value of I_D. I_D will then be used to calculate V_{DS}.

Menu for Bias Analysis
1. Assume a region of operation (Most often the saturation region)
2. Use circuit analysis to find V_{GS}
3. Use V_{GS} to calculate I_D, and I_D to determine V_{DS}
4. Check the validity of the operating region assumptions
5. Change assumptions and analyze again if necessary

DESIGN NOTE   **SATURATION BY CONNECTION!**

When making bias calculations for analysis or design, it is useful to remember that an NMOS *enhancement-mode* device that is operating with $V_{DS} = V_{GS}$ will always be in the pinch-off region. The same is true for an enhancement-mode PMOS transistor.

To demonstrate this result, it is easiest to keep the signs straight by considering an NMOS device with dc bias. For pinch-off, it is required that

$$V_{DS} \geq V_{GS} - V_{TN}$$

But if $V_{DS} = V_{GS}$, this condition becomes

$$V_{DS} \geq V_{DS} - V_{TN} \qquad \text{or} \qquad V_{TN} \geq 0$$

which is always true if V_{TN} is a positive number. $V_{TN} > 0$ corresponds to an NMOS enhancement-mode device. Thus an enhancement-mode device operating with $V_{DS} = V_{GS}$ is always in the saturation region! Similar arguments hold true for enhancement-mode PMOS devices operating with $V_{SD} = V_{SG}$.

---

[3] Remember $v_{GS} = V_{GS} + v_{gs}$

EXAMPLE **4.1(a)** CONSTANT GATE-SOURCE VOLTAGE BIAS

A basic bias circuit for the NMOS transistor is shown in Fig. 4.24, in which dc voltage source V_{GG} is used to establish a fixed gate-source bias for the MOSFET, source V_{DD} supplies drain current to the NMOS transistor through resistor R_D, and the value of R_D determines V_{DS}. This circuit is used to introduce a number of concepts related to biasing, but we shall find that it is not a very useful circuit in practical applications.

PROBLEM Find the quiescent operating point Q-point (I_D, V_{DS}) for the MOSFET in the fixed gate bias circuit in Fig. 4.24.

SOLUTION **Known Information and Given Data:** Circuit schematic in Fig. 4.24 with $V_{DD} = 10$ V, $V_{GG} = 10$ V, $R_1 = 300$ kΩ, $R_2 = 700$ kΩ, $R_D = 100$ kΩ, $V_{TN} = 1$ V, $K_n = 25\,\mu$A/V^2, $I_G = 0$, and $I_B = 0$

Unknowns: I_D, V_{DS}, and V_{GS}

Approach: We can find the Q-point using the mathematical model for the NMOS transistor. We must assume a region of operation, determine the Q-point, and then see if the resulting Q-point is consistent with the assumed region of operation.

Assumptions: We will assume that the MOSFET is pinched-off: $I_D = (K_n/2)(V_{GS} - V_{TN})^2$. Remember, we ignore λ in hand bias calculations. This assumption simplifies the mathematics because I_D is then modeled as being independent of V_{DS}.

Analysis: From the drain current expression and given data, we see that if we first find V_{GS}, then we can use it to find I_D. First label the variables in the circuit including I_D, V_{DS}, and V_{GS}. Then to simplify the analysis, we replace the gate-bias network consisting of V_{GG}, R_1, and R_2 with its Thévenin equivalent circuit as in Fig. 4.24(b) in which

$$V_{EQ} = \frac{R_1}{R_1 + R_2} V_{GG} = 3 \text{ V} \quad \text{and} \quad R_{EQ} = \frac{R_1 R_2}{R_1 + R_2} = 210 \text{ k}\Omega$$

We apply Kirchhoff's voltage law (KVL) to the loop containing the gate-source terminals of the device (referred to here as the input loop):

$$V_{EQ} = I_G R_{EQ} + V_{GS} \tag{4.42}$$

But, we know that $I_G = 0$ for the MOSFET, so that $V_{GS} = V_{EQ} = 3$ V. We can now find I_D using the transistor parameters from Fig. 4.24:

$$I_D = \frac{K_n}{2}(V_{GS} - V_{TN})^2 = \frac{25 \times 10^{-6} \text{ A}}{2 \text{ V}^2}(3 - 1)^2 \text{ V}^2 = 50\,\mu\text{A}$$

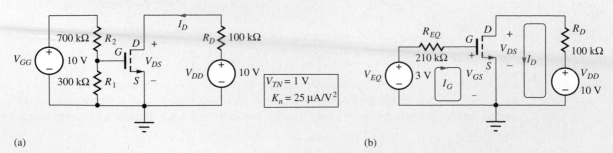

(a) (b)

Figure 4.24 (a) Constant gate-voltage bias using a voltage divider. (b) Simplified MOSFET bias circuit.

To determine V_{DS} we write a loop equation including the drain-source terminals of the device (referred to here as the output loop):

$$V_{DD} = I_D R_D + V_{DS} \tag{4.43}$$

Again substituting the values from Fig. 4.24,

$$V_{DS} = 10\,\text{V} - (50 \times 10^{-6}\,\text{A})(10^5\,\Omega) = 5.00\,\text{V}$$

Check of Results: We have $V_{DS} = 5$ V and $V_{GS} - V_{TN} = 2$ V. Since V_{DS} exceeds $V_{GS} - V_{TN}$, the transistor is indeed pinched-off and in the saturation region. Thus, the Q-point is (50.0 μA, 5.00 V) with $V_{GS} = 3$ V.

Discussion: Although this circuit introduces a number of concepts related to biasing, it is not a very useful circuit in practical applications because the Q-point is very sensitive to variations in the values of the transistor parameters. If the value of V_{GS} is fixed in the drain current expression, then I_D varies in direct proportion to K_n and depends on the square of changes in V_{TN}. The bias circuits that we will explore in Exs. 4.3 and 4.7 provide a much reduced sensitivity of the Q-point to changes in device parameters and are preferred methods of biasing the transistor.

EXERCISE: Find the Q-point for the circuit in Fig. 4.24 if R_D is changed to 50 kΩ.

ANSWER: (50.0 μA, 7.50 V)

EXERCISE: Find the Q-point for the circuit in Fig. 4.24 if $R_1 = 270$ kΩ, $R_2 = 750$ kΩ, and $R_D = 100$ kΩ.

ANSWER: (33.9 μA, 6.61 V)

EXERCISE: Suppose that $K_n = 30$ μA/V^2 instead of 25 μA/V^2 as in Ex. 4.1. What are the new values of V_{GS}, I_D, and V_{DS}?

ANSWER: (3 V, 60.0 μA, 4 V) (Note in this circuit that I_D is directly proportional to K_n.)

EXERCISE: Suppose that V_{TN} is 1.5 V instead of 1 V in Ex. 4.1. What are the new values of V_{GS}, I_D, and V_{DS}?

ANSWER: (3 V, 28.1 μA, 7.19 V) (We see that the current is also quite sensitive to the value of V_{TN}.)

EXAMPLE 4.1(b) ANALYSIS INCLUDING THE EFFECT OF CHANNEL-LENGTH MODULATION

It has been argued several times thus far that λ can be neglected in bias calculations. So, before we leave the constant gate bias circuit behind, let us repeat the bias calculation above with $\lambda = 0.02\,\text{V}^{-1}$.

PROBLEM Find the quiescent operating point Q-point for the MOSFET in the fixed gate bias circuit in Fig. 4.24 with $\lambda = 0.02\,\text{V}^{-1}$.

SOLUTION **Known Information and Given Data:** Simplified circuit schematic in Fig. 4.24 with $V_{DD} = 10$ V, $R_D = 100$ kΩ, $V_{TN} = 1$ V, $K_n = 25$ μA/V^2, $\lambda = 0.02\,\text{V}^{-1}$, $I_G = 0$, $I_B = 0$, and $V_{GS} = V_{EQ} = 3$ V

Unknowns: I_D, V_{DS}

Approach: We can make use of the information from the previous example. For this circuit, including a nonzero value of λ does not affect the equations describing the input loop. So $V_{GS} = 3\,\text{V}$, and we can directly reevaluate the drain current expression.

Assumptions: Assume that the MOSFET is in the pinch-off region. But now

$$I_D = \frac{K_n}{2}(V_{GS} - V_{TN})^2(1 + \lambda V_{DS}).$$

Analysis: To find V_{DS}, we still have

$$V_{DS} = V_{DD} - I_D R_D$$

Combining this equation with the expression for the drain current and substituting the values from Fig. 4.24 yields

$$V_{DS} = 10 - \frac{(25 \times 10^{-6})(10^5)}{2}(3 - 1)^2(1 + 0.02\,V_{DS})$$

in which the units have been eliminated for simplicity. Solving for V_{DS} yields $V_{DS} = 4.55\,\text{V}$. Using this value to calculate the drain current gives

$$I_D = \frac{25 \times 10^{-6}}{2}(3 - 1)^2[1 + 0.02(4.55)] = 54.5\,\mu\text{A}$$

Check of Results: We see that $V_{DS} = 4.55\,\text{V}$ exceeds $V_{GS} - V_{TN} = 2\,\text{V}$ so that the transistor is indeed pinched-off. Thus, the saturation region assumption is justified. The final Q-point is $(54.5\,\mu\text{A}, 4.55\,\text{V})$.

Discussion: We see that the Q-point values have each changed by approximately 10 percent from $(50\,\mu\text{A}, 5\,\text{V})$ to $(54.5\,\mu\text{A}, 4.55\,\text{V})$. From a practical point of view, the tolerances on circuit element and transistor parameter values will completely swamp out these small differences. Therefore we gain little from the additional complexity of including λ in our hand calculations. Note that, although this particular calculation including λ may have seemed relatively painless, the relative ease is an artifact of this particular circuit. Including λ in calculations for other bias circuits is considerably more difficult. On the other hand, if we use a circuit analysis program to perform the calculations, we might as well include λ.

EXERCISE: Repeat the channel length modulation calculation for $\lambda = 0.01\,\text{V}^{-1}$. What are the new values of I_D and V_{DS}?

ANSWER: $(52.4\,\mu\text{A}, 4.76\,\text{V})$

EXAMPLE 4.2   LOAD LINE ANALYSIS

The Q-point for the MOSFET circuit in Fig. 4.24 can also be found graphically with a load-line method very similar to the one used for analysis of diode circuits in Sec. 3.10. The graphical approach helps us visualize the operating point of the device and its location relative to the boundaries between the cutoff, triode and pinch-off regions of operation.

PROBLEM Use load line analysis to locate the Q-point for the MOSFET in the fixed gate bias circuit in Fig. 4.24.

SOLUTION

Known Information and Given Data: Circuit schematic in Fig. 4.24 with $V_{DD} = 10$ V, $V_{EQ} = 3$ V, $R_{EQ} = 210\,\text{k}\Omega$, $R_D = 100\,\text{k}\Omega$, $V_{TN} = 1$ V, $K_n = 25\,\mu\text{A/V}^2$, $I_G = 0$, and $I_B = 0$

Unknowns: Q-point $= (I_D, V_{DS})$

Approach: We need to find an equation for the load line, $I_D = f(V_{DS})$, so that it can be plotted on the output i-v characteristics. The Q-point can then be located on the output characteristics. Equation (4.43) represents the *load line* for this MOSFET circuit and is repeated here:

$$V_{DD} = I_D R_D + V_{DS}$$

Assumptions: We have already found $V_{GS} = 3$ V using the techniques in Ex. 4.1(a).

Analysis: For the values for the circuit in Fig. 4.24, the load line equation becomes

$$10 = 10^5 I_D + V_{DS}$$

Just as for the diode circuits in Sec. 3.10, the load line is constructed by finding two points on the line: for $V_{DS} = 0$, $I_D = 100\,\mu\text{A}$, and for $I_D = 0$, $V_{DS} = 10$ V. The resulting line is drawn on the output characteristics of the MOSFET in Fig. 4.25. The family of NMOS curves intersects the load line at many different points (actually infinitely many since each possible gate voltage corresponds to a different curve). The gate-source voltage is the parameter that determines which of the intersection points is the actual Q-point. In this circuit, we already found $V_{GS} = 3$ V; the Q-point is indicated by the circle in the Fig. 4.25. Reading the values from the graph yields $V_{DS} = 5$ V and $I_D = 50\,\mu\text{A}$.

Check of Results: This is the same Q-point that we found using our mathematical model for the MOSFET.

Discussion: From the graph, we can immediately see that the Q-point is in the saturation region of the transistor output characteristics. The Q-point is fairly well centered in the saturation region of operation, and the drain-source voltage is 3 V greater than that required to saturate the device.

Although we will seldom actually solve bias problems using graphical techniques, it is very useful to visualize the location of the Q-point in terms of the load line on the output characteristics as in Fig. 4.25. We can readily see if the device is operating in the triode or saturation regions as well as how far the operating point is from the boundaries between the various regions of operation.

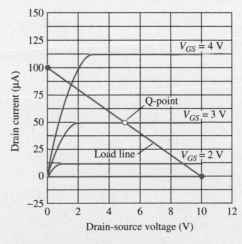

Figure 4.25 Load line for the circuit in Fig. 4.24.

EXERCISE: Draw the new load line and find the Q-point if R_D is changed to 66.7 kΩ.

ANSWER: (50 µA, 6.7 V)

EXAMPLE **4.3**   **FOUR-RESISTOR BIASING**

The circuit in Fig. 4.24 provides a fixed gate-source bias voltage to the transistor. Theoretically, this works fine. However, in practice the values of K_n, V_{TN}, and λ for the MOSFET will not be known with high precision. In addition, we must be concerned about resistor and power supply tolerances (you may wish to review Sec. 1.8) as well as component value drift with both time and temperature in an actual circuit.

The most general and important bias method that we will encounter is the **four-resistor bias** circuit in Fig. 4.26(a). The addition of the fourth resistor R_S helps stabilize the MOSFET Q-point in the face of many types of circuit parameter variations. This bias circuit is actually a form of *feedback circuit,* which will be studied in great detail in Chapters 12 and 18. Also observe that a single voltage source V_{DD} is now used to supply both the gate-bias voltage and the drain current. The four-resistor bias circuit is most often used to place the transistor in the saturation region of operation for use as an amplifier for analog signals.

PROBLEM   Find the Q-point = (I_D, V_{DS}) for the MOSFET in the four resistor bias circuit in Fig. 4.26.

SOLUTION   **Known Information and Given Data:** Circuit schematic in Fig. 4.26 with $V_{DD} = 10$ V, $R_1 = 1$ MΩ, $R_2 = 1.5$ MΩ, $R_D = 75$ kΩ, $R_S = 39$ kΩ, $K_n = 25$ µA/V^2, and $V_{TN} = 1$ V

Unknowns: Q-point = (I_D, V_{DS}), V_{GS}, and region of operation

Approach: We can find the Q-point using the mathematical model for the NMOS transistor. We assume a region of operation, determine the Q-point, and check to see if the resulting Q-point is consistent with the assumed region of operation.

Assumptions: The first step in our Q-point analysis of the equivalent circuit in Fig. 4.26 is to assume that the transistor is saturated (remember to use λ = 0):

$$I_D = \frac{K_n}{2}(V_{GS} - V_{TN})^2 \tag{4.44}$$

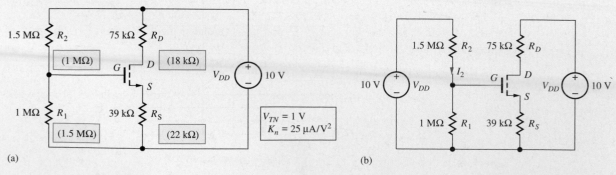

Figure 4.26 (a) Four-resistor bias network for a MOSFET. (b) Equivalent circuit with replicated sources. The shaded values in part (a) are used in Ex. 4.4.

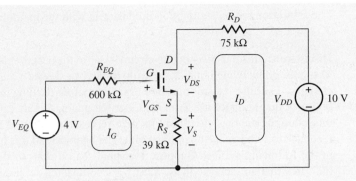

Figure 4.27 Equivalent circuit for the four-resistor bias network.

Also, $I_G = 0 = I_B$. Using the $\lambda = 0$ assumption simplifies the mathematics because I_D is then modeled as being independent of V_{DS}.

Analysis: To find I_D, the gate-source voltage must be determined, and we begin by simplifying the circuit. In the equivalent circuit in Fig. 4.26(b), the voltage source V_{DD} has been split into two equal-valued sources, and we recognize that the gate-bias voltage is determined by V_{EQ} and R_{EQ}, exactly as in Fig. 4.24. After the Thévenin transformation is applied to this circuit, the resulting equivalent circuit is given in Fig. 4.27 in which the variables have been clearly labeled. This is the final circuit to be analyzed.

Detailed analysis begins by writing the input loop equation containing V_{GS}:

$$V_{EQ} = I_G R_{EQ} + V_{GS} + (I_G + I_D)R_S \qquad \text{or} \qquad V_{EQ} = V_{GS} + I_D R_S \qquad (4.45)$$

because we know that $I_G = 0$. Substituting Eq. (4.44) into Eq. (4.45) yields

$$V_{EQ} = V_{GS} + \frac{K_n R_s}{2}(V_{GS} - V_{TN})^2 \qquad (4.46)$$

and we have a quadratic equation to solve for V_{GS}. For the values in Fig. 4.27 with $V_{TN} = 1 \text{ V}$ and $K_n = 25\ \mu\text{A/V}^2$,

$$4 = V_{GS} + \frac{(25 \times 10^{-6})(3.9 \times 10^4)}{2}(V_{GS} - 1)^2$$

and

$$V_{GS}^2 + 0.05\ V_{GS} - 7.21 = 0 \qquad \text{for which} \qquad V_{GS} = -2.71 \text{ V}, +2.66 \text{ V}$$

For $V_{GS} = -2.71 \text{ V}$, the MOSFET would be cut off because $V_{GS} < V_{TN}$. Therefore, $V_{GS} = +2.66 \text{ V}$ must be the answer we seek, and $I_D = 34.4\ \mu\text{A}$ is found using Eq. (4.44).

The second part of the Q-point, V_{DS}, can now be determined by writing the "output" loop equation including the drain-source terminals of the device:

$$V_{DD} = I_D R_D + V_{DS} + (I_G + I_D)R_S \qquad \text{or} \qquad V_{DD} = I_D(R_D + R_S) + V_{DS} \qquad (4.47)$$

Eq. (4.47) has been simplified since we know that $I_G = 0$. Substituting the values from the circuit gives

$$10 \text{ V} = (34.4\ \mu\text{A})(75\ \text{k}\Omega + 39\ \text{k}\Omega) + V_{DS} \qquad \text{or} \qquad V_{DS} = 6.08 \text{ V}$$

Check of Results: Checking the saturation region assumption, we have

$$V_{DS} = 6.08 \text{ V}, \quad V_{GS} - V_{TN} = 1.66 \text{ V} \qquad \text{and} \qquad V_{DS} > (V_{GS} - V_{TN}) \quad \checkmark$$

The saturation region assumption is consistent with the resulting Q-point: (34.4 μA, 6.08 V) with $V_{GS} = 2.66$ V.

Discussion: The four-resistor bias circuit is one of the best for biasing transistors in discrete circuits. The bias point is well stabilized with respect to device parameter variations and temperature changes. The four-resistor bias circuit is most often used to place the transistor in the saturation region of operation for use as an amplifier for analog signals, and as mentioned at the beginning of this example, the bias circuit in Fig. 4.26 represents a type of feedback circuit that uses negative feedback to stabilize the operating point. The operation of this feedback mechanism can be viewed in the following manner. Suppose for some reason that I_D begins to increase. Equation (4.45) indicates that an increase in I_D must be accompanied by a decrease in V_{GS} since V_{EQ} is fixed. But, this decrease in V_{GS} will tend to restore I_D back to its original value [see Eq. (4.44)]. This is negative feedback in action!

Note that this circuit uses the three-terminal representation for the MOSFET, in which it is assumed that the bulk terminal is tied to the source. If the bulk terminal is instead grounded, the analysis becomes more complex because the threshold voltage is then a function of the voltage developed at the source terminal of the device. This case will be investigated in more detail in Ex. 4.5. Let us now use the computer to explore the impact of neglecting λ in our hand analysis.

Computer-Aided Analysis: If we use SPICE to simulate the circuit using a LEVEL = 1 model and the parameters from our hand analysis (KP = 25 μA/V^2 and VTO = 1 V), we get exactly the same Q-point (34.4 μA, 6.08 V). If we add LAMBDA = 0.02 V^{-1}, SPICE yields a new Q-point of (35.9 μA, 5.91 V). The Q-point values change by less than 5 percent, a value that is well below our uncertainty in the device parameter and resistor values in a real situation.

EXERCISE: Suppose K_n increases to 30 μA/V^2 for the transistor in Fig. 4.27. What is the new Q-point for the circuit?

ANSWER: (36.8 μA, 5.81 V)

EXERCISE: Suppose V_{TN} changes from 1 V to 1.5 V for the MOSFET in Fig. 4.27. What is the new Q-point for the circuit?

ANSWER: (26.7 μA, 6.96 V)

EXERCISE: Find the Q-point in the circuit in Fig. 4.27 if R_S is changed to 62 kΩ.

ANSWER: (25.4 μA, 6.52 V)

DESIGN EXAMPLE 4.4   FOUR-RESISTOR BIAS REDESIGN

Let us redesign the four-resistor bias network in Ex. 4.3 to increase the current while keeping V_{DS} approximately the same; the new Q-point will be (100 μA, 6 V).

PROBLEM   Find the new values of R_S and R_D that will change the Q-point to (100 μA, 6 V).

SOLUTION   **Known Information and Given Data:** Simplified circuit schematic in Fig. 4.27 with $V_{EQ} = 4$ V, $R_{EQ} = 600$ kΩ, $K_n = 25$ μA/V^2, $V_{TN} = 1$ V, $I_G = 0$, $I_B = 0$, $I_D = 100$ μA, and $V_{DS} = 6$ V

Unknowns: V_{GS}, R_S, and R_D

Approach: The source and drain currents are controlled by the input loop and the value of R_S, which will be changed to achieve the desired value of I_D. The drain-source voltage can then be adjusted by changing the value of R_D.

Assumptions: Use the MOSFET saturation region model with $\lambda = 0$:

$$I_D = \frac{K_n}{2}(V_{GS} - V_{TN})^2$$

Analysis: Equation (4.45) can be rearranged to find the required value of R_S:

$$R_S = \frac{V_{EQ} - V_{GS}}{I_D} = \frac{V_S}{I_D} \tag{4.48}$$

but we must first find the new value of V_{GS}. (I_D is changed so V_{GS} must change.)

The gate-source voltage V_{GS} needed to establish $I_D = 100\,\mu\text{A}$ is found by rearranging the saturation region expression for the NMOS drain current, Eq. (4.44), with $\lambda = 0$:

$$V_{GS} = V_{TN} + \sqrt{\frac{2I_{DS}}{K_n}} = 1\,\text{V} + \sqrt{\frac{2(100\,\mu\text{A})}{25\,\dfrac{\mu\text{A}}{\text{V}^2}}} = 3.83\,\text{V} \tag{4.49}$$

Note that the positive root is used here since V_{GS} must exceed V_{TN} for conduction. Substituting this value in Eq. (4.49) yields

$$R_S = \frac{4\,\text{V} - 3.83\,\text{V}}{100\,\mu\text{A}} = \frac{0.17\,\text{V}}{100\,\mu\text{A}} = 1.7\,\text{k}\Omega$$

By rearranging the second expression in Eq. (4.47), we see that the sum of R_D and R_S in the bias network is determined by the desired Q-point values:

$$R_D + R_S = \frac{V_{DD} - V_{DS}}{I_D} = \frac{10\,\text{V} - 6\,\text{V}}{100\,\mu\text{A}} = 40\,\text{k}\Omega \quad \text{and} \quad R_D = (40 - 1.7)\,\text{k}\Omega - 38.3\,\text{k}\Omega$$

From Appendix A, the nearest standard 5 percent resistor values are $R_S = 1.6$ or $1.8\,\text{k}\Omega$ and $R_D = 39\,\text{k}\Omega$. Here we choose $R_S = 1.8\,\text{k}\Omega$ which will yield a drain current that is slightly lower than the design value, but, in the absence of any addition information, $1.6\,\text{k}\Omega$ would be an equally valid choice.

Since R_D and R_S are not exactly what we calculated, the values of V_{GS}, I_D, and V_{DS} will vary slightly from the design values. Using the approach in Ex. 4.3, the Q-point is found to be $(99.5\,\mu\text{A}, 5.95\,\text{V})$ with $V_{GS} = 3.82\,\text{V}$.

Check of Results: For this design, we now have

$$V_{DS} = 5.95\,\text{V} \qquad V_{GS} - V_{TN} = 2.82\,\text{V} \quad \text{and} \quad V_{DS} > (V_{GS} - V_{TN}) \quad ✔$$

The saturation region assumption is consistent with the solution.

Discussion: Note that although the value of R_S is 6 percent larger than the calculated value, I_D changed by less than 0.5 percent. Once again we see the effects of feedback in action!

Although $R_S = 1.8\,\text{k}\Omega$ represents a reasonable value of resistance, the voltage developed at the source of the MOSFET — only 0.17 V — is quite small and will be highly sensitive to changes in V_{DD}, V_{TN}, R_1, and R_2. If we pick a larger value of V_{EQ}, a greater value of V_S will appear across R_S, and the circuit design will be far less dependent on the device parameters. (See Prob. 4.94.)

So let us increase V_{EQ} from 4 to 6 V, which will directly increase V_S by 2 V — see Eq. (4.48). The new value of R_S and R_D are

$$R_S = \frac{6\,\text{V} - 3.83\,\text{V}}{100\,\mu\text{A}} = 21.7\,\text{k}\Omega \qquad \text{and} \qquad R_D = (40 - 21.2)\,\text{k}\Omega = 18.3\,\text{k}\Omega$$

From Appendix A, the nearest standard 5 percent resistor values are now $R_S = 22\,\text{k}\Omega$ and $R_D = 18\,\text{k}\Omega$.

The values of R_1 and R_2 must be modified to set V_{EQ} to 6 V. If we simply interchange the values of R_1 and R_2 in Fig. 4.26, we will have $V_{EQ} = 6\,\text{V}$, with R_{EQ} remaining $600\,\text{k}\Omega$. Our final design values, $R_1 = 1.5\,\text{M}\Omega$, $R_2 = 1\,\text{M}\Omega$, $R_S = 22\,\text{k}\Omega$, and $R_D = 18\,\text{k}\Omega$, are indicated by the shaded numbers in parentheses in Fig. 4.26(a).

EXERCISE: Show that the actual Q-point in the circuit in Fig. 4.26 for $R_1 = 1\,\text{M}\Omega$, $R_2 = 1.5\,\text{M}\Omega$, $R_S = 1.8\,\text{k}\Omega$, and $R_D = 39\,\text{k}\Omega$ is (99.5 μA, 5.95 V).

EXERCISE: Find the Q-point in the circuit in Fig. 4.26 for $R_1 = 1.5\,\text{M}\Omega$, $R_2 = 1\,\text{M}\Omega$, $R_S = 22\,\text{k}\Omega$, and $R_D = 18\,\text{k}\Omega$.

ANSWER: (99.1 μA, 6.04 V)

EXERCISE: Redesign the values of R_1 and R_2 to set the bias current to 2 μA while maintaining $V_{EQ} = 6\,\text{V}$. What is the value of R_{EQ}?

ANSWER: 3 MΩ, 2 MΩ, 1.2 MΩ

DESIGN NOTE
GATE VOLTAGE DIVIDER DESIGN

Resistors R_1 and R_2 in Fig. 4.26 are required to set the value of V_{EQ}, but the current in the resistors does not contribute directly to operation of the transistor. Thus we would like to minimize the current "lost" through R_1 and R_2. The sum $(R_1 + R_2)$ sets the current in the gate bias resistors. As a rule of thumb, $R_1 + R_2$ is usually chosen to limit the current to no more than a few percent of the value of the drain current. In Fig. 4.26, the value of current I_2 is 4 percent of the drain current $I_2 = 10\,\text{V}/(1\,\text{M}\Omega + 1.5\,\text{M}\Omega) = 4\,\mu\text{A}$.

EXAMPLE 4.5   ANALYSIS INCLUDING BODY EFFECT

The NMOS transistor in Fig. 4.27 was connected as a three-terminal device. This example explores how the Q-point is altered when the substrate is connected as shown in Fig. 4.28.

PROBLEM Find the Q-point $= (I_D, V_{DS})$ for the MOSFET in the four-resistor bias circuit in Fig. 4.28 including the influence of body effect on the transistor threshold.

SOLUTION **Known Information and Given Data:** The circuit schematic in Fig. 4.28 with $V_{EQ} = 6\,\text{V}$, $R_{EQ} = 600\,\text{k}\Omega$, $R_S = 22\,\text{k}\Omega$, $R_D = 18\,\text{k}\Omega$, $K_n = 25\,\mu\text{A/V}^2$, $V_{TO} = 1\,\text{V}$, and $\gamma = 0.5\,\text{V}^{-1}$

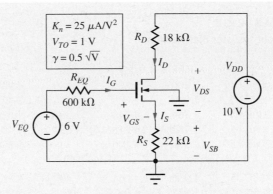

$K_n = 25\ \mu\text{A/V}^2$
$V_{TO} = 1\ \text{V}$
$\gamma = 0.5\ \sqrt{\text{V}}$

Figure 4.28 MOSFET with redesigned bias circuit.

Unknowns: I_D, V_{DS}, V_{GS}, V_{BS}, V_{TN}, and region of operation

Approach: In this case, the source-bulk voltage, $V_{SB} = I_S R_S = I_D R_S$, is no longer zero, and we must solve the following set of equations:

$$V_{GS} = V_{EQ} - I_D R_S \qquad V_{SB} = I_D R_S$$

$$V_{TN} = V_{TO} + \gamma\left(\sqrt{V_{SB} + 2\phi_F} - \sqrt{2\phi_F}\right) \qquad (4.50)$$

$$I_D = \frac{K_n}{2}(V_{GS} - V_{TN})^2$$

Although it may be possible to solve these equations analytically, it will be more expedient to find the Q-point by iteration using the computer with a spreadsheet, MATLAB®, MATHCAD®, or with a calculator.

Assumptions: Saturation region operation with $I_G = 0$, $I_B = 0$, and $2\phi_F = 0.6\ \text{V}$

Analysis: Using the assumptions and values in Fig. 4.28, Eq. set (4.50) becomes

$$V_{GS} = 6 - 22{,}000 I_D \qquad V_{SB} = 22{,}000 I_D$$

$$V_{TN} = 1 + 0.5\left(\sqrt{V_{SB} + 0.6} - \sqrt{0.6}\right) \qquad I_D' = \frac{25 \times 10^{-6}}{2}(V_{GS} - V_{TN})^2 \quad (4.51)$$

and the drain-source voltage is found from

$$V_{DS} = V_{DD} - I_D(R_D + R_S) = 10 - 40{,}000 I_D \qquad (4.52)$$

The expressions in Eq. (4.51) have been arranged in a logical order for an iterative solution:

1. Estimate the value of I_D.
2. Use I_D to calculate the values of V_{GS} and V_{SB}.
3. Calculate the resulting value of V_{TN} using V_{SB}.
4. Calculate I_D' using the results of steps 1 to 3, and compare to the original estimate for I_D.
5. If the calculated value of I_D' is not equal to the original estimate for I_D, then go back to step 1.

In this case, no specific method for choosing the improved estimate for I_D is provided (although the problem could be structured to use Newton's method), but it is easy to converge to the solution after a few trials, using the power of the computer to do the calculations. (Note that the SPICE circuit analysis program can also do the job for us.)

Table 4.3 shows the results of using a spreadsheet to iteratively find the solution to Eqs. (4.51) and (4.52) by trial and error. The first iteration sequence used by the author is shown; it converges

TABLE 4.3
Four-Resistor Bias Iteration

I_D	$I_D R_S$	V_{GS}	V_{TN}	I_D'	V_{DS}
1.000E-04	2.200	3.800	1.449	6.907E-05	6.000
9.000E-05	1.980	4.020	1.416	8.477E-05	6.400
8.000E-05	1.760	4.240	1.381	1.022E-04	6.800
8.100E-05	1.782	4.218	1.384	1.004E-04	6.760
8.200E-05	1.804	4.196	1.388	9.856E-05	6.720
⋮	⋮	⋮	⋮	⋮	⋮
8.800E-05	1.936	4.064	1.409	8.812E-05	6.480
8.805E-05	1.937	4.063	1.409	8.803E-05	6.478
8.804E-05	1.937	4.063	1.409	8.805E-05	6.478

to a drain current of 88.0 μA and drain-source voltage of 6.48 V. Care must be exercised to be sure that the spreadsheet equations are properly formulated to account for all regions of operation. In particular, $I_D = 0$ if $V_{GS} < V_{TN}$.

Check of Results: For this design, we now have

$$V_{DS} = 6.48 \text{ V}, V_{GS} - V_{TN} = 2.56 \text{ V} \quad \text{and} \quad V_{DS} > (V_{GS} - V_{TN}) \quad ✔$$

The saturation region assumption is consistent with the solution, and the Q-point is (88.0 μA, 6.48 V).

Discussion: Now that the analysis is complete, we see that the presence of body effect in the circuit has caused the threshold voltage to increase from 1 V to 1.41 V and the drain current to decrease by approximately 12 percent from 100 μA to 88 μA.

EXERCISE: Find the new drain current in the circuit in Fig. 4.28 if $\gamma = 0.75\sqrt{V}$.

ANSWER: 83.2 μA

DESIGN EXAMPLE 4.6   BIAS REDESIGN TO COMPENSATE FOR BODY EFFECT

The increase in threshold voltage of the MOSFET in Ex. 4.5 due to body effect has caused the drain current to be smaller than the original design value from Ex. 4.4. Let us change the design values for R_S and R_D to restore the Q-point to the original value.

PROBLEM Find the new values of R_S and R_D required to restore the Q-point to the original value of (100 μA, 6 V).

SOLUTION **Known Information and Given Data:** The circuit schematic in Fig. 4.28 with $V_{EQ} = 6$ V, $R_{EQ} = 600$ kΩ, $K_n = 25$ μA/V², $V_{TO} = 1$ V, $\gamma = 0.5$ V⁻¹, $I_D = 100$ μA, and $V_{DS} = 6$ V

Unknowns: V_{GS}, V_{BS}, V_{TN}, and region of operation

Approach: We have $I_S = I_D = 100$ μA. To find R_S, we must find the voltage across R_S. From the circuit we have

$$V_{EQ} - V_{GS} - V_S = 0 \quad \text{or} \quad V_{SB} = V_{EQ} - V_{GS} \tag{4.53}$$

However, V_{GS} is a function of V_{TN} and V_{TN} depends on V_{SB}.

$$V_{GS} = V_{TN} + \sqrt{\frac{2I_D}{K_n}} \quad \text{and} \quad V_{TN} = V_{TO} + \gamma\left(\sqrt{V_{SB} + 2\phi_F} - \sqrt{2\phi_F}\right)$$

We need a simultaneous solution to these equations.

Assumptions: Saturation region operation with $I_G = 0$, $I_B = 0$, and $2\phi_F = 0.6$ V.

Analysis: Since $V_{SB} = V_S$ and $V_{EQ} = 6$ V, the source-bulk voltage given by Eq. (4.53) is

$$V_{SB} = 6 - V_{GS}$$

The value of V_{GS} required to set $I_D = 100$ μA is

$$V_{GS} = V_{TN} + \sqrt{\frac{2I_D}{K_n}} = V_{TN} + \sqrt{\frac{2(100\,\mu\text{A})}{25\dfrac{\mu\text{A}}{\text{V}^2}}} = V_{TN} + 2.83\,\text{V} \tag{4.54}$$

Combining Eqs. (4.53) and (4.54) and adding the expression for V_{TN} from Eq. (4.51) gives

$$6 - \left[1 + 0.5\left(\sqrt{V_{SB} + 0.6} - \sqrt{0.6}\right) + 2.83\right] - V_{SB} = 0 \tag{4.55}$$

Rearranging and collecting terms in Eq. (4.55) yields a quadratic equation for V_{SB}:

$$V_{SB}^2 - 5.37V_{SB} + 6.40 = 0 \quad \text{or} \quad V_{SB} = 1.79\,\text{V}, 3.58\,\text{V}$$

The second value of V_{SB} is too large [$(V_{SB} + V_{GS}) = (2.83 + 3.58)$ V which exceeds $V_{EQ} = 6$ V], so $V_{SB} = 1.79$ V is selected as the valid answer. The new values of R_S and R_D required to bias the circuit to the Q-point of (100 μA, 6 V) are

$$R_S = \frac{V_{SB}}{I_D} = \frac{1.79\,\text{V}}{100\,\mu\text{A}} = 17.9\,\text{k}\Omega \quad \text{and} \quad R_D = 40\,\text{k}\Omega - R_S = 22.1\,\text{k}\Omega$$

From Appendix A, the nearest standard 5 percent resistor values are $R_S = 18$ kΩ and $R_D = 22$ kΩ.

Check of Results: Here again V_{DS} exceeds $(V_{GS} - V_{TN})$, 6 V > 2.83 V, so the transistor is pinched off as assumed in the design.

Discussion: Including the body effect in the analysis resulted in a 39 percent increase in threshold voltage ($V_{TN} - 1.39$ V) and required a significant change in the values of R_S and R_D to restore the operating point to the desired design value. We need to be aware of and be ready to account for body effect in many circuits.

EXERCISE: Find the new values of R_S and R_D needed to achieve the Q-point of (100 μA, 6 V) in Fig. 4.28 if $\gamma = 0.75\sqrt{\text{V}}$.

ANSWERS: 16.3 kΩ, 23.7 kΩ; nearest 5 percent values: 16 kΩ, 24 kΩ ($V_{SB} = 1.63$ V)

EXAMPLE 4.7   TWO-RESISTOR FEEDBACK BIAS

Another example of a feedback bias circuit is given in Fig. 4.29. This circuit requires only two resistors. R_D determines both the drain current and the drain-source voltage of the transistor. Resistor R_G provides a dc connection between the gate and drain, and also serves to isolate the two terminals when signals are applied (as will occur in analog amplifier applications for example).

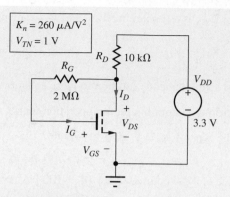

Figure 4.29 Two-resistor bias circuit.

PROBLEM   Find the Q-point (I_D, V_{DS}) for the MOSFET in the two-resistor bias circuit of Fig. 4.29.

SOLUTION   **Known Information and Given Data:** The circuit schematic in Fig. 4.29 with $V_{DD} = 3.3$ V, $R_D = 10$ kΩ, $R_G = 2$ MΩ, $K_n = 260$ μA/V², and $V_{TN} = 1$ V

Unknowns: I_D, V_{DS}, and V_{GS}

Assumptions: $I_G = 0$, $I_B = 0$. Note that the region of operation is actually known. For $I_G = 0$, there is no voltage drop across resistor R_G, and $V_{GS} = V_{DS}$. Since the transistor is an enhancement-mode device ($V_{TN} > 0$) it is pinched-off and in the saturation region — remember the Design Note in Sec. 4.8!

Approach: First find the value of V_{GS}; use V_{GS} to find I_D; use I_D to find V_{DS}.

Analysis: Writing the input loop equation for the value of V_{GS} yields

$$V_{GS} = V_{DS} - I_G R_G \qquad \text{or} \qquad V_{DS} = V_{GS} \text{ since } I_G = 0 \qquad (4.56)$$

Next, writing the output loop equation including V_{DS}:

$$V_{DS} = V_{DD} - (I_D + I_G)R_D = V_{DD} - I_D R_D \qquad (4.57)$$

Inserting saturation region Eq. (4.27) with $\lambda = 0$ into Eq. (4.57) yields

$$V_{GS} = V_{DD} - \frac{K_n R_D}{2}(\mathrm{V}_{GS} - \mathrm{V}_{TN})^2 \qquad (4.58)$$

Substituting the values from the circuit in Fig. 4.29 gives

$$V_{GS} = 3.3 - \frac{(2.6 \times 10^{-4})(10^4)}{2}(V_{GS} - 1)^2 \qquad \text{and} \qquad V_{GS} = -0.769 \text{ V}, +2.00 \text{ V}$$

Since $V_{TN} = 1$ V, $I_D = 0$ for the negative value of V_{GS}, and the answer must be $V_{GS} = 2.00$ V for which

$$I_D = 130 \text{ μA} \qquad \text{and} \qquad V_{DS} = 2.00 \text{ V}.$$

Check of Results: Since $V_{DS} = 2.00$ V and $V_{GS} - V_{TN} = 1$ V, the transistor is in the saturation region. The final Q-point is (130 μA, 2.00 V).

Evaluation and Discussion: The two-resistor bias circuit in Fig. 4.29 is another example of a circuit that uses negative feedback to stabilize the operating point. The negative feedback mechanism can be viewed in the following manner. Suppose for some reason that I_D begins to increase. An increase in I_D will cause a decrease in V_{DS} and hence a decrease in V_{GS} since $V_{GS} = V_{DS}$. The decrease in V_{GS} will cause I_D to decrease back toward its original value.

Computer-Aided Analysis: SPICE simulation with the LEVEL $= 1$ model gives precisely the same Q-point. Be sure to set KP $= 2.6 \times 10^{-4}$ A/V^2 and VTO $= 1$ V. If we add LAMBDA $= 0.02$ V^{-1}, the Q-point changes to (131 μA, 1.99 V), negligible shifts.

EXERCISE: Find the Q-point of the NMOS transistor in Fig. 4.29 if $V_{TN} = 1$ V and $K_n = 200$ μA/V^2.

ANSWER: (120 μA, 2.10 V) Note the small change in Q-point caused by the 15 percent change in K_n. This is a result of feedback.

EXERCISE: Find the Q-point of the NMOS transistor in this circuit if $V_{TN} = 1$ V and $K_n = 200$ μA/V^2.

ANSWER: (120 μA, 2.10 V)

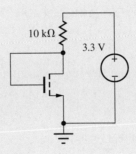

EXAMPLE 4.8 **ANALYSIS OF AN NMOS TRANSISTOR BIASED IN THE TRIODE REGION**

In all the previous circuit examples, we assumed and confirmed that the transistors were operating in the saturation region. But, what if our assumption for the region of operation is wrong? The circuit in Fig. 4.30 provides a simple example of such a circuit.

PROBLEM Determine the Q-point for the NMOSFET in Fig. 4.30.

SOLUTION **Known Information and Given Data:** The circuit schematic in Fig. 4.30 with $V_{DD} = 4$ V, $R_D = 1.6$ kΩ, $K_n = 250$ μA/V^2, and $V_{TN} = 1$ V

Unknowns: I_D, V_{DS}, V_{GS}

Approach: To find the Q-point using the mathematical model for the NMOS transistor, we will find V_{GS}, use it to find I_D and V_{DS} and then see if the resulting Q-point is consistent with the assumed region of operation.

Assumptions: $I_G = 0$ and $I_B = 0$. Assume that the transistor is operating in the saturation region as we have done in the past examples.

Analysis: In this circuit we immediately see that $V_{GS} = V_{DD} = 4$ V. Therefore, the MOSFET current is given by

$$I_D = \frac{250 \ \mu A}{2 \ V^2}(4 - 1)^2 = 1.13 \ \text{mA}$$

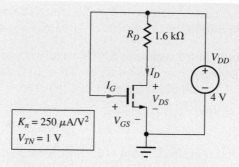

Figure 4.30 Bias circuit for example 4.8.

Writing the output-loop equation for V_{DS} in terms of I_D gives

$$4 = 1600\, I_D + V_{DS} \qquad (4.59)$$

and we find $V_{DS} = 2.19$ V using $I_D = 1.13$ mA.

Assumption Check: We have $V_{DS} = 2.19$ V. However, $V_{GS} - V_{TN} = 4 - 1 = 3$ V. Because $V_{GS} - V_{TN} > V_{DS}$ (3 V > 2.19 V), the assumption of saturation region operation is incorrect, and we must try again.

Analysis — Second Iteration: Substituting the triode region expression into Eq. (4.59) yields

$$4 - V_{DS} = 1600 K_n \left(V_{GS} - V_{TN} - \frac{V_{DS}}{2} \right) V_{DS} = 1600 \left(250 \frac{\mu A}{V^2} \right) \left(4 - 1 - \frac{V_{DS}}{2} \right) V_{DS} \quad (4.60)$$

After rearrangement we have

$$V_{DS}^2 - 11 V_{DS} + 20 = 0 \qquad \text{and} \qquad V_{DS} = 8.7 \text{ V} \qquad \text{or} \qquad 2.3 \text{ V}$$

The first voltage, 8.7 V, exceeds the magnitude of the power supply voltage and is not a possible result. So $V_{DS} = 2.3$ V, and

$$I_D = 250 \frac{\mu A}{V^2} \left(4 \text{ V} - 1 \text{ V} - \frac{2.3 \text{ V}}{2} \right) (2.3 \text{ V}) = 1.06 \text{ mA}$$

Check of Results: Checking the region of operation:

$$V_{GS} - V_{TN} = 4 \text{ V} - 1\text{V} = 3 \text{ V} \qquad \text{and} \qquad V_{GS} - V_{TN} > V_{DS} \quad ✔$$

The triode region is correct and the Q-point is (1.06 mA, 2.3 V).

Evaluation and Discussion: We have now found a Q-point consistent with the assumptions. We can use the value of I_D to double check our answer for V_{DS} from the "load line" equation: $V_{DS} = 4 - 1600 I_D = 2.30$ V.

In this case, we found that the circuit is biased in the triode region. However, the two and four resistor bias circuits of the previous examples are most often used to bias the transistor in the saturation region of operation for use as an amplifier.

EXERCISE: Find the values of I_D and V_{DS} in the circuit in Fig. 4.30 if $R_D = 1.8$ MΩ.

ANSWER: (2.22 µA, 2.96 mV)

Examples 4.1 through 4.8 of bias circuits represent but a few of the many possible ways to bias an NMOS transistor. Nevertheless, the examples have demonstrated the techniques that we need to analyze most of the circuits we will encounter. The four-resistor and two-resistor bias circuits are most often encountered in discrete design, whereas current sources and current mirrors, introduced in Chapter 15, find extensive application in integrated circuit design.

4.9 BIASING THE PMOS FIELD-EFFECT TRANSISTOR

CMOS technology, which uses a combination of NMOS and PMOS transistors, is the dominant IC technology in use today, and it is thus very important to know how to bias both types of devices. PMOS bias techniques mirror those used in the previous NMOS bias examples. In the circuits that follow, you will observe that the source of the PMOS transistor will be consistently drawn at the top of the device since the source of the PMOS device is normally connected to a potential that is higher than the drain. This is in contrast to the NMOS transistor in which the drain is connected to a more positive voltage than the source. The PMOS model equations were summarized in Sec. 4.3. Remember that the drain current I_D is positive when coming out of the drain terminal of the PMOS device, and the values of V_{GS} and V_{DS} will be negative.

EXAMPLE 4.9   **FOUR-RESISTOR BIAS FOR THE PMOS FET**

The four-resistor bias circuit in Fig. 4.31 functions in a manner similar to that used for the NMOS device in Ex. 4.3. In the circuit in Fig. 4.31(a), a single voltage source V_{DD} is used to supply both the gate-bias voltage and the source-drain current. R_1 and R_2 form the gate voltage divider circuit. R_S sets the source/drain current, and R_D determines the source-drain voltage.

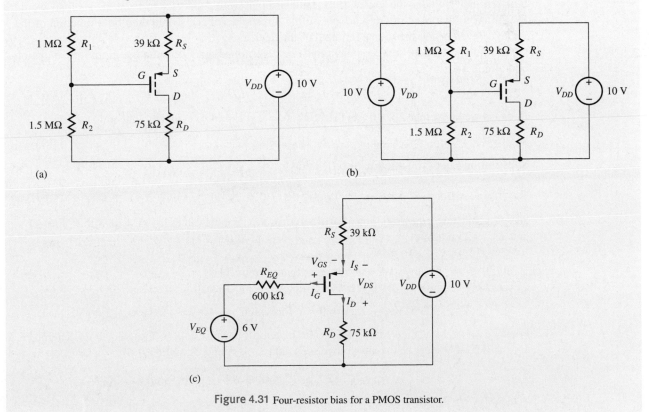

Figure 4.31 Four-resistor bias for a PMOS transistor.

PROBLEM   Find the quiescent operating point Q-point (I_D, V_{DS}) for the PMOS transistor in the four resistor bias circuit in Fig. 4.31.

SOLUTION   **Known Information and Given Data:** Circuit schematic in Fig. 4.31 with $V_{DD} = 10$ V, $R_1 = 1$ MΩ, $R_2 = 1.5$ MΩ, $R_D = 75$ kΩ, $R_S = 39$ kΩ, $K_P = 25$ μA/V^2, $V_{TP} = -1$ V, and $I_G = 0$

Unknowns: I_D, V_{DS}, V_{GS}, and the region of operation

Approach: We can find the Q-point using the mathematical model for the PMOS transistor. We assume a region of operation, determine the Q-point, and check to see if the Q-point is consistent with the assumed region of operation. First find the value of V_{GS}; use V_{GS} to find I_D; use I_D to find V_{DS}.

Assumptions: Assume that the transistor is operating in the saturation region (Once again, remember to use $\lambda = 0$)

$$I_D = \frac{K_p}{2}(V_{GS} - V_{TP})^2 \qquad (4.61)$$

Analysis: We begin by simplifying the circuit. In the equivalent circuit in Fig. 4.31(b), the voltage source has been split into two equal-valued sources, and in Fig. 4.31(c), the gate-bias circuit is replaced by its Thévenin equivalent

$$V_{EQ} = 10 \text{ V} \frac{1.5 \text{ M}\Omega}{1 \text{ M}\Omega + 1.5 \text{ M}\Omega} = 6 \text{ V} \qquad \text{and} \qquad R_{EQ} = 1 \text{ M}\Omega \| 1.5 \text{ M}\Omega = 600 \text{ k}\Omega$$

Figure 4.31(c) represents the final circuit to be analyzed (be sure to label the variables). Note that this circuit uses the three-terminal representation for the MOSFET, in which it is assumed that the bulk terminal is tied to the source. If the bulk terminal were connected to V_{DD}, the analysis would be similar to that used in Ex. 4.5 because the threshold voltage would then be a function of the voltage developed at the source terminal of the device.

To find I_D, the gate-source voltage must be determined, and we write the input loop equation containing V_{GS}:

$$V_{DD} = I_S R_S - V_{GS} + I_G R_G + V_{EQ} \qquad (4.62)$$

Because we know that $I_G = 0$ and therefore $I_S = I_D$, Eq. (4.62) can be reduced to

$$V_{DD} - V_{EQ} = I_D R_S - V_{GS} \qquad (4.63)$$

Substituting Eq. (4.61) into Eq. (4.63) yields

$$V_{DD} - V_{EQ} = \frac{K_p R_S}{2}(V_{GS} - V_{TP})^2 - V_{GS} \qquad (4.64)$$

and we again have a quadratic equation to solve for V_{GS}. For the values in Fig. 4.31 with $V_{TP} = -1$ V and $K_p = 25$ μA/V^2,

$$10 - 6 = \frac{(25 \times 10^{-6})(3.9 \times 10^4)}{2}(V_{GS} + 1)^2 - V_{GS}$$

and

$$V_{GS}^2 - 0.051 V_{GS} - 7.21 = 0 \qquad \text{for which} \qquad V_{GS} = +2.71 \text{ V}, -2.66 \text{ V}$$

For $V_{GS} = +2.71$ V, the PMOS FET would be cut off because $V_{GS} > V_{TP} (= -1$ V). Therefore, $V_{GS} = -2.66$ V must be the answer we seek, and I_D is found using Eq. (4.61):

$$I_D = \frac{25 \times 10^{-6}}{2}(-2.66 + 1)^2 = 34.4 \text{ μA}$$

The second part of the Q-point, V_{DS}, can now be determined by writing a loop equation including the source-drain terminals of the device:

$$V_{DD} = I_S R_S - V_{DS} + I_D R_D \qquad \text{or} \qquad V_{DD} = I_D(R_S + R_D) - V_{DS} \qquad (4.65)$$

Eq. (4.65) has been simplified since we know that $I_S = I_D$. Substituting the values from the circuit gives

$$10 \text{ V} = (34.4 \text{ μA})(39 \text{ kΩ} + 75 \text{ kΩ}) - V_{DS} \qquad \text{or} \qquad V_{DS} = -6.08 \text{ V}$$

Check of Results: We have

$$V_{DS} = -6.08 \text{ V} \qquad \text{and} \qquad V_{GS} - V_{TP} = -2.66 \text{ V} + 1 \text{ V} = -1.66 \text{ V}$$

and $|V_{DS}| > |V_{GS} - V_{TP}|$. Therefore the saturation region assumption is consistent with the resulting Q-point (34.4 μA, −6.08 V) with $V_{GS} = -2.66$ V.

Evaluation and Discussion: As mentioned in Ex. 4.3, the bias circuit in Fig. 4.31 uses negative feedback to stabilize the operating point. Suppose I_D begins to increase. Since V_{EQ} is fixed, an increase in I_D will cause a decrease in the magnitude of V_{GS} [see Eq. (4.63)], and this decrease will tend to restore I_D back to its original value.

EXERCISE: Find the Q-point in the circuit in Fig. 4.31 if R_S is changed to 62 kΩ.

ANSWER: (25.4 μA, −6.52 V)

 EXERCISE: (a) Use SPICE to find the Q-point in the circuit in Fig. 4.31. (b) Repeat if R_S is changed to 62 kΩ. (c) Repeat parts (a) and (b) with $\lambda = 0.02$.

ANSWERS: (a) (34.4 μA, −6.08 V); (b) (25.4 μA, −6.52 V); (c) (35.9 μA, −5.91 V), (26.3 μA, −6.39 V)

4.10 MOS TRANSISTOR SCALING

In Chapter 1, we discussed the phenomenal increase in integrated circuit density and complexity. These changes have been driven by our ability to aggressively scale the physical dimensions of the MOS transistor. A theoretical framework for MOSFET miniaturization was first provided by Dennard, Gaensslen, Kuhn, and Yu [6, 7]. The basic tenant of the theory is to require that the electrical fields be maintained constant within the device as the geometry is changed. Thus, if a physical dimension is reduced by a factor of α, then the voltage applied across that dimension must also be decreased by the same factor.

4.10.1 DRAIN CURRENT

These rules are applied to the transconductance parameter and triode region drain current expressions for the MOSFET in Eq. (4.66) in which the three physical dimensions, W, L, and T_{ox} are all reduced by the factor α, and each of the voltages including the threshold voltage is reduced by the same factor.

$$K_n^* = \mu_n \frac{\varepsilon_{ox}}{T_{ox}/\alpha} \frac{W/\alpha}{L/\alpha} = \alpha \mu_n \frac{\varepsilon_{ox}}{T_{ox}} \frac{W}{L} = \alpha K_n$$

$$i_D^* = \mu_n \frac{\varepsilon_{ox}}{T_{ox}/\alpha} \frac{W/\alpha}{L/\alpha} \left(\frac{v_{GS}}{\alpha} - \frac{V_{TN}}{\alpha} - \frac{v_{DS}}{2\alpha} \right) \frac{v_{DS}}{\alpha} = \frac{i_D}{\alpha} \qquad (4.66)$$

ELECTRONICS IN ACTION

Thermal Inkjet Printers

Inkjet printers have moved from a few niche applications in the 1960s to a widespread, mainstream consumer presence. Thermal inkjet technology was invented in 1979 at Hewlett-Packard Laboratories. Since that time, inkjet technology has evolved to the point where modern thermal inkjet printers deliver 10–20 picoliter droplets at rates of several KHz. Integration of the ink handling structures with microelectronics has been an important component of this evolution. Early versions of thermal inkjet printers had drive electronics that were separate from the ink delivery devices. Through the use of MEMS (micro-electro-mechanical system) technology, it has been possible to combine MOS transistors onto the same substrate with the ink handling structures.

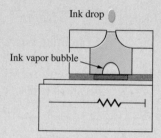

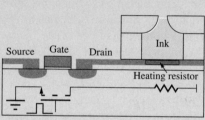

Simplified diagram of thermal inkjet structure integrated with MOS drive transistors. Voltage pulse on the gate causes I^2R heating in the resistor.

Heat from power dissipated in the resistor vaporizes a small amount of ink causing the ejection of an ink droplet out of the nozzle.

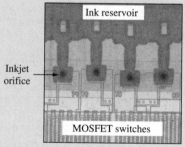

Photomicrograph of inkjet print head

This diagram is a simplified illustration of a merged thermal inkjet system. A MOSFET transistor is located in the left segment of the silicon substrate. A metal layer connects the drain of the transistor to the thin-film resistive heating material directly under the ink cavity. When the gate of the transistor is driven with a voltage pulse, current passes through the resistor leading to a rapid heating of the ink in the cavity. The temperature of the ink in contact with the resistor increases until a small portion of the ink vaporizes. The vapor bubble forces an ink drop to be ejected from the nozzle at the top of the ink cavity and onto the paper. (In practice, the drops are directed down onto the paper.) At the end of the gate drive pulse, the resistor cools and the vapor bubble collapses, allowing more ink to be drawn into the cavity from an ink reservoir.

Due to the high densities and resolutions made possible by the merging of control and drive electronics with the printing structures, inkjet printers are now capable of generating photo-quality images at reasonable costs. As we will see throughout this text, making high-technology affordable and widely available is a common trait of microelectronics-based systems.

We see that scaled transconductance K_n^* is increased by the scale factor α, whereas the scaled drain current is reduced from the original value by the scale factor.

4.10.2 GATE CAPACITANCE

In a similar manner, the total gate-channel capacitance of the device is also found to be reduced by α:

$$C_{GC}^* = (C_{ox}'')^* W^* L^* = \frac{\varepsilon_{ox}}{T_{ox}/\alpha} \frac{W/\alpha}{L/\alpha} = \frac{C_{GC}}{\alpha} \tag{4.67}$$

In Chapter 6 we will demonstrate that the delay of logic gates is limited by the transistor's ability to charge and discharge the capacitance associated with the circuit. Based on $i = C\, dv/dt$, an estimate of the delay of a scaled logic circuit is

$$\tau^* = C_{GC}^* \frac{\Delta V^*}{i_D^*} = \frac{C_{GC}}{\alpha} \frac{\Delta V/\alpha}{i_D/\alpha} = \frac{\tau}{\alpha} \tag{4.68}$$

We find that circuit delay is also improved by the scale factor α.

4.10.3 CIRCUIT AND POWER DENSITIES

As we scale down the dimensions by α, the number of circuits in a given area will increase by a factor of α^2. An important concern in scaling is therefore what happens to the power per circuit, and hence the power per unit area (power density) as dimensions are reduced. The total power supplied to a transistor circuit will be equal to the product of the supply voltage and the transistor drain current:

$$P^* = V_{DD}^* i_D^* = \left(\frac{V_{DD}}{\alpha}\right)\left(\frac{i_D}{\alpha}\right) = \frac{P}{\alpha^2}$$

and

$$\tag{4.69}$$

$$\frac{P^*}{A^*} = \frac{P^*}{W^* L^*} = \frac{P/\alpha^2}{(W/\alpha)(L/\alpha)} = \frac{P}{WL} = \frac{P}{A}$$

The result in Eq. (4.69) is extremely important. It indicates that the power per unit area remains constant if a technology is properly scaled. Even though we are increasing the number of circuits by α^2, the total power for a given size integrated circuit die will remain constant. Violation of the **scaling theory** over many years, by maintaining a constant 5-V power supply as dimensions were reduced, led to almost unmanagable power levels in many of today's integrated circuits. The power problem was finally resolved by changing from NMOS to CMOS technology, and then by reducing the power supply voltages.

4.10.4 POWER-DELAY PRODUCT

A useful figure of merit for comparing logic families is the **power-delay product** (PDP), which is discussed in more detail in Chapters 6 to 9. The product of power and delay time represents energy, and the power-delay product represents a measure of the energy required to perform a simple logic operation.

$$\text{PDP}^* = P^* \tau^* = \frac{P}{\alpha^2}\frac{\tau}{\alpha} = \frac{\text{PDP}}{\alpha^3} \tag{4.70}$$

The PDP figure of merit shows the full power of technology scaling. The power-delay product is reduced by the cube of the scaling factor.

Each new generation of lithography technology corresponds to a scale factor $\alpha = \sqrt{2}$. Therefore each new technology generation increases the potential number of circuits per chip by a factor of 2 and improves the PDP by a factor of almost 3. Table 4.4 summarizes the performance changes achieved with **constant electric field scaling.**

TABLE 4.4
Constant Electric Field Scaling Results

PERFORMANCE MEASURE	SCALE FACTOR	PERFORMANCE MEASURE	SCALE FACTOR
Area/circuit	$1/\alpha^2$	Circuit delay	$1/\alpha$
Transconductance parameter	α	Power/circuit	$1/\alpha^2$
Current	$1/\alpha$	Power/unit area (power density)	1
Capacitance	$1/\alpha$	Power-delay product (PDP)	$1/\alpha^3$

EXERCISE: A MOS technology is scaled from a 1-μm feature size to 0.25 μm. What is the increase in the number of circuits/cm^2? What is the improvement in the power-delay product?

ANSWERS: 16 times; 64 times

EXERCISE: Suppose that the voltages are not scaled as the dimensions are reduced by a factor of α? How does the drain current of the transistor change? How do the power/circuit and power density scale?

ANSWERS: $I_D^* = \alpha I_D$; $P^* = \alpha P$; $P^*/A^* = \alpha^3 P!!$

4.10.5 CUTOFF FREQUENCY

The ratio of transconductance g_m to gate-channel capacitance C_{GC} represents the highest useful frequency of operation of the transistor, and this ratio is called the cutoff frequency f_T of the device. The cutoff frequency represents the highest frequency at which the transistor can provide amplification. We can find f_T for the MOSFET by combining Eqs. (4.20) and (4.34):

$$f_T = \frac{1}{2\pi} \frac{g_m}{C_{GC}} = \frac{1}{2\pi} \frac{\mu_n}{L^2}(V_{GS} - V_{TN}) \tag{4.71}$$

Here we see clearly the advantage of scaling the channel length of MOSFET. The cutoff frequency improves with the square of the reduction in channel length.

EXERCISE: (a) A MOSFET has a mobility of 500 cm^2/V·s and channel length of 1 μm. What is its cutoff frequency if the gate voltage exceeds the threshold voltage by 1 V? (b) Repeat for a channel length of 0.25 μm.

ANSWERS: (a) 7.96 GHz; (b) 127 GHz

4.10.6 HIGH FIELD LIMITATIONS

Unfortunately the assumptions underlying constant-field scaling have often been violated due to a number of factors. For many years, the supply voltage was maintained constant at a standard level of 5 V, while the dimensions of the transistor were reduced, thus increasing the electric fields within the MOSFET. Increasing the electric field in the device can reduce long-term reliability and ultimately lead to breakdown of the gate oxide or *pn* junction.

High fields directly affect MOS transistor mobility in two ways. The first effect is a reduction in the mobility of the MOS transistor due to increasing carrier scattering at the channel oxide interface. The second effect of high electric fields is to cause a breakdown of the linear mobility-field relationship as discussed in Chapter 2. At low fields, carrier velocity is directly proportional to electric field, as assumed in Eq. (4.5), but for fields exceeding approximately 10^5 V/cm, the carriers reach a maximum velocity of approximately 10^7 cm/s called the saturation velocity v_{SAT} (see Fig. 2.5). Both mobility reduction and velocity saturation tend to linearize the drain current expressions for the MOSFET. The results of these effects can be incorporated into the drain current model for the MOSFET as indicated in Eqs. (4.72) and (4.73) in which the expression for carrier velocity is replaced with the maximum velocity limit v_{SAT}:

$$i_D = Q_n v_n = \frac{C''_{ox}W}{2}(v_{GS} - V_{TN})v_n \text{ and } v_n = \mu_n \frac{v_{DS}}{L} \to v_{SAT} \qquad (4.72)$$

This modification causes the square-law behavior to disappear from the saturation region equation:

$$\text{Saturation region:} \quad i_D = \frac{C''_{ox}W}{2}(v_{GS} - V_{TN})v_{SAT} \qquad (4.73)$$

EXERCISE: A MOSFET has a channel length of 1 μm. What value of V_{DS} will cause the electrons to reach saturation velocity? Repeat for a channel length of 0.1 μm.

ANSWERS: 10 V, 1 V

4.10.7 SUBTHRESHOLD CONDUCTION

In our discussion of the MOSFET thus far, we have assumed that the transistor turns off abruptly as the gate-source voltage drops below the threshold voltage. In reality, this is not the case. As depicted in Fig. 4.32, the drain current decreases exponentially for values of v_{GS} less than V_{TN} (referred to as the **subthreshold region**), as indicated by the region of constant slope in the graph. A measure of the rate of turn off of the MOSFET in the subthreshold region is specified as the reciprocal of the slope $(1/S)$ in mV/decade of current change. Typical values range from 60 to 120 mV/decade. The value depends on the relative magnitudes of C''_{ox} and C_d in Fig. 4.3(b).

From Eq. (4.66), we see that the threshold voltage of the transistor should be reduced as the dimensions are reduced. However, the subthreshold region does not scale properly, and the curve in Fig. 4.32 tends to shift horizontally as V_{TN} is decreased. The reduced threshold increases the leakage current in "off" devices, which ultimately limits data storage time in the dynamic memory cells (see Chapter 8) and can play an important role in limiting battery life in low-power portable devices.

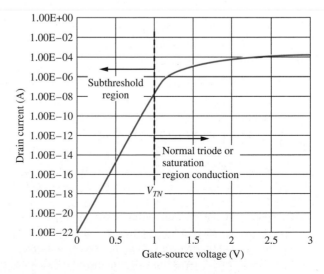

Figure 4.32 Subthreshold conduction in an NMOS transistor with $V_{TN} = 1$ V.

EXERCISE: (a) What is the leakage current in the device in Fig. 4.32 for $V_{GS} = 0.25$ V? (b) Suppose the transistor in Fig. 4.32 had $V_{TN} = 0.5$ V. What will be the leakage current for $V_{GS} = 0$ V? (c) A memory chip uses 10^9 of the transistors in part (b). What is the total leakage current if $V_{GS} = 0$ V for all the transistors?

ANSWERS: (a) $\cong 10^{-18}$ A; (b) $\cong 10^{-15}$ A; (c) $\cong 1$ μA

S U M M A R Y

- This chapter discussed the structures and $i\text{-}v$ characteristics of the metal-oxide-semiconductor FET, or MOSFET.

- At the heart of the MOSFET is the MOS capacitor, formed by a metallic gate electrode insulated from the semiconductor by an insulating oxide layer. The potential on the gate controls the carrier concentration in the semiconductor region directly beneath the gate; three regions of operation of the MOS capacitor were identified: accumulation, depletion, and inversion.

- A MOSFET is formed when two pn junctions are added to the semiconductor region of the MOS capacitor. The junctions act as the source and drain terminals of the MOS transistor and provide a ready supply of carriers for the channel region of the MOSFET. The source and drain junctions must be kept reverse-biased at all times in order to isolate the channel from the substrate.

- MOS transistors can be fabricated with either n- or p-type channel regions and are referred to as NMOS or PMOS transistors, respectively. In addition, MOSFETs can be fabricated as either enhancement-mode or depletion-mode devices.

  - For an enhancement-mode device, a gate-source voltage exceeding the threshold voltage must be applied to the transistor to establish a conducting channel between source and drain.

  - In the depletion-mode device, a channel is built into the device during its fabrication, and a voltage must be applied to the transistor's gate to quench conduction.

- The MOSFET is a symmetrical device. The source and drain terminals of the device are actually determined by the voltages applied to the terminals. For a given geometry and set of voltages, the n-channel transistor will conduct two to three times the current of the p-channel device because of the difference between the electron and hole mobilities in the channel.

- The MOSFET has three regions of operation.

  - In cutoff, a channel does not exist, and the terminal currents are zero.

  - In the triode region of operation, the drain current in the FET depends on both the gate-source and drain-source voltages of the transistor. For small values of drain-source voltage, the transistor exhibits an almost linear relationship between its drain current and drain-source voltage. In the triode region, the FET can be used as a voltage-controlled resistor, in which the on-resistance of the transistor is controlled by the gate-source voltage of the transistor. Because of this behavior, the name *transistor* was developed as a contraction of "transfer resistor."

  - For values of drain-source voltage exceeding the pinch-off voltage, the drain current of the FET becomes almost independent of the drain-source voltage. In this region, referred to variously as the pinch-off region, the saturation region, or the active region, the drain-source current exhibits a square-law dependence on the voltage applied between the gate and source terminals. Variations in drain-source voltage do cause small changes in drain current in saturation due to channel-length modulation.

- Mathematical models for the i-v characteristics of both NMOS and PMOS devices were presented. The MOSFET is actually a four-terminal device and has a threshold voltage that depends on the source-bulk voltage of the transistor.

  - Key parameters for the MOSFET include the transconductance parameters K_n or K_p, the zero-bias threshold voltage V_{TO}, body effect parameter γ, and channel-length modulation parameter λ as well as the width W and length L of the channel.

- A variety of examples of bias circuits were presented, and the mathematical model was used to find the quiescent operating point, or Q-point, for various types of MOSFETs. The Q-point represents the dc values of drain current and drain-source voltage: (I_D, V_{DS}).

- The i-v characteristics are often displayed graphically in the form of either the output characteristics, which plot i_D versus v_{DS}, or the transfer characteristics, which graph i_D versus v_{GS}. Examples of finding the Q-point using graphical load-line and iterative numerical analyses were discussed.

- The most important bias circuit in discrete design is the four-resistor circuit which yields a well-stabilized operating point.

- The gate-source, gate-drain, drain-bulk, source-bulk, and gate-bulk capacitances of MOS transistors were discussed, and the Meyer model [4] for the gate-source and gate-bulk capacitances was introduced. All the capacitances are nonlinear functions of the terminal voltages of the transistor.

- Complex models for MOSFETs are built into SPICE circuit analysis programs. These models contain many circuit elements and parameters in an attempt to model the true behavior of the transistor as closely as possible.

- Part of the IC designer's job often includes layout of the transistors based on a set of technology-specific ground rules that define minimum feature dimensions and spaces between features.

- Constant electric field scaling provides a framework for proper miniaturization of MOS devices in which the power density remains constant as the transistor density increases. In this case, circuit delay improves directly with the scale factor α, whereas the power-delay product improves with the cube of α.

- The cutoff frequency f_T of the transistor represents the highest frequency at which the transistor can provide amplification. Cutoff frequency f_T improves directly with the scale factor.

- The electric fields in small devices can become very high, and the carrier velocity tends to saturate at fields above 10 kV/cm. Subthreshold leakage current becomes increasingly important as devices are scaled to small dimension.

KEY TERMS

Accumulation	Capacitance per unit width
Accumulation region	Channel length L
Active region	Channel-length modulation
Alignment tolerance T	Channel-length modulation parameter λ
Body effect	Channel region
Body-effect parameter γ	Channel width W
Body terminal (B)	Constant electric field scaling
Bulk terminal (B)	Current sink
$C_{GS}, C_{GD}, C_{GB}, C_{DB}, C_{SB}, C_{ox}'', C_{GDO}, C_{GSO}$	Current source

Cutoff frequency

Depletion

Depletion-mode device

Depletion-mode MOSFETs

Depletion region

Design rules

Drain (D)

Electronic current source

Enhancement-mode device

Field-effect transistor (FET)

Four-resistor bias

Gate (G)

Gate-channel capacitance C_{GC}

Gate-drain capacitance C_{GD}

Gate-source capacitance C_{GS}

Ground rules

High field limitations

Inversion layer

Inversion region

KP

K'_n, K'_p

LAMDA, λ

Triode region

Metal-oxide-semiconductor field-effect transistor (MOSFET)

Minimum feature size F

Mirror ratio

MOS capacitor

n-channel MOS (NMOS)

n-channel MOSFET

n-channel transistor

NMOSFET

NMOS transistor

On-resistance (R_{on})

Output characteristics

Output resistance

Overlap capacitance

Oxide thickness

p-channel MOS (PMOS)

PHI

Pinch-off locus

Pinch-off point

Pinch-off region

PMOS transistor

Power delay product

Quiescent operating point

Q-point

Saturation region

Saturation voltage

Scaling theory

Small-signal output resistance

SPICE MODELS

Source (S)

Substrate sensitivity

Substrate terminal

Surface potential parameter $2\phi_F$

Subthreshold region

Threshold voltage V_{TN}, V_{TP}

Transconductance g_m

Transconductance parameter — K'_n, K'_p, KP

Transfer characteristic

Triode region

V_{TN}, V_{TP}, VT, VTO

Zero-substrate-bias value for V_{TN}

REFERENCES

1. U. S. Patent 1,900,018. Also see 1,745,175 and 1,877,140.

2. National Technology Road Map for Semiconductors, www.itrs.net.

3. Carver Mead and Lynn Conway, *Introduction to VLSI Systems,* Addison Wesley, Reading, Massachusetts: 1980.

4. J. E. Meyer, "MOS models and circuit simulations," *RCA Review,* vol. 32, pp. 42–63, March 1971.

5. B. M. Wilamowski and R. C. Jaeger, *Computerized Circuit Analysis Using SPICE Programs,* McGraw-Hill, New York: 1997.

6. R. H. Dennard, F. H. Gaensslen, L. Kuhn, and H. N. Yu, "Design of micron MOS switching devices," *IEEE IEDM Digest,* pp. 168–171, December 1972.

7. R. H. Dennard, F. H. Gaensslen, H-N. Yu, V. L. Rideout, E. Bassous and A. R. LeBlanc, "Design of ion-implanted MOSFET's with very small physical dimensions," *IEEE J. Solid-State Circuits,* vol. SC-9, no. 5, pp. 256–268, October 1974.

PROBLEMS

Use the parameters in Table 4.5 as needed in the problems here.

TABLE 4.5
MOS Transistor Parameters

	NMOS DEVICE	PMOS DEVICE
V_{TO}	+0.75 V	−0.75 V
γ	$0.75\sqrt{V}$	$0.5\sqrt{V}$
$2\phi_F$	0.6 V	0.6 V
K'	100 μA/V^2	40 μA/V^2

$\varepsilon_{ox} = 3.9\varepsilon_o$ and $\varepsilon_s = 11.7\varepsilon_o$ where $\varepsilon_o - 8.854 \times 10^{-14}$ F/cm

4.1 Characteristics of the MOS Capacitor

4.1. (a) The MOS capacitor in Fig. 4.1 has $V_{TN} = 1$ V and $V_G = 2$ V. To what region of operation does this bias condition correspond? (b) Repeat for $V_G = -3$ V. (c) Repeat for $V_G = 0.5$ V.

4.2. Calculate the capacitance of an MOS capacitor with an oxide thickness T_{ox} of (a) 50 nm, (b) 20 nm, (c) 10 nm, and (d) 5 nm.

4.3. The minimum value of the depletion-layer capacitance can be estimated using an expression similar to Eq. (3.18): $C_d = \varepsilon_S/x_d$ in which the depletion-layer width is $x_d \cong \sqrt{\frac{2\varepsilon_S}{qN_B}(0.75 \text{ V})}$ and N_B is the substrate doping. Estimate C_d for $N_B = 10^{-15}/\text{cm}^3$.

4.2 The NMOS Transistor
Triode (Linear) Region Characteristics

4.4. Calculate K'_n for an NMOS transistor with $\mu_n = 500$ cm^2/V · s for an oxide thickness of (a) 40 nm, (b) 20 nm, (c) 10 nm, and (d) 5 nm.

4.5. (a) What is the charge density (C/cm^2) in the channel if the oxide thickness is 25 nm and the oxide voltage exceeds the threshold voltage by 1 V? (b) Repeat for a 10-nm oxide and a bias 2 V above threshold.

4.6. (a) What is the electron velocity in the channel if $\mu_n = 500$ cm^2/V · s and the electric field is 2000 V/cm? (b) Repeat for $\mu_n = 400$ cm^2/V · s with a field of 4000 V/cm.

4.7. Equation (4.2) indicates that the charge/unit · length in the channel of a pinched-off transis-

tor decreases as one proceeds from source to drain. However, our text argued that the current entering the drain terminal is equal to the current exiting from the source terminal. How can a constant current exist everywhere in the channel between the drain and source terminals if the first statement is indeed true?

4.8. Calculate the drain current in an NMOS transistor for $V_{GS} = 0, 1$ V, 2 V, and 3 V, with $V_{DS} = 0.25$ V, if $W = 5$ μm, $L = 0.5$ μm, $V_{TN} = 0.80$ V, and $K'_n = 200$ μA/V^2. What is the value of K_n?

4.9. An NMOS transistor has $K'_n = 200$ μA/V^2. What is the value of K_n if $W = 60$ μm, $L = 3$ μm? If $W = 3$ μm, $L = 0.15$ μm? If $W = 10$ μm, $L = 0.25$ μm?

4.10. Calculate the drain current in an NMOS transistor for $V_{GS} = 0, 1$ V, 2 V, and 3 V, with $V_{DS} = 0.1$ V, if $W = 10$ μm, $L = 1$ μm, $V_{TN} = 1.0$ V, and $K'_n = 250$ μA/V^2. What is the value of K_n?

4.11. Identify the source, drain, gate, and bulk terminals and find the current I in the transistors in Fig. P4.11. Assume $V_{TN} = 0.70$ V.

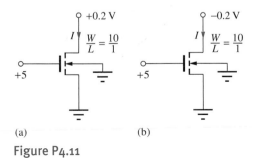

(a)   (b)

Figure P4.11

4.12. (a) What is the current in the transistor in Fig. P4.11(a) if the 0.2 V is changed to 0.5 V? Assume $V_{TN} = 0.75$ V. (b) If the gate voltage is changed to 3 V and the other voltage remains at 0.2 V?

4.13. (a) What is the current in the transistor in Fig. P4.11(b) if −0.2 V is changed to −0.5 V? Assume $V_{TN} = 0.75$ V. (b) If the gate voltage is changed to 3 V and the upper terminal voltage is replaced by −1 V?

4.14. (a) Design a transistor (choose W) to have $K_n = 4$ mA/V^2 if $L = 0.5$ μm. (See Table 4.5.) (b) Repeat for $K_n = 800$ μA/V^2.

On Resistance

4.15. What is the on-resistance of an NMOS transistor with $W/L = 100/1$ if $V_{GS} = 5$ V and $V_{TN} = 0.65$ V? (b) If $V_{GS} = 3.3$ V and $V_{TN} = 0.50$ V? (See Table 4.5.)

4.16. (a) What is the W/L ratio required for an NMOS transistor to have an on-resistance of 500 Ω when $V_{GS} = 5$ V and $V_{SB} = 0$? (b) Repeat for $V_{GS} = 3.3$ V.

4.17. Suppose that an NMOS transistor must conduct a current $I_D = 5$ A with $V_{DS} \leq 0.1$ V when it is on. What is the maximum on-resistance of M_S? If $V_G = 5$ V is used to turn on M_S and $V_{TN} = 2$ V, what is the minimum value of K_n required to achieve the required on-resistance?

Saturation of the i-v Characteristics

***4.18.** The output characteristics for an NMOS transistor are given in Fig. P4.18. What are the values of K_n and V_{TN} for this transistor? Is this an enhancement-mode or depletion-mode transistor? What is W/L for this device?

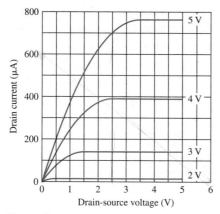

Figure P4.18

4.19. Add the $V_{GS} = 3.5$ V and $V_{GS} = 4.5$ V curves to the i-v characteristic of Fig. P4.18. What are the values of i_{DSAT} and v_{DSAT} for these new curves?

4.20. Calculate the drain current in an NMOS transistor for $V_{GS} = 0$, 1 V, 2 V, and 3 V, with $V_{DS} = 3.3$ V, if $W = 5$ μm, $L = 0.5$ μm, $V_{TN} = 1$ V, and $K'_n = 375$ μA/V^2. What is the value of K_n? Check the saturation region assumption.

4.21. Calculate the drain current in an NMOS transistor for $V_{GS} = 0$, 1 V, 2 V, and 3 V, with $V_{DS} = 4$ V, if $W = 10$ μm, $L = 1$ μm, $V_{TN} = 1.5$ V, and $K'_N = 200$ μA/V^2. What is the value of K_n? Check the saturation region assumption.

Regions of Operation

4.22. Find the region of operation and drain current in an NMOS transistor with $K'_n = 200$μA/V^2, $W/L = 10/1$, $V_{TN} = 0.75$ V and (a) $V_{GS} = 2$ V and $V_{DS} = 0.2$ V, (b) $V_{GS} = 2$ V and $V_{DS} = 2.5$ V, (c) $V_{GS} = 0$ V and $V_{DS} = 4$ V. (d) Repeat for $K'_n = 300$μA/V^2.

4.23. Identify the region of operation of an NMOS transistor with $K_n = 250$ μA/V^2 and $V_{TN} = 1$ V for (a) $V_{GS} = 5$ V and $V_{DS} = 6$ V, (b) $V_{GS} = 0$ V and $V_{DS} = 6$ V, (c) $V_{GS} = 2$ V and $V_{DS} = 2$ V, (d) $V_{GS} = 1.5$ V and $V_{DS} = 0.5$, (e) $V_{GS} = 2$ V and $V_{DS} = -0.5$ V, and (f) $V_{GS} = 3$ V and $V_{DS} = -6$ V.

4.24. Identify the region of operation of an NMOS transistor with $K_n = 400$ μA/V^2 and $V_{TN} = 0.7$ V for (a) $V_{GS} = 3.3$ V and $V_{DS} = 3.3$ V, (b) $V_{GS} = 0$ V and $V_{DS} = 3.3$ V, (c) $V_{GS} = 2$ V and $V_{DS} = 2$ V, (d) $V_{GS} = 1.5$ V and $V_{DS} = 0.5$, (e) $V_{GS} = 2$ V and $V_{DS} = -0.5$ V, and (f) $V_{GS} = 3$ V and $V_{DS} = -3$ V.

4.25. (a) Identify the source, drain, gate, and bulk terminals for the transistor in the circuit in Fig. P4.25. Assume $V_{DD} > 0$. (b) Repeat for $V_{DD} < 0$.

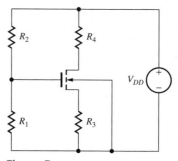

Figure P4.25

4.26. (a) Identify the source, drain, gate, and bulk terminals for each of the transistors in the circuit in Fig. P4.26(a). Assume $V_{DD} > 0$. (b) Repeat for the circuit in Fig. P4.26(b).

Transconductance

4.27. Calculate the transconductance for an NMOS transistor for $V_{GS} = 2$ V and 3.3 V, with $V_{DS} = 3.3$ V, if $W = 20$ μm, $L = 1$ μm, $V_{TN} = 0.7$ V, and $K'_n = 250$ μA/V^2. Check the saturation region assumption.

4.28. (a) Estimate the transconductance for the transistor in Fig. P4.18 for $V_{GS} = 4$ V and $V_{DS} = 4$ V. (*Hint:* $g_m \cong \Delta i_D / \Delta V_{GS}$.) (b) Repeat for $V_{GS} = 3$ V and $V_{DS} = 4.5$ V.

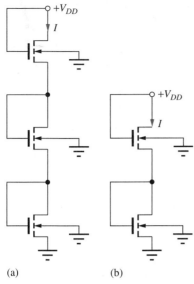

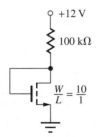

(a)   (b)

Figure P4.26

Channel-Length Modulation

4.29. (a) Calculate the drain current in an NMOS transistor if $K_n = 250$ μA/V^2, $V_{TN} = 0.75$ V, $\lambda = 0.025$ V^{-1}, $V_{GS} = 5$ V, and $V_{DS} = 6$ V. (b) Repeat assuming $\lambda = 0$.

4.30. (a) Calculate the drain current in an NMOS transistor if $K_n = 500$ μA/V^2, $V_{TN} = 1$ V, $\lambda = 0.02$ V^{-1}, $V_{GS} = 4$ V, and $V_{DS} = 5$ V. (b) Repeat assuming $\lambda = 0$.

4.31. (a) Find the drain current for the transistor in Fig. P4.31 if $\lambda = 0$. (b) Repeat if $\lambda = 0.025$ V^{-1}. (c) Repeat part (a) if the W/L ratio is changed to 25/1.

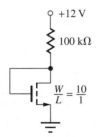

Figure P4.31

4.32. (a) Find the drain current for the transistor in Fig. P4.31 if $\lambda = 0$ and the W/L ratio is changed to 20/1. (b) Repeat if $\lambda = 0.020$ V^{-1}.

4.33. (a) Find the current I in Fig. P4.33 if $V_{DD} = 10$ V and $\lambda = 0$. Both transistors have $W/L = 10/1$.

(b) What is the current if both transistors have $W/L = 20/1$. (c) Repeat part (a) for $\lambda = 0.04$ V^{-1}.

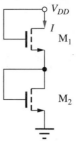

Figure P4.33

4.34. (a) Find the currents in the two transistors in Fig. P4.33 if $(W/L)_1 = 10/1$, $(W/L)_2 = 40/1$, and $\lambda = 0$ for both transistors. (b) Repeat for $(W/L)_2 = 40/1$ and $(W/L)_1 = 10/1$. (c) Repeat part (a) if $\lambda = 0.05$/V for both transistors.

4.35. (a) Find the currents in the two transistors in Fig. P4.33 if $(W/L)_1 = 25/1$, $(W/L)_2 = 12.5/1$ and $\lambda = 0$ for both transistors. (b) Repeat part (a) if $\lambda = 0.05$/V for both transistors.

Transfer Characteristics and the Depletion-Mode MOSFET

4.36. (a) Calculate the drain current in an NMOS transistor if $K_n = 250$ μA/V^2, $V_{TN} = -2$ V, $\lambda = 0$, $V_{GS} = 5$ V, and $V_{DS} = 6$ V. (b) Repeat assuming $\lambda = 0.03$ V^{-1}.

4.37. An NMOS depletion-mode transistor is operating with $V_{DS} = V_{GS} > 0$. What is the region of operation for this device?

4.38. (a) Calculate the drain current in an NMOS transistor if $K_n = 250$ μA/V^2, $V_{TN} = -3$ V, $\lambda = 0$, $V_{GS} = 0$ V, and $V_{DS} = 6$ V. (b) Repeat assuming $\lambda = 0.025$ V^{-1}.

4.39. (a) Find the Q-point for the transistor in Fig. P4.39 (a) if $V_{TN} = -2$ V. (b) Repeat for $R = 50$ kΩ and $W/L = 20/1$. (c) Repeat parts (a) & (b) for Fig. 4.39(b).

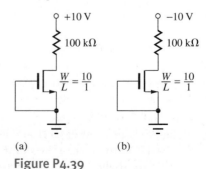

(a)   (b)

Figure P4.39

4.40. (a) Find the Q-point for the transistor in Fig. P4.39(a) if $V_{TN} = -1$ V and W/L is changed to 20/1. (b) Repeat for Fig. P4.39(b).

Body Effect or Substrate Sensitivity

4.41. Repeat Prob. 4.21 for $V_{SB} = 1.5$ V with the values from Table 4.5.

4.42. An NMOS transistor with $W/L = 16.8/1$ has $V_{TO} = 1.5$ V, $2\phi_F = 0.75$ V, and $\gamma = 0.5\sqrt{V}$. The transistor is operating with $V_{SB} = 4$ V, $V_{GS} = 2$ V, and $V_{DS} = 5$ V. What is the drain current in the transistor? (b) Repeat for $V_{DS} = 0.5$ V.

4.43. (a) An NMOS transistor with $W/L = 8/1$ has $V_{TO} = 1$ V, $2\phi_F = 0.6$ V, and $\gamma = 0.7\sqrt{V}$. The transistor is operating with $V_{SB} = 3$ V, $V_{GS} = 2.5$ V, and $V_{DS} = 5$ V. What is the drain current in the transistor? (b) Repeat for $V_{DS} = 0.5$ V.

4.44. A depletion-mode NMOS transistor has $V_{TO} = -1.5$ V, $2\phi_F = 0.75$ V, and $\gamma = 1.5\sqrt{V}$. What source-bulk voltage is required to change this transistor into an enhancement-mode device with a threshold voltage of $+0.85$ V?

*4.45. The measured body-effect characteristic for an NMOS transistor is given in Table 4.6. What are the best values of V_{TO}, γ, and $2\phi_F$ (in the least-squares sense — see Prob. 3.29) for this transistor?

TABLE 4.6

V_{SB} (V)	V_{TN} (V)
0	0.710
0.5	0.912
1.0	1.092
1.5	1.232
2.0	1.377
2.5	1.506
3.0	1.604
3.5	1.724
4.0	1.822
4.5	1.904
5.0	2.005

4.3 PMOS Transistors

4.46. Calculate K'_p for a PMOS transistor with $\mu_p = 200$ cm²/V · s for an oxide thickness of (a) 50 nm, (b) 20 nm, (c) 10 nm, and (d) 5 nm.

*4.47. The output characteristics for a PMOS transistor are given in Fig. P4.47. What are the values of K_p and V_{TP} for this transistor? Is this an enhancement-mode

or depletion-mode transistor? What is the value of W/L for this device?

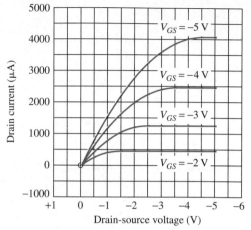

Figure P4.47

4.48. Add the $V_{GS} = -3.5$ V and $V_{GS} = -4.5$ V curves to the i-v characteristic of Fig. P4.47. What are the values of i_{DSAT} and v_{DSAT} for these new curves?

4.49. Find the region of operation and drain current in a PMOS transistor with $W/L = 20/1$ for $V_{BS} = 0$ V and (a) $V_{GS} = -1.1$ V and $V_{DS} = -0.2$ V and (b) $V_{GS} = -1.3$ V and $V_{DS} = -0.2$ V. (c) Repeat parts (a) and (b) for $V_{BS} = 1$ V.

4.50. (a) What is the W/L ratio required for a PMOS transistor to have an on-resistance of 1 Ω when $V_{GS} = -5$ V and $V_{SB} = 0$? Assume $V_{TP} = -0.70$ V. (b) Repeat for an NMOS transistor with $V_{GS} = +5$ V and $V_{BS} = 0$. Assume $V_{TN} = 0.70$ V.

4.51. (a) What is the W/L ratio required for an PMOS transistor to have an on-resistance of 2 kΩ when $V_{GS} = -5$ V and $V_{BS} = 0$? Assume $V_{TP} = -0.70$ V. (b) Repeat for an NMOS transistor with $V_{GS} = +5$ V and $V_{BS} = 0$. Assume $V_{TN} = 0.70$ V.

4.52. (a) Calculate the on-resistance for a PMOS transistor having $W/L = 200/1$ and operating with $V_{GS} = -5$ V and $V_{TP} = -0.75$ V. (b) Repeat for a similar NMOS transistor with $V_{GS} = 5$ V and $V_{TN} = 0.75$ V. (c) What W/L ratio is required for the PMOS transistor to have the same R_{on} as the NMOS transistor in (b)?

4.53. (a) Identify the source, drain, gate, and bulk terminals for the transistors in the two circuits in Fig. P4.53(a). Assume $V_{DD} = 10$ V. (b) Repeat for Fig. P4.53(b).

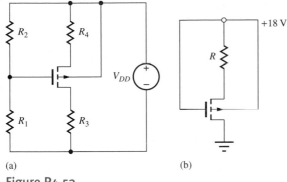

(a)   (b)

Figure P4.53

4.54. What is the on-resistance and voltage V_O for the parallel combination of the NMOS ($W/L = 10/1$) and PMOS ($W/L = 25/1$) transistors in Fig. P4.54 for $V_{IN} = 0$ V? (b) For $V_{IN} = 5$ V? This circuit is called a transmission-gate.

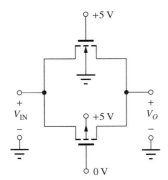

Figure P4.54

4.55. Suppose a PMOS transistor must conduct a current $I_D = 0.5$ A with $V_{SD} \leq 0.1$ V when it is on. What is the maximum on-resistance? If $V_G = 0$ V is used to turn on the transistor with $V_S = 10$ V and $V_{TP} = -2$ V, what is the minimum value of K_p required to achieve the required on-resistance?

4.56. A PMOS transistor is operating with $V_{BS} = 4$ V, $V_{GS} = -1.5$ V, and $V_{DS} = -4$ V. What are the region of operation and drain current in this device if $W/L = 25/1$?

4.57. A PMOS transistor is operating with $V_{BS} = 0$ V, $V_{GS} = -1.5$ V, and $V_{DS} = -0.5$ V. What are the region of operation and drain current in this device if $W/L = 40/1$?

4.4 MOSFET Circuit Symbols

4.58. The PMOS transistor in Fig. P4.53(a) is conducting current. Is $V_{TP} > 0$ or $V_{TP} < 0$ for this transistor?

Based on this value of V_{TP}, what type transistor is in the circuit? Is the proper symbol used in this circuit for this transistor? If not, what symbol should be used?

4.59. The PMOS transistor in Fig. P4.53(b) is conducting current. Is $V_{TP} > 0$ or $V_{TP} < 0$ for this transistor? Based on this value of V_{TP}, what type transistor is in the circuit? Is the proper symbol used in this circuit for this transistor? If not, what symbol should be used?

4.60. (a) Redraw the circuits in Fig. P4.53(a) with a three-terminal PMOS transistor with its body connected to its source. (b) Repeat for Fig. 4.53(b).

4.61. Redraw the circuit in Fig. 4.26 with a four-terminal NMOS transistor with its body connected to -3 V.

4.62. Redraw the circuit in Fig. 4.27 with a four-terminal NMOS transistor with its body connected to -5 V.

4.5 MOS Transistor Fabrication and Layout Design Rules

4.63. Layout a transistor with $W/L = 10/1$ similar to Fig. 4.18. What fraction of the total area does the channel represent?

4.64. Layout a transistor with $W/L = 5/1$ similar to Fig. 4.18 using $T = F = 2 \Lambda$. What fraction of the total area does the channel represent?

4.65. Layout a transistor with $W/L = 5/1$ similar to Fig. 4.18 but change the alignment so that mask 3 is aligned to mask 1. What fraction of the total area does the channel represent?

4.66. Layout a transistor with $W/L = 5/1$ similar to Fig. 4.18 but change the alignment so that masks 2, 3, and 4 are all aligned to mask 1. What fraction of the total area does the channel represent?

4.6 Capacitances in MOS Transistors

4.67. Calculate C''_{ox} and C_{GC} for an MOS transistor with $W = 20\,\mu\text{m}$ and $L = 2\,\mu\text{m}$ with an oxide thickness of (a) 50 nm, (b) 20 nm, (c) 10 nm, and (d) 5nm.

4.68. Calculate C''_{ox} and C_{GC} for an MOS transistor with $W = 5\,\mu\text{m}$ and $L = 0.5\,\mu\text{m}$ with an oxide thickness of 10 nm.

4.69. In a certain MOSFET, the value of C'_{OL} can be calculated using an effective overlap distance of $0.5\,\mu\text{m}$. What is the value of C'_{OL} for an oxide thickness of 10 nm.

4.70. What are the values of C_{GS} and C_{GD} for a transistor with $C''_{ox} = 1.4 \times 10^{-3}$ F/m^2 and

$C'_{OL} = 4 \times 10^{-9}$ F/m if $W = 10\,\mu$m and $L = 1\,\mu$m operating in (a) the triode region, (b) the saturation region, and (c) cutoff?

4.71. A large-power MOSFET has an effective gate area of $50 \times 10^6\,\mu$m^2. What is the value of C_{GC} if T_{ox} is 100 nm?

4.72. (a) Find C_{GS} and C_{GD} for the transistor in Fig. 4.18 for the triode region if $\Lambda = 0.5\,\mu$m, $T_{\text{ox}} = 150$ nm, and $C_{GSO} = C_{GDO} = 20$ pF/m. (b) Repeat for the saturation region. (c) Repeat for the cutoff region.

4.73. (a) Repeat Prob. 4.72 for a transistor similar to Fig. 4.18 but with $W/L = 10/1$. (b) With $W/L = 100/1$. Assume $L = 1\,\mu$m.

4.74. Find C_{SB} and C_{DB} for the transistor in Fig. 4.18 if $\Lambda = 0.5\,\mu$m, the substrate doping is 10^{16}/cm^3, the source and drain doping is 10^{20}/cm^3, and $C_{JSW} = C_J \times (5 \times 10^{-4}$/cm).

4.7 MOSFET Modeling in SPICE

4.75. What are the values of SPICE model parameters KP, LAMBDA, VTO, PHI, W, and L for a transistor with the following characteristics: $V_{TN} = 0.7$ V, $K_n = 175\,\mu$A/V^2, $W = 5\,\mu$m, $L = 0.25\,\mu$m, $\lambda = 0.02$ V^{-1}, and $2\phi_F = 0.8$ V?

4.76. (a) What are the values of SPICE model parameters VTO, PHI, and GAMMA for the transistor in Fig. 4.13? (b) Repeat for the transistor in Prob. 4.45.

4.77. What are the values of SPICE model parameters KP, LAMBDA, VTO, W and L for the transistor in Fig. 4.7 if $K'_n = 50\,\mu$A/V^2 and $L = 0.5\,\mu$m?

4.78. What are the values of SPICE model parameters KP, LAMBDA, VTO, W and L for the transistor in Fig. 4.8 if $K'_n = 10\,\mu$A/V^2 and $L = 0.6\,\mu$m?

4.79. What are the values of SPICE model parameters KP, LAMBDA, VTO, W and L, for the transistor in Fig. 4.14 if $K'_p = 10\,\mu$A/V^2 and $L = 0.5\,\mu$m?

4.80. What are the values of SPICE model parameters KP, LAMBDA, VTO, W and L, for the transistor in Fig. 4.23(b) if $K'_n = 25\,\mu$A/V^2 and $L = 0.6\,\mu$m?

4.8 Biasing the NMOS Field-Effect Transistor

Load Line Analysis

4.81. Draw the load line for the circuit in Fig. P4.81 on the output characteristics in Fig. P4.18 and locate the Q-point. Assume $V_{DD} = +4$ V. What is the operating region of the transistor?

4.82. Draw the load line for the circuit in Fig. P4.81 on the output characteristics in Fig.P4.18 and locate the Q-point. Assume $V_{DD} = +5$ V and the resistor is changed to 8.3 kΩ. What is the operating region of the transistor?

4.83. Draw the load line for the circuit in Fig. P4.83 on the output characteristics in Fig. P4.18 and locate the Q-point. Assume $V_{DD} = +6$ V. What is the operating region of the transistor?

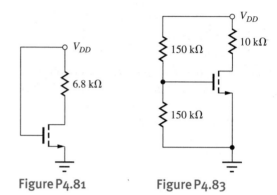

Figure P4.81   **Figure P4.83**

4.84. Draw the load line for the circuit in Fig. P4.83 on the output characteristics in Fig. P4.18 and locate the Q-point. Assume $V_{DD} = +8$ V. What is the operating region of the transistor?

Four-Resistor Biasing

4.85. (a) Find the Q-point for the transistor in Fig. P4.85 for $R_1 = 100$ kΩ, $R_2 = 220$ kΩ, $R_3 = 24$ kΩ, $R_4 = 12$ kΩ, and $V_{DD} = 12$ V. Assume that $V_{TO} = 1$ V, $\gamma = 0$, and $W/L = 5/1$. (b) Repeat for $W/L = 10/1$.

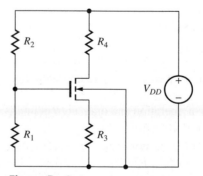

Figure P4.85

4.86. Repeat Prob. 4.85(a) if all resistor values are reduced by a factor of 10 and $W/L = 20/1$.

4.87. Repeat Prob. 4.85(a) if all resistor values are increased by a factor of 10.

4.88. Repeat Prob. 4.85 with $V_{DD} = 15$ V.

4.89. Find the Q-point for the transistor in Fig. P4.85 for $R_1 = 200$ kΩ, $R_2 = 430$ kΩ, $R_3 = 47$ kΩ, $R_4 = 24$ kΩ, and $V_{DD} = 12$ V. Assume that $V_{TO} = 1$ V, $\gamma = 0$, and $W/L = 5/1$. (b) Repeat for $W/L = 15/1$.

4.90. Use SPICE to simulate the circuit in Prob. 4.85 and compare the results to hand calculations.

4.91. Use SPICE to simulate the circuit in Prob. 4.88 and compare the results to hand calculations.

4.92. Use SPICE to simulate the circuit in Prob. 4.89 and compare the results to hand calculations.

4.93. The drain current in the circuit in Fig. 4.24 was found to be 50 μA. The gate bias circuit in the example could have been designed with many different choices for resistors R_1 and R_2. Some possibilities for (R_1, R_2) are (3 kΩ, 7 kΩ), (12 kΩ, 28 kΩ), (300 kΩ, 700 kΩ), and (1.2 MΩ, 2.8 MΩ). Which of these choices would be the best and why?

*4.94. Suppose the design of Ex. 4.4 is implemented with $V_{EQ} = 4$ V, $R_S = 1.7$ kΩ, and $R_D = 38.3$ kΩ. (a) What would be the Q-point if $K_n = 35$ μA/V^2? (b) If $K_n = 25$ μA/V^2 but $V_{TN} = 0.75$ V?

4.95. (a) Simulate the circuit in Ex. 4.3 and compare the results to the calculations. (b) Repeat for the circuit design in Ex. 4.4.

4.96. Design a four-resistor bias network for an NMOS transistor to give a Q-point of (100 μA, 4 V) with $V_{DD} = 12$ V and $R_{EQ} \cong 250$ kΩ. Use the parameters from Table 4.5.

4.97. Design a four-resistor bias network for an NMOS transistor to give a Q-point of (250 μA, 3 V) with $V_{DD} = 9$ V and $R_{EQ} \cong 250$ kΩ. Use the parameters from Table 4.5.

4.98. Design a four-resistor bias network for an NMOS transistor to give a Q-point of (500 μA, 5 V) with $V_{DD} = 15$ V and $R_{EQ} \cong 600$ kΩ. Use the parameters from Table 4.5.

Depletion-Mode Devices

4.99. What is the Q-point of the transistor in Fig. P4.85 if $R_1 = 1$ MΩ, $R_2 = \infty$, $R_3 = 10$ kΩ, $R_4 = 5$ kΩ, and $V_{DD} = 15$ V for $V_{TN} = -5$ V and $K_n = 1$ mA/V^2.

4.100. What is the Q-point of the transistor in Fig. P4.85 if $R_1 = 470$ kΩ, $R_2 = \infty$, $R_3 = 27$ kΩ, $R_4 = 51$ kΩ, and $V_{DD} = 12$ V for $V_{TN} = -4$ V and $K_n = 600$ μA/V^2.

4.101. Design a bias network for a depletion-mode NMOS transistor to give a Q-point of (250 μA, 5 V) with $V_{DD} = 15$ V if $V_{TN} = -5$ V and $K_n = 1$ mA/V^2.

*4.102. Design a bias network for a depletion-mode NMOS transistor to give a Q-point of (2 mA, 5 V) with $V_{DD} = 15$ V if $V_{TN} = -2$ V and $K_n = 250$ μA/V^2. (*Hint:* You may wish to consider the four-resistor bias network.)

Two-Resistor Biasing

4.103. (a) Find the Q-point for the transistor in the circuit in Fig. P4.103(a) if $V_{DD} = +12$ V. (b) Repeat for the circuit in Fig. P4.103(b).

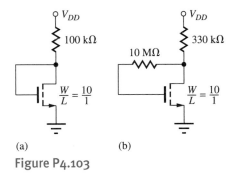

(a)　　　　　　　(b)

Figure P4.103

4.104. (a) Find the Q-point for the transistor in the circuit in Fig. P4.103(a) if $V_{DD} = +12$ V and W/L is changed to 20/1? (b) Repeat for the circuit in Fig. P4.103(b).

4.105. (a) Find the Q-point for the transistor in the circuit in Fig. P4.103(b) if $V_{DD} = +15$ V. (b) Repeat for $V_{DD} = +15$ V with W/L is changed to 25/1?

4.106. (a) Find the Q-point for the transistor in the circuit in Fig. P4.103(b) if $V_{DD} = +12$ V and the 330 kΩ resistor is increased to 470 kΩ. (b) Repeat if the 10 MΩ resistor is reduced to 2 MΩ.

Body Effect

4.107. Find the solution to Eq. set (4.51) using MATLAB. (b) Repeat for $\gamma = 0.75$ $\sqrt{V}$.

4.108. Find the solution to Eq. set (4.51) using a spreadsheet if $\gamma = 0.75$ $\sqrt{V}$. (b) Repeat for $\gamma = 1.25$ $\sqrt{V}$.

4.109. Find the Q-point for the transistor in Fig. P4.85 for $R_1 = 100$ kΩ, $R_2 = 220$ kΩ, $R_3 = 24$ kΩ, $R_4 = 12$ kΩ, and $V_{DD} = 12$ V. Assume that $V_{TO} = 1$ V, $\gamma = 0.6$ $\sqrt{V}$, and $W/L = 5/1$.

*4.110. (a) Repeat Prob. 4.109 with $\gamma = 0.75$ $\sqrt{V}$. (b) Repeat Prob. 4.109 with $R_4 = 24$ kΩ.

4.111. (a) Use SPICE to simulate the circuit in Prob. 4.109 and compare the results to hand calculations.

(b) Repeat for Prob. 4.110(a). (c) Repeat for Prob. 4.110(b).

4.112. Simulate the circuit in Prob. 4.85 using (a) $\gamma = 0$ and (b) $\gamma = 0.5$ V$^{-0.5}$ and $2\phi_F = 0.6$ V and compare the results. Does our neglect of body effect in hand calculations appear to be justified?

4.113. Simulate the circuit in Prob. 4.86 using (a) $\gamma = 0$ and (b) $\gamma = 0.5$ V$^{-0.5}$ and $2\phi_F = 0.6$ V and compare the results. Does our neglect of body effect in hand calculations appear to be justified?

4.114. Simulate the circuit in Prob. 4.87 using (a) $\gamma = 0$ and (b) $\gamma = 0.5$ V$^{-0.5}$ and $2\phi_F = 0.6$ V and compare the results. Does our neglect of body effect in hand calculations appear to be justified?

4.115. Simulate the circuit in Prob. 4.89 using (a) $\gamma = 0$ and (b) $\gamma = 0.5$ V$^{-0.5}$ and $2\phi_F = 0.6$ V and compare the results. Does our neglect of body effect in hand calculations appear to be justified?

General Bias Problems

4.116. (a) Find the current I in Fig. P4.116 if $V_{DD} = 5$ V assuming that $\gamma = 0$, $V_{TO} = 1$ V, and the transistors both have $W/L = 20/1$. (b) Repeat for $V_{DD} = 10$ V. *(c) Repeat part (a) with $\gamma = 0.5\sqrt{V}$.

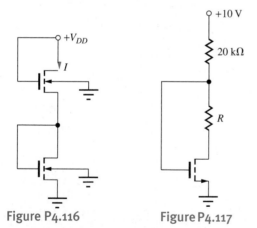

Figure P4.116   **Figure P4.117**

4.117. Find the Q-point for the transistor in Fig. P4.117 if $R = 20$ kΩ, $V_{TO} = 1$ V, and $W/L = 2/1$.

4.118. Find the Q-point for the transistor in Fig. P4.117 if $R = 10$ kΩ, $V_{TO} = 1$ V, and $W/L = 4/1$.

**4.119. (a) Find the current I in Fig. P4.119 assuming that $\gamma = 0$ and $W/L = 20/1$ for each transistor. (b) Repeat part (a) for $W/L = 50/1$. **(c) Repeat part (a) with $\gamma = 0.5\sqrt{V}$.

4.120. (a) Simulate the circuit in Fig. P4.119 using SPICE and compare the results to those of Prob.

4.119(a). (b) Repeat for Prob. 4.119(b). **(c) Repeat for Prob. 4.119(c).

4.121. What value of W/L is required to set $V_{DS} = 0.50$ V in the circuit in Fig. P4.121 if $V = 5$ V and $R = 82$ kΩ?

4.122. What value of W/L is required to set $V_{DS} = 0.25$ V in the circuit in Fig. P4.121 if $V = 3.3$ V and $R = 180$ kΩ?

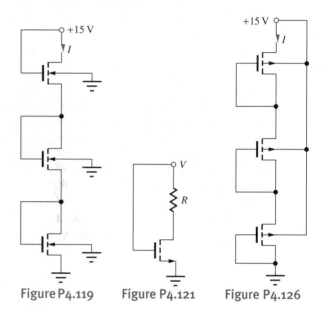

Figure P4.119   **Figure P4.121**   **Figure P4.126**

4.9 Biasing the PMOS Field-Effect Transistor

4.123. (a) Find the Q-point for the transistor in Fig. P4.123(a) if $V_{DD} = -15$ V, $R = 75$ kΩ, and $W/L = 1/1$. (b) Find the Q-point for the transistor in Fig. P4.123(b) if $V_{DD} = -15$ V, $R = 75$ kΩ, and $W/L = 1/1$.

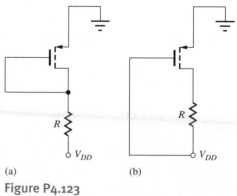

(a)   (b)

Figure P4.123

4.124. Simulate the circuits in Prob. 4.123 with $V_{DD} = -15$ V and compare the Q-point results to hand calculations.

*4.125. (a) Find current I and voltage V_O in Fig. P4.125 if $W/L = 20/1$ for both transistors and $V_{DD} = 10$ V. (b) What is the current if $W/L = 80/1$?

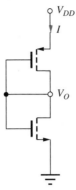

Figure P4.125

**4.126. (a) Find the current I in Fig. P4.126 (Page 204) assuming that $\gamma = 0$ and $W/L = 40/1$ for each transistor. (b) Repeat part (a) for $W/L = 75/1$. **(c) Repeat part (a) with $\gamma = 0.5\sqrt{V}$.

*4.127. (a) Simulate the circuit in Prob. 4.126(a) and compare the results to those of Prob. 4.126(a). (b) Repeat for Prob. 4.126(b). (c) Repeat for Prob. 4.126(c).

4.128. Draw the load line for the circuit in Fig. P4.128 on the output characteristics in Fig. P4.47 and locate the Q-point. What is the operating region of the transistor?

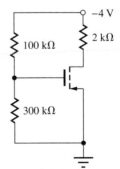

Figure P4.128

4.129. (a) Find the Q-point for the transistor in Fig. P4.129 if $R = 50$ kΩ. Assume that $\gamma = 0$ and $W/L = 20/1$. (b) What is the permissible range of values for R if the transistor is to remain in the saturation region?

4.130. Simulate the circuit of Prob. 4.129(a) and find the Q-point. Compare the results to hand calculations.

*4.131. (a) Find the Q-point for the transistor in Fig. P4.129 if $R = 43$ kΩ. Assume that $\gamma = 0.5 \sqrt{V}$ and $W/L = 20/1$. (b) What is the permissible range of values for R if the transistor is to remain in the saturation region?

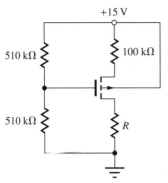

Figure P4.129

4.132. Simulate the circuit of Prob. 4.131(a) and find the Q-point. Compare the results to hand calculations.

4.133. (a) Find the Q-point for the transistor in Fig. P4.133 if $V_{DD} = 12$ V, $R = 100$ kΩ, $W/L = 10/1$, and $\gamma = 0$. (b) Repeat for $\gamma = 1 \sqrt{V}$.

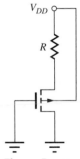

Figure P4.133

4.134. Find the Q-point current for the transistor in Fig. P4.129 if all resistors are reduced by a factor of 2. Assume saturation region operation. What value of R is needed to set $V_{DS} = -5$ V. Assume that $\gamma = 0$ and $W/L = 40/1$.

4.135. Repeat Prob. 4.134 if $\gamma = 0.5 \sqrt{V}$ and $W/L = 40/1$.

4.136. (a) Find the Q-point current for the transistor in Fig. P4.129 if the upper 510-kΩ resistor is changed to 270 kΩ. Assume that the transistor is saturated, $\gamma = 0$, and $W/L = 20/1$. (b) What is the permissible range of values for R if the transistor is to remain in the saturation region?

4.137. Repeat Prob. 4.136 if $\gamma = 0.5 \sqrt{V}$.

4.138. (a) Design a four-resistor bias network for a PMOS transistor to give a Q-point of (1 mA, −5 V) with $V_{DD} = -15$ V and $R_{EQ} \geq 100$ kΩ. Use the parameters from Table 4.5. (b) Repeat for an NMOS transistor with $V_{DS} = +6$ V and $V_{DD} = +15$ V.

4.139. (a) Design a four-resistor bias network for a PMOS transistor to give a Q-point of (500 μA, −3 V) with $V_{DD} = -9$ V and $R_{EQ} \geq 1$ MΩ. Use the parameters from Table 4.5. (b) Repeat for an NMOS transistor with $V_{DS} = +3$ V and $V_{DD} = +9$ V.

4.140. Find the Q-point for the transistor in Fig. P4.140 if $V_{TO} = +4$ V, $\gamma = 0$, and $W/L = 10/1$.

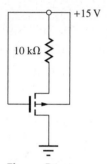

Figure P4.140

4.141. Find the Q-point for the transistor in Fig. P4.140 if $V_{TO} = +4$ V, $\gamma = 0.25 \sqrt{V}$, and $W/L = 10/1$.

4.142. Find the Q-point for the transistor in Fig. P4.142 if $V_{TO} = -1$ V and $W/L = 10/1$.

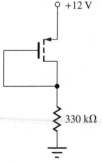

Figure P4.142

4.143. Find the Q-point for the transistor in Fig. P4.142 if $V_{TO} = -3$ V and $W/L = 30/1$.

4.144. What is the Q-point for each transistor in Fig. P4.144?

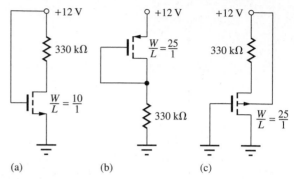

Figure P4.144

4.10 MOS Transistor Scaling

4.145. (a) A transistor has $T_{ox} = 40$ nm, $V_{TN} = 1$ V, $\mu_n = 500$ cm^2/V·s, $L = 2$ μm, and $W = 20$ μm. What are K_n and the saturated value of i_D for this transistor if $V_{GS} = 4$ V? (b) The technology is scaled by a factor of 2. What are the new values of T_{ox}, W, L, V_{TN}, V_{GS}, K_n, and i_D?

4.146. (a) A transistor has an oxide thickness of 20 nm with $L = 1$ μm and $W = 20$ μm. What is C_{GC} for this transistor? (b) The technology is scaled by a factor of 2. What are the new values of T_{ox}, W, L, and C_{GC}?

4.147. Show that the cutoff frequency of a PMOS device is given by $f_T = \frac{1}{2\pi} \frac{\mu_p}{L^2} |V_{GS} - V_{TP}|$.

4.148. (a) An NMOS device has $\mu_n = 400$ cm^2/V·s. What is the cutoff frequency for $L = 1$ μm if the transistor is biased at 1 V above threshold? What would be the cutoff frequency of a similar PMOS device if $\mu_p = 0.4 \mu_n$? (b) Repeat for $L = 0.1$ μm.

4.149. An NMOS transistor has $T_{ox} = 80$ nm, $\mu_n = 400$ cm^2/V·s, $L = 0.1$ μm, $W = 2$ μm, and $V_{GS} - V_{TN} = 2$ V. (a) What is the saturation region current predicted by Eq. (4.17)? (b) What is the saturation current predicted by Eq. (4.73) if we assume $v_{SAT} = 10^7$ cm/s?

4.150. The NMOS transistor in Fig. 4.32 is biased with $V_{GS} = 0$ V. What is the drain current? (b) What is the drain current if the threshold voltage is reduced to 0.5 V?

CHAPTER 5

BIPOLAR JUNCTION TRANSISTORS

John Bardeen, William Shockley, and Walter Brattain in Brattain's Laboratory in 1948.
Lucent Technologies Inc./Bell Labs

CHAPTER GOALS

- Explore the physical structure of the bipolar transistor
- Understand bipolar transistor action and the importance of carrier transport across the base region
- Study the terminal characteristics of the BJT
- Explore the differences between *npn* and *pnp* transistors
- Develop the Transport model for the bipolar device
- Define the four regions of operation of the BJT
- Explore model simplifications for each region of operation
- Understand the origin and modeling of the Early effect
- Present the SPICE model for the bipolar transistor
- Provide examples of worst-case and Monte Carlo analysis of bias circuits

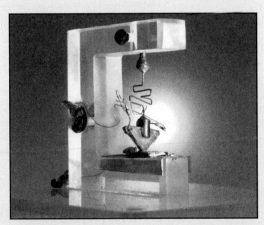

The first germanium bipolar transistor
Lucent Technologies Inc./Bell Labs

November 1997 was the 50th anniversary of the discovery of the bipolar transistor by John Bardeen and Walter Brattain at Bell Laboratories. In a matter of a few months, William Shockley managed to develop a theory describing the operation of the bipolar junction transistor. Only 10 years later in 1956, Bardeen, Brattain, and Shockley received the Nobel Prize in Physics for the discovery of the transistor.

In June 1948, Bell Laboratories held a major press conference to announce the discovery (which of course went essentially unnoticed by the public). Later in 1952, Bell Laboratories, operating under legal consent decrees, made licenses for the transistor available for the modest fee of $25,000 plus future royalty payments. About this time, Gordon Teal, another member of the solid-state group, left Bell Laboratories to work on the transistor at Geophysical Services Inc., which subsequently became Texas Instruments (TI). There he made the first silicon transistors, and

TI marketed the first all transistor radio. Another of the early licensees of the transistor was Tokyo Tsushin Kogyo which became the Sony Company in 1955. Sony subsequently sold a transistor radio with a marketing strategy based upon the idea that everyone could now have their own personal radio; thus was launched the consumer market for transistors. A very interesting account of these and other developments can be found in [1, 2] and their references.

Following its invention and demonstration in the late 1940s by Bardeen, Brattain, and Shockley at Bell Laboratories, the **bipolar junction transistor,** or **BJT,** became the first commercially successful three-terminal solid-state device. Its commercial success was based on its structure. In the structure of the BJT, the active base region of the transistor is below the surface of the semiconductor material, making it much less dependent on surface properties and cleanliness. Thus, it was initially easier to manufacture BJTs than MOS transistors. Commercial bipolar transistors were available in the late 1950s. The first integrated circuits, resistor-transistor logic gates, and operational amplifiers consisting of a few transistors and resistors appeared in the early 1960s.

While the FET has become the dominant device technology in modern integrated circuits, bipolar transistors are still widely used in both discrete and integrated circuit design. In particular, the BJT is still the preferred device in many applications that require high speed and/or high precision. Typical of these application areas are circuits for the growing families of wireless computing and communication products, and silicon-germanium (SiGe) BJTs offer the highest operating frequencies of any silicon transmitter.

The bipolar transistor is composed of a sandwich of three doped semiconductor regions and comes in two forms: the *npn* transistor and the *pnp* transistor. Performance of the bipolar transistor is dominated by *minority-carrier* transport via diffusion and drift in the central region of the transistor. Because carrier mobility and diffusivity are higher for electrons than holes, the *npn* transistor is an inherently higher-performance device than the *pnp* transistor. In Part III of this book, we will learn that the bipolar transistor typically offers a much higher voltage gain capability than the FET. On the other hand, the BJT input resistance is much lower, and a current must be supplied to the control electrode.

Our study of the BJT begins with a discussion of the *npn* transistor, followed by a discussion of the *pnp* device. The **transport model,** a simplified version of the Gummel-Poon model, is developed and used as our mathematical model for the behavior of the BJT. Four regions of operation of the BJT are defined and simplified models developed for each region. Examples of circuits that can be used to bias the bipolar transistor are presented. The chapter closes with a discussion of the worst-case and Monte Carlo analyses of the effects of tolerances on bias circuits.

5.1 PHYSICAL STRUCTURE OF THE BIPOLAR TRANSISTOR

The bipolar transistor structure consists of three alternating layers of *n*- and *p*-type semiconductor material. These layers are referred to as the **emitter (E), base (B),** and **collector (C)** of the transistor. Either an ***npn*** or a ***pnp* transistor** can be fabricated. The behavior of the device can be seen from the simplified cross section of the *npn* transistor in Fig. 5.1(a). During normal operation, a majority of the current enters the collector terminal, crosses the base region, and exits from the emitter terminal. A small current also enters the base terminal, crosses the base-emitter junction of the transistor, and exits the emitter.

The most important part of the bipolar transistor is the active base region between the dashed lines directly beneath the heavily doped ($n+$) emitter. Carrier transport in this region dominates the i-v characteristics of the BJT. Figure 5.1(b) illustrates the rather complex physical structure actually used to realize an *npn* transistor in integrated circuit form. Most of the structure in Fig. 5.1(b) is required to fabricate the external contacts to the collector, base, and emitter regions and to isolate one bipolar transistor from another. In the *npn* structure shown, collector current i_C and base current i_B enter the

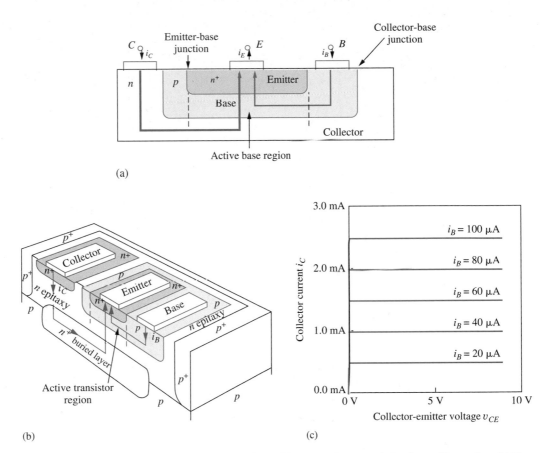

Figure 5.1 (a) Simplified cross section of an *npn* transistor with currents that occur during "normal" operation. (b) Three-dimensional view of an integrated *npn* bipolar junction transistor. (c) Output characteristics of an *npn* transistor.

collector (C) and base (B) terminals of the transistor, and emitter current i_E exits from the emitter (E) terminal. An example of the output characteristics of the bipolar transistor appears in Fig. 5.1(c), which plots collector current i_C versus collector-emitter voltage v_{CE} with base current as a parameter. The characteristics exhibit an appearance very similar to the output characteristics of the field-effect transistor. We find that a primary difference, however, is that a significant current must be supplied to the base of the device, whereas the dc-gate current of the FET is zero. In the sections that follow, a mathematical model is developed for these i-v characteristics for both *npn* and *pnp* transistors.

5.2 THE TRANSPORT MODEL FOR THE *npn* TRANSISTOR

Figure 5.2 is a conceptual model for the active region of the *npn* bipolar junction transistor structure. At first glance, the BJT appears to simply be two *pn* junctions connected back to back. However, the central region (the base) is very thin (0.1 to 100 μm), and the close proximity of the two junctions leads to coupling between the two diodes. This coupling is the essence of the bipolar device. The lower *n*-type region (the emitter) injects electrons into the *p*-type base region of the device. Almost all these injected electrons travel across the narrow base region and are removed (or collected) by the upper *n*-type region (the collector).

The three terminal currents are the **collector current i_C**, the **emitter current i_E,** and the **base current i_B.** The base-emitter voltage v_{BE} and the base-collector voltage v_{BC} applied to the two *pn* junctions in Fig. 5.2 determine the magnitude of these three currents in the bipolar transistor and

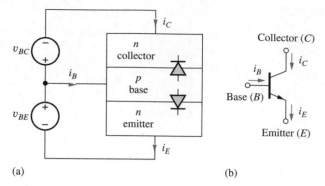

(a)   (b)

Figure 5.2 (a) Idealized *npn* transistor structure for a general-bias condition. (b) Circuit symbol for the *npn* transistor.

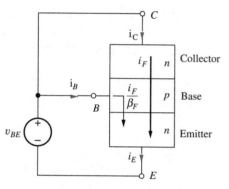

Figure 5.3 *npn* transistor with v_{BE} applied and $v_{BC} = 0$.

TABLE 5.1

Common-Emitter and Common-Base Current Gain Comparison

α_F or α_R	$\beta_F = \dfrac{\alpha_F}{1-\alpha_F}$ or $\beta_R = \dfrac{\alpha_R}{1-\alpha_R}$
0.1	0.11
0.5	1
0.9	9
0.95	19
0.99	99
0.998	499

are defined as positive when they forward-bias their respective *pn* junctions. The arrows indicate the directions of positive current in most *npn* circuit applications. The circuit symbol for the *npn* transistor appears in Fig. 5.2(b). The arrow identifies the emitter terminal and indicates that dc current normally exits the emitter of the *npn* transistor.

5.2.1 FORWARD CHARACTERISTICS

To facilitate both hand and computer analysis, we need to construct a mathematical model that closely matches the behavior of the transistor, and equations that describe the static *i-v* characteristics of the device can be constructed using a superposition of currents within the transistor structure.[1] In Fig. 5.3, an arbitrary voltage v_{BE} is applied to the base-emitter junction, and the voltage applied to the base-collector junction is set to zero. The base-emitter voltage establishes emitter current i_E, which equals the total current crossing the base-emitter junction. This current is composed of two components. The largest portion, the **forward-transport current i_F**, enters the collector, travels completely across the very narrow base region, and exits the emitter terminal. The collector current i_C is equal to i_F, which has the form of an ideal diode current

$$i_C = i_F = I_S \left[\exp\left(\frac{v_{BE}}{V_T} \right) - 1 \right] \tag{5.1}$$

---

[1] The differential equations that describe the internal physics of the BJT are linear second-order differential equations. These equations are linear in terms of the hole and electron concentrations; the currents are directly related to these carrier concentrations. Thus, superposition can be used with respect to the currents flowing in the device.

The parameter I_S is the **transistor saturation current**—that is, the saturation current of the bipolar transistor. I_S is proportional to the cross-sectional area of the active base region of the transistor, and can have a wide range of values:

$$10^{-18}\ \text{A} \le I_S \le 10^{-9}\ \text{A}$$

In Eq. (5.1), V_T should be recognized as the thermal voltage introduced in Chapter 2 and given by $V_T = kT/q = 0.025$ V at room temperature.

In addition to i_F, a second, much smaller component of current crosses the base-emitter junction. This current forms the base current i_B of the transistor, and it is directly proportional to i_F:

$$i_B = \frac{i_F}{\beta_F} = \frac{I_S}{\beta_F}\left[\exp\left(\frac{v_{BE}}{V_T}\right) - 1\right] \tag{5.2}$$

Parameter $\boldsymbol{\beta_F}$ is called the **forward** (or **normal**[2]) **common-emitter current gain.** Its value typically falls in the range

$$10 \le \beta_F \le 500$$

The emitter current i_E can be calculated by treating the transistor as a super node for which

$$i_C + i_B = i_E \tag{5.3}$$

Adding Eqs. (5.1) and (5.2) together yields

$$i_E = \left(I_S + \frac{I_S}{\beta_F}\right)\left[\exp\left(\frac{v_{BE}}{V_T}\right) - 1\right] \tag{5.4}$$

which can be rewritten as

$$i_E = I_S\left(\frac{\beta_F + 1}{\beta_F}\right)\left[\exp\left(\frac{v_{BE}}{V_T}\right) - 1\right] = \frac{I_S}{\alpha_F}\left[\exp\left(\frac{v_{BE}}{V_T}\right) - 1\right] \tag{5.5}$$

The parameter $\boldsymbol{\alpha_F}$ is called the **forward** (or **normal**[3]) **common-base current gain,** and its value typically falls in the range

$$0.95 \le \alpha_F < 1.0$$

The parameters α_F and β_F are related by

$$\alpha_F = \frac{\beta_F}{\beta_F + 1} \quad \text{or} \quad \beta_F = \frac{\alpha_F}{1 - \alpha_F} \tag{5.6}$$

Equations (5.1), (5.2), and (5.5) express the fundamental physics-based characteristics of the bipolar transistor. The three terminal currents are all exponentially dependent on the base-emitter voltage of the transistor. This is a much stronger nonlinear dependence than the square-law behavior of the FET.

For the bias conditions in Fig. 5.3, the transistor is actually operating in a region of high current gain, called the forward-active region[4] of operation, which is discussed more fully in Sec. 5.9. Two extremely useful auxiliary relationships are valid in the forward-active region. The first can be found

---

[2] β_N is sometimes used to represent the normal common-emitter current gain.

[3] α_N is sometimes used to represent the normal common-base current gain.

[4] Four regions of operation are fully defined in Sec. 5.6.

from the ratio of the collector and base current in Eqs. (5.1) and (5.2):

$$\frac{i_C}{i_B} = \beta_F \quad \text{or} \quad i_C = \beta_F i_B \quad \text{and} \quad i_E = (\beta_F + 1)i_B \qquad (5.7)$$

using Eq. (5.3). The second relationship is found from the ratio of the collector and emitter currents in Eqs. (5.1) and (5.5):

$$\frac{i_C}{i_E} = \alpha_F \quad \text{or} \quad i_C = \alpha_F i_E \qquad (5.8)$$

Equation (5.7) expresses an important and useful property of the bipolar transistor: The transistor "amplifies" (magnifies) its base current by the factor β_F. Because the current gain $\beta_F \gg 1$, injection of a small current into the base of the transistor produces a much larger current in both the collector and the emitter terminals. Equation (5.8) indicates that the collector and emitter currents are almost identical, because $\alpha_F \cong 1$.

5.2.2 REVERSE CHARACTERISTICS

Now consider the transistor in Fig. 5.4, in which voltage v_{BC} is applied to the base-collector junction, and the base-emitter junction is zero-biased. The base-collector voltage establishes the collector current i_C, now crossing the base-collector junction. The largest portion of the collector current, the reverse-transport current i_R, enters the emitter, travels completely across the narrow base region, and exits the collector terminal. Current i_R has a form identical to i_F:

$$i_R = I_S \left[\exp\left(\frac{v_{BC}}{V_T}\right) - 1 \right] \quad \text{and} \quad i_E = -i_R \qquad (5.9)$$

except the controlling voltage is now v_{BC}.

In this case, a fraction of the current i_R must also be supplied as base current through the base terminal:

$$i_B = \frac{i_R}{\beta_R} = \frac{I_S}{\beta_R} \left[\exp\left(\frac{v_{BC}}{V_T}\right) - 1 \right] \qquad (5.10)$$

Parameter $\boldsymbol{\beta_R}$ is called the **reverse** (or **inverse**[5]) **common-emitter current gain.**

In Chapter 4, we discovered that the FET was an inherently symmetric device. For the bipolar transistor, Eqs. (5.1) and (5.9) show the symmetry that is inherent in the current that traverses the base region of the bipolar transistor. However, the impurity doping levels of the emitter and collector regions of the BJT structure are quite asymmetric, and this fact causes the base currents in the forward

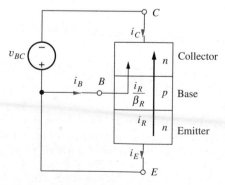

Figure 5.4 Transistor with v_{BC} applied and $v_{BE} = 0$.

---

[5] β_I is sometimes used to represent the inverse common-emitter current gain.

and reverse modes to be significantly different. For typical BJTs, β_R is in the range

$$0 < \beta_R \le 10 \quad \text{whereas} \quad 10 \le \beta_F \le 500$$

The collector current in Fig. 5.4 can be found by combining the base and emitter currents, as was done to obtain Eq. (5.5):

$$i_C = -\frac{I_S}{\alpha_R}\left[\exp\left(\frac{v_{BC}}{V_T}\right) - 1\right] \tag{5.11}$$

in which the parameter α_R is called the **reverse** (or **inverse**[6]) **common-base current gain:**

$$\alpha_R = \frac{\beta_R}{\beta_R + 1} \quad \text{or} \quad \beta_R = \frac{\alpha_R}{1 - \alpha_R} \tag{5.12}$$

Typical values of α_R fall in the range

$$0 < \alpha_R \le 0.95$$

Values of the common-base current gain α and the common-emitter current gain β are compared in Table 5.1 on page 210. Because α_F is typically greater than 0.95, β_F can be quite large. Values ranging from 10 to 500 are quite common for β_F, although it is possible to fabricate special-purpose transistors[7] with β_F as high as 5000. In contrast, α_R is typically less than 0.5, which results in values of β_R of less than 1.

EXERCISE: (a) What values of β correspond to $\alpha = 0.970, 0.993, 0.250$? (b) What values of α correspond to $\beta = 40, 200, 3$?

ANSWERS: (a) 32.3; 142; 0.333   (b) 0.976; 0.995; 0.750

5.2.3 THE COMPLETE TRANSPORT MODEL EQUATIONS FOR ARBITRARY BIAS CONDITIONS

Combining the expressions for the two collector, emitter, and base currents from Eqs. (5.1) and (5.11), (5.4) and (5.9), and (5.2) and (5.10) yields expressions for the total collector, emitter, and base currents for the *npn* transistor that are valid for the completely general-bias voltage situation in Fig. 5.2:

$$i_C = I_S\left[\exp\left(\frac{v_{BE}}{V_T}\right) - \exp\left(\frac{v_{BC}}{V_T}\right)\right] - \frac{I_S}{\beta_R}\left[\exp\left(\frac{v_{BC}}{V_T}\right) - 1\right]$$

$$i_E = I_S\left[\exp\left(\frac{v_{BE}}{V_T}\right) - \exp\left(\frac{v_{BC}}{V_T}\right)\right] + \frac{I_S}{\beta_F}\left[\exp\left(\frac{v_{BE}}{V_T}\right) - 1\right] \tag{5.13}$$

$$i_B = \frac{I_S}{\beta_F}\left[\exp\left(\frac{v_{BE}}{V_T}\right) - 1\right] + \frac{I_S}{\beta_R}\left[\exp\left(\frac{v_{BC}}{V_T}\right) - 1\right]$$

From this equation set, we see that three parameters are required to characterize an individual BJT: I_S, β_F, and β_R. (Remember that temperature is also an important parameter because $V_T = kT/q$.)

The first term in both the emitter and collector current expressions in Eqs. (5.13) is

$$i_T = I_S\left[\exp\left(\frac{v_{BE}}{V_T}\right) - \exp\left(\frac{v_{BC}}{V_T}\right)\right] \tag{5.14}$$

which represents the current being transported completely across the base region of the transistor. Equation (5.14) demonstrates the symmetry that exists between the base-emitter and base-collector voltages in establishing the dominant current in the bipolar transistor.

---

[6] α_I is sometimes used to represent the inverse common-base current gain.

[7] These devices are often called "super-beta" transistors.

Equations (5.13) actually represent a simplified version of the more complex **Gummel-Poon model** [3, 4] and form the heart of the BJT model used in the SPICE simulation program. The full Gummel-Poon model accurately describes the characteristics of BJTs over a wide range of operating conditions, and it has largely supplanted its predecessor, the **Ebers-Moll model** [5] (see Prob. 5.23).

EXAMPLE 5.1   TRANSPORT MODEL CALCULATIONS

The advantage of the full transport model is that it can be used to estimate the currents in the bipolar transistor for any given set of bias voltages.

PROBLEM   Use the transport model equations to find the terminal voltages and currents in the circuit in Fig. 5.5 in which an *npn* transistor is biased by two dc voltage sources.

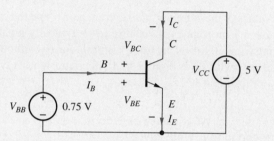

Figure 5.5 *npn* transistor circuit example: $I_S = 10^{-16}$ A, $\beta_F = 50$, $\beta_R = 1$.

SOLUTION   **Known Information and Given Data:** The *npn* transistor in Fig. 5.5 is biased by two dc sources $V_{BB} = 0.75$ V and $V_{CC} = 5.0$ V. The transistor parameters are $I_S = 10^{-16}$ A, $\beta_F = 50$, and $\beta_R = 1$.

Unknowns: Junction bias voltages V_{BE} and V_{BC}; emitter current I_E, collector current I_C, base current I_B

Approach: Determine V_{BE} and V_{BC} from the circuit. Use these voltages and the transistor parameters to calculate the currents using Eq. (5.13).

Assumptions: The transistor is modeled by the transport equations and is operating at room temperature with $V_T = 25.0$ mV.

Analysis: In this circuit, the base emitter voltage V_{BE} is set directly by source V_{BB}, and the base collector voltage is the difference between V_{BB} and V_{CC}:

$$V_{BE} = V_{BB} = 0.75 \text{ V}$$

$$V_{BC} = V_{BB} - V_{CC} = 0.75 \text{ V} - 5.00 \text{ V} = -4.25 \text{ V}$$

Substituting these voltages into Eqs. (5.13) along with the transistor parameters yields

$$i_C = 10^{-16} \text{ A} \left[\exp\left(\frac{0.75 \text{ V}}{0.025 \text{ V}}\right) - \overset{0}{\exp\left(\frac{-4.75 \text{ V}}{0.025 \text{ V}}\right)} \right] - \frac{10^{-16}}{1} \text{ A} \left[\overset{0}{\exp\left(\frac{-4.75 \text{ V}}{0.025 \text{ V}}\right)} - 1 \right]$$

$$i_E = 10^{-16} \text{ A} \left[\exp\left(\frac{0.75 \text{ V}}{0.025 \text{ V}}\right) - \overset{0}{\exp\left(\frac{-4.75 \text{ V}}{0.025 \text{ V}}\right)} \right] + \frac{10^{-16}}{50} \text{ A} \left[\exp\left(\frac{0.75 \text{ V}}{0.025 \text{ V}}\right) - 1 \right]$$

$$i_B = \frac{10^{-16}}{50} \text{ A} \left[\exp\left(\frac{0.75 \text{ V}}{0.025 \text{ V}}\right) - 1 \right] + \frac{10^{-16}}{1} \text{ A} \left[\overset{0}{\exp\left(\frac{-4.75 \text{ V}}{0.025 \text{ V}}\right)} - 1 \right]$$

and evaluating these expressions gives

$$I_C = 1.07 \text{ mA} \qquad I_E = 1.09 \text{ mA} \qquad I_B = 21.4 \text{ μA}$$

Check of Results: The sum of the collector and base currents equals the emitter current as required by KCL for the transistor treated as a super node. Also, the terminal currents range from microamperes to milliamperes, which are reasonable for most transistors.

Discussion: Note that the collector-base junction in Fig. 5.5 is reverse-biased, so the terms containing V_{BC} become negligibly small. In this example, the transistor is biased in the forward-active region of operation for which

$$\beta_F = \frac{I_C}{I_B} = \frac{1.07 \text{ mA}}{0.0214 \text{ mA}} = 50 \qquad \text{and} \qquad \alpha_F = \frac{I_C}{I_E} = \frac{1.07 \text{ mA}}{1.09 \text{ mA}} = 0.982$$

EXERCISE: Repeat the example problem for $I_S = 10^{-15}$ A, $\beta_F = 100$, $\beta_R = 0.50$, $V_{BE} = 0.70$ V, and $V_{CC} = 10$ V.

ANSWERS: $I_C = 1.45$ mA, $I_E = 1.46$ mA, and $I_B = 14.5$ μA

In Secs. 5.5 to 5.11 we completely define four different regions of operation of the transistor and find simplified models for each region. First, however, let us develop the transport model for the *pnp* transistor in a manner similar to that for the *npn* transistor.

5.3 THE *pnp* TRANSISTOR

In Chapter 4, we found we could make either NMOS or PMOS transistors by simply interchanging the *n*- and *p*-type regions in the device structure. One might expect the same to be true of bipolar transistors, and we can indeed fabricate *pnp* transistors as well as *npn* transistors.

The *pnp* transistor is fabricated by reversing the layers of the transistor, as diagrammed in Fig. 5.6. The transistor has been drawn with the emitter at the top of the diagram, as it appears in most circuit diagrams throughout this book. The arrows again indicate the normal directions of positive current in the *pnp* transistor in most circuit applications. The voltages applied to the two *pn* junctions are the emitter-base voltage v_{EB} and the collector-base voltage v_{CB}. These voltages are again positive when they forward-bias their respective *pn* junctions. Collector current i_C and base current i_B exit the transistor terminals, and the emitter current i_E enters the device. The circuit symbol for the *pnp*

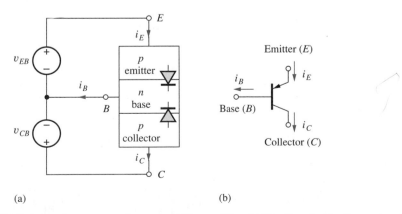

(a)   (b)

Figure 5.6 (a) Idealized *pnp* transistor structure for a general-bias condition. (b) Circuit symbol for the *pnp* transistor.

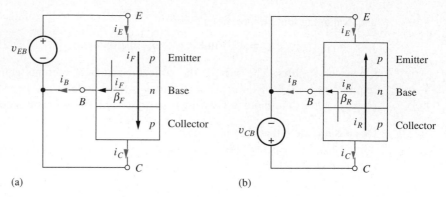

Figure 5.7 (a) *pnp* transistor with v_{EB} applied and $v_{CB} = 0$. (b) *pnp* transistor with v_{CB} applied and $v_{EB} = 0$.

transistor appears in Fig. 5.6(b). The arrow identifies the emitter of the *pnp* transistor and points in the direction of normal positive-emitter current.

Equations that describe the static *i-v* characteristics of the *pnp* transistor can be constructed using superposition of currents within the structure just as for the *npn* transistor. In Fig. 5.7(a), voltage v_{EB} is applied to the emitter-base junction, and the collector-base voltage is set to zero. The emitter-base voltage establishes forward-transport current i_F that traverses the narrow base region and base current i_B that crosses the emitter-base junction of the transistor:

$$i_C = i_F = I_S \left[\exp\left(\frac{v_{EB}}{V_T}\right) - 1 \right] \qquad i_B = \frac{i_F}{\beta_F} = \frac{I_S}{\beta_F} \left[\exp\left(\frac{v_{EB}}{V_T}\right) - 1 \right]$$

and

$$i_E = i_C + i_B = I_S \left(1 + \frac{1}{\beta_F} \right) \left[\exp\left(\frac{v_{EB}}{V_T}\right) - 1 \right]$$

(5.15)

In Fig. 5.7(b), a voltage v_{CB} is applied to the collector-base junction, and the emitter-base junction is zero-biased. The collector-base voltage establishes the reverse-transport current i_R and base current i_B:

$$-i_E = i_R = I_S \left[\exp\left(\frac{v_{CB}}{V_T}\right) - 1 \right]$$

$$i_B = \frac{i_R}{\beta_R} = \frac{I_S}{\beta_R} \left[\exp\left(\frac{v_{CB}}{V_T}\right) - 1 \right]$$

(5.16)

$$i_C = -I_S \left(1 + \frac{1}{\beta_R} \right) \left[\exp\left(\frac{v_{CB}}{V_T}\right) - 1 \right]$$

where the collector current is given by $i_C = i_E - i_B$.

For the general-bias voltage situation in Fig. 5.6, Eqs. (5.15) and (5.16) are combined to give the total collector, emitter, and base currents of the *pnp* transistor:

$$i_C = I_S \left[\exp\left(\frac{v_{EB}}{V_T}\right) - \exp\left(\frac{v_{CB}}{V_T}\right) \right] - \frac{I_S}{\beta_R} \left[\exp\left(\frac{v_{CB}}{V_T}\right) - 1 \right]$$

$$i_E = I_S \left[\exp\left(\frac{v_{EB}}{V_T}\right) - \exp\left(\frac{v_{CB}}{V_T}\right) \right] + \frac{I_S}{\beta_F} \left[\exp\left(\frac{v_{EB}}{V_T}\right) - 1 \right]$$

(5.17)

$$i_B = \frac{I_S}{\beta_F} \left[\exp\left(\frac{v_{EB}}{V_T}\right) - 1 \right] + \frac{I_S}{\beta_R} \left[\exp\left(\frac{v_{CB}}{V_T}\right) - 1 \right]$$

These equations represent the simplified Gummel-Poon or transport model equations for the *pnp* transistor and can be used to relate the terminal voltages and currents of the *pnp* transistor for any general-bias condition. Note that these equations are identical to those for the *npn* transistor except that v_{EB} and v_{CB} replace v_{BE} and v_{BC}, respectively, and are a result of our careful choice for the direction of positive currents in Figs. 5.2 and 5.6.

EXERCISE: Find I_C, I_E, and I_B for a *pnp* transistor if $I_S = 10^{-16}$ A, $\beta_F = 75$, $\beta_R = 0.40$, $V_{EB} = 0.75$ V, and $V_{CB} = +0.70$ V.

ANSWERS: $I_C = 0.563$ mA, $I_E = 0.938$ mA, $I_B = 0.376$ mA

5.4 EQUIVALENT CIRCUIT REPRESENTATIONS FOR THE TRANSPORT MODELS

For circuit simulation, as well as hand analysis purposes, the transport model equations for the *npn* and *pnp* transistors can be represented by the equivalent circuits shown in Fig. 5.8(a) and (b), respectively. In the *npn* model in Fig. 5.8(a), the total transport current i_T traversing the base is determined by I_S, v_{BE}, and v_{BC}, and is modeled by the current source i_T:

$$i_T = i_F - i_R = I_S \left[\exp\left(\frac{v_{BE}}{V_T}\right) - \exp\left(\frac{v_{BC}}{V_T}\right) \right] \tag{5.18}$$

The diode currents correspond directly to the two components of the base current:

$$i_B = \frac{I_S}{\beta_F} \left[\exp\left(\frac{v_{BE}}{V_T}\right) - 1 \right] + \frac{I_S}{\beta_R} \left[\exp\left(\frac{v_{BC}}{V_T}\right) - 1 \right] \tag{5.19}$$

Directly analogous arguments hold for the circuit elements in the *pnp* circuit model of Fig. 5.8(b).

EXERCISE: Find i_T if $I_S = 10^{-15}$ A, $V_{BE} = 0.75$ V, and $V_{BC} = -2.0$ V.

ANSWER: 10.7 mA

EXERCISE: Find the dc transport current I_T for the transistor in Example 5.1 on page 214.

ANSWER: $I_T = 1.07$ mA

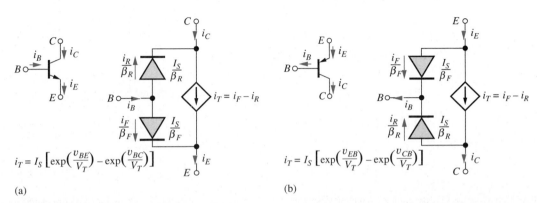

(a)   (b)

Figure 5.8 (a) Transport model equivalent circuit for the *npn* transistor. (b) Transport model equivalent circuit for the *pnp* transistor.

5.5 THE *i-v* CHARACTERISTICS OF THE BIPOLAR TRANSISTOR

Two complementary views of the *i-v* behavior of the BJT are represented by the device's **output and transfer characteristics.** (Remember that similar characteristics were presented for the FETs in Chapter 4.) The output characteristics represent the relationship between the collector current and collector-emitter or collector-base voltage of the transistor, whereas the transfer characteristic relates the collector current to the base-emitter voltage. A knowledge of both *i-v* characteristics is basic to understanding the overall behavior of the bipolar transistor.

5.5.1 OUTPUT CHARACTERISTICS

Circuits for measuring or simulating the **common-emitter output characteristics** are shown in Fig. 5.9. In these circuits, the base of the transistor is driven by a constant current source, and the output characteristics represent a graph of i_C vs. v_{CE} for the *npn* transistor (or i_C vs. v_{EC} for the *pnp*) with base current i_B as a parameter. Note that the Q-point (I_C, V_{CE}) or (I_C, V_{EC}) locates the BJT operating point on the output characteristics.

First, consider the *npn* transistor operating with $v_{CE} \geq 0$, represented by the first quadrant of the graph in Fig. 5.10. For $i_B = 0$, the transistor is nonconducting or cut off. As i_B increases above 0, i_C also increases. For $v_{CE} \geq v_{BE}$, the *npn* transistor is in the forward-active region, and collector current is independent of v_{CE} and equal to $\beta_F i_B$. Remember, it was demonstrated earlier that $i_C \cong \beta_F i_B$ in the forward-active region. For $v_{CE} \leq v_{BE}$, the transistor enters the saturation region of operation in which the total voltage between the collector and emitter terminals of the transistor is small.

It is important to note that the saturation region of the BJT does not correspond to the saturation region (active region) of the FET. The forward-active region (or just active region) of the BJT corresponds to the saturation region of the FET. When we begin our discussion of amplifiers in Part III, we will simply apply the term active region to both devices. The active region is the region most often used in transistor implementations of amplifiers.

In the third quadrant for $v_{CE} \leq 0$, the roles of the collector and emitter reverse. For $v_{BE} \leq v_{CE} \leq 0$, the transistor remains in saturation. For $v_{CE} \leq v_{BE}$, the transistor enters the reverse-active region, in which the *i-v* characteristics again become independent of v_{CE}, and now $i_C \cong -(\beta_R + 1)i_B$. The reverse-active region curves have been plotted for a relatively large value of reverse

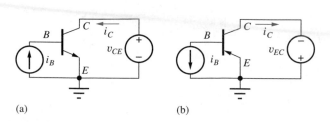

(a)　　　　　　　　(b)

Figure 5.9 Circuits for determining common-emitter output characteristics: (a) *npn* transistor, (b) *pnp* transistor.

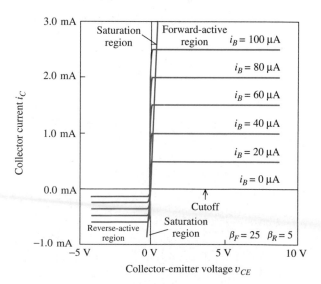

Figure 5.10 Common-emitter output characteristics for the bipolar transistor (i_C vs. v_{CE} for the *npn* transistor or i_C vs. v_{EC} for the *pnp* transistor).

common-emitter current gain, $\beta_R = 5$, to enhance their visibility. As noted earlier, the reverse-current gain β_R is often less than 1.

Using the polarities defined in Fig. 5.9(b) for the *pnp* transistor, the output characteristics will appear exactly the same as in Fig. 5.10, except that the horizontal axis will be the voltage v_{EC} rather than v_{CE}. Remember that $i_B > 0$ and $i_C > 0$ correspond to currents exiting the base and collector terminals of the *pnp* transistor.

Circuits for measuring or simulating the **common-base output characteristics** of the *npn* and *pnp* transistors are shown in Fig. 5.11. In these circuits, the emitter of the transistor is driven by a constant current source, and the output characteristics plot i_C vs. v_{CB} for the *npn* (or i_C vs. v_{BC} for the *pnp*), with the emitter-current i_E as a parameter. For $v_{CB} \geq 0$ V in Fig. 5.12, the transistor operates in the forward-active region with i_C independent of v_{CB}, and we saw earlier that $i_C \cong i_E$. For v_{CB} less than zero, the base-collector diode of the transistor becomes forward-biased, and the collector current grows exponentially (in the negative direction) as the base-collector diode begins to conduct.

Using the polarities defined in Fig. 5.11(b) for the *pnp* transistor, the output characteristics appear exactly the same as in Fig. 5.12, except that the horizontal axis is the voltage v_{BC} rather than v_{CB}. Again, remember that $i_B > 0$ and $i_C > 0$ correspond to currents exiting the emitter and collector terminals of the *pnp* transistor.

5.5.2 TRANSFER CHARACTERISTICS

The **common-emitter transfer characteristic** of the BJT defines the relationship between the collector current and the base-emitter voltage of the transistor. An example of the transfer characteristic for an *npn* transistor is shown in graphical form in Fig. 5.13, with both linear and semilog scales for

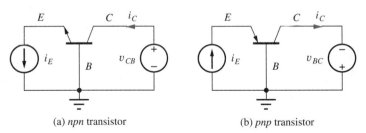

(a) *npn* transistor   (b) *pnp* transistor

Figure 5.11 Circuits to determine common-base output characteristics.

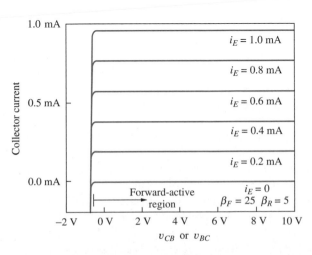

Figure 5.12 Common-base output characteristics for the bipolar transistor. (i_C vs. v_{CB} for the *npn* transistor or i_C vs. v_{BC} for the *pnp* transistor).

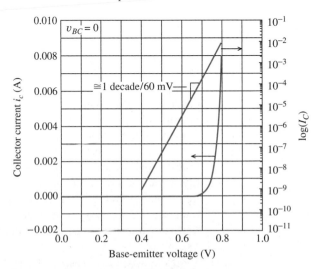

Figure 5.13 BJT transfer characteristic in the forward-active region.

TABLE 5.2
Regions of Operation of the Bipolar Transistor

BASE-EMITTER JUNCTION	BASE-COLLECTOR JUNCTION	
	Reverse Bias	Forward Bias
Forward Bias	Forward-active region (Normal-active region) (Good amplifier)	Saturation region* (Closed switch)
Reverse Bias	Cutoff region (Open switch)	Reverse-active region (Inverse-active region) (Poor amplifier)

* It is important to note that the saturation region of the bipolar transistor does *not* correspond to the saturation region of the FET. This unfortunate use of terms is historical in nature and something we just have to accept.

the particular case of $v_{BC} = 0$. The transfer characteristic is virtually identical to that of a *pn* junction diode. This behavior can also be expressed mathematically by setting $v_{BC} = 0$ in the collector-current expression in Eq. (5.13):

$$i_C = I_S \left[\exp \left(\frac{v_{BE}}{V_T} \right) - 1 \right] \tag{5.20}$$

Because of the exponential relationship in Eq. (5.20), the semilog plot exhibits the same slope as that for a *pn* junction diode. Only a 60-mV change in v_{BE} is required to change the collector current by a factor of 10, and for a fixed collector current, the base-emitter voltage of the silicon BJT will exhibit a -1.8-mV/°C temperature coefficient, just as for the silicon diode (see Sec. 3.5).

EXERCISE: What base-emitter voltage V_{BE} corresponds to $I_C = 100$ μA in an *npn* transistor at room temperature if $I_S = 10^{-16}$ A? For $I_C = 1$ mA?

ANSWERS: 0.691 V; 0.748 V

5.6 THE OPERATING REGIONS OF THE BIPOLAR TRANSISTOR

In the bipolar transistor, each *pn* junction may independently be forward-biased or reverse-biased, so there are four possible regions of operation, as defined in Table 5.2. The operating point establishes the region of operation of the transistor and can be defined by any two of the four terminal voltages or currents. The characteristics of the transistor are quite different for each of the four regions of operation, and in order to simplify our circuit analysis task, we need to be able to make an educated guess as to the region of operation of the BJT.

When both junctions are reverse-biased, the transistor is essentially nonconducting or *cut off* (**cutoff region**) and can be considered an open switch. If both junctions are forward-biased, the transistor is operating in the **saturation region**[8] and appears as a closed switch. Cutoff and saturation (colored in Table 5.2) are most often used to represent the two states in binary logic circuits implemented with BJTs. For example, switching between these two operating regions occurs in the transistor-transistor logic circuits that we shall study in Chapter 9 on bipolar logic circuits.

---

[8] It is important to note that the saturation region of the bipolar transistor does *not* correspond to the saturation region of the FET. This unfortunate use of terms is historical in nature and something we just have to accept.

In the **forward-active region** (also called the **normal-active region** or **just active region**), in which the base-emitter junction is forward-biased and the base-collector junction is reverse-biased, the BJT can provide high current, voltage, and power gains. The forward-active region is most often used to achieve high-quality amplification. In addition, in the fastest form of bipolar logic, called emitter-coupled logic, the transistors switch between the cutoff and the forward-active regions.

In the **reverse-active region** (or **inverse-active region**), the base-emitter junction is reverse-biased and the base-collector junction is forward-biased. In this region, the transistor exhibits low current gain, and the reverse-active region is not often used. However, we will see an important application of the reverse-active region in transistor-transistor logic circuits in Chapter 9. Reverse operation of the bipolar transistor has also found use in analog-switching applications.

The transport model equations describe the behavior of the bipolar transistor for any combination of terminal voltages and currents. However, the complete sets of equations in (5.13) and (5.17) are quite imposing. In subsequent sections, bias conditions specific to each of the four regions of operation will be used to obtain simplified sets of relationships that are valid for the individual regions. The Q-point for the BJT is (I_C, V_{CE}) for the *npn* transistor and (I_C, V_{EC}) for the *pnp*.

> **EXERCISE:** What is the region of operation of (a) an *npn* transistor with $V_{BE} = 0.75$ V and $V_{BC} = -0.70$ V? (b) A *pnp* transistor with $V_{CB} = 0.70$ V and $V_{EB} = 0.75$ V?
>
> **ANSWERS:** Forward-active region; saturation region

5.7 TRANSPORT MODEL SIMPLIFICATIONS

The complete sets of Transport Model Equations developed in Sections 5.2 and 5.3 describe the behavior of the *npn* and *pnp* transistors for any combination of terminal voltages and currents, and these equations are indeed the basis for the models used in SPICE circuit simulation. However, the full set of equations is quite imposing. Now we will explore simplifications that can be used to reduce the complexity of the model descriptions for each of the four different regions of operation identified in Table 5.2.

5.7.1 SIMPLIFIED MODEL FOR THE CUTOFF REGION

The easiest region to understand is the cutoff region, in which both junctions are reverse-biased. For an *npn* transistor, the cutoff region requires $v_{BE} \leq 0$ and $v_{BC} \leq 0$. Let us further assume that

$$v_{BE} < -\frac{4kT}{q} \qquad \text{and} \qquad v_{BC} < -4\frac{kT}{q} \qquad \text{where } -4\frac{kT}{q} = -0.1\text{V}$$

These two conditions allow us to neglect the exponential terms in Eqs. (5.13), yielding the following simplified equations for the *npn* terminal currents in cutoff:

$$i_C = I_S \left[\exp\left(\frac{v_{BE}}{V_T}\right)^{0} - \exp\left(\frac{v_{BC}}{V_T}\right)^{0} \right] - \frac{I_S}{\beta_R} \left[\exp\left(\frac{v_{BC}}{V_T}\right)^{0} - 1 \right]$$

$$i_E = I_S \left[\exp\left(\frac{v_{BE}}{V_T}\right)^{0} - \exp\left(\frac{v_{BC}}{V_T}\right)^{0} \right] + \frac{I_S}{\beta_F} \left[\exp\left(\frac{v_{BE}}{V_T}\right)^{0} - 1 \right] \qquad (5.21)$$

$$i_B = \frac{I_S}{\beta_F} \left[\exp\left(\frac{v_{BE}}{V_T}\right)^{0} - 1 \right] + \frac{I_S}{\beta_R} \left[\exp\left(\frac{v_{BC}}{V_T}\right)^{0} - 1 \right]$$

or

$$i_C = +\frac{I_S}{\beta_R} \qquad i_E = -\frac{I_S}{\beta_F} \qquad i_B = -\frac{I_S}{\beta_F} - \frac{I_S}{\beta_R}$$

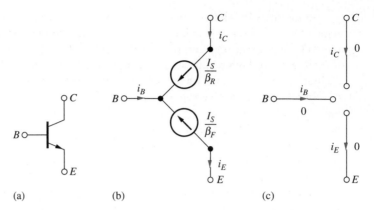

Figure 5.14 Modeling the *npn* transistor in cutoff: (a) *npn* transistor, (b) constant leakage current model, (c) open-circuit model.

In cutoff, the three terminal currents—i_C, i_E, and i_B—are all constant and smaller than the saturation current I_S of the transistor. The simplified model for this situation is shown in Fig. 5.14(b). In cutoff, only very small leakage currents appear in the three transistor terminals. In most cases, these currents are negligibly small and can be assumed to be zero.

We usually think of the transistor operating in the cutoff region as being "off" with essentially zero terminal currents, as indicated by the three-terminal open-circuit model in Fig. 5.14(c). The cutoff region represents an open switch and is used as one of the two states required for binary logic circuits.

EXAMPLE 5.2   **A BJT BIASED IN CUTOFF**

Cutoff represents the "off state" in switching applications, so an understanding of the magnitudes of the currents involved is important. In this example, we explore how closely the "off state" approaches zero.

PROBLEM Figure 5.15 is an example of a circuit in which the transistor is biased in the cutoff region. Estimate the currents using the simplified model in Fig. 5.14, and compare to calculations using the full transport model.

SOLUTION **Known Information and Given Data:** From the figure, $I_S = 10^{-16}$ A, $\alpha_F = 0.95$, $\alpha_R = 0.25$, $V_{BE} = 0$ V, $V_{BC} = -5$ V

Unknowns: I_C, I_B, I_E

Approach: First analyze the circuit using the simplified model of Fig. 5.14. Then, compare the results to calculations using the voltages to simplify the transport equations.

Assumptions: $V_{BE} = 0$ V, so the "diode" terms containing V_{BE} are equal to 0. $V_{BC} = -5$ V, which is much less than $-4kT/q = -100$ mV, so the transport model equations can be simplified.

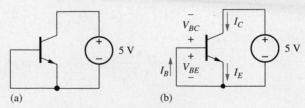

Figure 5.15 (a) *npn* transistor bias in the cutoff region. (For calculations, use $I_S = 10^{-16}$ A, $\alpha_F = 0.95$, $\alpha_R = 0.25$.) (b) Normal current directions.

Analysis: The voltages $V_{BE} = 0$ and $V_{BC} = -5$ V are consistent with the definition of the cutoff region. If we use the open-circuit model in Fig. 5.14(c), the currents I_C, I_E, and I_B are all predicted to be zero.

To obtain a more exact estimate of the currents, we use the transport model equations. For the circuit in Fig. 5.15, the base-emitter voltage is exactly zero, and $V_{BC} \ll 0$. Therefore, Eqs. (5.13) reduce to

$$I_C = I_S \left(1 + \frac{1}{\beta_R} \right) = \frac{I_S}{\alpha_R} = \frac{10^{-16} \text{ A}}{0.25} = 4 \times 10^{-16} \text{ A}$$

$$I_E = I_S = 10^{-16} \text{ A} \quad \text{and} \quad I_B = -\frac{I_S}{\beta_R} = -\frac{10^{-16} \text{ A}}{\frac{1}{3}} = -3 \times 10^{-16} \text{ A}$$

The calculated currents in the terminals are very small but nonzero. Note, in particular, that the base current is not zero and that small currents exit both the emitter and base terminals of the transistor.

Check of Results: As a check on our results, we see that Kirchhoff's current law is satisfied for the transistor treated as a super node: $i_C + i_B = i_E$.

Discussion: The voltages $V_{BE} = 0$ and $V_{BC} = -5$ V are consistent with the definition of the cutoff region. Thus, we expect the currents to be negligibly small. Here again we see an example of the use of different levels of modeling to achieve different degrees of precision in the answer [$(I_C, I_E, I_B) = (0, 0, 0)$ or $(4 \times 10^{-16} \text{ A}, 10^{-16} \text{ A}, -3 \times 10^{-16} \text{ A})$].

EXERCISE: Calculate the values of the currents in the circuit in Fig. 5.15(a) if the value of the voltage source is changed to 10 V and (b) if the base-emitter voltage is set to −3 V using a second voltage source.

ANSWERS: (a) No change; (b) 0.300 fA, 5.26 aA, −0.305 fA

5.7.2 MODEL SIMPLIFICATIONS FOR THE FORWARD-ACTIVE REGION

Arguably the most important region of operation of the BJT is the forward-active region, in which the emitter-base junction is forward-biased and the collector-base junction is reverse-biased. In this region, the transistor can exhibit high voltage and current gains and is useful for analog amplification. From Table 5.2, we see that the forward-active region of an *npn* transistor corresponds to $v_{BE} \geq 0$ and $v_{BC} \leq 0$. In most cases, the forward-active region will have

$$v_{BE} > 4\frac{kT}{q} = 0.1 \text{ V} \quad \text{and} \quad v_{BC} < -4\frac{kT}{q} = -0.1 \text{ V}$$

and we can assume that $\exp(-v_{BC}/V_T) \ll 1$ just as we did in simplifying Eq. Set (5.21). This simplification yields:

$$i_C = I_S \exp\left(\frac{v_{BE}}{V_T} \right) + \frac{I_S}{\beta_R}$$

$$i_E = \frac{I_S}{\alpha_F} \exp\left(\frac{v_{BE}}{V_T} \right) + \frac{I_S}{\beta_F} \tag{5.22}$$

$$i_B = \frac{I_S}{\beta_F} \exp\left(\frac{v_{BE}}{V_T} \right) - \frac{I_S}{\beta_F} - \frac{I_S}{\beta_R}$$

The exponential term in each of these expressions is usually huge compared to the other terms. By neglecting the small terms, we find the most useful simplifications of the BJT model for the forward-active region:

$$i_C = I_S \exp\left(\frac{v_{BE}}{V_T}\right) \qquad i_E = \frac{I_S}{\alpha_F} \exp\left(\frac{v_{BE}}{V_T}\right) \qquad i_B = \frac{I_S}{\beta_F} \exp\left(\frac{v_{BE}}{V_T}\right) \tag{5.23}$$

In these equations, the fundamental, exponential relationship between all the terminal currents and the base-emitter voltage v_{BE} is once again clear. In the forward-active region, the terminal currents all have the form of diode currents in which the controlling voltage is the base-emitter junction potential. It is also important to note that the currents are all independent of the base-collector voltage v_{BC}. The collector current i_C can be modeled as a voltage-controlled current source that is controlled by the base-emitter voltage and independent of the collector voltage.

By taking ratios of the terminal currents in Eq. (5.23), the two important auxiliary relationships for the forward-active region are found:

$$i_C = \alpha_F i_E \qquad \text{and} \qquad i_C = \beta_F i_B \tag{5.24}$$

Observing that $i_E = i_C + i_B$ and using Eq. (5.24) yields a third important result:

$$i_E = (\beta_F + 1)i_B \tag{5.25}$$

The results from Eqs. (5.24) and (5.25) are placed in a circuit context in the next two examples from Fig. 5.16.

DESIGN NOTE   FORWARD-ACTIVE REGION

Operating points in the forward-active region are normally used for linear amplifiers. Our dc model for the forward-active region is quite simple:

$$I_C = \beta_F I_B \qquad \text{and} \qquad I_E = (\beta_F + 1)I_B \quad \text{with} \quad V_{BE} \cong 0.7 \text{ V}.$$

Forward-active operation requires $V_{BE} > 0$ and $V_{CE} \geq V_{BE}$.

EXAMPLE 5.3   FORWARD-ACTIVE REGION OPERATION WITH EMITTER CURRENT BIAS

Current sources are widely utilized for biasing in circuit design, and such a source is used to set the Q-point current in the transistor in Fig. 5.16(a).

PROBLEM   Find the emitter, base and collector currents, and base-emitter voltage for the transistor biased by a current source in Fig. 5.16(a).

SOLUTION   **Known Information and Given Data:** An *npn* transistor biased by the circuit in Fig. 5.16(a) with $I_S = 10^{-16}$ A and $\alpha_F = 0.95$. From the circuit, $V_{BC} = V_B - V_C = -5$ V and $I_E = +100 \, \mu\text{A}$.

Unknowns: I_C, I_B, V_{BE}

Approach: Show that the transistor is in the forward-active region of operation and use Eqs. (5.22) to (5.25) to find the unknown currents and voltage.

Assumptions: Room temperature operation with $V_T = 25.0$ mV

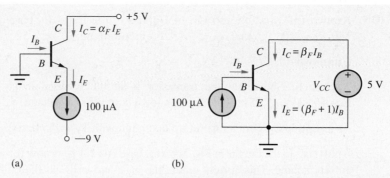

Figure 5.16 Two *npn* transistors operating in the forward-active region ($I_S = 10^{-16}$ A and $\alpha_F = 0.95$ are assumed for the example calculations).

Analysis: From the circuit, we observe that the emitter current is forced by the current source to be $I_E = +100 \, \mu$A, and the current source will forward-bias the base-emitter diode. Study of the mathematical model in Eq. (5.13) also confirms that the base-emitter voltage must be positive (forward bias) in order for the emitter current to be positive. Thus, we have $V_{BE} > 0$ and $V_{BC} < 0$, which correspond to the forward-active region of operation for the *npn* transistor.

The base and collector currents can be found using Eqs. (5.24) and (5.25) with $I_E = 100 \, \mu$A:

$$I_C = \alpha_F I_E = 0.95 \cdot 100 \, \mu\text{A} = 95 \, \mu\text{A}$$

Solving for β_F gives $\beta_F = \dfrac{\alpha_F}{1 - \alpha_F} = \dfrac{0.95}{1 - .95} = 19 \qquad \beta_F + 1 = 20$

and

$$I_B = \frac{I_E}{\beta_F + 1} = \frac{100 \, \mu\text{A}}{20} = 5 \, \mu\text{A}$$

The base-emitter voltage is found from the emitter current expression in Eq. (5.23):

$$V_{BE} = V_T \ln \frac{\alpha_F I_E}{I_S} = (0.025 \text{ V}) \ln \frac{0.95(10^{-4}\text{A})}{10^{-16}\text{A}} = 0.690 \text{ V}$$

Check of Results: As a check on our results, we see that Kirchhoff's current law is satisfied for the transistor treated as a super node: $i_C + i_B = i_E$. Also we can check V_{BE} using both the collector and base current expressions in Eq. (5.23).

Discussion: We see that most of the current being forced or "pulled" out of the emitter by the current source comes directly through the transistor from the collector. This is the common-base mode in which $i_C = \alpha_F i_E$ with $\alpha_F \cong 1$.

EXERCISE: Calculate the values of the currents and base-emitter voltage in the circuit in Fig. 5.16(a) if (a) the value of the voltage source is changed to 10 V. (b) The transistor's common-emitter current gain is increased to 50.

ANSWERS: (a) No change; (b) 100 μA, 1.96 μA, 98.0 μA, 0.690 V

EXAMPLE **5.4**   **FORWARD-ACTIVE REGION OPERATION WITH BASE CURRENT BIAS**

A current source is used to bias the transistor into the forward-active region in Fig. 5.16(b).

PROBLEM   Find the emitter, base and collector currents, and base-emitter and base-collector voltages for the transistor biased by the base current source in Fig. 5.16(b).

SOLUTION **Known Information and Given Data:** An *npn* transistor biased by the circuit in Fig. 5.16(b) with $I_S = 10^{-16}$ A and $\alpha_F = 0.95$. From the circuit, $V_C = +5$ V and $I_B = +100$ μA.

Unknowns: I_C, I_B, V_{BE}, V_{BC}

Approach: Show that the transistor is in the forward-active region of operation and use Eqs. (5.23) to (5.25) to find the unknown currents and voltage.

Assumptions: Room temperature operation with $V_T = 25.0$ mV

Analysis: In the circuit in Fig. 5.16(b), base current I_B is now forced to equal 100 μA by the ideal current source. This current enters the base and will exit the emitter, forward-biasing the base-emitter junction. From the mathematical model in Eq. (5.13), we see that positive base current can occur for positive V_{BE} and positive V_{BC}. However, we have $V_{BC} = V_B - V_C = V_{BE} - V_C$. Since the base-emitter diode voltage will be approximately 0.7 V, and $V_C = 5$ V, V_{BC} will be negative (e.g., $V_{BC} \cong 0.7 - 5.0 = -4.3$ V). Thus we have $V_{BE} > 0$ and $V_{BC} < 0$, which corresponds to the forward-active region of operation for the *npn* transistor, and the collector and emitter currents can be found using Eqs. (5.24) and (5.25) with $I_B = 100$ μA:

$$I_C = \beta_F I_B = 19 \cdot 100 \text{ μA} = 1.90 \text{ mA}$$

$$I_E = (\beta_F + 1)I_B = 20 \cdot 100 \text{ μA} = 2.00 \text{ mA}$$

The base-emitter voltage can be found from the collector current expression in Eq. (5.23):

$$V_{BE} = V_T \ln \frac{I_C}{I_S} = (0.025 \text{ V}) \ln \frac{1.9 \times 10^{-3} \text{ A}}{10^{-16} \text{ A}} = 0.764 \text{ V}$$

$$V_{BC} = V_B - V_C = V_{BE} - V_C = 0.764 - 5 = -4.24 \text{ V}$$

Check of Results: As a check on our results, we see that Kirchhoff's current law is satisfied for the transistor treated as a super node: $i_C + i_B = i_E$. Also we can check the value of V_{BE} using either the emitter or base current expressions in Eq. (5.23). The calculated values of V_{BE} and V_{BC} correspond to forward-active region operation.

Discussion: A large amplification of the current takes place when the current source is injected into the base terminal in Fig. 5.16(b) in contrast to the situation when the source is connected to the emitter terminal in Fig. 5.16(a).

EXERCISE: Calculate the values of the currents and base-emitter voltage in the circuit in Fig. 5.16(b) if (a) the value of the voltage source is changed to 10 V. (b) The transistor's common-emitter current gain is increased to 50.

ANSWERS: (a) No change; (b) 5.00 mA, 100 μA, 5.10 mA, 0.789 V, −4.21 V

EXERCISE: What is the minimum value of V_{CC} that corresponds to forward-active region bias in Fig. 5.16(b)?

ANSWERS: $V_{BE} = 0.764$ V

As illustrated in Examples 5.3 and 5.4, Eqs. (5.24) and (5.25) can often be used to greatly simplify the analysis of circuits operating in the forward-active region. However, remember this caveat well: **The results in Eqs. (5.24) and (5.25) are valid *only* for the forward-active region of operation!**

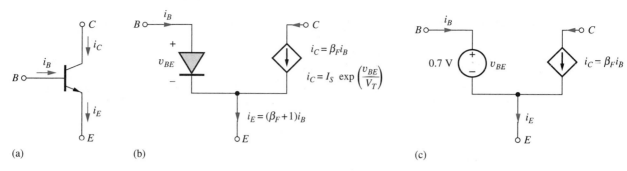

Figure 5.17 (a) *npn* transistor. (b) Simplified model for the forward-active region. (c) Further simplification for the forward-active region using the CVD model for the diode.

Based on Eq. (5.24), the BJT is often considered a current-controlled device. However, from Eqs. (5.23), we see that the fundamental physics-based behavior of the BJT in the forward-active region is that of a (nonlinear) voltage-controlled current source. The base current should be considered as an unwanted defect current that must be supplied to the base in order for the transistor to operate. In an ideal BJT, β_F would be infinite, the base current would be zero, and the collector and emitter currents would be identical, just as for the FET. (Unfortunately, it is impossible to fabricate such a BJT.)

Equations (5.23) lead to the simplified circuit model for the forward-active region shown in Fig. 5.17. The current in the base-emitter diode is amplified by the common-emitter current gain β_F and appears in the collector terminal. However, remember that the base and collector currents are exponentially related to the base-emitter voltage. Because the base-emitter diode is forward-biased in the forward-active region, the transistor model of Fig. 5.17(b) can be further simplified to that of Fig. 5.17(c), in which the diode is replaced by its constant voltage drop (CVD) model, in this case $V_{BE} = 0.7$ V. The dc base and emitter voltages differ by the 0.7-V diode voltage drop in the forward-active region.

EXAMPLE 5.5   **FORWARD-ACTIVE REGION BIAS USING TWO POWER SUPPLIES**

Analog circuits frequently operate from a pair of positive and negative power supplies so that bipolar input and output signals can easily be accommodated. The circuit in Fig. 5.18 provides one possible circuit configuration in which the resistor and −9-V source replace the current source utilized in Fig. 5.16(a). Collector resistor R_C has been added to reduce the collector-emitter voltage.

PROBLEM   Find the Q-point for the transistor in the circuit in Fig. 5.18.

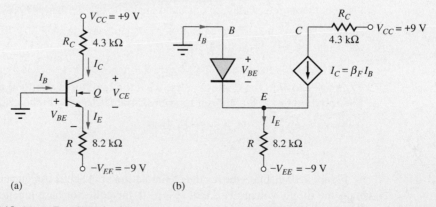

Figure 5.18 (a) *npn* Transistor circuit (assume $\beta_F = 50$ and $\beta_R = 1$). (b) Simplified model for the forward-active region.

SOLUTION   **Known Information and Given Data:** npn transistor in the circuit in Fig. 5.18(a) with $\beta_F = 50$ and $\beta_R = 1$

Unknowns: Q-point (I_C, V_{CE})

Approach: In this circuit, the base-collector junction will tend to be reverse-biased by the 9-V source. The combination of the resistor and the –9-V source will force a current out of the emitter and forward-bias the base-emitter junction. Thus, the transistor appears to be biased in the forward-active region of operation.

Assumptions: Assume forward-active region operation; since we do not know the saturation current, assume $V_{BE} = 0.7$ V; use the simplified model for the forward-active region to analyze the circuit as in Fig. 5.18(b).

Analysis: The currents can now be found by using KVL around the base-emitter loop:

$$V_{BE} + 8200 I_E - V_{EE} = 0$$

For $V_{BE} = 0.7$ V, $0.7 + 8200 I_E - 9 = 0$ or $I_E = \dfrac{8.3 \text{ V}}{8200 \ \Omega} = 1.01$ mA

At the emitter node, $I_E = (\beta_F + 1)I_B$, so

$$I_B = \frac{1.02 \text{ mA}}{50 + 1} = 19.8 \ \mu\text{A} \qquad \text{and} \qquad I_C = \beta_F I_B = 0.990 \text{ mA}$$

Because all the currents are positive, the assumption of forward-active region operation was correct. The collector-emitter voltage is equal to

$$V_{CE} = V_{CC} - I_C R_C - (-V_{BE}) = 9 - .990 \text{ mA}(4.3 \text{ k}\Omega) + 0.7 = 5.44 \text{ V}$$

The Q-point is (0.990 mA, 9.70 V).

Check of Results: We see that KVL is satisfied around the output loop containing the collector-emitter voltage: $+9 - V_{RC} - V_{CE} - V_R - (-9) = 9 - 4.3 - 5.4 - 8.3 + 9 = 0$. We must check the forward-active region assumption: $V_{CE} = 5.4$ V which is greater than $V_{BE} = 0.7$ V. Also, $I_C + I_B = I_E$.

Discussion: In this circuit, the combination of the resistor and the -9-V source replace the current source that was used to bias the transistor in Fig. 5.16(a).

Computer-Aided Analysis: SPICE contains a built-in model for the bipolar transistor that will be discussed in detail in Sec. 5.10. SPICE simulation with the default npn transistor model yields a Q-point that agrees well with our hand analysis: (0.993 mA, 9.77 V).

EXERCISE: (a) Find the Q-point in Ex. 5.5 if the resistor is changed to 5.6 kΩ. (b) What value of R is required to set the current to approximately 100 μA in the original circuit?

ANSWERS: (a) (1.45 mA, 3.5 V); (b) 82 kΩ.

Figure 5.19 displays the results of simulation of the collector current of the transistor in Fig. 5.18 versus the supply voltage V_{CC}. For $V_{CC} > 0$, the collector-base junction will be reverse-biased, and the transistor will be in the forward-active region. In this region, the circuit behaves essentially as a 1-mA ideal current source in which the output current is independent of V_{CC}. Note that the circuit

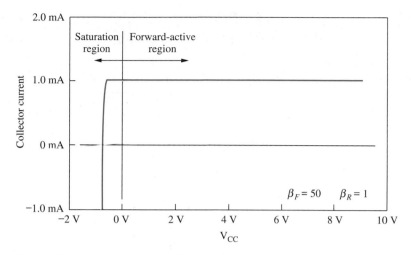

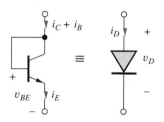

Figure 5.19 Simulation of output characteristics of circuit of Fig. 5.18(a).

Figure 5.20 Diode-connected transistor.

actually behaves as a current source for V_{CC} down to approximately -0.5 V. By the definitions in Table 5.2, the transistor enters saturation for $V_{CC} < 0$, but the transistor does not actually enter heavy saturation until the base-collector junction begins to conduct for $V_{BC} \geq +0.5$ V.

EXERCISE: Find the three terminal currents in the transistor in Fig. 5.18 if the 8.2 kΩ resistor value is changed to 5.6 kΩ.

ANSWER: 1.48 mA, 29.1 μA, 1.45 mA.

EXERCISE: What are the actual values of V_{BE} and V_{CE} for the transistor in Fig. 5.18(a) if $I_S = 5 \times 10^{-16}$ A? (Note that an iterative solution is necessary.)

ANSWERS: 0.708 V, 5.44 V

5.7.3 DIODES IN BIPOLAR INTEGRATED CIRCUITS

In integrated circuits, we often want the characteristics of a diode to match those of the BJT as closely as possible. In addition, it takes about the same amount of area to fabricate a diode as a full bipolar transistor. For these reasons, a diode is usually formed by connecting the base and collector terminals of a bipolar transistor, as shown in Fig. 5.20. This connection forces $v_{BC} = 0$.

Using the transport model equations for BJT with this boundary condition yields an expression for the terminal current of the "diode":

$$i_D = (i_C + i_B) = \left(I_S + \frac{I_S}{\beta_F} \right) \left[\exp\left(\frac{v_{BE}}{V_T} \right) - 1 \right] = \frac{I_S}{\alpha_F} \left[\exp\left(\frac{v_D}{V_T} \right) - 1 \right] \quad (5.26)$$

The terminal current has an i-v characteristic corresponding to that of a diode with a reverse saturation current that is determined by the BJT parameters. This technique is often used in both analog and digital circuit design; we will see many examples of its use in the analog designs in Part III.

EXERCISE: What is the equivalent saturation current of the diode in Fig 5.20 if the transistor is described by $I_S = 2 \times 10^{-14}$ A and $\alpha_F = 0.95$?

ANSWER: 21 fA

The Bipolar Transistor PTAT Cell

The diode version of the PTAT cell, that generates an output voltage **proportional to absolute** temperature, was introduced back in Chapter 3. We can also easily implement the PTAT cell using two bipolar transistors as shown in the figure here in which two identical bipolar transistors are biased in the forward-active region by current sources with a 10:1 current ratio.

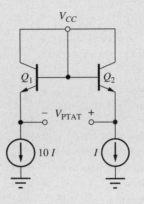

The PTAT voltage is given by

$$V_{\text{PTAT}} = V_{E2} - V_{E1} = (V_{CC} - V_{BE2}) - (V_{CC} - V_{BE1}) = V_{BE1} - V_{BE2}$$

$$V_{\text{PTAT}} = V_T \ln\left(\frac{10I}{I_S}\right) - V_T \ln\left(\frac{I}{I_S}\right) = \frac{kT}{q}\ln(10)$$

The bipolar PTAT cell is the circuit most commonly used in electronic thermometry.

5.7.4 SIMPLIFIED MODEL FOR THE REVERSE-ACTIVE REGION

In the reverse-active region, also called the inverse-active region, the roles of the emitter and collector terminals are reversed. The base-collector diode is forward-biased and the base-emitter junction is reverse-biased, and we can assume that $\exp(v_{BE}/V_T) \ll 1$ for $v_{BE} < -0.1$ V just as we did in simplifying Eq. Set (5.21). Applying this approximation to Eq. (5.13) and neglecting the -1 terms relative to the exponential terms yields the simplified equations for the reverse-active region:

$$i_C = -\frac{I_S}{\alpha_R}\exp\left(\frac{v_{BC}}{V_T}\right) \qquad i_E = -I_S\exp\left(\frac{v_{BC}}{V_T}\right) \qquad i_B = \frac{I_S}{\beta_R}\exp\left(\frac{v_{BC}}{V_T}\right) \qquad (5.27)$$

Ratios of these equations yield $i_E = -\beta_R i_B$ and $i_E = \alpha_R i_C$.

Equations (5.27) lead to the simplified circuit model for the reverse-active region shown in Fig. 5.21. The base current in the base-collector diode is amplified by the reverse common-emitter current gain β_R and enters the emitter terminal.

In the reverse-active region, the base-collector diode is now forward-biased, and the transistor model of Fig. 5.21(b) can be further simplified to that of Fig. 5.21(c), in which the diode is replaced by its CVD model with a voltage of 0.7 V. The base and collector voltages differ only by one 0.7-V diode drop in the reverse-active region.

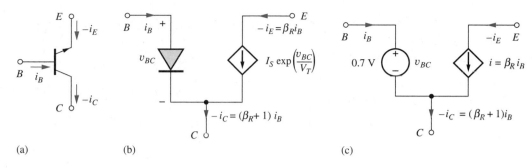

Figure 5.21 (a) *npn* transistor in the reverse active region. (b) Simplified circuit model for the reverse-active region. (c) Further simplification in the reverse-active region using the CVD model for the diode.

EXAMPLE 5.6   REVERSE-ACTIVE REGION ANALYSIS

Although the reverse active region is not often used, one does encounter it fairly frequently in the laboratory. If the transistor is inadvertently plugged in upside down, for example, the transistor will be operating in the reverse-active region. On the surface, the circuit will seem to be working but not very well. It is useful to be able to recognize when this error has occurred.

PROBLEM   The collector and emitter terminals of the *npn* transistor in Fig. 5.18 have been interchanged in the circuit in Fig. 5.22 (perhaps the transistor was plugged into the circuit backwards by accident). Find the new Q-point for the transistor in the circuit in Fig. 5.22.

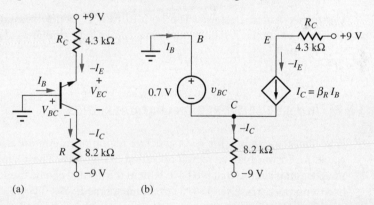

Figure 5.22 (a) Circuit of Fig. 5.18 with *npn* transistor orientation reversed. (b) Circuit simplification using the model for the reverse-active region. (Analysis of the circuit uses $\beta_F = 50$ and $\beta_R = 1$.)

SOLUTION   **Known Information and Given Data:** *npn* transistor in the circuit in Fig. 5.22 with $\beta_F = 50$ and $\beta_R = 1$

Unknowns: Q-point (I_C, V_{CE})

Approach: In this circuit, the base-emitter junction is reverse-biased by the 9-V source ($V_{BE} = V_B - V_E = -9$ V). The combination of the 8.2-kΩ resistor and the -9-V source will pull a current *out of* the collector and forward-bias the base-collector junction. Thus, the transistor appears to be biased in the reverse-active region of operation.

Assumptions: Assume reverse-active region operation; since we do not know the saturation current, assume $V_{BC} = 0.7$ V; use the simplified model for the reverse-active region to analyze the circuit as in Fig. 5.22(b).

Analysis: The current exiting from the collector $(-I_C)$ is now equal to

$$(-I_C) = \frac{-0.7 \text{ V} - (-9 \text{ V})}{8200 \, \Omega} = 1.01 \text{ mA}$$

The current through the 8.2-kΩ resistor is unchanged compared to that in Fig. 5.18. However, significant differences exist in the currents in the base terminal and the +9-V source. At the collector node, $(-I_C) = (\beta_R + 1)I_B$, and at the emitter, $(-I_E) = \beta_R I_B$:

$$I_B = \frac{1.01 \text{ mA}}{2} = 0.505 \text{ mA} \qquad \text{and} \qquad -I_E = (1)I_B = 0.505 \text{ mA}$$

$$V_{EC} = 9 - 4300(.505 \text{ mA}) - (-0.7 \text{ V}) = 7.5 \text{ V}$$

Check of Results: We see that KVL is satisfied around the output loop containing the collector-emitter voltage: $+9 - V_{CE} - V_R - (-9) = 9 - 9.7 - 8.3 + 9 = 0$. Also, $I_C + I_B = I_E$, and the calculated current directions are all consistent with the assumption of reverse-active region operation. Finally $V_{EB} = 9 - 43 \text{ k}\Omega \ (0.505 \text{ mA}) = 6.8 \text{ V}$. $V_{EB} > 0 \text{ V}$, and the reverse active assumption is correct.

Discussion: Note that both the base current is much larger than expected, whereas the current entering the upper terminal of the device is much smaller than would be expected if the transistor were in the circuit as originally drawn in Fig. 5.18. These significant differences in current often lead to unexpected shifts in voltage levels at the base and collector terminals of the transistor in more complicated circuits.

Computer-Aided Design: The built-in SPICE model is valid for any operating region, and simulation with the default model gives results very similar to hand calculations.

DESIGN NOTE   **REVERSE-ACTIVE REGION CHARACTERISTICS**

Note that the currents for reverse-active region operation are usually very different from those found for forward-active region operation in Fig. 5.18. These drastic differences are often useful in debugging circuits that we have built in the lab and can be used to discover transistors that have been improperly inserted into a circuit breadboard.

EXERCISE: Find the three terminal currents in the transistor in Fig. 5.22 if the resistor value is changed to 5.6 kΩ.

ANSWER: 1.48 mA, 0.741 mA, 0.741 mA

5.7.5 MODELING OPERATION IN THE SATURATION REGION

The fourth and final region of operation is called the saturation region. In this region, both junctions are forward-biased, and the transistor typically operates with a small voltage between collector and emitter terminals. In the saturation region, the dc value of v_{CE} is called the **saturation voltage** of the transistor: v_{CESAT} for the *npn* transistor or v_{ECSAT} for the *pnp* transistor.

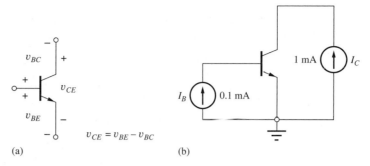

Figure 5.23 (a) Relationship between the terminal voltages of the transistor. (b) Circuit for Example 5.8.

In order to determine v_{CESAT}, we assume that both junctions are forward-biased so that i_C and i_B from Eqs. (5.13) can be approximated as

$$i_C = I_S \exp\left(\frac{v_{BE}}{V_T}\right) - \frac{I_S}{\alpha_R} \exp\left(\frac{v_{BC}}{V_T}\right)$$

$$i_B = \frac{I_S}{\beta_F} \exp\left(\frac{v_{BE}}{V_T}\right) + \frac{I_S}{\beta_R} \exp\left(\frac{v_{BC}}{V_T}\right) \tag{5.28}$$

Simultaneous solution of these equations using $\beta_R = \alpha_R/(1 - \alpha_R)$ yields expressions for the base-emitter and base-collector voltages:

$$v_{BE} = V_T \ln \frac{i_B + (1 - \alpha_R)i_C}{I_S\left[\frac{1}{\beta_F} + (1 - \alpha_R)\right]} \quad \text{and} \quad v_{BC} = V_T \ln \frac{i_B - \frac{i_C}{\beta_F}}{I_S\left[\frac{1}{\alpha_R}\right]\left[\frac{1}{\beta_F} + (1 - \alpha_R)\right]} \tag{5.29}$$

By applying KVL to the transistor in Fig. 5.23, we find that the collector-emitter voltage of the transistor is $v_{CE} = v_{BE} - v_{BC}$, and substituting the results from Eqs. (5.29) into this equation yields an expression for the saturation voltage of the *npn* transistor:

$$v_{\text{CESAT}} = V_T \ln\left[\left(\frac{1}{\alpha_R}\right) \frac{1 + \frac{i_C}{(\beta_R + 1)i_B}}{1 - \frac{i_C}{\beta_F i_B}}\right] \quad \text{for } i_B > \frac{i_C}{\beta_F} \tag{5.30}$$

This equation is important and highly useful in the design of saturated digital switching circuits. For a given value of collector current, Eq. (5.30) can be used to determine the base current required to achieve a desired value of v_{CESAT}.

Note that Eq. (5.30) is valid only for $i_B > i_C/\beta_F$. This is an auxiliary condition that can be used to define saturation region operation. The ratio i_C/β_F represents the base current needed to maintain transistor operation in the forward-active region. If the base current exceeds the value needed for forward-active region operation, the transistor will enter saturation. The actual value of i_C/i_B is often called the **forced beta β_{FOR}** of the transistor, where $\beta_{\text{FOR}} \leq \beta_F$.

EXAMPLE 5.7   **SATURATION VOLTAGE CALCULATION**

The BJT saturation voltage is important in many switching applications. Here we find an example of the value of the saturation voltage for a forced beta of 10.

PROBLEM   Calculate the saturation voltage for an *npn* transistor with $I_C = 1$ mA, $I_B = 0.1$ mA, $\beta_F = 50$, and $\beta_R = 1$.

SOLUTION   **Known Information and Given Data:** An *npn* transistor is operating with $I_C = 1$ mA, $I_B = 0.1$ mA, $\beta_F = 50$, and $\beta_R = 1$

Unknowns: Collector-emitter voltage of the transistor

Approach: Because $I_C/I_B = 10 < \beta_F$, the transistor will indeed be saturated. Therefore we can use Eq. (5.30) to find the saturation voltage.

Assumptions: Room temperature operation with $V_T = 0.025$ V

Analysis: Using $\alpha_R = \beta_R/(\beta_R + 1) = 0.5$ and $I_C/I_B = 10$ yields

$$v_{\text{CESAT}} = (0.025 \text{ V}) \ln \left[\left(\frac{1}{0.5} \right) \frac{1 + \dfrac{1 \text{ mA}}{2(0.1 \text{ mA})}}{1 - \dfrac{1 \text{ mA}}{50(0.1 \text{ mA})}} \right] = 0.068 \text{ V}$$

Check of Results: A small, nearly zero, value of saturation voltage is expected; thus the calculated value appears reasonable.

Discussion: We see that the value of V_{CE} in this example is indeed quite small. However, it is nonzero even for $i_C = 0$ [see Prob. 5.64]! It is impossible to force the forward voltages across both *pn* junctions to be exactly equal, which is a consequence of the asymmetric values of the forward and reverse current gains. The existence of this small voltage "offset" is an important difference between the BJT and the MOSFET. In the case of the MOSFET, the voltage between drain and source becomes zero when the drain current is zero.

Computer-Aided Analysis: We can simulate the situation in this example by driving the base of the BJT with one current source and the collector with a second. (This is one of the few circuit situations in which we can force a current into the collector using a current source.) SPICE yields $V_{\text{CESAT}} = 0.070$ V. The default temperature in SPICE is 27°C, and the slight difference in V_T accounts for the difference between SPICE result and our hand calculations.

EXERCISE: What is the saturation voltage in Ex. 5.7 if the base current is reduced to 40 μA?

ANSWER: 99.7 mV

EXERCISE: Use Eqs. (5.29) to find V_{BESAT} and V_{BCSAT} for the transistor in Ex. 5.7 if $I_S = 10^{-15}$ A.

ANSWERS: 0.694 V, 0.627 V

Figure 5.24 shows the simplified model for the transistor in saturation in which the two diodes are assumed to be forward-biased and replaced by their respective on-voltages. The forward voltages

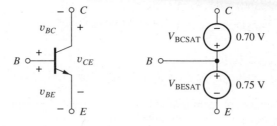

Figure 5.24 Simplified model for the *npn* transistor in saturation.

ELECTRONICS IN ACTION

Optical Isolators

The optical isolator drawn schematically here represents a highly useful circuit that behaves much like a single transistor, but provides a very high breakdown voltage and low capacitance between its input and output terminals. Input current i_{IN} drives a light emitting diode (LED) whose output illuminates the base region of an *npn* transistor. Energy lost by the photons creates hole–electron pairs in the base of the *npn*. The holes represent base current that is then amplified by the current gain β_F of the transistor, whereas the electrons simply become part of the collector current.

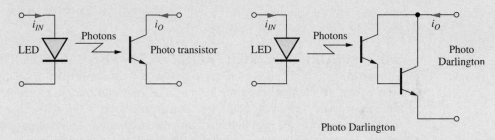

Photo Darlington

The output characteristics of the optical isolator are very similar to those of a BJT operating in the active region in Fig. 5.10. However, the conversion of photons to hole–electron pairs is not very efficient in silicon, and the current transfer ratio, $\beta_F = i_O/i_{IN}$, of the optical isolator is often only around unity. The "Darlington connection" of two transistors (see Prob. 15.48), is often used to improve the overall current gain of the isolator. In this case, the output current is increased by the current gain of the second transistor.

The dc isolation provided by such devices can exceed a thousand volts and is limited primarily by the spacing of the pins and the characteristics of the circuit board that the isolator is mounted upon. ac isolation is limited to the low picofarad range by stray capacitance between the input and outputs pins.

of both diodes are normally higher in saturation than in the forward-active region, as indicated in the figure by $V_{\text{BESAT}} = 0.75$ V and $V_{\text{BCSAT}} = 0.7$ V. In this case, V_{CESAT} is 50 mV. In saturation, the terminal currents are determined by the external circuit elements; no simplifying relationships exist between i_C, i_B, and i_E other than $i_C + i_B = i_E$.

5.8 NONIDEAL BEHAVIOR OF THE BIPOLAR TRANSISTOR

As with all devices, the BJT characteristics deviate from our ideal mathematical models in a number of ways. The emitter-base and collector-base diodes that form the bioplar transistor have finite reverse breakdown voltages (See Section 3.6.2) that we must carefully consider when choosing a transistor or the power supplies for our circuits. There are also capacitances associated with each of the diodes, and these capacitances place limitations on the high frequency response of the transistor. We also know that holes and electrons in semiconductor materials have finite velocities. Thus, it takes time for the carriers to move from the emitter to the collector, and this time delay places an additional limit on the upper frequency of operation of the bipolar transistor. Finally, the output characteristics of the BJT exhibit a dependence on collector-emitter voltage similar to the channel-length modulation effect that occurs in the MOS transistor (Section 4.2.7). This section considers each of these limitations in more detail.

5.8.1 JUNCTION BREAKDOWN VOLTAGES

The bipolar transistor is formed from two back-to-back diodes, each of which has a Zener breakdown voltage associated with it. If the reverse voltage across either *pn* junction is too large, the corresponding diode will break down. In the transistor structure in Fig. 5.1, the emitter region is the most heavily doped region and the collector is the most lightly doped region. These doping differences lead to a relatively low breakdown voltage for the base-emitter diode, typically in the range of 3 to 10 V. On the other hand, the collector-base diode can be designed to break down at much larger voltages. Transistors can be fabricated with collector-base breakdown voltages as high as several hundred volts.

Transistors must be selected with breakdown voltages commensurate with the reverse voltages that will be encountered in the circuit. In the forward-active region, for example, the collector-base junction is operated under reverse bias and must not break down. In the cutoff region, both junctions are reverse-biased, and the relatively low breakdown voltage of the emitter-base junction must not be exceeded.

5.8.2 MINORITY-CARRIER TRANSPORT IN THE BASE REGION

Current in the BJT is predominantly determined by the transport of *minority carriers* across the base region. In the *npn* transistor in Fig. 5.25, transport current i_T results from the diffusion of minority carriers—electrons in the *npn* transistor or holes in the *pnp*—across the base. Base current i_B is composed of hole injection back into the emitter and collector, as well as a small additional current I_{REC} needed to replenish holes lost to recombination with electrons in the base. These three components of base current are shown in Fig. 5.25(a).

An expression for the transport current i_T can be developed using our knowledge of carrier diffusion and the values of base-emitter and base-collector voltages. It can be shown from device physics (beyond the scope of this text) that the voltages applied to the base-emitter and base-collector junctions define the minority-carrier concentrations at the two ends of the base region through these relationships:

$$n(0) = n_{bo} \exp\left(\frac{v_{BE}}{V_T}\right) \qquad \text{and} \qquad n(W_B) = n_{bo} \exp\left(\frac{v_{BC}}{V_T}\right) \tag{5.31}$$

in which $\boldsymbol{n_{bo}}$ is the **equilibrium electron density** in the *p*-type base region.

The two junction voltages establish a minority-carrier concentration gradient across the base region, as illustrated in Fig. 5.25(b). For a narrow base, the minority-carrier density decreases linearly across the base, and the diffusion current in the base can be calculated using the diffusion current

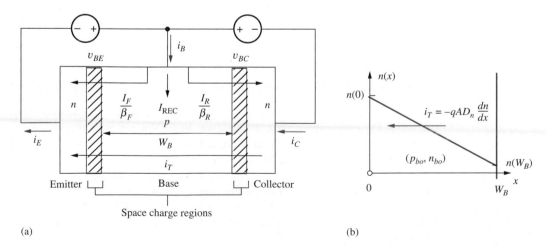

(a)

(b)

Figure 5.25 (a) Currents in the base region of an *npn* transistor. (b) Minority-carrier concentration in the base of the *npn* transistor.

expression in Eq. (2.14):

$$i_T = -qAD_n \frac{dn}{dx} = +qAD_n \frac{n_{bo}}{W_B} \left[\exp\left(\frac{v_{BE}}{V_T}\right) - \exp\left(\frac{v_{BC}}{V_T}\right) \right] \tag{5.32}$$

where $A =$ cross-sectional area of base region

$W_B =$ **base width**

Because the carrier gradient is negative, electron current i_T is directed in the negative x direction, exiting the emitter terminal (positive i_T).

Comparing Eqs. (5.32) and (5.19) yields a value for the bipolar transistor saturation current I_S:

$$I_S = qAD_n \frac{n_{bo}}{W_B} = \frac{qAD_n n_i^2}{N_{AB} W_B} \tag{5.33a}$$

where $N_{AB} =$ doping concentration in base of transistor

$n_i =$ intrinsic-carrier concentration ($10^{10}/\text{cm}^3$)

$n_{bo} = n_i^2 / N_{AB}$ using Eq. (2.12)

The corresponding expression for the saturation current of the *pnp* transistor is

$$I_S = qAD_p \frac{p_{bo}}{W_B} = \frac{qAD_p n_i^2}{N_{DB} W_B} \tag{5.33b}$$

Remembering from Chapter 2 that mobility μ, and hence diffusivity $D = (kT/q)\mu$ (cm^2/s), is larger for electrons than holes ($\mu_n > \mu_p$), we see from Eqs. (5.33) that the *npn* transistor will conduct a higher current than the *pnp* transistor for a given set of applied voltages.

EXERCISE: (a) What is the value of D_n at room temperature if $\mu_n = 500$ cm^2/V · s? (b) What is I_S for a transistor with $A = 50$ μm^2, $W = 1$ μm, $D_n = 12.5$ cm^2/s and $N_{AB} = 10^{18}/\text{cm}^3$?

ANSWERS: 12.5 cm^2/s; 10^{-18} A

5.8.3 BASE TRANSIT TIME

To turn on the bipolar transistor, minority-carrier charge must be introduced into the base to establish the carrier gradient in Fig. 5.25(b). The **forward transit time** τ_F represents the time constant associated with storing the required charge Q in the base region and is defined by

$$\tau_F = \frac{Q}{I_T} \tag{5.34}$$

Figure 5.26 depicts the situation in the neutral base region of an *npn* transistor operating in the forward-active region with $v_{BE} > 0$ and $v_{BC} = 0$. The area under the triangle represents the excess minority charge Q that must be stored in the base to support the diffusion current. For the dimensions in Fig. 5.26 and using Eq. (5.31),

$$Q = qA[n(0) - n_{bo}]\frac{W_B}{2} \qquad qAn_{bo}\left[\exp\left(\frac{v_{BE}}{V_T}\right) - 1\right]\frac{W_B}{2} \tag{5.35}$$

For the conditions in Fig. 5.26(a),

$$i_T = \frac{qAD_n}{W_B}n_{bo}\left[\exp\left(\frac{v_{BE}}{V_T}\right) - 1\right] \tag{5.36}$$

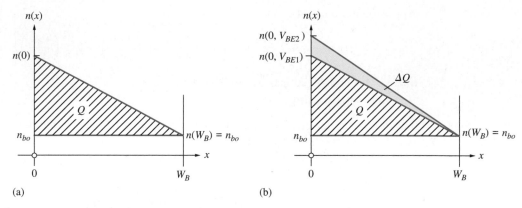

Figure 5.26 (a) Excess minority charge Q stored in the bipolar base region. (b) Stored charge Q changes as v_{BE} changes.

Substituting Eqs. (5.35) and (5.36) into Eq. (5.34), the forward transit time for the *npn* transistor is found to be

$$\tau_F = \frac{W_B^2}{2D_n} = \frac{W_B^2}{2V_T \mu_n} \tag{5.37a}$$

The corresponding expression for the transit time of the *pnp* transistor is

$$\tau_F = \frac{W_B^2}{2D_p} = \frac{W_B^2}{2V_T \mu_p} \tag{5.37b}$$

The base transit time can be viewed as the average time required for a carrier emitted by the emitter to arrive at the collector. Hence, one would not expect the transistor to be able to reproduce frequencies with periods that are less than the transit time, and the base transit time in Eq. (5.37) places an upper limit on the useful operating frequency f of the transistor,

$$f \leq \frac{1}{2\pi \tau_F} \tag{5.38}$$

From Eq. (5.37), we see that the transit time is inversely proportional to the minority-carrier mobility in the base, and the difference between electron and hole mobility leads to an inherent frequency and speed advantage for the *npn* transistor. Thus, an *npn* transistor may be expected to be 2 to 2.5 times as fast as a *pnp* transistor for a given geometry and doping. Equation (5.37) also indicates the importance of shrinking the base width W_B of the transistor as much as possible. Early transistors had base widths of 10 μm or more, whereas the base width of transistors in research laboratories today is 0.025 μm (25 nm) or less.

EXAMPLE 5.8   SATURATION CURRENT AND TRANSIT TIME

Device physics has provided us with expressions that can be used to estimate transistor saturation current and transit time based on a knowledge of physical constants and structural device information. Here we find representative values of I_S and τ_F for a bipolar transistor.

PROBLEM   Find the saturation current and base transit time for an *npn* transistor with a 100 μm × 100 μm emitter region, a base doping of $10^{17}/\text{cm}^3$, and a base width of 1 μm. Assume $\mu_n = 500$ cm²/V · s.

SOLUTION   **Known Information and Given Data:** Emitter area $= 100$ μm × 100 μm, $N_{AB} = 10^{17}/\text{cm}^3$, $W_B = 1$ μm, $\mu_n = 500$ cm²/V · s

Unknowns: Saturation current I_S; transit time τ_F

Approach: Evaluate Eqs. (5.33) and (5.37) using the given data

Assumptions: Room temperature operation with $V_T = 0.025$ V and $n_i = 10^{10}/\text{cm}^3$

Analysis: Using Eq. (5.33) for I_S,

$$I_S = \frac{qAD_nn_i^2}{N_{AB}W_B} = \frac{(1.6 \times 10^{-19}\ \text{C})(10^{-2}\ \text{cm})^2 \left(0.025\ \text{V} \times 500\ \dfrac{\text{cm}^2}{\text{V}\cdot\text{s}}\right)\left(\dfrac{10^{20}}{\text{cm}^6}\right)}{\left(\dfrac{10^{17}}{\text{cm}^3}\right)(10^{-4}\ \text{cm})} = 2 \times 10^{-15}\ \text{A}$$

in which $D_n = (kT/q)\mu_n$ has been used [remember Eq. (2.15)].
   Using Eq. (5.37),

$$\tau_F = \frac{W_B^2}{2V_T\mu_n} = \frac{(10^{-4}\ \text{cm})^2}{2(0.025\ \text{V})\left(500\ \dfrac{\text{cm}^2}{\text{V}\cdot\text{s}}\right)} = 4 \times 10^{-10}\ \text{s}$$

Check of Results: The calculations appear correct, and the value of I_S is within the range given in Sec. 5.2.

Discussion: Operation of this particular transistor is limited to frequencies below $f = 1/(2\pi\tau_F) = 400$ MHz.

5.8.4 DIFFUSION CAPACITANCE

Capacitances are circuit elements that represent the high-frequency limitations of MOS and bipolar devices.

For the base-emitter voltage and hence the collector current in the BJT to change, the charge stored in the base region also must change, as illustrated in Fig. 5.26(b). This change in charge with v_{BE} can be modeled by a capacitance C_D, called the **diffusion capacitance,** placed in parallel with the forward-biased base-emitter diode as defined by

$$C_D = \frac{dQ}{dv_{BE}}\bigg|_{Q-\text{point}} = \frac{1}{V_T}\frac{qAn_{bo}W_B}{2}\exp\left(\frac{V_{BE}}{V_T}\right) \tag{5.39}$$

This equation can be rewritten as

$$C_D = \frac{1}{V_T}\left[\frac{qAD_nn_{bo}}{W_B}\exp\left(\frac{V_{BE}}{V_T}\right)\right]\left(\frac{W_B^2}{2D_n}\right) \cong \frac{I_T}{V_T}\tau_F \tag{5.40}$$

Because the transport current actually represents the collector current in the forward-active region, the expression for the diffusion capacitance is normally written as

$$C_D = \frac{I_C}{V_T}\tau_F \tag{5.41}$$

From Eq. (5.41), we see that the diffusion capacitance C_D is directly proportional to current and inversely proportional to temperature T. For example, a BJT operating at a current of 1 mA with $\tau_F = 4 \times 10^{-10}$ s has a diffusion capacitance of

$$C_D = \frac{I_C}{V_T}\tau_F = \frac{10^{-3}\ \text{A}}{0.025\ \text{V}}\left(4 \times 10^{-10}\ \text{s}\right) = 16 \times 10^{-12}\ \text{F} = 16\ \text{pF}$$

This is a substantial capacitance, but it can be even larger if the transistor is operating at significantly higher currents.

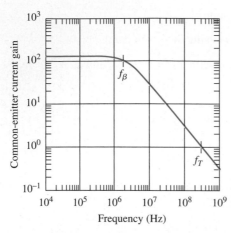

Figure 5.27 Magnitude of the common-emitter current gain β vs. frequency.

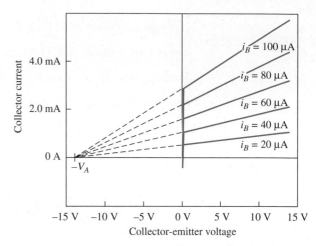

Figure 5.28 Transistor output characteristics identifying the Early voltage V_A.

EXERCISE: Calculate the value of the diffusion capacitance for a power transistor operating at a current of 10 A and a temperature of 100°C if $\tau_F = 4$ nS.

ANSWERS: 1.24 μF—a significant capacitance!

5.8.5 FREQUENCY DEPENDENCE OF THE COMMON-EMITTER CURRENT GAIN

The forward-biased diffusion and reverse-biased pn junction capacitances of the bipolar transistor cause the current gain of the transistor to be frequency-dependent. An example of this dependence is given in Fig. 5.27. At low frequencies, the current gain has a constant value β_F, but as frequency increases, the current gain begins to decrease. The **unity-gain frequency** f_T is defined to be the frequency at which the magnitude of the current gain is equal to 1. The behavior in the graph is described mathematically by

$$\beta(f) = \frac{\beta_F}{\sqrt{1 + \left(\dfrac{f}{f_\beta}\right)^2}} \tag{5.42}$$

where $f_\beta = f_T/\beta_F$ is the **β-cutoff frequency.** For the transistor in Fig. 5.27, $\beta_F = 125$ and $f_T = 300$ MHz.

EXERCISE: What is the β-cutoff frequency for the transistor in Fig. 5.27?

ANSWER: 2.4 MHz

5.8.6 THE EARLY EFFECT AND EARLY VOLTAGE

In the transistor output characteristics in Figs. 5.11 and 5.13, the current saturated at a constant value in the forward-active region. However, in a real transistor, there is actually a positive slope to the characteristics, as shown in Fig. 5.28. The collector current is not truly independent of v_{CE}. Note that this situation is the same as that found for the MOSFET in saturation. The slope of the output characteristics is greater than zero, not zero as predicted by the basic analysis.

It has been observed experimentally that when the output characteristic curves are extrapolated back to the point of zero collector current, the curves all intersect at a common point, $v_{CE} = -V_A$.

This phenomenon is called the **Early effect** [4], and the voltage V_A is called the **Early voltage** after James Early from Bell Laboratories, who first identified the source of the behavior. A relatively small value of Early voltage (14 V) has been used in Fig. 5.28 to exaggerate the characteristics. Values for the Early voltage more typically fall in the range

$$15 \text{ V} \le V_A \le 150 \text{ V}$$

5.8.7 MODELING THE EARLY EFFECT

The dependence of the collector current on collector-emitter voltage is easily included in the simplified mathematical model for the forward-active region of the BJT by modifying Eqs. (5.23) as follows:

$$i_C = I_S \left[\exp\left(\frac{v_{BE}}{V_T} \right) \right] \left[1 + \frac{v_{CE}}{V_A} \right]$$

$$\beta_F = \beta_{FO} \left[1 + \frac{v_{CE}}{V_A} \right] \tag{5.43}$$

$$i_B = \frac{I_S}{\beta_{FO}} \left[\exp\left(\frac{v_{BE}}{V_T} \right) \right]$$

β_{FO} represents the value of β_F extrapolated to $V_{CE} = 0$. In these expressions, the collector current and current gain now have the same dependence on v_{CE}, but the base current remains independent of v_{CE}. This is consistent with Fig. 5.28, in which the separation of the constant-base-current curves in the forward-active region increases as v_{CE} increases, indicating that the current gain β_F is increasing with v_{CE}.

> **EXERCISE:** A transistor has $I_S = 10^{-15}$ A, $\beta_{FO} = 75$, and $V_A = 50$ V and is operating with $V_{BE} = 0.7$ V and $V_{CE} = 10$ V. What are I_B, β_F, and I_C? What would be β_F and I_C if $V_A = \infty$?
>
> **ANSWERS:** 19.3 μA, 90, 1.74 mA; 75, 1.45 mA

5.8.8 ORIGIN OF THE EARLY EFFECT

Modulation of the base width W_B of the transistor by the collector-base voltage is the cause of the Early effect. As the reverse bias across the collector-base junction increases, the width of the collector-base depletion layer increases, and the width W_B of the base decreases. This mechanism, termed **base-width modulation,** is depicted in Fig. 5.29, in which the collector-base space charge

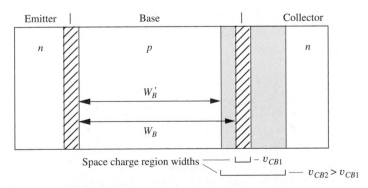

Figure 5.29 Base-width modulation, or Early effect.

region width is shown for two different values of collector-base voltage corresponding to effective base widths of W_B and W_B'. Equation (5.32) demonstrated that collector current is inversely proportional to the base width W_B, so a decrease in W_B results in an increase in transport current i_T. This decrease in W_B as V_{CB} increases is the cause of the Early effect.

The Early effect reduces the output resistance of the bipolar transistor and places an important limit on the amplification factor of the BJT. These limitations are discussed in detail in Part III, Chapter 13.

Note that both the Early effect in the BJT and channel-length modulation in the MOSFET are similar in the sense that the nonzero slope of the output characteristics is related to changes in a characteristic length within the device as the voltage across the output terminals of the transistor changes.

5.9 TRANSCONDUCTANCE

The important transistor parameter, **transconductance** g_m, was introduced during our study of the MOSFET in Chapter 4. For the bipolar transistor, g_m relates changes in i_C to changes in v_{BE} and is defined by

$$g_m = \frac{di_C}{dv_{BE}}\bigg|_{Q-\text{point}} \tag{5.44}$$

For Q-points in the forward-active region, Eq. (5.44) can be evaluated using the collector-current expression from Eq. (5.23):

$$g_m = \frac{d}{dv_{BE}}\left\{ I_S \exp\left(\frac{v_{BE}}{V_T}\right)\right\}\bigg|_{Q-\text{point}} = \frac{1}{V_T} I_S \exp\left(\frac{V_{BE}}{V_T}\right) = \frac{I_C}{V_T} \tag{5.45}$$

Equation (5.45) represents the fundamental relationship for the transconductance of the bipolar transistor, in which we find g_m is directly proportional to collector current. This is an important result that is used many times in bipolar circuit design. It is worth noting that the expression for the transit time defined in Eq. (5.41) can be rewritten as

$$\tau_F = \frac{C_D}{g_m} \quad \text{or} \quad C_D = g_m \tau_F \tag{5.46}$$

DESIGN NOTE   BIPOLAR TRANSCONDUCTANCE

$$g_m = \frac{I_C}{V_T}$$

The BJT transconductance is substantially higher than that of the FET for a given operating current. This difference will be discussed in more detail in Chapters 13 and 14.

DESIGN NOTE   TRANSIT TIME

$$\tau_F = \frac{C_D}{g_m}$$

Transit time τ_F places an upper limit on the frequency response of the bipolar device.

> **EXERCISE:** What is the value of the BJT transconductance g_m at $I_C = 100$ μA and $I_C = 1$ mA? What is the value of the diffusion capacitance for each of these currents if the base transit time is 25 psec?
>
> **ANSWERS:** 4 mS; 40 mS; 0.1 pF; 1.0 pF

5.10 BIPOLAR TECHNOLOGY AND SPICE MODEL

In order to create a comprehensive simulation model of the bipolar transistor, our knowledge of the physical structure of the transistor is coupled with the transport model expressions and experimental observations. We typically start with a circuit representation of our mathematical model that describes the intrinsic behavior of the transistor, and then add additional elements to model parasitic effects introduced by the actual physical structure. Remember, in any case, that our SPICE models represent only lumped element equivalent circuits for the distributed structure that we actually fabricate.

Although we will seldom use the equations that make up the simulation model in hand calculations, awareness and understanding of the equations can help when SPICE generates unexpected results. This can happen when we attempt to use a device in an unusual way, or the simulator may produce a circuit result that does not fit within our understanding of the device behavior. Understanding the internal model is SPICE will help us interpret whether our knowledge of the device is wrong or if the simulation has some built-in assumptions that may not be consistent with a particular application of the device.

5.10.1 QUALITATIVE DESCRIPTION

A detailed cross section of the classic *npn* structure from Fig. 5.1 is given in Fig. 5.30(a), and the corresponding SPICE circuit model appears in Fig. 5.30(b). Circuit elements i_C, i_B, C_{BE}, and C_{BC} describe the intrinsic transistor behavior that we have discussed thus far. Current source i_C represents the current transported across the base from collector to emitter, and current source i_B models the total base current of the transistor. **Base-emitter** and **base-collector** capacitances C_{BE} and C_{BC} include

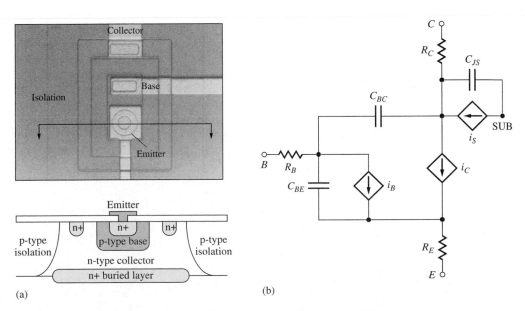

(a)   (b)

Figure 5.30 (a) Top view and cross section of a junction-isolated transistor. (b) SPICE model for the *npn* transistor.

models for the diffusion capacitances and the junction capacitances associated with the base-emitter and base-collector diodes.

Additional circuit elements are added to account for nonideal characteristics of the real transistor. The physical structure has a large-area *pn* junction that isolates the collector from the substrate of the transistor and separates one transistor from the next. The primary components related to this junction are diode current i_S and capacitance C_{JS}. Base resistance R_B accounts for the resistance between the external base contact and the intrinsic base region of the transistor. Similarly, collector current must pass through R_C on its way to the active region of the collector-base junction, and R_E models any extrinsic emitter resistance present in the device.

5.10.2 SPICE MODEL EQUATIONS

The SPICE models are comprehensive but quite complex. Even the model equations presented below represent simplified versions of the actual models. Table 5.3 defines the SPICE parameters that are used in these expressions. More complete descriptions can be found in [7].

The collector and base currents are given by

$$ i_C = \frac{(i_F - i_R)}{\text{KBQ}} - \frac{i_R}{\text{BR}} - i_{RG} \qquad \text{and} \qquad i_B = \frac{i_F}{\text{BF}} + \frac{i_R}{\text{BR}} + i_{FG} + i_{RG} $$

TABLE 5.3

Bipolar Device Parameters for Circuit Simulation (*npn/pnp*)

PARAMETER	NAME	DEFAULT	TYPICAL VALUES
Saturation current	**IS**	10^{-16} A	3×10^{-17} A
Forward current gain	**BF**	100	100
Forward emission coefficient	NF	1	1.03
Forward Early voltage	**VAF**	∞	75 V
Forward knee current	IKF	∞	0.05 A
Reverse knee current	IKR	∞	0.01 A
Reverse current gain	BR	1	0.5
Reverse emission coefficient	NR	1	1.05
Base resistance	RB	0	250 Ω
Collector resistance	RC	0	50 Ω
Emitter resistance	RE	0	1 Ω
Forward transit time	TF	0	0.15 nS
Reverse transit time	TR	0	15 nS
Base-emitter leakage saturation current	ISE	0	1 pA
Base-emitter leakage emission coefficient	NE	1.5	1.4
Base-emitter junction capacitance	CJE	0	0.5 pF
Base-emitter junction potential	PHIE	0.8 V	0.8 V
Base-emitter grading coefficient	ME	0.5	0.5
Base-collector leakage saturation current	ISC	0	1 pA
Base-collector leakage emission coefficient	NC	1.5	1.4
Base-collector junction capacitance	CJC	0	1 pF
Base-collector junction potential	PHIC	0.75 V	0.7 V
Base-collector grading coefficient	MC	0.33	0.33
Substrate saturation current	ISS	0	1 fA
Substrate emission coefficient	NS	1	1
Collector-substrate junction capacitance	CJS	0	3 pF
Collector-substrate junction potential	VJS	0.75 V	0.75 V
Collector-substrate grading coefficient	MJS	0	0.5

in which the forward and reverse components of the transport current are

$$i_F = \text{IS} \cdot \left[\exp\left(\frac{v_{BE}}{\text{NF} \cdot V_t} \right) - 1 \right] \qquad \text{and} \qquad i_R = \text{IS} \cdot \left[\exp\left(\frac{v_{BC}}{\text{NR} \cdot V_T} \right) - 1 \right] \qquad (5.47)$$

Base current i_B includes two added terms to model additional space-charge region currents associated with the base-emitter and base-collector junctions:

$$i_{FG} = \text{ISE} \cdot \left[\exp\left(\frac{v_{BE}}{\text{NE} \cdot V_T} \right) - 1 \right] \qquad \text{and} \qquad i_{RG} = \text{ISC} \cdot \left[\exp\left(\frac{v_{BC}}{\text{NC} \cdot V_T} \right) - 1 \right]$$

Another new addition is the KBQ term that includes voltages VAF and VAR to model the Early effect in both the forward and reverse modes, as well as "knee current" parameters IKF and IKR that model current gain fall-off at high operating currents. This phenomenon is discussed in more detail in Chapter 13.

$$\text{KBQ} = \left(\frac{1}{2} \right) \frac{1 + \left[1 + 4\left(\dfrac{i_F}{\text{IKF}} + \dfrac{i_R}{\text{IKR}} \right) \right]^{NK}}{1 + \dfrac{v_{CB}}{\text{VAF}} + \dfrac{v_{EB}}{\text{VAR}}}$$

Note as well that the Early effect is cast in terms of v_{BC} rather than v_{CE} as we have used in Eq. (5.43).

The substrate junction current is expressed as

$$i_S = \text{ISS} \cdot \left[\exp\left(\frac{v_{\text{SUB-C}}}{\text{NS} \cdot V_T} \right) - 1 \right]$$

The three device capacitances in Fig. 5.30(b) are represented by

$$C_{BE} = \frac{i_F}{\text{NE} \cdot V_T} \text{TF} + \frac{\text{CJE}}{\left(1 - \dfrac{v_{BE}}{\text{PHIE}} \right)^{\text{MJE}}} \qquad \text{and} \qquad C_{BC} = \frac{i_R}{\text{NC} \cdot V_T} \text{TR} + \frac{\text{CJC}}{\left(1 - \dfrac{v_{BC}}{\text{PHIC}} \right)^{\text{MJC}}}$$

$$C_{JS} = \frac{\text{CJS}}{\left(1 + \dfrac{v_{\text{SUB-C}}}{\text{VJS}} \right)^{\text{MJS}}} \qquad (5.48)$$

C_{BE} and C_{BC} consist of two terms representing the diffusion capacitance (modeled by TF and NE or TR and NC) and depletion-region capacitance (modeled by CJE, PHIE, and MJE or CJC, PHIC, and MJC). The substrate diode is normally reverse biased, so it is modeled by just the depletion-layer capacitance (CJS, VJS, and MJS). The base, collector, and emitter series resistances are RB, RC, and RE, respectively.

The SPICE model for the *pnp* transistor is similar to that presented in Fig. 5.30(b) except for reversal of the current sources and of the positive polarity for the transistor currents and voltages.

5.10.3 HIGH-PERFORMANCE BIPOLAR TRANSISTORS

Modern transistors designed for high-speed switching and analog RF applications use combinations of sophisticated shallow and deep trench isolation processes to reduce the device capacitances and minimize the transit times. These devices typically utilize polysilicon emitters, have extremely narrow bases, and may incorporate SiGe base regions. A drawing and cross section of a very high frequency, trench-isolated SiGe bipolar transistor appears in Fig. 5.31. In the research laboratory, SiGe transistors have already exhibited cutoff frequencies in excess of 300 GHz.

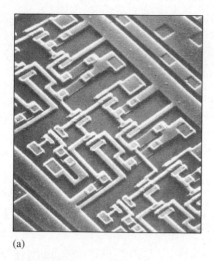
(a)

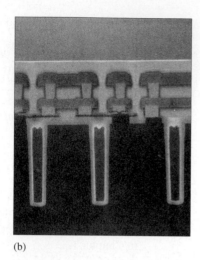
(b)

Figure 5.31 (a) Top view of a high-performance trench-isolated integrated circuit. (b) Cross section of a high-performance trench-isolated bipolar transistor. Copyright ©1995, IEEE. Reprinted with permission from [8].

EXERCISE: A bipolar transistor has a current gain of 80, a collector current of 350 μA for $V_{BE} = 0.68$ V, and an Early voltage of 70 V. What are the values of SPICE parameters BF, IS, and VAF? Assume $T = 27°C$.

ANSWERS: 80, 1.35 fA, 70 V

5.11 PRACTICAL BIAS CIRCUITS FOR THE BJT

The goal of biasing is to establish a known **quiescent operating point,** or **Q-point** that represents the initial operating region of the transistor. In the bipolar transistor, the Q-point is represented by the dc values of the collector-current and collector-emitter voltage (I_C, V_{CE}) for the *npn* transistor, or emitter-collector voltage (I_C, V_{EC}) for the *pnp*.

Logic gates and linear amplifiers use very different operating points. For example, the circuit in Fig. 5.32(a) can be used as either a logic inverter or a linear amplifier depending upon our choice of operating points. The voltage transfer characteristic (VTC) for the circuit appears in Fig. 5.33(a). For low values of v_{BE}, the transistor is nearly cut off, and the output voltage is 5 V, corresponding to a binary "1" in a logic applications. As v_{BE} increases above 0.6 V, the output drops quickly and reaches its "on-state" voltage of 0.18 V in for v_{BE} greater than 0.8 V. The BJT is now operating in its saturation region, and the small "on-voltage" would correspond to a "0" in binary logic. These two logic states are also shown on the transistor output characteristics in Fig. 5.33(b). When the transistor is "on," it conducts a substantial current, and v_{CE} falls to 0.18 V. When the transistor is off, v_{CE} equals 5 V. We study the design of logic gates in detail in Chapters 6–9.

For amplifier applications, the Q-point is located in the region of high slope (high gain) of the voltage transfer characteristic, also indicated in Fig. 5.33(a). At this operating point, the transistor is operating in the forward-active region, the region in which high voltage, current and/or power gain can be achieved. To establish this Q-point, a **dc bias** V_{BE} is applied to the base as in Fig. 5.32(b), and a small ac signal v_{be} is added to vary the base voltage around the bias value.[9] The variation in total base-emitter voltage v_{BE} causes the collector current to change, and an amplified replica of the

---

[9] Remember $v_{BE} = V_{BE} + v_{be}$

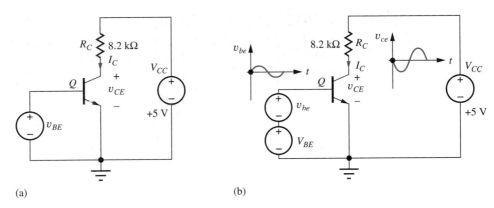

(a)   (b)

Figure 5.32 (a) Circuit for a logic inverter. (b) The same transistor used as a linear amplifier.

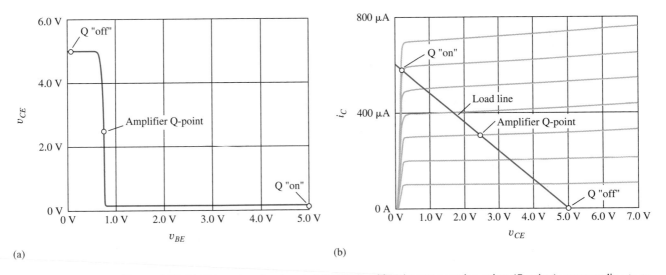

(a)   (b)

Figure 5.33 (a) Voltage transfer characteristic (VTC) with quiescent operating points (Q-points) corresponding to an "on-switch," an amplifier, and an "off switch." (b) The same three operating points located on the transistor output characteristics.

ac input voltage appears at the collector. Our study of the design of transistor amplifiers begins in Chapter 13 of this text.

In Secs. 5.6 to 5.10, we presented simplified models for the four operating regions of the BJT. In general, we will not explicitly insert the simplified circuit models for the transistor into the circuit but instead will use the mathematical relationships that were derived for the specific operating region of interest. For example, in the forward-active region, the results $V_{BE} = 0.7$ V and $I_C = \beta_F I_B$ will be utilized to directly simplify the circuit analysis.

In the dc biasing examples that follow, the Early voltage is assumed to be infinite. In general, including the Early voltage in bias circuit calculations substantially increases the complexity of the analysis but typically changes the results by less than 10 percent. In most cases, the tolerances on the values of resistors and independent sources will be 5 to 10 percent, and the transistor current-gain β_F may vary by a factor of 4:1 to 10:1. For example, the current gain of a transistor may be specified to be a minimum of 50 with a typical value of 100 but no upper bound specified. These tolerances will swamp out any error due to neglect of the Early voltage. Thus, basic hand design will be done ignoring the Early effect, and if more precision is needed, the calculations can be refined through SPICE analysis.

5.11.1 FOUR-RESISTOR BIAS NETWORK

One of the best circuits for stabilizing the Q-point of a transistor is the four-resistor bias network in Fig. 5.34. R_1 and R_2 form a resistive voltage divider across the power supplies (12 V and 0 V) and attempt to establish a fixed voltage at the base of transistor Q_1. R_E and R_C are used to define the emitter current and collector-emitter voltage of the transistor.

Our goal is to find the Q-point of the transistor: (I_C, V_{CE}). The first steps in analysis of the circuit in Fig. 5.34(a) are to split the power supply into two equal voltages, as in Fig. 5.34(b), and then to simplify the circuit by replacing the base-bias network by its Thévenin equivalent circuit, as shown in Fig. 5.34(c). V_{EQ} and R_{EQ} are given by

$$V_{EQ} = V_{CC} \frac{R_1}{R_1 + R_2} \qquad\qquad R_{EQ} = \frac{R_1 R_2}{R_1 + R_2} \qquad\qquad (5.49)$$

For the values in Fig. 5.34(c), $V_{EQ} = 4$ V and $R_{EQ} = 12$ kΩ.

Detailed analysis begins by assuming a region of operation in order to simplify the BJT model equations. Because the most common region of operation for this bias circuit is the forward-active region, we will assume it to be the region of operation. Using Kirchhoff's voltage law around loop 1:

$$V_{EQ} = R_{EQ} I_B + V_{BE} + R_E I_E \qquad \text{or} \qquad 4 = 12{,}000 I_B + V_{BE} + 16{,}000 I_E \qquad (5.50)$$

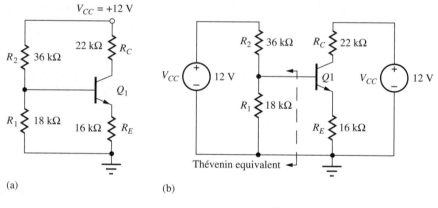

(a)

(b)

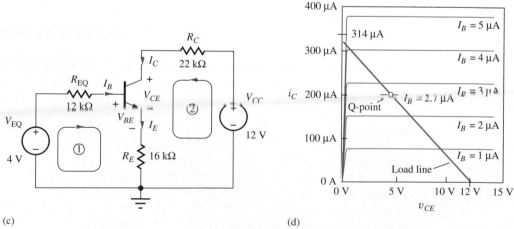

(c)

(d)

Figure 5.34 (a) The four-resistor bias network. (Assume $\beta_F = 75$ for analysis.) (b) Four-resistor bias circuit with replicated sources. (c) Thévenin simplification of the four-resistor bias network. (Assume $\beta_F = 75$.) (d) Load line for the four-resistor bias circuit.

Because we are assuming forward-active region operation, we have $V_{BE} = 0.7$ V and $I_E = (\beta_F + 1)I_B$, and Eq. (5.50) becomes

$$4 = 12{,}000 I_B + 0.7 + 16{,}000(\beta_F + 1)I_B \tag{5.51}$$

Using $\beta_F = 75$ and solving for I_B yields

$$I_B = \frac{4\text{ V} - 0.7\text{ V}}{1.23 \times 10^6\ \Omega} - 2.68\ \mu\text{A} \qquad I_C = \beta_F I_B = 201\ \mu\text{A} \qquad I_E = (\beta_F + 1)I_B = 204\ \mu\text{A}$$

To find V_{CE}, loop 2 is used:

$$V_{CE} = V_{CC} - R_C I_C - R_E I_E = V_{CC} - \left(R_C + \frac{R_E}{\alpha_F} \right) I_C \tag{5.52}$$

because $I_E = I_C/\alpha_F$. For the values in the circuit,

$$V_{CE} = 12 - 38{,}200 I_C = 12\text{ V} - 7.68\text{ V} = 4.32\text{ V} \tag{5.53}$$

All the calculated currents are greater than zero, and using the result in Eq. (5.53), $V_{BC} = V_{BE} - V_{CE} = 0.7 - 4.32 = -3.62$ V. Thus, the base-collector junction is reverse-biased, and the assumption of forward-active region operation was correct. The Q-point resulting from our analysis is $(205\ \mu\text{A}, 4.17\text{ V})$.

Before leaving this bias example, let us draw the load line for the circuit and locate the Q-point on the output characteristics. The load-line equation for this circuit already appeared as Eq. (5.52):

$$V_{CE} = V_{CC} - \left(R_C + \frac{R_E}{\alpha_F} \right) I_C = 12 - 38{,}200 I_C \tag{5.54}$$

Two points are needed to plot the load line. Choosing $I_C = 0$ yields $V_{CE} = 12$ V, and picking $V_{CE} = 0$ yields $I_C = 314\ \mu\text{A}$. The resulting load line is plotted on the transistor common-emitter output characteristics in Fig. 5.34(d). The base current was already found to be 2.7 μA, and the intersection of the $I_B = 2.7$-μA characteristic with the load line defines the Q-point. In this case we must estimate the location of the $I_B = 2.7$-μA curve.

EXERCISE: Find the Q-point for the circuit in Fig. 5.34(d) if $R_1 = 180$ kΩ and $R_2 = 360$ kΩ.

ANSWERS: $(185\ \mu\text{A}, 4.93\text{ V})$

5.11.2 DESIGN OBJECTIVES FOR THE FOUR-RESISTOR BIAS NETWORK

Now that we have analyzed a circuit involving the four-resistor bias network, let us explore the design objectives of this bias technique by solving for the emitter current from Eq. (5.50):

$$I_E = \frac{V_{EQ} - V_{BE} - R_{EQ}I_B}{R_E} \cong \frac{V_{EQ} - V_{BE}}{R_E} \qquad \text{for } R_{EQ}I_B \ll (V_{EQ} - V_{BE}) \tag{5.55}$$

The value of the Thévenin equivalent resistance R_{EQ} is normally designed to be small enough to neglect the voltage drop caused by the base current flowing through R_{EQ}. Under these conditions, I_E is set by the combination of V_{EQ}, V_{BE}, and R_E. In addition, V_{EQ} is normally designed to be large enough that small variations in the assumed value of V_{BE} will not materially affect the value of I_E.

In the original bias circuit reproduced in Fig. 5.35, the assumption that the voltage drop $I_B R_{EQ} \ll (V_{EQ} - V_{BE})$ is equivalent to assuming $I_B \ll I_2$ so that $I_1 \cong I_2$. For this case, the base current of Q_1 does not disturb the voltage divider action of R_1 and R_2. Using the approximate expression in Eq. (5.55) estimates the emitter current in the circuit in Fig. 5.34 to be

$$I_E = \frac{4\text{ V} - 0.7\text{ V}}{16{,}000\ \Omega} = 206\ \mu\text{A}$$

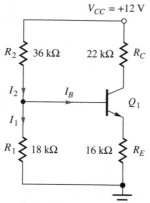

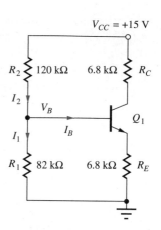

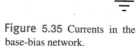

Figure 5.35 Currents in the base-bias network.

Figure 5.36 Final bias circuit design for a Q-point of (750 μA, 5 V).

which is essentially the same as the result that was calculated using the more exact expression. This is the result that should be achieved with a proper bias network design. If the Q-point is independent of I_B, it will also be independent of current gain β (a poorly controlled transistor parameter). The emitter current will then be approximately the same for a transistor with a current gain of 50 or 500.

Generally, a very large number of possible combinations of R_1 and R_2 will yield the desired value of V_{EQ}. An additional constraint is needed to finalize the design choice. A useful choice is to limit the current used in the base-voltage-divider network by choosing $I_2 \leq I_C/5$. This choice ensures that the power dissipated in bias resistors R_1 and R_2 is less than 17 percent of the total quiescent power consumed by the circuit and at the same time ensures that $I_2 \gg I_B$ for $\beta \geq 50$.

EXERCISE: Show that choosing $I_2 = I_C/5$ is equivalent to setting $I_2 = 10I_B$ when $\beta_F = 50$.

EXERCISE: Find the Q-point for the circuit in Fig. 5.34(a) if β_F is 500.

ANSWERS: (206 μA, 4.18 V)

DESIGN EXAMPLE 5.9   FOUR-RESISTOR BIAS DESIGN

Here we explore the design of the network most commonly utilized to bias the BJT—the four-resistor bias circuit.

PROBLEM   Design a four resistor bias circuit to give a Q-point of (750 μA, 5 V) using a 15-V supply with an *npn* transistor having a minimum current gain of 100.

SOLUTION   **Known Information and Given Data:** The bias circuit in Fig. 5.35 with $V_{CC} = 15$ V; the *npn* transistor has $\beta_F = 100$, $I_C = 750$ μA, and $V_{CE} = 5$ V.

Unknowns: Base voltage V_B, voltages across resistors R_E and R_C; values for R_1, R_2, R_C, and R_E

Approach: First, partition V_{CC} between the collector-emitter voltage of the transistor and the voltage drops across R_C and R_E. Next, choose currents I_1 and I_2 for the base bias network. Finally, use the assigned voltages and currents to calculate the unknown resistor values.

Assumptions: The transistor is to operate in the forward-active region. The base-emitter voltage of the transistor is 0.7 V. The Early voltage is infinite.

Analysis: To calculate values for the resistors, we must know the voltage across the emitter and collector resistors and the voltage V_B. V_{CE} is designed to be 5 V. One common choice is to divide the remaining power supply voltage $(V_{CC} - V_{CE}) = 10$ V equally between R_E and R_C. Thus, $V_E = 5$ V and $V_C = 5 + V_{CE} = 10$ V. The values of R_C and R_E are then given by

$$R_C = \frac{V_{CC} - V_C}{I_C} = \frac{5 \text{ V}}{750 \text{ }\mu\text{A}} = 6.67 \text{ k}\Omega \qquad \text{and} \qquad R_E = \frac{V_E}{I_E} = \frac{5 \text{ V}}{758 \text{ }\mu\text{A}} = 6.60 \text{ k}\Omega$$

The base voltage is given by $V_B = V_E + V_{BE} = 5.7$ V. For forward-active region operation, we know that $I_B = I_C/\beta_F = 750 \text{ }\mu\text{A}/100 = 7.5 \text{ }\mu\text{A}$. Now choosing $I_2 = 10I_B$, we have $I_2 = 75 \text{ }\mu\text{A}$, $I_1 = 9I_B = 67.5 \text{ }\mu\text{A}$, and R_1 and R_2 can be determined:

$$R_1 = \frac{V_B}{9I_B} = \frac{5.7 \text{ V}}{67.5 \text{ }\mu\text{A}} = 84.4 \text{ k}\Omega \qquad R_2 = \frac{V_{CC} - V_B}{10I_B} = \frac{15 - 5.7 \text{ V}}{75 \text{ }\mu\text{A}} = 124 \text{ k}\Omega \quad (5.56)$$

Check of Results: We have $V_{BE} = 0.7$ V and $V_{BC} = 5.7 - 10 = -4.3$ V, which are consistent with the forward-active region assumption.

Discussion: The values calculated above should yield a Q-point very close to the design goals. However, if we were going to build this circuit in the laboratory, we must use standard values for the resistors. In order to complete the design, we refer to the table of resistor values in Appendix A. There we find that the closest available values are $R_1 = 82$ kΩ, $R_2 = 120$ kΩ, $R_E = 6.8$ kΩ, and $R_C = 6.8$ kΩ.

Computer-Aided Analysis: SPICE can now be used as a tool to check our design. The final design using these values appears in Fig. 5.36 for which SPICE (with IS $= 2 \times 10^{-15}$ A) predicts the Q-point to be (734 μA, 4.97 V), with $V_{BE} = 0.65$ V. We neglected the Early effect in our hand calculations, but SPICE represents an easy way to check this assumption. If we set VAF $= 75$ V in SPICE, keeping the other parameters the same, the new Q-point is (737 μA, 4.93 V). Clearly, the changes caused by the Early effect are negligible.

EXERCISE: Redesign the four resistor bias circuit to yield $I_C = 75$ μA and $V_{CE} = 5$ V.

ANSWERS: (66.7 kΩ, 66.0 kΩ, 844 kΩ, 1.24 MΩ) → (68 kΩ, 68 kΩ, 820 kΩ, 1.20 MΩ)

DESIGN NOTE FOUR-RESISTOR BIAS DESIGN

1. Choose the Thévenen equivalent base voltage V_{EQ}: $\qquad \dfrac{V_{CC}}{4} \le V_{EQ} \le \dfrac{V_{CC}}{2}$

2. Select R_1 to set $I_1 = 9I_B$: $\qquad R_1 = \dfrac{V_{EQ}}{9I_B}$

3. Select R_2 to set $I_2 = 10I_B$: $\qquad R_2 = \dfrac{V_{CC} - V_{EQ}}{10I_B}$

4. R_E is determined by V_{EQ} and the desired collector current: $\qquad R_E \cong \dfrac{V_{EQ} - V_{BE}}{I_C}$

5. R_C is determined by the desired collector-emitter voltage: $\qquad R_C \cong \dfrac{V_{CC} - V_{CE}}{I_C} - R_E$

EXAMPLE 5.10 ANALYSIS OF A TRANSISTOR OPERATING IN SATURATION

This example demonstrates an analysis in which the assumption of forward-active region operation is discovered to be incorrect, and a second analysis iteration is required. We explore the impact of changing the collector resistor in the circuit of Fig. 5.34(a) from 22 kΩ to 56 kΩ, as shown in Fig. 5.37. (Perhaps the resistor color code was misread by the builder of the circuit.)

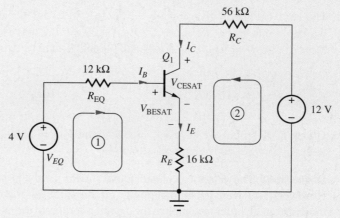

Figure 5.37 Bias circuit with collector resistor R_C increased to 56 kΩ ($\beta_F = 75$).

PROBLEM Find the Q-point for the transistor in Fig. 5.37.

SOLUTION **Known Information and Given Data:** Simplified equivalent circuit in Fig. 5.37 with $V_{EQ} = 4$ V, $R_{EQ} = 12$ kΩ, $V_{CC} = 12$ V, $R_E = 16$ kΩ, and $R_C = 56$ kΩ

Unknowns: I_C, V_{CE}

Approach: Assume a region of operation and calculate the Q-point; check answer to see if it is consistent with the assumptions. Analyze the input loop to find I_B, I_C, and I_E. Use currents in the output loop to find V_{CE}.

Assumptions: Forward-active region of operation with $V_{BE} = 0.7$ V; $V_A = \infty$

Analysis: The analysis starts by analyzing input loop 1 in Fig. 5.37, which is identical to that in Fig. 5.34(c). Therefore I_B is determined by Eq. (5.51) with $\beta_F = 75$:

$$I_B = \frac{4 \text{ V} - 0.7 \text{ V}}{1.21 \times 10^6 \text{ } \Omega} = 2.73 \text{ } \mu\text{A} \qquad I_C = \beta_F I_B = 205 \text{ } \mu\text{A}$$

$$I_E = (\beta_F + 1)I_B = 208 \text{ } \mu\text{A}$$

Using loop 2 to determine an expression for V_{CE} as in Eq. (5.52) yields:

$$V_{CE} = V_{CC} - \left(R_C + \frac{R_E}{\alpha_F} \right) I_C = 12 - 72,200 I_C = -2.80 \text{ V—Oops!}$$

The calculated Q-point is $(-2.80$ V, 205 μA$)$.

Check of Results: The calculated value of V_{CE} is negative, which violates the assumption of forward-active region operation that requires $V_{CE} \geq V_{BE}$. (In addition, it is physically impossible for V_{CE} to become negative in this circuit.) Therefore, we must choose a new region of operation and reanalyze the circuit.

Analysis—Second Iteration: Because V_{CE} was found to be negative, our second analysis attempt will assume that Q_1 is saturated (V_{CE} as small as possible. We will need to assume a value for V_{CESAT}.) Writing a new set of equations for loops 1 and 2:

$$4 = 12,000 I_B + V_{BESAT} + 16,000 I_E$$
$$12 = 56,000 I_C + V_{CESAT} + 16,000 I_E$$

$$(5.57)$$

If we substitute assumed values of $V_{BESAT} = 0.75$ V and $V_{CESAT} = 0.05$ V, and use $I_E = I_B + I_C$, then simultaneous solution of Eqs. (5.57) gives

$$I_C = 160 \ \mu A \qquad I_B = 24 \ \mu A \qquad \text{and} \qquad I_E = I_C + I_B = 184 \ \mu A$$

The Q-point is (0.05 V, 160 μA).

Check of Results: The three terminal currents are all positive, and $I_C/I_B < \beta_F$ (that is, $\beta_{FOR} < \beta_F$). Therefore, the assumption of saturation region operation is correct. The values of V_{BESAT} and V_{CESAT} can be calculated using Eqs. (5.57) as a check on the hand analysis and are found to be close to the assumed values: $V_{BESAT} = 0.77$ V and $V_{CESAT} = 0.096$ V.

Discussion: This problem provides an example in which the initial assumed region of operation was incorrect, and a second analysis iteration was required to find the correct Q-point.

Computer-Aided Analysis: This problem is another good place to use SPICE analysis to check our hand calculations. SPICE simulation yields

$$I_C = 160 \ \mu A \qquad I_B = 28 \ \mu A \qquad I_E = 188 \ \mu A$$

The slight discrepancies are caused by the differences in V_{BESAT} and V_{CESAT} between our hand analysis and SPICE.

EXERCISE: What is the largest value for resistor R_C that can be used in the circuit in Fig. 5.37 if the transistor is to remain biased in the forward-active region ($V_{CE} = 0.7$ V)?

ANSWER: 38.9 kΩ

EXERCISE: Substitute I_C, I_B, and I_E into Eqs. (5.57) and verify the values of V_{BESAT} and V_{CESAT}.

EXAMPLE 5.11   **TWO-RESISTOR BIASING**

The two-resistor feedback bias circuit, that was first introduced for the MOSFET in Ex. 4.7, can also be used to bias the bipolar transistor. In this example, we find an analysis of the two-resistor circuit applied to the problem of biasing a *pnp* transistor.

PROBLEM   Find the Q-point for the *pnp* transistor in the two-resistor bias circuit in Fig. 5.38. Assume $\beta_F = 50$.

SOLUTION   **Known Information and Given Data:** Two-resistor bias circuit in Fig. 5.38 with a *pnp* transistor with $\beta_F = 50$

Unknowns: I_C, V_{CE}

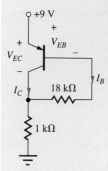

Figure 5.38 Two-resistor bias circuit with a *pnp* transistor.

Approach: Assume a region of operation and analyze the circuit to determine the Q-point; check answer to see if it is consistent with the assumptions.

Assumptions: Forward-active region operation with $V_{EB} = 0.7$ V and $V_A = \infty$

Analysis: The voltages and currents are first carefully labeled as in Fig. 5.38. To find the Q-point, an equation is written involving V_{EB}, I_B, and I_C:

$$9 = V_{EB} + 18{,}000I_B + 1000(I_C + I_B) \tag{5.58}$$

Applying the assumption of forward-active region operation with $\beta_F = 50$ and $V_{EB} = 0.7$ V,

$$9 = 0.7 + 18{,}000I_B + 1000(51)I_B \tag{5.59}$$

and

$$I_B = \frac{9\text{ V} - 0.7\text{ V}}{69{,}000\ \Omega} = 120\ \mu\text{A} \qquad I_C = 50I_B = 6.01\text{ mA} \tag{5.60}$$

The emitter-collector voltage is given by

$$V_{EC} = 9 - 1000(I_C + I_B) = 2.88\text{ V} \qquad \text{and} \qquad V_{BC} = 2.18\text{ V} \tag{5.61}$$

The Q-point is $(I_C, V_{EC}) = (6.01\text{ mA}, 2.88\text{ V})$.

Check of Results: Because I_B, I_C, and V_{BC} are all greater than zero, the assumption of forward-active region operation is valid, and the Q-point is correct.

Computer-Aided Analysis: For this circuit, SPICE simulation yields (6.04 mA, 2.95 V), which agrees with the Q-point found from our hand calculations.

EXERCISE: What is the Q-point if the 18 kΩ resistor is increased to 36 kΩ?

ANSWER: (4.77 mA, 4.13 V)

EXERCISE: Draw the two-resistor bias circuit (a "mirror image" of Fig. 5.38) that would be used to bias an *npn* transistor from a single +9-V supply using the same two resistor values as in Fig. 5.38.

ANSWER: See circuit topology in Fig. P5.93.

The bias circuit examples that have been presented in this section have only scratched the surface of the possible techniques that can be used to bias *npn* and *pnp* transistors. However, the analysis techniques have illustrated the basic approaches that need to be followed in order to determine the Q-point of any bias circuit.

5.12 TOLERANCES IN BIAS CIRCUITS

When a circuit is actually built in discrete form in the laboratory or fabricated as an integrated circuit, the components and device parameters all have tolerances associated with their values. Discrete resistors can easily be purchased with 10 percent, 5 percent, or 1 percent tolerances, whereas typical resistors in ICs can exhibit even wider variations (±30 percent). Power supply voltage tolerances are often 5 to 10 percent.

For a given bipolar transistor type, parameters such as current gain may cover a range of 5:1 to 10:1, or may be specified with only a nominal value and lower bound. The BJT (or diode) saturation current may vary by a factor varying from 10:1 to 100:1, and the Early voltage may vary by ±20 percent. In FET circuits, the values of threshold voltage and the transconductance parameter can vary widely, and in op-amp circuits all the op-amp parameters (for example, open-loop gain, input resistance, output resistance, input bias current, unity gain frequency, and the like) typically exhibit wide specification ranges.

In addition to these initial value uncertainties, the values of the circuit components and parameters change as temperature changes and the circuit ages. It is important to understand the effect of these variations on our circuits and be able to design circuits that will continue to operate correctly in the face of these element variations. Worst-case analysis and Monte Carlo analysis, introduced in Chapter 1, are two approaches that can be used to quantify the effects of tolerances on circuit performance.

5.12.1 WORST-CASE ANALYSIS

Worst-case analysis is often used to ensure that a design will function under an expected set of component variations. In Q-point analysis, for example, the values of components are simultaneously pushed to their various extremes in order to determine the worst possible range of Q-point values. Unfortunately, a design based on worst-case analysis is usually an unnecessary overdesign and economically undesirable, but it is important to understand the technique and its limitations.

EXAMPLE 5.12   **WORST-CASE ANALYSIS OF THE FOUR-RESISTOR BIAS NETWORK**

Now we explore the application of worst-case analysis to the four-resistor bias network with a given set of tolerances assigned to the elements. In Ex. 5.13, the bounds generated by the worst-case analysis will be compared to a statistical sample of the possible network realizations using Monte Carlo analysis.

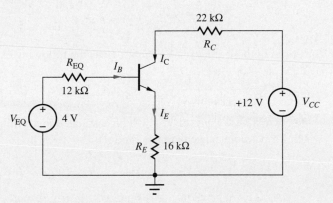

Figure 5.39 Simplified four-resistor bias circuit of Fig. 5.34(c) assuming nominal element values.

PROBLEM   Find the worst-case values of I_C and V_{CE} for the transistor in Fig. 5.39. The circuit in Fig. 5.39 is the simplified version of the four-resistor bias circuit in Figs. 5.33. Assume that the 12-V power supply has a 5 percent tolerance and the resistors have 10 percent tolerances. Also, assume that the transistor current gain has a nominal value of 75 with a 50 percent tolerance.

SOLUTION   **Known Information and Given Data:** Simplified version of the four-resistor bias circuit in Fig. 5.39; 5 percent tolerance on V_{CC}; 10 percent tolerance for each resistor; current $\beta_{FO} = 75$ with a 50 percent tolerance

Unknowns: Minimum and maximum values of I_C and V_{CE}

Approach: Find the worst-case values of V_{EQ} and R_{EQ}; use the results to find the extreme values of the base and collector current; use the collector current values to find the worst-case values of collector-emitter voltage

Assumptions: To simplify the analysis, assume that the voltage drop in R_{EQ} can be neglected and β_F is large so that I_C is given by

$$I_C \cong I_E = \frac{V_{EQ} - V_{BE}}{R_E} \tag{5.62}$$

Assume V_{BE} is fixed at 0.7 V.

Analysis: To make I_C as large as possible, V_{EQ} should be at its maximum extreme and R_E should be a minimum value. To make I_C as small as possible, V_{EQ} should be minimum and R_E should be a maximum value. Variations in V_{BE} are assumed to be negligible but could also be included if desired.

The extremes of R_E are $0.9 \times 16 \text{ k}\Omega = 14.4 \text{ k}\Omega$, and $1.1 \times 16 \text{ k}\Omega = 17.6 \text{ k}\Omega$. The extreme values of V_{EQ} are somewhat more complicated:

$$V_{EQ} = V_{CC} \frac{R_1}{R_1 + R_2} = \frac{V_{CC}}{1 + \dfrac{R_2}{R_1}} \tag{5.63}$$

To make V_{EQ} as large as possible, the numerator of Eq. (5.63) should be large and the denominator small. Therefore, V_{CC} and R_1 must be as large as possible and R_2 as small as possible. Conversely, to make V_{EQ} as small as possible, V_{CC} and R_1 must be small and R_2 must be large. Using this approach, the maximum and minimum values of V_{EQ} are

$$V_{EQ}^{\max} = \frac{12 \text{ V}(1.05)}{1 + \dfrac{36 \text{ k}\Omega(0.9)}{18 \text{ k}\Omega(1.1)}} = 4.78 \text{ V} \qquad \text{and} \qquad V_{EQ}^{\min} = \frac{12 \text{ V}(.95)}{1 + \dfrac{36 \text{ k}\Omega(1.1)}{18 \text{ k}\Omega(0.9)}} = 3.31 \text{ V}$$

Substituting these values in Eq. (5.62) gives the following extremes for I_C:

$$I_C^{\max} = \frac{4.78 \text{ V} - 0.7 \text{ V}}{14,400 \, \Omega} = 283 \text{ μA} \qquad \text{and} \qquad I_C^{\min} = \frac{3.31 \text{ V} - 0.7 \text{ V}}{17,600 \, \Omega} = 148 \text{ μA}$$

The worst-case range of V_{CE} will be calculated in a similar manner, but we must be careful to watch for possible cancellation of variables:

$$V_{CE} = V_{CC} - I_C R_C - I_E R_E \cong V_{CC} - I_C R_C - \frac{V_{EQ} - V_{BE}}{R_E} R_E \tag{5.64}$$

$$V_{CE} \cong V_{CC} - I_C R_C - V_{EQ} + V_{BE}$$

The maximum value of V_{CE} in Eq. (5.64) occurs for minimum I_C and minimum R_C and vice versa. Using (5.91), the extremes of V_{CE} are

$$V_{CE}^{\max} \cong 12 \text{ V}(1.05) - (148 \text{ μA})(22 \text{ k}\Omega \times 0.9) - 3.31 \text{ V} + 0.7 \text{ V} = 7.06 \text{ V} \quad ✔$$

$$V_{CE}^{\min} \cong 12 \text{ V}(0.95) - (283 \text{ μA})(22 \text{ k}\Omega \times 1.1) - 4.78 \text{ V} + 0.7 \text{ V} = 0.471 \text{ Saturated!}$$

Check of Results: The transistor remains in the forward-active region for the upper extreme, but the transistor saturates (weakly) at the lower extreme. Because the forward-active region assumption is violated in the latter case, the calculated values of V_{CE} and I_C would not actually be correct for this case.

Discussion: Note that the worst-case values of I_C differ by a factor of almost 2:1! The maximum I_C is 38 percent greater than the nominal value of 210 μA, and the minimum value is 37 percent below the nominal value. The failure of the bias circuit to maintain the transistor in the desired region of operation for the worst-case values is evident.

5.12.2 MONTE CARLO ANALYSIS

In a real circuit, the parameters will have some statistical distribution, and it is unlikely that the various components will all reach their extremes at the same time. Thus, the worst-case analysis technique will overestimate (often badly) the extremes of circuit behavior. A better approach is to attack the problem statistically using the method of Monte Carlo analysis.

As discussed in Chapter 1, **Monte Carlo analysis** uses randomly selected versions of a given circuit to predict its behavior from a statistical basis. For Monte Carlo analysis, values for each parameter in the circuit are selected at random from the possible distributions of parameters, and the circuit is then analyzed using the randomly selected element values. Many random parameter sets are generated, and the statistical behavior of the circuit is built up from analysis of the many test cases.

In Ex. 5.13, an Excel spreadsheet will be used to perform a Monte Carlo analysis of the four-resistor bias circuit. As discussed in Chapter 1, Excel contains the function RAND( ), which generates random numbers uniformly distributed between 0 and 1, but for Monte Carlo analysis, the mean must be centered on R_{nom} and the width of the distribution set to $(2\varepsilon) \times R_{\text{nom}}$:

$$R = R_{\text{nim}}[1 + 2\varepsilon(\text{RAND}() - 0.5)] \tag{5.65}$$

EXAMPLE 5.13   **MONTE CARLO ANALYSIS OF THE FOUR-RESISTOR BIAS NETWORK**

Now, let us compare the worst-case results from Ex. 5.12 to a statistical sample of 500 randomly generated realizations of the transistor embedded in the four-resistor bias network.

PROBLEM Perform a Monte Carlo analysis to determine statistical distributions for the collector current and collector-emitter voltage for the four-resistor circuit in Figs. 5.34 and 5.39 with a 5 percent tolerance on V_{CC}, 10 percent tolerances for each resistor and a 50 percent tolerance on the current gain $\beta_{FO} = 75$.

SOLUTION **Known Information and Given Data:** Circuit in Fig. 5.34(a) as simplified in Fig. 5.39; 5 percent tolerance on the 12-V power supply V_{CC}; 10 percent tolerance on each resistor; current $\beta_{FO} = 75$ with a 50 percent tolerance

Unknowns: Statistical distributions of I_C and V_{CE}

Approach: To perform a Monte Carlo analysis of the circuit in Fig. 5.34, random values are assigned to V_{CC}, R_1, R_2, R_C, R_E, and β_F and then used to determine I_C and V_{CE}. A spreadsheet is used to make the repetitive calculations. Since the computer is performing the calculations, the most exact formulas will be used in the analyses.

Assumptions: V_{BE} is fixed at 0.7 V. Random values are statistically independent of each other.

Computer-Aided Analysis: Using the tolerances from the worst-case analysis, the power supply, resistors, and current gain are represented as

$$1. \quad V_{CC} = 12(1 + 0.1(\text{RAND}(\,) - 0.5))$$

$$2. \quad R_1 = 18{,}000(1 + 0.2(\text{RAND}(\,) - 0.5))$$

$$3. \quad R_2 = 36{,}000(1 + 0.2(\text{RAND}(\,) - 0.5)) \qquad (5.66)$$

$$4. \quad R_E = 16{,}000(1 + 0.2(\text{RAND}(\,) - 0.5))$$

$$5. \quad R_C = 22{,}000(1 + 0.2(\text{RAND}(\,) - 0.5))$$

$$6. \quad \beta_F = 75(1 + (\text{RAND}(\,) - 0.5))$$

Remember, each variable evaluation must invoke a separate call of the function RAND( ) so that the random values will be independent of each other.

In the spreadsheet results presented in Fig. 5.40, the random elements in Eqs. (5.66) are used to evaluate the equations that characterize the bias circuit:

$$7. \quad V_{EQ} = V_{CC} \frac{R_1}{R_1 + R_2} \qquad\qquad 10. \quad I_C = \beta_F I_B$$

$$8. \quad R_{EQ} = \frac{R_1 R_2}{R_1 + R_2} \qquad\qquad 11. \quad I_E = \frac{I_C}{\alpha_F} \qquad (5.67)$$

$$9. \quad I_B = \frac{V_{EQ} - V_{BE}}{R_{EQ} + (\beta_F + 1)R_E} \qquad 12. \quad V_{CE} = V_{CC} - I_C R_C - I_E R_E$$

Because the computer is doing the work, the complete expressions rather than the approximate relations for the various calculations are used in Eqs. (5.67).[10] Once Eqs. (5.66) and (5.67) have been entered into one row of the spreadsheet, that row can be copied into as many additional rows as the number of statistical cases that are desired. The analysis is automatically repeated for the random selections to build up the statistical distributions, with each row representing one analysis of the circuit. At the end of the columns, the mean and standard deviation can be calculated using built-in spreadsheet functions, and the overall spreadsheet data can be used to build histograms for the circuit performance.

An example of a portion of the spreadsheet output for 25 cases of the circuit in Fig. 5.39 is shown in Fig. 5.40, whereas the full results of the analysis of 500 cases of the four-resistor bias circuit are given in the histograms for I_C and V_{CE} in Fig. 5.41. The mean values for I_C and V_{CE} are 207 μA and 4.06 V, respectively, which are close to the values originally estimated from the nominal circuit elements. The standard deviations are 19.6 μA and 0.64 V, respectively.

Check of Results and Discussion: The worst-case calculations from Sec. 5.12.1 are indicated by the arrows in the figures. It can be seen that the worst-case values of V_{CE} lie well beyond the edges of the statistical distribution, and that saturation does not actually occur for the worst statistical case evaluated. If the Q-point distribution results in the histograms in Fig. 5.41 were not sufficient to meet the design criteria, the parameter tolerances could be changed and the Monte Carlo simulation redone. For example, if too large a fraction of the circuits failed to be within some specified limits, the tolerances could be tightened by specifying more expensive, higher accuracy resistors.

---

[10] Note that V_{BE} could also be treated as a random variable.

Monte Carlo Spreadsheet

Case #	V_{CC} (1)	R_1 (2)	R_2 (3)	R_E (4)	R_C (5)	β_F (6)	V_{EQ} (7)	R_{EQ} (8)	I_B (9)	I_C (10)	V_{CE} (12)
1	12.277	16827	38577	15780	23257	67.46	3.729	11716	2.87E-06	1.93E-04	4.687
2	12.202	18188	32588	15304	23586	46.60	4.371	11673	5.09E-06	2.37E-04	2.891
3	11.526	16648	35643	14627	20682	110.73	3.669	11348	1.87E-06	2.07E-04	4.206
4	11.658	17354	33589	14639	22243	44.24	3.971	11442	5.00E-06	2.21E-04	3.420
5	11.932	19035	32886	16295	20863	62.34	4.374	12056	3.61E-06	2.25E-04	3.500
6	11.857	18706	32615	15563	21064	60.63	4.322	11888	3.83E-06	2.32E-04	3.286
7	11.669	18984	39463	17566	21034	42.86	3.790	12818	4.07E-06	1.75E-04	4.859
8	12.222	19291	37736	15285	22938	63.76	4.135	12765	3.53E-06	2.25E-04	3.577
9	11.601	17589	34032	17334	23098	103.07	3.953	11596	1.85E-06	1.90E-04	3.873
10	11.533	17514	33895	17333	19869	71.28	3.929	11547	2.63E-06	1.88E-04	4.505
11	11.436	19333	34160	15107	22593	68.20	4.133	12346	3.34E-06	2.28E-04	2.797
12	11.962	18810	33999	15545	22035	53.69	4.261	12110	4.25E-06	2.28E-04	3.330
13	11.801	19610	37917	14559	21544	109.65	4.023	12925	2.11E-06	2.31E-04	3.426
14	12.401	17947	34286	15952	21086	107.84	4.261	11780	2.09E-06	2.26E-04	4.002
15	11.894	16209	35321	17321	23940	45.00	3.741	11111	3.89E-06	1.75E-04	4.607
16	12.329	16209	37873	16662	23658	112.01	3.695	11351	1.63E-06	1.83E-04	4.923
17	11.685	19070	35267	15966	21864	64.85	4.101	12377	3.29E-06	2.13E-04	3.559
18	11.456	18096	37476	15529	20141	91.14	3.730	12203	2.17E-06	1.98E-04	4.370
19	12.527	18752	38261	15186	21556	69.26	4.120	12584	3.26E-06	2.26E-04	4.180
20	12.489	17705	36467	17325	20587	83.95	4.082	11919	2.35E-06	1.97E-04	4.979
21	11.436	18773	34697	16949	21848	65.26	4.015	12182	3.01E-06	1.96E-04	3.768
22	11.549	16830	38578	16736	19942	109.22	3.508	11718	1.57E-06	1.71E-04	5.247
23	11.733	16959	39116	15944	21413	62.82	3.548	11830	2.86E-06	1.80E-04	4.965
24	11.738	18486	35520	17526	20455	70.65	4.018	12158	2.70E-06	1.90E-04	4.457
25	11.679	18908	38236	15160	21191	103.12	3.864	12652	2.05E-06	2.12E-04	3.958
Mean	11.848	18014	35102	15973	21863	67.30	4.024	11885	3.44E-06	2.09E-04	3.880
Std. Dev.	0.296	958	2596	1108	1309	23.14	0.264	520	1.14E-06	2.18E-05	0.657

(X) = Equation number in text

Figure 5.40 Example of a Monte Carlo analysis using a spreadsheet.

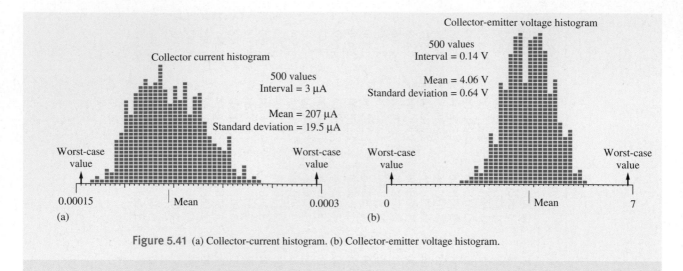

Figure 5.41 (a) Collector-current histogram. (b) Collector-emitter voltage histogram.

Some implementations of the SPICE circuit analysis program—PSPICE, for example—actually contain a Monte Carlo option in which a full circuit simulation is automatically performed for any number of randomly selected test cases. These programs are a powerful tool for performing much more complex statistical analysis than is possible by hand. Using these programs, statistical estimates of delay, frequency response, and so on of circuits with large numbers of transistors can be performed.

SUMMARY

- The bipolar junction transistor (BJT) was invented in the late 1940s at the Bell Telephone Laboratories by Bardeen, Brattain, and Shockley and became the first commercially successful three-terminal solid-state device.

- Although the FET has become the dominant device technology in modern integrated circuits, bipolar transistors are still widely used in both discrete and integrated circuit design. In particular, the BJT is still the preferred device in many applications that require high speed and/or high precision such as op-amps, A/D and D/A converters, and wireless communication products.

- The basic physical structure of the BJT consists of a three-layer sandwich of alternating p- and n-type semiconductor materials and can be fabricated in either *npn* or *pnp* form.

- The emitter of the transistor injects carriers into the base. Most of these carriers traverse the base region and are collected by the collector. The carriers that do not completely traverse the base region give rise to a small current in the base terminal.

- A mathematical model called the transport model (a simplified Gummel-Poon model) characterizes the i-v characteristics of the bipolar transistor for general terminal voltage and current conditions. The transport model requires three unique parameters to characterize a particular BJT: the saturation current I_S and the forward and reverse common-emitter current gains β_F and β_R.

- β_F is a relatively large number, ranging from 20 to 500, and characterizes the significant current amplification capability of the BJT. Practical fabrication limitations cause the bipolar transistor structure to be inherently asymmetric, and the value of β_R is much smaller than β_F, typically between 0 and 2.

- SPICE circuit analysis programs contain a comprehensive built-in model for the transistor that is an extension of the transport model.

- Four regions of operation—cutoff, forward-active, reverse-active, and saturation—were identified for the BJT based on the bias voltages applied to the base-emitter and base-collector junctions. The transport model can be simplified for each individual region of operation.

- The cutoff and saturation regions are most often used in switching applications and logic circuits. In cutoff, the transistor approximates an open switch, whereas in saturation, the transistor represents a closed switch. However, in contrast to the "on" MOSFET, the saturated bipolar transistor has a small voltage, the collector-emitter saturation voltage V_{CESAT}, between its collector and emitter terminals, even when operating with zero collector current.

- In the forward-active region, the bipolar transistor can provide high voltage and current gain for amplification of analog signals. The reverse-active region finds limited use in some analog- and digital-switching applications.

- The i-v characteristics of the bipolar transistor are often presented graphically in the form of the output characteristics, i_C versus v_{CE} or v_{CB}, and the transfer characteristics, i_C versus v_{BE} or v_{EB}.

- In the forward-active region, the collector current increases slightly as the collector-emitter voltage increases. The origin of this effect is base-width modulation, known as the Early effect, and it can be included in the model for the forward-active region through addition of the parameter called the Early voltage V_A.

- The collector current of the bipolar transistor is determined by minority-carrier diffusion across the base of the transistor, and expressions were developed that relate the saturation current and base transit time of the transistor to physical device parameters. The base width plays a crucial role in determining the base transit time and the high-frequency operating limits of the transistor.

- Minority-carrier charge is stored in the base of the transistor during its operation, and changes in this stored charge with applied voltage result in diffusion capacitances being associated with forward-biased junctions. The value of the diffusion capacitance is proportional to the collector current I_C.

- The capacitances of the bipolar transistor cause the current gain to be frequency-dependent. At the beta cutoff frequency f_β, the current gain has fallen to 71 percent of its low frequency value, whereas the value of the current gain is only 1 at the unity-gain frequency f_T.

- The transconductance g_m of the bipolar transistor in the forward-active region relates differential changes in collector current and base-emitter voltage and was shown to be directly proportional to the dc collector current I_C.

- A number of biasing circuits were analyzed to determine the Q-point of the transistor. Design of the four-resistor network was investigated in detail. The four-resistor bias circuit provides highly stable control of the operating point and is the most important bias circuit for discrete design.

- Techniques for analyzing the influence of element tolerances on circuit performance include the worst-case analysis and statistical Monte Carlo analysis methods. In worst-case analysis, element values are simultaneously pushed to their extremes, and the resulting predictions of circuit behavior are often overly pessimistic. The Monte Carlo method analyzes a large number of randomly selected versions of a circuit to build up a realistic estimate of the statistical distribution of circuit performance. Random number generators in high-level computer languages, spreadsheets, or MATLAB can be used to randomly select element values for use in the Monte Carlo analysis. Some circuit analysis packages such as PSPICE provide a Monte Carlo analysis option as part of the program.

KEY TERMS

Base

Base current

Base width

Base-collector capacitance

Base-emitter capacitance

Base-width modulation

β-cutoff frequency f_β

Bipolar junction transistor (BJT)

Collector

Collector current

Common-base output characteristic

Common-emitter output characteristic

Common-emitter transfer characteristic

Cutoff region

Diffusion capacitance

Early effect

Early voltage V_A

Ebers-Moll model

Emitter

Emitter current

Equilibrium electron density

Forced beta

Forward-active region

Forward common-emitter current gain β_F

Forward common-base current gain α_F

Forward transit time τ_F

Forward transport current

Gummel-Poon model

Inverse-active region

Inverse common-emitter current gain

Inverse common-base current gain

Monte Carlo analysis

Normal-active region

Normal common-emitter current gain

Normal common-base current gain

npn transistor

Output characteristic

pnp transistor

Quiescent operating point

Q-point

Reverse-active region

Reverse common-base current gain α_R

Reverse common-emitter current gain β_R

Saturation region

Saturation voltage

Small-signal output resistance

SPICE model parameters BF, IS, VAF

Transconductance

Transfer characteristic

Transistor saturation current

Transport model

Unity-gain frequency f_T

Worst-case analysis

REFERENCES

1. William F. Brinkman, "The transistor: 50 glorious years and where we are going," *IEEE International Solid-State Circuits Conference Digest,* vol. 40, pp. 22–26, February 1997.

2. William F. Brinkman, Douglas E. Haggan, William W. Troutman, "A history of the invention of the transistor and where it will lead us," *IEEE Journal of Solid-State Circuits,* vol. 32, pp. 1858–1865, December 1997.

3. H. K. Gummel and H. C. Poon, "A compact bipolar transistor model," *ISSCC Digest of Technical Papers,* pp. 78, 79, 146, February 1970.

4. H. K. Gummel, "A charge control relation for bipolar transistors," *Bell System Technical Journal,* January 1970.

5. J. J. Ebers and J. L. Moll, "Large signal behavior of junction transistors," *Proc. IRE.,* pp. 1761–1772, December 1954.

6. J. M. Early, "Effects of space-charge layer widening in junction transistors," *Proc. IRE.,* pp. 1401–1406, November 1952.

7. B. M. Wilamowski and R. C. Jaeger, *Computerized Circuit Analysis Using SPICE Programs,* McGraw-Hill, New York: 1997.

8. J. D. Cressler, "Reengineering silicon: S_iG_e heterojunction bipolar technology," *IEEE Spectrum,* pp. 49–55, March 1995.

PROBLEMS

If not otherwise specified, use $I_S = 10^{-15}$ A, $V_A = 50$ V, $\beta_F = 100$, $\beta_R = 1$, and $V_{BE} = 0.70$ V.

5.1 Physical Structure of the Bipolar Transistor

5.1. Figure P5.1 is a cross section of an *npn* bipolar transistor similar to that in Fig. 5.1. Indicate the letter (*A* to *G*) that identifies the base contact, collector contact, emitter contact, *n*-type emitter region, *n*-type collector region, and the active or intrinsic transistor region.

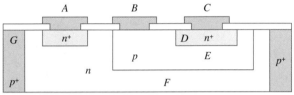

Figure P5.1

5.2 The Transport Model for the *npn* Transistor

5.2. (a) Label the collector, base, and emitter terminals of the transistor in the circuit in Fig. P5.2. (b) Label the base-emitter and base-collector voltages, V_{BE} and V_{BC}, respectively. (c) If $V = 0.640$ V, $I_C = 275\,\mu$A, and $I_B = 4\,\mu$A, find the values of I_S, β_F, and β_R for the transistor if $\alpha_R = 0.5$.

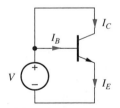

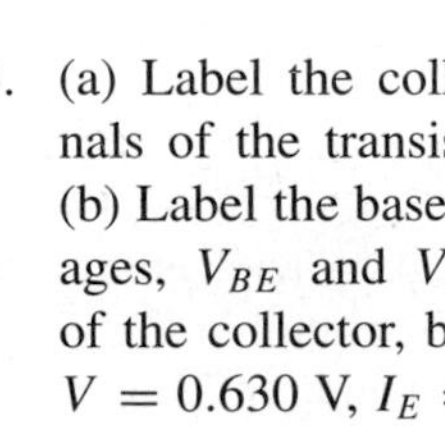

Figure P5.2

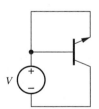

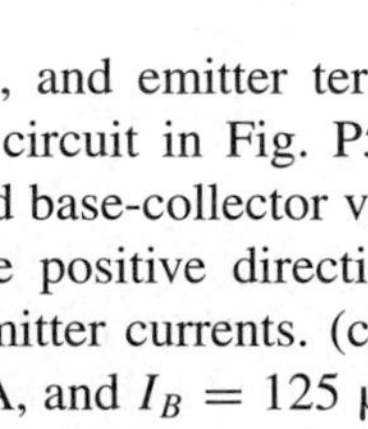

Figure P5.3

5.3. (a) Label the collector, base, and emitter terminals of the transistor in the circuit in Fig. P5.3. (b) Label the base-emitter and base-collector voltages, V_{BE} and V_{BC}, and the positive directions of the collector, base, and emitter currents. (c) If $V = 0.630$ V, $I_E = -275\,\mu$A, and $I_B = 125\,\mu$A, find the values of I_S, β_F, and β_R for the transistor if $\alpha_F = 0.975$.

5.4. Fill in the missing entries in Table 5.P1.

TABLE 5.P1

α	β
	0.200
0.400	
0.750	
	10.0
0.980	
	200
	1000
0.9998	

5.5. (a) Find the current I_{CBS} in Fig. P5.5(a). (Use the parameters specified at the beginning of the problem set.) (b) Find the current I_{CBO} and the voltage V_{BE} in Fig. P5.5(b).

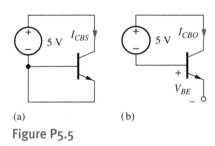

(a)   (b)

Figure P5.5

5.6. For the transistor in Fig. P5.6, $I_S = 2 \times 10^{-15}$ A, $\beta_F = 100$, and $\beta_R = 0.25$. (a) Label the collector, base, and emitter terminals of the transistor. (b) What is the transistor type? (c) Label the base-emitter and base-collector voltages, V_{BE} and V_{BC}, respectively, and label the normal directions for I_E, I_C, and I_B. (d) What is the relationship between V_{BE} and V_{BC}? (e) Write the simplified form of the transport model equations that apply to this particular circuit configuration. Write an expression for I_E/I_B. Write an expression for I_E/I_C. (f) Find the values of I_E, I_C, I_B, V_{BC}, and V_{BE}.

Figure P5.6

5.7. For the transistor in Fig. P5.7, $I_S = 2 \times 10^{-15}$ A, $\beta_F = 100$, and $\beta_R = 0.25$. (a) Label the collector, base, and emitter terminals of the transistor. (b) What is the transistor type? (c) Label the base-emitter and base-collector voltages, V_{BE} and V_{BC}, and the normal directions for I_E, I_C, and I_B. (d) Find the values of I_E, I_C, I_B, V_{BC}, and V_{BE} if $I = 175$ μA.

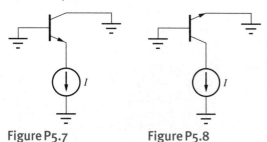

Figure P5.7   Figure P5.8

5.8. For the transistor in Fig. P5.8, $I_S = 2 \times 10^{-15}$ A, $\beta_F = 100$, and $\beta_R = 0.25$. (a) Label the collector, base, and emitter terminals of the transistor. (b) What is the transistor type? (c) Label the base-emitter and base-collector voltages, V_{BE} and V_{BC}, and label the normal directions for I_E, I_C, and I_B. (d) Find the values of I_E, I_C, I_B, V_{BC}, and V_{BE} if $I = 175$ μA.

5.9. The *npn* transistor is connected in a "diode" configuration in Fig. P5.9(a). Use the transport model equations to show that the i-v characteristics of this connection are similar to those of a diode as defined by Eq. (3.11). What is the reverse saturation current of this "diode" if $I_S = 2 \times 10^{-15}$ A, $\beta_F = 100$, and $\beta_R = 0.25$?

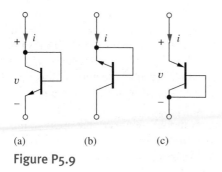

(a)   (b)   (c)

Figure P5.9

5.10. The *npn* transistor is connected in an alternate "diode" configuration in Fig. P5.9(b). Use the transport model equations to show that the i-v characteristics of this connection are similar to those of a diode as defined by Eq. (3.11). What is the reverse saturation current of this "diode" if $I_S = 5 \times 10^{-15}$ A, $\beta_F = 60$, and $\beta_R = 3$?

5.11. Calculate i_T for an *npn* transistor with $I_S = 10^{-16}$ A for (a) $V_{BE} = 0.75$ V and $V_{BC} = -3$ V and (b) $V_{BC} = 0.75$ V and $V_{BE} = -3$ V.

5.12. Calculate i_T for an *npn* transistor with $I_S = 10^{-15}$ A for (a) $V_{BE} = 0.70$ V and $V_{BC} = -3$ V and (b) $V_{BC} = 0.70$ V and $V_{BE} = -3$ V.

5.3 The *pnp* Transistor

5.13. Figure P5.13 is a cross section of a *pnp* bipolar transistor similar to the *npn* transistor in Fig. 5.1. Indicate the letter (A to G) that represents the base contact, collector contact, emitter contact, *p*-type emitter region, *p*-type collector region, and the active or intrinsic transistor region.

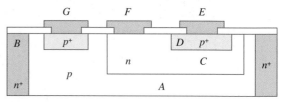

Figure P5.13

5.14. For the transistor in Fig. P5.14(a), $I_S = 10^{-15}$ A, $\alpha_F = 0.985$, and $\alpha_R = 0.25$. (a) What type of transistor is in this circuit? (b) Label the collector, base, and emitter terminals of the transistor. (c) Label the emitter-base and collector-base voltages, and label the normal direction for I_E, I_C, and I_B. (d) Write the simplified form of the transport model equations that apply to this particular circuit configuration. Write an expression for I_E/I_C. Write an expression for I_E/I_B. (e) Find the values of I_E, I_C, I_B, β_F, β_R, V_{EB}, and V_{CB}.

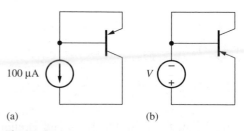

(a)   (b)

Figure P5.14

5.15. (a) Label the collector, base and, emitter terminals of the transistor in the circuit in Fig. P5.14(b). (b) Label the emitter-base and collector-base voltages, V_{EB} and V_{CB}, and the normal directions for I_E, I_C, and I_B. (c) If $V = 0.640$ V, $I_C = 300$ μA, and $I_B = 4$ μA, find the values of I_S, β_F, and β_R for the transistor if $\alpha_R = 0.2$.

5.16. Repeat Prob. 5.9 for the "diode-connected" *pnp* transistor in Fig. P5.9(c).

5.17. For the transistor in Fig. P.5.17, $I_S = 2 \times 10^{-15}$ A, $\beta_F = 75$, and $\beta_R = 4$. (a) Label the collector, base, and emitter terminals of the transistor. (b) What is the transistor type? (c) Label the emitter-base and collector-base voltages, and label the normal direction for I_E, I_C, and I_B. (d) Write the simplified form of the transport model equations that apply to this particular circuit configuration. Write an expression for I_E/I_B. Write an expression for I_E/I_C. (e) Find the values of I_E, I_C, I_B, V_{CB}, and V_{EB}.

Figure P5.17

5.18. For the transistor in Fig. P5.18(a), $I_S = 4 \times 10^{-15}$ A, $\beta_F = 100$, and $\beta_R = 5$. (a) Label the collector, base, and emitter terminals of the transistor. (b) What is the transistor type? (c) Label the emitter-base and collector-base voltages, V_{EB} and V_{CB}, and the normal directions for I_E, I_C, and I_B. (d) Find the values of I_E, I_C, I_B, V_{CB}, and V_{EB} if $I = 300$ μA.

and emitter terminals of the transistor. (b) What is the transistor type? (c) Label the emitter-base and collector-base voltages, V_{EB} and V_{CB}, and label the normal directions for I_E, I_C, and I_B. (d) Find the values of I_E, I_C, I_B, V_{CB}, and V_{EB} if $I = 300$ μA.

5.20. Calculate i_T for a *pnp* transistor with $I_S = 5 \times 10^{-16}$ A for (a) $V_{EB} = 0.70$ V and $V_{CB} = -3$ V and (b) $V_{CB} = 0.70$ V and $V_{EB} = -3$ V.

5.4 Equivalent Circuit Representations for the Transport Models

5.21. Calculate the values of i_T and the two diode currents for the equivalent circuit in Fig. 5.8(a) for an *npn* transistor with $I_S = 4 \times 10^{-15}$ A, $\beta_F = 80$, and $\beta_R = 2$ for (a) $V_{BE} = 0.73$ V and $V_{BC} = -3$ V and (b) $V_{BC} = 0.73$ V and $V_{BE} = -3$ V.

5.22. Calculate the values of i_T and the two diode currents for the equivalent circuit in Fig. 5.8(b) for a *pnp* transistor with $I_S = 6 \times 10^{-15}$A, $\beta_F = 60$, and $\beta_R = 3$ for (a) $V_{EB} = 0.68$ V and $V_{CB} = -3$ V and (b) $V_{CB} = 0.68$ V and $V_{EB} = -3$ V.

5.23. The Ebers-Moll model was one of the first mathematical models used to describe the characteristics of the bipolar transistor. Show that the *npn* Transport Model equations can be transformed into the Ebers-Moll equations below. (*Hint:* Add and subtract 1 from the collector and emitter current expressions in Eqs. (5.13).)

$$i_E = I_{ES}\left[\exp\left(\frac{v_{BE}}{V_T}\right) - 1\right] - \alpha_R I_{CS}\left[\exp\left(\frac{v_{BC}}{V_T}\right) - 1\right]$$

$$i_C = \alpha_F I_{ES}\left[\exp\left(\frac{v_{BE}}{V_T}\right) - 1\right] - I_{CS}\left[\exp\left(\frac{v_{BC}}{V_T}\right) - 1\right]$$

$$i_B = (1 - \alpha_F)I_{ES}\left[\exp\left(\frac{v_{BE}}{V_T}\right) - 1\right] + (1 - \alpha_R)I_{CS}\left[\exp\left(\frac{v_{BC}}{V_T}\right) - 1\right]$$

$$\alpha_F I_{ES} = \alpha_R I_{CS}$$

5.24. What are the values of α_F, α_R, I_{ES} and I_{CS} for an *npn* transistor with $I_S = 2 \times 10^{-15}$ A, $\beta_F = 100$ and $\beta_R = 0.5$? Show that $\alpha_F I_{ES} = \alpha_R I_{CS}$.

5.25. The Ebers-Moll model was one of the first mathematical models used to describe the characteristis of the bipolar transistor. Show that the *pnp* Transport Model equations can be transformed into the Ebers-Moll equations that follow. (*Hint*: Add and subtract 1 from the collector and emitter current expressions in Eqs. (5.17).)

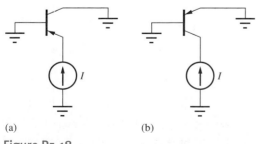

(a) (b)

Figure P5.18

5.19. For the transistor in Fig. P5.18(b), $I_S = 5 \times 10^{-15}$ A, $\beta_F = 75$, and $\beta_R = 1$. (a) Label the collector, base,

$$i_E = I_{ES} \left[\exp\left(\frac{v_{EB}}{V_T}\right) - 1 \right] - \alpha_R I_{CS} \left[\exp\left(\frac{v_{CB}}{V_T}\right) - 1 \right]$$

$$i_C = \alpha_F I_{ES} \left[\exp\left(\frac{v_{EB}}{V_T}\right) - 1 \right] - I_{CS} \left[\exp\left(\frac{v_{CB}}{V_T}\right) - 1 \right] \qquad \alpha_F I_{ES} = \alpha_R I_{CS}$$

$$i_B = (1 - \alpha_F) I_{ES} \left[\exp\left(\frac{v_{EB}}{V_T}\right) - 1 \right] + (1 - \alpha_R) I_{CS} \left[\exp\left(\frac{v_{CB}}{V_T}\right) - 1 \right]$$

5.5 The i-v Characteristics of the Bipolar Transistor

*5.26. The common-emitter output characteristics for an *npn* transistor are given in Fig. P5.26. What are the values of β_F at (a) $I_C = 5$ mA and $V_{CE} = 5$ V? (b) $I_C = 7$ mA and $V_{CE} = 7.5$ V? (c) $I_C = 10$ mA and $V_{CE} = 14$ V?

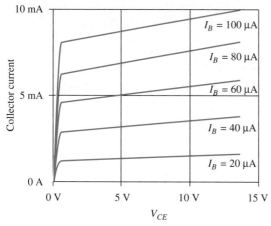

Figure P5.26

5.27. Plot the common-emitter output characteristics for an *npn* transistor having $I_S = 1$ fA, $\beta_{FO} = 75$, and $V_A = 50$ V for six equally spaced base current steps ranging from 0 to 200 μA and V_{CE} ranging from 0 to 10 V.

5.28. Use SPICE to plot the common-emitter output characteristics for the *npn* transistor in Prob. 5.27.

5.29. Use SPICE to plot the common-base output characteristics for an *npn* transistor having $I_S = 1$ fA, $\beta_{FO} = 75$, and $V_A = 50$ V for six equally spaced emitter current steps ranging from 0 to 2 mA and V_{CB} ranging from 0 to 10 V.

5.30. Plot the common-emitter output characteristics for a *pnp* transistor having $I_S = 1$ fA, $\beta_{FO} = 75$, and $V_A = 50$ V for six equally spaced base current steps ranging from 0 to 250 μA and V_{EC} ranging from 0 to 10 V.

5.31. Use SPICE to plot the common-emitter output characteristics for the *pnp* transistor in Prob. 5.30.

5.32. Use SPICE to plot the common-base output characteristics for an *pnp* transistor having $I_S = 1$ fA, $\beta_{FO} = 75$, and $V_A = 50$ V for six equally spaced emitter current steps ranging from 0 to 2 mA and V_{BC} ranging from 0 to 10 V.

5.33. What is the reciprocal of the slope (in mV/decade) of the logarithmic transfer characteristic for an *npn* transistor in the common-emitter configuration at a temperature of (a) 200 K, (b) 250 K, (c) 300 K and (d) 350 K?

Junction Breakdown Voltages

*5.34. In the circuits in Fig. P5.9, the Zener breakdown voltages of the collector-base and emitter-base junctions of the transistors are 50 V and 6 V, respectively. What is the Zener breakdown voltage for each "diode" connected transistor configuration?

5.35. In the circuits in Fig. P5.35, the Zener breakdown voltages of the collector-base and emitter-base junctions of the *npn* transistors are 75 V and 6.3 V, respectively. What is the current in the resistor in each circuit? (*Hint:* The equivalent circuits for the transport model equations may help in visualizing the circuit.)

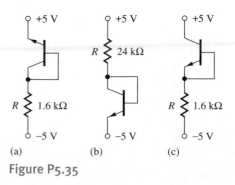

Figure P5.35

5.36. An *npn* transistor is biased as indicated in Fig. 5.9(a). What is the largest value of V_{CE} that can be applied without junction breakdown if the breakdown voltages of the collector-base and emitter-base junctions of the *npn* transistors are 65 V and 6 V, respectively?

*5.37. (a) For the circuit in Fig. P5.37, what is the maximum value of I according to the transport model equations if $I_S = 1 \times 10^{-15}$ A, $\beta_F = 50$, and $\beta_R = 0.5$? (b) Suppose that $I = 1$ mA. What happens to the transistor? (*Hint:* The equivalent circuits for the transport model equations may help in visualizing the circuit.)

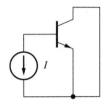

Figure P5.37

5.6 The Operating Regions of the Bipolar Transistor

5.38. Indicate the region of operation in the following table for an *npn* transistor biased with the indicated voltages.

BASE-EMITTER VOLTAGE	BASE-COLLECTOR VOLTAGE	
	0.7 V	−5.0 V
−5.0 V		
0.7 V		

5.39. (a) What are the regions of operation for the transistors in Fig. P5.9? (b) In Fig. P5.46(a)? (c) In Fig. P5.49? (d) In Fig. P5.62?

5.40. (a) What is the region of operation for the transistor in Fig. P5.5(a)? (b) In Fig. P5.5(b)?

5.41. (a) What is the region of operation for the transistor in Fig. P5.6? (b) In Fig. P5.7? (c) In Fig. P5.8?

5.42. Indicate the region of operation in the following table for a *pnp* transistor biased with the indicated voltages.

EMITTER-BASE VOLTAGE	COLLECTOR-BASE VOLTAGE	
	0.7 V	−0.65 V
0.7 V		
−0.65 V		

5.43. (a) What is the region of operation for the transistor in Fig. P5.2? (b) In Fig. P5.3?

5.44. (a) What is the region of operation for the transistor in Fig. P5.14(a)? (b) In Fig. P5.14(b)?

5.45. (a) What is the region of operation for the transistor in Fig. P5.17? (b) In Fig. P5.18(a)? (c) In Fig. P5.18(b).

5.7 Transport Model Simplifications
Cutoff Region

5.46. (a) What are the three terminal currents I_E, I_B, and I_C in the transistor in Fig. P5.46(a) if $I_S = 1 \times 10^{-15}$ A, $\beta_F = 75$, and $\beta_R = 4$? (b) Repeat for Fig. P5.46(b).

**5.47. An *npn* transistor with $I_S = 1 \times 10^{-16}$ A, $\alpha_F = 0.95$ and $\alpha_R = 0.5$ is operating with $V_{BE} = 0.3$ V and $V_{BC} = -5$ V. This transistor is not truly operating in the region defined to be cutoff, but we still say the transistor is off. Why? Use the transport model equations to justify your answer. In what region is the transistor actually operating according to our definitions?

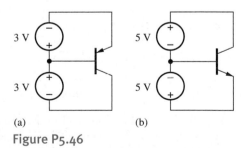

(a) (b)

Figure P5.46

Forward-Active Region

5.48. What are the values of β_F and I_S for the transistor in Fig. P5.48?

5.49. What are the values of β_F and I_S for the transistor in Fig. P5.49?

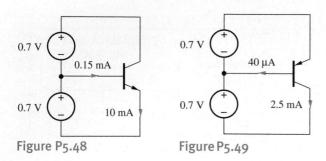

Figure P5.48   Figure P5.49

5.50. What are the emitter, base, and collector currents in the circuit in Fig. 5.18 if $V_{EE} = 3.3$ V, $R = 47$ kΩ, and $\beta_F = 80$?

**5.51. A transistor has $f_T = 500$ MHz and $\beta_F = 75$. (a) What is the β-cutoff frequency f_β of this transistor? (b) Use Eq. (5.42) to find an expression for the frequency dependence of α_F—that is, $\alpha_F(f)$. [*Hint:* Write an expression for $\beta(s)$.] What is the α-cutoff frequency for this transistor?

*5.52. (a) Start with the transport model equations for the *pnp* transistor, Eqs. (5.17), and construct the simplified version of the *pnp* equations that apply to the forward-active region [similar to Eqs. (5.23)]. (b) Draw the simplified model for the *pnp* transistor similar to the *npn* version in Fig. 5.21(c).

Reverse-Active Region

5.53. What are the values of β_R and I_S for the transistor in Fig. P5.53?

5.54. What are the values of β_R and I_S for the transistor in Fig. P5.54?

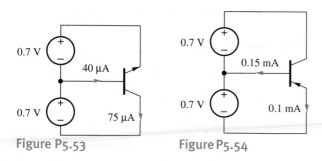

Figure P5.53   Figure P5.54

5.55. Find the emitter, base, and collector currents in the circuit in Fig. 5.22 if the negative power supply is -3.3 V, $R = 56$ kΩ, and $\beta_R = 0.75$.

Saturation Region

5.56. What is the saturation voltage of an *npn* transistor operating with $I_C = 1$ mA and $I_B = 1$ mA if $\beta_F = 50$ and $\beta_R = 2$? What is the forced β of this transistor? What is the value of V_{BE} if $I_S = 10^{-15}$ A?

5.57. Derive an expression for the saturation voltage V_{ECSAT} of the *pnp* transistor in a manner similar to that used to derive Eq. (5.30).

5.58. (a) What is the collector-emitter voltage for the transistor in Fig. P5.58(a) if $I_S = 5 \times 10^{-15}$ A, $\alpha_F = 0.99$, and $\alpha_R = 0.5$? (b) What is the emitter-collector voltage for the transistor in Fig. P5.59(b) for the same transistor parameters?

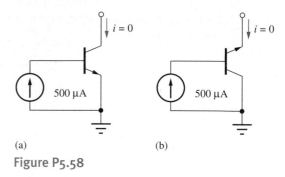

(a)   (b)

Figure P5.58

5.59. Repeat Prob. 5.58 for $\alpha_F = 0.95$ and $\alpha_R = 0.33$.

5.60. (a) What base current is required to achieve a saturation voltage of $V_{CESAT} = 0.1$ V in an *npn* power transistor that is operating with a collector current of 20 A if $\beta_F = 15$ and $\beta_R = 0.9$? What is the forced β of this transistor? (b) Repeat for $V_{CESAT} = 0.04$ V.

**5.61. An *npn* transistor with $I_S = 1 \times 10^{-16}$ A, $\alpha_F = 0.975$, and $\alpha_R = 0.5$ is operating with $V_{BE} = 0.70$ V and $V_{BC} = 0.50$ V. By definition, this transistor is operating in the saturation region. However, in the discussion of Fig. 5.19 it was noted that this transistor actually behaves as if it is still in the forward-active region even though $V_{BC} > 0$. Why? Use the transport model equations to justify your answer.

5.62. The current I in both circuits in Fig. P5.62 is 175 μA. Find the value of V_{BE} for both circuits if $I_S = 1 \times 10^{-16}$ A, $\beta_F = 50$, and $\beta_R = 0.5$. What is V_{CESAT} in Fig. P5.62(b)?

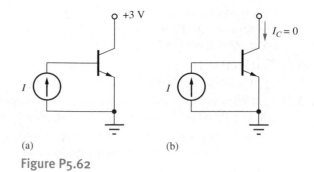

(a)   (b)

Figure P5.62

Diodes in Bipolar Integrated Circuits

5.63. Derive the result in Eq (5.26) by applying the circuit constraints to the transport equations.

5.64. What is the reverse saturation current of the diode in Fig. 5.20 if the transistor is described by $I_S = 10^{-15}$ A, $\alpha_R = 0.20$, and $\alpha_F = 0.98$?

5.65. The two transistors in Fig. P5.65 are identical. What is the collector current of Q_2 if $I = 25$ μA and $\beta_F = 25$?

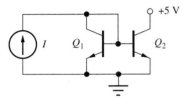

Figure P5.65

5.8 Nonideal Behavior of the Bipolar Transistor

5.66. Calculate the diffusion capacitance of a bipolar transistor with a forward transit time $\tau_F = 50$ ps for collector currents of (a) 2 μA, (b) 200 μA, (c) 20 mA.

5.67. (a) What is the forward transit time τ_F for an *npn* transistor with a base width $W_B = 1$ μm and a base doping of $10^{18}/cm^3$? (b) Repeat the calculation for a *pnp* transistor.

5.68. A transistor has a dc current gain of 200 and a current gain of 10 at 75 MHz. What are the unity-gain and beta-cutoff frequencies of the transistor?

5.69. A transistor has $f_T = 900$ MHz and $f_\beta = 5$ MHz. What is the dc current gain of the transistor? What is the current gain of the transistor at 50 MHz?

5.70. What is the saturation current for a transistor with a base doping of $6 \times 10^{18}/cm^3$, a base width of 0.4 μm, and a cross-sectional area of 25 μm²?

5.71. An *npn* transistor is needed that will operate at a frequency of at least 5 GHz. What base width is required for the transistor if the base doping is $5 \times 10^{18}/cm^3$?

The Early Effect and Early Voltage

5.72. An *npn* transistor is operating in the forward-active region with a base current of 3 μA. It is found that $I_C = 240$ μA for $V_{CE} = 5$ V and $I_C = 265$ μA for $V_{CE} = 10$ V. What are the values of β_{FO} and V_A for this transistor?

5.73. An *npn* transistor with $I_S = 10^{-16}$ A, $\beta_F = 100$, and $V_A = 65$ V is biased in the forward-active region with $V_{BE} = 0.72$ V and $V_{CE} = 10$ V. (a) What is the collector current I_C? (b) What would be the collector current I_C if $V_A = \infty$? (c) What is the ratio of the two answers in parts (a) and (b)?

5.74. The common-emitter output characteristics for an *npn* transistor are given in Fig. P5.26. What are the values of β_{FO} and V_A for this transistor?

5.75. (a) Recalculate the currents in the transistor in Fig. 5.16 if $I_S = 5 \times 10^{-15}$ A, $\beta_{FO} = 19$, and $V_A = 50$ V. What is V_{BE}? (b) What was V_{BE} for $V_A = \infty$?

5.76. Recalculate the currents in the transistor in Fig. 5.18 if $\beta_{FO} = 50$ and $V_A = 50$ V.

5.9 Transconductance

5.77. What is the transconductance of an *npn* transistor operating at 300 K and a collector current of (a) 10 μA, (b) 100 μA, (c) 1 mA, and (d) 10 mA? (e) Repeat for a *pnp* transistor.

5.78. What is the diffusion capacitance for an npn transistor with TF = 10 psec if it is operating at 300 K with a collector currents of 1 μA, 1 mA, and 10 mA?

5.10 Bipolar Technology and SPICE Model

5.79. (a) Find the default values of the following parameters for the generic *npn* transistor in the version of SPICE that you use in class: IS, BF, BR, VAF, VAR, TF, TR, NF, NE, RB, RC, RE, ISE, ISC, ISS, IKF, IKR, CJE, CJC. (*Note:* The values in Table 5.P1 may not agree exactly with your version of SPICE.) (b) Repeat for the generic *pnp* transistor.

5.80. A SPICE model for a bipolar transistor has a forward knee current IKF = 10 mA and NK = 0.5. How much does the KBQ factor reduce the collector current of the transistor in the forward-active region if i_F is (a) 1 mA? (b) 10 mA? (c) 50 mA?

5.81. Plot a graph of KBQ versus i_F for an *npn* transistor with IKF = 40 mA and NK = 0.5. Assume forward-active region operation with VAF = ∞.

5.11 Practical Bias Circuits for the BJT
Four-Resistor Biasing

5.82. (a) Find the Q-point for the circuit in Fig. P5.82(a). Assume that $\beta_F = 50$ and $V_{BE} = 0.7$ V. (b) Repeat the calculation if all the resistor values are decreased by a factor of 5. (c) Repeat (a) for Fig. 5.82(b). (d) Repeat (b) for Fig. P5.82(b).

5.83. (a) Find the Q-point for the circuit in Fig. 5.82(a) if the 33-kΩ resistor is replaced with a 22-kΩ resistor. Assume that $\beta_F = 75$. (b) Repeat (a) for the circuit in Fig. P5.82(b).

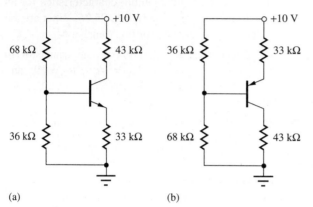

(a)   (b)

Figure P5.82

5.84. (a) Simulate the circuits in Fig. P5.82 and compare the SPICE results to your hand calculations of the Q-point. Use $I_S = 1 \times 10^{-16}$ A, $\beta_F = 50$, $\beta_R = 0.25$, and $V_A = \infty$. (b) Repeat for $V_A = 60$ V. (c) Repeat (a) for the circuit in Fig. 5.34(c). (d) Repeat (b) for the circuit in Fig. 5.34(c).

5.85. Find the Q-point in the circuit in Fig. 5.34 if $R_1 = 6.2$ kΩ, $R_2 = 12$ kΩ, $R_C = 5.1$ kΩ, $R_E = 7.5$ kΩ, $\beta_F = 100$, and the positive power supply voltage is 10 V.

5.86. Find the Q-point in the circuit in Fig. 5.34 if $R_1 = 120$ kΩ, $R_2 = 240$ kΩ, $R_E = 100$ kΩ, $R_C = 150$ kΩ, $\beta_F = 100$, and the positive power supply voltage is 15 V.

5.87. (a) Design a four-resistor bias network for an *npn* transistor to give $I_C = 1$ mA, $V_{CE} = 5$ V, and $V_E = 2$ V if $V_{CC} = 12$ V and $\beta_F = 100$. (b) Replace your exact values with the nearest values from the resistor table in Appendix C and find the resulting Q-point.

5.88. (a) Design a four-resistor bias network for an *npn* transistor to give $I_C = 10$ μA and $V_{CE} = 6$ V if $V_{CC} = 18$ V and $\beta_F = 75$. (b) Replace your exact values with the nearest values from the resistor table in Appendix C and find the resulting Q-point.

5.89. (a) Design a four-resistor bias network for a *pnp* transistor to give $I_C = 850$ μA, $V_{EC} = 2$ V, and $V_E = 1$ V if $V_{CC} = 5$ V and $\beta_F = 60$. (b) Replace

your exact values with the nearest values from the resistor table in Appendix C and find the resulting Q-point.

5.90. (a) Design a four-resistor bias network for a *pnp* transistor to give $I_C = 11$ mA and $V_{EC} = 5$ V if $V_{RE} = 1$ V, $V_{CC} = -15$ V and $\beta_F = 50$. (b) Replace your exact values with the nearest values from the resistor table in Appendix C and find the resulting Q-point.

Load Line Analysis

*5.91. Find the Q-point for the circuit in Fig. P5.91 using the graphical load-line approach. Use the characteristics in Fig. P5.26.

*5.92. Find the Q-point for the circuit in Fig. P5.92 using the graphical load-line approach. Use the characteristics in Fig. P5.26, assuming that the graph is a plot of i_C vs. v_{EC} rather than i_C vs. v_{CE}.

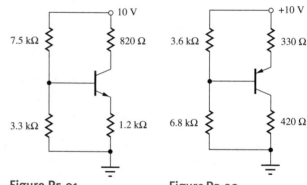

Figure P5.91   **Figure P5.92**

Two-Resistor Biasing

5.93. Find the Q-point for the circuit in Fig. P5.93 for (a) $\beta_F = 30$, (b) $\beta_F = 100$, (c) $\beta_F = 250$, (d) $\beta_F = \infty$.

*5.94. Write the load-line expression for the circuit in Fig. P5.93. Draw the load line on the characteristics in Fig. P5.26. Find the Q-point by drawing a curve that plots I_B vs. V_{CE}.

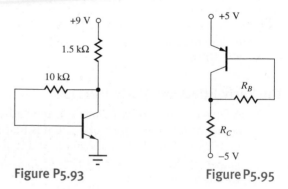

Figure P5.93   **Figure P5.95**

5.95. Design the bias circuit in Fig. P5.95 to give a Q-point of $I_C = 10$ mA and $V_{EC} = 3$ V if the transistor current gain $\beta_F = 60$. What is the Q-point if the current gain of the transistor is actually 40?

5.96. Design the bias circuit in Fig. P5.96 to give a Q-point of $I_C = 20$ μA and $V_{CE} = 0.90$ V if the transistor current gain is $\beta_F = 50$ and $V_{BE} = 0.65$ V. What is the Q-point if the current gain of the transistor is actually 125?

*5.97. Find the Q-point for the circuit in Fig. P5.97 if the Zener diode has $V_Z = 7$ V and $R_C = 500$ Ω. Use $\beta_F = 100$.

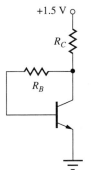

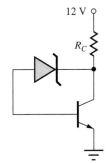

Figure P5.96   Figure P5.97

Bias Circuit Applications

5.98. The Zener diode in Fig. P5.98 has $V_Z = 6$ V and $R_Z = 100$ Ω. What is the output voltage if $I_L = 20$ mA? Use $I_S = 1 \times 10^{-16}$ A, $\beta_F = 50$, and $\beta_R = 0.5$ to find a precise answer.

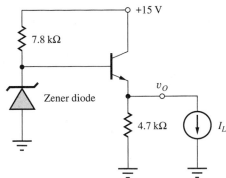

Figure P5.98

*5.99. Create a model for the Zener diode and simulate the circuit in Prob. P5.98. Compare the SPICE results to your hand calculations. Use $I_S = 1 \times 10^{-16}$ A, $\beta_F = 50$, and $\beta_R = 0.5$.

**5.100. The circuit in Fig. P5.100 has $V_{EQ} = 7$ V and $R_{EQ} = 100$ Ω. What is the output resistance R_o of this circuit for $i_L = 20$ mA if R_o is defined as $R_o = -dv_O/di_L$? Assume $\beta_F = 50$.

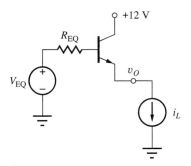

Figure P5.100

5.101. What is the output voltage in Fig. P5.101 if the op-amp is ideal? What are the values of the emitter current and the total current supplied by the 15-V source? Assume $\beta_F = 60$.

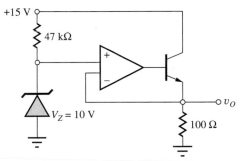

Figure P5.101

5.102. What is the output voltage in Fig. P5.102 if the op-amp is ideal? What are the values of the emitter current and the total current supplied by the 15-V source? Assume $\beta_F = 40$.

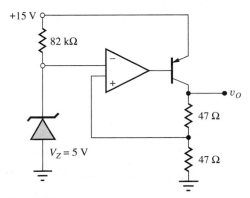

Figure P5.102

5.12 Tolerances in Bias Circuits

5.103. Find the worst-case values of the collector current and collector-emitter voltage for the circuit in Fig. 5.36 if the resistor and power supply tolerances are 5 percent.

*5.104. Perform a Monte Carlo analysis of the circuit in Fig. 5.36 assuming that the resistor and power supply tolerances are 5 percent.

*5.105. (a) Perform a worst-case analysis of the circuit in Fig. 5.35 assuming the resistor tolerances are 5 percent and $\beta_F = 100 \pm 50$ percent. (b) Perform a 100-case Monte Carlo analysis of the same circuit and compare the results to part (a).

*5.106. Repeat the Monte Carlo analysis of the circuit in Fig. 5.35 for resistor tolerances of 5 percent.

**5.107. The collector current of the circuit described by the histograms in Fig. 5.41 must be 210 μA ± 40 μA. (a) What percentage of the circuits will have to be discarded or rebuilt because they do not meet this specification? (b) Approximately what percentage of the circuits fail to meet a specification of $V_{CE} = 4.0$ V ± 20 percent?

**5.108. Choose the resistor tolerances in the circuit in Fig. 5.34 so that 99 percent of the circuits will meet the specification $I_C = 210$ μA ± 5 percent. Demonstrate your success using Monte Carlo analysis. Assume that the resistors all have the same tolerance. Do not change the tolerance on β and V_{CC}.

*5.109. Perform a worst-case analysis of the circuit in Fig. 5.35 assuming $\beta_F = 100 \pm 50$ percent, the resistors have a 20 percent tolerance, and V_{CC} has a 5 percent tolerance.

**5.110. (a) Perform a Monte Carlo analysis of the circuit in Fig. 5.35 assuming $\beta_{FO} = 100 \pm 50$ percent, $V_A = 75$ V ± 33 percent, the resistors have a 20 percent tolerance, and V_{CC} has a 5 percent tolerance. (b) Compare the results to the worst-case analysis in Prob. 5.109.

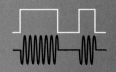

PART TWO
DIGITAL ELECTRONICS

INTRODUCTION TO DIGITAL ELECTRONICS

CHAPTER GOALS

- Introduce binary digital logic concepts
- Explore the voltage transfer characteristics of ideal and nonideal inverters
- Define logic levels and logic states at the input and output of logic gates
- Present goals for logic gate design
- Understand the need for noise rejection and the concept of noise margin
- Introduce measures of dynamic performance of logic gates including rise time, fall time, propagation delay, and power-delay product
- Review Boolean algebra and the NOT, OR, AND, NOR, and NAND functions
- Explore simple transistor implementations of the inverter
- Explore the design of MOS logic gates employing single transistor types—either NMOS or PMOS transistors (known as single-channel technology)
- Learn basic inverter design; discover why transistors are used in place of resistors
- Understand design and performance differences between saturated load, linear load, depletion-mode, and pseudo NMOS load circuits
- Present examples of noise margin calculations
- Learn to design multiinput NAND and NOR gates
- Learn to design complex logic gates including sum-of-products representations
- Develop expressions and approximation techniques for calculating rise time, fall time, and propagation delay of the various single-channel logic families

Digital electronics has had a profound effect on our lives through the pervasive application of microprocessors and microcontrollers in consumer and industrial products. The microprocessor chip forms the heart of personal computers and workstations, and digital signal processing is the basis of modern telecommunications. Microcontrollers are found in everything from CD/MP3 players to refrigerators to washing machines to vacuum cleaners, and in today's luxury automobiles often more than 50 microprocessors work together to control the vehicle. In fact, as much as 40 to 50 percent of the total cost of luxury cars is projected to come from electronics in the near future.

The digital electronics market is dominated by far by **complementary MOS,** or **CMOS, technology.** However, as pointed out in previous chapters, the first successful manufacturing processes were developed for bipolar devices, and the first integrated circuits utilized bipolar transistors. The rapid advance in the application of digital electronics was facilitated by circuit designers who developed early bipolar logic families called resistor-transistor logic (RTL) and diode-transistor logic (DTL). These families were subsequently replaced with highly robust bipolar logic families including **transistor-transistor logic (TTL)** and **emitter-coupled logic (ECL)** that could be easily interconnected to form highly reliable digital systems. High-performance forms of TTL and ECL remain in use today.

It took almost a decade to develop viable MOS manufacturing processes. The first high-density MOS integrated circuits utilizing PMOS technology appeared around 1970. The landmark development of the microprocessor is attributed to Ted Hoff who convinced Intel to develop the 4-bit 4004 microprocessor chip containing approximately of 2300 transistors that was introduced in 1971. As with many advances, work on single-chip processors advanced

Intel Founders Andy Grove, Robert Noyce, and Gordon Moore with rubylith layout of 8080 microprocessor.
Photo Courtesy of Intel Corporation

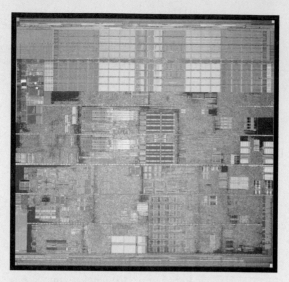

Intel® Pentium® 4 Processor
© *2006 Intel Corporation. All Rights Reserved.*

rapidly in research and development laboratories around the world. In the following 30 years, the industry went on to develop microprocessor chips of incredible complexity. As this edition is written, chips employing more than one billion transistors have been introduced, and the ITRS projections in Chapter 1 predict microprocessors with more than 10 billion transistors will appear by the year 2018.

By the mid 1970s, PMOS was being rapidly replaced by the higher-performance NMOS technology. The Intel 8080, 8085, and 8086 were all implemented in NMOS logic. A significant advance in NMOS circuit performance was achieved with the introduction of the depletion-mode load device, and this work was formally recognized when Dr. Toshiaki Masuhara of Hitachi received the 1990 IEEE Solid-State Circuits Award for this work.

But by the mid 1980s, power dissipation levels associated with NMOS microprocessors had reached unmanagable levels, and the industry made a transition to CMOS technology almost overnight. CMOS has remained the dominant technology since that time.

In this chapter, we begin our study of digital logic circuits with the introduction of a number of important concepts and definitions related to logic circuits. Then the chapter looks in detail at the design of MOS logic circuits built using only a single transistor type—either NMOS or PMOS—referred to as "single channel technology." Pseudo NMOS utilizes a PMOS load transistor and provides a bridge to modern Complementary MOS (CMOS) logic that uses both NMOS and PMOS transistors, as discussed in Chapter 7. MOS memory and storage circuits are introduced in Chapter 8, and bipolar logic circuits are discussed in Chapter 9.

This chapter explores the requirements and general characteristics of digital logic gates. The chapter then investigates the detailed implementation of logic gates in MOS technologies. The initial discussion in this chapter focuses on the characteristics of the inverter. Important logic levels associated with binary logic are defined, and the concepts of the voltage transfer characteristic and noise margin are introduced. Later, the temporal behavior and time delays of the gates are addressed. A short review of Boolean algebra, used for representation and analysis of logic functions, is included. NMOS inverters with various types of load elements are studied in detail, including static design

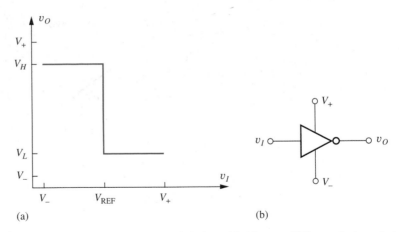

Figure 6.1 (a) Voltage transfer characteristic for an ideal inverter. (b) Inverter logic symbol.

and behavior in the time domain. In integrated circuits, transistors replace resistors as load devices in order to minimize circuit area. NAND, NOR, and complex gate implementations are based upon the basic inverter designs.

6.1 IDEAL LOGIC GATES

We begin our discussion of logic gates by considering the characteristics of the ideal logical inverter. Although we cannot achieve the ideal behavior, the concepts and definitions form the basis for our study of actual circuit implementations of MOS and bipolar logic families in Chapters 6–9.

In the discussions in this book, we limit consideration to binary logic, which requires only two discrete states for operation. In addition, the positive logic convention will be used throughout: The higher voltage level will correspond to a logic 1, and the lower voltage level will correspond to a logic 0.

The logic symbol and **voltage transfer characteristic (VTC)** for an ideal inverter are given in Fig. 6.1. The positive and negative power supplies, shown explicitly as V_+ and V_-, respectively, are not included in most logic diagrams. For input voltages v_I below the **reference voltage V_{REF}**, the output v_o will be in the **high logic level at the gate output V_H**. As the input voltage increases and exceeds V_{REF}, the output voltage changes abruptly to the **low logic level at the gate output V_L**. The output voltages corresponding to V_H and V_L generally fall between the supply voltages V_+ and V_- but may not be equal to either voltage. For an input equal to V_+ or V_-, the output does not necessarily reach either V_- or V_+. The actual levels depend on the individual logic family, and the reference voltage V_{REF} is determined by the internal circuitry of the gate.

In most digital designs, the power supply voltage is predetermined either by technology constraints or system-level power supply criteria. For example, $V_+ = 5.0$ V (with $V_- = 0$) represented the standard power supply for logic for many years. However, because of the power-dissipation, heat-removal, and breakdown-voltage limitations of advanced technology, many ICs now operate from supply voltages of 1.8 to 3.3 V, and many low-power systems must be designed to operate from battery voltages as low as 1.0 to 1.5 V.

6.2 LOGIC LEVEL DEFINITIONS AND NOISE MARGINS

Now, let us look at electronic implementations of the inverter. Conceptually, the basic inverter circuit consists of a load resistor and a switch controlled by the input voltage v_I, as indicated in Fig. 6.2(b). When closed, the switch forces v_o to V_L, and when open, the resistor sets the output to V_H. In Fig. 6.2(b), for example, $V_L = 0$ V and $V_H = V_+$.

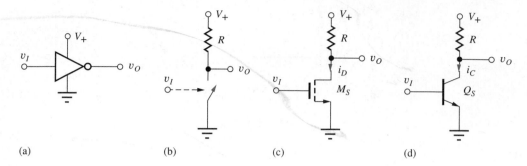

Figure 6.2 (a) Inverter operating with power supplies of 0 V and V_+. (b) Simple inverter circuit comprising a load resistor and switch. (c) Inverter with NMOS transistor switch. (d) Inverter with BJT switch.

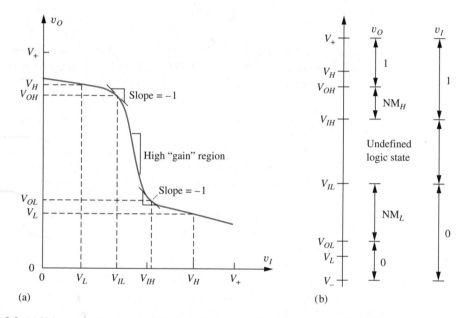

Figure 6.3 (a) Voltage transfer characteristic for the inverters in Fig. 6.2 with $V_- = 0$. (b) Voltage levels and logic state relationships for positive logic.

The voltage-controlled switch can be realized by either the MOS transistor in Fig. 6.2(c) or the bipolar transistor in Fig. 6.2(d). Transistors M_S and Q_S switch between two states: nonconducting or "off," and conducting or "on". Load resistor R sets the output voltage to $V_H = V_+$ when switching transistor M_S or Q_S is off. If the input voltage exceeds the threshold voltage of M_S or the turn-on voltage of the base-emitter junction of Q_S, the transistors conduct a current that causes the output voltage to drop to V_L. When transistors are used as switches, as in Figs. 6.2(c) and (d), $V_L \neq 0$ V. Detailed discussion of the design of these circuits appears later in this chapter and in Chapter 9.

In the inverter circuit, the transition between V_H and V_L does not occur in the abrupt manner indicated in Fig. 6.1 but is more gradual, as indicated by the more realistic transfer characteristic shown in Fig. 6.3(a). A single, well-defined value of V_{REF} does not exist. Instead, several additional input voltage levels are important.

When the input v_I is below the **input low-logic-level V_{IL}**, the output is defined to be in the high-output or 1 state. As the input voltage increases, the output voltage v_o falls until it reaches the low output or 0 state as v_I exceeds the voltage of the **input high-logic-level V_{IH}**. The input voltages V_{IL} and V_{IH} are defined by the points at which the slope of the voltage transfer characteristic equals -1.

Voltages below V_{IL} are reliably recognized as logic 0s at the input of a logic gate, and voltages above V_{IH} are recognized reliably as logic 1s at the input. Voltages corresponding to the region between V_{IL} and V_{IH} do not represent valid logic input levels and generate logically indeterminate output voltages. The transition region of high negative slope between these two points[1] represents an undefined logic state. The voltages labeled as V_{OL} and V_{OH} represent the gate output voltages at the -1 slope points and correspond to input levels of V_{IH} and V_{IL}, respectively.

In Part III of this book, we will find that the region of the VTC with a high negative slope between V_{IL} and V_{IH} corresponds to a large "voltage gain," and we actually use this region for amplification of analog signals. The gain is the slope of the voltage transfer characteristic. The higher the gain, the narrower will be the voltage range corresponding to the undefined logic state in Fig. 6.3.

An alternate representation of the voltages and voltage ranges appears in Fig. 6.3(b), along with quantities that represent the voltage noise margins. The various terms are defined more fully next.

6.2.1 LOGIC VOLTAGE LEVELS

V_L The nominal voltage corresponding to a low-logic state at the output of a logic gate for $v_I = V_H$. Generally, $V_- \leq V_L$.

V_H The nominal voltage corresponding to a high-logic state at the output of a logic gate for $v_I = V_L$. Generally, $V_H \leq V_+$.

V_{IL} The maximum input voltage that will be recognized as a low input logic level.

V_{IH} The minimum input voltage that will be recognized as a high input logic level.

V_{OH} The output voltage corresponding to an input voltage of V_{IL}.

V_{OL} The output voltage corresponding to an input voltage of V_{IH}.

For subsequent discussions of MOS logic, V_- will usually be taken to be 0 V, and V_+ will be either 2.5 V or 3.3 V. Five volts was commonly used in bipolar logic. However, other values are possible. For example, emitter-coupled logic, discussed in Chapter 9 has historically used $V_+ = 0$ V and $V_- = -5.2$ V or -4.5 V, and low-power ECL gates have been developed to operate with a total supply voltage span of only 2 V.

6.2.2 NOISE MARGINS

The **noise margin in the high state NM$_H$** and the **noise margin in the low state NM$_L$** represent "safety margins" that prevent the gate from producing erroneous logic decisions in the presence of noise sources. The noise margins are needed to absorb voltage differences that may arise between the outputs and inputs of various logic gates due to a variety of sources. These may be extraneous signals coupled into the gates or simply parameter variations between gates in a logic family.

Figure 6.4 shows several interconnected inverters and illustrates why noise margin is important. The signal and power interconnections on a printed circuit board or integrated circuit, which we most often draw as zero resistance wires (or short circuits), really consist of distributed *RLC* networks. In Fig. 6.4 the output of the first inverter, v_{O1}, and the input of the second inverter, v_{I2}, are not necessarily equal. As logic signals propagate from one logic gate to the next, their characteristics become degraded by the resistance, inductance, and capacitance of the interconnections. Rapidly switching signals may induce transient voltages and currents directly onto nearby signal lines through capacitive and inductive coupling indicated by C_c and M. In an RF environment, the interconnections may even act as small antennae that can couple additional extraneous signals into the logic circuitry. Similar problems occur in the power distribution network. Both direct current and transient currents during gate switching generate voltage drops across the various components (R_p, L_p, C_p) of the power distribution network.

---

[1] This region corresponds to a region of relatively high voltage gain. See Probs. 6.6 and 6.7.

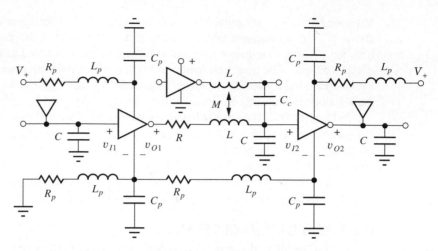

Figure 6.4 Inverters embedded in a signal and power and distribution network.

Noise margins also help absorb parameter variations that occur between individual logic gates. During manufacture, there will be unavoidable variations in device and circuit parameters, and variations will occur in the power supply voltages and operating temperature during application of the logic circuits. Normally, the logic manufacturer specifies worst-case values for V_H, V_L, V_{IL}, V_{OL}, V_{IH}, and V_{OH}. In our analysis, however, we will generally restrict ourselves to finding nominal values of these voltages.

There are a number of different ways to define the noise margin [1–3] of a logic gate. In this text, we will use a definition based on the input and output voltages at the -1 slope points on the inverter voltage transfer characteristic, as identified in Fig. 6.3:

NM_L The noise margin associated with a low input level is defined by

$$NM_L = V_{IL} - V_{OL} \qquad (6.1)$$

NM_H The noise margin associated with a high input level is defined by

$$NM_H = V_{OH} - V_{IH} \qquad (6.2)$$

The noise margins represent the voltages necessary to upset the logic levels in a long chain (actually an infinite chain) of inverters, or in the cross-coupled flip-flop storage elements that we explore in Chapter 8. The definitions in Eqs. (6.1) and (6.2) can be shown [1–3] to maximize the sum of the two noise margins. These definitions provide a reasonable metric for comparing the noise margins of different logic families and are relatively easy to understand and calculate.

EXERCISE: A certain TTL gate has the following values for its logic levels: $V_{OH} = 3.6$ V, $V_{OL} = 0.4$ V, $V_{IH} = 2.0$ V, $V_{IL} = 0.8$ V. What are the noise margins for this TTL gate?

ANSWERS: $NM_H = 1.6$ V; $NM_L = 0.4$ V

6.2.3 LOGIC GATE DESIGN GOALS

As we explore the design of logic gates, we should keep in mind a number of goals.

1. From Fig. 6.1, we see that the ideal logic gate is a highly nonlinear device that attempts to quantize the input signal into two discrete output levels. In the actual gate in Figs. 6.2 and 6.3, we should strive to minimize the width of the undefined input voltage range, and the noise margins should generally be as large as possible.

2. Logic gates should be unidirectional in nature. The input should control the output to produce a well-defined logic function. Voltage changes at the output of a gate should not affect the input side of the circuit.

3. The logic levels must be regenerated as the signal passes through the gate. In other words, the voltage levels at the output of one gate must be compatible with the input voltage levels of the same or similar logic gates.

4. The output of one gate should also be capable of driving the inputs of more than one gate. The number of inputs that can be driven by the output of a logic gate is called the **fan-out** capability of that gate. The term **fan in** refers to the number of input signals that may be applied to the input of a gate.

5. In most design situations, the logic gate should consume as little power (and area in an IC design) as needed to meet the speed requirements of the design.

6.3 DYNAMIC RESPONSE OF LOGIC GATES

In today's environment, even the general public is familiar with the seemingly endless increase in logic performance as we are bombarded with marketing of the latest microprocessors in terms of their clock frequencies, 1 GHz, 2 GHz, 3 GHz, and so on. The clock rate of a processor is ultimately set by the dynamic performance of the individual logic circuits. In engineering terms, the time domain performance of a logic family is cast in terms of its average propagation delay, rise time, and fall time as defined in this section.

Figure 6.5 shows idealized time domain waveforms for an inverter. The input and output signals are switching between the two static logic levels V_L and V_H. Because of capacitances in the circuits, the waveforms exhibit nonzero rise and fall times, and propagation delays occur between the switching times of the input and output waveforms.

6.3.1 RISE TIME AND FALL TIMES

The **rise time** t_r for a given signal is defined as the time required for the signal to make the transition from the **"10 percent" point** to the **"90 percent" point** on the waveform, as indicated in Fig. 6.5. The **fall time** t_f is defined as the time required for the signal to make the transition between the 90 percent point and the 10 percent point on the waveform. The voltages corresponding to the 10 percent and 90 percent points are defined in terms of V_L and V_H and the logic swing ΔV:

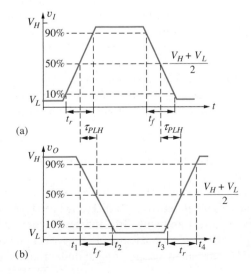

(a)

(b)

$$V_{10\%} = V_L + 0.1\,\Delta V$$
$$V_{90\%} = V_L + 0.9\,\Delta V = V_H - 0.1\,\Delta V \qquad (6.3)$$
$$\Delta V = V_H - V_L$$

Figure 6.5 Switching waveforms for an *idealized* inverter: (a) input voltage signal, (b) output voltage waveform.

where $\Delta V = V_H - V_L$. Rise and fall times usually have unequal values; the characteristic shapes of the input and output waveforms also differ.

6.3.2 PROPAGATION DELAY

Propagation delay is measured as the difference in time between the input and output signals reaching the **"50 percent" points** in their respective transitions. The 50 percent point is the voltage level corresponding to one-half the total transition between V_H and V_L:

$$V_{50\%} = \frac{V_H + V_L}{2} \tag{6.4}$$

The **propagation delay** on the **high-to-low output transition** is τ_{PHL} and that of the **low-to-high transition** is τ_{PLH}. In the general case, these two delays will not be equal, and the **average propagation delay** τ_P is defined by

$$\tau_P = \frac{\tau_{PLH} + \tau_{PHL}}{2} \tag{6.5}$$

Average propagation delay is one figure of merit that is commonly used to compare the performance of different logic families. In Chapters 6, 7, and 9, we explore the propagation delays for various MOS and bipolar logic circuits.

EXERCISE: Suppose the waveforms in Fig. 6.5 are those of an ECL gate with $V_L = -2.6$ V and $V_H = -0.6$ V, and $t_1 = 100$ ns, $t_2 = 105$ ns, $t_3 = 150$ ns, and $t_4 = 153$ ns. What are the values of $V_{10\%}$, $V_{90\%}$, $V_{50\%}$, t_r, and t_f?

ANSWERS: -2.4 V; -0.8 V; -1.6 V; 3 ns; 5 ns

6.3.3 POWER-DELAY PRODUCT

The overall performance of a logic family is ultimately determined by how much energy is required to change the state of the logic circuit. The traditional metric for comparing various logic families is the power-delay product, which tells us the amount of energy that is required to perform a basic logic operation.

Figure 6.6 shows the behavior of the average propagation delay of a general logic gate versus the average power supplied to the gate. The power consumed by a gate can be changed by increasing or decreasing the sizes of the transistors and resistors in the gate or by changing the power supply voltage. At low power levels, gate delay is dominated by intergate wiring capacitance, and the delay decreases as power increases. As device size and power are increased further, circuit delay becomes limited by the inherent speed of the electronic switching devices, and the delay becomes independent of power. In bipolar logic technology, the properties of the transistors begin to degrade at even higher power levels, and the delay can actually become worse as power increases further, as indicated in Fig. 6.6.

In the low power region, the propagation delay decreases in direct proportion to the increase in power. This behavior corresponds to a region of constant **power-delay product (PDP),**

$$\text{PDP} = P\tau_P \tag{6.6}$$

in which P is the average power dissipated by the logic gate. The PDP represents the energy (Joules) required to perform a basic logic operation.

Early logic families had power-delay products of 10 to 100 pJ (1 pJ $= 10^{-12}$ J), whereas many of today's IC logic families now have PDPs in the 10 to 100 fJ range (1 fJ $= 10^{-15}$ J). It has been estimated that the minimum energy required to reliably differentiate two logic states is on the order of $(\ln 2)kT$, which is approximately 4×10^{-20} J at room temperature [4]. Thus even today's best logic families have power-delay products that are many orders of magnitude from the ultimate limit [5].

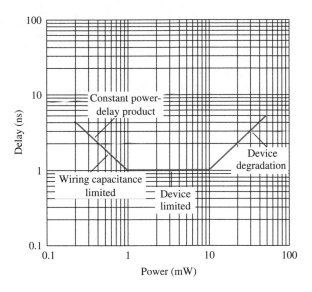

Figure 6.6 Logic gate delay versus power dissipation.

TABLE 6.1
Basic Boolean Operations

OPERATION	BOOLEAN REPRESENTATION
NOT	$Z = \overline{A}$
OR	$Z = A + B$
AND	$Z = A \cdot B = AB$
NOR	$Z = \overline{A + B}$
NAND	$Z = \overline{A \cdot B} = \overline{AB}$

TABLE 6.2
NOT, OR, AND Gate Truth Tables

B	A	NOT (INVERTER) $Z = A$	OR GATE $Z = A + B$	AND GATE $Z = AB$
0	0	0	0	0
0	1	1	1	0
1	0	0	1	0
1	1	1	1	1

TABLE 6.3
NOR and NAND Gate Truth Table

A	B	NOR $Z = \overline{A + B}$	NAND $Z = \overline{AB}$
0	0	1	1
0	1	0	1
1	0	0	1
1	1	0	0

EXERCISE: (a) What is the power-delay product at low power for the logic gate characterized by Fig. 6.6? (b) What is the PDP at $P = 3$ mW? (c) At 20 mW?

ANSWERS: 1 pJ; 3 pJ; 40 pJ

6.4 REVIEW OF BOOLEAN ALGEBRA

In order to be able to effectively deal with logic system analysis and design, we need a mathematical representation for networks of logic gates. Fortunately, way back in 1849, G. Boole [6] presented a powerful mathematical formulation for dealing with logical thought and reasoning, and the formal algebra we use today to manipulate binary logic expressions is known as **Boolean algebra.** Tables 6.1 to 6.3 and the following discussion summarize Boolean algebra.

Table 6.1 lists the basic logic operations that we need. The logic function at the gate output is represented by variable Z and is a function of logical input variables A and B: $Z = f(A, B)$. To perform general logic operations, a logic family must provide logical inversion (NOT) plus at least one other function of two input variables, such as the OR or AND functions. We will find in Chapter 7 that NMOS logic can easily be used to implement **NOR gates** as well as **NAND gates,** and in Chapter 9 we will see that the basic TTL gate provides a NAND function whereas OR/NOR logic is provided by the basic ECL gate. Note in Table 6.1 that the NOT function is equivalent to the output of either a single input NOR gate or NAND gate.

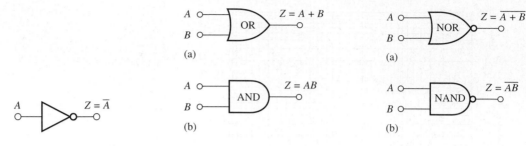

Figure 6.7 Inverter symbol.

Figure 6.8 (a) OR gate symbol. (b) AND gate symbol.

Figure 6.9 (a) NOR gate symbol. (b) NAND gate symbol.

TABLE 6.4
Useful Boolean Identities

$A + 0 = A$	$A \cdot 1 = A$	Identity operation
$A + B = B + A$	$AB = BA$	Commutative law
$A + (B + C) = (A + B) + C$	$A(BC) = (AB)C$	Associative law
$A + BC = (A + B)(A + C)$	$A(B + C) = AB + AC$	Distributive law
$A + \overline{A} = 1$	$A \cdot \overline{A} = 0$	Complements
$A + A = A$	$A \cdot A = A$	Idempotency
$A + 1 = 1$	$A \cdot 0 = 0$	Null elements
$\overline{A} + \overline{B} = \overline{AB}$	$\overline{AB} = \overline{A} + \overline{B}$	DeMorgan's theorem

Truth tables and logic symbols for the five functions in Table 6.1 appear in Tables 6.2 and 6.3 and Figs. 6.7 to 6.9. The truth table presents the output Z for all possible combinations of the input variables A and B. The inverter, $Z = \overline{A}$, has a single input, and the output represents the logical inversion or complement of the input variable, as indicated by the overbar (Table 6.2; Fig. 6.7).

Table 6.2 presents the truth tables for a two-input **OR gate** and a two-input **AND gate,** respectively, and the corresponding logic symbols appear in Fig. 6.8. The OR operation is indicated by the $+$ symbol; its output Z is a 1 when either one or both of the input variables A or B is a 1. The output is a 0 only if both inputs are 0. The AND operation is indicated by the $\cdot$ symbol, as in $A \cdot B$, or in a more compact form as simply AB, and the output Z is a 1 only if both the input variables A and B are in the 1 state. If either input is 0, then the output is 0. We shall use AB to represent A AND B throughout the rest of this text.

Table 6.3 gives the truth tables for the two-input NOR gate and the two-input NAND gate, respectively, and the logic symbols appear in Fig. 6.9. These functions represent the complements of the OR and AND operations—that is, the OR or AND operations followed by logical inversion. The NOR operation is represented as $Z = \overline{A + B}$, and its output Z is a 1 only if both inputs are 0. For the NAND operation, $Z = \overline{AB}$, output Z is a 1 except when both the input variables A or B are in the 1 state.

In this chapter and Chapter 8, we will find that a major advantage of MOS logic is its capability to readily form more complex logic functions, particularly logic expressions represented in a complemented **sum-of-products** or **AND-OR-INVERT** (AOI) form:

$$Z = \overline{AB + CD + E} \quad \text{or} \quad Z = \overline{ABC + DE} \quad (6.7)$$

The Boolean identities that are shown in Table 6.4 can be very useful in finding simplified logic expressions, such as those expressions in Eq. (6.7). This table includes the identity operations as well as the basic commutative, associative, and distributive laws of Boolean algebra.

EXAMPLE 6.1   **LOGIC EXPRESSION SIMPLIFICATION**

Here is an example of the use of Boolean identities to simplify a logic expression.

PROBLEM Use the Boolean relationships in Table 6.4 to show that the expression

$$Z = A\overline{B}C + ABC + \overline{A}BC \quad \text{can be reduced to} \quad Z = (A + B)C.$$

SOLUTION **Known Information and Given Data:** Two expressions for Z just given; Boolean identities in Table 6.4.

Unknowns: Proof that Z is equivalent to $(A + B)C$

Approach: Apply various identities from Table 6.4 to simplify the formula for Z

Assumptions: None

Analysis:

$$Z = A\overline{B}C + ABC + \overline{A}BC$$
$$Z = A\overline{B}C + ABC + ABC + \overline{A}BC \quad \text{using } ABC = ABC + ABC$$
$$Z = A(\overline{B} + B)C + (\overline{A} + A)BC \quad \text{using distributive law}$$
$$Z = A(1)C + (1)BC \quad \text{using } (\overline{B} + B) = (B + \overline{B}) = 1$$
$$Z = AC + BC \quad \text{since } A(1)C = AC(1) = AC$$
$$Z = (A + B)C \quad \text{using distributive law}$$

Check of Results: We have reached the desired answer. A double check indicates the sequence of steps appears valid.

EXERCISE: Simplify the logic expression $Z = (A + B)(B + C)$

ANSWER: $Z = B + AC$

6.5 NMOS LOGIC DESIGN

The rest of Chapter 6 focuses on understanding the design of MOS logic gates that use n-channel MOS transistors (NMOS logic) and p-channel MOS transistors (PMOS logic). Study of these circuits provides a background for understanding many important logic circuit concepts as well as the improvements gained by going to CMOS circuitry, which is the topic for Chapter 7. The discussion begins by investigating the design of the MOS inverter in order to gain an understanding of its voltage transfer characteristic and noise margins. Inverters with four different NMOS load configurations are considered: the resistor load, saturated load, linear load, and depletion-mode load circuits. Pseudo NMOS is a modern extension of classic NMOS logic that uses a PMOS transistor as a load device. NOR, NAND, and more complex logic gates can be easily designed as simple extensions of the reference inverter designs. Later, the rise time, fall time, and propagation delays of the gates are analyzed.

The drain current of the MOS device depends on its gate-source voltage v_{GS}, drain-source voltage v_{DS}, and source-bulk voltage v_{SB}, and on the device parameters, which include the transconductance parameter K'_n, threshold voltage V_{TN}, and width-to-length or **W/L ratio.** The power supply voltage constrains the range of v_{GS} and v_{DS}, and the technology sets the values of K'_n and V_{TN}. Thus, the

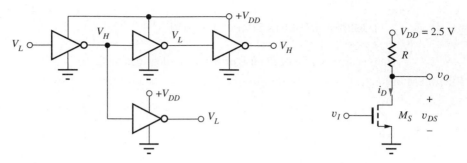

Figure 6.10 A network of inverters.

Figure 6.11 NMOS inverter with resistive load.

circuit designer's job is to choose the circuit topology and the W/L ratios of the MOS transistors to achieve the desired logic function.

In most logic design situations, the power supply voltage is predetermined by either technology reliability constraints or system-level criteria. For example, as mentioned in Sec. 6.1, V_{DD}[2] $= 5.0$ V was the standard power supply for logic for many years. However, 1.8–3.3 V power supply levels are gaining widespread use. In addition, many portable low-power systems, such as cell phones and PDAs, must operate from battery voltages as low as 1.0 to 1.5 V.

We begin our study of MOS logic circuit design by considering the detailed design of the NMOS inverter with the resistor load that was introduced in Chapter 5. Although we will seldom use this exact circuit, it provides a good basis for understanding operation of the basic logic gate. In integrated logic circuits, the load resistor occupies too much silicon area, and is replaced by a second MOS transistor. NMOS "load devices" can be connected in three different configurations called the saturated load, linear load, and depletion-mode load circuits, whereas pseudo NMOS uses a PMOS load device. We will explore the design of the NMOS load configurations in detail in this and Secs. 6.6 through 6.7.

6.5.1 NMOS INVERTER WITH RESISTIVE LOAD

Complex digital systems can consist of millions of logic gates, and it is helpful to remember that each individual logic gate is generally interconnected in a larger network. The output of one logic gate drives the input of another logic gate, as shown schematically by the four inverters in Fig. 6.10. Thus, a gate has $v_O = V_H$ when an input voltage $v_I = V_L$ is applied to its input, and vice versa.

The basic inverter circuit shown in Fig. 6.11 consists of an NMOS switching device M_S designed to force v_O to V_L and a **resistor load** element to "pull" the output up toward the power supply V_{DD}. The NMOS transistor is designed to switch between the triode region for $v_I = V_H$ and the cutoff (nonconducting) state for $v_I = V_L$. The circuit designer must choose the values of the load resistor R and the W/L ratio of **switching transistor** M_S so the inverter meets a set of design specifications. In this case, these two design variables permit us to choose the V_L level and set the total power dissipation of the logic gate.

Let us explore the inverter operation by considering the requirements for the design of such a logic gate. Writing the equation for the output voltage, we find

$$v_O = v_{DS} = V_{DD} - i_D R \tag{6.8}$$

When the input voltage is at a low state, $v_I = V_L$, M_S should be cut off, with $i_D = 0$, so that

$$v_O = V_{DD} = V_H \tag{6.9}$$

Thus, in this particular logic circuit, the value of V_H is set by the power supply voltage $V_{DD} = 2.5$ V.

---

[2] V_{DD} and V_{SS} have traditionally been used to denote the positive and negative power supply voltages in MOS circuits.

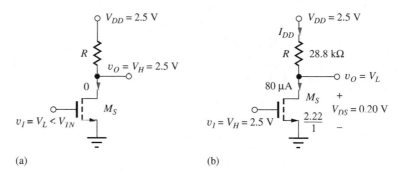

Figure 6.12 Inverters in the (a) $v_I = V_L(0)$ and (b) $v_I = V_H(1)$ logic states.

To ensure that transistor M_S is cut off when the input is equal to V_L, as in Fig. 6.12(a), the gate-source voltage of M_S ($v_{GS} = V_L$) must be less than its threshold voltage V_{TN}. For $V_{TN} = 0.6$ V, a normal design point would be for V_L to be in the range of 25 to 50 percent of V_{TN} or 0.15 to 0.30 V to ensure adequate noise margins. Let us assume a design value of $V_L = 0.20$ V.

DESIGN NOTE DESIGN OF V_L

To ensure that switching transistor M_S is cut off when the input is in the low logic state, V_L is designed to be 25 to 50 percent of the threshold voltage of switch M_S. This choice also provides a reasonable value for noise margin NM_L.

6.5.2 DESIGN OF THE W/L RATIO OF M_S

The value of W/L required to set $V_L = 0.20$ V can be calculated if we know the parameters of the MOS device. For now, the values $V_{TN} = 0.6$ V and $K'_n = 100 \times 10^{-6}$ A/V^2 will be used. In addition, we need to know a value for the desired operating current of the inverter. The current is determined by the permissible power dissipation of the NMOS gate when $v_o = V_L$. Using $P = 0.20$ mW (see Probs. 6.1 and 6.2),[3] the current in the gate can be found from $P = V_{DD} \times I_{DD}$. For our circuit,

$$0.20 \times 10^{-3} = 2.5 \times I_{DD} \quad \text{or} \quad I_{DD} = 80\ \mu\text{A} = i_D$$

Now we can determine the value for the W/L ratio of the NMOS switching device from the MOS drain current expression using the circuit conditions in Fig. 6.12(b). In this case, the input is set equal to $V_H = 2.5$ V, and the output of the inverter should then be at V_L. The expression for the drain current in the triode region of the device is used because $v_{GS} - V_{TN} = 2.5\ \text{V} - 0.6\ \text{V} = 1.9$ V, and $v_{DS} = V_L = 0.20$ V, yielding $v_{DS} < v_{GS} - V_{TN}$.

$$i_D = K'_n \left(\frac{W}{L}\right)_S (v_{GS} - V_{TN} - 0.5 v_{DS})\, v_{DS} \tag{6.10}$$

or

$$8 \times 10^{-5}\ \text{A} = \left(100 \times 10^{-6}\ \frac{\text{A}}{\text{V}^2}\right)\left(\frac{W}{L}\right)_S (2.5\ \text{V} - 0.6\ \text{V} - 0.10\ \text{V})(0.20\ \text{V})$$

Solving Eq. (6.10) for $(W/L)_S$ gives $(W/L)_S = 2.22/1$.

---

[3] It would be worth exploring these problems before continuing.

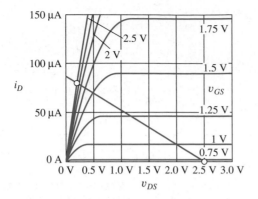

Figure 6.13 MOSFET output characteristics and load line.

6.5.3 LOAD RESISTOR DESIGN

The value of the load resistor R is chosen to limit the current when $v_O = V_L$ and is found from

$$R = \frac{V_{DD} - V_L}{i_D} = \frac{(2.5 - 0.20)\ \text{V}}{8 \times 10^{-5}\ \text{A}} = 28.8\ \text{k}\Omega \tag{6.11}$$

These design values are shown in the circuit in Fig. 6.12(b).

EXERCISE: Redesign the logic gate in Fig. 6.12 to operate at a power of 0.4 mW while maintaining $V_L = 0.20$ V.

ANSWER: $(W/L)_S = 4.44/1$; $R = 14.4\ \text{k}\Omega$

6.5.4 LOAD-LINE VISUALIZATION

An important way to visualize the operation of the inverter is to draw the load line on the MOS transistor output characteristics as in Fig. 6.13. Equation (6.8), repeated here, represents the equation for the load line:

$$v_{DS} = V_{DD} - i_D R$$

When the transistor is cut off, $i_D = 0$ and $v_{DS} = V_{DD} = 2.5$ V, and when the transistor is on, the MOSFET is operating in the triode region, with $v_{GS} = V_H = 2.5$ V and $v_{DS} = v_O = V_L = 0.20$ V. The MOSFET switches between the two operating points on the load line, as indicated by the circles in Fig. 6.13. At the right-hand end of the load line, the MOSFET is cut off. At the Q-point near the left end of the load line, the MOSFET represents a relatively low resistance, and the current is determined primarily by the load resistance. (Note how the Q-point is nearly independent of v_{GS}.)

DESIGN EXAMPLE 6.2   **DESIGN OF AN INVERTER WITH RESISTIVE LOAD**

Design a resistively loaded NMOS inverter to operate from a 3.3-V power supply.

PROBLEM   Design an inverter with a resistive load for $V_{DD} = 3.3$ V and $P = 0.1$ mW with $V_L = 0.2$ V. Assume $K'_n = 60\ \mu\text{A/V}^2$ and $V_{TN} = 0.75$ V.

SOLUTION   **Known Information and Given Data:** Circuit topology in Fig. 6.11; $V_{DD} = 3.3$ V, $P = 0.1$ mW, $V_L = 0.2$ V, $K'_n = 60\ \mu\text{A/V}^2$, and $V_{TN} = 0.75$ V

Unknowns: Value of load resistor R; W/L ratio of switching transistor M_S

Approach: Use the power dissipation specification to find the current I_{DD} for $v_O = V_L$. Use V_{DD}, V_L, and I_{DD} to calculate R. Determine V_H. Use V_H, V_L, and I_{DD} to find $(W/L)_S$.

Assumptions: M_S is off for $v_I = V_L$; M_S is in the triode region for $v_O = V_L$.

Analysis: Using the power specification with the inverter circuit in Fig. 6.11, we have

$$I_{DD} = \frac{P}{V_{DD}} = \frac{10^{-4} \text{ W}}{3.3 \text{ V}} = 30.3 \text{ μA} \qquad R = \frac{V_{DD} - V_L}{I_{DD}} = \frac{3.3 - 0.2}{30.3} \frac{\text{V}}{\text{μA}} = 102 \text{ kΩ}$$

For $v_I = V_L = 0.2$ V, the MOSFET will be off since 0.2 V is less than the threshold voltage, and the output high level will be $V_H = V_{DD} = 3.3$ V. The triode region expression for the MOSFET drain current with $v_{GS} = v_I = V_H$ and $v_{DS} = v_O = V_L$ is

$$I_D = K_n' \left(\frac{W}{L}\right)_S \left(V_H - V_{TN} - \frac{V_L}{2}\right) V_L$$

Equating this expression to the drain current yields

$$30.3 \text{ μA} = (60 \times 10^{-6}) \left(\frac{W}{L}\right)_S \left(3.3 - 0.75 - \frac{0.2}{2}\right) 0.2 \rightarrow \left(\frac{W}{L}\right)_S = \frac{1.03}{1}$$

Thus our completed design values are $R = 102$ kΩ and $(W/L)_S = 1.03/1$.

Check of Results: We should check the triode region assumption for the MOSFET for $v_O = V_L$: $V_{GS} - V_{TN} = 3.3 - 0.75 = 2.55$ V, which is indeed greater than $V_{DS} = 0.2$ V. Let us also double check the value of W/L by using it to calculate the drain current:

$$I_D = (60 \times 10^{-6}) \left(\frac{1.03}{1}\right) \left(3.3 - 0.75 - \frac{0.2}{2}\right) 0.2 = 30.3 \text{ μA} \quad ✔$$

Discussion: This new design for a reduced voltage and reduced power requires a larger value of load resistor to limit the current, but a smaller device to conduct the reduced level of current.

Computer-Aided Analysis: Let us verify our design values with SPICE. The circuit drawn with a schematic capture tool is given below. The NMOS transistor uses the LEVEL = 1 model with KP = 6.0E-5, VTO = 1, W = 1.03U, and L = 1U. The Q-point of the transistor is (30.4 μA, 0.201 V), which agrees with the design specifications.

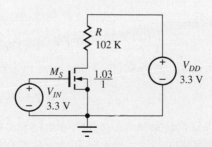

EXERCISE: (a) Redesign the inverter in Ex. 6.2 to have $V_L = 0.1$ V with $R = 102$ kΩ. (b) Verify your design with SPICE.

ANSWER: $(W/L)_S = 2.09/1$

6.5.5 ON-RESISTANCE OF THE SWITCHING DEVICE

When the logic gate output is in the low state, the output voltage can also be calculated from a resistive voltage divider formed by the load resistor R and the **on-resistance** R_{on} of the MOSFET, as in Fig. 6.14.

$$V_L = V_{DD} \frac{R_{on}}{R_{on} + R} = V_{DD} \frac{1}{1 + \dfrac{R}{R_{on}}} \tag{6.12}$$

where

$$R_{on} = \frac{v_{DS}}{i_D} = \frac{1}{K'_n \dfrac{W}{L}\left(v_{GS} - V_{TN} - \dfrac{v_{DS}}{2}\right)} \tag{6.13}$$

R_{on} must be much smaller than R in order for V_L to be small. It is important to recognize that R_{on} represents a nonlinear resistor because the value of R_{on} is dependent on v_{DS}, the voltage across the resistor terminals. All the NMOS gates that we study in this chapter demonstrate **"ratioed" logic**— that is, designs in which the on-resistance of the switching transistor must be much smaller than that of the load resistor in order to achieve a small value of V_L ($R_{on} \ll R$).

EXAMPLE 6.3   **ON-RESISTANCE CALCULATION**

Find the on-resistance for the MOSFET in the completed inverter design in Fig. 6.12(b).

PROBLEM What is the value of the on-resistance for the NMOS FET in Fig. 6.12 when the output voltage is at V_L?

SOLUTION **Known Information and Given Data:** $K'_n = 100\ \mu\text{A/V}^2$, $V_{TN} = 0.60\ \text{V}$, $W/L = 2.22/1$, $V_{DS} = V_L = 0.20\ \text{V}$

Unknowns: On-resistance of the switching transistor

Approach: Use the known values to evaluate Eq. (6.13).

Assumptions: The transistor is in the triode region of operation.

Analysis: R_{on} can be found using Eq. (6.13).

$$R_{on} = \frac{1}{\left(100 \times 10^{-6} \dfrac{\text{A}}{\text{V}^2}\right)\left(\dfrac{2.22}{1}\right)\left(2.5 - 0.60 - \dfrac{0.20}{2}\right)\text{V}} = 2.50\ \text{k}\Omega$$

Check of Results: We can check this value by using it to calculate V_L:

$$V_L = V_{DD} \frac{R_{on}}{R_{on} + R} = 2.5\ \text{V} \frac{2.5\ \text{k}\Omega}{2.5\ \text{k}\Omega + 28.8\ \text{k}\Omega} = 0.20\ \text{V}$$

$R_{on} = 2.5\ \text{k}\Omega$ does indeed give the correct value of V_L. Note that $R_{on} \ll R$. Checking the triode region assumption: $V_{GS} - V_{TN} = 2.5 - 0.6 = 1.9\ \text{V}$ and $V_{DS} = V_L = 0.20\ \text{V}$. ✔

EXERCISE: What value of R_{on} is needed to set $V_L = 0.15\ \text{V}$ in Ex. 6.3? What is the new value of W/L needed for the MOSFET to achieve this value of R_{on}?

ANSWERS: 1.84 kΩ; 2.98/1

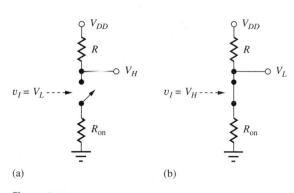

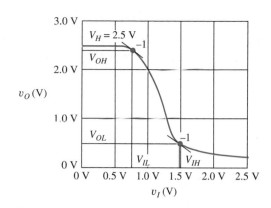

Figure 6.14 Simplified representation of an inverter: (a) the off or nonconducting state, (b) the on or conducting state.

Figure 6.15 Simulated voltage transfer characteristic of an NMOS inverter with resistive load.

EXERCISE: What is the value of R_{on} for the MOSFET in Ex. 6.2? Use R_{on} to find V_L.

ANSWERS: 6.61 kΩ; 0.201 V

6.5.6 NOISE MARGIN ANALYSIS

Figure 6.15 is a SPICE simulation of the voltage transfer function for the completed inverter design from Fig. 6.12. Now we are in a position to find the values of V_{IL}, V_{OL}, V_{IH}, and V_{OH} that correspond to the points at which the slope of the voltage transfer characteristic for the inverter is equal to -1, as defined in Sec. 6.2.

6.5.7 CALCULATION OF V_{IL} AND V_{OH}

Our analysis begins with the expression for the load line, repeated here from Eq. (6.8):

$$v_O = V_{DD} - i_D R \tag{6.14}$$

Referring to Fig. 6.15 with $v_I = V_{IL}$, v_{GS} is small and v_{DS} is large, so we expect the MOSFET to be operating in saturation, with drain current given by

$$i_D = (K_n/2)(v_{GS} - V_{TN})^2 \quad \text{where } K_n = K_n'(W/L) \text{ and } v_{GS} = v_I$$

Substituting this expression for i_D in load-line Eq. (6.14),

$$v_O = V_{DD} - \frac{K_n}{2}(v_I - V_{TN})^2 R \tag{6.15}$$

and taking the derivative of v_O with respect to v_I results in

$$\frac{dv_O}{dv_I} = -K_n(v_I - V_{TN})R \tag{6.16}$$

Setting this derivative equal to -1 for $v_I = V_{IL}$ yields

$$V_{IL} = V_{TN} + \frac{1}{K_n R} \quad \text{with} \quad V_{OH} = V_{DD} - \frac{1}{2K_n R} \tag{6.17}$$

We see that the value of V_{IL} is slightly greater than V_{TN}, since the input must exceed V_{TN} for M_S to begin conduction, and V_{OH} is slightly less than V_{DD}. The $1/K_n R$ terms represent the ratio of the transistor's transconductance parameter to the value of the load resistor. As K_n increases for a given value of R, V_{IL} decreases and V_{OH} increases.

EXERCISE: Show that $(1/K_n R)$ has the units of voltage.

6.5.8 CALCULATION OF V_{IH} AND V_{OL}

For $v_I = V_{IH}$, v_{GS} is large and v_{DS} is small, so we now expect the MOSFET to be operating in the triode region with drain current given by $i_D = K_n[v_{GS} - V_{TN} - (v_{DS}/2)]v_{DS}$. Substituting this expression for i_D into Eq. (6.14) and realizing that $v_O = v_{DS}$ yields

$$v_O = V_{DD} - K_n R \left(v_I - V_{TN} - \frac{v_O}{2}\right) v_O \qquad \text{or} \qquad \frac{v_O^2}{2} - v_O\left[v_I - V_{TN} + \frac{1}{K_n R}\right] + \frac{V_{DD}}{K_n R} = 0 \tag{6.18}$$

Solving for v_O and then setting $dv_O/dv_I = -1$ for $v_I = V_{IH}$ yields

$$V_{IH} = V_{TN} - \frac{1}{K_n R} + 1.63\sqrt{\frac{V_{DD}}{K_n R}} \qquad \text{with} \qquad V_{OL} = \sqrt{\frac{2V_{DD}}{3K_n R}} \tag{6.19}$$

Combining the results from Eqs. (6.17) and (6.19) yields expressions for the noise margins:

$$\text{NM}_H = V_{DD} - V_{TN} + \frac{1}{2K_n R} - 1.63\sqrt{\frac{V_{DD}}{K_n R}} \quad \text{and} \quad \text{NM}_L = V_{TN} + \frac{1}{K_n R} - \sqrt{\frac{2V_{DD}}{3K_n R}} \tag{6.20}$$

The product $K_n R$ compares the drive capability of the MOSFET to the resistance of the load resistor, and the noise margins increase as $K_n R$ increases for typical values of $K_n R$ greater than two.

EXAMPLE 6.4   **NOISE MARGIN CALCULATION FOR THE RESISTIVE LOAD INVERTER**

Find the noise margins associated with the inverter design in Fig. 6.12(b).

PROBLEM   Calculate the noise margins for the inverter in Fig. 6.12(b).

SOLUTION   **Known Information and Given Data:** The NMOS inverter circuit with resistor load in Fig. 6.11 with $R = 28.8\ \text{k}\Omega$, $(W/L)_S = 2.22/1$, $K_n' = 100\ \mu\text{A/V}^2$, and $V_{TN} = 0.60\ \text{V}$

Unknowns: The values of V_{IL}, V_{OH}, V_{IH}, V_{OL}, NM_L, and NM_H

Approach: Use the given data to evaluate Eqs. (6.17) and (6.18). Use the results to find the noise margins: $\text{NM}_H = V_{OH} - V_{IH}$ and $\text{NM}_L = V_{IL} - V_{OL}$.

Assumptions: Equation (6.17) assumes saturation region operation; Eq. (6.18) assumes triode region operation.

Analysis: For the inverter design in Fig. 6.12(b),

$$V_{TN} = 0.6\ \text{V} \qquad K_n'\frac{W}{L} = 100\left(\frac{2.22}{1}\right)\frac{\mu\text{A}}{\text{V}^2} = 222\frac{\mu\text{A}}{\text{V}^2} \qquad R = 28.8\ \text{k}\Omega$$

Evaluating Eq. (6.17),

$$V_{IL} = 0.6 + \frac{1}{(222\ \mu\text{A})(28.8\ \text{k}\Omega)} = 0.756\ \text{V} \quad \text{and} \quad V_{OH} = 2.5 - \frac{1}{2(222\ \mu\text{A})(28.8\ \text{k}\Omega)} = 2.42\ \text{V}$$

and Eq. (6.18),

$$V_{IH} = 0.6 - \frac{1}{(222\ \mu\text{A})(28.8\ \text{k}\Omega)} + 1.63\sqrt{\frac{2.5}{(222\ \mu\text{A})(28.8\ \text{k}\Omega)}} = 1.46\ \text{V}$$

$$V_{OL} = \sqrt{\frac{2(2.5)}{3(222\ \mu\text{A})(28.8\ \text{k}\Omega)}} = 0.51\ \text{V}$$

The noise margins are found to be

$$NM_H = 2.42 - 1.46 = 0.96 \text{ V} \qquad \text{and} \qquad NM_L = 0.76 - 0.51 = 0.25 \text{ V}$$

Check of Results: The values of V_{IL}, V_{OH}, V_{IH}, and V_{OL} all agree well with the simulation results in Fig. 6.15. Equation (6.17) is based on the assumption of saturation region operation. We should check to see if this assumption is consistent with the results in Eq. (6.17): $v_{DS} = 2.42$ and $v_{GS} - V_{TN} = 0.76 - 0.6 = 0.16$. Because $v_{DS} > (v_{GS} - V_{TN})$, our assumption was correct. Similarly, Eq. (6.18) is based on the assumption of triode region operation. Checking this assumption, we have $v_{DS} = 0.51$ and $v_{GS} - V_{TN} = 1.46 - 0.6 = 0.86$. Since $v_{DS} < (v_{GS} - V_{TN})$, our assumption was correct.

Discussion: Our analysis indicates that a long chain of inverters can tolerate electrical noise and process variations equivalent to 0.25 V in the low-input state and 0.96 V in the high state. Note that it is common for the values of the two noise margins to be unequal, as illustrated here.

EXERCISE: (a) Find the noise margins for the inverter in Ex. 6.2. (b) Verify your results with SPICE.

ANSWERS: $NM_L = 0.32$ V; $NM_H = 1.45$ V ($V_{IL} = 0.090$ V, $V_{OH} = 3.22$ V, $V_{IH} = 1.77$ V, $V_{OL} = 0.591$ V)

As mentioned earlier, V_{IL}, V_{OL}, V_{IH}, and V_{OH}, as specified by a manufacturer, actually represent guaranteed specifications for a given logic family and take into account the full range of variations in technology parameters, temperature, power supply, loading conditions, and so on. In Ex. 6.4, we have computed only V_{IL}, V_{OL}, V_{IH}, and V_{OH} and the noise margins under nominal conditions at room temperature.

6.5.9 LOAD RESISTOR PROBLEMS

The NMOS inverter with resistive load has been used to introduce the concepts associated with static logic gate design. Although a simple discrete component logic gate could be built using this circuit, IC realizations do not use resistive loads because the resistor would take up far too much area.

To explore the load resistor problem further, consider the rectangular block of semiconductor material in Fig. 6.16 with a resistance given by

$$R = \frac{\rho L}{t W} \tag{6.21}$$

where $\rho =$ resistivity, and L, W, $t =$ length, width, and thickness of the resistor, respectively.

In an integrated circuit, a resistor might typically be fabricated with a thickness of 1 μm in a silicon region with a resistivity of 0.001 Ω · cm. For these parameters, the 28.8-kΩ load resistor in

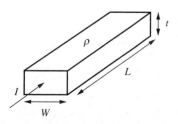

Figure 6.16 Geometry for a simple rectangular resistor.

the previous section would require the ratio of L/W to be

$$\frac{L}{W} = \frac{Rt}{\rho} = \frac{(2.88 \times 10^4 \ \Omega)(1 \times 10^{-4} \ \text{cm})}{0.001 \ \Omega \cdot \text{cm}} = \frac{2880}{1}$$

If the resistor width W were made a minimum line width of 1 μm, which we will call the **minimum feature size F,** then the length L would be 2880 μm, and the area would be 2880 μm^2.

For the switching device M_S, W/L was found to be 2.22/1. If the device channel length is made equal to the minimum feature size of 1 μm, then the gate area of the NMOS device is only 2.22 μm^2. Thus, the load resistor would consume more than 1000 times the area of the switching transistor M_S. This is simply not an acceptable utilization of area in IC design. The solution to this problem is to replace the load resistor with a transistor.

6.6 TRANSISTOR ALTERNATIVES TO THE LOAD RESISTOR

Six different alternatives for replacing the load resistor with a three-terminal MOSFET are shown in Fig. 6.17. When we replace the load resistor with a transistor, we are replacing the two terminal resistor with a three-terminal (or actually four-terminal) MOSFET, and we must decide where to connect the extra terminals. Current in the NMOS transistor goes from drain to source, so these terminals attach to the terminals where the resistor was removed. However, there are a number of possibilities for the gate terminal as indicated in the figure.

One possibility is to connect the gate to the source as in Fig. 6.17(a). However, for this case $v_{GS} = 0$, and MOSFET M_L will be non-conducting, assuming it is an enhancement-mode device with $V_{TN} > 0$. A similar problem exists if the gate is grounded as in Fig. 6.17(b). Here again, the connection forces $v_{GS} \leq 0$, and the load device is always turned off. Neither of these two

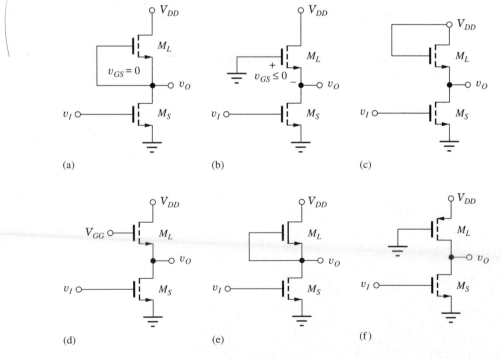

Figure 6.17 NMOS inverter load device options: (a) NMOS inverter with gate of the load device connected to its source, (b) NMOS inverter with gate of the load device grounded, (c) saturated load inverter, (d) linear load inverter, (e) depletion load inverter, and (f) pseudo NMOS inverter. Note that (a) and (b) are not useful.

alternatives work because an enchancement-mode NMOS device can never conduct current under these conditions.

The next three sections present an overview of the behavior of the circuits in Figs. 6.17(c–e). Saturated load logic played an important role in the history of electronic circuits. This form of logic was used in the design of early microprocessors, first in PMOS and then in NMOS technology. We briefly explore its static design in the next section. The characteristics of the linear load and depletion load technologies are outlined in Sections 6.6.2 and 6.6.3. The pseudo NMOS circuit in Fig. 6.17(d) is often encountered today in CMOS design, and a detailed discussion of its design appears in Section 6.6.4.

6.6.1 THE NMOS SATURATED LOAD INVERTER

The first workable circuit alternative, used in NMOS (and earlier in PMOS) logic design for many years, appears in Fig. 6.17(c). Here $v_{DS} = v_{GS}$, and the load device will operate in the saturation region because $v_{GS} - V_{TN} = v_{DS} - V_{TN} < v_{DS}$ for $V_{TN} > 0$. Because the connection forces the load transistor to always operate in the saturation region, we refer to this circuit as the **saturated load inverter.**

Figure 6.18(a) shows the circuit diagram for the saturated load inverter, and Fig. 6.18(b) gives the cross section of the inverter implementation in integrated circuit form. Here we see a very important aspect of the structure. The substrate is common to both transistors; thus, the substrate voltage must be the same for both M_S and M_L in the inverter, and the substrate terminal of M_L cannot be connected to its source as originally indicated in Fig. 6.17(c). This extra substrate terminal is most commonly connected to ground (0 V) (although voltages of -5 V and -8 V have been used in the past). For a substrate voltage of 0 V, v_{SB} for the switching device is always zero, but v_{SB} for the load device M_L changes as v_O changes. In fact, $v_{SB} = v_O$, as indicated in Fig. 6.18(a). The threshold voltages of transistors M_S and M_L will no longer be the same, and we will indicate the different values by V_{TNS} and V_{TNL}, respectively.

For the design of the saturated load inverter, we use the same circuit conditions that were used for the case of the resistive load ($I_{DD} = 80$ μA with $V_{DD} = 2.5$ V and $V_L = 0.20$ V). We first choose the W/L ratio of M_L to limit the operating current and power in the inverter. Because M_L is forced to operate in saturation by the circuit connection, its drain current is given by

$$i_D = \frac{K'_n}{2} \left(\frac{W}{L} \right)_L (v_{GS} - V_{TNL})^2 \tag{6.22}$$

For the circuit conditions in Fig. 6.19, load device M_L has $v_{GS} = 2.30$ V when $v_O = 0.20$ V.

Before we can calculate W/L, we must find the value of threshold voltage V_{TNL}, which is determined by the body effect relation represented by Eq. (4.23) in Chapter 4:

$$V_{TN} = V_{TO} + \gamma \left(\sqrt{v_{SB} + 2\phi_F} - \sqrt{2\phi_F} \right) \tag{6.23}$$

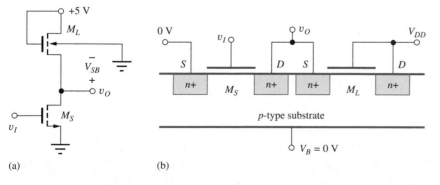

(a)   (b)

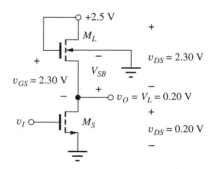

Figure 6.18 (a) Saturated load inverter. (b) Cross section of two integrated MOSFETs forming an inverter.

Figure 6.19 Saturated load inverter with $v_O = V_L$.

TABLE 6.5

	NMOS ENHANCEMENT-MODE DEVICE PARAMETERS	NMOS DEPLETION-MODE DEVICE PARAMETERS	PMOS ENHANCEMENT-MODE DEVICE PARAMETERS
V_{TO}	0.6 V	-1 V	-0.6 V
γ	$0.5 \sqrt{V}$	$0.5 \sqrt{V}$	$0.75\sqrt{V}$
$2\phi_F$	0.6 V	0.6 V	0.70 V
K'_n	100 $\mu A/V^2$	100 $\mu A/V^2$	40 $\mu A/V^2$

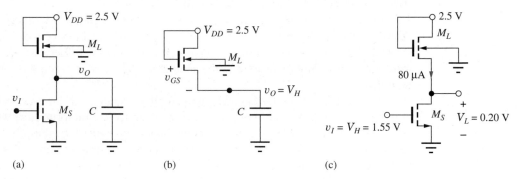

Figure 6.20 (a) Inverter with load capacitance. (b) High output level is reached when $v_I = V_L$ and M_S is off. (c) Bias conditions used to determine $(W/L)_S$.

where $V_{TO} =$ zero bias value of V_{TN} (V)

$\qquad \gamma =$ body effect parameter ($\sqrt{V}$)

$\qquad 2\phi_F =$ surface potential parameter (V)

For the rest of the discussion in this chapter, we use the set of device parameters given in Table 6.5. For the load transistor, we have $v_{SB} = v_S - v_B = 0.25$ V $- 0$ V $= 0.25$ V, and

$$V_{TNL} = 0.6 + 0.5\left(\sqrt{0.20 + 0.6} - \sqrt{0.6}\right) = 0.660 \text{ V}$$

Now, we can find the W/L ratio for the load transistor:

$$\left(\frac{W}{L}\right)_L = \frac{2i_D}{K'_n(v_{GS} - V_{TN})^2} = \frac{2 \cdot 80 \text{ } \mu A}{100\dfrac{\mu A}{V^2}(2.30 - 0.66)^2} = \frac{1}{1.68} \tag{6.24}$$

Note that the length of this load device is larger than its width. In most digital IC designs, one of the two dimensions will be made as small as possible corresponding to the minimum feature size in one direction. The W/L ratio is usually written with the smallest number normalized to unity. For $L = 1$ μm, the gate area of M_L is now only 1.68 μm^2, which is comparable to the area of M_S.

Calculation of V_H

Unfortunately, the use of the saturated load device has a detrimental effect on other characteristics of the logic gate. The value of V_H will no longer be equal to V_{DD}. In order to understand this effect, it is helpful to imagine a capacitive load attached to the logic gate, as in Fig. 6.20. Consider the logic gate with $v_I = V_L$ so that M_S is turned off. When M_S turns off, load device M_L charges capacitor C until the current through M_L becomes zero, which occurs when $v_{GS} = V_{TN}$:

$$v_{GS} = V_{DD} - V_H = V_{TN} \qquad \text{or} \qquad V_H = V_{DD} - V_{TN} \tag{6.25}$$

Thus, for the NMOS saturated load inverter, the output voltage reaches a maximum value equal to one threshold voltage drop below the power supply voltage V_{DD}. Without body effect, the output voltage in Fig. 6.20 would reach $V_H = 2.5 - 0.6 = 1.9$ V, which represents a substantial degradation in V_H compared to the resistive load inverter with $V_H = 2.5$ V.

However, body effect makes the situation even worse. As the output voltage increases toward V_H, v_{SB} increases, the threshold voltage increases above V_{TO} (see Eq. 6.23), and the steady-state value of V_H is degraded further. When v_O reaches V_H, Eq. (6.26) must be true because $v_{SB} = V_H$:

$$V_H = V_{DD} - V_{TNL} = V_{DD} - \left[V_{TO} + \gamma \left(\sqrt{V_H + 2\phi_F} - \sqrt{2\phi_F} \right) \right] \tag{6.26}$$

Using Eq. (6.26) with the parameters from Table 6.5 and $V_{DD} = 2.5$ V, we can solve for V_H, which yields the following equation:

$$\left(V_H - 1.9 - 0.5\sqrt{0.6} \right)^2 = 0.25(V_H + 0.6)$$

We can find the value of V_H using the solver in our calculator or by rearranging this equation into a quadratic equation. Either method yields $V_H = 1.55$ V or $V_H = \text{3.27 V.}$ In this circuit, the steady-state value of V_H cannot exceed the power supply voltage V_{DD} (actually it cannot exceed $V_{DD} - V_{TNL}$), so the answer must be $V_H = 1.55$ V. We can check our result for V_H by computing the threshold voltage of the load device using Eq. (6.23):

$$V_{TNL} = 0.6 \text{ V} + 0.5 \sqrt{V} \left(\sqrt{(1.55 + 0.6) \text{ V}} - \sqrt{0.6 \text{ V}} \right) = 0.95 \text{ V}$$

and

$$V_H = V_{DD} - V_{TNL} = 2.5 - 0.95 = 1.55 \text{ V} \quad ✔$$

which checks with the previous calculation of V_H.

> **EXERCISE:** Use your solver to find the two roots of Eq. (6.26) for the values used above.

Calculation of $(W/L)_S$

Now we are in a position to complete the inverter design by calculating W/L for the switching transistor. The bias conditions for $v_O = V_L$ appear in Fig. 6.20(c) in which the drain current of M_S must equal the design value of 80 μA. For $V_{GS} = 1.55$ V, $V_{DS} = 0.20$ V, and $V_{TNS} = 0.6$ V, the switching transistor is operating in the triode region. Therefore,

$$i_D = K'_n \left(\frac{W}{L} \right)_S \left(v_{GS} - V_{TNS} - \frac{v_{DS}}{2} \right) v_{DS}$$

$$80 \text{ μA} = 100 \frac{\text{μA}}{\text{V}^2} \left(\frac{W}{L} \right)_S \left(1.55 - 0.6 - \frac{0.20}{2} \right) 0.20 \text{ V}^2$$

$$\left(\frac{W}{L} \right)_S = \frac{4.71}{1}$$

The final inverter design appears in Fig. 6.21 in which $(W/L)_S = 4.71/1$ and $(W/L)_L = 1/1.68$. Note that the size of M_S has increased because of the reduction in the value of V_H.

> **EXERCISE:** Find V_H for the inverter in Fig. 6.18(a) if $V_{TO} = 0.75$ V. Assume the other parameters remain constant.
>
> **ANSWER:** 1.43 V
>
> **EXERCISE:** (a) What value of $(W/L)_S$ is required to achieve $V_L = 0.15$ V in Fig. 6.20? Assume that $i_D = 80$ μA. What is the new value of V_{TNL} for $v_O = V_L$? What value of $(W/L)_L$ is required to set $i_D = 80$ μA for $V_L = 0.15$ V? (b) Repeat for $V_L = 0.10$ V.
>
> **ANSWER:** (a) 6.10/1, 0.646 V, 1/1.82; (b) 8.89/1, 0.631 V, 1/1.96

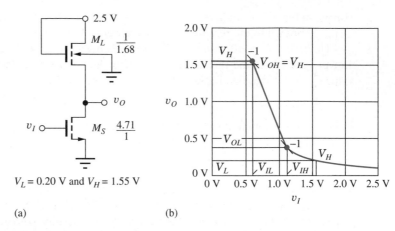

$V_L = 0.20$ V and $V_H = 1.55$ V

(a) (b)

Figure 6.21 (a) Completed inverter design with saturated load devices. (b) SPICE simulation of the voltage transfer function for the NMOS inverter with saturated load.

Figure 6.21 shows the results of SPICE simulation of the voltage transfer function for the final design. For low values of input voltage, the output is constant at 1.55 V. As the input voltage increases, the slope of the transfer function abruptly changes at the point at which the switching transistor begins to conduct, as the input voltage exceeds the threshold voltage of M_S. As the input voltage continues to increase, the output voltage decreases rapidly and ultimately reaches the design value of 0.20 V for an input of 1.55 V.

DESIGN NOTE

STATIC LOGIC INVERTER DESIGN STRATEGY

1. Given design values of V_{DD}, V_L, and power level, find I_{DD} from V_{DD} and the power.
2. Assume switching transistor M_S is off, and find the high output voltage level V_H.
3. Apply V_H to the inverter input and calculate $(W/L)_S$ of the switching transistor based upon design values of V_L and I_{DD}.
4. Calculate load resistor value or $(W/L)_L$ for the load transistor based on design values of V_L and I_{DD}.
5. Check operating region assumptions for M_S and M_L for $v_O = V_L$.
6. Check overall design with SPICE simulation.

DESIGN EXAMPLE 6.5

DESIGN OF AN INVERTER EMPLOYING A SATURATED LOAD DEVICE

Now let's design a saturated load inverter to operate from a 3.3-V supply including the influence of body effect on the transistor design.

PROBLEM Design a saturated load inverter similar to that of Fig. 6.21 with $V_{DD} = 3.3$ V and $V_L = 0.2$ V. Assume $I_{DD} = 60$ μA, $K'_n = 50$ μA/V^2, $V_{TN} = 0.75$ V, $\gamma = 0.5 \sqrt{V}$, and $2\phi_F = 0.6$ V.

SOLUTION **Known Information and Given Data:** Circuit topology in Fig. 6.21; $V_{DD} = 3.3$ V, $I_{DD} = 60$ μA, $V_L = 0.2$ V, $K'_n = 50$ μA/V^2, $V_{TO} = 0.75$ V, $\gamma = 0.5 \sqrt{V}$, and $2\phi_F = 0.6$ V

Unknowns: W/L ratios of the load and switching transistors M_S and M_L

Approach: First determine V_H including the influence of body effect on the load transistor threshold voltage by evaluating Eq. (6.26). Use V_H and the specified values of V_L and I_D to find $(W/L)_S$. Use I_D and the voltages in the circuit to find $(W/L)_L$.

Assumptions: M_S is off for $v_I = V_L$. For $v_O = V_L$, M_S is in the triode region, and M_L is in the saturation region.

Analysis: For the values associated with this technology, Eq. (6.26) becomes

$$V_H = 3.3 - \left[0.75 + 0.5\left(\sqrt{V_H + 0.6} - \sqrt{0.6}\right)\right]$$

and rearranging this equation gives

$$V_H^2 - 6.125V_H + 8.476 = 0 \quad \text{for which} \quad V_H = 2.11 \text{ V}, \; \cancel{4.01 \text{ V}}$$

Since V_H cannot exceed V_{DD}, the correct choice must be $V_H = 2.11$ V. Note that an extra digit was included in the calculation to increase the precision of the result.

The transistor operating conditions for the load and switching transistors appear in the (a) part of circuit below for $v_O = V_L$. The triode region expression for the switching transistor drain current with $v_I = V_H$ and $v_O = V_L$ is

$$I_{DS} = K_n'\left(\frac{W}{L}\right)_S \left(V_H - V_{TN} - \frac{V_L}{2}\right)V_L$$

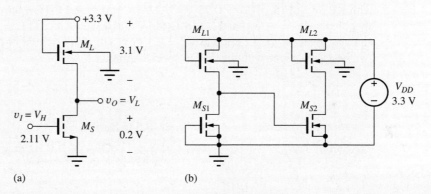

(a)   (b)

Equating this expression to the drain current yields

$$60 \; \mu\text{A} = (50 \times 10^{-6})\left(\frac{W}{L}\right)_S \left(2.11 - 0.75 - \frac{0.2}{2}\right)0.2 \rightarrow \left(\frac{W}{L}\right)_S = \frac{4.76}{1}$$

To find the W/L ratio for the load device, the saturation region expression is evaluated at a drain current of 60 μA. We must recalculate the threshold voltage since the body voltage of the load is 0.2 V when $v_O = V_L = 0.2$ V.

$$I_{DL} = \frac{K_n'}{2}\left(\frac{W}{L}\right)_L (V_{GSL} - V_{TNL})^2$$

$$V_{TNL} = 0.75 + 0.5\left(\sqrt{0.2 + 0.6} - \sqrt{0.6}\right) = 0.81 \text{ V}$$

$$60 \; \mu\text{A} = \frac{50\frac{\mu\text{A}}{\text{V}^2}}{2}\left(\frac{W}{L}\right)_L (3.3 - 0.2 - 0.81)^2 \rightarrow \left(\frac{W}{L}\right)_L = \frac{1}{2.19}$$

Our completed design values are $(W/L)_S = 4.76/1$ and $(W/L)_L = 1/2.19$.

Check of Results: We must check the triode and saturation region assumptions for the two MOSFETs: For the switch, $V_{GS} - V_{TN} = 2.11 - 0.75 = 1.36$ V, which is greater than $V_{DS} = 0.2$ V, and the triode region assumption is correct. For the load device, $V_{GS} - V_{TN} = 3.1 - 0.81 = 2.29$ V and $V_{DS} = 3.1$ V, which is consistent with the saturation region of operation. We can double check our V_H calculation by using it to find the threshold of M_L:

$$V_{TNL} = 0.75 + 0.5(\sqrt{2.11 + 0.6} - \sqrt{0.6}) = 1.19 \text{ V}$$

This is correct since $V_H + V_{TNL} = 2.11 + 1.19 = 3.3$ V, which must equal the value of V_{DD}.

Let us also double check the values of W/L by using them to recalculate the drain currents:

$$I_{DS} = (50 \times 10^{-6})\left(\frac{4.76}{1}\right)\left(2.11 - 0.75 - \frac{0.2}{2}\right)0.2 = 60.0 \text{ μA} \quad ✔$$

$$I_{DL} = \frac{50\dfrac{\text{μA}}{\text{V}^2}}{2}\left(\frac{1}{2.19}\right)(3.3 - 0.2 - 0.81)^2 = 59.9 \text{ μA} \quad ✔$$

Both results agree within round off error.

Computer-Aided Analysis: To verify our design with SPICE, we draw the circuit with a schematic capture tool, as in part (b) of the figure on the previous page. Two inverters are cascaded in order to get both V_H and V_L with one simulation. The NMOS transistors use the LEVEL = 1 model with KP = 5.0E-5, VTO = 0.75 V, GAMMA = 0.5, and PHI = 0.6 V. The transistor sizes are specified as W = 4.76 U and L = 1 U for M_S, and W = 1 U and L = 2.19 U for M_L. SPICE dc analysis gives $V_H = 2.11$ V and $V_L = 0.196$ V. The drain current of transistor M_{S2} is 60.1 μA. All the values agree with the design specifications.

EXERCISE: Redesign the inverter in Ex. 6.5 to have $V_L = 0.1$ V.

ANSWER: $(W/L)_S = 9.16/1$; $(W/L)_L = 1/2.44$ (Note $V_{TNL} = 0.781$ V)

EXAMPLE 6.6 **LOGIC LEVEL ANALYSIS FOR THE SATURATED LOAD INVERTER**

Finding the logic levels associated with someone else's design involves a somewhat different thought process than that used in designing our own inverter. Here we find V_H and V_L for a specified inverter design.

PROBLEM Find the high and low logic levels and the power supply current for a saturated load inverter with $(W/L)_S = 10/1$ and $(W/L)_L = 2/1$. The inverter operates with $V_{DD} = 2.5$ V. Assume $K_n' = 100$ μA/V², $V_{TO} = 0.60$ V, $\gamma = 0.5\sqrt{\text{V}}$, and $2\phi_F = 0.6$ V.

SOLUTION **Known Information and Given Data:** Circuit topology in Fig. 6.18(a); $V_{DD} = 2.5$ V, $(W/L)_S = 10/1$, $(W/L)_L = 2/1$, $K_n' = 100$ μA/V², $V_{TO} = 0.60$ V, $\gamma = 0.5\sqrt{\text{V}}$, and $2\phi_F = 0.6$ V

Unknowns: V_H, V_L, and I_{DD} for both logic states

Approach: First, determine V_H. Include the influence of body effect on the load transistor threshold voltage by solving Eq. (6.26). Use V_H and the specified transistor parameters to find V_L by equating the drain currents in the switching and load transistors. Use V_L to find the I_{DS}.

Assumptions: M_S is off for $v_I = V_L$. For $v_O = V_L$, M_S operates in the triode region, and M_L is in the saturation region.

Analysis: First we find V_H, and then we use it to find V_L. For the values associated with this technology, Eq. (6.26) becomes

$$V_H = 2.5 - \left[0.60 + 0.5\left(\sqrt{V_H + 0.6} - \sqrt{0.6}\right)\right]$$

Rearranging this equation gives

$$V_H^2 - 4.824V_H + 5.082 = 0 \qquad \text{for which} \qquad V_H = \cancel{3.27 \text{ V}} \text{ or } 1.55 \text{ V}$$

Since, V_H cannot exceed V_{DD}, the correct choice must be $V_H = 1.55$ V. Note that an extra digit was included in the calculation to increase the precision of the result. Since M_S is off, there is no path for current from V_{DD} and $I_{DD} = 0$ for $v_O = V_H$.

At this point we should check our result to avoid propagation of errors in our calculations. We can use V_H to find V_{TNL} and see if it is consistent with the value of V_H:

$$V_{TNL} = 0.60 + 0.5\left(\sqrt{1.55 + 0.6} - \sqrt{0.6}\right) = 0.946 \text{ V}$$

$$V_H = 2.5 - 0.946 = 1.55 \text{ V}$$

We see that the value of V_H is correct.

To find V_L, we use the condition that I_{DS} must equal I_{DL} in the steady state. The load transistor is saturated by connection, and we expect the switching transistor to be in the triode region since its drain-source voltage should be small. ($V_{DS} = V_L$.)

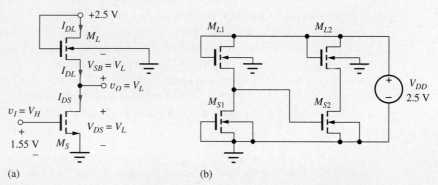

(a)   (b)

For $I_{DS} = I_{DL}$, we have $K_n'\left(\dfrac{10}{1}\right)\left(V_{GSS} - V_{TNS} - \dfrac{V_L}{2}\right)V_L = \dfrac{K_n'}{2}\left(\dfrac{2}{1}\right)(2.5 - V_L - V_{TNL})^2$

where

$$V_{TNL} = 0.60 + 0.5\left(\sqrt{V_L + 0.6} - \sqrt{0.6}\right)$$

From the circuit shown, $V_{GSS} = 1.55$ V and $V_{TNS} = 0.60$ V, since there will be no body effect in M_S. Unfortunately, V_{TNL} is a function of the unknown voltage V_L, since the source-bulk voltage of M_L is equal to V_L.

Approach 1: Since we expect V_L to be small, its effect on V_{TNL} will also be small, and one approach to finding V_L is to simply ignore body effect in the load transistor. For this case, equating I_{DS} and I_{DL} gives

$$K_n'\left(\frac{10}{1}\right)\left(1.55 - 0.6 - \frac{V_L}{2}\right)V_L = \frac{K_n'}{2}\left(\frac{2}{1}\right)(2.5 - V_L - 0.6)^2$$

which can be rearranged to yield a quadratic equation for which $V_L = 1.80$ V or 0.33 V. We must choose $V_L = 0.33$ V since the other root is not consistent with the assumed regions of operation of the transistors. For this value of V_L, the current in M_S is

$$I_{DS} = \left(100\frac{\mu A}{V^2}\right)\left(\frac{10}{1}\right)\left(1.55 - 0.6 - \frac{0.33}{2}\right)(0.33) \text{ V}^2 = 259 \text{ }\mu A$$

Approach 2: For a more exact result, we can find the simultaneous solution to the drain current and threshold voltage equations with the solver in a calculator, with a spreadsheet, or by direct iteration. The result is $V_L = 0.290$ V with $V_{TNL} = 0.68$ V. Using the value of V_L, we can find the current in M_S:

$$I_{DS} = \left(100\frac{\mu A}{V^2}\right)\left(\frac{10}{1}\right)\left(1.55 - 0.6 - \frac{0.29}{2}\right)(0.29)\ V^2 = 234\ \mu A$$

The approximate values in Approach 1 overestimate the more exact values in Approach 2 by approximately 10 percent. In most cases, this would be a negligible error.

Check of Results: Note that we double checked the value of V_H earlier. For V_L, we should check our triode and saturation region assumptions for the two MOSFETs: For the switching transistor, $V_{GS} - V_{TN} = 1.55 - 0.6 = 0.96$ V, which is greater than $V_{DS} = 0.29$ V, and the triode region assumption is correct. For the load device, $V_{GS} - V_{TN} = 2.5 - 0.29 - 0.68 = 1.53$ V and $V_{DS} = 2.5 - 0.29 = 2.21$, which are consistent with the saturation region of operation. We can further check our results by finding the drain current in M_L:

$$I_{DL} = \left(\frac{100\ \mu A}{2\ V^2}\right)\left(\frac{2}{1}\right)(2.5 - 0.29 - 0.68)^2 = 234\ \mu A$$

This value agrees with I_{DS} within round-off error.

Computer-Aided Analysis: To verify our design with SPICE, we draw the circuit with a schematic capture tool, as in part (b) of the figure on the previous page. Two inverters are cascaded in order to get both V_H and V_L with one simulation. The gate of MS1 is grounded to force MS1 to be off. The NMOS transistors use the LEVEL = 1 model with KP = 10E-5, VTO = 0.60 V, GAMMA = 0.5, and PHI = 0.6 V. The transistor sizes are specified as W = 10 U and L = 1 U for M_S, and W = 2 U and L = 1 U for M_L. SPICE dc analysis gives $V_H = 1.55$ V and $V_L = 0.289$ V. The current in V_{DD} is 234 μA. All the values agree with the hand calculations.

EXERCISE: Use the "Solver" on your calculator to find V_H in Ex. 6.6.

EXERCISE: Repeat the calculations with $\gamma = 0$. Check your results with SPICE.

ANSWERS: 1.90 V; 0 A; 0.235 V; 278 μA

Noise Margin Analysis

Detailed analysis of the noise margins for saturated load inverters operating from low power supply voltages is very tedious and results in expressions that yield little additional insight into the behavior of the inverter. So here we explore the values of V_{IL}, V_{OH}, V_{IH}, and V_{OL} based upon the SPICE simulation results presented in Fig. 6.21. Remember that these voltages are defined by the points in the voltage transfer characteristic at which the slope is -1. Looking at Fig. 6.21, we see that the slope of the VTC abruptly changes at the point where M_S just begins to conduct. This occurs for $v_I = V_{TN}$ and defines V_{IL} and V_{OH}. Therefore, $V_{IL} = V_{TNS} = 0.6$ V, and $V_{OH} = V_H = 1.55$ V. The values of V_{IH} and V_{OL} are found from the graph at the second point where the slope is -1. Reading the values from the graph yields $V_{IH} \cong 1.12$ V and $V_{OL} \cong 0.38$ V.[4]

The noise margins for this saturated load inverter are

$$\text{NM}_H = V_{OH} - V_{IH} = 1.55 - 1.12 = 0.33\ \text{V}$$
$$\text{NM}_L = V_{IL} - V_{OL} = 0.60 - 0.38 = 0.22\ \text{V}$$

---

[4] Note that we can have SPICE estimate the derivative of the VTC numerically by plotting the output $D(VO)/D(VI)$ for example, and we can quite accurately locate the points for which the slope is -1.

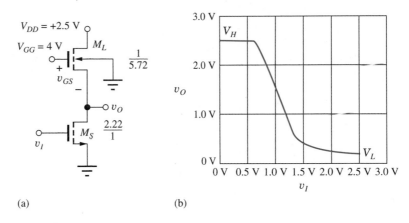

Figure 6.22 (a) Linear load inverter design. (b) Linear load inverter VTC.

Let's compare these values to those of the resistive load inverter ($\text{NM}_H = 0.96$ V, $\text{NM}_L = 0.25$ V). The reduction in V_H caused by the saturated load device has significantly reduced the value of NM_H, whereas the value of NM_L is very similar since M_S has been redesigned to maintain the same value of V_L.

6.6.2 NMOS INVERTER WITH A LINEAR LOAD DEVICE

Figure 6.17(d) provides a second workable choice for the load transistor M_L. In this case, the gate of the load transistor is connected to a separate voltage V_{GG} as in Fig. 6.22(a). V_{GG} is normally chosen to be at least one threshold voltage greater than the supply voltage V_{DD}:

$$V_{GG} \geq V_{DD} + V_{TNL}$$

For this value of V_{GG}, the output voltage in the high output state V_H is equal to V_{DD}.

The region of operation of M_L in Fig. 6.22 can be found by comparing $V_{GS} - V_{TNL}$ to V_{DS}. For the load device with its output at v_O and $V_{GG} \geq V_{DD} + V_{TNL}$:

$$
\begin{aligned}
v_{GS} - V_{TNL} &= V_{GG} - v_O - V_{TNL} \\
&\geq V_{DD} + V_{TNL} - v_O - V_{TNL} \\
&\geq V_{DD} - v_O
\end{aligned}
\tag{6.27}
$$

So $v_{GS} - V_{TNL} \geq V_{DD} - v_O$, but $v_{DS} = V_{DD} - v_O$, which demonstrates that the load device always operates in the triode (linear) region.

The W/L ratios for M_S and M_L can be calculated using methods similar to those in the previous sections; the results are shown in Fig. 6.22. Because V_H is now equal to $V_{DD} = 2.5$ V, M_S is again 2.22/1. However, for $v_O = V_L$, v_{GS} of M_L is large, and $(W/L)_L$ must be set to (1/5.72) in order to limit the current to the desired level. (Verification of these values is left for Prob. 6.64.)

Introduction of the additional power supply voltage V_{GG} overcomes the reduced output voltage problem associated with the saturated load device. However, the cost of the additional power supply level, as well as the increased wiring congestion introduced by distribution of the extra supply voltage to every logic gate, cause this form of load topology to rarely be used.

EXERCISE: Estimate the values of V_{IL}, V_{OH}, V_{IH}, V_{OL}, NM_H and NM_L for the linear load inverter using the graph in Fig. 6.22(b).

ANSWERS: 0.64 V, 2.42 V, 1.46 V, 0.52 V, 0.12 V, 0.96 V.

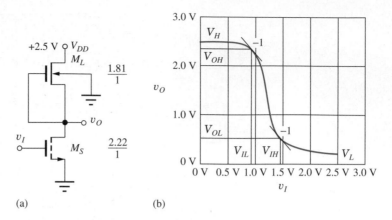

Figure 6.23 (a) NMOS inverter with depletion-mode load. (b) SPICE simulation results for the voltage transfer function of the NMOS depletion-load inverter of part (a).

6.6.3 NMOS INVERTER WITH A DEPLETION-MODE LOAD

The saturated load and linear load circuits were developed for use in integrated circuits because all the devices had the same threshold voltages in early NMOS and PMOS technologies. However, once ion-implantation technology was perfected, it became possible to selectively adjust the threshold of the load transistors to alter their characteristics to become those of NMOS depletion-mode devices with $V_{TN} < 0$, and the use of the circuit in Fig. 6.23(a) became feasible.

The circuit topology for the NMOS inverter with a depletion-mode load device is shown in Fig. 6.23(a). Because the threshold voltage of the NMOS depletion-mode device is negative, a channel exists even for $v_{GS} = 0$, and the load device conducts current until its drain-source voltage becomes zero. When the switching device M_S is off ($v_I = V_L$), the output voltage rises to its final value of $V_H = V_{DD}$.

For $v_I = V_H$, the output is low at $v_O = V_L$. In this state, current is limited by the depletion-mode load device, and it is normally designed to operate in the saturation region, requiring:

$$v_{DS} \geq v_{GS} - V_{TNL} = 0 - V_{TNL} \quad \text{or} \quad v_{DS} \geq -V_{TNL}$$

Design of the W/L Ratios of M_L

As an example of inverter design, if we assume $V_{DD} = 2.5$ V, $V_L = 0.20$ V, and $V_{TNL} = -1$ V, then the operating voltage for the load device with $v_O = V_L$ is $V_{DS} = 4.75$ V, which is greater than $-V_{TNL} = 1$ V, and the MOSFET operates in the saturation region. The drain current of the depletion-mode load device operating in the saturation region with $V_{GS} = 0$ is given by

$$i_{DL} = \frac{K_n'}{2} \left(\frac{W}{L}\right)_L (v_{GSL} - V_{TNL})^2 = \frac{K_n'}{2} \left(\frac{W}{L}\right)_L (V_{TNL})^2 \tag{6.28}$$

Just as for the case of the saturated load inverter, body effect must be taken into account in the depletion-mode MOSFET, and we must calculate V_{TNL} before $(W/L)_L$ can be properly determined. For depletion-mode devices, we use the parameters in Table 6.5, and

$$V_{TNL} = -1 \text{ V} + 0.5 \sqrt{V} \left(\sqrt{(0.20 + 0.6) \text{ V}} - \sqrt{0.6 \text{ V}}\right) = -0.94 \text{ V}$$

Using our previous design current of 80 μA with $K_n' = 100$ μA/V^2 and the depletion-mode threshold voltage of -0.94 V, we find $(W/L)_L = 1.81/1$.

Design of the W/L Ratio of M_S

When $v_I = V_H = V_{DD}$, the switching device once again has the full supply voltage applied to its gate, and its W/L ratio will be identical to the design of the NMOS logic gate with resistor load: $(W/L)_S = 2.22/1$. The completed depletion-mode load inverter design appears in Fig. 6.23, and the logic levels of the final design are $V_L = 0.20$ V and $V_H = 2.5$ V.

Figure 6.23 shows the results of SPICE simulation of the voltage transfer function for the final inverter design with the depletion-mode load. For low values of input voltage, the output is 2.5 V. As the input voltage increases, the slope of the transfer function gradually changes as the switching transistor begins to conduct for an input voltage exceeding the threshold voltage. As the input voltage continues to increase, the output voltage decreases rapidly and ultimately reaches the design value of 0.20 V for an input of 2.5 V.

Noise Margin Analysis

As for the saturated load inverter, detailed analysis of the noise margins for depletion load inverters operating from low power supply voltages is very tedious. So here we explore the values of V_{IL}, V_{OH}, V_{IH} and V_{OL} based upon the SPICE simulation results presented in Fig. 6.23. Remember that these voltages are defined by the points in the voltage transfer characteristic at which the slope is -1. Reading values from Fig. 6.23, we estimate $V_{IL} = 0.93$ V and $V_{OH} = 2.35$ V, and $V_{IH} \cong 1.45$ V and $V_{OL} \cong 0.50$ V.

The noise margins for this saturated load inverter are

$$\text{NM}_H = V_{OH} - V_{IH} = 2.35 - 1.45 = 0.90 \text{ V}$$

$$\text{NM}_L = V_{IL} - V_{OL} = 0.93 - 0.50 = 0.43 \text{ V}$$

Compared to the noise margins of the resistive load inverter ($\text{NM}_H = 0.96$ V, $\text{NM}_L = 0.25$ V), we see that NM_H is similar and NM_L has actually improved.

DESIGN EXAMPLE 6.7   **NMOS INVERTER WITH DEPLETION-MODE LOAD**

Now we will redesign the depletion-load inverter for operation with 3.3-V power supply voltage.

PROBLEM   Design the inverter with depletion-mode load of Fig. 6.23 for operation with $V_{DD} = 3.3$ V. Assume $V_{TO} = 0.7$ V for the switching transistor and $V_{TO} = -1.5$ V for the depletion-mode load. Keep the other design parameters the same (i.e., $V_L = 0.20$ V and $P = 0.20$ mW).

SOLUTION   **Known Information and Given Data:** Circuit topology in Fig. 6.23; $V_{DD} = 3.3$ V, $P = 0.20$ mW, $V_L = 0.20$ V, $K'_n = 100$ μA/V^2, $V_{TOS} = 0.60$ V, $V_{TOL} = -1$ V, $\gamma = 0.5 \sqrt{\text{V}}$, and $2\phi_F = 0.6$ V for both transistor types

Unknowns: Power supply current I_{DD}, W/L ratios of the load and switching transistors M_S and M_L

Approach: Find V_H. Use V_H, I_{DD}, and the specified value of V_L to find $(W/L)_S$. Calculate V_{TNL}. Use I_{DD}, V_{TNL}, and the voltages in the circuit to find $(W/L)_L$.

Assumptions: M_S is off for $v_I = V_L$. For $v_O = V_L$, M_S is in the triode region, and M_L is in the saturation region.

Analysis: First, we need to know the power supply current for $v_O = V_L$ in order to calculate the W/L ratios of both transistors.

$$I_{DD} = \frac{P}{V_{DD}} = \frac{0.20 \text{ mW}}{3.3 \text{ V}} = 60.6 \text{ μA}$$

The value of V_H will be equal to V_{DD} as long as the threshold of the depletion-mode device remains negative for $v_O = V_{DD}$. Checking the value of V_{TNL}:

$$V_{TNL} = -1 + 0.5\left(\sqrt{3.3 + 0.6} - \sqrt{0.6}\right) = -0.40 \text{ V} \quad ✔$$

Therefore, $V_H = V_{DD} = 3.3$ V. Now the size of the switching transistor can be determined. The transistor has $V_{GS} = V_H = 3.3$ V and $V_{DS} = V_L = 0.20$ V, as shown in the figure.

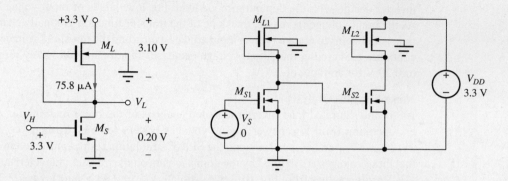

$$60.6\ \mu A = 100\ \mu A\ \left(\frac{W}{L}\right)_S\left(3.3 - 0.6 - \frac{0.20}{2}\right)0.20 \rightarrow \left(\frac{W}{L}\right)_S = \frac{1.17}{1}$$

In order to design the load transistor, we calculate its threshold voltage with $v_O = V_L = 0.20$ V, and then use V_{TNL} to find W/L (note that $V_{SB} = V_L = 0.20$ V):

$$V_{TNL} = -1 + 0.5\left(\sqrt{0.20 + 0.6} - \sqrt{0.6}\right) = -0.940\ V$$

$$60.6\ \mu A = \frac{100\ \mu A}{2}\left(\frac{W}{L}\right)_L(-0.94)^2 \rightarrow \left(\frac{W}{L}\right)_L = \frac{1.37}{1}$$

Check of Results: We must check the triode and saturation region assumptions for the two MOSFETs. For the switch, $V_{GS} - V_{TN} = 3.3 - 0.60 = 2.7$ V, which is greater than $V_{DS} = 0.20$ V, and the triode region assumption is correct. For the load device, $V_{GS} - V_{TN} = 0 - (-0.93) = 0.93$ V, and $V_{DS} = 3.3 - 0.20 = 3.10$ V, which are consistent with the saturation region of operation. Let us also double check the values of W/L by directly calculating the drain currents:

$$I_{DS} = (100 \times 10^{-6})\left(\frac{1.17}{1}\right)\left(3.3 - 0.60 - \frac{0.20}{2}\right)0.20 = 60.8\ \mu A \quad ✔$$

$$I_{DL} = \frac{100\frac{\mu A}{V^2}}{2}\left(\frac{1.37}{1}\right)[0 - (-0.94)]^2 = 60.5\ \mu A \quad ✔$$

Both results agree within round-off error.

Computer-Aided Analysis: Let us verify our design with SPICE. Here again, two inverters are cascaded in order to get both V_H and V_L with one simulation. The enhancement-mode transistors use the LEVEL = 1 model with KP = 1E-4, VTO = 0.60 V, GAMMA = 0.5, and PHI = 0.6 V. For the depletion mode devices, VTO is changed to VTO = −1.0 V. The transistor sizes are specified as W = 1.17 U and L = 1 U for M_S, and W = 1.37 U and L = 1 U for M_L. SPICE gives $V_H = 3.30$ V and $V_L = 0.20$ V with $I_D = 60.6\ \mu A$ for transistor M_{S2}. All the values confirm our design calculations.

6.6.4 STATIC DESIGN OF THE PSEUDO NMOS INVERTER

It is also possible to replace the load resistor with a PMOS transistor with its source connected to V_{DD}, its drain is connected to the output node, and its gate connected to ground, as in Fig. 6.24. This circuit has become known as **pseudo NMOS** since circuit operation is very similar to that of NMOS logic even though it is usually found embedded in CMOS designs that we will study in detail in Chapter 7.

In order to design the circuit, we use the same circuit conditions that were used for the case of the resistive load. ($I_{DD} = 80$ μA, $V_{DD} = 2.5$ V and $V_L = 0.20$ V). First we choose the W/L ratio of the PMOS load device to limit the operating current in the inverter. Then we calculate the size of M_S required to achieve the specified value of V_L. (Note that neither transistor suffers from any body effect since the bulk terminals of both transistors are connected to their respective sources. This is an important advantage of the PMOS load transistor in comparison to NMOS load devices.)

Calculation of $(W/L)_P$ and $(W/L)_S$

For the PMOS device in Fig. 6.24, we see that $V_{GS} = -V_{DD}$, and the transistor will be in the conducting state. Since $V_{DS} = 0.2 - 2.5 = -2.3$ V and $V_{GS} - V_{TP} = -2.5 - (-0.6) = -1.9$ V, the transistor will be saturated ($|V_{DS}| > |V_{GS} - V_{TP}|$—see Section 4.2). We need to find the value of W/L that sets the PMOS drain current to 80 μA:

$$i_D = \frac{K'_p}{2}\left(\frac{W}{L}\right)_P (V_{GS} - V_{TP})^2 \quad \text{or} \quad 80\ \mu A = \frac{1}{2}\left(40\frac{\mu A}{V^2}\right)\left(\frac{W}{L}\right)_P [-2.5 - (-0.6)]^2 V^2$$

which gives $\left(\dfrac{W}{L}\right)_P = \dfrac{1.11}{1}$.

Calculation of V_H and $(W/L)_S$

In order to calculate $(W/L)_S$, we need to determine the high output level V_H, since this is the voltage that is used to drive switching transistor M_S to achieve $v_O = V_L$. As shown in Fig. 6.24(b), the PMOS load transistor has a fixed value of $V_{GS} = -2.5$ V. Thus it will always be in the conducting state. With M_S off, current will flow through the PMOS device to charge the output node until the drain-source voltage V_{DS} of the transistor collapses to zero. Thus, $V_H = V_{DD}$, just as for the inverter with the resistor load.

Now, the conditions on switching transistor M_S for $v_O = V_L$ in Fig. 6.24(a) are $V_{GS} = V_H = 2.5$ V and $V_{DS} = V_L = 0.20$ V with $i_D = 80$ μA. These are identical to those of the switching

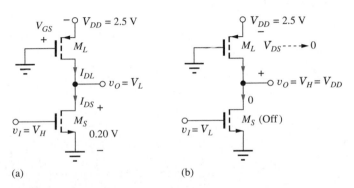

(a)   (b)

Figure 6.24 Pseudo NMOS Inverter with (a) $v_I = V_H$ and (b) $v_I = V_L$.

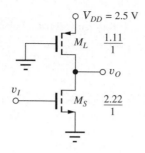

Figure 6.25 Completed pseudo NMOS inverter design.

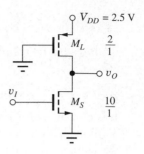

Figure 6.26 Pseudo NMOS inverter used in Ex. 6.8.

transistor in the resistor load inverter in Section 6.5.2. Thus, $(W/L)_S = 2.22/1$. The completed pseudo NMOS inverter design appears in Fig. 6.25.

> **EXERCISE:** Verify the value of $(W/L)_S$ by calculating the drain current of M_S.

EXAMPLE 6.8   LOGIC LEVEL ANALYSIS FOR THE PSEUDO NMOS INVERTER

Finding the logic levels associated with someone else's inverter design involves a different thought process than that required to design the inverter. Here we find V_H and V_L for a specified inverter design.

PROBLEM Find the high and low logic levels and the power supply current for the pseudo NMOS inverter in Fig. 6.26 with $(W/L)_S = 10/1$ and $(W/L)_L = 2/1$. The inverter operates with $V_{DD} = 2.5$ V. Assume $K'_n = 100\ \mu\text{A/V}^2$, $V_{TN} = 0.60$ V, $K'_p = 40\ \mu\text{A/V}^2$, $V_{TP} = -0.60$ V.

SOLUTION **Known Information and Given Data:** Circuit topology in Fig. 6.26; $V_{DD} = 2.5$ V, $(W/L)_N = 10/1$, $(W/L)_P = 2/1$, $K'_n = 100\ \mu\text{A/V}^2$, $V_{TN} = 0.60$ V, $K'_p = 40\ \mu\text{A/V}^2$, and $V_{TP} = -0.60$ V.

Unknowns: V_H, V_L, I_{DD} for both logic states

Approach: First determine V_H. Use V_H and the specified transistor parameters to find V_L by equating the drain currents in the switching and load transistors. Use V_L to find power supply current I_{DD} which is equal to switching transistor drain current I_{DS}.

Assumptions: M_S is off for $v_I = V_L$. For $v_O = V_L$, M_S operates in the triode region, and M_L is in the saturation region.

Analysis: First we find V_H, and then we use it to find V_L. For the pseudo NMOS logic gate, $V_H = V_{DD}$. Thus, for our circuit, $V_H = 2.5$ V. To find V_L, we use the condition that the two transistor drain currents must be equal in the steady state: $I_{DS} = I_{DL}$. For $v_O = V_L$, we expect that the load transistor will be saturated since the magnitude of its drain-source voltage is large ($V_{DS} = V_L - V_{DD}$), and we expect the switching transistor to be in the triode region since its drain-source voltage will be small. ($V_{DS} = V_L$). For $I_{DS} = I_{DL}$, we have

$$K'_n \left(\frac{10}{1}\right)\left(V_{GSN} - V_{TN} - \frac{V_L}{2}\right)V_L = \frac{K'_p}{2}\left(\frac{2}{1}\right)(V_{GSP} - V_{TP})^2$$

For the circuit in Fig. 6.24, $V_{GSN} = 2.5$ V and $V_{GSP} = -2.5$ V, and

$$100\frac{\mu\text{A}}{\text{V}^2}\left(\frac{10}{1}\right)\left(2.5 - 0.6 - \frac{V_L}{2}\right)V_L = \frac{40\ \mu\text{A}}{2\ \text{V}^2}\left(\frac{2}{1}\right)(-2.5 - (-0.6))^2$$

which can be rearranged to yield a quadratic equation:

$$12.5V_L^2 - 47.5V_L + 3.61 = 0 \qquad \text{for which} \qquad V_L = 0.0776 \text{ V}, \cancel{3.72 \text{ V}}.$$

$V_L = 3.72$ V exceeds the 2.5-V power supply, so that answer must be discarded. Hence the answer must be $V_L = 0.776$ V. For this value of V_L, the current in M_S is

$$I_{DS} = \left(100\frac{\mu A}{V^2}\right)\left(\frac{10}{1}\right)\left(2.5 - 0.6 - \frac{0.0776}{2}\right)(0.0776)V^2 = 144 \ \mu A$$

Check of Results: For V_L, we should check our triode and saturation region assumptions for the two MOSFETs: For the switching transistor, $V_{GS} - V_{TN} = 2.5 - 0.6 = 1.90$ V which is greater than $V_{DS} = 0.078$ V, and the triode region assumption is correct. For the load device, $V_{GS} - V_{TN} = -2.5 - (-0.6) = -1.9$ V, and $V_{DS} = 0.078 - 2.5 = -2.42$, which are consistent with the saturation region of operation. We can further check our results by finding the drain current in M_L:

$$I_{DL} = \left(\frac{40 \ \mu A}{2 \ V^2}\right)\left(\frac{2}{1}\right)(-2.5 + 0.6)^2 = 144 \ \mu A \quad \text{which agrees with } I_{DS}.$$

Computer-Aided Analysis: To verify our design with SPICE, we draw the circuit with a schematic capture tool. Two inverters are cascaded in order to get both V_H and V_L with one simulation. The gate of MS1 is grounded to force MS1 to be off. The NMOS transistor uses the LEVEL = 1 model with KP = 10E-5, VTO = 0.60 V, GAMMA = 0.5 and PHI = 0.6 V, and the PMOS parameters are KP = 4E-5, VTO = -0.60 V, GAMMA = 0.5 and PHI = 0.6 V. The transistor sizes are specified as W = 10 U and L = 1 U for M_S, and W = 2 U and L = 1 U for M_L. SPICE gives $V_H = 2.50$ V and $V_L = 0.0776$ V. The current in V_{DD} is 144 μA. All the values agree with the hand calculations.

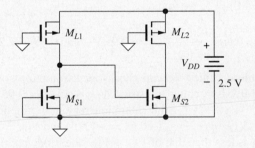

EXERCISE: Use the "Solver" in your calculator to check the value of V_L found in Section 6.7.2.

EXERCISE: Repeat the calculations with $(W/L)_S = 5/1$. Check your results with SPICE.

ANSWERS: 2.50 V, 0.159 V, 144 μA.

Noise Margin Analysis for the Pseudo NMOS Inverter

Let us now find the noise margins for the pseudo NMOS inverter. We need to calculate the values of V_{IL}, V_{OL}, V_{IH}, and V_{OH} and remember these voltages are defined by the points on the voltage transfer characteristic at which the slope $dv_O/dv_I = -1$, as indicated on the graph in Fig. 6.27.

First let us find V_{IL} and V_{OH}. We need to find a relationship between v_I and v_O that we can differentiate. Remember that the drain currents in the switching and load devices must be equal at all points on the static VTC. Also, at $v_I = V_{IL}$ the input will be at a relatively low voltage, and the output will be a relatively high voltage. Thus, we guess that M_S will be operating in the saturation region and that M_L will operate in the triode region. Setting $i_{DS} = i_{DL}$ yields

$$\frac{K_S}{2}(v_I - V_{TN})^2 = K_L \left(-V_{DD} - V_{TP} - \frac{v_O - V_{DD}}{2} \right)(v_O - V_{DD})$$

with

$$K_S = K'_n \left(\frac{W}{L} \right)_S \quad \text{and} \quad K_L = K'_p \left(\frac{W}{L} \right)_L \tag{6.29}$$

The point of interest is $\frac{\partial v_O}{\partial v_I} = -1$, but solving for the value of v_O would be quite tedious. Since we expect the derivatives to be smooth, continuous, and nonzero, we will assume that $\frac{\partial v_O}{\partial v_I} = \left(\frac{\partial v_I}{\partial v_O} \right)^{-1}$ and solve for v_I in terms of v_O:

$$v_I = V_{TN} + \frac{1}{\sqrt{K_R}} \sqrt{[2(V_{DD} + V_{TP}) - (V_{DD} - v_O)](V_{DD} - v_O)} \quad \text{where} \quad K_R = \frac{K_S}{K_L}$$

Evaluating the derivative is still quite tedious, so only the results are given here:[5]

$$V_{IL} = V_{TN} + \frac{(V_{DD} + V_{TP})}{\sqrt{K_R^2 + K_R}} \quad \text{and} \quad V_{OH} = V_{DD} - (V_{DD} + V_{TP})\left(1 - \sqrt{\frac{K_R}{K_R + 1}} \right) \tag{6.30}$$

These values appear reasonable. The input must exceed the threshold voltage of the NMOS transistor before it begins to conduct, so V_{IL} should be somewhat larger than V_{TN}, and the value of V_{OH} should be somewhat below V_{DD} as in Fig. 6.27.

For the inverter design of Fig. 6.26 with $V_{DD} = 2.5$ V, $V_{TP} = -0.6$ V and $K_R = (2.22)(100)/(1.11)(40) = 5$, we find

$$V_{IL} = 0.6 + \frac{(2.5 - 0.6)}{\sqrt{(5)^2 + 5}} = 0.95 \text{ V} \quad \text{and} \quad V_{OH} = 2.5 - (2.5 - 0.6)\left(1 - \sqrt{\frac{5}{5 + 1}} \right) = 2.33 \text{ V}$$

With these values we can check our assumptions of the operating regions of M_S and M_L. For the NMOS switching transistor, $V_{GS} - V_{TN} = 0.95 - 0.6 = 0.35$ V and $V_{DS} = 2.33$ V. Since $V_{DS} > V_{GS} - V_{TN}$, the saturation region assumption was correct. For the PMOS load device, $V_{GS} - V_{TP} = -2.5 - (-0.6) = -1.9$ V and $V_{DS} = 2.33 - 2.5 = -0.17$ V. Since the magnitude of V_{DS} is less than that of $V_{GS} - V_{TP}$, the triode region assumption was correct.

A similar process is used to find V_{IH} and V_{OL}. We again observe that the drain currents in the switching and load devices must be equal. At $v_I = V_{IH}$, the input will be at a relatively high voltage, and the output will be at a relatively low voltage. Thus, we guess that M_S will operate in the triode region and M_L will be in the saturation region. Equating drain currents in the switching and load transistors yields

$$K_S \left(v_I - V_{TN} - \frac{v_O}{2} \right)v_O = \frac{K_L}{2}(-V_{DD} - V_{TP})^2 \tag{6.31}$$

We again assume that $\frac{\partial v_O}{\partial v_I} = \left(\frac{\partial v_I}{\partial v_O} \right)^{-1}$ and solve for v_I in terms of v_O:

$$v_1 = V_{TN} + \frac{v_O}{2} + \frac{(V_{DD} + V_{TP})^2}{2K_R}\left(\frac{1}{v_O} \right) \quad \text{where} \quad K_R = \frac{K_S}{K_L} \tag{6.32}$$

---

[5] The details of the derivation can be found on the MCD website.

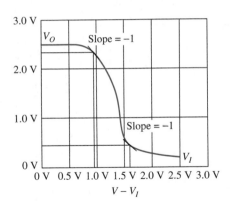

Figure 6.27 PSPICE simulation of the voltage transfer function for the pseudo NMOS inverter.

Psuedo NMOS Inverter Noise Margins

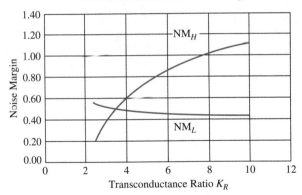

Figure 6.28 Noise margins versus transconductance ratio K_R for the pseudo NMOS inverter with $V_{DD} = 2.5$ V, $V_{TN} = 0.6$ V and $V_{TP} = -0.6$ V.

Taking the derivative

$$\frac{\partial v_I}{\partial v_O} = \frac{1}{2} - \frac{1}{2K_R}\frac{(V_{DD} + V_{TP})^2}{v_O^2} \tag{6.33}$$

and setting it equal to -1 at $v_O = V_{OL}$ yields

$$-1 = \frac{1}{2} - \frac{1}{2K_R}\frac{(V_{DD} + V_{TP})^2}{V_{OL}^2} \quad \text{or} \quad V_{OL} = \frac{V_{DD} + V_{TNP}}{\sqrt{3K_R}}$$

Substituting this result in Eq. (6.32) with $v_I = V_{IH}$ gives

$$V_{IH} = V_{TN} + \frac{2(V_{DD} + V_{TP})}{\sqrt{3K_R}} = V_{TN} + 2V_{OL} \tag{6.34}$$

For the inverter design of Fig. 6.26,

$$V_{OL} = \frac{V_{DD} + V_{TP}}{\sqrt{3K_R}} = \frac{(2.5 - 0.6)\text{ V}}{\sqrt{3(5)}} = 0.491\text{ V} \quad \text{and} \quad V_{IH} = 0.6 + 2(0.49) = 1.58\text{ V} \tag{6.35}$$

With these values we should again check our assumptions of the operating regions of M_S and M_L. For the NMOS switching transistor, $V_{GS} - V_{TN} = 1.58 - 0.6 = 0.98$ V and $V_{DS} = 0.491$ V. Since $V_{DS} < V_{GS} - V_{TN}$, the triode region assumption was correct. For the PMOS load device, $V_{GS} - V_{TP} = -2.5 - (-0.6) = -1.9$ V and $V_{DS} = 0.491 - 2.5 = -2.01$ V. Since the magniude of V_{DS} exceeds that of $V_{GS} - V_{TP}$, the saturation region assumption was correct. In Fig. 6.25, it can be seen that these calculated values of V_{IL}, V_{OL}, V_{IH} and V_{OH} all agree well with SPICE simulation results.

The noise margins for this pseudo NMOS inverter are

$$NM_H = V_{OH} - V_{IH} = 2.33 - 1.58 = 0.75\text{ V}$$

$$NM_L = V_{IL} - V_{OL} = 0.95 - 0.49 = 0.46\text{ V}$$

With Eqs. (6.30–6.35), we can easily explore the dependence of the noise margins on transconductance ratio K_R, and the results are plotted in Fig. 6.28. High-state noise margin NM_H increases monotonically as the drive capacity of switching transistor M_S, and hence K_R, increases, whereas NM_L gradually decreases.

6.7 NMOS INVERTER SUMMARY AND COMPARISON

Figure 6.29 and Table 6.6 summarize the NMOS inverter designs discussed in Secs. 6.5 and 6.6. The gate with the resistive load takes up too much area to be implemented in IC form. The saturated load configuration is the simplest circuit, using only NMOS transistors. However, it has a disadvantage

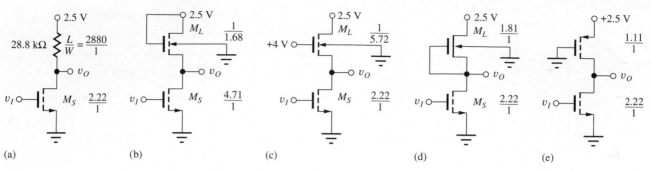

Figure 6.29 Comparison of various NMOS inverter designs: (a) Inverter with resistor load, (b) saturated load inverter, (c) linear load inverter, (d) inverter with depletion-mode load, (e) pseudo NMOS inverter.

TABLE 6.6
Inverter Characteristics

	INVERTER WITH RESISTOR LOAD	SATURATED LOAD INVERTER	LINEAR LOAD INVERTER	INVERTER WITH DEPLETION-MODE LOAD	PSEUDO NMOS INVERTER
V_H	2.50 V	1.55 V	2.50 V	2.50 V	2.50 V
V_L	0.20 V	0.20 V	0.20 V	0.20 V	0.20 V
N_{ML}	0.25 V	0.22 V	0.12 V	0.43 V	0.46 V
N_{MH}	0.96 V	0.33 V	0.96 V	0.90 V	0.75 V
Area (μm^2)	2880	6.39	7.94	4.03	3.33

that the high logic state no longer reaches the power supply. Also, in Sec. 6.11, the speed of the saturated load gate will be demonstrated to be poorer than that of other circuit implementations. The linear load circuit solves the logic level and speed problems but requires an additional costly power supply voltage that causes wiring congestion problems in IC designs.

Following successful development of the ion-implantation process and invention of depletion-mode load technology, NMOS circuits with depletion-mode load devices quickly became the circuit of choice. From Fig. 6.29 and Table 6.6, we see that the additional process complexity is traded for a simple inverter topology that gives $V_H = V_{DD}$ with the smallest overall transistor sizes. At the same time, the depletion-load gate yields the best combination of noise margins. At the end of the chapter, we will find that the depletion load gate also yields the highest speed of the four circuit configurations. The depletion-mode load in Sec. 6.11 tends to act as a current source during most of the output transition, and it offers high speed with significantly reduced area compared to the other purely NMOS inverter circuits. In pseudo NMOS, the PMOS load transistor acts as a current source during much of the output transition, and it offers the best speed with smallest area. We will refer to the gate designs of Fig. 6.29 as our **reference inverter designs** and use these circuits as the basis for more complex designs in subsequent sections.

Because of its many advantages, depletion-mode NMOS logic was the dominant technology for many years in the design of microprocessors. However, the large static power dissipation inherent in NMOS logic eventually limited further increases in IC chip density, and a rapid shift took place to the more complex CMOS technology, which is discussed in detail in the next chapter.

6.8 NMOS NAND AND NOR GATES

A complete logic family must provide not only the logical inversion function but also the ability to form some combination of at least two input variables such as the AND or OR function. In NMOS logic, an additional transistor can be added to the simple inverter to form either a NOR or

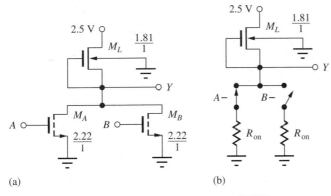

Figure 6.30 (a) Two-input NMOS NOR gate: $Y = \overline{A + B}$. (b) Simplified model with switching transistor A on.

TABLE 6.7		
NOR Gate Truth Table		
A	**B**	**$Y = \overline{A + B}$**
0	0	1
0	1	0
1	0	0
1	1	0

a NAND logic gate. The NOR gate represents the combination of an OR operation followed by inversion, and the NAND function represents the AND operation followed by inversion. One of the advantages of MOS logic is the ease with which both the NOR and NAND functions can be implemented. The switching devices inherently provide the inversion operation, whereas series and parallel combinations of transistors produce the AND and OR operations, respectively.

In the following discussion, remember that we use the positive logic convention to relate voltage levels to logic variables: a high logic level corresponds to a logical 1 and a low logic level corresponds to a logical 0:

$$V_H \equiv 1 \quad \text{and} \quad V_L \equiv 0$$

6.8.1 NOR GATES

In Fig. 6.30, switching transistor M_S of the inverter has been replaced with two devices, M_A and M_B, to form a two-input NOR gate. If either, or both, of the inputs A and B is in the high logic state, a current path will exist through at least one of the two switching devices, and the output will be in the low logic state. Only if inputs A and B are both in the low state will the output of the gate be in the high logic state. The truth table for this gate, Table 6.7, corresponds to that of the NOR function $Y = \overline{A + B}$.

We will pick the size of the devices in our logic gates based on the reference inverter design defined at the end of Sec. 6.7 [Fig. 6.29(d)]. The size of the various transistors must be chosen to ensure that the gate meets the desired logic level and power specifications under the worst-case set of logic inputs.

Consider the simplified schematic for the two-input NOR gate in Fig. 6.30(b). The worst-case condition for the output low state occurs when either M_A or M_B is conducting alone, so the on-resistance R_{on} of each individual transistor must be chosen to give the desired low output level. Thus, $(W/L)_A$ and $(W/L)_B$ should each be equal to the size of M_S in the reference inverter (2.22/1). If M_A and M_B both happen to be conducting ($A = 1$ and $B = 1$), then the combined on-resistance will be equivalent to $R_{on}/2$, and the actual output voltage will be somewhat lower than the original design value of $V_L = 0.20$ V.

When either M_A or M_B is conducting alone, the current is limited by the load device, and the voltages are exactly the same as in the reference inverter.[6] Thus, the W/L ratio of the load device is the same as in the reference inverter (1.81/1). The completed NOR gate design is given in Fig. 6.30(a).

---

[6] Actually, the worst-case situation for current in the load device occurs when M_A and M_B are both on because the voltage is slightly higher across the load device, and its value of V_{SB} is smaller. However, this effect is small enough to be neglected. See Prob. 6.81.

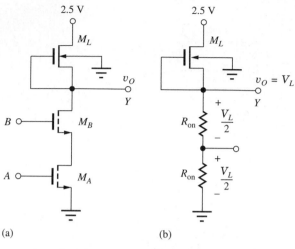

Figure 6.31 Two-input NMOS NAND gate: $Y = \overline{AB}$.

TABLE 6.8
NAND Gate Truth Table

A	B	$Y = \overline{AB}$
0	0	1
0	1	1
1	0	1
1	1	0

EXERCISE: Draw the schematic of a three-input NOR gate. What are the W/L ratios for the transistors based on Fig. 6.30?

ANSWERS: 1.81/1; 2.22/1; 2.22/1; 2.22/1

6.8.2 NAND GATES

In Fig. 6.31(a), a second NMOS transistor has been added in series with the original switching device of the basic inverter to form a two-input NAND gate. Now, if inputs A and B are *both* in a high logic state, a current path exists through the series combination of the two switching devices, and the output is in a low logic state. If either input A or input B is in the low state, then the conducting path is broken and the output of the gate is in the high state. The truth table for this gate, Table 6.8, corresponds to that of the NAND function $Y = \overline{AB}$.

Selecting the Sizes of the Switching Transistors
The sizes of the devices in the NAND logic gate are again chosen based on the reference inverter design from Fig. 6.29(d). The W/L ratios of the various transistors must be selected to ensure that the gate still meets the desired logic level and power specifications under the worst-case set of logic inputs. Consider the simplified schematic for the two-input NAND gate in Fig. 6.31(b). The output low state occurs when both M_A and M_B are conducting. The combined on-resistance will now be equivalent to $2R_{\text{on}}$, where R_{on} is the on-resistance of each individual transistor conducting alone. In order to achieve the desired low level, $(W/L)_A$ and $(W/L)_B$ must both be approximately twice as large as the W/L ratio of M_S in the reference inverter because the on-resistance of each device in the triode region is inversely proportional to the W/L ratio of the transistor:

$$R_{\text{on}} = \frac{v_{DS}}{i_D} = \frac{1}{K_n' \dfrac{W}{L}\left(v_{GS} - V_{TN} - \dfrac{v_{DS}}{2}\right)} \tag{6.36}$$

A second way to approach the choice of device sizes is to look at the voltage across the two switching devices when v_O is in the low state. For our design, $V_L = 0.20$ V. If we assume that one-half of this voltage is dropped across each of the switching transistors and that $(v_{GS} - V_{TN}) \gg v_{DS}/2$,

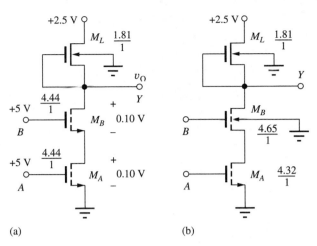

Figure 6.32 NMOS NAND gate: $Y = \overline{AB}$: (a) approximate design, (b) corrected design.

then it can be seen from

$$i_D = K_n' \left(\frac{W}{L}\right)_S (v_{GS} - V_{TN} - 0.5v_{DS})v_{DS} \cong K_n' \left(\frac{W}{L}\right)_S (v_{GS} - V_{TN})v_{DS} \qquad (6.37)$$

that the W/L of the transistors must be approximately doubled in order to keep the current at the same value. Figure 6.32(a) shows the NAND gate design based on these arguments.

Two approximations have crept into this analysis. First, the source-bulk voltages of the two transistors are not equal, and therefore the values of the threshold voltages are slightly different for M_A and M_B. Second, $V_{GSA} \neq V_{GSB}$. From Fig. 6.32(a), $V_{GSA} = 2.5$ V, but $V_{GSB} = 2.4$ V. The results of taking these two effects into account are shown in Fig. 6.32(b). (Verification of these W/L values is left for Prob. (6.82). The corrected device sizes have changed by only a small amount. The values in Fig. 6.32(a) represent an adequate level of design for most purposes.

Choosing the Size of the Load Device
When both M_A and M_B are conducting, the current is limited by the load device, but the voltages applied to the load device are exactly the same as those in the reference inverter design. Thus, the W/L ratio of the load device is the same as in the reference inverter. The completed NAND gate design, based on the simplified device sizing, is given in Fig. 6.32(a).

> **EXERCISE:** Draw the schematic of a three-input NAND gate. What are the W/L ratios for the transistors based on Fig. 6.32(a)?
>
> **ANSWERS:** $1.81/1; 6.66/1; 6.66/1; 6.66/1$

6.8.3 NOR AND NAND GATE LAYOUTS IN NMOS DEPLETION-MODE TECHNOLOGY

Sample layouts for two-input NOR and two-input NAND gates appear in Fig. 6.33 based on ground rules similar to those discussed in Chapter 4. The metal overlap has been reduced in the layout to make the figure clearer.

The NOR gate has the sources and drains of switching transistors A and B connected in parallel using the n^+ layer. The source of the load device is also connected to the common drain region of the switching transistors using the n^+ layer. The gate of the load device is connected to the switching transistors using the metal layer, which also is the output terminal.

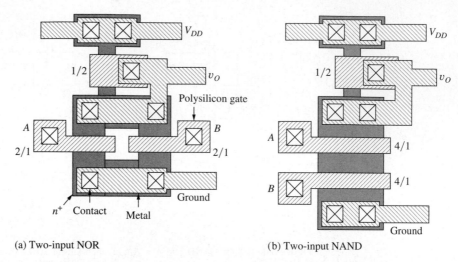

(a) Two-input NOR   (b) Two-input NAND

Figure 6.33 Possible layouts for (a) two-input NOR gate and (b) two-input NAND gate.

Input transistors A and B are stacked above each other in the NAND gate layout. Note that the source of transistor A and the drain of transistor B are the same n^+ region; no contacts are required between the transistors. The widths of transistors A and B have been made twice as wide to maintain the desired low output level, whereas the size of the load transistor remains unchanged.

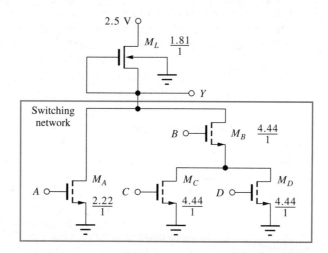

Figure 6.34 Complex NMOS logic gate: $Y = \overline{A + BC + BD}$.

6.9 COMPLEX NMOS LOGIC DESIGN

A major advantage of MOS logic over various forms of bipolar logic comes through the ability to directly combine NAND and NOR gates into more complex configurations. Three examples of **complex logic gate** design are discussed in this section.

Consider the circuit in Fig. 6.34. The output Y will be in a low state whenever a conducting path is developed through the switching transistor network. For this circuit, the output voltage will be low if any one of the following paths is conducting: A or BC (B and C) or BD (B and D). The output Y is represented logically as

$$Y = \overline{A + BC + BD} \quad \text{or} \quad Y = \overline{A + B(C + D)}$$

Silicon Art

Successful integrated circuit designers are typically a very creative group of people. In the course of a large chip design project, engineers generate numerous innovations. The process involves many long hours leading up to the release of the chip layout data to manufacturing.

A small herd of buffalo added to a Hewlett-Packard 64-bit combinatorial divider created by HP engineer Dick Vlach.

A train found on an analog shift register from a LeCroy MVV200 integrated circuit.

A roadrunner drawn in aluminum on silicon by Dan Zuras of Hewlett-Packard.

A compass placed on a prototype optical navigation chip by Hewlett-Packard Labs designer Travis Blalock.

As the end of the design process nears, exhausted designers often want to add a more personal imprint on their work. Traditionally this has taken the form of using patterns in the metal layers of the chip layout to create graphical images relating to the chip's internal code name.

Sadly, most modern IC foundries are now forbidding designers to express themselves in this way over concerns about design rule violations and potential processing problems. Designers tell us that this has forced them to become covert with their doodles and they are sometimes embedding the graphics directly into functional design structures.

which directly implements a complemented **sum-of-products logic function**. This logic gate is most commonly referred to as the **AND-OR-INVERT** or **AOI** gate, and it is widely used as one of the basic building blocks in chips such as field programmable logic arrays (FPGAs). The AND terms (A[7], BC, BD) are formed by vertical stacking of two transistors. These paths are then placed in parallel to form the OR function, and the logic gate inherently provides the logical inversion.

[7] $A = A \cdot 1$

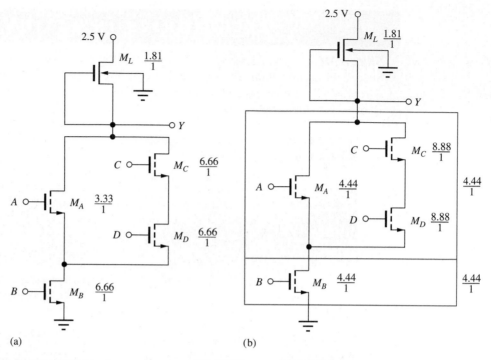

(a)   (b)

Figure 6.35 (a) NMOS implementation of $Y = \overline{AB + CDB}$ or $Y = \overline{(A + CD)B}$. (b) An alternate transistor sizing for the logic gate in (a).

In the final minimum size version in Fig. 6.34, it is recognized that transistor B need not be replicated.

Device sizing will again be based on the worst-case logic state situations. Referring to the reference inverter design, device M_A must have $W/L = 2.22/1$ because it must be able to maintain the output at 0.20 V when it is the only device that is conducting. In the other two paths, M_B will appear in series with either M_C or M_D. Thus, in the worst case, there will be two devices in series in this path, and the simplest choice will be $M_B = M_C = M_D = 4.44/1$. The load device size remains unchanged.

The circuit in Fig. 6.35 provides a second example of transistor sizing in complex logic gates. There are two possible conducting paths through the switching transistor network: AB (A and B) or CDB (C and D and B). The output will be low if either path is conducting, resulting in

$$Y = \overline{AB + CDB} \qquad \text{or} \qquad Y = \overline{(A + CD)B}$$

Transistor sizing can be done in two ways. In the first method, we find the worst-case path in terms of transistor count. For this example, path CDB has three transistors. By making each transistor three times the size of the reference switching transistor, the CDB path will have an on-resistance equivalent to that of M_S in the reference inverter. Thus, each of the three transistors should have $W/L = 6.66/1$.

The second path contains transistors M_A and M_B. In this path, we want the sum of the on-resistances of the devices to be equal to the on-resistance of M_S in the reference inverter:

$$\frac{R_{\text{on}}}{\left(\dfrac{W}{L}\right)_A} + \frac{R_{\text{on}}}{\left(\dfrac{W}{L}\right)_B} = \frac{R_{\text{on}}}{\left(\dfrac{W}{L}\right)_S} \tag{6.38}$$

In Eq. (6.38), R_{on} represents the on-resistance of a transistor with $W/L = 1/1$. Because $(W/L)_B$ has already been chosen,

$$\frac{R_{on}}{\left(\dfrac{W}{L}\right)_A} + \frac{R_{on}}{6.66} = \frac{R_{on}}{2.22} \tag{6.39}$$

Solving for $(W/L)_A$ yields a value of 3.33/1. Because the operating current of the gate is to be the same as the reference inverter, the geometry of the load device remains unchanged. The completed design values appear in Fig. 6.35(a).

A slightly different approach is used to determine the transistor sizes for the same logic gate in Fig. 6.35(b). The switching circuit can be partitioned into two sub-networks connected in series: transistor B in series with the parallel combination of A and CD. We make the equivalent on-resistance of these two subnetworks equal. Because the two subnetworks are in series, $(W/L)_B = 2(2.22/1) = 4.44/1$. Next, the on-resistance of each path through the $(A + CD)$ network should also be equivalent to that of a 4.44/1 device. Thus $(W/L)_A = 4.44/1$ and $(W/L)_C = (W/L)_D = 8.88/1$. These results appear in Fig. 6.35(b).

6.9.1 SELECTING BETWEEN THE TWO DESIGNS

If the unity dimension corresponds to the minimum feature size F, then the total gate area of the switching transistors for the design in Fig. 6.35(b) is $28.5F^2$. The previous implementation of Fig. 6.35(a) had a total gate area of $25.1F^2$. With this yardstick, the second design requires 13 percent more area than the first. Minimum area utilization is often a key consideration in IC design, and the device sizes in Fig. 6.35(a) would be preferred over those in Fig. 6.35(b).

DESIGN EXAMPLE 6.9   TRANSISTOR SIZING IN COMPLEX LOGIC GATES

Choose the transistor sizes for a complex logic gate based on a given reference inverter design.

PROBLEM   Find the logic expression for the gate in Fig. 6.36. Design the W/L ratios of the transistors based on the pseudo NMOS reference inverter in Fig. 6.29(e).

SOLUTION   **Known Information and Given Data:** Logic circuit diagram in Fig. 6.36; reference inverter design in Fig. 6.29(e) with $(W/L)_S = 2.22/1$ and $(W/L)_L = 1.11/1$.

Unknowns: Logic expression for output Y; W/L ratios for all the transistors

Approach: Identify the conducting paths that force the output low; output Y can be represented as a complemented sum-of-products function of the conducting path descriptions. Size the transistors in each path to yield the same on-resistance as the reference inverter.

Assumptions: Neglect the effects of the small source-bulk voltages on the switching transistors. Neglect V_{GS} differences among the switching transistors.

Analysis: Comparing the circuit in Fig. 6.36 to that in 6.35, we see that a fifth transistor has been added to the switching network. Now there are *four* possible conducting paths through the switching transistor network: AB or CDB or CE or ADE. The output will be low when any one of these paths is conducting, resulting in

$$Y = \overline{AB + CDB + CE + ADE}$$

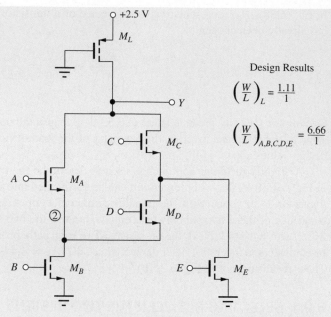

Design Results

$$\left(\frac{W}{L}\right)_L = \frac{1.11}{1}$$

$$\left(\frac{W}{L}\right)_{A,B,C,D,E} = \frac{6.66}{1}$$

Figure 6.36 NMOS implementation of $Y = \overline{AB + CDB + CE + ADE}$.

We desire the current and power to be the same in the circuit when the output is in the low state. Thus the load device will be identical to that of the inverter. The switching transistor network cannot be broken into series and parallel branches, and transistor sizing will follow the worst-case path approach. Path CDB has three transistors in series, so each W/L will be set to three times that of the switching transistor in the reference inverter, or 6.66/1. Path ADE also has three transistors in series, and, because D has $(W/L) = 6.66/1$, the W/L ratios of A and E can also be 6.66/1. All transistors are now 6.66/1 devices.

Check of Results: The remaining paths, AB and CE, must be checked to ensure that the low output level will be properly maintained. Each has two transistors with $W/L = 6.66/1$ in series for an equivalent $W/L = 3.33/1$. Because the W/L of 3.33/1 is greater than 2.22/1, the low output state will be maintained at $V_L < 0.20$ V when paths AB or CE are conducting.

Discussion: Note that the current traverses transistor D in one direction when path CDB is conducting, but in the opposite direction when path ADE is active! Remember from the device cross section in Fig. 6.18(b) that the MOS transistor is a symmetrical device. The only way to actually tell the drain terminal from the source terminal is from the values of the applied potentials. For the NMOS transistor, the drain terminal will be the terminal at the higher voltage, and the source terminal will be the terminal at the lower potential. This bidirectional nature of the MOS transistor is a key to the design of high-density dynamic random access memories (DRAMs), which are discussed in Chapter 8.

Computer-Aided Design: Now we can use SPICE to find the actual values of V_L for different input combinations, including the influence of body effect and nonzero source voltages on the operation of the gate. For the circuit below with VTO = 0.60, KP = 100E-6, GAMMA = 0.5, PHI = 0.6, W = 6.66 U, and L = 1 U for the switching devices and VTO = −0.6, KP = 40E-6, GAMMA = 0.5, PHI = 0.6, W = 1.11 U, and L = 1 U for the load device, SPICE gives these results:

ABCDE	Y (mV)	NODE 2 (mV)	NODE 3 (mV)	I_{DD} (μA)
11000	132	64.4	0	80.1
01110	203	64.4	132	80.1
00101	132	0	64.4	80.1
11111	64.6	31.9	31.9	80.1

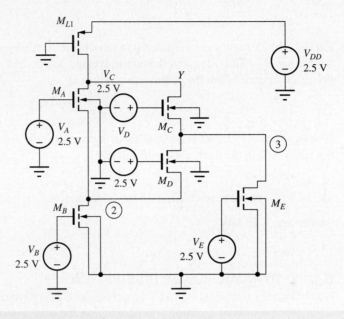

EXERCISE: (a) Calculate the power supply current I_{DD} if the voltage at node Y is 202 mV. (b) Repeat for 132 mV. (c) Repeat for 76.6 mV.

ANSWERS: (a) 80.1 μA; (b) 80.1 μA; (c) 80.1 μA

EXERCISE: Make a complete table for node voltages Y, 2, and 3 and I_{DD} for all 32 possible combinations of inputs for the circuit in Ex. 6.9. Fill in the table entries based on the SPICE simulation results presented in the example.

6.10 POWER DISSIPATION

In this section we consider the two primary contributions to power dissipation in NMOS inverters. The first is the steady-state power dissipation that occurs when the logic gate output is stable in either the high or low states. The second is power that is dissipated in order to charge and discharge the total equivalent load capacitance during dynamic switching of the logic gate.

6.10.1 STATIC POWER DISSIPATION

The overall **static power dissipation** of a logic gate is the average of the power dissipations of the gate when its output is in the low state and the high state. The power supplied to the logic gate is expressed as $P = V_{DD}i_{DD}$, where i_{DD} is the current provided by the source V_{DD}. In the circuits considered so far, i_{DD} is equal to the current through the load device, and the total power supplied by source V_{DD} is dissipated in the load and switching transistors. The average power dissipation depends on the fraction of time that the output spends in the two logic states. If we assume that the average logic gate spends one-half of the time in each of the two output states (a 50 percent duty cycle), then the average power dissipation is given by

$$P_{\text{av}} = \frac{V_{DD}I_{DDH} + V_{DD}I_{DDL}}{2} \tag{6.40}$$

where $I_{DDH} = $ current in gate for $v_O = V_H$

$I_{DDL} = $ current for $v_O = V_L$

For the NMOS logic gates considered in this chapter, the current in the gate becomes zero when the v_O reaches V_H. Thus, $I_{DDH} = 0$, and the average power dissipation becomes equal to one-half the power dissipation when the output is low, given by

$$P_{\text{av}} = \frac{V_{DD}I_{DDL}}{2} \tag{6.41}$$

If some other duty factor is deemed more appropriate (for example, 33 percent), it simply changes the factor of 2 in the denominator of Eq. (6.41).

> **EXERCISE:** What is the average power dissipation of the gates in Fig. 6.29?
>
> **ANSWER:** 0.10 mW

6.10.2 DYNAMIC POWER DISSIPATION

A second, very important source of power dissipation is **dynamic power dissipation,** which occurs during the process of charging and discharging the load capacitance of a logic gate. Consider the simple circuit in Fig. 6.37(a), in which a capacitor is being charged toward positive voltage V_{DD} through a nonlinear resistor (such as an MOS load device).

Let us assume the capacitor is initially discharged; at $t = 0$ the switch closes, and the capacitor then charges toward its final value. We also assume that the nonlinear element continues to deliver current until the voltage across it reaches zero (for example, a depletion-mode NMOS or PMOS load). The total energy E_D delivered by the source is given by

$$E_D = \int_0^\infty P(t)\,dt \tag{6.42}$$

The power $P(t) = V_{DD}i(t)$, and because V_{DD} is a constant,

$$E_D = V_{DD}\int_0^\infty i(t)\,dt \tag{6.43}$$

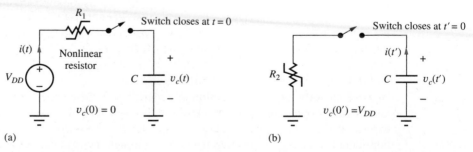

(a)

(b)

Figure 6.37 Simple circuit model for dynamic power calculation: (a) charging C, (b) discharging C.

The current supplied by source V_{DD} is also equal to the current in capacitor C, and so

$$E_D = V_{DD} \int_0^\infty C \frac{dv_C}{dt} \, dt = CV_{DD} \int_{V_C(0)}^{V_C(\infty)} dv_C \tag{6.44}$$

Integrating from $t = 0$ to $t = \infty$, with $V_C(0) = 0$ and $V_C(\infty) = V_{DD}$ results in

$$E_D = CV_{DD}^2 \tag{6.45}$$

We also know that the energy E_S stored in capacitor C is given by

$$E_S = \frac{CV_{DD}^2}{2} \tag{6.46}$$

and thus the energy E_L lost in the resistive element must be

$$E_L = E_D - E_S = \frac{CV_{DD}^2}{2} \tag{6.47}$$

Now consider the circuit in Fig. 6.37(b), in which the capacitor is initially charged to V_{DD}. At $t' = 0$, the switch closes and the capacitor discharges toward zero through another nonlinear resistor (such as an enhancement-mode MOS transistor). Again, we wait until the capacitor reaches its final value, $V_C = 0$. The energy E_S that was stored on the capacitor has now been completely dissipated in the resistor. The total energy E_{TD} dissipated in the process of first charging and then discharging the capacitor is equal to

$$E_{TD} = \frac{CV_{DD}^2}{2} + \frac{CV_{DD}^2}{2} = CV_{DD}^2 \tag{6.48}$$

Thus, every time a logic gate goes through a complete switching cycle, the transistors within the gate dissipate an energy equal to E_{TD}. Logic gates normally switch states at some relatively high frequency f (switching events/second), and the dynamic power P_D dissipated by the logic gate is then

$$P_D = CV_{DD}^2 f \tag{6.49}$$

In effect, an average current equal to $(CV_{DD}f)$ is supplied from source V_{DD}.

EXERCISE: What is the dynamic power dissipated by alternately charging and discharging a 1-pF capacitor between 2.5 V and 0 V at a frequency of 32 MHz?

ANSWER: 200 μW

Note that the power dissipation in the previous exercise is the same as the static power dissipation that we allocated to the $v_O = V_L$ state in our original NMOS logic gate design. In high-speed logic systems, the dynamic component of power can become dominant—we see in Chapter 7 that this is in fact the primary source of power dissipation in CMOS logic gates!

6.10.3 POWER SCALING IN MOS LOGIC GATES

During logic design in complex systems, gates with various power dissipations are often needed to provide different levels of drive capability and to drive different values of load capacitance at different speeds. For example, consider the saturated load inverter in Fig. 6.38(a). The static power dissipation is determined when $v_O = V_L$. M_S is operating in the linear region, M_L is saturated, and

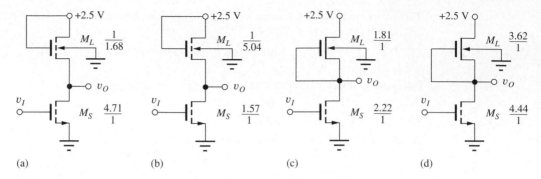

Figure 6.38 Inverter power scaling. The NMOS inverter of (b) operates at one-third the power of circuit (a), and the NMOS inverter of (d) operates at twice the power of circuit (c).

the drain currents of the two transistors are given by

$$i_{DL} = \frac{K_n'}{2} \left(\frac{W}{L} \right)_L (v_{GSL} - V_{TNL})^2$$

$$i_{DS} = K_n' \left(\frac{W}{L} \right)_S \left(v_{GSS} - V_{TNS} - \frac{v_{DSS}}{2} \right) v_{DSS}$$

(6.50)

in which the W/L ratios have been chosen so that $i_{DS} = i_{DL}$ for $v_O = V_L$. For fixed voltages, both drain currents are directly proportional to their respective W/L ratios. If we double the W/L ratio of the load device *and* the switching device, then the drain currents both double, with no change in operating voltage levels.

Or, if we reduce the W/L ratios of both the load device and the switching device by a factor of 3, then the drain currents are both reduced by a factor of 3, with no change in operating voltage levels. Thus, if the W/L ratios of M_L and M_S are changed by the same factor, the power level of the gate can easily be scaled up and down without affecting the values of V_H and V_L. With this technique, the inverter in Fig. 6.38(b) has been designed to operate at one-third the power of the inverter of Fig. 6.38(a) by reducing the value of W/L of each device by a factor of 3. This **power scaling** is a property of ratioed logic circuits. The power level can be scaled up or down without disturbing the voltage levels of the design.

Similar arguments can be used to scale the power levels of any of the NMOS gate configurations that we have studied, and the depletion-mode load inverter in Fig. 6.38(d) has been designed to operate at twice the power of the inverter of that of Fig. 6.38(c) by increasing the value of W/L of each device by a factor of 2. As we will see shortly, this same technique can also be used to scale the dynamic response time of the inverter to compensate for various capacitive load conditions.

EXERCISE: What are the new W/L ratios for the transistors in the gate in Fig. 6.38(a) for a power of 0.1 mW?

ANSWERS: 1/3.36 and 2.36/1

EXERCISE: What are the new W/L ratios for the transistors in the gate in Fig. 6.38(c) for a power of 4 mW?

ANSWERS: 36.2/1 and 44.4/1

EXERCISE: What are the W/L ratios of the transistors in the gate in Fig. 6.35(a) required to reduce the power by a factor of three while maintaining the same value of V_L?

ANSWERS: 1/1.66; 1.11/1; 2.22/1; 2.22/1; 2.22/1

6.11 DYNAMIC BEHAVIOR OF MOS LOGIC GATES

Thus far in this chapter the discussion has been concerned with only the static design of NMOS logic gates. The time domain response, however, plays an extremely important role in the application of logic circuits. There are delays between input changes and output transitions in logic circuits because every node is shunted by capacitance to ground and is not able to change voltage instantaneously. This section reviews the sources of capacitance in the MOS circuit and then explores the dynamic or time-varying behavior of logic gates. Calculations of rise time t_r, fall time t_f, and the average propagation delay τ_p (all defined in Sec. 6.3) are presented, and expressions are then developed for estimating the response time of various inverter configurations.

6.11.1 CAPACITANCES IN LOGIC CIRCUITS

Figure 6.29(a) shows two NMOS inverters including the various capacitances associated with each transistor. These capacitances were introduced in Sec. 4.6. Each device has capacitances between its gate-source, gate-drain, source-bulk, and drain-bulk terminals. Some of the capacitances do not appear in the schematic because they are shorted out by the various circuit connections (for example, C_{SB1}, C_{GS2}, C_{SB3}, C_{GS4}). In addition to the **MOS device capacitances,** the figure includes a wiring capacitance C_W, representing the capacitance of the electrical interconnection between the two logic gates. For simplicity in analyzing the delay times in logic circuits, the capacitances on a given node will be lumped together into a fixed effective nodal capacitance C, as indicated in Fig. 6.39(b), and our hand analysis will cast the behavior of circuits in terms of this effective capacitance C. The MOS

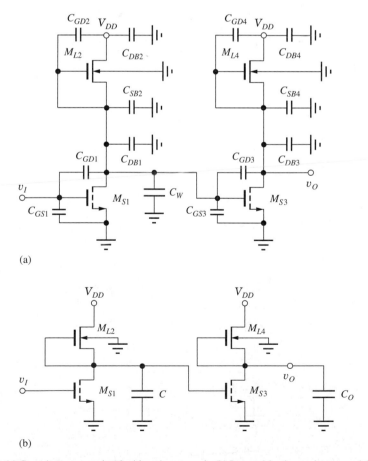

(a)

(b)

Figure 6.39 (a) Capacitances associated with an inverter pair. (b) Lumped-load capacitance model for inverters.

device capacitances are nonlinear functions of the various node voltages; they are highly dependent on circuit layout in an integrated circuit. We will not attempt to find a precise expression for C in terms of all the capacitances in Fig. 6.39(a), but we assume that we have an estimate for the value of C. Simulation tools exist that will extract values of C from a given IC layout, and more accurate predictions of time-domain behavior can be obtained using SPICE circuit simulations.

Fan-Out Limitations in NMOS Logic

Since no dc current needs to be supplied to the input of an NMOS logic gate, the fan-out of an MOS logic gate is not limited by static design constraints. (But this is not the case for bipolar design discussed later in Chapter 9.) However, as more and more gates are attached to a given output as in Figs. 6.39 or 6.10, the value of capacitor C increases, and as we shall see shortly, the temporal responses of the circuit will decrease accordingly. Thus the fan-out will be limited by how much degradation can be tolerated in the time delays of the circuit.

6.11.2 DYNAMIC RESPONSE OF THE NMOS INVERTER WITH A RESISTIVE LOAD

Figure 6.40 shows the circuit from our earlier discussion of the inverter with a resistive load. For hand analysis, the logic input signal is represented by an ideal step function, and we now calculate the rise time, fall time, and delay times for this inverter.

Calculation of t_r and τ_{PLH}

For analysis of the rise time, assume that the input and output voltages have reached their steady-state levels for $t < 0$: $v_I = V_H = 2.5$ V and $v_O = V_L = 0.20$ V. At $t = 0$, the input drops from $v_I = 2.5$ V to $v_I = 0.20$ V. Because the gate-source voltage of the switching transistor drops below V_{TNS}, the MOS transistor abruptly stops conducting. The output then charges from $v_O = V_L = 0.20$ V to $v_O = V_H = V_{DD} = 2.5$ V. In this case, the waveform is that of the simple RC network formed by the load resistor R and the load capacitor C. Using our knowledge of single-time constant circuits:

$$v_O(t) = V_F - (V_F - V_I)\exp\left(-\frac{t}{RC}\right) = V_F - \Delta V \exp\left(-\frac{t}{RC}\right) \tag{6.51}$$

where V_F is the final value of the capacitor voltage, V_I is the initial capacitor voltage, and $\Delta V = (V_F - V_I)$ is the change in the capacitor voltage. For the inverter in Fig. 6.40, $V_F = 2.5$ V, $V_I = 0.20$ V, and $\Delta V = 2.30$ V.

The rise time is determined by the difference between the time t_1 when $v_O(t_1) = V_I + 0.1\,\Delta V$ and the time t_2 when $v_O(t_2) = V_I + 0.9\,\Delta V$. Using Eq. (6.51),

$$V_I + 0.1\,\Delta V = V_F - \Delta V \exp\left(\frac{-t_1}{RC}\right) \qquad \text{yields} \qquad t_1 = -RC\ln 0.9 \tag{6.52}$$

$$V_I + 0.9\,\Delta V = V_F - \Delta V \exp\left(\frac{-t_2}{RC}\right) \qquad \text{yields} \qquad t_2 = -RC\ln 0.1 \tag{6.53}$$

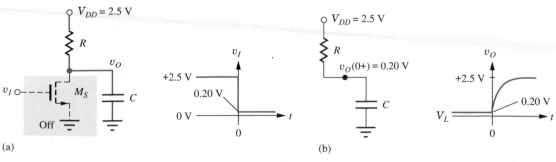

Figure 6.40 Model for rise time in resistively loaded inverter.

and

$$t_r = t_2 - t_1 = RC \ln 9 = 2.2RC \tag{6.54}$$

The delay time τ_{PLH} is determined by $v_O(\tau_{PLH}) = V_I + 0.5 \, \Delta V$, which yields

$$\tau_{PLH} = -RC \ln 0.5 = 0.69RC \tag{6.55}$$

Note that these expressions apply only to the simple RC network.

Equations (6.54) and (6.55) represent the classical expressions for the rise time and propagation delay for an RC network. Similar analyses show that $t_f = 2.2RC$ and $t_{PHL} = 0.69RC$. Remember that these expressions only apply to the simple RC network.

DESIGN NOTE

The rise and fall times and propagation delays for an RC network are given by

$$t_r = t_f = 2.2RC \qquad \tau_{PLH} = \tau_{PHL} = 0.69RC$$

EXERCISE: Find t_r and τ_{PLH} for the resistively loaded inverter with $C = 0.2$ pF and $R = 28.8$ kΩ.

ANSWERS: 12.7 ns; 3.97 ns

EXERCISE: Derive expressions for the fall time and high-to-low propagation delay for an RC network.

ANSWERS: $t_f = 2.2RC$; $\tau_{PHL} = 0.69RC$

Calculation of τ_{PHL} and t_f

Now consider the other switching situation, with $v_I = V_L = 0.20$ V and $v_O = V_H = 2.5$ V, as displayed in Fig. 6.41. At $t = 0$, the input abruptly changes from $v_I = 0.20$ V to $v_I = 2.5$ V. At $t = 0^+$, M_S has $v_{GS} = 2.5$ V and $v_{DS} = 2.5$ V, so it conducts heavily and discharges the capacitance until the value of v_O reaches V_L.

Figure 6.42 shows the currents i_R and i_D in the load resistor and switching transistor as a function of v_O during the transition between V_H and V_L. The current available to discharge the capacitor C is the difference in these two currents:

$$i_C = i_D - i_R$$

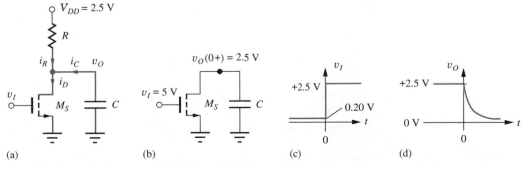

Figure 6.41 Simplified circuit for determining t_f and τ_{PHL}. $v_I(0^+) = V_H = V_{DD}$.

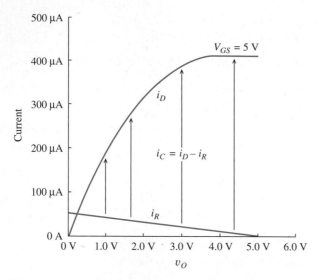

Figure 6.42 Drain current and resistor current versus v_O.

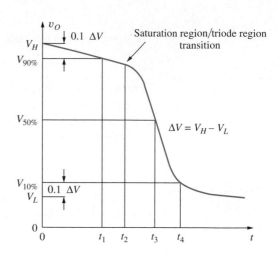

Figure 6.43 Times needed for calculation of τ_{PHL} and t_f for the inverter. Fall time $t_f = t_4 - t_1$; propagation delay $\tau_{PHL} = t_3$.

Because the load element is a resistor, the current in the resistor increases linearly as v_O goes from V_H to V_L. However, when M_S first turns on, a large drain current occurs, rapidly discharging the load capacitance C. V_L is reached when the current through the capacitor becomes zero and $i_R = i_D$. Note that the drain current is much greater than the current in the resistor for most of the period of time corresponding to τ_{PHL}. This leads to values of τ_{PHL} and t_f that are much shorter than τ_{PLH} and t_r associated with the rising output waveform. This behavior is characteristic of NMOS (or PMOS) logic circuits. Another way to visualize this difference is to remember that the on-resistance of the MOS transistor must be much smaller than R in order to force V_L to be a low value. Thus, the apparent "time constant" for the falling waveform will be much smaller than that of the rising waveform.

An exact calculation of t_f and τ_{PHL} is much more complicated than that for a fixed resistor charging the load capacitance because the NMOS transistor changes regions of operation during the voltage transition as shown in Fig. 6.43. At time t_2, the transistor moves from the saturation region of operation to the triode region, and the differential equation that models the V_H to V_L transition changes at that point. An example of these direct calculations can be found in the previous editions of this text or on the website. However, even those "exact" calculations only represent approximations because of the assumptions involved.

Rather than following this more involved approach, we can get very usable estimates for t_f and τ_{PHL} by defining an effective value for the on-resistance of the MOS transistor. Throughout the transient, the on-resistance of the transistor is continually changing as the drain-source voltage of the transistor changes. The effective on-resistance will be chosen to minimize the difference between the MOS and exponential transient curves, and it will then be used with Eqs. (6.54) and (6.55) to find t_f and τ_{PHL}.

First, let us simplify our model for the circuit. From Fig. 6.42, we can see that $i_D \gg i_R$ except for v_O very near V_L. Therefore, the current through the resistor will be neglected so that we can assume that all the drain current of the NMOS transistor is available to discharge the load capacitance, as in Fig. 6.41(b). The input signal v_I is assumed to be a step function changing to $v_I = 2.5$ V at $t = 0$. At $t = 0$, the output voltage V_C on the capacitor is $V_H = V_{DD} = 2.5$ V, and the gate voltage is forced to $V_G = 2.5$ V.

Figure 6.44 displays a SPICE simulation of the high-to-low transition for the resistor load inverter with $R = 28.8 \text{ k}\Omega$ and $(W/L)_S = 2.22/1$. Superimposed on this plot is the transient for the

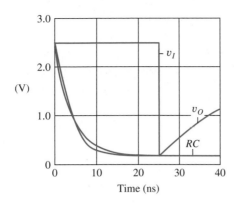

Figure 6.44 SPICE simulation of high-to-low output transient for the resistor load inverter with $C = 1$ pF and its effective constant on-resistance approximation.

exponential discharge of an RC network with a constant value of R. We see that the actual discharge curve is very similar to a purely exponential decay. The effective value of on-resistance used in this simulation is

$$R = R_{eff} = 1.7 R_{onS} \qquad \text{where} \qquad R_{onS} = \frac{1}{K_n(V_H - V_{TNS})} \qquad (6.56)$$

where the factor of 1.7 minimizes the integral of the magnitude of the errors between the MOS and exponential transient curves. R_{onS} represents the on-resistance of the switching transistor as originally defined in Eq. (4.16) with $v_{GS} = V_H$. Substituting $R = R_{eff}$ into the equations for t_f and τ_{PHL} (see the design note below Eq. (6.55)) yields

$$\tau_{PHL} = 0.69(1.7R_{onS})C \cong 1.2R_{onS}C \quad \text{and} \quad t_f = 2.2(1.7R_{onS})C \cong 3.7R_{onS}C \qquad (6.57)$$

EXAMPLE 6.10   DYNAMIC PERFORMANCE OF THE INVERTER WITH RESISTOR LOAD

Find numerical values for the dynamic performance measures of the reference inverter in Fig. 6.29(a).

PROBLEM   Find $t_f, t_r, \tau_{PHL}, \tau_{PHL}$, and τ_p for the resistively loaded inverter in Fig. 6.29 with $C = 0.5$ pF and $R = 28.8$ kΩ.

SOLUTION   **Known Information and Given Data:** Basic resistively loaded inverter circuit in Fig. 6.29; $R = 28.8$ kΩ, $C = 0.5$ pF, $V_{DD} = 2.5$ V, $W/L = 2.22/1$, $V_H = 2.5$ V, $V_L = 0.20$ V, and $K_S = (2.22)(100 \times 10^{-6} \text{ A/V}^2)$

Unknowns: $t_f, t_r, \tau_{PHL}, \tau_{PHL}$, and τ_P

Approach: Find t_r and τ_{PLH} using Eqs. (6.54) and (6.55); calculate R_{onS} and use it to evaluate Eq. (6.57); $\tau_P = (\tau_{PHL} + \tau_{PHL})/2$

Assumptions: None

Analysis: For the resistive load inverter, the rise time and low-to-high propagation delay are

$$t_r = 2.2RC = 2.2(28.8 \text{ kΩ})(0.5 \text{ pF}) = 31.7 \text{ ns}$$

$$\tau_{PLH} = 0.69RC = 0.69(28.8 \text{ kΩ})(0.5 \text{ pF}) = 9.94 \text{ ns}$$

To find t_f and τ_{PHL}, we first calculate the value of R_{onS}:

$$R_{\text{onS}} = \frac{1}{K_S(V_H - V_{TNS})} = \frac{1}{(2.22)\left(100\dfrac{\mu A}{V^2}\right)(2.5 - 0.6)\,V} = 2.37\ k\Omega$$

Substituting the data values into Eq. (6.57):

$$\tau_{PHL} = 1.2 R_{\text{onS}} C = 1.2(2.37\ k\Omega)(0.5\ pF) = 1.42\ ns$$

$$t_f = 3.7 R_{\text{onS}} C = 3.7(2.37\ k\Omega)(0.5\ pF) = 4.39\ ns$$

$$\tau_P = \frac{\tau_{PHL} + \tau_{PLH}}{2} = \frac{9.94 + 1.42}{2}\ ns = 5.68\ ns$$

Check of Results: A double check of the arithmetic indicates our calculations are correct. We see the expected asymmetry in the rise and fall times as well as in the two propagation delay values.

Discussion: Remember that the asymmetry in the rise and fall times and in τ_{PHL} and τ_{PHL} will occur in all NMOS (or PMOS) logic gates because the switching device must have a much smaller on-resistance than that of the load in order to produce the desired value of V_L. We see that τ_{PLH} is approximately 7 times τ_{PHL} and that t_r is more than 7 times t_f!

Computer-Aided Analysis: Let us check our hand calculations using the SPICE transient simulation capability. In the circuit schematic, VI is a pulse source with an initial value of 0, peak value of 2.5 V, zero delay time, 0.1-ns rise time, 0.1-ns fall time, 24.9-ns pulse width, and a 100-ns period. Note the pulse width is chosen so that the rise time plus the pulse width add up to a convenient value of 25 ns. The rise and fall times for VIN are chosen to be much smaller than those expected for the inverter. The transient simulation parameters are a start time of zero, stop time of 100 ns and a time step of 0.025 ns.

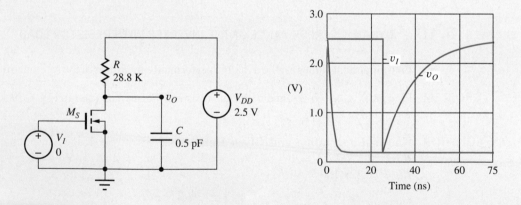

The SPICE results yield values that are very similar to our hand calculations: $t_f = 3.9$ ns, $\tau_{PHL} = 1.6$ ns, $t_r = 31$ ns, and $\tau_{PLH} = 10$ ns. (Note: In order to extract these values from the simulation one must expand the scale for the falling portion of the waveform.)

EXERCISE: Recalculate the values of t_f, t_r, τ_{PHL}, τ_{PLH}, and τ_P if C is decreased to 0.25 pF.

ANSWERS: 2.20 ns; 15.8 ns; 0.71 ns; 4.97 ns; 2.84 ns

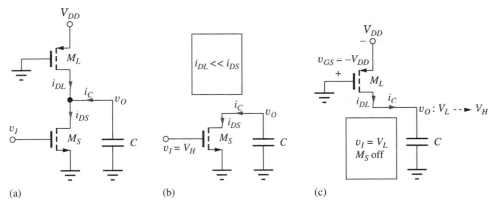

Figure 6.45 (a) Pseudo NMOS logic gate. (b) V_H to V_L transient. (c) V_L to V_H transient.

6.11.3 PSEUDO NMOS INVERTER

Because of its important relationship to CMOS design, we will develop estimates for the delays of the pseudo NMOS inverter. The conditions for the two switching transients appear in Fig. 6.45, and the time response for this inverter can quite easily be found based upon the results from the previous section. In Fig. 6.45(b), we assume that $i_{DL} \ll i_{DS}$, so that the full drain current of switching transistor M_S is available to discharge the capacitor from V_H to V_L, and the gate-source voltage of M_S is $v_{GS} = V_H = V_{DD}$. These conditions are exactly the same as those for the resistor load inverter depicted in Fig. 6.41. Thus the expressions for τ_{PHL} and t_f are the same as in Eq. (6.57):

$$\tau_{PHL} = 0.69(1.7R_{\text{on}S})C \cong 1.2R_{\text{on}S}C \qquad \text{and} \qquad t_f = 2.2(1.7R_{\text{on}S})C \cong 3.7R_{\text{on}S}C$$

with
$$R_{\text{on}S} = \frac{1}{K_n(V_H - V_{TNS})} = \frac{1}{K_n(V_{DD} - V_{TNS})} \tag{6.58}$$

The situation for the low-to-high transient is the same as given in Fig. 6.40(a–b) in which we assume a step change in the input from V_H to V_L at $t = 0$. The switching transistor turns off abruptly, and the load device charges the capacitor from V_L to V_H. We see that the operating conditions for the PMOS transistor are similar to those in Fig. 6.45(b): the source-gate voltage is equal to V_{DD}, and the source-drain voltage changes from a large voltage toward zero. Thus the expressions in Eq. (6.58) can be used to obtain τ_{PLH} and t_r with suitable changes in subscripts:

$$\tau_{PHL} = 0.69(1.7R_{\text{on}L})C \cong 1.2R_{\text{on}L}C \qquad \text{and} \qquad t_r = 2.2(1.7R_{\text{on}L})C \cong 3.7R_{\text{on}L}C$$

with
$$R_{\text{on}L} = \frac{1}{K_p|V_{DD} - V_{TNL}|} \tag{6.59}$$

Figure 6.46 presents SPICE simulation results for the pseudo NMOS Inverter from Fig. 6.29(e) with $(W/L)_S = 2.22/1$, $(W/L)_L = 1.11/1$ and $C = 1$ pF. Based upon the data from the SPICE output file, $\tau_{PHL} = 3.25$ ns, $t_f = 7.8$ ns, $\tau_{PLH} = 15.0$ ns, and $t_r = 35.0$ ns, whereas Eqs. 6.58 and 6.59 predict

$$R_{\text{on}S} = \frac{1}{K_n(V_{DD} - V_{TNS})} = \frac{1}{(100\ \mu\text{A/V}^2)(2.22/1)(2.5 - 0.6)\text{V}} = 2.37\ \text{k}\Omega$$

$$R_{\text{on}L} = \frac{1}{K_n|V_{DD} - V_{TNL}|} = \frac{1}{(40\ \mu\text{A/V}^2)(1.11/1)(2.5 - 0.6)\text{V}} = 11.9\ \text{k}\Omega$$

$$\tau_{PHL} = 1.2R_{\text{on}S}C = 1.2(2.37\ \text{k}\Omega)(1\ \text{pF}) = 2.84\ \text{ns} \qquad t_f = 3.7R_{\text{on}S}C = 8.77\ \text{ns}$$

$$\tau_{PLH} = 1.2R_{\text{on}L}C = 1.2(11.9\ \text{k}\Omega)(1\ \text{pF}) = 14.3\ \text{ns} \qquad t_f = 3.7R_{\text{on}L}C = 44.0\ \text{ns}$$

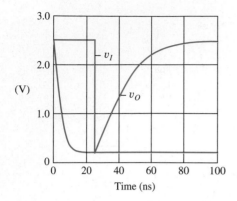

Figure 6.46 SPICE simulation of the pseudo NMOS Inverter with a 1 pF load.

TABLE 6.9
NMOS Inverter Time Delays*

	τ_{PHL}	τ_{PLH}	t_f	t_r
Resistor load	0.31 ns	2.3 ns	0.71 ns	9.9 ns
Pseudo NMOS	0.33 ns	1.5 ns	0.77 ns	3.7 ns
Depletion load	0.31 ns	2.5 ns	0.74 ns	4.6 ns
Saturated load	0.33 ns	1.7 ns	0.71 ns	13.7 ns
Linear load	0.31 ns	2.3 ns	0.72 ns	9.8 ns

*Circuits from Fig. 6.29 with $C = 0.1\ pF$

We see that the analytic expressions provide reasonable estimates for time response of the inverter. The largest error occurs from underestimating the long tail on the rise time.

> **EXERCISE:** Estimate $\tau_{PHL}, t_f, \tau_{PLH}$ and t_r from Fig. 6.46.
>
> **ANSWERS:** 3 ns; 7 ns; 15 ns; 35 ns

6.11.4 A FINAL COMPARISON OF NMOS INVERTER DELAYS

Rather than spend additional time developing closed-form expressions for the other less important NMOS logic gates, we will instead compare the delays of all five NMOS inverter configurations using the results of SPICE simulations that appear in Table 6.9. In this table we observe that the gates all have nearly the same values for both τ_{PHL} and t_f.

This occurs since the switching transistors have each been designed with equal on-resistances in order to achieve the same value of V_L. Note that if we crudely assume that the charging current from the switching transistor[8] is constant, we find

$$I_D = \left(\frac{100\ \mu A}{2\ V^2}\right)\left(\frac{2.22}{1}\right)(2.5 - 0.6)^2\ V^2 = 401\ \mu A$$

$$\tau_{PHL} \cong \frac{C}{I_D}\left(\frac{\Delta V}{2}\right) = \left(\frac{0.1\ pF}{40\ \mu A}\right)\left(\frac{2.5 - 0.2}{2}\right) = 0.29\ ns \tag{6.60}$$

which represents a slightly optimistic estimate for the propagation delays.

We also see that the values of τ_{PLH} are similar with the exception of the saturated load and pseudo NMOS inverters. As we shall see, the pseudo NMOS inverter has the best drive current by far, whereas the reduced logic swing ($\Delta V = 1.35$ V versus 2.3 V for the other inverters) is responsible for the faster response of the saturated load inverter. All the loads deliver the same charging current at the start of the low-high transient (i.e., 80 μA), and if we assume that the charging current from the load is constant during the τ_{PLH} time, we get

$$\tau_{PLH} \cong \frac{C}{I}\left(\frac{\Delta V}{2}\right) = \left(\frac{0.1\ pF}{80\ \mu A}\right)\left(\frac{2.5 - 0.2}{2}\right) = 1.44\ ns\ \text{(pseudo NMOS)}$$

$$\tau_{PLH} \cong \frac{C}{I}\left(\frac{\Delta V}{2}\right) = \left(\frac{0.1\ pF}{80\ \mu A}\right)\left(\frac{1.55 - 0.2}{2}\right) = 0.84\ ns\ \text{(saturated load inverter)} \tag{6.61}$$

---

[8] The switching transistors are all saturated at the start of the transition.

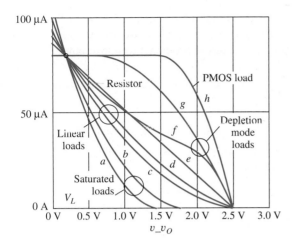

Figure 6.47 A comparison of NMOS inverter load device characteristics with current normalized to 80 μA for $v_o = V_L = 0.20$ V. (a) Saturated load including body effect, (b) saturated load without body effect, (c) linear load with body effect, (d) linear load without body effect, (e) 28.8-kΩ load resistor, (f) depletion-mode load with body effect, (g) depletion-mode load with no body effect, (h) PMOS load transistor for the pseudo NMOS inverter.

TABLE 6.10
NMOS Inverter Time Delays

	τ_{PHL}	τ_{PLH}	t_f	t_r
Resistor load	$1.2R_{\mathrm{on}S}C$	$0.69RC$	$3.7R_{\mathrm{on}S}C$	$2.2RC$
Pseudo NMOS	$1.2R_{\mathrm{on}S}C$	$1.2R_{\mathrm{on}L}C$	$3.7R_{\mathrm{on}S}C$	$3.7R_{\mathrm{on}L}C$
Depletion load	$1.2R_{\mathrm{on}S}C$	$3.6R_{\mathrm{on}L}C$	$3.7R_{\mathrm{on}S}C$	$8.1R_{\mathrm{on}L}C$
Saturated load	$1.2R_{\mathrm{on}S}C$	$3.0R_{\mathrm{on}L}C$	$3.7R_{\mathrm{on}S}C$	$11.9R_{\mathrm{on}L}C$
Linear load	$1.2R_{\mathrm{on}S}C$	$0.69R_{\mathrm{on}L}C$	$3.7R_{\mathrm{on}S}C$	$3.7R_{\mathrm{on}L}C$

$$R_{\mathrm{on}S} = \frac{1}{K_S(V_H - V_{TNS})} \qquad R_{\mathrm{on}L} = \frac{1}{K_L|V_{GS} - V_{TNL}|}$$

These simple calculations again represent an approximation that underestimates the simulated values of the propagation delays.

The largest differences in the delay times appear in the rise time results and are due to the widely differing current carrying capability of the various load transistors as the output rises from V_L to V_H as depicted in Fig. 6.47. The transistor sizes have been chosen so that the current in each device is 80 μA when v_O is at the output low level of 0.20 V, and the drive current of each device decreases as the output voltage rises. The PMOS load provides the largest current whereas the saturated load is the weakest. As a reference for comparison, curve (e) is the straight line corresponding to the constant 28.8-kΩ load resistor. Note that body effect significantly degrades the characteristics of the saturated, linear, and depletion-mode load devices.

Both saturated load characteristics (a and b in Fig. 6.47) deliver substantially less current than the resistor throughout the full output voltage transition. Thus, we expect gates with saturated loads to have the poorest values of t_r. We also can observe that the load current of the saturated load devices goes to zero before the output reaches 2.5 V. The linear loads c and d are an improvement over the saturated load devices but still provide less current than the resistive load. The depletion-mode load devices f and g provide the largest current throughout the transition, and thus they should exhibit the smallest value of t_r. The ideal depletion-mode load g would deliver substantially more current than the resistor load. However, body effect significantly degrades the current source characteristics of the depletion-mode device. We can clearly see the substantial current advantage offered by the PMOS load device that leads to the improvements in t_r and τ_{PLH} in Table 6.9.

Table 6.10 presents the inverter delay equations derived earlier in the chapter as well as similar approximations for the temporal responses of the other inverters. Each of the time responses is determined by an RC product consisting of the load capacitance and the effective on-resistance of either the switching transistor or the load device. The constants multiplying the RC product are determined by the nonlinear behavior of the transistors.

As we should expect from our knowledge of RC networks, all the delays are directly proportional to the load capacitance C and to the effective resistance of the switching and load transistors. Note that the time responses are inversely proportional to carrier mobility since the on-resistances contain mobility in K_S and K_L. Thus, we expect NMOS logic to be approximately 2.5 times faster than PMOS circuits operating at equivalent voltages. This fact was the main reason for the rapid switch to NMOS technology from PMOS as soon as the manufacturing problems were overcome.

DESIGN EXAMPLE 6.11 PROPAGATION DELAY DESIGN FOR AN INVERTER

The logic gates that drive signals off an integrated-circuit chip are typically loaded by relatively large capacitances. This example determines the size of an inverter that is required to drive a large load capacitor with a small value of propagation delay.

PROBLEM Design a pseudo NMOS inverter to provide an average propagation delay of 2 ns when driving a 5-pF capacitor. Use the reference inverter design in Fig. 6.29 as described by the equations in Table 6.10. Find the rise and fall times for the logic gate.

Known Information and Given Data: Pseudo NMOS reference inverter design in Fig. 6.29(e) with $C = 5$ pF, $V_{DD} = 2.5$ V, $K'_n = 100$ μA/V^2, $V_{TNS} = 0.6$ V, $V_{TNL} = -0.6$ V, $V_H = 2.5$ V, $V_L = 0.20$ V, $(W/L)_S = 2.22/1$, and $(W/L)_L = 1.11/1$

Unknowns: $(W/L)_S$, $(W/L)_L$, t_f, and t_r

Approach: Use the results in Table 6.10 to find the required values of R_{onL} and R_{onS}. Determine the W/L ratios from the on-resistance values and the reference inverter design.

Assumptions: None

SOLUTION Using the equations in Table 6.10 for the pseudo NMOS inverter, we can write an expression for the average propagation delay.

$$\tau_P = \frac{1.2 R_{onS} C + 1.2 R_{onL} C}{2} = \frac{0.6 C}{K_n} \left[\frac{1}{(V_{DD} - V_{TN})} + \frac{1}{\frac{K_p}{K_n} | - V_{DD} - V_{TP}|} \right]$$

Using the parameters from the reference inverter design, we can solve for the W/L ratios of the two transistors:

$$\left(\frac{W}{L} \right)_S = \frac{0.6(5 \text{ pF})}{2 \text{ ns } (10^{-4} \text{ A/V}^2)} \left[\frac{1}{(2.5 - 0.6)} + \frac{1}{\frac{1.11(40)}{2.22(100)} | - 2.5 - (-0.6)|} \right] = \frac{47.4}{1}$$

$$\left(\frac{W}{L} \right)_L = \frac{1}{2} \left(\frac{W}{L} \right)_S = \frac{23.7}{1}$$

The rise and fall times can be estimated using the equations in Table 6.10.

$$t_f = 3.7 R_{onL} C = 3.7 \frac{5 \text{ pF}}{47.4(100 \text{ μA/V}^2)(2.5 - 0.6) \text{ V}} = 2.05 \text{ ns}$$

$$t_r = 3.7 R_{onL} C = 3.7 \frac{5 \text{ pF}}{23.7(40 \text{ μA/V}^2)(2.5 - 0.6) \text{ V}} = 10.3 \text{ ns}$$

Check of Results: We have found the unknowns, and the values appear reasonable. The best check will be to simulate the transient response of the new inverter design using SPICE.

Discussion: Large transistor sizes are required to achieve the low propagation delay with a large capacitive load. The input capacitance to this inverter will also be large and require another large driver stage. The optimum form of these "cascade" buffers will be described in Chapter 7.

Computer-Aided Analysis: Now we confirm the behavior of our design using SPICE transient simulation. In the circuit schematic, VIN is a pulse source with initial value of 0.2 V, peak value of 2.5 V, zero delay time, 0.10-ns rise time, 0.10-ns fall time, 14.9-ns pulse width, and a 40-ns period. The pulse width is chosen so that the rise time plus the pulse width add up to a convenient value of 15 ns. The rise and fall times for VIN are chosen to be much smaller than the expected values of the inverter rise and fall times. The transient simulation parameters are a start time of zero, a stop time of 40 ns, and a time step of 0.025 ns.

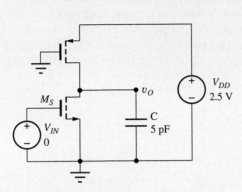
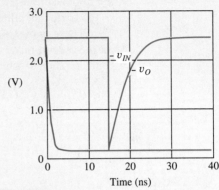

From the graph of the transient response, we find $\tau_{PHL} = 0.75$ ns and $\tau_{PLH} = 3.5$ ns, yielding $\tau_P = 2.1$ ns. The rise and fall times are 8.0 ns and 1.8 ns, respectively. The values all agree well with the design goals and estimates.

EXERCISE: What is the static power dissipation in the inverter in Ex. 6.11? What is the dynamic power dissipation if the inverter is switching on and off every 2 ns?

ANSWERS: 2.14 mW; 13.2 mW!

EXERCISE: What would be the transistor sizes in Ex. 6.11 if the inverter was required to drive 20 pF with an average propagation delay of 1 ns (a 1-GHz rate)? What is the dynamic power consumption of this inverter?

ANSWERS: 379/1; 189/1; 0.11 W

6.12 PMOS LOGIC

In the previous sections of this chapter, we have concentrated on understanding NMOS logic circuits. However, as already mentioned several times, PMOS logic historically preceeded NMOS logic, but was quickly replaced by the higher-performance NMOS logic as soon as NMOS technology could be reliably manufactured. This section presents a brief discussion of PMOS logic circuits.

6.12.1 PMOS INVERTERS

PMOS logic circuits mirror those presented for NMOS logic as exhibited in Fig. 6.48 which presents the PMOS equivalents of the inverter designs in Fig. 6.29. In these circuits, the power supply has been changed to −5 V and each NMOS transistor has been replaced with a PMOS device.

Each circuit has been designed to have the same power level as the equivalent NMOS circuit: $P = 0.20$ mW. Note that for the circuit in Fig. 6.48(a), $V_L = -2.5$ V and $V_H = -0.20$ V. In the saturated load circuit in Fig. 6.48(b), $V_L = -1.55$ V and $V_H = -0.20$ V, assuming the value of $V_{TP} = -0.6$ V. The W/L ratios have been found by simply scaling the W/L ratios of the NMOS inverters by the mobility ratio $\mu_n/\mu_p = 2.5$, not by going through detailed calculations.

ELECTRONICS IN ACTION

MEMS-Based Computer Projector

Microelectromechanical systems (MEMS) devices are becoming increasingly important in a variety of applications. MEMS allow one to design mechanical devices controllable by electrical signals and micromachined using tools and techniques similar to those used in the microelectronics industry. MEMS devices are used in applications such as air-bag accelerometers, radio frequency filters, and electrically addressable light modulators.

These figures show a MEMS-based technology developed at Texas Instruments and used at the core of many computer-driven projectors. The device has a large array of 12 to 16 μm mirrors, each of which is controlled by a CMOS SRAM memory cell located underneath each

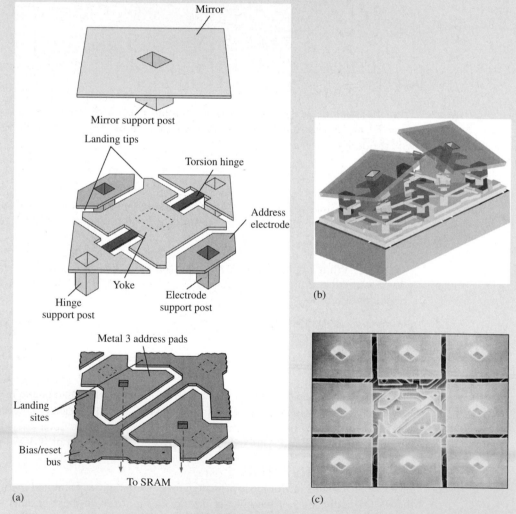

(a) Details of the micro-mirror pixel structure. (b) Three-dimensional view of two adjacent pixels. (c) Magnified view of 9 micro-mirror pixels. The middle mirror and yoke have been removed.

Device photos and drawings are courtesy of Texas Instruments, Inc.

mirror. The mirror rotates about the torsion hinge shown. Depending on the digital value stored in each cell, voltages are driven onto the address electrodes creating electrostatic forces that rotate the mirror into one of two positions. Combined with the appropriate optics, light is directed onto the device and reflected onto a projection screen. Writing appropriate data values in each pixel allows any arbitrary pattern to be created on the screen.

This scheme allows one to turn on or off each pixel. To also include grayscale, the ON time of each pixel is varied by writing different data into the cells during a single display interval. If a mirror is held in the ON position for 20 percent of a display interval, the human eye will perceive a pixel with an intensity of 20 percent. Color is created by a sequence of red, green, and blue illumination sources used in rapid sequence to create three sequential color frames. Again, the eye integrates the three color signals and creates the perception of a vast palette of colors. A few projection systems use three separate display chips to create the three-color image frames simultaneously, allowing for greater total optical power to be projected onto very large screens.

The combination of microscale movable mirrors with microelectronics has made possible a new technique for the projection of high resolution and high quality digital images and video. The integration of MEMS and microelectronics is enabling new applications in fields as diverse as medicine, science, transportation, and consumer electronics.

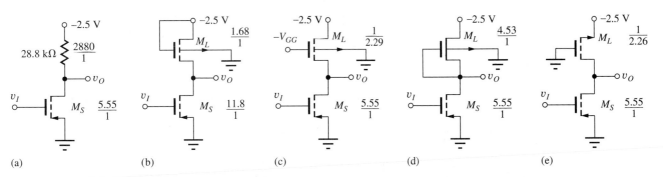

Figure 6.48 PMOS inverters: (a) resistive load, (b) saturated load, (c) linear load, (d) depletion-mode load, and (e) pseudo PMOS.

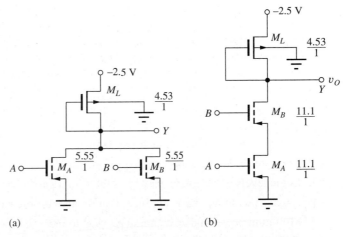

Figure 6.49 Two-input PMOS gates: (a) NOR gate and (b) NAND gate.

6.12.2 NOR AND NAND GATES

The PMOS NOR and NAND gates in Fig. 6.49 mirror the NMOS circuits in Figs. 6.30 and 6.31. The power supply has been changed to -2.5 V, and each NMOS transistor has been replaced with a PMOS device. The W/L ratios are scaled by the mobility ratio of 2.5. Complex logic gates are built up in a manner analogous to the NMOS case. As noted previously, NMOS logic will have a $2.5 \times$ speed advantage over PMOS logic for a given capacitance and gate size. The various delay times can be calculated using the formulas presented in Table 6.10.

SUMMARY

Chapter 6 introduced a number of concepts and definitions that form a basis for logic gate design. NMOS technology was then used as a vehicle to explore detailed logic circuit design. The geometry of the load device, $(W/L)_L$, is designed to limit the current and power dissipation of the logic gate to the desired level, whereas the geometry of the switching device, $(W/L)_S$, is chosen to provide the desired value of V_L. Transistors are usually designed with either W or L set equal to the minimum feature size achievable in a given technology.

- *Boolean algebra:* Boolean algebra, developed by G. Boole in the mid-1800s, is a powerful mathematical tool for manipulating binary logic expressions. Basic logic gates provide some combination of the NOT, AND, OR, NAND, or NOR logic functions. A complete logic family must provide at least the NOT function and either the AND or OR functions.

- *Binary logic states:* Binary logic circuits use two voltage levels to represent the Boolean logic variables 1 and 0. In the positive logic convention used throughout this book, the more positive voltage represents a 1, and the more negative level represents a 0. The output of an ideal logic gate would take on only two possible voltage levels: V_H corresponding to the 1 state, and V_L corresponding to the 0 state.

- *Logic state transitions:* The output of the ideal gate would abruptly change state as the input crossed through a fixed reference voltage V_{REF}. However, such an abrupt transition cannot be achieved (it requires infinite gain devices). Logic gates implemented with electronic circuits can only approximate this ideal behavior. The transition between states as the input voltage changes is much more gradual, and a precise reference voltage is not defined. V_{IL} and V_{IH} are defined by the input voltages at which the slope of the voltage transfer characteristic is equal to -1, and these voltages define the boundaries of the transition region between the logical 1 and 0 levels.

- *Noise margins:* Noise margins are very important in logic gates and represent a measure of the gate's ability to reject extraneous signals. The high-state and low-state noise margins are defined by $\text{NM}_H = V_{OH} - V_{IH}$ and $\text{NM}_L = V_{IL} - V_{OL}$, respectively. Voltages V_{OL} and V_{OH} represent the gate output voltages at the -1 slope points and correspond to input levels V_{IH} and V_{IL}, respectively. The unwanted signals can be voltages or currents coupled into the circuit from adjacent logic gates, from the power distribution network, or by electromagnetic radiation. The noise margins must also absorb manufacturing process tolerance variations and power supply voltage variations.

- *Logic design goals:* Keep in mind a number of logic gate design goals.
  1. The logic gate should quantize the input signal into two discrete output levels and minimize the width of the undefined input voltage range.
  2. The gate should be unidirectional in nature.
  3. Logic levels must be regenerated as the signal passes through the gate.
  4. Logic gates should have significant fan-in and fan-out capability.
  5. Minimum power and area should be used to meet the speed requirements of the design. Noise margins generally should be as large as possible.

- *Logic delays:* In the time domain, the transition between logic states cannot occur instantaneously. Capacitances exist in any real circuit and slow down the state transitions, thereby degrading the logic signals. Rise time t_r and fall time t_f characterize the time required for a given signal to change between the 0 and 1 or 1 and 0 states, respectively, and the average propagation delay τ_P characterizes the time required for the output of a given gate to respond to changes in its input signals. The propagation delays on the high-to-low (τ_{PHL}) and low-to-high (τ_{PLH}) transitions are typically not equal, and τ_P is equal to the average of these two values.

- *Power-delay product:* The power-delay product PDP, expressed in picojoules (pJ) or femtojoules (fJ), is an important figure of merit for comparing logic families. At low power levels, the power-delay product is a constant, and the propagation delay of a given logic family decreases as power is increased. At intermediate power levels, the propagation delay becomes independent of power level, and at high power levels, the propagation delay of bipolar logic families actually degrades as power is increased.

- *NMOS inverter with resistor load:* Basic inverter design was introduced by considering the static behavior of an inverter using an NMOS switching transistor and a resistor load. Although simple in concept, the resistor is not feasible for use as a load element in ICs because it consumes too much area.

- *IC inverters:* In integrated circuits, the resistor load in the logic gate is replaced with a second MOS transistor, and four possibilities were investigated in detail: the saturated load device, the linear load device, the depletion-mode load device, and the pseudo NMOS inverter.

- *Saturated load:* The saturated load device is the most economical configuration because it does not require any modification to the basic MOS fabrication process.

- *Linear load:* The linear load configuration offers improved performance but requires an additional power supply voltage, which is both expensive and causes substantial wiring congestion in ICs.

- *Depletion-mode load:* Depletion-mode load circuits require additional processing in order to create MOSFETs with a second value of threshold voltage. However, substantial performance improvement can be obtained, and NMOS depletion-mode load technology was the workhorse of the microprocessor industry for many years.

- *Pseudo NMOS Inverter:* A PMOS transistor is used as the load device in this logic family. Pseudo NMOS is a high-performance form of NMOS logic that is most commonly found embedded in today's CMOS IC chip designs.

- *NOR and NAND gates:* Multi-input NOR and NAND gates can both easily be implemented in MOS logic. The NOR gate is formed by placing additional transistors in parallel with the switching transistor of the basic inverter, whereas the NAND gate is formed by several switching devices connected in series.

- *Complex logic gates:* An advantage of MOS logic is its ability to implement complex sum-of-products and product-of sums logic equations in a single logic gate, by utilizing both parallel and series connections of the switching transistors. A single load device is required for each logic gate, and one switching transistor is required for each logic input variable.

- *Reference inverter based design:* Once the reference inverter for a logic family is designed, NAND, NOR, and complex gates can all be designed by applying simple scaling rules to the geometry of the reference inverter.

- *MOS body effect:* The influence of the MOSFET body effect cannot be avoided in integrated circuits, and it plays an important role in the design of NMOS (or PMOS) logic gates. Body effect reduces the value of V_H in saturated load logic and generally degrades the current delivery capability of all the load device configurations, thereby increasing the delay of all the logic gates. The MOS body effect has a minor influence on the design of the W/L ratios of the switching transistors in complex logic gates.

- *Rise time, fall time, and propagation delay:* Equations were developed for the rise time, fall time, and propagation delays of the various types of NMOS logic gates, and it was shown that all the time delays of MOS logic circuits are directly proportional to the total equivalent capacitance connected to the output node of the gate. The total effective capacitance is a complicated function of operating point and is due to the capacitance of the interconnections between gates as well as the capacitances of the MOS devices, which include the gate-source (C_{GS}), gate-drain (C_{GD}), drain-bulk (C_{DB}), and source-bulk (C_{SB}) capacitances.

- *Static and dynamic power dissipation:* Power dissipation of a logic gate has a static component and a dynamic component. Dynamic power dissipation is proportional to the switching frequency of the logic gate, the total capacitance, and the square of the logic voltage swing. At low switching frequencies, static power dissipation is most important, but at high switching rates the dynamic component becomes dominant. For a given load capacitance, the power and speed of a logic gate can be changed by proportionately scaling the geometry of the load and switching transistors. For example, doubling the W/L ratios of all devices doubles the power of the gate without changing the static voltage levels of the design. This behavior is characteristic of "ratioed" MOS logic.

- *PMOS logic:* PMOS logic gates are mirror images of the NMOS gates. In order to equal the performance of NMOS, the size of the transistors must be increased in order to compensate for the lower mobility of holes compared to electrons.

KEY TERMS

AND gate
Average propagation delay (τ_p)
Boolean algebra
Complementary MOS (CMOS) technology
Complex logic gates
Depletion load inverter
Dynamic power dissipation
Emitter-coupled logic (ECL)
Fall time t_f
Fan in
Fan out
High logic level at the gate input (V_{IH})
High logic level at the gate output (V_H)
Linear load inverter
Load transistor
Low logic level at the gate input (V_{IL})
Low logic level at the gate output (V_L)
Minimum feature size (F)
MOS device capacitances
NAND gate
Noise margin in high state (NM$_H$)
Noise margin in low state (NM$_L$)
NOR gate
On-resistance
OR gate
Output high logic level (V_H)
Output low logic level (V_L)
Power-delay product (PDP)

Power scaling
Propagation delay—high-to-low transition (τ_{PHL})
Propagation delay—low-to-high transition (τ_{PLH})
Pseudo NMOS Inverter
Ratioed logic
Reference inverter design
Reference voltage (V_{REF})
Resistor load inverter
Rise time t_r
Saturated load inverter
Single-channel technology
Static power dissipation
Sum-of-products logic function
Switching transistor
Transistor-transistor logic (TTL)
Truth table
V_{OH}—The output voltage corresponding to an input voltage of V_{IL}
V_{OL}—The output voltage corresponding to an input voltage of V_{IH}
Voltage transfer characteristic (VTC)
10 percent point
50 percent point
90 percent point
W/L ratio

REFERENCES

1. J. R. Houser, "Noise margin criteria for digital logic circuits," *IEEE Trans. on Education,* vol. 36, no. 4, pp. 363–368, November 1993.
2. C. F. Hill, "Noise margin and noise immunity in logic circuits," *Microelectronics,* vol. 1, pp. 16–21, April 1968.
3. J. Lohstroh, E. Seevinck, and J. Degroot, "Worst-case static noise margin criteria for logic circuits and their mathematical equivalence," *IEEE J. of Solid-State Circuits,* vol. SC-18, no. 6, pp. 803–806, December 1983.
4. J. D. Meindl and J. A. Davis, "The fundamental limit on binary switching energy for terascale integration (TSI)," *IEEE J. of Solid-State Circuits,* vol. 35, no. 10, pp. 1515–1516, October 2000.
5. R. M. Swanson and J. D. Meindl, "Ion-implanted complementary MOS transistors in low-voltage circuits," *IEEE J. of Solid-State Circuits,* vol. SC-7, no. 2, pp. 146–153, April 1972.
6. G. Boole, *An Investigation of the Laws of Thought, on Which Are Founded the Mathematical Theories of Logic and Probability,* 1849. Reprinted by Dover Publications, Inc., New York: 1954.

ADDITIONAL READING

Nelson, V. P., et al. *Analysis and Design of Logic Circuits.* Prentice-Hall, Englewood Cliffs, N.J.: 1995.

PROBLEMS

Use $K'_n = 60 \, \mu\text{A/V}^2$, $K'_p = 25 \, \mu\text{A/V}^2$, $V_{TN} = 0.6$ V, and $V_{TP} = -0.6$ V unless otherwise indicated.

General Introduction

6.1. Integrated circuit chips packaged in plastic can typically dissipate only 1 W per chip. Suppose we have an IC design that must fit on one chip and requires 100,000 logic gates. (a) What is the average power that can be dissipated by each logic gate on the chip? (b) If a supply voltage of 2.5 V is used, how much current can be used by each gate?

6.2. A high-performance microprocessor design requires 20 million logic gates and is placed in a package that can dissipate 100 W. (a) What is the average power that can be dissipated by each logic gate on the chip? (b) If a supply voltage of 2.5 V is used, how much current can be used by each gate? (c) What is the total current required by the IC chip?

Ideal Gates, Logic Level Definitions, and Noise Margins

6.3. (a) The ideal inverter in Fig. 6.2(b) has $R = 100$ kΩ and $V_+ = 2.5$ V. What are V_H and V_L? What is the power dissipation of the gate for $v_O = V_H$ and $v_O = V_L$? (b) Repeat for $V_+ = 3.3$ V.

6.4. Plot a graph of the voltage transfer characteristic for an ideal inverter with $V_+ = 3.3$ V, $V_- = 0$ V, and $V_{\text{REF}} = 1.1$ V. Assume $V_H = V_+$ and $V_L = V_-$.

6.5. (a) Plot a graph of the overall voltage transfer function for two cascaded ideal inverters if each individual inverter has a voltage transfer characteristic as defined in Prob. 6.4. (b) What is the overall logic expression $Z = f(A)$ for the two cascaded inverters?

6.6. Plot a graph of the voltage gain A_v of the ideal inverter in Fig. 6.1 as a function of input voltage v_I. ($A_v = dv_O/dv_I$)

6.7. The voltage transfer characteristic for an inverter is given in Fig. P6.7. What are V_H, V_L, V_{IH}, V_{IL}, and the voltage gain A_v of the inverter in the transition region? ($A_v = dv_O/dv_I$)

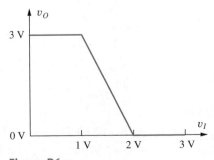

Figure P6.7

6.8. Plot a graph of the overall voltage transfer characteristic for two cascaded inverters if each individual inverter has the voltage transfer function defined in Fig. P6.7.

6.9. Suppose $V_H = 5$ V, $V_L = 0$ V, and $V_{REF} = 2.0$ V for the ideal logic gate in Fig. 6.1. What are the values of V_{IH}, V_{OL}, V_{IL}, V_{OH}, NM_H, and NM_L?

6.10. Suppose $V_H = 3.3$ V and $V_L = 0$ V for the ideal logic gate in Fig. 6.1. Considering noise margins, what would be the best choice of V_{REF}, and why did you make this choice?

6.11. The static voltage transfer characteristic for a practical CMOS inverter is given in Fig. P6.11. What are the values of V_H, V_L, V_{OH}, V_{OL}, V_{IH}, V_{IL}, NM_H, and NM_L?

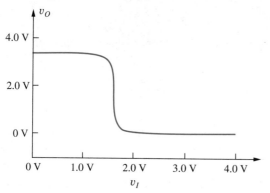

Figure P6.11

6.12. The graph in Fig. P6.12 gives the results of a SPICE simulation of an inverter. What are V_H and V_L for this gate?

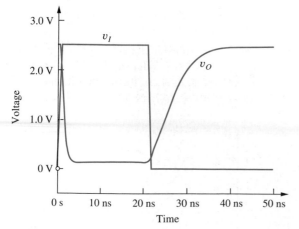

Figure P6.12

6.13. The graph in Fig. P6.13 gives the results of a SPICE simulation of an inverter. What are V_H and V_L for this gate?

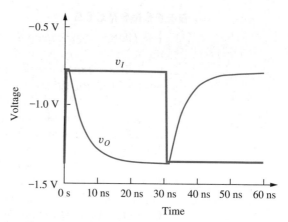

Figure P6.13

6.14. An ECL gate exhibits the following characteristics: $V_{OH} = -0.8$ V, $V_{OL} = -2.0$ V, and $NM_H = NM_L = 0.5$ V. What are the values of V_{IH} and V_{IL}?

Dynamic Response of Logic Gates

6.15. A logic family has a power-delay product of 100 fJ. If a logic gate consumes a power of 100 μW, estimate the propagation delay of the logic gate.

6.16. Integrated circuit chips packaged in plastic can typically dissipate only 1 W per chip. Suppose we have an IC design that must fit on one chip and requires 250,000 logic gates. (a) What is the average power that can be dissipated by each logic gate on the chip? (b) If a supply voltage of 2.5 V is used, how much current can be used by each gate? (c) If the average gate delay for these circuits must be 2 ns, what is the power-delay product required for the circuits in this design?

6.17. A high-performance microprocessor design requires 100 million logic gates and is placed in a package that can dissipate 100 W. (a) What is the average power that can be dissipated by each logic gate on the chip? (b) If a supply voltage of 2.5 V is used, how much current can be used by each gate? (c) If the average gate delay for these circuits must be 1 ns, what is the power-delay product required for the circuits in this design?

6.18. Plot the power-delay product versus power for the logic gate with the power-delay characteristic depicted in Fig. 6.6.

*6.19. (a) Derive an expression for the rise time of the circuit in Fig. P6.19(a) in terms of the circuit time constant. Assume that $v(t)$ is a 1-V step function, changing state at $t = 0$. (b) Derive a similar expression for the fall time of the capacitor voltage in Fig. P6.19(b) if the capacitor has an initial voltage of 1 V at $t = 0$.

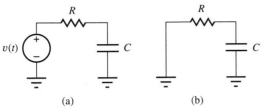

(a)   (b)

Figure P6.19

6.20. The graph in Fig. P6.12 gives the results of a SPICE simulation of an inverter. (a) What are V_H and V_L for this gate? (b) What are the rise and fall times for v_I and v_O? (c) What are the values of τ_{PHL} and τ_{PLH}? (d) What is the average propagation delay for this gate?

6.21. The graph in Fig. P6.13 gives the results of a SPICE simulation of an inverter. (a) What are V_H and V_L for this gate? (b) What are the rise and fall times for v_I and v_O? (c) What are the values of τ_{PHL} and τ_{PLH}? (d) What is the average propagation delay for this gate?

6.4 Review of Boolean Algebra

6.22. Use the results in Table 6.2 to prove that $(A + B) \cdot (A + C) = A + BC$.

6.23. Use the results in Table 6.2 to simplify the logic expression $Z = AB\overline{C} + ABC + \overline{A}BC$.

6.24. Make a truth table for the expression in Prob. 6.23.

6.25. Use the results in Table 6.2 to simplify the logic expression $Z = \overline{A}\overline{B}C + ABC + \overline{A}BC + A\overline{B}C$.

6.26. Make a truth table for the expression in Prob. 6.25.

6.27. Make a truth table and write an expression for the logic function in Fig. P6.27.

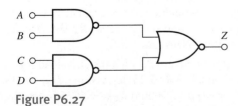

Figure P6.27

6.28. Make a truth table and write an expression for the logic function in Fig. P6.28.

6.29. (a) What is the fan out of the NAND gate in Fig. P6.28? (b) Of each NAND gate in Fig. P6.27?

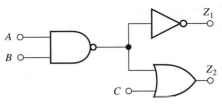

Figure P6.28

General Problems

6.30. A high-speed microprocessor must drive a 64-bit data bus in which each line has a capacitive load C of 40 pF, and the logic swing is 3.3 V. The bus drivers must discharge the load capacitance from 3.3 V to 0 V in 1 ns, as depicted in Fig. P6.30. Draw the waveform for the current in the output of the bus driver as a function of time for the indicated waveform. What is the peak current in the microprocessor chip if all 64 drivers are switching simultaneously?

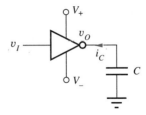

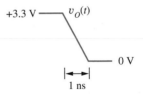

Figure P6.30   Bus driver and switching waveform.

6.31. Repeat Prob. 6.30 for a processor with a 1-GHz clock. Assume that the fall time must be 0.1 ns instead of 1 ns, as depicted in Fig. P6.30.

*6.32. A particular interconnection between two logic gates in an IC chip runs one-half the distance across a 7.5-mm-wide die. The interconnection line is insulated from the substrate by silicon dioxide. If the line is 1.5 μm wide and the oxide ($\varepsilon_{ox} = 3.9\varepsilon_o$ and $\varepsilon_o = 8.85 \times 10^{-14}$ F/cm) beneath the line is 1 μm thick, what is the total capacitance of this line assuming that the capacitance is three times that predicted by the parallel plate capacitance formula? Assume that the silicon beneath the oxide represents a conducting ground plane.

6.33. Ideal constant-electric-field scaling of a MOS technology reduces all the dimensions and voltages by the same factor α. Assume that the circuit delay ΔT can be estimated from

$$\Delta T = C \frac{\Delta V}{I}$$

in which the capacitance C is proportional to the total gate capacitance of the MOS transistor, $C = C''_{\text{ox}} WL$, ΔV is the logic swing, and I is the MOSFET drain current in saturation. Show that constant-field scaling results in a reduction in delay by a factor of α and a reduction in power by a factor of α^2 so that the PDP is reduced by a factor of α^3. Show that the power density actually remains constant under constant-field scaling.

6.34 For many years, MOS devices were scaled to smaller and smaller dimensions without changing the power supply voltage. Suppose that the width W, length L, and oxide thickness T_{ox} of a MOS transistor are all reduced by a factor of 2. Assume that V_{TN}, v_{GS}, and v_{DS} remain the same. (a) Calculate the ratio of the drain current of the scaled device to that of the original device. (b) By what factor has the power dissipation changed? (c) By what factor has the value of the total gate capacitance changed? (d) By what factor has the circuit delay ΔT changed? (Use the delay formula in Prob. 6.33)

6.5 NMOS Logic Design

6.35. Integrated circuit chips packaged in plastic DIPs (dual-in-line packages) can typically dissipate 1 W per chip. Suppose that we have an IC design that must fit on one chip and requires two million logic gates. Assume that one-half the logic gates on the chip are conducting current at any given time. (a) What is the average power that can be dissipated by each logic gate on the chip? (b) If a supply voltage of 1.8 V is used, how much current can be used by each gate?

6.36. A high-performance microprocessor design requires 20 million logic gates and will be placed in a package that can dissipate 20 W. (a) What is the average power that can be dissipated by each logic gate on the chip? (b) If a supply voltage of 1.8 V is used, how much current can be used by each gate? Assume that two-thirds of the logic gates on the chip are in the conducting state at any given time.

6.37. Design a resistive load inverter to operate from a 2.5-V power supply with a power dissipation of 50 μW.

6.38. (a) Find V_H, V_L, and the power dissipation (for $v_O = V_L$) for the logic inverter with resistor load in Fig. P6.38(a). (b) Repeat for Fig. P6.38(b).

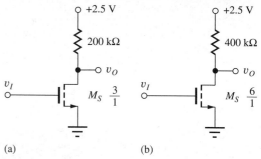

Figure P6.38

6.39. A manufacturing problem caused $V_{TN} = 0.8$ V instead of 0.6 V for the inverter in Fig. P6.38(a). (a) What are the values of V_H, V_L, and power dissipation? (b) Repeat for $V_{TN} = 0.4$ V.

6.40. (a) What are the noise margins for the circuit in Fig. P6.38(a)? (b) Fig. P6.38(b)?

6.41. (a) Find V_H, V_L, and the power dissipation (for $v_O = V_L$) for the logic inverter with resistor load in Fig. P6.38(b). (b) A manufacturing problem caused $V_{TN} = 0.5$ V instead of 0.6 V. What are the new values of V_H, V_L, and power dissipation? (c) Repeat for $V_{TN} = 0.7$ V.

6.42. (a) What are the noise margins for the circuit in Fig. P6.38(b)? (b) What percentage increase in K_S will result in $N_{ML} = 0$ for the resistive load inverter in Fig. P6.38(b)?

6.43. The resistive load inverter in Fig. 6.12 is to be redesigned for $V_L = 0.5$ V. (a) What are the new values of R and $(W/L)_S$ assuming that the power dissipation remains the same? (b) What are the values of NM_L and NM_H?

6.44. (a) Redesign the resistive load inverter of Fig. 6.12 for operation at a power level of 0.25 mW with $V_{DD} = 3.3$ V. Assume $V_{TO} = 0.7$ V. Keep the other design parameters the same. What is the new size of M_S and the value of R? (b) What are the values for NM_H and NM_L?

6.45. Design an inverter with a resistive load for $V_{DD} = 3$ V and $V_L = 0.25$ V. Assume $I_{DD} = 33$ μA, $K'_n = 60$ μA/V^2, and $V_{TN} = 0.75$ V. (b) Confirm the validity of your design with SPICE.

6.46. (a) Design an inverter with a resistive load for $V_{DD} = 2.0$ V and $V_L = 0.15$ V. Assume $I_{DD} = 10$ μA, $K'_n = 75$ μA/V^2, and $V_{TN} = 0.6$ V. (b) Confirm the validity of your design with SPICE.

6.6 Transistor Alternatives to Resistor Load

6.47. (a) Calculate the on-resistance of an NMOS transistor with $W/L = 10/1$ for $V_{GS} = 5$ V, $V_{SB} = 0$, $V_{TO} = 1$ V, and $V_{DS} = 0$ V. (b) Calculate the on-resistance of a PMOS transistor with $W/L = 10/1$ for $V_{SG} = 5$ V, $V_{SB} = 0$, $V_{TO} = -1$ V, and $V_{SD} = 0$ V. (c) What do we mean when we say that a transistor is "on" even though I_D and $V_{DS} = 0$? (d) What must be the W/L ratios of the NMOS and PMOS transistors if they are to have the same on-resistance as parts (a) and (b) with $|V_{GS}| = 3.0$ V?

6.48. Find V_H for an NMOS logic gate with a saturated load if $V_{TO} = 0.75$ V, $\gamma = 0.75 \sqrt{\text{V}}$, $2\phi_F = 0.7$ V, and $V_{DD} = 3.3$ V.

6.49. Find V_H for an NMOS logic gate with a saturated load if $V_{TO} = 0.6$ V, $\gamma = 0.6 \sqrt{\text{V}}$, $2\phi_F = 0.6$ V, and $V_{DD} = 3.3$ V.

6.50. Find V_H for an NMOS logic gate with a saturated load if $V_{TO} = 0.5$ V, $\gamma = 0.85 \sqrt{\text{V}}$, $2\phi_F = 0.6$ V, and $V_{DD} = 2.5$ V.

6.51. Find V_H, V_L, and the power dissipation (for $v_O = V_L$) for the logic inverter with the saturated load in Fig. P6.51. Assume $\gamma = 0$.

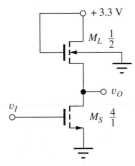

Figure P6.51

6.52. A manufacturing problem caused $V_{TN} = 0.8$ V instead of 0.6 V in the inverter in Fig. P6.51. What are the new values of V_H, V_L, and power dissipation? (c) Repeat for $V_{TN} = 0.4$ V.

*6.53. What are the noise margins for the circuit in Fig. P6.51?

6.54. (a) Find V_H, V_L, and the power dissipation (for $v_o = V_L$) for the logic inverter with the saturated load in Fig. P6.51 if the transistor sizes are changed to $(W/L)_S = 8/1$ and $(W/L)_L = 1/1$. (b) What are the noise margins for the circuit? (c) A manufacturing problem caused $V_{TN} = 0.7$ V instead of 0.6 V. What are the new values of V_H, V_L, and power dissipation?

6.55. (a) Redesign the saturated load inverter of Fig. 6.21 for operation at a power level of 0.25 mW with $V_{DD} = 3.3$ V. Assume $V_{TO} = 0.7$ V. Keep the other design parameters the same. What are the new sizes of M_L and M_S? (b) What are the new values for NM_H and NM_L?

6.56. Redesign the NMOS logic gate with saturated load of Fig. 6.21 to give $V_L = 0.3$ V and $P = 0.4$ mW for $v_O = V_L$.

6.57. (a) Design a saturated load inverter similar to that of Fig. 6.17(c) with $V_{DD} = 3.3$ V and $V_L = 0.2$ V. Assume $I_{DD} = 75$ μA. (b) Recalculate the values of W/L including body effect, as in Fig. 6.21.

6.58. (a) Design a saturated load inverter similar to that of Fig. 6.17(c) with $V_{DD} = 2.0$ V and $V_L = 0.15$ V. Assume $I_{DD} = 25$ μA and $V_{TN} = 0.6$ V. (b) Recalculate the values of W/L including body effect, as in Fig. 6.21 if $V_{TO} = 0.6$ V, $\gamma = 0.6 \sqrt{\text{V}}$, and $2\phi_F = 0.6$ V. (c) Confirm the validity of your designs with SPICE.

6.59. The logic input of the saturated load inverter of Fig. 6.21 is connected to 2.5 V. What is v_O for this input voltage? (This problem will probably require an iterative solution.)

6.60. The saturated load inverter of Fig. 6.29(b) was designed using $K'_n = 100$ μA/V^2, but due to process variations during fabrication, the value actually turned out to be $K'_n = 80$ μA/V^2. What will be the new values of V_H, V_L, and the power dissipation in the gate for this new value of K'_n?

6.61. The saturated load inverter of Fig. 6.29(b) was designed using $K'_n = 100$ μA/V^2, but due to process variations during fabrication, the value actually turned out to be $K'_n = 120$ μA/V^2. What will be the new values of V_H, V_L, and the power dissipation in the gate for this new value of K'_n?

**6.62. Plot the noise margins for the saturated load inverter similar to the design of Fig. 6.29(b) versus $K_R = K_S/K_L$ (see the graph in Fig. 6.28). Note that V_L will be changing.

6.63. The inverter designs in Fig. 6.29 assume $\lambda = 0$. (a) Does V_H depend upon the value of λ? (b) Use SPICE to find I_{DD} and V_L for the saturated load inverter in Fig. 6.29(b) if $\lambda = 0.02, 0.05$, and 0.1 V^{-1}.

NMOS Inverter with a Linear Load Device

6.64. Calculate the W/L ratio for the linear load device using the circuit and device parameters that apply to Sec. 6.8, and show that the values presented in Fig. 6.22 are correct.

6.65. What is the minimum value of V_{GG} required for linear region operation of M_L in Fig. 6.29(c) if $V_{TO} = 0.8$ V, $\gamma = 0.5 \sqrt{V}$, and $2\phi_F = 0.6$ V.

6.66. Find V_H, V_L, and the power dissipation (for $v_O = V_{OL}$) for the linear load inverter in Fig. P6.66.

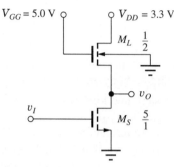

Figure P6.66

6.67. What is the minimum value of V_{GG} in the circuit in Fig. P6.66 if $V_{TO} = 0.6$ V, $\gamma = 0.6 \sqrt{V}$, and $2\phi_F = 0.6$ V.

6.68. (a) Design a linear load inverter similar to that of Fig. P6.66 with $V_{DD} = 3.3$ V, $V_L = 0.20$ V, and $P = 300 \ \mu$W. Assume $V_{TO} = 0.6$ V, $\gamma = 0.6 \sqrt{V}$, $2\phi_F = 0.6$ V. (b) Confirm the validity of your design using SPICE.

NMOS Inverter with a Depletion-Mode Load

6.69. We know that body effect deteriorates the behavior of NMOS logic gates with depletion-mode loads. Assume that the depletion-mode load device has $V_{TO} = -1$ V and is operating in an inverter circuit with $V_{DD} = 2.5$ V. What is the largest value of the body-effect parameter γ that still will allow $V_{OH} = V_{DD}$? Assume $2\phi_F = 0.6$ V.

6.70. (a) The depletion load inverter of Fig. 6.29(d) was designed using $K'_n = 100 \ \mu$A/V^2, but due to process variations during fabrication, the value actually turned out to be $K'_n = 80 \ \mu$A/V^2. What will be the new values of V_H, V_L, and the power dissipation in the gate for this new value of K'_n? (b) Repeat for $K'_n = 120 \ \mu$A/V^2.

6.71. (a) Redesign the inverter with depletion-mode load of Fig. 6.29(d) for operation with $V_{DD} = 3.3$ V. Assume $V_{TO} = 0.6$ V for the switching transistor and $V_{TO} = -1$ V, $\gamma = 0.5 \sqrt{V}$, and $2\phi_F = 0.6$ V for the depletion-mode load. Design for $V_L = 0.20$ V and $P = 0.25$ mW.

6.72. (a) Design a depletion-load inverter to operate with $V_{DD} = 3.3$ V, $V_L = 0.20$ V, and $P = 250 \ \mu$W. Assume $V_{TO} = -2$ V, $\gamma = 0.5 \sqrt{V}$, and $2\phi_F = 0.6$ V for the load transistor and $V_{TO} = 0.6$ V for M_S. (b) Confirm the validity of your design using SPICE.

6.73. The inverter designs in Fig. 6.29 assume $\lambda = 0$. (a) Does V_H depend upon the value of λ? (b) Use SPICE to find I_{DD} and V_L for the depletion-load inverter in Fig. 6.29(d) if $\lambda = 0.02$, 0.05, and 0.1 V^{-1}.

Pseudo NMOS Inverter

6.74. (a) The inverter in Fig. 6.29(e) is to be redesigned to have $V_L = 0.25$ V. What is the new value of $(W/L)_S$? (b) What are the noise margins for the new design?

6.75. (a) Due to process variations, the NMOS threshold voltage for the inverter in Fig. 6.29(e) was found to be $V_{TN} = 0.5$ V instead of 0.6 V. What are the new values of V_L and I_{DD}? (b) What are the new values of the noise margins? (c) Repeat parts (a) and (b) for $V_{TN} = 0.7$ V.

6.76. (a) Due to process variations, the value of K'_n for the NMOS transistor in Fig. 6.29(e) was found to be $K'_n = 120 \ \mu$A/V^2 instead of 100 μA/V^2. What are the new values of V_L and I_{DD}? (b) What are the new values of the noise margins? (c) Repeat parts (a) and (b) for $K'_n = 80 \ \mu$A/V^2.

6.77. (a) Due to process variations, the PMOS threshold voltage for the inverter in Fig. 6.29(e) was found to be $V_{TP} = -0.5$ V instead of -0.6 V. What are the new values of V_L and I_{DD}? (b) What are the new values of the noise margins? (c) Repeat parts (a) and (b) for $V_{TP} = -0.7$ V.

6.78. (a) Due to process variations, the value of K'_p for the PMOS transistor in Fig. 6.29(e) was found to be $K'_p = 50 \ \mu$A/V^2 instead of 40 μA/V^2. What are the new values of V_L and I_{DD}? (b) What are the new values of the noise margins? (c) Repeat parts (a) and (b) for $K'_p = 30 \ \mu$A/V^2.

6.79. (a) Design a pseudo NMOS inverter to operate from $V_{DD} = 1.8$ V with $V_L = 0.2$ V and a power of 100 μW. Assume $V_{TN} = 0.5$ V and $V_{TP} = -0.5$ V. (b) Find the noise margins for your design.

6.80. (a) Design a pseudo NMOS inverter to operate from $V_{DD} = 3.0$ V with $V_L = 0.3$ V and a power of 200 μW. (b) Find the noise margins for your design.

6.8 NMOS NAND and NOR Gates

6.81. (a) What is the value of V_L in the two-input NOR gate in Fig. 6.30(a) when both $A = 1$ and $B = 1$? (b) What is the current in V_{DD} for this input condition?

6.82. Calculate the W/L ratios of the switching devices in the NAND gate in Fig. 6.32(b) and verify that they are correct.

**6.83. The two-input NAND gate in Fig. 6.32 was designed with equal values of R_{on} (approximately equal voltage drops) in the two series-connected switching transistors, but an infinite number of other choices are possible. Show that the equal R_{on} design requires the minimum total gate area for the switching transistors.

6.84. Draw the schematic for a four-input NOR gate with a saturated load device. What are the W/L ratios of all the transistors, based on the reference inverter in Fig. 6.29?

6.85. Draw the schematic for a four-input NAND gate with a depletion-mode load device. What are the W/L ratios of all the transistors, based on the reference inverter in Fig. 6.32?

6.86. (a) Draw the schematic for a three-input pseudo NMOS NOR gate. Choose the device sizes based upon the reference inverter in Fig. 6.29. (b) What is V_L if all the logic inputs are equal to 1?

6.87. (a) Draw the schematic for a three-input pseudo NMOS NAND gate. Choose the device sizes based upon the reference inverter in Fig. 6.29. Ignore body effect in your design. (b) What is V_L if all the logic inputs are equal to 1? Do not ignore body effect. (c) Redesign the sizes of the three switching transistors to account for body effect.

6.88. Draw the layout of a two-input NOR gate similar to that in Fig. 6.33(a) but use a saturated load device. Be sure to scale the transistor sizes properly based on Fig. 6.29(b).

6.89. (a) Draw the layout of a three-input NOR gate similar to that in Fig. 6.33(a). Be sure to scale the transistor sizes properly. (b) Draw the layout of a three-input NAND gate similar to that in Fig. 6.33(b). Be sure to scale the transistor sizes properly.

6.9 Complex NMOS Logic

6.90. (a) What is the logic function that is implemented by the gate in Fig. P6.90? (b) What are the W/L ratios for the transistors, based on the reference inverter design of Fig. 6.29(d)?

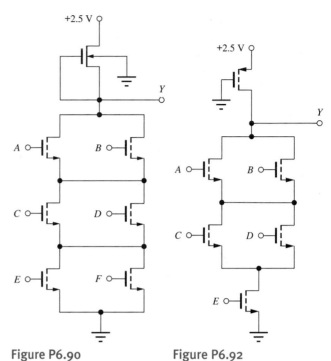

Figure P6.90   **Figure P6.92**

6.91. (a) Redraw the circuit in Fig. P6.90 using a saturated load transistor. (b) What is the logic function of the new circuit? (c) What are the W/L ratios of the transistors based upon the reference inverter design in Fig. 6.29(b)?

6.92. (a) What is the logic function that is implemented by the gate in Fig. P6.92? (b) What are the W/L ratios for the transistors, based on the reference inverter design of Fig. 6.29(d)?

6.93. (a) Redraw the circuit in Fig. P6.92 using a saturated load transistor. (b) What is the logic function of the new circuit? (c) What are the W/L ratios of the transistors based upon the reference inverter design in Fig. 6.29(b)?

6.94. (a) What is the logic function that is implemented by the gate in Fig. P6.94? (b) What are the W/L ratios for the transistors if the gate is to dissipate three times as much power as the reference inverter design of Fig. 6.29(d)?

6.95. (a) Redraw the circuit in Fig. P6.94 using a saturated load transistor. (b) What is the logic function of the new circuit? (c) What are the W/L ratios of the transistors based on the reference inverter design in Fig. 6.29(b)?

6.96. Design a depletion-load gate that implements the logic function $Y = \overline{A[B + C(D + E)]}$, based on the reference inverter design of Fig. 6.29(d).

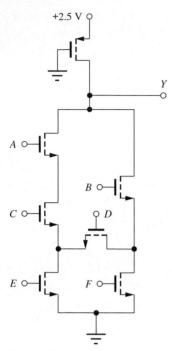

Figure P6.94

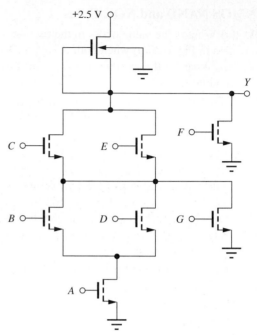

Figure P6.100

6.97. Design a pseudo NMOS gate that implements the logic function $Y = \overline{A(BC + DE)}$ and consumes one-half the power of the reference inverter design of Fig. 6.29(e).

6.98. Design a saturated-load gate that implements the logic function $Y = \overline{A(BC + DE)}$, based on the reference inverter design of Fig. 6.29(b).

6.99. Design a pseudo NMOS gate that implements the logic function $Y = \overline{A(B + CD) + E}$, based on the reference inverter design of Fig. 6.29(e).

6.100. What is the logic function for the gate in Fig. P6.100? What are the W/L ratios of the transistors that form the gate if the gate is to consume twice as much power as the reference inverter in Fig. 6.29(d)?

6.101. (a) Design a depletion-load gate that implements the logic function $Y = \overline{A(B + CD) + E}$, based on the reference inverter design of Fig. 6.29(d). (b) Redesign the W/L ratios of this gate to account for body effect and differences in values of V_{DS} for the various transistors.

6.102. Recalculate the W/L ratios of the transistors in the gate in Fig. 6.34 to account for the body effect and differences in the V_{DS} of the various transistors.

*6.103. Recalculate the W/L ratios of the transistors in the gate in Fig. 6.35(a) to account for the body effect and differences in the V_{DS} of the various transistors.

*6.104. Recalculate the W/L ratios of the transistors in the gate in Fig. 6.35(b) to account for the body effect and differences in the V_{DS} of the various transistors.

*6.105. Recalculate the W/L ratios of the transistors in the gate in Fig. 6.36 to account for the body effect and differences in the V_{DS} of the various transistors.

**6.106. (a) What is the truth table for the logic function Y for the gate in Fig. P6.106? (b) Write an expression for the logical output of this gate. (c) What are the sizes of the transistors M_S and M_P in order for $V_L \leq 0.20$ V? (d) Qualitatively describe how the sizes of M_S and M_P will change if body effect is included in the models for the transistors. (e) What is the name for this logic function?

6.10 Power Dissipation

6.107. Scale the sizes of the resistors and transistors in the five inverters in Fig. 6.29 to change the power dissipation level to 1 mW.

6.108. What are the W/L ratios of the transistors in the gate in Fig. P6.100 if the gate is to consume four times as much power as the reference inverter in Fig. 6.29(d)?

6.109. What are the W/L ratios for the transistors in Fig. P6.90 if the gate is to dissipate one-quarter as much power as the reference inverter design of Fig. 6.29(d)?

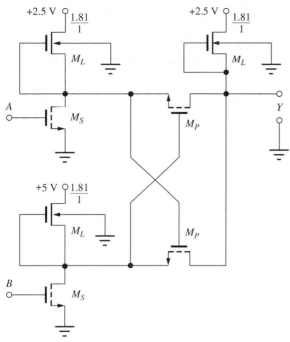

Figure P6.106

6.110. (a) Scale the transistor sizes in Fig. 6.35(a) to increase the gate power by a factor of three. (b) Scale the transistor sizes in Fig. 6.35(a) to decrease the gate power by a factor of five.

6.111. (a) Scale the transistor sizes in Fig. 6.35(b) to decrease the gate power by a factor of 10. (b) Scale the transistor sizes in Fig. 6.35(b) to increase the gate power by a factor of 2.5.

6.112. (a) Scale the transistor sizes in Fig. 6.36 to quadruple the gate power. (b) Scale the transistor sizes in Fig. 6.36 to decrease the gate power by a factor of three.

*6.113. For many years, MOS devices were scaled to smaller and smaller dimensions without changing the power supply voltage. Suppose that the width W, length L, and oxide thickness t_{ox} are all reduced by a factor of 2. Assume that V_{TN}, v_{GS}, and v_{DS} remain the same. Calculate the ratio of the drain current of the scaled device to that of the original device. How has the power dissipation changed?

6.114. A high-speed NMOS microprocessor has a 64-bit address bus and performs a memory access every 50 ns. Assume that all address bits change during every memory access, and that each bus line represents a load of 10 pF. (a) How much power is being dissipated by the circuits that are driving these signals if the power supply is 2.5 V? (b) 3.3 V?

6.11 Dynamic Behavior of MOS Logic Gates

**6.115. The capacitive load on a logic gate becomes dominated by the channel capacitance as the transistors are made wider and wider. Assume that W is very large and show that τ_{PHL} and τ_{PLH} become independent of W and are both proportional to (L^2/μ), where L is the channel length. This result shows the importance of achieving as small a channel length as possible.

6.116. A logic family has a power-delay product of 100 fJ. If a logic gate consumes a power of 100 μW, what is the expected propagation delay of the logic gate?

6.117. The graph in Fig. P6.117 gives the results of a SPICE simulation of an inverter. (a) What are the rise and fall times for v_I and v_O? (b) What are the values of τ_{PHL} and τ_{PLH}? (c) What is the average propagation delay for this gate?

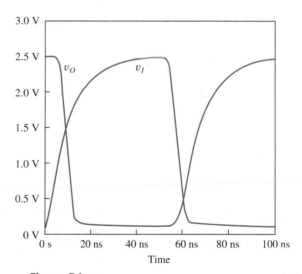

Figure P6.117

6.118. One method to estimate the average propagation delay of an inverter is to construct a long ring of inverters, as shown in the Fig. P6.118. This circuit is called a *ring oscillator,* and the output of any inverter in the chain will be similar to a square wave. (a) Suppose that the chain contains 301 inverters and the average propagation delay of an inverter is 100 ps. What will be the period of the square wave generated by the oscillator? (b) Why should the number of inverters be odd? What could happen if an even number of inverters were used in the ring oscillator?

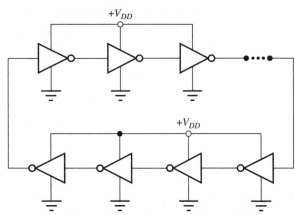

Figure P6.118 Ring oscillator formed from a chain of an *odd* number of inverters.

Resistor Load

6.119. What are the rise time, fall time, and average propagation delay of the NMOS gate in Fig. 6.12(b) if a load capacitance $C = 0.5$ pF is attached to the output of the gate?

6.120. What are the rise time, fall time, and average propagation delay of the NMOS gate in Fig. 6.12(b) if a load capacitance $C = 0.5$ pF is attached to the output of the gate and V_{DD} is increased to 3.3 V?

6.121. Design an NMOS inverter with resistor load ($V_{DD} = 2.5$ V, $V_L = 0.20$ V) to have an average propagation delay of 2.5 ns with a capacitive load of 1 pF. What is the average static power dissipation of this gate?

6.122. Repeat the simulation of Ex. 6.10 with $\lambda = 0.04$/V. Compare the new values rise time, fall time, and propagation delays with those of the example.

Saturated Load

6.123. What are the rise and fall times and average propagation delays of the NMOS gate in Fig. P6.123 if

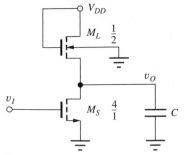

Figure 6.123

$C = 0.5$ pF and $V_{DD} = 2.5$ V? Use the estimates in Table 6.10.

6.124. What are the rise and fall times and average propagation delays of the NMOS gate in Fig. P6.123 if $C = 0.3$ pF and $V_{DD} = 3.3$ V? Use Table 6.10.

6.125. Design an NMOS saturated load inverter ($V_{DD} = 2.5$ V, $V_L = 0.25$ V) to have an average propagation delay of 2 ns with a capacitive load of 1 pF. What is the average static power dissipation of this gate? Make use of Table 6.10.

Linear Load

6.126. What are the rise and fall times and average propagation delay for the linear load inverter in Fig. 6.29(c) with a load capacitance of 0.7 pF? Use Table 6.10.

6.127. Use SPICE to determine the characteristics of the NMOS inverter with a linear load device for the design given in Fig. 6.29. (a) Simulate the voltage transfer function. (b) Determine t_r, t_f, τ_{PHL}, and τ_{PLH} for this inverter with a square wave input and $C = 0.15$ pF.

Depletion-Mode Load

6.128. What are the sizes of the transistors in the NMOS depletion-mode load inverter if it must drive a 1-pF capacitance with an average propagation delay of 3 ns? Assume $V_{DD} = 3.0$ V and $V_L = 0.25$ V. What are the rise and fall times for the inverter? Use $V_{TNL} = -3$ V ($\gamma = 0$). Make use of Table 6.10.

6.129. Design an NMOS depletion load inverter ($V_{DD} = 3.3$ V, $V_L = 0.20$ V, $V_{TNS} = 0.75$ V, $V_{TNL} = -2$ V, $\gamma = 0$) to have an average propagation delay of 1 ns with a capacitive load of 0.2 pF. What is the average static power dissipation of this gate? Use Table 6.10.

A Final Comparison of Load Devices

6.130. Currents in the various load devices are shown in Fig. 6.47. The resistor load has a value of 28.8 kΩ. The W/L ratios of the devices were chosen to set the current in each load device to 80 μA when $v_O = V_L = 0.20$ V. Calculate the values of the W/L ratios of the load devices that were used in the figure for the: (a) saturated load device including body effect, (b) saturated load device with no body effect, (c) linear load device including body effect, (d) linear load device with no body effect, (e) depletion-mode load device including body effect, (f) depletion-mode load with no body effect (g) pseudo NMOS load.

6.12 PMOS Problems

6.131. Design five PMOS logic gates similar to the ones in Fig. 6.29. Do not do a complete mathematical redesign, but design the PMOS circuits by scaling the W/L ratios in Fig. 6.29, assuming that $K'_p = 40\ \mu\text{A/V}^2$.

*6.132. What are the values of V_L and V_H for the inverter in Fig. P6.132?

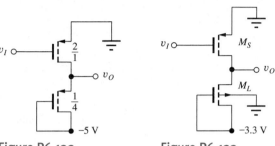

Figure P6.132   Figure P6.133

6.133. Design the transistors in the inverter of Fig. P6.133 so that $V_H = -0.33$ V and the power dissipation $= 0.1$ mW. Use the values in Table 6.5 on page 296.

6.134. What are the values of V_H and V_L for the inverter of Fig. P6.134? Use the values in Table 6.5 on page 296.

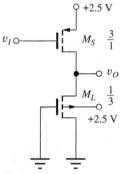

Figure P6.134

6.135. What is the logic function Y for the gate in Fig. P6.135?

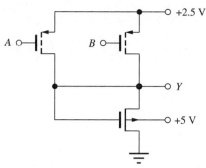

Figure P6.135

6.136. What is the logic function Y for the gate in Fig. P6.136?

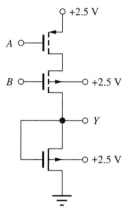

Figure P6.136

6.137. Simulate the voltage transfer characteristic for the PMOS gate in Fig. P6.132, and compare the results to those of Prob. 6.132.

6.138. Simulate the voltage transfer characteristic for the PMOS gate in Fig. P6.133, and compare the results to those of Prob. 6.133.

6.139. Simulate the delay of the PMOS gate in Fig. P6.132 with a load capacitance of 1 pF, and determine the rise time, fall time, and average propagation delay.

6.140. Simulate the delay of the PMOS gate in Fig. P6.133 with a load capacitance of 1 pF, and determine the rise time, fall time, and average propagation delay.

CHAPTER 7

COMPLEMENTARY MOS (CMOS) LOGIC DESIGN

Chapter Outline

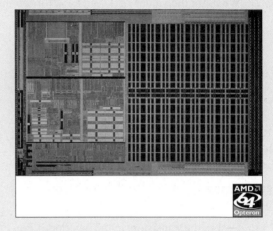

AMD (Advanced Micro Devices, Inc.) Operton™ die plot.
Copyright © 2006 Advanced Micro Devices, Inc.

Chapter Goals

From Chapter 7, we shall gain a basic understanding of the design of CMOS logic and memory circuits.

- Introduce CMOS Logic Gates
- Explore the voltage transfer characteristics of CMOS inverters
- Learn to design CMOS inverter, NOR, and NAND gates
- Learn to design complex CMOS logic gates
- Explore sources of static and dynamic power dissipation in CMOS logic
- Develop expressions for the rise time, fall time, propagation delay, and power-delay product of CMOS logic
- Develop expressions for CMOS noise margins
- Introduce the concept of dynamic logic and domino CMOS logic techniques
- Learn to design "cascade buffers" for driving high load capacitances
- Explore layout of CMOS inverters and logic gates
- Understand the problem of "latchup" in CMOS technology

Today, **Complementary MOS,** or CMOS, is by far the dominant integrated circuit technology. CMOS logic was first described by Wanlass and Sah in 1963 [1], but for many years, CMOS technology was only available in unit-logic form, with several gates packaged together in a single dual in-line package (DIP). CMOS circuitry requires both NMOS and PMOS transistors to be built into the same substrate, and this increase in process complexity leads to higher-cost circuits. In addition, standard CMOS gates require more transistors than the corresponding NMOS or PMOS logic gates, and thereby require more silicon area to implement. Larger silicon area contributes further to increased cost.

The increased cost that results from CMOS fabrication and circuit complexities represented the primary reason why CMOS technology was not widely used in integrated circuits. CMOS was utilized in applications that required extremely low power and where the increased cost could be justified.

However, as mentioned in Chapter 6, power levels associated with NMOS microprocessors reached crisis levels by the mid 1980s, with power dissipations as high as 50 W per chip or more. To solve the static power dissipation problem, the microprocessor industry rapidly moved to CMOS technology. Today, CMOS is the industrywide standard technology, and this important circuit innovation was recognized by the presentation of the 1991 IEEE Solid-State Circuits Society Field Award to Frank Wanlass for his invention of CMOS logic.

In Chapter 7, we investigate the design of CMOS logic circuits, starting with characterization of the CMOS inverter, and followed by a discussion of the design of NOR, NAND, and complex gates based on a CMOS reference inverter. CMOS gate design is shown to be determined primarily by logic delay considerations. CMOS noise margins and power-delay product are discussed, and the **transmission gate** is also introduced.

The physical structure of CMOS technology is presented, and parasitic bipolar transistors are shown to exist within the integrated CMOS structure. If these bipolar devices become active, a potentially destructive phenomenon called latchup can occur.

7.1 CMOS INVERTER TECHNOLOGY

As noted, CMOS requires a fabrication technology that can produce both NMOS and PMOS transistors together on one IC substrate. The basic IC structure used to accomplish this task is shown in Fig. 7.1. In this cross section, NMOS transistors are shown fabricated in a normal manner in a p-type silicon substrate. A lightly doped n-type region, called the n-well, is formed in the p-type substrate, and PMOS transistors are then fabricated in the n-well region, which becomes the body of the PMOS device. Note that a large-area diode exists between the n-well and p-type substrate. This pn diode must always be kept reverse-biased by the proper connection of V_{DD} and V_{SS} (for example, 2.5 V and 0 V).

The connections between the transistors needed to form a CMOS inverter are shown in Fig. 7.1 and correspond to the circuit schematic for the CMOS inverter in Fig. 7.2. The inverter circuit consists of one NMOS and one PMOS transistor similar to pseudo NMOS introduced in Chapter 6. The NMOS transistor functions in a manner identical to that of the switching device in NMOS gates. However, in CMOS, the PMOS load device is also switched on and off under control of the logic input signal v_I.

The CMOS logic gate can be conceptually modeled by the circuit in Fig. 7.2(b), in which the position of the two switches is controlled by the input voltage v_I. In an ideal inverter, there would never be a conducting path between the positive and negative power supplies under steady-state conditions. When the NMOS transistor is on, the PMOS transistor is off; if the PMOS transistor is on, the NMOS device is off. In an actual inverter, input v_I cannot abruptly jump from V_L to V_H or V_H to V_L, and there will be a period of time during the input voltage transition when both transistors are on and a conducting path exists between V_{DD} and ground.

In the CMOS inverter of Fig. 7.2, the source of the PMOS transistor is connected to V_{DD}, the source of the NMOS transistor is connected to V_{SS} (0 V in this case), and the drain terminals of the two MOSFETs are connected together to form the output node. Also, the substrates of both the NMOS and PMOS transistors are connected to their respective sources, and so body effect is eliminated in both devices.

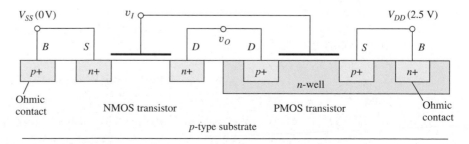

Figure 7.1 n-well CMOS structure for forming both NMOS and PMOS transistors in a single silicon substrate.

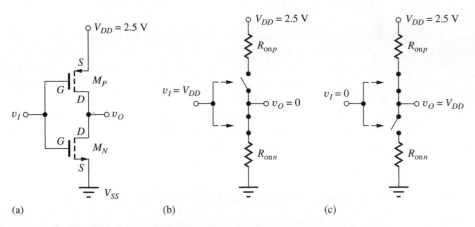

Figure 7.2 (a) A CMOS inverter uses one NMOS and one PMOS transistor. (b) A simplified model of the inverter for a high input level. The output is forced to zero through the on-resistance of the NMOS transistor. (c) Simplified model of the inverter for a low input level. The output is pulled to V_{DD} through the on-resistance of the PMOS transistor.

TABLE 7.1
CMOS Transistor Parameters

	NMOS DEVICE	PMOS DEVICE
V_{TO}	0.6 V	−0.6 V
γ	$0.50 \sqrt{\text{V}}$	$0.75 \sqrt{\text{V}}$
$2\phi_F$	0.60 V	0.70 V
K'	100 μA/V^2	40 μA/V^2

Simplified models for the CMOS inverter operation appear in Fig. 7.2(b) and 7.2(c). The input signal controls the state of the two switches that effectively work as a single-pole double-throw switch. In Fig. 7.2(b), the input is at a high input level ($v_I = V_{DD}$), and the output is connected to ground through the on-resistance of the NMOS transistor. In Fig. 7.2(c), the input is at a low input level ($v_I = 0$), and the output is connected to V_{DD} through the on-resistance of the PMOS transistor.

Before we explore the design of the CMOS inverter, we need parameters for the CMOS devices, as given in Table 7.1. The NMOS parameters are the same as those given in Chapter 6. In CMOS technology, the transistors are normally designed to have equal and opposite threshold voltages: for example, $V_{TON} = +0.6$ V and $V_{TOP} = -0.6$ V. Remember that $K'_n = \mu_n C''_{ox}$ and $K'_p = \mu_p C''_{ox}$ and that hole mobility in the channel of the PMOSFET is approximately 40 percent of the electron mobility in the NMOSFET channel. The values in Table 7.1 reflect this difference because the value of K' for the NMOS device is shown as 2.5 times that for the PMOS transistor. Processing differences are also reflected in the different values for γ and $2\phi_F$ in the two types of transistors.

Remember that the threshold voltage of the NMOS transistor is denoted by V_{TN} and that of the PMOS transistor by V_{TP}:

$$V_{TN} = V_{TON} + \gamma_N \left(\sqrt{V_{SBN} + 2\phi_{FN}} - \sqrt{2\phi_{FN}} \right)$$

and

$$V_{TP} = V_{TOP} - \gamma_P \left(\sqrt{V_{BSP} + 2\phi_{FP}} - \sqrt{2\phi_{FP}} \right) \tag{7.1}$$

For $v_{SBN} = 0$, $V_{TN} = V_{TON}$, and for $v_{BSP} = 0$, $V_{TP} = V_{TOP}$.

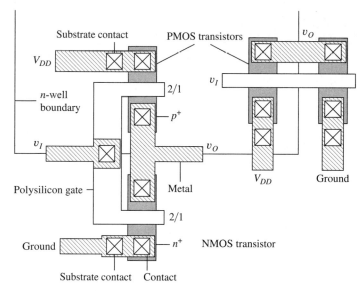

Figure 7.3 Layout of two CMOS inverters.

EXERCISES: (a) What are the values of K_p and K_n for transistors with $W/L = 20/1$? (b) An NMOS transistor has $V_{SB} = 2.5$ V. What is the value of V_{TN}? (c) A PMOS transistor has $V_{BS} = 2.5$ V. What is the value of V_{TP}?

ANSWERS: (a) 800 μA/V², 2 mA/V²; (b) 1.09 V; (c) −1.31 V

7.1.1 CMOS INVERTER LAYOUT

Two possible layouts for CMOS inverters appear in Fig. 7.3. In the left-hand layout, the lower transistor is the NMOS device in the p-type substrate, whereas the upper transistor is the PMOS device, which is within the boundary of the n-well. The polysilicon gates of the two transistors are connected together to form the input v_I, and the drain diffusions of the two transistors are connected together by the metallization to form output v_O. The sources of the NMOS and PMOS devices are at the bottom and top of the drawing, respectively, and are connected to the ground and power supply buss metallization. Each device has a local substrate contact connected to the source of the transistor. In this particular layout, each transistor has a W/L of 2/1. In the right-hand layout, the two transistors appear side by side with a common polysilicon gate running through both. Various options for the design of the W/L ratios of the transistors are a major focus of the rest of this chapter.

7.2 STATIC CHARACTERISTICS OF THE CMOS INVERTER

We now explore the static behavior of CMOS logic gates. First, consider the CMOS inverter with an input $v_I = +2.5$ V (1 state), as shown in Fig. 7.4(a). For this input condition, $v_{GS} = +2.5$ V for the NMOS transistor, and $v_{GS} = 0$ V for the PMOS transistor. For the NMOS device, $v_{GS} > V_{TN}$ (0.6 V), so a channel exists in the NMOS transistor, but the PMOS transistor is off because $v_{GS} = 0$ V for the PMOS device. Thus, load capacitor C discharges through the NMOS transistor, and v_O reaches 0 V. Because the PMOS transistor is off, a dc current path does not exist through M_N and M_P! A simplified equivalent circuit for the inverter for a high input level appears in Fig. 7.4(b). The output capacitance C is discharged to zero through the on-resistance of the NMOS transistor.

If v_I is now set to 0 V (0 state), as in Fig. 7.4(c), v_{GS} becomes 0 V for the NMOS transistor, and it is cut off. For the PMOS transistor, $v_{GS} = -2.5$ V, a channel exists in the PMOS transistor, and the

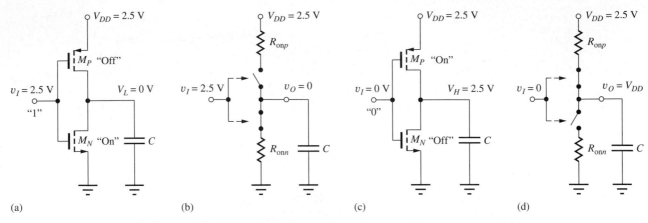

Figure 7.4 (a) CMOS inverter with the input high. M_N is on and M_P is off. (b) Simplified model of the inverter for a high input level. Output capacitance C is discharged to zero through the on-resistance of the NMOS transistor. (c) CMOS inverter with the input low. M_N is off and M_P is on. (d) Simplified model of the inverter for a low input level. Output capacitance C is charged to V_{DD} through the on-resistance of the PMOS transistor.

load capacitor C charges to the positive power supply voltage V_{DD} (2.5 V). Once a steady-state condition is reached, the currents in M_N and M_P must both be zero because the NMOS transistor is off. The corresponding simplified equivalent circuit for the inverter for the low input level appears in Fig. 7.4(d). In this case, we see that the capacitance C is charged to V_{DD} through the on-resistance of the PMOS transistor.

Several important characteristics of the CMOS inverter are evident. The values of V_H and V_L are equal to the positive and negative power supply voltages, and the logic swing ΔV is equal to the full power supply span. For the circuit in Fig. 7.4, $V_H = 2.5$ V, $V_L = 0$ V, and the logic swing $\Delta V = 2.5$ V. Of even greater importance is the observation that the static power dissipation is zero because $i_D = 0$ in both logic states!

7.2.1 CMOS VOLTAGE TRANSFER CHARACTERISTICS

Figure 7.5 shows the result of simulation of the voltage transfer characteristic (VTC) of a **symmetrical CMOS inverter,** designed with $K_P = K_N$. The VTC can be divided into five different regions, as shown in the figure and summarized in Table 7.2. For an input voltage less than $V_{TN} = 0.6$ V in region 1, the NMOS transistor is off, and the output is maintained at $V_H = 2.5$ V by the PMOS device.

Similarly, for an input voltage greater than $(V_{DD} - |V_{TP}|)$ (1.9 V) in region 5, the PMOS device is off, and the output is maintained at $V_L = 0$ V by the NMOS transistor. In region 2, the NMOS transistor is saturated, and the PMOS transistor is in the triode region. For the input voltage near $V_{DD}/2$ (region 3), both transistors are operating in the saturation region. The boundary between regions 2 and 3 is defined by the boundary between the saturation and triode regions of operation for the PMOS transistor. Saturation of the PMOS device requires $|v_{DS}| \geq |v_{GS} - V_{TP}|$:

$$(2.5 - v_O) \geq (2.5 - v_I) - 0.6 \qquad \text{or} \qquad v_O \leq v_I + 0.6 \tag{7.2}$$

In a similar manner, the boundary between regions 3 and 4 is defined by saturation of the NMOS device:

$$v_{DS} \geq v_{GS} - V_{TN} \qquad \text{or} \qquad v_O \geq v_I - 0.6 \tag{7.3}$$

In region 4, the voltages place the NMOS transistor in the triode region, and the PMOS transistor remains saturated.

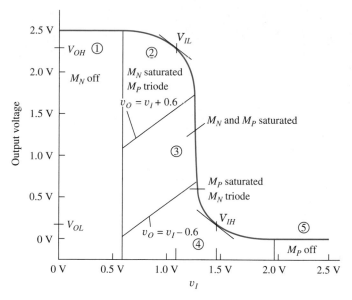

Figure 7.5 CMOS voltage transfer characteristic may be broken down into the five regions outlined in Table 7.2.

TABLE 7.2
Regions of Operation of Transistors in a Symmetrical CMOS Inverter

REGION	INPUT VOLTAGE v_I	OUTPUT VOLTAGE v_O	NMOS TRANSISTOR	PMOS TRANSISTOR
1	$v_I \leq V_{TN}$	$V_H = V_{DD}$	Cutoff	Triode
2	$V_{TN} < v_I \leq v_O + V_{TP}$	High	Saturation	Triode
3	$v_I \cong V_{DD}/2$	$V_{DD}/2$	Saturation	Saturation
4	$v_O + V_{TN} < v_I \leq (V_{DD} - |V_{TP}|)$	Low	Triode	Saturation
5	$v_I \geq (V_{DD} - |V_{TP}|)$	$V_L = 0$	Triode	Cutoff

EXERCISE: Suppose $v_I = 1$ V for the CMOS inverter in Fig. 7.4. (a) What is the range of values of v_O for which M_N is saturated and M_P is in the triode region? (b) For which values are both transistors saturated? (c) For which values is M_P saturated and M_N in the triode region?

ANSWERS: (1.6 V $\leq v_O \leq$ 2.5 V); (0.4 V $\leq v_O \leq$ 1.6 V); (0 V $\leq v_O \leq$ 0.4 V)

EXERCISE: The $(W/L)_N$ of M_N in Fig. 7.4 is 10/1. What is the value of $(W/L)_P$ required to form a symmetrical inverter?

ANSWER: 25/1

Figure 7.6 shows the results of simulation of the voltage transfer characteristics for a CMOS inverter with a symmetrical design ($K_P = K_N$) for several values of V_{DD}. Note that the output voltage levels V_H and V_L are always determined by the two power supplies. As the input voltage rises from 0 to V_{DD}, the output remains constant for $v_I < V_{TN}$ and $v_I > (V_{DD} - |V_{TP}|)$. For this symmetrical design case, the transition between V_H and V_L is centered at $v_I = V_{DD}/2$. The straight line on the graph represents $v_O = v_I$, which occurs for $v_I = V_{DD}/2$ for the symmetrical inverter.

If $K_p \neq K_n$, then the transition shifts away from $V_{DD}/2$. To simplify notation, a parameter K_R is defined: $K_R = K_n/K_p$. K_R represents the relative current drive capability of the NMOS

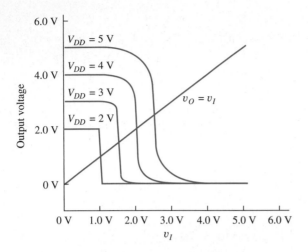

Figure 7.6 Voltage transfer characteristics for a symmetrical CMOS inverter ($K_R = 1$) with $V_{DD} = 5$ V, 4 V, 3 V, and 2 V.

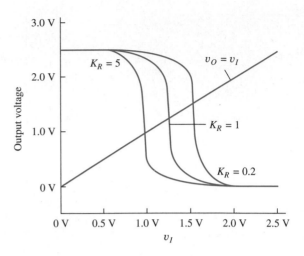

Figure 7.7 CMOS voltage transfer characteristics for $K_R = 5$, 1, and 0.2 for $V_{DD} = 2.5$ V.

and PMOS devices in the inverter. Voltage transfer characteristics for inverters with $K_R = 5$, 1, and 0.2 are shown in Fig. 7.7. For $K_R > 1$, the NMOS current drive capability exceeds that of the PMOS transistor, so the transition region shifts to $v_I < V_{DD}/2$. Conversely, for $K_R < 1$, PMOS current drive capability is greater than that of the NMOS device, and the transition region occurs for $v_I > V_{DD}/2$.

As discussed briefly in Chapter 4, FETs do not actually turn off abruptly as indicated in Eq. (4.9), but conduct small currents for gate-source voltages below threshold. This characteristic enables a CMOS inverter to function at very low supply voltages. In fact, it has been shown that the minimum supply voltage for operation of CMOS is only $[2V_T \ln (2)]$ V [2, 3]. At room temperature, this voltage is less than 40 mV!

EXERCISES: **Equate the expressions for the drain currents of M_N and M_P to show that $v_O = v_I$ occurs for a voltage equal to $V_{DD}/2$ in a symmetrical inverter. What voltage corresponds to $v_O = v_I$ in an inverter with $K_R = 10$ and $V_{DD} = 4$ V?**

ANSWER: 1.27 V

7.2.2 NOISE MARGINS FOR THE CMOS INVERTER

Because of the importance of the CMOS logic family, we explore the noise margins of the inverter in some detail. V_{IL} and V_{IH} are identified graphically in Figs. 7.5 and 7.8 as the points at which the voltage transfer characteristic has a slope of -1. First, we will find V_{IH}.

For v_I near V_{IH}, v_{DS} of M_P will be large and that of M_N will be small. Therefore, we assume that the PMOS device is saturated, and the NMOS device is in its triode region. The two drain currents must be equal, so $i_{DN} = i_{DP}$, and

$$K_n \left(v_I - V_{TN} - \frac{v_O}{2} \right)(v_O) = \frac{K_p}{2}(v_I - V_{DD} - V_{TP})^2 \tag{7.4}$$

For M_N, $v_{GS} = v_I$ and $v_{GS} = v_O$. For M_P, $v_{GS} = v_I - V_{DD}$ and $v_{DS} = v_O - V_{DD}$. Now

$$K_R(2v_I - 2V_{TN} - v_O)(v_O) = (v_I - V_{DD} - V_{TP})^2 \tag{7.5}$$

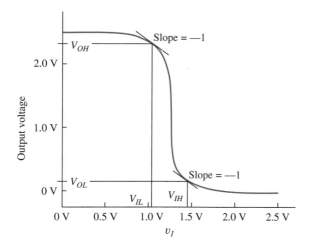

Figure 7.8 CMOS voltage transfer characteristic, with V_{IL} and V_{IH} indicated.

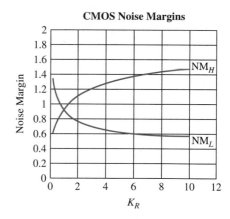

Figure 7.9 Noise margins versus K_R for the CMOS inverter with $V_{DD} = 2.5$ V and $V_{TN} = -V_{TP} = 0.6$ V.

in which $K_R = K_n/K_p$. Solving for v_O yields

$$v_O = (v_I - V_{TN}) \pm \sqrt{(v_I - V_{TN})^2 - \frac{(v_I - V_{DD} - V_{TP})^2}{K_R}} \qquad (7.6)$$

Taking the derivative with respect to v_I and setting it equal to -1 at $v_I = V_{IH}$ is quite involved (and is most easily done with a symbolic algebra package[1] on the computer), but eventually yields this result:

$$V_{IH} = \frac{2K_R(V_{DD} - V_{TN} + V_{TP})}{(K_R - 1)\sqrt{1 + 3K_R}} - \frac{(V_{DD} - K_R V_{TN} + V_{TP})}{K_R - 1} \qquad (7.7)$$

The value of V_{OL} corresponding to V_{IH} is

$$V_{OL} = \frac{(K_R + 1)V_{IH} - V_{DD} - K_R V_{TN} - V_{TP}}{2K_R} \qquad (7.8)$$

For the special case of $K_R = 1$,

$$V_{IH} = \frac{5V_{DD} + 3V_{TN} + 5V_{TP}}{8} \qquad \text{and} \qquad V_{OL} = \frac{V_{DD} - V_{TN} + V_{TP}}{8} \qquad (7.9)$$

V_{IL} can be found in a similar manner using $i_{DN} = i_{DP}$. For v_I near V_{IL}, the v_{DS} of M_P will be small and that of M_N will be large, so we assume that the NMOS device is saturated and the PMOS device is in its linear region. Equating drain currents:

$$K_p\left(v_I - V_{DD} - V_{TP} - \frac{v_O - V_{DD}}{2}\right)(v_O - V_{DD}) = \frac{K_n}{2}(v_I - V_{TN})^2$$

or

$$(2v_I - V_{DD} - 2V_{TP} - v_O)(v_O - V_{DD}) = K_R(v_I - V_{TN})^2 \qquad (7.10)$$

Again, we solve for v_O, take the derivative with respect to v_I, and set the result equal to -1 at $v_I = V_{IL}$. This process yields

$$V_{IL} = \frac{2\sqrt{K_R}(V_{DD} - V_{TN} + V_{TP})}{(K_R - 1)\sqrt{K_R + 3}} - \frac{(V_{DD} - K_R V_{TN} + V_{TP})}{K_R - 1} \qquad (7.11)$$

---

[1] For example, Mathematica, MAPLE, Macsyma, and so on.

The value of V_{OH} corresponding to V_{IL} is given by

$$V_{OH} = \frac{(K_R + 1)V_{IL} + V_{DD} - K_R V_{TN} - V_{TP}}{2} \tag{7.12}$$

For the special case of $K_R = 1$,

$$V_{IL} = \frac{3V_{DD} + 5V_{TN} + 3V_{TP}}{8} \qquad \text{and} \qquad V_{OH} = \frac{7V_{DD} + V_{TN} - V_{TP}}{8} \tag{7.13}$$

The noise margins for $K_R = 1$ are found using the results in Eqs. (7.9) and (7.13):

$$\text{NM}_H = \frac{V_{DD} - V_{TN} - 3V_{TP}}{4} \qquad \text{and} \qquad \text{NM}_L = \frac{V_{DD} + 3V_{TN} + V_{TP}}{4} \tag{7.14}$$

For $V_{DD} = 2.5$ V, $V_{TN} = 0.6$ V, and $V_{TP} = -0.6$ V, the noise margins are both equal to 0.93 V.

Figure 7.9 is a graph of the CMOS noise margins versus K_R from Eqs. (7.7–7.12) for the particular case $V_{DD} = 5$ V, $V_{TN} = 1$ V, and $V_{TP} = -1$ V. For a symmetrical inverter design ($K_R = 1$), the noise margins are both equal to 0.93 V, in agreement with the results already presented.

EXERCISE: What is the value of K_R for a CMOS inverter in which $(W/L)_P = (W/L)_N$? Calculate the noise margins for this value of K_R with $V_{DD} = 2.5$ V, $V_{TN} = 0.6$ V, and $V_{TP} = -0.6$ V.

ANSWERS: 2.5; $\text{NM}_L = 0.728$ V, $\text{NM}_H = 1.15$ V

7.3 DYNAMIC BEHAVIOR OF THE CMOS INVERTER

Static power dissipation and the values of V_H and V_L do not really represent design parameters in CMOS circuits. Instead, the choice of sizes of the NMOS and PMOS transistors is dictated by the dynamic behavior of the logic gate—namely, by the desired average **propagation delay τ_p**.

7.3.1 PROPAGATION DELAY ESTIMATE

We can get an estimate of the propagation delay in the CMOS inverter by studying the circuit in Fig. 7.10, in which the inverter is driven by an ideal step function. For $t < 0$, the NMOS transistor is off and the PMOS transistor is on, forcing the output into the high state with $v_O = V_H = V_{DD}$. At $t = 0$, the input abruptly changes from 0 V to 2.5 V, and for $t = 0^+$, the NMOS transistor is on ($v_{GS} = +2.5$ V) and the PMOS transistor is off ($v_{SG} = 0$ V). Thus, the circuit simplifies to that in Fig. 7.10(b). The capacitor voltage is equal to V_{DD} at $t = 0^+$, and the capacitor voltage begins to fall as C is discharged through the NMOS transistor. The NMOS device starts conduction in the saturation region with $v_{DS} = v_{GS} = V_{DD}$ and enters the triode region of operation when $v_{DS} = (v_{GS} - V_{TN}) = (V_{DD} - V_{TN})$. The NMOS device continues to discharge C until its v_{DS} becomes zero. Therefore, $V_L = 0$ V.

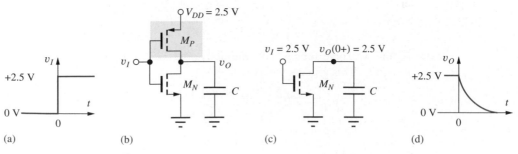

Figure 7.10 High-to-low output transition in a CMOS inverter.

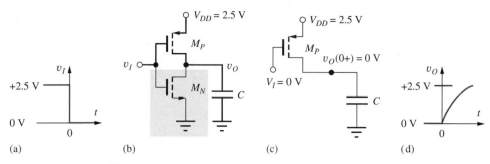

Figure 7.11 Low-to-high output transition in a CMOS inverter.

A significant amount of effort can be saved by realizing that the set of operating conditions in Fig. 7.10 is exactly the same as was used to determine τ_{PHL} for the NMOS inverter with a resistive load. Using Eq. (6.57),

$$\tau_{PHL} = 1.2 R_{\text{on}n} C \qquad \text{where } R_{\text{on}n} = \frac{1}{K_n(V_H - V_{TN})} \tag{7.15}$$

For the CMOS inverter with $V_H = 2.5$ V and $V_{TN} = 0.6$ V, τ_{PHL} becomes

$$\tau_{PHL} = 1.2 R_{\text{on}n} C = \frac{0.63C}{K_n} \tag{7.16}$$

Now consider the inverter driven by a step function that switches from $+2.5$ V to 0 V at $t = 0$, as in Fig. 7.11. For $t < 0$, the PMOS transistor is off and the NMOS transistor is on, forcing the output into the low state with $v_O = V_L = 0$. At $t = 0$, the input abruptly changes from 2.5 V to 0 V. For $t = 0^+$, the PMOS transistor will be on ($v_{GS} = -2.5$ V) and the NMOS transistor will be off ($v_{GS} = 0$ V). Thus, the circuit simplifies to that in Fig. 7.11(b). The voltage on the capacitor at $t = 0^+$ is equal to zero, and it begins to rise toward V_{DD} as charge is supplied through the PMOS transistor. The PMOS device begins conduction in the saturation region and subsequently enters the triode region of operation. This device continues to conduct until v_{DS} becomes zero, when $v_O = V_H = V_{DD}$. The same set of equations that was used to arrive at Eqs. (7.15) and (7.16) applies to this circuit, and for the CMOS inverter with $V_{DD} = 2.5$ V and $V_{TP} = -0.6$ V, τ_{PLH} becomes

$$\tau_{PLH} = 1.2 R_{\text{on}p} C \qquad \text{for } R_{\text{on}p} = \frac{1}{K_p(V_H + V_{TP})} \tag{7.17}$$

$$\tau_{PLH} = \frac{0.63C}{K_p} \tag{7.18}$$

The only difference between Eqs. (7.16) and (7.18) is the value of $R_{\text{on}n}$ and $R_{\text{on}p}$. From Table 7.1, we expect K_n' to be approximately 2.5 times the value of K_p'. In CMOS, a "symmetrical" inverter with $\tau_{PLH} = \tau_{PHL}$ can be designed if we set $(W/L)_P = (K_n'/K_p')(W/L)_N = 2.5(W/L)_N$ in order to compensate for the difference in mobilities. Because of the layout design rules in many MOS technologies, it is often convenient to design the smallest transistor with $(W/L) = (2/1)$. We use the symmetrical inverter design in Fig. 7.12, which has $(W/L)_N = (2/1)$ and $(W/L)_P = (5/1)$ as our **CMOS reference inverter** in the subsequent design of more complex CMOS logic gates.

Because we have designed this gate to have $\tau_{PLH} = \tau_{PHL}$, the average propagation delay is given by

$$\tau_P = \frac{\tau_{PHL} + \tau_{PLH}}{2} = \tau_{PHL} = 1.2 R_{\text{on}n} C \tag{7.19}$$

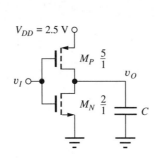

Figure 7.12 Symmetrical reference inverter design.

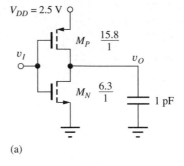

(a)

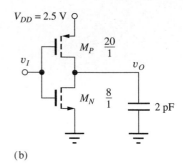

(b)

Figure 7.13 Scaled inverters: (a) $\tau_p = 1$ ns, (b) $\tau_p = 1.6$ ns.

EXERCISE: Calculate the average propagation delay of the reference inverter in Fig. 7.12 for $C = 1$ pF.

ANSWER: 3.16 ns

7.3.2 RISE AND FALL TIMES

The rise and fall times for the CMOS gate can also be found using our results for the NMOS inverter. Based upon Eq. (6.57), we see that the fall time is approximately three times τ_{PHL}, and because of the symmetry of the CMOS gate, the rise time is three times τ_{PLH}:

$$t_f = 3\tau_{PHL} \qquad \text{and} \qquad t_r = 3\tau_{PLH} \tag{7.20}$$

Based on the results of the preceding exercise for the 1-pF load capacitance, the expected rise and fall times are 9.5 ns.

To increase or decrease the speed of the CMOS inverter, the sizes of the transistors are changed. The inverter in Fig. 7.13(a) has a delay of 1 ns because the W/L ratios of both transistors are 3.16 times larger than those in the reference inverter design in Fig. 7.12. The inverter in Fig. 7.13(b) has a delay of 1.6 ns because its transistors are four times larger than those in the reference inverter, but it is driving twice as much capacitance:

$$\tau_P = (3.16 \times 2/4) \text{ ns} = 1.6 \text{ ns}$$

EXERCISE: An inverter must drive a 5-pF load capacitance with $\tau_p = 1$ ns. Scale the reference inverter to achieve this delay.

ANSWER: $(W/L)_P = 78.8/1$ and $(W/L)_N = 31.5/1$

DESIGN EXAMPLE 7.1   REFERENCE INVERTER DESIGN

In this example, we design a reference inverter to achieve a desired value of delay.

DESIGN SPECIFICATIONS   Design a reference inverter to achieve a delay of 250 ps when driving a 0.2-pF load using a 3.3-V power supply. Assume that the threshold voltages of the CMOS technology are $V_{TN} = -V_{TP} = 0.75$ V.

SOLUTION   **Specifications and Known Information:** $V_{DD} = 3.3$ V, $C = 0.2$ pF, $\tau_P = 250$ ps, $V_{TN} = -V_{TP} = 0.75$ V

Unknowns: $(W/L)_n$, $(W/L)_p$

Approach: Use Eq. (7.15) to find the value of R_{onn} and $(W/L)_n$; $(W/L)_p = 2.5(W/L)_p$.

Assumptions: A symmetrical inverter design is desired; the K' values are the same as Table 7.1: $K'_n = 100 \ \mu\text{A/V}^2$, $K'_p = 40 \ \mu\text{A/V}^2$

Analysis: Using Eq. (7.15),

$$R_{onn} = \frac{\tau_{PHL}}{1.2C} = \frac{250 \times 10^{-12}\text{sec}}{1.2(0.2 \times 10^{-12} \ \text{F})} = 1040 \ \Omega$$

$$\left(\frac{W}{L}\right)_n = \frac{1}{R_{onn}K'_n(V_{DD} - V_{TN})} = \frac{1}{(1040 \ \Omega)\left(100\dfrac{\mu\text{A}}{\text{V}^2}\right)(3.3 - 0.75) \ \text{V}} = \frac{3.77}{1}$$

$$\left(\frac{W}{L}\right)_p = \frac{K'_n}{K'_p}\left(\frac{W}{L}\right)_n = 2.5\left(\frac{W}{L}\right)_n = \frac{9.43}{1}$$

Check of Results: Let us double check the value of on-resistance for the PMOS device:

$$R_{onp} = \frac{1}{K_p(V_{DD} + V_{TP})} = \frac{1}{(9.43)\left(40\dfrac{\mu\text{A}}{\text{V}^2}\right)(3.3 - 0.75) \ \text{V}} = 1040 \ \Omega \quad ✔$$

Discussion: The on-resistances are equal, so the rise and fall times should be the same. The $R_{on}C$ time constant is 208 ps, which appears consistent with our design goal.

Computer-Aided Analysis: Our design is checked in the SPICE circuit below. Source VI is a pulse source, and the output waveform is the voltage across the capacitor. From the output waveforms, we find symmetrical propagation delays of approximately 280 ps. Our approximation in Eq. (7.15) is slightly optimistic in its estimate of τ_p.

PULSE SOURCE DESCRIPTION (VS)		SIMULATION PARAMETERS	
Initial voltage	0	Start time	0
Peak voltage	3.3 V	Stop time	6 ns
Delay time	0		
Rise time	50 ps		
Fall time	50 ps		
Pulse width	2.95 ns		
Pulse period	6 ns		

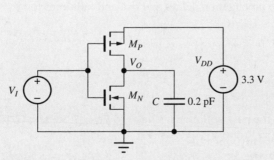

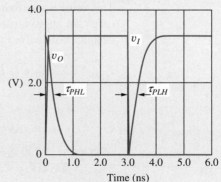

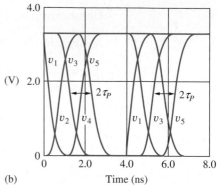

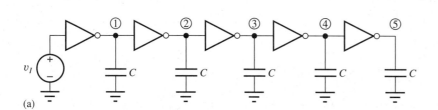

(a)

(b)

Figure 7.14 (a) Cascade of five identical inverters. (b) Output waveforms for five inverters using design from Ex. 7.1. V_S is a step function, $V_{DD} = 3.3$ V and $C = 0.2$ pF.

EXERCISE: Based upon the simulation results in Ex. 7.1, what W/L ratios are required to actually achieve an average propagation delay of 250 psec?

ANSWERS: 4.22/1, 10.6/1

EXERCISE: What would be the W/L ratios in Ex. 7.1 if the threshold voltages of the transistors were +0.5 V and −0.5 V?

ANSWERS: 3.43/1, 8.59/1

7.3.3 DELAY OF CASCADED INVERTERS

Obviously, the ideal step-function input waveform used in the analysis leading to Eq. (7.15) cannot be achieved in real circuits. The waveforms will have nonzero rise and fall times, and the propagation delay estimate given by Eq. (7.15) will be more optimistic than the delay encountered in actual circuits. Let us use SPICE to help improve the propagation delay design equation through simulation of the cascade of five identical inverters shown in Fig. 7.14(a). The first inverter is driven by a pulse with a 5-ns width, a 10-ns period, and rise and fall times of 2 ps.

The outputs of the five inverters are shown in the waveforms in Fig. 7.14(b). The propagation time between the step input and the output of the first inverter is approximately 0.28 ns, in agreement with the design of Ex. 7.1. However, the delay times of the other four inverters are significantly slower because of the rise and fall times of the waveforms driving each successive inverter. Looking at every other inverter, we see that the waveforms are very similar and quickly reach a periodic state. The average delay time through each inverter pair is 1 ns, corresponding to $\tau_{PHL} = \tau_{PLH} = 0.5$ ns. This value is approximately two times that predicted by Eq. (7.15). We also find that the rise and fall times are approximately two times the propagation delays. Based on these results, we modify our design estimate for the propagation delays and rise and fall times to be

$$\tau_{PHL} \cong 2.4 R_{onn} C \quad R_{onn} = \frac{1}{K_n(V_{DD} - V_{TN})} \quad \tau_{PLH} \cong 2.4 R_{onp} C \quad R_{onp} = \frac{1}{K_p(V_{DD} + V_{TP})}$$

$$t_r = 2\tau_{PLH} \qquad t_f = 2\tau_{PHL} \tag{7.21}$$

EXERCISE: What are the new equations [similar to Eqs. (7.16) and (7.18)] for τ_{PHL} and τ_{PLH} for $V_{DD} = 2.5$ V, $V_{TN} = 0.6$ V, and $V_{TP} = -0.6$ V?

ANSWERS: $\tau_{PHL} = 2.4 R_{onn} C = 1.26C/K_n$, $\tau_{PLH} = 2.4 R_{onp} C = 1.26C/K_p$

EXERCISE: What are the new equations [similar to Eqs. (7.16) and (7.18)] for τ_{PHL} and τ_{PLH} for $V_{DD} = 3.3$ V, $V_{TN} = 0.75$ V, and $V_{TP} = -0.75$ V?

ANSWERS: $\tau_{PHL} = 2.4 R_{onn} C = 0.94 C/K_n$, $\tau_{PLH} = 2.4 R_{onp} C = 0.94 C/K_p$

7.4 POWER DISSIPATION AND POWER DELAY PRODUCT IN CMOS

7.4.1 STATIC POWER DISSIPATION

CMOS logic is often considered to have zero static power dissipation. When the CMOS inverter is resting in either logic state, no direct current path exists between the two power supplies (V_{DD} and ground). However, in very low power applications or in IC designs with extremely large numbers of gates, static power dissipation can be important. The actual static power dissipation is nonzero due to the leakage currents associated with the reverse-biased drain-to-substrate junctions of the NMOS and PMOS transistors, as well as with the large area of reverse-biased n-well (or p-well) regions. The leakage current of these pn junctions flows between the supplies and contributes directly to static power dissipation. Power dissipation resulting from leakage can be a serious concern, particularly for battery-powered applications.

As this edition is being prepared, new circuit techniques are continually being invented to minimize static power dissipation. One straight-forward example appears in Fig. 7.15 in which large PMOS transistors are added to control the power to logic blocks. Power to inactive blocks can be turned off under either hardware or software control. The logic block can be a small number of gates or a much larger function, such as a multiplier or arithmetic logic unit (ALU) or even a CPU in chips containing multiple processor cores. When the PMOS switch is off, it is the only device that contributes to leakage and static power dissipation.

7.4.2 DYNAMIC POWER DISSIPATION

Dynamic power dissipation represents the power associated with switching a logic gate between states. There are two components of dynamic power dissipation in CMOS logic gates. As the gate charges and discharges load capacitance C at frequency f, the power dissipation is equal to that given by Eq. (6.49) in Chapter 6:

$$P_D = C V_{DD}^2 f \tag{7.22}$$

Power P_D is usually the largest component of power dissipation in CMOS gates operating at high frequency.

A second mechanism for power dissipation also occurs during switching of the CMOS logic gate and can be explored by referring to Fig. 7.16, which shows the current through a symmetrical CMOS inverter (with $C = 0$) as a function of the input voltage v_I. The current is zero for $v_I < V_{TN}$ and $v_I > (V_{DD} - |V_{TP}|)$ because either the NMOS or PMOS transistor is off for these conditions. However, in a logic gate we realize that input v_I does not make an abrupt jump between V_H and V_L,

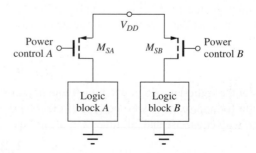

Figure 7.15 Static power reduction can be achieved by selectively turning power off to inactive logic blocks.

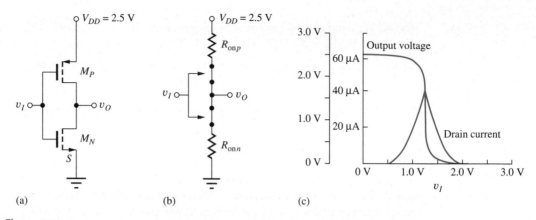

Figure 7.16 (a) CMOS inverter. (b) Switch and on-resistance model for the inverter with $V_{TN} \leq v_I \leq V_{DD} - |V_{TP}|$. (c) Supply current versus input voltage for a symmetrical CMOS inverter.

but instead makes a smooth transition between the two input states. During the input transition when $V_{TN} \leq v_I \leq V_{DD} - |V_{TP}|$, both transistor switches are on as in Fig. 7.16(b), and a current path exists through both the NMOS and PMOS devices. The current reaches a peak for $v_I = v_O = V_{DD}/2$. As v_I increases further, the current decreases back to zero. In the time domain, a pulse of current occurs between the power supplies as the output switches state, as shown in Fig. 7.17. In very high-speed CMOS circuits, this second component of dynamic power dissipation can approach 20 to 30 percent of that given by Eq. (7.22).

In the technical literature, this current pulse between supplies is often referred to as a "short circuit current." This reference is a misnomer, however, because the current is limited by the device characteristics, and a short circuit does not actually exist between the power supplies.

EXERCISE: Calculate the value of the maximum current similar to that in Fig. 7.16 for the inverter design in Fig. 7.12.

ANSWER: 42.3 μA

7.4.3 POWER-DELAY PRODUCT

The **power-delay product (PDP)**, defined in Sec. 6.3, is an important figure of merit for comparing various logic technologies:

$$\text{PDP} = P_{\text{av}} \tau_P \tag{7.23}$$

For CMOS operating at high frequency, the power consumed when charging and discharging the load capacitance C is usually the dominant source of power dissipation. For this case, $P_{\text{av}} = C V_{DD}^2 f$. The switching frequency $f = (1/T)$ can be related to the rise time t_r, fall time t_f, and the propagation delay τ_P of the CMOS waveform by referring to Fig. 7.18, in which we see that the period T must satisfy

$$T \geq t_r + t_a + t_f + t_b \tag{7.24}$$

For the highest possible switching frequency, times t_a and t_b approach 0, and the rise and fall times account for approximately 80 percent of the total transition time. Assuming a symmetrical inverter design with equal rise and fall times and using Eqs. (7.21) permits Eq. (7.24) to be approximated by

$$T \geq \frac{2t_r}{0.8} = \frac{2(2\tau_P)}{0.8} = 5\tau_P \tag{7.25}$$

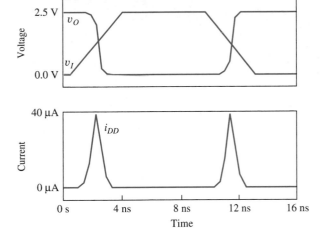

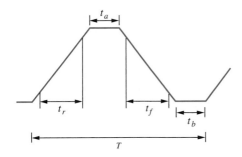

Figure 7.17 SPICE simulation of the transient current pulses between the power supplies during switching of a CMOS inverter.

Figure 7.18 CMOS switching waveform at high frequency.

A lower bound on the power-delay product for CMOS is then given by

$$\text{PDP} \geq \frac{CV_{DD}^2}{5\tau_P}\tau_P = \frac{CV_{DD}^2}{5} \tag{7.26}$$

The importance of using lower power supply voltages is obvious in Eq. (7.26), from which we see that the PDP is reduced in proportion to the square of any reduction in power supply. Moving from a power supply voltage of 5 V to one of 3.3 V reduces the power-delay product by a factor of 2.5. The importance of reducing the capacitance is also clear in Eq. (7.26). The lower the effective load capacitance, the smaller the power-delay product.

> **EXERCISE:** (a) What is the power-delay product for the symmetrical reference inverter in Fig. 7.12 operating from $V_{DD} = 2.5$ V and driving an average load capacitance of 100 fF? (b) Repeat for a 3.3-V supply.
>
> **ANSWERS:** 0.13 pJ; 0.22 pJ

7.5 CMOS NOR AND NAND GATES

The next several sections explore the design of CMOS gates, including the NOR gate, the NAND gate, and complex CMOS gates. The structure of a general static CMOS logic gate is given in Fig. 7.19 and consists of an NMOS transistor-switching network and a PMOS transistor-switching network. For each logic input variable in a CMOS gate, there is one transistor in the NMOS network *and* one transistor in the PMOS network. Thus, a CMOS logic gate has two transistors for every input variable. When a static conducting path exists through the NMOS network, a path must not exist through the PMOS network and vice versa. In other words, the conducting paths must represent logical complements of each other. CMOS forms an unusually powerful logic family because it is easy to realize both NOR and NAND gates as well as much more complex logic functions in a single gate. In Secs. 7.5.1 and 7.5.2, we will look at the NOR and NAND gate design. Then more complex gate design will be discussed.

7.5.1 CMOS NOR GATE

The realization of a two-input CMOS **NOR gate** is shown in Fig. 7.19 (b) in which the output should be low when either input A or input B is high. Thus the NMOS portion of the gate is identical to

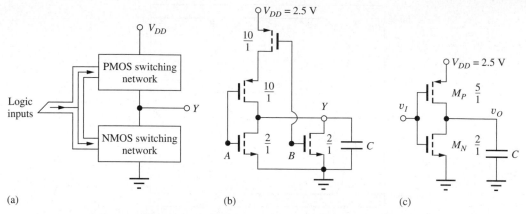

Figure 7.19 (a) Basic CMOS logic gate structure. (b) Two-input CMOS NOR gate and reference inverter.

TABLE 7.3
CMOS NOR Gate Truth Table and Transistor States

A	B	$Y = \overline{A+B}$	NMOS-A	NMOS-B	PMOS-A	PMOS-B
0	0	1	Off	Off	On	On
0	1	0	Off	On	On	Off
1	0	0	On	Off	Off	On
1	1	0	On	On	Off	Off

that of the NMOS NOR gate. However, in the CMOS gate, we must ensure that a static current path does not exist through the logic gate, and this requires the use of two PMOS transistors in series in the PMOS transistor network.

The complementary nature of the conducting paths can be seen in Table 7.3. A conducting path exists through the NMOS network for $A = 1$ or $B = 1$, as indicated by the highlighted entries in the table. However, a path exists through the PMOS network only when both $A = 0$ and $B = 0$ (no conducting path through the NMOS network).

In general, a parallel path in the NMOS network corresponds to a series path in the PMOS network, and a series path in the NMOS network corresponds to a parallel path in the PMOS network. (Bridging paths correspond to bridging paths in both networks.) We will study a rigorous method for implementing these two networks in Sec. 7.6.

Transistor Sizing

We can determine the sizes of the transistors in the two-input NOR gate by using our knowledge of NMOS gate design. In the CMOS case, one approach is to maintain the delay times equal to the reference inverter design under the worst-case input conditions. For the NMOS network with $AB = 10$ or 01, transistor A or transistor B must individually be capable of discharging load capacitance C, so each must be the same size as the NMOS device of the reference inverter. The PMOS network conducts only when $AB = 00$ and there are two PMOS transistors in series. To maintain the same **on-resistance** as the reference inverter, each PMOS device must be twice as large: $(W/L)_P = 2(5/1) = 10/1$. The resulting W/L ratios are indicated in Fig. 7.19(b).

Body Effect

In the preceding design, we ignored the influence of body effect. In the series-connected PMOS network, the source of the interior PMOS transistor cannot actually be connected to its substrate.

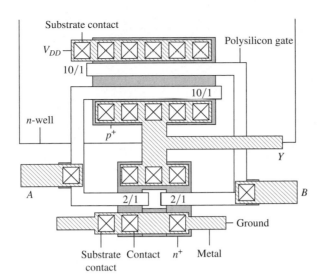

Figure 7.20 Layout of symmetrical two-input CMOS NOR gate. Note the large area of the PMOS transistors relative to the NMOS devices.

During switching, its threshold voltage changes as its source-to-bulk voltage changes. However, once the steady-state condition is reached, with $v_O = V_H$, for example, all the PMOS source and drain nodes will be at a voltage equal to V_{DD}. Thus, the total on-resistance of the PMOS transistors is not affected by the body effect once the final logic level is reached. During the transient response, however, the threshold voltage changes as a function of time, and $|V_{TP}| > |V_{TOP}|$, which slows down the rise time of the gate slightly. An investigation of this effect is left for Probs. 7.55 to 7.58.

Two-Input NOR Gate Layout

A possible layout for the two-input NOR gate is shown in Fig. 7.20. The two NMOS transistors are formed in the substrate, and each has a $2/1$ W/L ratio. The PMOS transistors, each with $W/L = 10/1$, are located in a common n-well. Note that the drain of the upper PMOS device is merged with the source of the lower PMOS device to form the connection between these two transistors. No contacts or metal are required to make this connection.

The gates of the NMOS and PMOS transistors are connected together with the metal level at inputs A and B, and the drains of three transistors are connected together by the metal layer to form output Y. Local substrate contacts are provided next to the sources of the NMOS transistors and the upper PMOS device. Note the much larger area taken up by the PMOS devices caused by the symmetrical gate design specification.

Three-Input NOR Gate

Figure 7.21(a) is the schematic for a three-input version of the NOR gate, and Table 7.4 is the truth table. As in the case of the NMOS gate, the output is low when NMOS transistor A or transistor B or transistor C is conducting. The only time that a conducting path exists through the PMOS network is for $A = B = C = 0$. The NMOS transistors must each individually be able to discharge the load capacitance in the desired time, so the W/L ratios are each $2/1$ based on our reference inverter design in Fig. 7.19. The PMOS network now has three devices in series, so each must be three times as large as the reference inverter device ($W/L = 15/1$).

It is possible to find the PMOS network directly from the truth table. From Table 7.3 for the two-input NOR gate, we see that $Y = \overline{A}\,\overline{B}$ and, using the identities in Table 6.4, the expression for $\overline{Y} = A + B$. For the three-input NOR gate described by Table 7.4, $Y = \overline{A}\,\overline{B}\,\overline{C}$ and $\overline{Y} = A + B + C$. The PMOS network directly implements the logic function $Y = \overline{A}\,\overline{B}\,\overline{C}$, written in terms of the complements of the input variables, and the NMOS network implements the logic function $\overline{Y} = A + B + C$, written in terms of the uncomplemented input variables. In effect, the PMOS transistors directly complement the input variables for us! The two functions $\overline{Y}$ and Y for the NMOS and PMOS

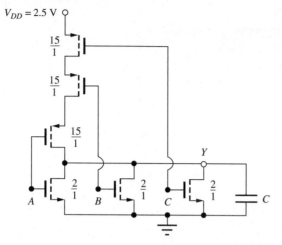

Figure 7.21 Three-input CMOS NOR gate.

TABLE 7.4 Three-Input NOR Gate Truth Table			
A	B	C	$Y = \overline{A + B + C}$
0	0	0	1
0	0	1	0
0	1	0	0
0	1	1	0
1	0	0	0
1	0	1	0
1	1	0	0
1	1	1	0

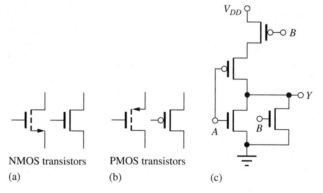

Figure 7.22 Shorthand notation for (a) NMOS and (b) PMOS transistors. (c) Two-input CMOS NOR gate using shorthand transistor symbols.

networks need to be written in minimum form in order to have the minimum number of transistors in the two networks.

The complementing effect of the PMOS devices has led to a shorthand representation for CMOS logic gates that is used in many VLSI design texts. This notation is shown by the right-hand symbol in each transistor pair in Fig. 7.22. The NMOS and PMOS transistor symbols differ only by the circle at the input of the PMOS gate. This circle identifies the PMOS transistor and indicates the logical inversion operation that is occurring to the input variable. In this book, however, we will continue to use the standard symbols.

7.5.2 CMOS NAND GATES

The NAND gate is also easy to realize in CMOS, and the structure and truth table for a two-input static CMOS **NAND gate** are given in Fig. 7.23 and Table 7.5, respectively. From the truth table for the NAND gate, we see that $Y = \overline{AB}$, and the output should be low only when both input A and input B are high. Thus, the NMOS switching network is identical to that of the NMOS NAND gate with transistors A and B in series.

Expanding the equation for Y in terms of complemented variables, we have $Y = \overline{A} + \overline{B}$. If either input A or input B is low, the output must be pulled high through a PMOS transistor, resulting in two transistors in parallel in the PMOS switching network. Once again, there is one transistor in the NMOS network and one transistor in the PMOS network for each logic input variable.

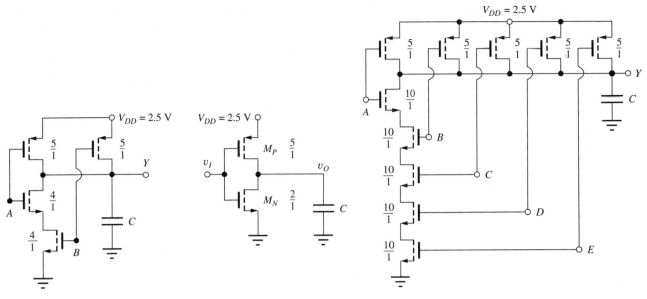

Figure 7.23 Two-input CMOS NAND gate and reference inverter.

Figure 7.24 Five-input CMOS NAND gate: $Y = \overline{ABCDE}$.

TABLE 7.5
CMOS NAND Gate Truth Table and Transistor States

A	B	$Y = \overline{AB}$	NMOS-A	NMOS-B	PMOS-A	PMOS-B
0	0	1	Off	Off	On	On
0	1	1	Off	On	On	Off
1	0	1	On	Off	Off	On
1	1	0	On	On	Off	Off

Transistor Sizing

We determine the sizes of the transistors in the two-input NAND gate by again using our knowledge of NMOS design. There are two transistors in series in the NMOS network, so each should be twice as large as in the reference inverter or 4/1. In the PMOS network, each transistor must individually be capable of discharging load capacitance C, so each must be the same size as the PMOS device in the reference inverter. These W/L ratios are indicated in Fig. 7.23.

The Multi-Input NAND Gate

As another example, the circuit for a five-input NAND gate is given in Fig. 7.24. The NMOS network consists of a series stack of five transistors with one MOS device for each input variable. The PMOS network consists of a group of PMOS devices in parallel, also with one transistor for each input. To maintain the speed on the high-to-low transition in the five-input gate, the NMOS transistors must each be five times larger than that of the reference inverter, whereas the PMOS transistors are each identical to that of the reference inverter.

EXERCISE: Draw a four-input NAND gate similar to the five-input gate in Fig. 7.24. What are the W/L ratios of the transistors?

ANSWERS: 8/1; 5/1

7.6 DESIGN OF COMPLEX GATES IN CMOS

Just as with NMOS design, the real power of CMOS is not realized if the designer uses only NANDs, NORs, and inverters. The ability to implement **complex logic gates** directly in CMOS is an advantage for CMOS design, just as it is in NMOS. This section investigates complex gate design through Design Examples 7.2 and 7.3.

DESIGN EXAMPLE 7.2   COMPLEX CMOS LOGIC GATE DESIGN

This example presents the design of a CMOS logic gate with the same logic function as the NMOS design from Fig. 6.34 in Chapter 6. A graphical approach for relating NMOS and PMOS transistor networks is introduced as part of the example.

DESIGN SPECIFICATIONS   Design a CMOS logic gate logic that implements the function $Y = \overline{A + BC + BD}$.

SOLUTION   **Specifications and Known Information:** Logic function $Y = \overline{A + BC + BD}$ or $\overline{Y} = A + BC + BD$. The NMOS network in Fig. 7.25(a) for $\overline{Y}$ is exactly the same as that of the corresponding NMOS gate in Fig. 6.34. (Note that $\overline{Y}$ is represented as a sum-of-products form.)

Unknowns: Topology of the PMOS network; W/L ratios of all the transistors

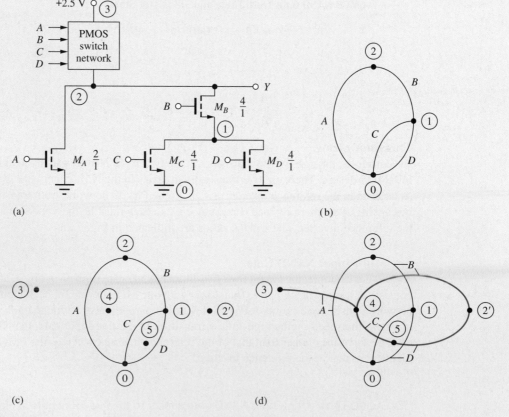

(a)

(b)

(c)

(d)

Figure 7.25 Steps in constructing graphs for NMOS and PMOS networks. (a) NMOS network. (b) NMOS graph. (c) NMOS graph with new nodes added. (d) Graph with PMOS arcs added.

Approach: In this case, we already have the NMOS network (Fig. 6.34) but need to construct the corresponding PMOS network. A new graphical method for finding the PMOS network will be introduced for this design. Once the PMOS circuit topology is found, then the transistor W/L values can be determined.

Assumptions: Use a symmetrical design based on the reference inverter in Fig. 7.12.

Analysis: In this case, we have been given the NMOS network. First we construct a graph of the NMOS network, which is shown in Fig. 7.25(b). Each node in the NMOS network corresponds to a node in the graph, including node 0 for ground and node 2 for the output. Each NMOS transistor is represented by an arc connecting the source and drain nodes of the transistors and is labeled with the logical input variable.

Next, we construct the PMOS network directly from the graph of the NMOS network. First, place a new node inside of every enclosed path [nodes 4 and 5 in Fig. 7.25(c)] in the NMOS graph. In addition, two exterior nodes are needed: one representing the output and one representing V_{DD} [nodes 2′ and 3 in Fig. 7.25(c)]. An arc, ultimately corresponding to a PMOS transistor, is then added to the graph for each arc in the NMOS graph. The new arcs cut through the NMOS arcs and connect the pairs of nodes that are separated by the NMOS arcs. A given PMOS arc has the same logic label as the NMOS arc that is intersected. This construction results in a minimum PMOS logic network, which has only one PMOS transistor per logic input. The completed graph is given in Fig. 7.25(d), and the corresponding PMOS network is shown in Fig. 7.26. A transistor is added to the PMOS switching network corresponding to each arc in the PMOS graph. Note that nodes 2 and 2′ represent the same output node.

To complete the design of this CMOS gate, we must choose the W/L ratios of the transistors. In the NMOS switching network, the worst-case path contains two transistors in series, and transistors B, C, and D should each be twice as large as those of the reference inverter. Transistor

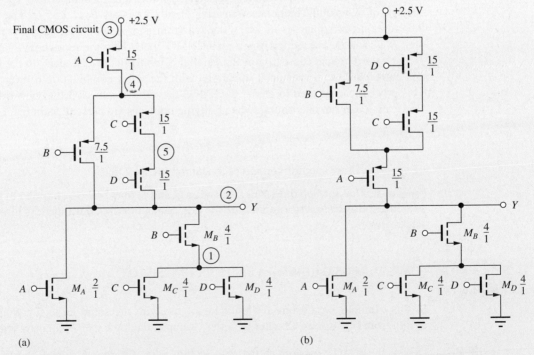

(a)

(b)

Figure 7.26 (a) CMOS implementation of $Y = \overline{A + BC + BD}$. (b) Circuit equivalent to gate in Fig. 7.26(a).

A should be the same size as the NMOS transistor in the reference inverter because it must be able to discharge the load capacitance by itself. In the PMOS switching network, the worst-case path has three transistors in series, and PMOS transistors A, C, and D should be three times as large as the PMOS device in the reference inverter. The size of transistor B is determined using the on-resistance method from Eq. (6.38), in which R_{on} represents the on-resistance of a reference transistor with $W/L = 1/1$:

$$\frac{R_{on}}{\left(\frac{15}{1}\right)} + \frac{R_{on}}{\left(\frac{W}{L}\right)_B} = \frac{R_{on}}{\left(\frac{5}{1}\right)} \qquad \text{or} \qquad \left(\frac{W}{L}\right)_B = \frac{7.5}{1}$$

The completed design appears in Fig. 7.26.

Figure 7.26(b) gives a second implementation of the same logic network; here nodes 2 and 3 in the PMOS graph have been interchanged. The conducting paths through the PMOS network are identical, and hence the logic function is also the same.

Check of Results: Another approach can be used to check the PMOS network topology. Note that the PMOS network implements the function obtained by expanding Y as a product of sums:

$$Y = \overline{A + B(C + D)} = \overline{A} \cdot (\overline{B} + \overline{C}\,\overline{D})$$

The inversion of each variable is provided directly by the PMOS transistor (remember the transistor symbol introduced in Fig. 7.22). A conducting path is formed through the PMOS network when A is conducting, and either B is conducting, or C and D are both conducting. The network structure appears to be correct. Thus, in summary, the NMOS network implements Y as a sum-of-products, whereas the PMOS network implements the function Y as a product-of-sums in the complemented variables.

Discussion: Note that the PMOS network in Fig. 7.26 could also be obtained from the NMOS network by successive application of the series/parallel transformation rule. The NMOS network consists of two parallel branches: transistor A and transistors B, C, and D. The PMOS network, therefore, has two branches in series, one with transistor A and a second representing the branch with B, C, and D. The second branch in the NMOS network is the series combination of transistor B and a third branch consisting of the parallel combination of C and D. In the PMOS network, this corresponds to transistor B in parallel with the series combination of transistors C and D. This process can be used to create the PMOS network from the NMOS network, or vice versa. However, it can run into trouble when bridging branches are present, as in Ex. 7.3.

EXERCISE: Draw another version of the circuit in Fig. 7.26(b).

ANSWER: The position of PMOS transistors C and D may be interchanged in Fig. 7.26(a) or (b). Another possibility is to interchange the position of the NMOS transistors forming the $B(C+D)$ subnetwork.

DESIGN EXAMPLE 7.3 COMPLEX CMOS GATE WITH A BRIDGING TRANSISTOR

In this example, we design a CMOS logic gate with the same logic function as the NMOS design from Fig. 6.36 in Chapter 6. In this example, the logic gate contains a bridging transistor.

DESIGN SPECIFICATIONS Design the CMOS version of the gate in Fig. 6.36, Chapter 6, which implements the logic function $Y = \overline{AB + CE + ADE + CDB}$.

SOLUTION **Specifications and Known Information:** $Y = \overline{AB + CE + ADE + CDB}$ or $\overline{Y} = AB + CE + ADE + CDB$. The NMOS network implementation of $\overline{Y}$ is exactly the same as that of the NMOS gate in Fig. 6.36.

Unknowns: Topology of the PMOS network; W/L ratios of all the transistors

Approach: We have the NMOS network and need to construct the corresponding PMOS network. Use the graphical method introduced in Ex. 7.2 to find the PMOS network. Once the PMOS circuit topology is found, the transistor W/L values can be determined.

Assumptions: Use a symmetrical design based upon the reference inverter in Fig. 7.12.

Analysis: Figure 7.27(b) is the graph of the NMOS network. The corresponding graph for the PMOS network is constructed in Fig. 7.27(c): new nodes 5 and 6 are added inside the enclosed paths formed by arcs ACD and BED in Fig. 7.27(b), and nodes 2' and 4 are added to represent the output and +5-V power supply. New arcs are added for each transistor. The resulting CMOS gate design is given in Fig. 7.28. In this unusual case, the NMOS and PMOS network topologies are identical. Transistor D is a bridging transistor in both networks. The worst-case path in each network contains three devices in series, and all the transistors in the CMOS gate are three times the size of the corresponding transistors in the reference inverter.

Check of Results: A recheck of our work indicates that it is correct.

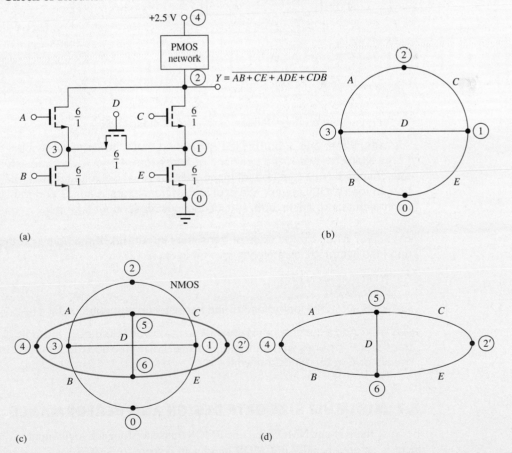

Figure 7.27 (a) NMOS portion of CMOS gate. (b) Graph for NMOS network. (c) Construction of the PMOS graph. (d) Resulting graph for the PMOS network.

Discussion: For this case, it is not easy to check our work using the series/parallel transformations because of the bridging transistor in the network.

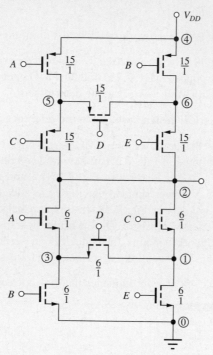

Figure 7.28 CMOS implementation of $Y = \overline{AB + CE + ADE + CDB}$.

EXERCISE: Draw another version of the circuit in Fig. 7.28.

ANSWER: Either one or both of the NMOS and PMOS switching networks can be rearranged. In the NMOS network, the positions of transistors *A* and *B* and transistors *C* and *E* can be interchanged. Note that both of these changes must be done to retain the correct logic function. In the PMOS network, the positions of transistors *A* and *C* and transistors *B* and *E* can be interchanged. Again, both sets of changes must occur together.

EXERCISE: What are the sizes of transistors in part (b) of the figure on the next page based upon the symmetrical reference inverter in Fig. 7.12?

ANSWER: 4/1, 2/1, 15/1

EXERCISE: Draw the logic diagram and transistor implementation for a (2-2-2) AOI (see page 377).

ANSWER: Add a second input labeled *F* to the lower AND gate in the logic diagram. Add NMOS transistor *F* in series with transistor *E* in the NMOS tree and PMOS transistor *F* in parallel with transistor *E* in the PMOS network.

7.7 MINIMUM SIZE GATE DESIGN AND PERFORMANCE

Because there is one NMOS and one PMOS transistor for each logic input variable in static CMOS, there is an area penalty in CMOS logic with respect to NMOS logic that requires only a single-load device regardless of the number of input variables. In addition, series paths in the PMOS or NMOS switching networks require large devices in order to maintain logic delay. (See Fig. 7.20.)

ELECTRONICS IN ACTION

And-Or-Invert Gates in a Standard Cell Library

The Standard Cell design approach utilizes a library of predesigned logic blocks that are placed and interconnected on a chip by the circuit designer or more often by automated design tools. The library contains a large number of cells ranging from basic logic gates to multipliers, ALU's, shifters, CPU cores, and memories, for example. The resulting integrated circuit chips have a highly regular appearance, as can be observed in the photograph below.

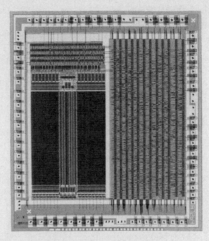

Die photograph courtesy of Professor C. E. Stroud, Auburn University, Auburn, AL.

The And-Or-Invert or AOI gate introduced in Chapter 6 (page 317) represents a basic logic block that is widely used in Standard Cell designs. The logic diagram and its transistor implementation for a (2-2-1) AOI gate that implements $\overline{Y = AB + CD + E}$ appear below. The 2-2-1 notation represents the input configuration for the gate: two 2-input AND's and one 1-input AND. In the transistor realization, the NMOS switching tree is the same as would be used in the corresponding NMOS gate. The PMOS tree implements $Y = (\overline{A} + \overline{B})(\overline{C} + \overline{D})(\overline{E})$. The three parallel branches in the NMOS network are replaced by three series branches in the PMOS network, and the AND branches in the NMOS network become OR branches in the PMOS network.

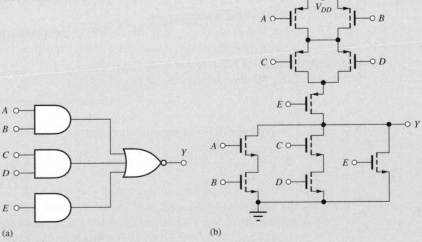

(a) Logic diagram for a (2-2-1) AOI gate. (b) Transistor implementation of $Y = \overline{AB + CD + E}$.

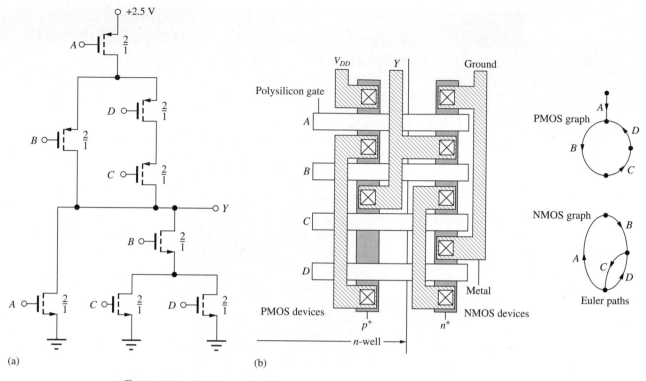

Figure 7.29 (a) Minimum size implementation of a complex CMOS gate. (b) Layout of complex CMOS gate in (a).

However, in most logic designs, only the critical logic delay paths need to be scaled to maintain maximum performance. If gate delay is not the primary concern, then considerable area savings can be achieved if all transistors are made of minimum geometry. For example, the gate from Fig. 7.26 is shown implemented with minimum size transistors in Fig. 7.29 . Here, area is being traded directly for increased logic delay. The total gate area for Fig. 7.29 is $16F^2$, where F is the minimum feature size, whereas that of Fig. 7.26 is $66.5F^2$, an area four times larger.

Let us estimate the worst-case propagation delay of this minimum size logic gate relative to our reference inverter design. In the NMOS switching network, the worst-case path contains two transistors in series, with $W/L = 2/1$, which is equivalent to the R_{on} of a 1/1 device, as compared to the reference inverter in which $W/L = 2/1$. Thus, the high-to-low propagation delay for this gate is twice that of the symmetrical reference inverter $\tau_{PHLI} : \tau_{PHL} = 2\tau_{PHLI}$. The worst-case path in the PMOS switching network contains three minimum geometry transistors in series, yielding an effective W/L ratio of 2/3, versus 5/1 for the reference inverter. The low-to-high propagation delay is related to that of the reference inverter τ_{PLHI} by

$$\tau_{PLH} = \frac{\left(\frac{5}{1}\right)}{\left(\frac{2}{3}\right)}\tau_{PLHI} = 7.5\tau_{PLHI}$$

The average propagation delay of the minimum size logic gate is

$$\tau_p = \frac{(\tau_{PLH} + \tau_{PHL})}{2} = \frac{(2\tau_{PLHI} + 7.5\tau_{PLHI})}{2} = \frac{9.5\tau_{PLHI}}{2} = 4.75\tau_{PLHI}$$

and the propagation delay of the minimum size gate will be 4.75 times slower than the reference inverter design when driving the same load capacitance and will be highly assymetrical.

Layout of the gate in Fig. 7.29 is implemented by two vertical linear arrays of transistors with common horizontal polysilicon gate stripes. The diffusions between various transistors are interconnected on the metal level to create the desired circuit. This layout strategy requires identification of an "Euler path" in the graphs of the PMOS and NMOS switching trees. The **Euler path** connects all the transistors in the graph of the NMOS or PMOS network, but must pass through each transistor once and only once. The path may go through a given node more than once. To create the layout in Fig. 7.29, the Euler paths in both the NMOS and PMOS networks must have the same transistor order—in this case the paths have the common order *ABCD*.

7.8 DYNAMIC DOMINO CMOS LOGIC

Dynamic logic circuits were first developed as a means of reducing power in PMOS and NMOS logic. **Dynamic logic** uses separate **precharge** and **evaluation phases** governed by a system clock signal to eliminate the dc current path that exists in single-channel static logic gates. However, the logical outputs are now valid only during a portion of the evaluation phase of the clock. In addition, static power represents only one component of power dissipation in high-speed systems, and the power required to drive the clock signals that must connect to every dynamic logic gate can be significant. Early PMOS and NMOS dynamic circuits used complicated two-phase and four-phase clocking techniques. However, a more recently developed form of dynamic CMOS logic called **domino CMOS** is based on a single clock signal.

The circuit in Fig. 7.30 represents a general gate in a dynamic domino CMOS. Operation of the domino CMOS circuit begins with the clock signal in the low state. M_{NC} is off, which disables the current path to ground through the logic function block F. The same clock signal turns on M_{PC}, pulling node 4 high to V_{DD} and forcing the inverter output low.

When the clock signal goes high, the capacitance at node 4 is selectively discharged. If a conducting path exists through the logic network F (that is, if $F = 1$), then node 4 is discharged to zero, and the output of the inverter rises to a 1 level. If $F = 0$, no discharge path exists, the voltage at node 4 remains high, and the output of the inverter remains at 0.

Simulated waveforms for the single-input domino CMOS gate in Fig. 7.31(a) are shown in Fig. 7.31(b) for the case of $A = 1$. After the clock goes high, the output rises following the delay to discharge the capacitances at nodes 3 and 4 plus the delay through the inverter. The output drops back to zero following the clock signal's return to zero.

The inputs to a given domino CMOS gate, generated by other domino gates, make only low-to-high transitions following the clock transition, and the functional evaluation during the positive clock phase ripples through the gates like a series of dominos falling over—hence the name domino logic. Figure 7.31(c) shows the ripple-through effect for the case of three domino gates in series. Output O_1 follows the rising edge of the clock and drives the input of O_2, and output O_3 responds to the change in O_2. Note that the outputs are all reset to zero at the same time, following the falling edge of the clock.

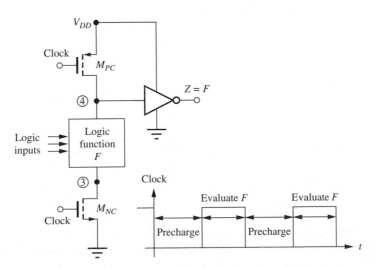

Figure 7.30 Dynamic domino CMOS logic.

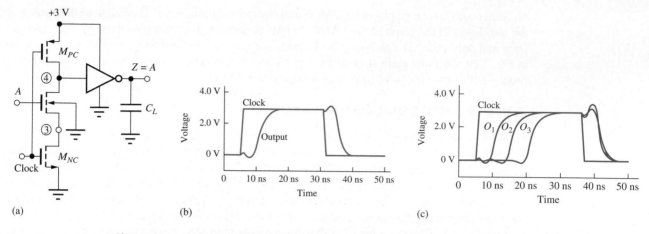

Figure 7.31 (a) Domino CMOS gate with a single input. (b) Waveforms for the clocked domino circuit. (c) Outputs of three cascaded domino CMOS gates.

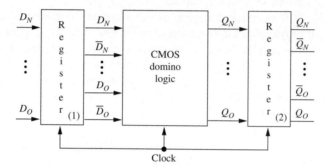

Figure 7.32 Section of logic in a complex digital system.

External inputs that are not from other domino gates should remain stable during the evaluation phase of the clock. As with most CMOS circuits, power dissipation is set by the dynamic power consumed in charging and discharging the capacitances at the various nodes. A major advantage of the domino CMOS circuit is the requirement for only two PMOS transistors per logic stage, no matter how complex the function F.

The observant reader has probably noticed that domino CMOS gates do not form a complete logic family because only true output functions are available. However, this does not represent a problem in system designs in which a register transfer structure exists, as in Fig. 7.32. In this logic structure, data are stored in a static register (1), which can be designed to produce both true and complemented data values at its outputs. Domino CMOS performs the combinational logic functions on the positive phase of the clock signal, and the results of the logic operations are clocked into register (2) on the falling edge of the clock signal.

EXERCISE: (a) Draw the circuit schematic for a domino CMOS gate for $Z = AB + C$. (b) What power is required in a domino CMOS clock driver circuit that drives a 50-pF load at 10 MHz if $V_{DD} = 5$ V?

ANSWER: Transistor A in Fig. 7.31 is replaced with a three-transistor NMOS structure for $AB + C$. 12.5 mW

7.9 CASCADE BUFFERS

In today's high-density ICs, the input capacitance of a logic gate may be in the range of only 10 to 100 fF, and the propagation delay of a CMOS gate driving such a small load capacitance can be well below 1 ns. However, there are many cases in which a much higher load capacitance is

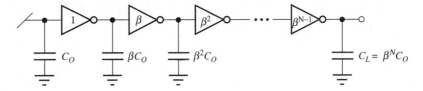

Figure 7.33 Cascade buffer for driving large capacitive loads.

encountered. For example, the wordlines in RAMs and ROMs represent relatively large capacitances; long interconnection lines and internal data buses in microprocessors also represent significant load capacitances. In addition, the circuits that drive off-chip data buses may encounter capacitances as large as 10 to 50 pF, a capacitance 1000 times larger than the internal load capacitance. We know that the propagation delay of a CMOS gate is proportional to its load capacitance, so if a minimum size inverter is used to drive such a large capacitance, then the delay will be extremely long. If the inverter is scaled up in size to reduce its own delay, then its input capacitance increases, slowing down the propagation delay of the previous stage.

7.9.1 CASCADE BUFFER DELAY MODEL

It has been shown [4, 5] that a minimum overall delay can be achieved by using an optimized cascade of several inverter stages, as depicted in Fig. 7.33. The W/L ratios of the transistors are increased by the **taper factor** β in each successive inverter stage in order to drive the large load capacitance. In this analysis, the load capacitance is $C_L = \beta^N C_o$, in which C_o is the input capacitance of the normal reference inverter stage and N is the number of stages in the **cascade buffer.** The analysis is simplified by assuming that the total capacitance at a given node is dominated by the input capacitance of the next inverter. Thus, the capacitive load on the first inverter is βC_o, the load on the second inverter is $\beta^2 C_o$, and so on. If the propagation delay of the unit size inverter driving a load capacitance of C_o is τ_o, then the delay of each inverter stage will be $\beta \tau_o$, and the total propagation delay of the N-stage buffer will be

$$\tau_B = N\beta\tau_o \tag{7.27}$$

in which

$$\beta^N = \frac{C_L}{C_o} \qquad \text{or} \qquad \beta = \left(\frac{C_L}{C_o}\right)^{1/N} \tag{7.28}$$

Substituting this result into Eq. (7.27) yields an expression for the total propagation delay of the buffer:

$$\tau_B = N\left(\frac{C_L}{C_o}\right)^{1/N}\tau_o \tag{7.29}$$

In Eq. (7.29), N increases as additional stages are added to the buffer, but the value of the capacitance ratio term decreases with increasing N. The opposite behavior of these two factors leads to the existence of an optimum value of N for a given capacitance ratio.

7.9.2 OPTIMUM NUMBER OF STAGES

The number of stages that minimizes the overall buffer delay can be found by differentiating Eq. (7.29) with respect to N and setting the derivative equal to zero. The optimization can be simplified by taking logarithms of both sides of Eq. (7.29) before taking the derivative:

$$\ln \tau_B = \ln N + \frac{1}{N}\ln\left(\frac{C_L}{C_o}\right) + \ln \tau_o \tag{7.30}$$

Taking the derivative with respect to N and setting it equal to zero gives the optimum value of N:

$$\frac{d\ln\tau_B}{dN} = \frac{1}{N} - \frac{1}{N^2}\ln\left(\frac{C_L}{C_o}\right) \qquad \text{and} \qquad N_{\text{opt}} = \ln\left(\frac{C_L}{C_o}\right) \tag{7.31}$$

Substituting N_{opt} into Eqs. (7.28) into (7.29) yields the optimum value of the taper factor β_{opt} and the optimum buffer delay $\tau_{B\text{opt}}$:

$$\beta_{\text{opt}} = \left(\frac{C_L}{C_o}\right)^{[1/\ln(C_L/C_o)]} = \varepsilon \quad \text{and} \quad \tau_{B\text{opt}} = \varepsilon \tau_o \ln\left(\frac{C_L}{C_o}\right) \qquad (7.32)$$

The optimum value of the taper factor is equal to the natural base $\varepsilon \cong 2.72$.

DESIGN EXAMPLE 7.4   CASCADE BUFFER DESIGN

In digital chip designs, such as microprocessors and memory chips, we often design circuits that must drive high-capacitance loads. This example designs a buffer to drive a 50-pF load.

DESIGN SPECIFICATIONS   Design a cascade buffer to drive a load capacitance of 50 pF if $C_o = 50$ fF. Find the overall delay for the buffer design for a 3.3-V supply with $V_{TN} = 0.75$ V and $V_{TP} = -0.75$ V.

SOLUTION   **Specifications and Known Information:** $C_L = 50$ pF; technology has $C_o = 50$ fF.

Unknowns: The number of stages N, taper factor β, and relative size of each inverter in the cascade buffer; overall buffer delay

Approach: Select the number of stages N needed using the result in Eq. (7.31). Calculate the taper factor using the selected value of N. Scale the reference inverter using N.

Assumptions: The value of C_o (50 fF) is known for the symmetrical reference inverter in Fig. 7.12.

Analysis: The optimum value of N is $N_{\text{opt}} = \ln C_L/C_o = \ln(1000) = 6.91$, and the optimum delay is $\tau_{B\text{opt}} = 6.91 \times (2.72) \times t_o = 18.8 t_o$. N must be an integer, but either $N = 6$ or $N = 7$ can be used because the delay minimum is quite broad, as illustrated by these numeric results. Using these two values of N yields

$$N = 6: \quad \tau_B = 6(1000^{1/6})\tau_o = 19.0\tau_o$$

$$N = 7: \quad \tau_B = 7(1000^{1/7})\tau_o = 18.8\tau_o$$

Little speed is lost by using the smaller value of N. The choice between $N = 6$ and $N = 7$ will usually be made based on buffer area. For $N = 6$, the taper factor β is found using Eq. (7.28):

$$\beta = \left(\frac{C_L}{C_o}\right)^{1/N} = \left(\frac{50 \text{ pF}}{50 \text{ fF}}\right)^{1/6} = (1000)^{1/6} = \sqrt{10} = 3.16$$

Each successive stage of the cascade buffer is 3.16 times larger than the preceding stage, and the result appears in Fig. 7.34. The transistor sizes (NMOS, PMOS) for the six inverters are (2/1, 5/1), (6.33/1, 15.8/1), (20/1, 50/1), (63.3/1, 158/1), (200/1, 500/1), and (633/1, 1580/1).

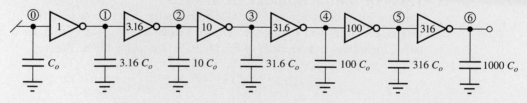

Figure 7.34   Optimum buffer design.

For $N = 6$, we already found $\tau_B = 19.0\tau_o$, and we need the value of τ_o. For the symmetrical inverter designs,

$$\tau_o \cong \frac{2.4C}{K_n(V_{DD} - V_{TN})} = \frac{2.4(50 \text{ fF})}{2(100 \text{ }\mu\text{A/V}^2)(3.3 - 0.75)} = 0.24 \text{ ns}$$

and the estimated buffer delay is $\tau_B = 4.5$ ns.

Check of Results: A recheck of calculations indicates the work is correct. We now check the overall buffer delay using SPICE.

Computer-Aided Analysis: The output waveforms for the cascade buffer in Fig. 7.34 appear below. The first inverter is driven by a pulse having a 6.25-ns width, a 12.5-ns period, and rise and fall times of 2 ps. The propagation delay of the full buffer is 4 ns. This is slightly shorter than our estimate because the delay of the first inverter is shortened by the fast rise and fall times of the input signal. The delay between the output of inverters 1 and 6 should be $15.8\tau_o = 3.8$ ns, and SPICE also yields an identical result.

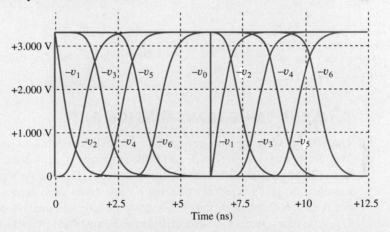

Discussion: The buffer optimization just discussed is based on a set of very simple assumptions for the change in nodal capacitance with buffer size. Many refinements to this analysis have been published in the literature, and those more complex analyses indicate that the optimum tapering factor lies between 3 and 4. The results in Eqs. (7.28), (7.29), and (7.31) should be used as an initial guide, with the final design determined using circuit simulation based on a given set of device and technology parameters.

EXERCISE: Suppose we try to drive the 50-pF capacitance in Ex. 7.4 by adding only one additional buffer stage, i.e., $N = 2$. What is the relative size of the buffer stage? What is the overall delay?

ANSWERS: 31.6, $63.2\tau_o$

EXERCISE: Prove that $z^{1/\ln z} = \varepsilon$.

EXERCISE: What would be the relative inverter sizes for the cascade buffer example if N were chosen to be 7? Compare the total buffer area for $N = 6$ and $N = 7$.

ANSWERS: 1, 2.68, 7.20, 19.3, 51.8, 139, 372; 462 times the unit inverter area versus 593 times

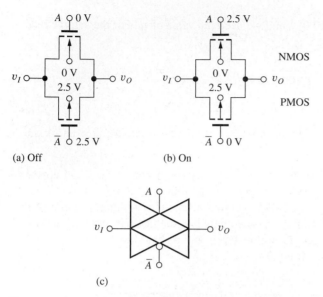

(a) Off   (b) On

(c)

Figure 7.35 CMOS transmission gate in (b) on state and (a) off state. (c) Special circuit symbol for the transmission gate.

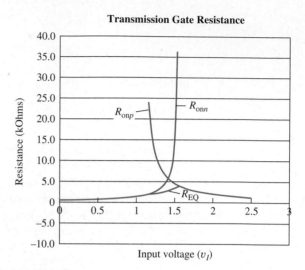

Figure 7.36 On-resistance of a transmission gate versus input voltage v_I including body effect using the values from Table 7.1 and $(W/L)_N = (W/L)_P = 10/1$. The maximum value of R_{EQ} is approximately 4 kΩ.

7.10 THE CMOS TRANSMISSION GATE

The **CMOS transmission gate** in Fig. 7.35 is a circuit useful in both analog and digital design. The circuit consists of NMOS and PMOS transistors, with source and drain terminals connected in parallel and gate terminals driven by opposite phase logic signals indicated by A and $\overline{A}$. The transmission gate is used so often that it is given the special circuit symbol shown in Fig. 7.35(c). For $A = 0$, both transistors are off, and the transmission gate represents an open circuit.

When the transmission gate is in the conducting state ($A = 1$), the input and output terminals are connected together through the parallel combination of the on-resistances of the two transistors, and the transmission gate represents a *bidirectional* resistive connection between the input and output terminals. The individual on-resistances R_{onp} and R_{onn}, as well as the equivalent on-resistance R_{EQ}, all vary as a function of the input voltage v_I, as shown in Fig. 7.36. The value of R_{EQ} is equal to the parallel combination of R_{onp} and R_{onn}:

$$R_{EQ} = \frac{R_{onp} R_{onn}}{R_{onp} + R_{onn}} \tag{7.33}$$

In Fig. 7.36, the PMOS transistor is cut off ($R_{onp} = \infty$) for $v_I \leq 1.05$ V, and the NMOS transistor is cut off for $v_I \geq 1.56$ V. For the parameters used in Fig. 7.36, the equivalent on-resistance is always less than 4 kΩ, but it can be made smaller by increasing the W/L ratios of the n- and p-channel transistors.

> **EXERCISE:** What W/L ratios are required in the transmission gate in Fig. 7.36 to ensure that $R_{EQ} \leq 1$ kΩ if the W/L ratios of both transistors are the same?
>
> **ANSWER:** $W/L = 40/1$

7.11 CMOS LATCHUP

The basic CMOS structure has a potentially destructive failure mechanism called **latchup** that was a major concern in early implementations and helped delay adoption of the technology. By the mid-1980s, technological solutions were developed that effectively surpress the latchup phenomenon.

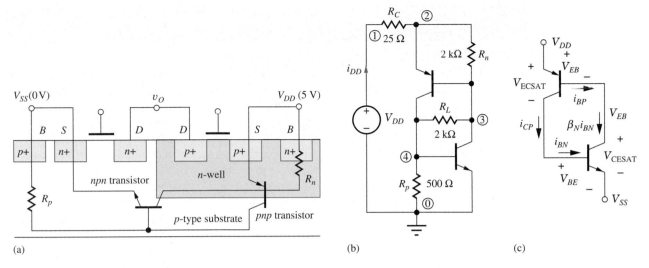

(a) (b) (c)

Figure 7.37 (a) CMOS structure with parasitic *npn* and *pnp* transistors identified. (b) Circuit including shunting resistors R_n and R_p for SPICE simulation. See text for description of R_L. (c) Regenerative structure formed by parasitic *npn* and *pnp* transistors.

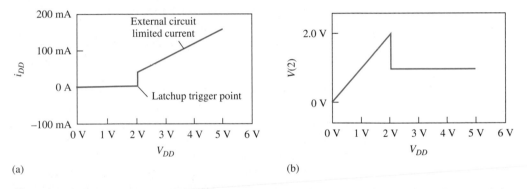

(a) (b)

Figure 7.38 SPICE simulation of latchup in the circuit of Fig. 7.37(a): (a) current from V_{DD}, (b) voltage at node 2.

However, it is important to understand the source of the problem that arises from the complex nature of the CMOS integrated structure that produces **parasitic bipolar transistors.** These bipolar transistors are normally off but can conduct under some transient fault conditions.

In the cross section of the CMOS structure in Fig. 7.37(a), a *pnp* transistor is formed by the source region of the PMOS transistor, the *n*-well, and the *p*-type substrate, and an *npn* transistor is formed by the source region of the NMOS transistor, the *p*-type substrate, and the *n*-well. The physical structure connects the *npn* and *pnp* transistors together in the equivalent circuit shown in Fig. 7.38(b). R_n and R_p model the series resistances existing between the external power supply connections and the internal base terminals of the bipolar transistors.

If the currents in R_n and R_p in the circuit model in Fig. 7.37(b) are zero, then the base-emitter voltages of both bipolar transistors are zero, and both devices are off. The total supply voltage ($V_{DD} - V_{SS}$) appears across the reverse-biased well-to-substrate junction that forms the collector-base junction of the two parasitic bipolar transistors. However, if a current should develop in the base of either the *npn* or the *pnp* transistor, latchup can be triggered, and high currents can destroy the structure. In the latchup state, the current is limited only by the external circuit components.

The problem can be more fully understood by referring to Fig. 7.37(c), in which R_n and R_p have been neglected for the moment. Suppose a base current i_{BN} begins to flow in the base of the *npn* transistor. This base current is amplified by the *npn* current gain β_N and must be supplied from

CMOS–The Enabler for Handheld Technologies

Starting with science fiction stories of the 1940s, moving on to space-going movies of the 50s and 60s, and eventually finding its way into the Star Trek voyages, powerful handheld sensing, computing, and communication technologies have been a dream of many an author. Enabling those dreams has also been the goal of many microelectronic circuit designers in the last 30 years. The beginnings of the integrated circuit era in the 1960s gave us a peek at the possibilities, but the realization was still elusive due to a lack of functional density and power consumption too high to enable battery powered operation.

In the1980s, CMOS technologies became widely available. Finally, we had a technology that promised to combine high computational power and low power consumption in a way that would lead to the handheld devices envisioned by authors years ago. As CMOS transistors continued to be scaled to smaller and smaller sizes during the 1990s, handheld devices such as cell phones, GPS receivers, and PDAs were introduced and rapidly improved from year to year.

Cell phones are highly complex devices and typically include a central processing unit (CPU), memory, input/output circuits, RF transceiver, power management, touchscreen, liquid-crystal display, and various optional modules such as Bluetooth radio, infrared, and other interfaces. In total, a cell phone will typically contain millions of transistors. The extremely low standby current of CMOS logic enables the designer to build complex capabilities that only dissipate significant power when activated.

Throughout this chapter, we have assumed that a CMOS transistor has zero current when the gate-source voltage is less than the threshold voltage. This property is responsible for the extremely low standby-power dissipation of CMOS logic. However, in practice, the current is not exactly zero; a small sub-threshold current (approximately $10^{-11} - 10^{-9}$ amps) flows when the device is OFF (See Fig. 4.32). This would seem to be an insignificant current, but when multiplied by gate counts of several million, it can add up to a sizable limitation on battery lifetime. Modern sub 100 nm gate-length processes have such a large sub-threshold leakage current that designers are being forced to invent new logic forms to reduce the leakage cur-rents, and processes are being modified to incorporate multiple thresholds to better control the leakage current power dissipation. These new CMOS logic forms build on the logic topologies developed in this chapter and are a topic of advanced VLSI design courses.

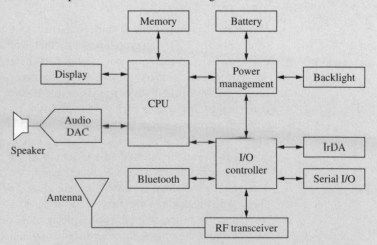

(a) Verizon Wireless Treo 700p Application Screen. Copyright © 2006 Palm, Inc. All Rights Reserved.

(b) PDA Block Diagram

the base of the *pnp* transistor. The *pnp* base current is then amplified further by the current gain β_P of the *pnp* transistor, yielding a collector current equal to

$$i_{CP} = \beta_P i_{BP} = \beta_P(\beta_N i_{BN}) \tag{7.34}$$

However, the *pnp* collector current i_{CP} is also equal to the *npn* base current i_{BN}. If the product of the two current gains $\beta_P \beta_N$ exceeds unity, then all the currents will grow without bound. This situation is called *latchup*. Once the circuit has entered the latchup state, both transistors enter their low impedance state, and the voltage across the structure collapses to one diode drop plus one emitter collector voltage:

$$V = V_{EB} + V_{\text{CESAT}} = V_{BE} + V_{\text{ECSAT}} \tag{7.35}$$

Shunting resistors R_n and R_p shown in Fig. 7.37 actually play an important role in determining the latchup conditions in a real CMOS structure. As mentioned before, latchup would not occur in an ideal structure for which $R_n = 0 = R_p$, and modern CMOS technology uses special substrates and processing to minimize the values of these two resistors.

The results of SPICE simulation of the behavior of the circuit in Fig. 7.37(b) for representative circuit elements are presented in Fig. 7.38. Resistor R_L is added to the circuit to provide a leakage path across the collector-base junctions to initiate the latchup phenomenon in the simulation. Prior to latchup in Fig. 7.38, all currents are very small, and the voltage at node 2 is directly proportional to the input voltage V_{DD}. At the point that latchup is triggered, the voltage across the CMOS structure collapses to approximately 0.8 V, and the current increases abruptly to $(V_{DD} - 0.8)/R_C$. The current level is limited only by the external circuit component values. Large currents cause high power dissipation that can rapidly destroy most CMOS structures.

Under normal operating conditions, latchup does not occur. However, if a fault or transient occurs that causes one of the source or drain diffusions to momentarily exceed the power supply voltage levels, then latchup can be triggered. Ionizing radiation or intense optical illumination are two other possible sources of latchup initiation.

Note that this section actually introduced another form of modeling. Figure 7.37(a) is a cross section of a complex three-dimensional distributed structure, whereas Figs. 7.37(b) and 7.37(c) are attempts to represent or model this complex structure using a simplified network of discrete transistors and resistors. Note, too, that Fig.7.37(b) is only a crude model of the real situation, so significant deviations between model predictions and actual measurements should not be surprising. It is easy to forget that circuit schematics generally represent only idealized models for the behavior of highly complex circuits.

SUMMARY

- *CMOS inverters:* In this chapter, we discussed the design of CMOS logic circuits beginning with the design of a reference inverter. The shape of the voltage transfer characteristic (VTC) of the CMOS inverter is almost independent of power supply voltage, and the noise margins of a symmetrical inverter can approach one-half the power supply voltage. The design of the W/L ratios of the transistors in a CMOS gate is determined primarily by the desired propagation delay τ_P, which is related directly to the device parameters K'_n, V_{TN}, K'_p, and V_{TP}, and the total load capacitance C.
- *CMOS logic gates:* In CMOS logic, each gate contains both an NMOS and a PMOS switching network, and every logical input is connected to at least one NMOS and one PMOS transistor. NAND gates, NOR gates, and complex CMOS logic gates can all be designed using the reference inverter concept, similar to that introduced in Chapter 6. As for NMOS circuitry, complex CMOS gates can directly implement Boolean logic equations expressed in a sum-of-products form. In contrast to NMOS logic, which has highly asymmetric rise and fall times, symmetrical inverters

in which t_f and t_r are equal can easily be designed in CMOS, although there can be a significant area penalty. A number of examples of styles for the layout of CMOS inverters and more complex logic gates were presented.

- *Body effect:* Body effect has a smaller influence on CMOS design than on NMOS design because the source-bulk voltages of all the transistors in a CMOS gate become zero in the steady state. However, the source-bulk voltages are nonzero during switching transients, and the body effect degrades the rise and fall times and propagation delays of CMOS logic.

- *Dynamic power dissipation and power delay product:* Except for very low power applications, CMOS power dissipation is determined by the energy required to charge and discharge the effective load capacitance at the desired switching frequency. A simple expression for the power-delay product of CMOS was developed. For a given capacitive load, the power and delay of the CMOS gate may be scaled up or down by simply modifying the W/L ratios of the NMOS and PMOS transistors.

- *Static power dissipation:* For low-power applications, particularly where battery life is important, leakage current from the reverse-biased wells and drain-substrate junctions can become an important source of power dissipation. This leakage current places a lower bound on the power required to operate a CMOS circuit.

- *"Short circuit" current:* During switching of the CMOS logic gate, a pulse of current occurs between the positive and negative power supplies. This current causes an additional component of power dissipation in the CMOS gate that can be as much as 20 to 30 percent of the dissipation resulting from charging and discharging the load capacitance.

- *Cascade buffers:* High capacitance loads are often encountered in logic design, and cascade buffers are used to minimize the propagation delay associated with driving these large capacitance values. Cascade buffers are widely used in wordline drivers and for on-chip and off-chip bus driver applications.

- *Dynamic logic:* Dynamic logic circuits, such as domino CMOS, operate on two phases—a precharge phase and a logic evaluation phase. This circuit family requires only a single PMOS transistor in each gate (plus an output inverter), thus reducing the silicon area overhead traditionally associated with static CMOS logic circuits. Dynamic circuits are also used to reduce power consumption in many applications.

- *The CMOS transmission gate:* A bidirectional circuit element, the CMOS transmission gate that utilizes the parallel connection of an NMOS and a PMOS, transistor, was introduced. When the transmission gate is on, it provides a low-resistance connection between its input and output terminals over the entire input voltage range. We will find the transmission gate used in circuit implementations of both the D latch and the master-slave D flip-flop in Chapter 8.

- *Latchup:* An important potential failure mechanism in CMOS is the phenomenon called latchup, which is caused by the existence of parasitic *npn* and *pnp* bipolar transistors in the CMOS structure. A lumped circuit model for latchup was developed and used to simulate the latchup behavior of a CMOS inverter. Special substrates and IC processing are used to minimize the possibility of latchup in modern CMOS technologies.

KEY TERMS

Cascade buffer	Dynamic logic
CMOS transmission gate	Euler path
CMOS reference inverter	Evaluation phase
Complementary MOS (CMOS)	Fall time
Complex logic gate	Latchup
Domino CMOS	NAND gate

NOR gate
On-resistance
Parasitic bipolar transistors
Power-delay product
Precharge phase

Propagation delay
Rise time
Symmetrical CMOS inverter
Taper factor
Transmission gate

REFERENCES

1. F. M. Wanlass and C. T. Sah, "Nanowatt logic using field-effect metal-oxide-semiconductor triodes," *IEEE International Solid-State Circuits Conference Digest,* vol. VI, pp. 32–33, February 1963.
2. J. D. Meindl and J. A. Davis, "The fundamental limit on binary switching energy for terascale integration (TSI)," *IEEE Journal of Solid-State Circuits,* vol. 35, no. 10, pp. 1515–1516, October 2000.
3. R. M. Swanson and J. D. Meindl, "Ion-implanted complementary MOS transistors in low-voltage circuits," *IEEE Journal of Solid-State Circuits,* vol. SC-7, no. 2, pp. 146–153, April 1972.
4. H. C. Lin and L. W. Linholm, "An optimized output stage for MOS integrated circuits," *IEEE Journal of Solid-State Circuits,* vol. 10, pp. 106–109, April 1975.
5. R. C. Jaeger, "Comments on An optimized output stage for MOS integrated circuits," *IEEE Journal of Solid-State Circuits,* vol. 10, pp. 185–186, June 1975.

PROBLEMS

Use $K_n' = 100$ μA/V^2, $K_p' = 40$ μA/V^2, $V_{TN} = 0.6$ V, and $V_{TP} = -0.6$ V unless otherwise indicated. *For simulation purposes, use the values in Appendix B.*

7.1 CMOS Inverter Technology

7.1. Calculate the values of K_n' and K_p' for NMOS and PMOS transistors with a gate oxide thickness of 100 Å. Assume that $\mu_n = 500$ cm^2/V · s, $\mu_p = 200$ cm^2/V · s, and the relative permittivity of the gate oxide is 3.9. ($\varepsilon_0 = 8.854 \times 10^{-14}$ F/cm).

7.2. Draw a cross section similar to that in Fig. 7.1 for a CMOS process that uses a p-well instead of an n-well. Show the connections for a CMOS inverter, and draw an annotated version of the corresponding circuit schematic. (*Hint:* Start with an n-type substrate and interchange all the n- and p-type regions.)

*7.3. (a) The n-well in a CMOS process covers an area of 1 cm × 0.5 cm, and the junction saturation current density is 500 pA/cm^2. What is the total leakage current of the reverse-biased well? (b) Suppose the drain and source regions of the NMOS and PMOS transistors are 2 μm × 5 μm, and the saturation current density of the junctions is 100 pA/cm^2. If the chip has 20 million inverters, what is the

total leakage current due to the reverse-biased junctions when $v_O = 2.5$ V? Assume $V_{DD} = 2.5$ V and $V_{SS} = 0$ V. (c) When $v_O = 0$ V?

*7.4. A particular interconnection between two logic gates in an IC chip runs one-half the distance across a 10-mm-wide die. If the line is 1 μm wide and the oxide ($\varepsilon_r = 3.9$, $\varepsilon_0 = 8.854 \times 10^{-14}$ F/cm) beneath the line is 1 μm thick, what is the total capacitance of this line, assuming that the capacitance is three times that predicted by the parallel plate capacitance formula? Assume that the silicon beneath the oxide represents a conducting ground plane.

7.5. The CMOS inverter in Fig. P7.5 has $V_{DD} = 2.5$ V and $V_{SS} = 0$ V. What are the values of V_H and V_L for this inverter? (b) Repeat for $V_{DD} = 1.8$ V.

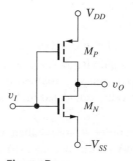

Figure P7.5

7.6. The CMOS inverter in Fig. P7.5 has $V_{DD} = 2.5$ V, $V_{SS} = 0$ V, $(W/L)_N = 4/1$, and $(W/L)_P = 10/1$. What are the values of V_H and V_L for this inverter? (b) Repeat for $(W/L)_N = 4/1$ and $(W/L)_P = 4/1$.

7.7. The CMOS inverter in Fig. P7.5 has $V_{DD} = 3.3$ V, $V_{SS} = 0$ V, $(W/L)_N = 4/1$, and $(W/L)_P = 10/1$. What are the values of V_H and V_L for this inverter? (b) Repeat for $(W/L)_N = 6/1$ and $(W/L)_P = 15/1$.

7.8. The CMOS inverter in Fig. P7.5 has $V_{DD} = 2.5$ V and $V_{SS} = 0$ V. If $V_{TN} = 0.60$ V and $V_{TP} = -0.60$ V, what are the regions of operation of the transistors for (a) $V_I = V_L$? (b) $V_I = V_H$? (c) $V_I = V_O = 1.25$ V?

7.9. The CMOS inverter in Fig. P7.5 has $V_{DD} = 3.3$ V and $V_{SS} = 0$ V. If $V_{TN} = 0.75$ V and $V_{TP} = -0.75$ V, what are the regions of operation of the transistors for (a) $V_I = V_L$? (b) $V_I = V_H$? (c) $V_I = V_O = 1.65$ V?

7.10. (a) The CMOS inverter in Fig. P7.5 with $(W/L)_N = 20/1$ and $(W/L)_P = 50/1$ is operating with $V_{DD} = 0$ V and $-V_{SS} = -5.2$ V. What are V_L and V_H? (b) If $(W/L)_N = 10/1$ and $(W/L)_P = 10/1$?

7.11. (a) Calculate the voltage at which $v_O = v_I$ for a CMOS inverter with $K_n = K_p$. (*Hint:* Always remember that $i_{DN} = i_{DP}$.) Use $V_{DD} = 2.5$ V, $V_{TN} = 0.6$ V, $V_{TP} = -0.6$ V. (b) What is the current I_{DD} from the power supply for $v_O = V_I$ if $(W/L)_N = 2/1$? (c) Repeat the calculation in (a) for a CMOS inverter with $K_n = 2.5K_p$. (d) What is the current I_{DD} from the power supply for $v_O = V_I$ if $(W/L)_N = 2/1$?

7.12. (a) Repeat Prob. 7.11 for $V_{DD} = 3.3$ V, $V_{TN} = 0.75$ V, and $V_{TP} = -0.75$ V. (b) Repeat Prob. 7.11 for $V_{DD} = 2.5$ V, $V_{TN} = 0.60$ V, and $V_{TP} = -0.50$ V.

7.13. (a) Repeat Prob. 7.11 for $V_{DD} = 1.8$ V, $V_{TN} = 0.5$ V and $V_{TP} = -0.5$ V. (b) Repeat Prob. 7.11 for $V_{DD} = 2.5$ V, $V_{TN} = 0.75$ V and $V_{TP} = -0.65$ V. (c) Repeat Prob. 7.11 for $V_{DD} = 2.5$ V, $V_{TN} = 0.65$ V and $V_{TP} = -0.75$ V.

7.2 Static Characteristics of the CMOS Inverter

7.14. Simulate the VTC for a CMOS inverter with $K_n = 2.5K_p$. Find the input voltage for which $v_O = v_I$ and compare to the value calculated by hand. Use $V_{DD} = 2.5$ V.

7.15. (a) The CMOS gate in Fig. P7.15 is called pseudo-NMOS. Find V_H and V_L for this gate. (b) Repeat for $V_{DD} = 2.5$ V.

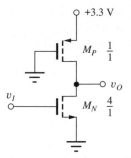

Figure P7.15

**7.16. A CMOS inverter is to be designed to drive a single TTL inverter (which will be studied in Chapter 9). When $v_O = V_L$, the CMOS inverter must sink a current of 1.5 mA and maintain $V_L \leq 0.6$ V. When $v_O \geq V_H$, the CMOS inverter must source a current of 60 μA and maintain $V_H \geq 2.4$ V. What are the minimum W/L ratios of the NMOS and PMOS transistors required to meet these specifications? Assume $V_{DD} = 5$ V.

7.17. The outputs of two CMOS inverters are accidentally tied together, as shown in Fig. P7.17. What is the voltage at the common output node if the NMOS and PMOS transistors have W/L ratios of 20/1 and 40/1, respectively? What is the current in the circuit?

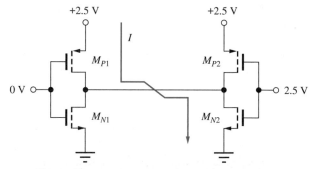

Figure P7.17

7.18. What are the noise margins of a minimum size CMOS inverter in which both W/L ratios are 2/1 and $V_{DD} = 2.5$ V and $V_{TN} = -V_{TP} = 0.6$ V?

7.19. What are the noise margins for a symmetrical CMOS inverter operating with $V_{DD} = 3.3$ V and $V_{TN} = -V_{TP} = 0.75$ V? (b) Repeat for a CMOS inverter having $(W/L)_N = (W/L)_P$ operating with $V_{DD} = 3.3$ V and $V_{TN} = -V_{TP} = 0.75$ V.

7.20. Use SPICE to plot the VTC for a CMOS inverter with $(W/L)_N = 2/1$, $(W/L)_P = 5/1$, $V_{DD} = 3.3$ V, $V_{SS} = 0$ V, $V_{TN} = 0.75$ V, and

$V_{TP} = -0.75$ V. Repeat if the threshold voltages are mismatched with values $V_{TN} = 0.85$ V and $V_{TP} = 0.65$ V. Repeat for $(W/L)_N = 2/1$ and $(W/L)_P = 4/1$ with the original threshold voltages. Plot the three curves on one graph.

7.21. (a) Plot a graph of the noise margins of the CMOS inverter (similar to Fig. 7.9) for $V_{DD} = 3.3$ V, $V_{TN} = 0.75$ V, and $V_{TP} = -0.75$ V. (b) Repeat for $V_{DD} = 2.0$ V, $V_{TN} = 0.50$ V, and $V_{TP} = -0.50$ V.

7.3 Dynamic Behavior of the CMOS Inverter

7.22. (a) What are the rise time, fall time, and average propagation delay for a symmetrical CMOS inverter with $(W/L)_N = 2/1$, $(W/L)_P = 5/1$, $V_{DD} = 2.5$ V, and $C = 0.25$ pF? (b) Repeat for $V_{DD} = 2.0$ V. (c) Repeat for $V_{DD} = 1.8$ V.

7.23. What are the rise time, fall time, and average propagation delay for a minimum size CMOS inverter in which both W/L ratios are 2/1? Assume a load capacitance of 0.5 pF and $V_{DD} = 2.5$ V.

7.24. What are the rise time, fall time, and average propagation delay for a symmetrical CMOS inverter with $(W/L)_N = 2/1$, $(W/L)_P = 5/1$, $C = 0.15$ pF, $V_{DD} = 2.5$ V, $V_{TN} = 0.60$ V, and $V_{TP} = -0.60$ V?

7.25. What are the rise time, fall time, and average propagation delay for a symmetrical CMOS inverter with $(W/L)_N = 2/1$, $(W/L)_P = 5/1$, $C = 0.20$ pF, $V_{DD} = 3.3$ V, $V_{TN} = 0.75$ V, and $V_{TP} = -0.75$ V?

7.26. What are the sizes of the transistors in the CMOS inverter if it must drive a 1-pF capacitance with an average propagation delay of 3 ns? Design the inverter for equal rise and fall times. Use $V_{DD} = 2.5$ V, $V_{TN} = 0.6$ V, $V_{TP} = -0.6$ V.

7.27. Design a symmetrical CMOS reference inverter to provide a delay of 1 ns when driving a 10-pF load. (a) Assume $V_{DD} = 2.5$ V. (b) Assume $V_{DD} = 3.3$ V and $V_{TN} = -V_{TP} = 0.75$ V.

7.28. Design a symmetrical CMOS reference inverter to provide a propagation delay of 200 ps for a load capacitance of 100 fF. Use $V_{DD} = 1.5$ V, $V_{TN} = 0.50$ V, and $V_{TP} = -0.50$ V.

7.29. Design a symmetrical CMOS reference inverter to provide a propagation delay of 400 ps for a load capacitance of 100 fF. Use $V_{DD} = 2.5$ V, $V_{TN} = 0.60$ V, and $V_{TP} = -0.60$ V.

*7.30. Use SPICE to determine the characteristics of the CMOS inverter for the design given in Fig. 7.12 if

$C = 100$ fF. (a) Simulate the voltage transfer function. (b) Determine t_r, t_f, τ_{PHL}, and τ_{PLH} for this inverter with a square wave input. What must be the total effective load capacitance C based on the propagation delay formula developed in the text?

**7.31. Use SPICE to simulate the behavior of a chain of five CMOS inverters similar to those in Fig. 7.13(b). The input to the first inverter should be a square wave with 0.1-ns rise and fall times and a period of 100 ns. (a) Calculate t_r, t_f, τ_{PHL}, and τ_{PLH} using the input and output waveforms from the first inverter in the chain and compare your results to the formulas developed in the text. (b) Determine t_r, t_f, τ_{PHL}, and τ_{PLH} from the waveforms at the input and output of the fourth inverter in the chain, and compare your results to the formulas developed in the text. (c) Discuss the differences between the results in (a) and (b).

7.4 Power Dissipation and Power Delay Product in CMOS

7.32. A high-performance CMOS microprocessor design requires 100 million logic gates and will be placed in a package that can dissipate 100 W. (a) What is the average power that can be dissipated by each logic gate on the chip? (b) If a supply voltage of 1.8 V is used, what is the average current that must be supplied to the chip?

7.33. A certain packaged IC chip can dissipate 5 W. Suppose we have a CMOS IC design that must fit on one chip and requires 2 million logic gates. What is the average power that can be dissipated by each logic gate on the chip? If the average gate must switch at 5 MHz, what is the maximum capacitive load on a gate for $V_{DD} = 3.3$ V? For $V_{DD} = 2.5$ V?

7.34. (a) The n-well in a CMOS process covers an area of 5 mm × 10 mm, and the saturation current density of the junction is 400 pA/cm^2. What is the total leakage current of the reverse-biased well? (b) Suppose the drain and source regions of the NMOS and PMOS transistors are each 1 μm × 2.5 μm in size, and the saturation current density of the junctions is 150 pA/cm^2. If the chip has 75 million inverters, what is the total leakage current when $v_O = 2.5$ V? Assume $V_{DD} = 2.5$ V. (b) Repeat for $v_O = 0$ V.

7.35. A high-speed CMOS microprocessor has a 64-bit address bus and performs a memory access every 10 ns. Assume that all address bits change during every memory access and that each bus line represents a load of 25 pF. (a) How much power is being dissipated by the circuits that are driving these signals if the power supply is 2.5 V? (b) Repeat for 3.3 V.

*7.36. (a) A CMOS inverter has $(W/L)_N = 20/1$, $(W/L)_P = 20/1$, and $V_{DD} = 3.3$ V. What is the peak current in the logic gate and at what input voltage does it occur? (b) Repeat for $V_{DD} = 2.5$ V.

*7.37. (a) A CMOS inverter has $(W/L)_N = 2/1$, $(W/L)_P = 5/1$, and $V_{DD} = 3.3$ V. Assume $V_{TN} = -V_{TP} = 0.7$ V. What is the peak current in the logic gate and at what input voltage does it occur? (b) Repeat for $V_{DD} = 2.0$ V with $V_{TN} = -V_{TP} = 0.5$ V.

7.38. (a) Repeat Prob. 7.37(a) for $V_{DD} = 2.5$ V. (b) Repeat for $V_{DD} = 2.5$ V, $V_{TN} = 0.65$ V, and $V_{TP} = -0.55$ V.

7.39. (a) Repeat Prob. 7.37(a) for $V_{DD} = 2.0$ V, $V_{TN} = 0.55$ V, and $V_{TP} = -0.45$ V. (b) Repeat Prob. 7.37(a) for $V_{DD} = 2.0$ V, $V_{TN} = 0.45$ V, and $V_{TP} = -0.55$ V.

7.40. What is the power-delay product for the inverter in Prob. 7.22? How much power does the inverter dissipate if it is switching at a frequency of 100 MHz?

7.41. (a) What is the power-delay product for the inverter in Prob. 7.25? (b) Estimate the maximum switching frequency for this inverter. (c) How much power does the inverter dissipate if it is switching at the frequency found in (b)?

7.42. (a) What is the power-delay product for the inverter in Prob. 7.24? (b) Estimate the maximum switching frequency for this inverter. (c) How much power does the inverter dissipate if it is switching at the frequency found in (b)?

7.43. Plot the power-delay characteristic for the CMOS inverter family based on an inverter design in which $(W/L)_N = (W/L)_P$. Assume the load capacitance $C = 0.2$ pF. Use $V_{DD} = 2.5$ V and vary the power by changing the W/L ratios.

**7.44. Ideal constant-electric-field scaling of a MOS technology reduces all the dimensions and voltages by the same factor α. Assume that the capacitor C in Eq. (7.26) is proportional to the total gate capacitance of the MOS transistor: $C = C''_{ox} WL$, and show that constant-field scaling results in a reduction of the PDP by a factor of α^3.

**7.45. For many years, MOS technology was scaled by reducing all the dimensions by the same factor α, but keeping the voltages constant. Assume that

the capacitor C in Eq. (7.26) is proportional to the total gate capacitance of the MOS transistor: $C = C''_{ox} WL$, and show that this geometry scaling results in a reduction of the PDP by a factor of α.

**7.46. Use SPICE to simulate the behavior of a chain of five CMOS inverters with the same design as in Fig. 7.12 with $C = 0.25$ pF. The input to the first inverter should be a square wave with 0.1-ns rise and fall times and a period of 30 ns. (a) Calculate t_r, t_f, τ_{PHL}, and τ_{PLH} using the input and output waveforms from the first inverter in the chain, and compare your results to the formulas developed in the text. (b) Determine t_r, t_f, τ_{PHL}, and τ_{PLH} from the waveforms at the input and output of the fourth inverter in the chain, and compare your results to the formulas developed in the text. (c) Discuss the differences between the results in (a) and (b).

**7.47. Use SPICE to simulate the behavior of a chain of five CMOS inverters with the same design as in Fig. 7.12 with $C = 1$ pF. The input to the first inverter should be a square wave with 0.1-ns rise and fall times and a period of 40 ns. (a) Determine t_r, t_f, τ_{PHL}, and τ_{PLH} from the waveforms at the input and output of the fourth inverter in the chain, and compare your results to the formulas developed in the text. (b) Repeat the simulation for $(W/L)_P = (W/L)_N = 2/1$, and compare the results to those obtained in (a).

7.5 CMOS NOR and NAND Gates

7.48. (a) Draw the circuit schematic for a four-input NAND gate. What are the W/L ratios of the transistors based on the reference inverter design in Fig. 7.12? (b) What should be the W/L ratios if the NOR gate must drive three times the load capacitance with the same delay as the reference inverter?

7.49. (a) Draw the circuit schematic for a four-input NOR gate. What are the W/L ratios of the transistors based on the reference inverter design in Fig. 7.12? (b) What should be the W/L ratios if the NOR gate must drive twice the load capacitance with the same delay as the reference inverter?

7.50. Draw the circuit schematic of a four-input NOR gate. Suppose the PMOS transistors are chosen to have $(W/L)_P = 2/1$. What are the corresponding W/L ratios of the NMOS devices, if the gate is to have symmetrical delay characteristics?

7.51. Draw the circuit schematic of a three-input NOR gate. Suppose the PMOS transistors are chosen to have $(W/L)_P = 2/1$. What are the corresponding W/L ratios of the NMOS devices, if the gate is to have symmetrical delay characteristics?

7.52. Draw the circuit schematic of a three-input NAND gate. Suppose the NMOS transistors are chosen to have $(W/L)_N = 2/1$. What are the corresponding W/L ratios of the PMOS devices, if the gate is to have symmetrical delay characteristics?

7.53. Draw the circuit schematic of a four-input NAND gate. Suppose the NMOS transistors are chosen to have $(W/L)_N = 2/1$. What are the corresponding W/L ratios of the PMOS devices, if the gate is to have symmetrical delay characteristics?

7.54. Design a circuit to multiply two one-bit numbers. (*Hint:* Construct a truth table for output bit M based on two inputs A and B.) Choose the W/L ratios based on the inverter in Fig. 7.12.

**7.55. Use SPICE to determine the characteristics of the two-input CMOS NOR gate given in Fig. 7.19 with a load capacitance of 1 pF. Assume that $\gamma = 0$ for all transistors. (a) Simulate the voltage transfer function by varying the voltage at input A with the voltage at input B fixed at 2.5 V. (b) Repeat the simulation in (a) but now vary the voltage at input B with the voltage at input A fixed at 2.5 V. Plot the results from (a) and (b) and note any differences. (c) Determine t_r, t_f, τ_{PHL}, and τ_{PLH} for this inverter with a square wave input at input A with the voltage at input B fixed at 2.5 V. (d) Determine t_r, t_f, τ_{PHL}, and τ_{PLH} for this inverter with a square wave input at input B with the voltage at input A fixed at 2.5 V. (e) Compare the results from (c) and (d). (f) Determine t_r, t_f, τ_{PHL}, and τ_{PLH} for this inverter with a single square wave input applied to both inputs A and B. Compare the results to those in (c) and (d).

**7.56. Repeat (a) and (b), Prob. 7.55, using the nonzero values for the parameter γ from the device parameter tables.

**7.57. Use SPICE to determine the characteristics of the two-input CMOS NAND gate given in Fig. 7.23 with a load capacitance of 1 pF. Assume that $\gamma = 0$ for all transistors. (a) Simulate the voltage transfer function by varying the voltage at input A with the voltage at input B fixed at 2.5 V. (b) Repeat the simulation in (a) but now vary the voltage at input B with the voltage at input A fixed at 2.5 V. Plot the results from (a) and (b) and note any differences. (c) Determine t_r, t_f, τ_{PHL}, and τ_{PLH} for this inverter with a square wave input at input A with the voltage

at input B fixed at 2.5 V. (d) Determine t_r, t_f, τ_{PHL}, and τ_{PLH} for this inverter with a square wave input at input B with the voltage at input A fixed at 2.5 V. (e) Compare the results from (c) and (d). (f) Determine t_r, t_f, τ_{PHL}, and τ_{PLH} for this inverter with a single square wave input applied to both inputs A and B. Compare the results to those in (c) and (d).

**7.58. Repeat (a) and (b), Prob. 7.57, using the nonzero values for the parameter γ from the device parameter tables.

7.6 Design of Complex Gates in CMOS

7.59. What are the worst case rise and fall times and average propagation delays of the CMOS gate in Fig. 7.26(b) for a load capacitance of 1.25 pF?

**7.60. (a) How many transistors are needed to implement the CMOS gate in Fig. 7.29 using depletion-mode NMOS? (b) Compare the total gate area of the CMOS and NMOS designs if they are both designed for a 10-ns average propagation delay for a load capacitance of 1 pF.

7.61. (a) What is the logic function implemented by the gate in Fig. P7.61? (b) Design the PMOS transistor network. Select the device sizes for both the NMOS and PMOS transistors to give a delay similar to that of the CMOS reference inverter.

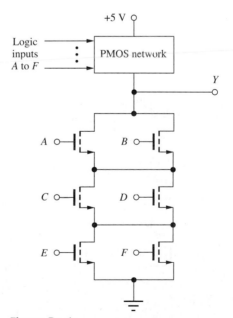

Figure P7.61

7.62. (a) What is the logic function implemented by the gate in Fig. P7.62? (b) Design the PMOS transistor network. Select the device sizes for both the NMOS and PMOS transistors to give a delay of approximately one-third the delay of the CMOS reference inverter.

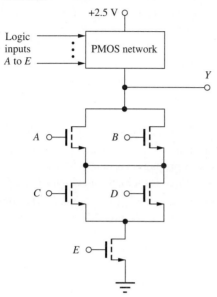

Figure P7.62

7.63. (a) What is the logic function implemented by the gate in Fig. P7.63? (b) Design the PMOS transistor network. Select the device sizes for both the NMOS and PMOS transistors to give a delay of approximately one-half the delay of the CMOS reference inverter.

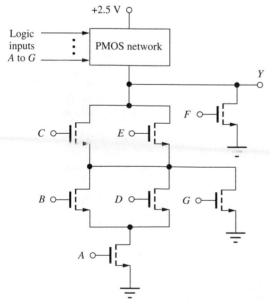

Figure P7.63

7.64. (a) What is the logic function implemented by the gate in Fig. P7.64? (b) Design the NMOS transistor network. Select the device sizes for both the NMOS and PMOS transistors to give a delay similar to that of the CMOS reference inverter.

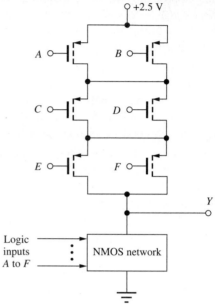

Figure P7.64

7.65. (a) What is the logic function implemented by the gate in Fig. P7.65? (b) Design the NMOS transistor network. Select the device sizes for both the NMOS and PMOS transistors to give a delay of approximately one-third the delay of the CMOS reference inverter.

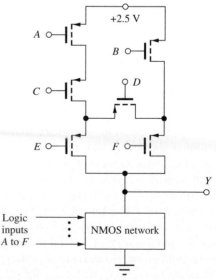

Figure P7.65

7.66. Draw the logic diagram and transistor implementation for a (2-3-1) AOI gate. Use the graphical approach to design the PMOS network. Choose the size of the transistors based upon the reference inverter in Fig. 7.12.

7.67. Draw the logic diagram and transistor implementation for a (3-2-3-1) AOI gate. Use the graphical approach to design the PMOS network. Choose the size of the transistors based upon the reference inverter in Fig. 7.12.

7.68. (a) Draw the NMOS and PMOS graphs for the (2-2-1) AOI in the Electronics in Action figure on page 377. (b) Find an Euler path for this circuit if it exists. (c) Draw the NMOS and PMOS graphs for a (2-2-2) AOI. (d) Find an Euler path for part (d) if it exists.

7.69. Redraw Fig. 7.28 and highlight the conducting path(s) for the following sets of inputs for ABCDE: (a) 10011, (b) 10001, (c) 11101, (d) 00010.

7.70. Draw the circuit for Prob. 7.61 and highlight the conducting path(s) for the following sets of inputs for ABCDEF: (a) 100110, (b) 011001, (c) 010101, (d) 110011.

7.71. Design a CMOS logic gate that implements the logic function $Y = \overline{ABC + DE}$, based on the CMOS reference inverter. Select the transistor sizes to give the same delay as that of the reference inverter if the load capacitance is the same as that of the reference inverter.

7.72. Design a CMOS logic gate that implements the logic function $Y = \overline{A(BC + DE)}$ and is twice as fast as the CMOS reference inverter when loaded by a capacitance of $2C$.

7.73. Design a CMOS logic gate that implements the logic function $Y = \overline{A(B + C(D + E))}$, based on the CMOS reference inverter. Select the transistor sizes to give the same delay as that of the reference inverter if the load capacitance is the same as that of the reference inverter.

7.74. Design a CMOS logic gate that implements the logic function $Y = \overline{A(B + CD) + E}$ and has the same logic delay as the CMOS reference inverter when driving a capacitance of $4C$.

7.75. Design a complex gate implementation of a one-bit half adder for which the sum bit is described by $S = X \oplus Y$, and the carry bit is given by $C = A \cdot B$. Choose the W/L ratios based on the reference inverter design in Fig. 7.12. Assume that true and complement values of each variable are available as inputs. (*Note:* Two gate designs are needed, one for S and one for C.)

7.76. Design a complex gate implementation of a 1-bit full adder for which the ith sum bit is described by $S_i = X_i \oplus Y_i \oplus C_{i-1}$, and the ith carry bit is given by $C_i = X_i \cdot Y_i + X_i \cdot C_{i-1} + Y_i \cdot C_{i-1}$. Choose the W/L ratios based upon the reference inverter design in Fig. 7.12. Assume that true and complement values of each variable are available as inputs. (*Note:* Two gate designs are needed, one for S_i and one for C_i.)

7.77. Design a complex gate implementation of a 2-bit parallel multiplier. [*Note:* The circuit should produce a 4-bit output (e.g., $11_2 \times 11_2 = 1001_2$), and a separate circuit should be designed for each output bit.] Choose the W/L ratios based on the inverter in Fig. 7.12.

7.7 Minimum Size Gate Design and Performance

7.78. The three-input NOR gate in Fig. 7.21 is implemented with transistors all having $W/L = 2/1$. What is the propagation delay for this gate for a load capacitance $C = 400$ fF? Assume $V_{DD} = 2.5$ V. What would be the delay of the reference inverter for $C = 400$ fF?

7.79. The five-input NAND gate in Fig. 7.24 is implemented with transistors all having $W/L = 2/1$. What is the propagation delay for this gate for a load capacitance $C = 180$ fF? Assume $V_{DD} = 2.5$ V. What would be the delay of the reference inverter for $C = 180$ fF?

7.80. A (2-2-2) AOI is implemented with transistors all having $W/L = 2/1$. What are the worst-case values of τ_{PLH} and τ_{PHL} if $V_{DD} = 2.5$ V and $C = 250$ fF?

7.81. A (2-3-1) AOI is implemented with transistors all having $W/L = 2/1$. What are the worst-case values of τ_{PLH} and τ_{PHL} if $V_{DD} = 2.5$ V and $C = 400$ fF?

7.82. What are the worst-case values of τ_{PHL} and τ_{PLH} for the gate in Fig. 7.28 when it is implemented using only 2/1 transistors and drives a load capacitance of 1 pF? Assume $V_{DD} = 2.5$ V.

7.83. What is the worst-case value of τ_{PHL} for the gate in Fig. P7.61 when it is implemented using only 2/1 transistors and drives a load capacitance of 1 pF? Assume $V_{DD} = 2.5$ V.

7.84. (a) Use a transient simulation in SPICE to find the average propagation delay of a cascade connection of 10 minimum size inverters ($W/L = 2/1$) in series. Assume each has a capacitive load C of 200 fF and $V_{DD} = 2.5$ V. (b) Repeat for a cascade of 10 symmetrical reference inverters with the same design as in Fig. 7.12, and compare the average propagation delays.

7.8 Dynamic Domino CMOS Logic

7.85. (a) Draw the circuit schematic for a two-input domino CMOS OR gate. Assume that true and complement values of each variable are available as inputs. (b) Repeat for a two-input domino CMOS AND gate.

7.86. (a) Draw the circuit schematic for a two-input domino CMOS NOR gate. Assume that true and complement values of each variable are available as inputs. (b) Repeat for a two-input domino CMOS NAND gate.

7.87. (a) Draw the circuit schematic for a three-input domino CMOS OR gate. Assume that true and complement values of each variable are available as inputs. (b) Repeat for a three-input domino CMOS AND gate.

7.88. (a) Draw the circuit schematic for a three-input domino CMOS NOR gate. Assume that true and complement values of each variable are available as inputs. (b) Repeat for a three-input domino CMOS NAND gate.

7.89. Draw the circuit schematic for a (2-2-2) AOI in domino CMOS.

7.90. Draw the circuit schematic for a (3-2-1) AOI in domino CMOS.

*7.91. (a) Suppose that inputs A_0, A_1, and A_2 are all 0 in the domino CMOS gate in Fig. P7.91, and the clock has just changed to the evaluate phase. If A_0 now changes to a 1, what happens to the voltage at node B if $C_1 = 2C_2$? (*Hint:* Charge sharing occurs between C_1 & C_2.) (b) Now A_1 changes to a 1. What happens to the voltage at node B if $C_3 = C_2$? (c) If the output inverter is a symmetrical design, what is the minimum ratio of C_1/C_2 (assume $C_3 = C_2$) for which the gate maintains a valid output? $V_{DD} = 2.5$ V

7.92. Draw the mirror image of the gate in Fig. P7.91 by replacing NMOS transistors with PMOS transistors and vice versa. Assume the logic inputs remain the same and write an expression for the logic function Z.

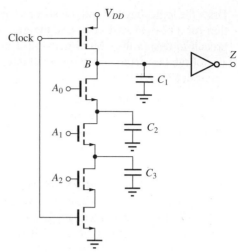

Figure P7.91

7.93. Draw the circuit schematic for a domino CMOS gate that implements the product of sums (POS) logic function $Z = (A + B)(C + D)$. Assume that true and complement values of each variable are available as inputs.

7.94. Draw the circuit schematic for a domino CMOS gate that implements the sum of products (SOP) logic function $Z = AB + CD$. Assume that true and complement values of each variable are available as inputs.

7.95. Draw the circuit schematic for a domino CMOS gate that implements the product-of-sums (POS) logic function $Z = (A + B)(C + D)(E + F)$. Assume that true and complement values of each variable are available as inputs. (Remember DeMorgan's theorem.)

7.96. Draw the circuit schematic for a domino CMOS gate that implements the sum-of-products (SOP) logic function $Z = AB + CD + EF$. Assume that true and complement values of each variable are available as inputs. (Remember DeMorgan's theorem.)

7.9 Cascade Buffers

7.97. Design an optimized cascade buffer to drive a load capacitance of $4000C_o$. (a) What is the optimum number of stages? (b) What are the relative sizes of each inverter in the chain (see Fig. 7.33)? (c) What is the delay of the buffer in terms of τ_o?

7.98. Design an optimized cascade buffer to drive a load capacitance of 10 pF if the capacitance of the symmetrical reference inverter is 100 fF. What is the optimum number of stages? What are the relative sizes of each inverter in the chain? What is the total delay of the buffer for $V_{DD} = 2.5$ V?

7.99. Design an optimized cascade buffer to drive a load capacitance of 40 pF if the capacitance of a symmetrical reference inverter is 50 fF. What is the optimum number of stages? What are the relative sizes of each inverter in the chain? What is the total delay of the buffer for $V_{DD} = 2.5$ V?

**7.100. Assume that the area of each inverter in a cascade buffer is proportional to the taper factor β and that the unit size inverter has as area A_o. Write an expression for the total area of an N-stage cascade buffer. In the example in Fig. 7.34, buffers with $N = 6$ and $N = 7$ have approximately the same delay. Compare the area of these two buffer designs using your formula.

7.10 The CMOS Transmission Gate

7.101. (a) Calculate the on-resistance of an NMOS transistor with $W/L = 20/1$ for $V_{GS} = 2.5$ V, $V_{SB} = 0$ V, and $V_{DS} = 0$ V. (b) Calculate the on-resistance of a PMOS transistor with $W/L = 20/1$ for $V_{SG} = 2.5$ V, $V_{SB} = 0$ V, and $V_{SD} = 0$ V. (c) What do we mean when we say that a transistor is "on" even though I_D and $V_{DS} = 0$?

7.102. (a) What is the largest value of the on-resistance of a transmission gate with $W/L = 10/1$ for both transistors if the input voltage range is $0 \leq v_I \leq 1$ V and the power supply is 2.5 V? At what input voltage does it occur? (b) Repeat for $0 \leq v_I \leq 2.5$ V.

7.103. A certain analog multiplexer application requires the equivalent on-resistance R_{EQ} of a transmission gate to always be less than 250 Ω for $0 \leq v_I \leq 2.5$ V. What are the minimum values of W/L for the NMOS and PMOS transistors if $V_{TON} = 0.75$ V, $V_{TOP} = -0.75$ V, $\gamma = 0.5\sqrt{V}$, $2\phi_F = 0.6$ V, $K'_p = 40$ μA/V², and $K'_n = 100$ μA/V²?

7.104. (a) What are the voltages at the nodes in the pass-transistor networks in Fig. P7.104. For NMOS transistors, use $V_{TO} = 0.75$ V, $\gamma = 0.55\sqrt{V}$, and $2\phi_F = 0.6$ V. For PMOS transistors, $V_{TO} = -0.75$ V. (b) What would be the voltages if transmission gates were used in place of each transistor?

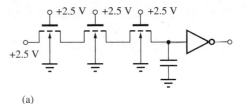

(a)

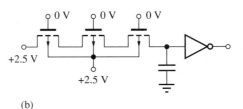

(b)

Figure P7.104

7.11 CMOS Latchup

7.105. Simulate CMOS latchup using the circuit in Fig. 7.37(b) and plot graphs of the voltages at nodes 2, 3, and 4 as well as the current supplied by V_{DD}. Discuss the behavior of the voltages and identify important voltage levels, current levels, and slopes on the graphs.

7.106. Repeat Prob. 7.105 if the values of R_n and R_p are reduced by a factor of 10.

7.107. Draw the cross section and equivalent circuit, similar to Fig. 7.37, for a p-well CMOS technology.

Additional Problems

7.108. (a) Verify Eq. (7.9). (b) Verify Eq. (7.13).

**7.109. (a) Calculate the sensitivity $S^{\tau_p}_{K_n} = (K_n/\tau_p)(d\tau_p/dK_n)$ of the propagation delay τ_p in Eq. (7.19) to changes in K_n. If the IC processing causes K_n to be 25 percent below its nominal value, what will be the percentage change in τ_p? (b) Calculate the sensitivity $S^{\tau_p}_{V_{TN}} = (V_{TN}/\tau_p)(d\tau_p/dV_{TN})$ of the propagation delay τ_p in Eq. (7.19) to changes in V_{TN}. If the IC processing causes V_{TN} to change from a nominal value of 0.75 V to 0.85 V, what will be the percentage change in τ_p?

7.110. Calculate logic delay versus input signal rise time for a minimum size inverter with a load capacitance of 1 pF for 0.1 ns $\leq t_r \leq 5$ ns.

CHAPTER 8

MOS MEMORY AND STORAGE CIRCUITS

CHAPTER OUTLINE

CHAPTER GOALS

From Chapter 8, we shall gain a basic understanding of the design of computer memory and storage circuits including

- Overall memory chip organization
- Static memory circuits using the six-transistor cell
- Dynamic memory circuits including the one-transistor and four-transistor cells
- Sense amplifier circuits required to detect the information stored in the memory cells
- Row and address decoders used to select cells from large memory arrays
- The implementation of various types of flip-flops used in CPU registers
- Pass transistor logic
- Read only memory

Robert H. Dennard, inventor of the 1-transistor DRAM cell
Courtesy of IBM Archives

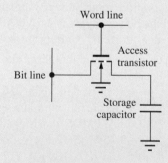

1-T DRAM cell.

For many years, high-density memory has served as the IC industry's vehicle for driving technology to ever smaller dimensions. In the mid-1960s, the first random-access memory (RAM) chip using MOS technology [1] was discussed at the IEEE International Solid-State Circuits Conference (ISSCC) [2], and in 1974 the first commercial 1024-bit (1-Kb) memory was introduced [3]. By 2000, experimental 1-gigabit (Gb) chips had been described at the ISSCC, and the technology for future 1-Gb memories had been discussed at the IEEE International Electron Devices Meeting (IEDM)[4]. Thus, just 30 years after the introduction of the first commercial MOS RAM chips, chips have been

demonstrated with more that 1 million times the storage capacity of the original RAMs.

The circuit that made these incredible memory chips possible is called the **one-transistor** dynamic RAM **cell** or 1-T DRAM. This elegant circuit, which requires only one transistor and one capacitor to store a single bit of information [1], was invented in 1966 by Robert H. Dennard of the IBM Thomas J. Watson Research Center. In this circuit, patented in 1968, the binary information is temporarily stored as a charge on the capacitor, and the data must be periodically refreshed in order to prevent information loss. In addition, the process of reading the data out of most DRAM circuits destroys the information, and the data must be put back into memory as part of the read operation. At the time of the invention, Dennard and a few of his colleagues were probably the only ones that believed the circuit could be made to actually work. Today, there are arguably more 1-T DRAM bits in the world than any other electronic circuit.

8.1 RANDOM ACCESS MEMORY

Thus far, our study of logic circuits concentrated on understanding the design of basic inverters and combinational logic circuits. In addition to logic, however, digital systems generally require data storage capability in the form of high-speed registers, high-density **random-access memory (RAM),** and **read-only memory (ROM).**

In digital systems, the term RAM is used to refer to memory with both read and write capability. This is the type of memory used when information needs to be changed with great frequency. The data in any given storage location in RAM can be directly read or altered just as quickly as the information at any other location.

Early memory designs were **static RAM** or **SRAM** circuits in which the information remains stored in memory as long as the power supply voltage is maintained. The SRAM cell requires the equivalent of six transistors per memory bit and features nondestructive readout of its stored information.

In the **dynamic RAM** or **DRAM** circuit, information is temporarily stored as a charge on a capacitor, and the data must be periodically refreshed in order to prevent information loss. The process of reading the data out of most DRAMs destroys the information, and the data must be put back into memory as part of the read operation.

Because the SRAM cell takes up considerably more area than the DRAM cell, an SRAM memory chip typically has only one-fourth the number of bits as a DRAM memory of the same technology generation. For example, using the same IC technology, it would be possible to fabricate a 64-Mb DRAM and a 16-Mb SRAM. The majority of RAM chips with densities below 4 Mb have provided a single output bit, but because the capacity of recent memory chips has become so large, the external interface to many memory chips is now designed to be four, eight, or more bits wide.

Read-only memories or ROMs, also sometimes called **read-only storage,** or **ROS,** represent another important class of memory. In these memories, data is permanently stored within the physical structure of the array. However, ROM technology can also be used to perform logic using the programmable logic array, or PLA, structure. Digital systems also require high-speed storage in the form of individual flip-flops and registers, and this chapter concludes with a discussion of the basic circuits used to realize RS and D flip-flops.

8.1.1 RANDOM ACCESS MEMORY (RAM) ARCHITECTURE

Random-access memory (RAM) provides the high-speed temporary storage used in digital computers, and digital systems have an almost insatiable demand for RAM. Today's high-function word processing and publishing software often require many tens of megabytes of RAM for operation, and the operating systems may require hundreds of megabytes. Thus, it is common even for a personal computer to have a gigabyte (GB) (1 byte = 8 bits) or more of RAM. In contrast, high-end computer mainframes contain multigigabytes of RAM.

It is mind-boggling to realize that a single 1-Gb memory chip contains 128 MB of storage and that the chips contain more than 2 billion electronic components that must all be working! Only the very regular repetitive structure of the memory array permits the design and realization of such complex IC chips. This section explores the basic structure of an IC memory; subsequent sections look at individual memory cells and subcircuits in more detail.

8.1.2 A 256-MB MEMORY CHIP

Figure 8.1(a) is a microphotograph of a 256-Mb memory chip [8] and its block structure. Internally, the 256-Mb array is divided into eight 32-Mb subarrays. To select a group of bits within the large array, the memory address is selected by **column** and **row address decoders.** In Fig. 8.1, the column decoders occupy the center of the die and separate it into upper and lower halves. Row decoders and **wordline drivers** bisect each 32-Mb subarray. Each 32-Mb subarray is further subdivided into 16 2-Mb sections, each of which contains 16 blocks of 128 kilobits (Kb). Thus, the 128-Kb array represents the basic building block of this 256-Mb memory.

Figure 8.2 is a block diagram of a basic memory array that could correspond to one of the 128-Kb (2^{17}-bit) subarrays in Fig. 8.1. The array contains 2^{M+N} storage locations, and the address is split into M bits of row address plus N bits of column address. Each $(M + N)$-bit address corresponds to a single storage location or memory cell within the array. For the 128-Kb memory segment in Fig. 8.1, $M = 10$ and $N = 7$. When a given bit is addressed, information can be written into the storage cell

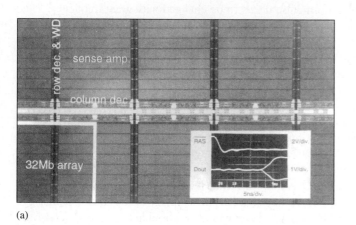

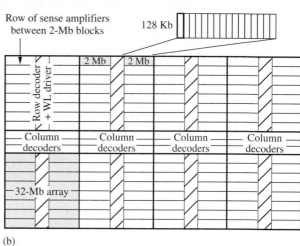

Figure 8.1 256-Mb RAM chip: (a) RAM micrograph and measured output waveforms, (b) functional identification of areas of the chip. (Micrograph from Mikio Asakura et al., *1994 ISSCC Digest of Technical Papers,* February 1994, vol. 37, © 1994 IEEE.)

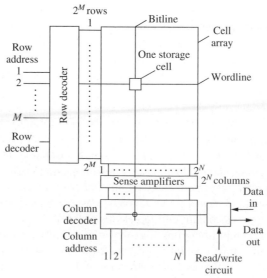

Figure 8.2 Block diagram of a basic memory array.

or the contents of the cell may be read out. Each 128-Kb array has a set of **sense amplifiers** to read and write the information of the selected memory cells.

When a row is selected, the contents of a 128-bit-wide word (2^7) are actually accessed in parallel. These horizontal rows are normally referred to as **wordlines (WL).** The lines running in the vertical direction contain the cell data and are called **bitlines (BL).** One or two bitlines may run through each cell, and bitlines can be shared between adjacent cells. The 7-bit column address is used to select the individual bit or group of bits that is actually transferred to the output during a **read operation,** or modified during a **write operation.**

In addition to the storage array, memory chips require several other types of peripheral circuits. Address decoder circuits are obviously required to select the desired row and column. In addition, the wordlines present heavy capacitive loads to the address decoders, and special wordline driver circuits are needed to drive these lines. Also, during a read operation the signal coming from the cell can be quite small, and sense amplifiers are required to detect the state of the memory cell and restore the signal to a full logic level for use in the external interface. The next several sections explore the individual circuits used to implement static and dynamic memory cells as well as sense amplifier and address decoder circuits. Both static and dynamic decoder circuits are discussed.

EXERCISE: (a) How many 128-Kb segments form the 256-Mb memory? (b) A 1-Gb memory made by doubling the dimensions of the main arrays in Fig. 8.1 (32 Mb → 128 Mb, 2 Mb → 8 Mb, 128 Kb → 512 Kb). How many 512 Kb segments are required in the 1-Gb memory?

ANSWERS: 2048; 2048

8.2 STATIC MEMORY CELLS

The basic electronic storage element consists of two inverters in series, with the output of the second inverter connected back to the input of the first, as shown in Fig. 8.3. If the input of the first inverter is a 0, as in Fig. 8.3(a), then its output will be a 1 and the output of the second inverter will be a 0. In Fig. 8.3(b), an alternate representation of the circuit in Fig. 8.3(a), the input of the first inverter is a 1, its output is a 0, and the output of the second inverter is a 1. For both cases, the output of the second inverter is connected back to the input of the first inverter to form a logically stable configuration. These circuits have two stable states and are termed **bistable circuits.** The pair of **cross-coupled inverters** is also often called a **latch.** The latch is a circuit that we also often use to study the noise margin of logic gates [16, 17].

The behavior of the circuit can be understood more completely by looking at the voltage transfer characteristic (VTC) in Fig. 8.4 for two cascaded inverters. A line with unity slope has been drawn on the figure, indicating three possible operating points with $v_O = v_I$. The points with $v_O = V_L$ and $v_O = V_H$ are the two stable Q-points already noted and represent the two data states of the binary latch.

However, the third point, corresponding to the midpoint of the VTC ($v_O \cong 1.5$ V), represents an **unstable equilibrium point.** It is unstable in the sense that any disturbance to the voltages in the circuit will cause the latch to quickly make a transition to one of its two stable operating points. For example, suppose the inverter is operating with $v_I = v_O = 1.5$ V, and then the input increases slightly. The output will immediately move toward V_H due to the large positive gain of the circuit. A small negative change from the 1.5-V equilibrium point would drive the output immediately to V_L. Using nonlinear analysis techniques beyond the scope of this text, it can actually be shown that any imbalance in the voltages between the two output nodes of the latch will be reinforced; the node at the higher potential will become a logic 1, and the node at the lower potential will become a logic 0.

The two stable points in the VTC are obviously useful for storing binary data. However, in Sec. 8.4 we will see that the latch can be forced to operate at the unstable equilibrium point and find that this third point is highly useful in designing sense amplifiers.

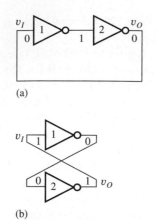

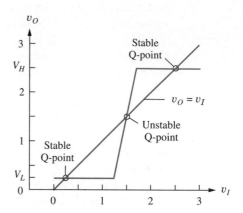

Figure 8.3 Two inverters forming a static storage element or latch.

Figure 8.4 VTC for two inverters in series, indicating the three possible operating points for a latch with $v_O = v_I$.

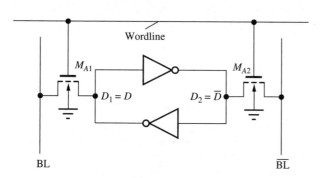

Figure 8.5 Basic memory cell formed from the two-inverter latch and two access transistors M_{A1} and M_{A2}.

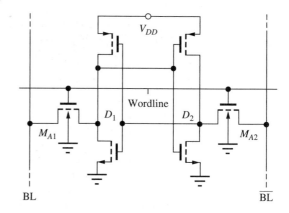

Figure 8.6 Six-transistor CMOS memory cell.

8.2.1 MEMORY CELL ISOLATION AND ACCESS—THE 6-T CELL

The cross-coupled inverter pair is the basic storage element needed in Fig. 8.2 to build a static memory. In Fig. 8.5, two additional transistors are added to the latch to isolate it from other memory cells and to provide a path for information to be written to and read from the memory cell. In NMOS or CMOS technology, each inverter requires two transistors, so the memory circuit in Fig. 8.5 is usually referred to as the **six-transistor (6-T) SRAM cell.** Note that the 6-T cell provides both true and complemented data outputs, D and $\overline{D}$.

Figure 8.6 is a 6-T CMOS cell implementation. The advantage of CMOS inverters is that only very small leakage currents exist in the cell because a static current path does not exist through either inverter. Because of higher mobility and lower on-resistance for a given W/L ratio, the access devices M_{A1} and M_{A2} are shown as NMOS transistors in all the circuits in this chapter. However, PMOS transistors can successfully be used in some designs.

The 6-T cell presents an interesting set of conflicting design requirements. During the read operation, the state of the memory cell must be determined through the access transistors without upsetting the data in the cell. However, during a write operation, the data in the cell must be forced to the desired state using the same access devices. The design of these cells is explored in more detail in the next two subsections. In the following discussion, a 0 in the memory cell will correspond to a low level (0 V) on the left-hand data storage node (D_1) and a high level (V_{DD}) on the right-hand

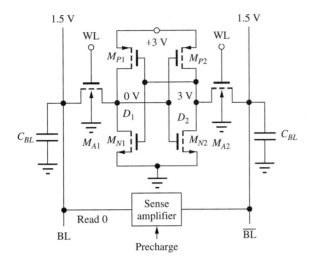

Figure 8.7 Reading data from a 6-T cell with a 0 stored in the cell.

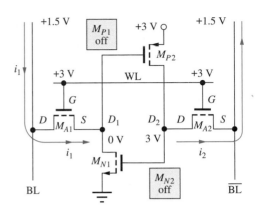

Figure 8.8 Conditions immediately following activation of the wordline.

data node (D_2); a 1 in the memory cell will correspond to a high level (V_{DD}) at D_1 and a low level (0 V) at D_2.

EXERCISES: (a) How many storage cells are actually in a 256-Mb memory? (b) Suppose a 256-Mb memory is to use the cells in Fig. 8.6, and the total static power consumption of the memory must be ≤ 50 mW with a 3.3-V power supply. What is the permissible leakage current in each cell?

ANSWERS: 268,435,456; 56.4 pA

EXERCISE: Draw a version of the storage cell in Fig. 8.6 with PMOS access transistors.

ANSWER: Simply reverse the direction of the substrate arrows in the access devices (M_{A1} and M_{A2}), and connect the substrates to V_{DD}.

8.2.2 THE READ OPERATION

Figure 8.7 is a 6-T CMOS memory cell in the 0 state, in which V_{DD} has been chosen to be 3 V. Although a number of different strategies can be used to read the state of the cell, we will assume that both bitlines are initially precharged to approximately one-half V_{DD} by the sense amplifier circuitry, while M_{A1} and M_{A2} are turned off by holding the wordline WL at 0 V. The exact precharge level is determined by the sense amplifiers and is discussed in the next section. Precharge levels equal to V_{DD}, $\frac{1}{2}V_{DD}$, and $\frac{2}{3}V_{DD}$ have all been proposed for memory design.

Once the bitline voltages have been precharged to the desired level, cell data can be accessed through transistors M_{A1} and M_{A2} by raising the wordline voltage to a high logic level (3 V). The conditions immediately following initiation of such a read operation are shown in Fig. 8.8, in which the substrate terminals of the access transistors have been omitted for clarity.[1] M_{P1} and M_{N2} are off. M_{A1} will be operating in the triode region (for typical values of V_{TN}) because $V_{GS} = 3$ V and $V_{DS} = 1.5$ V, and current i_1 enters the cell from the bitline into the cell. M_{A2} is saturated because both V_{GS} and V_{DS} are equal to 1.5 V, and the current i_2 exits cell into $\overline{\text{BL}}$.

---

[1] Note that the capacitances at nodes D_1 and D_2 prevent the voltages at these nodes from changing at $t = 0^+$.

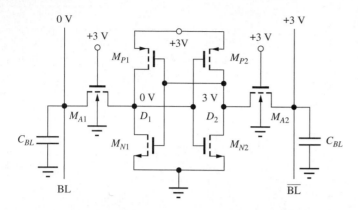

Figure 8.9 Final state after the sense amplifier has reached steady-state.

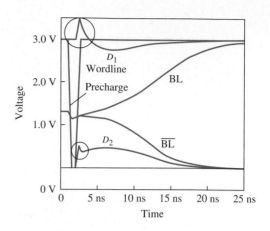

Figure 8.10 SPICE memory cell waveforms during a read operation.

As current increases through M_{A1} and M_{A2}, the voltage on data node D_1 tends to rise, and the voltage at D_2 tends to fall. For the data stored in the cell not to be disturbed, a conservative design ensures that the voltage at D_1 remains below the threshold voltage of M_{N2}, and that the voltage at D_2 remains high enough ($> 3 - |V_{TP}|$) to maintain M_{P1} off. Currents i_1 and i_2 in the two bitlines cause the sense amplifier to rapidly assume the same state as the data stored in the cell, and the BL and $\overline{BL}$ voltages become 0 V and 3 V, respectively.

The voltages in the circuit after the sense amplifier reaches steady-state are shown in Fig. 8.9. The bitline voltages match the original cell voltages, and the bistable latch in the storage cell has restored the cell voltages to the original full logic levels. In the final steady-state condition, both M_{A1} and M_{A2} will be on in the triode region but not conducting current because $V_{DS} = 0$.

It is important to note in Fig. 8.8 that the source terminal of M_{A1} is connected to the cell, whereas the source of M_{A2} is connected to the bitline. This is an example of the bidirectional nature of the FET. Remember that the source and drain of the FET are always determined by the relative polarities of the voltages in the circuit.

Rather than try to analyze the details of this complex circuit by hand, the waveforms resulting from a SPICE simulation of the 6-T circuit are presented in Fig. 8.10. The simulation assumes a total capacitance on each bitline of 500 fF, and the W/L ratios of all transistors in the memory cell are 1/1. In Fig. 8.10, the bitlines can be observed to be precharged to slightly less than one-half V_{DD}, approximately 1.3 V. At $t = 1$ ns, the precharge signal is turned off, and at $t = 2$ ns the wordline begins its transition from 0 V to 3 V. As the two access transistors turn on, BL and $\overline{BL}$ begin to diverge as the sense amplifier responds and reinforces the data stored in the cell. As the bitlines increase further, the cell voltages at D_1 and D_2 return to the full 3 V and 0 V levels. The state of the memory cell is disturbed, but not destroyed, during the read operation. Thus the 6-T cell provides data storage with nondestructive readout.

Note that the time delay from the midpoint of the wordline transition to the point when the bitlines reach full logic levels is approximately 20 ns. Also, the two rapid positive transients at D_1 and D_2 (in the circles) result from direct coupling of the rapid transition of the wordline signal through the MOSFET gate capacitances to the internal nodes of the latch. This capacitive coupling of the wordline signal causes the initial transients both to be in the same direction. D_1 actually goes above the 3-V supply.

Reading a 1 stored in the memory cell in Fig. 8.11 simply reverses the conditions in Figs. 8.7 to 8.10. The two cell currents i_1 and i_2 reverse directions, and the sense amplifier flips to the opposite state. Note that the source and drain terminals and direction of current in the two access transistors have all reversed.

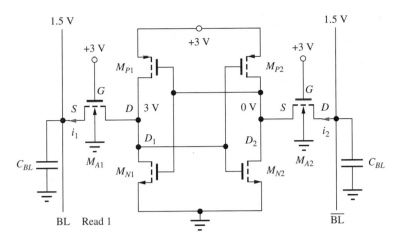

Figure 8.11 Reading data from 6-T cell with a 1 stored in the cell.

EXAMPLE 8.1 **CURRENTS IN THE 6-T STATIC MEMORY CELL**

This example demonstrates calculation of the currents in the access transistors in the six-transistor memory cell during a read operation.

PROBLEM Calculate bitline currents i_1 and i_2 in Fig. 8.8 immediately following activation of the wordline. Assume $W/L = 1/1$ for all devices and use $K'_n = 60 \ \mu\text{A/V}^2$, $V_{TO} = 0.7$ V, $\gamma = 0.5 \ \text{V}^{1/2}$, and $2\phi_F = 0.6$ V.

SOLUTION **Known Information and Given Data:** The circuit is the six-transistor memory cell in Figs. 8.7 and 8.8 with 0 V initially stored at D_1 and 3 V at D_2. All W/L values are 1/1; $K'_n = 60 \ \mu\text{A/V}^2$, $V_{TO} = 0.7$ V, $\gamma = 0.5 \ \text{V}^{1/2}$, and $2\phi_F = 0.6$ V.

Unknowns: Find currents i_1 and i_2 just after the wordline is activated.

Approach: "Activation" of the wordline means the wordline has just stepped from 0 V to 3 V. Current is supplied to the drain of M_{A1} from the left-hand bitline BL, and, at the right-hand side, current exits the source of M_{A2} onto $\overline{\text{BL}}$. The currents are set by these two transistors. Note that the drain-source voltages of M_{N1} and M_{P2} are both zero, so that the drain currents in these two devices must be zero. First we find the terminal voltages for M_{A1} and M_{A2} from Fig. 8.8. Then we use the terminal voltages to find the region of operation of each device. Once the region of operation has been identified, the drain currents are found using the equation appropriate for the region of operation.

Assumptions: The wordline voltage change is a step function that occurs at $t = 0$.

Analysis: At $t = 0^+$, the wordline has just stepped from 0 V to 3 V, and current goes from left-hand bitline BL into the drain of M_{A1}. Referring to the conditions in Fig. 8.8, we see that the source of M_{A1} is at 0 V, the drain is at 1.5 V, and the gate is at 3 V. Since the source is at 0 V, $V_{TN} = V_{TO}$, and we have $V_{GS} - V_{TN} = (3 - 0) - 0.7 = 2.3$ V, and $V_{DS} = 1.5 - 0 = 1.5$ V. Thus, the device is in the triode region of operation, and the drain current is given by

$$i_1(0^+) = K'_n \left(\frac{W}{L}\right) \left(V_{GS} - V_{TN} - \frac{V_{DS}}{2}\right) V_{DS} = \left(\frac{60 \ \mu\text{A}}{\text{V}^2}\right) \left(\frac{1}{1}\right) \left(3 - 0.7 - \frac{1.5}{2}\right) 1.5 \ \text{V}^2 = 140 \ \mu\text{A}$$

At the right-hand bitline, current exits the source of M_{A2} onto $\overline{BL}$. Referring again to the conditions in Fig. 8.8, the drain of M_{A2} is at 3 V, the source is at 1.5 V, and the gate is at 3 V. Since the source of M_{A2} is not at 0 V, we must find its threshold voltage using Eq. (4.20).

$$V_{TN} = V_{TO} + \gamma\left(\sqrt{v_{SB} + 2\phi_F} - \sqrt{2\phi_F}\right) = 0.7 + 0.5\left(\sqrt{1.5 + 0.6} - \sqrt{0.6}\right) = 1.04 \text{ V}$$

To find the region of operation, we have $V_{GS} - V_{TN} = (3 - 1.5) - 1.04 = 0.46$ V, and $V_{DS} = 3 - 1.5 = 1.5$ V. Thus, M_{A2} is in the saturation region of operation for which the drain current is given by

$$i_2(0^+) = \frac{K_n'}{2}\left(\frac{W}{L}\right)(V_{GS} - V_{TN})^2 = \left(\frac{60 \text{ μA}}{2 \text{ V}^2}\right)\left(\frac{1}{1}\right)(1.5 - 1.04)^2 = 6.35 \text{ μA}$$

We have found the two required currents $i_1(0^+) = 140$ μA and $i_1(0^+) = 6.35$ μA. Note that the current on the left-hand side is more than 20 times that on the right. Thus, the sense amplifier will have a significant current difference on which to make a decision.

Check of Results: We have found the two required currents, and the values appear reasonable (they are both in the μA to mA range).

Discussion: Note that the terminals identified as the drain and source of M_{A1} and M_{A2} are different in the two devices and are determined by the potentials at the various nodes. If the opposite state were stored in the memory cell, then the source and drain terminal identifications would be reversed. At this point, we should be puzzled about where the currents that we have calculated are actually going. Since the voltage across M_{N1} and M_{P2} are zero at $t = 0^+$, the drain-source currents are zero in these devices. The initial currents in M_{A1} and M_{A2} begin to charge and discharge the capacitances at nodes D_1 and D_2, respectively. As the voltages change at nodes D_1 and D_2, transistors M_{N1} and M_{P2} begin to conduct current.

EXERCISE: Find bitline currents i_1 and i_2 in Fig. 8.8 immediately following activation of the wordline. Assume W/L = 1/1 for all devices, and use V_{DD} = 5 V, WL = 5 V, the bitline voltages are 2.5 V, K_n' = 60 μA/V², V_{TO} = 1 V, γ = 0.6 V$^{1/2}$, and $2\phi_F$ = 0.6 V.

ANSWERS: 413 μA; 24.7 μA (V_{TN} = 1.592 V)

8.2.3 WRITING DATA INTO THE 6-T CELL

For a write operation, the bitlines are initialized with the data that is to be written into the cell. In Fig. 8.12, a zero is being written into a cell that already contains a zero. It can be seen that the access transistors both have $V_{DS} = 0$. The currents i_1 and i_2 are zero, and virtually nothing happens, except for the transients that occur due to internode coupling of the wordline signal through the MOS transistor capacitances (see Prob. 8.8).

The more interesting case is shown in Fig. 8.13, in which the state of the cell must be changed. When the wordline is raised to 3 V, access transistor M_{A1} conducts current in the saturation region, with $V_{GS} = 3$ V and $V_{DS} = 3$ V, and the voltage at D_1 tends to discharge toward 0 through M_{A1}. Access transistor M_{A2} is also in saturation, with $V_{GS} = 3$ V and $V_{DS} = 3$ V, and the voltage at D_2 initially tends to charge toward a voltage of $(3 - V_{TN})$ V. As soon as the voltage at D_2 exceeds that at D_1, positive feedback takes over, and the cell rapidly completes the transition to the new desired state, with $D_1 = 0$ V and $D_2 = 3$ V.

Figure 8.14 shows waveforms from a SPICE simulation of this write operation. As the wordline transition begins at $t = 0.5$ ns, the fixed levels on the bitlines are transferred to nodes D_1 and D_2 through the two access transistors. Minimum area transistors are normally used throughout the

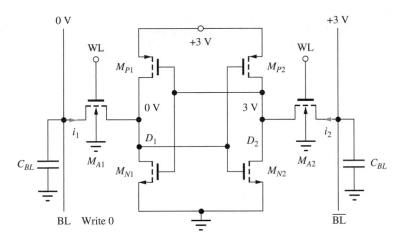

Figure 8.12 A memory cell set up for a write 0 operation with a 0 already stored in the cell.

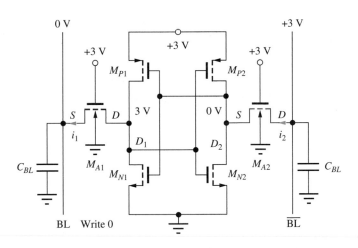

Figure 8.13 A memory cell set up for a write 0 operation with a 1 previously stored in the cell.

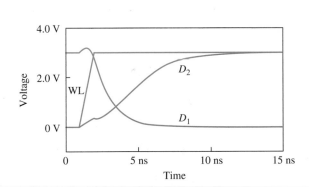

Figure 8.14 SPICE bitline and data node waveforms as a 0 is written into the 6-T cell in Fig. 8.13.

memory cell array, and the capacitances on the memory cell nodes are quite small. This small nodal capacitance is the reason why the voltages at D_1 and D_2 reach the desired state in approximately 10 ns in this simulation.

In the simulation results in Fig. 8.14, the bitlines were connected to ideal voltage sources. However, in a real memory implementation, the two bitlines will be driven by logic buffers, and the current driving capability of the buffer must exceed that of the inverters in the RAM cell in order for data to be written into the cell. The buffer must "overpower" the state of the memory cell.

EXAMPLE 8.2   **INITIAL MEMORY CELL CURRENTS DURING A WRITE OPERATION**

This example calculates of the currents in a 6-T memory cell when the state of the cell is being changed.

PROBLEM   Find bitline currents i_1 and i_2 in Fig. 8.13 immediately following activation of the wordline. Assume $W/L = 1/1$ for all devices and use $K'_n = 60 \ \mu A/V^2$, $V_{TO} = 0.7$ V, $\gamma = 0.5 \ V^{1/2}$, and $2\phi_F = 0.6$ V.

SOLUTION   **Known Information and Given Data:** The circuit is the six-transistor memory cell in Fig. 8.13 with 3 V stored at D_1 and 0 V at D_2. The state of the cell is to be changed, so the left-hand bitline is set to 0 V and the right-hand bitline is set to 3 V prior to activation of the wordline. All W/L values are 1/1; $K'_n = 60$ μA/V^2, $V_{TO} = 0.7$ V, $\gamma = 0.5$ V$^{1/2}$, and $2\phi_F = 0.6$ V. "Activation" of the wordline means that it changes from 0 V to +3 V.

Unknowns: Find currents i_1 and i_2 just after the wordline is activated.

Approach: At $t = 0^+$, the wordline has just completed a step from 0 V to 3 V, and current will exit the cell through M_{A1} and enter the cell through M_{A2}. At $t = 0^+$, the drain currents in M_{N1}, M_{N2}, M_{P1}, and M_{P2} are all zero since either the gate-source voltage is zero or the drain-source voltage is zero for all four of these devices. The initial cell currents are set by transistors M_{A1} and M_{A2}. First we find the terminal voltages for M_{A1} and M_{A2} from Fig. 8.13, and then we use the terminal voltages to find the region of operation of each device. Once the region of operation has been identified, we calculate the drain currents using the equation appropriate for the region of operation.

Assumptions: The wordline voltage change is a step function that occurs at $t = 0$.

Analysis: At $t = 0^+$, the wordline has just stepped from 0 V to 3 V. In Fig. 8.13, we see that the drain terminals of M_{A1} and M_{A2} are both at 3 V, the source terminals are both at 0 V and the gate terminals are both at 3 V. The bias conditions for both transistors are identical. Since the source terminal of each transistor is at 0 V, $V_{TN} = V_{TO}$, and we have $V_{GS} - V_{TN} = (3-0) - 0.7 = 2.3$ V and $V_{DS} = 3 - 0 = 3$ V. Thus, the devices are in saturation, and the drain currents are given by

$$i_1(0^+) = i_2(0^+) = \frac{K'_n}{2}\left(\frac{W}{L}\right)(V_{GS} - V_{TN})^2 = \left(\frac{60\ \mu A}{2\ V^2}\right)\left(\frac{1}{1}\right)(3 - 0.7)^2 = 158\ \mu A$$

Check of Results: We have found the two unknown currents, and the values appear reasonable (they are both in the μA to mA range).

Discussion: Note once again that the terminals identified as the drain and source of M_{A1} and M_{A2} are different in the two devices and are determined solely by the potentials at the various nodes. Again, we may be puzzled about where the currents that we have calculated are actually going. The initial currents in M_{A1} and M_{A2} begin to discharge and charge the capacitances at nodes D_1 and D_2, respectively, to begin the process of changing the information stored in the cell.

EXERCISE: **Find bitline currents i_1 and i_2 in Fig. 8.13 immediately following activation of the wordline. Assume $W/L = 1/1$ for all devices, and use $V_{DD} = 5$ V, WL $= 5$ V, bitline voltages $= 2.5$ V, $K'_n = 60$ μA/V^2, $V_{TO} = 1$ V, $\gamma = 0.6$ V$^{1/2}$, and $2\phi_F = 0.6$ V.**

ANSWERS: 480 μA; 480 μA

8.3 DYNAMIC MEMORY CELLS

As long as power is applied to static memory cells, the information stored in the cells should be retained. In addition, static cells feature nondestructive readout of data from the cell. Although voltage levels in the cell are disturbed during the read operation, the cross-coupled latch automatically restores the levels once the access transistors are turned off.

However, much smaller memory cells can be built if the requirement for static data storage is relaxed. These memory cells are referred to as dynamic memory, and the most important dynamic random-access memory cell is the one-transistor cell. The operation of the 1-T cell is explored in the next several subsections.

 **ELECTRONICS IN ACTION**

Field Programmable Gate Arrays (FPGAs)

Field programmable gate arrays (FPGAs) are widely used in electronic prototype development as well as in many completed products in which the volume and/or time schedule cannot justify the cost of custom integrated circuit development. FPGAs consist of a two-dimensional array of programmable logic blocks (PLBs), programmable input/output cells, and programmable interconnect that are controlled by the contents of a configuration memory that defines the logic functions of the PLBs, the I/O cells, and the interconnections between the PLBs and I/O cells. The configuration memory is loaded with the bit patterns required to define and connect the desired functions. Changing the configuration memory data changes the system function and, in some FPGAs, can even be done while the FPGA is operating without destroying the contents of RAMs and other memory elements on the chip.

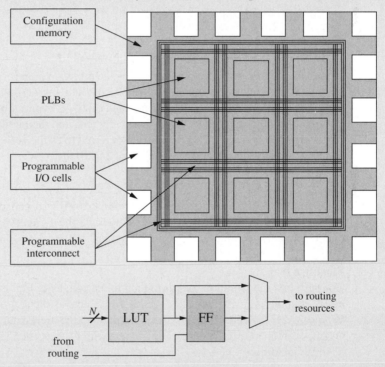

A typical PLB consists of Look-Up-Tables (LUTs), flip-flops (FF), and multiplexers. The LUTs store truth tables for combinational logic functions, and flip-flops provide memory elements for sequential logic functions. The programmable interconnect consists of a large number of wire segments that can be interconnected with MOS transistor switches to form longer wires, cross points, and other interconnect structures. An I/O cell contains latches or flip-flops and bi-direction data buffers.

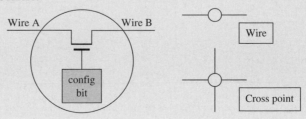

FPGAs have grown from relatively simple devices to extremely complex system chips as the density of integrated circuit technology has increased. Some chips contain general purpose microprocessor and DSP cores.

Current FPGA Characteristics

5 M—100 M transistors
32 Kb to 50 Mb of configuration memory
100—22,000 PLBs per FPGA
50—400 wire segments/PLB
80—2400 switches/PLB
50—750 I/O cells per FPGA

Drawings and data courtesy of Professor Charles E. Stroud, ECE Department, Auburn University.

8.3.1 THE ONE-TRANSISTOR CELL

In the 1-T cell in Fig. 8.15, data is stored as the presence or absence of charge on **cell capacitor** C_C. Because leakage currents exist in the drain-bulk and source-bulk junctions of the transistor and in the transistor channel, the information stored on C_C is eventually corrupted. To prevent this loss of information, the state of the cell is periodically read and then written back into the cell to reestablish the desired cell voltages. This operation is referred to as the **refresh operation.** Each storage cell in a DRAM typically must be refreshed every 2 to 10 ms.

8.3.2 DATA STORAGE IN THE 1-T CELL

In the analysis that follows, a 0 will be represented by 0 V on capacitor C_C, and a 1 will be represented by a high level on C_C. These data are written into the 1-T cell by placing the desired voltage level on the single bitline and turning on access transistor M_A.

Storing a 0

Consider first the situation for storing a 0 in the cell, as in Fig. 8.16. In this case, the bitline is held at 0 V, and the bitline terminal of the MOSFET acts as the source of the FET. The gate is raised to $V_{DD} = 3$ V. If the cell voltage is already zero, then the drain-source voltage of the MOSFET is zero, and the current is zero. If the cell contains a 1 with $v_C > 0$, then the MOSFET completely discharges C_C, also yielding $v_C = 0$.

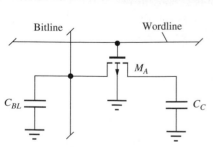

Figure 8.15 One-transistor (1-T) storage cell in which binary data is represented by the presence or absence of charge on C_C.

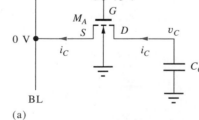

(a)

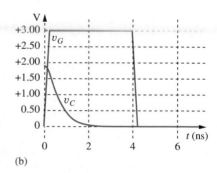

(b)

Figure 8.16 (a) Writing a 0 into the 1-T cell. (b) Waveform during WRITE operation.

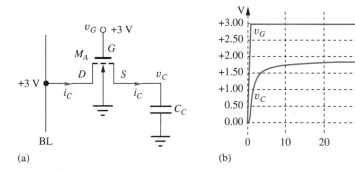

Figure 8.17 (a) Conditions for writing a 1 into the 1-T cell. (b) Waveform during WRITE operation.

The cell voltage waveform resulting from writing a zero into a cell containing a one is given in Fig. 8.17(b). The initial capacitor voltage, calculated in the next section, is rapidly discharged by the access transistor.

EXERCISE: (a) What is the cell current i_C in Fig. 8.16 just after access transistor M_A is turned on if $V_C = 1.9$ V, $K_n = 60$ μA/V^2, and $V_{TO} = 0.7$ V? (b) Estimate the fall time of the voltage on the capacitor using Eqs. (7.15) and (7.20), with $C_C = 50$ fF.

ANSWERS: 154 μA; 1.30 ns

Storing a 1

Now consider the case of writing a 1 into the 1-T cell in Fig. 8.17. The bitline is first set to V_{DD} (3 V), and the wordline is then raised to V_{DD}. The bitline terminal of M_A acts as its drain, and the cell capacitance terminal acts as the FET source. Because $V_{DS} = V_{GS}$, and M_A is an enhancement mode device, M_A will operate in the saturation region. If a full 1 level already exists in the cell, then the current is zero in M_A. However, if V_C is less than a full 1 level, current through M_A will charge up the capacitor to a potential one threshold voltage below the gate voltage.

We see from this analysis that the voltage levels corresponding to 0 and 1 in the 1-T cell are 0 V and $V_G - V_{TN}$. The threshold voltage must be evaluated for a source-bulk voltage equal to V_C:

$$V_C = V_G - V_{TN} = V_G - \left[V_{TO} + \gamma\left(\sqrt{V_C + 2\phi_F} - \sqrt{2\phi_F}\right)\right] \tag{8.1}$$

Equation (8.1) is identical to Eq. (6.19) used to determine V_H for the saturated load NMOS logic circuit.

Note once again the important use of the bidirectional characteristics of the MOSFET. Charge must be able to flow in both directions through the transistor in order to write the desired data into the cell. To read the data, current must also be able to change directions.

The waveform for writing a one into the 1-T cell appears in Fig. 8.17(b). Note the relatively long time required to reach final value. This is exactly the same situation as the long transient tail observed on the low-to-high transition in NMOS saturated load logic. However, access transistor M_A could be turned off at the 10-ns point without significant loss in cell voltage.

EXERCISE: Find the cell voltage V_C if $V_{DD} = 3$ V, $V_{TO} = 0.7$ V, $\gamma = 0.5 \sqrt{V}$, and $2\phi_F = 0.6$ V. What is V_C if $\gamma = 0$?

ANSWERS: 1.89 V; 2.3 V

EXERCISE: If a cell is in a 1 state, how many electrons are stored on the cell capacitor if $C_C = 25$ fF?

ANSWER: 2.95×10^5 electrons

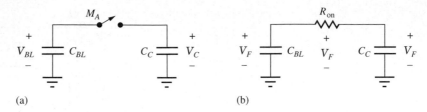

Figure 8.18 Model for charge sharing between the 1-T storage cell capacitance and the bitline capacitance: (a) circuit model before activation of access transistor, (b) circuit following closure of access switch.

The results in the preceding exercises are typical of the situation for the 1 level in the cell. A significant part of the power supply voltage is lost because of the threshold voltage of the MOSFET, and the body effect has an important role in further reducing the cell voltage for the 1 state. If there were no body effect in the first exercise, then V_C would increase to 2.3 V.

8.3.3 READING DATA FROM THE 1-T CELL

To read the information from the 1-T cell, the bitline is first precharged (**bitline precharge**) to a known voltage, typically V_{DD} or one-half V_{DD}. The access transistor is then turned on, and the cell capacitance is connected to the bitline through M_A. A phenomenon called **charge sharing** occurs. The total charge, originally stored separately on the **bitline capacitance** C_{BL} and cell capacitance C_C, is shared between the two capacitors following the switch closure, and the voltage on the bitline changes slightly. The magnitude and sign of the change are related to the stored information.

Detailed behavior of data readout can be understood using the circuit model in Fig. 8.18. Before access transistor M_A is turned on, the switch is open, and the total initial charge Q_I on the two capacitors is

$$Q_I = C_{BL}V_{BL} + C_C V_C \qquad (8.2)$$

After access transistor M_A is activated, corresponding to closing the switch, current through the on-resistance of M_A equalizes the voltage on the two capacitors. The final value of the stored charge Q_F is given by

$$Q_F = (C_{BL} + C_C)V_F \qquad (8.3)$$

Because no mechanism for charge loss exists, Q_F must equal Q_I. Equating Eqs. (8.2) and (8.3) and solving for V_F yields

$$V_F = \frac{C_{BL}V_{BL} + C_C V_C}{C_{BL} + C_C} \qquad (8.4)$$

The signal to be detected is the change in the voltage on the bitline from its initial precharged value:

$$\Delta V = V_F - V_{BL} = \frac{C_C}{C_{BL} + C_C}(V_C - V_{BL}) = \frac{(V_C - V_{BL})}{\dfrac{C_{BL}}{C_C} + 1} \qquad (8.5)$$

Equation (8.5) can be used to guide our selection of the precharge voltage. If V_{BL} is set midway between the 1 and 0 levels, then ΔV will be positive if a 1 is stored in the cell and negative if a 0 is stored. Study of Eq. (8.5) also shows that the signal voltage ΔV can be quite small. If there are 128 rows in our memory array, then there will be 128 access transistors connected to the bitline, and the ratio of bitline capacitance to cell capacitance can be quite large.

If we assume that $C_{BL} \gg C_C$, Eq. (8.4) shows that the final voltage on the bitline and cell is

$$V_F = \frac{C_{BL}V_{BL} + C_C V_C}{C_{BL} + C_C} \cong V_{BL} \qquad (8.6)$$

Thus, the content of the cell is destroyed during the process of reading the data from the cell—the 1-T cell is a cell with destructive readout. To restore the original contents following a read operation, the data must be written back into the cell.

Except for the case of an ideal switch, charge transfer cannot occur instantaneously. If the on-resistance were constant, the voltages and currents in the circuit in Fig. 8.18 would change exponentially with a time constant τ determined by R_{on} and the series combination of C_{BL} and C_C:

$$\tau = R_{\mathrm{on}}\frac{C_C C_{BL}}{C_C + C_{BL}} \cong R_{\mathrm{on}}C_C \qquad \text{for } C_{BL} \gg C_C \tag{8.7}$$

EXERCISE: (a) Find the change in bitline voltage for a memory array in which $C_{BL} = 49\, C_C$ if the bitline is precharged midway between the voltages corresponding to a 1 and a 0. Assume that 0 V corresponds to a 0 and 1.9 V corresponds to a 1. (b) What is τ if $R_{\mathrm{on}} = 5\, k\Omega$ and $C_C = 25\, fF$?

ANSWERS: +19.0 mV, −19.0 mV; 0.125 ns

The preceding exercise reinforces the fact that the voltage change that must be detected by the sense amplifier for a 1-T cell is quite small. Designing a sense amplifier to rapidly detect this small change is one of the major challenges of DRAM design. Also, note that the **charge transfer** occurs rapidly.

8.3.4 THE FOUR-TRANSISTOR CELL

We saw earlier that the 6-T static cell provides a large signal current to drive the sense amplifiers. This is one reason why static memory designs generally provide shorter access times than dynamic memories. The **four-transistor (4-T) cell** in Fig. 8.19 is a compromise between the 6-T and 1-T cells. The load devices of the 6-T cell are eliminated, and the information is stored on the capacitances at the interior nodes. The cross-coupled transistors provide high current for sensing, as well as both true and complemented outputs. If BL, $\overline{\mathrm{BL}}$, and the wordline are all forced high, the two access transistors temporarily act as load devices for the 4-T cell, and the cell levels are automatically refreshed.

The conditions for writing information into the 4-T cell using a 3-V power supply are shown in Fig. 8.20. Following wordline activation, node D charges up to $3 - V_{TN} = 1.9$ V through access transistor M_{A1}, and node $\overline{D}$ discharges to 0 V through M_{A2} and M_{N2}. The regenerative nature of the two cross-coupled transistors enhances the speed of the write operation.

Figure 8.21 shows conditions during a read operation, in which the bitline capacitances have been precharged to 1.5 V. The voltages stored in the cell initially force M_{N1} to be off and M_{N2} on. When the wordline is raised to 3 V, charge sharing occurs on BL, and the D node drops to approximately 1.5 V. However, the voltage on $\overline{\mathrm{BL}}$ rapidly divides between M_{A2} and M_{N2}. In a conservative design, the W/L ratios of M_{N2} and M_{A2} keep the voltage at $\overline{D}$ from exceeding the threshold voltage of M_{N1} to ensure that M_{N1} remains off. As the sense amplifier responds and drives the bitlines to 0 and 3 V, the 1 level within the cell is also restored to its original value through the access transistors. When the wordline drops, the cross-coupled transistors fully discharge the 0 node to 0 V.

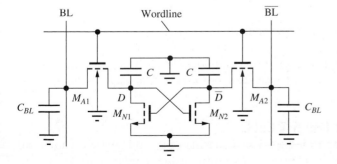

Figure 8.19 Four-transistor (4-T) dynamic memory cell.

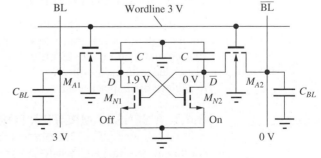

Figure 8.20 Writing data into the 4-T memory cell.

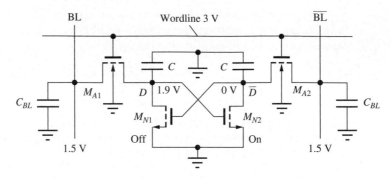

Figure 8.21 Reading data from the 4-T cell.

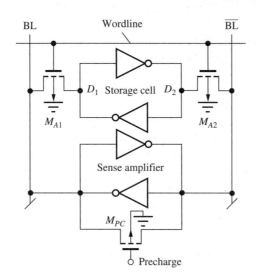

Figure 8.22 Memory array that includes a sense amplifier.

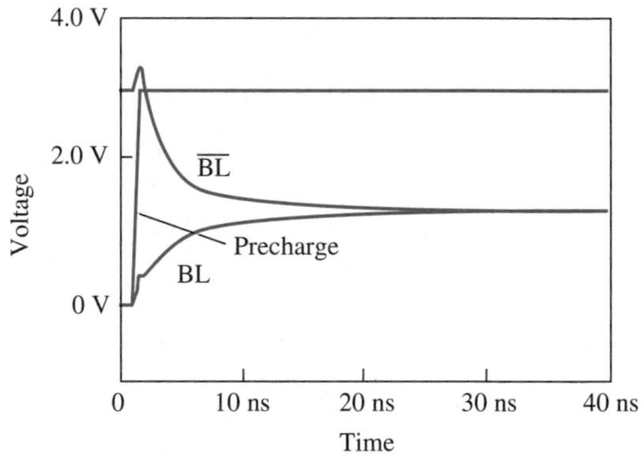

Figure 8.23 Results of SPICE simulation of the bitline voltage waveforms during the precharge operation.

EXERCISE: What is the drain current of M_{A2} in Fig. 8.21 just after the wordline is raised to 3 V if all the devices have $W/L = 2/1$? Use $K_n' = 60$ μA/V^2, $V_{TO} = 0.7$ V, $\gamma = 0.5 \sqrt{V}$, and $2\phi_F = 0.6$ V.

ANSWER: 280 μA

8.4 SENSE AMPLIFIERS

The sense amplifiers for the cells discussed in the previous sections must detect the small currents that run through the access transistors of the cell or the small voltage difference that arises from charge sharing and then rapidly amplify the signal up to full on-chip logic levels. One sense amplifier is associated with each bitline or bitline pair. The regenerative properties of the latch circuit are used to achieve high-speed sensing.

8.4.1 A SENSE AMPLIFIER FOR THE 6-T CELL

A basic sense amplifier that can be used with the 6-T cell consists of a two-inverter latch plus an additional **precharge transistor,** as shown in Fig. 8.22. Transistor M_{PC} is used to force the latch to operate at the unstable equilibrium point, originally noted in Fig. 8.4 with equal voltages at BL and

$\overline{BL}$. When the precharge device is on, it operates in the triode region and represents a low-resistance connection between the two bitlines. As long as transistor M_{PC} is on, the two nodes of the sense amplifier are forced to remain at equal voltages.

Figure 8.23 shows waveforms from a SPICE simulation of the precharge operation. The voltages on BL and $\overline{BL}$ begin at 0 V and 3 V. These levels are arbitrary, but they result from a preceding read or write operation. At $t = 1$ ns, the precharge signal turns on, forcing the two bitlines toward the same potential. The time required for the latch to reach the $v_O = v_I$ state in the simulation, approximately 30 ns, is limited by the W/L ratio of the precharge device, the capacitance of the bitlines, and the current drive capability of the inverters in the latch. In the simulation in Fig. 8.23, the precharge transistor is a 50/1 device, $C_{BL} = 500$ fF, and all devices in the sense amplifier have $W/L = 50/1$. One problem with this simple sense amplifier is its relatively slow precharge of the bitlines. Precharge must remain active until the bitline voltages are equal, or sensing errors may occur.

Once equilibrium is reached, the precharge transistor can be turned off, and the precharge level will be maintained temporarily on the bitline capacitances. The wordline can then be activated to read the cell, as demonstrated previously in Fig. 8.10.

EXAMPLE 8.3   CURRENT AND POWER IN THE PRECHARGED SENSE AMPLIFIER

This example evaluates the currents in a sense amplifier during the precharge phase, when significant power can be consumed.

PROBLEM   Find the currents in the transistors in the latch in Fig. 8.22 when the precharge transistor is turned on and the circuit has reached a steady-state condition. Use $V_{DD} = 3$ V and assume $W/L = 2/1$ for all devices; for the NMOS transistors, $K'_n = 60$ μA/V^2, $V_{TO} = 0.7$ V, $\gamma = 0.5$ V$^{1/2}$, and $2\phi_F = 0.6$ V; for the PMOS devices, $K'_p = 25$ μA/V^2, $V_{TO} = -0.7$ V, $\gamma = 0.75$ V$^{1/2}$, and $2\phi_F = 0.6$ V.

SOLUTION   **Known Information and Given Data:** The sense amplifier utilizes a two-inverter latch with a precharge transistor, as in Fig. 8.22, with $V_{DD} = 3$ V. All W/L values are 2/1; for the NMOS transistors, $K'_n = 60$ μA/V^2, $V_{TO} = 0.7$ V, $\gamma = 0.5$ V$^{1/2}$, and $2\phi_F = 0.6$ V; for the PMOS devices, $K'_p = 25$ μA/V^2, $V_{TO} = -0.7$ V, $\gamma = 0.7$ V$^{1/2}$, and $2\phi_F = 0.6$ V.

Unknowns: Find the drain currents in the transistors and the power dissipated by the sense amplifier.

Approach: The four-transistor latch is forced to the unstable equilibrium point by the precharge transistor M_{PC}. Because of the circuit symmetry, the two output voltages will be the same and the current through M_{PC} will be zero when the steady-state condition is reached. The output voltages are found by equating the drain currents of the two transistors. We must identify the region of operation of the devices and can then calculate the drain currents using the equation appropriate for the region of operation. Once the output voltages are determined, they will be used to find the drain currents.

Assumptions: Each inverter forming the latch is composed of the two-transistor CMOS inverter in Fig. 7.2(a).

Analysis: Since the output voltages will be the same on both sides of the latch, $V_{GS} = V_{DS}$ for the NMOS devices, and $V_{SG} = V_{SD}$ for the PMOS devices. Hence, we immediately know that all the transistors are saturated! We also note that all the source-body voltages are zero, allowing us to disregard the body effect in the calculation. Since the drain current of M_{PC} is zero due to its zero drain-source voltage, the drain currents of the PMOS and NMOS transistors must be identical on both sides of the latch. The source gate voltage of the NMOS device is V_O and that of the PMOS device is $3\,V - V_O$. Equating the two drain currents yields a single quadratic equation that can be solved for V_O:

$$\frac{K'_p}{2}\left(\frac{W}{L}\right)(V_{SG} + V_{TP})^2 = \frac{K'_n}{2}\left(\frac{W}{L}\right)(V_{GS} - V_{TN})^2$$

$$\frac{1}{2}\left(\frac{25\,\mu A}{V^2}\right)\left(\frac{2}{1}\right)(3 - V_O - 0.7)^2 = \frac{1}{2}\left(\frac{60\,\mu A}{V^2}\right)\left(\frac{2}{1}\right)(V_O - 0.7)^2$$

$$35V_O^2 + 31V_O - 102.9 = 0 \rightarrow V_O = 1.33\,V$$

The NMOS drain current is then

$$i_D = \frac{1}{2}\left(\frac{60\,\mu A}{V^2}\right)\left(\frac{2}{1}\right)(1.33 - 0.7)^2 = 23.6\,\mu A$$

The currents on both sides of the cell are equal, so the power dissipation is

$$P = 2i_D V_{DD} = 2(23.5\,\mu A)(3\,V) = 0.140\,mW$$

Check of Results: We have found the two required currents and the power dissipation. As a check, we can independently calculate the PMOS drain current

$$i_D = \frac{1}{2}\left(\frac{25\,\mu A}{V^2}\right)\left(\frac{2}{1}\right)(3 - 1.33 - 0.7)^2 = 23.5\,\mu A$$

which agrees with the NMOS calculation for the NMOS device.

Discussion: Although the power in this simple sense amplifier appears to be small, we must realize that thousands of these sense amplifiers can be simultaneously active in a large memory chip, so that the power dissipation is extremely important. For example, if $2^{10} = 1024$ of these sense amplifiers are on at once, then the power would be 144 mW. To minimize this power, the amount of time that these latches remain in the precharge state must be minimized. More complex clocked sense amplifiers have been developed to minimize this component of power dissipation.

EXERCISE: Repeat Ex. 8.3 if the transistors all have $W/L = 5/1$.

ANSWERS: 59.0 μA; 0.350 mW

EXERCISE: Repeat Ex. 8.3 if V_{DD} is changed to 2.5 V with $V_{TON} = 0.6$ V and $V_{TOP} = -0.6$ V.

ANSWERS: 15.6 μA; 78.0 μW ($V_O = 1.11$ V)

8.4.2 A SENSE AMPLIFIER FOR THE 1-T CELL

The two-inverter latch that was used with the 6-T static memory cell in Fig. 8.22 can also be used as the sense amplifier for the 1-T cell. The 1-T cell and a latch with precharge transistor M_{PC} are shown attached to the bitline in Fig. 8.24, and the waveforms associated with the sensing operation are shown in Fig. 8.25. Because the cross-coupled latch is highly sensitive, it remains connected across two bitlines in order to balance the capacitive load on the two nodes of the latch and to share the sense amplifier between two columns of cells.

In the circuit in Fig. 8.24, the 1-T cell is shown connected to BL. In most designs, a dummy cell with its own access transistor and storage capacitor would be connected to $\overline{BL}$ to try to balance the

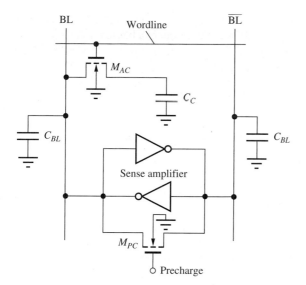

Figure 8.24 Simple sense amplifier for the 1-T cell.

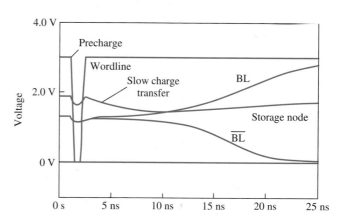

Figure 8.25 Single-ended sensing of the 1-T cell.

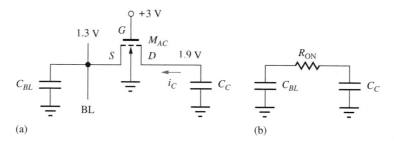

Figure 8.26 Voltages on the access transistor immediately following activation of the wordline.

switching transients on the two sides of the latch, as well as to improve the response time of the sense amplifier. Use of the dummy cell is discussed in more detail in the section on clocked sense amplifiers.

During the **precharge phase** of the circuit, as shown by the waveforms in Fig. 8.25, the bitlines are forced to a level determined by the relative W/L ratios of the NMOS and PMOS transistors in the sense amplifier. Following turnoff of the precharge transistor, the data in the storage cells is accessed by raising the wordline, enabling charge sharing between cell capacitance C_C and bitline capacitance C_{BL}. In the simulation, the sense amplifier amplifies the small difference and generates almost the full 3-V logic levels on the two bitlines in approximately 25 ns.

A closer inspection of the bitline waveform indicates that the desired charge sharing is actually not occurring in this circuit. The voltage on the storage node does not drop instantaneously because of the large on-resistance of the access transistor. Let us explore this problem further by looking at the voltages applied to the access transistor immediately following activation of the wordline, as in Fig. 8.26.

For M_{AC}:
$$V_{TN} = 0.70 + 0.5\left(\sqrt{1.3 + 0.6} - \sqrt{0.6}\right) = 1.0 \text{ V}$$

$$V_{GS} = 3 - 1.3 = 1.7 \text{ V} \quad \text{and} \quad V_{DS} = 1.9 - 1.3 = 0.6 \text{ V}$$

Because $V_{GS} > V_{DS}$, the transistor is operating in the triode region as desired, but for these voltages, the initial current through the MOSFET is quite small. Assuming $W/L = 1/1$,

$$i_D = K_n'\frac{W}{L}\left(v_{GS} - V_{TN} - \frac{v_{DS}}{2}\right)v_{DS} = (60 \times 10^{-6})\left(\frac{1}{1}\right)\left(1.7 - 1.0 - \frac{0.6}{2}\right)0.6 = 14.4 \text{ μA}$$

A measure of the charge-sharing response time is the initial discharge rate of the cell capacitance, given by $\Delta v / \Delta t = i_D / C_C$. In the simulation, $C_C = 25$ fF and the initial discharge rate is 0.24 V/ns. This relatively low discharge rate is responsible for the incomplete charge transfer and shallow slope on the storage node waveform. Even though charge sharing is not complete in this circuit, the small initial current drawn from the storage cell is still enough to unbalance the sense amplifier and cause it to reach the proper final state.

> **EXERCISE:** What are the initial values of R_{on} and $\tau = R_{on}C_C$ in the circuit in Fig. 8.26?
>
> **ANSWERS:** 23.8 kΩ, 0.59 ns

8.4.3 THE BOOSTED WORDLINE CIRCUIT

In high-speed memory circuits, every fraction of a nanosecond is precious, and some DRAM designs use a separate voltage level for the wordline. The additional level raises the voltage corresponding to a 1 level in the 1-T cell and substantially increases i_D during cell access. The waveforms for the circuit of Fig. 8.24 are repeated in Fig. 8.27 for the case in which the wordline is driven to $+5$ V instead of $+3$ V (referred to as a **boosted wordline**). In this case the cell voltage becomes 3.7 V and the initial current from the cell is increased to 216 μA, 15 times larger (see Prob. 8.14)! A much more rapid charge transfer is evident in the storage node waveform in Fig. 8.27, where the sense amplifier has developed a 1.5-V difference between the two bitlines approximately 10 ns after the wordline is raised. In the original case in Fig. 8.25, approximately 15 ns were required to reach the same bitline differential.

8.4.4 CLOCKED CMOS SENSE AMPLIFIERS

In the sense amplifiers in Figs. 8.22 and 8.24, there is considerable current between the two power supplies during the precharge phase. In addition, a relatively long time is required for precharge. Because a large number of sense amplifiers will be active simultaneously—128 in the 256-Mb memory chip example—minimizing power dissipation in the individual sense amplifier is an important design consideration. By introducing a more sophisticated clocking scheme, sense amplifier power dissipation can be reduced.

Clocked sense amplifiers were originally introduced in NMOS technology to reduce power dissipation in saturated load and depletion-mode load sense circuits. The same techniques are also routinely used in CMOS sense amplifiers; an example of such a circuit is shown in Fig. 8.28. For sensing 1-T cells, a dummy cell is used that has one-half the capacitance of the 1-T cell. In Fig. 8.28, the bottom plate of all the capacitors has been connected to V_{DD} instead of to 0 V. This change represents another design alternative but does not alter the basic theory of operation.

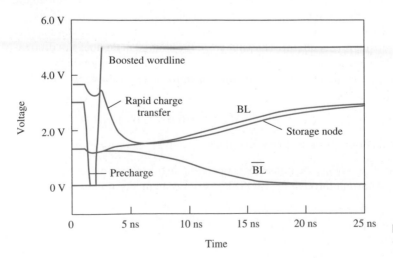

Figure 8.27 1-T sensing with wordline voltage boosted to 5 V.

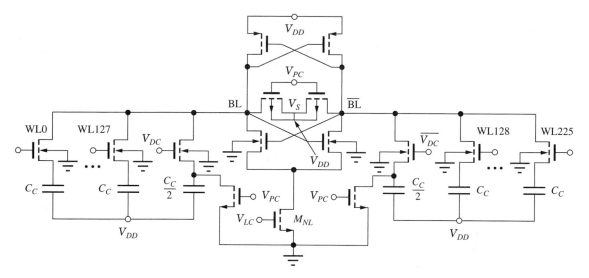

Figure 8.28 Clocked CMOS sense amplifier showing an array of 1-T memory cells and a dummy cell on each side of the sense amplifier. The right-hand dummy cell ($C_C/2$) is used with cells 0 to 127 and the left-hand dummy cell is used with cells 128 to 255.

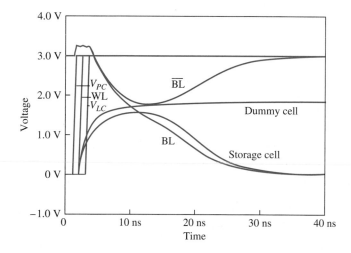

Figure 8.29 Simulated behavior of the clocked CMOS sense amplifier.

During precharge, V_{PC} is held at 0 V, and the bitlines are precharged to V_{DD} through the two PMOS transistors. After the precharge signal is removed, the access transistor of the addressed cell is activated, and the corresponding bitline drops slightly. The magnitude of the change depends on the data stored in the cell. A relatively large change occurs if cell voltage is 0, and a small change occurs if the cell voltage is $V_{DD} - V_{TN}$.

A dummy cell is required in this circuit to ensure that a voltage difference of the proper polarity will always develop between the two bitlines following cell access. Dummy cell capacitance is designed to be one-half the capacitance of the data storage cell, and cell voltage is always preset to 0 V by V_{PC}. During charge sharing, the selected dummy cell causes the corresponding bitline to drop by an amount equal to one-half the voltage drop that occurs when a 0 is stored in the 1-T cell. Thus, a positive voltage difference exists between BL and $\overline{BL}$ if a 1 is stored in the 1-T cell, and a negative difference exists if a 0 is stored in the cell.

Following charge sharing, the lower part of the CMOS latch is activated by turning on M_{NL}, and the small difference between the two bitline voltages is amplified by the full cross-coupled CMOS latch. Simulated waveforms for the clocked CMOS sense amplifier are shown in Fig. 8.29. For clarity, the three clock signals, precharge V_{PC}, wordline WL, and latch clock V_{LC} have each been staggered by 1 ns in the simulation. Note that both BL and $\overline{BL}$ are driven above 3 V by the coupling of the clock signal to the bitlines, but the voltage difference is maintained, and the latch responds properly.

No static current paths exist through the latch during the precharge period, and hence only transient switching currents exist in the sense amplifier. Although there are many variants and refinements to the circuit in Fig. 8.28, the basic circuit ideas presented here form the heart of the sense amplifiers in most dynamic RAMs.

To achieve high-speed precharge and sensing with minimum size transistors, the bitline capacitance C_{BL} in either static or dynamic RAMs must be kept as small as possible. This requirement restricts the number of cells that can be connected to each bitline. Many clever techniques have been developed to segment the bitlines in order to reduce the size of C_{BL}, and the interested reader is referred to the references. In particular, information can be found in the annual digests of the IEEE International Solid-State Circuits Conference [2], the Custom Integrated Circuits Conference (CICC) [9], and the Symposium on VLSI Circuits [10], as well as in the yearly Special Issues of the *IEEE Journal of Solid State Circuits* [11]. New information on memory-cell technology is discussed yearly at the IEEE International Electron Devices Meeting [4], the Symposium on VLSI Technology [10], and in the *IEEE Transactions on Electron Devices* [12].

8.5 ADDRESS DECODERS

Two additional major blocks in design of a memory are the *row address* and *column address decoders* shown in the block diagram in Fig. 8.2. The row address circuits decode the row address information to determine the single wordline that is to be activated. The decoded column address information is then used to select a bit or group of bits from the selected word. This section first explores NOR and NAND row address decoders and then discusses the use of an NMOS pass-transistor tree decoder for selecting the desired data from the wide internal memory word. Dynamic logic techniques for implementing these decoders are also introduced.

8.5.1 NOR DECODER

Figure 8.30 is the schematic for a 2-bit **NOR decoder.** The circuit must fully decode all possible combinations of the input variables and is equivalent to at least four 2-input NOR gates (2^N N-input gates in the general case). In the circuit, true and complemented address information is fed through an array of NMOS transistors. Each row of the decoder contains two FETs, with each gate connected to one of the desired address bits or its complement and the two drains connected in parallel to the output line being enabled. At the end of each row is a depletion-mode load device to pull the row output high, which occurs only if all the inputs to that row are low. Only one output line will be high for any given combination of input variables; the rest will be low. Each row corresponds to one possible address combination.

> **EXERCISE:** What are the sizes of the transistors for the NOR decoder in Fig. 8.30 based on the reference inverter in Fig. 6.29(d)? What is the static power dissipation of this decoder?
>
> **ANSWERS:** Depletion-mode load devices: 1.81/1; NMOS switching devices: 2.22/1; 1.00 mW

8.5.2 NAND DECODER

Figure 8.31 is a NAND version of the same decoder. Again, true and complemented address information is fed through the array. For the **NAND decoder,** all outputs are high except for the single row in which the transistor gates are all at a 1 level. Because additional driver circuits are normally required between the decoder and the highly capacitive wordline, the logical inversion that is needed to actually drive wordlines in a memory array is easily accommodated.

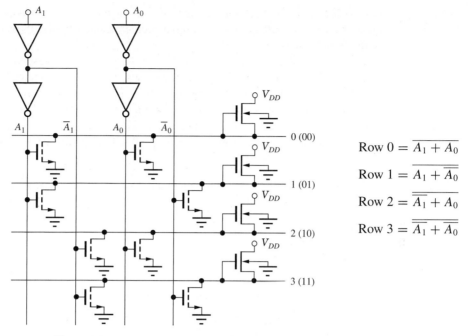

Figure 8.30 NMOS static NOR address decoder.

$$\text{Row } 0 = \overline{A_1 + A_0}$$

$$\text{Row } 1 = \overline{A_1 + \overline{A_0}}$$

$$\text{Row } 2 = \overline{\overline{A_1} + A_0}$$

$$\text{Row } 3 = \overline{\overline{A_1} + \overline{A_0}}$$

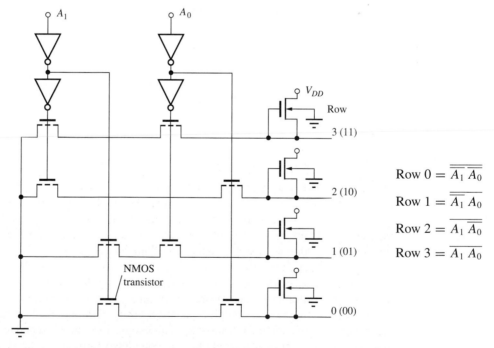

Figure 8.31 NMOS NAND decoder circuit.

$$\text{Row } 0 = \overline{\overline{A_1}\ \overline{A_0}}$$

$$\text{Row } 1 = \overline{\overline{A_1}\ A_0}$$

$$\text{Row } 2 = \overline{A_1\ \overline{A_0}}$$

$$\text{Row } 3 = \overline{A_1\ A_0}$$

As for standard NOR and NAND gates, the stacked series structure of the NAND gate tends to be slower than the corresponding parallel NOR structure, particularly if minimum-size devices are used throughout the decoder.

The NMOS static decoder circuits in Figs. 8.30 and 8.31 cause power consumption problems in high-density memories. In the memory array, 1 wordline will be high for a given address, and $2^N - 1$

will be low. For static NMOS circuits, $2^N - 1$ of the individual NOR gates in the NOR decoder dissipate power simultaneously. A similar problem occurs if NMOS inverters are used at the output of the NAND decoder.

Standard CMOS circuits offer low power dissipation but cause layout and area problems when there are a large number of inputs, because each input must be connected to both an NMOS and a PMOS transistor, which doubles the number of devices in the array. A full CMOS decoder can be more efficiently implemented as a combination of an NMOS NOR array and a PMOS NAND array. However, because memories are generally used in clocked systems, they can use dynamic decoders, which consume low power and require only a few PMOS transistors. We introduce these next.

EXERCISE: **What are the sizes of the load devices in the NAND decoder in Fig. 8.31 if the** W/L **ratios of the switching devices are all 2/1? Base your design on the reference inverter in Fig. 6.29(d).**

ANSWER: **1.63/1**

8.5.3 DECODERS IN DOMINO CMOS LOGIC

Domino CMOS, introduced in the last chapter in Sec. 7.8, is highly useful in the design of compact decoder circuits for memory circuits. As an example, Fig. 8.32 is the schematic for one row of a domino CMOS NAND decoder for a 3-bit address. In this circuit, a discharge path exists through the function block only for address $A_0 A_1 A_2 = 111$, and only the single-addressed wordline makes a low-to-high transition at the output of the inverter. This structure ensures that the voltage changes on only one of the high-capacitance wordlines during a given row access. The output inverter can be designed to drive the high-capacitance load by utilizing the cascade buffer techniques that we studied in Sec. 7.9 of the last chapter.

Figure 8.33 is the circuit schematic for a full 3-bit address decoder using the NAND decoder array of Fig. 8.31; here the load devices in Fig. 8.32 have been replaced with clocked PMOS transistors, and an NMOS clock transistor has been added to the beginning of each row. A CMOS inverter is connected to each output line to complete the domino CMOS implementation.

8.5.4 PASS-TRANSISTOR COLUMN DECODER

The column address decoder of the memory in Fig. 8.2 must choose a group of data bits—usually 1, 4, or 8 bits—from the much wider word that has been selected by the row address decoder. Another form of data selection circuit using NMOS **pass-transistor logic** is shown in Fig. 8.34. For a large number of data bits, the pass-transistor circuit technique requires far fewer transistors than would the more direct approach using standard NOR and NAND gates.

The pass-transistor decoder structure is quite similar to an analog multiplexer in which the transistors behave as switches. "On" switches pass the data from one node to the next, whereas "off" switches prevent the data transfer. In the pass-transistor implementation, true and complement address information is fed through a transistor array, with one level of the array corresponding to each address bit. Although all transistors with a logic 1 on their respective gates will be on, the tree structure ensures that only a single path is completed through the array for each combination of inputs. In Fig. 8.34, an address of 101 is provided to the array, and the transistors indicated in blue all have a logic 1 on their respective gates, creating a conducting channel region in each. In this case, a completed conducting path connects input column 5 to the data output.

Examples of the conducting paths through a three-level pass-transistor array for input data of 0 and 1 are shown in Fig. 8.35 for the case of $V_{DD} = 3$ V. For a 0 input equal to 0 V, the output node capacitor C is discharged to zero through the series combination of the three pass transistors. However, for a 1 input voltage of 3 V, the output of the first pass transistor is one threshold voltage

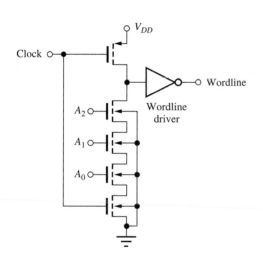

Figure 8.32 One row of a 3-bit NAND decoder in domino CMOS logic.

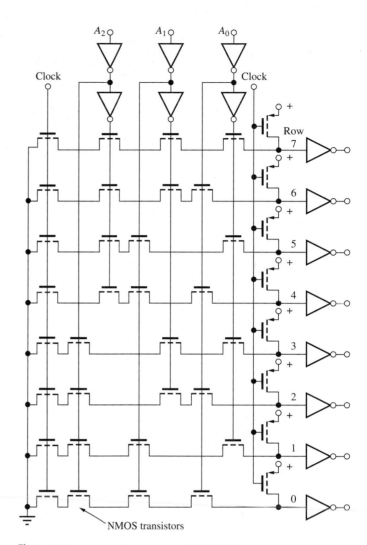

Figure 8.33 Complete 3-bit domino CMOS NAND address decoder.

below the gate voltage of 3 V. Using the NMOS parameters from earlier in the chapter, we find that the output voltage is

$$V_O = V_G - V_{TN} = 3 - 1.1 = 1.9 \text{ V}$$

The other node voltages can reach this same potential, so the output capacitance will charge to $V_G - V_{TN} = 1.9$ V, regardless of the number of levels in the array. The data buffer at the output must be designed to have a switching threshold below 1.9 V in order to properly restore the logic level at the output.

Full logic levels can also be achieved in pass-transistor logic for both 1 and 0 inputs by replacing each NMOS transistor with a CMOS transmission gate. However, this significantly increases the area and complexity of the design with little actual benefit because the data buffer can easily be designed to compensate for the loss in signal level through the NMOS (or PMOS) pass-transistor array. In addition to doubling the number of transistors in the array, the full CMOS version requires distribution of both true and complement address information to each transmission gate. However, design techniques to simplify the layout of CMOS transmission gate arrays do exist [13].

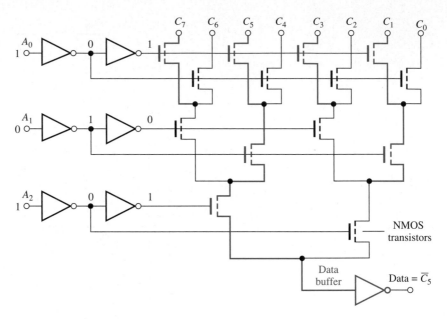

Figure 8.34 3-bit column data selector using pass-transistor logic.

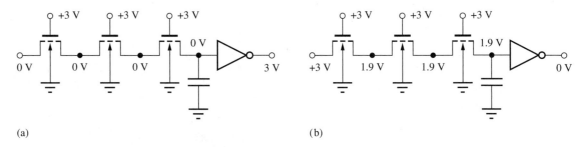

(a)   (b)

Figure 8.35 Data transmission through the pass-transistor decoder: (a) 0 input data, (b) 1 input data.

EXERCISE: What are the voltage levels at the nodes in Fig. 8.35(a) and 8.35(b) if the gate voltages are 5 V instead of 3 V? Assume $V_{TO} = 0.70$ V, $\gamma = 0.5 \sqrt{V}$, and $2\phi_F = 0.6$ V. What is the largest γ for which the output of the pass-transistor is at least 3 V?

ANSWERS: 0 V, 3.00 V; $1.16 \sqrt{V}$

EXERCISE: What would be the actual voltage at the output of the CMOS inverter in Fig. 8.35(b) if the inverter utilized a symmetrical design operating from a 3-V supply and $V_{TN} = -V_{TP} = 0.7$ V?

ANSWER: 68.6 mV

8.6 READ-ONLY MEMORY (ROM)

Read-only memory (ROM) is another form of memory often required in digital systems, and many common applications exist. Many microprocessors use microcoded instruction sets that reside in ROM, and a portion of the operating system for personal computers usually resides in ROM. The fixed programs for microcontrollers, often called *firmware,* are also typically stored in ROM, and

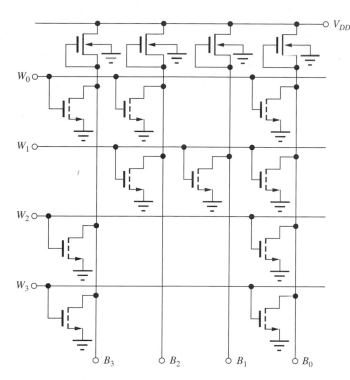

Figure 8.36 Basic structure of an NMOS static ROM with four 4-bit words:
$W_0 = 0010$; $W_1 = 1000$; $W_2 = 0110$; $W_3 = 0110$.

TABLE 8.1
Contents of ROM in Fig. 8.36

WORD	DATA
0	0010
1	1000
2	0110
3	0110

cartridges for home video games are simply ROMs that contain the firmware programs that define the characteristics of the particular game. Plug-in modules for hand-held scientific calculators are another application of read-only memory.

Most ROMs are organized as an array of 2^N words, where each word contains the number of bits required by the intended application. Common values are 4, 8, 12, 16, 18, 24, 32, 48, 64, and so on. Figure 8.36 shows the structure of a static NMOS ROM using depletion-mode load devices. This particular ROM contains four 4-bit words. Each column corresponds to 1 bit of the stored word, and an individual word W_i is selected when the corresponding wordline is raised to a 1 state by an address decoder similar to that in RAM. For example, this ROM could be driven by a NOR decoder circuit similar to Fig. 8.30.

An NMOS transistor can be placed at the intersection of each row and column within the array. In this ROM, the gate of the transistor is tied to the wordline and the drain is connected to the output data line. If connections to an NMOS transistor exist at a given array site, then the corresponding output data line is pulled low when the word is selected. If no FET exists, then the data line is maintained at a high level by the load device. Thus, the presence of an FET corresponds to a 0 stored in the array and the absence of an FET corresponds to a 1 stored in ROM. The particular data pattern stored in the array is often referred to as the **array personalization.** Table 8.1 contains the contents of the ROM array in Fig. 8.36.

Information can be personalized in the ROM array in many ways; we mention three possibilities here. Suppose an FET is fabricated at every possible site within the array. One method of storing the desired data is to eliminate the contact between the drain and data line wherever a 1 bit is to be stored. This design yields a high capacitance on the wordline because a gate is connected to the wordline at every possible site, but a low capacitance exists on the output line because only selected drain diffusions are connected to the output lines. A second technique is to use ion implantation to alter the MOSFET threshold voltage of the FETs wherever a 1 is to be stored. If the threshold is raised

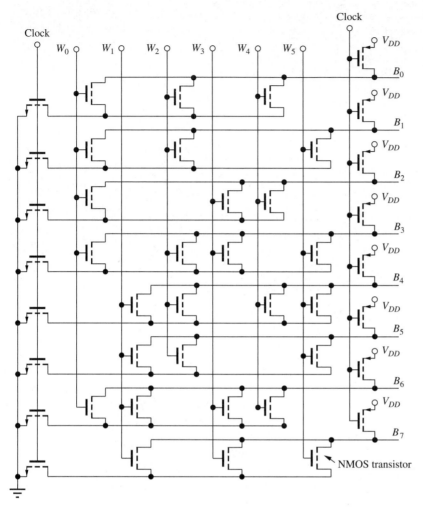

Figure 8.37 Domino CMOS ROM containing six 8-bit words. The array contains NMOS transistors in which substrate connections have been eliminated for clarity.

TABLE 8.2
Contents of ROM in Fig. 8.37

WORD	DATA
0	10110000
1	00001111
2	11000100
3	00110011
4	10101010
5	01000101

high enough, then the MOSFET cannot be turned on, and therefore cannot pull the corresponding data line low. A third method is to personalize the array by eliminating gate contacts instead of drain contacts.

Standard NMOS ROM circuits exhibit substantial static power dissipation. To eliminate this power dissipation, ROMs can be designed using dynamic circuitry such as domino CMOS, as demonstrated by the ROM in Fig. 8.37, which contains six 8-bit words. When the clock signal is low, the capacitance on the output data lines is precharged high. As the clock is raised to a high level, the PMOS transistors turn off, and the data bits of the addressed word are selectively discharged to zero if a transistor connection exists at a given intersection point in the array. A full domino implementation will add an inverter to each output bitline. Table 8.2 lists the contents of the ROM personalization in Fig. 8.37.

The ROMs in Figs. 8.36 and 8.37 use the NOR gate structure and are often called NOR arrays. It is also possible to use a NAND array structure, as shown in Fig. 8.38. In the NAND array, all the wordlines except the desired word are raised high. Thus, all the MOSFETs in the array are turned on except for the unselected row. If a MOSFET exists at a given cross-point in the unselected row, then

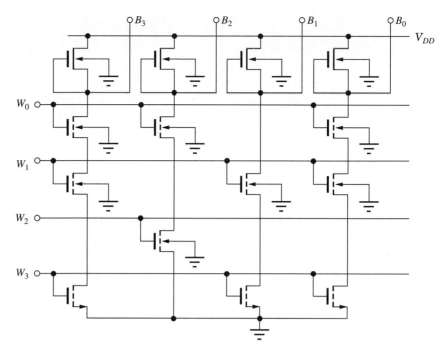

Figure 8.38 ROM based on a NAND array.

the conducting path is broken, and the data for that column will be a 1. If the MOSFET has been replaced by a connection (possibly resistive)—by making the MOSFET a depletion-mode device for example—then the corresponding data bit will be pulled low. Note that the NAND ROM array could be driven directly by the NAND decoder in Fig. 8.31.

EXERCISE: **Draw the schematic of an additional row in the ROM in Fig. 8.37 with contents of 11001101. What are the contents of the ROM in Fig. 8.38?**

ANSWERS: **NMOS transistors connected to B_5, B_4, and B_1; (0010, 0100, 1011, 0100)**

The ROMs already mentioned are all personalized at the mask level, which must be done during IC design and subsequent fabrication. If a design error occurs, the IC must be redesigned and the complete fabrication process repeated. To solve this problem, **programmable read-only memories (PROMs),** which can be programmed once from the external terminals, have been developed. **Erasable programmable read-only memories (EPROMs)** are another type of ROM. These can be erased using intense ultraviolet light and reprogrammed many times. **Electrically erasable read-only memories (EEROMs)** can be both erased and reprogrammed from the external terminals. High-density **flash memories** allow selective electrical erasure and reprogramming of large blocks of cells.

8.7 FLIP-FLOPS

Temporary storage in the form of high-speed registers is another requirement in most digital systems. The external data interface to these registers may be in either parallel or serial form and includes various **flip-flops (FFs)** and shift registers. There are many different types of flip-flops and shift registers, and this section presents several examples of circuits that can be used in static parallel registers. However, an exhaustive discussion of the various possibilities is not attempted.

ELECTRONICS IN ACTION

Flash Memory

The ever increasing number of handheld, low-power devices has given rise to a new class of memory devices known as flash memory. Cell phones maintain phone number lists as well as configuration information, digital cameras store pictures on flash memory cards, digital audio players use flash memory for storage, and most computer users carry flash memory drives in their pockets. All of these devices need to maintain stored information with no applied power but can be both read and written when powered.

Flash memory is a type of electrically eraseable, programmable ROM (EEPROM). As seen below, the basic storage cell is similar to a MOSFET with two gates, but one gate is electrically floating. The *write* process involves placing negative charge on the floating gate by driving a large positive voltage on the control gate and a positive drain-source voltage across the device. The positive control gate voltage inverts the MOSFET channel, and the large drain-source voltage accelerates electrons through the channel to the drain. The electrons gain so much kinetic energy that they are known as 'hot' electrons; they have the same energy as they would have if the device temperature were much higher. The vertical field created by the large control voltage captures some of the 'hot' electrons, and they tunnel through the thin oxide to the floating gate.

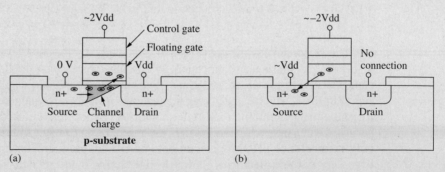

Flash memory cell structure and control voltages during write operation (left) and read operation.

Erasure occurs by removing the negative charge with a process known as Fowler-Nordheim tunneling.[1] The source is driven positive, and the control gate is set to a large negative voltage. This creates an electric field that causes electrons to slowly tunnel from the floating gate to the source. Unlike the *write* operation, the *erase* operation for flash memory is performed on a large number of cells simultaneously. This is done to amortize the rather slow erasure time over several cells, reducing the effective per cell erasure time.

The presence or absence of charge on the floating gate changes the effective threshold voltage of the control gate. Reading the state of the memory cell is done by driving the control gate with an appropriate voltage and measuring the resulting drain-source current.

Due to the proliferaton of handheld digital devices, the market for flash memory products will grow from about $50 million in 1987 to more than $25 billion dollars in 2009. This is another example of the microelectronics industry inventing a solution to a market need and rapidly creating an entirely new market segment.

---

[1] Lenzlinger, M. and Snow, E. H., "Fowler-Nordheim Tunneling into Thermally Grown SiO_2," *Journal of Applied Physics*, vol. 40, No. 1, January 1967, pp. 278–283.

8.7.1 RS FLIP-FLOP

The **RS (reset-set) flip-flop (RS-FF)** can be formed in a straightforward manner by using either two NOR gates or two NAND gates to replace the inverters in the simple latch. The desired state is stored by setting ($Q = 1$) or resetting ($Q = 0$) the flip-flop with the RS control inputs.

Figure 8.39 is the circuit for an RS-FF constructed using two-input CMOS NOR gates and corresponding to the truth table in Table 8.3. If the R and S inputs are low, they are both inactive, and the previously stored state of the flip-flop is maintained. However, if S is high and R is low, output $\overline{Q}$ is forced low, and Q then becomes high, setting the latch. If R is high and S is low, node Q is low, and $\overline{Q}$ is then forced high, resetting the latch. Finally, if both R and S are high, both output nodes are forced low, and the final state is determined by the input that is maintained high for the longest period of time. The RS = 11 state is usually avoided in logic design.

The RS-FF can also be implemented using two-input NAND gates, as shown in Fig. 8.39 and Table 8.4. In this case, the latch maintains its state as long as both the $\overline{R}$ and $\overline{S}$ inputs are high, thus maintaining a conducting channel in both NMOS transistors. If the $\overline{R}$ input is set to 0 and $\overline{S}$ is a 1, then the $\overline{Q}$ output becomes a 1, causing the Q output to be reset to 0. If $\overline{R}$ returns to a 1, the reset condition is maintained within the latch. If the $\overline{S}$ input becomes a 0, and $\overline{R}$ remains a 1, the Q output is set to a 1 and the $\overline{Q}$ output becomes 0. If $\overline{S}$ returns to a 1, the latch remains "set." In the NAND implementation, both outputs are forced to 1 when $\overline{R}$ and $\overline{S}$ are both 0.

The flip-flops in Figs. 8.39 and 8.40 utilize full CMOS implementations of the NOR and NAND gates. If static power dissipation can be tolerated when the R and S inputs are both active, then the simplified implementation of Fig. 8.41 can be used. This is essentially the two-inverter latch used in the static memory circuits, with R and S transistors added to force either the Q or $\overline{Q}$ output low. If both R and S are 1, both outputs will be 0, and both load devices and the R and S transistors will conduct current. PMOS transistors can also be used to replace the NMOS RS transistors to pull Q or $\overline{Q}$ toward V_{DD}.

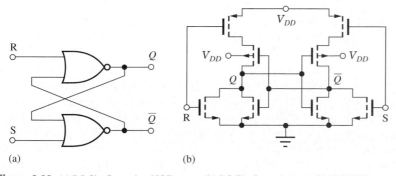

TABLE 8.3			
NOR RS Flip-Flop			
R	S	Q	$\overline{Q}$
0	0	Q	$\overline{Q}$
0	1	1	0
1	0	0	1
1	1	0	0

(a)   (b)

Figure 8.39 (a) RS flip-flop using NOR gates. (b) RS flip-flop using two CMOS NOR gates.

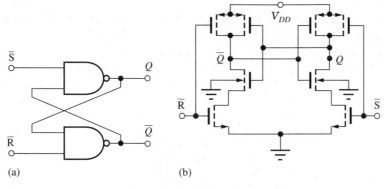

Figure 8.40 (a) A NAND $\overline{\text{RS}}$ flip-flop. (b) $\overline{\text{RS}}$ flip-flop implemented with two CMOS NAND gates.

TABLE 8.4
NAND RS Flip-Flop

$\overline{R}$	$\overline{S}$	Q	$\overline{Q}$
1	1	Q	$\overline{Q}$
0	1	0	1
1	0	1	0
0	0	1	1

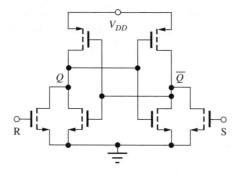

Figure 8.41 Simplified RS flip-flop using two NOR gates.

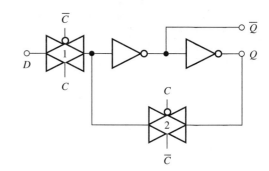

Figure 8.42 D latch.

8.7.2 THE D-LATCH USING TRANSMISSION GATES

The CMOS transmission gate introduced in Sec. 7.10 is used to implement another basic form of storage element called the **D latch,** as shown in Fig. 8.42. When the clock input $C = 1$, transmission gate 1 is on and transmission gate 2 is off. The state of the D input is stored on the capacitance at the input of the first inverter and transferred through the inverter pair to the $\overline{Q}$ and Q outputs. The Q output equals the D input as long as $C = 1$. When the clock changes state to $C = 0$, the D input is disabled, and the state of the inverter pair is latched through transmission gate 2. The state at Q and $\overline{Q}$ remains constant as long as $C = 0$.

8.7.3 A MASTER-SLAVE D FLIP-FLOP

The **master-slave D flip-flop (D-FF)** in Fig. 8.43 is a storage element in which the data is stable during both **phases of the clock.** Master-slave D-FFs can be directly cascaded to form a shift register. This D-FF is formed by a cascade connection of two D-type latches operating on opposite phases of the clock.

When the clock $C = 1$, transmission gates 1 and 4 are closed and 2 and 3 are open, resulting in the simplified circuit in Fig. 8.43(b). The D input is connected to the input of the first inverter pair, and the D input data appears at the output of the second inverter. The second pair of inverters is connected as a latch, holding the information previously placed on the input to the second inverter pair.

When the clock changes states and $C = 0$, as depicted in Fig. 8.43(c), transmission gate 1 disables the D input, and transmission gate 2 latches the information that was on the D input just before the clock state change. During the clock transition, data from the D input is maintained temporarily on the nodal capacitances associated with the first two inverters. Transmission gate 3

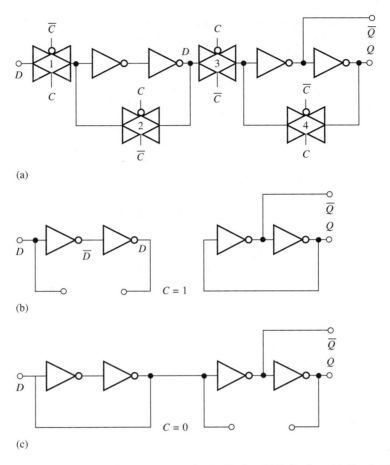

(a)

(b)

(c)

Figure 8.43 Master-slave D flip-flop. (a) Complete master-slave D flip-flop. (b) D flip-flop with $C = 1$. (c) D flip-flop with $C = 0$.

propagates the stored data onto the $\overline{Q}$ and Q outputs. Q is now equal to the data originally on the D input when C was equal to 1.

Note that the data at the output of the master-slave flip-flop is constant during both phases of the clock, except for the time required for the latches to change state. A path should never exist completely from the D input to the Q or $\overline{Q}$ outputs.

SUMMARY

- *Memory organization:* In Chapter 8, we have explored basic MOS memory circuits, including both random-access memory (RAM) and read-only memory (ROM), sometimes called read-only storage (ROS). The internal organization of IC memories was presented, and examples of the major building blocks of a memory chip, including row and column address decoders, sense amplifiers, wordline drivers, and output buffers were investigated.

- *Static RAM cells:* A six-transistor cell is normally used as the storage element in the SRAM, and the integrity of the stored data is maintained as long as power is applied to the circuit.

- *Dynamic RAM cells:* The one-transistor cell is utilized in most high-density dynamic RAM (DRAM) designs. Dynamic memory circuits store the information temporarily as the charge on a capacitor, and the data must be periodically refreshed to prevent loss of the information. Read operations also destroy the data in the cell, and it must be written back into the cell as part of the read cycle. Dynamic memories can also be based on four-transistor dynamic memory cells.

- *Decoder design:* Static NAND and NOR array structures are often utilized in address decoder circuits. Domino CMOS, introduced in Chapter 7 can also be used effectively to reduce power consumption in many applications; a domino CMOS decoder was presented.

- *Pass transistors:* Pass-transistor logic was introduced as one method for simplifying the design of the column decoding circuitry in a RAM.

- *ROMs:* The structure of read only memory or ROM is very similar to that of RAM, but the data is embedded in the physical design of the circuitry.

- *Register elements:* Bistable storage elements based on the cross-coupled inverter pair were introduced, including the RS flip-flop, the dynamic D flip-flop or D latch, and the master-slave flip-flop. Flip-flops use the two stable equilibrium points of a cross-coupled pair of inverters to represent the binary data.

- *Sense amplifiers:* The bistable latch also forms the heart of many sense amplifier circuits, and the unstable equilibrium point of the latch plays a key role in the design of high-speed sensing circuits. Both static and clocked dynamic sense amplifiers can be used in memory designs.

KEY TERMS

Array personalization
Bistable circuit
Bitline
Bitline precharge
Bitline capacitance
Boosted wordline
Cell capacitor
Charge sharing
Charge transfer
Clocked sense amplifier
Clock phases
Column address decoder
Cross-coupled inverter
Domino CMOS
Dynamic random-access memory (DRAM)
D latch
Electrically erasable read-only memory (EEROM)
Erasable programmable read-only memory (EPROM)
Flash memory
Flip-flop (FF)
Four-transistor (4-T) cell

Latch
Master-slave D flip-flop (D-FF)
NAND decoder
NOR decoder
One-transistor (1-T) cell
Pass-transistor logic
Precharge phase
Precharge transistor
Programmable read-only memory (PROM)
Random-access memory (RAM)
Read-only memory (ROM)
Read-only storage (ROS)
Read operation
Refresh operation
Row address decoder
RS flip-flop (RS-FF)
Sense amplifier
Six-transistor (6-T) SRAM cell
Static random-access memory (SRAM)
Unstable equilibrium point
Wordline
Wordline driver
Write operation

REFERENCES

1. J. Wood and R. G. Wood, "The use of insulated-gate field-effect transistors in digital storage systems," *ISSCC Digest of Technical Papers,* pp. 82–83, February 1965.

2. *Digest of Technical Papers of the IEEE International Solid-State Circuits Conference* (ISSCC), February of each year.

3. W. M. Regitz and J. A. Karp, "A three-transistor cell, 1024-bit, 500 ns MOS RAM," *ISSCC Digest of Technical Papers*, pp. 36–39, February 1970.

4. *Digest of the IEEE International Electron Devices Meeting* (IEDM), December of each year.

5. Robert H. Dennard; patent 3,387,286 assigned to the IBM Corporation.

6. H. C. Lin and L. W. Linholm, "An optimized output stage for MOS integrated circuits," *IEEE JSSC,* vol. SC-10, pp. 106–109, April 1975.

7. R. C. Jaeger, "Comments on 'An optimized output stage for MOS integrated circuits,'" *IEEE JSSC,* vol. SC-10, pp. 185–186, June 1975.

8. Mikio Asakura et al., "A 34 ns 256 Mb DRAM with boosted sense-ground scheme," *ISSCC Digest of Technical Papers*, pp. 140–141, 324, February 1994.

9. *Digest of the IEEE Custom Integrated Circuits Conference,* April of each year.

10. *Digests of the Symposium on VLSI Circuits and the Symposium on VLSI Technology,* June of each year.

11. *IEEE Journal of Solid-State Circuits,* monthly.

12. *IEEE Transactions on Electron Devices,* monthly.

13. Carver Mead and Lynn Conway, *Introduction to VLSI Systems,* Addison-Wesley, Reading, MA: 1980.

14. R. M. Swanson and J. D. Meindl, "Ion-implanted complementary MOS transistors in low-voltage circuits," *IEEE Journal of Solid-State Circuits,* vol. SC-7, no. 2, pp. 146–152, April 1972.

15. J. D. Meindl and J. A. Davis, "The fundamental limit on binary switching energy for terascale integration," *IEEE Journal of Solid-State Circuits,* vol. SC-35, no. 10, pp. 1515–1516, October 2000.

16. J. R. Houser, "Noise margin criteria for digital logic circuits," *IEEE Trans. on Education,* vol. 36, no. 4, pp. 363–368, November 1993.

17. J. Lohstroh, E. Seevinck, and J. Degroot, "Worst-case static noise margin criteria for logic circuits and their mathematical equivalence," *IEEE Journal of Solid-State Circuits,* vol. SC-18, no. 6, pp. 803–806, December 1983.

PROBLEMS

Unless otherwise specified, use $K'_n = 100$ μA/V^2, $K'_p = 40$ μA/V^2, $V_{TON} = 0.7$ V, $V_{TOP} = -0.7$ V, $\gamma = 0.5 \sqrt{V}$, $2\phi_F = 0.6$ V. For simulation, use the models in Appendix B.

8.1 Random-Access Memory (RAM)

8.1. (a) How many bits are actually in a 256-Mb memory chip? In a 1-Gb chip? (b) How many 128-Kb blocks must be replicated to form the 256-Mb memory in Fig. 8.1?

8.2. How much leakage is permitted per memory cell in a 256-Mb static CMOS memory chip if the total standby current of the memory is to be less than 1 mA? (b) Repeat for a 4-Gb memory.

8.3. Suppose a memory chip has a 64-bit-wide external memory bus. What is the power dissipated driving the memory bus at a 1-GHz data rate if each bus line has 10 pF of capacitance, and the voltage is 3.3 V? (b) Repeat for 3 GHz and 2.5 V.

8.4. Suppose that each cell in a 1-Gbit memory chip must be refreshed every 12 ms. What is the power dissipated in refreshing the chip if the cell capacitance is 100 fF and the cell voltage is 2.5 V? Assume that 50 percent of the cells have 1 bits stored and that the cell voltage is completely discharged and restored during the refresh operation.

8.2 Static Memory Cells

8.5. Find the voltages corresponding to D and $\overline{D}$ in an NMOS memory cell with resistor loads in place of the PMOS transistors in Fig. 8.6 if $R = 10^{10}$ Ω, $V_{DD} = 3$ V, and $W/L = 2/1$. Use $V_{TO} = 0.75$ V, $\gamma = 0.5 \sqrt{V}$, and $2\phi_F = 0.6$ V.

*8.6. Simulate the response time of the 6-T cell in Fig. 8.6 from an initial condition of $D_1 = 1.55$ V and

$D_2 = 1.45$ V with the access transistors off. How long does it take for the cell voltages to reach 90 percent of their final values? Use $V_{DD} = 3$ V and a symmetrical cell design, with W/L of the NMOS transistors $= 2/1$. Use the SPICE models from Appendix B.

*8.7. Assume that the two bitlines are fixed at 1.5 V in the circuit in Figs. 8.7 and 8.8 and that a steady-state condition has been reached, with the wordline voltage equal to 3 V. Assume that the inverter transistors all have $W/L = 1/1$, $V_{TN} = 0.7$ V, $V_{TP} = -0.7$ V, and $\gamma = 0$. What is the largest value of W/L for M_{A1} and M_{A2} (use the same value) that will ensure that the voltage at $D_1 \leq 0.7$ V and the voltage at $D_2 \geq 2.3$ V?

8.8. Simulate and plot a graph of the transients that occur when writing a 0 into a cell containing a 0, as in Fig. 8.12. Discuss the results.

8.3 Dynamic Memory Cells

8.9. The 1-T cell in Fig. P8.9 uses a bitline voltage of 2.5 V and a wordline voltage of 2.5 V. (a) What are the cell voltages stored on C_C for a 1 and 0 if $V_{TO} = 0.6$ V, $\gamma = 0.5 \sqrt{V}$, and $2\phi_F = 0.6$ V? (b) What would be the minimum wordline voltage needed in order for the cell voltage to reach 2.5 V for a 1?

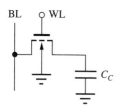

Figure P8.9

8.10. Substrate leakage currents usually tend to destroy only one of the two possible states in the 1-T cell. For the circuit in Fig. P8.9, which level is the most sensitive to leakage currents and why?

*8.11. The gate-source and drain-source capacitances of the MOSFET in Fig. P8.9 are each 100 fF, and $C_C = 75$ fF. The bitline and wordline have been stable at 2.5 V for a long time. The wordline signal is shown in Fig. P8.11. What is the voltage stored on C_C before the wordline drops? Estimate the drop in voltage on the C_C due to coupling of the wordline signal through the gate-source capacitance. Use $V_{TO} = 0.70$ V, $\gamma = 0.5 \sqrt{V}$, and $2\phi_F = 0.6$ V.

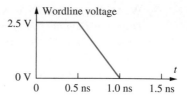

Figure P8.11

8.12. A 1-T cell has $C_C = 60$ fF and $C_{BL} = 7.5$ pF. (a) If the bitlines are precharged to 2.5 V, and the cell voltage is 0 V, what is the change in bitline voltage ΔV following cell access? (b) What is the final voltage in the cell?

8.13. A 1-T cell memory can be fabricated using PMOS transistors in the array shown in Fig. P8.13. (a) What are the voltages stored on the capacitor corresponding to logic 0 and 1 levels for a technology using $V_{DD} = 3.3$ V? (b) Repeat for $V_{DD} = 2.5$ V.

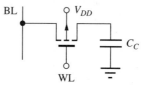

Figure P8.13

8.14. (a) Calculate the cell voltage for the boosted wordline version of the 1-T cell in Fig. 8.24 and show that the value in the text is correct. (b) Verify that the value of the current entering the sense amplifier node from the 1-T cell immediately following activation of the wordline is 216 μA.

8.15. The bottom electrode of the storage capacitor in the 1-T cell is often connected to a voltage V_{PP} rather than ground, as shown in Fig. P8.15. Suppose that $V_{PP} = 5$ V. (a) What are the voltages stored in the cell at node V_C for $0 = 0$ V on the bitline and $1 = 3$ V on the bitline? Assume the wordline can be driven to 3 V. (b) Which level will deteriorate due to leakage in this cell?

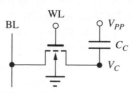

Figure P8.15

*8.16. The 1-T cell in Fig. P8.16 uses bitline and wordline voltages of 0 V and 5 V. (a) What are the cell voltages stored on C_C for a 1 and 0 if $V_{TO} = -0.80$ V, $\gamma = 0.65$ V$^{0.5}$, and $2\phi_F = 0.6$ V? (b) What would be the minimum wordline voltage needed for the stored cell voltage to reach 0 V for a 0 state?

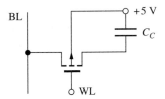

Figure P8.16

*8.17. In the discussion of the 1-T cell in the text, an improvement factor of 15 was stated for current drive from the boosted wordline cell compared to the normal cell. How much of this factor of 15 is attributable to the increased V_{DS} across the access transistor, and what portion is attributable to the increased gate voltage?

8.18. Suppose that each cell in a 1-Gbit memory chip must be refreshed every 10 ms. What is the power dissipated in refreshing the chip if the cell capacitance is 100 fF, and the cell voltage is 2.5 V? Assume that 50 percent of the cells have a 1 stored and that the cell voltage is completely discharged and restored during the refresh operation.

8.19. Simulate the refresh operation of the 4-T dynamic cell in Fig. P8.19. For initial conditions, assume that node D has decreased to 1 V, and node $\overline{D}$ is at 0 V. Use $BL = 3$ V, $\overline{BL} = 3$ V, $W/L = 2/1$ for all transistors, and the bitline capacitance is 500 fF.

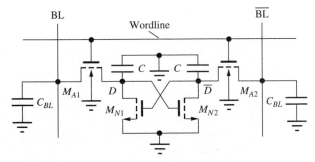

Figure P8.19

**8.20. Simulate the read access operation of the 4-T cell in Fig. P8.20 and discuss the waveforms that you

obtain. What is the access time of the cell from the time the wordline is activated until the data is valid at the output of the sense amplifier? Use $W/L = 2/1$ for all devices, and assume $C_{BL} = 1$ pF, with $V_{DD} = 3$ V.

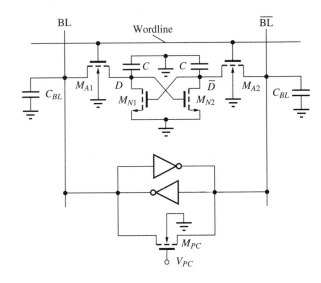

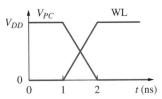

Figure P8.20

8.4 Sense Amplifiers

8.21. A simple CMOS sense amplifier is shown in Fig. P8.21. Suppose $V_{DD} = 3.3$ V and the W/L ratios of all the NMOS and PMOS transistors are 5/1 and 10/1, respectively. What is the total current through the sense amplifier when the precharge transistor is on? How much power will be consumed

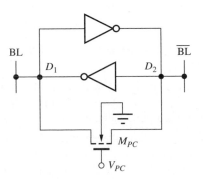

Figure P8.21

by 1024 of these sense amplifiers operating simultaneously?

**8.22. A transient drop can be observed in the waveforms for the two bitlines in Fig. 8.25 due to capacitive coupling of the precharge signal through the gate capacitance of the precharge device. Calculate the expected voltage change ΔV on the bitlines due to this coupling and compare to the simulation results in the figure. The BL capacitances are each 500 fF. See Appendix B for transistor models.

**8.23. The two bitlines in Fig. 8.29 are driven above V_{DD} by capacitive coupling of the precharge signal through the gate capacitance of the precharge devices. (a) Calculate the expected voltage change ΔV on the bitlines due to this coupling and compare to the simulation results in the figure. (b) What is the largest possible value of ΔV? See Appendix B for transistor models. Use $C_{BL} = 500$ fF.

**8.24. Figure P8.24 shows the basic form of a charge-transfer sense amplifier that can be used for amplifying the output of a 1-T cell. Assume that the switch closes at $t = 0$, that capacitor C_C is initially discharged, and that C_L is initially charged to +3 V. Also assume that charge sharing between C_C and C_{BL} occurs instantaneously. Find the total change in the output voltage Δv_O that occurs once the circuit returns to steady-state conditions following the switch closure. Assume $C_C = 50$ fF, $C_{BL} = 1$ pF, $C_L = 100$ fF, and $W/L = 50/1$. (*Hint:* The MOSFET will restore the BL potential to the original value, and the total charge that flows out of the source of the FET must be supplied from the drain.)

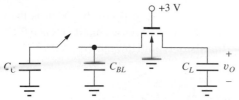

Figure P8.24 Charge transfer sense amplifier.

8.25. Simulate the circuit in Fig. P8.24 using a MOSFET ($W/L = 4/1$) for the switch and compare the results to your hand calculations.

**8.26. Convince yourself of the statement that any voltage imbalance in the cross-coupled latch will be reinforced by simulating the CMOS latch of Fig. P8.26 using the following initial conditions: (a) $D_1 = $

1.45 V and $D_2 = 1.55$ V, (b) $D_1 = 1$ V and $D_2 = 1.25$ V, (c) $D_1 = 2.75$ V and $D_2 = 2.70$ V. Assume all W/L ratios are 2/1 and $V_{DD} = 3.3$ V, and use bitline capacitances of 1 pF.

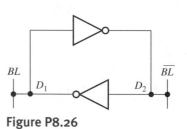

Figure P8.26

*8.27. The W/L ratios of the NMOS and PMOS transistors are 2/1 and 4/1, respectively, in the CMOS inverters in Fig. P8.27. The bitline capacitances are 400 fF, W/L of M_{PC} is 10/1, and $V_{DD} = 3$ V. (a) Simulate the switching behavior of the symmetrical latch and explain the behavior of the voltages at nodes D_1 and D_2. (b) Now suppose that a design error occurred and the W/L ratio of M_{N2} is 2.2/1 instead of 2/1. Simulate the latch again and explain any changes in the behavior of the voltages at nodes D_1 and D_2.

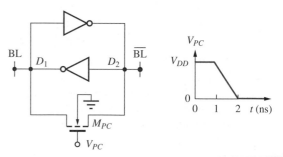

Figure P8.27

*8.28. Simulate the response of the NMOS clocked sense amplifier in Fig. P8.28 if $V_{DD} = 3$ V. What are the final voltage values on the two bitlines? How long does it take the sense amplifier to develop a difference of 1.5 V between the two bitlines? Assume that all clock signals have amplitudes equal to V_{DD} and rise or fall times of 1 ns. Assume that the three signals are delayed successively by 0.5 ns in a manner similar to Fig. 8.29.

*8.29. Repeat Prob. 8.28 for $V_{DD} = 5$ V.

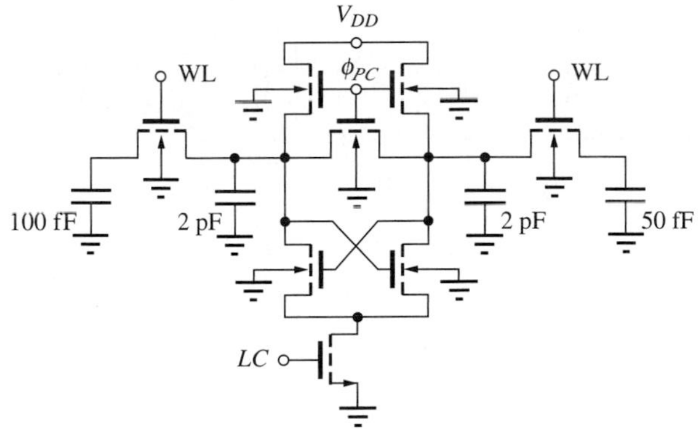

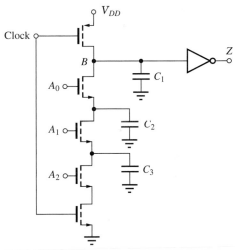

Figure P8.28

8.30. Simulate the transfer function of two cascaded CMOS inverters with all 2/1 devices and find the three equilibrium points. Use $V_{DD} = 3$ V.

8.5 Address Decoders

8.31. (a) How many transistors are required to implement a full 12-bit NOR address decoder similar to that of Fig. 8.30? (b) How many transistors are required to implement a full 12-bit NAND address decoder similar to that of Fig. 8.31?

8.32. Calculate the number of transistors required to implement a 7-bit column decoder using (a) NMOS pass-transistor logic and (b) standard NOR logic.

8.33. Draw the schematic of a 3-bit OR address decoder using domino CMOS.

8.34. What are the voltages at the nodes in the pass-transistor networks in Fig. P8.34? For NMOS transistors, use $V_{TO} = 0.75$ V, $\gamma = 0.55 \sqrt{V}$,

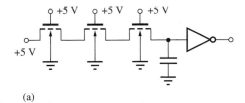

(a)

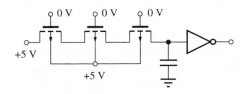

(b)

Figure P8.34

and $2\phi_F = 0.6$ V. For PMOS transistors, $V_{TO} = -0.75$ V.

*8.35. (a) Suppose that inputs A_0, A_1, and A_2 are all 0 in the domino CMOS gate in Fig. P8.35, and the clock has just changed to the evaluate phase. If A_0 now changes to a 1, what happens to the voltage at node B if $C_1 = 2C_2$? (*Hint:* Remember the charge-sharing phenomena.) (b) Now A_1 changes to a 1—what happens to the voltage at node B if $C_3 = C_2$? (c) If the output inverter is a symmetrical design, what is the minimum ratio of C_1/C_2 (assume $C_3 = C_2$) for which the gate maintains a valid output? Assume $V_{DD} = 5$ V.

Figure P8.35

8.36. Draw the mirror image of the gate in Fig. P8.35 by replacing NMOS transistors with PMOS transistors and vice versa. Assume the logic inputs remain the same and write an expression for the logic function Z.

8.6 Read-Only Memory (ROM)

8.37. What are the six output data words for the ROM in Fig. P8.37?

8.38. Identify and simulate the worst-case delay path in the ROM in Fig. P8.37.

8.39. What are the contents of the ROM in Fig. P8.39? (All FETs are NMOS.)

8.40. What are the contents of the ROM in Fig. P8.40? (All FETs are NMOS.)

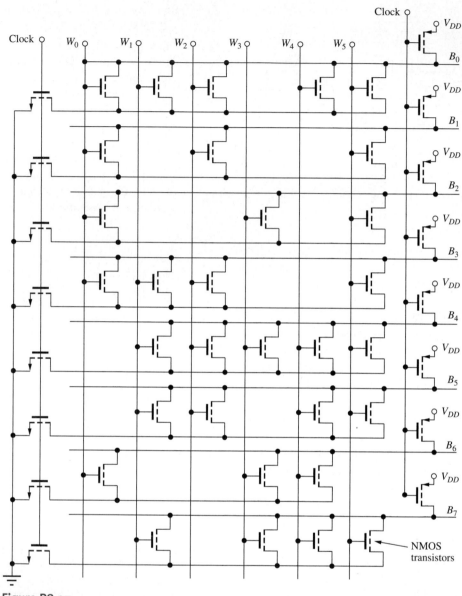

Figure P8.37

8.7 Flip-Flops

8.41. What is the minimum size of the transistors connected to the R and S inputs in Fig. P8.41 that will ensure that the latch can be forced to the desired state? Do not be concerned with speed of the latch.

8.42. What are the logic functions of inputs 1 and 2 in the flip-flop in Fig. P8.42?

8.43. Simulate the propagation delay through the D latch to Q and $\overline{Q}$ in Fig. 8.42. Assume that D is stable and the clock signal is a square wave. Assume the transistors all have $W/L = 2/1$ and use $V_{DD} = 2.5$ V. Use the transistor models on Appendix B.

8.44. Simulate the master-slave D-flip-flop with the slowly rising clock ($T = 20$ μs) in Fig. P8.44(a). Assume all $W/L = 2/1$. What happens to data on the D input? Use the transistor models in Appendix B.

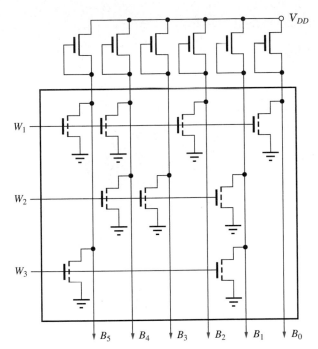

Figure P8.39

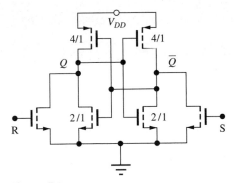

Figure P8.41

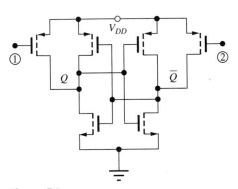

Figure P8.42

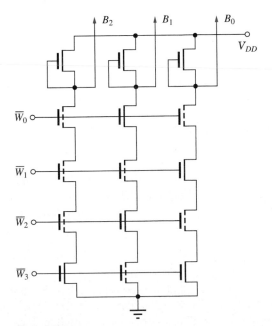

Figure P8.40

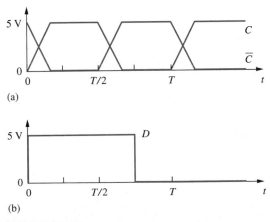

Figure P8.44

CHAPTER 9

BIPOLAR LOGIC CIRCUITS

CHAPTER OUTLINE

Photo of a group of TTL unit logic.

CHAPTER GOALS

From Chapter 9, we shall gain a basic appreciation of the switching characteristics of the bipolar transistor and for the design of the most important bipolar logic circuit families. The material in this chapter includes

- Bipolar current switch circuits
- Emitter-coupled logic (ECL)
- Behavior of the bipolar transistor as a saturated switch
- Transistor-transistor logic (TTL)
- Schottky clamping techniques for preventing saturation
- Operation of the transistor in the inverse-active region
- Voltage reference design
- BiCMOS logic circuits

As mentioned in Chapters 1 and 2, bipolar transistors were the first three-terminal solid-state devices to reach high-volume manufacturing, and in fact the first integrated circuits were bipolar logic circuits. The earliest logic family, built with just resistors and transistors, was called resistor-transistor logic or RTL. Later it was discovered that improved logic characteristics could be obtained by adding input diodes to the logic gate, creating a family called diode-transistor logic or DTL. Shortly thereafter, it was realized that the diodes in DTL could be merged and replaced by a multi-input transistor, and this circuit became the basis of transistor-transistor logic—TTL or T^2L.

TTL evolved into an extremely robust logic family that was very easy to use, and TTL was the dominant logic technology from the mid-1960s through the mid-1980s. A tremendous number of different types of logic gates and system components were made available by the manufacturers. TTL is still widely used today in prototype systems and as "glue" logic in most digital systems.

The second form of bipolar logic that found wide use is emitter-coupled logic or ECL. ECL has traditionally represented the highest-speed form of logic that is available, and it was the technology of choice for large mainframe computers and supercomputers for many many years. ECL unit logic families also offer an extremely wide range of logic gates and system building blocks.

Low-voltage and low-power versions of both TTL and ECL were eventually developed for VLSI applications, but still suffered from relatively high levels of power consumption compared to CMOS technology. Today, submicron CMOS now offers logic delay performance approaching that of emitter-coupled logic but with much higher circuit density and lower power consumption. Yet, the high transconductance of the bipolar transistor is still a significant advantage, and a number of semiconductor companies have developed complex BiCMOS processes that add bipolar transistors (often both *npn* and *pnp* devices) to a full CMOS technology.

Chapter 9 explores details of the two bipolar circuit families that have been used extensively in logic circuit design since the mid-1960s. **Emitter-coupled logic (ECL)** historically has been the fastest form of logic available. The ECL circuit uses bipolar transistors operating in a differential circuit that is often called a current switch. For binary logic operation, two states are needed, and in ECL, the transistors operate in the forward-active region with either a relatively large collector current or a very small collector current, actually near cutoff. The transistors avoid saturation and an attendant delay time that substantially slows down BJT switching speed.

Transistor-transistor logic (TTL or T²L) was the dominant logic family for systems designed through the mid-1980s, when CMOS began to replace it. TTL was the family that established 5 V as the standard power supply level. The main transistors in TTL switch between the **forward-active—** but essentially nonconducting—and **saturation regions** of operation. In the TTL circuit, we find one of the few actual applications for the **reverse-active region** of operation of the BJT. Because various transistors in the TTL circuit enter saturation, TTL delays tend to be poorer than those that can be achieved with ECL. However, an improved circuit, Schottky-clamped TTL, is substantially faster than standard TTL or can achieve delays similar to standard TTL at much less power dissipation.

9.1 THE CURRENT SWITCH (EMITTER-COUPLED PAIR)

We begin our study of bipolar logic circuits with emitter-coupled logic, or ECL. At the heart of an ECL gate is the **current switch circuit** in Fig. 9.1, consisting of two identical transistors, Q_1 and Q_2, two matched-load resistors, R_C, and current source I_{EE}. This circuit is also known as an **emitter-coupled pair.** The input logic signal v_I is applied to the base of Q_1 and is compared to the **reference voltage** V_{REF}, which is connected to the base of transistor Q_2. If v_I is greater than V_{REF} by a few hundred millivolts, then the current from source I_{EE} is supplied through the emitter of Q_1. If v_I is less than V_{REF} by a few hundred millivolts, then the current from source I_{EE} is supplied by the emitter of Q_2. Thus input voltage v_I "switches" the current from source I_{EE} back and forth between Q_1 and Q_2. This behavior is conceptually illustrated in Fig. 9.2, in which transistors Q_1 and Q_2 have been replaced by a single-pole, double-throw switch whose position is controlled by the input v_I.

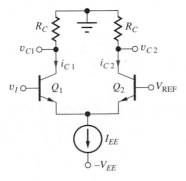

Figure 9.1 Current switch circuit used in an ECL gate.

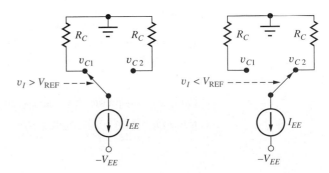

Figure 9.2 Conceptual representation of the current switch.

9.1.1 MATHEMATICAL MODEL FOR STATIC BEHAVIOR OF THE CURRENT SWITCH

The behavior of the current switch can be understood in more detail using the transport model for the forward-active region from Chapter 5, in which the collector currents in the two transistors are represented by

$$i_{C1} = I_S \exp\left(\frac{v_{BE1}}{V_T}\right) \qquad \text{and} \qquad i_{C2} = I_S \exp\left(\frac{v_{BE2}}{V_T}\right) \tag{9.1}$$

It is assumed in Eq. (9.1) that $v_{BE} > 4V_T$. If the transistors are identical, so that the saturation currents are the same, then the ratio of these two collector currents can be written as

$$\frac{i_{C2}}{i_{C1}} = \frac{I_S \exp\left(\dfrac{v_{BE2}}{V_T}\right)}{I_S \exp\left(\dfrac{v_{BE1}}{V_T}\right)} = \exp\left(\frac{v_{BE2} - v_{BE1}}{V_T}\right) = \exp\left(-\frac{\Delta v_{BE}}{V_T}\right) \tag{9.2}$$

Now, suppose that v_{BE2} exceeds v_{BE1} by 300 mV: $\Delta v_{BE} = (v_{BE1} - v_{BE2}) = -0.3$ V. For $V_T = 0.025$ V, we find that i_{C2} is approximately 1.6×10^5 times bigger than i_{C1}.[1] However, if v_{BE1} exceeds v_{BE2} by 300 mV, then i_{C1} will be 1.6×10^5 times larger than i_{C2}. Thus, the assumption that all the current from the source I_{EE} is switched from one side to the other appears justified for a few-hundred-millivolt difference in v_{BE}.

A useful expression for the normalized difference in collector currents can be derived from Eq. (9.1):

$$\frac{i_{C1} - i_{C2}}{i_{C1} + i_{C2}} = \frac{\exp\left(\dfrac{v_{BE1}}{V_T}\right) - \exp\left(\dfrac{v_{BE2}}{V_T}\right)}{\exp\left(\dfrac{v_{BE1}}{V_T}\right) + \exp\left(\dfrac{v_{BE2}}{V_T}\right)} = \tanh\left(\frac{v_{BE1} - v_{BE2}}{2V_T}\right) \tag{9.3}$$

Using Kirchhoff's current law at the emitter node yields

$$i_{E1} + i_{E2} = I_{EE} \qquad \text{so that} \qquad i_{C1} + i_{C2} = \alpha_F I_{EE} \tag{9.4}$$

since $i_C = \alpha_F i_E$ in the forward-active region, and assuming matched devices with identical current gains. Combining Eqs. (9.3) and (9.4) gives the desired result for the collector current difference in terms of the difference in base-emitter voltages:

$$i_{C1} - i_{C2} = \alpha_F I_{EE} \tanh\left(\frac{v_{BE1} - v_{BE2}}{2V_T}\right) \tag{9.5}$$

Figure 9.3 plots a graph of Eq. (9.5) in normalized form, showing that only a small voltage change is required to switch the current from one collector to the other. Ninety-nine percent of the current switches for $|\Delta v_{BE}| > 4.6V_T$ (130 mV)! We see that a relatively small voltage change is required to completely switch the current from one side to the other in the current switch. This small voltage change directly contributes to the high speed of ECL logic gates.

EXERCISE: Calculate the ratio i_{C2}/i_{C1} for $(v_{BE2} - v_{BE1}) = 0.2$ V and 0.4 V.

ANSWERS: 2.98×10^3; 8.89×10^6

---

[1] Remember that one decade of current change in a BJT corresponds to approximately a 60-mV change in v_{BE}, so a factor of 10^5 is precisely the change we expect for a ΔV_{BE} of 300 mV.

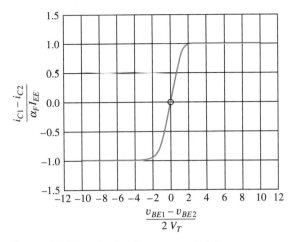

Figure 9.3 Normalized collector current difference versus $\Delta v_{BE}/2V_T$ for the bipolar current switch.

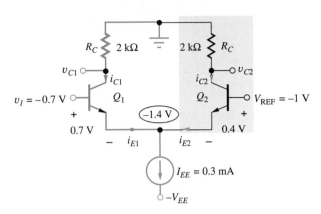

Figure 9.4 Current switch circuit with $v_I > V_{REF}$. Q_1 is conducting; Q_2 is "off."

9.1.2 CURRENT SWITCH ANALYSIS FOR $V_I > V_{REF}$

Now let us explore the actual current switch circuit for the case of $v_I = V_{REF} + 0.3 \text{ V} = -0.7 \text{ V}$, as in Fig. 9.4. From Fig. 9.3 we expect 0.3 V to be more than enough to fully switch the current. In this design, V_{REF} has been selected to be -1.0 V (the reasons for this choice will become clear shortly). Because $v_I > V_{REF}$, we assume that Q_2 is off ($i_{C2} = 0$) and Q_1 is conducting in the forward-active region with $V_{BE1} = 0.7$ V. Applying Kirchhoff's voltage law to the circuit in Fig. 9.4:

$$v_I - v_{BE1} + v_{BE2} - V_{REF} = 0$$

$$(V_{REF} + 0.3 \text{ V}) - (0.7) + v_{BE2} - V_{REF} = 0 \quad \text{and} \quad v_{BE2} = 0.4 \text{ V} \quad (9.6)$$

The base-emitter voltage difference is given by $v_{BE1} - v_{BE2} = 300$ mV, so essentially all the current I_{EE} switches to the emitter of Q_1, and Q_2 is nearly cut off. [However, Q_2 is actually still in the forward-active region by our strict definition of the regions of operation for the bipolar transistor ($v_{BE} \geq 0$, $v_{BC} \leq 0$)].

At the emitter node, we have $i_{E1} \cong I_{EE}$ because $i_{E2} \cong 0$ and the output voltages v_{C1} and v_{C2} at the two collectors are given by

$$v_{C1} = -i_{C1}R_C = -\alpha_F i_{E1} R_C \cong -\alpha_F I_{EE} R_C \quad (9.7)$$

$$v_{C2} = -i_{C2}R_C = -\alpha_F i_{E2} R_C \cong 0 \quad (9.8)$$

in which $i_C = \alpha_F i_E$ in the forward-active region. For $\alpha_F \cong 1$, the two output voltages become

$$v_{C1} = -i_{C1}R_C \cong -I_{EE}R_C \quad \text{and} \quad v_{C2} = -i_{C2}R_C = 0 \quad (9.9)$$

For the circuit in Fig. 9.4, $I_{EE} = 0.3$ mA and $R_C = 2$ kΩ, and

$$v_{C1} = -0.6 \text{ V} \quad \text{and} \quad v_{C2} = 0 \text{ V} \quad (9.10)$$

Check of Forward-Active Region Assumptions

Now we can check our assumptions concerning the forward-active region of operation. For Q_1, $v_{BC1} = v_{B1} - v_{C1} = -0.7 \text{ V} - (-0.6 \text{ V}) = -0.1$ V, and the collector-base junction is indeed reverse-biased. We assumed that the emitter-base junction was forward-biased, so the assumption of forward-active region is consistent with our circuit analysis. For Q_2, $V_{BC2} = -1.0 \text{ V} - (0 \text{ V}) = -1.0$ V and $v_{BE2} = 0.4$ V, so Q_2 is also in the forward-active region, although, it is conducting a negligibly small current.

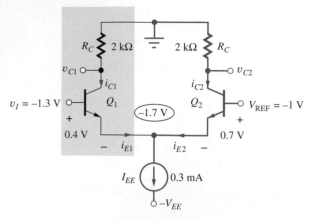

Figure 9.5   Current switch circuit with $v_I < V_{REF}$. Q_2 is conducting; Q_1 is "off."

TABLE 9.1
Current Switch Voltage Levels

v_I ($V_{REF} = -1.0$ V)	v_{C_1}	v_{C_2}
$V_{REF} + 0.3$ V $= -0.7$ V	-0.6 V	0 V
$V_{REF} - 0.3$ V $= -1.3$ V	0 V	-0.6 V

9.1.3 CURRENT SWITCH ANALYSIS FOR $V_I < V_{REF}$

The second logic state occurs for $v_I = V_{REF} - 0.3$ V $= -1.3$ V, as in Fig. 9.5. Because $v_I < V_{REF}$, we assume that Q_2 is conducting in the forward-active region, with $v_{BE2} = 0.7$ V. Kirchhoff's voltage law again requires

$$v_I - v_{BE1} + v_{BE2} - V_{REF} = 0$$

which yields

$$(V_{REF} - 0.3 \text{ V}) - v_{BE1} + (0.7) - V_{REF} = 0 \qquad \text{and} \qquad v_{BE1} = 0.4 \text{ V}$$

v_{BE1} is much less than v_{BE2}, so now $i_{E1} \cong 0$, and $i_{E2} \cong I_{EE}$. The output voltages at v_{C1} and v_{C2} are given by

$$v_{C1} = -i_{C1}R_C = -\alpha_F i_{E1} R_C \cong 0 \qquad\qquad (9.11)$$
$$v_{C2} = -i_{C2}R_C = -\alpha_F i_{E2} R_C \cong -\alpha_F I_{EE} R_C = -0.6 \text{ V}$$

for $\alpha_F \cong 1$.

The results from Eqs. (9.10) and (9.11) are combined in Table 9.1. We see there are two discrete voltage levels at the two outputs, 0 V and −0.6 V, that can correspond to a logic 1 and a logic 0, respectively. Note, however, that the voltages at the outputs of the current switch do not match the input voltages used at v_I. Thus, this current switch circuit fails to meet one of the important criteria for logic gates set forth in Sec. 6.2.3: The logic levels must be restored as the signal goes through the gate. That is, the voltage levels at the output of a logic gate must be compatible with the levels used at the input of the gate.

EXERCISE: Redesign the circuit in Fig. 9.4 to reduce the power by a factor of five while maintaining the same voltage levels.

ANSWER: $R_C = 10$ kΩ, $I_{EE} = 60$ μA

9.2 THE EMITTER-COUPLED LOGIC (ECL) GATE

We observe in Table 9.1 that the high and low logic levels at the input and output of the current switch differ by exactly one base-emitter voltage drop (0.7 V), which leads to the circuit of the complete ECL inverter in Fig. 9.6. Two transistors, Q_3 and Q_4, have been added, and their base-emitter junctions

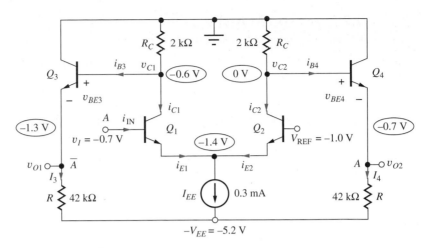

Figure 9.6 Emitter-coupled logic circuit with $v_I = V_H$.

are used to shift the output voltages down by one base-emitter drop. These transistors act as **level shifters** in the circuit and are usually called emitter followers.

9.2.1 ECL GATE WITH $V_I = V_H$

To understand how the level-shifting operation takes place in the ECL circuit, consider the case for $v_I = V_H = -0.7$ V indicated in Fig. 9.6. Equations for the output voltages v_{O1} and v_{O2} can be written as

$$v_{O1} = -(i_{C1} + i_{B3})R_C - v_{BE3} \quad \text{and} \quad v_{O2} = -(i_{C2} + i_{B4})R_C - v_{BE4} \quad (9.12)$$

The base currents are given by

$$i_{B3} = \frac{I_3}{\beta_F + 1} \quad \text{and} \quad i_{B4} = \frac{I_4}{\beta_F + 1}$$

In a typical digital IC technology, $\beta_F \geq 20$ and $i_B R_C$ is designed to be much less than v_{BE}. Then

$$v_{O1} \cong -i_{C1}R_C - v_{BE3} = -0.6 - 0.7 = -1.3 \text{ V}$$

and

$$v_{O2} \cong -i_{C2}R_C - v_{BE4} = 0 - 0.7 = -0.7 \text{ V}$$

(9.13)

For sufficiently large β_F, the addition of Q_3 and Q_4 does not change the voltage at v_{C1} or v_{C2}.

The base-collector voltages of Q_3 and Q_4 will be -0.6 V and 0 V, respectively, and the two emitter resistors R set up an average current of 0.1 mA in the emitters of Q_3 and Q_4:

$$I_{E3} = \frac{-1.3 - (5.2)}{42} \frac{\text{V}}{\text{k}\Omega} = 92.9 \text{ μA} \quad \text{and} \quad I_{E4} = \frac{-0.7 - (-5.2)}{42} \frac{\text{V}}{\text{k}\Omega} = 107 \text{ μA} \quad (9.14)$$

Thus, both Q_3 and Q_4 are in the forward-active region, so $v_{BE3} = v_{BE4} = 0.7$ V has been used.

EXERCISE: What are the base currents i_{B3} and i_{B4} in Fig. 9.6 if $\beta_F = 20$? Compare $i_B R_C$ to V_{BE}.

ANSWERS: 4.42 μA, 5.10 μA, 8.84 mV $\ll$ 0.7 V, 10.2 mV $\ll$ 0.7 V

9.2.2 ECL GATE WITH $V_I = V_L$

For $v_I = V_L = -1.3$ V, the outputs change state, and:

$$v_{O1} \cong -i_{C1} R_C - v_{BE3} = 0 - 0.7 = -0.7 \text{ V}$$

$$v_{O2} \cong -i_{C2} R_C - v_{BE4} = -0.6 - 0.7 = -1.3 \text{ V} \tag{9.15}$$

9.2.3 INPUT CURRENT OF THE ECL GATE

In NMOS and CMOS logic circuits, the inputs are normally connected to FET gates, and the static input current to the logic gate is zero. In bipolar logic circuits, however, there is a nonzero current in the input. For the ECL gate in Fig. 9.6, the input current i_{IN} is equal to the base current of Q_1. When Q_1 is conducting ($v_I = -0.7$ V), the input current is given by

$$i_{IN} = i_{B1} = \frac{i_{E1}}{\beta_F + 1} = \frac{0.3 \text{ mA}}{21} = 14.3 \text{ μA} \tag{9.16}$$

and $i_{IN} \cong 0$ when Q_1 is off ($v_I = -1.3$ V). Thus, a circuit that is providing an input to an ECL gate must be capable of supplying 14.3 μA to each input that it drives.

9.2.4 ECL SUMMARY

Table 9.2 summarizes the behavior of the basic ECL inverter in Fig. 9.6. The requirement for level compatibility between the input and output voltages is now met. For this ECL gate design,

$$V_H = -0.7 \text{ V}, \ V_L = -1.3 \text{ V}, \text{ and } \Delta V = V_H - V_L = 0.6 \text{ V} \tag{9.17}$$

To provide symmetrical noise margins, the reference voltage V_{REF} is normally centered midway between the two logic levels:

$$V_{\text{REF}} = \frac{V_H + V_L}{2} = -1.0 \text{ V} \tag{9.18a}$$

In the design in Fig. 9.6, the logic signal swings symmetrically above and below V_{REF} by one-half the logic swing, or 0.3 V. Note that the logic swing ΔV is just equal to the voltage drop developed across the load resistor R_C:

$$\Delta V = I_{EE} R_C \tag{9.18b}$$

Several important observations can be made at this point. If the input at v_I is now defined as the logic variable A, then the output at v_{O1} corresponds to $\overline{A}$, but the output at v_{O2} corresponds to A! A complete ECL gate generates both true and complement outputs for a given logic function. Having both true and complement outputs available can often reduce the total number of logic gates required to implement a given logic function.

A second observation relates to the speed of emitter-coupled logic. The transistors remain in the forward-active region at all times. The "off" transistor is actually conducting current but at a very low level, and it is ready to switch rapidly into high conduction for a base-emitter voltage change of only a few tenths of a volt. The transistors avoid the saturation region, which substantially slows down the switching speed of the bipolar transistor. (A detailed discussion of this problem is in Sec. 9.9.) The reduced logic swing of ECL also contributes to its high speed, and the small ΔV reduces the dynamic power required to charge and discharge the load capacitances.

TABLE 9.2

ECL Voltage Levels and Input Current

v_I	v_{O1}	v_{O2}	i_{IN}
$V_{\text{REF}} + 0.3 \text{ V} = -0.7 \text{ V}$	-1.3 V	-0.7 V	$+14.3$ μA
$V_{\text{REF}} - 0.3 \text{ V} = -1.3 \text{ V}$	-0.7 V	-1.3 V	0

Another benefit of ECL is the nearly constant power supply current maintained by the current sources in Fig. 9.6, regardless of the gate's logic state. This constant supply current reduces noise in the power distribution network.

> **EXERCISE:** Find V_H, V_L, V_{REF}, and ΔV for the ECL gate in Fig. 9.6 if I_{EE} is changed to 0.2 mA.
>
> **ANSWERS:** −0.7 V, −1.1 V, −0.9 V, 0.4 V

9.3 NOISE MARGIN ANALYSIS FOR THE ECL GATE

The simulated VTC for the ECL gate is given in Fig. 9.7, in which both outputs switch between the two logic levels specified in Table 9.2. The two outputs remain constant until the input comes within approximately 100 mV of the reference voltage, and then they rapidly change states as the input voltage changes by an additional 200 mV. The approach to finding the values of V_{IH} and V_{IL} and the noise margins is similar to that used for NMOS and CMOS circuits, but the algebra is simpler.

9.3.1 V_{IL}, V_{OH}, V_{IH}, AND V_{OL}

V_{IH} and V_{IL} are defined by the points at which $\partial v_{O1}/\partial v_I = -1$ for the inverting output or $\partial v_{O2}/\partial v_I = +1$ for the noninverting output. Writing the expression for v_{O1} in Fig. 9.6 yields

$$v_{O1} = v_{C1} - v_{BE3} = -(i_{C1} + i_{B3})R_C - v_{BE3} \tag{9.19}$$

and taking the derivative with respect to v_I gives

$$\frac{\partial v_{O1}}{\partial v_I} = -R_C \frac{\partial i_{C1}}{\partial v_I} \tag{9.20}$$

The base current and base-emitter voltage of Q_3 are constant because $i_{E3} = I_3$, a constant current. An expression for i_{C1} in terms of v_I can be obtained using the same procedure as that used to derive Eqs. (9.2) and (9.3):

$$\frac{i_{C1}}{i_{C1} + i_{C2}} = \frac{1}{1 + \exp\left(\dfrac{v_{BE2} - v_{BE1}}{V_T}\right)} \quad \text{and} \quad i_{C1} = \frac{\alpha_F I_{EE}}{1 + \exp\left(\dfrac{v_{BE2} - v_{BE1}}{V_T}\right)} \tag{9.21}$$

because $i_{C1} + i_{C2} = \alpha_F I_{EE}$. Equation (9.6) can be rearranged to yield a relationship between the input voltage and the base-emitter voltages:

$$(v_{BE2} - v_{BE1}) = V_{REF} - v_I \tag{9.22}$$

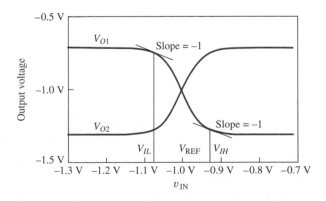

Figure 9.7 SPICE simulation results for ECL voltage transfer function.

Rewriting i_{C1} from Eq. (9.21):

$$i_{C1} = \frac{\alpha_F I_{EE}}{1 + \exp\left(\dfrac{V_{\text{REF}} - v_I}{V_T}\right)} \tag{9.23}$$

Taking the derivative and substituting the result in Eq. (9.20) yields

$$\frac{\partial v_{O1}}{\partial v_I} = -R_C \frac{(-\alpha_F I_{EE}) \exp\left(\dfrac{V_{\text{REF}} - v_I}{V_T}\right)\left(\dfrac{-1}{V_T}\right)}{\left[1 + \exp\left(\dfrac{V_{\text{REF}} - v_I}{V_T}\right)\right]^2} = \frac{-\left(\dfrac{1}{V_T}\right) i_{C1} R_C}{\left[1 + \exp\left(\dfrac{v_I - V_{\text{REF}}}{V_T}\right)\right]} \tag{9.24}$$

At $v_I = V_{IL}$, $v_I < V_{\text{REF}}$, $\exp[(v_I - V_{\text{REF}})/V_T] \ll 1$, and Eq. (9.24) simplifies to

$$\frac{\partial v_{O1}}{\partial v_I} \cong -\left(\frac{1}{V_T}\right) i_{C1} R_C = -1 \qquad \text{or} \qquad i_{C1} = \frac{V_T}{R_C} \tag{9.25}$$

Using Eq. (9.2),

$$V_{\text{REF}} - V_{IL} = V_T \ln\left(\frac{I_{EE} - i_{C1}}{i_{C1}}\right) = V_T \ln\left(\frac{I_{EE} - \dfrac{V_T}{R_C}}{\dfrac{V_T}{R_C}}\right) \tag{9.26}$$

Solving for V_{IL} yields

$$V_{IL} = V_{\text{REF}} - V_T \ln\left(\frac{I_{EE} R_C}{V_T} - 1\right) = V_{\text{REF}} - V_T \ln\left(\frac{\Delta V}{V_T} - 1\right) \tag{9.27}$$

and $$V_{OH} = V_H - i_{C1} R_C = V_H - V_T$$

Using symmetry, or a similar analysis, the value of V_{IH} is

$$V_{IH} = V_{\text{REF}} + V_T \ln\left(\frac{I_{EE} R_C}{V_T} - 1\right) = V_{\text{REF}} + V_T \ln\left(\frac{\Delta V}{V_T} - 1\right) \tag{9.28}$$

and $$V_{OL} = V_L + i_{C1} R_C = V_L + V_T$$

9.3.2 NOISE MARGINS

The noise margins are found using Eqs. (9.27) and (9.28):

$$\text{NM}_L = V_{IL} - V_{OL} = V_{\text{REF}} - V_T \ln\left(\frac{\Delta V}{V_T} - 1\right) - \left(V_{\text{REF}} - \frac{\Delta V}{2} + V_T\right)$$

and

$$\text{NM}_L = \frac{\Delta V}{2} - V_T \left[1 + \ln\left(\frac{\Delta V}{V_T} - 1\right)\right] \tag{9.29}$$

By symmetry,

$$\text{NM}_H = \frac{\Delta V}{2} - V_T \left[1 + \ln\left(\frac{\Delta V}{V_T} - 1\right)\right] \tag{9.30}$$

Using the values from Fig. 9.6, we find

$$\text{NM}_H = \text{NM}_L = \frac{0.6\ \text{V}}{2} - 0.025\ \text{V}\left[1 + \ln\left(\frac{0.6}{0.025} - 1\right)\right] = 0.197\ \text{V}$$

V_{IL} occurs at an input voltage approximately 78 mV below V_{REF}, and V_{IH} occurs at an input voltage approximately 78 mV above V_{REF}. These numbers are in excellent agreement with the circuit simulation results in Fig. 9.7, and the high and low state noise margins are both equal to approximately 0.20 V.

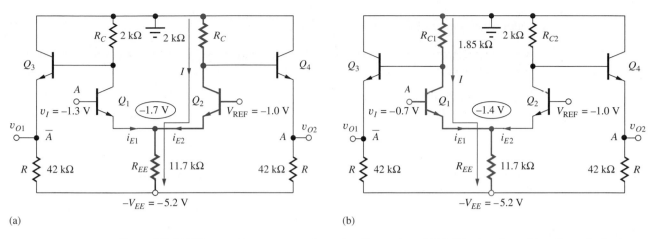

Figure 9.8 (a) ECL gate with current source I_{EE} replaced by a resistor. (b) Modification of one of the collector-load resistors.

> **EXERCISE:** What are the noise margins for the circuit in Fig. 9.6 if I_{EE} is changed to 0.2 mA?
>
> **ANSWER:** 0.107 V

9.4 CURRENT SOURCE IMPLEMENTATION

Source I_{EE} in Fig. 9.6 is often replaced by a resistor, as in Fig. 9.8(a). For $v_I = -1.3$ V, the voltage at the emitters of Q_1 and Q_2 is the same as that in Fig. 9.5, -1.7 V, and the value of R_{EE} required to set the emitter current i_{E2} to 0.3 mA is

$$R_{EE} = \frac{[-1.7 - (-5.2)] \text{ V}}{0.3 \text{ mA}} = 11.7 \text{ k}\Omega$$

The use of resistor R_{EE} is normally accompanied by a slight modification in the value of the resistor connected to the collector of Q_1. Referring to Fig. 9.8(b) for the case of $v_I = -0.7$ V, we find that the voltage at the emitters of Q_1 and Q_2 is -1.4 V, and hence the emitter current has changed slightly due to the voltage change across resistor R_{EE}:

$$i_E = \frac{[-1.4 - (-5.2)] \text{ V}}{11.7 \text{ k}\Omega} = 0.325 \text{ mA}$$

Because the emitter current increases, the value of R_{C1} must be decreased to maintain a constant logic swing ΔV. The new value of the resistor R_{C1} in the collector of Q_1 is

$$R_{C1} = \frac{0.6 \text{ V}}{0.325 \text{ mA}} = 1.85 \text{ k}\Omega$$

The corrected design values appear in the circuit in Fig. 9.8(b).

DESIGN EXAMPLE 9.1 ECL GATE DESIGN

In this example, we design the resistors in the ECL gate so that the gate can operate from a reduced supply voltage of -3.3 V.

PROBLEM Redesign the ECL gate in Fig. 9.8(b) to work from a -3.3-V supply.

SOLUTION **Known Information and Given Data:** Circuit topology in Fig. 9.8(b).

Unknowns: Values of resistors R_{EE}, R_{C1}, R_{C2}, and R

Approach: Determine the new voltage and current in each resistor, and then calculate the required value of resistance.

Assumptions: Maintain the same currents and values of V_H and V_L as in the design in Fig. 9.8(b): $V_H = -0.7$ V, $V_L = -1.3$ V, $\Delta V = 0.6$ V, $V_{REF} = -1.0$ V, $I_{E2} = 0.3$ mA, and the average emitter follower current is 0.1 mA.

Analysis: For $v_I = -1.3$ V as in Fig. 9.8(a), Q_1 is off and Q_2 is conducting. The voltage at the emitter of Q_2 is $V_{E2} = -1 - 0.7 = -1.7$ V, and the value of R_{EE} is given by

$$R_{EE} = \frac{-1.7 - (-3.3)}{0.3}\frac{V}{mA} = 5.33 \text{ k}\Omega$$

The value of R_{C2} is given by

$$R_{C2} = \frac{\Delta V}{I_{C2} + I_{B4}} = \frac{\Delta V}{I_{C2} + I_{B2} + (I_{B4} - I_{B2})} \cong \frac{\Delta V}{I_{E2}} = \frac{0.6 \text{ V}}{0.3}\frac{V}{mA} = 2.00 \text{ k}\Omega$$

in which the difference in the base currents between Q_4 and Q_2 is neglected.

For $v_I = -0.7$ V, as in Fig 9.8(b), Q_2 is off and Q_1 is conducting. The voltage at the emitter of Q_2 is $V_{E2} = -0.7 - 0.7 = -1.4$ V, and the value of I_{E1} is

$$I_{E1} = \frac{-1.4 - (-3.3)}{2.00}\frac{V}{k\Omega} = 357 \text{ }\mu\text{A}$$

The value of R_{C1} is given by

$$R_{C1} = \frac{\Delta V}{I_{C1} + I_{B3}} \cong \frac{\Delta V}{I_{E1}} = \frac{0.6 \text{ V}}{0.357}\frac{V}{mA} = 1.68 \text{ k}\Omega$$

The value of R is determined by the mean output voltage and current:

$$R = \frac{\dfrac{V_H + V_L}{2} - (-V_{EE})}{I_{E3}} \cong \frac{-1 + 3.3}{0.1}\frac{V}{mA} = 23 \text{ k}\Omega$$

Check of Results: We have found the four resistor values required to complete the design. Let us check the results using alternate methods of calculation. Since the voltage and current in R_{C2} have not changed, the value should be unchanged, as we calculated. For $v_I = V_H$, R_{C1} and R_{EE} are conducting approximately the same current. So, the voltage across R_{C1} should be equal to

$$\Delta V = R_{C1}\frac{V_{R_{EE}}}{R_{EE}} = 1.68 \text{ k}\Omega \frac{1.9 \text{ V}}{5.33 \text{ k}\Omega} = 0.599 \text{ V}$$

which is correct (0.6 V) within round-off error. The resistors R should be proportional to the voltage across them

$$R = \frac{-1 - (3.3)}{-1 - (5.2)}(42 \text{ k}\Omega) = 23 \text{ k}\Omega$$

which again agrees with the earlier calculation.

Discussion and Computer-Aided Analysis: The simulated outputs for our new design are given below using IS $= 0.3$ fA, BF $= 40$, and BR $= 0.25$. By design, we expect the VTC results to be essentially the same as those in Fig. 9.7, and the two graphs appear very similar. However, we should immediately note that asymmetries exist. The transfer function in Fig. 9.7 was generated with an ideal current source bias, whereas in this circuit the current in R_{EE} is different for the two logic states. In this circuit, the values of R_{C1} and R_{C2} are different, which causes the two transition slopes to be different. The circuit asymmetry causes the intersection of the two curves to shift slightly away from the $v_I = -1$-V point. The values of V_H and V_L are -0.694 and -1.27 V, respectively.

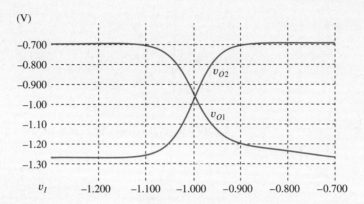

EXERCISE: Calculate the power dissipation and noise margins for the new ECL design.

ANSWERS: 1.65 mW, 1.84 mW; 0.20 V, 0.20 V

EXERCISE: What are the values of V_{IL} and V_{IH} from the VTC simulation in Design Ex. 9.1? What are the noise margins based on these values?

ANSWERS: -1.08 V, -0.91 V; 0.19 V, 0.21 V

EXERCISE: Redesign the ECL gate in Design Ex. 9.1 to reduce the power by a factor of 3.

ANSWERS: 6.00 kΩ, 5.04 kΩ, 16.0 kΩ, 69 kΩ

EXERCISE: Simulate the circuit in Design Ex. 9.1 with R_{EE} replaced by a 0.3-mA current source. Then change R_{C1} to 2 kΩ and simulate the circuit again. Note the differences in the various VTCs.

EXERCISE: What are the new values of R_{EE} and R_{C1} for the circuit in Fig. 9.8(b) if I_{EE} is changed to 0.2 mA and $R_{C2} = 2$ kΩ?

ANSWERS: 18.0 kΩ, 1.89 kΩ

9.5 THE ECL OR-NOR GATE

The ECL inverter becomes an OR-NOR gate through the addition of transistors in parallel with the original input transistor of the inverter, as in Fig. 9.9(a). If any one of the inputs (A or B or C) is at a high input level ($v_I > V_{\text{REF}}$), then the current from source I_{EE} will be switched into collector node v_{C1}, output Y_1 will drop to a low level, and output Y_2 will rise to a high level. Y_1 therefore represents a NOR output, and Y_2 is an OR output:

$$Y_1 = \overline{A + B + C} \qquad \text{and} \qquad Y_2 = A + B + C$$

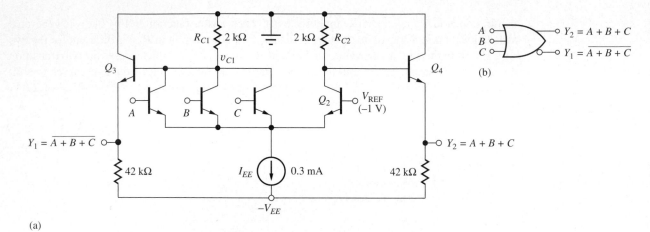

Figure 9.9 (a) Three-input ECL OR-NOR gate and (b) logic symbol.

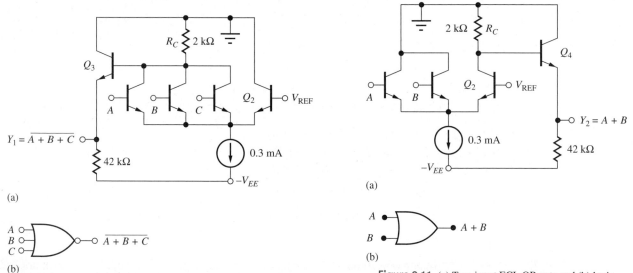

Figure 9.10 (a) Three-input ECL NOR gate and (b) logic symbol.

Figure 9.11 (a) Two-input ECL OR gate and (b) logic symbol.

The logic symbol in Fig. 9.9(b) is used for the dual output OR-NOR gate. The NOR output is marked by the small circle, which indicates the complemented or inverted output. The full ECL gate produces both true and complemented outputs. However, not every gate need be implemented this way. For example, the three-input logic gate in Fig. 9.10 has been designed with only the NOR output available; the two-input gate in Fig. 9.11 provides only the OR function. Note that the resistor in the collector of Q_2 is not needed in Fig. 9.10 and has been eliminated. Similarly, the left-hand collector resistor is eliminated from the circuit in Fig. 9.11.

EXERCISES: What are the new values of R_{C1} and R_{C2} for the circuit in Fig. 9.9 if I_{EE} is replaced by a 11.7-kΩ resistor? Assume $V_{REF} = -1$ V and $V_{EE} = -5.2$ V. What is the new value of R_C for the circuit in Fig. 9.10 if I_{EE} is replaced by a 11.7-kΩ resistor? Assume $V_{REF} = -1$ V and $V_{EE} = -5.2$ V.

ANSWERS: 1.85 kΩ, 2.00 kΩ; 1.85 kΩ

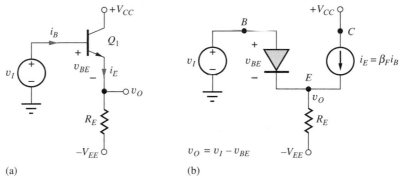

Figure 9.12 (a) Emitter follower and (b) transport model for the forward-active region.

EXERCISE: **What is the new value of R_C for the circuit in Fig. 9.11 if I_{EE} is replaced by a 11.7-kΩ resistor? Assume $V_{REF} = -1$ V and $V_{EE} = -5.2$ V.**

ANSWER: **2.00 kΩ**

9.6 THE EMITTER FOLLOWER

Let us now look in more detail at the operation of the emitter followers that provide the level-shifting function in the ECL gate. An emitter follower is shown biased by emitter resistor R_E in Fig. 9.12(a). For $v_I \leq V_{CC}$, Q_1 operates in the forward-active region because its base-collector voltage is negative, and current comes out of the emitter through resistor R_E. The behavior of the emitter follower can be better understood by replacing Q_1 with its model for the forward-active region, as in Fig. 9.12(b).

Using Kirchhoff's voltage law:

$$v_O = v_I - v_{BE} \tag{9.31}$$

Although the emitter current of the transistor changes as the output voltage changes,

$$i_E = \frac{v_O - (-V_{EE})}{R_E} = \frac{v_O + V_{EE}}{R_E} \tag{9.32}$$

v_{BE} does not change significantly[2] because of the logarithmic dependence of v_{BE} on i_E, as obtained from Eq. (9.1):

$$v_{BE} = V_T \ln \left(\frac{\alpha_F i_E}{I_S} \right) \tag{9.33}$$

So the difference between the input and output voltages is approximately constant,

$$v_O = v_I - v_{BE} \cong v_I - 0.7 \text{ V}$$

Thus, the voltage at the emitter "follows" the voltage at the input, but with a fixed offset equal to one base-emitter diode voltage. This is also clearly evident in Fig. 9.12(b), in which the base and emitter terminals are connected together through the base-emitter diode.

The voltage transfer characteristic for the emitter follower appears in Fig. 9.13. The output voltage v_O at the emitter follows the input voltage with a slope of $+1$ and a fixed offset voltage equal to $V_{BE} \cong 0.7$ V. For positive inputs, the output follows the input voltage until the BJT begins to enter saturation at the point when $v_I > V_{CC}$. The maximum output voltage occurs when the transistor saturates with $v_O = V_{CC} - V_{CESAT}$ and $v_I = V_{CC} - v_{CESAT} + v_{BE}$. At this point, the input voltage

---

[2] Remember that a factor of 10 change in i_E requires only a 60-mV change in v_{BE} at room temperature.

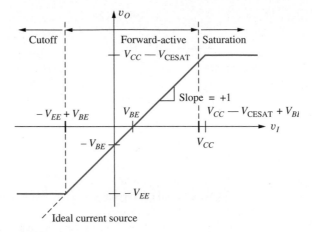

Figure 9.13 Voltage transfer characteristic for the emitter follower.

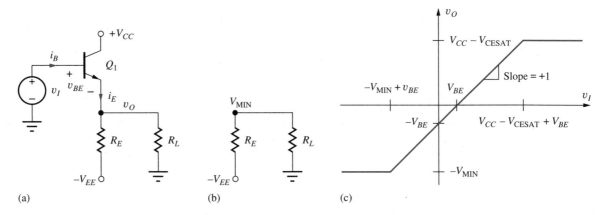

Figure 9.14 (a) Emitter follower with load resistor R_L added; (b) circuit with Q_1 cut off; (c) voltage transfer function for emitter follower with load resistor.

is approximately one diode-drop above V_{CC}! Any further increase in v_I will destroy the bipolar transistor.

The minimum output voltage is set by the negative power supply. The total emitter current i_E cannot become negative, so Q_1 turns off as input v_I falls below $(-V_{EE} + 0.7 \text{ V})$, and the output becomes $-V_{EE}$.

EXERCISE: What value of R_E is required to set $i_E = 0.3$ mA for $v_I = 0$ in the emitter follower if $-V_{EE} = -5.2$ V?

ANSWER: 15.0 kΩ

9.6.1 EMITTER FOLLOWER WITH A LOAD RESISTOR

An external load resistor is often connected to an emitter follower, as shown by R_L in Fig. 9.14(a). The addition of R_L sets a new limit V_{MIN} on the negative output swing of the emitter follower. V_{MIN} represents the Thévenin equivalent voltage at v_O for $i_E = 0$, as in Fig. 9.14(b):

$$V_{\text{MIN}} = \frac{R_L}{R_L + R_E}(-V_{EE}) \tag{9.34}$$

For $V_I > V_{\text{MIN}} + v_{BE}$, the behavior of the emitter follower is the same as discussed previously. However, for v_O to drop below V_{MIN}, emitter current i_E has to be negative, which is impossible in this circuit. The modified VTC for the emitter follower is shown in Fig. 9.14(c), in which the minimum output voltage is now V_{MIN}.

> **EXERCISE:** If $R_E = 15$ kΩ, $V_{CC} = 0$ V, and $-V_{EE} = -5.2$ V, what is the minimum output voltage of the emitter follower in Fig. 9.14(a) if $R_L = \infty$? If $R_L = 10$ kΩ?
>
> **ANSWERS:** −5.20 V; −2.08 V

DESIGN EXAMPLE 9.2

EMITTER FOLLOWER DESIGN

Emitter followers are widely used to buffer analog signals such as sine waves, and this example investigates the design of a circuit in such an application.

PROBLEM An emitter follower has an input voltage $v_I = 3 \sin 2000\pi t$ V. Design an emitter follower to deliver this signal to a 5-kΩ load resistor. The available power supplies are ±5 V.

SOLUTION **Known Information and Given Data:** An emitter follower circuit is specified; $v_I = 3 \sin 2000\pi t$ V; power supply voltages are +5 V and −5 V; the load resistor is 5 kΩ.

Unknowns: Bias circuit and operating current for the transistor

Approach: The simplest circuit implementation is that of Fig. 9.14(a) in which we need to choose only the value of R_E. First determine the required output voltage range; then calculate the required value of emitter resistance.

Assumptions: Use the emitter follower circuit biased with resistor R_E as in Fig. 9.14.

Analysis: The minimum value of v_I is −3 V, and the minimum value of v_O will be one base emitter diode drop (0.7 V) below this voltage or −3.7 V. Using Eq. (9.34):

$$-3.7\text{ V} = \frac{5\text{ k}\Omega}{5\text{ k}\Omega + R_E}(-5\text{ V}) \rightarrow R_E = 1.76\text{ k}\Omega$$

The nearest 5 percent resistor value is 1.8 kΩ, but this value is too large since 1.76 kΩ represents the maximum allowable value for R_E. So we choose $R_E = 1.6$ kΩ, the nearest smaller value.

Check of Results: We have found the resistor required for the design. Let us use the value to check the minimum output voltage.

$$V_{\text{MIN}} = \frac{5\text{ k}\Omega}{5\text{ k}\Omega + 1.6\text{ k}\Omega}(-5\text{ V}) = -3.79\text{ V} \quad \checkmark$$

We have not checked the maximum input condition and really need to be sure that the transistor is in the forward-active region. The maximum input voltage is +3 V, and the collector voltage is fixed at +5 V. Therefore the base-collector junction remains reverse-biased throughout the full input signal range.

Discussion: Let us explore the currents in the transistor at the Q-point and the extremes of the input voltage. The current in the emitter follower is given by

$$i_E = \frac{v_O - (-V_{EE})}{R_E} - \frac{v_O}{R_L} = \frac{v_O + V_{EE}}{R_E} - \frac{v_O}{R_L}$$

The Q-point current is defined as the current for $v_I = 0$ for which $v_O = -0.7$ V:

$$i_E = \frac{-0.7 + 5 \text{ V}}{1.6 \text{ k}\Omega} - \frac{0.7 \text{ V}}{5 \text{ k}\Omega} = 2.55 \text{ mA}$$

The minimum and maximum emitter follower currents are

$$i_E^{min} = \frac{-3.7 \text{ V} - (-5 \text{ V})}{1.6 \text{ k}\Omega} - \frac{3.7 \text{ V}}{5 \text{ k}\Omega} = 72.5 \text{ }\mu\text{A} \quad \text{and} \quad i_E^{max} = \frac{+2.3 \text{ V} + 5 \text{ V}}{1.6 \text{ k}\Omega} - \frac{2.3 \text{ V}}{5 \text{ k}\Omega} = 4.10 \text{ mA}$$

Use of the 1.6-kΩ emitter resistor causes a nonzero emitter current to flow in the transistor for $v_I = -3$ V so that the transistor remains in the forward-active region rather than entering cutoff at the minimum value of output voltage. This is the preferred situation in the design of the emitter follower. However, the minimum current is a relatively small fraction of the Q-point current. We might consider decreasing the value of R_E somewhat further to increase this current (e.g., 1.5 kΩ or 1.3 kΩ) to provide a design safety margin.

Computer-Aided Analysis: The results of two simulations are given here. For the design input, $v_I = 3 \sin 4000\pi t$ V, the output follows the input as desired. Note that the input and output are simply shifted by the base-emitter voltage drop (approximately 0.7 V). However, if the input is increased above the 3 V design limit, the waveform becomes distorted. The second graph presents the results for $v_I = 5 \sin 4000\pi t$ V. The follower circuit is unable to replicate the input as it becomes more negative than -3 V.

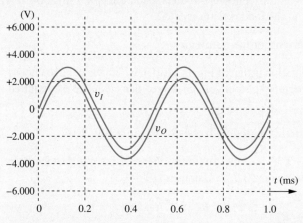

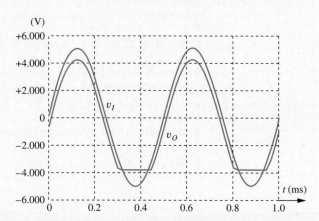

9.7 "EMITTER DOTTING" OR "WIRED-OR" LOGIC

In most logic circuits including NMOS, CMOS, and TTL, the outputs of two logic gates cannot be directly connected together. (See Prob. 7.17.) However, it is possible to tie the outputs of emitter followers together, and this capability provides a powerful enhancement to the logic function of ECL.

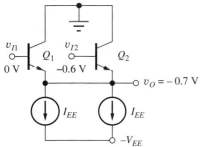

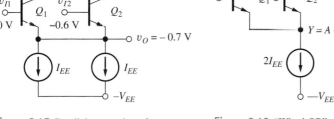

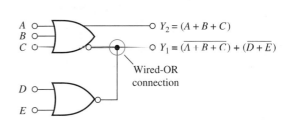

Figure 9.15 Parallel connection of two emitter followers.

Figure 9.16 "Wired-OR" connection of emitter followers.

Figure 9.17 "Wired-OR" connection of two ECL logic gates.

9.7.1 PARALLEL CONNECTION OF EMITTER-FOLLOWER OUTPUTS

Consider the circuit of Fig. 9.15, with the input voltages as shown. The output follows the most positive input voltage (minus the 0.7-V base-emitter offset), whereas the transistor with the lower input voltage operates near cutoff. For the specific example in Fig. 9.15, the input voltage to Q_2 is at -0.6 V. If Q_2 were conducting, then its emitter would be one diode drop below its base at -1.3 V. However, the input to Q_1 is 0 V, and its emitter can only drop down to -0.7 V. Thus, the output is at -0.7 V. The base-emitter voltage of transistor Q_2 is forced to become -0.1 V, and Q_2 is cut off. Note that because Q_2 is cut off, the emitter of Q_1 must now supply the current of both current sources, $i_{E1} = 2I_{EE}$.

9.7.2 THE WIRED-OR LOGIC FUNCTION

Now, suppose that one emitter-follower input corresponds to the logic variable A, and the second input corresponds to logic variable B, as in Fig. 9.16. The output will be high if either A or B is high, whereas the output will be low only if both A and B are low. This corresponds to the OR function: $Y = A + B$. The logical OR function can be obtained by simply connecting the outputs of two ECL gates. This is a powerful additional feature provided by the ECL logic family.

A simple example is provided in the logic circuit in Fig. 9.17, in which the outputs of two ECL gates are connected to provide the logic function $Y_1 = \overline{(A + B + C)} + \overline{(D + E)}$. The upper gate also provides a second output, $Y_2 = A + B + C$.

The **wired-OR logic** function in ECL represents an example of the reason we need to understand the internal circuitry associated with various logic families. We cannot arbitrarily connect the outputs of logic gates together. In many logic families, the wired-OR function is not permitted. If we connect the outputs, the logic levels will not be valid, and the logic gate may even be destroyed in some cases.

9.8 ECL POWER-DELAY CHARACTERISTICS

As pointed out in the chapters on MOS logic design, the power-delay product (PDP) is an important figure of merit for comparing logic families. In this section, we first explore the power dissipation of the ECL gate and then characterize its delay at low power. The results are then combined to form the power-delay product.

9.8.1 POWER DISSIPATION

The static power dissipation of the ECL gate is easily calculated based on our original inverter circuit, shown in Fig. 9.18. The total current I in the inverter is independent of the logic state within the gate: $I = I_{EE} + I_3 + I_4$. Thus, the average ECL power dissipation $\langle P \rangle$ is independent of logic state and is equal to

$$\langle P \rangle = V_{EE}(I_{EE} + I_3 + I_4) \tag{9.35}$$

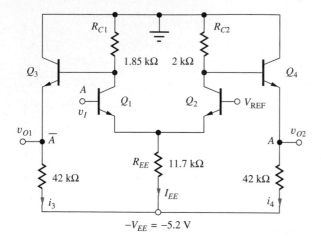

Figure 9.18 ECL gate with resistor biasing.

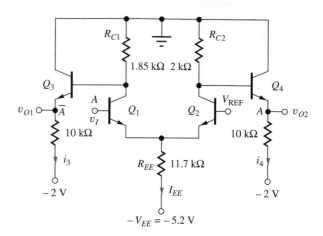

Figure 9.19 ECL circuit with reduced power in the emitter followers.

For the circuit in Fig. 9.18, we remember that the sum $i_3 + i_4 = 0.2$ mA is a constant regardless of input state, and the average value of the current I_{EE} is

$$\langle I_{EE} \rangle = \frac{0.300 + 0.325}{2} \text{ mA} = 0.313 \text{ mA} \tag{9.36}$$

based on a 50 percent logic state duty cycle. Thus, the average power dissipation is $\langle P \rangle = (5.2 \text{ V})(0.200 + 0.313 \text{ mA}) = 2.7$ mW.

EXERCISE: Scale the resistors in the ECL gate in Fig. 9.18 to reduce the power by a factor of 10.

ANSWERS: 20 kΩ, 18.5 kΩ, 117 kΩ, 420 kΩ

Power Reduction

Note that 40 percent of the power in the circuits in Fig. 9.18 is dissipated in the emitter-follower stages. Two techniques have been used to reduce the power consumption in more advanced ECL gates. The first is to simply return the emitter-follower resistors to a second, less negative power supply, such as the −2-V supply in Fig. 9.19. The resistors in the emitters of Q_3 and Q_4 have been changed to 10 kΩ to keep the currents in Q_3 and Q_4 equal to 0.1 mA. The power dissipation in this circuit is reduced by 33 percent to

$$P = 5.2(0.313) + 2(0.1) = 1.8 \text{ mW} \tag{9.37}$$

This method, however, requires the cost of another power supply and its associated wiring for power distribution.

Another power-reduction technique is illustrated in Fig. 9.20, in which the resistors that supply current to the emitter followers are now connected between each input and the −5.2-V supply. In Fig. 9.15, we saw that one emitter follower would have to supply the current of multiple current sources when the wired-OR function was used. Using the repartitioned circuit in Fig. 9.20, the emitter follower current is always equal to the current of only one of the original emitter followers. This redesign significantly reduces the overall power consumption in large logic networks. However, any output that does not drive the input of another logic gate needs to have an external termination resistor connected from its output to the negative power supply.

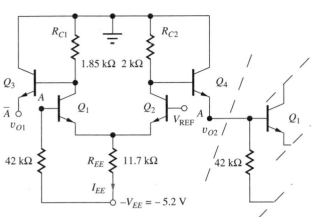

Figure 9.20 Repartitioned ECL gate.

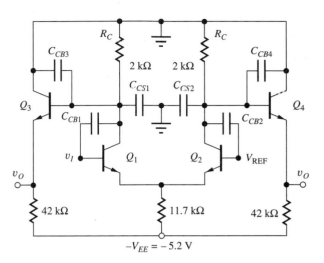

Figure 9.21 ECL inverter with capacitances at the collector nodes of the current switch.

9.8.2 GATE DELAY

The capacitances that dominate the delay of the ECL inverter at low power levels have been added to the circuit in Fig. 9.21. The symbol C_{CB} represents the capacitance of the reverse-biased collector-base junction, and C_{CS} represents the capacitance between the collector and the substrate of the transistor. Transistors Q_1 and Q_2 switch the current I_{EE} back and forth very rapidly in response to the input v_I. The emitter followers can supply large amounts of current to quickly charge any load capacitances connected to the two outputs.

At low power, the speed of the ECL gate is dominated by the $R_C - C_L$ time constant at the collectors of Q_1 and Q_2, and the response of the inverter can be modeled by the simple RC circuit in Fig. 9.22, in which R_C is the collector-load resistance and C_L is the effective load capacitance at the collector node of Q_2, given by

$$C_L = C_{CS2} + C_{CB2} + C_{CB4} \tag{9.38}$$

The load capacitance consists of the base-collector capacitances of Q_2 and Q_4 plus the collector-substrate capacitance of Q_2.

For the negative-going transient, the capacitor is initially discharged, and current $-I_{EE}$ is switched into the node at $t = 0$. The voltage at node v_{C2} is described by

$$v_{C2}(t) = -I_{EE}R_C \left[1 - \exp\left(-\frac{t}{R_C C_L} \right) \right] \tag{9.39}$$

The collector-node voltage exponentially approaches the final value of $-I_{EE}R_C$. The actual output voltage at v_O is level-shifted down by one 0.7-V drop by the emitter follower.

The propagation-delay time is the time required for the output to make 50 percent of its transition:

$$v_{C2}(\tau_{PHL}) = -\frac{I_{EE}R_C}{2} \quad \text{and} \quad \tau_{PHL} = 0.69 R_C C_L \tag{9.40}$$

For the positive-going transition, the capacitor is initially charged to the negative voltage $-I_{EE}R_C$. At $t = 0$, the current source is switched off, and the capacitor simply discharges through R_C:

$$v_{C2}(t) = -I_{EE}R_C \exp\left(-\frac{t}{R_C C_L} \right) \tag{9.41}$$

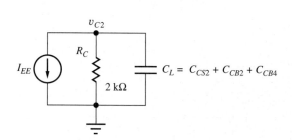

Figure 9.22 Simplified model for the dynamic response of an ECL gate.

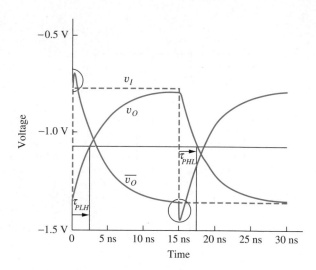

Figure 9.23 Simulated switching waveforms for the ECL inverter of Fig. 9.21.

For this case, the propagation delay is

$$v_{C2}(\tau_{PLH}) = -\frac{I_{EE}R_C}{2} \qquad \text{and} \qquad \tau_{PLH} = 0.69 R_C C_L \tag{9.42}$$

Using Eqs. (9.40) and (9.42), the average propagation delay of the ECL gate is

$$\tau_P = \frac{\tau_{PHL} + \tau_{PLH}}{2} = 0.69 R_C C_L \tag{9.43}$$

Figure 9.23 shows the results of simulation of the switching behavior of the ECL gate in Fig. 9.21. In this case, the transistor capacitances are $C_{CB} = 0.5$ pF and $C_{CS} = 1.0$ pF. For $R_C = 2$ kΩ, the two propagation-delay times from Eqs. (9.40) and (9.42) are estimated to be 2.8 ns. This prediction agrees very well with the waveforms in Fig. 9.23.

Note the two transient "spikes" (circled in Fig. 9.23) that show up on the $\overline{v_O}$ output coinciding with the switching points on the input waveform. These spikes do not show up on the v_O output. The transients are in the same direction as the input signal change and result from the coupling of the input waveform directly through capacitance C_{BC1} to the inverting output. A similar path to the noninverting output does not exist. This is another good illustration of detailed simulation results that one should always try to understand. Such unusual observations should be studied to determine if they are real effects or some artifact of the simulation tool, as well as to understand how they might affect the performance of the circuit.

9.8.3 POWER-DELAY PRODUCT

Using the values calculated from Eqs. (9.35) and (9.43) for the gate in Fig. 9.21, the power-delay product is 2.6 mW × 2.8 ns, or 7.3 pJ. From Eq. (9.43), we see that the propagation delay of the inverter for a given capacitance is directly proportional to the choice of R_C, and we know R_C is related to the logic swing and current I_{EE} by $R_C = \Delta V / I_{EE}$. Assuming that we want to keep the logic swing and the noise margins constant, then we can reduce R_C only if we increase the current I_{EE} and hence the power of the gate. This illustrates the direct power-delay trade-off involved in gate design because I_{EE} accounts for most of the power in the ECL logic gate.

The analysis presented in the previous section was valid for operation in the region of constant power-delay product. However, as power is increased, the effect of charge storage in the BJT (discussed in detail in Sec. 9.9) becomes more and more important, and the delay of the ECL gate enters

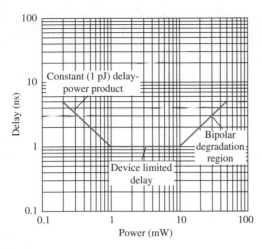

Figure 9.24 Delay versus power behavior for bipolar logic.

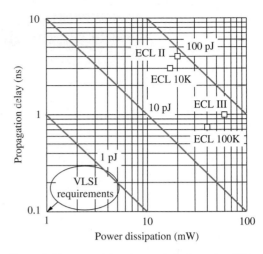

Figure 9.25 Power-delay characteristics for various ECL gates. VLSI requirements are less than 100 fJ.

a region in which it becomes independent of power. Finally, at even higher power levels, the delay starts to degrade as the f_T of the BJT falls. These three regions are shown in Fig. 9.24.

Figure 9.25 summarizes the power-delay characteristics of a number of commercial ECL unit logic gates as well as the requirements for high-performance circuits for use in high-density IC chips. The more recent ECL unit logic families offer subnanosecond performance but consume relatively large amounts of power in order to reliably drive the large off-chip capacitances associated with printed circuit board mounting and interconnect. The large power-delay products are not usable for VLSI circuit densities. Much lower power-delay products are associated with state-of-the-art on-chip logic circuits that benefit from smaller capacitive loads as well as significantly improved bipolar device technology.

EXERCISE: What are the delay and power-delay product for the ECL gate in Fig. 9.21 if I_{EE} is changed to 0.5 mA, but the logic swing is maintained the same?

ANSWERS: 1.4 ns, 5.0 pJ

9.9 THE SATURATING BIPOLAR INVERTER

At the heart of many bipolar logic gates is the simple saturating bipolar inverter circuit in Fig. 9.26, which uses a single BJT switching device Q_2 to pull v_O down to V_L and a resistor load element to pull the output up to the power supply V_{CC}. The input voltage v_I and the base current i_B supplied to the base of the bipolar transistor will be designed to switch Q_2 between the saturation and nonconducting states.

For analysis and design of saturating BJT circuits, we use the transistor parameters in Table 9.3. The base-emitter or base-collector voltages in the forward- or reverse-active regions are assumed to be 0.7 V. However, when a transistor saturates, its base-emitter voltage increases slightly, and 0.8 V will be used for the base-emitter voltage of a saturated transistor (V_{BESAT}). Our BJT circuits will be designed to have a worst-case $V_{\text{CESAT}} = 0.15$ V. Transistors in most logic technologies are optimized for speed, and current gain is often compromised, as indicated by the relatively low value of β_F in Table 9.3. β_R typically ranges between 0.1 and 2. Up to now, we have not found a use for the reverse-active region of operation, but as we shall see in Sec. 9.10, it plays a very important role in TTL circuits.

ELECTRONICS IN ACTION

Electronics for Optical Communications

Optical fiber communication systems provide the backbone of today's high bandwidth Internet and cellular communication systems. For example, in the OC-192 and OC-768 links, digital information is modulated onto optical carriers generated by solid-state lasers operating at data rates of 10 Gb/s and 40 Gb/s, respectively. Extremely high speed electronic circuits are required at both ends of the optical fiber link to convert from electrical to optical (E/O) and optical to electrical (O/E) form. These interface circuits include the data multiplexers and de-multiplexers, modulators, detectors, preamplifiers and clock recovery circuits shown in the accompanying figure. Because of the very high speed requirements, the circuit implementations are typically based on emitter-coupled logic (ECL) circuits using bipolar transistors with f_T's in the 50-200 GHz range. Versions of the circuits have been implemented using silicon BJTs,[1] Indium Phosphide (InP) heterojunction bipolar transistors (HBTs),[2] and HBTs in Silicon Germanium (SiGe—"Siggy").[3]

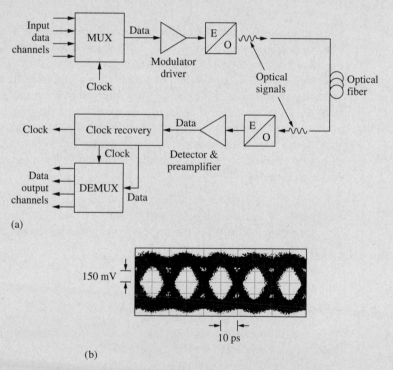

(a)

(b)

(a) Block diagram of a 40 Gb/s optical fiber communication system. (b) Eye diagram at data output of 2/1 MUX running at 50 Gb/s. Reprinted with permission from [1]. © 1996 IEEE.

[1] Alfred Felder, Michael Möller, Josef Popp, Josef Böck, and Hans-Martin Rein, "46 Gb/s DEMUX, 50 Gb/s MUX, and 30 GHz Static Frequency Divider in Silicon Bipolar Technology, IEEE Journal Of Solid-State Circuits, vol. 31, no. 4, pp. 481–486, April 1996.

[2] Mario Reinhold, Claus Dorschky, Eduard Rose, Rajasekhar Pullela, Peter Mayer, Frank Kunz, Yves Baeyens, Thomas Link, and John-Paul Mattia, "A Fully Integrated 40-Gb/s Clock and Data Recovery IC With 1:4 DEMUX in SiGe Technology, IEEE Journal Of Solid-State Circuits, vol. 36, no. 12, pp. 1937–1945, December, 2001.

[3] Y. Baeyens, G. Georgiou, J. S. Weiner, A. Leven, V. Houtsma, P. Paschke, Q. Lee, R. F. Kopf, Y. Yang, L. Chua, C. Chen, C. T. Liu, and Y. Chen, "InP D-HBT ICs for 40-Gb/s and higher bitrate lightwave tranceivers," IEEE Journal of Solid-State Circuits, vol. 37, no. 9, pp. 1152–1159, September 2002.

On the receiver side of the optical fiber system, the digital signal appearing at the output of the detector and preamplifier contains a combination of both data and clock information. The clock is first separated from the data and then used to synchronize the data recovery process. The clock recovery circuit shown represents one example of the use of ECL circuitry. The differential input signal is buffered by two-stages of emitter-followers. The output of the second pair of emitter followers drives the input of an ECL flip-flop divider formed from three current switch circuits. The output of the current switches is buffered and level-shifted by additional pairs of emitter followers and fed back to the upper pair of current switches to form the flip-flop. The clock recovery circuit produces two differential clock signals with a 90° phase separation (quadrature) at the C0/$\overline{\text{C0}}$ and C90/$\overline{\text{C90}}$ outputs. See the Electronics in Action topics in Chapters 12 and 16 for a further discussion of detector and amplifier interface circuitry for optical communications.

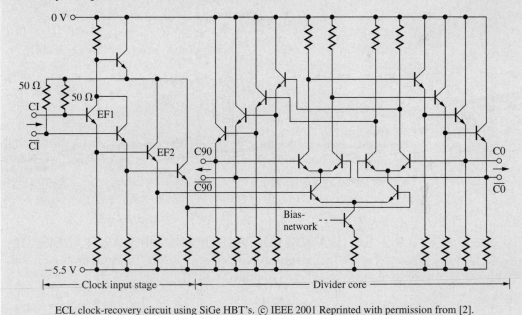

ECL clock-recovery circuit using SiGe HBT's. © IEEE 2001 Reprinted with permission from [2].

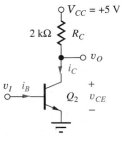

Figure 9.26 Single transistor bipolar inverter and device parameters.

TABLE 9.3 BJT Parameters	
I_S	10^{-15} A
β_F	40
β_R	0.25
V_{BE}	0.70 V
V_{BESAT}	0.80 V
V_{CESAT}	0.15 V

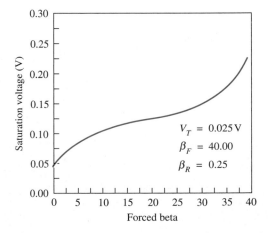

Figure 9.27 Saturation voltage V_{CESAT} versus forced beta i_C/i_B.

9.9.1 STATIC INVERTER CHARACTERISTICS

Writing the equation for the output voltage of the inverter in Fig. 9.26, we find

$$v_O = V_{CC} - i_C R_C \tag{9.44}$$

If the base-emitter voltage ($v_{BE} = v_I$) is several hundred millivolts less than the normal turn-on voltage of the base-emitter junction (0.6 to 0.7 V), then Q_2 will be nearly cut off, with $i_C \cong 0$. Equivalently, if the input base current i_B is zero, then Q_2 will also be near cutoff. For either case,

$$v_O = V_H \cong V_{CC} \tag{9.45}$$

In this particular logic circuit, the value of V_H is set by the power supply voltage: $V_H = 5$ V.

The low-state output level is set by the saturation voltage of the bipolar transistor, $V_L = V_{CESAT}$. To ensure saturation, as discussed in Sec. 5.7.5, we require that

$$i_B > \frac{i_{CMAX}}{\beta_F} \quad \text{where } i_{CMAX} = \frac{V_{CC} - V_{CESAT}}{R_C} \cong \frac{V_{CC}}{R_C} \tag{9.46}$$

For the circuit in Fig. 9.26, $i_{CMAX} = 2.43$ mA, assuming $V_{CESAT} = 0.15$ V. Using $\beta_F = 40$, the transistor will be saturated for

$$i_B > \frac{2.43 \text{ mA}}{40} = 60.8 \text{ } \mu\text{A}$$

> **EXERCISE:** Estimate the static power dissipation of the inverter in Fig. 9.26 for $v_O = V_H$ and $v_O = V_L$. What value of R_C is required to reduce the power dissipation by a factor of 10?
>
> **ANSWERS:** 0, 12.1 mW; 20 kΩ

9.9.2 SATURATION VOLTAGE OF THE BIPOLAR TRANSISTOR

An expression for the **saturation voltage** of the BJT in terms of its base and collector currents was derived in Eq. (5.43) and is repeated here:

$$V_{CESAT} = V_T \ln \left[\left(\frac{1}{\alpha_R} \right) \frac{1 + \frac{i_C}{(\beta_R + 1)i_B}}{1 - \frac{i_C}{\beta_F i_B}} \right] = V_T \ln \left[\left(\frac{1}{\alpha_R} \right) \frac{1 + \frac{\beta_{FOR}}{(\beta_R + 1)}}{1 - \frac{\beta_{FOR}}{\beta_F}} \right] \quad \text{for } i_B > \frac{i_C}{\beta_F} \tag{9.47}$$

Recall that the ratio i_C/i_B is the forced beta β_{FOR} and that $i_C/i_B = \beta_F$ is true only for the forward-active region. Note that Eq. (9.47) becomes infinite for $i_C/i_B = \beta_F$.

Equation (9.47) is plotted as a function of $\beta_{FOR} = i_C/i_B$ in Fig. 9.27. As i_B becomes very large or i_C becomes very small, β_{FOR} approaches zero, and V_{CESAT} reaches its minimum value:

$$V_{CESAT}^{MIN} = V_T \ln \left(\frac{1}{\alpha_R} \right) \tag{9.48}$$

The minimum value of V_{CESAT} is approximately 40 mV for the transistor depicted in Fig. 9.27.

For the more general case, i_{CMAX} is known from Eq. (9.46), and Eq. (9.47) can be used to ensure that our circuit design supplies enough base current to saturate the transistor to the desired level. As assumed earlier, $V_{CESAT} \leq 0.15$ V. Solving Eq. (9.47) for i_C/i_B,

$$\frac{i_C}{i_B} \leq \beta_F \left[\frac{1 - \frac{1}{\alpha_R \Gamma}}{1 + \frac{\beta_F}{\beta_R \Gamma}} \right] \quad \text{where } \Gamma = \exp(V_{CESAT}/V_T) \tag{9.49}$$

Using Eq. (9.49) with $i_C = 2.43$ mA and the values from Table 9.3, reaching $V_{CESAT} = 0.15$ V requires $i_C/i_B \leq 28.3$ or $i_B \geq 86$ µA.

DESIGN EXAMPLE 9.3 — BIPOLAR TRANSISTOR SATURATION VOLTAGE

In this example, we will choose the base current required to achieve a desired saturation voltage in a power transistor used in a switching application.

PROBLEM
A bipolar power transistor has a forward current gain of 20 and an inverse current gain of 0.1. How much base current is required to achieve a saturation voltage of 0.1 V at a collector current of 10 A?

SOLUTION
Known Information and Given Data: Bipolar transistor with $\beta_F = 20$, $\beta_R = 0.1$, and $I_C = 10$ A

Unknowns: Base current I_B required to achieve the desired "on-voltage"

Approach: Find Γ and calculate I_B based on Eq. (9.49).

Assumptions: Current gain values are independent of current; room temperature operation with $V_T = 25$ mV

Analysis: Let us first check to be sure that $V_{CESAT} = 0.1$ V is possible:

$$V_{CEMIN} = V_T \ln \frac{1}{\alpha_R} = V_T \ln \left(\frac{\beta_R + 1}{\beta_R} \right) = 0.025 \text{ V} \ln \left(\frac{0.1 + 1}{0.1} \right) = 0.025 \text{ V} \ln(11) = 0.06 \text{ V}$$

Since 60 mV is less than the required saturation voltage of 0.1 V, we can proceed. Using Eq. (9.49) to find Γ yields

$$\Gamma = \exp \left(\frac{V_{CESAT}}{V_T} \right) = \exp \left(\frac{0.1 \text{ V}}{0.025 \text{ V}} \right) = 54.6$$

and solving the same equation set for I_B yields

$$I_B \geq \frac{I_C}{\beta_F} \left[\frac{1 + \dfrac{\beta_F}{\beta_R \Gamma}}{1 - \dfrac{1}{\alpha_R \Gamma}} \right] = \frac{10 \text{ A}}{20} \left[\frac{1 + \dfrac{20}{0.1(54.6)}}{1 - \dfrac{11}{54.6}} \right] = 2.92 \text{ A}$$

A minimum base current of 2.92 A is required to achieve the saturation voltage of 100 mV. Note that an additional safety margin should be used in the design of the base current drive circuit.

Check of Results: The forced beta is

$$\beta_{FOR} = \frac{I_C}{I_B} = \frac{10}{2.92} = 3.43$$

and the collector-emitter voltage of the saturated transistor will be

$$V_{CESAT} = V_T \ln \left[\left(\frac{1}{\alpha_R} \right) \frac{1 + \dfrac{\beta_{FOR}}{\beta_R + 1}}{1 - \dfrac{\beta_{FOR}}{\beta_F}} \right] = 0.025 \text{ V} \ln \left[(11) \frac{1 + \dfrac{3.43}{0.1 + 1}}{1 - \dfrac{3.43}{20}} \right] = 0.100 \text{ V} \quad ✔$$

Computer-Aided Analysis and Discussion: SPICE simulation uses the circuit shown here in which current source I_B is swept from 0 to 4 A. VCC is chosen arbitrarily to be 5 V, and RC is selected so that the collector current will be 10 A when $v_C = 0.1$ V. The default *npn* model is used with the parameters set as BF = 20, BR = 0.1, and IS = 10 fA. The output voltage doesn't reach 0.1 V until IB = 3.37 A. We should ask ourselves why there is such a discrepancy. The answer lies in our choice of the value of the thermal voltage. Remember that the temperature in SPICE defaults to $T = 27°C$. If we recalculate the required base current using $V_T = 25.8$ mV, we find the base current should be at least 3.33 A.

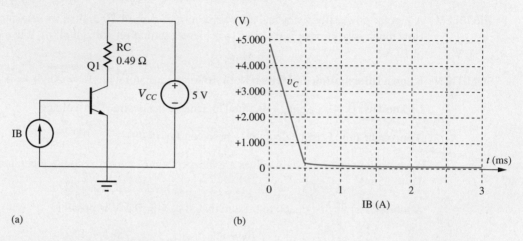

(a)   (b)

Note that the transistor enters saturation as the base current exceeds approximately 0.5 A, but the base current must increase to 3.4 A to force V_{CESAT} to reach 0.1 V. At a forced beta of 15 $(I_B = 0.667$ A$)$, V_{CESAT} is 0.16 V.

EXERCISE: Recalculate the minimum value of I_B and β_{FOR} assuming $V_T = 25.8$ mV.

ANSWER: 3.33 A; 6.00

EXERCISE: What is the minimum base current required in Design Ex. 9.3 if β_R of the transistor changes to 0.2?

ANSWER: 1.59 A

EXERCISE: What minimum base current is required to achieve a saturation voltage of 0.15 in Design Ex. 9.3?

ANSWER: 0.769 A

EXERCISE: What is the minimum saturation voltage of a power transistor having $\beta_R = 0.05$ and operating at a junction temperature of 150°C?

ANSWER: 111 mV

9.9.3 LOAD-LINE VISUALIZATION

As presented in Chapter 6, an important way of visualizing inverter operation is to look at the load line, Eq. (9.44), drawn on the BJT transistor output characteristics, as in Fig. 9.28. The BJT switches

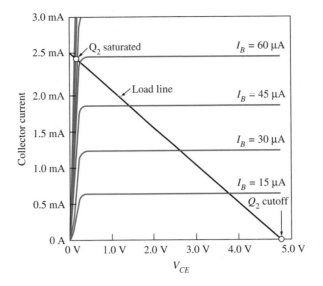

Figure 9.28 BJT output characteristics and load line.

between the two operating points on the load line indicated by the circles in Fig. 9.28. At the right-hand end of the load line, the BJT is cut off, with $i_C = 0$ and $v_{CE} = 5$ V. At the Q-point near the left end of the load line, the BJT represents a low resistance in the saturation region, with $v_O = V_{CESAT}$. Note that the current in saturation is limited primarily by the load resistance and is nearly independent of the base current.

EXERCISE: A transistor must reach a saturation voltage ≤ 0.1 V with $I_C = 10$ mA. What are the maximum value of β_{FOR} and the minimum value of the base current? Use the transistor parameters from Table 9.3.

ANSWERS: 9.24, 1.08 mA

9.9.4 SWITCHING CHARACTERISTICS OF THE SATURATED BJT

A very important change occurs in the switching characteristics of bipolar transistors when they saturate. The excess base current that drives the transistor into saturation causes additional charge storage in the base region of the transistor. This charge must be removed before the transistor can turn off, and an extra delay time, termed the **storage time t_S,** appears in the switching characteristic of the saturated BJT.

We illustrate the problem through the circuit simulation results presented in Fig. 9.30 for the inverter circuit in Fig. 9.29, in which the BJT is driven by a current source that forces current I_{BF} into the base to turn the transistor on and pulls current I_{BR} out of the base to turn the transistor back off. Negative base current can only occur in the *npn* transistor during the transient that removes charge from the base. Diode D_1 is added to the circuit to provide a steady-state path for the negative source current because we know that the *npn* transistor cannot support a large negative steady-state base current without junction breakdown.

Referring to Figs. 9.29 and 9.30 at $t = 0$, we see that the current source forces a base current $I_{BF} = 1$ mA into the transistor, rapidly charging the base-emitter capacitance and supplying charge to the base. The BJT turns on and saturates within approximately 5 ns. At $t = 15$ ns, the direction of the current source reverses, and simulation results are shown for three different values of reverse current. For a small reverse current (0.01 mA), electron–hole recombination in the base is the only

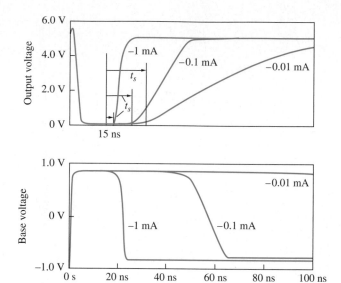

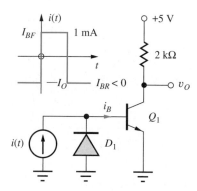

Figure 9.29 Bipolar inverter with current source drive.

Figure 9.30 Switching behavior for the BJT inverter for three values of reverse base current: $I_{BR} = -1$ mA, -0.1 mA, and -0.01 mA.

mechanism available to remove the excess base charge. The transistor turns off very slowly because of the slow decay of the charge stored on the base-emitter junction capacitance. As the reverse base current is increased, both storage time and rise time are substantially reduced.

Observe the behavior of the voltage at the base of the transistor. The base voltage remains constant at $+0.85$ V until the excess **stored base charge** has been removed from the base. Then the base voltage can drop, and the transistor turns off. Note that the base voltage is still above 0.8 V at $t = 100$ ns for the smallest reverse base current!

The storage time t_S can be calculated from this formula [1]:

$$ t_S = \tau_S \ln\left(\frac{I_{BF} - I_{BR}}{\dfrac{i_{\text{CMAX}}}{\beta_F} - I_{BR}}\right) \qquad \text{with} \qquad \tau_S = \frac{\alpha_F(\tau_F + \alpha_R \tau_R)}{1 - \alpha_F \alpha_R} \tag{9.50} $$

in which τ_S is called the **storage time constant.** I_{BF} and I_{BR} are the forward and reverse base currents defined in Fig. 9.29, and α_F and α_R are the forward and reverse common-base current gains. Note that the value of I_{BR} is negative.

The constants τ_F and τ_R are called the **forward** and **reverse transit times** for the transistor and determine the amount of charge stored in the base in the forward and reverse active modes of operation—Q_F and Q_R, respectively:

$$ Q_F = i_F \tau_F \qquad \text{and} \qquad Q_R = i_R \tau_R \tag{9.51} $$

In Eq. (9.51), i_F and i_R are the forward and reverse current components from the transport model.

The storage time constant quantifies the amount of excess charge, over and above that needed to support the actual collector current, stored in the base when the bipolar transistor enters saturation:

$$ Q_{XS} = \tau_S \left(i_B - \frac{i_{\text{CMAX}}}{\beta_F}\right) \tag{9.52} $$

The storage time t_S represents a significant degradation in the speed of the BJT when it tries to come out of saturation, as can be seen in Fig. 9.30 and Ex. 9.1.

EXAMPLE **9.4**   STORAGE TIME CALCULATIONS

In this example, we calculate the impact of the reverse turnoff current on the transistor storage time.

PROBLEM   Calculate the storage time constant and storage times for the three currents used in Figs. 9.29 and 9.30 if $\alpha_F = 0.976$, $\alpha_R = 0.20$, $\tau_F = 0.25$ ns, and $\tau_R = 25$ ns.

SOLUTION   **Known Information and Given Data:** Circuit in Fig. 9.29; transistor parameters $\alpha_F = 0.976$, $\alpha_R = 0.20$, $\tau_F = 0.25$ ns, and $\tau_R = 25$ ns; a forward current (I_{BF}) of 1 mA and reverse turnoff currents (I_{BR}) of -1 mA, -0.1 mA, and -0.01 mA

Unknowns: Storage time constant τ_S and storage times for the three turnoff conditions

Approach: Use Eq. (9.50) to find the storage time constant and then to calculate the storage time for the three different turnoff currents.

Assumptions: None

Analysis: Equation (9.50) gives $\tau_S = \dfrac{0.976(0.25 \text{ ns} + 0.20(25 \text{ ns}))}{1 - 0.976(0.20)} = 6.4 \text{ ns}$

In order to evaluate Eq. (9.50) we need values for i_{CMAX} and β_F. Equation (9.46) defines the value of i_{CMAX},

$$i_{CMAX} \cong \frac{V_{CC}}{R_C} = \frac{5 \text{ V}}{2 \text{ k}\Omega} = 2.5 \text{ mA} \qquad \text{and} \qquad \beta_F = \frac{\alpha_F}{1 - \alpha_F} = \frac{0.976}{1 - 0.976} = 40.7$$

Using Eq. (9.50) with $\tau_S = 6.4$ ns, $I_{BR} = 1$ mA, and $I_{BR} = -0.01$ mA, -0.1 mA, and -1.0 mA yields

$$t_S = (6.4 \text{ ns}) \ln \left[\frac{1 - (-0.01) \ \frac{\text{mA}}{\text{mA}}}{\frac{2.5}{40.7} - (-0.01) \ \text{mA}} \right] = 17.0 \text{ ns} \qquad t_S = (6.4 \text{ ns}) \ln \left[\frac{1 - (-0.1) \ \frac{\text{mA}}{\text{mA}}}{\frac{2.5}{40.7} - (-0.1) \ \text{mA}} \right] = 12.3 \text{ ns}$$

$$t_S = (6.4 \text{ ns}) \ln \left[\frac{1 - (-1) \ \frac{\text{mA}}{\text{mA}}}{\frac{2.5}{40.7} - (-1) \ \text{mA}} \right] = 4.06 \text{ ns}$$

Check of Results: A double check of the calculations indicates they are correct, and the values agree well with the storage times that can be observed in Fig. 9.30.

EXERCISE: What value of I_{BR} is required to achieve a storage time of 1 ns in the circuit in Fig. 9.29? Use the transistor parameters from Ex. 9.4.

ANSWER: 5.49 mA

EXERCISE: For Ex. 9.4, calculate the excess charge stored in the base and compare to Q_F.

ANSWERS: 6.01 pC, 0.625 pC, $Q_{XS} \gg Q_F$

9.10 A TRANSISTOR-TRANSISTOR LOGIC (TTL) PROTOTYPE

Now that we have studied the characteristics of the saturating transistor inverter, we have the knowledge in place to understand the behavior of transistor-transistor logic or TTL. For years, TTL has been a workhorse technology for implementing digital functions and for providing "glue logic"

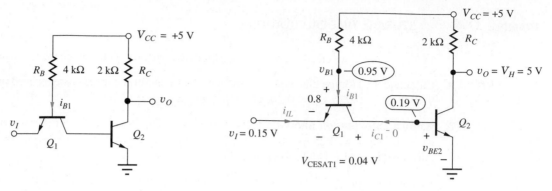

Figure 9.31 TTL inverter prototype.

Figure 9.32 TTL gate with input $v_I = V_L$.

necessary in microprocessor system design. TTL is interesting from another point of view since it is the only circuit that we shall encounter that makes use of transistors operating in all four regions of operation—forward-active, reverse-active, saturation, and cutoff. In this section and Secs. 9.11 to 9.14, we will first explore a simplified TTL prototype, and then move to analysis of the full standard TTL logic gate, as well as other members of the TTL logic family.

Figure 9.31 is the prototype for a low power TTL inverter. Transistor Q_1 has been added to the inverter of Fig. 9.26 to control the supply of base current to Q_2. Input voltage v_I causes the current i_{B1} to switch between either the base-emitter diode or the base-collector diode of Q_1. We first explore circuit behavior for $v_I = V_L$, find V_H, and then use it to study the circuit for $v_I = V_H$.

9.10.1 TTL INVERTER FOR $V_I = V_L$

The input to the TTL gate in Fig. 9.32 is set to $V_L = V_{CESAT} = 0.15$. The +5-V supply tends to force current i_{B1} down through the 4-kΩ resistor and into the base of Q_1, turning on the base-emitter diode of Q_1. Transistor Q_1 attempts to pull a collector current $i_{C1} = \beta_F i_{B1}$ out of the base of Q_2, but only leakage currents can flow in the reverse direction from the base. Therefore, $i_{C1} \cong 0$. Because Q_1 is operating with $\beta_F i_B > i_C$, it saturates, with both junctions being forward-biased. Thus the voltage at the base of Q_2 is given by

$$v_{BE2} = v_I + V_{CESAT1} = 0.15 + 0.04 = 0.19 \text{ V} \tag{9.53}$$

where $V_{CESAT1} = 0.04$ V has been used because $i_{C1} \cong 0$ (see Fig. 9.27). Since v_{BE2} is only 0.19 V, Q_2 does not conduct any substantial collector current (although it technically remains in the forward-active region—see Prob. 9.56), and the output voltage will be $v_O = V_{CC} = 5$ V.

Base current i_{B1} is given by

$$i_{B1} = \frac{5 - v_{B1}}{4 \text{ k}\Omega} \qquad \text{where } v_{B1} = v_1 + V_{BESAT} = 0.95 \text{ V for } V_{BESAT} = 0.8 \text{ V} \tag{9.54}$$

For the gate in Fig. 9.32, we find that $i_{B1} = 1.01$ mA. This current enters the base and exits the emitter of Q_1 because $i_{C1} \cong 0$. For this TTL gate, the input current for the low input state is $i_{IL} = -i_{E1}$, or

$$-i_{IL} = i_{E1} = (i_{B1} + i_{C1}) \cong i_{B1} = \frac{5 - v_{B1}}{4 \text{ k}\Omega} = \frac{5 - 0.95}{4 \text{ k}\Omega} = 1.01 \text{ mA} \tag{9.55}$$

This is a logic gate characteristic that we have not seen before. In TTL, there is a relatively large current that the input signal source v_I must absorb. We shall see shortly how this current limits the fanout of the TTL gate.

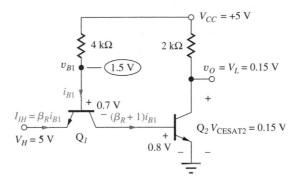

Figure 9.33 TTL gate with input $v_I = V_H$.

EXERCISE: What are the values of V_{BE2} and i_{IL} in Fig. 9.32 if $\beta_{R1} = 2$?

ANSWERS: 0.19 V, −1.01 mA

9.10.2 TTL INVERTER FOR $V_I = V_H$

$V_H = 5$ V is applied as the input to the inverter in Fig. 9.33. Because the emitter of Q_1 is at 5 V, base current i_{B1} cannot enter the base-emitter diode; it instead forward-biases the base-collector diode. Transistor Q_1 enters the reverse-active region with the base-collector junction forward-biased and the base-emitter junction reverse-biased. Base current i_{B1} causes a current $(\beta_R + 1)i_{B1}$ to *exit* the collector terminal, and this current becomes the base current of Q_2. In addition, a current $\beta_R i_{B1}$ *enters* the emitter terminal, and this current represents the input current i_{IH} for the input high state. A small value of β_R is desired to keep i_{IH} low.

$$i_{B2} = (\beta_R + 1)i_{B1} \quad \text{and} \quad i_{IH} = +\beta_R i_{B1} \tag{9.56}$$

For proper circuit operation, $i_{B2} = (\beta_R + 1)i_{B1}$ is designed to be much greater than i_{C2}/β_F, Q_2 saturates, and its base-emitter voltage becomes 0.8 V. The voltage $v_{B1} = v_{BC1} + V_{BESAT2} = 0.7 + 0.8 = 1.5$ V. The base current of Q_1 is now

$$i_{B1} = \frac{(5 - 1.5)\ \text{V}}{4\ \text{k}\Omega} = 0.875\ \text{mA} \quad \text{and} \quad i_{IH} = \beta_R i_{B1} = 0.25(0.875\ \text{mA}) = 0.219\ \text{mA} \tag{9.57}$$

Evaluating Eq. (9.56), we find that the base current of Q_2 is

$$i_{B2} = (\beta_R + 1)i_{B1} = (1.25)0.875\ \text{mA} = 1.09\ \text{mA} \tag{9.58}$$

Current i_{B2} is much greater than the 86 μA required to saturate Q_2, as we calculated immediately following Eq. (9.49). The forced beta in this circuit is $\beta_{FOR} = 2.43\ \text{mA}/1.09\ \text{mA} = 2.22$, and Eq. (9.47) indicates that Q_2 actually has $V_{CESAT2} = 67$ mV. The additional base current just calculated is needed to ensure that the V_{CESAT} specification is met when the inverter must drive a large fanout.

EXERCISE: What would be the values of v_{BE2} and i_{IH} in Fig. 9.33 if $\beta_{R1} = 2$?

ANSWERS: 0.80 V, 1.75 mA

9.10.3 POWER IN THE PROTOTYPE TTL GATE

Now we explore the power dissipation in the TTL gate for the two logic states. For $v_I = 5$ V and $v_O = V_L$, as in Fig. 9.34(a), the total power being consumed by the gate is

$$P_L = V_{CC}i_{CC} + v_I i_{IH} = V_{CC}(i_{C2} + i_{B1}) + v_I i_{IH} \tag{9.59}$$

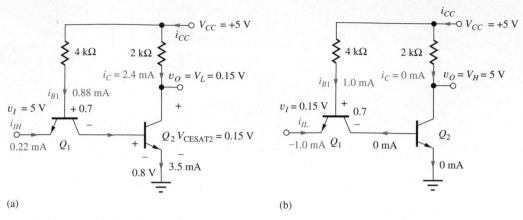

Figure 9.34 (a) Currents in the prototype inverter for $v_O = V_L$. (b) Currents in the TTL gate for $v_I = V_L$.

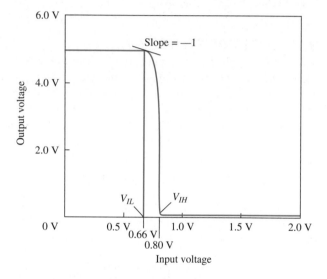

Figure 9.35 SPICE simulation of the VTC of the prototype TTL gate.

For the circuit in Fig. 9.34(a), $P_L = 5 \text{ V}(2.4 \text{ mA} + 0.88 \text{ mA} + 0.22 \text{ mA}) = 17.6 \text{ mW}$.

For $v_I = V_L$ and $v_O = V_H$, as in Fig. 9.34(b), the power being consumed by the gate is

$$P_H = V_{CC}i_{CC} + v_I i_{IL} = V_{CC}i_{B1} + v_I i_{IL} \tag{9.60}$$

For the values in Fig. 9.34(b), $P_H = 5 \text{ V}(1.0 \text{ mA}) + 0.15 \text{ V}(-1.0 \text{ mA}) = 4.85 \text{ mW}$. Assuming that the gate spends 50 percent of the time in each state (a 50 percent duty cycle), the average power dissipation $\langle P \rangle$ is

$$\langle P \rangle = \frac{17.6 + 4.85}{2} \text{ mW} = 11.2 \text{ mW}$$

9.10.4 V_{IH}, V_{IL}, AND NOISE MARGINS FOR THE TTL PROTOTYPE

Figure 9.35 shows the results of circuit simulation of the VTC for the TTL inverter of Figs. 9.31 to 9.33. As expected, V_H is equal to V_{CC}. For $v_O = V_L$, the output transistor is heavily saturated, with $V_L < 0.1 \text{ V}$. The transition region between V_H and V_L is quite narrow, which is a result of the exponential characteristics of the BJT.

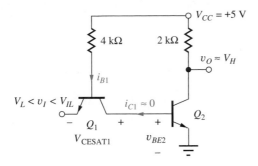

Figure 9.36 TTL gate with input v_I below V_{IL}.

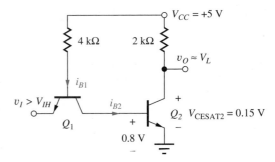

Figure 9.37 TTL gate with input V_I above V_{IH}.

As can be observed in the VTC, the difference between V_{IH} and V_{IL} is only slightly larger than 0.1 V (although large enough to change i_C by a factor of more than 50!). Calculating the exact input voltages for which the slope of the VTC equals -1 is complex, but a simplified analysis of the circuit in Fig. 9.36 yields the approximate location of the V_{IH} and V_{IL} transition points.

For an input voltage near V_L, Q_1 is saturated and the voltage at the base of Q_2 is given by

$$v_{BE2} = v_I + V_{CESAT1} = v_I + 0.04 \text{ V} \tag{9.61}$$

using the actual value of saturation voltage from Fig. 9.27 for $i_{C1} = 0$. Because of the exponential turn-on of the transistor, the slope of the VTC changes rapidly at the point at which Q_2 just begins to conduct as v_{BE2} reaches 0.7 V, and this point marks V_{IL}:

$$V_{IL} \cong 0.7 - V_{CESAT1} = 0.66 \text{ V} \tag{9.62}$$

Near V_{IL}, Q_1 is saturated, and the collector voltage of Q_1 follows the voltage applied to the emitter.

Calculation of the value of V_{OH} corresponding to V_{IL} is very similar to the method used to obtain Eq. (9.27), and V_{OH} is very close to V_H:

$$V_{OH} \cong V_H - V_T \cong V_H = 5 \text{ V} \tag{9.63}$$

We see that this value agrees well with the simulation results in Fig. 9.35.

Similar arguments yield an approximate value of V_{IH}. In Fig. 9.37, the input voltage is above V_{IH}, and the base-emitter voltage of Q_2 is given by $V_{BESAT2} = 0.8$ V. As the input decreases, the output remains at V_{CESAT} until current begins to be diverted away from the base of Q_2. This occurs approximately as the base-emitter and base-collector voltages of Q_1 become equal, or $v_I = 0.8$ V (see Prob. 9.58). Thus, we expect V_{IH} to be given approximately by

$$V_{IH} \cong V_{BESAT2} = 0.8 \text{ V} \tag{9.64}$$

Calculation of the value of V_{OL} is more complex and will not be attempted here. However, we see from the results in Fig. 9.35 that V_{OL} is very close to V_L, and we will incur little error if we use V_L for our estimate of V_{OL}:

$$V_{OL} \cong V_L = V_{CESAT2} = 0.15 \text{ V} \tag{9.65}$$

These approximate values for V_{IH} and V_{IL}, based on Eqs. (9.62) and (9.64), agree well with the simulation results in Fig. 9.35. These estimates are accurate because of the exponential behavior of the BJT. Less than a 60-mV change in v_{BE} changes i_C by a factor of 10, so the estimates of V_{IH} and V_{IL} should not be in error by more than ± 60 mV.

Using Eqs. (9.62) and (9.64) with the values of V_{OH} and V_{OL} yields

$$\text{NM}_L \cong 0.66 - 0.15 = 0.51 \text{ V} \quad \text{and} \quad \text{NM}_H \cong 5.0 - 0.8 = 4.2 \text{ V} \tag{9.66}$$

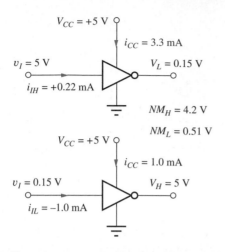

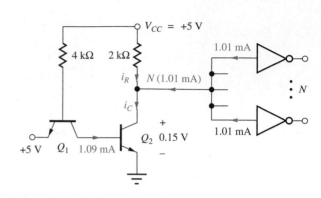

Figure 9.38 Summary of TTL prototype inverter characteristics.

Figure 9.39 Fanout conditions for $v_O = V_L$.

9.10.5 PROTOTYPE INVERTER SUMMARY

Figure 9.38 is a summary of the static characteristics of the prototype TTL inverter. V_H is equal to the power supply voltage of 5 V, and an input of this value produces an output low state of $V_L \leq 0.15$ V. The highly asymmetrical noise margins are $\text{NM}_L = 0.51$ V and $\text{NM}_H = 4.2$ V. When the input is high, a current of 0.22 mA enters the TTL input, and when the input is low, a current of 1.0 mA exits from the input.

9.10.6 FANOUT LIMITATIONS OF THE TTL PROTOTYPE

For NMOS, CMOS, and ECL logic, fanout restrictions were not investigated in detail because the input current to the various logic gates was zero, as for the case of NMOS and CMOS, or it was a very small base current, as for the case of ECL. However, a substantial current exists in the input terminal of the TTL inverter for both input states, as summarized in Fig. 9.38. This input current limits the number of gates that can be connected to the output of an individual TTL logic gate. Both logic states must be checked to see which set of conditions actually limits the fanout of the gate.

Fanout Limit for $v_O = V_L$

N inverters are shown connected to the output of one TTL inverter in Fig. 9.39. The maximum value of N is termed the **fanout** capability of the TTL gate. N can be determined from the conditions required to maintain Q_2 in saturation with $V_{\text{CESAT2}} \leq 0.15$ V.

Referring to Fig. 9.39, the collector current of Q_2 is given by

$$i_C = i_R + N(1.01 \text{ mA}) = 2.43 \text{ mA} + N(1.01 \text{ mA}) \tag{9.67}$$

In Sec. 9.9 we found that a forced beta $\beta_{\text{FOR}} \leq 28.3$ was required to maintain $V_{\text{CESAT}} \leq 0.15$ V. Also, the base current of Q_2 was previously found to be 1.09 mA for $v_I = 5$ V, so the collector current must satisfy

$$i_C \leq 28.3 i_B = 28.3 \times 1.09 \text{ mA} = 30.8 \text{ mA}$$

or

$$\tag{9.68}$$

$$2.43 \text{ mA} + N(1.01 \text{ mA}) \leq 30.8 \text{ mA} \qquad \text{and} \qquad N \leq 28.1$$

Because the number of gates must be an integer, the fanout capability of this gate is $N = 28$ for the output in the low state. This result indicates that the prototype TTL gate can drive a valid low output into 28 gates, a relatively large fanout capability.

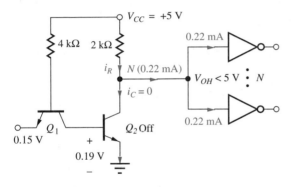

Figure 9.40 Fanout conditions for $v_O = V_H$.

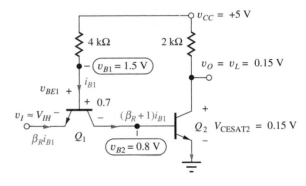

Figure 9.41 Circuit for determining V_H limitations.

Fanout Limit for $v_O = V_H$

The circuit conditions for $v_I = 0.15$ V and $v_O = V_H$ are given in Fig. 9.40. At the output node, the collector current of Q_2 is zero, but the input currents of the N inverters must be supplied through the 2-kΩ load resistor. The resulting expression for v_O becomes:

$$v_O = 5 - 2000 i_R = 5 - N(2000\ \Omega)(0.22\ \text{mA}) = 5 - N(0.44)\ \text{V} \qquad (9.69)$$

Each added fanout connection causes the output voltage to drop an additional 0.44 V below 5 V.

To determine N, we must understand how far v_O can drop without having a detrimental effect on the inverters connected to the output. The inverter circuit with $v_I \cong V_{IH}$ is redrawn in Fig. 9.41. Q_1 is operating in the reverse-active region, and its collector current, $i_C = -(\beta_R + 1)i_{B1}$, is independent of the base-emitter voltage as long as the base-emitter junction is reverse-biased. Therefore, the base current of Q_2 will be constant as long as $v_{BE1} = (v_{B1} - v_I) \leq 0$, which requires $v_I \geq 1.5$ V. Combining this limit with Eq. (9.69) enables us to determine the fanout N:

$$5 - N(2000\ \Omega)(0.22\ \text{mA}) \geq 1.5 \qquad \text{or} \qquad N \leq 7.95 \qquad (9.70)$$

For the output voltage in the high state, the fanout N must not exceed 7.95. Because N must once again be an integer, the maximum fanout is limited to 7. Further, because the overall fanout specification for the TTL gate must be the smallest N that works properly regardless of logic state, the fanout for the prototype TTL inverter must be specified as $N = 7$.

EXAMPLE 9.5 **FANOUT LIMITATIONS**

Here we explore the effects of changes in V_{CESAT} and β_R on the fanout of our prototype TTL gate.

PROBLEM (a) What is the fanout limit in the prototype TTL gate if V_{CESAT2} is required to be less than 0.1 V?
(b) What are the input current i_{IH} and fanout limit for $v_I = V_{OH}$ for the TTL prototype inverter circuit if $\beta_{R1} = 2$?

SOLUTION— **Known Information and Given Data:** Simplified TTL gate in Figs. 9.31–9.37; $V_{CESAT2} = 0.1$ V,
PART (a) $\beta_F = 40$, and $\beta_R = 0.25$

Unknowns: Fanout limit N for the new value of V_{CESAT}

Approach: We must recalculate the forced beta needed to achieve V_{CESAT} of 0.1 V and see if this changes the value of N for either logic state. The new value of V_{CESAT} also changes the value of i_{IN}.

Assumptions: $V_{\text{CESAT1}} = 0.04$ V, $V_{\text{BESAT}} = 0.8$ V, and $V_{BE} = 0.7$ V

Analysis: First we will determine N for $v_O = V_L$, and we begin by reevaluating Eqs. (9.55) and (9.67) for $v_I = 0.1$ V.

$$i_{IL} = -\frac{5 - 0.8 - 0.1}{4}\frac{\text{V}}{\text{k}\Omega} = 1.03 \text{ mA}$$

$$i_C = i_R + N(1.03 \text{ mA}) = \frac{5 - 0.1}{2}\frac{\text{V}}{\text{k}\Omega} + N(1.03 \text{ mA}) = 2.45 \text{ mA} + N(1.03 \text{ mA})$$

To find the maximum value of i_C, we must find the new value of forced beta using Eq. (9.49).

$$\Gamma = \exp\left(\frac{0.1 \text{ V}}{0.025 \text{ V}}\right) = 54.6 \qquad \text{and} \qquad \beta_{\text{FOR}} \leq 40\frac{1 - \dfrac{1}{0.25(54.6)}}{1 + \dfrac{20}{(0.333)(54.6)}} = 17.7$$

The base current drive to transistor Q_2 is the same as in Fig. 9.39, $i_B = 1.09$ mA. Therefore, the collector current must be no greater than $\beta_{\text{FOR}}i_B = 17.7(1.09 \text{ mA}) = 19.3$ mA, and the fanout for $v_O = V_L$ is found to be

$$2.45 \text{ mA} + N(1.03 \text{ mA}) \leq 19.3 \text{ mA} \rightarrow N = 16$$

For $v_O = V_H$, the conditions are unchanged from those in Fig. 9.40, and the value of fanout computed from Eq. (9.70) was $N = 7$. So the fanout of the gate remains limited to $N = 7$ by the $v_O = V_H$ condition.

Check of Results: We have evaluated N for both logic states and found $N = 7$.

Discussion: The reduction of V_{CESAT} by 50 mV did not change the overall value of N, which is still limited by the circuit conditions for $v_O = V_H$. However, it should be noted that this small change did substantially alter the value of N for the $v_O = V_L$ state (N was reduced from 28 to 16). The second part of this example shows that we must maintain good control of β_R.

SOLUTION — PART (b) **Known Information and Given Data:** Simplified TTL gate in Figs. 9.31–9.37; $V_{\text{CESAT2}} = 0.1$ V, $\beta_F = 40$, and $\beta_R = 2$

Unknowns: Current i_{IH} and fanout N for $v_I = V_{OH}$

Approach: Find the new value of input current for $\beta_R = 2$ and then find the updated value of N from Eq. (9.70).

Assumptions: $V_{\text{CESAT1}} = 0.04$ V, $V_{\text{BESAT}} = 0.8$ V, and $V_{BE} = 0.7$ V

Analysis: The voltage conditions are the same as in Fig. 9.41. The new value of input current is

$$i_{IH} = \beta_R i_{B1} = 2\frac{5 - 0.8 - 0.7}{4}\frac{\text{V}}{\text{k}\Omega} = 1.75 \text{ mA}$$

Substituting this value of base current into Eq. (9.70) gives

$$5 \text{ V} - N(2 \text{ k}\Omega)(1.75 \text{ mA}) \geq 1.5 \text{ V} \rightarrow N = 1$$

Check of Results: N seems low, but the calculations appear to be correct. We have found all the required information.

Discussion: The fanout is barely one. Variations due to circuit tolerances would certainly cause N to drop below 1. From this answer, we see that we must be concerned with control of β_R during transistor manufacture or the fanout will deteriorate rapidly. The full TTL gate discussed in Sec. 9.11 adds an emitter follower output stage to the TTL prototype, and thereby solves the fanout problem associated with the $v_O = V_H$ state.

EXERCISE: What value is the maximum value of β_R in the TTL prototype gate if we desire $N = 5$? $N = 10$?

ANSWERS: 0.4; 0.2

EXERCISE: How much base current is available to saturate Q_2 if $\beta_R = 2$ in Fig. 9.41? What is the fanout capability of the TTL prototype for the $v_O = V_L$ state for $V_{CESAT} = 0.15$ V as in Fig. 9.41 if $\beta_R = 2$?

ANSWERS: 2.63 mA; 71

9.11 THE STANDARD 7400 SERIES TTL INVERTER

The TTL gate described thus far has been a prototype for understanding the more complex gate that forms standard TTL logic. This basic circuit is also useful for low-level on-chip applications. However, with an overdriven transistor Q_2 to pull the output down, but only the 2-kΩ load resistor to pull the output up, the dynamic response of this gate will be highly asymmetric, as was the case for NMOS logic. In addition, we found in Sec. 9.10 that the prototype circuit has a limited fanout capability and is quite sensitive to the value of β_R.

The circuit for the classical TTL inverter shown in Fig. 9.42 solves these problems. This circuit is typically found in TTL unit logic in which several identical gates are packaged together in a single **dual-in-line package,** or **DIP.** Input transistor Q_1 operates in exactly the same manner as Q_1 in the prototype gate, and Q_2 forces the output low to V_{CESAT2}. The load resistor is replaced with an **active**

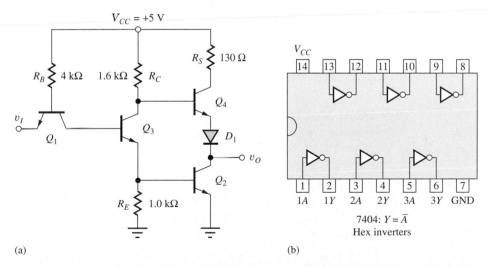

(a)

(b)

Figure 9.42 (a) Standard TTL inverter. (b) 7404 hex inverter.

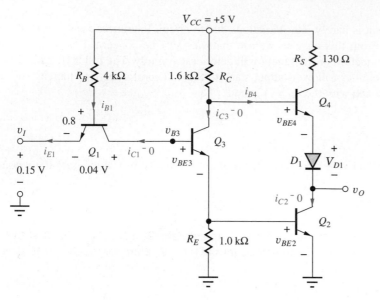

Figure 9.43   Standard TTL inverter with $v_I = 0.15$ V.

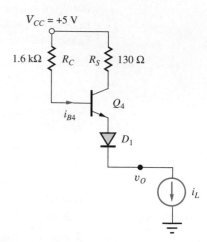

Figure 9.44   Simplified TTL output stage for $v_O = V_H$.

pull-up circuit formed by transistor Q_4 and diode D_1. Inverter Q_3 and D_1 are required to ensure that Q_4 is turned off when Q_2 is turned on and vice versa.

9.11.1 ANALYSIS FOR $V_I = V_L$

Figure 9.43 is the full TTL circuit with an input voltage of 0.15 V. Base current i_{B1} causes Q_1 to saturate, with $i_{C1} \cong 0$ and $V_{CESAT1} = 0.04$ V. The emitter current is equal to the base current given by

$$i_{E1} \cong i_{B1} = \frac{(5 - 0.8 - 0.15)\ \text{V}}{4\ \text{k}\Omega} = 1.0\ \text{mA} \tag{9.71}$$

The voltage v_{B3} at the base of Q_3 is approximately $0.15\ \text{V} + 0.04\ \text{V} = 0.19$ V, which keeps both Q_2 and Q_3 off because 0.19 V is less than one base-emitter voltage drop, and two are required to turn on both Q_2 and Q_3 ($v_{BE2} + v_{BE3} = 1.4$ V).

Because Q_2 and Q_3 are off with i_{C2} and i_{C3} approximately zero, the output portion of the gate may be simplified to the circuit of Fig. 9.44. In this circuit, base current is supplied to Q_4 through resistor R_C, and the output reaches

$$v_O = 5 - i_{B4} R_C - v_{BE4} - v_{D1} \tag{9.72}$$

For normal values of load current i_L, modeled by the current source in Fig. 9.44, the voltage drop in R_C is usually negligible, and the nominal value of V_{OH} is

$$V_H \cong 5 - v_{BE4} - v_{D1} = 5 - 0.7 - 0.7 = 3.6\ \text{V} \tag{9.73}$$

The 130-Ω resistor R_S is added to the circuit to protect transistor Q_4 from accidental short circuits to ground. Resistor R_S allows Q_4 to saturate and limits the power dissipation in the transistor. For example, if v_O is connected directly to ground, then the current through Q_4 and D_1 will be limited to approximately

$$i_{C4} = \frac{V_{CC} - V_{CESAT4} - V_{D1}}{R_S} \quad \text{and} \quad i_{C4} \le \frac{(5 - 0 - 0.7)\ \text{V}}{130\ \Omega} = 33.1\ \text{mA} \tag{9.74}$$

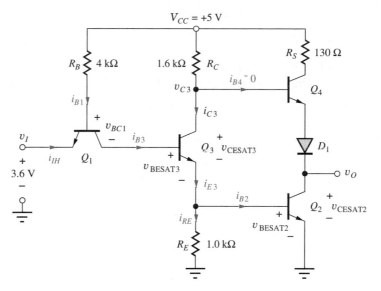

Figure 9.45 Standard TTL inverter with $v_I = V_H$.

EXERCISES: What is i_{C4} if Q_4 remains in the forward-active region when v_O is shorted to ground? Use $\beta_F = 40$. What is the maximum value of i_L for which $v_O \geq 3$ V if $\beta_{F4} = 40$? Is Q_4 in the forward-active region at this value of i_C?

ANSWERS: 92.3 mA; 15.4 mA; yes

9.11.2 ANALYSIS FOR $V_I = V_H$

V_H is now applied as the input to the TTL circuit in Fig. 9.45. This input level exceeds the voltage required at the base of Q_3 to turn on both Q_2 and Q_3, ($v_{B3} \geq v_{BE3} + v_{BE2}$), and the base current to Q_3 becomes

$$i_{B3} = (\beta_R + 1)i_{B1} = (\beta_R + 1)\frac{V_{CC} - v_{BC1} - v_{BESAT3} - v_{BESAT2}}{R_B}$$

or

$$i_{B3} = (0.25 + 1)\frac{(5 - 0.7 - 0.8 - 0.8)\ \text{V}}{4\ \text{k}\Omega} = 0.84\ \text{mA}$$

(9.75)

Both Q_2 and Q_3 are designed to saturate, so both base-emitter voltages are assumed to be 0.8 V in Eq. (9.75). For this case, the input current entering the emitter is

$$i_{IH} = -i_{E1} = \beta_R i_{B1} = 0.25(0.68\ \text{mA}) = 0.17\ \text{mA}$$

(9.76)

The base current of Q_2 ultimately determines the fanout limit of this TTL gate; it is given by

$$i_{B2} = i_{E3} - i_{RE} = i_{C3} + i_{B3} - i_{RE}$$

(9.77)

Currents i_{RE} and i_{C3} are given by

$$i_{RE} = \frac{V_{BESAT2}}{R_E} = \frac{0.8\ \text{V}}{1.0\ \text{K}} = 0.80\ \text{mA}$$

$$i_{C3} = \frac{5 - V_{CESAT3} - V_{BESAT2}}{1.6\ \text{K}} = \frac{5 - 0.15 - 0.8}{1.6\ \text{K}} = 2.53\ \text{mA}$$

(9.78)

and

$$i_{B2} = (2.53 + 0.84 - 0.80) \text{ mA} = 2.57 \text{ mA} \tag{9.79}$$

in which it has been assumed that Q_4 is off with $i_{B4} = 0$.

The assumption that Q_4 is off can be checked by calculating the voltage $v_{C3} - v_O$:

$$v_{C3} - v_O = (V_{\text{CESAT3}} + V_{\text{BESAT2}}) - (V_{\text{CESAT2}}) \tag{9.80}$$

$$= 0.80 + 0.15 - 0.15 = 0.80 \text{ V}$$

This voltage, 0.8 V, must be shared by the base-emitter junction of Q_4 and diode D_1, but it is not sufficient to turn on the series combination of the two. However, if D_1 were not present, the full 0.8 V would appear across the base-emitter junction of Q_4, and it would saturate. Thus, D_1 must be added to the circuit to ensure that Q_4 is off when Q_2 and Q_3 are saturated.

9.11.3 POWER CONSUMPTION

Figure 9.46 summarizes the voltages and currents in the TTL gate. Assuming a 50 percent duty cycle, the average power consumed by the TTL gate is

$$\langle P \rangle = \frac{P_{OL} + P_{OH}}{2} \tag{9.81}$$

$$\langle P \rangle = \frac{[5 \text{ V}(0.92 \text{ mA}) + 3.6 \text{ V}(0.17 \text{ mA})] + [5 \text{ V}(1.0 \text{ mA}) + 0.15 \text{ V}(-1.0 \text{ mA})]}{2}$$

$$\langle P \rangle = 5.03 \text{ mW}$$

9.11.4 TTL PROPAGATION DELAY AND POWER-DELAY PRODUCT

Analysis of the propagation delay of the TTL inverter is fairly complex because several saturating transistors are involved. Therefore, we investigate the behavior by looking at the results of simulation. From Fig. 9.47, the average propagation delay is approximately

$$\tau_P = \frac{\tau_{PHL} + \tau_{PLH}}{2} = \frac{6 + 14}{2} \text{ ns} = 10 \text{ ns} \tag{9.82}$$

This value represents the nominal delay of standard TTL.

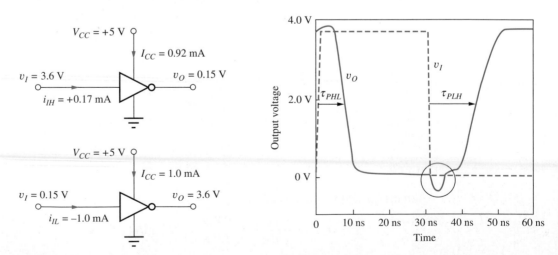

Figure 9.46 Voltage and current summary for standard TTL.

Figure 9.47 SPICE simulation of the full TTL inverter propagation delay.

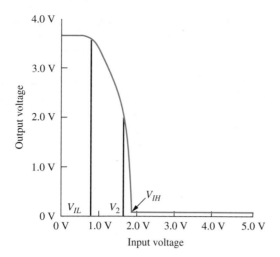

Figure 9.48 SPICE simulation of the VTC for the TTL gate in Fig. 9.42.

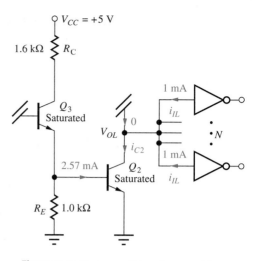

Figure 9.49 Fanout conditions for $v_O = V_{OL}$.

On the high-to-low transition, transistor Q_1 must come out of saturation and enter the reverse-active mode, while Q_2 and Q_3 must go from near cutoff to saturation. For the low-to-high transition, Q_2 and Q_3 must both come out of saturation. Thus, we may expect one storage time delay for the first case and two storage time delays for the second. Thus, τ_{PLH} should be greater than τ_{PHL}, as in the simulation results.

Using these simulation results and the power calculated from Eq. (9.81), we estimate the rather large power-delay product for the standard TTL gate to be

$$PDP = (5.0 \text{ mW})(10 \text{ ns}) = 50 \text{ pJ} \tag{9.83}$$

9.11.5 TTL VOLTAGE TRANSFER CHARACTERISTIC AND NOISE MARGINS

Figure 9.48 gives the results of simulation of the VTC for the TTL inverter. The various break points in the characteristic can be easily identified in a manner similar to that used for the TTL prototype gate. As the input voltage increases to become equal to 0.7 V, base current begins to enter Q_3. The emitter voltage of Q_3 starts to rise, and its collector voltage begins to fall. Q_4 functions as an emitter follower, and the output drops as the collector voltage of Q_3 falls. As Q_3 turns on, the slope changes abruptly, giving $V_{IL} \cong 0.7$ V.

As the input voltage (and the voltage at v_{B3}) approaches 1.5 V, there is enough voltage to both saturate Q_3 and turn on Q_2. As Q_2 turns on and Q_3 saturates, turning off Q_4, the output voltage drops more abruptly, causing break point V_2 in the curve. In Fig. 9.48 it can be seen that $V_{IH} \cong 1.8$ V, which is slightly larger than $V_{\text{BESAT3}} + V_{\text{BESAT2}}$, the voltage at v_{B3} for which both Q_2 and Q_3 are heavily saturated. For this voltage, Q_1 is coming out of heavy saturation and starting to operate in the reverse-active mode.

Using $V_{IL} = 0.7$ V, $V_{OL} = 0.15$ V, $V_{IH} = 1.8$ V, and $V_{OH} = 3.5$ V yields

$$NM_L = 0.7 \text{ V} - 0.15 \text{ V} = 0.55 \text{ V} \qquad \text{and} \qquad NM_H = 3.5 \text{ V} - 1.8 \text{ V} = 1.7 \text{ V} \tag{9.84}$$

9.11.6 FANOUT LIMITATIONS OF STANDARD TTL

The active pull-up circuit can supply relatively large amounts of current with the output changing very little (see Probs. 9.80 to 9.82), so the fanout becomes limited by the current sinking capability of Q_2. For $v_O = V_L$, Q_4 and D_1 are off, and the collector current of Q_2 is equal to the input currents of the N gates connected to the output, as shown in Fig. 9.49:

$$Ni_{IL} \leq \beta_{FOR}i_{B2} \quad \text{or} \quad N(1 \text{ mA}) \leq 28.3(2.57 \text{ mA}) \quad \text{and} \quad N \leq 72.7 \tag{9.85}$$

Using the transistor parameters in Table 9.3 with this circuit yields a fanout limit of 72. However, from Eq. (9.77), we see that the fanout is sensitive to the actual values of β_R, R_B, and R_E. The parameters of the transistors in the standard TTL IC process are somewhat different from those in Table 9.3, and the specifications must be guaranteed over a wide range of temperature, supply voltages, and IC process variations. Thus, the fanout of standard TTL is actually specified to be $N \leq 10$.

9.12 LOGIC FUNCTIONS IN TTL

Now, we will explore multi-input gates but will begin our discussion with the prototype TTL gate in order to simplify the discussion. The TTL inverter becomes a two-input gate with the addition of a second transistor in parallel with transistor Q_1, as drawn in the schematic in Fig. 9.50. If either the emitter of Q_{1A} or the emitter of Q_{1B} is in a low state, then base current i_B will be diverted out of the corresponding emitter terminal, the base current of Q_2 will be negligibly small, and the output will be high. Base current will be supplied to Q_2, and the output will be low, only if both inputs A and B are high. Table 9.4 is the truth table for this gate, which corresponds to a two-input NAND gate.

9.12.1 MULTI-EMITTER INPUT TRANSISTORS

Standard TTL logic families provide gates with eight or more inputs. An eight-input gate conceptually has eight transistors in parallel. However, because the input transistors all have common base and collector connections, these devices are actually implemented as a single multi-emitter transistor, which is usually drawn as shown in the two-input NAND gate diagram in Fig. 9.51(a).

The concept of the **merged transistor structure** for a multi-emitter transistor appears in Fig. 9.51(b), in which the two-emitter transistor with merged base and collector regions takes up far less area than two individual transistors would require.

9.12.2 TTL NAND GATES

A complete standard three-input TTL NAND gate is shown in the schematic in Fig. 9.52. If any one of the three input emitters is low, then the base current to transistor Q_3 will be zero, and the output will be high, yielding $Y = \overline{ABC}$. The behavior of the rest of the gate is identical to that described in the discussion of Figs. 9.42 to 9.46.

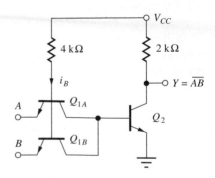

Figure 9.50 Two-input TTL NAND gate: $Y = \overline{AB}$.

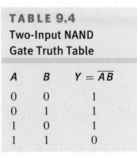

TABLE 9.4

Two-Input NAND Gate Truth Table

A	B	$Y = \overline{AB}$
0	0	1
0	1	1
1	0	1
1	1	0

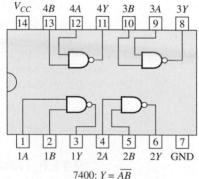

7400: $Y = \overline{AB}$
Quadruple two-input NAND gates

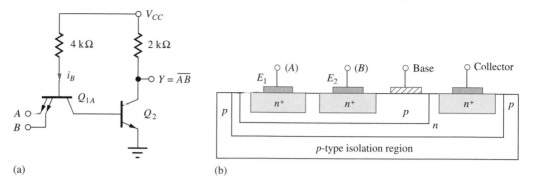

Figure 9.51 (a) Multi-emitter realization of the two-input NAND gate. (b) Merged structure for two-emitter bipolar transistor.

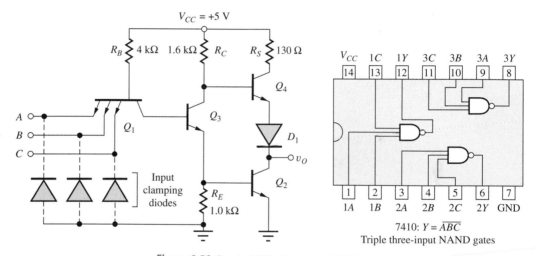

Figure 9.52 Standard TTL three-input NAND gate.

Although the basic TTL gate provides the NAND function, other logic operations can be implemented with the addition of more transistors. One example is shown in Fig. 9.53, in which the input circuitry of Q_1 and Q_3 is replicated to provide the **AND-OR-Invert,** or **AOI,** logic function (a complemented **sum-of-products function**). This five-input gate provides the logic function $Y = \overline{ABC + DE}$.

TTL also has several power options, including standard, high-power, and low-power versions, in which the resistor values are modified to change the power level and hence the gate delay. **Low-power TTL** has a delay of approximately 30 ns, and the **high-power TTL** series (54H/74H) has a delay of approximately 7 ns.

9.12.3 INPUT CLAMPING DIODES

In Fig. 9.47, we observe a negative-going transient on the output signal near $t = 15$ ns, which resulted from the rapid input signal transition. Another source of such transients is "ringing," which results from high-speed signals exciting the distributed L-C interconnection network between logic gates, as diagrammed in Fig. 6.4. To prevent large excursions of the inputs below ground level, which can damage the TTL input transistors, a diode is usually added to each input to clamp the input signal to no more than one diode-drop below ground potential. These input **clamping diodes** are added to the TTL NAND gate schematic in Fig. 9.52.

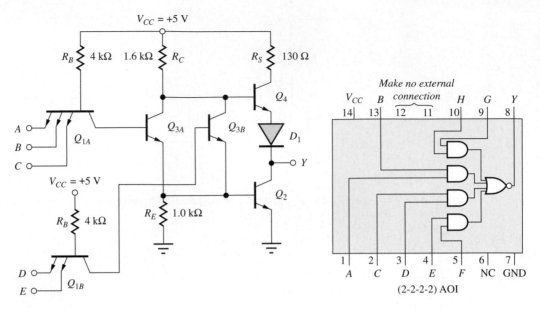

Figure 9.53 TTL AND-OR-Invert (AOI) gate: $Y = \overline{ABC + DE}$.

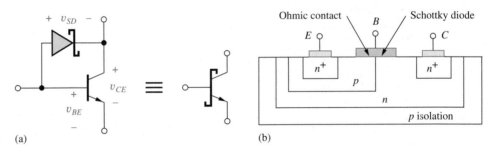

Figure 9.54 (a) Schottky-clamped transistor. (b) Cross section of the merged Schottky diode and bipolar transistor structure.

9.13 SCHOTTKY-CLAMPED TTL

As discussed in Sec. 9.11, saturation of the transistors in TTL logic substantially slows down the dynamic response of the logic gates. The **Schottky-clamped transistor** drawn in Fig. 9.54(a) was developed to solve this problem. The Schottky-clamped transistor consists of a metal semiconductor **Schottky barrier diode** in parallel with the collector-base junction of the bipolar transistor.

When conducting, the forward voltage drop of the Schottky diode is designed to be approximately 0.45 V, so it will turn on before the collector-base diode of the bipolar transistor becomes strongly forward-biased. Referring to Fig. 9.54(a), we see that

$$v_{CE} = v_{BE} - v_{SD} = 0.70 - 0.45 = 0.25 \text{ V} \tag{9.86}$$

The Schottky diode prevents the BJT from going into deep saturation by diverting excess base current through the Schottky diode and around the BJT. Because the BJT is prevented from entering heavy saturation, 0.7 V has been used for the value of v_{BE} in Eq. (9.86).

A cross section of the structure used to fabricate the Schottky transistor is given in Fig. 9.54(b). Conceptually, an aluminum base contact overlaps the collector-base junction, forming an ohmic contact to the p-type base region and a Schottky diode to the more lightly doped n-type collector

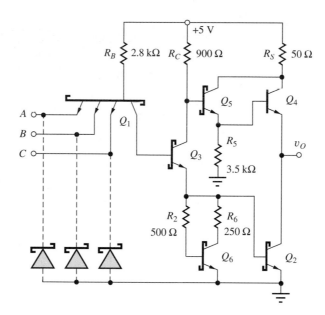

Figure 9.55 Schottky TTL NAND gate.

region. (Remember that aluminum is a p-type dopant in silicon.) This is another example of the novel merged structures that can be fabricated using IC technology.

Invention of this circuit required a good understanding of the exponential dependence of the BJT collector current on base-emitter voltage as well as knowledge of the differences between Schottky and pn junction diodes. Successful manufacture of the circuit relies on tight process control to maintain the desired difference between the forward drops of the base-collector and Schottky diodes.

Figure 9.55 is the schematic for a full three-input Schottky TTL NAND gate. Each saturating transistor in the original gate—Q_1, Q_2, and Q_3—is replaced with a Schottky transistor. Q_6, R_2, and R_6 replace the resistor R_E in the original TTL gate and eliminate the first "knee" voltage corresponding to V_{IL} in the VTC in Fig. 9.48. Thus, the transition region for the Schottky TTL is considerably narrower than for the original TTL circuit (see Prob. 9.101). Q_5 provides added drive to emitter follower Q_4 and eliminates the need for the series output diode D_1 in the original TTL gate. Q_4 cannot saturate in this circuit because the smallest value for v_{CB4} is the positive voltage V_{CESAT5}, so it is not a Schottky transistor. The input clamp diodes are replaced with Schottky diodes to eliminate charge storage delays in these diodes.

The use of Schottky transistors substantially improves the speed of the gate, reducing the nominal delay for the standard Schottky TTL series gate (54S/74S) to 3 ns at a power dissipation of approximately 15 mW. An extremely popular TTL family is **low-power Schottky TTL** (54LS/74LS), which provides the delay of standard TTL (10 ns) but at only one-fifth the power. The resistor values are increased to decrease the power, but speed is maintained at lower power by eliminating the storage times associated with saturating transistors. This family is widely used to replace standard TTL because it offers the same delay at substantially less power. As IC technology has continued to improve, the complexity and performance of TTL has also continued to increase. Advanced Schottky logic (ALS) and advanced low-power Schottky logic families were introduced with improved power-delay characteristics.

9.14 COMPARISON OF THE POWER-DELAY PRODUCTS OF ECL AND TTL

Figure 9.56 is a comparison of the power versus delay characteristics of a number of the ECL and TTL unit-logic families that have been produced. ECL, with its nonsaturating transistors and low logic swing, is generally faster than TTL, although one can see the performance overlap between

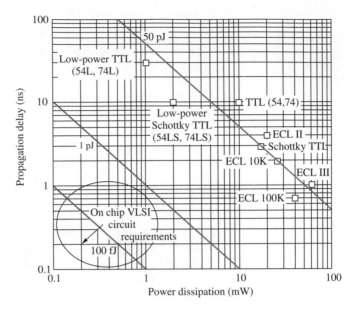

Figure 9.56 Power-delay products for various commercial unit-logic families.

generations that occurred for ECL II versus Schottky TTL, for example. A new generation of IC technology and better circuit techniques achieve higher speed circuits for a given power level. However, for high-density integrated circuits, significant improvements in power and power-delay products are required, as indicated by the circle in the lower left-hand corner of Fig. 9.56.

9.15 BiCMOS LOGIC

There have been numerous attempts to combine various bipolar and FET technologies in order to realize the advantages of both types of transistors in a single process. For example, the BiFET process combines BJTs and JFETs and has often been used in analog circuits. BiCMOS processes combine the *n*- and *p*-channel transistors from CMOS with bipolar transistors, and the most sophisticated forms of BiCMOS include both *npn* and *pnp* transistors. BiCMOS is a complex technology, but it permits one to use the bipolar and MOS transistors wherever they provide the most advantage. In BiCMOS logic gates, MOS transistors are typically used to provide high-impedance inputs with the power of MOS NAND, NOR, and complex gate implementations. Bipolar transistors are then used to provide high-output-current capacity for driving high-capacitance loads. BiCMOS is also highly useful in "mixed signal" designs that combine both analog and digital signal processing.

For many years, BiCMOS was considered a niche technology. There was strong debate whether the performance advantages were real and worth the additional process complexity and cost. However, the demand for ever higher microprocessor performance as well as the requirements for mixed-signal system-on-a-chip components has led to the development of BiCMOS processes by many companies, including IBM, Intel, and Texas Instruments, to name a few. Today, BiCMOS is used in high-performance Pentium processors, in mixed-signal ICs, and in advanced silicon-germanium (SiGe) processes. Silicon-germanium BiCMOS is the highest performance silicon-based technology available today.

With fully complementary MOS and bipolar devices available, the circuit design possibilities become almost endless. One need only search through the topic of BiCMOS in the *IEEE Journal of Solid-State Circuits,* for example, to discover the wide variety of novel circuits proposed for use with BiCMOS technology. Here we will attempt to get a flavor for how BJT and MOS devices can be combined to achieve improved logic performance. In Part III of this book we explore some of the possibilities that BiCMOS offers the analog circuit designer.

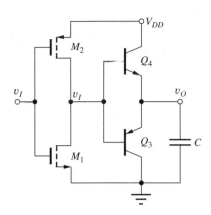

Figure 9.57 BiCMOS buffer employing complementary emitter followers.

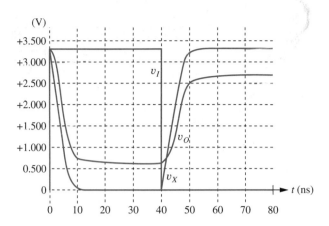

Figure 9.58 Simulation results for the BiCMOS gate in Fig. 9.57 with $C = 2$ pF and $W/L = 6/1$ and $15/1$ for the NMOS and PMOS transistors, respectively. $V_{DD} = 3.3$ V.

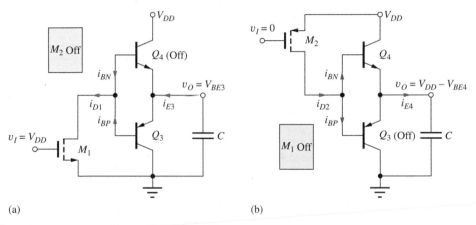

(a)   (b)

Figure 9.59 BiCMOS buffer (a) following low-to-high input transition and (b) after high-to-low input transition.

9.15.1 BiCMOS BUFFERS

Let us start by seeing how we can add bipolar transistors to improve performance of the CMOS inverter. The bipolar transistor offers high current gain and higher transconductance than the FET for a given operating current and can therefore rapidly charge and discharge large capacitances. On the other hand, the very high input impedance of the MOS device requires very little driving power, and CMOS gates provide ease of implementation of complex logic functions. Thus BJTs are used in the output stage of the on- and off-chip buffers used to drive high-capacitance busses. These buffers replace the large-area cascade buffer designs that are required to drive large capacitances in standard CMOS technology.

A basic BiCMOS buffer is shown in Fig. 9.57 in which a complementary *npn/pnp* pair of emitter followers, Q_4 and Q_3, is driven by a standard CMOS inverter. Operation of the circuit for the two logic states is outlined in Fig. 9.59. Figure 9.59(a) shows the circuit following a low-to-high transition of input v_I. PMOS transistor M_2 is off, and NMOS transistor M_1 turns on, providing a path for the base current i_{BP} of *pnp* Q_3. The base current is amplified by the *pnp* current gain ($\beta_F + 1$), and the resulting emitter current rapidly discharges the load capacitance to a voltage approximately equal to V_{EB3}. At the beginning of the transition, the NMOS transistor also provides a reverse base current path to quickly turn off *npn* transistor Q_4 and discharge the equivalent capacitance at the bases of the BJTs.

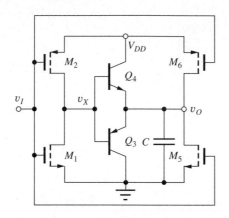

Figure 9.60 BiCMOS buffer with an auxiliary inverter to restore full logic swing.

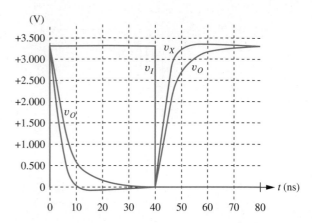

Figure 9.61 Simulation results for buffer in Fig. 9.60 with $W/L = 12/1$ and $30/1$ for M_5 and M_6, respectively. $V_{DD} = 3.3$ V.

Circuit operation for the opposite input transition is shown in Fig. 9.59(b). When v_I returns to zero, NMOS transistor M_1 is cut off. PMOS transistor M_2 turns on and supplies forward base current to *npn* transistor Q_4 as well as reverse base current to the *pnp* transistor. Q_3 turns off, and the *npn* device rapidly charges load capacitor C to a high logic level approximately equal to $V_{DD} - V_{BE4}$.

In order to charge or discharge the load capacitance, the CMOS inverter must supply only base current to the bipolar transistors and current to the equivalent capacitance of the bases of Q_3 and Q_4. Note in this circuit that the base-emitter junctions of Q_3 and Q_4 are connected in parallel. When the *npn* base-emitter junction is forward biased, the *pnp* base-emitter junction is reverse-biased and vice versa. So, the *pnp* must be off when the *npn* is on, and the *npn* must be off when the *pnp* is on. Furthermore, neither collector-base junction can become forward biased, so the bipolar transistors can never saturate, and storage time delays are eliminated, enabling high-speed operation.

We see that the overall circuit behaves as an inverter. The input is a standard CMOS inverter, and the bipolar output configuration is a noninverting follower. Thus the two-stage combination forms an inverter. Note, however, that the logic levels have deteriorated. The high logic level is now one base emitter voltage drop below the power supply, $V_H \cong V_{DD} - V_{BE4}$, and the low logic level is $V_L \cong V_{EB3}$. This loss of logic swing ($\Delta V \cong V_{DD} - 1.4$ V) is undesirable, particularly as power-supply voltages are reduced.

Figure 9.58 presents the results of SPICE simulations for the BiCMOS circuit in Fig. 9.57 with $V_{DD} = 3.3$ V and $C = 2$ pF. After the initial transient, V_L and V_H are 0.62 V and 2.68 V, respectively. The propagation delay is approximately 6 ns.

The full logic swing can be restored using various techniques. One approach is shown in Fig. 9.60 in which the output of an auxiliary inverter, composed of transistors M_5 and M_6, is connected in parallel with the output of the emitter followers. The bipolar transistors charge and discharge the load capacitance rapidly through most of the transition, and then the CMOS devices take over near the end of the transition and force the capacitor to charge fully to V_{DD} or discharge completely to ground potential. The results of simulation of the circuit in Fig. 9.60 appear in Fig. 9.61. The auxiliary CMOS inverter restores the full logic swing.

A second method is to simply place a resistor in parallel with the base-emitter junctions of the emitter followers as shown in Fig. 9.62. The resistor path permits the final voltage on the capacitor to reach 0 or V_{DD}.

9.15.2 BiNMOS INVERTERS

Simplified BiCMOS process implementations include high-performance *npn* devices but not *pnp* transistors. This places restrictions on the circuitry that can be designed, and the resulting BiCMOS circuits are sometimes referred to as BiNMOS rather than BiCMOS. BiNMOS gates use *npn*

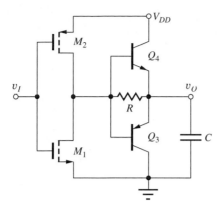

Figure 9.62 BiCMOS buffer using a resistor to increase logic swing.

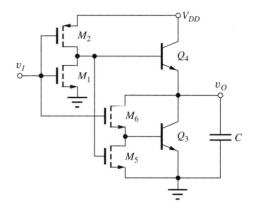

Figure 9.63 BiNMOS buffer.

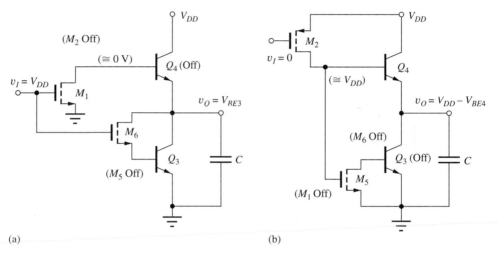

(a)   (b)

Figure 9.64 BiNMOS buffer (a) following low-to-high input transition (M_2 and M_5 are cut off) and (b) after high-to-low input transition (M_1 and M_6 are cut off).

transistors to drive the load capacitance in a manner that is similar to the "totem-pole" output stage of the TTL gate, as shown in the schematic in Fig. 9.63. Here *npn* transistor Q_4 functions as an emitter follower to assist in charging load capacitor C toward V_{DD}, and transistor Q_3 is used to discharge C.

The CMOS drive circuits must provide forward base-current drive to only one of the *npn* transistors at a time. With the input high as in Fig. 9.64(a), the CMOS inverter output will be low, and transistors M_2 and M_5 will both be off. NMOS transistor M_6 provides a base-current path from the output node to the base of Q_3, and Q_3 discharges the load capacitance. At the same time, M_1 provides a path for reverse base current to speed up the turn off of transistor Q_4. With M_6 connected between the collector and base of Q_3, the BJT cannot saturate, and the output low level is approximately V_{BE3}.

For the low-input state in Fig. 9.64(b), the CMOS inverter output will be high, and transistors M_1 and M_6 will be cut off. PMOS transistor M_2 provides base current to Q_4. This current is amplified, and the emitter follower action of Q_4 charges up capacitor C to the high logic level: $V_H \cong V_{DD} - V_{BE4}$. Transistor M_6 turns off, removing the path for forward base current to Q_3. Although the circuit would function logically without transistor M_5, it is added to provide a reverse base-current path to ensure rapid turn off of Q_3.

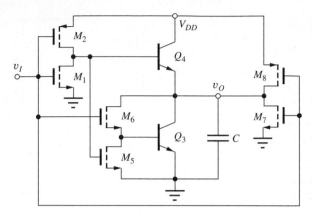

Figure 9.65 Full swing BiNMOS inverting buffer.

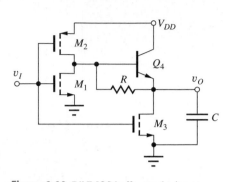

Figure 9.66 BiNMOS buffer employing a single *npn* transistor.

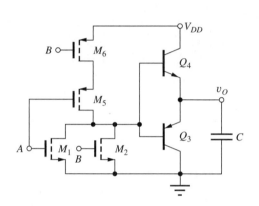

Figure 9.67 Two-input BiCMOS NOR gate.

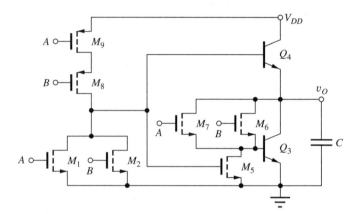

Figure 9.68 Two-Input BiNMOS NOR gate.

This BiNMOS gate suffers from the reduced logic swing problems just as the circuit in Fig. 9.57. However, the solution to this problem is similar. Figure 9.65 shows the addition of an auxiliary inverter (M_6 and M_7) that forces V_H and V_L to equal the full power supply levels.

A simplified BiNMOS gate appears in Fig. 9.66. Here NMOS device M_3 is used to discharge the load capacitance, and the *npn* transistor is only used to assist in the low-to-high transition. Resistor R is added to increase the value of V_H. The resistor can be implemented by a MOS transistor biased in the linear region.

9.15.3 BiCMOS LOGIC GATES

Although this section has focused on BiCMOS inverter circuitry, additional logic function may easily be added by converting the CMOS inverters to more complex gates. Examples of two-input NOR gates appear in Figs. 9.67 and 9.68. The BiCMOS NOR gate simply adds the complementary bipolar follower stage to the standard CMOS NOR gate as in Fig. 9.67. A NAND gate is constructed in a similar manner.

The BiNMOS NOR gate in Fig. 9.68 requires a little more effort since base current must be provided to Q_3 when either input A or input B is high. Thus the two transistors designated as M_6 and M_7 are required.

> EXERCISE: (a) Draw the circuit for a two-input BiCMOS NAND gate. (b) Draw the schematic for a two-input BiNMOS NAND gate.

SUMMARY

The two commercially most important forms of bipolar logic are emitter-coupled logic (ECL) and transistor-transistor logic (TTL or T^2L). In this chapter, we examined simple prototype circuits for the ECL and TTL gates, and then we explored the full gate structures.

- *ECL:* ECL logic has traditionally operated from a negative supply voltage, typically -5.2 V, and V_H and V_L are therefore negative. The voltage transfer characteristic for the ECL gate was investigated, and the ECL logic swing ΔV is relatively small, ranging between 0.2 V and 0.8 V with noise margins approaching $\Delta V/2$. The ECL gate introduced two new circuit techniques, the current switch and the emitter-follower circuit, and also requires a reference voltage circuit. ECL logic gates generate both true and complement outputs, and the basic ECL gate provides the OR-NOR logic functions. Standard ECL unit-logic families provide delays in the 0.25- to 5-ns range with a power-delay product of approximately 50 pJ.

- *Current switches:* The current switch consists of two matched BJTs and a current source. This circuit rapidly switches the bias current back and forth between the two transistors, based on a comparison of the logic input signal with a reference voltage. In the ECL gate, the transistors actually switch between two points in the forward-active region, which is one reason why ECL is the highest speed form of bipolar logic. A second factor is the relatively small logic swing, typically in the 0.4- to 0.8-V range. The low ECL logic swing results in noise margins of a few tenths of a volt. ECL is somewhat unusual compared to the other logic families that have been studied in that it is typically designed to operate from a single negative power supply, historically -5.2 V, and V_H and V_L are both negative voltages.

- *Emitter followers:* In the emitter-follower circuit, the output signal replicates the input signal except for a fixed offset equal to one base-emitter diode voltage, approximately 0.7 V. In ECL, this fixed-voltage offset is used to provide the level-shifting function needed to ensure that the logic levels at the input and output of the gates are the same. The emitter followers permit additional logic power through the use of the "wired-OR" circuit technique.

- *TTL circuits:* Classical TTL circuits operate from a single 5-V supply and provide a logic swing of approximately 3.5 V, with noise margins exceeding 1 V. During operation, the transistors in standard TTL circuits switch between the cutoff and saturation regions of operation. Basic TTL gates realize multi-input NAND functions; however, more complex gates can be used to realize almost any desired logic function. Standard TTL unit-logic families provide delays in the 3- to 30-ns range, with a power-delay product of approximately 50 pJ. Schottky diodes are used to prevent BJT saturation and speed up the TTL logic circuits.

- *BJT saturation region:* The collector-emitter saturation voltage of the BJT is controlled by the value of the forced beta, defined as $\beta_{\text{FOR}} = i_C/i_B$. The transistor enters saturation if the base current exceeds the value needed to support the collector current (that is, $i_B > i_C/\beta_F$ so that $\beta_{\text{FOR}} < \beta_F$). An undesirable result of saturation is storage of excess charge in the base region of the transistor. The time needed to remove the excess charge can cause the BJT to turn off slowly. This delayed turnoff response is characterized by the storage time t_S and is proportional to the value of the storage time constant τ_S, which determines the magnitude of the excess charge stored in the base during saturation.

- *Schottky-clamped transistors:* The Schottky-clamped transistor merges a standard bipolar transistor with a Schottky diode and was developed as a way to prevent saturation in bipolar transistors. The Schottky diode diverts excess base current around the base- collector diode of the BJT and prevents heavy saturation of the device. Schottky TTL circuits offer considerable improvement in speed compared to standard TTL for a given power dissipation because storage time delays are eliminated.

- *Inverse operation:* The input transistors in a TTL gate operate in the reverse-active mode when the input is in the high state. This is the only use of this mode of operation that we encounter in this text.

- *Fanout:* The TTL gate has relatively large input currents for both high- and low-input voltages. The input current is positive for high-input levels and negative for low-input levels. This input current limits the fanout capability of the gate, and the fanout capability of TTL was analyzed in detail. At the output of the TTL gate, another emitter follower can be found. The emitter follower provides the high-current drive needed to support large fanouts as well as to rapidly pull up the output.

- *TTL family members:* TTL gates are available in many forms, including standard, low-power, high-power, Schottky, low-power Schottky, advanced Schottky, and advanced low-power Schottky versions. Standard TTL has essentially been replaced by low-power Schottky (54LS/74LS) TTL, which provides similar delay but at reduced power. Schottky TTL (54S/74S) provides a high-speed alternative for circuits with critical speed requirements.

- *Power delay products:* The standard ECL and TTL unit-logic families have relatively large power-delay products (20 to 100 pJ), which are not suitable for high-density VLSI chip designs. VLSI circuit densities require subpicojoule power-delay products; simplified circuit designs with much lower values of power-delay product are required for VLSI applications. Low-voltage forms of TTL and ECL have been designed for VLSI applications, but for the most part they have been replaced by CMOS circuitry.

- *BiCMOS:* BiCMOS is a highly complex integrated circuit technology, but it provides the advantages of both bipolar and MOS transistors. Full BiCMOS technologies provide NMOS, PMOS, *npn,* and *pnp* transistors. Thus the circuit designer has maximum flexibility to choose the best device for each place in a circuit. In BiCMOS logic gates, MOS transistors are typically used to provide high-impedance inputs with the simplicity of MOS NAND, NOR, and complex gate implementations. Bipolar transistors are used to provide high-output-current capacity for driving large load capacitances. BiCMOS is also highly useful for "mixed-signal" designs that combine both analog and digital signal processing. Simplified BiCMOS technologies add *npn* transistors to the CMOS process, and the resulting circuits are often referred to as BiNMOS instead of BiCMOS.

KEY TERMS

Active pull-up circuit
AND-OR-Invert (AOI) logic
Base-emitter saturation voltage (V_{BESAT})
Collector-emitter saturation voltage (V_{CESAT})
Clamping diodes
Current switch circuit
Dual-in-line package (DIP)
Emitter-coupled logic (ECL)
Emitter-coupled pair
Emitter dotting
Emitter follower
Fanout
Forward-active region
Forward transit time
High-power TTL
Level shifter
Low-power TTL
Low-power Schottky TTL

Merged transistor structure
Reference voltage
Reverse-active region
Reverse transit time
Saturation region
Saturation voltage
Schottky barrier diode
Schottky-clamped transistor
Schottky TTL inverter
Stored base charge
Storage time t_S
Storage time constant τ_S
Sum-of-products logic function
Temperature compensation
Transistor-transistor logic (TTL, T^2L)
V_{BESAT}
Wired-OR logic

REFERENCE

1. Hodges, D. A. and H. G. Jackson, *Analysis and Design of Digital Integrated Circuits,* 2d ed. McGraw-Hill, New York: 1988.

ADDITIONAL READING

Haznedar, H. *Digital Microelectronics.* Benjamin/Cummings, Redwood City, CA: 1991.

Glasford, G. M. *Digital Electronic Circuits.* Prentice-Hall, Englewood Cliffs, NJ: 1988.

J. N. Rabaey, A. Chandrakkasan and B. Nikolic', *Digital Integrated Circuits—A Design Perspective, Second Edition,* Prentice Hall, Upper Saddle River, NJ: 2003.

Sedra, A. S. and K. C. Smith. *Microelectronic Circuits. 3d ed.* Saunders, Philadelphia, PA: *1990.*

PROBLEMS

For SPICE simulations, use the device parameters in Appendix B. For hand calculations, use the values in Table 9.3 on page 463.

9.1 The Current Switch (Emitter-Coupled Pair)

9.1. What are the voltages at v_{C1} and v_{C2} in the circuit in Fig. P9.1 for $v_I = -1.6$ V if $I_{EE} = 2.0$ mA, $R_C = 350\ \Omega$, and $V_{REF} = -1.25$ V?

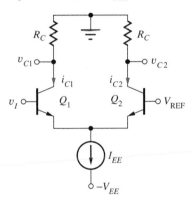

Figure P9.1

9.2. (a) What value of v_I is required in Fig. P9.1 to switch 99.5 percent of the current I_{EE} into transistor Q_1 if $V_{REF} = -1.25$ V? What value of v_I will switch 99.5 percent of I_{EE} into transistor Q_2? (b) Repeat part (a) if $V_{REF} = -2$ V.

9.3. What are the voltages at v_{C1} and v_{C2} in the circuit in Fig. P9.1 for $v_I = -1.6$ V, $I_{EE} = 2.5$ mA, $R_C = 700\ \Omega$, and $V_{REF} = -2$ V?

9.4. What are the voltages at v_{C1} and v_{C2} in the circuit in Fig. P9.4 for $v_I = -1.7$ V, $I_{EE} = 0.3$ mA, $R_1 = 3.33$ kΩ, $R_C = 2$ kΩ, and $V_{REF} = -2$ V?

*9.5. A bipolar transistor is operating with $v_{BE} = +0.7$ V and $v_{BC} = +0.3$ V. By the strict definitions given in the chapter on bipolar transistors, this

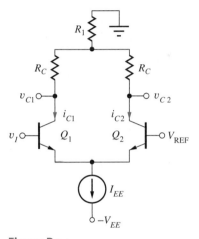

Figure P9.4

transistor is operating in the saturation region. Use the transport equations to demonstrate that it actually behaves as if it is still in the forward-active region. Discuss this result. (You may use $I_S = 10^{-15}$ A, $\alpha_F = 0.98$, and $\alpha_R = 0.2$.)

**9.6. A low-voltage current switch is shown in Fig. P9.6. (a) What are the voltage levels corresponding to V_H and V_L at v_{O2}? (b) Do these voltage levels appear to be compatible with the levels needed at the

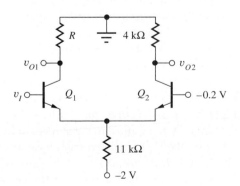

Figure P9.6

input v_I? Why? (c) What is the value of R? (d) For $v_I = V_H$ from part (a), what are the regions of operation of transistors Q_1 and Q_2? (e) For $v_I = V_L$ from part (a), what are the regions of operation of transistors Q_1 and Q_2? (f) Your answers in parts (c) and (d) should have involved regions other than the forward-active region. Discuss whether this appears to represent a problem in this circuit.

9.7.   (a) What is the average power in the current switch in Fig. P9.6? Assume v_I spends 50 percent of the time in each logic state. (b) Redesign the resistor values to reduce the power by a factor of five.

9.2 The Emitter-Coupled Logic (ECL) Gate

9.8.   The values of I_{EE} and R_C in Fig. 9.6 are changed to 5 mA and 200 Ω, respectively. What are the new values of V_H, V_L, V_{REF}, and ΔV?

9.9.   The values of I_{EE}, R_C, and I_3 and I_4 in Fig. 9.6 are changed to 1 mA, 600 Ω, and 0.3 mA, respectively. What are the new values of V_H, V_L, V_{REF}, and ΔV?

9.10.   Redesign the values of resistors and current sources in Fig. 9.6 to increase the power consumption by a factor of 4. The values of V_H and V_L should remain constant.

*9.11.   Redesign the circuit of Fig. 9.6 to have a logic swing of 0.8 V. Use the same currents. (a) What are the new values of V_H, V_L, V_{REF}, and the resistors? (b) What are the values of the noise margins? (c) What is the minimum value of V_{CB} for Q_1 and Q_2? Do the values of V_{CB} represent a problem?

9.12.   Emitter followers are added to the outputs of the circuit in Fig. P9.4 in the same manner as in Fig. 9.6. (a) Draw the new circuit. (b) If $I_{EE} = 0.3$ mA, $R_1 = 3.33$ kΩ, $R_C = 2$ kΩ, and $V_{REF} = -2$ V, what are the values of V_H, V_L, and the logic swing ΔV? (c) Are the input and output levels of this gate compatible with each other?

9.13.   Emitter followers are added to the outputs of the circuit in Fig. P9.4 in the same manner as in Fig. 9.6. (a) Draw the new circuit. (b) What are the values of R_C, V_H, V_L, and V_{REF} if $I_{EE} = 1.5$ mA, $R_1 = 800$ Ω, and $\Delta V = 0.4$ V?

**9.14.   Calculate the fanout for the ECL inverter in Fig. P9.6 at room temperature for $\beta_F = 30$. Define the fanout N to be equal to the number of inverters for which the V_H level deteriorates by no more than one V_T. (*Hint:* At v_{O2}, $\Delta V_H = \Delta V_{BE4} + \Delta i_{B4} R_C$.) Do we need to consider the case for $v_O = V_L$? Why?

9.3–9.4 V_{IH}, V_{IL}, and Noise Margin Analysis for the ECL Gate and Current Source Implementation

9.15.   Change the values of resistors and current sources in Fig. 9.8(b) to reduce the power consumption by a factor of 8. The values of V_H and V_L should remain constant.

9.16.   (a) Redesign the values of resistors and current sources in Fig. 9.8(b) to increase the power consumption by a factor of 8. The values of V_H and V_L should remain constant. (b) Change the values of resistors and current sources in Fig. 9.8(b) to reduce the power consumption by a factor of 5. The values of V_H and V_L should remain constant.

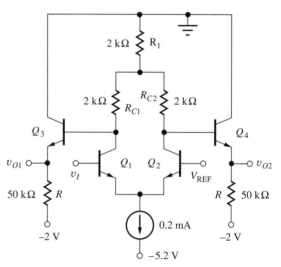

Figure P9.16

9.17.   What are the values of V_H, V_L, V_{REF}, ΔV, the noise margins, and the average power dissipation for the circuit in Fig. P9.16?

9.18.   What is the minimum logic swing ΔV required for an ECL gate to have a noise margin of 0.1 V at room temperature?

*9.19.   Suppose the values of resistors in Fig. 9.8(a) all increase in value by 20 percent. (a) How do the values of V_H, V_L, and the noise margins change? (b) Repeat part (a) for the circuit of Fig. 9.8(b).

9.20.   Redesign the circuit of Fig. 9.8(b) to have a logic swing of 0.8 V. Use the same currents. (a) What are the new values of V_H, V_L, V_{REF}, and the resistors? (b) What are the values of the noise margins? (c) What is the minimum value of V_{CB} for Q_1 and Q_2? Do the values of V_{CB} represent a problem?

9.21. Suppose that an ECL gate is to be designed to operate over the $-55°C$ to $+75°C$ temperature range and must maintain a minimum noise margin of 0.1 V over that full range. What is the value of the logic swing of this ECL gate at room temperature?

9.22. Replace the current source in Fig. P9.16 with a resistor. Assume $V_{REF} = -1.3$ V. Do any of the three collector resistors need to be changed? If so, what are the new values?

9.23. The current source for a current switch is implemented with a transistor and three resistors, as in Fig. P9.23. What is the current I_{EE}? What is the minimum permissible value of V_{REF} if transistor Q_S is to remain in the forward-active region? Assume that β_F is large. (See Section 5.11.)

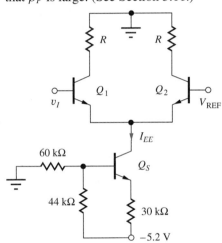

Figure P9.23

9.24. Design a 0.2-mA electronic current source to replace the ideal source in Fig. P9.16. Use the circuit topology from Fig. P9.23 (e.g., Q_S and its three bias resistors). The base bias network should not use more than 20 μA of current.

9.25. Simulate the voltage transfer characteristic of the ECL inverter in Fig. 9.8(b) at $T = -55°C$, $25°C$, and $85°C$. Use a constant voltage source for V_{REF}. What are V_H, V_L, and the noise margins at the three temperatures?

9.5 The ECL OR-NOR Gate

9.26. Draw the schematic of a five-input ECL OR gate.

9.27. Draw the schematic of a four-input ECL NOR gate.

9.28. $V_{CC} = +1$ V and $V_{EE} = -2.5$ V in the ECL gate in Fig. P9.28. (a) Find V_H and V_L at the emitter of

Q_5. (b) What is the value of R required to give the same voltage levels at the emitter of Q_1?

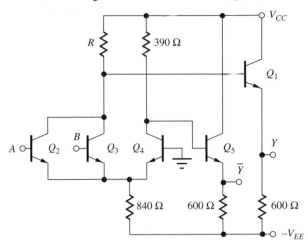

Figure P9.28

9.29. $V_{CC} = +1.3$ V and $V_{EE} = -3.2$ V in the ECL gate in Fig. P9.28 (a) Find V_H and V_L at the emitter of Q_5. (b) What value of R is required to give the same voltage levels at the emitter of Q_1?

9.30. (a) Only the OR output is needed in the circuit of Fig. P9.28. Redraw the circuit, eliminating the unneeded components. Use $V_{CC} = +1.3$ V and $V_{EE} = -3.2$ V. (b) Repeat part (a) if only the NOR output is needed.

9.6 The Emitter Follower

9.31. In the circuit in Fig. P9.31, $\beta_F = 50$, $V_{CC} = +5$ V, $-V_{EE} = -5$ V, $I_{EE} = 2.5$ mA, and $R_L = 1.2$ kΩ. What is the minimum voltage at v_O? What are the emitter, base, and collector currents if $v_I = +4$ V?

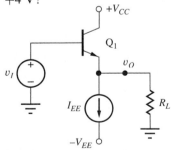

Figure P9.31

9.32. The input voltage v_I for the circuit in Fig. P9.31 is a symmetrical 1-kHz triangular signal ranging between $+3$ V and -3 V. (a) Sketch the output voltage v_O if $\beta_F = 40$, $V_{CC} = +4$ V, $-V_{EE} = -4$ V, $I_{EE} = 4$ mA, and $R_L = 1$ kΩ. (b) Repeat for

$I_{EE} = 2$ mA. (c) What is the minimum value of I_{EE} needed to insure that the output voltage is an undistorted replica of the input voltage?

9.33. Simulate the circuit in Prob. 9.32 using SPICE.

9.34. The input voltage for the circuit in Fig. P9.31 is $v_I = -1 + \sin 2000\pi t$ V. (a) Write an expression for the output voltage v_O if $\beta_F = 40$, $V_{CC} = 0$ V, $-V_{EE} = -3$ V, $I_{EE} = 0.2$ mA, and $R_L = 20$ kΩ. (b) What is the minimum value of I_{EE} needed to insure that the output voltage is an undistorted replica of the input voltage?

9.35. Simulate the circuit in Prob. 9.34 using SPICE.

9.36. In the circuit in Fig. P9.31, $\beta_F = 40$, $V_{CC} = +1.5$ V, $-V_{EE} = -1.5$ V, $I_{EE} = 0.5$ mA, and $R_L = 1$ kΩ. (a) What range of values is permitted for v_I if Q_1 is to stay within the forward-active region of operation? (b) What is the minimum value of I_{EE} required for the circuit to work properly for -1.5 V $\leq v_I \leq +1.5$ V.

9.37. In the circuit in Fig. P9.31, $\beta_F = 40$, $V_{CC} = 0$ V, $-V_{EE} = -15$ V, and $R_L = 1$ kΩ. What is the minimum value of I_{EE} required for the circuit to work properly for -10 V $\leq v_I \leq 0$ V?

9.38. In the circuit in Fig. P9.38, $R_L = 2$ kΩ, $V_{CC} = 15$ V, and $-V_{EE} = -15$ V. (a) What is the maximum value of R_E that can be used if v_O is to reach -12 V? (b) What is the emitter current of Q_1 when $v_O = +12$ V?

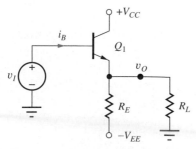

Figure P9.38

9.39. In the circuit in Fig. P9.38, $\beta_F = 100$, $V_{CC} = 0$ V, $-V_{EE} = -15$ V, and $R_L = 4.7$ kΩ. (a) What is the maximum value of R_E that can be used for the circuit to work properly for -10 V $\leq v_I \leq 0$ V? (b) What is the emitter current of Q_1 when $v_I = 0$ V? (c) For $v_I = -10$ V?

9.40. In the circuit in Fig. P9.38, $\beta_F = 100$, $V_{CC} = 0$ V, $-V_{EE} = -6$ V, $R_E = 1.3$ kΩ, $R_L = 4.7$ kΩ,

and $v_I = -1.5 + 1.5 \sin 2000\pi t$ V. (a) Plot the output voltage versus time. (b) Write an expression for the output voltage. (c) What is the maximum value of emitter current and for what value of v_I does it occur? (d) What is the minimum value of emitter current, and for what value of v_I does it occur? (e) What is the maximum value of R_E that can be used for the output voltage to be an undistorted replica of the input voltage?

9.41. (a) Simulate the circuit in Prob. 9.40 for $0 \leq t \leq 3$ ms. (b) Repeat the simulation if R_E is changed to 4.7 kΩ.

9.7 "Emitter Dotting" or "Wired-OR" Logic

9.42. The two outputs of the inverter in Fig. 9.6 are accidentally connected together. What will be the output voltage for $v_I = -0.7$ V? For $v_I = -1.3$ V? What will be the currents in transistors Q_3 and Q_4 for each case?

9.43. What are the logic functions for Y and Z in Fig. P9.43?

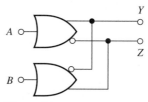

Figure P9.43

*9.44. Draw the full circuit schematic for an ECL implementation of the logic function $Y = A + B + \overline{(C + D)}$ using a wired-OR connection of two ECL gates.

9.8 ECL Power-Delay Characteristics

9.45. Suppose the ECL inverter in Figs. 9.21 and 9.23 must operate at a power of 20 μW. If the current sources are scaled by the same factor, and the capacitances and voltage levels remain the same, what is the new propagation delay of the inverter?

9.46. The logic swing in the inverter in Fig. 9.21 is reduced by a factor of 2 by reducing the value of R_C and changing V_{REF}. What is the new value of V_{REF}? What is the new value of the power-delay product?

9.47. The logic swing in the inverter in Fig. P9.21 is reduced by a factor of 2 by reducing the value of all the current sources by a factor of 2 and changing V_{REF}. What is the new value of V_{REF}? What is the new value of the power-delay product?

*9.48. (a) The logic circuit in Fig. P9.48 represents an alternate form of an ECL gate. If $V_{EE} = -3.3$ V, $V_{REF} = -1.0$ V, $R_B = 3.2$ kΩ, and $R_E = 1.6$ kΩ, find the values of V_L, V_H, and the power consumption in the gate. What are the values of R_{C1} and R_{C2}? (b) What are the logic function descriptions for the two outputs? (c) Compare the number of transistors in this gate with a standard ECL gate providing the same logic function.

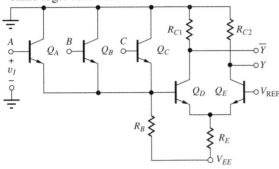

Figure P9.48

**9.49. Assume that you need 0.6 V across R_E to properly stabilize the current in the modified ECL gate in Fig. P9.48. Design the resistors in the gate for a logic swing of 0.4 V and an average current of 1 mA through R_B and R_E. What are the minimum values of V_{EE} and the value of V_{REF}?

9.50. Use SPICE and the values in Prob. P9.48 to find the propagation delay of the gate in Fig. P9.48.

9.51. Redesign the ECL inverter in Fig. P9.18 to change the average power dissipation to 1 mW. Scale the power in the current switch and the emitter followers by the same factor.

9.52. The power supply $-V_{EE}$ in Fig. 9.19 is changed to -2 V. What are the new values of R_{EE} and R_{C1} required to keep the logic levels and logic swing unchanged? What is the new power dissipation?

*9.53. What type of logic gate is the circuit in Fig. P9.53? What logic function is provided at the output Y?

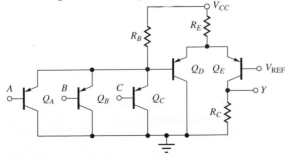

Figure P9.53

*9.54. Design the circuit in Fig. P9.53 to provide a logic swing of 0.6 V from a power supply of $+3$ V. The power consumption should average 1 mW, with 90% of the power consumed by Q_D and Q_E.

9.55. (a) Use SPICE to find the propagation delay of the gate in Prob. 9.54. (b) In Prob. 9.4. (c) In Fig. P9.16.

9.9 The Saturating Bipolar Inverter

9.56. Calculate the collector and base currents in a bipolar transistor operation with $v_{BE} = 0.20$ V and $v_{BC} = -4.8$ V. Use the BJT parameters from Table 9.3.

9.57. What is the value of V_{CESAT} of transistor Q_1 in Fig. 9.33 based on Eq. (9.47) and the BJT parameters from Table 9.3.

9.58. (a) What is the ratio i_C/i_E in Fig. P9.58 for $v_I = 0.8$ V? If needed, you may use the parameters in Table 9.3. (b) For $v_I = 0.6$ V? (c) For $v_I = 1.0$ V? (d) What is the dc input voltage v_I required for $i_C/i_E = -1$?

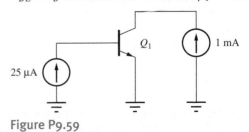

Figure P9.58

9.59. (a) What is the base-emitter voltage in the circuit in Fig. P9.59? Use the transport equations and the transistor parameters from Table 9.3. (b) What is the base-emitter voltage if $\beta_F = 80$? (c) What is V_{BE} if I_B is increased to 1 mA and $\beta_F = 40$?

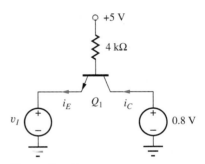

Figure P9.59

9.60. What is V_{CESAT} for the transistor in Fig. P9.59 if $\beta_F = 60$ and $\beta_R = 1$? (b) If $\beta_F = 60$, $\beta_R = 1$, and the 25-μA current source is increased to 40 μA?

9.61.   What is V_{CESAT} for the transistor in Fig. P9.59 if $\beta_F = 50$ and $\beta_R = 2$? (b) If $\beta_F = 100$ and $\beta_R = 2$?

9.62.   (a) What base current is required to reach $V_{CESAT} = 0.2$ V in the circuit in Fig. 9.26? (b) To reach $V_{CESAT} = 0.1$ V?

9.63.   What base current is required to reach $V_{CESAT} = 0.1$ V in the circuit in Fig. 9.26 if R_C is changed to 3.6 kΩ?

9.64.   Calculate the value of v_{CE} for the two circuits in Fig. P9.64.

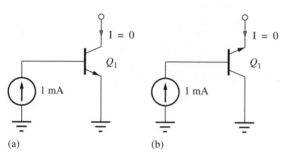

(a)   (b)

Figure P9.64

9.65.   Calculate the storage time for the inverter in Fig. 9.29 if $I_{BF} = 2$ mA, $I_{BR} = -0.5$ mA, $\tau_F = 0.4$ ns, and $\tau_R = 12$ ns. Use Table 9.3.

9.10 A Transistor-Transistor Logic (TTL) Prototype

*9.66. Suppose the TTL circuit in Fig. 9.31 is operated with $V_{CC} = +3$ V. What are the new values of V_H, V_L, V_{IH}, V_{IL}, and fanout for this gate?

9.67.   What are the worst-case minimum and maximum values of the power consumed by the gate in Fig. 9.31 if the 2-kΩ and 4-kΩ resistors have a tolerance of ±20 percent?

**9.68. A fixed value for V_L was assumed in the analysis of the prototype TTL gate in Fig. 9.33. However, an exact value can be found by the simultaneous solution of Eqs. (9.44) and (9.47) if we assume that $i_{B2} = 1.09$ mA is constant. Find the actual V_L level for the circuit in Fig. 9.33 using an iterative numerical solution of these two equations.

9.69.   A low-power TTL gate is shown in Fig. P9.69. Find V_H, V_L, V_{IL}, V_{IH}, and the noise margins for this circuit.

*9.70. A low-power TTL gate similar to the gate in Fig. P9.69 is needed for a VLSI design. These supply voltages are being considered: 0.5, 1.0, 1.5, 2.0,

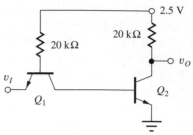

Figure P9.69

and 2.5 V. Which of these voltages represents the minimum supply needed for the circuit to operate properly? Why?

*9.71. The TTL prototype in Fig. 9.31 is operated from $V_{DD} = 3.0$ V. (a) Based on the results in Fig. 9.35, draw the new VTC. (b) What are approximate values of the new V_{IL} and V_{IH}? (c) What are the new values of the noise margins?

9.72.   Scale the resistors in the TTL prototype in Fig. P9.31 to change the power of the gate to 1.0 mW.

9.73.   (a) Use SPICE to determine the average propagation delay for the TTL prototype of Fig. 9.31. (b) Repeat for the scaled gate design from Prob. 9.72.

*9.74. The inverter in Fig. P9.74(a) is a member of the diode-transistor logic (DTL) family that was used prior to the invention of TTL. (a) Find V_H and V_L for the circuit in Fig. P9.74(a). What are the input currents in the two logic states? (b) What is the fanout limit for the DTL inverter? (c) Compare the values from parts (a) and (b) to those for the circuit in Fig. P9.74(b).

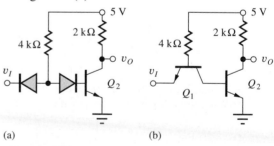

(a)   (b)

Figure P9.74

*9.75. The inverter in Fig. P9.74(a) is a member of the diode-transistor logic (DTL) family that was used prior to the invention of TTL. Sketch the VTC and compare to the VTC for the circuit in Fig. P9.74(b).

*9.76. The inverter in Fig. P9.74(a) is a member of the diode-transistor logic (DTL) family that was used before the introduction of TTL. Simulate the VTC of the DTL inverter and compare to that of the circuit in Fig. P9.74(b). Discuss the location of the break points in the characteristic.

*9.77. Simulate the propagation delay of the two inverters in Fig. P9.74. Discuss the reasons for any differences that are observed. (*Hint:* Calculate the values of I_{BR} available to bring Q_2 out of saturation and calculate the storage times in the two inverters.)

**9.78. The circuits in Fig. P9.78 are members of the diode-transistor logic (DTL) family that was used before the invention of TTL. (a) Simulate the propagation delay of the two DTL inverters in Fig. P9.78. Discuss the reasons for any differences that are observed. What is the function of the 1-kΩ resistor in Fig. P9.78(b), and why is it important? (*Hint:* Calculate the values of I_{BR} available to bring Q_2 out of saturation and calculate the storage times in the two inverters.)

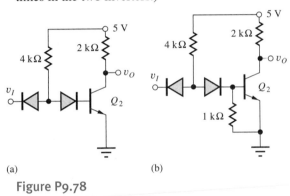

(a)   (b)

Figure P9.78

*9.79. Sketch the VTC for the simplified TTL gate in Fig. P9.79. Discuss the relationship between the observed break points in the VTC to the switching points of the various transistors. Estimate the noise margins.

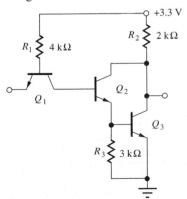

Figure P9.79

9.80. Simulate the VTC for the simplified TTL gate in Fig. P9.79. Discuss the relationship between the observed break points in the VTC to the switching points of the various transistors. What are the noise margins?

Fanout Limitations of the TTL Prototype

9.81. A fabrication process control problem causes $\beta_F = 25$ and $\beta_R = 1$ in the transistors used in the TTL circuit in Fig. 9.31. What is the new value of the fanout?

*9.82. What is the minimum value of the fanout specification for the gate in Fig. 9.31 if the 2-kΩ and 4-kΩ resistors have a tolerance of ±20 percent? Assume the resistors in each gate track each other.

9.83. The fanout for the circuit in Fig. 9.31 was calculated to be 7. Redesign the value of R_B to increase the fanout to 10.

9.11 The Standard 7400 Series TTL Inverter

9.84. Suppose the output in Fig. P9.84 is accidentally shorted to ground. (a) Calculate the emitter current in the circuit if $R_S = 0$ and $\beta_F = 100$. (b) Repeat for $R_S = 130\ \Omega$.

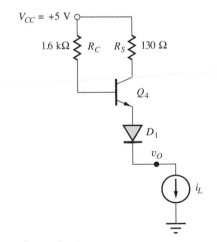

Figure P9.84

9.85. Calculate the power dissipation in the circuit in both parts of Prob. 9.84.

9.86. Use SPICE to plot v_O versus I_L for the circuit in Fig. P9.84.

*9.87. Simulate the voltage transfer characteristic for the modified TTL gate in Fig. P9.87. Discuss why the first "knee" voltage at V_2 in Fig. 9.48 has been eliminated.

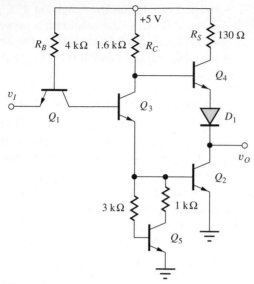

Figure P9.87

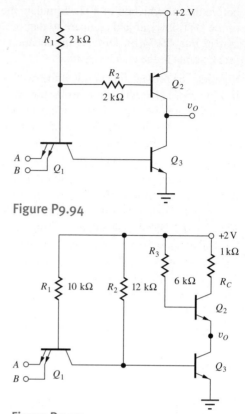

Figure P9.94

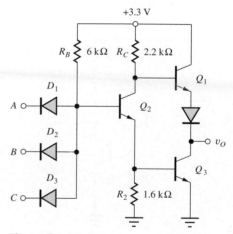

Figure P9.95

9.88. Calculate the currents in transistors Q_1 to Q_4 in a low-power version of the TTL gate in Fig. 9.42 in which $R_B = 20$ kΩ, $R_C = 8$ kΩ, $R_E = 5$ kΩ, and $R_S = 650$ Ω for (a) $v_I = V_H$ and (b) $v_I = V_L$.

9.89. Simulate the voltage transfer characteristic for the low-power TTL gate in Prob. 9.88 for $T = 25°C$.

9.90. Simulate the voltage transfer characteristic for the low-power TTL gate in Prob. 9.88 for $T = -55°C$, $+25°C$, and $+85°C$.

Fanout Limitations of Standard TTL

*9.91. Calculate the fanout limit for $v_O = V_H$ for the standard TTL gate in Fig. 9.52. Assume that V_H must not drop below 2.4 V.

*9.92. Calculate the fanout limit for the standard TTL gate in Fig. 9.52 if $R_B = 5$ kΩ, $R_C = 2$ kΩ, $R_E = 1.25$ kΩ, $\beta_R = 0.05$, and $\beta_F = 20$. Assume that V_H must not drop below 2.4 V.

*9.93. Plot the fanout of the standard TTL gate of Fig. 9.52 versus β_R for β_R ranging between 0 and 5.

*9.94. (a) Calculate V_H and V_L for the inverter in Fig. P9.94. (b) What are the input currents in the two logic states? (c) What is the fanout capability of the gate?

*9.95. (a) Calculate V_H and V_L for the inverter in Fig. P9.95. (b) What are the input currents in the two logic states? (c) What is the fanout capability of the gate?

9.12 Logic Functions in TTL

*9.96. The circuit in Fig. P9.96 can be considered a member of the diode-transistor logic (DTL) family. (a) What is the logic function of this gate? (b) Calculate V_H and V_L for this DTL inverter. (c) What are the input currents in the two logic states? (d) Sketch the VTC and compare to the VTC for the standard TTL circuit.

Figure P9.96

*9.97. For the circuit in Fig. P9.97, $R_1 = 4$ kΩ, $R_2 = 4$ kΩ, $R_3 = 4.3$ kΩ, $R_4 = 10$ kΩ, $R_5 = 5$ kΩ, and $R_6 = 5$ K. (a) What is the logic function Y? (b) What are the values of V_L and V_H? (c) What are the input currents in the high- and low-input states?

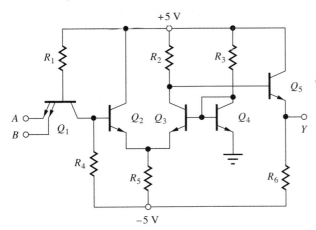

Figure P9.97

9.13 Schottky-Clamped TTL

9.98. (a) Find V_H and V_L for the Schottky DTL gate in Fig. P9.98. (b) What are the input currents in the two logic states? (c) What is the fanout of the gate?

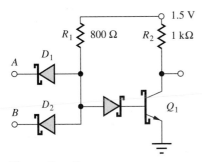

Figure P9.98

9.99. Prior to the availability of Schottky diodes, the *Baker clamp* circuit in Fig. P9.99 was used to prevent saturation. What is the collector-emitter

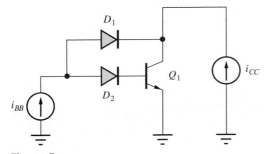

Figure P9.99

voltage of the transistor in this circuit assuming that $i_B > i_C/\beta_F$? What are i_{D1}, i_{D2}, and i_C if $i_{BB} = 250\,\mu$A, $i_{CC} = 1$ mA, $\beta_F = 20$, and $\beta_R = 2$?

*9.100. Calculate the currents in Q_1 to Q_6 in the Schottky TTL gate in Fig. 9.55 for (a) all inputs at V_H and (b) all inputs at V_L.

9.101. Simulate the voltage transfer characteristic and propagation delay for the Schottky TTL gate in Fig. 9.55.

*9.102. What is the logic function of the gate in Fig. P9.102. Find V_H, V_L, and V_{REF} for the circuit, assuming $V_{EE} = -3$ V, $R_C = 0.54$ kΩ, and $R_E = 0.75$ kΩ. Assume $V_{BE} = 0.7$ and $V_D = 0.4$ V for the Schottky diode.

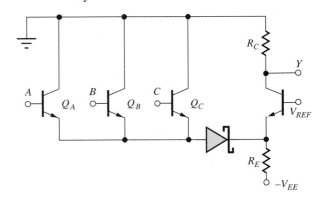

Figure P9.102

9.103. What is the logic function of the gate in Fig. P9.103. Find V_H, V_L, and V_{REF} for the circuit, assuming $V_{EE} = -3$ V, $R_C = 3.3$ kΩ, and $R_E = 2.4$ kΩ. Assume $V_{BE} = 0.7$ and $V_D = 0.4$ V for the Schottky diode.

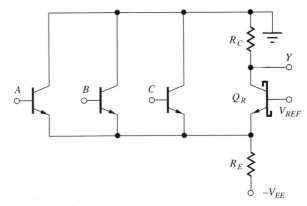

Figure P9.103

9.104. Estimate i_B and i_C of the Schottky transistor in Fig. P9.104 if the external collector terminal is open. Assume the forward voltage of the Schottky diode is 0.45 V.

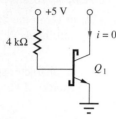

Figure P9.104

9.14 Comparison of Power-Delay Products of ECL and TTL

9.105.   The power dissipation of a particular IC chip is 50 W, and the chip will contain 50 million logic gates. The gates must have an average propagation delay of 1 ns. (a) What is the power-delay product of the logic family? (b) What is the lowest PDP that can be plotted on the graph in Fig. 9.56?

9.106.   The power dissipation of a particular IC chip is limited to 100 W, and the chip will contain 200 million logic gates. The gates must have an average propagation delay of 0.25 ns. (a) What is the power-delay product of the logic family? (b) What is the lowest PDP that can be plotted on the graph in Fig. 9.56?

9.107.   A low-power ECL gate has a 0.5-pJ PDP. (a) What will be the gate delay for a gate operating at a power level of 0.3 mW? (b) What power is required for the gate to achieve a delay of 1 ns?

9.108.   The 74LS gate in Fig. 9.56 is redesigned to operate at a power of 5 mW. (a) What is the gate delay of the new design? Assume a constant power-delay product. (b) What power is required to achieve a delay of 0.3 ns?

9.109.   The ECL-100K gate in Fig. 9.56 is redesigned to operate at a power of 10 mW. (a) What is the gate delay of the new design? Assume a constant power-delay product. (b) What power is required to achieve a delay of 0.2 ns?

9.15 BiCMOS Logic

9.110.   (a) Simulate the VTC for the BiNMOS buffer in Fig. 9.63 if the W/L ratios of all the transistors are 10/1. (b) Use SPICE to find the propagation delays for $C = 2$ pF. Use the BJT parameters from the simulation for Fig. 9.23.

9.111.   Add MOS transistors to the circuit in Fig. 9.62 to create a two-input BiCMOS NAND gate.

9.112.   Add MOS transistors to the circuit in Fig. 9.63 to create a two-input BiCMOS NAND gate.

9.113.   (a) Simulate the VTC for the BiNMOS buffer in Fig. 9.66 if the W/L ratios of all the transistors are 10/1. (b) Use SPICE to find the propagation delays for $C = 2$ pF. Use the BJT parameters from the simulation for Fig. 9.23.

9.114.   (a) Add MOS transistors to the circuit in Fig. 9.66 to create a two-input BiNMOS NOR gate. (b) Add MOS transistors to the circuit in Fig. 9.66 to create a two-input BiCMOS NAND gate.

9.115.   Simulate the VTC for the BiCMOS NOR gate in Fig. 9.67 if the W/L ratios of the NMOS transistors are 4/1 and those of the PMOS transistors are 10/1. (b) Use SPICE fo find the propagation delays for $C = 2$ pF. Use the BJT parameters from the simulation for Fig. 9.23.

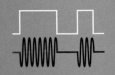

PART THREE
ANALOG CIRCUIT DESIGN

ANALOG SYSTEMS

Lee deForest and the Audion.
Courtesy of © Bettmann/Corbis.

CHAPTER GOALS

Chapter 10 begins our study of the circuits used for analog electronic signal processing. We will develop an understanding of this group of concepts related to linear amplification:

- Voltage gain, current gain, and power gain
- Gain conversion to decibel representation
- Input resistance and output resistance
- Transfer functions and Bode plots
- Cutoff frequencies and bandwidth
- Low-pass, high-pass, band-pass, and band-reject amplifiers
- Biasing for linear amplification
- Distortion in amplifiers
- Two-port representations of amplifiers
- Use of transfer function (TF) analysis in SPICE

Invention of the Audion tube by Lee DeForest in 1906 was a milestone event in electronics as it represented the first device that provided amplification with reasonable isolation between the input and output [1, 2, 3]. Amplifiers today, most often in solid-state form, play a key role in the multitude of electronic devices that we encounter in our daily activities, even in devices that we often think are digital in nature. Examples include cell phones, disk drives, digital audio and DVD players, and global positioning systems. All these devices utilize amplifiers to transform very small analog signals to levels where they can be reliably converted into digital form. Analog circuit technology also lies at the heart of the interface between the analog and digital portions of these devices in the form of analog-to-digital (A/D) and digital-to-analog (D/A) converters. Every day the world is becoming connected through an increasing variety of communications links. Optical fiber systems, cable modems, digital subscriber lines, and wireless communications technologies rely on amplifiers to both generate and then detect extremely small signals containing the transmitted information.

In Part II, we explored the analysis and design of circuits used to manipulate information in discrete form—that is, binary data. However, much information about the world around us, such as temperature, humidity, pressure, velocity, light intensity, sound, and so on, is "analog" in nature, may take on any value within some continuous range, and can be represented by the analog signal in Fig. 10.1. In electrical form, these signals may be the output of transducers that measure pressure, temperature, or flow rate, or the audio signal from a microphone or stereo amplifier. The characteristics of these signals are most often manipulated using linear amplifiers, which change the amplitude and/or phase of a signal without affecting its spectral content.

Today we recognize it is frequently advantageous to do as much signal processing as possible in the digital domain. Because of noise and dynamic range considerations, most A/D and D/A converters have full-scale ranges of 1–5 volts, whereas signals in sensor, transducer, communications, and

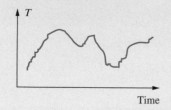

Figure 10.1 **Temperature vs. time—a continuous analog signal.**

many other applications are typically at much lower levels. For example, temperature sensor outputs may be less than 1 mV/°C, and cell phones and satellite radios require sensitivities in the microvolt range. Thus, we require the use of amplifiers to increase the voltage, current, and/or power levels of these signals. At the same time, amplifiers are used to limit (filter) the frequency content of the signals.

Part III of this text explores the design of amplifiers that are required in all these applications. Most of these devices mentioned in the previous paragraphs employ "mixed signal" designs that require a knowledge of both analog and digital circuitry as well as the A/D and D/A conversion interfaces between the two.

10.1 AN EXAMPLE OF AN ANALOG ELECTRONIC SYSTEM

We begin exploring some of the uses for analog amplifiers by examining a familiar electronic system, the FM stereo receiver, shown schematically in Fig. 10.2. This figure is representative of the FM receiver in our automobiles, as well as that in our home audio systems. At the receiving antenna are **very high frequency,** or **VHF,**[1] radio signals in the 88- to 108-MHz range that contain the information for at least two channels of stereo music.[2] In our FM receiver, these signals may have amplitudes as small as 1 μV and often reach the receiver input through a 50- or 75-Ω coaxial cable. At the output of the receiver are audio amplifiers that develop the voltage and current necessary to deliver 100 W of power to the 8-Ω speakers in the 50- to 15,000-Hz audio frequency range.

This receiver is a complex analog system that provides many forms of analog signal processing, some linear and some nonlinear (see Table 10.1). For example, the amplitude of the signal must be increased at **radio** and **audio frequencies** (**RF** and **AF**, respectively). Large overall voltage, current,

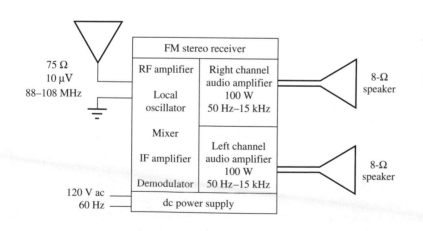

Stereo receiver
Copyright © Pioneer Electronics (USA) Inc.

Satellite radio receiver
Copyright © Pioneer Electronics (USA) Inc.

Figure 10.2 **FM stereo receiver.**

---

[1] The radio spectrum is traditionally divided into different frequency bands: RF, or radio frequency (0.5–50 MHz); VHF, or very high frequency (50–150 MHz); UHF, or ultra high frequencies (150–1000 MHz); and so on. Today, however, RF is commonly used to refer to the whole radio spectrum from 0.5 MHz to 10 GHz and higher.

[2] A satellite radio receiver is very similar except the input frequency is approximately 2.3 GHz.

TABLE 10.1
FM Stereo Receiver

LINEAR CIRCUIT FUNCTIONS	NONLINEAR CIRCUIT FUNCTIONS
Radio frequency amplification	
Audio frequency amplification	dc power supply (rectification)
Frequency selection (tuning)	Frequency conversion (mixing)
Impedance matching (75-Ω input)	Detection/demodulation
Tailoring audio frequency response	
Local oscillator	

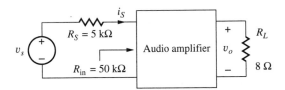

Figure 10.3 Audio amplifier channel from FM receiver.

and power gains are required to go from the very small signal received from the antenna to the 100-W audio signal delivered to the speaker. The input of the receiver is often designed to match the 75-Ω impedance of the coaxial transmission line coming from the antenna.

In addition, we usually want only one station to be heard at a time. The desired signal must be selected from the multitude of signals appearing at the antenna, and the receiver requires circuits with high frequency selectivity at its input. An adjustable frequency signal source, called the *local oscillator,* is also needed to tune the receiver. The electronic implementations of all these functions are based on **linear amplifiers.**

In most receivers, the incoming signal frequency is changed, through a process called *mixing,* to a lower **intermediate frequency (IF),**[3] where the audio information can be readily separated from the RF carrier through a process called *demodulation.* Mixing and demodulation are two basic examples of nonlinear analog signal processing. But even these nonlinear circuits are based on linear amplifier designs. Finally, the dc voltages needed to power the system are obtained using the nonlinear rectifier circuits described in Chapter 3.

10.2 AMPLIFICATION

Linear amplifiers are an extremely important class of circuits, and most of Part III discusses various aspects of their analysis and design. As an introduction to amplification, let us concentrate on one of the channels of the audio portion of the FM receiver in Fig. 10.3. In this figure, the input to the stereo amplifier channel is represented by the Thévenin equivalent source v_s and source resistor R_S. The speaker at the output is modeled by an 8-Ω resistor.

Based on Fourier analysis, we know that a complex periodic signal v_s can be represented as the sum of many individual sine waves:

$$v_s = \sum_{i=1}^{\infty} V_i \sin(\omega_i t + \phi_i) \tag{10.1}$$

where V_i = amplitude of ith component of signal, ω_i is the radian frequency, and ϕ_i is the phase.

If the amplifier is linear, the principle of superposition applies, so that each signal component can be treated individually and the results summed to find the complete signal. For simplicity in our

---

[3] Common IF frequencies are 11.7 MHz, 455 kHz, and 262 kHz.

analysis, we will consider only one component of the signal, with frequency ω_s and amplitude V_s:

$$v_s = V_s \sin \omega_s t \tag{10.2}$$

For this example, we assume $V_s = 0.001$ V, 1 mV. Because this signal serves as our reference input, we can assume $\phi_s = 0$ without loss of generality.

The output of the linear amplifier is a sinusoidal signal at the same frequency but with a different amplitude V_o and phase θ:

$$v_o = V_o \sin(\omega_s t + \theta) \tag{10.3}$$

The amplifier output power is

$$P_o = \left(\frac{V_o}{\sqrt{2}}\right)^2 \frac{1}{R_L} \tag{10.4}$$

(Remember from circuit theory that the quantity $V_o/\sqrt{2}$ in this equation represents the rms value of the sinusoidal voltage signal.) For an amplifier delivering 100 W to the 8-Ω load, the amplitude of the output voltage is

$$V_o = \sqrt{2 P_o R_L} = \sqrt{2 \times 100 \times 8} = 40 \text{ V}$$

This output power level also requires an output current

$$i_o = I_o \sin(\omega_s t + \theta) \tag{10.5}$$

with an amplitude

$$I_o = \frac{V_o}{R_L} = \frac{40 \text{ V}}{8 \text{ }\Omega} = 5 \text{ A}$$

Note that because the load element is a resistor, i_o and v_o have the same phase.

10.2.1 VOLTAGE GAIN

For sinusoidal signals, the **voltage gain** A_v of an amplifier is defined in terms of the **phasor representations** of the input and output voltages. Using $\sin \omega t = \text{Im}[\varepsilon^{j\omega t}]$ as our reference, the phasor representation of v_s is $\boldsymbol{v_s} = V_s \angle 0°$ and that for v_O is $\boldsymbol{v_o} = V_o \angle \theta$. Similarly, $\boldsymbol{i_s} = I_s \angle 0°$ and $\boldsymbol{i_o} = I_o \angle \theta$. The voltage gain is then expressed by the phasor ratio:

$$A_v = \frac{\boldsymbol{v_o}}{\boldsymbol{v_s}} = \frac{V_o \angle \theta}{V_s \angle 0} = \frac{V_o}{V_s} \angle \theta \quad \text{or} \quad |A_v| = \frac{V_o}{V_s} \quad \text{and} \quad \angle A_v = \theta \tag{10.6}$$

For the audio amplifier in Fig. 10.3, the magnitude of the required voltage gain is

$$|A_v| = \frac{V_o}{V_s} = \frac{40 \text{ V}}{10^{-3} \text{ V}} = 4 \times 10^4$$

We will find that the amplifier building blocks studied in the next several chapters have either $\theta = 0°$ or $\theta = 180°$ for frequencies in the "midband" range of the amplifier. (Midband will be defined shortly in Section 10.7.4)

We will find in subsequent chapters that achieving this level of voltage gain usually requires several stages of amplification. Be sure to note that the magnitude of the gain is defined by the amplitudes of the signals and is a constant; it is *not* a function of time! For the rest of this section, we concentrate on the magnitudes of the gains, saving a more detailed consideration of amplifier phase for Sec. 10.7.

10.2.2 CURRENT GAIN

The audio amplifier in our example requires a substantial increase in current level as well. The input current is determined by the **source resistance** R_S and the **input resistance** R_{in} of the amplifier.

When we write the input current as $i_s = I_s \sin \omega_s t$, the amplitude of the current is

$$I_s = \frac{V_s}{R_S + R_{\text{in}}} = \frac{10^{-3} \text{ V}}{5 \text{ k}\Omega + 50 \text{ k}\Omega} = 1.82 \times 10^{-8} \text{ A} \tag{10.7}$$

The phase $\phi = 0$ because the circuit is purely resistive.

The **current gain** is defined as the ratio of the phasor representations of i_o and i_s:

$$A_i = \frac{i_o}{i_s} = \frac{I_o \angle\theta}{I_s \angle 0} = \frac{I_o}{I_s} \angle\theta \tag{10.8}$$

The magnitude of the overall current gain is equal to the ratio of the amplitudes of the output and input currents:

$$|A_i| = \frac{I_o}{I_s} = \frac{5 \text{ A}}{1.82 \times 10^{-8} \text{ A}} = 2.75 \times 10^8$$

This level of current gain also requires several stages of amplification.

10.2.3 POWER GAIN

The power delivered to the amplifier input is quite small, whereas the power delivered to the speaker is substantial. Thus, the amplifier also exhibits a very large power gain. **Power gain** A_P is defined as the ratio of the output power P_o delivered to the load, to the power P_s delivered from the source:

$$A_P = \frac{P_o}{P_s} = \frac{\dfrac{V_o}{\sqrt{2}} \dfrac{I_o}{\sqrt{2}}}{\dfrac{V_s}{\sqrt{2}} \dfrac{I_s}{\sqrt{2}}} = \frac{V_o}{V_s} \frac{I_o}{I_s} = |A_v||A_i| \tag{10.9}$$

Note from Eq. (10.9) that either rms or peak values of voltage and current may be used to define power gain as long as the choice is applied consistently at the input and output of the amplifier. (This is also true for A_v and A_i.) For our ongoing example, we find the power gain to be a very large number:

$$A_P = \frac{40 \times 5}{10^{-3} \times 1.82 \times 10^{-8}} = 1.10 \times 10^{13}$$

EXERCISE: (a) Verify that $|A_P| = |A_v||A_i|$. (b) An amplifier must deliver 20 W to a 16-Ω speaker. The sinusoidal input signal source can be represented as a 5-mV source in series with a 10-kΩ resistor. If the input resistance of the amplifier is 20 kΩ, what are the voltage, current, and power gains required of the overall amplifier?

ANSWERS: 5060, 9.49×10^6, 4.80×10^{10}

10.2.4 THE DECIBEL SCALE

The various gain expressions often involve some rather large numbers, and it is customary to express the values of voltage, current, and power gain in terms of the **decibel,** or **dB** (one-tenth of a Bel):

$$A_{P\text{dB}} = 10 \log A_P \qquad A_{v\text{dB}} = 20 \log |A_v| \qquad A_{i\text{dB}} = 20 \log |A_i| \tag{10.10}$$

The number of decibels is 10 times the base 10 logarithm of the arithmetic power ratio, and decibels are added and subtracted just like logarithms to represent multiplication and division. Because power is proportional to the square of both voltage and current, a factor of 20 appears in the expressions for $A_{v\text{dB}}$ and $A_{i\text{dB}}$.

Table 10.2 has a number of useful examples. From this table, we can see that an increase in voltage or current gain by a factor of 10 corresponds to a change of 20 dB, whereas a factor of 10 increase in power gain corresponds to a change of 10 dB. A factor of 2 corresponds to a 6-dB change

TABLE 10.2
Expressing Gain in Decibels

	|GAIN|	$A_{v\text{dB}}$ or $A_{i\text{dB}}$	$A_{P\text{dB}}$
	1000	60 dB	30 dB
	500	54 dB	27 dB
	300	50 dB	25 dB
$A_{v\text{dB}} = 20\log\|A_v\|$	100	40 dB	20 dB
$A_{i\text{dB}} = 20\log\|A_i\|$	20	26 dB	13 dB
$A_{P\text{dB}} = 10\log A_P$	10	20 dB	10 dB
	$\sqrt{10} = 3.16$	10 dB	5 dB
	2	6 dB	3 dB
	1	0 dB	0 dB
	0.5	-6 dB	-3 dB
	0.1	-20 dB	-10 dB

in voltage or current gain or a 3-dB change in power gain. In the chapters that follow, the various gains routinely are expressed interchangeably in terms of arithmetic values or dB, so it is important to become comfortable with the conversions in Eqs. (10.10) and Table 10.2.

EXERCISE: Express the voltage gain, current gain, and power gain in the exercise at the end of Sec. 10.2.3 in dB.

ANSWERS: 74.1 dB, 140 dB, 107 dB

EXERCISE: Express the voltage gain, current gain, and power gain of the amplifier in Fig. 10.3 in dB.

ANSWERS: 92.0 dB, 169 dB, 130 dB

10.3 AMPLIFIER BIASING FOR LINEAR OPERATION

The amplifiers we discuss in subsequent chapters are constructed from solid-state devices that we know from our study of the large-signal models in Chapters 3 through 5 to be highly nonlinear. In addition, the output voltage range of a real amplifier is limited by its power supply voltages. Amplifiers must be carefully biased to ensure linear operation.

As an example, consider the voltage transfer characteristic (VTC) of a particular amplifier in Fig. 10.4. In this amplifier, the output voltage is restricted to $2\text{ V} \leq v_O \leq 16\text{ V}$. In addition, the input-output relationship is linear only over certain portions of the characteristic. Using our standard notation, described in Chapter 1, the total input voltage v_I is represented as the sum of two components:

$$v_I = V_I + v_i \tag{10.11}$$

in which V_I represents the dc value of v_I, and v_i is the time-varying component of the input signal. For the amplifier to provide linear amplification of a time-varying signal v_i, the total input signal to the amplifier must be **biased** by the dc voltage V_I into the desired region of the characteristic.

The voltage gain A_v of an amplifier describes the relation between changes in the output signal voltage

$$v_O = V_O + v_o \tag{10.12}$$

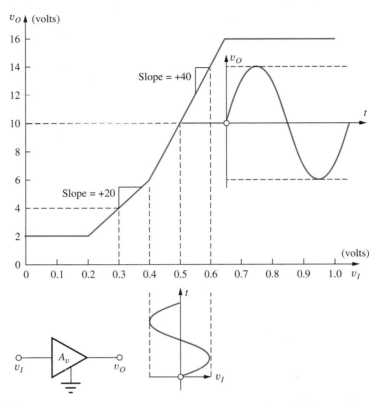

Figure 10.4 Voltage transfer characteristic for a hypothetical noninverting amplifier.

and changes in the input signal v_I and is defined by the slope of the amplifier's VTC, evaluated for an input voltage equal to the dc bias voltage V_I:

$$A_v = \frac{\partial v_O}{\partial v_I}\bigg|_{v_I = V_I} \tag{10.13}$$

The slope of the VTC in Fig. 10.4 is everywhere ≥ 0, so the amplifier input and output are in phase; this amplifier is a **noninverting amplifier.** If the slope had been negative, then the input and output signals would be 180° out of phase, and the amplifier would be characterized as an **inverting amplifier.**

The voltage gain of the amplifier in Fig. 10.4 depends on the bias point. If the amplifier input is biased at $V_I = 0.5$ V, for example, the voltage gain will be +40 for input signals v_i, satisfying $|v_i| \leq 0.1$ V. If the input signal exceeds this value, then the output signal will be distorted due to the change in amplifier slope. The amplifier could also be biased with $V_I = 0.3$ V, for which the voltage gain is +20. Note that $V_I = 0.4$ should not be chosen as a bias point for this amplifier because the gain is different for positive and negative values of v_i, and the output signal would be a distorted version of the input signal. Be sure to understand that the gain is *not* equal to the ratio of the bias point values: $A_v \neq V_O/V_I$! This represents a common point of misunderstanding.

Figure 10.5(a) shows the signals at the output of the amplifier in Fig. 10.4 for a 1 kHz sine wave input signal having an amplitude of 50 mV and biased at two different points: $V_I = 0.3$ V and $V_I = 0.5$ V. The input signals for these two bias points can be expressed as

$$v_{IA} = (0.3 + 0.05 \sin 2000\pi t) \text{ V} \qquad \text{and} \qquad v_{IB} = (0.5 + 0.05 \sin 2000\pi t) \text{ V}$$

and the corresponding output signals are described by

$$v_{OA} = (4 + 1 \sin 2000\pi t) \text{ V} \qquad \text{and} \qquad v_{OB} = (10 + 2 \sin 2000\pi t) \text{ V}$$

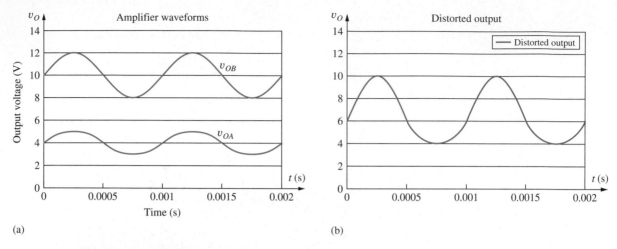

Figure 10.5 (a) Amplifier outputs with a gain of 20 (v_{OA}) and a gain of 40 (v_{OB}); (b) Distorted output caused by improper choice of bias point ($V_I = 0.4$ V).

The gain at the first bias point is 20, whereas the gain at the second bias point is 40. Both inputs produce undistorted sine waves at the amplifier output. Notice that the lower output varies about a dc level of 4 V, and the upper curve varies about 10 V.

EXERCISE: (a) What bias point should be chosen for the amplifier in Fig. 10.4 to provide the maximum possible linear input signal magnitude? What is the maximum input signal amplitude? (b) What is the voltage gain if the amplifier is biased at $V_I = 0.8$ V?

ANSWERS: 0.525 V, $|v_i| \leq 0.125$ V; 0

EXERCISE: Write an expression for $v_O(t)$ for the amplifier in Fig. 10.4 if $v_i(t) = (0.6 + 0.01 \sin 1000\pi t)$ V. What dc bias appears as part of the output voltage?

ANSWER: $(14 + 0.4 \sin 1000\pi t)$ V; 14 V

10.4 DISTORTION IN AMPLIFIERS

As mentioned in Sec. 10.3, if the input signal is biased at 0.4 V, then the output waveform will be significantly distorted since the gain for positive values of the input signal will be different from the gain for negative values. The resulting output is shown in Fig. 10.5(b). The downward excursions of the waveform appear "flattened," and there is a slope discontinuity in the waveform.

A measure of the distortion in such a signal is given by its **total harmonic distortion** (THD), which compares the undesired harmonic content of a signal to the desired component. If we expand the Fourier series representation for a signal $v(t)$ as given in Eq. (10.1), we have

$$v(t) = \underset{\text{dc}}{V_O} + \underset{\substack{\text{desired} \\ \text{output}}}{V_1 \sin(\omega_o t + \phi_1)} + \underset{\substack{\text{2nd harmonic} \\ \text{distortion}}}{V_2 \sin(2\omega_o t + \phi_2)} + \underset{\substack{\text{3rd harmonic} \\ \text{distortion}}}{V_3 \sin(3\omega_o t + \phi_3)} + \cdots$$

The signal at frequency ω_o is the desired output that has the same frequency as the input signal. The terms at $2\omega_o$, $3\omega_o$, etc. represent second-, third-, and higher-order harmonic distortion. The percent

THD is defined by

$$\text{THD}_\% = 100\% \times \frac{\sqrt{\displaystyle\sum_{2}^{\infty} V_n^2}}{V_1} \tag{10.14}$$

The numerator of this expression combines the amplitudes of the individual distortion terms in rms form, whereas the denominator contains only the desired component. Normally, only the first few terms are important in the numerator. For example, MATLAB yields this representation for the signal in Fig. 10.5(b),

$$v(t) = 6.635 + 3.000\sin(2000\pi t) - 0.426\sin(4000\pi t + 90°)$$
$$- 0.0852\sin(8000\pi t + 90°) - 0.0364\sin(12000\pi t + 90°) + \cdots$$

for which the total distortion is approximately

$$\text{THD} \cong 100\% \times \frac{\sqrt{0.426^2 + 0.852^2 + 0.0364^2}}{3} = 14.5\%$$

This value of THD represents a large amount of distortion, which is clearly visible in Fig. 10.5(b). Good distortion levels are well below 1 percent and are not readily apparent to the eye.

EXERCISE: Use MATLAB or Mathcad to plot both the signal in Fig. 10.5(b) and its reconstruction described by $v(t)$.

ANSWER:
```
w = 2000*pi*linspace(0,.001,512);
v = 6+4*sin(w).*(sin(w)>=0)+2*sin(w).*(sin(w)<0);
f = 6.635−3*sin(w)−.426*cos(2*w)−0.0852*cos(4*w)−.0364*cos(6*w);
plot(t,v,t,f)
```

(Note the close match between the two curves with only a few components of the series.)

EXERCISE: Use MATLAB to find the Fourier representation for $v(t)$.

ANSWER:
```
w = 2000**pi*linspace(0,.001,512);
v = 6+4*sin(w).*(sin(w)>=0)+2*sin(w).*(sin(w)<0);
s = fft(v)/512;
vmag=sqrt(s.*conj(s));
vmag(1:10)
```

(Note that the fft function in MATLAB generates the coefficients for the complex Fourier series.)

10.5 TWO-PORT MODELS FOR AMPLIFIERS

The **two-port network** in Fig. 10.6(a) is very useful for modeling the behavior of amplifiers in complex systems. We can use the two-port to provide a relatively simple representation of a much more complicated circuit. Thus, the two-port helps us hide or encapsulate the complexity of the circuit so we can more easily manage the overall analysis and design. One important limitation must be remembered, however. The two-ports we use are linear network models, and are valid under small-signal conditions that will be fully discussed in Chapter 13.

From network theory, we know that two-port networks can be represented in terms of **two-port parameters:** the g-, h-, y-, z-, s- and $abcd$-parameters. Note in these two-port representations that (v_1, i_1) and (v_2, i_2) represent the signal components of the voltages and currents at the two ports of

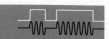

ELECTRONICS IN ACTION

Laptop Computer Touchpad

An essential element of a graphical user interface is a pointing device. This was clear to Douglas Engelbart in the late 1960s during his experimentation with graphical computer interfaces. In order to provide the feeling that a user was directly manipulating objects on the screen, Engelbart invented the computer mouse in 1968. It did not move into the computing mainstream until the introduction of the Apple Macintosh in 1984. As integrated circuit technology advanced and made possible the creation of laptop computers, it became necessary to develop a pointing device that was contained within the form factor of the laptop computer but maintained the 'connective' feel between the user and the computer interface. Trackballs were used in early machines, but they didn't allow the intuitive x-y hand displacement feedback of the mouse. Trackballs were also prone to accumulation of dirt and other debris which reduced their robustness.

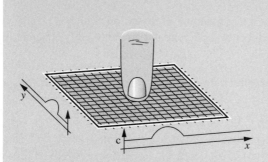

Touch screens were available in the early 1980s, but they required non-robust resistive membranes and/or expensive fabrication techniques. In the early 1990s, Synaptics Corporation developed the capacitive sensing touchpad. A simplified drawing of a capacitive touchpad is shown above. A thin insulating surface covers an x-y grid of wires. When a finger is placed on or near the surface, the capacitance of the wires directly underneath is changed. By measuring the capacitance between the wires and ground, it is possible to detect the presence of an object. If the capacitance measurement is performed with each of the wires in sequence, a capacitance versus position profile is developed. Calculating the centroid of the broad profile allows the system to form a precise indication of finger position over the touchpad.

The measurement itself can be done in a number of ways. The capacitance could be part of a tuned circuit to control the frequency of an oscillator. One could drive the capacitance with a sinusoid current and measure the peak-to-peak value of the resulting voltage. Or, as is the case with most touchpads, a step voltage is driven onto the wire and the resulting charging current is integrated. The magnitude of the integral is proportional to capacitance. Once again, integrated circuit technology made the device practical and inexpensive. A significant number of wires is required to achieve adequate resolution. If implemented with discrete components, the switches, signal routing, and signal processing would be large and expensive. A single mixed-signal CMOS integrated circuit, integrating precision analog circuits and digital processing, was designed to provide all of the necessary functionality, as well as to provide a digital interface that is easily incorporated into a computer. Bridging the gap between real world analog information and digital computers is an important and recurring theme in analog microelectronics.

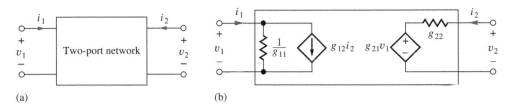

Figure 10.6 (a) Two-port network representation. (b) Two port g-parameter representation.

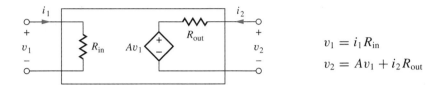

$$v_1 = i_1 R_{\text{in}}$$
$$v_2 = A v_1 + i_2 R_{\text{out}}$$

Figure 10.7 Simplified two-port with more intuitive notation, and "g_{12}" = 0.

the network. We will focus on the g-parameters description. The other parameter sets are discussed on the website for this text.

10.5.1 THE g-PARAMETERS

The **g-parameter** description is one of the most commonly used two-port representations for a voltage amplifier:

$$\mathbf{i_1} = g_{11}\mathbf{v_1} + g_{12}\mathbf{i_2}$$
$$\mathbf{v_2} = g_{21}\mathbf{v_1} + g_{22}\mathbf{i_2} \tag{10.15}$$

Figure 10.6(b) is a network representation of these equations.

The g-parameters are determined from a given network using a combination of **open-circuit** ($i = 0$) and **short-circuit** ($v = 0$) **termination** conditions by applying these parameter definitions:

$$g_{11} = \left.\frac{\mathbf{i_1}}{\mathbf{v_1}}\right|_{\mathbf{i_2}=0} = \textbf{open-circuit input conductance}$$

$$g_{12} = \left.\frac{\mathbf{i_1}}{\mathbf{i_2}}\right|_{\mathbf{v_1}=0} = \text{reverse } \textbf{short-circuit current gain}$$

$$g_{21} = \left.\frac{\mathbf{v_2}}{\mathbf{v_1}}\right|_{\mathbf{i_2}=0} = \text{forward } \textbf{open-circuit voltage gain} \tag{10.16}$$

$$g_{22} = \left.\frac{\mathbf{v_2}}{\mathbf{i_2}}\right|_{\mathbf{v_1}=0} = \textbf{short-circuit output resistance}$$

Unfortunately, the classic g-parameter notation doesn't provide much support for our intuition, so a more descriptive simplified representation is included in Eq. (10.17) and Fig. 10.7.

$$v_1 = i_1 R_{\text{in}}$$
$$v_2 = A v_1 + i_2 R_{\text{out}} \tag{10.17}$$

R_{in} represents the input resistance to the amplifier, A is the voltage gain when there is no external load on the amplifier, and R_{out} is the output resistance of the amplifier. In a normal amplifier design, we desire the forward gain (g_{21}) to be much larger than the reverse gain (g_{12}), that is, $g_{21} \gg g_{12}$, and Eq. (10.17) and Fig. 10.7 show the simplified two-port representation in which the reverse gain g_{12} is assumed to be zero.

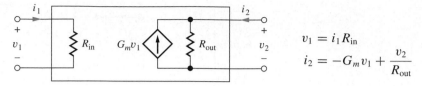

Figure 10.8 Norton transformation of the circuit in Fig. 10.7 in which $G_m = \dfrac{A}{R_{\text{out}}}$.

$$v_1 = i_1 R_{\text{in}}$$

$$i_2 = -G_m v_1 + \frac{v_2}{R_{\text{out}}}$$

Figure 10.8 presents an alternate two-port representation that we shall encounter frequently in our transistor circuits. In this equivalent circuit, the output port components have been found using Norton's theorem that yields $G_m = A_{v1}/R_{\text{out}}$.

EXAMPLE 10.1   FINDING A SET OF g-PARAMETERS

This example calculates a set of g-parameters for a network containing a dependent current source. We encounter this type of circuit often in analog circuit analysis and design because our models for both bipolar and field-effect transistors contain dependent current sources.

PROBLEM Find the g-parameters for the circuit shown here. Include g_{12} for completeness, and compare it to g_{21}.

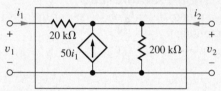

SOLUTION **Known Information and Given Data:** Circuit as given in the problem statement including element values; g-parameter definitions in Eq. (10.16)

Unknowns: Values of the four g-parameters

Approach: Apply the boundary conditions specified for each g-parameter and use circuit analysis to find the values of the four parameters. Note that each set of boundary conditions applies to two parameters.

Assumptions: None

Analysis—g_{11} and g_{21}: Looking at the definitions of the g-parameters,

$$G_{\text{in}} = g_{11} = \left. \frac{\mathbf{i_1}}{\mathbf{v_1}} \right|_{\mathbf{i_2}=0} \qquad \text{and} \qquad A = g_{21} = \left. \frac{\mathbf{v_2}}{\mathbf{v_1}} \right|_{\mathbf{i_2}=0}$$

we see that g_{11} and g_{21} use the same boundary conditions. We apply voltage $\mathbf{v_1}$ to the input port, and the output port is open circuited (i.e., $\mathbf{i_2}$ is set to zero), as in the figure here.

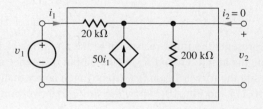

g_{11}: Writing an equation around the input loop and applying KCL at the output node yields

$$\mathbf{v_1} = (2 \times 10^4)\mathbf{i_1} + (\mathbf{i_1} + 50\mathbf{i_1})(200 \text{ k}\Omega)$$

$$G_{\text{in}} = \frac{\mathbf{i_1}}{\mathbf{v_1}} = \frac{1}{2 \times 10^4 \ \Omega + 51(200 \text{ k}\Omega)} = \frac{1}{10.2 \text{ M}\Omega} = 9.79 \times 10^{-8} \text{ S}$$

g_{21}: Since the external port current $\mathbf{i_2}$ is zero, the voltage $\mathbf{v_2}$ is given by

$$\mathbf{v_2} = (\mathbf{i_1} + 50\mathbf{i_1})(200 \text{ k}\Omega) = \mathbf{i_1}(51)(200 \text{ k}\Omega)$$

and $\mathbf{i_1}$ can be related to $\mathbf{v_1}$ using g_{11}:

$$\mathbf{v_2} = (g_{11}\mathbf{v_1})(51)(200 \text{ k}\Omega)$$

$$A = \frac{\mathbf{v_2}}{\mathbf{v_1}} = g_{11}(51)(200 \text{ k}\Omega) = (9.79 \times 10^{-8} \text{ S})(51)(200 \text{ k}\Omega) = +0.998$$

Analysis—g_{12} and g_{22}: Looking again at the definitions of the g-parameters, we see that g_{12} and g_{22} use the same boundary condition.

$$g_{12} = \left.\frac{\mathbf{i_1}}{\mathbf{i_2}}\right|_{\mathbf{v_1}=0} \qquad \text{and} \qquad R_{\text{out}} = g_{22} = \left.\frac{\mathbf{v_2}}{\mathbf{i_2}}\right|_{\mathbf{v_1}=0}$$

A current source $\mathbf{i_2}$ is applied to the input port, and the input port is short-circuited (i.e., $\mathbf{v_1}$ is set to zero) as shown in this figure:

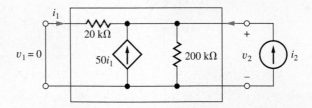

g_{22}: With $\mathbf{v_1} = 0$, we see that the network is just a single-node circuit. Writing a nodal equation for $\mathbf{v_2}$ yields

$$(\mathbf{i_2} + 50\mathbf{i_1}) = \frac{\mathbf{v_2}}{200 \text{ k}\Omega} + \frac{\mathbf{v_2}}{20 \text{ k}\Omega}$$

But, $\mathbf{i_1}$ can be written directly in terms of $\mathbf{v_2}$ as $i_1 = -v_2/20 \text{ k}\Omega$. Combining these two equations yields the short-circuit output resistance g_{22}:

$$\mathbf{i_2} = \frac{\mathbf{v_2}}{200 \text{ k}\Omega} + \frac{\mathbf{v_2}}{20 \text{ k}\Omega} + 50\frac{\mathbf{v_2}}{20 \text{ k}\Omega} \qquad \text{and} \qquad R_{\text{out}} = \frac{\mathbf{v_2}}{\mathbf{i_2}} = \frac{1}{\dfrac{1}{200 \text{ k}\Omega} + \dfrac{51}{20 \text{ k}\Omega}} = 391 \ \Omega$$

g_{12}: The reverse short-circuit current gain g_{12} can be found using the preceding results:

$$\mathbf{i_1} = -\frac{\mathbf{v_2}}{20 \text{ k}\Omega} = -\frac{R_{\text{out}}\mathbf{i_2}}{20 \text{ k}\Omega} \qquad \text{and} \qquad g_{12} = \frac{\mathbf{i_1}}{\mathbf{i_2}} = -\frac{391 \ \Omega}{20 \text{ k}\Omega} = -0.0196$$

The final g-parameter equations for the network are

$$\mathbf{i_1} = 9.79 \times 10^{-8}\mathbf{v_1} - 1.96 \times 10^{-2}\mathbf{i_2}$$

$$\mathbf{v_2} = 0.998\mathbf{v_1} + 3.91 \times 10^2\mathbf{i_2}$$

Check of Results: The results are confirmed below using SPICE.

Discussion: Note that the values of $R_{in} = 10.2$ MΩ and $R_{out} = 391$ Ω differ greatly from any of the resistor values in the network. This is a result of the action of the dependent current source and is an important effect that we will see throughout the analysis of analog transistor circuits. Here we see that g_{12} is indeed small and that $g_{12} \ll A$. The simplified mathematical and two-port models for the circuit become

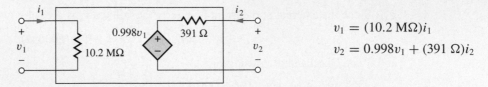

$$v_1 = (10.2 \text{ MΩ})i_1$$
$$v_2 = 0.998v_1 + (391 \text{ Ω})i_2$$

We will make use of this observation when we study feedback in Chapter 17.

Computer-Aided Analysis: Numerical values for two-port parameters can easily be found using the transfer function (TF) analysis capability of SPICE. In order to find the g-parameters for the circuit in this example, we drive the network with voltage source V1 at the input and current source I2 at the output, as in the figure here. These choices correspond to the boundary conditions in the definitions of the g-parameters.

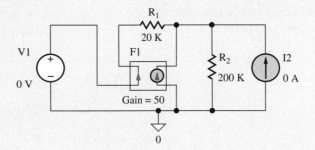

Both independent sources are assigned zero values. The TF analysis calculates how variables change in response to changes in an independent source. Therefore, a starting point of zero is fine. The zero value sources directly satisfy the boundary conditions required to calculate the g-parameters.

Two TF analyses are used—one to find g_{11} and g_{21} and a second to find g_{12} and g_{22}. The first analysis requests calculation of the transfer function from source V1 to the voltage at the output node, and SPICE will calculate the value of the transfer function, resistance at the input source node, and resistance at the output node. The SPICE results are transfer function = 0.998, input resistance = 10.2 MΩ, and output resistance = 391 Ω. Parameter g_{21} is the open-circuit voltage gain, which agrees with the hand calculations, and g_{11} is the input conductance equal to the reciprocal of 10.2 MΩ, again in agreement with our hand calculations.

The second analysis requests the transfer function from source I2 to (the current in) source V1. The results from SPICE are transfer function = 0.0196 and input resistance = 391 Ω. In this case, the output resistance (at V1) cannot be calculated because V1 represents a short at the input. Note that parameter g_{12} is the negative of the TF value. The sign difference arises from the sign convention assumed by SPICE in which positive current is directed downward through source V1. Parameter g_{22} is the 391-Ω resistance presented to source I2, which is the "input resistance" in this calculation. We find precise agreement with our hand calculations. It is important to remember that the SPICE TF analysis is a form of dc analysis and must be used very carefully in networks containing capacitors and inductors.

DESIGN NOTE

Remember, the transfer function analysis in SPICE is a dc analysis and should not be used in circuits containing capacitors and inductors!

10.6 MISMATCHED SOURCE AND LOAD RESISTANCES

In introductory circuit theory, the maximum power transfer theorem is usually discussed. Maximum power transfer occurs when the source and load resistances are matched (equal in value). In most voltage and **current amplifier** applications, however, the opposite situation is desired. A completely mismatched condition is used at both the input and output ports of the amplifier.

To understand the statement above, consider the **voltage amplifier** in Fig. 10.9. The input to the two-port is a Thévenin equivalent representation of the input source, and the output is connected to a load represented by resistor R_L. To find the voltage gain, voltage division is applied to each loop:

$$\mathbf{v_o} = A\mathbf{v_1}\frac{R_L}{R_{\text{out}} + R_L} \quad \text{and} \quad \mathbf{v_1} = \mathbf{v_s}\frac{R_{\text{in}}}{R_S + R_{\text{in}}} \tag{10.18}$$

Combining these two equations yields an expression for the magnitude of the voltage gain A_v:

$$|A_v| = \frac{V_o}{V_s} = A\frac{R_{\text{in}}}{R_S + R_{\text{in}}}\frac{R_L}{R_{\text{out}} + R_L} \tag{10.19}$$

To achieve maximum voltage gain, the resistors should satisfy $R_{\text{in}} \gg R_S$ and $R_{\text{out}} \ll R_L$. For this case,

$$|A_v| = A \tag{10.20}$$

The situation described by these two equations is a totally mismatched condition at both the input and the output ports. An **ideal voltage amplifier** satisfies the conditions in Eq. (10.19) by having $R_{\text{in}} = \infty$ and $R_{\text{out}} = 0$.

The magnitude of the current gain of the amplifier in Fig. 10.19 can be expressed as

$$|A_i| = \frac{I_o}{I_1} = \frac{\dfrac{V_o}{R_L}}{\dfrac{V_s}{R_S + R_{\text{in}}}} = \frac{V_o}{V_s}\frac{R_S + R_{\text{in}}}{R_L} \quad \text{or} \quad |A_i| = |A_v|\frac{R_S + R_{\text{in}}}{R_L} \tag{10.21}$$

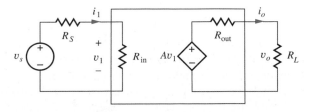

Figure 10.9 Two-port representation of an amplifier with source and load connected.

EXERCISE: Write an expression for the power gain of the amplifier in Fig. 10.9 in terms of the voltage gain.

ANSWER: $A_P = A^2 \dfrac{R_S + R_{in}}{R_L}$

EXERCISE: Suppose the audio amplifier in Fig. 10.3 can be modeled by $R_{in} = 50 \text{ k}\Omega$ and $R_{out} = 0.5 \ \Omega$. What value of open-circuit gain A is required to achieve an output power of 100 W if $v_s = 0.001 \sin 2000\pi t$? How much power is being dissipated in R_{out}? What is the current gain?

ANSWERS: 46,800 (93.4 dB); 6.25 W; 2.75×10^8 (169 dB)

EXERCISE: Repeat the preceding exercise if the input and output ports are matched to the source and load respectively (that is, $R_{in} = 5 \text{ k}\Omega$ and $R_{out} = 8 \ \Omega$). (It should become clear why we don't design R_{out} to match the load resistance.)

ANSWERS: 160,000 (104 dB); 100 W; 5×10^7 (153 dB)

10.7 AMPLIFIER TRANSFER FUNCTIONS AND FREQUENCY RESPONSE

The gain expressions evaluated thus far have been constants. However, real amplifiers cannot provide uniform amplification at all frequencies. As frequency increases, shunt capacitances between all circuit elements and the ground node always cause the gain to decrease to zero at very high frequencies. At low frequencies, any series capacitors in the circuit limit the amplifier's ability to provide gain, and circuits called *dc-coupled amplifiers* must be used if amplification is to be provided at dc. Because of these differences, amplifiers are classified into several general categories, including low-pass, high-pass, band-pass, **band-reject,** and **all-pass,** based on the characteristics of their **transfer functions** in the frequency domain.

The frequency-dependent voltage gain of an amplifier is characterized by the voltage transfer function $A_v(s)$, which is the ratio of the Laplace transforms $V_o(s)$ and $V_s(s)$ of the input and output voltages of the amplifier, where $s = \sigma + j\omega$ is the complex frequency variable and $j = \sqrt{-1}$:

$$A_v(s) = \frac{V_o(s)}{V_s(s)} \tag{10.22}$$

The frequency-dependent current gain is characterized in a similar manner by its transfer function $A_i(s)$:

$$A_i(s) = \frac{I_o(s)}{I_s(s)} \tag{10.23}$$

The circuits we study will be modeled entirely by lumped circuit elements (R, L, C, and so on), and the numerator $N(s)$ and denominator $D(s)$ of the transfer functions will be polynomials in s which can also be represented in factored form:

$$A_v(s) = \frac{N(s)}{D(s)} = \frac{a_m s^m + \cdots + a_1 s + a_o}{b_n s^n + \cdots + b_1 s + b_o} = K \frac{(s + z_1)(s + z_2) \cdots (s + z_m)}{(s + p_1)(s + p_2) \cdots (s + p_n)} \tag{10.24}$$

The frequencies $(-z_1, -z_2, \ldots, -z_m)$ for which the transfer function becomes zero are called the *zeros* of the transfer function, and the frequencies $(-p_1, -p_2, \ldots, -p_n)$, for which the transfer function becomes infinite are called the *poles* of the transfer function. In general, the values of the poles and zeros are complex numbers, although the majority of amplifiers we study will have only real values for the poles and zeros. Some of the zeros of the transfer function may either be at zero or infinite frequency.

EXERCISE: Find the poles and zeros of the voltage transfer function

$$A_v(s) = \frac{300s}{s^2 + 5100s + 500000}$$

ANSWERS: Poles: -100, -5000; zeros: 0 (and ∞)

10.7.1 BODE PLOTS

Transfer functions of general amplifiers can be quite complicated, having many poles and zeros, but their overall behavior can be broken down into the categories mentioned in the introduction: low-pass, high-pass, band-pass, band-reject, and all-pass amplifiers. The basic forms of each category are discussed in the next several subsections.

When we explore the characteristics of amplifiers, we are usually interested in the behavior of the transfer function for physical frequencies ω—that is, for $s = j\omega$, and the transfer function can then be represented in polar form by its **magnitude $|A_v(j\omega)|$** and **phase angle $\angle A_v(j\omega)$,** which are both functions of frequency:

$$A_v(j\omega) = |A_v(j\omega)| \angle A_v(j\omega) \tag{10.25}$$

It is often convenient to display this information separately in a graphical form called a **Bode plot.** The Bode plot displays the magnitude of the transfer function in dB and the phase in degrees (or radians) versus a logarithmic frequency scale. Examples will be given as the various amplifier types are discussed.

10.7.2 THE LOW-PASS AMPLIFIER

Circuits that amplify signals over a range of frequencies including dc are an extremely important class of circuits and are referred to as low-pass amplifiers. For instance, most operational amplifiers are designed as low-pass amplifiers. The simplest low-pass amplifier circuit is described by the single-pole[4] transfer function

$$A_v(s) = \frac{A_o \omega_H}{s + \omega_H} = \frac{A_o}{1 + \dfrac{s}{\omega_H}} \tag{10.26}$$

in which A_o is the low-frequency gain and ω_H represents the cutoff frequency of this low-pass amplifier. Let us first explore the behavior of the magnitude of $A_v(s)$ and then look at the phase response.

Magnitude Response

Substituting $s = j\omega$ into Eq. (10.26) and finding the magnitude of the function $A_v(j\omega)$ yields

$$|A_v(j\omega)| = \left| \frac{A_o \omega_H}{j\omega + \omega_H} \right| = \frac{|A_o \omega_H|}{\sqrt{\omega^2 + \omega_H^2}} \tag{10.27}$$

The Bode magnitude plot is given in terms of dB:

$$|A_v(j\omega)|_{dB} = 20 \log |A_o \omega_H| - 20 \log \sqrt{\omega^2 + \omega_H^2} \tag{10.28}$$

For a given set of numeric values, Eq. (10.28) can be easily evaluated and plotted using a package such as MATLAB or a spreadsheet, and results in the graph in Fig. 10.10.

---

[4] A general low-pass amplifier may have many poles. The single-pole version is the simplest approximation to the ideal low-pass characteristic described in Chapter 1.

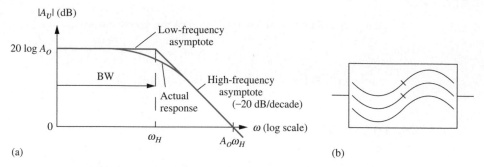

Figure 10.10 (a) Low-pass amplifier: BW $= \omega_H$. (b) Low-pass filter symbol.

For the general case, the graph is conveniently plotted in terms of its asymptotic behavior at low and high frequencies. For low frequencies, $\omega \ll \omega_H$, the magnitude is approximately constant:

$$\left. \frac{A_o \omega_H}{\sqrt{\omega^2 + \omega_H^2}} \right|_{\omega \ll \omega_H} \cong \frac{A_o \omega_H}{\sqrt{\omega_H^2}} = A_o \qquad \text{or} \qquad (20 \log A_o) \text{ dB} \qquad (10.29)$$

At frequencies well below ω_H, the gain of the amplifier is constant and equal to A_o, which corresponds to the horizontal asymptote in Fig. 10.10. Signals at frequencies below ω_H are amplified by the gain A_o. In fact, the gain of this amplifier is constant down to dc ($\omega = 0$)!

However, as ω exceeds ω_H, the gain of the amplifier begins to decrease (high-frequency roll-off). For sufficiently high frequencies, $\omega \gg \omega_H$, the magnitude can be approximated by

$$\left. \frac{A_o \omega_H}{\sqrt{\omega^2 + \omega_H^2}} \right|_{\omega \gg \omega_H} \cong \frac{A_o \omega_H}{\sqrt{\omega^2}} = \frac{A_o \omega_H}{\omega} \qquad (10.30)$$

and converting Eq. (10.30) to dB yields

$$|A_v(j\omega)|_{dB} \cong \left(20 \log A_o - 20 \log \frac{\omega}{\omega_H} \right) \text{ dB} \qquad (10.31)$$

For frequencies much greater than ω_H, the transfer function decreases at a rate of 20 dB per decade increase in frequency, as indicated by the high-frequency asymptote in Fig. 10.10. Obviously, ω_H plays an important role in characterizing the amplifier; this critical frequency is called the **upper-cutoff frequency** of the amplifier. At $\omega = \omega_H$, the gain of the amplifier is

$$|A_v(j\omega_H)| = \frac{A_o \omega_H}{\sqrt{\omega_H^2 + \omega_H^2}} = \frac{A_o}{\sqrt{2}} \qquad \text{or} \qquad [(20 \log A_o) - 3] \text{ dB} \qquad (10.32)$$

and ω_H is sometimes referred to as the **upper −3-dB frequency** of the amplifier. ω_H is also often termed the **upper half-power point** of the amplifier because the output power of the amplifier, which is proportional to the square of the voltage, is reduced by a factor of 2 at $\omega = \omega_H$. Note that when the expressions for the two asymptotes given in Eqs. (10.29) and (10.30) are equated, they intersect precisely at $\omega = \omega_H$. Also, the gain becomes unity (0 dB) at $\omega = A_o \omega_H$ which is known as the **gain-bandwidth product** of the amplifier.

Bandwidth
The gain of the amplifier in Fig. 10.10 is approximately uniform (it varies by less than 3 dB) for all frequencies below ω_H. This is called a **low-pass amplifier.** The **bandwidth (BW)** of an amplifier is defined by the range of frequencies in which the amplification is approximately constant; it is expressed in either radians/second or Hz. For the low-pass amplifier,

$$\text{BW} = \omega_H \text{ (rad/s)} \qquad \text{or} \qquad \text{BW} = f_H = \frac{\omega_H}{2\pi} \text{ Hz} \qquad (10.33)$$

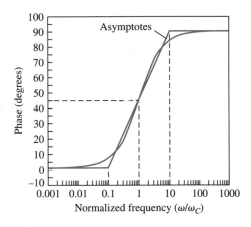

TABLE 10.3

Inverse Tangent

ω	$\tan^{-1}\dfrac{\omega}{\omega_C}$
$0.01\,\omega_C$	$0.057°$
$0.1\,\omega_C$	$5.7°$
ω_C	$45°$
$10\,\omega_C$	$84.3°$
$100\,\omega_C$	$89.4°$

Figure 10.11 Phase versus normalized frequency (ω/ω_C) resulting from a single inverse tangent term $+\tan^{-1}(\omega/\omega_C)$. The straight-line approximation is also given.

EXERCISE: Find the midband gain, cutoff frequency, bandwidth, and gain-bandwidth product of the low-pass amplifier with the following transfer function:

$$A_v(s) = -\frac{2\pi \times 10^6}{(s + 5000\pi)}$$

ANSWERS: −400, 2.5 kHz, 2.5 kHz, 1 MHz

Phase Response

The phase behavior versus frequency is also of interest in many applications and later will be found to be of great importance to the stability of feedback amplifiers. Again substituting $s = j\omega$ in Eq. (10.26), the phase response of the low-pass amplifier is found to be

$$\angle A_v(j\omega) = \angle \frac{A_o}{1 + j\dfrac{\omega}{\omega_H}} = \angle A_o - \tan^{-1}\left(\frac{\omega}{\omega_H}\right) \qquad (10.34)$$

The phase angle of A_o is $0°$ if A_o is positive and $180°$ if A_o is negative.

The frequency-dependent phase term associated with each pole or zero in a transfer function involves the evaluation of the inverse tangent function, as in Eq. (10.34). Important values appear in Table 10.3, and a graph of the complete inverse tangent function is given in Fig. 10.11. At the pole or zero frequency indicated by critical frequency ω_C, the magnitude of the phase shift is $45°$. One decade below ω_C, the phase is $5.7°$, and one decade above ω_C, the phase is $84.3°$. Two decades away from ω_C, the phase approaches its asymptotic limits of $0°$ and $90°$. Note that the phase response can also be approximated by the three straight-line segments in Fig. 10.11, in a manner similar to the asymptotes of the magnitude response.

The phase of more complex transfer functions with multiple poles and zeros is simply given by the appropriate sum and differences of inverse tangent functions. However, they are most easily evaluated with a computer or calculator.

EXAMPLE 10.2 TRANSFER FUNCTION EVALUATION

This example is included to help refresh our memory on calculations involving complex numbers.

PROBLEM Find the magnitude and phase of this voltage transfer function for $\omega = 0$ and $\omega = 3$ rad/s.

$$A_v(s) = 50\frac{s^2 + 4}{s^2 + 2s + 2}$$

SOLUTION   **Known Information and Given Data:** Transfer function describing the voltage gain

Unknowns: Magnitude and phase of $A_v(j0)$ and $A_v(j3)$

Approach: Substitute $s = j\omega$ into the expression for $A_v(s)$ and simplify. Substitute $\omega = 0$ and $\omega = 3$ into the resulting expressions. Then find magnitude and phase of the resulting complex numbers.

Assumptions: We remember how to do arithmetic with complex numbers.

Analysis: Inserting $s = j\omega$ into $A_v(s)$ and rearranging yields

$$A_v(j\omega) = 50\frac{\omega^2 - 4}{(\omega^2 - 2) - j(2\omega)}$$

The magnitude and phase of this expression are

$$|A_v(j\omega)| = 50\frac{|\omega^2 - 4|}{\sqrt{(\omega^2 - 2)^2 + 4\omega^2}} \quad \text{and} \quad \angle A_v(j\omega) = \angle(\omega^2 - 4) - \tan^{-1}\left(\frac{-2\omega}{\omega^2 - 2}\right)$$

Substituting $\omega = 0$ gives

$$|A_v(j\omega)| = \frac{200}{\sqrt{4}} = 100 \quad \text{or} \quad 40.0\,\text{dB}$$

$$\angle A_v(j\omega) = \angle(200) - \tan^{-1}(-0) = 0°$$

Substituting $\omega = 3$ gives

$$|A_v(j3)| = \frac{250}{\sqrt{49 + 36}} = 27.1 \quad \text{or} \quad 28.7\,\text{dB}$$

$$\angle A_v(j\omega) = \angle(250) - \tan^{-1}\left(\frac{-6}{7}\right) = 0° - (-40.6) = 40.6°$$

Check of Results: We can easily check the results using MATLAB or a calculator. With MATLAB,

$$h = \text{freqs}([50\ 0\ 200], [1\ 2\ 2], [0\ 3]);$$

$$\text{abs}(h)$$

$$\text{angle}\,(h) * 180/\text{pi}$$

The results confirm the preceding analysis.

EXERCISE: Find the magnitude and phase of the voltage gain in Ex. 10.2 for $\omega = 1$ rad/s and $\omega = 5$ rad/s.

ANSWERS: 36.5 dB, −63.4°; 32.4 dB, 23.5°

EXERCISE: Find the magnitude and phase of the following transfer function for $\omega = 0.95$, 1.0, and 1.10.

$$A_v(s) = 20\,\frac{s^2 + 1}{s^2 + 0.1s + 1}$$

ANSWERS: 14.3, −44.3°; 0, −90°; 17.7, +27.6°

EXERCISE: Make a Bode plot of the following $A_v(s)$ using MATLAB: $A_v(s) = -\dfrac{2\pi \times 10^6}{(s + 5000\pi)}$

ANSWER: $w = \text{logspace}(2, 6, 100)$
$\text{bode}(2*pi*1e6,[1\ 5000*pi],w)$

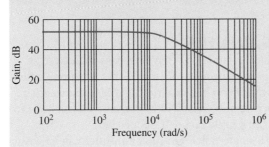
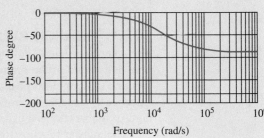

EXAMPLE 10.3   THE *RC* LOW-PASS FILTER

An *RC* low-pass network is a simple but important passive circuit that we will encounter in the upcoming chapters.

PROBLEM   Find the voltage transfer function $\mathbf{V_o}/\mathbf{V_s}$ for the low-pass network in the figure below.

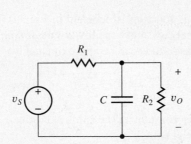

$$V_o = V_s \frac{\dfrac{R_2/sC}{R_2 + 1/sC}}{R_1 + \dfrac{R_2/sC}{R_2 + 1/sC}}$$

$$\frac{V_o}{V_s} = \left(\frac{R_2}{R_1 + R_2}\right)\left(\frac{1}{1 + \dfrac{s}{\omega_H}}\right)$$

SOLUTION   **Known Information and Given Data:** Circuit as given above in the problem statement.

Unknowns: Voltage transfer function $\mathbf{V_o}/\mathbf{V_s}$

Approach: Find the transfer function by applying voltage division in the frequency domain (*s* domain). Remember that the impedance of the capacitor is $1/sC$.

Assumptions: None

Analysis: Direct application of voltage division yields the equations next to the schematic, where the upper cutoff frequency is

$$\omega_H = \frac{1}{(R_1 \| R_2)C}$$

Check of Results: For $s \ll \omega_H$, the gain through the network is $R_2/(R_1 + R_2)$, which is correct.

Discussion: The cutoff frequency ω_H occurs at the frequency for which the reactance of the capacitor equals the parallel combination of resistors R_1 and R_2: $1/\omega_H C = R_1 \| R_2$. $R_1 \| R_2$ represents the Thévenin equivalent resistance present at the capacitor terminals. For $\omega \ll \omega_H$, the impedance of the capacitor is negligible with respect to the resistance in the circuit.

EXERCISE: What is the cutoff frequency for the low-pass circuit in Ex. 10.3 if $R_1 = 1$ kΩ, $R_2 = 100$ kΩ, and $C = 200$ pF?

ANSWER: 804 kHz

10.7.3 THE HIGH-PASS AMPLIFIER

The other basic single-pole transfer function is the high-pass characteristic, which includes a single pole and a zero at the origin. We will most often find this function combined with the low-pass function to form **band-pass amplifiers.** In fact a true high-pass characteristic is impossible to obtain since we will see that it requires infinite bandwidth. The best we can hope to do is approximate the high-pass characteristic over some finite range of frequencies.

The transfer function for a single-pole **high-pass amplifier** can be written as

$$A_v(s) = \frac{A_o s}{s + \omega_L} = \frac{A_o}{1 + \dfrac{\omega_L}{s}} \tag{10.35}$$

and for $s = j\omega$ the magnitude of Eq. (10.35) is

$$|A_v(j\omega)| = \left| \frac{A_o j\omega}{j\omega + \omega_L} \right| = \frac{A_o \omega}{\sqrt{\omega^2 + \omega_L^2}} = \frac{A_o}{\sqrt{1 + \left(\dfrac{\omega_L}{\omega} \right)^2}} \tag{10.36}$$

The Bode magnitude plot for this function is depicted in Fig. 10.12. In this case, the gain of the amplifier is constant for all frequencies above the **lower-cutoff frequency ω_L.** At frequencies high enough to satisfy $\omega \gg \omega_L$, the magnitude can be approximated by

$$\frac{A_o \omega}{\sqrt{\omega^2 + \omega_L^2}} \bigg|_{\omega \gg \omega_L} \cong \frac{A_o \omega}{\sqrt{\omega^2}} = A_o \qquad \text{or} \qquad (20 \log A_o) \text{ dB} \tag{10.37}$$

For ω exceeding ω_L, the gain is constant at the midband gain $A_{\text{mid}} = A_o$. At frequencies well below ω_L,

$$\frac{A_o \omega}{\sqrt{\omega^2 + \omega_L^2}} \bigg|_{\omega \ll \omega_L} \cong \frac{A_o \omega}{\sqrt{\omega_L^2}} = \frac{A_o \omega}{\omega_L} \tag{10.38}$$

Converting Eq. (10.38) to dB yields

$$|A_v(j\omega)| \cong (20 \log A_o) + 20 \log \frac{\omega}{\omega_L} \tag{10.39}$$

At frequencies below ω_L, the gain increases at a rate of 20 dB per decade increase in frequency.

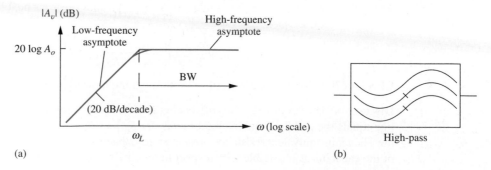

Figure 10.12 (a) High-pass amplifier. (b) High-pass filter symbol.

At **critical frequency** $\omega = \omega_L$,

$$|A_v(j\omega_L)| = \frac{A_o\omega_L}{\sqrt{\omega_L^2 + \omega_L^2}} = \frac{A_o}{\sqrt{2}} \qquad \text{or} \qquad [(20\log A_o) - 3]\ \text{dB} \qquad (10.40)$$

The gain is again 3 dB below its midband value. Besides being called the lower-cutoff frequency, ω_L is referred to as the **lower −3-dB frequency** or the **lower half-power point.**

The high-pass amplifier provides approximately uniform gain at all frequencies above ω_L, and its bandwidth is therefore infinite:

$$\text{BW} = \infty - \omega_L = \infty \qquad (10.41)$$

The phase dependence of the high-pass amplifier is found by evaluating the phase of $A_v(j\omega)$ from Eq. (10.35):

$$\text{and} \qquad \angle A_v(j\omega) = \angle\frac{A_o j\omega}{j\omega + \omega_L} = \angle A_o + 90° - \tan^{-1}\left(\frac{\omega}{\omega_L}\right) \qquad (10.42)$$

This phase expression is similar to that of the low-pass amplifier, except for a $+90°$ shift due to the s term in the numerator.

EXERCISE: Find the midband gain, cutoff frequency, and bandwidth of the amplifier with this transfer function:

$$A_v(s) = \frac{250s}{(s + 250\pi)}$$

ANSWERS: 250; 125 Hz; ∞

EXERCISE: Use MATLAB to produce a Bode plot of this transfer function.

ANSWER: w = logspace(1,5,100)
bode([250 0],[1 250*pi],w)

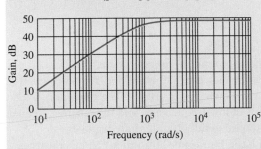

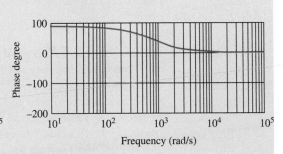

EXAMPLE 10.4   THE *RC* HIGH-PASS FILTER

The *RC* high-pass network is another important passive circuit that we will encounter in the upcoming chapters.

PROBLEM   Find the voltage transfer function $\mathbf{V_o}/\mathbf{V_s}$ for the high-pass network in the figure below.

$$V_o = V_s \frac{R_2}{R_1 + \dfrac{1}{sC} + R_2}$$

$$\frac{V_o}{V_s} = \left(\frac{R_2}{R_1 + R_2}\right)\left(\frac{s}{s + \omega_L}\right)$$

SOLUTION **Known Information and Given Data:** Circuit as given in the problem statement.

Unknowns: Voltage transfer function V_o/V_s

Approach: Find the transfer function by applying voltage division in the frequency domain (s domain). Remember that the impedance of the capacitor is $1/sC$.

Assumptions: None

Analysis: Direct application of voltage division yields the equation next to the circuit schematic, where the lower cutoff frequency is

$$\omega_L = \frac{1}{(R_1 + R_2)C}$$

Check of Results: For $s \gg \omega_L$, the gain through the network is $R_2/(R_1 + R_2)$, which is correct.

Discussion: Cutoff frequency ω_L occurs at the frequency for which the reactance of the capacitor equals the sum of resistors R_1 and R_2, which represents the Thévenin equivalent resistance at the capacitor terminals: $1/\omega_L C = R_1 + R_2$. For $\omega \gg \omega_L$, the impedance of the capacitor is negligible with respect to the resistance in the circuit.

EXERCISE: What is the cutoff frequency for the high-pass circuit in Ex. 10.4 if $R_1 = 1$ kΩ, $R_2 = 100$ kΩ, and $C = 0.1$ μF?

ANSWER: 15.8 Hz

10.7.4 BAND-PASS AMPLIFIERS

In certain cases, we may not need or want to amplify signals at dc, and capacitors are added to the amplifier circuit to produce an "ac-coupled" amplifier. An ac-coupled amplifier has a band-pass characteristic similar to that shown in Fig. 10.13. The capacitors added to the circuit cause the frequency response to roll off at low frequencies, and the inherent frequency limitations of the solid-state devices cause the response to fall off at high frequencies.

The transfer function for a basic **band-pass amplifier** can be constructed from the product of the low-pass and high-pass transfer functions from Eqs. (10.26) and (10.35):

$$A_v(s) = \frac{A_o s \omega_H}{(s + \omega_L)(s + \omega_H)} = A_o \frac{s}{(s + \omega_L)} \frac{1}{\left(\dfrac{s}{\omega_H} + 1\right)} \tag{10.43}$$

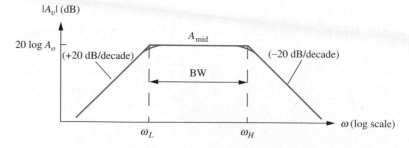

Figure 10.13 Band-pass amplifier.

Figure 10.13 is a graph of the magnitude of this transfer function. The concept of midband gain should be much clearer in this figure. The **midband** range of frequencies is defined by $\omega_L \leq \omega \leq \omega_H$, for which

$$|A_v(j\omega)| \cong A_o \tag{10.44}$$

A_o represents the gain in this midband region and is called the **midband gain**: $A_{\text{mid}} = A_o$.

The mathematical expression for the magnitude of $A_v(j\omega)$ is

$$|A_v(j\omega)| = \left| \frac{A_o j\omega\, \omega_H}{(j\omega + \omega_L)(j\omega + \omega_H)} \right| = \frac{A_o\, \omega\, \omega_H}{\sqrt{(\omega^2 + \omega_L^2)(\omega^2 + \omega_H^2)}} \tag{10.45}$$

or

$$|A_v(j\omega)| = \frac{A_{\text{mid}}}{\sqrt{\left(1 + \dfrac{\omega_L^2}{\omega^2}\right)\left(1 + \dfrac{\omega^2}{\omega_H^2}\right)}} \tag{10.46}$$

The expression in Eq. (10.46) has been written in a form that exposes the gain in the midband region. At both ω_L and ω_H, it is easy to show, assuming $\omega_L \ll \omega_H$, that

$$|A_v(j\omega_L)| = \frac{A_o}{\sqrt{2}} = |A(j\omega_H)| \quad \text{or} \quad [(20 \log A_o) - 3]\, \text{dB} \tag{10.47}$$

The gain is 3 dB below the midband gain at both critical frequencies. The region of approximately uniform gain (that is, the region of less than 3 dB variation) extends from ω_L to ω_H (f_L to f_H), and hence the bandwidth of the band-pass amplifier is

$$\text{BW} = f_H - f_L = \frac{\omega_H - \omega_L}{2\pi} \tag{10.48}$$

Evaluating the phase of $A_v(j\omega)$ from Eq. (10.43),

$$\angle A_v(j\omega) = \angle A_o + 90° - \tan^{-1}\left(\frac{\omega}{\omega_L}\right) - \tan^{-1}\left(\frac{\omega}{\omega_H}\right) \tag{10.49}$$

An example of this phase response is in the next exercise.

The graph in Fig. 10.13 is a representative of a relatively wide-band amplifier. In this figure, $f_H \gg f_L$, and

$$\text{BW} \cong f_H \tag{10.50}$$

The next section explores transfer functions for band-pass amplifiers with much narrower bandwidths.

EXERCISE: Find the midband gain, lower- and upper-cutoff frequencies, and bandwidth of the amplifier with the following transfer function:

$$A_v(s) = -\frac{2 \times 10^7 s}{(s + 100)(s + 50000)}$$

ANSWERS: 52 dB; 15.9 Hz; 7.96 kHz; 7.94 kHz

EXERCISE: Write an expression for the phase of this transfer function. What is the phase shift for $w = 0, 100, 50{,}000$, and ∞?

ANSWERS: $\angle A_v(j\omega) = -90° - \tan^{-1}\left(\dfrac{\omega}{100}\right) - \tan^{-1}\left(\dfrac{\omega}{50000}\right)$; $-90°, -135°, -225°, -270°$

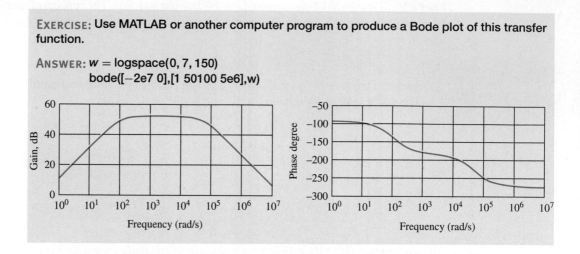

EXERCISE: Use MATLAB or another computer program to produce a Bode plot of this transfer function.

ANSWER: **w** = logspace(0, 7, 150)
bode([−2e7 0],[1 50100 5e6],w)

10.7.5 NARROW-BAND OR HIGH-Q BAND-PASS AMPLIFIERS

In Fig. 10.13 and the last exercise, the values of ω_H and ω_L were widely separated—that is, $\omega_H \gg \omega_L$. In applications such as the radio frequency (RF) amplifiers mentioned in Sec. 10.1, for example, a band-pass amplifier with a very narrow bandwidth is desired, as depicted in Fig. 10.14. For example, such a circuit is used to select one radio station from another in radio receiver applications.

In this case, the maximum gain A_o occurs at the **center frequency** ω_o of the amplifier and decreases rapidly by 3 dB at ω_H and ω_L. The bandwidth is again defined by

$$\text{BW} = f_H - f_L = \frac{\omega_H - \omega_L}{2\pi} \tag{10.51}$$

but now the bandwidth is a small fraction of center frequency ω_o. The width of this region is determined by the amplifier's Q, defined as

$$Q = \frac{\omega_o}{\omega_H - \omega_L} = \frac{f_o}{f_H - f_L} = \frac{f_o}{\text{BW}} \qquad \text{or} \qquad \text{BW} = \frac{f_o}{Q} \tag{10.52}$$

The values of ω_H and ω_L in the transfer function described by Eq. (10.43) are tacitly assumed to be real numbers. However, to achieve high Q, the poles of the band-pass amplifier are complex,

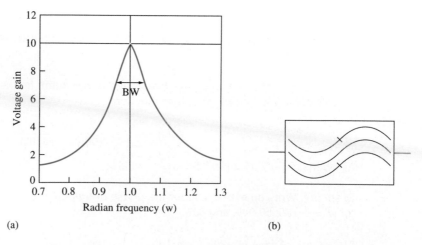

(a) (b)

Figure 10.14 (a) Band-pass amplifier with a high Q. For this graph, $A_o = 10$, $\omega_o = 1$, $\omega_L = 0.95$, $\omega_H = 1.05$, and $Q = 10$. Note the use of linear scales on both axes. (b) Band-pass filter symbol.

and the amplifier transfer function is described by

$$A_v(s) = A_o \frac{s\dfrac{\omega_o}{Q}}{s^2 + s\dfrac{\omega_o}{Q} + \omega_o^2} \tag{10.53}$$

The graph in Fig. 10.14 plots the magnitude of this transfer function versus frequency for $A_o = 10$, $\omega_o = 1$, $\omega_L = 0.95$, $\omega_H = 1.05$, and $Q = 10$ and clearly exhibits a sharp peak centered at $\omega = \omega_o$.

The phase of this transfer function changes rapidly near the center frequency ω_o. Writing the expression for the phase yields

$$\angle A_v(j\omega) = \angle A_o + 90° - \tan^{-1}\left(\frac{1}{Q}\frac{\omega\,\omega_o}{\omega_o^2 - \omega^2}\right) \tag{10.54}$$

If we assume that A_o is positive, then $\angle A_v(j\omega) = 90°$ for $\omega \ll \omega_o$. The phase is $0°$ at $\omega = \omega_o$, and reaches $-90°$ for $\omega \gg \omega_o$.

EXERCISE: Find the midband gain, center frequency, bandwidth, and **Q** of the amplifier with the following transfer function:

$$A_v(s) = \frac{6 \times 10^{13}s}{s^2 + 2 \times 10^5 s + 10^{14}}$$

ANSWERS: 30; 1.59 MHz; 31.8 kHz; 50

10.7.6 BAND-REJECTION AMPLIFIERS

In some cases, we need an amplifier that can reject a narrow band of frequencies, as depicted in Fig. 10.15. For example, such a circuit is useful in eliminating an interfering signal that is near a desired signal in communications receiver applications. This transfer function is referred to as a **band-reject characteristic** or as a **notch filter** and exhibits a sharp null at the center frequency ω_o. At frequencies well removed from ω_o, the gain approaches A_o. To achieve a sharp null, the transfer function has a pair of zeros on the $j\omega$ axis at the notch frequency ω_o, and the poles of the amplifier are complex:

$$A_v(s) = A_o \frac{s^2 + \omega_o^2}{s^2 + s\dfrac{\omega_o}{Q} + \omega_o^2} \tag{10.55}$$

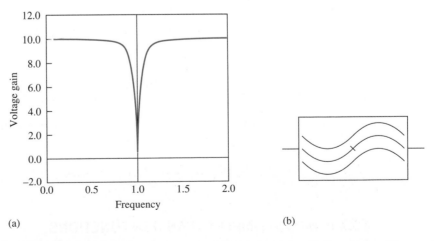

(a)

(b)

Figure 10.15 (a) Band-rejection transfer function for $A_o = 10$, $\omega_o = 1$, and $Q = 10$. Note the use of linear scales on both axes. (b) Band-rejection filter symbol.

Figure 10.15 is the Bode magnitude plot for Eq. (10.55) for the case $A_o = 10$, $\omega_o = 1$, and $Q = 10$. Note the sharp null at $\omega = \omega_o$. Details of the phase response of this transfer function are left for the next exercise. However, it will be discovered to change abruptly by $180°$ near $\omega = \omega_o$.

EXERCISE: Write the equation for the transfer function for the notch filter in Fig. 10.15.

ANSWER: $A_v(s) = 10\dfrac{s^2 + 1}{s^2 + 0.1s + 1}$

EXERCISE: Write expressions for the magnitude and phase of Eq. (10.55). Plot a graph of the phase versus frequency.

ANSWERS:

$$|A_v(j\omega)| = |A_o|\frac{\left|1 - \left(\dfrac{\omega}{\omega_o}\right)^2\right|}{\sqrt{\left(1 - \left(\dfrac{\omega}{\omega_o}\right)^2\right)^2 + \left(\dfrac{\omega}{Q\,\omega_o}\right)^2}}$$

$$\angle A_v(j\omega) = \angle A_o + \angle\left(\omega_o^2 - \omega^2\right) - \tan^{-1}\left(\frac{1}{Q}\frac{\omega\,\omega_o}{\omega_o^2 - \omega^2}\right)$$

10.7.7 THE ALL-PASS FUNCTION

As may be inferred from the name, an all-pass **network** transfer function has a uniform magnitude response at all frequencies. This unusual function can be used to tailor the phase characteristics of a signal. A simple example of the all-pass transfer function is

$$A_v(s) = A_o\frac{s - \omega_o}{s + \omega_o} \tag{10.56}$$

For positive A_o, $|A_v(j\omega)| = A_o$ and $\angle A_v(j\omega) = -2\tan^{-1}\dfrac{\omega}{\omega_o}$.

EXERCISE: Verify the results above for $|A_v(j\omega)|$ and $\angle A_v(j\omega)$.

10.7.8 MORE COMPLEX TRANSFER FUNCTIONS

Most amplifiers have much more complex transfer functions than those indicated in Figs. 10.10 to 10.15, but these can be built up as products of the basic functions given in earlier sections. An

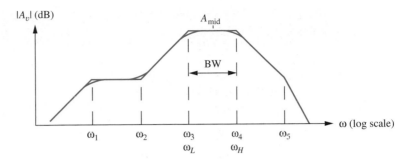

Figure 10.16 An amplifier transfer function with multiple poles and zeros.

example of an amplifier with four poles and four zeros (two at infinity) is

$$A_v(s) = \frac{Ks(s+\omega_2)}{(s+\omega_1)(s+\omega_3)(s+\omega_4)(s+\omega_5)} = \frac{A_{mid}s(s+\omega_2)}{(s+\omega_1)(s+\omega_3)\left(\dfrac{s}{\omega_4}+1\right)\left(\dfrac{s}{\omega_5}+1\right)}$$

(10.57)

which has the Bode plot in Fig. 10.16.

As can be observed, this amplifier has two frequency ranges in which the gain is approximately constant. The midband region is always defined as the region of highest gain, and the cutoff frequencies are defined in terms of the midband gain:

$$|A_v(j\omega_L)| = \frac{A_{mid}}{\sqrt{2}} = |A_v(j\omega_H)|$$

(10.58)

In Fig. 10.16, the poles and zeros are widely spaced, so $\omega_L \cong \omega_3$, $\omega_H \cong \omega_4$, and the bandwidth is BW $\cong f_4 - f_3 = (\omega_4 - \omega_3)/2\pi$.

If the critical frequencies are not widely spaced, then the poles and zeros interact and the process for determining the upper- and lower-cutoff frequencies can become substantially more complicated. As a simple example, consider a low-pass transfer function with two identical poles:

$$A_v(s) = \frac{A_o\omega_1^2}{(s+\omega_1)^2} \qquad \text{for which } A_v(0) = A_o$$

(10.59)

The upper-cutoff frequency is defined by

$$|A_v(j\omega_H)| = \frac{A_o}{\sqrt{2}} \qquad \text{or} \qquad \frac{A_o}{\left[\sqrt{1+\left(\dfrac{\omega_H}{\omega_1}\right)^2}\right]^2} = \frac{A_o}{\sqrt{2}}$$

(10.60)

Solving for ω_H yields

$$\omega_H = 0.644\omega_1$$

(10.61)

The cutoff frequency of the two-pole transfer function is only 64 percent that of the single-pole function. This "bandwidth shrinkage" in multipole amplifiers is discussed in more detail in Chapter 12. Approximation techniques for finding ω_L and ω_H in more complex cases are discussed in Chapter 17.

EXERCISE: Find the midband gain, lower- and upper-cutoff frequencies, and bandwidth of the amplifier with the following transfer function:

$$A_v(s) = \frac{6.4 \times 10^{12}\pi^2}{(s+200\pi)(s+80000\pi)^2}$$

ANSWERS: 1000; 100 Hz; 25.6 kHz; 25.5 kHz

ELECTRONICS IN ACTION

Player Characteristics

The headphone amplifier in a personal music player represents an everyday example of a basic audio amplifier. The traditional audio band spans the frequencies from 20 Hz to 20 kHz, a range that extends beyond the hearing capability of most individuals at both the upper and lower ends.

Apple iPods
Copyright © 2006
Apple Computer, Inc.
All Rights Reserved.

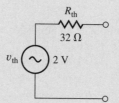

Thévenin equivalent
circuit for output stage

The characteristics of the Apple iPod in the accompanying figure are representative of a high quality audio output stage in an MP3 player or a computer sound card. The output can be represented by a Thévenin equivalent circuit with $v_{th} = 2$ V and $R_{th} = 32$ ohms, and the output stage is designed to deliver a power of approximately 15 mW into each channel of a headphone with a matched impedance of 32 ohms. The output power is approximately constant over the 20 Hz–20 kHz frequency range. At the lower and upper cutoff frequencies, f_L and f_H, the output power will be reduced by 3 dB, a factor of 2.

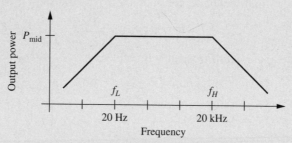

Power versus frequency for an audio amplifier

The distortion characteristics of the amplifier are also important, and this is an area that often distinguishes one sound card or MP3 player from another. A good audio system will have a total harmonic distortion (THD) specification of less than 0.1 percent at full power.

SUMMARY

- *Analog amplifiers:* This chapter introduced important amplifier characteristics including voltage gain A_v, current gain A_i, power gain A_P, input resistance, and output resistance. Gains are expressed in terms of the phasor representations of sinusoidal signals or as transfer functions using Laplace transforms. The magnitudes of the gains are often expressed in terms of the logarithmic decibel or dB scale.

- *Biasing:* It was demonstrated that bias must be provided to an amplifier to ensure that it operates in its linear region. The choice of bias point of the amplifier, its Q-point, can affect both the gain of the amplifier and the size of the input signal range for which linear amplification will occur. Improper choice of bias point can lead to nonlinear operation of an amplifier and distortion of the signal. One measure of linearity of a signal is its percent total harmonic distortion (THD).

- *Two-port representations:* Linear amplifiers can be conveniently modeled using two-port representations. The *g*-parameters are of particular interest for describing amplifiers in this text. In most of the amplifiers we consider, the 1–2 parameter (i.e., g_{12}) will be neglected. These networks were recast in terms of input resistance R_{in}, output resistance R_{out}, and open-circuit voltage gain A. Ideal voltage amplifiers have $R_{\mathrm{in}} = \infty$ and $R_{\mathrm{out}} = 0$.

- *Frequency response:* Linear amplifiers can be used to tailor the magnitude and/or phase of sinusoidal signals and are often characterized by their frequency response. Low-pass, high-pass, band-pass, band-reject (or notch), and all-pass characteristics were discussed. The characteristics of these amplifiers are conveniently displayed in graphical form as a Bode plot, which presents the magnitude (in dB) and phase (in degrees) of a transfer function versus a logarithmic frequency scale. Bode plots can be created easily using MATLAB.

  In an amplifier, the midband gain A_{mid} represents the maximum gain of the amplifier. At the upper- and lower-cutoff frequencies—f_H and f_L, respectively—the voltage gain is equal to $A_{\mathrm{mid}}/\sqrt{2}$ and is 3 dB below its midband value ($20 \log |A_{\mathrm{mid}}|$). The bandwidth of the amplifier extends from f_L to f_H and is defined as BW $= f_H - f_L$. Narrow band-pass amplifiers are characterized in terms of $Q = f_o/\mathrm{BW}$, in which f_o is the center frequency of the band-pass circuit.

KEY TERMS

All-pass network	Lower half-power point
Audio frequency (AF)	Lower -3 dB frequency
Band-pass amplifier	Magnitude
Band-reject amplifier	Midband gain
Bandwidth (BW)	Noninverting amplifier
Bias	Notch filter
Bode plot	Open-circuit input conductance
Center frequency	Open-circuit input resistance
Critical frequency	Open-circuit termination
Current amplifier	Open-circuit voltage gain
Current gain (A_i)	Phase angle
Decibel (dB)	Phasor representation
Gain-bandwidth product	Power gain (A_P)
g-parameters	Q
High-pass amplifier	Radio frequency (RF)
Ideal current amplifier	Short-circuit output conductance
Ideal voltage amplifier	Short-circuit output resistance
Input resistance (R_{in})	Short-circuit termination
Intermediate frequency (IF)	Source resistance (R_S)
Inverting amplifier	Total harmonic distortion
Linear amplifier	Transfer function
Low-pass amplifier	Two-port network
Lower-cutoff frequency	Upper-cutoff frequency

Upper half-power point
Upper -3-dB frequency
Very high frequency (VHF)

Voltage amplifier
Voltage gain (A_v)

REFERENCES

1. Tom Lewis, *Empire on the Air: The Men Who Made Radio,* Harper Collins: 1991.
2. James A. Hijiya, *Lee de Forest and the Fatherhood of Radio,* Lehigh University Press: 1992.
3. Thomas H. Lee, "A Non Linear History of Radio," Chapter 1 in *The Design of CMOS Radio-Frequency Integrated Circuits,* Cambridge University Press: 1998.
4. National Geographic Society, *Those Inventive Americans,* (Publisher and Editor, 1971). pp. 182–187 (Lee de Forest by Howard J. Lewis).

PROBLEMS

10.1 An Example of an Analog Electronic System

10.1. In addition to those given in the introduction, list 15 physical variables in your everyday life that can be represented as continuous analog signals.

10.2 Amplification

10.2. Convert the following to decibels: (a) voltage gains of 120, -60, 50,000, $-100,000$, 0.90; (b) current gains of 600, 3000, -10^6, 200,000, 0.95; (c) power gains of 2×10^9, 4×10^5, 6×10^8, 10^{10}.

10.3. Suppose the input and output voltages of an amplifier are given by

$$v_S = 1 \sin(1000\pi t) + 0.333 \sin(3000\pi t)$$
$$+ 0.200 \sin(5000\pi t) \text{ V}$$

and

$$v_O = 2 \sin\left(1000\pi t + \frac{\pi}{6}\right)$$
$$+ \sin\left(3000\pi t + \frac{\pi}{6}\right)$$
$$+ \sin\left(5000\pi t + \frac{\pi}{6}\right) \text{ V}$$

(a) Plot the input and output voltage waveforms of v_S and v_O for $0 \le t \le 4$ ms. (b) What are the amplitudes, frequencies, and phases of the individual signal components in v_S? (c) What are the amplitudes, frequencies, and phases of the individual signal components in v_O? (d) What are the voltage gains at the three frequencies? (e) Is this a linear amplifier?

10.4. What are the voltage gain, current gain, and power gain required of the amplifier in Fig. 10.3 if $V_s = 2.5$ mV and the desired output power is 40 W?

10.5. What are the voltage gain, current gain, and power gain required of the amplifier in Fig. 10.3 if $V_s = 10$ mV, $R_S = 2$ kΩ, and the output power is 20 mW?

10.6. The output of a PC sound card was set to be a 1-kHz sine wave with an amplitude of 1 V using MATLAB®. The outputs were monitored with an oscilloscope and ac voltmeter. (a) For the left channel, the rms value of the open-circuit output voltage at 1 kHz was measured to be 0.768 V, and it dropped to 0.721 V with a 430-Ω load resistor attached. Draw the Thévenin equivalent circuit representation for the left output of the sound card (i.e., What are v_{th} and R_{th}?). (b) For the right channel, the rms value of the open-circuit output voltage at 1 kHz was measured to be 0.760 V, and it dropped to 0.740 V with a 1040-Ω load resistor attached. Draw the Thévenin equivalent circuit representation for the output of the right channel. (c) What were the values of the measured amplitudes of the two open-circuit output voltages? What percent error was observed between the actual voltage and the desired voltage as defined by MATLAB®?

10.7. Go to the lab and determine the Thévenin equivalent output voltage and resistance for the sound card in your laptop PC (see Prob. 10.6).

10.8. Suppose that each output channel of a computer's sound card can be represented by a 1-V ac source in series with a 32-Ω resistor. Each channel of the amplifier in the external speakers has an input resistance of 20 kΩ, and must deliver 20 W into a 8-Ω speaker. (a) What are the voltage gain, current gain, and power gain required of the amplifier? (b) What would be a reasonable dc power supply voltage for this amplifier?

10.9. The amplifier in a battery-powered device is being designed to deliver 0.1 W to a set of headphones. The impedance of the headphones can be chosen to be 8 Ω, 24 Ω, or 1000 Ω. Calculate the voltage and current required to deliver 0.1 W to each of these resistances. Which resistance seems the best choice for a battery powered application?

10.3 Amplifier Biasing for Linear Operation

10.10. The circuit inside the box in Fig. P10.10 contains only resistors and diodes. The terminal V_O is connected to some point in the circuit inside the box. (a) Is the largest possible value of V_O most nearly 0 V, −6 V, +6 V, or +15 V? Why? (b) Is the smallest possible value of V_O most nearly 0 V, −9 V, +6 V, or +15 V? Why?

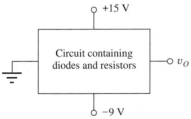

+15 V

Circuit containing diodes and resistors — v_O

−9 V

Figure P10.10

10.11. (a) The input voltage applied to the amplifier in Fig. P10.11 is $v_I = V_B + V_M \sin 1000t$. What is the voltage gain of the amplifier for small values of V_M if $V_B = 0.6$ V? What is the maximum value of V_M that can be used and still have an undistorted sinusoidal signal at v_O? (b) Write expressions for $v_I(t)$ and $v_O(t)$.

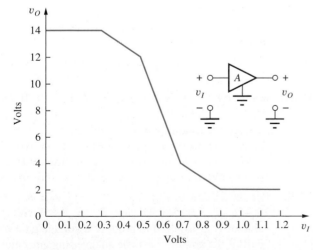

Figure P10.11

10.12. (a) Repeat Prob. 10.11 for $V_B = 0.5$ V. (b) For $V_B = 1.1$ V.

10.13. (a) Repeat Prob. 10.11 for $V_B = 0.8$ V. (b) For $V_B = 0.2$ V.

10.14. The input voltage applied to the amplifier in Fig. P10.11 is $v_I = (0.6 + 0.1 \sin 1000t)$ V. (a) Write expressions for the output voltage. (b) Draw a graph of two cycles of the output voltage. (c) Calculate the first five spectral components of this signal. You may use MATLAB or other computer analysis tools.

*10.15. The input voltage applied to the amplifier in Fig. P10.11 is $v_I = (0.5 + 0.1 \sin 1000t)$ V. (a) Write expressions for the output voltage. (b) Draw a graph of two cycles of the output voltage. (c) Calculate the first five spectral components of this signal. You may use MATLAB or other computer analysis tools.

10.4 Distortion in Amplifiers

10.16. The input signal to an audio amplifier is a 1-kHz sine wave with an amplitude of 4 mV, and the output is described by $v_O = (5 \sin 2000\pi t + 0.25 \sin 6000\pi t + 0.10 \sin 10,000\pi t)$ V. What is the voltage gain of the amplifier? What order harmonics are present in the signal? What is the total harmonic distortion in the output signal?

10.17. The input signal to an audio amplifier is described by $v_S = (0.5 + 0.25 \sin 1200\pi t)$ V, and the output is described by $v_O = (2 + 4 \sin 1200\pi t + 0.4 \sin 2400\pi t + 0.2 \sin 3600\pi t)$ V. What is the voltage gain of the amplifier? What order harmonics are present in the signal? What is the total harmonic distortion in the output signal?

10.18. (a) Use the FFT capability of MATLAB to find the Fourier series representation of $v(t)$ in Fig. 10.5(b). (b) Use MATLAB to find the coefficients of the first three terms of the Fourier series for $v(t)$ by evaluating the integral expression for the coefficients.

10.19. MATLAB limits the output of a sound signal to unity (1 V). Any signal value above this limit will be clipped (set to 1). (a) Use MATLAB to plot the following waveform: $y = \max(-1, \min(1, 1.5 \sin(1400\pi t)))$. (b) Use MATLAB to find the total harmonic distortion in waveform y. (c) Use the sound output on your computer to listen to and compare the following signals: $y = 1 \sin 1400\pi t$, $y = 1.5*\sin 1400\pi t$, and

$y = \max(-1, \min(1, 1.5\sin(1400\pi t)))$. Describe what you hear.

10.5 Two-Port Models for Amplifiers

10.20. Calculate the g-parameters for the circuit in Fig. P10.20.

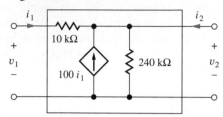

Figure P10.20

10.21. (a) Use SPICE transfer function analysis to find the g-parameters for the circuit in Fig. P10.20.

10.22. Calculate the g-parameters for the circuit in Fig. P10.22.

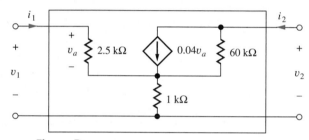

Figure P10.22

10.23. Calculate the g-parameters for the circuit in Fig. P10.23.

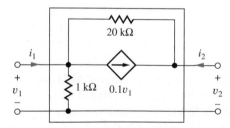

Figure P10.23

10.24. Use SPICE transfer function analysis to find the g-parameters for the circuit in Fig. P10.23.

10.6 Mismatched Source and Load Resistances

10.25. An amplifier connected in the circuit in Fig. P10.25 has the two-port parameters listed below, with $R_S = 1\ k\Omega$ and $R_L = 16\ \Omega$. (a) Find the overall voltage gain A_v, current gain A_i, and power gain A_P for the amplifier and express the results in dB. (b) What is

the amplitude V_s of the sinusoidal input signal v_s needed to deliver 1 W to the 16-Ω load resistor? (c) How much power is dissipated in the amplifier when 1 W is delivered to the load resistor?

$$\text{Input resistance } R_{\text{in}} = 1\ M\Omega$$
$$\text{Output resistance } R_{\text{out}} = 0.5\ \Omega$$
$$A = 54\ dB$$
$$v_s = V_s \sin \omega t$$

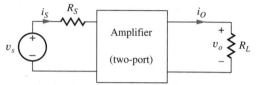

Figure P10.25

10.26. Suppose that the amplifier of Fig. P10.25 has been designed to match the source and load resistances with the parameters below. (a) What is the amplitude of the input signal v_s needed to deliver 1 W to the 16-Ω load resistor? (b) How much power is dissipated in the amplifier when 1 W is delivered to the load resistor?

$$\text{Input resistance } R_{\text{in}} = 1\ k\Omega$$
$$\text{Output resistance } R_{\text{out}} = 16\ \Omega$$
$$A = 54\ dB$$

10.27. The headphone amplifier in a battery-powered device has an output resistance of 28 Ω and is designed to deliver 0.1 W to the headphones. If the resistance of the headphones is 24 Ω, calculate the voltage and current required from the dependent voltage source (A_v in our model) to deliver 0.1 W to the headphones. How much power is delivered from dependent source? How much power is lost in the output resistance?

10.28. Repeat Prob. 10.27 if the headphones have a resistance of 1000 Ω.

10.29. For the circuit in Fig. 10.9, $R_S = 1\ k\Omega$, $R_L = 16\ \Omega$, and $A = -2000$. What values of R_{in} and R_{out} will produce maximum power in the load resistor R_L? What is the maximum power that can be delivered to R_L if v_s is a sine wave with an amplitude of 10 mV? What is the power gain of this amplifier?

10.30. For the circuit in Fig. 10.9, $R_S = 1\ k\Omega$, $R_{\text{in}} = 20\ k\Omega$, $R_{\text{out}} = 100\ \Omega$, and $R_L = 2\ k\Omega$. What value of A is required to produce a voltage gain of 77 dB if the amplifier is to be an inverting amplifier ($\theta = 180°$)?

10.31. The circuit in Fig. P10.31 represents a two-port model for a current amplifier. Write expressions for input current i_1, output current i_o and the current gain $A_i = I_o/I_s$. What values of R_{in} and R_{out} provide maximum magnitude for the current gain?

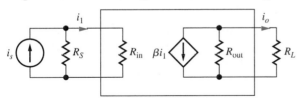

Figure P10.31

10.32. For the circuit in Fig. P10.31, $R_S = 200$ kΩ, $R_{\text{in}} = 10$ kΩ, $R_{\text{out}} = 300$ kΩ, and $R_L = 47$ kΩ. What value of β is required to produce a current gain of 200?

10.33. For the circuit in Fig. P10.31, $R_S = 100$ kΩ, $R_L = 10$ kΩ, and $\beta = 5000$. What values of R_{in} and R_{out} will produce maximum power in the load resistor R_L? What is the maximum power that can be delivered to R_L if i_s is a sine wave with an amplitude of 1 μA? What is the power gain of this amplifier?

10.34. Two amplifiers are connected in series, or cascaded, in the circuit in Fig. P10.34. If $R_S = 1$ kΩ, $R_{\text{in}} = 5$ kΩ, $R_{\text{out}} = 500$ Ω, $R_L = 100$ Ω, and $A = -1200$, what are the voltage gain, current gain, and power gain of the overall amplifier?

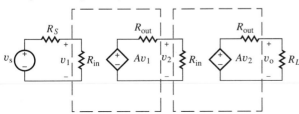

Figure P10.34

*10.35. For the circuit in Fig. 10.9, show that

$$A_{P\text{dB}} = A_{v\text{dB}} - 10\log\left(\frac{R_L}{R_S + R_{\text{in}}}\right)$$

and

$$A_{P\text{dB}} = A_{i\text{dB}} + 10\log\left(\frac{R_L}{R_S + R_{\text{in}}}\right)$$

10.7 Amplifier Transfer Functions and Frequency Response

*10.36. Find the poles and zeros of the following transfer functions:

(a) $A_i(s)$

$$= -\frac{3 \times 10^9 s^2}{(s^2 + 51s + 50)(s^2 + 13{,}000s + 3 \times 10^7)}$$

*(b) $A_v(s)$

$$= -\frac{10^5(s^2 + 51s + 50)}{s^5 + 1000s^4 + 50{,}000s^3 + 20{,}000s^2 + 13{,}000s + 3 \times 10^7}$$

10.37. Find the midband gain in dB and the upper cutoff frequency for the low-pass filter in Ex. 10.3 if $R_1 = 1$ kΩ, $R_2 = 1.5$ kΩ, and $C = 0.01$ μF.

10.38. Find the midband gain in dB and the upper cutoff frequency for the low-pass filter in Ex. 10.3 if $R_1 = 10$ kΩ, $R_2 = 100$ kΩ, and $C = 0.01$ μF.

10.39. (a) Design a low-pass filter using the circuit in Ex. 10.3 to provide a loss of no more than 0.5 dB at low frequencies and a cutoff frequency of 20 kHz if $R_1 = 560$ Ω. (b) Pick standard values from the tables in Appendix A.

10.40. Find the midband gain in dB and the upper cutoff frequency for the high-pass filter in Ex. 10.4 if $R_1 = 8.2$ kΩ, $R_2 = 20$ kΩ, and $C = 0.01$ μF.

10.41. Find the midband gain in dB and the upper cutoff frequency for the high-pass filter in Ex. 10.4 if $R_1 = 10$ kΩ, $R_2 = 78$ kΩ, and $C = 0.01$ μF.

10.42. (a) Design a high-pass filter using the circuit in Ex. 10.4 to provide a loss of no more than 0.5 dB at high frequencies and a cutoff frequency of 20 kHz if $R_1 = 330$ Ω. (b) Pick standard values from the tables in Appendix A.

10.43. What are A_{mid} in dB, f_H, f_L, and the BW in Hz for the amplifier described by

$$A_v(s) = \frac{2\pi \times 10^7 s}{(s + 20\pi)(s + 2\pi \times 10^4)}$$

What type of amplifier is this?

10.44. What are A_{mid} in dB, f_H, f_L, and the BW in Hz for the amplifier described by

$$A_v(s) = \frac{10^4 s}{s + 200\pi}$$

What type of amplifier is this?

10.45. What are A_{mid} in dB, f_H, f_L, and the BW in Hz for the amplifier described by

$$A_v(s) = \frac{2\pi \times 10^6}{s + 200\pi}$$

What type of amplifier is this?

10.46. What are A_{mid} in dB, f_H, f_L, the BW in Hz, and the Q for the amplifier described by

$$A_v(s) = -\frac{10^7 s}{s^2 + 10^5 s + 10^{14}}$$

What type of amplifier is this?

10.47. What are A_mid in dB, f_H, f_L, and the BW in Hz for the amplifier described by

$$A_v(s) = -20\frac{s^2 + 10^{12}}{s^2 + 10^4 s + 10^{12}}$$

What type of amplifier is this?

*10.48. What are A_mid in dB, f_H, f_L, and the BW in Hz for the amplifier described by

$A_v(s)$

$$= \frac{4\pi^2 \times 10^{14} s^2}{(s + 20\pi)(s + 50\pi)(s + 2\pi \times 10^5)(s + 2\pi \times 10^6)}$$

What type of amplifier is this?

10.49. Use MATLAB, a spreadsheet, or other computer program to generate a Bode plot of the magnitude and phase of the transfer function in Prob. 10.43.

10.50. Use MATLAB, a spreadsheet, or other computer program to generate a Bode plot of the magnitude and phase of the transfer function in Prob. 10.44.

10.51. Use MATLAB, a spreadsheet, or other computer program to generate a Bode plot of the magnitude and phase of the transfer function in Prob. 10.45.

10.52. Use MATLAB, a spreadsheet, or other computer program to generate a Bode plot of the magnitude and phase of the transfer function in Prob. 10.46.

10.53. Use MATLAB, a spreadsheet, or other computer program to generate a Bode plot of the magnitude and phase of the transfer function in Prob. 10.47.

10.54. Use MATLAB, a spreadsheet, or other computer program to generate a Bode plot of the magnitude and phase of the transfer function in Prob. 10.48.

10.55. The voltage gain of an amplifier is described by the transfer function in Prob. 10.43 and has an input $v_s = 0.002 \sin \omega t$ V. Write an expression for the amplifier's output voltage at a frequency of (a) 5 Hz, (b) 500 Hz, (c) 50 kHz.

10.56. The voltage gain of an amplifier is described by the transfer function in Prob. 10.44 and has an input $v_s = 0.3 \sin \omega t$ mV. Write an expression for the amplifier's output voltage at a frequency of (a) 1 Hz, (b) 50 Hz, (c) 5 kHz.

10.57. The voltage gain of an amplifier is described by the transfer function in Prob. 10.44 and has an input $v_s = 10 \sin \omega t$ μV. Write an expression for the amplifier's output voltage at a frequency of (a) 2 Hz, (b) 2 kHz, (c) 200 kHz.

10.58. The voltage gain of an amplifier is described by the transfer function in Prob. 10.46 and has an input $v_s = 0.004 \sin \omega t$ V. Write an expression for the amplifier's output voltage at a frequency of (a) 1.59 MHz, (b) 1 MHz, (c) 5 MHz.

10.59. The voltage gain of an amplifier is described by the transfer function in Prob. 10.47 and has an input $v_s = 0.25 \sin \omega t$ V. Write an expression for the amplifier's output voltage at a frequency of (a) 159 kHz, (b) 50 kHz, (c) 200 kHz.

10.60. The voltage gain of an amplifier is described by the transfer function in Prob. 10.48 and has an input $v_s = 0.002 \sin \omega t$ V. Write an expression for the amplifier's output voltage at a frequency of (a) 5 Hz, (b) 500 Hz, (c) 50 kHz.

10.61. (a) Write an expression for the transfer function of a low-pass voltage amplifier with a gain of 26 dB and $f_H = 5$ MHz. (b) Repeat if the amplifier exhibits a phase shift of $180°$ at $f = 0$.

10.62. (a) Write an expression for the transfer function of a voltage amplifier with a gain of 40 dB, $f_L = 200$ Hz, and $f_H = 100$ kHz. (b) Repeat if the amplifier exhibits a phase shift of $180°$ at $f = 0$.

10.63. Make a Bode plot of the transfer function in Eq. (10.59) if $A_o = -1000$ and $\omega_1 = 50{,}000\pi$. What are A_mid and ω_H? What is the slope of the magnitude plot for $\omega \gg \omega_H$ in dB/dec?

10.64. (a) What is the bandwidth of the low-pass amplifier described by

$$A_v(s) = A_o \left(\frac{\omega_1}{s + \omega_1}\right)^3$$

if $A_o = -2000$ and $\omega_1 = 50{,}000\pi$. (b) Make a Bode plot of this transfer function. What is the slope of the magnitude plot for $\omega \gg \omega_H$ in dB/dec?

*10.65. The input to an all-pass amplifier with a gain of 10 dB is

$$v_S = 1 \sin(1000\pi t) + 0.333 \sin(3000\pi t)$$
$$+ 0.200 \sin(5000\pi t)\ \text{V}$$

(a) If the phase shift of the amplifier at 500 Hz is $10°$, what must be the phase shift at the other two frequencies if the shape of the output waveform is to be the same as that of the input waveform? Write an expression for the output signal. (b) Use the computer to check your answer by plotting the input and output waveforms.

CHAPTER 11

IDEAL OPERATIONAL AMPLIFIERS

CHAPTER OUTLINE

CHAPTER GOALS

In Chapter 11, we hope to achieve these goals:

- Understand behavior and characteristics of ideal differential and operational amplifiers

- Demonstrate techniques used to analyze circuits containing both ideal op amps

- Understand the techniques used to determine voltage gain, input resistance, and output resistance of general amplifier circuits

- Fully characterize classic op amp circuits, including the inverting, noninverting, summing, and instrumentation amplifiers, the voltage follower, and the integrator

- Learn many of the factors that must be considered in the design of circuits using operational amplifiers

- Provide an introduction to active filters

- Explore applications of op amps in nonlinear circuits including precision rectifiers

- Provide examples of multivibrator circuits that employ positive feedback

- Demonstrate ac analysis capability of SPICE

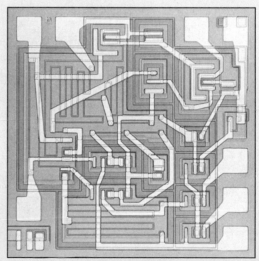

μA709 operational amplifier die photograph.
(Photo courtesy of Fairchild Semiconductor International.)

The **operational amplifier,** or **op amp,** is a fundamental building block of analog circuit design. The name "operational amplifier" originates from the use of this type of amplifier to perform specific electronic circuit functions or operations, such as scaling, summation, and integration, in analog computers.

Integrated circuit operational amplifiers evolved rapidly following development of the first bipolar integrated circuit processes in the 1960s. Although early IC amplifier designs offered little if any performance improvements over tube-type designs and discrete semiconductor realizations and were somewhat "delicate," they offered significant advantages in physical size, cost, and power consumption. The μA709, introduced by Fairchild Semiconductor in 1965, was one of the first widely used general-purpose IC operational amplifiers. IC op amp circuits improved quickly, and the now-classic Fairchild μA741 amplifier design, which appeared in the late 1960s, is a robust amplifier with excellent characteristics for general-purpose applications. The internal circuit design of these op amps used 20 to 50 bipolar transistors. Later designs improved performance in most specification areas. Today there is an almost overwhelming array of operational amplifiers from which to choose.

The chapter explores the characteristics of the ideal operational amplifier and ideal op amp circuits. A number of basic circuit applications are discussed, including inverting and noninverting amplifiers, the summing and instrumentation amplifiers, the integrator, and basic filters. Limitations caused by the nonideal behavior of the operational amplifier are discussed in Chapter 12, including finite gain, bandwidth, input and output resistances, common-mode rejection, offset voltage, and bias current.

11.1 THE DIFFERENTIAL AMPLIFIER

Differential amplifiers that respond to the difference of two input signals (and hence are sometimes referred to as difference amplifiers) are an extremely useful class of circuits. For example, they are used as error amplifiers in almost all electronic feedback and control systems, and operational amplifiers themselves are in fact very high performance versions of the differential amplifier. Thus, we begin our study of op amps by exploring the characteristics of the basic differential amplifier shown in schematic form in Fig. 11.1. The amplifier has two inputs, to which the input signals v_+ and v_- are connected, and a single output v_O, all referenced to the common (ground) terminal between the two power supplies V_{CC} and V_{EE}. In most applications, $V_{CC} \geq 0$ and $-V_{EE} \leq 0$, and the voltages are often symmetric—that is, ± 5 V, ± 12 V, ± 15 V, ± 18 V, ± 22 V, and so on. These power supply voltages limit the output voltage range: $-V_{EE} \leq v_O \leq V_{CC}$.

For simplicity, the amplifier is most often drawn without explicitly showing the power supplies, as in Fig. 11.2(a), or the ground connection, as in Fig. 11.2(b)—but we must remember that the power and ground terminals are always present in the implementation of a real circuit.

11.1.1 DIFFERENTIAL AMPLIFIER MODEL

For purposes of signal analysis, the differential amplifier can be represented by its input resistance R_{id}, output resistance R_o, and controlled voltage source Av_{id}, as in Fig. 11.3. This is the two-port

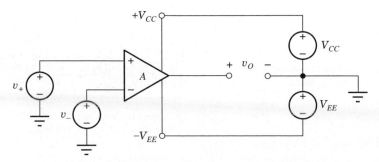

Figure 11.1 The differential amplifier, including power supplies.

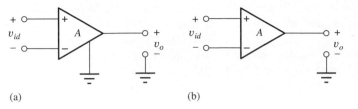

Figure 11.2 (a) Amplifier without power supplies explicitly included. (b) Differential amplifier with implied ground connection.

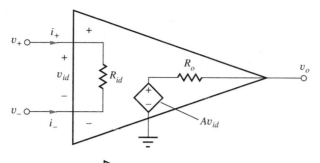

Figure 11.3 Differential amplifier.

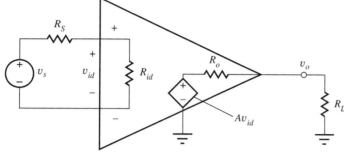

Figure 11.4 Amplifier with source and load attached.

representation introduced to Section 10.5.

$$A = \text{voltage gain (open-circuit voltage gain)}$$
$$v_{id} = (v_+ - v_-) = \text{differential input signal voltage}$$
$$R_{id} = \text{amplifier input resistance}$$
$$R_o = \text{amplifier output resistance}$$

$$(11.1)$$

The signal voltage developed at the output of the amplifier is in phase with the voltage applied to the $+$ input terminal and $180°$ out of phase with the signal applied to the $-$ input terminal. The v_+ and v_- terminals are therefore referred to as the **noninverting input** and **inverting input,** respectively.

In a typical application, the amplifier is driven by a signal source having a Thévenin equivalent voltage v_s and resistance R_S and is connected to a load represented by the resistor R_L, as in Fig. 11.4. For this simple circuit, the input voltage v_{id} and the output voltage can be written in terms of the circuit elements as

$$\mathbf{v_{id}} = \mathbf{v_s}\frac{R_{id}}{R_{id} + R_S} \qquad \text{and} \qquad \mathbf{v_o^*} = A\mathbf{v_{id}}\frac{R_L}{R_o + R_L} \qquad (11.2)$$

Combining Eqs. (11.2) yields an expression for the overall voltage gain of the amplifier circuit in Fig. 11.4 for arbitrary values of R_S and R_L:

$$A_v = \frac{\mathbf{v_o}}{\mathbf{v_s}} = A\frac{R_{id}}{R_S + R_{id}}\frac{R_L}{R_o + R_L} \qquad (11.3)$$

Operational amplifier circuits are most often **dc-coupled amplifiers,** and the signals v_o and v_s may in fact have a dc component that represents a dc shift of the input away from the Q-point. The op amp amplifies not only the ac components of the signal but also this dc component. We must remember that the ratio needed to find A_v, as indicated in Eq. (11.3), is determined by the amplitude and phase of the individual signal components and is not a time-varying quantity, but $\omega = 0$ is a valid signal frequency! Recall from Chapters 1 and 10 that v_s, v_o, i_2 and so on represent our signal voltages and currents and are generally functions of time: $v_s(t), v_o(t), i_2(t)$. But whenever we do algebraic calculations of voltage gain, current gain, input resistance, output resistance, and so on, we must use **phasor** representations of the individual signal components in our calculations: $\mathbf{v_s, v_o, i_2}$. Signals $v_s(t), v_o(t), i_2(t)$ and so on may be composed of many individual signal components, one of which may be a dc shift away from the Q-point value.

EXAMPLE 11.1   VOLTAGE GAIN ANALYSIS

Find the gain of a differential amplifier including the effects of load and source resistance.

PROBLEM   Calculate the voltage gain for an amplifier with the following parameters: $A = 100$, $R_{id} = 100\,\text{k}\Omega$, and $R_o = 100\,\Omega$, with $R_S = 10\,\text{k}\Omega$ and $R_L = 1000\,\Omega$. Express the result in dB.

SOLUTION   **Known Information and Given Data:** $A = 100$, $R_{id} = 100\,\text{k}\Omega$, $R_o = 100\,\Omega$, $R_S = 10\,\text{k}\Omega$, and $R_L = 1000\,\Omega$

Unknown: Voltage gain A_v

Approach: Evaluate the expression in Eq. (11.3). Convert answer to dB.

Assumptions: None

Analysis: Using Eq. (11.3),

$$A_v = 100 \left(\frac{100\,\text{k}\Omega}{10\,\text{k}\Omega + 100\,\text{k}\Omega} \right) \left(\frac{1000\,\Omega}{100\,\Omega + 1000\,\Omega} \right) = 82.6$$

$$A_{vdB} = 20 \log |A_v| = 20 \log |82.6| = 38.3\ \text{dB}$$

Check of Results: We have found the only unknown requested.

Discussion: The amplifier's internal voltage gain capability is $A = 100$, but an overall gain of only 82.6 is being realized because a portion of the signal source voltage ($\cong 9$ percent) is being dropped across R_S, and part of the internal amplifier voltage (Av_{id}) (also $\cong 9$ percent) is being lost across R_o.

Computer-Aided Analysis: The SPICE circuit is shown here, and a transfer function analysis from source VS to the output node is used to characterize the amplifier in this example.

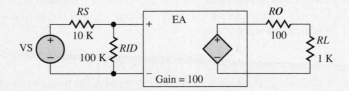

The SPICE results are transfer function $= 82.6$, input resistance $= 110\,\text{k}\Omega$, and output resistance $= 90.9\,\Omega$. A_v equals the value of the transfer function, the resistance at the terminals of VS is the input resistance, and the output resistance represents the total resistance at the output node. The voltage gain agrees with our hand analysis.

11.1.2 THE IDEAL DIFFERENTIAL AMPLIFIER

An ideal differential amplifier would produce an output that depends only on the voltage difference v_{id} between its two input terminals, and this voltage would be independent of source and load resistances. Referring to Eq. (11.3), we see that this behavior can be achieved if the input resistance of the amplifier is infinite and the output resistance is zero (as pointed out previously in Sec. 10.6).

For this case, Eq. (11.3) reduces to

$$\mathbf{v_o} = A\mathbf{v_{id}} \qquad \text{or} \qquad A_v = \frac{\mathbf{v_o}}{\mathbf{v_{id}}} = A \qquad (11.4)$$

and the full amplifier gain is realized. A is referred to as either the **open-circuit voltage gain** or **open-loop gain** of the amplifier and represents the maximum voltage gain available from the device.

As introduced in Chapter 10, we often want to achieve a completely mismatched resistance condition in voltage amplifier applications ($R_{id} \gg R_S$ and $R_o \ll R_L$), so that maximum voltage gain in Eq. (11.4) can be achieved. For the mismatched case, the overall amplifier gain is independent of the source and load resistances, and multiple amplifier stages can be cascaded without concern for interaction between stages.

11.2 THE IDEAL OPERATIONAL AMPLIFIER

As noted earlier, the term "operational amplifier" grew from use of these high-performance amplifiers to perform specific electronic circuit functions or operations, such as scaling, summation, and integration, in analog computers. The operational amplifier used in these applications is an ideal differential amplifier with an additional property: infinite voltage gain. Although it is impossible to realize the **ideal operational amplifier,** its conceptual use allows us to understand the basic performance to be expected from a given analog circuit and serves as a model to help in circuit design. Once the properties of the ideal amplifier and its use in basic circuits are understood, then various ideal assumptions can be removed in order to understand their effect on circuit performance.

The ideal operational amplifier is a special case of the ideal difference amplifier in Fig. 11.3, in which $R_{id} = \infty$, $R_o = 0$, and, most importantly, voltage gain $A = \infty$. Infinite gain leads to the first of two assumptions used to analyze circuits containing ideal op amps. Solving for $\mathbf{v_{id}}$ in Eq. (11.4),

$$\mathbf{v_{id}} = \frac{\mathbf{v_o}}{A} \qquad \text{and} \qquad \lim_{A \to \infty} \mathbf{v_{id}} = 0 \qquad (11.5)$$

If A is infinite, then the input voltage v_{id} will be zero for any finite output voltage. We will refer to this condition as Assumption 1 for ideal op-amp circuit analysis.

An infinite value for the input resistance R_{id} forces the two input currents i_+ and i_- to be zero, which will be Assumption 2 for analysis of ideal op amp circuits. These two results, combined with Kirchhoff's voltage and current laws, form the basis for analysis of all ideal op amp circuits.

EXERCISE: Suppose an amplifier is operating with $v_o = +10$ V. What is the input voltage v_{id} if (a) $A = 100$? (b) $A = 10,000$? (c) $A = 120$ dB?

ANSWERS: (a) 100 mV; (b) 1.00 mV; (c) 10.0 μV

11.2.1 ASSUMPTIONS FOR IDEAL OPERATIONAL AMPLIFIER ANALYSIS

As just described, the two primary assumptions used for analysis of circuits containing ideal op amps are:

$$
\begin{aligned}
&\text{1. Input voltage difference is zero: } v_{id} = 0 \\
&\text{2. Input currents are zero: } i_+ = 0 \text{ and } i_- = 0
\end{aligned}
\qquad (11.6)
$$

Infinite gain and infinite input resistance are the explicit characteristics that lead to Assumptions 1 and 2. However, the ideal operational amplifier actually has quite a number of additional implicit properties, but these assumptions are seldom clearly stated. They are

- Infinite common-mode rejection
- Infinite power supply rejection
- Infinite output voltage range (not limited by $-V_{EE} \le v_O \le V_{CC}$)

- Infinite output current capability
- Infinite open-loop bandwidth
- Infinite slew rate
- Zero output resistance
- Zero input-bias currents and offset current
- Zero input-offset voltage

These terms may be unfamiliar at this point, but they will all be defined and discussed in detail in Chapter 12.

11.3 ANALYSIS OF CIRCUITS CONTAINING IDEAL OPERATIONAL AMPLIFIERS

This section describes a number of classic operational amplifier circuits, including the basic inverting and noninverting amplifiers; the unity-gain buffer, or voltage follower; the summing, difference, and instrumentation amplifiers; the low-pass filter; the integrator; and the differentiator. Analysis of these various circuits demonstrates use of the two ideal op amp assumptions in combination with Kirchhoff's voltage and current laws (KVL and KCL, respectively).

11.3.1 THE INVERTING AMPLIFIER

An **inverting-amplifier** circuit is built by grounding the positive input of the operational amplifier and connecting resistors R_1 and R_2, called the **feedback network,** between the inverting input and the signal source and amplifier output node, respectively, as in Fig. 11.5. We wish to find a set of two-port parameters that characterize the overall circuit, including the open-circuit voltage gain $\mathbf{A_v}$, input resistance R_{in}, and output resistance R_{out}.

Inverting Amplifier Voltage Gain

We begin by determining the voltage gain. To find A_v, we need a relationship between $\mathbf{v_s}$ and $\mathbf{v_o}$, which we can find by writing an equation for the single loop shown in Fig. 11.6.

$$\mathbf{v_s} - \mathbf{i_s}R_1 - \mathbf{i_2}R_2 - \mathbf{v_o} = 0 \tag{11.7}$$

Applying KCL at the inverting input to the amplifier yields a relationship between $\mathbf{i_s}$ and $\mathbf{i_2}$

$$\mathbf{i_s} = \mathbf{i_-} + \mathbf{i_2} \qquad \text{or} \qquad \mathbf{i_s} = \mathbf{i_2} \tag{11.8}$$

since Assumption 2 states that $\mathbf{i_-}$ must be zero. Equation (11.7) then becomes

$$\mathbf{v_s} - \mathbf{i_s}R_1 - \mathbf{i_s}R_2 - \mathbf{v_o} = 0 \tag{11.9}$$

Now, current $\mathbf{i_s}$ can be written in terms of $\mathbf{v_s}$ as

$$\mathbf{i_s} = \frac{\mathbf{v_s} - \mathbf{v_-}}{R_1} \tag{11.10}$$

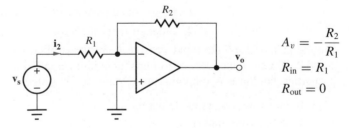

$$A_v = -\frac{R_2}{R_1}$$
$$R_{\text{in}} = R_1$$
$$R_{\text{out}} = 0$$

Figure 11.5 Inverting-amplifier circuit.

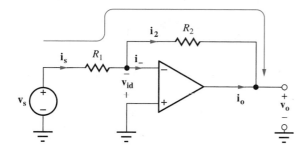

Figure 11.6 Inverting-amplifier circuit.

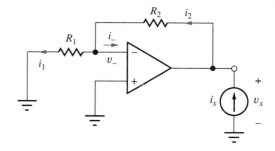

Figure 11.7 Test current applied to the amplifier to determine the output resistance: $R_{\text{out}} = \mathbf{v_x}/\mathbf{i_x}$.

where $\mathbf{v_-}$ is the voltage at the inverting input (negative input) of the op amp. But, Assumption 1 states that the input voltage $\mathbf{v_{id}}$ must be zero, so $\mathbf{v_-}$ must also be zero because the positive input is grounded:

$$\mathbf{v_{id}} = \mathbf{v_+} - \mathbf{v_-} = 0 \qquad \text{but} \qquad \mathbf{v_+} = 0 \qquad \text{so} \qquad \mathbf{v_-} = 0$$

Because $\mathbf{v_-} = 0$, $\mathbf{i_s} = \mathbf{v_s}/R_1$, and Eq. (11.9) reduces to

$$-\mathbf{v_s}\frac{R_2}{R_1} - \mathbf{v_o} = 0 \qquad \text{or} \qquad \mathbf{v_o} = -\mathbf{v_s}\frac{R_2}{R_1} \tag{11.11}$$

The voltage gain is given by

$$A_v = \frac{\mathbf{v_o}}{\mathbf{v_s}} = -\frac{R_2}{R_1} \tag{11.12}$$

Referring to Eq. (11.12), we should note several things. The voltage gain is negative, indicative of an inverting amplifier with a 180° phase shift between dc or sinusoidal input and output signals. In addition, the gain can be greater than or equal to 1 if $R_2 \geq R_1$ (the most common case), but it can also be less than 1 for $R_1 > R_2$.

In the amplifier circuit in Figs. 11.5 and 11.6, the inverting-input terminal of the operational amplifier is at ground potential, 0 V, and is referred to as a **virtual ground.** The ideal operational amplifier adjusts its output to whatever voltage is necessary to force the differential input voltage to zero. However, although the inverting input represents a virtual ground, it is *not* connected directly to ground (there is no direct dc path for current to reach ground). Shorting this terminal to ground for analysis purposes is a common error that must be avoided.

EXERCISE: Find A_v, v_o, i_s, and i_o for the amplifier in Fig. 11.6 if $R_1 = 68$ kΩ, $R_2 = 360$ kΩ, and $v_s = 0.5$ V.

ANSWERS: −5.29, −2.65 V, 7.35 μA, −7.35 μA

Input and Output Resistances of the Ideal Inverting Amplifier

The input resistance R_{in} of the overall amplifier is found directly from Eq. (11.10) with $v_- = 0$ (virtual ground).

$$R_{\text{in}} = \frac{\mathbf{v_s}}{\mathbf{i_s}} = R_1 \tag{11.13}$$

The output resistance R_{out} is the Thévenin equivalent resistance at the output terminal; it is found by applying a test signal current (or voltage) source to the output of the amplifier circuit and determining the voltage (or current), as in Fig. 11.7. All other *independent* voltage and current sources in the circuit must be turned off, and so v_s is set to zero in Fig. 11.7.

The output resistance of the overall amplifier is defined by

$$R_{\text{out}} = \frac{\mathbf{v_x}}{\mathbf{i_x}}$$

(11.14)

Writing a single-loop equation for Fig. 11.7 gives

$$\mathbf{v_x} = \mathbf{i_2} R_2 + \mathbf{i_1} R_1$$

(11.15)

but $\mathbf{i_1} = \mathbf{i_2}$ because $\mathbf{i_-} = 0$ based on op-amp Assumption 2. Therefore,

$$\mathbf{v_x} = \mathbf{i_1}(R_2 + R_1)$$

(11.16)

However, $\mathbf{i_1}$ must be zero because Assumption 1 tells us that $\mathbf{v_-} = 0$. Thus, $\mathbf{v_x} = 0$ independent of the value of $\mathbf{i_x}$, and

$$R_{\text{out}} = 0$$

(11.17)

DESIGN NOTE

For the **ideal inverting amplifier**, the closed-loop voltage gain A_v, input resistance R_{in}, and output resistance R_{out} are:

$$A_v = -\frac{R_2}{R_1} \qquad R_{\text{in}} = R_1 \qquad R_{\text{out}} = 0$$

DESIGN EXAMPLE 11.2   INVERTING AMPLIFIER DESIGN

Design an op amp inverting amplifier to meet a pair of specifications.

PROBLEM   Design an inverting amplifier (i.e., choose the values of R_1 and R_2) to have an input resistance of 20 kΩ and a gain of 40 dB.

SOLUTION   **Known Information and Given Data:** In this case, we are given the values for the gain and input resistance, and the amplifier circuit configuration has also been specified: op amp inverting amplifier topology; voltage gain = 20 dB; $R_{\text{in}} = 20$ kΩ.

Unknowns: Values of R_1 and R_2 required to achieve the specifications

Approach: Based on Eqs. (11.12) and (11.13), we see that the input resistance is controlled by R_1, and the voltage gain is set by R_2/R_1. First find the value of R_1; then use it to find the value of R_2.

Assumptions: The op amp is ideal so that Eqs. (11.12) and (11.13) apply.

Analysis: We must convert the gain from dB before we use it in the calculations:

$$|A_v| = 10^{40\,\text{dB}/20\,\text{dB}} = 100 \qquad \text{and} \qquad A_v = -100$$

The minus sign is added since an inverting amplifier is specified. Using Eqs. (11.13) and Eq. (11.12):

$$R_1 = R_{\text{in}} = 20 \text{ k}\Omega \qquad \text{and} \qquad A_v = -\frac{R_2}{R_1} \rightarrow R_2 = 100 R_1 = 2 \text{ M}\Omega$$

Check of Results: We have found all the answers requested.

Evaluation and Discussion: Looking at Appendix A, we find that 20 kΩ and 2 MΩ represent standard 5 percent resistor values, and our design is complete. (Murphy has been on our side for a change.) Note in this example, that we have two design constraints and two resistors to choose.

Computer-Aided Analysis: In the SPICE circuit shown here, the op amp is modeled by VCVC E1. In SPICE, we cannot set the gain of E1 to infinity. To approximate the ideal op amp, a value of 1E9 is assigned to E1. Remember that R2 = 2 MEG, not 2M = 0.002 Ω! A transfer function analysis from source VS to the output node is used to characterize the gain of the amplifier. A transient analysis gives the output voltage. VS is defined to have zero voltage offset, a 10 mV amplitude and a frequency of 1000 Hz (VS = 0.01 sin 2000πt). The transient solution starts at T = 0, stops at T = 0.003 s and uses a time step of 1 μs.

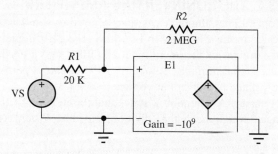

The SPICE results are: transfer function = −100, input resistance = 20 kΩ, and output resistance = 0. These values confirm our design, and the output signal is an inverted 1-V, 1000-Hz sine wave, as expected. Note that the small input signal is present but hard to see on the graph because of the scale.

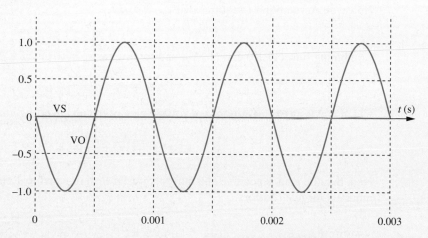

EXERCISE: If V_S = 2 V, R_1 = 4.7 kΩ, and R_2 = 24 kΩ, find I_S, I_2, and V_O in Fig. 11.6. Why is the symbol V_S being used instead of v_s, and so on?

ANSWERS: 20 kΩ, 2 MΩ; 0.426 mA, 0.426 mA, −10.2 V; the problem is stated specifically in terms of dc values.

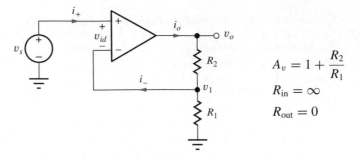

Figure 11.8 Noninverting amplifier configuration.

11.3.2 THE NONINVERTING AMPLIFIER

The operational amplifier can also be used to construct a **noninverting amplifier** with the circuit indicated in the schematic in Fig. 11.8. The input signal is applied to the positive or (noninverting) input terminal of the operational amplifier, and a portion of the output signal is fed back to the negative input terminal. Analysis of the circuit is performed by relating the voltage at v_1 to both input voltage v_s and output voltage v_o.

Because Assumption 2 states that the input current i_- is zero, v_1 can be related to the output voltage through the voltage divider formed by R_1 and R_2:

$$\mathbf{v_1} = \mathbf{v_o}\frac{R_1}{R_1 + R_2} \tag{11.18}$$

Writing an equation around the loop including v_s, v_{id}, and v_1 yields a relation between $\mathbf{v_1}$ and $\mathbf{v_s}$:

$$\mathbf{v_s} - \mathbf{v_{id}} = \mathbf{v_1} \tag{11.19}$$

However, Assumption 1 requires that $\mathbf{v_{id}} = 0$, so

$$\mathbf{v_s} = \mathbf{v_1} \tag{11.20}$$

Combining Eqs. (11.18) and (11.20) and solving for $\mathbf{v_o}$ in terms of $\mathbf{v_s}$ gives

$$\mathbf{v_o} = \mathbf{v_s}\frac{R_1 + R_2}{R_1} \tag{11.21}$$

which yields an expression for the voltage gain of the noninverting amplifier:

$$A_v = \frac{\mathbf{v_o}}{\mathbf{v_s}} = \frac{R_1 + R_2}{R_1} = 1 + \frac{R_2}{R_1} \tag{11.22}$$

Note that the gain is positive and must be greater than or equal to 1 because R_1 and R_2 are positive numbers for real resistors.

EXAMPLE 11.3   **NONINVERTING AMPLIFIER ANALYSIS**

Determine the characteristics of a noninverting amplifier with feedback resistors specified.

PROBLEM Find the voltage gain A_v, output voltage v_o, and output current i_o for the amplifier in Fig. 11.8 if $R_1 = 3\ \text{k}\Omega$, $R_2 = 43\ \text{k}\Omega$, and $v_s = +0.1\ \text{V}$.

SOLUTION **Known Information and Given Data:** Noninverting amplifier circuit with $R_1 = 3\ \text{k}\Omega$, $R_2 = 43\ \text{k}\Omega$, and $v_s = +0.1\ \text{V}$

Unknowns: Voltage gain A_v, output voltage $\mathbf{v_o}$, and output current $\mathbf{i_o}$

Approach: Use Eq. (11.22) to find the voltage gain. Use the gain to calculate the output voltage. Use the output voltage and KCL to find $\mathbf{i_o}$.

Assumptions: The op amp is ideal.

Analysis: Using Eq. (11.22),

$$A_v = 1 + \frac{R_2}{R_1} = 1 + \frac{43\text{ k}\Omega}{3\text{ k}\Omega} = +15.3 \qquad \text{and} \qquad \mathbf{v_o} = A_v \mathbf{v_s} = (15.3)(0.1\text{ V}) = 1.53\text{ V}$$

Since the current $i_- = 0$,

$$\mathbf{i_o} = \frac{\mathbf{v_o}}{R_2 + R_1} = \frac{1.53\text{ V}}{43\text{ k}\Omega + 3\text{ k}\Omega} = 33.3\ \mu\text{A}$$

Check of Results: We have found all the answers requested. SPICE is used to check the results.

Computer-Aided Analysis: The amplifier in the figure is characterized using a combination of an operating point analysis and a transfer function analysis. The gain of E1 is set to 10^9 to model the ideal op amp. The transfer function analysis results are: transfer function $= +15.3$, input resistance $= 10^{20}\ \Omega$, and output resistance $= 0$. The dc output voltage is 1.53 V, and the current in source E1 is $-33.3\ \mu\text{A}$. These values agree with our hand analysis. Note that 10^{20} is the representation of infinity in this particular version of SPICE, and the current in E1 is negative because SPICE uses the passive sign convention which assumes that positive current is entering the positive terminal of E1.

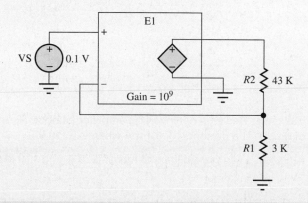

EXERCISE: What are the voltage gain A_v, output voltage v_o, and output current i_o for the amplifier in Fig. 11.8 if $R_1 = 2\text{ k}\Omega$, $R_2 = 36\text{ k}\Omega$, and $v_s = -0.2$ V?

ANSWERS: +19.0; −3.80 V; −100 μV

Input and Output Resistances of the Noninverting Amplifier

Using Assumption 2, $i_+ = 0$, we find that the input resistance of the noninverting amplifier is given by

$$R_{\text{in}} = \frac{\mathbf{v_s}}{\mathbf{i_+}} = \infty \tag{11.23}$$

To find the output resistance, a test current is applied to the output terminal, and the source v_s is set to 0 V. The resulting circuit is identical to that in Fig. 11.7, so the output resistance of the noninverting amplifier is also zero.

$$R_{\text{out}} = 0 \qquad\qquad (11.24)$$

DESIGN NOTE

For the **ideal noninverting amplifier,** the closed-loop voltage gain A_v, input resistance R_{in}, and output resistance R_{out} are:

$$A_v = 1 + \frac{R_2}{R_1} \qquad R_{\text{in}} = \infty \qquad R_{\text{out}} = 0$$

EXERCISE: Draw the circuit used to determine the output resistance of the noninverting amplifier and convince yourself that it is indeed the same as Fig. 11.7.

EXERCISE: What are the voltage gain in dB and the input resistance of the amplifier shown here? If $v_s = 0.25$ V, what are the values of v_o and i_o?

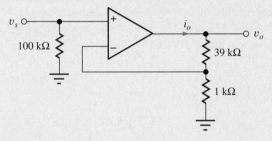

ANSWERS: 32.0 dB, 100 kΩ; +10.0 V, 0.250 mA

EXERCISE: Design a noninverting amplifier (choose R_1 and R_2 from Appendix A) to have a gain of 54 dB. The current $i_o \leq 0.1$ mA when $v_o = 10$ V.

ANSWERS: Two possibilities of many: (220 Ω and 110 kΩ) or (200 Ω and 100 kΩ)

11.3.3 THE UNITY-GAIN BUFFER, OR VOLTAGE FOLLOWER

A special case of the noninverting amplifier, known as the **unity-gain buffer,** or **voltage follower,** is shown in Fig. 11.9, in which the value of R_1 is infinite and that of R_2 is zero. Substituting these values in Eq. (11.22) yields $A_v = 1$. An alternative derivation can be obtained by writing a single-loop equation for Fig. 11.9:

$$\mathbf{v_s} - \mathbf{v_{id}} = \mathbf{v_o} \qquad \text{or} \qquad \mathbf{v_o} = \mathbf{v_s} \qquad \text{and} \qquad A_v = 1 \qquad (11.25)$$

because the ideal operational amplifier forces $\mathbf{v_{id}}$ to be zero.

Why is such an amplifier useful? The ideal unity-gain buffer provides a gain of 1 with infinite input resistance and zero output resistance and therefore provides a tremendous impedance-level transformation while maintaining the level of the signal voltage. Many transducers represent high-source impedances and cannot supply any significant current to drive a load. The ideal unity-gain buffer does not require any input current, yet can drive any desired load resistance without loss of signal voltage. Thus, the unity-gain buffer is found in many sensor and data acquisition applications.

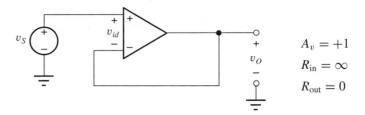

Figure 11.9 Unity-gain buffer (voltage follower).

TABLE 11.1
Summary of the Ideal Inverting and Noninverting Amplifier

	INVERTING AMPLIFIER	NONINVERTING AMPLIFIER
Voltage gain A_v	$-\dfrac{R_2}{R_1}$	$1 + \dfrac{R_2}{R_1}$
Input resistance R_{in}	R_1	∞
Output resistance R_{out}	0	0

Summary of Ideal Inverting and Noninverting Amplifier Characteristics

Table 11.1 summarizes the properties of the ideal inverting and noninverting amplifiers; the properties are recapitulated here. The gain of the noninverting amplifier must be greater than or equal to 1, whereas the inverting amplifier can be designed with a gain magnitude greater than or less than unity (as well as exactly 1). The gain of the inverting amplifier is negative, indicating a 180° phase inversion between input and output voltages.

The input resistance represents an additional major difference between the two amplifiers. R_{in} is extremely large for the noninverting amplifier but is relatively low for the inverting amplifier, limited by the value of R_1. The output resistance of both ideal amplifiers is zero.

EXAMPLE 11.4   INVERTING AND NONINVERTING AMPLIFIER COMPARISON

Compare the characteristics of the inverting and noninverting amplifier configurations.

PROBLEM   Explore the differences between the inverting and noninverting amplifiers in the figure here. Each amplifier is designed to have a gain of 40 dB.

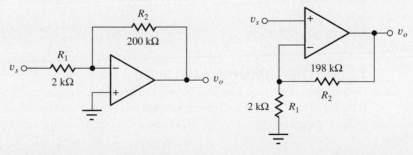

SOLUTION   **Known Information and Given Data:** Inverting amplifier topology with $R_1 = 2\ \text{k}\Omega$ and $R_2 = 200\ \text{k}\Omega$; noninverting amplifier topology with $R_1 = 2\ \text{k}\Omega$ and $R_2 = 198\ \text{k}\Omega$.

Unknowns: Voltage gains, input resistances and output resistances for the two amplifier circuits

Approach: Use the given data to evaluate the amplifier formulas that have already been derived for the two topologies.

Assumptions: The operational amplifiers are ideal.

Analysis:

$$\text{Inverting amplifier: } A_v = -\frac{200 \text{ k}\Omega}{2 \text{ k}\Omega} = -100 \text{ or } 40 \text{ dB}$$

$$R_{\text{in}} = 2 \text{ k}\Omega \quad \text{and} \quad R_{\text{out}} = 0 \text{ }\Omega$$

$$\text{Noninverting amplifier: } A_v = 1 + \frac{198 \text{ k}\Omega}{2 \text{ k}\Omega} = +100 \text{ or } 40 \text{ dB}$$

$$R_{\text{in}} = \infty \quad \text{and} \quad R_{\text{out}} = 0 \text{ }\Omega$$

Check of Results: We have indeed found all the answers requested. A double check indicates the calculations are correct.

TABLE 11.2
Numeric Comparison of the Ideal Inverting and Noninverting Amplifier

	INVERTING AMPLIFIER	NONINVERTING AMPLIFIER
Voltage gain A_v	-100 (40 dB)	$+100$ (40 dB)
Input resistance R_{in}	2 kΩ	∞
Output resistance R_{out}	0	0

Evaluation and Discussion: Table 11.2 lists the characteristics of the two amplifier designs. In addition to the sign difference in the gain of the two amplifiers, we see that the input resistance of the inverting amplifier is only 2 kΩ, whereas that of the noninverting amplifier is infinite. Note that the noninverting amplifier achieves our ideal voltage amplifier goals with $R_{\text{in}} = \infty$ and $R_{\text{out}} = 0 \text{ }\Omega$.

EXERCISE: What are the voltage gain A_v, input resistance R_{in}, output voltage v_o, and output current i_o for the amplifiers in Ex. 11.4 if $R_1 = 1.5$ kΩ, $R_2 = 30$ kΩ, and $v_s = 0.15$ V?

ANSWERS: -20.0, 1.5 kΩ, -3.00 V, -100 μA; $+21$, ∞, $+3.15$ V, $+100$ μA

EXERCISE: Use SPICE transfer function analysis to confirm the analysis in Ex. 11.4.

11.3.4 THE SUMMING AMPLIFIER

Operational amplifiers can also be used to combine signals using the **summing-amplifier** circuit depicted in Fig. 11.10. Here, two input sources v_1 and v_2 are connected to the inverting input of the amplifier through resistors R_1 and R_2. Because the negative amplifier input represents a virtual ground,

$$\mathbf{i_1} = \frac{\mathbf{v_1}}{R_1} \qquad \mathbf{i_2} = \frac{\mathbf{v_2}}{R_2} \qquad \mathbf{i_3} = -\frac{\mathbf{v_o}}{R_3} \tag{11.26}$$

Because $\mathbf{i_-} = \mathbf{0}$, $\mathbf{i_3} = \mathbf{i_1} + \mathbf{i_2}$, and substituting Eq. (11.26) into this expression yields

$$\mathbf{v_o} = -\frac{R_3}{R_1}\mathbf{v_1} - \frac{R_3}{R_2}\mathbf{v_2} \tag{11.27}$$

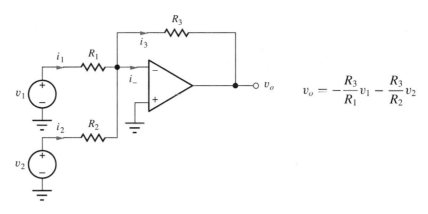

Figure 11.10 The summing amplifier.

$$v_o = -\frac{R_3}{R_1}v_1 - \frac{R_3}{R_2}v_2$$

The output voltage sums the scaled replicas of the two input voltages, and the scale factors for the two inputs may be independently adjusted through the choice of resistors R_1 and R_2. These two inputs can be scaled independently because of the virtual ground maintained at the inverting-input terminal of the op amp.

The inverting-amplifier input node is also commonly called the **summing junction** because the currents i_1 and i_2 are "summed" at this node and forced through the feedback resistor R_3. Although the amplifier in Fig. 11.10 has only two inputs, any number of inputs can be connected to the summing junction through additional resistors. A simple digital-to-analog converter can be formed in this way (see the EIA below and Probs. 11.25 and 11.52).

EXERCISE: What is the summing amplifier output voltage v_o in Fig. 11.10 if $v_1 = 2\sin 1000\pi t$ V, $v_2 = 4\sin 2000\pi t$ V, $R_1 = 1$ kΩ, $R_2 = 2$ kΩ, and $R_3 = 3$ kΩ? What are the input resistances presented to sources v_1 and v_2? What is the current supplied by the op amp output terminal?

ANSWERS: $(-6\sin 1000\pi t - 6\sin 2000\pi t)$ V; 1 kΩ, 2 kΩ; $(-2\sin 1000\pi t - 2\sin 2000\pi t)$ mA

ELECTRONICS IN ACTION

Digital-to-Analog Converter (DAC) Circuits

One of the simplest digital-to-analog converter (DAC) circuits, known as the **weighted resistor DAC,** is based upon the summing amplifier concept that we just encountered in Section 11.3.4. The DAC utilizes a binary-weighted resistor network, a reference voltage V_{REF}, and a group of single-pole, double-throw switches that are usually implemented using MOS transistors. Binary input data controls the switches, with a logical 1 indicating that the switch is connected to V_{REF} and a logical 0 corresponding to a switch connected to ground. Successive resistors are weighted progressively by a factor of 2, thereby producing the desired binary weighted contributions to the output:

$$v_O = (b_1 2^{-1} + b_2 2^{-2} + \cdots + b_n 2^{-n})V_{\text{REF}} \qquad \text{for } b_i \in \{1, 0\}$$

Bit b_1 has the highest weight and is referred to as the **most significant bit (MSB),** whereas bit b_n has the smallest weight and is referred to as the **least significant bit (LSB).**

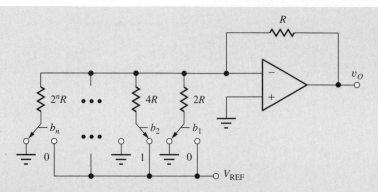

An n-bit weighted-resistor DAC.

Several problems arise in building a DAC using the weighted-resistor approach. The primary difficulty is the need to maintain accurate resistor ratios over a very wide range of resistor values (for example, 4096 to 1 for a 12-bit DAC). Linearity and gain errors occur when the resistor ratios are not perfectly maintained. In addition, because the switches are in series with the resistors, their on-resistance must be very low, and they should have zero offset voltage. The designer can meet these last two requirements by using good MOSFETs (or JFETs) as switches, and the (W/L) ratios of the FETs can be scaled with bit position to equalize the resistance contributions of the switches. However, the wide range of resistor values is not suitable for monolithic converters of moderate to high resolution. We should also note that the current drawn from the voltage reference varies with the binary input pattern. This varying current causes a change in voltage drop in the Thévenin equivalent source resistance of the voltage reference and can lead to data-dependent errors sometimes called **superposition errors.**

The **R-2R ladder** avoids the problem of a wide range of resistor values. It is well-suited to integrated circuit realization because it requires matching of only two resistor values, R and $2R$. The value of R typically ranges from 2 kΩ to 10 kΩ. By forming successive Thévenin equivalents proceeding from left to right at each node in the ladder, we can show that the contribution of each bit is reduced by a factor of 2 going from the MSB to LSB. Like the weighted-resistor DAC, this network requires switches with low on-resistance and zero offset voltage, and the current drawn from the reference still varies with the input data pattern.

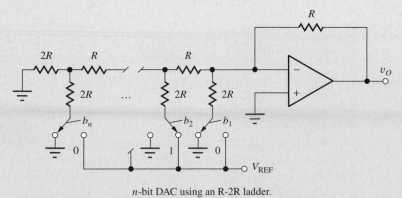

n-bit DAC using an R-2R ladder.

11.3.5 THE DIFFERENCE AMPLIFIER

Except for the summing amplifier, all the circuits thus far have had a single input. However, the operational amplifier may itself be used in a **difference amplifier** configuration, which amplifies the difference between two input signals as shown schematically in Fig. 11.11. Our analysis begins by

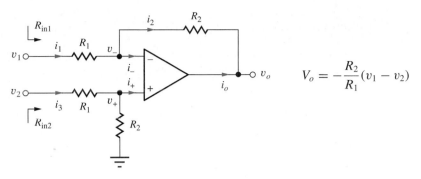

Figure 11.11 Circuit for the difference amplifier.

relating the output voltage to the voltage at v_- as

$$\mathbf{v_o} = \mathbf{v_-} - \mathbf{i_2}R_2 = \mathbf{v_-} - \mathbf{i_1}R_2 \tag{11.28}$$

because $\mathbf{i_2} = \mathbf{i_1}$ since $\mathbf{i_-}$ must be zero. The current $\mathbf{i_1}$ can be written as

$$\mathbf{i_1} = \frac{\mathbf{v_1} - \mathbf{v_-}}{R_1} \tag{11.29}$$

Combining Eqs. (11.28) and (11.29) yields

$$\mathbf{v_o} = \mathbf{v_-} - \frac{R_2}{R_1}(\mathbf{v_1} - \mathbf{v_-}) = \left(\frac{R_1 + R_2}{R_1}\right)\mathbf{v_-} - \frac{R_2}{R_1}\mathbf{v_1} \tag{11.30}$$

Because the voltage between the op amp input terminals must be zero, $\mathbf{v_-} = \mathbf{v_+}$, and the current $\mathbf{i_+}$ is also zero, $\mathbf{v_+}$ can be written using the voltage division formula as

$$\mathbf{v_+} = \frac{R_2}{R_1 + R_2}\mathbf{v_2} \tag{11.31}$$

Substituting Eq. (11.31) into Eq. (11.30) yields the final result

$$\mathbf{v_o} = \left(-\frac{R_2}{R_1}\right)(\mathbf{v_1} - \mathbf{v_2}) \tag{11.32}$$

Thus, the circuit in Fig. 11.11 amplifies the difference between v_1 and v_2 by a factor that is determined by the ratio of resistors R_2 and R_1. For $R_2 = R_1$,

$$\mathbf{v_o} = -(\mathbf{v_1} - \mathbf{v_2}) \tag{11.33}$$

This particular circuit is sometimes called a **differential subtractor.**

The input resistance of this circuit is limited by resistors R_2 and R_1. Input resistance R_{in2}, presented to source v_2, is simply the series combination of R_2 and R_1 because i_+ is zero. For $v_2 = 0$, input resistance R_{in1} equals R_1 because the circuit reduces to the inverting amplifier under this condition. However, for the general case, the input current i_1 is a function of both v_1 and v_2.

EXAMPLE 11.5   **DIFFERENCE AMPLIFIER ANALYSIS**

Here we find the various voltages and currents within the single op amp difference amplifier circuit with a specific set of input voltages.

PROBLEM   Find the values of V_O, V_+, V_-, I_1, I_2, I_3, and I_O for the difference amplifier in Figure 11.11 with $V_1 = 5$ V, $V_2 = 3$ V, $R_1 = 10$ kΩ, and $R_2 = 100$ kΩ.

SOLUTION   **Known Information and Given Data:** The input voltages, resistor values, and circuit topology are specified.

Unknowns: V_O, V_+, V_-, I_1, I_2, I_3, and I_O

Approach: We must use circuit analysis (KCL and KVL) coupled with the ideal op amp assumptions to determine the various voltages and currents, but we must find the node voltages in order to find the currents.

Assumptions: Since the op amp is ideal, we know $I_+ = 0 = I_-$, and $V_+ = V_-$.

Analysis: Since $I_+ = 0$, V_+ can be found directly by voltage division,

$$V_+ = V_2 \frac{R_2}{R_1 + R_2} = 3 \text{ V} \frac{100 \text{ k}\Omega}{10 \text{ k}\Omega + 100 \text{ k}\Omega} = 2.73 \text{ V} \quad \text{and} \quad V_- = 2.73 \text{ V}$$

V_O can be related to V_1 using Kirchhoff's voltage law:

$$V_1 - I_1 R_1 - I_2 R_2 - V_O = 0$$

We know that $I_- = 0$, and $I_2 = I_1$. We can find I_1 since we know the values of V_1, V_-, and R_1:

$$I_1 = \frac{V_1 - V_-}{R_1} = \frac{5 \text{ V} - 2.73 \text{ V}}{10 \text{ k}\Omega} = 227 \text{ }\mu\text{A} \quad \text{and} \quad I_2 = 227 \text{ }\mu\text{A}$$

Then the output voltage can be found

$$V_O = V_1 - I_1 R_1 - I_2 R_2 = V_1 - I_1(R_1 + R_2)$$

$$V_O = 5 \text{ V} - (227 \text{ }\mu\text{A})(110 \text{ k}\Omega) = -20.0 \text{ V}$$

The op amp output current is $I_O = -I_2 = -227 \text{ }\mu\text{A}$.

Check of Results: We have indeed found all the answers requested. The values of the voltages and currents all appear reasonable. This circuit is a difference amplifier that should amplify the difference in its inputs by the gain of $-R_2/R_1 = -10$. The output should be $-10(5 - 3) = -20$ V. ✔

Computer-Aided Analysis: SPICE can be used to check our calculations using the circuit below. The ideal op amp is modeled by VCVS E1 with a gain of 120 dB. An operating point analysis produces voltages that agree with our hand analysis: $V_+ = V_- = 2.73$ V, $V_O = -20$ V, $I_1 = 227 \text{ }\mu\text{A}$, and $I_O = -227 \text{ }\mu\text{A}$, (I(E1) = 227 μA).

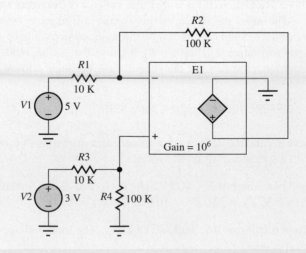

EXERCISE: What is the current exiting the positive terminal of source V_2 in Ex. 11.5?

ANSWER: 27.3 μA

EXERCISE: What are the voltage gain A_v, output voltage V_O, output current I_O, and the current in source V_2 for the amplifier in Ex. 11.5 if $V_1 = 3$ V and $V_2 = 5$ V?

ANSWERS: −10; 20.0 V; +154 μA, 45.5 μA

EXERCISE: What are the voltage gain A_v, output voltage V_O and output current I_O for the amplifier in Fig. 11.11 if $R_1 = 2$ kΩ, $R_2 = 36$ kΩ, $V_1 = 8$ V, and $V_2 = 8.25$ V?

ANSWERS: +18.0; 4.50 V; −92.0 μV

11.3.6 THE INSTRUMENTATION AMPLIFIER

We often need to amplify the difference in two signals but cannot use the difference amplifier in Fig. 11.11 because its input resistance is too low. In such a case, we can combine two noninverting amplifiers with a difference amplifier to form the high-performance composite **instrumentation amplifier** depicted in Fig. 11.12.

In this circuit, op amp 3, with resistors R_3 and R_4, forms a difference amplifier. Using Eq. (11.32), the output voltage $\mathbf{v_o}$ is

$$\mathbf{v_o} = \left(-\frac{R_4}{R_3} \right) (\mathbf{v_a} - \mathbf{v_b}) \tag{11.34}$$

in which voltages $\mathbf{v_a}$ and $\mathbf{v_b}$ are the outputs of the first two amplifiers. Because the $\mathbf{i_-}$ input currents to amplifiers 1 and 2 must be zero, voltages $\mathbf{v_a}$ and $\mathbf{v_b}$ are related to each other by

$$\mathbf{v_a} - \mathbf{i}R_2 - \mathbf{i}(2R_1) - \mathbf{i}R_2 = \mathbf{v_b} \qquad \text{or} \qquad \mathbf{v_a} - \mathbf{v_b} = 2\mathbf{i}(R_1 + R_2) \tag{11.35}$$

Because the voltage across the inputs of both op amps 1 and 2 must also be zero, the voltage difference $(\mathbf{v_1} - \mathbf{v_2})$ appears directly across the resistor $2R_1$, and

$$\mathbf{i} = \frac{\mathbf{v_1} - \mathbf{v_2}}{2R_1} \tag{11.36}$$

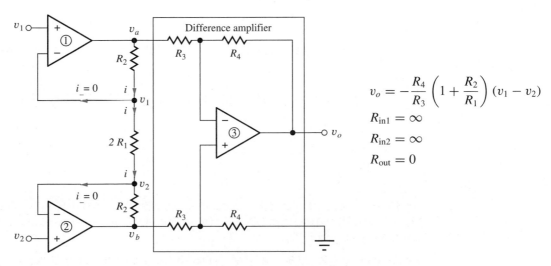

$$v_o = -\frac{R_4}{R_3} \left(1 + \frac{R_2}{R_1} \right) (v_1 - v_2)$$

$$R_{in1} = \infty$$

$$R_{in2} = \infty$$

$$R_{out} = 0$$

Figure 11.12 Circuit for the instrumentation amplifier.

Combining Eqs. (11.34), (11.35), and (11.36) yields a final expression for the output voltage of the instrumentation amplifier:

$$\mathbf{v_o} = -\frac{R_4}{R_3}\left(1 + \frac{R_2}{R_1}\right)(\mathbf{v_1} - \mathbf{v_2}) \tag{11.37}$$

The ideal instrumentation amplifier amplifies the difference in the two input signals and provides a gain that is equivalent to the product of the gains of the noninverting and difference amplifiers. The input resistance presented to both input sources is infinite because the input current to both op amps is zero, and the output resistance is forced to zero by the difference amplifier.

EXAMPLE **11.6**   INSTRUMENTATION AMPLIFIER ANALYSIS

The three op amp output voltages are calculated for a specific set of dc input voltages in this example.

PROBLEM   Find the values of V_O, V_A, and V_B for the instrumentation amplifier in Fig. 11.12 if $V_1 = 2.5$ V, $V_2 = 2.25$ V, $R_1 = 15$ kΩ, $R_2 = 150$ kΩ, $R_3 = 15$ kΩ, and $R_4 = 30$ kΩ.

SOLUTION   **Known Information and Given Data:** $V_1 = 2.5$ V, $V_2 = 2.25$ V, $R_1 = 15$ kΩ, $R_2 = 150$ kΩ, $R_3 = 15$ kΩ, and $R_4 = 30$ kΩ for the circuit configuration in Fig. 11.12.

Unknowns: The values of V_O, V_A, and V_B

Approach: All the values are specified to permit direct use of Eq. (11.37)

Assumptions: The op amps are ideal. Therefore $I_+ = 0 = I_-$ and $V_+ = V_-$ for each op amp.

Analysis: Using Eq. (11.37) with dc values, we find the output voltage is

$$V_O = -\frac{R_4}{R_3}\left(1 + \frac{R_2}{R_1}\right)(V_1 - V_2) = -\frac{30\text{ k}\Omega}{15\text{ k}\Omega}\left(1 + \frac{150\text{ k}\Omega}{15\text{ k}\Omega}\right)(2.5 - 2.25) = -5.50\text{ V}$$

Since the op amp input currents are zero, V_A and V_B can be related directly to the two input voltages and current i

$$V_A = V_1 + I R_2 \qquad \text{and} \qquad V_B = V_2 - I R_2$$

$$I = \frac{V_1 - V_2}{2R_1} = \frac{2.5\text{ V} - 2.25\text{ V}}{2(15\text{ k}\Omega)} = 8.33\ \mu\text{A}$$

$$V_A = 2.5 + (8.33\ \mu\text{A})(150\text{ k}\Omega) = +3.75\text{ V} \qquad V_B = 2.25 - (8.33\ \mu\text{A})(150\text{ k}\Omega) = 1.00\text{ V}$$

Check of Results: The unknowns have all been determined. Let us check to see if these voltages are consistent with the difference amplifier that should amplify its input by a factor of -2:

$$V_O = -\frac{R_4}{R_3}(V_A - V_B) = -\frac{30\text{ k}\Omega}{15\text{ k}\Omega}(3.75 - 1.00)\text{ V} = -5.50\text{ V} \quad ✔$$

EXERCISE: Suppose v_1 and v_2 are dc voltages with $V_1 = 5.001$ V, $V_2 = 4.999$ V, $R_1 = 1$ kΩ, $R_2 = 49$ kΩ, $R_3 = 10$ kΩ, and $R_4 = 10$ kΩ in Fig. 11.12. Write expressions for V_A and V_B. What are the values of V_O, V_A, V_B, and I?

ANSWERS: $V_A = V_1 + I R_2$, $V_B = V_2 - I R_2$; -0.100 V, 5.05 V, 4.95 V, $1.00\ \mu$A

11.3.7 AN ACTIVE LOW-PASS FILTER

Although the operational-amplifier circuit examples thus far have used only resistors in the feedback network, other passive elements or even solid-state devices can be part of the feedback path. The general case of the inverting configuration with passive feedback is shown in Fig. 11.13, in which resistors R_1 and R_2 have been replaced by general impedances $Z_1(s)$ and $Z_2(s)$, which may now be a function of frequency. (Note that resistive feedback is just one special case of the amplifier in Fig. 11.13.)

The gain of the amplifier in this figure is obtained in a manner identical to that in the resistive-feedback case. Replacing R_1 by Z_1 and R_2 by Z_2 in Eq. (11.12) yields the transfer function $A_v(s)$:

$$A_v(s) = \frac{\mathbf{V_o}(s)}{\mathbf{V_s}(s)} = -\frac{Z_2(s)}{Z_1(s)} \tag{11.38}$$

One useful circuit involving frequency-dependent feedback is the single-pole, **low-pass filter** in Fig. 11.14, for which

$$Z_1(s) = R_1 \quad \text{and} \quad Z_2(s) = \frac{R_2 \dfrac{1}{sC}}{R_2 + \dfrac{1}{sC}} = \frac{R_2}{sCR_2 + 1} \tag{11.39}$$

Substituting the results from Eq. (11.39) into Eq. (11.38) yields an expression for the voltage transfer function for the amplifier in Fig. 11.14(a).

$$A_v(s) = -\frac{R_2}{R_1} \frac{1}{(1 + sR_2C)} = -\frac{R_2}{R_1} \frac{1}{\left(1 + \dfrac{s}{\omega_H}\right)} \quad \text{where} \quad \omega_H = 2\pi f_H = \frac{1}{R_2C} \tag{11.40}$$

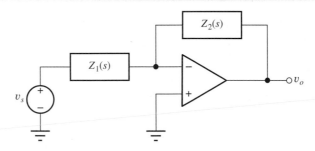

Figure 11.13 Generalized inverting-amplifier configuration.

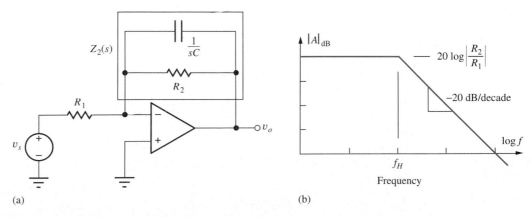

(a)   (b)

Figure 11.14 (a) Inverting amplifier with frequency-dependent feedback. (b) Bode plot of the voltage gain of the low-pass filter.

Figure 11.14(b) is the asymptotic Bode plot of the magnitude of the gain in Eq. (11.40). The transfer function exhibits a low-pass characteristic with a single pole at frequency ω_H, the upper-cutoff frequency (-3 dB point) of the low-pass filter. At frequencies below ω_H, the amplifier behaves as an inverting amplifier with gain set by the ratio of resistors R_2 and R_1; at frequencies above ω_H, the amplifier response rolls off at a rate of -20 dB/decade.

Note from Eq. (11.40) that the low-frequency gain and the cutoff frequency can be set independently in this **low-pass filter.** Indeed, because there are three elements—R_1, R_2, and C—the input resistance ($R_{in} = R_1$) can be a third design parameter. Since the inverting input terminal is at 0 V (remember it is a virtual ground), $R_{in} = R_1$.

DESIGN EXAMPLE 11.7   ACTIVE LOW-PASS FILTER DESIGN

Design a single-pole low-pass filter using the single op-amp circuit in Fig. 11.14(a) to meet a given cutoff frequency specification.

PROBLEM   Design an active low-pass filter (choose the values of R_1, R_2, and C) with $f_H = 2$ kHz, $R_{in} = 5$ kΩ, and $A_v = 40$ dB.

SOLUTION   **Known Information and Given Data:** In this case, we are given the values for the bandwidth, gain and input resistance ($f_H = 2$ kHz, $R_{in} = 5$ kΩ, and $A_v = 40$ dB), and the amplifier circuit configuration has also been specified. However, we must convert the gain from dB to purely numeric form before we use it in the calculations:

$$|A_v| = 10^{40\,\text{dB}/20\,\text{dB}} = 100$$

Unknowns: Find the values of R_1, R_2, and C.

Approach: Use the single-pole low-pass filter circuit in Fig. 11.14(a). The three specifications should uniquely determine the three unknowns. We will use R_{in} to determine R_1, R_1 to find R_2, and R_2 to find C.

Assumptions: The op amp is ideal. Note that the specified gain actually represents the low-frequency gain of the amplifier and that a gain of either $+100$ or -100 will satisfy the gain specification.

Analysis: Since the inverting input represents a virtual ground, the input resistance is set directly by R_1 so that

$$R_1 = R_{in} = 5 \text{ k}\Omega$$

and

$$|A_v| = \frac{R_2}{R_1} \rightarrow R_2 = 100R_1 = 500 \text{ k}\Omega$$

The value of C can now be determined from the f_H specification:

$$C = \frac{1}{2\pi f_H R_2} = \frac{1}{2\pi (2 \text{ kHz})(500 \text{ k}\Omega)} = 1.59 \times 10^{-10} \text{ F} = 159 \text{ pF}$$

Looking at Appendix A, we find the nearest values for R_1 and R_2 are 5.1 kΩ and 510 kΩ. In most applications, an input resistance of 5.1 kΩ (set by R_1) would be acceptable since it is only 2 percent higher than the design specification of 5 kΩ. Recalculating the value of C using the new value of R_2 yields

$$C = \frac{1}{2\pi f_H R_2} = \frac{1}{2\pi (2 \text{ kHz})(510 \text{ k}\Omega)} = 156 \text{ pF}$$

The closest capacitor value is 160 pF, which will lower f_H to 1.95 kHz. A second choice would be 150 pF for which $f_H = 2.08$ kHz.

Final Design: $R_1 = 5.1\,\text{k}\Omega$, $R_2 = 510\,\text{k}\Omega$, and $C = 160\,\text{pF}$, yielding a slightly smaller bandwidth than the design specification.

Check of Results: We have found the three required values. The SPICE analysis here confirms the design.

Discussion: A third but more costly option would be to use a parallel combination of two capacitors, 100 pF and 56 pf. In a similar vein, R_1 and R_2 could be synthesized from the series combination of two resistors (e.g., $R_1 = 4.7\,\text{k}\Omega + 300\,\Omega$). It might be preferable to just use more expensive 1 percent resistors with $R_1 = 4.99\,\text{k}\Omega$ and $R_2 = 499\,\text{k}\Omega$. In order to select between these options, one would need to know more details about the application. Note that trying to use an exact values of R and C doesn't provide much benefit if the resistor and capacitor tolerances are 5, 10, or 20 percent.

Computer-Aided Analysis: An ac analysis of the low-pass filter circuit is performed using the equivalent circuit below. The op amp is modeled by VCVS E1 whose gain is set to $-1\text{E}6$ so that the feedback is negative. The frequency response parameters are Start Frequency $= 10$ Hz, Stop Frequency $= 100$ KHz with 10 frequency points per decade. From the graph, $A_v = 40$ dB and $f_H = 1.95$ kHz as designed.

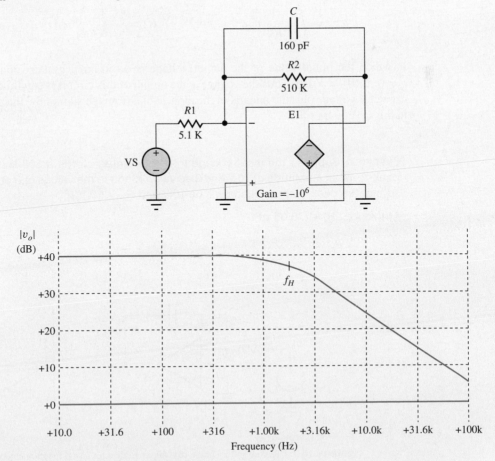

EXERCISE: Design an active low-pass filter (choose the values of R_1, R_2, and C) with $f_H = 3$ kHz, $R_{in} = 10$ kΩ, and $A_v = 26$ dB.

ANSWERS: Calculated values: 10 kΩ, 200 kΩ, 265 pF; Appendix A values: 10 kΩ, 200 kΩ, 270 pF

11.3.8 THE INTEGRATOR

The **integrator** is another highly useful building block formed from an operational amplifier with frequency-dependent feedback. In the integrator circuit, feedback-resistor R_2 is replaced by capacitor C, as in Fig. 11.15. This circuit provides an opportunity to explore op amp circuit analysis in the time domain (for frequency-domain analysis, see Prob. 11.42(a)).

Because the inverting-input terminal represents a virtual ground,

$$i_s = \frac{v_s}{R} \qquad \text{and} \qquad i_c = -C\frac{dv_o}{dt} \qquad (11.41)$$

For an ideal op amp, $i_- = 0$, so i_c must equal i_s. Equating the two expressions in Eq. (11.41) and integrating both sides of the result yields

$$\int dv_o = \int -\frac{1}{RC}v_s \, d\tau \qquad \text{or} \qquad v_o(t) = -\frac{1}{RC}\int_0^t v_s(\tau)\,d\tau + v_o(0) \qquad (11.42)$$

in which the initial value of the output voltage is determined by the voltage on the capacitor at $t = 0$: $v_o(0) = V_C(0)$. Thus the voltage at the output of this circuit at any time t represents the initial capacitor voltage plus the integral of the input signal from the start of the integration interval, chosen in this case to be $t = 0$.

EXERCISE: Suppose the input voltage $v_s(t)$ to an integrator is a 500-Hz square wave with a peak-to-peak amplitude of 10 V and 0 dc value. Choose the values of R and C in the integrator so that the peak output voltage will be 10 V and $R_{in} = 10$ kΩ.

ANSWERS: 10 kΩ, 0.05 μF

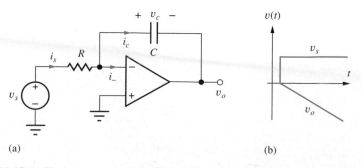

(a) (b)

Figure 11.15 (a) The integrator circuit. (b) Output voltage for a step-function input with $v_C(0) = 0$.

ELECTRONICS IN ACTION

Dual-Ramp or Dual-Slope Analog-to-Digital Converters (ADCs)

The dual-ramp or dual-slope analog-to-digital converter is widely used as the ADC in data acquisition systems, digital multimeters, and other precision instruments. The heart of the dual-ramp converter is the integrator circuit discussed in Section 11.3.8. As illustrated in the circuit schematic below, the conversion cycle consists of two separate integration intervals.

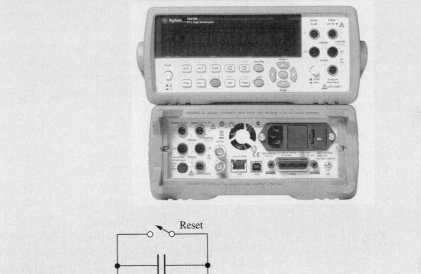

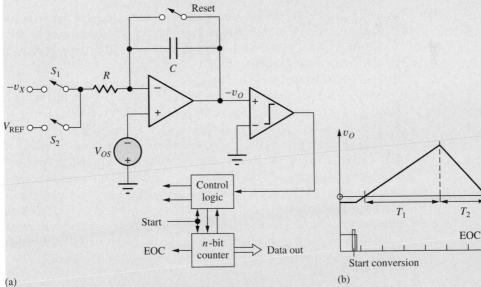

(a) Dual-ramp ADC and (b) timing diagram.
© Agilent Technologies 2006. All Rights Reserved.

First, unknown voltage v_X is integrated for a known period of time T_1. The value of this integral is then compared to that of a known reference voltage V_{REF}, which is integrated for a variable length of time T_2.

At the start of conversion the counter is reset, and the integrator is reset to a slightly negative voltage. The unknown input v_X is connected to the integrator input through switch S_1. Unknown voltage v_X is integrated for a fixed period of time $T_1 = 2^n T_C$, which begins when the integrator output crosses through zero. At the end of time T_1, the counter overflows, causing S_1 to be opened and the reference input V_{REF} to be connected to the integrator input through S_2. The integrator output then decreases until it crosses back through zero, and the comparator changes state, indicating the end of the conversion. The counter continues to accumulate pulses during the down ramp, and the final number in the counter represents the quantized value of the unknown voltage v_X.

Circuit operation forces the integrals over the two time periods to be equal:

$$\frac{1}{RC} \int_0^{T_1} v_X(t)\, dt = \frac{1}{RC} \int_{T_1}^{T_1 + T_2} V_{REF}\, dt$$

T_1 is set equal to $2^n T_C$ because the unknown voltage v_X was integrated over the amount of time needed for the n-bit counter to overflow. Time period T_2 is equal to $N T_C$, where N is the number accumulated in the counter during the second phase of operation.

Recalling the mean-value theorem from calculus, we have

$$\frac{1}{RC} \int_0^{T_1} v_X(t)\, dt = \frac{\langle v_X \rangle}{RC} T_1 \qquad \text{and} \qquad \frac{1}{RC} \int_{T_1}^{T_1 + T_2} V_{REF}(t)\, dt = \frac{V_{REF}}{RC} T_2$$

because V_{REF} is a constant. Equating these last two results, we find the average value of the input $\langle v_x \rangle$ to be

$$\frac{\langle v_X \rangle}{V_{REF}} = \frac{T_2}{T_1} = \frac{N}{2^n}$$

assuming that the RC product remains constant throughout the complete conversion cycle. The absolute values of R and C no longer enter directly into the relation between v_X and V_{FS}. The digital output word represents the average value of v_X during the first integration phase. Thus, v_X can change during the conversion cycle of this converter without destroying the validity of the quantized output value.

The conversion time T_T requires 2^n clock periods for the first integration period, and N clock periods for the second integration period. Thus the conversion time is variable and given by $T_T = (2^n + N)T_C \leq 2^{n+1} T_C$ because the maximum value of N is 2^n.

The dual ramp is a widely used converter. Although much slower than successive approximation converters, the dual-ramp converter offers excellent linearity. By combining its integrating properties with careful design, one can obtain accurate conversion at resolutions exceeding 20 bits, but at relatively low conversion rates. In a number of recent converters and instruments, the basic dual-ramp converter has been modified to include extra integration phases for automatic offset voltage elimination. These devices are often called *quad-slope* or *quad-phase converters.* Another converter, the *triple ramp,* uses coarse and fine down ramps to greatly improve the speed of the integrating converter (by a factor of $2^{n/2}$ for an n-bit converter).

Normal-Mode Rejection

As mentioned before, the quantized output of the dual-ramp converter represents the average of the input during the first integration phase. The integrator operates as a low-pass filter with the normalized transfer function shown in the accompanying figure. Sinusoidal input signals, whose frequencies are exact multiples of the reciprocal of the integration time T_1, have integrals of zero value and do not appear at the integrator output. This property is used in many digital multimeters, which are equipped with dual-ramp converters having an integration time that is some multiple of the period of the 50- or 60-Hz power-line frequency. Noise sources with frequencies at multiples of the power-line frequency are therefore rejected by these integrating ADCs. This property is usually termed **normal-mode rejection.**

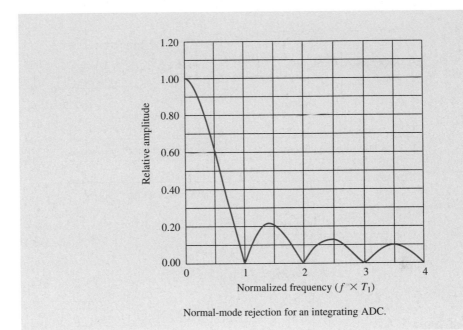

Normal-mode rejection for an integrating ADC.

11.3.9 THE DIFFERENTIATOR

The derivative operation can also be provided by an op amp circuit. The **differentiator** is obtained by interchanging the resistor and capacitor of the integrator as drawn in Fig. 11.16. The circuit is less often used than the integrator because the derivative operation is an inherently "noisy" operation; that is, the high-frequency components of the input signal are emphasized. Analysis of the circuit is similar to that of the integrator. Since the inverting-input terminal represents a virtual ground,

$$i_s = C\frac{dv_s}{dt} \qquad \text{and} \qquad i_R = -\frac{v_o}{R} \tag{11.43}$$

Since $i_- = 0$, the currents i_s and i_R must be equal, and

$$v_o = -RC\frac{dv_s}{dt} \tag{11.44}$$

The output voltage is a scaled version of the derivative of the input voltage.

EXERCISE: What is the output voltage of the circuit in Fig. 11.16 if $R = 20$ kΩ, $C = 0.02$ μF, and $v_s = 2.5 \sin 2000\pi t$ V?

ANSWER: $-6.28 \cos 2000\pi t$ V

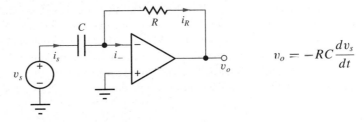

$$v_o = -RC\frac{dv_s}{dt}$$

Figure 11.16 Differentiator circuit.

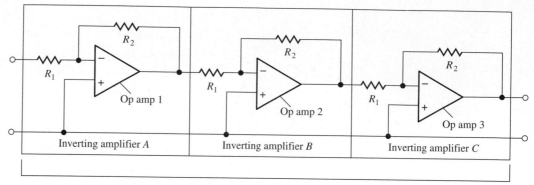

(a)

Three-stage amplifier

(b) A_v, R_{in}, R_{out}

(c) Closed-loop feedback amplifier

Figure 11.17 (a) Three-stage amplifier cascade. (b) Inverting amplifier using an operational amplifier. (c) Two-port representation of the overall amplifier.

11.3.10 CASCADED AMPLIFIERS

Often, a set of design specifications cannot be met using a single amplifier, but may be achieved by connecting several amplifiers in cascade[1] as indicated by the three-stage cascade in Fig. 11.17. In this situation, the output of one amplifier stage is connected to the input of the next. If the output resistance of one amplifier is much less than the input resistance of the next, $R_{\text{out}A} \ll R_{\text{in}B}$ and $R_{\text{out}B} \ll R_{\text{in}C}$, then the amplifiers don't interact and the overall voltage gain is simply the product of the voltage gains of the individual stages. In order to understand this behavior more fully, we will represent the amplifiers using their simplified two-port models discussed next.

Two-Port Representations

At each level in Fig. 11.17, we can represent the "amplifier" as a **two-port model** with a value of voltage gain, input resistance, and output resistance, defined as in Fig. 11.17(b) and (c). Each amplifier stage—A, B, and C—is built using an operational amplifier that has a gain A, input resistance R_{id}, and output resistance R_o. These quantities are usually called the open-loop parameters of the operational amplifier: *open-loop gain, open-loop input resistance,* and *open-loop output resistance.* They describe the op amp as a two-port by itself with no external elements connected.

Each single-stage amplifier built from an operational amplifier and the feedback network consisting of R_1 and R_2 is termed a **closed-loop amplifier.** We use A_v, R_{in}, and R_{out} for each closed-loop amplifier, as well as for the overall composite amplifier. Table 11.3 summarizes this terminology.

---

[1] In fact, the instrumentation amplifier is an example that we already encountered.

TABLE 11.3
Feedback-Amplifier Terminology Comparison

	VOLTAGE GAIN	INPUT RESISTANCE	OUTPUT RESISTANCE
Open-Loop Amplifier	A	R_{id}	R_o
Closed-Loop Amplifier	A_v	R_{in}	R_{out}

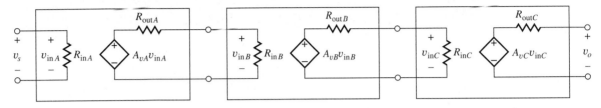

Figure 11.18 Two-port representation for three-stage cascaded amplifier.

Two-Port Model for the Three-Stage Cascade Amplifier

In Fig. 11.18 each individual amplifier has been replaced by its two-port model. By proceeding through the amplifier from left to right, the overall gain expression can be written as

$$\mathbf{v_o} = A_{vA}\mathbf{v_s}\left(\frac{R_{inB}}{R_{outA}+R_{inB}}\right)A_{vB}\left(\frac{R_{inC}}{R_{outB}+R_{inC}}\right)A_{vC} \qquad (11.45)$$

For the ideal amplifiers considered so far, $R_{out} = 0$, so the impedance mismatch requirement is always met (see Sec. 10.6), and the overall gain of the cascade amplifier is

$$A_v = \frac{\mathbf{v_o}}{\mathbf{v_s}} = A_{vA}\cdot A_{vB}\cdot A_{vC} \qquad (11.46)$$

If a test source $\mathbf{v_x}$ is applied to the input and the input current $\mathbf{i_x}$ is calculated, we find that R_{in} of the overall amplifier is determined solely by the input resistance of the first amplifier. In this case, $R_{in} = \mathbf{v_x}/\mathbf{i_x} = R_{inA}$. Similarly, if we apply a test source $\mathbf{v_x}$ at the output and find the input current $\mathbf{i_x}$, we find that R_{out} of the overall amplifier is determined only by the output resistance of the last amplifier. In this case, $R_{out} = R_{outC} = 0$.

 **DESIGN NOTE**

A very common mistake is to expect that the input resistance of a cascade of amplifiers results from some combination of the input resistances of the individual amplifiers, or that the output resistance is a function of R_{outA}, R_{outB}, and R_{outC}.

EXERCISE: The amplifier in Fig. 11.17 has $R_2 = 68$ kΩ and $R_1 = 2.7$ kΩ. What are the values of A_{vA}, A_{vB}, A_{vC}, R_{inA}, R_{inB}, R_{inC}, and R_{outA}, R_{outB}, R_{outC}, for the amplifier equivalent circuit in Fig. 11.18?

ANSWERS: −25.2; −25.2; −25.2; 2.7 kΩ; 2.7 kΩ; 2.7 kΩ; 0; 0; 0

EXERCISE: What are the gain A_v, input resistance, and output resistance of the three stage amplifier in Fig. 11.17(a) if $R_2 = 68$ kΩ and $R_1 = 2.7$ kΩ?

ANSWERS: $(-25.2)^3 = -1.60 \times 10^4$; 2.7 k$\Omega$; 0

EXERCISE: Suppose the three output resistances in the amplifier in the previous exercise are not zero. What is the largest value of R_{out} that can be permitted if the gain is not to be reduced by more than 1 percent? Assume that the three output resistance values are the same.

ANSWER: 13.5 Ω

ELECTRONICS IN ACTION

Fiber Optic Receiver

Interface circuits for optical communications were introduced in the Electronics in Action feature in Chapter 9. One of the important electronic blocks on the receiver side of such a fiber optic communication link is the circuit that performs the optical-to-electrical (O/E) signal conversion, and a common approach is shown in the accompanying figure. Light exiting the optical fiber is incident upon a photodiode (see Sec. 3.18) and generates photocurrent i_{ph} as modeled by the current source in the figure. This photocurrent flows through feedback resistor R and generates a signal voltage at the output given by $v_o = i_{ph} R$. The voltage V_{BIAS} can be used to provide reverse bias to the photodiode. In this case, the total output voltage is $v_O = V_{BIAS} + i_{ph} R$.

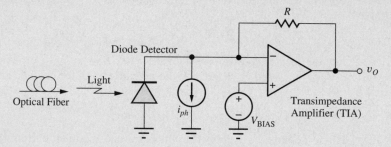

Optical-to-electrical interface for fiber optic data transmission.

Since the input to the amplifier is a current and the output is a voltage, the gain $A_{tr} = v_o / i_{ph}$ has the units of resistance, and the amplifier is referred to as a transresistance or (more generally) a transimpedance amplifier (TIA). The operational amplifier shown in the circuit must have an extremely wideband and linear design. The requirements are particularly stringent in OC-768 systems in which 40-GHz signals coming from the optical fiber must be amplified without the addition of any significant phase distortion.

11.3.11 AMPLIFIER TERMINOLOGY REVIEW

Now that we have analyzed a number of amplifier configurations, let us step back and review the terminology being used. Amplifier terminology is often a source of confusion because the portion of the circuit that is being called an *amplifier* must often be determined from the context of the discussion.

In Fig. 11.17, for example, an overall amplifier (the three-stage amplifier) is formed from the cascade connection of three inverting amplifiers (*A*, *B*, *C*), and each inverting amplifier has been implemented using an operational amplifier (op amp 1, 2, 3). Thus, we can identify at least seven different "amplifiers" in Fig. 11.17: operational amplifiers 1, 2, 3; inverting amplifiers *A*, *B*, *C*; and the composite three-stage amplifier *ABC*.

11.4 ACTIVE FILTERS

Filters come in many forms. We have looked at the characteristics of low-pass, high-pass, band-pass, and band-reject filters in Chapters 1 and 10. The simplest implementation uses passive components, resistors, capacitors, and inductors. In integrated circuits, however, inductors are difficult to fabricate,

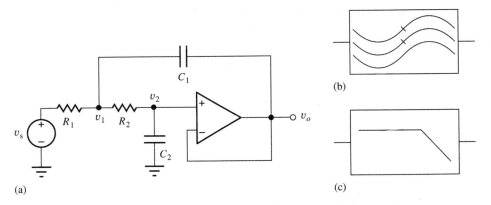

Figure 11.19 (a) A two-pole low-pass filter. (b) Low-pass filter symbol. (c) Alternate low-pass filter symbol.

take up significant area, and can only be made with very small values of inductance. With the advent of low-cost high-performance op amps, new circuits were invented that could realize the desired filter characteristics without the use of inductors. These filters utilizing op amps are referred to as **active filters,** and this section discusses examples of active low-pass, high-pass, and band-pass filters. A simple active low-pass filter was discussed in Sec. 11.3.7, but this circuit produced only a single pole. Many of the filters described in this section are more efficient in the sense that the circuits achieve two poles of filtering per op amp. The interested reader can explore the material further in many texts that deal exclusively with active-filter design.

11.4.1 LOW-PASS FILTER

A basic two-pole low-pass filter configuration is shown in Fig. 11.19 and is formed from an op amp with two resistors and two capacitors. In this particular circuit, the op amp operates as a voltage follower, which provides unity gain over a wide range of frequencies. The filter uses positive feedback through C_1 at frequencies above dc to realize complex poles without the need for inductors.

Let us now find the transfer function describing the voltage gain of this filter. The ideal op amp forces $\mathbf{V_o}(s) = \mathbf{V_2}(s)$, so there are only two independent nodes in the circuit. Writing nodal equations for $\mathbf{V_1}(s)$ and $\mathbf{V_2}(s)$ yields

$$\begin{bmatrix} G_1 \mathbf{V_s}(s) \\ 0 \end{bmatrix} = \begin{bmatrix} sC_1 + G_1 + G_2 & -(sC_1 + G_2) \\ -G_2 & sC_2 + G_2 \end{bmatrix} \begin{bmatrix} \mathbf{V_1}(s) \\ \mathbf{V_2}(s) \end{bmatrix} \tag{11.47}$$

and the determinant of this system of equations is

$$\Delta = s^2 C_1 C_2 + sC_2(G_1 + G_2) + G_1 G_2 \tag{11.48}$$

Solving for $\mathbf{V_2}(s)$ and remembering that $\mathbf{V_o}(s) = \mathbf{V_2}(s)$ yields

$$\mathbf{V_o}(s) = \mathbf{V_2}(s) = \frac{G_1 G_2}{\Delta} \mathbf{V_s}(s) \tag{11.49}$$

which can be rearranged as

$$A_{LP}(s) = \frac{\mathbf{V_o}(s)}{\mathbf{V_s}(s)} = \frac{\dfrac{1}{R_1 R_2 C_1 C_2}}{s^2 + s\dfrac{1}{C_1}\left(\dfrac{1}{R_1} + \dfrac{1}{R_2}\right) + \dfrac{1}{R_1 R_2 C_1 C_2}} \tag{11.50}$$

Equation (11.50) is most often written in standard form as

$$A_{LP}(s) = \frac{\omega_o^2}{s^2 + s\dfrac{\omega_o}{Q} + \omega_o^2} \tag{11.51}$$

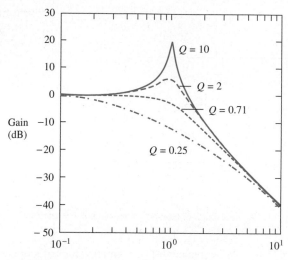

Figure 11.20 Low-pass filter response[2] for $\omega_o = 1$ and four values of Q.

in which

$$\omega_o = \frac{1}{\sqrt{R_1 R_2 C_1 C_2}} \qquad \text{and} \qquad Q = \sqrt{\frac{C_1}{C_2}} \frac{\sqrt{R_1 R_2}}{R_1 + R_2} \qquad (11.52)$$

The frequency ω_o is referred to as the cutoff frequency of the filter, although the exact value of the cutoff frequency, based on the strict definition of ω_H, is equal to ω_o only for $Q = 1/\sqrt{2}$. At low frequencies—that is, $\omega \ll \omega_o$—the filter has unity gain, but for frequencies well above ω_o, the filter response exhibits a two-pole roll-off, falling at a rate of 40 dB/decade. At $\omega = \omega_o$, the gain of the filter is equal to Q.

Figure 11.20 shows the response of the filter for $\omega_o = 1$ and four values of Q: 0.25, $1/\sqrt{2}$, 2, and 10. $Q = 1/\sqrt{2}$ corresponds to the **maximally flat magnitude** response of a **Butterworth filter,** which gives the maximum bandwidth without a peaked response. For a Q larger than $1/\sqrt{2}$, the filter response exhibits a peaked response that is usually undesirable, whereas a Q below $1/\sqrt{2}$ does not take maximum advantage of the filter's bandwidth capability. Because the voltage follower must accurately provide a gain of 1, ω_o should be designed to be one to two decades below the unity-gain frequency of the op amp.

From a practical point of view, a much wider selection of resistor values than capacitor values exists, and the filters are often designed with $C_1 = C_2 = C$. Then ω_o and Q are adjusted by choosing different values of R_1 and R_2. For the equal capacitor design,

$$\omega_o = \frac{1}{C\sqrt{R_1 R_2}} \qquad \text{and} \qquad Q = \frac{\sqrt{R_1 R_2}}{R_1 + R_2} \qquad (11.53)$$

Another practical consideration concerns the op amp bias currents that we will discuss in the next chapter. In order to operate properly, the active filter circuits must provide dc paths for the op amp bias currents. In the circuit in Fig. 11.19, the dc current for the noninverting input is supplied from the dc-referenced signal source through R_1 and R_2. The dc current in the inverting input is supplied from the op amp output.

DESIGN NOTE

In order for an op amp circuit to operate properly, the feedback network must provide a dc path for the amplifier's input bias currents.

---

[2] Using MATLAB: Bode(1,[1 0.1 1]), for example.

DESIGN EXAMPLE 11.8

LOW-PASS FILTER DESIGN

Determine the capacitor and resistor values required to meet a cutoff frequency specification in a two-pole active low-pass filter.

PROBLEM Design a low-pass filter using the circuit in Fig. 11.19 with an upper cutoff frequency of 5 kHz and a maximally flat response.

SOLUTION **Known Information and Given Data:** Second-order active low-pass filter circuit in Fig. 11.20; maximally flat design with $f_H = 5$ kHz

Unknowns: R_1, R_2, C_1, and C_2

Approach: As mentioned in Sec. 11.4.1, the maximally flat response for the transfer function in Eq. (11.51) is achieved for $Q = 1/\sqrt{2}$. For this case, we also find that $f_H = f_o$. Unfortunately, based on Eq. (11.52), the simple equal capacitor design cannot achieve this Q. We will need to explore another design option.

Assumptions: The operational amplifier is ideal.

Analysis: From Eq. (11.52), we see that one workable choice for the element values is $C_1 = 2C_2 = 2C$ and $R_1 = R_2 = R$. For these values,

$$ R = \frac{1}{\sqrt{2}\,\omega_o C} \qquad \text{and} \qquad Q = \frac{1}{\sqrt{2}} $$

but we still have two values to select and only one design constraint. We must call on our engineering judgment to make the design choice. Note that $1/\omega_o C$ represents the reactance of C at the frequency ω_o, and R is 30 percent smaller than this value. Thus, the impedance level of the filter is set by the choice of C (or R). If the impedance level is too low, the op amp will not be able to supply the current needed to drive the feedback network.

At 5 kHz, a 0.01-μF capacitor has a reactance of 3.18 kΩ:

$$ \frac{1}{\omega_o C} = \frac{1}{10^4 \pi (10^{-8})} = 3180 \ \Omega $$

This is a readily available value of capacitance, and so

$$ R = \frac{3180 \ \Omega}{\sqrt{2}} = 2250 \ \Omega $$

Referring to the precision resistor table in Appendix A, we find that the nearest 1 percent resistor value is 2260 Ω. The completed design values are

$$ R_1 = R_2 = 2.26 \text{ k}\Omega, \ C_1 = 0.02 \ \mu\text{F}, \ C_2 = 0.01 \ \mu\text{F} $$

Check of Results: Using the design values yields $f_o = 4980$ Hz and $Q = 0.707$.

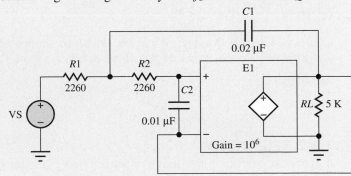

Discussion and Computer Aided Analysis: The frequency response of the filter is simulated using the circuit below. The op amp is modeled with VCVS E1 with the gain set to -10^6 so that the feedback is negative. An ac analysis is performed with VS as the source with FSTART = 10 Hz, FSTOP = 10 MHz, and 10 simulations points per frequency decade. The gain to the output node is 0 dB and $f_H = 5$ kHz, in agreement with the design specification.

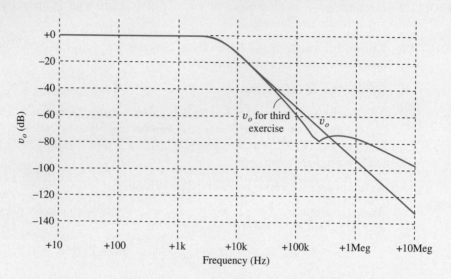

EXERCISE: What is $A_v(0)$ for the filter design in the above example? Show that $f_H = f_o$ for the maximally flat design with $Q = 1/\sqrt{2}$.

ANSWER: +1.00

EXERCISE: Redesign the filter in Ex. 11.8 to have an upper cutoff frequency of 10 kHz with a maximally flat response. Keep the impedance level of the filter the same.

ANSWERS: 0.01 µF, 0.005 µF, 2260 Ω, 2260 Ω.

EXERCISE: Starting with Eq. (11.51), show that $|A_{LP}(j\omega_o)| = Q$.

EXERCISE: Change the cutoff frequency of this filter to 2 kHz by changing the values of R_1 and R_2. Do not change the Q.

ANSWERS: $R_1 = R_2 = 5.62$ kΩ

EXERCISE: Use the Q expression in Eq. (11.53) to show that $Q = 1/\sqrt{2}$ cannot be realized using the equal capacitance design. What is the maximum Q for $C_1 = C_2$?

ANSWER: 0.5

11.4.2 SENSITIVITY

An important concern in the design of active filters is the **sensitivity** of ω_o and Q to changes in passive element values and op amp parameters. The sensitivity of design parameter P to changes in circuit parameter Z is defined mathematically as

$$S_Z^P = \frac{\dfrac{\partial P}{P}}{\dfrac{\partial Z}{Z}} = \frac{Z}{P}\frac{\partial P}{\partial Z} \tag{11.54}$$

Sensitivity S represents the fractional change in parameter P due to a given fractional change in the value of Z. For example, evaluating the sensitivity of ω_o with respect to the values of R and C using Eq. (11.52) yields

$$S_R^{\omega_o} = S_C^{\omega_o} = -\frac{1}{2} \tag{11.55}$$

A 2 percent increase in the value of R or C will cause a 1 percent decrease in the frequency ω_o.

EXERCISE: Calculate $S_{C_1}^Q$ and $S_{R_2}^Q$ for the low-pass filter using Eq. (11.52) and the values in the example.

ANSWERS: $+0.5$; 0

11.4.3 A HIGH-PASS FILTER WITH GAIN

A **high-pass filter** can be achieved with the same topology as Fig. 11.19 by interchanging the position of the resistors and capacitors, as shown in Fig. 11.21. In many applications, filters with gain in the midband region are preferred, and the voltage follower in the low-pass filter has been replaced with a noninverting amplifier with a gain of K in the filter of Fig. 11.21. Gain K provides an additional degree of freedom in the design of the filter elements. Note that dc paths exist for both op amp input bias currents through resistor R_2 and the two feedback resistors.

The analysis is virtually identical to that of the low-pass filter. Nodes v_1 and v_2 are the only independent nodes because $v_o = +Kv_2$, and writing the two nodal equations yields this system of equations:

$$\begin{bmatrix} sC_1\mathbf{V_s}(s) \\ 0 \end{bmatrix} = \begin{bmatrix} s(C_1 + C_2) + G_1 & -(sC_2 + KG_1) \\ -sC_2 & sC_2 + G_2 \end{bmatrix}\begin{bmatrix} \mathbf{V_1}(s) \\ \mathbf{V_2}(s) \end{bmatrix} \tag{11.56}$$

The system determinant is

$$\Delta = s^2C_1C_2 + s(C_1 + C_2)G_2 + sC_2G_1(1 - K) + G_1G_2 \tag{11.57}$$

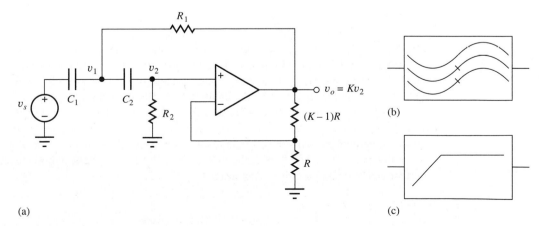

(a)

(b)

(c)

Figure 11.21 (a) A high-pass filter with gain. (b) High-pass filter symbol. (c) Alternate high-pass filter symbol.

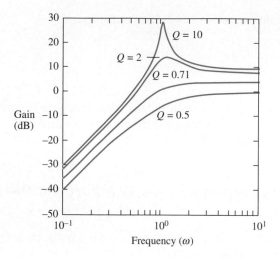

Figure 11.22 High-pass filter response[3] for $\omega_o = 1$ and four values of Q.

and the output voltage is

$$\mathbf{V_o}(s) = K\mathbf{V_2}(s) = K\frac{s^2 C_1 C_2 \mathbf{V_s}(s)}{\Delta} \tag{11.58}$$

Combining Eqs. (11.57) and (11.58) yields the filter transfer function that can be written in standard form as

$$A_{HP}(s) = \frac{\mathbf{V_o}(s)}{\mathbf{V_s}(s)} = K\frac{s^2}{s^2 + s\dfrac{\omega_o}{Q} + \omega_o^2} \tag{11.59}$$

in which

$$\omega_o = \frac{1}{\sqrt{R_1 R_2 C_1 C_2}} \qquad \text{and} \qquad Q = \left[\sqrt{\frac{R_1}{R_2}}\frac{C_1 + C_2}{\sqrt{C_1 C_2}} + (1-K)\sqrt{\frac{R_2 C_2}{R_1 C_1}}\right]^{-1} \tag{11.60}$$

For the case $R_1 = R_2 = R$ and $C_1 = C_2 = C$, Eqs. (11.59) and (11.60) can be simplified to

$$A_{HP}(s) = K\frac{s^2}{s^2 + s\dfrac{3-K}{RC} + \dfrac{1}{R^2 C^2}} \qquad \omega_o = \frac{1}{RC} \qquad \text{and} \qquad Q = \frac{1}{3-K} \tag{11.61}$$

For this design choice, ω_o and Q can be adjusted independently.

Figure 11.22 shows the high-pass filter responses for a filter with $\omega_o = 1$ and four values of Q. The parameter ω_o corresponds approximately to the lower-cutoff frequency of the filter, and $Q = 1/\sqrt{2}$ again represents the maximally flat, or Butterworth filter, response.

The noninverting amplifier circuit in Fig. 11.21 must have $K \geq 1$. Note in Eq. (11.61) that $K = 3$ corresponds to infinite Q. This situation corresponds to the poles of the filter being exactly on the imaginary axis at $s = j\omega_o$ and results in sinusoidal oscillation. (Oscillators are discussed in detail in Chapter 17.) For $K > 3$, the filter poles will be in the right-half plane, and values of $K \geq 3$ correspond to unstable filters. Therefore, $1 \leq K < 3$.

[3] Using MATLAB: Bode([(3-sqrt(2)) 0 0],[1 sqrt(2) 1]), for example.

EXERCISE: What is the gain at $\omega = \omega_o$ for the filter described by Eq. (11.61)?

ANSWER: $\dfrac{K}{3-K}$

EXERCISE: The high-pass filter in Fig. 11.21 has been designed with $C_1 = 0.0047\ \mu F$, $C_2 = 0.001\ \mu F$, $R_1 = 10\ k\Omega$, and $R_2 = 20\ k\Omega$, and the amplifier gain is 2. What are f_o and Q for this filter?

ANSWERS: 5.19 kHz, 0.828

EXERCISE: Derive an expression for the sensitivity of Q with respect to the closed-loop gain K for the high-pass filter in Fig. 11.22. What is the value of sensitivity if $Q = 1/\sqrt{2}$?

ANSWERS: $S_K^Q = Q(3-Q)$; 1.62

11.4.4 BAND-PASS FILTER

A **band-pass filter** can be realized by combining the low-pass and high-pass characteristics of the previous two filters. Figure 11.23 is one possible circuit for such a band-pass filter. In this case, the op amp is used in its inverting configuration; this circuit is sometimes called an "infinite-gain" filter because the full open-loop gain of the op amp, ideally infinity, is utilized. Resistor R_3 is added to provide an extra degree of design freedom so that gain, center frequency, and Q can be set with a minimum of interaction. Note again that dc paths exist for both op amp input bias currents.

Analysis of the circuit in Fig. 11.23(b) can be reduced to a one-node problem by using op amp theory to relate $\mathbf{V_o}(s)$ directly to $\mathbf{V_1}(s)$:

$$sC_2\mathbf{V_1}(s) = -\frac{\mathbf{V_o}(s)}{R_2} \qquad \text{or} \qquad \mathbf{V_1}(s) = -\frac{\mathbf{V_o}(s)}{sC_2R_2} \tag{11.62}$$

Using KCL at node v_1,

$$G_{\text{th}}\mathbf{V_{\text{th}}} = [s(C_1 + C_2) + G_{\text{th}}]\mathbf{V_1}(s) - sC_1\mathbf{V_o}(s) \tag{11.63}$$

Combining Eqs. (11.62) and (11.63) yields

$$\frac{\mathbf{V_o}(s)}{\mathbf{V_{\text{th}}}(s)} = \frac{-\dfrac{s}{R_{\text{th}}C_1}}{s^2 + s\dfrac{1}{R_2}\left(\dfrac{1}{C_1} + \dfrac{1}{C_2}\right) + \dfrac{1}{R_{\text{th}}R_2C_1C_2}} \tag{11.64}$$

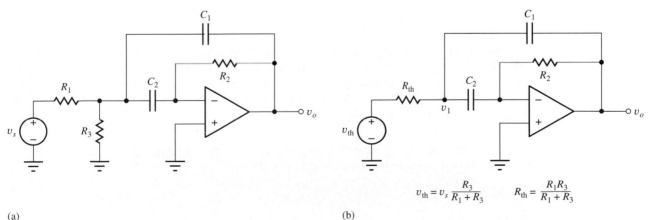

(a)

(b)

$$v_{\text{th}} = v_s\frac{R_3}{R_1 + R_3} \qquad R_{\text{th}} = \frac{R_1R_3}{R_1 + R_3}$$

Figure 11.23 (a) Band-pass filter using inverting op amp configuration. (b) Simplified band-pass filter circuit.

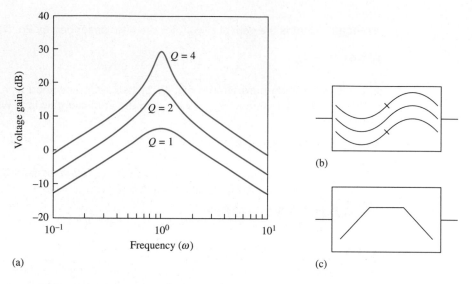

(a)

(b)

(c)

Figure 11.24 (a) Band-pass filter response[4] for $\omega_o = 1$ and three values of Q assuming $C_1 = C_2$ with $R_3 = \infty$. (b) Band-pass filter symbol. (c) Alternate band-pass filter symbol.

The band-pass output can now be expressed as

$$A_{BP}(s) = -\frac{V_o(s)}{V_s(s)} = -\sqrt{\frac{R_3}{R_1 + R_3}\frac{R_2 C_2}{R_1 C_1}}\frac{s\omega_o}{s^2 + s\dfrac{\omega_o}{Q} + \omega_o^2} \tag{11.65}$$

with

$$\omega_o = \frac{1}{\sqrt{R_{\text{th}} R_2 C_1 C_2}} \qquad \text{and} \qquad Q = \sqrt{\frac{R_2}{R_{\text{th}}}}\frac{\sqrt{C_1 C_2}}{C_1 + C_2} \tag{11.66}$$

If C_1 is set equal to $C_2 = C$, then

$$\omega_o = \frac{1}{C\sqrt{R_{\text{th}} R_2}} \qquad Q = \frac{1}{2}\sqrt{\frac{R_2}{R_{\text{th}}}} \qquad \text{BW} = \frac{2}{R_2 C} \tag{11.67}$$

$$A_{BP}(s) = -\left(\frac{2Q}{1 + \dfrac{R_1}{R_3}}\right)\left(\frac{s\omega_o}{s^2 + s\dfrac{\omega_o}{Q} + \omega_o^2}\right) \qquad A_{BP}(\omega_o) = -\frac{1}{2}\left(\frac{R_2}{R_1}\right)$$

The response of the band-pass filter is shown in Fig. 11.24 for $\omega_o = 1$, $C_1 = C_2$, $R_3 = \infty$, and three values of Q. Parameter ω_o now represents the center frequency of the band-pass filter. The response peaks at ω_o, and the gain at the center frequency is equal to $2Q^2$. At frequencies much less than or much greater than ω_o, the filter response corresponds to a single-pole high- or low-pass filter, changing at a rate of 20 dB/decade.

EXERCISE: The filter in Fig. 11.23 is designed with $C_1 = C_2 = 0.02$ μF, $R_1 = 2$ kΩ, $R_3 = 2$ kΩ, and $R_2 = 82$ kΩ. What are the values of f_o and Q?

ANSWERS: 879 Hz, 4.5

---

[4] Using MATLAB: Bode ([4 0],[1 .5 1]), for example.

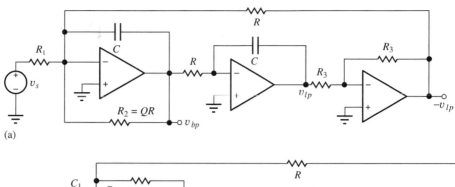

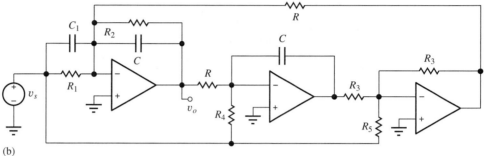

Figure 11.25 (a) Tow-Thomas filter with band-pass and low-pass outputs. (b) Full Tow-Thomas biquad.

11.4.5 THE TOW-THOMAS BIQUAD

Many single-amplifier filters do not permit independent design of ω_o, Q, and midband gain. Multi-amplifier filters trade increased component cost for ease of design and low sensitivity and can be used to realize the general biquadratic transfer function defined by

$$T(s) = \frac{a_2 s^2 + a_1 s + a_0}{s^2 + \dfrac{\omega_o}{Q} s + \omega_o^2} \tag{11.68}$$

Low-pass, high-pass, band-pass, **all-pass,** and **notch filters** can all be represented by this transfer function with appropriate choices of the numerator coefficients. Compare Eq. (11.68) with Eqs. (11.67), (11.59), and (11.51). Multi-op-amp filters are also used to achieve higher Qs than are practical with single op amp designs. One example of a two-pole filter using several op amps is the **Tow-Thomas biquad** in Fig. 11.25(a). The Tow-Thomas biquad inherently produces band-pass and low-pass outputs at v_{bp} and v_{lp}, respectively, but the other filter functions can easily be obtained with the addition of a few passive components, as in Fig. 11.25(b).

The filter response can be found by treating the first op amp as a multi-input integrator and noting that the third op amp simply forms an inverter with $V_o(s) = -V_{lp}(s)$. Using superposition, the output of the first integrator can be written as

$$\mathbf{V_{bp}}(s) = -\frac{1}{s R_1 C} \mathbf{V_s}(s) - \frac{1}{s R C}[-\mathbf{V_{lp}}(s)] - \frac{1}{s R_2 C} \mathbf{V_{bp}}(s) \tag{11.69}$$

$\mathbf{V_{lp}}(s)$ and $\mathbf{V_{bp}}(s)$ are related to each other by the second integrator:

$$\mathbf{V_{lp}}(s) = -\frac{1}{s R C} \mathbf{V_{bp}}(s) \tag{11.70}$$

Combining Eqs. (11.69) and (11.70) and solving for $\mathbf{V_{bp}}(s)$ gives

$$A_{BP}(s) = \frac{\mathbf{V_{bp}}(s)}{\mathbf{V_s}(s)} = -\frac{\left(\dfrac{R}{R_1}\right)\dfrac{s}{RC}}{s^2 + s\left(\dfrac{R}{R_2}\right)\dfrac{1}{RC} + \dfrac{1}{R^2 C^2}} = -K\frac{s\omega_o}{s^2 + s\dfrac{\omega_o}{Q} + \omega_o^2} \tag{11.71}$$

in which
$$K = \left(\frac{R}{R_1} \right) \qquad \omega_o = \frac{1}{RC} \qquad Q = \frac{R_2}{R} \qquad BW = \frac{1}{R_2 C}$$

The low-pass output is obtained using Eqs. (11.70) and (11.71):

$$A_{LP}(s) = K \frac{\dfrac{1}{R^2 C^2}}{s^2 + s\dfrac{1}{R_2 C} + \dfrac{1}{R^2 C^2}} = K \frac{\omega_o^2}{s^2 + s\dfrac{\omega_o}{Q} + \omega_o^2} \tag{11.72}$$

It can be observed that the center frequency ω_o, Q, and gain K of the filter are each controlled by separate element values and can all be adjusted independently.

Figure 11.25(b) shows the addition of the components needed to achieve other forms of the biquadratic transfer function. If v_O is defined as the output of the first amplifier, then

$$A_v(s) = \frac{\mathbf{V_o}(s)}{\mathbf{V_s}(s)} = - \frac{s^2 \left(\dfrac{C_1}{C} \right) + \dfrac{s}{C} \left(\dfrac{1}{R_1} - \dfrac{R_3}{RR_5} \right) + \dfrac{1}{RR_4 C^2}}{s^2 + s \left(\dfrac{R}{R_2} \right) \dfrac{1}{RC} + \dfrac{1}{R^2 C^2}} \tag{11.73}$$

DESIGN EXAMPLE 11.9 — TOW-THOMAS FILTER DESIGN

In this example, we will design a band-pass filter using the Tow-Thomas circuit.

PROBLEM Design a band-pass filter using the Tow-Thomas circuit to meet the following specifications: center frequency = 2000 Hz, bandwidth = 200 Hz, and midband gain = 20.

SOLUTION **Known Information and Given Data:** Tow-Thomas filter circuit with a center frequency $f_o = 2000$ Hz, bandwidth $BW = 200$ Hz, and midband gain $|A_v(j\omega_o)| = 20$

Unknowns: R, R_1, R_2, R_3, and C

Approach: We are given the circuit topology and the values of center frequency, bandwidth, and midband gain. Also, the Q of this filter is $Q = f_o / BW = 10$. In equation set (11.71), we see we have four circuit values to choose (R, R_1, R_2, C), but only three constraints have been specified. This is a situation usually encountered in design. In addition, we must pick a value for R_3 that does not directly appear in the band-pass filter equations.

Assumptions: The op amps are ideal.

Analysis: Let us set up the equations in a logical flow so that one value may be chosen, and it will then determine the others. Based on the values listed in Appendix A, we see that the choices for C are much more limited than those for the resistors, so we will choose C first. Then,

$$R = \frac{1}{\omega_o C} \qquad R_2 = QR \qquad R_1 = -\frac{R_2}{A_v(0)}$$

For the values in this design,

$$R = \frac{1}{4000\pi C} \qquad R_2 = 10R \qquad R_1 = \frac{R_2}{20} = \frac{R}{2}$$

An additional consideration is the "impedance level" of the filter. Looking at Fig. 11.25(a), we note that the input resistance to the filter is set by R_1. Also note that the magnitude of the reactance of the capacitor at the center frequency is given by

$$X_C = \frac{1}{\omega_o C} = R = 2R_1$$

As will be discussed in Chapter 12, op amps have limited output current drive capability (e.g., $R_L > 2$ kΩ for the LF155 and AD741 amplifiers—see website). The first op amp in the filter must supply ac signal current to the parallel combination of R, R_2, and C, the second op amp must drive the parallel combination of R_3 and C, and the third must drive R_3 in parallel with R. A spread sheet will help us visualize the design choices.

C	R	R_2	R_1	$R \parallel R_2 \parallel X_C$
1.00E − 09	7.96E + 04	7.96E + 05	3.98E + 04	19894
2.00E − 09	3.98E + 04	3.98E + 05	1.99E + 04	9947
2.20E − 09	3.62E + 04	3.62E + 05	1.81E + 04	9043
2.70E − 09	**2.95E + 04**	**2.95E + 05**	**1.47E + 04**	**7368**
3.30E − 09	2.41E + 04	2.41E + 05	1.21E + 04	6029

The bold row appears to be one reasonable choice. Based on Appendix A, our design can use $C = 2700$ pF, $R = 29.4$ kΩ, $R_2 = 294$ kΩ, and $R_1 = 14.7$ kΩ. The input resistance of the filter will be 14.7 kΩ, and the load impedance on the first op amp is on the order of 7.37 kΩ. Finally, the choice of R_3 is arbitrary as long as it is large enough not to load down the second and third op amps. $R_3 = 49.9$ kΩ is a reasonable choice.

Check of Results: Using the standard values selected:

$$A_v(0) = -20.0 \qquad f_o = \frac{1}{2\pi (29.4 \text{ k}\Omega)(2700 \text{ pF})} = 2005 \text{ Hz}$$

$$BW = \frac{1}{2\pi (294 \text{ k}\Omega)(2700 \text{ pF})} = 200.5 \text{ Hz}$$

Computer Aided Design: The filter response is simulated using the circuit below with these parameters for the op amps: input resistance = 2 MΩ, gain = 106 dB, and output resistance = 50 Ω. An ac analysis is performed with VS as the input source with FSTART = 100 Hz, FSTOP = 100 kHz, and 10 simulations points per frequency decade. The SPICE results are $f_o = 1.99$ kHz, $BW = 200$ Hz, and gain = 26.7 dB. Note that there is a slight gain enhancement and center frequency shift resulting from the nonideal op amp parameters.

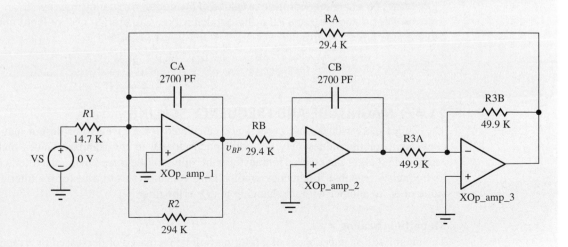

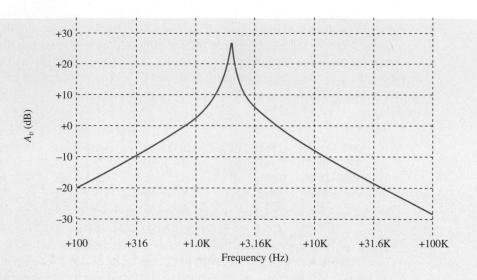

EXERCISE: Simulate the filter in Ex. 11.9 with op amps having an open loop gain of 80 dB. What is the filter gain at the center frequency?

ANSWER: 31.5 dB.

EXERCISE: Redesign the filter in Ex. 11.9 to have a gain of 20 dB at the center frequency.

ANSWER: Change R_1 to 29.4 kΩ.

EXERCISE: What standard resistor values would be used if the $C = 2000$ pF design were utilized in Design Ex. 11.9? What are the center frequency, bandwidth, and voltage gain for that filter design?

ANSWERS: $R = 40.2$ kΩ, $R_2 = 402$ kΩ, $R_1 = 20.0$ kΩ, $R_3 = 49.9$ kΩ can be used; 1980 Hz, 198 Hz, −20.1

EXERCISE: What are the worst-case values of the center frequency, bandwidth and voltage gain for the filter design in Ex. 11.9 if the capacitor has a tolerance of 2 percent and the resistor tolerances are all 1 percent?

ANSWERS: 1946 Hz; 2067 Hz; 195 Hz; 207 Hz; −19.6; −20.4

11.4.6 MAGNITUDE AND FREQUENCY SCALING

The values of resistance and capacitance calculated for a given filter design may not always be convenient, or the values may not correspond closely to the standard values that are available. Magnitude scaling can be used to transform the values of the impedances of a filter without changing its frequency response. Frequency scaling, however, allows us to transform a filter design from one value of ω_o to another without changing the Q of the filter.

Magnitude Scaling

The magnitude of impedances of a filter may all be increased or decreased by a **magnitude scaling** factor K_M without changing ω_o or Q of the filter. To scale the magnitude of the impedance of the filter elements, the value of each resistor[5] is multiplied by K_M and the value of the capacitor is divided by K_M:

$$R' = K_M R \qquad \text{and} \qquad C' = \frac{C}{K_M} \qquad \text{so that} \qquad |Z'_C| = \frac{1}{\omega C'} = \frac{K_M}{\omega C} = K_M |Z_C| \quad (11.74)$$

In all the filters discussed in Sec. 11.4, Q is determined by ratios of capacitor values and/or resistor values whereas ω_o always has the form $\omega_o = 1/\sqrt{R_1 R_2 C_1 C_2}$. Applying magnitude scaling to the low-pass filter described by Eq. (11.52) yields

$$\omega'_o = \frac{1}{\sqrt{K_M R_1 (K_M R_2) \dfrac{C_1}{K_M} \dfrac{C_2}{K_M}}} = \frac{1}{\sqrt{R_1 R_2 C_1 C_2}} = \omega_o$$

and

$$Q' = \sqrt{\frac{\dfrac{C_1}{K_M}}{\dfrac{C_2}{K_M}}} \, \frac{\sqrt{K_M R_1 (K_M R_2)}}{K_M R_1 + K_M R_2} = \sqrt{\frac{C_1}{C_2}} \, \frac{\sqrt{R_1 R_2}}{R_1 + R_2} = Q \qquad (11.75)$$

Thus, both Q and ω_o are independent of the magnitude scaling factor K_M.

EXERCISE: The filter in Fig. 11.23 is designed with $R_1 = R_2 = 2.26$ kΩ, $R_3 = \infty$, $C_1 = 0.02$ μF, and $C_2 = 0.01$ μF. What are the new values of C_1, C_2, R_1, R_2, f_o, and Q if the impedance magnitude is scaled by a factor of (a) 5 and (b) 0.885?

ANSWERS: (a) 0.004 μF, 0.002 μF, 11.3 kΩ, 11.3 kΩ, 4980 Hz, 0.707; (b) 0.0226 μF, 0.0113 μF, 2.00 kΩ, 2.00 kΩ, 4980 Hz, 0.707

Frequency Scaling

The cutoff or center frequencies of a filter may be scaled by a **frequency scaling** factor K_F without changing the Q of the filter if each capacitor value is divided by K_F, and the resistor values are left unchanged.

$$R' = R \qquad \text{and} \qquad C' = \frac{C}{K_F}$$

Once again, using the low-pass filter as an example yields:

$$\omega'_o = \frac{1}{\sqrt{R_1 R_2 \dfrac{C_1}{K_F} \dfrac{C_2}{K_F}}} = \frac{K_F}{\sqrt{R_1 R_2 C_1 C_2}} = K_F \omega_o$$

and

$$Q' = \sqrt{\frac{\dfrac{C_1}{K_F}}{\dfrac{C_2}{K_F}}} \, \frac{\sqrt{R_1 R_2}}{R_1 + R_2} = \sqrt{\frac{C_1}{C_2}} \, \frac{\sqrt{R_1 R_2}}{R_1 + R_2} = Q \qquad (11.76)$$

In this case, we see that the value of ω_o is increased by the factor K_F, but Q remains unaffected.

EXERCISE: The filter in Fig. 11.23 is designed with $C_1 = C_2 = 0.02$ μF, $R_1 = 2$ kΩ, $R_3 = 2$ kΩ, and $R_2 = 82$ kΩ. (a) What are the values of f_o and Q? (b) What are the new values of C_1, C_2, R_1, R_2, f_o, and Q if the frequency is scaled by a factor of 4?

ANSWERS: 880 Hz, 4.5; 0.005 μF, 0.005 μF, 1 kΩ, 82 kΩ, 3.5 kHz, 4.5.

---

[5] In *RLC* filters, each inductor value is also increased by K_M: $L' = K_M L$ so $|Z'_L| = K_M |Z_L|$.

ELECTRONICS IN ACTION

Band-pass Filters in BFSK Reception

Binary frequency shift keying (BFSK) is a basic form of modulation that is studied in communications classes and represents a type of communications that is commonly used for radio teletype transmissions in the high-frequency or "short wave" radio bands (3–30 MHz). The signal transmitting the data shifts back and forth between two closely spaced radio frequencies (for example, 18,080,000 Hz and 18,080,170 Hz). In the block diagram in the accompanying figure, a communications receiver that is receiving this transmission produces an audio signal at its output that shifts between 2125 Hz and 2295 Hz (or some other convenient frequency pair separated by a frequency shift of 170 Hz). In the analog signal processing circuit here, six-pole filters are used to separate these two audio tones. Each filter bank consists of a cascade of three, two-pole active band-pass filters as described in Sec. 11.4.4. The outputs of the two filter banks are rectified and filtered by an RC network to form a simple frequency discriminator. The output of the discriminator then drives circuitry that recovers the original data transmission.

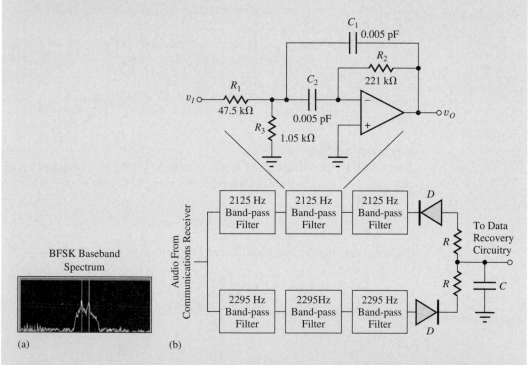

(a) (b)

These same functions can be performed in the digital domain using digital signal processing (DSP) if the audio signal from the communications receiver is first digitized by an analog-to-digital (A/D) converter.

11.5 NONLINEAR CIRCUIT APPLICATIONS

Up to this point, we have considered only operational amplifier circuits that use passive linear-circuit elements in the feedback network. But many interesting and useful circuits can be constructed using nonlinear elements, such as diodes and transistors in the feedback network. This section explores several examples of such circuits.

In addition, our op amp circuits thus far have involved only negative feedback configurations, but a number of important nonlinear circuits employ positive feedback. Section 11.5 looks at this important class of circuits, including op amp implementations of the astable and monostable multivibrators and the Schmitt trigger circuit.

11.5.1 A PRECISION HALF-WAVE RECTIFIER

An op amp and diode are combined in Fig. 11.26 to form a **precision half-wave rectifier** circuit. The output v_O represents a rectified replica of the input signal v_S without loss of the voltage drop encountered with a normal diode rectifier circuit. The op amp tries to force the voltage across its input terminals to be zero. For $v_S > 0$, v_O equals v_S, and $i > 0$. Because current i_- must be zero, diode current i_D is equal to i, diode D is forward-biased, and the feedback loop is closed through the diode. However, for negative output voltages, currents i and i_D would be less than zero, but negative current cannot go through D_1. Thus, the diode cuts off ($i_D = 0$), the feedback loop is broken (inactive), and $v_O = 0$ because $i = 0$.

The resulting voltage transfer function for the precision rectifier is shown in Fig. 11.27. For $v_S \geq 0$, $v_O = v_S$, and for $v_S \leq 0$, $v_O = 0$. The rectification is precise; for $v_S \geq 0$, the operational amplifier adjusts its output v_1 to exactly absorb the forward voltage drop of the diode:

$$v_1 = v_O + v_D = v_S + v_D \tag{11.77}$$

This circuit provides accurate rectification even for very small input voltages and is sometimes called a **superdiode.** The primary sources of error are gain error due to the finite gain of the op amp, as well as an offset error due to the offset voltage of the amplifier. These errors are discussed in Chapter 12.

A practical problem occurs in this circuit for negative input voltages. Although the output voltage is zero, as desired for the rectifier, the voltage across the op amp input terminals is now negative, and the output voltage v_1 is saturated at the negative supply limit. Most modern op amps provide input voltage protection and will not be damaged by a large voltage across the input. However, unprotected op amps can be destroyed if the magnitude of the input voltage is larger than a few volts. The saturated output of the op amp is not usually harmful to the amplifier, but it does take time for the internal circuits to recover from the saturated condition, thus slowing down the response time of the circuit. It is preferable to prevent the op amp from saturating, if possible.

EXERCISE: Suppose diode D_1 in Fig. 11.26 has an "on-voltage" of 0.6 V, and the op amp is operating with ±10-V power supplies. What are the voltages v_O and v_1 for the circuit if $v_S = +1$ V? For $v_S = -1$ V? What is the minimum Zener breakdown voltage for the diode?

ANSWERS: +1 V, +1.6 V; 0 V, −10 V; 10 V

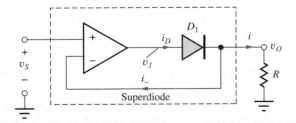

Figure 11.26 Precision half-wave rectifier circuit (or "superdiode").

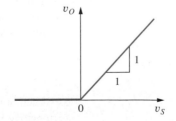

Figure 11.27 Voltage transfer characteristic for the precision rectifier.

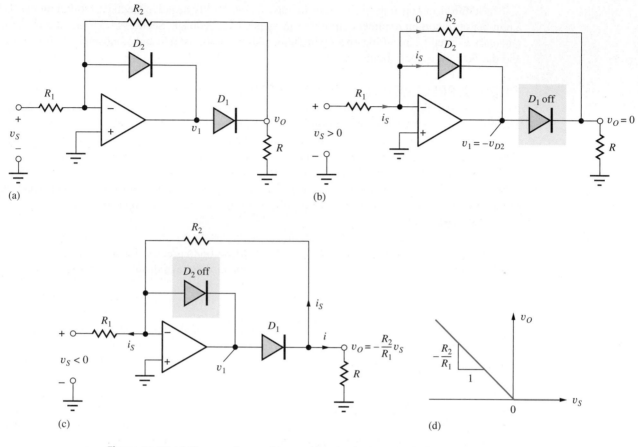

Figure 11.28 (a) Nonsaturating precision-rectifier circuit. (b) Active feedback elements for $v_S \geq 0$. (c) Active feedback elements for $v_S < 0$. (d) Improved rectifier voltage transfer characteristic.

11.5.2 NONSATURATING PRECISION-RECTIFIER CIRCUIT

The saturation problem can be solved using the circuit given in Fig. 11.28. An inverting-amplifier configuration is used instead of the noninverting configuration, and diode D_2 is added to keep the feedback loop closed when the output of the rectifier is zero.

For positive input voltages depicted in Fig. 11.28(b), the op amp output voltage v_1 becomes negative, forward-biasing diode D_2 so that current i_S passes through diode D_2 and into the output of the op amp. The inverting input is at virtual ground, the current in R_2 is zero, and the output remains at zero. Diode D_1 is reverse-biased.

For $v_S < 0$ in Fig. 11.28(c), diode D_1 turns on and supplies current i_S and load current i, and D_2 is off. The circuit behaves as an inverting amplifier with gain equal to $-R_2/R_1$. Thus, the overall voltage transfer characteristic can be described by

$$v_O = 0 \text{ for } v_S \geq 0 \quad \text{and} \quad v_O = -\frac{R_2}{R_1} v_S \text{ for } v_S \leq 0 \qquad (11.78)$$

as shown in Fig. 11.28(d). The output voltage of the op amp itself, v_1, is one diode-drop below zero for positive input voltages and one diode above the output voltage for negative input voltages. The inverting input is a virtual ground in both cases, and the negative feedback loop is always active: through D_1 and R_2 for $v_S < 0$ and through D_2 for $v_S > 0$.

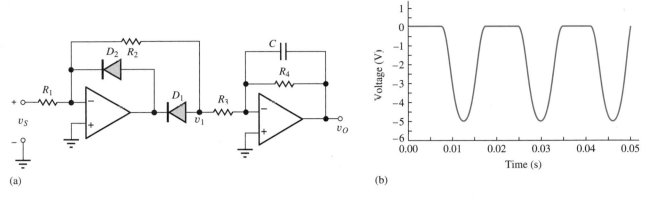

Figure 11.29 (a) AC voltmeter circuit consisting of a half-wave rectifier and low-pass filter. (b) Voltage waveform at rectifier output v_1 for $v_S = (-5 \sin 120\pi t)$ V and $R_2 = R_1$.

EXERCISE: Suppose the diodes in Fig. 11.28 have "on-voltages" of 0.6 V, and the op amp is operating with ±15-V power supplies. What are the voltages v_O and v_1 for the circuit if $R_1 = 22$ kΩ, $R_2 = 68$ kΩ, and $v_S = +2$ V? For $v_S = -2$ V? Estimate the most negative input voltage for which the circuit will operate properly. What is the minimum Zener breakdown voltage specification for the diodes assuming they are both the same?

ANSWERS: 0 V, −0.6 V; +6.18 V, +6.78 V; −4.66 V; 15 V

11.5.3 AN AC VOLTMETER

The half-wave rectifier circuit can be combined with a low-pass filter to form a basic ac voltmeter circuit, as in Fig. 11.29. For a sinusoidal input signal with an amplitude V_M at a frequency ω_o, the output voltage v_1 is a rectified sine wave that can be described by its Fourier series as:

$$v_1(t) = -\left(\frac{R_2}{R_1}\right)\left(\frac{V_M}{\pi}\right)\left[1 + \frac{\pi}{2}\sin\omega_o t - \sum_{n=2}^{\infty}\frac{1 + \cos n\pi}{(n^2 - 1)}\cos n\omega_o t\right] \qquad (11.79)$$

If the cutoff frequency of the low-pass filter is chosen such that $\omega_C \ll \omega_o$, then the output voltage v_O will consist primarily of the dc voltage component (see Prob. 11.93) given by

$$v_O = \frac{R_4}{R_3}\left[\frac{R_2}{R_1}\frac{V_M}{\pi}\right] \qquad (11.80)$$

The voltmeter range (scale factor) can be adjusted through the choice of the four resistors.

EXERCISE: What is the dc output voltage of the circuit in Fig. 11.29 if $R_1 = 3.24$ kΩ, $R_2 = 10.2$ kΩ, $R_3 = 20$ kΩ, $R_4 = 20$ kΩ, and $V_M = 2$ V?

ANSWER: 2.00 V

11.6 CIRCUITS USING POSITIVE FEEDBACK

Up to now, all our circuits except the active filters have used negative feedback: A voltage or current proportional to the output signal was returned to the inverting-input terminal of the operational amplifier. However, positive feedback can also be used to perform a number of useful nonlinear functions, and we investigate several possibilities in this final section, including the comparator, Schmitt trigger, and multivibrator circuits.

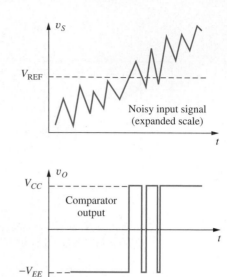

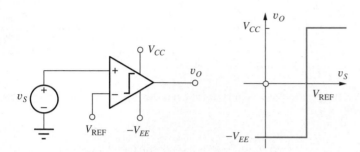

Figure 11.30 Comparator circuit using an infinite-gain amplifier.

Figure 11.31 Comparator response to noisy input signal.

11.6.1 THE COMPARATOR AND SCHMITT TRIGGER

It is often useful to compare a voltage to a known reference level. This can be done electronically using the **comparator** circuit in Fig. 11.30. We want the output of the comparator to be a logic 1 when the input signal exceeds the reference level and a logic 0 when the input is less than the reference level. The basic comparator is simply a very high gain amplifier without feedback, as indicated in Fig. 11.30. For input signals exceeding the reference voltage V_{REF}, the output saturates at V_{CC}; for input signals less than V_{REF}, the output saturates at $-V_{EE}$, as indicated in the voltage transfer characteristic in Fig. 11.30.[6] Amplifiers built for use as comparators are specifically designed to be able to saturate at the two voltage extremes without incurring excessive internal time delays.

However, a problem occurs when high-speed comparators are used with noisy signals, as indicated in Fig. 11.31. As the input signal crosses through the reference level, multiple transitions may occur due to noise present on the input. In digital systems, we often want to detect this threshold crossing cleanly by generating only a single transition, and the **Schmitt-trigger** circuit in Fig. 11.32 helps solve this problem.

The Schmitt trigger uses a comparator whose reference voltage is derived from a voltage divider across the output. The input signal is applied to the inverting-input terminal, and the reference voltage is applied to the noninverting input (positive feedback). For positive output voltages, $V_{REF} = \beta V_{CC}$, but for negative output voltages, $V_{REF} = -\beta V_{EE}$, where $\beta = R_1/(R_1 + R_2)$. Thus, the reference voltage changes when the output switches state.

Consider the case for an input voltage increasing from below V_{REF}, as in Fig. 11.33. The output is at V_{CC} and $V_{REF} = \beta V_{CC}$. As the input voltage crosses through V_{REF}, the output switches state to $-V_{EE}$, and the reference voltage simultaneously drops, reinforcing the voltage across the comparator input. In order to cause the comparator to switch states a second time, the input must now drop below $V_{REF} = -\beta V_{EE}$, as depicted in Fig. 11.34.

Now consider the situation as v_S decreases from a high level, as in the voltage transfer characteristic in Fig. 11.34. The output is at $-V_{EE}$ and $V_{REF} = -\beta V_{EE}$. As the input voltage crosses through V_{REF}, the output switches state to V_{CC}, and the reference voltage simultaneously increases, again reinforcing the voltage across the comparator input.

The voltage transfer characteristics from Figs. 11.33 and 11.34 are combined to yield the overall voltage transfer characteristic for the Schmitt trigger given in Fig. 11.35. The arrows indicate the

---

[6] In this section, we assume that the output can reach the supply voltages.

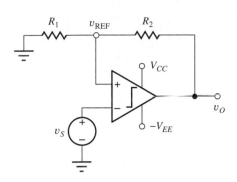

Figure 11.32 Schmitt-trigger circuit.

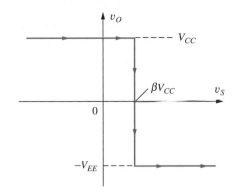

Figure 11.33 Voltage transfer characteristic for the Schmitt trigger as v_S increases from below $V_{\text{REF}} = +\beta V_{CC}$.

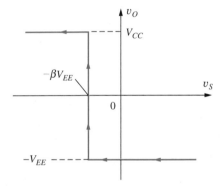

Figure 11.34 Voltage transfer characteristic for the Schmitt trigger as v_S decreases from above $V_{\text{REF}} = -\beta V_{EE}$.

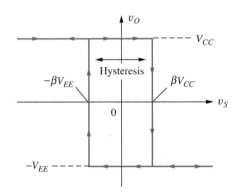

Figure 11.35 Complete voltage transfer characteristic for the Schmitt trigger.

portion of the characteristic that is traversed for increasing and decreasing values of the input signal. The Schmitt trigger is said to exhibit **hysteresis** in its VTC, and will not respond to input noise that has a magnitude V_n smaller than the difference between the two threshold voltages:

$$V_n < \beta[V_{CC} - (-V_{EE})] = \beta(V_{CC} + V_{EE}) \tag{11.81}$$

The Schmitt trigger with positive feedback is an example of a circuit with two stable states: a **bistable circuit,** or **bistable multivibrator.** Another example of a bistable circuit is the digital storage element usually called the flip-flop (see Chapter 8).

EXERCISE: If $V_{CC} = +10$ V $= -V_{EE}$, $R_1 = 1$ kΩ, and $R_2 = 9.1$ kΩ, what are the values of the switching thresholds for the Schmitt-trigger circuit in Figs. 11.32 through 11.35 and the magnitude of the hysteresis?

ANSWERS: $+0.99$ V; -0.99 V; 1.98 V

11.6.2 THE ASTABLE MULTIVIBRATOR

Another type of multivibrator circuit employs a combination of positive and negative feedback and is designed to oscillate and generate a rectangular output waveform. The output of the circuit in Fig. 11.36 has no stable state and is referred to as an **astable multivibrator.**

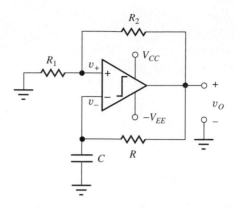

Figure 11.36 Operational amplifier in an astable multivibrator circuit.

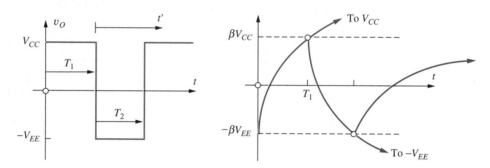

Figure 11.37 Waveforms for the astable multivibrator.

Operation of the astable multivibrator circuit can best be understood by referring to the waveforms in Fig. 11.37. The output voltage switches periodically (oscillates) between the two output voltages V_{CC} and $-V_{EE}$. Let us assume that the output has just switched to $v_O = V_{CC}$ at $t = 0$. The voltage at the inverting-input terminal of the op amp charges exponentially toward a final value of V_{CC} with a time constant $\tau = RC$. The voltage on the capacitor at the time of the output transition is $v_C = -\beta V_{EE}$. Thus, the expression for the voltage on the capacitor can be written as

$$v_C(t) = V_{CC} - (V_{CC} + \beta V_{EE}) \exp\left(-\frac{t}{RC}\right) \qquad (11.82)$$

The comparator changes state again at time T_1 when $v_C(t)$ just reaches βV_{CC}:

$$\beta V_{CC} = V_{CC} - (V_{CC} + \beta V_{EE}) \exp\left(-\frac{T_1}{RC}\right) \qquad (11.83)$$

Solving for time T_1 yields

$$T_1 = RC \ln \frac{1 + \beta\left(\dfrac{V_{EE}}{V_{CC}}\right)}{1 - \beta} \qquad (11.84)$$

During time interval T_2, the output is low and the capacitor discharges from an initial voltage of βV_{CC} toward a final voltage of $-V_{EE}$. For this case, the capacitor voltage can be expressed as

$$v_C(t') = -V_{EE} + (V_{EE} + \beta V_{CC}) \exp\left(-\frac{t'}{RC}\right) \qquad (11.85)$$

in which $t' = 0$ at the beginning of the T_2 interval. At $t' = T_2$, $v_C = -\beta V_{EE}$,

$$-\beta V_{EE} = -V_{EE} + (V_{EE} + \beta V_{CC}) \exp\left(-\frac{T_2}{RC}\right) \qquad (11.86)$$

and T_2 is equal to

$$T_2 = RC \ln \frac{1 + \beta \left(\dfrac{V_{CC}}{V_{EE}} \right)}{1 - \beta} \tag{11.87}$$

For the common case of symmetrical power supply voltages, $V_{CC} = V_{EE}$, and the output of the astable multivibrator represents a square wave with a period T given by

$$T = T_1 + T_2 = 2RC \ln \frac{1 + \beta}{1 - \beta} \tag{11.88}$$

EXERCISE: What is the frequency of oscillation of the circuit in Fig. 11.36 if $V_{CC} = +5$ V, $-V_{EE} = -5$ V, $R_1 = 6.8$ kΩ, $R_2 = 6.8$ kΩ, $R = 10$ kΩ, and $C = 0.001$ μF?

ANSWER: 45.5 kHz

ELECTRONICS IN ACTION

Function Generators
Analog Function Generators
The instrumentation in most introductory electronics laboratories includes some type of low frequency function generator that produces elementary waveforms including square, triangle, and sine wave outputs at frequencies up to a few MHz. For many years, inexpensive versions of these function generators utilized the astable multivibrator to generate the square wave signal as shown in the accompanying figure. The frequency of the multivibrator is varied by changing either R_3 or C_3. C_3 is often changed in decade steps; R_3 may be varied continuously using a potentiometer. The output of the astable multivibrator drives an op amp integrator circuit to produce a triangular waveform. The output of the integrator can then be passed through a low-pass filter or piecewise linear shaping circuit to produce a low-distortion sine wave.

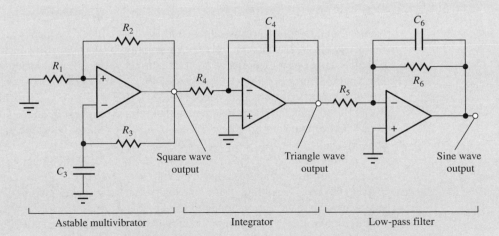

Simple function generator using an astable multivibrator, integrator, and low-pass filter.

Function generator.
© Agilent Technologies 2006. All Rights Reserved.

Direct Digital Synthesis

Today, traditional analog function generators have been largely replaced with instruments based upon direct digital synthesis (DDS) as depicted in the block diagram below. In a DDS, the signal waveform is constructed in the digital domain, and the analog output signal is produced using a digital-to-analog converter followed by a low-pass filter.

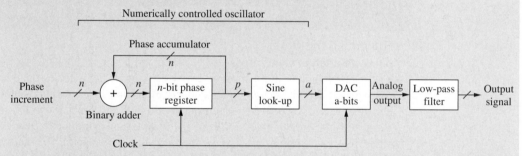

The digital portion of the DDS consists of an n-bit phase accumulator and a sine look-up table. To generate a sine wave, an n-bit phase increment is added to the accumulator at each clock cycle. A full phase register (2^n) corresponds to 2π radians or 1 cycle of the output sine wave. If the phase is incremented by one at each clock interval, the maximum period T_{max} of the output waveform, corresponding to minimum output frequency f_{min}, will be

$$T_{max} = 2^n T_{clk} \qquad \text{or} \qquad f_{min} = \frac{f_{clk}}{2^n}$$

where T_{clk} is the period of the clock, and f_{clk} is the clock frequency. This minimum output frequency also represents the frequency resolution of the DDS. To generate higher frequency signals, a larger phase increment N is added to the phase at each clock cycle, and $f_o = N f_{min}$. For example, for $f_{clk} = 20$ MHz and $n = 24$, $f_{min} \approx 1.192$ Hz. In order to generate a 10 kHz sine wave, a phase increment of $8389(10,000/1.192)$ would be added to the phase at each clock cycle. Based upon the Nyquist sampling theorem, the highest frequency that can be generated is one-half of the clock frequency (using $N = 2^{n/2}$), since f_{clk} is the update rate of the D/A converter.

In order to reduce the size of the look-up table, only the upper p bits of the phase accumulator are used to address the sine table, and a number of ROM compression techniques are also utilized. The output of the sine table is an a-bit representation of the amplitude of the sine wave where "a" corresponds to the number of bits of resolution of the D/A converter. Finite resolution in the representation of both the signal phase and amplitude lead-to distortion in the output waveform. The low-pass filter helps to remove distortion and the high frequency content related to the update rate of the DAC (f_{clk}).

Direct digital synthesis provides a great deal of flexibility since the digital hardware can create highly complex waveforms in the digital domain. For example, sine and cosine waves can be generated with very precise 90-degree phase relationships, and square and triangular waveforms can be generated by changing the phase update information as a function of time. AM or FM modulated waveforms can also readily be generated with only minor additional complexity.

11.6.3 THE MONOSTABLE MULTIVIBRATOR OR ONE SHOT

A third type of multivibrator operates with one stable state and is used to generate a single pulse of known duration following application of a trigger signal. The circuit rests quiescently in its stable state, but can be "triggered" to generate a single transient pulse of fixed duration T. Once the time T is past, the circuit returns to the stable state to await another **triggering** pulse. This **monostable circuit** is variously called a **monostable multivibrator, a single shot,** or a **one shot.**

An example of a comparator-based monostable multivibrator circuit is given in Fig. 11.38. Diode D_1 has been added to the astable multivibrator in Fig. 11.36 to couple the triggering signal v_T into the circuit, and clamping diode D_2 has been added to limit the negative voltage excursion on capacitor C.

The circuit rests in its quiescent state with $v_O = -V_{EE}$. If the trigger signal voltage v_T is less than the voltage at node 2,

$$v_T < -\frac{R_1}{R_1 + R_2} V_{EE} = -\beta V_{EE} \tag{11.89}$$

diode D_1 is cut off. Capacitor C discharges through R until diode D_2 turns on, clamping the capacitor voltage at one diode-drop V_D below ground potential. In this condition, the differential-input voltage v_{ID} to the comparator is given by

$$v_{ID} = -\beta V_{EE} - (-V_D) = -\beta V_{EE} + V_D \tag{11.90}$$

As long as the value of the voltage divider is chosen so that

$$v_{ID} < 0 \quad \text{or} \quad \beta V_{EE} > V_D \quad \text{where } \beta = \frac{R_1}{R_1 + R_2} \tag{11.91}$$

then the output of the circuit will have one stable state.

Triggering the Monostable Multivibrator

The monostable multivibrator can be triggered by applying a positive pulse to the trigger input v_t, as shown in the waveforms in Fig. 11.39. As the trigger pulse level exceeds a voltage of $-\beta V_{EE}$, diode D_1 turns on and subsequently pulls the voltage at node 2 above that at node 3. At this point, the comparator output changes state, and the voltage at the noninverting-input terminal rises abruptly to

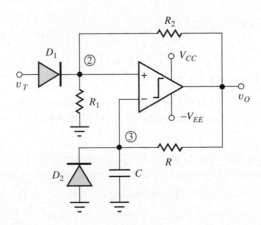

Figure 11.38 Example of an operational-amplifier monostable-multivibrator circuit.

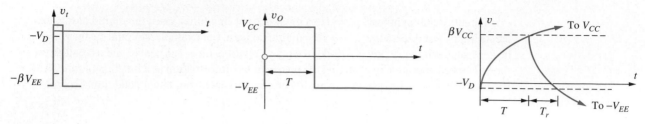

Figure 11.39 Monostable multivibrator waveforms.

a voltage equal to $+\beta V_{CC}$. Diode D_1 cuts off, isolating the comparator from any further changes on the trigger input.

The voltage on the capacitor now begins to charge from its initial voltage $-V_D$ toward a final voltage of V_{CC} and can be expressed mathematically as

$$v_C(t) = V_{CC} - (V_{CC} + V_D) \exp\left(-\frac{t}{RC}\right) \tag{11.92}$$

where the time origin ($t = 0$) coincides with the start of the trigger pulse. However, the comparator changes state again when the capacitor voltage reaches $+\beta V_{CC}$. Thus, the pulse width T is given by

$$\beta V_{CC} = V_{CC} - (V_{CC} + V_D) \exp\left(-\frac{T}{RC}\right) \quad \text{or} \quad T = RC \ln \frac{1 + \left(\dfrac{V_D}{V_{CC}}\right)}{1 - \beta} \tag{11.93}$$

The output of the circuit consists of a positive pulse with a fixed duration T set by the values of R_1, R_2, R, and C.

For a well-defined pulse width to be generated, this circuit should not be retriggered until the voltages on the various nodes have all returned to their quiescent steady-state values. Following the return of the output to $-V_{EE}$, the capacitor voltage charges from a value of βV_{CC} toward $-V_{EE}$, but reaches steady state when diode D_2 begins to conduct. Thus, the recovery time can be calculated from

$$-V_D = -V_{EE} + (V_{EE} + \beta V_{CC}) \exp\left(-\frac{T_r}{RC}\right) \quad \text{and} \quad T_r = RC \ln \frac{1 + \beta\left(\dfrac{V_{CC}}{V_{EE}}\right)}{1 - \left(\dfrac{V_D}{V_{EE}}\right)} \tag{11.94}$$

EXERCISE: For the monostable multivibrator circuit in Fig. 11.38, $V_{CC} = +5\,\text{V} = V_{EE}$, $R_1 = 22\,\text{k}\Omega$, $R_2 = 18\,\text{k}\Omega$, $R = 11\,\text{k}\Omega$, and $C = 0.002\,\mu\text{F}$. What is the pulse width of the one shot? What is the minimum time between trigger pulses for this circuit?

ANSWERS: 20.4 μs; 33.4 μs

SUMMARY

The introduction to operational amplifiers in Chapter 11 began with a discussion of the ideal op amp and then explored the behavior of circuits containing ideal op amps. Key points and topics are outlined here.

- Ideal operational amplifiers are assumed to have infinite gain and zero input current, and circuits containing these amplifiers were analyzed using two primary assumptions:

   1. The differential input voltage is zero: $v_{id} = 0$.

   2. The input currents are zero: $i_+ = 0$ and $i_- = 0$.

- Assumptions 1 and 2, combined with Kirchhoff's voltage and current laws, are used to analyze the ideal behavior of circuit building blocks based on operational amplifiers. Constant feedback with resistive voltage dividers is used in the inverting and noninverting amplifier configurations, the voltage follower, the difference amplifier, the summing amplifier, and the instrumentation amplifier, whereas frequency-dependent feedback is used in the integrator, low-pass filter, and differentiator circuits.

- Infinite gain and input resistance are the explicit characteristics that lead to Assumptions 1 and 2. However, many additional properties are implicit characteristics of ideal operational amplifiers; these assumptions are seldom clearly stated, though. They are

  - Infinite common-mode rejection
  - Infinite power supply rejection
  - Infinite output voltage range
  - Infinite output current capability
  - Infinite open-loop bandwidth
  - Infinite slew rate
  - Zero output resistance
  - Zero input-bias currents
  - Zero input-offset voltage

- Active RC filters including low-pass, high-pass, and band-pass circuits were introduced. These designs use RC feedback networks and operational amplifiers to replace bulky inductors that would normally be required in RLC filters designed for the audio range. Single-amplifier active filters employ a combination of negative and positive feedback to realize second-order low-pass, high-pass, and band-pass transfer functions.

- Sensitivity of filter characteristics to passive component and op amp parameter tolerances is an important design consideration. Multiple op amp filters offer low sensitivity and ease of design, compared to their single op amp counterparts.

- Magnitude and frequency scaling can be used to change the impedance level and ω_o of a filter without affecting its Q.

- Many circuits employ comparators to compare an unknown input voltage with a precision reference voltage. The comparator can be considered to be a high-gain, high-speed op amp designed to operate without feedback.

- Nonlinear circuit applications of operational amplifiers were also introduced including several precision-rectifier circuits.

- Multivibrator circuits are used to develop various forms of electronic pulses. The bistable Schmitt-trigger circuit has two stable states and is often used in place of the comparator in noisy environments. The monostable multivibrator, or one shot, is used to generate a single pulse of known duration, whereas the astable multivibrator has no stable state and oscillates continuously, producing a square wave output.

KEY TERMS

Active filters
All-pass filter
Astable multivibrator
Band-pass filter
Bias current compensation resistor

Bistable circuit
Bistable multivibrator
Butterworth filter
Closed-loop amplifier
Closed-loop gain

Comparator
dc-coupled amplifier
Difference amplifier
Differential amplifier
Differential-mode gain
Differential-mode input resistance
Differential-mode input voltage
Differential subtractor
Differentiator
Digital-to-analog converter
 (DAC or D/A converter)
Dual-ramp (dual slope) ADC
Feedback amplifier
Feedback factor (β)
Feedback network f_M
Frequency scaling
High-pass filter
Hysteresis
Ideal operational amplifier
Instrumentation amplifier
Integrator
Inverted R-2R ladder
Inverting amplifier
Inverting input
Least significant bit (LSB)
Loop gain ($A\beta$) (Loop transmission T)
Low-pass filter
Magnitude scaling
Maximally flat magnitude
Monostable circuit

Monostable multivibrator
Most significant bit (MSB)
Noninverting amplifier
Noninverting input
Normal mode rejection
Notch filter
One shot
Open-circuit voltage gain
Open-loop gain
Operational amplifier (op amp)
Precision half-wave rectifier
R-2R ladder
Reference voltage
Schmitt trigger
Sensitivity
Series feedback
Shunt feedback
Single-pole frequency response
Single shot
Summing amplifier
Summing junction
Superdiode
Tow-Thomas biquad
Triggering
Two-port model
Unity-gain buffer
Virtual ground
Voltage follower
Weighted-resistor DAC

REFERENCES

1. Connelly, J. A. and P. Choi, *Macromodeling with SPICE,* Prentice Hall, Englewood Cliffs, NJ: 1992.

2. Boyle, G. R., B. M. Cohen, D. O. Pederson, and J. E. Soloman, "Macromodeling of integrated circuit operational amplifiers," *IEEE Journal of Solid-State Circuits,* vol. SC-9, no. 6, pp. 353–364, December 1974.

3. M. S. Ghausi and K. R. Laker. *Modern Filter Design—Active RC and Switched Capacitor.* Prentice-Hall, Englewood Cliffs, NJ: 1981.

4. L. P. Huelsman and P. E. Allen. *Introduction to Theory and Design of Active Filters.* McGraw-Hill, New York: 1980.

ADDITIONAL READING

Franco, Sergio, *Design with Operational Amplifiers and Analog Integrated Circuits,* Third Edition, McGraw-Hill, New York: 2001.

Gray, P. R., P. J. Hurst, S. H. Lewis, and R. G. Meyer, *Analysis and Design of Analog Integrated Circuits,* Fourth Edition, John Wiley and Sons, New York: 2001.

Kennedy, E. J. *Operational Amplifier Circuits—Theory and Applications.* Holt, Rinehart and Winston, New York: 1988.

Roberge, James K. *Operational Amplifiers—Theory and Practice.* Wiley, New York: 1975.

Rogers, J., C. Plett, and F. Dai, Integrated Circuit Design for High Speed Frequency Synthesis, Artech House, 2006, Chapters 2 and 10.

PROBLEMS

11.1 The Differential Amplifier

11.1. A differential amplifier connected in the circuit in Fig. P11.1 has the parameters listed below with $R_S = 5$ kΩ and $R_L = 1$ kΩ. (a) Find the overall voltage gain A_v, current gain A_i, and power gain A_P for the amplifier, and express the results in dB. (b) What is the amplitude V_S of the sinusoidal input signal needed to develop a 10-V peak-to-peak signal at v_o?

$$\text{Input resistance } R_{id} = 1 \text{ M}\Omega$$
$$\text{Output resistance } R_o = 0.5 \text{ }\Omega$$
$$A = 60 \text{ dB}$$
$$v_s = V_S \sin \omega t$$

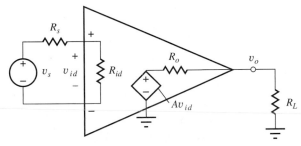

Figure P11.1

11.2. Suppose that the amplifier in Fig. P11.1 has been designed to match the source and load resistances in Prob. 11.1 with the parameters below. (a) What is the amplitude of the input signal v_s needed to develop a 20-V peak-to-peak signal at v_o? (b) How much power is dissipated in the amplifier when 0.5 W is delivered to the load resistor?

$$\text{Input resistance } R_{id} = 5 \text{ k}\Omega$$
$$\text{Output resistance } R_o = 1 \text{ k}\Omega$$
$$A = 30 \text{ dB}$$

11.3. The input to an amplifier comes from a transducer that can be represented by a 1-mV voltage source in series with a 50-kΩ resistor. What input resistance is required of the amplifier for $v_{id} \geq 0.99$ mV?

11.4. An amplifier has a sinusoidal output signal and is delivering 100 W to a 50-Ω load resistor. What output resistance is required if the amplifier is to dissipate no more than 5 W in its own output resistance?

11.2 The Ideal Operational Amplifier

11.5. An almost ideal op amp has an open-circuit output voltage $v_o = 10$ V and a gain $A = 100$ dB. What is the input voltage v_{id}? How large must the voltage gain be to make $v_{id} \leq 1$ μV?

11.6. Suppose a differential amplifier has $A = 120$ dB, and it is operating in a circuit with an open-circuit output voltage $v_o = 15$ V. What is the input voltage v_{id}? How large must the voltage gain be to make $v_{id} \leq 1$ μV? What is the input current i_+ if $R_{id} = 1$ MΩ?

11.3 Analysis of Circuits Containing Ideal Operational Amplifiers

Inverting Amplifiers

11.7. (a) What are the voltage gain, input resistance, and output resistance of the amplifier in Fig. P11.7 if $R_1 = 4.7$ kΩ and $R_2 = 220$ kΩ? Express the voltage gain in dB. (b) Repeat for $R_1 = 47$ kΩ and $R_2 = 2.2$ MΩ.

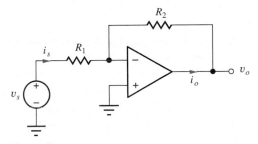

Figure P11.7

11.8. What are the voltage gain, input resistance, and output resistance of the amplifier in Fig. P11.7 if $R_1 = 12$ kΩ and $R_2 = 120$ kΩ? (b) Repeat for $R_1 = 140\ \text{k}\Omega$ and $R_2 = 330\ \text{k}\Omega$? (c) Repeat for $R_1 = 4.3\ \text{k}\Omega$ and $R_2 = 240\ \text{k}\Omega$?

11.9. Write an expression for the output voltage $v_o(t)$ of the circuit in Fig. P11.7 if $R_1 = 750\ \Omega$, $R_2 = 8.2\ \text{k}\Omega$, and $v_s(t) = (0.05 \sin 4638t)$ V. Write an expression for the current $i_s(t)$.

11.10. $R_1 = 22\ \text{k}\Omega$ and $R_2 = 110\ \text{k}\Omega$ in the amplifier circuit in Fig. P11.7. (a) What is the output voltage if $v_s = 0$? (b) What is the output voltage if a dc signal $V_S = 0.22$ V is applied to the circuit? (c) What is the output voltage if an ac signal $v_s = 0.15 \sin 2500\,\pi t$ V is applied to the circuit? (d) What is the output voltage if the input signal is $v_S = 0.22 - 0.15 \sin 2500\,\pi t$ V? (e) What is the input current i_S for parts (b), (c), and (d)? (f) What is the op amp output current i_o for the input signals in parts (b), (c), and (d)? (g) What is the voltage at the inverting input of the op amp for the input signal in part (d)?

11.11. Find an expression for the output voltage v_O in Fig. P11.11.

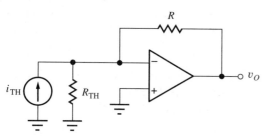

Figure P11.11

11.12. Design an inverting amplifier with an input resistance of 1.5 kΩ and a gain of 40 dB. Choose values from the 1-percent resistor table in Appendix A.

11.13. Design an inverting amplifier with an input resistance of 30 kΩ and a gain of 26 dB. Choose values from the 1-percent resistor table in Appendix A.

11.14. Design an inverting amplifier with an input resistance of 100 kΩ and a gain of 12 dB. Choose values from the 1-percent resistor table in Appendix A.

Noninverting Amplifiers

11.15. What are the voltage gain, input resistance, and output resistance of the amplifier in Fig. P11.15 if $R_1 = 8.2\ \text{k}\Omega$ and $R_2 = 750\ \text{k}\Omega$? Express the voltage gain in dB.

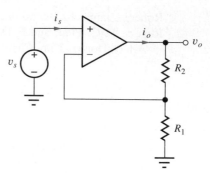

Figure 11.15

11.16. What are the voltage gain, input resistance, and output resistance of the amplifier in Fig. P11.15 if $R_1 = 24\ \text{k}\Omega$ and $R_2 = 120\ \text{k}\Omega$? (b) Repeat for $R_1 = 15\ \text{k}\Omega$ and $R_2 = 300\ \text{k}\Omega$. (c) Repeat for $R_1 = 4.3\ \text{k}\Omega$ and $R_2 = 360\ \text{k}\Omega$.

11.17. Write an expression for the output voltage $v_o(t)$ of the circuit in Fig. P11.15 if $R_1 = 910\ \Omega$, $R_2 = 8.2\ \text{k}\Omega$, and $v_s(t) = (0.05 \sin 9125t)$ V.

11.18. $R_1 = 22\ \text{k}\Omega$ and $R_2 = 110\ \text{k}\Omega$ in the amplifier circuit in Fig. P11.11. (a) What is the output voltage if $v_S = 0$? (b) What is the output voltage if a dc signal $V_S = 0.33$ V is applied to the circuit? (c) What is the output voltage if an ac signal $v_s = 0.18 \sin 3250\,\pi t$ V is applied to the circuit? (d) What is the output voltage if the input signal is $v_S = 0.33 - 0.18 \sin 3250\pi t$ V? (e) Write an expression for the input current i_S for parts (b), (c), and (d). (f) Write an expression for the op amp output current i_o for the input signals in parts (b), (c) and (d). (g) What is the voltage at the inverting input of the op amp for the input signal in part (d)?

*11.19. (a) What are the gain, input resistance, and output resistance of the amplifier in Fig. P11.19 if $R_1 = 180\ \Omega$ and $R_2 = 47\ \text{k}\Omega$? Express the gain in dB. (b) If the resistors have 10 percent tolerances, what are the worst-case values (highest and lowest) of gain that could occur? (c) What are the resulting positive and negative tolerances on the voltage gain with respect to the ideal value? (d) What is the ratio of the largest to the smallest voltage gain? (e) Perform a 500-case Monte Carlo analysis of this circuit. What percentage of the circuits has a gain within ±5 percent of the nominal design value?

11.20. Design a noninverting amplifier with a gain of 32 dB. Choose values from the 1-percent resistor table in Appendix A, and use values that are no smaller than 2 kΩ.

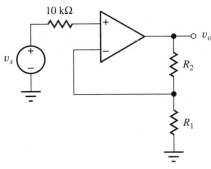

Figure P11.19

11.21. Design a noninverting amplifier with a gain of 26 dB. Choose values from the 1-percent resistor table in Appendix A, and use values that are no smaller than 1 kΩ.

11.22. Design a noninverting amplifier with an input resistance of 100 kΩ and a gain of 6 dB. Choose values from the 1-percent resistor table in Appendix A, and use values that are no smaller than 2 kΩ.

11.23. What are the gain, input resistance, and output resistance for the three circuits in Fig. P11.23?

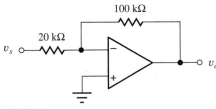

(a)

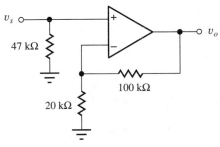

(b)

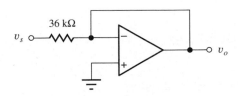

(c)

Figure P11.23

Summing Amplifiers

11.24. Write an expression for the output voltage $v_o(t)$ of the circuit in Fig. P11.24 if $R_1 = 1\ \text{k}\Omega$, $R_2 = 2\ \text{k}\Omega$, $R_3 = 51\ \text{k}\Omega$, $v_1(t) = (0.01 \sin 3770t)$ V, and $v_2(t) = (0.04 \sin 10000t)$ V. Write an expression for the voltage appearing at the summing junction (v_-).

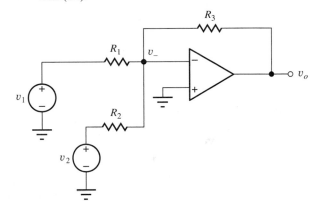

Figure P11.24

11.25. The summing amplifier can be used as a digitally controlled volume control using the circuit in Fig. P11.25. The individual bits of the 4-bit binary input word $(b_1 b_2 b_3 b_4)$ are used to control the position of the switches with the resistor connected to 0 V if $b_i = 0$ and connected to the input signal v_S if $b_i = 1$. (a) What is the output voltage v_O with the data input 0110 if $v_S = 1 \sin 4000\pi t$ V? (b) Suppose the input changes to 1011. What will be the new output voltage? (c) Make a table giving the output voltages for all 16 possible input combinations.

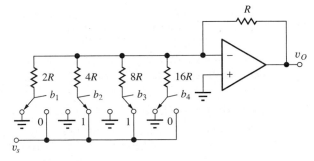

Figure P11.25

11.26. The switches in Fig. P11.25 can be implemented using MOSFETs, as shown in Fig. P11.26. What are the W/L ratios of the transistors if the on-resistance of the transistor is to be less than 1 percent of the resistor $2R = 10\ \text{k}\Omega$? Use $V_{\text{REF}} = 3.0$ V. Assume that the voltage applied to the gate of the MOSFET is 5 V when $b_1 = 1$ and 0 V when $b_1 = 0$. For the MOSFET, $V_{TN} = 1$ V, $K'_n = 50$ μA/V^2, $2\phi_F = 0.6$ V, and $\gamma = 0.5\ \sqrt{V}$.

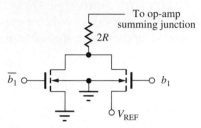

Figure P11.26

Difference Amplifier

11.27. (a) What is the gain of the circuit in Fig. P11.27 if $A_v = \mathbf{v_o}/(\mathbf{v_1} - \mathbf{v_2})$ and $R = 10\,\text{k}\Omega$? (b) What is the input resistance presented to v_2? (c) What is the input resistance at terminal v_1? (d) What is the output voltage if $v_1 = 3$ V and $v_2 = 1.5\cos 8300\pi t$ V? (e) What is the output voltage if $v_1 = [3 - 1.5\cos 8300\pi t]$ V and $v_2 = 1.5\sin 8300\pi t$ V?

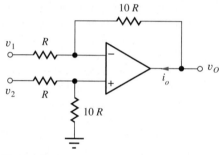

Figure P11.27

11.28. (a) What are the voltages at all the nodes in the difference amplifier in Fig. P11.27 if $V_1 = 3.2$ V, $V_2 = 3.1$ V, and $R = 100\,\text{k}\Omega$? (b) What is amplifier output current I_O? (c) What are the currents entering the circuit from v_1 and v_2?

Instrumentation Amplifier

11.29. What is the voltage gain of the instrumentation amplifier in Fig. 11.12 if $R_1 = 2\,\text{k}\Omega$, $R_2 = 100\,\text{k}\Omega$, $R_3 = 10\,\text{k}\Omega$, and $R_4 = 10\,\text{k}\Omega$. What is the output voltage if $v_1 = (2 + 0.1\sin 2000\pi t)$ V and $v_2 = 2.1$ V?

11.30. What is the voltage gain of the instrumentation amplifier in Fig. 11.12 if $R_1 = 20\,\text{k}\Omega$, $R_2 = 100\,\text{k}\Omega$, $R_3 = 10\,\text{k}\Omega$, and $R_4 = 20\,\text{k}\Omega$. What is the output voltage if $v_1 = (4 - 0.1\sin 4000\pi t)$ V and $v_2 = 3.5$ V?

Integrator

11.31. The input voltage to the integrator circuit in Fig. 11.15 is given by $v_s = 0.1\sin 2000\pi t$ V. What is the output voltage if $R = 10\,\text{k}\Omega$, $C = 0.005$ μF, and $v_o(0) = 0$?

11.32. The input voltage to the integrator circuit in Fig. 11.15 is a rectangular pulse with an amplitude of 5 V and a width 1 ms. Draw the waveform at the output of the integrator if the pulse starts at $t = 0$, $R = 10\,\text{k}\Omega$, and $C = 0.1$ μF. Assume $v_o = 0$ for $t \le 0$.

Low-Pass Filter

11.33. (a) What are the low-frequency voltage gain (in dB) and cutoff frequency f_H for the amplifier in Fig. 11.14 if $R_1 = 2\,\text{k}\Omega$, $R_2 = 10\,\text{k}\Omega$, and $C = 0.001$ μF? (b) Repeat for $R_1 = 2.7\,\text{k}\Omega$, $R_2 = 56\,\text{k}\Omega$, and $C = 100$ pF?

11.34. (a) Design a low-pass amplifier (i.e., choose R_1, R_2, and C) to have a low-frequency input resistance of $10\,\text{k}\Omega$, a midband gain of 20 dB, and a bandwidth of 20 kHz. (b) Choose element values from the tables in Appendix A.

Differentiator

11.35. What is the transfer function $T(s) = V_O(s)/V_S(s)$ for the differentiator circuit in Fig. 11.16?

11.36. What is the output voltage of the differentiator circuit in Fig. 11.16 if $v_S(t) = 2\cos 3000\pi t$ V with $C = 0.02$ μF and $R = 100\,\text{k}\Omega$?

11.37. What is the transfer function $A_v(s) = V_o(s)/V_s(s)$ for the circuit in Fig. P11.37?

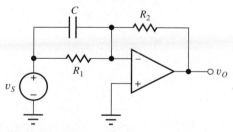

Figure P11.37

Ideal Op Amp Problems

11.38. Find the voltage gain, input resistance, and output resistance for the three circuits in Fig. P11.38.

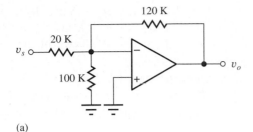

(a)

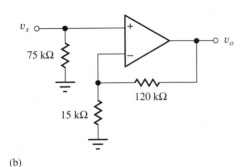

(b)

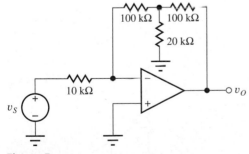

(c)

Figure P11.38

11.39. Find the voltage gain, input resistance, and output resistance for the circuits in Fig. P11.39.

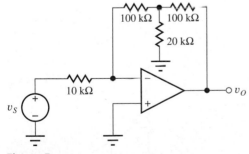

Figure P11.39

*11.40. (a) What is the output current I_O in the circuit of Fig. P11.40 if $-V_{EE} = -10$ V and $R = 10$ Ω? Assume that the MOSFET is saturated. (b) What is the minimum voltage V_{DD} needed to saturate the MOSFET if $V_{TN} = 2.5$ V and $K_n = 0.25$ A/V². (c) What must be the power dissipation rating of resistor R?

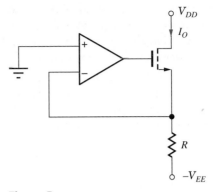

Figure P11.40

*11.41. (a) What is the output current I_O in the circuit in Fig. P11.41 if $-V_{EE} = -15$ V and $R = 30$ Ω? Assume that the BJT is in the forward-active region and $\beta_F = 30$. (b) What is the voltage at the output of the operational amplifier if the saturation current I_S of the BJT is 10^{-13} A? (c) What is the minimum voltage V_{CC} needed for forward-active region operation of the bipolar transistor? (d) Find the power dissipation rating of the resistor R. How much power is dissipated in the transistor if $V_{CC} = 15$ V?

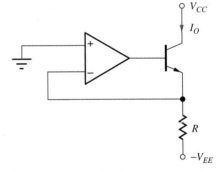

Figure P11.41

**11.42. (a) What is the voltage transfer $V_o(s)/V_s(s)$ function for the integrator in Fig. 11.15. (b) What is the voltage transfer function for the circuit in Fig. P11.42?

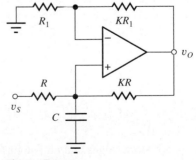

Figure P11.42

11.43. The resistors in the circuit in Fig. P11.42 will differ from their ideal values due to tolerances. Suppose $KR = 10R$, but $KR_1 = 9R_1$. What is the transfer function for the circuit in Fig. P11.42?

11.44. The circuit in Fig. P11.44 represents a current source that forces a current through the floating load impedance Z_L. (a) Find the relationship between the current i_o and the input voltages v_1 and v_2. (b) What is the output impedance of the current source as seen at the terminals of Z_L?

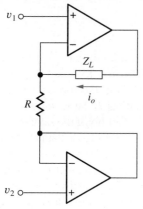

Figure P11.44

*11.45. The circuit in Fig. P11.45 is a current source that operates with a grounded load Z_L. (a) Find the relationship between the current i_o and the input voltages v_1 and v_2 if $R_1 = R_2 = R_3 = R_4$. (b) What is the output impedance of the current source as seen at the terminals of Z_L for $R_1 = R_2 = R_3 = R_4$?

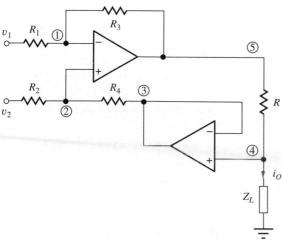

Figure P11.45

*11.46. Find the five node voltages and output current for the current source circuit in Fig. P11.45 if $v_1 = 4$ V,

$v_2 = 6$ V, $R = 10$ kΩ, $R_1 = 5$ kΩ, $R_2 = 5$ kΩ, $R_3 = 5.01$ kΩ, $R_4 = 4.99$ kΩ, and $Z_L = 10$ kΩ. (b) What is the output resistance of the current source?

11.47. Derive an expression for the current I_L in terms of I_S, R_1, and R_2 for the circuit in Fig. P11.47. Calculate the input and output resistance of this circuit if the op amp is ideal.

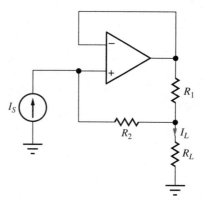

Figure P11.47

11.48. What are the input and output resistances of the circuit in Fig. P11.47 if $R_1 = 10$ kΩ, $R_2 = 1$ kΩ, and $Z_L = 3.6$ kΩ?

11.49. Find expressions for the voltage v_{o1} and v_{o2} in terms of v_s, R_1, R_2, and R_3 for the circuit in Fig. P11.49. The op amps are ideal.

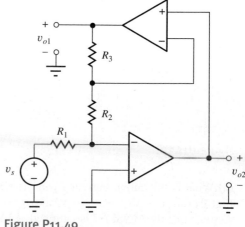

Figure P11.49

11.50. What is the input resistance seen by source v_S in the circuit in Fig. P11.50?

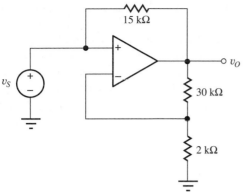

Figure P11.50

D/A and A/D Converters

11.51. How many resistors are needed to realize a 10-bit weighted-resistor DAC? What is the ratio of the largest resistor to the smallest resistor?

11.52. (a) What is the output voltage for the 4-bit DAC in Fig. P11.52, as shown with input data of 0110 if $V_{REF} = 3.0$ V? (b) Suppose the input data changes to 1001. What is the new output voltage? (c) Make a table giving the output voltages for all 16 possible input data combinations.

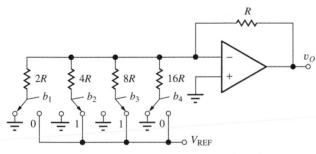

Figure P11.52

11.53. Use Thévenin equivalent circuits for the R-2R ladder network in Fig. P11.53 to find the output voltage for the four input combinations 0001, 0010, 0100, and 1000 if $V_{REF} = 5.0$ V.

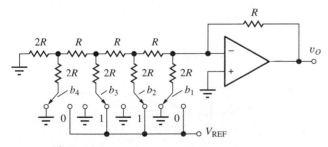

Figure P11.53

11.54. A 14-bit ADC with $V_{FS} = 5.12$ V has an output code of 10101010110010. What is the possible range of input voltages?

11.55. A 20-bit ADC has $V_{FS} = 2$ V. (a) What is the value of the LSB? (b) What is the ADC output code for an input voltage of 1.630000 V? (c) What is the ADC output code for an input voltage of 0.997003 V?

*11.56. A 20-bit dual-ramp converter is to have an integration time $T_1 = 0.2$ s. How rapidly must the unknown and reference voltage switches change state if this timing uncertainty is to be equivalent to less than 0.1 LSB?

**11.57. Derive the transfer function for the integrator in the dual-ramp converter, and show that it has the functional form of $|\sin x/x|$.

Amplifier Terminology Review

11.58. Seven amplifiers were identified in Fig. 11.17. Find two more possibilities.

11.59. An amplifier is formed by cascading two operational-amplifier stages, as shown in Fig. P11.59(a). (a) Replace each amplifier stage with its two-port representation. (b) Use the circuit model from part (a) to find the overall two-port representation (A_v, R_{in}, R_{out}) for the complete two-stage amplifier. (c) Draw the circuit of the two-port corresponding to the complete two-stage amplifier.

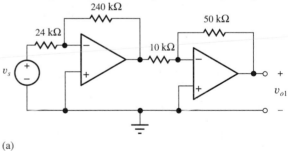

(a)

(b)

Figure P11.59   (*Continued on next page*)

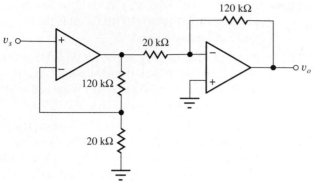

(c)

Figure P11.59   *(Continued)*

11.60. Two inverting op amp stages are cascaded as shown in Fig. P11.59(a). What are the values of A_v, R_{in}, and R_{out} for the overall two-stage amplifier?

11.61. Two noninverting op amp stages are cascaded as shown in Fig. P11.59(b). What are the values of A_v, R_{in}, and R_{out} for the overall two-stage amplifier?

11.62. Two op amp stages are cascaded as shown in Fig. P11.59(c). (a) What are the values of A_v, R_{in}, and R_{out} for the overall two-stage amplifier? (b) What are the values of A_v, R_{in}, and R_{out} for the overall two-stage amplifier if the two amplifier stages are interchanged?

11.63. An amplifier is formed by cascading three identical operational-amplifier stages, as shown in

Fig. P11.63. (a) Replace each op amp circuit with its two-port representation. (b) Use the circuit model from part (a) to find the overall two-port representation (A_v, R_{in}, R_{out}) for the complete three-stage amplifier. (c) Draw the two-port circuit corresponding to the complete three-stage amplifier.

11.64. What are the values of A_v, R_{in}, and R_{out} for the overall three-stage amplifier in Fig. P11.63 if the 2-kΩ resistors are replaced with 3.9-kΩ resistors?

11.65. The 2-kΩ resistors in Fig. P11.63 are to be replaced with a value that gives an overall gain of 40 dB. What is the new resistor value? What is the new value of R_{in}?

11.66. The op amps in Fig. P11.66 are ideal. (a) What are the voltage gain, input resistance, and output resistance of the overall amplifier? (b) If the input voltage $v_I = 10$ mV, what are the voltages at each of the eight nodes in the amplifier circuit?

11.67. The op amps in Fig. P11.66 are ideal. What are the nominal, minimum, and maximum values of the voltage gain, input resistance, and output resistance of the overall amplifier if all the resistors have 5 percent tolerances?

11.68. The op amps in Fig. P11.68 are ideal (a) What are the voltage gain, input resistance, and output resistance of the overall amplifier? (b) If the input voltage $v_I = 0.002$ V, what are the voltages at each of the eight nodes in the amplifier circuit?

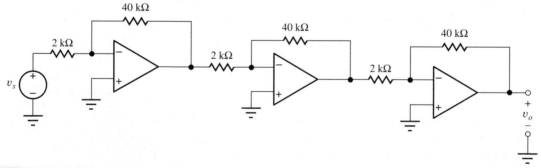

Figure P11.63

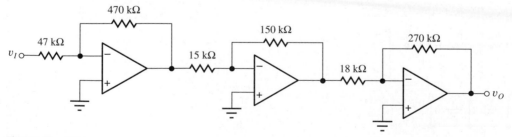

Figure P11.66

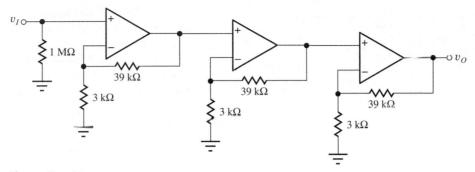

Figure P11.68

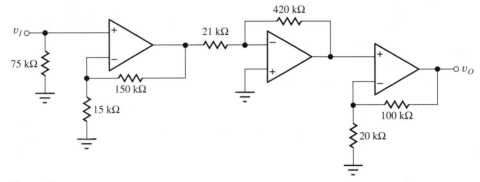

Figure P11.71

11.69. Repeat Prob. 11.68 if the 3-kΩ resistors are all re-placed with 1-kΩ resistors, and the 1-MΩ resistor is replaced with a 2-MΩ resistor.

11.70. The op amps in Fig. P11.68 are ideal. What are the nominal, minimum, and maximum values of the voltage gain, input resistance, and output resistance of the overall amplifier if all the resistors have 2 percent tolerances?

11.71. The op amps in Fig. P11.71 are ideal. (a) What are the voltage gain, input resistance, and output resistance of the overall amplifier? (b) If the input voltage $v_I = 0.005$ V, what are the voltages at each of the eight nodes in the amplifier circuit?

11.72. The op amps in Fig. P11.71 are ideal. What are the nominal, minimum, and maximum values of the voltage gain, input resistance, and output resistance of the overall amplifier if all the resistors have 1 percent tolerances?

11.4 Active Filters

11.73. (a) Repeat Design Example 11.8 for a max-imally flat second-order low-pass filter with $f_o = 20$ kHz, using the circuit in Fig. 11.19. Assume

$C = 0.005$ μF. What is the filter bandwidth? (b) Use frequency scaling to change f_o to 40 kHz.

*11.74. (a) Use MATLAB or other computer tool to make a Bode plot for the response of the filter in Prob. 11.73, assuming the op amp is ideal. (b) Use SPICE to simulate the characteristics of the filter in Prob. 11.73 using a 741 op amp. (c) Discuss any disagreement between the SPICE results and the ideal response.

11.75. Derive an expression for the input impedance of the filter in Fig. 11.19.

11.76. Use MATLAB or another computer tool to plot the input impedance of the low-pass filter in Fig. 11.19 versus frequency for $R_1 = R_2 = 2.26$ kΩ, $C_1 = 0.02$ μF, and $C_2 = 0.01$ μF.

*11.77. (a) What is the transfer function for the low-pass filter in Fig. P11.77? (b) What is S_K^Q for this filter if $R_1 = R_2$ and $C_1 = C_2$?

11.78. What are the expressions for $S_{R_1}^{\omega_o}$ and $S_{C_1}^{\omega_o}$ for the high-pass filter of Fig. 11.21?

11.79. What is $S_Q^{\omega_o}$ for the band-pass filter of Fig. 11.23 for $C_1 = C_2$? What is the value for $f_o = 10$ kHz and $Q = 10$?

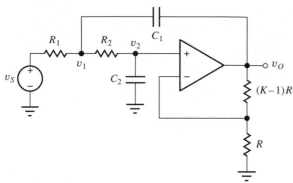

Figure P11.77

11.80. Design a maximally flat second-order low-pass filter with a bandwidth of 1 kHz using the circuit in Fig. 11.19.

11.81. Design a high-pass filter with a lower-half power frequency of 20 kHz and $Q = 1$ using the circuit in Fig. 11.21.

11.82. (a) Calculate f_o, Q, and the bandwidth for the band-pass filter in Fig. 11.23 if $R_{th} = 1$ kΩ, $R_2 = 200$ kΩ, and $C_1 = C_2 = 220$ pF. (b) Use magnitude scaling to change the element values so that $R_{th} = 3.3$ kΩ. (c) Use frequency scaling to double f_o for the filter in part (a).

11.83. (a) Design a band-pass filter with a center frequency of 1 kHz and $Q = 5$ using the circuit in Fig. 11.23 with $R_3 = \infty$. What is the filter bandwidth? (b) Use frequency scaling to change f_o to 2.25 kHz.

*11.84. (a) Use MATLAB or another computer tool to make a Bode plot for the response of the filter in Prob. 11.82(a), assuming the op amp is ideal. (b) Use SPICE to simulate the characteristics of the filter in Prob. 11.82(a) using a 741 op amp. (c) Discuss any disagreement between the SPICE results and the ideal response.

11.85. (a) Two identical band-pass filters having $\omega_o = 1$ and $Q = 3$ are designed using the circuit in Fig. 11.23 with $C_1 = C_2$ and $R_3 = \infty$. If the filters are cascaded, what are the center frequency, Q, and bandwidth of the overall filter? (b) Write the transfer function for the composite filter.

11.86. Use MATLAB or other computer tool to produce a Bode plot for the two-stage filter in Prob. 11.85.

*11.87. The first stage of a two-stage filter consists of a band-pass filter with $f_o = 5$ kHz and $Q = 5$. The second stage is also a band-pass filter, but it has $f_o = 6$ kHz and $Q = 5$. If the filters use Fig. 11.23 with $C_1 = C_2$ and $R_3 = \infty$, what are the center frequency, Q, and bandwidth of the overall filter?

11.88. Use MATLAB or another computer tool to produce a Bode plot for the two-stage filter in Prob. 11.87.

11.89. Design a band-pass filter with 20-dB gain at $f_o = 600$ Hz, $Q = 5$, and $R_{in} = 10$ kΩ using the Tow-Thomas biquad in Fig. 11.25.

11.90. Calculate S_C^Q for the Tow-Thomas biquad in Fig. 11.25.

11.5 Nonlinear Circuit Applications

Nonlinear Feedback

11.91. Draw the output voltage waveform for the circuit in Fig. P11.91 for the triangular input waveform shown.

*11.92. The signal v_S in Fig. P11.91 is used as the input voltage to the circuit in Fig. 11.29. What will be the dc component of the voltage waveform at v_O if $R_1 = 2.7$ kΩ, $R_2 = 8.2$ kΩ, $R_3 = 10$ kΩ, $R_4 = 10$ kΩ, $C = 0.22$ μF, and $T = 1$ ms?

**11.93. What must be the cutoff frequency of the low-pass filter in Fig. 11.29 if the rms value of the total ac component in the output voltage must be less than 1 percent of the dc voltage? Assume $R_1 = R_2$ and $v_S = -5 \sin 120\pi t$ V.

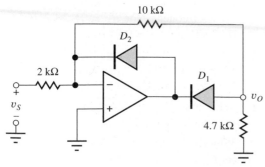

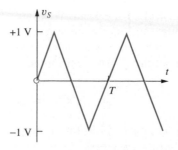

Figure P11.91

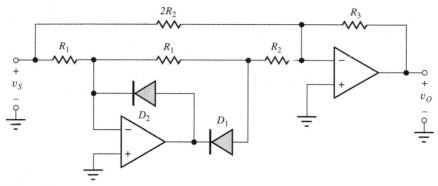

Figure P11.94

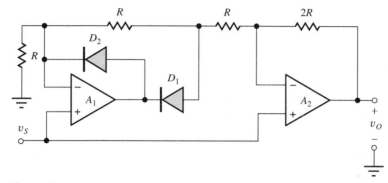

Figure P11.95

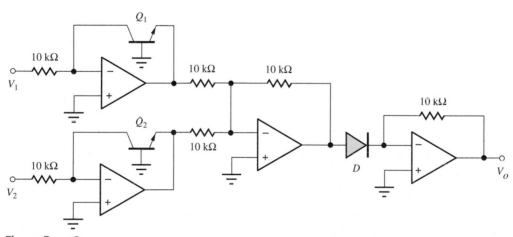

Figure P11.98

11.94. The triangular waveform in Fig. P11.91 is applied to the circuit in Fig. P11.94. Draw the corresponding output waveform for $R_3 = R_2$.

11.95. The triangular waveform in Fig. P11.91 is applied to the circuit in Fig. P11.95. Draw the corresponding output waveform.

11.96. Simulate the circuit in Prob. 11.94 using $R_1 = 10\,k\Omega$, $R_2 = 10\,k\Omega$, $R_3 = 10\,k\Omega$, and $R_4 = 10\,k\Omega$. Use op amps with $A_o = 100$ dB.

11.97. Simulate the circuit in Prob. 11.95 for $R = 10\,k\Omega$. Use op amps with $A_o = 100$ dB.

*11.98. Write an expression for the output voltage in terms of the input voltage for the circuit in Fig. P11.98. Diode D is formed from a diode-connected transistor, and all three transistors are identical.

11.6 Circuits Using Positive Feedback

11.99. What are the values of the two switching thresholds and hysteresis in the Schmitt-trigger circuit in Fig. P11.99?

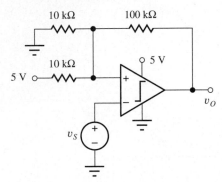

Figure P11.99

11.100. What are the switching thresholds and hysteresis for the Schmitt-trigger circuit in Fig. P11.100?

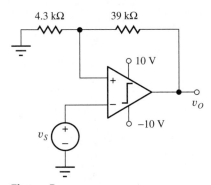

Figure P11.100

11.101. What are the switching thresholds and hysteresis for the Schmitt-trigger circuit in Fig. P11.101?

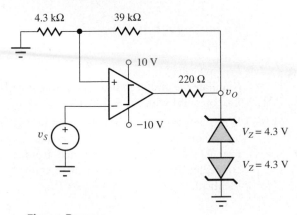

Figure P11.101

11.102. Design a Schmitt trigger to have its switching thresholds centered at 1 V with a hysteresis of ±0.05 V, using the circuit topology in Fig. P11.99.

11.103. What is the frequency of oscillation of the astable multivibrator in Fig. P11.103?

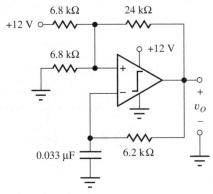

Figure P11.103

**11.104. Draw the waveforms for the astable multivibrator in Fig. P11.104. What is its frequency of oscillation? (Be careful—think before you calculate!)

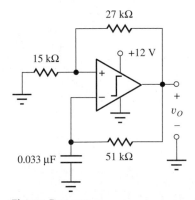

Figure P11.104

11.105. (a) Design an astable multivibrator to oscillate at a frequency of 1 kHz. Use the circuit in Fig. 11.36 with symmetric supplies of ±5 V. Assume that the total current from the op amp output must never exceed 1 mA. (b) If the resistors have ±5 percent tolerances and the capacitors have ±10 percent tolerances, what are the worst-case values of oscillation frequency? (c) If the power supplies are actually +4.75 and −5.25 V, what is the oscillation frequency for the nominal resistor and capacitor design values?

11.106. The function generator circuit in the EIA on page 591 has been designed to generate a sine wave output voltage with an amplitude of 5 V at a frequency of 1 kHz. The low-pass filter has been designed to have a low-frequency gain of -1 and a cutoff frequency of 1.5 kHz. What are the magnitudes of the undesired frequency components in the output waveform at frequencies of 2 kHz, 3 kHz, and 5 kHz?

11.107. Two diodes are added to the circuit in Fig. P11.104 to convert it to a monostable multivibrator similar to the circuit in Fig. 11.38, and the power supplies are changed to ± 10 V. What are the pulse width and recovery time of the monostable circuit?

11.108. Design a monostable multivibrator to have a pulse width of 10 μs and a recovery time of 5 μs. Use the circuit in Fig. 11.38 with ± 5 V supplies.

CHAPTER 12

CHARACTERISTICS AND LIMITATIONS OF OPERATIONAL AMPLIFIERS

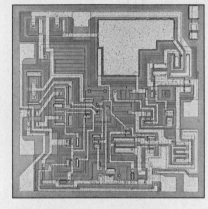

uA741 Die Photograph (Courtesy of Fairchild Semiconductor)

CHAPTER GOALS

- Study nonideal operational amplifier behavior
- Demonstrate techniques used to analyze circuits containing nonideal op amps
- Determine the voltage gain, input resistance, and output resistance of general amplifier circuits
- Explore common-mode rejection limitations and the effect of common-mode input resistance
- Learn how to model dc errors including offset voltage, input bias current, and input offset current
- Explore limits imposed by power supply voltages and finite output current capability
- Model amplifier limitations due to limited bandwidth and slew rate of the op amp
- Perform SPICE simulation of nonideal op amp circuits

- Finite open loop gain
- Finite input resistance
- Nonzero output resistance
- Offset voltage
- Input bias and offset currents
- Limited output voltage range
- Limited output current capability
- Finite common-mode rejection
- Finite power supply rejection
- Limited bandwidth
- Limited slew rate

Chapter 11 explored the characteristics of circuits employing ideal operational amplifiers having infinite gain, zero input current, and zero output resistance. Real operational amplifiers, on the other hand, do not exhibit any of these ideal characteristics. In fact, they have a significant number of additional limitations as tabulated in the next column.

There are literally hundreds of commercial hybrid and integrated circuit operational amplifiers available to the engineer for use in circuit design. The only way to choose among this large set of options is to fully understand the characteristics and limitations of real operational amplifiers. Thus, this chapter explores the impact of each of these limitations in detail and demonstrates the approaches used to analyze circuits employing nonideal op amps. Generally, we look at the effect of each of the nonideal characteristics independently, while assuming the others are still ideal. Then we can combine the results to understand how the circuits behave in general.

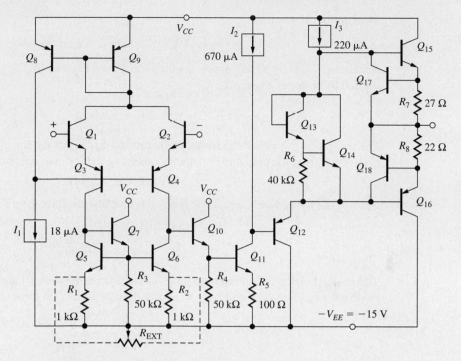

Circuit schematic for μA741 op amp die on preceding page

12.1 GAIN, INPUT RESISTANCE, AND OUTPUT RESISTANCE

12.1.1 FINITE OPEN-LOOP GAIN

A real operational amplifier provides a large but noninfinite gain. Op amps are commercially available with minimum open-loop gains of 80 dB (10,000) to over 120 dB (1,000,000). The finite open-loop gain contributes to deviations of the **closed-loop gain,** input resistance, and output resistance from those presented for the ideal amplifiers in Chapter 11.

Evaluation of the closed-loop gain for the noninverting amplifier of Fig. 12.1 provides our first example of amplifier calculations involving nonideal amplifiers. In Fig. 12.1, the output voltage of

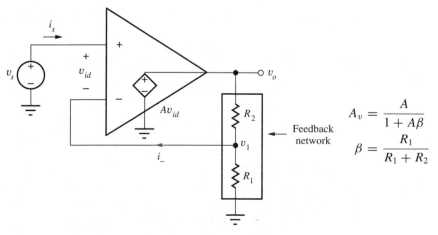

$$A_v = \frac{A}{1 + A\beta}$$

$$\beta = \frac{R_1}{R_1 + R_2}$$

Figure 12.1 Operational amplifier with finite open-loop gain A.

the amplifier is given by

$$\mathbf{v_o} = A\mathbf{v_{id}} \qquad \text{where} \qquad \mathbf{v_{id}} = \mathbf{v_s} - \mathbf{v_1} \tag{12.1}$$

Because $i_- = 0$ by ideal op-amp Assumption 2 (see Eq. (11.6)), v_1 is set by the voltage divider formed by resistors R_1 and R_2:

$$\mathbf{v_1} = \frac{R_1}{R_1 + R_2}\mathbf{v_o} = \beta\mathbf{v_o} \qquad \text{where} \qquad \beta = \frac{R_1}{R_1 + R_2} \tag{12.2}$$

The parameter β is called the **feedback factor** and represents the fraction of the output voltage that is fed back from the output to the input. Combining the last two equations gives

$$\mathbf{v_o} = A(\mathbf{v_s} - \beta\mathbf{v_o}) \tag{12.3}$$

and solving for $\mathbf{v_o}$ yields the classic **feedback amplifier** voltage-gain formula in Eq. (12.4).

$$A_v = \frac{\mathbf{v_o}}{\mathbf{v_s}} = \frac{A}{1 + A\beta} \tag{12.4}$$

The product $A\beta$ is called the **loop gain** (or **loop transmission T**) and plays an important role in feedback amplifiers. For $A\beta \gg 1$, A_v approaches the ideal gain expression found previously:

$$A_{\text{ideal}} = \frac{1}{\beta} = 1 + \frac{R_2}{R_1} \tag{12.5}$$

The meaning of the loop gain is explored more fully in Chapter 17, which is dedicated to the study of feedback systems.

The voltage $\mathbf{v_{id}}$ across the input is given by

$$\mathbf{v_{id}} = \mathbf{v_s} - \mathbf{v_1} = \mathbf{v_s} - \beta\mathbf{v_o} = \mathbf{v_s} - \frac{A\beta}{1 + A\beta}\mathbf{v_s} = \frac{\mathbf{v_s}}{1 + A\beta} \tag{12.6}$$

Although $\mathbf{v_{id}}$ is no longer zero, it is small for large values of $A\beta$. Thus, when we apply an input voltage $\mathbf{v_s}$, only a small portion of it appears across R_{in} (remember $\mathbf{v_{id}} \to 0$ as $A \to \infty$) so the input current is very small.

EXERCISE: Suppose $A = 10^5$, $\beta = 1/100$, and $v_s = 100$ mV. What are A_{ideal}, A_v, v_o, and v_{id}?

ANSWERS: 100, 99.9, 10.0 V, 100 μV (v_{id} is small but nonzero)

EXERCISE: What are the nominal, minimum, and maximum values of the open-loop gain at 25°C for an OP-27 operational amplifier?

ANSWERS: With 15-V supplies: 1,000,000; 1,800,000; no maximum value specified

EXERCISE: What value of open-loop gain is guaranteed for the OP-27 op amp over the full temperature range with a load resistance of at least 2 kΩ?

ANSWER: 600,000 with 15-V power supplies

12.1.2 GAIN ERROR

In many applications it is important to know, or to control by design, just how far the actual gain in Eq. (12.4) deviates from the ideal gain expression in Eq. (12.5). The **gain error (GE)** is defined as the difference between the ideal gain and the actual gain:

$$\text{GE} = (\text{ideal gain}) - (\text{actual gain}) \tag{12.7}$$

This error is more often expressed as a fractional error or percentage error, and the **fractional gain error (FGE)** is defined as

$$\text{FGE} = \frac{(\text{ideal gain}) - (\text{actual gain})}{(\text{ideal gain})} \qquad (12.8)$$

For the noninverting amplifier in Fig. 12.1, the GE and FGE are given by

$$\text{GE} = \frac{1}{\beta} - \frac{A}{1 + A\beta} = \frac{1}{\beta(1 + A\beta)} \qquad (12.9)$$

and

$$\text{FGE} = \frac{\dfrac{1}{\beta} - \dfrac{A}{1 + A\beta}}{\dfrac{1}{\beta}} = \frac{1}{1 + A\beta} \cong \frac{1}{A\beta} \quad \text{for } A\beta \gg 1 \qquad (12.10)$$

For $A\beta \gg 1$, we see that the value of the FGE is determined by the reciprocal of the loop gain $A\beta$.

EXERCISE: **Show that FGE $= 1/(1 + A\beta)$ for the inverting amplifier.**

EXAMPLE 12.1   GAIN ERROR ANALYSIS

Characterize the gain and gain error of a noninverting amplifier implemented with a finite gain operation amplifier.

PROBLEM A noninverting amplifier is designed to have a gain of 200 (46 dB) and is built using an operational amplifier with an open-loop gain of 80 dB. Find the values of the ideal gain, the actual gain, and the gain error. Express the gain error in percent.

SOLUTION **Known Information and Given Data:** Noninverting amplifier circuit with closed-loop gain of 46 dB. The open-loop gain of the op amp is 10,000 (80 dB).

Unknowns: The values of the ideal gain, the actual gain, and the gain error in percent

Approach: First, we need to clarify the meaning of some terminology. We normally design an amplifier to produce a given value of ideal gain, and then determine the deviations to be expected from the ideal case. So, when it is said that this amplifier is designed to have a gain of 200, we set $\beta = \frac{1}{200}$. We do not normally try to adjust the design values of R_1 and R_2 to try to compensate for the finite open-loop gain of the amplifier. One reason is that we do not know the exact value of the gain A but generally only know its lower bound. Also, the resistors we use have tolerances, and their exact values are also unknown.

Assumptions: The op amp is ideal except for its finite open-loop gain.

Analysis: The ideal gain of the circuit is 200. The actual gain and FGE are given by

$$A_v = \frac{A}{1 + A\beta} = \frac{10^4}{1 + \dfrac{10^4}{200}} = 196 \quad \text{and} \quad \text{FGE} = \frac{200 - 196}{200} = 0.02 \text{ or 2 percent}$$

Check of Results: The three unknown values have been found. The value of A_v is slightly less than $1/\beta$ and therefore appears to be a reasonable result.

Discussion: The actual gain is 196, representing a 2 percent error from the ideal design gain of 200. Note that this gain error expression does not include the effects of resistor tolerances, which are an additional source of gain error in an actual circuit. If the gain must be more precise, a higher-gain op amp must be used, or the resistors can be replaced by a potentiometer so the gain can be adjusted manually.

Computer Aided Analysis: The circuit in Ex. 11.3 can be used to check the results of this example by setting R1 = 1 kΩ, R2 = 199 kΩ, and the gain of E1 to 10,000. The transfer function analysis gives A_v = 196, in agreement with our hand calculations.

EXERCISE: A noninverting amplifier is designed with R_1 = 1 kΩ, R_2 = 39 kΩ, and an op amp with an open-loop gain of 80 dB. What is the fractional gain error of the amplifier?

ANSWER: 0.4 percent

12.1.3 NONZERO OUTPUT RESISTANCE

The next effect we explore is the influence of a nonzero output resistance on the characteristics of the inverting and noninverting closed-loop amplifiers. In this case, we assume that the op amp has a nonzero output resistance R_o as well as a finite open-loop gain A. (As we shall see, finite gain must also be assumed; otherwise, we would get the same output resistance as for the idealcase.)

To determine the (Thévenin equivalent) output resistances of the two amplifiers in Fig. 12.2, each output terminal is driven with a test signal source v_x (a current source could also be used), and the current i_x is calculated; all other independent sources in the network must be turned off. The output resistance is then given by

$$R_{\text{out}} = \frac{\mathbf{v_x}}{\mathbf{i_x}} \tag{12.11}$$

From Fig. 12.2 we observe that the two amplifier circuits are identical for the output resistance calculation. Thus, analysis of the circuit in Fig. 12.3 gives the expression for R_{out} for both the inverting and noninverting amplifiers.

Analysis begins by expressing currents $\mathbf{i_x}$ and $\mathbf{i_o}$ as

$$\mathbf{i_x} = \mathbf{i_o} + \mathbf{i_2} \qquad \text{and} \qquad \mathbf{i_o} = \frac{\mathbf{v_x} - A\mathbf{v_{id}}}{R_o} \tag{12.12}$$

Current $\mathbf{i_2}$ can be found from

$$\mathbf{v_x} = \mathbf{i_2}R_2 + \mathbf{i_1}R_1 \qquad \text{or} \qquad \mathbf{i_2} = \frac{\mathbf{v_x}}{R_1 + R_2} \tag{12.13}$$

because $\mathbf{i_1} = \mathbf{i_2}$ due to op amp Assumption 2: $\mathbf{i_-} = 0$. The input voltage $\mathbf{v_{id}}$ is equal to $-\mathbf{v_1}$, and because $\mathbf{i_-} = 0$,

$$\mathbf{v_1} = \frac{R_1}{R_1 + R_2}\mathbf{v_x} = \beta\mathbf{v_x} \tag{12.14}$$

Combining Eqs. (12.12) through (12.14) yields

$$\frac{1}{R_{\text{out}}} = \frac{\mathbf{i_x}}{\mathbf{v_x}} = \frac{1 + A\beta}{R_o} + \frac{1}{R_1 + R_2} \tag{12.15}$$

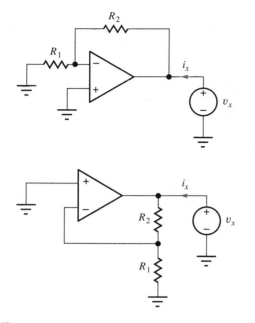

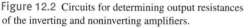

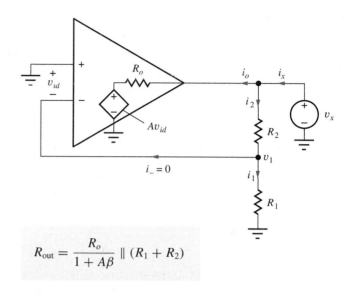

$$R_{out} = \frac{R_o}{1 + A\beta} \,\|\, (R_1 + R_2)$$

Figure 12.2 Circuits for determining output resistances of the inverting and noninverting amplifiers.

Figure 12.3 Circuit explicitly showing amplifier with A and R_o.

Equation (12.15) represents the output conductance of the amplifier and corresponds to the sum of the conductances of two parallel resistors. Thus, the output resistance can be expressed as

$$R_{out} = \frac{R_o}{1 + A\beta} \,\|\, (R_1 + R_2) \qquad (12.16)$$

The output resistance in Eq. (12.15) represents the series combination of R_1 and R_2 in parallel with a resistance $R_o/(1 + A\beta)$ that represents the output resistance of the operational amplifier including the effects of feedback. In almost every practical situation, the value of $R_o/(1 + A\beta)$ is much less than that of $(R_1 + R_2)$, and the output resistance expression in Eq. (12.16) simplifies to

$$R_{out} \cong \frac{R_o}{1 + A\beta} \qquad (12.17)$$

An example of the degree of dominance of the resistance term in Eq. (11.67) is given in Example 12.2.

Note that the output resistance would be zero if A were assumed to be infinite in Eqs. (12.16) or (12.17). This is the reason why the analysis must simultaneously account for both finite A and nonzero R_o.

EXERCISE: What are the nominal, minimum, and maximum values of the open-loop gain and output resistance for an OP-77E operational amplifier (see MCD website)?

ANSWERS: 12,000,000 (142 dB); 5,000,000 (134 dB); no maximum value specified; 60 Ω; no minimum or maximum value specified

---

EXAMPLE **12.2**   OP AMP OUTPUT RESISTANCE

Perform a numeric calculation of the output resistance of a noninverting amplifier implemented using an op amp with a finite open-loop gain and nonzero output resistance.

PROBLEM   A noninverting amplifier is constructed with $R_1 = 1$ kΩ and $R_2 = 39$ kΩ using an operational amplifier with an open-loop gain of 80 dB and an output resistance of 50 Ω. Find the output resistance of the noninverting amplifier.

SOLUTION   **Known Information and Given Data:** Noninverting op amp amplifier circuit with $R_1 = 1$ kΩ, $R_2 = 39$ kΩ, $A = 10{,}000$, and $R_o = 50$ Ω.

Unknowns: Output resistance of the overall amplifier

Approach: Use known values to evaluate Eq. (12.16)

Assumptions: The op amp is ideal except for finite gain and nonzero output resistance.

Analysis: Evaluating Eq. (12.16):

$$1 + A\beta = 1 + A\frac{R_1}{R_1 + R_2} = 1 + 10^4 \frac{1 \text{ k}\Omega}{1 \text{ k}\Omega + 39 \text{ k}\Omega} = 251$$

and

$$R_{\text{out}} = \frac{50 \text{ } \Omega}{251} \| (40 \text{ k}\Omega) = 0.199 \text{ } \Omega \| 40 \text{ k}\Omega = 0.198 \text{ } \Omega$$

Check of Results: The unknown value of output resistance has been calculated. The value is much smaller than the value of R_o, which is expected.

Evaluation and Discussion: We see that the effect of the feedback in the circuits in Fig. 12.2 is to reduce the output resistance of the closed-loop amplifier far below that of the individual op amp itself. In fact, the output resistance is quite small and represents a good practical approximation to that of an ideal amplifier ($R_{\text{out}} = 0$). This is a characteristic of shunt feedback at the output port, in which the feedback network is in parallel with the port. Shunt feedback tends to lower the resistance at a port, whereas feedback in series with a port, termed series feedback, tends to raise the resistance at that port. The properties of series and shunt feedback are explored in greater detail in Chapter 17.

Computer-Aided Design: The output resistance of the non-inverting amplifier can be simulated by adding the output resistance RO to the circuit of Ex. 11.3 as indicated in the figure. The gain of E1 is set to 10,000. A transfer function analysis from VS to output node v_o gives a gain of $+39.8$ and an output resistance of 0.199 Ω.

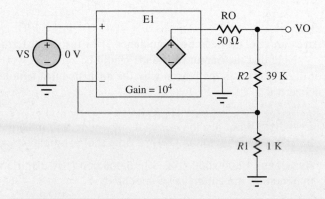

EXERCISE: Calculate the value of closed-loop gain in Ex. 12.2 and verify that the simulation result is correct.

EXERCISE: Suppose the resistors in Ex. 12.2 both have 5 percent tolerances. What are the worst-case (highest and lowest) values of gain that can be expected if the open-loop gain were infinite? What is the gain error for each of these two cases?

ANSWERS: 44.1, 36.3, +10.3 percent, −9.25 percent

We see that the effect of the feedback in the circuits in Fig. 12.2 is to reduce the output resistance of both closed-loop amplifiers far below that of the individual op amp itself. In fact, the output resistance is quite small and represents a good practical approximation to that of an ideal amplifier ($R_{\text{out}} = 0$). This is a characteristic of **shunt feedback** at the output port, in which the feedback network is in parallel with the port. Shunt feedback tends to lower the resistance at a port, whereas feedback in series with a port, termed **series feedback,** tends to raise the resistance at that port. The properties of series and shunt feedback are explored in greater detail in Chapter 17.

DESIGN EXAMPLE 12.3 OPEN-LOOP GAIN DESIGN

In this example, we find the value of open-loop gain required to meet an amplifier output resistance specification.

PROBLEM Design a noninverting amplifier to have a closed-loop gain of 35 dB and an output resistance of no more than 0.2 Ω. The only op amp available has an output resistance of 250 Ω. What is the minimum open-loop gain of the op amp that will meet the design requirements?

SOLUTION **Known Information and Given Data:** For the noninverting amplifier: closed-loop gain = 35 dB, closed-loop output resistance = 0.2 Ω. For the operational amplifier used to realize the noninverting amplifier: open-loop output resistance = 250 Ω.

Unknowns: Open-loop gain A of the op amp

Approach: The required value of operational amplifier gain can be found using Eq. (12.17), in which all the variables are known except A.

Assumptions: The operational amplifier is ideal except for finite open-loop gain and nonzero output resistance.

Analysis: The closed-loop output resistance is given by

$$R_{\text{out}} = \frac{R_o}{1 + A\beta} \leq 0.2\ \Omega$$

R_o and R_{out} are given, and β is determined by the desired gain:

$$R_o = 250\ \Omega \qquad R_{\text{out}} = 0.2\ \Omega \qquad \beta = \frac{1}{|A_v|}$$

We must convert the gain from dB to purely numeric form before we use it in the calculations:

$$|A_v| = 10^{35\,\text{dB}/20\,\text{dB}} = 56.2 \qquad \text{and} \qquad \beta = \frac{1}{|A_v|} = \frac{1}{56.2}$$

The minimum value of the open-loop gain A can now be determined from the R_{out} specification:

$$A \geq \frac{1}{\beta}\left(\frac{R_o}{R_{\text{out}}} - 1\right) = 56.2\left(\frac{250}{0.2} - 1\right) = 7.03 \times 10^4$$

$$A_{\text{dB}} = 20\log(7.03 \times 10^4) = 96.9 \text{ dB}$$

Check of Results: We have found the required unknown value.

Discussion: By exploring the world wide web, we see that op amps are available with 100 dB gain. So the value required by our design is achievable.

Computer-Aided Analysis: If we change the parameter values in the circuit in Ex. 12.2 and rerun the simulation, we will see if we meet the output resistance specification. Using R1 = 10 kΩ, R2 = 552 kΩ, and RO = 250 Ω with the gain of E1 set to 7.03E4, the SPICE transfer function analysis yields $A_v = 56.2$ and $R_{\text{out}} = 0.200$ Ω. The values of R1 and R2 were deliberately chosen to be large so that they would not materially affect the output resistance. We see from the voltage gain result that we have chosen the correct value for the ratio R2/R1.

EXERCISE: A noninverting amplifier must have a closed-loop gain of 40 dB and an output resistance of less than 0.1 Ω. The only op amp available has an output resistance of 200 Ω. What is the minimum open-loop gain of the op amp that will meet the requirements?

ANSWER: 106 dB

12.1.4 FINITE INPUT RESISTANCE

Next we explore the effect of the finite input resistance of the operational amplifier on the open-loop input resistances of the noninverting and inverting amplifier configurations. In this case, we shall find that the results are greatly different for the two amplifiers.

Let us first consider the noninverting amplifier circuit in Fig. 12.4, in which test source $\mathbf{v_x}$ is applied to the input. To find R_{in}, we must calculate the current $\mathbf{i_x}$ given by

$$\mathbf{i_x} = \frac{\mathbf{v_x} - \mathbf{v_1}}{R_{id}} \tag{12.18}$$

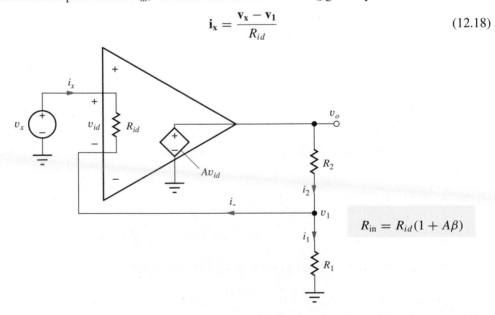

$$R_{\text{in}} = R_{id}(1 + A\beta)$$

Figure 12.4 Input resistance of the noninverting amplifier.

Voltage $\mathbf{v_1}$ is equal to

$$\mathbf{v_1} = \mathbf{i_1}R_1 = (\mathbf{i_2} - \mathbf{i_-})R_1 \cong \mathbf{i_2}R_1 \tag{12.19}$$

which has been simplified by assuming that the input current $\mathbf{i_-}$ to the op amp can still be neglected with respect to $\mathbf{i_2}$. We will check this assumption shortly. The assumption is equivalent to saying that $\mathbf{i_1} \cong \mathbf{i_2}$ and permits the voltage $\mathbf{v_1}$ to again be written in terms of the resistive voltage divider as

$$\mathbf{v_1} \cong \frac{R_1}{R_1 + R_2}\mathbf{v_o} = \beta\mathbf{v_o} = \beta(A\mathbf{v_{id}}) = A\beta(\mathbf{v_x} - \mathbf{v_1}) \tag{12.20}$$

Solving for $\mathbf{v_1}$ in terms of $\mathbf{v_x}$ yields

$$\mathbf{v_1} = \frac{A\beta}{1 + A\beta}\mathbf{v_x} \tag{12.21}$$

and substituting this result into Eq. (12.18) yields an expression for R_{in}

$$\mathbf{i_x} = \frac{\mathbf{v_x} - \dfrac{A\beta}{1 + A\beta}\mathbf{v_x}}{R_{id}} = \frac{\mathbf{v_x}}{(1 + A\beta)R_{id}} \quad \text{and} \quad \boxed{R_{\text{in}} = R_{id}(1 + A\beta)} \tag{12.22}$$

Note from Eq. (12.22) that the input resistance can be very large—much larger than that of the op amp itself. R_{id} is often large (1 MΩ to 1 TΩ) to start with, and it is multiplied by the loop gain $A\beta$, which is typically designed to be much greater than 1.

EXERCISE: What are the nominal, minimum, and maximum values of the open-loop gain and input resistance for an AD745 operational amplifier (see MCD website)? Repeat for the input resistance of the OP-27.

ANSWERS: 132 dB; 120 dB; no maximum value specified; 10^{10} Ω; minimum and maximum values not specified; 6 MΩ; 1.3 MΩ; no maximum specified

EXAMPLE 12.4 NONINVERTING AMPLIFIER INPUT RESISTANCE

Find a numeric value for the input resistance of a noninverting feedback amplifier circuit.

PROBLEM The noninverting amplifier in Fig. 12.4 is built with an op amp having an input resistance of 2 MΩ and an open-loop gain of 90 dB. What is the amplifier input resistance if $R_1 = 20$ kΩ and $R_2 = 510$ kΩ?

SOLUTION **Known Information and Given Data:** A noninverting feedback amplifier circuit is built with feedback resistors $R_1 = 20$ kΩ and $R_2 = 510$ kΩ. For the op amp: $A = 90$ dB, $R_{id} = 2$ MΩ

Unknown: Closed-loop amplifier input resistance R_{in}

Approach: In this case, we are given the values necessary to directly evaluate Eq. (12.22) including A, R_{id}, and the two feedback resistors.

Assumptions: The operational amplifier is ideal except for finite open-loop gain and finite input resistance.

Analysis: In order to evaluate Eq. (12.22) we must find β, which is determined by the feedback resistors

$$\beta = \frac{R_1}{R_1 + R_2} = \frac{20 \text{ k}\Omega}{20 \text{ k}\Omega + 510 \text{ k}\Omega} = \frac{1}{26.5}$$

We must also convert the gain from dB to purely numeric form before we use it in the calculations:

$$R_{id} = 2 \text{ M}\Omega \qquad \text{and} \qquad A = 10^{90\,\text{dB}/20\,\text{dB}} = 31{,}600$$

The closed-loop input resistance is given by

$$R_{\text{in}} = R_{id}(1 + A\beta) = 2 \text{ M}\Omega \left[1 + \frac{31600}{26.5} \right] = 2.39 \times 10^9 \ \Omega = 2.39 \text{ G}\Omega$$

Check of Results: We have found the only unknown value. The value is large as expected from our analysis of the noninverting amplifier.

Discussion: The calculated input resistance of the noninverting amplifier is very large (although not infinite as for case of an ideal op amp). In fact, the calculated value of R_{in} is so large that we must consider other factors that may limit the actual input resistance. These include surface leakage of the printed circuit board in which the op amp is mounted as well as common-mode input resistance of the op amp itself, which we discuss in Sec. 12.2.3.

Computer-Aided Analysis: To check our result with SPICE, we add RID = 2 MEG to the noninverting amplifier circuit model as shown below (with the gain of E1 = 31600) and perform a transfer function analysis. The results are $A_v = 26.5$ and $R_{\text{in}} = 2.39$ GΩ confirming our hand calculations.

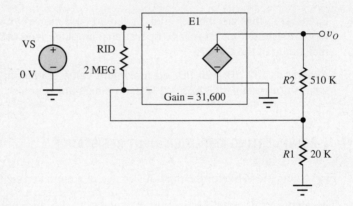

For the numbers in the preceding exercise, it is easy to see that the current $\mathbf{i}_-$, which equals $-\mathbf{i}_x$, is small compared to the current through R_2 and R_1 (see Prob. 12.10). Thus, our simplifying assumption that led to Eqs. (12.19) and (12.20) is well justified.

Input Resistance for the Inverting-Amplifier Configuration
The input resistance of the inverting amplifier can be determined using the circuit in Fig. 12.5(a). The input resistance is

$$R_{\text{in}} = \frac{\mathbf{v}_x}{\mathbf{i}_x} \qquad\qquad (12.23)$$

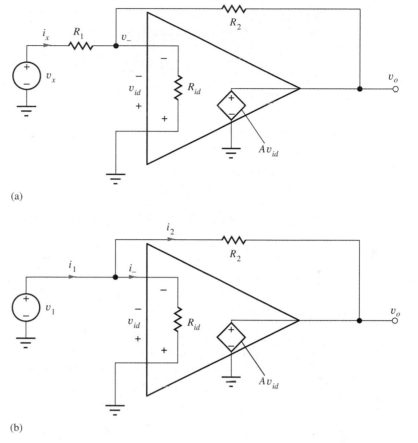

(a)

(b)

Figure 12.5 Inverting amplifier input-resistance calculation: (a) complete amplifier, (b) amplifier with R_1 removed.

where test signal $\mathbf{v_x}$ can be expressed as

$$\mathbf{v_x} = \mathbf{i_x} R_1 + \mathbf{v_-} \qquad \text{and} \qquad R_{\text{in}} = R_1 + \frac{\mathbf{v_-}}{\mathbf{i_x}} \tag{12.24}$$

The total input resistance R_{in} is equal to R_1 plus the resistance looking into the inverting terminal of the operational amplifier, which can be found using the circuit in Fig. 12.5(b). The input current in Fig. 12.5(b) is

$$\mathbf{i_1} = \mathbf{i_-} + \mathbf{i_2} = \frac{\mathbf{v_1}}{R_{id}} + \frac{\mathbf{v_1} - \mathbf{v_o}}{R_2} = \frac{\mathbf{v_1}}{R_{id}} + \frac{\mathbf{v_1} + A\mathbf{v_1}}{R_2} \tag{12.25}$$

Using this result, the input conductance can be written as

$$G_1 = \frac{\mathbf{i_1}}{\mathbf{v_1}} = \frac{1}{R_{id}} + \frac{1 + A}{R_2} \tag{12.26}$$

which represents the sum of two conductances. Thus, the equivalent resistance looking into the inverting-input terminal is the parallel combination of two resistors,

$$R_{id} \left\| \left(\frac{R_2}{1 + A} \right) \right. \tag{12.27}$$

and the overall input resistance of the inverting amplifier becomes

$$R_{\text{in}} = R_1 + R_{id} \left\| \left(\frac{R_2}{1 + A} \right) \right. \tag{12.28}$$

TABLE 12.1
Inverting and Noninverting Amplifier Summary

$\beta = \dfrac{R_1}{R_1+R_2}$	INVERTING AMPLIFIER	NONINVERTING AMPLIFIER
Voltage Gain A_v	$-\dfrac{R_2}{R_1}\left(\dfrac{A\beta}{1+A\beta}\right) \cong -\dfrac{R_2}{R_1}$	$\dfrac{A}{1+A\beta} \cong \dfrac{1}{\beta} = 1 + \dfrac{R_2}{R_1}$
Input Resistance R_{in}	$R_1 + \left(R_{id}\left\|\dfrac{R_2}{1+A}\right.\right) \cong R_1$	$R_{id}(1+A\beta) \cong R_{id}A\beta$
Output Resistance R_{out}	$\dfrac{R_O}{1+A\beta} \cong \dfrac{R_O}{A\beta}$	$\dfrac{R_O}{1+A\beta} \cong \dfrac{R_O}{A\beta}$

Normally, R_{id} will be large and Eq. (12.28) can be approximated by

$$R_{\text{in}} \cong R_1 + \left(\frac{R_2}{1+A}\right) \tag{12.29}$$

For large A and common values of R_2, the input resistance approaches the ideal result $R_{\text{in}} \cong R_1$. In other words, we see that the input resistance is usually dominated by R_1 connected to the quasi virtual ground at the op amp input. (Remember, v_{id} is no longer zero for a finite-gain amplifier.)

EXERCISE: Find the input resistance R_{in} of an inverting amplifier that has $R_1 = 1$ kΩ, $R_2 = 100$ kΩ, $R_{id} = 1$ MΩ, and $A = 100$ dB. What is the deviation of R_{in} from its ideal value?

ANSWERS: 1001 Ω; 1 Ω out of 1000 Ω or 0.1 percent

12.1.5 SUMMARY OF NONIDEAL INVERTING AND NONINVERTING AMPLIFIERS

Table 12.1 is a summary of the simplified expressions for the closed-loop voltage gain, input resistance, and output resistance of the inverting and noninverting amplifiers. These equations are most often used in the design of these basic amplifier circuits.

12.2 COMMON-MODE REJECTION AND INPUT RESISTANCE

12.2.1 FINITE COMMON-MODE REJECTION RATIO

Unfortunately, the output voltage of the real amplifier in Fig. 12.6 contains components in addition to the scaled replica of the input voltage (Av_{id}). In particular, a real amplifier also responds to the signal that is in common to both inputs, called the **common-mode input voltage** v_{ic} defined as

$$v_{ic} = \left(\frac{v_1 + v_2}{2}\right) \tag{12.30}$$

The common-mode input signal is amplified by the **common-mode gain** A_{cm} to give an overall output voltage expressed by

$$v_o = A(v_1 - v_2) + A_{cm}\left(\frac{v_1 + v_2}{2}\right) \qquad \text{or} \qquad v_o = A(v_{id}) + A_{cm}(v_{ic}) \tag{12.31}$$

where A (or A_{dm}) = **differential-mode gain**

A_{cm} = common-mode gain

$v_{id} = (v_1 - v_2)$ = differential-mode input voltage

$v_{ic} = \left(\dfrac{v_1 + v_2}{2}\right)$ = common-mode input voltage

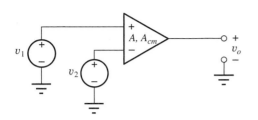

Figure 12.6 Operational amplifier with inputs v_1 and v_2.

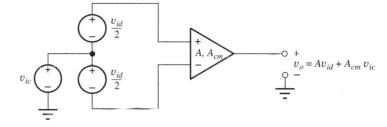

Figure 12.7 Operational amplifier with common-mode and differential-mode inputs shown explicitly.

Simultaneous solution of these last two equations allows voltages v_1 and v_2 to be expressed in terms of v_{ic} and v_{id} as

$$v_1 = v_{ic} + \frac{v_{id}}{2} \qquad \text{and} \qquad v_2 = v_{ic} - \frac{v_{id}}{2} \qquad (12.32)$$

and the amplifier in Fig. 12.6 can be redrawn in terms of v_{ic} and v_{id}, as in Fig. 12.7.

An ideal amplifier would amplify the **differential-mode input voltage** v_{id} and totally reject the common-mode input signal ($A_{cm} = 0$), as has been tacitly assumed thus far. However, an actual amplifier has a nonzero value of A_{cm}, and Eq. (12.31) is often rewritten in a slightly different form by factoring out A:

$$v_o = A\left[v_{id} + \frac{A_{cm}v_{ic}}{A}\right] = A\left[v_{id} + \frac{v_{ic}}{\text{CMRR}}\right] \qquad (12.33)$$

In this equation, **CMRR** is the **common-mode rejection ratio,** defined by the ratio of A and A_{cm}

$$\text{CMRR} = \left|\frac{A}{A_{cm}}\right| \qquad (12.34)$$

CMRR is often expressed in dB as

$$\text{CMRR}_{\text{dB}} = 20\log\left|\frac{A}{A_{cm}}\right| \text{ dB} \qquad (12.35)$$

An ideal amplifier has $A_{cm} = 0$ and therefore infinite CMRR. Actual amplifiers usually have $A \gg A_{cm}$, and the CMRR typically falls in the range

$$60 \text{ dB} \le \text{CMRR}_{\text{dB}} \le 120 \text{ dB}$$

A value of 60 dB is a relatively poor level of common-mode rejection, whereas achieving 120 dB (or even higher) is possible but difficult. Generally, the sign of A_{cm} is unknown ahead of time. In addition, CMRR specifications represent a lower bound. An illustration of the problems that can be caused by finite common-mode rejection is given in Example 12.5.

EXERCISE: What are the nominal, minimum, and maximum values of CMRR for the op-27 operational amplifier. (See MCD website for specification sheets.) Repeat for the AD745.

ANSWERS: 126 dB, 114 dB, no maximum value specified; 95 dB, 80 dB, no maximum value specified

12.2.2 WHY IS CMRR IMPORTANT?

The common-mode signal concept may initially seem obscure, but we actually encounter common-mode signals quite often. In digital systems, capacitive coupling of high frequency signals between signal lines on a bus or backplane can induce the same signal on more than one line. This induced signal often appears as a common-mode signal. Many high-speed computer buses utilize differential

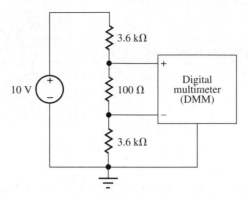

Figure 12.8 Common-mode input in a digital multimeter application.

signaling so that the undesired common-mode signals can be eliminated by amplifiers with good CMRR.

Probably the most common time that we encounter common-mode signals is when we use instruments to make measurements. Consider the circuit in Fig. 12.8 in which we are trying to measure the voltage across the 100-Ω resistor with a digital multimeter (DMM). The dc voltage difference across the DMM input terminals (its differential-mode input V_{DM}) is easily found by voltage division:

$$V_{DM} = V_+ - V_- = 10\ \text{V}\left(\frac{100\ \Omega}{7300\ \Omega}\right) = 0.137\ \text{V}$$

However, there is also a dc common-mode input to the DMM:

$$V_{CM} = \frac{V_+ + V_-}{2} = \frac{1}{2}\left[10\ \text{V}\left(\frac{3700\ \Omega}{7300\ \Omega}\right) + 10\ \text{V}\left(\frac{3600\ \Omega}{7300\ \Omega}\right)\right] = 4.5\ \text{V}$$

Thus our DMM must accurately measure the 0.137 V differential input in the presence of a 4.5 V common-mode input, and this requires the digital multimeter to have good common-mode rejection capability. If we want the common-mode input to produce an error of less that 0.1% in the measurement of the 0.137-V input, we need

$$\frac{4.5}{\text{CMRR}} \leq 10^{-3} \times (0.137\ \text{V}) \quad \text{or} \quad \text{CMRR} \geq 3.28 \times 10^4$$

which represents a CMRR of more than 90 dB.

A similar measurement problem occurs when an oscilloscope is used in its differential mode. In this case, the differential- and common-mode inputs may have very high frequency signal components in addition to dc. Unfortunately, good common-mode rejection at high frequencies is difficult to obtain.

EXAMPLE 12.5   **COMMON-MODE ERROR CALCULATION**

Calculate the error in a differential amplifier with nonideal values of gain and common-mode rejection.

PROBLEM   Suppose the amplifier in Fig. 12.6 has a differential-mode gain of 2500 and a CMRR of 80 dB. What is the output voltage if $v_1 = 5.001$ V and $v_2 = 4.999$ V? What is the error introduced by the finite CMRR?

SOLUTION   **Known Information and Given Data:** For the amplifier in Fig. 12.6: $A = 2500$, CMRR = 80 dB, $v_1 = 5.001$ V, and $v_2 = 4.999$ V.

Unknowns: Output voltage v_o; common-mode contribution to the error

Approach: Use the known values to evaluate Eq. (12.33)

Assumptions: The op amp is ideal except for finite gain and CMRR. The CMRR specification of 80 dB corresponds to CMRR $= \pm 10^4$. Let us assume CMRR $= +10^4$ for this example.

Analysis: The differential- and common-mode input voltages are

$$v_{id} = 5.001 \text{ V} - 4.999 \text{ V} = 0.002 \text{ V} \qquad \text{and} \qquad v_{ic} = \frac{5.001 + 4.999}{2} \text{ V} = 5.000 \text{ V}$$

$$v_o = A \left[v_{id} + \frac{v_{ic}}{\text{CMRR}} \right] = 2500 \left[0.002 + \frac{5.000}{10^4} \right] \text{ V}$$

$$= 2500[0.002 + 0.0005] \text{ V} = 6.25 \text{ V}$$

The error introduced by the common-mode input is 25 percent of the differential input voltage.

Check for Results: We have found the required unknowns. The output voltage is a reasonable value for power supplies normally used with integrated circuit op amps.

Evaluation and Discussion: An ideal amplifier would amplify only v_{id} and produce an output voltage of 5.00 V. For this particular situation, the output voltage is in error by 25 percent due to the finite common-mode rejection of the amplifier. Common-mode rejection is often important in measurements of small voltage differences in the presence of large common-mode voltages, as in the example shown here. Note in this case that

$$A_{cm} = \frac{A}{\text{CMRR}} = \frac{2500}{10000} = 0.25 \text{ or } -12 \text{ dB}$$

Computer-Aided Analysis: Let's build a model to simulate this example. The output of the amplifier can be rewritten as

$$v_o = A_{dm} v_{id} + \frac{A_{cm}}{2} v_1 + \frac{A_{cm}}{2} v_2 = 2500 v_{id} + 0.125 v_1 + 0.125 v_2$$

which is implemented below using three voltage-controlled voltage sources. EDM depends on the voltage difference V1–V2, ECM1 depends on the voltage V1, and ECM2 depends on the voltage V2. An operating point analysis confirms our hand analysis with $v_o = 6.25$ V. RL is added to have two connections at the output node and does not affect the calculation.

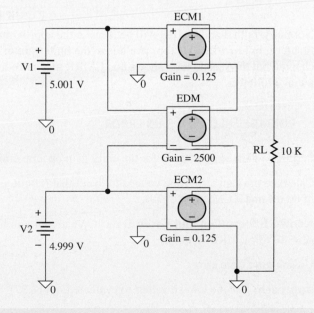

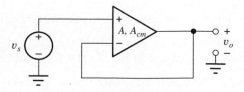

Figure 12.9 CMRR error in the voltage follower.

12.2.3 VOLTAGE-FOLLOWER GAIN ERROR DUE TO CMRR

Finite CMRR can also play an important role in determining the gain error in the voltage-follower circuit in Fig. 12.9, for which

$$\mathbf{v_{id}} = \mathbf{v_s} - \mathbf{v_o} \qquad \text{and} \qquad \mathbf{v_{ic}} = \frac{\mathbf{v_s} + \mathbf{v_o}}{2}$$

Using Eq. (12.31)

$$\mathbf{v_o} = A\left[(\mathbf{v_s} - \mathbf{v_o}) + \frac{\mathbf{v_s} + \mathbf{v_o}}{2\,\text{CMRR}}\right] \tag{12.36}$$

Solving this equation for $\mathbf{v_o}$ yields

$$A_v = \frac{\mathbf{v_o}}{\mathbf{v_s}} = \frac{A\left[1 + \dfrac{1}{2\,\text{CMRR}}\right]}{1 + A\left[1 - \dfrac{1}{2\,\text{CMRR}}\right]} \tag{12.37}$$

The ideal gain for the voltage follower is unity, so the gain error is equal to

$$\text{GE} = 1 - A_v = \frac{1 - \dfrac{A}{\text{CMRR}}}{1 + A\left[1 - \dfrac{1}{2\,\text{CMRR}}\right]} \cong \frac{1}{A} - \frac{1}{\text{CMRR}} \tag{12.38}$$

Normally, both A and CMRR will be $\gg 1$, so the approximation in Eq. (12.38) to usually valid. The first term in Eq. (12.38) is the error due to the finite gain of the amplifier, as discussed earlier in this chapter, but the second term shows that CMRR may introduce an error of even greater import in the voltage follower.

EXAMPLE **12.6**   **VOLTAGE FOLLOWER GAIN ERROR**

Perform a gain error analysis for the unity gain op amp circuit.

PROBLEM   Calculate the gain error for a voltage follower that is built using an op amp with an open-loop gain of 80 dB and a CMRR of 60 dB.

SOLUTION   **Known Information and Given Data:** Operational amplifier configured as a voltage follower; $A = 80$ dB; CMRR $= 60$ dB

Unknowns: Gain error

Approach: Use the known values to evaluate Eq. (12.37).

Assumptions: The op amp is ideal except for finite open-loop gain and CMRR. The CMRR specification of 60 dB corresponds to CMRR = ±1000. Let us assume CMRR = +1000 for this example. Since both A and CMRR are much greater than one, we will use the approximate form of Eq. (12.38).

Analysis: Equation (12.38) gives a gain error of

$$\text{GE} \cong \frac{1}{10^4} - \frac{1}{10^3} = -9.00 \times 10^{-4} \qquad \text{or} \qquad -0.090 \text{ percent}$$

Check of Results: We have found the desired gain error. However, the sign is negative, which may seem a bit unusual. We better explore this result further.

Discussion: In this calculation, the error due to finite CMRR is ten times larger than that due to finite gain. As pointed out above, the gain error is negative, which corresponds to a gain that is greater than 1! Finite open-loop gain alone always causes A_v to be slightly less than 1. However, for this case,

$$A_v = \frac{A\left[1 + \dfrac{1}{2\,\text{CMRR}}\right]}{1 + A\left[1 - \dfrac{1}{2\,\text{CMRR}}\right]} = \frac{10^4\left[1 + \dfrac{1}{2(1000)}\right]}{1 + 10^4\left[1 - \dfrac{1}{2(1000)}\right]} = 1.001$$

We must be aware of errors related to CMRR whenever we are trying to perform precision amplification and measurement.

Computer-Aided Analysis: The amplifier model from the previous example is reconnected in the circuit below as a voltage follower with V1 = 0. The gains of EDM, ECM1, and ECM2 are set to 10,000, 5, and 5, respectively. A SPICE transfer function analysis gives a voltage gain of +1.001.

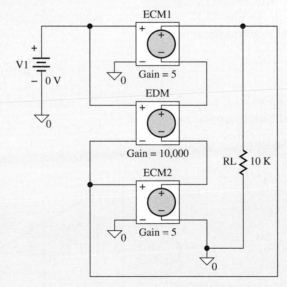

Discussion of CMRR often focuses on amplifier behavior at dc. However, CMRR can be an even greater problem at higher frequencies. **Common-mode rejection** decreases rapidly as frequency increases, typically with a slope of at least −20 dB/decade increase in frequency. This roll-off of the CMRR can begin at frequencies below 100 Hz. Thus, common-mode rejection at 60 or 120 Hz can be much worse than that specified for dc.

EXERCISE: A voltage follower is to be designed to provide a gain error of less than 0.005 percent. Develop a set of minimum required specifications on open-loop gain and CMRR.

ANSWER: Several possibilities: A = 92 dB, CMRR = 92 dB; A = 100 dB, CMRR = 88 dB; CMRR = 100 dB, A = 88 dB.

12.2.4 COMMON-MODE INPUT RESISTANCE

Up to now, the discussion of the input resistance of an op amp has been limited to the resistance R_{id}, which is actually the approximate resistance presented to a purely differential-mode input voltage v_{id}. In Fig. 12.10, two new resistors with value $2\,R_{ic}$ have been added to the circuit to model the finite common-mode input resistance of the amplifier.

When a purely common-mode signal v_{ic} is applied to the input of this amplifier, as depicted in Fig. 12.11, with $v_{id} = 0$, the input current is nonzero even though R_{id} is shorted out. In this situation, the total resistance presented to source v_{ic} is the parallel combination of the two resistors with value $2R_{ic}$, which thus equals R_{ic}. Therefore, R_{ic} is the equivalent resistance presented to the common-mode source; it is called the **common-mode input resistance** of the op amp. The value of R_{ic} is often much greater than that of the **differential-mode input resistance** R_{id}, typically in excess of 10^9 Ω (1 GΩ).

From Fig. 12.12, we see that a purely differential-mode input signal actually sees an input resistance equivalent to

$$R_{\text{in}} = R_{id} \parallel 4R_{ic} \qquad (12.39)$$

As mentioned, however, R_{ic} is typically much greater than R_{id}, and the differential-mode input resistance is approximately equal to R_{id}.

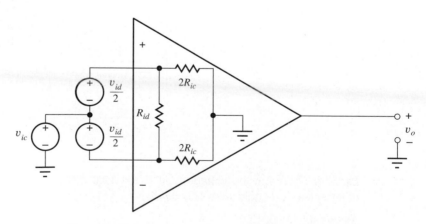

Figure 12.10 Op amp with common-mode input resistances added.

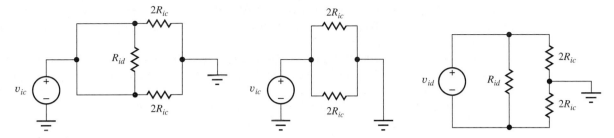

Figure 12.11 Amplifier with only a common-mode input signal present.

Figure 12.12 Amplifier input for a purely differential-mode input.

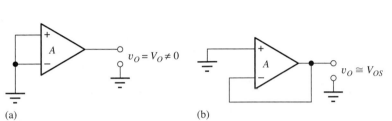

(a)   (b)

Figure 12.13 (a) Amplifier with zero input voltage but nonzero output voltage. (*Note:* The offset voltage cannot be measured in this manner.) (b) Circuit for measuring offset voltage.

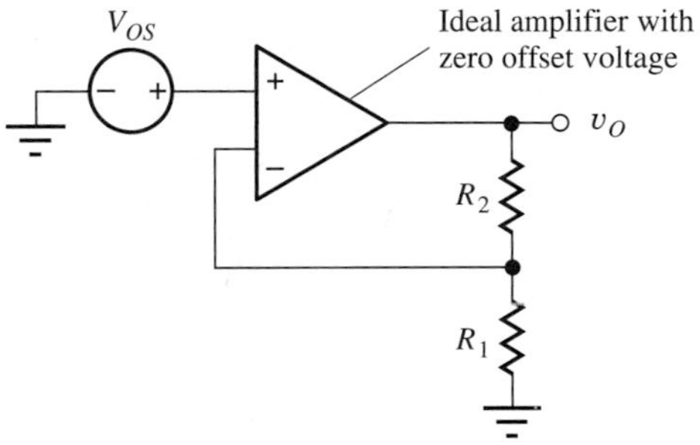

Figure 12.14 Offset voltage can be modeled by a voltage source V_{OS} in series with the amplifier input.

12.3 DC ERROR SOURCES AND OUTPUT RANGE LIMITATIONS

Another important class of error sources results from the need to bias the internal circuits that form the operational amplifier and from mismatches between pairs of solid-state devices in these circuits. These dc error sources include the input-offset voltage V_{OS}, the input-bias currents I_{B1} and I_{B2}, and the input-offset current I_{OS}. Power supply rejection is also considered.

12.3.1 INPUT-OFFSET VOLTAGE

When the inputs of the amplifier in Fig. 12.13 are both zero, the output of the amplifier is not truly zero but is resting at some nonzero dc voltage level. A small dc voltage seems to have been applied to the input of the amplifier, which is being amplified by the gain.[1] The equivalent dc **input-offset voltage** V_{OS} is defined as

$$V_{OS} = \left.\frac{V_O}{A}\right|_{v_1=o=v_2} \tag{12.40}$$

Equation (12.33) can be modified to include the effects of this offset voltage by adding the V_{OS} term:

$$v_O = A\left[v_{id} + \frac{v_{ic}}{\text{CMRR}} + V_{OS}\right] \tag{12.41}$$

The first term in brackets represents the desired differential input signal to the amplifier, whereas the second and third terms represent the common-mode and offset-voltage error sources that corrupt the desired signal.

As in the case of CMRR, the actual sign of the offset voltage is not known, and only the magnitude of the worst-case offset voltage is specified. Most commercial operational amplifiers

---

[1] The voltage arises mainly from mismatches in the transistors in the input stage of the operational amplifier. We will explore this problem in Chapter 15.

have offset-voltage specifications of less than 10 mV, and op amps can easily be purchased with V_{OS} specified to be less than a few mV. For additional cost, internally trimmed op amps are available with $V_{OS} < 0.25$ mV.

The offset voltage usually cannot be measured with the operational amplifier connected as depicted in Fig. 12.13(a) because of the high gain of the amplifier. However, the circuit in Fig. 12.13(b) can be used. Here the amplifier is connected as a voltage follower, and the output voltage is equal to the offset voltage of the amplifier (except for the small gain error of the amplifier since $A \neq \infty$.)

In Example 12.7, the effect of offset voltage on circuit performance is found using the model in Fig. 12.14, in which the offset voltage is represented by a source in series with the input to an otherwise ideal amplifier. V_{OS} is amplified just as any input signal source, and the dc output voltage of the amplifier in Fig. 12.14 is

$$v_O = \left(1 + \frac{R_2}{R_1}\right) V_{OS} \tag{12.42}$$

EXERCISE: What are the nominal, minimum, and maximum values of offset voltage for the AD745 operational amplifier at 25°C? (See MCD website for specification sheets.) Repeat for the OP77E.

ANSWERS: 0.25 mV, no minimum value specified, 1.0 mV; 10 μV, no minimum value specified, 25 μV

EXAMPLE 12.7   OFFSET VOLTAGE ANALYSIS

Practice calculating the output voltage of an op amp circuit caused by the offset voltage of the op amp.

PROBLEM Suppose the amplifier in Fig. 12.14 has $|V_{OS}| \leq 3$ mV and R_2 and R_1 are 99 kΩ and 1.2 kΩ, respectively. What is the quiescent dc voltage at the amplifier output?

SOLUTION **Known Information and Given Data:** Noninverting amplifier configuration with $R_1 = 1.2$ kΩ and $R_2 = 99$ kΩ. The amplifier has and equivalent input voltage of $|V_{OS}| \leq 3$ mV.

Unknowns: Amplifier dc output voltage V_O

Approach: Use the known values to evaluate Eq. (12.42).

Assumptions: The op amp is ideal except for the specified value of nonzero offset voltage.

Analysis: Using Eq. (12.42), we find that the output voltage is

$$|V_O| \leq \left(1 + \frac{99 \text{ k}\Omega}{1.2 \text{ k}\Omega}\right)(0.003) = 0.25 \text{ V}$$

Check of Results: We have found the value of the only unknown, and the value appears reasonable for standard IC power supplies.

Discussion: We do not actually know the sign of V_{OS} since the V_{OS} specification represents an upper bound. Therefore we actually know only that

$$-0.25 \text{ V} \leq V_O \leq 0.25 \text{ V}$$

EXERCISE: Repeat the calculation in Ex. 12.7 if the noninverting amplifier gain is set to 50 and the offset voltage is 2 mV?

ANSWER: 100 mV

12.3.2 OFFSET-VOLTAGE ADJUSTMENT

Addition of a potentiometer allows the **offset voltage** of most IC op amps to be manually adjusted to zero. Commercial amplifiers typically provide two terminals to which the potentiometer can be connected, as in Fig. 12.15. The third terminal of the potentiometer is connected to the positive or negative power supply voltage. The potentiometer value depends on the internal design of the amplifier.

12.3.3 AN ALTERNATE INTERPRETATION OF CMRR

If the differential input voltage v_{id} is set to zero in Eq. (12.41), then any residual output voltage is due to two equivalent input error voltage contributions:

$$v_O = A\left(V_{OS} + \frac{v_{ic}}{\text{CMRR}}\right) = A(v_{OS}) \tag{12.43}$$

We can view the CMRR as being a measure of how the total offset voltage v_{OS} changes from its dc value V_{OS} when a common-mode voltage is applied. We may find CMRR as

$$\text{CMRR} = \frac{v_{OS}}{v_{ic}} \qquad \text{or} \qquad \text{CMRR}_{\text{dB}} = 20\,\log\left|\frac{v_{os}}{v_{ic}}\right|^{-1} \tag{12.44}$$

where the first form has units of μV/V.

Power Supply Rejection Ratio

A parameter closely related to CMRR is the **power supply rejection ratio,** or **PSRR.** When power supply voltages change due to long-term drift or the existence of noise on the supplies, the equivalent input-offset voltage changes slightly. PSRR is a measure of the ability of the amplifier to reject these power supply variations.

In a manner similar to the CMRR, the power supply rejection ratio indicates how the offset voltage changes in response to a change in the power supply voltages.

$$\text{PSRR}_+ = \frac{\Delta V_{OS}}{\Delta V_{CC}} \qquad \text{and} \qquad \text{PSRR}_- = \frac{\Delta V_{OS}}{\Delta V_{EE}} \quad \text{usually expressed in} \quad \frac{\mu V}{V} \tag{12.45}$$

PSRR is also often expressed in dB as $\text{PSRR}_{\text{dB}} = 20\log|1/\text{PSRR}|$.

Generally the PSRR is different for changes in V_{CC} and V_{EE}, and the op amp PSRR specification usually represents the poorer of these two values. PSRR values are similar to those of CMRR with typical values ranging from 60 to 120 dB at dc. It is important to note that both CMRR and PSRR fall rapidly as frequency increases.

EXERCISE: What are the nominal, minimum, and maximum values of PSRR and CMRR for the OP77E operational amplifier (see MCD website for specification sheets)? Repeat for the AD741C.

ANSWERS: 123 dB, 120 dB, no maximum value specified; 140 dB, 120 dB, no maximum value specified; 90 dB, 76 dB, no maximum value specified; 90 dB, 70 dB, no maximum value specified

12.3.4 INPUT-BIAS AND OFFSET CURRENTS

For the transistors that form the operational amplifier to operate, a small but nonzero dc bias current must be supplied to each input terminal of the amplifier. These currents represent base currents in

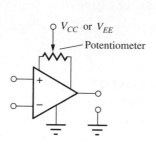

Figure 12.15 Offset-voltage adjustment of an operational amplifier.

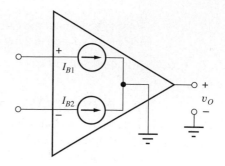

Figure 12.16 Operational amplifier with input-bias currents modeled by current sources I_{B1} and I_{B2}.

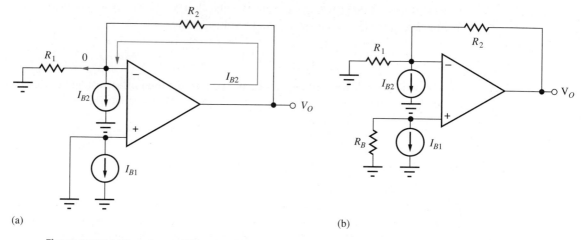

(a)

(b)

Figure 12.17 (a) Inverting amplifier with input-bias currents modeled by current sources I_{B1} and I_{B2}. (b) Inverting amplifier with bias current compensation resistor R_B.

an amplifier built with bipolar transistors or gate currents in one designed with MOSFETs or JFETs. As we shall see, this dc current, although small, is an additional source of error.

The bias currents can be modeled by two current sources I_{B1} and I_{B2} connected to the noninverting and inverting inputs of the amplifier, as in Fig. 12.16. The values of I_{B1} and I_{B2} are similar but not identical, and the actual direction of the currents depends on the details of the internal amplifier circuit (*npn*, *pnp*, NMOS, PMOS, and so on). The difference between the two bias currents is called the **offset current I_{OS}**.

$$I_{OS} = I_{B1} - I_{B2} \quad \text{and} \quad |I_{OS}| \le I_{MAX} \tag{12.46}$$

The offset-current specification for an op amp is normally expressed as an upper bound I_{MAX} on the magnitude of I_{OS}, and the actual sign of I_{OS} for a given op amp is not known.

In an operational amplifier circuit, the **input-bias currents** produce an undesired voltage at the amplifier output. Consider the inverting amplifier in Fig. 12.17(a) as an example. In this circuit, I_{B1} is shorted out by the direct connection of the noninverting input to ground and does not affect the circuit. However, because the inverting input represents a virtual ground, the current in R_1 must be zero, forcing I_{B2} to be supplied by the amplifier output through R_2. Thus, the dc output voltage is equal to

$$V_O = I_{B2}R_2 \tag{12.47}$$

The output-voltage error in Eq. (12.47) can be reduced by placing a **bias current compensation resistor R_B** in series with the noninverting input of the amplifier, as in Fig. 12.17(b). Using analysis

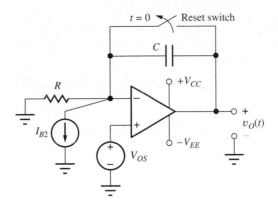

Figure 12.18 Example of dc offset-voltage and bias current errors in an integrator.

by superposition, the output due to I_{B1} acting alone is

$$V_O = -I_{B1}R_B \left(1 + \frac{R_2}{R_1}\right) \tag{12.48}$$

The total output voltage is the sum of Eqs. (12.47) and (12.48):

$$V_O^T = I_{B2}R_2 - I_{B1}R_B \left(1 + \frac{R_2}{R_1}\right) \tag{12.49}$$

If R_B is set equal to the parallel combination of R_1 and R_2, then the expression for the output-voltage error reduces to

$$V_O^T = (I_{B2} - I_{B1})R_2 = -I_{OS}R_2 \qquad \text{for} \qquad R_B = \frac{R_1 R_2}{R_1 + R_2} \tag{12.50}$$

The value of the offset current is typically a factor of 5–10 smaller than either of the individual bias currents, so the dc output-voltage error can be substantially reduced by using bias current compensation techniques.

Another example of the problems associated with offset-voltage and bias currents occurs in the integrator circuit in Fig. 12.18. A reset switch has been added to the integrator and is kept closed for $t < 0$. With the switch closed, the circuit is equivalent to a voltage follower, and the output voltage v_O is equal to the offset voltage V_{OS}. However, when the switch opens at $t = 0$, the circuit begins to integrate its own offset-voltage and bias current. Again using superposition analysis, it is easy to show (see Prob. 12.35) that the output voltage becomes

$$v_O(t) = V_{OS} + \frac{V_{OS}}{RC}t + \frac{I_{B2}}{C}t \qquad \text{for} \qquad t \geq 0 \tag{12.51}$$

The output voltage becomes a ramp with a constant slope determined by the values of V_{OS} and I_{B2}. Eventually, the integrator output saturates at a limit set by one of the two power supplies, as discussed in more detail in Sec. 12.3.5. If an integrator is built in the laboratory without a reset switch, the output is normally found to be resting near one of the power supply voltages.

EXERCISE: What are the nominal, minimum, and maximum values of the input bias and offset currents for the µA741C operational amplifier (see MCD website for specification sheets)? Repeat for the AD745J.

ANSWERS: 80 nA, no minimum value, 500 nA; 20 nA, no minimum value, 200 nA; 150 pA, no minimum value, 400 pA; 40 pA, no minimum value, 150 pA

EXERCISE: An inverting amplifier is designed with $R_1 = 1$ kΩ and $R_2 = 39$ kΩ. What value of resistance should be placed in series with the noninverting input terminal for bias current compensation?

ANSWERS: 975 Ω; Note that 1 kΩ is the closest 5 percent resistor value.

EXERCISE: An integrator has $R = 10$ kΩ, $C = 100$ pF, $V_{OS} = 1.5$ mV, and $I_{B2} = 100$ nA. How long will it take v_O to saturate (reach V_{CC} or V_{EE}) after the power supplies are turned on if $V_{CC} = V_{EE} = 15$ V?

ANSWER: $t = 6.0$ ms

12.3.5 OUTPUT VOLTAGE AND CURRENT LIMITS

An actual operational amplifier has a limited range of voltage and current capability at its output. For example, the voltage at the output of the amplifier in Fig. 12.19 cannot exceed V_{CC} or be more negative than $-V_{EE}$. In fact, for many real op amps, the output-voltage range is limited to several volts less than the power supply span. For example, the output-voltage limits for a particular op amp might be specified as

$$(-V_{EE} + 1) \text{ V} \le v_O \le (V_{CC} - 2) \text{ V} \tag{12.52}$$

Commercial operational amplifiers also contain circuits that restrict the magnitude of the current in the output terminal in order to limit power dissipation in the amplifier and to protect the amplifier from accidental short circuits. The current-limit specification is often given in terms of the minimum load resistance that an amplifier can drive with a given voltage swing. For example, an amplifier may be guaranteed to deliver an output of ± 10 V only for a total load resistance ≥ 5 kΩ. This is equivalent to saying that the total output current i_O is limited to

$$|i_O| \le \frac{10 \text{ V}}{5 \text{ k}\Omega} = 2 \text{ mA} \tag{12.53}$$

The output-current specification not only affects the size of load resistor that can be connected to the amplifier, it also places lower limits on the value of the feedback resistors R_1 and R_2. The total output current i_O in Fig. 12.20 is given by $\mathbf{i_O} = \mathbf{i_L} + \mathbf{i_F}$, and since the current into the inverting input is zero,

$$\mathbf{i_O} = \frac{\mathbf{v_O}}{R_L} + \frac{\mathbf{v_O}}{R_2 + R_1} = \frac{\mathbf{v_O}}{R_{\text{EQ}}} \tag{12.54}$$

The amplifier output must supply current not only to the load but also to its own feedback network! From Eq. (12.54), we see that the resistance that the noninverting amplifier must drive is equivalent

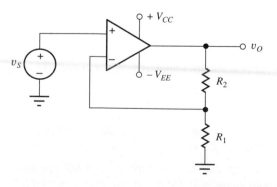

Figure 12.19 Amplifier with power supply voltages indicated.

ELECTRONICS IN ACTION

Sample-and-Hold Circuits

Sample-and-hold (S/H) circuits are used throughout sampled data systems and are needed ahead of many types of analog-to-digital converters in order to prevent the ADC input signal from changing during the conversion time. Several op amp based S/H circuits are described here[1] including one based upon an op amp integrator.

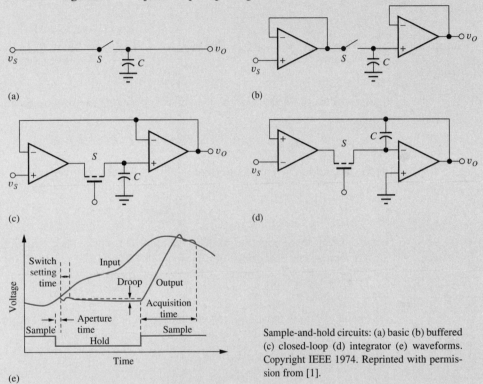

Sample-and-hold circuits: (a) basic (b) buffered (c) closed-loop (d) integrator (e) waveforms. Copyright IEEE 1974. Reprinted with permission from [1].

The basic sample-and-hold in (a) of the figure includes a sampling switch S and a capacitor C that stores the sampled voltage. However, this simple circuit can incur errors due to loading of the signal being sampled. Circuit (b) utilizes voltage followers to solve the problem by buffering both the input to, and the output from, sampling capacitor C. The closed-loop sample-and-hold circuits in (c) and (d) place C within a global feedback loop to improve circuit performance. The integrator circuit in (d) greatly increases the effective value of the sampling capacitor. If we apply our ideal op amp assumptions to each of the three S/H circuits, we find that both the capacitor and output voltages are always forced to be equal to the input voltage v_S.

The graph in part (e) illustrates some of design issues associated with sample-and-hold operation. The aperture time represents the time required for the switching devices to change state between the sample and hold modes. A settling time is then required for the feedback circuits to recover from the switching transients. During the hold mode, the voltage stored on the capacitor can change slightly due to switch leakage and op amp bias currents. This change is referred to as "droop." Finally, an acquisition time is required for the circuit to catch back up to the input voltage after the circuit switches from hold mode back to sample mode.

[1] K. R. Stafford, P. R. Gray, and R. A. Blanchard, "A complete monolithic sample-and-hold," *IEEE Journal of Solid-State Circuits,* vol. SC-9, no. 6, pp. 381–387, December 1974.

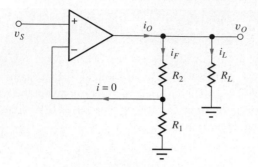

Figure 12.20 Output-current limit in the noninverting amplifier.

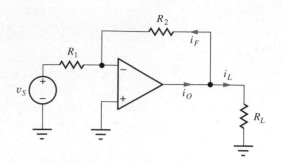

Figure 12.21 Output-current limit in the inverting-amplifier circuit.

to the parallel combination of the load resistance and the series combination of R_1 and R_2:

$$R_{EQ} = R_L \parallel (R_1 + R_2) \tag{12.55}$$

For the case of the inverting amplifier in Fig. 12.21, R_{EQ} is given by

$$R_{EQ} = R_L \parallel R_2 \tag{12.56}$$

because the inverting-input terminal of the amplifier represents a virtual ground.

The output-current constraint represented by Eqs. (12.55) and (12.56) often helps us choose the size of the feedback resistors during the design process.

EXERCISE: What is the maximum guaranteed value for the output current of the OP-27A operational amplifier (see MCD website for specification sheets)?

ANSWER: 12 V/2 kΩ = 6 mA

DESIGN EXAMPLE 12.8 INVERTING AMPLIFIER DESIGN WITH OUTPUT CURRENT LIMITS

Here we explore op amp circuit design including constraints on the output current capability of the op amp.

PROBLEM The amplifier in Fig. 12.21 is to be designed to have a gain of 20 dB and must develop a peak output voltage of at least 10 V when connected to a minimum load resistance of 5 kΩ. The op amp output current specification states that the output current must be less than 2.5 mA. Choose acceptable values of R_1 and R_2 from the table of 5 percent resistor values in Appendix A.

SOLUTION **Known Information and Given Data:** Inverting amplifier configuration; $A_v = 20$ dB, $|v_O| \leq 10$ V with $R_L \geq 5$ kΩ. The magnitude of the op amp output current must not exceed 2.5 mA.

Unknowns: Feedback resistors R_1 and R_2. Choose real values from the tables in Appendix A.

Approach: The op amp must supply current to both the load resistor and the feedback networks. We must account for both.

Assumptions: The op amp is ideal except for its limited output current capability.

Analysis: The equivalent load resistance on the amplifier must be greater than 4 kΩ:

$$R_{EQ} \geq \frac{10 \text{ V}}{2.5 \text{ mA}} = 4 \text{ k}\Omega \qquad \text{or} \qquad R_L \| R_2 \geq 4 \text{ k}\Omega$$

Because the minimum value of R_L is 5 kΩ, the feedback resistor R_2 must satisfy $R_2 \geq 20$ kΩ, and we also have $R_2/R_1 = 10$ because the gain was specified as 20 dB. We should allow some safety margin in the value of R_2. For example, a 27-kΩ resistor with a 5 percent tolerance will have a minimum value of 25.6 kΩ and would be satisfactory. A 22-kΩ resistor would have a minimum value of 20.9 kΩ and would also meet the specification. A wide range of choices still exists for R_1 and R_2.

Several acceptable choices would be

$$R_2 = 22 \text{ k}\Omega \qquad \text{and} \qquad R_1 = 2.2 \text{ k}\Omega$$
$$R_2 = 27 \text{ k}\Omega \qquad \text{and} \qquad R_1 = 2.7 \text{ k}\Omega$$
$$R_2 = 47 \text{ k}\Omega \qquad \text{and} \qquad R_1 = 4.7 \text{ k}\Omega$$
$$R_2 = 100 \text{ k}\Omega \qquad \text{and} \qquad R_1 = 10 \text{ k}\Omega$$

Let us select the last choice: $R_1 = 10$ kΩ and $R_2 = 100$ kΩ to provide an input resistance of 10 kΩ.

Check of Results: The gain is $-R_2/R_1 = -10$, which is correct. The maximum output current will be

$$i_o \leq \frac{10 \text{ V}}{100 \text{ k}\Omega} + \frac{10 \text{ V}}{5 \text{ k}\Omega} = 2.1 \text{ mA}$$

which is less than 2.5 mA (2.2 mA if we include 5 percent tolerances).

Discussion: Note that an input resistance specification would help us decide on a value for R_1.

Computer-Aided Design: SPICE can be used to check our design using the circuit below. The gain of E1 is set to 1E6 to approximate an ideal op amp. VS is set to -1 V to produce an output of $+10$ V. Operating point and transfer function analyses yield $V_O = 10$ V, $I_O = 2.1$ mA, $A_v = -10$, and $R_{in} = 10$ kΩ, all in agreement with our theory.

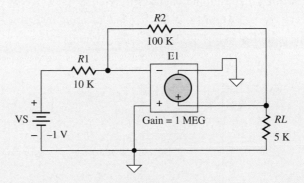

EXERCISE: What is the maximum guaranteed value for the output current of the AD745J operational amplifier (see MCD website for specification sheets)?

ANSWER: 12 V/2 kΩ = 6 mA

EXERCISE: Design a noninverting amplifier to have a gain of 20 dB and to develop a peak output voltage of at least 20 V when connected to a load resistance of at least 5 kΩ. The op amp output current specification states that the output current must be less than 5 mA. Choose acceptable values of R_1 and R_2 from the table of 5 percent resistor values in Appendix A.

ANSWER: Some possibilities: 27 kΩ and 3 kΩ; 270 kΩ and 30 kΩ; 180 kΩ and 20 kΩ; *but not* 18 kΩ and 2 kΩ because of tolerances.

12.4 FREQUENCY RESPONSE AND BANDWIDTH OF OPERATIONAL AMPLIFIERS

Most general-purpose operational amplifiers are low-pass amplifiers designed to have high gain at dc and a **single-pole frequency response** described by

$$A(s) = \frac{A_o \omega_B}{s + \omega_B} = \frac{\omega_T}{s + \omega_B} \qquad (12.57)$$

in which A_o is the open-loop gain at dc, ω_B is the open-loop bandwidth of the op amp, and ω_T is called the **unity-gain frequency,** the frequency at which $|A(j\omega)| = 1$ (0 dB). The magnitude of Eq. (12.57) versus frequency can be expressed as

$$|A(j\omega)| = \frac{A_o \omega_B}{\sqrt{\omega^2 + \omega_B^2}} = \frac{A_o}{\sqrt{1 + \dfrac{\omega^2}{\omega_B^2}}} \qquad (12.58)$$

An example is depicted graphically in the Bode plot in Fig. 12.22. For $\omega \ll \omega_B$, the gain is constant at the dc value A_o. The bandwidth of the open-loop amplifier, the frequency at which the gain is 3 dB below A_o, is ω_B (or $f_B = \omega_B/2\pi$). In Fig. 12.22, $A_o = 10,000\,(80\,\text{dB})$ and $\omega_B = 1000\,\text{rad/s}\,(159\,\text{Hz})$.

At high frequencies, $\omega \gg \omega_B$, the transfer function can be approximated by

$$|A(j\omega)| \cong \frac{A_o \omega_B}{\omega} = \frac{\omega_T}{\omega} \qquad (12.59)$$

Using Eq. (12.59), we see that the magnitude of the gain is indeed unity at $\omega = \omega_T$:

$$|A(j\omega)| = \frac{\omega_T}{\omega} = 1 \qquad \text{for } \omega = \omega_T \qquad (12.60)$$

Rewriting the result in Eq. (12.60),

$$|A(j\omega)|\,\omega \cong \omega_T \qquad \text{and dividing by } 2\pi \qquad |A(j\omega)|\,f \cong f_T \qquad (12.61)$$

The amplifier in Fig. 12.22 has $\omega_T = 10^7$ rad/s.

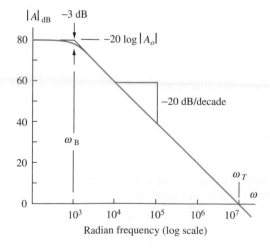

Figure 12.22 Voltage gain vs. frequency for an operational amplifier.

ELECTRONICS IN ACTION

CMOS Navigation Chip Prototype for Optical Mice

Agilent Technologies has sold over 100 million optical navigation mouse sensors, the devices at the core of most optical mice sold today. However, as is often the case in the engineering world, the navigation technology was originally developed for a different application. In 1993, a group of engineers at Hewlett-Packard Laboratories led by Ross Allen envisioned a handheld, battery powered document scanner that could be moved across a page in a freehand motion and still accurately recover the text. To help make this vision a reality, Travis Blalock and Dick Baumgartner at HP Labs designed a CMOS integrated circuit to optically measure movement of the scanner across the paper. The chip, known as "Magellan" within HP, is shown below.

Similar to digital cameras, the prototype contains a photo-receiver array to acquire images of the scanned surface, which is illuminated and positioned under the chip. The images are then transferred from the photo-receiver array to a computation array. The computation array always contains a reference image and a current image. Two-dimensional cross-correlations are then computed between the two images. The cross-correlation results can then be used to calculate the physical movement between the reference image and current image.

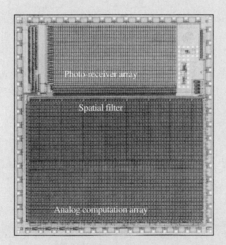 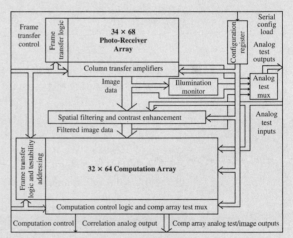

Optical navigation chip photo and block diagram. (Courtesy of Travis N. Blalock)

The Magellan optical navigation chip contains over 6000 operational amplifiers and sample-and-hold circuits, over 2000 photo-transistor amplifiers, and acquires 25,000 images per second. The chip calculates the cross-correlations at a rate of over 1.5 billion computations per second.

After a successful technology demonstration, the prototype was transferred to a product division and modified to create a commercial product. At some point in this process, it was recognized that the optical navigation architecture could be used as the basis for an optical mouse. The navigation chip design was again modified and became the basis of an optical navigation module sold by Agilent Technologies (a spinoff of the Hewlett-Packard company) and is used as the basis of most of the available optical mice on the market today.

Equation (12.61) states that, for any frequency $\omega \gg \omega_B$, the product of the magnitude of amplifier gain and frequency has a constant value equal to the unity-gain frequency ω_T. For this reason, the parameter ω_T (or f_T) is often referred to as the **gain-bandwidth product (GBW)** of the amplifier. The important result in Eq. (12.61) is a property of *single-pole* amplifiers that can be represented by transfer functions of the form of Eq. (12.57).

EXAMPLE 12.9   **OP AMP TRANSFER FUNCTION**

Determine an op amp transfer function from a given Bode plot.

PROBLEM   Write the transfer function that describes the frequency-dependent voltage gain of the amplifier in Fig. 12.22.

SOLUTION   **Known Information and Given Data:** From the figure we see that the amplifier has a single-pole response as modeled by Eq. (12.57).

Unknowns: To evaluate the transfer function, we need to find A_o and ω_B.

Approach: The values must be found from the graph and converted into proper form before insertion into Eq. (12.57)

Assumptions: The amplifier can be modeled by the single pole formula.

Analysis: At low frequencies, the gain asymptotically approaches 80 dB, which must be converted back from dB:

$$A_o = 10^{80\,\text{dB}/20\,\text{dB}} = 10^4$$

The cutoff frequency ω_B can also be read directly from the graph and is already in radian form: $\omega_B = 10^3$ rad/s. Substituting the values of A_o and ω_B into Eq. (12.57) yields the desired transfer function.

$$A_v(s) = \frac{A_o \omega_B}{s + \omega_B} = \frac{10^4(10^3)}{s + 10^3} = \frac{10^7}{s + 10^3}$$

Check of Results: We have found the unknown transfer function. We can check the answer for consistency by observing that the numerator value represents the unity-gain frequency ω_T. From the graph we see that ω_T is indeed 10^7 rad/s.

Discussion: Note that we often express the frequency values in Hz and that $A_o f_B = f_T$.

$$f_B = \frac{\omega_B}{2\pi} = 159 \text{ Hz} \qquad \text{and} \qquad f_T = \frac{\omega_T}{2\pi} = 1.59 \text{ MHz}$$

EXERCISE: An op amp has a gain of 100 dB at dc and a unity-gain frequency of 5 MHz. What is f_B? Write the transfer function for the gain of the op amp.

ANSWER: 50 Hz; $A(s) = \dfrac{10^7 \pi}{s + 100\pi}$

EXERCISE: What are the nominal values of open-loop gain and unity-gain frequency for the AD745 operational amplifier (see MCD website for specification sheets)? Write a transfer function for the op amp.

ANSWERS: 200,000; 1 MHz; $A_v(s) = \dfrac{A_o \omega_B}{s + \omega_B} = \dfrac{\omega_T}{s + \omega_B} = \dfrac{2\pi \times 10^6}{s + 10\pi}$

12.4.1 FREQUENCY RESPONSE OF THE NONINVERTING AMPLIFIER

We now use the frequency-dependent op-amp gain expression to study the closed-loop frequency response of the noninverting and inverting amplifiers. The closed-loop gain for the noninverting amplifier was found previously to be

$$A_v = \frac{A}{1 + A\beta} \quad \text{where} \quad \beta = \frac{R_1}{R_1 + R_2} \tag{12.62}$$

The algebraic derivation of this gain expression actually placed no restrictions on the functional form of A. Up to now, we have assumed A to be a constant, but we can explore the frequency response of the closed-loop feedback amplifier by replacing A in Eq. (12.62) by the frequency-dependent voltage-gain expression for the op amp, Eq. (12.57):

$$A_v(s) = \frac{A(s)}{1 + A(s)\beta} = \frac{\dfrac{A_o\omega_B}{s + \omega_B}}{1 + \dfrac{A_o\omega_B}{s + \omega_B}\beta} = \frac{A_o\omega_B}{s + \omega_B(1 + A_o\beta)} \tag{12.63}$$

Dividing by the factor $(1 + A_o\beta)\omega_B$, Eq. (12.63) can be written as

$$A_v(s) = \frac{\dfrac{A_o}{1 + A_o\beta}}{\dfrac{s}{(1 + A_o\beta)\omega_B} + 1} = \frac{A_v(0)}{\dfrac{s}{\omega_H} + 1} \tag{12.64}$$

where the upper-cutoff frequency is

$$\omega_H = \omega_B(1 + A_o\beta) = \omega_T \frac{(1 + A_o\beta)}{A_o} = \frac{\omega_T}{A_v(0)} \tag{12.65}$$

The closed-loop amplifier also has a single-pole response of the same form as Eq. (12.57), but its dc gain and bandwidth are given by

$$A_v(0) = \frac{A_o}{1 + A_o\beta} \quad \text{and} \quad \omega_H = \frac{\omega_T}{A_v(0)} \tag{12.66}$$

For $A_o\beta \gg 1$, Eq. (12.65) reduces to

$$A_v(0) \cong \frac{1}{\beta} \quad \text{and} \quad \omega_H \cong \beta\omega_T \tag{12.67}$$

Note that the gain-bandwidth product of the closed-loop amplifier is constant:

$$A_v(0)\,\omega_H = \omega_T$$

From Eq. (12.66), we see that the gain must be reduced in order to increase ω_H, or vice versa. We will explore this in more detail shortly.

The loop gain $A(s)\beta$ is now also a function of frequency depicted as in Fig. 12.23. At frequencies for which $|A(j\omega)\beta| \gg 1$, Eq. (12.63) reduces to $1/\beta$, the constant value derived previously for low frequencies. At frequencies for which $|A(j\omega)\beta| \ll 1$, Eq. (12.63) becomes $A_v \cong A(j\omega)$. At low frequencies, the gain is set by the feedback, but at high frequencies, we find that the gain must follow the gain of the amplifier. We should not expect a (negative) feedback amplifier to produce more gain than is available from the open-loop operational amplifier by itself.

These results are indicated graphically by the bold lines in Fig. 12.23 for an amplifier with $1/\beta = 35$ dB. The loop gain can be expressed as

$$A\beta = \frac{A}{\left(\dfrac{1}{\beta}\right)} \quad \text{and (in dB)} \quad |A\beta|_{dB} = |A|_{dB} - \left|\frac{1}{\beta}\right|_{dB} \tag{12.68}$$

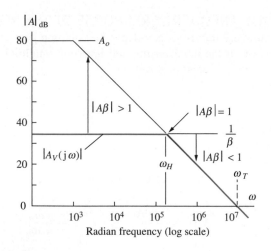

Figure 12.23 Graphical interpretation of operational amplifier with feedback.

At any given frequency, the magnitude of the loop gain is equal to the difference between A_{dB} and $(1/\beta)_{dB}$ on the graph. The upper half-power frequency $\omega_H = \beta\omega_t$ corresponds to the frequency at which $(1/\beta)$ intersects $|A(j\omega)|$ corresponding to $|A\beta| = 1$ (actually $A\beta \cong -j1 = 1\angle-90°$). For the case in Fig. 12.23, $\beta = 0.0178$ (-35 dB) and $\omega_H = 0.0178 \times 10^7 = 178 \times 10^3$ rad/s.

EXAMPLE 12.10 NONINVERTING AMPLIFIER FREQUENCY RESPONSE

Characterize the frequency response of a noninverting amplifier built with a nonideal op amp having limited gain and bandwidth.

PROBLEM An op amp has a dc gain of 100 dB and a unity-gain frequency of 10 MHz. (a) What is the cutoff frequency of the op amp? (b) If the op amp is used to build a noninverting amplifier with a closed-loop gain of 60 dB, what is the bandwidth of the feedback amplifier? (c) Write an expression for the transfer function of the op amp. (d) Write an expression for the transfer function of the noninverting amplifier.

SOLUTION **Known Information and Given Data:** We are given $A_o = 10^5$ (100 dB) and $f_T = 10^7$ Hz. The desired closed-loop gain is $A_v = 1000$ (60 dB).

Unknowns: (a) Bandwidth f_B of the operational amplifier, (b) Bandwidth f_H of the closed-loop amplifier, (c) op amp transfer function, (d) noninverting amplifier transfer function

Approach: Evaluate Eqs. (12.62)–(12.67), which model the behavior of the noninverting amplifier.

Assumptions: Since we have been given values for A_o and f_T, we will assume that the amplifier is described by a single-pole transfer function.

Analysis:

(a) The cutoff frequency of the op amp is f_B, and its -3-dB frequency is

$$f_B = \frac{f_T}{A_o} = \frac{10^7 \text{ Hz}}{10^5} = 100 \text{ Hz}$$

(b) Using Eq. (12.65), the bandwidth of the noninverting amplifier is

$$f_H = f_B(1 + A_o\beta) = 100(1 + 10^5 \cdot 10^{-3}) = 10.1 \text{ kHz}$$

in which the feedback factor β is determined by the desired closed-loop gain.

$$\beta = \frac{1}{A_v(0)} = \frac{1}{1000} = 10^{-3}$$

(c) Substituting the values of A_o and ω_B into Eq. (12.57) yields the op amp transfer function.

$$A_v(s) = \frac{A_o\omega_B}{s + \omega_B} = \frac{10^5(2\pi)(10^2)}{s + (2\pi)(10^2)} = \frac{2\pi \times 10^7}{s + 200\pi}$$

(d) Evaluating Eq. (12.63) yields the noninverting amplifier transfer function.

$$A_v(s) = \frac{A_o\omega_B}{s + \omega_B(1 + A_o\beta)} = \frac{10^5(2\pi)(10^2)}{s + (2\pi)(10^2)[1 + 10^5(10^{-3})]} = \frac{2\pi \times 10^7}{s + 2.02\pi \times 10^4}$$

Check of Results: We have found each of the requested answers. The numerators of the transfer functions should be equal to $\omega_T = 2\pi f_T$ and are correct.

12.4.2 INVERTING AMPLIFIER FREQUENCY RESPONSE

The frequency response for the inverting-amplifier configuration can be found in a manner similar to that for the noninverting case by substituting the frequency-dependent op amp gain expression, Eq. (12.57), into the equation for the closed-loop gain of the inverting amplifier.

$$A_v = \left(-\frac{R_2}{R_1}\right)\frac{A(s)\beta}{1 + A(s)\beta} \qquad \text{where} \qquad \beta = \frac{R_1}{R_1 + R_2}$$

or

(12.69)

$$A_v(s) = \left(-\frac{R_2}{R_1}\right)\frac{\dfrac{A_o\omega_B}{s + \omega_B}\beta}{1 + \dfrac{A_o\omega_B}{s + \omega_B}\beta} = \frac{\left(-\dfrac{R_2}{R_1}\right)\dfrac{A_o\beta}{1 + A_o\beta}}{\dfrac{s}{\omega_B(1 + A_o\beta)} + 1} = -\frac{R_2}{R_1}\frac{A_o\beta\omega_B}{s + \omega_B(1 + A_o\beta)}$$

For $A_o\beta \gg 1$, these equations reduce to

$$A_v = \frac{\left(-\dfrac{R_2}{R_1}\right)\dfrac{A_o\beta}{(1 + A_o\beta)}}{\dfrac{s}{\omega_H} + 1} \cong \frac{\left(-\dfrac{R_2}{R_1}\right)}{\dfrac{s}{\omega_H} + 1} \qquad \text{and} \qquad \omega_H = \frac{\omega_t}{\dfrac{A_o}{(1 + A_o\beta)}} \cong \beta\omega_T \qquad (12.70)$$

where the approximate values hold for $A_o\beta \gg 1$. This expression again represents a single-pole transfer function. The gain at low frequencies, $A_v(0)$, is set by the resistor ratio $(-R_2/R_1)$, and the bandwidth expression is identical to that of the noninverting amplifier, $\omega_H = \beta\omega_T$.

TABLE 12.2

Inverting and Noninverting Amplifier Frequency Response Comparison

$\beta = \dfrac{R_1}{R_1+R_2}$	NONINVERTING AMPLIFIER	INVERTING AMPLIFIER
dc gain	$A_v(0) = 1 + \dfrac{R_2}{R_1}$	$A_v(0) = -\dfrac{R_2}{R_1}$
Feedback factor	$\beta = \dfrac{1}{A_v(0)}$	$\beta = \dfrac{1}{1+\lvert A_v(0)\rvert}$
Bandwidth	$f_B = \beta f_T$	$f_B = \beta f_T$
Input resistance	$R_{ic}\lVert R_{id}(1 + A\beta)$	$R_1 + \left(R_{ID}\lVert \dfrac{R_2}{1+A}\right)$
Output resistance	$\dfrac{R_o}{1 + A\beta}$	$\dfrac{R_o}{1 + A\beta}$

The frequency response characteristics of the inverting and noninverting amplifiers are summarized in Table 12.2, in which the expressions have been recast in terms of the ideal value of the gain at low frequencies. The expressions are quite similar. However, for a given value of dc gain, the noninverting amplifier will have slightly greater bandwidth than the inverting amplifier because of the difference in the relation between β and $A_v(0)$. The difference is significant only for amplifier stages designed with low values of closed-loop gain.

EXAMPLE 12.11 INVERTING AMPLIFIER FREQUENCY RESPONSE

Characterize the frequency response of an inverting amplifier built using a nonideal op amp having limited gain and bandwidth.

PROBLEM An op amp has a dc gain of 200,000 and a unity-gain frequency of 500 kHz. (a) What is the cutoff frequency of the op amp? (b) If the op amp is used to build an inverting amplifier with a closed-loop gain of 40 dB, what is the bandwidth of the feedback amplifier? (c) Write an expression for the transfer function of the op amp. (d) Write an expression for the transfer function of the inverting amplifier.

SOLUTION **Known Information and Given Data:** For the op amp; $A_o = 2 \times 10^5$, $f_T = 5 \times 10^5$ Hz; for the inverting amplifier, $A_v = -100$ (40 dB)

Unknowns: (a) Op amp cutoff frequency, (b) inverting amplifier bandwidth, (c) op amp transfer function, (d) inverting amplifier transfer function

Approach: Evaluate Eq. (12.57) for the op amp. Evaluate Eqs. (12.69) and (12.70), which model the behavior of the inverting amplifier.

Assumptions: The op amp has a single-pole frequency response. Otherwise it is ideal.

Analysis:

(a) The cutoff frequency of the op amp is f_B, its -3-dB frequency:

$$f_B = \frac{f_T}{A_o} = \frac{5 \times 10^5 \text{ Hz}}{2 \times 10^5} = 2.5 \text{ Hz}$$

(b) Using Eq. (12.70), the bandwidth of the inverting amplifier is

$$f_H = f_B(1 + A_o\beta) = 2.5 \text{ Hz}\left(1 + \frac{2 \times 10^5}{101}\right) = 4.95 \text{ kHz}$$

in which the feedback factor β is determined by the desired closed-loop gain (see Table 12.2).

$$\beta = \frac{1}{1 + |A_v(0)|} = \frac{1}{101}$$

(c) Substituting the values of A_o and ω_B into Eq. (12.57) yields the op amp transfer function.

$$A_v(s) = \frac{A_o \omega_B}{s + \omega_B} = \frac{\omega_T}{s + \omega_B} = \frac{(2\pi)(5 \times 10^5)}{s + (2\pi)(2.5)} = \frac{10^6 \pi}{s + 5\pi}$$

(d) Evaluating Eq. (12.63) yields the inverting amplifier transfer function:

$$A_v(s) = \left(-\frac{R_2}{R_1}\right) \frac{A_o \beta \omega_B}{s + \omega_B(1 + A_o \beta)} = (-100) \frac{(2 \times 10^5)\left(\dfrac{1}{101}\right)(2\pi)(2.5)}{s + (2\pi)(2.5)\left(1 + \dfrac{2 \times 10^5}{101}\right)}$$

$$= -\frac{9.90 \times 10^5 \pi}{s + 9.91 \times 10^3 \pi}$$

Check of Results: We have found the answers to all the unknowns. We can also double check the last transfer function by evaluating its dc gain and bandwidth.

$$A_v(0) = -\frac{9.90 \times 10^5 \pi}{9.91 \times 10^3 \pi} = -99.9 \quad \text{and} \quad f_H = \frac{9.91 \times 10^3 \pi}{2\pi} = 4.96 \text{ kHz}$$

The values agree within round-off error.

EXERCISE: An op amp has a dc gain of 90 dB and a unity-gain frequency of 5 MHz. What is the cutoff frequency of the op amp? If the op amp is used to build an inverting amplifier with a closed-loop gain of 50 dB, what is the bandwidth of the feedback amplifier? Write an expression for the transfer function of the op amp. Write an expression for the transfer function of the inverting amplifier.

ANSWERS: 158 Hz; 15.8 kHz; $A(s) = \dfrac{10^7 \pi}{s + 316\pi}$; $A(s) = \dfrac{10^7 \pi}{s + 3.16\pi \times 10^5}$

EXERCISE: If the amplifier in Ex. 12.11 is used in a voltage follower, what is its bandwidth? If the amplifier is used in an inverting amplifier with $A_v = -1$, what is its bandwidth?

ANSWERS: 10 MHz; 5 MHz

12.4.3 FREQUENCY RESPONSE OF CASCADED AMPLIFIERS

When several amplifiers are connected in cascade, as in Fig. 12.24 for example, the overall transfer function can be written as the product of the transfer functions of the individual stages:

$$A_v(s) = \frac{V_{oN}(s)}{V_s(s)} = \frac{V_{o1}}{V_s} \frac{V_{o2}}{V_{o1}} \cdots \frac{V_{oN}}{V_{o(N-1)}} = A_{v1}(s) A_{v2}(s) \cdots A_{vN}(s) \tag{12.71}$$

It is extremely important to remember that this product representation implicitly assumes that the stages do not interact with each other, which can be achieved with $R_{out} = 0$ or $R_{in} = \infty$ (that is, the interconnection of the various amplifiers must not alter the transfer function of any of the amplifiers).

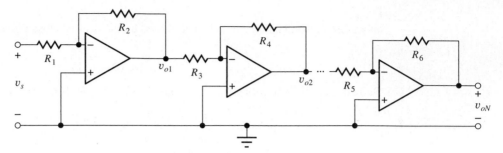

Figure 12.24 Multistage amplifier cascade.

In the general case, each amplifier has a different value of dc gain and bandwidth, and the overall transfer function becomes

$$A_v(s) = \frac{A_{v1}(0)}{\left(1 + \dfrac{s}{\omega_{H1}}\right)} \frac{A_{v2}(0)}{\left(1 + \dfrac{s}{\omega_{H2}}\right)} \cdots \frac{A_{vN}(0)}{\left(1 + \dfrac{s}{\omega_{HN}}\right)} \tag{12.72}$$

The gain at low frequencies ($s = 0$) is

$$A_v(0) = A_{v1}(0)A_{v2}(0) \cdots A_{vN}(0) \tag{12.73}$$

The overall bandwidth of the cascade amplifier is defined to be the frequency at which the voltage gain is reduced by a factor of $1/\sqrt{2}$ or -3 dB from its low-frequency value. Stated mathematically,

$$|A_v(j\omega_H)| = \frac{|A_{v1}(0)A_{v2}(0) \cdots A_{vN}(0)|}{\sqrt{2}} \tag{12.74}$$

In the general case, hand calculation of ω_H based on Eq. (12.74) can be quite tedious, and approximate techniques for estimating ω_H will be developed in Chapter 16. With the aid of a computer or calculator, solver routines or iterative trial-and-error can be used directly to find ω_H. Example 12.12 uses direct algebraic evaluation of Eq. (12.74) for the case of two amplifiers.

EXAMPLE 12.12 TWO-AMPLIFIER CASCADE

Calculate the gain and bandwidth of a two-stage amplifier.

PROBLEM Two amplifiers with transfer functions $A_{v1}(s)$ and $A_{v2}(s)$ are connected in cascade. What are the dc gain and bandwidth of the overall two-stage amplifier?

$$A_{v1} = \frac{500}{1 + \dfrac{s}{2000}} \quad \text{and} \quad A_{v2} = \frac{250}{1 + \dfrac{s}{4000}}$$

SOLUTION **Known Information and Given Data:** A cascade connection of two amplifiers; the individual transfer functions are specified for the two amplifiers.

Unknowns: $A_v(0)$ and f_H for the overall two-stage amplifier

Approach: The transfer function for the cascade is given by $A_v = A_{v1} \times A_{v2}$. Find $A_v(0)$. Apply the definition of bandwidth to find f_H.

Assumptions: The amplifiers are ideal except for their frequency dependencies and can be cascaded without interaction—i.e., the overall gain is equal to the product of the individual transfer functions.

Analysis: The overall transfer function is

$$A_v(s) = \left(\frac{500}{1 + \dfrac{s}{2000}} \right) \left(\frac{250}{1 + \dfrac{s}{4000}} \right) = \frac{125{,}000}{\left(1 + \dfrac{s}{2000}\right)\left(1 + \dfrac{s}{4000}\right)}$$

Calculating the dc gain $A_v(0)$:

$$A_v(0) = (500)(250) = 125{,}000 \text{ or } 102 \text{ dB}$$

Note that $A_v(0)$ is equal to the product of the dc gains of the two individual amplifiers.
The magnitude of the frequency response for $s = j\omega$ is

$$|A_v(j\omega)| = \frac{1.25 \times 10^5}{\sqrt{1 + \dfrac{\omega^2}{2000^2}} \sqrt{1 + \dfrac{\omega^2}{4000^2}}}$$

and we remember that ω_H is defined by

$$|A(j\omega_H)| = \frac{A_{\text{mid}}}{\sqrt{2}} = \frac{A_v(0)}{\sqrt{2}} = \frac{1.25 \times 10^5}{\sqrt{2}}$$

Equating the denominators of these two equations and squaring both sides yields

$$\left(1 + \frac{\omega_H^2}{2000^2}\right)\left(1 + \frac{\omega_H^2}{4000^2}\right) = 2$$

which can be rearranged into the following quadratic equation in terms of ω_H^2:

$$\left(\omega_H^2\right)^2 + 2.00 \times 10^7 \left(\omega_H^2\right) - 6.40 \times 10^{13} = 0$$

Using the quadratic formula or our calculator's root-finding routine gives these values for ω_H^2

$$\omega_H^2 = 2.81 \times 10^6 \quad \text{or} \quad -4.56 \times 10^7$$

The value of ω_H must be real, so the only acceptable answer is

$$\omega_H = 1.68 \times 10^3 \quad \text{or} \quad f_H = 267 \text{ Hz}$$

Check of Results: The bandwidth of the composite amplifier should be less than that of either individual amplifier:

$$f_{H1} = \frac{2000}{2\pi} = 318 \text{ Hz} \quad \text{and} \quad f_{H2} = \frac{4000}{2\pi} = 637 \text{ Hz}$$

The bandwidth we have calculated is indeed less than for either individual amplifier.

EXERCISE: An amplifier is formed by cascading two amplifiers with these transfer functions. What is the gain at low frequencies? What is the gain at f_H? What is f_H?

$$A_{v1}(s) = \frac{50}{1 + \dfrac{s}{10{,}000\pi}} \quad \text{and} \quad A_{v2}(s) = \frac{25}{1 + \dfrac{s}{20{,}000\pi}}$$

ANSWERS: 1250; 884; 8380 Hz

Cascade of Identical Amplifier Stages

For the special case in which a cascade-amplifier configuration is composed of identical amplifiers, then a simple result can be obtained for the bandwidth of the overall amplifier. For N identical stages,

$$A_v(s) = \left[\frac{A_{v1}(0)}{1 + \dfrac{s}{\omega_{H1}}}\right]^N = \frac{[A_{v1}(0)]^N}{\left(1 + \dfrac{s}{\omega_{H1}}\right)^N} \qquad \text{and} \qquad A_v(0) = [A_{v1}(0)]^N \qquad (12.75)$$

TABLE 12.3
Bandwidth Shrinkage Factor

N	$\sqrt{2^{1/N} - 1}$
1	1
2	0.644
3	0.510
4	0.435
5	0.386
6	0.350
7	0.323

in which $A_{v1}(0)$ and ω_{H1} are the closed-loop gain and bandwidth of each individual amplifier stage. The bandwidth ω_H of the overall cascade amplifier is determined from

$$|A_v(j\omega_H)| = \frac{[A_{v1}(0)]^N}{\left(\sqrt{1 + \dfrac{\omega_H^2}{\omega_{H1}^2}}\right)^N} = \frac{[A_{v1}(0)]^N}{\sqrt{2}} \qquad (12.76)$$

Solving for ω_H in terms of ω_{H1} for the cascaded-amplifier bandwidth yields

$$\omega_H = \omega_{H1}\sqrt{2^{1/N} - 1} \qquad \text{or} \qquad f_H = f_{H1}\sqrt{2^{1/N} - 1} \qquad (12.77)$$

The bandwidth of the cascade is less than that of the individual amplifiers. Sample values of the **bandwidth shrinkage factor** $\sqrt{2^{1/N} - 1}$ are given in Table 12.3.

Although most amplifier designs do not actually cascade identical amplifiers, Eq. (12.77) can be used to help guide the design of a multistage amplifier or, in some cases, to estimate the bandwidth of a portion of a more complex amplifier. (Additional useful results appear in Probs. 12.69 and 12.70.)

DESIGN EXAMPLE 12.13   A CASCADE AMPLIFIER DESIGN

In this example, a spreadsheet is used to assist in the design of a fairly complex multistage amplifier.

PROBLEM Design an amplifier to meet these specifications: $A_v \geq 100$ dB, bandwidth ≥ 50 kHz, $R_{\text{out}} \leq 0.1\ \Omega$, and $R_{\text{in}} \geq 20\ \text{k}\Omega$. Use an operational amplifier with these specifications: $A_o = 100$ dB, $f_T = 1$ MHz, $R_{id} = 1$ GΩ, and $R_o = 50\ \Omega$.

SOLUTION   **Known Information and Given Data:** Op amp and overall amplifier specifications as already tabulated.

Unknowns: Choice between inverting and noninverting configurations; gain and bandwidth of each amplifier; feedback resistor values

Approach: Because the required value of R_{in} is relatively low and can be met by a resistor, both the inverting and noninverting amplifier stages should be considered. More than one stage will be required because a single op amp by itself cannot simultaneously meet the specifications for A_v, f_H, and R_{out}. For example, if we were to use the open-loop op amp by itself, it would provide a gain of 100 dB (10^5) but have a bandwidth of only $f_t/10^5 = 10$ Hz. Thus, we must reduce the gain of each stage in order to increase the bandwidth (i.e., we must trade gain for bandwidth).

For simplicity in the design, we assume that the amplifier will be built from a cascade of N identical amplifier stages. The design formulas will be set up in a logical order so that we can choose one design variable, and the rest of the equations can then be evaluated based on that single design choice. For this particular design, the gain and bandwidth are the most difficult specifications to achieve, whereas the required input and output resistance specifications are easily met. We can initially force our design to meet either the gain or the bandwidth specification and then find the number of stages that will be required to achieve the other specifications.

Assumptions: The design must have the minimum number of stages required to meet the specifications in order to achieve minimum cost.

Analysis: In this example, we force the cascade amplifier to meet the gain specification, and then find the number of stages needed to meet the bandwidth by repeated trial and error.

To meet the gain specification, we set the gain of each stage to

$$A_v(0) = \sqrt[N]{10^5}$$

Based on this choice, we can then calculate the other characteristics of the amplifier using this process:

1. Choose N.
2. Calculate the gain required of each stage $A_v(0) = \sqrt[N]{10^5}$.
3. Find β using the numerical result from step 2.
4. Calculate the bandwidth f_{H1} of each stage.
5. Calculate the bandwidth of N stages using Eq. (12.77).
6. Using $A\beta$, calculate R_{out} and R_{in}.
7. See if specifications are met. If not, go back to step 1 and try a new value of N.

The formulas for the noninverting and inverting amplifiers are slightly different, as summarized in Table 12.4. These formulas have been used for the results tabulated in the spreadsheet in Table 12.5.

From Table 12.5, we see that a cascade of six noninverting amplifiers meets all the specifications, whereas seven inverting amplifier stages are required. This occurs because the inverting amplifier has a slightly smaller bandwidth than the noninverting amplifier for a given value of closed-loop gain. We are usually interested in the most economical design, so the six-stage amplifier will be chosen. Note that the R_{out} requirement is met with $N > 2$ for both amplifiers.

To complete the design, we must choose values for R_1 and R_2. From Table 12.5, the gain of each stage must be at least 6.81, requiring the resistor ratio R_2/R_1 to be 5.81. Because we will probably not be able to find two 5 percent resistors that give a ratio of exactly 5.81, we need to

TABLE 12.4

N-Stage Cascades of Noninverting and Inverting Amplifiers

$\beta = \dfrac{R_1}{R_1 + R_2}$	**NONINVERTING AMPLIFIER**	**INVERTING AMPLIFIER**
2. Single-stage gain $A_v(0) = \sqrt[N]{10^5}$	$A_v(0) = 1 + \dfrac{R_2}{R_1}$	$A_v(0) = -\dfrac{R_2}{R_1}$
3. Feedback factor	$\beta = \dfrac{1}{A_v(0)}$	$\beta = \dfrac{1}{1 + |A_v(0)|}$
4. Bandwidth of each stage	$f_{H1} = \dfrac{f_T}{\dfrac{A_o}{1 + A_o\beta}}$	$f_{H1} = \dfrac{f_T}{\dfrac{A_o}{1 + A_o\beta}}$
5. N-stage bandwidth	$f_H = f_{H1}\sqrt{2^{1/N} - 1}$	$f_B = f_{H1}\sqrt{2^{1/N} - 1}$
6. Input resistance	$R_{id}(1 + A_o\beta)$	R_1
Output resistance	$\dfrac{R_o}{1 + A_o\beta}$	$\dfrac{R_o}{1 + A_o\beta}$

TABLE 12.5

Design of Cascade of N Identical Operational Amplifier Stages

CASCADE OF IDENTICAL NONINVERTING AMPLIFIERS

NUMBER OF STAGES	$A_V(0)$ GAIN PER STAGE $1/\beta$	f_{H1} SINGLE STAGE $\beta \times f_T$	f_H N STAGES	R_{in}	R_{out}
1	1.00E + 05	1.00E + 01	1.000E + 01	2.00E + 09	2.50E + 01
2	3.16E + 02	3.16E + 03	2.035E + 03	3.17E + 11	1.58E − 01
3	4.64E + 01	2.15E + 04	1.098E + 04	2.16E + 12	2.32E − 02
4	1.78E + 01	5.62E + 04	2.446E + 04	5.62E + 12	8.89E − 03
5	1.00E + 01	1.00E + 05	3.856E + 04	1.00E + 13	5.00E − 03
6	**6.81E + 00**	**1.47E + 05**	**5.137E + 04**	**1.47E + 13**	**3.41E − 03**
7	5.18E + 00	1.93E + 05	6.229E + 04	1.93E + 13	2.59E − 03
8	4.22E + 00	2.37E + 05	7.134E + 04	2.37E + 13	2.11E − 03

CASCADE OF IDENTICAL INVERTING AMPLIFIERS

NUMBER OF STAGES	$A_V(0)$ $(1/\beta) - 1$	f_{H1} SINGLE STAGE	f_H N STAGES	R_{in}	R_{out}
1	1.00E + 05	1.00E + 01	1.00E + 01	R_1	2.50E + 01
2	3.16E + 02	3.15E + 03	2.03E + 03	R_1	1.58E − 01
3	4.64E + 01	2.11E + 04	1.08E + 04	R_1	2.32E − 02
4	1.78E + 01	5.32E + 04	2.32E + 04	R_1	8.89E − 03
5	1.00E + 01	9.09E + 04	3.51E + 04	R_1	5.00E − 03
6	6.81E + 00	1.28E + 05	4.48E + 04	R_1	3.41E − 03
7	**5.18E + 00**	**1.62E + 05**	**5.22E + 04**	R_1	**2.59E − 03**
8	4.22E + 00	1.92E + 05	5.77E + 04	R_1	2.11E − 03

explore the acceptable range for the ratio now that we know we need six stages. In Table 12.6, a spreadsheet is again used to study six-stage amplifier designs having gains ranging from 6.81 to 7.01. As the single-stage gain is increased, the overall bandwidth decreases. From Table 12.6, we see that the specifications will be met for resistor ratios falling between 5.81 and 5.99.

TABLE 12.6
Cascade of Six Identical Noninverting Amplifiers

NUMBER OF STAGES	$A_V(0)$ GAIN PER STAGE $1/\beta$	N STAGE GAIN	f_{H1} SINGLE STAGE $\beta \times f_T$	f_H N STAGES	R_{in}	R_{out}
6	6.81E + 00	1.00E + 05	1.47E + 05	5.137E + 04	1.47E + 13	3.41E − 03
6	6.83E + 00	1.02E + 05	1.46E + 05	5.121E + 04	1.46E + 13	3.42E − 03
6	6.85E + 00	1.04E + 05	1.46E + 05	5.107E + 04	1.46E + 13	3.43E − 03
6	6.87E + 00	1.05E + 05	1.45E + 05	5.092E + 04	1.46E + 13	3.44E − 03
6	6.89E + 00	1.07E + 05	1.45E + 05	5.077E + 04	1.45E + 13	3.45E − 03
6	6.91E + 00	1.09E + 05	1.45E + 05	5.062E + 04	1.45E + 13	3.46E − 03
6	6.93E + 00	1.11E + 05	1.44E + 05	5.048E + 04	1.44E + 13	3.47E − 03
6	6.95E + 00	1.13E + 05	1.44E + 05	5.033E + 04	1.44E + 13	3.48E − 03
6	6.97E + 00	1.15E + 05	1.43E + 05	5.019E + 04	1.43E + 13	3.49E − 03
6	6.99E + 00	1.17E + 05	1.43E + 05	5.004E + 04	1.43E + 13	3.50E − 03
6	7.01E + 00	1.19E + 05	1.43E + 05	4.990E + 04	1.43E + 13	3.51E − 03

Many acceptable resistor ratios can be found in the table of 5 percent resistors in Appendix C. Picking $A_v(0) = 6.91$, a value near the center of the range of acceptable gain and bandwidth, two possible resistors sets are

$$\text{(i)} \quad R_1 = 22\text{ k}\Omega,\ R_2 = 130\text{ k}\Omega$$

which gives

$$1 + \frac{R_2}{R_1} = 6.91,\ A(0) = 101\text{ dB},\ f_H = 50.6\text{ kHz},\ R_{out} = 3.46\text{ m}\Omega$$

and

$$\text{(ii)} \quad R_1 = 5.6\text{ k}\Omega,\ R_2 = 33\text{ k}\Omega$$

which yields

$$1 + \frac{R_2}{R_1} = 6.89,\ A(0) = 101\text{ dB},\ f_H = 50.8\text{ kHz},\ R_{out} = 3.45\text{ m}\Omega$$

The overall size of these resistors has been chosen so that the feedback resistors do not heavily load the output of the op amp. For example, the resistor pairs $R_1 = 220\ \Omega$ and $R_2 = 1.3\text{ k}\Omega$ or $R_1 = 56\ \Omega$ and $R_2 = 330\ \Omega$, although providing acceptable resistor ratios, would not be desirable choices for a final design.

Check of Results: Based on the spreadsheet results, a design has been found that meets the specifications.

Discussion: This example has explored the design of a fairly complex multistage amplifier. Economical design requires the use of the minimum number of amplifier stages. In this case, spreadsheets were used to explore the design space, and the calculations indicated that the specifications could be met with a cascade of six identical noninverting amplifiers. The design was completed through the choice of feedback resistors from the set of available discrete resistor values.

Computer-Aided Analysis: With the level of complexity in this example, it would obviously be quite useful to use SPICE to check our final design. We shall do this after we look at macro models in Secs. 12.4.5 and 12.4.6.

TABLE 12.7
Cascade of Six Identical Noninverting Amplifiers—Worse-Case Analysis

R VALUES	ONE-STAGE GAIN	SIX-STAGE GAIN	f_{H1}	f_H	R_{in}	R_{out}
Nominal	6.91E + 00	1.09E + 05	1.45E + 05	5.065E + 04	1.45E + 13	3.45E − 03
Max	7.53E + 00	1.82E + 05	1.33E + 05	**4.647E + 04**	1.33E + 13	3.77E − 03
Min	6.35E + 00	**6.53E + 04**	1.58E + 05	5.514E + 04	1.58E + 13	3.17E − 03

The Influence of Tolerances on Design

Now that we have completed Example 12.13, let us explore the effects of the resistor tolerances on our design. We have chosen resistors that have 5 percent tolerances; Table 12.7 presents the results of calculating the worst-case specifications, in which

$$A_v^{nom} = 1 + \frac{130 \text{ k}\Omega}{22 \text{ k}\Omega} = 6.91$$

$$A_v^{max} = 1 + \frac{130 \text{ k}\Omega (1.05)}{22 \text{ k}\Omega (0.95)} = 7.53$$

$$A_v^{min} = 1 + \frac{130 \text{ k}\Omega (0.95)}{22 \text{ k}\Omega (1.05)} = 6.35$$

The nominal design values easily meet both specifications, with a margin of 9 percent for the gain but only 1.3 percent for the bandwidth. When the resistor tolerances are set to give the largest gain per stage, the gain specification is easily met, but the bandwidth shrinks below the specification limit. At the opposite extreme, the gain of the six stages fails to meet the required specification. This analysis gives us an indication that there may be a problem with the design. Of course, assuming that all the amplifiers reach the worst-case gain and bandwidth limits at the same time is an extreme conclusion. Nevertheless, the nominal bandwidth does not exceed the specification limit by very much.

A Monte Carlo analysis would be much more representative of the actual design results. Such an analysis for 10,000 cases of our six-stage amplifier indicates that if this circuit is built with 5 percent resistors, more than 30 percent of the amplifiers will fail to meet either the gain or bandwidth specification. (The details of this calculation and exact results are left for Prob. 12.71.)

12.4.4 LARGE-SIGNAL LIMITATIONS—SLEW RATE AND FULL-POWER BANDWIDTH

Up to this point, we have tacitly assumed that the internal circuits that form the operational amplifier can respond instantaneously to changes in the input signal. However, the internal amplifier nodes all have an equivalent capacitance to ground, and only a finite amount of current is available to charge these capacitances. Thus, there will be some limit to the rate of change of voltage on the various nodes. This limit is described by the **slew-rate (SR)** specification of the operational amplifier. The slew rate defines the maximum rate of change of voltage at the output of the operational amplifier. Typical values of slew rate for general-purpose op amps fall into the range

$$0.1 \text{ V/}\mu\text{s} \le \text{SR} \le 10 \text{ V/}\mu\text{s}$$

although much higher values are possible in special designs. An example of a slew-rate limited signal at an amplifier output is shown schematically in Fig. 12.25.

For a given frequency, slew rate limits the maximum amplitude of a signal that can be amplified without distortion. Consider a sinusoidal output signal $v_o = V_M \sin \omega t$, for example. The maximum

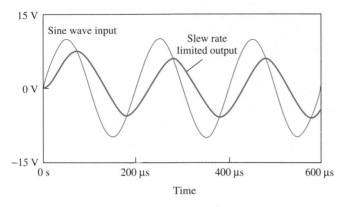

Figure 12.25 An example of a slew-rate limited output signal.

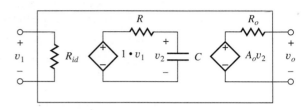

Figure 12.26 Simple macro model for an operational amplifier.

rate of change of this signal occurs at the zero crossings and is given by

$$\frac{dv_O}{dt}\bigg|_{\max} = V_M\omega \cos\omega t|_{\max} = V_M\omega \tag{12.78}$$

For no signal distortion, this maximum rate of change must be less than the slew rate:

$$V_M\omega \leq \text{SR} \quad \text{or} \quad V_M \leq \frac{\text{SR}}{\omega} \tag{12.79}$$

The **full-power bandwidth** f_M is the highest frequency at which a full-scale signal can be developed. Denoting the amplitude of the full-scale output signal by V_{FS}, the full-power bandwidth can be written as

$$f_M \leq \frac{\text{SR}}{2\pi V_{FS}} \tag{12.80}$$

> **EXERCISE:** Suppose that an op amp has a slew rate of 0.5 V/μs. What is the largest sinusoidal signal amplitude that can be reproduced without distortion at a frequency of 20 kHz? If the amplifier must deliver a signal with a 10-V maximum amplitude, what is the full-power bandwidth corresponding to this signal?
>
> **ANSWERS:** 3.98 V; 7.96 kHz

12.4.5 MACRO MODEL FOR OPERATIONAL AMPLIFIER FREQUENCY RESPONSE

The actual internal circuit of an operational amplifier may contain from 20 to 100 bipolar and/or field-effect transistors. If the actual circuits were used for each op amp, simulations of complex circuits containing many op amps would be very slow. Simplified circuit representations, called **macro models,** have been developed to model the terminal behavior of the op amp. The two-port model that we used in this chapter is one simple form of macro model. This section introduces a model that can be used for SPICE simulation of the frequency response of circuits utilizing operational amplifiers.

To model the single-pole roll-off, an auxiliary loop consisting of the voltage-controlled voltage source with value v_1 in series with R and C is added to the interior of the original two-port, as depicted in Fig. 12.26. The product of R and C is chosen to give the desired -3 dB point for the open-loop amplifier. If a voltage source is applied to the input, then the open-circuit voltage gain $(R_L = \infty)$ is

$$A_v(s) = \frac{\mathbf{V_o}(s)}{\mathbf{V_1}(s)} = \frac{A_o\omega_B}{s + \omega_B} \quad \text{where} \quad \omega_B = \frac{1}{RC} \tag{12.81}$$

This interior loop represents a "dummy" circuit added just to model the frequency response; the individual values of R and C are arbitrary. For example, $R = 1\,\Omega$ and $C = 0.0159\,\text{F}$, $R = 1000\,\Omega$ and $C = 15.9\,\mu\text{F}$, or $R = 1\,\text{M}\Omega$ and $C = 0.0159\,\mu\text{F}$ may all be used to model a cutoff frequency of 10 Hz.

EXERCISE: Create a macro model for the OP-27 based on Fig. 12.26 (see MCD website for specification sheets). Use the nominal specification values.

ANSWERS: $R_{id} = 6\,\text{M}\Omega$, $R = 2000\,\Omega$, $C = 17.9\,\mu\text{F}$, $A_o = 1.8 \times 10^6$, $R_o = 70\,\Omega$. The individual values of R and C are arbitrary as long as $RC = (1/8.89\,\pi)$s.

12.4.6 COMPLETE OP AMP MACRO MODELS IN SPICE

Most versions of SPICE contain sophisticated macro models for op amps [1, 2] including descriptions for many commercial op amps. These macro models include all of the nonideal limitations we have discussed in this chapter and contain a large number of parameters that can be adjusted to model op amp behavior. Both three- and five-terminal op amps may be included as shown in Fig. 12.27. An example parameter set is given in Table 12.8. In addition to those discussed previously in this chapter, we see parameters to model multiple poles and a zero in the op amp frequency response, to describe the input capacitance, and to set the input transistors to *npn* or *pnp* devices. This choice determines the direction of the input bias current; the bias current goes into the op amp terminals for an *n*-type input and comes out of the terminals for a *p*-type input.

12.4.7 EXAMPLES OF COMMERCIAL GENERAL-PURPOSE OPERATIONAL AMPLIFIERS

Now that we have explored the theory of circuits using ideal and nonideal operational amplifiers, let us look in more detail at the characteristics of a general-purpose operational amplifier. A portion of the specification sheets for one such commercial op amp, the AD745 series from Analog Devices

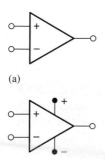

Figure 12.27 (a) Three-terminal op amp. (b) Five-terminal op amp.

TABLE 12.8

Typical Op Amp Macro Model Parameter Set

PARAMETER	TYPICAL VALUE
Differential-mode gain (dc)	106 dB
Differential-mode input resistance	2 MΩ
Input capacitance	1.5 pF
Common-mode rejection ratio	90 dB
Common-mode input resistance	2 GΩ
Output resistance	50 Ω
Input offset voltage	1 mV
Input bias current	80 nA
Input offset current	20 nA
Positive slew rate	0.5 V/μs
Negative slew rate	0.5 V/μs
Maximum output source current	25 mA
Maximum output sink current	25 mA
Input type (*n*- or *p*-type)	*n*-type
Frequency of first pole	5 Hz
Frequency of zero	5 MHz
Frequency of second pole	2 MHz
Frequency of third pole	20 MHz
Frequency of fourth pole	100 MHz
Power Supply Voltage (3-pin model)	15 V

Corporation, can be found on the MCD website. These amplifiers are fabricated with an IC technology that has both bipolar transistors and JFETs.

Note that many of the specifications are stated in terms of typical values plus either upper or lower bounds. For example, the voltage gain for the AD745J at $T = 25°C$ with ± 15-V power supplies is typically 132 dB but has a minimum value of 120 dB and no upper bound. The offset voltage is typically 0.25 mV with an upper bound of 1 mV, but the AD745K version is also available with a typical offset voltage of 0.1 mV and an upper bound of 0.5 mV. The input stage of this amplifier contains JFETs, so the input-bias current is very small at room temperature, and the nominal input resistance is very large.

The minimum common-mode rejection ratio (at dc) is 80 dB, and PSRR and CMRR specifications are the same. With ± 15-V power supplies, the amplifier can handle input signals with a common-mode range of $+13.3$ and -10.7 V, and the amplifier is guaranteed to develop an output-voltage swing of $+12$ V with a 2-kΩ load resistance.

The AD745J has a minimum gain-bandwidth product (unity-gain frequency f_T) of 20 MHz, and a slew rate of 12.5 V/μs. Considerable additional information is included concerning the performance of the amplifier family over a large range of power supply voltages and temperatures.

DESIGN EXAMPLE 12.14　**MACRO MODEL APPLICATION**

Use a SPICE op amp macro model to simulate the frequency response of a multistage amplifier.

PROBLEM　Use simulation to verify the frequency response of the six-stage amplifier designed in Ex. 12.13.

SOLUTION　**Known Information and Given Data:** The six-stage noninverting amplifier cascade design from Ex. 12.13 with $R_1 = 22$ kΩ and $R_2 = 130$ kΩ. The op amp specifications are $A_o = 100$ dB, $f_T = 1$ MHz, $R_{id} = 1$ GΩ, and $R_o = 50$ Ω.

Unknowns: A Bode plot of the amplifier frequency response; the values of $A_v(0)$ and f_H

Approach: Use a SPICE macro model for the op amp and use it to simulate the frequency response of the six-stage amplifier. Use SPICE subcircuits to simplify the analysis.

Assumptions: The amplifier is a single-pole amplifier. Symmetrical 15-V power supplies are available.

Analysis: After drawing the circuit with the schematic editor, we need to set the parameters of the SPICE op amp model to agree with our specifications. The differential-mode gain and input resistance and the output resistance are given. We need to calculate the frequency of the first pole: $f_\beta = f_T/A_o = 10$ Hz.

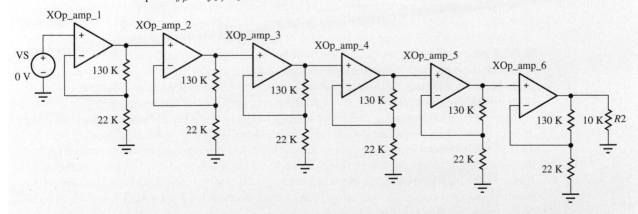

VS is an 1-V ac source with a dc value of 0 V. Since the amplifier is dc-coupled, a transfer function analysis from source VS to the output node will give the low-frequency gain, input resistance, and output resistance. An ac analysis using FSTART = 100 Hz, FSTOP = 1 MHz, and 10 frequency points per decade will produce the Bode plot needed to find the bandwidth. The simulation results yield a gain of 100.7 dB, $R_{in} = 28.9$ TΩ, $R_{out} = 3.52$ mΩ, and the bandwidth is 54.8 kHz.

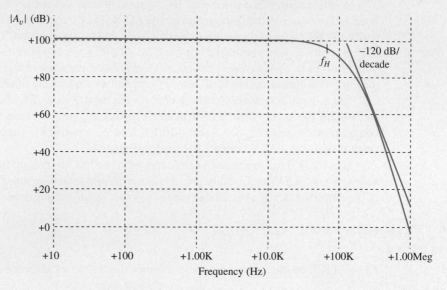

Check of Results: We see some discrepancies. The gain and output resistance agree with our calculations, but the bandwidth is larger than expected, and the input resistance is far too small. We should immediately be concerned about our simulation results. Indeed, a closer examination of the Bode plot also shows that the high-frequency roll-off is exceeding the $6(20 \text{ dB/decade}) = 120$ dB/decade slope that we should expect.

Discussion: The problems are buried in the macro model in which all of the unspecified parameters have default values. If we look at the op amp model in the version of SPICE used here, we find the default values are the same as given in Table 12.8: common-mode input resistance $= 2$ GΩ, second pole frequency $= 2$ MHz, offset voltage $= 1$ mV, input bias current $= 80$ nA, input offset current $= 20$ nA, etc. The input resistance cannot exceed the value set by R_{cm}, and the bandwidth and roll-off enhancements are actually caused by the second op amp pole at 2 MHz. If we change R_{cm} to 10^{15} Ω and set the higher-order pole frequencies all to 200 MHz, SPICE yields an input resistance of 13.2 TΩ and a bandwidth of 50.4 kHz, close to the expected values. In addition, the high-frequency roll-off rate is 120 dB/decade.

An additional problem was encountered in this simulation. In the initial simulation attempts, very small values of voltage gain were generated. An operating point analysis indicated that several of the op amp output voltages were at large values. Here again, the default parameter settings were causing the problem. This amplifier has very high overall gain, and a 1-mV offset voltage at the input of the first amplifier multiplied by the gain of 100,000 should produce 100 V at the output of the sixth amplifier! In order for the simulation to work, the offset voltage, input bias current, and input offset currents must all be set to zero in our op amp model!

The results discussed in the previous paragraph are also of significant practical interest! If we attempt to build this amplifier, we will encounter exactly the same problem. The offset voltages and input bias currents of the amplifier will cause the individual op amp outputs to saturate against the power supply levels. An overall feedback loop could be used to stabilize the dc operating point. However, the high gain and large phase shift through the amplifier will most likely cause stability problems on a breadboard, as well as with the overall feedback loop. (Feedback loop stability will be discussed in Chapter 17.)

EXERCISE: Simulate the amplifier with the dc value of VS set to 1 mV. What are the op amp output voltages?

ANSWERS: 6.91 mV, 47.7 mV, 330 mV, 2.28 V, 15 V, 15 V; the last two are saturated.

SUMMARY

The ideal operational amplifier was introduced previously in Chapter 11. Chapter 12 discussed removal of the ideal op amp assumptions, and quantified the effects and limitations caused by the nonideal op amp behavior. The nonideal behavior considered included:

- Finite open loop gain
- Finite differential-mode input resistance
- Nonzero output resistance
- Offset voltage
- Input bias and offset currents
- Limited output voltage range
- Limited output current capability
- Finite common-mode rejection
- Finite common-mode input resistance
- Finite power supply rejection
- Limited bandwidth
- Limited slew rate

- The effect of removing the various ideal operational amplifier assumptions was explored in detail. Expressions were developed for the gain, gain error, input resistance, and output resistance of the closed-loop inverting and noninverting amplifiers, and it was found that the loop gain $A\beta$ plays an important role in determining the value of these closed-loop amplifier parameters.

- The dc error sources, including offset voltage, bias current, and offset current, all limit the dc accuracy of op amp circuits. Real op amps also have limited output voltage and current ranges as well as a finite rate of change of the output voltage called the slew rate. Circuit design options are constrained by these factors.

- The frequency response of basic single-pole operational amplifiers is characterized by two parameters: the open-loop gain A_o and the gain-bandwidth product ω_T. Analysis of the gain and bandwidth of the inverting and noninverting amplifier configurations demonstrated the direct trade-off between the closed-loop gain and the closed-loop bandwidth of these amplifiers. The gain-bandwidth product is constant, and the closed-loop gain must be reduced in order to increase the bandwidth, or vice versa.

- The bandwidth of multistage amplifiers is less than the bandwidth of any of the single amplifiers acting alone. An expression was developed for the bandwidth of a cascade of N identical amplifiers and was cast in terms of the bandwidth shrinkage factor.

- A comprehensive example of the design of a multistage amplifier was presented in which a computer spreadsheet was used to explore the design space. The influence of resistor tolerances on this design was also explored.

- Simplified macro models are often used for simulation of circuits containing op amps. Simple macro models can be constructed in SPICE using controlled sources, and most SPICE libraries contain comprehensive macro models for a wide range of commercial operational amplifiers.

KEY TERMS

Bandwidth shrinkage factor
Bias current compensation resistor
Closed-loop amplifier
Closed-loop gain
Common-mode gain
Common-mode input resistance
Common-mode input voltage
Common-mode rejection
Common-mode rejection ratio (CMRR)
dc-coupled amplifier
Differential-mode gain
Differential-mode input resistance
Differential-mode input voltage
Feedback amplifier
Feedback factor (β)
Feedback network
Fractional gain error (FGE)
Full-power bandwidth f_M

Gain-bandwidth product (GBW)
Gain error (GE)
Ideal operational amplifier
Input-bias current
Input offset current
Input-offset voltage
Inverting amplifier
Loop gain ($A\beta$) (Loop transmission T)
Macro model
Noninverting amplifier
Offset current (I_{OS})
Offset voltage (V_{OS})
Open-loop gain (A_o)
Power supply rejection ratio (PSRR)
Single-pole frequency response
Slew rate (SR)
Unity-gain frequency ($\omega_T = 2\pi f_T$)

REFERENCES

1. Connelly, J. A. and P. Choi, *Macromodeling with SPICE,* Prentice Hall, Englewood Cliffs, NJ: 1992.
2. Boyle, G. R., B. M. Cohen, D. O. Pederson, and J. E. Soloman, "Macromodeling of integrated circuit operational amplifiers," *IEEE Journal of Solid-State Circuits,* vol. SC-9, no. 6, pp. 353–364, December 1974.

ADDITIONAL READING

Franco, Sergio, *Design with Operational Amplifiers and Analog Integrated Circuits,* Third Edition, McGraw-Hill, New York: 2001.

Gray, P. R., P. J. Hurst, S. H. Lewis, and R. G. Meyer, *Analysis and Design of Analog Integrated Circuits,* Fourth Edition, John Wiley and Sons, New York: 2001.

Kennedy, E. J. *Operational Amplifier Circuits—Theory and Applications.* Holt, Rinehart and Winston, New York: 1988.

Roberge, James K. *Operational Amplifiers—Theory and Practice.* Wiley, New York: 1975.

PROBLEMS

12.1 Gain, Input Resistance, and Output Resistance

12.1. A noninverting amplifier is built with $R_1 = 12$ kΩ and $R_2 = 150$ kΩ using an op amp with an open-loop gain of 86 dB. What are the closed-loop gain, the gain error, and the fractional gain error for this amplifier? (b) Repeat if R_1 is changed to 1.2 kΩ.

12.2. A noninverting amplifier is built with $R_2 = 47$ kΩ and $R_1 = 5.6$ kΩ using an op amp with an open-loop gain of 100 dB. What are the closed-loop gain, the gain error, and the fractional gain error for this amplifier? (b) Repeat if the open-loop gain is changed to 94 dB.

12.3. An inverting amplifier is built with $R_1 = 22$ kΩ and $R_2 = 220$ kΩ using an op amp with an open-loop gain of 92 dB. What are the closed-loop gain, the gain error, and fractional gain error for this amplifier? (b) Repeat if R_1 is changed to 1.1 kΩ.

12.4. An inverting amplifier is built with $R_2 = 47\,\text{k}\Omega$ and $R_1 = 4.7\,\text{k}\Omega$ using an op amp with an open-loop gain of 94 dB. What are the closed-loop gain, the gain error, and fractional gain error for this amplifier? (b) Repeat if the open-loop gain is changed to 100 dB.

12.5. An inverting amplifier is being built with a closed-loop gain of 46 dB. What op amp gain is required to have the gain error below 0.1 percent?

12.6. A voltage follower is built using an operational amplifier and must have a gain error ≤ 0.01 percent. What is the minimum open-loop gain specification for the op amp?

12.7. A noninverting amplifier is being designed to have a closed-loop gain of 32 dB. What op-amp gain is required to have the gain error less than 0.2 percent?

12.8. A noninverting amplifier is built with 0.01 percent precision resistors and designed with $R_2 = 99R_1$. What are the nominal and worst-case values of voltage gain if the op amp is ideal? What open-loop gain is required for the op amp if the gain-error due to finite op amp gain is to be less than 0.01 percent?

12.9. Repeat the derivation of the output resistance in Fig. 12.3 using a test current source rather than a test voltage source.

12.10. Calculate the currents i_1, i_2, and i_- for the amplifier in Fig. 12.4 if $v_x = 0.1\,\text{V}$, $R_1 = 1\,\text{k}\Omega$, $R_2 = 47\,\text{k}\Omega$, $R_{id} = 1\,\text{M}\Omega$, and $A = 10^5$.

12.11. A noninverting amplifier is built with $R_1 = 12\,\text{k}\Omega$ and $R_2 = 150\,\text{k}\Omega$ using an op amp with an open-loop gain of 86 dB, an input resistance of $250\,\text{k}\Omega$, and an output resistance of $250\,\Omega$. What are the closed-loop gain, input resistance, and output resistance for this amplifier? (b) Repeat if R_1 is changed to $1.2\,\text{k}\Omega$.

12.12. A noninverting amplifier is built with $R_2 = 47\,\text{k}\Omega$ and $R_1 = 5.6\,\text{k}\Omega$ using an op amp with an open-loop gain of 100 dB, an input resistance of $400\,\text{k}\Omega$, and an output resistance of $200\,\Omega$. What are the closed-loop gain, input resistance, and output resistance for this amplifier? (b) Repeat if the open-loop gain is changed to 94 dB.

12.13. An inverting amplifier is built with $R_2 = 47\,\text{k}\Omega$ and $R_1 = 5.6\,\text{k}\Omega$ using an op amp with an open-loop gain of 100 dB, an input resistance of $400\,\text{k}\Omega$, and an output resistance of $200\,\Omega$. What are the closed-loop gain, input resistance, and output resistance for this amplifier? (b) Repeat if the open-loop gain is changed to 94 dB.

12.14. An inverting amplifier is built with $R_2 = 47\,\text{k}\Omega$ and $R_1 = 4.7\,\text{k}\Omega$ using an op amp with an open-loop gain of 94 dB, an input resistance of $700\,\text{k}\Omega$, and an output resistance of $100\,\Omega$. What are the closed-loop gain, input resistance, and output resistance for this amplifier? (b) Repeat if the open-loop gain is changed to 100 dB.

12.15. What are the actual values of the two input resistances R_{in1} and R_{in2} and the output resistance R_{out} of the instrumentation amplifier in Fig. 11.12 if it is constructed using operational amplifiers with $A = 4 \times 10^4$, $R_{id} = 500\,\text{k}\Omega$, and $R_o = 75\,\Omega$? Assume $R_2 = 49R_1$ and $R_3 = R_4 = 10\,\text{k}\Omega$.

12.16. An op amp has $R_{id} = 500\,\text{k}\Omega$, $R_o = 35\,\Omega$, and $A = 5 \times 10^4$. You must decide if a single-stage amplifier can be built that meets all of the specifications below. (a) Which configuration (inverting or noninverting) must be used and why? (b) Assume that the gain specification must be met and show which of the other specifications can or cannot be met.

$$|A_v| = 200 \quad R_{in} \geq 2 \times 10^8\,\Omega \quad R_{out} \leq 0.2\,\Omega$$

12.17. An op amp has $R_{id} = 1\,\text{M}\Omega$, $R_o = 100\,\Omega$, and $A = 1 \times 10^4$. Can a single-stage amplifier be built with this op amp that meets all of the following specifications? Show which specifications can be met and which cannot.

$$|A_v| = 200 \quad R_{in} \geq 10^8\,\Omega \quad R_{out} \leq 0.2\,\Omega$$

12.18. The overall amplifier circuit in Fig. P12.18 is a two-terminal network. $R_1 = 6.8\,\text{k}\Omega$ and $R_2 = 110\,\text{k}\Omega$. What is its Thévenin equivalent circuit if the operational amplifier has $A = 5 \times 10^4$, $R_{id} = 1\,\text{M}\Omega$, and $R_o = 250\,\Omega$?

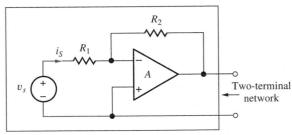

Figure P12.18

12.19. The circuit in Fig. P12.19 is a two-terminal network. $R_1 = 390\,\Omega$ and $R_2 = 56\,\text{k}\Omega$. What is its Thévenin equivalent circuit if the operational amplifier has $A = 1 \times 10^4$, $R_{id} = 100\,\text{k}\Omega$, and $R_o = 200\,\Omega$?

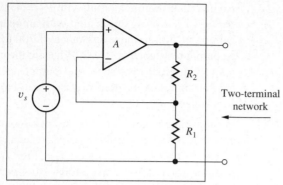

Figure P12.19

12.20. An inverting amplifier is to be designed to have a closed-loop gain of 60 dB. The only op amp that is available has an open-loop gain of 106 dB. What is the tolerance of the feedback resistors if the total gain error must be ≤ 1 percent? Assume that the resistors all have the same tolerances.

12.21. A noninverting amplifier is to be designed to have a closed-loop gain of 54 dB. The only op amp that is available has an open-loop gain of 40,000. What must be the tolerance of the feedback resistors if the total gain error must be ≤ 2 percent? Assume that the resistors all have the same tolerances.

12.22. The resistors in the difference amplifier in Fig. P12.22 are slightly mismatched due to their tolerances. What is the dc output voltage of the amplifier for $v_1 = 3$ V and $v_2 = 3$ V? What would be the output voltage if the resistor pairs were exactly matched?

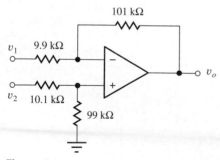

Figure P12.22

12.23. The resistors in the difference amplifier in Fig. P12.22 are slightly mismatched due to their tolerances. What is the amplifier output voltage for $v_1 = 3.95$ V and $v_2 = 4.05$ V? What would be the (ideal) output voltage if the resistor pairs were properly matched? What is the error in amplifying $(v_1 - v_2)$?

*12.24. The op amp in the amplifier circuit in Fig. P12.22 is ideal and

$$v_1 = (10 \sin 120\pi t + 0.25 \sin 5000\pi t) \text{ V}$$

$$v_2 = (10 \sin 120\pi t - 0.25 \sin 5000\pi t) \text{ V}$$

(a) What are the differential-mode and common-mode input voltages to this amplifier? (b) What are the differential-mode and common-mode gains of this amplifier? (c) What is the common-mode rejection ratio of this amplifier? (d) Find v_o.

12.25. The multimeter in Fig. P12.25 has a common-mode rejection specification of 80 dB. What possible range of voltages can be indicated by the meter?

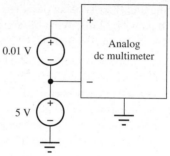

Figure P12.25

12.26. (a) We would like to measure the voltage ($V = V_1 - V_2$) appearing across the 20-kΩ resistor in Fig. P12.26 with a voltmeter. What is the value of V? What is the common-mode voltage associated with V ($V_{CM} = (V_1 + V_2)/2$)? What CMRR is required of the voltmeter if we are to measure V with an error of less than 0.01 percent? (b) Repeat if the 20-kΩ resistor is changed to 200 Ω.

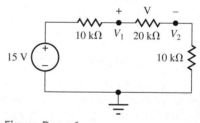

Figure P12.26

12.2 Common-Mode Rejection and Input Resistance

12.27. The common-mode rejection ratio of the difference amplifier in Fig. P12.27 is most often limited by the mismatch in the resistor pairs and not by the CMRR of the amplifier itself. Suppose that the nominal value of R is 10 kΩ and its tolerance is 0.05 percent. What is the worst-case value of the CMRR in dB?

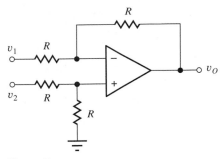

Figure P12.27

12.28. What are the actual values of the two input resistances R_{in1} and R_{in2} and the output resistance R_{out} of the instrumentation amplifier in Fig. 11.12 if it is constructed using operational amplifiers with $A = 7.5 \times 10^4$, $R_{id} = 1$ MΩ, $R_{ic} = 500$ MΩ, and $R_o = 100$ Ω? Assume $R_1 = 1$ kΩ, $R_2 = 24$ kΩ, and $R_3 = R_4 = 10$ kΩ.

12.29. In the instrumentation amplifier in Fig. P12.29, $v_a = 4.99$ V and $v_b = 5.01$ V. Find the values of node voltages v_1, v_2, v_3, v_4, v_5, v_6, v_o, and currents i_1, i_2, and i_3. What are the values of the common-mode gain, differential-mode gain, and CMRR of the amplifier? The op amps are ideal.

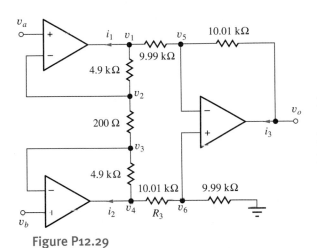

Figure P12.29

12.30. Find the values of v_1, v_2, v_3, v_4, v_5, v_6, v_o, i_1, i_2, and i_3 in the instrumentation amplifier in Fig. P12.29 if $v_a = 3$ V and $v_b = 3$ V.

12.3 DC Error Sources and Output Range Limitations

12.31. Calculate the worst-case output voltage for the circuit in Fig. P12.31 if $V_{OS} = 1$ mV, $I_{B1} = 100$ nA, and $I_{B2} = 95$ nA. What is the ideal output voltage.

What is the total error in this circuit? Is there a better choice for the value of R_1? If so, what is the value?

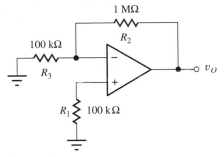

Figure P12.31

12.32. Repeat Prob. P12.31 if $V_{OS} = 10$ mV, $I_{B1} = 200$ nA, $I_{B2} = 250$ nA, and $R_2 = 510$ kΩ?

12.33. The voltage transfer characteristic for an operational amplifier is given in Fig. P12.33. What are the values of gain and offset voltage for this op amp?

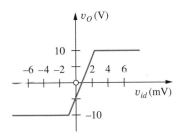

Figure P12.33

12.34. Plot voltage gain A versus v_{id} for the amplifier voltage transfer characteristic given in Fig. P12.33.

12.35. Use superposition to derive the result in Eq. (12.51).

12.36. The amplifier in Fig. P12.36 is to be designed to have a gain of 40 dB. What values of R_1 and R_2 should be used in order to meet the gain specification and minimize the effects of bias current errors?

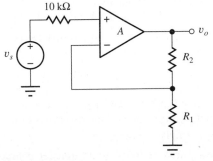

Figure P12.36

*12.37. The op amp in the circuit of Fig. P12.37 has an open-loop gain of 10,000, an offset voltage of 1 mV, and an input-bias current of 100 nA. (a) What is the output voltage for an ideal op amp? (b) What is the actual output voltage for the worst-case polarity of offset voltage? (c) What is the percentage error in the output voltage compared to the ideal output voltage?

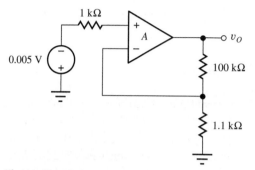

Figure P12.37

Voltage and Current Limits

12.38. The output-voltage range of the op amp in Fig. P12.38 is equal to the power supply voltages. What are the values of V_O and V_- for the amplifier if the dc input V_S is (a) 1 V and (b) -3 V?

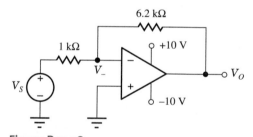

Figure P12.38

12.39. Plot a graph of the voltage transfer characteristic for the amplifier in Fig. P12.38.

12.40. The 6.2-kΩ resistor in Fig. P12.38 is replaced by a 10-kΩ resistor. What are the values of V_o and V_- if the dc input V_S is (a) 0.5 V and (b) 1.2 V?

12.41. Plot a graph of the voltage transfer characteristic for the amplifier in Prob. 12.40.

12.42. What are the voltages V_O and V_{ID} in the op amp circuit in Fig. P12.42 for dc input voltages of (a) $V_S = 250$ mV and (b) $V_S = 500$ mV if the output-voltage range of the op amp is limited to the power supply voltages.

12.43. Plot a graph of the voltage transfer characteristic for the amplifier in Fig. P12.42.

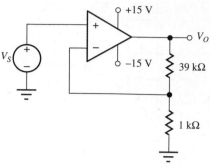

Figure P12.42

12.44. Repeat Prob. 12.42 if the 1-kΩ resistor in Fig. P12.42 is replaced by a 910-Ω resistor.

12.45. Design a noninverting amplifier with $A_v = 40$ dB that can deliver a ±10-V signal to a 10-kΩ load resistor. Your op amp can supply only 1.5 mA of output current. Use standard resistor 5 percent values in your design.

12.46. Design an inverting amplifier with $A_v = 46$ dB that can deliver ±15 V to a 5-kΩ load resistor. Your op amp can supply only 4 mA of output current. Use standard resistor 5 percent values in your design.

12.47. What is the minimum value of R in the circuit in Fig. P12.47 if the maximum op amp output current is 5 mA and the current gain of the transistor is $\beta_F \geq 50$?

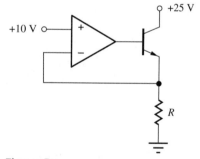

Figure P12.47

*12.48. (a) Design a single-stage inverting amplifier with a gain of 46 dB using an operational amplifier. The input resistance should be as low as possible while achieving the op amp output drive capability mentioned here. The amplifier must be able to produce the signal $v_o = (10 \sin 1000t)$ V at its output when an external load resistance $R_L \geq 5$ kΩ is connected to the output of the amplifier. You have an operational amplifier available whose output is guaranteed to deliver ±10 V into a 4-kΩ load resistance. Otherwise, the amplifier is ideal. (b) If the amplifier input signal is $v_s = V \sin 1000t$, what is the largest acceptable value for the input signal amplitude V? (c) What is the input resistance of your amplifier?

Cascaded Amplifiers

12.49. An amplifier is formed by cascading two operational amplifier stages, as shown in Fig. P12.49. For the op amps, $A = 10^5$, $R_{id} = 500$ kΩ, and $R_o = 100$ Ω. What are the closed-loop gain, input resistance, and output resistance for this amplifier?

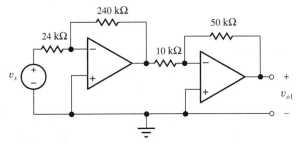

Figure P12.49

12.50. An amplifier is formed by cascading two operational amplifier stages, as shown in Fig. P12.50. Assume $A = 10^5$, $R_{id} = 250$ kΩ, and $R_o = 200$ Ω for each op amp. What are the closed-loop gain, input resistance, and output resistance for this amplifier?

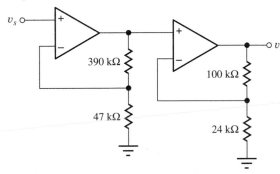

Figure P12.50

12.51. What are the closed-loop gain, input resistance, and output resistance for the amplifier in Fig. P12.51 if the op amps have an open-loop gain of 100 dB, an input resistance of 250 kΩ, and an output resistance of 200 Ω.

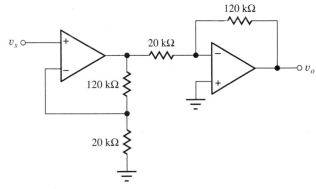

Figure P12.51

12.52. What are the closed-loop gain, input resistance, and output resistance for the amplifier in Fig. P12.52 if the op amps have an open-loop gain of 94 dB, an input resistance of 500 kΩ, and an output resistance of 300 Ω.

**12.53. A cascade amplifier is to be designed to meet these specifications:

$$A_v = 5000 \quad R_{in} \geq 10 \text{ M}\Omega \quad R_{out} \leq 0.1 \text{ }\Omega$$

How many amplifier stages will be required if the stages must use an op amp below? Because of bandwidth requirements, assume that no individual stage can have a gain greater than 50.

$$\text{Op amp specifications: } A = 85 \text{ dB}$$
$$R_{id} = 1 \text{ M}\Omega$$
$$R_o = 100 \text{ }\Omega$$
$$R_{ic} \geq 1 \text{ G}\Omega$$

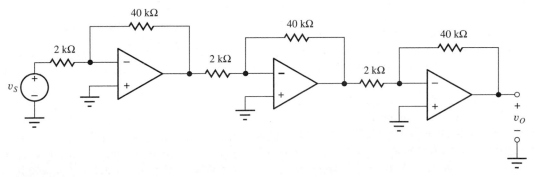

Figure P12.52

12.4 Frequency Response and Bandwidth of Operational Amplifiers

12.54. What is the transfer function for the voltage gain of the amplifier in Fig. P12.54?

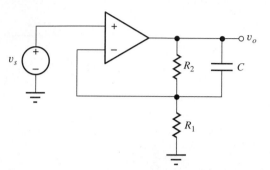

Figure P12.54

*12.55. The low-pass filter in Fig. P12.55 has $R_1 = 10$ kΩ, $R_2 = 330$ kΩ, and $C = 100$ pF. If the tolerances of the resistors are ± 10 percent and that of the capacitor is $+20$ percent/-50 percent, what are the nominal and worst-case values of the low-frequency gain and cutoff frequency?

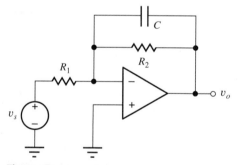

Figure P12.55 Low-pass filter

*12.56. Design a multistage low-pass filter to have a gain of 1000, a bandwidth of 20 kHz, and a response that rolls off at high frequencies at a rate of -60 dB/decade. Each stage should use the circuit in Fig. P12.55. (Questions to ask: How many stages are required? What are the gain and bandwidth required of each stage? What is the bandwidth shrinkage factor?)

*12.57. Derive an expression for the output impedance $Z_{out}(s)$ of the inverting and noninverting amplifiers in Fig. 12.2, assuming that the op amp has a transfer function given by Eq. (12.58).

*12.58. Use MATLAB, a spreadsheet, or other computer tool to plot the output impedance in Prob. 12.57

as a function of frequency if $R_1 = 10$ kΩ, $R_2 = 100$ kΩ, and the op amp has an open-loop gain $A_o = 100$ dB, $R_o = 100$ Ω, and $f_T = 1$ MHz.

*12.59. Derive an expression for the input impedance $Z_{in}(s)$ of the inverting amplifier in Fig. 12.5, assuming that the amplifier has a transfer function given by Eq. (12.58).

12.60. Use MATLAB, a spreadsheet, or other computer tool to produce a Bode plot for $A_{v1}(s)$, $A_{v2}(s)$, and the composite $A_v(s)$ for Example 12.12.

12.61. Use MATLAB, a spreadsheet, or other computer tool to draw a Bode plot for the low-pass filter circuit in Fig. 11.14 with $R_1 = 4.3$ kΩ, $R_2 = 82$ kΩ, and $C = 200$ pF if the op amp is not ideal but has an open-loop gain $A_o = 100$ dB and $f_T = 5$ MHz.

*12.62. (a) Derive an expression for the transfer function of the ideal integrator in Fig. 11.15 in the frequency domain. (b) Repeat if the transfer function of the op amp is given by Eq. (12.58).

12.63. Use MATLAB, a spreadsheet, or other computer tool to draw a Bode plot for the integrator circuit in Fig. 11.15 with $R_1 = 10$ kΩ, and $C = 470$ pF if the op amp is not ideal but has an open-loop gain $A_o = 100$ dB and $f_T = 5$ MHz.

12.64. (a) What are the gain and bandwidth of the individual amplifier stages in Fig. P12.52 if the op amps have $A_o = 10^5$ and $f_T = 3$ MHz? (b) What are the overall gain and bandwidth of the three-stage amplifier?

**12.65. Use these op amp parameters to design a multistage amplifier that meets the specifications below.

$$A_v = 86 \text{ dB} \pm 1 \text{ dB} \qquad R_{in} \geq 10 \text{ k}\Omega$$

$$R_{out} \leq 0.01 \text{ }\Omega \qquad f_H \geq 75 \text{ kHz}$$

The amplifier should use the minimum number of op amp stages that will meet the requirements. (A spreadsheet or simple computer program will be helpful in finding the solution.)

$$\text{Op amp specifications:} \quad A_o = 10^5$$
$$R_{id} = 10^9 \text{ }\Omega$$
$$R_o = 50 \text{ }\Omega$$
$$\text{GBW} = 1 \text{ MHz}$$

**12.66. (a) Design the amplifier in Prob. 12.65, including values for the feedback resistors in each stage. (b) What is the bandwidth of your amplifier if the op amps have $f_T = 5$ MHz?

12.67. A cascade amplifier is to be designed to meet these specifications:

$$A_v = 60 \text{ dB} \pm 1 \text{ dB} \qquad R_{in} = 27 \text{ k}\Omega$$

$$R_{out} \leq 0.1 \text{ }\Omega \qquad \text{Bandwidth} = 20 \text{ kHz}$$

How many amplifier stages will be required if the stages must use these op amp specifications?

Op amp specifications: $A_o = 85 \text{ dB}$

$$f_T = 5 \text{ MHz}$$
$$R_o = 100 \text{ }\Omega$$
$$R_{id} = 1 \text{ M}\Omega$$
$$R_{ic} \geq 1 \text{ G}\Omega$$

12.68. Design the amplifier in Prob. 12.67, including values for the feedback resistors in each stage.

**12.69. (a) Perform a Monte Carlo analysis of the six-stage cascade amplifier design resulting from the example in Tables 12.5 and 12.6, and determine the fraction of the amplifiers that will not meet either the gain or bandwidth specifications. Assume the resistors are uniformly distributed between their limits.

$$A_v \geq 100 \text{ dB} \qquad \text{and} \qquad f_H \geq 50 \text{ kHz}$$

(b) What tolerance must be used to ensure that less than 0.1 percent of the amplifiers fail to meet both specifications?

The equation here can be used to estimate the location of the half-power frequency for N closely spaced poles, where $\overline{f_{H1}}$ is the average of the individual cutoff frequencies of the N stages and f_{H1}^i is the cutoff frequency of the ith individual stage.

$$f_H = \overline{f_{H1}} \sqrt{2^{1/N} - 1}$$

where $\overline{f_{H1}} = \dfrac{1}{N} \sum_{i=1}^{N} f_{H1}^i$

**12.70. (a) Show that the number of stages that optimizes the bandwidth of a cascade of identical noninverting amplifier stages having a total gain G is given by

$$N = \dfrac{\ln 2}{\ln \left[\dfrac{\ln G}{\ln G - \ln \sqrt{2}} \right]}$$

(b) Calculate N for the amplifier in Example 12.13.

12.71. What are the gain and bandwidth of the amplifier in Fig. P12.71 for the nominal and worst-case values of the feedback resistors if $A = 50,000$ and $f_T = 1 \text{ MHz}$?

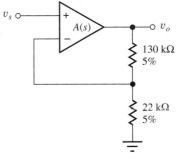

Figure P12.71

**12.72. Perform a Monte Carlo analysis of the circuit in Fig. P12.71. (a) What are the three sigma limits on the gain and bandwidth of the amplifier if $A_o = 50,000$ and $f_T = 1 \text{ MHz}$? (b) Repeat if A_o is uniformly distributed in the interval $[5 \times 10^4, 1.5 \times 10^5]$ and f_T is uniformly distributed in the interval $[10^6, 3 \times 10^6]$.

Large Signal Limitations—Slew Rate and Full-Power Bandwidth

12.73. An audio amplifier is to be designed to develop a 30-V peak-to-peak sinusoidal signal at a frequency of 20 kHz. What is the slew-rate specification of the amplifier?

12.74. An amplifier has a slew rate of 10 V/μs. What is the full-power bandwidth for signals having an amplitude of 10 V?

12.75. An amplifier must reproduce the output waveform in Fig. P12.75. What is its slew rate?

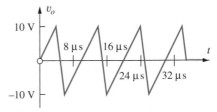

Figure P12.75

Macro Model for the Operational Amplifier Frequency Response

12.76. A single-pole op amp has these specifications:
$$A_o = 80,000 \qquad f_T = 5 \text{ MHz}$$
$$R_{id} = 250 \text{ k}\Omega \qquad R_o = 50 \text{ }\Omega$$
(a) Draw the circuit of a macro model for this operational amplifier. (b) Draw the circuit of a macro model for this operational amplifier if the op amp also has $R_{ic} = 500 \text{ M}\Omega$.

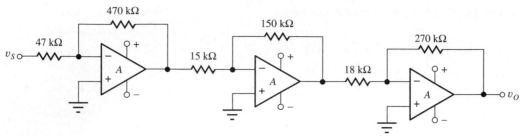

Figure P12.84

12.77. Draw a macro model for the amplifier in Prob. 12.76, including the additional elements necessary to model $R_{ic} = 100$ MΩ, $I_{B1} = 105$ nA, $I_{B2} = 95$ nA, and $V_{OS} = 1$ mV.

*12.78. A two-pole operational amplifier can be represented by the transfer function

$$A(s) = \frac{A_o \omega_1 \omega_2}{(s + \omega_1)(s + \omega_2)}$$

where

$$A_o = 80{,}000$$
$$f_1 = 1 \text{ kHz}$$
$$f_2 = 100 \text{ kHz}$$
$$R_{id} = 400 \text{ k}\Omega$$
$$R_o = 75 \text{ }\Omega$$

Create a macro model for this amplifier. (*Hint:* Consider using two "dummy" loops.)

12.79. Simulate the frequency response of the nominal design of the six-stage cascade amplifier from Table 12.7. Use the macro model in Fig. 12.26 to represent the op amp.

12.80. Simulate the frequency response of the six-stage cascade amplifier from Table 12.7 using the μA741 op amp macro model in SPICE.

12.81. Use the Monte Carlo analysis capability in PSPICE to simulate 1000 cases of the behavior of the six-stage amplifier in Table 12.7. Assume that all resistors and capacitors have 5 percent tolerances and the open-loop gain and bandwidth of the op amps each has a 50 percent tolerance. Use uniform statistical distributions. What are the lowest and highest observed values of gain and bandwidth for the amplifier?

Commercial General-Purpose Operational Amplifiers

12.82. (a) What are the element values for the macro model in Fig. 12.26 for the AD745J op amp? Use $R = 1$ kΩ and the nominal specifications.

12.83. What are the worst-case values (minimum or maximum, as appropriate) of the following parameters of the AD745 op amp: open-loop gain, CMRR, PSRR, V_{OS}, I_{B1}, I_{B2}, I_{OS}, R_{ID}, slew rate, gain-bandwidth product, and power supply voltages?

12.84. The op amps in Fig. P12.84 are described by $A_o = 100$ dB, $R_{ID} = 1$ MΩ, $R_O = 250$ Ω, and $f_T = 2$ MHz, and the power supplies are ± 18 V. (a) What are the voltage gain, input resistance, output resistance, and bandwidth of the overall amplifier? (b) If the input voltage $v_I = 50$ mV, what are the voltages (three significant digits) at each of the 10 nodes in the amplifier circuit?

12.85. What are the nominal, minimum, and maximum values of the voltage gain, input resistance, output resistance, and bandwidth of the overall amplifier in Prob. 12.84 if the resistors all have 5 percent tolerances?

12.86. The op amps in Fig. P12.86 are described by $A_o = 106$ dB, $R_{id} = 500$ kΩ, $R_O = 300$ Ω, and $f_T = 5$ MHz, and the power supplies are ± 12 V. (a) What are the voltage gain, input resistance, output resistance, and bandwidth of the overall amplifier? (b) If the input voltage $v_I = 5$ mV, what are the voltages (three significant digits) at each of the 10 nodes in the amplifier circuit?

12.87. Repeat Prob. 12.86 if the 3-kΩ resistors are all replaced with 1.5-kΩ resistors, and the 1-MΩ resistor is replaced with a 1.5-MΩ resistor.

12.88. What are the nominal, minimum, and maximum values of the voltage gain, input resistance, output resistance, and bandwidth of the overall amplifier in Prob. 12.86 if the resistors all have 2 percent tolerances?

12.89. The op-amps in Fig. P12.89 are described by $A_o = 80$ dB, $R_{id} = 300$ kΩ, $R_O = 200$ Ω, and $f_T = 3$ MHz, and the power supplies are ± 15 V. (a) What are the voltage gain, input resistance, output resistance, and bandwidth of the overall amplifier?

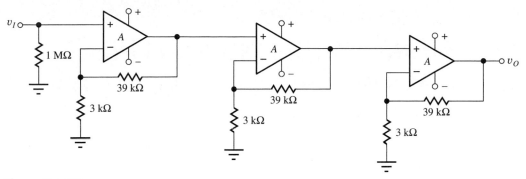

Figure P12.86

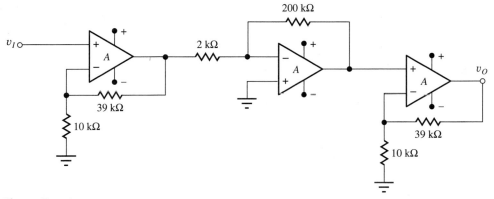

Figure P12.89

(b) Assume the offset voltage of each op-amp is equivalent to $+10$ mV at the positive input of the op amp. If the input voltage $v_I = 0$ V, what are the voltages (three significant digits) at each of the 10 nodes in the amplifier circuit?

12.90. What are the nominal, minimum, and maximum values of the voltage gain, input resistance, output resistance, and bandwidth of the overall amplifier in Prob. 12.89 if the resistors all have 10 percent tolerances?

SMALL-SIGNAL MODELING AND LINEAR AMPLIFICATION—INVERTING AMPLIFIERS

CHAPTER OUTLINE

CHAPTER GOALS

In Chapter 13, we develop a basic understanding of the following concepts related to linear amplification:

- Transistors as linear amplifiers
- dc and ac equivalent circuits
- Use of coupling and bypass capacitors and inductors to modify the dc and ac equivalent circuits
- The concept of small-signal voltages and currents
- Small-signal models for diodes and transistors
- Identification of common-emitter and common-source amplifiers
- Amplifier characteristics including voltage gain, input resistance, output resistance, and linear signal range
- Rule-of-thumb estimates for voltage gain of common-emitter and common-source amplifiers
- Improvement of our understanding of the use and differences between the ac small-signal transfer function, and transient analysis capabilities of SPICE

Chapter 13 begins our study of basic amplifier circuits that are used in the design of complex analog components and systems such as high-performance operational amplifiers, analog-to-digital and digital-to-analog converters, audio equipment, compact disk players, wireless devices, cellular telephones, and so on. At first glance, the operational amplifier schematic in the figure here represents an overwhelming interconnection of transistors and passive components. With this chapter, we begin our quest to understand and design a wide variety of such circuits. We will learn to simplify our job by separating the dc and ac circuit analyses.

In order to predict the detailed behavior of the circuit, we must also be able to build mathematical models that describe the circuit. This chapter develops these models. Then over the next several chapters, we become familiar with the basic subcircuits that serve as our electronic tool kit for building more complicated electronic systems. With practice over time, we will be able to spot these basic building blocks in more complex electronic circuits and use our knowledge of the subcircuits to understand the full system.

This chapter introduces the general techniques for employing individual transistors as amplifiers and then studies in detail the operation of the common-emitter bipolar transistor circuit. This is followed by analysis of common-source amplifiers employing MOSFETs. Circuits containing these devices are compared, and expressions are developed for the voltage gain and input and output resistances of the various amplifiers. The advantages and disadvantages of each are discussed in detail.

To simplify the analysis and design processes, the circuits are split into two parts: a dc equivalent circuit used to find the Q-point of the transistor, and an ac equivalent circuit used for analysis of the circuit's response to signal sources. As a by-product of this approach, we discover how capacitors and inductors are used to change the ac and dc circuit topologies.

The ac analysis is based on linearity and requires the use of "small-signal" models that exhibit a linear relation between their terminal voltages and currents. The concept of a small signal is developed, and small-signal models for

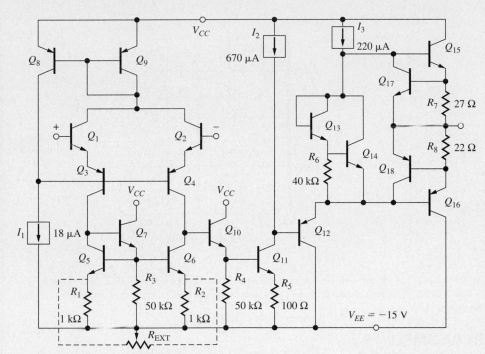

Circuit schematic for μA741 op amp

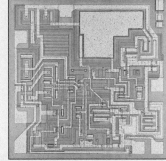

uA741 Die Photograph (Courtesy of Fairchild Semiconductor)

the diode, bipolar transistor, and MOSFET are all discussed in detail.

Examples of the complete analysis of common-emitter and common-source amplifiers are included in this chapter. The relationships between the choice of operating point and the small-signal characteristics of the amplifier are developed, as is the relationship between Q-point design and output signal voltage swing.

13.1 THE TRANSISTOR AS AN AMPLIFIER

As mentioned in Part I, the bipolar junction transistor is an excellent amplifier when it is biased in the forward-active region of operation; field-effect transistors should be operated in the saturation or pinch-off region in order to be used as amplifiers. For simplicity, we will now refer to bipolar transistors operating in the forward-active region and FETs in the saturation region as simply being in the "active region" where they can be used as linear amplifiers. In these regions of operation, the transistors have the capacity to provide high voltage, current, and power gains. This chapter focuses on determining voltage gain, input resistance, and output resistance. We will find that we need the input and output resistance information in order to calculate the lower and upper cutoff frequencies of the amplifiers. Evaluation of current gain and power gain are addressed in later chapters.

We must provide bias to the transistor in order to stabilize the operating point in the active region of operation. Once the dc operating point has been established, we can then use the transistor as an amplifier. The Q-point controls many other amplifier characteristics, including

- Small-signal parameters of the transistor
- Voltage gain, input resistance, and output resistance
- Maximum input and output signal amplitudes
- Power consumption

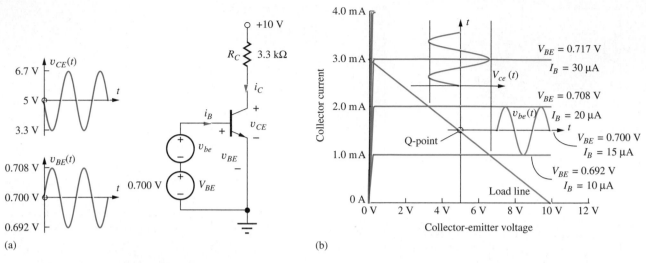

Figure 13.1 (a) BJT biased in the forward-active region by the voltage source V_{BE}. A small sinusoidal signal voltage v_{be} is applied in series with V_{BE} and generates a similar bit larger amplitude waveform at the collector. (b) Load-line Q-point and signals for circuit of Fig. 13.1(a).

13.1.1 THE BJT AMPLIFIER

To get a clearer understanding of how a transistor can provide linear amplification, let us assume that a bipolar transistor is biased in the active region by the dc voltage source V_{BE} shown in Fig. 13.1. For this particular transistor, the fixed base-emitter voltage of 0.700 V sets the Q-point to be $(I_C, V_{CE}) = (1.5 \text{ mA}, 5 \text{ V})$ with $I_B = 15$ μA, as indicated in Fig. 13.1(b). Both I_B and V_{BE} have been shown as parameters on the output characteristics in Fig. 13.1(b) (usually only I_B is given).

To provide amplification, a signal must be injected into the circuit in a manner that causes the transistor voltages and currents to vary with the applied signal. For the circuit in Fig. 13.1, the base-emitter voltage is forced to vary about its Q-point value by signal source v_{be} placed in series with V_{BE}, so the total base-emitter voltage becomes

$$v_{BE} = V_{BE} + v_{be} \tag{13.1}$$

In Fig. 13.1(b), we see that the 8-mV peak change in base-emitter voltage produces a 5-μA change in base current and hence a 500-μA change in collector current $(i_c = \beta_F i_b)$.

The collector-emitter voltage of the BJT in Fig. 13.1 can be expressed as

$$v_{CE} = 10 - i_C R_C = 10 - 3300 i_C \tag{13.2}$$

The change in collector current develops a time-varying voltage across the load resistor R_C and at the collector terminal of the transistor. The 500 μA change in collector current develops a 1.65-V change in collector-emitter voltage. Equation (13.2) represents the equation for the load line for this transistor.

If these changes in operating currents and voltages are all small enough ("small signals"), then the collector current and collector-emitter voltage waveforms will be undistorted replicas of the input signal. Small-signal operation is device-dependent; it will be precisely defined for the BJT when the small-signal model for the bipolar transistor is introduced.

In Fig. 13.1 we see that a small input voltage change at the base is causing a large voltage change at the collector. The voltage gain for this circuit is defined in terms of the frequency domain (phasor) representation of the signals:

$$A_v = \frac{\mathbf{v_{ce}}}{\mathbf{v_{be}}} = \frac{1.65 \angle 180}{0.008 \angle 0} = 206 \angle 180 = -206 \tag{13.3}$$

The magnitude of the collector-emitter voltage is 206 times larger than the base-emitter signal amplitude; this represents a voltage gain of 206. It is also important to note in Fig. 13.1 that the output signal voltage decreases as the input signal increases, indicating a 180° phase shift between the input and the output signals. This 180° phase shift is represented by the minus sign in Eq. (13.3).

EXERCISE: The common-emitter current gain β_F of the bipolar transistor is defined by $\beta_F = I_C/I_B$. (a) What is the value of β_F for the transistor in Fig. 13.1? (b) The dc collector current of the BJT in the (forward) active region is given by $I_C = I_S \exp V_{BE}/V_T$. Use the Q-point data to find the saturation current I_S of the transistor in Fig. 13.1. (Remember $V_T = 0.025$ V.) (c) The ratio of v_{be}/i_b represents the small-signal input resistance R_{in} of the BJT. What is its value for the transistor in Fig. 13.1? (d) Does the BJT remain in the active region during the full range of signal voltages at the collector? Why?

ANSWERS: $\beta_F = 100$; $I_S = 1.04 \times 10^{-15}$ A; $R_{in} = 1.6$ kΩ; yes, $v_{CE}^{MIN} > V_{BE}$: 3.4 V > 0.708 V.

13.1.2 THE MOSFET AMPLIFIER

The amplifier circuit using a MOSFET in Fig. 13.2 is directly analogous to the BJT amplifier circuit in Fig. 13.1. Here the gate-to-source voltage is forced to vary about its Q-point value ($V_{GS} = 3.5$ V) by signal source v_{gs} placed in series with dc bias source V_{GS}. In this case, the total gate-source voltage is

$$v_{GS} = V_{GS} + v_{gs}$$

The resulting signal voltages are superimposed on the MOSFET output characteristics in Fig. 13.2(b). $V_{GS} = 3.5$ V sets the Q-point (I_D, V_{DS}) at (1.56 mA, 4.8 V), and the 1-V p-p change in v_{GS} causes a 1.25-mA p-p change in i_D and a 4-V p-p change in v_{DS}.

EXERCISE: (a) Write an expression for the drain-source voltage (the load-line equation) for the MOSFET in Fig. 13.2. (b) What is the approximate voltage gain for the amplifier in Fig. 13.2? (c) Does the MOSFET in Fig. 13.2 remain in the active region of operation during the full-output signal swing? (d) If the dc drain current of the MOSFET in the active region is given by $I_D = (K_n/2)(V_{GS} - V_{TN})^2$, what are the values of the parameter K_n and threshold voltage V_{TN} for the transistor in Fig. 13.2?

ANSWERS: $v_{DS} = 10 - 3300i_D$; $A_V = -4$; no, not near the positive peak of v_{GS}, corresponding to the peak negative excursion of v_{DS}; $K_n = 5 \times 10^4$ A/V², $V_{TN} = 1$ V

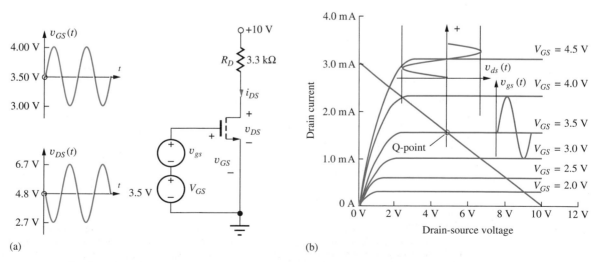

(a)

(b)

Figure 13.2 (a) A MOSFET common-source amplifier. (b) Q-point, load line, and signals for the circuit of Fig. 13.2.(a).

13.2 COUPLING AND BYPASS CAPACITORS

The constant base-emitter or gate-source voltage biasing techniques used in Figs. 13.1 and 13.2 are not very desirable methods of establishing the Q-point for a bipolar transistor or FET because the operating point is highly dependent on the transistor parameters. As discussed in detail in Chapters 4 and 5, bias circuits, such as the four-resistor network in Fig. 13.3, are much preferred for establishing a stable Q-point for the transistor.

To use the transistor as an amplifier, ac signals need to be introduced into the circuit, but application of these ac signals must not disturb the dc Q-point that has been established by the bias network. One method of injecting an input signal and extracting an output signal without disturbing the Q-point is to use **ac coupling** through capacitors. The values of these capacitors are chosen to have negligible impedances in the frequency range of interest, but at the same time, the capacitors provide open circuits at dc so the Q-point is not disturbed. When power is first applied to the amplifier circuit, transient currents charge the capacitors, but the final steady-state operating point is not affected.

Figure 13.4 is an example of the use of capacitors; the transistor is biased by the same four-resistor network shown in Fig. 13.3. An input signal v_i is *coupled* onto the base node of the transistor through capacitor C_1, and the signal developed at the collector is coupled to load resistor R_3 through capacitor C_3. C_1 and C_3 are referred to as **coupling capacitors,** or **dc blocking capacitors.**

For now, the values of C_1 and C_3 are assumed to be very large, so their reactance $(1/\omega C)$ at the signal frequency ω will be negligible. This assumption is indicated in the figure by $C \rightarrow \infty$. Calculation of more exact values of the capacitors is left until the discussion of amplifier frequency response in Chapters 14 and 17.

Figure 13.4 also shows the use of capacitor C_2, called a **bypass capacitor.** In many circuits, we want to force signal currents to go around elements of the bias network. Capacitor C_2 provides a low impedance path for ac current to "bypass" emitter resistor R_4. Thus R_4, which is required for good Q-point stability, can be effectively removed from the circuit when ac signals are considered.

Simulation results of the behavior of this circuit are shown in Fig. 13.5. A 5-mV sine wave signal at a frequency of 1 kHz is applied to the base terminal of transistor Q through coupling capacitor C_1; this signal produces a sinusoidal signal at the collector node with an amplitude of approximately 1.1 V, centered on the Q-point value of $V_C \cong 5.8$ V. Note once again that there is a 180° phase shift between the input and output voltage signals. These values indicate that the amplifier is providing a

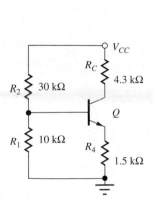

Figure 13.3 Transistor biased in the forward-active region using the four-resistor bias network (see Sec. 5.11 for an example).

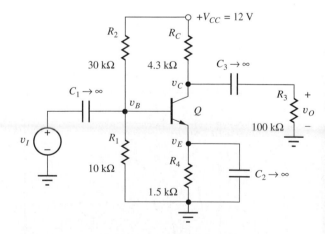

Figure 13.4 Common-emitter amplifier stage built around the four-resistor bias network. C_1 and C_3 function as coupling capacitors, and C_2 is a bypass capacitor.

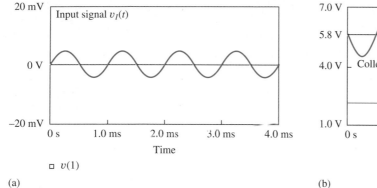

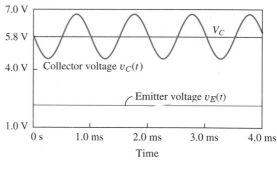

(a)

(b)

Figure 13.5 SPICE simulation results for v_S, v_C, and v_E for the amplifier in Fig. 13.4 with $v_I = 0.005 \sin 2000\pi t$ V.

voltage gain of

$$A_v = \frac{\mathbf{v_c}}{\mathbf{v_i}} = \frac{1.1\angle 180°}{.005\angle 0°} = 220\angle 180° = -220. \qquad (13.4)$$

In Fig. 13.5, we should also observe that the voltage at the emitter node remains constant at its Q-point value of slightly more than 2 V. The very low impedance of the bypass capacitor prevents any signal voltage from being developed at the emitter. We say that the bypass capacitor maintains an "ac ground" at the emitter terminal. In other words, zero signal voltage appears at the emitter, and the emitter voltage remains constant at its dc Q-point value.

EXERCISE: Calculate the Q-point for the bipolar transistor in Fig. 13.3. Use $\beta_F = 100$, $V_{BE} = 0.7$ V, and $V_A = \infty$.

ANSWER: (1.45 mA, 3.41 V)

EXERCISE: Write expressions for $v_C(t)$, $v_E(t)$, and $i_c(t)$ based on the waveforms shown in Fig. 13.5.

ANSWERS: $v_C(t) = (5.8 - 1.1 \sin 2000\pi t)$ V, $v_E(t) = 2.1$ V, $i_c(t) = -0.25 \sin 2000\pi t$ mA

EXERCISE: Suppose capacitor C_2 is 500 μF. What is its reactance at a frequency of 1000 Hz?

ANSWER: 0.318 Ω

13.3 CIRCUIT ANALYSIS USING dc AND ac EQUIVALENT CIRCUITS

To simplify the circuit analysis and design tasks, we break the amplifier into two parts, performing separate dc and ac circuit analyses. We find the Q-point of the circuit using the **dc equivalent circuit**—the circuit that is appropriate for steady-state dc analysis. To construct the dc equivalent circuit, we assume that capacitors are open circuits and inductors are short circuits.

Once we have found the Q-point, we determine the response of the circuit to the ac signals using an **ac equivalent circuit.** In constructing the ac equivalent circuit, we assume that the reactance of the coupling and bypass capacitors is negligible at the operating frequency ($|Z_C| = 1/\omega C = 0$), and we replace the capacitors by short circuits. Similarly, we assume the impedance of any inductors in the circuit is extremely large ($|Z_L| = \omega L \to \infty$), so we replace inductors by open circuits. Because the voltage at a node connected to a dc voltage source cannot change, these points represent grounds in the ac equivalent circuit (i.e., no ac voltage can appear at such a node: $v_{ac} = 0$). Furthermore, the current through a dc current source does not change even if the voltage across the source changes ($i_{ac} = 0$), so we replace dc current sources with open circuits in the ac equivalent circuit.

DESIGN NOTE

dc power supply nodes represent grounds and dc current sources represent open circuits in the ac equivalent circuit!

13.3.1 MENU FOR dc AND ac ANALYSIS

To summarize, our analysis of amplifier circuits is performed using the two-part process listed here:

dc Analysis

1. Find the dc equivalent circuit by replacing all capacitors with open circuits and inductors by short circuits.
2. Find the Q-point from the dc equivalent circuit using the appropriate large-signal model for the transistor.

ac Analysis

3. Find the ac equivalent circuit by replacing all capacitors by short circuits and all inductors by open circuits. dc voltage sources are replaced by ground connections, and dc current sources are replaced by open circuits in the ac equivalent circuit.
4. Replace the transistor by its small-signal model.
5. Analyze the ac characteristics of the amplifier using the small-signal ac equivalent circuit from step 4.
6. If desired, combine the results from steps 2 and 5 to yield the total voltages and currents in the network.

Since we are most often interested in determining the ac behavior of the circuit, we seldom actually perform this final step of combining the dc and ac results.

EXAMPLE 13.1   **CONSTRUCTING ac AND dc EQUIVALENT CIRCUITS FOR A BJT AMPLIFIER**

As has been stated several times, we usually split a circuit into its dc and ac equivalents, in order to simplify the analysis and design problem. This is a critical step, since we cannot get the correct answer if the equivalent circuits are improperly constructed.

PROBLEM Draw the dc and ac equivalent circuits (menu steps 1 and 3) for the common-emitter amplifier in Fig. 13.6(a). The circuit topology is similar to that in Fig. 13.4 except the original emitter resistor R_E has been split into two parts now labeled R_E and R_4, and resistor R_1, representing the Thévenin equivalent resistance of the signal source, has been added to the circuit. The resistor values have been changed to establish a new operating point.

SOLUTION **Known Information and Given Data:** The circuit with element values appears in Fig. 13.6(a).

Unknowns: dc equivalent circuit; ac equivalent circuit

Approach: Replace capacitors by open circuits to obtain the dc equivalent circuit. Replace capacitors and dc voltage sources by short circuits to obtain the ac equivalent circuit. Combine and simplify resistor combinations wherever possible.

Assumptions: Capacitor values are large enough that they can be treated as short circuits in the ac equivalent circuit.

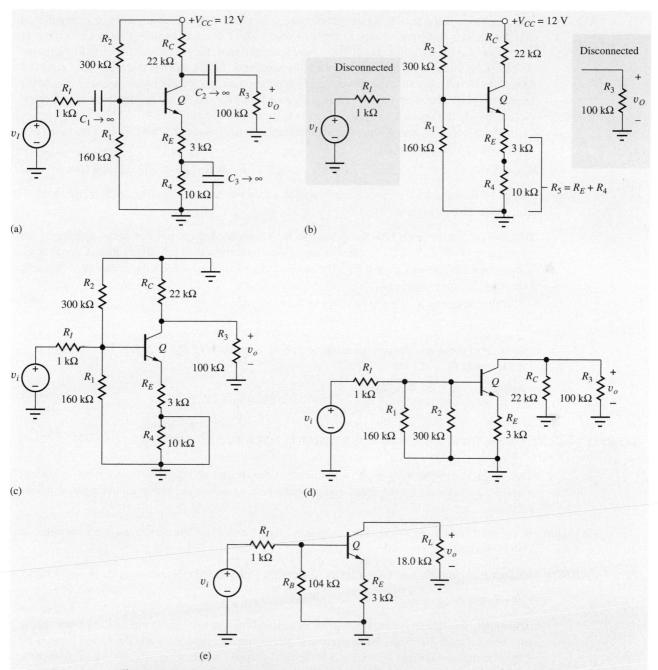

Figure 13.6 (a) Complete ac-coupled amplifier circuit. (b) Simplified equivalent circuit for dc analysis. (c) Circuit after step 3. Note that the input and output are now v_i and v_o. (d) Redrawn version of Fig. 13.6(c). (e) Continued simplification of the ac circuit.

ANALYSIS **dc Equivalent Circuit:** The dc equivalent circuit is found by open circuiting all the capacitors in the circuit. We find that the resulting dc equivalent circuit in Fig. 13.6(b) is identical to the four-resistor bias circuit of Fig. 13.3 (also see Sec. 5.11). Opening capacitors C_1 and C_2 disconnects v_I, R_I, and R_3 from the circuit.

ac Equivalent Circuit: To construct the ac equivalent circuit, we replace the capacitors by short circuits. Also, the dc voltage source becomes an ac ground in Fig. 13.6(c). In the ac equivalent circuit, source resistance R_I is now connected directly to the base node, and the external load resistor R_3 is connected directly to the collector node. Figure 13.6(d) is a redrawn version of Fig. 13.6(c). Although these two figures may look different, they are the same circuit! Note that resistor R_4 is shorted out by the presence of bypass capacitor C_3 and has therefore been removed from Fig. 13.6(d). Because the power supply represents an "ac ground," bias resistors R_1 and R_2 appear in parallel between the base node and ground, and R_C and R_3 are in parallel at the collector. Note that only the signal v_i is included in the ac equivalent circuit.

In Fig. 13.6(e), R_1 and R_2 have been combined into the resistor R_B, and R_C and R_3 have been combined into the resistor R_L:

$$R_B = R_1 \| R_2 = 160\,\text{k}\Omega \| 300\,\text{k}\Omega = 104\,\text{k}\Omega \quad \text{and} \quad R_L = R_C \| R_3 = 22\,\text{k}\Omega \| 100\,\text{k}\Omega = 18.0\,\text{k}\Omega$$

Check of Results: In this case, the best way to verify our results is to double check our work—everything seems correct.

Discussion: Notice again how the capacitors have been used to modify the circuit topologies for dc and ac signals. C_1 and C_2 isolate the source and load from the bias circuit at dc. Capacitor C_3 causes the node between R_4 and R_E to be connected directly to ground in the ac circuit, effectively removing R_4 from the circuit.

EXERCISE: What are the values of R_B and R_L in Fig. 13.6(e) if $R_1 = 20\,\text{k}\Omega$, $R_2 = 62\,\text{k}\Omega$, $R_C = 8.2\,\text{k}\Omega$, and $R_E = 2.7\,\text{k}\Omega$?

ANSWERS: 15.1 kΩ; 7.58 kΩ

EXAMPLE 13.2   CONSTRUCTING ac AND dc EQUIVALENTS FOR A MOSFET AMPLIFIER

This second example showing how to construct the dc and ac equivalent circuits of an overall amplifier circuit includes the use of a split-supply biasing technique, and an inductor has also been included in the circuit.

PROBLEM   Draw the dc and ac equivalent circuits (menu steps 1 and 3) for the common-source amplifier in Fig. 13.7(a).

SOLUTION   **Known Information and Given Data:** The circuit with labeled elements appears in Fig. 13.7(a).

Unknowns: dc equivalent circuit; ac equivalent circuit

Approach: Replace capacitors by open circuits and inductors by short circuits to obtain the dc equivalent circuit. To obtain the ac equivalent circuit, replace capacitors and dc voltage sources by short circuits, and current sources and inductors by open circuits. Combine and simplify resistor combinations wherever possible.

Assumptions: Capacitor values are large enough that they can be treated as short circuits in the ac equivalent circuit. The inductor value is large enough that it can be treated as an open circuit in the ac equivalent circuit.

ANALYSIS   **dc Equivalent Circuit:** The dc equivalent circuit is found by replacing the capacitors by open circuits and the inductor by a short circuit, resulting in the circuit in Fig. 13.7(b). Capacitors C_1 and C_2 again disconnect v_I, R_I, and R_3 from the circuit, and the shorted inductor connects the drain of the transistor directly to V_{DD}.

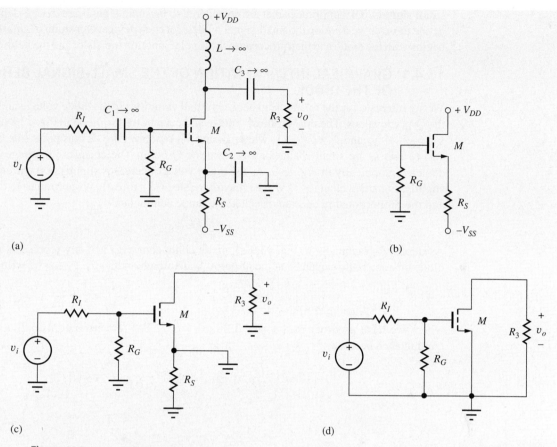

Figure 13.7 (a) An amplifier biased by two power supplies. (b) dc Equivalent circuit for Q-point analysis. (c) First step in generating the ac equivalent circuit. (d) Simplified ac equivalent circuit.

ac Equivalent Circuit: The ac equivalent circuit in Fig. 13.7(c) is obtained by replacing the capacitors by short circuits and the inductor by an open circuit. Figure 13.7(c) has been redrawn in final simplified form in Fig. 13.7(d). Only signal component v_i appears in the ac equivalent circuit.

Check of Results: For this case, the best way to verify our results is to double check our work—all seems correct.

Discussion: Here again, the designer has used the capacitors and inductor to achieve very different circuit topologies for the dc and ac equivalent circuits. Compare Figs. 13.7(b) and 13.7(d).

EXERCISE: Redraw the dc and ac equivalent circuits in Fig. 13.7(b) and (d) if C_2 were eliminated from the circuit.

ANSWERS: (b) No change, (d) resistor R_S appears between the MOSFET source and ground (See R_E in Fig 13.6(e).).

13.4 INTRODUCTION TO SMALL-SIGNAL MODELING

For ac analysis, we would like to be able to use our wealth of linear circuit analysis techniques. For this approach to be valid, the signal currents and voltages must be small enough to ensure that the ac circuit behaves in a linear manner. Thus, we must assume that the time-varying signal components are

small signals. The amplitudes that are considered to be small signals are device-dependent; we will define these as we develop the small-signal models for each device. Our study of **small-signal models** begins with the diode and then proceeds to the bipolar junction transistor and the field-effect transistor.

13.4.1 GRAPHICAL INTERPRETATION OF THE SMALL-SIGNAL BEHAVIOR OF THE DIODE

We are interested in the relationship between small variations in the diode voltage and current around the Q-point values. The total terminal voltage and current for the diode in Fig. 13.8 can be written as $v_D = V_D + v_d$ and $i_D = I_D + i_d$ where I_D and V_D represent the dc bias point (the Q-point) values, and v_d and i_d are small changes away from the Q-point. The changes in voltage and current are depicted graphically in Fig. 13.9. As the diode voltage increases slightly, the current also increases slightly. For small changes, i_d will be linearly related (i.e., directly proportional) to the change in v_d, and the proportionality constant is called the diode conductance g_d:

$$i_d = g_d v_d \tag{13.5}$$

As depicted graphically in Fig. 13.9(a), diode conductance g_d actually represents the slope of the diode characteristic evaluated at the Q-point. Stated mathematically, g_d can be written as

$$g_d = \left. \frac{\partial i_D}{\partial v_D} \right|_{\text{Q-point}} = \left. \frac{\partial}{\partial v_d} \left\{ I_S \left[\exp\left(\frac{v_D}{V_T} \right) - 1 \right] \right\} \right|_{\text{Q-point}} = \frac{I_S}{V_T} \exp\left(\frac{V_D}{V_T} \right) = \frac{I_D + I_S}{V_T} \tag{13.6}$$

where we have used our mathematical model for i_D. For forward bias with $I_D \gg I_S$, the diode conductance becomes

$$g_d \cong \frac{I_D}{V_T} \quad \text{or} \quad g_d \cong \frac{I_D}{0.025 \text{ V}} = 40 I_D \tag{13.7}$$

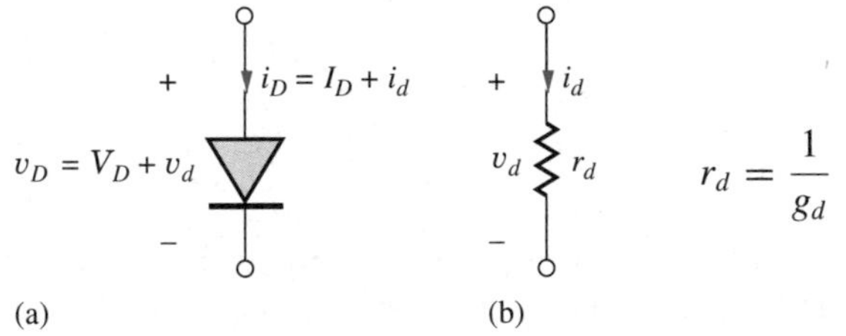

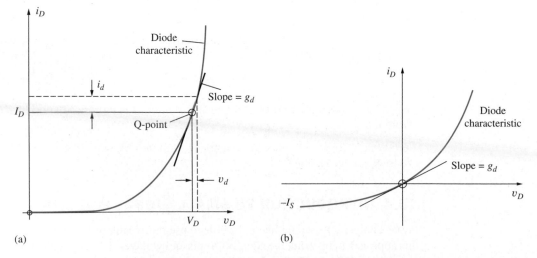

Figure 13.8 (a) Total diode terminal voltage and current. (b) Small signal model for the divide.

Figure 13.9 (a) The relationship between small increases in voltage and current above the diode operating point (I_D, V_D). For small changes $i_d = g_d v_d$. (b) The diode conductance is not zero for $I_D = 0$.

at room temperature. Note that g_d is small but not zero for $I_D = 0$ because the slope of the diode equation is nonzero at the origin, as depicted in Fig. 13.9(b).

13.4.2 SMALL-SIGNAL MODELING OF THE DIODE

Now we will use the diode equation to more fully explore the small-signal behavior of the diode and to actually define how large v_d and i_d can become before Eq. (13.5) breaks down. A relationship between the ac and dc quantities can be developed directly from the diode equation introduced in Chapter 3:

$$i_D = I_S \left[\exp\left(\frac{v_D}{V_T} \right) - 1 \right] \tag{13.8}$$

Substituting $v_D = V_D + v_d$ and $i_D = I_D + i_d$ into Eq. (13.8) yields

$$I_D + i_d = I_S \left[\exp\left(\frac{V_D + v_d}{V_T} \right) - 1 \right] = I_S \left[\exp\left(\frac{V_D}{V_T} \right) \exp\left(\frac{v_d}{V_T} \right) - 1 \right] \tag{13.9}$$

Expanding the second exponential using Maclaurin's series and collecting the dc and signal terms together,

$$I_D + i_d = I_S \left[\exp\left(\frac{V_D}{V_T} \right) - 1 \right] + I_S \exp\left(\frac{V_D}{V_T} \right) \left[\frac{v_d}{V_T} + \frac{1}{2}\left(\frac{v_d}{V_T} \right)^2 + \frac{1}{6}\left(\frac{v_d}{V_T} \right)^3 + \cdots \right] \tag{13.10}$$

We recognize the first term on the right-hand side of Eq. (13.10) as the dc diode current I_D:

$$I_D = I_S \left[\exp\left(\frac{V_D}{V_T} \right) - 1 \right] \qquad \text{and} \qquad I_S \exp\left(\frac{V_D}{V_T} \right) = I_D + I_S \tag{13.11}$$

Subtracting I_D from both sides of the equation yields an expression for i_d in terms of v_d:

$$i_d = (I_D + I_S) \left[\frac{v_d}{V_T} + \frac{1}{2}\left(\frac{v_d}{V_T} \right)^2 + \frac{1}{6}\left(\frac{v_d}{V_T} \right)^3 + \cdots \right] \tag{13.12}$$

We want the signal current i_d to be a linear function of the signal voltage v_d. Using only the first two terms of Eq. (13.12), we find that linearity requires

$$\frac{v_d}{V_T} \gg \frac{1}{2}\left(\frac{v_d}{V_T} \right)^? \qquad \text{or} \qquad v_d \ll 2V_T = 0.05 \text{ V} \tag{13.13}$$

If the relationship in Eq. (13.13) is met, then Eq. (13.12) can be written as

$$i_d = (I_D + I_S)\left(\frac{v_d}{V_T} \right) \qquad \text{or} \qquad i_d = g_d v_d \qquad \text{and} \qquad i_D = I_D + g_d v_d \tag{13.14}$$

in which g_d is the **small-signal conductance** of the diode originally given in Eq. (13.6). Equation (13.14) states that the total diode current is the dc current I_D (at the Q-point) plus a small change in current ($i_d = g_d v_d$) that is linearly related to the small voltage change v_d across the diode.

The values of the **diode conductance g_d**, or the equivalent **diode resistance r_d**, are determined by the operating point of the diode as defined in Eq. (13.7):

$$g_d = \frac{I_D + I_S}{V_T} \cong \frac{I_D}{V_T} = 40 I_D \qquad \text{and} \qquad r_d = \frac{1}{g_d} \tag{13.15}$$

The diode and its corresponding small-signal model, represented by resistor r_d, are given in Fig. 13.8.

Equation (13.13) defines the requirement for small-signal operation of the diode. The shift in diode voltage away from the Q-point value must be much less than 50 mV. Choosing a factor of 10 as adequate to satisfy the inequality yields $v_d \leq 0.005$ V for small-signal operation. This is indeed a small voltage change.

Note, however, that the maximum small-signal change in diode voltage represents a significant change in diode current:

$$i_d = g_d v_d = 0.005 \text{ V} \frac{I_D}{0.0025 \text{ V}} = 0.2 I_D \tag{13.16}$$

The 5-mV change in diode voltage corresponds to a 20 percent change in the diode current! This large change results from the exponential relationship between voltage and current in the diode.

DESIGN NOTE

Small changes in diode current and voltage are related by the small-signal conductance of the diode

$$i_d = g_d v_d \qquad \text{where} \qquad g_d \cong \frac{I_D}{V_T} \cong 40 I_D$$

at room temperature. For small-signal operation,

$$|v_d| \leq 0.005 \text{ V} \qquad \text{and} \qquad |i_d| \leq 0.20 I_D$$

EXERCISE: Calculate the values of the diode resistance r_d for a diode with $I_S = 1$ fA operating at $I_D = 0$, 50 μA, 2 mA, and 3 A.

ANSWERS: 25.0 TΩ, 500 Ω, 12.5 Ω, 8.33 mΩ

EXERCISE: What is the small-signal diode resistance r_d at room temperature for $I_D = 1.5$ mA? What is the small-signal resistance of this diode at $T = 100°$C?

ANSWERS: 16.7 Ω, 21.4 Ω

13.5 SMALL-SIGNAL MODELS FOR BIPOLAR JUNCTION TRANSISTORS

Now that the concept of small-signal modeling has been introduced, we shall develop the small-signal model for the more complicated bipolar transistor. The BJT is a three-terminal device, and its small-signal model is based on the two-port network representation[1] shown in Fig. 13.10 for which the input port variables are v_{be} and i_b, and the output port variables are v_{ce} and i_c. A set of two-port equations in terms of these variables can be written as

$$\mathbf{i_b} = g_\pi \mathbf{v_{be}} + g_r \mathbf{v_{ce}}$$
$$\mathbf{i_c} = g_m \mathbf{v_{be}} + g_o \mathbf{v_{ce}} \tag{13.17}$$

The port variables in Fig. 13.10 can be considered to represent either the time-varying portion of the total voltages and currents or small changes in the total quantities away from the Q-point values.

$$v_{BE} = V_{BE} + v_{be} \qquad v_{CE} = V_{CE} + v_{ce}$$
$$i_B = I_B + i_b \qquad i_C = I_C + i_c$$

or

$$v_{be} = \Delta v_{BE} = v_{BE} - V_{BE} \qquad v_{ce} = \Delta v_{CE} = v_{CE} - V_{CE}$$
$$i_b = \Delta i_B = i_B - I_B \qquad i_c = \Delta i_C = i_C - I_C \tag{13.18}$$

---

[1] These equations actually represent a *y*-parameter two-port network.

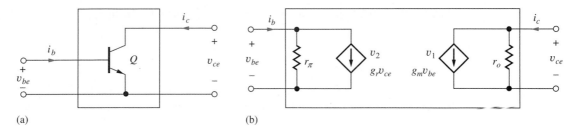

Figure 13.10 (a) Two-port representation of the *npn* transistor. (b) Two-port representation for the transistor in Fig. 13.10(a).

We can write the y-parameters in terms of small-signal voltages and currents or in terms of derivatives of the complete port variables, as in Eq. (13.19):

$$g_\pi = \left.\frac{\mathbf{i_b}}{\mathbf{v_{be}}}\right|_{\mathbf{v_{ce}}=0} = \left.\frac{\partial i_B}{\partial v_{BE}}\right|_{Q\text{-point}} \qquad g_r = \left.\frac{\mathbf{i_b}}{\mathbf{v_{ce}}}\right|_{\mathbf{v_{be}}=0} = \left.\frac{\partial i_B}{\partial v_{CE}}\right|_{Q\text{-point}}$$

$$g_m = \left.\frac{\mathbf{i_c}}{\mathbf{v_{be}}}\right|_{\mathbf{v_{ce}}=0} = \left.\frac{\partial i_C}{\partial v_{BE}}\right|_{Q\text{-point}} \qquad g_o = \left.\frac{\mathbf{i_c}}{\mathbf{v_{ce}}}\right|_{\mathbf{v_{be}}=0} = \left.\frac{\partial i_C}{\partial v_{CE}}\right|_{Q\text{-point}} \tag{13.19}$$

Because we have the transport model, Eq. (5.44), which expresses the BJT terminal currents in terms of the terminal voltages, as repeated in Eq. (13.20) for the forward-active region, we use the derivative formulation to determine the y-parameters for the transistor:

$$i_C = I_S\left[\exp\left(\frac{v_{BE}}{V_T}\right)\right]\left[1 + \frac{v_{CE}}{V_A}\right] \qquad i_B = \frac{i_C}{\beta_F} = \frac{I_S}{\beta_{FO}}\left[\exp\left(\frac{v_{BE}}{V_T}\right)\right] \tag{13.20}$$

$$\beta_F = \beta_{FO}\left[1 + \frac{v_{CE}}{V_A}\right]$$

Evaluating the various derivatives[2] of Eq. (13.20) yields the y-parameters for the BJT:

$$g_r = \left.\frac{\partial i_B}{\partial v_{CE}}\right|_{Q\text{-point}} = 0$$

$$g_m = \left.\frac{\partial i_C}{\partial v_{BE}}\right|_{Q\text{-point}} = \frac{I_S}{V_T}\left[\exp\left(\frac{v_{BE}}{V_T}\right)\right]\left[1 + \frac{v_{CE}}{V_A}\right]_{Q\text{-point}} = \frac{I_C}{V_T} \tag{13.21}$$

$$g_o = \left.\frac{\partial i_C}{\partial v_{CE}}\right|_{Q\text{-point}} = \frac{I_S}{V_A}\left[\exp\left(\frac{V_{BE}}{V_T}\right)\right] = \frac{I_C}{V_A + V_{CE}}$$

Calculation of g_π has been saved until last because it requires a bit more effort and the use of some new information. The current gain of a BJT is actually operating point-dependent, and this dependence should be included when evaluating the derivative needed for g_π:

$$g_\pi = \left.\frac{\partial i_B}{\partial v_{BE}}\right|_{Q\text{-point}} = \left[\frac{1}{\beta_F}\frac{\partial i_C}{\partial v_{BE}} - \frac{i_C}{\beta_F^2}\frac{\partial \beta_F}{\partial v_{BE}}\right]_{Q\text{-point}} = \left[\frac{1}{\beta_F}\frac{\partial i_C}{\partial v_{BE}} - \frac{i_C}{\beta_F^2}\frac{\partial \beta_F}{\partial i_C}\frac{\partial i_C}{\partial v_{BE}}\right]_{Q\text{-point}} \tag{13.22}$$

Factoring out the first term:

$$g_\pi = \frac{1}{\beta_F}\frac{\partial i_C}{\partial v_{BE}}\left[1 - \frac{i_C}{\beta_F}\frac{\partial \beta_F}{\partial i_C}\right]_{Q\text{-point}} = \frac{I_C}{\beta_F V_T}\left[1 - \left(\frac{i_C}{\beta_F}\frac{\partial \beta_F}{\partial i_C}\right)_{Q\text{-point}}\right] \tag{13.23}$$

---

[2] We could equally well have taken the direct approach used in analysis of the diode.

Finally, Eq. (13.23) can be simplified by defining a new parameter β_o:

$$g_\pi = \frac{I_C}{\beta_o V_T} \qquad \text{where } \beta_o = \frac{\beta_F}{\left[1 - I_C \left(\frac{1}{\beta_F} \frac{\partial \beta_F}{\partial i_C}\right)_{\text{Q-point}}\right]} \qquad (13.24)$$

β_o represents the **small-signal common-emitter current gain** of the bipolar transistor.

13.5.1 THE HYBRID-PI MODEL

The standard representation of the basic hybrid-pi small-signal model appears in Fig. 13.11, and the expressions for the model elements are given in Eq. (13.25). These results will be used throughout the rest of the text and should be committed to memory.

$$\text{Transconductance:} \quad g_m = \frac{I_C}{V_T} \cong 40 I_C$$

$$\text{Input resistance:} \quad r_\pi = \frac{\beta_o V_T}{I_C} = \frac{\beta_o}{g_m} \qquad (13.25)$$

$$\text{Output resistance:} \quad r_o = \frac{1}{g_o} = \frac{V_A + V_{CE}}{I_C} \cong \frac{V_A}{I_C}$$

Arguably the most important small-signal parameter is transconductance g_m. The transconductance characterizes how the collector current changes in response to a change in the base-emitter voltage, thereby modeling the forward gain of the device. Here again we see the fundamental voltage-controlled current behavior of the bipolar transistor. At room temperature, $V_T \cong 0.025$ mV, and transconductance $g_m \cong 40 I_C$. Also, collector-emitter voltage V_{CE} is typically much less than Early voltage V_A, so we can simplify the expression for the output resistance: $r_o \cong V_A/I_C$.

When we change the base-emitter voltage and hence the collector current, we must also supply a change in base current, and resistor r_π characterizes the relationship between changes in i_B and V_{BE}. Similarly, when the collector-emitter voltage changes slightly, the collector current changes, and resistor r_o characterizes the relationship between changes in i_c and v_{ce}.

The two-port representation in Fig. 13.11 using these symbols shows the intrinsic low-frequency **hybrid-pi small-signal model** for the bipolar transistor. Additional elements will be added to model frequency dependencies in Chapter 16.

From Eq. (13.25), we see that the values of the small-signal parameters are controlled explicitly by our choice of Q-point. Transconductance g_m is directly proportional to the collector current of the bipolar transistor, whereas input resistance r_π and output resistance r_o are both inversely proportional to the collector current. The output resistance exhibits a weak dependence on collector-emitter voltage (generally $V_{CE} \ll V_A$). Note that these parameters are independent of the geometry of the BJT. For example, small high-frequency transistors or large-geometry power devices all have the same value of g_m for a given collector current.

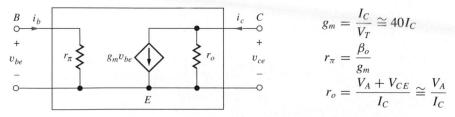

Figure 13.11 Hybrid-pi small-signal model for the intrinsic bipolar transistor.

13.5.2 GRAPHICAL INTERPRETATION OF THE TRANSCONDUCTANCE

Figure 13.12 depicts the diode-like exponential relation between total collector current i_C and total base-emitter voltage v_{BE} in the bipolar transistor. Transconductance g_m represents the slope of the $i_C - v_{BE}$ characteristic at the given operating point (Q-point). For a small increase v_{be} above the Q-point voltage V_{BE}, a small corresponding increase i_c occurs above the Q-point current I_C. When the small-signal condition $v_{be} \leq 5$ mV is met, these two changes are linearly related by the transconductance: $i_c = g_m v_{be}$.

13.5.3 SMALL-SIGNAL CURRENT GAIN

Two important auxiliary relationships also exist between the small-signal parameters. It can be seen in Eq. (13.25) that the parameters g_m and r_π are related by the small-signal current gain β_o:

$$\beta_o = g_m r_\pi \qquad (13.26)$$

As mentioned before, the dc current gain in a real transistor is not constant but is a function of operating current, as indicated in Fig. 13.13. From this figure, we see that

$$\frac{\partial \beta_F}{\partial i_C} > 0 \text{ for } i_C < I_M \qquad \text{and} \qquad \frac{\partial \beta_F}{\partial i_C} < 0 \text{ for } i_C > I_M$$

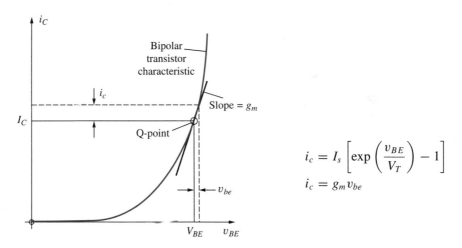

$$i_c = I_s \left[\exp\left(\frac{v_{BE}}{V_T}\right) - 1 \right]$$
$$i_c = g_m v_{be}$$

Figure 13.12 The relationship between small increases in base-emitter voltage and collector current above the BJT operating point (I_C, V_{CE}). For small changes $i_c = g_m v_{be}$.

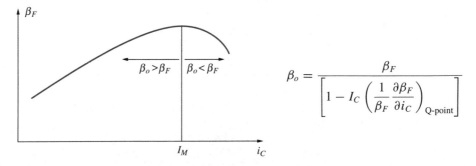

$$\beta_o = \frac{\beta_F}{\left[1 - I_C \left(\dfrac{1}{\beta_F} \dfrac{\partial \beta_F}{\partial i_C} \right)_{\text{Q-point}} \right]}$$

Figure 13.13 dc current gain versus current for the BJT.

where I_M is the collector current at which β_F is maximum. Thus, for the **small-signal current gain** defined by

$$\beta_o = \frac{\beta_F}{\left[1 - I_C \left(\frac{1}{\beta_F} \frac{\partial \beta_F}{\partial i_C} \right)_{\text{Q-point}} \right]} \tag{13.27}$$

$\beta_o > \beta_F$ for $i_C < I_M$, and $\beta_o < \beta_F$ for $i_C > I_M$. That is, the ac current gain β_o exceeds the dc current gain β_F when the collector current is below I_M and is smaller than β_F when I_C exceeds I_M. In practice, the difference between β_F and β_o is usually ignored, and β_F and β_o are commonly assumed to be the same.

13.5.4 THE INTRINSIC VOLTAGE GAIN OF THE BJT

The second important auxiliary relationship is given by the **intrinsic voltage gain** μ_f, which is equal to the product of g_m and r_o:

$$\mu_f = g_m r_o = \frac{V_A + V_{CE}}{V_T} \cong \frac{V_A}{V_T} \qquad \text{for} \qquad V_{CS} \ll V_A \tag{13.28}$$

From Eq. (13.28), the BJT amplification factor can be seen to be almost independent of operating point for $V_{CE} \ll V_A$. At room temperature $\mu_f \cong 40 V_A$.

We shall find that the amplification factor μ_f plays an important role in circuit design, and it appears often in the analysis of amplifier circuits. Parameter μ_f represents the maximum voltage gain that the individual transistor can provide and is also referred to as the **amplification factor** of the device. For V_A ranging from 25 V to 100 V, μ_f ranges from 1000 to 4000.

DESIGN NOTE

Remember, the voltage gain of a single transistor amplifier cannot exceed the transistor's intrinsic voltage gain μ_f, which ranges from 1000 to 4000 for the bipolar transistor.

$$\mu_f = \frac{V_A + V_{CE}}{V_T} \cong \frac{V_A}{V_T}$$

Table 13.1 displays examples of the variation of the small-signal parameters with operating point. The values of g_m, r_π, and r_o can each be varied over many orders of magnitude by changing the value of the dc collector current corresponding to the Q-point. Note that μ_f does not change with the choice of operating point. As we see later in this chapter, this is a very significant difference between the BJT and FET.

TABLE 13.1

BJT Small-Signal Parameters versus Current: $\beta_o = 100$, $V_A = 75$ V, $V_{CE} = 10$ V

I_C	g_M	r_π	r_o	μ_f
1 μA	4×10^{-5} S	2.5 MΩ	85 MΩ	3400
10 μA	4×10^{-4} S	250 kΩ	8.5 MΩ	3400
100 μA	0.004 S	25.0 kΩ	850 kΩ	3400
1 mA	0.04 S	2.5 kΩ	85 kΩ	3400
10 mA	0.40 S	250 Ω	8.5 kΩ	3400

It is important to realize that although we developed the small-signal model of the BJT based on analysis of the transistor oriented in the common-emitter configuration in Fig. 13.10, the resulting hybrid-pi model can actually be used in the analysis of any circuit topology. This point becomes clearer in Chapter 14.

EXERCISE: Calculate the values of g_m, r_π, r_o, and μ_f for a bipolar transistor with $\beta_o = 75$ and $V_A = 60$ V operating at a Q-point of (50 μA, 5 V).

ANSWERS: 2.00 mS, 37.5 kΩ, 1.30 MΩ, 2600

EXERCISE: Calculate the values of g_m, r_π, r_o, and μ_f for a bipolar transistor with $\beta_o = 50$ and $V_A = 75$ V operating at a Q-point of (250 μA, 15 V).

ANSWERS: 10.0 mS, 5.00 kΩ, 360 kΩ, 3600

EXERCISE: Use graphical analysis to find values of β_{FO}, g_m, β_o, and r_o at the Q-point for the transistor in Fig. 13.1(b). Calculate the value of r_π.

ANSWERS: 100, 62.5 mS, 100, ∞; 1.60 kΩ

13.5.5 EQUIVALENT FORMS OF THE SMALL-SIGNAL MODEL

The small-signal model in Fig. 13.14 includes the voltage-controlled current source $g_m v_{be}$. It is often useful in circuit analysis to transform this model into a current-controlled source. Recognizing that the voltage $\mathbf{v_{be}}$ in Fig. 13.13 can be written in terms of the current $\mathbf{i_b}$ as $\mathbf{v_{be}} = \mathbf{i_b} r_\pi$, the current-controlled source can be rewritten as

$$g_m \mathbf{v_{be}} = g_m r_\pi \mathbf{i_b} = \beta_o \mathbf{i_b} \qquad \text{where} \qquad \beta_o = g_m r_\pi \qquad (13.29)$$

Figure 13.14(a) and (b) shows the two equivalent forms of the small-signal BJT model. The model in Fig. 13.14(a) recognizes the fundamental voltage-controlled current source nature of the transistor that is explicit in the transport model. From the second model, Fig. 13.14(b), we see that

$$\mathbf{i_c} = \beta_o \mathbf{i_b} + \frac{\mathbf{v_{ce}}}{r_o} \cong \beta_o \mathbf{i_b} \qquad (13.30)$$

which demonstrates the auxiliary relationship that $\mathbf{i_c} \cong \beta_o \mathbf{i_b}$ in the active region of operation. For the most typical case, $\mathbf{v_{ce}}/r_o \ll \beta_o \mathbf{i_b}$. Thus, the basic relationship $i_C = \beta i_B$ is useful for both ac and dc analysis when the BJT is operating in the forward-active region. We will find that sometimes circuit analysis is more easily performed using the model in Fig. 13.14(a), and at other times more easily performed using the model in Fig. 13.14(b).[3]

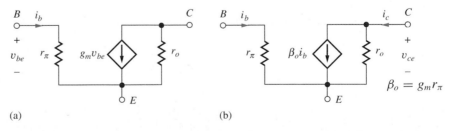

(a)   (b)

Figure 13.14 Two equivalent forms of the BJT small-signal model: (a) voltage-controlled current source model, and (b) current-controlled current source model.

---

[3] An alternative model, called the T-model, is in Prob. 13.59.

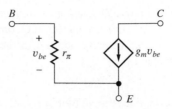

Figure 13.15 Simplified hybrid pi model in which r_o is neglected.

13.5.6 SIMPLIFIED HYBRID PI MODEL

As we investigate circuit behavior in more detail, we will find that output resistance r_o often has a relatively minor effect on circuit performance, especially on voltage gain, and we can often greatly simplify our circuit analysis if we neglect the output resistance in our model as shown in Fig. 13.15. Generally, we can make this simplification if the voltage gain of the circuit is much less than the intrinsic voltage gain μ_f. So our approach will be to neglect r_o, then calculate the voltage gain, and see if the result is consistent with the assumption that the voltage gain is much less than μ_f. However, r_o can have a much greater impact on amplifier output resistance calculations, and we must often add it back into the model in order to get nontrivial results. We will see examples of this as we proceed through Chapters 13 and 14.

DESIGN NOTE

Output resistance r_o can be neglected in calculations of voltage gain A_v as long as $A_v \ll \mu_f$.

13.5.7 DEFINITION OF A SMALL SIGNAL FOR THE BIPOLAR TRANSISTOR

For small-signal operation, we want the relationship between changes in voltages and currents to be linear. We can find the constraints on the BJT corresponding to small-signal operation using the simplified transport model for the total collector current of the transistor in the active region:

$$i_C = I_S \left[\exp \left(\frac{v_{BE}}{V_T} \right) \right] = I_S \left[\exp \left(\frac{V_{BE} + v_{be}}{V_T} \right) \right] \tag{13.31}$$

Rewriting the exponential as a product,

$$i_C = I_C + i_c = \left[I_S \exp \left(\frac{V_{BE}}{V_T} \right) \right] \left[\exp \left(\frac{v_{be}}{V_T} \right) \right] = I_C \left[\exp \left(\frac{v_{be}}{V_T} \right) \right] \tag{13.32}$$

in which it has been recognized that the collector current I_C is given by

$$I_C = I_S \exp \left(\frac{V_{BE}}{V_T} \right) \tag{13.33}$$

Now, expanding the remaining exponential in Eq. (13.32), its Maclaurin's series yields

$$i_C = I_C \left[1 + \frac{v_{be}}{V_T} + \frac{1}{2} \left(\frac{v_{be}}{V_T} \right)^2 + \frac{1}{6} \left(\frac{v_{be}}{V_T} \right)^3 + \cdots \right] \tag{13.34}$$

Recognizing $i_c = i_C - I_C$ yields

$$i_c = I_C \left[\frac{v_{be}}{V_T} + \frac{1}{2} \left(\frac{v_{be}}{V_T} \right)^2 + \frac{1}{6} \left(\frac{v_{be}}{V_T} \right)^3 + \cdots \right] \tag{13.35}$$

Linearity requires that i_c be proportional to v_{be}, so we must have

$$\frac{1}{2} \left(\frac{v_{be}}{V_T} \right)^2 \ll \frac{v_{be}}{V_T} \qquad \text{or} \qquad v_{be} \ll 2V_T \tag{13.36}$$

From Eq. (13.36), we see that small-signal operation requires the signal applied to the base-emitter junction to be much less than twice the thermal voltage, which is 50 mV at room temperature. In this book, we assume that a factor of 10 satisfies the condition in Eq. (13.36), and

$$|v_{be}| \leq 0.005 \text{ V} \tag{13.37}$$

is our definition of a **small signal for the BJT.** If the condition in Eq. (13.36) is met, then Eq. (13.34) can be approximated as

$$i_C \cong I_C \left[1 + \frac{v_{be}}{V_T} \right] = I_C + \frac{I_C}{V_T} v_{be} = I_C + g_m v_{be} \tag{13.38}$$

and the change in i_C is directly proportional to the change in v_{BE} (i.e., $i_c = g_m v_{be}$). The constant of proportionality is the transconductance g_m. Note that the quadratic, cubic, and higher-order powers of v_{be} in Eq. (13.35) are sources of the harmonic distortion that was discussed in Sec. 10.4.

From Eq. (13.37), the signal developed across the base-emitter junction must be no larger than 5 mV to qualify as a small signal! This is indeed small. But note well: We must not infer that signals at other points in the circuit need be small. Referring back to Fig. 13.1, we can see that a 16-mV p-p signal v_{be} generates a 3.3-V p-p signal at the collector. This is fortunate because we often want linear amplifiers to develop signals that are many volts in amplitude.

Let us now explore the change in collector current i_c that corresponds to small-signal operation. Using Eq. (13.38),

$$\frac{i_c}{I_C} = \frac{g_m v_{be}}{I_C} = \frac{v_{be}}{V_T} \leq \frac{0.005}{0.025} = 0.200 \tag{13.39}$$

A 5-mV change in v_{BE} corresponds to a 20 percent deviation in i_C from its Q-point value as well as a 20 percent change in i_E because $\alpha_F \cong 1$. Some authors prefer to permit $|v_{be}| \leq 10$ mV, which corresponds to a 40 percent change in i_C from the Q-point value. In either case, relatively large changes in voltage can occur at the collector and/or emitter terminals of the transistor when the signal currents i_c and i_e flow through resistors external to the transistor.

The strict small-signal guidelines introduced above are frequently violated in practice. The designer must accept the trade-off between a larger signal amplitude and a higher level of distortion. As we move beyond the small-signal limit, our small-signal analysis becomes more approximate. However, our hand analysis still represents a useful estimate of circuit performance that we can then refine with detailed transient simulation.

DESIGN NOTE

The small-signal limit for the bipolar transistor is set by

$$|v_{be}| \leq 0.005 \text{ V} \qquad \text{and} \qquad |i_c| \leq 0.2 I_C$$

EXERCISE: Does the amplitude of the signal in Figs. 13.1(a) and 13.1(b) satisfy the requirements for small-signal operation?

ANSWER: No, $v_{be} = 8$ mV exceeds our definition of a small signal.

13.5.8 SMALL-SIGNAL MODEL FOR THE *pnp* TRANSISTOR

The small-signal model for the *pnp* transistor is identical to that of the *npn* transistor. At first glance, this fact is surprising to most people because the dc currents flow in opposite directions. The circuits in Fig. 13.16 will be used to help explain this situation.

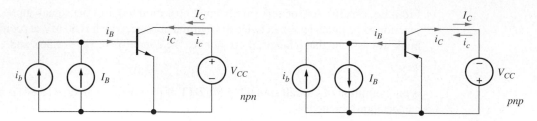

Figure 13.16 dc bias and signal currents for *npn* and *pnp* transistors.

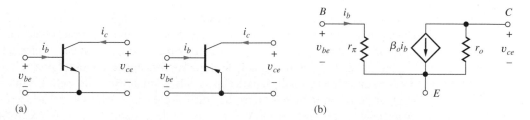

Figure 13.17 (a) Two-port notations for *npn* and *pnp* transistors. (b) The small-signal models are identical.

In Fig. 13.16, the *npn* and *pnp* transistors are each biased by dc current source I_B, establishing the Q-point current $I_C = \beta_F I_B$. In each case a signal current i_b is also injected *into* the base. For the *npn* transistor, the total base and collector currents are (for $\beta_o = \beta_F$):

$$i_B = I_B + i_b \quad \text{and} \quad i_C = I_C + i_c = \beta_F I_B + \beta_F i_b \qquad (13.40)$$

An increase in base current of the *npn* transistor causes an increase in current entering the collector terminal.

For the *pnp* transistor,

$$i_B = I_B - i_b \quad \text{and} \quad i_C = I_C - i_c = \beta_F I_B - \beta_F i_b \qquad (13.41)$$

The signal current injected into the base of the *pnp* transistor causes a decrease in the total collector current, which is again equivalent to an increase in the signal current entering the collector. Thus, for both the *npn* and *pnp* transistors, a signal current injected into the base causes a signal current to enter the collector, and the polarities of the current-controlled source in the small-signal model are identical, as in Fig. 13.17.

13.5.9 AC ANALYSIS VERSUS TRANSIENT ANALYSIS IN SPICE

Differences between the ac and transient analysis modes in SPICE are a constant source of confusion to new users of electronic simulation tools. ac analysis mirrors our hand calculations with small-signal models. In the SPICE calculations, the transistors are automatically replaced with their small-signal models, and a linear circuit analysis is then performed. On the other hand, SPICE transient analysis provides a time-domain representation similar to what we will see when we build the circuit and look at waveforms with an oscilloscope. The built-in models in SPICE attempt to fully account for the nonlinear behavior of the devices. If the small-signal limits are violated, distorted waveforms will result.

Once the circuit is converted to a linearized version, the magnitudes of the sources applied have no small-signal constraints. We typically use a value of 1 V or 1 A for convenience in ac analysis. On the other hand, this large a signal would cause significant distortion in many transient simulations.

13.6 THE COMMON-EMITTER (C-E) AMPLIFIER

Now we are in a position to analyze the small-signal characteristics of the complete **common-emitter (C-E) amplifier** shown in Fig. 13.18(a). The ac equivalent circuit of Fig. 13.18(b) is constructed earlier (Ex. 13.1) by assuming that the capacitors all have zero impedance at the signal frequency

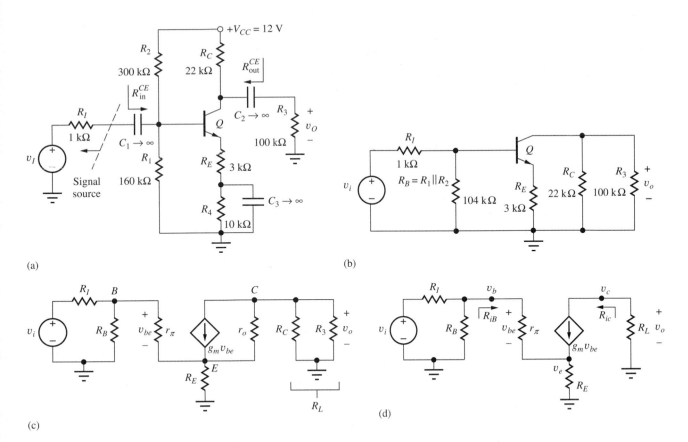

Figure 13.18 (a) Common-emitter amplifier circuit employing a bipolar transistor. (b) ac Equivalent circuit for the common-emitter amplifier in part (a). The common-emitter connection should now be evident. (c) ac Equivalent circuit with the bipolar transistor replaced by its small-signal model. (d) Final equivalent circuit for ac analysis of the common-emitter amplifier.

and the dc voltage source represents an ac ground. For simplicity, we assume that we have found the Q-point and know the values of I_C and V_{CE}. In Fig. 13.18(b), resistor R_B represents the parallel combination of the two base bias resistors R_1 and R_2,

$$R_B = R_1 \| R_2 \tag{13.42}$$

and R_4 is eliminated by bypass capacitor C_3.

Before we can develop an expression for the voltage gain of the amplifier, the transistor must be replaced by its small-signal model as in Fig. 13.18(c). A final simplification appears in Fig. 13.18(d), in which the resistor R_L represents the total equivalent load resistance on the transistor, the parallel combination of R_C and R_3:

$$R_L = R_C \| R_3 \tag{13.43}$$

Note that transistor output resistance r_o has been neglected in the final circuit in Fig. 13.18(d) as discussed in Section 13.5.5.

In Fig. 13.18(b) through (d), the reason why this amplifier configuration is called a common-emitter amplifier is apparent. The emitter terminal represents the common connection between the amplifier input and output ports. The input signal is applied to the transistor's base, the output signal appears at the collector, and both the input and output signals are referenced to the (common) emitter terminal (through R_E).

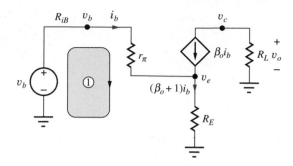

Figure 13.19 Simplified circuit model for finding the common-emitter terminal voltage gain A_{vt}^{CE} and input resistance R_{iB}.

13.6.1 TERMINAL VOLTAGE GAIN

Now we are ready to develop an expression for the overall gain of the amplifier from signal source v_i to the output voltage across resistor R_3. The voltage gain can be written as

$$A_v^{CE} = \frac{v_o}{v_i} = \left(\frac{v_o}{v_b}\right)\left(\frac{v_b}{v_i}\right) = A_{vt}^{CE}\left(\frac{v_b}{v_i}\right) \qquad \text{where} \qquad A_{vt}^{CE} = \left(\frac{v_o}{v_b}\right) \tag{13.44}$$

A_{vt}^{CE} represents the voltage gain between the base and collector terminals of the transistor, the **"terminal gain."** We will first find expressions for terminal gain A_{vt}^{CE} as well as the input resistance at the base of the transistor. Then we can relate v_b to v_i to find the overall voltage gain.

In Fig. 13.19, the BJT is replaced with its small-signal model, and the base terminal of the transistor is driven by test source v_b. Note that the small-signal model has been changed to its current-controlled form, and r_o is neglected as discussed before. Output voltage v_o is given by

$$v_o = -\beta_o i_b R_L \tag{13.45}$$

We can relate i_b to base voltage v_b by writing an equation around loop 1:

$$v_b = i_b r_\pi + (i_b + \beta_o i_b)R_E = i_b\left[r_\pi + (\beta_o + 1)R_E\right] \tag{13.46}$$

Solving for i_b and substituting the result in Eq. (13.45) yields

$$A_{vt}^{CE} = \frac{v_o}{v_b} = -\frac{\beta_o R_L}{r_\pi + (\beta_o + 1)R_E} \cong -\frac{g_m R_L}{1 + g_m R_E} \tag{13.47}$$

in which the approximation assumes $\beta_o \gg 1$ and uses $\beta_o = g_m r_\pi$.

The minus sign indicates that the common-emitter stage is an inverting amplifier in which the input and output are 180° out of phase. The gain is proportional to the product of the transistor transconductance g_m and load resistor R_L. This product places an upper bound on the gain of the amplifier, and we will encounter the $g_m R_L$ product over and over again as we study transistor amplifiers. We will explore gain expression (Eq. 13.47) in more detail shortly.

13.6.2 INPUT RESISTANCE

The resistance looking into the base terminal R_{iB} in Fig. 13.19 can easily be found by rearranging Eq. (13.46). The input resistance is simply the ratio of v_b and i_b,

$$R_{iB} = \frac{v_b}{i_b} = r_\pi + (\beta_o + 1)R_E \cong r_\pi(1 + g_m R_E) \tag{13.48}$$

in which the final approximation again assumes $\beta_o \gg 1$ and uses $\beta_o = g_m r_\pi$. The input resistance looking into the base of the transistor is equal to r_π plus the resistance reflected into the base by R_E. The effective value of R_E is increased by the current gain $(\beta_0 + 1)$.

13.6.3 SIGNAL SOURCE VOLTAGE GAIN

The overall voltage gain A_v^{CE} of the amplifier, including the effect of source resistance R_1, can now be found using the input resistance and terminal gain expressions. Voltage v_b at the base of the bipolar transistor in Fig. 13.18(d) is related to v_i by

$$v_b = v_i \frac{R_B \| R_{iB}}{R_I + (R_B \| R_{iB})} \tag{13.49}$$

Combining Eqs. (13.44), (13.47), and (13.49), yields a general expression for the overall voltage gain of the common-emitter amplifier:

$$A_v^{CE} = A_{vt}^{CE} \left(\frac{v_b}{v_i} \right) = -\left(\frac{g_m R_L}{1 + g_m R_E} \right) \left[\frac{R_B \| R_{iB}}{R_I + (R_B \| R_{iB})} \right] \tag{13.50}$$

In this expression, we see that the overall voltage gain is equal to the terminal gain A_{vt}^{CE} reduced by the voltage division between R_I and the equivalent resistance at the base of the transistor. Terminal gain A_{vt}^{CE} places an upper limit on the voltage gain since the voltage division factor will be less than one.

13.7 IMPORTANT LIMITS AND MODEL SIMPLIFICATIONS

We now explore the limits to the voltage gain of common-emitter amplifiers using model simplifications for large emitter resistance and zero emitter resistance. First, we will assume that the source resistance is small enough that $R_I \ll R_B \| R_{iB}$ so that

$$A_v^{CE} \cong A_{vt}^{CE} = -\frac{g_m R_L}{1 + g_m R_E} \qquad \text{for } R_I \ll R_B \| R_{iB} \tag{13.51}$$

This approximation is equivalent to saying that the total input signal appears at the base of the transistor.

13.7.1 ZERO RESISTANCE IN THE EMITTER

In order to achieve as large a gain as possible, we need to make the denominator in Eq. (13.51) as small as possible, and this is achieved by setting $R_E = 0$. The gain is then

$$A_v^{CE} \cong -g_m R_L = -g_m (R_C \| R_3) \tag{13.52}$$

Equation (13.52) states that the terminal voltage gain of the common-emitter stage is equal to the product of the transistor's transconductance g_m and load resistance R_L, and the minus sign indicates that the output voltage is "inverted" or 180° out of phase with respect to the input. Equation (13.52) places an upper limit on the gain we can achieve from a common-emitter amplifier with an external load resistor. The approximations that led to Eq. (13.52) are equivalent to saying that the total input signal appears across r_π as shown in Fig. 13.20.

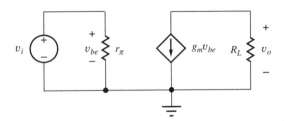

Figure 13.20 Simplified circuit corresponding to Eq. (13.52) with $R_E = 0$.

EXERCISE: What is the terminal voltage gain A_{vt} ($-g_m R_L$) for the amplifier in Ex. 13.3? The actual gain of the amplifier was only -130. Where is most of this gain being lost?

ANSWER: -222; A significant portion, 42 percent, of the input signal is lost by voltage division between the source resistance R_I and the amplifier input resistance.

13.7.2 A DESIGN GUIDE FOR THE COMMON-EMITTER AMPLIFIER WITH $R_E = 0$

During design of most amplifiers, we try to achieve $R_3 \gg R_C$. For these conditions, the load resistance on the collector of the transistor is approximately equal to R_C, and Eq. (13.52) can be reduced to

$$A_v^{CE} \cong A_{vt}^{CE} = -g_m R_C = -\frac{I_C R_C}{V_T} \tag{13.53}$$

The $I_C R_C$ product represents the dc voltage dropped across the collector resistor R_C. This voltage typically ranges between one-fourth and three-fourths of V_{CC}. Assuming $I_C R_C = \zeta V_{CC}$ with $0 \leq \zeta \leq 1$, and remembering that the reciprocal of V_T is 40 V^{-1}, Eq. (13.53) can be rewritten as

$$A_v^{CE} \cong -\frac{I_C R_C}{V_T} \cong -40\zeta V_{CC} \quad \text{with} \quad 0 \leq \zeta \leq 1 \tag{13.54}$$

A common design allocates 1/3 of the power supply voltage across the collector resistor. For this case, $\zeta = 1/3$, $I_C R_C = V_{CC}/3$, and Eq. (13.54) becomes $A_v \cong 13V_{CC}$. To further account for the approximations that led to this result and produce a number that is easy to remember, we will use this expression for our voltage gain estimate:

$$A_v^{CE} \cong -10V_{CC} \quad \text{for} \quad R_E = 0 \tag{13.55}$$

Equation (13.55) represents our basic rule-of-thumb for the design of resistively loaded common-emitter amplifiers; that is, the magnitude of the voltage gain is approximately equal to 10 times the power supply voltage.[4] We need to know only the supply voltage to make a rough prediction of the gain of the common-emitter amplifier. For a C-E amplifier operating from a 15-V power supply, we estimate the gain to be -150 or 44 dB; a C-E amplifier with a 10-V supply would be expected to produce a gain of approximately -100 or 40 dB. Note well: A nonzero value for R_E will reduce the gain below the $10 V_{CC}$ estimate.

DESIGN NOTE

The magnitude of the voltage gain of a resistively loaded common-emitter amplifier <u>with zero emitter resistance</u> is approximately equal to 10 times the power supply voltage.

$$A_v^{CE} \cong -10V_{CC} \quad \text{for} \quad R_E = 0$$

This result represents an excellent way to quickly check the validity of more detailed calculations. Remember that the rule-of-thumb estimate is not going to be exact, but will predict the order of magnitude of the gain, typically within a factor of two or so.

---

[4] For dual power supplies, the corresponding estimate would be $A_V = -10(V_{CC} + V_{EE})$.

13.7.3 COMMON-EMITTER VOLTAGE GAIN FOR LARGE EMITTER RESISTANCE

The presence of a nonzero value of emitter resistor R_E reduces the gain below that given by Eq. (13.52), and another very useful simplification occurs when R_E is large enough so that the $g_m R_E$ product is much larger than one:

$$A_{vt}^{CE} = -\frac{g_m R_L}{1 + g_m R_E} \cong -\frac{R_L}{R_E} \qquad \text{for} \qquad g_m R_E \gg 1 \qquad (13.56)$$

The gain is now set by the ratio of the load resistor R_L and emitter resistor R_E. This is an extremely useful result because the gain is now independent of the transistor characteristics that vary widely from device to device. The result in Eq. (13.56) is very similar to the one obtained for the op-amp inverting amplifier circuit and is a result of feedback introduced by resistor R_E.

Achieving the simplification in Eq. (13.56) requires $g_m R_E \gg 1$. We can relate this product to the dc bias voltage developed across R_E:

$$g_m R_E = \frac{I_C R_E}{V_T} = \alpha_F \frac{I_E R_E}{V_T} \cong \frac{I_E R_E}{V_T} \qquad \text{and we need} \qquad I_E R_E \gg V_T \qquad (13.57)$$

$I_E R_E$ represents the dc voltage drop across emitter resistor R_E and must be much greater than 25 mV, for example 0.250 V, a value that is easily achieved.

13.7.4 SMALL-SIGNAL LIMIT FOR THE COMMON-EMITTER AMPLIFIER

An important additional benefit of adding resistor R_E to the circuit is to increase the allowed size of the input signal v_b at the base. For small-signal operation, the magnitude of the base-emitter voltage v_{be}, developed across r_π in the small-signal model, must be less than 5 mV (you may wish to review Sec. 13.5.6). This voltage can be found using the input current $\mathbf{i}_b$ from Eq. (13.46):

$$\mathbf{v_{be}} = \mathbf{i}_b r_\pi = v_b \frac{r_\pi}{r_\pi + (\beta_o + 1) R_E} \cong \frac{\mathbf{v_b}}{1 + g_m R_E} \qquad (13.58)$$

The approximation requires $\beta_o \gg 1$. Requiring $|v_{be}|$ in Eq. (13.58) to be less than 5 mV gives

$$|v_b| \leq 0.005(1 + g_m R_E) \text{ V} \qquad (13.59)$$

If $g_m R_E \gg 1$, then v_b can be increased well beyond the 5-mV limit.

EXAMPLE 13.3   **VOLTAGE GAIN OF A COMMON-EMITTER AMPLIFIER**

In this example, we find the small-signal parameters of the bipolar transistor and then calculate the voltage gain of a common-emitter amplifier.

PROBLEM   Calculate the voltage gain of the common-emitter amplifier in Fig. 13.18 if the transistor has $\beta_F = 100$, $V_A = 75\text{V}$, $\lambda = 0.0133 \text{ V}^{-1}$, and the Q-point is (0.245 mA, 3.39 V). What is the maximum value of v_i that satisfies the small-signal assumptions?

SOLUTION   **Known Information and Given Data:** Common-emitter amplifier with its ac equivalent circuit given in Fig. 13.18; $\beta_F = 100$ and $V_A = 75$ V; the Q-point is (0.245 mA, 3.39 V); $R_1 = 1 \text{ k}\Omega$, $R_1 = 160 \text{ k}\Omega$, $R_2 = 300 \text{ k}\Omega$, $Rc = 22 \text{ k}\Omega$, $R_E = 3 \text{ k}\Omega$, $R_4 = 100 \text{ k}\Omega$ and $R_3 = 100 \text{ k}\Omega$.

Unknowns: Small-signal parameters of the transistor; voltage gain A_v; small-signal limit for the value of v_i

Approach: Use the Q-point information to find r_π. Use the calculated and given values to evaluate the voltage gain expression in Eq. (13.50).

Assumptions: The transistor is in the active region, and $\beta_o = \beta_F$. The signal amplitudes are low enough to be considered as small signals. Assume r_o can be neglected.

Analysis: To evaluate Eq. (13.50),

$$A_v = -\left(\frac{g_m R_L}{1 + g_m R_E}\right)\left[\frac{R_B \| R_{iB}}{R_I + (R_B \| R_{iB})}\right] \quad \text{with} \quad R_B = R_1 \| R_2 \quad \text{and} \quad R_{iB} = r_\pi + (\beta_o + 1)R_E$$

the values of the various resistors and small-signal model parameters are required. We have

$$g_m = 40 I_C = 40(0.245 \text{ mA}) = 9.80 \text{ mS} \qquad r_\pi = \frac{\beta_o V_T}{I_C} = \frac{100(0.025 \text{ V})}{0.245 \text{ mA}} = 10.2 \text{ k}\Omega$$

$$r_o = \frac{V_A + V_{CE}}{I_C} = \frac{75 \text{ V} + 3.39 \text{ V}}{0.245 \text{ mA}} \qquad R_{iB} = r_\pi + (\beta_o + 1)R_{EI} = 313 \text{ k}\Omega$$

$$R_B = R_1 \| R_2 = 104 \text{ k}\Omega \qquad R_L = R_c \| R_3 = 18.0 \text{ k}\Omega$$

Using these values,

$$A_v = -\left(\frac{9.80 \text{ mS}(18.0 \text{ k}\Omega)}{1 + 9.80 \text{ mS}(3.0 \text{ k}\Omega)}\right)\left[\frac{104 \text{ k}\Omega \| 313 \text{ k}\Omega}{1 \text{ k}\Omega + (104 \text{ k}\Omega \| 313 \text{ k}\Omega)}\right] = -5.80(0.987) = -5.72$$

Thus, the common-emitter amplifier in Fig. 13.18 provides a small-signal voltage gain $A_v = -5.72$ or 15.1 dB.

Small-signal operation requires $|v_{be}| \leq 0.005$ V. Based on Fig. 13.18, the base-emitter signal voltage can be related to v_i by

$$v_{be} = v_b \frac{r_\pi}{r_\pi + (\beta_o + 1)R_E} = v_i \left[\frac{R_B \| R_{iB}}{R_I + R_B \| R_{iB}}\right]\left[\frac{r_\pi}{r_\pi + (\beta_o + 1)R_E}\right]$$

so that

$$v_i \leq (0.005 \text{ V})\left[\frac{R_I + (R_B \| R_{iB})}{R_B \| R_{iB}}\right]\left[\frac{r_\pi + (\beta_o + 1)R_E}{r_\pi}\right]$$

$$v_i \leq (0.005 \text{ V})\left[\frac{1 \text{ k}\Omega + (104 \text{ k}\Omega \| 313 \text{ k}\Omega)}{104 \text{ k}\Omega \| 313 \text{ k}\Omega}\right]\left[\frac{10.2 \text{ k}\Omega + 101(3 \text{ k}\Omega)}{10.2 \text{ k}\Omega}\right] = 0.155 \text{ V}$$

Check of Results: We have found the required information. The amplification factor is $\mu_f = g_m r_o = (9.80 \text{ mS})(320 \text{ k}\Omega) = 3140$. The magnitude of the voltage gain of a single-transistor amplifier cannot exceed this value. Using the result from Eq. (13.56), we estimate the gain to be $A_v^{CE} = -18 \text{ k}\Omega/3 \text{ k}\Omega = -6$. Our answer satisfies both these checks.

Discussion: Note that the value of the voltage gain ($A_v = -5.72$) is much less than the intrinsic voltage gain ($\mu_f = 3140$), so neglecting r_o in the calculation is valid. Note also that the value of r_o is much greater than the load resistance connected to the collector terminal of the amplifier (18 kΩ). This is also consistent with our being able to neglect r_o in the voltage gain calculation. The maximum allowed input signal is increased significantly to 0.155 V due to the presence of R_E.

Computer-Aided Analysis: Now, let us can check our hand analysis using SPICE. We will simulate the circuit in Fig. 13.18(a), and must set the transistor parameters to be consistent with our hand analysis in order to achieve a similar Q-point: BF = 100, VAF = 75 V, and IS = 1 fA. We can use an ac analysis to find the voltage gain and will sweep from 1000 Hz to 100 kHz with five frequency points per decade. Several decades are simulated so we can be sure we are in a region where the effects of the capacitors are negligible. The capacitor values must be set to a large number, say 100 μF, so that they will have very small reactance at the simulation frequencies. The SPICE results are: Q-point = (0.248 mA, 3.30 V) and $A_v = -5.67$. [Note that an alternative method to check our calculations is to use SPICE to perform an ac analysis of the small-signal equivalent circuit in Fig. 13.18(d).]

The graph here shows the time-domain response of the amplifier output with an input of 0.15 V obtained with a transient simulation with TSTOP = 0.3 MS. In the graph, we observe good linearity with a gain of −5.7.

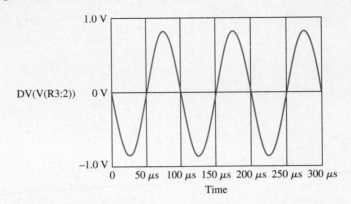

EXERCISE: (a) Suppose resistors R_C, R_E, and R_3 have 10 percent tolerances. What are the worst-case values of voltage gain for this amplifier? (b) What is the voltage gain in the original circuit if $\beta_o = 125$? (c) Suppose the Q-point current in the original circuit increases to 0.275 mA. What are the new values of V_{CE} and the voltage gain?

ANSWERS: (a)−4.75, −6.99; (b) −5.74; (c) 2.34 V, −5.76

EXAMPLE 13.4 COMMON-EMITTER VOLTAGE GAIN WITH BYPASSED EMITTER

Now we will find the voltage gain of the amplifier in Ex. 13.3 with bypass capacitor C_3 connected between ground and the emitter terminal of the BJT.

PROBLEM (a) Find the voltage gain of the amplifier in Ex. 13.3 with bypass capacitor C_3 connected between ground and the emitter terminal of the BJT. (b) Compare the result in (a) to the common-emitter "rule-of-thumb" gain estimate and the amplification factor of the transistor. (c) Find the value of v_i that corresponds to the small-signal limit.

SOLUTION **Known Information and Given Data:** Common-emitter amplifier in Fig. 13.18 with emitter terminal bypassed to ground. From Ex. 13.3, the Q-point = (0.245 mA, 3.39 V), $g_m = 9.80$ mS, $r_\pi = 10.2$ kΩ, and $r_o = 320$ kΩ.

Unknowns: Actual voltage gain, rule-of-thumb estimate, amplification factor of the transistor

Approach: (a) Evaluate the A_v^{CE} expression with $R_E = 0$ (See ac equivalent circuit on next page.). (b) Estimate the voltage gain using $A_v^{CE} \cong -10V_{CC}$; calculate $\mu_f = g_m r_o$

Assumptions: The bipolar transistor is operating in the active region. Signal amplitudes correspond to small-signal conditions. Transistor output resistance r_o can be neglected.

Analysis:

(a) With the emitter terminal bypassed to ground, $R_E = 0$:

$$R_{iB} = r_\pi + (\beta_o + 1)R_E = r_\pi \qquad \text{and} \qquad R_B = R_1 \| R_2$$

$$A_v^{CE} = -\left(\frac{g_m R_L}{1 + g_m R_E}\right)\left[\frac{R_B \| R_{iB}}{R_I + (R_B \| R_{iB})}\right] = -g_m R_L \frac{R_B \| r_\pi}{R_I + (R_B \| r_\pi)}$$

$$A_v^{CE} = -9.80\,\text{mS}\,(18\,\text{k}\Omega)\frac{104\,\text{k}\Omega \| 10.2\,\text{k}\Omega}{1\,\text{k}\Omega + (104\,\text{k}\Omega \| 10.2\,\text{k}\Omega)} = -159 \text{ or } 44.0\,\text{dB}$$

(b) $A_v^{CE} \cong -10(12) = -120$ and $\mu_f = 9.80\,\text{mS}\,(320\,\text{k}\Omega) = 3140$

(c) With the emitter bypassed, v_{be} is given by

$$v_{be} = v_i\left[\frac{R_B \| R_{iB}}{R_I + (R_B \| R_{iB})}\right] = v_i \frac{R_B \| r_\pi}{R_I + (R_B \| r_\pi)} = v_i \frac{104\,\text{k}\Omega \| 10.2\,\text{k}\Omega}{1\,\text{k}\Omega + (104\,\text{k}\Omega \| 10.2\,\text{k}\Omega)} = 0.903 v_i$$

and the small-signal limit for v_i is

$$|v_i| \leq \frac{0.005\,\text{V}}{0.903} = 5.53\,\text{mV}$$

Check of Results: The calculated voltage gain is similar to the rule-of-thumb estimate so our calculation appears correct. Remember, the rule-of-thumb formula is meant to only be a rough estimate; it will not be exact. The gain is much less than the amplification factor, so the neglect of r_o is valid.

Computer-Aided Analysis: SPICE simulation yields the Q-point (0.248 mA, 3.30 V) that is consistent with the assumed value. The small difference results from V_A being included in the SPICE simulation and not in our hand calculations. An ac sweep from 10 Hz to 100 kHz with 10 frequency points/decade is used to find the region in which the capacitors are acting as short circuits, and the gain is observed to be constant at 43.4 dB above a frequency of 1 kHz. The voltage gain is slightly less than our calculated value because r_o was neglected in our calculations. A transient simulation was performed with a 5-mV, 10-kHz sine wave. The output exhibits reasonably good linearity, but the positive and negative amplitudes are slightly different, indicating some waveform distortion. Enabling the Fourier analysis capability of SPICE yields THD = 3.9%.

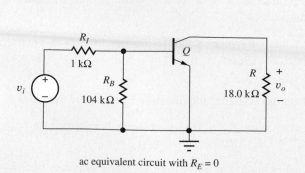

ac equivalent circuit with $R_E = 0$

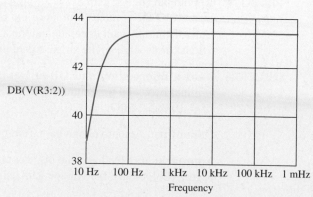

DB(V(R3:2))

Frequency

EXERCISE: A common-emitter amplifier similar to Fig. 13.18 is operating from a single $+20$-V power supply, and the emitter terminal is bypassed by capacitor C_3. The BJT has $\beta_F = 100$ and $V_A = 50$ V and is operating at a Q-point of (100 μA, 10 V). The amplifier has $R_I = 5$ kΩ, $R_B = 150$ kΩ, $R_C = 100$ kΩ, and $R_3 = \infty$. What is the voltage gain predicted using our rule of thumb estimate? What is the actual voltage gain? What is the value of μ_f for this transistor?

ANSWERS: -200; -278; 2400

DESIGN NOTE

Remember, the amplification factor μ_f places an upper bound on the voltage gain of a single-transistor amplifier. We can't do better than μ_f! For the BJT,

$$\mu_f \cong 40V_A$$

For 25 V $\le V_A \le 100$ V, we have $1000 \le \mu_f \le 4000$.

EXERCISE: (a) What is the voltage gain A_v of the amplifier in Ex. 13.4 if R_E is changed to 1 kΩ? Assume the Q-point does not change. (b) What is the new value of R_{E2} required to maintain the Q-points unchanged in the amplifier?

ANSWERS: -15.6, 12 kΩ

EXERCISE: (a) What is the voltage gain A_v of the amplifier in Ex. 13.4 if the upper terminal of C_2 is connected to the emitter of the transistor in Fig. 13.18?

ANSWERS: -145

EXERCISE: What value of saturation current I_S must be used in SPICE to achieve $V_{BE} = 0.7$ V for $I_C = 245$ μA? Assume a default temperature of 27°C.

ANSWER: 0.422 fA

EXERCISE: Calculate R_{in} for the C-E amplifier in Fig. 13.18. How does R_{in}^{CE} compare to r_π?

ANSWERS: 313 kΩ $\cong \beta_o R_E$; $R_{in}^{CE} \gg r_\pi$ for the C-E amplifier

13.7.5 RESISTANCE AT THE COLLECTOR OF THE BIPOLAR TRANSISTOR

The resistance looking into the collector of the transistor, R_{iC}, can be found with the aid of the equivalent circuit in Fig. 13.21 in which input source v_i has been set to zero, and test source v_x is applied to the collector of the transistor.

R_{ic} equals the ratio of $\mathbf{v_x}$ to $\mathbf{i_x}$, where $\mathbf{i_x}$ represents the current through independent source v_x. To find $\mathbf{i}$, we write an expression for $\mathbf{v_e}$:

$$\mathbf{v_e} = (\beta_o + 1)\mathbf{i}R_E \qquad \text{and} \qquad \mathbf{i_x} = \beta_o\mathbf{i} \qquad (13.60)$$

and realize that the current $\mathbf{i}$ can also be written directly in terms of $\mathbf{v_e}$:

$$\mathbf{i} = -\frac{\mathbf{v_e}}{R_{th} + r_\pi} \qquad (13.61)$$

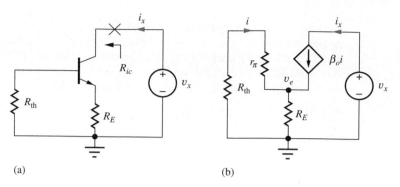

(a)   (b)

Figure 13.21 Circuits for calculating the resistance at the collector of the transistor.

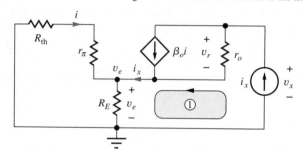

Figure 13.22 Collector resistance with r_o included.

Combining Eqs. (13.60) and (13.61) yields

$$\mathbf{v_e}\left[1 + \frac{(\beta_o + 1)R_E}{r_\pi + R_{th}}\right] = 0 \qquad \text{and} \qquad \mathbf{v_e} = 0 \qquad (13.62)$$

Because $\mathbf{v_e} = 0$, Eq. (13.61) requires that $\mathbf{i}$ equal zero as well. Hence, $\mathbf{i_x} = 0$, and the output resistance of this circuit is infinite!

On the surface, this result may seem acceptable. However, a red flag should go up. We must be suspicious of the results that indicate resistances are infinite (or zero). Using the simplified circuit model in Fig. 13.21(b), in which r_o is neglected, has led to an unreasonable result.

We improve our analysis by moving to the next level of model complexity, as shown in Fig. 13.22. For this analysis, the circuit is driven by the test current i_x, and the voltage v_x must be determined in order to find R_{out}.[5]

Summing the voltages around loop 1 and applying KCL at the output node,

$$\mathbf{v_x} = \mathbf{v_r} + \mathbf{v_e} = (\mathbf{i_x} - \beta_o\mathbf{i})r_o + \mathbf{v_e} \qquad (13.63)$$

Current i_x is forced through the parallel combination of $(R_{th} + r_\pi)$ and R_E, so that v_e can be expressed as

$$\mathbf{v_e} = \mathbf{i_x}[(R_{th} + r_\pi)\|R_E] = \mathbf{i_x}\frac{(R_{th} + r_\pi)R_E}{R_{th} + r_\pi + R_E} \qquad (13.64)$$

At the emitter node, current division can be used to find $\mathbf{i}$ in terms of $\mathbf{i_x}$:

$$\mathbf{i} = -\mathbf{i_x}\frac{R_E}{R_{th} + r_\pi + R_E} \qquad (13.65)$$

---

[5] The upcoming sequence of equations has been developed by the author as an "easy" way to derive this result; this approach is not expected to be obvious. Alternatively, the circuit in Fig. 13.22 can be formulated as a two-node problem by combining R_{th} and r_π.

Combining Eqs. (13.63) through (13.65) yields a somewhat messy expression for the output resistance of the C-E amplifier:

$$R_{iC} = r_o \left(1 + \frac{\beta_o R_E}{R_{\text{th}} + r_\pi + R_E}\right) + (R_{\text{th}} + r_\pi)\|R_E \cong r_o \left(1 + \frac{\beta_o R_E}{R_{\text{th}} + r_\pi + R_E}\right) \qquad (13.66)$$

If we now assume that $(r_\pi + R_E) \gg R_{\text{th}}$ and $r_o \gg R_E$ and remember that $\beta_o = g_m r_\pi$, we reach the approximate results that should be remembered:

$$R_{iC} \cong r_o[1 + g_m(R_E\|r_\pi)] \cong \mu_f(R_E\|r_\pi) \qquad (13.67)$$

Note that Eq. (13.67) reduces to the proper result for $R_E = 0$; that is, $R_{\text{out}} = r_o$. Now we can feel comfortable that our level of modeling is sufficient to produce a meaningful result.

Equation (13.67) tells us that the output resistance of the common-emitter amplifier is equal to the output resistance r_o of the transistor itself plus the equivalent resistance $(R_E\|r_\pi)$ multiplied by the amplification factor of the transistor. For $g_m(R_E\|r_\pi) \gg 1$, $R_{\text{out}} \gg r_o$, we find that the resistance at the collector can be designed to be much greater than the output resistance of the transistor itself!

Important Limit for the Bipolar Transistor

The finite current gain of the bipolar transistor places an upper bound on the size of R_{in}^c that can be achieved. Referring back to Fig. 13.22, we see that r_π appears in parallel with R_E when we neglect R_{th}. If we let $R_E \to \infty$ in Eq. (13.67), we find that the maximum value of output resistance is $R_{ic} \cong \mu_f r_\pi = \beta_o r_o$.

DESIGN NOTE

A quick design estimate for the resistance at the collector of a bipolar transistor with an unbypassed resistor R_E in the emitter is

$$R_{iC} \cong r_o[1 + g_m(r_\pi\|R_E)] \cong \mu_f(r_\pi\|R_E)$$

13.7.6 OUTPUT RESISTANCE OF THE OVERALL COMMON-EMITTER AMPLIFIER

The output resistance of the overall common-emitter amplifier is defined as the resistance looking into the circuit at input coupling capacitor C_2 in Fig. 13.18(a). Thus R_{out}^{CE} equals the parallel combination of collector resistor R_C and the resistance looking into the collector of the transistor itself, R_{iC}, as defined in Fig. 13.18(c):

$$R_{\text{out}}^{CE} = R_C\|R_{iC} = R_C\|r_o \left(1 + \frac{\beta_o R_E}{R_{\text{th}} + r_\pi + R_E}\right) \qquad (13.68)$$

EXAMPLE 13.5   COMMON-EMITTER OUTPUT RESISTANCE

Let us calculate the output resistance for the common-emitter amplifier in Fig. 13.18.

PROBLEM   Calculate R_{out}^{CE} for the C-E amplifier in Fig. 13.18.

SOLUTION **Known Information and Given Data:** The circuit with element values appears in Fig. 13.18. The Q-point and small-signal values appear in the table in Ex. 13.3.

Unknowns: Output resistance R_{out}^{CE} for the C-E amplifier

Approach: Substitute element values from the circuit into Eq. (13.66).

Assumptions: Use the parameter values tabulated in Ex. 13.3. Small-signal conditions apply.

Analysis: The resistance looking into the collector of the transistor is

$$R_{iC} = r_o \left[1 + \frac{\beta_o R_E}{R_{th} + r_\pi + R_E} \right] = 320 \, \text{k}\Omega \left[1 + \frac{100(3.00 \, \text{k}\Omega)}{(1 \, \text{k}\Omega \| 104 \, \text{k}\Omega) + 10.2 \, \text{k}\Omega + 3 \, \text{k}\Omega} \right] = 7.09 \, \text{M}\Omega$$

and the overall output resistance is

$$R_{out}^{CE} = 22 \, \text{k}\Omega \| 7.09 \, \text{M}\Omega = 21.9 \, \text{k}\Omega$$

Check of Results: We can quickly check these answers using the approximation $R_{iC} \cong \mu_f R_E$ from Eq. (13.67) for the C-E transistor:

$$R_{iC} \cong 3140 \, (3 \, \text{k}\Omega) = 9.42 \, \text{M}\Omega \qquad \text{and} \qquad 7.09 \, \text{M}\Omega < 9.42 \, \text{M}\Omega$$

Discussion: The detailed calculation is similar to the approximation, although not exact. The estimate of R_{iC} is somewhat high because R_{th} and R_E cannot be neglected in Eq. (13.66).

www **Computer-Aided Analysis[6]:** We can check the output resistance using SPICE. [Be sure to include β_F and V_A (BF = 100, VAF = 75 V) in your transistor model.] We place an ac current source i_o in parallel with resistor R_3 and perform an operating point analysis followed by an ac analysis with v_O as the output voltage. Make all the capacitor values large, say 100 μF, and sweep the frequency to find the midband range of frequencies (e.g., FSTART = 1 Hz and FSTOP = 100 kHz with 10 frequency points per decade.)

The (frequency dependent) output resistance of the common-emitter amplifier is equal to the voltage at the output node divided by the current entering the amplifier through coupling capacitor C_2: $R_{out}^{CE} = \text{V(IO)/I(C2)}$. At frequencies above 1 kHz, the SPICE output becomes constant at 21.9 kΩ, in agreement with our hand calculations.

The resistance looking into the collector of the transistor itself can be found from SPICE as $R_{iC} = \text{VC(Q1)/IC(Q1)} = 6.89 \, \text{M}\Omega$. The slight disagreement is due to differences in the calculated value of the Q-point and small-signal parameters in SPICE.

EXERCISE: What is the value of R_{out} for the common-emitter amplifier in Ex. 13.5 if R_E is changed to 2 kΩ? Assume the Q-point does not change. Compare the result to the new value of $\mu_f R_E$.

ANSWERS: 3.31 MΩ, which is less than 4.30 MΩ

EXERCISE: Show that the maximum output resistance for the common-emitter amplifier is $R_{out} \cong (\beta_o + 1)r_o$ by taking the limit as $R_E \to \infty$ in Eq. (13.67).

---

[6] See the MCD website for help with this circuit.

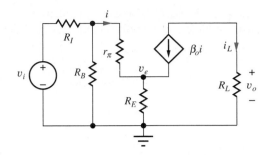

Figure 13.23 Circuit for calculating C-E current gain.

13.7.7 TERMINAL CURRENT GAIN FOR THE COMMON-EMITTER AMPLIFIER

The **terminal current gain** A_{it} is defined as the ratio of the current delivered to the load resistor R_L to the current being supplied to the base terminal. For the C-E amplifier in Fig. 13.23, the current in R_L is equal to i amplified by the current gain β_o, yielding a current gain equal to $-\beta_o$.

$$A_{it}^{CE} = -\beta_o \tag{13.69}$$

13.8 SMALL-SIGNAL MODELS FOR FIELD-EFFECT TRANSISTORS

We now turn our attention to the small-signal model for the field-effect transistor and then use it in Sec. 13.9 to analyze the behavior of the common-source amplifier stage that is the FET version of the common-emitter amplifier. First, we consider the MOSFET as a three-terminal device; we then explore the changes necessary when the MOSFET is operated as a four-terminal device.

13.8.1 SMALL-SIGNAL MODEL FOR THE MOSFET

The small-signal model of the MOSFET is based on the two-port network representation in Fig. 13.24 with the input port variables defined as v_{gs} and i_g and the output port variables defined as v_{ds} and i_d. Rewriting Eq. (13.17) in terms of these variables yields

$$\mathbf{i_g} = g_\pi \mathbf{v_{gs}} + g_r \mathbf{v_{ds}}$$
$$\mathbf{i_d} = g_m \mathbf{v_{gs}} + g_o \mathbf{v_{ds}} \tag{13.70}$$

Remember that the port variables in Fig. 13.24(a) can be considered to represent either the time-varying portion of the total voltages and currents or small changes in the total quantities.

$$v_{GS} = V_{GS} + v_{gs} \qquad v_{DS} = V_{DS} + v_{ds}$$
$$i_G = I_G + i_g \qquad i_{DS} = I_D + i_d \tag{13.71}$$

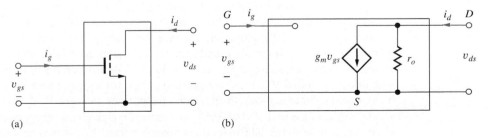

(a)   (b)

Figure 13.24 (a) The MOSFET represented as a two-port network. (b) Small-signal model for the three-terminal MOSFET.

The parameters in Eq. (13.70) can be written in terms of the small-signal variations or in terms of derivatives of the complete port variables, as in Eq. (13.72):

$$g_\pi = \left.\frac{\mathbf{i_g}}{\mathbf{v_{gs}}}\right|_{\mathbf{v_{ce}}=0} = \left.\frac{\partial i_G}{\partial v_{GS}}\right|_{\text{Q-point}} \qquad g_r = \left.\frac{\mathbf{i_g}}{\mathbf{v_{ds}}}\right|_{\mathbf{v_{be}}=0} = \left.\frac{\partial i_G}{\partial v_{DS}}\right|_{\text{Q-point}}$$

$$g_m = \left.\frac{\mathbf{i_d}}{\mathbf{v_{gs}}}\right|_{\mathbf{v_{ce}}=0} = \left.\frac{\partial i_{DS}}{\partial v_{GS}}\right|_{\text{Q-point}} \qquad g_o = \left.\frac{\mathbf{i_d}}{\mathbf{v_{ds}}}\right|_{\mathbf{v_{be}}=0} = \left.\frac{\partial i_{DS}}{\partial v_{DS}}\right|_{\text{Q-point}}$$

(13.72)

We can evaluate these parameters by taking appropriate derivatives of the large-signal model equations for the drain current of the active region MOS transistor, as developed in Chapter 4, and repeated here in Eq. (13.73).

$$i_D = \frac{K_n}{2}(v_{GS} - V_{TN})^2(1 + \lambda v_{DS}) \tag{13.73}$$

for $v_{DS} \geq v_{GS} - V_{TN}$ and $i_G = 0$, where $K_n = \mu_n C_{\text{ox}}(W/L)$.

$$g_\pi = \left.\frac{\partial i_G}{\partial v_{GS}}\right|_{v_{DS}} = 0 \qquad \text{and} \qquad g_r = \left.\frac{\partial i_G}{\partial v_{DS}}\right|_{v_{GS}} = 0$$

$$g_m = \left.\frac{\partial i_{DS}}{\partial v_{GS}}\right|_{\text{Q-point}} = K_n(V_{GS} - V_{TN})(1 + \lambda V_{DS}) = \frac{2I_D}{V_{GS} - V_{TN}}$$

(13.74)

$$g_o = \left.\frac{\partial i_{DS}}{\partial v_{DS}}\right|_{\text{Q-point}} = \lambda \frac{K_n}{2}(V_{GS} - V_{TN})^2 = \frac{\lambda I_D}{1 + \lambda V_{DS}} = \frac{I_D}{\dfrac{1}{\lambda} + V_{DS}}$$

Because i_G is always zero and therefore independent of v_{GS} and v_{DS}, g_π and g_r are both zero. Remembering that the gate terminal is insulated from the channel by the gate oxide, we can reasonably expect that the input resistance $(1/g_\pi)$ of the transistor is infinite.

As for the bipolar transistor, g_m is called the transconductance, and $1/g_o$ represents the output resistance of the transistor.

Transconductance: $\quad g_m = \dfrac{I_D}{\dfrac{V_{GS} - V_{TN}}{2}}$

Output resistance: $\quad r_o = \dfrac{1}{g_o} = \dfrac{\dfrac{1}{\lambda} + V_{DS}}{I_D} \cong \dfrac{1}{\lambda I_D}$

(13.75)

The small-signal circuit model for the MOSFET resulting from Eqs. (13.74) and (13.75) appears in Fig. 13.24(b).

From Eq. (13.75), we see that the values of the small-signal parameters are directly controlled by the design of the Q-point. The form of the equations for g_m and r_o of the MOSFET directly mirrors that of the BJT. However, one-half the internal gate drive $(V_{GS} - V_{TN})/2$ replaces the thermal voltage V_T in the transconductance expression, and $1/\lambda$ replaces the Early voltage in the output resistance expression. The value of $V_{GS} - V_{TN}$ is often a volt or more in MOSFET circuits, whereas $V_T = 0.025$ V at room temperature. Thus, for a given operating current, the MOSFET can be expected to have a much smaller transconductance than the BJT. However, the value of $1/\lambda$ is similar to V_A, so the output resistances are similar for a given operating point $(I_D, V_{DS}) = (I_C, V_{CE})$. Here, and similar to the BJT case, drain-source voltage V_{DS} is typically much less than $1/\lambda$, so we can simplify the expression for the output resistance to $r_o \cong 1/\lambda I_D$.

The actual dependence of transconductance g_m on current is not shown explicitly by Eq. (13.75) because I_D is a function of $(V_{GS} - V_{TN})$. Rewriting the expression for g_m from Eq. (13.74) yields

$$g_m = K_n(V_{GS} - V_{TN})(1 + \lambda V_{DS}) = \sqrt{2K_n I_D(1 + \lambda V_{DS})}$$

$$g_m \cong K_n(V_{GS} - V_{TN}) \qquad \text{or} \qquad g_m \cong \sqrt{2K_n I_D} \tag{13.76}$$

where the simplifications require $\lambda V_{DS} \ll 1$.

Here we see two other important differences between the MOSFET and BJT. The MOSFET transconductance increases only as the square root of drain current, whereas the BJT transconductance is directly proportional to collector current. In addition, the MOSFET transconductance is dependent on the geometry of the transistor because $K_n \propto W/L$, whereas the transconductance of the BJT is geometry-independent. It is also worth noting that the current gain of the MOSFET is infinite. Because the value of $r_\pi = (1/g_\pi)$ is infinite for the MOSFET, the "current gain" $\beta_o = g_m r_\pi$ equals infinity as well.

13.8.2 INTRINSIC VOLTAGE GAIN OF THE MOSFET

Another important difference between the BJT and MOSFET is the variation of the intrinsic voltage gain μ_f with operating point. Using Eq. (13.75) for g_m and r_o, we find that the intrinsic voltage gain becomes

$$\mu_f = g_m r_o = \frac{\dfrac{1}{\lambda} + V_{DS}}{\left(\dfrac{V_{GS} - V_{TN}}{2}\right)} \qquad \text{and} \qquad \mu_f \cong \frac{2}{\lambda(V_{GS} - V_{TN})} \cong \frac{1}{\lambda}\sqrt{\frac{2K_n}{I_D}} \tag{13.77}$$

for $\lambda V_{DS} \ll 1$.

The value of μ_f of the MOSFET decreases as the operating current increases. Thus the larger the operating current of the MOSFET, the smaller its voltage gain capability. In contrast, the intrinsic gain of the BJT is independent of operating point. This is an extremely important difference to keep in mind, particularly during the design process.

Table 13.2 displays examples of the values of the MOSFET small-signal parameters for a variety of operating points. Just as for the bipolar transistor, the values of g_m and r_o can each be varied over many orders of magnitude through the choice of Q-point. By comparing the results in Tables 13.1 and 13.2 we see that g_m, r_o, and μ_f of the MOSFET are all similar to those of the bipolar transistor at low currents. However, as the drain current increases, the value of g_m of the MOSFET does not grow as rapidly as for the bipolar transistor, and μ_f drops significantly. This particular MOSFET has a significantly lower intrinsic gain than the BJT for currents greater than a few tens of microamperes.

TABLE 13.2
MOSFET Small-Signal Parameters versus Current: $K_n = 1\,\text{mA/V}^2$,
$\lambda = 0.0133\,\text{V}^{-1}$, $V_{DS} = 10\,\text{V}$

I_D	g_m	r_π	r_o	μ_f
1 μA	4.76×10^{-5} S	∞	85.2 MΩ	4060
10 μA	1.51×10^{-4} S	∞	8.52 MΩ	1280
100 μA	4.76×10^{-4} S	∞	852 kΩ	406
1 mA	1.51×10^{-3} S	∞	85.2 kΩ	128
10 mA	4.76×10^{-3} S	∞	8.52 kΩ	40

13.8.3 DEFINITION OF SMALL-SIGNAL OPERATION FOR THE MOSFET

The limits of linear operation of the MOSFET can be explored using the simplified drain-current expression ($\lambda = 0$) for the MOSFET in the active region:

$$i_D = \frac{K_n}{2}(v_{GS} - V_{TN})^2 \qquad \text{for} \qquad v_{DS} \geq v_{GS} - V_{TN} \tag{13.78}$$

Expanding this expression using $v_{GS} = V_{GS} + v_{gs}$ and $i_D = I_D + i_d$ gives

$$I_D + i_d = \frac{K_n}{2}\left[(V_{GS} - V_{TN})^2 + 2v_{gs}(V_{GS} - V_{TN}) + v_{gs}^2\right] \tag{13.79}$$

Recognizing that the dc drain current is equal to $I_D = (K_n/2)(V_{GS} - V_{TN})^2$ and subtracting this term from both sides of Eq. (13.79) yields an expression for signal current i_d:

$$i_d = \frac{K_n}{2}\left[2v_{gs}(V_{GS} - V_{TN}) + v_{gs}^2\right] \tag{13.80}$$

For linearity, i_d must be directly proportional to v_{gs}, which is achieved for

$$v_{gs}^2 \ll 2v_{gs}(V_{GS} - V_{TN}) \qquad \text{or} \qquad v_{gs} \ll 2(V_{GS} - V_{TN}) \tag{13.81}$$

Using a factor of 10 to satisfy the inequality gives

$$v_{gs} \leq 0.2(V_{GS} - V_{TN}) \tag{13.82}$$

Because the MOSFET can easily be biased with $(V_{GS} - V_{TN})$ equal to several volts, we see that it can handle much larger values of v_{gs} than the values of v_{be} corresponding to the bipolar transistor. This is another fundamental difference between the MOSFET and BJT and can be very important in circuit design, particularly in RF amplifiers, for example.

Now let us explore the change in drain current that corresponds to small-signal operation. Using Eq. (13.82),

$$\frac{i_d}{I_D} = \frac{g_m v_{gs}}{I_D} = \frac{0.2(V_{GS} - V_{TN})}{\dfrac{V_{GS} - V_{TN}}{2}} \leq 0.4 \tag{13.83}$$

A change of 0.2 $(V_{GS} - V_{TN})$ in v_{GS} corresponds to a 40 percent deviation in the drain and source currents from their Q-point values.

13.8.4 BODY EFFECT IN THE FOUR-TERMINAL MOSFET

When the body terminal of the MOSFET is not connected to the source terminal, as in Fig. 13.25(a), an additional controlled source must be introduced into the small-signal model. Referring to the

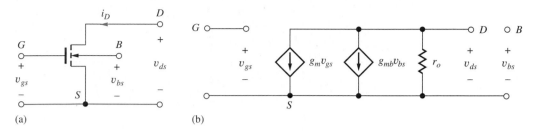

Figure 13.25 (a) MOSFET as a four-terminal device. (b) Small-signal model for the four-terminal MOSFET.

simplified drain-current expression for the MOSFET from Sec. 4.8:

$$i_D = \frac{K_n}{2}(v_{GS} - V_{TN})^2 \quad \text{and} \quad V_{TN} = V_{TO} + \gamma\left(\sqrt{v_{SB} + 2\phi_F} - \sqrt{2\phi_F}\right) \qquad (13.84)$$

We recognize that the drain current is dependent on the threshold voltage, and the threshold voltage changes as v_{SB} changes. Thus, a **back-gate transconductance** can be defined:

$$g_{mb} = \left.\frac{\partial i_D}{\partial v_{BS}}\right|_{\text{Q-point}} = -\left.\frac{\partial i_D}{\partial v_{SB}}\right|_{\text{Q-point}} = -\left(\frac{\partial i_D}{\partial V_{TN}}\right)\left.\left(\frac{\partial V_{TN}}{\partial v_{SB}}\right)\right|_{\text{Q-point}} \qquad (13.85)$$

Evaluating the derivative terms in brackets,

$$\left.\frac{\partial i_D}{\partial V_{TN}}\right|_{\text{Q-point}} = -K_n(V_{GS} - V_{TN}) = -g_m \quad \text{and} \quad \left.\frac{\partial V_{TN}}{\partial v_{SB}}\right|_{\text{Q-point}} = \frac{\gamma}{2\sqrt{V_{SB} + 2\phi_F}} = \eta \qquad (13.86)$$

in which η represents the **back-gate transconductance parameter.** Combining Eqs. (13.86) yields

$$g_{mb} = -(-g_m)\eta \quad \text{or} \quad g_{mb} = +\eta g_m \qquad (13.87)$$

for typical values of γ and V_{SB}, $0 \leq \eta \leq 1$.

We also need to explore the question of whether there is a conductance connected from the bulk terminal to the other terminals. However, the bulk terminal represents a reverse-biased diode between the bulk and channel. Using our small-signal model for the diode, Eq. (13.15), we see that

$$\left.\frac{\partial i_B}{\partial v_{BS}}\right|_{\text{Q-point}} = \frac{I_D + I_S}{V_T} \cong 0 \qquad (13.88)$$

because $I_D \cong -I_S$ for the reverse-biased diode. Thus, there is no conductance indicated between the bulk and source or drain terminals in the small-signal model.

The resulting small-signal model for the four-terminal MOSFET is given in Fig. 13.25(b), in which a second voltage-controlled current source has been added to model the back-gate transconductance g_{mb}.

EXERCISE: Calculate the values of η for a MOSFET transistor with $\gamma = 0.75$ V$^{0.5}$ and $2\phi_F = 0.6$ V for $V_{SB} = 0$ and $V_{SB} = 3$ V.

ANSWERS: 0.48, 0.20

13.8.5 SMALL-SIGNAL MODEL FOR THE PMOS TRANSISTOR

Just as was the case for the *pnp* and *npn* transistors, the small-signal model for the PMOS transistor is identical to that of the NMOS device. The circuits in Fig. 13.26 should help reinforce this result.

In Fig. 13.26, the NMOS and PMOS transistors are each biased by the dc voltage source V_{GG}, establishing Q-point current I_D. In each case, a signal voltage v_{gg} is added in series with V_{GG} so

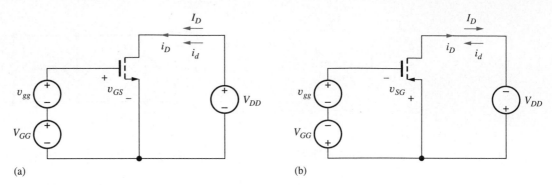

Figure 13.26 dc Bias and signal currents for (a) NMOS and (b) PMOS transistors.

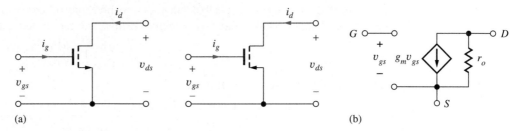

Figure 13.27 (a) NMOS and PMOS transistors. (b) The small-signal models are identical.

that a positive value of v_{gg} causes the gate-to-source voltage of each transistor to increase. For the NMOS transistor, the total gate-to-source voltage and drain current are

$$v_{GS} = V_{GG} + v_{gg} \qquad \text{and} \qquad i_{DS} = I_D + i_d \qquad (13.89)$$

and an increase in v_{gg} causes an increase in current into the drain terminal. For the PMOS transistor,

$$v_{SG} = V_{GG} - v_{gg} \qquad \text{and} \qquad i_D = I_D - i_d \qquad (13.90)$$

A positive signal voltage v_{gg} reduces the source-to-gate voltage of the PMOS transistor and causes a decrease in the total current exiting the drain terminal. This reduction in total current is equivalent to an increase in the signal current entering the drain.

Thus, for both the NMOS and PMOS transistors, an increase in the value of v_{GS} causes an increase in current into the drain, and the polarities of the voltage-controlled current source in the small-signal model are identical, as depicted in Fig. 13.27.

13.9 SUMMARY AND COMPARISON OF THE SMALL-SIGNAL MODELS OF THE BJT AND FET

Table 13.3 is a side-by-side comparison of the small-signal models of the bipolar junction transistor and the field-effect transistor; the table has been constructed to highlight the similarities and differences between the two types of devices.

The transconductance of the BJT is directly proportional to operating current, whereas that of the FET increases only with the square root of current. Both can be represented as the drain current divided by a characteristic voltage: V_T for the BJT and $(V_{GS} - V_{TN})/2$ for the MOSFET.

TABLE 13.3
Small-Signal Parameter Comparison

PARAMETER	BIPOLAR TRANSISTOR	MOSFET
Transconductance g_m	$\dfrac{I_C}{V_T}$	$\dfrac{2I_D}{V_{GS} - V_{TN}} \cong \sqrt{2K_n I_D}$
Input resistance	$r_\pi = \dfrac{\beta_o}{g_m} = \dfrac{\beta_o V_T}{I_C}$	∞
Output resistance r_o	$\dfrac{V_A + V_{CE}}{I_C} \cong \dfrac{V_A}{I_C}$	$\dfrac{\frac{1}{\lambda} + V_{DS}}{I_D} \cong \dfrac{1}{\lambda I_D}$
Intrinsic voltage gain μ_f	$\dfrac{V_A + V_{CE}}{V_T} \cong \dfrac{V_A}{V_T}$	$\dfrac{2\left(\frac{1}{\lambda} + V_{DS}\right)}{V_{GS} - V_{TN}} \cong \dfrac{1}{\lambda}\sqrt{\dfrac{2K_n}{I_D}}$
Small-signal requirement	$v_{be} \le 0.005$ V	$v_{gs} \le 0.2(V_{GS} - V_{TN})$

dc i-v active region expressions for use with Table 13.3:

BJT:
$$I_C = I_S\left[\exp\left(\frac{V_{BE}}{V_T}\right) - 1\right]\left[1 + \frac{V_{CE}}{V_A}\right] \qquad V_T = \frac{kT}{q}$$

MOSFET:
$$I_D = \frac{K_n}{2}(V_{GS} - V_{TN})^2(1 + \lambda V_{DS}) \qquad K_n = \mu_n C_{ox}\frac{W}{L}$$

The input resistance of the bipolar transistor is set by the value of r_π, which is inversely proportional to the Q-point current and can be quite small at even moderate currents (1 to 10 mA). On the other hand, the input resistance of the FETs is extremely high, approaching infinity.

The expressions for the output resistances of the transistors are almost identical, with the parameter $1/\lambda$ in the FET taking the place of the Early voltage V_A of the BJT. The value of $1/\lambda$ is similar to V_A, so the output resistances can be expected to be similar in value for comparable operating currents.

The intrinsic voltage gain of the BJT is nearly independent of operating current and has a typical value of several thousand at room temperature. In contrast, μ_f of the FET is inversely proportional to the square root of operating current and decreases as the Q-point current is raised. At very low currents, μ_f of the FET can be similar to that of the BJT, but in normal operation it is often much smaller and can even fall below 1 for high currents (see Prob. 13.77).

Small-signal operation is dependent on the size of the base-emitter voltage of the BJT or gate-source voltage of the field-effect transistor. The magnitude of voltage that corresponds to small-signal operation can be significantly different for these two devices. For the BJT, v_{be} must be less than 5 mV. This value is indeed small, and it is independent of Q-point. In contrast, the FET requirement is $v_{gs} \le 0.2(V_{GS} - V_{TN})$ or $0.2(V_{GS} - V_P)$, which is dependent on bias point and can be designed to be as much as a volt or more.

This discussion highlighted the similarities and differences between the bipolar and field-effect transistors. An understanding of Table 13.3 is extremely important to the design of analog circuits. As we study single and multistage amplifier design in the coming sections and chapters, we will note the effect of these differences and relate them to our circuit designs.

ELECTRONICS IN ACTION

Noise in Electronic Circuits

The linear signal-level limitations of transistors that we have introduced in this chapter may seem small, but we often deal with signal levels that are far below even the 5-mV v_{be} limit for the BJT. For example, the radio frequency signals from antennas on our cell phones can be in the microvolt range, and high frequency communications receivers often have minimum detectable signals of less than 0.1 μV! The minimum detectable signals are set by the noise in the RF amplifiers connected to the antennas. These amplifiers are often referred to as low noise amplifiers, or LNAs, in which the noise actually comes from the transistors and resistors that make up the circuit.

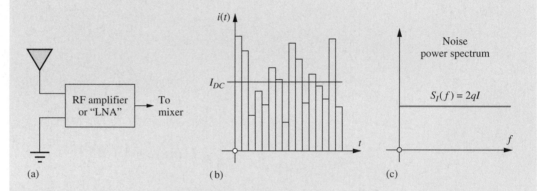

We often think that the dc voltages and currents that we calculate or measure with a dc voltmeter are constants, but they really represent averages of noisy signals. For example, the currents that we encounter in this text are made up of very large numbers of small current pulses due to individual electrons (e.g., 1 μA = 6.3×10^{12} electrons/sec). The current is constantly fluctuating or varying about the dc value as shown in the graph here, and these fluctuations represent one of the sources of noise in electronic devices. If we somehow listened to this current, it would sound much like rain on a tin roof. The background "din" from the rain is actually made up of the noise from a huge number of individual drops. This form of electronic noise is termed "shot" noise.

We model the noise in electron devices by adding noise voltage and current generators to our circuits. The noise generators represent random signals with zero mean and are therefore characterized by either their rms or mean square values. For example, both the base and collector currents in the bipolar transistor produce shot noise, and the noise is modeled by

$$\overline{i_{cn}^2} = \overline{[i_C(t) - I_C]^2} = 2qI_C B \qquad \text{and} \qquad \overline{i_{bn}^2} = \overline{[i_B(t) - I_B]^2} = 2qI_B B$$

These sources are referred to as "white noise" sources in which the noise power spectrum is independent of frequency. The mean square value of the noise current is directly proportional to the dc current and depends upon the bandwidth B (Hz) associated with the measurement. For instance, the rms value of the collector shot noise for $I_C = 1$ mA and $B = 1$ kHz would be

$$\sqrt{\overline{i_{cn}^2}} = \sqrt{2(1.6 \times 10^{-19})(10^{-3})(10^3)} = 0.566 \, \text{nA}$$

In addition to shot noise, resistors and other resistive elements in electronic circuits exhibit noise due to the thermal agitation of electrons in the resistor. This "thermal" noise or "Johnson" noise is modeled by a noise voltage source in series with the resistor as shown for the base resistance of the BJT in the figure below (also see Chapter 16). The mean square value of the noise voltage associated with a resistor R is given by

$$\overline{v_{rn}^2} = 4kTRB$$

where k is Boltzmann's constant, T is absolute temperature, and B is the bandwidth of interest. For a 1-kΩ resistor operating at 300 K with $B = 1$ kHz,

$$\sqrt{\overline{v_{rn}^2}} = \sqrt{4(1.38 \times 10^{-23})(300)(10^3)(10^3)} = 0.129 \, \mu V$$

Remembering that the channel region of the MOSFET is really a voltage-controlled resistor, we can model the thermal noise of the channel by a (Norton equivalent) current source whose mean square value is

$$\overline{i_{dn}^2} = \frac{8}{3}kTg_m B$$

The final figure presents our basic transistor noise models. For the BJT, current sources are added to model the shot noise of both the base and collector currents, and the thermal noise

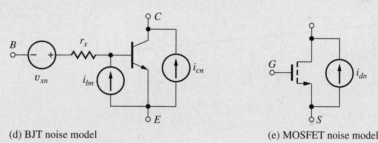

(d) BJT noise model   (e) MOSFET noise model

of the base resistance is also included. For the MOSFET, the thermal noise of the channel is modeled by an equivalent noise current source. These noise models are built into SPICE, and NOISE is one of the analysis options available. For further information on how to make noise calculations and use the noise analysis capability in SPICE, see the MCD website.

13.10 THE COMMON-SOURCE AMPLIFIER

Now we are in a position to analyze the small-signal characteristics of the **common-source (C-S) amplifier** shown in Fig. 13.28(a), which uses an enhancement-mode n-channel MOSFET ($V_{TN} > 0$) in a four-resistor bias network. The ac equivalent circuit of Fig. 13.28(b) is constructed by assuming that the capacitors all have zero impedance at the signal frequency and that the dc voltage sources represent ac grounds. Bias resistors R_1 and R_2 appear in parallel and are combined into gate resistor R_G, and R_L represents the parallel combination of R_D and R_3. In Fig. 13.28(c), the transistor has been replaced with its small-signal model. In subsequent analysis, we will assume that the voltage gain is much less than the intrinsic voltage gain of the transistor so we can neglect transistor output resistance r_o. For simplicity at this point, we assume that we have found the Q-point and know the values of I_D and V_{DS}.

In Fig. 13.28(b) through (d), the common-source nature of this amplifier should be apparent. The input signal is applied to the transistor's gate terminal, the output signal appears at the drain, and both the input and output signals are referenced to the (common) source terminal. Note that the small-signal models for the MOSFET and BJT are virtually identical at this step, except that r_π is replaced by an open circuit for the MOSFET.

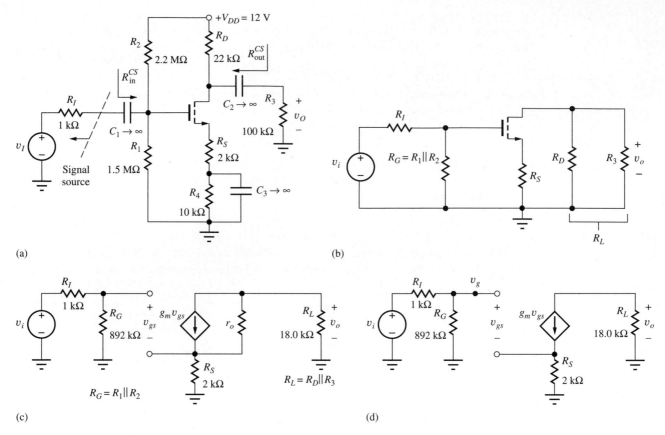

Figure 13.28 (a) Common-source amplifier circuit employing a MOSFET. (b) ac Equivalent circuit for common-source amplifier in part (a). The common-source connection is now apparent. (c) ac Equivalent circuit with the MOSFET replaced by its small-signal model. (d) Final equivalent circuit for ac analysis of the common-source amplifier in which r_o is neglected.

Our first goal is to develop an expression for the voltage gain A_v^{CS} of the circuit in Fig. 13.28(a) from the source v_s to the output v_o. As with the BJT, we will first find the terminal voltage gain A_{vt}^{CS} between the gate and drain terminals of the transistor. Then, we will use the terminal gain expression to find the gain of the overall amplifier.

13.10.1 COMMON-SOURCE TERMINAL VOLTAGE GAIN

Starting with the circuit in Fig. 13.28(d), the terminal voltage gain is defined as

$$A_{vt}^{CS} = \frac{v_d}{v_g} = \frac{v_o}{v_g} \qquad \text{where} \qquad v_o = -g_m v_{gs} R_L \tag{13.91}$$

We can relate v_{gs} to v_g by applying KVL at the gate of the FET:

$$v_g = v_{gs} + g_m v_{gs} R_S \qquad \text{or} \qquad v_{gs} = \frac{v_g}{1 + g_m R_S} \tag{13.92}$$

Combining Eqs. (13.91) and (13.92) yields an expression for the terminal gain.

$$A_{vt}^{CS} = -\frac{g_m R_L}{1 + g_m R_S} \tag{13.93}$$

13.10.2 SIGNAL SOURCE VOLTAGE GAIN FOR THE COMMON-SOURCE AMPLIFIER

Now we can find the overall gain from source v_i to the voltage across R_L. The overall gain can be written as

$$A_v^{CS} = \frac{v_o}{v_i} = \left(\frac{v_o}{v_g}\right)\left(\frac{v_g}{v_i}\right) = A_{vt}^{CS}\left(\frac{v_g}{v_i}\right) \qquad \text{where} \qquad v_g = v_i\frac{R_G}{R_G + R_I} \qquad (13.94)$$

in which v_g is related to v_i by the voltage divider formed by R_G and R_I. Combining Eqs. (13.93) and (13.94) yields a general expression for the voltage gain of the common-source amplifier:

$$A_v^{CS} = -\frac{g_m R_L}{1 + g_m R_S}\left(\frac{R_G}{R_G + R_I}\right) \qquad (13.95)$$

We now explore the limits to the voltage gain of common-source amplifiers using model simplifications for zero and large values of resistance R_S. First, we will assume that the signal source resistance R_I is much less than R_G so that

$$A_v^{CS} \cong A_{vt}^{CS} = -\frac{g_m R_L}{1 + g_m R_S} \qquad \text{for} \qquad R_I \ll R_G \qquad (13.96)$$

This approximation is equivalent to saying that the total input signal appears at the gate terminal of the transistor.

13.10.3 COMMON-SOURCE VOLTAGE GAIN FOR LARGE VALUES OF R_S

A very useful simplification occurs when R_S is large enough so that the $g_m R_S \gg 1$:

$$A_{vt}^{CS} = -\frac{g_m R_L}{1 + g_m R_S} \cong -\frac{R_L}{R_S} \qquad \text{for} \qquad g_m R_{EI} \gg 1 \qquad \text{and} \qquad R_G \gg R_I \qquad (13.97)$$

The gain is now set by the ratio of the load resistor R_L and source resistor R_S. This is an extremely useful result because the gain is now independent of the transistor characteristics that vary widely from device to device. The result in Eq. (13.97) is very similar to the one that we obtained for the op-amp inverting amplifier circuit and is a result of feedback introduced by resistor R_S.

Achieving the simplification in Eq. (13.97) requires $g_m R_S \gg 1$. We can relate this product to the dc bias voltage developed across R_S:

$$g_m R_S = \frac{2}{(V_{GS} - V_{TN})}I_D R_S \quad \text{and we need} \quad I_D R_S \gg \frac{V_{GS} - V_{TN}}{2} \qquad (13.98)$$

$I_D R_S$ represents the dc voltage drop across source resistor R_S and must be much greater than half the gate drive of the transistor. This inequality can be achieved, but not as easily as for the case of the BJT.

13.10.4 ZERO RESISTANCE IN THE SOURCE

In order to achieve as large a gain as possible, we need to make the denominator in Eq. (13.96) as small as possible, and this is achieved by setting $R_S = 0$. The gain is then

$$A_v^{CS} \cong -g_m R_L = -g_m(R_D \| R_3) \qquad \text{for} \qquad R_S = 0 \qquad (13.99)$$

Equation (13.99) places an upper limit on the gain we can achieve from a common-source amplifier with an external load resistor. Equation (13.99) states that the terminal voltage gain of the common-source stage is equal to the product of the transistor's transconductance g_m and load resistance R_L, and the minus sign indicates that the output voltage is "inverted" or 180° out of phase with respect to the input. The approximations that led to Eq. (13.99) are equivalent to saying that the total input signal appears across v_{gs} as shown in Fig. 13.29.

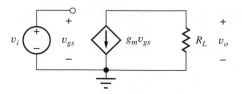

Figure 13.29 Simplified circuit for $R_G \gg R_1$ and $R_s = 0$.

EXAMPLE **13.6**   VOLTAGE GAIN OF A COMMON-SOURCE AMPLIFIER

In this example, we find the small-signal parameters of the MOSFET and then calculate the voltage gain of a common-source amplifier.

PROBLEM   Calculate the gain of the common-source amplifier in Fig. 13.28 if the transistor has $K_n = 0.500$ mA/V^2, $V_{TN} = 1$ V, and $\lambda = 0.0133$ V^{-1}, and the Q-point is (0.241 mA, 3.81 V). What is the largest value of v_i that does not violate the small-signal assumption?

SOLUTION   **Known Information and Given Data:** Common-source amplifier with its ac equivalent circuit given in Fig. 13.28; $K_n = 0.500$ mA/V^2, $V_{TN} = 1$ V, and $\lambda = 0.0133$ V^{-1}; the Q-point is (0.241 mA, 3.64 V); $R_I = 1$ kΩ, $R_1 = 1.5$ MΩ, $R_2 = 2.2$ MΩ, $R_D = 22$ kΩ, $R_3 = 100$ kΩ, $R_S = 2$ kΩ.

Unknowns: Small-signal parameters of the transistor; voltage gain A_v; small-signal limit for the value of v_i

Approach: Use the Q-point information to find g_m and r_o. Use the calculated and given values to evaluate the voltage gain expression in Eq. (13.95).

Assumptions: The transistor is in the active region of operation, and the signal amplitudes are below small-signal limit for the MOSFET.

Analysis: We need to evaluate Eq. (13.95):

$$A_v^{CS} = -\frac{g_m R_L}{1 + g_m R_S}\left(\frac{R_G}{R_G + R_I}\right)$$

Calculating the values of the various resistors and small-signal model parameters yields

$$g_m = \sqrt{2K_n I_{DS}(1 + \lambda V_{DS})}$$

$$= \sqrt{2\left(5 \times 10^{-4}\frac{A}{V^2}\right)(0.241 \times 10^{-3}\ A)\left(1 + \frac{0.0133}{V}3.81\ V\right)} = 0.503\ \text{mS}$$

$$r_o = \frac{\dfrac{1}{\lambda} + V_{DS}}{I_D} = \frac{\left(\dfrac{1}{0.0133} + 3.81\right)\ V}{0.241 \times 10^{-3}\ A} = 328\ \text{k}\Omega$$

$$R_G = R_1 \| R_2 = 892\ \text{k}\Omega \qquad R_L = R_D \| R_3 = 18.0\ \text{k}\Omega$$

$$g_m R_L = 9.05 \quad g_m R_S = 1.01 \quad A_v^{CS} = -\frac{9.05}{1 + 1.01}\left(\frac{892\ \text{k}\Omega}{892\ \text{k}\Omega + 1\ \text{k}\Omega}\right) = -4.50$$

Thus the common-source amplifier in Fig. 13.29 provides a small-signal voltage gain $A_v = -4.50$ or 13.1 dB.

Based on Eq. (13.82) for small-signal operation, we require

$$v_i \leq 0.2(V_{GS} - V_{TN})(1 + g_m R_S) = 0.2(0.982 \text{ V})(2.01) = 0.395 \text{ V}$$

Thus, the input signal amplitude must not exceed 0.40 V for small-signal operation.

Check of Results: We have found all the requested values. The amplification factor for this transistor is $\mu_f = g_m r_o = 165$. Our calculated voltage gain is much less than $\mu_f = 165$, so neglect of r_o is justified. With nonzero R_S, our estimate for the gain is $-R_L/R_S = -18 \text{ k}\Omega/2 \text{ k}\Omega = -9.00$. Our gain is lower than this prediction because the $g_m R_S$ product is not large compared to one. (See Ex. 13.7.) Checking the active region assumption: $V_{GS} - V_{TN} = 0.982 \text{ V}$ and $V_{DS} = 3.81 \text{ V}$. ✔

Discussion: Note that this C-S amplifier has been designed to operate at nearly the same Q-point as the C-E amplifier in Fig. 13.18, and R_S has been chosen to give about the same gain.

Computer-Aided Analysis: A SPICE operating point analysis (KP = 0.5 mA/V², VTO = 1 V, LAMBDA = 0.0133/V) yields the Q-point of (0.242 mA, 3.77 V). The slight variations result from including a nonzero value of λ. ac analysis yields a small-signal gain of −4.39. SPICE transient simulation results are given in the graphs below at a frequency of 10 kHz with TSTART = 0, TSTOP = 0.2 MS and TSTEP = 0.1 US. The first graph shows the results of an ac sweep from 0.1 Hz to 100 kHz with a 1-V input signal to identify the region (midband) where the capacitors are effectively short circuits. From the graph, we find that the gain is constant at −4.39 frequencies above 10 Hz. The second graph shows the result from the transient simulation with a 0.4-V, 10-kHz sine wave as the input. Although this amplitude is at the small signal limit, we do not visually observe any significant distortion in the waveform.

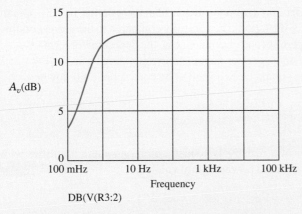

DB(V(R3:2))

Frequency response (as sweep) for a 1-V ac input signal.

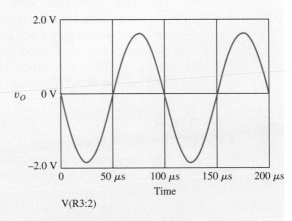

V(R3:2)

Transient response with $v_i = 0.4 \sin(20000\pi t)$ V.

EXERCISE: Calculate the Q-point for the transistor in Fig. 13.28.

EXERCISE: Convert the voltage gain in Ex. 13.6 to dB.

ANSWER: 13.1 dB

13.10.5 A DESIGN GUIDE FOR THE COMMON-SOURCE AMPLIFIER WITH $R_S = 0$

When a resistive load is used with the common-source amplifier, we often try to achieve $R_3 \gg R_D$. For these conditions, the total load resistance on the collector of the transistor is approximately equal to R_D, and Eq. (13.99) can be reduced to

$$A_v^{CS} \cong -g_m R_D = -\frac{I_D R_D}{\left(\dfrac{V_{GS} - V_{TN}}{2}\right)} \tag{13.100}$$

using the expression for g_m from Eq. (13.74).

The product $I_D R_D$ represents the dc voltage drop across drain resistor R_D. This voltage is usually in the range of one-fourth to three-fourths the power supply voltage V_{DD}. Assuming $I_D R_D = V_{DD}/2$ and $V_{GS} - V_{TN} = 1$ V, we can rewrite Eq. (13.99) as

$$A_v^{CS} \cong -\frac{V_{DD}}{V_{GS} - V_{TN}} \cong -V_{DD} \tag{13.101}$$

Equation (13.101) is a basic rule of thumb for the design of the resistively loaded common-source amplifier; its form is very similar to that for the BJT in Eq. (13.54). The magnitude of the gain is approximately the power supply voltage divided by the internal gate drive ($V_{GS} - V_{TN}$) of the MOSFET. For a common-source amplifier operating from a 12-V power supply, Eq. (13.101) predicts the voltage gain to be -12.

Note that this estimate is an order of magnitude smaller than the gain for the BJT operating from the same power supply. Equation (13.100) should be carefully compared to the corresponding expression for the BJT, Eq. (13.53). Except in special circumstances, the denominator term $(V_{GS} - V_{TN})/2$ in Eq. (13.100) for the MOSFET is much greater than the corresponding term $V_T = 0.025$ V for the BJT, and the MOSFET voltage gain should be expected to be correspondingly lower.

DESIGN NOTE

The magnitude of the voltage gain of a resistively loaded common-source amplifier with zero source resistance is approximately equal to power supply voltage:

$$A_v^{CS} \cong -V_{DD} \qquad \text{for} \qquad R_S = 0$$

This result represents an excellent way to quickly check the validity of more detailed calculations.

13.10.6 SMALL-SIGNAL LIMIT FOR THE COMMON-SOURCE AMPLIFIER

The presence of resistor R_S in series with the source of the MOSFET in the common-source amplifier increases the signal handling capability of the amplifier, just as for the case of the BJT. Using Eqs. (13.92), (13.94) and (13.82) and assuming $R_G \gg R_I$,

$$v_{gs} = \frac{v_i}{1 + g_m R_S} \le 0.2(V_{GS} - V_{TN}) \quad \text{or} \quad v_i \le 0.2(V_{GS} - V_{TN})(1 + g_m R_S) \tag{13.102}$$

The permissible input signal is increased by the factor $(1 + g_m R_S)$, which can be much larger than 1.

EXAMPLE **13.7**   COMMON-SOURCE VOLTAGE GAIN WITH BYPASSED SOURCE

Now we will find the voltage gain of the amplifier in Ex. 13.6 with bypass capacitor C_3 connected between ground and the source terminal of the FET.

PROBLEM   (a) Find the voltage gain of the amplifier in Ex. 13.6 with bypass capacitor C_3 connected between ground and the source terminal of the FET. (b) Compare the result in (a) to the common-source "rule-of-thumb" gain estimate and the amplification factor of the transistor. (c) What is the largest value of v_i that can be considered to be a small-signal?

SOLUTION   **Known Information and Given Data:** Common-source amplifier in Fig. 13.28 with source terminal bypassed. From Ex. 13.6, Q-point = (0.241 mA, 371 V), g_m = 0.503 mS, r_o = 328 kΩ.

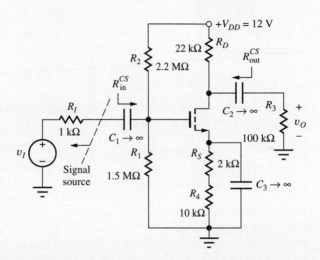

Unknowns: Actual voltage gain, rule-of-thumb estimate, amplification factor of the transistor, small-signal limit for v_i

Approach: Evaluate the A_v^{CS} expression with $R_S = 0$; estimate the voltage gain using Eq. (13.101); calculate $\mu_f = g_m r_o$.

Assumptions: The transistor is operating in the active region. Signal amplitudes correspond to small-signal conditions. Transistor output resistance r_o can be neglected.

Analysis:

(a)   With the source bypassed, $R_S = 0$ and

$$A_v^{CS} = -g_m R_L \frac{R_G}{R_G + R_I} = -0.503 \text{ mS}(18.0 \text{ k}\Omega)\frac{892 \text{ k}\Omega}{892 \text{ k}\Omega + 1 \text{ k}\Omega} = -9.04 \quad \text{or} \quad 19.1 \text{ dB}$$

(b)   Our "rule-of-thumb" estimate for the voltage gain is $A_v = -V_{DD} = -12$, which somewhat overestimates the actual gain.

For the given Q-point,

$$V_{GS} - V_{TN} \cong \sqrt{\frac{2I_{DS}}{K_n}} = \sqrt{\frac{2 \times 0.241 \times 10^{-3} \text{ A}}{5 \times 10^{-4} \dfrac{A}{V^2}}} = 0.982 \text{ V}$$

and our simple estimate for the gain is

$$A_v^{CS} \cong -\frac{V_{DD}}{V_{GS} - V_{TN}} = -\frac{12 \text{ V}}{0.982 \text{ V}} = -12.2$$

Our more detailed calculation is consistent with this estimate.

The amplification factor of the MOSFET is equal to

$$\mu_f = \frac{\dfrac{1}{\lambda} + V_{DS}}{\dfrac{V_{GS} - V_{TN}}{2}} = \frac{(75.2 + 3.71) \text{ V}}{0.491} = 161$$

With the source bypassed, essentially all of the input signal appears directly across the gate-source terminals of the transistor. The small-signal limit on the input signal is therefore

$$|v_{gs}| \leq 0.2(V_{GS} - V_{TN}) = 0.2(0.982 \text{ V}) = 0.196 \text{ V} \qquad \text{so} \qquad |v_{gs}| \leq 0.196 \text{ V}$$

Check of Results: The rule-of-thumb estimates are in reasonable agreement with the actual gains. The voltage gain is much less than the amplification factor, so neglect of r_o is valid.

Discussion: The rule-of-thumb produces a good estimate for the gain of this amplifier. Although the amplification factor for this MOSFET is much smaller than that for the BJT, the gain of this resistively loaded amplifier circuit is still not limited by amplification factor μ_f.

Computer-Aided Analysis: SPICE simulation yields a Q-point of (0.242 mA, 3.77 V) that is consistent with the assumed value. An ac sweep from 0.1 Hz to 100 kHz with 10 frequency points/decade is used to find the region in which the capacitors are acting as short circuits, and the gain is observed to be constant at 18.7 dB above a frequency of 10 Hz. The voltage gain is slightly less than our calculated value because r_o was neglected in our calculations. A transient simulation was performed with a 0.15-V, 10-kHz sine wave. The output exhibits reasonably good linearity, but note that the positive and negative amplitudes are slightly different, indicating some waveform distortion.

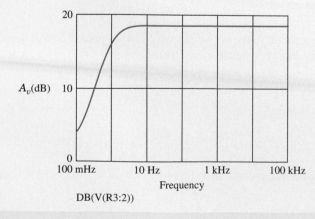

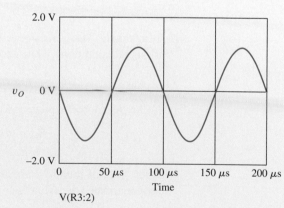

> **EXERCISE:** Draw the small-signal ac equivalent circuit for the amplifier in Ex. 13.7 including the transistor output resistance. What is the total load resistance on the transistor? What is the new value of the voltage gain?
>
> **ANSWERS:** $R_L = r_o \| R_D \| R_3 = 328 \text{ k}\Omega \| 22 \text{ k}\Omega \| 100 \text{ k}\Omega = 17.1 \text{ k}\Omega$
>
> $$A_v^{CS} = -0.503 \text{ mS } (17.1 \text{ k}\Omega)\frac{892 \text{ k}\Omega}{892 \text{ k}\Omega + 1 \text{ k}\Omega} = -8.59 \text{ or } 18.7 \text{ dB}$$
>
> **EXERCISE:** Suppose we increase the transconductance parameter of the transistor to $K_n = 2 \times 10^{-3}$ A/V^2 by increasing the W/L ratio of the device. If the drain current is kept the same, find a new estimate for the voltage gain in Ex. 13.7. By what factor was the W/L ratio increased?
>
> **ANSWERS:** -12.5; 4

13.10.7 INPUT RESISTANCES OF THE COMMON-EMITTER AND COMMON-SOURCE AMPLIFIERS

If the voltage gain of the MOSFET amplifier is generally much lower than that of the BJT, there must be other reasons for using the MOSFET. One of the reasons was mentioned earlier: A small signal can be much larger for the MOSFET than for the BJT. Another important difference is in the relative size of the input impedance of the amplifiers. This section explores the input resistances of the common-emitter and common-source amplifiers.

The input resistance R_in to the common-emitter and common-source amplifiers is defined in Figs. 13.30(a) and (b) to be the total resistance looking into the amplifier at coupling capacitor C_1. R_in represents the total resistance presented to the signal source represented by v_I and R_I. The input resistance definition is repeated in Fig. 13.31, in which the amplifiers have been reduced to their ac equivalent circuits.

Common-Emitter Input Resistance

Let us first calculate the input resistance for the common-emitter stage. In Fig. 13.32, the BJT has been replaced by its small-signal model, and input resistance $R_\text{in} = v_x/i_x$, where i_x has two components:

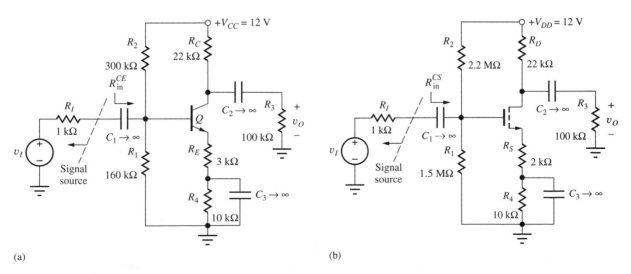

(a)

(b)

Figure 13.30 (a) Input resistance definition for the common-emitter amplifier. (b) Input resistance definition for the common-source amplifier.

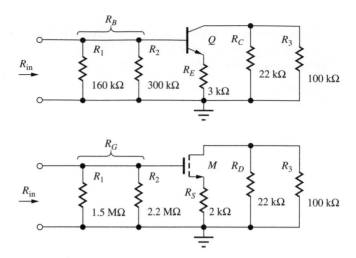

Figure 13.31 ac Equivalent circuits for the input resistance for the common-emitter and common-source amplifiers of Fig. 13.30.

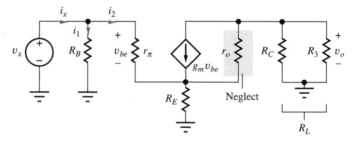

Figure 13.32 Input resistance for the common-emitter amplifier.

current i_1 going down through R_B and current i_2 going into the base of the transistor. The input resistance is the parallel combination of R_B and R_{iB}. More formally,

$$G_{\text{in}}^{CE} = \frac{1}{R_{\text{in}}^{CE}} = \frac{i_x}{v_x} = \frac{i_i}{v_x} + \frac{i_2}{v_x} = \frac{1}{R_B} + \frac{1}{R_{iB}}$$

The input conductance is the sum of two components. Thus, the input resistance is the parallel combination of base bias resistor R_B and R_{iB}, the resistance looking into the base of the common-emitter stage from Eq. (13.103):

$$R_{\text{in}}^{CE} = R_B \| R_{iB} = R_B \| [r_\pi + (\beta_o + 1)R_E] \tag{13.103}$$

EXAMPLE 13.8   INPUT RESISTANCE OF THE COMMON-EMITTER AMPLIFIER

Let us calculate R_{in} for the amplifier in Fig. 13.32 for a given Q-point.

PROBLEM   (a) Find the input resistance for the common-emitter amplifier in Figs. 13.31 and 13.32. The Q-point is (0.245 mA, 3.39 V). (b) Repeat the calculation if the bypass capacitor is connected between the transistor's emitter and ground.

SOLUTION   **Known Information and Given Data:** The small-signal circuit topology appears in Fig. 13.32. The Q-point is given as (0.245 mA, 3.39 V). From Fig. 13.30, we have $R_1 = 160\,\text{k}\Omega$, $R_2 = 300\,\text{k}\Omega$, $R_3 = 100\,\text{k}\Omega$, $R_E = 3\,\text{k}\Omega$.

Unknowns: Input resistance looking into the common-emitter amplifier (a) With emitter resistor $R_E = 3\,\text{k}\Omega$, (b) With $R_E = 0$.

Approach: Find r_π and use Eq. (13.103) to find the input resistance.

Assumptions: Small-signal conditions apply, $\beta_o = 100$, $V_T = 25$ mV

Analysis: The values of R_B and r_π are

$$R_B = R_1 \| R_2 = 160\,\text{k}\Omega \| 300\,\text{k}\Omega = 104\,\text{k}\Omega \quad \text{and} \quad r_\pi = \frac{\beta_o V_T}{I_C} = \frac{100(0.025)}{0.245\ \text{mA}} = 10.2\,\text{k}\Omega$$

(a) $R_{\text{in}}^{CE} = R_B \| [r_\pi + (\beta_o + 1) R_E] = 104\,\text{k}\Omega \| [10.2\,\text{k}\Omega + (101)3\,\text{k}\Omega] = 78.1\,\text{k}\Omega$

(b) If the emitter of the transistor is bypassed, then R_E is zero in the expression for R_{iB}, and the input resistance to the transistor would be much smaller:

$$R_{\text{in}}^{CE} = R_B \| R_{iB} = R_B \| r_\pi = 104\,\text{k}\Omega \| 10.2\,\text{k}\Omega = 9.29\,\text{k}\Omega$$

Check of Results: The input resistance must be smaller than any one of the resistors R_1, R_2, or R_{in}^{CE}, since they all appear in parallel. The calculated value of input resistance is consistent with this observation.

Discussion: With the emitter terminal bypassed, the input resistance to the amplifier, 9.29 kΩ, is quite low and is dominated by r_π. By inserting $R_E = 3$ kΩ in series with the emitter, we give up voltage gain in return for a significantly higher input resistance.

Computer-Aided Analysis: (a) We may use an ac analysis of the circuit from Fig. 13.30(a) to determine R_{in} by finding the signal current in source v_I. (Note that a TF analysis cannot be used because of the presence of capacitors in the network.) According to SPICE, the current produced by a 1-V sinusoidal input at 100 kHz is 12.7 μA, which yields an input resistance of 78.7 kΩ. R_I is in series with the input, and subtracting 1 kΩ yields an input resistance of 77.7 kΩ, in reasonable agreement with our hand calculations. The input resistance is also equal to the base voltage divided by the current entering the base terminal through C_1. From SPICE, VB(Q1)/I(C1) = 77.7 kΩ at frequencies above 1 kHz. (b) With the emitter bypassed, SPICE yields VB(Q1)I(C1) = 9.80 kΩ, which is 5 percent higher than our calculations. This discrepancy results from the values of ac current gain β_o and thermal voltage V_T used by SPICE, since both differ slightly from our hand calculations.

EXERCISE: What are the two values of R_{in} if the Q-point is changed to (0.725 mA, 3.86 V)?

ANSWER: (a) 77.6 kΩ; (b) 3.34 kΩ

Common-Source Input Resistance

Now let us compare the input resistance of the common-source amplifier to that of the common-emitter stage. In Fig. 13.33, the MOSFET in Fig. 13.30 has been replaced by its small-signal model. This circuit is similar to that in Fig. 13.32 except that $r_\pi \to \infty$. Because the gate terminal of the MOSFET itself represents an open circuit, the input resistance of the circuit is simply limited

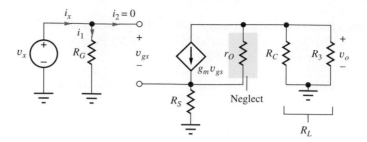

Figure 13.33 Input resistance for the common-source amplifier.

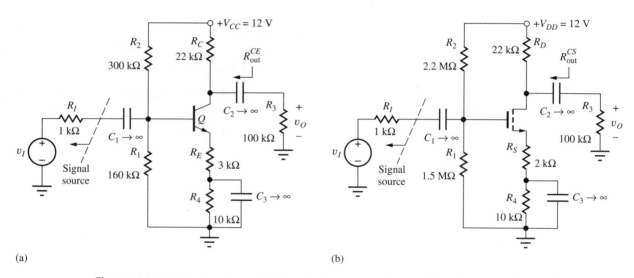

(a)

(b)

Figure 13.34 (a) Output resistance definition for the common-emitter amplifier. (b) Output resistance definition for the common-source amplifier.

by our value of R_G:

$$\mathbf{v_x} = \mathbf{i_x} R_G \qquad \text{and} \qquad R_{in}^{CS} = R_G \qquad (13.104)$$

In the C-S amplifier in Figs. 13.30, $R_G = 2.2\,\text{M}\Omega \| 1.5\,\text{M}\Omega = 892\,\text{k}\Omega$, so $R_{in}^{CS} = 892\,\text{k}\Omega$. We see that the input resistance of the C-S amplifier can easily be much larger than that of the corresponding C-E stage.

> **EXERCISE:** What is the input resistance of the common-source amplifier in Fig. 13.30(b) if $R_2 = 1.0\,\text{M}\Omega$ and $R_1 = 680\,\text{k}\Omega$? Is the Q-point of the amplifier changed?
>
> **ANSWERS:** 405 kΩ; no, the Q-point remains the same because $I_G = 0$ and the dc voltage at the gate is unchanged.

13.10.8 COMMON-EMITTER AND COMMON-SOURCE OUTPUT RESISTANCES

The output resistances of the C-E and C-S amplifiers are defined in Figs. 13.34(a) and (b) as the total equivalent resistance looking into the output of the amplifier at coupling capacitor C_3. The definition of the output resistance is repeated in Fig. 13.35, in which the two amplifiers have been reduced to their ac equivalent circuits. For the output resistance calculation, input source v_I is set to zero.

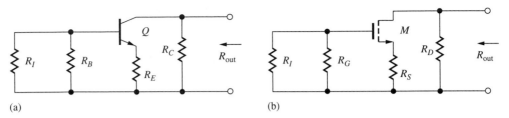

Figure 13.35 Output resistance definition for (a) common-emitter and (b) common-source amplifiers.

Output Resistance of the Common-Emitter Amplifier

From Fig. 13.35(a) we see that the total output resistance R_{out}^{CE} is equal to R_C in parallel with resistance R_{iC} looking into the collector of the biploar transistor (see Eqs. 13.66 and 13.67):

$$R_{\text{out}}^{CE} = R_C \| R_{iC} \cong R_C \| r_o[1 + g_m(r_\pi \| R_E)] \tag{13.105}$$

For nonzero R_E, R_{iC} is typically much larger than r_o. If R_E is zero, then Eq. (13.105) reduces to

$$R_{\text{out}}^{CE} = R_C \| r_o \quad \text{for} \quad R_E = 0 \tag{13.106}$$

Let us compare the values of r_o and R_C by multiplying each by I_C:

$$I_C r_o = I_C \frac{V_A + V_{CE}}{I_C} \cong V_A \quad \text{and} \quad I_C R_C \cong \frac{V_{CC}}{2} \tag{13.107}$$

As discussed previously, the dc voltage drop across R_C is typically a small fraction of the power supply voltage, whereas we see that the apparent voltage across r_o is the Early voltage. Thus, we expect $r_o \gg R_C$, and Eqs. (13.105) and (13.106) are both dominated by R_C:

$$R_{\text{out}}^{CE} \cong R_C \tag{13.108}$$

Output Resistance of the Common-Source Amplifier

The situation is the same in Fig. 13.35(b). The total output resistance R_{out}^{CS} is equal to R_D in parallel with the resistance R_{iD} looking into the drain of the FET:

$$R_{\text{out}}^{CS} = R_D \| R_{iD} \tag{13.109}$$

The small-signal model for the FET is the same as the bipolar transistor, except that $r_\pi = \infty$ so that

$$R_{iD} = r_o(1 + g_m R_S) \quad \text{and} \quad R_{\text{out}}^{CS} = R_D \| R_{iD} \cong R_D \| r_o(1 + g_m R_S) \tag{13.110}$$

Here again $r_o \gg R_D$, and the total output resistance of the common-source amplifier is determined by R_D:

$$R_{\text{out}}^{CS} \cong R_D \tag{13.111}$$

13.11 EXAMPLES OF COMMON-EMITTER AND COMMON-SOURCE AMPLIFIERS

This section presents examples of analysis of common-emitter and common-source amplifiers to provide additional examples of methods used to establish transistor Q-points. The circuits use symmetrical power supplies.

13.11.1 A COMMON-EMITTER AMPLIFIER

EXAMPLE 13.9   **A COMMON-EMITTER AMPLIFIER BIASED WITH DUAL POWER SUPPLIES**

The common-emitter stage shown schematically in Fig. 13.36 is biased by symmetrical positive and negative power supplies. The use of the second supply permits elimination of one of the bias resistors in the base circuit and is very common in both discrete and integrated circuit design. The remaining 100-kΩ resistor is required to isolate the base node from ground so that an input-signal voltage can be applied to the base.

PROBLEM Find the input resistance, output resistance, and overall voltage gain for the common-emitter amplifier in Fig. 13.36.

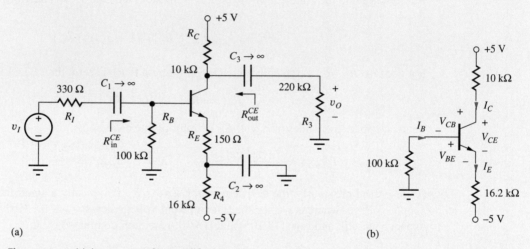

Figure 13.36 (a) A common-emitter amplifier biased by two supplies. The BJT parameters are $\beta_F = 65$ and $V_A = 50$ V. (b) dc Equivalent circuit for Fig. 13.36(a).

SOLUTION **Known Information and Given Data:** The circuit topology with element values appears in Fig. 13.36. The transistor parameters are specified in the figure caption to be $\beta_F = 65$ and $V_A = 50$ V.

Unknowns: Q-point (I_C, V_{CE}), small-signal parameters, R_{in}^{CE}, R_{out}^{CE}, and A_v^{CE}

Approach: To analyze the circuit, we first draw the dc equivalent circuit and find the Q-point. We then develop the ac equivalent circuit, find the small-signal model parameters, and characterize the small-signal properties of the amplifier.

Assumptions: Active region operation with $V_{BE} = 0.7$ V; V_A can be ignored in dc bias calculations; small-signal operating conditions apply; $V_T = 25$ mV

Q-Point Analysis: The first step is to draw the dc equivalent circuit. Opening the three capacitors yields the simplified circuit in Fig. 13.36(b). The Q-point can be found by first calculating the base current, then the collector current, and finally the collector-emitter voltage. For the input loop containing the base-emitter junction using $I_E = (\beta_F + 1)I_B$,

$$10^5 I_B + V_{BE} + (\beta_F + 1)I_B(1.62 \times 10^4) = 5 \tag{13.112}$$

or

$$10^5 I_B + 0.7 + 66 I_B(1.62 \times 10^4) = 5 \tag{13.113}$$

Solving for the base current yields

$$I_B = \frac{(5 - 0.7) \text{ V}}{10^5 \ \Omega + 1.06 \times 10^6 \ \Omega} = 3.67 \ \mu\text{A}$$

Then   (13.114)

$$I_C = 65 I_B = 239 \ \mu\text{A} \qquad \text{and} \qquad I_E = 66 I_B = 243 \ \mu\text{A}$$

Writing an equation for the output loop containing V_{CE},

$$5 - 10^4 I_C - V_{CE} - 1.62 \times 10^4 I_E - (-5) = 0 \qquad (13.115)$$

and solving for V_{CE} yields:

$$V_{CE} = 10 - 10^4 I_C - 1.62 \times 10^4 I_E = 3.67 \text{ V} \qquad (13.116)$$

Check of Results and Discussion: In writing Eqs. (13.112) and (13.115), we have assumed that the transistor is in the active region of operation, and the dc analysis is not complete until we check this assumption. Writing an expression for V_{CB} yields

$$5 - 10^4 I_C - V_{CB} + 10^5 I_B = 0 \qquad \text{or} \qquad V_{CB} = 5 - 2.41 + 0.371 = 2.22 > 0 \quad (13.117)$$

Because $V_{CB} > 0$ ($V_{BC} < 0$), the transistor is indeed in the active region and the Q-point is

$$(I_C, V_{CE}) = (0.239 \text{ mA}, 3.67 \text{ V}) \qquad (13.118)$$

It is important to note in this dc analysis that we used the simplified form of the transport model, in which $V_A = \infty$. As discussed in previous chapters, we want to use the lowest complexity model that provides reasonable answers. Using V_{CE} from our analysis, we see that $V_{CE}/V_A = 3.67 \text{ V}/50 \text{ V} = 0.0734$. Including the Early voltage term would change our answers by less than 10 percent but would considerably complicate the dc analysis.

ac ANALYSIS The next step is to draw the ac equivalent circuit and simplify it before beginning the detailed analysis. For the ac analysis, we replace all capacitors by short circuits and the dc voltage sources with ground connections. This step appears in Fig. 13.37(a).

The circuit in Fig. 13.37(a) is redrawn in simplified form in Fig. 13.37(b), in which the 16-kΩ resistor has been removed because it is "shorted out" by ground connections at both ends, and the parallel connection of the two resistors attached to the collector is shown more explicitly. Once we become skilled with this process, we can skip directly to Fig. 13.37(b) without the intermediate step.

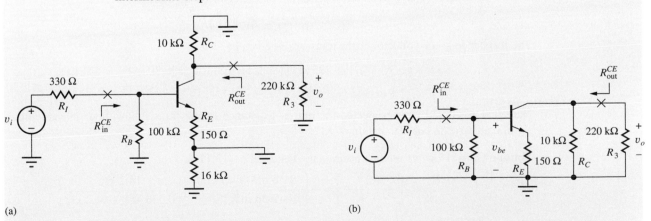

(a)   (b)

Figure 13.37 (a) First step in construction of the ac equivalent circuit. (b) Redrawn circuit of Fig. 13.37(a).

SMALL-SIGNAL ANALYSIS We wish to find the voltage gain from v_i to v_o for the amplifier in Fig. 13.37. The output voltage at the collector terminal is related to the voltage at the base by the terminal gain in Eq. (13.47), where R_L is the total load resistance at the collector terminal and equal to R_{out}^{CE} in parallel with external load resistor R_3. Base emitter voltage v_{be} is related to v_i through voltage division between the source resistance R_I and input resistance R_{in}^{CE}.

Combining these results yields the expression for the overall voltage gain from Eq.(13.50):

$$A_v^{CE} = A_{vt}^{CE} \left(\frac{v_b}{v_i} \right) = -\left(\frac{g_m R_L}{1 + g_m R_E} \right) \left[\frac{R_{in}^{CE}}{R_I + R_{in}^{CE}} \right] \tag{13.119}$$

To evaluate this expression, we will first find R_{in}^{CE} and R_{out}^{CE} and use the values to find A_v.

The final step prior to mathematical analysis is to find the small-signal model parameters. Using the Q-point values,

$$g_m = 40 I_C = \frac{40}{V}(2.39 \times 10^{-4}\ A) = 9.56 \times 10^{-3}\ S$$

$$r_\pi = \frac{\beta_o V_T}{I_C} = \frac{65(0.025\ V)}{2.39 \times 10^{-4}\ A} = 6.80\ k\Omega$$

$$r_o = \frac{V_A + V_{CE}}{I_C} = \frac{(50 + 3.67)\ V}{2.39 \times 10^{-4}\ A} = 225\ k\Omega$$

Input Resistance: The input resistance is defined looking into the amplifier at the position of coupling capacitor C_1 in Figs. 13.36(a) and 13.37(a). The simplified ac model used to calculate R_{in}^{CE} appears in Fig. 13.38(a). The input resistance is

$$R_{in}^{CE} = R_B \| R_{iB} = R_B \| r_\pi (1 + g_m R_E) = 100\ k\Omega \| 6.80\ k\Omega [1 + 9.5\ mS(150\ \Omega)] = 14.2\ k\Omega$$

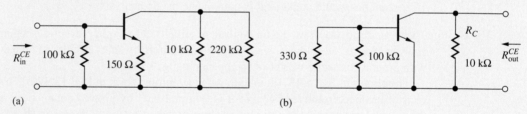

(a)   (b)

Figure 13.38 (a) ac Model for calculating R_{in}. (b) ac circuit model for calculating R_{out}.

Load Resistance: The total load resistance on the collector of the transistor is given by

$$R_L = R_{out}^{CE} \| R_3 = R_{iC} \| R_C \| R_3$$

The resistance at the collector of the transistor is given by

$$R_{iC} \cong r_o[1 + g_m(R_E \| r_\pi)] = 225\ k\Omega\ [1 + 9.56\ mS(150\ \Omega \| 6.80\ k\Omega)] = 541\ k\Omega$$

$$R_L = R_{iC} \| R_C \| R_3 = 541\ k\Omega \| 10\ k\Omega \| 220\ k\Omega = 9.40\ k\Omega$$

Note that we could have neglected the output resistance of the transistor without introducing significant error in our claculation of R_L.

Voltage Gain: Now we are in a position to evaluate Eq. (13.119).

$$A_v^{CE} = -\left(\frac{g_m R_L}{1 + g_m R_E} \right) \left[\frac{R_{in}^{CE}}{R_I + R_{in}^{CE}} \right] = -\left[\frac{9.56\ mS(9.40\ k\Omega)}{1 + 9.56\ mS(150\ \Omega)} \right] \left[\frac{14.2\ k\Omega}{0.330\ k\Omega + 14.2\ k\Omega} \right] = -36.1$$

Check of Results: We have found the answers requested in the problem. Our first-order estimate for the voltage gain would be $A_v^{CE} = -R_C/R_E = -10\ k\Omega/0.15\ k\Omega = -66.7$ so the calculated gain appears reasonable. Looking at the circuit in Fig. 13.37(a), we also see that the input and output resistances cannot exceed 100 kΩ and 10 kΩ, respectively, which agree with our more detailed calculations. In summary, the BJT amplifier of Fig. 13.36(a) has the following characteristics:

$$A_v = -36.1 \qquad R_{in} = 14.2 \text{ k}\Omega \qquad R_{out}^{CE} = R_{iC} \| R_C = 541 \text{ k}\Omega \| 10 \text{ k}\Omega = 9.82 \text{ k}\Omega$$

Computer-Aided Analysis: Our dc analysis can easily be confirmed by SPICE simulation of the circuit below, which yields a Q-point of (231 μA, 3.90 V). Note that the bias circuit stabilizes the Q-point with respect to variations in device parameters, and the changes in Q-point are actually even less than predicted by the V_{CE}/V_A estimate. In addition, these differences in Q-point values are much smaller than the uncertainty in either the bias resistor values or the device parameter values for a real circuit, so neglecting V_A does not introduce significant additional uncertainty.

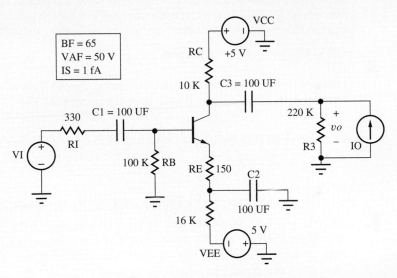

For ac analysis, the capacitors are set to large values so their impedances are small at the frequencies of interest. In this case, 100-μF capacitors are used. In Chapter 14, we will find how to choose the values for these capacitors. Source IO is an ac source added to determine the output resistance. An ac analysis with FSTART = 1000 Hz, FSTOP = 100 kHz, five frequency points per decade, a 1-V value for source ac VI, and a zero value for I_O yields $A_v = -34.9$. The input resistance to the amplifier is given by the ratio of the voltage at the base of transistor Q_1 to the current entering the base node through C_1: R_{in}^{CE} = VB(Q1)/I(C1). At 100 kHz, this value is 14.8 kΩ. The output resistance can be found by driving the output with a 1-A ac current source IO with VI set to zero. The output resistance of the amplifier is then found by dividing the voltage at the collector of transistor Q_1 by the current entering the collector node through C_3: R_{out}^{CE} = VC(Q1)/I(C3). At 100 kHz, this value is 9.81 kΩ. Our hand analysis results are confirmed. The small discrepancies result from the values of ac current gain β_o(69.2) and thermal voltage V_T (25.83 mV) used by SPICE, since both differ slightly from our hand calculations. Remember in SPICE, β_o = BF*(1+VCB/VA) = 69.2, and T defaults to 300 K.

EXERCISE: What would be the voltage gain of the amplifier in Ex. 13.9 if the output resistance of the transistor were neglected in the calculation (i.e., $V_A = \infty$)?

ANSWERS: −36.7, a negligible change

EXERCISE: What is the amplification factor of the BJT characterized by the parameters in Fig. 13.38? How does A_v compare to μ_f?

ANSWERS: 2220; $|A_v| \ll \mu_f$

EXERCISE: What is the largest value of v_i that corresponds to a small signal for the BJT amplifier in Fig. 13.38? What is the largest value of v_o that corresponds to a small signal in this amplifier?

ANSWERS: $5\,\text{mV}(2.43) = 12.2\,\text{mV}; (12.2\,\text{mV})(36.1) = 0.439\,\text{V}$

EXERCISE: Perform a transient simulation of the amplifier in Ex. 13.9 with a 12.2 mV sinusoidal input signal at a frequency of 20 kHz. Does the output appear to be an undistorted sine wave? Use the Fourier analysis capability of SPICE (part of the transient analysis) to find the total harmonic distortion (the results are in the output file).

ANSWERS: Yes; THD = 1.8 percent

EXERCISE: What would be the input resistance, output resistance, and voltage gain of the amplifier in Ex. 13.9 if capacitor C_2 were connected between the emitter of the transistor and ground?

ANSWERS: $100\,\text{k}\Omega \| 9.8\,\text{k}\Omega = 6.37\,\text{k}\Omega$, $10\,\text{k}\Omega \| 225\,\text{k}\Omega = 9.57\,\text{k}\Omega$, $-(9.56\,\text{mS})(9.18\,\text{k}\Omega)(0.951) = -83.5$

13.11.2 ac VERSUS TRANSIENT ANALYSIS IN SPICE—ANOTHER VISIT

In Ex. 13.9, you may have noted that the ac simulation utilized a 1-V value for the ac voltage source at the input and a 1-A value for the ac current source at the output. These values are far too large to be considered small-signals in this amplifier! On the other hand, in transient analyses we are careful to use small input signals. So what is going on?

The ac analysis assumes that the network is linear and uses the small-signal model for the amplifier. Since it is linear, any convenient value can be used for the signal source amplitudes, hence the choice of 1-V and 1-A sources. On the other hand, transient simulations utilize the full large-signal nonlinear models of the transistors. If we desire linear behavior, all signals must satisfy the small-signal constraints. It is extremely important to be sure to understand the subtle differences between the conditions used for ac analysis and transient analysis in SPICE!

13.11.3 A MOSFET COMMON-SOURCE AMPLIFIER

EXAMPLE **13.10** A MOSFET COMMON-SOURCE AMPLIFIER

A MOSFET in the common-source configuration is shown in Fig. 13.39. Bias is provided by pair of symmetrical power supplies. Capacitors C_1 and C_3 couple the ac signal into and out of the amplifier, respectively. Bypass capacitor C_2 provides an ac ground between resistors R_S and R_4.

PROBLEM Find the input resistance, output resistance, and voltage gain for the common-source amplifier in Fig. 13.39.

SOLUTION **Known Information and Given Data:** The circuit topology with element values appears in the circuit Fig. 13.39. The transistor parameters are specified in the figure to be $K_n = 500\,\mu\text{A/V}^2$, $V_{TN} = 1$ V, and $\lambda = 0.0133\,\text{V}^{-1}$.

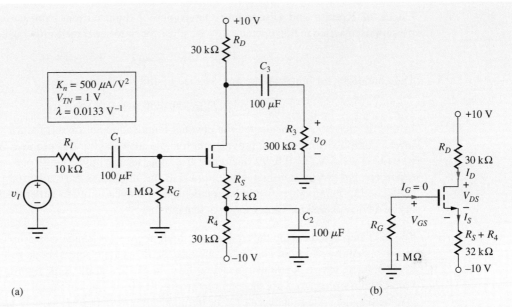

$$K_n = 500\ \mu A/V^2$$
$$V_{TN} = 1\ V$$
$$\lambda = 0.0133\ V^{-1}$$

Figure 13.39 (a) MOSFET common-source amplifier. (b) dc equivalent circuit.

Unknowns: Q-point (I_D, V_{DS}), small-signal parameters, R_{in}, R_{out}, and A_v

Approach: To analyze the circuit, we first draw the dc equivalent circuit and find the Q-point. Then we develop the ac equivalent circuit, find the small-signal model parameters, and characterize the small-signal properties of the amplifier.

Assumptions: Active region operation for the MOSFET; λ can be ignored in dc bias calculations; small-signal operating conditions apply; $T = 300$ K

Q-Point Analysis: Analysis begins by constructing the dc equivalent circuit in Fig. 13.39(b), which is obtained by replacing the capacitors by open circuits. To find the Q-point, we first find an expression for the drain-source voltage of the MOSFET.

$$10 - 3 \times 10^4 I_D - V_{DS} - 3 \times 10^4 I_S - (-10) = 0 \quad \text{or} \quad V_{DS} = 20 - 6.2 \times 10^4 I_D$$

since the drain and source currents are equal. The drain current is given by

$$I_D = \frac{K_n}{2}(V_{GS} - V_{TN})^2 \quad \text{for} \quad V_{DS} \geq V_{GS} - V_{TN}$$

and a second relationship can be found between I_D and V_{GS} using the gate-source loop.

$$10^6 I_G + V_{GS} + 3.2 \times 10^4 I_s - 10 = 0 \quad \text{or} \quad 10 - V_{GS} = 3.2 \times 10^4 I_D$$

in which we have used $I_G = 0$ and $I_S = I_D$. Substituting the drain current expression for I_D yields a quadratic equation in V_{GS}.

$$8V_{GS}^2 - 15V_{GS} - 2 = 0 \quad \text{and} \quad V_{GS} = -0.125\ V \text{ or } + 2.00\ V.$$

V_{GS} must exceed the 1-V threshold voltage for the transistor to be conducting, so V_{GS} must equal 2.00 V. Therefore,

$$I_D = \frac{500\ \mu A}{2\ V^2}(2V - 1V)^2 = 250\ \mu A \quad \text{and} \quad V_{DS} = 20 - 6.2 \times 10^4(2.5 \times 10^{-4}) = 4.50\ V$$

Check of Results and Discussion: Our choice of drain current expression for the MOSFET presupposed active region operation, so we must check to see if the following equation is satisfied:

$$V_{DS} \geq V_{GS} - V_{TN} \quad \text{and} \quad 4.5 > 1 \quad ✔$$

The conditions for pinch-off are met, so the Q-point in this circuit is

$$(I_D, V_{DS}) = (250 \ \mu A, 4.5 \ V)$$

In this dc analysis, we neglected the channel-length modulation term since we want to use the lowest complexity model that provides reasonable answers. For this problem, we see that λV_{DS} is $(0.0133 \ V^{-1})(5 \ V) = 0.0599$. Including the λV_{DS} term would change our answers by less than 6 percent but would considerably complicate the dc analysis. In addition, any differences in calculated Q-point values would be smaller than the uncertainty in either the bias resistor values or the device parameter values for an actual circuit.

ac ANALYSIS The next step is to develop the equivalent circuit needed for ac analysis. Replacing the capacitors in Fig. 13.39 by short circuits and grounding the dc voltage source node results in the circuit in Fig. 13.40(a), which is redrawn in clearer form in part (b) of the same figure. We wish to find the voltage gain from v_i to v_o for the amplifier in Fig. 13.40(b). The output voltage at the drain terminal is related to the voltage at the gate by the terminal gain in Eq. (13.93), $v_o = -g_m R_L v_{gs}/(1+g_m R_S)$, where R_L is the total load resistance at the drain terminal and equal to R_{out}^{CS} in parallel with external load resistor R_3, $R_L = R_{out}^{CS} \| R_3$. Neglecting r_o, $R_2 = R_D \| R_3$. Gate-source voltage v_{gs} is related to v_i through voltage division between the source resistance R_I and input resistance R_{in}^{CS} where $R_{in}^{CS} = R_G$. Combining these results yields an expression for the overall voltage gain:

$$A_v^{CS} = \frac{v_o}{v_i} = -\frac{g_m(R_D \| R_3)}{1 + g_m R_S} \left(\frac{R_G}{R_I + R_G} \right) \tag{13.120}$$

Using the Q-point values to find the small-signal parameters,

$$g_m = \sqrt{2K_n I_{DS}(1 + \lambda V_{DS})}$$

$$= \sqrt{2 \left(5 \times 10^{-4} \frac{A}{V^2} \right) (2.50 \times 10^{-4} \ A) \left(1 + \frac{0.0133}{V} 4.5 \ V \right)} = 5.15 \times 10^{-4} \ S$$

$$r_o = \frac{\left(\frac{1}{\lambda} \right) + V_{DS}}{I_D} = \frac{(60 + 4.5) \ V}{2.50 \times 10^{-4} \ A} = 258 \ k\Omega$$

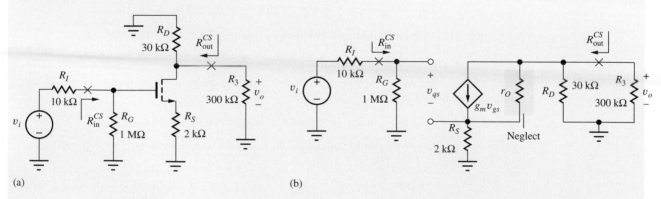

(a)   (b)

Figure 13.40 (a) Initial step in constructing the ac equivalent circuit. (b) Redrawn equivalent of circuit in (a).

Voltage Gain: The voltage gain is found to be

$$A_v^{CS} = \frac{0.515 \text{ mS } (30 \text{ k}\Omega \| 300 \text{ k}\Omega)}{1 + 0.515 \text{ mS } (2 \text{ k}\Omega)} \left(\frac{1 \text{ M}\Omega}{10 \text{ k}\Omega + 1 \text{M}\Omega} \right) = -6.85$$

Input Resistance: As mentioned above, the input resistance simply equals the value of gate resistance R_G.

$$R_{\text{in}}^{CS} = R_G = 1 \text{ M}\Omega$$

Output Resistance: R_{out}^{CS} represents the Thévenin equivalent resistance looking into coupling capacitor C_3 in Fig. 13.39, as redrawn in Fig. 13.41. The expression for R_{out}^{CS} was calculated earlier in Eq. 13.110:

$$R_{iD} = r_o(1 + g_m R_S) = 258 \text{ k}\Omega[1 + 5.15 \times 10^{-4} S(2000 \text{ }\Omega)] = 524 \text{ k}\Omega$$

$$R_{\text{out}}^{CS} = R_D \| R_{iD} = 30 \text{ k}\Omega \| 524 \text{ k}\Omega = 28.4 \text{ k}\Omega$$

As we expect, the resistance presented at the drain terminal of the transistor is much larger than the value of R_D. Thus, we could have safely neglected the effect of r_o on the overall output resistance.

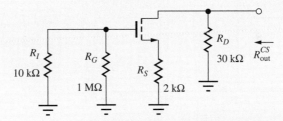

Figure 13.41 Simplified circuit for output resistance calculation.

Check of Results: We have found the answers requested in the problem. Our basic rule-of-thumb estimate for the voltage gain would be $A_v^{CS} \cong -R_D/R_S = -15$. The actual gain is less than this limit since we do not satisfy the requirement that $g_m R_S \gg 1$. Looking at the circuit in Fig. 13.50, we quickly see that the input and output resistances should not exceed $R_G = 1$ MΩ and $R_D = 30$ kΩ, respectively, which also agree with our more detailed calculations. In summary, the MOSFET amplifier in Fig. 13.39 has the following characteristics:

$$A_v^{CS} = -6.85 \qquad R_{\text{in}}^{CS} = 1 \text{ M}\Omega \qquad R_{\text{out}}^{CS} = 28.4 \text{ k}\Omega$$

Computer-Aided Analysis: A SPICE operating point analysis of the circuit below gives the Q-point (251 μA, 4.45 V). For ac analysis, the capacitors are set to large values so their impedances are small at the frequencies of interest. In this case, 100-μF capacitors are used. In Chapters 14 and 16, we will find how to choose the values for these capacitors. An ac analysis (DEC, FSTART = 0.1 Hz, FSTOP = 100 kHz, and 10 points/decade) with 1-V value for ac source VI (and IO = 0) yields $A_v = -6.56$. The input resistance can be found as VG(M1)/I(C1) and is equal to 1.00 MΩ at high frequencies where the impedance of the capacitors is negligible. The output resistance is found by driving the output with a 1-A ac source and plotting V(IO+)/I(C3). At high frequencies, SPICE yields $R_{\text{out}}^{CS} = 28.7$ kΩ. Our hand analysis results are once again confirmed.

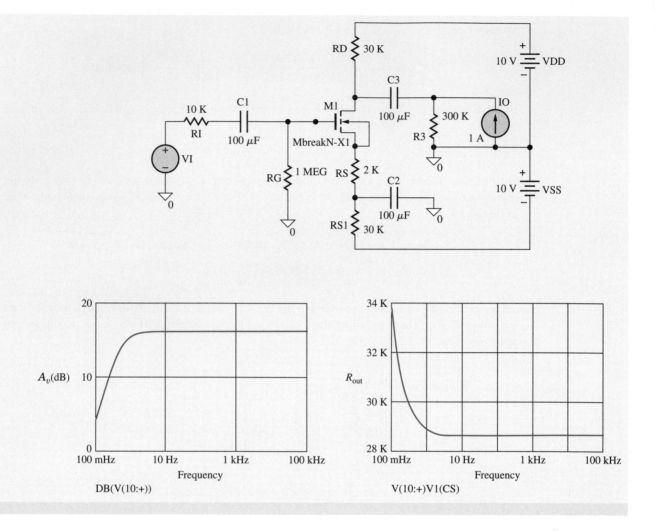

DB(V(10:+))

V(10:+)V1(CS)

EXERCISE: Verify the output resistance correction stated at the end of Ex. 13.10.

EXERCISE: What is the amplification factor of the MOSFET characterized by the parameters in Ex. 13.10? How does A_v compare to μ_f?

ANSWERS: 133; $|A_v| \ll \mu_f$

EXERCISE: What is the largest value of v_i that corresponds to a small signal for the amplifier in Fig. 13.39? What is the largest value of v_o that corresponds to a small signal in this amplifier?

ANSWERS: 406 mV; 2.78 V

EXERCISE: Verify the dc and ac analysis using SPICE. Use SPICE to find the voltage gain if C_2 is removed from the circuit.

ANSWER: −0.787

13.11.4 COMPARISON OF THE TWO AMPLIFIER EXAMPLES

Tables 13.4 and 13.5 compare the numerical results for the amplifiers analyzed in Exs. 13.9 and 13.10. The amplifiers have all been designed to have similar Q-points, as indicated in Table 13.4. In this table, we see that the BJT yields a much higher voltage gain than the FET circuit. However, all the voltage gains are well below the value of the amplification factor, which is characteristic of amplifiers with resistive loads in which the gain is limited by the external resistors (that is, in which $r_o \gg R_C$ or R_D).

Table 13.5 compares the input and output resistances. We see that the bipolar input resistance, in this case dominated by the value of r_π, is orders of magnitude smaller than that of the FET. On the other hand, R_{in} of the FET stage is limited by the choice of gate-bias resistor R_G. The output resistances are limited by the external resistors and are of similar magnitude.

13.11.5 COMMON-EMITTER AND COMMON-SOURCE AMPLIFIER SUMMARY

Table 13.6 presents a comparison of the ac small-signal characteristics of the common-emitter (C-E) and common-source (C-S) amplifiers based on the analyses presented in this chapter. The voltage gain expressions collapse to the same symbolic form, but the values will differ because the value of g_m for the BJT is usually much larger than that of the FET for a given operating current. The input resistance of the C-S stages is limited only by the design value of R_G and can be quite large, whereas the values of R_B and r_π limit the input resistance of the C-E amplifier to much smaller values. For a given operating point, the output resistances of the C-E and C-S stages are similar because R_{out} is limited by the collector- or drain-bias resistors R_C or R_D.

TABLE 13.4
Comparison of Two Amplifier Voltage Gains

AMPLIFIER	Q-POINT	A_V	μ_f	GAIN ESTIMATES
BJT	(239 μA, 3.67 V)	−36.1	2150	−66.7
MOSFET	(250 μA, 4.50 V)	−6.85	133	−15

TABLE 13.5
Comparison of Input and Output Resistances

AMPLIFIER	R_{in}	R_B or R_G	r_π	R_{out}	R_C or R_D	r_o
BJT	14.2 kΩ	104 kΩ	6.80 kΩ	9.40 kΩ	10 kΩ	225 kΩ
MOSFET	1.00 MΩ	1.00 MΩ	∞	28.4 kΩ	30 kΩ	258 kΩ

TABLE 13.6
Common-Emitter/Common-Source Amplifier Characteristics

	COMMON-EMITTER (C-E) AMPLIFIER	COMMON-SOURCE (C-S) AMPLIFIER
Terminal Voltage Gain A_{vt}	$-\dfrac{g_m R_L}{1 + g_m R_E}$	$-\dfrac{g_m R_L}{1 + g_m R_S}$
Signal Source Voltage Gain	$A_v = \dfrac{v_o}{v_i} = A_{vt}\dfrac{R_{\text{in}}}{R_I + R_{\text{in}}}$	
Rule-of-Thumb estimate for $g_m R_L$	$-10(V_{CC} + V_{EE})$	$-(V_{DD} + V_{SS})$
Input Resistance R_{in}	$R_B \| r_\pi (1 + g_m R_E)$	R_G
Output Resistance R_{out}	$R_C \| r_o (1 + g_m R_E)$	$R_D \| r_o (1 + g_m R_S)$
Input Signal Range	$0.005(1 + g_m R_E)$ V	$0.2(V_{GS} - V_{TN})(1 + g_m R_S)$
Terminal Current Gain	β_o	∞

13.11.6 FEEDBACK IN THE INVERTING AMPLIFIERS

In the common-emitter circuit, resistor R_E adds feedback to the amplifier that reduces the voltage gain by the factor $(1 + g_m R_E)$ but increases the input resistance, output resistance, and input signal range by the same amount. Resistor R_S has a similar impact on the voltage gain, output resistance, and input signal range of the common-source amplifier. Since the resistance at the gate terminal of the FET is already infinite, the overall input resistance of the C-S amplifier is not affected by R_S.

Let us relate these effects to the results we saw for the op-amp feedback circuits in Chapter 11. First consider the terminal voltage gain expression that can be rewritten as

$$A_{vt} = -\frac{g_m R_L}{1 + g_m R_L \left(\dfrac{R_E}{R_L}\right)} = \frac{A}{1 + A\beta} \qquad \text{with} \qquad A = -g_m R_L \text{ and } \beta = -\frac{R_E}{R_L} \quad (13.121)$$

The gain without feedback ($R_E = 0$) is $A = -g_m R_L$, whereas the gain is reduced by the factor $(1 + A\beta)$ when feedback is present. For large loop gain ($A\beta \gg 1$), the gain is set by the resistor ratio:

$$A_{vt} \cong \frac{1}{\beta} = -\frac{R_L}{R_E} \qquad \text{for} \qquad A\beta = g_m R_E \gg 1 \quad (13.122)$$

The feedback factor β can be interpreted in the following way. The output voltage across load resistor R_L is $v_o = i_c R_L$, whereas the voltage feed back to the input loop is $v_e = i_e R_E$. β represents the ratio of these two voltages: $\beta = v_e/v_o \cong -R_E/R_L$ since $i_c \cong i_e$.

The resistance at the base terminal of the transistor without feedback is r_π, whereas the input resistance with feedback is increased by $(1 + A\beta)$:

$$R_{iB} = r_\pi(1 + g_m R_E) = r_\pi(1 + A\beta) \quad (13.123)$$

Compare this result to $R_{in} = R_{id}(1 + A\beta)$ for the noninverting op-amp circuit.

Similarly, the resistance at the collector terminal of the transistor without feedback is r_o, whereas the resistance with feedback in this case is increased by $(1 + A\beta)$:

$$R_{iC} = r_o(1 + g_m R_E) = r_o(1 + A\beta) \quad (13.124)$$

Note that this is different from the op-amp circuits that we studied in Chapter 11 where the output resistance was reduced by the presence of feedback. This result occurs because of the differences between "series feedback" and "shunt feedback" that will be discussed in detail in Chapter 17.

13.11.7 GUIDELINES FOR NEGLECTING THE TRANSISTOR OUTPUT RESISTANCE

In all these amplifier examples, we found that the transistor's own output impedance did not greatly affect the results of the various calculations. The following question naturally arises: Why not just neglect r_o altogether, which will simplify the analysis? The answer is: The resistance r_o must be included whenever it makes a difference. We use the following rule: The transistor output resistance r_o can be neglected in voltage gain calculations as long as the computed value of $A_v \ll \mu_f$. However, in Thévenin equivalent resistance calculations r_o can play a very important role and one must be careful not to overlook limitations due to r_o. If r_o is neglected, and an input or output resistance is calculated that is similar to or much larger than r_o, then the calculation should be rechecked with r_o included in the circuit. At this point, this procedure may sound mysterious, but in the next several chapters we shall find circuits in which r_o is very important.

DESIGN NOTE

You can neglect the transistor output resistance in voltage gain calculations as long as the computed value is much less than the transistor's intrinsic gain μ_f! When the output resistance is included in a calculation, we often do not know V_{CE} or V_{DS}, and it is perfectly acceptable to use the simplified expressions for the output resistances:

$$r_o = \frac{V_A}{I_C} \quad \text{or} \quad r_o = \frac{1}{\lambda I_D}$$

13.12 AMPLIFIER POWER AND SIGNAL RANGE

We explored a number of examples showing how the selection of Q-point affects the value of the small-signal parameters of the transistors and hence affects the voltage gain, input resistance, and output resistance of common-emitter and common-source amplifiers. For the FET, the choice of Q-point also determines the value of v_{gs} that corresponds to small-signal operation. Two additional characteristics that are set by Q-point design are discussed in this section. The choice of operating point determines the level of power dissipation in the transistor and overall circuit, and it also determines the maximum linear signal range at the output of the amplifier.

13.12.1 POWER DISSIPATION

The static power dissipation of the amplifiers can be determined from the dc equivalent circuits used earlier. The power that is supplied by the dc sources is dissipated in both the resistors and transistors. For the amplifier in Fig. 13.42(a), for example, the power P_D dissipated in the transistor is the sum of the power dissipation in the collector-base and emitter-base junctions:

$$P_D = V_{CB}I_C + V_{BE}(I_B + I_C) = (V_{CB} + V_{BE})I_C + V_{BE}I_B$$

or (13.125)

$$P_D = V_{CE}I_C + V_{BE}I_B \quad \text{where} \quad V_{CE} = V_{CB} + V_{BE}$$

The total power P_S supplied to the amplifier is determined by the currents in the two power supplies:

$$P_S = V_{CC}I_C + V_{EE}I_E \tag{13.126}$$

Similarly for the MOSFET circuit in Fig. 13.42(b), the power dissipated in the transistor is given by

$$P_D = V_{DS}I_D + V_{GS}I_G = V_{DS}I_D \tag{13.127}$$

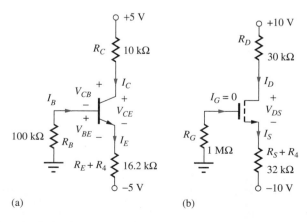

(a) (b)

Figure 13.42 dc Equivalent circuits for the (a) BJT and (b) MOSFET amplifiers from Figs. 13.36(a) and 13.39(a).

because the gate current is zero. The total power being supplied to the amplifier is equal to:

$$P_S = V_{DD}I_D + V_{SS}I_S = (V_{DD} + V_{SS})I_D \tag{13.128}$$

EXERCISE: What power is being dissipated by the bipolar transistor in Fig. 13.42(a)? Assume $\beta_F = 65$. What is the total power being supplied to the amplifier? Use the Q-point information already calculated (239 µA, 3.67 V).

ANSWERS: 880 µW; 2.41 mW

EXERCISE: What power is being dissipated by the MOSFET in Fig. 13.42(b)? What is the total power being supplied to the amplifier? Use the Q-point information already calculated (250 µA, 9.5 V).

ANSWERS: 1.13 mW; 5.00 mW

13.12.2 SIGNAL RANGE

We next discuss the relationship between the Q-point and the amplitude of the signals that can be developed at the output of the amplifier. Consider the amplifier in Fig. 13.43 with $V_{CC} = 12$ V, and the corresponding waveforms, which are given in Fig. 13.44. The collector and emitter voltages at the operating point are 5.9 V and 2.10 V, respectively, and hence the value of V_{CE} at the Q-point is 3.8 V.

Because the bypass capacitor at the emitter forces the emitter voltage to remain constant, the total collector-emitter voltage can be expressed as

$$v_{CE} = V_{CE} - V_M \sin \omega t \tag{13.129}$$

in which $V_M \sin \omega t$ is the signal voltage being developed at the collector. The bipolar transistor must remain in the active region at all times, which requires that the collector-emitter voltage remain larger than base-emitter voltage V_{BE}:

$$v_{CE} \geq V_{BE} \quad \text{or} \quad v_{CE} \geq 0.7 \text{ V} \tag{13.130}$$

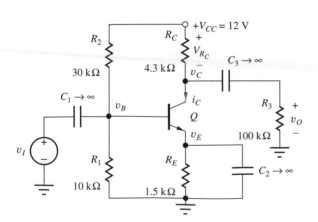

Figure 13.43 Common-emitter amplifier stage.

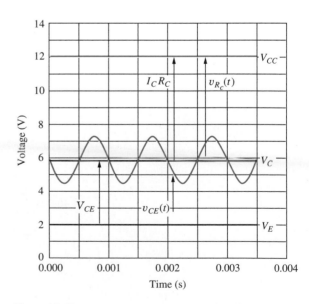

Figure 13.44 Waveforms for the amplifier in Fig. 13.43.

Thus the amplitude of the signal at the collector must satisfy

$$V_M \leq V_{CE} - V_{BE} \tag{13.131}$$

The positive power supply presents an additional limit to the signal swing. Writing an expression for the voltage across resistor R_C,

$$v_{R_c}(t) = I_C R_C + V_M \sin \omega t \geq 0 \tag{13.132}$$

In this circuit, the voltage across the resistor cannot become negative; that is, the voltage V_C at the transistor collector cannot exceed the power supply voltage V_{CC}. Equation (13.132) indicates that the amplitude V_M of the ac signal developed at the collector must be smaller than the voltage drop across R_C at the Q-point:

$$V_M \leq I_C R_C \tag{13.133}$$

Thus, the signal swing at the collector is limited by the smaller of the two limits expressed in Eqs. (13.131) or (13.133):

$$V_M \leq \min[I_C R_C, (V_{CE} - V_{BE})] \tag{13.134}$$

Similar expressions can be developed for field-effect transistor circuits. We must require that the MOSFET remains pinched off, or v_{DS} must always remain larger than $v_{GS} - V_{TN}$.

$$v_{DS} = V_{DS} - V_M \sin \omega t \geq V_{GS} - V_{TN} \tag{13.135}$$

in which it has been assumed that $v_{gs} \ll V_{GS}$. In direct analogy to Eq. (13.133) for a BJT circuit, the signal amplitude in the FET case also cannot exceed the dc voltage drop across R_D:

$$V_M \leq I_D R_D \tag{13.136}$$

So, for the case of the MOSFET, V_M must satisfy:

$$V_M \leq \min[I_D R_D, (V_{DS} - (V_{GS} - V_{TN}))] \tag{13.137}$$

EXERCISES: **(a) What is V_M for the bipolar transistor amplifier in Fig. 13.36(a)? (b) For the MOSFET amplifier in Fig. 13.39(a)?**

ANSWERS: 3.2 V; 3.50 V

ELECTRONICS IN ACTION

Electric Guitar Distortion Circuits
For most of this chapter we have focused on small-signal models and gain calculations. However, in some applications, it is desirable to intentionally violate small-signal constraints and generate a distorted waveform. In particular, electric guitars, the mainstay of rock music, intentionally use distortion to enrich the sound. The early Marshall and Fender tube amps, through substantial over-design and the natural characteristics of vacuum tube circuits, generated a rich soft-clipped sound when driven into overload. When excited with the right chords, the tube amplifier distortion can actually generate harmonics that are in-tune and add a great deal to the character of the electric guitar sound.

Modern guitar players use 'pedal' boxes to produce distortion and other effects without the excessive power levels required to produce the overdrive sound. Typical forms of these circuits are shown below. The first is an op-amp circuit with a pair of diodes in the feedback network. R_2 is 50 to 200 times larger than R_1, so the circuit has a large gain. As the voltage across the amplifier exceeds the diode turn-on voltage, the diodes begin to conduct. Since the diode impedance is much less than R_2, the gain is reduced during diode conduction. The resulting 'soft' clipped waveform is shown below.

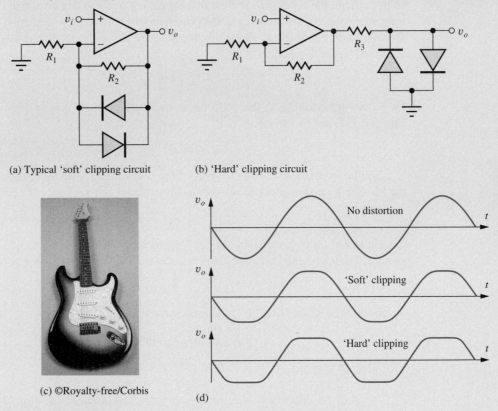

(a) Typical 'soft' clipping circuit   (b) 'Hard' clipping circuit

(c) ©Royalty-free/Corbis

(d)

Another form of distortion circuit is the 'hard' clipping circuit. The amplifier gain is again set to be quite large, and resistor R_3 is typically a few kilohms. As v_o exceeds the diode turn-on voltage, the output is clipped to the diode voltage. In this case, the diode current is limited by R_3, so v_o changes very little once the diode turns on. This results in a 'hard' clipped waveform as seen above. Typically, practical circuits also include some frequency shaping.

From Fourier analysis, we know that any cyclical waveform shape other than an ideal sine wave is composed of a possibly infinite set of harmonics or sine and cosine waves, each at frequencies which are multiples of the fundamental frequency. The sharper the transitions in a waveform, the more harmonic content it contains. The soft clipping circuit creates a waveform with smaller amplitude of harmonics than the hard clipping circuit.

There are also additional tones created by the intermodulation of the incoming frequencies. In these nonlinear clipping circuits, the incoming frequencies mix and give rise to sum and difference frequencies. This is an additional audible effect of the distortion circuits. There are many variations on these simple circuits which produce a wide range of sounds. The guitarist must select between a variety of different distortion and effects devices to create the sound that optimally presents their musical ideas. Additional information can be found through the MCD website.

SUMMARY

Chapter 13 initiated our study of the basic amplifier circuits used in the design of more complex analog components and systems such as operational amplifiers, audio amplifiers, and RF communications equipment. The chapter began with an introduction to the use of the transistor as an amplifier, and then explored the detailed operation of the BJT common-emitter (C-E) and FET common-source (C-S) amplifiers. Expressions were developed for the voltage gain and input and output resistances of these amplifiers. Several examples of the complete analysis of common-emitter and common-source amplifiers were included near the end of the chapter. The relationships between Q-point design and the small-signal characteristics of the amplifier were fully developed.

POINTS TO REMEMBER

- The common-emitter amplifier can provide good voltage gain but has only a low-to-moderate input resistance.

- In contrast, the FET stage can have very high input resistance but typically provides relatively modest values of voltage gain.

- The output resistances of both C-E and C-S circuits tend to be determined by the resistors in the bias network and are similar for comparable operating points.

- A two-step approach is used to simplify the analysis and design of amplifiers. Circuits are split into two parts: a dc equivalent circuit used to find the Q-point of the transistor, and an ac equivalent circuit used for analysis of the response of the circuit to signal sources. The design engineer often must respond to competing goals in the design of the dc and ac characteristics of the amplifier, and coupling capacitors, bypass capacitors, and inductors are used to change the ac and dc circuit topologies.

- Our ac analyses were all based on linear small-signal models for the transistors. The small-signal models for the diode, bipolar transistor (the hybrid-pi model), and MOSFET were all discussed in detail. The expressions relating the transconductance g_m, output resistance r_o, and input resistance r_π to the Q-point were all found by evaluating derivatives of the large-signal model equations developed in earlier chapters.

- The small-signal model for the diode is simply a resistor that has a value given by $r_d = V_T/I_D$.

- The results in Table 13.3 on page 707 for the three-terminal devices are extremely important. The structure of the models is similar. The transconductance of the BJT is directly proportional to current, whereas that of the FET increases only in proportion to the square root of current. Resistances r_π and r_o are inversely proportional to Q-point current. Resistor r_π is infinite for the case of the FET, so it does not actually appear in the small-signal model. It was discovered that each device pair, the *npn* and *pnp* BJTs, and the NMOS and PMOS FETs, has the same small-signal model.

- The small-signal current gain of the BJT was defined as $\beta_o = g_m r_\pi$, and its value generally differs from that of the large-signal current gain β_F. Because β_F for the FET is infinite, the FET exhibits an infinite small-signal current gain at low frequency.

- The intrinsic voltage gain, also known as the amplification factor of the transistor, is defined as $\mu_f = g_m r_o$ and represents the maximum gain available from the transistor in the C-E and C-S amplifiers. Expressions were evaluated for the intrinsic gain of the BJT and FETs. Parameter μ_f was found to be independent of Q-point for the BJT, but for the FET, the amplification factor decreases as operating current increases. For usual operating points, μ_f for the BJT will be several thousand, whereas that for the FET ranges between tens and hundreds.

- The definition of a small signal was found to be device-dependent. The signal voltage v_d developed across the diode must be less than 5 mV in order to satisfy the requirements of a small signal. Similarly, the base-emitter signal voltage v_{be} of the BJT must be less than 5 mV for small-signal operation. However, FETs can amplify much larger signals without distortion. For the MOSFET, $v_{gs} \leq 0.2(V_{GS} - V_{TN})$ represent the small-signal limits, respectively, and can be designed to range from 100 mV to more than 1 V.

- Common-emitter and common-source amplifiers were analyzed in detail. Table 13.4 on page 731 is another extremely important table. It summarizes the overall characteristics of these two amplifiers. The rule-of-thumb estimates in Table 13.4 were developed to provide quick predictions of the voltage gain of the C-E and C-S stages.

- The chapter closed with a discussion of the relationship between operating point design and the power dissipation and output signal swing of the amplifiers. The amplitude of the signal voltage at the output of the amplifier is limited by the smaller of the Q-point value of the collector-base or drain-gate voltage of the transistor, and by the Q-point value of the voltage across the collector or drain-bias resistors R_C or R_D.

- It is extremely important to understand the difference between ac analysis and transient analysis in SPICE. ac analysis assumes that the network is linear and uses small-signal models for the transistors and diodes. Since the circuit is linear, any convenient value can be used for the signal source amplitudes, hence the common choice of 1-V and 1-A sources. In contrast, transient simulations utilize the full large-signal non-linear models of the transistors. If we desire linear behavior in a transient simulation, all signals must satisfy the small-signal constraints.

KEY TERMS

ac coupling
ac equivalent circuit
Amplification factor
Analysis by superposition
Back-gate transconductance
Back-gate transconductance parameter
Bypass capacitor
Common-emitter (C-E) amplifier
Common-source (C-S) amplifier
Coupling capacitor
dc blocking capacitor
dc equivalent circuit
Diode conductance

Diode resistance
Hybrid-pi small-signal model
Input resistance
Intrinsic voltage gain μ_f
Output resistance r_o
Small signal
Small-signal common-emitter current gain
Small-signal conductance
Small-signal current gain
Small-signal models
Terminal voltage gain
Transconductance g_m

PROBLEMS

Figures P13.3 through P13.13 are used in a variety of problems in this chapter. Assume all capacitors and inductors have infinite value unless otherwise noted. Assume $V_{BE} = 0.7$ V and $\beta_F = \beta_o$ unless otherwise specified.

13.1 The Transistor as an Amplifier

13.1. (a) Suppose $v_{be}(t) = 0.005 \sin 2000\pi t$ V in the bipolar amplifier in Fig. 13.1. Write expressions for $v_{BE}(t)$, $v_{ce}(t)$, and $v_{CE}(t)$. (b) What is the

maximum value of I_C that corresponds to the active region of operation?

13.2. (a) Suppose $v_{gs}(t) = 0.25 \sin 2000\pi t$ V in the MOSFET amplifier in Fig. 13.2. Write expressions for $v_{GS}(t)$, $v_{ds}(t)$, and $v_{DS}(t)$. (b) What is the maximum value of I_D that corresponds to the active region of operation?

13.2 Coupling and Bypass Capacitors

13.3. (a) What are the functions of capacitors C_1, C_2, and C_3 in Fig. P13.3? (b) What is the magnitude of the signal voltage at the top of C_3?

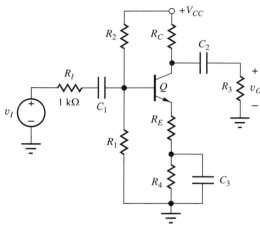

Figure P13.3

13.4. (a) What are the functions of capacitors C_1, C_2, and C_3 in Fig. P13.4? (b) What is the magnitude of the signal voltage at the base of Q_1?

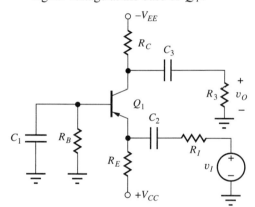

Figure P13.4

13.5. (a) What are the functions of capacitors C_1, C_2, and C_3 in Fig. P13.5? (b) What is the magnitude of the signal voltage at the emitter of Q_1?

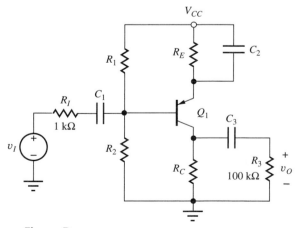

Figure P13.5

13.6. (a) What are the functions of capacitors C_1, C_2, and C_3 in Fig. P13.6? (b) What is the magnitude of the signal voltage at the source of M_1?

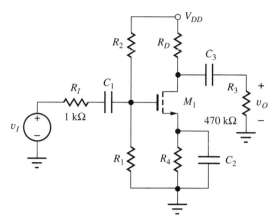

Figure P13.6

13.7. What are the functions of capacitors C_1 and C_2 in Fig. P13.7?

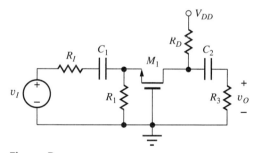

Figure P13.7

13.8. (a) What are the functions of capacitors C_1, C_2, and C_3, in Fig. P13.8? (b) What is the magnitude of the signal voltage at the source of M_1?

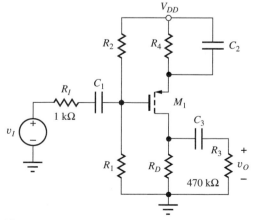

Figure P13.8

13.9. What are the functions of capacitors C_1, C_2, and C_3, in Fig. P13.9? What is the magnitude of the signal voltage at the emitter of Q_1?

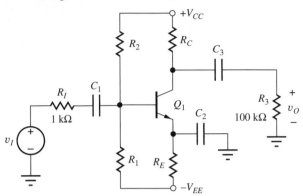

Figure P13.9

13.10. What are the functions of capacitors C_1, C_2, and C_3 in Fig. P13.10? What is the magnitude of the signal voltage at the collector of Q_1?

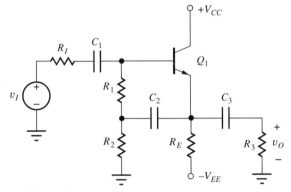

Figure P13.10

13.11. Describe the functions of capacitors C_1, C_2, and C_3 in Fig. P13.11. What is the magnitude of the signal voltage at the upper terminal of C_2?

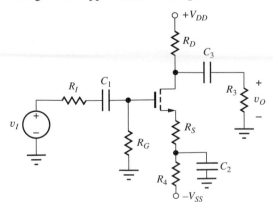

Figure P13.11

13.12. What are the functions of capacitors C_1 and C_2 in Fig. P13.12?

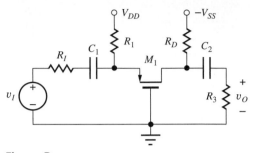

Figure P13.12

13.13. What are the functions of capacitors C_1 and C_2 in Fig. P13.13?

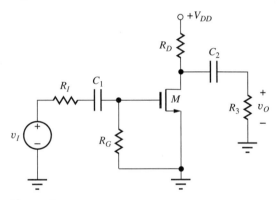

Figure P13.13

13.14. The phrase "dc voltage sources represent ac grounds" is used several times in the text. Use your own words to describe the meaning of this statement.

13.3 Circuit Analysis Using dc and ac Equivalent Circuits

13.15. Draw the dc equivalent circuit and find the Q-point for the amplifier in Fig. P13.13. Assume $\beta_F = 90$, $V_{CC} = 18$ V, $R_I = 2$ kΩ, $R_1 = 360$ kΩ, $R_2 = 750$ kΩ, $R_C = 270$ kΩ, $R_E = 8.2$ kΩ, $R_4 = 220$ kΩ, and $R_3 = 910$ kΩ.

13.16. (a) Use SPICE to find the Q-point for the circuit in Prob. 13.15. Assume $V_A = \infty$ and $I_S = 5$ fA. (b) Repeat with $V_A = 80$ V and $I_S = 5$ fA.

13.17. Draw the dc equivalent circuit and find the Q-point for the amplifier in Fig. P13.9. Assume $\beta_F = 75$, $V_{CC} = 12$ V, $-V_{EE} = -12$ V, $R_I = 1$ kΩ, $R_1 = 5$ kΩ, $R_2 = 10$ kΩ, $R_3 = 24$ kΩ, $R_E = 4$ kΩ, and $R_C = 6$ kΩ.

13.18. Use SPICE to find the Q-point for the circuit in Prob. 13.17. Compare the results to the hand calculations in Prob. 13.17.

13.19. Draw the dc equivalent circuit and find the Q-point for the amplifier in Fig. P13.4. Assume $\beta_F = 65$, $V_{CC} = 7.5$ V, $-V_{EE} = -7.5$ V, $R_I = 0.47$ kΩ, $R_B = 3$ kΩ, $R_C = 33$ kΩ, $R_E = 68$ kΩ, and $R_3 = 120$ kΩ.

13.20. Use SPICE to find the Q-point for the circuit in Prob. 13.19. Compare the results to the hand calculations in Prob. 13.19.

13.21. Draw the dc equivalent circuit and find the Q-point for the amplifier in Fig. P13.5. Assume $\beta_F = 135$ and $V_{CC} = 9$ V, $R_1 = 20$ kΩ, $R_2 = 62$ kΩ, $R_C = 13$ kΩ, and $R_E = 3.9$ kΩ.

13.22. Use SPICE to find the Q-point for the circuit in Prob. 13.21. Compare the results to the hand calculations in Prob. 13.21.

13.23. Draw the dc equivalent circuit and find the Q-point for the amplifier in Fig. P13.6. Assume $K_n = 250$ μA/V^2, $V_{TN} = 1$ V, $V_{DD} = 15$ V, $R_I = 1$ kΩ, $R_1 = 1$ MΩ, $R_2 = 2.7$ MΩ, $R_D = 82$ kΩ, and $R_4 = 27$ kΩ.

13.24. Use SPICE to find the Q-point for the circuit in Prob. 13.23. Compare the results to the hand calculations in Prob. 13.23.

13.25. Draw the dc equivalent circuit and find the Q-point for the amplifier in Fig. P13.7. Assume $K_n = 500$ μA/V^2, $V_{TN} = -2$ V, $V_{DD} = 15$ V, $R_I = 1$ kΩ, $R_1 = 3.9$ kΩ, $R_D = 4.3$ kΩ, and $R_3 = 51$ kΩ.

13.26. Use SPICE to find the Q-point for the circuit in Prob. 13.25. Compare the results to the hand calculations in Prob. 13.25.

13.27. Draw the dc equivalent circuit and find the Q-point for the amplifier in Fig. P13.8. Assume $K_p = 400$ μA/V^2, $V_{TP} = -1$ V, $V_{DD} = 18$ V, $R_1 = 3.3$ MΩ, $R_2 = 3.3$ MΩ, $R_D = 24$ kΩ, and $R_4 = 22$ kΩ.

13.28. Use SPICE to find the Q-point for the circuit in Prob. 13.27. Compare the results to the hand calculations in Prob. 13.27.

13.29. Draw the dc equivalent circuit and find the Q-point for the amplifier in Fig. P13.10. Assume $\beta_F = 100$, $V_{CC} = 9$ V, $-V_{EE} = -9$ V, $R_I = 1$ kΩ, $R_1 = 43$ kΩ, $R_2 = 43$ kΩ, $R_3 = 24$ kΩ, and $R_E = 82$ kΩ.

13.30. Use SPICE to find the Q-point for the circuit in Prob. 13.29. Compare the results to the hand calculations in Prob. 13.29.

13.31. Draw the dc equivalent circuit and find the Q-point for the amplifier in Fig. P13.12. Assume $K_p = 200$ μA/V^2, $V_{TP} = +1$ V, $V_{DD} = 12$ V, $-V_{SS} = -12$ V, $R_1 = 33$ kΩ, $R_D = 22$ kΩ, $R_I = 500$ Ω, and $R_3 = 100$ kΩ.

13.32. Use SPICE to find the Q-point for the circuit in Prob. 13.31. Compare the results to the hand calculations in Prob. 13.31.

13.33. Draw the dc equivalent circuit and find the Q-point for the amplifier in Fig. P13.11. Assume $K_n = 400$ μA/V^2, $V_{TN} = -5$ V, $V_{DD} = 18$ V, $R_G = 10$ MΩ, $R_D = 3.9$ kΩ, $R_I = 10$ kΩ, $R_1 = 2$ kΩ, $R_S = 1$ kΩ, $R_4 = 1$ kΩ, and $R_3 = 36$ kΩ.

13.34. Use SPICE to find the Q-point for the circuit in Prob. 13.33. Compare the results to the hand calculations in Prob. 13.33.

13.35. Draw the dc equivalent circuit and find the Q-point for the amplifier in Fig. P13.13. Assume $V_{DD} = 15$ V, $K_n = 225$ μA/V^2, $V_{TN} = -3$ V, $R_G = 2.2$ MΩ, $R_D = 7.5$ kΩ, $R_I = 10$ kΩ, and $R_3 = 220$ kΩ.

13.36. Use SPICE to find the Q-point for the circuit in Prob. 13.35. Compare the results to the hand calculations in Prob. 13.35.

13.37. (a) Draw the equivalent circuit used for ac analysis of the circuit in Fig. P13.3. (Use transistor symbols for this part.) Assume all capacitors have infinite value. (b) Redraw the ac equivalent circuit, replacing the transistor with its small-signal model. (c) Identify the function of each capacitor in the circuit (bypass or coupling).

13.38. (a) Repeat Prob. 13.37 for the circuit in Fig. P13.4. (b) Repeat Prob. 13.37 for the circuit in Fig. P13.5.

13.39. (a) Repeat Prob. 13.37 for the circuit in Fig. P13.9. (b) Repeat Prob. 13.37 for the circuit in Fig. P13.11.

13.40. (a) Repeat Prob. 13.37 for the circuit in Fig. P13.6. (b) Repeat Prob. 13.37 for the circuit in Fig. P13.7.

13.41. (a) Repeat Prob. 13.37 for the circuit in Fig. P13.8. (b) Repeat Prob. 13.37 for the circuit in Fig. P13.11.

13.42. (a) Repeat Prob. 13.37 for the circuit in Fig. P13.12. (b) Repeat Prob. 13.37 for the circuit in Fig. P13.13.

13.43. Describe the function of each of the resistors in the circuit in Fig. P13.3.

13.44. Describe the function of each of the resistors in the circuit in Fig. P13.6.

13.45. Describe the function of each of the resistors in the circuit in Fig. P13.9.

13.4 Introduction to Small-Signal Modeling

13.46. (a) Calculate r_d for a diode with $V_D = 0.6$ V if $I_S = 10$ fA. (b) What is the value of r_d for $V_D = 0$ V? (c) At what voltage does r_d exceed 10^{15} Ω?

13.47. What is the value of the small-signal diode resistance r_d of a diode operating at a dc current of 1 mA at temperatures of (a) 75 K, (b) 100 K, (c) 200 K, (d) 300 K, and (e) 400 K?

13.48. (a) Compare $[\exp(v_d/V_T) - 1]$ to v_d/V_T for $v_d = +5$ mV and -5 mV. How much error exists between the linear approximation and the exponential? (b) Repeat for $v_d = \pm10$ mV.

13.5 Small-Signal Models for Bipolar Junction Transistors

13.49. (a) What collector current is required for a bipolar transistor to achieve a transconductance of 30 mS? (b) Repeat for a transconductance of 250 μS. (c) Repeat for a transconductance of 50 μS.

13.50. At what Q-point current will $r_\pi = 10$ kΩ for a bipolar transistor with $\beta_o = 75$? What are the approximate values of g_m and r_o if $V_A = 100$ V?

13.51. Repeat Prob. 13.50 for $r_\pi = 2$ MΩ with $\beta_o = 125$ and $V_A = 75$ V.

13.52. Repeat Prob. 13.50 for $r_\pi = 250$ kΩ with $\beta_o = 100$.

13.53. At what Q-point current will $r_\pi = 1$ MΩ for a bipolar transistor with $\beta_o = 75$? What are the values of g_m and r_o if $V_A = 100$ V?

13.54. The following table contains the small-signal parameters for a bipolar transistor. What are the values of β_F and V_A? Fill in the values of the missing entries in the table if $V_{CE} = 10$ V.

Bipolar Transistor Small-Signal Parameters				
I_C (A)	g_m (S)	r_π (Ω)	r_o (Ω)	μ_f
0.002			40,000	
	0.12	500		
		480,000		

13.55. (a) Compare $[\exp(v_{be}/V_T) - 1]$ to v_{be}/V_T for $v_{be} = +5$ mV and -5 mV? How much error exists between the linear approximation and the exponential? (b) Repeat for $v_{be} = \pm7.5$ mV. (c) Repeat for $v_{be} = \pm2.5$ mV.

13.56. The output characteristics of a bipolar transistor appear in Fig. P13.143. (a) What are the values of β_F and β_o at $I_B = 4$ μA and $V_{CE} = 10$ V?

(b) What are the values of β_F and β_o at $I_B = 8$ μA and $V_{CE} = 10$ V?

13.57. (a) Suppose that a BJT is operating with a total collector current given by

$$i_C(t) = 0.001 \exp\left(\frac{v_{be}(t)}{V_T}\right) \text{ Amps}$$

and $v_{be}(t) = V_M \sin 2000\pi t$ with $V_M = 5$ mV. What is the value of the dc collector current? Plot the collector current using MATLAB. Use FFT capability of MATLAB to find the amplitude of i_c at 1000 Hz? At 2000 Hz? At 3000 Hz? (b) Repeat for $V_M = 50$ mV.

13.58. (a) Use SPICE to find the Q-point of the circuit in Fig. P13.9 using the element values in Prob. 13.17. Use the Q-point information from SPICE to calculate the values of the small-signal parameters of transistor Q_1. Compare the values with those printed out by SPICE and discuss the source of any discrepancies. (b) Repeat part (a) for the circuit in Fig. P13.5 with the element values from Prob. 13.21.

*13.59. Another small-signal model, the T-model in Fig. P13.59, is of historical interest and quite useful in certain situations. Show that this model is equivalent to the hybrid-pi model if the emitter resistance $r_e = r_\pi/(\beta_o + 1) = \alpha_o/g_m = V_T/I_E$. (Hint: Calculate the short-circuit input admittance y_{11} for both models assuming $\beta_F = \beta_o$.)

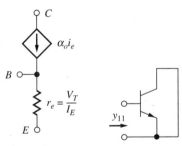

Figure P13.59

13.6 The BJT Common-Emitter (C-E) Amplifier

13.60. The ac equivalent circuit for an amplifier is shown in Fig. P13.60. Assume the capacitors have infinite value, $R_S = 750$ Ω, $R_B = 100$ kΩ, $R_C = 100$ kΩ, and $R_3 = 100$ kΩ. Calculate the voltage gain for the amplifier if the BJT Q-point is (50 μA, 10 V). Assume $\beta_o = 100$ and $V_A = 75$ V.

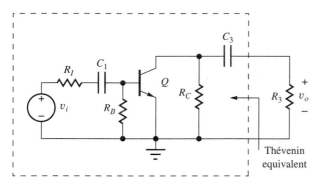

Figure P13.60

13.61. What are the worst-case values of voltage gain for the amplifier in Prob. 13.60 if β_o can range from 60 to 100? Assume that the Q-point is fixed.

13.62. The ac equivalent circuit for an amplifier is shown in Fig. P13.60. Assume the capacitors have infinite value, $R_I = 50\ \Omega$, $R_B = 4.7\ k\Omega$, $R_C = 4.3\ k\Omega$, and $R_3 = 10\ k\Omega$. Calculate the voltage gain for the amplifier if the BJT Q-point is (2.5 mA, 7.5 V). Assume $\beta_o = 75$ and $V_A = 50$ V.

13.63. The ac equivalent circuit for an amplifier is shown in Fig. P13.63. Assume the capacitors have infinite value, $R_I = 10\ k\Omega$, $R_B = 5\ M\Omega$, $R_C = 1.5\ M\Omega$ and $R_3 = 3.3\ M\Omega$. Calculate the voltage gain for the amplifier if the BJT Q-point is (1 μA, 1.5 V). Assume $\beta_o = 40$ and $V_A = 50$ V.

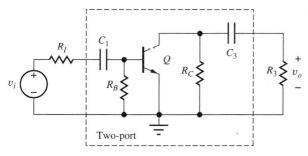

Figure P13.63

13.64. Simulate the behavior of the BJT common-emitter amplifier in Fig. 13.18 and compare the results to the calculations in Ex. 13.3. Use 100 μF for all capacitor values and perform the ac analysis at a frequency of 1000 Hz.

*13.65. An amplifier is required with a voltage gain of 25,000 and will be designed using a cascade of several C-E amplifier stages operating from a single 9-V power supply. Estimate the minimum number of amplifier stages that will be required to achieve this gain.

13.66. (a) Use SPICE to simulate the dc and ac characteristics of the amplifier in Prob. 13.21. What is the Q-point? What is the value of the small-signal voltage gain? Use 100 μF for all capacitor values and perform the ac analysis at a frequency of 1000 Hz. (b) Compare the results to hand calculations.

13.7 Important Limitations and Model Specifications

13.67. A C-E amplifier is operating from a single 12-V supply. Estimate its voltage gain.

*13.68. A C-E amplifier is operating from symmetrical ±15-V power supplies. Estimate its voltage gain.

13.69. A battery-powered amplifier must be designed to provide a gain of 50. Can a single-stage amplifier be designed to meet this goal if it must operate from two ±1.5-V batteries?

13.70. A battery-powered C-E amplifier is operating from a single 1.5-V battery. Estimate its voltage gain. What will the gain be if the battery voltage drops to 1 V?

*13.71. The common-emitter amplifier in Fig. P13.71 must develop a 10-V peak-to-peak sinusoidal signal across the 10-kΩ load resistor R_L. (a) What is the minimum collector current I_C that will satisfy the requirements of small-signal operation of the transistor? (b) What is the minimum power supply voltage V_{CC}?

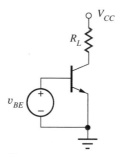

Figure P13.71

*13.72. A common-emitter amplifier has a voltage gain of 40 dB. What is the amplitude of the largest output signal voltage at the collector that corresponds to small-signal operation?

*13.73. A common-emitter amplifier has a gain of 50 dB and is developing a 15-V peak-to-peak ac signal at its output. Is this amplifier operating within its small-signal region? If the input signal to this amplifier is a sine wave, do you expect the output to be distorted? Why or why not?

13.74. (a) What is the voltage gain of the common-emitter amplifier in Fig. P13.9? (b) What is the voltage gain of the common-emitter amplifier in Fig. P13.5? Assume $\beta_F = 135$, $V_{CC} = V_{EE} = 9$ V, $R_1 = 20$ kΩ, $R_2 = 62$ kΩ, $R_C = 13$ kΩ, and $R_E = 3.9$ kΩ for part (a) and $V_{CC} = 18$ V for part (b).

13.8 Small-Signal Models for Field-Effect Transistors

13.75. The following table contains the small-signal parameters for a MOS transistor. What are the values of K_n and λ? Fill in the values of the missing entries in the table if $V_{DS} = 6$ V and $V_{TN} = 1$ V.

MOSFET Small-Signal Parameters				
I_{DS}	g_m (S)	r_o (Ω)	μ_f	SMALL-SIGNAL LIMIT V_{gs} (V)
0.8 mA		40,000		
50 μA	0.0002			
10 mA				

13.76. What value of W/L is required to achieve $\mu_f = 250$ in a MOSFET operating at a drain current of 200 μA if $K'_n = 50$ μA/V^2 and $\lambda = 0.02$/V? What is the value of $V_{GS} - V_{TN}$?

13.77. An n-channel MOSFET has $K_n = 250$ μA/V^2, $V_{TN} = 1$ V, and $\lambda = 0.02$ V^{-1}. At what drain current will the MOSFET no longer be able to provide any voltage gain (that is, $\mu_F \leq 1$)?

13.78. A MOSFET is needed with $g_m = 5$ mS at $V_{GS} - V_{TN} = 0.5$ V. What is W/L if $K'_n = 40$ μA/V^2?

13.79. Compare $[1 + v_{gs}/(V_{GS} - V_{TN})]^2 - 1$ to $[2v_{gs}/(V_{GS} - V_{TN})]$ for $v_{gs} = 0.2 (V_{GS} - V_{TN})$. How much error exists between the linear approximation and the quadratic expression? Repeat for $v_{gs} = 0.4 (V_{GS} - V_{TN})$.

13.80. Use SPICE to find the Q-point of the circuit in Prob. 13.23. Use the Q-point information from SPICE to calculate the values of the small-signal parameters of transistor M_1. Compare the values with those printed out by SPICE and discuss the source of any discrepancies.

13.81. Repeat Prob. 13.80 for the circuit in Prob. 13.27.

13.82. At approximately what Q-point will $R_{\text{out}} = 100$ kΩ in a common-source amplifier if the transistor has $\lambda = 0.02$ V^{-1} and the power supply is 18 V?

*13.83. At approximately what Q-point can we achieve an input resistance of $R_{\text{in}} = 2$ MΩ in a common-source amplifier if the transistor has $K_n = 500$ μA/V^2, $V_{TN} = 1$ V, $\lambda = 0.02$ V^{-1}, and the power supply is 18 V?

**13.84. Figure P13.84 gives the device characteristics and schematic of an amplifier circuit including a "new"[7] electronic device called a *triode* vacuum tube. (a) Write the equation for the load line for the circuit. (b) What is the Q-point (I_P, V_{PK})? Assume

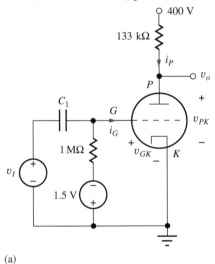

(a)

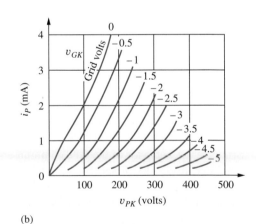

(b)

Figure P13.84   "New" electron device—the triode vacuum tube. (b) Triode output characteristics: G = grid, P = plate, K = cathode.

---

[7] New to us at least.

$i_G = 0$. (c) Using the following definitions, find the values of g_m, r_o, and μ_f. (d) What is the voltage gain of the circuit?

$$g_m = \left.\frac{\Delta i_P}{\Delta v_{GK}}\right|_{Q\text{-point}}$$

$$r_o = \left(\left.\frac{\Delta i_P}{\Delta v_{PK}}\right|_{Q\text{-point}}\right)^{-1} \qquad \mu_f = g_m r_o$$

13.9 Summary and Comparison of the Small-Signal Models of the BJT and FET

13.85. A circuit requires the use of a transistor with a transconductance of 0.5 S. A bipolar transistor with $\beta_F = 60$ and a MOSFET with $K_n = 25$ mA/V^2 are available. Which transistor would be preferred and why?

13.86. A circuit is to be biased at a current of 10 mA and achieve an input resistance of at least 1 MΩ. Should a BJT or FET be chosen for this circuit and why?

13.87. A BJT has $V_A = 25$ V and a MOSFET has $K_n = 25$ mA/V^2 and $\lambda = 0.02$ V^{-1}. At what current level is the amplification factor of the MOSFET equal to that of the BJT if $V_{DS} = V_{CE} = 10$ V? What is μ_f for the BJT?

13.88. A BJT has $V_A = 50$ V, and a MOSFET has $\lambda = 0.02$/V with $V_{GS} - V_{TN} = 0.5$ V. What are the amplification factors of the two transistors? What are the transconductances if the transistors are both operating at a current of 200 μA?

13.89. An amplifier circuit is needed with an input resistance of 75 Ω. Should a BJT or MOSFET be chosen for this circuit? Discuss.

13.90. (a) We need to amplify a 0.25-V signal by 26 dB. Would a BJT or FET amplifier be preferred? Why? (b) RF amplifiers must often amplify microvolt signals in the presence of many other interfering signals with amplitudes of 100 mV or more. Does an FET or BJT seem most appropriate for this application? Why?

13.10 The Common-Source Amplifier

13.91. A C-S amplifier is operating from a single 12-V supply with $V_{GS} - V_{TN} = 1$ V. Estimate its voltage gain.

13.92. A common-source amplifier has a gain of 15 dB and is developing a 15-V peak-to-peak ac signal at its output. Is this amplifier operating within its small-signal region? Discuss.

13.93. A C-S amplifier is operating from a single 9-V supply. What is the maximum value of $V_{GS} - V_{TN}$ that can be used if the amplifier must have a gain of at least 30?

13.94. A C-S amplifier is operating from a single 15-V supply. The MOSFET has $K_n = 1$ mA/V^2. What is the Q-point current required for a voltage gain of 30?

13.95. A MOSFET common-source amplifier must amplify a sinusoidal ac signal with a peak amplitude of 0.1 V. What is the minimum value of $V_{GS} - V_{TN}$ for the transistor? If a voltage gain of 35 dB is required, what is the minimum power supply voltage?

13.96. A MOSFET common-source amplifier must amplify a sinusoidal ac signal with a peak amplitude of 0.5 V. What is the minimum value of $V_{GS} - V_{TN}$ for the transistor? If a voltage gain of 20 dB is required, what is the minimum power supply voltage?

13.97. An amplifier is required with a voltage gain of 1000 and will be designed using a cascade of several C-S amplifier stages operating from a single 10-V power supply. Estimate the minimum number of amplifier stages required to achieve this gain.

13.98. What is the voltage gain of the amplifier in Fig. P13.98? Assume $K_n = 0.500$ mA/V^2, $V_{TN} = 1$ V, and $\lambda = 0.0133$ V^{-1}.

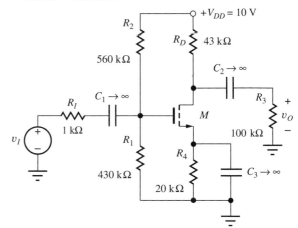

Figure P13.98

13.99. The ac equivalent circuit for an amplifier is shown in Fig. P13.99. Assume the capacitors have infinite value, $R_I = 100$ kΩ, $R_G = 6.8$ MΩ, $R_D = 50$ kΩ, and $R_3 = 120$ kΩ. Calculate the voltage gain for the amplifier if the MOSFET Q-point is (100 μA, 5 V). Assume $K_n = 500$ μA/V^2 and $\lambda = 0.02$ V^{-1}.

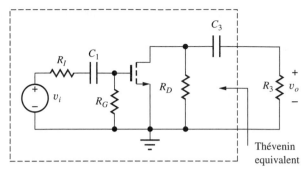

Figure P13.99

13.100. What are the worst-case values of voltage gain for the amplifier in Prob. 13.99 if K_n can range from 300 μA/V^2 to 700 μA/V^2? Assume the Q-point is fixed.

13.101. The ac equivalent circuit for an amplifier is shown in Fig. P13.99. Assume the capacitors have infinite value, $R_I = 100$ kΩ, $R_G = 10$ MΩ, $R_D = 560$ kΩ, and $R_3 = 2.2$ MΩ. Calculate the voltage gain for the amplifier if the MOSFET Q-point is (10 μA, 5 V). Assume $K_n = 100$ μA/V^2 and $\lambda = 0.02$ V^{-1}.

13.102. The ac equivalent circuit for an amplifier is shown in Fig. P13.102. Assume the capacitors have infinite value, $R_I = 10$ kΩ, $R_G = 1$ MΩ, $R_D = 3.9$ kΩ, and $R_3 = 270$ kΩ. Calculate the voltage gain for the amplifier if the MOSFET Q-point is (2 mA, 7.5 V). Assume $K_n = 1$ mA/V^2 and $\lambda = 0.015$ V^{-1}.

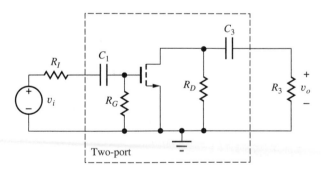

Figure P13.102

13.103. Use SPICE to simulate the dc and ac characteristics of the amplifier in Prob. 13.23. What is the Q-point? What are the values of the small-signal voltage gain, input resistance, and output resistance of the amplifier? Use 100 μF for all capacitor values and perform the ac analysis at a frequency of 1000 Hz.

13.104. Use SPICE to simulate the dc and ac characteristics of the amplifier in Prob. 13.27. What is the Q-point? What are the values of the small-signal voltage gain, input resistance, and output resistance of the amplifier? Use 100 μF for all capacitor values and perform the ac analysis at a frequency of 1000 Hz.

13.105. Use SPICE to simulate the dc and ac characteristics of the amplifier in Prob. 13.33. What is the Q-point? What are the values of the small-signal voltage gain, input resistance, and output resistance of the amplifier? Use 100 μF for all capacitor values and perform the ac analysis at a frequency of 1000 Hz.

13.106. Use SPICE to simulate the dc and ac characteristics of the amplifier in Prob. 13.35. What is the Q-point? What are the values of the small-signal voltage gain, input resistance, and output resistance of the amplifier? Use 100 μF for all capacitor values and perform the ac analysis at a frequency of 1000 Hz.

Input and Output Resistances of the Common-Emitter and Common-Source Amplifiers

13.107. The ac equivalent circuit for an amplifier is shown in Fig. P13.60. Assume the capacitors have infinite value, $R_I = 750$ Ω, $R_B = 100$ kΩ, $R_C = 100$ kΩ, and $R_3 = 100$ kΩ. Calculate the input resistance and output resistance for the amplifier if the BJT Q-point is (50 μA, 10 V). Assume $\beta_o = 100$ and $V_A = 75$ V.

13.108. What are the worst-case values of input resistance and output resistance for the amplifier in Prob. 13.60 if β_o can range from 60 to 100? Assume that the Q-point is fixed.

13.109. The ac equivalent circuit for an amplifier is shown in Fig. P13.63. Assume the capacitors have infinite value, $R_I = 10$ kΩ, $R_B = 5$ MΩ, $R_C = 1.5$ MΩ, and $R_3 = 3.3$ MΩ. Calculate the input resistance and output resistance for the amplifier if the BJT Q-point is (1 μA, 1.5 V). Assume $\beta_o = 40$ and $V_A = 50$ V.

13.110. The ac equivalent circuit for an amplifier is shown in Fig. P13.60. Assume the capacitors have infinite value, $R_I = 50$ Ω, $R_B = 4.7$ kΩ, $R_C = 4.3$ kΩ, and $R_3 = 10$ kΩ. Calculate the input resistance and output resistance for the amplifier if the BJT Q-point is (2.5 mA, 7.5 V). Assume $\beta_o = 75$ and $V_A = 50$ V.

13.111. What are the input resistance and output resistance of the amplifier in Prob. 13.98?

13.112. Calculate the input and output resistances for the amplifier in Prob. 13.99.

13.113. What are the worst-case values of the input and output resistances for the amplifier in Prob. 13.99 if K_n can range from 300 μA/V^2 to 700 μA/V^2? Assume the Q-point is fixed.

13.114. Calculate the input and output resistances for the amplifier in Prob. 13.101.

13.115. Calculate the input and output resistances for the amplifier in Prob. 13.102.

13.116. Calculate the Thévenin equivalent representation for the amplifier in Prob. 13.60.

13.117. Calculate the Thévenin equivalent representation for the amplifier in Prob. 13.62.

13.118. Calculate the Thévenin equivalent representation for the amplifier in Prob. 13.99.

13.119. Calculate the Thévenin equivalent representation for the amplifier in Prob. 13.101.

13.11 Examples of Common-Emitter and Common-Source Amplifiers

13.120. Simulate the behavior of the BJT common-emitter amplifier in Fig. 13.36 and compare the results to the calculations in the example. Use 100 μF for all capacitor values and perform the ac analysis at a frequency of 1000 Hz.

13.121. The amplifier in Fig. P13.121 is the bipolar amplifier in Fig. 13.36 with currents increased by a factor of approximately 10. What are the voltage gain and input resistance and output resistance of this C-E stage? Compare the gain to that of Fig. 13.36. Did you expect this result? Why? Assume $\beta_F = 65$ and $V_A = 50$ V.

13.122. Simulate the behavior of the BJT common-emitter amplifier in Fig. P13.121 and compare the results to the calculations in Prob. 13.121. Use 100 μF for all capacitor values and perform the ac analysis at a frequency of 10,000 Hz.

13.123. The amplifier in Fig. P13.123 is the bipolar amplifier in Fig. P13.36 with currents reduced by a factor of approximately 10. What are the voltage gain and input resistance and output resistance of this amplifier? Compare to that in Fig. 13.36, and discuss the reasons for any differences in gain.

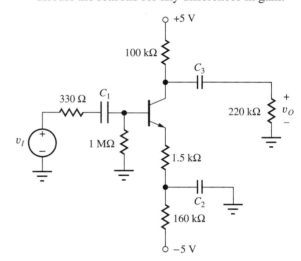

Figure P13.123

13.124. Simulate the behavior of the BJT common-emitter amplifier in Fig. P13.123 and compare the results to the calculations in Prob. 13.123. Use 100 μF for all capacitor values and perform the ac analysis at a frequency of 1000 Hz.

13.125. Use SPICE to simulate the behavior of the MOSFET common-source amplifier in Fig. 13.39 and compare the results to the calculations in the example. Use 100 μF for all capacitor values and perform the ac analysis at a frequency of 1000 Hz.

13.126. Use SPICE to simulate the voltage gain and input resistance and output resistance of the amplifier in Prob. 13.98. Use 100 μF for all capacitor values and perform the ac analysis at a frequency of 1000 Hz.

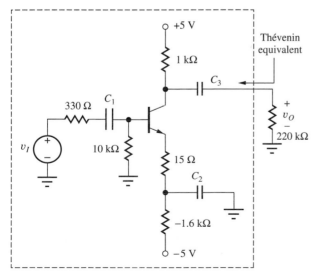

Figure P13.121

13.12 Amplifier Power and Signal Range

13.127. Calculate the dc power dissipation in each element in the circuit in Fig. 13.42(a) if $\beta_F = 65$. Compare the result to the total power delivered by the sources.

13.128. Calculate the dc power dissipation in each element in the circuit in Fig. 13.42(b). Compare the result to the total power delivered by the sources.

13.129. Calculate the dc power dissipation in each element in the circuit in Prob. 13.17. Compare the result to the total power delivered by the sources.

13.130. Repeat Prob. 13.129 for the circuit in Prob.13.19.

13.131. Repeat Prob. 13.129 for the circuit in Prob.13.23.

13.132. Repeat Prob. 13.129 for the circuit in Prob.13.27.

13.133. Repeat Prob. 13.129 for the circuit in Prob.13.33.

*13.134. A common bias point for a transistor is shown in Fig. P13.134. What is the maximum amplitude signal that can be developed at the collector terminal that will satisfy the small-signal assumptions (in terms of V_{CC})?

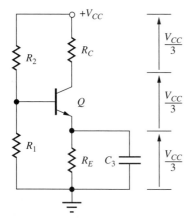

Figure P13.134

*13.135. The MOSFET in Fig. P13.135 has $K_n = 500$ μA/V² and $V_{TN} = -1.5$ V. What is the largest permissible signal voltage at the drain that will satisfy the requirements for small-signal operation if $R_D = 15$ kΩ? What is the minimum value of V_{DD}?

*13.136. The simple C-E amplifier in Fig. P13.136 is biased with $V_{CE} = V_{CC}/2$. Assume that the transistor can saturate with $V_{CESAT} = 0$ V and still be operating linearly. What is the amplitude of the largest sine wave that can appear at the output? What is the ac signal power P_{ac} being dissipated

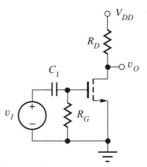

Figure P13.135

in the load resistor R_L? What is the total dc power P_S being supplied from the power supply? What is the efficiency ε of this amplifier if ε is defined as $\varepsilon = 100\% \times P_{ac}/P_S$?

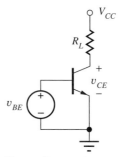

Figure P13.136

13.137. What is the amplitude of the largest ac signal that can appear at the collector of the transistor in Fig. P13.5 that satisfies the small-signal limit? Use the parameter values from Prob. 13.21.

13.138. What is the amplitude of the largest ac signal that can appear at the drain of the transistor in Fig. P13.6 that satisfies the small-signal limit? Use the parameter values from Prob. 13.23.

13.139. What is the amplitude of the largest ac signal that can appear at the drain of the transistor in Fig. P13.8 that satisfies the small-signal limit? Use the parameter values from Prob. 13.27.

13.140. What is the amplitude of the largest ac signal that can appear at the collector of the transistor in Fig. P13.9 that satisfies the small-signal limit? Use the parameter values from Prob. 13.17.

13.141. What is the amplitude of the largest ac signal that can appear at the drain of the transistor in Fig. P13.11 that satisfies the small-signal limit? Use the parameter values from Prob. 13.33.

13.142. What is the amplitude of the largest ac signal that can appear at the drain of the transistor in Fig. P13.13 that satisfies the small-signal limit? Use the parameter values from Prob. 13.35.

13.143. Draw the load line for the circuit in Fig. 13.1 on the output characteristics in Fig. P13.143 for $V_{CC} = 20$ V and $R_C = 20$ kΩ. Locate the Q-point for $I_B = 2$ μA. Estimate the maximum output voltage swing from the characteristics. Repeat for $I_B = 5$ μA.

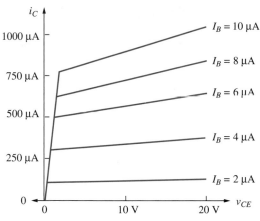

Figure P13.143

SINGLE-TRANSISTOR AND MULTISTAGE ac-COUPLED AMPLIFIERS

CHAPTER OUTLINE

CHAPTER GOALS

In Chapter 14, we fully explore the small-signal characteristics of three families of single-stage amplifiers. We will discover why certain transistor terminals are preferred for signal input whereas others are used for signal outputs. The results define three broad classes of amplifiers.

- Inverting amplifiers—the common-emitter and common-source configurations—that provide high voltage gain with a 180° phase shift
- Followers—the common-collector and common-drain configurations—that provide nearly unity gain similar to the op amp voltage follower
- Noninverting amplifiers—the common-base and common-gate configurations—that provide high voltage gain with no phase shift

For each type of amplifier, we discuss the detailed design of

- Voltage gain and input voltage range
- Current gain
- Input and output resistances
- Coupling and bypass capacitor design and lower cutoff frequency

The results become our design toolkit and are used to solve a number of examples of design problems.

As in most chapters, we will continue to increase our understanding of SPICE simulation and interpretation of SPICE results. In particular, we try to understand the differences between

- SPICE ac (small-signal), transient (large signal), and transfer function analysis modes

Chapter 13 introduced the common-emitter and common-source amplifiers, in which the input signal was applied to the base and gate terminals of the BJT and MOSFET, respectively, and the output signal was taken from the collector and drain. However, bipolar and field-effect transistors are three-terminal devices, and this chapter explores the use of other terminals for signal input and output. Three useful amplifier configurations are identified, each using a different terminal as the common or reference terminal. When implemented using bipolar transistors, these are called the common-emitter, common-collector, and common-base amplifiers; the corresponding names for the FET implementations are the common-source, common-drain, and common-gate amplifiers. Each amplifier category provides a unique set of characteristics in terms of voltage gain, input resistance, output resistance, and current gain.

The chapter reviews the characteristics of the common-emitter and common-source amplifiers, i.e., the inverting amplifiers that were developed in Chapter 13, and then looks in depth at the followers and noninverting amplifiers, focusing on the limits solid-state devices place on individual amplifier performance. Expressions are presented for the properties of each amplifier, and their similarities and differences are discussed in detail in order to build the understanding needed for the circuit design process. The transistor-level results are used throughout this book to analyze and design more complex single-stage and multi-stage amplifiers. We also explore amplifier frequency response at low frequencies and develop design equations useful for choosing coupling and bypass capacitors.

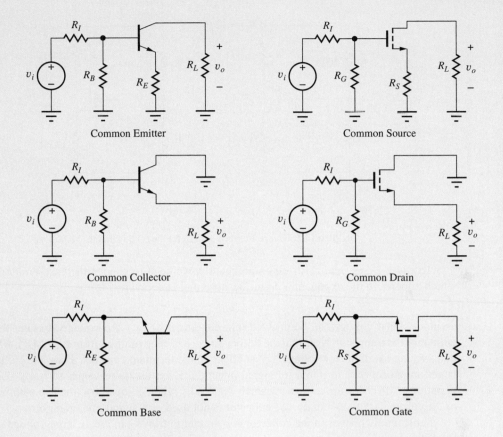

Much discussion is devoted to single-transistor ampli-fiers because they are the heart of analog design. These single-stage amplifiers are an important part of the basic "tool set" of analog circuit designers, and a good under-standing of their similarities and differences is a prerequisite for more complex amplifier design.

14.1 AMPLIFIER CLASSIFICATION

In Chapter 13, the input signal was applied to the base or gate of the transistor, and the output signal was taken from the collector or drain. However, the transistor has three separate terminals that may possibly be used to inject a signal for amplification: the base, emitter, and collector for the BJT; the gate, source, and drain for the FET. We will see shortly that only the base and emitter, or gate and source, are useful as signal insertion points; the collector and emitter, or drain and source, are useful points for signal removal. The examples we use in this chapter of the various amplifier configurations all use the same four-resistor bias circuits shown in Fig. 14.1. Coupling and bypass capacitors are then used to change the signal injection and extraction points and modify the ac characteristics of the amplifiers.

14.1.1 SIGNAL INJECTION AND EXTRACTION—THE BJT

For the BJT in Fig. 14.1(a), the large-signal transport model provides guidance for proper location of the input signal. In the forward-active region of the BJT,

$$i_C = I_S \left[\exp \left(\frac{v_{BE}}{V_T} \right) \right] \quad i_B = \frac{i_C}{\beta_F} = \frac{I_S}{\beta_{FO}} \left[\exp \left(\frac{v_{BE}}{V_T} \right) \right] \quad i_E = \frac{I_S}{\alpha_F} \left[\exp \left(\frac{v_{BE}}{V_T} \right) \right] \quad (14.1)$$

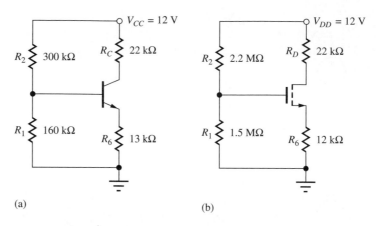

Figure 14.1 Four-resistor bias circuits for the (a) BJT and (b) MOSFET.

To cause i_C, i_E, and i_B to vary significantly, we need to change the base-emitter voltage v_{BE}, which appears in the exponential term. Because v_{BE} is equivalent to

$$v_{BE} = v_B - v_E \qquad (14.2)$$

an input signal voltage can be injected into the circuit to vary the voltage at either the base or the emitter of the transistor. Note that the Early voltage has been omitted from Eq. (14.1), which indicates that varying the collector voltage has no effect on the terminal currents. Thus, the collector terminal is not an appropriate terminal for signal injection. Even for finite values of Early voltage, current variations with collector voltage are small, especially when compared to the exponential dependence of the currents on v_{BE}—again, the collector is not used as a signal injection point.

Substantial changes in the collector and emitter currents can create large voltage signals across the collector and emitter resistors R_C and R_6 in Fig. 14.1. Thus, signals can be removed from the amplifier at the collector or emitter terminals. However, because the base current i_B is a factor of β_F smaller than either i_C or i_E, the base terminal is not normally used as an output terminal.

 **DESIGN NOTE**

> The input signal can be applied to the base or emitter terminal of the bipolar transistor, and the output signal can be taken from the collector or emitter. The collector is not used as an input terminal, and the base is not used as an output.

14.1.2 SIGNAL INJECTION AND EXTRACTION—THE FET

A similar set of arguments can be used for the FET in Fig. 14.1(b), based on the expression for the n-channel MOSFET drain current in pinchoff:

$$i_S = i_D = \frac{K_n}{2}(v_{GS} - V_{TN})^2 \qquad \text{and} \qquad i_G = 0 \qquad (14.3)$$

To cause i_D and i_S to vary significantly, we need to change the gate-source voltage v_{GS}. Because v_{GS} is equivalent to

$$v_{GS} = v_G - v_S \qquad (14.4)$$

an input signal voltage can be injected so as to vary either the gate or source voltage of the FET. Varying the drain voltage has only a minor effect (for $\lambda \neq 0$) on the terminal currents, so the drain terminal is not an appropriate terminal for signal injection. As for the BJT, substantial changes in the drain or source currents can develop large voltage signals across resistors R_D and R_6 in Fig. 14.1(b). However, the gate terminal is not used as an output terminal because the gate current is zero.

In summary, effective amplification requires a signal to be injected into either the base/emitter or gate/source terminals of the transistors in Fig. 14.1; the output signal can be taken from the collector/emitter or drain/source terminals. We do not inject a signal into the collector or drain or extract a signal from the base or gate terminals. These constraints yield three families of amplifiers: the **common-emitter/common-source (C-E/C-S)** circuits that we studied in Chapter 13, the **common-base/common-gate (C-B/C-G)** circuits, and the **common-collector/common-drain (C-C/C-D)** circuits.

These amplifiers are classified in terms of the structure of the ac equivalent circuit; each is discussed in detail in the next several sections. As noted earlier, the circuit examples all use the same four-resistor bias circuits in Fig. 14.1 in order to establish the Q-point of the various amplifiers. Coupling and bypass capacitors are then used to change the ac equivalent circuits. We will find that the ac characteristics of the various amplifiers are significantly different.

EXERCISE: Find the Q-points for the transistors in Fig. 14.1 and calculate the small-signal model parameters for the BJT and MOSFET. Use $\beta_F = 100$, $V_A = 50$ V, $K_n = 500$ μA/V^2, $V_{TN} = 1$ V, and $\lambda = 0.02$ V^{-1}. What are the values of μ_f? What is the value of $V_{GS} - V_{TN}$ for the MOSFET?

ANSWERS:

	I_C / I_D	V_{CE} / V_{DS}	$V_{GS} - V_{TN}$	g_m	r_π	r_o	μ_f
BJT	245 μA	3.64 V	...	9.80 mS	10.2 kΩ	219 kΩ	2150
FET	241 μA	3.81 V	0.982 V	0.491 mS	∞	223 kΩ	110

DESIGN NOTE

The input signal can be applied to the gate or source terminal of the FET, and the output signal can be taken from the drain or source. The drain is not used as an input terminal, and the gate is not used as an output.

14.1.3 COMMON-EMITTER (C-E) AND COMMON-SOURCE (C-S) AMPLIFIERS

The circuits in Fig. 14.2 are the common-emitter and common-source amplifiers discussed in Chapter 13. In these circuits, resistor R_6 in Fig. 14.1 has been split into two parts, with only resistor R_4 bypassed by capacitor C_2. As we discovered in the last chapter, we gain considerable flexibility in setting the voltage gain, input resistance, and output resistance of the amplifier by not bypassing all of the resistance in the transistor's emitter or source. In the C-E circuit in Fig. 14.2(a), the signal is injected into the base and taken out of the collector of the BJT. The emitter is the common terminal between the input and output ports. In the C-S circuit in Fig. 14.2(b), the signal is injected into the gate and taken out of the drain of the MOSFET; the source is the common terminal between the input and output ports.

The simplified ac equivalent circuits for these amplifiers appear in Figs. 14.2(c) and (d). We see that these network topologies are identical. Resistors R_E and R_S, connected between the emitter or source and ground, represent the unbypassed portion of the original bias resistor R_6. The presence of R_E and R_S in the ac equivalent circuits gives an added degree of freedom to the designer, and allows gain to be traded for increased input resistance, output resistance, and input signal range. Our comparative analysis will show that the C-E and C-S circuits can provide moderate-to-high values of voltage, current gain, input resistance, and output resistance.

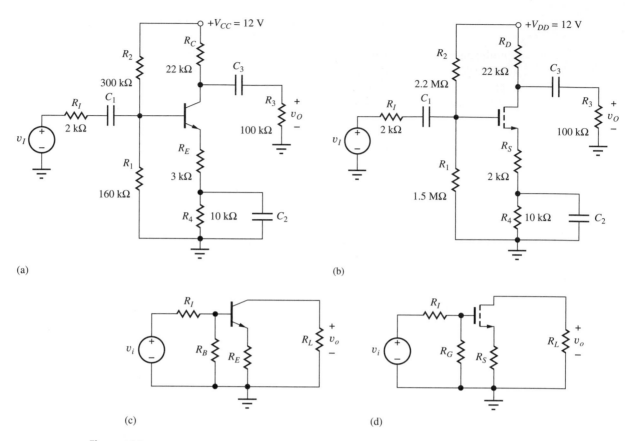

(a) (b) (c) (d)

Figure 14.2 Generalized versions of the (a) common-emitter (C-E) and (b) common-source (C-S) amplifiers. (c) Simplified ac equivalent circuit of the C-E amplifier in (a). (d) Simplified ac equivalent circuit of the C-S amplifier in (b).

EXERCISE: Construct the ac equivalent circuit for the C-E and C-S amplifiers in Fig. 14.2, and show that the ac models are correct. What are the values of R_B or R_G, R_E or R_S, and R_L?

ANSWERS: 104 kΩ, 3.00 kΩ, 18.0 kΩ; 892 kΩ, 2.00 kΩ, 18.0 kΩ

14.1.4 COMMON-COLLECTOR (C-C) AND COMMON-DRAIN (C-D) TOPOLOGIES

The C-C and C-D circuits are shown in Fig. 14.3. Here the signal is injected into the base [Fig. 14.3(a)] or gate [Fig. 14.3(b)] and extracted from the emitter or source of the transistors. The collector and drain are bypassed directly to ground by the capacitors C_2 and represent the common terminals between the input and output ports. Once again, the ac equivalent circuits in Figs. 14.3(c) and (d) are identical in structure; the only differences are the resistor and transistor parameter values. Analysis will show that the C-C and C-D amplifiers provide a voltage gain of approximately 1, a high input resistance and a low output resistance. In addition, the input signals to the C-C and C-D amplifiers can be quite large without exceeding the small-signal limits. These amplifiers, often called emitter or source followers, are the single-transistor equivalents of the op amp voltage-follower circuit that we studied in Chapter 11.

EXERCISE: Construct the ac equivalent circuit for the C-C and C-D amplifiers in Fig. 14.3, and show that the ac models are correct. Verify the values of R_B, R_G, and R_L.

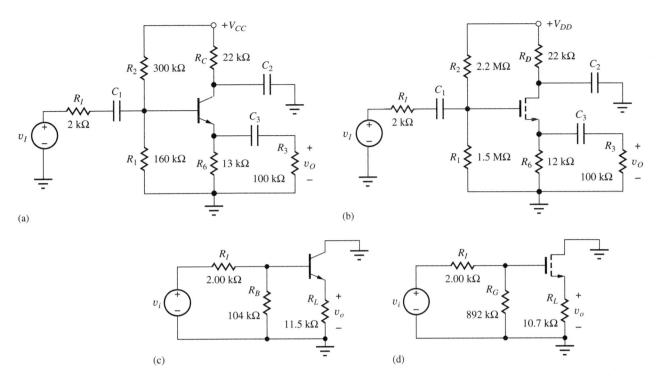

Figure 14.3 (a) Common-collector (C-C) amplifier. (b) Common-drain (C-D) amplifier. (c) Simplified ac equivalent circuit for the C-C amplifier. (d) Simplified ac equivalent circuit for the C-D amplifier.

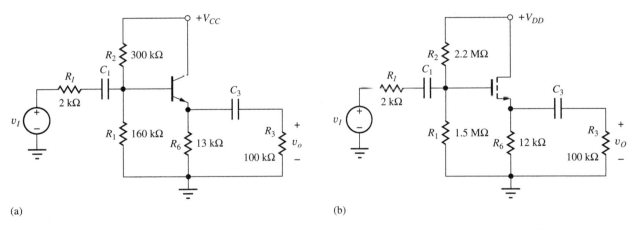

Figure 14.4 Simplified follower circuits with C_2, R_C and R_D eliminated. (a) Common-collector amplifier and (b) common-drain amplifier.

Circuit Simplification

For economy of design, we certainly do not want to include unneeded components, and the circuits in Fig. 14.3 can actually be simplified. The purpose of capacitor C_2 in the C-C and C-D amplifiers is to provide an ac ground at the collector or drain terminals of the two transistors, and since we do not wish to develop a signal voltage at either of these terminals, there is no reason to have resistors R_C or R_D in the circuits. We can achieve the desired ac ground by simply connecting the collector and drain terminals directly to V_{CC} and V_{DD}, respectively, which eliminates components R_C, R_D, and C_2 from the circuits, as shown in Fig. 14.4.

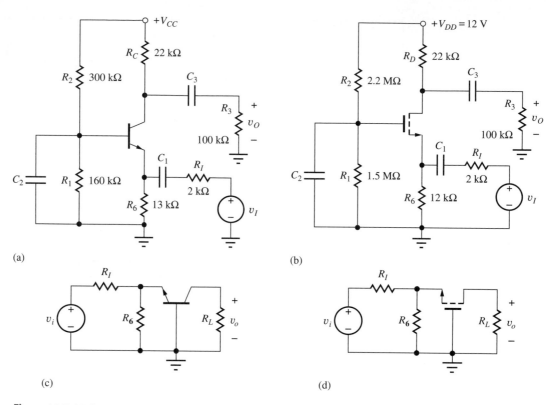

Figure 14.5 (a) Common-base (C-B) amplifier. (b) Common-gate (C-G) amplifier. (c) Simplified ac equivalent circuit for the C-B amplifier. (d) Simplified ac equivalent circuit for the C-G amplifier.

14.1.5 COMMON-BASE (C-B) AND COMMON-GATE (C-G) AMPLIFIERS

The third class of amplifiers contains the C-B and C-G circuits in Fig. 14.5. ac signals are injected into the emitter and source and extracted from the collector and drain of the transistors. The base and gate terminals are connected to signal ground through bypass capacitors C_2; these terminals are the common connections between the input and output ports. The resulting ac equivalent circuits in Figs. 14.5(c) and (d) are again identical in structure. Analysis will show that the C-B and C-G amplifiers provide a voltage gain and output resistance very similar to those of the C-E and C-S amplifiers, but they have a much lower input resistance.

Analyses in the next several sections involve the simplified ac equivalent circuits given in Figs. 14.2(c), (d), 14.3(c), (d), and 14.5(c), (d). We assume for purposes of analysis that the circuits have been reduced to these "standard amplifier prototypes." These circuits are used to delineate the limits that the devices place on performance of the various circuit topologies. The results from these simplified circuits will then be used to analyze and design complete amplifiers.

The circuits in Figs. 14.2 to 14.5 showed only the BJT and MOSFET. The small-signal model of the JFET is identical to that of the three-terminal MOSFET, and the results obtained for the MOSFET amplifiers apply directly to those for the JFETs as well. JFETs can replace the MOSFETs in many circuits. (Information on JFET can be found on the website.)

EXERCISE: **Construct the ac equivalent circuit for the C-B and C-G amplifiers in Fig. 14.5, and show that the ac models are correct. What are the values of** R_I, R_6, **and** R_L**?**

ANSWERS: 2 kΩ, 13.0 kΩ, 18.0 kΩ; 2 kΩ, 12.0 kΩ, 18.0 kΩ

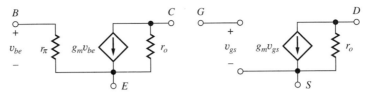

Figure 14.6 Small-signal models for the BJT and MOSFET.

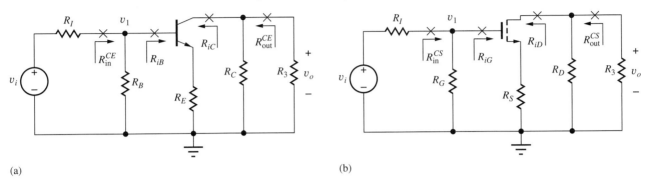

(a)   (b)

Figure 14.7 (a) ac Equivalent circuit for the C-E amplifier. (b) ac Equivalent circuit for the C-S amplifier.

TABLE 14.1

SMALL-SIGNAL PARAMETER	BJT	MOSFET
g_m	$\dfrac{I_C}{V_T} \cong 40 I_C$	$\dfrac{2I_D}{V_{GS} - V_{TN}} \cong \sqrt{2K_n I_D}$
r_π	$\dfrac{\beta_o}{g_m}$	∞
r_o	$\dfrac{V_A + V_{CE}}{I_C} \cong \dfrac{V_A}{I_C}$	$\dfrac{(1/\lambda) + V_{DS}}{I_D} \cong \dfrac{1}{\lambda I_D}$
β_o	$g_m r_\pi$	∞
$\mu_f = g_m r_o$	$\dfrac{V_A + V_{CE}}{V_T} \cong 40 V_A$	$\dfrac{2}{\lambda(V_{GS} - V_{TN})} \cong \dfrac{1}{\lambda}\sqrt{\dfrac{2K_n}{I_D}}$

14.1.6 SMALL-SIGNAL MODEL REVIEW

The small-signal models for the BJT and MOSFET appear in Fig. 14.6, and the formulae relating the small-signal model parameters to the Q-point are summarized in Table 14.1.

Again, we recognize that the topologies are very similar, except for the finite value of r_π for the BJT. Due to these similarities, we begin the analyses with that for the bipolar transistor because it has the more general small-signal model; we obtain results for the FET cases from the BJT expressions by taking limits as r_π and β_o approach infinity. In subsequent sections, expressions for the voltage gain, input resistance, output resistance, and current gain are developed for each of the single-transistor amplifiers based upon the small-signal models in Fig. 14.6.

14.2 INVERTING AMPLIFIERS—COMMON-EMITTER AND COMMON-SOURCE CIRCUITS

We begin our comparative analysis of the various amplifier families with the common-emitter and common-source amplifiers that we studied in detail in Chapter 13. The ac equivalent circuits are repeated in Fig. 14.7. Here again we note that the topologies are identical. Performance differences arise because of the differences in the parameters of the transistors used in the circuit.

TABLE 14.2
Common-Emitter/Common-Source Amplifier Summary

	COMMON-EMITTER (C-E) AMPLIFIER	COMMON-SOURCE (C-S) AMPLIFIER
Terminal voltage gain	$A_{vt}^{CE} = \dfrac{v_o}{v_1} = -\dfrac{g_m R_L}{1 + g_m R_E}$	$A_{vt}^{CS} = \dfrac{v_o}{v_1} = -\dfrac{g_m R_L}{1 + g_m R_S}$
Signal source voltage gain	$A_v^{CE} = \dfrac{v_o}{v_i} = A_{vt}^{CE}\dfrac{R_B \| R_{iB}}{R_I + R_B \| R_{iB}}$	$A_v^{CS} = \dfrac{v_o}{v_i} = A_{vt}^{CS}\dfrac{R_G}{R_I + R_G}$
Rule-of-thumb estimate for $g_m R_L$	$10(V_{CC} + V_{EE})$	$(V_{DD} + V_{SS})$
Input terminal resistance	$R_{iB} = r_\pi(1 + g_m R_E)$	$R_{iG} = \infty$
Output terminal resistance	$R_{iC} = r_o(1 + g_m R_E)$	$R_{iD} = r_o(1 + g_m R_S)$
Input signal range	$0.005(1 + g_m R_E)$ V	$0.2(V_{GS} - V_{TN})(1 + g_m R_S)$
Terminal current gain	β_o	∞

14.2.1 COMMON-EMITTER AND COMMON-SOURCE AMPLIFIER CHARACTERISTICS

Table 14.2 summarizes the results for the C-E and C-S amplifiers developed in Chapter 13. In the common-emitter circuit, resistor R_E adds feedback to the amplifier that reduces the voltage gain by the factor $(1 + g_m R_E)$, but increases the input resistance, output resistance, and input signal range by the same amount. Resistor R_S has a similar impact on the voltage gain, output resistance, and input signal range of the common-source amplifier. Since the resistance at the gate terminal of the FET is already infinite, the overall input resistance of the C-S amplifier is not affected by R_S.

EXAMPLE 14.1   **INVERTING AMPLIFIER VOLTAGE GAIN CALCULATIONS**

This example uses the results in Table 14.2 and compares the characteristics of the BJT and MOSFET amplifiers in Fig. 14.2.

PROBLEM   Find the voltage gain $A_v = v_o/v_i$ for the common-emitter and common-source amplifiers in Fig. 14.2. Use the Q-point information from Sec. 14.1. What are the maximum values of v_i for the two amplifiers? Find the resistance at the transistor output terminals.

SOLUTION   **Known Information and Given Data:** The common-emitter and common-source amplifier circuits appear in Figs. 14.2(a) and 14.2(b). The transistor Q-point values and small-signal parameter values were calculated in this chapter in Sec. 14.1 and are repeated here for convenience.

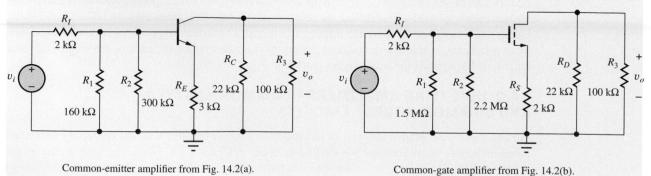

Common-emitter amplifier from Fig. 14.2(a).   Common-gate amplifier from Fig. 14.2(b).

	I_C or I_D	V_{CE} or V_{DS}	$V_{GS} - V_{TN}$	g_m	r_π	r_o	μ_f
BJT	245 μA	3.64 V	$\cdots$	9.80 mS	10.2 kΩ	219 kΩ	2150
FET	241 μA	3.81 V	0.982 V	0.491 mS	∞	223 kΩ	110

For Fig. 14.2(a), $R_I = 2\text{ k}\Omega$, $R_1 = 160\text{ k}\Omega$, $R_2 = 300\text{ k}\Omega$, $R_C = 22\text{ k}\Omega$, $R_E = 3\text{ k}\Omega$, and $R_3 = 100\text{ k}\Omega$. For Fig. 14.2(b), $R_I = 2\text{ k}\Omega$, $R_1 = 1.5\text{ M}\Omega$, $R_2 = 2.2\text{ M}\Omega$, $R_D = 22\text{ k}\Omega$, $R_S = 2\text{ k}\Omega$, and $R_3 = 100\text{ k}\Omega$.

Unknowns: Voltage gain $A_v = v_o/v_i$ for the common-emitter and common-source amplifiers, maximum input signals, resistance at the output terminal.

Approach: Calculate the values of R_B, R_{in}, and R_L. Substitute the transistor and circuit parameter values into the voltage gain expressions, in Table 14.2.

Assumptions: The transistors are operating under small-signal conditions

Analysis for the Common-Emitter Amplifier: In order to evaluate the gain expression that describes the equivalent circuit in Fig. 14.2(c), we need to find R_B, R_{iB}, R_L, and A_{vt}:

$$R_B = R_1 \| R_2 = 160\text{ k}\Omega \| 300\text{ k}\Omega = 104\text{ k}\Omega$$

$$R_{iB} = r_\pi(1 + g_m R_E) = 10.2\text{ k}\Omega[1 + 9.80\text{ mS}(3\text{ k}\Omega)] = 310\text{ k}\Omega$$

$$R_L = R_C \| R_3 = 22\text{ k}\Omega \| 100\text{ k}\Omega = 18.0\text{ k}\Omega$$

Now,

$$A_{vt}^{CE} = -\frac{g_m R_L}{1 + g_m R_E} = -\frac{9.80\text{ mS}(18.0\text{ k}\Omega)}{1 + 9.80\text{ mS}(3\text{ k}\Omega)} = -5.80$$

$$A_v^{CE} = A_{vt}^{CE} \left[\frac{R_B \| R_{iB}}{R_I + (R_B \| R_{iB})} \right] = -5.80 \left[\frac{104\text{ k}\Omega \| 310\text{ k}\Omega}{2.00\text{ k}\Omega + (104\text{ k}\Omega \| 310\text{ k}\Omega)} \right] = -5.65$$

The input resistance looking into coupling capacitor C_1 of the common-emitter amplifier is

$$R_{in}^{CE} = R_B \| R_{iB} = 104\text{ k}\Omega \| 310\text{ k}\Omega = 77.9\text{ k}\Omega$$

The resistance at the transistor's collector terminal and the overall output resistance to the common-emitter amplifier are

$$R_{iC} \cong r_o(1 + g_m R_E) = 219\text{ k}\Omega[1 + 9.80\text{ mS}(3\text{ k}\Omega)] = 6.68\text{ M}\Omega$$

$$R_{out}^{CE} = R_C \| R_{iC} = 22\text{ k}\Omega \| 6.68\text{ M}\Omega = 21.9\text{ k}\Omega$$

The maximum input signal that satisfies the small-signal limit is

$$|v_i| \leq 0.005(1 + g_m R_E)\text{ V} = 0.005\text{ V}[1 + 9.80\text{ mS}(3\text{ k}\Omega)] = 0.152\text{ V}$$

Note that more accurate estimate for the resistance R_{iC} at the collector terminal is given by Eq. (13.64):

$$R_{iC} = r_o\left(1 + \frac{\beta_o R_E}{R_{th} + r_\pi + R_E}\right) = 219\text{ k}\Omega\left[1 + \frac{100(3\text{ k}\Omega)}{1.96\text{ k}\Omega + 10.2\text{ k}\Omega + 3\text{ k}\Omega}\right] = 4.55\text{ M}\Omega$$

$$R_{out}^{CE} = R_C \| R_{iC} = 22\text{ k}\Omega \| 4.55\text{ M}\Omega = 21.9\text{ k}\Omega$$

Analysis for the Common-Source Amplifier: In order to evaluate the gain expression that describes the equivalent circuit in Fig. 14.2(d), we need to find R_G, R_L, and A_{vt}:

$$R_G = R_1 \| R_2 = 1.5 \text{ M}\Omega \| 2.2 \text{ M}\Omega = 892 \text{ k}\Omega$$

$$R_L = R_D \| R_3 = 22 \text{ k}\Omega \| 100 \text{ k}\Omega = 18.0 \text{ k}\Omega$$

$$A_{vt}^{CS} = -\frac{g_m R_L}{1 + g_m R_S} = -\frac{(0.491 \text{ mS})(18.0 \text{ k}\Omega)}{1 + (0.491 \text{ mS})(2.00 \text{ k}\Omega)} = -4.46$$

$$A_v^{CS} = A_{vt}^{CS} \left(\frac{R_G}{R_I + R_G} \right) = -4.46 \left(\frac{892 \text{ k}\Omega}{2 \text{ k}\Omega + 892 \text{ k}\Omega} \right) = -4.45$$

The input resistance looking into coupling capacitor C_1 of the common-source amplifier is

$$R_{in}^{CS} = R_G \| R_{iG} = 892 \text{ k}\Omega \| \infty = 892 \text{ k}\Omega$$

The resistance at the transistor's drain terminal and the overall output resistance to the common-source amplifier are

$$R_{iD} = r_o(1 + g_m R_S) = 223 \text{ k}\Omega[1 + 0.491 \text{ mS}(2 \text{ k}\Omega)] = 442 \text{ k}\Omega$$

$$R_{out}^{CD} = R_D \| R_{iD} = 22 \text{ k}\Omega \| 442 \text{ k}\Omega = 21.0 \text{ k}\Omega$$

The maximum input signal that satisfies the small-signal limit is

$$|v_i| \leq 0.2(V_{GS} - V_{TN})(1 + g_m R_S) = 0.2(0.982V)[1 + 0.491 \text{ mS}(2 \text{ k}\Omega)] = 0.389 \text{ V}$$

Check of Results: Since both amplifiers have a nonzero value of the emitter and source resistors, we can use Eq. (13.56) to check these calculations. For the BJT,

$$A_{vt}^{CE} \cong -\frac{R_L}{R_E} = -\frac{18 \text{ k}\Omega}{3 \text{ k}\Omega} = -6.00 \quad \checkmark$$

and for the MOSFET,

$$A_{vt}^{CS} \cong -\frac{R_L}{R_S} = -\frac{18 \text{ k}\Omega}{2 \text{ k}\Omega} = -9.00 \quad \checkmark$$

Discussion: The actual gain of the two amplifiers is very similar and is the result of including R_E and R_S in each circuit. For the common-emitter amplifier, $g_m R_E \gg 1$, and the terminal gain should closely approach the estimate, as it does. However, for the MOSFET, the $g_m R_S$ product is not much greater than one, and the gain is less than the resistor ratio estimate, which represents an upper bound on the gain of the circuit with a nonzero value of R_S. Note that the value of R_S was reduced to 2 kΩ in the common-source design to produce a gain similar to that of the common-emitter amplifier. Note that the signal handling capability of the circuits is similar because of the large $g_m R_E$ product for the C-E circuit.

Computer-Aided Analysis: We can check the voltage gains using SPICE by performing an operating point analysis followed by an ac analysis with v_I as the input and v_O as the output voltage. Be sure to set the transistor parameters to match our hand calculations (BF = 100, VAF = 50 V). Make all the capacitor values large, say 100 μF, and sweep the frequency to find the midband range of frequencies (e.g., FSTART = 1 Hz, FSTOP = 100 kHz, with 10 frequency points per decade.) Analysis of the two circuits yields -5.60 for the gain of the common-emitter amplifier and -4.35 for the gain of the common-source stage. These values agree closely with our hand calculations. Because we have specified VAF, the transistor output resistance r_o is included in the simulations and is the reason for the slightly smaller gain in SPICE.

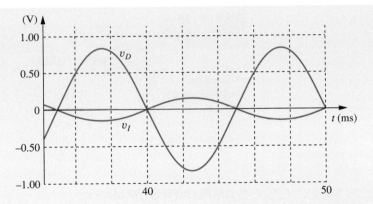

The graph here presents the results of a transient simulation (TSTART $= .034$ S, TSTOP $= .05$ S, and TSTEP $= 1$ US) for the common-emitter amplifier with an input signal of $v_I = 0.150 \sin 2000\pi t$ V. The output $v_O = -0.838 \sin 2000\pi t$ V and exhibits no visible sign of distortion with the input signal at the small signal limit. The start time in the graph is chosen to be 34 ms, rather than zero, to allow the transient simulation to reach a periodic steady-state condition.

EXERCISE: What is the voltage gain A_v^{CE} for the C-E amplifier based on the transient simulation results in the graph above?

ANSWER: -5.59

EXERCISE: (a) What is the voltage gain A_v of the two amplifiers in Ex. 14.1 if R_E and R_S are changed to 1 kΩ? Assume the Q-points do not change. (b) What are the new values of R_4 required to maintain the Q-points unchanged in the two amplifiers?

ANSWERS: -15.6, -5.9; 12 kΩ, 11 kΩ

EXERCISE: (a) What is the voltage gain A_v of the two amplifiers in Ex. 14.1 if the upper terminal of C_2 is connected to the emitter of the transistor in Fig. 14.2(a) and to the source of the transistor in Fig. 14.2(b)?

ANSWERS: -145, -8.82

EXERCISE: What value of saturation current I_S must be used in SPICE to achieve $V_{BE} = 0.7$ V for $I_C = 245$ μA? Assume a default temperature of 27°C.

ANSWER: 0.422 fA

EXERCISE: Evaluate $-g_m R_L$ and $-R_L/R$ for the C-E and C-S amplifiers in Ex. 14.1, and compare the magnitudes to the exact calculations in Ex. 14.1. ($R = R_E$ or R_S)

ANSWERS: -176, -6.00; -8.84, -9.00; $5.65 < 6.00$; $4.46 < 8.84$

EXERCISE: Calculate R_{in} for the C-E and C-S amplifiers in Fig. 14.5. How does R_{in}^{CE} compare to r_π?

ANSWERS: 313 k$\Omega \cong \beta_o R_E$, ∞; $R_{in}^{CE} \gg r_\pi$ for the C-E amplifier

TABLE 14.3
Common-Emitter/Common-Source Amplifier Comparison

	C-E AMPLIFIER	C-S AMPLIFIER
Voltage gain	−5.61	−4.45
Input resistance	310 kΩ	∞
Output resistance	4.55 MΩ	442 kΩ
Input signal range	152 mV	389 mV
Terminal current gain	−100	∞

14.2.2 C-E/C-S AMPLIFIER SUMMARY

The numeric results for the two specific amplifier cases are presented in Table 14.3. The common-emitter and common-source amplifiers have similar voltage gains. The C-E amplifier approaches the R_L/R_E limit (-6) more closely because $g_m R_E = 29.4$ for the BJT case, but only 0.982 for the MOSFET. The C-S amplifier provides extremely high input resistance, but that of the BJT amplifier is also substantial due to the $\mu_f R_E$ term. The output resistance of the C-E amplifier is also much higher than the C-S amplifier because μ_f is much larger for the BJT than for the FET. The input signal levels have been increased above the R_S or $R_E = 0$ case—again by a substantial amount in the BJT case. The terminal current gains are identical to those of the individual transistors.

14.2.3 EQUIVALENT TRANSISTOR REPRESENTATION OF THE GENERALIZED C-E/C-S TRANSISTOR

The equations in Table 14.1 can actually provide us with a way to "absorb" resistor R into the transistor. This action can often simplify our circuit analysis or help provide insight into the operation of a circuit that we haven't seen before. The process is depicted in Fig. 14.8, in which the original transistor Q and resistor R are replaced by a new equivalent transistor Q'. The small-signal parameters of the new transistor are given by

$$g'_m = \frac{g_m}{1 + g_m R} \qquad r'_\pi = r_\pi(1 + g_m R) \qquad r'_o = r_o(1 + g_m R) \qquad (14.5)$$

Here we see the direct trade-off between reduced transconductance and increased input and output resistance. It is also important to note, however, that current gain and amplification factor of the transistor are conserved—we cannot exceed the limitations of the transistor itself!

$$\beta'_o = g'_m r'_\pi = \beta_o \qquad \text{and} \qquad \mu'_f = g'_m r'_o = \mu_f \qquad (14.6)$$

Similar results apply to the FET except that the current gain and input resistance are both infinite.

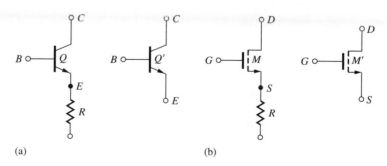

| (a) | (b) |

Figure 14.8 Composite transistor representation of (a) transistor Q and resistor R. (b) Transistor M and resistor R.

14.3 FOLLOWER CIRCUITS—COMMON-COLLECTOR AND COMMON-DRAIN AMPLIFIERS

We now consider a second class of amplifiers, the common-collector (C-C) and common-drain (C-D) amplifiers, as represented by the ac equivalent circuits in Fig. 14.9. We will see that the follower circuits provide high input resistance and low output resistance with a gain of approximately one. The BJT circuit is analyzed first, and then the MOSFET circuit is treated as a special case with $r_\pi \to \infty$.

14.3.1 TERMINAL VOLTAGE GAIN

To find the terminal gain in Fig. 14.9(a), the bipolar transistor is replaced by its small-signal model in Fig. 14.10 (r_o is again neglected). The output voltage v_o now appears across load resistor R_L connected to the emitter of the transistor and is equal to

$$\mathbf{v_o} = +(\beta_o + 1)\mathbf{i_b} R_L \qquad \text{where} \qquad R_L = R_3 \| R_6 \qquad (14.7)$$

The input current is related to applied voltage v_b by

$$\mathbf{v_b} = \mathbf{i_b} r_\pi + (\beta_o + 1)\mathbf{i_b} R_L = \mathbf{i_b}[r_\pi + (\beta_o + 1)R_L] \qquad (14.8)$$

Combining Eqs. (14.7) and (14.8) yields an expression for the terminal gain of the common-collector amplifier:

$$A_{vt}^{CC} = +\frac{(\beta_o + 1)R_L}{r_\pi + (\beta_o + 1)R_L} \cong \frac{g_m R_L}{1 + g_m R_L} \qquad (14.9)$$

where the approximation holds for large β_o.

Letting r_π (and β_o) approach infinity in Eq. (14.9) yields the corresponding terminal gain for the FET follower in Fig. 14.9(b):

$$A_{vt}^{CD} = \frac{g_m R_L}{1 + g_m R_L} \qquad (14.10)$$

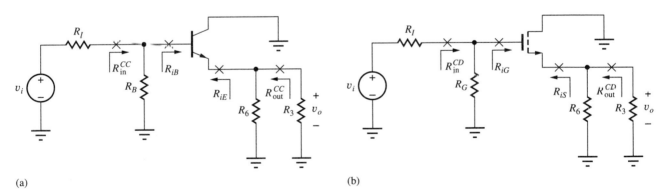

(a)   (b)

Figure 14.9 (a) ac equivalent circuit for the C-C amplifier. (b) ac equivalent circuit for the C-D amplifier.

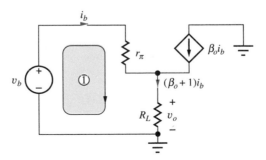

Figure 14.10 Small-signal model for the C-C amplifier. $R_L = R_3 \| R_6$.

In most common-collector and common-drain designs, $g_m R_L \gg 1$, and Eqs. (14.9) and (14.10) reduce to

$$A_{vt}^{CC} \cong A_{vt}^{CD} \cong 1 \tag{14.11}$$

The C-C and C-D amplifiers both have a gain that approaches 1. That is, the output voltage follows the input voltage, and the C-C and C-D amplifiers are often called **emitter followers** and **source followers,** respectively. In most cases, the BJT does a better job of achieving $g_m R_L \gg 1$ than does the FET, and the BJT gain is closer to unity than that of the FET. However, in both cases the value of voltage gain typically falls in the range of

$$0.75 \le A_{vt} \le 1 \tag{14.12}$$

Obviously, A_{vt} is much less than the amplification factor μ_f, so neglecting r_o in the model of Fig. 14.10 is valid. Note, however, that r_o appears in parallel with R_L, and its effect can be included by replacing R_L with $(R_L \| r_o)$ in the equations.

DESIGN NOTE

The terminal gain of single transistor voltage followers is given by

$$A_{vt}^{FO} = +\frac{g_m R_L}{1 + g_m R_L} \qquad \text{and typically} \qquad 0.75 < A_{vt}^{FO} < 1$$

14.3.2 RESISTANCE

The input resistance at the base terminal of the BJT is simply equal to the last term in brackets in Eq. (14.8):

$$R_{iB} = \frac{\mathbf{v_b}}{\mathbf{i_b}} = r_\pi + (\beta_o + 1) R_L \cong r_\pi (1 + g_m R_L) \qquad \text{and} \qquad R_{iG} = \infty \tag{14.13}$$

letting r_π (and β_o) approach infinity for the MOSFET. The input resistance of the emitter follower is equal to r_π plus an amplified replica of load resistor R_L, and can be made quite large. Of course, we see that the input resistance of the source follower is very large.

The overall input resistance R_{in}^{CC} to the common-collector amplifier in Fig. 14.9(a) is equal to the parallel combination of bias resistor and the equivalent resistance at the base of the BJT:

$$R_{in}^{CC} = R_B \| R_{iB} = R_B \| r_\pi (1 + g_m R_L) \qquad \text{where} \qquad R_L = R_4 \| R_3 \tag{14.14}$$

The overall input resistance R_{in}^{CD} to the common-drain amplifier in Fig. 14.9(b) is equal to the parallel combination of bias resistor and the equivalent resistance at the gate of the FET:

$$R_{in}^{CD} = R_G \| R_{iG} = R_G \| \infty = R_G \tag{14.15}$$

14.3.3 SIGNAL SOURCE VOLTAGE GAIN

The overall voltage gains from source v_i in Fig. 14.9 to the output are found using the terminal gain and input resistance expressions

$$A_v^{CC} = \frac{\mathbf{v_o}}{\mathbf{v_i}} = \left(\frac{\mathbf{v_o}}{\mathbf{v_b}}\right) \left(\frac{\mathbf{v_b}}{\mathbf{v_i}}\right) = A_{vt}^{CC} \left(\frac{\mathbf{v_b}}{\mathbf{v_i}}\right)$$

Voltage v_b at the base of the bipolar transistor in Fig. 14.9 is related to v_i by

$$\mathbf{v_b} = \mathbf{v_i} \frac{R_B \| R_{iB}}{R_I + \left(R_B \| R_{iB}\right)}$$

for $R_B = R_1 \| R_2$. Combining these expressions,

$$A_v^{CC} = A_{vt}^{CC} \left[\frac{R_B \| R_{iB}}{R_I + (R_B \| R_{iB})} \right] \tag{14.16}$$

For the common-source case with infinite input resistance, Eq. (14.16) reduces to

$$A_v^{CD} = A_{vt}^{CD} \left(\frac{R_G}{R_I + R_G} \right) \tag{14.17}$$

EXAMPLE 14.2   FOLLOWER VOLTAGE GAIN CALCULATIONS

The characteristics of the common-collector and common-drain amplifiers in Fig. 14.4 are calculated using the expressions derived in this section.

PROBLEM   Calculate the overall gain A_v, input resistances, output resistances, and signal handling capability of the C-C and C-D amplifiers using the results from section 14.3 and the parameter values from Ex. 14.1.

SOLUTION   **Known Information and Given Data:** The equivalent circuit with element values is redrawn below. The Q-point and small-signal values appear in the Table below from Ex. 14.1.

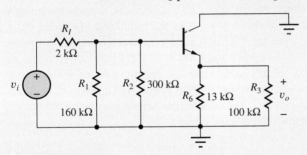

Common-Collector Amplifier from Fig. 14.4.

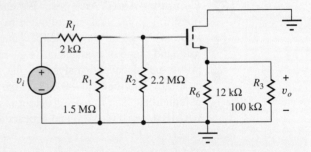

Common-Drain Amplifier from Fig. 14.4.

	I_C or I_D	V_{CE} or V_{DS}	$V_{GS} - V_{TN}$	g_m	r_π	r_o	μ_f
BJT	245 μA	3.64 V	—	9.80 mS	10.2 kΩ	219 kΩ	2150
FET	241 μA	3.81 V	0.982 V	0.491 mS	∞	223 kΩ	110

Unknowns: Voltage gains for the C-C and C-D amplifiers

Approach: Substitute element values from the two circuits into Eqs. (14.16) and (14.17).

Assumptions: Use the parameter values tabulated in Ex. 14.1.

Analysis: For the C-C amplifier in Fig. 14.3, load resistor $R_L = R_6 \| R_3 = 11.5$ kΩ, and bias resistor $R_B = R_1 \| R_2 = 104$ kΩ. The input resistances and terminal gain are

$$R_{iB} \cong r_\pi (1 + g_m R_L) = 10.2 \text{ k}\Omega[1 + 9.80 \text{ mS}(11.5 \text{ k}\Omega)] = 1.16 \text{ M}\Omega$$

$$R_{in}^{CC} = R_B \| R_{iB} = 104 \text{ k}\Omega \| 1.16 \text{ M}\Omega = 95.4 \text{ k}\Omega$$

$$A_{vt}^{CC} \cong \frac{g_m R_L}{1 + g_m R_L} = \frac{9.80 \text{ mS}(11.5 \text{ k}\Omega)}{1 + 9.80 \text{ mS}(11.5 \text{ k}\Omega)} = 0.991$$

Using Eq. (14.16), we find the overall gain to be

$$A_v^{CC} = A_{vt}^{CC} \left[\frac{R_B \| R_{iB}}{R_I + (R_B \| R_{iB})} \right] = 0.991 \left[\frac{104 \text{ k}\Omega \| 116 \text{ k}\Omega}{2.00 \text{ k}\Omega + (104 \text{ k}\Omega \| 116 \text{ k}\Omega)} \right] = 0.956$$

For the C-D amplifier, load resistor $R_L = R_6 \| R_3 = 10.7 \text{ k}\Omega$, and $R_G = R_1 \| R_2 = 892 \text{ k}\Omega$.

$$A_{vt}^{CD} = \frac{g_m R_L}{1 + g_m R_L} = \frac{0.491 \text{ mS}(10.7 \text{ k}\Omega)}{1 + 0.491 \text{ mS}(10.7 \text{ k}\Omega)} = 0.840$$

and

$$A_v^{CD} = A_{vt}^{CD} \left(\frac{R_G}{R_I + R_G} \right) = 0.840 \left(\frac{892 \text{ k}\Omega}{2 \text{ k}\Omega + 892 \text{ k}\Omega} \right) = 0.838$$

The overall input resistance for the common-drain amplifier is

$$R_{\text{in}}^{CD} = R_G \| R_{iG} = 892 \text{ k}\Omega \| \infty = 892 \text{ k}\Omega$$

Check of Results: Both voltage gains are approximately $+1$, as expected for a voltage follower. Both results are in the range specified in Eq. (14.12).

Discussion: The C-C amplifier has a gain much closer to 1 because $g_m R_L$ is much larger than it is for the C-D case. The C-C amplifier will normally have a gain closer to one than will the C-D stage.

 Computer-Aided Analysis:[1] We can check the voltage gains using SPICE by performing an operating point analysis followed by an ac analysis with v_I as the input and v_O as the output voltage. Make all the capacitor values large, say 100 μF, and sweep the frequency to find the midband range of frequencies (e.g., FSTART = 1 Hz and FSTOP = 100 kHz with 10 frequency points per decade). Analysis of the two circuits yields $+0.970$ for the gain of the common-collector amplifier and $+0.837$ for the gain of the common-drain stage. Both agree well with hand calculations. The transistor output resistance r_o is included in the simulations (VAF = 50 V or LAMBDA = 0.02 V^{-1}) and appears to have only a small effect.

EXERCISE: How large must R_L be for the common-drain amplifier to achieve the same value of gain as the common-collector amplifier in Ex. 14.2?

ANSWER: 73.1 kΩ

EXERCISE: What is the voltage gain for the two amplifiers in Fig. 14.4 if R_3 is removed ($R_3 \to \infty$)?

ANSWERS: 0.971, 0.853

EXERCISE: Redraw the circuit in Fig. 14.10 including r_o and show that it can easily be included in the analysis by changing the value of R_L.

ANSWER: Resistor r_o appears directly in parallel with R_L in Fig. 14.10; hence we simply replace R_L with a new value of load resistance in all the equations: $R_L' = R_L \| r_o$

EXERCISE: Compare the values of $g_m R_L$ for the C-C and C-D amplifiers in Ex. 14.3.

ANSWER: $113 \gg 5.25$

---

[1] See the MCD website for help with this circuit.

14.3.4 FOLLOWER SIGNAL RANGE

Because the emitter- and source-follower circuits have a gain approaching unity, only a small portion of the input signal actually appears across the base-emitter or gate-source terminals. Thus, these circuits can be used with relatively large input signals without violating their respective small-signal limits. The voltage developed across r_π in the small-signal model must be less than 5 mV for small-signal operation of the BJT. An expression for v_{be} is found in a manner identical to that used to derive Eq. (14.8):

$$\mathbf{v_{be}} = \mathbf{i_b} r_\pi = \mathbf{v_b} \frac{r_\pi}{r_\pi + (\beta_o + 1) R_L} \tag{14.18}$$

Requiring the amplitude of voltage v_{be} to be less than 5 mV gives

$$|v_b| \le 0.005(1 + g_m R_L) \text{ V} \tag{14.19}$$

for large β_o. Normally, $g_m R_L \gg 1$, and the magnitude of v_b can be increased well beyond the 5-mV limit.

For the case of the FET (letting $r_\pi \to \infty$), the corresponding expression becomes

$$|v_{gs}| = \frac{|v_g|}{1 + g_m R_L} \le 0.2(V_{GS} - V_{TN}) \tag{14.20}$$

and

$$|v_g| \le 0.2(V_{GS} - V_{TN})(1 + g_m R_L) \tag{14.21}$$

which also increases the permissible range for v_i.

DESIGN NOTE

An unbypassed resistor R in series with the emitter or source of a transistor increases the signal handling capability of the amplifier by a factor of approximately $(1 + g_m R)$.

EXERCISE: What are the largest values of v_{th} that correspond to small-signal operation of the amplifiers in Fig. 14.3?

ANSWERS: 0.569 V; 1.23 V

14.3.5 RESISTANCE AT THE EMITTER TERMINAL

The resistance looking into the output of the C-C circuit at the emitter terminal can be calculated based on the circuit in Fig. 14.11, in which test source v_x is applied directly to the emitter terminal.

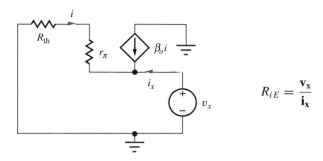

Figure 14.11 C-C/C-D output resistance calculation.

Using KCL at the emitter node yields

$$\mathbf{i_x} = -\mathbf{i} - \beta_o \mathbf{i} = \frac{\mathbf{v_x}}{r_\pi + R_{\text{th}}} - \beta_o \left(-\frac{\mathbf{v_x}}{r_\pi + R_{\text{th}}} \right) \tag{14.22}$$

Collecting terms and rearranging gives

$$R_{iE} = \frac{r_\pi + R_{\text{th}}}{\beta_o + 1} \cong \frac{1}{g_m} + \frac{R_{\text{th}}}{\beta_o} \tag{14.23}$$

for $\beta_o \gg 1$. Because the current gain is infinite for the FET,

$$R_{iS} = \frac{1}{g_m} \tag{14.24}$$

From Eqs. (14.23) and (14.24), it can be observed that the output resistance is primarily determined by the reciprocal of the transconductance of the transistor. This is an extremely important result to remember. For the BJT case, an additional term is added, but it is usually small, unless R_{th} is very large. The value of the output resistance for the C-C and C-D circuits can be quite low. For instance, at a current of 5 mA, the g_m of the bipolar transistor is $40 \times 0.005 = 0.2$ S, and $1/g_m$ is only 5 Ω.

DESIGN NOTE

The equivalent resistance looking into the emitter or source of a transistor is approximately $1/g_m$!

EXAMPLE 14.3 **COMMON-COLLECTOR AND COMMON-DRAIN OUTPUT RESISTANCES**

The output resistances for the common-collector and common-drain amplifiers in Fig. 14.3 are calculated using Eq. (14.23) and (14.24).

PROBLEM Calculate the output resistances of these C-C and C-D amplifiers using the parameter values from Ex. 14.1.

SOLUTION **Known Information and Given Data:** Equivalent circuit with element values appear in Fig. 14.14. The Q-point and small-signal values appear in the table in Ex. 14.1.

Unknowns: Output resistances R_{out} for the C-C and C-D amplifiers

Approach: Substitute element values from the two circuits into Eqs. (14.23) and (14.24).

Assumptions: Use the parameter values tabulated in Ex. 14.1.

Analysis: The overall output resistance of the common-collector amplifier R_{out}^{CC} is defined looking into the amplifier at coupling capacitor C_3 in Figs. 14.4 and 14.9(a) and is equal to resistor R_6 in parallel with the resistance at the emitter of the transistor:

$$R_{\text{out}}^{CC} = R_6 \| R_{iE} \cong R_6 \| \left(\frac{1}{g_m} + \frac{R_{\text{th}}}{\beta_o} \right) = 13\,\text{k}\Omega \| \left(\frac{1}{9.80\,\text{mS}} + \frac{1.96\,\text{k}\Omega}{100} \right) = 121\ \Omega$$

For the common-drain case, R_{out}^{CD} is defined looking into the amplifier at coupling capacitor C_3 in Figs. 14.4 and 14.9(b) and is equal to resistor R_6 in parallel with the resistance at the source of the transistor:

$$R_{out}^{CD} = R_6 \| R_{IS} = R_6 \| \frac{1}{g_m} = 12\,\text{k}\Omega \| \frac{1}{0.491\,\text{mS}} = 1.74\,\text{k}\Omega$$

Check of Results: The resistance looking into the emitter or source of the transistor should be relatively low, and the results agree with this expectation. Both results are approximately equal to $(1/g_m)$, which is the expected result.

Discussion: The output resistance of the C-C amplifier is much lower because of its larger transconductance at the given Q-point. Note that the term due to R_{th} increases the output resistance of the C-C amplifier by 19 percent above $(1/g_m)$ in this example.

[www] **Computer-Aided Analysis:**[2] We can check the output resistance using SPICE. Place an ac current source i_o in parallel with resistor R_3 and perform an operating point analysis followed by an ac analysis with v_O as the output voltage. Make all the capacitor values large, say $100\,\mu\text{F}$, and sweep the frequency to find the midband range of frequencies (e.g., FSTART $= 1$ Hz and FSTOP $= 100$ kHz with 10 frequency points per decade). In order to find the resistance looking into the emitter or source terminal of the transistor, we must subtract out the effect of the resistors R_3 and R_6 that appear in parallel with the transistor terminal. The SPICE analyses of the C-C and C-D stages yield a total resistance of $121.1\,\Omega$ and $1.643\,\text{k}\Omega$ at the two output nodes, respectively. Removing the effect of the $100\text{-}\text{k}\Omega$ and $13\text{-}\text{k}\Omega$ or $12\text{-}\text{k}\Omega$ resistors from the circuits yields $122\,\Omega$ for the output resistance of the common-collector amplifier and $1.94\,\text{k}\Omega$ for that of the common-drain stage. These values agree well with our hand calculations.

EXERCISE: What drain current is required to achieve $R_{out} = 120\,\Omega$ in the common-drain amplifier if $K_n = 500\,\mu\text{A/V}^2$? What would be the value of $(V_{GS} - V_{TN})$?

ANSWERS: 69.4 mA; 16.7 V!

Let us further interpret the two terms in Eq. (14.23) by injecting a current into the emitter of the BJT, as in Fig. 14.12. Multiplying **i** by the input resistance gives the voltage that must be developed at the emitter:

$$\mathbf{v_e} = \frac{\alpha_o \mathbf{i}}{g_m} + \frac{\mathbf{i}}{\beta_o + 1} R_{th} \tag{14.25}$$

Current $(\alpha_o \mathbf{i})$ comes out of the collector and must be supported by the emitter-base voltage $\mathbf{v_{eb}} = \alpha_o \mathbf{i}/g_m$, represented by the first term in Eq. (14.24). Base current $\mathbf{i_b} = -\mathbf{i}/\beta_o + 1$ creates a voltage drop in resistance R_{th} and yields the second term. In the FET case, only the first term exists because $i_g = 0$.

EXERCISE: Drive the emitter node in Fig. 14.11 with a test current source i_x, and verify the output resistance results in Eq. (14.23).

---

[2] See the MCD website for help with this circuit.

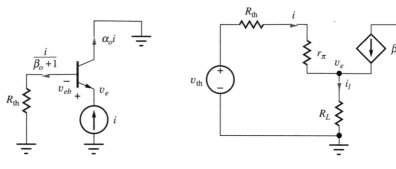

Figure 14.12 Circuit to aid in interpreting Eq. (14.25).

Figure 14.13 Circuit for calculating C-C/C-D terminal current gain.

TABLE 14.4
Common-Collector/Common-Drain Amplifier Summary

	COMMON-COLLECTOR (C-C) AMPLIFIER	COMMON-DRAIN (C-D) AMPLIFIER
Terminal voltage gain	$A_{vt}^{CC} = \dfrac{v_o}{v_1} = +\dfrac{g_m R_L}{1 + g_m R_L} \cong 1$	$A_{vt}^{CD} = \dfrac{v_o}{v_1} = +\dfrac{g_m R_L}{1 + g_m R_L} \cong 1$
Signal source voltage gain	$A_v^{CC} = \dfrac{v_o}{v_i} = A_{vt}^{CC}\dfrac{R_B \| R_{iB}}{R_I + R_B \| R_{iB}}$	$A_v^{CD} = \dfrac{v_o}{v_i} = A_{vt}^{CD}\dfrac{R_G}{R_I + R_G}$
Rule-of-thumb estimate for $g_m R_L$	$10(V_{CC} + V_{EE})$	$(V_{DD} + V_{SS})$
Input terminal resistance	$R_{iB} = r_\pi(1 + g_m R_L)$	$R_{iG} = \infty$
Output terminal resistance	$R_{iE} \cong \dfrac{1}{g_m} + \dfrac{R_{th}}{\beta_o}$	$R_{iS} = \dfrac{1}{g_m}$
Input signal range	$0.005(1 + g_m R_L)$ V	$0.2(V_{GS} - V_{TN})(1 + g_m R_L)$
Terminal current gain	$\beta_o + 1$	∞

14.3.6 CURRENT GAIN

Terminal current gain A_{it} is the ratio of the current delivered to the load element to the current being supplied from the Thévenin source. In Fig. 14.13, the current i plus its amplified replica ($\beta_o i$) are combined in load resistor R_L, yielding a current gain equal to ($\beta_o + 1$). For the FET, r_π is infinite, i is zero, and the current gain is infinite. Thus, for the C-C/C-D amplifiers,

$$A_{it}^{CC} = \frac{\mathbf{i_1}}{\mathbf{i}} = \beta_o + 1 \qquad \text{and} \qquad A_{it}^{CD} = \infty \tag{14.26}$$

14.3.7 C-C/C-D AMPLIFIER SUMMARY

Table 14.4 summarizes the results that have been derived for the common-collector and common-drain amplifiers in Fig. 14.14. As before, the FET results in the table can always be obtained from the BJT results by letting r_π and $\beta_o \to \infty$. The numeric results from the two specific amplifiers in Fig. 14.14 are gathered together in Table 14.5.

In Tables 14.4 and 14.5, the similarity between the characteristics of the C-C and C-D amplifiers should be readily apparent. Both amplifiers provide a gain approaching unity, a high input resistance,

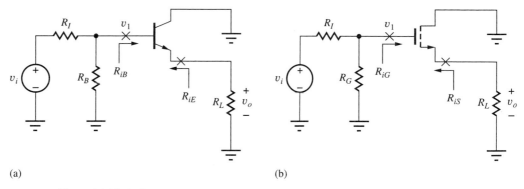

Figure 14.14 (a) Common-collector and (b) common-drain amplifiers for use with Table 14.4.

TABLE 14.5
Common-Collector/Common-Drain Amplifier Comparison

	C-C AMPLIFIER	C-D AMPLIFIER
Voltage gain	0.956	0.838
Input resistance	1.17 MΩ	∞
Output resistance	121 Ω	2.04 kΩ
Input signal range	0.575 V	1.23 V
Current gain	101	∞

and a low output resistance. The differences arise because of the finite value of r_π and β_o of the BJT. The FET can more easily achieve very high values of input resistance because of the infinite resistance looking into its gate terminal, whereas the C-C amplifier can more easily reach very low levels of output resistance because of its higher transconductance for a given operating current. Both amplifiers can be designed to handle relatively large input signal levels. The current gain of the FET is inherently infinite, whereas that of the BJT is limited by its finite value of β_o.

14.4 NONINVERTING AMPLIFIERS—COMMON-BASE AND COMMON-GATE CIRCUITS

The final class of amplifiers to be analyzed consists of the common-base and common-gate amplifiers represented by the two ac equivalent circuits in Fig. 14.15. From our analyses, we will find that the noninverting amplifiers provide a voltage gain and output resistance similar to that of the C-E/C-S stages but with much lower input resistance. As in Secs. 14.2 and 14.3, we analyze the BJT circuit first and treat the MOSFET in Fig. 14.15(b) as a special case of Fig. 14.15(a).

14.4.1 TERMINAL VOLTAGE GAIN AND INPUT RESISTANCE

The bipolar transistor is replaced by its small-signal model in Fig. 14.16(a). Because the amplifier has a resistor load, the circuit model is simplified by neglecting r_o, as redrawn in Fig. 14.16(b). In addition, the polarities of $\mathbf{v_{be}}$ and the dependent current source $g_m\mathbf{v_{be}}$ have both been reversed.

For the common-base circuit, output voltage v_o appears at the collector across resistor R_L and is equal to

$$\mathbf{v_o} = +g_m\mathbf{v_{eb}}R_L = +g_m R_L \mathbf{v_e} \tag{14.27}$$

and the terminal gain for the common-base transistor is

$$A_{vt}^{CB} = \frac{\mathbf{v_o}}{\mathbf{v_e}} = +g_m R_L \tag{14.28}$$

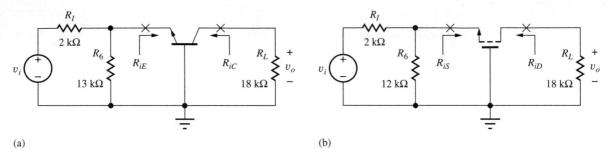

Figure 14.15 ac Equivalent circuits for the (a) C-B and (b) C-G amplifiers.

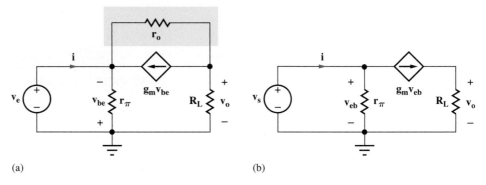

Figure 14.16 (a) Small-signal model for the common-base amplifier. (b) Simplified model neglecting r_o and reversing the direction of the controlled source.

which is the same as that for the common-emitter stage except for the sign. The input current i and input resistance at the emitter are given by

$$i = \frac{\mathbf{v_e}}{\mathbf{r}_\pi} + g_m \mathbf{v_e} \qquad \text{and} \qquad R_{iE} = \frac{\mathbf{v_e}}{\mathbf{i}} = \frac{r_\pi}{\beta_o + 1} \cong \frac{1}{g_m} \qquad (14.29)$$

assuming $\beta_o \gg 1$.

The corresponding expressions for the common-gate stage $(r_\pi \rightarrow \infty)$ are

$$A_{vt}^{CG} = \frac{\mathbf{v_o}}{\mathbf{v_e}} = +g_m R_L \qquad \text{and} \qquad R_{iS} = \frac{1}{g_m} \qquad (14.30)$$

14.4.2 SIGNAL SOURCE VOLTAGE GAIN

The overall gains for the amplifiers in Fig. 14.15 can now be expressed as

$$A_v^{CB} = \frac{\mathbf{v_o}}{\mathbf{v_i}} = \left(\frac{v_o}{v_e} \right) \left(\frac{v_e}{v_i} \right) = A_{vt}^{CB} \left[\frac{R_6 \| R_{iE}}{R_I + (R_4 \| R_{iE})} \right] \qquad (14.31)$$

and substituting $R_{iE} = 1/g_m$ yields

$$A_v^{CB} = \frac{g_m R_L}{1 + g_m (R_{\text{th}})} \left(\frac{R_6}{R_I + R_6} \right) \qquad \text{and} \qquad A_v^{CG} = \frac{g_m R_L}{1 + g_m (R_{\text{th}})} \left(\frac{R_6}{R_I + R_6} \right) \qquad (14.32)$$

where $R_{\text{th}} = R_6 \| R_I$. If we assume that $R_6 \gg R_I$, then the gain expressions in Eq. (14.32) become

$$A_v^{CB,CG} \cong \frac{g_m R_I}{1 + g_m R_I} \qquad \text{for} \qquad R_6 \gg R_I \qquad (14.33)$$

Because of the low input resistance of the common-base and common-gate amplifiers, the voltage gain A_v from the signal source to the output can be substantially less than the terminal gain. Note that the final expressions in Eq. (14.33) have a similar form to the overall gains for the inverting amplifiers and followers. We will explore this result more fully later in the chapter.

Note that the gain expressions in Eqs. (14.30) and (14.31) are positive, indicating that the output signal is in phase with the input signal. Thus, the C-B and C-G amplifiers are classified as noninverting amplifiers.

DESIGN NOTE

The terminal voltage gain of the noninverting amplifiers is given by

$$A_{vt}^{NI} \cong +g_m R_L$$

DESIGN NOTE

An estimate for the overall gain of the noninverting amplifiers is

$$A_v^{NI} \cong +\frac{g_m R_L}{1 + g_m R_{\text{th}}} \cong +\frac{R_L}{R_{\text{th}}} \quad \text{for} \quad g_m R_{\text{th}} \gg 1 \quad \text{and} \quad R_{\text{th}} = R_6 \| R_I$$

DESIGN NOTE

The equivalent resistance R looking into the emitter or source of a transistor is approximately $R = 1/g_m$.

Important Limits

As for the C-E/C-S amplifiers, two limiting conditions are of particular importance (see Prob. 14.38). The upper bound occurs for $g_m R_I \ll 1$, for which Eq. (14.33) reduces to

$$A_v^{CB} \cong +g_m R_L \qquad \text{and} \qquad A_v^{CG} \cong +g_m R_L \qquad (14.34)$$

Equation (14.34) represents the upper bound on the gain of the C-B/C-G amplifiers and is the same as that for the C-E/C-S amplifiers, except the gain is noninverting.

However, if $g_m R_{\text{th}} \gg 1$, then Eq. (14.34) reduces to

$$A_v^{CB} = A_v^{CG} \cong +\frac{R_L}{R_{\text{th}}} \qquad (14.35)$$

For this case, the C-B and C-G amplifiers both have a gain that approaches the ratio of the value of the load resistor to that of the Thévenin source resistance ($R_{\text{th}} = R_6 \| R_I$) and is independent of the transistor parameters. For resistor loads, the limit in Eq. (14.35) is much less than the amplification factor μ_f, so neglecting r_o is valid.

14.4.3 INPUT SIGNAL RANGE

The relationship between $\mathbf{v_{eb}}$ and $\mathbf{v_i}$ in Fig. 14.15(a) is given by

$$\mathbf{v_{eb}} = \mathbf{v_i} \frac{R_6 \| R_{iE}}{R_I + (R_6 \| R_{iE})} = \frac{\mathbf{v_i}}{1 + g_m(R_I \| R_6)} \left(\frac{R_6}{R_I + R_6} \right) \quad \text{and} \quad \mathbf{v_i} \cong \mathbf{v_{eb}}(1 + g_m R_I)$$

$$(14.36)$$

for $R_6 \gg R_I$.

The small-signal limit requires

$$|\mathbf{v_i}| \leq 0.005(1 + g_m R_I) \text{ V} \tag{14.37}$$

For the FET case, replacing $\mathbf{v_{eb}}$ by $\mathbf{v_{sg}}$ yields $\mathbf{v_i} = \mathbf{v_{sg}}(1 + g_m R_I)$ and

$$|\mathbf{v_i}| \leq 0.2(V_{GS} - V_{TN})(1 + g_m R_I) \tag{14.38}$$

The relative size of R_I and g_m will determine the signal-handling limits.

EXERCISE: Calculate the maximum values of v_i for the C-B and C-G amplifiers in Fig. 14.15 based on Eqs. (14.37) and (14.38).

ANSWERS: 103 mV; 389 mV

14.4.4 RESISTANCE AT THE COLLECTOR AND DRAIN TERMINALS

The resistance at the output terminal of the C-B/C-G transistors can be calculated for the circuit in Fig. 14.17, in which a test source v_x is applied to the collector terminal. The desired resistance is that looking into the collector with the base grounded and resistor R_{th} in the emitter. If the circuit is redrawn as shown in Fig. 14.17(b), we should recognize it to be the same as the C-E circuit in Fig. 13.21, repeated in Fig. 14.17(c), except that the resistance R_{th}^{CE} in the base is zero, and resistor R_E has been relabeled R_{th}.

Thus, the resistance at the output for the C-B device can be found using the results from the common-emitter amplifier, Eq. (13.66), without further detailed calculation, by substituting $R_{\text{th}}^{CE} = 0$ and replacing R_E with R_{th}:

$$R_{iC} = r_o\left(1 + \frac{\beta_o R_E}{R_{\text{th}}^{CE} + r_\pi + R_E}\right) = r_o\left(1 + \frac{\beta_o R_{\text{th}}}{r_\pi + R_{\text{th}}}\right) \tag{14.39}$$

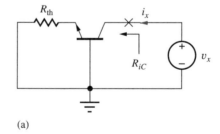

(a)

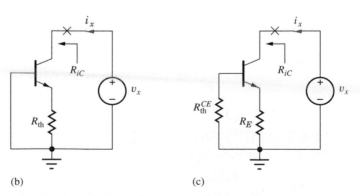

(b)   (c)

Figure 14.17 (a) Circuit for calculating the C-B output resistance. (b) Redrawn version of the circuit in (a). (c) Circuit used in common-emitter analysis (see Fig. 13.21).

Using $\beta_o = g_m r_\pi$

$$R_{iC} \cong r_o[1 + g_m(R_{th}\|r_\pi)] \quad \text{and} \quad R_{iD} = r_o(1 + g_m R_{th}) \quad (14.40)$$

DESIGN NOTE

A quick design estimate for the output resistance of an inverting or noninverting amplifier with an unbypassed resistor R in the emitter or source is

$$R_o \cong r_o[1 + g_m(R\|r_\pi)] \cong \mu_f(R\|r_\pi)$$

EXERCISE: Calculate the output resistances of the C-B and C-G amplifiers.

ANSWERS: 3.93 MΩ; 410 kΩ

14.4.5 CURRENT GAIN

The terminal current gain A_{it} is the ratio of the current through the load resistor to the current being supplied to the emitter. If a current i_e is injected into the emitter of the C-B transistor in Fig. 14.18, then the current $i_l = \alpha_o i_e$ comes out of the collector. Thus, the common-base current gain is simply α_o.

For the FET, α_o is exactly 1, and we have

$$A_{it}^{CB} = \frac{i_l}{i_e} = +\alpha_o \cong +1 \quad \text{and} \quad A_{it}^{CG} = +1 \quad (14.41)$$

14.4.6 OVERALL INPUT AND OUTPUT RESISTANCES FOR THE NONINVERTING AMPLIFIERS

The overall input and output resistances, R_{in}^{CB}, R_{in}^{CG}, R_{out}^{CB}, and R_{out}^{CG} of the common-base and common-gate amplifiers are defined looking into the input (C_1) and output (C_3) coupling capacitors in Fig. 14.5, as redrawn in the midband ac models in Fig. 14.19. The overall input resistance of the common-base or common-gate amplifiers equals the parallel combination of resistor R_6 and the resistance looking

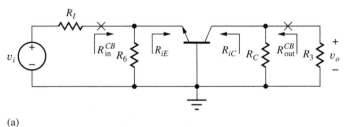

(a)

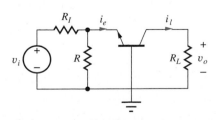

Figure 14.18 Common-base current gain.

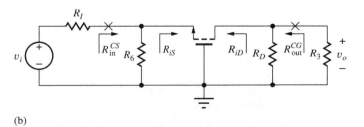

(b)

Figure 14.19 Midband ac equivalent circuits for the common-base and common gate amplifiers.

into the emitter or source terminal of the transistor:

$$R_{\text{in}}^{CB} = R_6 \| R_{iE} \cong R_6 \| \frac{1}{g_m} \qquad \text{and} \qquad R_{\text{in}}^{CG} = R_6 \| R_{iS} = R_6 \| \frac{1}{g_m} \qquad (14.42)$$

Similarly, the overall output resistance of the common-base or common-gate amplifiers equals the parallel combination of resistors R_C or R_D and the resistance looking into the collector or drain terminal of the transistor:

$$R_{\text{out}}^{CB} = R_C \| R_{iC} = R_C \| r_o \left[1 + g_m (R_6 \| R_I \| r_\pi) \right]$$

$$\text{and} \qquad R_{\text{out}}^{CD} = R_D \| R_{iD} = R_D \| r_o \left[1 + g_m (R_6 \| R_1) \right] \qquad (14.43)$$

EXAMPLE 14.4   NONINVERTING AMPLIFIER CHARACTERISTICS

A comparison of the characteristics of the common-base and common-gate amplifiers is provided by this example.

PROBLEM Calculate the signal-source voltage gains, input resistances, output resistances, and signal handling capability for the C-B and C-G amplifiers in Fig. 14.5.

SOLUTION **Known Information and Given Data:** The equivalent circuit with element values appear below. Q-point and small-signal values appear in the accompanying table.

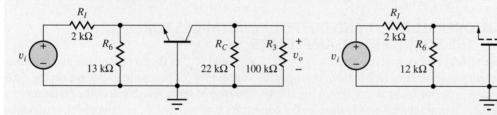

Common-Base Amplifier from Fig. 14.5.   Common-Gate Amplifier from Fig. 14.5.

	I_C or I_D	V_{CE} or V_{DS}	$V_{GS} - V_{TN}$	g_m	r_π	r_o	μ_f
BJT	245 μA	3.64 V	—	9.80 mS	10.2 kΩ	219 kΩ	2150
FET	241 μA	3.81 V	0.982 V	0.491 mS	∞	223 kΩ	110

Unknowns: Voltage gains, input resistances, output resistances, and maximum input signal amplitudes for the common-base and common-gate amplifiers

Approach: Verify the value of R_L, and substitute element values from the two circuits the appropriate equations from Sections 14.4.1–14.4.6.

Assumptions: Use the parameter values tabulated in Ex. 14.1.

Analysis: *For the C-B amplifier:* $R_I = 2\,\text{k}\Omega$, $R_6 = 13\,\text{k}\Omega$, $R_L = R_3 \| R_C = 18.0\,\text{k}\Omega$. The terminal input resistance and gain are

$$R_{iE} \cong \frac{1}{g_m} = \frac{1}{9.8\,\text{mS}} = 102\,\Omega \qquad \text{and} \qquad A_{vt}^{CB} = +g_m R_L = 9.80\,\text{mS}(18.0\,\text{k}\Omega) = +176$$

and the overall voltage gain is

$$A_v^{CB} = \frac{A_{vt}}{1 + g_m(R_6 \| R_I)} \left(\frac{R_6}{R_I + R_6} \right) = \frac{176}{1 + 9.8 \text{ mS}(1.73 \text{ k}\Omega)} \left(\frac{13 \text{ k}\Omega}{2 \text{ k}\Omega + 13 \text{ k}\Omega} \right) = +8.48$$

The input resistance of the common-base amplifier is found using Eq. (14.42)

$$R_{in}^{CB} = R_6 \| R_{iE} = 13 \text{ k}\Omega \| 102 \text{ }\Omega = 101 \text{ }\Omega$$

and the output resistance of the common-base amplifier is calculated using Eq. (14.43)

$$R_{iC} = r_o[1 + g_m(R_6 \| R_I \| r_\pi)] = 219 \text{ k}\Omega[1 + 9.80 \text{ mS}(13 \text{ k}\Omega \| 2 \text{ k}\Omega \| 10.2 \text{ k}\Omega)] = 3.40 \text{ M}\Omega$$

$$R_{out}^{CB} = R_C \| R_{iC} = 22 \text{ k}\Omega \| 3.40 \text{ M}\Omega = 21.9 \text{ k}\Omega$$

The maximum input signal amplitude is computed using Eq. (14.37).

$$|v_i| \leq 0.005V[1 + g_m(R_6 \| R_I)] \frac{R_I + R_6}{R_6}$$

$$|v_i| = 0.005V[1 + 9.80 \text{ mS}(13 \text{ k}\Omega \| 2 \text{ k}\Omega)] \frac{2 \text{ k}\Omega + 13 \text{ k}\Omega}{13 \text{ k}\Omega} = 104 \text{ mV}$$

For the C-G amplifier: $R_I = 2 \text{ k}\Omega$, $R_6 = 12 \text{ k}\Omega$, $R_L = R_3 \| R_D = 18.0 \text{ k}\Omega$. We have

$$R_{iS} = \frac{1}{g_m} = \frac{1}{0.491 \text{ mS}} = 2.04 \text{ k}\Omega \quad \text{and} \quad A_{vt}^{CG} = +g_m R_L = 0.491 \text{ mS}(18.0 \text{ k}\Omega) = +8.84$$

and

$$A_v^{CG} = \frac{A_{vt}^{CG}}{1 + g_m(R_6 \| R_I)} \left(\frac{R_6}{R_I + R_6} \right) = \frac{8.84}{1 + 0.491 \text{ mS}(1.71 \text{ k}\Omega)} \left(\frac{12 \text{ k}\Omega}{2 \text{ k}\Omega + 12 \text{ k}\Omega} \right) = +4.11$$

The input resistance of the common-gate amplifier is found using Eq. (14.42)

$$R_{in}^{CG} = R_6 \| R_{iS} = 12 \text{ k}\Omega \| 2.04 \text{ k}\Omega = 1.74 \text{ k}\Omega$$

and the output resistance of the common-gate amplifier is calculated using Eq. (14.43)

$$R_{iD} = r_o[1 + g_m(R_6 \| R_I)] = 223 \text{ k}\Omega[1 + 0.491 \text{ mS}(12 \text{ k}\Omega \| 2 \text{ k}\Omega)] = 411 \text{ k}\Omega$$

$$R_{out}^{CG} = R_D \| R_{iD} = 22 \text{ k}\Omega \| 411 \text{ k}\Omega = 20.9 \text{ k}\Omega$$

The maximum input signal amplitude is computed using Eq. (14.37).

$$|v_i| \leq 0.2(V_{GS} - V_{TN})[1 + g_m(R_6 \| R_I)] \frac{R_I + R_6}{R_6}$$

$$|v_i| = 0.2(0.982)[1 + 0.491 \text{ mS}(12 \text{ k}\Omega \| 2 \text{ k}\Omega)] \frac{2 \text{ k}\Omega + 12 \text{ k}\Omega}{12 \text{ k}\Omega} = 422 \text{ mV}$$

Check of Results: Both values are similar to and do not exceed the design estimate given by

$$A_v^{NI} \cong +\frac{R_L}{R_I} = \frac{18 \text{ k}\Omega}{2 \text{ k}\Omega} = +9.00$$

Discussion: Note that the overall gain of the common-base amplifier is much less than its terminal gain because significant signal loss occurs due to the low input resistance of the transistor relative to the source resistance:

$$A_v^{CB} = A_{vt}^{CB} \left(\frac{R_6 \| R_{iE}}{R_I + R_6 \| R_{iE}} \right) = 176 \left(\frac{13 \text{ k}\Omega \| 102 \text{ }\Omega}{2 \text{ k}\Omega + 13 \text{ k}\Omega \| 102 \text{ }\Omega} \right) = 176(0.0482) = +8.48$$

For the common-gate case, the loss factor is less,

$$A_v^{CG} = A_{vt}^{CG} \left(\frac{R_6 \| R_{iS}}{R_I + R_6 \| R_{iS}} \right) = 8.84 \left(\frac{12 \text{ k}\Omega \| 2.04 \text{ k}\Omega}{2.00 \text{ k}\Omega + 12 \text{ k}\Omega \| 2.04 \text{ k}\Omega} \right)$$

$$= 8.84(0.466) = +4.12$$

Once again, we see that the overall C-G gain differs from the simple design estimate by more than that of the C-B stage because of the lower transconductance (and $g_m R_{th}$ product) of the FET. The gains are both well below the value of μ_f, so neglecting r_o in Fig. 14.16 is valid.

Computer-Aided Analysis:[3] We can check the voltage gains using SPICE by performing an operating point analysis followed by an ac analysis with v_I as the input and v_O as the output voltage. Make all the capacitor values large, say 100 μF, and sweep the frequency to find the midband region (e.g., FSTART = 1 Hz and FSTOP = 100 kHz with 10 frequency points per decade). Analysis of the two circuits yields +8.38 for the gain of the common-base amplifier and +4.05 for the gain of the common-gate stage. These values agree closely with our hand calculations. The transistor output resistance r_o is included in the simulations (VAF = 50 V or LAMBDA = 0.02 V^{-1}) and appears to have a negligible effect.

EXERCISE: Show that Eq. (14.31) can be reduced to Eq. (14.32).

EXERCISE: What are the open circuit voltage gains ($R_3 = \infty$) for these two amplifiers?

ANSWERS: 11.9; 5.87

EXERCISE: Compare the gains of the C-B and C-G amplifiers calculated in Ex. 14.4 to the two limits developed in Eqs. (14.34) and (14.35).

ANSWERS: 6.92 < 8.48 < 176; 4.42 < 8.84 < 10.5

14.4.7 C-B/C-G AMPLIFIER SUMMARY

Table 14.6 summarizes the results derived for the common-base and common-gate amplifiers in Fig. 14.20, and the numeric results for the specific amplifiers in Fig. 14.15 are collected together in Table 14.7. Table 14.6 again displays the symmetry between the various characteristics of the common-base and common-gate amplifiers. The voltage gain and current gain are very similar. Numeric differences occur because of differences in the parameter values of the BJT and FET at similar operating points.

Both amplifiers can provide significant voltage gain, low input resistance, and high output resistance. The higher amplification factor of the BJT gives it an advantage in achieving high output resistance; the C-B amplifier can more easily reach very low levels of input resistance because of the BJT's higher transconductance for a given operating current. The FET amplifier can inherently handle larger signal levels.

14.5 AMPLIFIER PROTOTYPE REVIEW AND COMPARISON

Sections 14.2 to 14.4 compared the three individual classes of BJT and FET circuits: the C-E/C-S, C-C/C-D, and C-B/C-G amplifiers. In this section, we review these results and compare the three BJT and FET amplifier configurations.

---

[3] See the MCD website for help with this circuit.

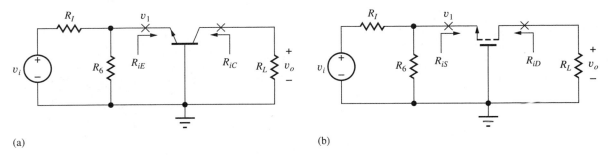

Figure 14.21 Circuits for use with summary Table 14.5. (a) Common-base amplifier, (b) common-gate amplifier.

TABLE 14.6

Common-Base/Common-Gate Amplifier Summary

	C-B AMPLIFIER	C-G AMPLIFIER
Terminal voltage gain $A_{vt} = \dfrac{v_o}{v_1}$	$+g_m R_L$	$+g_m R_L$
Signal-source voltage gain $A_v = \dfrac{v_o}{v_i} \qquad R_{th} = (R_I \| R_6)$	$\dfrac{g_m R_L}{1 + g_m R_{th}}\left(\dfrac{R_6}{R_I + R_6}\right)$	$\dfrac{g_m R_L}{1 + g_m R_{th}}\left(\dfrac{R_6}{R_I + R_6}\right)$
Input terminal resistance	$\dfrac{1}{g_m}$	$\dfrac{1}{g_m}$
Output terminal resistance	$r_o(1 + g_m R_{th}) = r_o + \mu_f R_{th}$	$r_o(1 + g_m R_{th}) = r_o + \mu_f R_{th}$
Input signal range	$0.005(1 + g_m R_{th})$	$0.2(V_{GS} - V_{TN})(1 + g_m R_{th})$
Terminal current gain	$\alpha_o \cong +1$	$+1$

TABLE 14.7

Common-Base/Common-Gate Amplifier Comparison

	C-B AMPLIFIER	C-G AMPLIFIER
Voltage gain	+8.48	+4.11
Input resistance	102 Ω	2.04 kΩ
Output resistance	3.40 MΩ	411 kΩ
Input signal range	104 mV	422 mV
Terminal current gain	1	1

14.5.1 THE BJT AMPLIFIERS

Table 14.8 collects the results of analysis of the three BJT amplifiers in Fig. 14.21; Table 14.9 gives approximate results.

A very interesting and important observation can be made from review of Table 14.8. If we assume the voltage loss across the source resistance is small, the signal-source gains of the three amplifiers have exactly the same form:

$$|A_v| \cong \frac{g_m R_L}{1 + g_m R_E} \qquad \text{or} \qquad |A_v| \cong \frac{R_L}{\dfrac{1}{g_m} + R_E} = \frac{R_L}{R_{EQ}} \qquad \text{for} \qquad R_{EQ} = R_E + \frac{1}{g_m} \quad (14.44)$$

in which R_E is the external resistance in the emitter of the transistor (R_E, R_L, or $R_I \| R_6$, respectively). We really only need to commit one formula to memory to get a good estimate of amplifier gain!

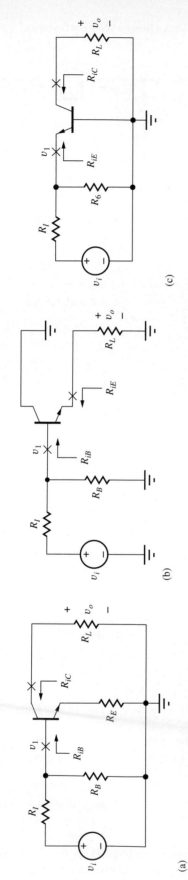

Figure 14.21 The three BJT amplifier configurations: (a) common-emitter amplifier, (b) common-collector amplifier, and (c) common-base amplifier.

TABLE 14.8
Single-Transistor Bipolar Amplifiers

	COMMON-EMITTER AMPLIFIER	COMMON-COLLECTOR AMPLIFIER	COMMON-BASE AMPLIFIER
Terminal voltage gain $A_{vt} = \dfrac{v_o}{v_1}$	$\cong -\dfrac{g_m R_L}{1+g_m R_E}$	$\cong +\dfrac{g_m R_L}{1+g_m R_L} \cong +1$	$+g_m R_L$
Signal-source voltage gain $A_v = \dfrac{v_o}{v_i}$	$-\dfrac{g_m R_L}{1+g_m R_E}\left[\dfrac{R_B\|R_{\text{in}}}{R_I+(R_B\|R_{\text{in}})}\right]$	$+\dfrac{g_m R_L}{1+g_m R_L}\left[\dfrac{R_B\|R_{\text{in}}}{R_I+(R_B\|R_{\text{in}})}\right] \cong +1$	$+\dfrac{g_m R_L}{1+g_m(R_I\|R_4)}\left(\dfrac{R_6}{R_I+R_6}\right)$
Input terminal resistance	$r_\pi+(\beta_o+1)R_E$ $\cong r_\pi(1+g_m R_E)$	$r_\pi+(\beta_o+1)R_L$ $\cong r_\pi(1+g_m R_L)$	$\dfrac{\alpha_o}{g_m}\cong \dfrac{1}{g_m}$
Output terminal resistance	$r_o(1+g_m R_E)$	$\dfrac{\alpha_o}{g_m}+\dfrac{R_{\text{th}}}{\beta_o+1}$	$r_o[1+g_m(R_I\|R_4)]$
Input signal range	$\cong 0.005(1+g_m R_E)$	$\cong 0.005(1+g_m R_L)$	$\cong 0.005[1+g_m(R_I\|R_6)]$
Terminal current gain	$-\beta_o$	β_o+1	$\alpha_o \cong +1$

TABLE 14.9
Simplified Characteristics of Single BJT Amplifiers

	COMMON-EMITTER $(R_E = 0)$	COMMON-EMITTER WITH EMITTER RESISTOR R_E	COMMON-COLLECTOR	COMMON-BASE
Terminal voltage gain	$-g_m R_L \cong -10 V_{CC}$	$-\dfrac{R_L}{R_E}$	1	$+g_m R_L \cong +10 V_{CC}$
$A_{vt} = \dfrac{v_o}{v_1}$	(high)	(moderate)	(low)	(high)
Input terminal resistance	r_π (moderate)	$\beta_o R_E$ (high)	$\beta_o R_L$ (high)	$1/g_m$ (low)
Output terminal resistance	r_o (moderate)	$\mu_f R_E$ (high)	$1/g_m$ (low)	$\mu_f (R_I \| R_4)$ (high)
Current gain	$-\beta_o$ (moderate)	$-\beta_o$ (moderate)	$\beta_o + 1$ (moderate)	1 (low)

In addition, the same symmetry exists in the expressions for input signal range:

$$|v_{be}| \le 0.005(1 + g_m R_E) \text{ V} \tag{14.45}$$

Note as well the similarity in the expressions for the input resistances of the C-E and C-C amplifiers, the input resistance of the C-B amplifier and the output resistance of the C-C amplifier, and the output resistances of the C-E and C-B amplifiers. Carefully review the three amplifier topologies in Fig. 14.21 to fully understand why these symmetries occur.

The second form of Eq. (14.44) deserves further discussion. The magnitude of the terminal gain of all three BJT stages can be expressed as the ratio of total resistance R_L at the collector to the total resistance R_{EQ} in the emitter loop! R_{EQ} is the sum of the external resistance R_E [i.e., R_E, R_L, or $(R_I \| R_6)$, as appropriate] plus the resistance $(1/g_m)$ found looking back into the emitter of the transistor itself. This is an extremely important conceptual result.

Table 14.9 is a simplified comparison. The common-emitter amplifier provides moderate-to-high levels of voltage gain, and moderate values of input resistance, output resistance, and current gain. The addition of emitter resistor R_E to the common-emitter circuit gives added design flexibility and allows a designer to trade reduced voltage gain for increased input resistance, output resistance, and input signal range. The common-collector amplifier provides low voltage gain, high input resistance, low output resistance, and moderate current gain. Finally, the common-base amplifier provides moderate to high voltage gain, low input resistance, high output resistance, and low current gain.

14.5.2 THE FET AMPLIFIERS
Tables 14.10 and 14.11 are similar summaries for the three FET amplifiers shown in Fig. 14.22. The signal source voltage gain and signal range of all three amplifiers can again be expressed approximately as

$$|A_v| \cong \frac{g_m R_L}{1 + g_m R} = \frac{R_L}{\dfrac{1}{g_m} + R} \tag{14.46}$$

and

$$|v_{gs}| \le 0.2(V_{GS} - V_{TN})(1 + g_m R) \text{ V} \tag{14.47}$$

in which R is the resistance in the source of the transistor (R_S, R_L, or $(R_I \| R_6)$, respectively). Note the symmetry between the output resistances of the C-S and C-G amplifiers. Also, the input resistance of the C-G amplifier and output resistance of the C-D amplifier are identical. Review the three amplifier topologies in Fig. 14.22 carefully to fully understand why these symmetries occur. The addition of resistor R_S to the common-source circuit allows the designer to trade reduced voltage gain for increased output resistance and input signal range.

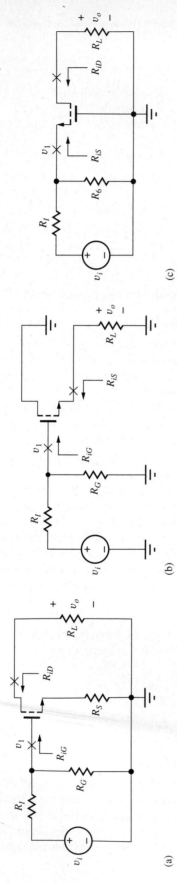

(a)

(b)

Figure 14.22 The three FET amplifier configurations: (a) common-source, (b) common-drain, and (c) common-gate.

(c)

TABLE 14.10
Single-Transistor FET Amplifiers

	COMMON-SOURCE AMPLIFIER	COMMON-DRAIN AMPLIFIER	COMMON-GATE AMPLIFIER
Terminal voltage gain $A_{vt} = \dfrac{v_o}{v_1}$	$-\dfrac{g_m R_L}{1 + g_m R_S}$	$+\dfrac{g_m R_L}{1 + g_m R_L} \cong +1$	$+g_m R_L$
Signal-source voltage gain $A_v = \dfrac{v_o}{v_i}$	$-\dfrac{g_m R_L}{1 + g_m R_S}\left(\dfrac{R_G}{R_I + R_G}\right)$	$+\dfrac{g_m R_L}{1 + g_m R_L}\left(\dfrac{R_G}{R_I + R_G}\right) \cong +1$	$+\dfrac{g_m R_L}{1 + g_m (R_I\|R_6)}\left(\dfrac{R_6}{R_I + R_6}\right)$
Input terminal resistance	∞	∞	$1/g_m$
Output terminal resistance	$r_o(1 + g_m R_S)$	$1/g_m$	$r_o[1 + g_m(R_I\|R_6)]$
Input signal range	$0.2(V_{GS} - V_{TN})(1 + g_m R_S)$	$0.2(V_{GS} - V_{TN})(1 + g_m R_L)$	$0.2(V_{GS} - V_{TN})[1 + g_m(R_I\|R_6)]$
Terminal current gain	∞	∞	$+1$

TABLE 14.11
Simplified Characteristics of Single FET Amplifiers

	COMMON-SOURCE ($R_S = 0$)	COMMON-SOURCE WITH SOURCE RESISTOR R_S	COMMON-DRAIN	COMMON-GATE
Terminal voltage gain $A_{vt} = \dfrac{v_o}{v_1}$	$-g_m R_L \cong -V_{DD}$ (moderate)	$-\dfrac{R_L}{R_S}$ (moderate)	1 (low)	$+g_m R_L \cong +V_{DD}$ (moderate)
Input terminal resistance	∞ (high)	∞ (high)	∞ (high)	$1/g_m$ (low)
Output terminal resistance	r_o (moderate)	$\mu_f R_S$ (high)	$1/g_m$ (low)	$\mu_f (R_I \| R_6)$ (high)
Current gain	∞ (high)	∞ (high)	∞ (high)	1 (low)

In a manner similar to the BJT amplifiers, the magnitude of the terminal gain of all three FET stages can be expressed as the ratio of total resistance R_L at the drain terminal to the total resistance R_{SQ} in the source loop. R_{SQ} represents the sum of the external resistance R_X [i.e., R_S, R_L, or $(R_I \| R_6)$, as appropriate] plus the resistance $(1/g_m)$ found looking back into the source of the transistor itself. Thus, when properly interpreted, the gain expressions for the single stage BJT and FET amplifier stages can all be considered as identical!

Table 14.11 is a relative comparison of the FET amplifiers. The common-source amplifier provides moderate voltage gain and output resistance but high values of input resistance and current gain. The common-drain amplifier provides low voltage gain and output resistance, and high input resistance and current gain. Finally, the common-gate amplifier provides moderate voltage gain, high output resistance, and low input resistance and current gain. Tables 14.8 to 14.11 are very useful in the initial phase of amplifier design, when the engineer must make a basic choice of amplifier configuration to meet the design specifications.

DESIGN NOTE

The magnitude of the overall gain of the single-stage amplifiers can all be expressed approximately by

$$|A_v| \cong \frac{g_m R_L}{1 + g_m R_X} = \frac{R_L}{\dfrac{1}{g_m} + R_X}$$

in which R_X is the external resistance in the emitter or source loop of the transistor.

Now we have a toolbox full of amplifier configurations that we can use to solve circuit design problems. Design Ex. 14.5 demonstrates how to use our understanding to make design choices between the various configurations.

DESIGN EXAMPLE 14.5   SELECTING AN AMPLIFIER CONFIGURATION

One of the first things we must do to solve a circuit design problem is to decide on the circuit topology to be used. A number of examples are given here.

PROBLEMS What is the preferred choice of amplifier configuration for each of these applications?

(a) A single-transistor amplifier is needed that has a gain of approximately 80 dB and an input resistance of 100 kΩ.

(b) A single-transistor amplifier is needed that has a gain of 52 dB and an input resistance of 250 kΩ.

(c) A single-transistor amplifier is needed that has a gain of 30 dB and an input resistance of 5 MΩ.

(d) A single-transistor amplifier is needed that has a gain of approximately 0 dB and an input resistance of 20 MΩ with a load resistor of 10 kΩ.

(e) A follower is needed that has a gain of at least 0.98 and an input resistance of at least 250 kΩ with a load resistance of 5 kΩ.

(f) A single-transistor amplifier is needed that has a gain of +10 and an input resistance of 2 kΩ.

(g) An amplifier is needed with an output resistance of 25 Ω.

SOLUTION **Known Information and Given Data:** In each case, we see that a minimum amount of information is provided, typically only a voltage gain and resistance specification.

Unknowns: Circuit topologies

Approach: Use our estimates of voltage gain, input resistance, and output resistance for the various configurations to make a selection.

Assumptions: Typical values of current gain, Early voltage, power supply voltage, and so on will be assumed as necessary: $\beta_o = 100$, $0.25\ \text{V} \leq V_{GS} - V_{TN} \leq 1\ \text{V}$, $V_T = 0.025\ \text{V}$, $V_A \leq 150\ \text{V}$.

Analyses:

(a) The required voltage gain is $A_v = 10^{80/20} = 10{,}000$. This value of voltage gain exceeds the intrinsic voltage gain of even the best BJTs:

$$A_v \leq \mu_f = 40 V_A = 40(150) = 6000$$

An FET typically has a much lower value of intrinsic gain and is at an even worse disadvantage. Thus, such a large gain requirement cannot be met with a single-transistor amplifier.

(b) For the second set of specifications, we have $R_{\text{in}} = 250\ \text{k}\Omega$ and $A_v = 10^{52/20} \cong 400$. We require both large gain and relatively large input resistance, which point us toward the common-emitter amplifier. For the C-E stage, $A_v = 10\ V_{CC} \rightarrow V_{CC} = 40\ \text{V}$, which is somewhat large. However, we know that the $10\ V_{CC}$ estimate for the voltage gain is conservative and can easily be off by a factor of 2 or 3, so we can probably get by with a smaller power supply, say 20 V. Achieving the input of resistance requirement requires r_π to exceed 250 kΩ:

$$r_\pi = \frac{\beta_o V_T}{I_C} \geq 250\ \text{k}\Omega \rightarrow I_C \leq \frac{100(0.025\ \text{V})}{2.5 \times 10^5\ \Omega} = 10\ \mu\text{A}$$

which is small but acceptable. Achieving the gain specification with an FET would be much more difficult. For example, even with a small gate overdrive,

$$A_v = \frac{V_{DD}}{V_{GS} - V_{TN}} \cong \frac{V_{DD}}{0.25\ \text{V}} \rightarrow V_{DD} = 100\ \text{V}$$

which is unreasonably large for most solid-state designs. Note that the sign of the gain was not specified, so either positive or negative gain would be satisfactory, based on our limited specifications. However, the input resistance of the noninverting (C-B or C-G) amplifiers is low, not high.

(c) In this case, we require $R_{in} = 5\,M\Omega$ and $A_v = 10^{30/20} \cong 31.6$—large input resistance and moderate gain. These requirements can easily be met by a common-source amplifier:

$$A_v = \frac{V_{DD}}{V_{GS} - V_{TN}} = \frac{15\,V}{0.5\,V} = 30$$

The input resistance is set by our choice of gate bias resistors (R_1 and R_2 in Fig. 14.2), and $5\,M\Omega$ can be achieved with standard resistor values.

Since the gain is moderate, a C-E stage with emitter resistor could probably achieve the required high input resistance, although the values of the base bias resistor could become a limiting factor. For example, the input resistance and voltage gain could be met approximately with

$$R_{in} \cong \beta_o R_E \geq 5\,M\Omega \rightarrow R_E \geq \frac{5\,M\Omega}{100} = 50\,k\Omega \quad \text{and} \quad |A_v| = \frac{R_L}{R_E} \rightarrow R_L = 1.5\,M\Omega$$

(d) Zero-dB gain corresponds to a follower. For an emitter follower, $R_{in} \cong \beta_o R_L \cong 100(10\,k\Omega) = 1\,M\Omega$, so the BJT will not meet the input resistance requirement. On the other hand, a source follower provides a gain of approximately one and can easily achieve the required input resistance.

(e) A gain of 0.98 and an input resistance of $250\,k\Omega$ should be achievable with either a source follower or an emitter follower. For the MOSFET,

$$A_v = \frac{g_m R_L}{1 + g_m R_L} = 0.98 \qquad \text{requires} \qquad g_m R_L = \frac{2I_D R_L}{V_{GS} - V_{TN}} = 49$$

which can be satisfied with $I_D R_L = 12.3\,V$ for $V_{GS} - V_{TN} = 0.5\,V$.

The BJT can achieve the required gain with a much lower supply voltage and still meet the input resistance requirement: $R_{in} \cong \beta_o R_L \cong 100(5\,k\Omega) = 500\,k\Omega$.

$$g_m R_L = \frac{I_C R_L}{V_T} = 49 \rightarrow I_C R_L = 49(0.025\,V) = 1.23\,V$$

(f) A noninverting amplifier with a gain of 10 and an input resistance of $2\,k\Omega$ should be achievable with either a common-base or common-gate amplifier with proper choice of operating point. The gain of 10 is easily achieved with either the MOSFET or BJT design estimate: $A_v = V_{DD}/(V_{GS} - V_{TN})$ or $A_v = 10V_{CC}$. $R_{in} \cong 1/g_m = 2\,k\Omega$ is within easy reach of either device.

(g) Twenty-five ohms represents a small value of output resistance. The follower stages are the only choices that provide low output resistances. For the followers, $R_{out} = 1/g_m$, and so we need $g_m = 40\,mS$.

$$\text{For the BJT:} \quad I_C = g_m V_T = 40\,mS(25\,mV) = 1\,mA$$

$$\text{For the MOSFET:} \quad I_D = \frac{g_m(V_{GS} - V_{TN})}{2} = \frac{40\,mS(0.5\,V)}{2} = 10\,mA$$

$$K_n = \frac{g_m^2}{2I_D} = \frac{(40\,mS)^2}{2(10\,mA)} = 0.08\frac{A}{V^2} \quad \text{and}$$

$$\frac{W}{L} = \frac{K_n}{K_n'} = \frac{80\,mA/V^2}{50\,\mu A/V^2} = \frac{1600}{1}$$

The 25-Ω requirement can be met with either device, but the BJT requires an order of magnitude less current. In addition, the MOSFET requires a large W/L ratio.

Discussion: The options developed here represent our first attempts, and there is no guarantee that we will actually be able to fully achieve the desired specifications. After attempting a full design, we may have to change the circuit choice or use more than one transistor in a more complex amplifier configuration.

EXERCISE: Suppose the BJT amplifier in part (b) of Design Ex. 14.5 will be designed with symmetric 15-V supplies using a circuit similar to the one in Figure 13.36(a). Choose a collector current.

ANSWER: 5 μA, (10 μA does not account for the effect of R_B)

EXERCISE: Estimate the collector current needed for a BJT to achieve the input resistance specification in part (f) of Design Ex. 14.5.

ANSWER: 12.5 μA

14.6 COUPLING AND BYPASS CAPACITOR DESIGN

Up to this point, we have assumed that the impedances of coupling and bypass capacitors are negligible, and have concentrated on understanding the properties of the single transistor building blocks in their "midband" region of operation. However, since the impedance of a capacitor increases with decreasing frequency, the coupling and bypass capacitors generally reduce amplifier gain at low frequencies. In this section, we discover how to pick the values of these capacitors to ensure that our midband assumption is valid. Each of the three classes of amplifiers will be considered in succession. The technique we use is related to the "short-circuit" time constant (SCTC) method that we shall study in greater detail in Chapter 16. In this method, each capacitor is considered separately with all the others replaced by short circuits ($C \to \infty$).

14.6.1 COMMON-EMITTER AND COMMON-SOURCE AMPLIFIERS

Let us start by choosing values for the capacitors for the C-E and C-S amplifiers in Fig. 14.2. For the moment, assume that C_2 is still infinite in value, thus shorting the bottom of R_E and R_S to ground, as drawn in Fig. 14.23(a) and (b).

Coupling Capacitors C_1 and C_3

First, consider C_1. In order to be able to neglect C_1, we require the magnitude of the impedance of the capacitor (its capacitive reactance) to be much smaller than the equivalent resistance that appears at its terminals. Referring to Fig. 14.23, we see that the resistance looking to the left from capacitor C_1 (with $v_i = 0$) is R_I, and that looking to the right is R_{in}. Thus, design of C_1 requires

$$\frac{1}{\omega C_1} \ll (R_I + R_{\text{in}}) \qquad \text{or} \qquad C_1 \gg \frac{1}{\omega(R_I + R_{\text{in}})} \tag{14.48}$$

Frequency ω is chosen to be the lowest frequency for which midband operation is required in the given application.

For the common-emitter stage, bias resistor R_B appears in parallel (shunt) with the input resistance of the transistor, so $R_{\text{in}} = R_B \| R_{iB}$. For the common-source stage, bias resistor R_G shunts the input resistance of the transistor, and $R_{\text{in}} = R_G \| R_{iG} = R_G$.

ELECTRONICS IN ACTION

Revisiting the CMOS Imager Circuitry

In the first Electronics in Action feature in Chapter 4, we introduced the CMOS imager circuit presented on the next page. The chip contains 1.3 million pixels in a 1280×1024 imaging array. A typical photodiode based imaging pixel consists of a photo diode with sensing and access circuitry. Let us revisit this sensor circuit in light of what we have learned about single transistor amplifiers.

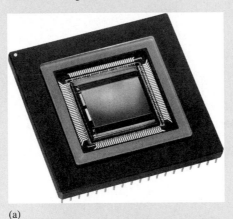

(a)

1.3 MegaPixel CMOS active-pixel image sensor.[1]

Typical photo diode pixel architecture.

M_1 is a reset switch, and after the $\overline{\text{RESET}}$ signal is asserted, the storage capacitor is fully charged to V_{DD}. The reset signal is then removed, and light incident on the photodiode generates a photo current that discharges the capacitor. Different light intensities produce different voltages on the capacitor at the end of the light integration time. Transistor M_2 is a source follower that buffers the photo-diode node. The source follower transfers the signal voltage at the photo-diode node to the output with nearly unity gain, and M_2 does not disturb the voltage at the photo diode output since it has an infinite dc input resistance. The voltage at the source of M_2 is then transferred to the output column via switch M_3. The source follower provides a low output resistance to drive the capacitance of the output column. The W/L ratio of switch M_3 must be chosen carefully so it does not significantly degrade the overall output resistance.

---

[1] The chip pictured above is a Micron Technology MI-MV13 image sensor and is reprinted here with permission from Micron Technology, Inc.

A similar analysis applies to C_3. We require the reactance of the capacitor to be much smaller than the equivalent resistance that appears at its terminals. Referring to Fig. 14.23(b), the resistance looking to the left from capacitor C_3 is R_{out}, and that looking to the right is R_3. Thus, C_3 must satisfy

$$\frac{1}{\omega C_3} \ll (R_{\text{out}} + R_3) \qquad \text{or} \qquad C_3 \gg \frac{1}{\omega(R_{\text{out}} + R_3)} \tag{14.49}$$

For the common-emitter stage, the collector resistor R_C appears in parallel with the output resistance of the transistor, and $R_{\text{out}} = R_C \| R_{iC}$. For the common-source stage, drain resistor R_D shunts the output resistance of the transistor, so $R_{\text{out}} = R_D \| R_{iD}$.

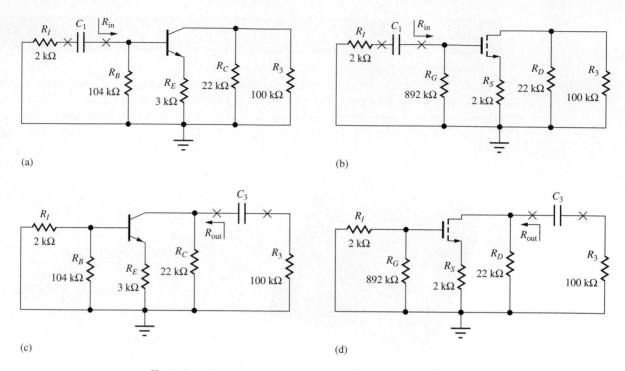

Figure 14.23 Coupling capacitors in the common-emitter and common-source amplifiers.

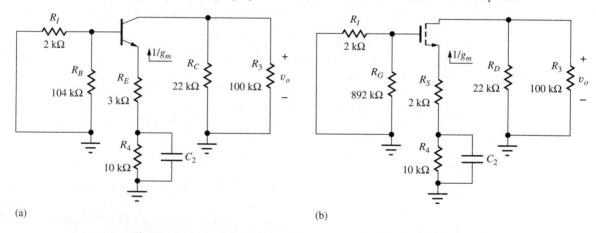

Figure 14.24 Bypass capacitors in the common-emitter and common-source amplifiers.

Bypass Capacitor C_2

The formula for C_2 is somewhat different. Figure 14.24 depicts the circuit assuming we can neglect the impedance of capacitors C_1 and C_3. At the terminals of C_2 in Fig. 14.24(a), the equivalent resistance is equal to R_4 in parallel with the sum $(R_E + 1/g_m)$,[4] the resistance looking up toward the emitter of the transistor. Thus, for the C-E and C-S amplifiers, C_2 must satisfy

$$C_2 \gg \frac{1}{\omega \left[R_4 \middle\| \left(R_E + \dfrac{1}{g_m} \right) \right]} \quad \text{or} \quad C_2 \gg \frac{1}{\omega \left[R_4 \middle\| \left(R_S + \dfrac{1}{g_m} \right) \right]} \tag{14.50}$$

---

[4] For the BJT case, we are neglecting the $R_{th}/(\beta_o + 1)$ term. Since the additional term will increase the equivalent resistance, its neglect makes Eq. (14.50) a conservative estimate.

In order to satisfy the inequalities in Eqs. (14.48) through (14.50), we will set the capacitor value to be approximately 10 times that calculated in the equations.

DESIGN EXAMPLE 14.6   CAPACITOR DESIGN FOR THE C-E AND C-S AMPLIFIERS

In this example, we select capacitor values for the three capacitors in both inverting amplifiers in Figs. 14.2, 14.23, and 14.24.

PROBLEM Choose values for the coupling and bypass capacitors for the amplifiers in Fig. 14.2 so that the presence of the capacitors can be neglected at a frequency of 1 kHz (1 kHz represents an arbitrary choice in the audio frequency range).

SOLUTION **Known Information and Given Data:** Frequency $f = 1000$ Hz; for the C-E stage described in Fig. 14.2 and Table 14.3, $R_{in}^{CE} = 310$ kΩ, $R_{out}^{CE} = 4.55$ MΩ, $R_I = 2$ kΩ, $R_B = 104$ kΩ, $R_C = 22$ kΩ, $R_E = 3$ kΩ, $R_4 = 10$ kΩ, and $R_3 = 100$ kΩ; for the C-S stage from Table 14.3, $R_{in}^{CS} = \infty$, $R_{out}^{CS} = 442$ kΩ, $R_I = 2$ kΩ, $R_G = 892$ kΩ, $R_D = 22$ kΩ, $R_S = 2$ kΩ, $R_4 = 10$ kΩ, and $R_3 = 100$ kΩ

Unknowns: Values of capacitors C_1, C_2, and C_3 for the common-emitter and common-source amplifier stages.

Approach: Substitute known values in Eq. (14.48) through (14.50). Choose nearest values from the appropriate table in Appendix A.

Assumptions: Small-signal operating conditions are valid, $V_T = 25$ mV.

Analysis: For the common-emitter amplifier,

$$R_{in} = R_B \| R_{iB} = 104 \text{ k}\Omega \| 310 \text{ k}\Omega = 77.9 \text{ k}\Omega$$

$$C_1 \gg \frac{1}{\omega(R_I + R_{in})} = \frac{1}{2000\pi(2 \text{ k}\Omega + 77.9 \text{ k}\Omega)} = 1.99 \text{ nF} \rightarrow C_1 = 0.02 \text{ }\mu\text{F} \text{ (20 nF)}^5$$

$$C_2 \gg \frac{1}{\omega\left[R_4 \left\|\left(R_E + \frac{1}{g_m}\right)\right.\right]} = \frac{1}{2000\pi\left[10 \text{ k}\Omega \left\|\left(3 \text{ k}\Omega + \frac{1}{9.80 \text{ mS}}\right)\right.\right]}$$
$$= 67.2 \text{ nF} \rightarrow C_2 = 0.68 \text{ }\mu\text{F}$$

$$C_3 \gg \frac{1}{\omega(R_{out} + R_3)} = \frac{1}{2000\pi(21.9 \text{ k}\Omega + 100 \text{ k}\Omega)} = 1.31 \text{ nF} \rightarrow C_3 = 0.015 \text{ }\mu\text{F} \quad (15 \text{ }\mu\text{F})$$

For the common-source stage, $R_{in} = R_G$ since the input resistance at the gate of the transistor is infinite,

$$C_1 \gg \frac{1}{\omega(R_I + R_{in})} = \frac{1}{2000\pi(2 \text{ k}\Omega + 892 \text{ k}\Omega)} = 178 \text{ pF} \rightarrow C_1 = 1800 \text{ pF}$$

$$C_2 \gg \frac{1}{\omega\left[R_4 \left\|\left(R_S + \frac{1}{g_m}\right)\right.\right]} = \frac{1}{2000\pi\left[10 \text{ k}\Omega \left\|\left(2 \text{ k}\Omega + \frac{1}{0.491 \text{ mS}}\right)\right.\right]}$$
$$= 55.3 \text{ nF} \rightarrow C_2 = 0.56 \text{ }\mu\text{F}$$

$$C_3 \gg \frac{1}{\omega(R_{out} + R_3)} = \frac{1}{2000\pi(21.5 \text{ k}\Omega + 100 \text{ k}\Omega)} = 1.31 \text{ nF} \rightarrow C_3 = 0.015 \text{ }\mu\text{F} \quad (15 \text{ nF})$$

---

5 We are using $C_1 = 10(1.99 \text{ nF})$ to satisfy the inequality.

Check of Results: A double check of the calculations indicates they are correct. This would be a good place to check the analysis with simulation.

Discussion: We have chosen each capacitor to have negligible reactance at the frequency of 1 kHz and would expect the lower cutoff frequency of the amplifier to be well below this frequency. The choice of frequency in this example was arbitrary and depends upon the lowest frequency of interest in the application.

Computer-Aided Analysis: The graph below gives SPICE simulation results for the common-emitter amplifier with the capacitors as designed here. The midband gain is 15.0 dB and the lower cutoff frequency is 195 Hz. Note the two-pole roll-off at low frequencies indicated by the 40-dB/decade slope in the magnitude characteristic. The slope indicates that there are two zeros at dc, which are associated with capacitors C_1 and C_3. A signal cannot pass through either capacitor at dc, hence the frequency response exhibits a double zero at the origin. We have ended up with an amplifier that has three low frequency poles at approximately 100 Hz (1 kHz/10), and bandwidth shrinkage (Sec. 12.4.3) causes the resulting lower cutoff frequency f_L to increase to 195 Hz.

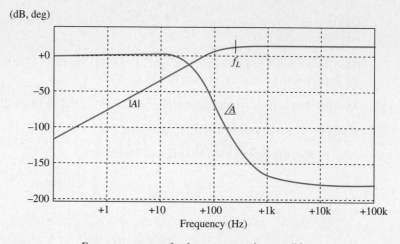

Frequency response for the common-emitter amplifier.

14.6.2 COMMON-COLLECTOR AND COMMON-DRAIN AMPLIFIERS

The simplified C-C and C-D amplifiers in Fig. 14.4 have only two coupling capacitors. In order to be able to neglect C_1, the reactance of the capacitor must be much smaller than the equivalent resistance that appears at its terminals. Referring to Fig. 14.25, we see that the resistance looking to the left

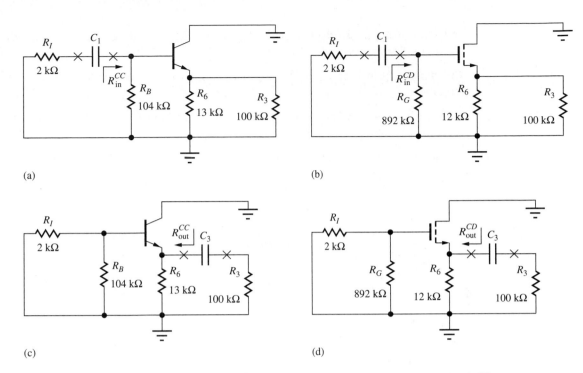

Figure 14.26 Coupling capacitors in the common-collector and common-drain amplifiers.

from C_1 is R_I, and that looking to the right is R_{in}. Thus, design of C_1 is the same as Eq. (14.48):

$$\frac{1}{\omega C_1} \ll (R_I + R_{\text{in}}) \qquad \text{or} \qquad C_1 \gg \frac{1}{\omega(R_I + R_{\text{in}})} \qquad (14.51)$$

Be sure to note that the values of the input and output resistances will be different in Eq. (14.51) from those in Eq. (14.48)! For the common-collector stage, bias resistor R_B shunts the input resistance of the transistor, so $R_{\text{in}} = R_B \| R_{\text{in}}^{CC}$. For the common-drain stage, gate bias resistor R_G appears in parallel with the input resistance of the transistor, and $R_{\text{in}} = R_G \| R_{\text{in}}^{CD}$.

For C_3, the resistance looking to the left from capacitor C_3 is R_{out}, and that looking to the right is R_3. Thus, design of C_3 requires

$$\frac{1}{\omega C_3} \ll (R_{\text{out}} \mid R_3) \qquad \text{or} \qquad C_3 \gg \frac{1}{\omega(R_{\text{out}} + R_3)} \qquad (14.52)$$

where $R_{\text{out}} = R_6 \| R_{\text{out}}^{CC,CD}$, because resistor R_6 appears in parallel with the output resistance of the transistor. Note again that the value of R_{out} in Eq. (14.52) differs from that in Eq. (14.49).

DESIGN EXAMPLE 14.7   **CAPACITOR DESIGN FOR THE C-C AND C-D AMPLIFIERS**

This example selects capacitor values for the followers in Figs. 14.4 and 14.25.

PROBLEM   Choose values for the coupling and bypass capacitors for the amplifiers in Fig. 14.4 and 14.25 so that the presence of the capacitors can be neglected at a frequency of 2 kHz.

SOLUTION   **Known Information and Given Data:** Frequency $f = 2000\,\text{Hz}$; for the C-C stage from Fig. 14.4 and Table 14.4, $R_{iB} = 1.17\,\text{M}\Omega$, $R_{iC} = 0.121\,\text{k}\Omega$, $R_6 = 13\,\text{k}\Omega$, $R_I = 2\,\text{k}\Omega$, $R_B = 104\,\text{k}\Omega$, and $R_3 = 100$ kΩ; for the C-S stage, $R_{iG} = \infty$, $R_{iS} = 2.04$ kΩ, $R_6 = 12$ kΩ, $R_I = 2$ kΩ, $R_G = 892$ kΩ, and $R_3 = 100$ kΩ

Unknowns: Values of capacitors C_1 and C_3 for the common-collector and common-drain amplifiers.

Approach: Substitute known values in Eqs. (14.51) and (14.52). Choose the nearest values from the capacitor table in Appendix A.

Assumptions: Small-signal operating conditions are valid.

Analysis: For the common-collector amplifier,

$$R_{in} = R_B \| R_{iB} = 104 \text{ k}\Omega \| 1.17 \text{ M}\Omega = 95.5 \text{ k}\Omega$$

$$C_1 \gg \frac{1}{\omega(R_I + R_{in})} = \frac{1}{4000\pi(2 \text{ k}\Omega + 95.5 \text{ k}\Omega)} = 816 \text{ pF} \rightarrow C_1 = 8200 \text{ pF}^6$$

$$R_{out} = R_6 \| R_{iC} = 13 \text{ k}\Omega \| 121 \text{ }\Omega = 120 \text{ }\Omega$$

$$C_3 \gg \frac{1}{\omega(R_{out} + R_3)} = \frac{1}{4000\pi(120 \text{ }\Omega + 100 \text{ k}\Omega)} = 795 \text{ pF} \rightarrow C_3 = 8200 \text{ pF}$$

and for the common-drain stage,

$$R_{in} = R_G \| R_{iG} = 892 \text{ k}\Omega \| \infty = 892 \text{ k}\Omega$$

$$C_1 \gg \frac{1}{\omega(R_I + R_{in})} = \frac{1}{4000\pi(2 \text{ k}\Omega + 892 \text{ k}\Omega)} = 89.0 \text{ pF} \rightarrow C_1 = 1000 \text{ pF}$$

$$R_{out} = R_6 \| R_{iS} = 12 \text{ k}\Omega \| 2.04 \text{ k}\Omega = 1.74 \text{ k}\Omega$$

$$C_3 \gg \frac{1}{\omega(R_{out} + R_3)} = \frac{1}{4000\pi(1.74 \text{ k}\Omega + 100 \text{ k}\Omega)} = 782 \text{ pF} \rightarrow C_3 = 8200 \text{ pF}$$

Check of Results: A double check of the calculations indicates they are correct. This represents a good place to check the analysis with simulation.

Discussion: We have chosen each capacitor to have negligible reactance at the frequency of 2 kHz and would expect the lower cutoff frequency of the amplifier to be well below this frequency.

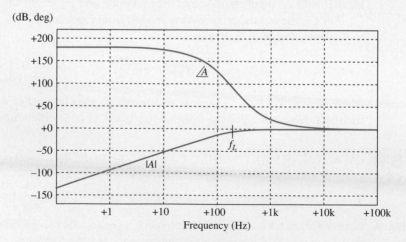

Emitter follower frequency response.

---

[6] $C_1 = 10(816 \text{ pF})$ is used to satisfy the inequality.

The choice of frequency in this example was arbitrary and depends upon the lowest frequency of interest in the application.

Computer-Aided Analysis: The graph on the next page shows SPICE simulation results for the common-emitter amplifier with the capacitors as designed above. The midband gain is -0.262 dB (0.970) and the lower cutoff frequency is 310 Hz. Note the two-pole roll off at low frequencies indicated by the 40-dB/decade slope in the magnitude characteristic. As in Design Ex. 14.6, a dc signal cannot pass through capacitor C_1 or C_3, and the amplifier transfer function is characterized by a double zero at the origin.

EXERCISE: Reevaluate the capacitor values for the two amplifiers in Ex. 14.7 if the frequency is 250 Hz and the values of R_I and R_3 are changed to 1 kΩ and 82 kΩ, respectively?

ANSWERS: 6.79 nF → 0.068 μF, 8.16 nF → 0.082 μF; 713 pF → 8200 pF, 7.98 nF → 0.082 μF

EXERCISE: Use SPICE to simulate the frequency response of the common-drain amplifier and find the midband gain and lower cutoff frequency.

ANSWERS: −1.54 dB; 293 Hz

14.6.3 COMMON-BASE AND COMMON-GATE AMPLIFIERS

For the C-B and C-G amplifiers, C_2 is first assumed to be infinite in value, thus shorting the base and gate of the transistors in Fig. 14.5 to ground as redrawn in Fig. 14.26. In order to neglect C_1 the magnitude of the impedance of the capacitor must be much smaller than the equivalent resistance that appears at its terminals. Referring to Fig. 14.26, the resistance looking to the left from the capacitor is R_I, and that looking to the right is R_{in}. Thus, design of C_1 is the same as Eq. (14.48):

$$\frac{1}{\omega C_1} \ll (R_I + R_{in}) \quad \text{or} \quad C_1 \gg \frac{1}{\omega (R_I + R_{in})} \tag{14.53}$$

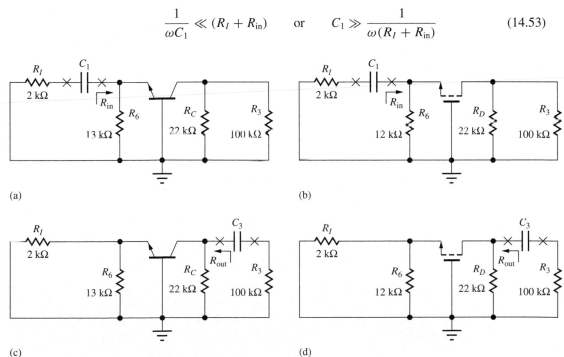

Figure 14.26 Coupling capacitors in the common-base and common-gate amplifiers.

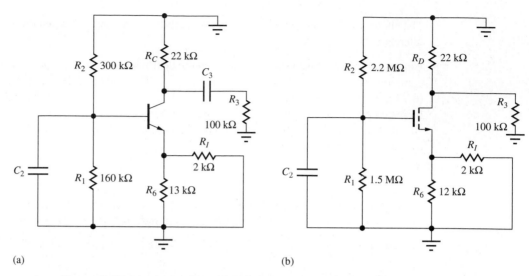

Figure 14.27 Bypass capacitors in the (a) common-collector and (b) common-drain amplifiers.

For the two amplifier stages, resistor R_6 appears in shunt with the input resistance of the transistor, so $R_{in} = R_6 \| R_{iE}$ or $R_{in} = R_6 \| R_{iS}$.

For C_3, we see that the resistance looking to the left from capacitor C_3 is R_{out}, and that looking to the right is R_3. Thus, design of C_3 requires

$$\frac{1}{\omega C_3} \ll (R_{out} + R_3) \qquad \text{or} \qquad C_3 \gg \frac{1}{\omega (R_{out} + R_3)} \tag{14.54}$$

For the amplifiers, resistor R_C or R_D appears in parallel with the output resistance of the transistor, so $R_{out} = R_C \| R_{iC}$ or $R_{out} = R_D \| R_{iD}$.

To be an effective bypass capacitor, the reactance of C_2 must be much smaller than the equivalent resistance at the base or gate terminal of the transistors in Fig. 14.5 with the other capacitors assumed to be infinite, as depicted in Fig. 14.27. The resistances at the base and gate nodes are

$$R_{eq}^{CB} = R_1 \| R_2 \| [r_\pi + (\beta_o + 1)(R_6 \| R_I)] \qquad \text{and} \qquad R_{eq}^{CG} = R_1 \| R_2 \tag{14.55}$$

respectively. The corresponding value of C_2 must satisfy

$$C_2 \gg \frac{1}{\omega R_{eq}^{CB,CG}}$$

DESIGN EXAMPLE 14.8 — CAPACITOR DESIGN FOR THE C-B AND C-G AMPLIFIERS

This example selects capacitor values for the noninverting amplifiers in Figs. 14.5 and 14.26.

PROBLEM Choose values for the coupling and bypass capacitors for the amplifiers in Figs. 14.5 and 14.26 so that the presence of the capacitors can be neglected at a frequency of 1 kHz.

SOLUTION **Known Information and Given Data:** Frequency $f = 1000$ Hz; for the C-B stage from Fig. 14.5 and Table 14.6, $R_{iE} = 102 \ \Omega$, $R_{iC} = 3.40 \ M\Omega$, $R_I = 2 \ k\Omega$, $R_1 = 160 \ k\Omega$, $R_2 = 300 \ k\Omega$, $R_C = 22 \ k\Omega$, and $R_6 = 13 \ k\Omega$; for the C-G amplifier, $R_{iS} = 2.04 \ k\Omega$, $R_{iD} = 411 \ k\Omega$, $R_I = 2 \ k\Omega$, Ω, $R_1 = 1.5 \ M\Omega$, $R_2 = 2.2 \ M\Omega$, $R_6 = 12 \ k\Omega$, and $R_D = 22 \ k\Omega$

Unknowns: Values of capacitors C_1, C_2, and C_3

Approach: Substitute known values in Eqs. (14.53) through (14.55). Choose the nearest values from the capacitor table in Appendix A.

Assumptions: Small-signal operating conditions are valid.

Analysis: For the common-base amplifier,

$$R_{\text{in}} = R_6 \| R_{iE} = 13 \text{ k}\Omega \| 102 \ \Omega = 100 \ \Omega$$

$$C_1 \gg \frac{1}{\omega(R_I + R_{\text{in}})} = \frac{1}{2000\pi(2 \text{ k}\Omega + 100 \ \Omega)} = 75.8 \text{ nF} \to C_1 = 0.82 \ \mu\text{F}^7$$

$$C_2 \gg \frac{1}{\omega(R_1 \| R_2 \| [r_\pi + (\beta_o + 1)(R_6 \| R_I)])}$$

$$= \frac{1}{2000\pi(160 \text{ k}\Omega \| 300 \text{ k}\Omega \| [10.2 \text{ k}\Omega + (101)(13 \text{ k}\Omega \| 2 \text{ k}\Omega)])}$$

$$= 2.38 \text{ nF} \to C_2 = 0.027 \ \mu\text{F}$$

$$R_{\text{out}} = R_C \| R_{iC} = 22 \text{ k}\Omega \| 3.40 \text{ M}\Omega = 21.9 \text{ k}\Omega$$

$$C_3 \gg \frac{1}{\omega(R_{\text{out}} + R_3)} = \frac{1}{2000\pi(21.9 \text{ k}\Omega + 100 \text{ k}\Omega)} = 1.31 \text{ nF} \to C_3 = 0.015 \ \mu\text{F} \quad (15 \text{ nF})$$

and for the common-gate stage,

$$R_{\text{in}} = R_6 \| R_{iS} = 12 \text{ k}\Omega \| 2.04 \ \Omega = 1.74 \text{ k}\Omega$$

$$C_1 \gg \frac{1}{\omega(R_I + R_{\text{in}})} = \frac{1}{2000\pi(2 \text{ k}\Omega + 1.74 \text{ k}\Omega)} = 42.6 \text{ nF} \to C_1 = 0.42 \ \mu\text{F}$$

$$C_2 \gg \frac{1}{\omega(R_1 \| R_2)} = \frac{1}{2000\pi(1.5 \text{ M}\Omega \| 2.2 \text{ M}\Omega)} = 178 \text{ pF} \to C_2 = 1800 \text{ pF}$$

$$R_{\text{out}} = R_6 \| R_{iD} = 22 \text{ k}\Omega \| 410 \text{ k}\Omega = 20.9 \text{ k}\Omega$$

$$C_3 \gg \frac{1}{\omega(R_{\text{out}} + R_3)} = \frac{1}{2000\pi(20.9 \text{ k}\Omega + 100 \text{ k}\Omega)} = 1.31 \text{ nF} \to C_3 = 0.015 \ \mu\text{F} \quad (15 \text{ nF})$$

Check of Results: A double check of the calculations indicates they are correct. This is a good place to check the analysis with simulation.

Discussion: We have chosen each capacitor to have negligible reactance at the frequency of 1 kHz and expect the lower cutoff frequency of the amplifier to be well below this frequency. The choice of frequency in this example was arbitrary and depends upon the lowest frequency of interest in the application.

Computer-Aided Analysis: The graph below shows SPICE simulation results for the common-base amplifier with the capacitors as just designed. The midband gain is 18.5 dB (8.41) and the lower cutoff frequency is 174 Hz. Note the two-pole roll-off at low frequencies indicated by the 40-dB/decade slope in the magnitude characteristic. Here again, since a dc signal cannot pass through capacitor C_1 or C_3, the amplifier transfer function exhibits a double zero at the origin.

---

[7] $C_1 = 10(75.8 \text{ nF})$ is used to satisfy the inequality.

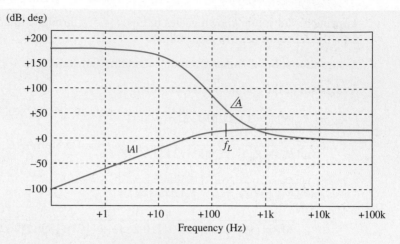

Common-base amplifier frequency response.

EXERCISE: Recalculate the capacitor values for the two amplifiers in Design Ex. 14.8 if the frequency is 250 Hz and the values of R_l and R_3 are changed to 1 kΩ and 82 kΩ, respectively.

ANSWERS: 0.578 μF → 6.8 μF, 6.37 nF → 0.068 μF, 9.54 nF → 0.10 μF; 0.209 μF → 2.2 μF, 6.44 nF → 0.068 μF, 714 pF → 8200 pF

EXERCISE: Use SPICE to simulate the frequency response of the common-gate amplifier and find the midband gain and lower cutoff frequency.

ANSWERS: 12.2 dB, 156 Hz

14.6.4 SETTING LOWER CUTOFF FREQUENCY f_L

In the previous sections, we have designed the coupling and bypass capacitors to have a negligible effect on the circuit at some particular frequency in the midband range of the amplifier. An alternative is to choose the capacitor values to set the lower cutoff frequency of the amplifier where we want it to be. Referring back to the high-pass filter analysis in Sec. 10.7.4, we see that the pole associated with the capacitor occurs at the frequency for which the capacitive reactance is equal to the resistance that appears at the capacitor terminals.

Multiple Poles and Bandwidth Shrinkage

In the circuits we have considered, there are several poles, and a bandwidth shrinkage occurs at low frequencies in a manner similar to that which was presented in Table 12.3 for high frequencies. A transfer function which exhibits n identical poles at a low frequency ω_o can be written as

$$T(s) = A_{\text{mid}} \frac{s^n}{(s + \omega_o)^n} \tag{14.56}$$

$$|T(j\omega)| = A_{\text{mid}} \frac{\omega^n}{\left(\sqrt{\omega^2 + \omega_o^2}\right)^n} \tag{14.57}$$

$$|T(j\omega_L)| = \frac{A_{\text{mid}}}{\sqrt{2}} \rightarrow \omega_L = \frac{\omega_o}{\sqrt{2^{1/n} - 1}} \quad \text{or} \quad f_L = \frac{f_o}{\sqrt{2^{1/n} - 1}} \tag{14.58}$$

The factor in the denominator of Eq. (14.58) is less than 1, so that the lower cutoff frequency is higher than the frequency corresponding to the individual poles. Table 14.12 gives the relationship between ω_o and ω_L for various values of n.

In Design Exs. 14.6, 14.7, and 14.8, we have effectively located three poles of each amplifier at a frequency of 1/10 of the midband frequency specified in the problem. For three identical poles, $f_L = 1.96 f_o$. In Design Ex. 14.5, the three poles were placed at a frequency of approximately 100 Hz (1000 Hz/10), which should yield a cutoff of 196 Hz based on the numbers in Table 14.12. The simulation results yielded $f_L = 195$ Hz. In Design Ex. 14.7, the poles were also placed at a frequency of approximately 100 Hz, which should yield a cutoff of 196 Hz. The simulation results yielded a slightly smaller value of f_L, 174 Hz.

The situation in Design Ex. 14.6 is slightly different. With capacitor C_3 eliminated from the circuit, the C-C and C-D amplifiers exhibit two poles at low frequencies. In this example, the two poles are at 200 Hz, which should yield a cutoff frequency of 310 Hz, and the simulation results agree with $f_L = 310$ Hz.

Setting f_L with a Dominant Pole

It is often easy and preferable to have the pole associated with just one of the capacitors determine the lower cutoff frequency, rather than have f_L set by the interaction of several poles. In this case, we set f_L with one of the capacitors, and then choose the other capacitors to have their pole frequencies much below f_L. This is referred to as a dominant pole design. In Design Exs. 14.6, 14.7, and 14.8, we see that the capacitor associated with the emitter or source portion of the circuit tends to be the largest (C_2 in Fig. 14.24, C_3 in Fig. 14.25, and C_1 in Fig. 14.26) because of the low resistance presented by the emitter or source terminal of the transistor. It is common to use these capacitors to set f_L, and then increase the value of the other capacitors by a factor of 10 to push their corresponding poles to much lower frequencies.

For the C-E stage in Design Ex. 14.6, we could set $f_L = 1000$ Hz by choosing $C_2 = 0.067 \, \mu\text{F}$ and leaving $C_1 = 0.02 \, \mu\text{F}$ and $C_3 = 0.015 \, \mu\text{F}$. In the C-D amplifier in Fig. 14.25(b), using $C_3 = 780$ pF with $C_1 = 1000$ pF sets the lower cutoff frequency to 2000 Hz. Finally, for the C-B amplifier in Design Ex. 14.8, choosing $C_1 = 0.082 \, \mu\text{F}$, $C_2 = 0.027 \, \mu\text{F}$, and $C_3 = 0.015 \, \mu\text{F}$ should set the cutoff frequency to approximately 1000 Hz.

EXERCISE: Use SPICE to find the values of f_L for the three designs presented in the preceding paragraph.

ANSWERS: C-E: 960 Hz; C-D: 2.04 kHz; C-B: 960 Hz

EXERCISE: (a) What value of capacitor C_2 should be used to set f_L to 1 kHz in the C-S amplifier in Design Ex. 14.6? (b) What value of capacitor C_3 should be used to set f_L to 2 kHz in the C-C amplifier in Design Ex. 14.7? (c) What value of capacitor C_1 should be used to set f_L to 1 kHz in the C-G amplifier in Design Ex. 14.8?

ANSWERS: (a) 0.056 μF; (b) 795 pF; (c) 0.039 μF

14.7 AMPLIFIER DESIGN EXAMPLES

Now that we have become "experts" in the characteristics of single-transistor amplifiers, we will use this knowledge to tackle several amplifier design problems. We should emphasize that no "cookbook" exists for design. Every design is a new, creative experience. Each design has its own unique set of constraints, and there may be more than one way to achieve the desired results. The examples presented here further illustrate the approach to design; they also underscore the interaction between the designer's choice of Q-point and the small-signal properties of the amplifiers.

DESIGN EXAMPLE 14.9   A FOLLOWER DESIGN

In this example, we will design a follower to meet a set of specifications.

PROBLEM Design an amplifier with a mid-band input resistance of at least 20 MΩ and a gain of at least 0.95 when driving an external load of at least 3 kΩ. Any capacitors present should not affect the performance of the circuit at frequencies above 50 Hz.

SOLUTION **Known Information and Given Data:** $A_v \geq 0.95$, $R_{\text{in}} \geq 20$ MΩ, $R_{\text{out}} \ll 3$ kΩ.

Unknowns: The circuit topology must be chosen, the Q-point must be selected, and the circuit element values must all be determined. The transistor parameters are unknown.

Approach: The gain is approximately one, a high input resistance is required, and the relatively small load resistance will require the amplifier to have a low output resistance. All three of these specifications lead us to consider a voltage follower. We must choose between the emitter-follower (C-C) and source-follower (C-D) configurations and then select the circuit values to meet the design specifications.

Assumptions: The transistors are operating in the active region. Small-signal operating conditions are satisfied, $V_T = 25$ mV.

Analysis: Reviewing Tables 14.9 and 14.11, we find that the input resistance of the C-D amplifier prototype is infinite, whereas that of the C-C amplifier is limited to $\beta_o R_L$. For a load resistance of 3 kΩ, a current gain β_o in excess of 6600 is required to meet the input-resistance specification. This current gain is beyond the range of normal bipolar transistors, so here we rule out the C-C amplifier. (However, be sure to watch for the Darlington circuit in Prob. 15.48.)

Figure 14.28 represents a basic source-follower circuit. In this amplifier, we recognize that R_{in} is set simply by the value of R_G, and we can pick $R_G = 22$ MΩ (± 5 percent) to meet the specification. The 22-MΩ value ensures that the design specifications are met when the effect of the tolerance is included.

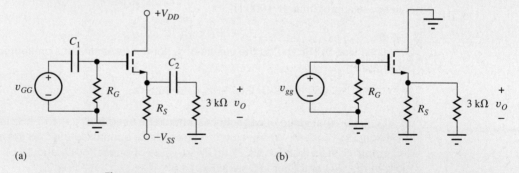

(a)   (b)

Figure 14.28 (a) Common-drain amplifier and (b) ac equivalent circuit.

The choices of source resistor R_S and power supply voltages are related to the voltage gain requirement:

$$\frac{g_m R_L}{1 + g_m R_L} \geq 0.95 \quad \text{or} \quad g_m R_L \geq 19 \quad \text{and} \quad R_L = R_S \| 3 \text{ k}\Omega \qquad (14.59)$$

The $g_m R_L$ product can be related to the drain current and device parameter K_n by using $g_m \cong \sqrt{2K_n I_D}$, and from Eq. (14.59),

$$\sqrt{2K_n I_D} \, R_L \geq 19 \quad \text{or} \quad \sqrt{K_n I_D} \geq \frac{19}{\sqrt{2} \, R_L} \qquad (14.60)$$

TABLE 14.13
Possible Solutions to Eq. (14.61)

I_D (mA)	K_n (mA/V²)	$(V_{GS} - V_{TN})$ (V)	V_{SS} (V)
3	10	0.78	$9.8 + V_{TN}$
5	10	1	$16 + V_{TN}$
8	10	1.27	$25.3 + V_{TN}$
5	20	0.71	$16.7 + V_{TN}$

In Fig. 14.28(b), the equivalent load resistor $R_L = R_S \| 3 \text{ k}\Omega \leq 3 \text{ k}\Omega$. As is often the case in design, one equation—here, Eq. (14.60)—contains more than one unknown. We must make a design decision. Let us choose $R_L \geq 1.5 \text{ k}\Omega$ (that is, $R_S \geq 3 \text{ k}\Omega$). Substituting this value into Eq. (14.60) yields

$$\sqrt{K_n I_D} \geq \frac{19/\sqrt{2}}{1.5 \text{ k}\Omega} = 8.96 \text{ mA} \tag{14.61}$$

Equation (14.61) indicates that the geometric mean of K_n and I_D must be at least 9 mA.

We can now attempt to select an FET and Q-point current. Here again, Eq. (14.61) contains two unknowns. We must make another design decision. Table 14.13 presents some possible solution pairs for Eq. (14.62), as well as their impact on the values of $(V_{GS} - V_{TN})$ and negative supply voltage V_{SS} since

$$I_D = \frac{K_n}{2}(V_{GS} - V_{TN})^2 \tag{14.62}$$

and

$$V_{SS} = I_D R_S + V_{GS} \tag{14.63}$$

based on analysis of the dc equivalent circuit in Fig. 14.29 (remember $I_G = 0$). The choice of $I_D = 5$ mA with $K_n = 20$ mA/V² seems to be reasonable, although the power supply voltage might be too large for some applications.

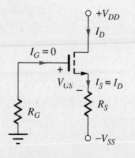

Figure 14.29 dc Equivalent circuit for the C-D amplifier.

Let us assume we have looked through our device catalogs and found a MOSFET with $V_{TN} = 1.5$ V and $K_n = 20$ mA/V². Evaluating Eq. (14.62) for this FET gives

$$V_{GS} = V_{TN} + \sqrt{\frac{2I_D}{K_n}} = 1.5 + \sqrt{\frac{2(0.005)}{0.02}} = 2.21 \text{ V} \tag{14.64}$$

Now we are finally in a position to find R_S using Eq. (14.64).

$$R_S = \frac{V_{SS} - V_{GS}}{I_D} = \frac{V_{SS} - 2.21}{0.005} \tag{14.65}$$

TABLE 14.14
Possible
Solutions to
Eq. (14.96)

V_{SS}	R_S
10 V	1.56 kΩ
15 V	2.56 kΩ
20 V	3.56 kΩ
25 V	4.56 kΩ

Values have been selected for V_{GS} and I_D but not for V_{SS}, and Eq. (14.65) is another equation with two unknowns. (The value in Table 14.13 represented only a lower bound.) Table 14.14 presents several possible solution pairs from which to make our design selection. Earlier in the design discussion, we assumed that $R_S \geq 3$ kΩ, so one acceptable choice is $V_{SS} = 20$ V and $R_S = 3.56$ kΩ.

Our final design decision is the choice of V_{DD}, which must be large enough to ensure that the MOSFET operates in the active region under all signal conditions:

$$v_{DS} \geq v_{GS} - V_{TN} \tag{14.66}$$

and

$$v_{DS} = v_D - v_S = V_{DD} + V_{GS} - v_s \tag{14.67}$$

for $v_S = V_S + v_s$ and $V_S = -V_{GS}$. Combining Eqs. (14.66) and (14.67) yields

$$v_{DD} + V_{GS} - v_s \geq V_{GS} - V_{TN} \qquad \text{or} \qquad V_{DD} \geq v_s - V_{TN} = v_s - 1.5 \text{ V} \tag{14.68}$$

The largest amplitude signal v_{gg} at the source that satisfies the small-signal requirements is

$$|v_{gg}| \leq 0.2(V_{GS} - V_{TN})(1 + g_m R_L) \frac{g_m R_L}{1 + g_m R_L} \leq 0.2(0.71)(19) = 2.70 \text{ V} \tag{14.69}$$

Thus, if we choose a V_{DD} of at least 1.2 V, then the MOSFET remains saturated for all signals that satisfy the small-signal criteria.

The final step in this design is to select values for the coupling capacitors. We desire the impedance of the capacitors at frequencies ≥ 50 Hz to be negligible with respect to the resistance that appears at their terminals. The resistance looking to the left from C_1 is zero, and that looking to the right is R_{in}^{CD}, which is 22 MΩ. Therefore,

$$\frac{1}{2\pi(50 \text{ Hz})C_1} \ll 22 \text{ M}\Omega \qquad \text{or} \qquad C_1 \gg 145 \text{ pF}$$

For C_2, the resistance looking back toward the source is

$$R_{out} = R_S \left\| \frac{1}{g_m} = 3.6 \text{ k}\Omega \right\| \frac{1}{\sqrt{2K_n I_D}} = 3.6 \text{ k}\Omega \left\| \frac{1}{\sqrt{2(20 \text{ mS})(5 \text{ mA})}} = 69.4 \ \Omega, \right.$$

and the resistance looking toward the right is 3 kΩ. Therefore,

$$\frac{1}{2\pi(50 \text{ Hz})C_2} \ll 3.097 \text{ k}\Omega \qquad \text{or} \qquad C_2 \gg 1.03 \ \mu\text{F}$$

Let us choose $C_1 = 1500$ pF and $C_2 = 10$ μF, which are standard values that exceed the minimum bound by a factor of approximately 10.

The final design appears in Fig. 14.30, in which the nearest 5 percent values have been used for the resistors and V_{DD} has been chosen to be a common power supply value of $+5$ V.

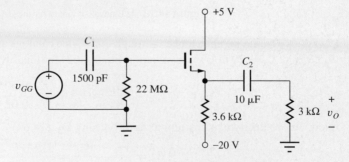

Figure 14.30 Completed source-follower design.

Check of Results: To check our design, we should now analyze the circuit and find the actual Q-point, input resistance, and voltage gain. This analysis is left as an exercise. Another approach at this point would be to check the analysis with SPICE.

Discussion: In this example, we see that even a problem that appears to be a relatively well specified problem takes considerable effort to achieve a design that meets the requirements and the design required a relatively large value of V_{SS}. Such is the situation in most real design situations. Most problems will be under-constrained with numerous choices to be made.

Computer-Aided Analysis: Simulation of the circuit in SPICE yields these results: Q-point: (4.94 mA, 7.20 V), $A_v = -0.369$ dB, and $f_L = 7.8$ Hz. With two poles at 5 Hz, the expected value of f_L is also 7.8 Hz.

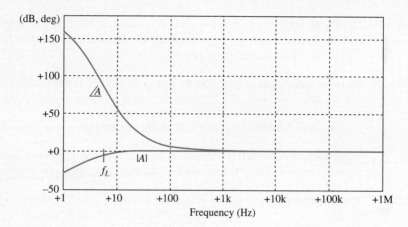

Source follower frequency response.

EXERCISE: (a) Create a two-port model (Fig. 10.7) for the midband region of the source follower in Fig. 14.37. (b) Use the model to calculate the voltage gain with the 3-kΩ load attached to the amplifier.

ANSWERS: (a) $R_{in} = 22$ MΩ, $A = +0.981$, $R_{out} = 69.4$ Ω; (b) 0.959

DESIGN EXAMPLE 14.10   A COMMON-BASE AMPLIFIER

The requirements of this design problem are even less specific than those in Design Ex. 14.9. A common-base amplifier will be found to be the most appropriate choice to meet the given design specifications.

PROBLEM   Design an amplifier to match a 75-Ω source resistance (for example, a coaxial transmission line) and to provide a voltage gain of 34 dB. Design the capacitors to have negligible impact on the circuit for RF frequencies above 500 kHz.

SOLUTION   **Known Information and Given Data:** Amplifier input resistance = 75 Ω; voltage gain = 50 (34 dB); capacitors should be negligible at a frequency of 500 kHz.

Unknowns: Amplifier topology; Q-point; circuit element values; transistor parameters

Approach: Use overall specifications to guide choice of circuit topology and transistor type; then choose circuit element values to meet numeric requirements

Assumptions: Active region operation with $V_{EB} = 0.7$ V; Small-signal conditions apply; $V_T = 25$ mV.

Analysis: Our first problem is to select a circuit configuration and transistor type. From the various examples in this and previous chapters, we realize that $A_v = 50$ (34 dB) is a moderate value of gain. At the same time, the required input resistance of 75 Ω is relatively low. Looking through our amplifier comparison charts in Tables 14.8 through 14.11, we find that the common-base and common-gate amplifiers most nearly meet these two requirements: good voltage gain and low input resistance. From past examples, we should recognize that it will probably be easier to achieve a gain of 50 with a BJT than with a FET, particularly since the matched input resistance requirement will increase the amplifier terminal gain requirement by a factor of 2! Thus, the common-base amplifier is the choice that seems to most nearly meet the problem specifications.

For simplicity, let us use the dual supply-bias circuit in Fig. 14.31, which requires only two bias resistors. In addition, to get some practice analyzing circuits using *pnp* devices, we have arbitrarily selected a *pnp* transistor. We happen to have a *pnp* transistor available with $\beta_F = 80$ and $V_A = 50$ V (e.g., a 2N3906—see MCD Web Resources).

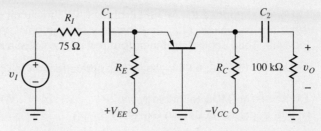

Figure 14.31   Common-base circuit topology.

Next, let us select the power supplies V_{CC} and V_{EE}. Remembering our rule-of-thumb from Chapter 13, $A_v = 10(V_{CC} + V_{EE})$. The matched input resistance situation causes a factor of two voltage loss between the signal source v_i and the emitter-base junction. Thus, an overall gain of 50 requires a value of $g_m R_L = 100$, and we estimate that a total supply voltage of 10 V is required. Using symmetrical supplies, we have $V_{CC} = V_{EE} = 5$ V.

Figures 14.32 and 14.33 are the dc and ac equivalent circuits needed to analyze the behavior of the amplifier in Fig. 14.31. Resistor R_E and the Q-point of the transistor can now be determined from the input resistance requirement. From Fig. 14.33, we recognize that the input resistance of the amplifier is equal to resistor R_E in parallel with the input resistance at the emitter of the transistor. From Table 14.6, $R_{iE} = 1/g_m$:

$$R_{\text{in}} = R_E \| R_{iE} = R_E \| \frac{1}{g_m} \tag{14.70}$$

Expanding Eq. (14.70) and using the expression for g_m yields

$$R_{\text{in}} = \frac{\dfrac{1}{g_m} R_E}{\dfrac{1}{g_m} + R_E} = \frac{R_E}{1 + g_m R_E} = \frac{R_E}{1 + 40 I_C R_E} \tag{14.71}$$

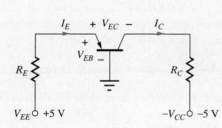

Figure 14.32 dc Equivalent circuit for common-base amplifier.

Figure 14.33 ac Equivalent circuits for the common-base amplifier.

Since $I_E \cong I_C$, the $I_C R_E$ product in Eq. (14.71) represents the dc voltage developed across the resistor R_E. Here again we see the direct coupling between the small-signal input resistance and the dc Q-point values. From the dc equivalent circuit in Fig. 14.32 and assuming $V_{EB} = 0.7$ V,

$$I_C R_E \cong I_E R_E = V_{EE} - V_{BE} = 5 - 0.7 = 4.3 \text{ V} \tag{14.72}$$

Combining Eqs. (14.71) and (14.72) with the input resistance specification,

$$75 = \frac{R_E}{1 + 40(4.3)} \quad \text{and} \quad R_E = 13.0 \text{ k}\Omega \tag{14.73}$$

I_C can now be found using Eq. (14.73):

$$I_C \cong I_E = \frac{4.3 \text{ V}}{13 \text{ k}\Omega} = 331 \text{ }\mu\text{A} \tag{14.74}$$

It is interesting to note that once V_{EE} was chosen for this circuit, R_E and I_C were both indirectly fixed.

The next step in the design is to choose collector resistor R_C. For the circuit in Fig. 14.33, the gain is

$$A_v^{CB} = g_m R_L \left(\frac{R_{\text{in}}}{R_I + R_{\text{in}}} \right) \tag{14.75}$$

For our circuit,

$$R_{\text{in}} = 75\ \Omega \qquad g_m = 40I_C = 40(331\ \mu\text{A}) = 13.2\ \text{mS} \tag{14.76}$$

$$R_L = R_C \| 100\ \text{k}\Omega$$

Solving for R_L in Eq. (14.75) yields

$$50 = (13.2\ \text{mS})\ R_L\left(\frac{75}{75 + 75}\right) \qquad \text{and} \qquad R_L = 7.58\ \text{k}\Omega \tag{14.77}$$

Since $R_L = R_C \| 100\ \text{k}\Omega$, $R_C = 8.20\ \text{k}\Omega$.

The next step is to finish checking the Q-point of the transistor by calculating V_{EC}. Using the circuit in Fig. 14.32,

$$V_{EB} = V_{EC} + I_C R_C - 5 \tag{14.78}$$

and solving for V_{EC} yields

$$V_{EC} = 5 + V_{EB} - I_C R_C = 5 + 0.7 - (331\ \mu\text{A})(8.20\ \text{k}\Omega) = 2.99\ \text{V} \tag{14.79}$$

V_{EC} is positive and greater than 0.7 V, so the *pnp* transistor is operating in the active region, as required.

The final step in this design is to select values for the coupling capacitors. We desire the impedance of the capacitors for frequencies of 500 kHz and above to be negligible with respect to the resistance that appears at their terminals. The resistance looking to the left from C_1 is 75 Ω, and the resistance looking to the right is R_{in}, which is also 75 Ω. Therefore,

$$\frac{1}{2\pi(500\ \text{kHz})C_1} \ll 150\ \Omega \qquad \text{or} \qquad C_1 \gg 2.12\ \text{nF}$$

For C_2, the resistance looking back toward the collector is at most 8.2 kΩ, and the resistance looking toward the right is 100 kΩ. Therefore,

$$\frac{1}{2\pi(500\ \text{kHz})C_2} \ll 108\ \text{k}\Omega \qquad \text{or} \qquad C_2 \gg 2.95\ \text{pF}$$

Let us choose $C_1 = 0.022\ \mu\text{F}$ and $C_2 = 33$ pF, which are standard values that are larger than the calculated values by a factor of at least 10.

The completed design is shown in Fig. 14.34, in which the nearest 5 percent values have been used for the resistors. This amplifier provides a gain of approximately 50 and an input resistance of approximately 75 Ω.

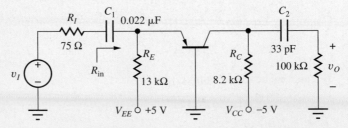

Figure 14.34 Final design for amplifier with $R_{\text{in}} = 75\ \Omega$ and $A_v = 50$.

One serious limitation of this amplifier design is its signal-handling ability. Only 5 mV can appear across the emitter-base junction, which sets a limit on the signal $\mathbf{v_i}$:

$$\mathbf{v_{eb}} = \mathbf{v_i}\,\frac{R_{\text{in}}}{R_I + R_{\text{in}}} = \mathbf{v_i}\,\frac{75\ \Omega}{75\ \Omega + 75\ \Omega} = \frac{\mathbf{v_i}}{2} \tag{14.80}$$

Thus, for small-signal operation to be valid, the magnitude of the input signal v_i must not exceed 10 mV.

Check of Results: At this point, an excellent way to check the design is to simulate the circuit in SPICE, which yields a Q-point of (323 μA, 3.09 V). The frequency response results appear in the figure here.

Discussion: In this design, we were lucky that we remembered to account for the factor of 2 loss in Eq. (14.80) due to the matched resistance condition at the input. Otherwise, our initial choice of power supplies might not have been sufficient to meet the gain specification, and a second design iteration could have been required. The signal handling capability of this stage is small. If an FET were used in place of the bipolar transistor, a much higher input range could be achieved.

Computer-Aided Analysis: The frequency response generated by SPICE with v_I as the input appears below. The simulation parameters are FSTART $= 1000$ Hz and FSTOP $= 10$ MHz with 10 frequency points per decade. The midband voltage gain is found to be 33.5 dB, and $f_L = 72$ kHz.

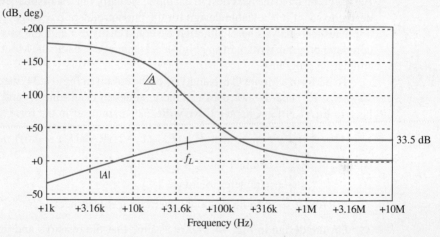

Common-base amplifier frequency response.

EXERCISE: Suppose the resistors and power supplies in the circuit in Fig. 14.34 all have 5 percent tolerances. Will the BJT remain in the active region in the worst-case situation? Repeat for tolerances of 10 percent. Do the values of current gain β_F or V_A have any significant effect on the design? Discuss.

ANSWERS: Yes; yes; no, not unless they become very small.

EXERCISE: (a) Create a two-port model (Fig. 10.7) for the midband region of the common-base amplifier in Fig. 14.34. (b) Use the model to calculate the voltage gain with the 100-kΩ load attached to the amplifier.

ANSWERS: (a) $R_{in} = 76.0\ \Omega$, $A = +107$, $R_{out} = 8200\ \Omega$; (b) $+49.8$

14.7.1 MONTE CARLO EVALUATION OF THE COMMON-BASE AMPLIFIER DESIGN

Before going on to the next design example, we carry out a statistical evaluation of the common-base design to see if it is a viable design for the mass production of large numbers of amplifiers. We use a spreadsheet analysis here, although we could easily evaluate the same equation set using a simple computer program written in any high-level language or using the Monte Carlo option in some circuit simulation programs.

To perform a Monte Carlo analysis of the circuit in Fig. 14.34, we assign random values to V_{CC}, V_{EE}, R_C, R_E, and β_F; we then use these values to determine I_C and V_{EC}, R_{in}, and A_v. Referring back to Eq. (1.50) in Chapter 1, we write each parameter in the form

$$P = P_{nom}(1 + 2\varepsilon(\text{RAND}(\ ) - 0.5)) \tag{14.81}$$

where $P_{nom} = $ nominal value of parameter

$\varepsilon = $ parameter tolerance

RAND( ) $ = $ random-number generator in spreadsheet

For the design in Fig. 14.34, we assume that the resistors and power supplies have 5 percent tolerances and the current gain has a ± 25 percent tolerance. As mentioned in Chapter 1 and Ex. 5.13, it is important that each variable invoke a separate evaluation of the random-number generator so that the random values are independent of each other. The random-element values are then used to characterize the Q-point, R_{in}, and A_v. The expressions for the Monte Carlo analysis are presented in a logical sequence for evaluation in Eqs. (14.82):

$$
\begin{aligned}
&1.\quad V_{CC} = 5(1 + 0.1(\text{RAND}(\ ) - 0.5))\\
&2.\quad V_{EE} = 5(1 + 0.1(\text{RAND}(\ ) - 0.5))\\
&3.\quad R_E = 13{,}000(1 + 0.1(\text{RAND}(\ ) - 0.5))\\
&4.\quad R_C = 8200(1 + 0.1(\text{RAND}(\ ) - 0.5))\\
&5.\quad \beta_F = 80(1 + 0.5(\text{RAND}(\ ) - 0.5))\\
&6.\quad I_C = \frac{V_{EE} - 0.7}{R_E}\\
&7.\quad V_{EC} = 0.7 + V_{CC} - I_C R_C\\
&8.\quad g_m = 40 I_C\\
&9.\quad R_{in} = R_E \left\|\frac{\alpha_o}{g_m}\right.\\
&10.\quad A_v = g_m R_L \frac{R_{in}}{R_I + R_{in}} \qquad \text{where } R_L = R_C \| 100\ \text{k}\Omega
\end{aligned}
\tag{14.82}
$$

Table 14.15 summarizes the results of a 1000-case analysis. The transistor is always in the active region. The mean collector current of 331 μA corresponds closely to the nominal value for

TABLE 14.15
Monte Carlo Analysis of the Common-Base Amplifier Design

CASE #	V_{CC} (1)	V_{EE} (2)	R_E (3)	R_C (4)	β_F (5)	I_C (6)	V_{EC} (7)	g_m (8)	R_{in} (9)	A_v (10)
1	4.932	5.090	13602	8461	96.02	3.23E-04	2.902	1.29E-02	76.2	50.8
2	4.951	5.209	12844	8208	93.01	3.51E-04	2.769	1.40E-02	70.1	51.4
3	4.844	4.759	13418	8440	98.33	3.03E-04	2.990	1.21E-02	81.3	49.0
4	4.787	5.162	13193	8294	72.82	3.38E-04	2.682	1.35E-02	72.5	50.9
5	5.073	5.181	12358	8542	79.30	3.63E-04	2.676	1.45E-02	67.7	54.2
⋮										
996	4.863	5.058	12453	8134	68.56	3.50E-04	2.716	1.40E-02	70.0	50.8
997	5.157	5.016	12945	8225	98.03	3.33E-04	3.115	1.33E-02	73.8	50.3
998	4.932	5.183	12458	8211	78.17	3.60E-04	2.677	1.44E-02	68.2	52.0
999	5.034	4.940	13444	7969	76.71	3.15E-04	3.221	1.26E-02	77.8	47.4
1000	5.119	5.002	12948	7892	95.25	3.32E-04	3.196	1.33E-02	74.0	48.3
Mean	5.006	4.997	12992	8205	79.95	3.31E-04	2.990	1.32E-02	74.29	49.88
std. dev.	0.143	0.146	381	239	11.27	1.44E-05	0.199	5.75E-04	3.22	1.74
min.	4.750	4.751	12351	7792	60.04	2.97E-04	2.409	1.19E-02	66.85	45.36
max.	5.248	5.250	13650	8609	99.98	3.67E-04	3.613	1.47E-02	82.54	54.63

(X) = equation number in Equation Set (14.82).

the standard 5 percent resistors that were selected for the final circuit. The mean values of R_{in} and A_v are 74.3 Ω and 49.9, respectively, and are also quite close to the design value. The 3σ limit corresponds to only slightly more than 10 percent deviation from the nominal design specification, and even the worst observed cases of R_{in} yield acceptable values of SWR (standing wave ratio) on the transmission line that the amplifier was designed to match. Overall, we should be able to mass produce this design and have few problems meeting the specifications.

DESIGN EXAMPLE 14.11   A COMMON-SOURCE AMPLIFIER

Let us now try to meet the requirements of the previous design using a C-E/C-S design.

PROBLEM   Design an amplifier to match a 75-Ω source resistance (for example, a coaxial transmission line) and to provide a voltage gain of 34 dB. Design the capacitors to have negligible impact on the circuit for frequencies above 500 kHz.

SOLUTION   **Known Information and Given Data:** Amplifier input resistance = 75 Ω; voltage gain = 50 (34 dB); frequency of application of amplifier is 500 kHz and above

Unknowns: Amplifier topology; Q-point; circuit element values; transistor parameters

Approach: Use overall specifications to guide choice of circuit topology and transistor type; then choose circuit element values to meet numeric requirements. Although the input resistance of the C-E and C-S amplifiers is usually considered in the moderate to high range, we can always limit it by reducing the size of the resistors in the bias network. For example, consider the common-source amplifier in Fig. 14.35. If the gate-bias resistor R_G is reduced to 75 Ω, then the input resistance of the amplifier will also be 75 Ω. (This design technique is sometimes referred to as **swamping** of the impedance level.) A BJT could also be used, but a MOSFET has been chosen because it offers the potential of a higher signal-handling capability and simple bias circuit design.

Assumptions: The transistor is in the active region. Small-signal conditions apply.

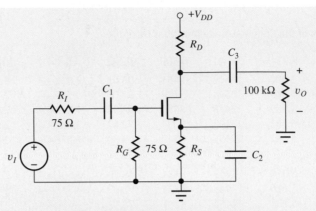

Figure 14.35 Common-source amplifier.

Analysis: If resistor R_S is bypassed, this amplifier yields the full gain $-g_m R_L$, but the matched input causes a loss of input signal by a factor of 2:

$$\mathbf{v_{gs}} = \mathbf{v_i} \frac{R_G}{R_I + R_G} = \mathbf{v_i} \frac{75\ \Omega}{75\ \Omega + 75\ \Omega} = \frac{\mathbf{v_i}}{2} \tag{14.83}$$

Thus, the prototype amplifier must deliver a gain of 100 for the overall amplifier to have a gain of 50. (This was also the case for the C-B amplifier designed in Design Ex. 14.10.) Referring back to Table 14.10 on page 782, we find that our design guide for the voltage gain of the common-source amplifier is

$$A_v = \frac{V_{DD}}{V_{GS} - V_{TN}} \tag{14.84}$$

Here again we have a single constraint equation with two variables; Table 14.16 presents some possible design choices. Let us choose the 20 V/0.2 V option.

Because $V_{GS} - V_{TN}$ must be small in order to achieve high gain, a MOSFET with a large K_n or K_p must be chosen if I_D is to be a reasonable current. Let us assume that we have found an n-channel depletion-mode MOSFET with $K_n = 10$ mS/V and $V_{TN} = -2$ V. With these parameters, the MOSFET drain current will be

$$I_D = \frac{K_n}{2}(V_{GS} - V_{TN})^2 = \frac{0.01}{2}(0.2)^2 = 0.200\text{ mA} \tag{14.85}$$

With reference to the dc equivalent circuit in Fig 14.36(a), we can now calculate the value of R_S. Because the gate current is zero for the FET, the voltage developed across R_S equals $-V_{GS}$:

$$R_S = \frac{-V_{GS}}{I_D} = \frac{-(V_{TN} + 0.2\text{ V})}{0.200\text{ mA}} = \frac{1.8\text{ V}}{0.200\text{ mA}} = 9.00\text{ k}\Omega \tag{14.86}$$

The gain of the amplifier is

$$A_v = \frac{v_{gs}}{v_i}(-g_m R_L) = -\frac{g_m R_L}{2} \qquad \text{where} \qquad R_L = R_D \| 100\text{ k}\Omega \tag{14.87}$$

Setting Eq. (14.87) equal to 50 and solving for R_L yields

$$R_L = \frac{2A_v}{g_m} = \frac{A_v(V_{GS} - V_P)}{I_D} = \frac{50(0.2\text{ V})}{0.2\text{ mA}} = 50\text{ k}\Omega \tag{14.88}$$

For $R_L = 50$ kΩ, R_D must be 100 kΩ.

Now we have encountered a problem. A drain current of 0.2 mA in $R_D = 100$ kΩ requires a voltage drop equal to the total power supply voltage of 20 V. Thus, the power supply voltage must be increased. For active region operation, $V_{DS} \geq V_{GS} - V_{TN}$, where

$$V_{DS} = V_{DD} - I_D R_D - I_D R_S \tag{14.89}$$

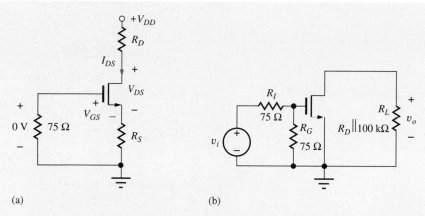

Figure 14.36 (a) dc and (b) ac equivalent circuits for the common-source amplifier.

Therefore,

$$V_{DD} - 20 - 1.8 \geq (-1.8) - (-2) \quad \text{or} \quad V_{DD} \geq 22 \text{ V} \tag{14.90}$$

is sufficient to ensure pinch-off operation. Let us choose $V_{DD} = 25$ V to provide additional design margin and room for additional signal voltage swing at the drain.

The final step in this design is to select values for the coupling capacitors. We desire the impedance of the capacitors for frequencies of 500 kHz and above to be negligible with respect to the resistance that appears at their terminals. The resistance looking to the left from C_1 is 75 Ω, and the input resistance looking to the right is also 75 Ω. Therefore,

$$\frac{1}{2\pi (500 \text{ kHz}) C_1} \ll 150 \ \Omega \quad \text{or} \quad C_1 \gg 2.12 \text{ nF}$$

For C_2, the resistance looking back toward the source is 9.1 kΩ in parallel with $(1/g_m)$ looking into the source of the transistor:

$$R_{eq} = 9.1 \text{ k}\Omega \left\| \frac{1}{g_m} = 9.1 \text{ k}\Omega \right\| \frac{1}{2 \text{ mS}} = 474 \ \Omega$$

Therefore,

$$\frac{1}{2\pi (500 \text{ kHz}) C_2} \ll 474 \ \Omega \quad \text{or} \quad C_2 \gg 644 \text{ pF}$$

For C_3, the resistance looking back toward the drain is 100 kΩ, and the resistance looking toward the right is also 100 kΩ. Therefore,

$$\frac{1}{2\pi (500 \text{ kHz}) C_3} \ll 200 \text{ k}\Omega \quad \text{or} \quad C_3 \gg 1.59 \text{ pF}$$

Let us choose $C_1 = 0.022 \ \mu\text{F}$, $C_2 = 0.0068 \ \mu\text{F}$, and $C_3 = 20$ pF, which are standard values that are larger than the calculated values by a factor of approximately 10. The circuit corresponding to the final amplifier design is in Fig. 14.37, where standard 5 percent resistor values have once again been selected.

Check of Results: At this point, an excellent way to check the design is to simulate the circuit in SPICE, which yields a Q-point of (198 μA, 3.41 V) with a gain of 33.9 dB and $f_L = 91.5$ kHz. The frequency response appears below.

Discussion: The designs in Design Exs. 14.10 and 14.11 demonstrate that usually more than one, often very different, design approaches can meet the specifications for a given problem. Choosing one design over another depends on many factors. For example, one criterion could be the use of power supply voltages that are already available in the rest of the system. Total power consumption

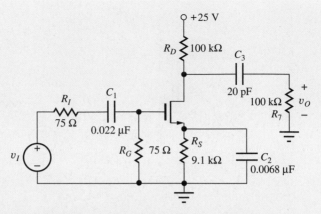

Figure 14.37 Final common-source amplifier design.

might be an important issue. Our common-base design uses a power of approximately 3.3 mW and uses two power supplies, whereas the common-source design consumes 5 mW from a single 25-V supply. In actuality, the design is somewhat of a struggle using the FET. A large power supply voltage is combined with a FET operating very near cutoff. It may be difficult to find a FET with $K_n = 10$ mA with $V_P = -2$ V.

Another important factor could be amplifier cost. The core of the FET amplifier requires three resistors, R_D, R_G, and R_S; bypass capacitor C_2; and the JFET. The common-base amplifier core requires resistors R_E and R_C and the BJT. The cost of the additional parts, plus the expense of inserting them into a printed circuit board (often more expensive than the parts themselves), will probably tilt the economic decision away from the C-S design toward the C-B amplifier. However, the maximum input signal capability of the JFET amplifier, $|v_i| = 2 \times 0.2(V_{GS} - V_P) = 0.08$ V can be of overriding importance in certain applications. Obviously, the final decision involves many factors.

Computer-Aided Analysis: As already noted, the frequency response generated by SPICE with v_S as the input appears in the graph. The simulation parameters are FSTART = 1000 Hz and FSTOP = 10 MHz with 10 frequency points per decade. The midband voltage gain is 33.9 dB and $f_L = 91.5$ kHz. Based on Table 14.12, three poles at 50 Hz are expected to produce $f_L = 97.5$ Hz.

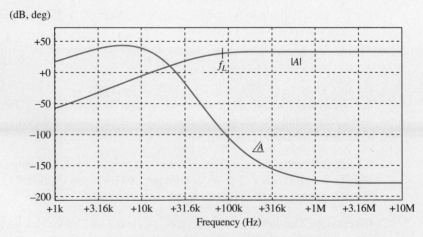

Frequency response of the common-source amplifier.

EXERCISE: Verify the results of the SPICE simulation of Design Ex. 14.11. What is the bandwidth predicted using the bandwidth shrinkage factor in Eq. (14.58)?

ANSWER: 89 Hz (using the average of the pole frequencies)

EXERCISE: Suppose the JFET chosen for the circuit in Fig. 14.37 also had $\lambda = 0.015$ V^{-1}. What are the values of r_o and the new voltage gain? (Use the Q-point values from the example.) Does neglecting the output resistance seem a reasonable thing to do?

ANSWERS: 333 kΩ, 43.5; No, r_o is important in this circuit! We will not achieve the desired gain with this FET.

EXERCISE: (a) Redesign the circuit using the 25 V/0.25 V case from Table 14.16. Use the same FET device parameters. (b) Verify your design with SPICE.

ANSWERS: 5.60 kΩ, 68 kΩ, $V_{DD} = 25$ V, 0.022 μF, 8200 pF, 20 pF

EXERCISE: (a) Create a two-port model (Fig. 10.7) for the midband region of the common-source amplifier in Fig. 14.37. (b) Use the model to calculate the voltage gain with the 100-kΩ load attached to the amplifier.

ANSWERS: (a) $R_{in} = 75$ Ω, $A = 200$, $R_{out} = 100$ kΩ; (b) 50.0

14.8 MULTISTAGE ac-COUPLED AMPLIFIERS

In most situations, a single-transistor amplifier cannot meet all the specifications of a given amplifier design. The required voltage gain often exceeds the amplification factor of a single transistor, or the required combination of voltage gain, input resistance, and output resistance cannot be met simultaneously. For example, consider the specifications of a good general purpose operational amplifier having an input resistance exceeding 1 MΩ, a voltage gain of 100,000, and an output resistance less than a few hundred ohms. It is clear from our investigation of amplifiers in this chapter that these requirements cannot be met with a single-transistor amplifier. A number of stages must be cascaded in order to create an amplifier that can meet all the requirements.

14.8.1 A THREE-STAGE ac-COUPLED AMPLIFIER

In this section, we study the three-stage ac-coupled amplifier in Fig. 14.38. Signals are coupled from one stage to the next through the use of coupling capacitors C_1, C_3, C_5 and C_6, whereas the same capacitors provide dc isolation between stages that permits independent design of the bias circuitry of the individual stages.

The function of the various stages can more readily be seen in the midband ac equivalent circuit for this amplifier in Fig. 14.39(a) in which all the capacitors have been replaced with short circuits. MOSFET M_1, operating in the common-source configuration, provides a high input resistance with modest voltage gain. Bipolar transistor Q_2 in the common-emitter configuration provides a second stage with high voltage gain. Q_3, an emitter follower, provides a low output resistance and buffers the high-gain stage, Q_2, from the relatively low load resistance (250 Ω). In Fig. 14.39(a), the base bias resistors have been replaced by $R_{B2} = R_1 \| R_2$ and $R_{B3} = R_3 \| R_4$.

In the amplifier in Fig. 14.38, the input and output of the overall amplifier are ac-coupled through capacitors C_1 and C_6. Bypass capacitors C_2 and C_4 are used to obtain maximum voltage gain from the two inverting amplifier stages. Interstage coupling capacitors C_3 and C_5 transfer the ac signals between the amplifiers but provide isolation at dc. Thus, the individual Q-points of the transistors are not affected by connecting the stages together. Figure 14.39(b) gives the dc equivalent circuit for the amplifier in which the capacitors have all been removed. The isolation of the three individual transistor amplifier stages is apparent in this figure.

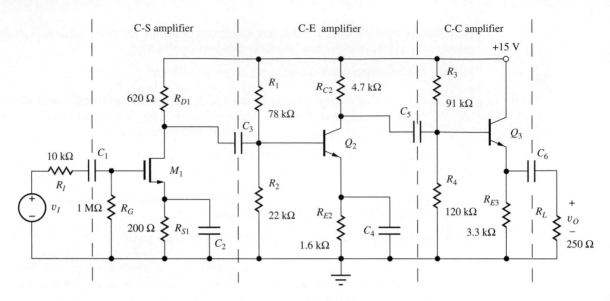

Figure 14.38 Three-stage ac-coupled amplifier.

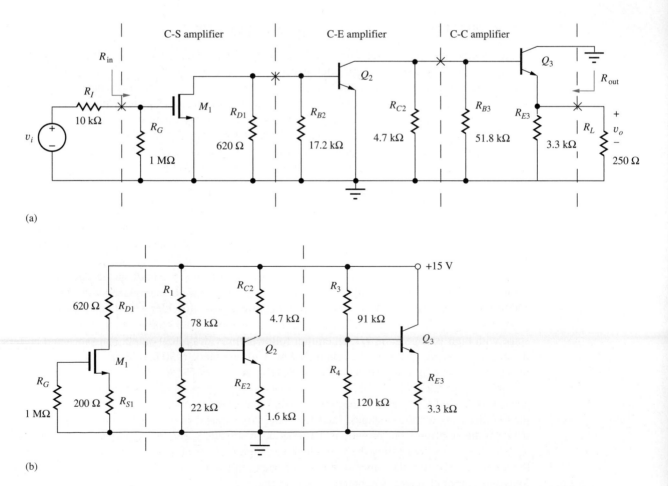

(a)

(b)

Figure 14.39 (a) Equivalent circuit for ac analysis. (b) dc Equivalent circuit for the three-stage ac-coupled amplifier.

TABLE 14.17
Transistor Parameters for Figs. 14.38–14.43

M_1	$K_n = 10 \text{ mA/V}^2$, $V_{TN} = -2 \text{ V}$, $\lambda = 0.02 \text{ V}^{-1}$
Q_2	$\beta_F = 150$, $V_A = 80 \text{ V}$, $V_{BE} = 0.7 \text{ V}$
Q_3	$\beta_F = 80$, $V_A = 60 \text{ V}$, $V_{BE} = 0.7 \text{ V}$

TABLE 14.18
Q-Points and Small-Signal Parameters for the Transistors in Fig. 14.39

	Q-POINT VALUES	SMALL-SIGNAL PARAMETERS
M_1	(5.00 mA, 10.9 V)	$g_{m1} = 10.0 \text{ mS}$, $r_{o1} = 12.2 \text{ k}\Omega$
Q_2	(1.57 mA, 5.09 V)	$g_{m2} = 62.8 \text{ mS}$, $r_{\pi2} = 2.39 \text{ k}\Omega$, $r_{o2} = 54.2 \text{ k}\Omega$
Q_3	(1.99 mA, 8.36 V)	$g_{m3} = 79.6 \text{ mS}$, $r_{\pi3} = 1.00 \text{ k}\Omega$, $r_{o3} = 34.4 \text{ k}\Omega$

We want to characterize this amplifier by determining its voltage, input and output resistances, current and power gains, and input signal range using the transistor parameters in Table 14.17. We will also estimate the lower cutoff frequency of the amplifier. First, the Q-points of the three transistors must be found. Each transistor stage in Fig. 14.39 is independently biased, and, for expediency, we assume that the Q-points listed in Table 14.18 have already been found using the dc analysis procedures developed in previous chapters. The details of these dc calculations are left for the next exercise.

EXERCISE: Verify the values of the Q-points and small-signal parameters in Table 14.18.

EXERCISE: Why can't a single transistor amplifier meet the op amp specifications mentioned in the introduction to this chapter?

14.8.2 VOLTAGE GAIN

The ac equivalent circuit for the three-stage amplifier example has been redrawn and is shown in simplified form in Fig. 14.40, in which the three sets of parallel resistors have been combined into the following: $R_{I1} = 620 \ \Omega \| 17.2 \text{ k}\Omega = 598 \ \Omega$, $R_{I2} = 4.7 \text{ k}\Omega \| 51.8 \text{ k}\Omega = 4.31 \text{ k}\Omega$, and $R_{L3} = 3.3 \text{ k}\Omega \| 250 \ \Omega = 232 \ \Omega$. The voltage gain of the overall amplifier can be expressed as

$$A_v = \frac{\mathbf{v_o}}{\mathbf{v_i}} = \left(\frac{\mathbf{v_o}}{\mathbf{v_3}} \right) \left(\frac{\mathbf{v_3}}{\mathbf{v_2}} \right) \left(\frac{\mathbf{v_2}}{\mathbf{v_1}} \right) \left(\frac{\mathbf{v_1}}{\mathbf{v_s}} \right) = A_{vt1} A_{vt2} A_{vt3} \left(\frac{\mathbf{v_1}}{\mathbf{v_i}} \right) \qquad (14.91)$$

where

$$\frac{\mathbf{v_1}}{\mathbf{v_i}} = \frac{R_{\text{in}}}{R_I + R_{\text{in}}} = \frac{R_G}{R_I + R_G} \qquad (14.92)$$

We see that the overall voltage gain is determined by the product of the individual terminal gains of the three amplifier stages, as well as the signal voltage loss across the source resistance.

We use our knowledge of single-transistor amplifiers, gained in Chapters 13 and 14, to determine expressions for the three voltage gains. The first stage is a common-source amplifier with a terminal gain

$$A_{vt1} = \frac{\mathbf{v_2}}{\mathbf{v_1}} = -g_{m1} R_{L1} \qquad (14.93)$$

in which R_{L1} represents the total load resistance[8] connected to the drain of M_1. From the ac circuit in Fig. 14.40(a) and the small-signal version in (b), we can see that R_{L1} is equal to the parallel

---

[8] The output resistances r_{o1}, r_{o2}, and r_{o3} are neglected because each amplifier has an external resistor as a load, and we expect $|A_v| \ll \mu_f$ for each stage.

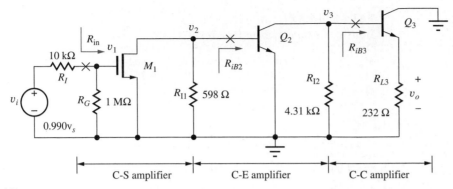

(a)

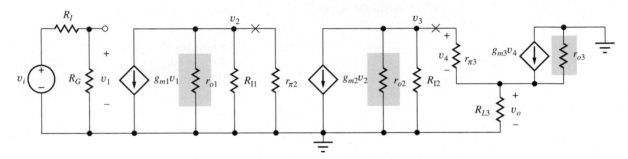

(b)

Figure 14.40 (a) Simplified ac equivalent circuit for the three-stage amplifier. (b) Small-signal equivalent circuit for the three-stage amplifier. Resistances r_{o1}, r_{o2}, and r_{o3} are neglected in the calculations.

combination of R_{I1} and R_{iB2}, the input resistance at the base of Q_2. Because Q_2 is a common-emitter stage with zero emitter resistance, $R_{iB2} = r_{\pi 2}$,

$$R_{L1} = 598\ \Omega \| r_{\pi 2} = 598\ \Omega \| 2390\ \Omega = 478\ \Omega \tag{14.94}$$

and the gain of the first stage is

$$A_{vt1} = \frac{\mathbf{v_2}}{\mathbf{v_1}} = -0.01\ \text{S} \times 478\ \Omega = -4.78 \tag{14.95}$$

The terminal gain of the second stage is that of a common-emitter amplifier:

$$A_{vt2} = \frac{\mathbf{v_3}}{\mathbf{v_2}} = -g_{m2}R_{L2} \tag{14.96}$$

in which R_{L2} represents the total load resistance connected to the collector of Q_2. In Fig. 14.41, R_{L2} is equal to the parallel combination of R_{I2} and R_{iB3}, where R_{iB3} represents the input resistance of Q_3. Q_3 is an emitter follower with $R_{iB3} = r_{\pi 3}(1 + g_{m3}R_{L3})$. Thus, R_{L2} is equal to

$$R_{L2} = R_{I2}\|[r_{\pi 3} + (\beta_{o3} + 1)R_{L3}] = 4310\ \Omega\|1000\ \Omega[1 + 79.6\ \text{mS}(232\ \Omega)] = 3.53\ \text{k}\Omega \tag{14.97}$$

and the gain of the second stage is

$$A_{vt2} = -62.8\ \text{mS} \times 3.53\ \text{k}\Omega = -222 \tag{14.98}$$

Finally, the terminal gain of the emitter follower stage is

$$A_{vt3} = \frac{\mathbf{v_o}}{\mathbf{v_3}} = \frac{g_{m3}R_{L3}}{1 + g_{m3}R_{L3}} = \frac{(79.6\ \text{mS})(232\ \Omega)}{1 + 79.6\ \text{mS}(232)\ \Omega} = 0.950 \tag{14.99}$$

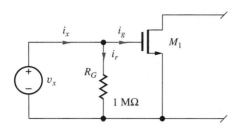

Figure 14.41 Input resistance of the three-stage amplifier.

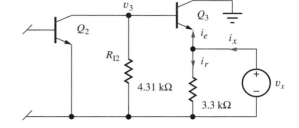

Figure 14.42 Output resistance of the three-stage amplifier.

Before we can complete the voltage gain calculation in Eq. (14.91), we must find input resistance R_{in} in order to evaluate the ratio v_1/v_i given in Eq. (14.92).

14.8.3 INPUT RESISTANCE

The input resistance R_{in} of this amplifier can be determined by referring to Figs. 14.39 through 14.41. Because the current i_g in Fig. 14.41 is zero, we see that the resistance presented to source v_x is simply $R_{\text{in}} = R_G = 1\ \text{M}\Omega$. Note that this result is independent of the circuitry connected to the source or drain of M_1.

14.8.4 SIGNAL SOURCE VOLTAGE GAIN

Substituting the voltage gains and resistance values into Eqs. (14.91) and (14.92) gives the voltage gain for the overall amplifier:

$$A_v = A_{vt1}A_{vt2}A_{vt3}\frac{R_{\text{in}}^{CS}}{R_I + R_{\text{in}}^{CS}} = (0.95)(-222)(-4.78)\left(\frac{1\ \text{M}\Omega}{10\ \text{k}\Omega + 1\ \text{M}\Omega}\right) = +998 \quad (14.100)$$

We find that the three-stage amplifier circuit realizes a noninverting amplifier with a voltage gain of approximately 60 dB and an input resistance of 1 MΩ. Because of the high input resistance, only a small portion (1 percent) of the input signal is lost across the source resistance.

EXERCISE: Recalculate A_v, including the influence of r_{o1}, r_{o2}, and r_{o3}.

ANSWER: 903 (59.1 dB)

EXERCISE: Estimate the gain of the amplifier in Fig. 14.38 using our simple design estimates if M_1 has $V_{GS} - V_{TN} = 1$ V. What is the origin of the discrepancy?

ANSWERS: $(-15)(-150)(1) = 2250$; only 3 V is dropped across R_{D1}, whereas the estimate assumes $V_{DD}/2 = 7.5$ V. Taking this difference into account, $(2250)(3/7.5) = 900$.

EXERCISE: What is the value of A_v if the interstage resistances R_{I1} and R_{I2} could be eliminated (made ∞)? Would r_{o1}, r_{o2}, and r_{o3} be required in this case?

ANSWERS: 28,200; r_{o2} would need to be included.

14.8.5 OUTPUT RESISTANCE

The output resistance R_{out} of the amplifier is defined looking back into the amplifier at the position of coupling capacitor C_6, as indicated in Figs. 14.38 and 14.39. To find R_{out}, test voltage v_x is applied to the amplifier output as in Fig. 14.42, and we see that the output resistance of the overall amplifier is determined by the output resistance of the emitter follower in parallel with the 3300-Ω resistor.

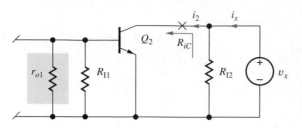

Figure 14.43 Thévenin equivalent source resistance for stage 3.

Writing this mathematically gives,

$$\mathbf{i_x} = \mathbf{i_r} + \mathbf{i_e} = \frac{\mathbf{v_x}}{3300} + \frac{\mathbf{v_x}}{R_{iE3}} \tag{14.101}$$

Using the results from Table 14.3, we find that the overall output resistance is

$$R_{\text{out}} = \frac{\mathbf{v_x}}{\mathbf{i_x}} = 3300 \| R_{iE3} \cong 3300 \left\| \left(\frac{1}{g_{m3}} + \frac{R_{\text{th}3}}{\beta_{o3}} \right) \right. \tag{14.102}$$

in which the Thévenin equivalent source resistance of stage 3, $R_{\text{th}3}$, must be found.

$R_{\text{th}3}$ can be determined with the aid of Fig. 14.43. The third stage Q_3 is removed, and test voltage v_x is applied to node v_3. Current i_x from the test source v_x is equal to

$$\mathbf{i_x} = \frac{\mathbf{v_x}}{R_{I2}} + \mathbf{i_2} = \frac{\mathbf{v_x}}{R_{I2}} + \frac{\mathbf{v_x}}{R_{iC}} \qquad \text{or} \qquad R_{\text{th}3} = \frac{\mathbf{v_x}}{\mathbf{i_x}} = R_{I2} \| R_{iC} = R_{I2} \| r_{o2} \tag{14.103}$$

$R_{\text{th}3}$ is equal to the parallel combination of interstage resistance R_{I2} and the resistance at the collector of Q_2, which we know is just equal to r_{o2}:

$$R_{\text{th}3} = 4310\ \Omega \| 54200\ \Omega = 3990\ \Omega$$

Evaluating Eq. (14.102) for the output resistance of the overall amplifier yields

$$R_{\text{out}} = 3300\ \Omega \left\| \left(\frac{1}{0.0796\ \text{S}} + \frac{3990\ \Omega}{80} \right) = 62.4\ \Omega \right. \tag{14.104}$$

14.8.6 CURRENT AND POWER GAIN

The input current delivered to the amplifier from source v_i in Fig. 14.39 is given by

$$\mathbf{i_i} = \frac{\mathbf{v_i}}{R_I + R_{\text{in}}} = \frac{\mathbf{v_i}}{10^4 + 10^6} = 9.90 \times 10^{-7} \mathbf{v_i} \tag{14.105}$$

and the current delivered to the load from the amplifier is

$$\mathbf{i_o} = \frac{\mathbf{v_o}}{250} = \frac{A_v \mathbf{v_i}}{250} = \frac{998 \mathbf{v_s}}{250} = 3.99 \mathbf{v_s} \tag{14.106}$$

Combining Eqs. (14.105) and (14.106) gives the current gain

$$A_i = \frac{\mathbf{i_o}}{\mathbf{i_i}} = \frac{3.99 \mathbf{v_i}}{9.90 \times 10^{-7} \mathbf{v_i}} = 4.03 \times 10^6 \tag{14.107}$$

Combining Eqs. (14.91) and (14.107) with the power gain expression from Chapter 10 yields a value for overall power gain of the amplifier:

$$A_P = \frac{P_o}{P_s} = \left| \frac{\mathbf{v_o}}{\mathbf{v_i}} \frac{\mathbf{i_o}}{\mathbf{i_i}} \right| = |A_v A_i| = 998 \times 4.03 \times 10^6 = 4.02 \times 10^9 \tag{14.108}$$

Because input resistance to the common-source stage is large, only a small input current is required to develop a large output current. Thus, current gain is large. In addition, the voltage gain of the amplifier is significant, and combining a large voltage gain with a large current gain yields a very substantial power gain.

14.8.7 INPUT SIGNAL RANGE

Our final step in characterizing this amplifier is to determine the largest input signal that can be applied to the amplifier. In a multistage amplifier, the small-signal assumptions must not be violated anywhere in the amplifier chain. The first stage of the amplifier in Figs. 15.39 and 15.40 is easy to check. Voltage source v_1 appears directly across the gate-source terminals of the MOSFET, and to satisfy the small-signal limit, $v_1 (= 0.990v_s)$ must satisfy

$$|v_1| \leq 0.2(V_{GS1} - V_{TN}) \quad \text{or} \quad |v_i| \leq \frac{0.2(-1+2)}{0.990} = 0.202 \text{ V} \tag{14.109}$$

The first stage limits the input signal to 202 mV.

To satisfy the small-signal requirements, the base-emitter voltage of Q_2 must also be less than 5 mV. In this amplifier, $v_{be2} = v_2$, and we have

$$|v_2| = |A_{vt1}v_1| \leq 5 \text{ mV}, \quad |v_1| \leq \frac{5 \text{ mV}}{A_{vt1}} = \frac{0.005}{4.78} = 1.05 \text{ mV} \tag{14.110}$$

and
$$|v_i| \leq \frac{1.05 \text{ mV}}{0.990} = 1.06 \text{ mV}$$

In this design, the small-signal requirements are violated at Q_2 if the amplitude of the input signal v_s exceeds 1.06 mV.

Finally, using Eq. (14.19) for the emitter-follower output stage (with $R_{th} = 0$),

$$\mathbf{v_{be3}} \cong \frac{\mathbf{v_3}}{1 + g_{m3}R_{L3}} = \frac{A_{vt1}A_{vt2}\mathbf{v_1}}{1 + g_{m3}R_{L3}} = \frac{A_{vt1}A_{vt2}(0.990\mathbf{v_s})}{1 + g_{m3}R_{L3}} \tag{14.111}$$

and requiring $|v_{be3}| \leq 5$ mV yields

$$|v_i| \leq \frac{1 + g_{m3}R_{L3}}{A_{vt1}A_{vt2}(0.990)}0.005 = \frac{1 + 0.0796 \text{ S}(232 \text{ }\Omega)}{(-4.78)(-222)(0.99)}0.005 \text{ V} = 92.7 \text{ }\mu\text{V} \tag{14.112}$$

To satisfy all the small-signal limitations, the maximum amplitude of the input signal to the amplifier must be no greater than the smallest of the three values computed in Eqs. (14.109), (14.110), and (14.112):

$$|v_i| \leq \min(202 \text{ mV}, 1.06 \text{ mV}, 92.7 \text{ }\mu\text{V}) = 92.7 \text{ }\mu\text{V} \tag{14.113}$$

In this design, output stage linearity limits the input signal amplitude to less than 93 μV. Note that the maximum output voltage that satisfies the small-signal limit is only

$$|v_o| \leq A_v(92.7 \text{ }\mu\text{V}) = 998(92.7 \text{ }\mu\text{V}) = 92.5 \text{ mV} \tag{14.114}$$

EXAMPLE 14.12 THREE-STAGE AMPLIFIER SIMULATION

Hand analysis of the three-stage amplifier is verified using SPICE simulation.

PROBLEM Use SPICE to find the midband voltage gain, input resistance, and output resistance of the amplifier in Fig. 14.38. Confirm the gain with both ac and transient analyses.

SOLUTION **Known Information and Given Data:** The original amplifier circuit appears in Fig. 14.38, and transistor parameters are given in Table 14.17.

Unknowns: A_v, R_{in}, and R_{out}

Approach: Use SPICE analysis to plot the frequency response and find the midband region. Then choose a midband frequency, and use ac analysis to find the voltage gain, input resistance, and output resistance. Assume large values for the capacitors. Verify the gain with a transient simulation.

Assumptions: The coupling and bypass capacitors are all arbitrarily set to 22 μF. Bipolar transistor parameters TF = 0.5 NS and CJC = 2 PF are added to the BJT models to cause the frequency response to roll off at high frequencies. These parameters are discussed in detail in Chapter 16.

Analysis: The circuit is created in SPICE using the schematic editor, as shown in the figure.

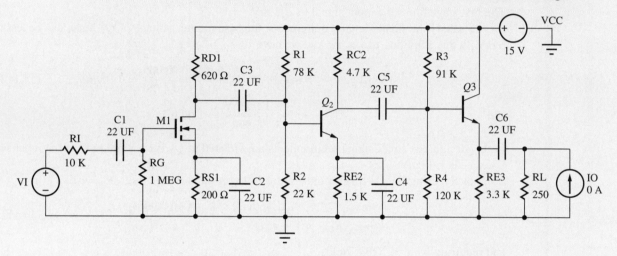

The MOSFET parameters are set to KP = 0.01 S/V, VTO = −2 V, and LAMBDA = 0.02 V^{-1}. The BJT parameters for Q_2 are set to BF = 150, VAF = 80 V, TF = 0.5 NS, and CJC = 2 PF. For Q_3, BF = 80, VAF = 60 V, TF = 0.5 NS, and CJC = 2 PF. As mentioned, TF and CJC are added to create a roll-off in the frequency response at high frequencies and will be discussed further in Chapter 16. Source VI is used for ac analysis of the voltage gain and input resistance. Source IO is an ac source used to find the output resistance.

First, we set VI = 1∠0° V and IO = 0∠0° A and perform an ac sweep from 10 Hz to 10 MHz with 20 points per decade in order to find the midband region. We obtain the response shown below.

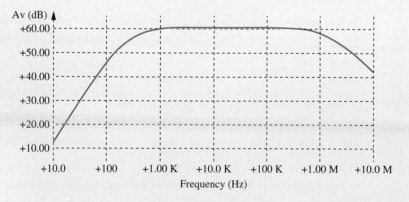

The midband region extends from approximately 500 Hz to 500 kHz. Choosing 20 kHz as a representative midband frequency, we find the gain is 60.1 dB ($A_v = 1010$), and the current in VI is −990 nA with a phase angle of 0°. The minus sign results from the sign convention in SPICE—positive current enters the positive terminal of an independent source. The input resistance presented to VI is 1 V/990 nA = 1.01 MΩ. Subtracting the 10-kΩ source resistance yields an amplifier input resistance of 1 MΩ. Both the gain and input resistance agree with our hand calculations.

The output resistance is found by setting VI = $0\angle 0°$ V and IO = $1\angle 0°$ A and finding the output voltage. The result yields $R = 45.6$ Ω. Removing the effect of the 250-Ω resistor in parallel with the output yields $R_{out} = 55.7$ Ω. The slight difference is caused by the value of current gain utilized in SPICE: $\beta_o =$ BF(1 + VCB/VA) = 80(1 + 7.6 V/60 V) = 90.1.

Check of Results: As a second check on the gain, we can run a transient simulation at $f = 20$ kHz, which we now know corresponds to a midband frequency. The graph here gives the output with an input amplitude of 100 μV, Start time = 0, Stop time = 100 US, and a time step of 0.01 US. The amplitude of the output is approximately 100 mV corresponding to a gain of 1000.

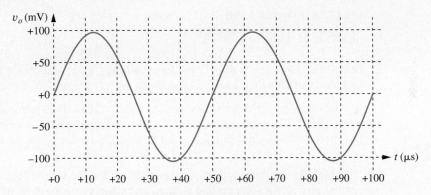

Simulation with undistorted output and gain of 1000.

Discussion: This input value in the transient simulation is just slightly above the small-signal limit that we calculated. The waveform looks like a good sine wave, and the Fourier analysis option in SPICE indicates that the total harmonic distortion in the waveform is less than 0.15 percent.

However, if one uses an input signal larger than about 650 μV in the transient solution, one discovers a new limitation. An example of the problem appears in the next figure. Because Q_3 is biased at a current of only 2 mA, the largest output signal that can be developed by the emitter follower is approximately 2 mA × 250 Ω or 0.5 V. The output will begin to show substantial distortion before this value is reached. In the figure, the amplitude of input v_I is 750 μV. The bottom of the output waveform is "clipped off," and the total harmonic distortion has increased to 8.2 percent. This output waveform is not desirable.

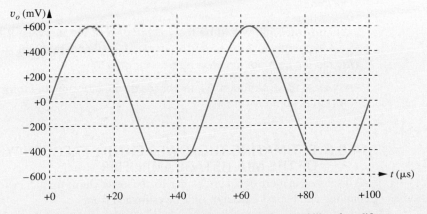

Distorted output with amplitude exceeding output voltage capability of amplifier.

TABLE 14.19
Three-Stage Amplifier Summary

	HAND ANALYSIS	SPICE RESULTS
Voltage gain	+998	+1010
Input signal range	92.7 μV	—
Input resistance	1 MΩ	1 MΩ
Output resistance	60.5 Ω	55.7 Ω
Current gain	$+4.03 \times 10^6$	—
Power gain	4.02×10^9	—

EXERCISE: Reevaluate Eq. (14.104) using a current gain of 90.1.

ANSWER: 55.3 Ω

EXERCISE: Find the output waveform, voltage gain, and total harmonic distortion if the amplitude of VI is increased to (a) 400 μV, (b) 600 μV, and (c) 1 mV.

ANSWERS: (a) Looks like a sine wave, $A_v = 826$, THD = 0.28 percent; (b) looks like a sine wave, $A_v = 790$, THD = 2.4 percent; (c) bottom of the waveform is clipped off, $A_v = 760$, THD = 18.3 percent. Note that the opponent voltage gain is dropping as the signal level increases.

Table 14.19 summarizes the characteristics for the three-stage amplifier in Fig. 14.38. The amplifier provides a noninverting voltage gain of approximately 60 dB, a high input resistance, and a low output resistance. The current and power gains are both quite large. The input signal must be kept below 92.7 μV in order to satisfy the small-signal limitations of the transistors.

14.8.8 IMPROVING AMPLIFIER VOLTAGE GAIN

We know that the gain of the C-S amplifier is inversely proportional to the square root of drain current. In this amplifier, there is no need to operate the first stage at a 5-mA bias current level, and the voltage gain of the amplifier could be increased by reducing I_{D1} while maintaining a constant voltage drop across R_{D1}. It should be possible to improve the signal range by increasing the current in the output stage and the voltage drop across R_{E3}. Another possibility is to replace Q_3 with a FET. Some gain loss might again occur in the third stage because the gain of a common-drain amplifier is typically less than that of a common-collector stage, but this could be made up by improving the gain of the first and second stages (see Probs. 14.101 and 14.105).

EXERCISE: (a) What would be the voltage gain of the amplifier if I_{D1} is reduced to 1 mA and R_{D1} is increased to 3 kΩ so that V_D is maintained constant? (b) The FET g_m decreases by $\sqrt{5}$. Why did the gain not increase by a factor of $\sqrt{5}$?

ANSWERS: 1840; although R_{D1} increases by a factor of 5, the total load resistance at the drain of M_1 does not.

14.8.9 ESTIMATING THE LOWER CUTOFF FREQUENCY OF THE MULTISTAGE AMPLIFIER

As discussed in more detail in Chapter 16, the lower cutoff frequency for an amplifier having multiple coupling and bypass capacitors can be estimated from

$$\omega_L \cong \sum_{i=1}^{n} \frac{1}{R_{iS}C_i} \tag{14.115}$$

Humbucker Guitar Pickup

Electric guitar pickups are devices that convert motion of a steel string into electrical signals. They function through an interesting interaction of materials and magnetic fields. A basic schematic of a pickup is shown below. The magnet induces a magnetization (aligning of the magnetic domains) in the steel string. When the string vibrates, this creates a moving magnetic field, which we know from Faraday's law induces a current in the wire coil located beneath the string. The signal from the wire coil is then amplified and sent to the rest of the amplification system. The coil is typically composed of extremely thin wire, with several hundred to over a thousand turns. The choice of magnet, wire material, and number of turns in the coil generates a set of compromises in frequency response and sensitivity. Guitar players often use acoustic feedback to generate a sustained note by placing the guitar near the amplifier speakers. The acoustic energy couples into the guitar, causing the string to vibrate, which generates more signal through the amplifier. Unlike the highly undesirable feedback that we often hear with poorly configured public address systems, a skilled guitarist can use acoustic feedback to create intentionally sustained notes.

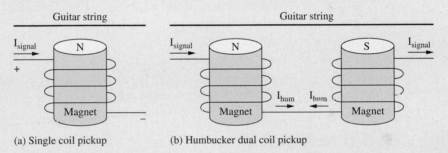

(a) Single coil pickup   (b) Humbucker dual coil pickup

Inherent with the use of the single coil pickup is sensitivity to extraneous magnetic fields. In particular, the 60 Hz power moving through most buildings gives rise to magnetic fields at the same frequency. As a result, the guitar pickup coil will generate the desired string vibration signal as well as a 60 Hz signal commonly referred to as hum.

To eliminate hum, one has to make two important observations: First, the polarity of the string vibration signal is a function of both the magnetization polarity of the string and the orientation of the coil relative to the string. Second, the polarity of the undesired hum signal is a function of only the orientation of the coil to the hum-producing external magnetic field.

Making use of these two observations, the humbucker pickup shown above was created. A second pickup coil has been added in series with the first. In the second coil, the orientation of the magnet has been reversed, resulting in the reversing of the string magnetization in the area above the coil. Additionally, the orientation of the second coil with respect to the string has also been reversed. The result is a system where the string vibration signal of the two coils has the same polarity and is additive, while the hum signal, dependent only on the coil orientation, is of opposite sign in the two coils and is cancelled out.

The humbucker coil is an example of excellent sensor design. Recognizing the unique characteristics of the desired versus the undesired signals allowed the designers to implement a sensor that rejects everything but the signal of interest. Rejecting unwanted signals at the sensor is almost always preferred to attempting to reject undesired signals in post-processing after detection and amplification.

Much of this material is attributable to the excellent guitar building website: http://groups.msn.com/GuitarBuilding/

in which R_{iS} represents the resistance at the terminals of the ith capacitor with all the other capacitors replaced by short circuits. The product $R_{iS}C_i$ represents the short circuit time constant associated with capacitor C_i. Let us now use this method to estimate the lower cutoff frequency of the three-stage amplifier in Ex. 14.12.

$$C_1: R_{1S} = R_I + R_G = 1.01\,\text{M}\Omega$$

$$C_2: R_{2S} = R_{S1}\|R_{iS1} = R_{S1}\|\frac{1}{g_{m1}} = 200\,\Omega\|\frac{1}{0.01S} = 66.7\,\Omega$$

$$C_3: R_{3S} = R_{D1} + R_{I1}\|R_{iB2} = R_{D1} + R_{I1}\|r_{\pi2} = 620\,\Omega + 17.2\,\text{k}\Omega\|2.39\,\text{k}\Omega = 2.72\,\text{k}\Omega$$

$$C_4: R_{4S} = R_{E2}\|R_{iE2} = R_{E2}\|\frac{r_{\pi2} + R_{th2}}{\beta_{o2} + 1} = 1.5\,\text{k}\Omega\|\frac{2.39\,\text{k}\Omega + (17.2\,\text{k}\Omega\|620\,\Omega)}{151} = 19.2\,\Omega$$

$$C_5: R_{3S} = R_{C2} + R_{I2}\|R_{iB3} = R_{C2} + R_{I2}\|r_{\pi3}(1 + g_{m3}R_{L3})$$
$$= 4.7\,\text{k}\Omega + 51.8\,\text{k}\Omega\|1.0\,\text{k}\Omega[1 + 0.0796S(232\,\Omega)] = 18.9\,\text{k}\Omega$$

$$C_6: R_{4S} = R_L + R_{E3}\|R_{iE3} = R_L + R_{E3}\|\frac{r_{\pi3} + R_{th3}}{\beta_{o3} + 1}$$
$$= 250\,\Omega + 3.3\,\text{k}\Omega\|\frac{1.0\,\text{k}\Omega + (51.8\,\text{k}\Omega\|4.7\,\text{k}\Omega)}{81} = 315\,\Omega$$

$$f_L \cong \frac{1}{2\pi}\left[\frac{1}{1.01\,\text{M}\Omega\,(22\,\mu\text{F})} + \frac{1}{66.7\,\Omega\,(22\,\mu\text{F})} + \frac{1}{2.72\,\text{k}\Omega\,(22\,\mu\text{F})} + \frac{1}{19.2\,\Omega\,(22\,\mu\text{F})}\right.$$
$$\left. + \frac{1}{18.9\,\text{k}\Omega\,(22\,\mu\text{F})} + \frac{1}{315\,\Omega\,(22\,\mu\text{F})}\right]$$

$$f_L \cong 511\,\text{Hz}$$

The $f_L = 511$ Hz estimate obtained using the short circuit time constant approach agrees very well with the SPICE simulation results presented in Ex. 14.12.

SUMMARY

This chapter presented an in-depth investigation of the characteristics of amplifiers implemented using single transistors.

- Of the three available device terminals of the BJT, only the base and emitter are useful as signal input terminals, whereas the collector and emitter are acceptable as output terminals. For the FET, the source and gate are useful as signal input terminals, and the drain and source are acceptable as output terminals. The collector or drain are not used as input terminals, and the base or gate are not used as output terminals.

- There are three basic classifications of amplifiers: inverting amplifiers—the common-emitter and common-source amplifiers; followers—the common-collector and common-drain amplifiers (also known as emitter followers or source followers); and the noninverting amplifiers—common-base and common-gate amplifiers.

- Detailed analyses of these three amplifier classes were performed using the small-signal models for the transistors. These analyses produced expressions for the voltage gain, current gain, input resistance, output resistance, and input signal range, which are summarized in a group of important tables:

Table 14.2 C-E/C-S Amplifier Summary	page 758
Table 14.4 C-C/C-D Amplifier Summary	page 770
Table 14.6 C-B/C-G Amplifier Summary	page 779
Table 14.8 Single-Transistor Bipolar Amplifiers	page 780
Table 14.10 Single-Transistor FET Amplifiers	page 782

The results summarized in these tables form the basic toolkit of the analog circuit designer. A thorough understanding of these results is a prerequisite for design and for the analysis of more complex analog circuits.

TABLE 14.20
Relative Comparison of Single-Transistor Amplifiers

	INVERTING AMPLIFIERS (C-E AND C-S)	FOLLOWERS (C-C AND C-D)	NONINVERTING AMPLIFIERS (C-B AND C-G)
Voltage gain	Moderate	Low ($\cong 1$)	Moderate
Input resistance	Moderate to high	High	Low
Output resistance	Moderate to high	Low	High
Input signal range	Low to moderate	High	Low to moderate
Current gain	Moderate	Moderate	Low ($\cong 1$)

- Inverting amplifiers (C-E and C-S amplifiers) can provide significant voltage and current gain, as well as high input and output resistance. If a resistor is included in the emitter or source of the transistor, the voltage gain is reduced but can be made relatively independent of the individual transistor characteristics. This reduction in gain is traded for increases in input resistance, output resistance, and input signal range. Because of its higher transconductance, the BJT more easily achieves higher values of voltage gain than the FET, whereas the infinite input resistance of the FET gives it the advantage in achieving high input resistance amplifiers. The FET also typically has a larger input signal range than the BJT.

- Emitter and source followers (C-C and C-D amplifiers) provide a voltage gain of approximately 1, high input resistance, and low output resistance. The followers provide moderate levels of current gain and achieve the highest input signal range. These C-C and C-D amplifiers are the single-transistor equivalents of the voltage-follower operational-amplifier configuration introduced in Chapter 11.

- The noninverting amplifiers (C-B and C-G amplifiers) provide voltage gain, signal range, and output resistances very similar to those of the inverting amplifiers but have relatively low input resistance and a current gain of less than one.

- All the amplifier classes provide at least moderate levels of either voltage gain or current gain (or both) and are therefore capable of providing significant power gain with proper design.

- Table 14.20 presents a relative comparison of these three amplifier classes.

- Design examples were presented for amplifiers using the inverting, noninverting, and follower configurations, and an example using Monte Carlo analysis to evaluate the effects of element tolerances on circuit performance was also given.

- The values of coupling and bypass capacitors can be chosen by setting the reactance of the capacitors to be much smaller than the Thévenin equivalent resistance that appears at the capacitor terminals. The reactance is calculated at the lowest frequency in the midband region of the amplifier's frequency response. The lower cutoff frequency f_L is determined by the frequency at which the capacitive reactance equals the equivalent resistance at the capacitor terminals. In the amplifiers in this chapter, there are two or three poles that interact to set f_L, and bandwidth shrinkage moves the cutoff frequency above that set by each individual capacitor acting alone.

KEY TERMS

Body effect
Common-base (C-B) amplifier
Common-collector (C-C) amplifier
Common-emitter (C-E) amplifier
Common-drain (C-D) amplifier

Common-gate (C-G) amplifier
Common-source (C-S) amplifier
Emitter follower
Input resistance
Output resistance

Signal range
Source follower
Swamping

Terminal current gain
Terminal voltage gain
Voltage gain

ADDITIONAL READING

P. R. Gray, P. J. Hurst, S. H. Lewis, and R. G. Meyer, *Analysis and Design of Analog Integrated Circuits,* 4th ed., John Wiley and Sons, New York: 2001.

A. S. Sedra and K. C. Smith, *Microelectronic Circuits,* 4th ed., Oxford University Press, New York: 1998.

M. N. Horenstein, *Microelectronic Circuits and Devices,* 2nd ed., Prentice-Hall, Englewood Cliffs, NJ: 1995.

C. J. Savant, M. S. Roden, and G. L. Carpenter, *Electronic Design—Circuits and Systems,* 2nd ed., Benjamin/Cummings, Redwood City, CA: 1990.

PROBLEMS

Assume all capacitors and inductors have infinite value unless otherwise indicated.

14.1 Amplifier Classification

14.1. Draw the ac equivalent circuits for, and classify (that is, as C-S, C-G, C-D, C-E, C-B, C-C, and not useful), the amplifiers in Figs. P14.1 (a) to (o).

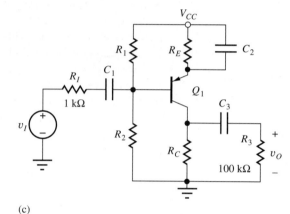

(c)

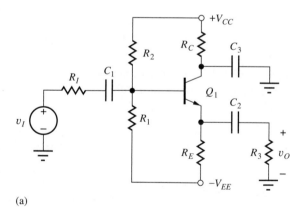

(a)

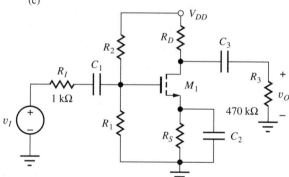

(d)

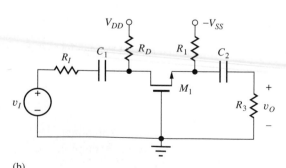

(b)

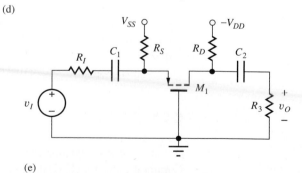

(e)

Figure P14.1   (a), (b)

Figure P14.1   (c), (d), (e)

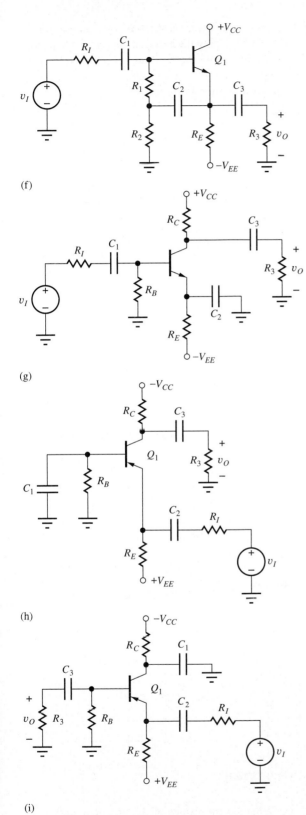

(f)

(g)

(h)

(i)

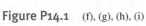

Figure P14.1   (f), (g), (h), (i)

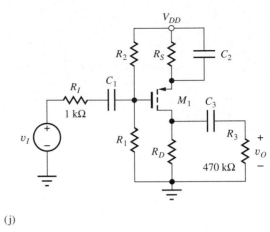

(j)

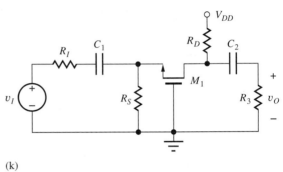

(k)

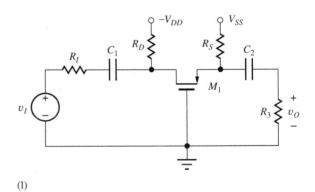

(l)

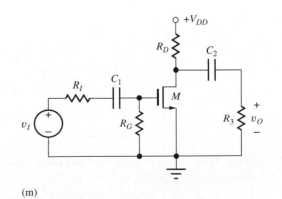

(m)

Figure P14.1   (j), (k), (l), (m)

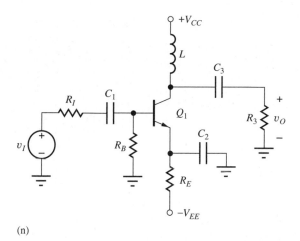

(n)

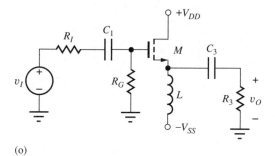

(o)

Figure P14.1 (n), (o)

14.2. A PMOS transistor is biased by the circuit in Fig. P14.2. Using the external source and load configurations in the figure, add coupling and bypass capacitors to the circuit to turn the amplifier into a common-gate amplifier.

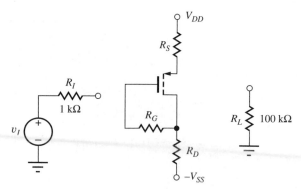

Figure P14.2

14.3. Repeat Prob. 14.2 to turn the amplifier into a common-drain amplifier.

14.4. Repeat Prob. 14.2 to turn the amplifier into a common-source amplifier.

14.5. An *npn* transistor is biased by the circuit in Fig. P14.5. Using the external source and load configurations in the figure, add coupling and bypass capacitors to the circuit to turn the amplifier into a common-emitter amplifier.

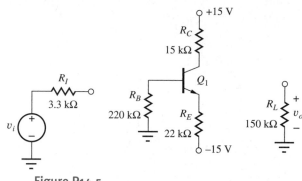

Figure P14.5

14.6. (a) Repeat Prob. 14.5 to turn the amplifier into a common-collector amplifier. (b) Redesign the circuit by deleting any unneeded component(s). Draw the new circuit.

14.7. (a) Repeat Prob. 14.5 to turn the amplifier into a common-base amplifier. (b) Eliminate R_B and any other unneeded components and draw the modified circuit.

14.2 Inverting Amplifiers—Common-Emitter and Common-Source Circuits

14.8. (a) What are the values of A_v, R_{in}, R_{out}, and $A_i = i_o/i_i$ for the common-source stage in Fig. P14.8 if $R_G = 2$ MΩ, $R_I = 75$ kΩ, $R_L = 2$ kΩ, and $R_S = 330$ Ω? Assume $g_m = 5$ mS and $r_o = 10$ kΩ. (b) What are the values of A_v, R_{in}, R_{out}, and A_i if R_S is bypassed by a capacitor?

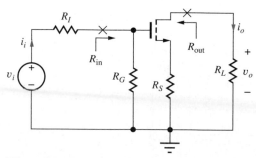

Figure P14.8

14.9. (a) What are the values of A_v, R_{in}, R_{out}, and $A_i = i_o/i_i$ for the common-emitter stage in Fig. P14.9 if

$g_m = 20$ mS, $\beta_o = 75$, $r_o = 100$ kΩ, $R_I = 500$ Ω, $R_B = 15$ kΩ, $R_L = 12$ kΩ, and $R_E = 300$ Ω? (b) What are the values if R_E is changed to 620 Ω?

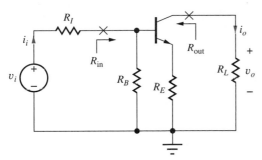

Figure P14.9

14.10. (a) Estimate the voltage gain of the inverting amplifier in Fig. P14.10. (b) Place a bypass capacitor in the circuit to change the gain to approximately -10. (c) Where should the bypass capacitor be placed to change the gain to approximately -20? (d) Where should the bypass capacitor be placed to achieve maximum gain? (e) Estimate this gain.

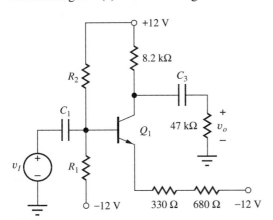

Figure P14.10

14.11. What values of R_E and R_L are required in the ac equivalent circuit in Fig. P14.11 to achieve $A_{vt} = -10$ and $R_{in} = 250$ kΩ? Assume $\beta_o = 75$.

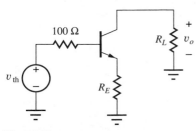

Figure P14.11

14.12. Assume that $R_E = 0$ in Fig. P14.11. What values of R_L and I_C are required to achieve $A_{vt} = -10$ and $R_{in} = 250$ kΩ? Assume $\beta_o = 75$.

14.13. Use nodal analysis to rederive the output resistance of the common-source circuit in Fig. P14.3, as expressed in Table 14.1.

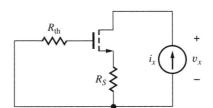

Figure P14.13

14.14. What are A_v, A_i, R_{in}, R_{out}, and the maximum amplitude of the signal source for the amplifier in Fig. P14.1(g) if $R_I = 500$ Ω, $R_E = 100$ kΩ, $R_B = 1$ MΩ, $R_3 = 500$ kΩ, $R_C = 39$ kΩ, $V_{CC} = 15$ V, $-V_{EE} = -15$ V? Use $\beta_F = 100$.

14.15. What are A_v, A_i, R_{in}, R_{out}, and the maximum amplitude of the signal source for the amplifier in Fig. P14.1(c) if $R_1 = 20$ kΩ, $R_2 = 62$ kΩ, $R_E = 3.9$ kΩ, $R_C = 8.2$ kΩ, and $V_{CC} = 9$ V? Use $\beta_F = 75$. Compare A_v to our rule-of-thumb estimate and discuss the reasons for any discrepancy.

14.16. What are A_v, A_i, R_{in}, R_{out}, and the maximum amplitude of the signal source for the amplifier in Fig. P14.1(d) if $R_1 = 500$ kΩ, $R_2 = 1.4$ MΩ, $R_S = 27$ kΩ, $R_D = 75$ kΩ, and $V_{DD} = 15$ V? Use $K_n = 250$ μA/V^2 and $V_{TN} = 1$ V. Compare A_V to our rule-of-thumb estimate and discuss the reasons for any discrepancy.

14.17. What are A_v, A_i, R_{in}, R_{out}, and the maximum amplitude of the signal source for the amplifier in Fig. P14.1(j) if $R_1 = 2.2$ MΩ, $R_2 = 2.2$ MΩ, $R_I = 22$ kΩ, $R_S = 22$ kΩ, $R_D = 18$ kΩ, and $V_{DD} = 22$ V? Use $K_p = 400$ μA/V^2 and $V_{TP} = -1$ V.

14.18. What are A_v, A_i, R_{in}, R_{out}, and the maximum value of the source voltage for the amplifier in Fig. P14.1(m) if $R_I = 5$ kΩ, $R_G = 10$ MΩ, $R_3 = 36$ kΩ, $R_D = 1.8$ kΩ, and $V_{DD} = 16$ V? Use $K_n = 0.4$ mS/V and $V_{TN} = -5$ V.

14.19. What are A_v, A_i, R_{in}, R_{out}, and the maximum value of the source voltage for the amplifier in Fig. P14.1(n) if $R_I = 250$ Ω, $R_B = 20$ kΩ, $R_3 = 1$ MΩ, $R_E = 9.1$ kΩ, $V_{CC} = 12$ V, and $V_{EE} = 12$ V? Use $\beta_F = 80$ and $V_A = 100$ V.

14.3 Follower Circuits—Common-Collector and Common-Drain Amplifiers

14.20. What are the values of A_v, R_{in}, R_{out}, and A_i for the common-collector stage in Fig. P14.20 if $R_I = 10$ kΩ, $R_B = 47$ kΩ, $R_L = 1$ kΩ, $\beta_o = 80$, and $g_m = 0.4$ S? ($A_i = i_o/i_i$).

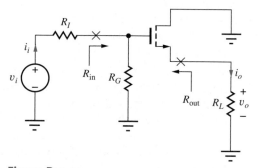

Figure P14.20

14.21. What are the values of A_v, R_{in}, R_{out}, and A_i for the common-drain stage in Fig. P14.21 if $R_G = 2$ MΩ, $R_I = 100$ kΩ, $R_L = 2$ kΩ, and $g_m = 10$ mS? ($A_i = i_o/i_i$).

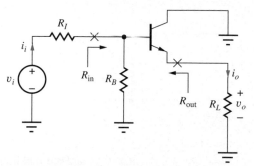

Figure P14.21

*14.22. The gate resistor R_G in Fig. P14.22 is said to be "bootstrapped" by the action of the source follower. (a) Assume that the FET is operating with $g_m = 3.54$ mS and r_o can be neglected. Draw the small-signal model and find A_v, R_{in}, and R_{out} for the amplifier. (b) What would R_{in} be if A_v were exactly $+1$?

14.23. What are A_v, R_{in}, R_{out}, and maximum input signal amplitude for the amplifier in Fig. P14.1(a) if $R_I = 500$ Ω, $R_1 = 100$ kΩ, $R_2 = 100$ kΩ, $R_3 = 24$ kΩ, $R_E = 4.7$ kΩ, $R_C = 2$ kΩ, and $V_{CC} = V_{EE} = 12$ V? Use $\beta_F = 125$ and $V_A = 50$ V.

14.24. What are A_v, R_{in}, R_{out}, and maximum input signal for the amplifier in Fig. P14.1(o) if $R_I = 10$ kΩ,

$R_G = 1$ MΩ, $R_3 = 100$ kΩ, and $V_{DD} = V_{SS} = 5$ V? Use $K_n = 500$ μA/V^2, $V_{TN} = 1.5$ V, and $\lambda = 0.02$ V^{-1}.

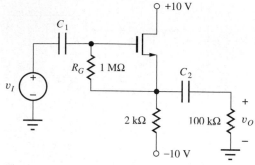

Figure P14.22

14.25. What are A_v, R_{in}, R_{out}, and the maximum input signal amplitude for the amplifier in Fig. P14.1(f) if $R_I = 500$ Ω, $R_1 = 500$ kΩ, $R_2 = 500$ kΩ, $R_3 = 500$ kΩ, $R_E = 430$ kΩ, and $V_{CC} = V_{EE} = 9$ V? Use $\beta_F = 100$ and $V_A = 60$ V.

*14.26. Recast the signal-range formula for the common-collector amplifier in Table 14.4 in terms of the dc voltage developed across the emitter resistor R_E in Fig. 14.3(a). Assume $R_3 = \infty$.

14.27. Rework Prob. 14.22(a) by using the formulas for the bipolar transistor by "pretending" that R_G makes the FET equivalent to a BJT with $r_\pi = R_G$.

*14.28. The input to a common-collector amplifier is a triangular input signal with a peak-to-peak amplitude of 10 V. (a) What is the minimum gain required of the C-C amplifier to meet the small-signal limit? (b) What is the minimum dc voltage required across the emitter resistor in this amplifier to satisfy the limit in (a)?

*14.29. Design the emitter-follower circuit in Fig. P14.29 to meet the small-signal requirements when $v_o = 5 \sin 2000\pi t$ V. Assume $C_1 = C_2 = \infty$ and $\beta_F = 50$.

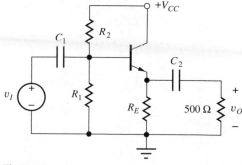

Figure P14.29

14.4 Noninverting Amplifiers—Common-Base and Common-Gate Circuits

14.30. What are the values of A_v, R_{in}, R_{out}, and A_i for the common-base stage in Fig. P14.30 operating with $I_C = 12.5$ μA, $\beta_o = 100$, $V_A = 60$ V, $R_I = 50$ Ω, $R_4 = 100$ kΩ and $R_L = 100$ kΩ? (b) What are the values if R_I is changed to 2.2 kΩ? ($A_i = i_o/i_i$).

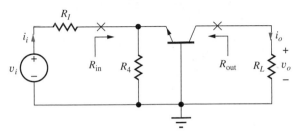

Figure P14.30

14.31. What are the values of A_v, R_{in}, R_{out}, and A_i for the common-gate stage in Fig. P14.31 operating with $g_m = 0.5$ mS, $R_I = 50$ Ω, $R_4 = 3$ kΩ and $R_L = 60$ kΩ? (b) What are the values if R_I is changed to 5 kΩ? ($A_i = i_o/i_i$).

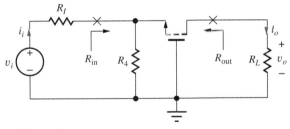

Figure P14.31

14.32. Estimate the voltage gain of the amplifier in Fig. P14.32. Explain your answer.

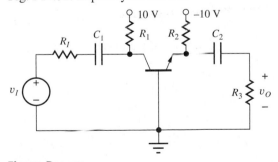

Figure P14.32

14.33. What are A_v, R_{in}, R_{out}, and the maximum input signal for the amplifier in Fig. P14.1(h) if $R_I = 500$ Ω, $R_B = 100$ kΩ, $R_3 = 100$ kΩ, $R_E = 82$ kΩ, $R_C = 39$ kΩ, and $V_{EE} = V_{CC} = 12$ V? Use $\beta_F = 50$ and $V_A = 50$ V.

14.34. What are A_v, R_{in}, R_{out}, and the maximum input signal for the amplifier in Fig. P14.1(h) if $R_I = 5$ kΩ, $R_B = 1$ MΩ, $R_3 = 1$ MΩ, $R_E = 820$ kΩ, $R_C = 390$ kΩ, and $V_{EE} = V_{CC} = 9$ V? Use $\beta_F = 50$ and $V_A = 50$ V.

14.35. What are A_v, R_{in}, R_{out}, and the maximum input signal for the amplifier in Fig. P14.1(k) if $R_I = 1$ kΩ, $R_S = 3.9$ kΩ, $R_3 = 51$ kΩ, $R_D = 20$ kΩ, and $V_{DD} = 16$ V? Use $K_n = 500$ μA/V^2 and $V_{TN} = -2$ V.

14.36. What are A_v, R_{in}, R_{out}, and the maximum input signal for the amplifier in Fig. P14.1(e) if $R_I = 250$ Ω, $R_S = 68$ kΩ, $R_3 = 200$ kΩ, $R_D = 43$ kΩ, and $V_{DD} = V_{SS} = 15$ V? Use $K_p = 200$ μA/V^2 and $V_{TP} = -1$ V.

14.37. What are A_v, R_{in}, R_{out}, and the maximum input signal amplitude for the amplifier in Fig. P14.1(e) if $R_I = 500$ Ω, $R_S = 33$ kΩ, $R_3 = 100$ kΩ, $R_D = 24$ kΩ, and $V_{DD} = V_{SS} = 10$ V? Use $K_p = 200$ μA/V^2 and $V_{TP} = -1$ V.

14.38. The gain of the common-gate and common-base stages can be written as $A_v = R_L/[(1/g_m) + R_{th}]$. When $R_{th} \ll 1/g_m$, the circuit is said to be "voltage driven," and when $R_{th} \gg 1/g_m$, the circuit is said to be "current driven." What are the approximate voltage gain expressions for these two conditions? Discuss the reason for the use of these adjectives to describe the two circuit limits.

14.39. What is the input resistance to the common-base stage in Fig. P14.39 if $I_C = 1$ mA and $\beta_F = 75$?

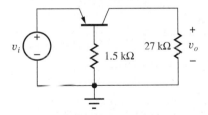

Figure P14.39

14.40. What is the input resistance to the common-gate stage in Fig. P14.40 if $I_D = 1$ mA, $K_p = 1.25$ mA/V^2, and $V_{TP} = 2$ V?

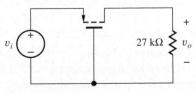

Figure P14.40

14.41. (a) Estimate the resistance looking into the collector of the transistor in Fig. P14.41 if $R_E = 143\text{ k}\Omega$, $V_A = 50\text{ V}$, $\beta_F = 100$, and $V_{EE} = 15\text{ V}$? (b) What is the minimum value of V_{CC} required to ensure that Q_1 is operating in the forward-active region? (c) Repeat parts (a) and (b) if $R_E = 15\text{ k}\Omega$.

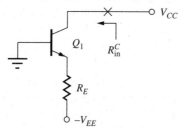

Figure P14.41

*14.42. What is the resistance looking into the collector terminal in Fig. P14.42 if $I_E = 50\text{ μA}$, $\beta_o = 125$, $V_A = 50\text{ V}$, and $V_{CC} = 10\text{ V}$? (*Hint:* r_o must be considered in this circuit. Otherwise $R_{\text{out}} = \infty$.)

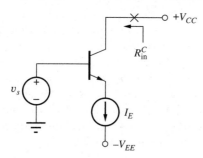

Figure P14.42

14.5 Amplifier Prototype Review and Comparison

14.43. A single-transistor amplifier is needed that has a gain of 43 dB and an input resistance of 350 Ω. What is the preferred choice of amplifier configuration? Discuss your reasons for your selection.

14.44. A single-transistor amplifier is needed that has a gain of 46 dB and an input resistance of 0.3 MΩ. What is the preferred choice of amplifier configuration? Discuss your reasons for making this selection.

14.45. A single-transistor amplifier is needed that has a gain of 26 dB and an input resistance of 10 MΩ. What is the preferred choice of amplifier configuration, and why did you make this selection?

14.46. A single-transistor amplifier is needed that has a gain of 58 dB and an input resistance of 50 kΩ. What is the preferred choice of amplifier configuration? Discuss your reasons for making this selection.

14.47. A single-transistor amplifier is needed that has a gain of −100 and an input resistance of 5 Ω. What is the preferred choice of amplifier configuration? Discuss your reasons for making this selection.

14.48. A single-transistor amplifier is needed that has a gain of approximately +20 and an input resistance of 5 kΩ. What is the preferred choice of amplifier configuration? Discuss your reasons for making this selection.

14.49. A single-transistor amplifier is needed that has a gain of approximately 0 dB and an input resistance of 20 MΩ with a load resistor of 20 kΩ. What is the preferred choice of amplifier configuration, and why did you make this selection?

14.50. A follower is needed that has a gain of at least 0.97 and an input resistance of at least 200 kΩ with a load resistance of 5 kΩ. What is the preferred choice of amplifier configuration? Discuss your reasons for making this selection.

14.51. A single-transistor amplifier is needed that has a gain of approximately 66 dB and an input resistance of 250 kΩ. What is the preferred choice of amplifier configuration, and why did you make this selection?

14.52. An inverting amplifier is needed that has an output resistance of at least 1 GΩ. What is the preferred choice of amplifier configuration? Discuss your reasons for making this selection. Estimate the Q-point current and emitter or source resistance required to achieve this specification.

14.53. A common-collector amplifier is being driven from a source having a resistance of 250 Ω. What is the minimum output resistance of this amplifier if the transistor has $\beta_o = 150$ and $V_A = 50\text{ V}$?

**14.54. Show that the emitter resistor R_E in Fig. P14.54 can be absorbed into the transistor by redefining the small-signal parameters of the transistor to be

$$g'_m \cong \frac{g_m}{1 + g_m R_E} \qquad r'_\pi \cong r_\pi(1 + g_m R_E)$$

$$r'_o \cong r_o(1 + g_m R_E)$$

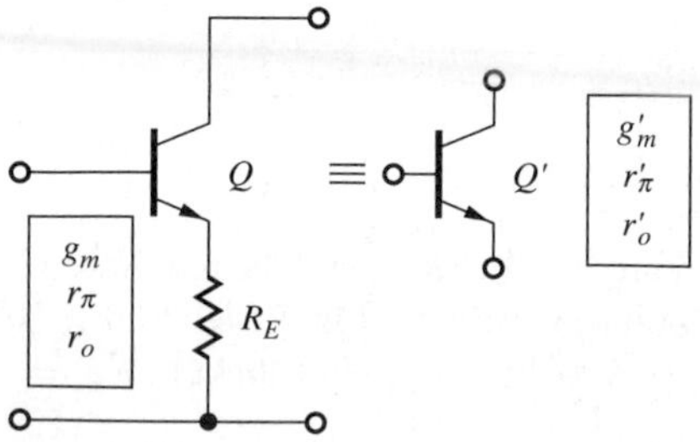

Figure P14.54

What is the expression for the common-emitter small-signal current gain β'_o for the new transistor? What is the expression for the amplification factor μ'_f for the new transistor?

*14.55. Perform a transient simulation of the behavior of the common-emitter amplifier in Fig. P14.55 for sinusoidal input voltages of 5 mV, 10 mV, and 15 mV at a frequency of 1 kHz. Use the Fourier analysis capability of SPICE to analyze the output waveforms. Compare the amplitudes of the 2-kHz and 3-kHz harmonics to the amplitude of the desired signal at 1 kHz. Assume $\beta_F = 100$ and $V_A = 70$ V.

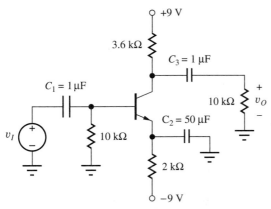

Figure P14.55

14.56. In the circuits in Fig. P14.56, $I_B = 10$ μA. Use SPICE to determine the output resistances of the two circuits by sweeping the voltage V_{CC} from 10 to 20 V. Use $\beta_F = 60$ and $V_A = 20$ V. Compare results to hand calculations using the small-signal parameter values from SPICE.

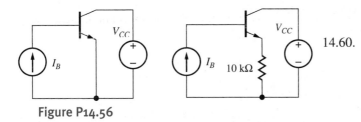

Figure P14.56

14.57. (a) What is the Thévenin equivalent representation for the amplifier in Fig. P14.57? (b) What are the values of v_{th} and R_{th} if $R_I = 270$ Ω, $\beta_o = 100$, $g_m = 2$ mS, and $r_o = 250$ kΩ?

14.58. (a) What is the Thévenin equivalent representation for the amplifier in Fig. P14.58? (b) What are the values of v_{th} and R_{th} if $R_I = 100$ kΩ, $R_S = 18$ kΩ, $g_m = 500$ μS, and $r_o = 250$ kΩ?

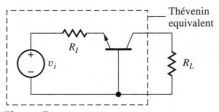

Figure P14.57

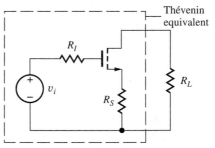

Figure P14.58

14.59. (a) What is the Thévenin equivalent representation for the amplifier in Fig. P14.59 if $R_I = 5$ kΩ, $R_L = 10$ kΩ, $\beta_o = 100$ and $g_m = 4$ mS?

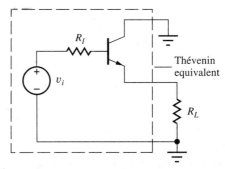

Figure P14.59

14.60. (a) An emitter follower is drawn as a two-port in Fig. P14.60. Calculate g_{21} and g_{12} for this amplifier in terms of the small-signal parameters. Compare the two results. (b) What are the values of g_{21} and g_{12} if $R_B = 150$ kΩ, $R_E = 2.4$ kΩ, $\beta_o = 100$, $g_m = 10$ mS, and $r_o = 250$ kΩ?

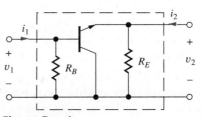

Figure P14.60

14.61. (a) A source follower is drawn as a two-port in Fig. P14.61. Calculate g_{21} and g_{12} for this amplifier in terms of the small-signal parameters. Compare the two results. (b) What are the values of g_{21} and g_{12} if $R_G = 1$ MΩ, $R_D = 50$ kΩ, $g_m = 400$ μS, and $r_o = 450$ kΩ?

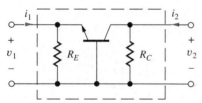

Figure P14.61

14.62. (a) A common-base amplifier is drawn as a two-port in Fig. P14.62. Calculate g_{21} and g_{12} for this amplifier in terms of the small-signal parameters. Compare the two results. (b) What are the values of g_{21} and g_{12} if $R_C = 18$ kΩ, $R_E = 3.9$ kΩ, $\beta_o = 100$, $g_m = 3$ mS, and $r_o = 800$ kΩ?

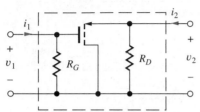

Figure P14.62

14.63. (a) A common-gate amplifier is drawn as a two-port in Fig. P14.63. Calculate g_{21} and g_{12} for this amplifier in terms of the small-signal parameters. Compare the two results. (b) What are the values of g_{21} and g_{12} if $R_S = 15$ kΩ, $R_D = 100$ kΩ, $g_m = 500$ μS, and $r_o = 500$ kΩ?

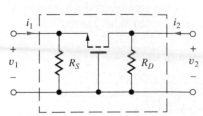

Figure P14.63

14.64. (a) A common-emitter amplifier is drawn as a two-port in Fig. P14.64. Calculate g_{21} and g_{12} for this amplifier in terms of the small-signal parameters.

Compare the two results. (b) What are the values of g_{21} and g_{12} if $R_B = 180$ kΩ, $R_E = 12$ kΩ, $R_C = 130$ kΩ, $\beta_o = 100$, $g_m = 2$ mS, and $r_o = 1$ MΩ?

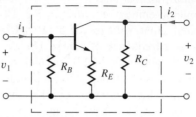

Figure P14.64

14.65. (a) A common-source amplifier is drawn as a two-port in Fig. P14.65. Calculate g_{21} and g_{12} for this amplifier in terms of the small-signal parameters. Compare the two results. (b) What are the values of g_{21} and g_{12} if $R_G = 1.5$ MΩ, $R_S = 12$ kΩ, $R_D = 130$ kΩ, $g_m = 750$ μS, and $r_o = 330$ kΩ?

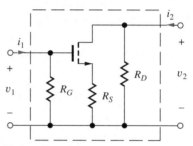

Figure P14.65

14.66. Our calculation of the input resistance of the common-gate and common-base amplifiers neglected r_o in the calculation. Calculate an improved estimate for R_{in} for the common-gate stage in Fig. P14.66.

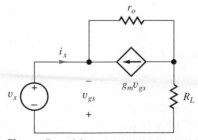

Figure P14.66

14.67. The circuit in Fig. P14.67 is called a phase inverter. Calculate the two gains $A_{v1} = \mathbf{v_{o1}}/\mathbf{v_i}$ and $A_{v2} = \mathbf{v_{o2}}/\mathbf{v_i}$. What is the largest ac signal that can be developed at output v_{O1} in this particular circuit? Assume $\beta_F = 100$.

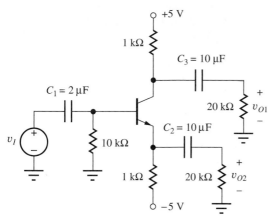

Figure P14.67

14.68. (a) Calculate the values of A_v, R_{in}, and R_{out} for the amplifier in Fig. P14.1(a) if $R_I = 600\ \Omega$, $R_1 = 100\ k\Omega$, $R_2 = 100\ k\Omega$, $R_3 = 24\ k\Omega$, $R_E = 4.7\ k\Omega$, $R_C = 2\ k\Omega$, and $V_{CC} = V_{EE} = 15\ V$. Use $\beta_F = 125$ and $V_A = 50\ V$. (b) Use SPICE to verify the results of your hand calculations. Assume $f = 10\ kHz$ and $C_1 = 10\ \mu F$, $C_2 = 47\ \mu F$, $C_3 = 10\ \mu F$.

14.69. (a) Calculate the values of A_v, R_{in}, and R_{out} for the amplifier in Fig. P14.1(g) if $R_I = 500\ \Omega$, $R_E = 68\ k\Omega$, $R_B = 1\ M\Omega$, $R_3 = 500\ k\Omega$, $R_C = 39\ k\Omega$, $V_{EE} = -10\ V$, and $V_{CC} = 10\ V$. Use $\beta_F = 80$ and $V_A = 75\ V$. (b) Use SPICE to verify the results of your hand calculations. Assume $f = 4\ kHz$, $C_1 = C_3 = 2.2\ \mu F$, and $C_2 = 47\ \mu F$.

14.70. (a) Calculate the values of A_v, R_{in}, and R_{out} for the amplifier in Fig. P14.1(c) if $R_1 = 20\ k\Omega$, $R_2 = 62\ k\Omega$, $R_E = 6.8\ k\Omega$, $R_C = 16\ k\Omega$, and $V_{CC} = 12\ V$. Use $\beta_F = 75$ and $V_A = 60\ V$. (b) Use SPICE to verify the results of your hand calculations. Assume $f = 5\ kHz$ and $C_1 = 2.2\ \mu F$, $C_2 = 47\ \mu F$, $C_3 = 10\ \mu F$.

14.71. (a) Calculate the values of A_v, R_{in}, and R_{out} for the amplifier in Fig. P14.1(d) if $R_1 = 500\ k\Omega$, $R_2 = 1.4\ M\Omega$, $R_S = 27\ k\Omega$, $R_D = 75\ k\Omega$, $V_{DD} = 18\ V$. Use $K_n = 500\ \mu A/V^2$, $\lambda = 0.02\ V^{-1}$, and $V_{TN} = 1\ V$. (b) Use SPICE to verify the results of your hand calculations. Assume $f = 5\ kHz$ and $C_1 = 2.2\ \mu F$, $C_2 = 47\ \mu F$, $C_3 = 10\ \mu F$.

14.72. (a) Calculate the values of A_v, R_{in}, and R_{out} for the amplifier in Fig. P14.1(e) if $R_I = 500\ \Omega$,

$R_S = 33\ k\Omega$, $R_3 = 100\ k\Omega$, $R_D = 24\ k\Omega$, and $V_{DD} = V_{SS} = 10\ V$. Use $K_p = 250\ \mu A/V^2$, $V_{TP} = -1\ V$, and $\lambda = 0.02\ V^{-1}$. (b) Use SPICE to verify the results of your hand calculations. Assume $f = 50\ kHz$, $C_1 = 10\ \mu F$, and $C_2 = 47\ \mu F$.

14.73. (a) Calculate the values of A_v, R_{in}, and R_{out} for the amplifier in Fig. P14.1(f) if $R_I = 500\ \Omega$, $R_1 = 500\ k\Omega$, $R_2 = 500\ k\Omega$, $R_3 = 500\ k\Omega$, $R_E = 330\ k\Omega$, and $V_{CC} = V_{EE} = 5\ V$. Use $\beta_F = 100$ and $V_A = 60\ V$. (b) Use SPICE to verify the results of your hand calculations. Assume $f = 10\ kHz$ and $C_1 = 10\ \mu F$, $C_2 = 47\ \mu F$, $C_3 = 10\ \mu F$.

14.74. (a) Calculate the values of A_v, R_{in}, and R_{out} for the amplifier in Fig. P14.1(h) if $R_I = 500\ \Omega$, $R_B = 100\ k\Omega$, $R_3 = 100\ k\Omega$, $R_E = 82\ k\Omega$, $R_C = 39\ k\Omega$, and $V_{EE} = V_{CC} = 12\ V$. Use $\beta_F = 50$ and $V_A = 50\ V$. (b) Use SPICE to verify the results of your hand calculations. Assume $f = 12\ kHz$ and $C_1 = 4.7\ \mu F$, $C_2 = 47\ \mu F$, $C_3 = 10\ \mu F$.

14.75. (a) Calculate the values of A_v, R_{in}, and R_{out} for the amplifier in Fig. P14.1(j) if $R_1 = 2.2\ M\Omega$, $R_2 = 2.2\ M\Omega$, $R_S = 110\ k\Omega$, $R_D = 90\ k\Omega$, and $V_{DD} = 18\ V$. Use $K_p = 400\ \mu A/V^2$, $\lambda = 0.02\ V^{-1}$, and $V_{TP} = -1\ V$. (b) Use SPICE to verify the results of your hand calculations. Assume $f = 7500\ Hz$ and $C_1 = 2.2\ \mu F$, $C_2 = 47\ \mu F$, $C_3 = 10\ \mu F$.

14.76. (a) Calculate the values of A_v, R_{in}, and R_{out} for the amplifier in Fig. P14.1(k) if $R_I = 1\ k\Omega$, $R_3 = 10.0\ k\Omega$, $R_S = 51\ k\Omega$, $R_D = 20\ k\Omega$, and $V_{DD} = 15\ V$. Use $K_n = 500\ \mu A/V^2$, $\lambda = 0.02\ V^{-1}$, and $V_{TN} = -2\ V$. (b) Use SPICE to verify the results of your hand calculations. Assume $f = 20\ kHz$ and $C_1 = 2.2\ \mu F$ and $C_2 = 47\ \mu F$.

14.77. (a) Calculate the values of A_v, R_{in}, and R_{out} for the amplifier in Fig. P14.1(m) if $R_I = 5\ k\Omega$, $R_G = 10\ M\Omega$, $R_3 = 36\ k\Omega$, $R_D = 1.8\ k\Omega$, and $V_{DD} = 16\ V$. Use $K_n = 10\ mS/V$, $V_{TN} = -5\ V$, and $\lambda = 0.02\ V^{-1}$. (b) Use SPICE to verify the results of your hand calculations. Assume $f = 3000\ Hz$ and $C_1 = 2.2\ \mu F$, $C_2 = 10\ \mu F$.

14.78. (a) Calculate the values of A_v, R_{in}, and R_{out} for the amplifier in Fig. P14.1(n) if $R_I = 250\ \Omega$, $R_B = 33\ k\Omega$, $R_3 = 1\ M\Omega$, $R_E = 7.8\ k\Omega$, $V_{CC} = 10\ V$, and $V_{EE} = 10\ V$. Use $\beta_F = 80$ and $V_A = 100\ V$. (b) Use SPICE to verify the results of your hand calculations. Assume $f = 500$ kHz and $C_1 = 4.7$ µF, $C_2 = 100$ µF, $C_3 = 1$ µF, and $L = 1$ H.

14.79. (a) Calculate the values of A_v, R_{in}, and R_{out} for the amplifier in Fig. P14.1(o) if $R_I = 10$ kΩ, $R_G = 2$ MΩ, $R_3 = 100$ kΩ, and $V_{DD} = V_{SS} = 6$ V. Use $K_n = 400$ µA/V^2, $V_{TN} = 1$ V, and $\lambda = 0.02$ V^{-1}. (b) Use SPICE to verify the results of your hand calculations. Assume $f = 1$ MHz and $C_1 = 2.2$ µF, $C_3 = 4.7$ µF, and $L = 100$ mH.

14.6 Coupling and Bypass Capacitor Design

14.80. (a) The amplifier in Fig. P14.1(d) has $R_1 = 500$ kΩ, $R_2 = 1.4$ MΩ, $R_S = 27$ kΩ, $R_D = 75$ kΩ, and $V_{DD} = 15$ V. Use $K_n = 400$ µA/V^2, $V_{TN} = 1$ V, and $\lambda = 0.02$ V^{-1}. Choose values for C_1, C_2, and C_3 so that they can be neglected at a frequency of 400 Hz. (b) Choose C_2 to set the lower cutoff frequency to 4 kHz assuming C_1 and C_3 remain unchanged.

14.81. (a) The amplifier in Fig. P14.1(c) has $R_1 = 20$ kΩ, $R_2 = 62$ kΩ, $R_C = 8.2$ kΩ, $R_E = 3.9$ kΩ, and $V_{CC} = 12$ V. Choose values for C_1, C_2, and C_3 so that they can be neglected at a frequency of 100 Hz. Use $\beta_F = 75$ and $V_A = 60$ V. (b) Choose C_2 to set the lower cutoff frequency to 1000 Hz assuming C_1 and C_3 remain unchanged.

14.82. Calculate the frequency for which each of the capacitors in Fig. P14.67 can be considered to have a negligible effect on the circuit.

14.83. Choose values of C_1 and C_2 in Fig. P14.83 so they will have negligible effect on the circuit at a frequency of 50 kHz. (b) Repeat for a frequency of 100 Hz.

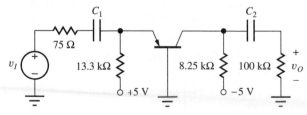

Figure P14.83

14.84. The amplifier in Fig. P14.1(a) has $R_I = 500$ Ω, $R_1 = 51$ kΩ, $R_2 = 100$ kΩ, $R_3 = 24$ kΩ, $R_E = 4.7$ kΩ, $R_C = 0$, and $V_{CC} = V_{EE} = 15$ V. Choose values for C_1 and C_2 so that they can be neglected at a frequency of 50 Hz. Use $\beta_F = 100$ and $V_A = 50$ V.

14.85. The amplifier in Fig. P14.1(k) has $R_I = 1$ kΩ, $R_1 = 3.9$ kΩ, $R_3 = 100$ kΩ, $R_D = 20$ kΩ, and $V_{DD} = 15$ V. Choose values for C_1, C_2, and C_3 so that they can be neglected at a frequency of 400 Hz. Use $K_n = 500$ µA/V^2, $V_{TN} = -2$ V, and $\lambda = 0.02$ V^{-1}.

14.86. (a) Use a dominant-pole approach to set the lower cutoff frequency of the C-S amplifier in Ex. 14.6 to 1000 Hz. Choose values for C_1, C_2, and C_3 based upon the values in the example. (b) Check your design with SPICE.

14.87. (a) Use a dominant-pole approach to set the lower cutoff frequency of the C-C amplifier in Ex. 14.7 to 2000 Hz. Choose values for C_1, and C_3 based upon the values in the example. (b) Use the dominant-pole approach to set the lower cutoff frequency of the C-G amplifier in Ex. 14.8 to 1000 Hz. Choose values for C_1, C_2, and C_3 based upon the values in the example. (c) Check your two designs with SPICE.

14.7 Amplifier Design Examples

14.88. Repeat the source-follower design in Design Ex. 14.5 for a MOSFET with $K_n = 30$ mA/V^2 and $V_{TN} = 2$ V. Assume $V_{GS} - V_{TN} = 0.5$ V.

14.89. A common-base amplifier was used in the design problem in Ex. 14.10 to match the 75-Ω input resistance. One could conceivably match the input resistance with a common-emitter stage (with $R_E = 0$). What collector current is required to set $R_{in} = 75$ Ω for a BJT with $\beta_o = 100$?

14.90. Rework Ex. 14.10 to achieve a 50-Ω input resistance.

*14.91. Redesign the bias network so that the common-base amplifier in Fig. 14.34 can operate from a single +10-V supply.

14.92. Redesign the amplifier in Fig. 14.34 to operate from symmetrical 9-V power supplies and achieve the same design specifications.

14.93. A common-gate amplifier is needed with an input resistance of 10 Ω. Two n-channel MOSFETs are available: one with $K_n = 5$ mA/V^2 and the other with $K_n = 500$ mA/V^2. Both are capable of providing the desired value of R_{in}. Which one would be preferred and why? (*Hint:* Find the required Q-point current for each transistor.)

14.94. $(1/g_m)$ is set to 50 Ω in a common-base design operating at 27°C. What are the values of $(1/g_m)$ at −40°C and +50°C?

*14.95. (a) Calculate worst-case estimates of the gain of the common-base amplifier in Fig. 14.34 if the resistors and power supplies all have 5 percent tolerances. (b) Compare your answers to the Monte Carlo results in Table 14.15.

**14.96. Use SPICE to perform a 1000-case Monte Carlo analysis of the common-base amplifier in Fig. 14.34 if the resistors and power supplies have 5 percent tolerances. Assume that the current gain β_F and V_A are uniformly distributed in the intervals [60, 100] and [50, 70], respectively. What are the mean and 3σ limits on the voltage gain predicted by these simulations? Compare the 3σ values to the worst-case calculations in Prob. 14.95. Compare your answers to the Monte Carlo results in Table 14.15. Use $C_1 = 47\,\mu F$, $C_2 = 4.7\,\mu F$, and $f = 10$ kHz.

**14.97. The common-base amplifier in Fig. P14.83 is the implementation of the design from Design Ex. 14.10 using the nearest 1 percent resistor values. (a) What are the worst-case values of gain and input resistance if the power supplies have ± 2 percent tolerances? (b) Use a computer program or spreadsheet to perform a 1000-case Monte Carlo analysis to find the mean and 3σ limits on the gain and input resistances. Compare these values to the worst-case estimates from part (a).

**14.98. Use SPICE to perform a 1000-case Monte Carlo analysis of the circuit in Fig. P14.83 assuming the resistors have 1 percent tolerances and the power supplies have ± 2 percent tolerances. Find the mean and 3σ limits on the gain and input resistance at a frequency of 10 kHz. Assume that the current gain β_F and V_A are uniformly

distributed in the intervals (60, 100) and (50, 70), respectively. Use $C_1 = 100\,\mu F$, $C_2 = 1\,\mu F$, and $f = 10$ kHz.

14.99. Suppose that we forgot about the factor of 2 loss in signal that occurs at the input of the common-base stage in Ex. 14.10 and selected $V_{CC} = V_{EE} = 2.5$ V. Repeat the design to see if the specifications can be met using these power supply values.

**14.100. (a) Use a spreadsheet or other computer tool to perform a Monte Carlo analysis of the design in Fig. 14.30. The resistors and power supplies have 5 percent tolerances. V_{TN} is uniformly distributed in the interval [1 V, 2 V], and K_n is uniformly distributed in the interval [10 mA/V², 30 mA/V²]. (b) Use the Monte Carlo option in PSPICE to perform the same analysis at a frequency of 10 kHz for $C_1 = 4.7\,\mu F$ and $C_2 = 68\,\mu F$. Compare the results.

Unless otherwise specified, use $\beta_F = 100$, $V_A = 70$ V, $K_p = K_n = 1$ mA/V², $V_{TN} = -V_{TP} = 1$ V, and $\lambda = 0.02$ V^{-1}.

14.8 Multistage ac-Coupled Amplifiers

14.101. What are the voltage gain, input resistance, and output resistance of the amplifier in Fig. 14.38 if bypass capacitors C_2 and C_4 are removed from the circuit?

14.102. Figure P14.102 is an "improved" version of the three-stage amplifier discussed in Sec. 14.8. Find the gain and input signal range for this amplifier. Was the performance actually improved?

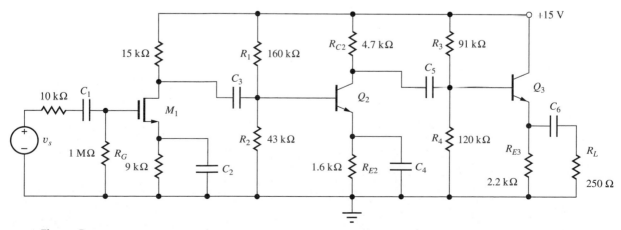

Figure P14.102

14.103. Use SPICE to simulate the amplifier in Fig. P14.102 at a frequency of 2 kHz, and determine the voltage gain, input resistance, and output resistance. Assume the capacitors all have a value of 22 μF.

14.104. Find the midband voltage gain and input resistance of the amplifier in Fig. P14.102 if capacitors C_2 and C_4 are removed from the circuit.

14.105. Use SPICE to determine the gain of the amplifier in Fig. P14.102 if C_2 and C_4 are removed from the circuit. Assume the capacitors all have a value of 22 μF.

*14.106. Figure P14.106 shows another "improved" design of the three-stage amplifier discussed in Sec. 14.8.

Find the gain and input signal range for this amplifier. Was the performance improved?

14.107. Use SPICE to simulate the amplifier in Fig. P14.106 at a frequency of 3 kHz and determine the voltage gain, input resistance, and output resistance. Assume the capacitors all have a value of 22 μF.

14.108. What is the gain of the amplifier in Fig. P14.106 if C_2 and C_4 are removed?

14.109. What are the midband voltage gain, input resistance, and output resistance of the amplifier in Fig. P14.109?

14.110. What are the voltage gain, input resistance, and output resistance of the amplifier in Fig. P14.109 if the bypass capacitors are removed?

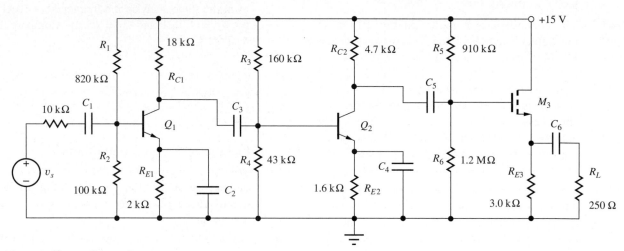

Figure P14.106

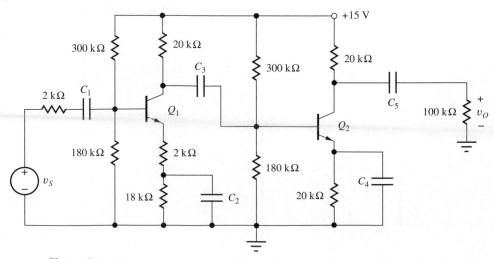

Figure P14.109

14.111. Use SPICE to simulate the amplifier in Fig. P14.109 at a frequency of 5 kHz and determine the voltage gain, input resistance, and output resistance. Assume the capacitors all have a value of 10 μF.

14.112. Find the midband voltage gain, input resistance, and output resistance of the amplifier in Fig. P14.109 if capacitor C_2 is connected between the emitter of Q_1 and ground?

14.113. What are the midband voltage gain, input resistance, and output resistance of the amplifier in Fig. P14.113 if $K_n = 50$ mA/V^2 and $V_{TN} = -2$ V?

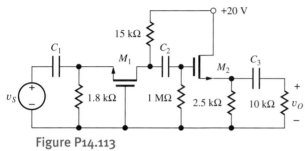

Figure P14.113

Lower Cutoff Frequency Estimates

14.114. Use the short circuit time constant technique to estimate the lower cutoff frequency for the amplifier in Prob. 14.68. Compare your answer to SPICE simulation.

14.115. Use the short circuit time constant technique to estimate the lower cutoff frequency for the amplifier in Prob. 14.69. Compare your answer to SPICE simulation.

14.116. Use the short circuit time constant technique to estimate the lower cutoff frequency for the amplifier in Prob. 14.70. Compare your answer to SPICE simulation.

14.117. Use the short circuit time constant technique to estimate the lower cutoff frequency for the amplifier in Prob. 14.71. Compare your answer to SPICE simulation.

14.118. Use the short circuit time constant technique to estimate the lower cutoff frequency for the amplifier in Prob. 14.72. Compare your answer to SPICE simulation.

14.119. Use the short circuit time constant technique to estimate the lower cutoff frequency for the amplifier in Prob. 14.74. Compare your answer to SPICE simulation.

14.120. Use the short circuit time constant technique to estimate the lower cutoff frequency for the amplifier in Prob. 14.76. Compare your answer to SPICE simulation.

14.121. Use the short circuit time constant technique to estimate the lower cutoff frequency for the amplifier in Prob. 14.77. Compare your answer to SPICE simulation.

14.122. Use the short circuit time constant technique to estimate the lower cutoff frequency for the amplifier in Prob. 14.113. Compare your answer to SPICE simulation. Use $C_1 = C_2 = C_3 = 1$ μF.

DIFFERENTIAL AMPLIFIERS AND OPERATIONAL AMPLIFIER DESIGN

CHAPTER OUTLINE

CHAPTER GOALS

In this chapter, we learn to work with dc-coupled amplifiers that contain several interconnected stages, and several important new amplifier concepts are introduced. Overall, we want to achieve these goals:

- Understand analysis and design of dc-coupled multistage amplifiers
- Explore the dc and ac properties of differential amplifiers
- Understand the basic three-stage operational amplifier circuit
- Explore the design of Class-A, Class-B, and Class AB output stages
- Discuss the characteristics and design of electronic current sources
- Understand bipolar and MOS current mirror operation and mirror ratio errors
- Explore the high output resistance cascode and Wilson current sources
- Add reference current circuit techniques to our kit of circuit building blocks
- Use current mirrors as active loads in differential and operational amplifiers

- Learn how to analyze the effects of device and component mismatch on the performance of symmetrical amplifier circuits
- Analyze the design of the classic μA741 operational amplifier
- Continue to increase our understanding of SPICE simulation techniques

In most situations, a single-transistor amplifier cannot meet all the given specifications. The required voltage gain often exceeds the amplification factor of a single transistor, or the combination of voltage gain, input resistance, and output resistance cannot be met simultaneously. For example, consider the specifications of a good general-purpose operational amplifier. Such an amplifier has an input resistance exceeding 1 MΩ, a voltage gain of 100,000, and an output resistance of less than 100 Ω. It should be clear from our investigation of amplifiers in Chapters 13 and 14 that these requirements cannot all be met simultaneously with a single-transistor amplifier. A number of stages must be cascaded in order to create an amplifier that can meet all these requirements.

Chapter 15 continues our study of combining single-transistor amplifier stages to achieve higher levels of overall performance. ac-coupled amplifiers discussed in Chapter 14 eliminate dc interactions between the various stages forming the amplifier, thus simplifying bias circuit design. On the other hand, in our work in Chapters 11 and 12, most of the operational amplifier circuits provided amplification of dc signals. To realize amplifiers of this type, coupling capacitors that block dc signal flow through the amplifier must be eliminated, which leads to the concept of direct-coupled or dc-coupled amplifiers that can satisfy the requirement for dc amplification. In the dc-coupled case, the operating point of one stage is dependent on the Q-point of the other stages, making the dc design somewhat more complex.

The most important dc-coupled amplifier is the symmetric two-transistor differential amplifier. Not only is the differential amplifier a key circuit in the design of operational amplifiers, it is also a fundamental building block in all analog IC design. In this chapter, we present the

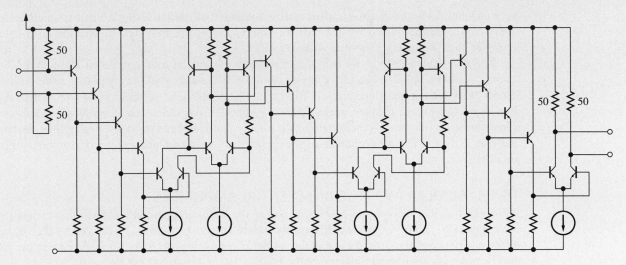

Schematic of a limiting amplifier in bipolar technology
(Copyright 2002 IEEE. Reprinted with permission.)

transistor-level implementation of BJT and FET differential amplifiers and explore how the differential-mode and common-mode gains, common-mode rejection ratio, differential-mode and common-mode input resistances, and output resistance of the amplifier are all related to transistor parameters.

Subsequently, a second gain stage and an output stage are added to the differential amplifier, creating the prototype for a basic operational amplifier. The definitions of class-A, class-B, and class-AB amplifiers are introduced, and the basic op amp design is further improved by adding class-B and class-AB output stages. In audio applications, these output stages often use transformer coupling.

Bias for analog circuits is most often provided by current sources. An ideal current source provides a fixed output current, independent of the voltage across the source; that is, the current source has an infinite output resistance. Electronic current sources cannot achieve infinite output resistance, but very high values are possible, and a number of basic current source circuits and techniques for achieving high output resistance are introduced and compared. Analysis of the various current sources uses the single-stage amplifier results from Chapters 13 and 14.

In this chapter, we also explore the design of the classic μA741 op amp. The μA741 circuit techniques spawned a broad range of follow-on designs that are still in use today.

Integrated circuit (IC) technology allows the realization of large numbers of virtually identical transistors. Although the absolute parameter tolerances of these devices are relatively poor, device characteristics can be matched to within 1 percent or better. The ability to build devices with nearly identical characteristics has led to the development of special circuit techniques that take advantage of the tight matching of the device characteristics.

We explore the use of matched transistors in the design of current sources, called **current mirrors,** in both MOS and bipolar technology. The cascode and Wilson current sources are subsequently added to our repertoire of high-output-resistance current source circuits. Circuit techniques that can be used to achieve **power supply independent biasing** are also introduced.

The current mirror is often used to bias analog circuits and to replace load resistors in differential and operational amplifiers. This active-load circuit can substantially enhance the voltage gain capability of many amplifiers, and a number of MOS and bipolar circuit examples are presented. The chapter then discusses circuit techniques used in IC operational amplifiers, including the classic 741 amplifier. This design was the first to provide a robust, high-performance, general-purpose operational amplifier with breakdown-voltage protection of the input stage and short-circuit protection of the output stage.

15.1 DIFFERENTIAL AMPLIFIERS

The coupling capacitors that were discussed in Chapter 14 limit the low-frequency response of the amplifiers and prevent their application as dc amplifiers. For an amplifier to provide gain at dc, capacitors in series with the signal path (e.g., C_1, C_3, C_5, and C_6 in Fig. 14.38) must be eliminated.

Such an amplifier is called a **dc-coupled** or **direct-coupled amplifier.** Using a direct-coupled design can also eliminate additional resistors that are required to bias the individual stages in an ac-coupled amplifier, thus producing a less expensive amplifier.

The dc-coupled differential amplifier represents one of the most important additions to our "toolkit" of basic building blocks for analog design. Differential amplifiers appear in some form in almost every analog integrated circuit! This circuit forms the heart of operational amplifier design as well as of most dc-coupled analog circuits. Although the differential amplifier contains two transistors in a symmetrical configuration, it is usually thought of as a single-stage amplifier, and our analyses will show that it has characteristics similar to those of common-emitter or common-source amplifiers.

15.1.1 BIPOLAR AND MOS DIFFERENTIAL AMPLIFIERS

Figure 15.1 shows bipolar and MOS versions of the differential amplifier. Each circuit has two input terminals, v_1 and v_2, and the **differential-mode output voltage** v_{OD} is defined by the voltage difference between the collectors or drains of the two transistors. Ground-referenced outputs can also be taken between either collector or drain—v_{C1}, v_{C2}, v_{D1}, or v_{D2}—and ground.

The symmetrical nature of the amplifier provides useful dc and ac properties. However, ideal performance is obtained from the differential amplifier only when it is perfectly symmetrical, and the best versions are built using IC technology in which the transistor characteristics can be closely matched. Two transistors are said to be **matched** if they have identical characteristics and parameter values; that is, the parameter sets (I_S, β_{FO}, V_A) or $(K_n, V_{TN}, \text{and } \lambda)$, Q-points, and temperatures of the two transistors are identical.

15.1.2 dc ANALYSIS OF THE BIPOLAR DIFFERENTIAL AMPLIFIER

The quiescent operating points of the transistors in the bipolar differential amplifier can be found by setting both input signal voltages to zero, as in Fig. 15.2. In this circuit, both bases are grounded and the two emitters are connected together. Therefore, $V_{BE1} = V_{BE2} = V_{BE}$. If bipolar transistors Q_1 and Q_2 are assumed to be matched, then the symmetry of the circuit also forces $V_{C1} = V_{C2} = V_C$, and the terminal currents of the two transistors are identical: $I_{C1} = I_{C2} = I_C$, $I_{E1} = I_{E2} = I_E$, and $I_{B1} = I_{B2} = I_B$.

The emitter currents can be found by writing a loop equation starting at the base of Q_1:

$$V_{BE} + 2I_E R_{EE} - V_{EE} = 0 \qquad \text{and} \qquad I_C = \alpha_F I_E = \alpha_F \frac{V_{EE} - V_{BE}}{2R_{EE}} \qquad (15.1)$$

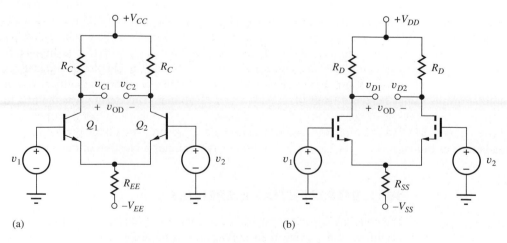

(a)   (b)

Figure 15.1 (a) Bipolar and (b) MOS differential amplifiers.

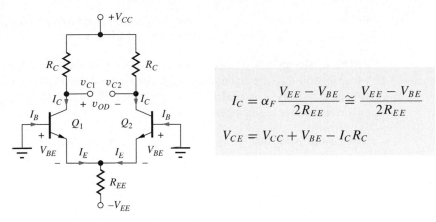

Figure 15.2 Circuit for dc analysis of the bipolar differential amplifier.

with

$$I_B = I_C / \beta_F$$

The voltages at the two collectors are equal to

$$V_{C1} = V_{C2} = V_{CC} - I_C R_C \qquad (15.2)$$

and $V_{CE1} = V_{CE2} = V_{CC} + V_{BE} - I_C R_C$. For the symmetrical amplifier, the dc output voltage is zero:

$$V_{OD} = V_{C1} - V_{C2} = 0 \text{ V} \qquad (15.3)$$

EXAMPLE 15.1 DIFFERENTIAL AMPLIFIER Q-POINT ANALYSIS

In this example, we determine the Q-point for an "emitter-coupled pair" of bipolar transistors.

PROBLEM Find the Q-points plus V_C and I_B for the differential amplifier in Fig. 15.1(a) if $V_{CC} = V_{EE} = 15$ V, $R_{EE} = 75$ kΩ, $R_C = 75$ kΩ, and $\beta_F = 100$.

SOLUTION **Known Information and Given Data:** Circuit topology appears in Fig. 15.1(a); symmetrical 15-V power supplies are used to operate the circuit; $R_C = R_{EE} = 75$ kΩ; $\beta_F = 100$.

Unknowns: I_C, V_{CE}, V_C, I_B for Q_1 and Q_2

Approach: Use the circuit element values and follow the analysis presented in Eqs. (15.1) through (15.3).

Assumptions: Active region operation with $V_{BE} = 0.7$ V; $V_A = \infty$

Analysis: Using Eqs. (15.1) and (15.2):

$$I_E = \frac{V_{EE} - V_{BE}}{2R_{EE}} = \frac{(15 - 0.7) \text{ V}}{2(75 \times 10^3 \text{ }\Omega)} = 95.3 \text{ }\mu\text{A}$$

$$I_C = \alpha_F I_E = \frac{100}{101} I_E = 94.4 \text{ }\mu\text{A} \qquad I_B = \frac{I_C}{\beta_F} = \frac{94.4 \text{ }\mu\text{A}}{100} = 0.944 \text{ }\mu\text{A}$$

$$V_C = 15 - I_C R_C = 15 \text{ V} - (9.44 \times 10^{-5} \text{ A})(7.5 \times 10^4 \text{ }\Omega) = 7.92 \text{ V}$$

$$V_{CE} = V_C - V_E = 7.92 \text{ V} - (-0.7 \text{ V}) = 8.62 \text{ V}$$

Because of the circuit symmetry, both transistors in the differential amplifier are biased at a Q-point of (94.4 μA, 8.62 V) with $I_B = 0.944$ μA and $V_C = 7.92$ V.

Check of Results: A double check of results indicates the calculations are correct. Note that when R_C and R_{EE} are equal, the voltage drop across R_C should be approximately one-half of the voltage across R_{EE}. Our calculations agree with this result. Also, $V_{CE} > V_{BE}$, so the assumption of forward-active region operation is correct.

Discussion: Note, that for $V_{EE} \gg V_{BE}$, I_E can be approximated by

$$I_E \cong \frac{V_{EE}}{2R_{EE}} = \frac{15 \text{ V}}{150 \text{ k}\Omega} = 100 \text{ μA}$$

This estimate represents only a 6 percent error compared to the more accurate calculation.

Computer-Aided Analysis: SPICE analysis with BF = 100 and IS = 5×10^{-16} A yields a Q-point of (94.6 μA, 8.57 V) with $V_{BE} = 0.672$ V. The collector voltage and base current values are 7.91 V and 0.946 μA, respectively, all in agreement with our hand calculations. We can also use SPICE to explore the effect of a nonzero Early voltage on the Q-point of the differential amplifier. A second simulation with VAF = 50 V yields Q-point values of (94.7 μA, 8.56 V). Almost no changes can be observed in the Q-point values! The collector voltage and base current values are now 7.90 V and 0.818 μA, respectively. Since the collector current has not changed, V_C also has not changed. However, the base current has been reduced by 14 percent. We should wonder why this has occurred. Remember that the current gain of the transistor in our transport model is given by $\beta_F = \beta_{FO}(1 + V_{CE}/V_A)$. Also, remember that a slightly different form is used in SPICE:

$$\beta_F = \beta_{FO}\left(1 + \frac{V_{CB}}{V_A}\right) = \beta_{FO}\left(1 + \frac{V_{CE} - V_{BE}}{V_A}\right) = 100\left(1 + \frac{7.90}{50}\right) = 116$$

and there is our 14 percent discrepancy!

EXERCISE: What is the Q-point if β_F is 60 instead of 100?

ANSWER: (93.7 μA, 8.67 V)

EXERCISE: What are the actual values of I_C and V_{BE} for the transistor if the transistor saturation current is 0.5 fA?

ANSWER: 0.649 V for $V_T = 25$ mV; 94.7 μA

EXERCISE: Draw a *pnp* version of the differential amplifier in Fig. 15.1(a).

ANSWER: See Fig. P15.14.

15.1.3 TRANSFER CHARACTERISTIC FOR THE BIPOLAR DIFFERENTIAL AMPLIFIER

The differential amplifier provides advantages in terms of signal range and distortion characteristics over that of a single bipolar transistor. We can explore these advantages using results already derived for the current switch in Sec. 9.1.1. The current switch simply represents a digital application of the differential amplifier, and the large-signal transfer characteristic of the differential amplifier is the same as that presented in Eq. (9.6) and repeated here (with $\alpha_F I_{EE} = 2I_C$).

$$i_{C1} - i_{C2} = 2I_C \tanh\left(\frac{v_{BE1} - v_{BE2}}{2V_T}\right) = 2I_C \tanh\left(\frac{v_{id}}{2V_T}\right)$$

$$G_m = \frac{d(i_{C1} - i_{C2})}{dv_{id}} = 2I_C \operatorname{sech}^2\left(\frac{v_{id}}{2V_T}\right)$$

(15.4)

for the symmetrical differential amplifier with $v_{BE1} = V_{BE} + \dfrac{v_{id}}{2}$ and $v_{BE2} = V_{BE} - \dfrac{v_{id}}{2}$.

Expanding the hyperbolic tangent using its Maclaurin series yields

$$I_{C1} - I_{C2} = 2I_C \left[\left(\frac{v_{id}}{2V_T}\right) - \frac{1}{3}\left(\frac{v_{id}}{2V_T}\right)^3 + \frac{2}{15}\left(\frac{v_{id}}{2V_T}\right)^5 - \frac{17}{315}\left(\frac{v_{id}}{2V_T}\right)^7 + \cdots\right] \quad (15.5)$$

First we see that subtraction of the two collector currents eliminates the even order distortion terms in Eq. (15.5). Second, for small-signal operation, we desire the linear term to be dominant. Setting the third-order term to be one-tenth of the linear term requires $v_{id} \leq 2V_T\sqrt{0.3}$ or $v_{id} \leq 27$ mV. On the surface, one would expect an increase by a factor of 2 (to 10 mV) since the input signal is shared equally by the two transistors of the differential pair. However, cancellation of the even-order distortion terms further increases the signal-handling capability of the differential pair! This expanded linear region of the transfer function can clearly be seen at the center of the plot of Eq. (15.4) that appears in Fig. 15.3.

The transconductance (G_m) of the differential pair defined in Eq. (15.4) as the derivative of the transfer characteristic is also plotted in Fig. 15.3 as a function of the normalized input voltage. The value of G_m peaks when the pair is balanced with $i_{C1} = i_{C2}$ and falls to nearly zero for $|\mathbf{v_{id}}| > 6V_T (150\,\text{mV})$.

15.1.4 ac ANALYSIS OF THE BIPOLAR DIFFERENTIAL AMPLIFIER

The ac analysis of the differential amplifier can be simplified by breaking input sources v_1 and v_2 into their equivalent differential-mode input $(\mathbf{v_{id}})$ and common-mode input $(\mathbf{v_{ic}})$ signal components, shown in Fig. 15.4, and defined by

$$\mathbf{v_{id}} = \mathbf{v_1} - \mathbf{v_2} \quad \text{and} \quad \mathbf{v_{ic}} = \frac{\mathbf{v_1} + \mathbf{v_2}}{2} \quad (15.6)$$

The input voltages can be written in terms of $\mathbf{v_{ic}}$ and $\mathbf{v_{id}}$ as

$$\mathbf{v_1} = \mathbf{v_{ic}} + \frac{\mathbf{v_{id}}}{2} \quad \text{and} \quad \mathbf{v_2} = \mathbf{v_{ic}} - \frac{\mathbf{v_{id}}}{2} \quad (15.7)$$

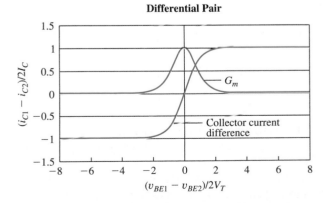

Figure 15.3 Large-signal transfer characteristic and transconductance for the bipolar differential pair.

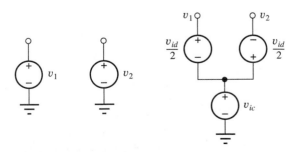

Figure 15.4 Definition of the differential-mode (v_{id}) and common-mode (v_{ic}) input voltages.

Circuit analysis can then be performed using superposition of the differential-mode and common-mode input signal components. This technique was originally used in our study of operational amplifiers in Chapter 12.

The differential-mode and common-mode output voltages, v_{od} and v_{oc}, are defined in a similar manner:

$$v_{od} = v_{c1} - v_{c2} \quad \text{and} \quad v_{oc} = \frac{v_{c1} + v_{c2}}{2} \tag{15.8}$$

For the general amplifier case, the voltages v_{od} and v_{oc} are functions of both v_{id} and v_{ic} and can be written as

$$\begin{bmatrix} v_{od} \\ v_{oc} \end{bmatrix} = \begin{bmatrix} A_{dd} & A_{cd} \\ A_{dc} & A_{cc} \end{bmatrix} = \begin{bmatrix} v_{id} \\ v_{ic} \end{bmatrix} \tag{15.9}$$

in which four gains are defined:

$$A_{dd} = \textbf{differential-mode gain}$$

$$A_{cd} = \textbf{common-mode (to differential-mode) conversion gain}$$

$$A_{cc} = \textbf{common-mode gain}$$

$$A_{dc} = \textbf{differential-mode (to common-mode) conversion gain}$$

For an ideal symmetrical amplifier with matched transistors, A_{cd} and A_{dc} are zero, and Eq. (15.9) reduces to

$$\begin{bmatrix} v_{od} \\ v_{oc} \end{bmatrix} = \begin{bmatrix} A_{dd} & 0 \\ 0 & A_{cc} \end{bmatrix} \begin{bmatrix} v_{id} \\ v_{ic} \end{bmatrix} \tag{15.10}$$

In this case, a differential-mode input signal produces a purely differential-mode output signal, and a purely common-mode input produces only a common-mode output.

However, when the differential amplifier is not completely balanced because of transistor or other circuit mismatches, A_{dc} or A_{cd} are no longer zero. In upcoming discussions, we assume that the transistors are identical unless stated otherwise.

EXERCISE: Measurement of a differential amplifier yielded the following sets of values:

$$v_{od} = 2.2 \text{ V and } v_{oc} = 1.002 \text{ V} \quad \text{for } v_1 = 1.01 \text{ V and } v_2 = 0.990 \text{ V}$$

$$v_{od} = 0 \text{ V and } v_{oc} = 5.001 \text{ V} \quad \text{for } v_1 = 4.995 \text{ V and } v_2 = 5.005 \text{ V}$$

What are v_{id} and v_{ic} for the two cases? What are the values of A_{dd}, A_{cd}, A_{dc}, and A_{cc} for the amplifier?

ANSWERS: 0.02 V, 1.00 V; −0.01 V, 5.00 V; 100, 0.200, 0.100, 1.00

15.1.5 DIFFERENTIAL-MODE GAIN AND INPUT RESISTANCE

Purely differential-mode input signals are applied to the differential amplifier in Fig. 15.5, and the two transistors are replaced with their small-signal models in Fig. 15.6. We want to find the gain for both differential and single-ended outputs as well as the input and output resistances. Because the transistors have resistor loads, the output resistances will be neglected in the calculations.

Summing currents at the emitter node in Fig. 15.6:

$$g_\pi v_3 + g_m v_3 + g_m v_4 + g_\pi v_4 = G_{EE} v_e \quad \text{or} \quad (g_m + g_\pi)(v_3 + v_4) = G_{EE} v_e \tag{15.11}$$

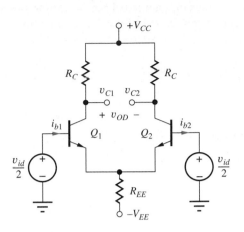

Figure 15.5 Differential amplifier with a differential-mode input signal.

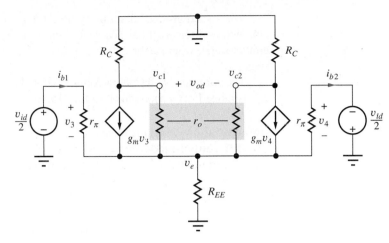

Figure 15.6 Small-signal model for differential-mode inputs. The output resistances are neglected in the calculations.

These equations have been simplified by representing resistances r_π and R_{EE} with their equivalent conductances g_π and G_{EE}. The base-emitter voltages are

$$\mathbf{v_3} = \frac{\mathbf{v_{id}}}{2} - \mathbf{v_e} \quad \text{and} \quad \mathbf{v_4} = -\frac{\mathbf{v_{id}}}{2} - \mathbf{v_e} \qquad (15.12)$$

giving $\mathbf{v_3} + \mathbf{v_4} = -2\mathbf{v_e}$. Combining Eq. (15.12) with Eq. (15.11) yields

$$\mathbf{v_e}(G_{EE} + 2g_\pi + 2g_m) = 0 \qquad (15.13)$$

which requires $\mathbf{v_e} = 0$.

For a purely differential-mode input voltage, the voltage at the emitter node is identically zero. This is an extremely important result. The "virtual ground" at the emitter node causes the differential amplifier to behave as a common-emitter (or common-source) amplifier.

DESIGN NOTE

The emitter node in the differential amplifier represents a **virtual ground** for differential-mode input signals.

Because the voltage at the emitter node is zero, Eq. (15.12) yields

$$\mathbf{v_3} = \frac{\mathbf{v_{id}}}{2} \quad \text{and} \quad \mathbf{v_4} = -\frac{\mathbf{v_{id}}}{2} \qquad (15.14)$$

and the output signal voltages are

$$\mathbf{v_{c1}} = -g_m R_C \frac{\mathbf{v_{id}}}{2} \qquad \mathbf{v_{c2}} = +g_m R_C \frac{\mathbf{v_{id}}}{2} \qquad \mathbf{v_{od}} = -g_m R_C \mathbf{v_{id}} \qquad (15.15)$$

The differential-mode gain A_{dd} for a **balanced output,** $\mathbf{v_{od}} = \mathbf{v_{c1}} - \mathbf{v_{c2}}$, is

$$A_{dd} = \left.\frac{\mathbf{v_{od}}}{\mathbf{v_{id}}}\right|_{\mathbf{v_{ic}}=0} = -g_m R_C \qquad (15.16)$$

If either $\mathbf{v_{c1}}$ or $\mathbf{v_{c2}}$ alone is used as the output, referred to as a **single-ended** (or ground-referenced) **outputs,** then

$$A_{dd1} = \left.\frac{\mathbf{v_{c1}}}{\mathbf{v_{id}}}\right|_{\mathbf{v_{ic}}=0} = -\frac{g_m R_C}{2} = \frac{A_{dd}}{2} \quad \text{or} \quad A_{dd2} = \left.\frac{\mathbf{v_{c2}}}{\mathbf{v_{id}}}\right|_{\mathbf{v_{ic}}=0} = +\frac{g_m R_C}{2} = -\frac{A_{dd}}{2} \quad (15.17)$$

depending on which output is selected.

The virtual ground condition at the emitter node causes the amplifier to behave as a single-stage common-emitter amplifier. The balanced differential output provides the full gain of a common-emitter stage, whereas the output at either collector provides a gain equal to one-half that of the C-E stage.

The common-mode output voltage, defined by Eq. (15.8), is zero since $v_{c2} = -v_{c1}$, and therefore A_{dc} is indeed zero, as assumed in Eq. (15.10).

Differential-Mode Input Resistance

The **differential-mode input resistance R_{id}** represents the small-signal resistance presented to the full differential-mode input voltage appearing between the two bases of the transistors. R_{id} is defined as

$$R_{id} = \frac{\mathbf{v_{id}}}{\mathbf{i_{b1}}} = 2r_\pi \qquad \text{because} \qquad \mathbf{i_{b1}} = \frac{\dfrac{\mathbf{v_{id}}}{2}}{r_\pi} \qquad (15.18)$$

If $\mathbf{v_{id}}$ is set to zero in Fig. 15.6, then $g_m\mathbf{v_3}$ and $g_m\mathbf{v_4}$ are zero, and the **differential-mode output resistance R_{od}** is equal to

$$R_{od} = 2(R_C \| r_o) \cong 2R_C \qquad (15.19)$$

For single-ended outputs,

$$R_{\text{out}} \cong R_C \qquad (15.20)$$

 **DESIGN NOTE**

The differential pair behaves the same as a common-emitter or common-source amplifier for differential-mode input signals.

15.1.6 COMMON-MODE GAIN AND INPUT RESISTANCE

Purely common-mode input signals are applied to the differential amplifier in Fig. 15.7. For this case, both sides of the amplifier are completely symmetrical. Thus, the two base currents, the emitter currents, the collector currents, and the two collector voltages must be equal. Using this symmetry as a basis, the output voltage can be developed by writing a loop equation including either base-emitter junction.

For the small-signal model in Fig. 15.8,

$$\mathbf{v_{ic}} = \mathbf{i_b}r_\pi + \mathbf{v_e} = \mathbf{i_b}[r_\pi + 2(\beta_o + 1)R_{EE}] \qquad \text{and} \qquad \mathbf{i_b} = \frac{\mathbf{v_{ic}}}{r_\pi + 2(\beta_o + 1)R_{EE}} \qquad (15.21)$$

The output voltage at either collector is given by

$$\mathbf{v_{c1}} = \mathbf{v_{c2}} = -\beta_o\mathbf{i_b}R_C = \frac{-\beta_o R_C}{r_\pi + 2(\beta_o + 1)R_{EE}}\mathbf{v_{ic}} \qquad (15.22)$$

and the voltage at the emitter is

$$\mathbf{v_e} = 2(\beta_o + 1)\mathbf{i_b}R_{EE} = \frac{2(\beta_o + 1)R_{EE}}{r_\pi + 2(\beta_o + 1)R_{EE}}\mathbf{v_{ic}} \cong \mathbf{v_{ic}} \qquad (15.23)$$

The common-mode output voltage $\mathbf{v_{oc}}$ is defined by Eq. (15.8), and A_{cc} is given by

$$A_{cc} = \left.\frac{\mathbf{v_{oc}}}{\mathbf{v_{ic}}}\right|_{\mathbf{v_{id}}=0} = -\frac{\beta_o R_C}{r_\pi + 2(\beta_o + 1)R_{EE}} \cong -\frac{R_C}{2R_{EE}} \qquad (15.24)$$

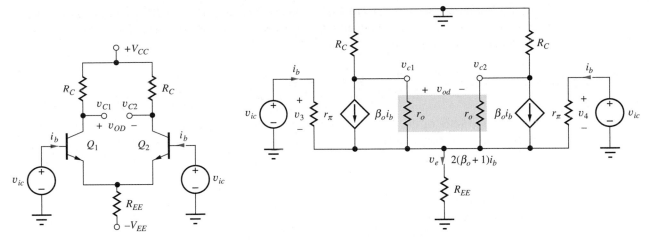

Figure 15.7 Differential amplifier with purely common-mode input.

Figure 15.8 Small-signal model with common-mode input. The output resistances are neglected in the calculations.

for $\beta_o \gg 1$. By multiplying and dividing Eq. (15.24) by collector current I_C, Eq. (15.24) can be rewritten as

$$A_{cc} = -\frac{I_C R_C}{2 I_C R_{EE}} \cong \frac{\dfrac{V_{CC}}{2}}{2 I_E R_{EE}} = \frac{V_{CC}}{2(V_{EE} - V_{BE})} \cong \frac{V_{CC}}{2 V_{EE}} \qquad (15.25)$$

where it is assumed that $\alpha_F = \alpha_o$ and $I_C R_C = V_{CC}/2$. In Eq. (15.25) we see that the common-mode gain A_{cc} is determined by the ratio of the two power supplies, and for symmetrical supplies, $A_{cc} = 0.5$. Note that the result in Eq. (15.25) only applies to the differential amplifier biased by resistor R_{EE}. We will shortly improve this result by replacing R_{EE} with an electronic current source.

The differential output voltage $\mathbf{v_{od}}$ is identically zero because the voltages are equal at the two collectors: $\mathbf{v_{od}} = \mathbf{v_{c1}} - \mathbf{v_{c2}} = 0$. Therefore, the common-mode conversion gain for a differential output is also 0, as assumed in Eq. (15.10):

$$A_{cd} = \left. \frac{\mathbf{v_{od}}}{\mathbf{v_{ic}}} \right|_{\mathbf{v_{id}}=0} = 0 \qquad (15.26)$$

The result in Eq. (15.24) indicates that the common-mode output voltage and A_{cc} tend toward zero as R_{EE} approaches infinity. This is another suspicious result, and it is in fact a direct consequence of neglecting the output resistances in the circuit in Fig. 15.8. If r_o is included, a small current $v_{ic}/\beta_o r_o$ results from the finite current gain of the BJT and appears in the collector terminal. A more accurate expression for the common-mode gain is

$$A_{cc} \cong R_C \left(\frac{1}{\beta_o r_o} - \frac{1}{2 R_{EE}} \right) \qquad (15.27)$$

Now for infinite R_{EE}, we find that A_{cc} is limited to $R_C/\beta_o r_o \cong V_{CC}/2\beta_o V_A$. It is also interesting to note that the sign difference allows a theoretical cancellation to occur. (See Prob. 15.125.)

Common-Mode Input Resistance

The **common-mode input resistance** is determined by the total signal current ($2i_b$) being supplied from the common-mode source and can be calculated using Eq. (15.21):

$$R_{ic} = \frac{\mathbf{v_{ic}}}{2\mathbf{i_b}} = \frac{r_\pi + 2(\beta_o + 1) R_{EE}}{2} = \frac{r_\pi}{2} + (\beta_o + 1) R_{EE} \qquad (15.28)$$

Equations (15.21), (15.22), (15.23), and the numerator of Eq. (15.28) should be recognized as those of a common-emitter amplifier with a resistor of value $2R_{EE}$ in the emitter. This observation is discussed in detail shortly.

DESIGN NOTE

The characteristics of the differential pair with a common-mode input are similar to those of a common-emitter (common-source) amplifier with a large emitter (source) resistor.

15.1.7 COMMON-MODE REJECTION RATIO (CMRR)

As defined in Chapter 11, the **common-mode rejection ratio,** or **CMRR,** characterizes the ability of an amplifier to amplify the desired differential-mode input signal and reject the undesired common-mode input signal. For a general differential amplifier stage characterized by Eq. (12.31), CMRR is defined in Eq. (12.32) as

$$\text{CMRR} = \left| \frac{A_{dm}}{A_{cm}} \right| \tag{15.29}$$

where A_{dm} and A_{cm} are the overall differential-mode and common-mode gains.

For the differential amplifier, CMRR is dependent on the designer's choice of output voltage. For a differential output v_{od}, the common-mode gain of the balanced amplifier is zero, and the CMRR is infinite. However, if the output is taken from either collector, we have

$$v_{c1} = v_{oc} + \frac{v_{od}}{2} = A_{cc}v_{ic} + \frac{A_{dd}}{2}v_{id} \qquad \text{and} \qquad v_{c2} = v_{oc} - \frac{v_{od}}{2} = A_{cc}v_{ic} - \frac{A_{dd}}{2}v_{id} \tag{15.30}$$

using Eqs. (15.8) and (15.10). Based upon Eqs. (15.29) with (15.17) and (15.27), the CMRR is given by

$$\text{CMMR} = \left| \frac{A_{dm}}{A_{cm}} \right| = \left| \frac{\dfrac{A_{dd}}{2}}{A_{cc}} \right| = \left| \frac{\dfrac{g_m R_c}{2}}{R_c \left(\dfrac{1}{\beta_o r_o} - \dfrac{1}{2R_{EE}} \right)} \right| = \left| \frac{1}{2 \left(\dfrac{1}{\beta_o \mu_f} - \dfrac{1}{2 g_m R_{EE}} \right)} \right| \tag{15.31}$$

For infinite R_{EE}, CMMR $\cong \beta_o \mu_f / 2$ and is limited by the $\beta_o \mu_f$ product of the transistor. On the other hand, if the term containing R_{EE} is dominant, we find the commonly quoted result:

$$\text{CMRR} \cong g_m R_{EE} \tag{15.32}$$

Let us explore Eq. (15.32) a bit further by writing g_m in terms of the collector current.

$$\text{CMRR} = 40 I_C R_{EE} = 20(2 I_E R_{EE}) = 20(V_{EE} - V_{BE}) \cong 20 V_{EE} \tag{15.33}$$

For the differential amplifier biased by resistor R_{EE}, CMRR is limited by the available negative power supply voltage V_{EE}. Also observe that the differential-mode gain is determined by the positive power supply voltage, that is $A_{dd} = -20V_{CC}$ based on our design guide from Chapter 13 with $I_C R_C = V_{CC}/2$.

EXERCISE: Estimate the differential-mode gain, common-mode gain, and CMRR for a differential amplifier with V_{EE} and $V_{CC} = 15$ V if the differential output is used and if the output v_{C2} is used.

ANSWERS: $-300, 0, \infty$; $+150, -0.5, 49.5$ dB (a poor CMRR)

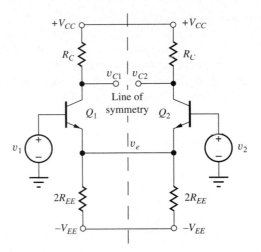

Figure 15.9 Circuit emphasizing symmetry of the differential amplifier.

Effects of Mismatches

Although the CMRR for an ideal differential amplifier with differential output is infinite, an actual amplifier will not be perfectly symmetrical because of mismatches in the transistors, and the two conversion gains A_{cd} and A_{dc} will not be zero. For this case, many of the errors will still be proportional to the result in Eq. (15.32) and will be of the form [1]:

$$\text{CMRR} \propto g_m R_{EE} \left(\frac{\Delta g}{g} \right) \tag{15.34}$$

in which the $\Delta g / g = 2(g_1 - g_2)/(g_1 + g_2)$ factor represents the fractional mismatch between the small-signal device parameters on the two sides of the differential amplifier (see Probs. 15.19 and 15.21). Therefore, maximizing the $g_m R_{EE}$ product is equally important to improving the performance of differential amplifiers with differential outputs.

15.1.8 ANALYSIS USING DIFFERENTIAL- AND COMMON-MODE HALF-CIRCUITS

We noted that the differential amplifier behaves much as the single-transistor common-emitter amplifier. The analogy can be carried even further using the **half-circuit** method of analysis, in which the symmetry of the differential amplifier is used to simplify the circuit analysis by splitting the circuit into **differential-mode** and **common-mode half-circuits.**

Half-circuits are constructed by first drawing the differential amplifier in a fully symmetric form, as in Fig. 15.9. To achieve full symmetry, the power supplies have been split into two equal value sources in parallel, and the emitter resistor R_{EE} has been separated into two equal parallel resistors, each of value $2R_{EE}$. It is important to recognize from Fig. 15.9 that these modifications have not changed any of the currents or voltages in the circuit.

Once the circuit is drawn in symmetrical form, two basic rules are used to construct the half-circuits: one for differential-mode signal analysis and one for common-mode signal analysis:

DESIGN NOTE   RULES FOR CONSTRUCTING HALF-CIRCUITS

Differential-mode signals   Points on the line of symmetry represent virtual grounds and can be connected to ground for ac analysis. (For example, remember that we found that $v_e = 0$ for differential-mode signals.)

Common-mode signals   Points on the line of symmetry can be replaced by open circuits. (No current flows through these connections.)

Differential-Mode Half Circuits

Applying the first rule to the circuit in Fig. 15.9 for differential-mode signals yields the circuit in Fig. 15.10(a). The two power supply lines and the emitter node all become ac grounds. (Of course, the power supply lines would become ac grounds in any case.) Simplifying the circuit yields the two differential-mode half-circuits in Fig. 15.10(b), each of which represents a common-emitter amplifier stage. The differential-mode behavior of the circuit, as described by Eqs. (15.15) to (15.20), can easily be found by direct analysis of the half-circuits:

and

$$\mathbf{v_{c1}} = -g_m R_C \frac{\mathbf{v_{id}}}{2} \qquad \mathbf{v_{c2}} = +g_m R_C \frac{\mathbf{v_{id}}}{2} \qquad \mathbf{v_o} = \mathbf{v_{c1}} - \mathbf{v_{c2}} = -g_m R_C \mathbf{v_{id}} \quad (15.35)$$

$$R_{id} = \frac{\mathbf{v_{id}}}{\mathbf{i_b}} = 2r_\pi \qquad \text{and} \qquad R_{od} = 2(R_C \| r_o) \qquad (15.36)$$

Common-Mode Half-Circuits

If the second rule is applied to the circuit in Fig. 15.9, all points on the line of symmetry become open circuits, and we obtain the circuit in Fig. 15.11. The common-mode half-circuits obtained from Fig. 15.11 are redrawn in Fig. 15.12. The dc circuit with V_{IC} set to zero in Fig. 15.12(a) is

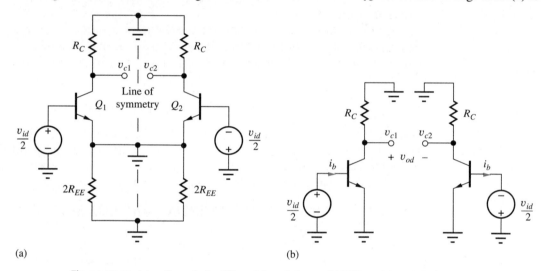

(a)   (b)

Figure 15.10   (a) ac Grounds for differential-mode inputs. (b) Differential-mode half-circuits.

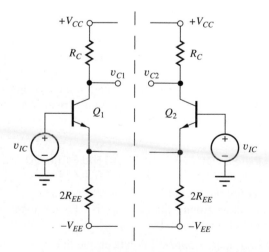

Figure 15.11   Construction of the common-mode half-circuit.

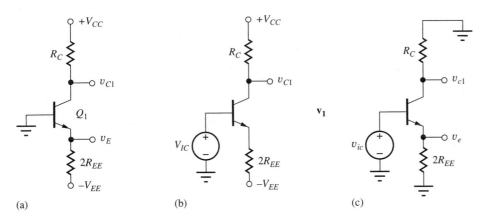

Figure 15.12 Common-mode half-circuits for (a) Q-point analysis, (b) dc common-mode input, and (c) common-mode signal analysis.

used to find the Q-point of the amplifier. The circuit in Fig. 15.12(b) should be used to find the operating point when a dc common-mode input is applied, and the ac circuit of Fig. 15.12(c) is used for common-mode signal analysis.

The common-mode half-circuit in Fig. 15.12(c) simply represents the common-emitter amplifier with an emitter resistor $2R_{EE}$, which was studied in great detail in Chapter 13. In addition, Eqs. (15.24), and (15.28) could have been written down directly using the results of our analysis from Chapter 13.

We can see that use of the differential-mode and common-mode half-circuits can greatly simplify the analysis of symmetric circuits. Half-circuit techniques are used shortly to analyze the MOS differential amplifier from Fig. 15.1.

Common-Mode Input Voltage Range

Common-mode input voltage range is another important consideration in the design of differential amplifiers. The upper limit to the dc common-mode input voltage V_{IC} in the circuit in Fig. 15.12(b) is set by the requirement that Q_1 remain in the forward-active region of operation. Writing an expression for the collector-base voltage of Q_1,

$$V_{CB} = V_{CC} - I_C R_C - V_{IC} \geq 0 \qquad \text{or} \qquad V_{IC} \leq V_{CC} - I_C R_C \tag{15.37}$$

in which

$$I_C = \alpha_F \frac{V_{IC} - V_{BE} + V_{EE}}{2R_{EE}} \tag{15.38}$$

Solving the preceding two equations for V_{IC} yields

$$V_{IC} \leq V_{CC} \frac{1 - \alpha_F \dfrac{R_C}{2R_{EE}} \dfrac{(V_{EE} - V_{BE})}{V_{CC}}}{1 + \alpha_F \dfrac{R_C}{2R_{EE}}} \tag{15.39}$$

For symmetrical power supplies, $V_{EE} \gg V_{BE}$, and with $R_C = R_{EE}$, Eq. (15.39) yields $V_{IC} \leq V_{CC}/3$.

Note from Eq. (15.38) that I_C changes as V_{IC} changes. The upper limit on V_{IC} is set by Eq. (15.39) and by the allowable shift in Q-point current as V_{IC} changes.

EXERCISE: Find the positive common-mode input voltage range for the differential amplifier in Fig. 15.7 if $V_{CC} = V_{EE} = 15$ V and $R_C = R_{EE}$.

ANSWER: $\cong 5.20$ V

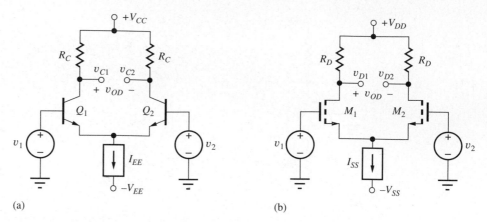

Figure 15.13 Differential amplifiers employing electronic current source bias.

Figure 15.14 Electronic current source and models.

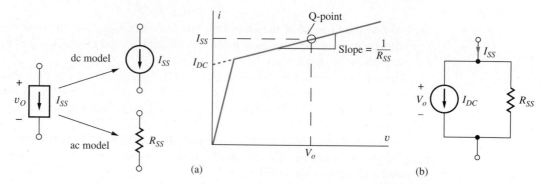

Figure 15.15 (a) i-v Characteristic for an electronic current source. (b) Proper SPICE representation of the electronic current source.

15.1.9 BIASING WITH ELECTRONIC CURRENT SOURCES

From Eqs. (15.1) and (15.2) we see that the Q-point of the differential amplifier is directly dependent on the value of the negative power supply, and from Eq. (15.31) we see that R_{EE} limits the CMRR. In order to remove these limitations, most differential amplifiers are biased using electronic current sources, which both stabilize the operating point of the amplifier and increase the effective value of R_{EE}. Electronic current source biasing of both the BJT and MOSFET differential amplifiers is shown in Fig. 15.13. In these circuits, the current source replaces resistor R_{EE} or R_{SS}.

The rectangular symbols in Figs. 15.13 and 15.14 denote an electronic current source with a finite output resistance, as shown graphically in the i-v characteristic in Fig. 15.15. The electronic source has a Q-point current equal to I_{SS} and an output resistance equal to R_{SS}.

For hand analysis using the dc equivalent circuit, we will replace the electronic current source with a dc current source of value I_{SS}. For ac analysis, the ac equivalent circuit is constructed by replacing the source with its output resistance R_{SS}. These substitutions are depicted symbolically in Fig. 15.14.

DESIGN NOTE

High common-mode rejection in differential amplifiers requires a large value of output resistance R_{SS} in the current source I_{SS} used for bias.

15.1.10 MODELING THE ELECTRONIC CURRENT SOURCE IN SPICE

Proper modeling of the electronic current source is slightly different in SPICE since the program must create its own dc and ac equivalent circuits. In order for SPICE to properly calculate the dc and ac behavior of a circuit, the network must contain both the dc current source and its output resistance R_{SS}. In the full circuit model in Fig. 15.15(b), a dc current will exist in the resistance R_{SS}, and the value of the current source in the SPICE circuit must be set to the value I_{DC} indicated in Fig. 15.15. I_{DC} represents the current in the equivalent circuit when voltage $V_o = 0$ and can be expressed as

$$I_{DC} = I_{SS} - \frac{V_o}{R_{SS}} \tag{15.40}$$

The equivalent circuit to be used in SPICE appears in Fig. 15.15(b). In cases where R_{SS} is very large, I_{DC} is approximately equal to I_{SS}.

> **EXERCISE:** Suppose an electronic current source has a current $I_{SS} = 100$ μA with an output resistance $R_{SS} = 750$ kΩ. (These values are representative of a single transistor current source operating at this current.) If $V_o = 15$ V, what is the value of I_{DC}?
>
> **ANSWER:** 80 μA

15.1.11 dc ANALYSIS OF THE MOSFET DIFFERENTIAL AMPLIFIER

MOSFETS provide very high input resistance and are often used in differential amplifiers implemented in CMOS and BiFET[1] technologies. In addition to high input resistance, op amps with FET inputs typically have a much higher slew rate than those with bipolar input stages.

The MOS version of the differential amplifier circuit appears in Fig. 15.13(b). We will use the MOSFET differential amplifier as our first direct application of half-circuit analysis. For dc analysis using half-circuits, the amplifier is redrawn in symmetrical form in Fig. 15.16(a). If the connections on the line of symmetry are replaced with open circuits, and the two input voltages are set to zero, we obtain in Fig. 15.16(b) the half-circuit needed for dc analysis.

It is immediately obvious from the dc half-circuit that the current in the source of the NMOS transistor must be equal to one-half of the bias current I_{SS}:

$$I_S = \frac{I_{SS}}{2} \tag{15.41}$$

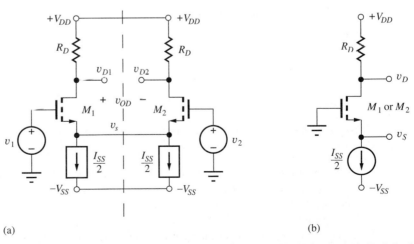

Figure 15.16 (a) Symmetric circuit representation of the MOS differential amplifier. (b) Half-circuit for dc analysis.

---

[1] BiFET technologies contain JFETs as well as bipolar transistors.

The gate-source voltage of the MOSFET can be determined directly from the drain-current expression for the transistor:

$$I_D = \frac{K_n}{2}(V_{GS} - V_{TN})^2 \quad \text{or} \quad V_{GS} = V_{TN} + \sqrt{\frac{2I_D}{K_n}} = V_{TN} + \sqrt{\frac{I_{SS}}{K_n}} \tag{15.42}$$

Note that $V_S = -V_{GS}$, and the voltages at both MOSFET drains are

$$V_{D1} = V_{D2} = V_{DD} - I_D R_D \quad \text{and} \quad V_O = 0 \tag{15.43}$$

Thus, the drain-source voltages are

$$V_{DS} = V_{DD} - I_D R_D + V_{GS} \tag{15.44}$$

EXAMPLE **15.2**   MOSFET DIFFERENTIAL AMPLIFIER ANALYSIS

A dc Q-point analysis is provided for the MOSFET differential amplifier in this example.

PROBLEM Find the Q-points for the MOSFETs in the differential amplifier in Fig. 15.13(b) if $V_{DD} = V_{SS} = 12$ V, $I_{SS} = 200$ μA, $R_{SS} = 500$ kΩ, $R_D = 62$ kΩ, $K_n = 5$ mA/V^2, $\lambda = 0.0133$ V^{-1}, and $V_{TN} = 1$ V. What is the maximum V_{IC} for which M_1 remains in the active region?

SOLUTION **Known Information and Given Data:** Circuit topology appears in Fig. 15.13(b); symmetrical 12-V power supplies are used to operate the circuit; $I_{SS} = 200$ μA, $R_D = 62$ kΩ, $V_{TN} = 1$ V, and $K_n = 5$ mA/V^2

Unknowns: I_D, V_{DS}, for M_1 and M_2, and maximum dc common-mode input voltage V_{IC}

Approach: Use the circuit element values and follow the analysis presented in Eq. (15.41) through (15.44).

Assumptions: Active region operation; ignore λ and R_{SS} for hand bias calculations

Analysis: Using Eqs. (15.41) through (15.44):

$$I_D = \frac{I_{SS}}{2} = 100 \text{ μA} \qquad V_{GS} = 1 + \sqrt{\frac{200 \text{ μA}}{5 \text{ mA/V}^2}} = 1.20 \text{ V}$$

$$V_{DS} = 12 \text{ V} - (100 \text{ μA})(62 \text{ kΩ}) + 1.2 \text{ V} = 7.00 \text{ V}$$

Thus, both transistors in the differential amplifier are biased at a Q-point of (100 μA, 7.00 V). Maintenance of pinch-off for M_1 for nonzero V_{IC} requires

$$V_{GD} = V_{IC} - (V_{DD} - I_D R_D) \leq V_{TN}$$

$$V_{IC} \leq V_{DD} - I_D R_D + V_{TN} = 6.8 \text{ V}$$

Check of Results: Checking for pinch-off, $V_{GS} - V_{TN} = 0.2$ V, and $V_{DS} \geq 0.2$. ✔

Discussion: Note that the drain currents are set by the current source and will be independent of device characteristics. This is demonstrated next using SPICE.

Computer-Aided Analysis: In our SPICE analysis, we can easily include λ and R_{SS} to see their impact on the Q-points of the transistors. Using Eq. (15.40) and the $V_{GS} = 1.2$ V as already calculated, the dc current source value for SPICE will be $200 - 21.6 = 178.4$ μA. We need to set up KP = 0.005 A/V^2, VTO = 1 V, and LAMBDA = 0.0133 V^{-1} in the SPICE device models. With these values, the Q-points from a SPICE operating point analysis are virtually the same as our hand calculations (100 μA, 6.99 V). Since the drain currents are locked by the current source, including λ causes only a small adjustment to occur in the value of gate-source voltage: $V_{GS} = 1.198$ V.

15.1.12 DIFFERENTIAL-MODE INPUT SIGNALS

The differential-mode and common-mode half-circuits for the differential amplifier in Fig. 15.16 are given in Fig. 15.17. In the differential-mode half-circuit, the MOSFET sources represent a virtual ground. In the common-mode circuit, the electronic current source has been modeled by twice its small-signal output resistance R_{SS}, representing the finite output resistance of the current source.

The differential-mode half-circuit represents a common-source amplifier, and the output voltages are given by

$$\mathbf{v_{d1}} = -g_m R_D \frac{\mathbf{v_{id}}}{2} \qquad \mathbf{v_{d2}} = +g_m R_D \frac{\mathbf{v_{id}}}{2} \qquad \mathbf{v_{od}} = -g_m R_D \mathbf{v_{id}} \tag{15.45}$$

The differential-mode gain is

$$A_{dd} = \left. \frac{\mathbf{v_{od}}}{\mathbf{v_{id}}} \right|_{\mathbf{v_{ic}}=0} = -g_m R_D \tag{15.46}$$

whereas taking the single-ended output between either drain and ground provides a gain of one-half A_{dd}:

$$A_{dd1} = \left. \frac{\mathbf{v_{d1}}}{\mathbf{v_{id}}} \right|_{\mathbf{v_{ic}}=0} = -\frac{g_m R_D}{2} = \frac{A_{dd}}{2} \qquad \text{and} \qquad A_{dd2} = \left. \frac{\mathbf{v_{d2}}}{\mathbf{v_{id}}} \right|_{\mathbf{v_{ic}}=0} = +\frac{g_m R_D}{2} = -\frac{A_{dd}}{2} \tag{15.47}$$

The differential-mode input and output resistances are infinite and $2R_D$, respectively:

$$R_{id} = \infty \qquad \text{and} \qquad R_{od} = 2R_D \tag{15.48}$$

The virtual ground at the source node causes the amplifier to again behave as a single stage inverting amplifier. A differential output provides the full gain of the common-source stage, whereas using the single-ended output at either drain reduces the gain by a factor of 2.

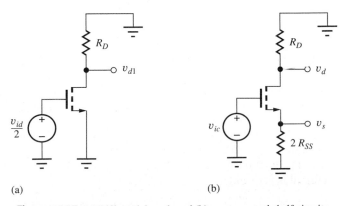

(a)   (b)

Figure 15.17 (a) Differential-mode and (b) common-mode half-circuits.

> **EXERCISE:** In a manner similar to the analysis of Fig. 15.8, derive directly from the full small-signal model the expressions for the differential-mode voltage gains of the MOS differential amplifier.

15.1.13 SMALL-SIGNAL TRANSFER CHARACTERISTIC FOR THE MOS DIFFERENTIAL AMPLIFIER

The MOS differential amplifier also provides improved linear input signal range and distortion characteristics over that of a single transistor. We can explore these advantages using the drain current expression for the MOSFET:

$$I_{D1} - I_{D2} = \frac{K_n}{2}[(v_{GS1} - V_{TN})^2 - (v_{GS2} - V_{TN})^2] \tag{15.49}$$

For the symmetrical differential amplifier with a purely differential-mode input, $v_{GS1} = V_{GS} + \dfrac{v_{id}}{2}$, $v_{GS2} = V_{GS} - \dfrac{v_{id}}{2}$, and

$$I_{D1} - I_{D2} = K_n(V_{GS} - V_{TN})v_{id} = g_m v_{id} \tag{15.50}$$

The second-order distortion product cancels out, and the output current expression is distortion free! As usual, we should question such a perfect result. In reality, MOSFETs are not perfect square-law devices, and some distortion will exist. There also will be distortion introduced through the voltage dependence of the output impedances of the transistors.

15.1.14 COMMON-MODE INPUT SIGNALS

The common-mode half-circuit in Fig. 15.17(b) is that of an inverting amplifier with a source resistor equal to $2R_{SS}$. Using the results from Chapter 14,

$$\mathbf{v_{d1}} = \mathbf{v_{d2}} = \frac{-g_m R_D}{1 + 2g_m R_{SS}}\mathbf{v_{ic}} \tag{15.51}$$

and the signal voltage at the source is

$$\mathbf{v_s} = \frac{2g_m R_{SS}}{1 + 2g_m R_{SS}}\mathbf{v_{ic}} \cong \mathbf{v_{ic}} \tag{15.52}$$

The differential output voltage is zero because the voltages are equal at the two drains:

$$\mathbf{v_{od}} = \mathbf{v_{d1}} - \mathbf{v_{d2}} = 0 \tag{15.53}$$

Thus, the common-mode conversion gain for a differential output is zero:

$$A_{cd} = \frac{\mathbf{v_{od}}}{\mathbf{v_{ic}}} = 0 \tag{15.54}$$

The common-mode gain is given by

$$A_{cc} = \frac{\mathbf{v_{oc}}}{\mathbf{v_{ic}}} = -\frac{g_m R_D}{1 + 2g_m R_{SS}} \cong -\frac{R_D}{2R_{SS}} \tag{15.55}$$

It should be noted, in contrast to the case of the BJT, that Eq. (15.55) is correct even if r_o is included because of the infinite current gain of the FET.

The common-mode input source is connected directly to the MOSFET gate. Thus, the input current is zero and

$$R_{ic} = \infty \tag{15.56}$$

Common-Mode Rejection Ratio (CMRR)

For a purely common-mode input signal, the output voltage of the balanced MOS amplifier is zero, and the CMRR is infinite. However, if a single-ended output is taken from either drain,

$$\text{CMRR} = \left| \frac{\dfrac{A_{dd}}{2}}{A_{cc}} \right| = \left| \frac{-\dfrac{g_m R_D}{2}}{-\dfrac{R_D}{2R_{SS}}} \right| = g_m R_{SS} \tag{15.57}$$

For high CMRR, a large value of R_{SS} is again desired. In Fig. 15.17, R_{SS} represents the output resistance of the current source in Fig. 15.13, and its value is much greater than resistor R_{EE}, which is used to bias the amplifier in Fig. 15.1. For this reason, as well as for Q-point stability, most differential amplifiers are biased by a current source, as in Fig. 15.13.

To compare the MOS amplifier more directly to the BJT analysis, however, let us assume for the moment that the MOS amplifier is biased by a resistor of value

$$R_{SS} = \frac{V_{SS} - V_{GS}}{I_{SS}} \tag{15.58}$$

Then Eq. (15.57) can be rewritten in terms of the circuit voltages, as was done for Eq. (15.33):

$$\text{CMRR} = \frac{2I_D R_{SS}}{V_{GS} - V_{TN}} = \frac{I_{SS} R_{SS}}{V_{GS} - V_{TN}} = \frac{(V_{SS} - V_{GS})}{V_{GS} - V_{TN}} \tag{15.59}$$

Using the numbers from the example,

$$\text{CMRR} = \frac{(V_{SS} - V_{GS})}{V_{GS} - V_{TN}} = \frac{(12 - 1.2)}{0.20} = 54 \tag{15.60}$$

—a paltry 35 dB. This is almost 10 dB worse than the result for the BJT amplifier. Because of the low values of CMRR in both the BJT and FET circuits, the use of current sources with much higher effective values of R_{SS} or R_{EE} is common in all differential amplifiers.

15.1.15 TWO-PORT MODEL FOR DIFFERENTIAL PAIRS

The ac analysis of circuits involving differential amplifiers can often be simplified by using the two-port small-signal model for the differential-pair appearing in Fig. 15.18. The two-port model can be substituted directly for the differential pair, or it can be used as a conceptual aid in simplifying circuits. The two current sources represent the signal currents generated by the two transistors in the pair. Resistors R_{oc} are the common-mode output resistances appearing at each collector or drain, D_1 and D_2, and R_{od} is the differential output resistance that appears between the two collectors or drains. (Remember for symmetrical differential-mode circuits, the node "x" will be a virtual ground.) For the pairs in Fig. 15.13, approximate expressions for the elements are

$$i_{dm} = g_m v_{dm} \qquad i_{cm} = \frac{g_m}{1 + 2g_m R_{EE}} v_{cm} \cong \frac{v_{cm}}{2R_{EE}} \tag{15.61}$$

$$R_{od} = 2r_o \qquad R_{oc} \cong 2\mu_f R_{EE}$$

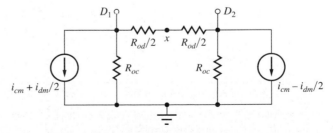

Figure 15.18 Two-port model for the differential pair.

Substitute R_{SS} for R_{EE} in these expressions for the FET case. We will make use of this two-port model later on in this and subsequent chapters.

EXERCISE: The bipolar differential amplifier in Fig. 15.13(a) is biased by a 75-μA current source with an output resistance of 1 MΩ. If the transistors have Early voltages of 60 V, estimate values of R_{od}, R_{oc}, i_{dm}, and i_{cm}.

ANSWERS: 3.2 MΩ; 4.9 GΩ; $1.50 \times 10^{-3} v_{dm}$; $5.00 \times 10^{-7} v_{cm}$

ELECTRONICS IN ACTION

Limiting Amplifiers for Optical Communications

Interface circuits for optical communications were introduced in the Electronics in Action features in both Chapters 9 and 11. Here we discuss the limiting amplifier (LA), another of the important electronic blocks on the receiver side of the fiber optic communication link. The LA amplifies the low level output voltage (e.g., 10 mV) of the transimpedance amplifier up to a level that can drive the clock and data recovery circuits (e.g., 250 mV).

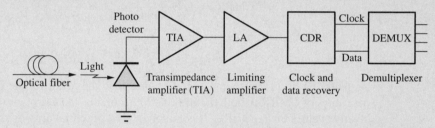

Optical fiber receiver block diagram

A typical limiting amplifier consists of a wide-band multistage dc-coupled amplifier similar to the one in the circuit schematic here [1–3]. The input signal from the transimpedance amplifier is buffered and level-shifted by two stages of emitter followers (2EF). This is followed

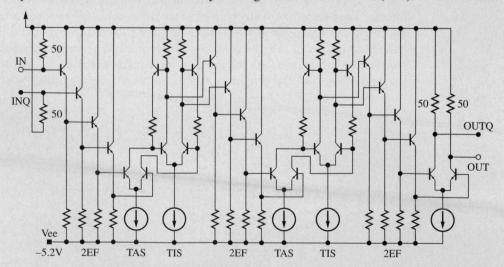

Schematic of a typical limiting amplifier in bipolar technology (Copyright 2002 IEEE. Reprinted with permission from [3]) (Note that this is a dc-coupled amplifier.)

by a transadmittance amplifier (TAS) that converts the voltage to a current and then drives a transimpedance amplifier (TIS) that converts the current back to a voltage. This TAS-TIS cascade was developed by Cherry and Hooper [4] and represents an important technique for realizing amplifiers with very wide bandwidth. The output is level-shifted by two more emitter followers and amplified by a second Cherry-and-Hooper stage. A third pair of emitter followers drives a differential amplifier with load resistors chosen to match a transmission line impedance of 50 Ω. Note that 50-Ω matching is used at the LA input as well.

We see that differential pairs are used throughout the limiting amplifier in the TAS and TIS stages, and in the gain stage at the output. Since these optical-to-electrical interface circuits typically push the state-of-the-art in speed, only *npn* transistors are used in the design. Remember that *npn* transistors are inherently faster than *pnp* transistors because of the mobility advantage of electrons over that of holes.

1. H-M. Rein, "Multi-gigabit-per-second silicon bipolar IC's for future optical-fiber transmission systems," *IEEE J. Solid-State Circuits,* vol. 23, no. 3, pp. 664–675, June 1988.
2. R. Reimann and H-M. Rein, "Bipolar high-gain limiting amplifier IC for optical-fiber receivers operating up to 4 Gbits/s," *IEEE J. Solid-State Circuits,* vol. 22, no. 4, pp. 504–510, August 1987.
3. Y. Baeyens et al., "InP D-HBT IC's for 40-Gb/s and higher bit rate lightwave transceivers," *IEEE J. Solid-State Circuits,* vol. 37, no. 9, pp. September 2002.
4. E. M. Cherry and D. E. Hooper, "The design of wide-band transistor feedback amplifiers," *Proc. Institute of Electrical Engineers,* vol. 110, pp. 375–389, February 1963.

EXAMPLE 15.3   DIFFERENTIAL AMPLIFIER DESIGN

Design a differential amplifier to meet a given set of specifications.

PROBLEM Design a differential amplifier stage with $A_{dd} = 40$ dB, $R_{id} \geq 250$ kΩ, and an input common-mode input range of at least ±5 V. Specify a current source to give CMRR of at least 80 dB for a single-ended output. MOSFETs are available with $K_n' = 50$ μA/V^2, $\lambda = 0.0133$ V^{-1}, and $V_{TN} = 1$ V. BJTs are available with $I_S = 0.5$ fA, $\beta_F = 100$, and $V_A = 75$ V.

SOLUTION **Known Information and Given Data:** Differential amplifier topologies appear in Fig. 15.13; $A_{dd} = 40$ dB, $R_{id} \geq 250$ kΩ, single-ended CMRR ≥ 80 dB, and $|V_{IC}| \geq 5$ V.

Unknowns: Power supply values, Q-points, R_C, bias source current and output resistance, M_1 and M_2, and maximum dc common-mode input voltage V_{IC}

Approach: Use theory developed in Ex. 15.2; choose transistor type and operating current based on A_{dm} and R_{id}; choose power supplies based on A_{dm}, V_{IC}, and small-signal range; choose current source output resistance to achieve desired CMRR.

Assumptions: Active region operation; symmetrical power supplies, $\beta_o = \beta_F$, $|v_{id}| \leq 30$ mV.

Analysis: 40 dB of gain corresponds to $A_{dd} = 100$. To achieve this gain with a resistively loaded amplifier, use of a BJT is indicated. For $A_{dd} = g_m R_C = 40 I_C R_C$, a gain of 100 can be achieved with a voltage drop of 2.5 V across the resistor R_C. For a bipolar differential amplifier, the input resistance $R_{id} = 2r_\pi$, so $r_\pi = 125$ kΩ, which requires

$$I_C \leq \frac{\beta_o V_T}{r_\pi} = \frac{100(0.025 \text{ V})}{125 \text{ kΩ}} = 20 \text{ μA}$$

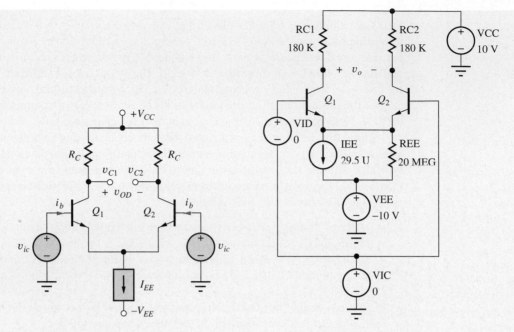

based on a current gain of 100. Let us choose $I_C = 15$ μA to provide some safety margin. Then, $R_C = 2.5$ V/15 μA = 167 kΩ. Choose $R_C = 180$ kΩ as the nearest value from the 5 percent resistor tables in Appendix A. (The larger value will also help compensate for our neglect of r_o in the gain calculation.)

A V_{IC} of 5 V requires the collector voltage of the BJT to be at least 5 V at all times. We do not know the signal level, but we know $|v_{id}| \leq 30$ mV for linearity in the differential pair. Thus, the ac component of the differential output voltage will be no greater than 100(0.03 V) = 3 V, half of which will appear at each collector. Thus the dc + ac signal across R_C will not exceed 4 V (2.5-V dc + 1.5-V ac), and the positive power supply must satisfy

$$V_{CC} \geq V_{IC} + 4 \text{ V} = 5 + 4 = 9 \text{ V}$$

Choosing $V_{CC} = 10$ V provides a design margin of 1 V. For symmetrical supplies, $-V_{EE} = -10$ V.
   The single-ended CMRR of 80 dB requires

$$R_{EE} \geq \frac{\text{CMRR}}{g_m} = \frac{10^4}{(40/\text{V})(15 \text{ μA})} = 16.7 \text{ MΩ}$$

A current source with $I_{EE} = 30$ μA and $R_{EE} \geq 20$ MΩ will provide some design margin.

Check of Results: Using the design values, $A_{dd} = 40(15 \text{ μA})(180 \text{ kΩ}) = 108$. CMRR = $40(15 \text{ μA})(20 \text{ MΩ}) = 12{,}000$ (81.6 dB), and $R_{id} = 2(2.5 \text{ V}/15 \text{ μA}) = 333$ kΩ. The bias voltages provide $V_{IC} = 6$ V. Thus the amplifier design should meet the specifications. We will check it further shortly with SPICE.

Discussion: Note that the drain currents are set by the current source and will be independent of device characteristics.

Computer-Aided Analysis: The SPICE schematic input appears in the figure above. Zero-value differential- and common-mode sources VID and VIC are for use in transfer function simulations. Requesting the transfer function from VID to output voltage v_O between the two collectors will produce values for A_{dd}, R_{id}, and R_{od}. Requesting the transfer function from VIC to either collector node produces values for A_{cc} and R_{ic}. In our SPICE analysis, we can easily include the Early voltage (set VAF = 75 V) and R_{EE} to see their impact on the Q-points of the transistors. Using Eq. (15.40) with $V_{BE} = (0.025 \text{ V}) \ln(15 \, \mu\text{A}/0.5 \text{ fA}) \cong 0.6 \text{ V}$, the dc current source value for SPICE will be $30 - 0.5 = 29.5 \, \mu\text{A}$.

With these values and REE = 20 MEG, the Q-points from a SPICE operating point analysis are virtually the same as our hand calculations (14.9 μA, 7.33 V). Using two transfer function analyses gives $A_{dd} = 100$, $R_{id} = 382 \text{ k}\Omega$, $R_{od} = 349 \text{ k}\Omega$, and $A_{cc} = 0.00416$, and we find the CMRR = 100/0.00416 = 24,000 or 87.6 dB.

The results of a transient simulation for the output voltage at the right-hand collector are given here for v_{id} equal to a 30-mV input sine wave at a frequency of 1 kHz and $V_{IC} = +5$ V. TSTART = 0, TSTOP = 0.002 s, and TSTEP = 0.001 ms (1 μs). As designed, we see an undistorted 1-kHz sine wave with an amplitude of 1.5 V biased at the Q-point level of 7.3 V.

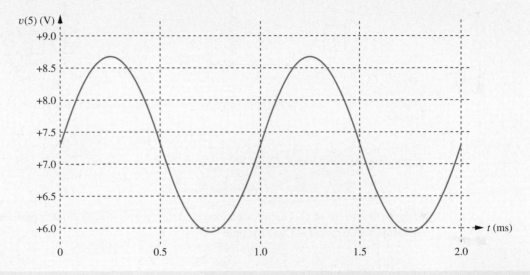

15.2 EVOLUTION TO BASIC OPERATIONAL AMPLIFIERS

One extremely important application of differential amplifiers is at the input stage of operational amplifiers. Differential amplifiers provide the desired differential input and common-mode rejection capabilities, and a ground-referenced signal is available at the output. However, an op amp usually requires higher voltage gain than is available from a single differential amplifier stage, and most op amps use two stages of gain. In addition, a third stage, the output stage, is added to provide low output resistance and high output current capability.

15.2.1 A TWO-STAGE PROTOTYPE FOR AN OPERATIONAL AMPLIFIER

To achieve a higher gain, a *pnp* common-emitter amplifier Q_3 has been connected to the output of a differential amplifier, Q_1–Q_2, to form the simple two-stage op amp depicted in Fig. 15.19a. Bias is provided by current source I_1.

dc Analysis

The dc equivalent circuit for the op amp is shown in Fig. 15.19(b) and will be used to find the Q-points of the three transistors. The emitter currents of Q_1 and Q_2 are each equal to one-half the bias current I_1: $I_{E1} = I_{E2} = I_1/2$. The voltage at the collector of Q_1 is equal to

$$V_{C1} = V_{CC} - I_{C1}R_C = V_{CC} - \alpha_{F1}\frac{I_1}{2}R_C \tag{15.62}$$

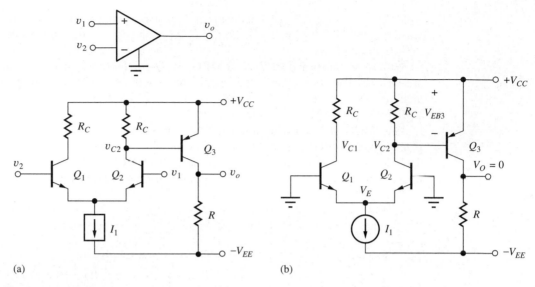

Figure 15.19 (a) A simple two-stage prototype for an operational amplifier. (b) dc equivalent circuit for the two-stage amplifier.

and that at the collector of Q_2 is

$$V_{C2} = V_{CC} - (I_{C2} - I_{B3})R_C = V_{CC} - \left(\alpha_{F2}\frac{I_1}{2} - I_{B3}\right)R_C \tag{15.63}$$

If the base current of Q_3 can be neglected, and the common-base current gains are approximately 1, then Eqs. (15.62) and (15.63) become

$$V_{C1} \cong V_{C2} \cong V_{CC} - \frac{I_1 R_C}{2} \tag{15.64}$$

and because $V_E = -V_{BE}$,

$$V_{CE1} \cong V_{CE2} \cong V_{CC} - \frac{I_1 R_C}{2} + V_{BE} \tag{15.65}$$

In this particular circuit, it is important to note that the voltage drop across R_C is constrained to be equal to the emitter-base voltage V_{EB3} of Q_3, or approximately 0.7 V.

The value of the collector current of Q_3 can be found by remembering that this circuit is going to represent an operational amplifier, and because both inputs in Fig. 15.19 are zero, V_O should also be zero. This is the situation that exists when the circuit is used in any of the negative feedback circuits discussed in Chapter 11.

Since $V_O = 0$, I_{C3} must satisfy

$$I_{C3} = \frac{V_{EE}}{R} \quad \text{and} \quad V_{EC3} = V_{CC} \tag{15.66}$$

We also know that V_{EB3} and I_{C3} are intimately related through the transport model relationship,

$$V_{EB3} = V_T \ln\left(1 + \frac{I_{C3}}{I_{S3}}\right) \tag{15.67}$$

in which I_{S3} is the saturation current of Q_3. For the offset voltage of this amplifier to be zero, the value of R_C must be carefully selected, based on Eqs. (15.66) and (15.68):

$$R_C = \frac{V_T}{\left(\alpha_{F2}\dfrac{I_1}{2} - \dfrac{I_{C3}}{\beta_{F3}}\right)} \ln\left(1 + \frac{I_{C3}}{I_{S3}}\right) \tag{15.68}$$

EXERCISE: Find the Q-points for the transistors in the amplifier in Fig. 15.19 if $V_{CC} = V_{EE} = 15$ V, $I_1 = 150$ μA, $R_C = 10$ kΩ, $R = 20$ kΩ, and $\beta_F = 100$. What is the value of I_{S3} if the output voltage is zero?

ANSWERS: (74.3 μA, 14.9 V), (74.3 μA, 15.01 V), (750 μA, 15.0 V); 1.90×10^{-15} A

dc Bias Sensitivity—A Word of Caution

It should be noted that the circuit in Fig. 15.19 cannot be operated open-loop without some form of feedback to stabilize the operating point of transistor Q_3, because the collector current of Q_3 is exponentially dependent on the value of its emitter-base voltage. If one attempts to build this circuit, or even simulate it with the default values in SPICE, the output will be found to be saturated at one of the power supply "rails" due to our $V_{EB} = 0.7$ V approximation. This sensitivity could be reduced by putting a resistance in series with the emitter of Q_3, at the expense of a significant loss in voltage gain.

EXERCISE: Simulate the circuit in Fig. 15.19 with $V_{CC} = V_{EE} = 15$ V, $I_1 = 150$ μA, $R_C = 10$ kΩ, and $R = 20$ kΩ using the default transistor parameters in SPICE. What are the transistor Q-points and output voltage v_O?

ANSWERS: (74.4 μA, 14.9 V), (74.4 μA, 14.9 V), (164 μA, 26.7 V), −11.7 V—not quite saturated against the negative rail. (The exact values will depend on the default parameters in your version of SPICE.) The problem can be solved by connecting the base of Q_1 to the output. Try the simulation again.

ac Analysis

The ac equivalent circuit for the two-stage op amp is shown in Fig. 15.20, in which bias source I_1 has been replaced by its equivalent ac resistance R_1. Analysis of the differential-mode behavior of the op amp can be determined from the simplified equivalent circuit in Fig. 15.21 based on the differential-mode half-circuit for the input stage.

It is important to realize that the overall two-stage amplifier in Fig. 15.20 no longer represents a symmetrical circuit. Thus, half-circuit analysis is not theoretically justified. However, we know that voltage variations at the collector of Q_2 (or at the drain of an FET) do not substantially alter the current in the transistor when it is operating in the forward-active region (or saturation region for the

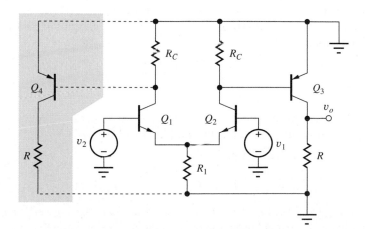

Figure 15.20 ac Equivalent circuit for the two-stage op amp.

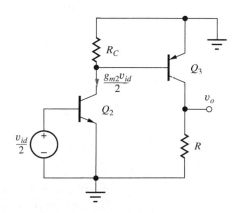

Figure 15.21 Simplified model using differential-mode half-circuit.

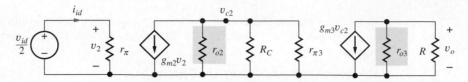

Figure 15.22 Small-signal model for Fig. 15.21.

FET). Thus, the emitters of the differential pair will remain approximately a virtual ground. One can also envision a fully symmetrical version of the amplifier with Q_3 and R replicated on the left-hand side of the circuit with the base of the additional transistor attached to the collector of Q_1. Thus, continuing to represent the differential amplifier by its half-circuit is a highly useful engineering approximation.

The approximate small-signal model corresponding to Fig. 15.21 appears in Fig. 15.22. In this analysis, output resistances r_{o2} and r_{o3} are neglected because they are in parallel with external resistors R_C and R. From Fig. 15.22, the overall differential-mode gain A_{dm} of this two-stage operational amplifier can be expressed as

$$A_{dm} = \frac{\mathbf{v_o}}{\mathbf{v_{id}}} = \frac{\mathbf{v_{c2}}}{\mathbf{v_{id}}} \frac{\mathbf{v_o}}{\mathbf{v_{c2}}} = A_{vt1} A_{vt2} \tag{15.69}$$

and the terminal gains A_{vt1} and A_{vt2} can be found from analysis of the circuit in the figure.

The first stage is a differential amplifier with the output taken from the inverting side,

$$A_{vt1} = \frac{\mathbf{v_{c2}}}{\mathbf{v_{id}}} = -\frac{g_{m2}}{2} R_{L1} = -\frac{g_{m2}}{2} \frac{R_C r_{\pi3}}{R_C + r_{\pi3}} \tag{15.70}$$

in which the load resistance R_{L1} is equal to the collector resistor R_C in parallel with the input resistance $r_{\pi3}$ of the second stage.

The second stage is also a resistively loaded common-emitter amplifier with gain

$$A_{vt2} = \frac{\mathbf{v_o}}{\mathbf{v_{C2}}} = -g_{m3} R \tag{15.71}$$

Combining Eqs. (15.69) to (15.71) yields the overall voltage gain for the two-stage amplifier:

$$A_{dm} = A_{vt1} A_{vt2} = \left(-\frac{g_{m2}}{2} \frac{R_C r_{\pi3}}{R_C + r_{\pi3}} \right) (-g_{m3} R) = \frac{g_{m2} R_C}{2} \frac{\beta_{o3} R}{R_C + r_{\pi3}} \tag{15.72}$$

Equation (15.72) appears to contain quite a number of parameters and is difficult to interpret. However, some thought and manipulation will help reduce this expression to its basic design parameters. Multiplying the numerator and denominator of Eq. (15.72) by g_{m3} and expanding the transconductances in terms of the collector currents yields

$$A_{dm} = \frac{1}{2} \frac{(40 I_{C2} R_C) \beta_{o3} (40 I_{C3} R)}{40 \frac{I_{C3}}{I_{C2}} I_{C2} R_C + \beta_{o3}} \tag{15.73}$$

If the base current of Q_3 is neglected, then $I_{C2} R_C = V_{BE3} \cong 0.7$ V, and $I_{C3} R = V_{EE}$, as pointed out during the dc analysis. Substituting these results into Eq. (15.73) yields

$$A_{dm} = \frac{1}{2} \frac{(28) \beta_{o3} (40 V_{EE})}{28 \left(\frac{I_{C3}}{I_{C2}} \right) + \beta_{o3}} = \frac{560 V_{EE}}{1 + \frac{28}{\beta_{o3}} \left(\frac{I_{C3}}{I_{C2}} \right)} \tag{15.74}$$

In the final result in Eq. (15.74), A_{dm} is reduced to its basics. Once the power supply voltage V_{EE} and transistor Q_3 (that is, β_{o3}) are chosen, the only remaining design parameter is the ratio of the collector currents in the first and second stages. An upper limit on I_{C2} and I_1 is usually set by the permissible dc bias current, I_{B1}, at the input of the amplifier, whereas the minimum value of I_{C3}

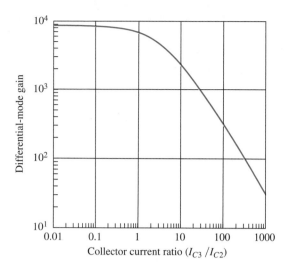

Figure 15.23 Differential-mode gain versus collector current ratio for $V_{EE} = 15$ V and $\beta_{o3} = 100$.

is determined by the current needed to drive the total load impedance connected to the output node. Generally, I_{C3} is several times larger than I_{C1}.

Figure 15.23 is a graph of Eq. (15.74), showing the variation of amplifier gain versus the collector current ratio. Observe that the gain starts to drop rapidly as I_{C3}/I_{C2} exceeds approximately 5. Such a graph is very useful as an aid in choosing the operating point during the design of the basic two-stage operational amplifier.

> **EXERCISE:** What is the maximum possible gain of the amplifier described by Eq. (15.74) for $V_{CC} = V_{EE} = 15$ V, $\beta_{o1} = 50$, and $\beta_{o3} = 100$? What is the maximum voltage gain for the amplifier if the input bias current to the amplifier must not exceed 1 μA, and $I_{C3} = 500$ μA? Repeat if $I_{C3} = 5$ mA.
>
> **ANSWERS:** 8400; 2210; 290
>
> **EXERCISE:** What is the maximum possible gain of the amplifier described by Eq. (15.74) if $V_{CC} = V_{EE} = 1.5$ V?
>
> **ANSWER:** 840

Input and Output Resistances

From the ac model of the amplifier in Figs. 15.21 and 15.22, the differential-mode input resistance of the simple op amp is equal to the input resistance of the differential amplifier given by

$$R_{id} = \frac{v_{id}}{i_{id}} = 2r_{\pi 2} = 2r_{\pi 1} \tag{15.75}$$

and the output resistance is given by

$$R_{\text{out}} = R \| r_{o3} \cong R \tag{15.76}$$

> **EXERCISE:** What are the input and output resistances for the two amplifier designs in the previous exercise?
>
> **ANSWERS:** 50 kΩ, 30 kΩ; 50 kΩ, 3 kΩ

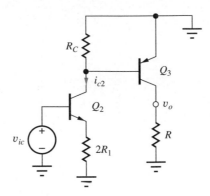

Figure 15.24 ac Equivalent circuit for common-mode inputs.

Before proceeding, we need to understand how the coupling and bypass capacitors have been eliminated from the two-stage op amp prototype. The virtual ground at the emitters of the differential amplifier allows the input stage to achieve the full inverting amplifier gain without the need for an emitter bypass capacitor. Use of the *pnp* transistor permits direct coupling between the first and second stages and allows the emitter of the *pnp* to be connected to an ac ground point. In addition, the *pnp* provides the voltage **level shift** required to bring the output back to 0 V. Thus, the need for any bypass or coupling capacitors is entirely eliminated, and $v_o = 0$ for $v_1 = 0 = v_2$.

CMRR

The common-mode gain and CMRR of the two-stage amplifier can be determined from the ac circuit model with common-mode input that is shown in Fig. 15.24, in which the half-circuit has again been used to represent the differential input stage. If Fig. 15.24 is compared to Fig. 15.21, we see that the circuitry beyond the collector of Q_2 is identical in both figures. The only difference in output voltage is therefore due to the difference in the value of the collector current i_{c2}. In Fig. 15.24, i_{c2} is the collector current of a C-E stage with emitter resistor $2R_1$:

$$\mathbf{i_{c2}} = \frac{\beta_{o2}\mathbf{v_{ic}}}{r_{\pi2} + 2(\beta_{o2} + 1)R_1} = \frac{g_{m2}\mathbf{v_{ic}}}{1 + 2\dfrac{g_{m2}}{\alpha_{o2}}R_1} \cong \frac{g_{m2}\mathbf{v_{ic}}}{1 + 2g_{m2}R_1} \tag{15.77}$$

whereas i_{c2} in Fig. 15.21 was

$$\mathbf{i_{c2}} = \frac{g_{m2}}{2}\mathbf{v_{id}} \tag{15.78}$$

Thus, the common-mode gain A_{cm} of the op amp is found from Eq. (15.72) by replacing the quantity $g_{m2}/2$ by $g_{m2}/(1 + 2g_{m2}R_1)$:

$$A_{cm} = \frac{g_{m2}R_C}{1 + 2g_{m2}R_1}\frac{\beta_{o3}R}{R_C + r_{\pi3}} = \frac{2A_{dm}}{1 + 2g_{m2}R_1} \tag{15.79}$$

From Eq. (15.79), the CMRR of the simple op amp is

$$\text{CMRR} = \left|\frac{A_{dm}}{A_{cm}}\right| = \frac{1 + 2g_{m2}R_1}{2} \cong g_{m2}R_1 \tag{15.80}$$

which is identical to the CMRR of the differential input stage alone.

EXERCISE: **What is the CMRR of the amplifier in Fig. 15.19 if $I_1 = 100\ \mu\text{A}$ and $R_1 = 750\ \text{k}\Omega$?**

ANSWER: 63.5 dB

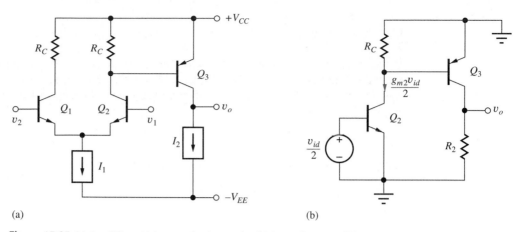

(a) (b)

Figure 15.25 (a) Amplifier with improved voltage gain. (b) Approximate ac differential-mode equivalent for op amp.

15.2.2 IMPROVING THE OP AMP VOLTAGE GAIN

From the previous several exercises, we can see that the prototype op amp has a relatively low overall voltage gain and a higher output resistance than is normally associated with a true operational amplifier. This section explores the use of an additional current source to improve the voltage gain; the next section adds an emitter follower to reduce the output resistance.

Figure 15.23 indicates that the overall amplifier gain decreases rapidly as the quiescent current of the second stage increases. In the exercise, the overall gain is quite low when $I_{C3} = 5$ mA. One technique that can be used to improve the voltage gain is to replace resistor R by a second current source, as shown in Fig. 15.25. The modified ac model is in Fig. 15.25(b). The small-signal model is the same as Fig. 15.22 except R is replaced by output resistance R_2 of current source I_2. The load on Q_3 is now the output resistance R_2 of the current source in parallel with the output resistance of Q_3 itself. In Sec. 15.7, we shall discover that it is possible to design a current source with $R_2 \gg r_{o3}$, and, by neglecting R_2, the differential-mode gain expression for the overall amplifier becomes

$$A_{dm} = A_{vt1} A_{vt2} = \left(-\frac{g_{m2}}{2} \frac{R_C r_{\pi 3}}{R_C + r_{\pi 3}} \right) (-g_{m3} r_{o3}) \tag{15.81}$$

We can reduce Eq. (15.81) to

$$A_{dm} = \frac{14 \mu_{f3}}{1 + \dfrac{28}{\beta_{o3}} \left(\dfrac{I_{C3}}{I_{C2}} \right)} \cong \frac{560 V_{A3}}{1 + \dfrac{28}{\beta_{o3}} \left(\dfrac{I_{C3}}{I_{C2}} \right)} \tag{15.82}$$

using the same steps that led to Eq. (15.74). This expression is similar to Eq. (15.74) except that power supply voltage V_{EE} has been replaced by the Early voltage of Q_3. For low values of the collector current ratio, excellent voltage gains, approaching $560 V_{A3}$, are possible from this simple two-stage amplifier. Also, note that the amplifier gain is no longer directly dependent on the choice of V_{CC} and V_{EE}.

Although adding the current source has improved the voltage gain, it also has degraded the output resistance. The output resistance of the amplifier is now determined by the characteristics of current source I_2 and transistor Q_3:

$$R_{\text{out}} = R_2 \| r_{o3} \cong r_{o3} \tag{15.83}$$

Because of the relatively high output resistance, this amplifier more nearly represents a transconductance amplifier with a current output ($A_{tc} = \mathbf{i_o}/\mathbf{v_{id}}$) rather than a true low output resistance voltage amplifier.

EXERCISE: Start with Eq. (15.81) and show that Eq. (15.82) is correct.

EXERCISE: What is the maximum possible voltage gain for the amplifier described by Eq. (15.82) for $V_{CC} = 15$ V, $V_{EE} = 15$ V, $V_{A3} = 75$ V, $\beta_{o1} = 50$, and $\beta_{o3} = 100$? What is the voltage gain if the input bias current to the amplifier must not exceed 1 μA, and $I_{C3} = 500$ μA? Repeat if $I_{C3} = 5$ mA.

ANSWERS: 42,000; 11,000; 1450

EXERCISE: What are the input and output resistances for the last two amplifier designs?

ANSWERS: 50 kΩ, 180 kΩ; 50 kΩ, 18kΩ

15.2.3 OUTPUT RESISTANCE REDUCTION

As mentioned earlier, the two-stage op amp prototype at this point more nearly represents a high-output resistance transconductance amplifier than a voltage amplifier with a low output resistance. A third stage, that maintains the amplifier voltage gain but provides a low output resistance, needs to be added to the amplifier. This sounds like the description of a follower circuit—unity voltage gain and low output resistance!

An emitter-follower (C-C) stage is added to the prototype amplifier in Fig. 15.26. In this case, the C-C amplifier is biased by a third current source I_3, and an external load resistance R_L has been connected to the output of the amplifier. The ac equivalent circuit is drawn in Fig. 15.26(b), in which the output resistances of I_2 and I_3 are assumed to be very large and will be neglected in the analysis. Based on the ac equivalent circuit, the overall gain of the three-stage operational amplifier can be expressed as

$$A_{dm} = \frac{\mathbf{v_2}}{\mathbf{v_{id}}} \frac{\mathbf{v_3}}{\mathbf{v_2}} \frac{\mathbf{v_o}}{\mathbf{v_3}} = A_{vt1} A_{vt2} A_{vt3} \tag{15.84}$$

The gain of the first stage is equal to the gain of the differential input pair (neglecting r_{o2}):

$$A_{vt1} = -\frac{g_{m2}}{2}(R_C \| r_{\pi3}) \tag{15.85}$$

The second stage is a common-emitter amplifier with a load resistance equal to the output resistance of Q_3 in parallel with the input resistance of emitter follower Q_4:

$$A_{vt2} = -g_{m3}\left(r_{o3} \| R_{iB4}\right) \qquad \text{where } R_{iB4} = r_{\pi4}(1 + gm_4 R_L) \tag{15.86}$$

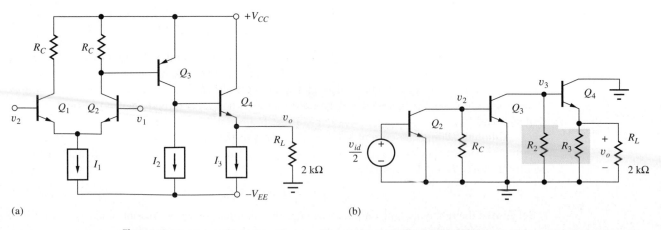

(a)   (b)

Figure 15.26 (a) Amplifier with common-collector stage Q_4 added. (b) Simplified ac equivalent circuit for the three-stage op amp.

Finally, the gain of emitter follower Q_4 is (neglecting r_{o4}):

$$A_{vt3} = \frac{g_{m4}R_L}{1 + g_{m4}R_L} \cong 1 \qquad (15.87)$$

The input resistance is set by the differential pair, and the output resistance of the amplifier is now determined by the resistance looking back into the emitter of Q_4:

$$R_{id} = 2r_{\pi 2} \qquad \text{and} \qquad R_{out} = \frac{1}{g_{m4}} + \frac{R_{th4}}{\beta_{o4} + 1} \qquad (15.88)$$

In this case, there is a relatively large Thévenin equivalent source resistance at the base of Q_4, $R_{th4} \cong r_{o3}$, and the overall output resistance is

$$R_{out} \cong \frac{1}{g_{m4}} + \frac{r_{o3}}{\beta_{o4}} = \frac{1}{g_{m4}}\left[1 + \frac{\mu_{f3}}{\beta_{o4}}\frac{I_{C4}}{I_{C3}}\right] \qquad (15.89)$$

EXAMPLE 15.4   **THREE-STAGE BIPOLAR OP AMP ANALYSIS**

Let us now determine the characteristics of a specific implementation of the three-stage op amp implemented with bipolar transistors.

PROBLEM   Find the differential-mode voltage gain, CMRR, input resistance, and output resistance for the amplifier in Fig. 15.27 if $V_{CC} = 15$ V, $V_{EE} = 15$ V, $V_{A3} = 75$ V, $\beta_{o1} = \beta_{o2} = \beta_{o3} = \beta_{o4} = 100$, $I_1 = 100$ μA, $I_2 = 500$ μA, $I_3 = 5$ mA, $R_1 = 750$ kΩ, and $R_L = 2$ kΩ. Assume R_2 and $R_3 = \infty$.

SOLUTION   **Known Information and Given Data:** Three-stage prototype op amp in Fig. 15.27 with $V_{CC} = 15$ V, $V_{EE} = 15$ V, $V_{A3} = 75$ V, $\beta_{o1} = \beta_{o2} = \beta_{o3} = \beta_{o4} = 100$, $I_1 = 100$ μA, $I_2 = 500$ μA, $I_3 = 5$ mA, $R_1 = 750$ kΩ, and $R_L = 2$ kΩ. Assume R_2 and $R_3 = \infty$.

Unknowns: Q-point values, R_C, A_{dm}, CMRR, R_{in}, and R_{out}

Approach: We need to evaluate the expressions in Eqs. (15.84) through (15.89). First, we must find the Q-point and then use it to calculate the small-signal parameters including g_{m2}, $r_{\pi 2}$, $r_{\pi 3}$, g_{m2}, r_{o3}, and $r_{\pi 4}$. The required Q-point information can be found from Fig. 15.27, in which v_1 and v_2 equal zero.

Assumptions: The Q-point is found with v_1 and v_2 set to zero, and output voltage v_o is also assumed to be zero for this set of input voltages. The transistors are all in the active region with V_{BE} or V_{EB} equal to 0.7 V.

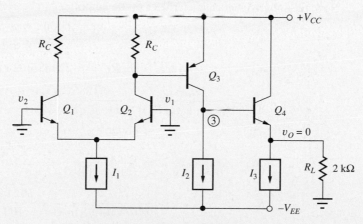

Figure 15.27 Operational amplifier with $v_1 = 0 = v_2$.

Analysis: The emitter current in the input stage is one-half the bias current source I_1 and

$$g_{m2} = 40I_{C2} = 40(\alpha_{F2}I_{E2}) = 40(0.99 \times 50 \text{ μA}) = 1.98 \text{ mS}$$

The collector of the second stage must supply the current I_2 plus the base current of Q_4:

$$I_{C3} = I_2 + I_{B4} = I_2 + \frac{I_{E4}}{\beta_{F4} + 1}$$

When the output voltage is zero, the current in load resistor R_L is zero, and the emitter current of Q_4 is equal to the current in source I_3. Therefore,

$$I_{C3} = I_2 + I_{B4} = I_2 + \frac{I_3}{\beta_{F4} + 1} = 5 \times 10^{-4} \text{ A} + \frac{5 \times 10^{-3} \text{ A}}{101} = 550 \text{ μA}$$

and

$$g_{m3} = 40I_{C3} = \frac{40}{\text{V}}(5.5 \times 10^{-4} \text{ A}) = 2.20 \times 10^{-2} \text{ S}$$

$$r_{\pi3} = \frac{\beta_{o3}}{g_{m3}} = \frac{100}{2.20 \times 10^{-2} \text{ S}} = 4.55 \text{ kΩ}$$

To find the output resistance of Q_3, V_{EC3} is needed. When properly designed, the dc output voltage of the amplifier will be zero when the input voltages are zero. Hence, the voltage at node 3 is one base-emitter voltage drop above zero, or $+0.7$ V, and $V_{EC3} = 15 - 0.7 = 14.3$ V. The output resistance of Q_3 is

$$r_{o3} = \frac{V_{A3} + V_{EC3}}{I_{C3}} = \frac{(75 + 14.3) \text{ V}}{5.50 \times 10^{-4} \text{ A}} = 162 \text{ kΩ}$$

Remembering that $I_{E4} = I_3$

$$I_{C4} = \alpha_{F4}I_{E4} = 0.990 \times 5 \text{ mA} = 4.95 \text{ mA}$$

and

$$g_{m4} = 40I_{C4} = 198 \text{ mS} \qquad r_{\pi4} = \frac{\beta_{o4}V_T}{I_{C4}} = \frac{100 \times 0.025 \text{ V}}{4.95 \times 10^{-3} \text{ A}} = 505 \text{ Ω}$$

Finally, the value of R_C is needed:

$$R_C = \frac{V_{EB3}}{I_{C2} - I_{B3}} = \frac{V_{EB3}}{I_{C2} - \dfrac{I_{C3}}{\beta_{F3}}} = \frac{0.7 \text{ V}}{\left(49.5 - \dfrac{550}{100}\right) \times 10^{-6} \text{ A}} = 15.9 \text{ kΩ}$$

Now, the small-signal characteristics of the amplifier can be evaluated:

$$A_{vt1} = -\frac{g_{m2}(R_C \| r_{\pi3})}{2} = -\frac{1.98 \text{ mS}(15.9 \text{ kΩ} \| 4.55 \text{ kΩ})}{2} = -3.50$$

$$A_{vt2} = -g_{m3}[r_{o3} \| r_{\pi4} + \beta_{o4}R_L] = -22 \text{ mS}(162 \text{ kΩ} \| 203 \text{ kΩ}) = -1980!$$

$$A_{vt3} = \frac{g_{m4}R_L}{r_{\pi4}(1 + g_{m4}R_L)} = \frac{0.198\text{S}(2 \text{ kΩ})}{1 + 0.198\text{S}(2 \text{ kΩ})} = 0.998 \cong 1$$

$$A_{dm} = A_{vt1}A_{vt2}A_{vt3} = +6920$$

$$R_{id} = 2r_{\pi2} = 2\frac{\beta_{o2}}{g_{m2}} = 2\frac{100}{(40/\text{V})(49.5 \text{ μA})} = 101 \text{ kΩ}$$

$$R_{out} \cong \frac{1}{g_{m4}} + \frac{r_{o3}}{\beta_{o4}} = \frac{1}{(40/\text{V})(4.95 \text{ mA})} + \frac{162 \text{ kΩ}}{100} = 1.62 \text{ kΩ}$$

$$\text{CMRR} = g_{m2}R_1 = (40/\text{V})(49.5 \text{ μA})(750 \text{ kΩ}) = 1490 \text{ or } 63.5 \text{ dB}$$

Check of Results: We can use our rules-of-thumb from Chapter 13 to estimate the voltage gain. The first stage should produce a gain of approximately $(1/2) \times 40 \times$ the voltage across the load resistor or $20(0.7) = 14$. The second stage should produce a gain of approximately $\mu_f = 40(75) = 3000$. The product is 42,000. Our detailed calculations give us about 1/6 of this value. Can we account for the discrepancies? We see the gain of the first stage is only 3.5 because $r_{\pi 3}$ is considerably smaller than R_C, and the gain of the second stage is approximately 2000 because the reflected loading of R_L is of the same order as r_{o3}. These two reductions account for the lower overall gain. The emitter follower produces a gain of one, as expected.

Discussion: This amplifier achieves a reasonable set of op amp characteristics for a simple circuit: $A_v = 6920$, $R_{id} = 101$ kΩ, and $R_{out} = 1.62$ kΩ. Note that the second stage, loaded by current source I_2 and buffered from R_L by the emitter follower, is achieving a gain that is a substantial fraction of Q_3's amplification factor. However, even with the emitter follower, the reflected load resistance $\beta_{o4} R_L$ is similar to the value of r_{o3} and is reducing the overall voltage gain by a factor of almost 2. Also, note that the output resistance is dominated by r_{o3}, present at the base of Q_4, and not by the reciprocal of g_{m4}. These last two factors point to a way to increase the performance of the amplifier by replacing Q_4 with an *npn* Darlington stage (See Prob. 15.48).

Computer-Aided Analysis: Since this amplifier is dc coupled, a transfer function analysis from an input source to the output node will automatically yield the voltage gain, input resistance, and output resistance. In order to force the output to be nearly zero (the normal operating point), we must determine the offset voltage of the amplifier, and then apply it as a dc input to the amplifier. This is done by first connecting the amplifier as a voltage follower with the input grounded (see the figure below). For this amplifier, the SPICE yields $V_{OS} = 0.437$ mV. Note that a current of approximately 20 μA will exist in R_1, the output resistance of current source I_1. Be sure to choose $I_1 = 80$ μA so that the bias currents in Q_1 and Q_2 will each be approximately 50 μA.

Next, the offset voltage is applied to the amplifier input with the feedback connection removed, and a transfer function analysis is requested from source V_{OS} to the output (Fig. b). The computed values are $A_{dm} = 8280$, $R_{in} = 105$ kΩ, and $R_{out} = 960$ Ω. The values are all somewhat higher than our hand calculations. Most of the differences can be traced to the higher temperature and hence higher value of V_T used in the simulations (T defaults to 27°C and $V_T = 25.9$ mV). R_{out} as calculated by SPICE includes the presence of R_L. Removing the 2-kΩ resistor from the SPICE result yields $R_{out} = [(1/960) - (1/2000)]^{-1} = 1.85$ kΩ, which agrees more closely with our hand calculations.

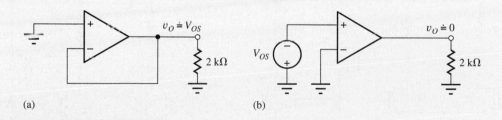

(a) (b)

EXERCISE: Suppose the output resistances of current sources R_2 and R_3 in the amplifier in Fig. 15.26 are 150 kΩ and 15 kΩ, respectively. (a) Recalculate the gain, input resistance, and output resistance. (b) Compare to SPICE simulation results. (c) What is the power consumption of the amplifier in Ex. 15.4?

ANSWERS: 4320, 101 kΩ, 776 Ω; 4480, 105 kΩ, 774 Ω; 168 mW

EXERCISE: Suppose the current gain β_F of all the transistors is 150 instead of 100. Recalculate the gain, input resistance, output resistance, and CMRR of the amplifier in Fig. 15.26.

ANSWERS: 11,000; 152 kΩ; 1.12 kΩ; 63.4 dB

EXERCISE: Suppose the Early voltage of Q_3 in the amplifier in Fig. 15.26 is 50 V instead of 75 V. Recalculate the gain, input resistance, output resistance, and CMRR.

ANSWERS: 5700; 101 kΩ; 1.16 kΩ; 63.5 dB

EXERCISE: The op amp in Ex. 15.4 is operated as a voltage follower. What are the closed-loop gain, input resistance, and output resistance?

ANSWERS: +0.99986, 699 MΩ, 0.233 Ω

15.2.4 A CMOS OPERATIONAL AMPLIFIER PROTOTYPE

Similar circuit design ideas have been used to develop the basic CMOS operational amplifier depicted in Fig. 15.28(a). A differential amplifier, formed by NMOS transistors M_1 and M_2, is followed by a PMOS common-source stage M_3 and NMOS source follower M_4. Current sources are again used to bias the differential input and source-follower stages and as a load for M_3. Referring to the ac equivalent circuit in Fig. 15.28(b), we see that the differential-mode gain is given by the product of the terminal gains of the three stages:

$$A_{dm} = A_{vt1} A_{vt2} A_{vt3} = \left(-\frac{g_{m2}}{2} R_D \right) [-g_{m3}(r_{o3} \| R_2)] \left[\frac{g_{m4}(R_3 \| R_L)}{1 + g_{m4}(R_3 \| R_L)} \right] \quad (15.90)$$

$$\cong \mu_{f3} \left(\frac{g_{m2}}{2} R_D \right) \left(\frac{g_{m4} R_L}{1 + g_{m4} R_L} \right) \quad (15.91)$$

in which we have assumed that $R_3 \gg R_L$ and $R_2 \gg r_{o3}$.

Equation (15.90) is relatively easy to construct using our single-stage amplifier formulas because the input resistance of each FET is infinite and the gain of one stage is not altered by the presence of the next.

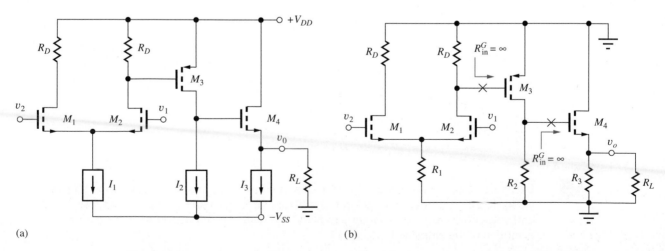

(a) (b)

Figure 15.28 (a) A CMOS operational amplifier prototype. (b) ac Equivalent circuit for the CMOS amplifier, in which the output resistances of current sources I_2 and I_3 have been neglected.

The overall differential-mode gain is approximately equal to the product of the voltage gain of the first stage and the amplification factor of the second stage.

Expanding g_{m2}, realizing that the product $I_{D2}R_D$ represents the voltage across R_D, which must equal V_{GS3}, and assuming that the source follower has a gain of nearly 1 yields

$$A_{dm} \cong A_{v1}A_{v2}(1) = \mu_{f3}\left(\frac{V_{SG3}}{V_{GS2} - V_{TN2}}\right) \tag{15.92}$$

Although Eq. (15.92) is a simple expression, we often prefer to have the gain expressed in terms of the various bias currents, and expanding μ_{f3}, V_{GS2}, and V_{SG3} yields

$$A_{dm} = \frac{1}{\lambda_3}\sqrt{\frac{K_{n2}}{I_{D2}}\frac{K_{p3}}{I_{D3}}}\left[\sqrt{\frac{2I_{D3}}{K_{p3}}} - V_{TP3}\right] \tag{15.93}$$

Because of the Q-point dependence of μ_f, there are more degrees of freedom in Eq. (15.93) than in the corresponding expression for the bipolar amplifier, Eq. (15.82). This is particularly true in the case of integrated circuits, in which the values of K_n and K_p can be easily changed by modifying the W/L ratios of the various transistors. However, the benefit of operating both gain stages of the amplifier at low currents is obvious from Eq. (15.93), and picking a transistor with a small value of λ for M_3 is also clearly important.

It is worth noting that because the gate currents of the MOS devices are zero, input-bias current does not place a restriction on I_{D1}, whereas it does place a practical upper bound on I_{C1} in the case of the bipolar amplifier. The input and output resistances of the op amp are determined by M_1, M_2 and M_4. From our knowledge of single-stage amplifiers,

$$R_{id} = \infty \qquad R_{out} = \frac{1}{g_{m4}}\bigg\|R_3 \qquad \text{CMRR} = g_{m2}R_1 \tag{15.94}$$

CMRR is once again determined by the differential input stage, where R_1 is the output resistance of current source I_1.

EXERCISE: For the CMOS amplifier in Fig. 15.28(a), $\lambda_3 = 0.01$ V, $K_{n1} = K_{n4} = 5.0$ mA/V^2, $K_{p3} = 2.5$ mA/V^2, $I_1 = 200$ μA, $I_2 = 500$ μA, $I_3 = 5$ mA, $R_1 = 375$ kΩ, and $V_{TP3} = -1$ V. What is the actual gain of the source follower if $R_L = 2$ kΩ? What are the voltage gain, CMRR, input resistance, and output resistance of the amplifier?

ANSWERS: 0.934; 2410, 51.5 dB, ∞, 141 Ω

EXERCISE: What is the quiescent power consumption of this op amp if $V_{DD} = V_{SS} = 12$ V?

ANSWER: 137 mW

15.2.5 BICMOS AMPLIFIERS

A number of integrated circuit processes exist that offer the circuit designer a combination of bipolar and MOS transistors or bipolar transistors and JFETs. These are commonly referred to as BiCMOS and BiFET technologies, respectively. The combination of BJTs and FETs offers the designer the ability to use the best characteristics of both devices to enhance the performance of the circuit. A simple BiCMOS op amp is shown in Fig. 15.29. In this case, a differential pair of PMOS transistors has been used as the input stage to demonstrate another design variation. The PMOS transistors at the input provide high input resistance and can be biased at relatively high input currents, since input current is not an issue. (We will discover later that this increased current improves the slew rate of the amplifier.) The second gain stage utilizes a bipolar transistor, which provides a superior amplification factor compared to the FET. Emitter resistor R_E increases the voltage across R_{D2} and hence, the voltage gain of the first stage without reducing the amplification factor of Q_1 (see Section 14.2.3).

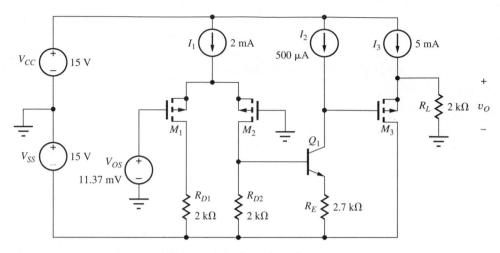

Figure 15.29 Basic BiCMOS op amp.

The follower stage uses another FET in order to maximize second-stage gain while maintaining a reasonable output resistance.

For the circuit shown, SPICE simulation uses VTO $= -1$ V, KP $= 25$ mA/V^2, VAF $= 75$ V, and BF $= 100$. SPICE is first used to find the offset voltage in the same manner as in Ex. 15.5. The value is found to be -11.37 mV, which is then applied to the input of the open-loop amplifier. A transfer function analysis from V_{OS} to the output yields infinite input resistance, a voltage gain of 13,200 and an output resistance of 61.4 Ω.

15.3 OUTPUT STAGES

The basic operational amplifier circuits discussed in Sec. 15.3 used followers for the output stages. The final stage of these amplifiers is designed to provide a low-output resistance as well as a relatively high current drive capability. However, because of this last requirement, the output stages of the amplifiers in the previous section consume approximately two-thirds or more of the total power.

Followers are **class-A amplifiers,** defined as circuits in which the transistors conduct during the full 360° of the signal waveform. The class-A amplifier is said to have a **conduction angle** $\theta_C = 360°$. Unfortunately, the maximum efficiency of the class-A stage is only 25 percent. Because the output stage must often deliver relatively large powers to the amplifier load, this low efficiency can cause high power dissipation in the amplifier. This section analyzes the efficiency of the class-A amplifier and then introduces the concept of the **class-B push-pull output stage.** The class-B push-pull stage uses two transistors, each of which conducts during only one-half, or 180°, of the signal waveform ($\theta_C = 180°$) and can achieve much higher efficiency than the class-A stage. Characteristics of the class-A and class-B stages can also be combined into a third category, the **class-AB amplifier,** which forms the output stage of most operational amplifiers.

15.3.1 THE SOURCE FOLLOWER—A CLASS-A OUTPUT STAGE

We analyzed the small-signal behavior of follower circuits in detail in Chapter 14 and found that they provide high input resistance, low output resistance, and a voltage gain of approximately 1. The large-signal operation of the emitter follower, biased by the **ideal current source,** was discussed in Chapter 10, so here we focus on the source-follower circuit in Fig. 15.30.

For $v_I \leq V_{DD} + V_{TN}$, M_1 will be operating in the saturation region (be sure to prove this to yourself). The current source forces a constant current I_{SS} to flow out of the source. Using Kirchhoff's

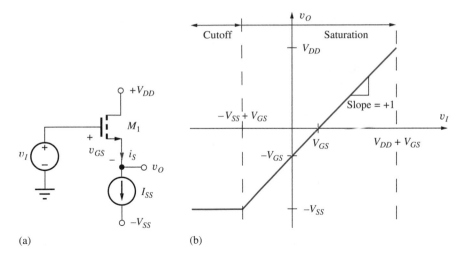

Figure 15.30 (a) Source-follower circuit. (b) Voltage transfer characteristic for the source follower.

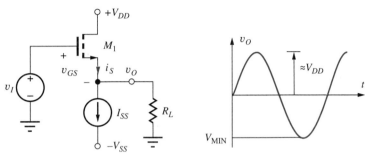

Figure 15.31 Source follower with external load resistor R_L.

voltage law, $v_O = v_I - v_{GS}$. Since the source current is constant, v_{GS} is also constant, and v_O is

$$v_O = v_I - V_{GS} = v_I - \left(V_{TN} + \sqrt{\frac{2I_{SS}}{K_n}} \right) \qquad (15.95)$$

The difference between the input and output voltages is fixed. Thus, from a large-signal perspective (as well as from a small-signal perspective), we expect the source follower to provide a gain of approximately 1.

The voltage transfer characteristic for the source follower appears in Fig. 15.30(b). The output voltage at the source follows the input voltage with a slope of +1 and a fixed offset voltage equal to V_{GS}. For positive inputs, M_1 remains in saturation until $v_I = V_{DD} + V_{TN}$. The maximum output voltage is $v_o = V_{DD}$ for $v_I = V_{DD} + V_{GS}$. Note that to actually reach this output, the input voltage must exceed V_{DD}.

The minimum output voltage is set by the characteristics of the current source. An ideal current source will continue to operate even with $v_o < -V_{SS}$, but most electronic current sources require $v_o \geq -V_{SS}$. Thus, the minimum possible value of the input voltage is $v_I = -V_{SS} + V_{GS}$.

Source Follower with External Load Resistor
When a load resistor R_L is connected to the output, as in Fig. 15.31, the output voltage range is restricted by a new limit. The total source current of M_1 is equal to

$$i_S = I_{SS} + \frac{v_O}{R_L} \qquad (15.96)$$

and must be greater than zero. In this circuit, current cannot go back into the MOSFET source, and the minimum output voltage occurs at the point at which transistor M_1 cuts off. In this situation,

$i_S = 0$ and $v_{MIN} = -I_{SS}R_L$. M_1 cuts off when the input voltage falls to one threshold voltage drop above V_{MIN}: $v_I = -I_{SS}R_L + V_{TN}$.

15.3.2 EFFICIENCY OF CLASS-A AMPLIFIERS

Now consider the emitter follower in Fig. 15.31 biased with $I_{SS} = V_{SS}/R_L$ and using symmetrical power supplies $V_{DD} = V_{SS}$. Assuming that V_{GS} is much less than the amplitude of v_I, then a sinusoidal output signal can be developed with an amplitude approximately equal to V_{DD},

$$v_O \cong V_{DD} \sin \omega t \tag{15.97}$$

The efficiency ζ of the amplifier is defined as the power delivered to the load at the signal frequency ω, divided by the average power supplied to the amplifier:

The average power P_{av} supplied to the source follower is

$$P_{av} = \frac{1}{T} \int_0^T \left[I_{SS}(V_{DD} + V_{SS}) + \left(\frac{V_{DD} \sin \omega t}{R_L} \right) V_{DD} \right] dt \tag{15.98}$$

$$= I_{SS}(V_{DD} + V_{SS}) = 2I_{SS}V_{DD}$$

where T is the period of the sine wave. The first term in brackets in Eq. (15.98) is the power dissipation due to the dc current source; the second term results from the ac drain current of the transistor. The last simplification assumes symmetrical power supply voltages. The average of the sine wave current is zero, and the sinusoidal current does not contribute to the value of the integral in Eq. (15.98).

Because the output voltage is a sine wave, the power delivered to the load at the signal frequency is

$$P_{ac} = \frac{\left(\dfrac{V_{DD}}{\sqrt{2}} \right)^2}{R_L} = \frac{V_{DD}^2}{2R_L} \tag{15.99}$$

Combining Eqs. (15.98) and (15.99) yields

$$\zeta = \frac{P_{ac}}{P_{av}} = \frac{\dfrac{V_{DD}^2}{2R_L}}{2I_{SS}V_{DD}} = \frac{1}{4} \qquad \text{or} \qquad 25\% \tag{15.100}$$

because $I_{SS}R_L = V_{SS} = V_{DD}$. Thus a follower, operating as a class-A amplifier, can achieve an efficiency of only 25 percent, at most, for sinusoidal signals (see Probs. 15.90 to 15.92). Equation (15.100) indicates that the low efficiency is caused by the Q-point current I_{SS} that flows continuously between the two power supplies.

15.3.3 CLASS-B PUSH-PULL OUTPUT STAGE

Class-B amplifiers improve the efficiency by operating the transistors at zero Q-point current, eliminating the quiescent power dissipation. A **complementary push-pull** (class-B) **output stage** using CMOS transistors is shown in Fig. 15.32, and the voltage and current waveforms for the composite output stage appear in Fig. 15.33. NMOS transistor M_1 operates as a source follower for positive input signals, and PMOS transistor M_2 operates as a source follower for negative inputs.

Consider the sinusoidal input in Fig. 15.33, for example. As the input voltage v_I swings positive, M_1 turns on supplying current to the load, and the output follows the input on the positive swing. When the input becomes negative, M_2 turns on sinking current from the load, and the output follows the input on the negative swing.

Each transistor conducts current for approximately $180°$ of the signal waveform, as shown in Fig. 15.33. Because the n- and p-channel gate-source voltages are equal in Fig. 15.32, only one of the two transistors can be on at a time. Also, the Q-point current for $v_O = 0$ is zero, and the efficiency can be high.

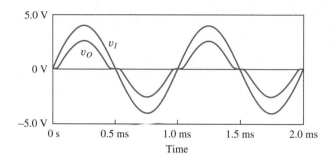

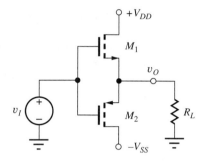

Figure 15.32 Complementary MOS class-B amplifier.

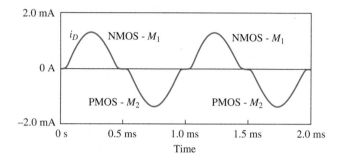

Figure 15.33 Cross-over distortion and drain currents in the class-B amplifier.

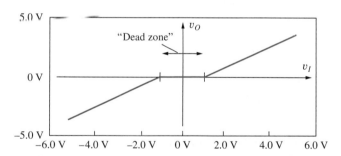

Figure 15.34 SPICE simulation of the voltage transfer characteristic for the complementary class-B amplifier.

However, although the efficiency is high, a distortion problem occurs in the class-B stage. Because V_{GS1} must exceed threshold voltage V_{TN} to turn on M_1, and V_{GS2} must be less than V_{TP} to turn on M_2, a "dead zone" appears in the push-pull class-B voltage transfer characteristic, shown in Fig. 15.34. Neither transistor is conducting for

$$V_{TP} \leq v_{GS} \leq V_{TN} \qquad (-1 \text{ V} \leq v_{GS} \leq 1 \text{ V in Fig. 15.34}) \qquad (15.101)$$

This **dead zone,** or **cross-over region,** causes distortion of the output waveform, as shown in the simulation results in Fig. 15.33. As the sinusoidal input waveform crosses through zero, the output voltage waveform becomes distorted. The waveform distortion in Fig. 15.33 is called **cross-over distortion.**

Class-B Efficiency

Simulation results for the currents in the two transistors are also included in Fig. 15.33. If cross-over distortion is neglected, then the current in each transistor can be approximated by a half-wave rectified sinusoid with an amplitude of approximately V_{DD}/R_L. Assuming $V_{DD} = V_{SS}$, the average power dissipated from each power supply is

$$P_{\text{av}} = \frac{1}{T} \int_0^{T/2} V_{DD} \frac{V_{DD}}{R_L} \sin \frac{2\pi}{T} t \, dt = \frac{V_{DD}^2}{\pi R_L} \qquad (15.102)$$

The total ac power delivered to the load is still given by Eq. (15.99), and ζ for the class-B output stage is

$$\zeta = \frac{\dfrac{V_{DD}^2}{2R_L}}{2\dfrac{V_{DD}^2}{\pi R_L}} = \frac{\pi}{4} \cong 0.785 \qquad (15.103)$$

By eliminating the quiescent bias current, the class-B amplifier can achieve an efficiency of 78.5 percent!

In closed-loop feedback amplifier applications such as those introduced in Chapter 11, the effects of cross-over distortion are reduced by the loop gain $A\beta$. However, an even better solution is to eliminate the cross-over region by operating the output stage with a small nonzero quiescent current. Such an amplifier is termed a class-AB amplifier.

15.3.4 CLASS-AB AMPLIFIERS

The benefits of the class-B amplifier can be maintained, and cross-over distortion can be minimized by biasing the transistors into conduction-but at a relatively low quiescent current level. The basic technique is shown in Fig. 15.35. A bias voltage V_{GG} is used to establish a small quiescent current in both output transistors. This current is chosen to be much smaller than the peak ac current that will be delivered to the load. In Fig. 15.35, the bias source is split into two symmetrical parts so that $v_O = 0$ for $v_I = 0$.

Because both transistors are conducting for $v_I = 0$, the cross-over distortion can be eliminated, but the additional power dissipation can be kept small enough that the efficiency is not substantially degraded. The amplifier in Fig. 15.35 is classified as a **class-AB amplifier.** Each transistor conducts for more than the 180° of the class-B amplifier but less than the full 360° of the class-A amplifier.

Figure 15.35(b) shows the results of circuit simulation of the voltage transfer characteristic of the class-AB output stage with a quiescent bias current of approximately 60 μA. The distorted cross-over region has been eliminated, even for this small quiescent bias current.

Figure 15.36(a) shows one method for generating the needed bias voltage that is consistent with the CMOS operational amplifier circuit of Fig. 15.28. Bias current I_G develops the required bias voltage for the output stage across resistor R_G. If we assume that $K_p = K_n$ and $V_{TN} = -V_{TP}$ for

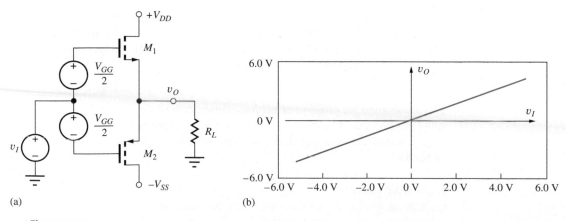

(a)

(b)

Figure 15.35 (a) Complementary output stage biased for class-AB operation. (b) SPICE simulation of voltage transfer characteristic for class-AB stage with $I_D \cong 60$ μA.

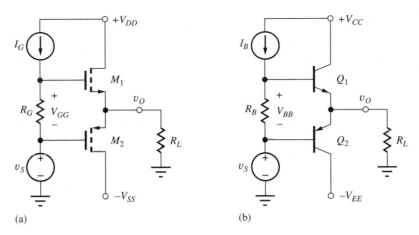

Figure 15.36 (a) Method for biasing the MOS class-AB amplifier. (b) Bipolar class-AB amplifier.

the MOSFETs, and $v_O = 0$, then the bias voltage splits equally between the gate-source terminals of the two transistors. The drain currents of the two transistors are both

$$I_D = \frac{K_n}{2} \left(\frac{V_{GG}}{2} - V_{TN} \right)^2 \tag{15.104}$$

The bipolar version of the class-AB push-pull output stage employs complementary *npn* and *pnp* transistors, as shown in Fig. 15.28(b). The principle of operation of the bipolar circuit is the same as that for the MOS case. Transistors Q_1 and Q_2 operate as emitter followers for the positive and negative excursions of the output signal, respectively. Current source I_B develops a bias voltage V_{BB} across resistor R_B, which is shared between the base-emitter junctions of the two BJTs.

For class-AB operation, voltage V_{BB} is designed to be approximately $2V_{BE} \cong 1.1$ V, so both transistors are conducting a small collector current. If we assume the saturation currents of the two transistors are equal, then the bias voltage V_{BB} splits equally between the base-emitter junctions of the two transistors, and the two collector currents are

$$I_C = I_S \exp \left(\frac{I_B R_B}{2V_T} \right) \tag{15.105}$$

Each transistor is biased into conduction at a low level to eliminate cross-over distortion.

A simplified small-signal model for the class-AB stage is a single follower transistor with a current gain equal to the average of the gains of Q_1 and Q_2 or with a transconductance parameter equal to the average of the values for M_1 and M_2.

A class-B version of the bipolar push-pull output stage is obtained by setting V_{BB} to zero. For this case, the output stage exhibits cross-over distortion for an input voltage range of approximately $2V_{BE}$.

EXERCISE: Find the bias current in the transistors in Fig. 15.36(a) for $v_O = 0$ if $K_n = K_p = 25$ mA/V^2, $V_{TN} = 1$ V, and $V_{TP} = -1$ V, $I_G = 500$ μA, and $R_G = 4.4$ kΩ.

ANSWER: 125 μA

EXERCISE: Find the bias current in the transistors in Fig. 15.36(b) for $v_O = 0$ if $I_S = 10$ fA, $I_B = 500$ μA, and $R_B = 2.4$ kΩ.

ANSWER: 265 μA

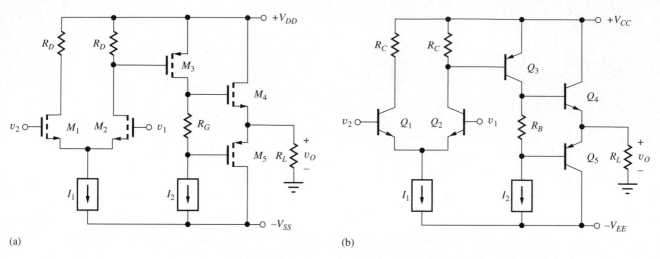

(a)

(b)

Figure 15.37 Class-AB output stages added to the (a) CMOS and (b) bipolar operational amplifiers.

15.3.5 CLASS-AB OUTPUT STAGES FOR OPERATIONAL AMPLIFIERS

In Figs. 15.37(a) and (b), the follower output stages of the prototype CMOS and bipolar op amps have been replaced with complementary class-AB output stages. Current source I_2, which originally provided a high impedance load to transistors Q_3 and M_3, is now also used to develop the dc bias voltage necessary for class-AB operation. The signal current is supplied by transistor M_3 or Q_3, respectively. The total quiescent power dissipation is greatly reduced in both these amplifiers.

15.3.6 SHORT-CIRCUIT PROTECTION

If the output of a follower circuit is accidentally shorted to ground, the transistor can be destroyed due to high current and high power dissipation, or, through direct destruction of the base-emitter junction of the BJT. To make op amps as "robust" as possible, circuitry is often added to the output stage to provide protection from short circuits.

In Fig. 15.38, transistor Q_2 has been added to protect emitter follower Q_1. Under normal operating conditions, the voltage developed across R is less than 0.7 V, transistor Q_2 is cut off, and Q_1 functions as a normal follower. However, if emitter current I_{E1} exceeds a value of

$$I_{E1} = \frac{V_{BE2}}{R} = \frac{0.7 \text{ V}}{R} \tag{15.106}$$

then transistor Q_2 turns on and shunts any additional current from R_1 down through the collector of Q_2 and away from the base of Q_1. Thus, the output current is limited to approximately the value in Eq. (15.106). For example, $R = 25 \text{ }\Omega$ will limit the maximum output current to 28 mA. Because R is directly in series with the output, however, the output resistance of the follower is increased by the value of R.

Figure 15.39(a) depicts the complementary bipolar output stage including **short-circuit protection**. *pnp* transistor Q_4 is used to limit the base current of Q_3 in a manner identical to that of Q_2 and Q_1. Similar **current-limiting circuits** can be applied to FET output stages, as shown in Fig. 15.39(b). Here, transistor M_2 steals the current needed to develop gate drive for M_1, and the output current is limited to

$$I_{S1} \cong \frac{V_{GS2}}{R} = \frac{V_{TN2} + \sqrt{\dfrac{2I_G}{K_{n2}}}}{R} \tag{15.107}$$

Transistor M_4 provides similar protection to M_3.

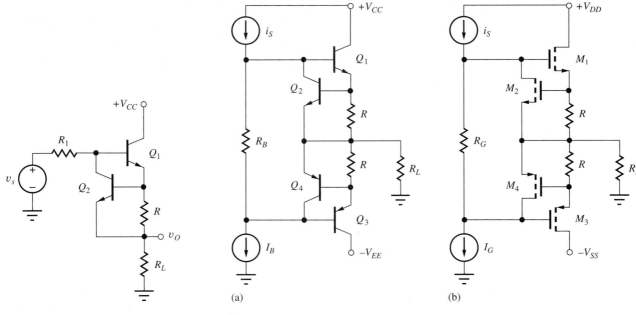

Figure 15.38 Short-circuit protection for an emitter follower.

Figure 15.39 Short-circuit protection for complementary output stages. ($i_S = I_B$ or I_G at the Q-point.)

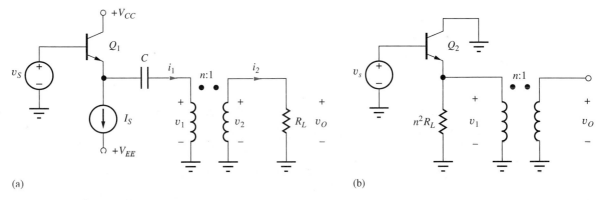

Figure 15.40 (a) Follower circuit using transformer coupling. (b) ac Equivalent circuit representation for the follower.

15.3.7 TRANSFORMER COUPLING

Designing amplifiers to deliver power to low impedance loads can be difficult. For example, loud-speakers typically have only an 8- or 16-Ω impedance. To achieve good voltage gain and efficiency in this situation, the output resistance of the amplifier needs to be quite low. One approach would be to use a feedback amplifier to achieve a low output resistance, as discussed in Chapter 11. An alternate approach to the problem is to use **transformer coupling.**

In Fig. 15.40, a follower circuit is coupled to load resistance R_L through an ideal transformer with a turns ratio of $n{:}1$. In this circuit, a coupling capacitor C is required to block the dc path through the primary of the transformer. (See Prob. 15.100 for an alternate approach.)

As defined in network theory, the terminal voltages and currents of the ideal transformer are related by

$$\mathbf{v_1} = n\mathbf{v_2} \qquad \mathbf{i_2} = n\mathbf{i_1} \qquad \frac{\mathbf{v_1}}{\mathbf{i_1}} = n^2\frac{\mathbf{v_2}}{\mathbf{i_2}} \qquad \text{or} \qquad Z_1 = n^2 Z_L \qquad (15.108)$$

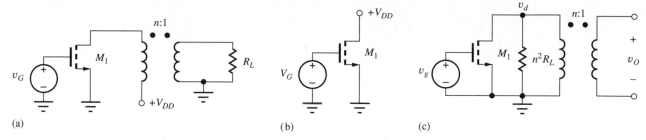

Figure 15.41 (a) Transformer-coupled inverting amplifier. (b) dc Equivalent circuit. (c) ac Equivalent circuit.

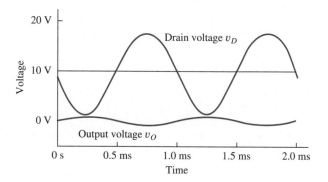

Figure 15.42 SPICE simulation of the transformer-coupled inverting amplifier stage for $n = 10$ with $V_{DD} = 10$ V.

Figure 15.43 Transformer-coupled class-B output stage.

The transformer provides an impedance transformation by the factor n^2. Based on these equations, the transformer and load resistor can be represented by the ac equivalent circuit in Fig. 15.40(b), in which the resistor has been moved to the primary side of the transformer and the secondary is now an open circuit. The effective resistance that the transistor must drive and the voltage at the transformer output are

$$R_{EQ} = n^2 R_L \qquad \text{and} \qquad \mathbf{v_o} = \frac{\mathbf{v_1}}{n} \tag{15.109}$$

Transformer coupling can reduce the problems associated with driving very low impedance loads. However, it is obviously restricted by the transformer to frequencies above dc.

Figure 15.41 is a second example of the use of a transformer, in which an inverting amplifier stage is coupled to the load R_L through the ideal transformer. The dc and ac equivalent circuits appear in Figs. 15.41(b) and (c), respectively. At dc, the transformer represents a short circuit, the full dc power supply voltage appears across the transistor, and the quiescent operating current of the transistor is supplied through the primary of the transformer. At the signal frequency, a load resistance equal to $n^2 R_L$ is presented to the transistor.

Results of simulation of the circuit in Fig. 15.41 are in Fig. 15.42 for the case $R_L = 8\ \Omega$, $V_{DD} = 10$ V, and $n = 10$. The behavior of this circuit is different from most that we have studied. The quiescent voltage at the drain of the MOSFET is equal to the full power supply voltage V_{DD}. The presence of the inductance of the transformer permits the signal voltage to swing symmetrically above and below V_{DD}, and the peak-to-peak amplitude of the signal at the drain can approach $2V_{DD}$.

Figure 15.43 is a final circuit example, which shows a transformer-coupled class-B output stage. Because the quiescent operating current in Q_1 and Q_2 are zero, the emitters may be connected directly to the primary of the transformer.

EXERCISE: Find the small-signal voltage gains

$$A_{v1} = \frac{v_d}{v_g} \quad \text{and} \quad A_{vo} = \frac{v_o}{v_g}$$

for the circuit in Fig. 15.41 if $V_{TN} = 1$ V, $K_n = 50$ mA/V^2, $V_G = 2$ V, $V_{DD} = 10$ V, $R_L = 8\ \Omega$, and $n = 10$. What are the largest values of v_g, v_d, and v_o that satisfy the small-signal limitations?

ANSWERS: -40, -4; 0.2 V, 8 V, 0.8 V

ELECTRONICS IN ACTION

Class-D Audio Amplifiers

As mentioned in the main body of the text, the efficiency of class-A, -B, and -AB amplifiers is limited to less than 80 percent. To achieve higher efficiencies, a number of forms of switching amplifiers have been developed for use in portable and other low-power electronic applications. One of these is the class-D amplifier shown here, in which the output is a pulse-width modulated (PWM) signal that switches rapidly between the positive and negative power supplies. High efficiency is achieved by using CMOS transistors as switches. In a manner similar to a CMOS inverter, the goal is to have only one transistor on at a given time.

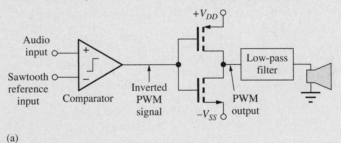

(a)

Conceptual implementation of a class-D audio amplifier.

A basic PWM signal can be generated by comparing the audio input signal to a sawtooth reference waveform. Referring to the sample waveforms, we see that the PWM output is

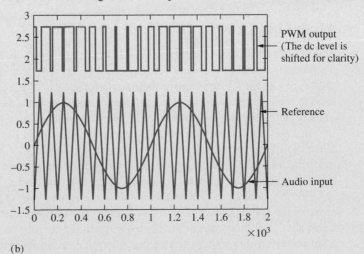

(b)

Illustration of PWM waveforms.

switched high to V_{DD} when the audio input exceeds the reference waveform, and the output is switched to $-V_{SS}$ when the reference input exceeds the analog input. In the waveform illustration, the sawtooth reference input is operating at a frequency that is 10 times that of the sinusoidal input. For an audio signal with a bandwidth of 20 Hz to 20 kHz, the reference frequency may range from 250 kHz to more than 1 MHz. Before being fed to the speaker, the PWM signal is passed through a low-pass filter to remove the unwanted high frequency content.

In order to achieve higher power levels with a given supply voltage, the load is often driven in a differential fashion using a complementary "H-bridge." In the CMOS version shown here, output voltage v_O equals $(V_{DD} + V_{SS})$ when input V is high, and v_O equals $-(V_{DD} + V_{SS})$ when input V is low. Thus, the total signal swing across the load is twice the sum of the power supply span, and the power that can be delivered to the load is four times that achieved without the H-bridge. A class-D amplifier using the H-bridge appears in the final figure in which the speaker is driven by the low-pass filtered output of the CMOS H-bridge.

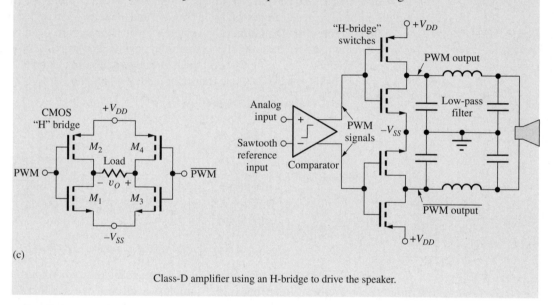

(c)

Class-D amplifier using an H-bridge to drive the speaker.

15.4 ELECTRONIC CURRENT SOURCES

The dc current source is clearly a fundamental and highly useful circuit component. In Sec. 15.3 we found that multiple current sources could be used to provide bias to the BJT and MOS op amp prototypes as well as to improve their ac performance. This section first explores the basic circuits used to realize electronic versions of ideal current sources and then explores current source design in more depth by looking at techniques specifically applicable to the design of integrated circuits.

In Fig. 15.44, the current-voltage characteristics of an ideal current source are compared with those of resistor and transistor current sources of Fig. 15.45. Current I_O through the ideal source is independent of the voltage appearing across the source, and the output resistance of the ideal source is infinite, as indicated by the zero slope of the current source i-v characteristic.

For the ideal source, the voltage across the source can be positive or negative, and the current remains the same. However, **electronic current sources** must be implemented with resistors and transistors, and their operation is usually restricted to only one quadrant of the total i-v space. In addition, electronic sources have a finite output resistance, as indicated by the nonzero slope of the i-v characteristic. We will find that the output resistance of the transistor is much greater than a resistor for an equivalent Q-point.

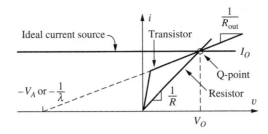

Figure 15.44 i-v Characteristics of basic electronic current sources.

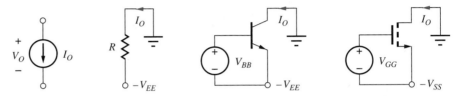

Figure 15.45 Ideal, resistor, BJT, and MOS current sources.

In normal use, the circuit elements in Fig. 15.44 will actually be *sinking* current from the rest of the network, and some authors prefer to call these elements **current sinks.** In this book, we use the generic term *current source* to refer to both sinks and sources.

15.4.1 SINGLE-TRANSISTOR CURRENT SOURCES

The simplest forms of electronic current sources are shown in Fig. 15.44. A resistor is often used to establish bias currents in many circuits—differential amplifiers, for example—but it represents our poorest approximation to an ideal current source. Individual transistor implementations of current sources generally operate in only one quadrant because the transistors must be biased in the forward-active or pinch-off regions in order to maintain high impedance operation. However, the transistor source can realize very high values of output resistance.

For simplicity, the transistors in Fig. 15.45 are biased into conduction by sources V_{BB} and V_{GG}. In these circuits, we assume that the collector-emitter and drain-source voltages are large enough to ensure operation in the forward-active or pinch-off (active) regions, as appropriate for each device.

15.4.2 FIGURE OF MERIT FOR CURRENT SOURCES

Resistor R in Fig. 15.45 will be used as a reference for comparing current sources. The resistor provides an output current and output resistance of

$$I_O = \frac{V_{EE}}{R} \qquad \text{and} \qquad R_{\text{out}} = R \qquad (15.110)$$

The product of the dc current I_O and output resistance R_{out} is the effective voltage V_{CS} across the current source, and we will use it as a **figure of merit (FOM)** for comparing various current sources:

$$V_{CS} = I_O R_{\text{out}} \qquad (15.111)$$

For a given Q-point current, V_{CS} represents the equivalent voltage that will be needed across a resistor for it to achieve the same output resistance as the given current source. The larger the value of V_{CS}, the higher the output resistance of the source. For the resistor itself, V_{CS} is simply equal to the power supply voltage V_{EE}.

If ac models are drawn for each source in Fig. 15.45, the base, emitter, gate, and source of each transistor will be connected to ground, and each transistor will be considered operating in either the common-source or common-emitter configuration. The output resistance therefore will be equal to

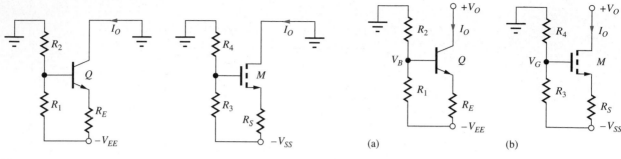

Figure 15.46 High-output resistance current sources.

Figure 15.47 (a) *npn* and (b) NMOS current source circuits.

r_o in all cases, and the figures of merit for these sources will be

$$\text{BJT:} \quad V_{CS} = I_O R_{\text{out}} = I_C r_o = I_C \frac{V_A + V_{CE}}{I_C} = V_A + V_{CE} \cong V_A$$

$$\text{FET:} \quad V_{CS} = I_O R_{\text{out}} = I_D r_o = I_D \frac{\dfrac{1}{\lambda} + V_{DS}}{I_D} = \frac{1}{\lambda} + V_{DS} \cong \frac{1}{\lambda} \tag{15.112}$$

V_{CS} for the C-E/C-S transistor current sources is approximately equal to either the Early voltage V_A or $1/\lambda$. We can expect that both these values generally will be at least several times the available power supply voltage. Therefore, any of the single transistor sources will provide an output resistance that is greater than that of a resistor.

15.4.3 HIGHER OUTPUT RESISTANCE SOURCES

From our study of single-stage amplifiers in Chapters 13 and 14, we know that placing a resistor in series with the emitter or source of the transistor, as in Fig. 15.46, increases the output resistance. Referring back to Eq. (13.66), we find that the output resistances for the circuits in Fig. 15.46 are

and

$$\text{BJT:} \quad R_{\text{out}} = r_o \left[1 + \frac{\beta_o R_E}{R_1 \| R_2 + r_\pi + R_E} \right] \leq (\beta_o + 1) r_o \tag{15.113}$$

$$\text{FET:} \quad R_{\text{out}} = r_o (1 + g_m R_S) \cong \mu_f R_S \tag{15.114}$$

The figures of merit are

$$\text{BJT: } V_{CS} \cong \beta_o (V_A + V_{CE}) \cong \beta_o V_A \qquad \text{and} \qquad \text{FET: } V_{CS} \cong \mu_f \frac{V_{SS}}{3} \tag{15.115}$$

where it has been assumed that $I_o R_S \cong V_{SS}/3$. Based on these figures of merit, the output resistance of the current sources in Fig. 15.46 can be expected to reach very high values, particularly at low current levels.[2] Table 15.1 compares V_{CS} for the various sources for typical device parameter values.

15.4.4 CURRENT SOURCE DESIGN EXAMPLES

This section provides examples of the design of current sources using the three-resistor bias circuits in Fig. 15.47. The computer (via a spreadsheet) is used to help explore the design space. The current source requirements are provided in the following design specifications.

---

[2] Because of its importance in analog circuit design, the $\beta_o V_A$ product is often used as a basic figure of merit for the bipolar transistor.

TABLE 15.1
Comparison of the Basic Current Sources $\beta_o = 100$, $V_A = 1/\lambda = 50$ V, $\mu_{f_{FET}} = 100$

TYPE OF SOURCE	R_{out}	V_{CS}	TYPICAL VALUES
Resistor	R	V_{EE}	15 V
Single transistor	r_o	V_A or $\dfrac{1}{\lambda}$	50–100 V
BJT with emitter resistor R_E	$\beta_o r_o$	$\cong \beta_o V_A$	5000 V
FET with source resistor R_S ($V_{SS} = 15$ V)	$\mu_f R_S$	$\cong \mu_f \dfrac{V_{SS}}{3}$	500 V or more

Design Specifications

Design a current source using the circuits in Fig. 15.47 with a nominal output current of 200 μA and an output resistance greater than 10 MΩ using a single −15-V power supply. The source must also meet the following additional constraints.

> Output voltage (compliance) range should be as large as possible while meeting the output resistance specification.
>
> The total current used by the source should be less than 250 μA.
>
> Bipolar transistors are available with (β_o, V_A) of (80, 100 V) or (150, 75 V). FETs are available with $\lambda = 0.01$ V^{-1}; K_n can be chosen as necessary.

When used in an actual application, the collector and drain of the current sources in Fig. 15.47 will be connected to some other point in the overall circuit, as indicated by the voltage $+V_O$ in the figure. For the current source to provide a high output resistance, the BJT must remain in the active region, with the collector-base junction reverse-biased ($V_O \geq V_B$), or the FET must remain in pinchoff ($V_O \geq V_G - V_{TN}$).

Specifications include the requirement that the output voltage range be as large as possible. Thus, the design goal is to achieve $I_O = 200$ μA and $R_{out} \geq 10$ MΩ with as low a voltage as possible at V_B or V_G. A range of designs is explored to see just how low a voltage can be used at V_B or V_G and still meet the I_O and R_{out} requirements. Investigating this design space is most easily done with the aid of the computer.

DESIGN EXAMPLE 15.5 DESIGN OF A BIPOLAR TRANSISTOR CURRENT SOURCE

Here we design a current source to meet a given set of design specifications using a bipolar transistor and the three resistor bias circuit.

PROBLEM Design a current source using the circuit in Fig. 15.47(a) with a nominal output current of 200 μA and an output resistance greater than 10 MΩ using a single −15-V power supply. The source must also meet the following additional constraints.

> Output voltage (compliance) range should be as large as possible while meeting the output resistance specification.
>
> The total current used by the source should be less than 250 μA.
>
> Bipolar transistors are available with (β_o, V_A) of (80, 100 V) or (150, 75 V).

SOLUTION **Known Information and Given Data:** Current source circuit in Fig. 15.47(a); $I_O = 200$ μA; $V_{EE} = 15$ V; $I_{EE} < 250$ μA; $R_{out} > 10$ MΩ; V_B as low as possible; BJTs are available with (β_o, V_A) of (80, 100 V) and (150, 75 V)

Unknowns: Values of resistors R_1, R_2, and R_E

Approach: Set up equations for analysis; using a computer program or spreadsheet, search for a set of bias conditions that satisfy the requirements; choose nearest resistor values from 1 percent resistor table in Appendix A.

Assumptions: Active region and small-signal operating conditions apply; $V_{BE} = 0.7$ V; $V_T = 0.025$ V; choose $V_O = 0$ V as a representative value for the output voltage.

Analysis: We start the design of the bipolar version of the current source with the expression for the output resistance of the source. Because we will use a computer to help in the design, we use the most complete expression for the output resistance:

$$R_{\text{out}} = r_o \left[1 + \frac{\beta_o R_E}{R_E + r_\pi + R_1 \| R_2} \right] \le \beta_o r_o \tag{15.116}$$

The figure of merit for this source is

$$V_{CS} = I_o R_{\text{out}} \le \beta_o V_A \tag{15.117}$$

and the design specifications require

$$\beta_o V_A = I_o R_{\text{out}} \ge (200\ \mu\text{A})(10\ \text{M}\Omega) = 2000\ \text{V} \tag{15.118}$$

Although both the specified transistors easily meet the requirement of Eq. (15.118), the denominator of Eq. (15.117) can substantially reduce the output resistance below that predicted by the $\beta_o r_o$ limit. Thus, it will be judicious to select the transistor with the higher $\beta_o V_A$ product—that is, (150, 75 V).

Having made this decision, the equations relating the dc Q-point design to the output resistance of the source can be developed. In Fig. 15.48, the three-resistor bias circuit is simplified using a $-V_{EE}$ referenced Thévenin transformation, for which

$$V_{BB} = 15 \frac{R_1}{R_1 + R_2} = 15 \frac{R_{BB}}{R_2} \quad \text{with} \quad R_{BB} = \frac{R_1 R_2}{R_1 + R_2} \tag{15.119}$$

The Q-point can be calculated using

$$I_B = \frac{V_{BB} - V_{BE}}{R_{BB} + (\beta_F + 1) R_E} \qquad I_O = I_C = \beta_F I_B$$

and

$$\tag{15.120}$$

$$V_{CE} = V_O + V_{EE} - (V_{BB} - I_B R_{BB} - V_{BE})$$

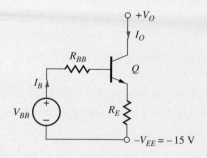

Figure 15.48 Equivalent circuit for the current source.

The small-signal parameters required for evaluating Eq. (15.116) are given by their usual formulas:

$$r_o = \frac{V_A + V_{CE}}{I_C} \quad \text{and} \quad r_\pi = \frac{\beta_o V_T}{I_C} \quad (15.121)$$

From Eq. (15.116), we can see that $R_{BB} = (R_1 \| R_2)$ should be made as small as possible in order to achieve maximum output resistance. From the design specifications, the complete current source must use no more than 250 μA. Because the output current is 200 μA, a maximum current of 50 μA can be used by the base bias network. The bias network current should be a factor of 5 to 10 times larger than the base current of the transistor which is 1.33 μA for the transistor with a current gain of 150. Thus, a bias network current of 20 μA is more than enough. However, in this case, we will trade increased operating current for a higher output resistance by picking a bias network current of 40 μA, which sets the sum of R_1 and R_2 to be (neglecting base current)

$$R_1 + R_2 \cong \frac{15 \text{ V}}{40 \text{ } \mu\text{A}} = 375 \text{ k}\Omega \quad (15.122)$$

Equations (15.116) to (15.122) provide the information necessary to explore the design space with the aid of a computer. These equations have been rearranged in order of evaluation in Eq. (15.123), with V_{BB} selected as the primary design variable.

Once V_{BB} is selected, R_1 and R_2 can be calculated. Then R_E and the Q-point can be determined, the small-signal parameters evaluated, and the output resistance determined from Eq. (15.113).

$$I_B = \frac{I_o}{\beta_F}$$
$$R_1 = (R_1 + R_2)\frac{V_{BB}}{15} = 375 \text{ k}\Omega\left(\frac{V_{BB}}{15}\right)$$
$$R_2 = (R_1 + R_2) - R_1 = 375 \text{ k}\Omega - R_1$$
$$R_{BB} = R_1 \| R_2$$
$$R_E = \alpha_F\left[\frac{V_{BB} - V_{BE} - I_B R_{BB}}{I_o}\right] \quad (15.123)$$
$$V_{CE} = V_{EE} - (V_{BB} - I_B R_{BB} - V_{BE})$$
$$r_o = \frac{V_A + V_{CE}}{I_o} \quad r_\pi = \frac{\beta_o V_T}{I_o}$$
$$R_{\text{out}} = r_o\left[1 + \frac{\beta_o R_E}{R_{BB} + r_\pi + R_E}\right]$$

Table 15.2 presents the results of using a spreadsheet to assist in evaluating these equations for a range of V_{BB}. The smallest value of V_{BB} for which the output resistance exceeds 10 MΩ

TABLE 15.2
Spreadsheet Results for Current Source Design

V_{BB}	R_1	R_2	R_{BB}	R_E	r_o	R_{out}
1.0	2.50E + 04	3.50E + 05	2.33E + 04	1.34E + 03	4.49E + 05	2.52E + 06
2.0	5.00E + 04	3.25E + 05	4.33E + 04	6.17E + 03	4.44E + 05	6.46E + 06
3.0	7.50E + 04	3.00E + 05	6.00E + 04	1.10E + 04	4.39E + 05	8.52E + 06
3.5	8.75E + 04	2.88E + 05	6.71E + 04	1.35E + 04	4.36E + 05	9.31E + 06
4.0	1.00E + 05	2.75E + 05	7.33E + 04	1.59E + 04	4.34E + 05	1.00E + 07
4.5	**1.13E + 05**	**2.63E + 05**	**7.88E + 04**	**1.84E + 04**	**4.32E + 05**	**1.07E + 07**
5.0	1.25E + 05	2.50E + 05	8.33E + 04	2.08E + 04	4.29E + 05	1.13E + 07
5.5	1.38E + 05	2.38E + 05	8.71E + 04	2.33E + 04	4.27E + 05	1.20E + 07
6.0	1.50E + 05	2.25E + 05	9.00E + 04	2.57E + 04	4.24E + 05	1.26E + 07

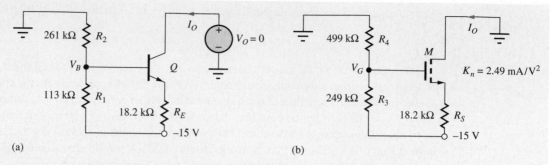

Figure 15.49 Final current source designs with $I_O = 200$ μA and $R_{out} \geq 10$ MΩ.

with some safety margin is 4.5 V. Note that this value of output resistance is achieved as

$$R_{out} = 432 \text{ k}\Omega \left[1 + \frac{150(18.4 \text{ k}\Omega)}{(78.8 + 18.8 + 18.4) \text{ k}\Omega} \right] = 10.7 \text{ M}\Omega \qquad (15.124)$$

Check of Results: Analysis of the circuit with the 1 percent resistor values in Fig. 15.49 yields $I_O = 203$ μA, $R_{out} = 10.4$ MΩ, and the supply current is 244 μA.

Discussion: For this design, the denominator in Eq. (15.124) reduces the output resistance by a factor of 6.3 below the $\beta_o r_o$ limit. So, it was a wise decision to choose the transistor with the largest $\beta_o V_A$ product. The final design appears in Fig. 15.49 using the nearest values from the 1 percent table in Appendix A.

Computer-Aided Analysis: Now we can check our hand design using SPICE with BF = 150, VAF = 75 V, and IS = 0.5 fA. (IS is selected to give $V_{BE} \cong 0.7$ V for a collector current or 200 μA.) In the circuit shown here, zero-value source V_O is added to directly measure the output current I_O as well as to provide a source that can be used to find R_{out} with a SPICE transfer function analysis. The results are $R_{out} = 11.4$ MΩ with $I_O = 205$ μA and $I_{EE} = 245$ μA, which meet all the design specifications. This could also be a good point to do a Monte Carlo analysis to explore the influence of tolerances on the design.

EXERCISE: What is the output resistance of the bipolar current source if the base were bypassed to ground with a capacitor?

ANSWER: 32.5 MΩ

EXERCISE: The current source is to be implemented using the nearest 5 percent resistor values. What are the best values? Are resistors with a 1/4-W power dissipation rating adequate for use in this circuit? What are the actual output current and output resistance of your current source, based on these 5 percent resistor values?

ANSWERS: 110 kΩ, 270 kΩ, 18 kΩ; yes; 195 μA, 10.6 MΩ

EXERCISE: Rework Design Ex. 15.5 using a bias network current of 20 μA. What are the new values of V_{BB}, R_1, R_2, R_E, and R_{out}?

ANSWERS: 9 V; 450 kΩ; 300 kΩ; 40.0 kΩ; 10.7 MΩ

DESIGN EXAMPLE 15.6   DESIGN OF A MOSFET CURRENT SOURCE

Now we design a current source to meet the same set of design specifications as in Ex. 15.5 but with a MOSFET replacing the BJT.

PROBLEM   Design a current source using the circuit in Fig. 15.50(b) with a nominal output current of 200 μA and an output resistance greater than 10 MΩ using a single −15-V power supply. The source must also meet the following additional constraints.

> Output voltage (compliance) range should be as large as possible while meeting the output resistance specification.
>
> The total current used by the source should be less than 250 μA.
>
> MOS transistors are available with $\lambda = 0.01$ V^{-1}. K_n can be chosen as required.

SOLUTION   **Known Information and Given Data:** Current source circuit in Fig. 15.50; $I_O = 200$ μA; $V_{SS} = 15$ V; $I_{SS} < 250$ μA; $R_{\text{out}} > 10$ MΩ; V_{GG} as low as possible; MOS transistors are available with $\lambda = 0.01$ V^{-1}. K_n can be chosen as required.

Unknowns: Values of resistors R_3, R_4, and R_S

Approach: Use $R_S = R_E$ and $V_S = V_E$ from the bipolar design in the previous example so the two designs can be easily compared. Find the amplification factor and value of K_n required to meet the output resistance requirement. Find V_{GS} and V_{GG}, and then choose R_3 and R_4 from the 1 percent resistor table in Appendix A.

Assumptions: Active region and small-signal operating conditions apply; $V_{TN} = 1$ V; choose $V_O = 0$ V as a representative value for the output voltage.

Analysis: We begin the design of the MOSFET current source by writing the expression for the transistor's output resistance. Because of the infinite current gain of the MOSFET, the expression for the output resistance of the current source is much less complex than that of the BJT source and is given by

$$R_{\text{out}} = r_o(1 + g_m R_S) \cong \mu_f R_S$$

If values of R_S and V_S are selected that are the same as those of the BJT source, 18 kΩ and −11.4 V, respectively, then the MOSFET must have an amplification factor of

$$\mu_f \geq \frac{10 \text{ M}\Omega}{18 \text{ k}\Omega} = 556 \gg 1$$

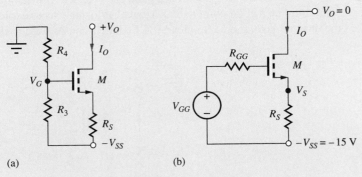

(a)   (b)

Figure 15.50 (a) MOSFET current source. (b) Equivalent circuit.

The amplification factor of the MOSFET is given by

$$\mu_f = \frac{1}{\lambda}\sqrt{\frac{2K_n}{I_D}}(1 + \lambda V_{DS})$$

and solving for K_n yields

$$K_n = \frac{I_D}{2}\left(\frac{\lambda \mu_f}{1 + \lambda V_{DS}}\right)^2 = 100 \ \mu\text{A} \left(\frac{\frac{0.01}{\text{V}}(556)}{1 + \frac{0.01}{\text{V}}(11.5 \ \text{V})}\right)^2 = 2.49 \ \frac{\text{mA}}{\text{V}^2}$$

This value of K_n is achievable using either discrete components or integrated circuits. In Fig. 15.50, the required gate voltage V_{GG} is

$$V_{GG} = I_D R_S + V_{GS} = 3.60 + V_{TN} + \sqrt{\frac{2I_D}{K_n}}$$

$$= 3.60 \ \text{V} + 1 \ \text{V} + \sqrt{\frac{2(0.2 \ \text{mA})}{\frac{2.49 \ \text{mA}}{\text{V}^2}}} = 5.00 \ \text{V}$$

If the current in the bias resistors is limited to 10 percent of the drain current, then

$$R_3 + R_4 = \frac{15 \ \text{V}}{20 \ \mu\text{A}} = 750 \ \text{k}\Omega \qquad \text{and} \qquad R_3 = \frac{5.00 \ \text{V}}{15 \ \text{V}} 750 \ \text{k}\Omega = 250 \ \text{k}\Omega$$

The nearest 1 percent values from Appendix A are $R_3 = 249 \ \text{k}\Omega$ and $R_4 = 499 \ \text{k}\Omega$ with $R_S = 18.2 \ \text{k}\Omega$. The final design appears in Fig. 15.49.

Check of Results: A recheck of the math indicates that our calculations are correct. SPICE can now be used to verify our design and the results appear below.

Discussion: For the MOS source, we can use a larger set of gate bias resistors, since the output resistance of the current source does not depend on R_{GG}.

Computer-Aided Analysis: Now we can check our hand design using SPICE with VTO = 1 V, KP = 2.49 mA/V^2, and LAMBDA = 0.01 V^{-1}. In the circuit shown here, zero-value source V_O is added to directly measure the output current I_O and to provide a source that can be used to find R_{out} with a SPICE Transfer Function analysis. The results using the 1 percent resistor values are $R_{\text{out}} = 11.3 \ \text{M}\Omega$ with $I_O = 198 \ \mu\text{A}$ and $I_{SS} = 219 \ \mu\text{A}$, which meet all the design specifications. This could be a good point to do a Monte Carlo analysis to explore the influence of tolerances on the design. Also, more complex SPICE models can be used to double check the design.

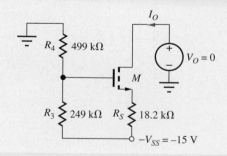

EXERCISE: What is the minimum drain voltage for which MOSFET *M* in the figure above remains saturated?

ANSWER: −9.34 V

EXERCISE: What W/L ratio is required for the preceding FET if $K_n' = 25$ μA/V^2?

ANSWER: 99.6/1

EXERCISE: What is the minimum collector voltage for which the BJT in Fig. 15.49(a) remains in the forward-active region?

ANSWER: −10.8 V

EXERCISE: The MOS current source is to be implemented using the nearest 5 percent resistor values. What are the best values? Are resistors with a 1/4-W power dissipation rating adequate for use in this circuit? What are the actual output current and output resistance of your current source based on these 5 percent resistor values?

ANSWERS: 510 kΩ, 240 kΩ, 18 kΩ; yes; 200 μA, 10.6 MΩ

15.5 CIRCUIT ELEMENT MATCHING

Integrated circuit (IC) technology allows the realization of large numbers of virtually identical transistors. Although the absolute parameter tolerances of these devices are relatively poor, device[1] characteristics can be matched to within 1 percent or better. The ability to build devices with nearly identical characteristics has led to the development of special circuit techniques that take advantage of the tight matching of device characteristics. Transistors are said to be **matched** when they have identical sets of device parameters: (I_S, β_{FO}, V_A) for the BJT, (V_{TN}, K', λ) for the MOSFET, or (I_{DSS}, V_P, λ) for the JFET. The planar geometry of the devices can easily be changed in integrated designs, and so the emitter area A_E of the BJT and the W/L ratio of the MOSFET become important circuit design parameters. (Remember from our study of MOS digital circuits in Part II that W/L represents a fundamental circuit design parameter.)

In integrated circuits, absolute parameter values may vary widely from fabrication process run to process run, with ±25 to 30 percent tolerances not uncommon (see Table 15.3). However, the matching between nearby circuit elements on a given IC chip is typically within a fraction of a percent. Thus, IC design techniques have been invented that rely heavily on **matched device** characteristics and resistor ratios rather than absolute parameter values. The circuits described in this chapter depend, for proper operation, on the tight device matching that can be realized through IC fabrication processes, and many will not operate correctly if built with mismatched discrete

TABLE 15.3
IC Tolerances and Matching [1]

	ABSOLUTE TOLERANCE, %	MISMATCH, %
Diffused resistors	30	≤ 2
Ion-implanted resistors	5	≤ 1
V_{BE}	10	≤ 1
I_S, β_F, V_A	30	≤ 1
V_{TN}, V_{TP}	15	≤ 1
K', λ	30	≤ 1

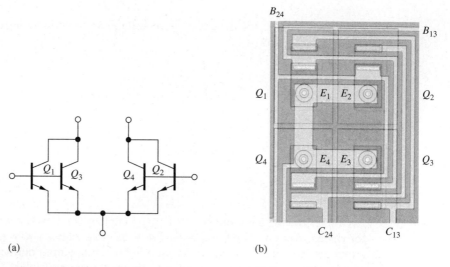

Figure 15.51 (a) Differential amplifier formed with a cross-connected quad of identical transistors. (b) Layout of the cross-coupled transistor quad in Fig. 15.51(a).

components. However, many of these circuits can be used in discrete circuit design if integrated transistor arrays are used in the implementation.

Figure 15.51 shows one example of the use of four matched transistors to improve the performance of the differential amplifier that we studied earlier in this chapter. The four devices are cross-connected to further improve the overall parameter matching and temperature tracking of the circuit. In Section 15.8, we explore the use of matched bipolar and MOS transistors in the design of IC current sources called **current mirrors.**

EXERCISE: An IC resistor has a nominal value of 10 kΩ and a tolerance of ±30 percent. A particular process run has produced resistors with an average value 20 percent higher than the nominal value, and the resistors are found to be matched within 2 percent. What range of resistor values will occur in this process run?

ANSWER: 11.88 kΩ–12.12 kΩ

15.6 CURRENT MIRRORS

Current mirror biasing is an extremely important technique in integrated circuit design. Not only is it heavily used in analog applications, it also appears routinely in digital circuit design as well. Figure 15.52 shows the circuits for basic MOS and bipolar current mirrors. In Fig. 15.52(a), MOSFETs M_1 and M_2 are assumed to have identical characteristics (V_{TN}, K_n', λ) and W/L ratios; in Fig. 15.52(b), the characteristics of Q_1 and Q_2 are assumed to be identical (I_S, β_{FO}, V_A). In both circuits, a **reference current** I_{REF} provides operating bias to the mirror, and the output current is represented by current I_O. These basic circuits are designed to have $I_O = I_{\text{REF}}$; that is, the output current mirrors the reference current—hence, the name "current mirror."

15.6.1 dc ANALYSIS OF THE MOS TRANSISTOR CURRENT MIRROR

In the MOS current mirror in Fig. 15.52(a), reference current I_{REF} goes through "diode-connected" transistor M_1, establishing gate-source voltage V_{GS}. V_{GS} is applied to transistor M_2, developing an identical drain current $I_{D2} = I_{\text{REF}}$. Detailed analysis of the current mirror operation follows in the paragraphs below.

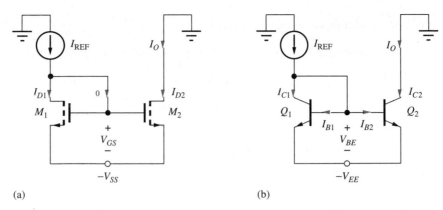

Figure 15.52 (a) MOS and (b) BJT current mirror circuits.

Because the gate currents are zero for the MOSFETs, reference current I_{REF} must flow into the drain of M_1, which is forced to operate in pinch-off by the circuit connection because $V_{DS1} = V_{GS1} = V_{GS}$. V_{GS} must equal the value required for $I_{D1} = I_{REF}$. Assuming matched devices:[3]

$$I_{REF} = \frac{K_n}{2}(V_{GS1} - V_{TN})^2(1 + \lambda V_{DS1}) \quad \text{or} \quad V_{GS1} = V_{TN} + \sqrt{\frac{2I_{REF}}{K_{n1}(1 + \lambda V_{DS1})}} \quad (15.125)$$

Current I_O is equal to the drain current of M_2:

$$I_O = I_{D2} = \frac{K_n}{2}(V_{GS2} - V_{TN})^2(1 + \lambda V_{DS2}) \quad (15.126)$$

However, the circuit connection forces $V_{GS2} = V_{GS1}$, and $V_{DS1} = V_{GS1}$. Substituting Eq. (15.125) into Eq. (15.126) yields

$$I_O = I_{REF}\frac{(1 + \lambda V_{DS2})}{(1 + \lambda V_{DS1})} \cong I_{REF} \quad (15.127)$$

For equal values of V_{DS}, the output current is identical to the reference current (that is, the output mirrors the reference current). Unfortunately, in most circuit applications, $V_{DS2} \neq V_{DS1}$, and there is a slight mismatch between the output current and the reference current, as demonstrated in Ex. 15.7.

For convenience, we define the ratio of I_O to I_{REF} to be the **mirror ratio** MR given by

$$MR = \frac{I_O}{I_{REF}} = \frac{(1 + \lambda V_{DS2})}{(1 + \lambda V_{DS1})} \quad (15.128)$$

EXAMPLE 15.7   **OUTPUT CURRENT OF THE MOS CURRENT MIRROR**

In this example, we find the output current for the standard current mirror configuration.

PROBLEM   Calculate the output current I_O for the MOS current mirror in Fig. 15.52(a) if $V_{SS} = 10$ V, $K_n = 250$ μA/V^2, $V_{TN} = 1$ V, $\lambda = 0.0133$ V^{-1}, and $I_{REF} = 150$ μA.

SOLUTION   **Known Information and Given Data:** Current mirror circuit in Fig. 15.52(a); $V_{SS} = 10$ V; transistor parameters are given as $K_n = 250$ μA/V^2, $V_{TN} = 1$ V, $\lambda = 0.0133$ V^{-1}, and $I_{REF} = 150$ μA

---

[3] Matching between elements in the current mirror is very important; this is a case in which the $(1 + \lambda V_{DS})$ term is included in the dc, as well as ac, calculations.

Unknowns: Output current I_O

Approach: Find V_{GS1} and V_{DS2} and then evaluate Eq. (15.52) to give the output current.

Assumptions: Transistors are identical and operating in the active region of operation

Analysis: We need to evaluate Eq. (15.127) and must find the value of V_{GS1} using Eq. (15.125). Since $V_{DS1} = V_{GS1}$, we can write

$$V_{DS1} = V_{TN} + \sqrt{\frac{2I_{REF}}{K_n}} = 1 + \sqrt{\frac{2(150 \ \mu A)}{250\frac{\mu A}{V^2}}} = 2.10 \ V$$

in which we have neglected the $(1 + \lambda V_{DS1})$ term to simplify the dc bias calculation. Substituting this value and $V_{DS2} = 10$ V in Eq. (15.127):

$$I_O = (150 \ \mu A)\frac{[1 + 0.0133(10)]}{[1 + 0.0133(2.10)]} = 165 \ \mu A$$

The ideal output current would be 150 μA, whereas the actual currents are mismatched by approximately 10 percent.

Check of Results: A double check shows the calculations to be correct. M_1 is pinched-off by connection, and M_2 will also be active as long as its drain-source voltage exceeds $(V_{GS1} - V_{TN})$, which is easily met in Fig. 15.52(a) since $V_{DS2} = 10$ V.

Discussion: We could attempt to improve the precision of our answer slightly by including the $(1 + \lambda V_{DS1})$ term in the evaluation of V_{GS1}. The solution then requires an iterative analysis that barely changes the value of I_O.

Computer-Aided Analysis: We can check our analysis directly with SPICE by setting the MOS transistor parameters to KP = 250 μA/V², VTO = 1 V, LEVEL = 1, and LAMBDA = 0. SPICE yields an output current of 150 μA with $V_{GS} = 2.095$ V. With nonzero λ, LAMBDA = 0.0133 V⁻¹, SPICE yields $I_O = 165$ μA with $V_{GS} = 2.081$ V. The values are in agreement with our hand calculations.

EXERCISE: Suppose we include the $(1 + \lambda V_{DS1})$ term in the evaluation of V_{GS1}. Show that the equation to be solved is

$$V_{DS1} = V_{TN} + \sqrt{\frac{2I_{REF}}{K_n(1 + \lambda V_{DS1})}}$$

Find the new value of V_{DS1} using the numbers in Ex. 15.7. What is the new value of I_O?

ANSWERS: 2.08 V; 165 μA

EXERCISE: Based on the numbers in Ex. 15.7, what is the minimum value of the drain voltage required to keep M_2 saturated in Fig. 15.52(a)?

ANSWER: −8.9 V

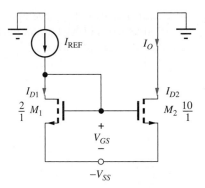

Figure 15.53 MOS current mirror with unequal (W/L) ratios.

15.6.2 CHANGING THE MOS MIRROR RATIO

The power of the current mirror is greatly increased if the mirror ratio can be changed from unity. For the MOS current mirror, the ratio can easily be modified by changing the W/L ratios of the two transistors forming the mirror. In Fig. 15.53, for example, remembering that $K_n = K_n'(W/L)$ for the MOSFET, the K_n values of the two transistors are given by

$$K_{n1} = K_n'\left(\frac{W}{L}\right)_1 \qquad \text{and} \qquad K_{n2} = K_n'\left(\frac{W}{L}\right)_2 \qquad (15.129)$$

Substituting these two different values of K_n in Eqs. (15.125) and (15.126) yields the mirror ratio given by

$$\text{MR} = \frac{\left(\dfrac{W}{L}\right)_2}{\left(\dfrac{W}{L}\right)_1} \frac{(1 + \lambda V_{DS2})}{(1 + \lambda V_{DS1})} \qquad (15.130)$$

In the ideal case ($\lambda = 0$) or for $V_{DS2} = V_{DS1}$, the mirror ratio is set by the ratio of the W/L values of the two transistors. For the particular values in Fig. 15.53, this design value of the mirror ratio would be 5, and the output current would be $I_O = 5I_{\text{REF}}$. However, the differences in V_{DS} will again create an error in the mirror ratio.

EXERCISE: (a) Calculate the mirror ratio for the MOS current mirrors in the figure here for $\lambda = 0$. (b) For $\lambda = 0.02$ V^{-1} if $V_{TN} = 1$ V, $K_n' = 25$ μA/V^2, and $I_{\text{REF}} = 50$ μA.

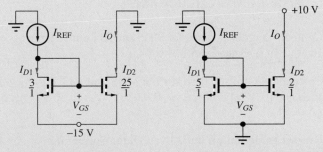

ANSWERS: 8.33, 0.400; 10.4, 0.462

15.6.3 dc ANALYSIS OF THE BIPOLAR TRANSISTOR CURRENT MIRROR

The operation of the bipolar current mirror in Fig. 15.52(b) is similar to that of the MOS circuit. Reference current I_{REF} goes through diode-connected transistor Q_1, establishing base-emitter voltage

V_{BE}. V_{BE} also biases transistor Q_2, developing an almost identical collector current at its output: $I_{C2} \cong I_{\text{REF}}$. Detailed analysis of the current mirror operation follows in the paragraphs below.

Analysis of the BJT current mirror in Fig. 15.52(b) is similar to that of the FET. Applying KCL at the collector of "diode-connected" transistor Q_1 yields

$$I_{\text{REF}} = I_{C1} + I_{B1} + I_{B2} \quad \text{and} \quad I_O = I_{C2} \tag{15.131}$$

The currents needed to relate I_O to I_{REF} can be found using the transport model, noting that the circuit connection forces the two transistors to have the same base-emitter voltage V_{BE}:

$$I_{C1} = I_S \exp\left(\frac{V_{BE}}{V_T}\right)\left(1 + \frac{V_{CE1}}{V_A}\right) \qquad I_{C2} = I_S \exp\left(\frac{V_{BE}}{V_T}\right)\left(1 + \frac{V_{CE2}}{V_A}\right)$$

$$\beta_{F1} = \beta_{FO}\left(1 + \frac{V_{CE1}}{V_A}\right) \qquad\qquad \beta_{F2} = \beta_{FO}\left(1 + \frac{V_{CE2}}{V_A}\right) \tag{15.132}$$

$$I_{B1} = \frac{I_S}{\beta_{FO}} \exp\left(\frac{V_{BE}}{V_T}\right) \qquad\qquad I_{B2} = \frac{I_S}{\beta_{FO}} \exp\left(\frac{V_{BE}}{V_T}\right)$$

Substituting Eq. (15.132) into Eq. (15.131) and solving for $I_O = I_{C2}$ yields

$$I_O = I_{\text{REF}} \frac{\left(1 + \dfrac{V_{CE2}}{V_A}\right)}{\left(1 + \dfrac{V_{CE1}}{V_A} + \dfrac{2}{\beta_{FO}}\right)} = I_{\text{REF}} \frac{\left(1 + \dfrac{V_{CE2}}{V_A}\right)}{\left(1 + \dfrac{V_{BE}}{V_A} + \dfrac{2}{\beta_{FO}}\right)} \tag{15.133}$$

If the Early voltage were infinite, Eq. (15.133) would give a mirror ratio of

$$\text{MR} = \frac{I_O}{I_{\text{REF}}} = \frac{1}{1 + \dfrac{2}{\beta_{FO}}} \tag{15.134}$$

and the output current would mirror the reference current, except for a small error due to the finite current gain of the BJT. For example, if $\beta_{FO} = 100$, the currents would match within 2 percent. As for the FET case, however, the collector-emitter voltage mismatch in Eq. (15.133) is generally more significant than the **current gain defect** term, as indicated in Ex. 15.8.

EXAMPLE 15.8   **MIRROR RATIO CALCULATIONS**

Compare the mirror ratios for MOS and BJT current mirrors operating with similar bias conditions and output resistances ($V_A = 1/\lambda$).

PROBLEM   Calculate the mirror ratio for the MOS and BJT current mirrors in Fig. 15.52 for $V_{GS} = 2$ V, $V_{DS2} = 10$ V $= V_{CE2}$, $\lambda = 0.02$ V^{-1}, $V_A = 50$ V, and $\beta_{FO} = 100$. Assume $M_1 = M_2$ and $Q_1 = Q_2$.

SOLUTION   **Known Information and Given Data:** Current mirror circuits in Fig. 15.52 with $M_2 = M_1$ and $Q_2 = Q_1$; $V_{SS} = 10$ V; operating voltages: $V_{GS} = 2$ V, $V_{DS2} = V_{CE2} = 10$ V and $V_{BE} = 0.7$ V; transistor parameters: $\lambda = 0.02$ V^{-1}, $V_A = 50$ V, and $\beta_{FO} = 100$

Unknowns: Mirror ratio MR for each current mirror

Approach: Use Eqs. (15.130) and (15.133) to determine the mirror ratios.

Assumptions: BJTs and MOSFETs are in the active region of operation, respectively. Assume $V_{BE} = 0.7$ V for the BJTs and the MOSFETs are enhancement-mode devices.

Analysis: For the MOS current mirror,

$$MR = \frac{(1 + \lambda V_{DS2})}{(1 + \lambda V_{DS1})} = \frac{\left[1 + \dfrac{0.02}{V}(10 \text{ V}) \right]}{\left[1 + \dfrac{0.02}{V}(2 \text{ V}) \right]} = 1.15$$

and for the BJT case

$$MR = \frac{\left(1 + \dfrac{V_{CE2}}{V_A} \right)}{\left(1 + \dfrac{2}{\beta_{FO}} + \dfrac{V_{CE1}}{V_A} \right)} = \frac{\left(1 + \dfrac{10 \text{ V}}{50 \text{ V}} \right)}{\left(1 + \dfrac{2}{100} + \dfrac{0.7 \text{ V}}{50 \text{ V}} \right)} = 1.16$$

Check of Results: A double check shows our calculations to be correct. M_1 is forced to be active by connection. M_2 has $V_{DS2} > V_{GS2}$ and will be pinched-off for $V_{TN} > 0$ (enhancement-mode transistor). Q_1 has $V_{CE} = V_{BE}$, so it is forced to be in the active region. Q_2 has $V_{CE2} > V_{BE2}$ and is also in the active region. The assumed regions of operation are valid.

Discussion: The FET and BJT mismatches are very similar—15 percent and 16 percent, respectively. The current gain error is a small contributor to the overall error in the BJT mirror ratio.

Computer-Aided Analysis: We can easily perform an analysis of the current mirrors using SPICE, which will be done shortly as part of Ex. 15.9.

EXERCISE: What is the actual value of V_{BE} in the bipolar current mirror in Ex. 15.8 if $I_S = 0.1$ fA and $I_{REF} = 100$ μA? What is the minimum value of the collector voltage required to maintain Q_2 in the active region in Fig. 15.52(b)?

ANSWERS: 0.691 V; $-V_{EE} + 0.691$ V

15.6.4 ALTERING THE BJT CURRENT MIRROR RATIO

In bipolar IC technology, the designer is free to modify the emitter area of the transistors, just as the W/L ratio can be chosen in MOS design. To alter the BJT mirror ratio, we use the fact that the saturation current of the bipolar transistor is proportional to its emitter area A_E and can be written as

$$I_S = I_{SO} \frac{A_E}{A} \tag{15.135}$$

I_{SO} represents the saturation current of a bipolar transistor with one unit of emitter area: $A_E = 1 \times A$. The actual dimensions associated with A are technology-dependent.

By changing the relative sizes of the emitters (**emitter area scaling**) of the BJTs in the current mirror, the IC designer can modify the mirror ratio. For the modified mirror in Fig. 15.54,

$$I_{C1} = I_{SO} \frac{A_{E1}}{A} \exp\left(\frac{V_{BE}}{V_T} \right) \left(1 + \frac{V_{CE1}}{V_A} \right) \qquad I_{C2} = I_{SO} \frac{A_{E2}}{A} \exp\left(\frac{V_{BE}}{V_T} \right) \left(1 + \frac{V_{CE2}}{V_A} \right)$$

$$\tag{15.136}$$

$$I_{B1} = \frac{I_{SO}}{\beta_{FO}} \frac{A_{E1}}{A} \exp\left(\frac{V_{BE}}{V_T} \right) \qquad\qquad I_{B2} = \frac{I_{SO}}{\beta_{FO}} \frac{A_{E2}}{A} \exp\left(\frac{V_{BE}}{V_T} \right)$$

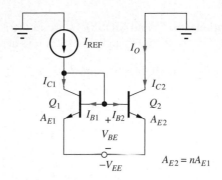

Figure 15.54 BJT current mirror with unequal emitter area.

Substituting these equations in Eq. (15.131) and then solving for I_O yields

$$I_O = nI_{REF} \frac{1 + \dfrac{V_{CE2}}{V_A}}{1 + \dfrac{V_{BE}}{V_A} + \dfrac{1+n}{\beta_{FO}}} \qquad \text{where} \qquad n = \frac{A_{E2}}{A_{E1}} \qquad (15.137)$$

In the ideal case of infinite current gain and identical collector-emitter voltages, the mirror ratio would be determined only by the ratio of the two emitter areas: $MR = n$.

However, for finite current gain,

$$MR = \frac{n}{1 + \dfrac{1+n}{\beta_{FO}}} \qquad \text{where} \qquad n = \frac{A_{E2}}{A_{E1}} \qquad (15.138)$$

For example, suppose $A_{E2}/A_{E1} = 10$ and $\beta_{FO} = 100$; then the mirror ratio would be 9.01. A relatively large error (10 percent) is occurring even though the effect of collector-emitter voltage mismatch has been ignored. For high mirror ratios, the current gain error term can become quite important because the total number of units of base current increases directly with the mirror ratio.

EXERCISE: (a) Calculate the ideal mirror ratio for the BJT current mirrors in the figure below if $V_A = \infty$ and $\beta_{FO} = \infty$. (b) If $V_A = \infty$ and $\beta_{FO} = 75$. (c) If $V_A = 60$ V, $\beta_{FO} = 75$, and $V_{BE} = 0.7$ V.

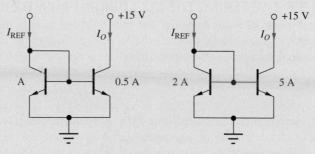

ANSWERS: 0.500, 2.50; 0.490, 2.39; 0.606, 2.95

15.6.5 MULTIPLE CURRENT SOURCES

Analog circuits often require a number of different current sources to bias the various stages of the design. A single reference transistor, M_1 or Q_1, can be used to generate multiple output currents using the circuits in Fig. 15.55. In Fig. 15.55 (a), the unusual connection of the gate terminals through

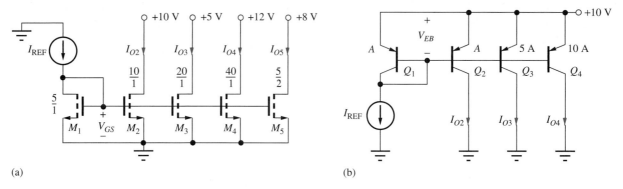

Figure 15.55 (a) Multiple MOS current sources generated from one reference voltage. (b) Multiple bipolar sources biased by one reference device.

the MOSFETs is being used as a "short-hand" method to indicate that all the gates are connected together. Circuit operation is similar to that of the basic current mirror. The reference current enters the **"diode-connected" transistor**—here, the MOSFET M_1—establishing gate-source voltage V_{GS}, which is then used to bias transistors M_2 through M_5, each having a different W/L ratio. Because there is no current gain defect in MOS technology, a large number of output transistors can be driven from one reference transistor.

EXERCISE: What are the four output currents in the circuit in Fig. 15.55(a) if $I_{REF} = 100$ μA and $\lambda = 0$ for all the FETs?

ANSWERS: 200 μA; 400 μA; 800 μA; 50.0 μA

EXERCISE: Recalculate the four output currents in the circuit in Fig. 15.55(a) if $\lambda = 0.02$ for all the FETs. Assume $V_{GS} = 2$ V.

ANSWERS: 231 μA; 423 μA; 954 μA; 55.8 μA

The situation is very similar in the *pnp* bipolar mirror in Fig. 15.55(b). Here again, the base terminals of the BJTs are extended through the transistors to simplify the drawing. In this circuit, reference current I_{REF} is supplied by diode-connected BJT Q_1 to establish the emitter-base reference voltage V_{EB}. V_{EB} is then used to bias transistors Q_2 to Q_4, each having a different emitter area relative to that of the reference transistor. Because the total base current increases with the addition of each output transistor, the base current error term gets worse as more transistors are added, which limits the number of outputs that can be used with the basic bipolar current mirror. The buffered current mirror in Sec. 15.6.6 was invented to solve this problem.

An expression for the output current from a given collector can be derived following the steps that led to Eq. (15.137):

$$I_{Oi} = n_i I_{REF} \frac{1 + \dfrac{V_{ECi}}{V_A}}{1 + \dfrac{V_{EB}}{V_A} + \dfrac{1 + \sum\limits_{i=2}^{m} n_i}{\beta_{FO}}} \qquad \text{where} \qquad n_i = \frac{A_{Ei}}{A_{E1}} \qquad (15.139)$$

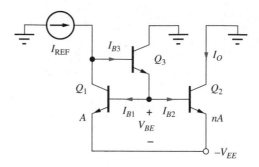

Figure 15.56 Buffered current mirror.

15.6.6 BUFFERED CURRENT MIRROR

The current gain defect in the bipolar current mirror can become substantial when a large mirror ratio is used or if many source currents are generated from one reference transistor. However, this error can be reduced greatly by using the circuit in Fig. 15.56, called a **buffered current mirror.** The current gain of transistor Q_3 is used to reduce the base current that is subtracted from the reference current. Applying KCL at the collector of transistor Q_1, and assuming that $V_A = \infty$ for simplicity, I_{C1} is expressed as

$$I_{C1} = I_{REF} - I_{B3} = I_{REF} - \frac{(1+n)\dfrac{I_{C1}}{\beta_{FO1}}}{\beta_{FO3} + 1} \tag{15.140}$$

and solving for the collector current yields

$$I_O = nI_{C1} = nI_{REF}\frac{1}{1 + \dfrac{(1+n)}{\beta_{FO1}(\beta_{FO3} + 1)}} \tag{15.141}$$

The current gain error term in the denominator has been reduced by a factor of $(\beta_{FO3} + 1)$ from the error in Eq. (15.138).

15.6.7 OUTPUT RESISTANCE OF THE CURRENT MIRRORS

Now that we have found the dc output current of the current mirror, we will focus on the second important parameter that characterizes the electronic current source—the output resistance. The output resistance of the basic current mirror can be found by referring to the ac model of Fig. 15.57. Diode-connected bipolar transistor Q_1 represents a simple two-terminal device, and its small-signal model is easily found using nodal analysis of Fig. 15.58:

$$\mathbf{i} = g_\pi \mathbf{v} + g_m \mathbf{v} + g_o \mathbf{v} = (g_m + g_\pi + g_o)\mathbf{v} \tag{15.142}$$

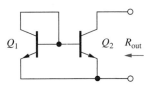

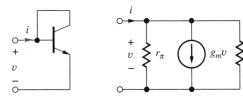

Figure 15.57 ac Model for the output resistance of the bipolar current mirror.

Figure 15.58 Model for "diode-connected" transistor.

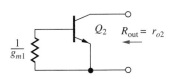

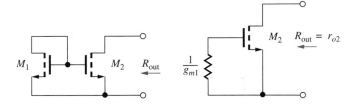

Figure 15.59 Simplified small-signal model for the bipolar current mirror.

Figure 15.60 Output resistance of the MOS current mirror.

By factoring out g_m, an approximate result for the diode conductance is

$$\frac{\mathbf{i}}{\mathbf{v}} = g_m \left[1 + \frac{1}{\beta_o} + \frac{1}{\mu_f} \right] \cong g_m \qquad \text{and} \qquad R \cong \frac{1}{g_m} \qquad (15.143)$$

for β_o and $\mu_f \gg 1$. The small-signal model for the diode-connected BJT is simply a resistor of value $1/g_m$. Note that this result is the same as the small-signal resistance r_d of an actual diode that was developed in Sec. 13.4.

Using this diode model simplifies the ac model for the current mirror to that shown in Fig. 15.59. This circuit should be recognized as a common-emitter transistor with a Thévenin equivalent resistance $R_{th} = 1/g_m$ connected to its base; the output resistance just equals the output resistance r_{o2} of transistor Q_2.

The equation describing the small-signal model for the two-terminal "diode-connected" MOS-FET is similar to that in Eq. (15.143) except that the current gain is infinite. Therefore, the two-terminal MOSFET is also represented by a resistor of value $1/g_m$, as in Fig. 15.60; the output resistance of the MOS current mirror is equal to r_{o2} of MOSFET M_2.

Thus, the output resistance and figure of merit (see Section 15.4.2) for the basic current mirror circuits are determined by output transistors Q_2 and M_2:

$$R_{out} = r_{o2} \qquad \text{and} \qquad V_{CS} \cong V_{A2} \qquad \text{or} \qquad \frac{1}{\lambda_2} \qquad (15.144)$$

EXERCISE: What are the output resistances of sources I_{O2} and I_{O3} in Fig. 15.55(a) for $I_{REF} = 100\ \mu A$ and Fig. 15.55(b) for $I_{REF} = 10\ \mu A$ if $V_A = 1/\lambda = 50$ V and $\beta_F = 100$?

ANSWERS: 260 kΩ, 130 kΩ; 6.77 MΩ, 1.35 MΩ

15.6.8 TWO-PORT MODEL FOR THE CURRENT MIRROR

We shall see shortly that the current mirror can be used not only as a dc current source but, in more complex circuits, as a current amplifier and active load. It will be useful to understand the small-signal behavior of the current mirror, redrawn as a two-port in Fig. 15.61.

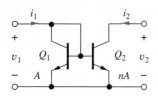

Figure 15.61 Current mirror as a two-port.

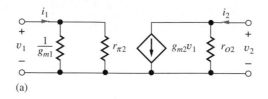 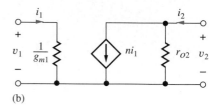

Figure 15.62 (a) Small-signal model for the current mirror. (b) Simplified small-signal model for the current mirror.

The small-signal model for the current mirror is in Fig. 15.62, in which diode-connected transistor Q_1 is represented in its simplified form by $1/g_{m1}$. From the circuit in Fig. 15.62(a),

$$R_{in} = \frac{v_1}{i_1}\bigg|_{v_2=0} = \frac{1}{(g_{m1} + g_{\pi 2})} = \frac{1}{g_{m1}\left(1 + \dfrac{n}{\beta_{o2}}\right)} \cong \frac{1}{g_{m1}}$$

$$\beta = \frac{i_2}{i_1}\bigg|_{v_2=0} = \frac{g_{m2}r_{\pi 2}}{1 + g_{m1}r_{\pi 2}} = \frac{\beta_{o2}}{1 + \dfrac{g_{m1}}{g_{m2}}\beta_{o2}} \cong \frac{g_{m2}}{g_{m1}} \cong \frac{I_{C2}}{I_{C1}} = n \qquad (15.145)$$

$$R_{out} = \frac{v_2}{i_2}\bigg|_{i_1=0} = r_{o2}$$

Figure 15.62(b) shows the final two-port model representation. The bipolar current mirror has an input resistance of $1/g_{m1}$, determined by diode Q_1 and an output resistance equal to r_{o2} of Q_2. The current gain is determined approximately by the emitter-area ratio $n = A_{E2}/A_{E1}$. Be sure to remember to use the correct values of I_{C1} and I_{C2} when calculating the values of the small-signal parameters.

Analysis of the MOS current mirror yields similar results [or by simply setting $\beta_{o2} = \infty$ in Eq. (15.145)]:

$$R_{in} = \frac{1}{g_{m1}} \qquad \beta = \frac{g_{m2}}{g_{m1}} \cong \frac{\left(\dfrac{W}{L}\right)_2}{\left(\dfrac{W}{L}\right)_1} \cong n \qquad R_{out} = r_{o2} \qquad (15.146)$$

In this case, the current gain β is determined by the W/L ratios of the two FETs rather than by the bipolar emitter-area ratio.

EXERCISE: What are the values of I_{C1} and I_{C2} and the small-signal parameters for the current mirror in Fig. 15.54 if $I_{REF} = 100\ \mu A$, $\beta_{FO} = 50$, $V_A = 50\ V$, $V_{BE} = 0.7\ V$, $V_{CE2} = 10\ V$, and $n = 5$?

ANSWERS: 89.4 μA; 529 μA; 280 Ω; 0; 5.92; 113 kΩ

EXAMPLE 15.9   **CALCULATING THE TWO-PORT PARAMETERS OF A CURRENT MIRROR USING SPICE**

Transfer function analysis is used to find the two-port parameters of the BJT current mirror.

PROBLEM   Use the transfer function capability of SPICE to find the two-port parameters of the BJT current mirror biased by a reference current of 100 μA and a power supply of +10 V.

SOLUTION **Known Information and Given Data:** A current mirror using bipolar transistors; $I_{\text{REF}} = 100\ \mu\text{A}$ and $V_{CC} = 10\ \text{V}$

Unknowns: Output current I_O, V_{BE}, R_{in}, β, and R_{out} for the current mirror

Approach: Construct the circuit using the schematic editor in SPICE. Use the transfer function analysis to find the forward transfer function from I_{REF} to $I(V_{CC})$ and reverse transfer function from V_{CC} to node 1. The SPICE transfer function analysis automatically calculates three values: the requested transfer function, the resistance at the input source node, and the resistance at the output source node. However, since the output node is connected to V_{CC}, the output resistance calculated at that node will be zero, and two analyses will be required to find all the two-port parameters.

Assumptions: Use the current mirror with a single positive supply V_{CC} biased by current source I_{REF}, as shown in the figure here. $V_A = 50\ \text{V}$, $\beta_{FO} = 100$, and $I_S = 0.1\ \text{fA}$.

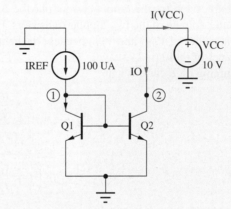

Analysis: First, we must set the BJT parameters to the desired values: BF = 100, VAF = 50 V, and IS = 0.1 fA. An operating point and two transfer function analyses are used in this example. The first asks for the transfer function from input source I_{REF} to output variable $I(V_{CC})$. The operating point analysis yields $V(1) = 0.719\ \text{V}$ and $I_O = 116\ \mu\text{A}$. The transfer function analysis gives input resistance $R_{\text{in}} = 259\ \Omega$ and current gain $\beta = +1.16$. The second analysis requests the transfer function from voltage source V_{CC} to node 1. SPICE analysis gives $R_{\text{out}} = 510\ \text{k}\Omega$.

Check of Results: Based on equation set (15.145) and the operating point results, we expect

$$R_{\text{in}} = 250\ \Omega \qquad \beta = +1.16 \qquad R_{\text{out}} = 517\ \text{k}\Omega$$

and we see that agreement with theory is very good.

Discussion: One should always try to understand and account for the differences between our theory and SPICE. In this example, the input resistance difference can be traced to the use of $V_T = 25.9\ \text{mV}$. Be careful not to make a sign error in interpreting the data for β. A negative sign appears in the SPICE output because of the assumed polarity of V_{CC} and $I(V_{CC})$. Finally, the SPICE model uses $r_o = (V_A + V_{CB})/I_C = 511\ \text{k}\Omega$, accounting for the small difference in the values of R_{out}.

15.6.9 THE WIDLAR CURRENT SOURCE

Resistor R in the **Widlar**[4] **current source** circuit shown in the schematic in Fig. 15.63 gives the designer an additional degree of freedom in adjusting the mirror ratio of the current mirror. In this circuit, the difference in the base-emitter voltages of transistors Q_1 and Q_2 appears across resistor R and determines the output current I_O. Transistor Q_3 buffers the mirror reference transistor in Fig. 15.63(b) to minimize the effect of finite current gain.

An expression for the output current may be determined from the standard expressions for the base-emitter voltage of the two bipolar transistors. In this analysis, we must accurately calculate the individual values of V_{BE1} and V_{BE2} because the behavior of the circuit depends on small differences in the values of these two voltages.

Assuming high current gain,

$$V_{BE1} = V_T \ln\left(1 + \frac{I_{\text{REF}}}{I_{S1}}\right) \cong V_T \ln \frac{I_{\text{REF}}}{I_{S1}}$$

and (15.147)

$$V_{BE2} = V_T \ln\left(1 + \frac{I_O}{I_{S2}}\right) \cong V_T \ln \frac{I_O}{I_{S2}}$$

The current in resistor R is equal to

$$I_{E2} = \frac{V_{BE1} - V_{BE2}}{R} = \frac{V_T}{R} \ln\left(\frac{I_{\text{REF}}}{I_O} \frac{I_{S2}}{I_{S1}}\right) \tag{15.148}$$

If the transistors are matched, then $I_{S1} = (A_{E1}/A)I_{SO}$ and $I_{S2} = (A_{E2}/A)I_{SO}$, and Eq. (15.148) can be rewritten as

$$I_O = \alpha_F I_{E2} \cong \frac{V_T}{R} \ln\left(\frac{I_{\text{REF}}}{I_O} \frac{A_{E2}}{A_{E1}}\right) \tag{15.149}$$

If I_{REF}, R, and the emitter-area ratio are all known, then Eq. (15.149) represents a transcendental equation that must be solved for I_O. The solution can be obtained by iterative trial and error, using Newton's method, or utilizing the solver in our calculators.

Widlar Source Output Resistance

The ac model for the Widlar source in Fig. 15.63(a) represents a common-emitter transistor with resistor R in its emitter and a small value of R_{th} ($= 1/g_{m1}$) from diode Q_1 in its base, as indicated in Fig. 15.64. In normal operation, the voltage developed across resistor R is usually small ($\leq 10V_T$). Referring to Table 14.1, or by simplifying Eq. (15.113) for this case, we can reduce the output

---

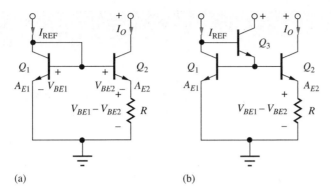

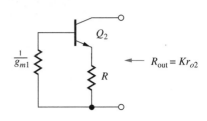

Figure 15.64 Widlar source output resistance – $K = 1 + \ln[(I_{\text{REF}}/I_{C2})(A_{E2}/A_{E1})]$.

Figure 15.63 (a) Basic Widlar current source and (b) buffered Widlar source.

resistance of the source to

$$R_{\text{out}} \cong r_{o2}[1 + g_{m2}R] = r_{o2}\left[1 + \frac{I_O R}{V_T}\right] \tag{15.150}$$

in which $I_O R$ can be found from Eq. (15.148):

$$R_{\text{out}} \cong r_{o2}\left[1 + \ln\frac{I_{\text{REF}}}{I_O}\frac{A_{E2}}{A_{E1}}\right] = Kr_{o2} \qquad \text{and} \qquad V_{CS} \cong KV_{A2} \tag{15.151}$$

where

$$K = \left[1 + \ln\frac{I_{\text{REF}}}{I_O}\frac{A_{E2}}{A_{E1}}\right]$$

For typical values, $1 < K < 10$.

EXERCISE: What value of R is required to set $I_O = 25\ \mu\text{A}$ if $I_{\text{REF}} = 100\ \mu\text{A}$ and $A_{E2}/A_{E1} = 5$? What are the values of output resistance and K in Eq. (15.151) for this source if $V_A + V_{CE} = 75\ \text{V}$?

ANSWERS: 3000 Ω; 12 MΩ, 4

EXERCISE: Find the output current in the Widlar source if $I_{\text{REF}} = 100\ \mu\text{A}$, $R = 100\ \Omega$, and $A_{E2} = 10A_{E1}$. What are the values of output resistance and K in Eq. (15.151) for this source if $V_A + V_{CE} = 75\ \text{V}$?

ANSWERS: 301 μA; 551 kΩ, 2.20

15.6.10 THE MOS VERSION OF THE WIDLAR SOURCE

Figure 15.65 is the MOS version of the Widlar source. In this circuit, the difference between the gate-source voltages of transistors M_1 and M_2 appears across resistor R, and I_O can be expressed as

$$I_O = \frac{V_{GS1} - V_{GS2}}{R} = \frac{\sqrt{\dfrac{2I_{\text{REF}}}{K_{n1}}} - \sqrt{\dfrac{2I_O}{K_{n2}}}}{R}$$

or $\tag{15.152}$

$$I_O = \frac{1}{R}\sqrt{\frac{2I_{\text{REF}}}{K_{n1}}}\left(1 - \sqrt{\frac{I_O}{I_{\text{REF}}}\frac{(W/L)_1}{(W/L)_2}}\right)$$

ELECTRONICS IN ACTION

The PTAT Voltage

The voltage developed across resistor R in Fig. 15.63 represents an extremely useful quantity because it is directly **proportional to absolute temperature** (referred to as PTAT). V_{PTAT} is equal to the difference in the two base-emitter voltages described by Eq. (15.147):

$$V_{\text{PTAT}} = V_{BE1} - V_{BE2} = V_T \ln\left(\frac{I_{C1}}{I_{C2}} \frac{A_{E2}}{A_{E1}}\right) = \frac{kT}{q} \ln\left(\frac{I_{C1}}{I_{C2}} \frac{A_{E2}}{A_{E1}}\right)$$

and the change of V_{PTAT} with temperature is

$$\frac{\partial V_{\text{PTAT}}}{\partial T} = +\frac{k}{q} \ln\left(\frac{I_{C1}}{I_{C2}} \frac{A_{E2}}{A_{E1}}\right) = +\frac{V_{\text{PTAT}}}{T}$$

For example, suppose $T = 300$ K, $I_{C1} = I_{C2}$ and $A_{E2} = 10A_{E1}$. Then $V_{\text{PTAT}} = 59.6$ mV with a temperature coefficient of slightly less than $+0.2$ mV/K.

The PTAT voltage developed in the Widlar cell, combined with an analog-to-digital converter, forms the heart of all of today's highly accurate electronic thermometers.

PTAT Voltage Based Digital Thermometry

The PTAT generator produces a well-defined output voltage that is used in many of today's digital thermometers, as in the block diagram on the next page. The output of the PTAT circuit is scaled, and the offset voltage is shifted to provide an output voltage that directly represents either the Fahrenheit or Celsius temperature scales. The analog voltage is converted to a digital representation by an A/D converter and the digital output is sent to an alphanumeric display. The scaling and offset shift can also be easily done in digital form after the A/D conversion operation is performed.

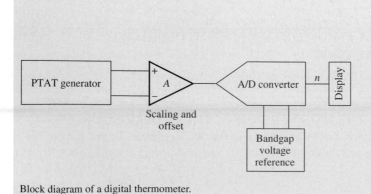

Block diagram of a digital thermometer.

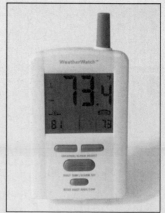

Wireless digital thermometer.

Dividing through by I_{REF},

$$\frac{I_O}{I_{\text{REF}}} = \frac{1}{R}\sqrt{\frac{2}{K_{n1} I_{\text{REF}}}}\left(1 - \sqrt{\frac{I_O}{I_{\text{REF}}} \frac{(W/L)_1}{(W/L)_2}}\right) \tag{15.153}$$

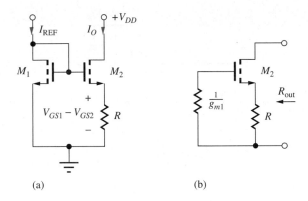

Figure 15.65 (a) MOS Widlar source and (b) small-signal model.

If I_O is known, then I_{REF} can be calculated directly from Eq. (15.152). If I_{REF}, R, and the W/L ratios are known, then Eq. (15.153) can be written as a quadratic equation in terms of $\sqrt{I_O/I_{REF}}$:

$$\left(\sqrt{\frac{I_O}{I_{REF}}}\right)^2 + \frac{1}{R}\sqrt{\frac{2}{K_{n1}I_{REF}}}\sqrt{\frac{(W/L)_1}{(W/L)_2}}\left(\sqrt{\frac{I_O}{I_{REF}}}\right) - \frac{1}{R}\sqrt{\frac{2}{K_{n1}I_{REF}}} = 0 \qquad (15.154)$$

MOS Widlar Source Output Resistance

In Fig. 15.65(b), the small-signal model for the MOS Widlar source is recognized as a common-source stage with resistor R in its source. Therefore, from Table 14.1,

$$R_{out} = r_{o2}(1 + g_{m2}R) \qquad (15.155)$$

EXERCISE: (a) Find the output current in Fig. 15.65(a) if $I_{REF} = 200$ μA, $R = 2$ kΩ, and $K_{n2} = 10K_{n1} = 250$ μA/V². (b) What is R_{out} if $\lambda = 0.02$/V and $V_{DS} = 10$ V?

ANSWERS: 764 μA; 176 kΩ

15.7 HIGH-OUTPUT-RESISTANCE CURRENT MIRRORS

In our introductory discussion of differential amplifiers in Section 15.2, we found that current sources with very high output resistances are needed to achieve good CMRR. The basic current mirrors discussed in the previous sections have a figure of merit V_{CS} equal to V_A or $1/\lambda$; that for the Widlar source is typically a few times higher. This section continues our introduction to current mirrors by discussing two additional circuits, the Wilson current source and the cascode current source, which enhance the value of V_{CS} to the order of $\beta_o V_A$ or μ_f/λ.

15.7.1 THE WILSON CURRENT SOURCES

The **Wilson current sources** [5] depicted in Fig. 15.66 use the same number of transistors as the buffered current mirror but achieve much higher output resistance; they are often used in applications requiring precisely matched current sources. In the MOS version, the output current is taken from the drain of M_3, and M_1 and M_2 form a current mirror. During circuit operation, the three transistors are all pinched-off and in the active region.

Because the gate current of M_3 is zero, I_{D2} must equal reference current I_{REF}. If the transistors all have the same W/L ratios, then

$$V_{GS3} = V_{GS1} = V_{GS} \qquad \text{because } I_{D3} = I_{D1}$$

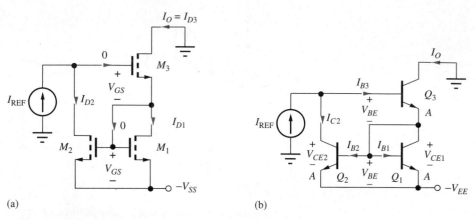

(a)   (b)

Figure 15.66 (a) MOS Wilson current source. (b) Original Wilson current source circuit using BJTs.

The current mirror requires

$$I_{D2} = I_{D1} \frac{1 + 2\lambda V_{GS}}{1 + \lambda V_{GS}}$$

and because $I_O = I_{D3}$ and $I_{D3} = I_{D1}$, the output current is given by

$$I_O = I_{REF} \frac{1 + \lambda V_{GS}}{1 + 2\lambda V_{GS}} \qquad \text{where} \qquad V_{GS} \cong V_{TN} + \sqrt{\frac{2I_{REF}}{K_n}} \qquad (15.156)$$

For small λ, $I_O \cong I_{REF}$. For example, if $\lambda = 0.02/V$ and $V_{GS} = 2$ V, then I_O and I_{REF} differ by 3.7 percent.

The Wilson source actually appeared first in bipolar form as drawn in Fig. 15.66(b). The circuit operates in a manner similar to the MOS source, except for the loss of current from I_{REF} to the base of Q_3 and the current gain error in the mirror formed by Q_1 and Q_2. Applying KCL at the base of Q_3, $I_{REF} = I_{C2} + I_{B3}$ in which I_{C2} and I_{B3} are related through the current mirror formed by Q_1 and Q_2:

$$I_{C2} = \frac{1 + \dfrac{2V_{BE}}{V_A}}{1 + \dfrac{V_{BE}}{V_A} + \dfrac{2}{\beta_{FO}}} I_{E3} = \frac{1 + \dfrac{2V_{BE}}{V_A}}{1 + \dfrac{V_{BE}}{V_A} + \dfrac{2}{\beta_{FO}}} (\beta_{FO} + 1) I_{B3} \qquad (15.157)$$

Note in Fig. 15.66(b) that $V_{CE1} = V_{BE}$ and $V_{CE2} = 2V_{BE}$.

Solving directly for $I_{C3} = \beta_F I_{B3}$ yields a messy expression that is difficult to interpret. However, if we assume the error terms are small, then we can eventually reduce (with considerable effort) the expression to the following approximate result:

$$I_O \cong I_{REF} \frac{1 + \dfrac{V_{BE}}{V_A}}{1 + \dfrac{2}{\beta_{FO}(\beta_{FO} + 2)} + \dfrac{2V_{BE}}{V_A}} \qquad (15.158)$$

For $\beta_{FO} = 50$, $V_A = 60$ V, and $V_{BE} = 0.7$ V, the mirror ratio is 0.988. The primary source of error results from the collector-emitter voltage mismatch between transistors Q_1 and Q_2. The base current error has been reduced to less than 0.1 percent of I_{REF}.

The errors due to drain-source voltage mismatch in Fig. 15.66(a), or collector-emitter voltage mismatch in Fig. 15.66(b), may still be too large for use in precision circuits, but this problem can be significantly reduced by adding one more transistor to balance the circuit as in Fig. 15.67. Transistor Q_4 reduces the collector-emitter voltage of Q_2 by one V_{BE} drop and balances the collector-emitter

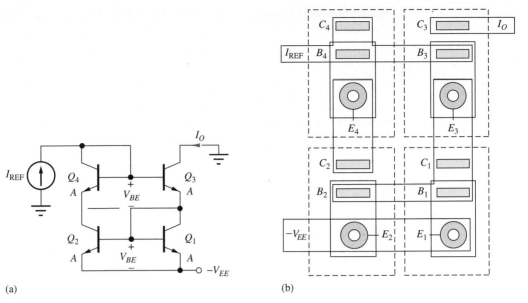

Figure 15.67 (a) Wilson source using balanced collector-emitter voltages. (b) Layout of Wilson source.

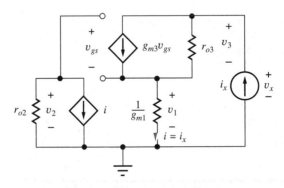

Figure 15.68 Small-signal model for the MOS version of the Wilson source.

voltages of Q_1 and Q_2:

$$V_{CE2} = V_{BE1} + V_{BE3} - V_{BE4} \cong V_{BE}$$

All four transistors are operating at approximately the same value of collector current, and the values of V_{BE} are all the same if the devices are matched with equal emitter areas.

EXERCISE: Draw a voltage-balanced version of the MOS Wilson source by adding one additional transistor to the circuit in Fig. 15.66(a).

ANSWER: See Fig. P15.169.

15.7.2 OUTPUT RESISTANCE OF THE WILSON SOURCE

The primary advantage of the Wilson source over the standard current mirror is its greatly increased output resistance. The small-signal model for the MOS version of the Wilson source is given in Fig. 15.68, in which test current i_x is applied to determine the output resistance.

The current mirror formed by transistors M_1 and M_2 is represented by its simplified two-port model assuming $n = 1$. Voltage v_x is determined from

$$\mathbf{v_x} = \mathbf{v_3} + \mathbf{v_1} = [\mathbf{i_x} - g_{m3}\mathbf{v_{gs}}]r_{o3} + \mathbf{v_1} \tag{15.159}$$

where

$$\mathbf{v_{gs}} = \mathbf{v_2} - \mathbf{v_1} \qquad \text{with} \qquad \mathbf{v_1} = \frac{\mathbf{i_x}}{g_{m1}} \qquad \text{and} \qquad \mathbf{v_2} = -\mu_{f2}\mathbf{v_1}$$

Combining these equations and recognizing that $g_{m1} = g_{m2}$ for $n = 1$ yields

$$R_{\text{out}} = \frac{\mathbf{v_x}}{\mathbf{i_x}} = r_{o3}\left[\mu_{f2} + 2 + \frac{1}{\mu_{f2}}\right] \cong \mu_{f2}r_{o3} \tag{15.160}$$

and

$$V_{CS} = I_{D3}\mu_{f2}\frac{1 + \lambda_3 V_{DS3}}{\lambda_3 I_{D3}} \simeq \frac{\mu_{f2}}{\lambda_3} \tag{15.161}$$

Analysis of the bipolar source is somewhat more complex because of the finite current gain of the BJT and yields the following result:

$$R_{\text{out}} \cong \frac{\beta_{o3}r_{o3}}{2} \qquad \text{and} \qquad V_{CS} \cong \frac{\beta_o V_A}{2} \tag{15.162}$$

Derivation of this equation is left for Prob. 15.167.

EXERCISE: Calculate R_{out} for the Wilson source in Fig. 15.66(b) if $\beta_F = 150$, $V_A = 50$ V, $V_{EE} = 15$ V, and $I_O = I_{REF} = 50$ μA. What is the output resistance of a standard current mirror operating at the same current?

ANSWER: 96.6 MΩ versus 1.30 MΩ

EXERCISE: Use SPICE to find the output current and output resistance of the Wilson source in the previous exercise.

ANSWERS: $I_O = 49.5$ μA; 118 MΩ

15.7.3 CASCODE CURRENT SOURCES

In this section, we learn that the output resistance of the cascode connection (C-E/C-B cascade) of two transistors is very high, approaching $\mu_f r_o$ for the FET case and $\beta_o r_o$ for the BJT circuit. Figure 15.69 shows the implementation of the MOS and BJT **cascode current sources** using current mirrors.

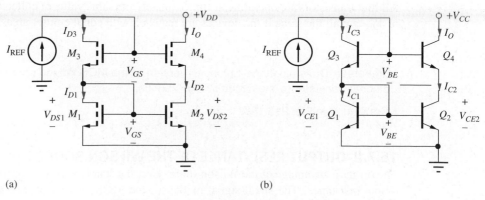

(a)   (b)

Figure 15.69 (a) MOS and (b) BJT cascode current sources.

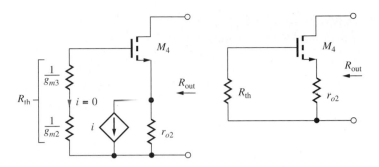

Figure 15.70 Small-signal model for the MOS cascode source.

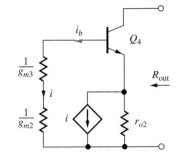

Figure 15.71 Small-signal model for the BJT cascode source.

In the MOS circuit, $I_{D1} = I_{D3} = I_{REF}$. The current mirror formed by M_1 and M_2 forces the output current to be approximately equal to the reference current because $I_O = I_{D4} = I_{D2}$. Diode-connected transistor M_3 provides a dc bias voltage to the gate of M_4 and balances V_{DS1} and V_{DS2}. If all transistors are matched with the same W/L ratios, then the values of V_{GS} are all the same, and V_{DS2} equals V_{DS1}:

$$V_{DS2} = V_{GS1} + V_{GS3} - V_{GS4} = V_{GS} \qquad \text{and} \qquad V_{DS1} = V_{GS}$$

Thus, the M_1-M_2 current mirror is precisely balanced, and $I_O = I_{REF}$.

The BJT source in Fig. 15.69(b) operates in the same manner. For $\beta_F = \infty$, $I_{REF} = I_{C3} = I_{C1}$ on the reference side of the source. Q_1 and Q_2 form a current mirror, which sets $I_O = I_{C4} = I_{C2} = I_{C1} = I_{REF}$. Diode Q_3 provides the bias voltage at the base of Q_4 needed to keep Q_2 in the active region and balances the collector-emitter voltages of the current mirror:

$$V_{CE2} = V_{BE1} + V_{BE3} - V_{BE4} = 2V_{BE} - V_{BE} = V_{BE} = V_{CE1}$$

15.7.4 OUTPUT RESISTANCE OF THE CASCODE SOURCES

Figure 15.70 shows the small-signal model for the MOS cascode source; the two-port model has been used for the current mirror formed of transistors M_1 and M_2. Because current i represents the gate current of M_4, which is zero, the circuit can be reduced to that on the right in Fig. 15.70, which should be recognized as a common-source stage with resistor r_{o2} in its source. Thus, its output resistance is

$$R_{out} = r_{o4}(1 + g_{m4}r_{o2}) \cong \mu_{f4}r_{o2} \qquad \text{and} \qquad V_{CS} \cong \frac{\mu_{f4}}{\lambda_2} \cong \frac{\mu_{f4}}{\lambda_4} \qquad (15.163)$$

Analysis of the output resistance of the BJT source in Fig. 15.71 is again more complex because of the finite current gain of the BJT. If the base of Q_4 were grounded, then the output resistance would be just equal to that of the cascode stage, $\beta_o r_o$. However, the base current i_b of Q_4 enters the current mirror, doubles the output current, and causes the overall output resistance to be reduced by a factor of 2:

$$R_{out} \cong \frac{\beta_{o4} r_{o4}}{2} \qquad \text{and} \qquad V_{CS} \cong \frac{\beta_{o4} V_{A4}}{2} \qquad (15.164)$$

Detailed calculation of this result is left as Prob. 15.191.

EXERCISE: Calculate the output resistance of the MOS cascode current source in Fig. 15.69(a) and compare it to that of a standard current mirror if $I_O = I_{REF} = 50 \ \mu A$, $V_{DD} = 15$ V, $K_n = 250 \ \mu A/V^2$, $V_{TN} = 0.8$ V, and $\lambda = 0.015 \ V^{-1}$.

ANSWER: 379 MΩ versus 1.63 MΩ including all λV_{DS} terms

TABLE 15.4
Comparison of the Basic Current Mirrors

TYPE OF SOURCE	R_{out}	V_{CS}	TYPICAL VALUES OF V_{CS}
Resistor	R	V_{EE}	15 V
Two-transistor mirror	r_o	V_A or $\dfrac{1}{\lambda}$	75 V
Cascode BJT	$\dfrac{\beta_o r_o}{2}$	$\dfrac{\beta_o V_A}{2}$	3750 V
Cascode FET	$\mu_f r_o$	$\dfrac{\mu_f}{\lambda}$	10,000 V
BJT Wilson	$\dfrac{\beta_o r_o}{2}$	$\dfrac{\beta_o V_A}{2}$	3750 V
FET Wilson	$\mu_f r_o$	$\dfrac{\mu_f}{\lambda}$	10,000 V

EXERCISE: Use SPICE to find the output current and output resistance of the cascode current source in the previous exercise.

ANSWERS: $I_O = 50.0$ μA; 382 MΩ

EXERCISE: Calculate the output resistance of the BJT cascode current source in Fig. 15.69(b) and compare it to that of a standard current mirror if $I_O = I_{REF} = 50$ μA, $V_{CC} = 15$ V, $\beta_o = 100$, and $V_A = 67$ V.

ANSWER: 81.3 MΩ versus 1.63 MΩ

15.7.5 CURRENT MIRROR SUMMARY

Table 15.4 is a summary of the current mirror circuits discussed in this chapter. The cascode and Wilson sources can achieve very high values of V_{CS} and often find use in the design of differential and operational amplifiers, as well as in many other analog circuits.

DESIGN EXAMPLE 15.10 **ELECTRONIC CURRENT SOURCE DESIGN**

Design an IC current source to meet a given set of specifications.

PROBLEM Design a 1:1 current mirror with a reference current of 25 μA and a mirror ratio error of less than 0.1 percent when the output is operating from a 20-V supply. Devices with these parameters are available: $\beta_{FO} = 100$, $V_A = 75$ V, $I_{SO} = 0.5$ fA; $K_n' = 50$ μA/V², $V_{TN} = 0.75$ V, and $\lambda = 0.02/$V.

SOLUTION **Known Information and Given Data:** $I_{REF} = 25$ μA. A mirror ratio error of less than 0.1 percent requires an output current of 25 μA ± 25 nA when the output voltage is 20 V. Either a bipolar or MOS realization is acceptable.

Unknowns: Current source configuration; transistor sizes

Approach: The specifications define the required values of R_{out} and V_{CS}. Use this information to choose a circuit topology. Complete the design by choosing device sizes based on the output resistance expressions for the selected circuit topology.

Assumptions: Room temperature operation; devices are in the active region of operation.

Analysis: The output resistance of the current source must be large enough that 20 V applied across the output does not change (increase) the current by more than 25 nA. Thus, the output resistance must satisfy $R_{\text{out}} \geq 20$ V/25 nA $= 800$ MΩ. Let us choose $R_{\text{out}} = 1$ GΩ to provide some safety margin. The effective current source voltage is then $V_{CS} = 25$ μA $(1$ GΩ$) = 25{,}000$ V! From Table 15.4 we see that either a cascode or Wilson source will be required to meet this value of V_{CS}. In fact, the source must be an MOS version, since our BJTs can at best reach $V_{CS} = 100(75$ V$)/2 = 3750$ V.

The choice between the Wilson and cascode sources is arbitrary at this point. Let us pick the cascode source, which does not involve an internal feedback loop. In order to achieve the small mirror error, a voltage-balanced version is required. Our final circuit choice is therefore the circuit shown in Fig. 15.69(a). Now we must choose the device sizes. In this case, the W/L ratios are all the same since we require MR $= 1$.

Again referring to Table 15.4, the required amplification factor for the transistor is

$$\mu_f = \lambda V_{CS} = \left(\frac{0.02}{\text{V}}\right)(25{,}000 \text{ V}) = 500$$

The MOS transistor's amplification factor is given approximately by

$$\mu_f = g_m r_o \cong \sqrt{2K_n I_D} \frac{1}{\lambda I_D}$$

Using $\mu_f = 500$, $\lambda = 0.02$/V, and $I_D = 25$ μA gives a value of $K_n = 1.25$ mS. Since $K_n = K_n'(W/L)$, we need a W/L ratio of 25/1 for the given technology. (This W/L ratio is easy to achieve in integrated circuit form.) In this circuit, all the transistors are operating at the same current, so the W/L ratios should all be the same size in order to maintain the required voltage balance.

Check of Results: Let us check the calculations by directly calculating the output resistance of the source.

$$R_{\text{out}} \cong g_{m4} r_{o4} r_{o2} \qquad g_{m4} = \sqrt{2K_n I_D(1 + \lambda V_{DS4})} \qquad r_o = \frac{(1/\lambda) + V_{DS}}{I_D}$$

We can either neglect the values of V_{DS} in these expressions, or we can calculate them. In order to best compare with simulation, let us find V_{DS} and the corresponding values of g_m and r_o.

$$V_{DS2} = V_{GS2} = V_{TN} + \sqrt{\frac{2I_D}{K_n}} = 0.75 + \sqrt{\frac{50 \text{ μA}}{1.25 \text{ mS}}} = 0.95 \text{ V}$$

$$V_{DS4} = 20 - V_{DS2} = 19.0 \text{ V}$$

$$g_{m4} = \sqrt{2K_n I_D(1 + \lambda V_{DS4})} = \sqrt{2(1.25 \text{ mA/V}^2)(25 \text{ μA})[1 + .02(19)]} = 0.294 \text{ mS}$$

$$r_{o2} = \frac{(1/\lambda) + V_{DS2}}{I_D} = \frac{51.0 \text{ V}}{25 \text{ μA}} = 2.04 \text{ MΩ}$$

$$r_{o4} = \frac{(1/\lambda) + V_{DS4}}{I_D} = \frac{69.0}{25 \text{ μA}} = 2.76 \text{ MΩ}$$

Multiplying the small-signal parameters together produces an output resistance estimate of 1.65 GΩ, which exceeds the design requirement that we originally calculated from the design specifications.

Discussion: Note that our ability to set the amplification factor of the MOS transistor was very important in achieving the design goals. In this case $\mu_{f4} = 811$. A possible layout for the cascode current source is presented in the figure. The four 25/1 NMOS transistors are stacked vertically. G_1 and G_2 are the gates of the current mirror transistors. Gates G_1 and G_3 are connected directly to their respective drains. The drain of M_1 and the source of M_3 are merged as are those of M_2 and M_4. However, there are no contacts required to the connection between the drain of M_2 and the source of M_4.

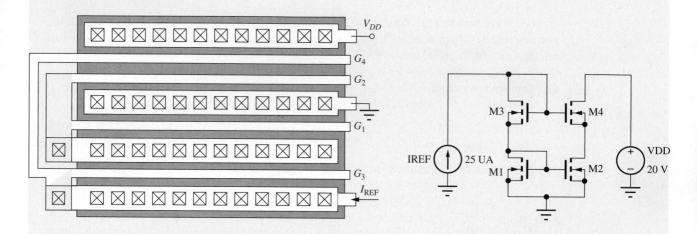

Computer-Aided Analysis: SPICE represents a good way to double check the results. First, we must set the MOS device parameters: KP = 50 μA/V², VTO = 0.75 V, LAMBDA = 0.02/V, W = 25 μm, and L = 1 μm. A dc simulation of the final circuit (shown next) with the given device parameters yields an output current of 25.014 μA. In addition, the voltages at the drains of M_1 and M_2 are 0.948 V and 0.976 V, respectively, indicating that the voltage balancing is working as desired.

A transfer function analysis from source V_{DD} to the output node yields an output resistance of 1.66 GΩ, easily meeting the specifications with a satisfactory safety margin. We also have good agreement with the value of R_{out} that we calculated by hand.

EXERCISE: In the SPICE results in Design Ex. 15.10, $I_O = 25.015$ μA at $V_{DD} = 20$ V. If $R_{\text{out}} = 1.66$ GΩ, what will be the output current at $V_{DD} = 10$ V?

ANSWER: 25.008 μA

EXERCISE: What is the minimum value of V_{DD} for which M_4 remains in the active region of operation?

ANSWER: 1.15 V

EXERCISE: Repeat the design in Design Ex. 15.10 for a current source with a mirror ratio of 2 ± 0.1 percent.

ANSWERS: $(W/L)_3 = (W/L)_1 = 25/1$; $(W/L)_4 = (W/L)_2 = 50/1$

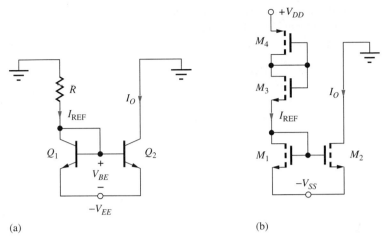

Figure 15.72 Reference current generation for current mirrors: (a) resistor reference and (b) series-connected MOSFETs.

15.8 REFERENCE CURRENT GENERATION

A **reference current** is required by all the current mirrors that have been discussed. The least complicated method for establishing this reference current is to use resistor R, as shown in Fig. 15.72(a).

However, the source's output current is directly proportional to the supply voltage V_{EE}:

$$I_{\text{REF}} = \frac{V_{EE} - V_{BE}}{R} \tag{15.165}$$

In MOS technology, the gate-source voltages of MOSFETs can be designed to be large, and several MOS devices can be connected in series between the power supplies to eliminate the need for large-value resistors. An example of this technique is given in Fig. 15.72(b), in which

$$V_{DD} + V_{SS} = V_{SG4} + V_{SG3} + V_{GS1}$$

and the drain currents must satisfy $I_{D1} = I_{D3} = I_{D4}$. However, any change in the supply voltages directly alters the values of the gate-source voltages of the three MOS transistors and again changes the reference current. Note that the series device technique is not usable in bipolar technology because of the small fixed voltage ($\cong 0.7$ V) developed across each diode, as well as the exponential relationship between voltage and current in the diode.

EXERCISE: What is the reference current in Fig. 15.72(a) if $R = 43$ kΩ and $V_{EE} = -5$ V? (b) If $V_{EE} = -7.5$ V?

ANSWERS: 100 μA; 158 μA

EXERCISE: What is the reference current in Fig. 15.72(b) if $K_n = K_p = 400$ μA/V^2, $V_{TN} = -V_{TP} = 1$ V, and $V_{SS} = -5$ V? (b) If $V_{SS} = -7.5$ V?

ANSWERS: 88.9 μA; 450 μA. (*Note:* the variation is worse than in the resistor bias case because of the square-law MOSFET characteristic.)

15.8.1 SUPPLY-INDEPENDENT BIASING

In most cases, the supply voltage dependence of I_{REF} is undesirable. For example, we would like to fix the bias points of the devices in general-purpose op amps, even though they must operate from power supply voltages ranging from ±3 V to ±22 V. In addition, Eq. (15.165) indicates that relatively large values of resistance are required to achieve small operating currents, and these resistors use significant area in integrated circuits, as was discussed in detail in Sec. 6.5.9. Thus, a number of circuit techniques that yield currents relatively independent of the power supply voltages have been invented.

A V_{BE}-Based Reference

One possibility is the V_{BE}-**based reference,** shown in Fig. 15.73, in which the output current is determined by the base-emitter voltage of Q_1. For high current gain, the collector current of Q_1 is equal to the current through resistor R_1,

$$I_{C1} = \frac{V_{EE} - V_{BE1} - V_{BE2}}{R_1} \cong \frac{V_{EE} - 1.4 \text{ V}}{R_1} \tag{15.166}$$

and the output current I_O is approximately equal to the current in R_2:

$$I_O = \alpha_{F2} I_{E2} = \alpha_{F2} \left(\frac{V_{BE1}}{R_2} + I_{B1} \right) \cong \frac{V_{BE1}}{R_2} \cong \frac{0.7 \text{ V}}{R_2} \tag{15.167}$$

Rewriting V_{BE1} in terms of V_{EE},

$$I_O \cong \frac{V_T}{R_2} \ln \frac{V_{EE} - 1.4 \text{ V}}{I_{S1} R_1} \tag{15.168}$$

A substantial degree of supply-voltage independence has been achieved because the output current is now only logarithmically dependent on changes in the supply voltage V_{EE}. However, the output current is temperature dependent due to the temperature coefficients of both V_{BE} and resistor R.

EXERCISE: (a) Calculate I_O in Fig. 15.73 for $I_S = 10^{-16}$ A, $R_1 = 39$ kΩ, $R_2 = 6.8$ kΩ, and $V_{EE} = -5$ V. Assume infinite current gains. (b) Repeat for $V_{EE} = -7.5$ V.

ANSWERS: 101 μA; 103 μA

The Widlar Source

Actually, we already discussed another source that achieves a similar independence from power supply voltage variations. The expression for the output current of the Widlar source given in Fig. 15.63 and Eq. (15.149) is

$$I_O = \alpha_F I_{E2} \cong \frac{V_T}{R} \ln \left(\frac{I_{REF}}{I_O} \frac{A_{E2}}{A_{E1}} \right) \tag{15.169}$$

Here again, the output current is only logarithmically dependent on the reference current I_{REF} (which may be proportional to V_{CC}).

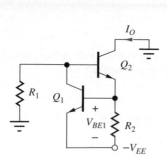

Figure 15.73 V_{BE}-based current source.

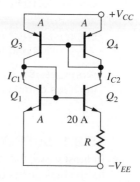

Figure 15.74 Power-supply-independent bias circuit using the Widlar source and a current mirror.

Power-Supply-Independent Bias Cell

Bias circuits with an even greater degree of power supply voltage independence can be obtained by combining the Widlar source with a standard current mirror, as indicated in the circuit in Fig. 15.74. Assuming high current gain, the *pnp* current mirror forces the currents on the two sides of the reference cell to be equal—that is, $I_{C1} = I_{C2}$. In addition, the emitter-area ratio of the Widlar source in Fig. 15.74 is equal to 20.

With these constraints, Eq. (15.169) can be satisfied by an operating point of

$$I_{C2} \cong \frac{V_T}{R} \ln(20) = \frac{0.0749 \text{ V}}{R} \tag{15.170}$$

In this example, a fixed voltage of approximately 75 mV is developed across resistor R, and this voltage is independent of the power supply voltages. Resistor R can then be chosen to yield the desired operating current.

Obviously, a wide range of mirror ratios and emitter-area ratios can be used in the design of the circuit in Fig. 15.74. Although the current, once established, is independent of supply voltage, the actual value of I_C still depends on temperature, as well as the absolute value of R and varies with run-to-run process variations.

Unfortunately, $I_{C1} = I_{C2} = 0$ is also a stable operating point for the circuit in Fig. 15.74. **Start-up circuits** must be included in IC realizations of this reference to ensure that the circuit reaches the desired operating point.

EXERCISE: Find the output current in the current source in Fig. 15.74 if $A_{E3} = 10A_{E4}$, $A_{E2} = 10A_{E1}$, and $R = 1 \text{ k}\Omega$.

ANSWER: 115 μA

EXERCISE: What is the minimum power supply voltage for proper operation of the supply-independent bias circuit in Fig. 15.74?

ANSWER: $2V_{BE} \cong 1.4$ V

Once the current has been established in the reference cell consisting of Q_1–Q_4 in Fig. 15.74, the base-emitter voltages of Q_1 and Q_4 can be used as reference voltages for other current mirrors, as shown in Fig. 15.75. In this figure, buffered current mirrors have been used in the reference cell to

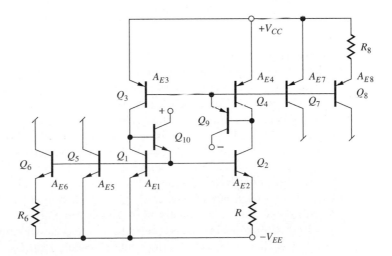

Figure 15.75 Multiple source currents generated from the supply-independent cell.

minimize errors associated with finite current gains of the *npn* and *pnp* transistors. Output currents are shown generated from basic mirror transistors Q_5 and Q_7 and from Widlar sources, Q_6 and Q_8.

15.8.2 A SUPPLY-INDEPENDENT MOS REFERENCE CELL

The MOS analog of the circuit in Fig. 15.74 appears in Fig. 15.76. In this circuit, the PMOS current mirror forces a fixed relationship between drain currents I_{D3} and I_{D4}. For the particular case in Fig. 15.76, $I_{D3} = I_{D4}$, and so $I_{D1} = I_{D2}$. Substituting this constraint into Eq. (15.153) yields an

ELECTRONICS IN ACTION

Legends of Analog Design

(a)   (b)

Legends of Analog Design (a) Robert J. Widlar. (b) Barrie Gilbert
(a) Courtesy of National Semiconductor. (b) Courtesy of Analog Devices

Two legends of analog integrated circuit design, Robert Widlar and Barrie Gilbert, made extremely clever and lasting contributions to the analog circuits field. Widlar developed the LM101 operational amplifier and contributed to many of the innovations that led to the design of the classic μA741 op amp. The μA741 circuit techniques spawned a broad range of follow-on designs that are still in use today. Widlar was also responsible for the bandgap reference that forms the heart of most precision voltage references and voltage regulator circuits [6], and is also used as a temperature sensor in digital thermometry.

Gilbert invented a four-quadrant analog multiplier circuit referred to today as the **Gilbert multiplier** (see website). Circuits related to the analog multiplier are used in RF mixers (the Gilbert mixer) and phase detectors in phase-locked loops. Gilbert mixers are utilized in the transceiver ICs found in virtually all of today's cell phones.

Both Gilbert and Widlar received the Solid-State Circuits Field Award from the IEEE Solid-State Circuits Society for their individual contributions to analog circuit design.

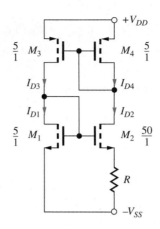

Figure 15.76 Supply-independent current source using MOS transistors.

equation for the value of R required to establish a given current I_{D2}:

$$R = \sqrt{\frac{2}{K_{n1} I_{D2}}} \left(1 - \sqrt{\frac{(W/L)_1}{(W/L)_2}} \right) \tag{15.171}$$

Based on Eq. (15.171), we see that the MOS source is independent of supply voltage but is a function of the absolute values of R and K_n'.

EXERCISE: **What value of R is required in the current source in Fig. 15.76 if I_{D2} is to be designed to be 100 μA and $K_n' = 25$ μA/V^2?**

ANSWER: **8.65 kΩ**

DESIGN EXAMPLE 15.11

REFERENCE CURRENT DESIGN

Design a supply-independent current source using bipolar technology.

PROBLEM Design a supply-independent current source to provide an output current of 45 μA at $T = 300$ K using the circuit topology in Fig. 15.74 with symmetrical 5-V power supplies. The circuit should use no more than 1 kΩ of resistance or 60 μA of total current. Use SPICE to determine the sensitivity of the design current to power supply voltage variations. Assume that a unit-area BJT has the following parameters: $\beta_{FO} = 100$, $V_A = 75$ V, and $I_{SO} = 0.1$ fA for both *npn* and *pnp* transistors.

SOLUTION **Known Information and Given Data:** Circuit topology in Fig. 15.74, $\beta_{FO} = 100$, $V_A = 75$ V, $I_{SO} = 0.1$ fA. Total current ≤ 60 μA.

Unknowns: R and the area ratio between Q_1 and Q_2

Approach: The current in the circuit is described by Eq. (15.169). Use the maximum resistance values to select the area ratio. Select a current ratio in the sides of the reference to satisfy the total supply current requirement.

Assumptions: Transistors operate in the active region. $I_{C2} = 45$ μA.

Analysis: At $T = 300$ K and $V_T = 25.88$ mV, and from Eq. (15.169), we have

$$\ln\left(\frac{I_{C1}}{I_{C2}}\frac{A_{E2}}{A_{E1}}\right) = \frac{I_{C2}R}{V_T} \leq \frac{(45\ \mu\text{A})(1\ \text{k}\Omega)}{25.88\ \text{mV}} = 1.739 \quad \text{or} \quad \frac{I_{C1}}{I_{C2}}\frac{A_{E2}}{A_{E1}} \leq 5.69$$

In addition, the maximum current specification requires

$$\frac{I_{C2}}{I_{C1}} \geq \frac{45\ \mu\text{A}}{15\ \mu\text{A}} = \frac{3}{1}$$

Let's choose $I_{C2} = 5I_{C1}$. Then $A_{E2}/A_{E1} \leq 28.45$. Choosing $A_{E2}/A_{E1} = 20$, we obtain

$$R = \frac{25.88\ \text{mV}\ \ln(4)}{45\ \mu\text{A}} = 797\ \Omega$$

The final design is $R = 797\ \Omega$, $A_{E1} = A$, $A_{E2} = 20$ A, $A_{E3} = A$, $A_{E4} = 5$ A with 35.88 mV across resistor R.

Check of Results: Since we need to use SPICE to find the power supply sensitivity, let us use it to also check our design.

Computer-Aided Analysis: The circuit shown is drawn using the schematic editor. Zero-valued sources V_{IC2} and V_{IC3} function as ammeters to measure the collector currents of transistors Q_2 and Q_3. First we must remember to set the *npn* and *pnp* BJT parameters to BF = 100, VAF = 75 V, IS = 0.1 fA, and TEMP = 27 C. We must also specify AREA = 1, AREA = 20, AREA = 1 and AREA = 5 for Q_1 through Q_4, respectively. SPICE then gives $I_{C2} = 49.6\ \mu$A and $I_{C3} = 10.94\ \mu$A with 39.89 mV across R. The currents and voltage are slightly higher than predicted, and this is primarily due to having neglected the mirror ratio error due to the different values of V_{EC4} and V_{EC3}. (Try the exercise after this example.) We can correct for this error by modifying the emitter area ratio:

$$A_{E4} = 5\left(1 + \frac{V_{EC3} - V_{EC4}}{V_A}\right) = 5\left(1 + \frac{9.34 - 0.65}{75}\right) = 5.58$$

SPICE now yields $I_{C2} = 45.9\ \mu$A, $I_{C3} = 9.08\ \mu$A and $V_{E2} = 36.9$ mV. A transfer function analysis from V_{CC} to V_{IC2} gives a total output resistance of 928 kΩ for the current source, and the sensitivity of I_{C2} to changes in V_{CC} is 0.808 μA/V.

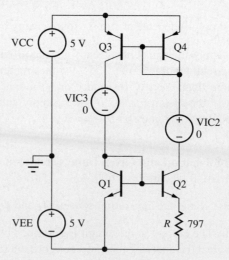

Discussion: The current source meets the specifications.

EXERCISE: Explore the errors caused by finite current gain and Early voltage by simulating the circuit with BF = 10,000 and VAF = 10,000 V. What are the new values of I_{C2}, I_{C3}, and the voltage developed across R?

ANSWERS: 45.0 μA; 9.01 μA; 35.88 mV

EXERCISE: What are the new design values if we choose $A_{E2}/A_{E1} = 25$?

ANSWERS: $R = 925\ \Omega$; $A_{E1} = A$; $A_{E2} = 25\ A$; $A_{E3} = A$; $A_{E4} = 5.57\ A$

ELECTRONICS IN ACTION

The Bandgap Reference

Precision **voltage references** need to not only be independent of power supply voltage, but also be independent of temperature. Although the circuits described in Sec. 15.8 can produce reference currents and voltages that are substantially independent of power supply voltage, they all still vary with temperature. Robert Widlar solved this problem with his invention of the elegant bandgap reference circuit, and today, the bandgap reference is the most common technique used to generate a precision voltage. It has supplanted Zener reference diodes in the majority of applications.

Based on his detailed understanding of bipolar transistor characteristics, Widlar realized that the negative temperature coefficient associated with the base-emitter junction could be canceled out by the positive temperature dependence of a scaled PTAT voltage as indicated conceptually in the figure here.

We desire the output voltage V_{BG} to have a zero temperature coefficient:

$$V_{BG} = V_{BE} + G V_{\text{PTAT}} \qquad \text{and} \qquad \frac{\partial V_{BG}}{\partial T} = \frac{\partial V_{BE}}{\partial T} + G \frac{\partial V_{\text{PTAT}}}{\partial T} = 0$$

The dependence of V_{BE} and V_{PTAT} on temperature were developed previously in Eq. (3.15) and

$$\frac{V_{BE} - V_{GO} - 3V_T}{T} + G \frac{V_{\text{PTAT}}}{T} = 0 \qquad \text{or} \qquad G V_{\text{PTAT}} = V_{GO} + 3V_T - V_{BE}$$

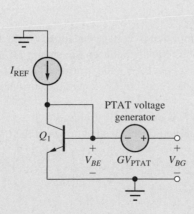

Concept for the bandgap reference.

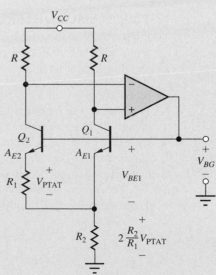

Brokaw version of the bandgap reference.

where V_{GO} is the silicon bandgap voltage at 0 K (1.12 V). Combining these results gives $V_{BG} = V_{GO} + 3V_T$. The output voltage at which zero temperature coefficient is achieved is slightly above the bandgap voltage of silicon. Hence, this circuit is referred to as a "bandgap reference." At room temperature, the output voltage is approximately 1.20 V.

The circuit realization of the bandgap reference shown here is attributed to another talented designer, Paul Brokaw of Analog Devices [7], and is easier to understand than the original circuit of Widlar. In this circuit, the output voltage is equal to the sum of the base-emitter voltage of Q_1 plus the voltage across resistor R_2, which is a scaled replica of the PTAT voltage being developed across resistor R_1. The scaling factor is controlled by the op amp and resistors R.

The ideal op amp forces the voltage across the two matched collector resistors to be the same, thereby setting $I_{C2} = I_{C1}$ and $I_{E2} = I_{E1}$. Thus, the PTAT voltage is equal to $V_T \ln(A_{E2}/A_{E1})$, and the emitter current of Q_2 equals V_{PTAT}/R_1. The current in R_2 is twice that in R_1, since $I_{E2} = I_{E1}$. Combining these results yields an expression for the output voltage V_{BG}:

$$V_{BG} = V_{BE1} + 2\frac{R_2}{R_1}V_T \ln \frac{A_{E2}}{A_{E1}}$$

For this circuit, the gain $G = 2R_2/R_1$, and the resistor ratio is given by

$$\frac{R_2}{R_1} = -\frac{1}{2}\frac{\dfrac{\partial V_{BE1}}{\partial T}}{\dfrac{\partial V_{PTAT}}{\partial T}} = \frac{V_{GO} + 3V_T - V_{BE1}}{2V_{PTAT}}$$

15.9 THE CURRENT MIRROR AS AN ACTIVE LOAD

One of the most important applications of the current mirror[5] is as a replacement for the load resistors of differential amplifier stages in IC operational amplifiers. This elegant application of the current mirror can greatly improve amplifier voltage gain while maintaining the operating-point balance necessary for good common-mode rejection and low offset voltage. When used in this manner, the current mirror is referred to as an **active load** because the passive load resistors have been replaced with active transistor circuit elements.

15.9.1 CMOS DIFFERENTIAL AMPLIFIER WITH ACTIVE LOAD

Figure 15.77 shows a CMOS differential amplifier with an active load; the load resistors have been replaced by a PMOS current mirror. Let us first study the quiescent operating point of this circuit and then look at its small-signal characteristics.

dc Analysis

Assume for the moment that the amplifier is voltage balanced (in fact, it will turn out that it *is* balanced). Then bias current I_{SS} divides equally between transistors M_1 and M_2, and I_{D1} and I_{D2} are each equal to $I_{SS}/2$. Current I_{D3} must equal I_{D1} and is mirrored as I_{D4} at the output of the PMOS current mirror. Thus, I_{D3} and I_{D4} are also equal to $I_{SS}/2$, and the current in the drain of M_4 is exactly the current required to satisfy M_2.

The mirror ratio set by M_3 and M_4 is exactly unity when $V_{SD4} = V_{SD3}$ and hence $V_{DS1} = V_{DS2}$. Thus, the differential amplifier is completely balanced at dc when the quiescent output voltage is

$$V_O = V_{DD} - V_{SD4} = V_{DD} - V_{SG3} = V_{DD} - \left(\sqrt{\frac{I_{SS}}{K_p}} - V_{TP}\right) \tag{15.172}$$

---

[5] In addition to its role as a current source.

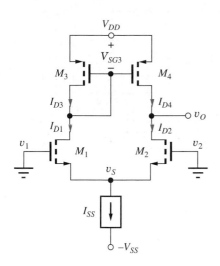

Figure 15.77 CMOS differential amplifier with PMOS active load.

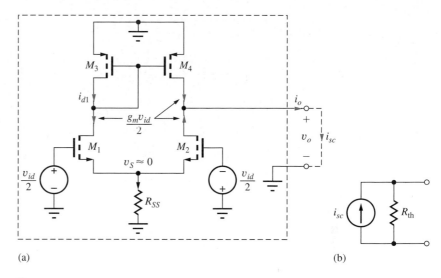

Figure 15.78 (a) CMOS differential amplifier with differential-mode input. (b) The circuit is a one port and can be represented by its Norton equivalent circuit.

Q-Points

The drain-source voltages of M_1 and M_2 are

$$V_{DS1} = V_O - V_S = V_{DD} - \left(\sqrt{\frac{I_{SS}}{K_p}} - V_{TP} \right) + \left(V_{TN} + \sqrt{\frac{I_{SS}}{K_n}} \right)$$

or

$$V_{DS1} = V_{DD} + V_{TN} + V_{TP} + \sqrt{\frac{I_{SS}}{K_n}} - \sqrt{\frac{I_{SS}}{K_p}} \cong V_{DD} \qquad (15.173)$$

and those of M_3 and M_4 are

$$V_{SD3} = V_{SG3} = \sqrt{\frac{I_{SS}}{K_p}} - V_{TP} \qquad (15.174)$$

(Remember that $V_{TP} < 0$ for p-channel enhancement-mode devices.)

The drain currents of all the transistors are equal:

$$I_{DS1} - I_{DS2} = I_{SD3} = I_{SD4} = \frac{I_{SS}}{2} \qquad (15.175)$$

Small-Signal Analysis

Now that we have found the operating points of the transistors, we can proceed to analyze the small-signal characteristics of the amplifier, including differential-mode gain, differential-mode input and output resistances, common-mode gain, CMRR, and common-mode input and output resistances.

Differential-Mode Signal Analysis

Analysis of the ac behavior of the differential amplifier begins with the differential-mode input applied in the ac circuit model in Fig. 15.78. Upon studying the circuit in Fig. 15.78, we realize that it is a two terminal network and can be represented by its Norton equivalent circuit consisting of the short-circuit output current and Thévenin equivalent output resistance. With the output terminals short circuited, the NMOS differential pair produces equal and opposite currents with amplitude $g_{m2}v_{id}/2$ at the drains of M_1 and M_2. Drain current i_{d1} is supplied by current mirror transistor M_3 and is replicated at the output of M_4. Thus, the total short circuit output current is

$$\mathbf{i_o} = 2\frac{g_{m2}\mathbf{v_{id}}}{2} = g_{m2}\mathbf{v_{id}} \qquad (15.176)$$

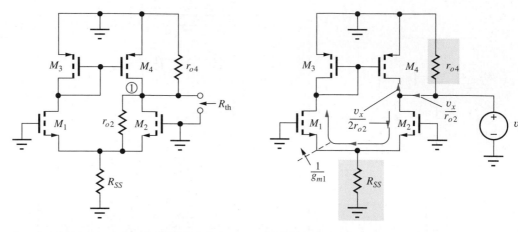

Figure 15.79 Simple CMOS op amp with active load in the first stage.

Figure 15.80 Output resistance component due to r_{o2}.

The current mirror provides a single-ended output but with a transconductance equal to the full value of the C-S amplifier.

The Thévenin equivalent output resistance will be found using the circuit in Fig. 15.79 in which the internal output resistances of M_2 and M_4 are shown next to their respective transistors. In the next sub-section, we will show that R_{th} is equal to the parallel combination of r_{o2} and r_{o4}:

$$R_{\text{th}} = r_{o2} \| r_{o4} \qquad (15.177)$$

The differential-mode voltage gain of the open-circuited differential amplifier is simply the product of i_{sc} and R_{th}:

$$A_{dm} = g_{m2}(r_{o2} \| r_{o4}) = \frac{\mu_{f2}}{1 + \dfrac{r_{o2}}{r_{o4}}} \cong \frac{\mu_{f2}}{2} \qquad (15.178)$$

Equation (15.178) indicates that the gain of the input stage of the amplifier approaches one-half the amplification factor of the transistors forming the differential pair. We are now within a factor of 2 of the theoretical voltage gain limit for a single-transistor amplifier!

Output Resistance of the Differential Amplifier
The origin of the output resistance expression in Eq. (15.177) can be thought of conceptually in the following (although technically incorrect) manner. At node 1 in Fig. 15.79, r_{o4} is connected directly to ac ground at the positive power supply, whereas r_{o2} appears connected to virtual ground at the sources of M_2 and M_1. Thus, r_{o2} and r_{o4} are effectively in parallel. Although this argument gives the correct answer, it is not precisely correct. Because the differential amplifier with active load no longer represents a symmetric circuit, the node at the sources of M_1 and M_2 *is not* truly a virtual ground.

Exact Analysis
A more precise analysis can be obtained from the circuit in Fig. 15.80. The output resistance r_{o4} of M_4 is indeed connected directly to ac ground and represents one component of the output resistance. However, the current from v_x due to r_{o2} is more complicated. The actual behavior can be determined from Fig. 15.80, in which R_{SS} is assumed to be negligible with respect to $1/g_{m1}$, and $R_{SS} \gg 1/g_{m1}$.

Transistor M_2 is operating as a common-gate transistor with an effective resistance in its source of $R_S = 1/g_{m1}$. Based on the results in Table 14.1, the resistance looking into the drain of M_2 is

$$R_{o2} = r_{o2}(1 + g_{m2}R_S) = r_{o2}\left(1 + g_{m2}\frac{1}{g_{m1}}\right) = 2r_{o2} \qquad (15.179)$$

Therefore, the drain current of M_2 is equal to $\mathbf{v_x}/2r_{o2}$. However, the current goes around the differential pair and into the current mirror at M_3. The current is replicated by the mirror to become the drain current of M_4. The total current from source v_x becomes $2(\mathbf{v_x}/2r_{o2}) = \mathbf{v_x}/r_{o2}$.

Combining this current with the current through r_{o4} yields a total current of

$$i_x^T = \frac{\mathbf{v_x}}{r_{o2}} + \frac{\mathbf{v_x}}{r_{o4}} \qquad \text{and} \qquad R_{od} = r_{o2}\|r_{o4} \qquad (15.180)$$

The equivalent resistance at the output node is, in fact, exactly equal to the parallel combination of the output resistances of M_2 and M_4.

EXERCISE: Find the Q-points of the transistors in Fig. 15.77 if $I_{SS} = 250\ \mu\text{A}$, $K_n = 250\ \mu\text{A/V}^2$, $K_p = 200\ \mu\text{A/V}^2$, $V_{TN} = -V_{TP} = 0.75\ \text{V}$, and $V_{DD} = V_{SS} = 5\ \text{V}$. What are the transconductance, output resistance, and voltage gain of the amplifier if $\lambda = 0.0133\ \text{V}^{-1}$?

ANSWERS: (125 μA, 4.88 V), (125 μA, 1.87 V); 250 μS, 314 kΩ, 78.5

Common-Mode Input Signals

Figure 15.81 is the CMOS differential amplifier with a common-mode input signal. The common-mode input voltage causes a common-mode current i_{oc} in both sides of the differential pair consisting of M_1 and M_2. The common-mode current (i_{oc}) in M_1 is mirrored at the output of M_4 with a small error since no current can appear in r_{o4} with the output shorted. In addition, the small voltage difference developed between the drains of M_1 and M_2 causes a current in the differential output resistance $(2r_{o2})$ of the pair that is then doubled by the action of the current mirror.

An expression for the short-circuit output current can be found using the small-signal model for the circuit in Fig. 15.81(b). The differential pair with common-mode input is represented by the two-port model from Sec. 15.1.15 with

$$i_{oc} \cong \frac{v_{ic}}{2R_{SS}} \qquad R_{od} = 2r_{o2} \qquad R_{oc} = 2\mu_f R_{SS} \qquad (15.181)$$

With the output short-circuited, we have a one-node problem. Solving for v_3,

$$v_3 = \frac{-i_{oc}}{g_{m3} + g_{o3} + \dfrac{g_{o2}}{2} + G_{oc}} \qquad \text{and} \qquad i_{sc} = -\left(i_{oc} + g_{m4}v_3 - \frac{g_{o2}}{2}v_3\right) \qquad (15.182)$$

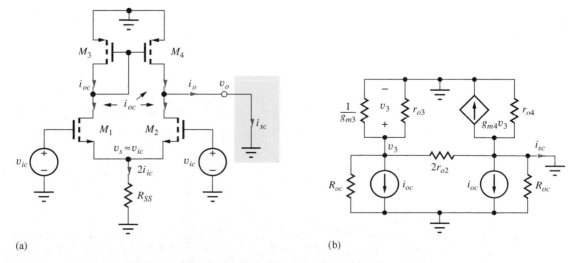

(a)   (b)

Figure 15.81 (a) CMOS differential amplifier with common-mode input. (b) Small-signal model.

which together with Eq. (15.181) yield

$$i_{sc} = -\frac{g_{o3} + g_{o2}}{g_{m3} + g_{o3} + \dfrac{g_{o2}}{2} + G_{oc}} i_{oc} \cong -\frac{1 + \dfrac{r_{o3}}{r_{o2}}}{\mu_{f3}} \left(\frac{v_{ic}}{2R_{SS}}\right) \qquad (15.183)$$

where it is assumed that $g_{m4} = g_{m3}$ and $G_{oc} \ll g_{m3}$. The Thévenin equivalent output resistance is exactly the same as found in the previous section, $R_{th} = r_{o2} \| r_{o4}$. Thus, the common-mode gain is

$$A_{cm} = \frac{i_{sc} R_{th}}{v_{ic}} = -\frac{\left(1 + \dfrac{r_{o3}}{r_{o2}}\right)}{2\mu_{f3} R_{SS}} (r_{o2} \| r_{o4}) \qquad (15.184)$$

where $\mu_{f3} \gg 1$ has been assumed. The common-mode rejection ratio is

$$\text{CMRR} = \left|\frac{A_{dm}}{A_{cm}}\right| = \frac{2\mu_{f3} g_{m2} R_{SS}}{\left(1 + \dfrac{r_{o3}}{r_{o2}}\right)} \cong \mu_{f3} g_{m2} R_{SS} \quad \text{for} \quad r_{o3} \cong r_{o2} \qquad (15.185)$$

which is improved by a factor of approximately μ_{f3} over that of the pair with a resistor load!

EXERCISE: Evaluate Eq. (15.185) for $K_p = K_n = 5$ mA/V^2, $\lambda = 0.0167$ V^{-1}, $I_{SS} = 200$ μA, and $R_{SS} = 10$ MΩ.

ANSWER: 6.00×10^6 or 136 dB

In the last exercise, we find that the CMRR predicted by Eq. (15.185) is quite large, whereas typical op amp specs are 80 to 100 dB. We need to look deeper. In reality, this level will not be achieved, but will be limited by mismatches between the devices in the circuit.

Mismatch Contributions to CMRR

In this section, we explore the techniques used to calculate the effects of device mismatches on CMRR. Figure 15.82 presents the small-signal model for the differential amplifier with mismatches

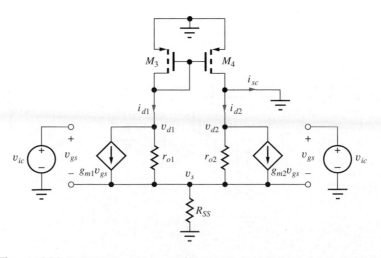

Figure 15.82 CMOS differential amplifier in which M_1 and M_2 are no longer matched.

in transistors M_1 and M_2 in which we assume

$$g_{m1} = g_m + \frac{\Delta g_m}{2} \qquad g_{m2} = g_m - \frac{\Delta g_m}{2} \qquad g_{o1} = g_o + \frac{\Delta g_o}{2} \qquad g_{o2} = g_o - \frac{\Delta g_o}{2} \qquad (15.186)$$

In this analysis, M_3 and M_4 are still identical. We desire to find the short circuit output current $i_{sc} = (i_{d1} - i_{d2})$ in which i_{d1} is replicated by the current mirror. Let us use our knowledge of the gross behavior of the circuit to simplify the analysis. We have $v_{d2} = 0$, since we are finding the short-circuit output current, and based on previous common-mode analyses, we expect the signal at v_{d1} to be small. So let us assume that $v_{d1} \cong 0$. With this assumption, and noting that the two gate-source voltages are identical,

$$i_{sc} = i_{d1} - i_{d2} = (g_{m1} - g_{m2})v_{gs} - (g_{o1} - g_{o2})v_s = \Delta g_m v_{gs} - \Delta g_o v_s \qquad (15.187)$$

To evaluate this expression, we need to find source voltage v_s and gate-source voltage v_{gs}. Writing a nodal equation for v_s with $v_{gs} = v_{ic} - v_s$, $v_{d1} = 0$ and $v_{d2} = 0$, yields

$$\left(g_m + \frac{\Delta g_m}{2} + g_m - \frac{\Delta g_m}{2} \right)(v_{ic} - v_s) = \left(g_o + \frac{\Delta g_o}{2} + g_o - \frac{\Delta g_o}{2} + G_{SS} \right) v_s$$

in which we may be surprised to see all the mismatch terms cancel out! Thus, for common-mode inputs, v_s and v_{gs} are not affected by the transistor mismatches:[6]

$$v_s \cong \frac{2g_m R_{SS}}{1 + 2g_m R_{SS}} v_{ic} \cong v_{ic}$$

and

$$v_{gs} \cong \frac{1 + 2g_o R_{SS}}{1 + 2g_m R_{SS}} v_{ic} \cong \left(\frac{1}{2g_m R_{SS}} + \frac{1}{\mu_f} \right) v_{ic} \qquad (15.188)$$

The short-circuit output current goes through the Thévenin output resistance $R_{th} = r_{o2} \| r_{o4}$ to produce the output voltage, and

$$A_{cm} = \frac{i_{sc} R_{th}}{v_{ic}} = \left[\Delta g_m \left(\frac{1}{2g_m R_{SS}} + \frac{1}{\mu_f} \right) - \Delta g_o \right] (r_{o2} \| r_{o4}) \qquad (15.189)$$

The CMRR is then

$$\mathrm{CMRR}^{-1} = \left| \frac{A_{cm}}{A_{dm}} \right| = \left| \frac{A_{cm}}{g_m (r_{o2} \| r_{o4})} \right|$$

$$= \left[\frac{\Delta g_m}{g_m} \left(\frac{1}{2g_m R_{SS}} + \frac{1}{\mu_f} \right) - \frac{\Delta g_o}{g_o} \frac{1}{\mu_f} \right] \qquad (15.190)$$

For very large R_{SS}, we see that CMRR is now limited by the transistor mismatches and value of the amplification factor. For example, a 1 percent mismatch with an amplification factor of 500 limits the individual terms in Eq. (15.190) to 2×10^{-5}. Since we cannot predict the signs on the $\Delta g/g$ terms, the expected CMRR is 2.5×10^4 or 88 dB. This is much more consistent with observed values of CMRR.

---

[6] An exact analysis without assuming that $v_{d1} = 0$ shows that a negligibly small change actually occurs.

ELECTRONICS IN ACTION

G_m-C Integrated Filters

The design of integrated circuit filters is complicated by the lack of well-controlled resistive components in most mainstream CMOS processes. One approach to overcome this is the use of G_m-C filter topologies based on the operational transconductance amplifier (OTA). The OTA is characterized by both a high input and high output impedance. A simple form of an OTA is shown below. The high impedance output is a small-signal current given by the product of the differential pair g_m and the differential input voltage v_{id}. Typically, commercial OTA designs include additional devices to improve output resistance and voltage swing.

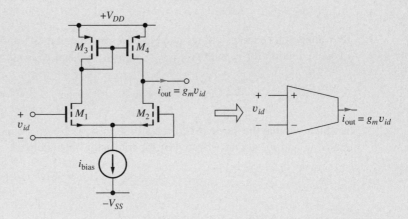

Equivalent schematic symbol for operational transconductance amplifier (OTA).

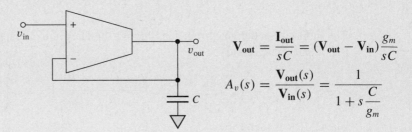

$$\mathbf{V_{out}} = \frac{\mathbf{I_{out}}}{sC} = (\mathbf{V_{out}} - \mathbf{V_{in}})\frac{g_m}{sC}$$

$$A_v(s) = \frac{\mathbf{V_{out}}(s)}{\mathbf{V_{in}}(s)} = \frac{1}{1 + s\dfrac{C}{g_m}}$$

Single pole g_m-C low-pass filter.

A simple low-pass filter formed with an OTA and a capacitor is shown above. The transfer characteristic is also included and indicates that the upper cutoff frequency occurs at $f_H = g_m/2\pi C$. One of the more useful characteristics of the g_m-C filter approach is the ease with which the characteristics can be tuned. Recalling that g_m is a function of the differential pair current, we see that the cutoff frequency of the filter is easily modified by adjusting the bias current.

A second-order version (a biquad topology) is shown below. This version allows for the adjustment of cutoff frequency with constant Q, and still requires no resistors. High-pass, band-pass, and band-reject are also readily derived from this basic form. Because of their compatibility with standard CMOS processes and excellent power efficiency, g_m-C filters have become prevalent in communication circuits, A/D converter anti-alias filters, noise shaping, and many other applications.

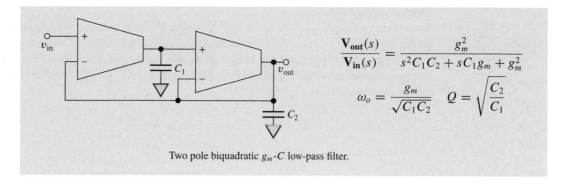

$$\frac{\mathbf{V_{out}}(s)}{\mathbf{V_{in}}(s)} = \frac{g_m^2}{s^2 C_1 C_2 + s C_1 g_m + g_m^2}$$

$$\omega_o = \frac{g_m}{\sqrt{C_1 C_2}} \qquad Q = \sqrt{\frac{C_2}{C_1}}$$

Two pole biquadratic g_m-C low-pass filter.

15.9.2 BIPOLAR DIFFERENTIAL AMPLIFIER WITH ACTIVE LOAD

The bipolar differential amplifier with an active load formed from a *pnp* current mirror is depicted in Fig. 15.83 with $v_1 = 0 = v_2$. If we assume that the circuit is balanced with $\beta_{FO} = \infty$, then the bias current I_{EE} divides equally between transistors Q_1 and Q_2, and I_{C1} and I_{C2} are equal to $I_{EE}/2$. Current I_{C1} is supplied by transistor Q_3 and is mirrored as I_{C4} at the output of *pnp* transistor Q_4. Thus, I_{C3} and I_{C4} are both also equal to $I_{EE}/2$, and the dc current in the collector of Q_4 is exactly the current required to satisfy Q_2.

If β_{FO} is very large, then the current mirror ratio is exactly 1 when $V_{EC4} = V_{EC3} = V_{EB}$, and the differential amplifier is completely balanced when the quiescent output voltage is

$$V_O = V_{CC} - V_{EB} \qquad (15.191)$$

Q-Points
The collector currents of all the transistors are equal:

$$I_{C1} = I_{C2} = I_{C3} = I_{C4} = \frac{I_{EE}}{2} \qquad (15.192)$$

The collector-emitter voltages of Q_1 and Q_2 are

$$V_{CE1} = V_{CE2} = V_C - V_E = (V_{CC} - V_{EB}) - (-V_{BE}) \cong V_{CC} \qquad (15.193)$$

and for Q_3 and Q_4,

$$V_{EC3} = V_{EC4} = V_{EB} \qquad (15.194)$$

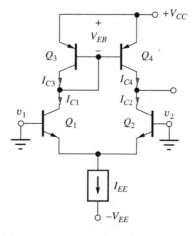

Figure 15.83 Bipolar differential amplifier with active load.

Finite Current Gain

The current gain defect in the current mirror upsets the dc balance of the circuit. However, as long as the transistors remain in the forward-active region, the collector current of Q_4 must equal the collector current of Q_2, and the collector-emitter voltage of Q_4 adjusts itself to make up for the current-gain defect of the current mirror. The required value of V_{EC4} can be found using the current mirror expression from Eq. (16.10):

$$I_{C4} = I_{C1} \frac{\left[1 + \dfrac{V_{EC4}}{V_A} \right]}{\left[1 + \dfrac{V_{EB}}{V_A} + \dfrac{2}{\beta_{FO4}} \right]} \tag{15.195}$$

However, because $I_{C4} = I_{C2}$ and $I_{C2} = I_{C1}$, the mirror ratio must be unity, which requires

$$V_{EC4} = V_{EB} + \frac{2V_A}{\beta_{FO4}} \tag{15.196}$$

For $\beta_{FO3} = 50$, $V_A = 60$ V, and $V_{EB} = 0.7$ V, $V_{EC4} = 3.10$ V.

This collector-emitter voltage difference represents a substantial offset at the amplifier output and translates to an equivalent input offset voltage of

$$V_{OS} = \frac{V_{EC4} - V_{EC3}}{A_{dd}} = \frac{V_{EC4} - V_{EB}}{A_{dd}} \tag{15.197}$$

V_{OS} represents the input voltage needed to force the output voltage differential to be zero. For $A_{dd} = 100$, V_{OS} would be 24.0 mV. To eliminate this error, a buffered current mirror is usually used as the active load, as shown in Fig. 15.84.

It should be noted that Eq. (15.196) actually overestimates the value of V_{EC4} because the increase in V_{EC4} decreases V_{CE2} and thereby reduces I_{C2}.

EXERCISE: Calculate the dc value of V_{EC4} if the circuit buffered current mirror replaces the active load in Fig. 15.83. What is V_{OS} if $A_{dd} = 100$?

ANSWERS: $V_{EC4} = 1.25$ V and $\Delta V_{EC} = 47$ mV; $V_{OS} = 0.47$ mV

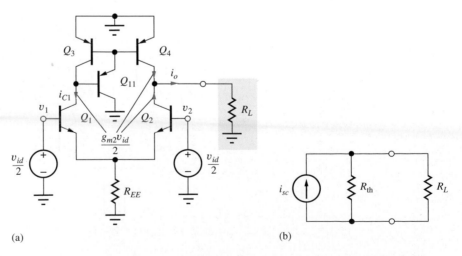

(a) (b)

Figure 15.84 (a) BJT differential amplifier with differential-mode input. (b) Equivalent circuit.

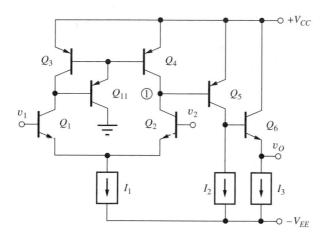

Figure 15.85 Bipolar op amp with active load in first stage.

Differential-Mode Signal Analysis

Analysis of the ac behavior of the differential amplifier begins with the differential-mode input applied in the ac circuit model in Fig. 15.84. The differential input pair produces equal and opposite currents with amplitude $g_{m2}v_{id}/2$ at the collectors of Q_1 and Q_2. Collector current i_{c1} is supplied by Q_3 and is replicated at the output of Q_4. Thus, the total short circuit output current is equal to

$$\mathbf{i_{sc}} = 2\frac{g_{m2}\mathbf{v_{id}}}{2} = g_{m2}\mathbf{v_{id}} \tag{15.198}$$

The output resistance is identical to Eq. (15.177) $R_{th} = r_{o2}\|r_{o4}$ and

$$A_{dd} = \frac{i_{sc}(R_L\|R_{th})}{v_{dm}} = g_{m2}(R_L\|r_{o2}\|r_{o4}) = -g_{m2}R_L \tag{15.199}$$

The current mirror provides a single-ended output but with a voltage equal to the full gain of the C-E amplifier, just as for the FET case. Here we have included R_L, which models the loading of the next stage in a multistage amplifier.

The power of the current mirror is again most apparent when additional stages are added, as in the prototype operational amplifier in Fig. 15.85. The resistance at the output of the differential input stage, node 1, is now equivalent to the parallel combination of the output resistances of transistors Q_2 and Q_4 and the input resistance of Q_5 ($R_L = r_{\pi5}$):

$$R_{eq} = r_{o2}\|r_{o4}\|r_{\pi5} \cong r_{\pi5} \tag{15.200}$$

and the gain of the differential input stage becomes

$$A_{dm} = g_{m2}R_{eq} \cong g_{m2}r_{\pi5} = \beta_{o5}\frac{I_{C2}}{I_{C5}} \tag{15.201}$$

EXERCISE: What is the approximate differential-mode voltage gain of the amplifier in Fig. 15.85 if $\beta_{FO} = 150$, $V_A = 75$ V, and $I_{C5} = 3\,I_{C2}$?

ANSWER: 50

Common-Mode Input Signals

The circuits in Fig. 15.86 represent the bipolar differential amplifier with current mirror load and a buffered current mirror load. The detailed analysis is quite involved and tedious, particularly for the buffered mirror, so here we will argue the result based on earlier analyses. The common-mode

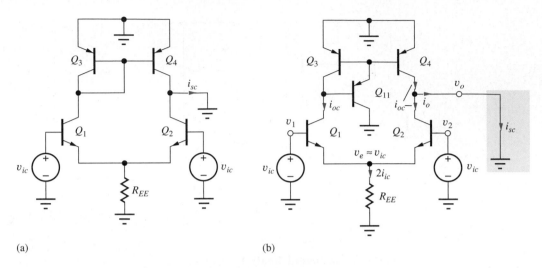

(a) (b)

Figure 15.86 Bipolar differential amplifiers with common-mode input.

current i_{oc} in Q_1 and Q_2 is found with the help of Eq. (15.27):

$$i_{oc} = \frac{A_{cc}v_{ic}}{R_C} = v_{ic}\left(\frac{1}{2R_{EE}} - \frac{1}{\beta_o r_o}\right) \tag{15.202}$$

The current from Q_1 is mirrored at the output of Q_4 with a mirror error of $2/\beta_o$. Thus, the short-circuit output current is

$$i_{sc} = v_{ic}\frac{2}{\beta_o}\left(\frac{1}{\beta_o r_o} - \frac{1}{2R_{EE}}\right) \tag{15.203}$$

In a manner similar to that of the FET pair, the voltage developed at the collector of Q_1, i_{oc}/g_{m3}, forces a current in the differential output resistance of the pair $(2r_{o2})$, which is doubled by the action of the current mirror:

$$i_{sc} = 2v_{ic}\left(\frac{1}{\beta_o r_o} - \frac{1}{2R_{EE}}\right)\frac{1}{g_{m3}(2r_{o2})} \cong \frac{v_{ic}}{\mu_{f2}}\left(\frac{1}{\beta_o r_o} - \frac{1}{2R_{EE}}\right) \tag{15.204}$$

Since $\mu_f \gg \beta_o$ for the BJT, the output will be dominated by Eq. (15.203), and the CMRR is

$$\text{CMRR} = \left|\frac{g_{m2}R_{th}}{i_{sc}R_{th}/v_{ic}}\right| \cong \left[\frac{2}{\beta_{o3}}\left(\frac{1}{\beta_{o2}\mu_{f2}} - \frac{1}{2g_{m2}R_{EE}}\right)\right]^{-1} \tag{15.205}$$

EXERCISE: Evaluate Eq. (15.205) for $\beta_F = 100$, $V_A = 75$ V, $I_{EE} = 200$ μA, and $R_{EE} = 10$ MΩ.

ANSWER: 5.45 × 10⁶ or 135 dB

The expression in Eq. (15.205) yields a very large CMRR that is almost impossible to achieve. The CMRR predicted for the buffered current mirror is even larger, since the mirror error is approximately $2/\beta_{o11}\beta_{o3}$. In both these circuits, however, the CMRR will actually be limited to much smaller levels by small mismatches between the various transistors:

$$\text{CMRR}^{-1} = \left[\left(\frac{\Delta g_m}{g_m} + \frac{\Delta g_\pi}{g_\pi}\right)\left(\frac{1}{2g_m R_{SS}} + \frac{1}{\mu_f}\right) - \frac{\Delta g_o}{g_o}\frac{1}{\mu_f}\right] \tag{15.206}$$

Equation (15.206) is similar to the results for the FET from Eq. (15.90) with the addition of the $\Delta g_\pi/g_\pi$ term. In an actual amplifier, the common-mode gain is determined by small imbalances in the bipolar transistors and overall symmetry of the amplifier.

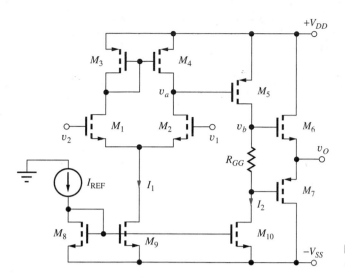

Figure 15.87 Complete CMOS op amp with current mirror bias.

15.10 ACTIVE LOADS IN OPERATIONAL AMPLIFIERS

Let us now explore more fully the use of active loads in MOS and bipolar operational amplifiers. Figure 15.87 shows a complete three-stage MOS operational amplifier. The input stage consists of NMOS differential pair M_1 and M_2 with PMOS current mirror load, M_3 and M_4, followed by a second common-source gain stage M_5 loaded by current source M_{10}. The output stage is a class-AB amplifier consisting of transistors M_6 and M_7. Bias currents I_1 and I_2 for the two gain stages are set by the current mirrors formed by transistors M_8, M_9, and M_{10}, and class-AB bias for the output stage is set by the voltage developed across resistor R_{GG}. At most, only two resistors are required: R_{GG} and one for the current mirror reference current.

15.10.1 CMOS OP AMP VOLTAGE GAIN

Assuming that the gain of the output stage is approximately 1, then the overall differential-mode gain A_{dm} of the three-stage operational amplifier is approximately equal to the product of the terminal gains of the first two stages:

$$A_{dm} = \frac{\mathbf{v_a}}{\mathbf{v_{id}}} \frac{\mathbf{v_b}}{\mathbf{v_a}} \frac{\mathbf{v_o}}{\mathbf{v_b}} = A_{vt1} A_{vt2}(1) \cong A_{vt1} A_{vt2} \tag{15.207}$$

As discussed earlier, the input stage provides a gain of

$$A_{vt1} = g_{m2}(r_{o2} \| r_{o4}) \cong \frac{\mu_{f2}}{2} \tag{15.208}$$

The terminal gain of the second stage is equal to

$$A_{vt2} = g_{m5}(r_{o5} \| (R_{GG} + r_{o10})) \cong g_{m5}(r_{o5} \| r_{o10}) \cong g_{m5}(r_{o5} \| r_{o5}) = \frac{\mu_{f5}}{2} \tag{15.209}$$

assuming that the output resistances of M_5 and M_{10} are similar in value and $R_{GG} \ll r_{o10}$. Combining the three equations above yields

$$A_{dm} \cong \frac{\mu_{f2}\mu_{f5}}{4} \tag{15.210}$$

The gain approaches one-quarter of the product of the amplification factors of the two gain stages.

The factor of 4 in the denominator of Eq. (15.210) can be eliminated by improved design. If a Wilson source is used in the first-stage active load, then the output resistance of the current mirror is much greater than r_{o2}, and A_{v1} becomes equal to μ_{f2}. The gain of the second stage can also be increased to the full amplification factor of M_5 if the current source M_{10} is replaced by a Wilson or

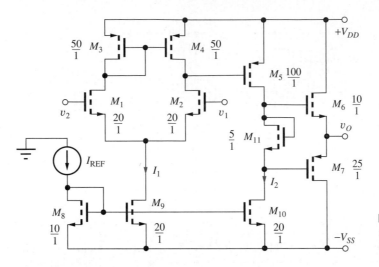

Figure 15.88 Op amp with current mirror bias of the class-AB output stage.

cascode source. If both these circuit changes are used (see Prob. 15.233), then the gain of the op amp can be increased to

$$A_{dm} \cong \mu_{f2}\mu_{f5} \qquad (15.211)$$

This discussion has only scratched the surface of the many techniques available for increasing the gain of the CMOS op amp. Several examples appear in the problems at the end of this chapter; further discussion can be found in the bibliography.

15.10.2 dc DESIGN CONSIDERATIONS

When the circuit in Fig. 15.87 is operating in a closed-loop op amp configuration, the drain current of M_5 must be equal to the output current I_2 of current source transistor M_{10}. For the amplifier to have a minimum offset voltage, the (W/L) ratio of M_5 must be carefully selected so the source-gate bias of M_5, $V_{SG5} = V_{SD4} = V_{SG3}$, is precisely the proper voltage to set $I_{D5} = I_2$. Usually the W/L ratio of M_5 is also adjusted to account for V_{DS} and λ differences between M_5 and M_{10}. R_{GG} and the (W/L) ratios of M_6 and M_7 determine the quiescent current in the class-AB output stage.

Even resistor R_{GG} has been eliminated from the op amp in Fig. 15.88 by using the gate-source voltage of FET M_{11} to bias the output stage. The current in the class-AB stage is determined by the W/L ratios of the output transistors and the matching diode-connected MOSFET M_{11}.

EXAMPLE 15.12 CMOS OP AMP ANALYSIS

Find the small-signal characteristics of a CMOS operational amplifier.

PROBLEM Find the voltage gain, input resistance, and output resistance of the amplifier in Fig. 15.88 if $K'_n = 25\ \mu\text{A}/\text{V}^2$, $K'_p = 10\ \mu\text{A}/\text{V}^2$, $V_{TN} = 0.75$ V, $V_{TP} = -0.75$ V, $\lambda = 0.0125$ V^{-1}, $V_{DD} = V_{SS} = 5$ V, and $I_{REF} = 100\ \mu\text{A}$.

SOLUTION **Known Information and Given Data:** The schematic for the operational amplifier appears in Fig. 15.88; $V_{DD} = V_{SS} = 5$ V, and $I_{REF} = 100\ \mu\text{A}$; device parameters are given as $K'_n = 25\ \mu\text{A}/\text{V}^2$, $K'_p = 10\ \mu\text{A}/\text{V}^2$, $V_{TN} = 0.75$ V, $V_{TP} = -0.75$ V, $\lambda = 0.0125$ V^{-1}.

Unknowns: Q-points, A_{dm}, R_{id}, and R_{out}

Approach: Find the Q-point currents and use the device parameters to evaluate Eq. (15.210) for A_{dm}. Since we have MOSFETS at the input, $R_{id} = R_{ic} = \infty$. R_{out} is set by M_6 and M_7: $R_{out} = (1/g_{m6}) \| (1/g_{m7})$.

Assumptions: MOSFETs operate in the active region.

Analysis: The gain can be estimated using Eq. (15.210).

$$A_{dm} \cong \frac{\mu_{f2}\mu_{f5}}{4} = \frac{1}{4}\left(\frac{1}{\lambda_2}\sqrt{\frac{2K_{n2}}{I_{D2}}}\right)\left(\frac{1}{\lambda_5}\sqrt{\frac{2K_{p5}}{I_{D5}}}\right)$$

For the amplifier in Fig. 15.88,

$$I_{D2} = \frac{I_1}{2} = \frac{2I_{REF}}{2} = 100 \ \mu A \qquad I_{D5} = I_2 = 2I_{REF} = 200 \ \mu A$$

$$K_{n2} = 20K_n' = 500\frac{\mu A}{V^2} \qquad\qquad K_{p5} = 100K_p' = 1000\frac{\mu A}{V^2}$$

and

$$A_{dm} \cong \frac{\mu_{f2}\mu_{f5}}{4} = \frac{1}{4}\left(\frac{1}{0.0125}\right)^2 V^2 \sqrt{\frac{2\left(500\frac{\mu A}{V^2}\right)}{100 \ \mu A}}\sqrt{\frac{2\left(1000\frac{\mu A}{V^2}\right)}{200 \ \mu A}} = 16,000$$

The input resistance is twice the input resistance of M_1, which is infinite: $R_{id} = \infty$. The output resistance is determined by the parallel combination of the output resistances of M_6 and M_7, which act as two source followers operating in parallel:

$$R_{out} = \frac{1}{g_{m6}} \,\Big\|\, \frac{1}{g_{m7}} = \frac{1}{\sqrt{2K_{n6}I_{D6}}} \,\Big\|\, \frac{1}{\sqrt{2K_{p7}I_{D7}}}$$

To evaluate this expression, the current in the output stage must be found. The gate-source voltage of M_{11} is

$$V_{GS11} = V_{TN11} + \sqrt{\frac{2I_{D11}}{K_{n11}}} = 0.75 \ V + \sqrt{\frac{2(200 \ \mu A)}{125\left(\frac{\mu A}{V^2}\right)}} = 2.54 \ V$$

In this design, $V_{TP} = -V_{TN}$ and the W/L ratios of M_6 and M_7 have been chosen so that $K_{p7} = K_{n6}$. Because I_{D6} must equal I_{D7}, $V_{GS6} = V_{SG7}$. Thus, both V_{GS6} and V_{SG7} are equal to one-half V_{GS11}, and

$$I_{D7} = I_{D6} = \frac{250 \ \mu A}{2 \ V^2}(1.27 \ V - 0.75 \ V)^2 = 33.7 \ \mu A$$

The transconductances of M_6 and M_7 are also equal,

$$g_{m7} = g_{m6} = \sqrt{2\left(2.50 \times 10^{-4}\frac{\mu A}{V^2}\right)(33.7 \times 10^{-6} \ \mu A)} = 1.30 \times 10^{-4} \ S$$

and the output resistance at the Q-point is $R_{out} = 3.85 \ k\Omega$.

Check of Results: A double check of our hand calculations indicates they are correct. Because of the complexity of the circuit, SPICE simulation represents an excellent check of hand calculations. The simulation results appear in the next exercise.

Discussion: Simulation of the open-loop characteristics of high-gain amplifiers in SPICE can be difficult. The open-loop gain will amplify the offset voltage of the amplifier and may saturate the output. One approach is to first determine the offset voltage and then to apply a compensating voltage to the amplifier input to bring the output near zero. The steps are outlined next. In very high gain cases, SPICE may still be unable to converge because numerical "noise" during the simulation steps is amplified just as an input voltage. The successive voltage and current injection method discussed in Chapter 17 solves this problem.

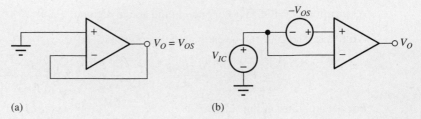

(a)   (b)

Figure 15.89 Op amp setups for SPICE simulation. (a) Offset voltage determination. (b) Circuit for open-loop analysis using SPICE transfer functions.

Computer-Aided Analysis: After drawing the circuit of Fig. 15.88 with the schematic editor, be sure to set the device parameters to the desired values. For the NMOS devices, KP = 25 μA/V², VTO = 0.75 V, and LAMBDA = 0.0125 V⁻¹. For the PMOS devices, KP = 10 μA/V², VTO = −0.75 V, and LAMBDA = 0.0125 V⁻¹. W and L must be specified for each individual transistor. For example, use $W = 5$ μm and $L = 1$ μm for a 5/1 device.

The first step in the simulation is to find the offset voltage by operating the op amp in a voltage-follower configuration for which $V_O = V_{OS}$, as in Fig. 15.89(a). V_{OS} is then applied as a differential input to the amplifier in Fig. 15.89(b) with a common-mode input $V_{IC} = 0$. If the value of V_{OS} is correct, an operating point analysis should yield a value of approximately 0 for V_O. A transfer function analysis from V_{OS} to the output will give values of A_{dm}, R_{id}, and R_{out}. A transfer function analysis from V_{IC} to the output will give A_{cm}, R_{ic}, and R_{out}. The SPICE results are given as the answers to the next exercise.

EXERCISE: Simulate the amplifier in Fig. 15.88 using SPICE and compare the results to the answers in Ex. 15.12. Which terminal is the noninverting input? What are the offset voltage, common-mode and differential-mode gains, CMRR, common-mode and differential-mode input resistances, and output resistance?

ANSWERS: v_1; 64.164 μV; 17,800; 0.52 V; 90.7 dB; ∞; ∞; 3.63 kΩ

15.10.3 BIPOLAR OPERATIONAL AMPLIFIERS

Active-load techniques can be applied equally well to bipolar op amps. In fact, most of the techniques discussed thus far were developed first for bipolar amplifiers and later applied to MOS circuits as NMOS and CMOS technologies matured. In the circuit in Fig. 15.90, a differential input stage with active load is formed by transistors Q_1 to Q_4. The first stage is followed by a high gain C-E amplifier formed of Q_5 and its current source load Q_8. Load resistance R_L is driven by the class-AB output stage, consisting of transistors Q_6 and Q_7 biased by current I_2 and diodes Q_{11} and Q_{12}. (The diodes will actually be implemented with BJTs, in this case with emitter areas five times those of Q_6 and Q_7.)

Based on our understanding of multistage amplifiers, the gain of this circuit is approximately $A_{dm} = A_{vt1} A_{vt2} A_{vt3}$ and

$$A_{dm} \cong [g_{m2}r_{\pi 5}][g_{m5}(r_{o5}\|r_{o8}\|(\beta_{o6} + 1)R_L)][1] \cong \frac{g_{m2}}{g_{m5}} g_{m5}r_{\pi 5}g_{m5}\frac{r_{o5}}{2} = \frac{I_{C2}}{I_{C5}}\beta_{o5}\frac{\mu_{f5}}{2} \qquad (15.212)$$

in which it has been assumed that the input resistance of the class-AB output stage is much larger than the parallel combination of r_{o5} and r_{o8}. Note that the upper limit to Eq. (15.212) is set by the $\beta_o V_A$ product of Q_5, because I_{C2} is typically less than or equal to I_{C5}.

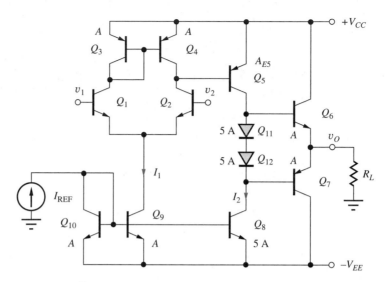

Figure 15.90 Complete bipolar operational amplifier.

EXERCISE: Estimate the voltage gain of the amplifier in Fig. 15.90 using Eq. (15.212) if $I_{REF} = 100\ \mu A$, $V_{A5} = 60\ V$, $\beta_{o1} = 150$, $\beta_{o5} = 50$, $R_L = 2\ M\Omega$, and $V_{CC} = V_{EE} = 15\ V$. What is the gain of the first stage? The second stage? What should be the emitter area of Q_5? What is R_{ID}? Which terminal is the inverting input?

ANSWERS: 7500; 5; 1500; 10 A; 150 kΩ; v_1

EXERCISE: Simulate the amplifier in the previous exercise using SPICE and determine the offset voltage, voltage gain, differential-mode input resistance, CMRR, and common-mode input resistance.

ANSWERS: 3.28 mV; 8440; 165 kΩ; 84.7 dB; 59.1 MΩ

15.10.4 INPUT STAGE BREAKDOWN

Although the bipolar amplifier designs discussed thus far have provided excellent voltage gain, input resistance, and output resistance, the amplifiers all have a significant flaw. The input stage does not offer **overvoltage protection** and can easily be destroyed by the large input voltage differences that can occur, not only under fault conditions but also during unavoidable transients during normal use of the amplifier. For example, the voltage across the input of an op amp can temporarily be equal to the total supply voltage span during slew-rate limited overload recovery.

Consider the worst-case fault condition applied to the differential pair in Fig. 15.91. Under the conditions shown, the base-emitter junction of Q_1 will be forward-biased, and that of Q_2 reverse-biased by a voltage of $(V_{CC} + V_{EE} - V_{BE1})$. If $V_{CC} = V_{EE} = 22$ V, the reverse voltage exceeds 41 V. Because of heavy doping in the emitter, the typical Zener breakdown voltage of the base-emitter junction of an *npn* transistor is only 5 to 7 V. Thus, any voltage exceeding this value by more than one diode drop may destroy at least one of the transistors in the differential input pair.

Early IC op amps required circuit designers to add external diode protection across the input terminals, as shown Fig. 15.91(b). The diodes prevent the differential input voltage from exceeding approximately 1.4 V, but this technique adds extra components and cost to the design. The two resistors limit the current through the diodes. The μA741 described in the next section was the first commercial IC op amp to solve this problem by providing a fully protected input, as well as output, stage.

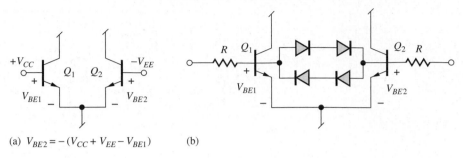

(a) $V_{BE2} = -(V_{CC} + V_{EE} - V_{BE1})$ (b)

Figure 15.91 (a) Differential input stage voltages under a fault condition. (b) Simple diode input protection circuit.

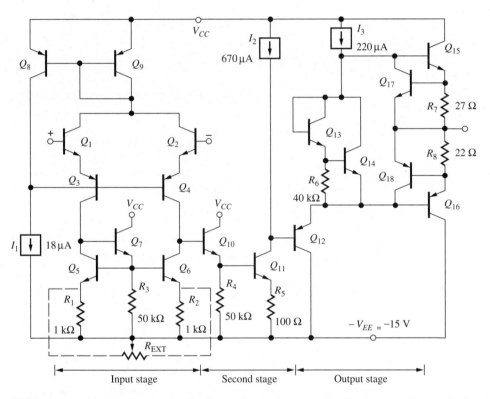

Figure 15.92 Overall schematic of the classic Fairchild μA741 operational amplifier (the bias network appears in Fig. 15.93).

15.11 THE μA741 OPERATIONAL AMPLIFIER

The now classic Fairchild **μA741** operational-amplifier design was the first to provide a highly robust amplifier from the application engineer's point of view. The amplifier provides excellent overall characteristics (high gain, input resistance and CMRR, low output resistance, and good frequency response) while providing overvoltage protection for the input stage and short-circuit current limiting of the output stage. The 741 style of amplifier design quickly became the industry standard and spawned many related designs. By studying the 741 design, we will find a number of new amplifier circuit design and bias techniques.

Figure 15.92 is a simplified schematic of the μA741 operational amplifier. The three bias sources shown in symbolic form are discussed in more detail following a description of the overall circuit. The op amp has two stages of voltage gain followed by a class-AB output stage. In the first stage, transistors Q_1 to Q_4 form a differential amplifier with a buffered current mirror active load, Q_5 to Q_7.

Practical operational amplifiers offer an offset voltage adjustment port, which is provided in the 741 through the addition of 1-kΩ resistors R_1 and R_2 and an external potentiometer R_{EXT}.

The second stage consists of emitter follower Q_{10} driving common-emitter amplifier Q_{11} with current source I_2 and transistor Q_{12} as load. Transistors Q_{13} to Q_{18} form a short-circuit protected class-AB push-pull output stage that is buffered from the second gain stage by emitter follower Q_{12}.

> **EXERCISE:** Reread this section and be sure you understand the function of each individual transistor in Fig. 15.92. Make a table listing the function of each transistor.

15.11.1 BIAS CIRCUITRY

The three current sources shown symbolically in Fig. 15.92 are generated by the bias circuitry in Fig. 15.93. The value of the current in the two diode-connected reference transistors Q_{20} and Q_{22} is determined by the power supply voltage and resistor R_5:

$$I_{\text{REF}} = \frac{V_{CC} + V_{EE} - 2V_{BE}}{R_5} = \frac{15 + 15 - 1.4}{39 \text{ k}\Omega} = 0.733 \text{ mA} \tag{15.213}$$

assuming ±15-V supplies. Current I_1 is derived from the Widlar source formed of Q_{20} and Q_{21}. The output current for this design is

$$I_1 = \frac{V_T}{5000} \ln\left[\frac{I_{\text{REF}}}{I_1}\right] \tag{15.214}$$

Using the reference current calculated in Eq. (15.213) and iteratively solving for I_1 in Eq. (15.214) yields $I_1 = 18.4 \ \mu\text{A}$.

The currents in mirror transistors Q_{23} and Q_{24} are related to the reference current I_{REF} by their emitter areas using Eq. (15.137). Assuming $V_O = 0$ and $V_{CC} = 15$ V, and neglecting the voltage drop across R_7 and R_8 in Fig. 15.92, $V_{EC23} = 15 + 1.4 = 16.4$ V and $V_{EC24} = 15 - 0.7 = 14.3$ V. Using these values with $\beta_F = 50$ and $V_A = 60$ V, the two source currents are

$$I_2 = 0.75(733 \ \mu\text{A}) \frac{1 + \dfrac{16.4 \text{ V}}{60 \text{ V}}}{1 + \dfrac{0.7 \text{ V}}{60 \text{ V}} + \dfrac{2}{50}} = 666 \ \mu\text{A}$$

$$I_3 = 0.25(733 \ \mu\text{A}) \frac{1 + \dfrac{14.4 \text{ V}}{60 \text{ V}}}{1 + \dfrac{0.7 \text{ V}}{60 \text{ V}} + \dfrac{2}{50}} = 216 \ \mu\text{A}$$

$$\tag{15.215}$$

and the two output resistances are

$$R_2 = \frac{V_{A23} + V_{EC23}}{I_2} = \frac{60 \text{ V} + 16.4 \text{ V}}{0.666 \text{ mA}} = 115 \text{ k}\Omega$$

$$R_3 = \frac{V_{A24} + V_{EC24}}{I_3} = \frac{60 \text{ V} + 14.3 \text{ V}}{0.216 \text{ mA}} = 344 \text{ k}\Omega$$

$$\tag{15.216}$$

> **EXERCISE:** What are the values of I_{REF}, I_1, I_2, and I_3 in the circuit in Fig. 15.93 for $V_{CC} = V_{EE} = 22$ V?
>
> **ANSWERS:** 1.09 mA, 20.0 μA, 1.08 mA, 351 μA

> **EXERCISE:** What is the output resistance of the Widlar source in Fig. 15.93 operating at 18.4 μA for $V_A = 60$ V and $V_{EE} = 15$ V?
>
> **ANSWER:** 18.8 MΩ

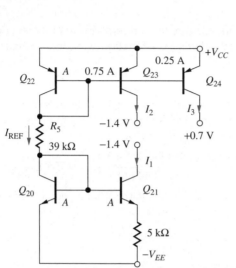

Figure 15.93 741 bias circuitry with voltages corresponding to $V_O = 0$ V.

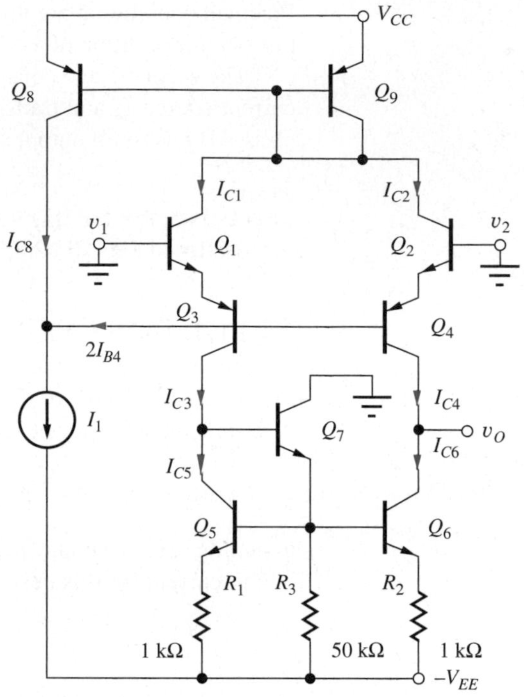

Figure 15.94 μA741 input stage.

15.11.2 dc ANALYSIS OF THE 741 INPUT STAGE

The input stage of the μA741 amplifier is redrawn in the schematic in Fig. 15.94. As noted earlier, Q_1, Q_2, Q_3, and Q_4 form a differential input stage with an active load consisting of the buffered current mirror formed by Q_5, Q_6, and Q_7. In this input stage, there are four base-emitter junctions between inputs v_1 and v_2, two from the *npn* transistors and, more importantly, two from the *pnp* transistors, and $(v_1 - v_2) = (V_{BE1} + V_{EB3} - V_{EB4} - V_{BE2})$.

In standard bipolar IC processes, *pnp* transistors are formed from lateral structures in which both junctions exhibit breakdown voltages equal to that of the collector-base junction of the *npn* transistor. This breakdown voltage typically exceeds 50 V. Because most general-purpose op amp specifications limit the power supply voltages to less than ±22 V, the emitter-base junctions of Q_3 and Q_4 provide sufficient breakdown voltage to fully protect the input stage of the amplifier, even under a worst-case fault condition, such as that depicted in Fig. 15.91(a).

Q-Point Analysis

In the 741 input stage in Fig. 15.94, the current mirror formed by transistors Q_8 and Q_9 operates with transistors Q_1 to Q_4 to establish the bias currents for the input stage. Bias current I_1 represents the output of the Widlar source discussed previously (18 μA) and must be equal to the collector current of Q_8 plus the base currents of matched transistors Q_3 and Q_4:

$$I_1 = I_{C8} + I_{B3} + I_{B4} = I_{C8} + 2I_{B4} \tag{15.217}$$

For high current gain, the base currents are small and $I_{C8} \cong I_1$.

The collector current of Q_8 mirrors the collector currents of Q_1 and Q_2, which are summed together in mirror reference transistor Q_9. Assuming high current gain and ignoring the collector-voltage mismatch between Q_7 and Q_8,

$$I_{C8} = I_{C1} + I_{C2} = 2I_{C2} \tag{15.218}$$

Combining Eqs. (15.217) and (15.218) yields the ideal bias relationships for the input stage

$$I_{C1} = I_{C2} \cong \frac{I_1}{2} \quad \text{and} \quad I_{C3} = I_{C4} \cong \frac{I_1}{2} \tag{15.219}$$

because the emitter currents of Q_1 and Q_3 and Q_2 and Q_4 must be equal. The collector current of Q_3 establishes a current equal to $I_1/2$ in current mirror transistors Q_5 and Q_6. Thus, transistors Q_1 to Q_6 all operate at a nominal collector current equal to one-half the value of source I_1.

Now that we understand the basic ideas behind the input stage bias circuit, let us perform a more exact analysis. Expanding Eq. (15.217) using the current mirror expression from Eq. (15.137),

$$I_1 = 2I_{C2} \frac{1 + \dfrac{V_{EC8}}{V_{A8}}}{1 + \dfrac{2}{\beta_{FO8}} + \dfrac{V_{EB8}}{V_{A8}}} + 2I_{B4} \tag{15.220}$$

I_{C2} is related to I_{B4} through the current gains of Q_2 and Q_4:

$$I_{C2} = \alpha_{F2}I_{E2} = \alpha_{F2}(\beta_{FO4} + 1)I_{B4} = \frac{\beta_{FO2}}{\beta_{FO2} + 1}(\beta_{FO4} + 1)I_{B4} \tag{15.221}$$

Combining Eqs. (15.220) and (15.221) and solving for I_{C2} yields

$$I_{C1} = \frac{I_1}{2} \times \left[\frac{1 + \dfrac{V_{EC8}}{V_{A8}}}{1 + \dfrac{2}{\beta_{FO8}} + \dfrac{V_{EB8}}{V_{A8}}} + \frac{1}{\dfrac{\beta_{FO2}}{\beta_{FO2} + 1}(\beta_{FO4} + 1)} \right]^{-1} \tag{15.222}$$

which is equal to the ideal value of $I_1/2$ but reduced by the nonideal current mirror effects because of finite current gain and Early voltage.

The emitter current of Q_4 must equal the emitter current of Q_2, and so the collector current of Q_4 is

$$I_{C4} = \alpha_{F4}I_{E4} = \alpha_{F4}\frac{I_{C2}}{\alpha_{F2}} = \frac{\beta_{FO4}}{\beta_{FO4} + 1}\frac{\beta_{FO2} + 1}{\beta_{FO2}}I_{C2} \tag{15.223}$$

The use of buffer transistor Q_7 essentially eliminates the current gain defect in the current mirror. Note from the full amplifier circuit in Fig. 15.92 that the base current of transistor Q_{10}, with its 50-kΩ emitter resistor R_4, is designed to be approximately equal to the base current of Q_7, and $V_{CE6} \cong V_{CE5}$ as well. Thus, the current mirror ratio is quite accurate and $I_{C5} = I_{C6} = I_{C3} \cong I_1/2$.

If 50-kΩ resistor R_3 were omitted, then the emitter current of Q_7 would be equal only to the sum of the base currents of transistors Q_5 and Q_6 and would be quite small. Because of the Q-point dependence of β_F, the current gain of Q_7 would be poor. R_3 increases the operating current of Q_7 to improve its current gain, as well as to improve the dc balance and transient response of the amplifier. The value of R_3 is chosen to approximately match I_{B7} to I_{B10}.

To complete the Q-point analysis, the various collector-emitter voltages must be determined. The collectors of Q_1 and Q_2 are $1V_{EB}$ below the positive power supply, whereas the emitters are $1V_{BE}$ below ground potential. Hence,

$$V_{CE1} = V_{CE2} = V_{CC} - V_{EB9} + V_{BE2} \cong V_{CC} \tag{15.224}$$

The collector and emitter of Q_3 are approximately $2V_{BE}$ above the negative power supply voltage and $1V_{BE}$ below ground, respectively:

$$V_{EC3} = V_{E3} - V_{C3} = -0.7 \text{ V} - (-V_{EE} + 1.4 \text{ V}) = V_{EE} - 2.1 \text{ V} \tag{15.225}$$

The buffered current mirror effectively minimizes the error due to the finite current gain of the transistors, and $V_{CE6} = V_{CE5} \cong 2V_{BE} = 1.4$ V, neglecting the small voltage drop (<10 mV) across R_1 and R_2. Finally, the collector of Q_8 is $2V_{BE}$ below zero, and the emitter of Q_7 is $1V_{BE}$ above $-V_{EE}$:

$$V_{EC8} = V_{CC} + 1.4 \text{ V} \qquad V_{CE7} = V_{EE} - 0.7 \text{ V} \qquad (15.226)$$

EXAMPLE **15.13**   µA741 INPUT STAGE BIAS CURRENTS

Find the currents in the 741 input stage.

PROBLEM   Calculate the bias currents in the 741 input stage if $I_1 = 18$ µA, $\beta_{FOnpn} = 150$, $V_{Anpn} = 75$ V, $\beta_{FOpnp} = 60$, $V_{Apnp} = 60$ V, and $V_{CC} = V_{EE} = 15$ V.

SOLUTION   **Known Information and Given Data:** µA741 input stage depicted in Fig. 15.94. $I_1 = 18$µA, $\beta_{FOnpn} = 150$, $V_{Anpn} = 75$ V, $\beta_{FOpnp} = 60$, $V_{Apnp} = 60$ V, and $V_{CC} = V_{EE} = 15$ V.

Unknowns: I_{C1}, I_{C2}, I_{C3}, I_{C4}, I_{C5}, and I_{C6}

Approach: Use given data to evaluate Eqs. (15.222) through (15.226).

Assumptions: Transistors are in the active region; use default values of I_S.

Analysis: From Fig. 15.94, we find that the emitter-collector voltage of Q_8 is equal to $V_{CC} + V_{BE1} + V_{EB3} \cong 16.4$ V. Substituting the known values into Eq. (15.222) gives

$$I_{C1} = I_{C2} = \frac{18 \text{ µA}}{2} \cdot \cfrac{1}{\cfrac{1 + \cfrac{16.4 \text{ V}}{60 \text{ V}}}{1 + \cfrac{2}{50} + \cfrac{0.7 \text{ V}}{60 \text{ V}}} + \cfrac{1}{\cfrac{150}{150 + 1}(60 + 1)}} = 7.32 \text{ µA}$$

Equation (15.223) yields

$$I_{C3} = I_{C4} = \alpha_{F4}\frac{I_{C2}}{\alpha_{F2}} = \frac{\beta_{FO4}}{\beta_{FO4} + 1}\left(\frac{\beta_{FO2} + 1}{\beta_{FO2}}\right)I_{C2} = \frac{60}{61}\left(\frac{151}{150}\right)I_{C2} = 7.25 \text{ µA}$$

$$I_{C5} \cong I_{C3} = 7.25 \text{ µA} \qquad \text{and} \qquad I_{C6} = I_{C4} = 7.25 \text{ µA}$$

Check of Results: The basic objective of the bias circuit would be to set all currents to 18 µA/2 or 9 µA. Our calculations are close to this value and appear correct.

Discussion: The actual bias currents are slightly greater than 7 µA, whereas the ideal value would be 9 µA. The dominant source of error arises from the collector-emitter voltage mismatch of the *pnp* current mirror.

Computer-Aided Analysis: We draw the circuit using the schematic editor and set the BJT parameters. For the *npn* devices, BF = 150 and VAF = 75 V. For the *pnp* transistors, BF = 60 and VAF = 60 V. Source V_O is added to balance the circuit by forcing the output voltage to the same voltage as that which will appear at the collector of Q_5. Otherwise, the voltage at the collectors of Q_4 and Q_6 will float to a value determined by the difference in overall output resistances of transistors Q_4 and Q_6. When balance is achieved, the current in source V_O will be nearly zero. Table 15.5 summarizes the Q-points based on these calculations and Eqs. (15.219) to (15.226) and compares them with the SPICE operating point simulation results.

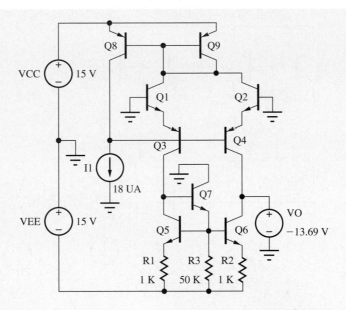

TABLE 15.5
Q-points of 741 Input Stage Transistors for $I_1 = 18$ μA and $V_{CC} = V_{EE} = 15$ V

TRANSISTORS	Q-POINT	SPICE RESULTS
Q_1 and Q_2	7.32 μA, 15 V	7.30 μA, 15.0 V
Q_3 and Q_4	7.25 μA, 12.9 V	7.24 μA, 13.0 V
Q_5 and Q_6	7.25 μA, 1.4 V	7.16 μA, 1.30 V
Q_7	12.2 μA, 14.3 V	13.1 μA, 14.3 V
Q_8	17.7 μA, 16.4 V	17.8 μA, 16.3 V
Q_9	14.0 μA, 0.7 V	14.1 μA, 0.66 V

EXERCISE: Remove V_O and simulate the 741 input stage amplifier. What are the new collector currents? What are the voltages at the collectors or Q_5 and Q_6?

ANSWERS: 7.31 μA, 7.28 μA, 7.25 μA, 7.22 μA, 7.18 μA, 7.22 μA, 13.1 μA, 17.8 μA, 14.1 μA; −13.7 V, −13.1 V

EXERCISE: Suppose buffer transistor Q_7 and resistor R_3 are eliminated from the amplifier in Fig. 15.94 and Q_5 and Q_6 were connected as a standard current mirror. What would be the collector-emitter voltage of Q_6 if $V_{BE6} = 0.7$ V, $\beta_{FO6} = 100$, and $V_{A6} = 60$ V? Use Eq. (15.196).

ANSWER: 1.90 V

15.11.3 ac ANALYSIS OF THE 741 INPUT STAGE

The 741 input stage is redrawn in symmetric form in Fig. 15.95, with its active load temporarily replaced by two resistors. From Fig. 15.95, we see that the collectors of Q_1 and Q_2 as well as the bases of Q_3 and Q_4, lie on the line of symmetry of the amplifier and represent virtual grounds for differential-mode input signals.

The corresponding differential-mode half-circuit shown in Fig. 15.96 is a common-collector stage followed by a common-base stage, a C-C/C-B cascade. The characteristics of the C-C/C-B cascade can be determined from Fig. 15.96 and our knowledge of single-stage amplifiers.

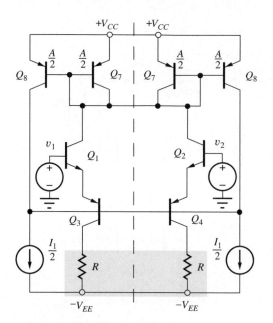

Figure 15.95 Symmetry in the 741 input stage.

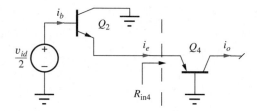

Figure 15.96 Differential-mode half-circuit for the 741 input stage.

The emitter current of Q_2 is equal to its base current i_b multiplied by $(\beta_{o2} + 1)$, and the collector current of Q_4 is α_{o4} times the emitter current. Thus, the output current can be written as

$$\mathbf{i_o} = \alpha_{o4}\mathbf{i_e} = \alpha_{o4}(\beta_{o2} + 1)\mathbf{i_b} \cong \beta_{o2}\mathbf{i_b} \tag{15.227}$$

The base current is determined by the input resistance to Q_2:

$$\mathbf{i_b} = \frac{\dfrac{\mathbf{v_{id}}}{2}}{r_{\pi 2} + (\beta_{o2} + 1)R_{\text{in4}}} = \frac{\dfrac{\mathbf{v_{id}}}{2}}{r_{\pi 2} + (\beta_{o2} + 1)\left(\dfrac{r_{\pi 4}}{\beta_{o4} + 1}\right)} = \frac{\dfrac{\mathbf{v_{id}}}{2}}{r_{\pi 2} + r_{\pi 4}} \cong \frac{\mathbf{v_{id}}}{4r_{\pi 2}} \tag{15.228}$$

in which $R_{\text{in4}} = r_{\pi 4}/(\beta_{o4} + 1)$ represents the input resistance of the common-base stage. Combining Eqs. (15.227) and (15.228) yields

$$\mathbf{i_o} \cong \beta_{o2}\frac{\mathbf{v_{id}}}{4r_{\pi 2}} = \frac{g_{m2}}{4}\mathbf{v_{id}} \tag{15.229}$$

Each side of the C-C/C-B input stage has a transconductance equal to one-half of the transconductance of the standard differential pair. From Eq. (15.228) we can also see that the differential-mode input resistance is twice the value of the corresponding C-E stage:

$$R_{id} = \frac{\mathbf{v_{id}}}{\mathbf{i_b}} = 4r_{\pi 2} \tag{15.230}$$

From Fig. 15.97, we can see that the output resistance is equivalent to that of a common-base stage with a resistor of value $1/g_{m2}$ in its emitter:

$$R_{\text{out}} \cong r_{o4}(1 + g_{m4}R) = r_{o4}\left(1 + g_{m4}\frac{1}{g_{m2}}\right) = 2r_{o4} \tag{15.231}$$

15.11.4 VOLTAGE GAIN OF THE COMPLETE AMPLIFIER

We now use the results from the previous section to analyze the overall ac performance of the op amp. We find a Norton equivalent circuit for the input stage and then couple it with a two-port model for the second stage.

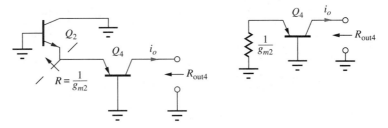

Figure 15.97 Output resistance of C-C/C-B cascade.

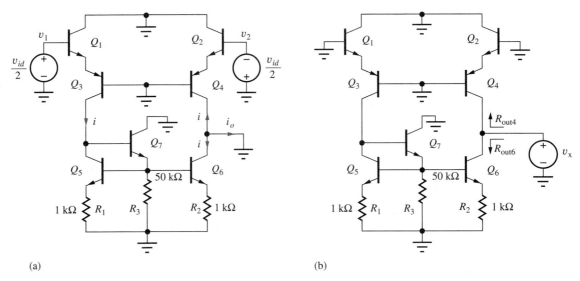

(a)

(b)

Figure 15.98 Circuits for finding the Norton equivalent of the input stage.

Norton Equivalent of the Input Stage

Figure 15.98 is the simplified differential-mode ac equivalent circuit for the input stage. We use Figure 15.98(a) to find the short-circuit output current of the first stage. Based on our analysis of Fig. 15.96, the differential-mode input signal establishes equal and opposite currents in the two sides of the differential amplifier where $\mathbf{i} = (g_{m2}/4)\mathbf{v_{id}}$. Current i, exiting the collector of Q_3, is mirrored by the buffered current mirror so that a total signal current equal to $2i$ flows in the output terminal:

$$\mathbf{i_o} = -2\mathbf{i} = -\frac{g_{m2}\mathbf{v_{id}}}{2} = (-20I_{C2})\mathbf{v_{id}}$$

$$= -\left(\frac{20}{V}7.32 \times 10^{-6}\ \text{A}\right)\mathbf{v_{id}} = (-1.46 \times 10^{-4}\ \text{S})\mathbf{v_{id}} \tag{15.232}$$

The Thévenin equivalent resistance at the output is found using the circuit in Fig. 15.98(b) and is equal to

$$R_{\text{th}} = R_{\text{out6}} \| R_{\text{out4}} \tag{15.233}$$

Because only a small dc voltage is developed across R_2, the output resistance of Q_6 can be calculated from

$$R_{\text{out6}} \cong r_{o6}[1 + g_{m6}R_2] \cong r_{o6}\left[1 + \frac{I_{C6}R_2}{V_T}\right] = r_{o6}\left[1 + \frac{0.0073\ \text{V}}{0.025\ \text{V}}\right] = 1.3r_{o6} \tag{15.234}$$

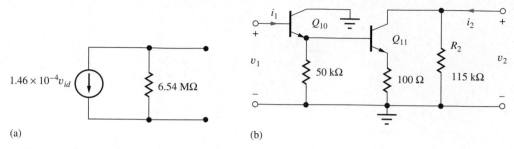

Figure 15.99 (a) Norton equivalent of the 741 input stage. (b) Two-port representation for the second stage.

The output resistance of Q_4 was already found in Eq. (15.231) to be $2r_{o4}$. Substituting the results from Eqs. (15.231) and (15.234) into Eq. (15.233),

$$R_{th} = 2r_{o4} \| 1.3r_{o6} = 0.79r_{o4} \cong 0.79 \frac{60 \text{ V}}{7.25 \times 10^{-6} \text{ A}} = 6.54 \text{ M}\Omega \qquad (15.235)$$

in which $r_{o4} = r_{o2}$ has been assumed for simplicity with $V_A + V_{CE} = 60$ V.

The resulting Norton equivalent circuit for the input stage appears in Fig. 15.99(a). Based on the values in this figure, the open-circuit voltage gain of the first stage is -955. SPICE simulations yield values very similar to those in Fig. 15.99(a): $(1.40 \times 10^{-4} \text{ S})\mathbf{v_{id}}$, 6.95 M$\Omega$, and $A_{dm} = -973$.

EXERCISE: Improve the estimate of R_{th} using the actual values of V_{CE6} and V_{CE4} if $V_{CC} = V_{EE} = 15$ V and $V_A = 60$ V. What are the values of R_{out4} and R_{out6}?

ANSWERS: 7.12 MΩ; 20.2 MΩ, 11.0 MΩ

Model for the Second Stage

Figure 15.99(b) is a two-port representation for the second stage of the amplifier. Q_{10} is an emitter follower that provides high input resistance and drives a common-emitter amplifier consisting of Q_{11} and its current source load represented by output resistance R_2. A y-parameter model is constructed for this network.

From Fig. 15.92 and the bias current analysis, we can see that the collector current of Q_{11} is approximately equal to I_2 or 666 μA. Calculating the collector current of Q_{10} yields

$$I_{C10} \cong I_{E10} = \frac{I_{C11}}{\beta_{F11}} + \frac{V_{B11}}{50 \text{ k}\Omega} = \frac{666 \text{ }\mu\text{A}}{150} + \frac{0.7 + (0.67 \text{ mA})(0.1 \text{ k}\Omega)}{50 \text{ k}\Omega} = 19.8 \text{ }\mu\text{A} \quad (15.236)$$

Using these values to find the small-signal parameters with ($\beta_{on} = 150$) gives

$$r_{\pi 10} = \frac{\beta_{o10} V_T}{I_{C10}} = \frac{3.75 \text{ V}}{19.8 \text{ }\mu\text{A}} = 189 \text{ k}\Omega \quad \text{and} \quad r_{\pi 11} = \frac{3.75 \text{ V}}{0.666 \text{ mA}} = 5.63 \text{ k}\Omega \quad (15.237)$$

Parameters y_{11} and y_{21} are calculated by applying a voltage v_1 to the input port and setting $v_2 = 0$, as in Fig. 15.100. The input resistance to Q_{11} is that of a common-emitter stage with a 100-Ω emitter resistor:

$$R_{in11} = r_{\pi 11} + (\beta_{o11} + 1)100 \cong 5630 + (151)100 = 20.7 \text{ k}\Omega \qquad (15.238)$$

This value is used to simplify the circuit, as in Fig. 15.100(b), and the input resistance to Q_{10} is

$$R_{in10} = r_{\pi 10} + (\beta_{o10} + 1)(50 \text{ k}\Omega \| R_{in11}) = 189 \text{ k}\Omega + (151)(50 \text{ k}\Omega \| 20.7 \text{ k}\Omega) = 2.40 \text{ M}\Omega \quad (15.239)$$

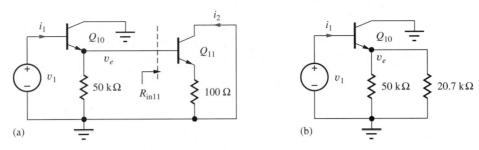

Figure 15.100 Network for finding y_{11} and y_{21}.

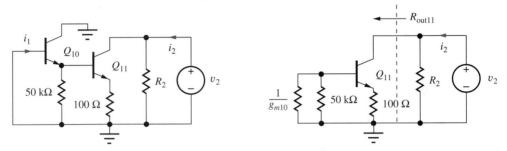

Figure 15.101 Network for finding y_{12} and y_{22}.

The gain of emitter follower Q_{10} is:

$$\mathbf{v_e} = \mathbf{v_1} \frac{(\beta_{o10} + 1)(50 \text{ k}\Omega \| R_{\text{in}11})}{r_{\pi 10} + (\beta_{o10} + 1)(50 \text{ k}\Omega \| R_{\text{in}11})} = \frac{(151)(50 \text{ k}\Omega \| 20.7 \text{ k}\Omega)}{189 \text{ k}\Omega + (151)(50 \text{ k}\Omega \| 20.7 \text{ k}\Omega)} = 0.921\mathbf{v_1} \tag{15.240}$$

The output current $\mathbf{i_2}$ in Fig. 15.99(a) is given by

$$\mathbf{i_2} = \frac{\mathbf{v_e}}{\dfrac{1}{g_{m11}} + 100 \ \Omega} = \frac{0.921\mathbf{v_1}}{\dfrac{1}{\dfrac{40}{\text{V}}(0.666 \text{ mA})} + 100 \ \Omega} = 0.00670\mathbf{v_1} \tag{15.241}$$

yielding a forward transconductance of

$$G_m = 6.70 \text{ mS} \tag{15.242}$$

Parameters y_{12} and y_{22} can be found from the network in Fig. 15.101. We assume that the reverse transconductance y_{12} is negligible and reserve its calculation for Prob. 15.247. The output conductance can be determined from Fig. 15.101(b).

$$G_o = i_2/v_2 = [R_2 \| R_{\text{out}11}]^{-1} \tag{15.243}$$

where $R_2 = 115 \text{ k}\Omega$ was calculated during the analysis of the bias circuit.

Because the voltage drop across the 100-Ω resistor is small, the output resistance of Q_{11} is approximately

$$R_{\text{out}11} = r_{o11}[1 + g_{m11}R_E] = \frac{V_{A11} + V_{CE11}}{I_{C11}} \left[1 + \frac{I_{C11}R_E}{V_T}\right] \tag{15.244}$$

$$= \frac{60 \text{ V} + 13.6 \text{ V}}{0.666 \text{ mA}} \left[1 + \frac{0.067 \text{ V}}{0.025 \text{ V}}\right] = 407 \text{ k}\Omega$$

and

$$R_o = 115 \text{ k}\Omega \| 407 \text{ k}\Omega = 89.1 \text{ k}\Omega \tag{15.245}$$

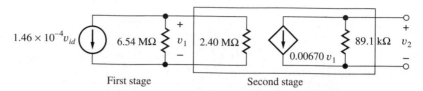

Figure 15.102 Combined model for first and second stages.

Figure 15.102 depicts the completed two-port model for the second stage, driven by the Norton equivalent of the input stage. Using this model, the open-circuit voltage gain for the first two stages of the amplifier is

$$\mathbf{v_2} = -0.00670(89.1 \text{ k}\Omega)\mathbf{v_1} = -597\mathbf{v_1}$$

$$\mathbf{v_1} = -1.46 \times 10^{-4}(6.54 \text{ M}\Omega \| 2.40 \text{ M}\Omega)\mathbf{v_{id}} = -256\mathbf{v_{id}} \qquad (15.246)$$

$$\mathbf{v_2} = -597(-256\mathbf{v_{id}}) = 153{,}000\mathbf{v_{id}}$$

Note from Eq. (15.245) that the 2.42-MΩ input resistance of Q_{10} reduces the voltage gain of the first stage by a factor of almost 4.

EXERCISE: What is the voltage gain of the input stage if transistor Q_{10} and its 50-kΩ emitter resistor are omitted so that the output of the first stage is connected directly to the base of Q_{11}? Use the small-signal element values already calculated.

ANSWER: −3.00

15.11.5 THE 741 OUTPUT STAGE

Figure 15.103 shows simplified models for the 741 output stage. Transistor Q_{12} is the emitter follower that buffers the high impedance node at the output of the second stage and drives the push-pull output stage composed of transistors Q_{15} and Q_{16}. Class-AB bias is provided by the sum of the base-emitter voltages of Q_{13} and Q_{14}, represented as diodes in Fig. 15.103(b). The 40-kΩ resistor is used to increase the value of I_{C13}. Without this resistor, I_{C13} would only be equal to the base current of Q_{14}. The short-circuit protection circuitry in Fig. 15.92 is not shown in Fig. 15.103 in order to simplify the diagram.

The input and output resistances of the class-AB output stage are actually complicated functions of the signal voltage because the operating current in Q_{15} and Q_{16} changes greatly as the output voltage changes. However, because only one transistor conducts strongly at any given time in the class-AB stage, separate circuit models can be used for positive and negative output signals. The model for positive signal voltages is shown in Fig. 15.104. (The model for negative signal swings is similar except npn transistor Q_{15} is replaced by pnp transistor Q_{16} connected to the emitter of Q_{12}.)

Let us first determine the input resistance of transistor Q_{12}. If R_{in12} is much larger than the 89-kΩ output resistance of the two-port in Fig. 15.104, then it does not significantly affect the overall voltage gain of the amplifier. Using single-stage amplifier theory,

$$R_{in12} = r_{\pi 12} + (\beta_{o12} + 1)R_{eq1} \qquad (15.247)$$

where

$$R_{eq1} = r_{d14} + r_{d13} + R_3 \| R_{eq2} \qquad \text{and} \qquad R_{eq2} = r_{\pi 15} + (\beta_{o15} + 1)R_L \cong (\beta_{o15} + 1)R_L \qquad (15.248)$$

The value of R_3 (344 kΩ) was calculated in the bias circuit section. For $I_{C12} = 216 \text{ μA}$, and assuming a representative collector current in Q_{15} of 2 mA,

$$R_{eq2} = r_{\pi 15} + (\beta_{o15} + 1)R_L = \frac{3.75 \text{ V}}{2 \text{ mA}} + (151)2 \text{ k}\Omega = 304 \text{ k}\Omega \qquad (15.249)$$

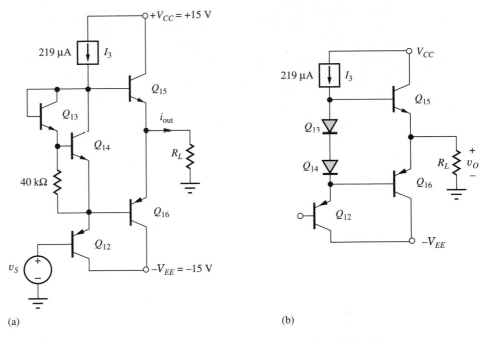

(a)   (b)

Figure 15.103 (a) 741 output stage without short-circuit protection. (b) Simplified output stage.

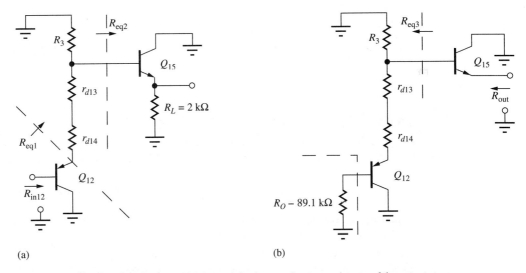

(a)   (b)

Figure 15.104 Circuits for determining input and output resistance of the output stage.

Note that the value of R_{eq2} is dominated by the reflected load resistance $\beta_{o15}R_L$. Resistor $r_{\pi 15}$ represents a small part of R_{eq2}, and knowing the exact value of I_{C15} is not critical.

$$R_{eq1} = r_{d14} + r_{d13} + R_3 \| R_{eq2} = 2\frac{0.025\ \text{V}}{0.216\ \text{mA}} + 344\ \text{k}\Omega \| 304\ \text{k}\Omega = 162\ \text{k}\Omega \qquad (15.250)$$

and

$$R_{in12} = r_{\pi 12} + (\beta_{o12} + 1)R_{eq1} = \frac{0.025\ \text{V}}{0.216\ \text{mA}} + (51)162\ \text{k}\Omega = 8.27\ \text{M}\Omega \qquad (15.251)$$

Because R_{in12} is approximately 100 times the output resistance R_o of the second stage, R_{in12} has little effect on the gain of the second stage. Although the value of R_{in12} changes for different values of load resistance, the overall op amp gain is not affected because the value of R_{in12} is so much larger than the value of R_o in Fig. 15.104.

Similar results are obtained for negative signal voltages. The values are slightly different because the current gain of the *pnp* transistor Q_{16} differs from that of the *npn* transistor Q_{15}.

15.11.6 OUTPUT RESISTANCE

The output resistance of the amplifier for positive output voltages is determined by transistor Q_{15}

$$R_o = \frac{r_{\pi 15} + R_{eq3}}{\beta_{o15} + 1} \tag{15.252}$$

in which

$$\begin{aligned} R_{eq3} &= R_3 \left\| \left[r_{d13} + r_{d14} + \frac{r_{\pi 12} + y_{22}^{-1}}{\beta_{o12} + 1} \right] \right. \\ &= 304\,\text{k}\Omega \left\| \left[2\frac{0.025\,\text{V}}{0.219\,\text{mA}} + \frac{5.71\,\text{k}\Omega + 89.1\,\text{k}\Omega}{51} \right] = 2.08\,\text{k}\Omega \right. \end{aligned} \tag{15.253}$$

From Fig. 15.95, we can see that the 27-Ω resistor R_7, which determines the short-circuit current limit, adds directly to the overall output resistance of the amplifier so that actual op-amp output resistance is

$$R_{out} = R_o + R_7 = \frac{1.88\,\text{k}\Omega + 2.08\,\text{k}\Omega}{151} + 27\,\Omega = 53\,\Omega \tag{15.254}$$

> **EXERCISE:** Repeat the calculation of R_{in12} and R_{out} if *pnp* transistor Q_{16} has a current gain of 50, $I_{C16} = 2$ mA, and $I_{C15} = 0$. Be sure to draw the new equivalent circuit of the output stage for negative output voltages.
>
> **ANSWERS:** 3.94 MΩ ($\gg$89.1 kΩ), 53 Ω + 22 Ω = 75 Ω

15.11.7 SHORT CIRCUIT PROTECTION

For simplicity, the output short-circuit protection circuitry was not shown in Fig. 15.103. Referring back to the complete op amp schematic in Fig. 15.92, we see that **short-circuit protection** is provided by resistors R_7 and R_8 and transistors Q_{17} and Q_{18}. The circuit is identical to the one presented in Fig. 15.39(a). Transistors Q_{17} and Q_{18} are normally off, but if the current in resistor R_7 becomes too high, then transistor Q_{17} turns on and steals the base current from Q_{15}. Likewise, if the current in resistor R_8 becomes too large, then transistor Q_{18} turns on and removes the base current from Q_{16}. The positive and negative short-circuit current levels will be limited to approximately V_{BE17}/R_7 and $-V_{EB18}/R_8$, respectively. As already mentioned, resistors R_7 and R_8 increase the output resistance of the amplifier since they appear directly in series with the output terminal.

> **EXERCISE:** Estimate the positive and negative short-circuit output current in the 741 op amp in Fig. 16.53.
>
> **ANSWERS:** 26 mA; −32 mA

15.11.8 SUMMARY OF THE μA741 OPERATIONAL AMPLIFIER CHARACTERISTICS

Table 15.6 is a summary of the characteristics of the μA741 operational amplifier. Column 2 gives our calculated values; column 3 presents values typically found in the actual commercial product.

TABLE 15.6
μA741 Characteristics

	CALCULATION	TYPICAL VALUES
Voltage gain	153,000	200,000
Input resistance	2.05 MΩ	2 MΩ
Output resistance	53 Ω	75 Ω
Input bias current	49 nA	80 nA
Input offset voltage	—	2 mV

The observed values depend on the exact values of current gain and Early voltage of the *npn* and *pnp* transistors and vary from process run to process run.

ELECTRONICS IN ACTION

Medical Ultrasound Imaging

Medical ultrasound imaging systems are widely used in clinical applications for many diagnostic procedures such as characterization of tumors, measurement of cardiac function, and monitoring of prenatal development. Ultrasound systems work by sending 1 to 20 MHz acoustic pulses into the body and then measuring the acoustic echo. Different types of tissue absorb different amounts of acoustic energy, so the acoustic return varies with tissue type and characteristic. In order to measure tissue properties at specific points within the body, a phased array technique is used to focus the transmit and receive pulses. For example, a simplified view of the receive process is depicted below. The acoustic propagation time of the reflected wave varies with the distance from a particular transducer element to the focus point, resulting in a set of received waves separated in time. By introducing the appropriate delays to the received waves, they can then be summed. Random noise will average out, but the signal of interest adds coherently. The transmit process is also focused by time-varying the pulses driven onto each of the transducer elements.

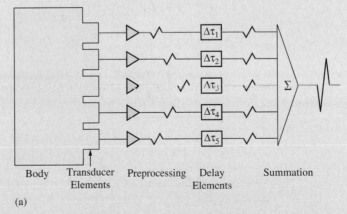

Simplified view of ultrasound receive focusing.

A more detailed look at the electronics of an ultrasound system is shown in the accompanying figures. Because of the lossy nature of the transducers and body tissue, the received ultrasound signal is extremely small, often on the order of microvolts. As a consequence, the analog preamp must be a very low noise, multistage amplifier. Total gain is about 100 dB. With such a high gain, it is important that the amplifier be either ac coupled or have some form

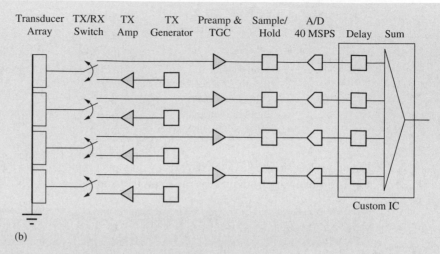

(b)

Block diagram of typical commercial ultrasound system transmit/receive electronics.

of offset-correction. For example, if the amplifier has an input offset of 5 mV, a gain of 100 dB would yield an output that is clipped. Another interesting aspect of ultrasound preamplifiers is the need for time gain control (TGC). As an ultrasound signal propagates through the body, it is heavily attenuated. The longer a signal propagates, the more it attenuates. This is compensated with a circuit that continuously varies the gain of the amplifier over a 60–80 dB range during the few microseconds required to receive an ultrasonic waveform.

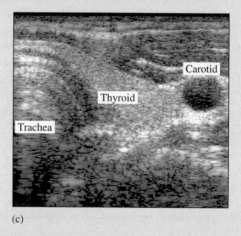

(c)

Ultrasound image of trachea, thyroid, and carotid artery*.

After the preamp, the signal is sampled and then digitized. In a typical 128-channel system with 40 MSample/sec 10-bit ADCs, the total data rate is 6.4 gigabytes/sec! This vast data pipeline is processed by several custom ASICs which digitally perform the real-time delay and summing operations, correcting for many non-idealities not described here.

Modern medical systems, such as the one shown here, are tremendous opportunities for innovative circuit design. As medical knowledge increases, it is increasingly important to accurately measure physiological responses and interactions to properly apply and utilize new understanding.

---

* Ultrasound image appears courtesy of William F. Walker, University of Virginia

SUMMARY

In most situations, the single-stage amplifiers discussed in Chapters 13 and 14 cannot simultaneously meet all the requirements of an application (e.g., high voltage gain, high input resistance, and low output resistance). Therefore, we must combine single-stage amplifiers in various ways to form multistage amplifiers that achieve higher levels of overall performance.

- Both ac- and dc-coupling (also called direct-coupling) methods are used in multistage amplifiers depending on the application. ac coupling allows the Q-point design of each stage to be done independently of the other stages, and bypass capacitors can be utilized to eliminate bias elements from the ac equivalent circuit of the amplifier. However, dc coupling can eliminate circuit elements, including both coupling capacitors and bias resistors, and can represent a more economical approach to design. In addition, direct coupling is required to achieve a low-pass amplifier that provides gain at dc, and dc-coupling is utilized in most op amp designs.

- The most important dc-coupled amplifier is the symmetric two-transistor differential amplifier. Not only is the differential amplifier a key circuit in the design of operational amplifiers, but it is also a fundamental building block of all analog circuit design. In this chapter, we studied BJT and MOS differential amplifiers in detail. Differential-mode gain, common-mode gain, common-mode rejection ratio (CMRR), and differential- and common-mode input and output resistances of the amplifier are all directly related to transistor parameters and, hence, Q-point design.

- Either a balanced or a single-ended output is available from the differential amplifier. The balanced output provides a voltage gain that is twice that of the single-ended output, and the CMRR of the balanced output is inherently much higher (infinity for the ideal case). A two-port model can be used to model the small-signal characteristics at the output of the differential pairs.

- One of the most important applications of differential amplifiers is to form the input stage of the operational amplifier. By adding a second gain stage plus an output stage to the differential amplifier, a basic op amp is created. The performance of differential and operational amplifiers can be greatly enhanced by the use of electronic current sources. Op amp designs usually require a number of current sources, and, for economy of design, these multiple sources are often generated from a single-bias voltage.

- An ideal current source provides a constant output current, independent of the voltage across the source; that is, the current source has an infinite output resistance. Although electronic current sources cannot achieve infinite output resistance, very high values are possible, and there are a number of basic current source circuits and techniques for achieving high output resistance.

- For a current source, the product of the source current and output resistance represents a figure of merit, V_{CS}, that can be used to compare current sources. A single-transistor current source can be built using the bipolar transistor in which V_{CS} can approach the $\beta_o V_A$ product of the BJT. For a very good bipolar transistor, this product can reach 10,000 V. For the FET case, V_{CS} can approach a significant fraction of $\mu_f V_{SS}$, in which V_{SS} represents the power supply voltage. Values well in excess of 1000 V are achievable with the FET source.

- The electronic current source can be modeled in SPICE as a dc current source in parallel with a resistor equal to the output resistance of the source. For greatest accuracy, the value of the dc source should be adjusted to account for any dc current existing in the output resistance.

- Class-A, Class-B, and Class-AB amplifiers are defined in terms of their conduction angles: 360° for Class-A, 180° for Class-B, and between 180° and 360° for Class-AB operation. The efficiency of the Class-A amplifier cannot exceed 25 percent for sinusoidal signals, whereas that of the Class-B amplifier has an upper limit of 78.5 percent. However, Class-B amplifiers suffer from cross-over distortion caused by a dead zone in the transfer characteristic.

- The Class-AB amplifier trades a small increase in quiescent power dissipation and a small loss in efficiency for elimination of the cross-over distortion. The efficiency of the Class-AB amplifier can approach that of the Class-B amplifier when the quiescent operating point is properly chosen. The basic op-amp design can be further improved by replacing the Class-A follower output stage with a Class-AB output stage. Class-AB output stages are often used in operational amplifiers and are usually provided with short-circuit protection circuitry.

- Amplifier stages may also employ transformer coupling. The impedance transformation properties of the transformer can be used to simplify the design of circuits that must drive low values of load resistances, such as loudspeakers.

- Integrated circuit (IC) technology permits the realization of large numbers of virtually identical transistors. Although the absolute parameter tolerances of these devices are relatively poor, device characteristics can actually be matched to within less than 1 percent. The availability of large numbers of such closely matched devices has led to the development of special circuit techniques that depend on the similarity of device characteristics for proper operation. These matched circuit design techniques are used throughout analog circuit design and produce high-performance circuits that require very few resistors.

- One of the most important of the IC techniques is the current mirror circuit, in which the output current replicates, or mirrors, the input current. Multiple copies of the replicated current can be generated, and the gain of the current mirror can be controlled by scaling the emitter areas of bipolar transistors or the W/L ratios of FETs. Errors in the mirror ratio of current mirrors are related directly to the finite output resistance and/or current gain of the transistors through the parameters λ, V_A, and β_F.

- In bipolar current mirrors, the finite current gain of the BJT causes an error in the mirror ratio, which the buffered current mirror circuit is designed to minimize. In both FET and BJT circuits, the ideal balance of the current mirror is disturbed by the mismatch in dc voltages between the input and output sections of the mirror. The degree of mismatch is determined by the output resistance of the current sources.

- The figure of merit V_{CS} for the basic current mirror is approximately equal to V_A for the BJT or $1/\lambda$ for the MOS version. However, the value of V_{CS} can be improved by up to two orders of magnitude through the use of either the cascode or Wilson current sources.

- Current mirrors can also be used to generate currents that are independent of the power supply voltages. The V_{BE}-based reference and the Widlar reference produce currents that depend only on the logarithm of the supply voltage. By combining a Widlar source with a current mirror, a reference is realized that exhibits first-order independence of the power supply voltages. The only variation is due to the finite output resistance of the current mirror and Widlar source used in the supply-independent cell.

- The Widlar cell produces a PTAT voltage (proportional to absolute temperature) which is used as the basic sensing element in most electronic thermometers.

- An extremely important application of the current mirror is as a replacement for the load resistors in differential and operational amplifiers. This active-load circuit can substantially enhance the voltage gain capability of most amplifiers while maintaining the operating-point balance necessary for low offset voltage and good common-mode rejection. Amplifiers with active loads can achieve single-stage voltage gains that approach the amplification factor of the transistor. Analysis of the ac behavior of circuits employing current mirrors can often be simplified using a two-port model for the mirror.

- Active current mirror loads are used to enhance the performance of both bipolar and MOS operational amplifiers. The classic μA741 operational amplifier, introduced in the late 1960s, was the first highly robust design combining excellent overall amplifier performance with input-stage breakdown-voltage protection and short-circuit protection of the output stage. Active loads are used to achieve a voltage gain in excess of 100 dB in an amplifier with two stages of gain. This operational amplifier design immediately became the industry standard op amp and spawned many similar designs.

KEY TERMS

ac-coupled amplifiers
Active load
Balanced output
Buffered current mirror
Cascode amplifier
Cascode current source
Class-A, class-B, and class-AB amplifiers
Class-B push-pull output stage
Common-mode conversion gain
Common-mode gain
Common-mode half-circuit
Common-mode input resistance
Common-mode input voltage range
Common-mode rejection ratio (CMRR)
Complementary push-pull output stage
Conduction angle
Cross-over distortion
Cross-over region
Current gain defect
Current-limiting circuit
Current mirror
Current sink
Darlington circuit
dc-coupled (direct-coupled) amplifiers
Dead zone
Differential amplifier
Differential-mode conversion gain
Differential-mode gain

Differential-mode half-circuit
Differential-mode input resistance
Differential-mode output resistance
Differential-mode output voltage
"Diode-connected" transistor
Electronic current source
Emitter area scaling
Figure of merit (FOM)
Half-circuit analysis
Ideal current source
Level shift
Matched (devices)
Matched transistors
μA741
Mirror ratio
Overvoltage protection
Power-supply independent biasing
Reference current
Short-circuit protection
Single-ended output
Start-up circuit
Transformer coupling
V_{BE}-based reference
Virtual ground
Voltage reference
Widlar current source
Wilson current source

REFERENCES

1. R. D. Thornton, et. al., *Multistage Transistor Circuits,* SEEC Volume 5, Wiley, New York: 1965.

2. P. R. Gray, P. J. Hurst, S. H. Lewis, and R. G. Meyer, *Analysis and Design of Analog Integrated Circuits,* 4th ed., John Wiley and Sons, New York: 2001.

3. R. J. Widlar, "Some circuit design techniques for linear integrated circuits," *IEEE Transactions on Circuit Theory,* vol. CT-12, no. 12, pp. 586–590, December 1965.

4. R. J. Widlar, "Design techniques for monolithic operational amplifiers," *IEEE Journal of Solid-State Circuits,* vol. SC-4, no. 4, pp. 184–191, August 1969.

5. G. R. Wilson, "A monolithic junction FET-NPN operational amplifier," *IEEE Journal of Solid-State Circuits,* vol. SC-3, no. 6, pp. 341–348, December 1968.

6. Robert J. Widlar, "New developments in IC voltage regulators," *IEEE Journal of Solid-State Circuits,* vol. SC-6, no. 1, pp. 2–7, January 1991.

7. A. Paul Brokaw, "A simple three-terminal IC bandgap reference," *IEEE Journal of Solid-State Circuits,* vol. SC-9, no. 6, pp. 388–393, December 1994.

ADDITIONAL READING

R. C. Jaeger, "A high output resistance current source." *IEEE JSSC,* vol. SC-9, pp. 192–194, August 1974.

R. C. Jaeger, "Common-mode rejection limitations of differential amplifiers." *IEEE JSSC,* vol. SC-11, pp. 411–417, June 1976.

R. C. Jaeger, and G. A. Hellwarth. "On the performance of the differential cascode amplifier." *IEEE JSSC,* vol. SC-8, pp. 169–174, April 1973.

PROBLEMS

Unless otherwise specified, use $\beta_F = 100$, $V_A = 70$ V, $K_p = K_n = 1$ mA/V^2, $V_{TN} = -V_{TP} = 1$ V, and $\lambda = 0.02$ V^{-1}.

15.1 Differential Amplifiers

BJT Amplifiers

15.1. (a) What are the Q-points for the transistors in the amplifier in Fig. P15.1 if $V_{CC} = 12$ V, $V_{EE} = 12$ V, $R_{EE} = 270$ kΩ, $R_C = 330$ kΩ, and $\beta_F = 100$? (b) What are the differential-mode gain, and differential-mode input and output resistances? (c) What are the common-mode gain, CMRR, and common-mode input resistance for a single-ended output?

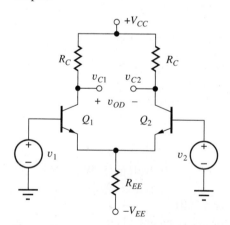

Figure P15.1

15.2. (a) What are the Q-points for the transistors in the amplifier in Fig. P15.1 if $V_{CC} = 1.5$ V, $V_{EE} = 1.5$ V, $\beta_F = 60$, $R_{EE} = 75$ kΩ, and $R_C = 100$ kΩ? (b) What are the differential-mode gain, common-mode gain, CMRR, and differential-mode and common-mode input and output resistances?

15.3. (a) Use SPICE to simulate the amplifier in Prob. P15.1 at a frequency of 1 kHz, and determine the differential-mode gain, common-mode gain, CMRR, and differential-mode and common-mode input resistances. (b) Apply a 25 mV, 1 kHz sine wave as an input signal, and plot the output signals using SPICE transient analysis. Use the SPICE distortion analysis capability to find the harmonic distortion in the output.

15.4. (a) What are the Q-points for the transistors in the amplifier in Fig. P15.1 if $V_{CC} = 18$ V, $V_{EE} = 18$ V, $R_{EE} = 47$ kΩ, $R_C = 100$ kΩ, and $\beta_F = 100$? (b) What are the differential-mode gain, common-mode gain, CMRR, and differential-mode and common-mode input and output resistances?

*15.5. (a) Use the common-mode gain to find voltages v_{C1}, v_{C2}, and v_{OD} for the differential amplifier in Fig. P15.1 if $V_{CC} = 12$ V, $V_{EE} = 12$ V, $R_{EE} = 270$ kΩ, $R_C = 390$ kΩ, $v_1 = 5.000$ V, and $v_2 = 5.000$ V. (b) Find the Q-points of the transistors directly with V_{IC} applied. Recalculate v_{c1} and v_{c2} and compare to the results in part (a). What is the origin of the discrepancy?

15.6. Design a differential amplifier to have a differential gain of 58 dB and $R_{ID} = 100$ kΩ using the topology in Fig. P15.1, with $V_{CC} = V_{EE} = 9$ V and $\beta_F = 120$. (Be sure to check feasibility of the design using our rule-of-thumb estimates from Chapter 13 before you move deeper into the design calculations.)

15.7. Design a differential amplifier to have a differential gain of 46 dB and $R_{ID} = 1$ MΩ using the topology in Fig. P15.1, with $V_{CC} = V_{EE} = 12$ V and $\beta_F = 100$. (Be sure to check feasibility of the design using our rule-of-thumb estimates from Chapter 13 before you move deeper into the design calculations.)

15.8. (a) What are the Q-points for the transistors in the amplifier in Fig. P15.8 if $V_{CC} = 12$ V, $V_{EE} = 12$ V, $I_{EE} = 400$ μA, $\beta_F = 100$, $R_{EE} = 200$ kΩ, $R_C = 39$ kΩ, $V_A = \infty$, and $\beta_F = 100$? (b) What are the differential-mode gain, common-mode gain, CMRR, and differential-mode and common-mode input and output resistances? (c) Repeat part (b) for $V_A = 50$ V.

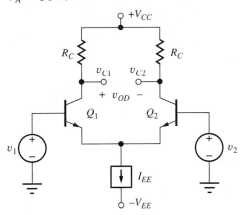

Figure P15.8

*15.9. What are the voltages v_{C1}, v_{C2}, and v_{OD} for the differential amplifier in Fig. P15.8 if $V_{CC} = 12$ V, $V_{EE} = 12$ V, $\beta_F = 75$, $I_{EE} = 400$ μA, $R_{EE} = 200$ kΩ, $R_C = 39$ kΩ, $v_1 = 2.005$ V, and $v_2 = 1.995$ V? What is the common-mode input range of this amplifier?

15.10. What is the value of the current I_{EE} required to achieve $R_{id} = 5$ MΩ in the circuit in Fig. P15.8 if $\beta_o = 100$? What output resistance R_{EE} is required for CMRR = 100 dB?

15.11. For the amplifier in Fig. P15.11, $V_{CC} = 10$ V, $V_{EE} = 10$ V, $\beta_F = 100$, $I_{EE} = 20$ μA, and $R_C = 910$ kΩ. (a) What are the output voltages v_o and V_O for the amplifier for $v_s = 0$ V and $v_s = 2$ mV? (b) What is the maximum value of v_s?

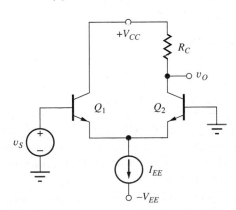

Figure P15.11

15.12. For the amplifier in Fig. P15.11, $V_{CC} = 12$ V, $V_{EE} = 12$ V, $\beta_F = 120$, $I_{EE} = 200$ μA, and

$R_C = 110$ kΩ. (a) What are the output voltages V_O and v_o for the amplifier for $v_s = 0$ V and $v_s = 1$ mV? (b) What is the maximum value of v_s?

15.13. (a) Use SPICE to simulate the amplifier in Prob. 15.12 at a frequency of 1 kHz, and determine the differential-mode gain, common-mode gain, CMRR, and differential-mode and common-mode input resistances. Use $V_A = 60$ V. (b) Apply a 25-mV, 1-kHz sine wave as an input signal and plot the output signal using SPICE transient analysis. Use the SPICE distortion analysis capability to find the harmonic distortion in the output.

15.14. (a) What are the Q-points for the transistors in the amplifier in Fig. P15.14 if $V_{CC} = 15$ V, $V_{EE} = 15$ V, $\beta_F = 150$, $R_{EE} = 150$ kΩ, and $R_C = 200$ kΩ? (b) What are the differential-mode gain, common-mode gain, CMRR, and differential-mode and common-mode input resistances?

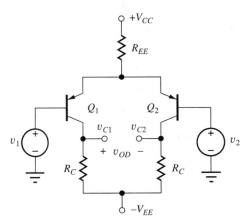

Figure P15.14

15.15. What are the voltages v_{C1}, v_{C2}, and v_{OD} for the differential amplifier in Fig. P15.14 if $V_{CC} = 10$ V, $V_{EE} = 10$ V, $\beta_F = 100$, $R_{EE} = 430$ kΩ, $R_C = 560$ kΩ, $v_1 = 1$ V, and $v_2 = 0.99$ V?

15.16. Use SPICE to simulate the amplifier in Prob. 15.15 at a frequency of 5 kHz, and determine the differential-mode gain, common-mode gain, CMRR, and differential-mode and common-mode input resistances.

15.17. (a) What are the Q-points for the transistors in the amplifier in Fig. P15.17 if $V_{CC} = 3$ V, $V_{EE} = 3$ V, $\beta_F = 80$, $I_{EE} = 10$ μA, $R_{EE} = 5$ MΩ, and $R_C = 390$ kΩ? (b) What are the differential-mode gain, common-mode gain, CMRR, differential-mode and common-mode input resistances, and common-mode input range?

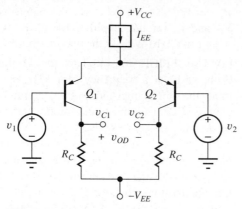

Figure P15.17

15.18. What are the voltages v_{C1}, v_{C2}, and v_{OD} for the differential amplifier in Fig. P15.17 if $V_{CC} = 22$ V, $V_{EE} = 22$ V, $\beta_F = 120$, $I_{EE} = 1$ mA, $R_{EE} = 500$ kΩ, $R_C = 15$ kΩ, $v_1 = 0.01$ V, and $v_2 = 0$ V?

*15.19. The differential amplifier in Fig. P15.19 has mismatched collector resistors. Calculate A_{dd}, A_{cd}, and the CMRR of the amplifier if the output is the differential output voltage v_{od}, and $R = 100$ kΩ, $\Delta R/R = 0.01$, $V_{CC} = V_{EE} = 15$ V, $R_{EE} = 100$ kΩ, and $\beta_F = 100$.

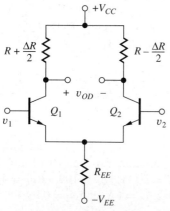

Figure P15.19

15.20. Use SPICE to simulate the amplifier in Prob. P15.19 at a frequency of 100 Hz, and determine the differential-mode gain, common-mode gain, and CMRR.

**15.21. The transistors in the differential amplifier in Fig. P15.21 have mismatched transconductances. Calculate A_{dd}, A_{cd}, and the CMRR of the amplifier if the output is the differential output voltage v_{OD}, and $R = 100$ kΩ, $g_m = 3$ mS, $\Delta g_m/g_m = 0.01$, $V_{CC} = V_{EE} = 15$ V, and $R_{EE} = 100$ kΩ.

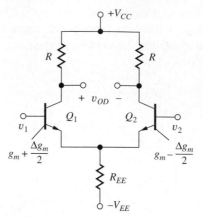

Figure P15.21

FET Differential Amplifiers

15.22. (a) What are the Q-points for the transistors in the amplifier in Fig. P15.22 if $V_{DD} = 12$ V, $V_{SS} = 12$ V, $R_{SS} = 220$ kΩ, and $R_D = 330$ kΩ? Assume $K_n = 400$ μA/V² and $V_{TN} = 1$ V. (b) What are the differential-mode gain, common-mode gain, CMRR, and differential-mode and common-mode input resistances?

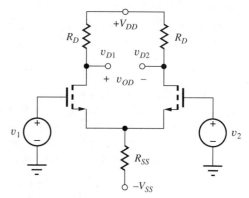

Figure P15.22

15.23. (a) What are the Q-points for the transistors in the amplifier in Fig. P15.22 if $V_{DD} = 15$ V, $V_{SS} = 15$ V, $R_{SS} = 62$ kΩ, and $R_D = 62$ kΩ? Assume $K_n = 400$ μA/V² and $V_{TN} = 1$ V. (b) What are the differential-mode gain, common-mode gain, CMRR, and differential-mode and common-mode input resistances?

15.24. (a) Use SPICE to simulate the amplifier in Prob. 15.22 at a frequency of 1 kHz, and determine the differential-mode gain, common-mode gain, CMRR, and differential-mode and common-mode input resistances. (b) Apply a 250-mV, 1-kHz sine wave as an input signal and plot the output signals using SPICE transient analysis. Use the SPICE distortion analysis capability to find the harmonic distortion in the output.

15.25. Design a differential amplifier to have a differential-mode output resistance of 5 kΩ and $A_{dm} = 20$ dB, using the circuit in Fig. P15.22 with $V_{DD} = V_{SS} = 5$ V. Assume $V_{TN} = 1$ V and $K_n = 25$ mA/V^2.

*15.26. (a) What are the Q-points for the transistors in the amplifier in Fig. P15.26 if $V_{DD} = 15$ V, $V_{SS} = 15$ V, $R_{SS} = 62$ kΩ, and $R_D = 62$ kΩ? Assume $K_n = 400$ μA/V^2, $\gamma = 0.75$ V$^{0.5}$, $2\phi_F = 0.6$ V, and $V_{TO} = 1$ V. (b) What are the differential-mode gain, common-mode gain, CMRR, and differential-mode and common-mode input resistances? (c) What would the Q-points be if $\gamma = 0$?

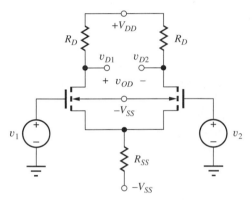

Figure P15.26

15.27. (a) Use SPICE to simulate the amplifier in Prob. 15.26 at a frequency of 1 kHz, and determine the differential-mode gain, common-mode gain, CMRR, and differential-mode and common-mode input resistances. (b) Apply a 250-mV, 1-kHz sine wave as an input signal and plot the output signal using SPICE transient analysis. Use the SPICE distortion analysis capability to find the harmonic distortion in the output.

*15.28. (a) What are the Q-points for the transistors in the amplifier in Fig. P15.26 if $V_{DD} = 12$ V, $V_{SS} = 12$ V, $R_{SS} = 220$ kΩ, and $R_D = 330$ kΩ? Assume $K_n = 400$ μA/V^2, $\gamma = 0.75$ V$^{0.5}$, $2\phi_F = 0.6$ V, and $V_{TO} = 1$ V. (b) What are the differential-mode gain, common-mode gain, CMRR, and differential-mode and common-mode input resistances? (c) What would the Q-points be if $\gamma = 0$?

15.29. (a) What are the Q-points for the transistors in the amplifier in Fig. P15.29 if $V_{DD} = 9$ V,

$V_{SS} = 9$ V, $I_{SS} = 40$ μA, $R_{SS} = 1.25$ MΩ, and $R_D = 300$ kΩ? Assume $K_n = 400$ μA/V^2 and $V_{TN} = 1$ V. (b) What are the differential-mode gain, common-mode gain, CMRR, and differential-mode and common-mode input resistances?

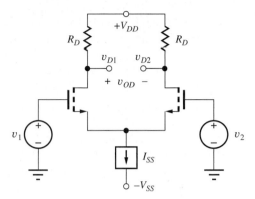

Figure P15.29

15.30. (a) What are the Q-points for the transistors in the amplifier in Fig. P15.29 if $V_{DD} = 15$ V, $V_{SS} = 15$ V, $I_{SS} = 300$ μA, $R_{SS} = 160$ kΩ, and $R_D = 75$ kΩ? Assume $K_n = 400$ μA/V^2 and $V_{TN} = 1$ V. (b) What are the differential-mode gain, common-mode gain, CMRR, and differential-mode and common-mode input resistances?

*15.31. (a) What are the Q-points for the transistors in the amplifier in Fig. P15.31 if $V_{DD} = 9$ V, $V_{SS} = 9$ V, $I_{SS} = 40$ μA, $R_{SS} = 1.25$ MΩ, and $R_D = 300$ kΩ? Assume $K_n = 400$ μA/V^2, $\gamma = 0.75$V$^{0.5}$, $2\phi_F = 0.6$ V, and $V_{TO} = 1$ V. (b) What are the differential-mode gain, common-mode gain, CMRR, and differential-mode and common-mode input resistances?

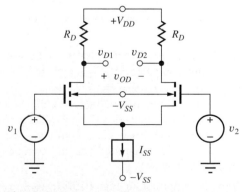

Figure P15.31

*15.32. (a) What are the Q-points for the transistors in the amplifier in Fig. P15.31 if $V_{DD} = 15$ V, $V_{SS} = 15$ V,

$I_{SS} = 300$ μA, $R_{SS} = 160$ kΩ, and $R_D = 75$ kΩ? Assume $K_n = 400$ μA/V^2, $\gamma = 0.75$ V$^{0.5}$, $2\phi_F = 0.6$ V, and $V_{TO} = 1$ V. (b) What are the differential-mode gain, common-mode gain, CMRR, and differential-mode and common-mode input resistances?

15.33. Design a differential amplifier to have a differential-mode gain of 30 dB, using the circuit in Fig. P15.29 with $V_{DD} = V_{SS} = 7.5$ V. The circuit should have the maximum possible common-mode input range. Assume $V_{TN} = 1$ V and $K_n = 5$ mA/V^2.

15.34. (a) What are the Q-points for the transistors in the amplifier in Fig. P15.34 if $V_{DD} = 18$ V, $V_{SS} = 18$ V, $R_{SS} = 56$ kΩ, and $R_D = 91$ kΩ? Assume $K_p = 200$ μA/V^2 and $V_{TP} = -1$ V. (b) What are the differential-mode gain, common-mode gain, CMRR, and differential-mode and common-mode input resistances?

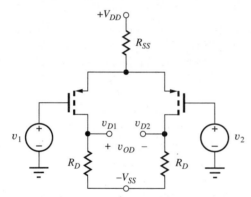

Figure P15.34

15.35. Use SPICE to simulate the amplifier in Prob. 15.34 at a frequency of 3 kHz, and determine the differential-mode gain, common-mode gain, CMRR, and differential-mode and common-mode input resistances.

*15.36. (a) What are the Q-points for the transistors in the amplifier in Fig. P15.36 if $V_{DD} = 10$ V, $V_{SS} = 10$ V, $I_{SS} = 40$ μA, $R_{SS} = 1.25$ MΩ, and $R_D = 300$ kΩ? Assume $K_p = 200$ μA/V^2, $\gamma = 0.6$ V$^{0.5}$, $2\phi_F = 0.6$ V, and $V_{TO} = -1$ V. (b) What are the differential-mode gain, common-mode gain, CMRR, and differential-mode and common-mode input resistances?

15.37. For the amplifier in Fig. P15.37, $V_{DD} = 12$ V, $V_{SS} = 12$ V, $I_{SS} = 20$ μA, and $R_D = 820$ kΩ. Assume $K_p = 1$ mA/V^2 and $V_{TP} = +1$ V. (a) What are the output voltages v_O for the amplifier

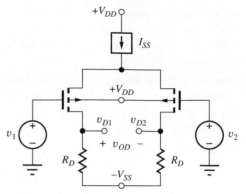

Figure P15.36

for $v_1 = 0$ V and $v_1 = 20$ mV? (b) What is the maximum permissible value of v_s?

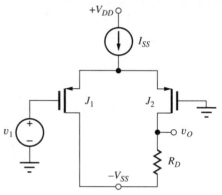

Figure P15.37

Half-Circuit Analysis

*15.38. (a) Draw the differential-mode and common-mode half-circuits for the differential amplifier in Fig. P15.38. (b) Use the half-circuits to find the Q-points, differential-mode gain, common-mode gain, and differential-mode input resistance

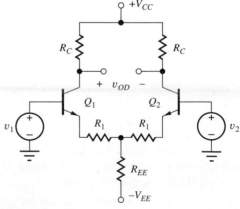

Figure 15.38

for the amplifier if $\beta_o = 150$, $V_{CC} = 22$ V, $V_{EE} = 22$ V, $R_{EE} = 200$ kΩ, $R_1 = 2$ kΩ, and $R_C = 200$ kΩ.

15.39. Use SPICE to simulate the amplifier in Prob. 15.38 at a frequency of 1 kHz, and determine the differential-mode gain, common-mode gain, and differential-mode input resistances.

*15.40. (a) Draw the differential-mode and common-mode half-circuits for the differential amplifier in Fig. P15.40. (b) Use the half-circuits to find the Q-points, differential-mode gain, common-mode gain, and differential-mode input resistance for the amplifier if $\beta_o = 100$, $V_{CC} = 20$ V, $V_{EE} = 20$ V, $I_{EE} = 100$ μA, and $R_{EE} = 600$ kΩ?

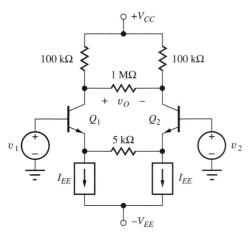

Figure P15.40

15.41. Use SPICE to simulate the amplifier in Prob. 15.40 at a frequency of 1 kHz, and determine the differential-mode gain, common-mode gain, and differential-mode input resistances.

*15.42. (a) Draw the differential-mode and common-mode half-circuits for the differential amplifier in Fig. P15.42. (b) Use the half-circuits to find the Q-points, differential-mode gain, common-mode gain, and differential-mode input resistance for the amplifier if $V_{CC} = 15$ V, $V_{EE} = 15$ V, $I_{EE} = 100$ μA, $R_D = 75$ kΩ, $R_{EE} = 600$ kΩ, $\beta_o = 100$, $K_n = 200$ μA/V^2, and $V_{TN} = -4$ V. (c) Show that Q_1 and Q_2 are in the active region.

**15.43. (a) Draw the differential-mode and common-mode half-circuits for the differential amplifier in Fig. P15.43. (b) Use the half-circuits to find the Q-points, differential-mode gain, common-mode gain, and differential-mode input resistance for the

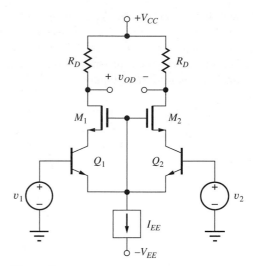

Figure P15.42

amplifier if $K_n = 1000$ μA/V^2, $V_{TN} = 0.75$ V, $K_p = 500$ μA/V^2, $V_{TP} = -0.75$ V, $I_1 = 200$ μA, $I_2 = 100$ μA, $V_{DD} = 6$ V, $V_{SS} = 6$ V, and $R_D = 30$ kΩ.

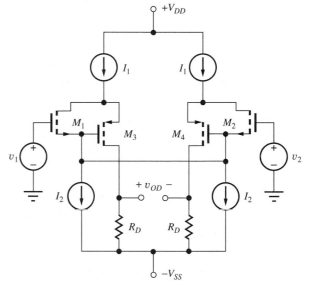

Figure P15.43

15.44. (a) Repeat Prob. 15.43 for $V_{DD} = 1.5$ V, $-V_{SS} = -1.5$ V, and $R_D = 10$ kΩ. (b) What is the common-mode input range for this amplifier?

15.2 Evolution to Basic Operational Amplifiers

15.45. (a) What are the Q-points of the transistors in the amplifier in Fig. P15.45 if $V_{CC} = 12$ V, $V_{EE} = 12$ V, $I_1 = 50$ μA, $R = 24$ kΩ, $\beta_o = 100$, and $V_A = 60$ V? (b) What are the differential-mode voltage gain and input resistance? (c) What is the amplifier output resistance? (d) What is the common-mode input resistance? (e) Which terminal is the noninverting input?

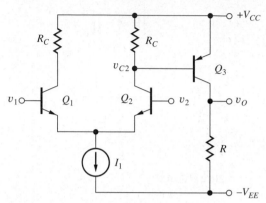

Figure P15.45

15.46. What is the common-mode input range for the amplifier in Prob. 15.45 if current source I_1 is replaced with an electronic current source that must have 0.75 V across it to operate properly?

15.47. Use SPICE to simulate the amplifier in Prob. 15.45 at a frequency of 1 kHz, and determine the differential-mode gain, CMRR, and differential-mode input resistance and output resistance.

15.48. The circuit in Fig. P15.48 is called a Darlington connection of two transistors. Assume the emitter is grounded and derive the expressions below.

$$I_{C1} = \beta_{F1}I_B \qquad I_{C2} = \beta_{F2}(\beta_{F1} + 1)I_B$$
$$I_C \cong \beta_{F1}\beta_{F2}I_B$$
$$g_{m2} = \beta_{o1}g_{m1} \qquad r_{\pi1} = \beta_{o1}r_{\pi2}$$
$$r_{o1} = \beta_{o1}r_{o2}$$
$$\beta_o = \frac{i_c}{i_B} \cong \beta_{o1}\beta_{o2}$$
$$G_m = \frac{i_c}{v_{be}} = \frac{g_{m1}}{2} + \frac{g_{m2}}{2} \cong \frac{g_{m2}}{2}$$
$$R_{iB} = \frac{v_{be}}{i_b} = r_{\pi1} + (\beta_{o1} + 1)r_{\pi2} \cong 2\beta_{o1}r_{\pi2}$$
$$R_{iC} = \frac{v_{ce2}}{i_c} \cong r_{o2}\|2\frac{r_{o1}}{\beta_{o2}} \cong \frac{2}{3}r_{o2}$$

15.49. Transistor Q_3 in Fig. P15.45 is replaced with a *pnp* Darlington circuit. Draw the new amplifier and repeat Prob. 15.45. (See Figs. P15.48 and P15.81.)

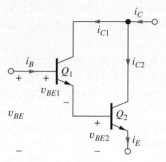

Figure P15.48

15.50. (a) What are the Q-points of the transistors in the amplifier in Fig. P15.50 if $V_{CC} = 15$ V, $V_{EE} = 15$ V, $I_1 = 200$ μA, $R_E = 2.4$ kΩ, $R = 50$ kΩ, $\beta_o = 80$, and $V_A = 70$ V? (b) What are the differential-mode voltage gain and input resistance? (c) What is the amplifier output resistance? (d) What is the common-mode input resistance? (e) Which terminal is the noninverting input?

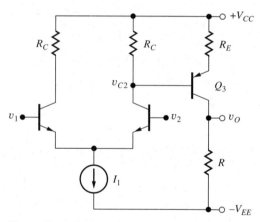

Figure P15.50

15.51. What is the common-mode input range for the amplifier in Prob. 15.50 if current source I_1 is replaced with an electronic current source that must have 0.75 V across it to operate properly?

15.52. (a) What are the Q-points of the transistors in the amplifier in Fig. P15.50 if $V_{CC} = 15$ V, $V_{EE} = 15$ V, $I_1 = 200$ μA, $R_E = 0$, $R = 50$ kΩ, $\beta_o = 80$, and $V_A = 70$ V? (b) What are the differential-mode voltage gain and input resistance? (c) What is the common-mode input resistance?

*15.53. Plot a graph of the differential-mode voltage gain of the amplifier in Prob. 15.52 versus the value of R_E. (The computer might be a useful tool.)

15.54. Design an amplifier to have $R_{out} = 1$ kΩ and $A_{dm} = 2000$, using the circuit in Fig. P15.45. Use $V_{CC} = V_{EE} = 9$ V, and $β_F = 100$.

15.55. (a) What are the Q-points of the transistors in the amplifier in Fig. P15.55 if $V_{CC} = 15$ V, $V_{EE} = 15$ V, $I_1 = 200$ μA, $I_2 = 300$ μA, $R_E = 2.4$ kΩ, $β_o = 80$, and $V_A = 70$ V? (b) What are the differential-mode voltage gain, input resistance, and output resistance?

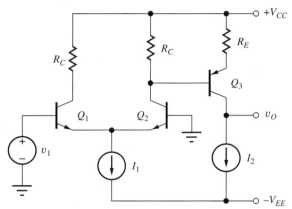

Figure P15.55

15.56. Use SPICE to simulate the amplifier in Prob. 15.55 and compare the results to hand calculations.

15.57. What are the Q-points of the transistors in the amplifier in Fig. P15.55 if $V_{CC} = 15$ V, $V_{EE} = 15$ V, $I_1 = 200$ μA, $I_2 = 300$ μA, $R_E = 0$, $β_o = 100$, and $V_A = 70$ V?

*15.58. Plot a graph of the differential-mode voltage gain of the amplifier in Prob. P15.55 versus the value of R_E. (The computer might be a useful tool.)

15.59. (a) What are the Q-points of the transistors in the amplifier in Fig. P15.59 if $V_{CC} = V_{EE} = 15$ V,

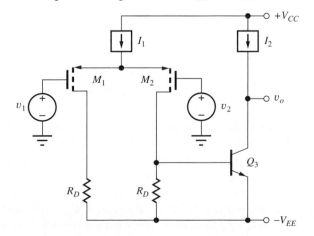

Figure P15.59

$I_1 = 500$ μA, $R_1 = 2$ MΩ, $I_2 = 500$ μA, and $R_2 = 2$ MΩ? Use $β_o = 80$, $V_A = 75$ V, $K_p = 5$ mA/V², and $V_{TP} = -1$ V. (b) What are the differential-mode voltage gain and input resistance and output resistance of the amplifier? (c) Which terminal is the noninverting input? (d) Which terminal is the inverting input?

*15.60. Use SPICE to simulate the amplifier in Prob. 15.59 at a frequency of 1 kHz, and determine the differential-mode gain, CMRR, and differential-mode input resistance and output resistance.

15.61. What is the voltage gain of the amplifier in Fig. P15.59 if $V_{CC} = V_{EE} = 5$ V, $I_1 = 500$ μA, $R_1 = 20$ MΩ, $I_2 = 100$ μA, $R_2 = 10$ MΩ, $β_o = 80$, $V_A = 75$ V, $K_p = 5$ mA/V², and $V_{TP} = -1$ V?

15.62. What is the common-mode input voltage range for the amplifier in Prob. 15.61 if current source I_1, must have a 0.75-V drop across it to operate properly?

15.63. (a) What are the Q-points of the transistors in the amplifier in Fig. P15.63 if $I_1 = 500$ μA, $R_1 = 1$ MΩ, $I_2 = 500$ μA, $R_2 = 1$ MΩ, and $V_{CC} = V_{EE} = 5$ V. Use $β_o = 80$, $V_A = 75$ V, $K_p = 5$ mA/V², and $V_{TP} = -1$ V. (b) What are the differential-mode voltage gain and input resistance and output resistance of the amplifier? (See Prob. 15.48.)

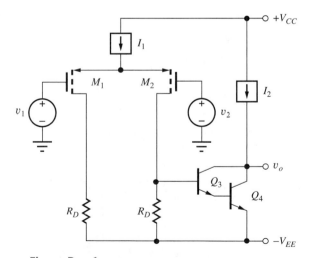

Figure P15.63

*15.64. Use SPICE to simulate the amplifier in Prob. 15.63 at a frequency of 1 kHz, and determine the differential-mode gain, CMRR, and differential-mode input resistance and output resistance.

15.65. (a) Redraw the op amp circuit in Fig. 15.26(a) with Q_4 replaced by the *npn* Darlington configuration from Prob. 15.48. (b) What are the new values of voltage gain, CMRR, input resistance, and output resistance? Use the circuit element values from Ex. 15.4, and compare your results to those of the example.

15.66. Simulate the circuit in Prob. 15.65 using SPICE and compare the results of the two problems.

15.67. (a) What are the Q-points of the transistors in the amplifier in Fig. P15.67 if $V_{CC} = 15$ V, $V_{EE} = 15$ V, $I_1 = 100$ μA, $I_2 = 350$ μA, $I_3 = 1$ mA, $\beta_F = 100$, and $V_A = 50$ V? (b) What are the differential-mode voltage gain and input resistance? (c) What is the amplifier output resistance? (d) What is the common-mode input resistance? (e) Which terminal is the noninverting input?

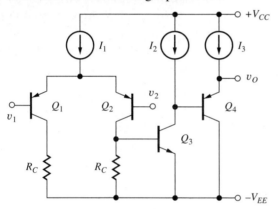

Figure P15.67

*15.68. Use SPICE to simulate the amplifier in Prob. 15.68 at a frequency of 1 kHz, and determine the differential-mode gain, CMRR, and differential-mode input resistance and output resistance.

15.69. (a) What are the Q-points of the transistors in the amplifier in Fig. P15.69 if $V_{DD} = 12$ V, $V_{SS} = 12$ V,

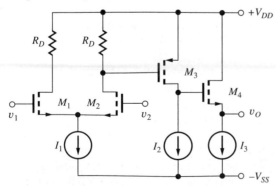

Figure P15.69

$I_1 = 500$ μA, $I_2 = 2$ mA, $I_3 = 5$ mA, $K_n = 5$ mA/V², $V_{TN} = 0.75$ V, $\lambda_n = 0.02$ V⁻¹, $K_p = 2$ mA/V², $V_{TP} = -0.75$ V, and $\lambda_p = 0.015$ V⁻¹? (b) What are the differential-mode voltage gain and input resistance and output resistance of the amplifier?

*15.70. Use SPICE to simulate the amplifier in Prob. 15.69 at a frequency of 1 kHz, and determine the differential-mode gain, CMRR, and differential-mode input resistance and output resistance.

15.71. (a) What are the Q-points of the transistors in the amplifier in Fig. P15.71 if $V_{DD} = 5$ V, $V_{SS} = 5$ V, $I_1 = 600$ μA, $I_2 = 500$ μA, $I_3 = 2$ mA, $K_n = 5$ mA/V², $V_{TN} = 0.70$ V, $\lambda_n = 0.02$ V⁻¹, $K_p = 2$ mA/V², $V_{TP} = -0.70$ V, and $\lambda_p = 0.015$ V⁻¹? (b) What are the differential-mode voltage gain and input resistance and output resistance of the amplifier?

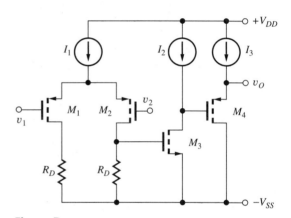

Figure P15.71

15.72. Simulate the circuit in Prob. 15.71 using SPICE and compare the results to hand calculations.

15.73. Transistor M_3 in Fig. P15.71 is replaced with an *npn* device with $\beta_o = 150$ and $V_A = 70$ V. What are the values of the differential-mode voltage gain and input resistance, and the output resistance of the new amplifier? Use the circuit element values from Prob. 15.71.

15.74. Simulate the circuit in Prob. 15.73 using SPICE and compare the results with hand calculations.

15.75. (a) What are the Q-points of the transistors in the amplifier in Fig. P15.75 if $V_{CC} = 5$ V, $V_{EE} = 5$ V, $I_1 = 200$ μA, $I_2 = 500$ μA, $I_3 = 2$ mA, $R_L = 2$ kΩ, $\beta_o = 100$, $V_A = 50$ V, $K_n = 5$ mA/V², and $V_{TN} = 0.70$ V? (b) What are the differential-mode voltage gain and input resistance and output resistance of the amplifier? (c) Use SPICE to simulate the amplifier in Fig. 15.116 at a frequency of 2 kHz, and determine the differential-mode gain, CMRR, and differential-mode input resistance and output resistance.

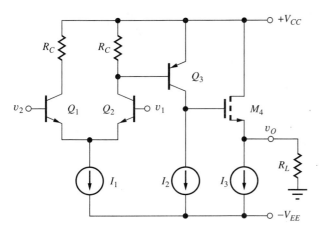

Figure P15.75

*15.76. (a) What are the Q-points of the transistors in the amplifier in Fig. P15.76 if $V_{CC} = 3$ V, $V_{EE} = 3$ V, $I_1 = 10$ μA, $I_2 = 50$ μA, $I_3 = 250$ μA, $R_{C1} = 300$ kΩ, $R_{C2} = 78$ kΩ, $R_L = 5$ kΩ, $\beta_{on} = 100$, $V_{AN} = 50$ V, $\beta_{op} = 50$, and $V_{AP} = 70$ V? (b) What are the differential-mode voltage gain and input resistance and output resistance of the amplifier? (c) Which terminal is the noninverting input? Which terminal is the inverting input? (d) What is the gain predicted by our rule-of-thumb estimate? What are the reasons for any discrepancy?

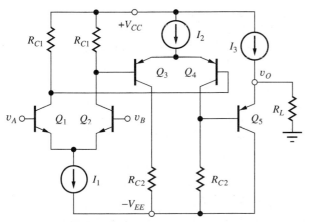

Figure P15.76

**15.77. (a) What are the Q-points of the transistors in the amplifier in Fig. P15.76 if $V_{CC} = V_{EE} = 18$ V,

$I_1 = 100$ μA, $I_2 = 200$ μA, $I_3 = 750$ μA, $R_{C1} = 120$ kΩ, $R_{C2} = 170$ kΩ, $R_L = 2$ kΩ, $\beta_{on} = 100$, $V_{AN} = 50$ V, $\beta_{op} = 50$, and $V_{AP} = 70$ V? (b) What are the differential-mode voltage gain and input resistance and output resistance of the amplifier? (c) What is the common-mode input range? (d) Estimate the offset voltage of this amplifier.

15.78. (a) What are the Q-points of the transistors in the amplifier in Fig. P15.78 if $V_{CC} = V_{EE} = 12$ V, $I_1 = 200$ μA, $R = 12$ kΩ, $\beta_F = 100$, and $V_A = 70$ V? (b) What are the differential-mode voltage gain and input resistance and output resistance of the amplifier?

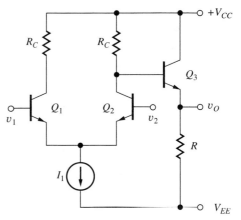

Figure P15.78

*15.79. Design an amplifier using the topology in Fig. P15.78 to have an input resistance of 300 kΩ and an output resistance of 100 Ω. Can these specifications all be met if $V_{CC} = V_{EE} = 12$ V, $\beta_{FO} = 100$, and $V_A = 60$ V? If so, what are the values of I_1, R_C, and R, and the voltage gain of the amplifier? If not, what needs to be changed?

*15.80. Design an amplifier using the topology in Fig. P15.78 to have an input resistance of 1 MΩ and an output resistance ≤ 2 Ω. Can these specifications all be met if $V_{CC} = V_{EE} = 9$ V, $\beta_{FO} = 100$, and $V_A = 60$ V? If so, what are the values of I_1, R_C, and R, and the voltage gain of the amplifier? If not, what needs to be changed?

**15.81. (a) What are the Q-points of the transistors in the amplifier in Fig. P15.81 if $V_{CC} = V_{EE} = 18$ V, $I_1 = 50$ μA, $I_2 = 500$ μA, $I_3 = 5$ mA, $\beta_{on} = 100$, $V_{AN} = 50$ V, $\beta_{op} = 50$, and $V_{AP} = 70$ V? (b) What are the differential-mode voltage gain and input resistance and output resistance of the amplifier?

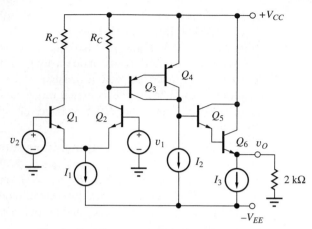

Figure P15.81

15.82. (a) What are the Q-points of the transistors in the amplifier in Fig. P15.81 if $V_{CC} = V_{EE} = 22$ V, $I_1 = 50$ μA, $I_2 = 500$ μA, $I_3 = 5$ mA, $\beta_{on} = 100$, $V_{AN} = 50$ V, $\beta_{op} = 50$, and $V_{AP} = 70$ V? (b) What are the differential-mode voltage gain and input resistance and output resistance of the amplifier?

15.3 Output Stages

15.83. What is the quiescent current in the class-AB stage in Fig. P15.83 if $K_p = K_n = 600$ μA/V^2 and $V_{TN} = -V_{TP} = 0.75$ V?

15.84. What is the quiescent current in the class-AB stage in Fig. P15.83 if $K_p = 400$ μA/V^2, $K_n = 600$ μA/V^2, $V_{TP} = -0.8$ V, and $V_{TN} = 0.7$ V?

15.85. What is the quiescent current in the class-AB stage in Fig. P15.85 if both transistors have $I_S = 2 \times 10^{-15}$ A?

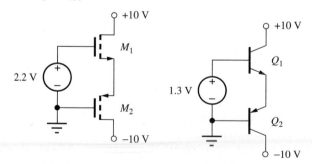

Figure P15.83   **Figure P15.85**

15.86. What is the quiescent current in the class-AB stage in Fig. P15.85 if $I_S = 10^{-15}$ A for the *pnp* transistor and $I_S = 4 \times 10^{-15}$ A for the *npn* transistor?

15.87. Draw a sketch of the voltage transfer characteristic for the circuit in Fig. P15.87. Label important voltages on the characteristic.

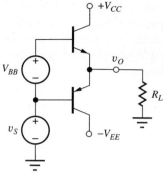

Figure P15.87

15.88. Use SPICE to plot the voltage transfer characteristic for the class-AB stage in Fig. P15.87 if $I_S = 10^{-15}$ A and $\beta_F = 50$ for the *pnp* transistor, $I_S = 5 \times 10^{-15}$ A and $\beta_F = 60$ for the *npn* transistor, $V_{BB} = 1.3$ V, and $R_L = 1$ kΩ.

15.89. What is the quiescent current in the class-AB stage in Fig. P15.89 if I_S for the *npn* transistor is 10^{-15} A, I_S for the *pnp* transistor is 10^{-16} A, $I_B = 250$ μA, and $R_B = 5$ kΩ? Assume $\beta_F = \infty$ and $v_O = 0$.

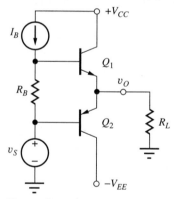

Figure P15.89

15.90. What is the quiescent current in the class-AB stage in Fig. P15.90 if $V_{TN} = 0.75$ V, $V_{TP} = -0.75$ V,

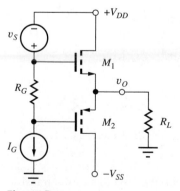

Figure P15.90

$K_n = 500$ μA/V^2, $K_p = 200$ μA/V^2, $I_G = 500$ μA, and $R_G = 4$ kΩ?

15.91. The source-follower in Fig. 15.31 has $V_{DD} = V_{SS} = 10$ V and $R_L = 1$ kΩ. If the amplifier is developing an output voltage of $5 \sin 2000\pi t$ V, what is the minimum value of I_{SS}? What are the maximum and minimum values of source current i_S that occur during the signal swing? What is the efficiency?

15.92. An ideal complementary class-B output stage is generating a square wave output signal across a 5-kΩ load resistor with a peak value of 5 V from ±5-V supplies. What is the efficiency of the amplifier?

*15.93. An ideal complementary class-B output stage is generating a triangular output signal across a 100-kΩ load resistor with a peak value of 10 V from ±10-V supplies. What is the efficiency of the amplifier?

**15.94. (a) Use the Fourier analysis capability of SPICE to find the amplitude of the first, second, third, fourth, and fifth harmonics of the input signal introduced by the cross-over region of the class-B amplifier in Fig. P15.87 if $V_{BB} = 0$, $V_{CC} = V_{EE} = 5$ V, $v_S = 4 \sin 2000\pi t$, and $R_L = 2$ kΩ. (b) Repeat for $V_{BB} = 1.3$ V.

Short-Circuit Protection

15.95. What is the current in the R_L circuit in Fig. 15.38 at the point when current just begins to limit ($V_{BE2} = 0.7$ V) if $R = 10$ Ω, $R_1 = 1$ kΩ, and $R_L = 250$ Ω? For what value of v_S does the output begin to limit current?

15.96. Use SPICE to simulate the circuit in Prob. 15.95, and compare the results to your hand calculations. Discuss the reasons for any discrepancies.

15.97. What would be the Q-point currents in M_4 and M_5 in the amplifier in Fig. 15.37(a) if $V_{DD} = V_{SS} = 15$ V, $I_2 = 250$ μA, $R_G = 7$ kΩ, $R_L = 2$ kΩ, and $V_{TN} = 0.75$ V, $V_{TP} = -0.75$ V, $K_n = 5$ mA/V^2, and $K_p = 2$ mA/V^2?

15.98. What would be the currents in Q_4 and Q_5 in the amplifier in Fig. 15.37(b) if $V_{CC} = V_{EE} = 15$ V, $I_2 = 500$ μA, $R_B = 2.4$ kΩ, $R_L = 2$ kΩ, and Q_3 is modeled by a voltage of $V_{CESAT} = 0.2$ V in series with a resistance of 50 Ω when it is saturated?

Transformer Coupling

15.99. Calculate the output resistance of the follower circuit (as seen at R_L) in Fig. 15.40(a) if $n = 10$ and $I_S = 10$ mA.

15.100. For the circuit in Fig. P15.100, $v_S = \sin 2000\pi t$, $R_E = 82$ kΩ, $R_B = 200$ kΩ, and $V_{CC} = V_{EE} = 9$ V. What value of n is required to deliver maximum power to R_L if $R_L = 10$ Ω? What is the power? Assume $C_1 = C_2 = \infty$.

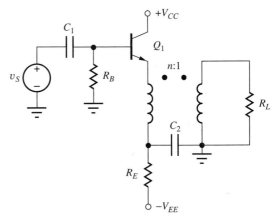

Figure P15.100

15.4 Electronic Current Sources

15.101. (a) What are the output current and output resistance of the current source in Fig. P15.101(a) if $V_{EE} = 12$ V, $R_1 = 2$ MΩ, $R_2 = 2$ MΩ, $R_E = 220$ kΩ, $\beta_o = 100$, and $V_A = 50$ V? (b) Repeat for the circuit in Fig. P15.101(b).

15.102. What are the output current and output resistance of the current source in Prob. 15.101(a) if node V_B is bypassed to ground with a capacitor?

15.103. (a) What are the output current and output resistance of the current source in Fig. P15.101(a) if $-V_{EE} = -9$ V, $R_1 = 270$ kΩ, $R_2 = 430$ kΩ, $R_E = 18$ kΩ, $\beta_o = 150$, and $V_A = 75$ V? (b) Repeat for the circuit in Fig. P15.101(b).

15.104. (a) What are the output current and output resistance of the current source in Fig. P15.101(a) if $-V_{EE} = -5$ V, $R_1 = 100$ kΩ, $R_2 = 200$ kΩ, $R_E = 15$ kΩ, $\beta_o = 100$, and $V_A = 75$ V? (b) Repeat for the circuit in Fig. 15.101(b).

15.105. Design a current source to provide an output current of 1 mA using the topology of Fig. P15.101(a). The current source should use no more than 1.2 mA and have an output resistance of at least 500 kΩ. Assume $V_{EE} = 12$ V. (b) Repeat for Fig. 15.101(b).

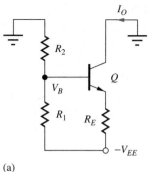

(a)

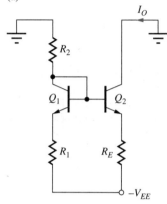

(b)

Figure P15.101

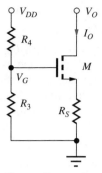

Figure P15.106

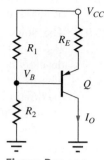

Figure P15.109

15.111. What are the output current and output resistance of the current source in Fig. P15.109 if $V_{CC} = 10$ V, $R_1 = 100$ kΩ, $R_2 = 300$ kΩ, $R_E = 18$ kΩ, $\beta_o = 90$, and $V_A = 75$ V?

15.112. What are the output current and output resistance of the current source in Fig. P15.112 if $V_{DD} = 6$ V, $R_4 = 200$ kΩ, $R_3 = 100$ kΩ, and $R_S = 16$ kΩ? Use the device parameters from Prob. 15.113.

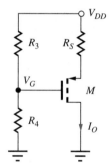

Figure P15.112

15.106. What are the output current and output resistance of the current source in Fig. P15.106 if $V_O = V_{DD} = 10$ V, $R_4 = 680$ kΩ, $R_3 = 330$ kΩ, $R_S = 33$ kΩ, $K_n = 500$ µA/V², $V_{TN} = 1$ V, and $\lambda = 0.01$ V^{-1}?

15.107. What are the output current and output resistance of the current source in Fig. P15.106 if $V_O = V_{DD} = 3$ V, $R_4 = 200$ kΩ, $R_3 = 68$ kΩ, and $R_S = 56$ kΩ? Use the device parameters from Prob. 15.106.

15.108. What are the output current and output resistance of the current source in Fig. P15.106 if $V_O = V_{DD} = 6$ V, $R_4 = 200$ kΩ, $R_3 = 100$ kΩ, and $R_S = 16$ kΩ? Use the device parameters from Prob. 15.106.

15.109. What are the output current and output resistance of the current source in Fig. P15.109 if $V_{CC} = 15$ V, $R_1 = 100$ kΩ, $R_2 = 200$ kΩ, $R_E = 47$ kΩ, $\beta_o = 75$, and $V_A = 50$ V?

15.110. What are the output current and output resistance of the current source in Fig. P15.109 if $V_{CC} = 5$ V, $R_1 = 10$ kΩ, $R_2 = 33$ kΩ, $R_E = 1.5$ kΩ, $\beta_o = 75$, and $V_A = 60$ V?

15.113. What are the output current and output resistance of the current source in Fig. P15.112 if $V_{DD} = 9$ V, $R_4 = 2$ MΩ, $R_3 = 1$ MΩ, $R_S = 120$ kΩ, $K_p = 750$ µA/V², $V_{TP} = -0.75$ V, and $\lambda = 0.01$ V^{-1}?

15.114. What are the output current and output resistance of the current source in Fig. P15.112 if $V_{DD} = 4$ V, $R_4 = 200$ kΩ, $R_3 = 62$ kΩ, and $R_S = 43$ kΩ? Use the device parameters from Prob. 15.113.

15.115. Design a current source to provide an output current of 175 µA using the topology in Fig. P15.112. The current source should use no more than 200 µA and have an output resistance of at least 2.5 MΩ. Assume $V_{DD} = 12$ V, $K_p = 200$ µA/V², $V_{TP} = -0.75$ V, and $\lambda = 0.02$ V^{-1}.

15.116. (a) What are the two output currents and output resistances of the current source in Fig. P15.116(a) if $V_{EE} = 12$ V, $\beta_o = 125$, $V_A = 50$ V, $R_1 = 33$ kΩ,

$R_2 = 68 \text{ k}\Omega$, $R_3 = 20 \text{ k}\Omega$, and $R_4 = 100 \text{ k}\Omega$?
(b) Repeat for the circuit in Fig. P15.116(b).

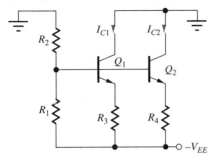

(a)

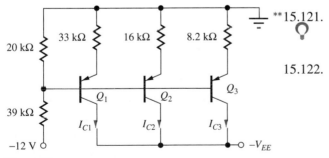

(b)

Figure P15.116

15.117. What are the three output currents and output resistances of the current source in Fig. P15.117 if $V_{EE} = 15$ V, $\beta_o = 75$, and $V_A = 60$ V?

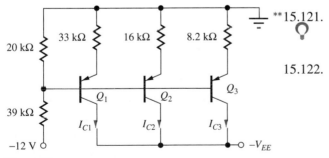

Figure P15.117

15.118. What are the two output currents and output resistances of the current sources in Fig. P15.118 if $V_{DD} = 12$ V, $K_p = 250$ μA/V², $V_{TP} = -1$ V, $\lambda = 0.02$ V⁻¹, $R_1 = 100$ kΩ, $R_2 = 470$ kΩ, $R_3 = 2$ MΩ, and $R_4 = 2$ MΩ?

15.119. Use SPICE to simulate the current source array in Prob. 15.118, and find the output currents and output resistances of the source. Use transfer function analysis to find the output resistances.

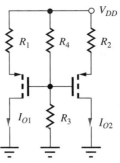

Figure P15.118

*15.120. The op amp in Fig. P15.120 is used in an attempt to increase the overall output resistance of the current source circuit. If $V_{REF} = 5$ V, $V_{CC} = 0$ V, $V_{EE} = 15$ V, $R = 50$ kΩ, $\beta_o = 120$, $V_A = 70$ V, and $A = 50{,}000$, what are the output current I_O and output resistance of the current source? Did the op amp help increase the output resistance? Explain why or why not.

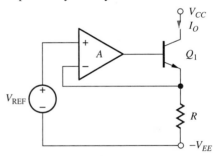

Figure P15.120

**15.121. How might the output resistance of the circuit in Fig. P15.120 be improved further using an additional bipolar transistor?

15.122. The op amp in Fig. P15.122 is used to increase the overall output resistance of current source M_1. If $V_{REF} = 5$ V, $V_{DD} = 0$ V, $V_{SS} = 15$ V, $R = 50$ kΩ, $K_n = 800$ μA/V², $V_{TN} = 0.8$ V, $\lambda = 0.02$ V⁻¹, and $A = 50{,}000$, what are the output current I_O and output resistance of the current source?

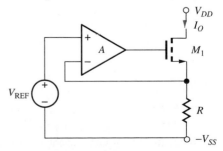

Figure P15.122

15.123. (a) What are the Q-points of the transistors in the amplifier in Fig. P15.123(a) if $\beta_o = 85$ and $V_A = 70$ V? (b) What are the differential-mode gain and CMRR of the amplifier? (c) Repeat for Fig. P15.123(b).

$V_{TN} = +1$ V, $\lambda = 0.02$ V^{-1}, $R_1 = 51$ kΩ, $R_2 = 100$ kΩ, $R_S = 7.5$ kΩ, and $R_D = 36$ kΩ? (b) What are the differential-mode gain and CMRR of the amplifier?

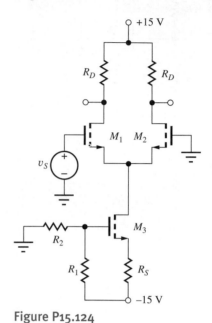

Figure P15.124

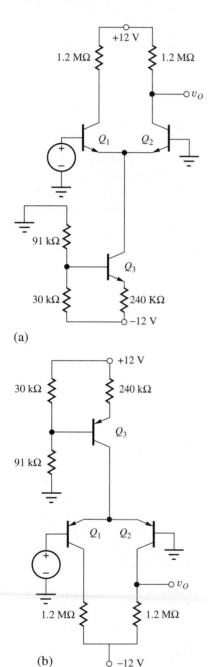

Figure P15.123

15.124. (a) What are the Q-points of the transistors in the amplifier in Fig. P15.124 if $K_n = 400$ μA/V^2,

15.125. (a) A current source with $R_{out} = \beta_o r_o$ is used to bias a standard bipolar differential amplifier. What is an expression for the CMRR of this amplifier for single-ended outputs?

15.126. The output resistance of the MOS current source in Fig. P15.124 is given by $R_{out} = \mu_f R_S$. How much voltage must be developed across R_S to achieve an output resistance of 5 MΩ at a current of 100 μA if $K_n = 500$ μA/V^2 and $\lambda = 0.02$ V^{-1}?

**15.127. Use PSPICE to perform a Monte Carlo analysis of the circuits in Fig. 15.49. Assume 5 percent resistors and a 5 percent power-supply tolerance. Find the nominal and 3σ limits on I_O and R_{out}.

15.5 Circuit Element Matching

15.128. An integrated circuit resistor has a nominal value of 4.02 kΩ. A given process run has produced resistors with a mean value 15 percent higher than the nominal value, and the resistors are found to be matched within 3 percent. What are the maximum and minimum resistor values that will occur?

15.129. (a) The emitter areas of two bipolar transistors are mismatched by 10 percent. What will be the base-emitter voltage difference between these two transistors when their collector currents are identical? (Assume $V_A = \infty$.) (b) Repeat for a 20 percent area mismatch. (c) What degree of matching is required for a base-emitter voltage difference of less than 1 mV?

15.130. The bipolar transistors in the differential pair in Fig. 15.1(a) are mismatched. (a) What will be the offset voltage if the current gains are mismatched by 5 percent? (b) If the saturation currents are mismatched by 5 percent? (c) If the Early voltages are mismatched by 5 percent? (d) If the collector resistors are mismatched by 5 percent? (Remember, the offset voltage is the input voltage required to force the differential output voltage to be zero.)

*15.131. The collector currents of two BJTs are equal when the base-emitter voltages differ by 2 mV. What is the fractional mismatch $\Delta I_S / I_S$ in the saturation current of the two transistors if $I_{S1} = I_S + \Delta I_S/2$ and $I_{S2} = I_S - \Delta I_S/2$? Assume that the collector-emitter voltages and Early voltages are matched. If $\Delta \beta_{FO} / \beta_{FO} = 5$ percent, what are the values of I_{B1} and I_{B2} for the transistors at a Q-point of (100 μA, 10 V)? Assume $\beta_{FO} = 100$ and $V_A = 50$ V.

15.132. What is the worst-case fractional mismatch $\Delta I_D / I_D$ in drain currents in two MOSFETs if $K_n = 250$ μA/V^2 ± 5 percent and $V_{TN} = 1$ V ± 25 mV for (a) $V_{GS} = 2$ V? (b) $V_{GS} = 4$ V? Assume $I_{D1} = I_D + \Delta I_D/2$ and $I_{D2} = I_D - \Delta I_D/2$.

15.133. (a) A layout design error causes the W/L ratios of the two NMOSFETs in a differential amplifier to differ by 10 percent. What will be the gate-source voltage difference between these two transistors when their drain currents are identical if the nominal value of $(V_{GS} - V_{TN}) = 0.5$ V? (Assume $V_{TN} = 1$ V, $\lambda = 0$ and identical values of K'_n). (b) What degree of matching is required for a gate-source voltage difference of less than 3 mV? (c) For 1 mV?

15.134. The MOS transistors in the differential pair in Fig. 15.1(b) are mismatched. The nominal value of $(V_{GS} - V_{TN}) = 0.75$ V. (a) What will be the offset voltage if the (W/L) ratios are mismatched by 5 percent? (b) If the threshold voltages are mismatched by 5 percent? (c) If the values of λ are mismatched by 5 percent? (d) If the drain resistors are mismatched by 5 percent? (Remember, the offset voltage is the input voltage required to force the differential output voltage to be zero.)

15.6 Current Mirrors

15.135. (a) What are the output currents and output resistances for the current sources in Fig. P15.135 if $I_{REF} = 30$ μA, $K'_n = 25$ μA/V^2, $V_{TN} = 0.75$ V and $\lambda = 0.015$ V^{-1}? (b) What are the currents if I_{REF} is changed to 50 μA? (c) What would be the values if $\lambda = 0$?

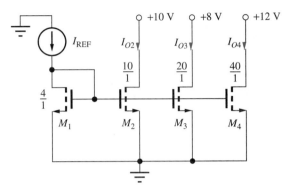

Figure P15.135

15.136. (a) What are the output currents for the circuit in Prob. 15.135 if the W/L ratio of M_1 is changed to 2.5/1? (b) If $I_{REF} = 20$ μA and $(W/L)_1 = 6/1$?

15.137. The current sources in Prob. 15.135 could represent the binary weighted currents needed for a 3-bit D/A converter. (a) What are the ideal values of the three output currents (i.e., $\lambda = 0$)? (b) Express the current errors from Prob. 15.135 in terms of LSBs.

15.138. Simulate the current source array in Fig. P15.135 and compare the results to the hand calculations in Prob. 15.135.

*15.139. What are the output currents and output resistances for the current sources in Fig. P15.139 if $R = 30$ kΩ, $K'_p = 15$ μA/V^2, $V_{TP} = -0.90$ V, and $\lambda = 0.01$ V^{-1}?

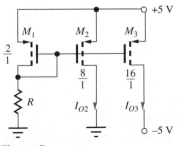

Figure P15.139

15.140. (a) What are the output currents for the circuit in Prob. 15.139 if the W/L ratio of M_1 is changed to 3.3/1? (b) $R = 50$ kΩ and $(W/L)_1 = 4/1$?

15.141. Simulate the current source array in Fig. P15.139 and compare the results to the hand calculations in Prob. 15.139.

15.142. What value of R is required in Fig. P15.139 to have $I_{O2} = 55$ µA? Use device data from Prob. 15.139.

15.143. (a) What are the output currents and output resistances for the current sources in Fig. P15.143(a) if $R = 75$ kΩ, $\beta_{FO} = 50$, and $V_A = 60$ V? (b) Repeat part (a) if the emitter areas of all the transistors are doubled. (c) Repeat for Fig. P15.143(b).

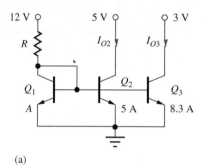

(a)

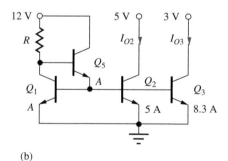

(b)

Figure P15.143

15.144. Simulate the current source array in Fig. P15.143(a) and compare the results to the hand calculations in Prob. 15.143. (b) Repeat for Fig. P15.143(b).

15.145. What value of R is required in Fig. P15.143(a) to have $I_{O3} = 150$ µA? What is the value of I_{O2}? Assume $\beta_{FO} = 50$ and $V_A = 60$ V. (b) Repeat for the circuit in Fig. P15.143(b).

15.146. (a) What are the output currents in the circuit in Fig. P15.143(a) if the area of transistor Q_1 is

changed to $2A$, and $R = 75$ kΩ? Use $\beta_{FO} = 125$ and $V_A = 75$ V. (b) Repeat for Fig. P15.143(b).

15.147. What are the output currents in the circuit in Fig. P15.143(b) if the area of transistor Q_1 is changed to $3A$, and $R = 100$ kΩ? Use $\beta_{FO} = 100$ and $V_A = 75$ V.

15.148. (a) What are the output currents in the circuit in Fig. P15.143(a) if $R = 140$ kΩ? Use $\beta_{FO} = 125$ and $V_A = 75$ V. (b) What value of R is required to produce the same output currents in Fig. P15.143(b).

15.149. (a) What are the output currents in Fig. P15.143(a) if $R = 100$ kΩ? (b) What are the output currents if the 5-V supply increases to 6 V? (c) What are the output currents if the 12-V supply decreases to 11 V? (d) Show that the change in I_{O2} in part (b) is equal to $g_{o2} \Delta V$.

15.150. What are the output currents and output resistances for the current sources in Fig. P15.150 if $R = 60$ kΩ, $\beta_{FO} = 50$, and $V_A = 60$ V?

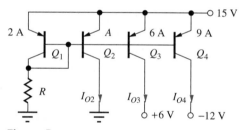

Figure P15.150

15.151. What value of R is required in the circuit in Fig. P15.150 to set $I_{O3} = 65$ µA? What are the values of I_{O2} and I_{O4}?

*15.152. Draw a buffered current mirror version of the source in Fig. P15.150 and find the value of R required to set $I_{REF} = 25$ µA if $\beta_{FO} = 50$ and $V_A = 60$ V. What are the values of the three output currents? What is the collector current of the additional transistor?

15.153. In Fig. P15.153, $R_2 = 5R_3$. What value of n is required to set I_{E3} to be equal to exactly $5I_{E2}$?

*15.154. What are the output currents and output resistances for the current sources in Fig. P15.153 if $R = 10$ kΩ, $R_1 = 10$ kΩ, $R_2 = 5$ kΩ, $R_3 = 2.5$ kΩ, $n = 4$, $\beta_{FO} = 75$, and $V_A = 60$ V?

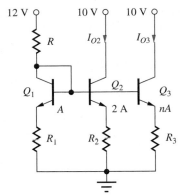

Figure P15.153

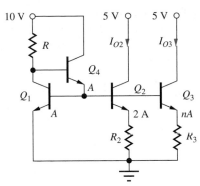

Figure P15.160

*15.155. What values of n and R_3 would be required in Prob. 15.154 so that $I_{O2} = 3I_{O3}$?

15.156. Repeat Prob. 15.154 if the area of transistor Q_1 is changed to 0.5 A and R_1 is changed to 20 kΩ.

15.157. What are the output current I_O and output resistance in the circuit in Fig. P15.157 if $-V_{EE} = -5$ V, $n = 7.2$, $K_n = 50$ μA/V², $V_{TN} = 0.75$ V, $I_{REF} = 15$ μA, $\beta_{FO} = 100$, and $V_A = 75$ V?

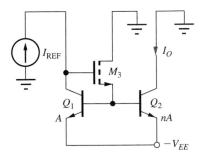

Figure P15.157

15.158. Use SPICE to simulate the circuit in Prob. 15.157 and compare the results to hand calculations.

15.159. (a) What is the input resistance presented to source I_{REF} at the gate of transistor M_3 in Fig. P15.157 if $n = 1$? Use the other parameters from Prob. 15.157. (b) Use transfer function analysis in SPICE to verify your result.

Widlar Sources

15.160. (a) What are the output current and output resistance for the Widlar current source I_{O2} in Fig. P15.160 if $R = R_2 = 10$ kΩ and $V_A = 60$ V? (b) For I_{O3} if $R_3 = 5$ kΩ and $n = 12$?

15.161. What value of R is required to set $I_{REF} = 75$ μA in Fig. P15.160? If $I_{REF} = 75$ μA, what value of R_2 is needed to set $I_{O2} = 5$ μA? If $R_3 = 2$ kΩ, what value of n is required to set $I_{O3} = 10$ μA?

15.162. Simulate the source of Prob. 15.161 and compare the results to hand calculations.

15.163. (a) What are the output current and output resistance for the Widlar current source I_{O2} in Fig. P15.163 if $R = 40$ kΩ and $R_2 = 5$ kΩ? Use $V_A = 70$ V and $\beta_F = 100$. (b) For I_{O3} if $R_3 = 2.5$ kΩ and $n = 20$?

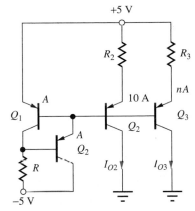

Figure P15.163

15.164. What value of R is required to set $I_{REF} = 50$ μA in Fig. P15.163. If $I_{REF} = 50$ μA, what value of R_2 is needed to set $I_{O2} = 10$ μA? If $R_3 = 2$ kΩ, what value of n is required to set $I_{O3} = 10$ μA?

15.7 High-Output-Resistance Current Mirrors

Wilson Sources

15.165. $I_{REF} = 50$ μA, $-V_{EE} = -5$ V, $\beta_{FO} = 125$, and $V_A = 40$ V in the Wilson source in Fig. P15.165.

(a) What are the output current and output resistance for $n = 1$? (b) For $n = 3$? (c) What is the value of V_{CS} for the current source in (b)? (d) What is the minimum value of V_{EE}?

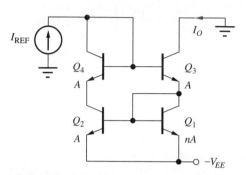

Figure P15.165

**15.166. Derive an expression for the output resistance of the BJT Wilson source in Fig. P15.165 and show that it can be reduced to Eq. (15.162), use $n = 1$. What assumptions were used in this simplification?

*15.167. Derive an expression for the output resistance of the Wilson source in Fig. P15.165 as a function of the area ratio n.

15.168. What is the minimum voltage that can be applied to the collector of Q_3 in Fig. P15.165 and have the transistor remain in the active region if $I_{REF} = 15\ \mu A$, $n = 5$, $\beta_{FO} = 125$, and $I_{SO} = 3$ fA? Calculate an exact value based on the value of I_{SO}.

15.169. $R = 30\ k\Omega$ in the Wilson source in Fig. P15.169. (a) What is the output current if $(W/L)_1 = 5/1$, $(W/L)_2 = 20/1$, $(W/L)_3 = 20/1$, $K'_n = 25\ \mu A/V^2$, $V_{TN} = 0.75$ V, $\lambda = 0$ V^{-1}, and $V_{SS} = -5$ V. What value of $(W/L)_4$ is required to balance the drain voltages of M_1 and M_2?

(*b) Repeat if $\lambda = 0.015$ V^{-1}. (c) Check your results in (b) with SPICE simulation.

*15.170. Derive an expression for the output resistance of the Wilson source in Fig. P15.169 as a function of $(W/L)_1$, $(W/L)_2$, $(W/L)_3$, $(W/L)_4$, and the reference current I_{REF}. Assume $R = \infty$.

15.171. (a) Derive an expression for the equivalent resistance presented to I_{REF} in the Wilson source in Fig. 15.66(a). (b) Derive an expression for the equivalent resistance presented to I_{REF} in the Wilson source in Fig. 15.66(b).

15.172. What is the minimum voltage required on the drain of M_3 to maintain it in pinch-off in the circuit in Fig. P15.169 if $I_{REF} = 150\ \mu A$, $(W/L)_1 = 5/1$, $(W/L)_2 = 20/1$, $(W/L)_3 = 20/1$, $K'_n = 25\ \mu A/V^2$, $V_{TN} = 0.75$ V, $\lambda = 0$ V^{-1}, and $-V_{SS} = -10$ V?

15.173. In Fig. P15.169, $(W/L)_3 = 5/1$, $(W/L)_4 = 5/1$, and $I_{REF} = 50\ \mu A$. What value of $(W/L)_2$ is required for $R_{out} = 250\ M\Omega$ if $K'_n = 25\ \mu A/V^2$, $V_{TN} = 0.75$ V, $\lambda = 0.0125$ V^{-1}. Assume $(W/L)_2 = (W/L)_1$, $R = \infty$, and $V_{SS} = 5$ V. Neglect V_{DS}.

**15.174. Redraw the equivalent circuit used to calculate the output resistance of the MOS Wilson source in Figs. 15.66(a) and 15.68 including a finite output resistance R_{REF} for the reference source. Based on this circuit, how large must R_{REF} be to keep from degrading the output resistance of the Wilson source? What type of current source could be used to implement I_{REF} to meet this requirement?

Cascode Current Sources

15.175. (a) What are the output current and output resistance for the cascode current source in Fig. P15.175 if $I_{REF} = 17.5\ \mu A$, $V_{DD} = 5$ V, $K_n = 75\ \mu A/V^2$, $V_{TN} = 0.75$ V, and $\lambda = 0.0125$ V^{-1}.

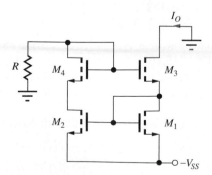

Figure P15.169

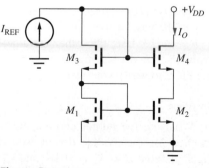

Figure P15.175

(b) What is the value of V_{CS} for this current source?
(c) What is the minimum value of V_{DD}?

15.176. Use SPICE to simulate the current source in Prob. 15.175 and compare the results to your calculations.

15.177. (a) A layout error causes the W/L ratio of M_2 to be 5 percent larger than that of M_1 in Prob. 15.175. What is the error in the output current I_O? (a) Repeat if $M_1 = M_2$, but M_4 is 5 percent larger than that of M_3.

15.178. In Fig. P15.175, $(W/L)_1 = 5/1$, $(W/L)_2 = 5/1$, $(W/L)_3 = 5/1$, and $I_{REF} = 50$ μA. What value of $(W/L)_4$ is required for $R_{out} = 250$ MΩ if $K'_n = 25$ μA/V^2, $V_{TN} = 0.75$ V, and $\lambda = 0.0125$ V^{-1}?

15.179. (a) Repeat Prob. 15.175 for $I_{REF} = 25$ μA. (b) Repeat Prob. 15.175 for $I_{REF} = 50$ μA.

15.180. What is the equivalent resistance presented to I_{REF} in the cascode current source in Prob. 15.175?

15.181. (a) What are the output current and output resistance for the cascode current source in Fig. P15.181 if $I_{REF} = 17.5$ μA, $\beta_{FO} = 110$, and $V_A = 50$ V? (b) What is the value of V_{CS} for this current source? (c) What is the minimum value of V_{CC}?

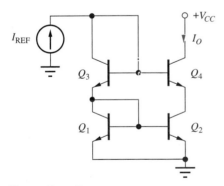

Figure P15.181

15.182. Simulate the current source in Prob. 15.181 and compare the results to hand calculations.

15.183. What is the equivalent resistance presented to I_{REF} in the cascode current source in Prob. 15.181?

15.8 Reference Current Generation

15.184. What are the output current and output resistance for the Widlar source in Fig. P15.184 if $I_{REF} = 80$ μA and $R_2 = 500$ Ω? (b) What is the new value of the output current if a layout error causes the area of Q_2 to be 5 percent larger than desired? (c) What are the new values or output current and resistance if the emitter area of Q_2 is reduced to $14A$?

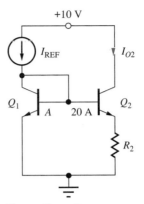

Figure P15.184

15.185. What are the output current and output resistance for the Widlar source in Fig. P15.184 if $I_{REF} = 35$ μA and $R_2 = 935$ Ω? (b) What are the new values if the emitter area of Q_1 is increased to $2A$?

15.186. $I_{REF} = 73$ μA in Fig. P15.184. (a) What value of R_2 is required to set $I_{O2} = 22$ μA? (b) To set $I_{O2} = 5.7$ μA? (c) To set $I_{O2} = 5.7$ μA if the area of Q_2 is changed to $10A$?

15.187. $I_{REF} = 62$ μA in Fig. P15.184. (a) What value of R_2 is required to set $I_{O2} = 12$ μA if the area Q_1 is changed to $2A$? (b) If the area of Q_2 is changed to $10A$?

15.188. Plot the variation of the output current vs. I_{REF} for the Widlar source in Fig. P15.184 for 50 μA < $I_{REF} \leq 5$ mA if $R_2 = 4$ kΩ and $\beta_{FO} = 100$.

15.189. (a) What is the output current of the V_{BE}-based reference in Fig. P15.189(a) if $I_S = 10^{-15}$ A, $\beta_F = \infty$, $R_1 = 10$ kΩ, $R_2 = 2.2$ kΩ, and $V_{EE} = 15$ V? (b) For $V_{EE} = 3.3$ V? (c) What is the output current of the V_{BE}-based reference in Fig. P15.189(b) if $R_1 = 10$ kΩ, $R_2 = 10$ kΩ, and $V_{CC} = 5$ V?

15.190. (a) Design the reference in Fig. P15.189(a) to produce an output current $I_O = 30$ μA. Assume $-V_{EE} = -3.3$ V and $I_S = 0.1$ fA and $\beta_{FO} = 130$ for both transistors. (b) Repeat for the circuit in Fig. P15.189(b) if $V_{CC} = 3.3$ V.

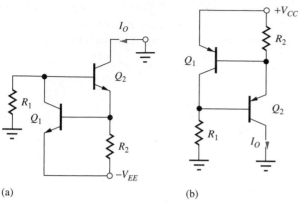

(a)

(b)

Figure P15.189

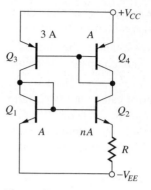

Figure P15.196

*15.191. Derive an expression for the output resistance of the cascode current source in Figs. 15.69(b) and 15.71.

*15.192. What is the output current of the NMOS reference in Fig. P15.192(a) if $R_1 = 10$ kΩ, $R_2 = 15$ kΩ, $K_n = 250$ μA/V^2, $V_{TN} = 0.75$ V, $\lambda = 0.017$ V^{-1}, and $V_{DD} = 10$ V?

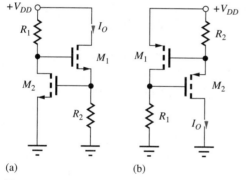

(a)

(b)

Figure P15.192

15.193. Design the reference in Fig. P15.192(a) to produce an output current $I_O = 75$ μA. Assume $V_{DD} = 6$ V and use the transistor parameters from Prob. 15.192.

*15.194. What is the output current of the PMOS reference in Fig. P15.192(b) if $R_1 = 10$ kΩ, $R_2 = 18$ kΩ, $K_p = 100$ μA/V^2, $V_{TP} = -0.75$ V, $\lambda = 0.02$ V^{-1}, and $V_{DD} = 5$ V?

15.195. Design the reference in Fig. P15.192(b) to produce an output current $I_O = 125$ μA. Assume $V_{DD} = 9$ V and use the transistor parameters from Prob. 15.194.

15.196. What are the collector currents in Q_1 and Q_2 in the reference in Fig. P15.196 if $V_{CC} = V_{EE} = 1.5$ V, $n = 20$, and $R = 2.2$ kΩ? Assume $\beta_{FO} = \infty$ and $V_A = \infty$.

15.197. Simulate the reference in Prob. 15.196 using SPICE, assuming $\beta_{FO} = 100$ and $V_A = 50$ V. Compare the currents to hand calculations and discuss the source of any discrepancies. Use SPICE to determine the sensitivity of the reference currents to power supply voltage changes.

15.198. What are the collector currents in the four transistors in Fig. P15.196 if $V_{CC} = V_{EE} = 3.3$ V, $n = 8$, and $R = 4$ kΩ?

15.199. What is the smallest value of n required for the circuit in Fig. P15.196 to operate properly (i.e., $V_{PTAT} > 0$)?

15.200. (a) What value of R is required to set $I_{C2} = 35$ μA in Fig. P15.196 if $n = 5$ and $T = 50°$C? (b) For $n = 10$ and $T = 0°$C?

15.201. What are the drain currents in M_1 and M_2 in the reference in Fig. P15.201 if $R = 5.1$ kΩ and $V_{DD} = V_{SS} = 5$ V? Use $K_n' = 25$ μA/V^2, $V_{TN} = 0.75$ V, $K_p' = 10$ μA/V^2, and $V_{TP} = -0.75$ V. Assume $\gamma = 0$ and $\lambda = 0$ for both transistor types.

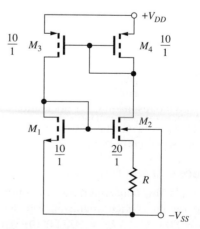

Figure P15.201

15.202. (a) Find the currents in both sides of the reference cell in Fig. P15.201 if $R = 10$ kΩ and $V_{DD} = V_{SS} = 5$ V, using $K'_n = 25$ µA/V^2, $V_{Ton} = 0.75$ V, $K'_p = 10$ µA/V^2, $V_{Top} = -0.75$ V, $\gamma_n = 0$ and $\gamma_p = 0$. Use $2\phi_F = 0.6$ V and $\lambda = 0$ for both transistor types. (b) Repeat for $\gamma_n = 0.5$ V$^{0.5}$ and $\gamma_p = 0.75$ V$^{0.5}$ and compare the results.

15.203. Simulate the references in Prob. 15.202(a) and (b) using SPICE with $\lambda = 0.017$ V^{-1}. Compare the currents to hand calculations (with $\gamma = 0$ and $\lambda = 0$) and discuss the source of any discrepancies. Use SPICE to determine the sensitivity of the reference currents to power supply voltage changes.

15.204. What are the collector currents in Q_1 to Q_8 in the reference in Fig. P15.204 if $V_{CC} = 0$ V, $V_{EE} = 3.3$ V, $R = 11$ kΩ, $R_6 = 3$ kΩ, $R_8 = 4$ kΩ, and $A_{E2} = 5$ A, $A_{E3} = 2$ A, $A_{E4} = A$, $A_{E5} = 2.5$ A, $A_{E6} = A$, $A_{E7} = 5$ A, and $A_{E8} = 3$ A?

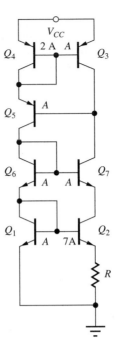

Figure P15.206

15.208. Repeat Prob. 15.206 assuming the emitter area of transistor Q_3 is changed to $2A$.

*15.209. (a) What are the drain currents in M_1 and M_2 in the reference in Fig. P15.209 if $R = 3300$ Ω, $V_{DD} = 15$ V, $K'_n = 25$ µA/V^2, $V_{TN} = 0.75$ V, $K'_p = 10$ µA/V^2, $V_{TP} = -0.75$ V, and $\lambda = 0$ for both transistor types? (b) Repeat part (a) if the W/L ratios of transistors M_5, M_6, and M_7 are all increased to 15/1.

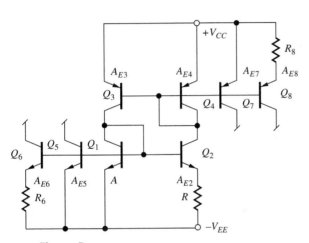

Figure P15.204

15.205. Repeat Prob. 15.204 if $A_{E2} = 10A$ and $A_{E3} = A$.

*15.206. (a) What are the collector currents in Q_1 to Q_7 in the reference in Fig. P15.206 if $V_{CC} = 5$ V and $R = 4300$ Ω? Assume $\beta_F = \infty = V_A$. (b) Repeat part (a) if the emitter areas of transistors Q_5, Q_6, and Q_7 are all changed to $2A$.

*15.207. (a) Simulate the reference in Prob. 15.206 using SPICE. Assume $\beta_{FOn} = 100$, $\beta_{FOp} = 50$, and both Early voltages $= 50$ V. Compare the currents to hand calculations and discuss the source of any discrepancies. Use SPICE to determine the sensitivity of the reference currents to power supply voltage changes.

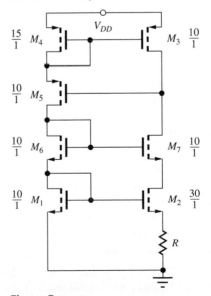

Figure P15.209

15.210. Simulate the reference in Prob. 15.209 with SPICE using $\lambda = 0.017$ V^{-1} for both transistor types. Compare the currents to those in Prob. 15.209 and discuss the source of any discrepancies. Use SPICE to determine the sensitivity of the reference currents to power supply voltage changes.

15.211. Repeat Prob. 15.209 assuming the W/L ratio of transistor M_3 is changed to 15/1.

15.9 The Current Mirror as an Active Load

15.212. What are the values of A_{dd}, A_{cd}, and CMRR for the amplifier in Fig. 15.77 if $I_{SS} = 200$ μA, $R_{SS} = 25$ MΩ, $K_n = K_p = 500$ μA/V^2, $V_{TN} = 1$ V, and $V_{TP} = -1$ V and $\lambda = 0.02$ V^{-1} for both transistors?

15.213. Use SPICE to simulate the amplifier in Prob. 15.212 and compare the results to the hand calculations. Use symmetrical 12-V supplies.

15.214. What are the values of A_{dd}, A_{cd}, and CMRR for the amplifier in Fig. 15.77 if $I_{SS} = 1$ mA, $R_{SS} = 10$ MΩ, $K_n = K_p = 500$ μA/V^2, $V_{TN} = -V_{TP} = 1$ V, and $\lambda = 0.015$/V for both transistors? What are the minimum power supply voltages if the common-mode input range must be ±5 V? Assume symmetrical supply voltages.

15.215. Use SPICE to simulate the amplifier in Prob. 15.214 and compare the results to hand calculations. Use symmetrical 12-V power supplies.

**15.216. (a) What are A_{dd} and A_{cd} for the bipolar differential amplifier in Fig. 15.83 ($R_L = \infty$) if $\beta_{op} = 70$, $\beta_{on} = 125$, $I_{EE} = 200$ μA, $R_{EE} = 25$ MΩ, and the Early voltages for both transistors are 60 V? What is the CMRR for $v_{C1} = v_{C2}$? (b) What are the minimum power supply voltages if the common-mode input range must be ±1.5 V? Assume symmetrical supply voltages.

15.217. Use SPICE to calculate A_{dd} and A_{cd} for the differential amplifier in Prob. 15.216. Compare the results to hand calculations.

15.218. (a) Repeat Prob. 15.216 if I_{EE} is changed to 50 μA, $R_{EE} = 100$ MΩ, and $V_A = 75$ V. (b) Repeat part (a) for $V_A = 100$ V.

15.219. Use SPICE to simulate the amplifier in Prob. 15.218 and compare the results to hand calculations. Use symmetrical 3-V power supplies.

*15.220. (a) Find the Q-points of the transistors in the CMOS differential amplifier in Fig. P15.220 if $V_{DD} = V_{SS} = 10$ V, $I_{SS} = 200$ μA, and $R_{SS} = 25$ mΩ. Assume $K_n' = 25$ μA/V^2, $V_{TN} = 0.75$ V, $K_p' = 10$ μA/V^2, $V_{TP} = -0.75$ V, and $\lambda = 0.017$ V^{-1} for both transistor types. (b) What is the voltage gain A_{dd} of the amplifier? (c) Compare this result to the gain of the amplifier in Fig. 15.77 if the Q-point and W/L ratios of M_1 to M_4 are the same.

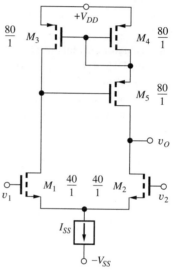

Figure P15.220

15.221. Use SPICE to simulate the amplifier in Prob. 15.220(a,b) and compare the results to hand calculations.

*15.222. Find the Q-points of the transistors in the folded-cascode CMOS differential amplifier in Fig. P15.222 if $V_{DD} = V_{SS} = 5$ V, $I_1 = 250$ μA, $I_2 = 250$ μA, $(W/L) = 40/1$ for all transistors, $K_n' = 25$ μA/V^2, $V_{TN} = 0.75$ V, $K_p' = 10$ μA/V^2, $V_{TP} = -0.75$ V, and $\lambda = 0.017$ V^{-1} for both transistor types. Draw the differential-mode half-circuit for transistors M_1 to M_4 and show that the circuit is in fact a cascode amplifier. What is the differential-mode voltage gain of the amplifier?

15.223. Use SPICE to simulate the amplifier in Prob. 15.222 and determine its voltage gain, output resistance, and CMRR. Compare to hand calculations.

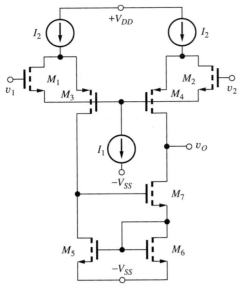

Figure P15.222

*15.224. Design a current mirror bias network to supply the three currents needed by the amplifier in Prob. 15.222.

Output Stages

15.225. What are the currents in Q_3 and Q_4 in the class-AB output stage in Fig. P15.225 if $R_1 = 20$ kΩ, $R_2 = 20$ kΩ, and $I_{S4} = I_{S3} = I_{S2} = 10^{-14}$ A. Assume $\beta_F = \infty$.

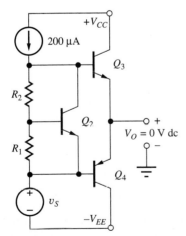

Figure P15.225

*15.226. (a) Show that the currents in Q_3 and Q_4 in the class-AB output stage in Fig. P15.226 are equal to $I_o = I_2\sqrt{(A_{E3}A_{E4})/(A_{E1}A_{E2})}$. (b) What are the currents in Q_3 and Q_4 if $A_{E1} = 3A_{E3}$, $A_{E2} = 3A_{E4}$, $I_2 = 300$ μA, $I_{SOpnp} = 4$ fA, and $I_{SOnpn} = 10$ fA?

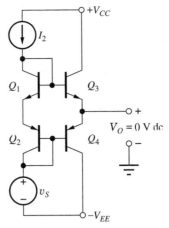

Figure P15.226

15.10 Active Loads in Operational Amplifiers

15.227. (a) Find the Q-points of the transistors in the CMOS op amp in Fig. 15.88 if $V_{DD} = V_{SS} = 5$ V, $I_{REF} = 250$ μA, $K_n' = 25$ μA/V^2, $V_{TN} = 0.75$ V, $K_p' = 10$ μA/V^2, and $V_{TP} = -0.75$ V. (b) What is the voltage gain of the op amp assuming the output stage has unity gain and $\lambda = 0.017$ V^{-1} for both transistor types? (c) What is the voltage gain if I_{REF} is changed to 500 μA?

15.228. Based on the example calculations and your knowledge of MOSFET characteristics, what will be the gain of the op amp in Ex. 15.11 if the I_{REF} is set to (a) 250 μA? (b) 20 μA? (*Note:* These should be short calculations.)

*15.229. What is the differential-mode gain of the amplifier in Fig. P15.229 if $V_{DD} = V_{SS} = 10$ V, $I_{REF} = 100$ μA, $K_n' = 25$ μA/V^2, $V_{TON} = 0.75$ V, $K_p' = 10$ μA/V^2, $V_{TOP} = -0.75$ V, $\gamma_n = 0$, and $\gamma_p = 0$. Use $\lambda = 0.017$ V^{-1} for both transistor types.

15.230. (a) Use SPICE to find the Q-points of the transistors of the amplifier in Prob. 15.229. (b) Repeat with $2\phi_F = 0.8$ V, $\gamma_n = 0.60$ V$^{0.5}$, and $\gamma_p = 0.75$ V$^{0.5}$, and compare the results to (a).

*15.231. Find the Q-points of the transistors in Fig. P15.229 if $V_{DD} = V_{SS} = 7.5$ V, $I_{REF} = 250$ μA, $(W/L)_{12} = 40/1$, $K_n' = 25$ μA/V^2, $V_{TN} = 0.75$ V, $K_p' = 10$ μA/V^2, and $V_{TP} = -0.75$ V. What is the differential-mode voltage gain of the op amp if $\lambda = 0.017$ V^{-1} for both transistor types?

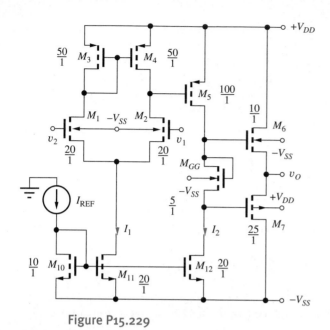

Figure P15.229

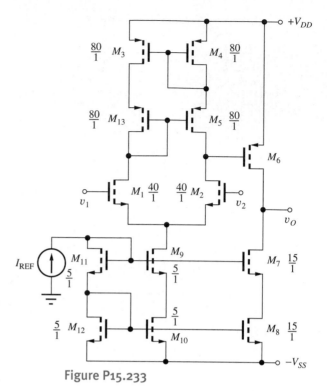

Figure P15.233

*15.232. (a) Estimate the minimum values of V_{DD} and V_{SS} needed for proper operation of the amplifier in Prob. 15.229. Use $K'_n = 25$ μA/V^2, $V_{TN} = 0.75$ V, $K'_p = 10$ μA/V^2, and $V_{TP} = -0.75$ V. (b) What are the minimum values of V_{DD} and V_{SS} needed to have at least a ± 5-V common-mode input range in the amplifier?

*15.233. (a) Find the Q-points of the transistors in Fig. P15.233 if $V_{DD} = V_{SS} = 10$ V, $I_{REF} = 250$ μA, $K'_n = 25$ μA/V^2, $V_{TN} = 0.75$ V, $K'_p = 10$ μA/V^2, and $V_{TP} = -0.75$ V. (b) What is the approximate value of the W/L ratio for M_6 of the CMOS op amp in order for the offset voltage to be zero? What is the differential-mode voltage gain of the op amp if $\lambda = 0.017$ V^{-1} for both transistor types?

*15.234. (a) Simulate the amplifier in Prob. 15.233 and compare its differential-mode voltage gain to the hand calculations in Prob. 15.233. (b) Use SPICE to calculate the offset voltage and CMRR of the amplifier.

15.235. Draw the amplifier that represents the mirror image of Fig. 15.88 by interchanging NMOS and PMOS transistors. Choose the W/L ratios of the NMOS and PMOS transistors so the voltage gain of the new amplifier is the same as the gain of the amplifier in Fig. 15.88. Maintain the operating currents the same and use the device parameter values from Ex. 15.12.

15.236. Draw the amplifier that represents the mirror image of Fig. 15.90 by interchanging *npn* and *pnp* transistors. If $\beta_{on} = 150$, $\beta_{op} = 60$, and $V_{AN} = V_{AP} = 60$ V, which of the two amplifiers will have the highest voltage gain? Why?

*15.237. What is the approximate emitter area of Q_{16} needed to achieve zero offset voltage in the amplifier in Fig. P15.237 if $I_B = 250$ μA and $V_{CC} = V_{EE} = 5$ V? What is the value of R_{BB} needed to set the quiescent current in the output stage to 75 μA? What are the voltage gain and input resistance of this amplifier? Assume $\beta_{on} = 150$, $\beta_{op} = 60$, $V_{AN} = V_{AP} = 60$ V, and $I_{SOnpn} = I_{SOpnp} = 15$ fA.

15.238. Use SPICE to simulate the characteristics of the amplifier in Prob. 15.237. Determine the offset voltage, voltage gain, input resistance, output resistance, and CMRR of the amplifier.

15.239. (a) What are the minimum values of V_{CC} and V_{EE} needed for proper operation of the amplifier in Fig. P15.237? (b) What are the minimum values of V_{CC} and V_{EE} needed to have at least a ± 1-V common-mode input range in the amplifier?

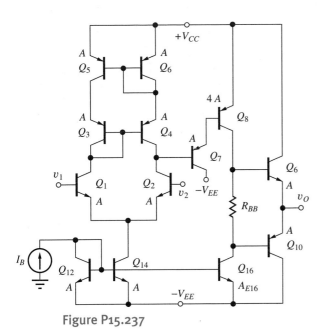

Figure P15.237

15.11 The μA741 Operational Amplifier

15.240. (a) What are the three bias currents in the source in Fig. P15.240 if $R_1 = 100$ kΩ, $R_2 = 4$ kΩ, and $V_{CC} = V_{EE} = 3$ V. (b) Repeat for $V_{CC} = V_{EE} = 22$ V. (c) Why is it important that I_1 in the μA741 be independent of power supply voltage but it does not matter as much for I_2 and I_3?

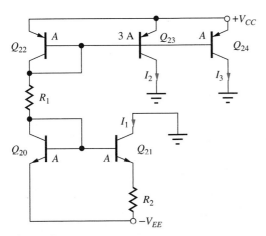

Figure P15.240

15.241. Choose the values of R_1 and R_2 in Fig. P15.240 to set $I_2 = 250$ μA and $I_1 = 50$ μA if $V_{CC} = V_{EE} = 12$ V. What is I_3?

15.242. Choose the values of R_1 and R_2 in Fig. P15.240 to set $I_3 = 300$ μA and $I_1 = 75$ μA if $V_{CC} = V_{EE} = 15$ V. What is I_2?

*15.243. (a) Based on the schematic in Fig. 15.92, what are the minimum values of V_{CC} and V_{EE} needed for proper operation of μA741 amplifier? (b) What are the minimum values of V_{CC} and V_{EE} needed to have at least a ±1-V common-mode input range in the amplifier?

15.244. What are the values of the elements in the Norton equivalent circuit in Fig. 15.99(a) if I_1 in Fig. 15.92 is increased to 50 μA?

15.245. Suppose Q_{23} in Fig. 15.93 is replaced by a cascode current source. (a) What is the new value of output resistance R_2? (b) What are the new values of the y-parameters of Fig. 15.99(b)? (c) What is the new value of A_{dm} for the op amp?

15.246. Draw a schematic for the cascode current source in Prob. 15.245.

15.247. Create a small-signal SPICE model for the circuit in Fig. 15.99(b) and verify the values of R_{in10}, G_m, and G_o.

**15.248. Figure P15.248 represents an op amp input stage that was developed following the introduction of the μA741. (a) Find the Q-points for all the transistors in the differential amplifier in Fig. P15.248 if $V_{CC} = V_{EE} = 15$ V and $I_{REF} = 100$ μA. (b) Discuss how this bias network operates to establish the Q-points. (c) Label the inverting and noninverting input terminals. (d) What are the transconductance and output resistance of this amplifier? Use $V_A = 60$ V.

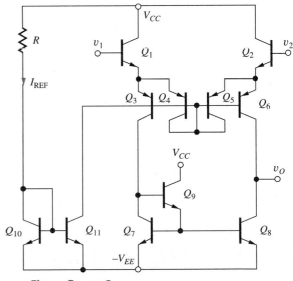

Figure P15.248

15.249. Figure P15.249 represents an op amp input stage that was developed following the introduction of the μA741. Find the Q-points for all the transistors in the differential amplifier in Fig. P15.237 if $V_{CC} = V_{EE} = 15$ V and $I_{REF} = 100$ μA. (b) Discuss how this bias network operates to establish the Q-points. (c) Label the inverting and noninverting input terminals. (d) What are the transconductance and output resistance of this amplifier? Use $V_A = 60$ V.

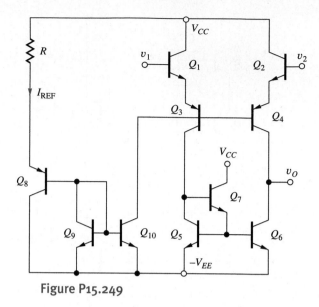

Figure P15.249

CHAPTER 16

FREQUENCY RESPONSE

CHAPTER GOALS

- Review transfer function analysis and determination of cutoff frequencies
- Understand dominant-pole approximations of amplifier transfer functions
- Learn to partition ac circuits into low-frequency and high-frequency equivalent circuits
- Learn the short-circuit time constant approach for estimating lower-cutoff frequency f_L
- Complete development of the small-signal models of both bipolar and MOS transistors with the addition of device capacitances
- Understand the unity-gain bandwidth product limitations of bipolar and field-effect transistors
- Learn the open-circuit time constant technique for estimating upper-cutoff frequency f_H
- Develop expressions for the upper-cutoff frequency of the inverting, noninverting, and follower configurations
- Demonstrate that the gain-bandwidth product limitations of the inverting, noninverting, and follower configurations approach the same upper limit

- Learn to apply the two time-constant approaches to the analysis of the frequency response of multistage amplifiers
- Explore bandwidth limitations of two-transistor circuits including current mirrors, cascode amplifiers, and differential pairs
- Understand the Miller effect
- Develop relationships between op amp unity-gain frequency and amplifier slew rate
- Understand the use of tuned circuits to produce narrow-band (high-Q) band-pass amplifiers
- Demonstrate the use of ac analysis in SPICE
- Demonstrate the use of MATLAB® to display frequency response information

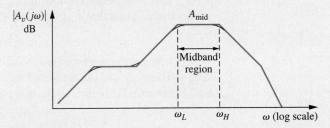

Chapters 13 to 15 discussed analysis and design of the midband characteristics of amplifiers. Low-frequency limitations due to coupling and bypass capacitors were discussed, but the internal capacitances of electronic devices, which limit the response at high frequencies, were neglected. This chapter completes the discussion of basic amplifier design with the introduction of methods used to tailor the frequency response of analog circuits at both low and high frequencies. As part of this discussion, the internal device capacitances of bipolar and field-effect transistors are discussed, and frequency-dependent small-signal models of the transistors are introduced. The unity-gain bandwidth product of the devices is expressed in terms of the small-signal parameters.

In order to complete our basic circuit-building block toolkit, expressions for the frequency responses of the single-stage inverting, noninverting, and follower configurations are each developed in detail. We show that the bandwidth of high-gain inverting and noninverting stages can be quite limited (although much wider than a typical op-amp stage of equal gain), whereas that of followers is normally very wide. Use of the cascode configuration is shown to significantly improve the frequency response of inverting amplifiers. Narrow-band (high-Q) band-pass amplifiers based on tuned circuits are also discussed.

Transfer functions for multistage amplifiers may have large numbers of poles and zeros, and direct circuit analysis, although theoretically possible, can be complex and unwieldy. Therefore, approximation techniques—the short-circuit and open-circuit time-constant methods—have been developed to estimate the upper- and lower-cutoff frequencies ω_H and ω_L.

The Miller effect is introduced, and the relatively low bandwidth associated with inverting amplifiers is shown to be caused by Miller multiplication of the collector-base or gate-drain capacitance of the transistor in the amplifier.

16.1 AMPLIFIER FREQUENCY RESPONSE

Figure 16.1 is the Bode plot for the magnitude of the voltage gain of a hypothetical amplifier. Regardless of the number of poles and zeros, the voltage transfer function $A_v(s)$ can be written as the ratio of two polynomials in s:

$$A_v(s) = \frac{N(s)}{D(s)} = \frac{a_0 + a_1 s + a_2 s^2 + \cdots + a_m s^m}{b_0 + b_1 s + b_2 s^2 + \cdots + b_n s^n} \tag{16.1}$$

In principle, the numerator and denominator polynomials of Eq. (16.1) can be written in factored form, and the poles and zeros can be separated into two groups. Those associated with the low-frequency response below the midband region of the amplifier can be combined into a function $F_L(s)$, and those associated with the high-frequency response above the midband region can be grouped into a function $F_H(s)$. Using F_L and F_H, $A_v(s)$ can be rewritten as

$$A_v(s) = A_{\text{mid}} F_L(s) F_H(s) \tag{16.2}$$

in which A_{mid} is the **midband gain**[1] of the amplifier in the region between the **lower-** and **upper-cutoff frequencies** (ω_L and ω_H, respectively). For A_{mid} to appear explicitly as shown in Eq. (16.2), $F_H(s)$ and $F_L(s)$ must be written in the two particular standard forms defined by Eqs. (16.3) and (16.4):

$$F_L(s) = \frac{\left(s + \omega_{Z1}^L\right)\left(s + \omega_{Z2}^L\right) \cdots \left(s + \omega_{Zk}^L\right)}{\left(s + \omega_{P1}^L\right)\left(s + \omega_{P2}^L\right) \cdots \left(s + \omega_{Pk}^L\right)} \tag{16.3}$$

$$F_H(s) = \frac{\left(1 + \dfrac{s}{\omega_{Z1}^H}\right)\left(1 + \dfrac{s}{\omega_{Z2}^H}\right) \cdots \left(1 + \dfrac{s}{\omega_{Zl}^H}\right)}{\left(1 + \dfrac{s}{\omega_{P1}^H}\right)\left(1 + \dfrac{s}{\omega_{P2}^H}\right) \cdots \left(1 + \dfrac{s}{\omega_{Pl}^H}\right)} \tag{16.4}$$

The representation of $F_H(s)$ is chosen so that its magnitude approaches a value of 1 at frequencies well below the upper-cutoff frequency ω_H,

$$|F_H(j\omega)| \to 1 \quad \text{for} \quad \omega \ll \omega_{Zi}^H, \omega_{Pi}^H \quad \text{for } i = 1 \ldots l \tag{16.5}$$

Thus, at low frequencies, the transfer function $A_v(s)$ becomes

$$A_L(s) \cong A_{\text{mid}} F_L(s) \tag{16.6}$$

The form of $F_L(s)$ is chosen so its magnitude approaches a value of 1 at frequencies well above ω_L:

$$|F_L(j\omega)| \to 1 \quad \text{for} \quad \omega \gg \omega_{Zj}^L, \omega_{Pj}^L \quad \text{for } j = 1 \ldots k \tag{16.7}$$

Thus, at high frequencies the transfer function $A_v(s)$ can be approximated by

$$A_H(s) \cong A_{\text{mid}} F_H(s) \tag{16.8}$$

---

[1] You may wish to review some of the frequency response definitions in Chapter 10.

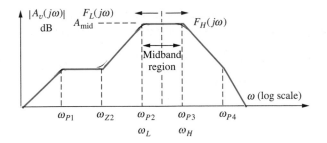

Figure 16.1 Bode plot for a general amplifier transfer function.

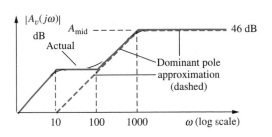

Figure 16.2 Bode plot for a complete transfer function and its dominant pole approximation.

16.1.1 LOW-FREQUENCY RESPONSE

In many designs, the zeros of $F_L(s)$ can be placed at frequencies low enough to not influence the lower-cutoff frequency ω_L. In addition, one of the low-frequency poles in Fig. 16.1, say ω_{P2}, can be designed to be much larger than the others. For these conditions, the low-frequency portion of the transfer function can be written approximately as

$$F_L(s) \cong \frac{s}{s + \omega_{P2}} \tag{16.9}$$

Pole ω_{P2} is referred to as the **dominant low-frequency pole** and the lower-cutoff frequency ω_L is approximately

$$\omega_L \cong \omega_{P2} \tag{16.10}$$

The Bode plot in Fig. 16.2 is an example of a transfer function and its dominant pole approximation. The overall transfer function $A_L(s)$ for this figure has two poles and two zeros.

16.1.2 ESTIMATING ω_L IN THE ABSENCE OF A DOMINANT POLE

If a dominant pole does not exist at low frequencies, then the poles and zeros interact to determine the lower-cutoff frequency, and a more complicated analysis must be used to find ω_L. As an example, consider the case of an amplifier having two zeros and two poles at low frequencies:

$$A_L(s) = A_{\text{mid}} F_L(s) = A_{\text{mid}} \frac{(s + \omega_{Z1})(s + \omega_{Z2})}{(s + \omega_{P1})(s + \omega_{P2})} \tag{16.11}$$

For $s = j\omega$,

$$|A_L(j\omega)| = A_{\text{mid}} |F_L(j\omega)| = A_{\text{mid}} \sqrt{\frac{\left(\omega^2 + \omega_{Z1}^2\right)\left(\omega^2 + \omega_{Z2}^2\right)}{\left(\omega^2 + \omega_{P1}^2\right)\left(\omega^2 + \omega_{P2}^2\right)}} \tag{16.12}$$

and remembering that ω_L is defined as the -3 dB frequency,

$$|A(j\omega_L)| = \frac{A_{\text{mid}}}{\sqrt{2}} \quad \text{and} \quad \frac{1}{\sqrt{2}} = \sqrt{\frac{\left(\omega_L^2 + \omega_{Z1}^2\right)\left(\omega_L^2 + \omega_{Z2}^2\right)}{\left(\omega_L^2 + \omega_{P1}^2\right)\left(\omega_L^2 + \omega_{P2}^2\right)}} \tag{16.13}$$

Squaring both sides and expanding Eq. (16.13),

$$\frac{1}{2} = \frac{\omega_L^4 + \omega_L^2\left(\omega_{Z1}^2 + \omega_{Z2}^2\right) + \omega_{Z1}^2\omega_{Z2}^2}{\omega_L^4 + \omega_L^2\left(\omega_{P1}^2 + \omega_{P2}^2\right) + \omega_{P1}^2\omega_{P2}^2} = \frac{1 + \dfrac{\left(\omega_{Z1}^2 + \omega_{Z2}^2\right)}{\omega_L^2} + \dfrac{\omega_{Z1}^2\omega_{Z2}^2}{\omega_L^4}}{1 + \dfrac{\left(\omega_{P1}^2 + \omega_{P2}^2\right)}{\omega_L^2} + \dfrac{\omega_{P1}^2\omega_{P2}^2}{\omega_L^4}} \tag{16.14}$$

If we assume that ω_L is larger than all the individual pole and zero frequencies, then the terms involving $1/\omega_L^4$ can be neglected, and the lower-cutoff frequency can be estimated from

$$\omega_L \cong \sqrt{\omega_{P1}^2 + \omega_{P2}^2 - 2\omega_{Z1}^2 - 2\omega_{Z2}^2} \tag{16.15}$$

For the more general case of n poles and n zeros, a similar analysis yields

$$\omega_L \cong \sqrt{\sum_n \omega_{Pn}^2 - 2\sum_n \omega_{Zn}^2} \tag{16.16}$$

EXERCISE: Use Eq. (16.15) to estimate f_L for the transfer functions

$$A_v(s) = \frac{200s(s+50)}{(s+10)(s+1000)} \quad \text{and} \quad A_v(s) = \frac{100s(s+500)}{(s+100)(s+1000)}$$

ANSWERS: 159 Hz, 114 Hz

EXAMPLE 16.1   **ANALYSIS OF A TRANSFER FUNCTION**

The midband gain, poles, zeros, and cutoff frequency are identified from a specified transfer function.

PROBLEM   Find the midband gain, $F_L(s)$, and lower-cutoff frequency f_L for

$$A_L(s) = 2000\frac{s\left(\dfrac{s}{100}+1\right)}{(0.1s+1)(s+1000)}$$

Identify the frequencies corresponding to the poles and zeros. Find a dominant pole approximation for the transfer function, if one exists.

SOLUTION   **Known Information and Given Data:** The transfer function is specified.

Unknowns: A_{mid}, $F_L(s)$, f_L, poles, zeros, dominant-pole approximation

Approach: Rearrange $A_L(s)$ into the form of Eqs. (16.6) and (16.3). Identify the pole and zero frequencies. Find the midband region and A_{mid}. Since the poles and zeros can all be found, use Eq. (16.16) to find f_L. If the poles and zeros are widely separated, find the dominant pole representation.

Assumptions: None

Analysis: To begin, we need to rearrange the transfer function by factoring 0.01 out of the numerator and 0.1 out of the denominator in order to have all the poles and zeros written as in Eq. (16.3):

$$A_L(s) = 200\frac{s(s+100)}{(s+10)(s+1000)}$$

Now, $A_L(s) = A_{\text{mid}}F_L(s)$ with $A_{\text{mid}} = 200$ and

$$F_L(s) = \frac{s(s+100)}{(s+10)(s+1000)}$$

Zeros occur at the values of s for which the numerator is zero: $s = 0$ and $s = -100$ rad/s. Poles occur at the frequencies s for which the denominator is zero: $s = -10$ rad/s and $s = -1000$ rad/s. Substituting these values in to Eq. (16.16) yields an estimate of f_L:

$$f_L = \frac{1}{2\pi}\sqrt{10^2 + 1000^2 - 2(0^2 + 100^2)^2} = \frac{990}{2\pi} = 158 \text{ Hz}$$

Note that these are all at low frequencies and are separated from one another by a decade of frequency. Thus, a dominant pole exists at $\omega = 1000$, and the lower-cutoff frequency is given approximately by $f_L \cong 1000/2\pi = 159$ Hz. For frequencies above a few hundred rad/s, the transfer function can be approximated by

$$A_L(s) \cong 200\frac{s}{s + 1000} \qquad \text{for} \qquad \omega > 200 \text{ rad/s}$$

Check of Results: The requested unknowns have been found. For $\omega \gg 1000$, the transfer original function reaches its largest value and becomes constant—the midband region:

$$A_L(s)|_{s \gg 1000} \cong 200\frac{s^2}{s^2} = 200$$

Thus, $A_{\text{mid}} = 200$ or 46 dB. We also see that the value of f_L predicted by Eq. (16.16) is the same as that of the dominant-pole model, indicating correctness of the dominant-pole approximation.

Discussion: Figure 16.2 graphs the original transfer function and its dominant-pole approximation. The midband region is clearly viable for $\omega > 1000$ rad/s, and the single pole roll-off is valid for frequencies down to approximately 200 rad/s.

Computer-Aided Analysis: We can easily visualize the transfer function with the aid of MATLAB®: bode ([200 20000 0], [1 1010 10000]). The resulting graph of the magnitude and phase of $A_L(s)$ appears in the figure. The alternating sequence of zeros and poles is apparent in both the magnitude and phase plots, and the gain approaches 46 dB at high frequencies.

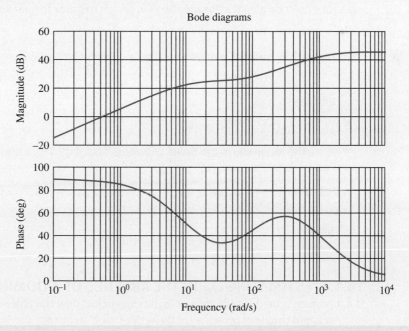

EXERCISE: For what range of frequencies does the approximation to $A_v(s)$ in Example 16.1 differ from the actual transfer function by less than 10 percent?

ANSWER: $\omega \geq 205$ rad/s

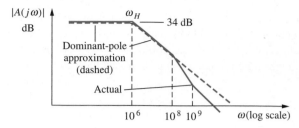

Figure 16.3 Bode plot for a complete transfer function and its dominant-pole approximation.

16.1.3 HIGH-FREQUENCY RESPONSE

In the region above midband, $A_v(s)$ can be represented by its high-frequency approximation:

$$A_H(s) \cong A_{\text{mid}} F_H(s) \tag{16.17}$$

Many of the zeros of $F_H(s)$ are often at infinite frequency, or high enough in frequency that they do not influence the value of $F_H(s)$ near ω_H. If, in addition, one of the **pole frequencies**—for example, ω_{P3} in Fig. 16.1—is much smaller than all the others, then a **dominant high-frequency pole** exists in the high-frequency response, and $F_H(s)$ can be represented by the approximation

$$F_H(s) \cong \frac{1}{1 + \dfrac{s}{\omega_{P3}}} \tag{16.18}$$

For the case of a dominant pole, the upper-cutoff frequency is given by $\omega_H \cong \omega_{P3}$. Figure 16.3 is an example of a Bode plot of a transfer function at high frequencies and its dominant-pole approximation.

EXERCISE: The transfer function for the amplifier in Fig. 16.3 is

$$A_H(s) = 50 \frac{\left(1 + \dfrac{s}{10^9}\right)}{\left(1 + \dfrac{s}{10^6}\right)\left(1 + \dfrac{s}{10^8}\right)}$$

What are the locations of the poles and zeros of $A_H(s)$? What are A_{mid}, $F_H(s)$ for the dominant-pole approximation, and f_H?

ANSWERS: $\omega_{Z1} = -10^9$ rad/s, $\omega_{P1} = -10^6$ rad/s, $\omega_{P2} = -10^8$ rad/s; 50, $F_H(s) = \dfrac{1}{\left(1 + \dfrac{s}{10^6}\right)}$,

159 kHz

16.1.4 ESTIMATING ω_H IN THE ABSENCE OF A DOMINANT POLE

If a dominant pole does not exist at high frequencies, then the poles and zeros interact to determine ω_H. An approximate expression for the upper-cutoff frequency can be found from the expression for F_H in a manner similar to that used to arrive at Eq. (16.16). Consider the case of an amplifier having two zeros and two poles at high frequencies:

$$A_H(s) = A_{\text{mid}} F_H(s) = A_{\text{mid}} \frac{\left(1 + \dfrac{s}{\omega_{Z1}}\right)\left(1 + \dfrac{s}{\omega_{Z2}}\right)}{\left(1 + \dfrac{s}{\omega_{P1}}\right)\left(1 + \dfrac{s}{\omega_{P2}}\right)} \tag{16.19}$$

and for $s = j\omega$,

$$|A_H(j\omega)| = A_{\text{mid}}|F_H(j\omega)| = A_{\text{mid}}\sqrt{\frac{\left(1 + \dfrac{\omega^2}{\omega_{Z1}^2}\right)\left(1 + \dfrac{\omega^2}{\omega_{Z2}^2}\right)}{\left(1 + \dfrac{\omega^2}{\omega_{P1}^2}\right)\left(1 + \dfrac{\omega^2}{\omega_{P2}^2}\right)}} \qquad (16.20)$$

At the upper-cutoff frequency $\omega = \omega_H$,

$$|A(j\omega_H)| = \frac{A_{\text{mid}}}{\sqrt{2}} \qquad \text{and} \qquad \frac{1}{\sqrt{2}} = \sqrt{\frac{\left(1 + \dfrac{\omega_H^2}{\omega_{Z1}^2}\right)\left(1 + \dfrac{\omega_H^2}{\omega_{Z2}^2}\right)}{\left(1 + \dfrac{\omega_H^2}{\omega_{P1}^2}\right)\left(1 + \dfrac{\omega_H^2}{\omega_{P2}^2}\right)}} \qquad (16.21)$$

By squaring both sides and expanding Eq. (16.21), and assuming ω_H is smaller than all the individual pole and zero frequencies, the upper-cutoff frequency can be found to be

$$\omega_H \cong \frac{1}{\sqrt{\dfrac{1}{\omega_{P1}^2} + \dfrac{1}{\omega_{P2}^2} - \dfrac{2}{\omega_{Z1}^2} - \dfrac{2}{\omega_{Z2}^2}}} \qquad (16.22)$$

The expression for the general case of n poles and n zeros can be found in a manner similar to Eq. (16.22), and the resulting approximation for ω_H is

$$\omega_H \cong \frac{1}{\sqrt{\displaystyle\sum_n \frac{1}{\omega_{Pn}^2} - 2\sum_n \frac{1}{\omega_{zn}^2}}} \qquad (16.23)$$

EXERCISE: Write the expression for the $A_H(s)$ below in standard form. What are the pole and zero frequencies? What are A_{mid}, $F_H(s)$, and f_H?

$$A_H(s) = \frac{2.5 \times 10^7 (s + 2 \times 10^5)}{(s + 10^5)(s + 5 \times 10^5)}$$

ANSWERS: $A_H(s) = 100 \dfrac{\left(1 + \dfrac{s}{2 \times 10^5}\right)}{\left(1 + \dfrac{s}{10^5}\right)\left(1 + \dfrac{s}{5 \times 10^5}\right)}$; -10^5 rad/s, -5×10^5 rad/s, -2×10^5 rad/s; ∞, 40 dB, 21.7 kHz

16.2 DIRECT DETERMINATION OF THE LOW-FREQUENCY POLES AND ZEROS—THE COMMON-SOURCE AMPLIFIER

To apply the theory in Sec. 16.1, we need to know the location of all the individual poles and zeros. In principle, the frequency response of an amplifier can always be calculated by direct analysis of the circuit in the frequency domain, so this section begins with an example of this form of analysis for the common-source amplifier. However, as circuit complexity grows, exact analysis by hand rapidly becomes intractable. Although SPICE analysis can always be used to study the characteristics of an amplifier for a given set of parameter values, a more general understanding of the factors that control the cutoff frequencies of the amplifier is needed for design. Because we are most often interested in the position of ω_L and ω_H, we subsequently develop approximation techniques that can be used to estimate ω_L and ω_H.

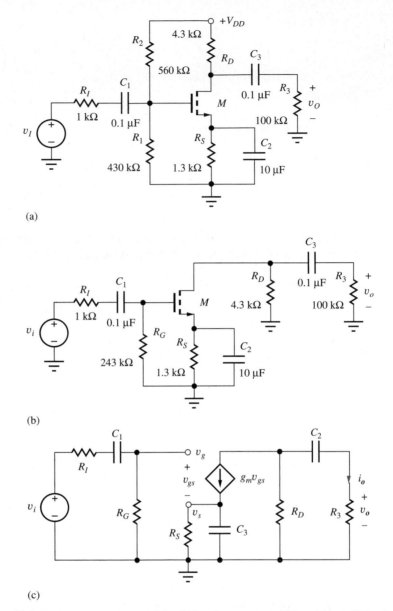

Figure 16.4 (a) A common-source amplifier, (b) low-frequency ac model, and (c) small-signal model.

The circuit for the common-source amplifier from Chapter 13 is repeated in Fig. 16.4(a) along with its ac equivalent circuit in Fig. 16.4(b). At low frequencies below midband, the impedance of the capacitors can no longer be assumed to be negligible, and they must be retained in the ac equivalent circuit. To determine circuit behavior at low frequencies, we replace transistor Q_1 by its low-frequency small-signal model, as in Fig. 16.4(c). Because the stage has an external load resistor, r_o is neglected in the circuit model.

In the frequency domain, output voltage $\mathbf{V_o}(s)$ can be found by applying current division at the drain of the transistor:

$$\mathbf{V_o}(s) = \mathbf{I_o}(s)R_3 \quad \text{where} \quad \mathbf{I_o}(s) = -g_m \mathbf{V_{gs}}(s)\dfrac{R_D}{R_D + \dfrac{1}{sC_3} + R_3}$$

and

$$\mathbf{V_o}(s) = -g_m \mathbf{V_{gs}}(s) \frac{R_D}{R_D + \dfrac{1}{sC_3} + R_3} R_3 = -g_m(R_3 \| R_D) \frac{s}{s + \dfrac{1}{C_3(R_D + R_3)}} \mathbf{V_{gs}}(s) \quad (16.24)$$

Next, we must find $\mathbf{V_{gs}}(s) = \mathbf{V_g}(s) - \mathbf{V}_s(s)$. Because the gate terminal in Fig. 16.4(c) represents an open circuit, $\mathbf{V_g}(s)$ can be determined using voltage division:

$$\mathbf{V_g}(s) = \mathbf{V_i}(s) \frac{R_G}{R_I + \dfrac{1}{sC_1} + R_G} = \mathbf{V_i}(s) \frac{sC_1 R_G}{sC_1(R_I + R_G) + 1} \quad (16.25)$$

and the voltage at the source of the FET can be found by writing a nodal equation for $\mathbf{V}_s(s)$:

$$g_m(\mathbf{V_g} - \mathbf{V_s}) - G_S \mathbf{V_s} - sC_2 \mathbf{V_s} = 0 \quad \text{or} \quad \mathbf{V_s} = \frac{g_m}{sC_2 + g_m + G_S} \mathbf{V_g} \quad (16.26)$$

and

$$\mathbf{V_{gs}}(s) = (\mathbf{V_g} - \mathbf{V_s}) = \mathbf{V_g} \left[1 - \frac{g_m}{sC_2 + g_m + G_S} \right] = \frac{sC_2 + G_S}{sC_2 + g_m + G_S} \mathbf{V_g} \quad (16.27)$$

By dividing through by C_3, Eq. (16.27) can be rewritten as

$$(\mathbf{V_g} - \mathbf{V_s}) = \frac{s + \dfrac{1}{C_2 R_S}}{s + \dfrac{1}{C_2 \left(\dfrac{1}{g_m} \| R_S \right)}} \mathbf{V_g}(s) \quad (16.28)$$

Finally, combining Eqs. (16.24), (16.25), and (16.28) yields an overall expression for the voltage transfer function:

$$A_v(s) = \frac{\mathbf{V_o}(s)}{\mathbf{V_i}(s)} = A_{\text{mid}} F_L(s)$$

$$= \left[-g_m(R_3 \| R_D) \frac{R_G}{(R_I + R_G)} \right] \frac{s^2 \left[s + \dfrac{1}{C_2 R_S} \right]}{\left[s + \dfrac{1}{C_1(R_I + R_G)} \right] \left[s + \dfrac{1}{C_2 \left(\dfrac{1}{g_m} \| R_S \right)} \right] \left[s + \dfrac{1}{C_3(R_D + R_3)} \right]}$$

$$(16.29)$$

In Eq. (16.29), $A_v(s)$ has been written in the form that directly exposes the midband gain and $F_L(s)$:

$$A_v(s) = A_{\text{mid}} F_L(s) \quad \text{where} \quad A_{\text{mid}} = -g_m(R_D \| R_3) \frac{R_G}{R_G + R_I} \quad (16.30)$$

A_{mid} should be recognized as the voltage gain of the circuit with the capacitors all replaced by short circuits.

Although the analysis in Eqs. (16.24) to (16.30) may seem rather tedious, we nevertheless obtain a complete description of the frequency response. In this example, the poles and zeros of the transfer function appear in factored form in Eq. (16.29). Unfortunately, this is an artifact of this particular FET circuit and generally will not be the case. The infinite input resistance of the FET and absence of r_o in the circuit have decoupled the nodal equations for v_g, v_s, and v_o. In most cases, the mathematical analysis is even more complex. For example, if a bipolar transistor were used in which both r_π and r_o were included, the analysis would require the simultaneous solution of three equations in three unknowns.

EXERCISE: Draw the midband ac equivalent circuit for the amplifier in Fig. 16.2 and derive the expression for A_{mid} directly from this circuit.

ANSWER: Eq. (16.30)

Let us now explore the origin of the poles and zeros of the voltage transfer function. Eq. (16.29) has three poles and three zeros, *one pole and one zero for each independent capacitor* in the circuit. Two of the zeros are at $s = 0$ (dc), corresponding to series capacitors C_1 and C_3, each of which blocks the propagation of dc signals through the amplifier. The third zero occurs at the frequency for which the impedance of the parallel combination of R_S and C_2 becomes infinite. At this frequency, propagation of signal current through the MOSFET is blocked, and the output voltage must be zero. Thus, the three zero locations are

$$s = 0, 0, -\frac{1}{R_S C_2} \tag{16.31}$$

From the denominator of Eq. (16.29), the three poles are located at frequencies of

$$s = -\frac{1}{(R_I + R_G)C_1}, \ -\frac{1}{(R_D + R_3)C_3}, \ -\frac{1}{\left(R_S \left\| \frac{1}{g_m} \right.\right)C_2} \tag{16.32}$$

These pole frequencies are determined by the time constants associated with the three individual capacitors. Because the input resistance of the FET is infinite, the resistance present at the terminals of capacitor C_1 is simply the series combination of R_I and R_G, and because the output resistance r_o of the FET has been neglected, the resistance associated with capacitor C_3 is the series combination of R_3 and R_D. The effective resistance in parallel with capacitor C_2 is the equivalent resistance present at the source terminal of the FET, which is equal to the parallel combination of resistor R_S and $1/g_m$. Section 16.3 has a more complete interpretation of these resistance expressions.

EXAMPLE 16.2   **DIRECT CALCULATION OF THE POLES AND ZEROS OF THE COMMON-SOURCE AMPLIFIER**

Analyze the low-frequency behavior of a common-source amplifier, including the effects of coupling and bypass capacitors.

PROBLEM   Find the midband gain, poles, zeros, and cutoff frequency for the common-source amplifier in Fig. 16.4. Assume $g_m = 1.23$ mS. Write a complete expression for the amplifier transfer function. Write a dominant-pole representation for the amplifier transfer function.

SOLUTION   **Known Information and Given Data:** The circuit with element values appears in Fig. 16.4, and $g_m = 1.23$ mS. Expressions for A_{mid} and the individual poles and zeros are given in Eqs. (16.29) and (16.30).

Unknowns: A_{mid}, poles, zeros, f_L, dominant-pole approximation, complete transfer function

Approach: Use the circuit element values to find A_{mid} and the poles and zeros from Eqs. (16.31) and (16.32). Use the pole and zero values to find f_L from Eq. (16.15).

Assumptions: Small-signal conditions apply; output resistance r_o can be neglected

Analysis: To begin, we will find A_{mid}:

$$A_{mid} = -(1.23 \text{ mS})(4.3 \text{ k}\Omega \| 100 \text{ k}\Omega)\frac{243 \text{ k}\Omega}{1.0 \text{ k}\Omega + 243 \text{ k}\Omega} = -5.05 \quad \text{or} \quad 14.1 \text{ dB}$$

From Eq. (16.29), the three zeros are

$$\omega_{Z1} = 0 \qquad \omega_{Z2} = 0 \qquad \omega_{Z3} = -\frac{1}{(10\ \mu\text{F})(1.3\ \text{k}\Omega)} = -76.9\ \text{rad/s}$$

and the three poles are

$$\omega_{P1} = -\frac{1}{(0.1\ \mu\text{F})(1\ \text{k}\Omega + 243\ \text{k}\Omega)} = -41.0\ \text{rad/s}$$

$$\omega_{P3} = -\frac{1}{(0.1\ \mu\text{F})(4.3\ \text{k}\Omega + 100\ \text{k}\Omega)} = -95.9\ \text{rad/s}$$

$$\omega_{P2} = -\frac{1}{(10\ \mu\text{F})\left(1.3\ \text{k}\Omega \left\|\dfrac{1}{1.23\ \text{mS}}\right.\right)} = -200\ \text{rad/s}$$

The lower-cutoff frequency is given by

$$f_L = \frac{1}{2\pi}\sqrt{41.0^2 + 95.9^2 + 200^2 - 2(0^2 + 0^2 + 76.9^2)} = \frac{197}{2\pi} = 31.5\ \text{Hz}$$

and the complete transfer function is

$$A_v(s) = -5.05\frac{s^2(s + 76.9)}{(s + 41.0)(s + 95.9)(s + 200)}$$

The dominant-pole estimate could be written either using the calculated value of f_L or the highest pole

$$A_v(s) \cong -5.05\frac{s}{s + 197} \qquad \text{or} \qquad A_v(s) \cong -5.05\frac{s}{s + 200}$$

Check of Results: A double check of our math indicates the calculations are correct. We see that A_{mid} is small, so that neglecting r_o should be reasonable.

Discussion: Although the poles and zeros are not widely spaced, the lower-cutoff frequency is surprisingly close to ω_{P3}. This occurs because of an approximate pole-zero cancellation that is taking place between ω_{Z3} and ω_{P2}.

Computer-Aided Analysis: SPICE simulation results for the common-source amplifier appear in the figure here. The simulation used $V_{DD} = 12$ V, FSTART = 0.01 Hz, and FSTOP = 10 kHz with 10 frequency points per decade. The values of A_{mid} and f_L agree with our hand calculations. The small discrepancies are related to our neglect of r_o in the calculations. We can also plot $A_v(s)$ by multiplying out the numerator and denominator and then using MATLAB®: bode (−5.05* [1 76.9 0 0],[1 336.9 31311.9 786380]) or by using the convolution function to multiply the polynomials for us: bode(−5.05*[1 76.9 0 0],[conv([1 41],conv([1 95.9],[1 200]))]).

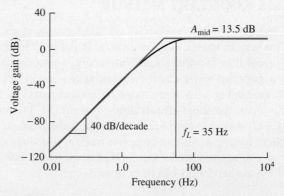

SPICE simulation results for the C-S amplifier in Fig. 16.4 ($V_{DD} = 12$ V).

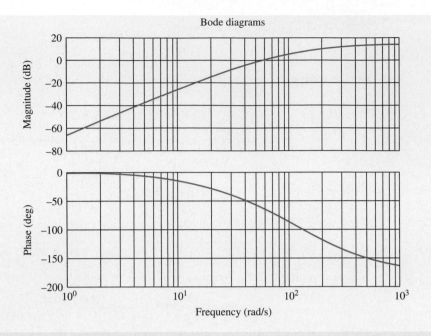

16.3 ESTIMATION OF ω_L USING THE SHORT-CIRCUIT TIME-CONSTANT METHOD

To use Eq. (16.16) or Eq. (16.23), the location of all the poles and zeros of the amplifier must be known. In most cases, however, it is not easy to find the complete transfer function, let alone represent it in factored form. Fortunately, we are most often interested in the values of A_{mid}, and the upper- and lower-cutoff frequencies ω_H and ω_L that define the bandwidth of the amplifier, as indicated in Fig. 16.5. Knowledge of the exact position of all the poles and zeros is not necessary. Two techniques, the **short-circuit time-constant (SCTC) method** and the **open-circuit time-constant (OCTC) method,** have been developed; these produce good estimates of ω_L and ω_H, respectively, without having to find the complete transfer function.

It can be shown theoretically [1] that the lower-cutoff frequency for a network having n coupling and bypass capacitors can be estimated from

$$\omega_L \cong \sum_{i=1}^{n} \frac{1}{R_{iS}C_i} \tag{16.33}$$

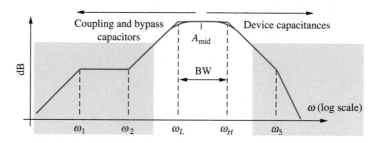

Figure 16.5 Midband region of primary interest in most amplifier transfer functions.

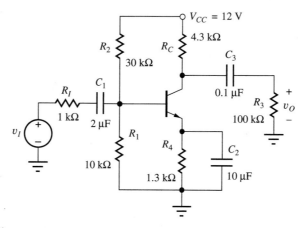

Figure 16.6 Common-emitter amplifier including finite capacitor values.

in which R_{iS} represents the resistance at the terminals of the ith capacitor C_i with all the other capacitors replaced by short circuits. The product $R_{iS}C_i$ represents the short-circuit time constant associated with capacitor C_i. We now use the SCTC method to find ω_L for the three classes of single-stage amplifiers.

16.3.1 ESTIMATE OF ω_L FOR THE COMMON-EMITTER AMPLIFIER

We use the C-E amplifier in Fig. 13.4, Chapter 13, as a first example of the SCTC method; this is redrawn in Fig. 16.6 and now includes finite values for the capacitors. R_4 has also been reduced slightly to shift the Q-point. The presence of r_π in the bipolar model causes direct calculation of the transfer function to be complex; including r_o leads to even further difficulty. Thus, the circuit is a good example of applying the method of short-circuit time constants to a network.

The ac model for the C-E amplifier in Fig. 16.7 contains three capacitors, and three short-circuit time constants must be determined in order to apply Eq. (16.33). The three analyses rely on the expressions for the midband input and output resistances of the BJT amplifier in Table 14.8.

R_{1S}

For C_1, R_{1S} is found by replacing C_2 and C_3 by short circuits, yielding the network in Fig. 16.8. R_{1S} represents the equivalent resistance present at the terminals of capacitor C_1. Based on Fig. 16.8,

$$R_{1S} = R_I + (R_B \| R_{iB}) = R_I + (R_B \| r_\pi) \tag{16.34}$$

R_{1S} is equal to the source resistance R_I in series with the parallel combination of the base bias resistor R_B and the input resistance r_π of the BJT.

The Q-point for this amplifier is found to be (1.66 mA, 2.70 V), and for $\beta_o = 100$ and $V_A = 75$ V,

$$r_\pi = 1.51 \text{ k}\Omega \quad \text{and} \quad r_o = 46.8 \text{ k}\Omega$$

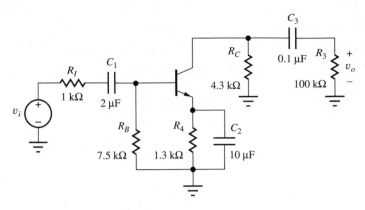

Figure 16.7 ac Model for the C-E amplifier in Fig. 16.6.

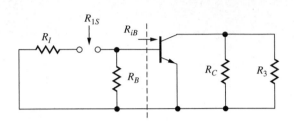

Figure 16.8 Circuit for finding R_{1S}.

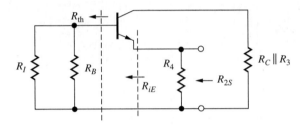

Figure 16.9 Circuit for finding R_{3S}.

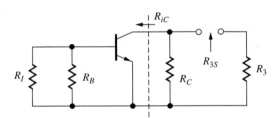

Figure 16.10 Circuit for finding R_{3S}.

Using these values and those of the other circuit elements,

$$R_{1S} = 1000\ \Omega + (7500\ \Omega\|1510\ \Omega) = 2260\ \Omega$$

and

$$\frac{1}{R_{1S}C_1} = \frac{1}{(2.26\ \text{k}\Omega)(2.00\ \mu\text{F})} = 222\ \text{rad/s} \qquad (16.35)$$

R_{2S}

The network used to find R_{2S} is constructed by shorting capacitors C_1 and C_3, as in Fig. 16.9, and

$$R_{2S} = R_4 \| R_{iE} = R_4 \left\| \frac{r_\pi + R_{\text{th}}}{\beta_o + 1} \right. \qquad \text{where} \qquad R_{\text{th}} = R_I \| R_B \qquad (16.36)$$

R_{2S} represents the combination of emitter resistance R_4 in parallel with the equivalent resistance at the emitter terminal of the BJT. For the values in this particular circuit,

$$R_{\text{th}} = R_I \| R_B = 1000\ \Omega \| 7500\ \Omega = 882\ \Omega$$

$$R_{2S} = 1300\ \Omega \left\| \frac{1510\ \Omega + 882\ \Omega}{101} \right. = 23.3\ \Omega$$

and

$$\frac{1}{R_{2S}C_2} = \frac{1}{(23.3\ \Omega)(10\ \mu\text{F})} = 4300\ \text{rad/s} \qquad (16.37)$$

R_{3S}

Finally, the network used to find R_{3S} is constructed by shorting capacitors C_1 and C_2, as in Fig. 16.10. For this network,

$$R_{3S} = R_3 + (R_C \| R_{iC}) = R_3 + (R_C \| r_o) \cong R_3 + R_C \qquad (16.38)$$

R_{3S} represents the combination of load resistance R_3 in series with the parallel combination of collector resistor R_C and the collector resistance r_o of the BJT. For the values in this particular circuit,

$$R_{3S} = 100 \text{ k}\Omega + (4.30 \text{ k}\Omega \| 46.8 \text{ k}\Omega) = 104 \text{ k}\Omega \tag{16.39}$$

and

$$\frac{1}{R_{3S}C_3} = \frac{1}{(104 \text{ k}\Omega)(0.100 \text{ μF})} = 96.1 \text{ rad/s} \tag{16.40}$$

The ω_L Estimate

Using the three time-constant values from Eqs. (16.35), (16.37), and (16.40) yields estimates for ω_L and f_L:

$$\omega_L \cong \sum_{i=1}^{3} \frac{1}{R_{iS}C_i} = 222 + 96.1 + 4300 = 4620 \text{ rad/s} \tag{16.41}$$

and

$$f_L = \frac{\omega_L}{2\pi} = 735 \text{ Hz}$$

The lower-cutoff frequency of the amplifier is approximately 735 Hz.

Note in this example that the time constant associated with emitter bypass capacitor C_2 is dominant; that is, the value of $R_{2S}C_2$ is more than an order of magnitude larger than the other two time constants so that $\omega_L \cong 1/R_{2S}C_2$ ($f_L \cong 4300/2\pi = 685$ Hz). This is a common situation and a practical approach to the design of ω_L. Because the resistance presented at the emitter or source of the transistor is low, the time constant associated with an emitter or source bypass capacitor is often dominant and can be used to set ω_L. The other two time constants can easily be designed to be much larger.

EXERCISE: Simulate the frequency response of the circuit in Fig. 16.6 using SPICE, and find the midband gain and lower-cutoff frequency. Use $\beta_o = 100$, $I_S = 1$ fA, and $V_A = 75$ V. What is the Q-point?

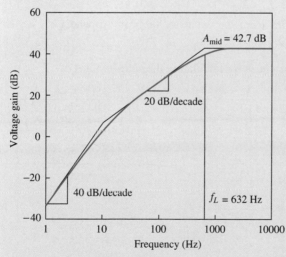

ANSWERS: 135, 635 Hz, (1.64 mA, 2.79 V)

SPICE simulation results.

EXERCISE: Find the short-circuit time constants and f_L for the common-emitter amplifier in Fig. 16.7 if $R_B = 75$ kΩ, $R_4 = 13$ kΩ, $R_C = 43$ kΩ, and $I_C = 175$ μA. Assume $\beta_o = 140$ and $V_A = 80$ V. The other values remain unchanged.

ANSWERS: 11.1 ms; 10.4 ms; 1.35 ms; 148 Hz

DESIGN EXAMPLE 16.3   LOWER-CUTOFF FREQUENCY DESIGN IN THE COMMON-EMITTER AMPLIFIER

Choose the coupling and bypass capacitors to set the value of f_L of the common-emitter to a specified value.

PROBLEM   Choose C_1, C_2, and C_3 to set $f_L = 2000$ Hz in the amplifier in Fig. 16.6.

SOLUTION   **Known Information and Given Data:** The circuit with resistor values appears in Fig. 16.6 with $\beta_o = 100$, $r_\pi = 1.51$ kΩ, and $r_o = 46.8$ kΩ. From Eqs. (16.34) through (16.39), we have $R_{1S} = 2.26$ kΩ, $R_{2S} = 23.3$ Ω, and $R_{3S} = 104$ kΩ.

Unknowns: C_1, C_2, and C_3

Approach: Because R_{3S} is much smaller than the other two resistors, its associated time constant can easily be designed to dominate the value of ω_L as occurred in Eq. (16.41). Thus, the approach taken here is to use C_3 to set f_L and to choose C_1 and C_2 so that their contributions are negligible.

Assumptions: Small-signal conditions apply. $V_T = 25.0$ mV.

Analysis: Choosing C_2 to set f_L yields

$$C_2 \cong \frac{1}{R_{2S}\omega_L} = \frac{1}{23.3\ \Omega\,(2\pi)(2000\ \text{Hz})} = 3.42\ \mu\text{F}$$

Let us choose C_1 and C_3 so that their individual time constants are each 100 times larger than that associated with C_2—that is, each capacitor will contribute a 1 percent error to f_L.

$$C_1 = 100\frac{R_{2S}C_2}{R_{1S}} = 100\frac{(23.2\ \Omega)(3.42\ \mu\text{F})}{2.26\ \text{k}\Omega} = 3.51\ \mu\text{F}$$

$$C_3 = 100\frac{R_{2S}C_2}{R_{3S}} = 100\frac{(23.2\ \Omega)(3.42\ \mu\text{F})}{104\ \text{k}\Omega} = 0.0763\ \mu\text{F}$$

Picking the nearest values from the capacitor table in Appendix A, we have $C_1 = 3.9$ μF, $C_2 = 3.9$ μF, and $C_3 = 0.082$ μF.

Check of Results: Let us check by calculating the actual values of f_L.

$$f_L = \frac{1}{2\pi}\left[\frac{1}{2.26\ \text{k}\Omega\,(3.9\ \mu\text{F})} + \frac{1}{23.2\ \Omega\,(3.9\ \mu\text{F})} + \frac{1}{104\ \text{k}\Omega\,(0.082\ \mu\text{F})}\right] = 1800\ \text{Hz}$$

Discussion: The cutoff frequency is approximately 10 percent lower than the design value because of the use of the 3.9-μF capacitor and the small contributions from C_1 and C_3. At additional cost, one could use two capacitors to make up the 3.5-μF value. However, the tolerances on typical capacitors are relatively large, and one would need to use a precision capacitor (and resistors) if a more accurate value of f_L is required. (See simulation results on previous page.)

Computer-Aided Analysis: The frequency response with the new capacitor values can be simulated using SPICE ac analysis with FSTART = 10 Hz and FSTOP = 10 MHz with 20 frequency points per decade. The transistor parameters were set to IS = 3 fA, BF = 100, and VAF = 75 V. SPICE simulation results for the new common-emitter design results yields $A_{mid} = -138$ and $f_L = 1610$ Hz. The value of f_L is approximately 10 percent less than our hand calculations. This discrepancy is due to differences in V_T and the Q-point current.

EXERCISE: Estimate the midband gain for the circuit in Fig. 16.6. What is the source of the error between this value and SPICE?

ANSWER: −157; Neglect of r_o accounts for most of the difference.

16.3.2 ESTIMATE OF ω_L FOR THE COMMON-SOURCE AMPLIFIER

Equations (16.34), (16.36), and (16.38) can be applied directly to the C-S FET amplifier in Fig. 16.11 by substituting infinity for the values of the transistor's input resistance and current gain. These equations reduce directly to:

$$R_{1S} = R_I + (R_G \| R_{iG}) = R_I + R_G$$

$$R_{2S} = R_S \| R_{iS} = R_S \left\| \frac{1}{g_m} \right. \tag{16.42}$$

$$R_{3S} = R_3 + (R_D \| R_{iD}) = R_3 + (R_D \| r_o) \cong R_3 + R_D$$

The three expressions in Eq. (16.42) represent the short-circuit time constants associated with the three capacitors in the circuit, as indicated in the ac circuit models in Figs. 16.12(a) to (c). Note that the three time constants are the same as those found by the direct approach that yielded Eq. (16.30).

EXERCISE: Find the short-circuit time constants and f_L for the common-source amplifier in Fig. 16.11 if $I_D = 1.5$ mA and $V_{GS} - V_{TN} = 0.5$ V. Assume $\lambda = 0.015/V$. The other values remain unchanged.

ANSWERS: 24.4 ms; 10.4 ms; 1.48 ms; 129 Hz

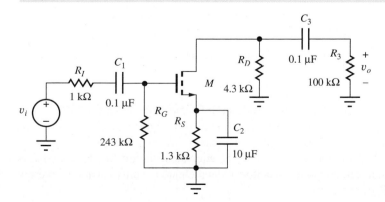

Figure 16.11 ac Model for common-source amplifier.

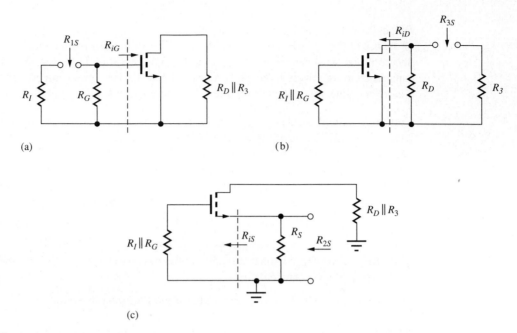

(a)

(b)

(c)

Figure 16.12 (a) Resistance R_{1S} at the terminals of C_1. (b) Resistance R_{2S} at the terminals of C_2. (c) Resistance R_{3S} at the terminals of C_3.

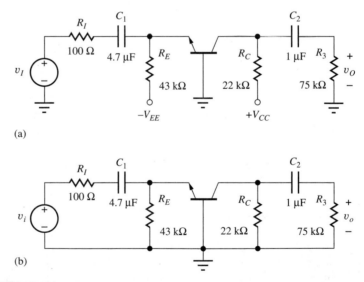

(a)

(b)

Figure 16.13 (a) Common-base amplifier. (b) Low-frequency ac equivalent circuit.

16.3.3 ESTIMATE OF ω_L FOR THE COMMON-BASE AMPLIFIER

Next, we apply the short-circuit time-constant technique to the common-base amplifier in Fig. 16.13. The results are also directly applicable to the common-gate case if β_o and r_π are set equal to infinity. Figure 16.13(b) is the low-frequency ac equivalent circuit for the common-base amplifier. In this particular circuit, coupling capacitors C_1 and C_2 are the only capacitors present, and expressions for R_{1S} and R_{2S} are needed.

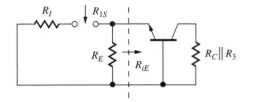

Figure 16.14 Equivalent circuit for determining R_{1S}.

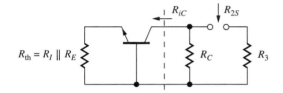

Figure 16.15 Equivalent circuit for determining R_{2S}.

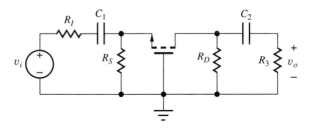

Figure 16.16 ac Circuit for common-gate amplifier.

R_{1S}

R_{1S} is found by shorting capacitor C_2, as indicated in the circuit in Fig. 16.14. Based on this figure,

$$R_{1S} = R_I + (R_E \| R_{iE}) \cong R_I + \left(R_E \left\| \frac{1}{g_m} \right. \right) \tag{16.43}$$

R_{2S}

Shorting capacitor C_1 yields the circuit in Fig. 16.15, and the expression for R_{2S} is

$$R_{2S} = R_3 + (R_C \| R_{iC}) \cong R_3 + R_C \tag{16.44}$$

because $R_{iC} \cong r_o(1 + g_m R_{th})$ is large.

EXERCISE: Find the short-circuit time constants and f_L for the common-base amplifier in Fig. 16.13 if $\beta_o = 100$, $V_A = 70$ V, and the Q-point is (0.1 mA, 5 V). What is A_{mid}?

ANSWERS: 1.64 ms, 97.0 ms, 98.7 Hz; 48.6

16.3.4 ESTIMATE OF ω_L FOR THE COMMON-GATE AMPLIFIER

The expressions for R_{1S} and R_{2S} for the common-gate amplifier in Fig. 16.16 are virtually identical to those of the common-base stage:

$$R_{1S} = R_I + (R_S \| R_{iS}) = R_I + \left(R_S \left\| \frac{1}{g_m} \right. \right) \tag{16.45}$$

$$R_{2S} = R_3 + (R_D \| R_{iD}) \cong R_3 + R_D \qquad \text{because} \qquad R_{iD} \cong \mu_f(R_S \| R_I)$$

EXERCISE: Draw the circuits used to find R_{1S} and R_{2S} for the common-gate amplifier in Fig. 16.16 and verify the results presented in Eq. (16.45).

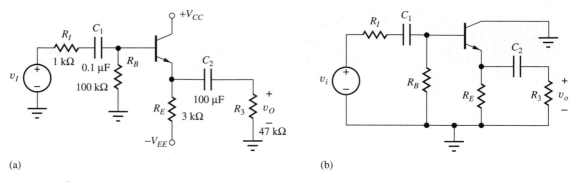

(a)

(b)

Figure 16.17 (a) Common-collector amplifier. (b) Low-frequency ac model for the common-collector amplifier.

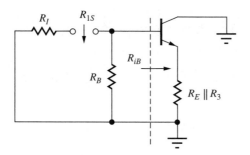

Figure 16.18 Circuit for finding R_{1S}.

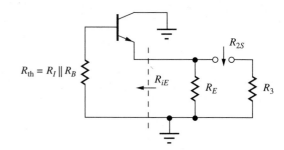

Figure 16.19 Circuit for finding R_{2S}.

EXERCISE: Find the short-circuit time constants and f_L for the common-gate amplifier in Fig. 16.16 if $R_I = 100\ \Omega$, $R_S = 1.3\ \mathrm{k\Omega}$, $R_D = 4.3\ \mathrm{k\Omega}$, $R_3 = 75\ \mathrm{k\Omega}$, $C_1 = 1\ \mu\mathrm{F}$, $C_2 = 0.1\ \mu\mathrm{F}$, $I_D = 1.5\ \mathrm{mA}$, and $V_{GS} - V_{TN} = 0.5\ \mathrm{V}$. Assume $\lambda = 0$.

ANSWERS: 0.248 ms; 7.93 ms; 663 Hz

16.3.5 ESTIMATE OF ω_L FOR THE COMMON-COLLECTOR AMPLIFIER

Figures 16.17(a) and (b) are schematics of an emitter follower and its corresponding low-frequency ac model, respectively. This circuit has two coupling capacitors, C_1 and C_2. The circuit for R_{1S} in Fig. 16.18 is constructed by shorting C_2, and the expression for R_{1S} is

$$R_{1S} = R_I + (R_B \| R_{iB}) = R_I + (R_B \| [r_\pi + (\beta_o + 1)(R_E \| R_3)]) \tag{16.46}$$

Similarly, the circuit used to find R_{2S} is found by shorting capacitor C_1, as in Fig. 16.19, and

$$R_{2S} = R_3 + (R_E \| R_{iE}) = R_3 + \left(R_E \left\| \frac{R_{\mathrm{th}} + r_\pi}{\beta_o + 1} \right. \right) \tag{16.47}$$

16.3.6 ESTIMATE OF ω_L FOR THE COMMON-DRAIN AMPLIFIER

The corresponding low-frequency ac model for the common-drain amplifier appears in Fig. 16.20. Taking the limits as β_o and r_π approach infinity, Eqs. (16.46) and (16.47) become

$$R_{1S} = R_I + (R_G \| R_{iG}) = R_I + R_G \qquad \text{because} \qquad R_{iG} = \infty$$

and $\tag{16.48}$

$$R_{2S} = R_3 + (R_S \| R_{iS}) = R_3 + \left(R_S \left\| \frac{1}{g_m} \right. \right)$$

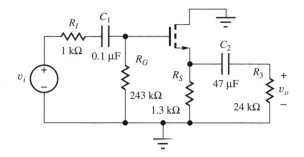

Figure 16.20 Low-frequency ac equivalent circuit for common-drain amplifier.

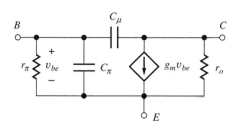

Figure 16.21 Capacitances in the hybrid-pi model of the BJT.

EXERCISE: Find the short-circuit time constants and f_L for the common-collector amplifier in Fig. 16.17(a) if $\beta_o = 100$, $V_A = 70$ V, and the Q-point $= (1$ mA, 5 V). What is A_{mid}?

ANSWERS: 7.52 ms, 4.70 s, 21.2 Hz; 0.978

EXERCISE: Find the short-circuit time constants and f_L for the common-drain amplifier in Fig. 16.20 if $g_m = 1$ mS. What is A_{mid}?

ANSWERS: 43.1 ms, 1.16 s, 3.83 Hz; 0.551

16.4 TRANSISTOR MODELS AT HIGH FREQUENCIES

To explore the upper limits of amplifier frequency response, the high-frequency limitations of the transistors, which we have ignored thus far, must be taken into account. All electronic devices have capacitances between their various terminals, and these capacitances limit the range of frequencies for which the devices can provide useful voltage, current, or power gain. This section develops the description of the frequency-dependent hybrid-pi model for the bipolar transistor, as well as the pi model for the field-effect transistor.

16.4.1 FREQUENCY-DEPENDENT HYBRID-PI MODEL FOR THE BIPOLAR TRANSISTOR

In the BJT, capacitances appear between the base-emitter and base-collector terminals of the transistor and are included in the small-signal hybrid-pi model in Fig. 16.21. The capacitance between the base and collector terminals, denoted by C_μ, represents the capacitance of the reverse-biased collector-base junction of the bipolar transistor and is related to the Q-point through an expression equivalent to Eq. (3.21), Chapter 3:

$$C_\mu = \frac{C_{\mu o}}{\sqrt{1 + \dfrac{V_{CB}}{\phi_{jc}}}} \qquad (16.49)$$

In Eq. (16.49), $C_{\mu o}$ represents the total collector-base junction capacitance at zero bias, and ϕ_j is the built-in potential of the collector-base junction, typically 0.6 to 1.0 V.

The internal capacitance between the base and emitter terminals, denoted by C_π, represents the diffusion capacitance associated with the forward-biased base-emitter junction of the transistor. C_π is related to the Q-point through Eq. (5.39) in Sec. 5.8:

$$C_\pi = g_m \tau_F \qquad (16.50)$$

in which τ_F is the forward transit-time of the bipolar transistor. In Fig. 16.21, C_π appears directly in parallel with r_π. For a given input signal current, the impedance of C_π causes the base-emitter voltage

v_{be} to be reduced as frequency increases, thereby reducing the current in the controlled source at the output of the transistor.

Shunt capacitances such as C_π are always present in electronic devices and circuits. At low frequencies, the impedance of these capacitances is usually very large and so has negligible effect relative to the resistances such as r_π. However, as frequency increases, the impedance of C_π becomes smaller and smaller, and v_{be} eventually approaches zero. Thus, transistors cannot provide amplification at arbitrarily high frequencies.

16.4.2 MODELING C_π AND C_μ IN SPICE

In SPICE, the values of C_π and C_μ are controlled by the forward transit time TF, the zero-bias value of the collector-junction capacitance CJC, the built-in potential VJC of the collector-base junction, and the grading factor MJC of the collector-base junction. In SPICE, C_π and C_μ are referred to as C_{BE} and C_{BC}, respectively.

$$C_{BE} = g_m \cdot \text{TF} \quad \text{and} \quad C_{BC} = \frac{\text{CJC}}{\left(1 + \dfrac{\text{VCB}}{\text{VJC}}\right)^{\text{MJC}}} \tag{16.51}$$

VJC defaults to 0.75 V, and MJC defaults to 0.33.

16.4.3 UNITY-GAIN FREQUENCY f_T

A quantitative description of the behavior of the transistor at high frequencies can be found by calculating the frequency-dependent short-circuit current gain $\beta(s)$ from the circuit in Fig. 16.22. For a current $\mathbf{I_b}(s)$ injected into the base, the collector current $\mathbf{I_c}(s)$ consists of two components:

$$\mathbf{I_c}(s) = g_m \mathbf{V_{be}}(s) - \mathbf{I_\mu}(s) \tag{16.52}$$

Because the voltage at the collector is zero, v_{be} appears directly across C_μ and $\mathbf{I_\mu}(s) = s C_\mu \mathbf{V_{be}}(s)$. Therefore,

$$\mathbf{I_c}(s) = (g_m - s C_\mu) \mathbf{V_{be}}(s) \tag{16.53}$$

Because the collector is connected directly to ground, C_π and C_μ appear in parallel in this circuit, and the base current flows through the parallel combination of r_π and $(C_\pi + C_\mu)$ to develop the base-emitter voltage:

$$\mathbf{V_{be}}(s) = \mathbf{I_b}(s) \frac{r_\pi \dfrac{1}{s(C_\pi + C_\mu)}}{r_\pi + \dfrac{1}{s(C_\pi + C_\mu)}} = \mathbf{I_b}(s) \frac{r_\pi}{s(C_\pi + C_\mu)r_\pi + 1} \tag{16.54}$$

By combining Eqs. (16.53) and (16.54), we reach an expression for the frequency-dependent current gain:

$$\beta(s) = \frac{\mathbf{I_c}(s)}{\mathbf{I_b}(s)} = \frac{\beta_o \left(1 - \dfrac{s C_\mu}{g_m}\right)}{s(C_\pi + C_\mu)r_\pi + 1} \tag{16.55}$$

A right-half-plane transmission zero occurs in the current gain at an extremely high frequency, $\omega_Z = +g_m/C_\mu$, and can almost always be neglected. Neglecting ω_Z results in the following simplified

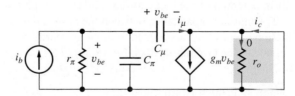

Figure 16.22 Finding the short-circuit current gain β of the BJT.

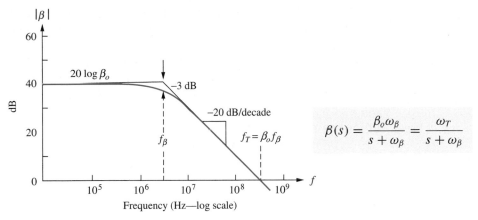

Figure 16.23 Common-emitter current gain versus frequency for the BJT.

expression for $\beta(s)$:

$$\beta(s) \cong \frac{\beta_o}{s(C_\pi + C_\mu)r_\pi + 1} = \frac{\beta_o}{\dfrac{s}{\omega_\beta} + 1} \tag{16.56}$$

in which ω_β represents the **beta-cutoff frequency,** defined by

$$\omega_\beta = \frac{1}{r_\pi(C_\pi + C_\mu)} \quad \text{and} \quad f_\beta = \frac{\omega_\beta}{2\pi} \tag{16.57}$$

Figure 16.23 is a Bode plot for Eq. (16.56). From Eq. (16.56) and this graph, we see that the current gain has the value of $\beta_o = g_m r_\pi$ at low frequencies and exhibits a single-pole roll-off at frequencies above f_β, decreasing at a rate of 20 dB/decade and crossing through unity gain at $f = f_T$. The magnitude of the current gain is 3 dB below its low-frequency value at the beta-cutoff frequency, f_β.

Equation (16.56) can be recast in terms of $\omega_T = \beta_o \omega_\beta$ as

$$\beta(s) = \frac{\beta_o \omega_\beta}{s + \omega_\beta} = \frac{\omega_T}{s + \omega_\beta} \tag{16.58}$$

where $\omega_T = 2\pi f_T$. Parameter f_T is referred to as the **unity gain-bandwidth product** of the transistor and characterizes one of the fundamental frequency limitations of the transistor. At frequencies above f_T, the transistor no longer offers any current gain and fails to be useful as an amplifier.

A relationship between the unity gain-bandwidth product and the small-signal parameters can be obtained from Eqs. (16.57) and (16.58):

$$\omega_T = \beta_o \omega_\beta = \frac{\beta_o}{r_\pi(C_\pi + C_\mu)} = \frac{g_m}{C_\pi + C_\mu} \tag{16.59}$$

Note that the transmission zero occurs at a frequency beyond ω_T:

$$\omega_Z = \frac{g_m}{C_\mu} > \frac{g_m}{C_\pi + C_\mu} = \omega_T \tag{16.60}$$

To perform numeric calculations, we determine the values of f_T and C_μ from a transistor's specification sheet and then calculate C_π by rearranging Eq. (16.59):

$$C_\pi = \frac{g_m}{\omega_T} - C_\mu \tag{16.61}$$

From Eq. (16.49) we can see that C_μ is only a weak function of operating point, but recasting g_m in Eq. (16.61) demonstrates that C_π is directly proportional to collector current:

$$C_\pi = \frac{40 I_C}{\omega_T} - C_\mu \qquad (16.62)$$

EXAMPLE 16.4 BIPOLAR TRANSISTOR MODEL PARAMETERS

Find a set of model parameters for a bipolar transistor from its specification sheet.

PROBLEM Find values of β_o, I_S, V_A, f_T, C_π, and C_μ for the CA-3096 *npn* transistors operating at a collector current of 1 mA using the specifications sheets on the MCD website.

SOLUTION **Known Information and Given Data:** CA-3096 specification sheets; $I_C = 1$ mA

Unknowns: β_o, I_S, V_A, f_T, C_π, and C_μ

Approach: We will use our definitions of, and relationships between, the large-signal and small-signal parameters to find the unknown values.

Assumptions: $T = 25°C$ and $V_{CE} = 5$ V, corresponding to the electrical specification sheets; active region operation; $\beta_o \cong \beta_F$; the built-in potential of the collector-base junction is 0.75 V.

Analysis: Based on the typical values in the specification sheets, we find $\beta_F = h_{FE} = 390$, $V_{BE} = 0.69$ V, $f_T = 280$ MHz, and $C_{CB} = 0.46$ pF at $V_{CB} = 3$ V. From the graph of output resistance versus current, we find $r_o = 80$ kΩ for $I_c = 1$ mA. For $T = 25°C$, $V_T = 26.0$ mV.

We find the current gain and Early voltage using the values of h_{FE} and r_o,

$$\beta_o \cong h_{FE} = 390 \qquad V_A = I_C r_o - V_{CE} = 75 \text{ V}$$

and I_S is found from I_C, V_{BE}, and V_T:

$$I_S = \frac{I_C}{\exp\left(\dfrac{V_{BE}}{V_T}\right)} \frac{1 \text{ mA}}{\exp\left(\dfrac{0.69 \text{ V}}{26.0 \text{ mV}}\right)} = 2.98 \text{ fA}$$

Capacitance C_μ is equal to the collector-base capacitance of the transistor, but it is specified at $V_{CB} = 3$ V. Using Eq. (16.49), we find $C_{\mu o}$, and then calculate C_μ for $V_{CB} = 5 - .69 = 4.31$ V.

$$C_{\mu o} \cong C_{CB}\sqrt{1 + \frac{V_{CB}}{\phi_{jc}}} = 0.46 \text{ pF}\sqrt{1 + \frac{3}{0.75}} = 1.03 \text{ pF}$$

$$C_\mu \cong \frac{C_{\mu o}}{\sqrt{1 + \dfrac{V_{CB}}{\phi_{jc}}}} = \frac{1.03 \text{ pF}}{\sqrt{1 + \dfrac{4.31}{0.75}}} = 0.397 \text{ pF}$$

Now we can find C_π:

$$C_\pi = \frac{g_m}{\omega_T} - C_\mu = \frac{1 \text{ mA}}{26.0 \text{ mV}} \frac{1}{2\pi(280 \text{ MHz})} - 0.40 \text{ pF} = 21.5 \text{ pF}$$

Check of Results: We have found the required values of β_o, I_S, V_A, f_T, C_π, and C_μ. The calculations appear correct and reasonable. The calculated value of C_μ agrees reasonably well with the graph of C_{CB} versus V_{CB} in the specification sheets.

Discussion: The values in the specification sheets must often be mapped into the parameters that we need, and the data supplied is often incomplete. Some may be presented in tabular form; others must be found from graphs. Note that the current gain peaks at a collector current of approximately 1 mA, whereas f_T peaks at approximately 4 mA.

Computer-Aided Analysis: Let us now attempt to create a SPICE model that has these parameters. We must set IS = 2.98 fA, BF = 390, and VAF = 75 V. Using Eq. (16.51), we also have TF = 559 ps, CJC = 1.03 pF, VJC = 0.75 V, and MJC = 0.5. Let us bias the transistor as in the circuit shown here, and request the device parameters as an output following an operating point analysis. The results are $I_C = 1$ mA, $V_{BE} = 0.685$ V, $V_{BC} = -5$ V, $g_m = 38.7$ mS, $\beta_o = g_m/g_\pi = 416$, $r_o = 1/g_o = 79.9$ kΩ, $C_\pi = 21.6$ pF, and $C_\mu = 0.372$ pF. Our set of device parameters appears to be correct. Note that $\beta_o = \text{BF}(1 + \text{VCB}/\text{VAF}) = 416$.

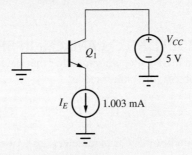

16.4.4 HIGH-FREQUENCY MODEL FOR THE FET

To model the FET at high frequencies, gate-drain and gate-source capacitances C_{GD} and C_{GS} are added to the small-signal model, as shown in Fig. 16.24. For the MOSFET, these two capacitors represent the gate oxide and overlap capacitances discussed previously in Sec. 4.6. At high frequencies, currents through these two capacitors combine to form a current in the gate terminal, and the signal current i_g can no longer be assumed to be zero. Thus, even the FET has a finite current gain at high frequencies.

The short-circuit current gain for the FET can be calculated in the same manner as for the BJT, as in Fig. 16.25:

$$\mathbf{I_d}(s) = (g_m - sC_{GD})\mathbf{V_{gs}}(s) = \mathbf{I_g}(s)\frac{(g_m - sC_{GD})}{s(C_{GS} + C_{GD})} \qquad (16.63)$$

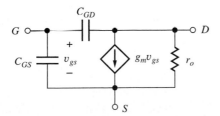

Figure 16.24 Pi model for the FET.

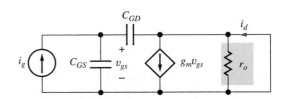

Figure 16.25 Circuit for calculating the short-circuit current gain of the FET.

and

$$\beta(s) = \frac{\mathbf{I_d}(s)}{\mathbf{I_g}(s)} = \frac{g_m\left(1 - \dfrac{sC_{GD}}{g_m}\right)}{s(C_{GS} + C_{GD})} = \frac{\omega_T}{s}\left(1 - \frac{s}{\omega_T\left(1 + \dfrac{C_{GS}}{C_{GD}}\right)}\right) \tag{16.64}$$

At dc, the current gain is infinite but falls at a rate of 20 dB/decade as frequency increases. The unity gain-bandwidth product ω_T of the FET is defined in a manner identical to that of the BJT,

$$\omega_T = \frac{g_m}{C_{GS} + C_{GD}} \tag{16.65}$$

and the FET current gain falls below 1 for frequencies in excess of ω_T, just as for the case of the bipolar transistor. The transmission zero now occurs at $\omega_Z = \omega_T(1 + C_{GS}/C_{GD})$, typically a few times ω_T.

16.4.5 MODELING C_{GS} AND C_{GD} IN SPICE

As discussed in Sec. 4.6, we remember that the gate-source and gate-drain capacitances in the active region (pinch-off) are expressed as

$$C_{GS} = C'_{OL}W + \frac{2}{3}C''_{ox}WL \qquad C_{GD} = C'_{OL}W \qquad C''_{ox} = \frac{\varepsilon_{ox}}{T_{ox}} \tag{16.66}$$

The corresponding SPICE parameters are oxide thickness TOX, gate width W, gate length L, gate-source overlap capacitance per unit length CGSO, and gate-drain overlap capacitance per unit length CGDO. Note that SPICE permits definition of different values of overlap capacitance for the source and drain regions of the transistor.

16.4.6 CHANNEL LENGTH DEPENDENCE OF f_T

The unity-gain bandwidth product of the MOSFET is strongly dependent on the channel length, and this fact represents one of the reasons for continuing to scale the technology to smaller and smaller dimensions. The basic expression for the intrinsic f_T ($C'_{OL} = 0$) of the MOSFET in terms of technology parameters can be found using Eqs. (16.65) and (16.66). If we remember that $g_m = K_n(V_{GS} - V_{TN})$, and assume $C'_{OL} = 0$:

$$f_T = \frac{\mu_n C''_{ox}\dfrac{W}{L}(V_{GS} - V_{TN})}{\dfrac{2}{3}C''_{ox}WL} = \frac{3}{2}\frac{\mu_n(V_{GS} - V_{TN})}{L^2} \tag{16.67}$$

The value of f_T is proportional to transistor mobility and inversely dependent upon the square of the channel length. Thus, an NMOS transistor will have a higher-cutoff frequency than a similar PMOS transistor for a given channel length and bias condition. Reducing the channel length by a factor of 10 results in an increase in f_T by a factor of 100!

EXAMPLE **16.5**   MOSFET MODEL PARAMETERS

Find a set of model parameters for a MOSFET from its specification sheet.

PROBLEM   Find values of V_{TN}, K_P, λ, C_{GS}, and C_{GD} for the NMOS transistors in the ALD-1116 transistor
www   operating at a drain current of 10 mA using the specifications sheets on the MCD website.

SOLUTION   **Known Information and Given Data:** ALD-1116 specification sheets; $I_D = 10$ mA

Unknowns: V_{TN}, K_n, λ, C_{GS}, and C_{GD}

Approach: We will use our definitions of the relationships between the large-signal and small-signal parameters to find the unknown values.

Assumptions: $T = 25°C$ and $V_{DS} = 5$ V, corresponding to the electrical specification sheets; our square-law transistor model applies to the device; the transistor is symmetrical.

Analysis: Based on the typical values in the specification sheets, we find $V_{TN} = 0.7$ V and $I_D = 4.8$ mA for $V_{GS} = 5$ V, output conductance $g_o = 200$ μS at 10 mA, and $C_{ISS} = 1$ pF. First, we can find λ using the output conductance value:

$$\lambda = \left(\frac{I_D}{g_o} - V_{DS} \right)^{-1} = \left(\frac{10 \text{ mA}}{0.2 \text{ mS}} - 5 \text{ V} \right)^{-1} = 0.0222 \text{ V}^{-1}$$

Now we can find K_n using the MOS drain current expression in the active region:

$$K_n = \frac{2I_D}{(V_{GS} - V_{TN})^2(1 + \lambda V_{DS})} = \frac{2(4.8 \text{ mA})}{(5 - 0.7)^2 \left(1 + \dfrac{5 \text{ V}}{45 \text{ V}} \right)} = 467 \frac{\mu A}{V^2}$$

From these results, we can set SPICE parameters to VTO $= 0.7$ V, KP $= 467$ μA/V², and LAMBDA $= 0.0222$ V^{-1}.

C_{ISS} is the short-circuit input capacitance of the transistor in the common-source configuration and is equal to the sum of C_{GS} and C_{GD}. Unfortunately, the test conditions are not specified, so we cannot be sure if the measurement was in the triode or active region of operation. If we assume active region operation and use Eq. (16.66), then

$$C_{GS} + C_{GD} = \frac{2}{3}C_{ox}''WL + 2C_{OL}'W = 1 \text{ pF}$$

However, we have no way of directly splitting the 1-pF capacitance between C_{GS} and C_{GD}. One approximation is to assume approximately that the oxide capacitance term is dominant. Then $C_{GS} \cong 1$ pF and $C_{GD} \cong 0$. As we shall see shortly, however, neglecting C_{GD} may cause significant errors in our calculations of the high-frequency response of amplifiers.

Check of Results: We have found the required values. The values of V_{TN} and λ appear reasonable. Let us see if the values of the amplification factor and f_T are reasonable.

$$g_m = \sqrt{2K_n I_D(1 + \lambda V_{DS})} = \sqrt{2(467 \text{ μA/V}^2)(10 \text{ mA})(1 + 5/45)} = 3.22 \text{ mS}$$

$$f_T = \frac{1}{2\pi} \frac{g_m}{C_{GS} + C_{GD}} = \frac{1}{2\pi} \frac{3.22 \text{ mS}}{1 \text{ pF}} = 513 \text{ MHz}$$

$$\mu_f = \frac{g_m}{g_o} = \frac{3.22}{0.2 \text{ mS}} = 16.1$$

The value of f_T is reasonable. The relatively low value of μ_f results from the 10-mA drain current condition. Although the value of g_m is larger than the typical value given in the table, it is reasonably consistent with the graph of the output characteristics: $g_m = \Delta I_D/\Delta V_{GS} = 4.5 \text{ mA}/2 \text{ V} = 2.25$ mS.

Discussion: The values in the specification sheets must often be mapped into the parameters that we need, and we often find that the data sheet information is incomplete and not necessarily self consistent. We often must contact the manufacturer for more information. The manufacturer may be able to supply a SPICE model for the device. As a last resort, we can directly measure the parameters for ourselves, or chose another device.

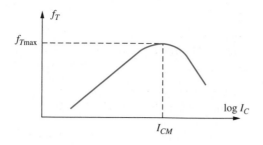

Figure 16.26 Current dependence of f_T.

EXERCISE: What are the values of C_{GS} and C_{GD} if C_{ISS} in Ex. 16.5 had been measured in the triode region?

ANSWER: 0.5 pF, 0.5 pF

EXERCISE: An NMOSFET has $f_T = 200$ MHz and $K_n = 10$ mA/V^2 and is operating at a drain current of 10 mA. Assume that $C_{GS} = 5C_{GD}$ and find the values of these two capacitors.

ANSWERS: $C_{GS} = 9.38$ pF, $C_{GD} = 1.88$ pF

16.4.7 LIMITATIONS OF THE HIGH-FREQUENCY MODELS

The pi-models of the transistor in Figs. 16.21 and 16.25 are good representations of the characteristics of the transistors for frequencies up to approximately $0.3 f_T$. Above this frequency, the behavior of the simple pi-models begins to deviate significantly from that of the actual device. In addition, our discussion has tacitly assumed that ω_T is constant. However, this is only an approximation. In an actual BJT, ω_T depends on operating current, as shown in Fig. 16.26.

For a given BJT, there will be a collector current I_{CM}, which yields a maximum value of $f_T = f_{T_{\max}}$. For the FET operating in the saturation region, C_{GS} and C_{GD} are independent of Q-point current so that $\omega_T \propto g_m \propto \sqrt{I_D}$. In the upcoming discussions, we assume that the specified value of f_T corresponds to the operating point being used.

EXERCISE: As an example of the problem of using a constant value for the transistor f_T, repeat the calculation of C_π and C_μ for a Q-point of (20 μA, 8 V) if $f_T = 500$ MHz, $C_{\mu o} = 2$ pF, and $\phi_{jc} = 0.6$ V.

ANSWERS: 0.551 pF, −0.296 pF. Impossible—C_π cannot have a negative value.

16.5 BASE RESISTANCE IN THE HYBRID-PI MODEL

One final circuit element, the **base resistance** r_x, completes the basic hybrid-pi description of the bipolar transistor. In the bipolar transistor cross section in Fig. 16.27, base current i_b enters the transistor through the external base contact and traverses a relatively high resistance region before actually entering the active area of the transistor. Circuit element r_x models the voltage drop between the base contact and the active region of the transistor and is included between the internal and external base nodes, B' and B, respectively, in the circuit model in Fig. 16.28. As discussed in the next section, the base resistance usually can be neglected at low frequencies. However, resistance r_x can represent an important limitation to the frequency response of the transistor in low-source resistance applications. Typical values of r_x range from a few ohms to a thousand ohms. In SPICE, BJT base resistance is modeled by parameter RB.

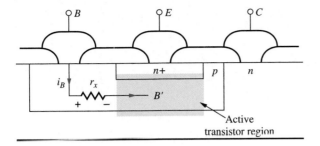

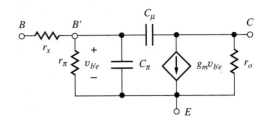

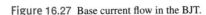

Figure 16.27 Base current flow in the BJT.

Figure 16.28 Completed hybrid-pi model, including the base resistance r_x.

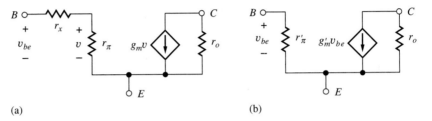

(a)   (b)

Figure 16.29 (a) Transistor model containing r_x. (b) Model transformation that "absorbs" r_x.

16.5.1 EFFECT OF BASE RESISTANCE ON MIDBAND AMPLIFIERS

Before considering the high-frequency response of single and multistage amplifiers, we explore the effect of base resistance on the midband gain expressions for single-stage amplifiers. Although the model used in deriving the midband voltage gain expressions in Chapters 13 and 14 did not include the effect of base resistance, the expressions can be easily modified to include r_x. A simple approach is to use the circuit transformation shown in Fig. 16.29, in which r_x is absorbed into an equivalent pi model. The current generator in the model in Fig. 16.29(a) is controlled by the voltage developed across r_π, which is related to the total base-emitter voltage through voltage division by

$$\mathbf{v} = \mathbf{v_{be}} \frac{r_\pi}{r_x + r_\pi} \tag{16.68}$$

and the current in the controlled source is

$$\mathbf{i} = g_m \mathbf{v} = g_m \frac{r_\pi}{r_x + r_\pi} \mathbf{v_{be}} = g_m' \mathbf{v_{be}} \qquad \text{where} \qquad g_m' = \frac{\beta_o}{r_x + r_\pi} \tag{16.69}$$

Equations (16.68) and (16.69) lead to the model in Fig. 16.29(b), in which the base resistance has been absorbed into r_π' and g_m' of an equivalent transistor Q' defined by

$$g_m' = g_m \frac{r_\pi}{r_x + r_\pi} = \frac{\beta_o}{r_x + r_\pi} \qquad \text{and} \qquad r_\pi' = r_x + r_\pi \tag{16.70}$$

Note that current gain is conserved during the transformation: $\beta_o' = \beta_o$.

Based on Eq. (16.70), the original expressions from Table 14.8 can be transformed to those in Table 16.1 for the three classes of amplifiers in Fig. 16.30 by simply substituting g_m' for g_m and r_π' for r_π. In many cases, particularly at bias points below a few hundred μA, $r_\pi \gg r_x$, and the expressions in Eq. (16.70) reduce to $g_m' \cong g_m$ and $r_\pi' \cong r_\pi$. The expressions in Table 16.1 then become identical to those in Table 14.8.

TABLE 16.1
Single-Stage Bipolar Amplifiers, Including Base Resistance

	COMMON-EMITTER AMPLIFIER	COMMON-COLLECTOR AMPLIFIER	COMMON-BASE AMPLIFIER
Terminal voltage gain $A_{vt} = \dfrac{v_o}{v_1}$ $r'_\pi = r_x + r_\pi$ $g'_m = \dfrac{\beta_o}{r'_\pi}$	$-\dfrac{\beta_o R_L}{r'_\pi + (\beta_o + 1)R_E}$ $\cong -\dfrac{g'_m R_L}{1 + g'_m R_E}$	$+\dfrac{\beta_o R_L}{r'_\pi + (\beta_o + 1)R_L}$ $\cong +\dfrac{g'_m R_L}{1 + g'_m R_L} \cong +1$	$+g'_m R_L$
Signal-source voltage gain $A_v = \dfrac{\mathbf{v_o}}{\mathbf{v_i}}$	$-\dfrac{g'_m R_L}{1 + g'_m R_E}\left(\dfrac{R_B\|R_{iB}}{R_I + R_B\|R_{iB}}\right)$	$+\dfrac{g'_m R_L}{1 + g'_m R_L}\left(\dfrac{R_B\|R_{iE}}{R_I + R_B\|R_{iE}}\right) \cong +1$	$+\dfrac{g'_m R_L}{1 + g'_m (R_I\|R_E)}\left(\dfrac{R_E}{R_I + R_E}\right)$
Input resistance	$r'_\pi + (\beta_o + 1)R_E$	$r'_\pi + (\beta_o + 1)R_L$	$\dfrac{1}{g'_m}$
Output resistance	$r_o(1 + g'_m R_E)$	$\dfrac{1}{g'_m} + \dfrac{R_{\text{th}}}{\beta_o + 1}$	$r_o[1 + g'_m (R_I\|R_E)]$
Input signal range	$\cong 0.005(1 + g'_m R_E)$	$\cong 0.005(1 + g'_m R_L)$	$\cong 0.005[1 + g'_m (R_I\|R_E)]$
Current gain	$-\beta_o$	$\beta_o + 1$	$\alpha_o \cong +1$

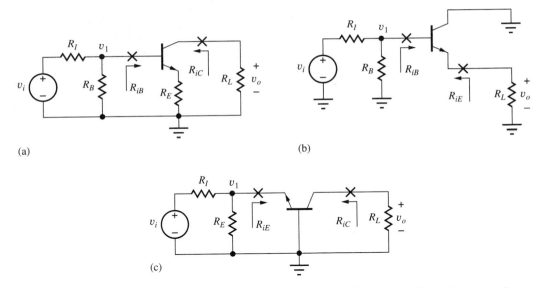

Figure 16.30 The three BJT amplifier configurations: (a) common-emitter; (b) common-collector; (c) common-base.

EXERCISE: Recalculate the midband gain for the circuit in Fig. 16.6, including base resistance $r_x = 250\ \Omega$. What was the value of A_{mid} with $r_x = 0$?

ANSWER: $-141;\ -157$

16.6 HIGH-FREQUENCY COMMON-EMITTER AND COMMON-SOURCE AMPLIFIER ANALYSIS

Now that the complete hybrid-pi model has been described, we can explore the high-frequency limitations of the three basic single-stage amplifiers. For each of the basic stages, we will develop expressions for the high-frequency poles at the input and output of each stage. This approach will allow us to easily extend our analysis to multistage amplifiers.

To begin our analysis, we will first review the high-frequency response of a single pole network with a high-frequency capacitor, shown in Figure 16.31. An expression for the high-frequency transfer characteristic for the RC circuit can be derived as

$$
\frac{\mathbf{V_x}}{\mathbf{V_i}} = \frac{R_2 \left\| \dfrac{1}{sC_1} \right.}{R_1 + R_2 \left\| \dfrac{1}{sC_1} \right.} = \frac{\dfrac{R_2}{1 + sR_2C_1}}{R_1 + \dfrac{R_2}{1 + sR_2C_1}} = \frac{R_2}{R_1 + R_2} \frac{1}{\left(1 + s\dfrac{R_1R_2}{R_1 + R_2}C_1\right)}
$$

$$
= \frac{R_2}{R_1 + R_2} \frac{1}{(1 + s[R_1 \| R_2]C_1)}
$$

(16.71)

Substituting $s = j2\pi f$ and using $f_p = 1/(2\pi[R_1 \| R_2]C_1)$,

$$
\frac{\mathbf{V_x}}{\mathbf{V_i}} = \frac{R_2}{R_1 + R_2} \frac{1}{\left(1 + j\dfrac{f}{f_p}\right)} = A_{mid}F_H(s)
$$

(16.72)

This expression has two parts, the midband gain, $R_2/(R_2 + R_1)$, and the high-frequency characteristic, $1/(1 + jf/f_p)$. Notice that the equivalent resistance, $R_1 \| R_2$ is the total equivalent resistance to ground at the output of the example network. If other branch connections are present, the equivalent

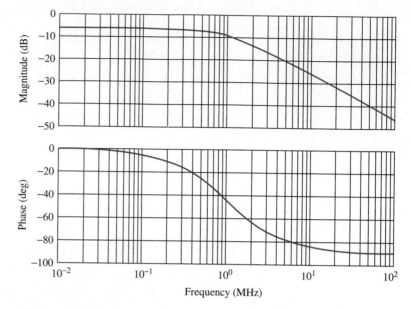

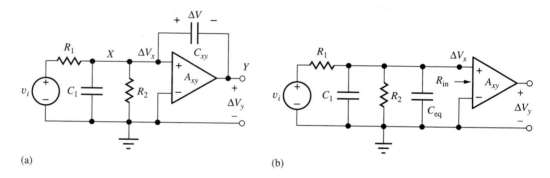

Figure 16.31 A two-resistor, one-capacitor circuit.

Figure 16.32 Magnitude and phase of single high-frequency pole with $R_1 = R_2$ and $f_p = 1$ MHz.

(a)   (b)

Figure 16.33 (a) Amplifier with a capacitance coupling its input and output. (b) The amplifier with the input-output capacitance replaced by an equivalent effective capacitance C_{eq} between the input and small-signal ground.

small-signal resistance of each branch will be added in parallel. Capacitance C_1 is the total equivalent capacitance to small-signal ground at the output. If other capacitors are present they too are simply added to find the total capacitance. The magnitude and phase of this single pole characteristic is shown in Figure 16.32.

16.6.1 THE MILLER EFFECT

Figure 16.33(a) shows a typical variation on the simple network of Fig. 16.31. Here a capacitor connected at node X is connected across an amplifier with a gain A_{xy} from node X to node Y.

We would like to find a method to convert the physical capacitor C_{xy} across the amplifier to an equivalent capacitance, C_{eq}, to small signal ground as shown in Fig. 16.33(b). We observe that a small signal capacitance can be defined as

$$C = \frac{\Delta Q}{\Delta V} \tag{16.73}$$

where ΔV is the voltage change across the capacitor and ΔQ is the charge required to develop that voltage change across the capacitor. For the case in Fig. 16.33(a), given a ΔV_x, a ΔV_y equal to

$A_{xy}\Delta V_x$ will appear at the output. The charge that must be delivered by the driving circuit can be calculated as

$$\Delta Q = C_{xy}(\Delta V_x - \Delta V_y) = C_{xy}(\Delta V_x - A_{xy}\Delta V_x) = C_{xy}\Delta V_x(1 - A_{xy}) \tag{16.74}$$

If we now consider the circuit in Fig. 16.33(b), one terminal of C_{eq} is connected to small-signal ground, so using our calculation of ΔQ from Eq. (16.74), C_{eq} can be written as

$$C_{eq} = \frac{C_{xy}\Delta V_x(1 - A_{xy})}{\Delta V_x} = C_{xy}(1 - A_{xy}) \tag{16.75}$$

The amplifier gain acts to produce an effective capacitance at its input which is scaled with respect to the physical capacitor by a factor $(1 - A_{xy})$. This is known as the **Miller effect**, or **Miller multiplication**, first described by John M. Miller in 1920.[2] Given our new equivalent capacitance, C_{eq}, based on our previous Eq. (16.70), we can now write an expression for the high-frequency transfer characteristic for the circuit in Fig. 16.33 as

$$\frac{\mathbf{V_x}}{\mathbf{V_i}} = \frac{R_2 \| R_{in}}{R_1 + R_2 \| R_{in}} \frac{1}{(1 + s[R_1 \| R_2 \| R_{in}][C_1 + C_{eq}])} \tag{16.76}$$

We see that the pole frequency is determined by a resistance of $R_1 \| R_2 \| R_{in}$, and a capacitance of $C_1 + C_{xy}(1 - A_{xy})$. As an example, consider the case with a gain $= -10$ V/V. The input capacitance due to C_{xy} will be $(1 - [-10])$ or eleven times larger than the physical capacitance C_{xy}. To understand this intuitively, consider a ΔV_x of 10 mV. With a gain of -10 V/V, the ΔV_y will be -100 mV. In other words, as the voltage at the input of the capacitor is increasing, the other terminal is rapidly decreasing in voltage, causing the driving circuit to deliver much more charge than would be expected given the actual value of the capacitor. On the other hand, for a gain of 0.9 V/V, the effective capacitance will be $(1 - 0.9)$, or 10 percent of the physical capacitance. For this gain value, the second terminal of the capacitor is approximately "following" the input terminal, leading to a much smaller delivery of charge from the driving circuit. Using the Miller effect allows us to separate capacitively coupled sections of a circuit into simpler RC circuits which are more easily analyzed.

In the following sections, we will generalize this approach to develop the high-frequency response of an amplifier as the product of the midband gain we have developed in previous chapters and a high-frequency transfer characteristic representing the effects of the high-frequency time constant at each node along the signal path.

16.6.2 COMMON-EMITTER AND COMMON-SOURCE AMPLIFIER HIGH-FREQUENCY RESPONSE

Figure 16.34 is a common-emitter amplifier with low-frequency coupling and bypass capacitors C_1, C_2, and C_3. In this section, we are concerned with the high-frequency response so we will consider the low-frequency capacitors to be open circuits at dc and short circuits at midband and high-frequencies. C_L represents the parasitic high-frequency load capacitance. We will use a simplified analysis approach similar to that presented in the previous section. We calculate the midband gain and then calculate a time constant at the input and output signal nodes. The Miller effect will be used to calculate an equivalent capacitance at the input.

The ac small-signal equivalent circuit is shown in Figure 16.35(a). The power supplies have been replaced with small-signal ground connections, and the low-frequency coupling and bypass capacitors are replaced with short circuits. The circuit has been further simplified in Fig. 16.35(c). The midband input gain is

$$A_i = \frac{\mathbf{v_b}}{\mathbf{v_i}} = \frac{R_{in}}{R_I + R_{in}} \cdot \frac{r_\pi}{r_x + r_\pi} = \frac{R_B \|(r_x + r_\pi)}{R_I + R_B \|(r_x + r_\pi)} \cdot \frac{r_\pi}{r_x + r_\pi} \tag{16.77}$$

---

[2] J. M. Miller, "Dependence of the input impedance of a three-electrode vacuum tube upon the load in the plate circuit," *Scientific Papers of the Bureau of Standards*, 15(351):367–385, 1920.

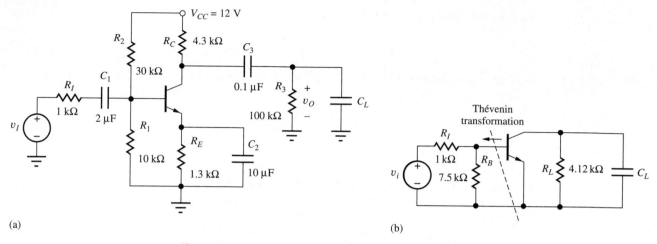

(a)

(b)

Figure 16.34 (a) Common-emitter amplifier. (b) High-frequency ac model for amplifier in (a).

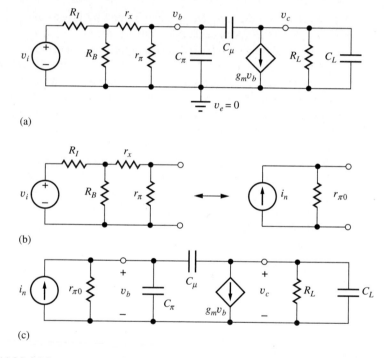

(a)

(b)

(c)

Figure 16.35 (a) Model for common-emitter stage at high frequencies. (b) Model used to determine the Norton source transformation for the CE amplifier. Resistor $r_{\pi 0}$ represents the equivalent resistance at the base node. (c) Simplified small-signal model for the high-frequency common-emitter amplifier.

where R_{in} is the parallel combination of R_1, R_2, and $(r_x + r_\pi)$, and R_L is the parallel combination of r_o, R_C, and R_3.

The terminal gain of the common-emitter amplifier (the effect of r_x was included in A_i) can be found as

$$A_{bc} = \frac{\mathbf{v_c}}{\mathbf{v_b}} = -g_m R_L \cong -g_m (R_C \| R_3) \tag{16.78}$$

We now use the Miller effect to calculate the input high frequency pole at the base based on the circuit in Fig. 16.35(a).

$$C_{eqB} = C_\mu(1 - A_{bc}) + C_\pi(1 - A_{be}) = C_\mu[1 - (-g_m R_L)] + C_\pi(1 - 0)$$
$$= C_\pi + C_\mu(1 + g_m R_L) \tag{16.79}$$

The equivalent small-signal resistance to ground at base node v_b is

$$R_{eqB} = r_\pi \| (r_x + R_B \| \beta_1) = r_{\pi 0} \tag{16.80}$$

Remember the voltage source v_i has a small-signal impedance of zero. The resulting time constant at the input is

$$\tau_B = R_{eqB} C_{eqB} \tag{16.81}$$

At the collector output node, R_{eqC} is found as

$$R_{eqC} = R_L = r_o \| R_C \| R_3 \cong R_C \| R_3 \tag{16.82}$$

We must now determine the equivalent capacitance at the collector, C_{eqC}. At first glance, we might expect to apply the Miller effect to model the equivalent capacitance at the output due to C_μ. However, the transistor does not operate in reverse: applying a signal at the collector does not result in a significant "output" signal at the base node. Therefore, the equivalent capacitance at the output only includes the physical capacitance C_μ as well as any additional load capacitance C_L.

$$C_{eqC} = C_\mu + C_L \tag{16.83}$$

The resulting time constant at the output node is

$$\tau_C = R_{eqC} C_{eqC} \tag{16.84}$$

If the input and output nodes are well isolated, we expect to calculate a separate pole frequency for the input and output time constants. However, in this case the input and output are coupled through C_μ. In addition, the input and output impedances are large, so we should expect the two time constants to interact. This interaction gives rise to a dominant pole equal to

$$\omega_{P1} = \frac{1}{r_{\pi 0}[C_\pi + C_\mu(1 + g_m R_L)] + R_L[C_\mu + C_L]} \tag{16.85}$$

We can rewrite this expression in terms of a single resistance and capacitance by scaling the second capacitive term so that when multiplied by the input resistance, $r_{\pi 0}$, the resulting time constant is identical to the second term in Eq. (16.85). We will label the resulting capacitance as C_T.

$$C_T = [C_\pi + C_\mu(1 + g_m R_L)] + \frac{R_L}{r_{\pi 0}}[C_\mu + C_L] \tag{16.86}$$

This substitution allows us to write the dominant pole frequency for the common-emitter as

$$\omega_{P1} = \frac{1}{r_{\pi 0} C_T} = \frac{1}{r_{\pi 0} \left([C_\pi + C_\mu(1 + g_m R_L)] + \frac{R_L}{r_{\pi 0}}[C_\mu + C_L] \right)} \tag{16.87}$$

16.6.3 DIRECT ANALYSIS OF THE COMMON-EMITTER TRANSFER CHARACTERISTIC

At this point, it is desirable to check our simplified analysis approach via a direct analysis of the common-emitter transfer function. Writing and simplifying the nodal equations in the frequency domain for the circuit in Fig. 16.35(c) yields

$$\begin{bmatrix} \mathbf{I}_n(s) \\ 0 \end{bmatrix} = \begin{bmatrix} s(C_\pi + C_\mu) + g_{\pi o} & -sC_\mu \\ -(sC_\mu - g_m) & s(C_\mu + C_L) + g_L \end{bmatrix} \begin{bmatrix} \mathbf{V}_b(s) \\ \mathbf{V}_c(s) \end{bmatrix} \tag{16.88}$$

An expression for the output voltage, node voltage $\mathbf{V}_2(s)$, can be found using Cramer's rule:

$$\mathbf{V}_c(s) = \mathbf{I}_n(s) \frac{(sC_\mu - g_m)}{\Delta} \tag{16.89}$$

in which Δ represents the determinant of the system of equations given by

$$\Delta = s^2[C_\pi(C_\mu + C_L) + C_\mu C_L] + s[C_\pi g_L + C_\mu(g_m + g_{\pi o} + g_L) + C_L g_{\pi o}] + g_L g_{\pi o} \tag{16.90}$$

From Eqs. (16.89) and (16.90), we see that the high-frequency response is characterized by two poles, one finite zero, and one zero at infinity. The finite zero appears in the right-half of the s-plane at a frequency

$$\omega_Z = +\frac{g_m}{C_\mu} > \omega_T \tag{16.91}$$

The zero given by Eq. (16.91) can usually be neglected because it appears at a frequency above ω_T (for which the model itself is of questionable validity). Unfortunately, the denominator appears in unfactored polynomial form, and the positions of the poles are more difficult to find. However, good estimates for both pole positions can be found using the approximate factorization technique shown below. Note that even though there are three capacitors, the circuit only has two poles. The three capacitors are connected in a "pi" configuration, and only two of the capacitor voltages are independent. Once we know two of the voltages, the third is also defined.

Approximate Polynomial Factorization
We estimate the pole locations based on a technique for approximate factorization of polynomials. Let us assume that the polynomial has two real roots a and b:

$$(s + a)(s + b) = s^2 + (a + b)s + ab = s^2 + A_1 s + A_0 \tag{16.92}$$

If we assume that a dominant root exists—that is, that $a \gg b$—then the two roots can be estimated directly from coefficients A_1 and A_0 using two approximations:

$$A_1 = a + b \cong a \qquad \text{and} \qquad \frac{A_0}{A_1} = \frac{ab}{a + b} \cong \frac{ab}{a} = b \tag{16.93}$$

so,

$$a \cong A_1 \qquad \text{and} \qquad b \cong \frac{A_0}{A_1}$$

Note in Eq. (16.92) that the s^2 term is normalized to unity. Also note that the approximate factorization technique can be extended to polynomials having any number of widely spaced real roots.

16.6.4 POLES OF THE COMMON-EMITTER AMPLIFIER
For the case of the common-emitter amplifier, the smallest root is the most important because it is the one that limits the high-frequency response of the amplifier. From Eq. (16.93) we see that the smaller root is given by the ratio of coefficients A_0 and A_1, resulting in the following expression for the first pole:

$$\omega_{P1} = \frac{1}{r_{\pi 0} C_T} = \frac{1}{r_{\pi 0}\left([C_\pi + C_\mu(1 + g_m R_L)] + \dfrac{R_L}{r_{\pi 0}}[C_\mu + C_L]\right)} \tag{16.94}$$

This result is identical to that of Eq. (16.87). The dominant pole is controlled by the combination of the input and output time constants set by the total equivalent capacitance and resistance at the input and output. Notice that if the driving resistance R_I is zero, $r_{\pi 0}$ reduces to approximately r_x, and the bandwidth is primarily limited by r_x.

There is also a second pole resulting from the normalized version of coefficient A_1:

$$\omega_{P2} = \frac{C_\pi g_L + C_\mu(g_m + g_{\pi 0} + g_L) + C_L g_{\pi 0}}{C_\pi(C_\mu + C_L) + C_\mu C_L} \tag{16.95}$$

or

$$\omega_{P2} \cong \frac{g_m}{C_\pi\left(1 + \dfrac{C_L}{C_\mu}\right) + C_L} \cong \frac{g_m}{C_\pi + C_L} \tag{16.96}$$

in which the $C_\mu g_m$ term has been assumed to be the largest term in the numerator, as is most often the case for C-E stages with reasonably high gain. We can interpret the last approximation in Eq. (16.96) in this manner, particularly when C_μ is large. At high frequencies, capacitor C_μ effectively shorts the collector and base of the transistor together so that C_L and C_π appear in parallel, and the transistor behaves as a diode with a small-signal resistance of $1/g_m$. Recall also that there is a right-half plane zero equal to $+g_m/C_\mu$. While the zero is typically quite high in frequency and can be neglected, we will see in Chapter 17 that in FET amplifiers it can be an important aspect of the negative feedback amplifier stability analysis.

EXAMPLE 16.6   **HIGH-FREQUENCY ANALYSIS OF THE COMMON-EMITTER AMPLIFIER**

Find the midband gain and upper-cutoff frequency of a common-emitter amplifier.

PROBLEM   Find the midband gain and upper-cutoff frequency of the common-emitter amplifier in Fig. 16.34 using the C_T approximation, assuming $\beta_o = 100$, $f_T = 500$ MHz, $C_\mu = 0.5$ pF, $r_x = 250\ \Omega$, and a Q-point of (1.60 mA, 3.00 V). Find the additional poles and zeros of the common-emitter amplifier. Assume $C_L = 0$, $C_1 = C_2 = 3.9\ \mu$F, $C_3 = 0.082\ \mu$F.

SOLUTION   **Known Information and Given Data:** Common-emitter amplifier circuit in Fig. 16.34; Q-point = (1.60 mA, 3.00 V); $\beta_o = 100$, $f_T = 500$ MHz, $C_\mu = 0.5$ pF, and $r_x = 250\ \Omega$; expressions for the gain, poles, and zeros are given in Eqs. (16.77) through (16.96).

Unknowns: Values for A_{mid}, f_H, ω_{Z1}, ω_{P1}, and ω_{P2}

Approach: Find the small-signal parameters for the transistor. Find the unknowns by substituting the given and computed values into the expressions developed in the text.

Assumptions: Small-signal operation in the active region; $V_T = 25.0$ mV; $C_L = 0$

Analysis: The common-emitter stage is characterized by Eqs. (16.77), (16.78), and (16.87).

$$A_{\text{mid}} = A_i A_{bc} \qquad A_i = \frac{R_{\text{in}}}{R_I + R_{\text{in}}}\left(\frac{r_\pi}{r_x + r_\pi}\right) \qquad A_{bc} = -g_m R_L$$

$$\omega_{P1} = \frac{1}{r_{\pi 0} C_T} \qquad \omega_{P2} = \frac{g_m}{C_\pi + C_L} \qquad \omega_Z = \frac{g_m}{C_\mu}$$

$$r_{\pi 0} = r_\pi \| (R_B \| R_I + r_x) \qquad C_T = C_\pi + C_\mu \left(1 + g_m R_L + \frac{R_L}{r_{\pi 0}}\right)$$

The values of the various small-signal parameters must be found:

$$g_m = 40 I_C = 40(0.0016) = 64.0\ \text{mS} \qquad r_\pi = \frac{\beta_o}{g_m} = \frac{100}{0.064} = 1.56\ \text{k}\Omega$$

$$C_\pi = \frac{g_m}{2\pi f_T} - C_\mu = \frac{0.064}{2\pi(5 \times 10^8)} - 0.5 \times 10^{-12} = 19.9\ \text{pF}$$

$$R_{\text{in}} = 10\ \text{k}\Omega \| 30\ \text{k}\Omega \| 1.81\ \text{k}\Omega = 1.46\ \text{k}\Omega$$

$$R_L = R_C \| R_3 = 4.3\ \text{k}\Omega \| 100\ \text{k}\Omega = 4.12\ \text{k}\Omega$$

$$R_{\text{th}} = R_B \| R_I = 7.5\ \text{k}\Omega \| 1\ \text{k}\Omega = 882\ \Omega$$

$$r_{\pi o} = r_\pi \| (R_{\text{th}} + r_x) = 1.56\ \text{k}\Omega \| (882\ \Omega + 250\ \Omega) = 656\ \Omega$$

Substituting these values into the expression for C_T ($C_L = 0$) yields

$$C_T = C_\pi + C_\mu \left(1 + g_m R_L + \frac{R_L}{r_{\pi o}}\right)$$

$$= 19.9 \text{ pF} + 0.5 \text{ pF}\left[1 + 0.064(4120) + \frac{4120}{656}\right]$$

$$= 19.9 \text{ pF} + 0.5 \text{ pF}(1 + \underline{264} + 6.28) = 156 \text{ pF}$$

and

$$f_{P1} = \frac{1}{2\pi r_{\pi o} C_T} = \frac{1}{2\pi (656 \text{ }\Omega)(156 \text{ pF})} = 1.56 \text{ MHz}$$

$$\omega_{P2} \cong \frac{g_m}{C_\pi + C_L} = \frac{0.064}{19.9 \text{ pF}} = 3.22 \times 10^9 \text{ rad/sec}$$

$$f_{P2} = \frac{\omega_{P2}}{2\pi} = 512 \text{ MHz}$$

$$f_z = \frac{g_m}{2\pi C_\mu} = \frac{0.064}{2\pi (0.5 \text{ pF})} = 20.4 \text{ GHz}$$

$$A_i = \frac{1.46 \text{ k}\Omega}{1.00 \text{ k}\Omega + 1.46 \text{ k}\Omega}\left(\frac{1.56 \text{ k}\Omega}{250 \text{ }\Omega + 1.56 \text{ k}\Omega}\right) = 0.512 \quad A_{bc} = -(0.064S)(4.12 \text{ k}\Omega) = -264$$

$$A_{\text{mid}} = 0.512(-264) = -135$$

Check of Results: We have found the desired information. By double checking, the calculations appear correct. Let us use the gain-bandwidth product as an additional check: $|A_{\text{mid}} f_{P1}| = 211 \text{ MHz}$, which does not exceed f_T.

Discussion: The dominant pole is located at a frequency $f_{P1} = 1.56 \text{ MHz}$, whereas f_{P2} and f_Z are estimated to be at frequencies above f_T (500 MHz). Thus, the upper-cutoff frequency f_H for this amplifier is determined solely by f_{P1}: $f_H \cong 1.56 \text{ MHz}$. Note that this value of f_H is less than 1 percent of the transistor f_T and is consistent with the concept of GBW product. We should expect f_H to be no more than $f_T/A_{\text{mid}} = 3.3 \text{ MHz}$ for this amplifier. Note also that f_{P1} and f_{P2} are separated by a factor of almost 1000, clearly satisfying the requirement for widely spaced roots that was used in the approximate factorization.

It is important to keep in mind that the most important factor in determining the value of C_T is the term in which C_μ is multiplied by $g_m R_L$. To increase the upper-cutoff frequency f_H of this amplifier, the gain ($g_m R_L$) must be reduced; a direct trade-off must occur between amplifier gain and bandwidth.

Computer-Aided Analysis: SPICE can be used to check our hand analysis, but we must define the device parameters that match our analysis. We set BF = 100 and IS = 5 fA, but let VAF default to infinity. The base resistance r_x must be added by setting SPICE parameter RB = 250 Ω. C_μ is determined in SPICE from the value of the zero-bias collector-junction capacitance CJC and the built-in potential ϕ_{jc}. The Q-point from SPICE gives $V_{CE} = 2.70 \text{ V}$, which corresponds to $V_{BC} = 2.0 \text{ V}$ if $V_{BE} = 0.7 \text{ V}$. In SPICE, VJC faults to 0.75 V, and MJC defaults to 0.33 (see Sec. 16.4). Therefore, to achieve $C_\mu = 0.5 \text{ pF}$, CJC is specified as

$$\text{CJC} = 0.5 \text{ pF}\left(1 + \frac{20 \text{ V}}{0.75 \text{ V}}\right)^{0.33} = 0.768 \text{ pF}$$

C_π is determined by the SPICE forward transit-time parameter TF, as defined by Eqs. (5.39) and (16.50):

$$\text{TF} = \frac{C_\pi}{g_m} = \frac{19.9 \text{ pF}}{64 \text{ mS}} = 0.311 \text{ ns}$$

After adding these values to the transistor model, we perform an ac analysis using FSTART = 100 Hz and FSTOP = 10 MHz with 20 frequency points per decade. The SPICE simulation results in the graph below yield $A_{\text{mid}} = -135$ (42.6 dB) and $f_H \cong 1.56$ MHz, which agree closely with our hand calculations. Checking the device parameters in SPICE, we also find $r_x(\text{RB}) = 250 \ \Omega$, $C_\pi(\text{CBE}) = 19.9$ pF, and $C_\mu(\text{CBC}) = 0.499$ pF, as desired.

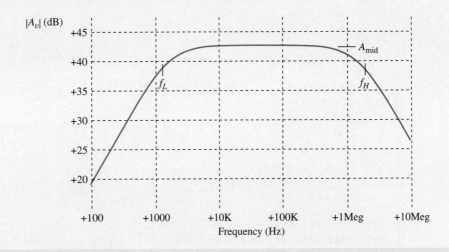

EXERCISE: Use SPICE to recalculate f_H for $V_A = 75$ V. For $V_A = 75$ V and $r_x = 0$?

ANSWER: 1.67 MHz; 1.96 MHz

EXERCISE: Repeat the calculations in Ex. 16.6 if a load capacitance $C_L = 3$ pF is added to the circuit.

ANSWERS: 1.39 MHz, 71.6 MHz (A significant decrease in f_{P2}!)

EXERCISE: Find the midband gain and the frequencies of the poles and zeros of the common-emitter amplifier in Ex. 16.6 if the transistor has $f_T = 500$ MHz, but $C_\mu = 1$ pF.

ANSWERS: −153, 835 kHz, 578 MHz, 10.2 GHz

16.6.5 DOMINANT POLE FOR THE COMMON-SOURCE AMPLIFIER

Analysis of the C-S amplifier in Fig. 16.36 mirrors that of the common-emitter amplifier. The small-signal model is similar to that for the C-E stage, except that both r_x and r_π are absent from the model. For Fig. 16.36(b),

$$R_{\text{th}} = R_I \| R_G \qquad R_L = R_D \| R_3 \qquad v_{\text{th}} = v_i \frac{R_G}{R_I + R_G} \qquad (16.97)$$

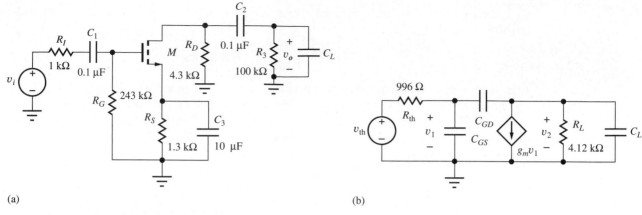

Figure 16.36 (a) Common-source amplifier. (b) The high-frequency small-signal model.

The expressions for the finite zero and poles of the C-S amplifier can be found by comparing Fig. 16.36(b) to Fig. 16.35:

$$\omega_{P1} = \frac{1}{R_{th}C_T} \quad \text{and} \quad C_T = C_{GS} + C_{GD}\left(1 + g_m R_L + \frac{R_L}{R_{th}}\right) + C_L \frac{R_L}{R_{th}}$$

(16.98)

$$\omega_{P2} = \frac{g_m}{C_{GS} + C_L} \qquad \omega_Z = \frac{+g_m}{C_{GD}}$$

EXERCISE: What is the upper-cutoff frequency for the amplifier in Fig. 16.36 if $C_{GS} = 10$ pF, $C_{GD} = 2$ pF, $C_L = 0$ pF, and $g_m = 1.23$ mS? What are the positions of the second pole and the zero? What is the f_T of this transistor?

ANSWERS: 5.26 MHz; 19.6 MHz, 97.9 MHz; 16.3 MHz

16.6.6 ESTIMATION OF ω_H USING THE OPEN-CIRCUIT TIME-CONSTANT METHOD

A technique also exists for estimating ω_H that is similar to the short-circuit time-constant method used to find ω_L. However, the upper-cutoff frequency ω_H is found by calculating the open-circuit time constants associated with the various device capacitances rather than the short-circuit time constants associated with the coupling and bypass capacitors. At high frequencies, the impedances of the coupling and bypass capacitors are negligibly small, and they effectively represent short circuits. The impedances of the device capacitances have now become small enough that they can no longer be neglected with respect to the internal resistances of the transistors. We will see shortly that the C_T approximation results can also be found using the OCTC method.

Although beyond the scope of this book, it can be shown theoretically[3] that the mathematical estimate for ω_H for a circuit having m capacitors is

$$\omega_H \cong \frac{1}{\displaystyle\sum_{i=1}^{m} R_{io}C_i}$$

(16.99)

---

[3] See [1]. The OCTC and SCTC methods represent dominant root factorizations similar to Eq. (16.92).

ELECTRONICS IN ACTION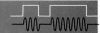

Graphic Equalizer

Graphic equalizers are used in audio applications to fine-tune the frequency response of an audio system. The equalizer is used to compensate for frequency-dependent absorption characteristics of a room, poor quality recordings, or just listener preferences. An example equalizer is shown in the figure below. This is the SG-9500 classic analog equalizer introduced by Pioneer in 1976 and features total harmonic distortion of only 0.05 percent. The unit sold for $300 and weighed about 15 pounds. It has a set of slider controls that set boost or cut levels for 10 different frequency bands within the audio frequency range. The term "graphic" is applied to equalizers where the physical position of the controls is representative of the boost or cut levels applied to the different bands.

Pioneer SG-9500 graphic equalizer.
(Image courtesy of Matt Francis.)

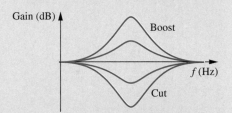

Typical graphic equalizer single band frequency response for different boost/cut settings.

A simplified schematic of a graphic equalizer is shown here. The circuit includes two summing amplifiers and a series of band-pass filters. The band-pass filters provide the frequency band selection. Resistor R_3 divides the output of the filters into two signals. The signal applied to the summing input of A_1 provides band reduction and the signal applied to the summing junction of A_2 provides boost for a particular frequency band.

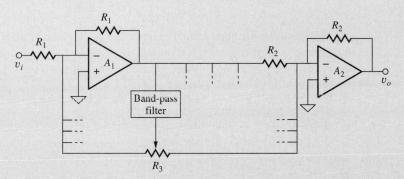

Typical graphic equalizer circuit.[1]

If potentiometer R_3 is set to the center point, the gain and boost signals are balanced, and no net signal is added or subtracted to the output. The slider controls in the picture above correspond to R_3 in the circuit diagram.

Graphic equalizers have been reduced greatly in size since the mid-1970s, and until recently, functioned similarly to the one shown above. However, with the advent of

---

[1] Dennis A. Bohn, "Constant-Q graphic equalizers." *J. Audio Eng. Soc.,* Vol. 34, No. 9, September 1986.

high-precision, low-cost A/D converters and low-cost high-performance digital signal processing (DSP), graphic equalizers have moved into the digital domain. DSP based equalizers with excellent accuracy and controllability are now available. This new class of equalizer has an A/D converter, followed by DSP circuits, with a digital-to-analog (D/A) converter at the output to move the signal back into the analog domain. The DSP allows the designer to generate complex transfer functions that account for non-idealities such as channel-to-channel interactions. DSP equalizers are commonly found in MP3 music players, for example. As integrated circuit process technology advances, it is always important to reevaluate the appropriate boundaries between analog and digital signal processing.

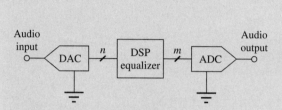

Graphic equalizer based upon digital signal processing.

Graphic equalizer screen from DSP-based audio software (Apple ITunes).

in which R_{io} represents the resistance measured at the terminals of capacitor C_i with the other capacitors open circuited. Because we already have results for the C-E stages, let us practice by applying the method to the high-frequency model for the C-E amplifier in Fig. 16.35.

If we assume $C_L = 0$, then two capacitors, C_π and C_μ, are present in Fig. 16.35, and $R_{\pi o}$ and $R_{\mu o}$ will be needed to evaluate Eq. (16.99). $R_{\pi o}$ can easily be determined from the circuit in Fig. 16.37, in which C_μ is replaced by an open circuit, and we see that

$$R_{\pi o} = r_{\pi o} \tag{16.100}$$

$R_{\mu o}$ can be determined from the circuit in Fig. 16.38, in which C_π is replaced by an open circuit. In this case, a bit more work is required. Test source i_x is applied to the network in Fig. 16.38(b), and v_x can be found by applying KVL around the outside loop:

$$\mathbf{v_x} = \mathbf{i_x} r_{\pi o} + \mathbf{i_L} R_L = \mathbf{i_x} r_{\pi o} + (\mathbf{i_x} + g_m \mathbf{v}) R_L \tag{16.101}$$

However, voltage $\mathbf{v}$ is equal to $\mathbf{i_x} r_{\pi o}$, and substituting this result into Eq. (16.101) yields

$$R_{\mu o} = \frac{\mathbf{v_x}}{\mathbf{i_x}} = r_{\pi o} + (1 + g_m r_{\pi o}) R_L = r_{\pi o} \left[1 + g_m R_L + \frac{R_L}{r_{\pi o}} \right] \tag{16.102}$$

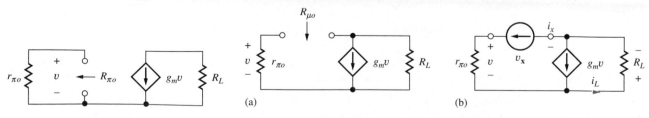

Figure 16.37 Circuit for finding $R_{\pi o}$.

Figure 16.38 (a) Circuit defining $R_{\mu o}$. (b) Test source applied.

which should look familiar [see Eq. (16.86)]. Substituting Eqs. (16.100) and (16.102) into Eq. (16.99) produces the estimate for ω_H:

$$\omega_H \cong \frac{1}{R_{\pi o}C_\pi + R_{\mu o}C_\mu} = \frac{1}{r_{\pi o}C_\pi + r_{\pi o}C_\mu\left(1 + g_m R_L + \dfrac{R_L}{r_{\pi o}}\right)} = \frac{1}{r_{\pi o}C_T} \qquad (16.103)$$

This is exactly the same result achieved from Eqs. (16.94) for $C_L = 0$ but with far less effort. (Remember, however, that this method does not produce an estimate for either the second pole or the zeros of the network.)

EXERCISE: Suppose that C_L is not zero in Fig. 16.34. What is R_{L0}?

ANSWER: R_L

16.6.7 COMMON-SOURCE AMPLIFIER WITH SOURCE DEGENERATION RESISTANCE

Figure 16.39(a) shows a common-source amplifier with unbypassed source resistance R_S. Figure 16.39(b) is the small-signal equivalent circuit. We find the input equivalent capacitance and resistance in the same manner as used for the common-emitter circuit. The midband gain is first calculated in two parts as before. The input gain expression is similar to that of the common-emitter

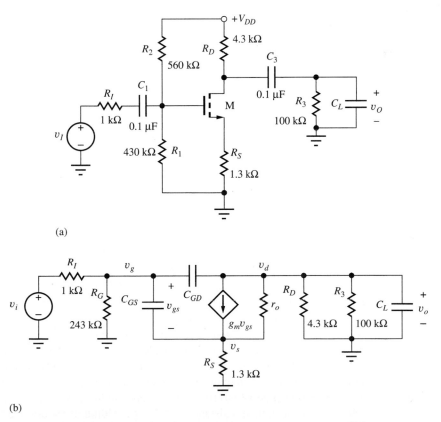

(a)

(b)

Figure 16.39 (a) Common-source amplifier with unbypassed source resistance. (b) Small-signal equivalent circuit.

except that r_π is not included since the impedance looking into the gate is infinite.

$$A_i = \frac{\mathbf{v_g}}{\mathbf{v_i}} = \frac{R_G}{R_I + R_G} = \frac{R_1 \| R_2}{R_I + (R_1 \| R_2)} \tag{16.104}$$

The terminal gain of the common-source amplifier is found as

$$A_{gd} = \frac{v_d}{v_g} = \frac{-g_m R_L}{1 + g_m R_S} = \frac{-g_m (R_{iD} \| R_D \| R_3)}{1 + g_m R_S} \cong \frac{-g_m (R_D \| R_3)}{1 + g_m R_S} \tag{16.105}$$

where

$$R_{iC} = r_o(1 + g_m R_S) \tag{16.106}$$

As seen in Eq. (16.106), R_{out} is typically quite large and can be neglected, and $R_L \cong R_D \| R_3$. We again use the Miller effect to calculate the input high-frequency time constant.

$$C_{eqG} = C_{GD}(1 - A_{gd}) + C_{GS}(1 - A_{gs})$$

$$= C_{GD}\left(1 - \frac{[-g_m R_L]}{1 + g_m R_S}\right) + C_{GS}\left(1 - \frac{g_m R_S}{1 + g_m R_S}\right) \tag{16.107}$$

$$= C_{GD}\left(1 + \frac{g_m (R_D \| R_3)}{1 + g_m R_S}\right) + \frac{C_{GS}}{1 + g_m R_S}$$

Note that we have used the expression for the gain of the common-drain amplifier to calculate the Miller multiplication of C_{GS}. Unlike the Miller effect with regard to C_{GD}, the effective capacitance of C_{GS} is reduced since A_{gs} will always be positive and less than 1. The unbypassed source resistance has also had the effect of reducing the effect of C_{GD} since the gate-to-drain gain has also been reduced by the $(1 + g_m R_S)$ term. The equivalent small-signal resistance to ground at the gate node is

$$R_{eqG} = R_G \| R_I = R_{th} \tag{16.108}$$

The equivalent capacitance and resistance at the output is similar to that of the common-emitter amplifier.

$$R_{eqD} = R_{iD} \| R_D \| R_3 \cong R_D \| R_3 \quad \text{and} \quad C_{eqD} = C_{GD} + C_L \tag{16.109}$$

Combining these results and using Eq. (16.98), we find the following general form for the poles and right-half plane zero.

$$\omega_{P1} = \frac{1}{R_{th}\left[\dfrac{C_{GS}}{1 + g_m R_S} + C_{GD}\left(1 + \dfrac{g_m R_L}{1 + g_m R_S}\right) + \dfrac{R_L}{R_{th}}(C_{GD} + C_L)\right]} \tag{16.110}$$

$$\omega_{P2} = \frac{g_m}{(1 + g_m R_S)(C_{GS} + C_L)} \tag{16.111}$$

$$\omega_z = \frac{+g_m}{(1 + g_m R_S)(C_{GD})} \tag{16.112}$$

Notice the gain bandwidth tradeoff indicated in Equations (16.105) and (16.110). The $(1 + g_m R_S)$ term decreases gain while increasing the frequency of the dominant pole ω_{P1}. In our study of op amps, we found that gain and bandwidth can be traded one for the other and the same relationship generally holds true in transistor circuits. Since the gain and bandwidth are inversely affected by the $(1 + g_m R_S)$ term, the gain-bandwidth product is held relatively constant, similar to what we found in our study of the gain-bandwidth characteristics of op amp based amplifiers.

The second pole and zero equations are modified to account for the degeneration of the effective g_m by the source resistance. Notice that although the dominant pole increases in frequency, the frequencies of second pole and zero are decreased. Increasing the gain of the stage increases

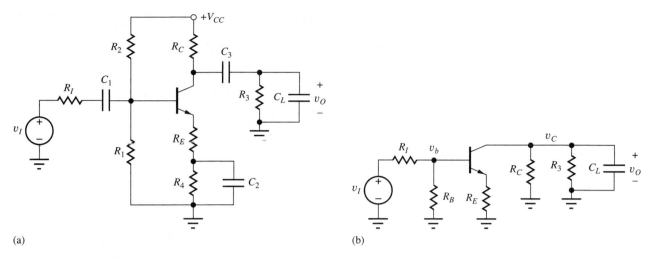

Figure 16.40 (a) Common-emitter amplifier with unbypassed emitter resistor R_E. (b) High-frequency ac equivalent circuit.

the frequency separation between ω_{P1} and ω_{P2}, resulting in what is often referred to as **pole-splitting.** Decreasing the gain moves the two poles closer in frequency, which can compromise the phase margin of feedback amplifiers.

If the small-signal source resistance R_S is reduced to zero, the equations for the common-source poles reduce to the simpler form of the common-source equations found previously. Likewise, if an unbypassed emitter resistance is included in the common-emitter amplifier, the pole equations can be modified in a manner similar to the common-source with source degeneration amplifier above.

16.6.8 POLES OF THE COMMON-EMITTER WITH EMITTER DEGENERATION RESISTANCE

The equations for the common-emitter with unbypassed source resistance are modified in a manner similar to those of the common-source. In Fig. 16.40 a portion of the emitter resistance R_E is unbypassed, and the input stage gain A_i is modified due to the increased impedance looking into the base.

$$A_i = \frac{\mathbf{v_b}}{\mathbf{v_i}} = \frac{(R_B \| R_{iB})}{R_I + (R_B \| R_{iB})} \cdot \frac{r_\pi + (\beta + 1)R_E}{r_x + r_\pi + (\beta + 1)R_E} \cong \frac{R_B \| R_{iB}}{R_I + R_B \| R_{iB}} \tag{16.113}$$

Recall the impedance looking into the base as

$$R_{iB} = r_x + r_\pi + (\beta + 1)R_E \tag{16.114}$$

The terminal gain of the common-emitter with unbypassed emitter resistance is found as

$$A_{bc} = \frac{v_c}{v_b} \cong \frac{-g_m(R_C \| R_3)}{1 + g_m R_E} \cong \frac{-g_m R_L}{1 + g_m R_E} \tag{16.115}$$

Including the effect of the emitter degeneration resistance R_E, the pole and zero equations are modified as follows:

$$\omega_{P1} = \frac{1}{r_{\pi 0}C_T} = \frac{1}{r_{\pi o}\left(\left[\dfrac{C_\pi}{1 + g_m R_E} + C_\mu\left(1 + \dfrac{g_m R_L}{1 + g_m R_E}\right)\right] + \dfrac{R_L}{r_{\pi o}}\left[C_\mu + C_L\right]\right)} \tag{16.116}$$

where $\quad r_{\pi o} = R_{eqB} = (R_{th} + r_x)\|[r_\pi + (\beta + 1)R_E] \quad$ with $R_{th} = R_B \| R_I$ $\qquad$ (16.117)

$$\omega_{P2} \cong \frac{g_m}{(1 + g_m R_E)(C_\pi + C_L)} \tag{16.118}$$

$$\omega_z = \frac{+g_m}{(1 + g_m R_E)(C_\mu)} \tag{16.119}$$

As with the common-source amplifier, the degeneration resistance causes a decrease in gain and an increase in the dominant pole frequency. The amplifier allows one to directly tradeoff gain and bandwidth, approximately maintaining a constant gain-bandwidth product.

EXAMPLE **16.7**   **COMMON-EMITTER AMPLIFIER WITH EMITTER DEGENERATION**

In this example, we explore the gain-bandwidth trade-off achieved by adding an unbypassed emitter resistor to the common-emitter amplifier from Ex. 16.6.

PROBLEM   Find the midband gain, upper-cutoff frequency, and gain-bandwidth product for the common-emitter amplifier in Fig. 16.34 if a 300-Ω portion of the emitter resistor is not bypassed. Assume $\beta_o = 100$, $f_T = 500$ MHz, $C_\mu = 0.5$ pF, $r_x = 250$ Ω, and the Q-point = (1.6 mA, 3.0 V).

SOLUTION   **Known Information and Given Data:** Common-emitter amplifier in Fig. 16.34 with a bypass capacitor placed around a 1000-Ω portion of the emitter resistor (in a manner similar to Fig. 14.2); $\beta_o = 100$, $f_T = 500$ MHz, $C_\mu = 0.5$ pF, $r_x = 250$ Ω; Q-point: (1.6 mA, 3.0 V).

Unknowns: A_{mid}, f_H, and GBW

Approach: Find A_{mid} and f_H using Eqs. (16.113–16.117). GBW $= A_{\text{mid}} \times f_H$.

Assumptions: $V_T = 25.0$ mV; small-signal operation in the active region

Analysis: Using the values from the analysis of Fig. 16.34 with $g_m = 40 I_C = 64$ mS:

$$r_\pi = \frac{\beta_o}{g_m} = \frac{100}{0.064} = 1.56 \,\text{k}\Omega \qquad R_{\text{th}} + r_x = 882\,\Omega + 250\,\Omega = 1130\,\Omega$$

$$R_{iB} = r_x + r_\pi + (\beta_o + 1)R_E = 250\,\Omega + 1560\,\Omega + (101)300\,\Omega = 32.1\,\text{k}\Omega$$

$$r_{\pi0} = R_{iB} \| (R_{\text{th}} + r_x) = 1.09\,\text{k}\Omega \qquad 1 + g_m R_E = 1 + 0.064(300) = 20.2$$

$$\omega_H \cong \cfrac{1}{r_{\pi0}\left[\cfrac{C_\pi}{1 + g_m R_E} + C_\mu\left(1 + \cfrac{g_m R_L}{1 + g_m R_E} + \cfrac{R_L}{r_{\pi0}}\right)\right]}$$

$$\cong \cfrac{1}{1090\left[\cfrac{19.9\,\text{pF}}{20.2} + 0.5\,\text{pF}\left(1 + \cfrac{264}{20.2} + \cfrac{4120}{1090}\right)\right]}$$

$$f_H \cong \frac{1}{2\pi}\,\frac{1}{1090\,\Omega\,(9.91\,\text{pF})} = 14.7\,\text{MHz}$$

$$A_i = \frac{R_1 \| R_2 \| R_{iB}}{R_1 + R_1 \| R_2 \| R_{iB}} = \frac{10\,\text{k}\Omega \| 30\,\text{k}\Omega \| 32.1\,\text{k}\Omega}{1\,\text{k}\Omega + 10\,\text{k}\Omega \| 30\,\text{k}\Omega \| 32.1\,\text{k}\Omega} = 0.859$$

$$A_{bc} = -\frac{g_m R_L}{1 + g_m R_E} = -\frac{0.064(4120\,\Omega)}{1 + 0.064(300\,\Omega)} = -13.0 \qquad \text{or} \qquad 22.3\,\text{dB}$$

$$A_{\text{mid}} = A_i A_{bc} = 0.859(-13.0) = -11.2 \qquad \text{GBW} = 11.2 \times 14.7\,\text{MHz} = 165\,\text{MHz}$$

Check of Results: A quick estimate for A_{mid} is $-R_L/R_E = -13.7$. Our more exact calculation is slightly less than this number, so it appears correct. The GBW product of the amplifier is 165 MHz, which is approximately 1/3 of f_T, also a reasonable result.

Discussion: Remember, the original C-E stage with no emitter resistance had $A_{\text{mid}} = -153$ and $f_H = 1.56$ MHz for GBW = 239 MHz. With $R_E = 300$ Ω, the gain has decreased by a factor of 14, and the bandwidth has increased by a factor of 8.9. The gain-bandwidth trade-off in the expression for ω_H is not exact because the effective resistance in the time-constant only increases from 882 Ω to 1130 Ω, as well as the $R_L C_\mu$ term that is not scaled by the $(1 + g_m R_E)$ factor.

Computer-Aided Analysis: SPICE can be used to check our hand analysis, but we must define the device parameters that match our analysis. The base resistance, collector-base capacitance, and forward transit-time were calculated in Ex. 16.6: RB = 250 Ω, CJC = 0.768 pF, and TF = 0.311 ns. After adding these values to the transistor model, we perform an ac analysis using FSTART = 10 Hz and FSTOP = 100 MHz with 20 frequency points per decade. SPICE yields $A_{\text{mid}} = -11.0$ and $f_H \cong 15.0$ MHz, which agree well with our hand calculations. Note that the lower-cutoff frequency has changed to 158 Hz. Use $C_1 = C_2 = 3.9$ µF, $C_3 = 0.082$ µF.

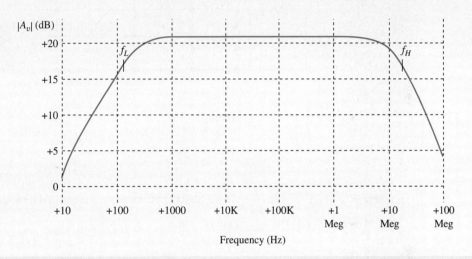

EXERCISE: Use SPICE to recalculate f_H for $V_A = 75$ V. For $V_A = 75$ V and $r_x = 0$?

ANSWERS: 14.8 MHz; 17.8 MHz

EXERCISE: Use the formulas to recalculate the midband gain, f_H, and GBW if the unbypassed portion of the emitter resistor is decreased to 100 Ω?

ANSWERS: −32.2; 6.10 MHz; 196 MHz

16.7 COMMON-BASE AND COMMON-GATE AMPLIFIER HIGH-FREQUENCY RESPONSE

We analyze the high-frequency response of the other single-stage amplifiers using the same approach we used in the previous section. At each node along the signal path, we determine an equivalent resistance to small-signal ground and an equivalent capacitance to small-signal ground. The resulting RC network gives rise to a high-frequency pole. We now apply this approach to the common-base amplifier shown in Figure 16.41(a). The high-frequency ac equivalent circuit is shown in Figure 16.41(b). Base resistance r_x has been neglected to simplify the analysis, as has output resistance r_o.

The input gain of the common base circuit is found as

$$A_i = \frac{\mathbf{v_e}}{\mathbf{v_i}} = \frac{R_{\text{in}}}{R_I + R_{\text{in}}} = \frac{R_E \| R_{iE}}{R_i + R_E \| R_{iE}} \tag{16.120}$$

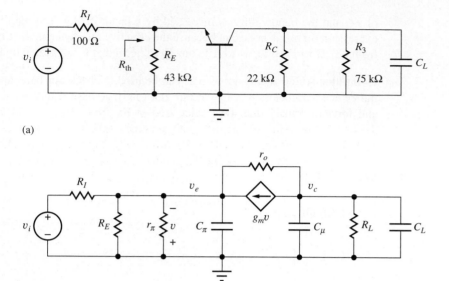

(a)

(b)

Figure 16.41 (a) High-frequency ac equivalent circuit for the common-base amplifier. (b) Small-signal model for the common-base amplifier.

where $R_{iE} = \dfrac{r_\pi}{\beta_o + 1} \cong \dfrac{1}{g_m}$. (To include the effect of r_x, we can add it to r_π.) Given that $R_{iE} \cong 1/g_m$ and if $1/g_m \ll R_E$, the input gain becomes

$$A_i \cong \frac{1}{1 + g_m R_I} \tag{16.121}$$

The terminal gain of the common-base amplifier is found as

$$A_{ec} = \frac{\mathbf{v_c}}{\mathbf{v_e}} = g_m(R_{iC} \| R_L) \cong +g_m R_L \tag{16.122}$$

where

$$R_{iC} = r_o[1 + g_m(r_\pi \| R_{\text{th}})] \tag{16.123}$$

The expression for R_{iC} is a simplified version of Eq. (15.183) with R_B equal to zero. Again, R_{iC} is typically much larger than the other resistances at the collector and can be neglected. The equivalent capacitance at the input is found as

$$C_{eqE} = C_\pi \tag{16.124}$$

An output capacitance associated with a driving stage would be added to C_π. To calculate the equivalent resistance at the emitter node, recall that due to the dependent generator, the resistance looking into the emitter is $R_{iE} \cong 1/g_m$. For the circuit in Fig. 16.41,

$$R_{eqE} = \frac{1}{g_m} \| R_E \| R_I \tag{16.125}$$

At the output, we determine the equivalent capacitance and resistance as

$$C_{eqC} = C_\mu + C_L \quad \text{and} \quad R_{eqC} = R_{iC} \| R_L \cong R_L \tag{16.126}$$

Since the input and output are well decoupled, we find the two poles for the common-base amplifier are

$$\omega_{P1} = \frac{1}{\left(\dfrac{1}{g_m} \| R_E \| R_I\right) C_\pi} \cong \frac{g_m}{C_\pi} \tag{16.127}$$

$$\omega_{P2} = \frac{1}{(R_{\text{out}} \| R_L)(C_\mu + C_L)} \cong \frac{1}{R_L(C_\mu + C_L)} \tag{16.128}$$

We should notice that the input pole of this stage has no Miller multiplication terms, and its equivalent resistance is dominated by the typically small $1/g_m$ term. As a result, the input pole of the common-base amplifier is typically a very high frequency, exceeding f_T. The bandwidth of the stage is dominated by the load resistance and capacitance, modeled with ω_{P2}.

EXERCISE: Find the midband gain and f_H using Eq. (16.128) for the common-base amplifier in Fig. 16.41 if the transistor has $\beta_o = 100$, $f_T = 500\,\text{MHz}$, $r_x = 250\,\Omega$, $C_\mu = 0.5\,\text{pF}$, and a Q-point (0.1 mA, 3.5 V). What is the gain-bandwidth product?

ANSWERS: $+48.0$; 10.7 MHz; 514 MHz

Figures 16.42(a) and (b) represent the high-frequency ac and small-signal equivalent circuits for a common-gate amplifier, and the analysis of the common-gate response is analogous to that of the common-base with R_4, C_{GS}, and C_{GD} replacing R_E, C_π, and C_μ.

$$\omega_{P1} = \frac{1}{\left(\dfrac{1}{g_m} \| R_4 \| R_I\right) C_{GS}} \cong \frac{g_m}{C_{GS}} \tag{16.129}$$

$$\omega_{P2} = \frac{1}{[R_{\text{out}} \| R_L][C_{GD} + C_L]} \cong \frac{1}{R_L[C_{GD} + C_L]} \tag{16.130}$$

EXERCISE: Find the midband gain and f_H for the common-gate amplifier in Fig. 16.42 if the transistor $C_{GS} = 10\,\text{pF}$, $C_{GD} = 1\,\text{pF}$, $g_m = 3\,\text{mS}$, and $C_L = 3\,\text{pF}$. What are the gain-bandwidth product and f_T?

ANSWERS: $+8.98$, 9.65 MHz; 86.7 MHz; 43.4 MHz

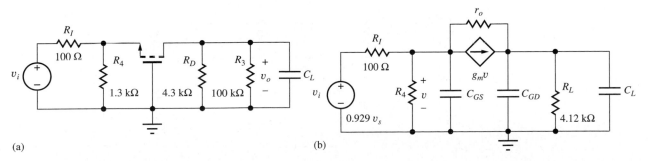

(a)   (b)

Figure 16.42 (a) High-frequency ac equivalent circuit for a common-gate amplifier. (b) Corresponding small-signal model.

16.8 COMMON-COLLECTOR AND COMMON-DRAIN AMPLIFIER HIGH-FREQUENCY RESPONSE

The high-frequency responses of the common-collector and common-drain amplifiers are found in a manner similar to the other single-stage amplifiers. Figure 16.43 illustrates a typical common-collector amplifier and its small-signal equivalents. (Note that r_o is included in R_L.)

The midband input gain looks very similar to that of the common-emitter amplifier.

$$A_i = \frac{\mathbf{v_b}}{\mathbf{v_i}} = \frac{R_{in}}{R_I + R_{in}} = \frac{R_B \| R_{iB}}{R_I + R_B \| R_{iB}} = \frac{R_B \| [r_x + r_\pi + (\beta_o + 1)R_L]}{R_I + R_B \| [r_x + r_\pi + (\beta_o + 1)R_L]} \qquad (16.131)$$

The base-to-emitter terminal voltage gain is

$$A_{be} = \frac{\mathbf{v_e}}{\mathbf{v_b}} = \frac{g_m R_L}{1 + g_m R_L} \qquad (16.132)$$

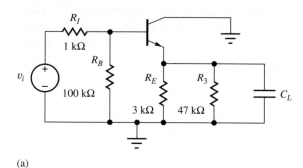

(a)

(b)

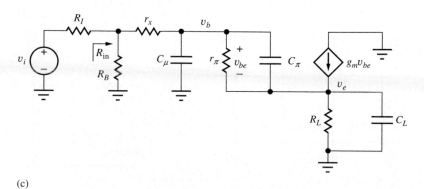

(c)

Figure 16.43 (a) Common-collector amplifier. (b) Small-signal model for the common-collector amplifier. (c) Simplification of the small-signal circuit for calculation of input and output high-frequency poles.

To calculate the high-frequency poles, we first evaluate the equivalent small-signal resistance to ground, R_{eqB} at node v_b.

$$R_{eqB} = [(R_I \| R_B) + r_x] \| [r_\pi + (\beta_o + 1)R_L] = (R_{th} + r_x) \| [r_\pi + (\beta_o + 1)R_L] \quad (16.133)$$

The equivalent capacitance is found using Miller multiplication as

$$C_{eqB} = C_\mu(1 - A_{bc}) + C_\pi(1 - A_{be}) = C_\mu(1 - 0) + C_\pi\left(1 - \frac{g_m R_L}{1 + g_m R_L}\right) = C_\mu + \frac{C_\pi}{1 + g_m R_L} \quad (16.134)$$

The equivalent small-signal resistance at the emitter can be found as

$$R_{eqE} = R_{iE} \| R_L = \left(\frac{r_\pi + R_{th} + r_x}{\beta_o + 1}\right) \| R_L \cong \frac{1}{g_m} + \frac{R_{th} + r_x}{\beta_o + 1} \quad (16.135)$$

where $R_{th} = R_B \| R_I$. The equivalent capacitance is found as the parallel combination of the load capacitance and the base-to-emitter capacitance.

$$C_{eqE} = C_\pi + C_L \quad (16.136)$$

Because of the low impedance at the output, the input and output time constants are relatively well decoupled, resulting in two poles for the common-collector amplifier.

$$\omega_{P1} = \frac{1}{([R_{th} + r_x] \| [r_\pi + (\beta_o + 1)R_L])\left(C_\mu + \dfrac{C_\pi}{1 + g_m R_L}\right)} \quad (16.137)$$

$$\omega_{P2} = \frac{1}{[R_{iE} \| R_L][C_\pi + C_L]} \cong \frac{1}{\left[\left(\dfrac{1}{g_m} + \dfrac{R_{th} + r_x}{\beta_o + 1}\right) \| R_L\right][C_\pi + C_L]} \quad (16.138)$$

Notice that the output pole of this stage is dominated by the typically small $1/g_m$ term. As a result, the output pole of the common-collector amplifier is typically a very high frequency, approaching f_T. The bandwidth of the stage is dominated by f_{p1}, the pole associated with the input section equivalent resistance and capacitance. We will typically ignore the high-frequency pole at the emitter. Because of the feed-forward high-frequency path through C_π, a common-collector stage also includes a high-frequency zero.

$$\omega_z \cong \frac{g_m}{C_\pi} \quad (16.139)$$

Notice that this zero is in the left-half plane. For low load capacitances, ω_z and ω_{P2} tend to cancel each other, so we should only include the effects of ω_{P2} when we also include ω_z.

EXERCISE: Find A_{mid} and f_H for the common-collector amplifier in Fig. 16.43 if the Q-point is (1.5 mA, 5 V), $\beta_o = 100$, $r_x = 150 \, \Omega$, $C_\mu = 0.5$ pF, and $f_T = 500$ MHz.

ANSWERS: 0.980, 229 MHz

A similar set of equations can be found for the common-drain amplifier of Fig. 16.44, making the appropriate changes for the different characteristics of the FET small-signal model.

$$\omega_{P1} = \frac{1}{R_{th}\left(C_{GD} + \dfrac{C_{GS}}{1 + g_m R_L}\right)} \qquad R_{th} = R_G \| R_I \quad (16.140)$$

$$\omega_{P2} = \frac{1}{[R_{iS} \| R_L][C_{GS} + C_L]} \cong \frac{1}{\left[\dfrac{1}{g_m} \| R_L\right][C_{GS} + C_L]} \quad (16.141)$$

$$\omega_z \cong \frac{g_m}{C_{GS}} \quad (16.142)$$

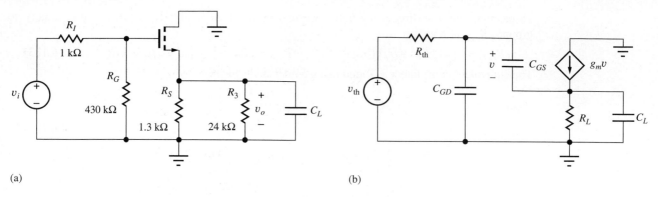

(a)

(b)

Figure 16.44 (a) High-frequency ac equivalent circuit for a source follower. (b) Corresponding high-frequency small-signal model.

Similar to the common-collector, the common-drain amplifier's high-frequency response is dominated by the input pole, f_{p1}, due to the small impedance, r_s, associated with the output pole and zero.

> **EXERCISE:** Find A_{mid} and f_H for the common-drain amplifier in Fig. 16.44 if $C_{GS} = 10\,\text{pF}$, $C_{GD} = 1\,\text{pF}$, and $g_m = 3\,\text{mS}$.
>
> **ANSWERS:** 0.785, 51.0 MHz

16.9 SINGLE-STAGE AMPLIFIER HIGH-FREQUENCY RESPONSE SUMMARY

Table 16.2 collects the expressions for the dominant poles of the three classes of single-stage amplifiers. The inverting amplifiers provide high voltage gain but with the most limited bandwidth. The noninverting stages offer improved bandwidth with voltage gains similar to those of the inverting amplifiers. Remember, however, that the input resistance of the noninverting amplifiers is relatively low. The followers provide unity gain with very wide bandwidth.

It is also worth noting at this point that both the C-E and C-B (or C-S and C-G) stages have a bandwidth that is always less than that set by the time constant of R_L and $(C_\mu + C_L)$ (or $C_{GD} + C_L$ and R_L) at the output node:

$$\omega_H < \frac{1}{R_L(C_\mu + C_L)} \qquad \text{or} \qquad \omega_H < \frac{1}{R_L(C_{GD} + C_L)}$$

16.9.1 AMPLIFIER GAIN-BANDWIDTH LIMITATIONS

The importance of the base resistance r_x (and gate resistance in high-frequency FETs) in ultimately limiting the frequency response of amplifiers should not be overlooked. Consider first the common-emitter amplifiers described by Table 16.2. If the Thévenin equivalent source resistance R_I were reduced to zero in an attempt to increase the bandwidth, then $r_{\pi\phi}$ would not become zero, but would be limited approximately to the value of r_x. If one assumes that the gain is large and the $g_m R_L C_\mu$ term is dominant in determining ω_H, then the gain bandwidth product for the common-emitter stage becomes $\text{GBW} \leq 1/r_x C_\mu$.

In the common-collector case, the gain is approximately one, and for $R_I = 0$ and large $g_m R_L$, the bandwidth becomes $1/r_x C_\mu$. Here again we find $\text{GBW} \leq 1/r_x C_\mu$. If we neglect C_π in Fig. 16.43, we can easily see that the bandwidth is set by r_x and C_μ, since the input resistance looking into r_π will be very large compared to r_x.

TABLE 16.2
Upper-Cutoff Frequency Estimates for the Single-Stage Amplifiers

$$\omega_H$$

Common-emitter

$$\frac{1}{r_{\pi0}C_T} = \frac{1}{r_{\pi0}\left[C_\pi + C_\mu(1 + g_m R_L) + (C_u + C_L)\dfrac{R_L}{r_{\pi0}}\right]} \qquad r_{\pi0} = r_\pi \,\|\, [r_x + (R_I\|R_B)]$$

Common-source

$$\frac{1}{R_{\text{th}}C_T} = \frac{1}{R_{\text{th}}\left[C_{GS} + C_{GD}(1 + g_m R_L) + (C_{GD} + C_L)\dfrac{R_L}{R_{\text{th}}}\right]} \qquad R_{\text{th}} = R_I\|R_G$$

Common-emitter with emitter resistor R_E

$$\frac{1}{r_{\pi0}\left[\dfrac{C_\pi}{1 + g_m R_E} + C_\mu\left(1 + \dfrac{g_m R_L}{1 + g_m R_E}\right) + (C_u + C_L)\dfrac{R_L}{r_{\pi0}}\right]} \qquad r_{\pi0} = r_\pi \,\|\, [r_x + (R_1\|R_B)]$$

Common-source with source resistor R_S

$$\frac{1}{R_{\text{th}}\left[\dfrac{C_{GS}}{1 + g_m R_S} + C_{GD}\left(1 + \dfrac{g_m R_L}{1 + g_m R_S}\right) + (C_{GD} + C_L)\dfrac{R_L}{R_{\text{th}}}\right]} \qquad R_{\text{th}} = R_I\|R_G$$

Common-base

$$\frac{1}{R_L(C_\mu + C_L)}$$

Common-gate

$$\frac{1}{R_L(C_{GD} + C_L)}$$

Common-collector

$$\frac{1}{[(R_I\|R_B) + r_x]\left(\dfrac{C_\pi}{1 + g_m R_L} + C_\mu\right)}$$

Common-drain

$$\frac{1}{(R_I\|R_G)\left(\dfrac{C_{GS}}{1 + g_m R_L} + C_{GD}\right)}$$

In order to simplify our analysis of the common-base amplifier, we neglected r_x. If r_x is included, it can be shown that the gain-bandwidth product is also limited by the $r_x C_\mu$ product in the C-B case as well. However, this limit is seldom reached since R_L is usually considerably larger than r_x.

Now we have found two important limits placed upon amplifier gain-bandwidth products by the characteristics of the transistor. The first was the unity-gain frequency for the current gain of the transistor, $f_T = g_m/(C_\pi + C_\mu)$. The second is set by GBW $\le 1/r_x C_\mu$. However, in typical amplifier designs, the GBW product will reach less than 60 percent of either of these bounds.

For transistors designed for very high-frequency operation, minimization of the $r_x C_\mu$ product is one of the main goals guiding the choice of the physical structure and the impurity profiles of the devices. As is often the case in engineering, tradeoffs are involved. The choices that minimize r_x increase C_μ, and vice-versa, and complex designs are utilized to optimize the $r_x C_\mu$ (or $R_{iG}C_{GD}$) product.

16.10 FREQUENCY RESPONSE OF MULTISTAGE AMPLIFIERS

The open- and short-circuit time-constant methods are not limited to single-transistor amplifiers but are directly applicable to multistage circuits as well; the power of the technique becomes more obvious as circuit complexity grows. This section uses the OCTC techniques to estimate the frequency response of several important two-stage dc-coupled amplifiers, including the differential amplifier, the cascode stage, and the current mirror. Because these amplifiers are direct-coupled, they have low-pass characteristics and only the OCTC method is needed to determine f_H. Following is an example of analysis of a general three-stage amplifier in which f_L and f_H are found using results from the SCTC and OCTC approaches.

16.10.1 DIFFERENTIAL AMPLIFIER

As pointed out several times, the differential amplifier is a key building block of analog circuits, and hence it is important to understand the frequency response of the differential pair. An important element, C_{EE}, has been included in the differential amplifier circuit in Fig. 16.45(a). C_{EE} represents the total capacitance at the emitter node of the differential pair. Analysis of the frequency response of the symmetrical amplifier in Fig. 16.45(a) is greatly simplified through the use of the half-circuits in (b) and (c).

Differential-Mode Signals

We recognize the differential-mode half-circuit in Fig. 16.45(b) as being equivalent to the standard common-emitter stage. Thus, the bandwidth for differential-mode signals is determined by the $r_{\pi o} C_T$ product that was developed in the analysis in Sec. 16.6, and we can expect amplifier gain-bandwidth products equal to a significant fraction of the f_T of the transistor. Because the emitter node is a virtual ground, C_{EE} has no effect on differential-mode signals.

Common-Mode Frequency Response

The important breakpoints in the Bode plot of the common-mode frequency response depicted in Fig. 16.46 can be determined from analysis of the common-mode half-circuit in Fig. 16.45(c). At very low frequencies, we know that the common-mode gain to either collector is small, given

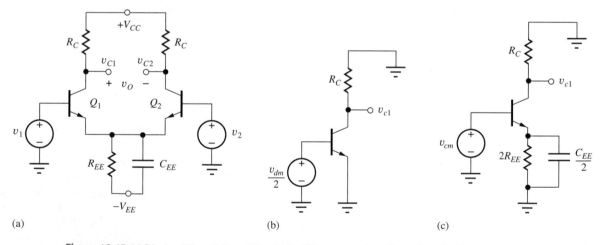

Figure 16.45 (a) Bipolar differential amplifier, (b) its differential-mode half-circuit, and (c) its common-mode half-circuit.

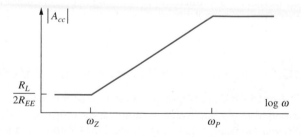

Figure 16.46 Bode plot for the common-mode gain of the differential pair.

approximately by

$$|A_{cc}(0)| \cong \frac{R_C}{2R_{EE}} \ll 1 \tag{16.143}$$

However, capacitance C_{EE} in parallel with emitter resistor R_{EE} introduces a transmission zero in the common-mode frequency response at the frequency for which the impedance of the parallel combination of R_{EE} and C_{EE} becomes infinite. This zero is given by

$$s = -\omega_z = -\frac{1}{R_{EE}C_{EE}} \tag{16.144}$$

and typically occurs at relatively low frequencies. Although C_{EE} may be small, resistance R_{EE} is normally designed to be large, often the output resistance of a very high impedance current source. The presence of this zero causes the common-mode gain to increase at a rate of $+20$ dB/decade for frequencies above ω_Z. The common-mode gain continues to increase until the dominant pole of the pair is reached at relatively high frequencies.

The common-mode half-circuit is equivalent to a common-emitter stage with emitter resistor $2R_{EE}$. The OCTC for $C_{EE}/2$ is just the resistance looking back into the emitter in Fig. 16.45(c):

$$R_{EEO} = 2R_{EE} \left\| \frac{r_x + r_\pi}{\beta_o + 1} \cong \frac{1}{g_m} \right. \tag{16.145}$$

Combining Eq. (16.145) with the equation in the third line of Table 16.2 for $R_{\text{th}} = 0$ yields the position of the pole in the common-mode response:

$$\omega_P \cong \frac{1}{r_x \left[\dfrac{C_\pi}{1 + 2g_m R_{EE}} + C_\mu \left(1 + \dfrac{g_m R_C}{1 + 2g_m R_{EE}} + \dfrac{R_C}{r_x} \right) \right] + \dfrac{C_{EE}}{2g_m}} \tag{16.146}$$

Because R_{EE} is usually designed to be very large, Eq. (16.146) can be reduced to

$$\omega_P \cong \frac{1}{C_\mu(r_x + R_C)} \tag{16.147}$$

EXERCISE: Find f_Z and f_P for the common-mode response of the differential amplifier in Fig. 16.45 if $r_x = 250\ \Omega$, $C_\mu = 0.5$ pF, $R_{EE} = 25$ MΩ, $C_{EE} = 1$ pF, and $R_C = 50$ kΩ.

ANSWERS: 6.37 kHz, 6.34 MHz

16.10.2 THE COMMON-COLLECTOR/COMMON-BASE CASCADE

Figure 16.47(a) is an unbalanced version of the differential amplifier. This circuit can also be represented as the cascade of a common-collector and common-base amplifier, as in Fig. 16.47(b). The poles of this two-stage amplifier are found by applying the OCTC approach, using the results of the previous single-stage amplifier analyses.

We assume the output resistance R_{EE} of the current source is very large and neglect it in the analysis because the resistances presented at the emitters of Q_1 and Q_2 in Fig. 16.48 are both small:

$$R_{iE}^{CC1} = \frac{r_{x1} + r_{\pi1}}{\beta_{o1} + 1} \cong \frac{1}{g_{m1}} \quad \text{and} \quad R_{iE}^{CB2} = \frac{r_{x2} + r_{\pi2}}{\beta_{o2} + 1} \cong \frac{1}{g_{m2}} \tag{16.148}$$

The high-frequency response of the circuit of Fig. 16.48 is found by utilizing the results of the common-collector and common-base stages. We will neglect the pole at the intermediate node

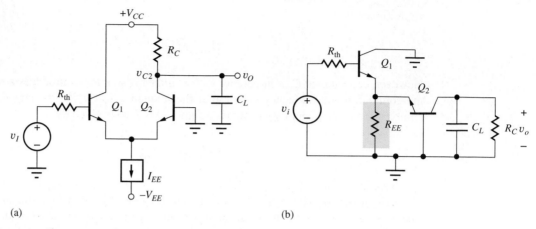

Figure 16.47 (a) The unbalanced differential amplifier and (b) its representation as a C-C/C-B cascade.

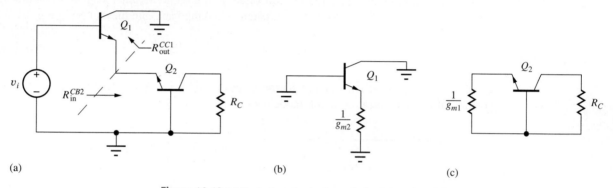

Figure 16.48 (a) Equivalent circuits for analysis of the poles of (b) Q_1 and (c) Q_2.

because it is expected to be very high frequency due to the low equivalent impedance to ground $(1/2g_m)$ at the node. We are left with the poles at the input and output of the two-stage amplifier. The pole at the input is that of a common-collector stage with $R_L = 1/g_m$.

$$\omega_{PB1} = \frac{1}{([R_{th} + r_x] \| [r_{\pi 1} + (\beta_{o1} + 1)R_L]) \left(C_\mu + \dfrac{C_\pi}{1 + g_m R_L} \right)}$$

$$= \frac{1}{([R_{th} + r_{x1}] \| [2r_{\pi 1}]) \left(C_{\mu 1} + \dfrac{C_{\pi 1}}{2} \right)} \tag{16.149}$$

Notice that if the source impedance is zero, the input pole response is set by r_x. The output pole is that of a common-base amplifier.

$$\omega_{PC2} \cong \frac{1}{R_C(C_{\mu 2} + C_L)} \tag{16.150}$$

Depending on the impedances in the circuit, both the input and output poles could contribute significantly to the high-frequency response.

16.10.3 HIGH-FREQUENCY RESPONSE OF THE CASCODE AMPLIFIER

The cascade of the common-emitter and common-base stages in Fig. 16.49 is referred to as the **cascode amplifier.** The cascode stage offers a midband gain and input resistance equal to that of the common-emitter amplifier but with a much improved upper-cutoff frequency f_H, as will be demonstrated by the forthcoming analysis.

The poles of the cascode stage follow directly from the analysis of the common-emitter and common-base stages using the model in Fig. 16.50. At the input to the cascode, we have the pole from our earlier analysis of this stage.

$$\omega_{PB1} = \frac{1}{r_{\pi\phi}C_T} = \frac{1}{r_{\pi\phi}\left([C_\pi + C_\mu(1 + g_m R_L)] + \dfrac{R_L}{r_{\pi\phi}}[C_\mu + C_L]\right)} \tag{16.151}$$

Since the load of the first stage is small $(1/g_m)$ we expect the second capacitive term in this expression to be insignificant, allowing us to simplify the expression. The bias current of the two transistors is equal so the g_m values of the two transistors are also equal.

$$\omega_{PB1} = \frac{1}{r_{\pi01}\left(\left[C_{\pi1} + C_{\mu1}\left(1 + \dfrac{g_{m1}}{g_{m2}}\right)\right] + \dfrac{1/g_{m2}}{r_{\pi01}}[C_{\mu1} + C_{\pi2}]\right)} \cong \frac{1}{r_{\pi\hat\phi1}(C_{\pi1} + 2C_{\mu1})} \tag{16.152}$$

The term due to Miller multiplication has been reduced from a very large factor (264 in the C-E example in Sec. 16.6) to only 2, and the $R_L/r_{\pi\phi}$ term has also been essentially eliminated. These reductions are the primary advantage of the cascode amplifier and can greatly increase the bandwidth of the overall amplifier.

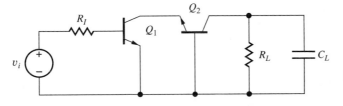

Figure 16.49 ac Model for the direct-coupled cascode amplifier.

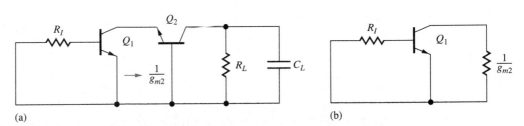

(a)   (b)

Figure 16.50 (a) Model for determining time constants associated with the two capacitances of Q_1. (b) Simplified model.

Similar to the previous circuit, the intermediate node impedance is quite low. At high frequencies, roughly equal to $1/2g_{m2}$ (recall the high-frequency shunting of Q_1 due to C_μ). Because of this, we may expect the pole at the intermediate node to be quite high frequency. The output pole is that of a common-base amplifier.

$$\omega_{PC2} \cong \frac{1}{R_L(C_{\mu2} + C_L)} \qquad (16.153)$$

Again, depending on the particular impedances in the circuit, both of these poles could be significant in determining the overall high-frequency response.

EXERCISE: Find the midband value of A_v and the poles of the cascode amplifier in Fig. 16.49 assuming $\beta_o = 100$, $f_T = 500$ MHz, $C_\mu = 0.5$ pF, $r_x = 250\,\Omega$, $R_I = 882\,\Omega$, $R_L = 4.12$ kΩ, $C_L = 5$ pF, and a Q-point of (1.60 mA, 3.00 V) for Q_2.

ANSWERS: -151, 22.7 MHz, 7.02 MHz

16.10.4 CUTOFF FREQUENCY FOR THE CURRENT MIRROR

As a final example of the analysis of direct-coupled amplifiers, let us find ω_H for the current mirror configuration in Fig. 16.51. The small-signal model in Fig. 16.51(b) represents the two-port model developed in Sec. 15.6 with the addition of the gate-source and gate-drain capacitances of M_1 and M_2. The gate-source capacitances of the two transistors appear in parallel, whereas the gate-drain capacitance of M_1 is shorted out by the circuit connection. The open-circuited load condition at the output represents a worst-case situation for estimating the current mirror bandwidth.

The circuit in Fig. 16.51(b) should be recognized as identical to the simplified model of the C-E stage in Fig. 16.35, and the results of the C_T approximation are directly applicable to the current mirror circuit with the following substitutions:

$$r_{\pi o} \to \frac{1}{g_{m1}} \qquad R_L \to r_{o2} \qquad C_\pi \to C_{GS1} + C_{GS2} \qquad C_\mu \to C_{GD2} \qquad (16.154)$$

Using the values from Eq. (16.154) in Eq. (16.93),

$$\omega_{P1} \cong \frac{1}{r_{\pi o}C_T} = \frac{1}{\dfrac{1}{g_{m1}}\left[C_{GS1} + C_{GS2} + C_{GD2}\left(1 + g_{m2}r_{o2} + \dfrac{r_{o2}}{\dfrac{1}{g_{m1}}}\right)\right]} \qquad (16.155)$$

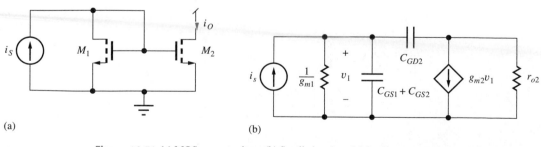

Figure 16.51 (a) MOS current mirror. (b) Small-signal model for the current mirror.

and for matched transistors with equal W/L ratios,

$$\omega_{P1} \cong \frac{1}{\dfrac{2C_{GS1}}{g_{m1}} + 2C_{GD2}r_{o2}} \cong \frac{1}{2C_{GD2}r_{o2}} \qquad (16.156)$$

The result in Eq. (16.156) indicates that the bandwidth of the current mirror is controlled by the time constant at the output of the mirror due to the output resistance and gate-drain capacitance of M_2. Note that the value of Eq. (16.156) is directly proportional to the Q-point current through the dependence of r_{o2}.

EXERCISE: (a) Find the bandwidth of the current mirror in Fig. 16.51 if $I_1 = 100\ \mu\text{A}$, $C_{GD} = 1\ \text{pF}$, and $\lambda = 0.02\ \text{V}^{-1}$. (b) If $I_1 = 25\ \mu\text{A}$.

ANSWERS: 318 kHz; 79.6 kHz

16.10.5 THREE-STAGE AMPLIFIER EXAMPLE

As an example of a more complex analysis, let us estimate the upper- and lower-cutoff frequencies for the multistage amplifier in Fig. 16.52 that was introduced in Chapter 14. We will use the method of short-circuit time constants to estimate the lower-cutoff frequency. In Chapter 17 we will need to

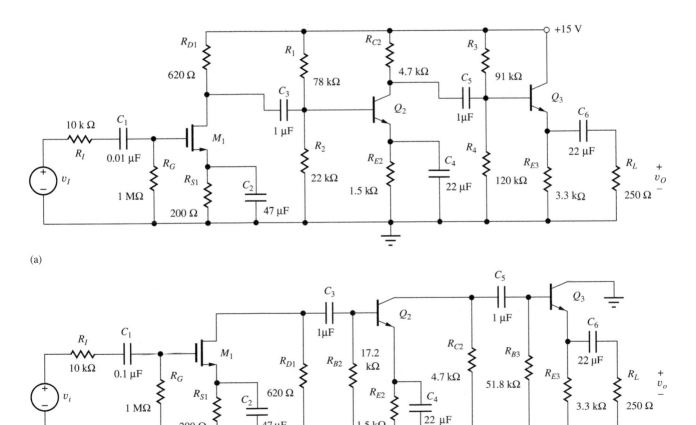

(a)

(b)

Figure 16.52 Three-stage amplifier and ac equivalent circuit.

know specific pole locations to accurately estimate feedback amplifier stability, so we will illustrate the calculation of high-frequency poles with this multistage example.

EXAMPLE 16.8   **MULTISTAGE AMPLIFIER FREQUENCY RESPONSE**

The time constant methods are used to find the upper- and lower-cutoff frequencies of a multistage amplifier.

PROBLEM   Use the direct calculation and the short-circuit time constant technique to estimate the upper- and lower-cutoff frequencies of a multistage amplifier.

SOLUTION   **Known Information and Given Data:** Three-stage amplifier circuit in Fig. 16.52; Q-points and small-signal parameters are given in Tables 14.18 and 16.3.

Unknowns: f_H, f_L, and bandwidth

Approach: The coupling and bypass capacitors determine the low-frequency response, whereas the device capacitances affect the high-frequency response. At low frequencies, the impedances of the internal device capacitances are very large and can be neglected. The coupling and bypass capacitors remain in the low-frequency ac equivalent circuit in Fig. 16.52(b), and an estimate for ω_L is calculated using the SCTC approach. An estimate for the upper-cutoff frequency is calculated based on the calculation of individual high-frequency poles from our single-stage analyses. At high frequencies, the impedances of the coupling and bypass capacitors are negligibly small, and we construct the circuit in Fig. 16.53 by replacing the coupling and bypass capacitors by short circuits.

We develop expressions for the various short- and open-circuit time constants using our knowledge of input and output resistances of single-stage amplifiers. Finally, the expressions can be evaluated using known values of circuit elements and small-signal parameters.

TABLE 16.3
Transistor Parameters

	g_m	r_π	r_o	β_0	C_{GS}/C_π	C_{GD}/C_μ	r_x
M_1	10 mS	∞	12.2 kΩ	∞	5 pF	1 pF	0 Ω
Q_2	67.8 mS	2.39 kΩ	54.2 kΩ	150	39 pF	1 pF	250 Ω
Q_3	79.6 mS	1.00 kΩ	34.4 kΩ	80	50 pF	1 pF	250 Ω

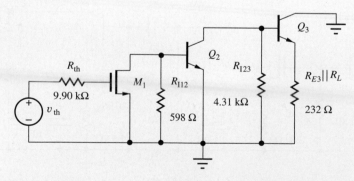

Figure 16.53 High-frequency ac model for three-stage amplifier in Fig. 16.52.

Assumptions: Transistors are in the active region. Small-signal conditions apply. $V_T = 25\,\text{mV}$.

ANALYSIS   **(a) SCTC Estimate for the Lower-Cutoff Frequency ω_L:** The circuit has six independent coupling and bypass capacitors; Fig. 16.54 gives the circuits for finding the six short-circuit time constants. The analysis proceeds using the small-signal parameters in Table 16.3. The low-frequency transistor parameter values in Table 16.3 are reproduced from Table 14.18.

R_{1S}: Because the input resistance to M_1 is infinite in Fig. 16.54(a), R_{1S} is given by

$$R_{1S} = R_I + R_G \| R_{iG} = 10\,\text{k}\Omega + 1\,\text{M}\Omega \| \infty = 1.01\,\text{M}\Omega \tag{16.157}$$

R_{2S}: R_{2S} represents the resistance present at the source terminal of M_1 in Fig. 16.54(b) and is equal to

$$R_{2S} = R_{S1} \left\| \frac{1}{g_{m1}} = 200\,\Omega \right\| \frac{1}{0.01\,\text{S}} = 66.7\,\Omega \tag{16.158}$$

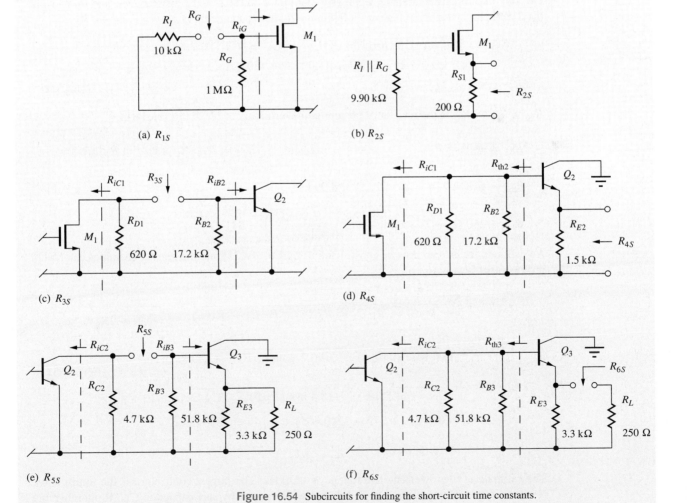

(a) R_{1S}

(b) R_{2S}

(c) R_{3S}

(d) R_{4S}

(e) R_{5S}

(f) R_{6S}

Figure 16.54 Subcircuits for finding the short-circuit time constants.

R_{3S}: Resistance R_{3S} is formed from a combination of four elements in Fig. 16.54(c). To the left, the output resistance of M_1 appears in parallel with the 620-Ω resistor R_{D1}, and on the right the 17.2-kΩ resistor R_{B2} is in parallel with the input resistance of Q_2:

$$R_{3S} = (R_{D1} \| R_{iC1}) + (R_{B2} \| R_{iB2}) = (R_{D1} \| r_{o1}) + (R_{B2} \| r_{\pi 2})$$

$$= (620\ \Omega \| 12.2\ \text{k}\Omega) + (17.2\ \text{k}\Omega \| 2.39\ \text{k}\Omega) = 2.69\ \text{k}\Omega \tag{16.159}$$

R_{4S}: R_{4S} represents the resistance present at the emitter terminal of Q_2 in Fig. 16.54(d) and is equal to

$$R_{4S} = R_{E2} \left\| \frac{R_{\text{th2}} + r_{\pi 2}}{(\beta_{o2} + 1)} \right. \qquad \text{where} \qquad R_{\text{th2}} = R_{B2} \| R_{D1} \| R_{iC1} = R_{B2} \| R_{D1} \| r_{o1}$$

$$R_{\text{th2}} = R_{B2} \| R_{D1} \| r_{o1} = 17.2\ \text{k}\Omega \| 620\ \Omega \| 12.2\ \text{k}\Omega = 571\ \Omega \tag{16.160}$$

$$R_{4S} = 1500\ \Omega \left\| \frac{571\ \Omega + 2390\ \Omega}{(150 + 1)} \right. = 19.4\ \Omega$$

R_{5S}: Resistance R_{5S} is also formed from a combination of four elements in Fig. 16.54(e). To the left, the output resistance of Q_2 appears in parallel with the 4.7-kΩ resistor R_{C2}, and to the right the 51.8-kΩ resistor R_{B3} is in parallel with the input resistance of Q_3:

$$R_{5S} = (R_{C2} \| R_{iC2}) + (R_{B3} \| R_{iB3}) = (R_{C2} \| r_{o2}) + (R_{B3} \| [r_{\pi 3} + (\beta_{o3} + 1)(R_{E3} \| R_L)])$$

$$= (4.7\ \text{k}\Omega \| 54.2) + 51.8\ \text{k}\Omega \| [1.00\ \text{k}\Omega + (80 + 1)(3.3\ \text{k}\Omega \| 250\ \Omega)]$$

$$= 18.4\ \text{k}\Omega \tag{16.161}$$

R_{6S}: Finally, R_{6S} is the resistance present at the terminals of C_6 in Fig. 16.54(f):

$$R_{6S} = R_L + \left(R_{E3} \left\| \frac{R_{\text{th3}} + r_{\pi 3}}{\beta_{o3} + 1} \right. \right) \qquad \text{where} \qquad R_{\text{th3}} = R_{B3} \| R_{C2} \| R_{iC2} = R_{B3} \| R_{C2} \| r_{o2}$$

$$R_{\text{th3}} = 51.8\ \text{k}\Omega \| 4.7\ \Omega \| 54.2\ \text{k}\Omega = 3.99\ \text{k}\Omega \tag{16.162}$$

$$R_{6S} = 250\ \Omega + \left(3.3\ \text{k}\Omega \left\| \frac{3.39\ \text{k}\Omega + 1.00\ \text{k}\Omega}{80 + 1} \right. \right) = 311\ \Omega$$

An estimate for ω_L can now be constructed using Eq. (16.33) and the resistance values calculated in Eqs. (16.157) to (16.162):

$$\omega_L \cong \sum_{i=1}^{n} \frac{1}{R_{iS} C_i} = \frac{1}{R_{1S} C_1} + \frac{1}{R_{2S} C_2} + \frac{1}{R_{3S} C_3} + \frac{1}{R_{4S} C_4} + \frac{1}{R_{5S} C_5} + \frac{1}{R_{6S} C_6}$$

$$\cong \frac{1}{(1.01\ \text{M}\Omega)(0.01\ \mu\text{F})} + \frac{1}{(66.7\ \Omega)(47\ \mu\text{F})} + \frac{1}{(2.69\ \text{k}\Omega)(1\ \mu\text{F})}$$

$$+ \frac{1}{(19.4\ \Omega)(22\ \mu\text{F})} + \frac{1}{(18.4\ \text{k}\Omega)(1\ \mu\text{F})} + \frac{1}{(311\ \Omega)(22\ \mu\text{F})} \tag{16.163}$$

$$\cong 99.0 + 319 + 372 + \underline{2340} + 54.4 + 146 = 3330\ \text{rad/s}$$

$$f_L = \frac{\omega_L}{2\pi} = 530\ \text{Hz}$$

The estimate of the lower-cutoff frequency is 530 Hz. The largest contributor is the fourth term, resulting from the time constant associated with emitter-bypass capacitor C_4. (Remember the design approach used in Design Ex. 16.3.)

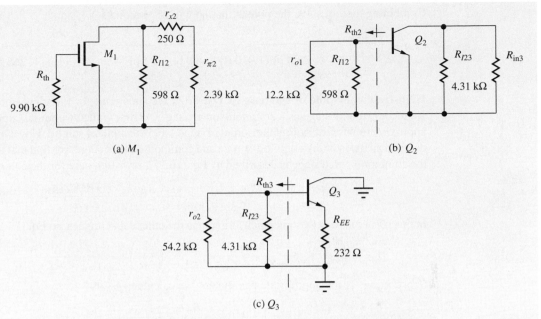

(a) M_1

(b) Q_2

(c) Q_3

Figure 16.55 Subcircuits for evaluating the OCTC for each transistor.

(b) Calculation of the Upper-Cutoff Frequency f_H: The upper-cutoff frequency can be found by calculating the high-frequency poles at each of the nodes within the high-frequency ac model of the amplifier in Fig. 16.52 and then applying Eq. 16.22. At high frequencies, the impedances of the coupling and bypass capacitors are negligibly small, and we construct the circuit in Fig. 16.53 by replacing the coupling and bypass capacitors with short circuits. The high-frequency poles can be calculated at each node based on our single-stage analyses in Table 16.2.

High-frequency pole at the gate of M_1: From the subcircuit for the transistor in Fig. 16.55(a), we recognize this stage as a common-source stage. Using the C_T approximation from Table 16.2,

$$f_{p1} = \left(\frac{1}{2\pi}\right) \frac{1}{R_{\text{th1}}[C_{GS1} + C_{GD1}(1 + g_{m1}R_{L1}) + \dfrac{R_{L1}}{R_{\text{th1}}}(C_{GD1} + C_{L1})]} \tag{16.164}$$

For this circuit, the unbypassed source resistance is zero, so we use a simpler form of the input pole frequency equation. In Eq. (16.164), the Thévenin source resistance is $9.9\,\text{k}\Omega$, and the load resistance is the parallel combination of resistances R_{I2}, $(r_{x2} + r_{\pi2})$, and r_{o1}:

$$R_{L1} = R_{I12}\|r_{\pi2}\|r_{o1} = 598\,\Omega\|(2.39\,\text{k}\Omega + 250\,\Omega)\|12.2\,\text{k}\Omega = 469\,\Omega \tag{16.165}$$

We use the Miller effect to evaluate C_{L1}, the capacitance seen looking into the second stage common-emitter amplifier:

$$C_{L1} = C_{\pi2} + C_{\mu2}(1 + g_{m2}R_{L2}) \tag{16.166}$$

From Fig. 16.55(b), we evaluate R_{L2} as

$$R_{L2} = R_{I23}\|R_{iB3}\|r_{o2} = R_{I23}\|[r_{x3} + r_{\pi3} + (\beta_{o3} + 1)(R_{E3}\|R_L\|r_{o2})]$$

$$= 4.31\,\text{k}\Omega\|[250 + 1\,\text{k}\Omega + (80 + 1)(3.3\,\text{k}\Omega\|250\,\Omega\|54.2\,\text{k}\Omega)] \tag{16.167}$$

$$= 3.33\,\text{k}\Omega$$

Using this result we find C_{L1} as

$$C_{L1} = 39\,\text{pF} + 1\,\text{pF}[1 + 67.8\,\text{mS}(3.33\,\text{k}\Omega)] = 266\,\text{pF} \tag{16.168}$$

Combining these results, the pole at the input of M_1 becomes

$$f_{p1} = \left(\frac{1}{2\pi}\right) \frac{1}{(9.9\,\text{k}\Omega)[1\,\text{pF}(1 + 0.01S(469\,\Omega)] + 5\,\text{pF} + \dfrac{469\,\Omega}{9.9\,\text{k}\Omega}(1\text{pF} + 266\,\text{pF})]} = 689\,\text{kHz}$$

(16.169)

High-frequency pole at the base of Q_2: From the subcircuit for the transistor in Fig 16.55(b), we recognize this stage as a common-source stage. At first glance we might expect to use the C_T approximation for the pole at the output of stage 1 and the input of stage 2. However, if we recall the detailed analysis of the common-source and common-emitter stage, we find that the output pole of the common-source stage is described by Eq. (16.95), rewritten here for the common-source case:

$$f_{p2} = \left(\frac{1}{2\pi}\right) \frac{C_{GS1}g_{L1} + C_{GD1}(g_{m1} + g_{th1} + g_{L1}) + C_{L1}g_{th1}}{[C_{GS1}(C_{GD1} + C_{L1}) + C_{GD1}C_{L1}]}$$

(16.170)

In this particular case, C_{L1} is much larger than the other capacitances, so Eq. (16.170) simplifies to

$$f_{p2} \cong \left(\frac{1}{2\pi}\right) \frac{C_{L1}g_{th1}}{[C_{GS1}C_{L1} + C_{GD1}C_{L1}]} \cong \left(\frac{1}{2\pi}\right) \frac{1}{R_{th1}(C_{GS1} + C_{GD1})}$$

(16.171)

Substituting for the appropriate parameters, we calculate f_{p2} as

$$f_{p2} = \left(\frac{1}{2\pi}\right) \frac{1}{(9.9\,\text{k}\Omega)(5\,\text{pF} + 1\,\text{pF})} = 2.68\,\text{MHz}$$

(16.172)

High-frequency pole at the base of Q_3: From the subcircuit for the transistor in Fig. 16.55(c), we recognize the third stage as a common-collector stage. Again, due to the pole splitting effect of the common-emitter second stage, we expect that the pole at the base of Q_3 will be set by Eq. (16.95). In this case, due to the small load capacitance and high g_{m2} of the second stage, the $g_{m2}C_\mu$ term simplification of Eq. (16.95) will dominate the numerator, and the first form of Eq. (16.95) will hold. However, since C_μ is quite small, we cannot simplify Eq. (16.95) to the form that applies when C_μ is large. As a consequence, we can expect the pole at the interstage node between Q_2 and Q_3 to be governed by

$$f_{p3} \cong \left(\frac{1}{2\pi}\right) \frac{g_{m2}}{\left[C_{\pi2}\left(1 + \dfrac{C_{L2}}{C_{\mu2}}\right) + C_{L2}\right]}$$

(16.173)

The load capacitance of Q_2 is the input capacitance for the common-collector output stage. This is calculated as

$$C_{L2} = C_{\mu3} + \frac{C_{\pi3}}{1 + g_{m3}(R_{E3}\|R_L)} = 1\,\text{pF} + \frac{50\,\text{pF}}{1 + 79.6\,\text{mS}(3.3\,\text{k}\Omega\|250\,\Omega)} = 3.55\,\text{pF}$$

(16.174)

To account for r_{x2}, we can use g_m as defined in Eq. (16.70) when evaluating f_{p3}.

$$f_{p3} \cong \left(\frac{1}{2\pi}\right) \frac{67.8\,\text{mS}[1\,\text{k}\Omega/(1\,\text{k}\Omega + 250\,\Omega)]}{\left[39\,\text{pF}\left(1 + \dfrac{3.55\,\text{pF}}{1\,\text{pF}}\right) + 3.55\,\text{pF}\right]} = 47.7\,\text{MHz}$$

(16.175)

There is an additional pole at the emitter of Q_3, but that will be at a very high frequency due to the relatively low equivalent resistance and capacitance at the output. The midband to high frequency response can now be written as

$$A(f) = \frac{A_{\text{mid}}}{\left(1 + j\dfrac{f}{f_{p1}}\right)\left(1 + j\dfrac{f}{f_{p2}}\right)\left(1 + j\dfrac{f}{f_{p3}}\right)}$$

$$= \frac{998\,\text{V/V}}{\left(1 + j\dfrac{f}{689\,\text{kHz}}\right)\left(1 + j\dfrac{f}{2.68\,\text{MHz}}\right)\left(1 + j\dfrac{f}{47.7\,\text{MHz}}\right)}$$

(16.176)

Applying Eq. (16.23), we estimate f_H as

$$f_H = \frac{1}{\sqrt{\dfrac{1}{f_{p1}^2} + \dfrac{1}{f_{p2}^2} + \dfrac{1}{f_{p3}^2}}} = 667 \text{ kHz} \qquad (16.177)$$

Check of Results: SPICE is an excellent method to check an analysis of this complexity. After drawing the circuit with the schematic editor, we need to set the MOSFET and BJT parameters. We can set up the device parameters by referring back to Tables 14.17, 14.18, and 16.3. For the depletion-mode MOSFET, KP = 10 mA/V^2, VTO = −2 V, and LAMDA = 0.02 V^{-1}. For this simulation, it is easiest to add external capacitors in parallel with the MOSFET to represent C_{GS} and C_{GD}. The values are 5 pF and 1 pF, respectively.

For the BJTs, RB = 250 Ω, BF = 150, and VAF = 80 V, and we can let IS take on its default value of 0.1 fA. The values of TF can also be found using the data in Table 16.3:

$$\text{TF}_2 = \frac{C_{\pi 2}}{g_{m2}} = \frac{39 \text{ pF}}{67.8 \text{ mS}} = 0.575 \text{ ns} \qquad \text{and} \qquad \text{TF}_3 = \frac{50 \text{ pF}}{79.6 \text{ mS}} = 0.628 \text{ ns}$$

The collector-emitter voltages from Table 14.18 are $V_{CE2} = 5.09$ V and $V_{CE3} = 8.36$ V. To achieve values of 1 pF for each C_μ, we must properly set the values of CJC:

$$\text{CJC2} = (1 \text{ pF}) \left(1 + \frac{5.09 - 0.7}{0.75} \right)^{0.33} = 1.89 \text{ pF}$$

and

$$\text{CJC3} = (1 \text{ pF}) \left(1 + \frac{8.36 - 0.7}{0.75} \right)^{0.33} = 2.22 \text{ pF}$$

Once the parameters are set, an ac analysis can be performed with FSTART = 10 Hz, FSTOP = 10 MHz, and 20 points per frequency decade. The resulting Bode magnitude plot appears next. We can also check the device parameters and see that the values of C_π and C_μ are approximately correct.

Discussion: Note that common-source stage M_1 and common-emitter stage Q_2 are both making substantial contributions to f_H, whereas follower Q_3 represents a negligible contribution. Based on our calculated results, the midband region of the amplifier extends from $f_L = 530$ Hz to $f_H = 667$ kHz for a bandwidth BW = 666 kHz.

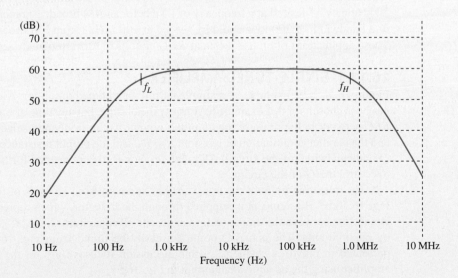

The SPICE results indicate that f_L and f_H are approximately 350 Hz and 675 kHz, respectively, and the midband gain is 60 dB. In this amplifier, we see that the SCTC method is overestimating the value of the lower-cutoff frequency. If we look at Eq. (16.163), we see that there is clearly a dominant time constant. If we use only this value, we get much better agreement with the SPICE results:

$$f_L \cong \frac{2340}{2\pi} = 372\,\text{Hz}$$

On the other hand, our estimate of f_H is in good agreement with simulation. We did have to be quite careful with our calculations to take into account the pole splitting behavior of common-emitter and common-source amplifiers. If we had not taken this into account, the estimate for f_H, based on dominant-pole calculations for each of the stages, would be less than 550 kHz. Of even more importance for feedback amplifier design, our analysis in this example also accurately characterizes the phase and magnitude response well beyond f_H.

EXERCISE: Calculate the reactance of $C_{\pi 2}$ at f_L and compare its value to $r_{\pi 2}$. Calculate the reactance of $C_{\mu 3}$ and compare it to $R_{B3} \| R_{\text{in}3}$ in Fig. 16.54(e).

ANSWERS: 7.7 MΩ ≫ 2.39 kΩ; 300 MΩ ≫ 14.3 kΩ

EXERCISE: Calculate the reactance of C_1, C_2, and C_3, in Fig. 16.52(b) at $f = f_H$, and compare the values to the midband resistances in the circuit at the terminals of the capacitors.

ANSWERS: 29.6 Ω ≪ 1.01 MΩ; 6.29 mΩ ≪ 66.7 Ω; 296 mΩ ≪ 2.69 kΩ

16.11 TUNED AMPLIFIERS

In radio-frequency (RF) applications, amplifiers with narrow bandwidths are needed in order to select one signal from the large number that may be present (from an antenna, for example). The frequencies of interest are typically well above the unity gain frequency of operational amplifiers so that RC active filters cannot be used. These amplifiers often have high Q; that is, f_H and f_L are very close together relative to the midband or center frequency of the amplifier. For example, a bandwidth of 20 kHz may be desired at a frequency of 1 MHz for an AM broadcast receiver application ($Q = 50$), or a bandwidth of 200 kHz could be needed at 100 MHz for an FM broadcast receiver ($Q = 500$). These applications often use resonant RLC circuits to form frequency selective **tuned amplifiers.**

16.11.1 SINGLE-TUNED AMPLIFIER

Figure 16.56 is an example of a simple narrow-band tuned amplifier. A depletion-mode MOSFET has been chosen for this example to simplify the biasing, but any type of transistor could be used. The RLC network in the drain of the amplifier represents the frequency-selective portion of the circuit, and the parallel combination of resistors R_D, R_3, and the output resistance r_o of the transistor set the Q and bandwidth of the circuit. Although resistor R_D is not needed for biasing, it is often included to control the Q of the circuit.

The operating point of the transistor can be found from analysis of the dc equivalent circuit in Fig. 16.56(b). Bias current is supplied through the inductor, which represents a direct short-circuit connection between the drain and V_{DD} at dc, and all capacitors C_1, C_2, C_S, and C have been replaced by open circuits. The actual Q-point can easily be found from Fig. 16.56(b) using the methods presented in previous chapters, so this discussion focuses only on the ac behavior of the tuned amplifier using the ac equivalent circuit in Fig. 16.57.

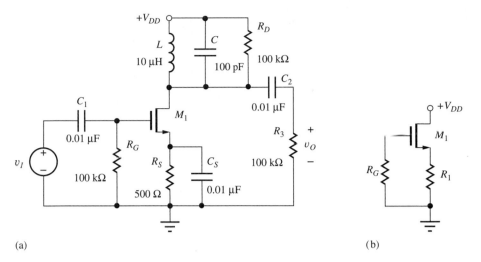

(a)   (b)

Figure 16.56 (a) Tuned amplifier using a depletion-mode MOSFET. (b) dc Equivalent circuit for the tuned amplifier in (a).

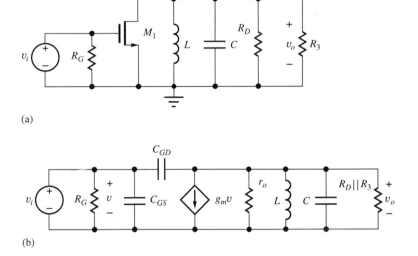

(a)

(b)

Figure 16.57 (a) High-frequency ac equivalent circuit and (b) small-signal model for the tuned amplifier in Fig. 16.56.

Writing a single nodal equation at the output node v_o of the circuit in Fig. 16.57(b) and observing that $v = v_i$ yields

$$(sC_{GD} - g_m)\mathbf{V_I}(s) = \mathbf{V_o}(s)\left[g_o + G_D + G_3 + s(C + C_{GD}) + \frac{1}{sL}\right] \qquad (16.178)$$

Making the substitution $G_P = g_o + G_D + G_3$, and then solving for the voltage transfer function:

$$A_v(s) = \frac{\mathbf{V_o}(s)}{\mathbf{V_I}(s)} = (sC_{GD} - g_m)R_P\frac{\dfrac{s}{R_P(C + C_{GD})}}{s^2 + \dfrac{s}{R_P(C + C_{GD})} + \dfrac{1}{L(C + C_{GD})}} \qquad (16.179)$$

If we neglect the right-half-plane zero, then Eq. (16.179) can be rewritten as

$$A_v(s) \cong A_{\text{mid}}\frac{s\dfrac{\omega_o}{Q}}{s^2 + s\dfrac{\omega_o}{Q} + \omega_o^2} \qquad \text{with} \qquad \omega_o = \frac{1}{\sqrt{L(C + C_{GD})}} \qquad (16.180)$$

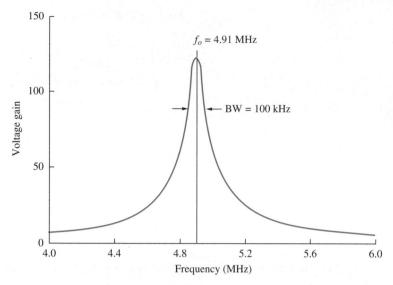

Figure 16.58 Simulated frequency response for the tuned amplifier in Fig. 16.56 with $C_{GS} = 50$ pF, $C_{GD} = 20$ pF, $V_{DD} = 15$ V, $K_n = 5$ mA/V^2, $V_{TN} = -2$ V, and $\lambda = 0.02$ V^{-1}.

In Eq. (16.180), ω_o is the **center frequency** of the amplifier, and the Q is given by

$$Q = \omega_o R_P (C + C_{GD}) = \frac{R_P}{\omega_o L}$$

The center or midband frequency of the amplifier is equal to the resonant frequency ω_o of the LC network. At the center frequency, $s = j\omega_o$, and Eq. (16.180) reduces to

$$A_v(j\omega_o) = A_{\text{mid}} \frac{j\omega_o \dfrac{\omega_o}{Q}}{(j\omega_o)^2 + j\omega_o \dfrac{\omega_o}{Q} + \omega_o^2} = A_{\text{mid}} \frac{j\omega_o \dfrac{\omega_o}{Q}}{-\omega_o^2 + j\omega_o \dfrac{\omega_o}{Q} + \omega_o^2} = A_{\text{mid}}$$

$$A_{\text{mid}} = -g_m R_P = -g_m (r_o \| R_D \| R_3) \tag{16.181}$$

For narrow bandwidth circuits—that is, high-Q circuits—the bandwidth is equal to

$$\text{BW} = \frac{\omega_o}{Q} = \frac{1}{R_P(C + C_{GD})} = \frac{\omega_o^2 L}{R_P} \tag{16.182}$$

A narrow bandwidth requires a large value of equivalent parallel resistance R_P, large capacitance, and/or small inductance. In this circuit, the maximum value of $R_P = r_o$. For this case, the Q is limited by the output resistance of the transistor and thus the choice of operating point of the transistor, and the midband gain A_{mid} equals the amplification factor μ_f.

An example of the frequency response of a tuned amplifier is presented in the SPICE simulation results in Fig. 16.58 for the amplifier in Fig. 16.56. This particular amplifier design has a center frequency of 4.91 MHz and a Q of approximately 50.

EXERCISE: What is the impedance of the 0.01-μF coupling and bypass capacitors in Fig. 16.56 at a frequency of 5 MHz?

ANSWERS: $-j3.18$ Ω (note that $X_C \ll R_G$ and $X_C \ll R_3$)

EXERCISE: Find the center frequency, bandwidth, Q, and midband gain for the amplifier in Fig. 16.56 using the parameters in Fig. 16.58, assuming $I_D = 3.20$ mA. (Remember that C_{GD} is Q-point dependent. Use $\phi_j = 0.9$ V.)

ANSWERS: 4.92 MHz, 105 kHz, 46.9, −116

EXERCISE: What is the new value of the center frequency if V_{DD} is reduced to 10 V?

ANSWER: 4.89 MHz

16.11.2 USE OF A TAPPED INDUCTOR—THE AUTO TRANSFORMER

The impedance of the gate-drain capacitance and output resistance of the transistor, C_{GD}, and r_o, can often be small enough in magnitude to degrade the characteristics of the tuned amplifier. The problem can be solved by connecting the transistor to a tap on the inductor instead of across the full inductor, as indicated in Fig. 16.59. In this case, the inductor functions as an auto transformer and changes the effective impedance reflected into the resonant circuit.

The n-turn auto transformer can be modeled by its total magnetizing inductance L_2 in parallel with an ideal transformer having a turns ratio of $(n - 1) : 1$. The ideal transformer has its primary and secondary windings interconnected, as in Fig. 16.60(b). Impedances are transformed by a factor

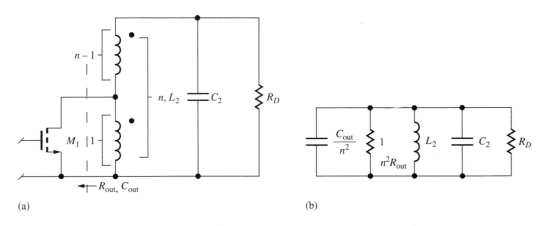

(a) (b)

Figure 16.59 (a) Use of a tapped inductor as an impedance transformer. (b) Transformed equivalent for the tuned circuit elements in Fig. 16.59(a). This circuit can be used to find ω_o and Q.

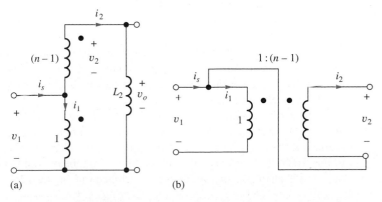

(a) (b)

Figure 16.60 (a) Tapped inductor and (b) its representation by an ideal transformer.

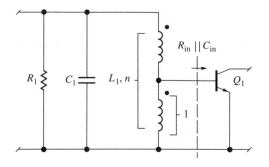

Figure 16.61 Use of an auto transformer at the input of transistor Q_1.

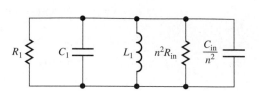

Figure 16.62 Transformed circuit model for the tuned circuit in Fig. 16.61.

of n^2 by the ideal transformer configuration:

$$\mathbf{V_o}(s) = \mathbf{V_2}(s) + \mathbf{V_1}(s) = (n-1)\mathbf{V_1}(s) + \mathbf{V_1}(s) = n\mathbf{V_1}(s)$$

$$\mathbf{I_s}(s) = \mathbf{I_1}(s) + \mathbf{I_2}(s) = (n-1)\mathbf{I_2}(s) + \mathbf{I_2}(s) = n\mathbf{I_2}(s) \tag{16.183}$$

and

$$\frac{\mathbf{V_o}(s)}{\mathbf{I_2}(s)} = \frac{n\mathbf{V_1}(s)}{\dfrac{\mathbf{I_s}(s)}{n}} = n^2\frac{\mathbf{V_1}(s)}{\mathbf{I_s}(s)} \qquad Z_s(s) = n^2 Z_p(s) \tag{16.184}$$

Thus, the impedance $Z_s(s)$ reflected into the secondary of the transformer is n^2 times larger than the impedance $Z_p(s)$ connected to the primary.

Using the result in Eq. (16.184), the resonant circuit in Fig. 16.59(a) can be transformed into the circuit representation in Fig. 16.59(b). L_2 represents the total inductance of the transformer. The equivalent output capacitance of the transistor is reduced by the factor of n^2, and the output resistance is increased by this same factor. Thus, a much higher Q can be obtained, and the center frequency is not shifted (detuned) significantly by changes in the value of C_{GD}.

A similar problem often occurs if the tuned circuit is placed at the input of the amplifier rather than the output, as in Fig. 16.61. For the case of the bipolar transistor in particular, the equivalent input impedance of Q_1 represented by R_{in} and C_{in} can be quite low due to r_π and the large input capacitance resulting from the Miller effect. The tapped inductor increases the impedance to that in Fig. 16.62, in which L_1 now represents the total inductance of the transformer.

16.11.3 MULTIPLE TUNED CIRCUITS—SYNCHRONOUS AND STAGGER TUNING

Multiple RLC circuits are often needed to tailor the frequency response of tuned amplifiers, as in Fig. 16.63, which has tuned circuits at both the amplifier input and output. The high-frequency ac equivalent circuit for the double-tuned amplifier appears in Fig. 16.63(b). The source resistor is bypassed by capacitor C_S, and C_C is a coupling capacitor. The **radio frequency choke (RFC)** is used for biasing and is designed to represent a very high impedance (an open circuit) at the operating frequency of the amplifier.

Two tuned circuits can be used to achieve higher Q than that of a single LC circuit if both are tuned to the same center frequency **(synchronous tuning),** or a broader band amplifier can be realized if the circuits are tuned to slightly different center frequencies **(stagger tuning),** as shown in Fig. 16.64. For the case of synchronous tuning, the overall bandwidth can be calculated using the bandwidth shrinkage factor that was developed in Chapter 12:

$$\text{BW}_n = \text{BW}_1\sqrt{2^{\frac{1}{n}} - 1} \tag{16.185}$$

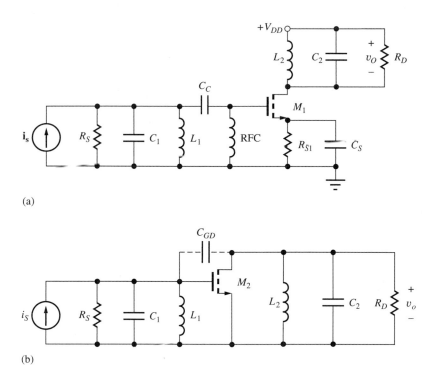

(a)

(b)

Figure 16.63 (a) Amplifier employing two tuned circuits. (b) High-frequency ac model for the amplifier employing two tuned circuits.

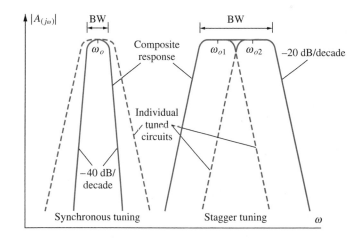

Figure 16.64 Examples of tuned amplifiers employing synchronous and stagger tuning of two tuned circuits.

in which n is the number of synchronous tuned circuits, and BW_1 is the bandwidth for the case of a single tuned circuit.

However, two significant problems can occur in the amplifier in Fig. 16.63, particularly for the case of synchronous tuning. First, alignment of the two tuned circuits is difficult because of interaction between the two tuned circuits due to the Miller multiplication of C_{GD}. Second, the amplifier can easily become an oscillator due to the coupling of signal energy from the output of the amplifier back to the input through C_{GD}. (A discussion of oscillators is in the next chapter.)

A technique called **neutralization** can be used to solve this feedback problem but is beyond the scope of this discussion. However, two alternative approaches are shown in Fig. 16.65, in which the

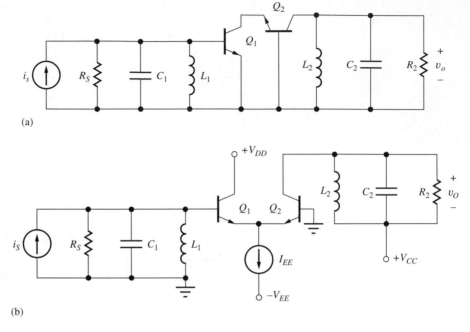

Figure 16.65 (a) Double-tuned cascode and (b) C-C/C-B cascade circuits that provide inherent isolation between input and output.

feedback path is eliminated. In Fig. 16.65(a), a cascode stage is used. Common-base transistor Q_2 effectively eliminates Miller multiplication and provides excellent isolation between the two tuned circuits. In Fig. 16.65(b), the C-C/C-B cascade is used to minimize the coupling between the output and input.

SUMMARY

- Amplifier frequency response can be determined by splitting the circuit into two models, one valid at low frequencies where coupling and bypass capacitors are most important, and a second valid at high frequencies in which the internal device capacitances control the frequency-dependent behavior of the circuit.

- Direct analysis of these circuits in the frequency domain, although usually possible for single-transistor amplifiers, becomes impractical for multistage amplifiers. In most cases, however, we are primarily interested in the midband gain and the upper- and lower-cutoff frequencies of the amplifier, and estimates of f_H and f_L can be obtained using the open-circuit and short-circuit time-constant methods. More accurate results can be obtained using SPICE circuit simulation.

- The frequency-dependent characteristics of the bipolar transistor are modeled by adding the base-emitter and base-collector capacitors C_π and C_μ and base resistance r_x to the hybrid-pi model. The value of C_π is proportional to collector current I_C, whereas C_μ is weakly dependent on collector-base voltage. The $r_x C_\mu$ product is one important figure of merit for the frequency limitations of the bipolar transistor.

- The frequency dependence of the FET is modeled by adding gate-source and gate-drain capacitances, C_{GS} and C_{GD}, to the pi-model of the FET. The values of C_{GS} and C_{GD} are independent of operating point when the FET is operating in the active region.

- Both the BJT and FET have finite current gain at high frequencies, and the unity gain-bandwidth product f_T for both devices is determined by the device capacitances and the transconductance of the transistor. In the bipolar transistor, the β-cutoff frequency f_β represents the frequency at which the current gain is 3 dB below its low-frequency value.

- In SPICE, the basic high-frequency behavior of the bipolar transistor is modeled using these parameters: forward transit-time TF, zero-bias collector-base junction capacitance CJC, collector junction built-in potential VJC, collector junction grading factor MJC, and base resistance RB.

- In SPICE, the high-frequency behavior of the MOSFET is modeled using the gate-source and gate-drain capacitances determined by the gate-source and gate-drain overlap capacitances CGSO and CGDO, as well as TOX, W, and L.

- If all the poles and zeros of the transfer function can be found from the low- and high-frequency equivalent circuits, then f_H and f_L can be accurately estimated using Eqs. (16.16) and (16.23). In many cases, a dominant pole exists in the low- and/or high-frequency responses, and this pole controls f_H or f_L. Unfortunately, the complexity of most amplifiers precludes finding the exact locations of all the poles and zeros except through numerical means.

- For design purposes, however, one needs to understand the relationship between the device and circuit parameters and f_H and f_L. The short-circuit time constant (SCTC) and open-circuit time constant (OCTC) approaches, as well as Miller effect, provide the needed information and were used to find detailed expressions for f_H and f_L for the three classes of single-stage amplifiers, the inverting, noninverting, and follower stages.

- The input impedance of an amplifier is decreased as a result of Miller multiplication, and the expression for the dominant pole of an inverting amplifier can be cast in terms of the Miller effect.

- It was found that the inverting amplifiers provide high gain but the most limited bandwidth. Noninverting amplifiers can provide improved bandwidth for a given voltage gain, but it is important to remember that these stages have a much lower input resistance. The follower configurations provide unity gain over a wide bandwidth. The three basic classes of amplifiers show the direct trade-offs that occur between voltage gain and bandwidth.

- The SCTC approach is used to estimate the value of the lower-cutoff frequency in multistage amplifiers, whereas the Miller effect and equivalent time constant approach is applied to the nodes in the signal path to find the upper cutoff frequency. The frequency responses of the differential pair, cascode amplifier, C-C/C-B cascade stage, and current mirror were all evaluated, as well as an example of calculations for a three-stage amplifier. The frequency response of another multistage amplifier is calculated in Chapter 17.

- Tuned amplifiers employing RLC circuits can be used to achieve narrow-band amplifiers at radio frequencies. Designs can use either single- or multiple-tuned circuits. If the circuits in a multiple-tuned amplifier are all designed to have the same center frequency, the circuit is referred to as synchronously tuned. If the tuned circuits are adjusted to different center frequencies, the circuit is referred to as stagger-tuned. Care must be taken to ensure that tuned amplifiers do not become oscillators, and the use of the cascode and C-C/C-B cascade configurations offers improved isolation between multiple-tuned circuits.

KEY TERMS

Base resistance	Dominant pole
Beta-cutoff frequency	Lower-cutoff frequency
Cascode amplifier	Midband gain
Center frequency	Miller compensation
Dominant high-frequency pole	Miller effect
Dominant low-frequency pole	Miller integrator

Miller multiplication	Stagger tuning
Neutralization	Synchronous tuning
Open-circuit time-constant (OCTC) method	Tuned amplifiers
Pole frequencies	Unity-gain-bandwidth product
Radio frequency choke (RFC)	Upper-cutoff frequency
Short-circuit time-constant (SCTC) method	

REFERENCE

1. P. E. Gray and C. L. Searle, *Electronic Principles,* Wiley, New York: 1969.

PROBLEMS

16.1 Amplifier Frequency Response

16.1. Find A_{mid} and $F_L(s)$ for this transfer function. Is there a dominant pole? If so, what is the dominant-pole approximation of $A_v(s)$? What is the cutoff frequency f_L of the dominant-pole approximation? What is the exact cutoff frequency using the complete transfer function?

$$A_v(s) = \frac{50s^2}{(s+2)(s+30)}$$

16.2. Find A_{mid} and $F_L(s)$ for this transfer function. Is there a dominant pole? If so, what is the dominant-pole approximation of $A_v(s)$? What is the cutoff frequency f_L of the dominant-pole approximation? What is the exact cutoff frequency using the complete transfer function?

$$A_v(s) = \frac{400s^2}{2s^2 + 1400s + 100{,}000}$$

16.3. Find A_{mid} and $F_L(s)$ for this transfer function. Is there a dominant pole? Use Eq. (16.16) to estimate f_L. Use the computer to find the exact cutoff frequency f_L.

$$A_v(s) = -\frac{150s(s+15)}{(s+12)(s+20)}$$

16.4. Find A_{mid} and $F_H(s)$ for this transfer function. Is there a dominant pole? If so, what is the dominant-pole approximation of $A_v(s)$? What is the cutoff frequency f_H of the dominant-pole approximation? What is the exact cutoff frequency using the complete transfer function?

$$A_v(s) = \frac{9 \times 10^{11}}{3s^2 + 3.3 \times 10^5 s + 3 \times 10^9}$$

16.5. Find A_{mid} and $F_H(s)$ for this transfer function. Is there a dominant pole? If so, what is the dominant-pole approximation of $A_v(s)$? What is the cutoff frequency f_H of the dominant-pole approximation? What is the exact cutoff frequency using the complete transfer function?

$$A_v(s) = \frac{(s + 3 \times 10^9)}{(s + 10^7)\left(1 + \dfrac{s}{10^9}\right)}$$

16.6. Find A_{mid} and $F_H(s)$ for this transfer function. Is there a dominant pole? Use Eq. (16.16) to estimate f_H. Use the computer to find the exact cutoff frequency f_H.

$$A_v(s) = \frac{2 \times 10^9 (s + 5 \times 10^5)}{(s + 1.5 \times 10^5)(s + 2 \times 10^6)}$$

16.7. Find A_{mid}, $F_L(s)$, and $F_H(s)$ for this transfer function. Is there a dominant pole at low frequencies? At high frequencies? Use Eqs. (16.16) and (16.23) to estimate f_L and f_H. Use the computer to find the exact cutoff frequencies and compare to the estimates.

$$A_v(s) = -\frac{6 \times 10^8 s^2}{(s+1)(s+2)(s+1000)(s+2000)}$$

*16.8. Find A_{mid}, $F_L(s)$ and $F_H(s)$ for this transfer function. Is there a dominant pole at low frequencies? At high frequencies? Use Eqs. (16.16) and (16.23) to estimate f_L and f_H. Use the computer to find the exact cutoff frequencies and compare to the estimates.

$$A_v(s) = \frac{10^{10} s^2 (s+1)(s+200)}{(s+3)(s+5)(s+7)(s+100)^2(s+300)}$$

16.2 Direct Determination of the Low-Frequency Poles and Zeros—The Common-Source Amplifier

16.9. (a) Draw the low-frequency and midband equivalent circuits for the common-source amplifier in Fig. P16.9 if $R_I = 2$ kΩ, $R_1 = 4.3$ MΩ,

$R_2 = 5.6$ MΩ, $R_S = 13$ kΩ, $R_D = 43$ kΩ, and $R_3 = 470$ kΩ. (b) What are the lower-cutoff frequency and midband gain of the amplifier if the Q-point $= (0.2$ mA, 5 V) and $V_{GS} - V_{TN} = 1$ V? (c) What is the value of V_{DD}?

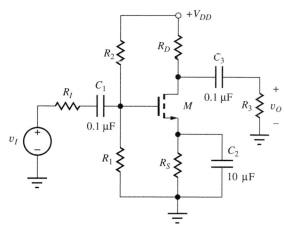

Figure P16.9

16.10. (a) Draw the low-frequency and midband equivalent circuits for the common-source amplifier in Fig. P16.9 if $R_I = 5$ kΩ, $R_1 = 430$ kΩ, $R_2 = 560$ kΩ, $R_S = 13$ kΩ, $R_D = 43$ kΩ, and $R_3 = 220$ kΩ. (b) What are the lower-cutoff frequency and midband gain of the amplifier if the Q-point $= (0.2$ mA, 5 V) and $V_{GS} - V_{TN} = 1$ V? (c) What is the value of V_{DD}?

16.11. (a) What is the value of C_2 required to set f_L to 50 Hz in the circuit in Prob. 16.9? (b) Choose the nearest standard value of capacitance from Appendix A. What is the value of f_L for this capacitor? (c) Repeat for the circuit in Prob. 16.10.

16.12. (a) Draw the low-frequency equivalent circuit for the common-gate amplifier in Fig. P16.12. (b) Write an expression for the transfer function of the amplifier and identify the location of the two low-frequency poles and two low-frequency zeros. Assume $r_o = \infty$ and $g_m = 5$ mS. (c) What are the lower-cutoff frequency and midband gain of the amplifier?

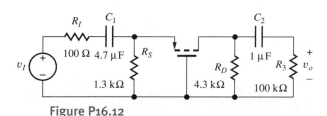

Figure P16.12

16.13. (a) What is the value of C_1 required to set f_L to 2000 Hz in the circuit in Prob. 16.12? (b) Choose the nearest standard value of capacitance from Appendix A. What is the value of f_L for this capacitor?

16.14. (a) Draw the low-frequency ac and midband equivalent circuits for the common-base amplifier in Fig. P16.14 if $R_I = 200$ Ω, $R_E = 4.3$ kΩ, $R_C = 2.2$ kΩ, $R_3 = 51$ kΩ, and $\beta_o = 100$. (b) Write an expression for the transfer function of the amplifier and identify the location of the two low-frequency poles and two low-frequency zeros. Assume $r_o = \infty$ and the Q-point $= (1$ mA, 5 V). (c) What are the midband gain and lower cutoff frequency of the amplifier? (d) What are the values of $-V_{EE}$ and V_{CC}? (e) What are the lower-cutoff frequency and midband gain of the amplifier if $R_E = 430$ kΩ, $R_C = 220$ kΩ, $R_3 = 510$ kΩ, and the Q-point is $(10$ μA, 5 V)? (f) What are the values of $-V_{EE}$ and V_{CC} in part (e)?

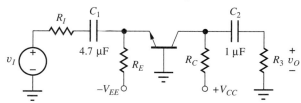

Figure P16.14

16.15. (a) What is the value of C_1 required to set f_L to 500 Hz in the circuit in Prob. 16.14(c)? (b) Choose the nearest standard value of capacitance from Appendix A. What is the value of f_L for this capacitor? (c) Repeat for the circuit in Prob. 16.14(e).

16.3 Estimation of ω_L Using the Short-Circuit Time-Constant Method

16.16. (a) The common-emitter circuit in Fig. 16.6 is redesigned with $R_1 = 100$ kΩ, $R_2 = 300$ kΩ, $R_E = 15$ kΩ, and $R_C = 43$ kΩ, and the Q-point is $(175$ μA, 2.3 V). The other values remain the same. Use the SCTC technique to find f_L. (b) Plot the frequency response of the amplifier with SPICE and find the value of f_L. (c) Calculate the Q-point for the transistor.

16.17. (a) What is the value of C_2 required to set f_L to 2500 Hz in the circuit in Fig. 16.6? (b) Choose the nearest standard value of capacitance from Appendix A. What is the actual value of f_L for this capacitor?

16.18. (a) Draw the low-frequency and midband equivalent circuits for the common-emitter amplifier in Fig. P16.18 if $R_I = 1$ kΩ, $R_1 = 100$ kΩ, $R_2 = 300$ kΩ, $R_E = 13$ kΩ, $R_C = 43$ kΩ, and $R_3 = 43$ kΩ. (b) What are the lower-cutoff frequency and midband gain of the amplifier assuming a Q-point of (0.164 mA, 2.79 V) and $\beta_o = 100$? (c) What is the value of V_{CC}?

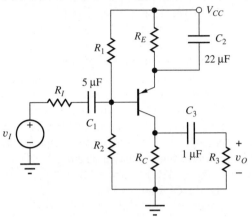

Figure P16.18

16.19. Use the SCTC technique to find the lower-cutoff frequency for the common-source amplifier in Fig. 16.11 if $R_G = 1$ MΩ, $R_3 = 68$ kΩ, $R_D = 22$ kΩ, $R_S = 6.8$ kΩ, and $g_m = 1.5$ mS. The other values remain unchanged.

16.20. Use the SCTC technique to find the lower-cutoff frequency for the common-source amplifier in Fig. 16.11 if $R_G = 500$ kΩ, $R_3 = 10$ kΩ, $R_D = 43$ kΩ, $R_S = 10$ kΩ and $g_m = 0.75$ mS. The other values remain unchanged.

16.21. (a) Draw the low-frequency and midband equivalent circuits for the common-gate amplifier in Fig. P16.21. (b) What are the lower-cutoff

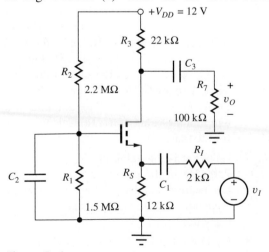

Figure P16.21

frequency and midband gain of the amplifier if the Q-point = (0.1 mA, 8.6 V), $V_{GS} - V_{TN} = 1$ V, $C_1 = 4.7$ μF, $C_2 = 0.1$ μF, and $C_3 = 0.1$ μF?

16.22. (a) Draw the low-frequency and midband equivalent circuits for the emitter follower in Fig. P16.22. (b) What are the lower-cutoff frequency and midband gain of the amplifier if the Q-point is (0.25 mA, 12 V), $\beta_o = 100$, $C_1 = 4.7$ μF, and $C_3 = 10$ μF?

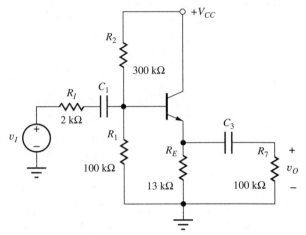

Figure P16.22

16.23. (a) Draw the low-frequency and midband equivalent circuits for the source follower in Fig. P16.23. (b) What are the lower-cutoff frequency and midband gain of the amplifier if the transistor is biased at 0.75 V above threshold with a Q-point = (0.1 mA, 8.8 V), $C_1 = 4.7$ μF, and $C_3 = 0.1$ μF? (c) What is the value of V_{DD}?

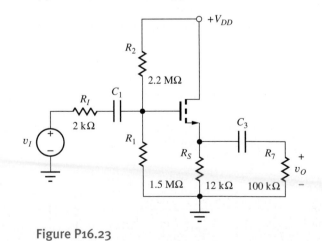

Figure P16.23

16.24. Redesign the value of C_2 in the C-S stage in Prob. 16.9 to set $f_L = 500$ Hz.

16.25. Redesign the value of C_1 in the C-G stage in Prob. 16.12 to set $f_L = 100$ Hz.

16.26. Redesign the value of C_2 in the C-E stage in Prob. 16.18 to set $f_L = 20$ Hz.

16.27. Redesign the value of C_1 in the C-G stage in Prob. 16.21 to set $f_L = 1$ Hz.

16.28. Redesign the value of C_3 in the C-C stage in Prob. 16.22 to set $f_L = 10$ Hz.

16.29. Redesign the value of C_3 in the C-D stage in Prob. 16.23 to set $f_L = 5$ Hz.

16.4 Transistor Models at High Frequencies

16.30. Fill in the missing parameter values for the BJT in the table if $r_x = 200\ \Omega$.

I_C	f_T	C_π	C_μ	$\dfrac{1}{2\pi r_x C_\mu}$
10 μA	50 MHz		0.50 pF	
100 μA	300 MHz	0.75 pF		
500 μA	1 GHz		0.25 pF	
10 mA		10 pF		1.59 GHz
1 μA		1 pF	1 pF	
	5 GHz	1 pF	0.5 pF	

16.31. A bipolar transistor with $f_T = 500$ MHz and $C_{\mu o} = 2$ pF is biased at a Q-point of (2 mA, 5 V). What is the forward-transit time τ_F if $\phi_{jc} = 0.9$ V?

16.32. Fill in the missing parameter values for the MOSFET in the table if $K_n = 2$ mA/V^2.

I_D	f_T	C_{GS}	C_{GD}
10 μA		1.5 pF	0.5 pF
250 μA		1.5 pF	0.5 pF
	250 MHz	1.5 pF	0.5 pF

16.33. (a) An n-channel MOSFET has a mobility of 600 cm^2/V · s and a channel length of 1 μm. What is the transistor's f_T if $V_{GS} - V_{TN} = 0.25$ V. (b) Repeat for a PMOS device with a mobility of 250 cm^2/V · s. (c) Repeat for transistors in a new technology with $L = 0.1$ μm. (d) Repeat for transistors in a technology with $L = 25$ nm.

16.5 Base Resistance in the Hybrid-Pi Model

16.34. (a) What is the midband gain for the common-emitter amplifier in Fig. P16.34 if $r_x = 500\ \Omega$, $I_C = 1$ mA, and $\beta_o = 125$? (b) If $r_x = 0$?

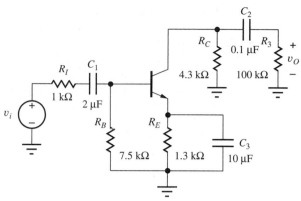

Figure P16.34

16.35. (a) What is the midband gain for the common-collector amplifier in Fig. P16.35 if $r_x = 350\ \Omega$, $I_C = 1$ mA, and $\beta_o = 125$? (b) If $r_x = 0$?

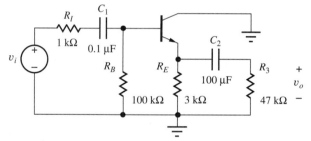

Figure P16.35

16.36. (a) What is the midband gain for the common-base amplifier in Fig. P16.36 if $r_x = 200\ \Omega$, $I_C = 0.1$ mA, and $\beta_o = 125$? (b) If $r_x = 0$?

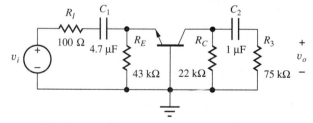

Figure P16.36

16.6 High-Frequency Response of the Common-Emitter and Common-Source Amplifiers

Factorization

16.37. Use dominant root factorization techniques to estimate the roots of these quadratic equations and compare the results to the exact roots: (a) $s^2 + 5100s + 500,000$; (b) $2s^2 + 700s + 30,000$; (c) $3s^2 + 3300s + 300,000$; (d) $0.5s^2 + 300s + 40,000$.

16.38. (a) Use dominant root factorization techniques to estimate the roots of this equation. (b) Compare the results to the exact roots.

$$s^3 + 1110s^2 + 111{,}000s + 1{,}000{,}000$$

16.39. Use Newton's method to help find the roots of this polynomial. (*Hint:* Find the roots one at a time. Once a root is found, factor it out to reduce the order of the polynomial. Use approximate factorization to find starting points for iteration.)

$$s^6 + 142s^5 + 4757s^4 + 58{,}230s^3$$
$$+ 256{,}950s^2 + 398{,}000s + 300{,}000$$

For Probs. 16.40 to 16.48, use $f_T = 500$ MHz, $r_x = 300\ \Omega$, $C_\mu = 0.75$ pF, $C_{GS} = C_{GD} = 2.5$ pF.

16.40. (a) What are the midband gain and upper-cutoff frequency for the common-emitter amplifier in Prob. 16.34(a) if $I_C = 1$ mA and $\beta_o = 100$? (b) What is the gain-bandwidth product for this amplifier?

16.41. Resistors R_1, R_2, R_E, and R_C in the common-emitter amplifier in Fig. 16.6 are all decreased in value by a factor of 2. (a) Draw the dc equivalent circuit for the amplifier, and find the new Q-point for the transistor. (b) Draw the ac small-signal equivalent circuit for the amplifier, and find the midband gain and upper-cutoff frequency for the amplifier. (c) What is the gain-bandwidth product for this amplifier?

16.42. The resistors in the common-emitter amplifier in Fig. 16.6 are all increased in value by a factor of 50. (a) Draw the dc equivalent circuit for the amplifier, and find the new Q-point for the transistor. (b) Draw the ac small-signal equivalent circuit for the amplifier, and find the midband gain and upper-cutoff frequency for the amplifier. (c) What is the gain-bandwidth product for this amplifier?

16.43. What are the midband gain and upper-cutoff frequency for the common-source amplifier in Prob. 16.9?

16.44. Simulate the frequency response of the amplifier in Prob. 16.9 and determine A_{mid}, f_L, and f_H.

16.45. In the common-source amplifier in Fig. 16.4, the value of R_S is changed to 3.9 kΩ and that of R_D to 10 kΩ. For the MOSFET, $K_n = 500\ \mu\text{A/V}^2$ and $V_{TN} = 1$ V. (a) Draw the dc equivalent circuit for the amplifier, and find the new Q-point for the transistor if $V_{DD} = 12$ V. (b) Draw the ac small-signal equivalent circuit for the amplifier, and find the midband gain and upper-cutoff frequency for the amplifier. (c) What is the gain-bandwidth product for this amplifier?

16.46. What are the midband gain and upper-cutoff frequency for the common-emitter amplifier in Prob. 16.18?

16.47. Simulate the frequency response of the amplifier in Prob. 16.18 and determine A_{mid}, f_L, and f_H.

16.48. The network in Fig. P16.48 models a common emitter stage with a load capacitor in parallel with R_L. (a) Write the two nodal equations and find the determinant of the system for the network in Fig. P16.48. (b) Use dominant root factorization to find the two poles. (c) There are three capacitors in the network. Why are there only two poles?

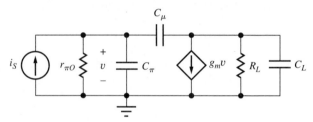

Figure P16.48

Miller Multiplication

16.49. What is the total input capacitance in the circuit in Fig. 16.35 if $C_\pi = 20$ pF, $C_\mu = 1$ pF, $I_C = 5$ mA, and $R_L = 1$ kΩ? What is the f_T of this transistor?

16.50. (a) What is the input capacitance of the circuit in Fig. P16.50 if Z is a 100-pF capacitor and the amplifier is an op amp with a gain of 100,000? (**b) What is the input impedance of the circuit in Fig. P16.50 at $f = 1$ kHz if element Z is a 100-kΩ resistor and $A(s) = 10^6/(s + 10)$? (c) At 50 kHz? (d) At 1 MHz?

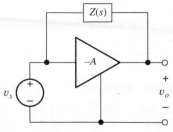

Figure P16.50

16.51. (a) Find the transfer function of the Miller integrator in Fig. 11.15 if $A(s) = 10A_o/(s + 10)$. The transfer function is really that of a low-pass amplifier. What is the cutoff frequency if $A_o = 10^5$? (b) For $A_o = 10^6$? (c) Show that the transfer function approaches that of the ideal integrator if $A_o \to \infty$.

16.52. Use Miller multiplication to calculate the impedance presented to v_i by the circuit in Fig. P16.52 at $f = 1$ kHz if $r_x = 250\ \Omega$, $r_\pi = 2.5$ kΩ, $g_m = 0.04$ S, $R_L = 2.5$ kΩ, $C_\pi = 15$ pF, and $C_\mu = 1$ pF. (b) At 50 kHz. (c) At 1 MHz. (d) Compare your results to SPICE.

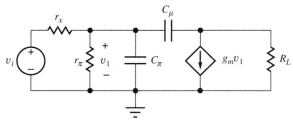

Figure P16.52

16.53. Use SPICE to find the midband gain, and upper- and lower-cutoff frequencies of the amplifier in Prob. 16.52.

16.54. (a) Estimate the upper-cutoff frequency for the common-emitter amplifier in Prob. 16.34(a) if $f_T = 500$ MHz and $C_\mu = 0.75$ pF. (b) Repeat for Prob. 16.34(b).

16.55. Resistors R_1, R_2, R_E, and R_C in the common-emitter amplifier in Fig. P16.34 are all increased in value by a factor of 10, and the collector current is reduced to 100 μA. (a) Draw the ac small-signal equivalent circuit for the amplifier, and find the midband gain and upper-cutoff frequency for the amplifier if $\beta_o = 100$, $r_x = 400\ \Omega$, $C_\mu = 0.75$ pF, and $f_T = 500$ MHz. (b) What is the gain-bandwidth product for this amplifier? Calculate the upper bound on GBW given by the $r_x C_\mu$ product.

16.56. Estimate the upper-cutoff frequency for the common-source amplifier in Prob. 16.9 if $C_{GS} = 5$ pF and $C_{GD} = 2$ pF. What is the gain-bandwidth product for this amplifier?

Estimation of ω_H for Inverting Amplifiers, Noninverting Amplifiers, and Followers Using the Open-Circuit Time-Constant Method

16.57. (a) Redesign the common-emitter amplifier in Fig. 16.34 to have an upper-cutoff frequency of

5 MHz by changing the value of the collector resistor R_C. What is the new value of the midband voltage gain? What is the gain-bandwidth product?

16.58. What are the values of (a) A_{mid}, f_L, and f_H for the common-emitter amplifier in Fig. P16.58 if $C_1 = 1$ μF, $C_3 = 0.1$ μF, $C_2 = 2.2$ μF, $R_3 = 100$ kΩ, $\beta_o = 100$, $f_T = 300$ MHz, $r_x = 300\ \Omega$, and $C_\mu = 0.5$ pF? (b) What is the gain-bandwidth product?

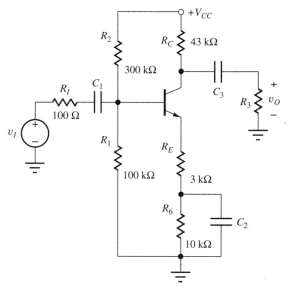

Figure P16.58

16.59. (a) Redesign the common-emitter amplifier in Fig. P16.58 to have an upper-cutoff frequency of 7.5 MHz by selecting new values for R_E and R_6. Maintain the sum $R_E + R_6 = 13$ kΩ. What is the new value of the midband voltage gain? What is the gain-bandwidth product?

16.60. Find (a) A_{mid}, (b) f_L, and (c) f_H for the amplifier in Fig. P16.60 if $\beta_o = 100$, $f_T = 200$ MHz, $C_\mu = 1$ pF, and $r_x = 350\ \Omega$.

16.61. Redesign the values of R_{E1} and R_{E2} in the amplifier in Prob. 16.60 to achieve $f_H = 12$ MHz. Do not change the Q-point.

*16.62. The network in Fig. P16.62 has two poles. (a) Estimate the lower-pole frequency using the short-circuit time-constant technique if $C_1 = 1$ μF, $C_2 = 10$ μF, $R_1 = 10$ kΩ, $R_2 = 1$ kΩ, and $R_3 = 1$ kΩ. (b) Estimate the upper-pole frequency. (c) Why do the positions of the poles seem to be backward? (d) Find the system determinant and compare its exact roots to those in (a) and (b).

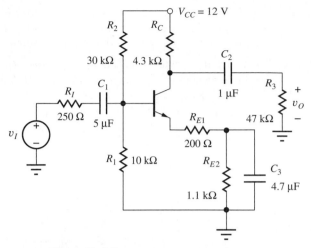

Figure P16.60

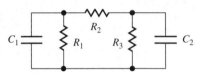

Figure P16.62

For Probs. 16.63 to 16.74, use $f_T = 500$ MHz, $r_x = 300$ Ω, $C_\mu = 0.60$ pF for the BJT, and $C_{GS} = 3$ pF and $C_{GD} = 0.60$ pF for the FET.

16.7 High-Frequency Response of Common-Base and Common-Gate Amplifiers

16.63. What are the midband gain and upper-cutoff frequency for the common-gate amplifier in Prob. 16.12?

16.64. Simulate the frequency response of the amplifier in Prob.16.12 and determine A_{mid}, f_L, and f_H.

16.65. What are the midband gain and upper-cutoff frequency for the common-base amplifier in Prob. 16.14(e)?

16.66. Simulate the frequency response of the amplifier in Prob. 16.14 with $V_{CC} = V_{EE} = 5$ V and determine A_{mid}, f_L, and f_H.

16.67. What are the midband gain and upper-cutoff frequency for the common-base amplifier in Prob. 16.14 if $V_{CC} = -V_{EE} = 10$ V?

16.68. What are the midband gain and upper-cutoff frequency for the amplifier in Prob. 16.21?

16.69. What are the midband gain and upper- and lower-cutoff frequencies for the amplifier in Prob. 16.21 if V_{DD} is increased to 18 V?

16.8 High-Frequency Response of Common-Collector and Common-Drain Amplifiers

16.70. (a) What are the midband gain and upper-cutoff frequency for the emitter follower in Prob. 16.22? (b) Simulate the frequency response of the amplifier in Prob. 16.22 with $V_{CC} = 15$ V and determine A_{mid}, f_L, and f_H.

16.71. What are the midband gain and upper-cutoff frequency for the common-collector amplifier in Fig. P16.22 if V_{CC} is 9 V?

16.72. (a) What are the midband gain and upper-cutoff frequency for the source follower in Prob. 16.23? (b) Simulate the frequency response of the amplifier in Prob. 16.23 with $V_{DD} = 10$ V and determine A_{mid}, f_L, and f_H.

16.73. What are the midband gain and upper-cutoff frequency for the common-drain amplifier in Prob. 16.23 if V_{DD} is 20 V?

**16.74. Derive an expression for the total input capacitance of the BJT in Fig. 16.43(c) looking into node v_b. Assume $C_L = 0$. Use it to interpret Eq. (16.137).

*16.75. Derive an expression for the total capacitance looking into the gate of the FET in Fig. 16.44(b). Use the expression to interpret Eq. (16.140).

16.9 Summary of the High-Frequency Response of Single-Stage Amplifiers

Gain-Bandwidth Product

16.76. A bipolar transistor must be selected for use in a common-emitter amplifier with a gain of 43 dB and a bandwidth of 6 MHz. What should be the minimum specification for the transistor's f_T? What should be the minimum $r_x C_\mu$ product? (Use a factor of 2 safety margin for each estimate.)

16.77. A bipolar transistor must be selected for use in a common-base amplifier with a gain of 40 dB and a bandwidth of 40 MHz. What should be the minimum specification for the transistor's f_T? What should be the minimum $r_x C_\mu$ product? (Use a factor of 2 safety margin for each estimate.)

17.78. A BJT will be used in a differential amplifier with load resistors of 100 kΩ. What are the maximum values of r_x and C_μ that can be tolerated if the gain and bandwidth are to be 100 and 1.8 MHz, respectively?

16.79. An FET with $C_{GS} = 7.5$ pF and $C_{GD} = 3$ pF will be used in a common-gate amplifier with a source resistance of 100 Ω, $A_{mid} = 20$, and a bandwidth of 25 MHz. Estimate the Q-point current needed to achieve these specifications if $K_n = 20$ mA/V^2 and $V_{DD} = 15$ V.

*16.80. An FET with $C_{GS} = 12$ pF and $C_{GD} = 5$ pF will be used in a common-source amplifier with a source resistance of 100 Ω and a bandwidth of 25 MHz. Estimate the minimum Q-point current needed to achieve this bandwidth if $K_n = 25$ mA/V^2 and $V_{GS} - V_{TN} \geq 0.25$ V.

16.81. What is the upper bound on the bandwidth of the circuit in Fig. P16.14 if $R_C = 12$ kΩ, $R_3 = 47$ kΩ, and $C_\mu = 2$ pF?

**16.82. (a) Estimate the cutoff frequency of the C-C/C-E cascade in Fig. P16.82(a). (b) Estimate the cutoff frequency of the Darlington stage in Fig. P16.82(b). Assume $I_{C1} = 0.1$ mA, $I_{C2} = 1$ mA, $\beta_o = 100$, $f_T = 300$ MHz, $C_\mu = 0.5$ pF, $V_A = 50$ V, $r_x = 300$ Ω, and $R_L = \infty$. (c) Which configuration offers better bandwidth? (d) Which configuration is used in the second stage in the μA741 amplifier in Chapter 15? Why do you think it was used?

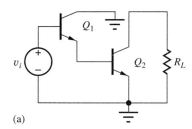

(a)

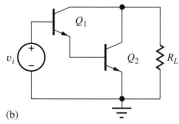

(b)

Figure P16.82

16.83. Draw a Bode plot for the common-mode rejection ratio for the differential amplifier in Fig. 16.45 if $I_C = 100$ μA, $R_{EE} = 10$ MΩ, $R_C = 6$ kΩ, $C_{EE} = 1$ pF, $\beta_o = 100$, $V_A = 50$ V, $f_T = 200$ MHz, $C_\mu = 0.3$ pF, $r_x = 175$ Ω, and $R_L = 100$ kΩ. R_L is connected between the collectors of transistors Q_1 and Q.

16.84. Use SPICE to plot the graph for Prob. 16.83.

16.10 Frequency Response of Multistage Amplifiers

16.85. What is the minimum bandwidth of the NMOS current mirror in Fig. P16.85 if $I_S = 100$ μA, $K_n' = 25$ μA/V^2, $\lambda = 0.02$ V^{-1}, $C_{GS1} = 3$ pF, $C_{GD1} = 0.5$ pF, and $(W/L)_1 = 5/1 = (W/L)_2$?

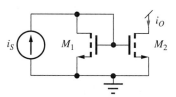

Figure P16.85

16.86. What is the minimum bandwidth of the MOS current mirror in Fig. P16.85 if $I_S = 200$ μA, $K_n' = 25$ μA/V^2, $\lambda = 0.02$ V^{-1}, $C_{GS1} = 3$ pF, $C_{GD1} = 1$ pF, $(W/L)_1 = 5/1$, and $(W/L)_2 = 25/1$?

16.87. What is the minimum bandwidth of the *npn* current mirror in Fig. P16.87 if $I_S = 100$ μA, $\beta_o = 100$, $V_A = 60$ V, $f_T = 600$ MHz, $C_\mu = 0.5$ pF, and $A_{E2} = 10\, A_{E1}$?

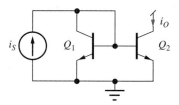

Figure P16.87

*16.88. What is the minimum bandwidth of the bipolar current mirror in Fig. P16.87 if $I_S = 250$ μA, $\beta_o = 100$, $V_A = 50$ V, $f_T = 500$ MHz, $C_\mu = 0.3$ pF, $r_x = 175$ Ω, and $A_{E2} = 4A_{E1}$?

16.89. What is the minimum bandwidth of the *pnp* current mirror in Fig. P16.89 if $I_S = 100$ μA, $\beta_o = 50$, $V_A = 60$ V, $f_T = 50$ MHz, $C_\mu = 2$ pF, and $A_{E2} = A_{E1}$?

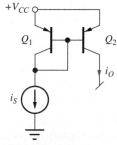

Figure P16.89

**16.90. Find the minimum bandwidth of the Wilson current mirror in Fig. P16.90 if $I_{REF} = 250$ μA, $K_n = 250$ μA/V², $V_{TN} = 0.75$ V, $\lambda = 0.02$ V⁻¹, $C_{GS} = 3$ pF, and $C_{GD} = 1$ pF.

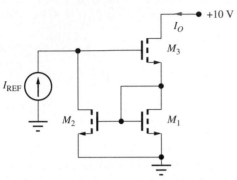

Figure P16.90

16.91. (a) The transistors in the differential amplifier in Fig. 16.45 are biased at a collector current of 15 μA, and $R_C = 430$ kΩ. The transistors have $f_T = 75$ MHz, $C_\mu = 0.5$ pF, and $r_x = 500$ Ω. What is the bandwidth of the differential amplifier? (b) Repeat if the collector current is increased to 50 μA and R_C is reduced to 140 kΩ.

16.92. (a) The transistors in the C-C/C-B cascade amplifier in Fig. 16.47 are biased with $I_{EE} = 250$ μA and $R_C = 62$ kΩ. The transistors have $f_T = 100$ MHz, $C_\mu = 1$ pF, and $r_x = 500$ Ω. What is the bandwidth of the amplifier? (b) Repeat if the current source is increased to 2 mA and R_C is reduced to 7.5 kΩ.

16.93. (a) The transistors in the cascode amplifier in Fig. 16.49 are biased at a collector current of 100 μA with $R_L = 75$ kΩ. The transistors have $f_T = 100$ MHz, $C_\mu = 1$ pF, and $r_x = 500$ Ω. What is the bandwidth of the amplifier if $R_{th} = 0$? (b) Repeat if the collector currents are increased to 1 mA and R_C is reduced to 7.5 kΩ.

16.94. The bias current in transistor Q_3 in Fig. 16.52(a) is doubled by reducing the value of resistors R_3, R_4, and R_{E3} by a factor of 2. What are the new values of midband gain, lower-cutoff frequency, and upper-cutoff frequency?

16.95. The bias current in transistor Q_2 in Fig. 16.52(a) is doubled by reducing the value of resistors R_1, R_2, R_{C2}, and R_{E2} by a factor of 2. What are the new values of midband gain, lower-cutoff frequency, and upper-cutoff frequency?

16.11 Tuned Amplifiers

16.96. What are the center frequency, Q, and midband gain for the amplifier in Fig. P16.96 if the FET has $C_{GS} = 50$ pF, $C_{GD} = 5$ pF, $\lambda = 0.0167$ V⁻¹, and it is biased at 2 V above threshold with $I_D = 10$ mA and $V_{DS} = 10$ V.

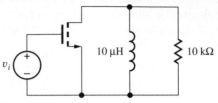

Figure P16.96

16.97. (a) What is the value of C required for $f_o = 10.7$ MHz in the circuit in Fig. P16.97 if $I_C = 10$ mA, $V_{CE} = 10$ V, $\beta_o = 100$, $C_\mu = 2$ pF, $f_T = 500$ MHz, and $V_A = 75$ V? (b) What is the Q of the amplifier? (c) Where should a tap be placed on the inductor to achieve a Q of 100? (d) What is the new value of C required to achieve $f_o = 10.7$ MHz?

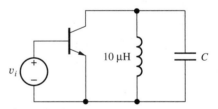

Figure P16.97

16.98. (a) Draw the dc and high-frequency ac equivalent circuits for the circuit in Fig. P16.98. (b) What is the resonant frequency of the circuit for $V_C = 0$ V if the diode is modeled by $C_{jo} = 20$ pF and $\phi_j = 0.9$ V? (c) For $V_C = 10$ V?

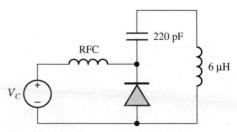

Figure P16.98

*16.99. (a) What are the center frequency, Q, and midband gain for the tuned amplifier in Fig. P16.99 if $L_1 = 5$ μH, $C_1 = 10$ pF, $C_2 = 10$ pF, $I_C = 1$ mA,

$C_\pi = 5$ pF, $C_\mu = 1$ pF, $R_L = 5$ kΩ, $r_\pi = 2.5$ kΩ, and $r_x = 0$ Ω? (b) What would be the answers if the base terminal of the transistor were connected to the top of the inductor?

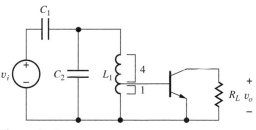

Figure P16.99

16.100. (a) What are the midband gain, center frequency, bandwidth, and Q for the circuit in Fig. P16.100(a) if $I_D = 20$ mA, $\lambda = 0.02$ V^{-1}, $C_{GD} = 5$ pF, and $K_n = 5$ mA/V^2? (b) Repeat for the circuit in Fig. P16.100(b).

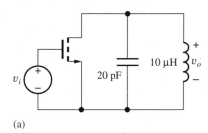

(a)

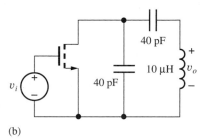

(b)

Figure P16.100

*16.101. Change the two capacitor values in the circuit in Fig. P16.100(a) to give the same center frequency as in Fig. P16.100(b). What are the Q and midband gain for the new circuit?

16.102. (a) Simulate the circuit in Prob. 16.100(a) and compare the results to the hand calculations in Prob. 16.100. (b) Simulate the circuit in Prob. 16.100(b) and compare the results to the hand calculations in Prob. 16.100. (c) Simulate the circuit in Prob. 16.101 and compare the results to the hand calculations in Prob. 16.101.

16.103. (a) What is the value of C_2 required to achieve synchronous tuning of the circuit in Fig. P16.103 if $L_1 = L_2 = 10$ μH, $C_1 = C_3 = 20$ pF, $C_{GS} = 20$ pF, $C_{GD} = 5$ pF, $V_{TN1} = -1$ V, $K_{n1} = 10$ mA/V^2, $V_{TN2} = -4$ V, $K_{n2} = 10$ mA/V^2, and $R_G = R_D = 100$ kΩ? (b) What are the Q, midband gain, and bandwidth of your design?

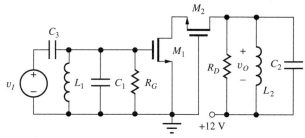

Figure P16.103

16.104. Simulate the frequency response of the circuit design in Prob. 16.103 and find the midband gain, center frequency, Q, and bandwidth of the circuit. Did you achieve synchronous tuning of your design?

**16.105. (a) What is the value of C_2 required to adjust the resonant frequency of the tuned circuit connected to the drain of M_2 to a frequency 2 percent higher than that connected at the gate of M_1 in Fig. P16.103 if $L_1 = L_2 = 10$ μH, $C_1 = C_3 = 20$ pF, $C_{GS} = 20$ pF, $C_{GD} = 5$ pF, $V_{TN1} = -1$ V, $K_{n1} = 10$ mA/V^2, $V_{TN2} = -4$ V, $K_{n2} = 10$ mA/V^2, and $R_G = R_D = 100$ kΩ? (b) What are the Q and bandwidth of your design?

*16.106. Simulate the frequency response of the circuit design in Prob. 16.105 and find the midband gain, center frequency, Q and bandwidth, and the Q of the circuit. Was the desired stagger tuning achieved?

*16.107. (a) Derive an expression for the high frequency input admittance at the base of the common-emitter circuit in Fig. 16.34(b) and show that the input capacitance and input resistance can be represented by the expressions below for $\omega C_\mu R_L \ll 1$.

$$C_{in} = C_\pi + C_\mu (1 + g_m R_L)$$

$$R_{in} = r_\pi \left\| \frac{R_L}{(1 + g_m R_L)(\omega C_\mu R_L)^2} \right.$$

(b) A MOSFET has $C_{GS} = 6$ pF, $C_{GD} = 2$ pF, $g_m = 5$ mS, and $R_L = 10$ kΩ. What are the values of C_{in} and R_{in} at a frequency of 5 MHz?

CHAPTER 17

FEEDBACK, STABILITY, AND OSCILLATORS

CHAPTER GOALS

- Review the concepts of negative and positive feedback
- Fully develop the two-port approach to the analysis of negative feedback amplifiers
- Understand the topologies and characteristics of the series-shunt, shunt-shunt, shunt-series, and series-series feedback configurations
- Discuss common errors that can occur in applying the two-port feedback theory
- Understand the effects of feedback on frequency response and feedback amplifier stability
- Learn to interpret feedback amplifier stability in terms of Nyquist and Bode plots
- Use SPICE ac and transfer function analyses to characterize feedback amplifiers
- Develop techniques to determine the loop-gain of closed-loop amplifiers using SPICE simulation or measurement

- Learn to design operational amplifier frequency compensation using Miller multiplication
- Develop relationships between op-amp unity gain frequency and slew rate.
- Discuss the Barkhausen criteria for oscillation
- Understand basic RC, LC, and crystal oscillator circuits
- Present the LCR model of the quartz crystal
- Discuss amplitude stabilization in oscillators

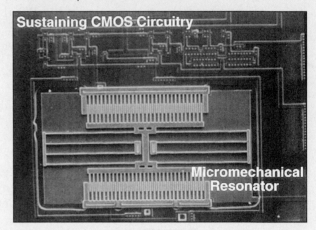

An oscillator employing a MEMS[1] frequency selective resonator. Copyright IEEE 1999. Reprinted with permission.

Examples of feedback systems abound in daily life. The thermostat that senses the temperature of a room and turns the air-conditioning system on and off is one example. Another is the remote control that we use to select a channel on the television or set the volume at an acceptable level. The heating and cooling system uses a simple temperature transducer to compare the temperature with a fixed set point. However, we are part of the TV remote control feedback system; we operate the control until our senses tell us that the audio and optical information is what we want.

The theory of negative feedback in electronic systems was first developed by Harold Black of the Bell Telephone System. In 1928, he invented the feedback amplifier to stabilize the gain of early telephone repeaters. Today, some form of feedback is used in virtually every electronic system. This chapter formally develops the concept of feedback, which is

---

[1] *Micro-Electro-Mechanical System.* C. T.-C. Nguyen and R. T. Howe, "An integrated micromechanical resonator high-Q oscillator," *IEEE J. Solid-State Circuits,* vol. 34, no. 4, pp. 440–445, April 1999.

an invaluable tool in the design of electronic systems. Valuable insight into the operation of many common electronic circuits can be gained by recasting the circuits as feedback amplifiers.

We already encountered **negative** (or **degenerative**) **feedback** in several forms. The four-resistor bias network uses negative feedback to achieve an operating point that is independent of variations in device characteristics. We also found that a source or emitter resistor can be used in an inverting amplifier to control the gain and bandwidth of the stage. Many of the advantages of negative feedback were actually uncovered during the discussion of operational amplifier circuit design. Generally, feedback can be used to achieve a trade-off between gain and many of the other properties of amplifiers:

- *Gain stability:* Feedback reduces the sensitivity of gain to variations in the values of transistor parameters and circuit elements.
- *Input and output impedances:* Feedback can increase or decrease the input and output resistances of an amplifier.

- *Bandwidth:* The bandwidth of an amplifier can be extended using feedback.
- *Nonlinear distortion:* Feedback reduces the effects of nonlinear distortion. (For example, feedback can be used to minimize the effects of the dead zone in a class-B amplifier stage.)

Feedback may also be **positive** (or **regenerative**), and we explore the use of positive feedback in sinusoidal **oscillator circuits** in this chapter. We encountered the use of a combination of negative and positive feedback in the discussion of *RC* active filters and multivibrator circuits in Chapter 11. Sinusoidal oscillators use positive feedback to generate signals at specific desired frequencies; they use negative feedback to stabilize the amplitude of the oscillations.

Positive feedback in amplifiers is usually undesirable. Excess phase shift in a feedback amplifier may cause the feedback to become regenerative and cause the feedback amplifier to break into oscillation. Remember that positive feedback was identified in Chapter 16 as a potential source of oscillation problems in tuned amplifiers.

17.1 CLASSIC FEEDBACK SYSTEMS

Figure 17.1 is the block diagram for a classic feedback system. This diagram may represent a simple feedback amplifier or a complex feedback control system. It consists of an amplifier with transfer function $A(s)$, referred to as the **open-loop amplifier,** a **feedback network** with transfer function $\beta(s)$, and a summing block indicated by Σ. The variables in this diagram are represented as voltages but could equally well be currents or even other physical quantities such as temperature, velocity, distance, and so on.

In Fig. 17.1, the input to the open-loop amplifier A is provided by the summing block, which develops the difference between the input signal v_i and the feedback signal v_f. In the frequency domain,

$$\mathbf{V_d}(s) = \mathbf{V_i}(s) - \mathbf{V_f}(s) \tag{17.1}$$

The output signal is equal to the product of the open-loop amplifier gain and the input signal to the amplifier:

$$\mathbf{V_o}(s) = A(s)\mathbf{V_d}(s) \tag{17.2}$$

The signal fed back to the input is given by

$$\mathbf{V_f}(s) = \beta(s)\mathbf{V_o}(s) \tag{17.3}$$

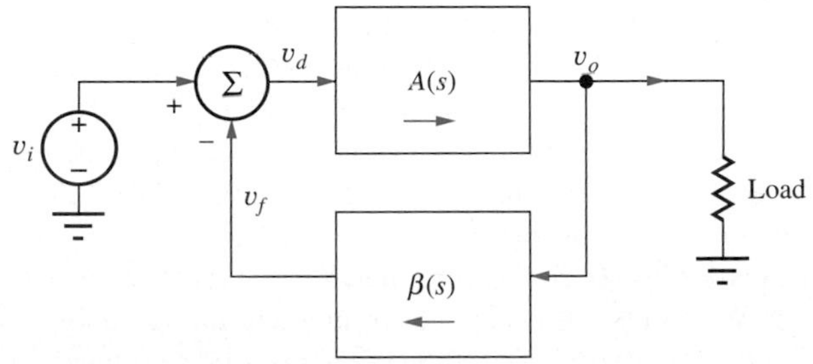

Figure 17.1 Classic block diagram for a feedback system.

Combining Eqs. (17.1) to (17.3) and solving for the overall voltage gain of the system yields the classic expression for the **closed-loop gain** of a feedback amplifier:

$$A_v(s) = \frac{\mathbf{V_o}(s)}{\mathbf{V_i}(s)} = \frac{A(s)}{1 + A(s)\beta(s)} = \frac{A(s)}{1 + T(s)} \tag{17.4}$$

Equation (17.4) was encountered in Chapter 11, where $A(s)$ was identified as the **open-loop gain,** and the product $T(s) = A(s)\beta(s)$ was defined as the **loop gain.** For the block diagram in Fig. 17.1, negative feedback requires $T(s) > 0$, whereas $T(s) < 0$ corresponds to the positive feedback condition.

A number of assumptions are implicit in this derivation. It is assumed that the blocks can be interconnected, as shown in Fig. 17.1, without affecting each other. That is, connecting the feedback network and the load to the output of the amplifier does not change the characteristics of the amplifier, nor does the interconnection of the summer, feedback network, and input of the open-loop amplifier modify the characteristics of either the amplifier or feedback network. In addition, it is tacitly assumed that signals flow only in the forward direction through the amplifier, and only in the reverse direction through the feedback network, as indicated by the arrows in Fig. 17.1.

Implementation of this block diagram with operational amplifiers having large input resistances, low output resistances, and essentially zero reverse-voltage gain is one method of satisfying these unstated assumptions. However, most general amplifiers and feedback networks do not necessarily satisfy these assumptions. The theory developed in the next several sections explores the analysis and design of more general feedback systems that do not satisfy the implicit restrictions just outlined.

17.2 FEEDBACK AMPLIFIER DESIGN USING TWO-PORT NETWORK THEORY

If we consider both the amplifier and feedback networks to each be represented as a two-port, then four basic feedback amplifier topologies can be defined: voltage, transresistance, current, and transconductance amplifiers. Figure 17.2 shows these four amplifiers in block diagram form; each is characterized by a specific combination of the input and output port connections. Port voltages are summed by connecting the ports in series, and port currents are summed by connecting the ports in parallel, usually termed a **shunt connection.**

In the voltage amplifier topology in Fig. 17.2(a), the input port of the feedback network is connected in series with the input port of the amplifier, and the output ports of the amplifier and feedback network form a parallel or shunt connection. Thus, this topology is referred to as a **series-shunt feedback** amplifier. The first adjective refers to the input port configuration, and the second refers to the output. As we know, high input resistance and low output resistance characterize a good voltage amplifier. In the series-shunt case, the series feedback increases the resistance at the input port, and the shunt feedback decreases the resistance at the output port. The op-amp based non-inverting amplifier is an example of series-shunt feedback.

In the transresistance topology in Fig. 17.2(b), shunt connections are used at both the input and output ports; thus this circuit is referred to as a **shunt-shunt feedback** amplifier. For the shunt-shunt configuration, feedback is used to lower the resistance at both ports. The current amplifier topology in Fig. 17.2(c) connects the input ports in parallel and the output ports in series and so is usually referred to as a **shunt-series feedback** amplifier. For the shunt-series case, shunt feedback lowers the resistance at the input port, and series feedback increases the resistance at the output port. Finally, for the transconductance amplifier topology in Fig. 17.2(d), the ports of the amplifier and feedback network are connected in series at both the input and output, so this topology is referred to as the **series-series feedback** amplifier. Series feedback provides high resistance at both ports.

We analyze the behavior of these amplifiers using two-port descriptions for the amplifiers and feedback networks. Based on the various port connections, analysis of each type uses a different set of two-port parameters.

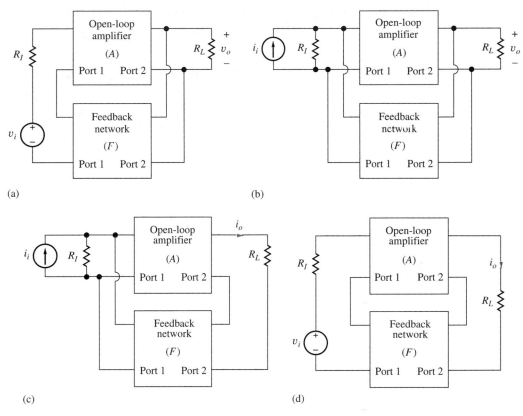

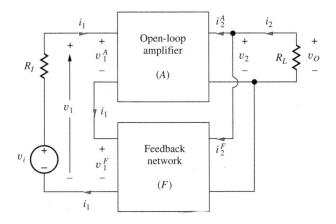

Figure 17.2 (a) Series-shunt feedback amplifier (h-parameters). Voltage amplifier topology: $A_v = \mathbf{v_o}/\mathbf{v_i}$. (b) Shunt-shunt feedback amplifier (y-parameters). Transresistance amplifier topology: $A_{tr} = \mathbf{v_o}/\mathbf{i_i}$. (c) Shunt-series feedback amplifier (g-parameters). Current amplifier topology: $A_i = \mathbf{i_o}/\mathbf{i_i}$. (d) Series- series feedback amplifier (z-parameters). Transconductance amplifier topology: $A_{tc} = \mathbf{i_o}/\mathbf{v_i}$.

Figure 17.3 Series-shunt feedback amplifier.

17.3 VOLTAGE AMPLIFIERS—SERIES-SHUNT FEEDBACK

Because of our study of op amp circuits, we are most familiar with **voltage amplifiers,** and we start our analysis with this configuration (see Fig. 17.3). As will become apparent during the analysis, the h-parameters are the appropriate two-port parameters for analyzing this configuration.

17.3.1 VOLTAGE GAIN CALCULATION

Analysis begins by describing the amplifier and feedback network by their individual h-parameter two-port descriptions:

$$\mathbf{v_1^A} = h_{11}^A \mathbf{i_1} + h_{12}^A \mathbf{v_2} \qquad \mathbf{v_1^F} = h_{11}^F \mathbf{i_1} + h_{12}^F \mathbf{v_2}$$

$$\text{and}$$

$$\mathbf{i_2^A} = h_{21}^A \mathbf{i_1} + h_{22}^A \mathbf{v_2} \qquad \mathbf{i_2^F} = h_{21}^F \mathbf{i_1} + h_{22}^F \mathbf{v_2} \tag{17.5}$$

in which the superscripts indicate the amplifier (A) and feedback network (F), respectively. Next, we proceed to find the two-port parameters of the overall feedback amplifier based on these individual h-parameter descriptions. Because of the series connection at the input, the overall input voltage v_1 of the feedback amplifier is just the sum of the input voltages of the individual two-ports and, because of the shunt connection at the output, the overall current i_2 into the output port is the sum of the currents at the output of the individual two-ports:

$$\mathbf{v_1} = \mathbf{v_1^A} + \mathbf{v_1^F} \qquad \text{and} \qquad \mathbf{i_2} = \mathbf{i_2^A} + \mathbf{i_2^F} \tag{17.6}$$

Substituting Eq. (17.5) into Eq. (17.6) yields a two-port description for the overall feedback amplifier:

$$\mathbf{v_1} = \left(h_{11}^A + h_{11}^F\right)\mathbf{i_1} + \left(h_{12}^A + h_{12}^F\right)\mathbf{v_2}$$

$$\mathbf{i_2} = \left(h_{21}^A + h_{21}^F\right)\mathbf{i_1} + \left(h_{22}^A + h_{22}^F\right)\mathbf{v_2} \tag{17.7}$$

Here we see the reasoning behind the choice of the h-parameters, which allows the two sets of network parameters to be conveniently added together. Because the corresponding parameters of both networks always appear together in Eq. (17.7), a more compact notation is achieved by defining

$$h_{ij}^T = h_{ij}^A + h_{ij}^F \tag{17.8}$$

so that the two-port equations become

$$\mathbf{v_1} = h_{11}^T \mathbf{i_1} + h_{12}^T \mathbf{v_2}$$

$$\mathbf{i_2} = h_{21}^T \mathbf{i_1} + h_{22}^T \mathbf{v_2} \tag{17.9}$$

Normally, the forward current gain of the amplifier far exceeds that of the feedback network, $h_{21}^A \gg h_{21}^F$, and the reverse voltage gain of the feedback network is much greater than that of the amplifier, $h_{12}^F \gg h_{12}^A$. Using these approximations to simplify Eq. (17.9) yields

$$\mathbf{v_1} = h_{11}^T \mathbf{i_1} + h_{12}^F \mathbf{v_2}$$

$$\mathbf{i_2} = h_{21}^A \mathbf{i_1} + h_{22}^T \mathbf{v_2} \tag{17.10}$$

An expression for the closed-loop gain of the feedback amplifier, including the effects of R_I and R_L, can now be found using Eq. (17.10). At the input port in Fig. 17.3, $\mathbf{v_1}$ and $\mathbf{i_1}$ are related by

$$\mathbf{v_1} = \mathbf{v_i} - \mathbf{i_1}R_I \tag{17.11}$$

and $\mathbf{v_2}$ and $\mathbf{i_2}$ at the output port are related by

$$\mathbf{i_2} = -\frac{\mathbf{v_2}}{R_L} = -G_L\mathbf{v_2} \tag{17.12}$$

Substituting Eqs. (17.11) and (17.12) into Eq. (17.10) yields

$$\mathbf{v_i} = \left(R_I + h_{11}^T\right)\mathbf{i_1} + h_{12}^F \mathbf{v_2}$$

$$0 = h_{21}^A \mathbf{i_1} + \left(h_{22}^T + G_L\right)\mathbf{v_2} \tag{17.13}$$

The source and load resistors, R_I and R_L, are absorbed into the expressions in Eq. (17.13) to develop a consistent set of equations for overall feedback amplifier gain, input resistance, and output resistance calculations.

The closed-loop voltage gain is found from Eq. (17.13) by solving for v_2 in terms of v_i:

$$A_v = \frac{v_2}{v_i} = \frac{h_{21}^A}{h_{21}^A h_{12}^F - \left(R_I + h_{11}^T\right)\left(h_{22}^T + G_L\right)} \tag{17.14}$$

By dividing numerator and denominator by the second denominator term, Eq. (17.14) can be rearranged into the standard form for a feedback system:

$$A_v = \frac{\dfrac{-h_{21}^A}{\left(R_I + h_{11}^T\right)\left(h_{22}^T + G_L\right)}}{1 + \dfrac{-h_{21}^A}{\left(R_I + h_{11}^T\right)\left(h_{22}^T + G_L\right)} h_{12}^F} = \frac{A}{1 + A\beta} \tag{17.15}$$

in which

$$A = -\frac{h_{21}^A}{\left(R_I + h_{11}^T\right)\left(h_{22}^T + G_L\right)} \quad \text{and} \quad \beta = h_{12}^F \tag{17.16}$$

Figure 17.4(a) and (b) provide an interpretation of Eqs. (17.15) and (17.16) and a general methodology for analyzing feedback amplifiers. Figure 17.4(a) shows the feedback amplifier with an explicit representation of the two-port parameters of the feedback network with $h_{21}^F = 0$. Equations (17.15) and (17.16) indicate that the gain of the amplifier A should be calculated including the effects of h_{11}^F,

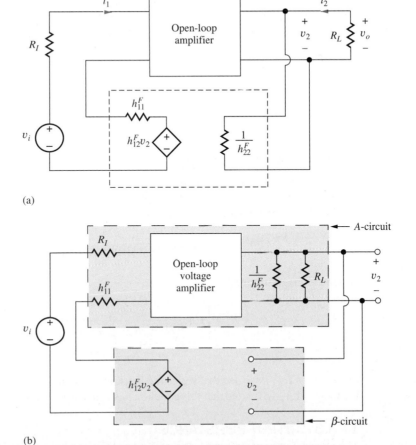

(a)

(b)

Figure 17.4 Schematic interpretation of the amplifier and feedback circuits described by Eqs. (17.15) and (17.16).

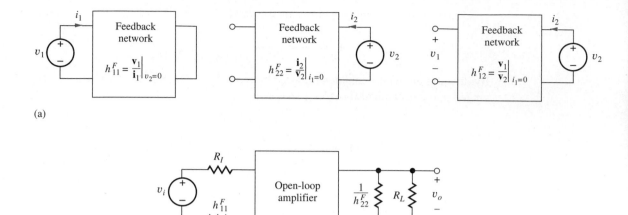

(a)

(b)

Figure 17.5 Subcircuits used for analysis of the series-shunt feedback amplifier: (a) circuits for determining the h-parameters of the feedback network, and (b) the A-circuit for finding the amplifier gain.

h_{22}^F, R_I, and R_L. A schematic representation of these equations is given by redrawing the circuit of Fig. 17.4(a) as in Fig. 17.4(b). Although the position of feedback elements h_{11}^F and h_{22}^F in the drawing has changed, the actual circuit remains the same. The amplifier circuit, the A-circuit, now includes h_{11}^F, h_{22}^F, R_I, and R_L, and the feedback network contains only h_{12}^F.

Figure 17.5 reinforces the analysis technique. The three required h-parameters of the feedback network are found based on their individual definitions. Then the voltage gain of the open-loop amplifier A is calculated from the circuit in Fig. 17.5(b), which includes the loading effects of h_{11}^F, h_{22}^F, R_I, and R_L. The gain is calculated directly from the A-circuit; we generally *do not* evaluate A using the mathematical two-port parameter description in Eq. (17.16) (although it can be done that way). Our first example helps make the overall analysis process clearer. But first, let us develop expressions for the input and output resistances of the feedback amplifier.

17.3.2 INPUT RESISTANCE

The **closed-loop input resistance** of the series-shunt feedback amplifier can be calculated from the two-port description of the overall amplifier in Fig. 17.4(b), described by Eq. (17.13). This time, solving the second equation for v_2 in terms of i_1 and substituting the result back into the first gives

$$\mathbf{v_i} = \left(R_I + h_{11}^T\right)\mathbf{i_1} + h_{12}^F \frac{-h_{21}^A}{\left(h_{22}^T + G_L\right)}\mathbf{i_1} \tag{17.17}$$

which can be rearranged as

$$R_{\text{in}} = \frac{\mathbf{v_i}}{\mathbf{i_1}} = \left(R_I + h_{11}^T\right)\left[1 + \frac{-h_{21}^A}{\left(R_I + h_{11}^T\right)\left(h_{22}^T + G_L\right)}h_{12}^F\right] = \left(R_I + h_{11}^T\right)[1 + A\beta] \tag{17.18}$$

or

$$R_{\text{in}} = R_{\text{in}}^A(1 + A\beta) \tag{17.19}$$

From Eq. (17.19), we see that series feedback at a port increases the input resistance at that port by the factor $(1 + A\beta)$. Note that this equation has exactly the same form as that obtained in Chapter 12.

17.3.3 OUTPUT RESISTANCE

The **closed-loop output resistance** of the amplifier can be calculated in a manner similar to the input resistance calculation, based on the circuit in Fig. 17.6. The signal source v_i is set to zero, a test

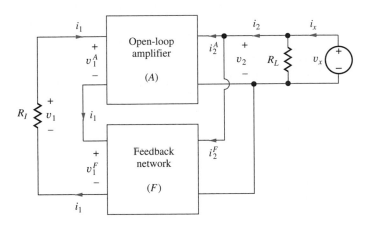

Figure 17.6 Output resistance of the series-shunt feedback network.

source v_x is applied to the output, and i_x must be determined. For Fig. 17.6,

$$\mathbf{v_1} = -\mathbf{i_1} R_I \qquad \mathbf{v_2} = \mathbf{v_x} \qquad \mathbf{i_2} = \mathbf{i_x} - G_L \mathbf{v_2} \tag{17.20}$$

Substituting these constraints into Eq. (17.10) yields

$$0 = \left(R_I + h_{11}^T \right) \mathbf{i_1} + h_{12}^F \mathbf{v_x}$$
$$\mathbf{i_x} = h_{21}^A \mathbf{i_1} + \left(h_{22}^T + G_L \right) \mathbf{v_x} \tag{17.21}$$

and solving for $\mathbf{i_x}$ in terms of $\mathbf{v_x}$ yields

$$\mathbf{i_x} = h_{21}^A \frac{-h_{12}^F}{\left(R_I + h_{11}^T \right)} \mathbf{v_x} + \left(h_{22}^T + G_L \right) \mathbf{v_x} \tag{17.22}$$

Rearranging Eq. (17.22) yields an expression for the output resistance of the overall amplifier, which again is in the same form as that derived in Chapter 12.

$$R_{\text{out}} = \frac{\mathbf{v_x}}{\mathbf{i_x}} = \frac{1}{\left(h_{22}^T + G_L \right) \left[1 + \dfrac{-h_{21}^A}{\left(R_I + h_{11}^T \right)\left(h_{22}^T + G_L \right)} h_{12}^F \right]} = \frac{\left(\dfrac{1}{h_{22}^T + G_L} \right)}{1 + A\beta} \tag{17.23}$$

$$R_{\text{out}} = \frac{R_{\text{out}}^A}{1 + A\beta} \tag{17.24}$$

The output resistance of the closed-loop amplifier is equal to the output resistance of the A-circuit, including the effects of R_I and R_L, reduced by the amount of feedback $(1 + A\beta)$. Shunt feedback reduces the impedance level at a port.

EXAMPLE 17.1   SERIES-SHUNT FEEDBACK AMPLIFIER ANALYSIS

Evaluate the closed-loop characteristics of a series-shunt feedback amplifier using the h-parameter two-port method.

PROBLEM   Find A, β, and the closed-loop voltage gain A_v for the series-shunt feedback amplifier in Fig. 17.7 if the op amp has an open-loop gain of 80 dB, a differential-mode input resistance of 25 kΩ, and an output resistance of 1 kΩ.

SOLUTION   **Known Information and Given Data:** Series-shunt feedback amplifier in Fig. 17.7 with 1-kΩ source resistance, 2-kΩ load resistance, and feedback network with $R_1 = 10$ kΩ and $R_2 = 91$ kΩ. Op amp parameters: $A = 10^4$, $R_{id} = 25$ kΩ, and $R_o = 1$ kΩ

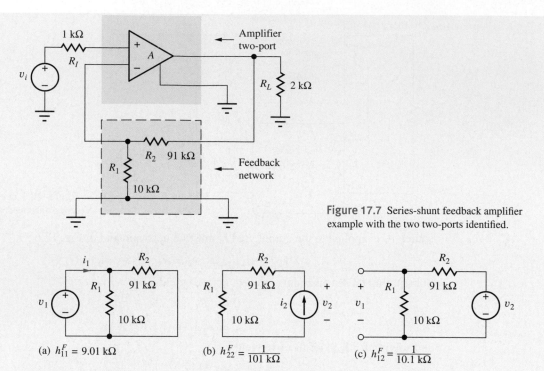

Figure 17.7 Series-shunt feedback amplifier example with the two two-ports identified.

(a) $h_{11}^F = 9.01 \text{ k}\Omega$

(b) $h_{22}^F = \dfrac{1}{101 \text{ k}\Omega}$

(c) $h_{12}^F = \dfrac{1}{10.1 \text{ k}\Omega}$

Figure 17.8 Circuits for determining the h-parameters of the feedback network.

Unknowns: Closed-loop voltage gain A_v, closed-loop input resistance R_{in}, and closed-loop output resistance R_{out}

Approach: Apply the h-parameter two-port method to the circuit. (1) Identify the amplifier and feedback network two ports; (2) find the h-parameters for the feedback network and determine the value of β; (3) form the augmented amplifier circuit and find its open-loop gain, input resistance, and output resistance; and (4) calculate the closed-loop values of A_v, R_{in}, and R_{out} using Eqs. (17.15), (17.19), and (17.24).

Assumptions: h_{12}^A and h_{21}^F can be neglected

Analysis: The first step in the analysis is to draw the amplifier as a pair of interconnected two-ports, as in Fig. 17.7, to (i) ensure that the theory can actually be applied to the amplifier configuration and (ii) clearly partition the circuit into the forward amplifier and the feedback network. The importance of this step must not be overlooked!

Feedback Network
In this case, the feedback network is formed by R_1 and R_2, and the values of h_{11}^F, h_{12}^F, and h_{22}^F, corresponding to the feedback network, are found from the three circuits in Fig. 17.8:

$$h_{11}^F = \left.\frac{\mathbf{v_1}}{\mathbf{i_1}}\right|_{\mathbf{v_2}=0} = R_1 \| R_2 = 10 \text{ k}\Omega \| 91 \text{ k}\Omega = 9.01 \text{ k}\Omega$$

$$h_{22}^F = \left.\frac{\mathbf{i_2}}{\mathbf{v_2}}\right|_{\mathbf{i_1}=0} = \frac{1}{R_1 + R_2} = \frac{1}{10 \text{ k}\Omega + 91 \text{ k}\Omega} = \frac{1}{101 \text{ k}\Omega}$$

$$h_{12}^F = \left.\frac{\mathbf{v_1}}{\mathbf{v_2}}\right|_{\mathbf{i_1}=0} = \frac{R_1}{R_1 + R_2} = \frac{10 \text{ k}\Omega}{10 \text{ k}\Omega + 91 \text{ k}\Omega} = 0.0990$$

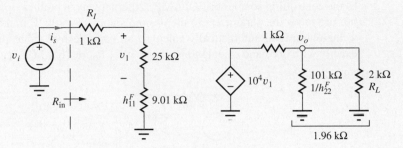

Figure 17.9 Augmented amplifier circuit (the A-circuit).

A-Circuit

The forward gain of the amplifier is found using the A-circuit in Fig. 17.9. R_I and h_{11}^F have been added in series with the amplifier input, and R_L and $1/h_{22}^F$ are placed in parallel with the amplifier output. The voltage gain of the A-circuit is

$$A = \frac{\mathbf{v_o}}{\mathbf{v_i}} = \frac{25 \text{ k}\Omega}{1 \text{ k}\Omega + 25 \text{ k}\Omega + 9.01 \text{ k}\Omega}(10^4)\frac{1.96 \text{ k}\Omega}{1.96 \text{ k}\Omega + 1.00 \text{ k}\Omega} = 4730$$

and the feedback factor is $\beta = h_{12}^F = 0.0990$. Note that A is less than the open-loop gain of the op amp because of the loading effects of R_I, R_L, h_{11}^F, and h_{22}^F.

Closed-Loop Gain

Using these results, the closed-loop gain is

$$A_v = \frac{A}{1 + A\beta} = \frac{4730}{1 + 4730(0.0990)} = 10.1$$

Note that the loop gain is large: $A\beta = 468$.

Closed-Loop Input Resistance

From Fig. 17.9 and Eq. (17.19),

$$R_{\text{in}} = R_{\text{in}}^A(1 + A\beta) = (1 \text{ k}\Omega + 25 \text{ k}\Omega + 9.01 \text{ k}\Omega)(469) = 16.4 \text{ M}\Omega$$

Closed-Loop Output Resistance

Using Eq. (17.24) and the A-circuit in Fig. 17.9,

$$R_{\text{out}} = \frac{R_{\text{out}}^A}{1 + A\beta} = \frac{2 \text{ k}\Omega\|101 \text{ k}\Omega\|1 \text{ k}\Omega}{469} = 1.41 \text{ } \Omega$$

Check of Results: We have found the three unknowns. In this case, the loop gain is much larger than 1, so A_v should be set by the feedback network:

$$A_v = \frac{A}{1 + A\beta} \cong \frac{1}{\beta} = 10.1$$

which agrees with the more detailed calculation. We also see that the overall input resistance is much larger than that of the amplifier itself, and the output resistance is much smaller than that of the amplifier. Both of these results are what we expect from the series-shunt configuration.

Discussion: This analysis demonstrates the proper method of including the effects of nonideal op amp characteristics in the noninverting amplifier configuration. We must be careful to include the loading effects of the feedback network and load and source resistors. The next exercise looks at the errors that occur if these effects are neglected.

Computer-Aided Analysis: The op amp can be modeled using two resistors, RID and RO and a voltage-controlled voltage source EOPAMP, similar to Fig. 17.9. Zero-valued source IOA is added as the controlling source for EOPAMP. (Alternatively, a built-in op amp model could be used, but be sure to set the parameters to the desired values.) A SPICE transfer function analysis from VI to the output will automatically calculate all three desired values, and the analysis results agree exactly with our hand calculations: $A_v = 10.1$, $R_{in} = 16.4 \ M\Omega$, and $R_{out} = 1.41 \ \Omega$.

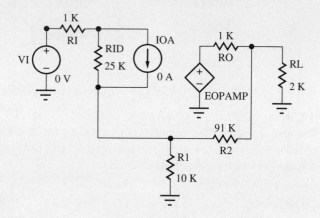

17.4 TRANSRESISTANCE AMPLIFIERS—SHUNT-SHUNT FEEDBACK

The **transresistance amplifier** is another important class of amplifier, widely used in optical communications systems to convert optical signals from a fiber into an electrical signal. For example, i_i and R_I are one model for a photodiode detector at the output of an optical fiber. The transresistance amplifier is formed using the shunt-shunt feedback configuration in Fig. 17.10, in which the amplifier and feedback network are connected in parallel with each other. The goal of the shunt-shunt feedback amplifier is to provide a low input resistance so that all of the current from source i_i enters the amplifier, as well as a low output resistance to drive external loads. Because the input port voltages are the same and the output port voltages are the same for the amplifier and feedback two-ports, the y-parameters are appropriate for analyzing this configuration.

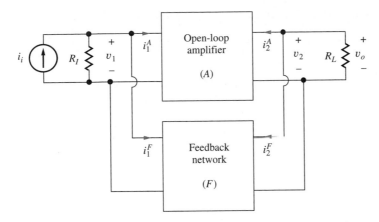

Figure 17.10 Shunt-shunt feedback amplifier.

17.4.1 TRANSRESISTANCE ANALYSIS

The analysis mirrors that in Sec. 17.3. In this case, the amplifier and feedback network are represented by their individual y-parameters:

$$\mathbf{i_1^A} = y_{11}^A \mathbf{v_1} + y_{12}^A \mathbf{v_2} \qquad \text{and} \qquad \mathbf{i_1^F} = y_{11}^F \mathbf{v_1} + y_{12}^F \mathbf{v_2}$$
$$\mathbf{i_2^A} = y_{21}^A \mathbf{v_1} + y_{22}^A \mathbf{v_2} \qquad\qquad \mathbf{i_2^F} = y_{21}^F \mathbf{v_1} + y_{22}^F \mathbf{v_2} \tag{17.25}$$

in which the superscripts again indicate the amplifier (A) and feedback network (F), respectively.

Based on the connections at the input and output ports, the overall input current $\mathbf{i_1}$ and output current $\mathbf{i_2}$ can be written as

$$\mathbf{i_1} = \mathbf{i_1^A} + \mathbf{i_1^F} \qquad \text{and} \qquad \mathbf{i_2} = \mathbf{i_2^A} + \mathbf{i_2^F} \tag{17.26}$$

Combining Eqs. (17.25) and (17.26) yields the two-port description for the overall shunt-shunt feedback amplifier:

$$\mathbf{i_1} = \left(y_{11}^A + y_{11}^F \right)\mathbf{v_1} + \left(y_{12}^A + y_{12}^F \right)\mathbf{v_2}$$
$$\mathbf{i_2} = \left(y_{21}^A + y_{21}^F \right)\mathbf{v_1} + \left(y_{22}^A + y_{22}^F \right)\mathbf{v_2} \tag{17.27}$$

Because the corresponding parameters of both networks again appear together in Eq. (17.27), a more compact notation is achieved by defining

$$y_{ij}^T = y_{ij}^A + y_{ij}^F \tag{17.28}$$

Once again assuming that $y_{21}^A \gg y_{21}^F$ and $y_{12}^F \gg y_{12}^A$, Eq. (17.28) is represented in simplified form as

$$\mathbf{i_1} = y_{11}^T \mathbf{v_1} + y_{12}^F \mathbf{v_2}$$
$$\mathbf{i_2} = y_{21}^A \mathbf{v_1} + y_{22}^T \mathbf{v_2} \tag{17.29}$$

The expression for the closed-loop gain of the shunt-shunt feedback amplifier, including the effects of R_I and R_L, can now be found with the aid of Eq. (17.29). In Fig. 17.10, $\mathbf{v_1}$ and $\mathbf{i_1}$, and $\mathbf{v_2}$ and $\mathbf{i_2}$, are related by

$$\mathbf{i_1} = \mathbf{i_i} - \mathbf{v_1} G_I \qquad \text{and} \qquad \mathbf{i_2} = -G_L \mathbf{v_2} \tag{17.30}$$

Substituting Eq. (17.30) into Eq. (17.29) yields

$$\mathbf{i_i} = \left(G_I + y_{11}^T \right)\mathbf{v_1} + y_{12}^F \mathbf{v_2}$$
$$0 = y_{21}^A \mathbf{v_1} + \left(y_{22}^T + G_L \right)\mathbf{v_2} \tag{17.31}$$

The closed-loop transresistance can now be found from Eq. (17.31) by solving for $\mathbf{v_2}$ in terms of $\mathbf{i_i}$:

$$A_{tr} = \frac{\mathbf{v_2}}{\mathbf{i_i}} = \frac{y_{21}^A}{y_{21}^A y_{12}^F - \left(G_I + y_{11}^T\right)\left(y_{22}^T + G_L\right)} \tag{17.32}$$

Rearranging Eq. (17.32) into the standard form for a feedback amplifier gives

$$A_{tr} = \frac{\mathbf{v_2}}{\mathbf{i_i}} = \frac{\dfrac{-y_{21}^A}{\left(G_I + y_{11}^T\right)\left(y_{22}^T + G_L\right)}}{1 + \dfrac{-y_{21}^A}{\left(G_S + y_{11}^T\right)\left(y_{22}^T + G_L\right)} y_{12}^F} = \frac{A}{1 + A\beta} \tag{17.33}$$

in which

$$A = \frac{\mathbf{v_o}}{\mathbf{i_i}} = -\frac{y_{21}^A}{\left(G_I + y_{11}^T\right)\left(y_{22}^T + G_L\right)} \quad \text{and} \quad \beta = y_{12}^F \tag{17.34}$$

Figure 17.11 provides the interpretation of Eqs. (17.33) and (17.34) for the case of shunt-shunt feedback. Figure 17.11(a) shows the feedback amplifier with an explicit representation of the two-port parameters of the feedback network with $y_{21}^F = 0$. Equations (17.33) and (17.34) indicate that the gain of the amplifier A should be calculated including the effects of y_{11}^F, y_{22}^F, R_I, and R_L, and a schematic representation of these equations is given by redrawing the amplifier, as in the circuit in Fig. 17.11(b). The positions of feedback circuit elements y_{11}^F and y_{22}^F have been changed, but the

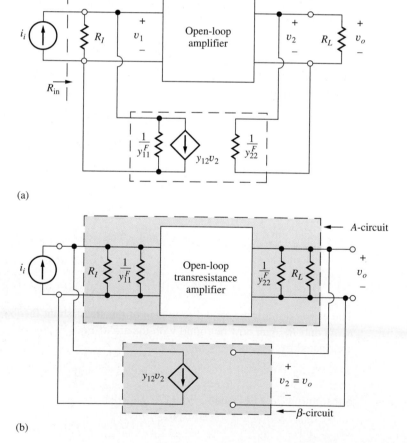

(a)

(b)

Figure 17.11 Schematic interpretation of the amplifier and feedback circuits described by Eqs. (17.33) and (17.34).

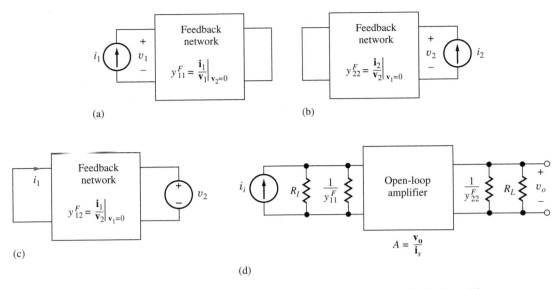

Figure 17.12 (a–c) Feedback circuits and (d) A-circuit for the shunt-shunt feedback amplifier.

overall circuit is once again the same. The amplifier A-circuit now includes y_{11}^F, y_{22}^F, R_I, and R_L, whereas the feedback network consists only of y_{12}^F.

Figure 17.12 reinforces the analysis technique. The three required y-parameters of the feedback network are found based on their individual definitions. Then the transresistance of the open-loop amplifier A is calculated from the circuit in Fig. 17.12(d), which includes the loading effects of y_{11}^F, y_{22}^F, R_I, and R_L. The gain is calculated directly from the A-circuit; remember, we normally *do not* evaluate A using the mathematical two-port parameter description given in Eq. (17.34).

17.4.2 INPUT RESISTANCE

The input resistance R_{in} of the closed-loop shunt-shunt feedback amplifier can be calculated for the overall amplifier in Fig. 17.11 using the two-port description in Eq. (17.31). The input resistance is

$$R_{in} = \frac{\mathbf{v_1}}{\mathbf{i_i}} \tag{17.35}$$

Solving Eq. (17.31) for $\mathbf{i_i}$ in terms of $\mathbf{v_1}$ gives

$$\mathbf{i_i} = \left(G_I + y_{11}^T\right)\mathbf{v_1} + y_{12}^F \frac{-y_{21}^A}{\left(y_{22}^T + G_L\right)}\mathbf{v_1} \tag{17.36}$$

which can be rearranged as

$$R_{in} = \frac{1}{\left(G_I + y_{11}^T\right)\left[1 + \dfrac{-y_{21}^A}{\left(G_S + y_{11}^T\right)\left(y_{22}^T + G_L\right)}y_{12}^F\right]} = \frac{\left(\dfrac{1}{G_I + y_{11}^T}\right)}{1 + A\beta} = \frac{R_{in}^A}{1 + A\beta} \tag{17.37}$$

Again, we see that shunt feedback reduces the resistance at the input port by the factor $(1 + A\beta)$. As the loop gain approaches infinity—for an ideal op amp, for example—the input resistance of the closed-loop transconductance amplifier approaches zero.

17.4.3 OUTPUT RESISTANCE

The output resistance of the closed-loop amplifier can be calculated in a manner similar to the input resistance calculation, but using the circuit in Fig. 17.13. We start with Eq. (17.29) and apply a test

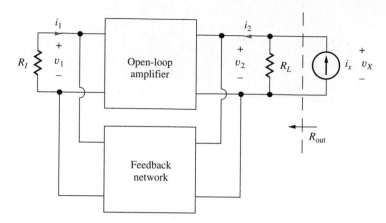

Figure 17.13 Output resistance of the shunt-shunt feedback amplifier.

source i_x to the output of the amplifier:

$$\mathbf{i_1} = y_{11}^T \mathbf{v_1} + y_{12}^F \mathbf{v_2}$$
$$\mathbf{i_2} = y_{21}^A \mathbf{v_1} + y_{22}^T \mathbf{v_2}$$

(17.38)

The voltage and current at the input and output ports are related by

$$\mathbf{i_1} = -\mathbf{v_1} G_I \qquad \text{and} \qquad \mathbf{i_2} = \mathbf{i_x} - G_L \mathbf{v_2}$$

(17.39)

Substituting Eq. (17.39) into Eq. (17.38) gives

$$0 = \left(G_I + y_{11}^T \right) \mathbf{v_1} + y_{12}^F \mathbf{v_x}$$
$$\mathbf{i_x} = y_{21}^A \mathbf{v_1} + \left(y_{22}^T + G_L \right) \mathbf{v_x}$$

(17.40)

and solving for $\mathbf{i_x}$ in terms of $\mathbf{v_x}$ yields

$$\mathbf{i_x} = y_{21}^A \frac{-y_{12}^F}{\left(G_I + y_{11}^T \right)} \mathbf{v_x} + \left(y_{22}^T + G_L \right) \mathbf{v_x}$$

(17.41)

Rearranging Eq. (17.41) yields an expression for the output resistance of the overall amplifier:

$$R_{\text{out}} = \frac{\mathbf{v_x}}{\mathbf{i_x}} = \frac{1}{\left(y_{22}^T + G_L \right) \left[1 + \dfrac{-y_{21}^A}{\left(G_I + y_{11}^T \right) \left(y_{22}^T + G_L \right)} y_{12}^F \right]} = \frac{\dfrac{1}{\left(y_{22}^T + G_L \right)}}{1 + A\beta} = \frac{R_{\text{out}}^A}{1 + A\beta}$$

(17.42)

The output resistance of the closed-loop amplifier is equal to the output resistance of the A-circuit decreased by the amount of feedback $(1 + A\beta)$. In the ideal case, the output resistance of the transresistance amplifier approaches zero as the loop gain approaches infinity.

EXAMPLE **17.2**   SHUNT-SHUNT FEEDBACK AMPLIFIER ANALYSIS

Analysis of the shunt-shunt feedback amplifier in Fig. 17.14 provides an example of application of the two-port theory to a practical single-transistor amplifier.

PROBLEM   Find A, β, the closed-loop transresistance A_{tr}, input resistance, and output resistance for the single-transistor shunt-shunt feedback amplifier in Fig. 17.14 assuming $\beta_F = 150$ and $V_A = 50$ V.

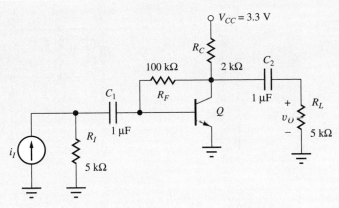

Figure 17.14 Shunt-shunt feedback amplifier.

SOLUTION **Known Information and Given Data:** Single transistor shunt-shunt feedback amplifier in Fig. 17.14; transistor parameters: $\beta_F = 150$ and $V_A = 50$ V

Unknowns: A, β, closed-loop transresistance, input resistance, and output resistance

Approach: Analyze the dc equivalent circuit to find the Q-point; construct midband ac equivalent circuits for A-circuit and feedback network; find A, β, A_{tr}, R_{in}, and R_{out} using equivalent circuits and Eqs. (17.35), (17.36), (17.37), and (17.42).

Assumptions: $V_{BE} = 0.7$ V; small-signal conditions apply, $V_T = 25$ mV; midband analysis is desired; $\beta_o = \beta_F$; $y_{21}^A \gg y_{21}^F$; $y_{12}^F \gg y_{12}^A$

dc Analysis: The analysis begins with determination of the Q-point for the dc equivalent circuit in Fig. 17.15. Writing a loop equation following the dashed line,

$$V_{CC} = (I_C + I_B)R_C + I_B R_F + V_{BE} \qquad \text{and} \qquad I_C = \beta_F I_B$$

Solving for the collector current yields

$$I_C = \frac{V_{CC} - V_{BE}}{R_C + \dfrac{R_C + R_F}{\beta_F}} = \frac{3.3 - 0.7}{2 \text{ k}\Omega + \dfrac{2 \text{ k}\Omega + 100 \text{ k}\Omega}{150}} = 0.970 \text{ mA}$$

The collector-emitter voltage is

$$V_{CE} = V_{CC} - (I_C + I_B)R_C = 3.3 \text{ V} - (0.977 \text{ mA})2 \text{ k}\Omega = 1.35 \text{ V}$$

For these Q-point values, the small-signal parameters are

$$g_m = 40(0.977 \text{ mA}) = 39.1 \text{ mS} \qquad r_\pi = \frac{150}{g_m} = 3.84 \text{ k}\Omega \qquad r_o = \frac{50 \text{ V} + 1.35 \text{ V}}{0.977 \text{ mA}} = 52.6 \text{ k}\Omega$$

Transresistance Analysis: The first step in the ac small-signal analysis is to make sure that the amplifier is indeed a shunt-shunt configuration. The midband ac equivalent circuit of the feedback amplifier is redrawn in Fig. 17.16, clearly identifying the two interconnected two-ports that compose the open-loop amplifier and feedback network.

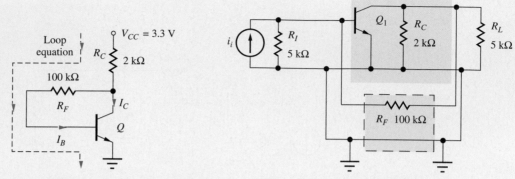

Figure 17.15 dc Equivalent circuit.

Figure 17.16 Amplifier decomposed into two-ports.

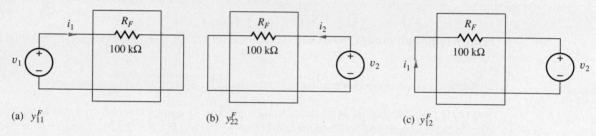

(a) y_{11}^F (b) y_{22}^F (c) y_{12}^F

Figure 17.17 Circuits for finding the y-parameters of the feedback network.

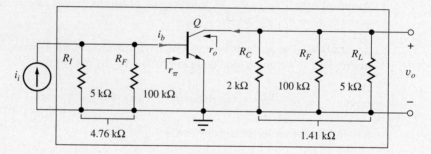

Figure 17.18 The augmented open-loop amplifier circuit (the A-circuit).

The parameters y_{11}^F, y_{21}^F, and y_{12}^F are next found by applying the y-parameter definitions to the feedback network, as in the circuits of Fig. 17.17.

$$A: \quad y_{11}^F = \left.\frac{\mathbf{i_1}}{\mathbf{v_1}}\right|_{\mathbf{v_2}=0} = \frac{1}{R_F} = 10^{-5}\text{ S}$$

$$B: \quad y_{22}^F = \left.\frac{\mathbf{i_2}}{\mathbf{v_2}}\right|_{\mathbf{v_1}=0} = \frac{1}{R_F} = 10^{-5}\text{ S}$$

$$C: \quad y_{12}^F = \left.\frac{\mathbf{i_1}}{\mathbf{v_2}}\right|_{\mathbf{v_1}=0} = -\frac{1}{R_F} = -10^{-5}\text{ S}$$

The A-circuit is constructed in Fig. 17.18 by placing R_I and $[y_{11}^F]^{-1} = R_F$ in parallel with the amplifier input and R_L and $[y_{22}^F]^{-1} = R_F$ in parallel with the amplifier output. The transresistance of the augmented open-loop amplifier is then found from the figure. Applying current division at the input,

$$\mathbf{i_b} = \mathbf{i_i} \frac{4.76 \text{ k}\Omega}{4.76 \text{ k}\Omega + r_\pi} \qquad \text{and} \qquad \mathbf{v_o} = -\beta_o \mathbf{i_b}(1.41 \text{ k}\Omega \| r_o)$$

Solving for A yields

$$A = \frac{\mathbf{v_o}}{\mathbf{i_i}} = -\frac{4.76 \text{ k}\Omega}{4.76 \text{ k}\Omega + 3.84 \text{ k}\Omega}(150)(1.41 \text{ k}\Omega \| 52.6 \text{ k}\Omega) = -114 \text{ k}\Omega$$

The feedback factor is $\beta = y_{12}^F = 10^{-5}$ S. Using these results, the closed-loop transresistance is

$$A_{tr} = \frac{A}{1 + A\beta} = \frac{-114 \text{ k}\Omega}{1 - 114 \text{ k}\Omega(-0.01 \text{ mS})} = -53.3 \text{ k}\Omega$$

Input and Output Resistances: We see from Eqs. (17.37) and (17.42) that we need to divide the input and output resistances of the augmented amplifier by $(1 + A\beta)$:

$$R_{\text{in}} = \frac{R_{\text{in}}^A}{1 + A\beta} \qquad \text{and} \qquad R_{\text{out}} = \frac{R_{\text{out}}^A}{1 + A\beta}$$

From Fig. 17.18 we have

$$R_{\text{in}}^A = R_I \| R_F \| r_\pi = 5 \text{ k}\Omega \| 100 \text{ k}\Omega \| 3.84 \text{ k}\Omega = 2.13 \text{ k}\Omega$$

and

$$R_{\text{out}}^A = R_L \| R_F \| R_C \| r_o = 5 \text{ k}\Omega \| 100 \text{ k}\Omega \| 2 \text{ k}\Omega \| 52.6 \text{ k}\Omega = 1.37 \text{ k}\Omega$$

The resulting values of overall input and output resistance are

$$R_{\text{in}} = \frac{2130 \ \Omega}{2.14} = 995 \ \Omega \qquad \text{and} \qquad R_{\text{out}} = \frac{1370 \ \Omega}{2.14} = 640 \ \Omega$$

Check of Results: The ideal value of transresistance is $A_{tr} = 1/\beta = R_F = 100 \text{ k}\Omega$. However, in this case, the loop gain is low, $A\beta = 1.14$, so the transresistance of the amplifier differs significantly from the ideal value of 100 kΩ. The reduced value is consistent with the low value of loop gain. Similarly, the closed-loop input and output resistances are reduced by only a factor of 2 compared to their open-loop values.

Discussion: The loop gain is low because the designer is trying to achieve too high a value of transresistance with this single-transistor amplifier.

Computer-Aided Analysis: After drawing the circuit in Fig. 17.13 with the SPICE schematic editor, we set the generic BJT parameters to BF = 150 and VAF = 50 V. We *cannot* use the

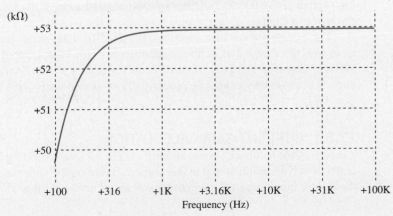

SPICE transfer function analysis capability to analyze this circuit since it is ac coupled! Instead, we must use an ac analysis to find the gain from source i_I to output voltage v_O (i.e., transresistance of the amplifier). For this calculation, the ac amplitude of current source i_I is set to 1 A, and the simulation parameters were chosen as FSTART = 100 Hz, FSTOP = 100 kHz, and 10 points per frequency decade so that we can be sure that we find the midband value of A_{tr}.

The Bode magnitude plot from SPICE yields a midband transresistance of 53.0 kΩ. The two coupling capacitors cause the transresistance to decrease for frequencies below 3 kHz. The minor disagreement with our hand calculation is due to the differences in V_T, V_{BE}, and I_C and their effect on g_m and r_π.

A similar plot of the input voltage yields an input resistance of 1028 Ω, which is slightly higher than our hand calculations. This value is sensitive to the exact value of r_π used in the calculations. The output resistance is found by driving the output node with an ac current source with an amplitude of 1 A. The output voltage then represents the output resistance and is found to be 645 Ω, in agreement with our hand calculations. Note that this value is much less sensitive to the transistor parameter values (i.e., r_o).

EXERCISE: Calculate y_{21}^A, y_{21}^F, and y_{12}^A for the shunt-shunt feedback amplifier in Fig. 17.13. Use the numeric results from the example. Compare y_{21}^A and y_{21}^F.

ANSWERS: −39.1 mS, −10^{-5} S, 0; $y_{21}^A \gg y_{21}^F$

EXERCISE: What are A, β, A_{tr}, R_{in}, and R_{out} for the transresistance amplifier in Ex. 17.2 if the values of R_C and V_{CC} are changed to 8.6 kΩ and 10 V, respectively? Assume the Q-point values do not change.

ANSWERS: −241 kΩ; −0.01 S; −70.7 kΩ, 623 Ω, 849 Ω

17.5 CURRENT AMPLIFIERS—SHUNT-SERIES FEEDBACK

Current amplifiers are another useful category of amplifiers; we encountered the most common application of open-loop current amplifiers in the form of current mirrors. By using feedback, we can produce a current amplifier that is more ideal than the basic current mirror. Shunt feedback at the input port produces a very low input resistance, and series feedback at the output achieves a very high output resistance. In fact, the high output resistances of the Wilson and cascode current mirrors, discussed in Chapter 15, are due to feedback in the circuit.

The g-parameters will be used for analysis of the shunt-series feedback amplifiers (see Sec. 10.5.1 to review the discussion of the g-parameters.) However, a word of caution is appropriate here. We will discover that it is often difficult to apply the shunt-series two-port theory to transistor circuits because they cannot be properly drawn as two-ports. This problem is addressed in detail in Sec. 17.7.

17.5.1 CURRENT GAIN CALCULATION

Analysis of the feedback current amplifier in Fig. 17.19 mirrors those presented in Secs. 17.3 and 17.4 and will only be summarized in this section. Based on the connections at the input and output ports, the overall input current $\mathbf{i_1}$ and output voltage $\mathbf{v_2}$ can be written as

$$\mathbf{i_1} = \mathbf{i_1^A} + \mathbf{i_1^F} \qquad \text{and} \qquad \mathbf{v_2} = \mathbf{v_2^A} + \mathbf{v_2^F} \tag{17.43}$$

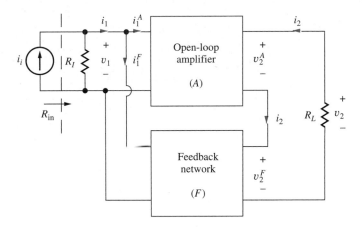

Figure 17.19 Shunt-series feedback amplifier.

For this case, the amplifier and feedback network are represented by their individual g-parameters. Assuming $g_{21}^A \gg g_{21}^F$ and $g_{12}^F \gg g_{12}^A$ yields

$$\mathbf{i_1} = g_{11}^T \mathbf{v_1} + g_{12}^F \mathbf{i_2}$$
$$\mathbf{v_2} = g_{21}^A \mathbf{v_1} + g_{22}^T \mathbf{i_2} \tag{17.44}$$

where

$$g_{ij}^T = g_{ij}^A + g_{ij}^F \tag{17.45}$$

Substituting $\mathbf{i_1} = \mathbf{i_i} - \mathbf{v_1} G_I$ and $\mathbf{v_2} = -\mathbf{i_2} R_L$ into Eq. (17.44) yields

$$\mathbf{i_i} = \left(G_I + g_{11}^T\right)\mathbf{v_1} + g_{12}^F \mathbf{i_2}$$
$$0 = g_{21}^A \mathbf{v_1} + \left(g_{22}^T + R_L\right)\mathbf{i_2} \tag{17.46}$$

The closed-loop current gain is found directly from Eq. (17.46):

$$A_i = \frac{\mathbf{i_2}}{\mathbf{i_i}} = \frac{g_{21}^A}{g_{21}^A g_{12}^F - \left(G_I + g_{11}^T\right)\left(g_{22}^T + R_L\right)} \tag{17.47}$$

Rearranging Eq. (17.47) into the standard form for a feedback amplifier gives

$$A_i = \frac{\dfrac{-g_{21}^A}{\left(G_I + g_{11}^T\right)\left(g_{22}^T + R_L\right)}}{1 + \dfrac{-g_{21}^A}{\left(G_I + g_{11}^T\right)\left(g_{22}^T + R_L\right)} g_{12}^F} = \frac{A}{1 + A\beta} \tag{17.48}$$

in which

$$A = -\frac{g_{21}^A}{\left(G_I + g_{11}^T\right)\left(g_{22}^T + R_L\right)} \quad \text{and} \quad \beta = g_{12}^F \tag{17.49}$$

Figure 17.20 presents the interpretation of Eqs. (17.48) and (17.49). The forward current gain of the amplifier should be calculated using the A-circuit in Fig. 17.20(b), in which the original amplifier is augmented by absorbing g_{11}^F and g_{22}^F of the feedback network as well as R_I and R_L, and the feedback factor is simply $\beta = g_{12}^F$. Based on the g-parameter definitions, the A-circuit and three g-parameters of the feedback network are found using the circuits in Fig. 17.21.

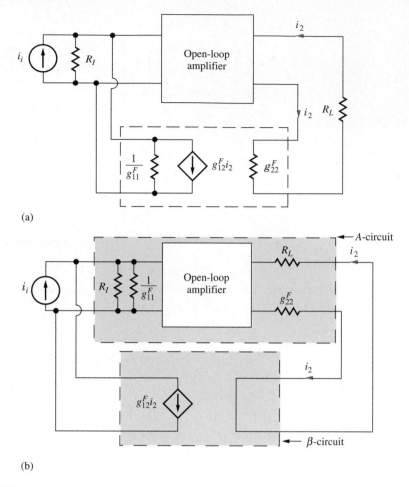

(a)

(b)

Figure 17.20 Schematic interpretation of the amplifier and feedback circuits described by Eqs. (17.48) and (17.49).

17.5.2 INPUT RESISTANCE

The input resistance of the closed-loop shunt-series feedback amplifier can be calculated from the two-port description in Eq. (17.46) for the overall amplifier in Fig. 17.19. Solving for $\mathbf{i_i}$ in terms of $\mathbf{v_1}$:

$$\mathbf{i_i} = \left(G_I + g_{11}^T\right)\mathbf{v_1} + g_{12}^F \frac{-g_{21}^A}{\left(g_{22}^T + R_L\right)}\mathbf{v_1} \tag{17.50}$$

which can be rearranged as

$$R_{\text{in}} = \frac{\mathbf{v_1}}{\mathbf{i_i}} = \frac{1}{\left(G_I + g_{11}^T\right)\left[1 + \dfrac{-g_{21}^A}{\left(G_I + g_{11}^T\right)\left(g_{22}^T + R_L\right)}g_{12}^F\right]} = \frac{\left(\dfrac{1}{G_I + g_{11}^T}\right)}{1 + A\beta} = \frac{R_{\text{in}}^A}{1 + A\beta} \tag{17.51}$$

Once again, we see that shunt feedback at a port reduces the resistance at the port by the factor $(1 + A\beta)$. For an ideal amplifier with loop gain approaching infinity, the input resistance of the closed-loop current amplifier approaches zero.

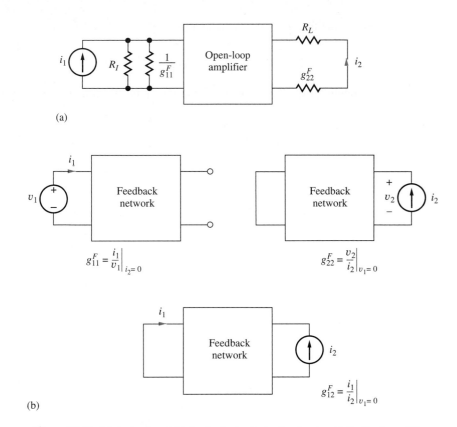

(a)

(b)

Figure 17.21 (a) A-circuit and (b) feedback circuits for the shunt-series feedback amplifier.

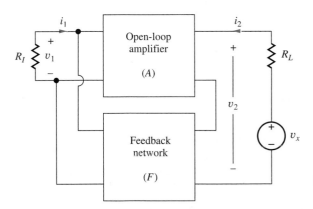

Figure 17.22 Output resistance of the shunt-series feedback amplifier.

17.5.3 OUTPUT RESISTANCE

The output resistance of the closed-loop amplifier can be calculated by applying a test source v_x to the output of the amplifier, as in Fig. 17.22. For this circuit, the port voltages and currents are related by

$$\mathbf{i_1} = -\mathbf{v_1} G_I \qquad \text{and} \qquad \mathbf{v_2} = \mathbf{v_x} - \mathbf{i_2} R_L \tag{17.52}$$

Substituting Eq. (17.52) into (17.44) gives

$$0 = \left(G_I + g_{11}^T\right)\mathbf{v_1} + g_{12}^F \mathbf{i_2}$$
$$\mathbf{v_x} = g_{21}^A \mathbf{v_1} + \left(g_{22}^T + R_L\right)\mathbf{i_2} \tag{17.53}$$

and solving for $\mathbf{v_x}$ in terms of $\mathbf{i_2}$ yields

$$\mathbf{v_x} = g_{21}^A \frac{-g_{12}^F}{\left(G_I + g_{11}^T\right)} \mathbf{i_2} + \left(g_{22}^T + R_L\right)\mathbf{i_2} \qquad (17.54)$$

Rearranging Eq. (17.54) yields an expression for the output resistance of the overall amplifier:

$$R_{\text{out}} = \frac{\mathbf{v_x}}{\mathbf{i_2}} = \left(g_{22}^T + R_L\right)\left[1 + \frac{-g_{21}^A}{\left(G_I + g_{11}^T\right)\left(g_{22}^T + R_L\right)} g_{12}^F\right] = \left(g_{22}^T + R_L\right)(1 + A\beta)$$

or

$$R_{\text{out}} = R_{\text{out}}^A(1 + A\beta) \qquad (17.55)$$

The output resistance of the closed-loop amplifier is equal to the output resistance of the A-circuit increased by the factor $(1 + A\beta)$. As the loop gain approaches infinity in an ideal amplifier, the output resistance of the current amplifier also approaches infinity.

Common errors occur in applying the two-port theory to transistor amplifiers, particularly in circuits that use series feedback at the output port. We delay presentation of examples until Sec. 17.7, following completion of the development of the mathematical description of the series-series feedback amplifier in Sec. 17.6.

17.6 TRANSCONDUCTANCE AMPLIFIERS—SERIES-SERIES FEEDBACK

The final configuration considered is the **transconductance amplifier,** which produces an output current proportional to the input voltage. Thus, it should have very high input resistance as well as very high output resistance. To achieve these characteristics, series feedback is utilized at both the input and output ports, as in Fig. 17.23. For this case, the input port currents are equal and the output port currents are equal for the amplifier and feedback two-ports; z-parameters are appropriate for analyzing this configuration.

17.6.1 TRANSCONDUCTANCE ANALYSIS

For the circuit in Fig. 17.23, the overall input voltage $\mathbf{v_1}$ and output voltage $\mathbf{v_2}$ can be written as

$$\mathbf{v_1} = \mathbf{v_1^A} + \mathbf{v_1^F} \qquad \text{and} \qquad \mathbf{v_2} = \mathbf{v_2^A} + \mathbf{v_2^F} \qquad (17.56)$$

and the z-parameter description of the overall circuit is

$$\mathbf{v_1} = z_{11}^T \mathbf{i_1} + z_{12}^T \mathbf{i_2}$$
$$\mathbf{v_2} = z_{21}^T \mathbf{i_1} + z_{22}^T \mathbf{i_2} \qquad (17.57)$$

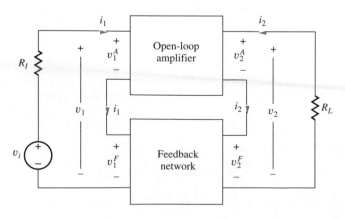

Figure 17.23 Series-series feedback amplifier (transconductance amplifier).

where
$$z_{ij}^T = z_{ij}^A + z_{ij}^F \tag{17.58}$$

Using $\mathbf{v_1} = \mathbf{v_i} - \mathbf{i_1}R_I$ and $\mathbf{v_2} = -\mathbf{i_2}R_L$, and assuming $z_{21}^A \gg z_{21}^F$ and $z_{12}^F \gg z_{12}^A$, yields the standard simplified form:

$$\mathbf{v_i} = \left(R_S + z_{11}^T\right)\mathbf{i_1} + z_{12}^F\mathbf{i_2}$$
$$0 = z_{21}^A\mathbf{i_1} + \left(z_{22}^T + R_L\right)\mathbf{i_2} \tag{17.59}$$

The closed-loop gain of the transconductance amplifier can be found from Eq. (17.59):

$$A_{tc} = \frac{\mathbf{i_2}}{\mathbf{v_i}} = \frac{\dfrac{-z_{21}^A}{\left(R_I + z_{11}^T\right)\left(z_{22}^T + R_L\right)}}{1 + \dfrac{-z_{21}^A}{\left(R_I + z_{11}^T\right)\left(z_{22}^T + R_L\right)}z_{12}^F} = \frac{A}{1 + A\beta} \tag{17.60}$$

in which

$$A = -\frac{z_{21}^A}{\left(R_I + z_{11}^T\right)\left(z_{22}^T + R_L\right)} \qquad \text{and} \qquad \beta = z_{12}^F \tag{17.61}$$

Figure 17.24 is the schematic interpretation of Eqs. (17.60) and (17.61). The forward transconductance of the amplifier should be calculated using the A-circuit in which the original amplifier is augmented by absorbing z_{11}^F and z_{22}^F of the feedback network as well as R_I and R_L, and the feedback factor is given by $\beta = z_{12}^F$.

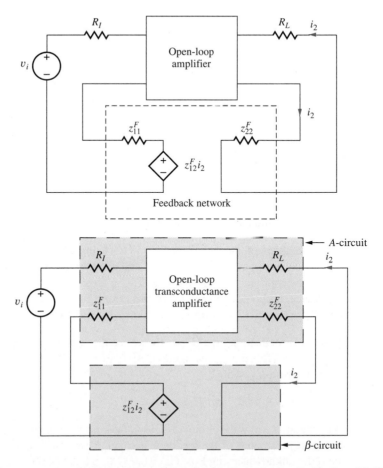

Figure 17.24 Schematic interpretation of the amplifier and feedback circuits described by Eqs. (17.60) and (17.61).

17.6.2 INPUT AND OUTPUT RESISTANCES

The input and output resistances of the closed-loop series-series feedback amplifier can be calculated from the two-port description for the overall amplifier in a manner similar to that used to derive results in the previous three sections:

$$R_{\text{in}} = \frac{\mathbf{v_i}}{\mathbf{i_1}} = \left(R_I + z_{11}^T\right)\left[1 + \frac{-z_{21}^A}{\left(R_I + z_{11}^T\right)\left(z_{22}^T + R_L\right)}z_{12}^F\right] \qquad (17.62)$$

$$R_{\text{in}} = \left(R_I + z_{11}^T\right)(1 + A\beta) = R_{\text{in}}^A(1 + A\beta) \qquad (17.63)$$

and

$$R_{\text{out}} = \frac{\mathbf{v_x}}{\mathbf{i_2}} = \left(z_{22}^T + R_L\right)\left[1 + \frac{-z_{21}^A}{\left(R_I + z_{11}^T\right)\left(z_{22}^T + R_L\right)}z_{12}^F\right] \qquad (17.64)$$

$$R_{\text{out}} = \left(z_{22}^T + R_L\right)(1 + A\beta) = R_{\text{out}}^A(1 + A\beta) \qquad (17.65)$$

As we should expect by now, series feedback increases the impedance levels at both ports by the factor $(1 + A\beta)$. For very large loop gain, the input and output resistances of the closed-loop transconductance amplifier both approach infinity.

17.7 COMMON ERRORS IN APPLYING TWO-PORT FEEDBACK THEORY

Great care must be exercised in applying the two-port theory to ensure that the amplifier and feedback networks can actually be represented as two-ports. This is particularly true for the case of amplifiers that appear to use series feedback at the output port. Many popular textbooks incorrectly apply the feedback theory to these amplifiers because a simple relationship seems to relate the output current to the feedback current. The best way to illustrate the problem is through an example that produces erroneous results.

EXAMPLE 17.3   *ERRONEOUS* APPLICATION OF TWO-PORT FEEDBACK THEORY

We use the basic implementation of the transconductance amplifier in Fig. 17.25 as an example of problems that can occur in the application of the two-port feedback theory, particularly in series-series and shunt-series feedback amplifiers.

PROBLEM Direct analysis of the circuit in Fig. 17.25 using ideal op amp theory indicates that the circuit develops an output current equal to (v_{REF}/R) with an infinite input resistance and an output resistance approaching $\beta_o r_o$ of the BJT. Use feedback theory to find the transresistance A_{tc}, input resistance, and output resistance of this circuit for an op amp with $A_o = 10,000$, $R_{id} = 25$ kΩ, and $R_o = 0$, and a BJT with $\beta_o = 100$ and $V_A = 50$ V.

SOLUTION **Known Information and Given Data:** Feedback amplifier circuit in Fig. 17.25 with $V_{\text{REF}} = 5$ V and $R = 5$ kΩ; op amp parameters: $A_o = 10,000$, $R_{id} = 25$ kΩ, and $R_o = 0$; BJT parameters: $\beta_o = 100$ and $V_A = 50$ V.

Unknowns: A_{tc}, R_{in}, and R_{out}

Approach: Redraw the circuit to identify the amplifier and feedback networks. Find the appropriate two-port parameters of the feedback network. Draw and analyze the augmented amplifier circuit. Find the unknowns using Eqs. (17.60), (17.63), and (17.65).

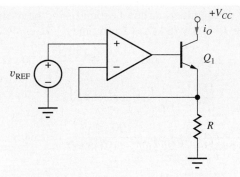

Figure 17.25 "Simple" series-series feedback amplifier ($V_{\text{REF}} = 5$ V, $R = 5$ kΩ).

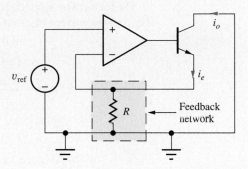

Figure 17.26 Feedback representation of the circuit in Fig. 17.25.

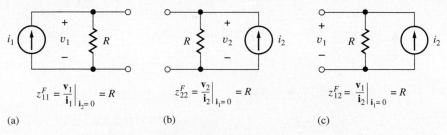

$$z_{11}^F = \left.\frac{\mathbf{v}_1}{\mathbf{i}_1}\right|_{\mathbf{i}_2=0} = R \qquad z_{22}^F = \left.\frac{\mathbf{v}_2}{\mathbf{i}_2}\right|_{\mathbf{i}_1=0} = R \qquad z_{12}^F = \left.\frac{\mathbf{v}_1}{\mathbf{i}_2}\right|_{\mathbf{i}_1=0} = R$$

(a) (b) (c)

Figure 17.27 z-Parameter calculations for the feedback network.

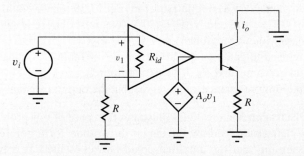

Figure 17.28 A-circuit, including the model for the op amp.

Assumptions: Series-series feedback theory can be applied to this amplifier. Q_1 is in the active region.

Analysis: As a first step, the amplifier in Fig. 17.25 is redrawn in Fig. 17.26 to emphasize the feedback network, which consists in this case of just resistor R. This appears to be a simple case of series-series feedback. A minor problem exists, however. The current sampled by the feedback network is i_e rather than the actual output current i_o, but because $\alpha_o \cong 1$ and $i_e \cong i_o$, this does not seem to be a major problem. But as we shall see, *it is!*

Let us go ahead and apply the series-series feedback theory to the amplifier. The required z-parameters for the feedback network are found from the three networks in Fig. 17.27, and the A-circuit is shown in Fig. 17.28, including the two-port representation of the operational amplifier.

The forward transconductance of the amplifier can now be found using the A-circuit in Fig. 17.28. The emitter current of Q_1 should be $I_E = 5 \text{ V}/5 \text{ k}\Omega = 1 \text{ mA}$, and

$$r_\pi \cong \frac{100(0.025 \text{ V})}{1 \text{ mA}} = 2.5 \text{ k}\Omega \qquad \text{and} \qquad r_o \cong \frac{50 \text{ V}}{1 \text{ mA}} = 50 \text{ k}\Omega$$

$$\mathbf{i_o} = \mathbf{v_i} \frac{R_{id}}{R_{id} + R} A_o \frac{\beta_o}{r_\pi + (\beta_o + 1)R} \qquad \text{neglecting } r_0$$

$$A = \frac{\mathbf{i_o}}{\mathbf{v_i}} = \frac{25 \text{ k}\Omega}{25 \text{ k}\Omega + 5 \text{ k}\Omega}(10^4)\frac{100}{2.5 \text{ k}\Omega + (101)5 \text{ k}\Omega} = 1.64 \text{ S}$$

$$R_{\text{in}}^A = R_{id} + R = 30 \text{ k}\Omega$$

$$R_{\text{out}}^A = r_o\left(1 + \frac{g_m R}{1 + \dfrac{R}{r_\pi}}\right) = 50 \text{ k}\Omega\left(1 + \frac{0.04 \text{ S}(5 \text{ k}\Omega)}{1 + \dfrac{5 \text{ k}\Omega}{2.5 \text{ k}\Omega}}\right) = 3.38 \text{ M}\Omega$$

Using these results, the closed-loop feedback amplifier predictions are

$$A_{tc} = \frac{\mathbf{i_o}}{\mathbf{v_i}} = \frac{A}{1 + A\beta} = \frac{1.64 \text{ S}}{1 + 1.64 \text{ S}(5 \text{ k}\Omega)} = \frac{1.64 \text{ S}}{8200} = 0.200 \text{ mS}$$

$$R_{\text{in}} = R_{\text{in}}^A(1 + A\beta) = 30 \text{ k}\Omega(8200) = 246 \text{ M}\Omega$$

$$R_{\text{out}} = R_{\text{out}}^A(1 + A\beta) = 3.38 \text{ M}\Omega(8200) = 27.7 \text{ G}\Omega$$

Check of Results Using Computer-Aided Analysis: Now, suppose we routinely check our hand calculations with SPICE. In this case, we can replace the op amp with the simple two-element (R_{id} and VCVS) circuit model in Fig. 17.28 (or we could use a built-in op amp component). We set BF = 100 and VAF = 50 V. We must choose a value of V_{CC} to keep the transistor in the forward-active region ($V_{CC} > 5.7$ V). $V_{CC} = 7.5$ V was used here. Two transfer function analyses are used to generate the SPICE results in Table 17.1. The first requests the TF from V_{REF} to V_{CC} and gives A_{tc} and R_{in}. The second from V_{CC} to V_{CC} yields R_{out}. We see that the transconductance and input resistance values agree, but the output resistance is off by a factor of nearly 5000!

Discussion: Figure 17.29 illustrates the cause of this problem. We (deliberately) violated our rule of being sure to draw the circuit in two-port form before proceeding with the analysis, blindly assuming that the amplifier could be represented as a two-port network. The problem occurs because the output of the op amp is referenced to ground, and the base current of the BJT escapes from the output port (and feedback loop). This base current loss limits the output resistance of the overall circuit to approximately

$$R_{\text{out}} \le \beta_o r_o = (100)50 \text{ k}\Omega = 5.00 \text{ M}\Omega^2$$

TABLE 17.1
Series-Series Feedback Amplifier

	TWO-PORT THEORY	SPICE
A_{tc}	0.200 mS	0.198 mS
R_{in}	246 MΩ	249 MΩ
R_{out}	27.7 GΩ	5.45 MΩ—*Oops!*

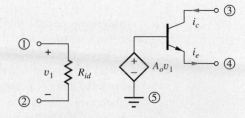

Figure 17.29 Five distinct terminals of the amplifier.

---

[2] The actual values used by SPICE are $\beta_o r_o = 112$ (56.7 kΩ) = 6.35 MΩ.

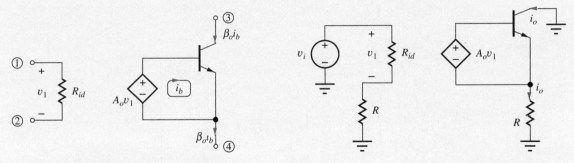

Figure 17.30 A different amplifier circuit that can be represented as a two-port.

Figure 17.31 The A-circuit using the amplifier in Fig. 17.30.

Terminals 3 and 4 do not represent valid terminals of a two-port network because the current entering terminal 3 is not equal to that exiting terminal 4. The amplifier has five distinct terminals, not four, and cannot be reduced to a two-port.

If the output of the op amp could somehow be referenced to the emitter of the transistor, as in the hypothetical A-circuit shown in Fig. 17.30, then the forward amplifier could be properly represented as a two-port. Figure 17.31 gives the new A-circuit for this hypothetical case; the new values for the A-circuit are

$$\mathbf{i_o} = \beta_o \mathbf{i_b} = v_i \frac{R_{id}}{R_{id} + R} \left(\frac{A_o}{r_\pi} \right) \beta_o$$

$$A = \frac{\mathbf{i_o}}{\mathbf{v_i}} = \frac{25 \text{ k}\Omega}{25 \text{ k}\Omega + 5 \text{ k}\Omega} \frac{10^4}{2.5 \text{ k}\Omega} 100 = 333 \text{ S}$$

$$R_{in}^A = R_{id} + R = 30 \text{ k}\Omega \quad \text{and} \quad R_{out}^A = r_o + R = 55 \text{ k}\Omega$$

Note the large changes in the values of the forward gain and output resistance. The closed-loop feedback amplifier parameters become

$$A_{tc} = \frac{\mathbf{i_o}}{\mathbf{v_i}} = \frac{A}{1 + A\beta} = \frac{333 \text{ S}}{1 + 333 \text{ S}(5 \text{ k}\Omega)} = \frac{333 \text{ S}}{1.67 \times 10^6} = 0.200 \text{ mS}$$

$$R_{in} = R_{in}^A (1 + A\beta) = 30 \text{ k}\Omega (1.67 \times 10^6) = 50.1 \text{ G}\Omega$$

$$R_{out} = R_{out}^A (1 + A\beta) = 55 \text{ k}\Omega (1.67 \times 10^6) = 91.9 \text{ G}\Omega$$

SPICE simulation of this circuit yields $A_{tc} = 0.2$ mS, $R_{in} = 42.3$ GΩ, and $R_{out} = 87.3$ GΩ. We see that the theory and SPICE simulation are now in reasonable agreement.

From the results in Table 17.1, the question of why A_{tc} and R_{in} are approximately correct naturally arises, and this is addressed in Fig. 17.32. As far as R_{in} and A_{tc} are concerned, the amplifier can be properly represented as the series-shunt feedback amplifier in Fig. 17.32, because the collector of Q_1 can be connected directly to ground for these calculations, and a valid two-port representation exists. Thus, the A_{tc} and R_{in} calculations are not in error. Because $i_o = \alpha_o i_e$, and $\alpha_o \cong 1$, the transconductance from input to i_e and from input to i_o have essentially the same value.

One final comment before we leave this example. If transistor Q_1 were replaced by a MOSFET, then the improperly applied feedback analysis would *appear* to work correctly. Consider the new A-circuit in Fig. 17.33. Because the current in the gate terminal is zero at dc, no current escapes through the ground terminal, and the calculations appear to be correct (however, they are still actually imprecise). Although the improper two-port analysis seems to give correct answers at dc,

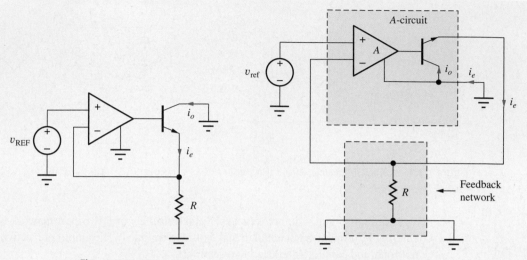

Figure 17.32 Circuit of Fig. 17.25 represented as a series-shunt feedback amplifier.

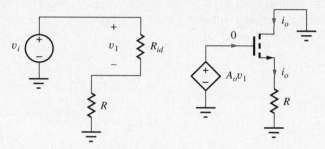

Figure 17.33 Transconductance amplifier with the BJT replaced by a MOSFET.

the analysis is still incorrect at higher frequencies for which the gate current is no longer zero due to currents through C_{GS} and C_{GD}. Significant errors can also occur if the MOSFET is modeled as a four-terminal device with the substrate connected to ac ground.

EXERCISE: Suppose resistor R in Fig. 17.25 is replaced by an ideal current source, and the op amp is ideal. Draw the small-signal equivalent circuit, and show that the output resistance approaches $R_{\text{out}} \cong \beta_o r_o$.

EXAMPLE **17.4** ANALYSIS OF THE SHUNT-SERIES FEEDBACK PAIR

A popular form of transconductance amplifier is the dc coupled shunt-series feedback pair in Fig. 17.34. In trying to apply our basic rule, however, we find that this is another example of an amplifier that cannot be drawn as a pair of two-ports in a shunt-series configuration.

PROBLEM Use feedback theory to analyze the shunt-series feedback pair in Fig. 17.34. Find the gain, input resistance, and output resistance. Use $\beta_o = 100$ and $V_A = 100$ V with these Q-points, Q_1: (0.66 mA, 2.3 V) and Q_2: (1.6 mA, 7.5 V).

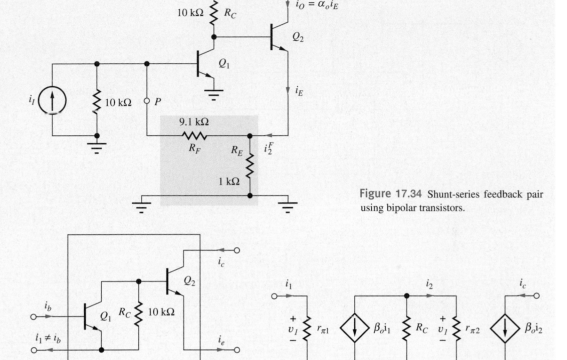

Figure 17.34 Shunt-series feedback pair using bipolar transistors.

Figure 17.35 (a) Failed attempt to represent the amplifier as a two-port. (b) Small-signal model showing common terminal between input and output.

SOLUTION

Known Information and Given Data: Shunt-series feedback amplifier circuit in Fig. 17.34; BJT parameters: $\beta_o = 100$ and $V_A = 100$ V; Q-point for Q_1: (0.66 mA, 2.3 V); Q-point for Q_2: (1.6 mA, 7.5 V)

Unknowns: A_{tc}, R_{in}, and R_{out}

Approach: Redraw the circuit to identify the amplifier and feedback networks. Find the appropriate two-port parameters of the feedback network. Draw and analyze the augmented amplifier circuit. Find the unknowns using the appropriate feedback equations.

Assumptions: Room temperature operation; active region operation; small-signal conditions apply.

Analysis: First we must carefully draw the ac-equivalent circuit as a pair of two-ports. As found in Ex. 17.3, however, we cannot draw the amplifier as a pair of two-ports in a shunt-series configuration. In attempting to define the ports as indicated in Fig. 17.35, we see that the currents entering the ports are not equal to those leaving the ports. The problem is even clearer in the small-signal model in Fig. 17.35(b).

However, the feedback pair can be properly represented as the shunt-shunt transresistance configuration in Fig. 17.36. Thus, we can use the y-parameter theory to find the transresistance and input resistance, but we cannot properly calculate the output resistance of the transconductance

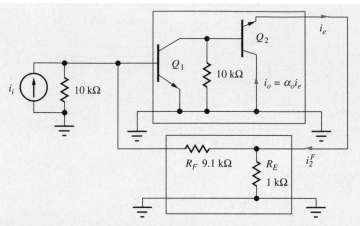

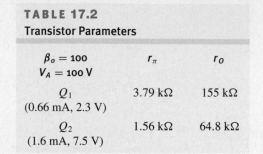

Figure 17.36 Successful representation of the circuit in Fig. 17.34 as a shunt-shunt feedback amplifier.

TABLE 17.2
Transistor Parameters

$\beta_o = 100$ $V_A = 100\text{ V}$	r_π	r_O
Q_1 (0.66 mA, 2.3 V)	3.79 kΩ	155 kΩ
Q_2 (1.6 mA, 7.5 V)	1.56 kΩ	64.8 kΩ

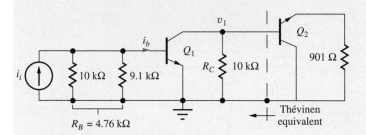

Figure 17.37 *A*-circuit for the shunt-series feedback pair.

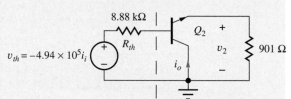

Figure 17.38 Simplified *A*-circuit using a Thévenin equivalent circuit.

configuration using the two-port theory when the output is defined at the collector. Therefore, let us partially analyze the circuit using the shunt-shunt feedback theory and compare the results to SPICE. SPICE and additional hand calculations can be used to find the output resistance.

The Q-points were given, and the corresponding small-signal parameters appear in Table 17.2. The *A*-circuit for the amplifier is given in Fig. 17.37, in which $[y_{11}^F]^{-1}$ and $[y_{22}^F]^{-1}$ appear in parallel with the input and output of the amplifier. We find the transresistance of the amplifier by finding Thévenin equivalent of the first stage, as indicated in Figs. 17.37 and 17.38. Cutting the circuit at the dashed line in Fig. 17.37 and finding the open-circuit voltage at v_1:

$$\mathbf{v_{th}} = -\beta_{o1}\mathbf{i_b}(r_{o1}\|R_C) \quad \text{and} \quad \mathbf{i_b} = \mathbf{i_i}\frac{R_B}{R_B + r_{\pi1}}$$

Combining these equations and evaluating:

$$\mathbf{v_{th}} = -\mathbf{i_i}\frac{R_B}{R_B + r_{\pi1}}\beta_{o1}(r_{o1}\|R_C) = -\mathbf{i_i}\frac{4.76\text{ k}\Omega}{4.76\text{ k}\Omega + 3.79\text{ k}\Omega}100(8.88\text{ k}\Omega) = -4.94\times10^5\mathbf{i_i}$$

The Thévenin equivalent resistance is $R_{th} = 10\text{ k}\Omega\|r_{o1} = 8.88\text{ k}\Omega$. The output voltage v_2 in Fig. 17.38 is expressed as

$$\mathbf{v_2} = \mathbf{v_{th}}\frac{(\beta_{o2}+1)(0.901\text{ k}\Omega)}{8.88\text{ k}\Omega + r_{\pi2} + (\beta_{o2}+1)(0.901\text{ k}\Omega)}$$

and the overall transresistance of the *A*-circuit is

$$A = \frac{\mathbf{v_2}}{\mathbf{i_i}} = -4.94\times10^5\frac{(101)(0.901\text{ k}\Omega)}{8.88\text{ k}\Omega + 1.56\text{ k}\Omega + (101)(0.901\text{ k}\Omega)} = -4.43\times10^5\ \Omega$$

The feedback factor and closed-loop transresistance are

$$\beta = y_{12}^F = -\frac{1}{9100} \text{ S} \quad \text{and} \quad (1 + A\beta) = 1 + \frac{-4.43 \times 10^5}{-9100} = 49.7$$

$$A_{tr} = \frac{A}{1 + A\beta} = \frac{-4.43 \times 10^5 \ \Omega}{49.7} = -8910 \ \Omega$$

and the closed-loop input resistance and the output resistance at the emitter of Q_2 are

$$R_{\text{in}} = \frac{R_{\text{in}}^A}{1 + A\beta} = \frac{R_B \| r_{\pi 1}}{1 + A\beta} = \frac{4.76 \text{ k}\Omega \| 3.79 \text{ k}\Omega}{49.7} = 42.5 \ \Omega$$

$$R_{\text{out}} = \frac{R_{\text{out}}^A}{1 + A\beta} = \frac{901 \ \Omega \left\| \dfrac{r_{\pi 2} + 8.88 \text{ k}\Omega}{\beta_{o2} + 1}\right.}{1 + A\beta} = \frac{901 \ \Omega \left\| \dfrac{1.56 \text{ k}\Omega + 8.88 \text{ k}\Omega}{101}\right.}{49.7} = 1.86 \ \Omega$$

In the original circuit, the desired output was actually the collector current of Q_2. The desired closed-loop current gain A_i can be found using the preceding results with a relationship between v_2 and i_o derived from the A-circuit:

$$A_i = \frac{i_o}{i_i} = \frac{\alpha_o i_e}{i_i} = \frac{\alpha_o \dfrac{v_2}{901 \ \Omega}}{i_i} = \frac{\alpha_o}{901 \ \Omega} \frac{v_2}{i_i} = \frac{\alpha_o}{901 \ \Omega} A_{tr}$$

$$A_i = \frac{0.99}{901}(-8910 \ \Omega) = -9.79$$

Check of Results: The ideal transresistance would be $-R_F = -9100 \ \Omega$, and the ideal current gain would be $1 + R_F / R_E = 10.1$. Our detailed calculations are consistent with these estimates.

Computer-Aided Analysis: The circuit shown below is drawn with the schematic editor. The BJT parameters are set to BF = 100 and VAF = 100 V. Saturation current IS defaults to 0.1 fA. VO and VCC are separated to facilitate the transfer function analyses discussed next. VIO is used to measure the emitter current of Q_2. Since the amplifier is dc-coupled, a SPICE transfer function analysis from current source II to the current in source VO can be used to find the gain, input resistance, and A_i in the circuit shown. R_{out} represents the resistance at the emitter of Q_2 and is found with a transfer function analysis from II to the emitter node. A third transfer function analysis from VO to any node can be used to find the output resistance presented to source VO. A transfer function analysis from II to zero-valued source VIO will give the current gain corresponding to the y-parameter calculations.

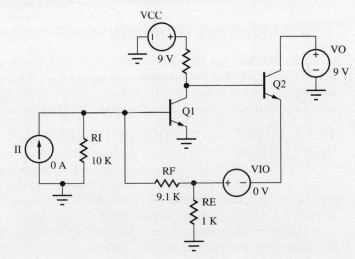

TABLE 17.3

Shunt-Series Feedback Pair Results

	SHUNT-SHUNT THEORY (y-PARAMETERS)	SPICE	SHUNT-SERIES THEORY (g-PARAMETERS)	SPICE
A_{tr}	−8910	−8925	—	—
R_{in}	42.5 Ω	41.9 Ω	41.5 Ω	41.9 Ω
R_{out}	1.86 Ω	1.70 Ω	18.0 MΩ	6.02 MΩ
A_i	9.79	9.91	9.90	9.81

Table 17.3 compares the results of our hand calculations with those of SPICE and includes similar calculations based on an *erroneous* analysis of the amplifier as a shunt-series configuration using g-parameters. The hand calculations using the shunt-shunt theory agree well with the SPICE results. We see that the g-parameter description would significantly overestimate the output resistance of the amplifier. In addition, there is a small but perceptible discrepancy in the calculated value of the current gain A_i.

Discussion: In summary, the key to analyzing feedback amplifiers using two-port theory is to be certain before beginning the analysis that the overall network can be properly represented as a pair of two-ports. When the theory is applied properly, the results will be correct. Note that the output resistance at the collector of bipolar transistor Q_2 is limited by $R_{out} \leq \beta_o r_o = 100(64.8 \text{ k}\Omega) = 6.48$ MΩ. The SPICE calculation is consistent with this bound.

EXERCISE: **Suppose the circuit in Figs. 17.34 and 17.36 is partitioned in a slightly different manner. Let R_E be part of the forward amplifier so that the shunt feedback network becomes just resistor R_F. Redraw the circuit in a manner similar to Fig. 17.36 and find the new values of A, β, and A_{tr}.**

ANSWERS: **No change: -4.43×10^5 Ω, $-(1/9100)$ S, -8910 Ω**

In Sec. 17.9, we will find that Blackman's theorem provides a way around the problems encountered with using two-port analysis. But first, we will explore methods of finding the loop gain in a feedback amplifier, since these techniques will be needed in calculations using Blackman's theorem.

17.8 FINDING THE LOOP GAIN

In the previous sections, we discovered the important role loop gain plays in feedback amplifiers. In Sec. 17.11, we will find that the stability of feedback amplifiers can be determined by analyzing the loop gain. Thus, it is important to know how to determine the loop gain directly from the circuit, not only theoretically, but computationally, using SPICE and actual circuit measurements.

17.8.1 DIRECT CALCULATION OF THE LOOP GAIN

Figure 17.39 illustrates a direct method of calculating the loop gain. The original input source v_s in Fig. 17.1 is first set to zero. Then, the feedback loop is opened at some arbitrary point, and a test source v_x is inserted into the loop. It is important to note that the output of the feedback network must be properly terminated. In Fig. 17.39, the loop has been broken at the output of the feedback network $\beta(s)$. In this case, the termination R_{IS} is equivalent to the input resistance of the summing circuit that would be connected to the feedback network when the loop is closed. (This is similar to the methods we learned using the two-port approach.)

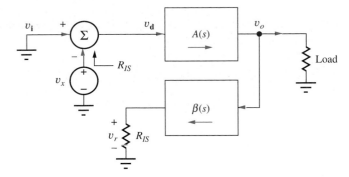

Figure 17.39 Direct calculation of the loop gain.

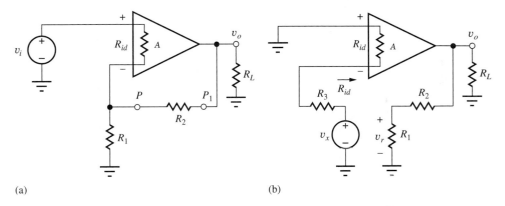

(a)   (b)

Figure 17.40 (a) Noninverting amplifier with (b) feedback loop broken at point P.

The loop gain T can now be calculated from Fig. 17.39 by finding the return voltage $\mathbf{v_r}$:

$$\mathbf{v_r} = \beta\mathbf{v_o} = \beta A(0 - \mathbf{v_x}) = -A\beta\mathbf{v_x} \quad \text{and} \quad T = A\beta = -\frac{\mathbf{v_r}}{\mathbf{v_x}} \tag{17.66}$$

From this equation, we can see that T is equal to the negative of the ratio of the voltage returned through the loop to the voltage applied. Note that the feedback loop can be broken at any point, and the answer will be the same.

As an example, consider the noninverting amplifier in Fig. 17.40. The feedback loop is broken at point P, and test source v_x is inserted into the loop. Resistor R_3 must be added in order to properly account for the loading of the feedback network on the amplifier. R_3 represents the input resistance of the feedback network and is equal to the parallel combination of R_1 and R_2:

$$R_3 = R_1 \| R_2 \tag{17.67}$$

Note that we have effectively constructed the feedback circuit representation in Fig. 17.4 for the series-shunt amplifier (with v_i set to zero) by including h_{11}^F and h_{22}^F in the circuit. For the circuit in Fig. 17.40(b),

$$\mathbf{v_r} = \frac{R_1}{R_2 + R_1}\mathbf{v_o} = \frac{R_1}{R_2 + R_1}\left(-A\mathbf{v_x}\frac{R_{id}}{R_{id} + R_3}\right) \tag{17.68}$$

and

$$T = -\frac{\mathbf{v_r}}{\mathbf{v_x}} = A\frac{R_1}{R_2 + R_1}\left(\frac{R_{id}}{R_{id} + R_3}\right) \tag{17.69}$$

EXERCISE: Break the feedback loop of the amplifier at point P_1 in Fig. 17.40 and show that loop gain can be written as

$$T = \frac{R_1 \| R_{id}}{R_2 + R_1 \| R_{id}} A$$

EXERCISE: Show that the loop gain expression in Eq. (17.69) and the expression in the previous exercise are equal.

17.8.2 FINDING THE LOOP GAIN USING SUCCESSIVE VOLTAGE AND CURRENT INJECTION

In many practical cases, particularly when the loop gain is large, the feedback loop cannot be opened to measure the loop gain because a closed loop is required to maintain a correct dc operating point. Another problem is electrical noise, which may cause an open-loop amplifier to saturate. A similar problem occurs in SPICE simulation of high-gain circuits, such as operational amplifiers, in which the circuit amplifies the numerical noise present in the calculations, and the analysis is unable to converge to a stable operating point. Fortunately, the method of **successive voltage and current injection** [1] can be used to measure the loop gain without opening the feedback loop, even in unstable systems.

Again consider the basic feedback amplifier in Fig. 17.40. To use the voltage and current injection method, an arbitrary point P within the feedback loop is selected, and a voltage source v_x is inserted into the loop, as in Fig. 17.41(a). The two voltages v_2 and v_1 on either side of the inserted source are measured, and T_v is calculated:

$$T_v = -\frac{\mathbf{v_2}}{\mathbf{v_1}} \tag{17.70}$$

Next, the voltage source is removed, a current i_x is injected into the same point P, and the ratio T_I of currents i_2 and i_1 is determined.

$$T_i = \frac{\mathbf{i_2}}{\mathbf{i_1}} \tag{17.71}$$

These two sets of measurements yield two equations in two unknowns: the loop gain T and the resistance ratio R_B/R_A. R_A represents the resistance seen looking to the left from test source v_x, and R_B represents the resistance seen looking to the right from the test source.

For the voltage injection case in Fig. 17.41(a),

$$\mathbf{v_1} = \mathbf{i} R_A = \frac{-A\mathbf{v_1} + \mathbf{v_x}}{R_A + R_B} R_A \tag{17.72}$$

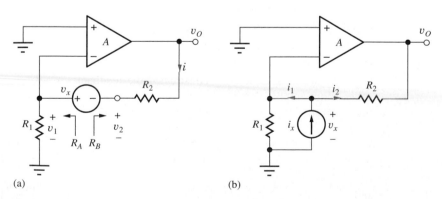

(a) (b)

Figure 17.41 (a) Voltage injection at point P and (b) current injection at point P. $R_{id} = \infty$.

Combining these two expressions yields

$$\mathbf{v_1} = \frac{\beta}{1 + A\beta}\mathbf{v_x} \qquad \text{where} \qquad \beta = \frac{R_A}{R_A + R_B} \tag{17.73}$$

After some algebra, voltage v_2 is found to be

$$\mathbf{v_2} = \mathbf{v_1} - \mathbf{v_x} = \frac{\beta - (1 + A\beta)}{1 + A\beta}\mathbf{v_x} \tag{17.74}$$

and T_v is equal to

$$T_v = \frac{1 + A\beta - \beta}{\beta} \tag{17.75}$$

We recognize the $A\beta$ product as the loop gain T, and using $1/\beta = 1 + R_B/R_A$, T_v can be rewritten as

$$T_v = T\left(1 + \frac{R_B}{R_A}\right) + \frac{R_B}{R_A} \tag{17.76}$$

The current injection circuit in Fig. 17.41(b) provides the second equation in two unknowns. Injection of current i_x causes a voltage v_x to develop across the current generator; currents i_1 and i_2 can each be expressed in terms of this voltage:

$$\mathbf{i_1} = \frac{\mathbf{v_x}}{R_A} \qquad \text{and} \qquad \mathbf{i_2} = \frac{\mathbf{v_x} - A\mathbf{v_x}}{R_B} = \mathbf{v_x}\frac{1 + A}{R_B} \tag{17.77}$$

Taking the ratio of these two expressions yields T_i

$$T_i = \frac{\mathbf{i_2}}{\mathbf{i_1}} = \frac{\dfrac{1 + A}{R_B}}{\dfrac{1}{R_A}} = (1 + A)\frac{R_A}{R_B} = \frac{R_A}{R_B} + A\frac{R_A}{R_B} \tag{17.78}$$

Multiplying the last term by β and again using $1/\beta = 1 + R_B/R_A$ yields

$$T_i = \frac{R_A}{R_B} + A\beta\frac{R_A}{R_B}\frac{1}{\beta} = \frac{R_A}{R_B} + T\left(1 + \frac{R_A}{R_B}\right) \tag{17.79}$$

Simultaneous solution of Eqs. (17.79) and (17.76) gives the desired result:

$$T = \frac{T_v T_i - 1}{2 + T_v + T_i} \qquad \text{and} \qquad \frac{R_B}{R_A} = \frac{1 + T_v}{1 + T_i} \tag{17.80}$$

Using this technique, we can find both the loop gain T and the resistance (or impedance) ratio at point P.

Although the resistance ratio would be dominated by R_2 and R_1 in the circuit in Fig. 17.41, R_B and R_A in the general case actually represent the two equivalent resistances that would be calculated looking to the right and left of the point P, where the loop is broken. This fact is illustrated more clearly by the SPICE analysis in Ex. 17.5.

EXAMPLE 17.5   LOOP GAIN AND RESISTANCE RATIO CALCULATION USING SPICE

We will use SPICE to find the loop gain for an amplifier using the successive voltage and current injection technique.

PROBLEM   Find the loop gain T and the resistance ratio for the shunt-series feedback pair of Fig. 17.34 using the method of successive voltage and current injection at point P.

SOLUTION **Known Information and Given Data:** Shunt-series feedback pair with element values given in Fig. 17.34. Apply the voltage and current injection at point P between the base of Q_1 and feedback resistor R_F.

Unknowns: Loop gain T; resistance ratio R_B/R_A

Approach: In the dc-coupled case, we can insert zero valued sources into the circuit and use the transfer function capability of SPICE to find the sensitivity of voltages v_1 and v_2 to changes in v_x and the sensitivity of i_1 and i_2 to changes in i_x.

Assumptions: $\beta_F = 100$ and $V_A = 100$ V

Analysis: The amplifier circuit is redrawn in Fig. 17.42 with sources v_{x1}, v_{x2}, and i_x added to the circuit. All three are zero-value sources, which do not affect the Q-point calculations. Source v_{x2} is added so that current i_2 can be determined by SPICE. The results of the four SPICE transfer function analyses are

$$\frac{v_7}{v_{x1}} = -0.995 \qquad \frac{v_1}{v_{x1}} = 4.56 \times 10^{-3}$$

$$\frac{i_2}{i_x} = 0.985 \qquad \frac{i_1}{i_x} = 0.0146$$

and the loop gain and resistance ratio calculated using these four values are

$$T_v = -\frac{-0.995}{4.56 \times 10^{-3}} = 218 \qquad T_i = \frac{0.985}{0.0146} = 67.4$$

$$T = \frac{T_v T_i - 1}{2 + T_v + T_i} = \frac{218(67.4) - 1}{2 + 218 + 67.4} = 51.1$$

$$\frac{R_B}{R_A} = \frac{1 + T_v}{1 + T_i} = \frac{1 + 218}{1 + 67.4} = 3.20$$

Check of Results: The value of T computed by hand in Ex. 17.4 was 48.7 $(1 + A\beta = 49.7)$ and compares well to the result based on SPICE. Resistances R_A and R_B associated with the open feedback loop are identified in Fig. 17.43. Calculating these resistances and their ratio by hand gives

$$R_A = 10 \text{ k}\Omega \| r_\pi = 10 \text{ k}\Omega \| 3.79 \text{ k}\Omega = 2.75 \text{ k}\Omega$$

$$R_B = 9.1 \text{ k}\Omega + 1 \text{ k}\Omega \left\| \frac{10 \text{ k}\Omega + r_{\pi 2}}{\beta_{o2} + 1} = 9.1 \text{ k}\Omega + 1 \text{ k}\Omega \| 116 \text{ }\Omega = 9.20 \text{ k}\Omega \right.$$

$$\frac{R_B}{R_A} = 3.35$$

Again, we find good agreement with SPICE.

Discussion: The small difference in the resistance ratio is caused by the slightly different values of r_π and β_o calculated by SPICE and by neglecting r_{o1}. If the exact values from the simulation are used, then the calculated resistance ratio is precisely 3.20. As an alternative to using transfer function analyses, v_x and i_x can be made 1-V and 1-A ac sources, and two ac analyses can be performed. The ac source method has the advantage that it can find the loop gain and impedance ratio as a function of frequency. We must know the loop gain as a function of frequency in order to determine the stability of a feedback amplifier. This topic is discussed in detail in Sec. 17.11.

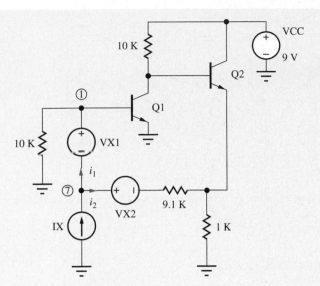

Figure 17.42 Shunt-series feedback pair with zero value sources added for SPICE analysis.

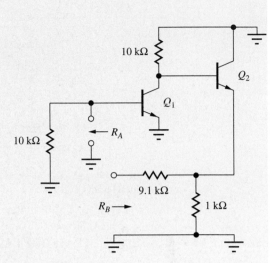

Figure 17.43 Definition of resistances R_A and R_B.

EXERCISE: Use SPICE to find the loop gain for the shunt-series feedback pair in Fig. 17.43 using successive voltage and current injection at a point between the 9.1-kΩ and 1-kΩ resistors.

ANSWERS: 51.2; Note: The loop gain should not change, but the exact SPICE models are important. It is best to compare your own results for both injection points.

EXERCISE: Use SPICE to find the loop gain T and resistance ratio for the amplifier in Ex. 17.1. Use the inverting input as the injection point. Compare to hand calculations.

ANSWERS: 467, 0.348; 467, 0.350

17.8.3 SIMPLIFICATIONS

Although analysis of the successive voltage and current injection method was performed using ideal sources, Middlebrook's analysis [1] shows that the technique is valid even if source resistances are included with both v_x and i_x. In addition, if point P is chosen at a position in the circuit where R_B is zero or R_A is infinite, then the equations can be simplified and T can be found from only one measurement. For example, if a point is found where R_A is infinite, then Eq. (17.76) reduces to $T = T_v$. In an ideal op amp circuit, such a point exists at the input of the op amp, as in Fig. 17.44(a).

Alternatively, if a point can be found where $R_B = 0$, then Eq. (17.76) also reduces to $T = T_v$. In an ideal op amp circuit, such a point exists at the output of the op amp, as in Fig. 17.44(b). A similar set of simplifications can be used for the current injection case. If $R_A = 0$ or R_B is infinite, then $T = T_I$.

In practice, the conditions $R_B \gg R_A$ or $R_A \gg R_B$ are sufficient to permit the use of the simplified expressions [2]. In the general case, where these conditions are not met, or we are not sure of the exact impedance levels, then the general method can always be applied.

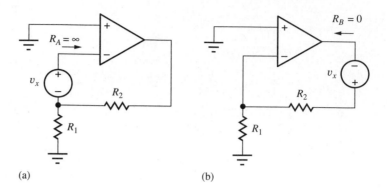

Figure 17.44 (a) Voltage injection at a point P_1, where $R_A = \infty$ and (b) voltage injection at a point P_2, where $R_B = 0$. (An ideal op amp is assumed.)

EXAMPLE 17.6   SIMPLIFIED ac LOOP-GAIN ANALYSIS IN SPICE

As we will see in the next section, we must be able to find the loop gain as a function of frequency in order to determine amplifier stability. Here we use SPICE to find the frequency-dependent loop gain for an op amp using the simplified approach outlined in Sec. 17.8.3.

PROBLEM Find the loop gain $T(j\omega)$ for the op amp in Figs. 15.85 and 17.62 using the simplified voltage injection method in Fig. 17.44. Add a 2-kW load resistor to the amplifier's output, and set $I_1 = 100\ \mu A$, $I_2 = 500\ \mu A$, and $I_3 = 5\ mA$. Use these transistor parameters: for the *npn*, BF = 100, VAF = 75 V, RB = 250 Ω, TF = 0.311 ns, and CJC = 0.768 pF; for the *pnp*, BF = 60, VAF = 75 V, RB = 250 Ω, TF = 3 ns, and CJC = 1 pF.

SOLUTION **Known Information and Given Data:** The op amp circuit from Fig. 17.62 appears here with the 2-kΩ load resistor attached to the output node. The transistor parameters are specified as BF = 100, VAF = 75 V, RB = 250 Ω, TF = 0.311 ns, and CJC = 0.768 pF for the *npn* device, and BF = 60, VAF = 75 V, RB = 250 Ω, TF = 3 ns, and CJC = 1 pF for the *pnp* device.

Unknowns: Magnitude and phase of the loop gain T as a function of frequency.

Approach: We close the feedback loop in the unity-gain configuration through voltage source V_X. V_X has 0 dc value, and its ac value is set to $1/\angle 0°$. The SPICE ac sweep analysis will be used to plot the loop gain as a function of frequency.

Assumptions: $T = 27°C$; small signal conditions apply

Analysis: The amplifier circuit from Fig. 17.62 is redrawn on the next page with source V_X connected from the output back to the inverting input. Load resistor R_L is connected from the output to ground.

   We expect the output resistance of the amplifier at the emitter of Q_6 ("R_B") to be small, and the input resistance at the base of Q_1 ("R_A") should be high. Thus, we have $R_A \gg R_B$, and the loop gain should be given by $T = T_v = -V_{E6}/V_{B1}$. We use the ac sweep capability in SPICE with FSTART = 1 Hz and FSTOP = 30 MHz with 20 analysis points per decade, and request SPICE to plot the magnitude in dB and the phase in degrees for the ratio of the two node voltages $-V(7)/V(1)$. The results appear here.

Check of Results: The graph of $|T(j\omega)|$ appears to match that given in Fig. 17.62(b).

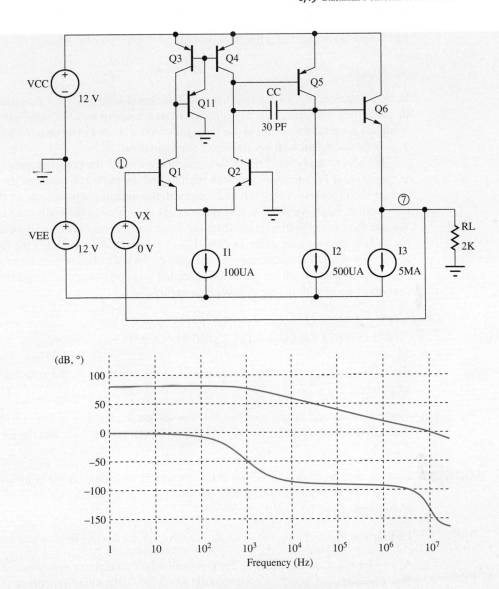

Discussion: This method enables us to find the open-loop gain of the op amp while keeping the feedback loop closed, so that the proper Q-points are maintained within the amplifier. As a by-product, we find that the dc offset voltage is approximately 3.83 mV.

EXERCISE: Use SPICE to verify the results in the Bode plot in Ex. 17.6.

17.9 BLACKMAN'S THEOREM TO THE RESCUE

Fortunately, Blackman's theorem provides a way to avoid the problems encountered using two-port analysis. Blackman was one of a group of individuals who first investigated the properties of feedback amplifiers at Bell Laboratories in the 1930s and 1940s [3–5]. One of his highly useful results is stated in Eq. (17.81) and provides us with an alternate approach for calculating the input and output resistance[3] of a feedback amplifier.

$$R_{CL} = R_D \frac{1 + |T_{SC}|}{1 + |T_{OC}|} \tag{17.81}$$

In this equation, R_{CL} is the resistance of the closed-loop feedback amplifier looking into one of its ports (any terminal pair), R_D is the resistance looking into the same pair of terminals with the feedback loop disabled, T_{SC} is the loop-gain with a short-circuit applied to the selected port, and T_{OC} is the loop gain with the same port open-circuited.

In order to apply Blackman's theorem, first we select the port terminals where we desire to find the resistance. For example, we often wish to find the input resistance or the output resistance of a closed-loop feedback amplifier, and the resistance appears between one of the amplifier terminals and ground. Next, we select one of the controlled sources in the equivalent circuit of the amplifier. We use this source to disable the feedback loop, and the source is also used as the reference source for finding the two loop gains T_{SC} and T_{OC}. Resistance R_D represents the driving-point resistance at the port of interest calculated with the gain of the controlled source set to zero, whereas T_{SC} and T_{OC} are calculated with the port short-circuited and open-circuited, respectively. This procedure is best understood with the aid of several examples.

EXAMPLE 17.7   RESISTANCE CALCULATIONS USING BLACKMAN'S THEOREM

Here we use Blackman's theorem to calculate the input and output resistances for the feedback amplifier in Ex. 17.3.

PROBLEM   Find the input and output resistance of the feedback amplifier in Fig. 17.25 using Blackman's theorem. From Ex. 17.3, $A_o = 10,000$, $R_{id} = 25$ kΩ, and $R_o = 0$, and the BJT has $\beta_o = 100$ and $V_A = 50$ V.

SOLUTION   **Known Information and Given Data:** Feedback amplifier circuit is given in Fig. 17.25 with $V_{REF} = 5$ V and $R = 5$ kΩ; op amp parameters: $A_o = 10,000$, $R_{id} = 25$ kΩ, and $R_o = 0$; BJT parameters: $\beta_o = 100$ and $V_A = 50$ V.

Unknowns: Closed-loop values of R_{out} and R_{in} for the overall feedback amplifier

Approach: Redraw the circuit including small-signal models for the transistor and amplifier. Find R_D, T_{SC}, and T_{OC}, and then substitute the results into Blackman's theorem to find R_{CL}.

Assumptions: We have the Q-point and small-signal parameters from Ex. 17.3: $g_m = 0.04$ S, $r_\pi = 2.5$ kΩ, and $r_o = 50$ kΩ.

Analysis: As a first step, the amplifier in Fig. 17.25 is redrawn on the next page. We must select the voltage-controlled voltage source (VCVS) associated with the op amp as the reference source for the calculations. It is the only source that is completely within the feedback loop, and the loop will be disabled when A_o is set to zero.

Output Resistance Calculation
We desire to find the equivalent resistance between the output and ground terminals of the amplifier. In order to find R_D, we disable the feedback loop by setting the gain of the VCVS to zero, resulting in the equivalent circuit in this figure. The resistance looking in the collector of the

---

[3] The results are also valid for general impedances.

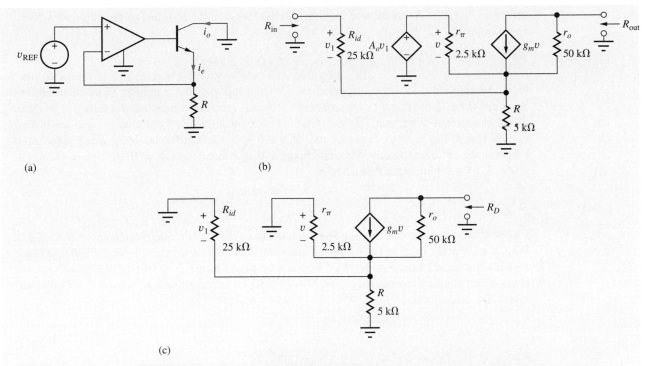

(a) Feedback circuit, (b) small-signal model, and (c) model for finding R_D for the output resistance calculation.

BJT can be found using Eq. (13.66):

$$R_D = r_o \left[1 + \frac{\beta_o(R \| R_{id})}{r_\pi + (R \| R_{id})} \right] = 50 \text{ k}\Omega \left[1 + \frac{100(5 \text{ k}\Omega \| 25 \text{ k}\Omega)}{2.5 \text{ k}\Omega + (5 \text{ k}\Omega \| 25 \text{ k}\Omega)} \right] = 3.18 \text{ M}\Omega$$

Next, we find loop gain T_{SC} with the output terminal shorted to ground by setting the value of the VCVS to 1 and finding the returned value of $A_o v_1$:

$$|T_{SC}| = \left| \frac{A_o v_1}{1} \right| = A_o \frac{(\beta_o + 1)(R \| R_{id} \| r_o)}{r_\pi + (\beta_o + 1)(R \| R_{id} \| r_o)} = 10^4 \frac{101(3.85 \text{ k}\Omega)}{2.5 \text{ k}\Omega + 101(3.85 \text{ k}\Omega)} = 9940$$

The calculation of T_{OC} is similar, except that a very interesting thing happens when the collector terminal of the transistor is opened. The transistor no longer exhibits a current gain because the collector current is zero, and the emitter current must now be equal to the base current. However, the loop gain does not become zero. We can obtain T_{OC} by setting β_o to zero in the expression for T_{SC}. Note that r_o is no longer in parallel with R.

$$|T_{OC}| = \frac{A_o v_1}{1} = A_o \frac{(R \| R_{id})}{r_\pi + (R \| R_{id})} = 10^4 \frac{(5 \text{ k}\Omega \| 25 \text{ k}\Omega)}{2.5 \text{ k}\Omega + (5 \text{ k}\Omega \| 25 \text{ k}\Omega)} = 6350$$

Now, Blackman's theorem gives the closed-loop output resistance as

$$R_{\text{out}} = 3.18 \text{ M}\Omega \left(\frac{1 + 9940}{1 + 6350} \right) = 5.06 \text{ M}\Omega$$

which is similar to the SPICE result in Table 17.1.

Input Resistance Calculation

Here we desire to find the equivalent resistance between the input terminal of the feedback amplifier and ground. For this case, we again disable the feedback loop by setting the gain of the VCVS to zero, and find R_D. The resistance looking into the noninverting terminal of the amplifier will be approximately

$$R_D = R_{id} + \left(R \left\| \frac{1}{g_m} \right. \right) = 25 \text{ k}\Omega + (5 \text{ k}\Omega \| 25 \text{ }\Omega) = 25.0 \text{ k}\Omega$$

Next, we make the loop gain calculations with the input of the amplifier short-circuited and open-circuited. (Note that the collector of the transistor remains connected to ground for all these calculations.) The loop gain T_{SC} with the input terminal shorted to ground is identical to that already found: $T_{SC} = 9940$. However, the value of T_{OC} is much different. When we open the input terminal, the current in R_{id} must be zero. So $v_1 = 0$, and therefore $T_{OC} = 0$. Blackman's theorem gives the closed-loop input resistance as

$$R_{\text{in}} = 25.0 \text{ k}\Omega \left(\frac{1 + 9940}{1 + 0} \right) = 249 \text{ M}\Omega$$

Check of Results: Both the input and output resistances agree with SPICE calculations in Ex. 17.3. We also see that the value of T_{SC} matches the value of $A\beta$ in the same example. The output resistance agrees with the BJT bound set by $(\beta_o + 1)r_o = 5.05 \text{ M}\Omega$. Since $T_{OC} = 0$ in the R_{in} calculation, we observe full amplification of R_D by the loop gain T_{SC}, whereas this amplification is not observed in the R_{out} calculation.

EXERCISE: Find R_I, T_{SC}, T_{OC}, and the closed-loop output resistance if the op amp output resistance is changed from 0 to 5 kΩ.

ANSWERS: 1.85 MΩ; 9830, 3570, 5.09 MΩ

EXAMPLE **17.8** APPLICATION OF BLACKMAN'S THEOREM TO EX. 17.4

As a second example, we use Blackman's theorem to calculate the input and output resistances for the shunt-series feedback pair in Ex. 17.4.

PROBLEM Find the input and output resistance of the feedback amplifier in Fig. 17.34 using Blackman's theorem. Use the transistor parameter and Q-point information from the example.

SOLUTION **Known Information and Given Data:** The shunt-series feedback pair appears in Fig. 17.34 with the transistor parameters given in Table. 17.2.

Unknowns: Closed-loop values of the input resistance at the base, and the output resistance at the collector of transistor Q_2.

Approach: Redraw the circuit including the small-signal models for the transistor and amplifier. Choose a controlled source as reference. Find R_D, T_{SC}, and T_{OC}, and then substitute the results into Blackman's theorem to find R_{CL}.

Assumptions: We have the Q-point and small-signal parameters from Ex. 17.4.

Analysis: As a first step, the small-signal model for the amplifier is drawn in the following figure. To simplify the circuit analysis, the feedback network has been replaced with its y-parameter equivalent with y_{21} assumed to be zero. Let us use source $y_{12}^f v_2$ as the reference source for the calculations. (We could also use the $g_{m1}v_1$ source of transistor Q_1; see the exercise at the end of this example.)

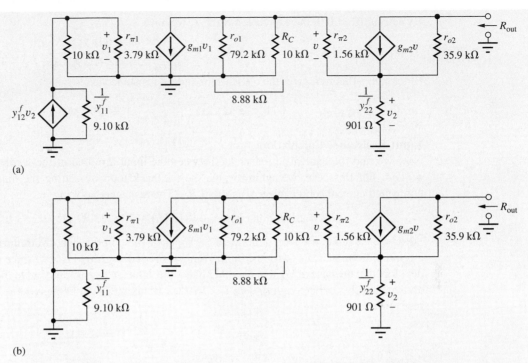

(a)

(b)

(a) Small-signal model for the shunt-series feedback pair of Fig. 17.34. (b) Shunt-series feedback pair with $y_{12}^f v_2 = 0$.

Output Resistance Calculation

We desire the equivalent resistance between the collector of Q_2 and ground. In order to find R_D, we disable the feedback loop by setting the gain of the controlled source to zero, resulting in the equivalent circuit in this figure. With the feedback disabled, the output resistance looking in the collector of the BJT can be found with the aid of Eq. (14.28):

$$R_D = r_{o2} \left[1 + \frac{\beta_o \left(\dfrac{1}{y_{22}^f} \right)}{(R_C \| r_{o1}) + r_{\pi 2} + \left(\dfrac{1}{y_{22}^f} \right)} \right]$$

$$= 35.9 \text{ k}\Omega \left[1 + \frac{100(0.901 \text{ k}\Omega)}{(10 \text{ k}\Omega \| 79.2 \text{ k}\Omega) + 1.56 \text{ k}\Omega + 0.901 \text{ k}\Omega} \right] = 321 \text{ k}\Omega$$

Next, we find loop gain T_{SC} with the output terminal shorted to ground by setting the value of the reference current source to 1 and finding the returned value of $y_{12}^f v_2$. The analysis is identical to that in Ex. 17.4 for Figs. 17.37 and 17.38 except that we must multiply v_2 by y_{12}^f:

$$|T_{SC}| = (9.1 \text{ k}\Omega \| 10 \text{ k}\Omega \| 3.79 \text{ k}\Omega)(g_{m1})(79 \text{ k}\Omega \| 10 \text{ k}\Omega \| [1.56 \text{ k}\Omega + (\beta_o + 1)0.901 \text{ k}\Omega])$$

$$\cdot \frac{(\beta_o + 1)0.901 \text{ k}\Omega}{1.56 \text{ k}\Omega + (\beta_o + 1)0.901 \text{ k}\Omega} y_{12}^f$$

$$|T_{SC}| = (2.11 \text{ k}\Omega)(40 \times 0.66 \text{ mA})(8.10 \text{ k}\Omega) \left(\frac{(101)0.901 \text{ k}\Omega}{1.56 \text{ k}\Omega + (101)0.901 \text{ k}\Omega} \right) \left(\frac{1}{9.1 \text{ k}\Omega} \right) = 48.7$$

The calculation of T_{OC} is similar to that in Ex. 17.4. When the collector terminal of the transistor is opened, the transistor no longer exhibits a current gain because the collector current is zero.

By setting $\beta_o = 0$ in the expression for T_{SC}, we have

$$|T_{SC}| = (2.11 \text{ k}\Omega)(40 \times 0.66 \text{ mA})(1.93 \text{ k}\Omega)\left(\frac{0.901 \text{ k}\Omega}{1.56 \text{ k}\Omega + 0.901 \text{ k}\Omega}\right)\left(\frac{1}{9.1 \text{ k}\Omega}\right) = 4.33$$

Blackman's theorem gives the closed-loop output resistance as

$$R_{\text{out}} = 321 \text{ k}\Omega\left(\frac{1 + 48.7}{1 + 4.33}\right) = 2.99 \text{ M}\Omega$$

Input Resistance Calculation

Now we find the equivalent resistance between the input terminal of the feedback amplifier and ground. For this case, we again disable the feedback loop by setting the gain of the voltage-controlled current source to zero, and find R_D. The resistance looking into the base of Q_1 is

$$R_D = 10 \text{ k}\Omega \| 9.1 \text{ k}\Omega \| r_{\pi 1} = 10 \text{ k}\Omega \| 9.1 \text{ k}\Omega \| 3.79 \text{ k}\Omega = 2.11 \text{ k}\Omega$$

Next, we calculate the loop gains with the input of the amplifier short-circuited and then open-circuited. Loop gain T_{SC} with the input terminal shorted to ground is zero since no voltage can be developed at the base of Q_1. On the other hand, the value of T_{OC} is identical to the loop gain found above with Q_2's collector grounded: $T_{OC} = 48.7$. Blackman's theorem gives the closed-loop input resistance as

$$R_{\text{in}} = 2.11 \text{ k}\Omega\left(\frac{1 + 0}{1 + 48.7}\right) = 42.5 \ \Omega$$

Check of Results: Both the input and output resistances agree with the SPICE calculations in Ex. 17.4.

EXERCISE: Repeat the calculations using source $g_{m1}v_1$ as the reference for the loop-gain calculations.

EXAMPLE 17.9   **WILSON SOURCE OUTPUT RESISTANCE**

As a final example, let us calculate the output resistance of the Wilson current source using Blackman's theorem.

PROBLEM   Find an expression for the output resistance of the BJT version of the Wilson current source in Fig. 15.66(b) using Blackman's theorem.

SOLUTION   **Known Information and Given Data:** The Wilson current source circuit is redrawn here.

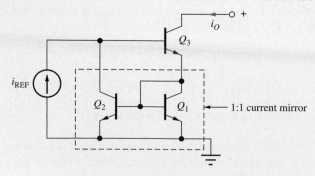

Wilson current source using bipolar transistors.

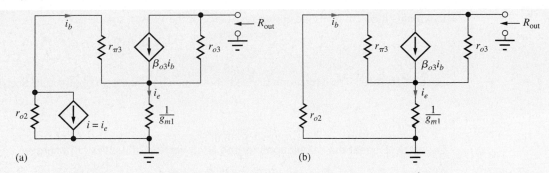

Figure 17.45 (a) Small-signal model for the Wilson current source. (b) Model with reference source i set to zero.

Unknowns: Output resistance looking into the collector of transistor Q_3.

Approach: Redraw the circuit including the small-signal model for transistor Q_3. Use the simplified small-signal two-port model for the current mirror from Sec. 15.6.8. Find R_D, T_{SC}, and T_{OC}, and then substitute the results into Blackman's theorem to find R_{out}.

Assumptions: The current mirror has a 1:1 mirror ratio. The transistors are all operating in the active region.

Analysis: A small-signal model for the Wilson current source is drawn in Fig. 17.45(a). To simplify the circuit analysis, the current mirror has been replaced with its equivalent circuit from Fig. 15.62. Current source i will be used as the reference source for the calculations.

Output Resistance Calculation

R_{out} is the equivalent resistance between the collector of Q_3 and ground. In order to find R_D, we disable the feedback loop by setting the gain of the controlled source to zero, resulting in the equivalent circuit in Fig. 17.45(b). The output resistance looking in the collector of the BJT can be found using Eq. (13.66).

$$R_D = r_{o3}\left[1 + \frac{\beta_{o3}\left(\dfrac{1}{g_{m1}}\right)}{r_{o2} + r_{\pi3} + \left(\dfrac{1}{g_{m1}}\right)}\right] = r_{o3}\left(1 + \frac{\beta_{o3}}{\mu_{f2} + \beta_{o3} + 1}\right) \cong r_{o3}$$

in which it is assumed that $g_{m1} = g_{m2} = g_{m3}$ and $\mu_f \gg \beta_o \gg 1$. Next, we find loop gain T_{SC} with the output terminal shorted to ground by setting the value of the reference current source i to 1 and then finding the returned value of $i = 1 \cdot i_e$. Because the current mirror has a gain of 1, i is equal to the current in the resistor $(1/g_m)$, and this current is equal to the emitter current of Q_3:

$$i = i_e = (\beta_{o3} + 1)i_b = -(\beta_{o3} + 1)(1)\frac{r_{o2}}{r_{o2} + r_{\pi3} + (\beta_{o3} + 1)\left(\dfrac{1}{g_{m1}}\right)} \cong -\frac{(\beta_o + 1)}{1 + \dfrac{2\beta_o + 1}{\mu_f}}$$

$$\cong -(\beta_o + 1)$$

Therefore, $|T_{SC}| = \beta_o + 1$. The calculation of T_{OC} is similar, except that the transistor's current gain becomes zero when the collector terminal of the transistor is opened. Modifying the last expression by setting $\beta_{o3} = 0$, we have

$$|T_{OC}| = \frac{i_e}{1} = (0 + 1)\frac{r_{o2}}{r_{o2} + r_{\pi3} + (0 + 1)\left(\dfrac{1}{g_{m1}}\right)} \cong \frac{1}{1 + \dfrac{\beta_o + 1}{\mu_f}} \cong 1$$

Blackman's theorem gives the output resistance of the Wilson source as

$$R_{\text{out}} = r_o \left(\frac{1 + \beta_o + 1}{1 + 1} \right) \cong \frac{\beta_o r_o}{2}$$

Check of Results: The expression derived above matches the result obtained in Sec. 15.7.2.

EXERCISE: Repeat the calculations for the MOS version of the Wilson source.

ANSWERS: $R_D = 2r_o$; $T_{SC} = \mu_f/2$; $T_{OC} = 0$; $R_{\text{out}} = \mu_f r_o$

In this section, we discovered that Blackman's theorem provides an elegant method for calculating resistances in closed-loop feedback amplifiers and that the method does not suffer from the limitations associated with the two-port analysis approach.

It is worth noting that the input and output resistance expressions that we derived earlier for the four two-port configurations are actually special cases of Eq. (17.81). If either $T_{SC} = 0$ or $T_{OC} = 0$, then Eq. (17.81) can be simplified to

$$R_{CL} = R_D(1 + |T_{SC}|) \quad \text{for} \quad T_{OC} = 0 \qquad \text{and} \qquad R_{CL} = \frac{R_D}{1 + |T_{OC}|} \quad \text{for} \quad T_{SC} = 0 \quad (17.82)$$

These expressions correspond to the results for series feedback [Eqs. (17.19), (17.55), (17.63), and (17.65)] and shunt feedback [Eqs. (17.24), (17.37), (17.42), and (17.51)].

17.10 USING FEEDBACK TO CONTROL FREQUENCY RESPONSE

In Secs. 17.3 to 17.6, we found that feedback can be used to stabilize the gain and improve the input and output resistances of an amplifier, and in Chapter 12 we found, in the discussion of operational amplifiers, that feedback can be used to trade reduced gain for increased bandwidth in low-pass amplifiers. A similar effect was noted in the discussion of the gain-bandwidth limitations of the inverting amplifiers. In this section, we extend the analysis to more general feedback amplifiers.

The closed-loop gain for all the feedback amplifiers in this chapter can be written as

$$A_v = \frac{A}{1 + A\beta} \qquad \text{or} \qquad A_v(s) = \frac{A(s)}{1 + A(s)\beta(s)} \qquad (17.83)$$

Up to now, we have worked with the midband value of A and assumed it to be a constant. However, we can explore the frequency response of the general closed-loop feedback amplifier by substituting a frequency-dependent voltage gain expression for A into Eq. (17.83).

Suppose that amplifier A is an amplifier with cutoff frequencies of ω_H and ω_L and midband gain A_o as described by

$$A(s) = \frac{A_o \omega_H s}{(s + \omega_L)(s + \omega_H)} \qquad (17.84)$$

Substituting Eq. (17.84) into Eq. (17.83) and simplifying the expression yields

$$A_v(s) = \frac{\dfrac{A_o \omega_H s}{(s + \omega_L)(s + \omega_H)}}{1 + \dfrac{A_o \omega_H s}{(s + \omega_L)(s + \omega_H)}\beta} = \frac{A_o \omega_H s}{s^2 + [\omega_L + \omega_H(1 + A_o\beta)]s + \omega_L\omega_H} \qquad (17.85)$$

Assuming that $\omega_H(1 + A_o\beta) \gg \omega_L$, then dominant-root factorization (see Sec. 16.6.3) yields these estimates of the upper- and lower-cutoff frequencies and bandwidth of the closed-loop feedback amplifier:

$$\omega_L^F \cong \frac{\omega_L\omega_H}{\omega_L + \omega_H(1 + A_o\beta)} \cong \frac{\omega_L}{1 + A_o\beta}$$

$$\omega_H^F \cong \omega_L + \omega_H(1 + A_o\beta) \cong \omega_H(1 + A_o\beta) \qquad (17.86)$$

$$\mathrm{BW}_F - \omega_H^F - \omega_L^F \cong \omega_H(1 + A_o\beta)$$

The upper- and lower-cutoff frequencies and bandwidth of the feedback amplifier are all improved by the factor $(1 + A_o\beta)$. Using the approximations in Eq. (17.86), we find that the transfer function in Eq. (17.85) can be rewritten approximately as

$$A_v(s) \cong \frac{\dfrac{A_o}{(1 + A_o\beta)}s}{\left(s + \dfrac{\omega_L}{(1 + A_o\beta)}\right)\left[1 + \dfrac{s}{\omega_H(1 + A_o\beta)}\right]} \qquad (17.87)$$

As expected, the midband gain is stabilized at

$$A_{\mathrm{mid}} = \frac{A_o}{1 + A_o\beta} \cong \frac{1}{\beta} \qquad (17.88)$$

It should once again be recognized that the gain-bandwidth product of the closed-loop amplifier remains constant:

$$\mathrm{GBW} = A_{\mathrm{mid}} \times \mathrm{BW}_F \cong \frac{A_o}{1 + A_o\beta}\omega_H(1 + A_o\beta) = A_o\omega_H \qquad (17.89)$$

These results are displayed graphically in Fig. 17.46 for an amplifier with $1/\beta = 20$ dB. The open-loop amplifier has $A_o = 40$ dB, $\omega_L = 100$ rad/s, and $\omega_H = 10,000$ rad/s, whereas the closed-loop amplifier has $A_v = 19.2$ dB, $\omega_L = 9.1$ rad/s, and $\omega_H = 110,000$ rad/s.

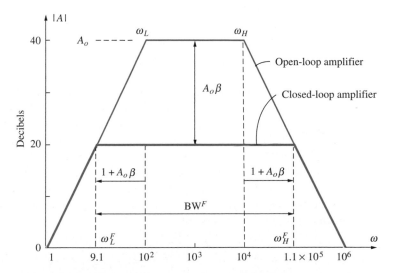

Figure 17.46 Graphical interpretation of feedback amplifier frequency response.

EXERCISE: An op amp has a dc gain of 100 dB and a unity-gain frequency of 10 MHz. What is the upper-cutoff frequency of the op amp itself? If the op amp is used to build a noninverting amplifier with a closed-loop gain of 60 dB, what is the bandwidth of the feedback amplifier? Write an expression for the transfer function of the op amp. Write an expression for the transfer function of the noninverting amplifier.

ANSWERS: 100 Hz; 10 kHz; $A(s) = 2\pi \times 10^7/(s + 200\pi)$; $A(s) = 2\pi \times 10^7/(s + 2\pi \times 10^4)$

17.11 STABILITY OF FEEDBACK AMPLIFIERS

Whenever an amplifier is embedded within a feedback network, a question of **stability** arises. Up to this point, it has tacitly been assumed that the feedback is negative. However, as frequency increases, the phase of the loop gain changes, and it is possible for the feedback to become positive at some frequency. If the gain is also greater than or equal to 1 at this frequency, then instability occurs, typically in the form of oscillation.

The locations of the poles of a feedback amplifier can be found by analysis of the closed-loop transfer function described by

$$A_v(s) = \frac{A(s)}{1 + A(s)\beta(s)} = \frac{A(s)}{1 + T(s)} \qquad (17.90)$$

The poles occur at the complex frequencies s, for which the denominator becomes zero:

$$1 + T(s) = 0 \qquad \text{or} \qquad T(s) = -1 \qquad (17.91)$$

The particular values of s that satisfy Eq. (17.91) represent the poles of $A_v(s)$. For amplifier stability, the poles must lie in the left half of the s-plane. Now we discuss two graphical approaches for studying stability using Nyquist and Bode plots.

17.11.1 THE NYQUIST PLOT

The **Nyquist plot** is a useful graphical method for qualitatively studying the locations of the poles of a feedback amplifier. The graph represents a mapping of the right half of the s-plane (RHP) onto the $T(s)$-plane, as in Fig. 17.47. Every value of s in the s-plane has a corresponding value of $T(s)$. The critical issue is whether any value of s in the RHP corresponds to $T(s) = -1$. However, checking every possible value of s would take a rather long time. Nyquist realized that to simplify the process, we need only plot $T(s)$ for values of s on the $j\omega$ axis

$$T(j\omega) = A(j\omega)\beta(j\omega) = |T(j\omega)| \angle T(j\omega) \qquad (17.92)$$

which represents the boundary between the RHP and LHP. $T(j\omega)$ is normally graphed using the polar coordinate form of Eq. (17.92). If the **−1 point** is enclosed by this boundary, then there must be some value of s for which $T(s) = -1$, a pole exists in the RHP, and the amplifier is not stable.[4] However, if −1 lies outside the interior of the Nyquist plot, then the poles of the closed-loop amplifier are all in the left half-plane, and the amplifier is stable.

Today, we are fortunate to have computer tools such as MATLAB, which can quickly construct the Nyquist plot for us. These tools eliminate the tedious work involved in creating the graphs so that we can concentrate on interpretation of the information. Let us consider examples of basic first-, second-, and third-order systems.

---

[4] If we mentally "walk" around the s-plane, keeping the shaded region on our right, then the corresponding region in the $T(s)$-plane will also be on our right as we "walk" in the $T(s)$-plane.

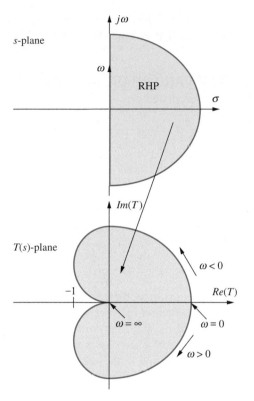

Figure 17.47 Nyquist plot as a mapping between the s-plane and the $T(s)$-plane.

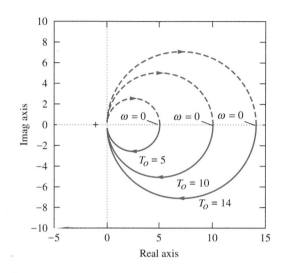

Figure 17.48 Nyquist plot for first-order $T(s)$ for $T_o = 5$, 10, and 14. (Nyquist plots are easily made using MATLAB. This figure is generated by three simple MATLAB statements: nyquist(14,[1 1]), nyquist(10,[1 1]), and nyquist(5,[1 1]).)

17.11.2 FIRST-ORDER SYSTEMS

In most of the feedback amplifiers we considered thus far, β was a constant and $A(s)$ was the frequency-dependent part of the loop gain $T(s)$. However, the important thing is the overall behavior of $T(s)$. The simplest case of $T(s)$ is that of a basic low-pass amplifier with a loop-gain described by

$$T(s) = \frac{A_o \omega_o}{s + \omega_o} \beta = \frac{T_o}{s + \omega_o} \tag{17.93}$$

For example, Eq. (17.93) might correspond to a single-pole operational amplifier with resistive feedback. The Nyquist plot for

$$T(j\omega) = \frac{T_o}{j\omega + 1} \tag{17.94}$$

is given in Fig. 17.48. At dc, $T(0) = T_o$, whereas for $\omega \gg 1$,

$$T(j\omega) \cong -j\frac{T_o}{\omega} \tag{17.95}$$

As frequency increases, the magnitude monotonically approaches zero, and the phase asymptotically approaches $-90°$.

From Eq. (17.93), we see that changing the feedback factor β scales the value of $T_o = T(0)$,

$$T(0) = A_o \omega_o \beta \tag{17.96}$$

but changing $T(0)$ simply scales the radius of the circle in Fig. 17.48, as indicated by the curves for $T_o = 5$, 10, and 14. It is impossible for the graph in Fig. 17.48 to ever enclose the $T = -1$ point, and the amplifier is stable regardless of the value of T_o. This is one reason why general-purpose op amps are often internally compensated to have a single-pole low-pass response. Single-pole op amps are stable for any fixed value of β.

17.11.3 SECOND-ORDER SYSTEMS AND PHASE MARGIN

A second-order loop-gain function can be described by

$$T(s) = \frac{A_o}{\left(1 + \dfrac{s}{\omega_1}\right)\left(1 + \dfrac{s}{\omega_2}\right)}\beta = \frac{T_o}{\left(1 + \dfrac{s}{\omega_1}\right)\left(1 + \dfrac{s}{\omega_2}\right)} \qquad (17.97)$$

An example appears in Fig. 17.49 for

$$T(s) = \frac{14}{(s + 1)^2} \qquad \text{and} \qquad T(j\omega) = \frac{14}{(j\omega + 1)^2} \qquad (17.98)$$

In this case, T_o is 14, but at high frequencies

$$T(j\omega) \cong (-j)^2 \frac{14}{\omega^2} = -\frac{14}{\omega^2} \qquad (17.99)$$

As frequency increases, the magnitude decreases monotonically from 14 toward 0, and the phase asymptotically approaches $-180°$. Again, it is theoretically impossible for this transfer function to encircle the -1 point. However, the second-order system can come arbitrarily close to this point, as indicated in Fig. 17.50, which is a blowup of the Nyquist plot in the region near the -1 point. The larger the value of T_o, the closer the curve will come to the -1 point. The curve in Fig. 17.50 is plotted for a T_o value of only 14, whereas an actual op amp circuit could easily have a T_o value of 1000 or more.

Although technically stable, the second-order system can have essentially zero **phase margin,** as defined in Fig. 17.51. Phase margin Φ_M represents the maximum increase in phase shift (phase lag) that can be tolerated before the system becomes unstable. Φ_M is defined as

$$\Phi_M = \angle T(j\omega_1) - (-180°) = 180° + \angle T(j\omega_1) \qquad \text{where} \qquad |T(j\omega_1)| = 1 \qquad (17.100)$$

To find Φ_M, we first must determine the frequency ω_1 for which the magnitude of the loop gain is unity, corresponding to the intersection of the Nyquist plot with the unit circuit in Fig. 17.51, and

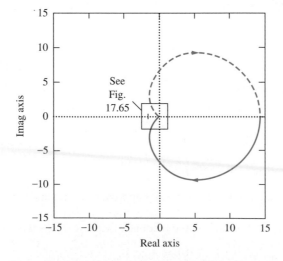

Figure 17.49 Nyquist plot for second-order $T(s)$. (Generated using MATLAB command: nyquist(14,[1 2 1]).)

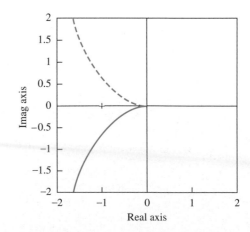

Figure 17.50 Blowup of Fig. 17.49 near the -1 point. The second-order system does not enclose the -1 point but may come arbitrarily close to doing so.

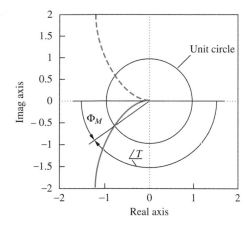

Figure 17.51 Definition of phase margin Φ_M.

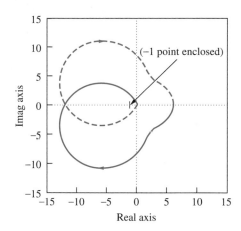

Figure 17.52 Nyquist plot for third-order $T(s)$. (Using MATLAB: nyquist(14,[1 1 3 2]).)

then determine the phase shift of T at this frequency. The difference between this angle and $-180°$ is Φ_M.

Small phase margin leads to excessive peaking in the closed-loop frequency response and undesirable ringing in the step response. In addition, any rotation of the Nyquist plot due to additional phase shift (from poles that may have been neglected in the model, for example) can lead to instability.

17.11.4 THIRD-ORDER SYSTEMS AND GAIN MARGIN

Third-order systems described by

$$T(s) = \frac{A_o}{\left(1 + \dfrac{s}{\omega_1}\right)\left(1 + \dfrac{s}{\omega_2}\right)\left(1 + \dfrac{s}{\omega_3}\right)} \qquad \beta = \frac{T_o}{\left(1 + \dfrac{s}{\omega_1}\right)\left(1 + \dfrac{s}{\omega_2}\right)\left(1 + \dfrac{s}{\omega_3}\right)} \qquad (17.101)$$

can easily have stability problems. Consider the example in Fig. 17.52, for

$$T(s) = \frac{14}{s^3 + s^2 + 3s + 2} \qquad (17.102)$$

For this case, $T(0) = 7$, and at high frequencies

$$T(j\omega) \cong (-j)^3 \frac{14}{\omega^3} = +j\frac{14}{\omega^3} \qquad (17.103)$$

At high frequencies, the polar plot asymptotically approaches zero along the positive imaginary axis, and the plot can enclose the critical -1 point under many circumstances. The particular case in Fig. 17.52 represents an unstable closed-loop system.

Gain margin is another important concept and is defined as the reciprocal of the magnitude of $T(j\omega)$ evaluated at the frequency for which the phase shift is $180°$:

$$\text{GM} = \frac{1}{|T(j\omega_{180})|} \qquad \text{where} \qquad \angle T(j\omega_{180}) = -180° \qquad (17.104)$$

Gain margin is often expressed in dB as $\text{GM}_{\text{dB}} = 20\log(\text{GM})$.

Equation (17.104) is interpreted graphically in Fig. 17.53. If the magnitude of $T(s)$ is increased by a factor equal to or exceeding the gain margin, then the closed-loop system becomes unstable, because the Nyquist plot then encloses the -1 point.

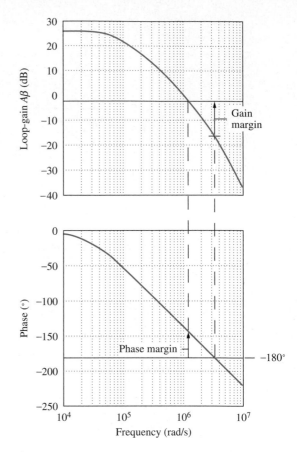

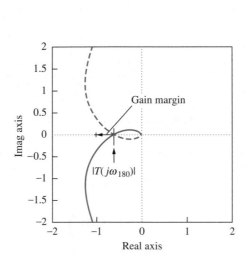

Figure 17.53 Nyquist plot showing gain margin of a third-order system. (Using MATLAB: nyquist(5,[1 3 3 1]).)

Figure 17.54 Phase and gain margin on the Bode plot. (Graph plotted using MATLAB: bode(2E19, [1 11.1E6 11.1 E12 1E18]).)

EXERCISE: Find the gain margin for the system in Fig. 17.53 described by

$$T(s) = \frac{5}{s^3 + 3s^2 + 3s + 1} = \frac{5}{(s+1)^3}$$

ANSWER: 4.08 dB

17.11.5 DETERMINING STABILITY FROM THE BODE PLOT

Phase and gain margin can also be determined directly from a **Bode plot** of the loop gain, as indicated in Fig. 17.54. This figure represents $A\beta$ for a third-order transfer function:

$$A\beta = \frac{2 \times 10^{19}}{(s + 10^5)(s + 10^6)(s + 10^7)} = \frac{2 \times 10^{19}}{s^3 + 11.1 \times 10^6 s^2 + 11.1 \times 10^{12} s + 10^{18}}$$

Phase margin is found by first identifying the frequency at which $|A\beta| = 1$ or 0 dB. For the case in Fig. 17.54, this frequency is approximately 1.2×10^6 rad/s. At this frequency, the phase shift is $-145°$, and the phase margin is $\Phi_M = 180° - 145° = 35°$. The amplifier can tolerate an additional phase shift of approximately $35°$ before it becomes unstable.

Gain margin is found by identifying the frequency at which the phase shift of the amplifier is exactly $180°$. In Fig. 17.54, this frequency is approximately 3.2×10^6 rad/s. The loop gain at this

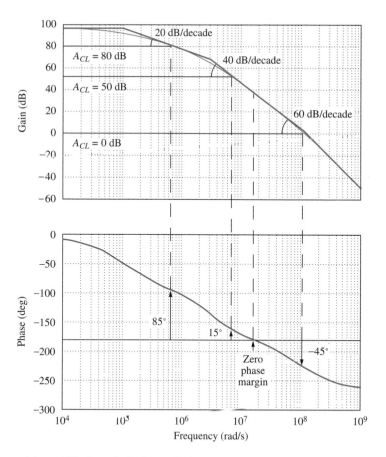

Figure 17.55 Determining stability from the Bode magnitude plot. Three values of closed-loop gain are indicated: 80 dB, 50 dB, and 0 dB. The corresponding phase margins are 85°, 15° (ringing and overshoot), and −45° (unstable). (Graph plotted using MATLAB: bode(2E24,[1 103.1E6 310.3E12 30E18]).)

frequency is −17 dB, and the gain margin is therefore +17 dB. The gain must increase by 17 dB before the amplifier becomes unstable.

Using a tool like MATLAB, we can easily construct the Bode plot for the gain of the amplifier and use it to determine the range of closed-loop gains for which the amplifier will be stable. Stability can be determined by properly interpreting the Bode magnitude plot. We use this mathematical approach:

$$20\log|A\beta| = 20\log|A| - 20\log\left|\frac{1}{\beta}\right| \tag{17.105}$$

Rather than plotting the loop gain $A\beta$ itself, the magnitude of the open-loop gain A and the reciprocal of the feedback factor β are plotted separately. (Remember, $A_v \cong 1/\beta$.) The frequency at which these two curves intersect is the point at which $|A\beta| = 1$, and the phase margin of the closed-loop amplifier can easily be determined from the phase plot.

Let us use the Bode plot in Fig. 17.55 as an example. In this case,

$$A(s) = \frac{2 \times 10^{24}}{(s + 10^5)(s + 3 \times 10^6)(s + 10^8)} \tag{17.106}$$

The asymptotes from Eq. (17.106) have also been included on the graph. For simplicity in this example, we assume that the feedback is independent of frequency (for example, a resistive voltage divider) so that $1/\beta$ is a straight line.

Three closed-loop gains are indicated. For the largest closed-loop gain, $(1/\beta) = 80$ dB, the phase margin is approximately 85°, and stability is not a problem. The second case corresponds to a closed-loop gain of 50 dB and has a phase margin of only 15°. Although stable, the amplifier operating at a closed-loop gain of 50 dB exhibits significant overshoot and "ringing" in its step response. Finally, if an attempt is made to use the amplifier as a unity gain voltage follower, the amplifier will be unstable (negative phase margin). We see that the phase margin is zero for a closed-loop gain of approximately 35 dB.

Relative stability can be inferred directly from the magnitude plot. If the graphs of A and $1/\beta$ intersect at a "rate of closure" of 20 dB/decade, then the amplifier will be stable. However, if the two curves intersect in a region of 40 dB/decade, then the closed-loop amplifier will have poor phase margin (in the best case) or be unstable (in the worst case). Finally, if the rate of closure is 60 dB or greater, the closed-loop system will be unstable. The closure rate criterion is equally applicable to frequency-dependent feedback as well.

EXERCISE: What is the phase margin for the amplifier depicted in Ex. 17.6?

ANSWER: Approximately 50°.

17.12 SINGLE-POLE OPERATIONAL AMPLIFIER COMPENSATION

General-purpose operational amplifiers often use internal **frequency compensation,** which forces the overall amplifier to have a single-pole frequency response, as discussed in Chapter 12. The voltage transfer functions of these amplifiers can be represented by Eq. (17.107):

$$A_v(s) = \frac{A_o \omega_B}{s + \omega_B} = \frac{\omega_T}{s + \omega_B} \tag{17.107}$$

This form of transfer function can be obtained by connecting a compensation capacitor C_C around the second gain stage of the basic operational amplifier, as depicted in Fig. 17.56.

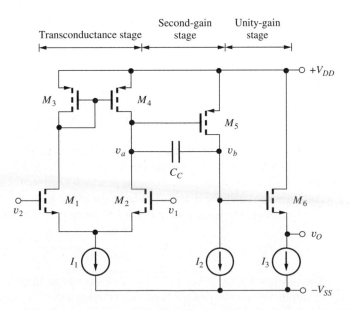

Figure 17.56 Frequency-compensation technique for single-pole operational amplifiers.

17.12.1 THREE-STAGE OP AMP ANALYSIS

Figure 17.57 is a simplified representation for the three-stage op amp. The input stage is modeled by its Norton equivalent circuit, represented by current source $G_m v_{dm}$ and output resistance R_o. The second stage provides a voltage gain $A_{v2} = g_{m5} r_{o5} = \mu_{f5}$, and the follower output stage is represented as a unity-gain buffer.

The circuit in Fig. 17.57 can be further simplified using the **Miller effect** relations. Feedback capacitor C_C is multiplied by the factor $(1 + A_{v2})$ and placed in parallel with the input of the second-stage amplifier, as in Fig. 17.58, and an expression for the output voltage can now be obtained from analysis of this figure. The output voltage $\mathbf{V_o}(s)$ must equal $\mathbf{V_b}(s)$ because the output buffer has a gain of 1. Also, $\mathbf{V_b}(s)$ equals $-A_{v2}\mathbf{V_a}(s)$.

Writing the nodal equation for $\mathbf{V_a}(s)$ assuming $i = 0$,

$$-G_m\mathbf{V_{dm}}(s) = \mathbf{V_a}(s)[sC_C(1 + A_{v2}) + G_o] \tag{17.108}$$

and

$$\frac{\mathbf{V_a}(s)}{\mathbf{V_{dm}}(s)} = \frac{-G_m R_o}{sR_o C_C(1 + A_{v2}) + 1} \tag{17.109}$$

Combining these results gives the overall gain of the op amp:

$$A_v(s) = \frac{\mathbf{V_o}(s)}{\mathbf{V_{dm}}(s)} = \frac{\mathbf{V_b}(s)}{\mathbf{V_{dm}}(s)} = \frac{-A_{v2}\mathbf{V_a}(s)}{\mathbf{V_{dm}}(s)} = \frac{G_m R_o A_{v2}}{1 + sR_o C_C(1 + A_{v2})} \tag{17.110}$$

Rewriting Eq. (17.110) in the form of (17.107) yields

$$A_v(s) = \frac{\dfrac{G_m A_{v2}}{C_C(1 + A_{v2})}}{s + \dfrac{1}{R_o C_C(1 + A_{v2})}} = \frac{\omega_T}{s + \omega_B} = \frac{A_o \omega_B}{s + \omega_B} \tag{17.111}$$

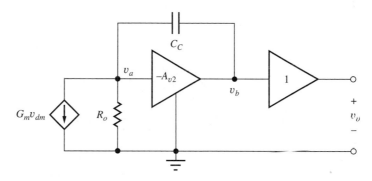

Figure 17.57 Simplified model for three-stage op amp.

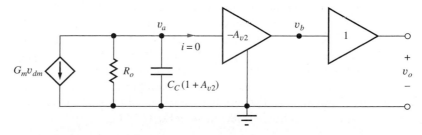

Figure 17.58 Equivalent circuit based on Miller multiplication.

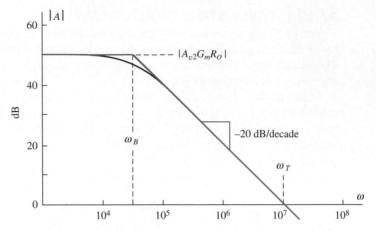

Figure 17.59 Gain magnitude plot for the ideal single-pole op amp.

Figure 17.59 is a Bode plot for this transfer function. At low frequencies the gain is $A_o = G_m R_o A_{v2}$, and the gain rolls off at 20 dB/decade above the frequency ω_B. Comparing Eq. (17.111) to (17.107),

$$\omega_B = \frac{1}{R_o C_C (1 + A_{v2})} \qquad \text{and} \qquad \omega_T = \frac{G_m A_{v2}}{C_C (1 + A_{v2})} \tag{17.112}$$

For large A_{v2},

$$\omega_T \cong \frac{G_m}{C_C} \tag{17.113}$$

Equation (17.113) is an extremely useful result. The unity gain frequency of the operational amplifier is set by the designer's choice of the values of the input stage transconductance and **compensation capacitor C_C.**

The single pole of the amplifier is at a relatively low frequency, as determined by the large values of the output resistance of the first stage and the Miller input capacitance of the second stage.

EXERCISE: What are the approximate values of G_m, R_o, f_T, and f_B for the op amp in Fig. 17.56 if $K_{n2} = 1$ mA/V^2, $K_{p5} = 1$ mA/V^2, $C_C = 20$ pF, $\lambda = 0.02$ V^{-1}, $I_1 = 100$ μA, and $I_2 = 500$ μA?

ANSWERS: 0.316 mS, 500 kΩ, 2.52 MHz, 158 Hz

17.12.2 TRANSMISSION ZEROS IN FET OP AMPS

Equation (17.113) presents an excellent method for controlling the frequency response of the operational amplifier with two gain stages. Unfortunately, however, we have overlooked a potential problem in the analysis of this amplifier: The simplified Miller approach does not take into account the finite transconductance of the second-stage amplifier.

The source of the problem can be understood by using the complete small-signal model for transistor M_5, as incorporated in Fig. 17.60. The previous analysis overlooked the zero that is determined by g_{m5} and the total feedback capacitance between the drain and gate of M_5. The circuit in Fig. 17.60 should once again look familiar. It is the same topology as the circuit for the simplified C-E amplifier, and we can use the results of the analysis in Eq. (16.93) by making the appropriate symbolic substitutions identified in Eq. (17.114):

$$r_{\pi o} \to R_o \qquad R_L \to r_{o5} \qquad C_\pi \to C_{GS5} \qquad C_\mu \to C_C + C_{GD5} \tag{17.114}$$

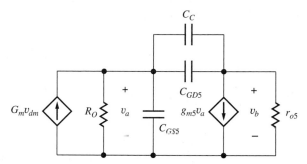

Figure 17.60 More complete model for op amp compensation.

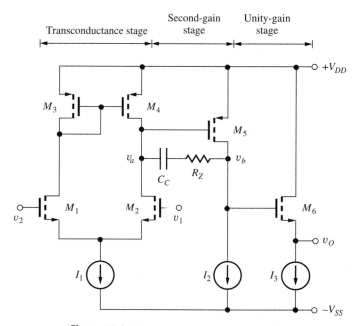

Figure 17.61 Zero cancellation using resistor R_Z.

With these transformations, the transfer function becomes

$$A_{vth}(s) = (-g_{m5}r_{o5})\frac{\left(1 - \dfrac{s}{\omega_Z}\right)}{\left(1 + \dfrac{s}{\omega_{P1}}\right)} \quad \text{in which} \quad \omega_Z = \frac{g_{m5}}{C_C + C_{GD5}} = \omega_T\frac{g_{m5}}{g_{m2}}$$

and

$$\omega_{P1} = \frac{1}{R_o C_T} \quad \text{where} \quad C_T = C_{GS5} + (C_C + C_{GD5})\left(1 + \mu_{f5} + \frac{r_{o5}}{R_o}\right)$$

(17.115)

In the case of many FET amplifier designs, ω_Z cannot be neglected because of the relatively low ratio of transconductances between FET M_5 and M_2. In bipolar designs, ω_Z can usually be neglected because of the much higher transconductance that is achieved for a given Q-point current. However, ω_Z can also be a problem in common-emitter amplifiers with emitter resistors that reduce the overall transconductance of the amplifier stage.

The problem can be overcome in FET amplifiers, however, through the addition of resistor R_Z in Fig. 17.61, which cancels the zero in Eq. (17.115). If we assume that $C_C \gg C_{GD}$, then the location of ω_Z in the numerator of Eq. (17.115) becomes

$$\omega_Z = \frac{\left(\dfrac{1}{g_{m5}}\right) - R_Z}{C_C} \tag{17.116}$$

and the zero can be eliminated by setting $R_Z = 1/g_{m5}$.

> **EXERCISE:** Find the approximate location of f_Z for the op amp in Fig. 17.61 using the values from the previous exercise. What value of R_Z is needed to eliminate f_Z?
>
> **ANSWERS:** 7.96 MHz; 1 kΩ

17.12.3 BIPOLAR AMPLIFIER COMPENSATION

The bipolar op amp in Fig. 17.62 is compensated in the same manner as the MOS amplifier. However, because the transconductance of the BJT is generally much higher than that of a FET for a given

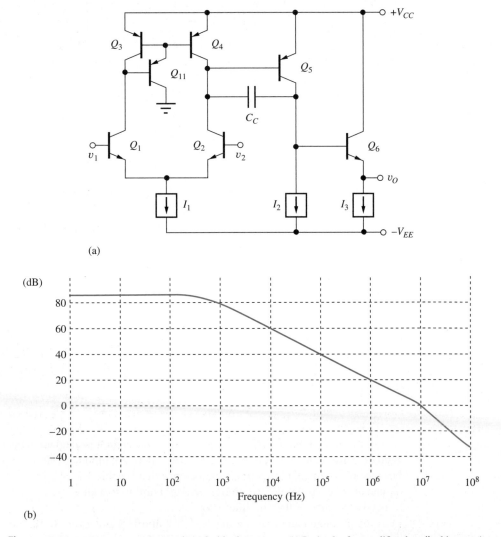

(a)

(b)

Figure 17.62 (a) Frequency compensation of a bipolar op amp. (b) Bode plot for amplifier described in exercise.

operating current, the transmission zero occurs at such a high frequency that it does not usually cause a problem. Applying Eq. (17.113) to the circuit in Fig. 17.62 yields an expression for the unity gain frequency of the two-stage bipolar amplifier:

$$\omega_T = \frac{g_{m2}}{C_C} = \frac{40I_{C2}}{C_C} = \frac{20I_1}{C_C} \qquad \text{and} \qquad \omega_Z = \frac{g_{m5}}{C_C} = \omega_T \left(\frac{I_{C5}}{I_{C2}}\right) \tag{17.117}$$

Because I_{C5} is 5 to 10 times I_{C2} in most designs, ω_Z is typically at a frequency of 5 to 10 times the unity gain frequency ω_T.

The simulated frequency response for the amplifier in Fig. 17.62(a) appears in Fig. 17.62(b) based on the values in the next exercise. The dominant pole, arising from the high resistance at the base of Q_5, occurs at approximately 565 Hz, and the unity-gain crossover occurs at 10 MHz. A second pole, due to the dominant pole of the *pnp* current mirror, causes the increased roll-off beyond 10 MHz.

> **EXERCISE:** Find the approximate locations of ω_T, ω_Z, and ω_B for the bipolar op amp in Fig. 17.62 if $C_C = 30$ pF, $V_A = 50$, $I_1 = 100$ μA, $I_2 = 500$ μA, and $I_3 = 5$ mA.
>
> **ANSWERS:** 10.6 MHz, 106 MHz, 565 Hz

17.12.4 SLEW RATE OF THE OPERATIONAL AMPLIFIER

Errors caused by slew-rate limiting of the output voltage of the amplifier were discussed in Chapter 12. Slew-rate limiting occurs because there is a limited amount of current available to charge and discharge the internal capacitors of the amplifier. For an internally compensated amplifier, C_C typically determines the **slew rate.** Consider the example of the CMOS amplifier with the large input signal (no longer a small signal) in Fig. 17.63. In this case, the voltages applied to the differential input stage cause current I_1 to switch completely to one side of the differential pair, in a manner directly analogous to the current switch discussed in Chapter 9.

Figure 17.64 is a simplified model for the amplifier in this condition. Because of the unity gain output buffer, output voltage v_O follows voltage v_B. Current I_1 must be supplied through compensation capacitor C_C, and the rate of change of the v_B, and hence v_O, must satisfy

$$I_1 = C_C \frac{d(v_B(t) - v_A(t))}{dt} = C_C \frac{d\left(v_B(t) + \dfrac{v_B(t)}{A_{v2}}\right)}{dt} \tag{17.118}$$

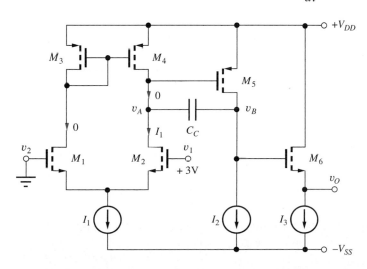

Figure 17.63 Operational amplifier with input stage overload.

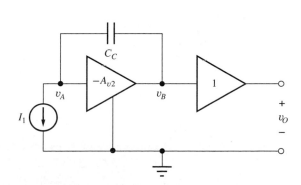

Figure 17.64 Simplified model for three-stage op amp.

If A_{v2} is assumed to be very large, then the amplifier will behave in a manner similar to an ideal integrator; that is, node voltage v_A represents a virtual ground, and Eq. (17.118) becomes

$$I_1 \cong C_C \frac{dv_B(t)}{dt} = C_C \frac{dv_O(t)}{dt} \tag{17.119}$$

The slew rate is the maximum rate of change of the output signal, and

$$SR = \left. \frac{dv_O(t)}{dt} \right|_{\text{max}} = \frac{I_1}{C_C} \tag{17.120}$$

The slew rate is determined by the total input stage bias current and the value of the compensation capacitor C_C. (It is seldom pointed out that this derivation tacitly assumes that the output of amplifier A_{v2} is capable of sourcing or sinking the current I_1. This requirement will be met as long as the amplifier is designed with $I_2 \geq I_1$.)

EXERCISE: Show that the slew rate is symmetrical in the CMOS amplifier in Fig. 17.63; that is, what is the current in capacitor C_C if $v_1 = 0$ V and $v_2 = +3$ V?

ANSWER: I_1

17.12.5 RELATIONSHIPS BETWEEN SLEW RATE AND GAIN-BANDWIDTH PRODUCT

Equation (17.120) can be related directly to the unity gain bandwidth of the amplifier using Eq. (17.113):

$$SR = \frac{I_1}{C_C} = \frac{I_1}{\left(\dfrac{G_m}{\omega_T}\right)} = \frac{\omega_T}{\left(\dfrac{G_m}{I_1}\right)} \tag{17.121}$$

For the simple CMOS amplifier in Fig. 17.56, the input stage transconductance is equal to that of transistors M_1 and M_2,

$$\left(\frac{G_m}{I_1}\right) = \frac{1}{I_1}\sqrt{2K_{n2}\frac{I_1}{2}} = \sqrt{\frac{2K_{n2}}{I_1}}$$

and

$$\tag{17.122}$$

$$SR = \omega_T \sqrt{\frac{I_1}{K_{n2}}}$$

For a given desired value of ω_T, the slew rate increases with the square root of the bias current in the input stage.

For the bipolar amplifier in Fig. 17.62,

$$\left(\frac{G_m}{I_1}\right) = \left(\frac{40\dfrac{I_1}{2}}{I_1}\right) = 20 \quad \text{and} \quad SR = \frac{\omega_T}{20} \tag{17.123}$$

In this case, the slew rate is related to the choice of unity gain frequency by a fixed factor.

EXERCISE: What is the slew rate of the CMOS amplifier in Fig. 17.56 if $K_{n2} = 1$ mA/V^2, $K_{p5} = 1$ mA/V^2, $C_C = 20$ pF, $\lambda = 0.02$ V^{-1}, $I_1 = 100$ μA, and $I_2 = 500$ μA?

ANSWER: 5.00 V/μS

EXERCISE: What is the slew rate of the bipolar amplifier in Fig. 17.62 if $C_C = 20$ pF, $I_1 = 100$ μA, and $I_2 = 500$ μA?

ANSWER: 5.00 V/μS

DESIGN EXAMPLE 17.10

OPERATIONAL AMPLIFIER COMPENSATION

In this example, we will choose the value of the compensation capacitor in a BJT op amp to give a desired value of phase margin.

PROBLEM Design the compensation capacitor in the BJT op amp circuit here to give a phase margin of 75°. Find the open-loop gain, bandwidth, and GBW product for the compensated op amp. For simplicity, assume that the *npn* and *pnp* transistors are described by the same set of SPICE parameters: BF = 100, VAF = 75 V, IS = 0.1 fA, RB = 250 Ω, TF = 0.75 ns, and CJC = 2 pF.

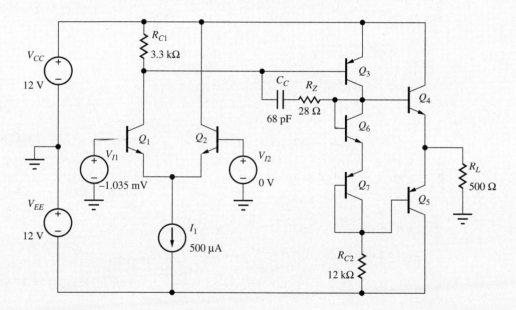

SOLUTION **Known Information and Given Data:** The three-stage op amp circuit appears here and consists of an *npn* differential input stage driving a common-emitter *pnp* gain stage. $R_{C1} = 3.3$ kΩ and $R_{C2} = 12$ kΩ. The output stage is a complementary *npn-pnp* emitter-follower stage. Transistor parameters are given as BF = 100, VAF = 75 V, IS = 0.1 fA, RB = 250 Ω, TF = 0.75 ns, and CJC = 2 pF, $\Phi_M = 75°$.

Unknowns: Value of C_C for 75° phase margin; the resulting open-loop gain and bandwidth, and unity-gain frequency; positions of the nondominant poles

Approach: Find the Q-points of the transistors and the small-signal parameters of the transistors. Assume that the dominant pole of the amplifier is set by compensation capacitor C_C around the *pnp* common-emitter gain stage. Find the nondominant poles of the amplifier resulting from the differential input stage and the emitter follower. Then choose C_C to give the unity-gain frequency required to achieve the desired phase margin.

Assumptions: The dominant pole of the op amp is set by compensation capacitor C_C and the *pnp* common-emitter stage; R_Z is included to remove the zero associated with C_C; $T = 27°C$; the *pnp* and *npn* transistors are identical; $V_{BE} = V_{EB} = 0.75$ V. The quiescent value of $V_O = 0$. Neglect all base currents in the Q-point analyses. VJC = 0.75 V and MJC = 0.33. Transistors Q_4 and Q_5 operate in parallel. The small-signal resistances of diode-connected transistors Q_6 and Q_7 can be neglected.

ANALYSIS **Q-Point:** Bias current I_1 splits equally between Q_1 and Q_2 so that $I_{C1} = I_{C2} = 250$ μA. For $V_O = 0$, the voltage across $R_{C2} = 12 - 0.75 = 11.3$ V, and the current in Q_3 is $I_{C3} = 11.3$ V/12 kΩ $= 938$ μA. Q_4 and Q_5 mirror the currents in Q_7 and Q_8, so $I_{C4} = I_{C5} = 938$ μA. For $V_O = 0$, $V_{CE4} = 12$ V, $V_{EC5} = 12$ V, and $V_{EC3} = 11.3$ V. For $V_I = 0$, $V_{CE2} = 12.8$ V and $V_{CE1} = 12 - 3300(0.25 \text{ mA}) + 0.75 = 11.9$ V.

Small-Signal Parameters: The small-signal parameters are found using these formulas cast in terms of the SPICE parameters:

$$\beta_o = \text{BF}\left(1 + \frac{V_{CE}}{\text{VAF}}\right) = 100\left(1 + \frac{V_{CE}}{75}\right) \qquad g_m = 40I_C \qquad r_\pi = \frac{\beta_o}{g_m}$$

$$r_o = \frac{\text{VAF} + V_{CE}}{I_C} = \frac{75 + V_{CE}}{I_C}$$

$$C_\pi = g_m\text{TF} = g_m(0.75 \times 10^{-9}) \qquad C_\mu = \frac{\text{CJC}}{\left(1 + \dfrac{V_{CB}}{\text{VJC}}\right)^{\text{MJC}}} = \frac{2 \text{ pF}}{\left(1 + \dfrac{V_{CB}}{0.75}\right)^{0.33}}$$

	I_C (μA)	V_{CE} (V)	β_o	g_m (S)	r_π (kΩ)	r_o (kΩ)	C_π (pF)	C_μ (pF)
Q_1	250	11.9	116	0.01	11.6	348	7.50	0.803
Q_2	250	12.8	117	0.01	11.7	351	7.50	0.784
Q_3	938	11.3	115	0.0375	3.07	92.0	28.1	0.818
Q_4	938	12.0	116	0.0375	3.09	92.8	28.1	0.801
Q_5	938	12.0	116	0.0375	3.09	92.8	28.1	0.801

Open Loop-Gain: $A_O = A_{vt1}A_{vt2}A_{vt3}$

$$A_{vt1} = \frac{g_{m1}}{2}(2r_{o1}\|R_{C1}\|r_{\pi3}) = \frac{0.01}{2}(696 \text{ kΩ}\|3.3 \text{ kΩ}\|3.07 \text{ kΩ}) = 7.93$$

$$A_{vt2} = g_{m2}\left[r_{o3}\|R_{C1}2\right\|\left(\frac{r_{\pi4}}{2} + (\beta_{o4} + 1)R_L\right)\right]$$

$$= 0.0375\left[92.0 \text{ kΩ}\|12 \text{ kΩ}\right\|\left(\frac{3.09 \text{ kΩ}}{2} + (117)500 \text{ Ω}\right)\right] = 338$$

$$A_{vt3} = \frac{(\beta_o + 1)R_L}{\dfrac{r_{\pi4}}{2} + (\beta_o + 1)R_L} = \frac{(117)500}{\dfrac{3090}{2} + (117)500} = 0.974$$

$$A_O = 2610$$

Compensation Capacitor Design: At the unity-gain frequency f_T, the dominant pole due to C_C will contribute a phase shift of 90°. The dominant poles of each of the other two stages will determine the phase margin. For a phase margin of 75°, the contributions of the additional poles can only be 15°. We expect these poles to be at frequencies above the op amp unity-gain frequency, typically 50 to 200 MHz.

Input Stage Pole

We are interested in the transfer function for the loop gain. In the feedback path, the input stage appears as a C-C/C-B cascade. Thus, we will use the equation from Table 16.2 for the pole at the input of a common-collector stage with $R_{L2} = 1/g_m$.

$$f_{pB2} = \left(\frac{1}{2\pi}\right) \frac{1}{([R_{th2} + r_{x2}] \| [r_{\pi2} + (\beta + 1)R_{L2}]) \left(C_{\mu2} + \dfrac{C_{\pi2}}{1 + g_{m2}R_{L2}}\right)}$$

$$= \left(\frac{1}{2\pi}\right) \frac{1}{([R_{th2} + r_{x2}] \| 2r_{\pi2}) \left(C_{\mu2} + \dfrac{C_{\pi2}}{2}\right)}$$

$$= \left(\frac{1}{2\pi}\right) \frac{1}{(250\,\Omega \| 2 \cdot 11.7\,\text{k}\Omega) \left(0.784\,\text{pF} + \dfrac{7.5\,\text{pF}}{2}\right)} = 142\,\text{MHz}$$

Gain Stage Pole

This pole will be dominated by the Miller effect capacitance associated with the compensation capacitance. The actual location of the pole will be calculated based on a desired phase margin.

Emitter-Follower Pole

The pole at the input to the emitter-follower stage will be affected by the pole-splitting action of the compensation capacitor placed across the gain stage. Assuming the compensation capacitor across the gain stage is much larger than the other capacitances in the circuit, the pole at the input to the follower stage is

$$f_{pB4} \cong \frac{g'_{m3}}{2\pi(C_{\pi3} + C_{L3})}$$

To account for r_x, we will use g'_{m3} as defined in Eq. (16.70). The C_π term represents the total equivalent capacitance to small-signal ground at the input to the gain stage, including the output capacitance of the differential pair. C_{L3} is the capacitance looking into the complementary pair follower stage. Assuming only one of the two devices in the complementary pair is carrying a signal at any instant in time, the complementary pair device is represented by a device with the same g_m, a current gain equal to the average of the two devices, and TF roughly equal to the average TF. Since C_μ is a junction parasitic capacitance, we will see the cumulative capacitance due to the C_μ of both devices. Given these conditions, the pole is calculated as

$$f_{pB4} \cong \left(\frac{1}{2\pi}\right) \frac{0.0375\,\text{mS}(3.07\,\text{k}\Omega / 3.32\,\text{k}\Omega)}{\left(0.8\,\text{pF} + 28.1\,\text{pF} + 2 \cdot 0.8\,\text{pF} + \dfrac{28.1\,\text{pF}}{1 + 0.0375\,\text{mS}\,(500\,\Omega)}\right)} = 173\,\text{MHz}$$

In addition to these terms, we should also expect to see a pole equal to approximately f_T, $[1/2\pi (\text{TF} + C_\mu/g_m)]$, at the emitter junction of the differential pair and at the output node since there is no additional output load capacitance. The f_T values for Q_1 and Q_4 are 192 MHz and 206 MHz, respectively.

We can now choose the unity-gain frequency, f_T, of the op amp to give the desired phase margin. At the unity-gain frequency, the primary pole of the op amp will contribute approximately 90° of phase shift. For a 75° phase margin, the remaining four poles can contribute an additional phase shift of 15°, which allows us to find the required value of f_T:

$$15° = \tan^{-1}\left(\frac{f_T}{142\,\text{MHz}}\right) + \tan^{-1}\left(\frac{f_T}{173\,\text{MHz}}\right) + \tan^{-1}\left(\frac{f_T}{192\,\text{MHz}}\right) + \tan^{-1}\left(\frac{f_T}{206\,\text{MHz}}\right)$$

Solving for the unity-gain frequency yields $f_T = 11.5$ MHz.

Using Eq. (17.113) from our op amp analysis,

$$(C_C + C_{\mu 3}) = \frac{G_{m1}}{\omega_T} = \left(\frac{g_{m1}}{2}\right)\left(\frac{1}{2\pi f_T}\right)$$

$$= \frac{0.005}{2\pi(11.5 \times 10^6)} = 69 \text{ pF}$$

since $C_{\mu 3}$ is approximately in parallel with C_C. To eliminate the unwanted zero associated with C_C, $R_Z = 1/g_{m3} = 27.5 \ \Omega$. Now we can also find the open-loop bandwidth:

$$f_B = \frac{f_T}{A_o} = \frac{11.5 \text{ MHz}}{2610} = 4.41 \text{ kHz}$$

Thus, our design values are $A_o = 68.3$ dB, $f_T = 11.5$ MHz, $f_B = 4.41$ kHz, and $\Phi_M = 75°$.

Check of Results: We will check our analysis using SPICE as outlined below.

Computer-Aided Analysis: In order to simulate the gain with the feedback loop open, we must first find the offset voltage of the amplifier. With the amplifier connected as a voltage follower, the offset voltage was found to be 1.035 mV. The Q-point collector currents for transistors Q_1 through Q_7 are 242 μA, 254 μA, 936 μA, 1.05 mA, 1.05 mA, 917 μA, and 917 μA, respectively.

The offset voltage was then applied to the input of the open-loop amplifier to set the output voltage to zero, and an ac sweep was performed from 1 Hz to 100 MHz with 20 simulation points per decade. The resulting open-loop gain is plotted below. The open-loop gain is 67.2 dB, the open-loop bandwidth is 4.52 kHz, the unity-gain frequency is 10.7 MHz, and the phase margin is 74°. These values all agree well with our design calculations. The phase margin is being affected by a zero that can be seen in the magnitude response above 30 MHz and was not included in our analysis.

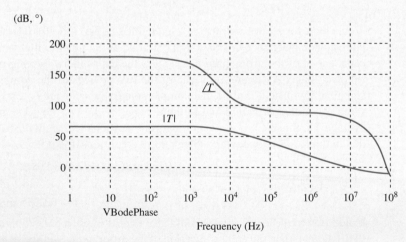

EXERCISE: Use the technique in Sec. 17.8.3 and Ex. 17.6 to verify the loop-gain plot in Ex. 17.10.

EXERCISE: Calculate the unity-gain frequency and phase margin for the amplifier in Ex. 17.10 if C_C is reduced to 50 pF.

ANSWER: 15.9 MHz, 27.6°

DESIGN
EXAMPLE 17.11

MOSFET OPERATIONAL AMPLIFIER COMPENSATION

In this example, we will choose the value of the compensation capacitor in a FET op amp to produce a desired value of phase margin for use in a unity-gain configuration.

PROBLEM Design the compensation capacitor in the FET op-amp circuit here to produce a phase margin of 70°. Find the open-loop gain, bandwidth, and GBW product for the compensated op amp. All of the NMOS FETS have SPICE parameters: KP = 10 mS/V, VTO = 1 V, LAMBDA = 0.01 V^{-1}. The PMOS FETS have SPICE parameters: KP = 4 mS/V, VTO = −1 V, LAMBDA = 0.01 V^{-1}. C_{GS} and C_{GD} are 5 pF and 1 pF, respectively, and will be added manually to the SPICE schematic. Consider M_5 to be the parallel combination of two PMOS FETs (or a PMOS with twice the W/L of the other PMOS FETs), KP = 8 mS/V, $C_{GS} = 10$ pF, and $C_{GD} = 2$ pF.

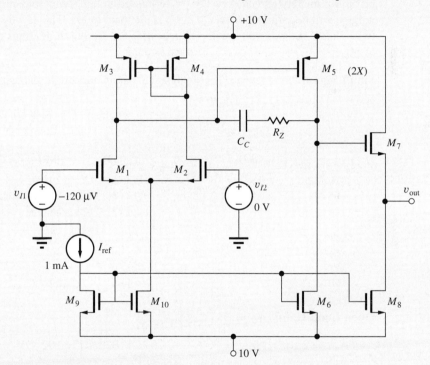

SOLUTION **Known Information and Given Data:** The three-stage op-amp circuit appears here and consists of an NMOS differential input stage with a PMOS current mirror load. The second stage is a PMOS common-source gain stage. The output stage is an NMOS source-follower stage. Transistor parameters given as KP = 10 mS/V (NMOS), KP = 4 mS/V (PMOS), VTO = −1 V, $C_{GD} = 1$ pF, and $C_{GS} = 5$ pF. Device M_5 is twice the width of the other PMOS FETs, so its KP, C_{GS}, and C_{GD} are doubled. $\Phi_M = 70°$.

Unknowns: Value of C_C for 70° phase margin; the resulting open-loop gain and bandwidth, and unity-gain frequency; positions of the nondominant poles.

Approach: Find the Q-points and small-signal parameters of the transistors. We initially assume that the dominant pole of the op amp is set by compensation capacitor C_C of the PMOS gain stage. Find the nondominant poles at the other nodes of the amplifier and use these to calculate the unity gain frequency required to achieve the desired phase margin.

Assumptions: The dominant pole is set by the compensation capacitor C_C and the C-S stage. We will include the appropriate value of R_Z to remove the right-half plane zero associated with the C-S gain stage. The circuit is operating at room temperature, and the circuit will be biased to produce a nominal output voltage of 0 volts. We will neglect the finite output impedance effects on device currents when calculating the operating points.

ANALYSIS **Q-Point:** The use of current mirror biasing and active loads greatly simplifies the calculation of the device operating currents. Given the reference current of 1 mA, we know that the bias currents for M_1–M_4 will all be 0.5 mA. M_6 and M_8 will nominally sink 1 mA. The V_{GS} of M_5 is of some interest. Because of the λ term in the FET current equation, we know that for I_{D3} and I_{D4} to be matched, they need to have the same V_{DS}. As a result, V_{GS} will nominally have the same value as the V_{GS} of M_3 and M_4. If M_5 is identical to M_4, their currents will therefore be approximately equal. However, M_6 is biased to sink twice the current of M_4, so the output voltage will be saturated near V_{SS} if M_5 is identical to M_4. This is why M_5 is specified as having twice the W/L of M_4, so it will produce twice the current of M_4, thus matching the current level of M_6.

Small-Signal Parameters: The small signal parameters are found using the following formulas:

$$r_o \cong \frac{1/\lambda + V_{DS}}{I_D} \qquad g_m = \sqrt{2KP \cdot I_D(1 + \lambda V_{DS})}$$

	I_D (mA)	V_{DS} (V)	g_m (mS)	r_o (kΩ)	C_{GD} (pF)	C_{GS} (pF)
M1, M2	0.5	9.8	3.46	120	1	5
M3, M4	0.5	1.5	2.03	103	1	5
M5	1	8.6	4.33	58.6	2	10
M6	1	11.4	4.96	61.4	1	5
M7	1	10	4.90	60	1	5
M8	1	10	4.90	60	1	5
M9	1	1.45	4.50	101	1	5
M10	1	8.55	4.66	109	1	5

Open Loop-Gain: $A_O = A_{vt1}A_{vt2}A_{vt3}$

$$A_{vt1} = g_{m1,2}(r_{o1}\|r_{o3}) = 3.46\,\text{mS}(120\,\text{k}\Omega\|103\,\text{k}\Omega) = 192\,\text{V/V}$$

$$A_{vt2} = -g_{m5}(r_{o5}\|r_{o6}) = 4.33\,\text{mS}(58.6\,\text{k}\Omega\|61.4\,\text{k}\Omega = -130\,\text{V/V}$$

$$A_{vt3} = \frac{g_{m7}R_{S7}}{1 + g_{m7}R_{S7}} = \frac{g_{m7}(r_{o7}\|r_{o8})}{1 + g_{m7}(r_{o7}\|r_{o8})} = \frac{4.90\,\text{mS}(60\,\text{k}\Omega\|60\,\text{k}\Omega)}{1 + 4.90\,\text{mS}(60\,\text{k}\Omega\|60\,\text{k}\Omega)} = 0.993\,\text{V/V}$$

$$A_O = -24{,}800\,\text{V/V} = 87.9\,\text{dB}$$

Compensation Capacitor Design: At f_T, the loop gain reaches 0 dB and the dominant pole will contribute approximately 90° of phase shift. To achieve a phase margin of 70°, the compensation capacitor is selected to set the unity-gain frequency such that the nondominant poles are contributing a total of 90 − 70 or 20° of phase shift (the inverting input contributes another 180°).

Input Stage Pole
We are interested in the transfer function for the loop gain, and in the feedback path, the input stage appears as a C-D/C-G cascade. Since we are driving our input with a zero impedance source, the pole at the gate of M_2 has infinite frequency. Since we are designing the op amp to be stable in a unity-gain configuration, we will include the M_2 input capacitance as an additional capacitive

load at the output in our calculations to model the capacitive loading seen by the output when the negative feedback is connected.

$$C_{in} = C_{GD} + \frac{C_{GS}}{1 + g_{m2}R_{S2}} = C_{GD} + \frac{C_{GS}}{1 + \dfrac{g_{m2}}{g_{m1}}} = 1\,\text{pF} + \frac{5\,\text{pF}}{2} = 3.5\,\text{pF}$$

Differential Pair Source Node Pole

There is a high-frequency pole at the differential pair source node. This pole is found as

$$f_{pS1} \cong \left(\frac{1}{2\pi}\right) \frac{1}{\left(\dfrac{1}{g_{m1}} \middle\| \dfrac{1}{g_{m2}}\right)(C_{GS1} + C_{GS2} + C_{GD10})}$$

$$= \left(\frac{1}{2\pi}\right) \frac{1}{\left(\dfrac{0.5}{3.46\,\text{mS}}\right)(5\,\text{pF} + 5\,\text{pF} + 1\,\text{pF})} = 100\,\text{MHz}$$

Gain Stage Pole

This pole will be dominated by the Miller effect capacitance at the input to the gain stage and associated with the compensation capacitance, C_C. The actual location of the pole will be calculated based on a desired phase margin.

Source Follower Input Pole

This pole at the input to the emitter-follower stage will be affected by the pole-splitting action of the compensation capacitor placed across the gain stage. Assuming the compensation capacitor across the gain stage is much larger than the other capacitances in the circuit, the pole at the input to the follower stage is

$$f_{pD5} \cong \frac{g_{m5}}{2\pi(C_{GS5} + C_{L5})}$$

As with our bipolar example, the C_{GS} term above represents the total equivalent capacitance to small-signal ground at the input to the gain stage, including the output capacitance of the differential pair.

$$C_{i5} = C_{GD1} + C_{GD3} + C_{GS5} = (1 + 1 + 10)\,\text{pF} = 12\,\text{pF}$$

C_{L3} is the capacitance looking into the C-C output stage plus the capacitance seen looking into the current source.

$$C_{L5} = C_{GD6} + C_{GD7} + \frac{C_{GS7}}{1 + g_{m7}(r_{o7}\|r_{o8})} = 1 + 1 + \frac{5}{1 + 4.9\,\text{mS}(30\,\text{k}\Omega)} = 2.03\,\text{pF}$$

Given these results, the pole is calculated as

$$f_{pD5} \cong \left(\frac{1}{2\pi}\right) \frac{4.33\,\text{mS}}{(12\,\text{pF} + 2.03\,\text{pF})} = 49.2\,\text{MHz}$$

Output Pole

The pole at the output will be set by finding the equivalent resistance and capacitance at the output node. As mentioned earlier, we will include the capacitance at the gate of M_2 to model the loading of the output when the output is fed back to the input.

$$C_{eqS7} = C_{GS7} + C_{GD8} + C_{in} = 5\,\text{pF} + 1\,\text{pF} + 3.5\,\text{pF} = 9.5\,\text{pF}$$

$$R_{eqS7} \cong 1/g_{m7} = 204\,\Omega$$

$$f_{pS7} \cong \left(\frac{1}{2\pi}\right) \frac{1}{(204)(9.5\,\text{pF})} = 82.1\,\text{MHz}$$

We can now choose the unity-gain frequency, f_T, of the op amp to give the desired phase margin. At the unity-gain frequency, the primary pole of the op amp will contribute approximately 90° of phase shift. For a 70° phase margin, the remaining two poles can contribute an additional phase shift of 20°, which allows us to find the required value of f_T:

$$20° = \tan^{-1}\left(\frac{f_T}{49.2\,\text{MHz}}\right) + \tan^{-1}\left(\frac{f_T}{82.1\,\text{MHz}}\right) + \tan^{-1}\left(\frac{f_T}{100\,\text{MHz}}\right) \rightarrow f_T \cong 8.5\,\text{MHz}$$

Using our single-pole op-amp compensation result from the previous section, we calculate the compensation capacitor as

$$(C_C + C_{GD5}) = \frac{g_{m1}}{2\pi f_T} \rightarrow C_C = \frac{3.46\,\text{mS}}{2\pi(8.5\,\text{MHz})} - 2\,\text{pF} = 63\,\text{pF}$$

To eliminate the unwanted right-half plane zero associated with C_C, $R_Z = 1/g_{m5} = 230\,\Omega$. The open-loop bandwidth is now calculated as a function of the midband gain and the unity-gain frequency.

$$f_B = \frac{f_T}{A_O} = \frac{8.5\,\text{MHz}}{24{,}800} = 343\,\text{Hz}$$

Our final design values are $A_O = 87.9$ dB, $f_T = 8.5\,\text{MHz}$, $f_B = 343\,\text{Hz}$, and $\Phi_M = 70°$.

Check of Results: In this case, results will be verified by SPICE simulation.

Simulation: With such a high gain amplifier, we should expect to have a significant offset at the output when the amplifier is operated in open loop. Any bias error at the input gets multiplied by the gain of the amplifier. As with the previous BJT example, we connect the amplifier in a follower configuration and then apply the opposite offset to the input to cancel the offset and allow us to perform open loop ac simulations while maintaining a 0 V bias at the output.

With the appropriate offset in place, the amplifier is simulated with an ac sweep from 1 Hz to 100 MHz. The first simulation is performed without R_Z to illustrate the stability problems created by the presence of the RHP zero in FET amplifiers.

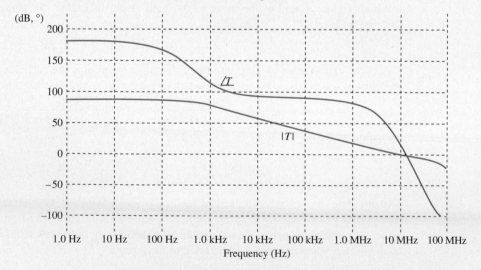

Loop gain without R_Z.

In the first simulation, the unity-gain frequency is 11 MHz but our phase margin is only 20°! The RHP zero is the worst of all conditions for stability. It simultaneously causes the magnitude response slope to decrease (increasing the 0 dB crossing frequency) while adding negative phase shift. If we resimulate with R_Z in place, we find the following loop gain response:

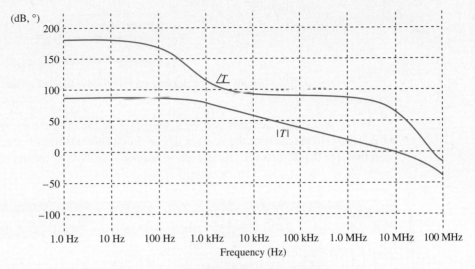

Loop gain with R_Z added to cancel RHP zero.

With the RHP zero cancelled by R_Z, our simulated result is quite close to the design values. The unity-gain frequency, f_T, is 8.4 MHz, and our phase margin is 69°. If we desire to increase the phase margin, R_Z can be increased to move the zero into the left-half plane and introduce some positive phase shift. The open-loop gain, A_O is 86.5 dB and the open-loop bandwidth, f_B, is approximately 410 Hz. These values are within the range of expected agreement with our design calculations.

EXERCISE: **What is the slew rate of the amplifier in Ex. 17.11?**

ANSWER: 15.4 V/μsec

EXERCISE: **(a) What value of compensation capacitor is required to achieve a 60° phase margin in the amplifier in Ex. 17.11? (b) Verify your results with SPICE simulation.**

ANSWER: 31.3 pF

17.13 OSCILLATORS

Oscillators are an important class of feedback circuits that are used for signal generation. We saw one form of oscillator, the nonlinear multivibrator, in Chapter 11. However, that circuit is limited to relatively low frequency operation by the characteristics of the operational amplifier. In Secs. 17.13.2 to 17.13.4, we consider **sinusoidal oscillators,** which are based on linear amplifiers suitable for signal generation at frequencies up to at least 0.5 to 1 GHz.

17.13.1 THE BARKHAUSEN CRITERIA FOR OSCILLATION

The oscillator can be described by a positive (or regenerative) feedback system using the block diagram in Fig. 17.65. A frequency-selective feedback network is used, and the oscillator is designed to produce an output even though the input is zero.

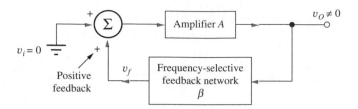

Figure 17.65 Block diagram for a positive feedback system.

For a sinusoidal oscillator, we want the poles of the closed-loop amplifier to be located at a frequency ω_o, precisely on the $j\omega$ axis. These circuits use positive feedback through the frequency-selective feedback network to ensure sustained oscillation at the frequency ω_o. Consider

ELECTRONICS IN ACTION

A MEMS Oscillator

Crystal oscillators have long been a mainstay for creating accurate, stable oscillators for clocks in watches and computer systems. Unlike oscillators based on integrated inductors and capacitors, crystal oscillators have very low equivalent series resistance, leading to low loss and high Q. However, conventional crystal oscillators are relatively bulky and are not easily integrated with CMOS processes. For this reason, researchers have developed microelectromechanical systems (MEMS) based resonant structures that can be integrated directly onto CMOS integrated circuits.

Illustrated below is a MEMS micromechanical resonator published in 1999 by Clark Nguyen and Roger Howe.[1] A photomicrograph of the device is shown below. The structure is an electrostatic comb-drive constructed from polysilicon material. A cross section of the MEMS post processing is also shown. The large polysilicon structure to the left is an example of the structures used to make the resonator structure. The structure makes electrical contact to a metal layer through a thin deposited polysilicon layer. Note that the horizontal beam in the left of the figure is actually suspended above the substrate. The structural polysilicon is deposited over a sacrificial phosphosilicate glass (PSG) that had been previously deposited and patterned. After the structural polysilicon is deposited and patterned, the PSG is chemically etched away, leaving the polysilicon beams suspended above the substrate.

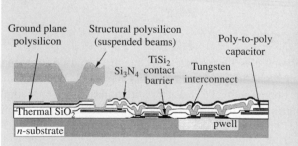

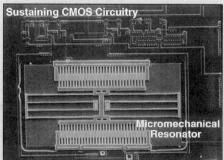

Copyright IEEE 1999. Reprinted with permission from [1].

[1] C. T.-C. Nguyen and R. T. Howe, "An integrated CMOS micromechanical resonator high-Q oscillator," *IEEE J. Solid-State Circuits*, vol. 34, no. 4, pp. 440–445, April 1999.

The physical structure of the comb drive is more clearly seen in the block diagram below. By driving the leftmost finger structure with a voltage, the suspended structure in the middle is pulled to the left. When the voltage is removed, the structure is pulled back to the right by the suspension. When the frequency of the drive voltage approaches the resonant frequency of the structure, sustainable oscillation begins. Similar to a quartz oscillator, the micromechanical resonator has a series RLC and parallel capacitance model. As the center structure oscillates back and forth, a displacement current is generated on the output port comb structure due to the changing capacitance as the comb fingers move in and out. In this design, the displacement current is sensed by a transresistance amplifier which amplifies the signal and drives the input. At the resonant frequency, the Barkhausen criteria is satisfied and the oscillation is sustained.

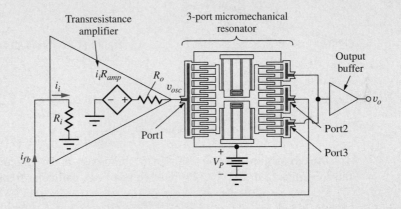

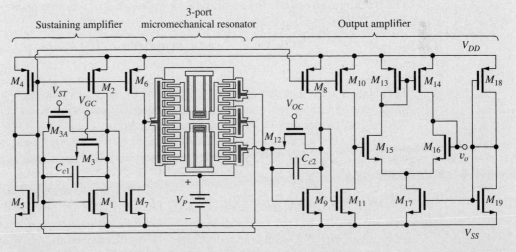

MEMS based devices are enabling fully integrated mixers, filters, and other resonator based structures. Because the structures are typically made from polysilicon material, they are compatible with conventional CMOS IC processing. The structure shown above has a resonant frequency in the tens of kilohertz, but researchers are exploring other resonator forms with demonstrated resonant frequencies in the hundreds of megahertz. The combination of MEMS and CMOS may soon enable highly efficient single-chip radio frequency transceivers that do not rely on the relatively lossy integrated capacitors and inductors used today.

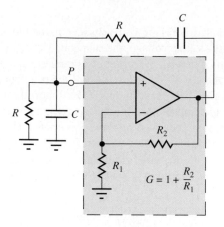

Figure 17.66 Wien-bridge oscillator circuit.

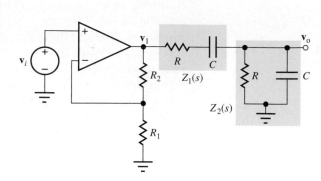

Figure 17.67 Circuit for finding the loop gain of the Wien-bridge oscillator.

the feedback system in Fig. 17.65, which is described by

$$A_v(s) = \frac{A(s)}{1 - A(s)\beta(s)} = \frac{A(s)}{1 - T(s)} \tag{17.124}$$

The use of positive feedback results in the minus sign in the denominator. For sinusoidal oscillations, the denominator of Eq. (17.124) must be zero for a particular frequency ω_o on the $j\omega$ axis:

$$1 - T(j\omega_o) = 0 \qquad \text{or} \qquad T(j\omega_o) = +1 \tag{17.125}$$

The **Barkhausen criteria for oscillation** are a statement of the two conditions necessary to satisfy Eq. (17.125):

1. $\angle T(j\omega_o) = 0°$ \qquad or even multiples of $360°$—$2n\pi$ rad

2. $|T(j\omega_o)| = 1$

$$\tag{17.126}$$

These two criteria state that the phase shift around the feedback loop must be zero degrees, and the magnitude of the loop gain must be unity. Unity loop gain corresponds to a truly sinusoidal oscillator. A loop gain greater than 1 causes a distorted oscillation to occur.

In Sec. 17.13.2 we look at several RC oscillators that are useful at frequencies below a few megahertz. Following that discussion, LC and crystal oscillators, both suitable for use at much higher frequencies, are presented.

17.13.2 OSCILLATORS EMPLOYING FREQUENCY-SELECTIVE RC NETWORKS

RC networks can be used to provide the required frequency-selective feedback at frequencies below a few megahertz. This section introduces two RC oscillator circuits: the Wien-bridge oscillator and the phase-shift oscillator. Another example, the quadrature oscillator, is in Prob. 17.96.

The Wien-Bridge Oscillator

The **Wien-bridge oscillator**[5] in Fig. 17.66 uses two RC networks to form the frequency-selective feedback network. The loop gain $T(s)$ for the Wien-bridge circuit can be found by breaking the loop at point P, as redrawn in Fig. 17.67. The operational amplifier is operating as a noninverting amplifier with a gain $G = V_1(s)/V_1(s) = 1 + R_2/R_1$. The loop gain can be found using voltage

---

[5] A version of this oscillator was the product that launched the Hewlett-Packard Company.

division between $Z_1(s)$ and $Z_2(s)$:

$$V_o(s) = V_1(s)\frac{Z_2(s)}{Z_1(s) + Z_2(s)}$$

(17.127)

$$Z_1(s) = R + \frac{1}{sC} = \frac{sCR + 1}{sC} \quad \text{and} \quad Z_2(s) = R\|\frac{1}{sC} = \frac{R}{sCR + 1}$$

Simplifying Eq. (17.127) yields the transfer function for the loop gain:

$$\mathbf{V_o}(s) = G\mathbf{V_1}(s)\frac{sRC}{s^2R^2C^2 + 3sRC + 1}$$

$$T(s) = \frac{\mathbf{V_o}(s)}{\mathbf{V_I}(s)} = \frac{sRCG}{s^2R^2C^2 + 3sRC + 1}$$

(17.128)

For $s = j\omega$,

$$T(j\omega) = \frac{j\omega RCG}{(1 - \omega^2 R^2 C^2) + 3jwRC}$$

(17.129)

Applying the first Barkhausen criterion, we see that the phase shift will be zero if $(1 - \omega_o^2 R^2 C^2) = 0$. At the frequency $\omega_o = 1/RC$,

$$\angle T(j\omega_o) = 0° \quad \text{and} \quad |T(j\omega_o)| = \frac{G}{3}$$

(17.130)

At $\omega = \omega_o$, the phase shift is zero degrees. If the gain of the amplifier is set to $G = 3$, then $|T(j\omega_o)| = 1$, and sinusoidal oscillations will be achieved.

The Wien-bridge oscillator is useful up to frequencies of a few megahertz, limited primarily by the characteristics of the amplifier. In signal generator applications, capacitor values are often switched by decade values to achieve a wide range of oscillation frequencies. The resistors can be replaced with potentiometers to provide continuous frequency adjustment within a given range.

The Phase-Shift Oscillator

A second type of RC oscillator is the **phase-shift oscillator** depicted in Fig. 17.68. A three-section RC network is used to achieve a phase shift of 180°, which, added to the 180° phase shift of the inverting amplifier, results in a total phase shift of 360°.

The phase-shift oscillator has many practical implementations. One possible implementation combines a portion of the phase-shift function with an op amp gain block, as in Fig. 17.69. The loop gain can be found by breaking the feedback loop at x–x' and calculating $\mathbf{V_o}(s)$ in terms of $\mathbf{V'_o}(s)$.

Writing the nodal equations for voltages $\mathbf{V_1}$ and $\mathbf{V_2}$,

$$\begin{bmatrix} sC\mathbf{V'_o}(s) \\ 0 \end{bmatrix} = \begin{bmatrix} (2sC + G) & -sC \\ -sC & (2sC + G) \end{bmatrix}\begin{bmatrix} \mathbf{V_1}(s) \\ \mathbf{V_2}(s) \end{bmatrix}$$

(17.131)

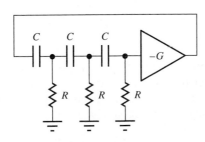

Figure 17.68 Basic concept for the phase-shift oscillator.

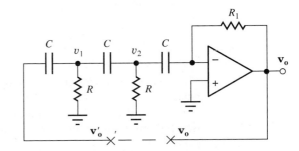

Figure 17.69 One possible realization of the phase-shift oscillator.

and using standard op amp theory:

$$\frac{\mathbf{V_o}(s)}{\mathbf{V_2}(s)} = -sCR_1 \tag{17.132}$$

Combining Eqs. (17.131) and (17.132) and solving for $\mathbf{V_o}(s)$ in terms of $\mathbf{V_o'}(s)$ yields

$$T(s) = \frac{\mathbf{V_o}(s)}{\mathbf{V_o'}(s)} = -\frac{s^3C^3R^2R_1}{3s^2R^2C^2 + 4sRC + 1} \tag{17.133}$$

and

$$T(j\omega) = -\frac{(j\omega)^3C^3R^2R_1}{(1 - 3\omega^2R^2C^2) + j4\omega RC} = \frac{j\omega^3C^3R^2R_1}{(1 - 3\omega^2R^2C^2) + j4\omega RC} \tag{17.134}$$

We can see from Eq. (17.134) that the phase shift of $T(j\omega)$ will be zero if the real term in the denominator is zero:

$$1 - 3\omega_o^2R^2C^2 = 0 \quad \text{or} \quad \omega_o = \frac{1}{\sqrt{3}RC} \tag{17.135}$$

and

$$T(j\omega_o) = +\frac{\omega_o^2C^2RR_1}{4} = +\frac{1}{12}\frac{R_1}{R} \tag{17.136}$$

For $R_1 = 12R$, the second Barkhausen criterion is met ($|T(j\omega_o)| = 1$).

Amplitude Stabilization in *RC* Oscillators

As power supply voltages, component values, and/or temperature change with time, the loop gain of an oscillator also changes. If the loop gain becomes too small, then the desired oscillation decays; if the loop gain is too large, waveform distortion occurs. Therefore, some form of **amplitude stabilization,** or gain control, is often used in oscillators to automatically control the loop gain and place the poles exactly on the $j\omega$ axis. Circuits will be designed so that, when power is first applied, the loop gain will be larger than the minimum needed for oscillation. As the amplitude of the oscillation grows, the gain control circuit reduces the gain to the minimum needed to sustain oscillation.

Two possible forms of amplitude stabilization are shown in Figs. 17.70 to 17.73. In the original Hewlett-Packard Wien-bridge oscillator, resistor R_1 was replaced by a nonlinear element, the light-bulb in Fig. 17.70. The small-signal resistance of the lamp is strongly dependent on the temperature of the filament of the bulb. If the amplitude is too high, the current is too large and the resistance of the lamp increases, thereby reducing the gain. If the amplitude is low, the lamp cools, the resistance decreases, and the loop gain increases. The thermal time constant of the bulb effectively averages the signal current, and the amplitude is stabilized using this clever technique.

In the Wien-bridge circuit in Fig. 17.71, diodes D_1 and D_2 and resistors R_1 to R_4 form an amplitude control network. For a positive output signal at node v_O, diode D_1 turns on as the voltage across R_3 exceeds the diode turn-on voltage. When the diode is on, resistor R_4 is switched in parallel with R_3, reducing the effective value of the loop gain. Diode D_2 functions in a similar manner on the negative peak of the signal. The values of the resistors should be chosen so that

$$\frac{R_2 + R_3}{R_1} > 2 \quad \text{and} \quad \frac{R_2 + R_3 \| R_4}{R_1} < 2 \tag{17.137}$$

The first ratio should be set to be slightly greater than 2, and the second to slightly less than 2. Thus, when the diodes are off, the op amp gain is slightly greater than 3, ensuring oscillation, but when one of the diodes is on, the gain is reduced to slightly less than 3.

An estimate for the amplitude of oscillation can be determined from the circuit in Fig. 17.72, in which diode D_1 is assumed to be conducting with an on-voltage equal to V_D. The current i can be expressed as

$$\mathbf{i} = \frac{\mathbf{v_O} - \mathbf{v_1}}{R_3} + \frac{\mathbf{v_O} - \mathbf{v_1} - V_D}{R_4} \tag{17.138}$$

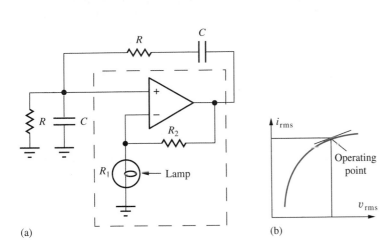

Figure 17.70 (a) Wien-bridge with amplitude stabilization. (b) Bulb i-v characteristic.

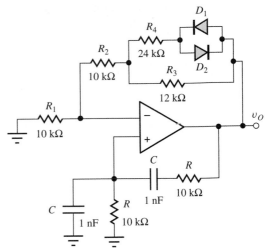

Figure 17.71 Diode amplitude stabilization of a Wien-bridge oscillator.

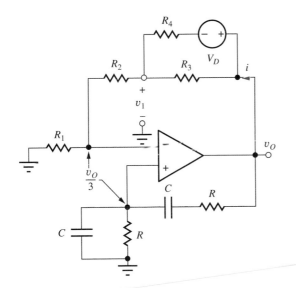

Figure 17.72 Equivalent circuit with diode D_1 on.

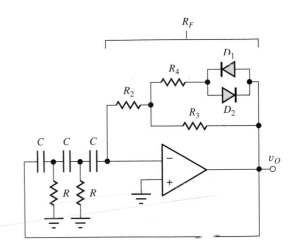

Figure 17.73 Diode amplitude stabilization of a phase-shift oscillator.

From Eq. (17.130) and ideal op amp behavior, we know that the voltages at both the inverting and noninverting input terminals are equal to one-third of the output voltage. Therefore,

$$\mathbf{v_1} = \frac{\mathbf{v_O}}{3}\left(1 + \frac{R_2}{R_1}\right) \tag{17.139}$$

Combining Eqs. (17.138) and (17.139) and solving for $\mathbf{v_O}$ yields

$$\mathbf{v_O} = \frac{3V_D}{\left(2 - \dfrac{R_2}{R_1}\right)\left(1 + \dfrac{R_4}{R_3}\right) - \dfrac{R_4}{R_1}} \qquad \text{where} \qquad \frac{R_2}{R_1} < 2 \tag{17.140}$$

Because the gain control circuit is actually a nonlinear circuit, Eq. (17.140) is only an estimate of the actual output amplitude; nevertheless, it does provide a good basis for circuit design.

A similar amplitude stabilization network is applied to the phase-shift oscillator in Fig. 17.73. In this case, conduction through the diodes adjusts the effective value of the total feedback resistance R_F, which determines the gain.

EXERCISE: What are the amplitude and frequency of oscillation for the Wien-bridge oscillator in Fig. 17.72? Assume $V_D = 0.6$ V.

ANSWERS: 9.95 kHz; 3.0 V

EXERCISE: Simulate the Wein-bridge oscillator using SPICE and find the frequency and amplitude of oscillation. Model the op amp using a macromodel with a gain of 100,000.

ANSWERS: 9.5 kHz; 3 V

17.13.3 *LC* OSCILLATORS

Individual transistors are used in oscillators designed for high-frequency operation, and the frequency-selective feedback network is formed from a high-Q LC network or a quartz crystal resonant element. Two classic forms of **LC oscillator** are introduced here: The Colpitts oscillator uses capacitive voltage division to adjust the amount of feedback, and the Hartley oscillator employs an inductive voltage divider. Crystal oscillators are discussed in Sec. 17.13.4.

The Colpitts Oscillator

Figure 17.74 shows the basic **Colpitts oscillator.** A resonant circuit is formed by inductor L and the series combination of C_1 and C_2; C_1, C_2, or L can be made variable elements in order to adjust the frequency of oscillation. The dc equivalent circuit is shown in Fig. 17.74(b). The gate of the FET is maintained at dc ground through inductor L, and the Q-point can be determined using standard techniques. In the small-signal model in Fig. 17.74(c), the gate-source capacitance C_{GS} appears in parallel with C_2, and the gate-drain capacitance C_{GD} appears in parallel with the inductor.

This circuit is used to illustrate another approach to finding the conditions for oscillation. The algebra in the analysis can be simplified by defining $G = 1/(R_S \| r_o)$ and $C_3 = C_2 + C_{GS}$. Writing nodal equations for $\mathbf{V_g}(s)$ and $\mathbf{V_s}(s)$ yields

$$\begin{bmatrix} 0 \\ 0 \end{bmatrix} = \begin{bmatrix} \left(s(C_3 + C_{GD}) + \dfrac{1}{sL}\right) & -sC_3 \\ -(sC_3 + g_m) & (s(C_1 + C_3) + g_m + G) \end{bmatrix} \begin{bmatrix} \mathbf{V_g}(s) \\ \mathbf{V_s}(s) \end{bmatrix} \qquad (17.141)$$

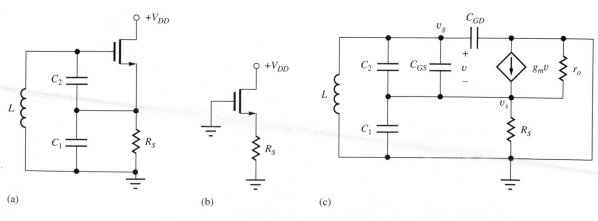

Figure 17.74 (a) Colpitts oscillator and (b) its dc and (c) small-signal models.

The determinant of this system of equations is

$$\Delta = s^2[C_1C_3 + C_{GD}(C_1 + C_3)] + s[(C_3 + C_{GD})G + GC_3] + \frac{(g_m + G)}{sL} + \frac{(C_1 + C_3)}{L} \tag{17.142}$$

Because the oscillator circuit has no external excitation, we must require $\Delta = 0$ for a nonzero output voltage to exist. For $s = j\omega$, the determinant becomes

$$\Delta = \left(\frac{(C_1 + C_3)}{L} - \omega^2[C_1C_3 + C_{GD}(C_1 + C_3)]\right)$$
$$+ j\left(\omega[(g_m + G)C_{GD} + GC_3] - \frac{(g_m + G)}{\omega L}\right) = 0 \tag{17.143}$$

after collecting the real and imaginary parts. Setting the real part equal to zero defines the frequency of oscillation ω_o,

$$\omega_o = \frac{1}{\sqrt{L\left(C_{GD} + \dfrac{C_1C_3}{C_1 + C_3}\right)}} = \frac{1}{\sqrt{LC_{TC}}} \qquad \text{where} \qquad C_{TC} = C_{GD} + \frac{C_1C_3}{C_1 + C_3} \tag{17.144}$$

and setting the imaginary part equal to zero yields a constraint on the gain of the FET circuit:

$$\omega^2 L \left[C_{GD} + \frac{G}{(g_m + G)}C_3\right] = 1 \tag{17.145}$$

At $\omega = \omega_o$, the gain requirement expressed by Eq. (17.145) can be simplified to yield

$$g_m R = \frac{C_3}{C_1} \qquad \left(g_m R \geq \frac{C_3}{C_1}\right) \tag{17.146}$$

From Eq. (17.144), we see that the frequency of oscillation is determined by the resonant frequency of the inductor L and the total capacitance C_{TC} in parallel with the inductor. The feedback is set by the capacitance ratio and must be large enough to satisfy the condition in Eq. (17.146). A gain that satisfies the equality places the oscillator poles exactly on the $j\omega$ axis. However, normally, more gain is used to ensure oscillation, and some form of amplitude stabilization is used.

The Hartley Oscillator

Feedback in the **Hartley oscillator** circuit in Fig. 17.75 is set by the ratio of the two inductors L_1 and L_2. The dc circuit for this case appears in Fig. 17.75(b). The conditions for oscillation can be found in a manner similar to that used for the Colpitts oscillator. For simplicity, the gate-source and gate-drain capacitances have been neglected, and no mutual coupling appears between the inductors. Writing the nodal equations for the small-signal model in Fig. 17.75(c):

$$\begin{bmatrix} 0 \\ 0 \end{bmatrix} = \begin{bmatrix} sC + \dfrac{1}{sL_2} & -\dfrac{1}{sL_2} \\ -\left(\dfrac{1}{sL_2} + g_m\right) & \dfrac{1}{sL_1} + \dfrac{1}{sL_2} + g_m + g_o \end{bmatrix} \begin{bmatrix} V_g(s) \\ V_s(s) \end{bmatrix} \tag{17.147}$$

The determinant of this system of equations is

$$\Delta = sC(g_m + g_o) + \frac{g_o}{sL_2} + \frac{1}{s^2L_1L_2} + C\left(\frac{1}{L_1} + \frac{1}{L_2}\right) \tag{17.148}$$

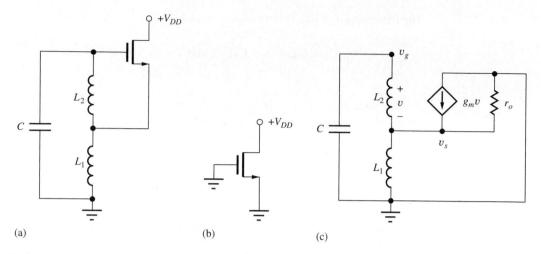

Figure 17.75 (a) Hartley oscillator using a JFET. (b) dc Equivalent circuit. (c) Small-signal model (C_{GS} and C_{GD} have been neglected for simplicity).

For oscillation, we require $\Delta = 0$. After collecting the real and imaginary parts for $s = j\omega$, the determinant becomes

$$\Delta = \left[C \left(\frac{1}{L_1} + \frac{1}{L_2} \right) - \frac{1}{\omega^2 L_1 L_2} \right] + j \left(\omega C (g_m + g_o) - \frac{g_o}{\omega L_2} \right) = 0 \qquad (17.149)$$

Setting the real part equal to zero again defines the frequency of oscillation ω_o,

$$\omega_o = \frac{1}{\sqrt{C(L_1 + L_2)}} \qquad (17.150)$$

and setting the imaginary part equal to zero yields a constraint on the amplification factor of the FET:

$$1 + g_m r_o = \frac{1}{\omega_2 C L_2} \qquad (17.151)$$

At $\omega = \omega_o$, the gain requirement expressed by Eq. (17.151) becomes

$$\mu_f = \frac{L_1}{L_2} \qquad \left(\mu_f \geq \frac{L_1}{L_2} \right) \qquad (17.152)$$

The frequency of oscillation is set by the resonant frequency of the capacitor and the total inductance, $L_1 + L_2$. The feedback is set by the ratio of the two inductors and must satisfy the condition in Eq. (17.152). For poles on the $j\omega$ axis, the amplification factor must be large enough to satisfy the equality. Generally, more gain is used to ensure oscillation, and some form of amplitude stabilization is used.

Amplitude Stabilization in *LC* Oscillators
The inherently nonlinear characteristics of the transistors are often used to limit oscillation amplitude. In JFET circuits for example, the gate diode can be used to form a peak detector that limits amplitude. In bipolar circuits, rectification by the base-emitter diode often performs the same function. In the Colpitts oscillator in Fig. 17.76, a diode and resistor are added to provide the amplitude-limiting function. The diode and resistor R_G form a rectifier that establishes a negative dc bias on the gate. The capacitors in the circuit act as the rectifier filter. In practical circuits, the onset of oscillation is accompanied by a slight shift in the Q-point values as the oscillator adjusts its operating point to limit the amplitude.

17.13.4 CRYSTAL OSCILLATORS
Oscillators with very high frequency accuracy and stability can be formed using quartz crystals as the frequency-determining element (**crystal oscillators**). The crystal is a piezoelectric device

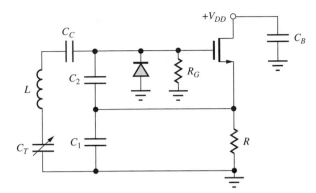

Figure 17.76 Tunable MOSFET version of the Colpitts oscillator with a diode rectifier for amplitude limiting.

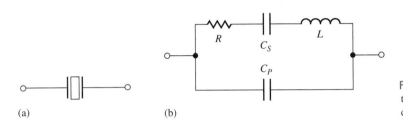

Figure 17.77 Symbol and electrical equivalent circuit for a quartz crystal.

(a)   (b)

that vibrates in response to electrical stimulus. Although the frequency of vibration of the crystal is determined by its mechanical properties, the crystal can be modeled electrically by a very high Q ($>10{,}000$) resonant circuit, as in Fig. 17.77.

L, C_S, and R characterize the intrinsic series resonance path through the crystal element itself, whereas the parallel capacitance C_P is dominated by the capacitance of the package containing the quartz element. The equivalent impedance of this network exhibits a series resonant frequency ω_S at which C_S resonates with L, and a parallel resonant frequency ω_P that is determined by L resonating with the series combination of C_S and C_P.

The impedance of the crystal versus frequency can easily be calculated using the circuit model in Fig. 17.77:

$$Z_C = \frac{Z_P Z_S}{Z_P + Z_S} = \frac{\dfrac{1}{sC_P}\left(sL + R + \dfrac{1}{sC_S}\right)}{\dfrac{1}{sC_P} + \left(sL + R + \dfrac{1}{sC_S}\right)} = \frac{1}{sC_P}\left(\frac{s^2 + s\dfrac{R}{L} + \dfrac{1}{LC_S}}{s^2 + s\dfrac{R}{L} + \dfrac{1}{LC_T}}\right) \qquad (17.153)$$

where $C_T = \dfrac{C_S C_P}{C_S + C_P}$

The figure accompanying Ex. 17.12 is an example of the variation of crystal impedance with frequency. Below ω_S and above ω_P, the crystal appears capacitive; between ω_S and ω_P, it exhibits an inductive reactance. As can be observed in the figure, the region between ω_S and ω_P is quite narrow. If the crystal is used to replace the inductor in the Colpitts oscillator, a well-defined frequency of oscillation will exist. In most crystal oscillators, the crystal operates between the two resonant points and represents an inductive reactance, replacing the inductor in the circuit.

EXAMPLE 17.12 QUARTZ CRYSTAL EQUIVALENT CIRCUIT

The values of L and C_S that represent the crystal have unusual magnitudes because of the extremely high Q of the crystal.

PROBLEM Calculate the equivalent circuit element values for a crystal with $f_S = 5$ MHz, $Q = 20,000$, $R = 50\ \Omega$, and $C_P = 5$ pF. What is the parallel resonant frequency?

SOLUTION **Known Information and Given Data:** The crystal parameters are specified as $f_S = 5$ MHz, $Q = 20,000$, $R = 50\ \Omega$, and $C_P = 5$ pF.

Unknowns: L and C_S

Approach: Use the definitions of Q and series resonant frequency to find the unknowns.

Assumptions: The equivalent circuit in Fig. 17.77 is adequate to model the crystal.

Analysis: Using Q, R, and f_S for a series resonant circuit,

$$L = \frac{RQ}{\omega_S} = \frac{50(20,000)}{2\pi(5 \times 10^6)} = 31.8\ \text{mH} \qquad C_S = \frac{1}{\omega_S^2 L} = \frac{1}{(10^7\pi)^2(0.0318)} = 31.8\ \text{fF}$$

Typical values of C_P fall in the range of 5 to 20 pF. For $C_P = 5$ pF, the parallel resonant frequency will be

$$f_P = \frac{1}{2\pi\sqrt{L\dfrac{C_S C_P}{C_S + C_P}}} = \frac{1}{2\pi\sqrt{(31.8\ \text{mH})(31.6\ \text{fF})}} = 5.02\ \text{MHz}$$

whereas

$$f_S = 5.00\ \text{MHz}$$

Check of Results: Let us use our values of L and C_S to calculate f_S.

$$f_S = \frac{1}{2\pi\sqrt{31.8\ \text{mH}\ (31.8\ \text{fF})}} = 5.00\ \text{MHz} \quad \checkmark$$

Discussion: Note that the two resonant frequencies differ by only 0.4 percent, and the high Q of the crystal results in a relatively large effective value for L and a small value for C_S.

Computer-Aided Analysis: The graph below presents results from a computer calculation of the reactance of the crystal versus frequency using the parameters calculated in Ex. 17.12.

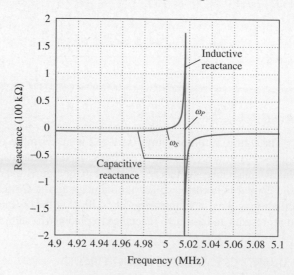

Reactance versus frequency for crystal parameters calculated in the example.

Below the series resonant frequency and above the parallel resonant frequency, the crystal exhibits capacitive reactance. Between f_S and f_P, the crystal appears inductive. In many oscillator circuits, the crystal behaves as an inductor and resonates with external capacitance. The oscillator frequency will therefore be between f_S and f_P.

EXERCISE: Calculate the parallel resonant frequency of the crystal if a 2-pF capacitor is placed in parallel with the crystal. Repeat for a 20-pF capacitor.

ANSWERS: 5.016 MHz; 5.008 MHz

Several examples of crystal oscillators are given in Figs. 17.78 to 17.81. Many variations are possible, but most of these oscillators are topological transformations of the Colpitts or Hartley oscillators. For example, the circuit in Fig. 17.78(a) represents a Colpitts oscillator with the source terminal chosen as the ground reference. The same circuit is drawn in a different form in Fig. 17.78(b). Figures 17.79 and 17.80 show Colpitts oscillators using bipolar and JFET devices.

The final crystal oscillator, shown in Fig. 17.81, represents a circuit that is often implemented using a CMOS logic inverter. The circuit forms yet another Colpitts oscillator, similar to Fig. 17.78(b). The inverter is initially biased into the middle of its operating region by feedback resistor R_F to ensure that the Q-point of the gate is in a region of high gain.

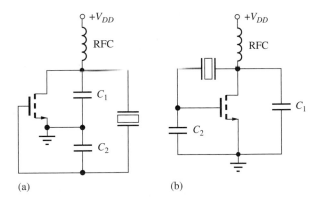

Figure 17.78 Two forms of the same Colpitts crystal oscillator.

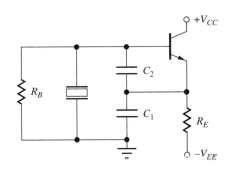

Figure 17.79 Crystal oscillator using a bipolar transistor.

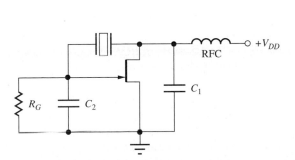

Figure 17.80 Crystal oscillator using a JFET.

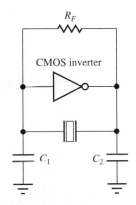

Figure 17.81 Crystal oscillator using a CMOS inverter as the gain element.

 **ELECTRONICS IN ACTION**

Numerically Controlled Oscillators and Direct Digital Synthesis

Modern D/A converter technology has advanced to the point that traditional analog feedback oscillators are being replaced with a direct digital synthesizer (DDS) that utilizes numerically controlled oscillators (NCOs) to synthesize the sinusoidal waveforms. The NCO can provide very small frequency step size and high-speed tuning. In the DDS, the signal waveform is constructed in the digital domain, and the analog output signal is produced using a digital-to-analog (DAC) converter followed by a low-pass filter.

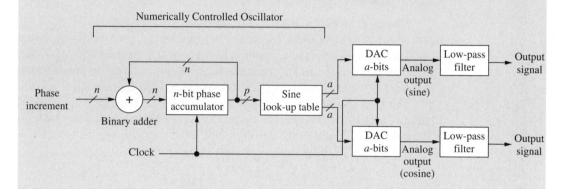

The digital NCO consists of an n-bit phase accumulator and a p-bit sine look-up table where $p \leq n$. To generate a sine wave, an n-bit phase increment is added to the accumulator during each clock cycle. A full counter (2^n counts) corresponds to 2π radians or 1 cycle of the output sine wave. If the counter is incremented by one at each clock interval, the maximum period T_{max} of the output waveform, corresponding to minimum output frequency f_{min}, will be

$$T_{max} = 2^n T_{clk} \qquad \text{or} \qquad f_{min} = \frac{f_{clk}}{2^n}$$

where T_{clk} is the period of the clock, and f_{clk} is the clock frequency. This minimum output frequency also represents the frequency resolution of the DDS. To generate higher frequency signals, a larger phase increment N is added to the phase accumulator at each clock cycle, and $f_O = N f_{min}$. For example, for $f_{clk} = 20$ MHz and $n = 24$, $f_{min} \approx 1.192$ Hz. In order to generate a 10-kHz sine wave, a phase increment of 8389 (10,000/1.192) would be added to the counter at each clock cycle. Based upon the Nyquist sampling theorem, the highest frequency that can be generated is one-half of the clock frequency (using $N = 2^{n/2}$), since f_{clk} is the update rate of the D/A converter.

In order to reduce the size of the look-up table, only the upper p bits of the phase accumulator are used to address the sine table. A number of ROM compression techniques are utilized to further reduce the size of the ROM. The output of the sine table is an a-bit representation of the amplitude of the sine wave where "a" corresponds to the number of bits of resolution of the D/A converter. Finite resolution in the representation of both the signal phase and amplitude lead to distortion in the output waveform. The low-pass filter helps to remove distortion and the high-frequency content related to the update rate of the DAC (f_{clk}). Many DDS chips provide two D/A outputs, producing sine and cosine waves with very precise 90-degree phase relationships for the in-phase (I) and quadrature (Q) channels in RF transceivers.

SUMMARY

- The characteristics of general feedback amplifiers can be expressed in terms of the two-port model parameters for the individual open-loop amplifier and feedback network. Analysis of each of the four different interconnections of the amplifier and feedback network uses a particular set of two-port parameters: series-shunt feedback uses the h-parameters; shunt-shunt feedback uses the y-parameters; shunt series feedback uses the g-parameters; and series-series feedback uses the z-parameters.

- Series feedback places ports in series and increases the overall impedance level at the series-connected port. Shunt feedback is achieved by placing ports in parallel and reduces the overall impedance level at the shunt-connected port.

- Before applying the methods, we must ensure that the amplifier and feedback networks can be properly represented as two-ports. Transistor realizations of series-shunt and shunt-shunt feedback amplifiers can readily be analyzed using h- and y-parameter descriptions, respectively. However, care must be exercised in the analysis of circuits that involve series feedback at the output port. Amplifiers in the shunt-series and series-series feedback circuits often cannot be represented as two-ports, particularly when we try to calculate output resistance.

- The loop gain $T(s)$ plays an important role in determining the characteristics of feedback amplifiers. For theoretical calculations, the loop gain can be found by breaking the feedback loop at some arbitrary point and directly calculating the voltage returned around the loop. However, both sides of the loop must be properly terminated before the loop-gain calculation is attempted.

- When using SPICE or making experimental measurements, it is often impossible to break the feedback loop. The method of successive voltage and current injection is a powerful technique for determining the loop gain without the need for opening the feedback loop.

- Whenever feedback is applied to an amplifier, stability becomes a concern. In most cases, a negative or degenerative feedback condition is desired. Stability can be determined by studying the characteristics of the loop gain $T(s) = A(s)\beta(s)$ of the feedback amplifier as a function of frequency, and stability criteria can be evaluated from either Nyquist diagrams or Bode plots.

- In the Nyquist case, stability requires that the plot of $T(j\omega)$ not enclose the $T = -1$ point.

- On the Bode plot, the asymptotes of the magnitudes of $A(j\omega)$ and $1/\beta(j\omega)$ must not intersect with a rate of closure exceeding 20 dB/decade.

- Phase margin and gain margin, which can be found from either the Nyquist or Bode plot, are important measures of stability.

- Miller multiplication represents a useful method for setting the unity-gain frequency of internally compensated operational amplifiers. This technique is often called Miller compensation. In these op amps, slew rate is directly related to the unity-gain frequency.

- In circuits called oscillators, feedback is actually designed to be positive or regenerative so that an output signal can be produced by the circuit without an input being present. The Barkhausen criteria for oscillation state that the phase shift around the feedback loop must be an even multiple of 360° at some frequency, and the loop gain at that frequency must be equal to 1.

- Oscillators use some form of frequency-selective feedback to determine the frequency of oscillation; RC and LC networks and quartz crystals can all be used to set the frequency.

- Wien-bridge and phase-shift oscillators are examples of oscillators employing RC networks to set the frequency of oscillation.

- Most *LC* oscillators are versions of either the Colpitts or Hartley oscillators. In the Colpitts oscillator, the feedback factor is set by the ratio of two capacitors; in the Hartley case, a pair of inductors determines the feedback.

- Crystal oscillators use a quartz crystal to replace the inductor in *LC* oscillators. A crystal can be modeled electrically as a very high Q resonant circuit, and when used in an oscillator, the crystal accurately controls the frequency of oscillation.

- For true sinusoidal oscillation, the poles of the oscillator must be located precisely on the $j\omega$ axis in the *s*-plane. Otherwise, distortion occurs. To achieve sinusoidal oscillation, some form of amplitude stabilization is normally required. Such stabilization may result simply from the inherent nonlinear characteristics of the transistors used in the circuit, or from explicitly added gain control circuitry.

KEY TERMS

Amplitude stabilization
Barkhausen criteria for oscillation
Blackman's theorem
Bode plot
Closed-loop gain
Closed-loop input resistance
Closed-loop output resistance
Colpitts oscillator
Crystal oscillator
Current amplifier
Degenerative feedback
Feedback amplifier stability
Feedback network
Gain margin (GM)
Hartley oscillator
LC oscillators
Loop gain
−1 Point
Negative feedback
Nyquist plot
Open-loop amplifier

Open-loop gain
Oscillator circuits
Oscillators
Phase margin
Phase-shift oscillator
Positive feedback
RC oscillators
Regenerative feedback
Series-series feedback
Series-shunt feedback
Shunt connection
Shunt-series feedback
Shunt-shunt feedback
Sinusoidal oscillator
Stability
Successive voltage and current injection
technique
Transconductance amplifier
Transresistance amplifier
Voltage amplifier
Wien-bridge oscillator

REFERENCES

1. R. D. Middlebrook, "Measurement of loop gain in feedback systems," *International Journal of Electronics,* vol. 38, no. 4, pp. 485–512, April 1975. Middlebrook credits a 1965 Hewlett-Packard Application Note as the original source of this technique.

2. R. C. Jaeger, S. W. Director, and A. J. Brodersen, "Computer-aided characterization of differential amplifiers," *IEEE JSSC,* vol. SC-12, pp. 83–86, February 1977.

3. R. B. Blackman, "Effect of feedback on impedance," *Bell System Technical Journal,* vol. 22, no. 3, 1943.

4. P. J. Hurst, "A comparison of two approaches to feedback circuit analysis," *IEEE Trans. on Education,* vol. 35, pp. 253–261, August 1992.

5. F. Corsi, C. Marzocca, and G. Matarrese, "On impedance evaluation in feedback circuits," *IEEE Trans. on Education,* vol. 45, no. 4, pp. 371–379, November 2002.

PROBLEMS

17.1 Classic Feedback Systems

17.1. A classic feedback amplifier in Fig. 17.1 has $\beta = 0.2$. What are the loop gain T, the closed-loop gain A_v, and the fractional gain error FGE (see Sec. 12.1.2) if $A = \infty$? (b) If $A = 80$ dB? (c) If $A = 10$?

17.2. The feedback amplifier in Fig. P17.2(a) has $R_1 = 1$ kΩ, $R_2 = 100$ kΩ, $R_I = 0$, and $R_L = 10$ kΩ. (a) What is $\beta(s)$? (b) If $A = 86$ dB, what are the loop gain T and the closed-loop gain A_v?

17.3. The inverting amplifier in Fig. 11.5 is implemented with an op amp with finite gain $A = 80$ dB. If $R_1 = 1$ kΩ and $R_2 = 100$ kΩ, what are $\beta(s)$, $T(s)$, and $A_v(s)$?

17.4. The integrator in Fig. 11.15 is implemented with an op amp with finite gain $A = 80$ dB. If $R = 20$ kΩ and $C = 0.01$ μF, What is $\beta(s)$? What is $T(s)$? What is $A_v(s)$?

17.5. An amplifier's closed-loop voltage gain A_v is described by Eq. (17.4). What is the minimum value of open-loop gain needed if the gain error is to be less than 0.01 percent for a voltage follower ($A_{CL} \cong 1$ with $\beta = 1$)?

17.6. Use SPICE to simulate and compare the transfer characteristics of the two class-B output stages in Fig. P17.6 if the op amp is described by $A_o = 1000$, $R_{id} = 100$ kΩ, and $R_o = 100$ Ω. Assume $V_I = 0$.

17.7. An amplifier's closed-loop voltage gain is described by Eq. (17.4). What is the minimum value of open-loop gain needed if the gain error is to be less than 0.2 percent for an ideal gain of 200?

17.8. (a) Calculate the sensitivity of the closed-loop gain A_v with respect to changes in open-loop gain A, $S_A^{A_v}$, using Eq. (17.4) and the definition of sensitivity originally presented in Chapter 11:

$$S_A^{A_v} = \frac{A}{A_v} \frac{\partial A_v}{\partial A}$$

(b) Use this formula to estimate the percentage change in closed-loop gain if the open-loop gain A changes by 10 percent for an amplifier with $A = 100$ dB and $\beta = 0.01$.

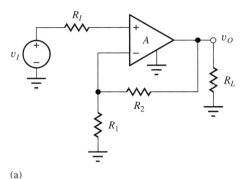

(a)

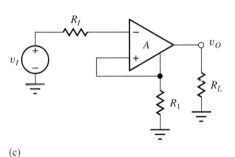

(b)

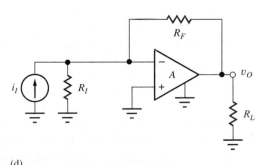

(c)

(d)

Figure P17.2 For each amplifier A: $A_o = 4000$, $R_{id} = 20$ kΩ, $R_o = 1$ kΩ.

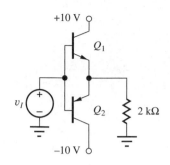

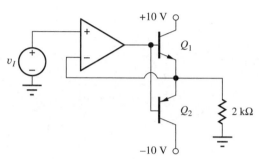

Figure P17.6

17.2 Feedback Amplifier Design Using Two-Port Network Theory

17.9. Identify the type of negative feedback that should be used to achieve these design goals: (a) high input resistance and high output resistance, (b) low input resistance and high output resistance, (c) low input resistance and low output resistance, (d) high input resistance and low output resistance.

17.10. Identify the type of feedback being used in the four circuits in Fig. P17.2.

17.11. Of the four circuits in Fig. P17.2, which tend to provide a high output resistance? (b) Which provide a relatively low output resistance?

17.12. Of the four circuits in Fig. P17.2, which tend to provide a high input resistance? (b) Which provide a relatively low input resistance?

17.13. An amplifier has an open-loop voltage gain of 90 dB, $R_{id} = 40$ kΩ, and $R_o = 1000$ Ω. The amplifier is used in a feedback configuration with a resistive feedback network. (a) What is the largest current gain that can be achieved with this feedback amplifier? (b) What is the largest transconductance that can be achieved with this feedback amplifier?

17.14. An amplifier has an open-loop voltage gain of 90 dB, $R_{id} = 40$ kΩ, and $R_o = 1000$ Ω. The amplifier is used in a feedback configuration with a resistive feedback network. (a) What is the largest value of input resistance that can be achieved in the feedback amplifier? (b) What is the smallest value of input resistance that can be achieved? (c) What is the largest value of output resistance that can be achieved? (d) What is the smallest value of output resistance that can be achieved?

17.3 Voltage Amplifiers—Series-Shunt Feedback

17.15. Draw the A and F circuits for the circuit in Fig. P17.2(a) and find the voltage gain, input resistance, and output resistance of the feedback amplifier. Assume $R_I = 1$ kΩ, $R_L = 5$ kΩ, $R_1 = 5$ kΩ, and $R_2 = 45$ kΩ.

*17.16. (a) Draw the amplifier in Fig. P17.2(a) as a *series-shunt* feedback amplifier. (b) Find h_{11}^T, h_{22}^T, h_{21}^A, and h_{12}^F. (c) Calculate $A = -h_{21}^A/(R_I + h_{11}^T)(h_{22}^T + G_L)$ and β using these values. (d) Find the closed-loop gain. (e) Compare the values of h_{12}^F to h_{12}^A and the value of h_{21}^A to h_{21}^F. $R_L = 5.6$ kΩ, $R_1 = 4.3$ kΩ, $R_2 = 39$ kΩ, and $R_I = 1$ kΩ.

17.17. Use the two-port approach to find the voltage gain, input resistance, and output resistance of the feedback amplifier in Fig. P17.17 if $R_1 = 1$ kΩ, $R_2 = 7.5$ kΩ, $\beta_o = 100$, $V_A = 50$ V, $I = 200$ μA, $V_{CC} = 10$ V, $A = 50$ dB, $R_{id} = 40$ kΩ, and $R_o = 1$ kΩ.

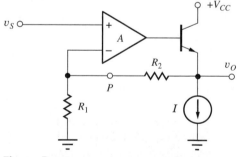

Figure P17.17

17.18. Draw the amplifier in Fig. P17.18 as a series-shunt feedback amplifier, and use two-port theory to find the voltage gain $A_v = v_o/v_{\text{ref}}$, input resistance, and output resistance. Use the results of these calculations to find the transconductance $A_{TC} = i_o/v_{\text{ref}}$. Assume $\beta_o = 100$, $V_A = 50$ V, $I = 200$ μA, $V_{\text{REF}} = 0$ V, and $R = 10$ kΩ.

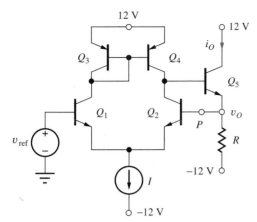

Figure P17.18

17.19. Rework Ex. 17.1 including R_L as part of the feedback network (i.e., the feedback circuit will be a three-resistor "pi-network," and there will not be an external load resistor. Draw the new A and F circuits and find the voltage gain, input resistance, and output resistance of the amplifier.

17.20. (a) Calculate the sensitivity of the closed-loop input resistance of the series-shunt feedback amplifier with respect to changes in open-loop gain A:

$$S_A^{R_{in}} = \frac{A}{R_{in}} \frac{\partial R_{in}}{\partial A}$$

(b) Use this formula to estimate the percentage change in closed-loop input resistance if the open-loop gain A changes by 10 percent for an amplifier with $A = 94$ dB and $\beta - 0.01$. (c) Calculate the sensitivity of the closed-loop output resistance of the series-shunt feedback amplifier with respect to changes in open-loop gain A:

$$S_A^{R_{out}} = \frac{A}{R_{out}} \frac{\partial R_{out}}{\partial A}$$

(d) Use this formula to estimate the percentage change in closed-loop output resistance if the open-loop gain A changes by 10 percent for an amplifier with $A = 100$ dB and $\beta = 0.01$.

17.4 Transresistance Amplifiers—Shunt-Shunt Feedback

17.21. Draw the A and F circuits for the circuit in Fig. P17.2(d) and find the transresistance, input resistance, and output resistance of the feedback amplifier. Assume $R_I = 100$ kΩ, $R_L = 5$ kΩ, and $R_F = 36$ kΩ.

17.22. (a) Draw the amplifier in Fig. P17.2(d) as a shunt-shunt feedback amplifier if $R_F = 10$ kΩ, $R_I = 100$ kΩ, and $R_L = 10$ kΩ. (Note that the amplifier parameters are in the figure.) (b) Find y_{11}^T, y_{22}^T, y_{21}^A, and y_{12}^F. (c) Calculate A and β using these values and find the closed-loop transresistance. (d) Compare the values of y_{12}^F to y_{12}^A and y_{21}^A to y_{21}^F.

17.23. The circuit in Fig. P17.23 is a shunt-shunt feedback amplifier. Use the two-port method to find the input resistance, output resistance, and transresistance of the amplifier if $R_I = 1$ kΩ, $R_E = 1$ kΩ, $\beta_o = 100$, $V_A = 50$ V, $R_L = 4.7$ kΩ, and $R_F = 36$ kΩ. What is the voltage gain of this amplifier? (*Note:* Represent v_i and R_I by a Norton equivalent circuit.)

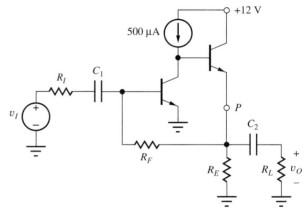

Figure P17.23

17.24. Use SPICE to find the input resistance, output resistance, and transresistance of the amplifier in Fig. P17.23 and compare the results to those in Prob. 17.23. $C_1 = 82$ μF and $C_2 = 47$ μF.

17.25. Use two-port analysis to find the midband transresistance, input resistance, and output resistance of the amplifier in Fig. P17.25 if $g_m = 2$ mS and $r_o = 40$ kΩ.

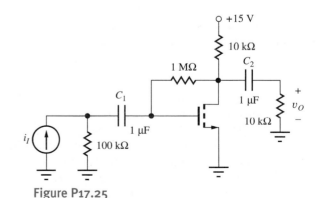

Figure P17.25

*17.26. Use two-port theory to derive an expression for the input impedance of the shunt-shunt feedback amplifier in Fig. P17.26.

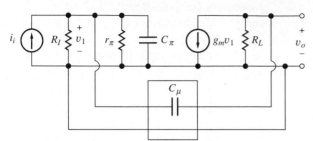

Figure P17.26

*17.27. Draw the Wilson current source in Fig. P17.27 as a shunt-shunt feedback amplifier and find the current gain i_o/i_{ref} and input resistance of the source. Use the two-port model (Fig. P17.27(b)) for the current mirror. For simplicity, assume all transistors have the same W/L ratio. Assume $g_m = 2$ mS and $r_o = 36$ kΩ.

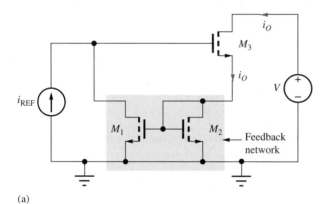

(a)

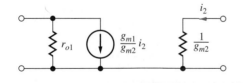

(b)

Figure P17.27

*17.28. Draw the Wilson current source in Fig. P17.28 as a shunt-shunt feedback amplifier and find the current gain i_o/i_{ref} and input resistance of the source. Use the two-port model, Fig. P17.27(b), for the current mirror. For simplicity, assume all transistors have the same emitter area with $\beta_o = 100$, $V_A = 50$, $g_m = 50$ mS, and $V_{CC} \ll V_A$.

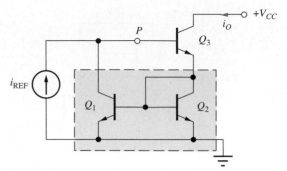

Figure P17.28

17.29. Use SPICE to simulate the Wilson BJT source in Fig. P17.28 with $i_{REF} = 100$ µA, $V_{CC} = 6$ V, and $V_A = 50$ V for current gains of 10^2, 10^4, and 10^6 and show that R_{out} goes from a limit of $\beta_o r_o/2$ to $\mu_f r_o$.

17.5 Current Amplifiers—Shunt-Series Feedback

17.30. Draw the A and F circuits for the circuit in Fig. P17.2(b) and find the current gain, input resistance, and output resistance of the feedback amplifier. Assume $R_I = 100$ kΩ, $R_L = 5$ kΩ, $R_1 = 10$ kΩ, and $R_2 = 1$ kΩ.

17.31. (a) Draw the amplifier in Fig. P17.2(b) as a shunt-series feedback amplifier if $R_1 = 20$ kΩ, $R_2 = 2$ kΩ, $R_I = 150$ kΩ, and $R_L = 10$ kΩ. (Note that the amplifier parameters are in the figure.) (b) Find g_{11}^T, g_{22}^T, g_{21}^A, and g_{12}^F. (c) Calculate A and β using these values and find the closed-loop current gain. (d) Compare the values of g_{12}^F to g_{12}^A and g_{21}^A to g_{21}^F.

17.32. Analyze the amplifier in Prob. 17.23 as a shunt-series feedback amplifier.

17.6 Transconductance Amplifiers—Series-Series Feedback

17.33. Draw the A and F circuits for the circuit in Fig. P17.2(c) and find the voltage gain, input resistance, and output resistance of the feedback amplifier. Assume $R_I = 2$ kΩ, $R_L = 5$ kΩ, and $R_1 = 5$ kΩ.

17.34. (a) Draw the amplifier in Fig. P17.2(c) as a series-series feedback amplifier if $R_1 = 5$ kΩ, $R_I = 2$ kΩ, and $R_L = 5$ kΩ. (Note that the amplifier parameters are in the figure.) (b) Find z_{11}^T, z_{22}^T, z_{21}^A, and z_{12}^F. (c) Calculate A and β using these values and find the closed-loop current gain. (d) Compare the values of z_{12}^F to z_{12}^A and z_{21}^A to z_{21}^F.

17.35. (a) Draw the emitter follower as a *series-series* feedback amplifier. (b) Draw the A-circuit and β-network. (c) Use these circuits to find expressions for the voltage gain and input resistance of the amplifier.

*17.36. In Chapter 14, the gain of the bipolar inverting amplifier was expressed as

$$A_v = -\frac{\beta_o R_L}{R_S + r_\pi + (\beta_o + 1)R_E}$$

(a) Show that this expression can be written as

$$A_v = \left(\frac{A}{1 + A\beta}\right)R_L$$

What are the expressions for A and β? (b) Show that the amplifier can be represented as a series-series connection of two-ports.

17.7 Common Errors in Applying Two-Port Feedback Theory

17.37. Draw the Wilson current source in Fig. P17.28 as a shunt-series feedback amplifier and find the current gain, input resistance, and *incorrect value* of the output resistance of the source. Use the two-port model for the current mirror.

17.38. (a) Draw the small-signal model for the "series-series feedback triple" in Fig. P17.38 and show that it cannot be drawn as a series-series feedback amplifier for two-port analysis.

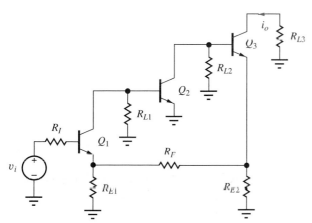

Figure P17.38

17.39. The circuit in Fig. P17.39 represents a high-resistance current source called a "regulated cascode" source. We wish to find the resistance presented by the circuit to supply voltage V_{DD}. (a) Draw the ac equivalent circuit. (b) Draw the

circuit as a shunt-shunt feedback amplifier with the output across r_{o2}. Identify the A and F circuits. (c) Can the circuit be represented as a shunt-series feedback amplifier with the output at the drain of the upper transistor? If so, draw the amplifier and identify the A and F circuits. If not, why not? (d) Find an expression for the resistance R_{in} presented to V_{CC} by direct analysis (not using the two-port approach). (e) What is the value of R_{in} if the two current sources are 100 μA, and the transistors are all described by $K_n = 750$ μA/V^2, $V_{TN} = 0.75$ V, $V_{DD} = 10$ V, and $\lambda = 0.02$/V. (f) Confirm the calculation with SPICE.

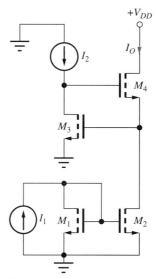

Figure P17.39

17.8 Finding the Loop Gain

Direct Calculation

*17.40. Break the feedback loop of the amplifier in Fig. P17.17 at point P and calculate the loop gain. Assume $R_1 = 1$ kΩ, $R_2 = 9.1$ kΩ, $\beta_o = 100$, $V_A = 50$ V, $I = 200$ μA, $V_{CC} = 10$ V, $A_o = 50$ dB, $R_{id} = 40$ kΩ, and $R_o = 1$ kΩ.

17.41. Break the feedback loop of the amplifier in Fig. P17.18 at point P and calculate the loop gain. Assume $\beta_o = 100$, $V_A = 50$ V, $I = 200$ μA, $V_{REF} = 0$ V, and $R = 10$ kΩ.

17.42. Break the feedback loop of the Wilson current source in Fig. P17.28 at point P and calculate the loop gain (current gain). Use the two-port model, Fig. P17.27(b), for the current mirror. Assume all transistors have the same emitter area with $\beta_o = 100$, $V_A = 50$, and $g_m = 50$ mS.

17.43. Break the feedback loop of the amplifier in Fig. P17.23 at point P and calculate the loop gain. Assume $R_I = 1$ kΩ, $R_E = 1$ kΩ, $\beta_o = 100$, $V_A = 50$ V, $R_L = 4.7$ kΩ, and $R_F = 36$ kΩ.

*17.44. Break the feedback loop at point P in the active low-pass filter in Fig. 17.44 and write an expression for the loop gain of the circuit.

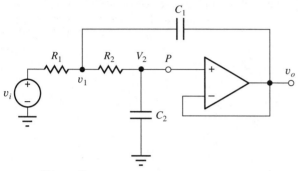

Figure P17.44

17.45. Break the feedback loop at point P in the active high-pass filter in Fig. P17.45 and write an expression for the loop gain of the circuit.

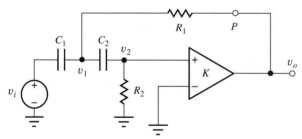

Figure P17.45

Voltage and Current Injection

17.46. Use the successive voltage and current injection technique at point P with SPICE to calculate the loop gain of the amplifier in Fig. P17.17. Assume $R_1 = 1$ kΩ, $R_2 = 7.5$ kΩ, $\beta_o = 100$, $V_A = 50$ V, $I = 200$ μA, $V_{CC} = 10$ V, $A_o = 50$ dB, $R_{id} = 40$ kΩ, and $R_o = 1$ kΩ.

17.47. Use the successive voltage and current injection technique at point P with SPICE to calculate the loop gain of the amplifier in Fig. P17.18. Assume $\beta_o = 100$, $V_A = 50$ V, $I = 200$ μA, $V_{REF} = 0$ V, and $R = 10$ kΩ.

17.48. Use the successive voltage and current injection technique at point P with SPICE to calculate the

loop gain of the amplifier in Fig. P17.17. Assume $R_1 = 40$ kΩ, $R_2 = 300$ kΩ, $\beta_o = 100$, $V_A = 50$ V, $I = 200$ μA, $V_{CC} = 10$ V, $A_{vo} = 50$ dB, $R_{id} = 40$ kΩ, and $R_o = 1$ kΩ.

17.49. Use the successive voltage and current injection technique at point P with SPICE to calculate the loop gain of the amplifier in Fig. P17.2(b). Assume $R_I = 1$ kΩ, $R_2 = 1$ kΩ, $R_L = 4.7$ kΩ, and $R_1 = 36$ kΩ.

17.50. Use the successive voltage and current injection technique at point P with SPICE to calculate the loop gain of the Wilson current source in Fig. P17.28. Assume all transistors have the same emitter area with $\beta_o = 100$, $V_A = 50$ V, $i_{REF} = 100$ μA, and $V_{CC} = 6$ V.

17.51. Use voltage injection at point P in Fig. P17.45 with SPICE to find the loop gain versus frequency for frequencies from 1 Hz to 1 MHz. Assume $C_1 = C_2 = 0.005$ μF, $R_1 = R_2 = 2$ kΩ, and the amplifier can be modeled by the transfer function

$$K(s) = \frac{10^7}{(s + 5 \times 10^6)}$$

17.9 Blackman's Theorem to the Rescue

17.52. (a) Use Blackman's theorem to find the input and output resistances of the amplifier in Fig. P17.2(a). Use the values from Prob. 17.15. (b) Use Blackman's theorem to find the input and output resistances of the amplifier in Fig. P17.2(b). Use the values from Prob. 17.30. (c) Use Blackman's theorem to find the input and output resistances of the amplifier in Fig. P17.2(c). Use the values from Prob. 17.33. (d) Use Blackman's theorem to find the input and output resistances of the amplifier in Fig. P17.2(d). Use the values from Prob. 17.21.

17.53. Use Blackman's Theorem to find the midband input and output resistances of the amplifier in Fig. P17.23. Use the element values from Prob. 17.23.

17.54. Use Blackman's theorem to find the input and output resistances of the amplifier in Fig. P17.17. Use the element values from Prob. P17.17.

17.55. Use Blackman's theorem to find the input and output resistances of the amplifier in Fig. P17.18. Use the element values from Prob. 17.18.

17.56. Use Blackman's theorem to find the midband input and output resistances of the amplifier in Fig. P17.25. Use the element values from Prob. 17.25.

17.57. Use Blackman's theorem to find the input and output resistances of the amplifier in Fig. P17.39. Use the element values from Prob. 17.39.

17.58. Use Blackman's theorem to find the input and output resistances of the regulated cascode current source in Fig. P17.39 if M_1 and M_2 are replaced with BJTs with $\beta_F = 100$ and $V_A = 75$ V. Use the other element values from Prob. 17.39.

17.10 Using Feedback to Control Frequency Response

17.59. The voltage gain of an amplifier is described by

$$A(s) = \frac{2\pi \times 10^{10}s}{(s + 2000\pi)(s + 2\pi \times 10^6)}$$

(a) What are the open-loop gain and upper- and lower-cutoff frequencies of this amplifier? (b) If this amplifier is used in a feedback amplifier with a closed-loop gain of 100, what are the upper- and lower-cutoff frequencies of the closed-loop amplifier? (c) Repeat for a closed-loop gain of 40.

17.60. Repeat Prob. 17.59 if the voltage gain of the amplifier is given by

$$A(s) = \frac{2 \times 10^{14}\pi^2}{(s + 2\pi \times 10^3)(s + 2\pi \times 10^5)}$$

17.61. Repeat Prob. 17.59 if the voltage gain of the amplifier is given by

$$A(s) = \frac{4\pi^2 \times 10^{18}s^2}{(s + 200\pi)(s + 2000\pi)(s + 2\pi \times 10^6)(s + 2\pi \times 10^7)}$$

17.62. (a) Find f_H and f_L for the A-circuit for the shunt-shunt feedback circuit in Fig. P17.25 if $g_m = 2$ mS, $r_o = 25$ kΩ, $C_{GD} = 3$ pF, and $C_{GS} = 10$ pF. (b) What are the closed-loop values of f_H and f_L?

17.63. (a) Find f_H, and f_L for the A-circuit for the BJT shunt-shunt feedback example in Fig. 17.14 if $f_T = 500$ MHz, $r_x = 0$, and $C_\mu = 0.75$ pF. (b) What are the closed-loop values of f_H and f_L?

17.64. (a) Find f_H for the A-circuit for the series-shunt feedback amplifier in Example 17.1 and Fig. 17.7 if the amplifier has $f_T = 10$ MHz. What is the closed-loop value of f_H? (*Hint:* Include the frequency dependence in the controlled voltage source.) (b) Repeat for $f_T = 1$ MHz.

17.65. (a) Calculate the sensitivity of the closed-loop bandwidth of a low-pass amplifier with respect to changes in open-loop gain A_o:

$$S_{A_o}^{\omega_H^F} = \frac{A}{\omega_H^F} \frac{\partial \omega_H^F}{\partial A_o}$$

(b) Use this formula to estimate the percentage change in closed-loop bandwidth if the open-loop gain A_o changes by 10 percent for an amplifier with $A = 100$ dB and $\beta = 0.01$.

17.11 Stability of Feedback Amplifiers

17.66. What are the phase and gain margins for the amplifier in Prob. 17.59?

17.67. What are the phase and gain margins for the amplifier in Prob. 17.60?

17.68. What are the phase and gain margins for the amplifier in Prob. 17.61?

17.69. The voltage gain of an amplifier is described by

$$A(s) = \frac{4 \times 10^{19}\pi^3}{(s + 2\pi \times 10^4)(s + 2\pi \times 10^5)^2}$$

(a) If resistive feedback is used, find the frequency at which the loop gain will have a phase shift of 180°. (b) At what value of closed-loop gain will the amplifier break into oscillation? (c) Is the amplifier stable for larger or smaller values of closed-loop gain?

*17.70. What is the maximum load capacitance C_L that can be connected to the output of the voltage follower in Fig. P17.70 if the phase margin of the amplifier is to be 60°? Assume that the amplifier voltage gain is described by this transfer function, and that it has an output resistance of $R_o = 500$ Ω:

$$A(s) = \frac{10^7}{(s + 50)}$$

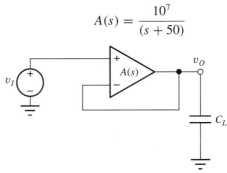

Figure P17.70

17.71 The voltage gain of an amplifier is described by

$$A(s) = \frac{4 \times 10^{13}\pi^2}{(s + 2\pi \times 10^3)(s + 2\pi \times 10^4)}$$

(a) Will this amplifier be stable for a closed-loop gain of 4? (b) What is the phase margin?

17.72. The voltage gain of an amplifier is described by

$$A(s) = \frac{2 \times 10^{14}\pi^2}{(s + 2\pi \times 10^3)(s + 2\pi \times 10^5)}$$

(a) Will this amplifier be stable for a closed-loop gain of 5? (b) What is the phase margin?

17.73. (a) Use MATLAB to make a Bode plot for the amplifier in Prob. 17.71 for a closed-loop gain of 5. Is the amplifier stable? What is the phase margin? (b) Repeat for the unity-gain case.

17.74. Find the loop gain for an integrator that uses a single-pole op amp with $A_o = 100$ dB and $f_T = 1$ MHz. Assume the integrator feedback elements are $R = 100$ kΩ and $C = 0.01$ μF. What is the phase margin of the integrator?

*17.75. Find the closed-loop transfer function of an integrator that uses a two-pole op amp with $A_o = 100$ dB, $f_{p1} = 1$ kHz, and $f_{p2} = 100$ kHz. Assume the integrator feedback elements are $R = 100$ kΩ and $C = 0.01$ μF. What is the phase margin of the integrator?

*17.76. (a) Write an expression for the loop gain $T(s)$ of the amplifier in Fig. P17.76 if $R_1 = 1$ kΩ, $R_2 = 20$ kΩ, $C_C = 0$, and the op amp transfer function is

$$A(s) = \frac{2 \times 10^{11}\pi^2}{(s + 2\pi \times 10^2)(s + 2\pi \times 10^4)}$$

(b) Use MATLAB to make a Bode plot of $T(s)$. What is the phase margin of this circuit? (c) Can compensation capacitor C_C be added to achieve a phase margin of $45°$? If so, what is the value of C_C?

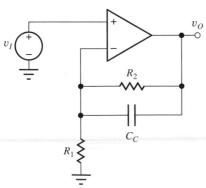

Figure P17.76

17.77. (a) Use MATLAB to make a Bode plot for the amplifier in Prob. 17.69. Find the frequency for which the phase shift is $180°$. (b) At what value of closed-loop gain will the amplifier break into oscillation?

17.78. Repeat Prob. 17.77 for the amplifier in Prob. 17.59.

17.79. Repeat Prob. 17.77 for the amplifier in Prob. 17.60.

17.80. Repeat Prob. 17.77 for the amplifier in Prob. 17.61.

17.81. Use MATLAB to make a Bode plot for the amplifier in Prob. 17.76 for a closed-loop gain of 100. Is the amplifier stable? What is the phase margin?

17.82. Use MATLAB to make a Bode plot for the loop gain of the active low-pass filter in Fig. P17.44 if the op amp can be modeled as a single-pole amplifier with $A_o = 100$ dB and $f_T = 1$ MHz. Assume $C_1 = 0.05$ μF, $C_2 = 0.01$ μF, and $R_1 = R_2 = 2$ kΩ. What is the phase margin of this circuit? What is the gain margin?

17.83. Use MATLAB to make a Bode plot for the loop gain of the active high-pass filter in Fig. P17.45 if the amplifier is modeled by $K(s)$ in Prob. 17.51. Assume $C_1 = C_2 = 0.005$ μF, and $R_1 = R_2 = 2$ kΩ. What is the phase margin of this circuit? What is the gain margin?

17.84. Use MATLAB to make a Bode plot for the integrator in Prob. 17.74. What is the phase margin of the integrator?

*17.85. Use MATLAB to make a Bode plot for the integrator in Prob. 17.75. What is the phase margin of the integrator?

17.86. The noninverting amplifier in Fig. P17.86 has $R_1 = 47$ kΩ, $R_2 = 390$ kΩ, and $C_S = 45$ pF. Find the phase margin of the amplifier if amplifier voltage gain is described by the following transfer function:

$$A(s) = \frac{10^7}{(s + 50)}$$

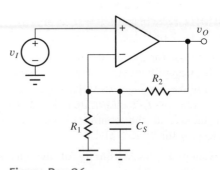

Figure P17.86

17.12 Single-Pole Operational Amplifier Compensation

17.87. (a) What are the unity-gain frequency and positive and negative slew rates for the CMOS amplifier in Fig. 17.56 if $I_1 = 250\ \mu A$, $I_2 = 500\ \mu A$, $K_{n1} = 1\ mA/V^2$, and $C_C = 7.5\ pF$? (b) If $I_1 = 500\ \mu A$, $I_2 = 250\ \mu A$, and $C_C = 10\ pF$?

17.88. Repeat Prob. 17.87(a) for $I_1 = 500\ \mu A$, $I_2 = 3\ mA$, and $C_C = 10\ pF$.

17.89. Simulate the frequency response of the CMOS amplifier in Fig. 17.61 for $R_Z = 0$ and for $R_Z = 1\ k\Omega$. Compare the values of the unity-gain frequency and phase shift of the amplifier at the unity-gain frequency. Use $I_1 = 250\ \mu A$, $I_2 = 500\ \mu A$, $I_3 = 2\ mA$, $(W/L)_1 = 20/1$, $(W/L)_3 = 40/1$, $(W/L)_5 = 160/1$, $(W/L)_6 = 60/1$, and $C_C = 7.5\ pF$. $V_{DD} = V_{SS} = 10\ V$. Use CMOS models from Appendix B.

17.90. (a) What are the unity-gain frequency and slew rate of the bipolar amplifier in Fig. 17.62 if $I_1 = 50\ \mu A$, $I_2 = 500\ \mu A$, and $C_C = 12\ pF$? (b) If $I_1 = 200\ \mu A$, $I_2 = 250\ \mu A$, and $C_C = 12\ pF$?

17.91. Repeat Prob. 17.90(a) for $I_1 = 500\ \mu A$, $I_2 = 3\ mA$, and $C_C = 10\ pF$.

17.92. (a) What are the positive and negative slew rates of the amplifier in Fig. P17.92 just after a 2-V step function is applied to input v_2 if $I_1 = 40\ \mu A$, $I_2 = 400\ \mu A$, $I_3 = 500\ \mu A$, and $C_C = 5\ pF$? Assume v_1 is grounded. (b) Check your answers with SPICE.

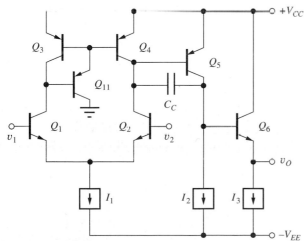

Figure P17.92

17.93. Repeat Prob. 17.92(a) for $I_1 = 100\ \mu A$, $I_2 = 1\ mA$, $I_3 = 2.5\ mA$, and $C_C = 8\ pF$.

*17.94. (a) Use SPICE to calculate the frequency response of the amplifier in Fig. P17.92 for $I_1 = 100\ \mu A$, $I_2 = 500\ \mu A$, $I_3 = 500\ \mu A$, and $C_C = 15\ pF$. (b) Repeat for $I_1 = 100\ \mu A$, $I_2 = 50\ \mu A$, and $C_C = 15\ pF$. Compare the unity-gain frequency to that predicted by Eq. (17.117). Discuss the reasons for any discrepancy. Use $V_{CC} = V_{EE} = 15\ V$.

17.95. (a) Use SPICE to simulate the amplifier in Ex. 17.10 and verify the results of the frequency response simulations presented in the example. (b) Use SPICE to simulate the amplifier in Ex. 17.11 and verify the results of the frequency response simulations presented in the example. Calculate the slew rate of the amplifier, and verify it with SPICE simulation. Are the positive and negative slew rates the same?

17.13 Oscillators

Frequency-Selective *RC* Networks

17.96. The circuit in Fig. P17.96 is called a quadrature oscillator. Derive an expression for its frequency of oscillation. What is the value of R_F required for sinusoidal oscillation?

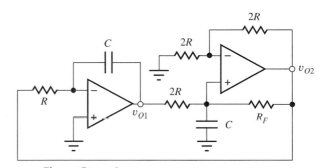

Figure P17.96

17.97. Derive an expression for the frequency of oscillation of the three-stage phase-shift oscillator in Fig. P17.97. What is the ratio R_2/R_1 required for oscillation?

17.98. Calculate the frequency and amplitude of oscillation of the Wien-bridge oscillator in Fig. 17.71 if $R = 5\ k\Omega$, $C = 500\ pF$, $R_1 = 10\ k\Omega$, $R_2 = 15\ k\Omega$, $R_3 = 6.2\ k\Omega$, and $R_4 = 10\ k\Omega$.

17.99. Use SPICE transient simulation to find the frequency and amplitude of the oscillator in Prob. 17.98. Start the simulation with a 1-V initial condition on the grounded capacitor C.

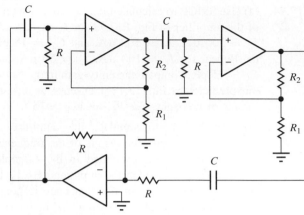

Figure P17.97

17.100. Calculate the frequency and amplitude of oscillation of the phase-shift oscillator in Fig. 17.73 if $R = 5$ kΩ, $C = 1000$ pF, $R_2 = 47$ kΩ, $R_3 = 15$ kΩ, and $R_4 = 68$ kΩ.

17.101. Use SPICE transient simulation to find the frequency and amplitude of the oscillator in Prob. 17.100. Start the simulation with a 1-V initial condition on the capacitor connected to the inverting input of the amplifier.

LC Oscillators

Colpitts Oscillators

**17.102. The ac equivalent circuit for a Colpitts oscillator is given in Fig. P17.102. (a) What is the frequency of oscillation if $g_m = 10$ mS, $\beta_o = 100$, $R_E = 1$ kΩ, $L = 5$ μH, $C_1 = 20$ pF, $C_2 = 100$ pF, $C_4 = 0.01$ μF, and $C_3 = $ infinity? Assume that the capacitances of the transistor can be neglected (see Prob. 17.103). (b) A variable capacitor C_3 is added to the circuit and has a range of 5–50 pF. What range of frequencies of oscillation can be achieved? (c) What is the minimum transconductance needed to ensure oscillation in part (a)? What is the minimum collector current required in the transistor?

17.103. The ac equivalent circuit for a Colpitts oscillator is given in Fig. P17.102. (a) What is the frequency of oscillation if $L = 20$ μH, $C_1 = 20$ pF, $C_2 = 100$ pF, $C_3 = $ infinity, $C_4 = 0.01$ μF, $f_T = 500$ MHz, $r_\pi = \infty$, $V_A = 50$ V, $r_x = 0$, $R_E = 1$ kΩ, $C_\mu = 3$ pF, and the transistor is operating at a Q-point of (5 mA, 5 V)? (b) What is the frequency of oscillation if the Q-point current is doubled?

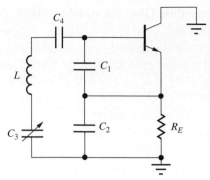

Figure P17.102

17.104. Design a Colpitts oscillator for operation at a frequency of 20 MHz using the circuit in Fig. 17.74(a). Assume $L = 3$ μH, $K_p = 1.25$ mA/V^2, and $V_{TN} = -4$ V. Ignore the device capacitances.

17.105. What is the frequency of oscillation of the MOSFET Colpitts oscillator in Fig. P17.105 if $L = 10$ μH, $C_1 = 50$ pF, $C_2 = 50$ pF, $C_3 = 0$ pF, $C_{GS} = 10$ pF, and $C_{GD} = 4$ pF? What is the minimum amplification factor of the transistor?

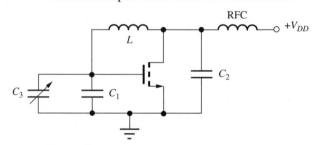

Figure P17.105

17.106. Capacitor C_3 is added to the Colpitts oscillator in Prob. 17.105 to allow tuning the oscillator. (a) Assume C_3 can vary from 5 to 50 pF and calculate the frequencies of oscillation for the two adjustment extremes. (b) What is the minimum value of amplification factor needed to ensure oscillation throughout the full tuning range?

17.107. A variable-capacitance diode is added to the Colpitts oscillator in Fig. P17.107 to form a voltage tunable oscillator. (a) The parameters of the diode are $C_{jo} = 20$ pF and $\phi_j = 0.8$ V [see Eq. (3.21)]. Calculate the frequencies of oscillation for $V_{TUNE} = 2$ V and 20 V if $L = 10$ μH, $C_1 = 75$ pF, and $C_2 = 75$ pF. Assume the RFC has infinite impedance and C_C has zero impedance. (b) What is the minimum value of voltage gain needed to ensure oscillation throughout the full tuning range?

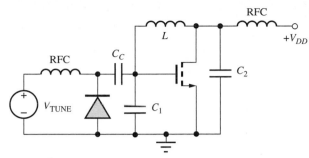

Figure P17.107

17.108. (a) Perform a SPICE transient simulation of the Colpitts oscillator in Fig. 17.74 and compare its frequency of oscillation to hand calculations if $V_{DD} = 10$ V, $K_p = 1.25$ mA/V^2, $V_{TN} = -4$ V, $R_S = 820$ Ω, $C_2 = 220$ pF, $C_1 = 470$ pF, and $L = 10$ μH. (b) Repeat if $C_2 = 470$ pF and $C_1 = 220$ pF.

17.109. Perform a SPICE transient simulation of the Colpitts oscillator in Fig. P17.105 if $L = 10$ μH, $C_1 = 50$ pF, $C_2 = 50$ pF, $C_3 = 0$ pF, RFC = 20 mH, $V_{DD} = 12$ V, $K_n = 10$ mA/V^2, $V_{TN} = 1$ V, $C_{GS} = 10$ pF, and $C_{GD} - 4$ pF. What are the amplitude and frequency of oscillation?

Hartley Oscillators

17.110. What is the frequency of oscillation of the Hartley oscillator in Fig. P17.110 if the diode is replaced by a short circuit and $L_1 = 10$ μH, $L_2 = 10$ μH, and $C = 20$ pF? Neglect C_{GS} and C_{GD}.

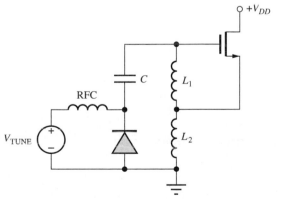

Figure P17.110

17.111. A variable-capacitance diode is added to the Hartley oscillator in Prob. 17.110 to form a voltage-tunable oscillator, and the value of C is

changed to 220 pF. (a) If the parameters of the diode are $C_{jo} = 20$ pF and $\phi_j = 0.8$ V [see Eq. (3.21)], calculate the frequencies of oscillation for $V_{TUNE} = 2$ V and 20 V. Assume the RFC has infinite impedance. (b) What is the minimum value of amplification factor of the FET needed to ensure oscillation throughout the full tuning range?

Crystal Oscillators

17.112. A crystal has a series resonant frequency of 10 MHz, series resistance of 40 Ω, Q of 25,000, and parallel capacitance of 10 pF. (a) What are the values of L and C_S for this crystal? (b) What is the parallel resonant frequency of the crystal? (c) The crystal is placed in an oscillator circuit in parallel with a total capacitance of 22 pF. What is the frequency of oscillation?

17.113. The crystal in the oscillator in Fig. P17.113 has $L = 15$ mH, $C_S = 20$ fF, and $R = 50$ Ω. (a) What is the frequency of oscillation if $R_E = 1$ kΩ, $R_B = 100$ kΩ, $V_{CC} = V_{EE} = 5$ V, $C_1 = 100$ pF, $C_2 = 470$ pF, and $C_3 = \infty$? Assume the transistor has $\beta_f = 100$, $V_A = 50$ V, and infinite f_T. (b) Repeat if $C_\mu = 5$ pF and $f_T = 250$ MHz.

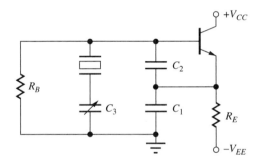

Figure P17.113

17.114. A variable capacitor C_3 is placed in series with the crystal in the oscillator in Prob. 17.113(a) to provide a calibration adjustment. Assume C_3 can vary from 1 pF to 35 pF and calculate the frequencies of oscillation for the two adjustment extremes.

17.115. Simulate the crystal oscillator in Fig. P17.113 and find the frequency of oscillation if $R_E = 1$ kΩ, $R_B = 100$ kΩ, $V_{CC} = V_{EE} = 5$ V, $C_1 = 100$ pF, $C_2 = 470$ pF, and $C_3 = \infty$. The crystal has $L = 15$ mH, $C_S = 20$ fF, $R = 50$ Ω, and $C_P = 20$ pF. Assume the transistor has $\beta_F = 100$, $V_A = 50$ V, $C_\mu = 5$ pF, and $\tau_F = 1$ ns.

APPENDIX A

Standard Discrete Component Values

Resistor Coding

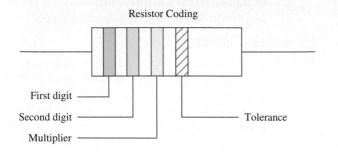

First digit

Second digit

Multiplier

Tolerance

Resistor Color Code			
COLOR	DIGIT	MULTIPLIER	TOLERANCE, %
Silver	· · ·	0.01	10
Gold	· · ·	0.1	5
Black	0	1	
Brown	1	10	
Red	2	10^2	
Orange	3	10^3	
Yellow	4	10^4	
Green	5	10^5	
Blue	6	10^6	
Violet	7	10^7	
Gray	8	10^8	
White	9	10^9	

Standard resistor values: All values available with a 5 percent tolerance. Bold values are available with 10 percent tolerance.

				OHMS				MEGOHMS	
1.0	**5.6**	**33**	**180**	**1000**	**5600**	**33000**	**180000**	**1.0**	**5.6**
1.1	6.2	36	200	1100	6200	36000	200000	1.1	6.2
1.2	**6.8**	**39**	**220**	**1200**	**6800**	**39000**	**220000**	**1.2**	**6.8**
1.3	7.5	43	240	1300	7500	43000	240000	1.3	7.5
1.5	**8.2**	**47**	**270**	**1500**	**8200**	**47000**	**270000**	**1.5**	**8.2**
1.6	9.1	51	300	1600	9100	51000	300000	1.6	9.1
1.8	**10**	**56**	**330**	**1800**	**10000**	**56000**	**330000**	**1.8**	**10**
2.0	11	62	360	2000	11000	62000	360000	2.0	11
2.2	**12**	**68**	**390**	**2200**	**12000**	**68000**	**390000**	**2.2**	**12**
2.4	13	75	430	2400	13000	75000	430000	2.4	13
2.7	**15**	**82**	**470**	**2700**	**15000**	**82000**	**470000**	**2.7**	**15**
3.0	16	91	510	3000	16000	91000	510000	3.0	16
3.3	**18**	**100**	**560**	**3300**	**18000**	**100000**	**560000**	**3.3**	**18**
3.6	20	110	620	3600	20000	110000	620000	3.6	20
3.9	**22**	**120**	**680**	**3900**	**22000**	**120000**	**680000**	**3.9**	**22**
4.3	24	130	750	4300	24000	130000	750000	4.3	
4.7	**27**	**150**	**820**	**4700**	**27000**	**150000**	**820000**	**4.7**	
5.1	30	160	910	5100	30000	160000	910000	5.1	

PRECISION (1%) RESISTORS

10.0	19.1	36.5	69.8	133	255	487	931	1.78K	3.40K	6.49K	12.4K	23.7K	45.3K	84.5K	158K	294K	549K
10.2	19.6	37.4	71.5	137	261	499	953	1.82K	3.48K	6.65K	12.7K	24.3K	46.4K	86.6K	162K	301K	562K
10.5	20.0	38.3	73.2	140	267	511	976	1.87K	3.57K	6.81K	13.0K	24.9K	47.5K	88.7K	165K	309K	576K
10.7	20.5	39.2	75.0	143	274	523	1.00K	1.91K	3.65K	6.98K	13.3K	25.5K	48.7K	90.9K	169K	316K	590K
11.0	21.0	40.2	76.8	147	280	536	1.02K	1.96K	3.74K	7.15K	13.7K	26.1K	49.9K	93.1K	174K	324K	604K
11.3	21.5	41.2	78.7	150	287	549	1.05K	2.00K	3.83K	7.32K	14.0K	26.7K	51.1K	95.3K	178K	332K	619K
11.5	22.1	42.2	80.6	154	294	562	1.07K	2.05K	3.92K	7.50K	14.3K	27.4K	52.3K	97.6K	182K	340K	634K
11.8	22.6	43.2	82.5	158	301	576	1.10K	2.10K	4.02K	7.68K	14.7K	28.0K	53.6K	100K	187K	348K	649K
12.1	23.2	44.2	84.5	162	309	590	1.13K	2.15K	4.12K	7.87K	15.0K	28.7K	54.9K	102K	191K	357K	665K
12.4	23.7	45.3	86.6	165	316	604	1.15K	2.21K	4.22K	8.06K	15.4K	29.4K	56.2K	105K	196K	365K	681K
12.7	24.3	46.4	88.7	169	324	619	1.18K	2.26K	4.32K	8.25K	15.8K	30.1K	57.6K	107K	200K	374K	698K
13.0	24.9	47.5	90.9	174	332	634	1.21K	2.32K	4.42K	8.45K	16.2K	30.9K	59.0K	110K	205K	383K	715K
13.3	25.5	48.7	93.1	178	340	649	1.24K	2.37K	4.53K	8.66K	16.5K	31.6K	60.4K	113K	210K	392K	732K
13.7	26.1	49.9	95.3	182	348	665	1.27K	2.43K	4.64K	8.87K	16.9K	32.4K	61.9K	115K	215K	402K	750K
14.0	26.7	51.1	97.6	187	357	681	1.30K	2.49K	4.75K	9.09K	17.4K	33.2K	63.4K	118K	221K	412K	768K
14.3	27.4	52.3	100	191	365	698	1.33K	2.55K	4.87K	9.31K	17.8K	34.0K	64.9K	121K	226K	422K	787K
14.7	28.0	53.6	102	196	374	715	1.37K	2.61K	4.99K	9.53K	18.2K	34.8K	66.5K	124K	232K	432K	806K
15.0	28.8	54.9	105	200	383	732	1.40K	2.67K	5.11K	9.76K	18.7K	35.7K	68.1K	127K	237K	442K	825K
15.4	29.4	56.2	107	205	392	750	1.43K	2.74K	5.23K	10.0K	19.1K	36.5K	69.8K	130K	243K	453K	845K
15.8	30.1	57.6	110	210	402	768	1.47K	2.80K	5.36K	10.2K	19.6K	37.4K	71.5K	133K	249K	464K	866K
16.2	30.9	59.0	113	215	412	787	1.50K	2.87K	5.49K	10.5K	20.0K	38.3K	73.2K	137K	255K	475K	887K
16.5	31.6	60.4	115	221	422	806	1.54K	2.94K	5.62K	10.7K	20.5K	39.2K	75.0K	140K	261K	487K	909K
16.9	32.4	61.9	118	226	432	825	1.58K	3.01K	5.76K	11.0K	21.0K	40.2K	76.8K	143K	267K	499K	931K
17.4	33.2	63.4	121	232	443	845	1.62K	3.09K	5.90K	11.3K	21.5K	41.2K	78.7K	147K	274K	511K	953K
17.8	34.0	64.9	124	237	453	866	1.65K	3.16K	6.04K	11.5K	22.1K	42.2K	80.6K	150K	280K	523K	976K
18.2	34.8	66.5	127	243	464	887	1.69K	3.24K	6.19K	11.8K	22.6K	43.2K	82.5K	154K	287K	536K	1.00M
18.7	35.7	68.1	130	249	475	909	1.74K	3.32K	6.34K	12.1K	23.2K	44.2K					

Standard Capacitor Values (Larger values are also available)

pF	pF	pF	pF	μF	μF	μF	μF	μF	μF	μF
1	10	100	1000	0.01	0.1	1	10	100	1000	10000
	12	120	1200	0.012	0.12	1.2	12	120	1200	12000
1.5	15	150	1500	0.015	0.15	1.5	15	150	1500	15000
	18	180	1800	0.018	0.18	1.8	18	180	1800	
	20	200	2000	0.020	0.20				2000	20000
2.2	22	220	2200	0.022	0.22	2.2	22	220	2200	22000
	27	270	2700	0.027	0.27	2.7	27	270	2700	
3.3	33	330	3300	0.033	0.33	3.3	33	330	3300	33000
	39	390	3900	0.039	0.39	3.9	39	390	3900	
4.7	47	470	4700	0.047	0.47	4.7	47	470	4700	47000
5.0	50	500	5000	0.050	0.50					50000
5.6	56	560	5600	0.056	0.56	5.6	56	560	5600	
6.8	68	680	6800	0.068	0.68	6.8	68	680	6800	68000
8.2	82	820	8200	0.082	0.82	8.2	82	820	8200	

Standard Inductor Values

μH	μH	μH	μH	mH	mH	mH
0.10	1.0	10	100	1.0	10	100
	1.1	11	110			
	1.2	12	120	1.2	12	120
0.15	1.5	15	150	1.5	15	
0.18	1.8	18	180	1.8	18	
	2.0	20	200			
0.22	2.2	22	220	2.2	22	
	2.4	24	240			
0.27	2.7	27	270	2.7	27	
0.33	3.3	33	330	3.3	33	
0.39	3.9	39	390	3.9	39	
	4.3	43	430			
0.47	4.7	47	470	4.7	47	
0.56	5.6	56	560	5.6	56	
	6.2	62	620			
0.68	6.8	68	680	6.8	68	
	7.5	75	750			
0.82	8.2	82	820	8.2	82	
	9.1	91	910			

APPENDIX B

Solid-State Device Models and SPICE Simulation Parameters

B.1 *pn* JUNCTION DIODES

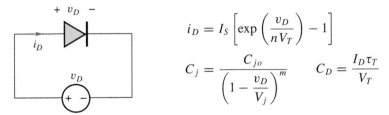

$$i_D = I_S \left[\exp \left(\frac{v_D}{n V_T} \right) - 1 \right]$$

$$C_j = \frac{C_{jo}}{\left(1 - \dfrac{v_D}{V_j} \right)^m} \qquad C_D = \frac{I_D \tau_T}{V_T}$$

Figure B.1 Diode with applied voltage v_D.

TABLE B.1
Diode Parameters for Circuit Simulation

PARAMETER	NAME	DEFAULT	TYPICAL VALUE
Saturation current	IS	1×10^{-14} A	3×10^{-17} A
Emission coefficient (ideality factor — n)	N	1	1
Transit time (τ_T)	TT	0	0.15 nS
Series resistance	RS	0	10 Ω
Junction capacitance	CJO	0	1.0 pF
Junction potential (V_j)	VJ	1 V	0.8 V
Grading coefficient (m)	M	0.5	0.5

B.2 MOS FIELD-EFFECT TRANSISTORS (MOSFETs)

A summary of the mathematical models for both the NMOS and PMOS transistors follows. The terminal voltages and currents are defined in Fig. B.2.

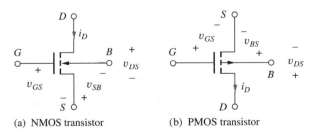

(a) NMOS transistor   (b) PMOS transistor

Figure B.2 NMOS and PMOS transistor circuit symbols.

NMOS TRANSISTOR MODEL SUMMARY

$$K_n = K_n' \frac{W}{L} = \mu_n C_{\text{ox}}'' \frac{W}{L}$$

$$i_G = 0 \text{ and } i_B = 0 \qquad \text{for all regions}$$

Cutoff Region

$$i_D = 0 \qquad \text{for } v_{GS} \leq V_{TN}$$

Triode Region

$$i_D = K_n \left(v_{GS} - V_{TN} - \frac{v_{DS}}{2} \right) v_{DS} \qquad \text{for } v_{GS} - V_{TN} \geq v_{DS} \geq 0$$

Saturation Region

$$i_D = \frac{K_n}{2} (v_{GS} - V_{TN})^2 (1 + \lambda v_{DS}) \qquad \text{for } v_{DS} \geq (v_{GS} - V_{TN}) \geq 0$$

Threshold Voltage

$$V_{TN} = V_{TO} + \gamma \left(\sqrt{v_{SB} + 2\phi_F} - \sqrt{2\phi_F} \right)$$

PMOS TRANSISTOR MODEL SUMMARY

$$K_p = K_p' \frac{W}{L} = \mu_p C_{\text{ox}}'' \frac{W}{L}$$

$$i_G = 0 \text{ and } i_B = 0 \qquad \text{for all regions}$$

Cutoff Region

$$i_D = 0 \qquad \text{for } v_{GS} \geq V_{TP}$$

Triode Region

$$i_D = K_p \left(v_{GS} - V_{TP} - \frac{v_{DS}}{2} \right) v_{DS} \qquad \text{for } v_{GS} - V_{TP} \leq v_{DS} \leq 0$$

Saturation Region

$$i_D = \frac{K_p}{2} (v_{GS} - V_{TP})^2 (1 + \lambda |v_{DS}|) \qquad \text{for } v_{DS} \leq (v_{GS} - V_{TP}) \leq 0$$

Threshold Voltage

$$V_{TP} = V_{TO} - \gamma \left(\sqrt{v_{BS} + 2\phi_F} - \sqrt{2\phi_F} \right)$$

TABLE B.2
Types of MOSFET Transistors

NMOS DEVICE	PMOS DEVICE	
Enhancement-Mode	$V_{TN} > 0$	$V_{TP} < 0$
Depletion-Mode	$V_{TN} \leq 0$	$V_{TP} \geq 0$

MOS TRANSISTOR PARAMETERS FOR CIRCUIT SIMULATION

For simulation purposes, use the LEVEL=1 models in SPICE with the following SPICE parameters in your NMOS and PMOS devices:

TABLE B.3

Representative MOS Device Parameters for SPICE Simulation (MOSIS 0.5-μm p-well process)

PARAMETER	SYMBOL	NMOS TRANSISTOR	PMOS TRANSISTOR
Threshold voltage	VTO	0.91 V	−0.77 V
Transconductance	KP	50 μA/V^2	20 μA/V^2
Body effect	GAMMA	0.99 $\sqrt{V}$	0.53 $\sqrt{V}$
Surface potential	PHI	0.7 V	0.7 V
Channel-length modulation	LAMBDA	0.02 V^{-1}	0.05 V^{-1}
Mobility	UO	615 cm^2	235 cm^2/s
Ohmic drain resistance	RD	0	0
Ohmic source resistance	RS	0	0
Junction saturation current	IS	0	0
Built-in potential	PB	0	0
Gate-drain capacitance per unit width	CGDO	330 pF/m	315 pF/m
Gate-source capacitance per unit width	CGSO	330 pF/m	315 pF/m
Gate-bulk capacitance per unit width	CGBO	395 pF/m	415 pF/m
Junction bottom capacitance per unit area	CJ	3.9×10^{-4} F/m^2	2×10^{-4} F/m^2
Grading coefficient	MJ	0.45	0.47
Sidewall capacitance	CJSW	510 pF/m	180 pF/m
Sidewall grading coefficient	MJSW	0.36	0.09
Source-drain sheet resistance	RSH	22 Ω/square	70 Ω/square
Oxide thickness	TOX	4.15×10^{-6} cm	4.15×10^{-6} cm
Junction depth	XJ	0.23 μm	0.23 μm
Lateral diffusion	LD	0.26 μm	0.25 μm
Substrate doping	NSUB	2.1×10^{16}/cm^3	5.9×10^{16}/cm^3
Critical field	UCRIT	9.6×10^5 V/cm	6×10^5 V/cm
Critical field exponent	UEXP	0.18	0.28
Saturation velocity	VMAX	7.6×10^7 cm/s	6.5×10^7 cm/s
Fast surface state density	NFS	9×10^{11}/cm^2	3×10^{11}/cm^2
Surface state density	NSS	1×10^{10}/cm^2	1×10^{10}/cm^2

INDEX

DIODE EQUATIONS

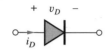

$$i_D = I_S \left[\exp\left(\frac{v_D}{nV_T}\right) - 1 \right] \qquad V_T = \frac{kT}{q} \qquad C_j = \frac{C_{jo}A}{\sqrt{1 - \dfrac{v_D}{\phi_j}}} \qquad C_D = \frac{I_D}{V_T}\tau_r$$

(FORWARD) ACTIVE REGION EQUATIONS—
npn TRANSISTOR ($v_{BE} > 0$ AND $v_{CE} \geq v_{BE}$)

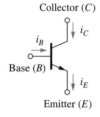

$$i_C = I_S \exp\left(\frac{v_{BE}}{V_T}\right)\left(1 + \frac{v_{CE}}{V_A}\right) \qquad i_C = \beta_F i_B \qquad i_E = (\beta_F + 1)i_B \qquad \beta_F = \beta_{FO}\left(1 + \frac{v_{CE}}{V_A}\right)$$

BJT SMALL-SIGNAL MODEL PARAMETER RELATIONSHIPS ($\beta_o \cong \beta_F$)

$$g_m = \frac{I_C}{V_T} \cong 40I_C \qquad \beta_o = g_m r_\pi \qquad r_o = \frac{V_A + V_{CE}}{I_C} \cong \frac{V_A}{I_C} \qquad \mu_f = g_m r_o \qquad \omega_T = \frac{g_m}{C_\pi + C_\mu}$$

LARGE SIGNAL MODEL EQUATIONS—NMOS TRANSISTOR

Triode (Linear) Region ($v_{GS} > V_{TN}$ and $v_{DS} \leq v_{GS} - V_{TN}$)

$$i_D = K_n\left(v_{GS} - V_{TN} - \frac{v_{DS}}{2}\right)^2 v_{DS} \qquad i_G = 0 \qquad i_S = i_D \qquad K_n = K_n'\frac{W}{L}$$

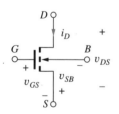

Active (Saturation) Region ($v_{GS} > V_{TN}$ and $v_{DS} \geq v_{GS} - V_{TN}$)

$$i_D = \frac{K_n}{2}(v_{GS} - V_{TN})^2(1 + \lambda v_{DS}) \qquad i_G = 0 \qquad i_S = i_D \qquad K_n = K_n'\frac{W}{L}$$

$$V_{TN} = V_{TO} + \gamma\left(\sqrt{v_{SB} + 2\phi_f} - \sqrt{2\phi_f}\right)$$

FET SMALL-SIGNAL MODEL PARAMETER RELATIONSHIPS

$$g_m = \frac{2I_D}{V_{GS} - V_{TN}} \cong \sqrt{2K_n I_D} \qquad r_o = \frac{1 + \lambda V_{DS}}{\lambda I_D} \cong \frac{1}{\lambda I_D} \qquad \mu_f = g_m r_o \qquad \omega_T = \frac{g_m}{C_{GS} + C_{GD}}$$